				Uranus' Known Satellites†				
Miranda	129,800	0.017	3.4	1.41349 days	16.5	—	$\sim 1 \times 10^{-6}$	—
Ariel	190,900	0.0028	~0	2.52038	14.4	—	$\sim 15 \times 10^{-6}$	—
Umbriel	266,000	0.0035	~0	4.14418	15.3	—	$\sim 6 \times 10^{-6}$	—
Titania	436,000	0.0024	~0	8.70587	14.0	—	$\sim 50 \times 10^{-6}$	—
Oberon	583,400	0.0007	~0	13.46325	14.2	—	$\sim 29 \times 10^{-6}$	—
				Neptune's Known Satellites				
Triton	355,550	0	159.945°	5.876844 days	13.5	4,400 ± 800	$1.3 \pm 0.3 \times 10^{-3}$	~3.0
Nereid	5,567,000	0.74934	27.71°	359.881	18.7	—	—	—
				Pluto's Known Satellite				
Charon[(g)]	~17,000	~0	~105°[(h)]	6.3875 days	16–17	—	—	—

[(a)] With respect to equatorial plane of planet.
[(b)] Known to be synchronously rotating with respect to primary.
[(c)] The eccentricities and inclinations for the regular satellites are slightly variable. Those for the irregular satellites are extremely variable.
[(d)] S.M.Larson (DPS, Honolulu meeting) suggests two satellites are required to satisfy the 1966 positions.
[(e)] To plane of ring.
[(f)] Varies from 17′ to 56′.
[(g)] Proposed name, not yet adopted by the International Astronomical Union.
[(h)] To plane of sky.
[(i)] Mass with respect to primary. Planet masses are listed in another table.

*The common nomenclature for the rings of Saturn is D, C, B, A, and D′ in order of increasing distance from the planet. The D and D′ rings are extremely tenuous. The inner and outer radii of the other three are 1.21 and 1.53, 1.53 and 1.95, and 2.03 and 2.29 equatorial radii of Saturn respectively. For details see "The Rings of Saturn", NASA SP-343, 1974.

†Uranus recently has been discovered to have at least nine narrow rings, currently designated 6, 5, 4, α, β, η, γ, δ, and ϵ. Their respective mean radii in km are 41,980, 42,360, 42,663, 44,844, 45,799, 47,323, 47,746, 48,423, and about 51,400. For details see Elliot *et al. Astron. J.* **83**, 980, 1978.

A recent collection of review papers on satellites is to be found in *Planetary Satellites*, J. Burns (ed.), University of Arizona Press. Tucson, 1977.

GEOCHIMICA ET COSMOCHIMICA ACTA

SUPPLEMENT 10

PROCEEDINGS
OF THE
NINTH LUNAR AND PLANETARY SCIENCE
CONFERENCE

Houston, Texas, March 13–17, 1978

COVER ILLUSTRATIONS

M. C. Escher described his work in the following words: "The ideas that are basic to [my work] often bear witness to my amazement and wonder at the laws of nature which operate in the world around us. He who wonders discovers that this is in itself a wonder. By keenly confronting the enigmas that surround us, and by considering and analyzing the observations that I had made, I ended up in the domain of mathematics" (1960, p. 8).

Many scientists active in research today could subscribe to this philosophy as a description of their feelings toward their own work. Because Escher's talent lent itself well to capturing this sense of amazement and wonder at the laws of nature, and because he often was able to capture as well the quirkiness of man as nature's student, we have chosen several of his prints to adorn our covers.

Volume One, Front Cover.	**Stars**, 1948.
Volume One, Back Cover.	**Dragon**, 1952.
Volume Two, Front Cover.	**Double Planetoid**, 1949.
Volume Two, Back Cover.	**Relativity**, 1953.
Volume Three, Front Cover.	**Concentric Rinds**, 1953.
Volume Three, Back Cover.	**Belvedere**, 1958.

We thank the Escher Foundation - Haags Gemeentemusem - The Hague for allowing us to reproduce these works and the Vorpal Gallery, San Francisco, for supplying reproducible copies.

Escher M. C. (1960) *The Graphic Work of M. C. Escher.* Meredith Press, N.Y.

GEOCHIMICA ET COSMOCHIMICA ACTA

Journal of The Geochemical Society and The Meteoritical Society

SUPPLEMENT 10

PROCEEDINGS OF THE NINTH LUNAR AND PLANETARY SCIENCE CONFERENCE

Houston, Texas, March 13–17, 1978

Compiled by the

Lunar and Planetary Institute

Houston, Texas

VOLUME 2

LUNAR AND PLANETARY SURFACES

PERGAMON PRESS

NEW YORK · OXFORD · TORONTO · SYDNEY · FRANKFURT

Pergamon Press Offices:

U.S.A.	Pergamon Press Inc., Maxwell House, Fairview Park, Elmsford, New York 10523, U.S.A.
U.K.	Pergamon Press Ltd., Headington Hill Hall, Oxford OX3 0BW, England
CANADA	Pergamon of Canada Ltd., 150 Consumers Road, Willowdale, Ontario M2J 1P9, Canada
AUSTRALIA	Pergamon Press (Aust) Pty. Ltd., P. O. Box 544, Potts Point, NSW 2011, Australia
FRANCE	Pergamon Press SARL, 24 rue des Ecoles, 75240 Paris, Cedex 05, France
FEDERAL REPUBLIC OF GERMANY	Pergamon Press GmbH, 6242 Kronberg/Taunus, Pferdstrasse 1, Federal Republic of Germany

---

First edition 1978

Library of Congress Cataloging in Publication Data

Lunar and Planetary Science Conference, 9th,
Houston, Tex., 1978.
Proceedings of the Ninth Lunar and Planetary
Science Conference, Houston, Texas, March 13-17,
1978.

(Geochimica et cosmochimica acta: Supplement;
10)
'A companion volume, Mare Crisium: the view from
Luna 24 . . . could be considered a fourth volume of
these Proceedings.'
CONTENTS: v. 1. Petrogenetic studies: the
Moon and meteorites.—v.2. Lunar and planetary
surfaces.—v. 3. The Moon and the inner solar
system.
1. Moon-Congresses. 2. Planets—Congresses.
3. Meteorites—Congresses. I. Lunar and Planetary
Institute. II. Series.
QB580.L83 1978 559.9'l 78-24332
ISBN 0-08-022966-2
Printed in the United States of America

TECHNICAL EDITOR

EDITORIAL STAFF

Contents
(Volume 2)

Regolith Studies

Soil Maturation and Agglutinates

RADIOMETRIC, COSMIC-RAY, AND TRACK CHRONOLOGIES FOR METEORITES AND LUNAR SAMPLES

GEOCHIMICA ET COSMOCHIMICA ACTA

SUPPLEMENT 10

PROCEEDINGS OF THE NINTH LUNAR AND PLANETARY SCIENCE CONFERENCE

Houston, Texas, March 13–17, 1978

Proc. Lunar Planet. Sci. Conf. 9th (1978), p. 1449–1458.
Printed in the United States of America

The floor of crater Le Monier: A study of Lunokhod 2 data

C. P. FLORENSKY, A. T. BASILEVSKY, N. N. BOBINA, G. A. BURBA,
N. N. GREBENNIK, R. O. KUZMIN, B. P. POLOSUKHIN, V. D. POPOVICH,
and A. A. PRONIN

V. I. Vernadsky Institute of Geochemistry and Analytical Chemistry
U.S.S.R. Academy of Sciences, Moscow, U.S.S.R.

L. B. RONCA

Department of Geology, Wayne State University
Detroit, Michigan 48202, U.S.A.

Abstract—The first part of the report consists of a brief summary of the observations obtained from the data of Lunokhod 2, previously published mainly in Russian. The second part presents new observations and interpretations.

The remotely controlled vehicle travelled extensively on the lava-flooded floor of crater Le Monier on the eastern side of Mare Serenitatis. The floor of Le Monier appears to be a typical mare surface, practically horizontal, with gentle undulations, and peppered with craters ranging from the limit of resolution (a few centimeters) to several hundred meters. Age of the flooding is estimated to be Late Imbrian or Early Eratosthenian. Boulders are relatively common, especially concentrated in and around craters. The edges of a linear rille, unofficially called Fossa Recta, show the presence of elongated boulder fields, interpreted as being not the usual erratic boulders found on the mare surface, but bedrock protuberances.

New observations and interpretations are presented concerning the nature of the uppermost regolith. The depth and characteristics of the vehicle wheel tracks indicate the presence on the uppermost regolith of a layer of material of considerably less bearing and traction strength than the lower regolith. This uppermost layer ranges from a few to 20 centimeters and is generally deeper at the foot of slopes. It seems to be responsible for terraces and furrows on inner crater slopes. The data for the X-ray fluorescence spectroscope, i.e., a systematic but gradual compositional change as a function of distance from the mare shore, can be included in these observations to suggest that considerable lateral movement of regolithic material may occur. Some of this movement may be due to micrometeorites, but some seems to be caused by soil creep, occurring on slopes having angles less than the angle of repose, and therefore of problematic origin.

INTRODUCTION

On January 18, 1973, the automatic spacecraft Luna 21 deposited Lunokhod 2 on the eastern side of Mare Serenitatis. The mission and results have been discussed in Russian by several papers (for example, Florensky *et al.*, 1976) and some observations also in English (Basilevsky *et al.*, 1977).

The first part of this report will present a brief summary of the mission and its results, most of which is unavailable to non-Russian readers, and the second part will discuss new observations and interpretations.

THE LUNOKHOD 2 MISSION

Luna 21 landed at coordinates 25°51′N, 30°27′E, in the southern part of the lava-flooded floor of crater Le Monier. The travels of Lunokhod 2 are schematically shown in Fig. 1. The construction and operational methods were basically similar to those of Lunokhod 1 (Academy of Sciences, U.S.S.R., 1971). The vehicle was equipped with a wide-angle panoramic camera, a "piloting" camera, tilmeter and other instruments.

The interpretation of the data collected by Lunokhod 2 was helped by analysing orbital photographs of the area, kindly supplied by the United States Geological Survey and the National Aeronautics and Space Administration.

The crater Le Monier has a diameter of 55 km and is a clear example of a polygenetic structure on the lunar surface. The rim of the crater, relatively poorly preserved, is much older than the lava covering the floor. This classifies the crater as belonging to Class IV of Ronca and Green's (1970) classification. Wilhelms

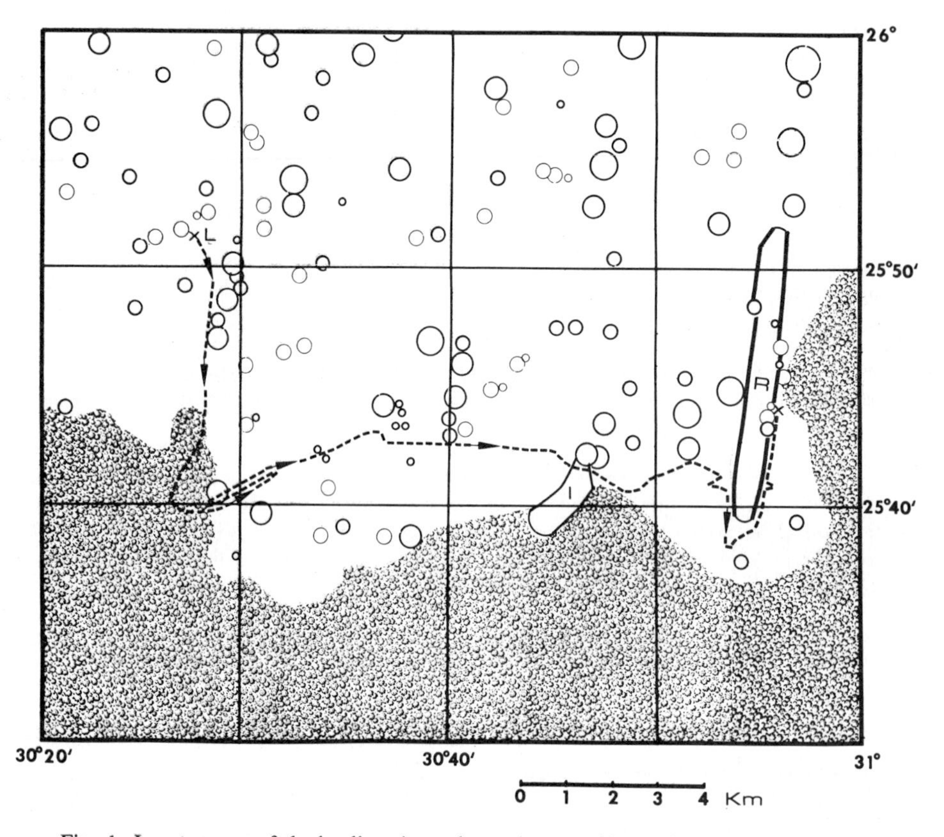

Fig. 1. Locator map of the landing site and travel route of Lunokhod 2. The patterned area represents highland terrains, white areas mare terrains. The elongated patterns marked with I and R represent, respectively, Fossa Incospicua and Fossa Recta (from Basilevsky *et al.*, 1977).

and McCauley (1971) considered the crater to be Early Imbrian in age, while the lava floor is Late Imbrian or Early Eratosthenian. It should be noted that the absolute age of mare basalts at Littrow-Taurus Valley is 3.7 to 3.8 × 10^9 years (Evensen *et al.*, 1973; Huneke *et al.*, 1973; Kirsten *et al.*, 1973), which corresponds to the first half of the Imbrian Period.

During the first lunar day the vehicle travelled southward toward the nearest highland point visible from the landing site. The second and third lunar days were spent investigating the highland immediately south of the flooded area. Then Lunokhod 2 returned to the mare surface and proceeded in an easterly direction toward a small mare bay, unofficially named Circle Harbour. On its way, it crossed a gentle linear depression, unofficially named Fossa Incospicua. Within Circle Harbour a more prominent linear feature unofficially named Fossa Recta was visited. The fourth and fifth lunar days were spent investigating the southern part of Fossa Recta.

The floor of Le Monier appears to be a typical mare surface, practically horizontal, with gentle undulations and peppered with small craters (Fig. 2). West of the area travelled by Lunokhod 2 a wrinkle ridge is located.

The craters range in diameter from a few centimeters (the limit of resolution) to several hundred meters. Craters smaller than about 50 meters in diameter are in steady state (D_s = 50 m) as defined by Shoemaker and Morris (1968). Most of the large craters belong to morphological class B, BC or C, as defined by

Fig. 2. Typical view of the floor of Crater Le Monier. The artifacts are the wheel tracks.

Fig. 3. Typical distribution of boulders.

Fig. 4. Typical boulders. Shapes range from irregular to prismatic with angular to rounded corners. Surface texture can be pitted or pristine.

Florensky *et al.* (1972). The most degraded craters (class C) are smaller than about 450 meters in diameter. Using Basilevsky's method (1974) the morphologies and sizes indicate that most craters are Copernican and Late and Middle Eratosthenian in age. This, together with the low D_s, suggests that the floor of Le Monier is early Eratosthenian.

In the scenery seen by Lunokhod 2, essentially a ground-level human-scale viewpiont, craters from a few centimeters to 20 meters in diameter are the easiest to see. The most common of these small craters are bowl-shaped, with gentle rims. Fresh-looking craters (classes A, AB and B), with steeper slopes and sharper rims, account for approximately 15 to 30 percent of all small craters. Occasionally, craters are found with slopes and bottoms covered with fragments of diameter 3 to 10 cm. They account for approximately 0.25% of all small craters. These craters are small, diameter between 0.5 and 3 meters, and shallow, with depth to diameter ratio about 1/10. They have been previously interpreted as low velocity impact craters (Florensky *et al.*, 1971).

The distribution of craters in the range between 5 cm and 10 meters was determined using stereopairs of panoramic pictures. It was found to follow $N_o = 10^{10.9} \cdot D^{-2}/10^6 km^2$, which is typical for mare surfaces (Shoemaker and Morris, 1968).

Boulders on mare surfaces are commonly related with the ejecta from craters or occur inside craters. As a rule, they are present only when related with craters of diameter larger than 10 meters. When boulders accompany a crater on a horizontal surface, they are symmetrically located around the crater. If a crater is on a slope, fewer boulders occur on the up-slope side. These characteristics leave little doubt that the great majority of boulders on a mare, if not all, are products of cratering impacts.

The boulders can be classified in a number of species on the basis of their form, texture and degree of destruction (Florensky *et al.*, 1971). Irregular and prismatic forms predominate (Figs. 3 and 4). From the degree of destruction, boulders can be classified as angular, angular-rounded and rounded. Surface texture ranges from highly pitted to pristine. It appears that boulders within craters are less destroyed and less pitted than boulders between craters. Also, boulders within craters tend to be partially buried and somewhat powdered by fine soil, while boulders between craters are not only more exposed, but often even are placed on a pedestal-like structure.

The approximate average density of the boulders on the mare surface is as follows: per 100 m^2, no boulder larger than 1 m, 3 larger than 20 cm, and 10 larger than 10 cm. It should be noted, however, that the route of Lunokhod 2 was at a distance from any large fresh crater and, therefore, this boulder distribution may be lower than a typical mare distribution.

The appearance of the surface in the vicinity of Fossa Recta drastically changes. A boulder field follows closely the edge of the fossa (Figs. 5 and 6). A previous publication in English (Basilevsky *et al.*, 1977) presents evidence that suggests that these boulders are not the usual "erratic" boulders found on a normal mare surface, but are bedrock protuberances.

Fig. 5. The edge of Fossa Recta. The boulder field follows the sudden increase in slope that can be considered to be the edge of the fossa.

Fig. 6. Another view of the boulder field at the edge of Fossa Recta.

Fig. 7. Typical appearance of the wheel tracks.

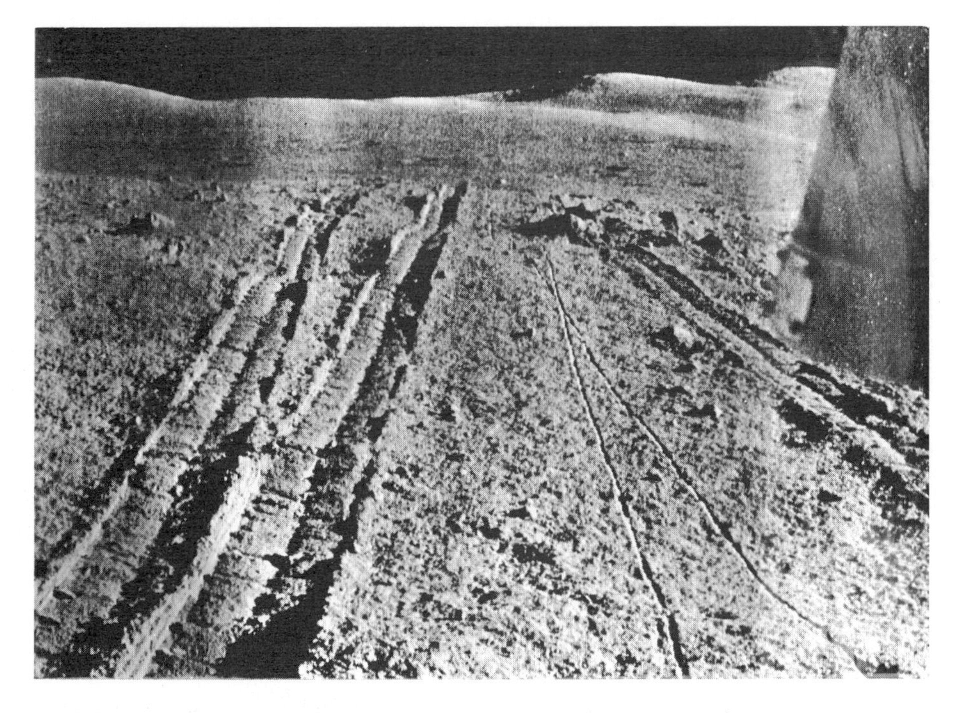

Fig. 8. Notice how the wheel tracks become deeper nearer the crater.

THE REGOLITH

The data collected by Lunokhod 2 present new information about the characteristics of the uppermost portion of the regolith.

Chemical composition of the surface of the regolith is revealed by the X-ray fluorescence spectroscope (Kocharov and Viktorov, 1974). Suprisingly, the composition is more similar to a highland regolith than to a mare regolith. At the landing point (6 km from the highland shores), the Al content is 9 ± 1%, (as compared to 7% for Mare Imbrium) and the Fe content is 6 ± 0.6% (as compared to 12% for Mare Imbrium). This must not be interpreted as being caused by the lava flows in Le Monier being of different composition than the flows of other maria, but by having a thin supply of highland material mixed in the upper layer of the mare regolith. This interpretation is suggested by the variation of the chemistry of the upper layer of the regolith incurred as the vehicle travelled. The nearer to the highland shores Lunokhod 2 arrived, the more "highlandic" the composition of the regolith. At a distance of 1.5 km, the Fe content was 4.9 ± 0.4%. When the travel was essentially parallel to the shores, the composition remained constant.

The above measurements are a strong indication that considerable lateral movement occurs in the regolith. Similar conclusions are suggested by looking at the wheel tracks of the vehicle. Figure 7 shows the most common appearance of the tracks. The regolith is relatively firm, ruts are a few centimeters deep and a few large fragments or lumps are visible. Figure 8 shows the wheel tracks

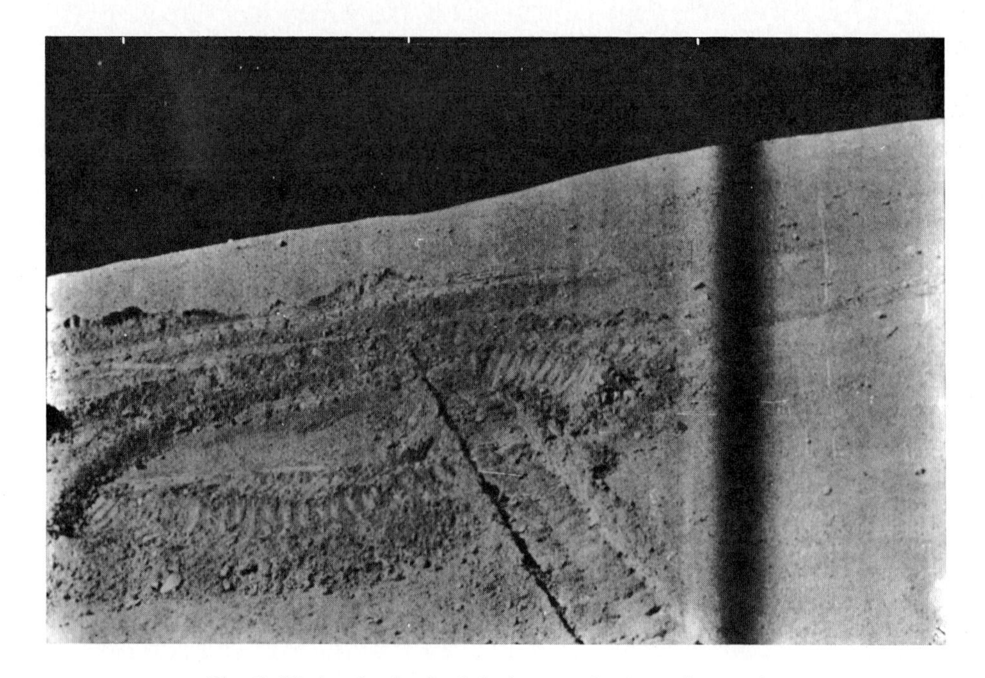

Fig. 9. Notice the depth of the low-traction layer (see text).

appearance when they near a crater. They become deeper, about 20 cm or more, and fragments or lumps are more common. This becomes even more so inside a crater, especially at the foot of the slope. Figures 9 and 10 show locations where Lunokhod 2 encountered difficulties in proceeding, because of the softness of the uppermost portion of the regolith.

Before these observations are analysed, other aspects of the regolith surface must be discussed. On areas relatively removed from craters, a vague pattern is apparent, consisting of slightly elevated zones making a rough cellular network. This pattern is very difficult to see in any of the photographs, but it was more obvious to one of the authors (A. T. Basilevsky) when the vehicle was transmitting "live" through the "piloting" camera. On areas within craters, both radial and concentric furrows or terraces are present.

The "highlandic" composition of the uppermost layer, the changes in firmness as a function of topography, the terraces and furrows, can be interpreted as being indicators of considerable lateral movement of the upper layer of the regolith. Some of this movement may be caused by the random ejections from the whole

Fig. 10. A location at the foot of a slope, where the low-traction layer is deeper than average.

mass spectrum of impacts, but a soil creep component may also be present.

Several oral presentations at the conference discussed the difficulty in making sure that a regolith core is actually complete to its real top. If such a poorly compacted uppermost layer is ubiquitous, it seems reasonable to expect difficulties in coring it.

One possible way of determining the amount and extension of such layers may be by studying infrared characteristics of the lunar surface. Mendell and Low (1975) and Mendell (1976) have shown the presence of "cold spots" which may be areas where soil creep or other processes have accumulated such uncompacted material (Mendell, comments during presentation).

Acknowledgments—Part of this research was performed while L. B. Ronca was at the Vernadsky Institute as a participant of an Academy of Sciences, U.S.A.-Academy of Sciences, U.S.S.R. exchange program.

REFERENCES

Academy of Sciences U.S.S.R. (1971) *A lunar travelling laboratory: Lunokhod 1.* Nauka, Moscow. 128 pp. (In Russian.)

Basilevsky A. T. (1974) Ages of small lunar craters. *ISVESTIA Acad. Nauk. SSR, Ser. Geol.* **8**, 139–142. (In Russian.)

Basilevsky A. T., Florensky C. P., and Ronca L. B. (1977) A possible lunar outcrop: A study of Lunokhod 2 data. *The Moon* **17**, 19–28.

Evensen N. M., Murthy V. R., and Coscio M. R. Jr. (1973) Rb-Sr ages of some mare basalts and isotopic and trace elements systematics in lunar fines. *Proc. Lunar Sci. Conf. 4th*, p. 1707–1724.

Florensky K. P., Basilevksy A. T., Bobina N. N., Burba G. A., Grebennic N. N., Kuzmin R. O., Polosuchin V. P., Popovich V. D., and Pronin A. A. (1976) Processes of reorganization on the surface of the moon in the region of Le Monier as indicated by a detailed study of Lunokhod 2 data. International Geological Congress, XXI Session, Planetology (Vinogradov chief editor), p. 205–234. Academy of Sciences, U.S.S.R. and Ministry of Geology, U.S.S.R. (In Russian.)

Florensky K. P., Basilevsky A. T., Gurshtein A. A., Zezin R. B., Pronin A. A., Polushin V. P., Popova Z. V., and Taborko I. M. (1972) On the problem of the structure of the surface of lunar maria. In *Modern Conceptions about the Moon*, p. 21–45. Nauka, Moscow. In Russian.

Florensky K. P., Basilevsky A. T., Pronin A. A., and Popova Z. V. (1971) Preliminary results of the geomorphological study of panoramas. In *A Lunar travelling laboratory: Lunokhod 1*, p. 96–115. Nauka, Moscow. (In Russian.)

Huneke J. C., Jessberger E. K., Podosek F. A., and Wasserberg G. J. (1973) $^{40}Ar/^{39}Ar$ measurements in Apollo 16 and 17 samples and the chronology of metamorphic and volcanic activity in the Taurus-Littrow region. *Proc. Lunar Sci. Conf. 4th*, p. 1725–1756.

Kirsten T., Horn P., and Heymann D. (1973) Chronology of the Taurus-Littrow region. *Earth Planet. Sci. Lett.* **20**, 55–76.

Kocharov G. E. and Victorov S. V. (1974) Chemical composition of the lunar surface in the area investigated by Lunokhod 2. *Dokl. Acad. Nauk. SSSR* **214**, 72–74.

Mendell W. W. (1976) The Apollo 17 infra-red scanning radiometer. Ph.D. Thesis, Rice University, Houston, Texas. 183 pp.

Mendell W. W. and Low F. J. (1975) Infra-red orbital mapping of lunar features. *Proc. Lunar Sci. Conf. 6th*, p. 2711–2719.

Ronca L. B. and Green R. R. (1970) Statistical geomorphology of the lunar surface. *Bull. Geol. Soc. Amer.* **81**, 337–352.

Shoemaker E. M. and Morris E. C. (1968) Craters. In *Surveyor Project Final Report, Pt. II.* J. P. L. Tech. Rep. N32–1265. Pasadena, California.

Wilhelms D. E. and McCauley J. F. (1971) Geologic map of the near side of the Moon. In *Geological Atlas of the Moon*, U. S. Geol. Surv. Map I-703.

Proc. Lunar Planet. Sci. Conf. 9th (1978), p. 1459–1471.
Printed in the United States of America

Recognition of lunar glass droplets produced directly from endogenous liquids: The evidence from S-ZN coatings.

PATRICK BUTLER, JR.,

SN2/NASA Johnson Space Center,
Houston, Texas 77058

Abstract—Concentrated deposits of ultramafic droplets of green glass at Apollo 15 and orange and black glass at Apollo 17 are generally accepted as having volcanic, and so endogenous, origins. Since these droplets have S + Zn surface coatings with characteristics that fit a volcanic origin, the presence of similar coatings on dispersed droplets are taken as a good criterion of their volcanic origin. This criterion in concert with other criteria (homogeneity, absence of shocked relicts) that distinguish between volcanic and impact origins, has led to identification of a new ultramafic endogenous liquid that was erupted at Apollo 15. The brown glass droplets representing this liquid have a higher TiO_2 content and are more mafic than any of the Apollo 15 basalts.

INTRODUCTION

Homogeneous glass droplets are present in every soil sample returned from the moon. Considerable interest attaches to these droplets because either they may represent endogenous igneous rocks, remelted and sprayed into droplets by meteorite impact, or they may be direct samples of endogenous melts sprayed from a lava fountain or splashed by meteorite impact from a lava lake. The use of the compositions of droplets produced by impacts on solids as representative of the compositions of primary rocks entails the further complication of establishing that the melt was not a mixture of rocks and/or soil (Reid *et al.*, 1972a). Wood (1975) showed that the cluster of Apollo 11 glass compositions most like the basalts resembles the bulk Apollo 11 soil more closely than the basalts. On the other hand, if a droplet can be established as having been produced from a liquid lava, its composition can be used directly as representative of material generated within the moon.

Figure 1 shows the compositional ranges defined by the basalts of the four mare landing sites. It also shows the average compositions for diffuse clusters of glasses with mare basalt compositions (high FeO and low Al_2O_3) at each landing site. For the individual landing sites, few of these averages lie within the ranges defined by the rock compositions (Reid *et al.*, 1972a and b). The ultramafic glasses, those with FeO +MgO >30% by weight, form much tighter compositional clusters at each landing site; the ranges defined by their compositions are shown in Fig. 2. However, there are no returned rock equivalents to these clusters either. (Irrespective of the origin of the glass droplets, by impact melting or from a lava, the absence of crystalline equivalents is a puzzle.)

Strong arguments have been made that droplets that have been found as rich

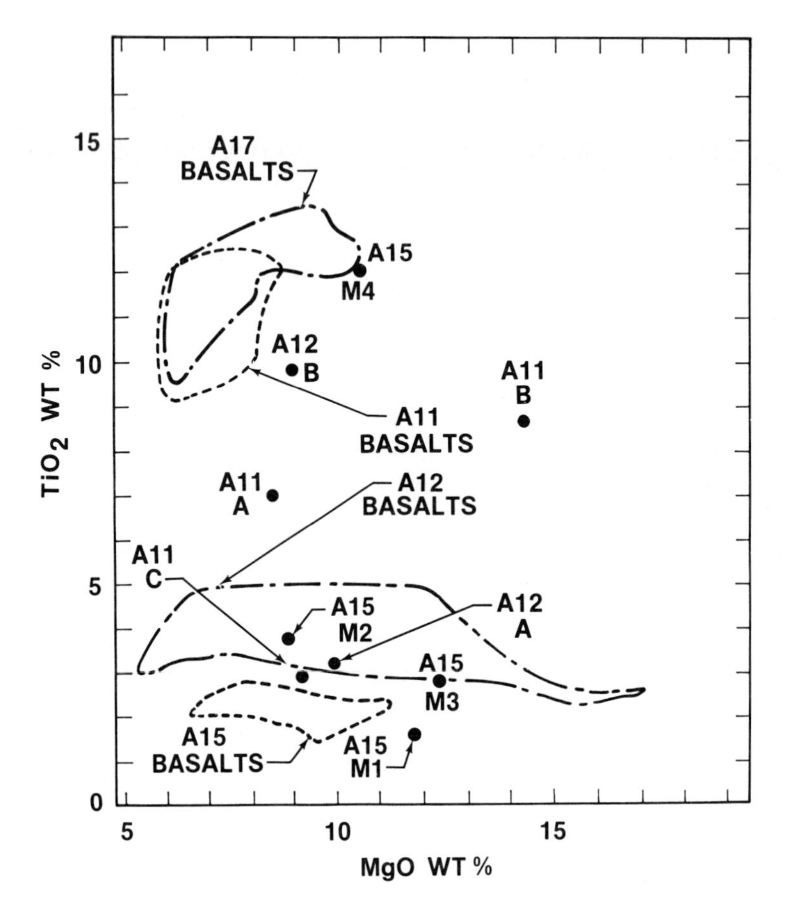

Fig. 1. TiO_2-MgO plot of the ranges in compositions defined by the mare basalt samples for each mission, and of the average compositions of clusters of mare glass droplets (designated by mission A11, A12, etc.—and by A, B, or M1, M2, etc., which are easily relatable to the designations of Reid *et al.* (1972a and 1972b) from whence these averages came).

concentrations (80 to 100% of droplets and fragments of them) of green glass at Apollo 15 (lower right corner of Fig. 2) and of orange glass at Apollo 17 were produced in lava fountains from endogenous liquids. If these arguments are accepted, are the other glass droplets in tight compositional clusters also the direct products of endogenous liquids? Further, might not some of the glass droplets in diffuse compositional clusters, or outside of clusters altogether, also be samples of endogenous melts? It is important to answer these questions because droplets that can be established as endogenous give critical information on the moon's mantle (Ridley *et al.*, 1973; Green and Ringwood 1973; Rhodes 1978; Ringwood *et al.*, 1978).

Some of the droplets are enriched in volatiles, such as Zn and Pb, and also in siderophiles (Ni, Au, Ir), two geochemical groups depleted in most lunar samples

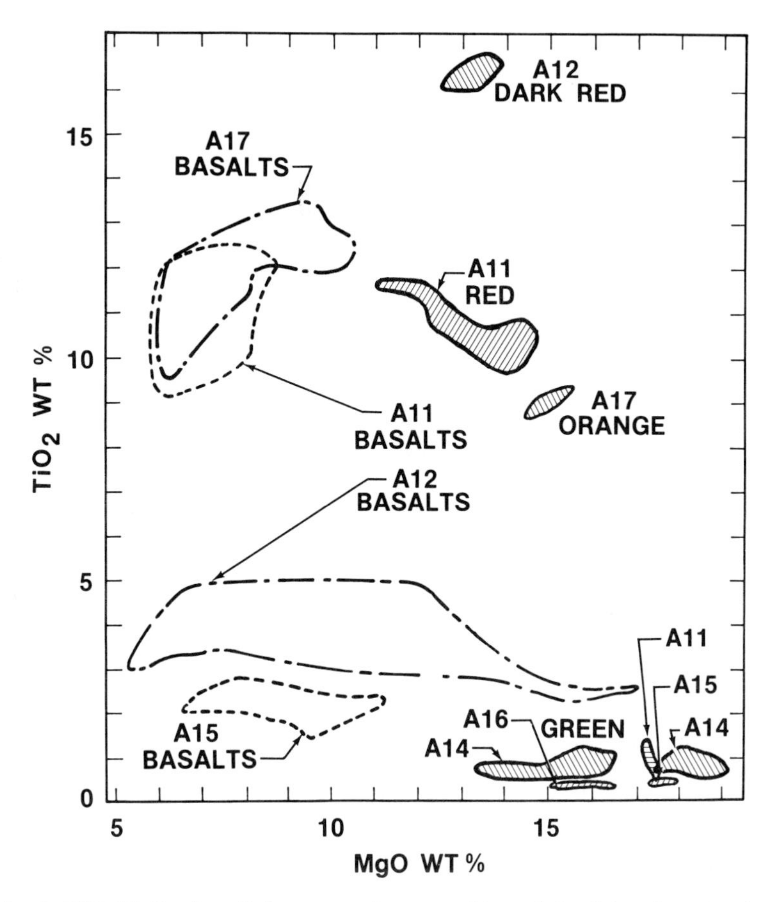

Fig. 2. TiO_2-MgO plot of the ranges in compositions defined by the mare basalt samples of each mission, and of the ranges in composition defined by ultramafic glasses (cross-hatched and designated by mission A11, A12, etc.—and by color). Ultramafic glass compositions are from: Bunch *et al.* (1972), Delano (1975) and J. W. Delano written communication (1977), Marvin (1971), Wood *et al.* (1971), Prinz *et al.* (1971), Ridley *et al.* (1973), Brown *et al.* (1971).

that are uncontaminated by meteorites (Wasson *et al.*, 1976). If the droplets are endogenous, the moon may not be as depleted in elements of these groups as would be concluded from other samples. The present report develops a logical method for distinguishing endogenous glass droplets from those produced by impact on solids. Preliminary use of this method has already allowed the identification of one endogenous liquid composition.

Considerable evidence has been marshalled in support of an origin for the concentrated deposits of droplets at Apollo 15 and Apollo 17 in a lava fountain (McKay *et al.*, 1973; Heiken *et al.*, 1974; Chou *et al.*, 1975; Meyer *et al.*, 1975). There are many lines of this evidence, much of which derives from analyses of bulk concentrates of the droplets for trace chemistry and age. For the present study, however, it is necessary to focus on evidence detectable in individual droplets, so that the criteria developed can be applied to single droplets occurring by themselves. Such evidence for formation from an endogenous melt would be homogeneity, the presence of euhedral phenocrysts, which must have formed in the melts before dispersal into droplets (Roedder and Weiblen, 1973; Prinz *et al.*, 1973; McKay and Heiken, 1973), and the absence of the features generally accepted to have been produced by impact such as schlieren, shock-damaged relicts, and Fe metal dust (Roedder and Weiblen, 1973). Since all of the droplets of the same composition in a deposit or at a landing site most likely had the same origin, each one of the droplets must be homogeneous and must be free of impact features in order that formation of any droplets of the compositional cluster directly from an endogenous liquid remain a viable hypothesis. Conversely, discovery of one droplet with a euhedral olivine phenocryst is very strong evidence that all droplets of the same composition formed directly from an endogenous melt.

None of the many Apollo 15 green droplets or of the Apollo 17 orange droplets studied has been reported to be inhomogeneous or to contain impact features. In addition, euhedral phenocrysts have been observed in 0.1% to 0.2% of the Apollo 17 orange droplets (Roedder and Weiblen, 1973; Prinz *et al.*, 1973), and several phenocrysts have been observed in Apollo 15 green droplets (Agrell *et al.*, 1973; D. S. McKay, pers. comm., 1978). All of this evidence indicates that the droplets of these two deposits formed directly from endogenous melts, either from a lava fountain or from a lava lake splashed by meteorites (Roedder and Weiblen, 1973).

Basis for use of the coatings as evidence for volcanic origin

The logic underlying the use of an S + Zn coating on an isolated glass droplet as evidence for its volcanic origin follows these steps:

1. As is generally accepted (but not universally, e.g., Wood and Ryder, 1977; Rhodes, 1978), the orange and green glass droplets in highly concentrated deposits are of volcanic origin.
2. These droplets in the highly concentrated deposits have coatings of S, Z, and other volatiles.
3. The coatings are cogenetic with the droplets (this point is examined further below) and are therefore also of volcanic origin.
4. Consequently, S + Zn coatings on glass droplets, whether or not the droplets occur as isolated individuals or in concentration, is *prima facie* evidence for their volcanic origin.

Approaches in identification of droplets from endogenous liquids

A number of approaches to the problem of identification of droplets from endogenous liquids are possible, if the criteria discussed above are accepted

(S + Zn coatings and euhedral olivine phenocrysts favor the endogenous origin; schlieren, shocked inclusions, Ni-Fe spherules are evidence against it). The approach used in the present study is:

1. Sieve soil sample to remove material too small to handle (<90 microns).
2. Pick out and mount all droplets. (Coatings are chiefly confined to the primary surfaces of droplets, as discussed below.)
3. Survey all droplets for S and Zn coatings by electron microprobe.
4. Study the surface characteristics of coated droplets by scanning electron microscope. Make energy dispersive X-ray analyses of surface features and look for correlations between elements in the coatings.
5. Make thin sections of coated droplets for petrography and quantitative electron microprobe analyses of their interiors.

An alternative approach would be to look for droplets with olivine phenocrysts in thin sections of grain mounts.

A third approach would be to look for compositional clusters of glass droplets. This approach has already been extensively used by many researchers (Essene *et al.*, 1970; Meyer *et al.*, 1971; Reid *et al.*, 1972a, 1972b; Wood, 1975), although generally not involving especially tight clusters, and never for the purpose of identifying endogenous liquids. What is needed now is to establish which tight clusters represent endogenous liquids by searching for a member-droplet of each with an S + Zn coating or an olivine phenocryst. On the other hand, a member-droplet showing one of the characteristics of an impact origin would suggest that all droplets of the cluster originated by impact onto similar solid material.

The pair of elements, S and Zn, were chosen as indicators of coatings on the basis of experience in studying the coatings on the Apollo 15 green droplets and the Apollo 17 orange and black droplets (Butler and Meyer, 1976). S is the prevailing element in coatings in which other elements were detected; furthermore, S correlates with most of these elements (Butler and Meyer, 1976). S is readily detected in electron microprobe X-ray analysis (EMX). FeS grains on the surface of some soil grains, unrelated to coatings, makes it necessary to use Zn, the second most abundant element of coatings, as an additional criterion for the presence of a coating. There are many other elements in these coatings, but they are less abundant and harder to detect, i.e., Pb.

Evidence that the S + Zn coatings are cogenetic with the droplets

Three lines of evidence emerge from the characteristics of the concentrated deposits of droplets and from the individual droplets themselves. *1.* The coatings are generally confined to the primary surfaces of the droplets, (Fig. 3 and also Fig. 3F of Butler and Meyer, 1976). Since these concentrations probably would have been diluted through mixing with other fines, the highly concentrated deposits of glass droplets could not have undergone much reworking to produce

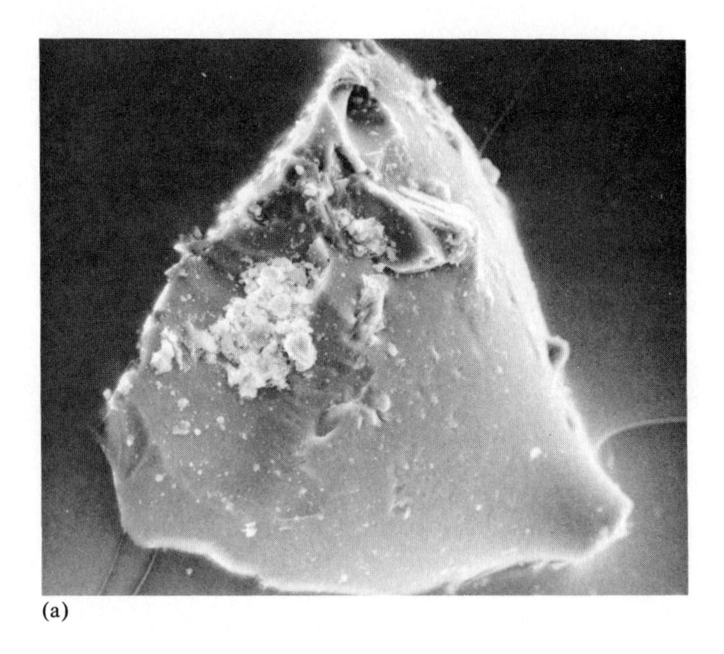

(a)

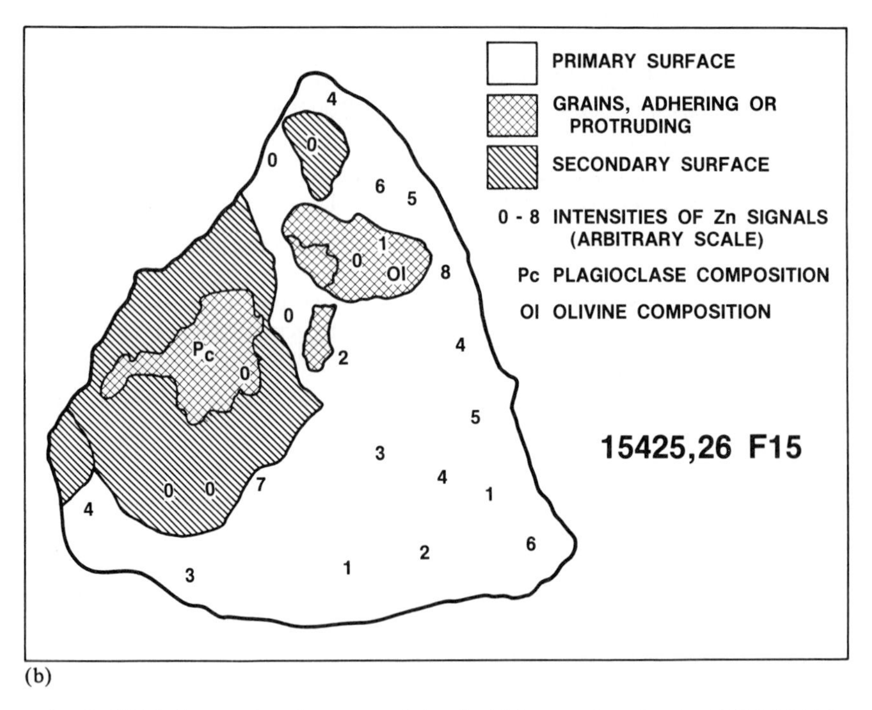

(b)

Fig. 3. SEM photo and surface map of Apollo 15 brown glass droplet 15425,26 F15. Map shows relative intensities of the Zn X-ray signals.

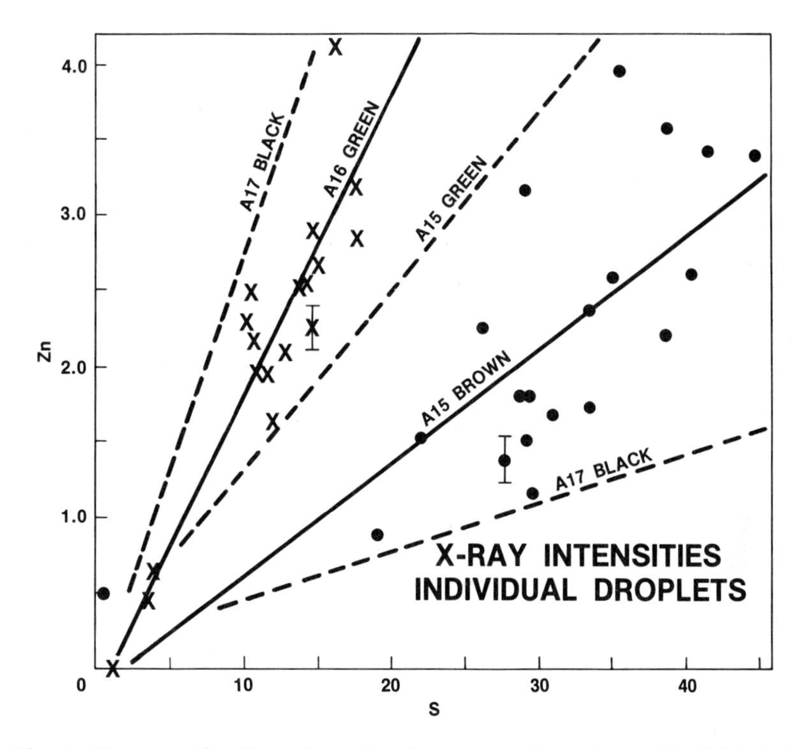

Fig. 4. Zinc vs. sulfur X-ray intensities for spots on the surfaces of individual glass droplets. The solid line "A16 GREEN" is a least squares fit to the X's, which apply to droplet 65701,21 B155. The solid line "A15 BROWN" is a least squares fit to the dots, which apply to droplet 15425,26 F15. Error bars show standard counting errors for Zn; the symbols are about the size of standard counting errors for S. The dashed lines "A17 BLACK" apply to the data for two black droplets from 74001 (Butler and Meyer, 1976, Fig. 6); they show the different S to Zn proportions that can occur between droplets from the same deposit.

breakage and so secondary surfaces after the original deposition. Therefore, most of the breaking and spalling to produce secondary surfaces must have occurred through mutual collisions while the droplets were still inflight and when they landed. McKay *et al.* (1973) found that most spalls on green glass droplets were from low velocity collisions, and that in many cases the projectile, also of green glass, apparently was still molten at the time of collision and either splashed over or stuck to the surface of the spalled droplet. The coatings must have been applied shortly after the droplets formed, while they were in flight and before most of the spalling occurred. *2.* The amount of coating varies greatly from droplet to droplet in the same deposit (see the droplet histograms for S and Zn, Fig. 4 in Butler and Meyer, 1976). *3.* The proportions of Zn to S are constant on the surfaces of individual droplets, but these proportions differ from droplet to droplet, (Fig. 4 and also Fig. 6 in Butler and Meyer, 1976). These observations are easily explained by a lava fountain hypothesis. The individual droplets in the same deposit would have had individual trajectories in the lava fountain and thus individual histories of cooling and exposure to gas of varying composition before their deposition together. Had the coating taken place after deposition, by volatiles moving through a permeable regolith, the coatings also would have covered the secondary surfaces, and the distribution and composition of the coatings probably would have been more uniform.

Mobility of some of the elements in the coatings has been established. For example, Cirlin and Housley (1978) showed that much of Pb in coatings is present as $PbCl_2$, and they found very similar surface Pb contents in 74220, in the upper part of core 74001, both composed of 100% orange plus black glass, and in 74241 with only 8% orange and black glass. The implication is strong that $PbCl_2$ migrated uniformly to the glass droplets and to the non-glass grains of 74241 after the deposits attained their present locations with respect to the exposed surface. Consistent with this implication of chloride mobility, Butler and Meyer (1976) reported much of the Cl detected is on secondary surfaces of glass from core 74001, unlike Zn and S which were found to be mainly confined to primary surfaces on these droplets.

Mobility of Zn from the orange and black soils into the adjacent gray soils 74240 and 74260 is suggested by the higher bulk concentrations of Zn that can be accounted for by the amount of orange and black droplets present in them (see Table 2 in Butler and Meyer, 1976). Consequently, an electron microprobe survey was made of other components of these soils to determine whether or not any had gotten coated. None of twenty ropy glass grains from 74241, or of 40 plagioclase grains and 50 grains of all kinds from 74260 showed a detectable S + Zn coating. Apparently, there has been no detectable migration of these two elements. (The high Zn contents of the gray soils may be due to more richly coated droplets than in the orange soils.)

A search for the location of the 91 ppm Zn reported by Haskin *et al.* (1973) for soil 65701 (most Apollo 16 soils have about 30 ppm Zn) led to an EMX survey of the surface of 1900 grains. A green glass droplet was the only one with a detectable S + Zn coating. The high bulk Zn concentration is obviously not from a few coated grains like this droplet. (Nevertheless, Zn may be on many grain surfaces in concentrations too low for detection by the EMX method employed.)

Preservation of coatings

The soils with high concentrations of these unusual glass droplets are among the least mature returned from the moon. In contrast to most lunar soils which show maturity index values (characteristic ferromagnetic resonance normalized to total iron content, I_s/FeO) in the range 30 to 100, the Apollo 15 green glass and the Apollo 17 orange glass soils have maturity index values of 1 or less (Morris, 1976). Since the maturation process in a soil must involve considerable gardening (or successive deposition of thin layers with long exposure times between each), it is not surprising that these soils, with their high concentration of original deposits of glass droplets, have not suffered much mixing. Maturation involves alterations of grain surfaces by solar wind, by cosmic rays, by micrometeorite impacts, and by accretion of debris from impacts (Morris, 1978). Since coatings on grain surfaces would be affected by these processes, the most complete preservation of primary surface coatings would be expected for the least mature soils. Indeed, coated droplets are abundant in the immature green and orange glass soils. On the other hand, there is a handicap in looking for surface coatings on droplets dispersed in mature soils because the coating once present on the droplets may have been degraded or destroyed. Even so, some few particles may be expected to have missed being exposed.

Does the volatile coating evidence discriminate among origins for glass droplets?

The presence of the coatings and their characteristics seem to strongly favor a fire fountain origin, first, because the volatiles that are essential elements of the fire fountain would be expected to contain a condensable portion. Furthermore, the variations in coating characteristics from droplet to droplet in the same deposit is reasonable in terms of the different trajectories followed by the droplets subsequently deposited together.

Conceivably, an impact on rock or soil, or an impact on a molten lava, could generate a transient volatile atmosphere that would coat the droplets. The average lunar rock has little Zn, Cd, Cl, ^{204}Pb, etc.; however, either a target material with a higher concentration of these elements than those found for any lunar sample, or a projectile such as a comet, rich in volatiles, could conceivably produce these elements (El Goresy *et al.*, 1973). The finding of coated droplets with impact characteristics would suggest either that there had been a transient atmosphere produced by the impact, or that the coating was secondary and not cogenetic with the droplet. There does not seem to be any evidence that would bear on whether or not the droplets were splashed from molten lava by impact.

The distorted forms of many of the droplets into discs, dumbbells, and tear drops strongly suggested to Roedder and Weiblen (1973) that vigorous rotations would have been necessary to overcome the surface tension on small droplets. They concluded that impact processes would be more likely to produce the high rates of shear necessary to generate rotations than a volcanic process. A survey of the shapes of all droplets, including droplets from mature soils, many of which must have been produced by impact melting of solids, should give an idea of the degree of association of nonspherical droplets with origin by impact.

RESULTS

The large X's in Fig. 5 show the compositions of the coated droplets studied to date. Many examples have been found for the Apollo 15 green glasses and the Apollo 17 orange glasses (Butler and Meyer, 1976); only one X in a central position is used for each group.

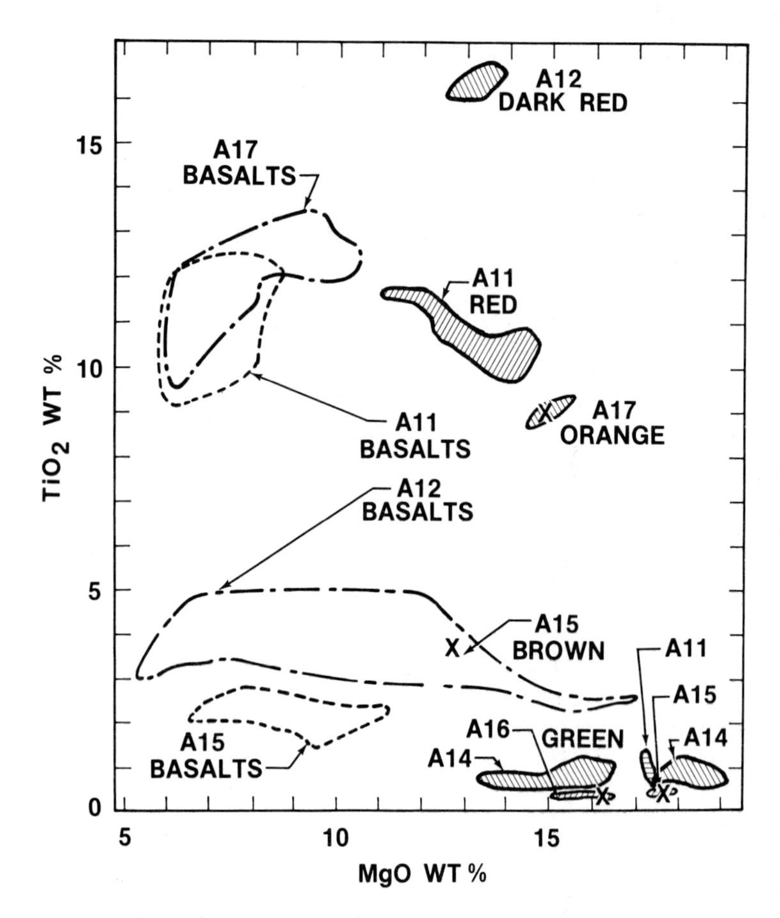

Fig. 5. TiO_2-MgO plot of compositions as in Fig. 2. X's show the compositions of ultramafic glasses that have S + Zn coatings. The "A15 BROWN" represents four droplets of virtually identical composition, "A16 GREEN" represents one droplet. The X's in the Apollo 17 orange field and in the Apollo 15 green field are generalized compositions representing many coated droplets. (Fig. 2 caption gives the source of the ultramafic glass compositions.)

In Apollo 16 soil 65701, out of about 1900 soil grains of all kinds surveyed for coatings, a green glass droplet was the only grain with detectable S + Zn. This Apollo 16 droplet shows the regular relation between the two elements (Fig 4), typical of the Apollo 15 green glass and the Apollo 17 orange and black glass droplets. The discovery of one green droplet with an S + Zn coating in an Apollo 16 soil adds a degree of validation to the endogenous origin for all members of the compositional cluster (Fig. 5) of these green glasses (Delano, 1975).

Four brown glass droplets with S + Zn coatings were found in green glass clod sample 15425. Their compositions are nearly identical (Table 1) and appear in Fig. 5 as a large X in the Apollo 12 basalt field. This result demonstrates the existence at the Apollo 15 site of an endogenic melt more mafic and richer in Ti

Table 1. Electron Microprobe X-Ray Analyses of Glass Droplets.

	15425,26 Brown Glasses	Apollo 16 Green Glasses*	65701 Green Glass	15426,25 Green Glass	15427 Green Glasses**
SiO_2	42.86(0.10)	43.5 (0.3)	42.8	43.9	45.38(0.32)
TiO_2	3.58(0.04)	0.36(0.05)	0.40	0.45	0.39(0.01)
Al_2O_3	8.48(0.04)	9.5 (0.6)	7.50	6.9	7.34(0.18)
Cr_2O_3	0.63(0.03)	0.47(0.06)	0.47	0.53	0.44(0.01)
FeO	21.97(0.01)	20.8 (0.2)	23.1	20.8	19.44(0.50)
MgO	12.63(0.10)	15.7 (0.6)	16.2	17.3	17.29(0.17)
CaO	8.40(0.03)	9.0 (0.2)	8.53	8.1	8.49(0.10)
Na_2O	0.36(0.01)	0.16	0.15	0.11	0.13(0.01)
K_2O	0.09(0.00)	0.04	0.04	0.04	0.02(0.01)
MnO	0.32(0.02)				
Total	99.32	99.53	99.19	98.13	98.92
Number of droplets	3***	5	1	1	11

Numbers in parentheses are standard deviations.

*Delano (1975). 3 droplets from 60003, 2 droplets from 61295.

**Ridley *et al.* (1973).

***Initial analyses showed 4 droplets to be virtually identical, but all gave low SiO_2 values. Only 3 droplets were reanalyzed.

than any of the Apollo 15 basalts. Since this study has not yet been exhaustive, more droplets with different ultramafic compositions and S + Zn coatings may occur at the Apollo 15 site.

An important aspect of this study is that the hypothesized relationship of glass droplets coated with S + Zn to an origin in a lava fountain is constantly tested. Discovery of a droplet with the same composition as a coated droplet but with any of the characteristics of impact origin would be evidence against the hypothesis.

Concluding Observations

These generalizations have been valid so far:

1. All droplets coated with S + Zn are ultramafic, but not all ultramafic droplets have (detectable) coatings.
2. All droplets that contain euhedral olivine phenocrysts are ultramafic, but not all ultramafic droplets contain phenocrysts.
3. All droplets that have schlieren and other features of impact origin are nonultramafic, but not all nonultramafic droplets have impact features.

Acknowledgments—I thank C. Meyer, J. L. Warner, C. M. Wai, and P. H. Warren for helpful reviews.

REFERENCES

Agrell S. O., Agrell J. E., and Arnold A. R. (1973) Observations on glass from 15425, 15426, 15427 (abstract). In *Lunar Science IV*, p. 12–14. The Lunar Science Institute, Houston.

Brown R. W., Reid A. M., Ridley W. I., Warner J. L., Jakeš P., Butler P., Williams R. J., and Anderson D. H. (1971) Microprobe analyses of glasses and minerals from Apollo 14 sample 14259. NASA Technical Memorandum TMX-58080. Johnson Space Center, Houston.

Bunch T. E., Quaide W., Prinz M., Keil K., and Dowty E. (1972) Lunar ultramafic glasses, chondrules and rocks. *Nature* **239**, 57–59.

Butler P. Jr, and Meyer C. Jr. (1976) Sulfur prevails in coatings on glass droplets: Apollo 15 green and brown glasses and Apollo 17 orange and black (devitrified) glasses. *Proc. Lunar Sci. Conf. 7th*, p. 1561–1581.

Chou C.-L., Boynton W. V., Sundberg L. L., and Wasson J. T. (1975). Volatiles on the surface of Apollo 15 green glass and trace element distributions among Apollo 15 soils. *Proc. Lunar Sci. Conf. 6th*, p. 1701–1727.

Cirlin E. H., and Housley R. M. (1978) Flameless atomic absorption studies of volatile trace metals in Apollo 17 samples (abstract). In *Lunar and Planetary Science IX*, p. 169–171. Lunar and Planetary Institute Houston.

Delano J. W. (1975) Petrology of the Apollo 16 mare component: Mare Nectaris. *Proc. Lunar Sci. Conf. 6th*, p. 15–47.

El Goresy A., Ramdohr P., Pavićević M., Medenbach O., Müller O., and Gentner W. (1973) Zinc, lead, chlorine and FeOOH-bearing assemblages in the Apollo 16 sample 66095: Origin by impact of a comet or a carbonaceous chondrite? *Earth Planet. Sci. Lett.* **18**, 411–419.

Essene E. J., Ringwood A. E., and Ware N. G. (1970) Petrology of the Lunar rocks from Apollo 11 landing site. *Proc. Apollo 11 Lunar Sci. Conf.* p. 385–397.

Green D. H. and Ringwood A. E. (1973) Significance of a primitive lunar basaltic composition present in Apollo 15 soils and breccias. *Earth Planet. Sci. Lett.* **19**, 1–8.

Haskin L. A., Helmke P. A., Blanchard D. P., Jacobs J. W., and Telander K. (1973) Major and trace element abundances in samples from the lunar highlands. *Proc. Lunar Sci. Conf. 4th*, p. 1275–1296.

Heiken G. H., McKay D. S., and Brown R. W. (1974) Lunar deposits of possible pyroclastic origin. *Geochim. Cosmochim. Acta* **38**, 1703–1718.

Marvin U. B. (1971) Two unique lunar microbreccias. EOS: *Trans. Amer. Geophys. Union* **52**, 272.

McKay D. S., Clanton U. S., and Ladle G. (1973) Scanning electron microscope study of Apollo 15 green glass. *Proc. Lunar Sci. Conf. 4th*, p. 225–238.

McKay D. S., and Heiken G. H. (1973) Petrography and scanning electron microscope study of Apollo 17 orange and black glass. EOS: *Trans. Amer. Geophys. Union* **54**, 599–600.

Meyer C. Jr., Brett R., Hubbard N. J., Morrison D. A., McKay D. S., Aiken F. K., Takeda H., and Schonfeld E. (1971) Mineralogy, chemistry, and origin of the KREEP component in soil samples from the Ocean of Storms. *Proc. Lunar Sci. Conf. 2nd*, p. 393–411.

Meyer C. Jr., McKay D. S., Anderson D. H., and Butler P. Jr. (1975) The source of sublimates on the Apollo 15 green and Apollo 17 orange glass samples. *Proc. Lunar Sci. Conf. 6th*, p. 1673–1699.

Morris R. V. (1976) Surface exposure indices of lunar soils: A comparative FMR study. *Proc. Lunar Sci. Conf. 7th*, p. 315–335.

Morris R. V. (1978) The maturity of lunar soils: Concepts and more values of I_s/FeO (abstract). In *Lunar and Planetary Science IX*, p. 760–762. Lunar and Planetary Institute, Houston.

Prinz M., Bunch T. E., and Keil K. (1971) Composition and origin of lithic fragments and glasses in Apollo 11 samples. *Contrib. Mineral. Petrol.* **32**, 211–230.

Prinz M., Dowty E., and Keil K. (1973) A model for the origin of orange and green glasses and the filling of mare basins. EOS: *Trans. Amer. Geophys. Union* **54**, 605–606.

Reid A. M., Warner J., Ridley W. I., Johnson D. A., Harmon R. S., Jakeš P., and Brown R. W. (1972a) The major element compositions of lunar rocks as inferred from glass compositions in the lunar soils. *Proc. Lunar Sci. Conf. 3rd*, p. 363–378.

Reid A. M., Warner J., Ridley W. I., and Brown R. W. (1972b) Major element compositions of glasses in three Apollo 15 soils. *Meteoritics* **7**, 395–415.

Rhodes J. M. (1978) Primary mare basalts and green glass (abstract). In *Lunar and Planetary Science IX*, p. 958–960. Lunar and Planetary Institute, Houston.

Ridley W. I., Reid A. M., Warner J. L., and Brown R. W. (1973) Apollo 15 green glasses. *Phys. Earth Planet. Inter.* **7**, 133–136.

Ringwood A. E., Kesson S. E., and Delano J. W. (1978) Geochemical constraints on the existence of a lunar core (abstract). In *Lunar and Planetary Science IX*, p. 964–966. Lunar and Planetary Institute, Houston.

Roedder E., and Weiblen P. W. (1973) Apollo 17 "orange soil" and meteorite impact on liquid lava. *Nature* **244**, 210–212.

Wasson J. T., Boynton W. V., Kallemeyn G. W., Sundberg L. L., and Wai C. M. (1976). Volatile compounds released during lunar lava fountaining. *Proc. Lunar Sci. Conf. 7th*, p. 1583–1595.

Wood J. A., Marvin U. B., Reid J. B., Taylor G. J., Bower J. F., Powell B. N., and Dickey J. S. (1971) Mineralogy and Petrology of the Apollo 12 lunar samples. Smithsonian Astrophysical Observatory, Spec. Rept. 333.

Wood J. A. (1975) Glass compositions as a clue to unsampled mare basalt lithologies. In *Papers Presented to the Conference on Origins of Mare Basalts and their Implications for Lunar Evolution*, p. 194–198. The Lunar Science Institute, Houston.

Wood J. A., and Ryder G. (1977) The Apollo 15 green clods and the green glass enigma (abstract). In *Lunar Science VIII*, p. 1026–1028. The Lunar Science Institute, Houston.

Proc. Lunar Planet. Sci. Conf. 9th (1978), p. 1473–1484.
Printed in the United States of America

Modeling lunar volcanic eruptions

R. M. HOUSLEY

Rockwell International Science Center, Thousand Oaks, California 91360

Abstract—We use simple physical arguments to show that basaltic volcanoes on different planetary bodies would fountain to the same height if the mole fraction of gas in the magma scaled with the acceleration of gravity. We then suggest that the actual eruption velocities and fountain heights are controlled by the velocities of sound in the two phase gas/liquid flows. These velocities are in turn determined by the gas contents in the magma. Predicted characteristics of Hawaiian volcanoes are in excellent accord with observations.

Assuming that the only gas in lunar volcanoes is the CO which would be produced if the observed Fe metal in lunar basalts resulted from graphite reduction, lunar volcanoes would fountain vigorously, but not as spectacularly, as their terrestrial counterparts. The volatile trace metals, halogens, and sulfur released would be transported over the entire moon by the transient atmosphere. Orange and black glass type pyroclastic materials would be transported in sufficient amounts to produce the observed dark mantle deposits.

INTRODUCTION

Heiken *et al.* (1974) assembled an impressive array of chemical, petrographic, and photogeologic evidence that led them to conclude that the Apollo 17 orange and black glasses and the Apollo 15 green glass are lunar pyroclastics, produced by volcanic fire fountaining similar to that currently observed in Hawaii.

Most subsequent workers have interpreted their chemical (Meyer *et al.*, 1975; Chou *et al.*, 1975; Wasson *et al.*, 1976; Reed *et al.*, 1977) and petrographic (Heiken and McKay, 1977) data in terms of this general picture. However, until recently, two problems, regarded as serious by some workers, have remained.

The one that appeared to be most severe was that the bulk of terrestrial pyroclastic material is pumiceous and bears little resemblance to the lunar glass balls. In addition, the relatively small number of balls found in terrestrial pyroclastic deposits are generally quite vesicular. O'Keefe (1976) pointed out that normal liquids will not support thin-walled or adjacent bubbles and that it is the terrestrial pumice which should be regarded as unusual. Thin-walled bubbles as in foam or pumice can only form if the surface tension increases as the film becomes thinner, so that material is drawn out from regions of bulk liquid as the bubbles grow.

In the terrestrial magma apparently water in solution lowers the surface tension. In the thin wall regions of bubbles, escape of this water increases the surface tension and permits the formation of a pumiceous texture. Thus, pumice would not be expected to form from a dry magma free of volatiles *in solution.*

The other problem has been to identify a plausible volatile component to drive the fountain, which is compatible with the dry, reduced chemistry of lunar basalts. The presence of abundant vesicles in some of the returned samples suggested that some solution to this problem must exist.

Sato (1976) has recently analyzed this problem and shown that all available data can be interpreted by assuming that about 100 ppm of finely divided graphite was present in the magma at depth. At depths shallower than about 3 km in the moon, this graphite would react with the melt to form bubbles containing CO and CO_2 and would leave finely divided grains of Fe metal. This avoids any problem associated with the nucleation of bubbles in a volatile poor magma and is consistent with the different physical characteristics of lunar and terrestrial pyroclastics.

A large amount of work has already been published on these lunar pyroclastics and, with the recent availability of samples from the whole 74001/74002 double drive tube, much more can be expected in the near future. We have devoted a considerable effort to a study of volatile trace metals in these samples using x-ray photoemission spectroscopy and atomic absorption techniques (Cirlin *et al.*, 1978).

In order to provide a framework in which to interpret these data and other lunar observations as fully as possible, it seems appropriate to develop as thorough models of lunar volcanism as possible.

As a step in this direction, we assume along with Heiken *et al.* (1974) that terrestrial Hawaiian volcanoes provide a plausible analog for the interior structure of possible lunar volcanoes, but then explicitly consider the possible effects of differences in magma viscosity, the acceleration of gravity g, magma volatile content, and the ambient atmosphere.

In addition to their original purpose, the results appear to shed fresh light on terrestrial volcanic phenomena and planetary volcanism in general. In particular, they allow us to formulate a correspondence principle to the effect that for a given internal structure volcanic eruptions would appear similar on planets with different g, if the mole fraction of volatiles in the magma scaled with g. We next derive this relationship. We then discuss some important aspects of gas/liquid two phase flows and use them quantitatively to predict characteristics of terrestrial and lunar eruptions.

GENERAL

We start with the common assumption that the ultimate driving force responsible for volcanic activity is the difference in pressure head between a magma column to the surface and the rock column through which it passes. Equilibrium between these columns, which constrains the gross size of volcanic features, depends only on the relative densities of the magma and the rocks, and is independent of g. Thus, as is well known, the great height of Olympus Mons on Mars forces one to conclude that its magma source is at considerable depth. On the other hand, the thick low density crust on the moon apparently constrained volcanoes there to very modest heights.

The eruption of a Hawaiian volcano generally begins when magma which has been accumulating in a fairly shallow reservoir breaks through to the surface along a tensional crack (McDonald and Abbott, 1970; Eaton and Murata, 1960). Within a short time the eruption becomes confined to one or a few well defined vents. To facilitate discussion a schematic representation of the volcano at this stage is presented in Fig. 1.

Pure Liquid Flow

We will begin by considering the expected behavior of a hypothetical volcano with gas free magma. The gravitational potential energy associated with the difference in average density between the magma column and the rock through which it passes will go largely to overcoming frictional losses along the conduit and will also supply the kinetic energy corresponding to the magma velocity at the point of eruption.

A numerical example will help us develop two important ideas. From observations of recent Hawaiian vents, we take 10 m as a reasonable estimate for the diameter d of a magma conduit below any near surface funnelling. For our purposes we take the conduit to be a long vertical pipe of this diameter with an average surface roughness ϵ of 0.5 m. We also assume the average rock density ρ_r is 10% greater than the liquid density ρ_l.

The steady state magma flow velocity v may now be obtained by solving the equation relating friction loss per unit length to pressure head drop

$$\frac{fv^2}{2d} = \frac{\rho_r - \rho_l}{\rho_l} g, \tag{1}$$

where f is a friction factor that depends on the roughness parameter ϵ/d and the magma viscosity μ through the Reynolds number

$$N = \frac{\rho_l dv}{\mu}.$$

Making use of a standard engineering compilation of f as a function of ϵ/d and N (Anderson, 1967) and taking $\rho_l = 3\ \mathrm{gm/cm^3}$, $\mu = 145$ poise for terrestrial basalt at 1200°C (Heiken *et al.*, 1974), and using the terrestrial g value, we find the self consistent solution

$$v = 16.5\ \mathrm{m/sec}\ \text{and}\ N = 3.4 \times 10^4.$$

From the same compilation, we see that f (≈ 0.07) and hence v depend only very weakly on N and hence μ between $N = 5 \times 10^3$ and $N = \infty$. The flow is highly turbulent over this whole range of N values.

Using the above v, we obtain a magma eruption rate of $4.7 \times 10^6\ \mathrm{m^3/hr}$. This is about an order of magnitude high in comparison with estimates of recent Kilauea eruption rates, but comfortably within the range of estimates for Mauna Loa eruption rates (MacDonald and Abbott, 1970).

Our assumed 10% average density contrast is unquestionably considerably on the high side. Real magma conduits also must vary considerably and possibly

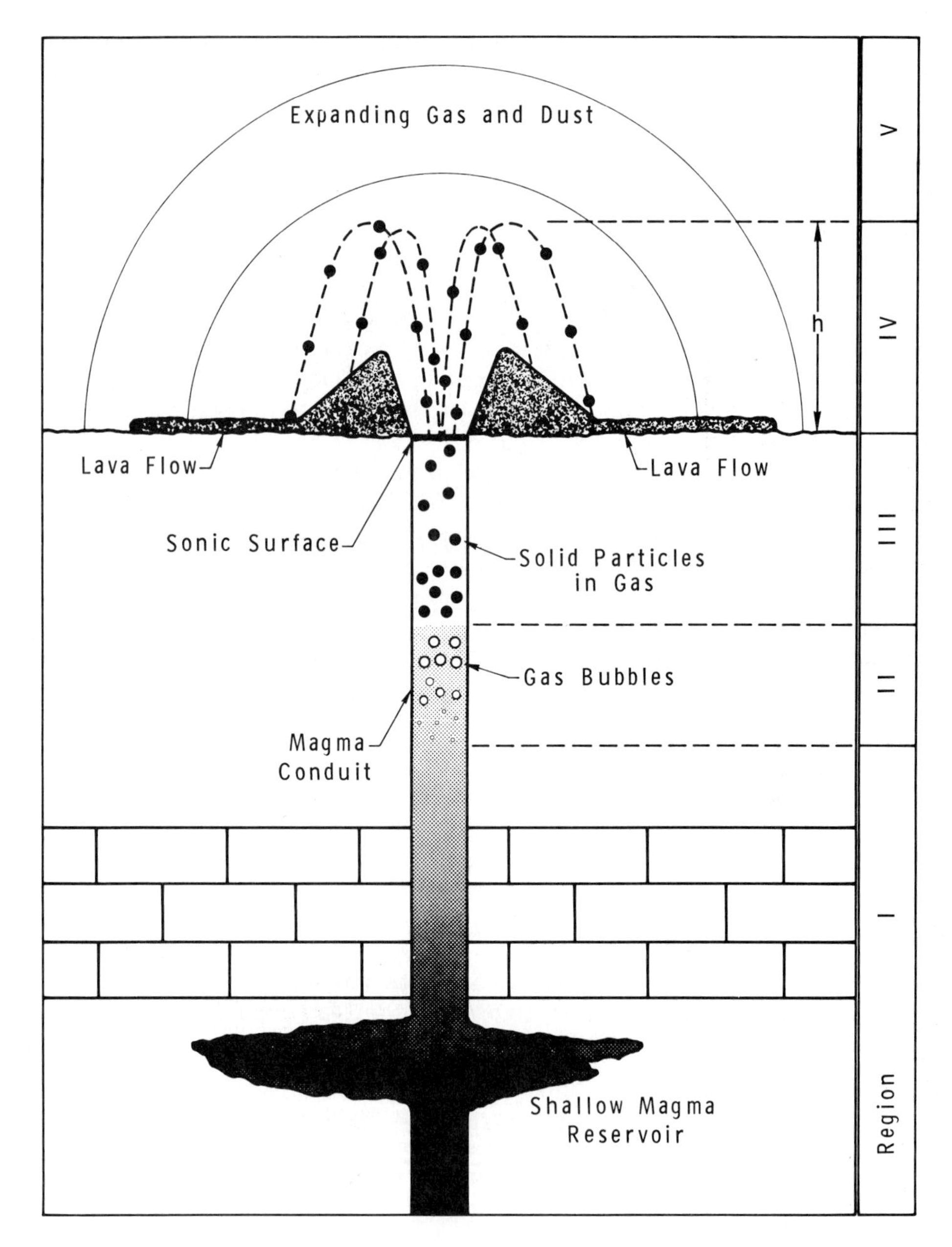

Fig. 1. Schematic diagram of the upper part of a basaltic volcano (not to scale). In the deep part designated by region I, gas volume is negligible and the magma can be regarded as a pure liquid. In general, gas bubbles become important at shallower depths and can exert a controlling influence on actual eruption characteristics as discussed in the text.

erratically in cross section with depth. The above results suggest as a working hypothesis, however, that in actual basaltic volcanoes the bulk of the frictional losses occur in constricted parts of the conduit where the flow is highly turbulent.

If this is the case, magma velocities would be insensitive to decreases or modest increases in the magma viscosity. In particular, the lower viscosities of lunar magmas in comparison to terrestrial ones would not lead to increased flow velocities.

Taking the above hypothesis to be correct, both the acceleration terms and the loss terms in the hydrodynamic equations describing magma flow depend on $\rho_l v^2$, while the driving term depends on g. Thus a change in g would simply lead to proportionate changes in v^2 and the pressure P at every point in the flow.

Volatiles and Similarity Principle

The height h to which lava will fountain during an eruption is given by $v_e^2/2g$ where v_e is the exit velocity at the surface. For the numerical example of the previous section, this comes out to be 14 m. Because in the gas-free case, v^2 scales with g, it would be the same on the moon where the magma velocity would be 6.8 m/sec.

It appears from direct visual observations of terrestrial volcanic eruptions that a large fraction of the volume of escaping material is gas at the instant of eruption. Hawaiian volcanoes are frequently observed to fountain to heights of 200 to 300 m (MacDonald and Abbott, 1970) and a record height of about 580 m has been recorded. Unreasonably large pressure heads would be required to produce the required exit velocities in a pure liquid. In any case, such high velocities could not be reconciled with the observed magma eruption rates without assuming that a large fraction of the volume at the time of eruption is gas. The appropriate description seems to be liquid blobs suspended in gas as indicated by region III in Fig. 1, although bubbles in liquid designated by region II is also possible.

While it is clear that volatiles play a dominant role in determining the eruption characteristics of volcanoes, it is surprisingly easy to specify conditions such that fountain heights would be the same on bodies with different g values. In the gas-free volcano considered in the previous section, P and v^2 scaled with g. If we now allow a small amount of gas to be present, but require the mole fraction to scale with g, we find that to a good approximation the gas volume V and average magma density at any depth are independent of g. Thus, the scaling of v^2 and P with g is maintained. This result can be stated as a simple correspondence principle. Volcanoes on different bodies will fountain to the same height if the mole fraction of volatiles in the magma scales with g. Higher fountaining would imply more volatiles and lower fountaining less volatiles than given by this relationship.

The presence of bubbles in the magma will decrease its effective viscosity while increasing the velocity, assuring that the flow will be highly turbulent with frictional losses proportional to v^2 as required.

While the above discussion provides a very interesting basis for comparing volcanic behavior on different bodies, it does not allow *a priori* predictions of eruption behavior as a function of volatile content to be made. That requires some understanding of the subtleties of two phase flow which will be discussed in the next section.

TWO PHASE FLOW

Gas/liquid two phase flows frequently differ dramatically in behavior from pure liquid flows primarily because the gas can make them highly compressible. The importance of this in volcanology was already apparent in the pioneering work of McGetchin and Ulrich (1973).

The presence of only a small volume fraction of bubbles can be sufficient to control the total compressibility of the fluid. On the other hand, in systems with a small mass fraction of gas, the liquid phase has most of the mass and keeps the average density much higher than that of the gas. Since the sound velocity is determined by the square root of the ratio of compressibility to density, it can be unexpectedly low (Kieffer, 1977). This can have important consequences to the flow behavior. In the static case the upper part of such a mixture expands to arbitrarily low density as a gas atmosphere would. In the following subsections, we will recall some important aspects of compressible gas flow and then derive relations for the sound velocity and density profile in two phase flows. These will be used later in semi-quantitative discussions of volcanic eruption behavior.

Ideal gas behavior

Several standard texts, for example, Leipmann and Roshko (1957) give thorough discussions of ideal gas behavior. The sound velocity v_s is given by $v_s^2 = \gamma P/\rho$ where γ is the ratio of the specific heat at constant pressure to that at constant volume. Since the density ρ is proportional to P for an ideal gas, v_s is a function of temperature only.

A gas expanding into vacuum or a sufficiently low pressure region will accelerate to supersonic velocities. At subsonic velocities, any disturbance in the flow will propagate as a pressure wave upstream and influence the approaching flow characteristics. At supersonic velocities, this cannot happen and the flow incident on the region of disturbance remains unchanged. The disturbance, of course, affects the downstream character of the flow in both cases.

From the above physical discussion, it is clear that if a gas flow is to become supersonic, the sonic surface across which it does so must be located at the position where it is most constricted. Almost paradoxically, higher pressures will not increase the flow velocity across this sonic surface, nor lower pressures decrease it. They will, however, increase or decrease the gas density and hence the mass flow rate. After passing the sonic surface, the gas can continue to accelerate until all of its thermal energy is converted into kinetic flow energy. These considerations are familiar in rocket technology.

Two phase sound velocities

Throughout this paper, we are assuming that the bubbles or droplets in our fluid are sufficiently small and uniformly distributed that it may be considered homogeneous and isotropic. The velocity of sound in a liquid containing bubbles has been carefully considered by Hsieh and Plesset (1961). The velocity for a gas containing liquid droplets is quoted by Rudinger (1964). Under conditions appropriate for our discussions both results are well approximated by

$$v_s^2 = m_g \frac{P}{\rho_g}, \tag{2}$$

where m_g is the mass fraction gas and ρ_g is the gas density. It is to be noted that γ does not occur in this expression because the assumptions of small bubble or droplet size and small gas fraction ensure that the gas compressions and rarefactions take place isothermally.

For an ideal gas (2) is equivalent to

$$v_s^2 = \frac{m_g}{M_g} RT, \tag{3}$$

where M_g is the molecular weight of the gas, R the gas constant per mole, and T the absolute temperature. Thus, v_s^2 scales with the mole fraction of the gas phase. It is worth explicitly mentioning that this result is in complete agreement with the correspondence principle developed in the previous section. If the mole fraction of gas scales with g, the ratio v/v_s at any point in the flow will be unchanged.

Pressure and density profiles

If we consider a static gas/liquid two phase fluid in a gravitational field, it is easy to derive the following expression for the pressure P versus height x

$$(P_o\text{-}P) - \alpha\rho_l \frac{RT}{M_g} ln \frac{P}{P_o} = g\rho_l (1 + \alpha)x, \tag{4}$$

where P_o is the pressure at the reference elevation $x = 0$ and α is mass ratio of gas to liquid.

At depths where the gas volume is considerably larger than the liquid volume, the first term on the left in (4) can be neglected and the resulting expression simplifies to

$$\frac{P}{P_o} = \frac{\rho}{\rho_o} = e^{-\frac{gx}{v_s^2}} \equiv e^{-\frac{x}{s}}. \tag{5}$$

Thus, the scale height s remains constant as long as v_s^2 varies linearly with g.

Lunar Volatiles

It is evident, from the highly vesicular nature of many returned lunar basalt samples, that some significant volatile phase was associated with lunar volcanism. Its nature and amount, however, are still strongly debated.

All lunar basalts contain a small amount, usually between 0.05 and 0.15 wt.%, of metallic Fe. This and the absence of evidence for any hydrated phases place strong constraints on the composition of a gas phase in equilibrium. Although a suitable ratio of H_2 and H_2O would be compatible at low pressure, such an equilibrium is very sensitive to pressure. Such a system would become reducing at high pressure and metallic Fe would form at depth and differentiate downwards. Then, during an eruption, the excess H_2O in the melt would be oxidizing.

Sato (1976) has carefully considered various suggested mechanisms for the formation of the Fe metal in lunar basalts. He showed that all available evidence can be interpreted by assuming that on the order of 100 ppm of graphite was present in a Fe^{++}-containing melt at depth. Above about 3 km in the moon, this graphite would react with the melt to form bubbles containing CO and CO_2 and leave finely divided grains of Fe metal in the melt. This idea is consistent with cosmochemical considerations since most chondritic meteorites contain small amounts of graphite, but no hydrated phases.

For the purpose of making an illustrative comparison between lunar and terrestrial volcanic activity, we will assume that the major lunar volatile is the CO that would correspond with the observed amount of Fe metal. This ranges from 250 to 750 ppm by weight. We take 0.5 to 1.0 wt.% H_2O to be representative of terrestrial Hawaiian basalts (MacDonald and Abbott, 1970).

Lunar and Terrestrial Eruptions

We now make use of the relationship and ideas developed in the last two sections to make semi-quantitative estimates of the typical eruption behavior expected for terrestrial and lunar basaltic volcanoes. The main results are presented in Table 1. We first note that the fountain heights h_s predicted for terrestrial volcanoes using the sonic velocity as the exit velocity at the surface are well within the range commonly observed (MacDonald and Abbott, 1970). The sonic velocity is also close to the exit velocities observed by McGetchin *et al.* (1974) for Northeast Crater, Mt. Etna ejecta.

The highest recorded fountaining of Kilaeua Iki, 580 m, would require either considerably more than 1% H_2O in the magma, which seems reasonable, or some further acceleration beyond sonic velocity. The latter is possible if the vent flairs out into a cone of suitable dimensions above the throat. Limiting velocities v_∞ and the corresponding heights h_∞ theoretically obtainable if all the thermal energy in the gas, at the throat, were converted into vertical motion energy are also presented in Table 1. It appears that significant post-sonic vertical acceleration must be uncommon.

Since the mole fractions of gas assumed in the lunar magma fall below linear scaling with g by factors of 0.2 and 0.6 assuming 0.5 wt.% H_2O in terrestrial

Table 1. Parameters describing the eruption behavior of terrestrial and lunar basaltic volcanoes.

	$v_s \left(\frac{m}{sec}\right)$	$v_\infty \left(\frac{m}{sec}\right)$	h_s (m)	h_∞(m)	s(m)
Terrestrial					
0.5 wt. % H_2O	58.3	142	174	1045	348
1.0 wt. % H_2O	82.5	202	348	2100	696
Lunar					
250 ppm CO	10.4	25.6	33.5	202	67
750 ppm CO	18.2	44.5	102	612	204

magma, the lunar eruptions are, as expected, predicted to be significantly less spectacular than their terrestrial counterparts. This appears to be in general agreement with reported observations of lunar cinder cones (McGetchin and Head, 1973; Wood and Head, 1977).

The last column of Table 1 gives the static scale heights calculated for each composition. We suggest that they may provide an order of magnitude estimate of the depth to which the magma conduit and reservoir system will empty during the late stages of an eruption. They seem adequate to provide for subsidence and flow of outgassed lava back down the conduit as is sometimes observed. With this view, even though the apparent density of a lava column might decrease with time, the actual mass ratio of gas to liquid would be unchanged as long as sonic velocity in the throat could be maintained.

Lunar Transient Atmosphere

In the fire fountain of a volcano, separation of the gas and liquid phases takes place, with most of the liquid falling back to form a lava flow containing relatively little gas, while most of the gas forms a radially expanding transient atmosphere carrying with it only a modest amount of condensed material.

Within a radial distance equivalent to a few times the fountain height, the gas will have accelerated to approach a limiting velocity corresponding to the conversion of all its thermal energy into kinetic energy. For CO starting at 1200°C, this limiting velocity is 1.76 km/sec, which is a large fraction of the lunar escape velocity of about 2.4 km/sec.

With a conduit diameter of 10 m and a density contrast of 10%, a hypothetical lunar eruption would bring magma to the surface at a rate of 1.9×10^6 m^3/hr of liquid. Assuming 750 ppm by weight of CO, the corresponding gas evolution rate m is an impressive 1.2 metric tons per second.

At radial distances r exceeding a few fountain heights, this gas can be assumed to be expanding nearly hemispherically with close to its limiting velocity. Then from the conservation of mass we can write

$$2\pi r^2 \rho_g v_\infty = m. \tag{6}$$

Using m and v_∞ estimated above, we find the density ρ_g as a function of radius to be given by

$$\rho_g = \frac{1.1 \times 10^{-4}}{r^2} \frac{gm}{cm^2}, \tag{7}$$

where r is in meters. The static pressure at 0°C which would give the same density is

$$P = \frac{8.7 \times 10^{-2}}{r^2} \text{ atmos, or } P = \frac{66}{r^2} \text{ torr.} \tag{8}$$

It seems clear from the above semi-quantitative description that a single volcanic eruption could distribute volatiles including trace metals, halogens, and sulfur over the entire moon. Thus, mare volcanism could easily be the primary source of excess volatiles observed in both mare and highlands regoliths and breccias.

The limited droplet size range found in the 74220 and 74001/74002 pyroclastic deposits strongly suggests sorting during atmospheric transport rather than ballistic deposition. The transient atmosphere described above has sufficient energy to transport a significant fraction of the liquid mass to distances comparable to the observed extent of dark mantle deposits (Lucchitta, 1973). For example, 100 times as much mass as the mass of gas could conceivably be accelerated to 175 m/sec velocities. Thus, 2.5 to 7.5% of the total mass could be carried as far as about 20 km. Actual gas transport is probably much less efficient than this and more detailed estimates remain to be made.

The above discussion shows that the observed characteristics of lunar pyroclastics and pyroclastic deposits are compatible with the normally expected behavior of lunar volcanoes. Special types of lunar volcanism such as one recently suggested by Sato (1977) do not appear to be required.

SUMMARY

We have shown that for normal basaltic volcanic eruptions a correspondence principle exists stating that volcanoes on different planetary bodies will fountain to the same height if the mole fraction of gas in the magma scales with g. Higher or lower fountaining would imply more or less gas than given by this relationship. This correspondence principle holds independent of decreases or modest increases in magma viscosity in comparison to terrestrial basalts.

We suggested that the sound velocity in the two gas/liquid magma fluid controls the eruption velocity of volcanoes. This allows semi-quantitative predictions to be made for ejecta velocities and fountain heights as a function of the gas content of the magma. Predictions for Hawaiian volcanoes proved to be in good accord with observations.

We were also able to estimate the depths to which magma reservoirs would empty during eruptions.

Assuming that the only gas phase in lunar magmas was the CO produced by graphite reduction of Fe^{++}, to form the Fe metal observed in lunar basalts, we found that lunar volcanoes would fountain, though somewhat less impressively than terrestrial ones. Volatile trace metals, halogens, and sulfur would be carried over the whole moon by the expanding transient atmosphere. Orange and black glass type pyroclastic material could be transported by the gas in amounts sufficient to account for observed dark mantle deposits.

Acknowledgments—We gratefully acknowledge the help of Dr. N. D. Malmuth who offered various suggestions on this work and who called the solution to the radially expanding gas cloud to our attention, and to Dr. V. Shankar who located appropriate references on two phase flow. Our appreciation of the importance of the sonic velocity in two phase flow developed during discussions with Dr. S. I. Pai and Dr. T. R. McGetchin during the 9th LPSC. This manuscript has benefited from reviews by Drs. Elaine Padovani, T. R. McGetchin, Dave Miller, and especially L. Wilson. This work was supported by NASA Contract No. NAS 9-11539.

REFERENCES

Anderson C. T. (1967) Mechanics of fluids. In *Standard Handbook for Mechanical Engineers* (T. Baumeister, ed.), p. 3-48–3-63. McGraw-Hill, N. Y.

Chou C.-L., Boynton W. V., Sundberg L. L. and Wasson J. T. (1975) Volatiles on the surface of Apollo 15 green glass and trace-element distributions among Apollo 15 soils. *Proc. Lunar Sci. Conf. 6th*, p. 1701–1727.

Cirlin E. H., Housley R. M. and Grant R. W. (1978) Studies of volatiles in Apollo 17 samples and their implications to vapor transport processes. *Proc. Lunar Planet. Sci. Conf. 9th*. This volume.

Eaton J. P. and Murata K. J. (1960) How volcanoes grow. *Science* **132**, 925–938.

Heiken G. and McKay D. S. (1977) A model for eruption behavior of a volcanic vent in eastern Mare Serenitatis. *Proc. Lunar Sci. Conf. 8th*, p. 3243–3255.

Heiken G. H., McKay D. S. and Brown R. W. (1974) Lunar deposits of possible pyroclastic origin. *Geochim. Cosmochim. Acta* **38**, 1703–1718.

Hseih D. Y. and Plesset M. S. (1961) On the propagation of sound in a liquid containing gas bubbles. *Phys. Fluids* **4**, 970–975.

Kieffer S. W. (1977) Sound speed in liquid-gas mixtures: Water-air and water-steam. *J. Geophys. Res.* **82**, 2895–2904.

Leipmann H. W. and Roshko A. (1957) *Elements of Gasdynamics*. Wiley, N. Y.

Lucchitta B. K. (1973) Geologic setting of the dark mantling material in the Taurus-Littrow region of the moon. Apollo 17 Prelim. Sci. Rep. NASA SP-330, p. 29-13–29-25.

MacDonald G. A. and Abbott A. T. (1970) *Volcanoes in the sea*. The University Press of Hawaii, Honolulu. 441 pp.

McGetchin T. R. and Head J. W. (1973) Lunar cinder cones. *Science* **180**, 68–71.

McGetchin T. R., Settle M. and Chouet B. A. (1974) Cinder cone growth modeled after Northeast crater Mount Etna, Sicily. *J. Geophys. Res.* **79**, 3257–3272.

McGetchin T. R. and Ullrich G. W. (1973) Xenoliths in maars and diatremes, with inferences for the Moon, Mars, and Venus. *J. Geophys. Res.* **78**, 1833–1853.

Meyer C. Jr., McKay D. S., Anderson D. H. and Butler P. Jr. (1975) The source of sublimates on Apollo 15 green and Apollo 17 orange glass samples. *Proc. Lunar Sci. Conf. 6th*, p. 1673–1699.

O'Keefe J. A. (1976) Volcanic ash: Terrestrial versus extraterrestrial. *Science* **194**, 190.

Reed G. W. Jr., Allen R. O. Jr. and Jovanovic S. (1977) Volatile metal deposits on lunar soils—relation to volcanism. *Proc. Lunar Sci. Conf. 8th*, p. 3917–3930.

Rudinger G. (1964) Some properties of shock relaxation in gas flows carrying small particles. *Phys. Fluids* **7**, 658–663.
Sato M. (1976) Oxygen fugacity and other thermochemical parameters of Apollo 17 high-Ti basalts and their implications on the reduction mechanism. *Proc. Lunar Sci. Conf. 7th*, p. 1323–1344.
Sato M. (1977) The driving mechanism of lunar pyroclastic eruptions inferred from the oxygen fugacity behavior of Apollo 17 orange glass. *EOS (Trans. Amer. Geophys. Union)* **58**, 425.
Wasson J. T., Boynton W. V., Kallemeyn G. W., Sundberg L. L. and Wai C. M. (1976) Volatile compounds released during lunar lava fountaining. *Proc. Lunar Sci. Conf. 7th*, p. 1583–1595.
Wood C. A. and Head J. W. (1977) Cinder cones on Earth, Moon, and Mars. *EOS (Trans. Amer. Geophys. Union)* **58**, 425.

Proc. Lunar Planet. Sci. Conf. 9th (1978), p. 1485–1508.
Printed in the United States of America

Similar explosive eruptions of lunar and terrestrial volcanoes

S. I. PAI and Y. HSU

Institute for Physical Science and Technology and Aerospace Engineering
Department, University of Maryland, College Park, Maryland 20742

JOHN A. O'KEEFE

Goddard Space Flight Center, Greenbelt, Maryland 20771

Abstract—A mathematical model of one-dimensional, steady duct flow of a mixture of gas and small solid particles (rock), has been applied to the earth and the moon under geometrically and dynamically similar conditions to simulate explosive eruption conditions. Numerical results for equilibrium two-phase flows for lunar and terrestrial explosive eruptions under similar conditions indicate that: (1) a lunar explosive vent is much larger than the corresponding terrestrial vent, (2) the exit velocity from the lunar explosive flow may be higher than the lunar escape velocity, but the exit velocity of terrestrial explosive flow is much less than that of the lunar case (this result supports the hypothesis that Australian tektites came from the moon as a stream of a mixture of rock and gas of extremely high speed), and (3) the thermal effects on the lunar explosive flows are much larger than those of the terrestrial case.

I. INTRODUCTION

There is a real need for physical and mathematical models of geologic processes, especially in conjunction with planetological studies, because we cannot observe planetological processes in action. To some extent, planetary structures may be explained by comparison with terrestrial structures, *after allowance for scaling*. The pi-theorem (Bridgeman, 1922) states that a physical law can be stated in terms of one or several non-dimensional parameters, each a product of a number of physical variables. If these parameters have the same values in two objects, one on planet A and the other on planet B, both produced by the same process, then it is reasonable to try to explain what is seen in B, in terms of what is known about A.

One of the most important problems of planetary structures is to understand the circular depressions which are the most conspicuous features of the surface of the moon and the surface of most of the planets and satellites which have been examined. These depressions are all called craters; but it is important to realize that the use of one name does not mean that they are all the same kind of object on the moon or the planets, any more than on the earth, where some are due to impact and others to volcanism.

The craters Maunder (Fig. 1) and Kopff (Fig. 2) in Mare Orientale on the moon are of roughly the same size (Maunder 55 km, Kopff 40 km), but they are fundamentally different in structure (McCauley, 1968). Kopff has a rim which is

Fig. 1. The lunar crater Maunder, 55 km in diameter, on Mare Orientale. Note the hummocky region behind it, and the very gentle slope of the outside wall. The slope of the inner wall has been diminished by slumping. A smaller crater, Hohmann, near the bottom of the figure, is analogous to Kopff.

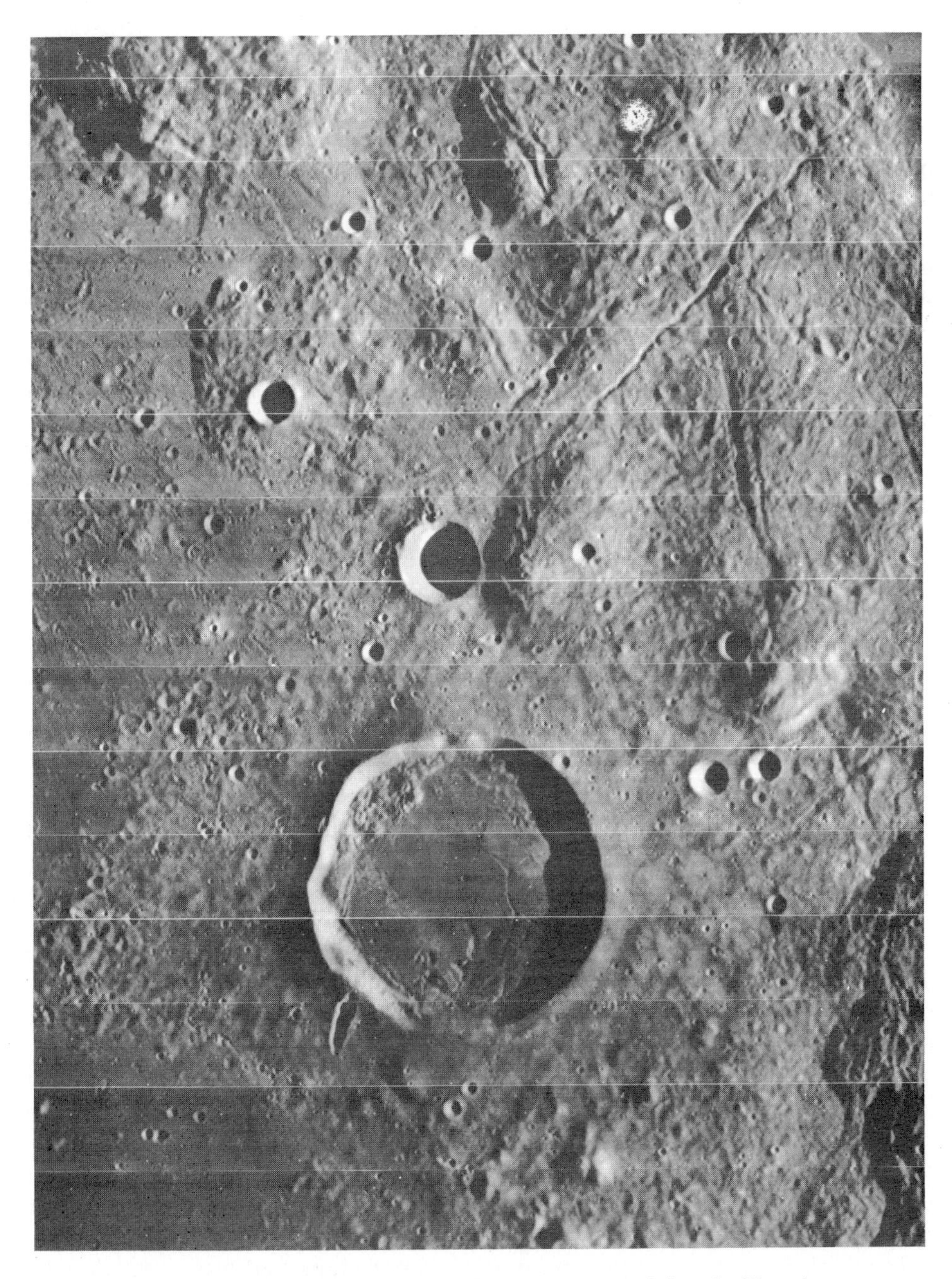

Fig. 2. The lunar crater Kopff, 40 km in diameter, in Mare Orientale. Note the steep outer wall, and the smooth region around it. Kopff is obviously a different kind of crater from Maunder, yet it is superposed on the characteristic Mare Orientale Terrain, which has not been reworked by cratering. It is thus not likely that Kopff is the worn-down remnant of a crater like Maunder.

polygonal in plan and Δ-shaped in cross-section, while the wall of Maunder is circular in plan and much deeper on the inside than the outside; Kopff has a smooth surrounding surface, while that around Maunder is hummocky; Kopff is shallow while Maunder is deep. Both craters are on Mare Orientale, which has a peculiar, wavy surface, clearly not degraded by any erosional process. The survival of this peculiar surface guarantees that neither crater could have been seriously modified by erosion; both must have been found in about their observed shape. It seems to follow that the moon has at least two radically different ways of forming craters. Since all impact craters tend to be similar, it follows that if Maunder is an impact crater, then Kopff is not. This example is important because it indicates that the moon has large craters of volcanic origin. The explosion craters considered in this paper are probably different from Kopff in morphology.

Similarly, Cameron and Padgett (1974) have drawn attention to regional variations in the frequency of what they call ringdikes, i.e., craters more or less like Kopff. C. R. Chapman *et al.* (1971) find that the most subtelescopic craters on the maria are of internal origin, because they are ten times as numerous as the subtelescopic highland craters, whereas telescopic mare craters are one tenth as numerous as telescopic highland craters.

In addition, the lunar origin of tektites, which seems especially probable for Australian and Javanese tektites, their flanges and ringwaves, seems to imply explosive lunar volcanism (O'Keefe 1976); the argument rests in part on studies of australite ablation, especially those by Chapman and Larson (1963). The calculations have been thoroughly checked by independent computations; they seem to rule out any non-lunar origin of tektites. The line of argument is supported by calculations of the size of the crater which would be needed for a terrestrial origin, and of the time needed to produce glass pieces of the necessary size, density, and homogeneity, as well as of the probability of hitting compositions as unusual as those of some of the components of tektite stream fields.

Most lunar craters resemble in shape terrestrial explosion pits, rather than terrestrial volcanoes (Pike, 1974). The exceptions are the maar-type craters and the tuff rings which are terrestrial, but are more like typical lunar craters than like the majority of terrestrial volcanoes. Pike rejects the maars and tuff rings as lunar crater analogs because the maximum size of these structures is about 5 km, while lunar craters range up to 250 to 300 km. Obviously the question of scale is crucial here.

It thus appears useful to calculate the physics of eruptive volcanism, with the earth and the moon as examples. (Martian eruptive volcanism will then represent an intermediate case.)

In comparing lunar and terrestrial volcanism we note that water, which is the principal driving gas in a terrestrial eruption, will be less important than hydrogen in a lunar case. This is a direct consequence of the low oxygen fugacity in a lunar lava. The equilibrium is governed by the usual relation

$$\frac{(H_2O)}{(H_2)(O_2)^{1/2}} = K, \qquad (1)$$

where (H_2O) is the partial pressure of steam, etc., and K is the equilibrium constant; from the JANAF tables (JANAF, 1971) we find, at 1500°K, $\log_{10}K = 3.225$ if the pressure is in N m^{-2} (Newton's per square meter). The fugacity of oxygen in lunar lava at this temperature is known to be about 10^{-8} N m^{-2}, this is the partial pressure of oxygen in contact with the lava. Substituting this in Eq. (1), we find that the ratio of the partial pressure of water and hydrogen is 0.17. Hence in our mathematical mode, we consider steam as the gas in terrestrial volcanoes while hydrogen as the gas in lunar volcanoes.

Although the moon is presently deficient in volatiles compared to the earth, the supply of volatiles earlier in lunar history may have been sufficient to drive explosive volcanism. Friedman *et al.* (1971) find that lunar basalts contain about 6 ppm of hydrogen. At that concentration the moon as a whole would contain some 4×10^{14} tons of hydrogen, which means a reservoir in the range of 3×10^9 megatons of energy. If this hydrogen reservoir were properly concentrated it would be sufficient to generate a dozen or so 100 km craters. In fact the hydrogen content found on the lunar surface may underestimate that found at depth. Also, potassium argon ages of lunar basalts indicate that they were devolatized at the time of eruption, indicating that the present hydrogen content may be an underestimate of the content available earlier in lunar history.

Hydrogen is the most important gas for our present study, not only because it is geochemically plausible, but also because its molecular height places it at one extreme of the possible gases, and so illustrates the effects of molecular weight in the most effective way.

II. Mathematical Model of Explosive Flow

To explain the mechanism of explosive eruption, whether lunar or terrestrial, we may consider the explosive eruption as a two-phase system consisting of the solid particles and a gas. We analyze the one-dimensional flow due to the eruption of such a two-phase system. From full observations (Fielder, 1971) and some preliminary theoretical calculations (McGetchin and Ullrich, 1973), the general explosive flow field is sketched in Fig. 3.

The general flow field may be divided into two regions: the lower part, region I, starts from a great depth in the planet; region I has a height H_1, of the order of 100 km in terrestrial case. At the base, due to chemical reactions, the rock is melting at a temperature of 1000°C. The original pressure is large enough to eliminate gas pockets (McGetchin and Ullrich, 1973). As a result, a mixture of molten rock with dissolved gas flows upward in a duct of almost constant cross-sectional area. In this region, since the pressure is very high, not only the molten rock but also the gas, if any, behaves like an incompressible fluid. From fluid dynamic point of view, the flow problem is very simple. The flow field should not deviate much from the simple analysis by McGetchin and Ullrich (1973) which is essentially a flow of incompressible fluid in a narrow pipe.

When the flow is near the surface of the planet, with $x \leq - H_o$, where x is the vertical coordinate, the area of the duct increases and the mixture expands as it

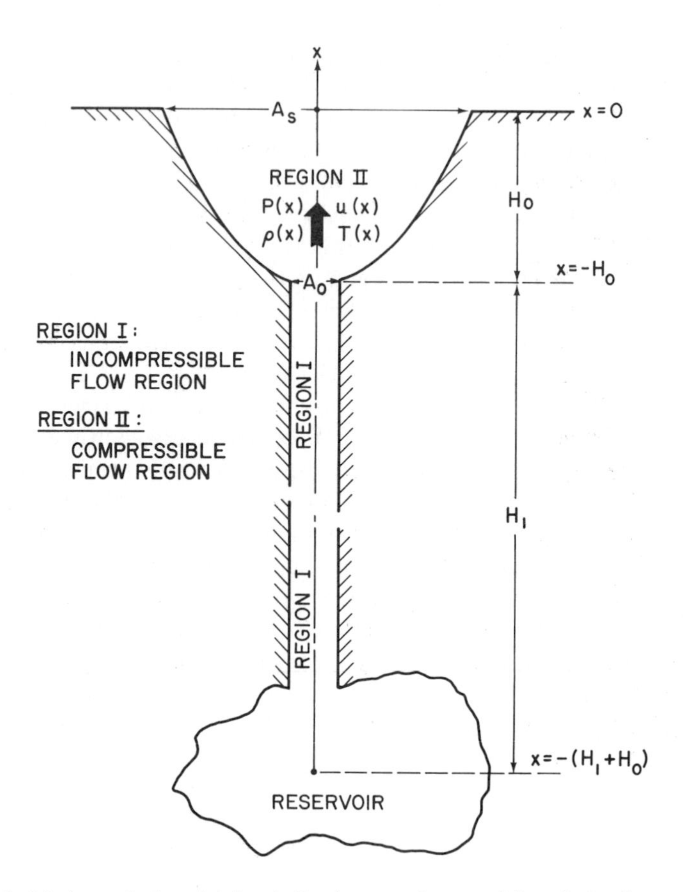

Fig. 3. Mathematical model for similar lunar and terrestrial explosive flows.

flows upward. We call this region II of lunar and terrestrial explosive flows. We are especially interested in the similar flow patterns of a terrestrial explosive flow with a corresponding lunar explosive flow.

III. Fundamental Equations of Explosive Flows and the Important Non-Dimensional Parameters

Since the flow of a mixture of solid particles and a gas comes a long way from the great depth of the planet to the initial section $x = -H_0$ of region II, it should reach an equilibrium condition such that the initial velocity of the rock and that of the gas are approximately equal and that the temperatures of the rock and the gas are also equal. This condition is known as equilibrium flow (Pai, 1977). In order to get some essential features of the explosive flow in region II, we assume for the present that the flow is in equilibrium in region II. We shall investigate the non-equilibrium effects on the flow field in region II in a later paper.

For equilibrium flow of the two-phase system in region II, we consider the following variables:

velocity of the mixture u
temperature of the mixture T
pressure of the mixture p, and
density of the mixture ρ.

The density of the rock ρ_r is assumed to be constant and the density of the mixture depends on the density of the gas ρ_g and the volume fraction of the rock in the mixture Z.

The values of the variables at the initial section $x = -H_o$ of cross-sectional area A_o are given from the calculation of region I as follows:

$$p = p_o,\ u = u_o,\ T = T_o,\ \rho = \rho_o. \tag{2}$$

Since we are interested in similar flows in the lunar and the terrestrial case, the relations between the initial values of these two cases may be obtained by similarity considerations as we shall show in section IV.

We define the non-dimensional quantities (with a bar) as follows:

$$\bar{p} = \frac{P}{P_o},\ \bar{T} = \frac{T}{T_o},\ \bar{\rho} = \frac{\rho}{\rho_o},\ \bar{u} = \frac{u}{u_o},\ \bar{x} = \frac{x}{L},\ \bar{A} = \frac{A}{A_o},$$

$$\bar{a} = \frac{a}{a_o},\ \bar{a}_M = \frac{a_M}{a_{Mo}}, \tag{3}$$

where L is the characteristic length of the flow field of region II which may be taken as H_o.

The sound speed of the gas a is given by the following formula:

$$a = \sqrt{\gamma \frac{R_u}{m} T}, \tag{4}$$

where γ is the ratio of specific heats c_p/c_v, R_u is the universal gas constant and m is the molecular weight of the gas.

The non-dimensional equations for one-dimensional steady equilibrium flows of a two phase mixture of solid particles and a gas (Pai, 1977) are as follows:

(i) Equation of state

$$\bar{p} = \frac{1 - Z_o}{1 - Z} \bar{\rho}\, \bar{T}, \tag{5}$$

where subscript o refers to values at initial section $x = -H_o$.

(ii) Equation of continuity

$$\bar{\rho}\, \bar{u}\, \bar{A} = 1 \tag{6}$$

(iii) Equation of motion

$$\frac{d\bar{u}}{d\bar{x}} = \frac{\bar{u}}{\bar{a}_M^2 - M_o^2 \bar{u}^2} \left(-\frac{\bar{a}_M^2}{\bar{A}} \frac{d\bar{A}}{d\bar{x}} + \frac{M_o^2}{Fr_o} + 2c_f M_o^2 \frac{\bar{u}^2}{\bar{D}} \right), \tag{7}$$

where the sound speed of the mixture a_M is given by the formula:

$$a_M^2 = \frac{\Gamma(1 - k_p)}{(1 - Z)^2} \left(\frac{R_u}{m}\right) T, \tag{8}$$

where

$$k_p = \frac{Z\rho_r}{\rho} \tag{9}$$

is the mass concentration of the solid particles in the mixture and Γ is the effective ratio of specific heats of the mixture, i.e.,

$$\Gamma = \gamma \frac{\left(1 + \frac{\eta\sigma}{\gamma}\right)}{1 + \eta\sigma} \tag{10}$$

where $\eta = (1 - k_P)/k_p$, $\delta = c_s/c_v$, c_s is the specific heat of the solid particles, c_v is the specific heat of the gas at constant volume and c_p is the specific heat of the gas at constant pressure.

The initial Mach number M_o is defined as follows:

$$M_o = \frac{u_o}{a_o} \tag{11}$$

and the initial Froude number Fr_o is defined as follows:

$$Fr_o = \frac{u_o^2}{gL}, \tag{12}$$

where g is the gravitational acceleration of the planet. The friction coefficient of the duct is c_f and $\bar{D} = D/L$ where D is the hydraulic diameter, A(x) is the cross-sectional area of the duct in region II.

(iv) The energy equation:

$$[b_1(1\text{-}Z)^2 + b_4 \frac{(1\text{-}Z)}{\bar{u}\bar{A}} + b_3(1\text{-}Z)^2]\, \bar{a}_M^2 + \frac{1}{2} M_o^2\bar{u}^2 + \frac{\bar{M}_o}{Fr_o} \bar{x} -$$

$$- \sum_{n=o}^{n} 2c_f M_o^2 \frac{\bar{u}_{av}}{\bar{D}_{av}} (\bar{x}_{n+1} - \bar{x}_n) =$$

$$= \left[\frac{k_p}{\Gamma_o} \left(\frac{\sigma_o}{\gamma_o - 1}\right) \frac{(1\text{-}Z_o)^2}{(1\text{-}k_p)} + \frac{k_p}{G_o} \frac{\rho_o}{\rho_{go}} \frac{(1\text{-}Z_o)}{\Gamma_o} + \frac{\gamma_o}{(\gamma_o - 1)} \frac{(1\text{-}Z_o)^2}{\Gamma_o}\right] + \frac{1}{2} M_o^2 - \frac{M_o^2}{Fr_o}, \tag{13}$$

where

$$\left.\begin{aligned} b_1 &= \frac{k_p}{(1 - k_p)} \frac{\delta}{(\gamma - 1)\Gamma} \\ b_3 &= \left(\frac{\gamma}{\gamma - 1}\right) \frac{1}{\Gamma} \\ b_4 &= \frac{k_p}{\Gamma} \frac{1}{G_o} \frac{\rho_o}{\rho_{go}} \quad , \end{aligned}\right\} \tag{14}$$

$$G_o = \frac{\rho_r}{\rho_{go}} \tag{15}$$

is the initial density ratio and D_{av} is the average hydraulic diameter between two steps and u_{av} is the average velocity between two steps (Hsu, 1976).

From the non-dimensional Eqs. (5), (6), (7) and (13), we have four important non-dimensional parameters: k_p (9); M_o (11); Fr_o (12) and G_o (15), which characterize the explosive flow. The mass concentration k_p indicates the mass fraction of the solid particles in the mixture, which is important in the determination of the velocity of the mixture. The initial density ratio G_o indicates the relative importance of the gas and the solid particles in the mixture. The Mach number M_o is a measure of the compressibility effect of the mixture due to high speed. The initial Froude number Fr_o indicates the effect of gravitational force on the flow field.

IV. Similarity Laws of Lunar and Terrestrial Explosive Flows

We would like to compare a terrestrial explosive flow with a similar lunar explosive flow. For the similarity of these two explosive flows, we should have both geometrical similarity and dynamic similarity (Sedov, 1959).

By geometrical similarity, we mean that there is a linear scaling of the region II of the terrestrial and the lunar explosive duct. We should compare these explosive flow fields with the same non-dimensional initial area A_o and same slope of the non-dimensional variation of the cross-sectional area in region II, i.e., the same dA/dx. These two ducts are geometrically similar. It should be noted that the ratio of the exit area A_s to the initial area A_o for these two ducts is not the same because we have to satisfy the conditions of dynamic similarity.

By dynamic similarity, we mean that the most important non-dimensional parameters of the fundamental equations of the explosive flows for these two cases should be equal. In other words, for dynamic similarity, the non-dimensional parameters: M_o, G_o, Fr_o and k_p for the similar terrestrial and lunar flows should be equal. From these conditions, we may determine the corresponding size and initial values of these two similar explosive flows as follows.

We assume that in the terrestrial flow, the gas is steam while in the lunar explosive flow the gas is hydrogen as we have discussed in section I. We have the following values of various characteristic quantities in two geometrically and dynamically similar terrestrial and lunar explosive flows:

(i) For geometrical similarity, we have in region II

$$\bar{A}_{om} = \bar{A}_{oe} = 1, \left(\frac{d\bar{A}}{d\bar{x}}\right)_m = \left(\frac{d\bar{A}}{d\bar{x}}\right)_e, \tag{16}$$

where subscript o refers to the value at the initial section $x = -H_o$ or $\bar{x} = -1$, subscript m refers to the value for lunar explosive flow while subscript e refers to the value for terrestrial explosive flow. We compare two similar explosive flows that satisfy Eq. (16).

(ii) We assume that the initial temperature T_o is the same in these two similar terrestrial and lunar explosive flows, i.e.,

$$T_{om} = T_{oe} = 1{,}000°\text{C}. \tag{17}$$

(iii) In the terrestrial explosive flow, the principal gas is steam while that in the lunar explosive is hydrogen. Hence at $T_o = 1000°\text{C}$, we have

$$m_e = 18.016;\ c_{pe} = 0.593\ \text{cal/gm °K};\ \gamma_e = \frac{c_{pe}}{c_{ve}} = 1.23\ \text{(for steam)} \tag{18a}$$

$$m_m = 2.016;\ c_{pm} = 3.721\ \text{cal/gm °K};\ \gamma_m = \frac{c_{pm}}{c_{vm}} = 1.36\ \text{(for hydrogen)}. \tag{18b}$$

The ratio of the initial sound speed in these two cases is

$$\frac{a_{om}}{a_{oe}} = \sqrt{\frac{\gamma_m m_e}{\gamma_e m_m}} = 3.126. \tag{19}$$

(iv) By dynamical similarity, the initial Mach number of these two similar flows should be the same, i.e.,

$$M_{om} = \frac{u_{om}}{a_{om}} = M_{oe} = \frac{u_{oe}}{a_{oe}}. \tag{20}$$

Hence,

$$u_{om} = 3.126\ u_{oe}. \tag{21}$$

For dynamical similarity, the initial value of velocity in the lunar case should be 3.126 times that of the terrestrial case.

(v) By dynamical similarity, the density ratio G_o of these two similar flows should be the same, i.e.,

$$G_{om} = \frac{\rho_r}{\rho_{gom}} = G_{oe} = \frac{\rho_r}{\rho_{goe}}. \tag{22}$$

Hence,

$$\rho_{gom} = \rho_{goe} \tag{23}$$

because we assume that the rock densities ρ_r in these two cases are the same.

Since the initial density of the gas is related to the initial pressure of the mixture by the following formula,

$$\rho_{go} = \frac{P_o}{\frac{R_u}{m} T_o}, \tag{24}$$

we have the relation of the two initial pressures for those two similar explosive flows as follows:

$$p_{om} = \frac{m_e}{m_m} p_{oe} \cong 9\, p_{oe}. \tag{25}$$

If we assume that the pressures in the reservoirs are the same and because the smaller gravitational forces, the drop in pressure as the flow ascends must be smaller in the lunar case. Hence it is possible that p_{om} is larger than p_{oe}.

(vi) By dynamical similarity, the initial Froude numbers Fr_o of these two similar flows should be the same, i.e.,

$$Fr_{om} = \frac{u_{om}^2}{g_m L_m} = Fr_{oe} = \frac{u_{oe}^2}{g_e L_e}. \tag{26}$$

Hence for dynamic similarity, the ratio of the two characteristic lengths of these two similar explosive flows is

$$\frac{L_m}{L_e} = \frac{u_{om}^2}{u_{oe}^2}\frac{g_e}{g_m} \cong 60. \tag{27}$$

We shall take $L = H_o$.

(vii) By dynamic similarity, the k_p should be the same in these two similar explosive flows. In our numerical calculations we calculate the flow field corresponding to various values of k_p and then compare the terrestrial and lunar explosive flow with the same k_p.

For simplicity in our numerical calculations, we shall assume that

$$\frac{d\bar{A}}{d\bar{x}} = K = \text{constant}. \tag{28}$$

We calculate the cases for K = 1, 3, and 5 and the corresponding ratio of the exit area A_s to the initial area A_o as

$$\begin{aligned} K &= 1, \quad 3, \quad 5 \\ (A_s/A_o)_e &= 2, \quad 4, \quad 6 \\ (A_s/A_o)_m &= 61, \ 131, \ 301 \end{aligned}$$

Since we are mainly interested in the flow variables at the exit of the duct, the ratio A_s/A_o is more important than the slope dA/dx which determines the flow inside the duct of the volcano. It is interesting to notice that the exit area, the crater area of the lunar case, is about 30 to 50 times that of the terrestrial crater if the initial areas of the two eruptions A_o are the same.

V. Numerical Solutions for Similar Lunar and Terrestrial Equilibrium Explosive Flows

We shall show later that the viscous effects in region II are negligible. Hence in our numerical calculations, we consider the equilibrium inviscid flow only so that the friction terms in Eqs. (7) and (13) are neglected.

The initial conditions in our numerical calculations are:

$$T_o = 1{,}000°C,\ M_o = 1.0,\ c_s = 0.3\ \text{cal/gm °K},\ \rho_r = 3.30\ \text{gr/cc}$$
$$p_{oe} = 100\ \text{atmospheres},\ L_e = H_{oe} = 1\ \text{km}$$
$$p_{om} = 900\ \text{atmospheres},\ L_m = H_{om} = 60\ \text{km}$$

and the values for the gases are given in Eq. (18).

We have calculated two cases.

(1) *Isothermal case*

In this case, we assume that the temperature T of the mixture remains at its initial temperature T_o. This is physically possible because when the mixture moves upward, some of the molten rock may be solidified and release sufficient heat so that the resultant temperature of the mixture remains unchanged during the region II. Furthermore, from our previous experience in ash flow calculations, we find that the isothermal approximation is a good approximation for the terrestrial case (Pai *et al.*, 1972). We would like to analyze this simpler case first.

(2) *Adiabatic case*

In this case we assume that there is no heat transfer between the flow and the wall of the duct, nor viscous dissipation. The temperature of the mixture is not a constant in region II but decreases according to the isentropic process as it ascends.

Detailed numerical results are given in the reference by Y. Hsu (1976). Here we just give some typical results in Figs. 4 to 14 to show the similarities and the essential differences between similar lunar and terrestrial explosive flows.

In Fig. 4, we compare the actual velocity in the isothermal similar lunar and terrestrial explosive flows with the mass concentration k_p as a variable parameter and the corresponding area ratios $(A_s/A_o)_e = 4$ and $(A_s/A_o)_m = 181$. The lunar crater area is about 45 times that of the similar terrestrial one in this case. The velocity in all cases increases as k_p decreases. The velocity in the lunar case is much larger than the corresponding value in the terrestrial case. When k_p is equal to or less than 0.8, the exit velocity of the lunar explosive flow is larger than the escape velocity on the moon which is 2,372 m/sec.

Figure 5 shows the effect of the area ratio (A_s/A_o) on the velocity distribution in the isothermal case. When A_s/A_o increases, the velocity increases too, but the effect of (A_s/A_o) on the velocity is smaller than that due to k_p. In the lunar case, the velocity increases rapidly in the first part of the duct and remains almost constant for the remaining portion of the duct.

Figure 6 shows the effect of k_p and (A_s/A_o) on the pressure distributions for the isothermal terrestrial explosive flow. The influence of k_p on the pressure distribution is small, particularly when (A_s/A_o) is large.

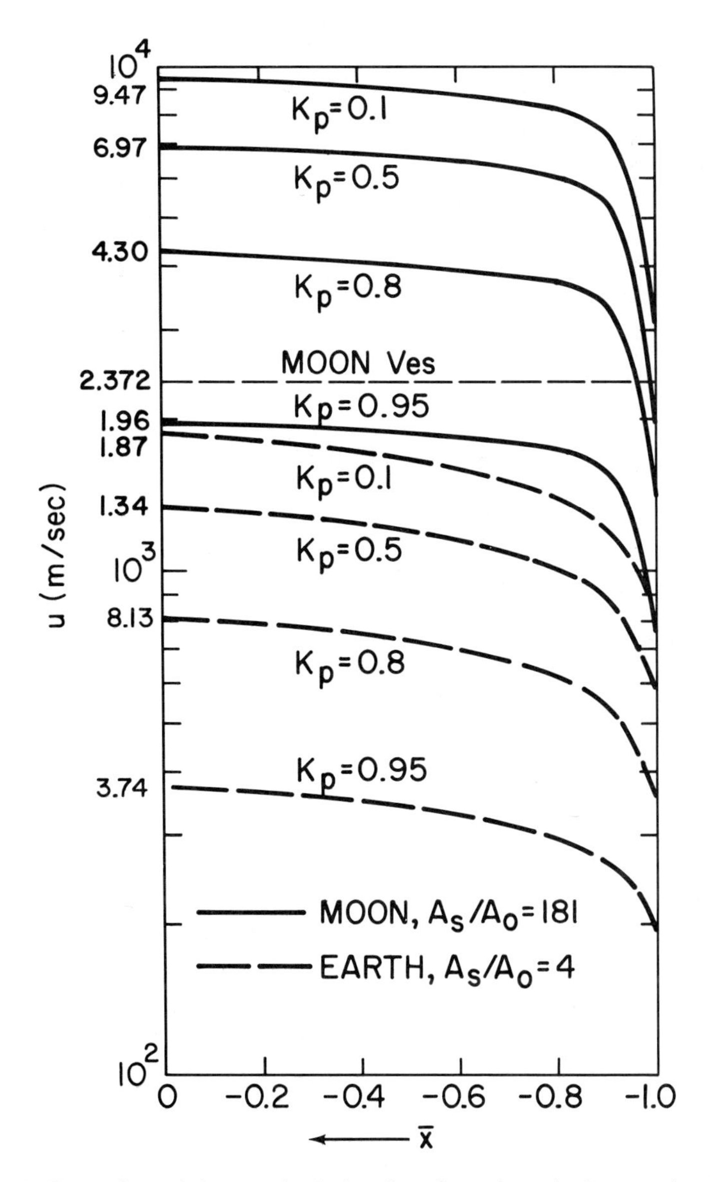

Fig. 4. Comparison of the actual velocity of *isothermal similar* lunar and terrestrial explosive flows along the duct with the k_p as a parameter, for the corresponding area ratio $(A_s/A_o)_e = 4$ and $(A_s/A_o)_m = 181$.

Figure 7 shows the effect of k_p and (A_s/A_o) on the pressure distributions for the isothermal lunar explosive flow. The influence of k_p on the pressure distributions is negligible.

Figures 8 to 13 show the results of adiabatic explosive flows. In Fig. 8, the

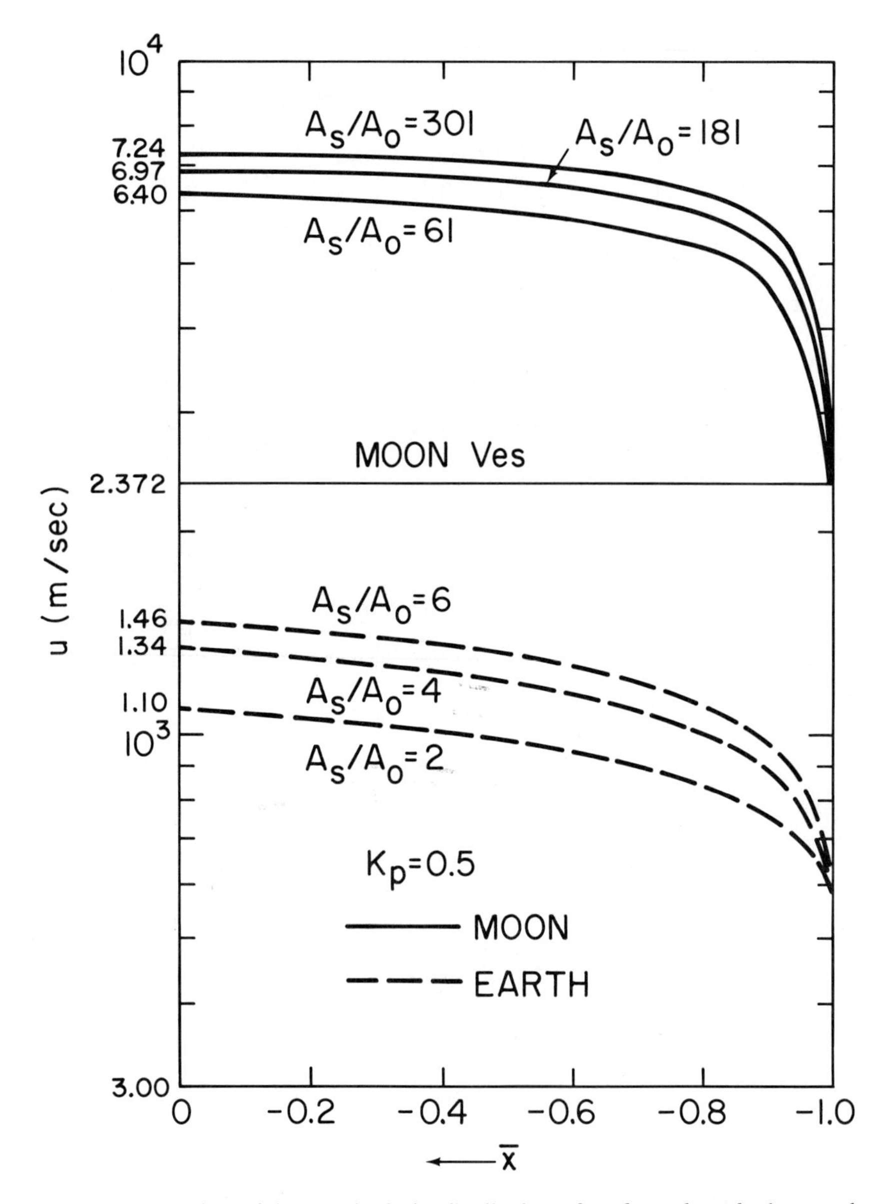

Fig. 5. Comparison of the actual velocity distributions of *isothermal similar* lunar and terrestrial explosive flows along the duct with the corresponding area ratio as a parameter for a fixed $k_p = 0.5$.

velocity distributions of the similar lunar and terrestrial explosive flows with corresponding area ratio $(A_s/A_o)_e = 4$, and $(A_s/A_o)_m = 181$ are given with k_p as a parameter. In general, these velocity distributions are similar to those of the isothermal case (Fig. 4), but the veolcity is less than the corresponding value of

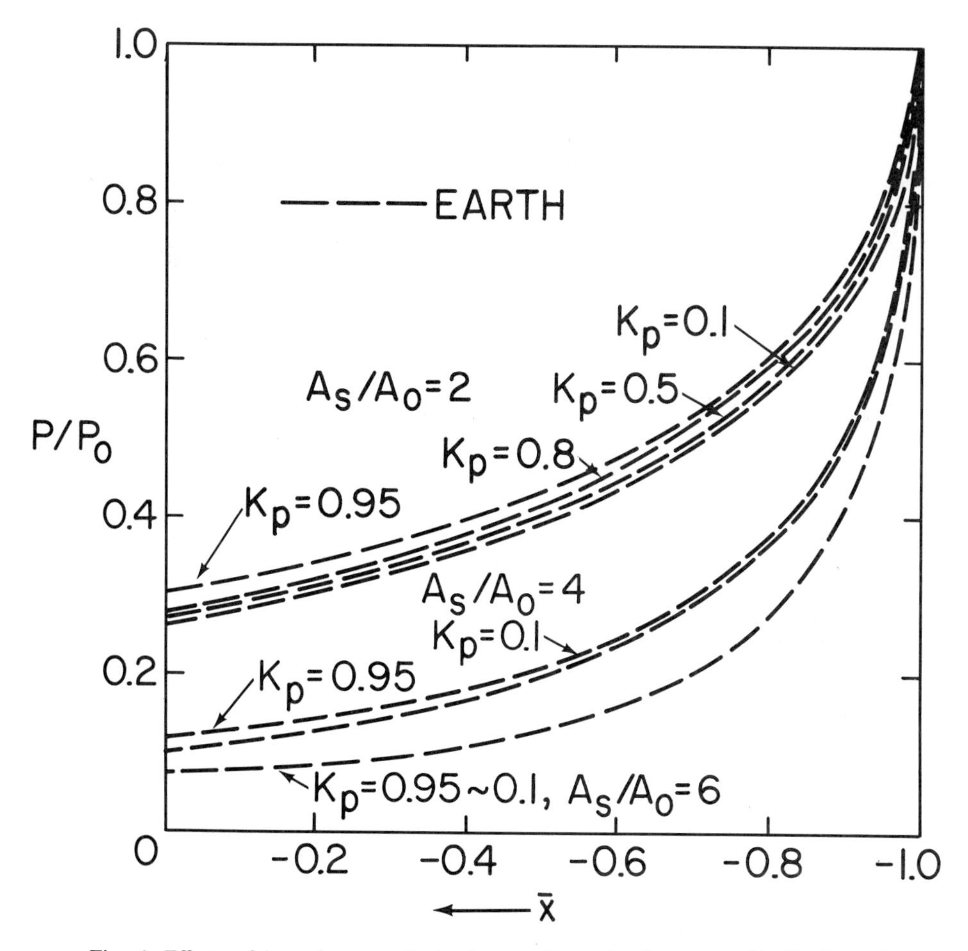

Fig. 6. Effects of k_p and area ratio to the non-dimensional pressure distributions of *isothermal terrestrial* explosive flow along the duct.

the isothermal case. The initial rate of increase of velocity of the lunar case is larger than that of the isothermal case. For small value of k_p, the velocity of the lunar explosive flow will reach its maximum possible value at a short distance above the initial section. From that point up, the lunar explosive flow will be a jet of a given width less than the width of the duct moving with the constant velocity of the value of maximum possible value.

Figure 9 shows the temperature distributions in the terrestrial adiabatic explosive flow. The temperature drops very fast initially, especially when k_p is small. In the major portion of the duct, however, the temperature is almost constant but at a smaller value than the initial temperature.

Figure 10 shows the effects of k_p and (A_s/A_o) on the temperature distributions of the adiabatic terrestrial explosive flow. For large k_p (>0.8), the isothermal

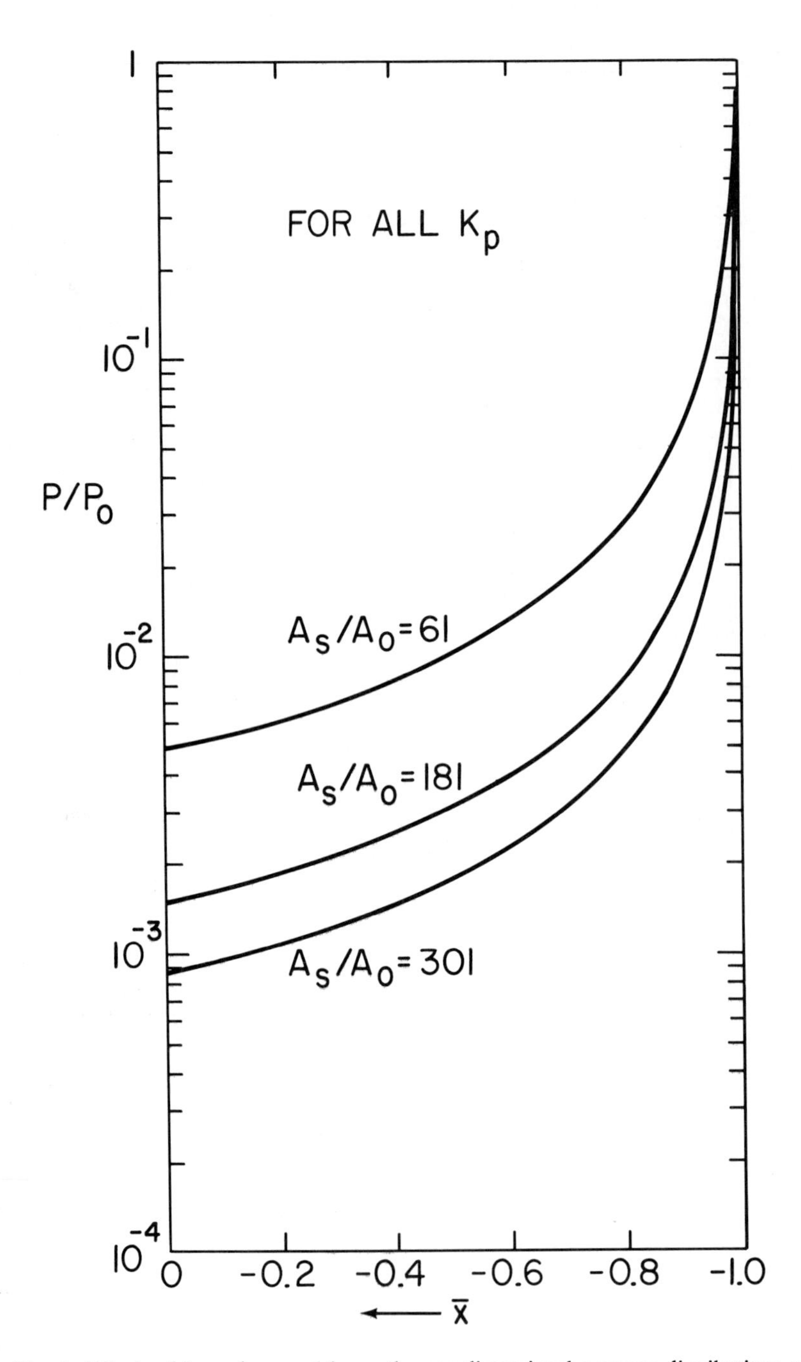

Fig. 7. Effects of k_p and area ratio on the non-dimensional pressure distributions of *isothermal lunar* explosive flow along the duct.

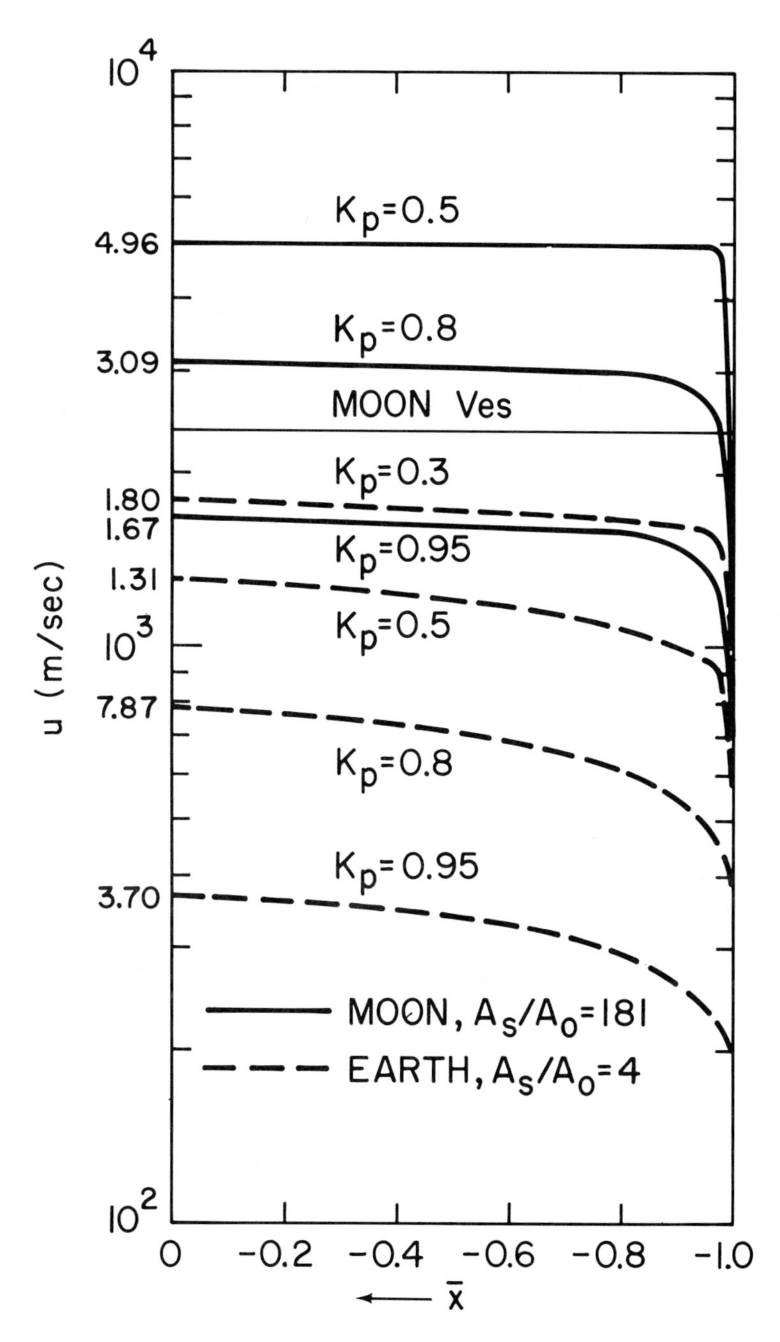

Fig. 8. Comparison of the actual velocity of *adiabatic similar* lunar and terrestrial explosive flows along the duct with the k_p as a parameter for the corresponding area ratio $(A_s/A_o)_e = 4$ and $(A_s/A_o)_m = 181$.

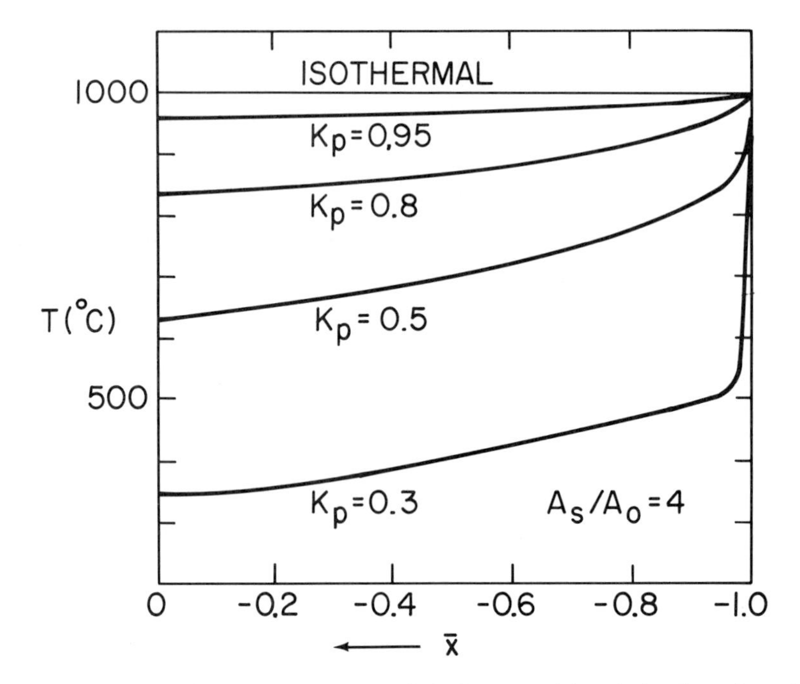

Fig. 9. Temperature distribution of the adiabatic terrestrial explosive flow along the duct with k_p as a parameter and $(A_s/A_o)_e = 4$.

approximation is a good one for terrestrial explosive flow.

Figure 11 shows the effects of k_p and (A_s/A_o) on the temperature distributions of the lunar adiabatic explosive flow. The thermal effect on the lunar explosive flow is larger than that of terrestrial flow. The isothermal approximation is not representative of adiabatic flow condition. The temperature of the lunar adiabatic flow drops very fast initially, especially when k_p is small. When k_p is less than 0.5, the condition of maximum possible velocity and zero temperature will be reached at a short distance from the initial section.

Figure 12 shows the effects of k_p and (A_s/A_o) on the pressure distributions for the adiabatic terrestrial explosive flows while Fig. 13 shows the corresponding pressure distributions of similar lunar adiabatic explosive flow. The pressure drops initially very fast when k_p is small but remains almost constant in the major portion of the duct. The pressure in the lunar case is much smaller than the corresponding value in the lunar case even though the initial pressure of the lunar case is large.

Finally, Fig. 14 shows the effect of friction on the velocity distribution on the isothermal terrestrial explosive flow. The effect of friction is negligible.

VI. Summary and Conclusions

From our theoretical study and numerical results, the following conclusions may be drawn.

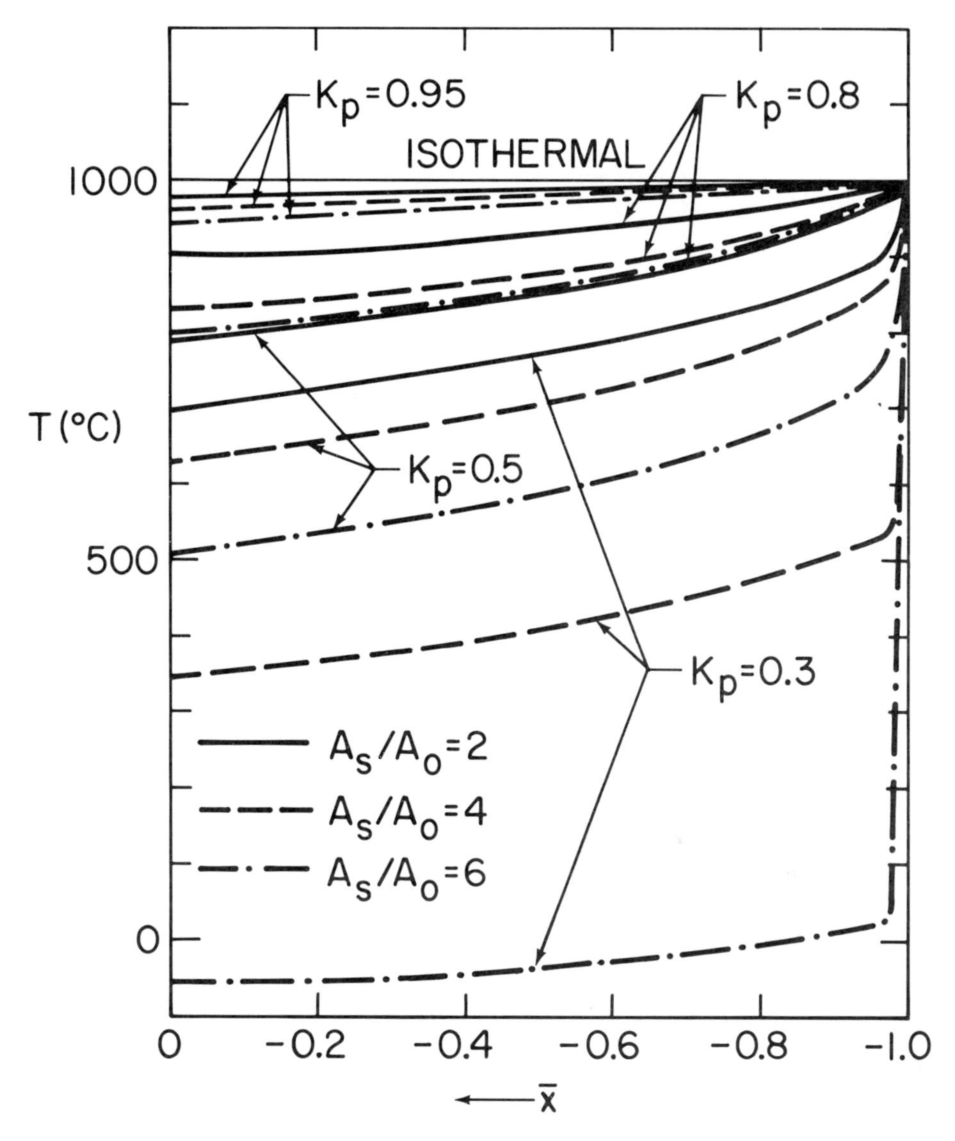

Fig. 10. Effect of k_p and area ratio on the temperature distribution of terrestrial explosive flow along the duct for adiabatic conditions.

1) The important non-dimensional parameters which characterize the explosive flow are (i) the Mach number, (ii) the Froude number, (iii) the density ratio between the rock and the gas, and (iv) the mass concentration of the rock in the explosive flow.
2) By dynamical similarity, the above four non-dimensional parameters should be the same. If hydrogen powers lunar explosive eruptions, this

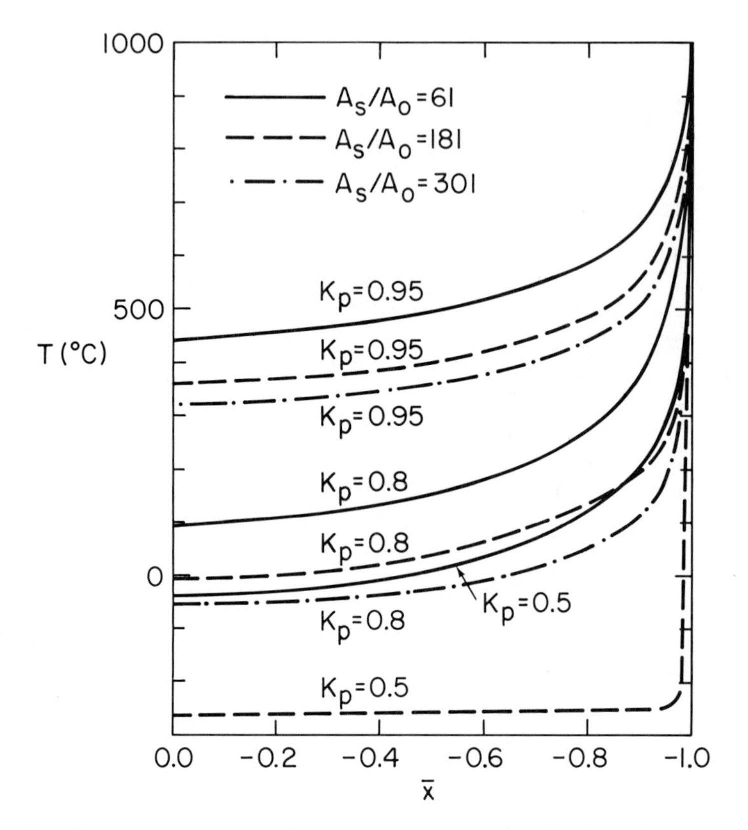

Fig. 11. Effects of k_p and area ratio on the temperature distribution of lunar explosive flow along the duct for adiabatic conditions.

suggests that terrestrial explosive eruption, with steam as the gas are to be compared (other things being equal) with lunar explosive eruption, 60 times larger in area. Hence the morphological similarity found by Pike (1974) between terrestrial maars and tuff rings, on the one hand, and the commonest kind of large lunar crater on the other is not necessarily invalid, despite the fact that the lunar craters are 50 times larger than the maars and tuff rings.

3) In the terrestrial cases, the exit velocity from the duct is supersonic, but its value is much smaller than the terrestrial escape velocity.
4) In lunar cases, the exit velocity from the duct may be higher than the lunar escape velocity when the mass concentration of the rock is less than $k_p = 0.8$.
5) In the adiabatic case, the lunar explosive flow may reach its maximum velocity before the exit and the explosive flow will then leave the duct as a

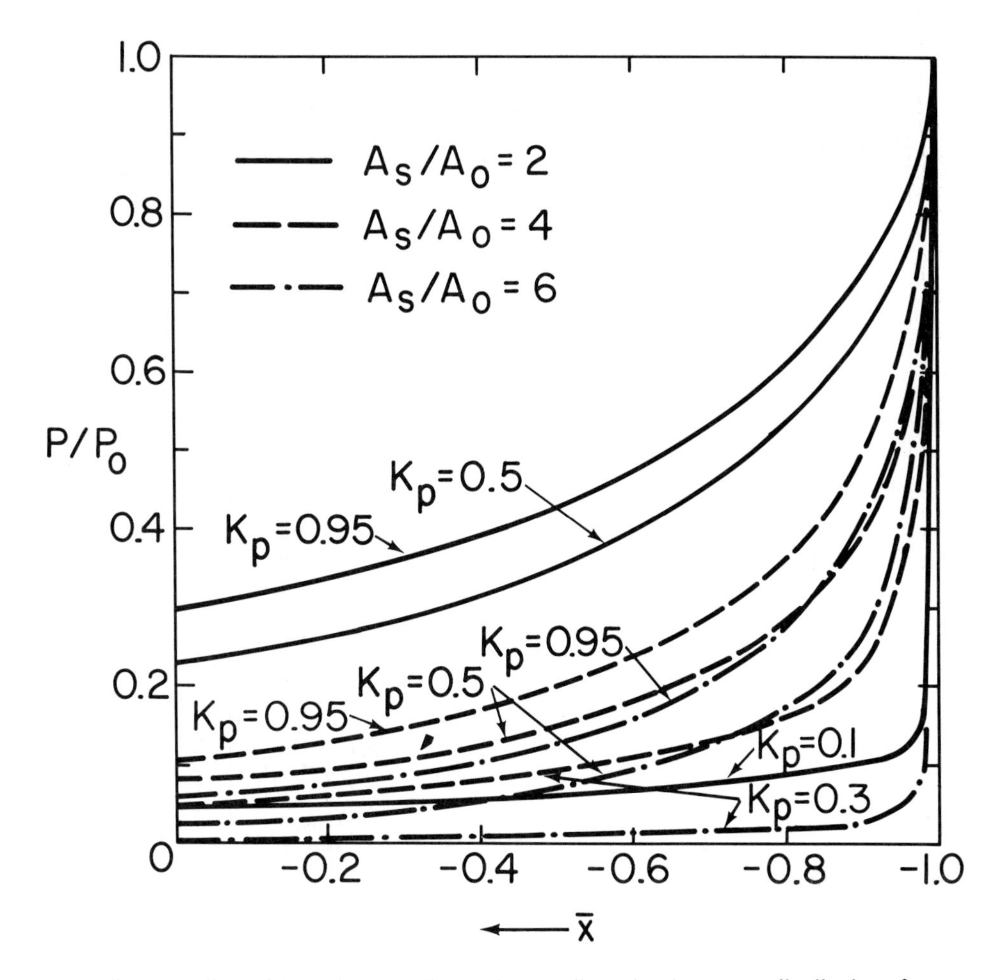

Fig. 12. Effect of k_p and area ratio on the non-dimensional pressure distribution of adiabatic terrestrial explosive flow along the duct.

jet stream with velocity greater than the lunar escape velocity. Thus the flow field agrees with the assumed flow pattern of a stream of tektites coming from the moon (Chapman and Larson, 1963).

6) When k_p is large and (A_s/A_o) is small, the temperature in the terrestrial cases is quite close to its initial value under adiabatic condition. When k_p is small, the initial drop of the temperature in both the terrestrial and the lunar cases is large; in the major portion of the duct, the temperature is almost constant but is less than the initial temperature.
7) The variation of both the velocity and the pressure in the terrestrial cases is gradual. For the lunar cases, the pressure drops very rapidly to a very

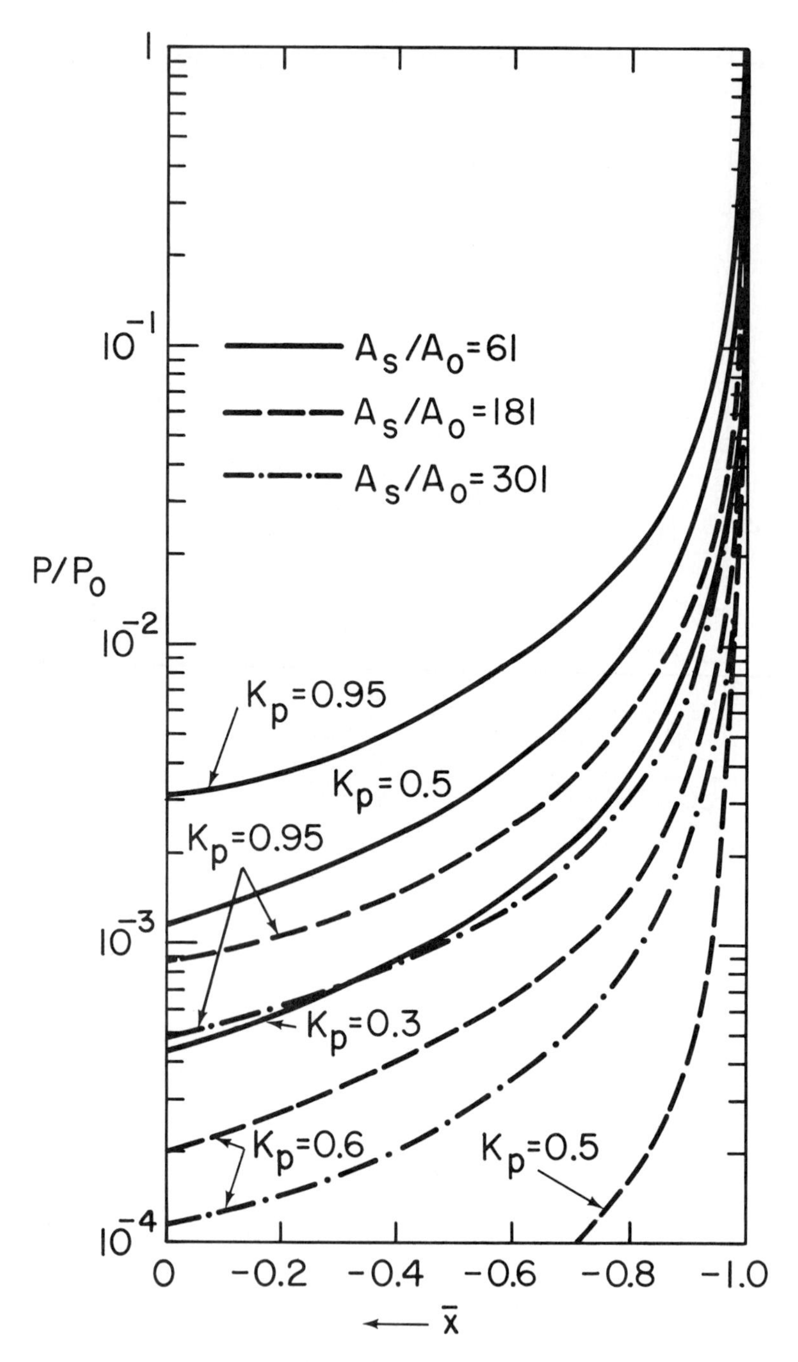

Fig. 13. Effects of k_p and area ratio on the non-dimensional pressure distribution of adiabatic lunar explosive flow along the duct.

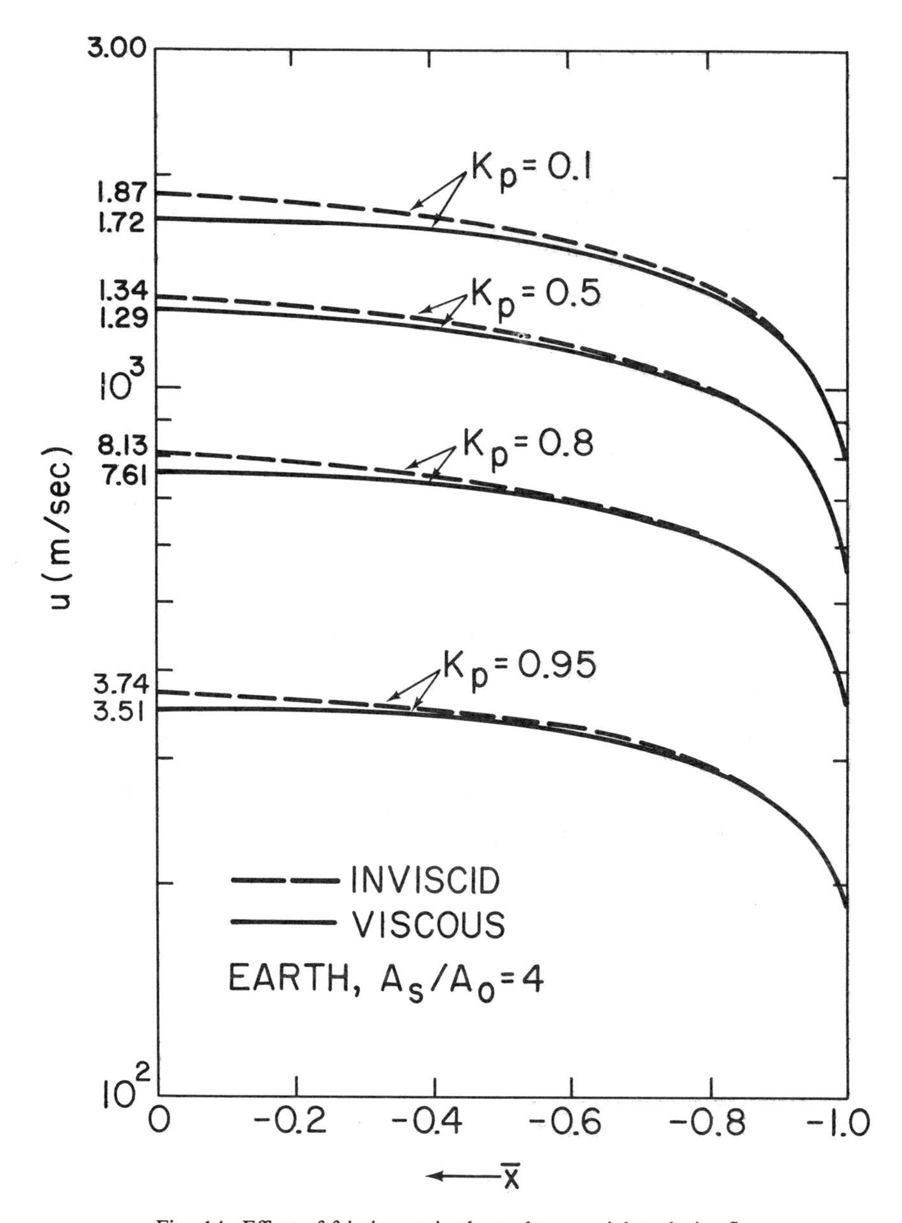

Fig. 14. Effect of friction on isothermal terrestrial explosive flow.

low value initially and then remains almost constant for the major portion of the duct. The variation of the pressure with k_p is always negligibly small. The velocity in the lunar case drops rapidly first and remains almost constant for the remaining portion of the duct.

Acknowledgments—This research is supported in part by the National Aeronautics and Space Administration under Grants No. NSG-5083 and No. NSG-7365 to the University of Maryland. Computer time was supported by NASA grant under NSG-398 to the Computer Science Center of the University of Maryland.

References

Bridgeman P. W. (1922) *Dimensional Analysis.* Yale University Press, New Haven. 112 pp.

Cameron W. S. and Padgett J. L. (1974) Possible lunar ring dikes. *The Moon* **9**, 249–294.

Chapman C. R., Mosher J. H. and Simmons G. (1970) Lunar cratering and erosion from Orbiter 5 photographs. *J. Geophys. Res.* **75**, 1445–1466.

Chapman D. R. and Larson H. K. (1963) The lunar origin of tegtites. *J. Geophys. Res.* **68**, 4305–4358.

Fielder G. (ed.) (1971) *Geology and Physics of the Moon.* Elsevier, N.Y. 159 pp.

Friedman I., O'Neil J. R., Gleason J. D. and Hardcastle K. (1971) The carbon and hydrogen content and isotopic composition of some Apollo 12 materials. *Proc. Lunar Sci. Conf. 2nd*, p. 1407–1415.

Hsu Y. (1976) Similarity laws of lunar and terrestrial volcanic flows. Ph. D. thesis, University of Maryland, College Park. 175 pp.

JANAF (1971) *JANAF Thermochemical Tables*, 2nd ed. (D. R. Stull and H. Prophet, eds.), NSRDS-NBS37, p. 1141.

McCauley J. F. (1967) Geologic results from lunar precursor probes. *Amer. Inst. Aeron. Astron. J.* **6**, 1991–1996.

McGetchin T. R. and Ullrich G. W. (1973) Deep-seated xenoliths observed terrestrial abundance, distribution and emplacement in some maars and diatremes with influences for the Moon, Mars, and Venus. *J. Geophys. Res.* **78**, 1833–1853.

O'Keefe J. A. (1976) *Tektites and their origin.* Elsevier, Amsterdam and New York. 244 pp.

Pai S. I. (1977) *Two Phase Flows*, Chapter V., p. 116–167. Vieweg, Weisbaden, Germany.

Pai S. I., Hsieh T. and O'Keefe J. A. (1972) Lunar ash flow, isothermal approximation. *J. Geophys. Res.* **77**, 3631–3649.

Pike R. J. (1974) Craters on Earth, Moon, and Mars: Multivariate analysis classification and mode of origin. *Earth Planet. Sci. Lett.* **22**, 245–255.

Sedov L. I. (1959) *Similarity and Dimensional Methods in Mechanics.* Academic Press, N.Y. 363 pp.

Proc. Lunar Planet. Sci. Conf. 9th (1978), p. 1509–1526.
Printed in the United States of America

A comparison of a lunar and a terrestrial volcanic section

J. STEWART NAGLE

Lunar Curatorial Laboratory (Northrop Services Inc.)
Johnson Space Center, Houston, Texas 77058

Abstract—A section through a terrestrial cinder cone was compared to the section in lunar core tubes 74002/1 to characterize the changes that could be expected in a localized volcanic deposit of compact particles, and thus evaluate the lunar section for completeness and overturning. Both the lunar and terrestrial sections were internally non-stratified and contained relatively fine-grained glassy particles at one end, and, at the other end, were stratified, coarse-grained and contained crystallized particles similar in form to the glassy particles. Changes in both sections were rapid but without discontinuities suggesting that the lunar section is essentially intact and complete. Trends in the lunar section were reversed from those of the terrestrial section, suggesting that the lunar section is overturned.

INTRODUCTION

Systematic and progressive changes in ultramafic droplet petrography were observed during dissection of lunar drive tubes 74002/1. In order to understand and interpret the lunar cores, the question was asked, "Are there comparable petrographic changes in terrestrial volcanic deposits?". Since a search of the literature and discussions with active volcanolgists indicated that little is known about variations and trends within mafic terrestrial pyroclastic deposits, observations were made, therefore, of material from cinder cones in the western United States and from Hawaiian cinder cones. In all cones observed to date, there are fine-grained glassy cinders at the bottom and large, crystallized cinders at the top. Whether these changes occur in most cinder cones is still under investigation, but trends opposite to those reported here have not been found, and enough is now known about sections through terrestrial cinder cones to make a tentative comparison to the lunar cores.

This paper reports on variations within one terrestrial cinder cone, representative of the Flagstaff, Arizona, volcanic field, and compares trends to the part of the lunar section that is best interpreted as pyroclastic. There is considerable evidence for considering the lunar section to be fire-fountain volcanic origin and for rejecting alternative hypotheses of origin (Heiken, 1976). Furthermore, McGetchin and Head (1973) have described and interpreted some features near the Apollo 17 landing site to be the lunar equivalent of cinder cones; if so, one could expect an analog to cinder cone pyroclastics at the landing site. Although much of the material in the drive tubes may be pyroclastic, the core itself was a section through a large clast of partially indurated material that is common in the ejecta blanket on the rim of Shorty Crater, an impact crater. Age dating of 74220 indicates that the orange and black pyroclastics are old (Hueneke *et al.*,

1973, give an age of 3.6 b.y.). They have been redeposited relatively recently on the rim of Shorty Crater (see Eugster *et al.*, 1977 or Nagle, 1978, for discussion of the age of Shorty Crater) probably by the impact that produced Shorty Crater. Nagle has mapped the megaclasts on the crater rim (Duke and Nagle, 1974, p. VII. 5.3) and Heiken *et al.* (1974) have identified the horizon near the bottom of Shorty Crater, where the orange and black soil probably originated. Post-Shorty activity has modified the upper part of 74002, and the upper 10 cm of that core does not continue the progressive changes seen lower in the core; rather, it is a mixture of orange and dark soil, which contains agglutinates near the top and is discussed as a detrital zone in another paper (Nagle, 1978). Assuming that the lunar orange and black droplets are pyroclastic and that trends seen in the terrestrial and lunar sections are comparable, questions to be dealt with are: (1) Is the lunar drive tube section intact or disrupted? (2) In what ways are the lunar and terrestrial sections comparable? (3) Is the lunar volcanic section upright or overturned?

DESCRIPTION OF TERRESTRIAL AND LUNAR VOLCANIC SECTIONS

TERRESTRIAL SECTION

Location and mode of occurrence

The terrestrial volcanic section sampled is on the property of the Flagstaff Cinder Company, 2 km East of Flagstaff, Arizona. Here, two quarries excavate a 50 m continuous section through both the upper and lower parts of the 50–60 m high cinder cone. The cone and associated basalt flows are interpreted by Moore *et al.* (1976) to be of Tappan age, deposited 0.2–0.7 m.y. ago. Because the cone was deeply quarried, it was possible to obtain samples from the base of the cone, very near the beginning of the deposit, through the entire cone (Table 1). Sections, or exposed portions of other cinder cones within the basaltic part of the San Francisco volcanic field, are similar to the sampled section, in showing dark,

Table 1. Location of terrestrial volcanic samples.

Grab samples of 100-500 gm mass were collected from sheer walls of the quarry as follows:

(1) 1 M above base, to sample black cinders in massive strata
(2) 3 M above base, to sample black cinders in massive strata
(3) 5 M above base, to sample mixed black-brown cinders
(4) 8 M above base, to sample mixed brown-black cinders
(5) 15 M above base, to sample part of massive section that contains the first red cinders, mixed with brown and black cinders
(6) 30 M above base, to sample weakly-stratified part of section that contains 100% red cinders
(7) 35 M above base of section, in stratified part of section, a suite of three samples was collected to sample the base, middle, and top of a graded bed, in red cinders.

unstratified, fine cinders at the base, and red, stratified, coarse cinders and lapilli at the top. Hence, the sampled section is believed to be representative (at least qualitatively) of the volcanic field.

Volcanics of Tappan age are weathered on the surface, but not dissected by erosion (Moore *et al.*, 1976), and the sampled cinder cone fits this category. A clayey-altered zone occurred approximately 5–10 m below the partially vegetated surface of the volcano, but cinders below the weathered zone were both fresh and unaltered, and appeared glassy or crystalline under the hand lens. Samples were collected a minimum of 10 m below the lowest visibly weathered material; sericite and alteration oxides were not seen in thin sections of the cinders, indicating weathered material was avoided.

Changes in stratification

The lowest 20 m of the section is black and massive in outcrop while the upper part is red and stratified. At the base, the deposit consists of black, glassy, cm-sized cinders mixed with similar-sized white limestone xenoliths, derived from local Paleozoic and Mesozoic sedimentary rocks (Moore *et al.*, 1976, p. 557). Xenoliths diminish in abundance upward and cinders become brownish and less vitreous. Still higher in the section cinders become reddish and crystalline. Above 20–25 m, all the cinders are red and the section becomes stratified, with layers being clearly present but with very irregular and crenulate boundaries. These layers appear to be graded beds 0.5–2 m thick, which dip away from the center of the cinder cone at approximately 30 degrees and show normal graded bedding, becoming noticeably finer upward in each bed. At the base of each bed are numerous flattish compound lapilli up to 0.5 m across, consisting of aggregates of dark reddish-gray (5R 3/1) cm-sized cinders containing 5–10% mm sized laths of plagioclase. Higher in each graded bed, bombs or cinder aggregates are smaller, more equant, and there are great quantities of single as well as compound cinders. At the top of each stratum, individual cinders are mm-sized, equant, fairly well-sorted, light reddish gray (7.5R 3/2), and contain 5–10% conspicuous mm-sized plagioclase laths. Strata are internally transitional, but contacts between strata, although crenulate, are very distinct and well-defined with darker, compound cinders showing re-entrants containing light-colored cinders from the top of the underlying stratum. Samples were taken at positions indicated in Table 1 and Fig. 1 in order to characterize the red, stratified cinders, the black, massive cinders, and the transition zone.

Changes in physical properties

Textural properties of grain size distribution and sorting were determined by dry sieve analysis of 0.5, 1, 2, 4, 10, 20 and 40 mm size fractions. Abundance of compound cinders and presence of rounded and compressed vesicles were determined under the binocular microscope. The location of samples is given in Fig. 1. The black cinders from the base of the section, in sample 1, averaged 5

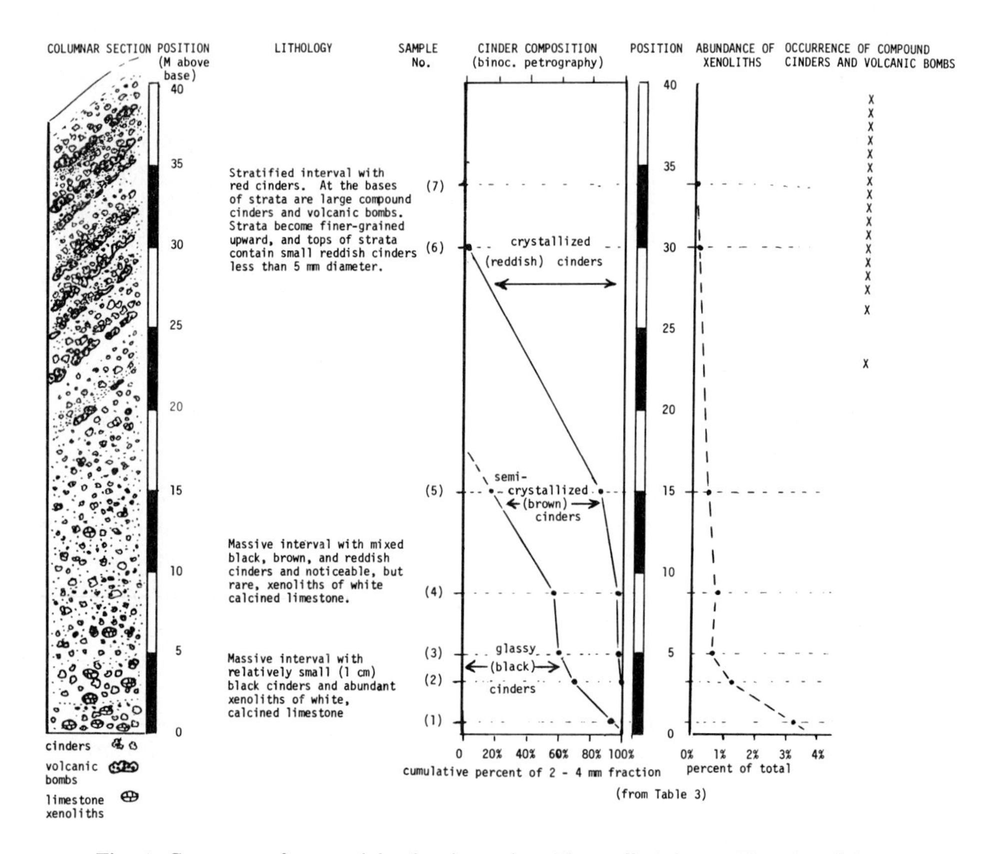

Fig. 1. Summary of terrestrial volcanic section, Flagstaff, Arizona. Figs. 1 and 2 are comparable in that following the columnar section, is general lithology, particle petrography, xenolith or possible xenolith abundance, and occurrence of very large particles.

mm in diameter, but the mean size of black, brown, and red cinders in samples 2–4 was a nearly constant 9–10 mm. There was a slight increase to 12 mm in partially stratified material represented by sample 5, primarily because of the occurrence of compound particles, composed of adhering cinders in this and stratigraphically higher samples. Mean particle size in the graded beds of the stratified part of the cinder cone was near 40 mm, because of the occurrence of the large aggregate masses composed of numerous cm-sized cinders that were welded together. However, mean size of individual, non-aggregated cinders remained constant throughout the entire section. In cross-section the compound cinders show flattened vesicles or vertically-aligned trains of vesicles whereas the black cinders all contain vesicles with circular cross-sections. Xenolith size distribution is the same as that of individual cinders. In summary, the most striking observations are the constancy of size of individual cinders throughout the section and the occurrence of large compound cinders only in the stratified upper part of the section.

Petrographic changes

Components in the Flagstaff volcano include red, brown and black cinders, and xenoliths. To the unaided eye the terrestrial cinders appear black and shiny when vitreous, dull and brownish when semivitreous, and cinders with crystallized matrix appear a dull earthy red. Xenoliths of white, calcined limestone of Paleozoic Age (Moore *et al.*, 1976, p. 557) are obvious against the black glass of the terrestrial section, but ultramafic and igneous xenoliths, if present, were not recognized. In thin section, black, glassy cinders have a matrix of brownish sideromelane, rounded vesicles, and randomly-aligned plagioclast phenocrysts; whereas, reddish cinders have a crystallized opaque matrix, vesicles are collapsed, and plagioclase laths are aligned. Brownish cinders show intermediate characteristics, with matrix consisting of patches of interspersed brownish glass and opaques, with some collapsed vesicles.

Glassy black cinders (Table 2; Fig. 1) are most abundant at the base of the cinder cone and are succeeded by semi-crystallized brown cinders which, in turn, are succeeded by the wholly crystallized red cinders which make up nearly 100% of the stratified part of the section. In contrast to the gradual succession of cinder types, xenoliths rapidly decrease in abundance up the section, although a few persist as high as 30 m above the base of the section.

LUNAR SECTION

Location and mode of occurrence

The Lunar section is preserved in two continuous cores, 74002 and its underlying companion core, 74001. Both were collected during the Apollo 17 mission to sample the orange soil on the rim of Shorty Crater. The orange soil deposit is now believed to have been excavated by the Shorty Crater event from more continuous material near the floor of the crater (Heiken *et al.*, 1974). It now occurs as discrete megaclasts on the rim of the Shorty Crater (Duke and Nagle, 1974, p. VII.5.1).

Changes in stratification

Presence of possible stratification in 74002/1 was first seen in X-radiographs of the cores (Duke and Nagle, 1974, VII.5.9), and evidence for subtle transverse layering was seen in the peel, taken after dissection. The double drive tube containing the section was carefully dissected during 1977 and was found to contain a relatively continuous succession from dark clastic material in the lower tube (74001) to orange clastic material in the upper part of 74002. In the

Table 2. Flagstaff Cinder Co., Data.

Sample	Category	Black (glassy) Cinders	Brown (Semi-glassy) Cinders	Red (Devi-trified) Cinders	Sum Of Interval	Compound Cinders (Inc. Red and Dk.)	Xenoliths (Wt.% of Tot. Samp. in Frac.)	Total Mass in Fraction	Binocular Petrography % of Glass In Matrix
1)	Number of Particles	236	8	0	252	0			
2–4 mm	Numerical Pct.	94%	3%	0%		0%			
	Weight of Particles	5.15 gm	0.20 gm	0.00 gm	5.55 gm		1.3 gm		
	Weight Pct.	93%	4%	0%			3.4%	38.1 gm	
1–2 mm	Number of Particles	158	5	0	165	0			
	Numerical Pct.	96%	3%	0%		0%			
	Weight of Particles	0.40 gm	.02 gm	0.00 gm	0.44 gm		0.4 gm		
	Weight Pct.	91%	4%	0%			1.9%	20.6 gm	
<1 mm	Number of Particles	588	0	0	648		4.3 (entire sample)		
	Numerical Pct.	91%	0%	0%			3.12%	137.9 gm	
2)	Number of Particles	156	63	0	222	0			7*
2–4 mm	Numerical Pct.	70%	28%	0%					60% (40–80%)
	Weight of Particles	2.35 gm	0.95 gm	0.00 gm	3.35 gm	0%	1.2 gm		
	Weight Pct.	70%	28%	0%			1.7%	71.7 gm	
1–2 mm	Number of Particles	359	168	0	540				
	Numerical Pct.	66%	31%	0%					
	Weight of Particles	0.75 gm	0.60 gm	0.00 gm	1.40 gm		1.8 gm		
	Weight Pct.	54%	43%	0%			2.8%	65.2 gm	
<1 mm	Number of Particles	173	111	0	287		5.1 (entire sample)		
	Numerical Pct.	60%	39%	0%			1.07%	478.2 gm	
3)	Number of Particles	94	59	3	157	0			11*
2–4 mm	Numerical Pct.	60%	38%	2%					50% (10–95%)
	Weight of Particles	1.75 gm	1.25 gm	0.1 gm	3.15 gm	0%	0.2 gm		
	Weight Pct.	56%	39%	3%			1.1%	17.9 gm	
1–2 mm	Number of Particles	109	49	9	172	0			
	Numerical Pct.	63%	29%	5%					
	Weight of Particles	0.40 gm	0.20 gm	0.03 gm	0.65 gm	0%	0.1 gm		
	Weight Pct.	61%	31%	5%			1.6%	6.2 gm	

Table 2. *(cont'd.)*

Sample	Category	Black (glassy) Cinders	Brown (Semi-glassy) Cinders	Red (Devi-trified) Cinders	Sum Of Interval	Compound Cinders (Inc. Red and Dk.)	Xenoliths (Wt.% of Tot. Samp. in Frac.)	Total Mass in Fraction	Binocular Petrography % of Glass In Matrix
<1 mm	Number of Particles	194	103	32	335	0	0.8 (entire sample)		
	Numerical Pct.	58%	31%	1%			0.6%	139.6 gm	
4) 2–4 mm	Number of Particles	125	90	7		2			7*
	Numerical Pct.	56%	40%	3%		2%			60% (10–80%)
	Weight of Particles	2.53 gm	2.36 gm	0.12 gm	5.01 gm		0.3 gm		
	Weight Pct.	56%	40%	2%			0.6%	53.1 gm	
1–2 mm	Number of Particles	163	107	9	281	2			
	Numerical Pct.	58%	38%	3%		1%			
	Weight of Particles	0.44 gm	0.40 gm	0.03 gm	0.88 gm		0.1 gm		
	Weight Pct.	50%	45%	3%			0.5%	21.9 gm	
<1 mm	Number of Particles	283	35	71	391		2.9 gm (entire sample)		
	Numerical Pct.	72%	9%	18%			0.94%	394.0 gm	
5) 2–4 mm	Number of Particles	66	228	37	342	4			10*
	Numerical Pct.	19%	67%	11%		4%			60% (10–90%)
	Weight of Particles	1.25 gm	4.32 gm	0.90 gm	6.69 gm		0.2 gm	9.0 gm	
	Weight Pct.	19%	65%	13%			2.2%		
1–2 mm	Number of Particles	57	285	56	408	10			
	Numerical Pct.	14%	70%	14%		2.5%			
	Weight of Particles	.15 gm	.82 gm	.18 gm	1.18 gm		0.1 gm	4.0 gm	
	Weight Pct.	13%	69%	15%			2.5%		
<1 mm	Number of Particles	86	362	84	543		1.1 (entire sample)		
	Numerical Pct.	16%	67%	15%			0.59%	186.5 gm	
6) 2–4 mm	Number of Particles	0	0	110	111	12			6*
	Numerical Pct.			99%		11%			1% (0–5%)
	Weight of Particles			1.4 gm	1.4 gm		0.16 gm	17.9 gm	
	Weight Pct.			99%			0.9%		
<1 mm	Number of Particles			284	286		.23 (entire sample)		
	Numerical Pct.			99%			0.14%	164.3 gm	

*Number of particles in slide.

Table 2. *(cont'd.)*

Sample	Category	Black (glassy) Cinders	Brown (Semi-glassy) Cinders	Red (Devi-trified) Cinders	Sum Of Interval	Compound Cinders (Inc. Red and Dk.)	Xenoliths (Wt.% of Tot. Samp. in Frac.)	Total Mass in Fraction	Binocular Petrography % of Glass In Matrix
7) M**	Number of Particles	0	0	81	81	12			4*
2-4 mm	Numerical Pct.			100%		5%			5% (0-1%)
	Weight of Particles			2.0 gm	2.0 gm		0		
	Weight Pct.			100%			0%		
	Number of Particles	0	0	286	286	16	0		
1-2 mm	Numerical Pct.			100%		6%	0%		
	Weight of Particles			1.0 gm	1.0 gm				
	Weight Pct.			100%					
<1 mm	Number of Particles						0% (entire sample)		
	Numerical Pct.								
7) U	Number of Particles	0	0	84	84	2			6*
2-4 mm	Numerical Pct.	0%		100%		2%			0%
	Weight of Particles	0.00 gm		2.0 gm	2.0 gm				
	Weight Pct.	0		100%					
	Number of Particles			262	262	5			
1-2 mm	Numerical Pct.			100%		2%			
	Weight of Particles			1.0 gm	1.0 gm		1		
	Weight Pct.			100%			Tr		
	Number of Particles								
	Numerical Pct.								

*Number of particles in slide.

**7 consists entirely of 20–40 cm flattish masses, 10% cpd.

Note that cinder petrography was determined by point count, and checked by weighing of aliquots from each size fraction; then all light-colored xenoliths were picked out of the entire fraction and weighed directly. To save space, xenolith abundance was left out of the cinder petrography tables; hence some weights and percentages do not add up to the sum of the interval.

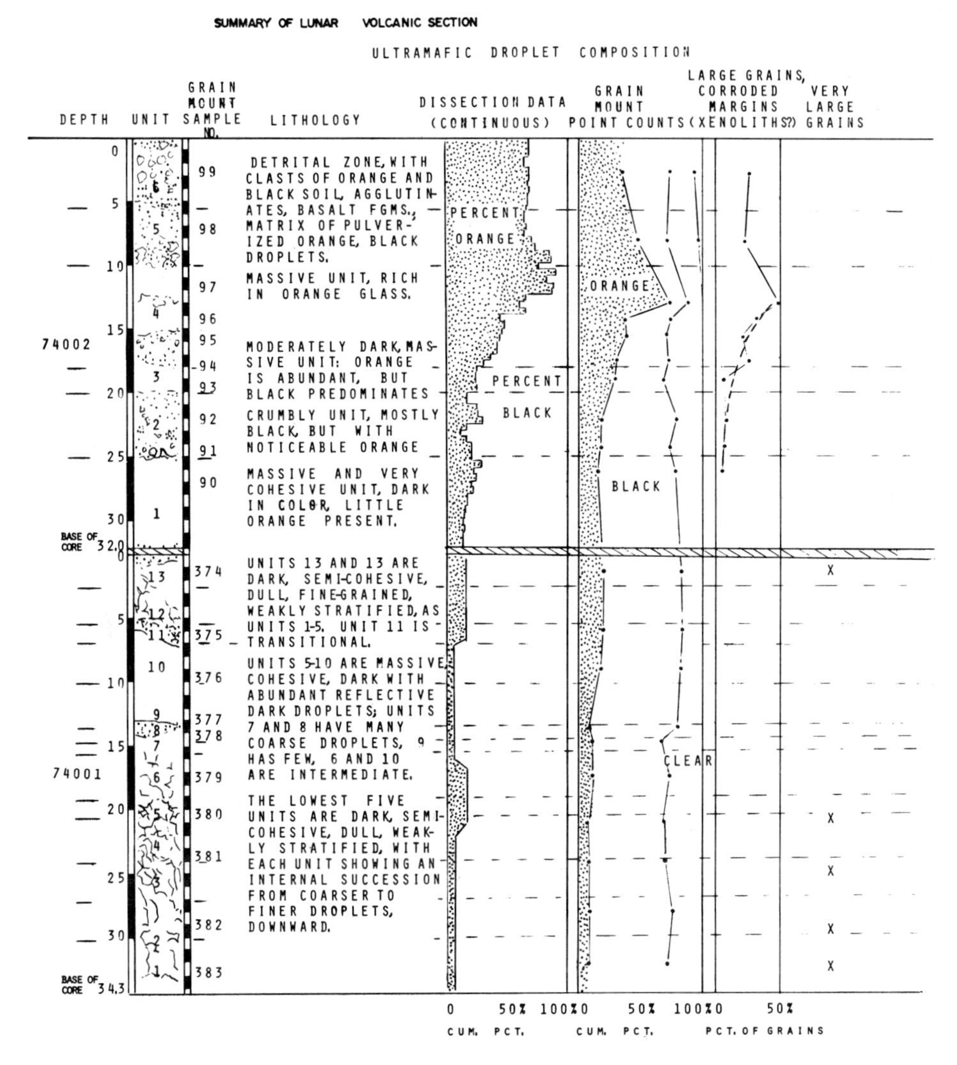

Fig. 2. Summary of Lunar volcanic section from Taurus-Littrow Valley, Drive Tubes 74002/1. Compare to Fig. 1.

dissection cabinet the double core appears black through all of 74001 to 25 cm below the top of 74002 and gradually becomes orange between 25 and 13 cm of 74002. (All subsequent locations are referenced to positions indicated in Fig. 2.) The most distinct orange color appears between 13 and 10 cm, above which the core becomes noticeably darker. Material in the core is alternately massive and crumbly, giving the appearance of stratification. Massive zones occur between 7 and 19 cm of 74001, and between 10 and 15, 18–20 and 25–32 cm of 74002. Abundance of orange and black particles was estimated over the entire length of

Table 3. Compositional summary for coarser size classes, grain mounts from drive tube 74001.

		Orange Glass		Black "Glass"		Clear Droplets			% Compound Orange		
Sample No.	Size Class	No. of Points	% of Total	No. of Points	% of Total	No. of Points	% of Total	Total points Counted	No. of Points	% of Total	Total points Counted
–,374	3*	49	16%	225	75%	27	9%	300			
	4	56	18%	188	62%	59	19%	303	37	18%	202
	5	28	14%	123	60%	53	26%	204			
–,375	3	36	18%	136	70%	23	12%	195			
	4	147	18%	533	63%	143	18%	823	32	16%	205
	5	76	18%	213	49%	142	33%	431			
–,376	3	19	9%	146	73%	36	18%	201			
	4	33	17%	124	63%	39	20%	196	43	21%	208
	5	45	18%	114	47%	85	35%	244			
–,377	3	4	4%	92	81%	18	15%	114			
	4	13	6%	159	72%	50	22%	222	32	22%	144
	5	19	9%	105	50%	86	40%	21			
–,378	3	3	2%	116	78%	30	20%	149			
	4	24	9%	156	56%	96	34%	276	28	18%	160
	5	22	9%	113	45%	115	46%	250			
–,379	3	14	7%	154	77%	32	17%	200			
	4	27	9%	190	63%	83	28%	300	37	18%	201
	5	28	14%	95	48%	75	37%	198			
–,380	3	7	4%	146	73%	46	24%	199			
	4	15	6%	156	63%	75	32%	246	48	24%	204
	5	26	13%	92	46%	82	41%	200			
–,381	3	28	6%	333	72%	99	22%	460			
	4	30	7%	251	61%	133	32%	414	44	22%	204
	5	36	12%	123	39%	158	50%	317			

-,382	3	34	8%	262	68%	91	23%	387			
	4	14	7%	122	58%	73	34%	209	47	21%	219
	5	32	15%	82	38%	101	47%	215			
-,383	3	13	6%	143	72%	44	22%	290			
	4	15	7%	143	62%	69	30%	228	67	27%	248
	5	18	9%	95	48%	85	43%	198			
-,97	3	141	70%	53	26%	6	3%	200			
	4	159	67%	48	20%	30	13%	237	40	8%	524
	5	200	73%	32	12%	41	15%	273			
-,96	3	112	53%	91	43%	7	3%	210			
	4	114	35%	122	38%	87	27%	322	19	6%	316
	5	171	42%	120	29%	116	28%	407			
-,95	3	97	43%	112	50%	16	7%	225			
	4	105	37%	94	33%	83	30%	282	19	7%	274
	5	143	44%	83	26%	96	29%	322			
-,94	3	92	44%	101	48%	15	7%	208			
	4	96	28%	154	44%	97	28%	347	58	12%	492
	5	170	35%	142	30%	170	35%	482			
-,93	3	68	32%	121	57%	22	10%	211			
	4	95	24%	175	44%	131	33%	401	34	12%	277
	5	104	26%	151	38%	146	37%	401			
-,92	3	49	25%	139	70%	12	6%	200			
	4	50	16%	187	61%	68	22%	305	33	13%	248
	5	55	18%	146	48%	103	34%	304			
-,91	3	51	15%	210	60%	89	25%	350			
	4	56	16%	199	56%	98	28%	353		poor slide	
	5	81	24%	158	47%	100	30%	339			
-,90	3	36	8%	325	73%	83	18%	444			
	4	67	17%	237	59%	51	24%	355	25	17%	145
	5	46	17%	174	64%	53	20%	273			

*Size classes are as follows: 2-.25-.12 mm, 3-.12-.06 mm, and 4-.06-.01 mm.

both cores during dissection (Data are in Duke and Nagle, 1974, p. VII.5.14). These data are limited by poor viewing conditions under the binocular microscope. To better document changes in petrographic characteristics, 1 mg of soil was extracted from several locations within the core and prepared as epoxy-stabilized grain mounts. (Table II in Duke and Nagle, 1974, p. VII.5.8., shows the location of these grain mounts and Table 3 of this paper indicates properties studied.) The volcanic part of the section occurs between 10 cm below the top of 74002 and the base of the core. Above 10 cm, structure changes from stratified to marbled. Agglutinates appear above 5.5 cm (Duke and Nagle, 1974, p. VII.5.32; Heiken *et al.*, 1978, p. 492), and Morris *et al.* (1978) p. 763, find an increase in the Is/FeO profile near the top of the core. This evidence indicates that the upper part of the core is detrital, that the surface was reworked by micrometeorites, and that it has undergone post-Shorty modification. Discussion of strata and structures of the upper 10 cm is not relevant to interpretation of the volcanic part of the section and is discussed in detail elsewhere (Nagle, 1978).

Changes in grain size

The best data on grain size distribution in the double core are from the sieve analyses of McKay and Waits (1978, p. 275). They report a decrease in size from the bottom to the top of the core, with samples from the lower portions of 74001 being 45 μm in mean diameter, and samples in 74002 being 34–38 μm. Furthermore, there is an upward decrease in the weight percentage in the coarsest sieve fraction of McKay and Waits (0.5 to 1 mm) from over 8% near the base of 74001 to less than 1% in 74002. These data agree with sizes noted during dissection, when all particles were passed through 1 mm sieve. During dissection a total of 5 particles over 1 mm were found between 10 cm of 74002 and the base of 74001. All occurred in 74001, and all were compound particles of dark, partially crystallized glass.

Changes in petrographic composition

Variations in 74002/1 were observed and quantified during dissection by point counting at least 100, and generally 150–200 particles in every 5 mm dissection interval. Although orange, black, and clear droplets were observed during dissection, comparison to grain mounts revealed that only the abundance of orange droplets had been unambiguously determined during binocular examination. To retain consistency with other data, clear and black droplets were considered to be a single class (Fig. 2).

During study of the grain mounts, 200–500 grains were counted in each size class to characterize the petrographic changes presented in Table 3 and Fig. 2. Particle types seen in the grain mounts include droplets, fragments of droplets, and large crystal fragments. Droplets range through an entire spectrum from orange glass through black, quench-crystallized glass to clear droplets consisting wholly of single, transparent crystals in optical continuity. Material classified here as orange glass ranges from transparent light orange to nearly opaque, blackish orange. Particles classified here as dark droplets are opaque or semi-opaque, rich in ilmenite, and contain relatively small transparent areas that are finely twinned and highly birefringent. Many of the dark droplets show large transparent areas, and others are almost completely clear, with no ilmenite. Droplets with more than 65% transparent minerals are classified as clear droplets (Table 3). Such particles were found to be much more abundant in grain mounts than during dissection, probably because of differences in viewing conditions. Although most clear droplets appear to be finely twinned, highly birefringent and irregular-fracturing, other particles with droplet morphology are more coarsely twinned, less birefringent, and show definite cleavage. The transparent mineral phase is mostly olivine, but the presence of other minerals, such as pyroxene, cannot be excluded. However, because it was not possible to obtain interference figures or index of refraction in the .1 mm-sized, epoxy-stabilized grain mounts, the term clear droplets is applied rather than a specific mineral designation.

The dark and mixed dark-clear particles appear to represent the quench-crystallized olivine-ilmenite droplets of Heiken and McKay (1977). Some clear droplets and droplet fragments are not as finely twinned as the darker droplets, and appear, at least in part, to represent the more coarsely crystallized, subequant skeletal forms reported by Heiken *et al.* (1978).

Within the cores, the orange glass gradually increases in abundance from less than 10% in lower 74001 to approximately 25% at 20 cm of 74002. From 20 cm to 10–12 cm of 74002, abundance of orange glass increases rapidly to 80–90%. Clear droplets are more abundant low in 74001, and show a minor abrupt decrease in mid-74001, but black, partially crystallized glass and compound particles decrease as orange glass increases.

In the grain mounts, crystal fragments appeared as a small, but persistent, quantity of large (0.25–1 mm) untwinned single crystals. Most are fragmented, anhedral grains of olivine; but blocky crystals with low birefringence and low relief also occur. In 74002, many, but not all, of these larger grains show irregular indentations less than 10 micrometers in diameter along the margins. These indentations give a moth-eaten appearance to the particles and are interpreted here as possible corroded margins. Such margins contrast to the margins of ultramafic droplets, which are smooth or cleanly-broken, or which have tiny droplets adhering to the surface. Most large crystal fragments in the lower core, 74001, do not show the irregular margins, but rather, have smooth or regular margins comparable to margins of the ultramafic droplets. The large crystal fragments with the possible corroded margins in the upper core, 74002, could represent xenocrysts in disequilibrium with the melt, or volcanic pheno-crysts. Their position in the core makes it unlikely that they are derived from the regolith. Abundance of these large crystal fragments in 74002 is given in Table 4 and Fig. 2. Although large crystal fragments were present throughout both drive tubes, those with possible corroded margins were common only in 74002, and their abundance decreases downward in the core, in the same way that orange glass decreases.

Table 4. Changes in clear grains, grain mounts from drive tube 74002.

		Corroded-margin Large Crystal Fragments		Crystallized Droplets and Fragments		Sharp-edged Large Crystal Fragments		
Sample No.	Size Class	No. of Points	% of Points	No. of Points	% of Points	No. of Points	% of Points	Total Points Counted
-,99	.25-.12 mm	3		2		0		5
	.12-.06 mm	18	29%	40	63%	5	8%	63
-,98	.25-.12 mm	5		9		0		14
	.12-.06 mm	14	22%	46	73%	3	5%	63
-,97	.25-.12 mm	12		3		0		15
	.12-.06 mm	32	53%	22	37%	6	10%	60
-,96	.25-.12 mm	3		2		0		5
	.12-.06 mm	15	31%	33	69%	0		48
-,95	.25-.12 mm	1		6		0		7
	.12-.06 mm	10	21%	34	72%	3	6%	47
-,94	.25-.12 mm	4		11		0		15
	.12-.06 mm	19	26%	51	71%	1	1%	72
-,93	.25-.12 mm	0		14		0		14
	.12-.06 mm	4	4%	85	92%	3	3%	92
-,92	.25-.12 mm	0		14		0		14
	.12-.06 mm	6	7%	79	91%	2	2%	87
-,91	.25-.12 mm	2	5%	35	95%	0		37
	.12-.06 mm	7	4%	186	95%	2	1%	195
-,90	.25-.12 mm	1	4%	25	96%	0		26
	.12-.06 mm	9	4%	203	94%	5	2%	217

DISCUSSION

Comparability of the lunar and terrestrial sections

The terrestrial section was selected for study and comparison to the lunar section because both are derived from relatively mafic, and thus relatively fluid lavas. As pyroclastics, such lavas form compact particles, not fragmented foams, and are emplaced ballistically into discrete localized deposits, not blanket deposits moved by gas clouds.

Some aspects of both sections are not closely comparable. For instance, a thick terrestrial cinder cone would be equivalent to a thin, blanket-like deposit on the moon. Based on the calculations of McGetchin *et al.* (1974), it is expected that the 55 cm lunar section in drive tubes 74002/1 is comparable to a 70 m terrestrial section. The studied terrestrial and lunar sections may be equivalent, although not comparable in thickness. The lunar section is nearly homogeneous, chemically (Butler and Meyer, 1976; Blanchard and Budhan, 1978), but no comparable chemical data are available for the terrestrial section. Color (and related mineralogy) of terrestrial and lunar glassy and crystallized particles differs in that terrestrial glass is black, lunar is orange, and terrestrial crystallized glass is reddish; whereas, lunar crystallized droplets are black to clear.

Many specific aspects of lunar and terrestrial deposits are only partially comparable. Cinder cone lavas are basaltic to basaltic andesite and moderately viscous (2×10^3 to 6×10^3 poises) and form vesicular bombs or scoria-like cinders; whereas, the lunar lavas are ultramafic and have a very low viscosity (3 to 10 poises) and form droplets (Heiken, 1976). However, the compact scoria-like cinders are more comparable to droplets than they are to widely dispersed pumice and foam-fragment shards produced by highly gaseous and viscous rhyolitic lavas (Heiken, 1974) (Table 1). Terrestrial particle size was much larger, with particles averaging approximately 1 cm in diameter with maximum size of 1 m, whereas lunar particles averaged <0.06 mm with maximum size slightly over 1 mm. Processing of materials into size classes was similar, but procedures were of necessity different, because terrestrial samples could be sieved and weighed; whereas, scarce lunar material was grain-mounted and had to be point-counted giving abundance, not weight data.

Although specific aspects of texture and composition are only partially comparable, trends in both the lunar and terrestrial sections seem to be closely analogous. Both sections show glassy particles at one end, crystallized particles at the other, and a relatively rapid transition between. Particle size reaches a maximum in the part of the section that contains crystallized material. In both sections, abundance of compound particles increases concurrently with abundance of crystallized particles. Indistinct bedding occurs in both sections only where there are large compound particles; otherwise the sections are massive.

In the terrestrial section, xenoliths decrease as glass decreases, and if the large eroded-margin single-crystal lunar grains are xenocrysts, they also decrease along with glass.

Is the lunar section complete or disrupted?

It is important to know whether a section is complete or disrupted because interpretation of the history of the section could be altered by an incomplete, reversed, or repeated section. Indications of disruption could include discontinuities in trends, repetition of trends, and the occurrence of spall fragments in crumbly units that match voids in massive units. On the other hand, presence of smooth and progressive trends can be taken as evidence for continuous section. Discontinuity in one trend, accompanied by continuity in others, can be an indication of changes in some, but not all, the conditions of deposition.

The section in 74002/1 shows differences in cohesiveness and units, that could be interpreted as evidence of disruption. There are no clear-cut trends in size patterns in both cores, so disruption cannot be proven or excluded on this basis. By matching fracture polygons in units 2 and 4 of 74002 to similar-shaped voids at the margins of units 1 and 3 (Fig. 2), it is possible to infer that there was some disruption of the fragmented, friable portions of 74002. Most grain-mount compositional parameters show a distinct, but minor, break in trends between units 7 and 8 in 74001 (Duke and Nagle, 1974, Fig. VII.5.11), suggesting some missing section. There is a distinct textural change between the friable and massive units of 74001, suggesting lack of section continuity.

Despite the appearance of matching voids and spall fragments in lower 74002 and major changes in texture in 74001, there are no corresponding discontinuities in compositional trends at unit boundaries, suggesting that any disruption of the section was minimal. The break in grain mount trends between units 7 and 8 in 74001 is distinct in most parameters, suggesting some missing section; however, the break is minor and trends are not reversed or repeated across it. The general progression and smooth continuity of trends through the rest of the core suggests that there is no repetition of section, nor are major parts missing.

Is the lunar section upright or overturned?

Earlier in the discussion it was inferred that trends in occurrence and abundance of glass and crystallized particles, compound grains, possible zenoliths, and glass-crystalline states of particles were analogous in the lunar and terrestrial section (cf., Figs. 1 and 2). Trends in the lunar section are opposite to trends in the terrestrial cinder cone section, suggesting the overturning of the Lunar section. Such overturning is common in crater ejecta (Gault *et al.*, 1968) and can be suspected in any crater rim deposit, but cannot be reliably inferred without additional independent evidence. Furthermore, Bogard and Hirsch (1978) find a ^{21}Ne profile that increases downward through much of the section, as might be expected from an overturned section.

On the basis of this evidence, any reconstruction of history of the lunar volcanic succession should use a starting point at a depth of 10 cm in 74002 (above that is a reworked, non-volcanic section) and progress to the base of 74001, in order to derive the succession of events which represent the history of the lunar eruption.

SUMMARY

The following tentative conclusions are offered:

1. Despite difference in thickness, lava composition, and lava viscosity, the terrestrial and lunar sections studied show similar internal trends. Both are glassy, relatively fine-grained and massive at one end, contain devitrified glass, are coarse-grained and stratified at the other end. The change from glassy to crystallized material is transitional but relatively rapid in both sections.
2. Trends in the lunar section are continuous and progressive from one end of

the section to the other, and show smooth changes except at 15 cm in 74001, where there is a minor step-wise increment. No reversals in trends were seen. These characteristics suggest that the lunar section is essentially complete.

3. Although they are similar, trends in the lunar section are upside-down compared to the terrestrial section. This indicates that the lunar section is overturned. Such an occurrence is commonplace in deposits on rims of impact craters.

Acknowledgments—The kind efforts of many people helped make this manuscript possible. T. McGetchin, L. P. I., provided volcanic particles from Hawaii. G. Heiken, T. McGetchin, D. Morris, R. Williams, and an unknown reviewer all contributed to the clarity and readability of the work. S. Goudie faithfully typed the many revisions, and NASA contract NAS9-15425 supported the work.

REFERENCES

Blanchard D. P. and Budhan J. R. (1978) Compositional variations among size fractions in a homogeneous, unmatured regolith: Drive core 74001/2 (abstract). In *Lunar and Planetary Science IX*, p. 100–102. Lunar and Planetary Institute, Houston.

Bogard D. D. and Hirsch W. C. (1978) Noble gas contents and irradiation history of orange-black glass in the 74001–74002 core (abstract). In *Lunar and Planetary Science IX*, p. 111–113. Lunar and Planetary Institute, Houston.

Butler P. Jr. and Meyer C. Jr. (1976) Sulfur prevails in coatings on glass droplets: Apollo 15 green and brown glasses and Apollo 17 orange and black (devitrified) glasses. *Proc. Lunar Sci. Conf. 7th*, p. 1561–1581.

Duke M. B. and Nagle J. S. (1974) Lunar Core Catalog, Jan. 1978 supplement NASA SP. 09252.

Eugster O., Eberhardt P., Geiss J., Grogler N., Jungck M. and Morgeli M. (1977) The cosmic exposure history of Shorty Crater samples: The age of Shorty Crater. *Proc. Lunar Sci. Conf. 8th*, p. 3059–3081.

Gault D. E., Quaide W. L. and Oberbeck V. R. (1968) Impact cratering mechanics and structures. In *Shock Metamorphism of Natural Materials*. Mono Book Corp., Baltimore. p. 87–99.

Heiken G. (1974) An atlas of volcanic ash. *Smithsonian Contrib. Earth Sci. 12*. Smithsonian Press, Washington, D.C. 101 pp.

Heiken G. H. (1978) Volcanic ash, its formation and characterization. In *Encyclopedia of Volcanoes and Volcanology* (Jack Green, ed.). Dowden, Hutchinson and Ross, Inc. In press.

Heiken G. H., McKay D. S. and Brown R. W. (1974) Lunar deposits of possible pyroclastic origin. *Geochim. Cosmochim. Acta* **38**, 1703–1718.

Heiken G. and McKay D. S. (1977) A model for eruption behavior of a volcanic vent in eastern Mare Serentatis. *Proc. Lunar Sci. Conf. 8th*, p. 3243–3255.

Heiken G., McKay D. S. and Gooley R. (1978) Petrology of a sequence of pyroclastic rocks from the Valley of Taurus-Littrow (Apollo 17 landing site) (abstract). In *Lunar and Planetary Sci. IX*, p. 491–493. Lunar and Planetary Institute, Houston.

Huneke J. C., Jessberger E. K., Pedosek F. A. and Wasserburg J. G. (1973) $^{40}Ar/^{30}Ar$ measurements in Apollo 16 and 17 samples and the chronology of metamorphic and volcanic activity in the Taurus-Littrow region. *Proc. Lunar Sci. Conf. 4th*, p. 1725–1756.

McGetchin R. G. and Head J. W. (1973) Lunar cinder cones. *Science* **180**, 68–71.

McGetchin T. F., Settle M. and Chouet B. A. (1974) Cinder cone growth modeled after Northeast Crater, Mount Etna, Sicily. *J. Geophys. Res.* **79**, No. 23, 3257–3272.

McKay D. and Waits G. (1978) Grain size distribution of samples from core 74001 and 74002 (abstract). In *Lunar and Planetary Science IX*, p. 723–725. Lunar and Planetary Institute, Houston.

Moore R. B., Wolfe E. W. and Ulrich G. E. (1976) Volcanic rocks of the eastern and northern parts of the San Francisco volcanic field, Arizona. *J. Res. U. S. Geol. Survey* **4**, 549–560.

Morris R. B., Gose W. A. and Lauer H. V. Jr. (1978) Depositional and reworking history of core 74001/2 (abstract). In *Lunar and Planetary Science IX*, p. 763–765. Lunar and Planetary Institute, Houston.

Nagle J. S. (1978) The detrital zone in the Shorty Crater cores. *Moon and Planets* **18**, 499–517.

Proc. Lunar Planet. Sci. Conf. 9th (1978), p. 1527–1536.
Printed in the United States of America

The formation kinetics of lunar glasses

D. R. Uhlmann, C. A. Handwerker, P. I. K. Onorato*, R. Salomaa and D. Goncz

Department of Materials Science and Engineering
Massachusetts Institute of Technology
Cambridge, Massachusetts 02139

Abstract—An extension of the kinetic treatment of glass formation is presented which permits the description of crystallization of a body initially cooled to the glassy state. The treatment is applied to the matrix composition of lunar breccia 67975, and excellent agreement is obtained between predicted and experimental values of the temperature of maximum crystallization rate. The analysis has also been applied to anorthite and suggests an exceptionally large barrier to crystal nucleation, in the range of 75 kT at $\Delta T_r = 0.2$. Measurements of the nucleation barrier in the 67975 matrix composition are presented, and lie in the range of 42–44 kT at $\Delta T_r = 0.2$. Use of this value in the kinetic treatment of glass formation gives predicted critical cooling rates of 1–2°C min^{-1} for a volume fraction crystallized of 10^{-2} which are in excellent ageement with the measured range of 1–3°C min^{-1}. The same value of the nucleation barrier was used in predicting the temperature of maximum crystallization rate.

Taken *in toto*, the present results provide strong support for the kinetic descriptions of glass formation and the development of partial crystallinity in cooled bodies. It is suggested that further confidence can be placed in the use of such descriptions in elucidating thermal histories.

I. Introduction

During the past few years, kinetic treatments have been developed to describe glass formation and the development of various degrees of partial crystallinity in bodies cooled at different rates. The analysis has been applied to a number of glasses and glass-containing bodies returned from the lunar surface, and has provided insight into the critical conditions required to form the bodies as glasses [e.g., Uhlmann *et al.* (1974) and Uhlmann (1977)]. When combined with treatments of viscous sintering, the analysis can be used to obtain information about the thermal histories of breccias whose matrices are not completely crystalline [e.g., Uhlmann *et al.* (1975) and Uhlmann *et al.* (1977)].

The treatment of crystallization and glass formation involves the construction of time-temperature-transformation (TTT) curves corresponding to the degree of crystallinity which is of interest. These curves are usually constructed using measured values of the crystal growth rate and calculated values of the nucleation frequency. For nucleation and growth rates which are independent of time, the volume fraction, V_c/V, crystallized in a time, t, can be related to the

*now with GTE Laboratories, Waltham, Massachusetts 02154

crystal growth rate, u, and the nucleation rate per unit volume, I_v, as:

$$V_c/V = 1 - \exp\left(-\frac{\pi}{3} I_v u^3 t^4\right). \qquad (1)$$

In the case of glass formation, a volume fraction crystallized of 10^{-6} has been identified as a just-detectable degree of crystallinity.

The results of studies of crystal nucleation on a variety of materials indicate barriers to homogeneous nucleation of about 50–60 kT at $\Delta T_r = 0.2$ (here $\Delta T_r = \Delta/T_E$, where the ΔT is the undercooling and T_E is the melting point or liquidus temperature). Such studies on two silicate melts indicate that the formation of crystals takes place predominantly by a homogeneous nucleation mechanism in the range of large undercoolings which are critical in the process of glass formation. In the discussion to follow, only homogeneous nucleation will be considered. This is in accord with the results obtained on a number of lunar glass-forming compositions in our laboratory. The analysis can, however, readily be extended to include the effects of nucleating heterogeneities [Onorato and Uhlmann (1976)].

The TTT curves indicate the time required at any temperature to reach the indicated fraction crystallized. In using such curves to estimate the cooling rate required for glass formation, use is made of continuous cooling (CT) curves. The CT curves are constructed from the TTT curves following the approach of Grange and Kiefer (1941) as elaborated and applied to the problem of glass formation by Onorato and Uhlmann (1976).

More detailed information about the state of crystallinity in a body can be obtained from the crystal distribution function ψ, introduced by Hopper *et al.* (1974). This function is defined such that the number dn of crystals in the volume dv at position r having a radius between R and R + dR at time t is given by

$$dn = \psi(r,t,R)\, dv\, dR. \qquad (2)$$

To date, the kinetic treatments have been usefully predictive and have yielded estimated cooling rates in general accord with estimates provided by other methods. In the present paper, we shall describe a test of the kinetic treatment of glass formation which directly compares the predicted cooling rates for glass formation with the results of controlled cooling experiments. Also to be described is a new theoretical development which provides a description of crystallization on reheating a material initially cooled to the glassy state (i.e., the process of devitrification). It is assumed that the sample is of uniform temperature so ψ is not a function of r. The predictions of this treatment will also be compared with experimental data on the same material used to test the preditions of critical cooling rates.

II. Treatment of Crystallization on Reheating

In treating the cooling of a body to the glassy state and crystallization on subsequent heating, consider a crystal nucleating at a time t_O ($t_O < t$), which has a radius R at time t of

$$R(t,t_o) = \int_{t_o}^{t} u[T(t')]dt' + R^*[T(t_o)]. \quad (3)$$

Here R^* is the radius the critical nucleus, which can be expressed from classical nucleation theory as:

$$R^* \approx \frac{0.5 V_M^{1/3} T_E^2}{N_o^{1/3} \Delta T\, T}, \quad (4)$$

where V_M is the molar volume and N_O is Avogadro's number. Hence the crystal size R is a function only of t and t_O.

It is assumed that if at any time and temperature the crystallite is smaller than the critical size corresponding to that temperature, it will melt completely and will not be included in any future calculations of the crystal distributions; that is,

$$\psi(t,t_O) = 0 \quad (5)$$

when $R(t',t_O) < R^*(t')$ for some t' on (t_O,t). There will be a range of $R(t,t_O)$ for which $R(t',t_O) > R^*(t')$, and the number of crystallites in that range can be calculated.

The number of crystals nucleating in the volume dv during the time interval between t_1 and t_2 for which Eq. (5) does not apply is

$$dn_{12} = dv \int_{t_1}^{t_2} I_v\, [T(t_o)]dt_o. \quad (6)$$

From Eq (3), one obtains

$$\frac{\partial R}{\partial t_o} = -\,u[T(t_o)] + \frac{dR^*}{dt_o}. \quad (7)$$

Combining Eqs. (6) and (7) and changing variables

$$dn_{12} = dv \int_{R_1}^{R_2} -\frac{I_v}{u - dR^*(t_o)/dt_o}\, dR, \quad (8)$$

where $R_1 = R(t,t_1)$ and $R_2 = R(t,t_2)$. From Eqs. (2) and (8), one obtains an expression for the crystal distribution function

$$\psi(t,R) = \frac{I_v\,\{T[t_o(t,R)]\}}{u\{T(t_o(t,R)]\} - \dfrac{dR^*[t_o(t,R)]}{dt_o(t,R)}}. \quad (9)$$

The inclusion of R^* in the expression for the crystallite size is significant when calculating the volume fraction crystallized. When the temperature is cycled below the glass transition, where the growth rate is essentially zero but dR^*/dt_O is not, as well as when a sample is cooled below and reheated above the glass transition, the expressions of Eqs. (3), (4), (5), and (9) are necessary to obtain reliable results.

An alternative method of calculating the volume fraction crystallized in a material with a known thermal history is to integrate the number of nuclei and their volume over time

$$F_v(t) = \int_o^t \frac{4}{3} \pi R(t,t_o)^3 \, I_v[T(t_o)] \, dt_o, \tag{10}$$

where the finite size of the critical nucleus must in general be considered.

The volume fraction crystallized has been calculated for several materials with thermal histories that include quenching from the melt to a temperature below the glass transition temperature and subsequent reheating until crystallization takes place. This is a method used in many laboratory measurements of the glass-forming ability of materials. Using differential thermal analysis (DTA) or differential scanning calorimetry (DSC), the temperature at which a sample crystallizes, T_{cr}, is determined. In detail, T_{cr} is usually taken as the temperature of maximum crystallization rate.

Hruby (1972) has formulated a parameter

$$K_{gl} = \frac{T_{cr} - T_g}{T_E - T_{cr}}, \tag{11}$$

as a measure of glass-forming ability, where T_g is the glass transition temperature and T_E is the liquidus temperature. This parameter provides a convenient reference for describing the process of crystallization on reheating a glass.

To determine the crystallization temperature of several materials, Eq. (7) was integrated numerically. The growth rate and viscosity for each material were determined experimentally. The nucleation barrier was varied in the range suggested by the results of previous workers.

$$I_v = N_v^\circ \, \nu \exp \frac{-.02048B \, T_E{}^5}{(\Delta T)^2 T^3}, \tag{12}$$

where N_v° is the number of molecules per unit volume; ν is the frequency of transport at the nucleus-matrix interface (which is inversely related to the viscosity); and $B = 40–75$, which corresponds to a nucleation barrier of BkT at a relative undercooling (ΔT_r) of 0.2. At every step in the integration, the crystallite sizes were compared with the critical nucleus size for that temperature; any crystallites that were smaller than the critical size were no longer included in calculations of the volume fraction crystallized.

The temperature range in which the greatest amount of crystallization takes place is taken as the crystallization temperature, T_{cr}. The predicted crystallization temperature can be compared with the temperature of the maximum exotherm in DTA or DSC runs. The temperature of crystallization is found to depend strongly on the heating rate, and also upon the cooling rate, used to form the glass. The results calculated for lunar composition 70019 are shown in Fig. 1. The results indicate that not all glasses are in the same state at the glass transition, but the state depends upon the cooling rate at which the glass was formed. It has also been found that the magnitude of the nucleation barrier has a pronounced effect on the crystallization temperature. This is shown in Fig. 2 for anorthite. From this figure, it is clear that one can estimate the nucleation barrier from the temperature of the crystallization peak as a function of heating rate.

In the following section, the analysis is applied to the matrix composition of lunar breccia 67975 and to anorthite. The predictions of the new analysis will be compared with experimental data obtained using DTA techniques; and the predictions of the critical cooling rate for glass formation will be compared with other kinetic data obtained from constant cooling rate experiments and from direct measurements of the nucleation barrier in the case of the lunar composition.

III. Experimental Studies

To provide a test of the ability of the kinetic treatment of glass formation to predict critical cooling rates and to apply the development outlined in Section II to evaluate the expected crystallization temperature as a function of heating rate, detailed experiments have been carried out on the matrix composition of lunar breccia 67975. The viscous flow and crystallization behavior of this composition have been determined previously [Uhlmann *et al.* (1977)]. The material was found to be an excellent glass former; it is, in fact, the best glass former of any lunar composition investigated to date in our laboratory. In the previous work, principal attention was directed to the cooling rate at which the degree of crystallinity observed in the matrix of the breccia would be formed. In evaluating this cooling rate, the nucleation barrier was taken as 60 kT at $\Delta T_r = 0.2$, in accord with experimental results obtained as of that time on other materials.

In the present work, the nucleation barrier in the 67975 matrix composition has been directly measured following the approach used by Klein and Uhlmann (1976). Samples were heat-treated at various temperatures for desired periods of time, and the samples then quenched in ice water. After removal from the quench bath, the central region of each sample was removed, examined optically, crushed and mounted on a glass slide for X-ray diffraction study. From the resultant diffraction patterns, the transition between glassy materials and materials with a sensible degree of crystallinity was determined; and the time required at each temperature to reach the indicated fraction crystallized was determined. Using these times together with measured crystal growth rates, the nucleation frequencies were evaluated from Eq. (1). The temperature dependence of the nucleation frequency is shown in Fig. 3 as a plot of $ln(I_v\eta)$ vs. $1/T_r^3\,\Delta T_r^2$, where $T_r = T/T_E$. In constructing this figure, use has been made of the viscosity data for the 67975 composition determined previously.

According to the classical theory of nucleation, the $ln(I_v\eta)$ vs. $1/T_r^3\,\Delta T_r^2$ relation should be a straight line of negative slope; such a straight line is indicated by the data in Fig. 3. The slope of the least squares line through the data indicates a nucleation barrier of about 43 kT at $\Delta T_r = 0.2$, and the intercept at $1/T_r^3\,\Delta T_r^2 = 0$ corresponds to a value of 10^{26} cm^{-3} sec^{-1} poise. The latter value is in good agreement with the value expected from the classical theory of homogeneous nucleation (about 10^{32} cm^{-3} sec^{-1} poise).

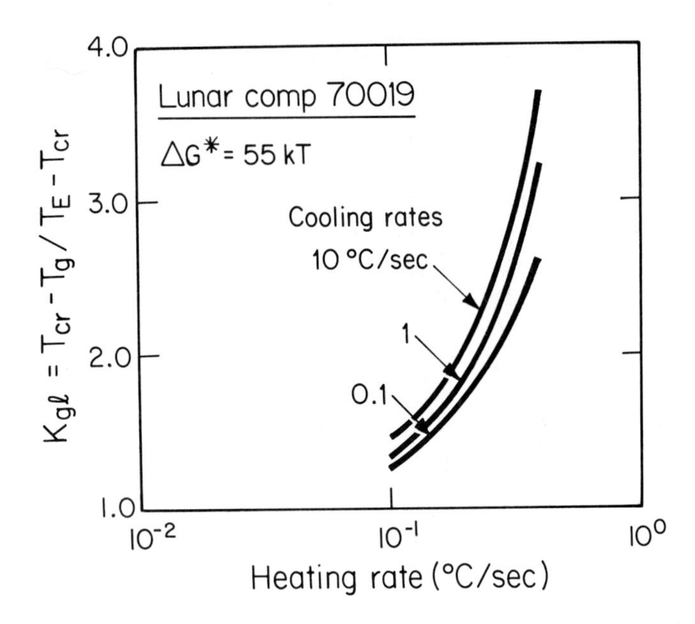

Fig. 1. Calculated variation of K_{gl} with heating rate and cooling rate for lunar composition 70019.

These results indicate that homogeneous nucleation represents the dominant contribution to the formation of crystals, at least over the range of undercooling covered by the investigation. In the range of large volume fractions crystallized, heterogeneous nucleation on the crystals which originally formed by homogeneous nucleation may also be significant; primary heterogeneous nucleation may well be important in the range of small undercoolings. The results obtained in the present investigation also indicate that the classical theory of nucleation provides a close description of the nucleation frequency and its temperature dependence.

Using this value of the nucleation barrier, TTT and CT curves corresponding to a detectable degree of crystallinity ($V_c/V = 10^{-2}$) have been constructed. These are shown in Fig. 4. From the CT analysis, the critical cooling rate required to form this material as a glass is estimated as about 1–2°C min^{-1}. To verify this estimate and to test the predictive power of the kinetic analysis, samples of the 67975 matrix composition were cooled at a variety of rates under controlled atmosphere conditions (the same as used in the nucleation and crystal growth studies) to determine the critical cooling rate for glass formation. It was found that sensibly crystalline bodies were produced at a cooling rate of 1°C min^{-1}, while detectably crystalline bodies were produced at a cooling rate of 3°C min^{-1}. This indicates a remarkable agreement between the predictions of the kinetic analysis and experimental data.

As a further test of the kinetic approach and as a direct attempt at verification of the new treatment of crystallization behavior on reheating a glass, samples of the 67975 composition were cooled rapidly to the glassy state and were subsequently reheated at a number of rates in a differential thermal analyzer (DTA). The predictions of the kinetic analysis are compared with the experimental data in Table 1. In obtaining the predicted temperatures of maximum crystallization rate, the nucleation barrier, crystal growth rate and viscosity data measured for this material were employed. The agreement between the predicted and experimental values of the temperature of maximum crystallization rate is seen to be quite good, as is the dependence of the temperature of maximum crystallization rate on heating rate.

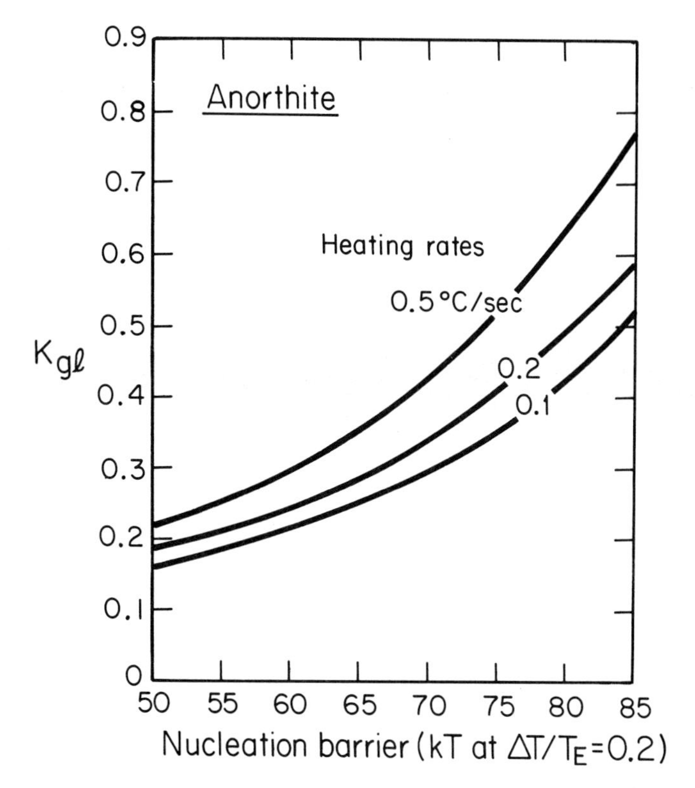

Fig. 2. Calculation variation of K_{gl} with nucleation barrier and heating rate for anorthite.

Differential thermal analysis runs were also carried out on samples of anorthite which had been rapidly quenched to the glassy state. While only limited data were obtained on each composition, the results shown in Fig. 5 indicate a very large barrier to nucleation, in the range of 75 kT at $\Delta T_r = 0.2$. These large nucleation barriers in anorthite would, if verified by subsequent direct measurements, explain why the materials are better glass formers than indicated by their crystallization and viscous flow behavior.

IV. Discussion and Conclusions

The present analysis of crystallization on reheating a sample initially cooled to the glassy state represents a further extention of the kinetic treatment of glass formation and the development of partial crystallinity in cooled bodies. The close agreement between the calculated and measured temperatures of maximum crystallization rate at various heating rates for 67975 matrix composition are quite gratifying. These results are of particular note since they support the validity of the critical assumption of the new analysis—viz., that if at any time and temperature a crystallite is smaller than the critical size corresponding to that temperature, it will melt completely and need not be included in any further calculations of the crystal distribution.

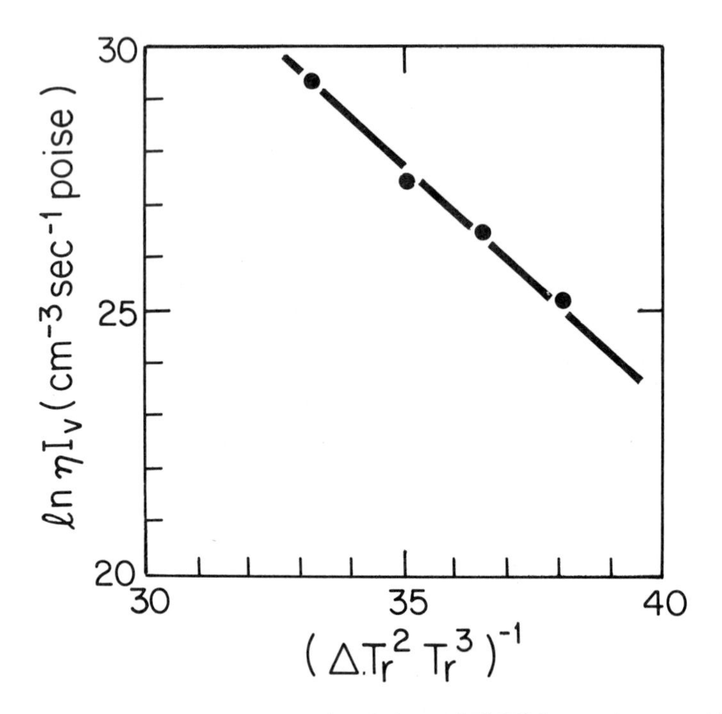

Fig. 3. Logarithm (nucleation rate × viscosity) vs. $1/T_r^3 \Delta T_r^2$ for matrix composition of lunar breccia 67975.

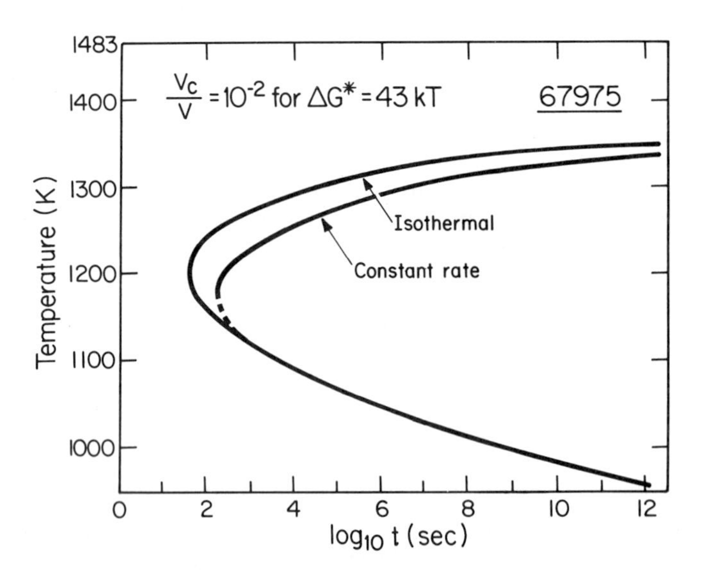

Fig. 4. Time-temperature-transformation (TTT) and continuous cooling (CT) curves for matrix composition of lunar breccia 67975. Nucleation barrier = 43 kT at $\Delta T_r = 0.2$.

Table 1. A comparison of T_{cr} values for 67975 calculated from the kinetic analysis ($\Delta G^* = 43$ kT) with T_{cr} values measured by DTA.

T_{cr} from kinetic analysis ($\Delta G^* = 43$ kT at $\Delta T_r = 0.2$)	Heating rate	T_{cr} from DTA
1204	5°/min.	1193°K
1220	10°/min.	1203°K
1240	20°/min.	1223°K
1270	50°/min.	1248°K

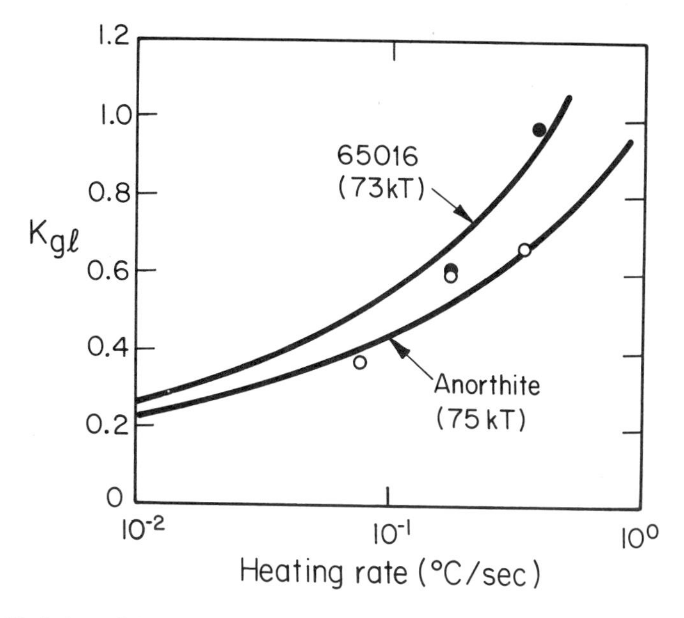

Fig. 5. Variation of K_{gl} with heating rate for anorthite. Points experimental data; curve = calculated relation for nucleation barrier of 75 kT at $\Delta T_r = 0.2$.

The results obtained in the DTA studies of anorthite indicate exceptionally large values of the barrier to crystal nucleation. If these values are correct, they present an interesting theoretical problem to explain the large nucleation barriers in these materials—barriers which are appreciably larger than those measured for sodium disilicate and for the matrix compositions of lunar breccias 67975 and 70019. Because of this interest and the general importance of anorthite, direct measurements of the nucleation barrier are presently underway in our laboratory.

The nucleation rates determined for 67975 composition are in close agreement with the predictions of classical nucleation theory. The magnitude of the barrier to crystal nucleation indicated for this material (about 42–44 kT at $\Delta T_r = 0.2$) is similar to that determined recently for sodium disilicate [Klein *et al.* (1977)], and is somewhat smaller than determined previously for the matrix composition of lunar breccia 70019 [Klein and Uhlmann (1976)].

More generally, the totality of the present results indicates that the kinetic treatments of glass formation gives good agreement with measured critical cooling rates. The quality of this agreement suggests that one can have greater confidence in applying the models to describe glass formation and the development of various degrees of partial crystallinity in cooled bodies. The new extension of the model permits a description of crystallization on heating a body after initially cooling into a glassy state. Excellent predictive power has been found in applying the new analysis to the 67975 matrix composition; the interesting prediction of a large barrier to crystal nucleation for anorthite should be the subject of close attention in the future.

Acknowledgments—Financial support for the present work was provided by the National Aeronautics and Space Administration. This support is gratefully acknowledged.

REFERENCES

Grange R. A. and Kiefer J. M. (1941) Transformation of austenite on continuous cooling and its relation to transformation at constant temperature. *Trans. Amer. Soc. Metals* **29**, 85–115.

Hopper R. W., Scherer G. and Uhlmann D. R. (1974) Crystallization statistics, thermal history and glass formation. *J. Non-Cryst. Solids* **15**, 45–62.

Hruby A. (1972) Glass transition. *Czech. J. Phys.* **B22**, 1187–1195.

Klein L. C., Handwerker C. A. and Uhlmann D. R. (1977) Nucleation kinetics of sodium disilicate. *J. Cryst. Growth* **42**, 47–51.

Klein L. C. and Uhlmann D. R. (1976) The kinetics of lunar glass formation, revisited. *Proc. Lunar Sci. Conf. 7th*, p. 1113–1121.

Onorato P. I. K. and Uhlmann D. R. (1976) Nucleating heterogeneities and glass formation. *J. Non-Cryst. Solids* **22**, 367–378.

Uhlmann D. R. (1977) Glass formation. *J. Non-Cryst. Solids* **25**, 43–85.

Uhlmann D. R., Klein L. C. and Handwerker C. A. (1977) Crystallization kinetics, viscous flow and thermal history of lunar breccia 67975. *Proc. Lunar Sci. Conf. 8th*, p. 2067–2078.

Uhlmann D. R., Klein L., Kritchevsky G. and Hopper R. W. (1974) The formation of lunar glasses. *Proc. Lunar Sci. Conf. 5th*, p. 2317–2331.

Uhlmann D. R., Klein L., Onorato P. I. K. and Hopper R. W. (1975). The formation of lunar breccias: Sintering and crystallization kinetics. *Proc. Lunar Sci. Conf. 6th*, p. 693–705.

Proc. Lunar Planet. Sci. Conf. 9th (1978), p. 1537–1550.
Printed in the United States of America

Meteoritic material in lunar highland samples from the Apollo 11 and 12 sites

MARIE-JOSÉE JANSSENS[1*], HERBERT PALME[1†], JAN HERTOGEN[1*]
ALFRED T. ANDERSON[2] and EDWARD ANDERS[1]

[1]Enrico Fermi Institute and Department of Chemistry
University of Chicago, Chicago, Illinois 60637
[2]Department of Geophysical Sciences
University of Chicago, Chicago, Illinois 60637

Abstract—Two samples of rock 12013 and 9 single fragments from coarse soils 12033 and 10085, including Luny Rock, were analyzed by radiochemical neutron activation for Ag, Au, Bi, Br, Cd, Cs, Ge, In, Ir, Ni, Os, Pd, Rb, Re, Sb, Te, Tl, U, and Zn; the 12033 separates were also studied petrographically. Most of the Apollo 12 samples have a 1L (Imbrium) meteoritic component, but some, typified by 12013, contain a unique component of low Re content. Others have a Group 2 component, derived from the Serenitatis projectile or a smaller body of similar composition. It is characterized by metal of unusual composition (Ni = 15–22%, Co = 0.9–1.7%).

The Apollo 11 anorthositic samples have components of higher Ir/Au ratio (3L, 5L, and 7), and their enrichment in meteoritic siderophiles parallels that of Fe, Mg, REE, and other KREEP elements; presumably this reflects progressive contamination of an initially anorthositic crust (Wänke *et al.*, 1976; Warner *et al.* 1977). Luny Rock falls in Group 2. The most pristine anorthosites tend to show only a small, Group 7 meteoritic component of very high Ir/Au ratio, which supports the view that this component is derived from the moon's original planetesimal population. Other tentative identifications strengthened by the new data are: 5H = Nectaris; 5L = small, Pre-Serenitatis basin in NE quadrant.

1. INTRODUCTION

Continuing our efforts to characterize the bodies that bombarded the lunar highlands (Hertogen *et al.*, 1977, and earlier papers cited therein), we have analyzed 11 samples from the Apollo 11 and 12 sites. Though located in maria, both sites have yielded small amounts of highland material, such as anorthosite and KREEP fragments in soils (Wood *et al.*, 1970; Chao *et al.*, 1970; King *et al.*, 1970; Short, 1970), and the felsite-KREEP breccia 12013 (Lunatic Asylum, 1970). We had previously analyzed several such samples, i.e., 2 anorthosite fractions from <1 mm and 1–10 mm soils (Laul *et al.*, 1971), 3 KREEPy fractions from 2–4 and <1 mm soils (Morgan *et al.*, 1972, 1973), and 7 samples of rock 12013 (Laul *et al.*, 1970). However, these earlier analyses gave rather

*Institut voor Nucleaire Wetenschappen, Rijksuniversiteit Gent, Proeftuinstraat, 86 B-9000 Gent, Belgium.

†Present address: Max-Planck Institut für Chemie, Saarstrasse 23, Postfach 3060, Mainz, Germany.

inadequate information on meteoritic components, because they were done on composites and included only 2–4 of the 8 siderophiles that we measure in our current procedure.

We therefore obtained a new suite of samples, the first 9 from the Lunar Sample Curator, and the last 3 from the Wasserburg Collection:

4 anorthositic fragments from 1–3 mm soil 10085,104;
5 KREEP fragments from 2–4 mm soil 12033,2;
1 sample of "Luny Rock" 10085,31 LR-1;
2 samples of 12013 (one granitic, one KREEPy).

Thin sections were prepared from portions of the five 12033 samples. Upon examination, it turned out that two of the putative KREEP samples actually were mare rocks. We therefore measured only one of them.

All 12 samples were sent to R. A. Schmitt for INAA (Ma and Schmitt, 1978a,b), prior to radiochemical analysis. Details of our procedure may be found in Hertogen *et al.* (1977).

2. SAMPLES

2.1 Apollo 11

Anorthosites. No petrographic information is available. However, Ma and Schmitt (1978a) conclude from their analyses that sample 773 is a gabbroic anorthosite, whereas 770, 771, and 772 are anorthositic gabbros.

Luny Rock. A single 2.6 mg fragment (10085,9015), produced during the original sampling of the rock, was selected by D. A. Papanastassiou. It appeared to be mostly interior material.

2.2 Apollo 12

Rock 12013. The light, felsitic lithology was represented by a single 5.9 mg fragment (12013,9037A; Caltech number 12013,10,12A). The dark, KREEPy part was represented by about a dozen fragments totaling 6.5 mg, chipped free of sawed surfaces by a tungsten chisel (12013,9037B; Caltech number 12013,10,01A). We are indebted to D. A. Papanastassiou for the preparation of this sample.

12033,390 (thin section 12033,406 B2). This rock is an olivine-rich gabbro (of mare origin, according to Ma and Schmitt, 1978b). It consists of about 20 volume percent of subhedral grains of olivine up to about 0.3 mm long, about 40 volume percent of subhedral to anhedral clinopyroxene up to 0.5 mm long, about 20 volume percent of orthopyroxene up to 0.6 mm long, 15 volume percent of anhedral interstitial plagioclase up to 0.6 mm long and about 5 percent of euhedral to subhedral Ti-Cr spinels(?). There is a trace of residuum into which grow several minor minerals too small to identify. One circular section of metal 60 μm in diameter is enclosed in a clinopyroxene which has an optic angle of about 30° (subcalcic augite), suggesting a temperature of crystallization above about 1000°C and a cooling rate sufficient to prevent microscopic exsolution.

Although this rock is submillimeter in grain size, it is mineralogically and texturally similar to accumulative rocks. Probably it is not a frozen liquid aliquot, but rather a mechanical concentration of comparatively early crystallizing minerals. It is deficient in incompatible elements and probably rich in refractory elements such as Mg,Cr compared to the parental liquid from which it formed.

12033,393 (thin section 12033,407 B2). This rock, which we did not analyze, is a partly brecciated gabbro (of mare origin, according to Ma and Schmitt, 1978b). It consists of subequal amounts of pyroxene and plagioclase as anhedral grains up to 2 mm long. The clinopyroxene grains have blebby lamellae up to 10–20 μm thick. Locally, the clinopyroxenes have curved cleavage and undulatory extinction. Most plagioclase grains have undulatory extinction and many grains consist of granular mosaics rich in voids. Ilmenite is rare and occurs only in one patch. There is about as much sulfide as ilmenite, but no metal.

Although the curved cleavage and mosaic plagioclase indicate shock metamorphism, there is no evidence of regrowth of pyroxene, such as the poikilitic texture of orthopyroxene grains typical of various metabreccias.

12033,396 (thin section 12033,408 B2). This rock is a recrystallized breccia. About 40 volume percent of the rock consists of irregular, anhedral angular crystals of plagioclase up to 0.4 mm long. Some of the smaller crystals have subround cores and subhedral to euhedral rims richer in albite which are up to 1–30 μm thick. The albitic zones are not evident on all grains—possibly an orientation effect. About 50 volume percent consists of pyroxenes up to 0.7 mm long. The orthopyroxenes are sieved with inclusions of subhedral plagioclase. Rarely there are inclusions of glass up to 15 μm diameter in the orthopyroxenes. Relict ophitic intergrowths of plagioclase laths in clinopyroxene are present in a few grains. There is about 10 volume percent of ilmenite. Metallic iron amounts to about 0.01 volume precent; there is about five times as much sulfide. The rock has about 5 volume percent of voids, many of which are partly filled with shattered grains.

The rims on the anhedral cores of the plagioclase grains appear to postdate brecciation and correspond to a recrystallization. The sieved orthopyroxenes possibly grew during the same interval of recrystallization, but it is not certain if the sodic rims on plagioclase are earlier or later than the sieved orthopyroxenes. Inclusions of glass in the sieved orthopyroxenes indicate growth in the presence of a melt.

12033,399 (thin section 12033,409 B2). This rock is a recrystallized polymict breccia. One part of the rock is a fragment of breccia containing about 40 volume percent of angular, anhedral grains of olivine up to about 0.1 mm across. The same fragment has about 10 percent porosity as subequant, subround voids up to about 0.02 mm diameter. In addition, the olivine-rich fragment contains 5 volume percent or less of ilmenite. Most of the rest of the olivine-rich fragment consists of plagioclase.

The remainder of the rock is less porous and contains about 50–60 volume percent of angular anhedral grains of plagioclase with rare rims of sodic plagioclase up to 10 μm thick. The plagioclase clasts are in a matrix of more plagioclase together with a few percent of intersertal glass and intergranular pyroxene and ilmenite grains 10–50 μm long. There is less than about 0.01 volume percent of metal but about 0.1 to 0.2 volume percent of zircon and several percent of phosphate. Pyroxene poikilitically encloses subhedral grains of plagioclase. This part of the rock probably is comparatively rich in residual elements like P and Zr, and appears to be more thoroughly indurated by metamorphic recrystallization than the part rich in olivine, perhaps because of a greater concentration of fluxes in the former part.

12033,402 (thin section 12033,410 B2). This rock is a recrystallized, probably polymict breccia.

Angular, anhedral pieces of plagioclase up to 1 mm across comprise about 60 percent of this rock. Some 0.1 mm size grains have 10–20 μm wide rims of relatively sodic plagioclase. Some grains either lack such rims or are in unfavorable orientation. Judging from variation in extinction angle, the compositional variation of the zoned plagioclases is no more than 10 atom percent in Na/Na + Ca. One angular grain of olivine 0.7 mm across is present.

Vugs up to 0.1 mm across comprise about 10 volume percent. The vugs are lined with crystals of various minerals including metal and troilite.

The matrix of the rock consists mostly of elongate grains of orthopyroxene, some of which have needle-like protrusions suggestive of skeletal growth. Subhedral grains of plagioclase are poikilitically enclosed in orthopyroxene and ilmenite. The ilmenite amounts to about 10 percent of the rock and has blebs and rare single lamellae of intergrown gray armalcolite(?). About 1 percent of phosphate is present; intersertal glass is rare and restricted to small areas where phosphate and perhaps sulfide are concentrated. There is about 0.1 volume percent of metal and the metal to sulfide ratio is near unity.

3. Results

Trace element data are given in Table 1. Accuracy and precision are

Table 1. Trace Elements in Lunar Samples.*

Sample	Description†	Ir ppb	Os ppb	Re ppb	Au ppb	Pd ppb	Ni ppm	Sb ppb	Ge ppb	Se ppb	Te ppb
10085,770	An gabbro	0.507	0.478	0.035	0.190	≤3	≤100	0.08	3.7	69	*2.5*
10085,771	An gabbro	1.05	0.90	0.034	0.028	≤9	≤300	0.16	12.3	92	10
10085,772	An gabbro	20.6	22.2	1.63	6.31	*18*	330	0.88	86.5	110	24
10085,773	Gabbroic an	0.26	≤0.21	0.019	0.056	≤5	≤60	0.14	4.0	27	≤17
10085,9015	Luny Rock	4.01	3.41	0.31	2.26	*21*	*120*	0.98	149	540	≤30
12013,9037A	Felsite	0.17	0.09	0.028	0.081	≤4.5	≤50	0.25	30.5	400	≤13
12013,9037B	KREEP	7.46	5.87	0.26	3.97	13	260	1.14	409	930‡	*15*
12033,390	Ol-r gb (m)	*0.006*	≤0.023	≤0.003	0.0049	≤1.3	120	0.079	30.7	111	≤2.4
12033,396	Recr bc	2.53	2.18	0.24	2.12	*5.7*	150	0.52	181	250	≤10
12033,399	Recr pol bc	5.36	5.17	0.50	3.40	9.2	240	0.86	362	260	20
12033,402	Recr pol bc	6.04	5.67	0.46	3.87	10.9	240	0.97	445	185	≤7.2
BCR-1	Terr basalt	0.0019	≤0.0008	0.77	2.22	≤0.053	*8*	440	1360	79.8	5.2

*Italicized values have errors greater than 20%.

†Abbreviations: An = anorthosite, bc = breccia, gb = gabbro, m = mare, ol-r = olivine-rich, pl = plagioclase, pol = polymict, recr = recrystallized.

‡Doubtful because of Ta contamination.

comparable to those of past analyses (Keays *et al.*, 1974), except for certain marginal elements where the small sample size led to significantly larger statistical counting errors: Pd, Ni, Te, Br, Bi, and Cd. Where the statistical errors exceeded 50%, the results are reported only as 2σ upper limits; where they fell between 20 and 50%, they are given in italics.

Electron microprobe analyses of 26 metal grains in 3 samples are shown in Table 3, later in the text. (We are indebted to Catherine A. Leitch for these analyses).

4. NON-METEORITIC ELEMENTS

4.1. *Highland samples*

Abundances of volatiles in highland samples are occasionally altered by fumarolic volcanism (thus far seen only at Apollo 16; Krähenbühl *et al.*, 1973), or by impact metamorphism. Both processes can, in principle, be recognized by changes in the ratio of two geochemically correlated elements of different volatility, such as Rb/U, Tl/U, Ge/Ir, etc. Of course, since these correlations are not perfect, some criterion is needed for distinguishing volatility effects from variations caused by igneous processes alone. We shall therefore compare 4 correlated elements whose volatilities differ greatly (Tl > Cs > Rb ≫ U), and check whether the observed trends are consistent with this volatility sequence.

Of the 11 samples in Table 1, 8 show anomalous ratios, which may signify volatility losses or enrichment (Table 2). Sample 10085,773 has high Tl (8×) but normal Rb,Cs. The remaining samples are depleted in Tl (up to 16×), and generally also in Rb and Cs, by smaller factors (2–4×). This trend certainly is

Table 1. *(cont'd.)*

Ag ppb	Br ppb	In ppb	Bi ppb	Zn ppm	Cd ppb	Tl ppb	Rb ppm	Cs ppb	U ppb	Weight mg	Sample
0.54	*9*	1.0	≤0.3	1.7	*5*	0.45	0.495	14.8	60	13.02	10085,770
0.70	*16*	1.3	≤0.8	1.2	*6*	≤0.3	0.14	7.3	122	4.98	10085,771
0.30	≤120	1.4	≤1.5	2.2	*7*	1.15	0.71	43.2	182	3.40	10085,772
0.88	≤19	1.1	≤1.1	3.4	*3*	1.43	0.43	14.4	60	5.28	10085,773
0.71	≤90	28	≤3	92	51	6.9	13.2	800	7510	1.95	10085,9015
1.76	90	28	≤0.9	2.5	48	22.4	55.5	2700	11150	5.85	12013,9037A
1.20	250	3.6	≤0.8	3.9	37	0.98	4.94	710	5710	5.99	12013,9037B
0.18	14	0.2	≤0.4	4.6	*1.3*	≤0.1	0.239	10.7	78	14.27	12033,390
0.90	168	0.4	≤0.7	3.9	17	1.92	13.6	625	7060	7.09	12033,396
0.90	205	2.2	≤0.3	4.2	17	1.58	8.42	470	7320	17.49	12033,399
0.78	123	0.6	≤0.5	3.6	9	1.48	23.1	1640	6390	12.64	12033,402
	66.6	99	47.1	134	≡150	281	48.1	1050	≡1750	105.96	BCR-1

consistent with volatilization, perhaps during the recrystallization episode inferred from the petrography. But since even larger variations in Tl/Rb, K/Rb ratio have been seen in terrestrial basalts where volatilization was not a factor (de Albuquerque *et al.*, 1972), it would seem premature to ascribe these trends to volatilization without further evidence. The abundances of other non-meteoritic volatiles (Br, Cd) do not correlate well with the putative losses.

4.2. Mare basalt

Olivine gabbro 12033,390 is very low in the siderophile and volatile elements Ir, Ag, Te, In, Tl, Rb, Cs, compared to typical Apollo 12 mare basalts (Anders *et al.*, 1971), but is higher in Ni and Zn. Qualitatively, this trend might be explained by a high content of olivine, because this mineral tends to take up Ni and Zn while excluding larger ions. However, the petrographically observed olivine content of 20% is too low to account for the difference. Volatilization might be invoked at least for Tl, Rb, Cs, because their ratios relative to U are low (Table 2). But as pointed out above, igneous processes alone can cause low ratios. Obviously, this problem requires more thorough study.

The Os content of our sample ≤0.023 ppb, is more than an order of magnitude lower than the values of 0.26 to 3.0 ppb reported by Jovanovic and Reed (1976) for 6 Apollo 12 basalts. The remaining samples in Table 1 continue to show an Os:Ir correlation in nearly 1:1 ratio (Hertogen and Janssens, 1977), and give no indication for a variable non-meteoritic Os component decoupled from other siderophiles, as proposed by Jovanovic and Reed (1977).

5. Ancient Meteoritic Components

All 10 highland samples are enriched in siderophiles over indigenous levels (Gros *et al.*, 1976), but in at least 2 cases (10085,773 and 12013,9037A) the enrichment is too slight for reliable characterization of the meteoritic component.

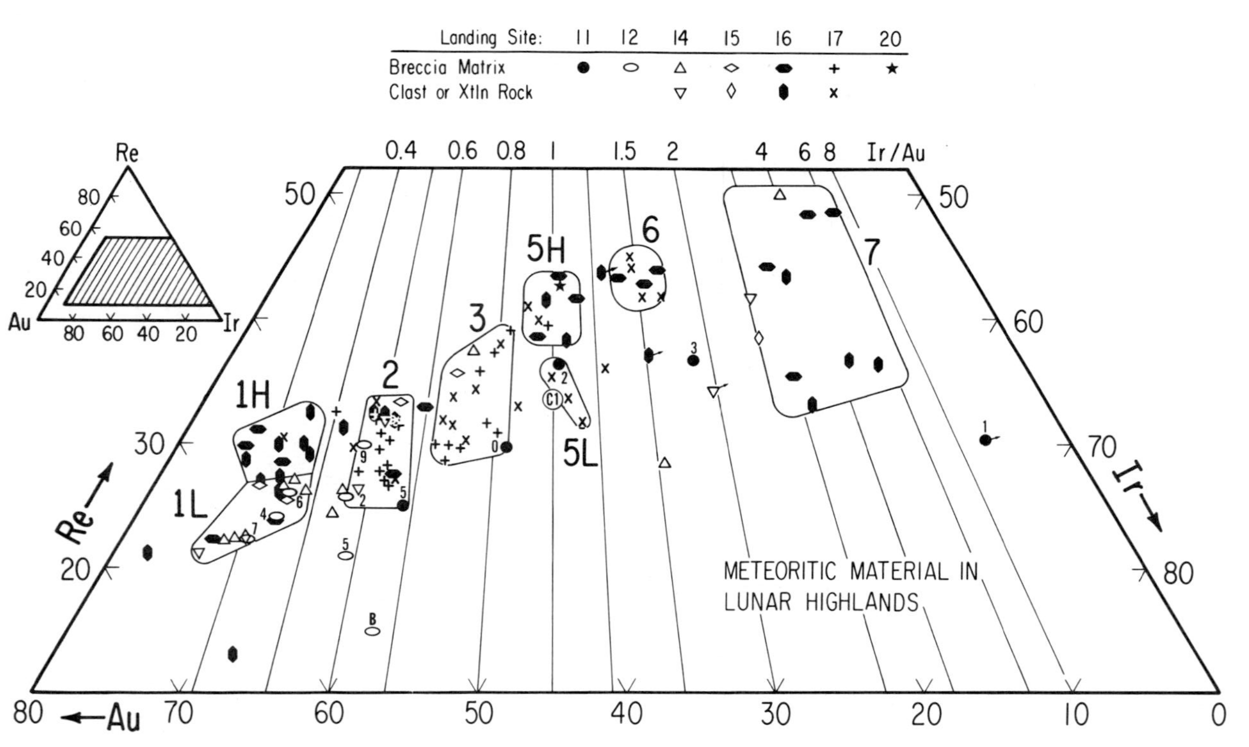

Fig. 1. Ancient meteoritic components in lunar highland samples can be divided into about 8 groups, according to the proportions of siderophile elements Au, Re, and Ir. Reality of groups has been confirmed by objective statistical tests, such as cluster, factor, and discriminant analysis (Higuchi and Morgan, 1975; Gros *et al.*, 1976). Some groups are heavily represented at one landing site, e.g., Groups 1H and 7 at Apollo 16 and Groups 2, 3 at Apollo 17.

Samples from the present study are identified by the last character of the sample number (Table 1). KREEP-rich samples from Apollo 12, as well as Luny Rock 10085,9015, generally have low Ir/Au ratios and fall in or near Groups 1L and 2. Anorthositic samples from Apollo 11 have higher Ir/Au ratios and favor higher Groups, 3 to 7.

Table 2. Possible Evidence for Volatilization.

	Rb/U	Cs/U	Tl/U × 100
Lunar Ave.	*5 ± 1*	*0.25*	*0.28*
10085,771	1.2	0.06	≤0.25
10085,773	7.2	0.24	2.3
10085,9015	1.8	0.11	0.09
12013,9037B	0.9	0.12	0.017
12033,390	3.1	0.14	≤0.13
12033,396	1.9	0.09	0.027
12033,399	1.2	0.06	0.022
12033,402	3.6	0.26	0.023

The remaining 8 samples, duly corrected for the indigenous component, are plotted on an AuReIr diagram (Fig. 1), along with most of our earlier rock analyses. Omitted were samples of mixed pedigree: all soil separates (except 12033, 67602,14-3, and 22006,3), all glasses, and 6 rocks known to be contaminated or hybrids (Hertogen *et al.*, 1977).* Group boundaries in Fig. 1 were slightly adjusted relative to the previous version, and Groups 3H and 3L were combined.

5.1. Meteoritic components at Apollo 11 and 12

Apollo 11. Sample 10085,772, with the highest content of siderophiles, falls in Group 5L. Though richer in a few volatiles (Se, Te, Zn, Tl), it resembles the other members of this group (72215,89; 72235,37; and 76235,9) to a remarkable degree. All are anorthositic breccias of similar Al_2O_3 contents (25.8–27.3%) corresponding to the feldspathic granulitic impactites of Warner *et al.* (1977), all have low U, Rb, and Cs contents (U = 0.18–0.50 ppm, Rb = 0.45–1.48 ppm), and 3 of the 4 have similarly high contents of siderophiles (Ir = 17.6–22.5 ppb). Though some other groups also show correlations between meteoritic component, rock type, and indigenous trace elements (Morgan *et al.*, 1974; Hertogen *et al.*, 1977), such correlations are rarely so strong.

Relative to other Apollo 11 anorthosites, 772 and its kin in Group 5L show enhancements in MgO (7.0–8.5%) and FeO (4.8–6.2%), which Wänke *et al.* (1976) have attributed to a late-accreted "primary component" that also contained siderophiles. However, among the 4 Group 5L members, there is little correlation between MgO-FeO and siderophiles.

Of the remaining 2 anorthositic samples, 771 has exceptionally high (Ir, Re, Os)/Au ratios, even without any indigenous correction, and falls on the Re,Ir-rich side of Group 7 (Fig. 1). This group is more disperse than any other, because

*The rocks omitted were: 15418,30,5D; 64455,27; 67015,104m; 67016,170; 76295,49; 76295,52.

most of its members have very low Au contents, so that sampling, analytical, and contamination problems are magnified. Indeed, we cannot rule out the possibility that the pure Group 7 component contains only the refractory siderophiles Ir, Re, Os, and no "normal" siderophiles such as Au. The variable Ir/Au ratios would then reflect variable contamination with Au from other sources. Such refractory-rich material must have been prominent during the moon's accretion, and may well have survived until the end (Gros *et al.*, 1976). Whatever the nature of this material, we find it expedient to classify all such samples of high Ir/Au into a single Group 7 (Hertogen *et al.*, 1977).

Sample 770 falls into the extreme SE corner of Group 3L, which is populated mainly by Apollo 17 samples. To determine whether this group represents a basin or a local crater, it is important to establish its presence at other sites. Sample 770 suggests such presence at Apollo 11, but in view of its very low siderophile-element content (Au = 0.19 ppb), one must consider the possibility that this sample was shifted leftward from Group 5 or 7 by a small admixture of material of low Ir/Au ratio. Present data, though not definitive, speak against this possibility. Material of low Ir/Au is rare at Apollo 11: bulk soil actually plots near the Cl chondrite point in Fig. 1 (center), and the only known candidate, Luny Rock, has too much Rb, Cs, U relative to Au to have served as the source of excess Au. Moreover, of the two Apollo 11 anorthosite fractions analyzed by Laul *et al.* (1971), 10085,107 (a composite of 5 fragments) gave an Ir/Au ratio of 0.71, in the middle of Group 3, coupled with a Rb content so low (0.80 ppm) as to preclude significant contamination with KREEPy material of low Ir/Au ratio.

Sample 773 has the lowest Ir content of the 4 anorthosites (0.26 vs 0.51–20.6 ppb). Ma and Schmitt (1978a) have shown that it is compositionally the most extreme of the 4, having the highest Al_2O_3 (30.6%) and Eu (31x chondritic), but lowest REE (0.6–0.18x chondritic), FeO (1.6%), as well as Mn, Cr, Sc, V, Co, and Hf. This trend fits the Warner *et al.* (1977) model, according to which an initial anorthositic crust through repeated impacts gradually became contaminated with mafics and the KREEP suite of elements. In terms of this model, sample 773 is the most pristine of the group, and fittingly has the smallest meteoritic component. The Ir/Au ratio of 1.8 is too high even for Group 6, and since a mixing line from the Au corner or Cl chondrite point through 773 passes into Group 7, it seems likely that 773 contains a Group 7 component with a small amount of extraneous Au (~0.02 ppb, or 1×10^{-13} g Au).

Luny Rock falls in the extreme SE corner of Group 2, and has the highest Ir/Au ratio in that group. It could have been excluded by redrawing the boundary, but since it seems to fit the trend defined by the Apollo 17 samples (increasing Ir/Au with decreasing Re/Au), we have left that portion of the Group 2 boundary unchanged.

The meteoritic component of Luny Rock probably can be accepted at face value, because the sample is not polymict and is fairly rich in siderophiles. Thus the presence of Group 2 at yet another site seems likely (though not certain, in view of the extreme position of Luny Rock within Group 2).

Apollo 12. Of the three 12033 separates, monomict breccia 396 falls in Group 1L, which is populated mainly by Apollo 14 and 15 KREEP-rich samples. The two polymict breccias, 399 and 402, lie at the western edge of Group 2. Both are richer in siderophiles than sample 396 (5.4 and 6.0 vs. 2.5 ppb Ir), and since both samples are polymict, it is possible that they were contaminated with a meteoritic component of higher Ir/Au ratio. There is no shortage of such components; ordinary chondrites or C-chondrites would do, as would the ancient component responsible for the high and variable Ir/Au ratios in breccia 14306 (Ganapathy *et al.*, 1974). On the other hand, data on polymict Apollo 16 breccias (Misra and Taylor, 1975; Higuchi and Morgan, 1975) show that the amount of alien metal often is too small to move the samples from one group into another.

To resolve this question, 26 metal particles from 396, 399, and 402 were analyzed for Fe, Ni, Co, P, and Si on the electron microprobe (Table 3). The data are shown on a conventional Co-Ni diagram (Fig. 2), with genetically distinct fields marked according to Goldstein *et al.* (1974).

The distribution is highly atypical. Only 7 of the 26 grains fall in the "meteoritic," high-Fe, or high-Co fields, where the vast majority of metal grains from lunar soils lie (Goldstein and Yakowitz, 1971; Goldstein *et al.*, 1972; 1974; Goldstein and Axon, 1973; Hewins and Goldstein, 1975; Wlotzka *et al.*, 1972; 1973). The majority (18) fall in the thinly populated "supermeteoritic" field; 10 of them in the nearly unoccupied high-Ni portion above 15% Ni. The closest match to this distribution is matrix 76295 from the Apollo 17, Station 6 boulder (Misra *et al.*, 1976), which shows a preponderance of supermeteoritic Co values in the range 3 to 14% Ni, and a few grains in the submeteoritic range.

Because Station 6 boulder matrix is a prototypical member of Group 2, this resemblance strengthens the assignment of 12033,399 and 12033,402 to Group 2. To be sure, the Co-Ni distribution is not a foolproof criterion, because it differs measurably among the 5 boulder matrix samples of Misra *et al.* (1976), although all 5 belong to Group 2. Nonetheless, Fig. 2 indirectly strengthens the assignment to Group 2, by eliminating two alternatives. First, the metal of 399 and 402 is not a hybrid of 396 metal and something else, because mixing lines from the mean composition of 396 through 399 and 402 point to empty regions of the graph: the required high-Ni, low-Co grains are absent in these samples and exceedingly rare in others. (This point can be checked most readily on a Co/Fe vs. Ni/Fe plot, which gives straight mixing lines). Second, the 399 and 402 metal probably was not made by remelting of the rock prior to brecciation. The data lie neither on observed trend lines of melt rocks such as 14310 or 68415, nor on calculated trend lines (Hewins and Goldstein, 1975). Thus it seems likely that 399 and 402 belong to Group 2.

We can compare the new samples with the older composites from Morgan *et al.* (1972): 12033,105 (1 glassy, 1 crystalline fragment), 12033,20,4 (glassy KREEP, +30 mesh), and 12033,20,7 (cindery KREEP, +30 mesh). They are plotted in Fig. 1, and are again identified by the last digit of the sample number. The two <1 mm fractions (4 and 7), which provide the best statistical average,

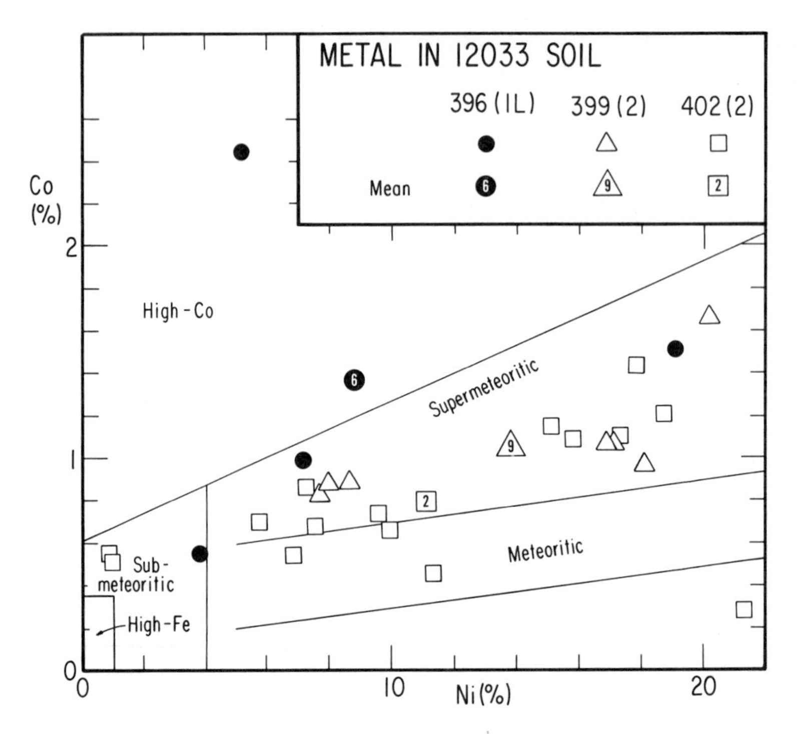

Fig. 2. Metal grains from 12033 KREEP separates have an unusual composition, and lie mainly in the "supermeteoritic" range. Samples 399 and 402, which have a Group 2 ancient meteoritic component, probably are not hybrids of 396 metal (Group 1L) and another component, because mixing lines drawn through their mean compositions require that this component lie in the nearly unpopulated lower right-hand corner of the graph.

fall in Group 1L, confirming the predominance of the 1L component at Apollo 12. The 2–4 mm composite (5), on the other hand, plots below Group 2; though it has the right Ir/Au ratio for that group, it has too little Re.

Interestingly, the KREEPy sample of rock 12013 (9037B) lies even lower, and since Re loss is out of the question, in view of the extreme involatility of this element, it would seem that a new meteoritic component, of very low Re content, is present at Apollo 12. 12013,9037B may be the endmember, and since a line from this sample through 12033,105 passes very close to the 1L field, it seems likely that the latter sample was a mixture of the 12013 component with a conventional 1L component. It may be significant that these unconventional components show up mainly in coarser material (12013 and 2–4 mm soils), not in <1 mm soil.

5.2. Distribution of meteoritic components among landing sites

An important clue to the origin of the ancient meteoritic components is their

Table 3. Composition of Metal Grains in 12033 KREEP Samples. (Analysis by Catherine A. Leitch).

Thin Section	Location or Description	Fe %	Co %	Ni %	P %	Si %	Sum %
408	Nr. voids, Il residuum	91.6	2.44	5.15	0.49	0.05	99.73
	Plagioclase	91.4	0.99	7.15	0.007	0.09	99.64
	Void	79.1	1.51	19.1	≤0.005	<0.01	99.71
	Rock	95.6	0.55	3.81	≤0.005	0.05	100.01
409	Rock; void-rich area	81.4	1.07	17.1	0.013	0.01	99.59
	w. Su; Me+Su-rich area	90.3	0.89	8.64	0.006	0.05	99.89
	w. Su; Me+Su-rich area	91.6	0.83	7.70	0.006	0.11	100.25
	Rock; void-rich area	90.7	0.88	7.94	0.008	0.14	99.67
	Edge of section	77.7	1.66	20.2	–	0.07	99.63
	Rock, Me-Su-rich area	79.9	0.97	18.1	0.025	0.13	99.13
	Edge of section	81.7	1.07	16.9	0.008	0.10	99.78
410	Large } 120 μm apart	98.3	0.55	0.82	0.005	0.05	99.73
	Large } 120 μm apart	78.5	0.28	21.3	0.004	0.05	100.13
	Small, w. Pl, Px	89.0	0.74	9.58	0.007	0.06	99.39
	Small, w. Pl, Px	90.9	0.68	7.52	–	0.08	99.18
	w. Pl in rock	80.1	1.44	17.8	–	0.07	99.41
	Large, in void	98.4	0.51	0.93	0.009	0.03	99.88
	Small, in void	88.8	0.66	9.91	0.012	0.07	99.45
	Large, in rock	91.7	0.54	6.82	0.016	0.04	99.12
	Large, in rock	93.6	0.70	5.76	0.009	0.06	100.13
	Large, edge of section	87.7	0.46	11.3	0.005	–	99.47
	Large, edge of section	79.6	1.21	18.7	0.006	0.01	99.53
	Edge of same void	82.2	1.09	15.8	0.006	0.07	99.17
	Edge of same void	91.3	0.86	7.22	–	0.06	99.44
	Edge of same void	80.9	1.10	17.3	–	0.06	99.36
	Vermicular Px	83.7	1.15	15.1	–	0.08	100.03

selenographic distribution. Components restricted to a single site are likely to come from a local crater, whereas those distributed over several sites may come either from a large basin or from several smaller impacts by bodies of similar composition.

The new data generally support our previous interpretations (Morgan *et al.*, 1974; Hertogen *et al.*, 1977), and in some cases help resolve ambiguities. We shall briefly comment on our latest assignments, as given in the last column of Table 4. For a complete discussion, we refer the reader to Hertogen *et al.* (1977).

Group 1L. This group has now been confirmed at Apollo 12, though its presence was already apparent from earlier data on bulk soils and KREEP separates. Its assignment to Imbrium seems secure.

Group 1H. No material belonging to this group has turned up at Apollo 11, and though this result needs to be confirmed with better statistics, it favors assignment of 1H to a local crater at Apollo 16 (Unnamed B) rather than the Nectaris basin.

Group 2. The presence of a tentative Group 2 sample at Apollo 11 (Luny Rock) supports the assignment of this group to Serenitatis. It would be interesting to look for chronological links between Luny Rock and other rocks assigned to this basin.

Table 4. Distribution of ancient meteoritic components among landing sites.

Group	A11	A12	A14	A15	A16	A17	L20	Assignment
1LL					1			
1L		3	6	3	3			Imbrium
1H					10	1?		Unnamed B
2	1	2?	2	1	4	15		Serenitatis
3	1?		1?	1?		15		Pre-Serenitatis basin or crater
5H					5	3	1	Nectaris?
5L	1					3		Old basin
6					4	4		Old basin or crater
7	2		4	2	9			Planetesimals?
U		2	2		2	4		
0			2	9	10	8		

Group 3. The discovery of a tentative Group 3 member at Apollo 11 does not resolve the question whether this group is derived from a basin or crater. Because this group lies in the middle of the AuReIr diagram, it is especially prone to intrusions by hybrids, and so its members must be authenticated with particular care. Though the assignment of sample 10085,790 is supported by other evidence (Sec. 4.1), it must remain tentative in view of the marginal position, low content of siderophiles, and lack of petrographic information. Since the remaining 2 samples at sites other than Apollo 17 are equally tentative, the case for a basin remains weak.

Group 5H. As pointed out by Hertogen *et al.* (1977), the assignment of this group is strongly coupled to that of 1H. Because the assignment of 1H to a local Apollo 16 crater has now been strengthened, the identification of 5H with Nectaris looks correspondingly more probable.

Group 5L. Now that this group has been found at Apollo 11, its assignment to a basin rather than a crater has been strengthened. The anorthositic mineralogy and the low Rb, Cs, U content of all its members suggest a high age, whereas the restricted occurrence suggests a small size, and a location in the NE quadrant of the moon.

Group 7. The two new cases at Apollo 11 fit the previous trends: anorthositic mineralogy, low Rb, Cs, U content, and very low siderophile-element abundance. As pointed out by Hertogen *et al.* (1977), these traits, as well as its ubiquity and moon-like composition suggest that this group represents the last few impacts by the moon's original planetesimal population.

Acknowledgments—We are greatly indebted to Catherine A. Leitch for the electron microprobe analyses. This work was supported in part by NASA Grant NGL 14-001-167. One of us (JH) is grateful to the Belgian National Science Foundation (NFWO) for granting leave of absence.

REFERENCES

de Albuquerque C. A. R., Muysson J. R. and Shaw D. M. (1972) Thallium in basalts and related rocks. *Chem. Geol.* **10**, 41–58.

Anders E., Ganapathy R., Keays R. R., Laul J. C. and Morgan J. W. (1971) Volatile and siderophile elements in lunar rocks: Comparison with terrestrial and meteoritic basalts. *Proc. Lunar Sci. Conf. 2nd*, p. 1021–1036.

Chao E. C. T., James O. B., Minkin J. A., Boreman J. A., Jackson E. D. and Raleigh C. B. (1970) Petrology of unshocked crystalline rocks and shock effects in lunar rocks and minerals. *Science* **167**, 644–647.

Ganapathy R., Morgan J. W., Higuchi H., Anders E. and Anderson A. T. (1974) Meteoritic and volatile elements in Apollo 16 rocks and in separated phases from 14306. *Proc. Lunar Sci. Conf. 5th*, p. 1659–1683.

Goldstein J. I. and Axon H. J. (1973) Composition, structure, and thermal history of metallic particles from 3 Apollo 16 soils, 65701, 68501, and 63501. *Proc. Lunar Sci. Conf. 4th*, p. 751–775.

Goldstein J. I., Axon H. J. and Yen C. F. (1972) Metallic particles in the Apollo 14 lunar soil. *Proc. Lunar Sci. Conf. 3rd*, p. 1037–1064.

Goldstein J. I., Hewins R. H. and Axon H. (1974) Metal-silicate relationships in Apollo 17 soils. *Proc. Lunar Sci. Conf. 5th*, p. 653–671.

Goldstein J. I. and Yakowitz H. (1971) Metallic inclusions and metal particles in the Apollo 12 lunar soils. *Proc. Lunar Sci. Conf. 2nd*, p. 177–191.

Gros J., Takahashi H., Hertogen J., Morgan J. W. and Anders E. (1976) Composition of the projectiles that bombarded the lunar highlands. *Proc. Lunar Sci. Conf. 7th*, p. 2403–2435.

Hertogen J. and Janssens M.-J. (1977) Is osmium chemically fractionated on the Moon? *Proc. Lunar Sci. Conf. 8th*, p. 47–52.

Hertogen J., Janssens M.-J., Takahashi H., Palme H. and Anders E. (1977) Lunar basins and craters: Evidence for systematic compositional changes of bombarding population. *Proc. Lunar Sci. Conf. 8th*, p. 17–45.

Hewins R. H. and Goldstein J. I. (1975) The provenance of metal in anorthositic rocks. *Proc. Lunar Sci. Conf. 6th*, p. 343–362.

Higuchi H. and Morgan J. W. (1975) Ancient meteoritic component in Apollo 17 boulders. *Proc. Lunar Sci. Conf. 6th*, p. 1625–1651.

Jovanovic S. and Reed G. W. Jr. (1976) Chemical fractionation of Ru and Os in the moon. *Proc. Lunar Sci. Conf. 7th*, p. 3437–3446.

Jovanovic S. and Reed G. W. Jr. (1977) Is Os chemically fractionated on the moon? A response. *Proc. Lunar Sci. Conf. 8th*, p. 53–56.

Keays R. R., Ganapathy R., Laul J. C., Krähenbühl U. and Morgan J. W. (1974) The simultaneous determination of 20 trace elements in terrestrial, lunar and meteoritic materials by radiochemical neutron activation analysis. *Anal. Chim. Acta* **72**, 1–29.

King E. A. Jr., Carman M. F. and Butler J. C. (1970) Mineralogy and petrology of coarse particulate material from the lunar surface at Tranquility Base. *Proc. Apollo 11 Lunar Sci. Conf.*, p. 599–606.

Krähenbühl U., Ganapathy R., Morgan J. W. and Anders E. (1973) Volatile elements in Apollo 16 samples: Implications for highland volcanism and accretion history of the moon. *Proc. Lunar Sci. Conf. 4th*, p. 1325–1348.

Laul J. C., Keays R. R. Ganapathy R. and Anders E. (1970) Abundance of 14 trace elements in lunar rock 12013,10. *Earth Planet. Sci. Lett.* **9**, 211–215.

Laul J. C., Morgan J. W., Ganapathy R. and Anders E. (1971) Meteoritic material in lunar samples: Characterization from trace elements. *Proc. Lunar Sci. Conf. 2nd*, p. 1139–1158.

Lunatic Asylum (1970) Mineralogic and isotopic investigations on lunar rock 12013. *Earth Planet. Sci. Lett.* **9**, 137–163.

Ma M.-S. and Schmitt R. A. (1978a) Chemistry of 10085 fragments and Apollo 11 basalts and breccias (abstract). In *Lunar and Planetary Science IX*, p. 678–680. Lunar and Planetary Institute, Houston.

Ma M.-S. and Schmitt R. A. (1978b) Chemistry of 12033 fragments (abstract). In *Lunar and Planetary Science IX*, p. 681–682. Lunar and Planetary Institute, Houston.

Misra K. C. and Taylor L. A. (1975) Characteristics of metal particles in Apollo 16 rocks. *Proc. Lunar Sci. Conf. 6th*, p. 615–639.

Misra K. C., Walker B. M. and Taylor L. A. (1976) Textures and compositions of metal particles in Apollo 17, Station 6 boulder samples. *Proc. Lunar Sci. Conf. 7th*, p. 2251–2266.

Morgan J. W., Ganapathy R., Laul J. C. and Anders E. (1973) Lunar crater Copernicus: Search for debris of impacting body. *Geochim. Cosmochim. Acta* **37**, 141–154.

Morgan J. W., Ganapathy R., Higuchi H., Krähenbühl U. and Anders E. (1974) Lunar basins: Tentative characterization of projectiles, from meteoritic elements in Apollo 17 boulders. *Proc. Lunar Sci. Conf. 5th*, p. 1703–1737.

Morgan J. W., Laul J. C., Krähenbühl U., Ganapathy R. and Anders E. (1972) Major impacts on the moon: Characterization from trace elements in Apollo 12 and 14 samples. *Proc. Lunar Sci. Conf. 3rd*, p. 1377–1395.

Short N. M. (1970) Evidence and implications of shock metamorphism in lunar samples. *Proc. Apollo 11 Lunar Sci. Conf.*, p. 865–872.

Wänke H., Palme H., Kruse H., Baddenhausen H., Cendales M., Dreibus G., Hofmeister H., Jagoutz E., Palme C., Spettel B. and Thacker R. (1976) Chemistry of lunar highland rocks: A refined evaluation of the composition of the primary matter. *Proc. Lunar Sci. Conf. 7th*, p. 3479–3499.

Warner J. L., Phinney W. C., Bickel C. E. and Simonds C. H. (1977) Feldspathic granulitic impactites and pre-final bombardment lunar evolution. *Proc. Lunar Sci. Conf. 8th*, p. 2051–2066.

Wlotzka F., Jagoutz E., Spettel B., Baddenhausen H., Balacescu A. and Wänke H. (1972) On lunar metallic particles and their contribution to the trace element content of Apollo 14 and 15 sites. *Proc. Lunar Sci. Conf. 3rd*, p. 1077–1084.

Wlotzka F., Spettel B. and Wänke H. (1973) On the composition of metal from Apollo 16 fines and the meteoritic component. *Proc. Lunar Sci. Conf. 4th*, p. 1483–1491.

Wood J. A., Dickey J. S. Jr., Marvin U. B. and Powell B. N. (1970) Lunar anorthosites and a geophysical model of the moon. *Proc. Apollo 11 Lunar Sci. Conf.*, p. 965–988.

Proc. Lunar Planet. Sci. Conf. 9th (1978), p. 1551–1570.
Printed in the United States of America

Ion microprobe analyses of aluminous lunar glasses: A test of the "rock type" hypothesis

CHARLES MEYER, JR.

SN7, Geochemistry Branch, NASA Johnson Space Center
Houston, Texas 77058

Abstract—Previous soil survey investigations found that there are natural groupings of glass compositions in lunar soils and that the average major element composition of some of these groupings is the same at widely separated lunar landing sites. This led soil survey enthusiasts to promote the hypothesis that the average composition of glass groupings represents the composition of primary lunar "rock types".

In this investigation the trace element composition of numerous aluminous glass particles was determined by the ion microprobe method as a test of the above mentioned "rock type" hypothesis. It was found that within any grouping of aluminous lunar glasses by major element content, there is considerable scatter in the refractory trace element content. In addition, aluminous glasses grouped by major elements were found to have different average trace element contents at different sites (Apollo 15, 16 and Luna 20). This evidence argues that natural groupings in glass compositions are determined by regolith processes and may *not* represent the composition of primary lunar "rock types".

INTRODUCTION

The comprehensive study of numerous small glass particles from the lunar regolith is thought to be a powerful method for studying the chemistry of the lunar surface. Certainly the wide range in glass compositions indicates a wide diversity of lunar rocks. On the other hand, statistical groupings of the compositions of glass particles may or may not indicate the relative abundance of various lunar materials at a given landing site. Likewise, the average composition of natural groupings of glass particles may or may not correspond to the average composition of primary "rock types". The problem is that most of the glass particles found in a lunar soil sample were undoubtedly produced by meteorite impact. For this reason the exact nature of the protolith of these glass particles is necessarily clouded by various regolith processes such as mixing, vaporization, and differential melting. Nevertheless, the large number of investigations that have reported on the compositions of glass particles testifies to the intrinsic scientific interest in these particles.

A proper interpretation of the study of small particles from lunar soils is a necessary prerequisite for planning the sampling of the surfaces of other planetary objects. The moon provides a natural laboratory where we can establish whatever linkage there is between separate data sets. First of all, there are the many small particles from soils and core tubes from a variety of landing sites. In addition, we have extensive, detailed chemical and petrological studies on numerous large samples for direct comparison. Finally, we have a partial data set from lunar orbit for a number of key elements. One wonders how accurate our interpretation would be if we had not collected the larger samples from the manned landing sites! Such will undoubtedly be the case for other planetary objects.

The purpose of this project was to measure the trace element composition of small (50–150 μm) particles of glass from lunar soil samples by means of the ion microprobe technique and to use this data along with major element data on the same particles to test and/or constrain the various models for the petrogenesis of these glasses and their parental rock types. The particles analyzed in this study are from soil 15101, 61221, Luna 20 (various sub-samples) and 74240 (ropy glass only). These soil samples differ in bulk composition and maturity, but they all contain numerous aluminous glasses suitable for ion probe analysis.

A limited number of ion microprobe analyses of lunar glass particles have previously been reported by Andersen *et al.* (1970), Fredriksson *et al.* (1971) and Prinz *et al.* (1973). It is important to continue development of this technique so that a larger number of elements can be determined in more particles and with more accuracy.

PROCEDURE

The ion microprobe (ARL) used in this investigation has been computer automated to allow reproducible data collection for a large number of particles. In this study a negative oxygen ion beam (15 kv, 2 nano-amps) was used to sputter a 20 μm spot on the surface of the sample. The sputtering rate for this size spot was about 1 μm per hour. The secondary ions that were sputtered off the sample were accelerated 1.5 kv and analyzed by a double focussing mass spectrometer. Slits set at α = .32 cm, β = .64 cm, and final = .05 cm gave a resolution $\frac{M}{\Delta M} = 150$. A Daly detector and photomultiplier with pulse counting electronics were used to detect secondary ions (Andersen and Hinthorne, 1972).

Mass spectra were obtained by repeatedly and reproducibly scanning the magnetic field of the secondary magnet using voltage supplied by the computer and storing the counts with respect to time in a multiscaler. Complete spectra (mass 4 to 150) were integrated in 4096 channels using 100 separate magnet scans with a scan rate of 10^{-2} sec/channel. A pause of 15 sec before each scan allowed the magnet to "flyback" to the beginning of the scan. In this way each spectrum took about an hour and a half to integrate during which time the ion beam erroded away a spot 20 μm diameter and 1.5 μm deep. In this way complete spectra were integrated in a reproducible manner for both standards and unknowns. Since the samples were carbon coated, each spot was first sputter cleaned for several minutes before integration. A correction for deadtime of 200 nano-seconds was applied to the large peaks.

Trace element concentrations were determined by comparing peak heights of Li 7, Be 9, B 11, F 19, K 39, Y 89, Zr 90, Ba 138, and Ce 140 relative to Si 30

Table 1. Analytical Parameters.

Element	Isotope	Interference	Count Rate on Standard cps	12033,* 97,1A ppm	CPS ppm	Minimum Detect-ability ppm	Reproduc-ibility percent	Interference level ppm
Li	7		3070	89	35	0.2	12–25	0.1
Be	9	Al^{+3}	56	(10)	5.6	1.3	10–30	2
B	11		52	(26)	2.0	3.6	6–12	2
F	19	H_3O^+	31	(250)	0.12	60	9–15	40
K	39	NaO	171,300	6490	26	0.28	16–29	1
Y	89	$Si_2O_2{}^+$	900	(270)	3.3	2.2	4–10	10
Zr	90	$Si_2O_2{}^+$	1200	(1200)	1.0	7.2	3–9	10
Ba	138	$ZrO_3{}^+$	3700	1210	3.06	2.3	5–10	10
La	139	?	180	105	1.71	4.2	5–10	10
Ce	140	?	325	265	1.23	5.8	5–10	10
Si	30		24,000					
Bkg.			1.44					

*Wiesmann and Hubbard (1975).

and comparing these ratios to those obtained from the spectra of a secondary standard measured in the same way. The secondary standard used was a thin section of KREEP glass (12033,97,1A). For Li, K, Ba, La and Ce, the concentration of this standard are known from isotope dilution analyses of an adjacent piece (Hubbard and Gast, 1971). Zr was determined by electron probe and the other elements were estimated using the bulk concentrations in 12033 soil and an assumed KREEP content of 68% (Meyer *et al.*, 1971). Other standards were used to verify the assumed values of this secondary standard by means of the relative sensitivities of various elements. Direct ratioing to a standard has the advantage of being simple and allows easy intercomparison of standards. Data reduction schemes based on assumed thermal equilibrium (Andersen and Hinthorne, 1973) are less satisfactory.

Results on a wide variety of materials are compared by plotting data as relative sensitivities (Fig. 1). The relative sensitivities are calculated to take into account isotopic abundances and atom percentages. The lack of agreement between different samples may be accounted for by one or more causes. First of all, there is a matrix effect which is interpreted as due to differences in the ionization conditions at the surface of different materials. It is observed that the ratio of single to double charge species of the same element varies widely from one mineral to another. There is also a corresponding difference in relative sensitivity for elements that differ in their ionization potentials. Presumably this is related to differences in how ions are neutralized as they leave the surface of the sample. In this investigation only aluminous glasses were analyzed and the Si^{+2}/Si^{+1} ratio was monitored throughout. There were no systematic variations of this ratio between standards and unknowns and this ratio was not found to be a function of iron contents (FeO, 0–10%) of the glasses. However, considerable scatter (± 10%) in the Si^{+2}/Si^{+1} ratios was observed and may be the cause of the relatively poor reproducibility as measured on the secondary standard.

Another problem is caused by the fact that ionization conditions and relative

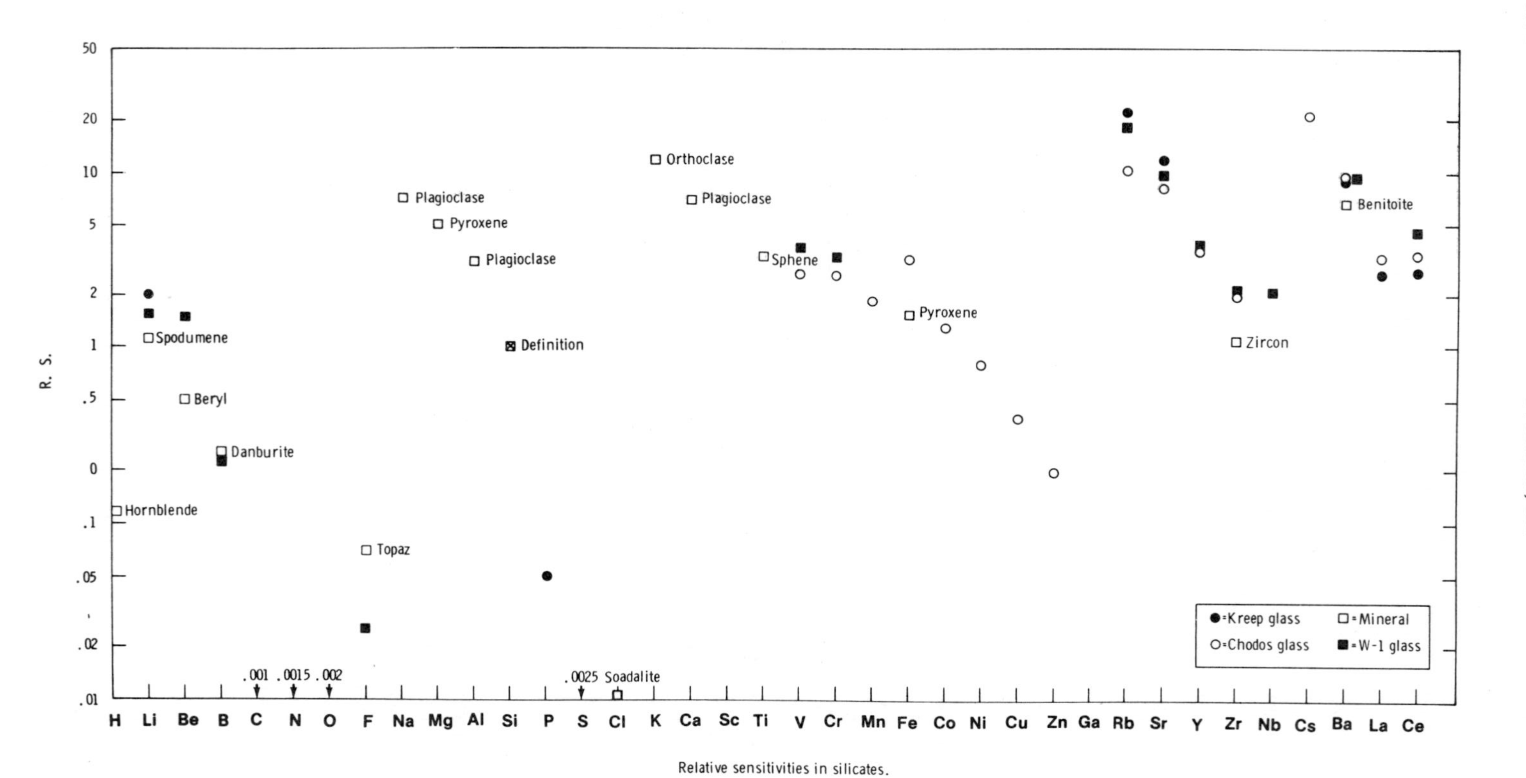

Fig. 1. The ratios of peak heights (corrected for deadtime and percent isotopic abundance) divided by the atom percentage concentration and relative to silicon are termed relative sensitivities (R.S.). These data were collected under constant operating conditions (15 kv, 2 nanoamps, same settings) on a wide variety of stoichiometric silicates and analyzed glasses.

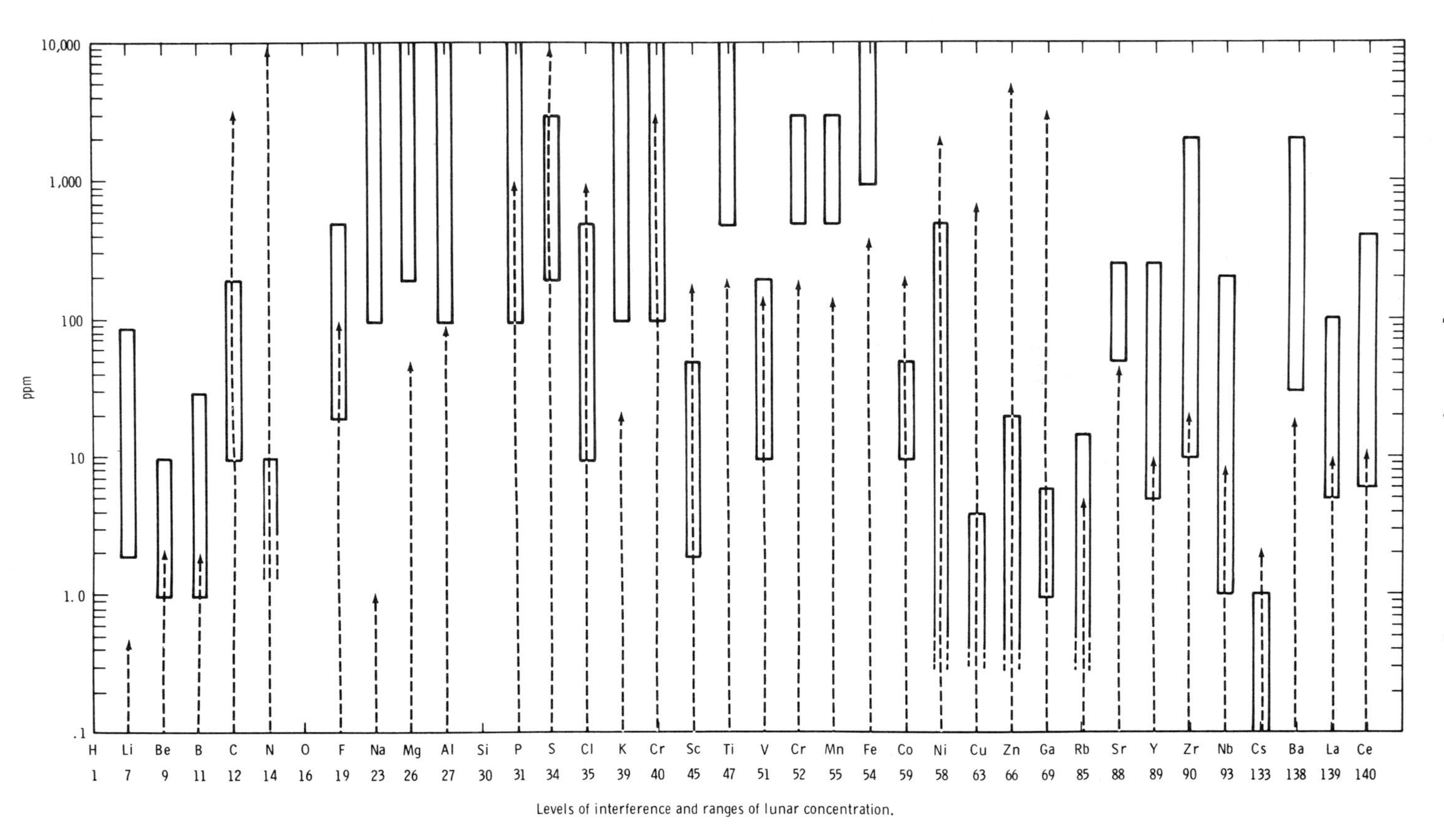

Fig. 2. Levels of interference and/or background expressed as concentrations using relative sensitivities derived from Fig. 1. The boxes indicate the ranges of concentrations of the various elements found in lunar samples by various other techniques.

sensitivities change rapidly during the first few minutes of sputtering. During this time the carbon coat is sputtered away and the silicate lattice becomes stuffed full of oxygen ions. A steady state condition is achieved when diffusion of elements into this zone is balanced by sputtering. This problem was not recognized until after this investigation, and not enough care was taken to assure steady state conditions before beginning each integration. Some of the scatter in Si^{+2}/Si^{+1} ratio and the poor reproducibility may be due to this effect.

The level of detectability of a given trace element by the ion probe method is determined by: 1) the relative sensitivity; 2) the relative isotopic abundance; 3) the background; and 4) the level of molecular ion interference in the matrix studied. If the relative sensitivities in Fig. 1 are directly applied to the peak

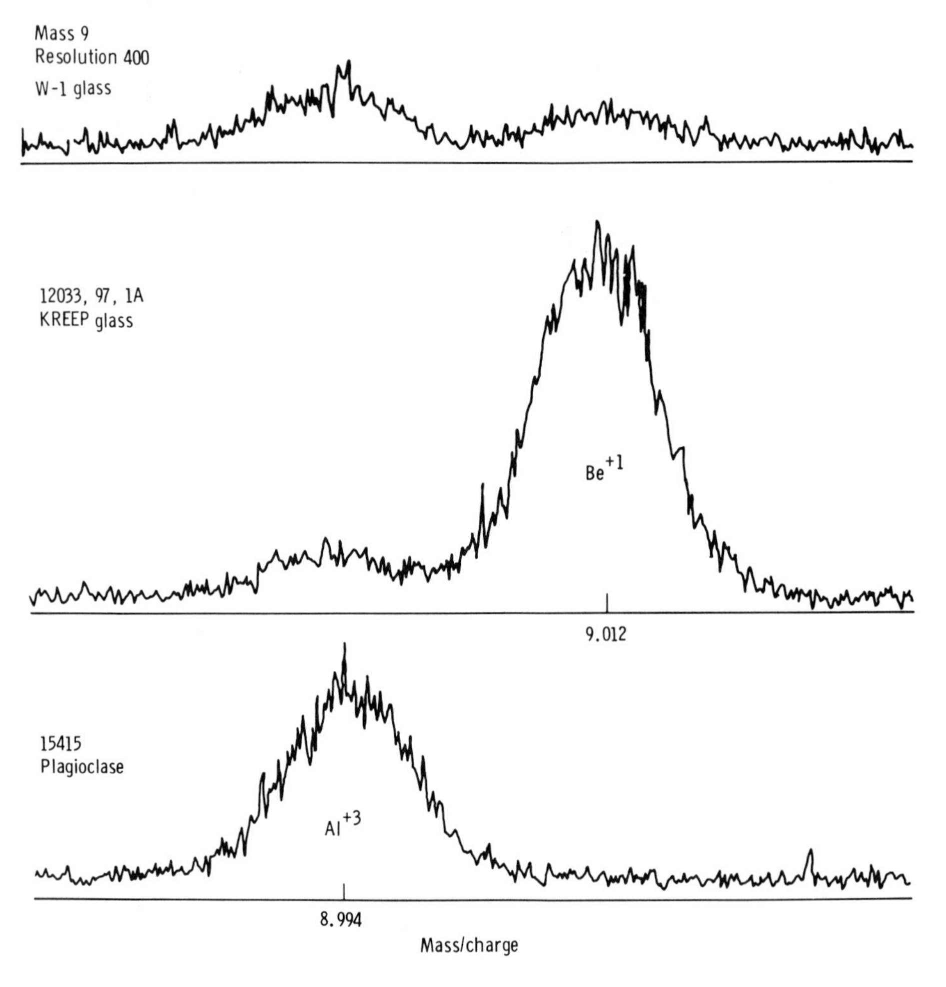

Fig. 3. Resolution of Al^{+3} and Be^{+1} peaks in region of mass 9. The 12033 sample has about 10 ppm Be and the 15415 plagioclase has none.

heights in an unknown spectra, one may calculate effective "concentrations". However, depending on the matrix these "concentrations" may be meaningless unless molecular ion interferences are carefully considered. For aluminous lunar glasses, Fig. 2 compares these molecular ion interferences (calculated as concentrations) with the ranges of trace element abundances usually found in lunar rock samples. From this plot one can determine which elements can be measured by a low resolution ion probe. Figure 2 shows that Li, Be, B, F, K, Y, Zr, Ba, and Ce are relatively free of interferences and these are the elements chosen in this study. Aluminous lunar glasses have less interferences than the more iron-rich lunar glasses. In addition, as the FeO content of aluminous lunar glasses increases, so does their inherent trace element content.

In the case of Be, the Al^{+3} interference can be resolved with the ARL ion probe

Table 2. Rucklidge Glass Standards Ion Microprobe Analyses.

	AGV-1		BCR-1		W-1		G-2		GSP-1	
	Flanagan	IMA	Flanagan	IMA	Flanagan	IMA	Flanagan	IMA	Flanagan	IMA
ppm										
Li	12	7.1	13	12	14.5 12.5**	10.4	35	21	32	27
Be	(3)	1.1	(2)	1.5	0.8 0.6**	0.65	2.6	2.4	1.5	1.1
B	(5)	8.6	(5)	10.6	15 9**	15	2.0	2.1	3.0	0.6
F	435 408*	211	470 502*	245	250	194	1290 1224*	285	3200	3833
Y	21	32	37	40	25	28	12	22	30	59
Zr	225	243	190	185	105	105	300	115	500	650
Ba	1208	1077	675	546	160	151	1870	1700	1300	1400
La	35	37	26	23	10	18	96	82	191	231
Ce	63	78	54	44	23	30	150	168	394	660
percent										
K_2O	2.89	2.26	1.70	1.50	0.64	0.54	4.51	—	5.53	—
SiO_2	59		54.5		52.6		69.1		67.4	

Flanagan, 1973.
**Eugster, 1971.
*Allen and Clarke, 1977.

using narrow slits (Fig. 3). However, data were normally collected at low resolution and the Al^{+3} interference was estimated using the height of the Al^{+2} peak and the Al^{+3}/Al^{+2} ratios as measured on 15415—which has no Be. The molecular ion interferences for Y, Zr, Ba, La and Ce were estimated from analyses of trace element-poor aluminous glasses and these corrections were applied at the end of the data reduction procedure.

The accuracy of this method is indicated by the analyses of glass standards made from USGS rock powders by Rucklidge (1969) (Table 2). The values reported by Flanagan (1973) are based on a wide variety of techniques applied to the rock powders. The IMA values are based on the 12033,97,A standard (Table I). Differences in these values are hard to explain and further improvement in this technique is desirable.

RESULTS

A total of 103 aluminous glass particles (Al_2O_3 >14%) were analyzed in several thin sections of 15101, Luna 20, 61221 and 74240 soils (50–150 μm). Each particle was analyzed for Li, Be, B, F, K, Y, Zr, Ba, and Ce by the ion probe as well as for major elements by the electron probe. The complete data set can be obtained from the author. In general, the trace elements measured were found to vary proportionally to Ba over a wide range (Fig. 4) and the elemental ratios are similar to those of larger samples. For this reason much of the discussion of this paper will refer to Ba contents of these particles without reference to the similar trends of the other trace elements. Figure 5 shows the variation of Ba (and other trace elements) with Al_2O_3 in individual glasses from these samples.

The major element composition of the glass particles, as determined by automated electron probe (MAC), were used to group the glasses according to their position on the low pressure, pseudo-quaternary anorthite-olivine-silica phase diagram (Walker *et al.*, 1973). These groupings are similar to those of the "soil survey" which were based on "cluster analysis" of a much larger set of glass data (Reid *et al.*, 1972c). The averages of these groupings are given in Table 3. Figure 6 illustrates these groups and gives the average value for Ba for each group. Note that the average value of the bulk soil is near or within a major grouping in each case. Although the center of each cluster lies close to the cotectic, it is important to notice that individual analyses show considerable scatter. Within each grouping the trace element data also show considerable scatter.

Since averages can be misleading, it is important to consider the composition of individual glasses. Some individuals in each group have very low concentrations of refractory trace elements (Table 4). These few individuals are probably the ones that best represent the original or primary "rock types."

Some elements, like F, B, K and Na, are volatile and should have been lost by volatilization during the glass forming event(s). However, no correlation could be found. Presumably this lack of regularity among the volatile elements is due to the multiple number of events that produced these glasses.

Aluminous glasses of anorthositic gabbro composition are quite numerous in the Luna 20 soil samples (Glass, 1973; Reid *et al.*, 1973). These glasses do not form a single natural grouping, however, but rather vary along the spinel-plagioclase cotectic. They become more trace element-rich with decreasing Al_2O_3 content.

The aluminous glasses from 74240 were of a ropy nature (Fruland *et al.*, 1977)

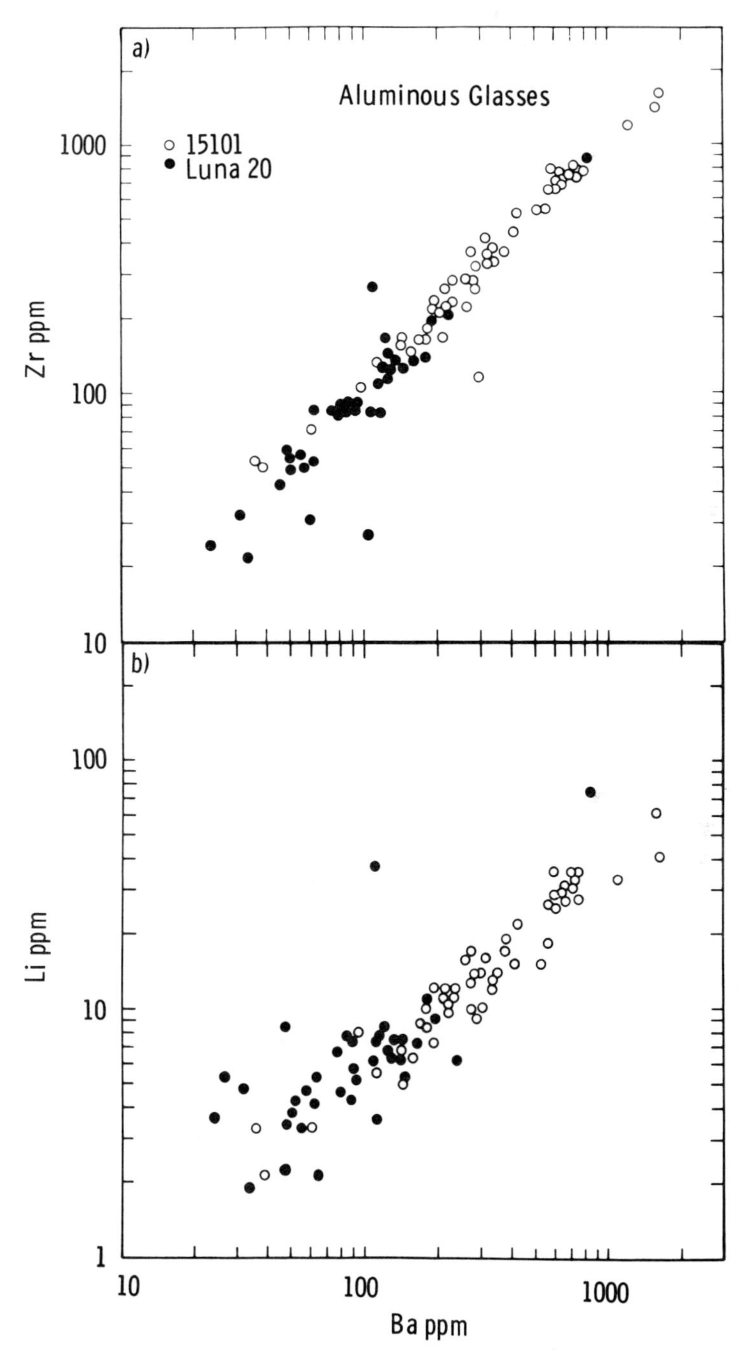

Fig. 4. Correlation of Zr, Li and Ba analyses of aluminous glass particles from Luna 20 and 15101 soils.

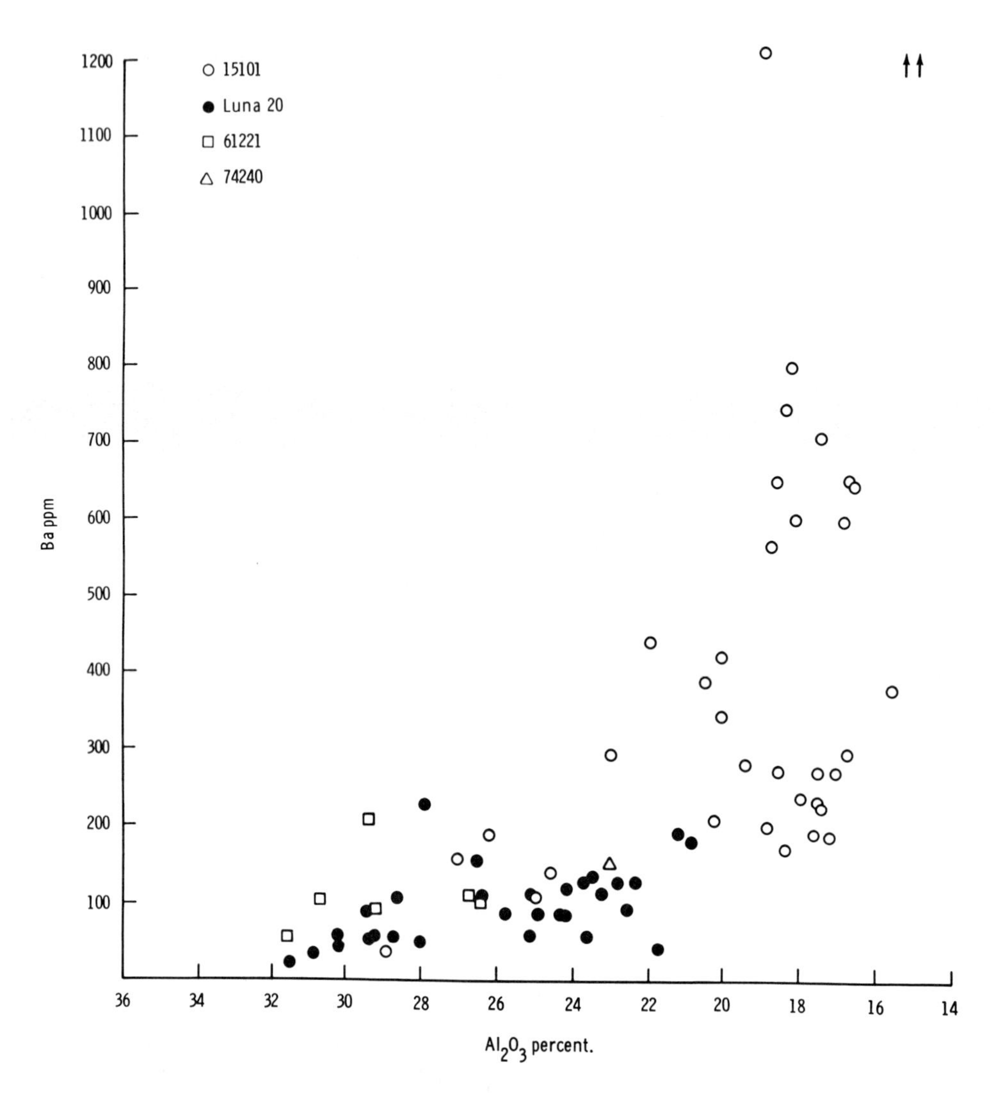

Fig. 5. Ba analyses of individual aluminous lunar glass particles from four different soils.

and were presumably formed during a single event. The trace element data obtained in this study support the interpretation that these glasses are mixtures of noritic and anorthositic breccias.

Discussion

A primary objective of lunar science is to discern the distribution of "rock types" on the lunar surface. A knowledge of this spatial distribution is needed to help constrain theories of the moon's origin and evolution. However, the

Table 3. Average composition of glass groupings ($S_x = \sqrt{(\Sigma x_i^2 - nx^2)/n\text{-}1}$).

	15101 HB	15101 LKFM	15101 KREEP	Luna 20 HBb	Luna 20 HBa	Luna 20 LKFM	61221 HB	74240 Ropy
Number	8	26	14	6	15	4	6	7
SiO_2	44.9 ± 1.2	46.7 ± 1.5	49.3 ± 0.9	43.4 ± 0.7	44.8 ± 0.6	45.8 ± 0.5	44.6 ± 0.5	44.4 ± 0.4
TiO_2	.32 ± .12	1.02 ± .27	1.46 ± .19	.24 ± .20	.43 ± .13	1.06 ± .27	.41 ± .10	.49 ± .19
Al_2O_3	26.5 ± 2.6	18.3 ± 1.9	16.7 ± 3.0	29.8 ± 1.2	24.2 ± 1.5	19.2 ± 2.1	27.0 ± 1.6	23.2 ± .6
Cr_2O_3	.12 ± .03	.25 ± .08	.24 ± .04	.04 ± .04	.14 ± .05	.28 ± .14	.10 ± .04	–
FeO	5.2 ± 1.7	9.8 ± 1.6	9.4 ± 0.8	3.6 ± 0.9	6.5 ± 1.1	10.3 ± 4.0	4.5 ± 0.9	6.8 ± 0.6
MgO	6.8 ± 2.1	11.1 ± 1.9	8.8 ± 0.8	6.5 ± 1.3	8.9 ± 1.0	9.2 ± 0.7	5.8 ± 1.5	8.9 ± 0.9
CaO	15.4 ± 1.7	11.5 ± 1.0	10.2 ± 2.4	16.4 ± 0.7	14.5 ± 0.5	13.1 ± 1.1	15.8 ± 0.8	14.4 ± 1.1
Na_2O	0.16 ± .09	0.44 ± .16	0.71 ± .14	0.30 ± .11	0.27 ± .08	0.34 ± .15	0.56 ± .19	0.24 ± .12
K_2O	0.05 ± .04	0.19 ± .08	0.49 ± .11	0.04 ± .01	0.07 ± .02	0.10 ± .01	0.10 ± .04	0.09 ± .03
P_2O_5	0.03 ± .03	0.11 ± .07	0.40 ± .18	–	–	–	0.03 ± .01	0.06 ± .02
Total	99.5	99.4	97.8	100.0	99.8	99.4	98.9	98.6
ppm								
K	677 ± 469	1720 ± 733	4565 ± 1232	233 ± 146	434 ± 163	742 ± 100	800 ± 360	607 ± 270
Li	5.1 ± 2.1	12.7 ± 3.5	29 ± 5	3.8 ± 0.9	6.0 ± 1.7	8.7 ± 1.8	6.8 ± 1.6	7.5 ± 1.1
Be	0.8 ± 0.8	2.1 ± 0.9	6.3 ± 1.4	0.5 ± 0.5	0.7 ± 0.6	1.2 ± 0.3	0.9 ± 0.9	1.2 ± 0.6
B	8 ± 6	13 ± 10	24 ± 7	6 ± 6	7 ± 6	7 ± 4	7 ± 5	8 ± 7
F	132 ± 76	137 ± 45	310 ± 160	138 ± 123	150 ± 70	166 ± 54	93 ± 33	109 ± 17
Y	27 ± 15	70 ± 26	176 ± 36	5 ± 5	19 ± 7	32 ± 11	21 ± 9	39 ± 6
Zr	106 ± 50	269 ± 99	738 ± 142	38 ± 24	82 ± 23	135 ± 49	81 ± 26	162 ± 20
Ba	99 ± 56	261 ± 79	686 ± 160	44 ± 23	90 ± 31	134 ± 55	113 ± 45	150 ± 23
Ce	33 ± 18	73 ± 27	172 ± 34	9 ± 7	20 ± 12	33 ± 10	23 ± 8	44 ± 6

HB = Highland Basalt (Anorthositic Gabbro).

Table 4. Individual glasses with low trace element contents.

	15101 14–29	15101 S4	15101 S27	22002,7 ,6	SA0525 ,7	22001,27 ,7	SA0529 ,8	22001,26 ,6
SiO_2	45.9	43.4	46.2	43.0	43.9	44.5	44.6	45.5
TiO_2	0.21	0.18	1.34	0.09	0.28	0.28	0.65	0.90
Al_2O_3	28.7	25.3	19.1	31.5	28.0	25.1	21.8	18.1
Cr_2O_3	0.09	–	–	0.00	0.09	0.10	0.26	–
FeO	4.5	6.5	11.0	4.1	5.0	5.2	9.0	12.1
MgO	5.6	6.6	10.7	4.9	6.9	8.9	7.3	9.2
CaO	15.6	15.8	11.9	16.7	16.0	14.7	14.3	12.6
Na_2O	0.26	0.10	0.39	0.40	0.13	0.32	0.16	0.26
K_2O	0.03	0.04	0.17	0.03	0.03	0.07	0.03	0.10
P_2O_5	0.01	0.00	0.09	–				0.05
ppm								
K	283	790	1130	150	123	522	106	600
Li	2.2	3.3	8.0	3.6	3.8	5.3	3.4	6.6
Be	2.3	0.8	0.0	0.0	0.0	0.2	0.8	1.6
B	–	20	13	1	6	7	4	5
F	283	70	87	55	49	93	58	235
Y	(6)	(9)	(24)	0	(5)	(14)	(14)	(19)
Zr	40	43	98	(14)	44	75	49	71
Ba	29	26	87	(14)	40	54	38	67
Ce	(8)	(14)	(21)	0	(4)	(15)	(7)	(20)
	HB	HB	LKFM	HB	HB	HB	HB	LKFM

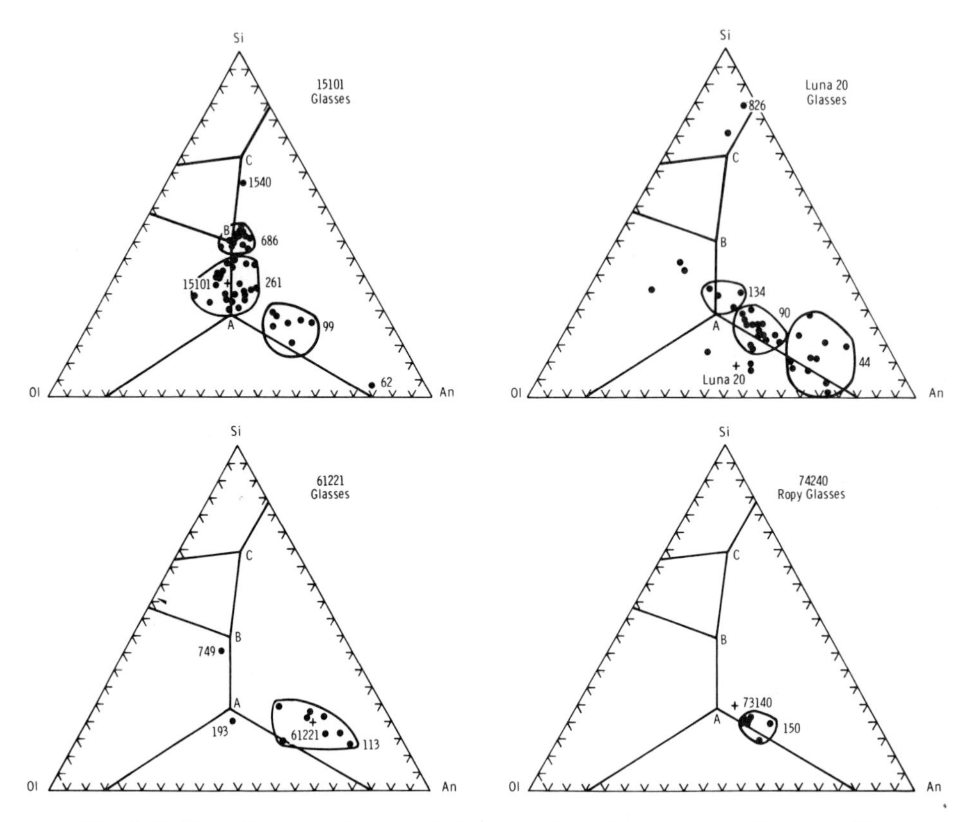

Fig. 6. Individual glass analyses plotted on pseudo-quaternary phase diagrams. The average Ba contents of natural groupings are indicated. The crosses indicate the compositions of bulk soil samples.

conversion of maps of chemical composition into maps of rock type is difficult at best and requires a general synthesis of the extensive data on lunar samples from 9 different landing sites, spectroscopic data from earth-based telescopes, and geochemical measurements from lunar orbit.

A necessary step in this synthesis is the proper recognition of the fundamental lunar "rock types" and their average composition. The larger lunar samples are a good starting point for this exercise, but it is recognized that only a few of the lunar rock types may be represented in this collection (Warren and Wasson, 1977).

Another method that has been used is to perform statistical surveys of small particles in the lunar soils (e.g., Apollo Soil Survey, 1971; Reid *et al.*, 1972c; Glass, 1973, and many others). It is known that the smaller particles at a landing site sample a wider area than do the larger samples. For example, samples of the lunar highlands were identified among the smaller particles from the Apollo 11 landing site. Perhaps the most useful particles in the smaller size range are the glasses. For the reason numerous investigators of lunar soils have concentrated their efforts on analyses of glass particles in the size range 50–150 μm. Using this large data base, some soil survey enthusiasts have promoted the hypothesis that statistical groupings of glass particles give the average major element compositions of primary "rock types." Several lines of reasoning have been used to support this interpretation (Ridley *et al.*, 1973), but searches for pristine examples of these rock types have yielded few good examples (Reid *et al.*, 1977; Warren and Wasson, 1977). Taylor *et al.* (1973) have extended the compositions of these "rock types" to trace elements by measurements on highlands breccias with major element compositions that are similar to the average composition of the glass clusters. However, this linkage is uncertain at best.

In the present investigation, trace element compositions were directly determined in individual glass particles to test the above mentioned "rock type hypothesis." Especially useful in this regard are trace element analyses of the aluminous glasses generally termed "Anorthositic Gabbro" and "Low K Fra Mauro Basalt." Previous investigators have reported that these glasses have remarkably uniform major element composition at all the landing sites studied (Reid *et al.*, 1972a). If the "rock type hypothesis" is valid, the trace element compositions should also be uniform for glasses with the same major element composition. On the other hand, if the trace element composition of these particles varies with bulk soil composition or with soil maturity, then one could argue that the glass particles could be the product of regolith processes and, as such, likely to be altered in both their major and trace element compositions. On the other hand, there remains the less likely possibility that those "rock types" could simply have different trace element contents at different places on the moon! A test of this second possibility would be to measure glasses in soils of different maturity from the same site.

The result of this investigation is that within any grouping of aluminous glasses by major element composition there is considerable scatter in the trace element analyses. Figure 7 shows the variation of Ba vs. Al_2O_3 contents of individual glasses for the "rock type" previously termed "Anorthositic Gabbro" (see Reid *et al.*, 1972a). In addition, there is a general increase in the average Ba content of each glass type with increasing Ba content of the bulk soil samples in which they are found (Table 5). This relationship is not exact and is complicated by the additional parameter of soil maturity. Presumably contamination of these aluminous glasses by Ba-rich materials in the soil causes both the variability in the individual analyses and the general increase in the more mature and Ba-rich soils.

Figure 8 compares these averages with some of the best analyzed larger lunar rock samples. However, these larger lunar rocks are polymict breccias containing meteoritic materials and presumably they have also been contaminated by Ba-rich components. Finally, Fig. 8 also compares these analyses with values of Ba estimated for large regions of the lunar highlands assuming a constant Ba/Th relationship. Supposedly these large regions also contain a significant foreign component. It is interesting that only a few of the individual glasses measured have Ba contents below those of the large rocks or as estimated from orbit. These

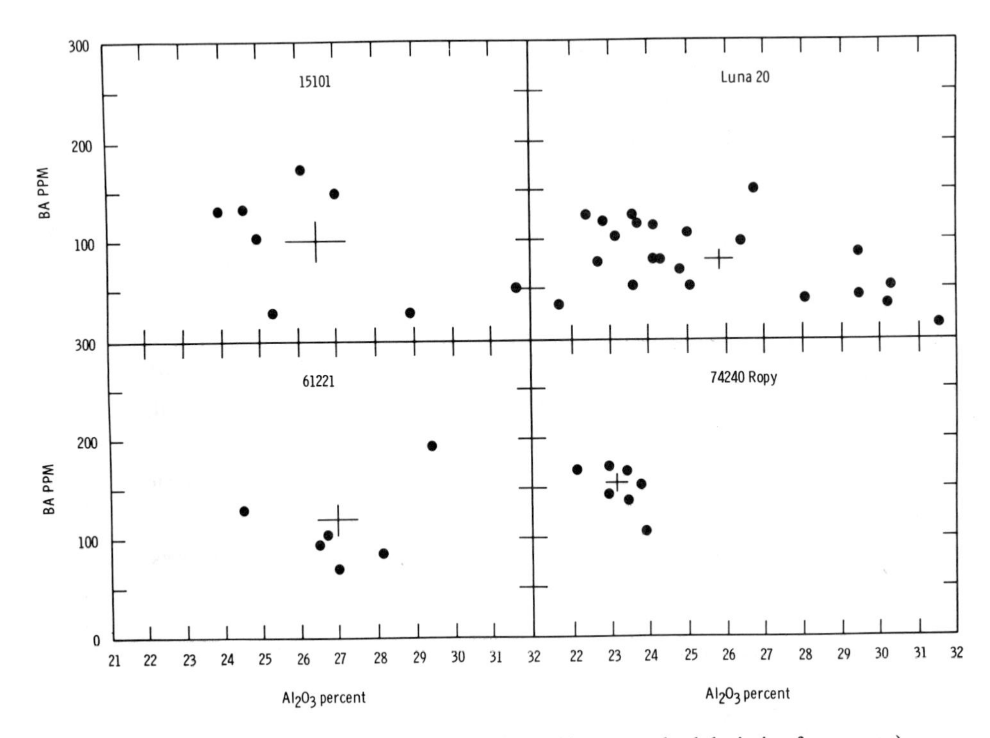

Fig. 7. Individual glass analyses and averages (1 sigma standard deviation from mean) for glasses with "anorthositic gabbro" composition.

few glasses (5–10%) are the only ones that may represent the composition of primary rocks.

Another rock type of problematical origin is the Low K Fra Mauro Basalt (LKFM) which makes up much of the glasses in the 15101 soil (Reid *et al.*, 1972b). Although these glasses contain more Fe and hence have larger interferences, the high Ba contents swamp out any such interference. Figure 9 compares individual Ba and Al_2O_3 analyses of LKFM particles with the data from large Apollo 15 and 17 rocks thought to represent this rock type. Again, the glasses

Table 5. Ba Concentrations in Glass Types and Bulk Soils (ppm).

	"Anorthositic Gabbro" Glass Type	LKFM Glass Type	Bulk Soil[a]	Maturity[b]
Luna 20	79 ± 30	134 ± 55	87	—
61221	113 ± 45	—	96	10
15101	99 ± 56	261 ± 79	203	70

[a]Wiesmann and Hubbard (1975).
[b]Morris (1976) (I_s/FeO).

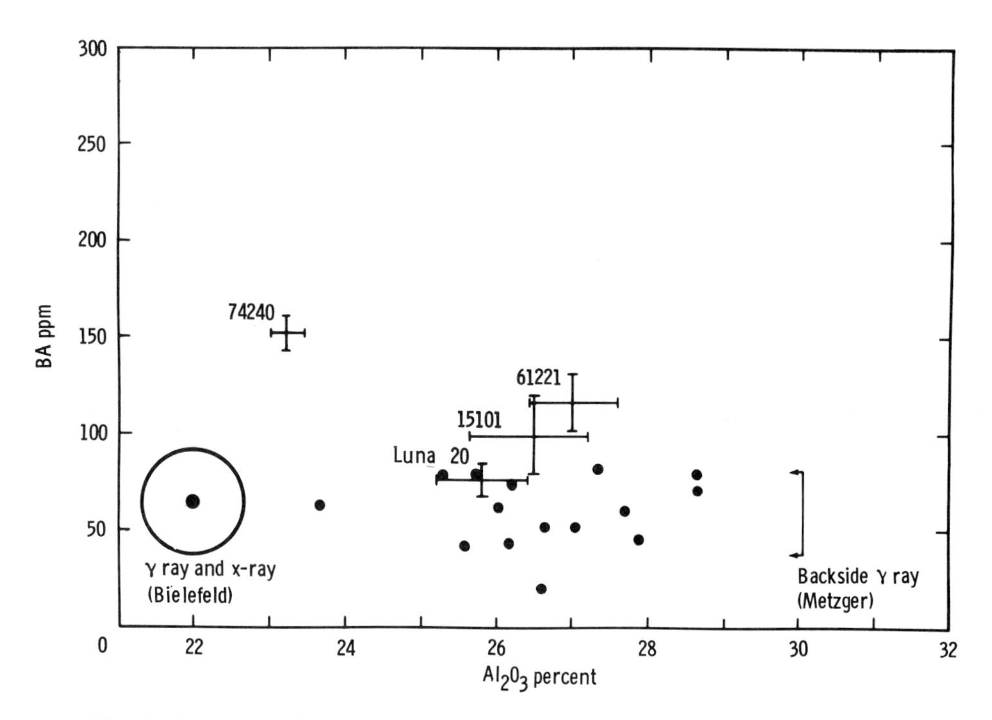

Fig. 8. Comparison of average glass analyses for four landing sites with analyses of large individual lunar samples (15418, 15455, 15445, 63549, 67955, 68415, 72215, 72235, 73215, 76230, 76315, 77017, 78155, 79215). Data from many sources. Also indicated are Ba contents of large areas of the lunar surface as inferred from Th measurement from lunar orbit and an assumed constant ratios of Ba/Th = 65 (from Bielefeld, 1977 and Metzger *et al.*, 1977).

within this group have a wide range of trace element contents and the average values of Ba are different for different soils and different from what previously had been identified as LKFM (Taylor, 1972; Ryder and Wood, 1977). Thus the data obtained in this study indicate that either LKFM has a different trace element content at different sites (Apollo 15, 16 and Luna 20) or LKFM is a myth (hypothesis 1, Reid *et al.*, 1977).

Grieve *et al.* (1974), in an important paper on lunar impact melts and terrestrial analogs, argue that the compositions represented by glass clusters (as in Ridley *et al.*, 1973) "could all be the product of impact mixing of heterogeneous targets containing a high-Ca-Al component (anorthosite) counterbalanced by a high Fe-Mg component (spinel troctolite or dunite) with perhaps minor amounts of a high Si-K-component (granite)." Mixing model calculations (such as those by Schonfeld, 1974) show that it is at least *possible* to make LKFM and Anorthositic Gabbro compositions by mixing various proportions of other components.

Let us examine more closely where glass particles in lunar soils come from and their probable origin. First of all there is the mechanism by lava fountaining, but

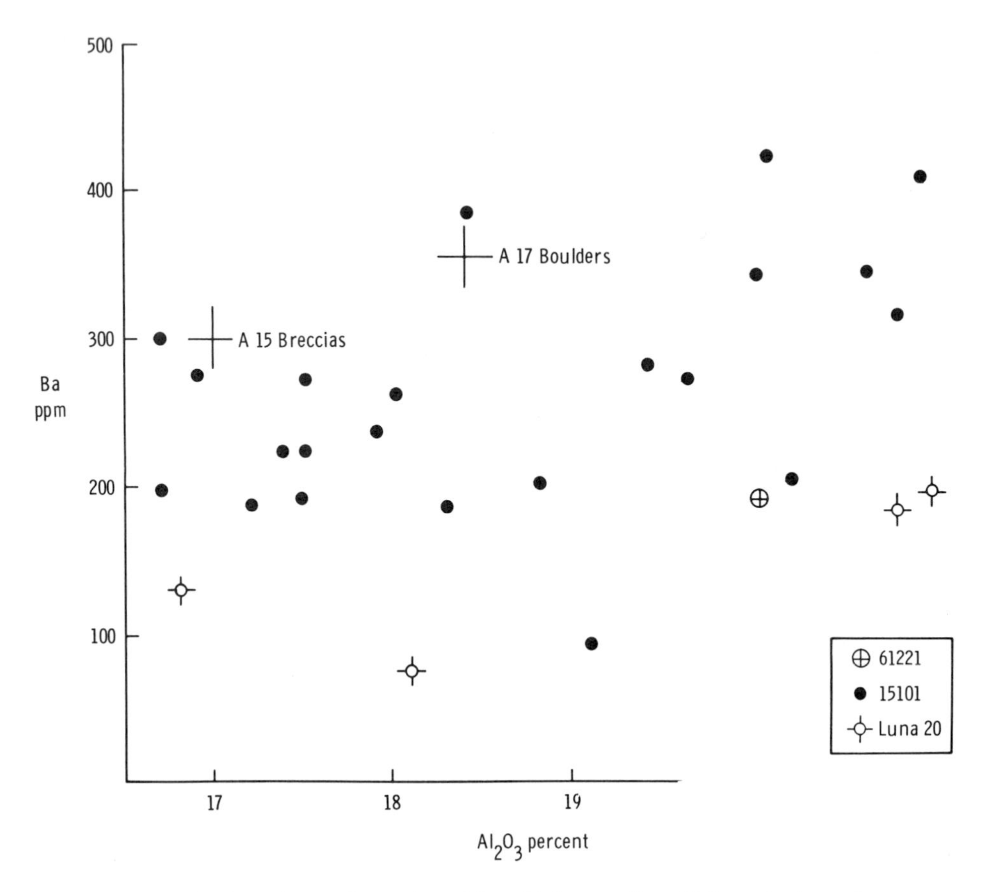

Fig. 9. Individual glass analyses of particles with intermediate Al_2O_3 contents previously termed Low K Fra Mauro Basalt (LKFM). Average values for Apollo 17 boulders (72215, 73215, 76315, 77115) and the Apollo 15 breccias (15445 and 15455) are also indicated.

this is restricted to glasses of mafic composition (Meyer *et al.*, 1975; Butler, 1978) because no volatile coatings have been reported on aluminous glass particles. This leaves meteorite bombardment as the principal mechanism. If one is to argue that these glass clusters represent primary rock types then one needs to depend on large impacts that punch through the regolith. However, the size distribution (see Neukum *et al.*, 1975) of lunar craters indicates that there are many more small craters into the regolith than through it. Consequently, it is reasonable to expect that most of the glass found in a lunar soil should be derived from the regolith itself. Another possibility is a two-stage mechanism whereby microcraters produce glass from rock samples that were dug up by the large impacts. The trouble with this argument is that glasses at mare sites do not closely represent the basalts which were returned (Wood, 1975). Furthermore, only a portion of the regolith surface contains exposed rock surfaces so that most of the small impacts are into soil. When soils melt they produce what are known as agglutinates, which are vesicular glasses filled with minute iron particles (Morris, 1976). As the regolith matures it should contain more and more glass particles, and the glasses from the most mature regoliths should include the highest percentage of mixed compositions.

A related type of glass is found as glass coatings on the surfaces of many lunar breccias (Winzer *et al.*, 1978) and at the bottoms of small lunar craters (Schaber *et al.*, 1972; Gold, 1969). Such glass is clearly made by meteorite impacts into the regolith (Greenwood and Heiken, 1970). Comminution of these glass splashes by further meteorite bombardment will presumably yield large numbers of the glass particles in the fine particles of the soil. Because of the mixed-up nature of the lunar regolith, no direct relationship of such glass to "primary rock types" should be expected.

What then are the major lunar rock types and to what extent can glass analyses be used to estimate their compositions? Although the trace element contents indicate contamination of the glasses by extraneous components, we are

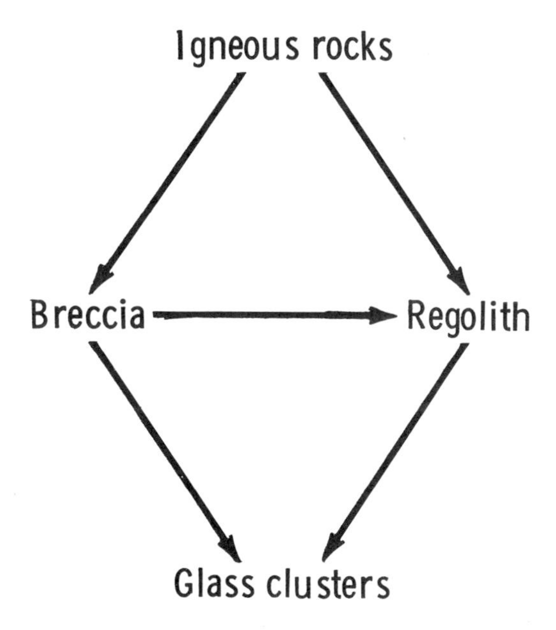

Fig. 10. Origin of glass clusters is via brecciation and regolith processes involving extensive mixing of various igneous rock types. Clusters in glass composition cannot originate directly from impact melting of single igneous rock types.

still left not knowing how much and in which direction the major element composition has been altered. If KREEP basalt is the major additive then only a few percent KREEP component is needed because of the extremely high trace element content of this component (Schonfeld and Meyer, 1972). This then would not significantly change the major element chemistry of LKFM and would only slightly alter the major element chemistry of Anorthositic Gabbro. However, the KREEP component in these soils is only part of the total (a very small part at Luna 20) and the other components must also get worked into the glass.

The Apollo Soil Survey (1972) noted the considerable spread in the major element compositions of individual glasses within each cluster and recognized that this spread might be due to mixing with other regolith materials. However, they assumed that these effects would cancel. This would, however, only be so for a component near the bulk soil composition. All other components should have their compositions altered *towards* the soil values (Schonfeld and Meyer, 1972).

CONCLUSIONS

It is important to continue development of techniques for the study of small particles from planetary objects. Especially useful will be techniques that can be used on a statistically large number of small particles. This study shows the ion probe can be used in this way to document the trace element chemistry of numerous small particles. An instrument with higher mass resolution, if it simultaneously had high sensitivity, would allow more elements to be measured.

This application of the ion probe to lunar glasses has shown that within groupings of aluminous glasses by major element compositions, there is considerable scatter in refractory trace element contents. Further, the trace element content of such groupings are found to vary proportional to the bulk soil chemistry, indicating that they have become contaminated by the trace element-rich components present in some soils.

Figure 10 illustrates the origin of glass particles via brecciation and regolith processes involving extensive mixing and contamination of trace element-poor rock types by the more trace element-rich components. For this reason clusters in glass compositions probably do *not* represent the composition of primary igneous rock types.

Acknowledgments—Dan Anderson, T. Rohrer, J. Bradley and F. Gibbons designed and built the computer interface that made this study possible. Special thanks go to A. Chodos and J. E. Rucklidge for the loan of their glass standards and to J. Wood for the loan of the Luna 20 sections from the SAO collection. E. Schonfeld and P. Butler, Jr. provided useful comments throughout the study.

REFERENCES

Allen R. O. and Clark P. J. (1977) Fluorine in meteorites. *Geochim. Cosmochim. Acta* **41**, 581–585.

Anderson C. A. and Hinthorne J. R. (1972) Ion microprobe mass analyzer. *Science* **175**, 853–860.

Andersen C. A. and Hinthorne J. R. (1973) Thermodynamic approach to the quantitative interpretation of sputtered ion mass spectra. *Anal. Chem.* **45**, 1421–1438.

Andersen C. A., Hinthorne J. R. and Fredriksson K. (1970) Ion microprobe analysis of lunar material from Apollo 11. *Proc. Apollo 11 Lunar Sci. Conf.*, p. 159–167.

Apollo Soil Survey (1971) Apollo 14: Nature and origin of rock types in soil from the Fra Mauro Formation. *Earth Planet. Sci. Lett.* **12**, 49–54.

Bielefeld M. J. (1977) Lunar surface chemistry of regions common to the orbital X-ray and gamma-ray experiments. *Proc. Lunar Sci. Conf. 8th*, p. 1131–1147.

Butler P. Jr. (1978) Recognition of lunar glass droplets produced directly from endogenous melts: The evidence from S-Zn coatings (abstract). In *Lunar and Planetary Science IX*, p. 143–145. Lunar and Planetary Institute, Houston.

Eugster O. (1971) Li, Be, and B abundances in fines from the Apollo 11, Apollo 12, Apollo 14 and Luna 16 mission. *Earth Planet. Sci. Lett.* **12**, 273–281.

Flanagan F. J. (1973) 1972 values for international geochemical reference samples. *Geochim. Cosmochim. Acta* **37**, 1189–1200.

Fredriksson K., Nelen J., Noonan A., Andersen C. A. and Hinthorne J. R. (1971) Glasses and sialic components in Mare Procellarum soil. *Proc. Lunar Sci. Conf. 2nd*, p. 727–735.

Fruland R. M., Morris R. V., McKay D. S. and Clanton U. S. (1977) Apollo 17 ropy glasses. *Proc. Lunar Sci. Conf. 8th*, p. 3095–3111.

Glass B. P. (1973) Major element compositions of Luna 20 glass particles. *Geochim. Cosmochim. Acta* **37**, 841–846.

Gold T. (1969) Apollo 11 observations of a remarkable glazing phenomenon on the lunar surface. *Science* **165**, 1345–1349.

Greenwood W. R. and Heiken G. (1970) Origin of glass deposits in lunar craters. *Science* **168**, 610–611.

Grieve R. A. F., Plant A. G. and Dence M. R. (1974) Lunar impact melts and terrestrial analogs: Their characteristics, formation and implications for lunar crustal evolution. *Proc. Lunar Sci. Conf. 5th*, p. 261–273.

Hubbard N. J. and Gast P. W. (1971) Chemical composition and origin of nonmare lunar samples. *Proc. Lunar Sci. Conf. 2nd*, p. 999–1020.

Meyer C. Jr., Brett R., Hubbard N. J., Morrison D. A., McKay D. S., Aiken F. K., Takeda H. and Schonfeld E. (1971) Mineralogy, chemistry, and origin of the KREEP component in soil samples from the Ocean of Storms. *Proc. Lunar Sci. Conf. 2nd*, p. 393–411.

Meyer C. Jr., McKay D. S., Anderson D. H. and Butler P. Jr. (1975) The source of sublimates on the Apollo 15 green and Apollo 17 orange glass samples. *Proc. Lunar Sci. Conf. 6th*, p. 1673–1699.

Metzger A. E., Haines E. L., Parker R. E. and Radocinski R. G. (1977) Thorium concentrations in the lunar surface. I: Regional values and crustal content. *Proc. Lunar Sci. Conf. 8th*, p. 949–999.

Morris R. V. (1976) Surface exposure indices of lunar soils: A comparative FMR study. *Proc. Lunar Sci. Conf. 7th*, p. 315–335.

Neukum G., Konig B. and Arkani-Hamed J. (1975) A study of lunar impact crater size-distribution. The *Moon* **12**, 201–229.

Prinz M., Dowty E., Keil K., Andersen C. A. and Hinthorne J. R. (1973) Ion microprobe study of high-alumina basaltic glasses (abstract). In *Lunar Science IV*, p. 603–605. The Lunar Science Institute, Houston.

Reid A. M., Ridley W. I., Harmon R. S., Warner J., Brett R., Jakes P. and Brown R. W. (1972a) Highly aluminous glasses in lunar soils and the nature of the lunar highlands. *Geochim. Cosmochim. Acta* **36**, 903–912.

Reid A. M., Warner J., Ridley W. I. and Brown R. W. (1972b) Major element composition of glasses in three Apollo 15 soils. *Meteoritics* **7**, 395–415.

Reid A. M., Warner J. L., Ridley W. I. and Brown R. W. (1973) Luna 20 soil: Abundance and composition of phases in the 45–125 micron fraction. *Geochim. Cosmochim. Acta* **37**, 1011–1030.

Reid A. M., Warner J. L., Ridley W. I., Johnston D. A., Harmon R. S., Jakes P. and Brown R. W. (1972c) The major element compositions of lunar rocks as inferred from glass composition in the lunar soils. *Proc. Lunar Sci. Conf. 3rd*, p. 363-378.

Reid A. M., Duncan A. R. and Richardson S. H. (1977) In search of LKFM. *Proc. Lunar Sci. Conf. 8th*, p. 2321–2338.

Ridley W. I., Reid A. M., Warner J. L., Brown R. W., Gooley R. and Donaldson C. (1973) Glass compositions in Apollo 16 soils 60501 and 61221. *Proc. Lunar Sci. Conf. 4th*, p. 309–321.

Rucklidge J. E. (1969) Rapid rock analysis by electron probe. *Geochim. Cosmochim. Acta* **34**, 243–247.

Ryder G. and Wood J. (1977) Serenitatis and Imbrium impact melts: Implications for large-scale layering in the lunar crust. *Proc. Lunar Sci. Conf. 8th*, p. 655–668.

Schaal R. B. and Hörz F. (1977) Shock metamorphism of lunar terrestrial basalts. *Proc. Lunar Sci. Conf. 8th*, p. 1697–1729.

Schaber G. G., Scott D. R. and Irwin J. B. (1972) Glass in the bottom of small lunar craters: An observation from Apollo 15. *Geol. Soc. Amer. Bull.* **83**, 1573–1578.

Schonfeld E. (1974) The contamination of lunar highland rocks by KREEP: Interpretation by mixing models. *Proc. Lunar Sci. Conf. 5th*, p. 1269–1286.

Schonfeld E. and Meyer C. Jr. (1972) The abundances of components of the lunar soils by a least-squares mixing model and the formation age of KREEP. *Proc. Lunar Sci. Conf. 3rd*, p. 1397–1420.

Taylor S. R. (1972) Geochemistry of the lunar highlands. The *Moon* **7**, 181–195.

Taylor S. R., Gorton M. P., Muir P., Nance W. B., Rudowski R. and Ware N. (1973) Composition of the Descartes region, lunar highlands. *Geochim. Cosmochim. Acta* **37**, 2665–2683.

Walker D., Grove T. L., Longhi J., Stolper E. and Hays J. F. (1973) Origin of lunar feldspathic rocks. *Earth Planet. Sci. Lett.* **20**, 325–336.

Wiesmann H. and Hubbard N. J. (1975) A compilation of the Lunar Sample Data generated by the Gast, Nyquist and Hubbard Lunar Sample PIships. Unpublished.

Warren P. H. and Wasson J. T. (1977) Pristine nonmare rocks and the nature of the lunar crust. *Proc. Lunar Sci. Conf. 8th*, p. 2215–2235.

Winzer S. R., Breen K., Ritter A., Meyerhoff M. and Schuhmann P. J. (1978) A study of glass coatings from some Apollo 15 breccias (abstract). In *Lunar and Planetary Science IX*, p. 1259–1261. Lunar and Planetary Institute, Houston.

Wood J. A. (1975) Glass compositions as a clue to unsampled mare basalt lithologies. In *Origins of Mare Basalts and Their Implications for Lunar Evolution*, p. 194–198. The Lunar Science Institute, Houston.

Proc. Lunar Planet. Sci. Conf. 9th (1978), p. 1571–1597.
Printed in the United States of America

Excess fission xenon at Apollo 16

T. J. BERNATOWICZ, C. M. HOHENBERG, B. HUDSON,
B. M. KENNEDY and F. A. PODOSEK

McDonnell Center for the Space Sciences, Washington University
St. Louis, Missouri 63130

Abstract—We report the results of stepwise heating analyses of Xe and Kr in four Apollo 16 low grade breccias: 60019, 60255, 60275 and 67455. The experiment is intended to determine whether Apollo 16 samples exhibit the "excess fission Xe" effect previously found in several Apollo 14 samples. All four Apollo 16 samples contain significant amounts of solar wind gases and at least three contain excess fission Xe (results for the fourth, 60255, are ambiguous). This observation indicates that the excess fission Xe effect is global rather than local, i.e., results from moon-wide redistribution of fission Xe rather than local redistribution in KREEP-rich Apollo 14 material. Accordingly, more confidence can be placed in the view that the excess fission Xe phenomenon represents a temporal variation in the amount of fission Xe available for redistribution, that it is prominent in samples from both highland sites simply because they are older than mare sites, and that samples from these sites experienced at least some of their surface exposure at an early epoch. At least two of the breccias (60019 and 60275) also appear to contain excess ^{129}Xe, presumably originating from decay of ^{129}I and redistributed in the same way as excess fission Xe.

INTRODUCTION

It is generally recognized that several Apollo 14 samples are characterized by what may be termed the "excess fission Xe" effect, as originally observed by Drozd *et al.* (1972). The defining feature of this effect is a concentration of fission Xe too large to be attributed to *in situ* spontaneous fission of ^{238}U and ^{244}Pu in the age of the sample. Associated with this excess and equally important in defining the excess fission Xe effect is the observation that the fission Xe in a given sample is more closely correlated with solar wind Xe than with an *in situ* Xe component such as cosmic-ray-induced spallation Xe; this is seen in both the thermal release pattern (in stepwise heating) of fission Xe and in its enrichment in fine grain sizes (Behrmann *et al.*, 1973; Basford *et al.*, 1973). Both of these features indicate that the excess fission Xe originates in a reservoir external to the sample; additional support for this conclusion is found in the failure of the fission Xe to correlate with *in situ* neutron-induced ^{235}U fission (Reynolds *et al.*, 1974) or with fossil fission tracks (Graf *et al.*, 1973). The excess fission Xe effect is thus one of redistribution, indicating mobilization of fission Xe from the reservoir in which it was formed and perhaps from intermediate reservoirs and finally, concentration in the samples in which it is now found.

It is convenient to describe the amount of excess fission Xe in terms of an excess factor, the ratio of observed fission ^{136}Xe to the amount expected from spontaneous fission of ^{238}U. For highland age samples an excess factor greater than unity is to be expected: thus, for an initial $^{244}Pu/^{238}U$ ratio~0.015 (Podosek, 1972) 4.5×10^9 yr ago, decay of ^{244}Pu in a sample of age 3.8×10^9 yr will produce 24% as much fission ^{136}Xe as will decay of ^{238}U, thus producing an excess factor of 1.24. There are several rocks of this type; a good example is 14321, with an age of 3.95×10^9 yr (Papanastassiou and Wasserburg, 1971) and a fission excess factor of 1.6 (Marti *et al.*, 1973). This kind of case is *not* what is meant by "excess fission Xe," and we will not consider such cases further in this paper. The excess fission Xe phenomenon is defined by much larger excess factors (in the range 10^1 to 10^2) and evidence that the fission is not generated *in situ*.

There are several other features associated with the effect. In general, the excess fission is not perfectly correlated with solar wind Xe, and is often enhanced, relative to solar wind, at very low gas extraction temperatures. In a few cases of particularly favorable enhancement it has been possible to determine the isotopic spectrum of the fission component and make a clear identification of ^{244}Pu as the parent (Behrmann *et al.*, 1973). Because ^{244}Pu is so short-lived (82 m.y.) the phenomenon has become associated with an ancient epoch of lunar history. This association is enhanced by the observation that some of the samples containing excess fission Xe also apparently contain excess ^{129}Xe, presumably originating in decay of ^{129}I (17 m.y.) and presumably redistributed in a fashion similar to that of the fission Xe. So far the effect has been observed only in Apollo 14 samples; it has been observed in all breccias (and one soil) which contain large amounts of solar wind noble gases, and is absent in breccias which do not contain large amounts of solar wind gases.

There have been several observations of the excess fission Xe effect and suggestions for its origin (Crozaz *et al.*, 1972; Drozd *et al.*, 1972, 1975, 1976; Basford *et al.*, 1973; Behrmann *et al.*, 1973; Reynolds *et al.*, 1974). On the whole, understanding of the phenomenon is not very satisfactory. The timing and mechanism of mobilization and transport are unclear, as is the means by which the fission Xe and ^{129}Xe are incorporated in the samples. Since the ^{244}Pu identification pertains only to very small gas amounts at low extraction temperatures, it is not even clear that the parent of most of the total amount of excess fission Xe is indeed ^{244}Pu. A recent review of the data observations and of models for origin of the effect is given by Drozd *et al.* (1976). In this paper we will focus attention on one specific aspect of the problem.

As noted, excess fission Xe has so far been identified only in Apollo 14 samples. Most authors have presumed that the distinguishing feature of the Apollo 14 site is its antiquity (relative to mare sites). As an alternative hypothesis, however, it might be supposed that the redistribution of fission Xe is a local effect, prominent at Apollo 14 because of the high local concentration of actinides. The obvious test of whether excess fission Xe is a local or a global phenomenon is examination of samples from a site of comparable antiquity but different chemistry. To this end we have examined Xe and Kr in a suite of four low-grade Apollo 16 breccias: 60019, 60255, 60275 and 67455. This paper presents the results and their bearing on the excess fission Xe problem.

Experimental Procedures

Samples

Rock 60019 was a 2 kg solitary block collected at Station 10′, about 70 m SW of the Lunar Module (LM). Warner *et al.* (1973) classify it as a glassy breccia similar in nature to Apollo 14 breccias, characterized by texturally and mineralogically unequilibrated matrix. It is classified as a Type I polymict breccia (light color and cataclastic, nearly glass-free matrix) by the Lunar Sample Preliminary Examination Team, 1972 (hereafter referred to as LSPET). Our sample (60019,105), has white clasts set in a dark, glass-rich matrix and is therefore more consistent with the classification of Warner *et al.* (1973). Compositional data are reported by Rose *et al.* (1975).

Both 60255 (0.9 kg) and 60275 (0.3 kg) were solitary blocks collected near the LM. Warner *et al.* (1973) classify them as glassy breccias, and our samples (60255,91 and 60275,49), macroscopically similar to 60019, reflect that classification. LSPET (1972) tentatively put 60255 in their Type I category, but our sample looks more like one of their Type IV, which is more or less equivalent to the Warner *et al.* glassy breccias. Boynton *et al.* (1975) report compositional data.

Rock 67455 (1 kg) was collected from a large (5 m long, 2 m high) white boulder on the SE rim of North Ray Crater; the rock was quite friable and broke into several pieces in transit (LSPET, 1972). Warner *et al.* (1973) classify 67455 as a glassy breccia to a light matrix breccia (Type II in LSPET, 1972). 67455,20 was the least coherent rock which we studied, consisting mostly of a light grey powder. The light color is due to a high plagioclase content, and the texture indicates that the rock has been extensively crushed but has remained unequilibrated (LSPET, 1972; Warner *et al.*, 1973). Compositional data are reported by Rose *et al.* (1973) and by Müller (1975).

Gas extraction and analysis

All samples were wrapped in Al foil. After loading in the vacuum system they were heated overnight to about 100°C under vacuum. In analysis, gases were removed from each sample in a series of one hour extractions in a tungsten crucible which had been previously degassed at 1700°C. Reactive gases were removed by three successive exposures to freshly deposited Ti films. Kr and Xe were trapped on charcoal cooled to dry ice temperature; Ne and Ar were then pumped away. Kr was released (and Xe retained) by warming the charcoal to the melting point of Hg; after ½ hour Kr was admitted into the spectrometer for analysis. Xe was then released from the charcoal by heating (100°C) and admitted into the spectrometer after completion of the Kr analysis. The mass spectrometer used was a 4.5 inch radius glass Reynolds-type instrument operated in static mode with source magnet and electron multiplier detector. A thorough description of the apparatus and data reduction is given by Drozd (1974). Instrumental sensitivity and mass discrimination were determined by following the same preparation procedure with pipetted aliquots of air. The results of the analyses are presented in Table 1.

Procedural blanks were determined by following the same analysis procedure prior to dropping the sample in the crucible, with the results noted in Table 1. Before the analysis described in the next section the data were corrected for blanks by subtraction of the appropriate blank amounts with an assumed uncertainty of 25%. Blank amounts were taken to be constant from 200°C to 1000°C and were linearly interpolated between 1000°C and 1500°C. In general, the blanks are small compared to sample gas concentrations and do not introduce significant uncertainty in the subsequent data analysis.

Component Resolution

In general terms the noble gases contained in a rock are a superposition of several different components, where a component is defined as some distinct source or reservoir of gas with a well-defined isotopic composition. An *in situ* component is one which is generated by nuclear transformation in the sample; a *trapped* component is one incorporated in the rock from some external source. For the samples under consideration here there are three major components. One is

Table 1a. Krypton[a,b] Analyses of Apollo 16 Breccias.

Temp. (°C)	[84] $cm^3STP/g \times 10^{-11}$	Isotopic Composition (^{84}Kr = 100) ^{78}Kr	^{80}Kr	^{81}Kr	^{82}Kr	^{83}Kr	^{86}Kr
				60019			
300	2.1	0.428±0.206	3.569±0.049		19.06±0.17	19.62±0.13	31.87±0.15
500	43.2	0.604±0.074	4.152±0.042		20.53±0.14	20.54±0.11	30.81±0.02
600	3.1	0.659±0.214	4.360±0.023	0.0149±0.0038	20.80±0.06	20.83±0.09	30.43±0.14
700	7.1	0.728±0.089	4.476±0.023	0.0159±0.0015	21.14±0.05	21.25±0.07	30.11±0.07
800	11.7	0.630±0.055	4.073±0.027	0.0064±0.0010	20.52±0.09	20.72±0.05	31.65±0.08
900	13.9	0.647±0.057	4.208±0.014	0.0100±0.0021	20.73±0.05	21.10±0.03	30.47±0.05
1000	23.6	0.730±0.040	4.419±0.018	0.0098±0.0008	21.17±0.03	21.36±0.04	30.39±0.04
1100	128	0.677±0.011	4.204±0.007	0.0036±0.0003	20.72±0.05	20.67±0.03	30.40±0.02
1200	1143	0.639±0.003	4.072±0.007	0.0012±0.0002	20.39±0.07	20.36±0.08	30.43±0.02
1300	125	0.645±0.012	4.155±0.015	0.0038±0.0008	20.55±0.04	20.54±0.02	30.42±0.02
1400	58.8	0.606±0.014	3.935±0.017	0.0029±0.0005	20.18±0.02	20.42±0.05	30.55±0.04
1500	2.6	0.713±0.041	3.868±0.051	0.0078±0.0042	19.75±0.09	20.34±0.07	31.34±0.17
Total	1560	0.642±0.004	4.094±0.006		20.44±0.05	20.43±0.06	30.45±0.02
				60275			
300	1.8	0.585±0.274	3.889±0.056		20.02±0.10	19.95±0.06	30.97±0.05
500	13.4	0.547±0.149	3.978±0.037		20.40±0.12	20.22±0.08	30.98±0.06
600	1.8	0.589±0.330	4.121±0.040	0.0499±0.0065	20.33±0.08	20.40±0.07	31.01±0.10
700	4.1	0.593±0.161	4.244±0.059	0.0279±0.0022	20.59±0.03	20.89±0.08	30.67±0.16
800	4.1	0.565±0.153	4.321±0.035	0.0201±0.0039	20.72±0.05	21.16±0.09	30.41±0.09
900	7.8	0.651±0.070	4.225±0.022	0.0162±0.0013	20.72±0.05	21.33±0.08	30.91±0.09
1000	13.1	0.730±0.056	4.467±0.026	0.0144±0.0014	21.10±0.04	21.58±0.04	30.28±0.05
1100	71.8	0.699±0.017	4.238±0.010	0.0070±0.0003	20.61±0.05	20.79±0.04	30.53±0.03
1200	838	0.644±0.002	4.058±0.006	0.0015±0.0002	20.33±0.06	20.32±0.06	30.47±0.02
1300	118	0.653±0.008	4.107±0.015	0.0037±0.0002	20.46±0.03	20.43±0.05	30.61±0.03
1400	40.0	0.666±0.017	4.179±0.016	0.0050±0.0004	20.77±0.05	20.71±0.02	30.43±0.02
1500	2.2	0.392±0.220	3.889±0.073	0.0008±0.0056	19.86±0.11	20.14±0.09	31.34±0.11
Total	1120	0.648±0.003	4.085±0.005		20.39±0.04	20.40±0.04	30.50±0.01
				67455			
300	0.8	0.939±0.479	4.355±0.096		20.37±0.11	20.34±0.18	30.37±0.28
500	241	0.646±0.030	4.056±0.015		20.27±0.06	20.34±0.04	30.90±0.04
600	2.0	0.804±0.108	4.935±0.054	0.0127±0.0036	21.51±0.17	22.07±0.27	30.21±0.10
700	0.7	1.079±0.653	7.306±0.171		24.12±0.19	27.21±0.66	31.03±0.24
800	1.7	1.103±0.286	5.900±0.171	0.0100±0.0045	24.37±0.61	23.80±0.19	29.23±0.22
900	1.6	0.865±0.373	5.095±0.053	0.0223±0.0083	21.57±0.16	22.99±0.13	30.71±0.14
1000	0.8	1.860±0.805	10.429±0.340		29.09±0.26	32.90±0.81	30.42±0.11
1100	1.6	3.192±0.429	14.497±0.313	0.0784±0.0096	34.55±0.70	40.51±0.84	28.54±0.15
1200	1.3	2.910±0.402	11.975±0.130	0.0550±0.0086	29.89±0.16	34.57±0.56	28.82±0.07
1300	0.6	2.544±0.949	12.297±0.527	0.0509±0.0102	32.02±0.91	36.92±1.46	29.01±0.17
1500	0.7	3.781±0.530	15.655±0.434	0.1005±0.0204	37.01±1.00	44.46±1.72	26.79±0.20
Total	253	0.699±0.029	4.271±0.015		20.57±0.05	20.76±0.04	30.84±0.04

Table 1a. *(cont'd.)*

60255							
300	4.9	0.617±0.099	4.265±0.047	20.48±0.05	20.30±0.07	30.37±0.09
500	23.5	0.529±0.244	4.065±0.032	20.01±0.02	20.17±0.05	30.93±0.07
600	8.2	0.575±0.322	4.423±0.028	20.76±0.11	20.80±0.08	30.59±0.05
700	25.5	0.709±0.180	4.829±0.042	20.94±0.09	21.40±0.07	30.91±0.11
800	17.3	1.246±0.067	6.588±0.028	23.34±0.05	24.24±0.05	29.75±0.03
900	22.5	1.058±0.056	5.812±0.039	22.51±0.02	23.25±0.05	30.09±0.05
1000	2129	0.807±0.005	4.797±0.032	23.00±0.28	23.14±0.28	32.78±0.31
1100	3932	0.681±0.002	4.184±0.019	20.48±0.14	20.64±0.14	30.57±0.03
1150	5292	0.649±0.002	4.064±0.031	20.21±0.19	20.36±0.19	30.27±0.02
1175	1900	0.682±0.002	4.195±0.017	20.57±0.12	20.58±0.12	30.61±0.03
1200	957	0.686±0.002	4.187±0.006	20.41±0.05	20.49±0.06	30.63±0.03
1250	420	0.675±0.002	4.211±0.005	20.66±0.04	20.77±0.03	30.46±0.03
1300	203	0.636±0.004	3.941±0.004	19.64±0.06	20.40±0.04	31.27±0.03
1400	216	0.675±0.005	4.197±0.012	20.51±0.03	20.89±0.03	30.85±0.03
1500	5.6	0.644±0.103	4.305±0.059	20.43±0.06	20.89±0.06	32.38±0.06
1600	1.9	0.512±0.237	4.209±0.075	20.12±0.06	20.39±0.05	31.17±0.11
Total	15200	0.688±0.001	4.234±0.013	20.75±0.09	20.89±0.09	30.79±0.04
BEOC-12[d]		0.593±0.005	3.885±0.020	20.05±0.07	20.09±0.07	30.50±0.07
AIR[e]		0.610±0.003	3.960±0.015	20.22±0.07	20.16±0.06	30.55±0.07

cosmogenic gas, produced by cosmic-ray-induced reactions on target nuclides somewhat heavier in mass than the gas of interest. In simple cases cosmogenic gas may be treated as a single component, although in general it must be recognized that different targets are not homogeneously distributed throughout the sample so that in some cases cosmogenic gas cannot be treated as a single component. The second major component is trapped gas. Two possible sources of trapped gas may be recognized; gas implanted in grains exposed on the surface of the moon (which we will assume to have the BEOC-12 composition reported by Eberhardt *et al.*, 1972) and terrestrial atmospheric gas which contaminates the samples during storage and processing. While air and BEOC-12 compositions are generally similar the difference between them is large enough to be significant in subsequent discussion. The third major component is fission gas (chiefly Xe). Like cosmogenic and trapped gas the fission component must also be considered composite; two plausible sources may be identified, arising from spontaneous fission of ^{238}U and ^{244}Pu. The ^{238}U and ^{244}Pu compositions are sufficiently like each other and unlike cosmogenic, air and BEOC-12 compositions, that it is often possible to identify a fission component with reasonable accuracy but still with insufficient precision to resolve the contributions of ^{238}U and ^{244}Pu.

In this section we discuss the resolution of the observed data (Table 1) into the various contributory components, as summarized in Table 2. The basic algebra for resolution of superposed components is straightforward, and is illustrated in three-isotope correlation diagrams (Figs. 1–3).

Cosmogenic Xenon

The light isotopes ^{124}Xe, ^{126}Xe, ^{128}Xe and ^{130}Xe may be regarded as a superposition of trapped and cosmogenic contributions, since they are not produced in fission or nuclear-specific reactions in significant quantity (a possible exception is ^{128}Xe which is produced by $^{127}I(n,\gamma)^{128}I \rightarrow {}^{128}Xe$). The cosmogenic Xe isotopes easiest to identify are ^{126}Xe and ^{124}Xe, which are major constituents of cosmogenic Xe but very deficient in trapped Xe; the enrichments of ^{126}Xe are easily seen in Figs. 1–3. We can estimate total ^{126}Xe contributions (Table 2) by assuming a two component superposition: BEOC-12 composition for the trapped component and a cosmogenic component with $(^{126}Xe/^{130}Xe)_c = 1.0 \pm 0.2$ for 60019 and 60725, $= 1.24 \pm 0.25$ for 67455 and 60255. Uncertainty in cosmogenic Xe composition is not very important in estimating $^{126}Xe_c$; it is, however, more important

Table 1b. Xenon[C] Analyses of Apollo 16 Breccias.

Temp. (°C)	[132] $cm^3STP/g \times 10^{-12}$	Isotopic Composition (^{132}Xe = 100)							
		^{124}Xe	^{126}Xe	^{128}Xe	^{129}Xe	^{130}Xe	^{131}Xe	^{134}Xe	^{136}Xe
					60019				
300	5.4	0.573±0.023	0.677±0.018	8.11±0.09	102.12±0.38	15.83±0.12	82.01±0.16	38.01±0.19	31.76±0.20
500	11.0	0.654±0.008	0.786±0.015	8.58±0.07	103.17±0.19	16.43±0.05	83.09±0.20	38.02±0.12	31.55±0.07
600	3.0	0.692±0.030	0.843±0.022	8.92±0.06	105.54±0.21	16.45±0.13	83.74±0.28	38.19±0.07	31.59±0.17
700	4.1	0.811±0.025	1.018±0.032	8.97±0.05	103.54±0.41	16.21±0.11	83.31±0.20	40.45±0.25	34.94±0.31
800	11.8	0.803±0.015	1.117±0.030	9.23±0.03	102.69±0.18	16.14±0.08	83.08±0.16	41.08±0.19	35.00±0.19
900	15.0	1.022±0.012	1.496±0.027	9.87±0.05	104.58±0.21	16.81±0.09	85.11±0.20	39.64±0.14	33.77±0.11
1000	26.4	1.025±0.016	1.503±0.021	10.01±0.08	105.13±0.14	16.78±0.09	85.44±0.17	39.64±0.07	33.43±0.05
1100	234	0.752±0.006	0.985±0.006	9.14±0.02	104.67±0.13	16.29±0.02	82.52±0.07	40.14±0.03	33.81±0.05
1200	2496	0.528±0.002	0.525±0.002	8.40±0.03	104.65±0.34	15.96±0.03	80.59±0.16	39.79±0.04	33.43±0.04
1300	283	0.561±0.005	0.601±0.006	8.50±0.02	104.28±0.07	16.04±0.01	80.82±0.07	40.06±0.04	33.80±0.06
1400	253	0.567±0.006	0.617±0.004	8.57±0.04	104.70±0.08	16.04±0.02	80.83±0.13	39.91±0.05	33.52±0.04
1500	14.2	0.583±0.015	0.627±0.024	8.51±0.06	103.99±0.47	16.09±0.12	81.34±0.33	39.57±0.12	33.11±0.12
Total	3360	0.558±0.002	0.587±0.002	8.50±0.02	104.61±0.25	16.01±0.03	80.85±0.12	39.84±0.03	33.49±0.03
					60275				
300	3.7	0.381±0.031	0.435±0.015	7.01±0.11	97.89±0.40	15.15±0.17	79.46±0.43	39.01±0.25	33.17±0.11
500	11.9	0.411±0.012	0.438±0.016	7.35±0.04	97.25±0.20	15.14±0.07	79.47±0.17	39.58±0.16	33.98±0.08
600	1.9	0.349±0.026	0.501±0.035	7.70±0.11	99.26±0.53	15.73±0.22	80.38±0.55	39.46±0.26	33.73±0.15
700	7.0	0.565±0.027	0.658±0.039	7.74±0.08	98.80±0.48	15.64±0.09	81.01±0.26	39.43±0.19	34.14±0.15
800	2.8	1.006±0.012	1.371±0.048	9.86±0.09	103.66±0.41	16.26±0.14	85.22±0.41	39.14±0.28	33.14±0.13
900	10.6	1.143±0.023	1.799±0.033	10.40±0.11	104.04±0.43	16.86±0.05	87.03±0.27	38.87±0.03	32.77±0.10
1000	19.4	1.120±0.019	1.774±0.020	10.49±0.08	106.81±0.39	17.09±0.11	87.17±0.28	39.26±0.16	32.46±0.08
1100	107	0.855±0.015	1.196±0.014	9.55±0.04	105.45±0.15	16.64±0.03	83.98±0.08	39.57±0.06	33.19±0.05
1200	1228	0.539±0.002	0.535±0.004	8.46±0.01	105.15±0.17	16.17±0.03	81.00±0.09	39.50±0.02	32.98±0.02
1300	656	0.556±0.002	0.569±0.004	8.58±0.02	105.52±0.11	16.15±0.02	81.24±0.08	39.28±0.03	32.68±0.01
1400	110	0.576±0.010	0.608±0.010	8.56±0.06	105.17±0.18	16.19±0.05	81.17±0.07	39.49±0.05	32.98±0.04
1500	7.6	0.583±0.017	0.568±0.018	8.44±0.06	103.61±0.07	15.94±0.09	80.53±0.26	38.98±0.18	32.54±0.17
Total	2170	0.570±0.002	0.600±0.003	8.58±0.01	105.20±0.10	16.19±0.02	81.30±0.06	39.43±0.02	32.90±0.01

Table 1b. *(cont'd.)*

67455									
300	1.8	0.614 ± 0.020	0.697 ± 0.031	8.45 ± 0.13	104.92 ± 0.40	16.27 ± 0.16	82.76 ± 0.22	38.37 ± 0.14	32.80 ± 0.21
500	474	0.671 ± 0.005	0.791 ± 0.003	8.84 ± 0.02	104.88 ± 0.15	16.59 ± 0.03	83.03 ± 0.07	37.76 ± 0.03	31.16 ± 0.03
600	8.8	0.692 ± 0.016	0.826 ± 0.009	9.03 ± 0.06	107.02 ± 0.34	16.75 ± 0.09	83.91 ± 0.25	38.65 ± 0.22	32.31 ± 0.26
700	1.7	0.756 ± 0.020	0.834 ± 0.031	8.86 ± 0.06	105.65 ± 0.54	16.80 ± 0.13	83.42 ± 0.43	38.06 ± 0.11	32.03 ± 0.14
800	1.7	0.646 ± 0.012	1.002 ± 0.065	8.99 ± 0.15	103.00 ± 0.50	16.51 ± 0.14	83.18 ± 0.18	37.88 ± 0.11	31.92 ± 0.07
900	2.8	0.882 ± 0.028	1.266 ± 0.092	9.61 ± 0.12	106.98 ± 0.44	17.10 ± 0.14	84.31 ± 0.47	37.03 ± 0.14	29.88 ± 0.25
1000	2.3	1.086 ± 0.054	1.512 ± 0.071	10.02 ± 0.16	106.53 ± 0.53	17.39 ± 0.11	85.89 ± 0.31	37.58 ± 0.19	31.16 ± 0.18
1100	5.1	1.780 ± 0.061	2.803 ± 0.149	11.71 ± 0.27	105.84 ± 0.50	18.21 ± 0.11	89.84 ± 0.42	37.80 ± 0.17	31.16 ± 0.25
1200	6.1	1.779 ± 0.074	2.538 ± 0.086	11.38 ± 0.13	104.20 ± 0.41	17.99 ± 0.13	88.66 ± 0.56	37.64 ± 0.18	31.86 ± 0.20
1300	2.2	1.799 ± 0.059	2.668 ± 0.147	11.30 ± 0.19	105.19 ± 0.16	17.83 ± 0.18	88.92 ± 0.82	39.01 ± 0.24	33.07 ± 0.25
1500	4.5	2.062 ± 0.112	3.391 ± 0.177	12.74 ± 0.27	102.83 ± 0.43	18.30 ± 0.21	91.64 ± 0.77	38.29 ± 0.17	32.44 ± 0.12
Total	511	0.716 ± 0.005	0.870 ± 0.004	8.96 ± 0.02	104.92 ± 0.14	16.65 ± 0.03	83.30 ± 0.06	37.78 ± 0.03	31.21 ± 0.03
60255									
300	6.7	0.469 ± 0.018	0.488 ± 0.014	7.72 ± 0.06	101.08 ± 0.28	15.65 ± 0.15	80.86 ± 0.39	38.14 ± 0.20	31.96 ± 0.08
500	50.0	0.520 ± 0.013	0.561 ± 0.009	7.89 ± 0.03	100.56 ± 0.22	15.67 ± 0.05	80.58 ± 0.18	39.11 ± 0.09	33.11 ± 0.07
600	6.6	0.646 ± 0.024	0.856 ± 0.017	8.23 ± 0.08	101.24 ± 0.39	15.84 ± 0.11	82.06 ± 0.29	38.44 ± 0.18	32.53 ± 0.12
700	45.8	0.537 ± 0.026	0.616 ± 0.019	7.72 ± 0.08	99.61 ± 0.38	15.58 ± 0.06	81.07 ± 0.18	39.18 ± 0.14	33.57 ± 0.13
800	9.7	1.133 ± 0.013	1.732 ± 0.032	10.22 ± 0.06	104.54 ± 0.19	17.20 ± 0.10	88.88 ± 0.20	37.72 ± 0.10	31.34 ± 0.08
900	15.2	0.941 ± 0.020	1.403 ± 0.039	9.30 ± 0.08	100.36 ± 0.25	16.56 ± 0.10	85.27 ± 0.17	38.22 ± 0.12	32.43 ± 0.07

Table 1b. *(cont'd.)*

60255									
1000	2044	1.034±0.002	1.470±0.003	10.02±0.03	106.02±0.30	17.45±0.04	87.10±0.14	36.47±0.03	29.46±0.02
1100	6678	0.841±0.002	1.125±0.006	9.37±0.07	105.23±0.67	17.02±0.08	85.39±0.37	36.62±0.04	29.69±0.03
1150	9418	0.644±0.002	0.759±0.005	8.78±0.09	105.02±0.84	16.73±0.10	83.71±0.45	36.65±0.07	29.72±0.05
1175	2775	0.632±0.001	0.727±0.003	8.82±0.04	105.25±0.43	16.72±0.05	83.45±0.18	36.79±0.03	29.85±0.02
1200	2771	0.644±0.003	0.749±0.002	8.87±0.03	105.58±0.40	16.77±0.05	83.52±0.19	36.75±0.03	29.75±0.02
1250	1016	0.677±0.004	0.805±0.004	8.95±0.02	104.91±0.17	16.79±0.02	83.61±0.10	36.96±0.02	30.06±0.02
1300	563	0.704±0.005	0.854±0.004	9.00±0.01	105.14±0.16	16.85±0.04	83.88±0.08	36.86±0.03	29.95±0.03
1400	590	0.778±0.003	0.995±0.007	9.26±0.02	105.58±0.11	17.03±0.03	84.54±0.09	36.74±0.03	29.80±0.02
1500	31.7	0.757±0.013	0.961±0.014	9.22±0.03	105.85±0.17	16.92±0.04	84.30±0.12	37.01±0.04	29.83±0.06
1600	17.7	0.704±0.020	0.838±0.008	8.91±0.08	104.56±0.39	16.56±0.08	83.35±0.16	36.85±0.11	30.45±0.03
Total	26000	0.729±0.001	0.914±0.002	9.06±0.04	105.22±0.35	16.87±0.04	84.37±0.19	36.68±0.03	29.75±0.02
BEOC-12[d]		0.478±0.012	0.427±0.015	8.31±0.05	104.82±0.39	16.50±0.04	82.28±0.28	36.94±0.16	29.99±0.13
AIR[f]		0.353±0.002	0.329±0.002	7.12±0.02	98.33±0.00	15.12±0.03	78.98±0.08	38.81±0.08	32.99±0.06

[a]Sample weights: 60019,105 (726 mg); 60255,91 (811 mg); 67455,20 (527 mg); 60275,49 (819 mg). Krypton and xenon isotopic ratios have been corrected for mass discrimination but not for discrimination uncertainty (about 0.2%/mass for Kr and 0.1%/mass for Xe). Uncertainty in concentrations is ±10%.

[b]Blank corrections have not been applied to the data; all Kr blanks were of atmospheric composition and in the following concentrations ($\times 10^{-12}$cm^{3}STP): 60019, 200° = 0.5, 1500° = 1.9; 67455, 200° = 0.8, 1500° = 1.3; 60255, 200° = 0.5, 1500° = 0.9; 60275, 200° = 0.8, 1500° = 2.3.

Corrections for hydrocarbon background at masses 78 and 81 have been made by using 78/77 and 81/79 ratios measured in blanks and the 77/78 and 79/81 ratios for each extraction temperature. The errors for ^{78}Kr and ^{81}Kr represent an estimated 10% error in the correction ratios compounded with the normal statistical error.

[c]Blank corrections have not been applied to the data; all Xe blanks were of atmospheric composition and in the following concentrations ($\times 10^{-12}$cm^{3}STP): 60019, 200° = 0.2, 1500° = 1.2; 67455, 200° = 0.2, 1500° = 1.1; 60255, 200° = 0.2, 1500° = 0.8; 60275, 200° = 0.2, 1500° = 1.1.

[d]Eberhardt *et al.*, 1972

[e]Nief, 1960

[f]Podosek *et al.*, 1971.

in correcting for cosmogenic contributions at other isotopes. This problem will be taken up in later subsections.

Cosmogenic Krypton

In identifying cosmogenic Kr we first make a small correction for fission ^{86}Kr by assuming $^{86}Kr_f = (0.02 \pm 0.02)^{136}Xe_f$, where $^{136}Xe_f$ is the fission component whose calculation is described later. Cosmogenic ^{83}Kr (Table 2) is then identified by partitioning ^{86}Kr and ^{83}Kr into a trapped component with BEOC composition and a cosmogenic component with $(^{86}Kr/^{83}Kr)_c = 0.015 \pm 0.015$. The uncertainties in fission and cosmogenic compositions are not important sources of error in estimation of $^{83}Kr_c$.

The only formal exposure age calculation we report here is a ^{81}Kr-^{83}Kr age (Marti, 1967) for 67455. This age (Table 2), calculated following the procedure of Drozd *et al.* (1974), is based on the 1100°C and 1200°C fractions, and is nearly identical to the ^{81}Kr-^{83}Kr age obtained by Morgan (1975) for 67455 and indistinguishable from several other ^{81}Kr-^{83}Kr ages found in rocks recovered from the rim of North Ray Crater (cf., Drozd *et al.*, 1974). The other rocks, in general, have considerably higher total $^{83}Kr_c$ and $^{126}Xe_c$ than 67455, and presumably have had longer and more complex histories of exposure to cosmic rays in the lunar regolith. We will not attempt to unravel these histories here.

Trapped gases

The resolution of the light isotopes described above also determines the amount of trapped ^{130}Xe and ^{86}Kr (Table 2). Since most of the observed gases are trapped, these abundances are insensitive to the details of the resolution and the only significant uncertainty is the 10% uncertainty in absolute instrumental sensitivity.

Table 2. Isotopic Decomposition of Noble Gases† in Apollo 16 Breccias.

	60019	60255	60275	67455
^{84}Kr trapped[b]	15,600	151,000	11,100	2,500
^{132}trapped[b]	3,220	25,700	2,100	490
^{83}Kr cosmogenic	61 ± 11	940 ± 180	36 ± 7	12 ± 2
^{81}Kr–^{83}Kr exposure age ($\times 10^6$ yr)				50.3 ± 5.6
^{126}Xe cosmogenic	5.9 ± 1.1	128 ± 19	4.0 ± 0.7	2.3 ± 0.3
^{136}Xe fission	157 ± 16	2 ± 24[c]	79 ± 8	8.5 ± 1.3[c]
^{129}Xe excess[a]	124 ± 12	67 ± 193	68 ± 13	5.9 ± 3.8

†All gas amounts are in units of $10^{-12}cm^3STP/g$.

[a]Amount of ^{129}Xe in excess of trapped and cosmogenic gas.

[b]Gas amounts uncertain by 10%, as dictated by uncertainty in spectrometer sensitivity.

[c]An alternative calculation, illustrated in Fig. 3, gives 35 ± 13 and 9.5 ± 1.1 for 60255 and 67455, respectively.

Although the total concentration of trapped Xe is not in doubt, the resolution of fission Xe described in the following subsection is very sensitive to whether the trapped composition is BEOC-12 or air. A distinction can be made on the basis of correlation among the light, fission-shielded isotopes. If only two components (cosmogenic and trapped) are present, we expect a linear correlation on a

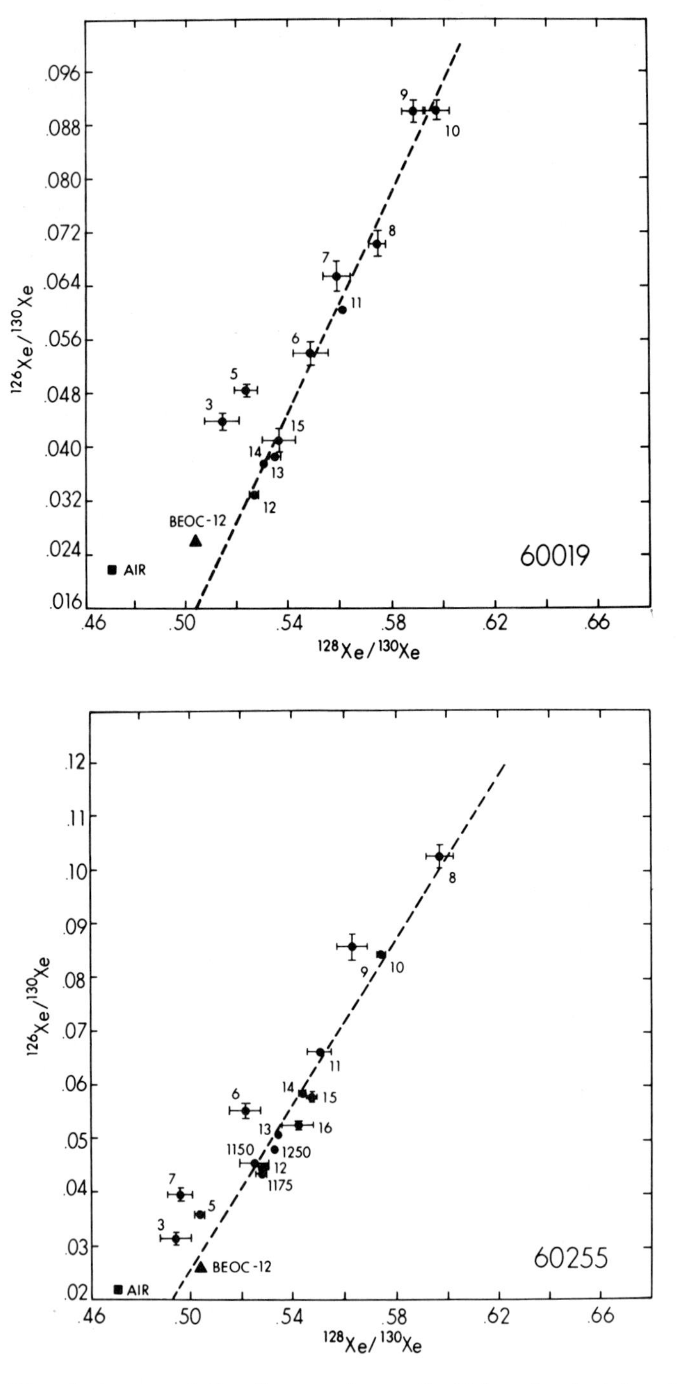

Fig. 1. Three isotope correlation diagrams for Apollo 16 breccias. In this and in the following figures the data for individual temperature release fractions are labeled in units of 100°C. For data where error bars are not shown the errors are smaller than the points drawn. Dashed lines in this figure represent least-squares fits to the data. The three isotopes here are not produced by fission, and are composed of trapped and

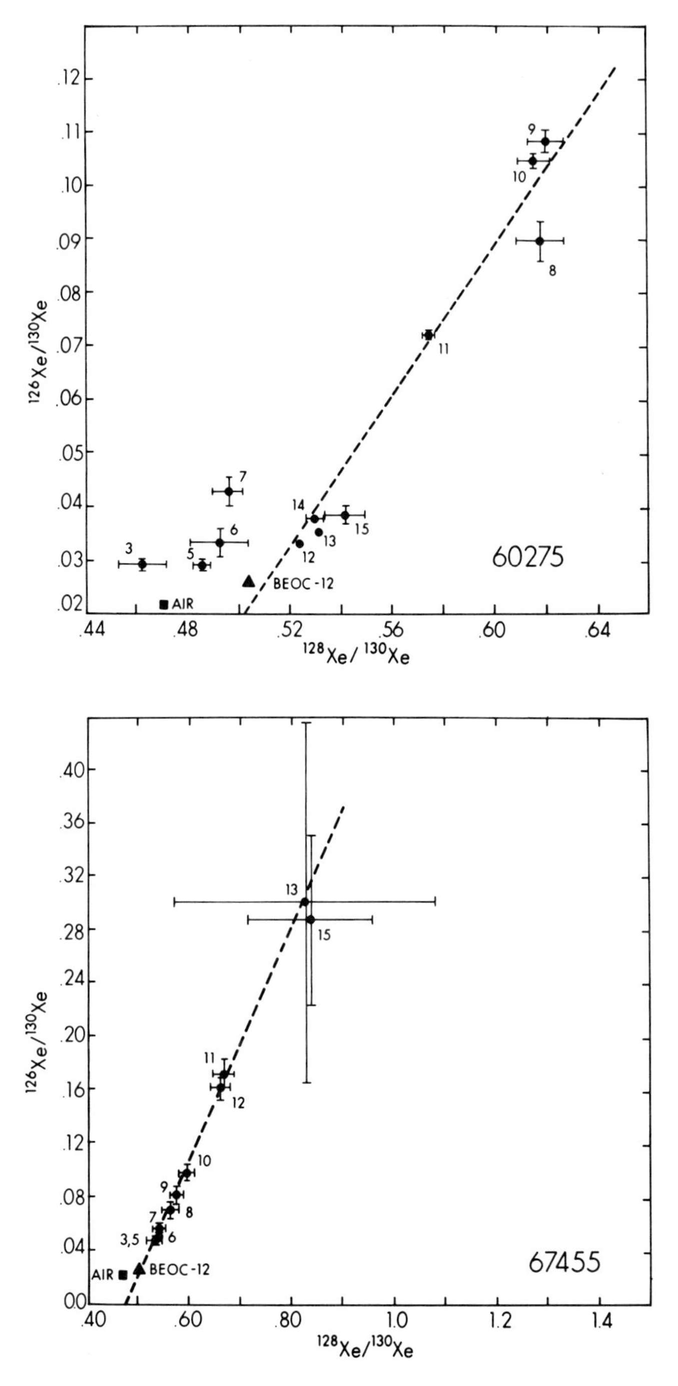

cosmogenic xenon. A good linear correlation indicates mixing of a single trapped component (BEOC-12) and a single cosmogenic component. As a result of admixture of adsorbed air, some of the low temperature points lie to the left of the line defined by the higher temperature data. In 60019 and 60275 a possible addition of ^{128}Xe from $^{127}I(n,\gamma)$ may have deflected the data points to the right.

three-isotope diagram, with both components constrained to lie on this line. The distinction between BEOC-12 and air compositions can be seen most clearly on $^{128}Xe/^{130}Xe$ and $^{126}Xe/^{130}Xe$ correlation diagrams (Fig. 1). 67455, for example, exhibits a good linear correlation consistent with two-component mixing and strongly indicates that the trapped composition is BEOC-12 rather than air. For the other rocks the data trends also indicate or are consistent with BEOC-12 compositions for most temperature fractions; the low temperature data in each case, however, indicate that air is the dominant trapped composition in the lowest extraction fractions (up to 500°C in 60019, 700°C in 60275, and 900°C in 60255). For 60019 and 60275 the deflection of some data points to the right of a plausible trapped-cosmogenic correlation line suggests a third component. A plausible interpretation is excess ^{128}Xe from $^{127}I(n,\gamma)$. For these two rocks a correlation between $^{124}Xe/^{130}Xe$ and $^{126}Xe/^{130}Xe$ is shown in Fig. 2. The separation between BEOC-12 and air is not as favorable in this diagram as it is in Fig. 1, but again it is indicated that except at the lowest extraction temperatures the dominant trapped component is BEOC-12 rather than air.

Fission Xenon

The heavy Xe isotopes also contain, in general, a contribution from spontaneous fission; this contribution is most prominent at ^{136}Xe. The effect of an added fission component can be most clearly illustrated in a three-isotope correlation diagram relating $^{126}Xe/^{130}Xe$ to $^{136}Xe/^{130}Xe$ (Fig. 3). In the simple case where only two components—cosmogenic and trapped—are present, the data must define a linear correlation between these compositions. Addition of a fission component causes displacement of data to the right of the tieline between cosmogenic and trapped compositions, since fission produces ^{136}Xe but not ^{126}Xe or ^{130}Xe.

If it is assumed that the trapped composition is known, resolution of the fission component is straightforward. Specifically, if we *assume* that the trapped composition is BEOC-12 and that there is a single cosmogenic composition, a lower limit to the amount of $^{136}Xe_f$ is readily determined. We can imagine the trapped-cosmogenic tieline in Fig. 3 to be pivoted at BEOC-12 but free to rotate in the lower left sector of the diagram, the rotation reflecting uncertainty in cosmogenic composition. All data points must lie to the right of the tieline, so the most extreme clockwise position of the tieline is set by the first data point it intersects as it rotates clockwise. If it is assumed that this extreme tieline is in fact the correct one (i.e., that this limiting data point contains no $^{136}Xe_f$), then the Xe_f in the other data points is determined by their positions to the right of this line. This assumption for the position of the tieline also determines the $(^{126}Xe/^{130}Xe)_c$ composition at its intersection with the ordinate, $(^{136}Xe/^{130}Xe)_c = 0$ (allowance for reasonable $^{136}Xe_c$, say $(^{136}Xe/^{130}Xe)_c < 0.1$, does not significantly affect the $(^{126}Xe/^{130}Xe)_c$ value). This is the construction shown for 67455 and 60255 in Fig. 3, and leads to the $(^{126}Xe/^{130}Xe)_c$ values of 0.8055 and 0.8060, respectively. These are the values used earlier in the resolution of $^{126}Xe_c$ in these rocks; they are also quite plausible compositions in view of compositions found in other lunar rocks in which cosmogenic gas is more dominant, and are presumably only fortuitously coincident.

The construction illustrated for 67455 and 60255 (Fig. 3) is easily seen to dictate an upper limit to $(^{126}Xe/^{130}Xe)_c$ and a lower limit to $^{136}Xe_f$. For 60019 and 60275 this procedure leads to unrealistic cosmogenic compositions and we have adopted $(^{126}Xe/^{130}Xe)_c = 1.0 \pm 0.2$ for resolution of $^{126}Xe_c$, since this range reflects the variations seen in similar rocks. The upper limit $(^{126}Xe/^{130}Xe)_c = 1.2$ (to which the tielines lead) then defines a plausible lower limit to $^{136}Xe_f$ in these two rocks.

Although the uncertainty in cosmogenic composition leads to corresponding uncertainty in the resolution of $^{136}Xe_f$, the ambiguity is not very severe, since fission effects are relatively large. If $(^{126}Xe/^{130}Xe)_c$ is indeed confined to the range 1.0 ± 0.2 (a range supported by other lunar data and cosmogenic systematics) and the underlying assumptions are valid, the lower limit $^{136}Xe_f$ values can be taken to be actual values with less systematic uncertainty than the statistical uncertainty inherent in the finite precision of measurement.

If the assumption of constant cosmogenic composition is dropped, the preceding analysis cannot be performed rigorously. Demonstrated cases of variable cosmogenic compositions are known, presumably because of differential release of cosmogenic Xe from differently sited target elements, and in some cases this greatly impedes resolution of fission components. Nevertheless, we will not further

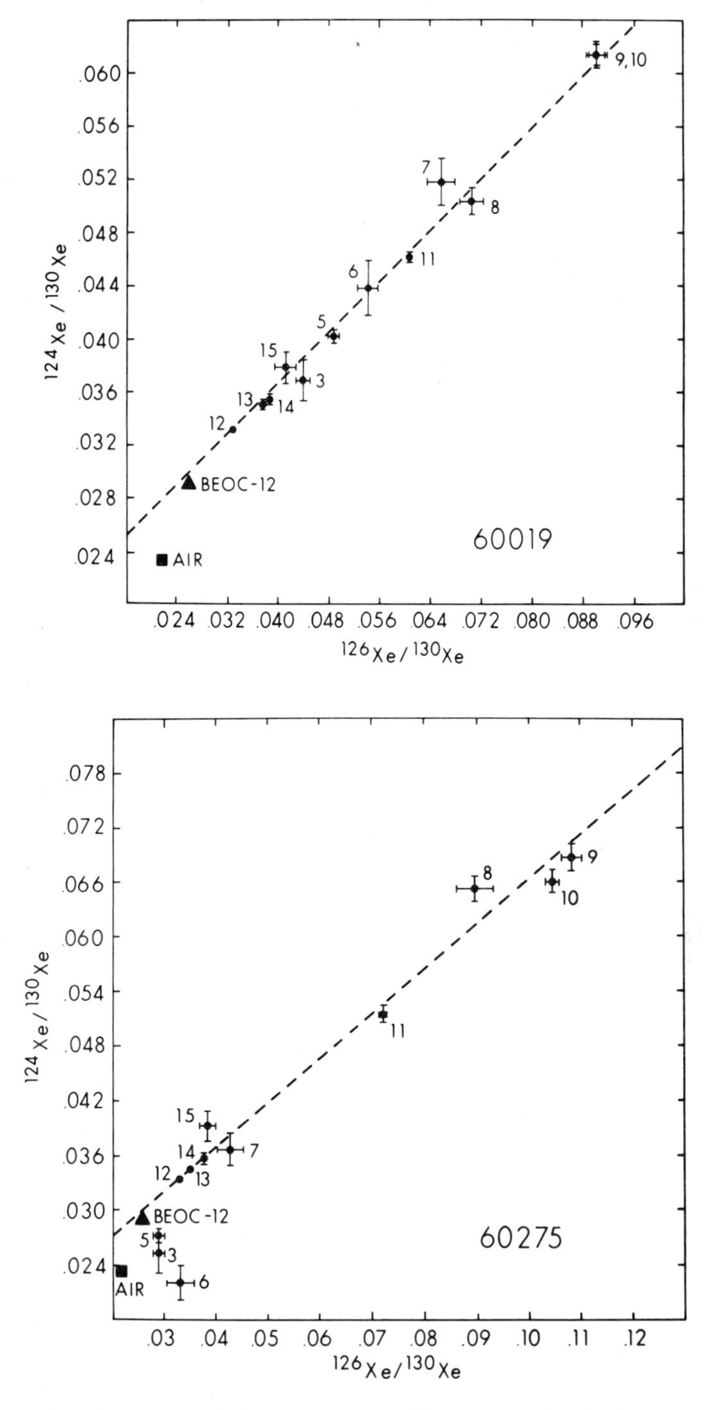

Fig. 2. Three isotope correlation diagrams for 60019 and 60275. The linear correlation of the data indicates a mixture of BEOC-12 and a single cosmogenic component. Deflection of some of the lowest temperature points below the correlation lines is due to air contamination. The dashed lines represent least-squares fits to the data.

consider this complication here, since inspection of Fig. 3 shows that sensibly modest variations in $(^{126}Xe/^{130}Xe)_c$ would not constitute a major perturbation on the calculation of $^{136}Xe_f$ contributions outlined above. The case for 60255 is an exception, as described later.

A much more serious problem arises through ambiguity in the trapped composition. A prevailing bias stipulates that large quantities of trapped gases in lunar samples reflect implantation of solar wind species in grains exposed at the surface of the lunar regolith. Strictly speaking, the BEOC-12 composition, and others like it, are computed in a manner which identifies surface-correlated trapped gas (Eberhardt *et al.*, 1970); such gas is presumably dominated by the solar wind but may also include nontrivial amounts of indigenous lunar gases. Several distinct surface-correlated compositions based on mare samples have been reported (e.g., Eberhardt *et al.*, 1970, 1972; Podosek *et al.*, 1971; Basford *et al.*, 1973); whether the differences reflect genuine variation or unrecognized experimental error is unclear. For our immediate purposes, it is not important which of the mare surface-correlated compositions is chosen as a lunar trapped component, since differences among them are small in comparison with the effects identified as "excess fission Xe." Similarly, the degree to which the BEOC-12 composition faithfully records solar wind composition is not relevant, and our study is primarily concerned with why trapped, surface-correlated gases characterized by excess fission Xe should be different from the trapped surface correlated gases in samples such as the soil from which BEOC-12 composition was determined.

The problem with trapped Xe is thus not ambiguity regarding BEOC-12 composition. The problem is possible contamination with terrestrial atmospheric Xe (adsorbed on or reacted with grain surfaces, as opposed to procedural blanks). Indeed, we have already noted (Figs. 1 and 2) the probability that trapped gas released at low extraction temperatures is dominated by air rather than BEOC-12 composition. These release fractions do not account for the major inferred $^{136}Xe_f$, however, so their contributions can be discounted, and the major concern is with the dominant gas amounts released at higher temperatures. It is frequently considered that gases released from samples at intermediate and high temperatures (say, 900°C and up) cannot be attributed to superficial contamination with air. This generalization is deceptive, however, and in at least some Apollo 16 samples gas released at high temperatures and initially thought to be lunar is apparently due to atmospheric contamination (see Niemeyer and Leich, 1976). Inspection of Fig. 3 indicates that the isotopic variations attributed to mixing of fission with BEOC-12 and cosmogenic components might, for most data, alternatively be attributed to admixture of air rather than or in addition to fission. As noted, at the high temperatures which account for most of the apparent $^{136}Xe_f$, atmospheric contamination is unexpected. We have also argued earlier that at these temperatures the data for the fission-shielded isotopes (Figs. 1 and 2) indicate BEOC-12 rather than air as the trapped composition. Nevertheless, some more quantitive evaluation seems appropriate.

Such a quantitative approach is possible through consideration of other isotopes besides ^{136}Xe. For each temperature fraction we have performed a series of calculations illustrated by the specimen data in Table 3. First, the measured (blank-corrected) data are corrected for cosmogenic gas by assuming $(^{126}Xe/^{130}Xe)_c = 1.0 \pm 0.2$ and BEOC-12 for the trapped composition (use of air instead of BEOC-12 makes no difference in the following argument). With this resolution the cosmogenic spectrum indicated in the first line of Table 3 is subtracted from the measured composition. This cosmogenic spectrum includes fairly wide error limits, and is intended to cover most of the range of compositions observed in lunar rocks. It is then presumed that the remaining, cosmogenic-free gas represents a superposition of BEOC-12 composition and an "excess" component which contributes no ^{130}Xe. The corresponding amounts of ^{130}Xe and other isotopes are then subtracted to give the spectrum of the excess component. If the isotopic variations are indeed due to addition of a component free of ^{130}Xe, e.g., a fission component, this calculation will produce the composition of the source, e.g., the ^{238}U or ^{244}Pu spontaneous fission compositions listed in the second and third lines in Table 3. If, on the other hand, the isotopic variations are actually due to admixture of air (which does contain ^{130}Xe), this same calculation will produce a characteristic apparent excess spectrum (fourth line). By comparison of actual calculation (e.g., fifth and sixth lines) with these expectations, we may make a more realistic evaluation of the likelihood of significant air contamination.

Thus, in Table 3 we can see that the dominant effect in the 300°C fraction of 60275 is apparently air contamination rather than fission, while in the 1200°C fraction the dominant effect is clearly addition of fission rather than air contamination. The 1200°C data are more suggestive of ^{238}U fission than of ^{244}Pu fission, but this distinction is rather more delicate than the distinction between fission and air and we are not prepared to defend this identification as a conclusion.

The comparison involving $(^{134}\text{Xe}/^{136}\text{Xe})_{ex}$ is not very helpful, since the value in a real fission spectrum is not very different from the value in the apparent excess spectrum produced by adding air to BEOC-12. In most cases the $(^{131}\text{Xe}/^{136}\text{Xe})_{ex}$ value is also not very helpful (although it is in the examples shown in Table 3) because of the large error assigned to $(^{131}\text{Xe}/^{126}\text{Xe})_c$ to cover the range of highly variable cosmogenic ^{131}Xe. For most of the data, then, the primary discriminant between fission and air addition is in the $(^{132}\text{Xe}/^{136}\text{Xe})_{ex}$ ratio, although the ^{131}Xe and ^{134}Xe criteria are compatible with the identification made on the basis of ^{132}Xe. In general terms, the use of this calculation to discriminate between fission addition and air contamination yields the same results as the more qualitative arguments described in reference to Figs. 1 and 2.

The total fission contents presented in Table 2 are calculated by the method described above, counting only those temperature fractions in which the comparison illustrated in Table 3 clearly indicates an actual fission addition rather than air contamination. These tabulated results are not significantly different than those obtained by the limit calculations described earlier in connection with Fig. 3; i.e., except for the very low extraction temperature fractions, which account for only a minor contribution to total gas amounts, air contamination is not a serious problem. The only exception is the case for 60255. For this rock the low temperature (≤900°C) fractions are dominated by air and in the high temperature fractions the isotopic effects are so small that the resolution procedure described in Table 3 does not afford an unambiguous distinction between fission and air. If it is *assumed* that air contamination is unimportant above 900°C *and* the cosmogenic composition is uniform, then the limit calculation described earlier (Fig. 3) is valid; this result is given in the note to Table 2. We have reservations about this figure, however, not only because of the possibility of air contamination but also because if the assumption of constant cosmogenic composition is dropped and cosmogenic composition allowed to vary in the range $(^{126}\text{Xe}/^{130}\text{Xe})_c = 1.0 \pm 0.2$, the data are consistent with essentially *no* $^{136}\text{Xe}_f$.

We conclude that at least three of the Apollo 16 rocks studied here—60019, 60275 and 67455—have significant contents of fission Xe. As seen in Table 4, these fission contents are too high to be supported by *in situ* fission. Thus, these three samples are members of the set of lunar rocks characterized by the excess fission xenon phenomenon. With a simple set of assumptions the same statement applies to 60255, but relaxation of these assumptions obviates this conclusion, so we cannot unambiguously state that 60255 has excess fission Xe. Unfortunately, we also cannot argue for the contrary proposition that 60255 does *not* have excess fission Xe, which, in view of the present results, would be a more interesting circumstance. Our results for 60255 are thus best regarded as inconclusive.

Table 3. Compositional Data for Trapped. Fission and Cosmogenic Xenon.

	^{126}Xe	^{129}Xe	^{130}Xe	^{131}Xe	^{132}Xe	^{134}Xe	^{136}Xe
Spallation	1.00 ± 0.20	1.75 ± 0.50	≡1.00	6.00 ± 3.00	0.75 ± 0.25	0.05 ± 0.05	≡0
^{244}Pu fission[a]			≡0	0.25 ± 0.02	0.89 ± 0.03	0.94 ± 0.01	≡1.00
^{238}U fission[b]			≡0	0.08 ± 0.00	0.60 ± 0.01	0.83 ± 0.01	≡1.00
BEOC + air[c]		0.41 ± 0.09	≡0	0.65 ± 0.07	1.52 ± 0.07	0.90 ± 0.05	≡1.00
60275–300°C	≡0	0.32 ± 0.11	≡0	0.68 ± 0.11	1.47 ± 0.06	0.91 ± 0.06	≡1.00
60275–1200°C	≡0	0.78 ± 0.14	≡0	0.07 ± 0.14	0.69 ± 0.07	0.94 ± 0.06	≡1.00

[a]Lewis (1975).

[b]Tabulated in Hyde (1971).

[c]Apparent "excess" composition produced by admixture of air and BEOC-12 compositions and subsequently assuming that all ^{130}Xe was contributed by BEOC-12 component.

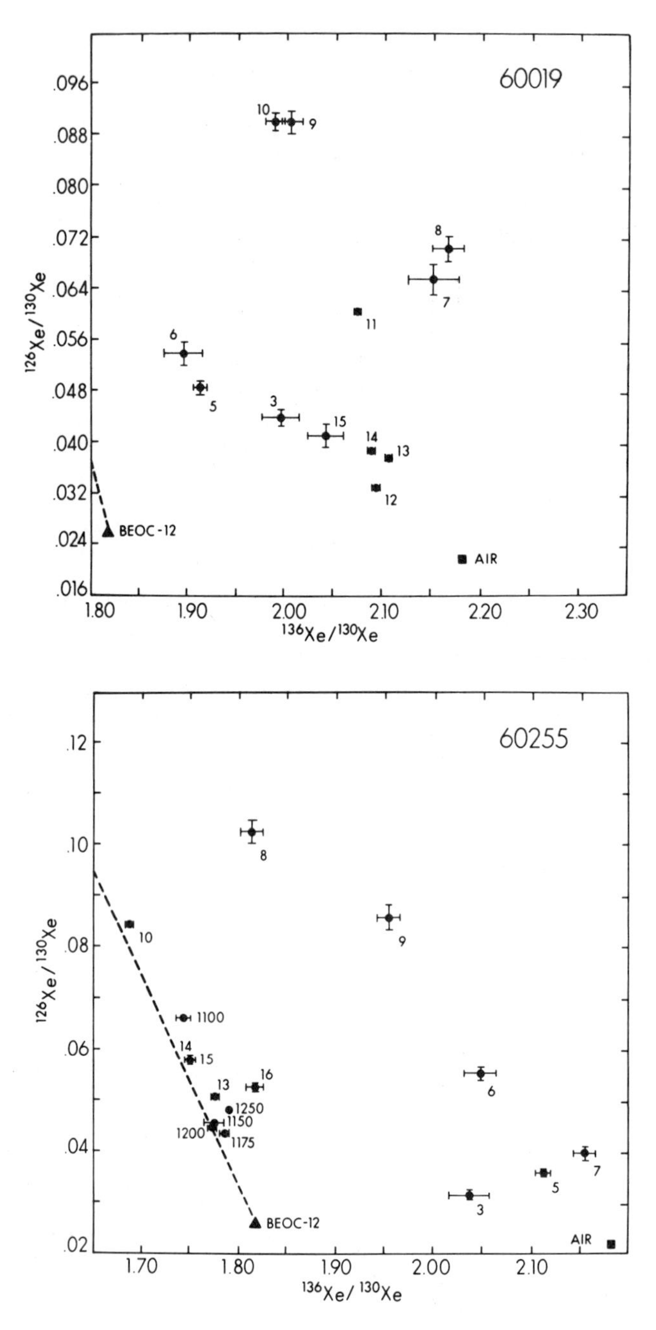

Fig. 3. Three isotope correlation diagrams for Apollo 16 breccias. ^{126}Xe and ^{130}Xe contain both trapped and cosmogenic gas, but no fission gas. ^{136}Xe has trapped and fission components only, since $^{136}Xe_c \approx 0$. Assuming a single cosmogenic composition, a maximum value for $(^{126}Xe/^{130}Xe)_c$ and a minimum value for $^{136}Xe_f$ is obtained by rotating clockwise a tie line pinned at the BEOC-12 datum until it meets the first data point; then extrapolation of the tie line to the axis $^{136}Xe/^{130}Xe = 0$ defines the

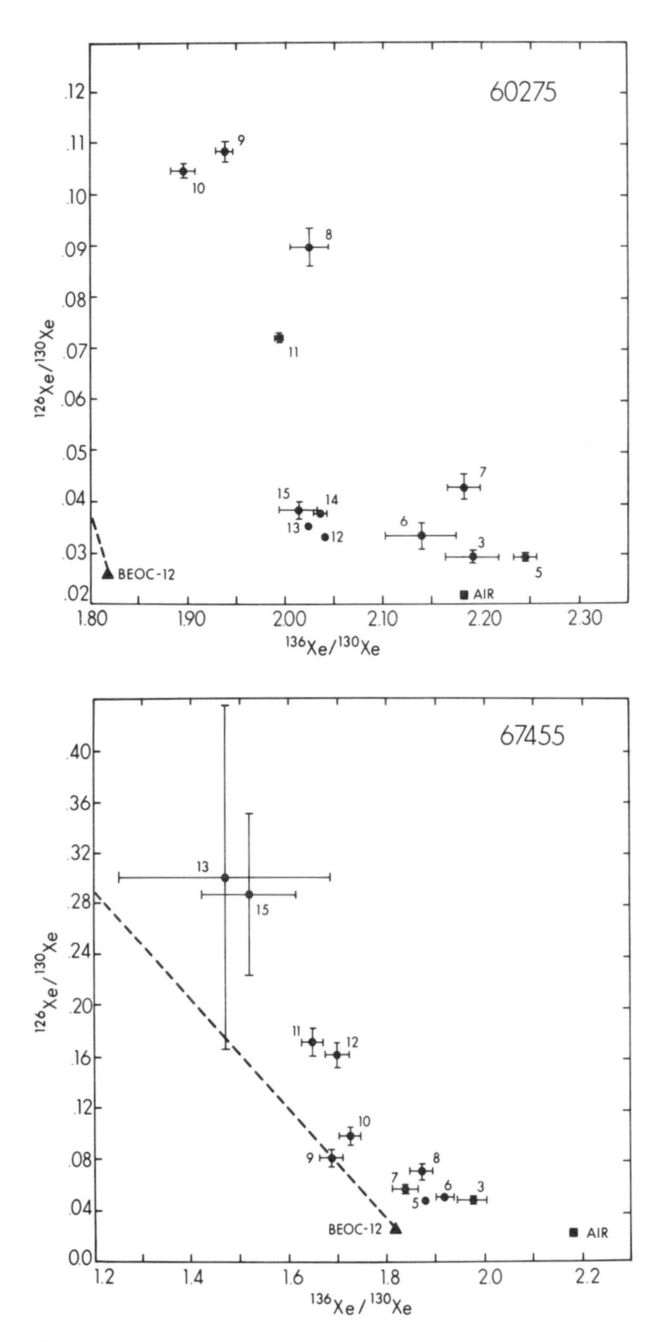

maximum $(^{126}Xe/^{130}Xe)_c$. Mixtures of BEOC-12 and cosmogenic xenon should plot on the tie line, and additions of ^{136}Xe due to fission will deflect the data points directly to the right. This construction for 67455 and 60255 gives $(^{126}Xe/^{130}Xe)_c \geq 0.806$. Because all of the temperature fractions for 60275 and 60019 appear to contain significant fission ^{136}Xe, no such construction is possible, and the tie line is arbitrarily drawn to a plausible upper limit of $(^{126}Xe/^{130}Xe)_c = 1.2$.

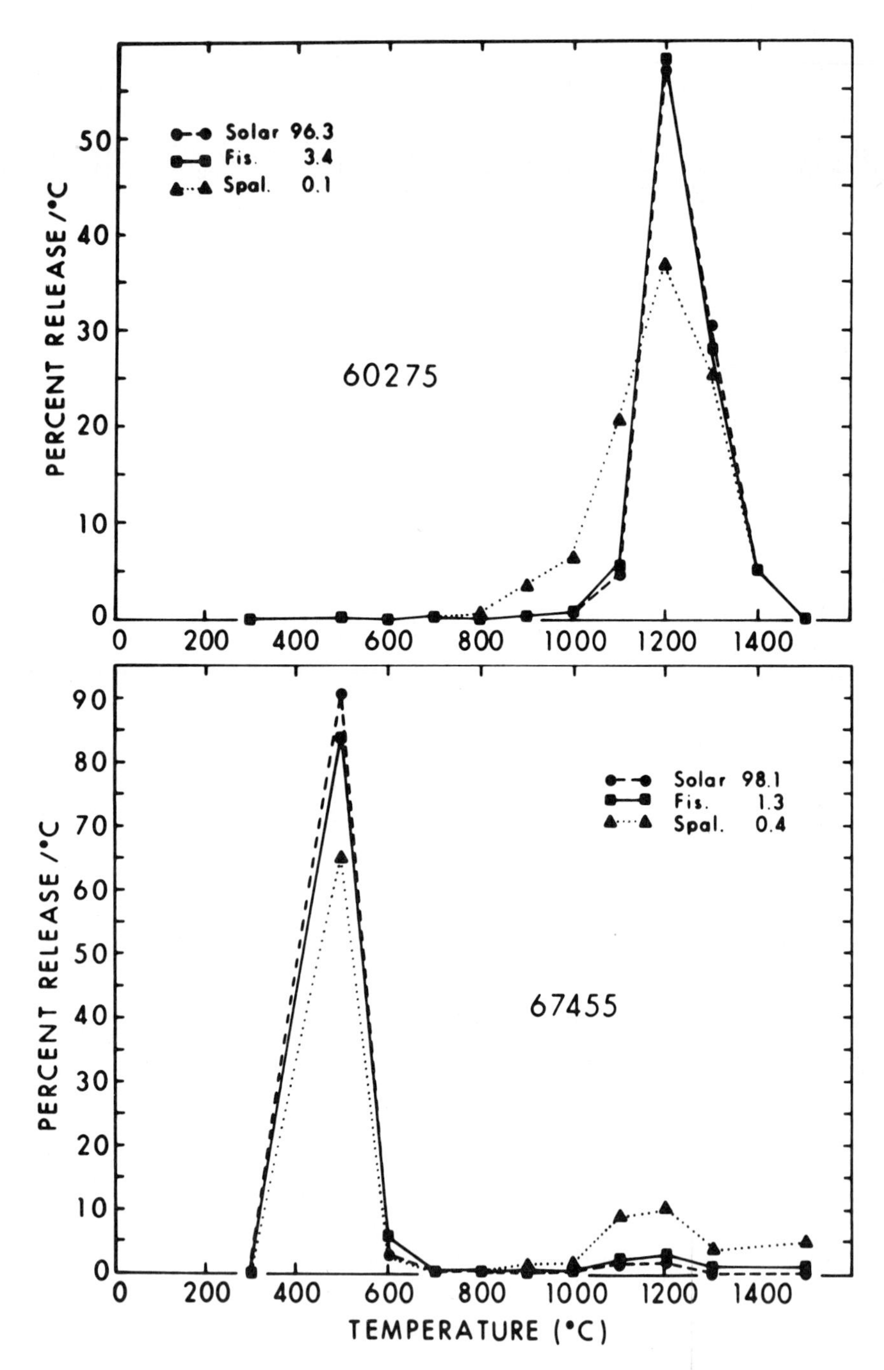

Fig. 4. Temperature release patterns for solar wind, cosmogenic, and fission xenon in Apollo 16 breccias. Each temperature fraction is resolved into "solar" (BEOC-12), fission and cosmogenic components as described in text; the data plotted represent the fraction of the total of each component which is released at a given temperature. Fission

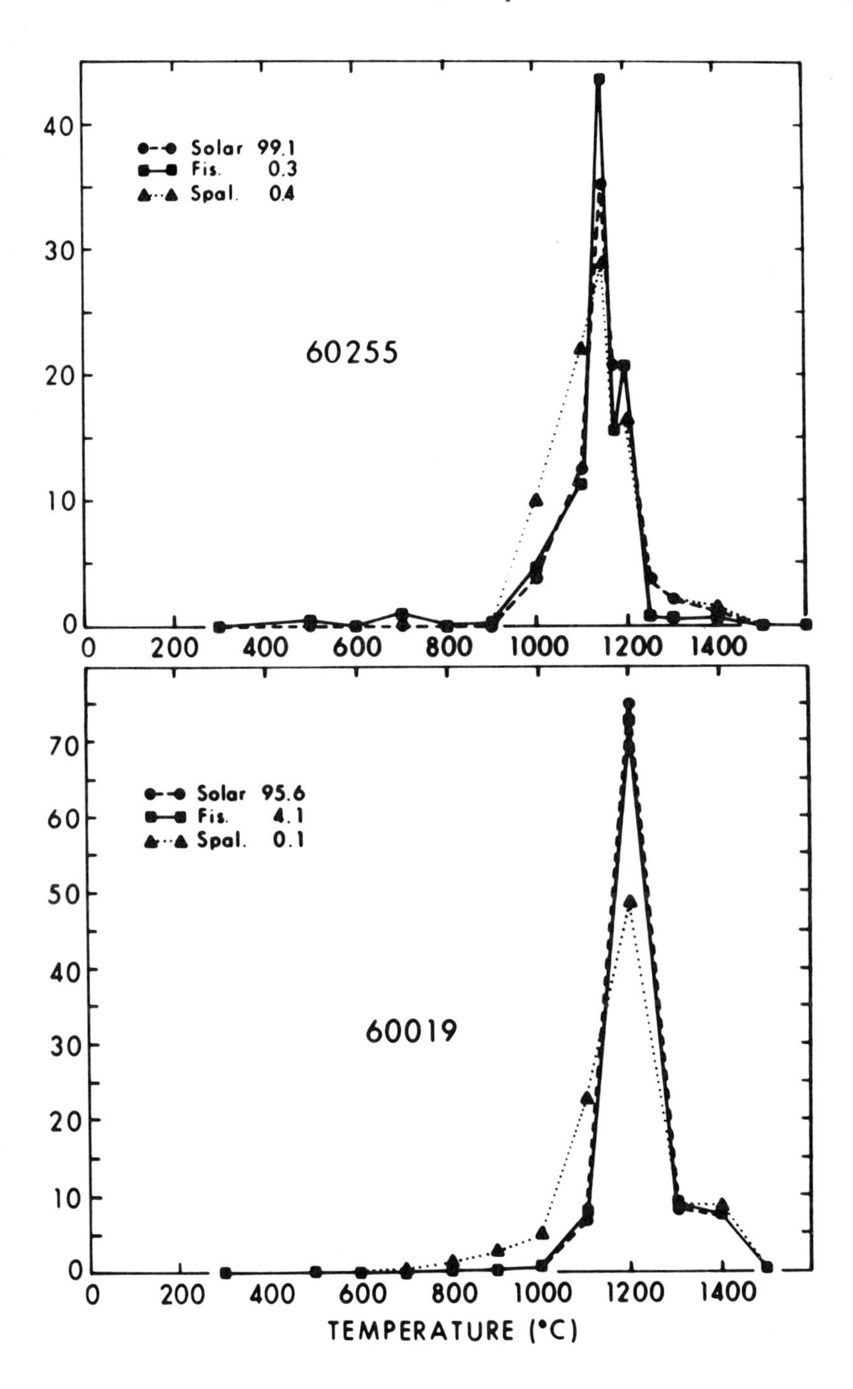

amounts are those computed from deflection of data to the right of the tie line in Fig. 3; as described in text these amounts are not the same as those in Table 2, although only for 60255 is the difference significant.

We have implicitly assumed that the source of the fission Xe is lunar, but the point is worthy of explicit consideration. As an alternative hypothesis, it might be supposed that the effect results from the import of fission Xe in meteorite influx. Reynolds *et al.* (1974) have discussed the meteorite influx hypothesis, and consider it implausible. We may also make a simple mass balance calculation. If meteorite input has an approximately chondritic U content of about 10 ppb, and corresponding ^{244}Pu, a characteristic excess fission ^{136}Xe content of a few times 10^{-10}cm^{3}STP/g requires transfer of fission Xe from about 10^2 g of meteorite to one g of lunar breccia (assuming perfect efficiency in redistribution). If the total regolith is about 10^{-2} meteoritic material, this requires that the meteoritic fission Xe be concentrated in about 10^{-4} of the regolith. We may thus conclude that meteoritic import is an unlikely source of the excess fission Xe, and that the source must be the moon itself.

Release patterns

One of the earmarks of excess fission xenon is that its thermal release pattern closely parallels that of trapped gas. Figure 4 shows the fractional release of fission, cosmogenic and trapped gas as a function of extraction temperature. One of the release patterns (67455) is distinctly different from the others, but in all cases the release of fissiogenic xenon tracks that of solar wind xenon. A possible explanation of this style of gas release can be found in the petrology of the samples. 60019, 60255 and 60275 are all petrologically similar; they consist of clasts and appreciable matrix glass (Warner *et al.*, 1973). If surficially sited solar wind xenon becomes more tenaciously trapped by the processes of glass formation, then the high temperature release of solar wind xenon in these three rocks can be attributed to their glassy constituents. 67455, on the other hand, has only minor matrix glass (LSPET, 1972) and the gas release pattern for this rock is clearly bimodal (Fig. 4). The lower temperature release presumably represents xenon bound in surficial sites (clast surfaces, interstices, etc.), while the higher temperature release may represent xenon associated with minor matrix glass or more tightly bound in clasts. It is noteworthy that the higher temperature release in 67455 occurs in roughly the same temperature range as that for the major gas release in the other three rocks.

Excess ^{129}Xe

The principles already discussed for resolution of excess heavy isotopes are, of course, more generally applicable to all isotopes. The case of ^{129}Xe is of particular interest, since it can be produced by decay of ^{129}I and excesses of ^{129}Xe have already been identified in a few of the Apollo 14 samples which exhibit excess fission Xe. If we resolve the Table 1 data into trapped (BEOC-12) and cosmogenic contributions for $(^{126}\text{Xe}/^{130}\text{Xe})_c = 1.0 \pm 0.2$ and allow for cosmogenic ^{129}Xe according to $(^{129}\text{Xe}/^{130}\text{Xe}) = 1.75 \pm 0.50$, we do indeed find ^{129}Xe in excess of the trapped and cosmogenic contributions. As with ^{136}Xe, some caution is necessary in interpreting such excesses, since admixture of air can also produce an apparent excess of ^{129}Xe (Table 3). Nevertheless, applying the same criterion as above in counting only those fractions in which the isotopic effects cannot be attributed to air contamination (as explained earlier in the construction of Table 3), we obtain the amounts of excess ^{129}Xe given in Table 2. In two cases—60019 and 60275—excesses of ^{129}Xe are quite definite. For 67455, the apparent excess of ^{129}Xe is only slightly larger than the (one standard deviation) uncertainty and accordingly may not be significant. For 60255 the isotopic excesses are small and there is no evidence for excess ^{129}Xe; the absolute gas abundance in 60255 is so high, however, that the uncertainty in the calculations allows, within error limits, as much excess ^{129}Xe as in the other samples. Our evaluation of the possible presence or absence of excess ^{129}Xe in 67455 and 60255 must thus be considered inconclusive.

DISCUSSION

The principal results of the data analysis of the previous section, as summarized in Table 2, are determinations of the amounts of trapped Xe and excesses of

^{136}Xe and ^{129}Xe. These are restated in Table 4. Only for the case of 67455 is there a U measurement available for comparison; for the other three Apollo 16 breccias studied here we will assume 1 ppm as what should be a generous upper limit of U concentration. For at least three samples—60019, 60275 and 67455—we conclude that the "excess fission factor" is $\gg 1$, i.e., that the amount of fission Xe in these samples is far too large to be attributed to *in situ* decay of ^{238}U and ^{244}Pu. Indeed, in 60019 and 60275 we might realistically expect the actual U concentrations to be substantially lower than 1 ppm, so that the fission excess factors can plausibly be taken to be substantially larger than expected even for samples approximately 4.5×10^9 yr old (excess factor about 70–80 for an initial ratio ^{244}Pu/^{238}U $\simeq 0.015$). This underscores the conclusion that the large amounts of "excess" fission in these samples result from gas redistribution: a concentration in these samples of Xe produced (by fission) elsewhere. As noted earlier, we cannot unambiguously reach the same conclusion for 60255, but neither can we argue for the opposite conclusion.

We will return to consideration of release patterns shortly. For now we note that while the release of fission Xe is not (except for 67455) particularly peaked

Table 4. Selected Data for Apollo 14, 15 and 16 Samples.[a]

Sample	$^{132}Xe_{tr}$	$^{136}Xe_f$	Excess[(c)] ^{129}Xe	Excess[(d)] Fission Factor	[U] ppm	Excess ^{129}Xe / Excess $^{136}Xe_f$	Excess $^{136}Xe_f$ / $^{132}Xe_{tr}$	Excess Fission Xenon?	Excess ^{129}Xe?	Solar Wind Rich?	Ref
60019	32	1.5	1.2	>72	<1[(e)]	0.8	0.05	yes	yes	yes	(1)
60255	257				<1[(e)]			?	?	yes	(1)
60275	21	0.8	0.7	>36	<1[(e)]	0.9	0.04	yes	yes	yes	(1)
67455	5	0.1	0.06	80	0.05[(f)]	0.6	0.02	yes	?	yes	(1)
67455	0.1							no	no	no	(7)
15205	1.4				3.4			no	no	no	(5)
15405	1.8				3.9			no	no	no	(5)
14047	317	1.7		25	3.4		0.005	yes	yes	yes	(4)
14055	363	2.0		27	3.7		0.006	yes	no	yes	(4)
14063	0.1				2			no	no	no	(5)
14082	0.3				1.9			no	no	no	(5)
14149	67	0.5	0.02	8	3.2	0.04	0.007	yes	yes	yes	(4)
14301	53	1.3	0.2	19	3.6	0.2	0.02	yes	yes	yes	(2)
14307	824	4.0	27	80	3.3	6.8	0.005	yes	yes	yes	(6)
14313	313	4.8		76	3.2		0.015	yes	no	yes	(3)
14318	33	2.2		29	3.8		0.07	yes	no	yes	(3)

(a) All gas amounts are given in units of 10^{-10}cm^{3}STP/g; all samples are breccias except trench soil 14149.
(b) Amount of $^{136}Xe_f$ in excess of that expected from *in situ* spontaneous fission of ^{238}U. Ages of 3.85 AE for Apollo 16 samples and 3.7 AE for Apollo 14 samples were arbitrarily chosen.
(c) Amount of ^{129}Xe in excess of amount attributable to cosmogenic and trapped.
(d) Ratio of total ^{136}Xe to amount expected from *in situ* ^{238}U spontaneous fission.
(e) U contents of these rocks have not been measured; 1 ppm is assumed as a plausible upper limit.
(f) Wänke *et al.* (1973); a value of 0.04 ppm is given by Müller (1975).
(1) This work.
(2) Drozd *et al.* (1972).
(3) Behrmann *et al.* (1973).
(4) Drozd *et al.* (1975).
(5) Drozd *et al.* (1976).
(6) Bernatowicz *et al.* (1977) (calculated from matrix data).
(7) Drozd (1974).

at low temperatures, as expected for a superficial component, neither is the release of "solar wind" Xe. The release of fission Xe correlates as well or better with solar wind gas than with volume-correlated cosmogenic gas (Fig. 4). In consideration of both the amounts of fission Xe and the association with solar wind gas we reach our principal conclusion and the answer to the question which prompted this study: Apollo 16 breccias *do* exhibit the excess fission Xe effect in the same fashion as do Apollo 14 breccias.

A number of generalizations about the excess fission Xe phenomenon and models for its origin have been reviewed by Drozd *et al.* (1976). We will not give another review here, but will instead concentrate on how the present results bear most closely on this problem.

So far the excess fission Xe effect has been found only in lunar highland samples—Apollo 14 and now Apollo 16—which at one time were part of the regolith: low to medium grade breccias (and one soil). It is not found in higher grade breccias or clasts which were heated and metamorphosed prior to or during brecciation. Table 4 lists pertinent quantities for all samples in which the effect has been found, as well as for similar samples (low grade highland breccias) in which it has been sought but *not* found. Consideration of the Apollo 16 samples substantially strengthens a generalization already noted on the basis of the data previously available: excess fission is found *only* in samples which are solar wind rich, i.e., which experienced significant exposure to the solar wind, and (with due allowance for ambiguity in 60255), excess fission Xe is found in *every* breccia which is solar wind rich. Altogether, the association of excess fission Xe with /olar wind exposure is quite clear and indicates that for the effect to be present it is not sufficient for the sample to have been part of a regolith; exposure at the surface of the regolith seems to have been necessary also.

An independent and compelling argument for global scale rather than local scale redistribution is the simple observation that the effect is present at Apollo 16 as well as at Apollo 14. Accordingly, the source of the fission Xe cannot be associated with the high actinide concentration of the local rocks at Apollo 14. Whatever the source of the original reservoir (i.e., the fissioning nuclides), at least the mechanism for redistribution must be global in scale. Together with the apparent requirement for direct surface exposure, this observation very strongly favors models in which redistribution is accomplished by transport in a transient lunar atmosphere.

Viewed in one perspective, the excess fission Xe effect is an interesting phenomenon, but an isolated one whose relationship to other lunar phenomena is not very clear. Such a viewpoint is implicit in the previous section in which we have attempted to resolve excess fission Xe (and excess ^{129}Xe) from normal "trapped" or "solar wind" Xe. This perspective is encouraged by consideration of the data which led to the initial recognition of the effect (cf., Drozd *et al.*, 1972): in a few of the Apollo 14 breccias the very low temperature release fractions exhibit quite prominent fission enhancements relative to the more nearly normal "solar wind" compositions released from these same samples at high temperatures. In this perspective there is a tendency to consider the excess fission Xe effect as a discrete phenomenon: Apollo 14, and now Apollo 16, samples have it and mare samples do not. While we have stressed the observation that acquisition of excess fission Xe, like that of solar wind gases, apparently required direct surface exposure, the two components—solar wind and excess fission Xe—are clearly different. They differ not only in location of source reservoir but also in means of incorporation, since in some cases they can still be separated by stepwise heating. Also, while there is an evident qualitative correlation between solar wind and excess fission Xe, there is an equally evident lack of quantitative correlation; as discussed by Drozd *et al.* (1976) the ratio of $^{136}Xe_f$ to $^{132}Xe_{tr}$ is by no means constant (Table 4), nor is there any perceptible quantitative correlation between excess fission Xe and any other index of soil maturity.

It is also possible to view the problem from a different perspective, one which may be more fruitful in terms of relating excess fission Xe to other lunar phenomena. This alternative perspective focuses on the *similarity* between "solar wind" and excess fission Xe; this is not a difference of interpretation but simply one of emphasis. Like solar wind, it appears that excess fission Xe (and excess ^{129}Xe) is a trapped component, one whose nuclear identity originates outside the sample in which it is now found. While direct evidence (Basford *et al.*, 1973) is still meager, it seems plausible that excess fission Xe is surface-correlated to the same extent that solar wind Xe is. In practice, in many samples neither solar wind nor excess fission Xe can be considered superficially sited (cf., Fig. 4) or truly surface correlated, presumably because of microscale redistribution processes such as diffusion and agglutinate formation (cf., Bogard, 1977). In many cases excess fission Xe has been at least partially homogenized with solar wind Xe. Thus, for the Apollo 16 data reported here, the data analysis described in the previous section can be considered in large part to be a resolution of and correction for cosmogenic and terrestrial contamination components, leading to identification of trapped lunar Xe. Within this composition excess fission Xe is identified less because it is *separable* from a solar wind composition than simply because it is *different* from the "solar wind" surface-correlated composition found in mare soils, as represented by the BEOC-12 composition. It is noteworthy that this generalization also applies to the excess ^{129}Xe seen in these Apollo 16 samples, as well as to most of the excess fission Xe found in the Apollo 14 samples, including those in which there *is* significant isotopic separation in the very low gas amounts seen at low extraction temperatures.

Viewed in this light, the excess fission Xe effect can be considered simply a variation in the net composition of Xe available for trapping at the surface of the lunar regolith, where this variation consists of a variable relative amount of a fission component added to a base (solar wind) component. The data presented in this paper allow us to infer with greater confidence that the principal parameter controlling this variation is time, since this is the element that the Apollo 14 and Apollo 16 sites have in common with each other and by which they differ from mare sites. Specifically, we can imagine a changing composition of lunar trapped Xe as fission Xe becomes less available with time.

A corollary inference of this reasoning is that since the presence of excess fission Xe in highland samples exposed to the solar wind is nearly universal (Table 4), most highland soil breccias must have experienced at least some of their exposure very early in lunar history, before the beginning of comparable exposure on mare surfaces (otherwise the distinction between highland and mare samples would be lost). Thus, at least some of the solar wind gases, and presumably also the other results of surface exposure such as agglutinates or impact pits, reflect an ancient epoch in lunar history. This is not particularly surprising, but direct support for this thesis is mostly nonexistent and we know of only one case where this proposition has been supported by experimental evidence, that of 14307 (as argued, by a completely different line of reasoning, by Bernatowicz *et al.*, 1977).

This alternative perspective also encourages a less discrete view of the presence or absence of excess fission Xe, and one which is probably closer to the truth. Rather than consider the effect simply present in highland samples and absent in mare samples, the notion of a changing composition of lunar trapped Xe suggests that the highland trapped compositions simply incorporate more fission Xe (of lunar origin) than do the mare compositions. This accounts in a straightforward way for the conclusion that even the mare surface-correlated compositions contain some contribution of indigenous lunar gas (most perceptibly fission) and do not accurately measure real solar wind composition, as has already been concluded (Podosek *et al.*, 1971; Eberhardt *et al.*, 1972; Basford *et al.*, 1973) on the basis of comparisons with meteoritic and terrestrial Xe compositions. The extent to which such Xe compositions incorporate an indigenous lunar contamination of solar wind Xe cannot be quantitatively assessed. As a possible extreme case, if real solar wind Xe resembles the primordial composition deduced from meteorite analyses by Pepin and Phinney (1976), we would have to conclude that both mare and highland Xe compositions contain unexpectedly large (several percent) heavy isotope contributions, so that the highland samples would have to be viewed as containing just a little more excess fission Xe than mare samples. In this context it is worth reemphasizing that while the excess fission Xe effect is commonly associated with ^{244}Pu, this identification is based only on very small gas amounts in a few samples, and that for most of the excess fission Xe the identity of the parent must still be considered an open question (cf., Table 3). On the other hand, if a small fraction of the excess fission xenon has precisely the isotopic composition of ^{244}Pu it is highly suggestive that the bulk of the excess is due to this nuclide in these samples.

In principle at least, within the broader perspective we must also consider the possibility that other indigenous lunar components besides fission could be mobilized and redistributed. An interesting example is suggested by the release pattern for 67455 (Fig. 4), in which not only the solar wind and fission Xe but also the major portion of the cosmogenic Xe is released at a very low temperature. This is unexpected for an *in situ* component, so we should consider the possibility that the low temperature release of spallation Xe represents cosmogenic gas produced elsewhere and incorporated in 67455 as a trapped component. Actually, in this specific instance this seems not to be the case. For

40 ppm Ba (Wänke *et al.*, 1973) and the cosmogenic production rate advocated by Bogard *et al.* (1971) we expect that a 50 m.y. exposure to cosmic rays will produce $2.6 \times 10^{-12} cm^3 STP/g$ cosmogenic ^{126}Xe, essentially identical to the total amount observed in 67455 (Table 2). Thus, cosmogenic Xe in 67455 is *not* present in excess of expectations for *in situ* production. Unless small grain size can be blamed we have no plausible explanation for the anomalously low temperature release of cosmogenic Xe. While it seems unlikely that cosmogenic Xe in 67455 is a trapped rather than an *in situ* component, this case is an interesting example of the possibility of variation in trapped gas composition not usually considered.

An important reason for interest in the excess fission Xe phenomenon is its potential use as a chronometer. This application seems more reasonable in light of the present results which strengthen the view that it represents a global phenomenon rather than merely a local effect. If the composition of Xe available for trapping on the surface of the regolith varies with time, specifically if this variation can be characterized as on the average monotonically decreasing relative (to the solar wind flux) availability of fission Xe, and this composition is characteristic of the whole moon, then the amount of excess fission Xe observed in a given sample serves as a stratigraphic index of time of exposure. We have in fact already made such an interpretation by arguing that most highland breccias received at least some of their surface exposures very early. In principle such a stratigraphy, ordered by the ratio of ^{136}Xe to ^{132}Xe or by the ratio of "excess" ^{129}Xe to "excess" ^{136}Xe, could be extended to finer detail as suggested by Behrmann *et al.* (1973), but such inference is sensitively model-dependent and not readily testable. As seen in Table 4, both of these potential indices vary widely and with no apparent order. This could plausibly be due to time of exposure, but could equally plausibly be due to multiple surface exposure or to erratic temporal variation of ^{136}Xe and ^{129}Xe available for trapping. At present we feel that the only meaningful distinction which can be made is that between highland samples and mare samples, but the situation can hopefully be expected to improve with further elaboration of systematic variations in trapped Xe compositions.

Acknowledgments—We gratefully acknowledge the assistance of Claire Morgan in the mass spectrometric analyses of the Apollo 16 samples. The authors would also like to acknowledge Barbara Wilcox and Lou Ross for their help in preparation of the manuscript. This work was supported by NASA through grants NGL 26-008-065 and NSG 07016.

REFERENCES

Basford J. R., Dragon J. C., Pepin R. O., Coscio M. R. Jr. and Murthy V. R. (1973) Krypton and xenon in lunar fines. *Proc. Lunar Sci. Conf. 4th*, p. 1915–1955.

Behrmann C. J., Drozd R. J. and Hohenberg C. M. (1973) Extinct lunar radio-activities: xenon from ^{244}Pu and ^{129}I in Apollo 14 breccias. *Earth Planet. Sci. Lett.* **17**, 446–455.

Bernatowicz T., Drozd R. J., Hohenberg C. M., Lugmair G., Morgan C. J. and Podosek F. A. (1977) The regolith history of 14307. *Proc. Lunar Sci. Conf. 8th*, p. 2763–2783.

Bogard D. D. (1977) Effects of soil maturation on grain size-dependence of trapped solar gases. *Proc. Lunar Sci. Conf. 8th*, p. 3705–3718.

Bogard D. D., Funkhouser J. G., Schaeffer O. A. and Zähringer J. (1971) Noble gas abundances in lunar material—cosmic ray spallation products and radiation ages from the Sea of Tranquility and the Ocean of Storms. *J. Geophys. Res.* **76**, 2757–2779.

Boynton W. V., Baedecker P. A., Chou C.-L., Robinson K. L. and Wasson J. T. (1975) Mixing and transport of lunar surface materials: evidence obtained by the determination of lithophile, siderophile, and volatile elements. *Proc. Lunar Sci. Conf. 6th*, p. 2241–2259.

Crozaz G., Drozd R., Graf H., Hohenberg C. M., Monnin M., Ragan D., Ralston C., Seitz M., Shirck J., Walker R. M. and Zimmerman J. (1972) Uranium and extinct Pu^{244} effects in Apollo 14 materials. *Proc. Lunar Sci. Conf. 3rd*, p. 645–659.

Drozd R. J. (1974) Krypton and xenon in lunar and terrestrial samples, Ph.D. Thesis, Washington University, St. Louis.

Drozd R. J., Hohenberg C. M. and Morgan C. J. (1975) Krypton and xenon in Apollo 14 samples: fission and neutron capture effects in gas-rich samples. *Proc. Lunar Sci. Conf. 6th*, p. 1857–1877.

Drozd R. J., Hohenberg C. M., Morgan C. J. and Ralston C. E. (1974) Cosmic ray exposure at the Apollo 16 and other lunar sites: lunar surface dynamics. *Geochim. Cosmochim. Acta* **38**, 1625–1642.

Drozd R., Hohenberg C. M. and Ragan D. (1972) Fission xenon from extinct ^{244}Pu in 14301. *Earth Planet. Sci. Lett.* **15**, 338–346.

Drozd R. J., Kennedy B. M., Morgan C. J., Podosek F. A. and Taylor G. J. (1976) The excess fission xenon problem in lunar samples. *Proc. Lunar Sci. Conf. 7th*, p. 599–623.

Eberhardt P., Geiss J., Graf H., Grögler N., Krähenbühl U., Schwaller H., Schwartzmüller J. and Stettler A. (1970) Trapped solar wind noble gases, exposure age and K/Ar-age in Apollo 11 lunar fine material. *Proc. Apollo 11 Lunar Sci. Conf.*, p. 1037–1070.

Eberhardt P., Geiss J., Graf H., Grögler N., Mendia M. D., Mörgeli M., Schwaller H. and Stettler A. (1972) Trapped solar wind noble gases in Apollo 12 lunar fines 12001 and Apollo 11 breccia 10046. *Proc. Lunar Sci. Conf. 3rd*, p. 1821–1856.

Graf H., Shirck J., Sun S. and Walker R. M. (1973) Fission track astrology of three Apollo 14 gas-rich breccias. *Proc. Lunar Sci. Conf. 4th*, p. 2145–2155.

Hyde E. K. (1971) *The Nuclear Properties of the Heavy Elements 3.* Dover, N. Y., 1636 pp.

Lewis R. S. (1975) Rare gases in separated whitlockite from the St. Severin chondrite: xenon and krypton from fission of extinct ^{244}Pu. *Geochim. Cosmochim. Acta* **39**, 417–432.

LSPET (Lunar Sample Preliminary Examination Team) (1972) Preliminary Examination of Lunar Samples. Part A. *Apollo 16 Preliminary Science Report.* NASA SP 315, p. 7-1–7-24.

Marti K. (1967) Mass-spectrometric detection of cosmic ray produced ^{81}Kr in meteorites and the possibility of Kr-Kr dating. *Phys. Rev. Lett.* **18**, 264–266.

Marti K., Lightner B. D. and Osborn T. W. (1973) Krypton and xenon in some lunar samples and the age of North Ray Crater. *Proc. Lunar Sci. Conf. 4th*, p. 2037–2048.

Morgan C. J. (1975) Exposure age dating of lunar features: lunar heavy rare gases. Ph.D. Thesis, Washington University, St. Louis.

Müller O. (1975) Lithophile trace and major elements in Apollo 16 and 17 lunar samples. *Proc. Lunar Sci. Conf. 6th*, p. 1303–1311.

Nief G. (1960) Isotopic Abundance Ratios Reported for Reference Samples Stocked by the National Bureau of Standards, (F. Mohler, ed.), NBS Technical Note 51.

Niemeyer S. and Leich D. A. (1976) Atmospheric rare gases in lunar rock 60015. *Proc. Lunar Sci. Conf. 7th*, p. 587–597.

Papanastassiou D. A. and Wasserburg G. J. (1971) Rb-Sr ages of igneous rocks from Apollo 14 mission and the age of the Fra Mauro formation. *Earth Planet. Sci. Lett.* **12**, 36–48.

Pepin R. O. and Phinney D. (1976) The formation interval of the earth (abstract). In *Lunar Science VII*, p. 682–684. The Lunar Science Institute, Houston.

Podosek F. A. (1972) Gas retention chronology of Petersburg and other meteorites. *Geochim. Cosmochim. Acta* **36**, 755–772.

Podosek F. A., Huneke J. C., Burnett D. S. and Wasserburg G. J. (1971) Isotopic composition of xenon and krypton in the lunar soil and in the solar wind. *Earth Planet. Sci. Lett.* **10**, 199–216.
Reynolds J. H., Alexander E. C., Davis P. K. and Srinivasan B. (1974) Studies of K-Ar dating and xenon from extinct radioactivities in breccia 14318; implications for early lunar history. *Geochim. Cosmochim. Acta* **38**, 401–417.
Rose H. J. Jr., Baedecker P. A., Berman S., Christian R. P., Dwornik E. J., Finkelman R. B. and Schnepfe M. M. (1975) Chemical composition of rocks and soils returned by the Apollo 15, 16, and 17 missions. *Proc. Lunar Sci. Conf. 6th*, p. 1363–1373.
Rose H. J. Jr., Cuttitta F., Berman S., Carron M. K., Christian R. P., Dwornik E. J., Greenland L. P. and Ligon D. T. Jr. (1973) Compositional data for twenty-two Apollo 16 samples. *Proc. Lunar Sci. Conf. 4th*, p. 1149–1158.
Wänke H., Baddenhausen H., Dreibus G., Jagoutz E., Kruse H., Palme H., Spettel B. and Teschke F. (1973) Multielement analyses of Apollo 15, 16, and 17 samples and the bulk composition of the moon. *Proc. Lunar Sci. Conf. 4th*, p. 1461–1481.
Warner J. L., Simonds C. H. and Phinney W. C. (1973) Apollo 16 rocks: classification and petrogenetic model. *Proc. Lunar Sci. Conf. 4th*, p. 481–504.

Proc. Lunar Planet. Sci. Conf. 9th (1978), p. 1599–1617.
Printed in the United States of America

Direct measurement of surface carbon concentrations for lunar soil breccias

C. Filleux, R. H. Spear, T. A. Tombrello and D. S. Burnett

California Institute of Technology, Pasadena, California 91125

Abstract—We report measurements of the depth distribution of C on grain surfaces for Apollo 11, 15, 16 and 17 soil breccias using a nuclear reaction depth profiling technique. The depth resolution of the ^{12}C(d,p) reaction utilized permits only determination of "surface" (0–0.5 microns) and "volume" (0–2.5 microns) components. Bombardment with 2 MeV F ions removes an adsorbed layer of terrestrial CO/CO_2 and permits measurement of a stable residual C surface layer which we identify as lunar. The spectral distortion of the surface C proton spectrum by a rough lunar sample surface has been evaluated experimentally using difference spectra for varying amounts of adsorbed CO/CO_2. It is possible to calculate the equivalent distribution for the volume component. This distortion is important because it increases the calculated amounts of surface C by about a factor of 4 from those calculated using a spectrum from a smooth surface. The soil breccia spectra are described well by the surface and volume model, but the spectrum of a <125 micron sample of the 14259 soil is anomalous. For all samples the surface C concentration lies between 2 and 8x10^{15} C atoms/cm^2, showing no obvious correlation with the volume C, which varies over an order of magnitude. The variation probably represents the presence of unexposed mineral grains on the surfaces (1-4 mm^2 area) analyzed; thus our results are best interpreted as indicating a steady state surface C concentration of 5-8x10^{15}/cm^2, which accumulates on lunar soil grains on a time scale fast compared to the production of volume correlated C. Approximately ⅓–½ of the total C in a lunar soil appears to be surface-correlated. Because of the possibility of surface C of meteoritic origin, our results are only an upper limit for directly implanted solar wind C, but they permit this to be an important source.

Introduction

Although most workers have ascribed the carbon found in lunar soils to the solar wind, contributions from the impact vaporization of carbonaceous meteorites and from the early outgassing of the moon cannot be excluded. New measurements of carbon concentrations for Apollo 15 and Apollo 17 mare basalts by DesMarais (1978) yield values (about 5ppm) that are much lower than previously believed. The DesMarais results together with the low C concentrations (3.5 ppm, Epstein and Taylor, 1973; 11 ppm, Chang *et al.*, 1974) for the Apollo 17 orange soil (which is enriched in other lunar volatile elements) suggest that the contribution from indigenous lunar carbon is small, although the origin of the gas phase producing vesiculation in mare basalts (Goldberg *et al.*, 1976) remains a problem.

The mineralogical and physical (i.e., surface vs. volume) distributions of C and $^{13}C/^{12}C$ in lunar soils provide a better set of data from which to evaluate the sources and sinks of C to the regolith than do the corresponding bulk soil values. It is only recently that details of the accumulation and distribution of carbon among different particle types and sizes have been investigated (DesMarais *et al.*, 1975, total C; Gardiner *et al.*, 1977, hydrolyzable C, CH_4). The earlier work by DesMarais *et al.* (1973) provided the first estimates of the amounts of surface-correlated C; at the same time these authors recognized that the reworking of the lunar regolith by micrometeorite bombardment redistributes significant amounts of carbon into the interiors of composite particles (agglutinates, microbreccias). Noble gas studies by Bogard (1977), Signer *et al.* (1977) and Schultz *et al.* (1977) show that the bulk of the trapped solar wind gases are also concentrated, probably as a volume-correlated component, in microbreccias and agglutinates. Surface C concentrations have been obtained from the slope of total C vs. inverse grain size plots. However, as discussed by Bogard (1977) and Schultz *et al.* (1977), the value of the slope depends critically on the distribution of volume correlated carbon. For this reason direct experimental measurements of surface C concentrations are desirable and have already been reported by Zinner *et al.* (1976) and Filleux *et al.* (1977). In the present study we report surface and volume correlated C concentrations in soil breccias originating from both mare and highland sites and in a sample of Apollo 14 fines.

The average surface C concentrations for lunar soils are also important in assessing the relative rates of erosion (e.g., by solar wind sputtering) and deposition (e.g., of impact volatized material) on grain surfaces; however, this problem is obviously closely related to the question of the origin of the surface C, and it is not desirable to attempt to infer erosion/deposition rates directly from the surface C data, (Filleux *et al.*, 1977; Jull and Pillinger, 1977).

Sample description and preparation

Because they had never been exposed to fluorocarbon packing material, partially-glass-coated soil breccia fragments from the Apollo 16 double drive tube were analyzed as received. In our laboratory these samples were handled under N_2 at all times prior to irradiation. Sample 70019,17 is a highly vitrified breccia from which three different fresh surfaces were analyzed. Two surfaces for each of the soil breccias 10048, 10068, 15299 and 79135 were analyzed. For these samples fresh surfaces were obtained in a N_2 atmosphere by breaking new fragments with cleaned quartz or steel chisels. An Apollo 14 soil sample was also irradiated in the horizontal beam line; the soil was held in a cylindrical tube designed to maintain a soil layer slumping at the angle of repose. This sample was a size fraction (less than 125μm) of 14259 which was skimmed off the upper 1 cm of the regolith. A surface glass chip from the unpitted lunar bottom of rock 15059 was analyzed. Although this sample was exposed to fluorocarbon packing material, it was analyzed to check the spectrum produced by a smooth lunar sample surface (see below) and to measure the volume carbon concentration for an impact-produced glass. Inspection of the sample surfaces with a stereomicroscope after irradiation showed that 70019 and the Apollo 11 soil breccias were typical regolith samples, which we visually estimate to contain 10–25% of mineral and rock fragments with 10048 having a slightly larger proportion than 10068. The 15299 and 79135 breccias, on the other hand, contained large (2–10 mm) angular rock clasts with only small amounts of visible glass in the surrounding matrix. Thus, the latter samples cannot be realistically compared with soil samples.

Nuclear reaction technique for ^{12}C analysis

The nuclear reaction technique used in this work has been described by Filleux *et al.* (1977). The samples are mounted in an ultra-high vacuum chamber and irradiated with a 1.07 MeV deuteron beam. The d beam spot was adjusted to have an area from 1 to 4 mm^2, depending on the size of the samples. The protons produced by the reaction of the deuterons with ^{12}C nuclei are measured by a silicon surface barrier detector positioned at an angle of 160°. Elastically scattered deuterons or alpha particles from (d,α) reactions are absorbed by a 3.4 mg/cm^2 Al foil placed in front of the detector. Protons produced by reactions of deuterons with ^{12}C at the surface will have maximum energy. The deuterons will slow down as they penetrate the sample, and, similarly, the emitted protons will lose energy as they emerge from within the sample. Therefore, protons originating from below the surface will have lower energies. The measured proton energy spectrum, i.e., the proton counting rate vs. proton energy, can thus be related to the concentration and depth distribution of C in the sample.

The previous analysis (Filleux *et al.*, 1977) assumed a flat background beneath the proton distribution from $^{12}C(d,p_o)$. To investigate and correct for possible structure in this background, we have measured reference thick target spectra for possible interfering elements: O, F, Na, Mg, Al, Si and Ca. To calculate the background contribution due to a particular interfering element, the reference spectrum is normalized to some characteristic feature of the contaminant which is identifiable in the sample spectrum. So far it has been found that such an accurate background subtraction is important for Mg and to a lesser extent, for Al and Si. Figure 1 shows a typical uncorrected sample spectrum (10068) with its appropriate background spectra. The present background-corrected spectra (see e.g., Fig. 5) show an almost monotonic decrease with decreasing energy. The principal change from our previous spectra (Filleux *et al.*, 1977) is the disappearance of a relatively sharp break in the lower energy range (about channel 480).

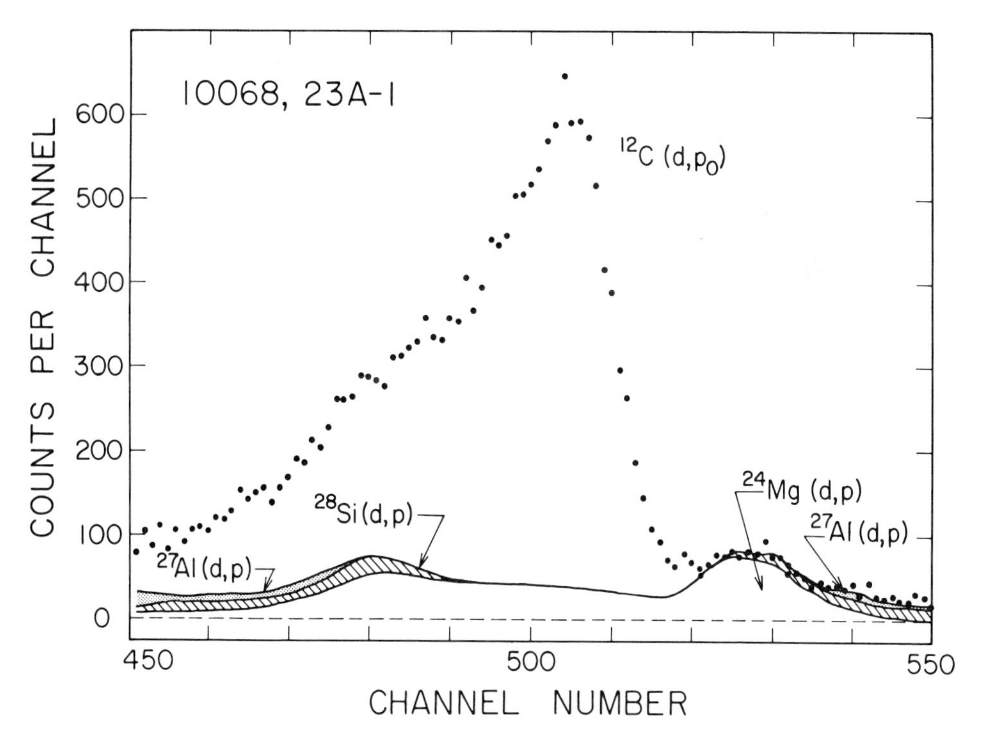

Fig. 1. The raw proton spectrum obtained for sample 10068. The smooth curves shown are the proton spectra obtained for pure targets of ^{24}Mg, ^{27}Al, and ^{28}Si that have been normalized to features characteristic of these nuclides in the spectrum for 10068. The remaining counts are from the ^{12}C in the sample.

Results

Because our depth resolution is limited by range straggling of both the protons and deuterons and by surface roughness effects, we are only able to perform a simple decomposition of our background-corrected spectra f(E) into "surface" (0–0.5 microns) and "volume" (0–2.5 microns) components:

$$f(E) = a_s g_s(E) + a_v g_v(E)$$

The functions $g_s(E)$ and $g_v(E)$ are the spectral shapes for a thin surface layer and a uniform (depth-independent) C concentration, respectively. The spectral shapes are based on measured spectra for a thin C film and for calcite but have been modified for the effect of a rough lunar sample surface as discussed below. The spectral shapes are the input to a linear regression fit of f(E) to give a_s and a_v. The coefficients a_s and a_v are the surface and volume concentrations, which we express in C atoms/cm^2 and ppm, respectively. The quality of the fit to our observed spectra by a superposition of $g_s(E)$ and $g_v(E)$ is estimated by a "reduced χ^2" value (see, e.g., Bevington, 1969).

Last year we showed that our handling did not introduce hydrocarbon or fluorocarbon contamination (Filleux *et al.*, 1977). We also showed that we were able to remove adsorbed CO/CO_2 by a surface desorption (sputtering?) cleaning procedure using a 2 MeV fluorine beam. The principal source of the adsorbed CO or CO_2 is the residual gas in the ultra-high vacuum system (10^{-10}torr), which builds up a monolayer in about two weeks. (It cannot be ruled out that some surface-adsorbed CO_2 is also introduced during sample preparation in the glove box). Prolonged F bombardment produces a progressive decrease in the surface C concentration up to a dose of about 3×10^{16} F/cm^2 (Fig. 2), but essentially no change with increasing dose beyond this value. We conclude from these data that sputter-desorption essentially removes completely the laboratory adsorbed contaminants leaving behind a stable residual lunar surface C concentration which is insensitive to further bombardment.

Comparison of C and H surface concentrations (Leich *et al.*, 1973) allows us to argue strongly against the possibility of a very mobile lunar C component that would be lost along with the adsorbed terrestrial CO/CO_2. Figure 3 shows the H profile for 10048,23 before sputtering and after two sputtering doses of 1×10^{16}F/cm^2. The H profiles shown on Fig. 3 are typical for all samples and are consistent with those observed by Leich *et al.* (1974) and ascribed to solar wind H. The decrease in H content observed after both sputtering bombardments corresponds only to about 20% of the total hydrogen in the outer 0.4 micron. Some diffusive loss of H under bombardment is consistent with the mobility of solar wind implanted H observed by Leich *et al.* (1974). An upper limit to the amount of C loss can be calculated if we assume that all the hydrogen escaped as methane, the most volatile C species observed in hydrolysis experiments. This calculation shows that no more than about 5% of the measured surface carbon would be lost from the surface. We conclude that diffusive loss of carbon is negligible for the F doses ($2–3 \times 10^{16}$/cm^2) and beam intensities normally used in the sputtering phase of our experiments. Extended F bombardment, to levels of $\sim 6 \times 10^{16}/cm^2$, does result in extensive (factor of 2–3) H loss; however, Fig. 2 shows that this loss is not accompanied by a decrease in surface C. The mechanism for the H loss is not clear. It may be associated with the formation of a thick amorphous layer (compared to that due to solar wind ions) at the higher doses, followed by H loss.

In Table 1 we report surface C concentrations before and after sputtering. On the average the amount of CO or CO_2 removed is 2.5×10^{15} atoms of C/cm^2. The variation in the amount of CO/CO_2 removed is not especially large and can be explained by the considerable variation in the elapsed time from the breaking of a fresh surface in the N_2 glove box until the first C analysis. Chemical and grain size effects may contribute to the variation too. For comparison, Table 1 also includes data for an interior fragment of rock 65315 (which has no known solar wind exposure) and terrestrial SiO_2 glass control samples. The amount of contamination observed on anorthosite 65315,7 is well within the range of contamination observed on our soil breccias. The terrestrial quartz samples, which were treated with perchloric acid prior to handling in the N_2 glove box, yield somewhat lower C concentrations, which are readily explained by the difference in adsorptivity between SiO_2 glass and soil breccia surfaces.

Microscopic examination of the samples after analysis shows that the sputtered areas are much "cleaner" than unsputtered areas in that most of the superficial adhering dust on the larger grains (both glasses and minerals) has been removed, presumably by electrostatic hopping. On the assumption that welded-on accretionary particles cannot be electrostatically removed, loss of superficial dust during sputtering should not produce any bias in our measured C distributions, particularly since some of the adhering material is a residue of chiseling operations and, in reality, a contaminant on the surface analyzed. An additional argument against any bias is that the dust is probably removed in the first deuteron bombardment, and comparison of the C spectra obtained during the first and second halves of the initial sample irradiation have consistently been indistinguishable. The sputtered surfaces show no effects of radiation damage in that plagioclase and other clear mineral grains are not darkened. Similarly, no effects of melting (e.g., of glass features) by beam heating are observed. We have devised an experimental method to correct for distortion of the surface and volume C spectral shapes (g_s and g_v) due to surface roughness effects. Protons produced from surface C atoms in a re-entrant location or atoms located within the outer 0.5 micron on grain surfaces subparallel to the deuteron beam will have slightly lower energies from passing through adjoining material before entering the detector and thus will appear to have originated from a greater depth. This effect will be important primarily for roughness on a 1–10 micron scale. We previously analyzed our data in terms of "surface" (0–0.5 μ) and uniform "volume" (0–2.5 μ) components using experimental spectra from a uniform thick target (calcite) and a thin C film. Both of these end-member spectra will be distorted by a rough surface which would give an under-estimation of the surface and an overestimation of the volume concentration. However, surface roughness

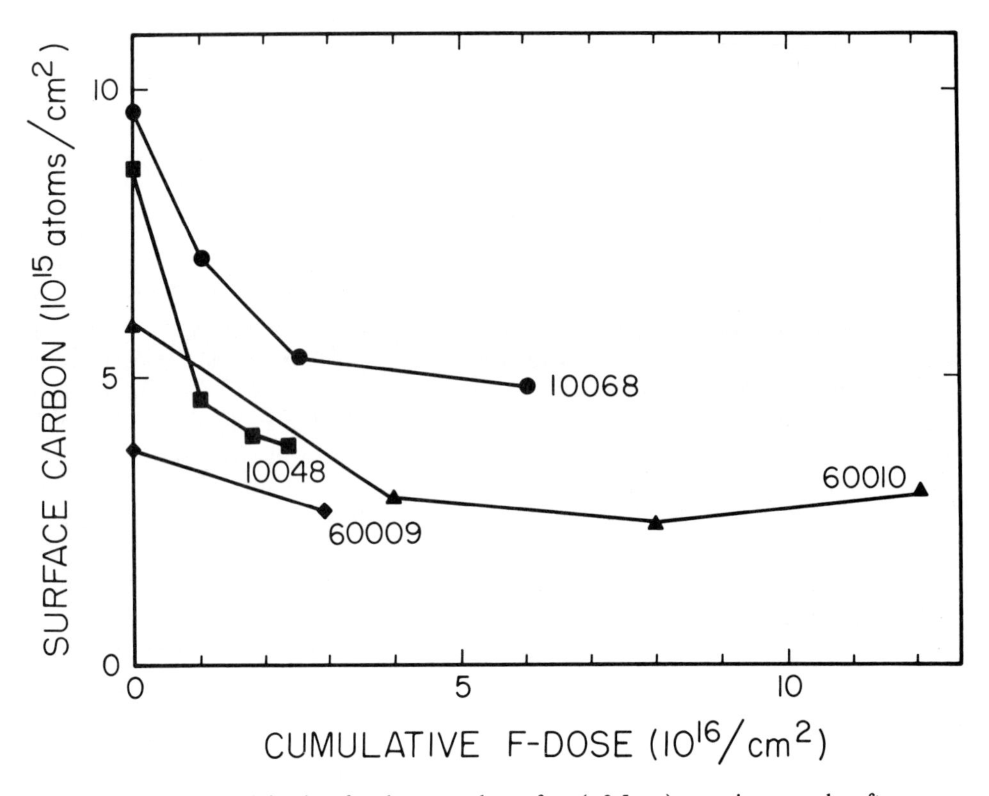

Fig. 2. The areal density of carbon near the surface ($\leq 0.5\ \mu m$) ov various samples after bombardment by varying fluences of 2 MeV F ions. Although carbon is removed quite readily at first, after bombardment by $\sim 3 \times 10^{16}$ F ions/cm^2 there is no further change in the observed concentration of surface carbon. These data indicate that the sputter-desorption induced by the F-beam removes only the laboratory adsorbed contaminents. (Refer to the entries for 65315,7 and the quartz samples in Table 1.)

will also affect the measured spectrum of the adsorbed CO and CO_2. Subtracting spectra having varying amounts of adsorbed CO/CO_2 yields "difference spectra" (data points, Fig. 4a), which are considerably distorted compared to that from a thin C film on a smooth surface (almost a pure Gaussian, Fig. 4a). Only a small amount of distortion was found for CO/CO_2 on an interior surface of anorthosite 65315; thus no explicit corrections were made in our first paper (Filleux *et al.*, 1977). The soil breccia samples, including those from Apollo 16 reported previously, appear to be considerably rougher than 65315. Two types of difference spectra are available: (a) from the spectra before and after sputterng and (b) from spectra obtained after different exposure times to the residual gas in the ultra-high vacuum system, with no intermediate sputtering. Most of our difference spectra are of type a. In three cases where both types of difference spectra were available, the difference spectra obtained from (a) and (b) were in agreement. This shows that possible electrostatic loss of dust upon sputtering

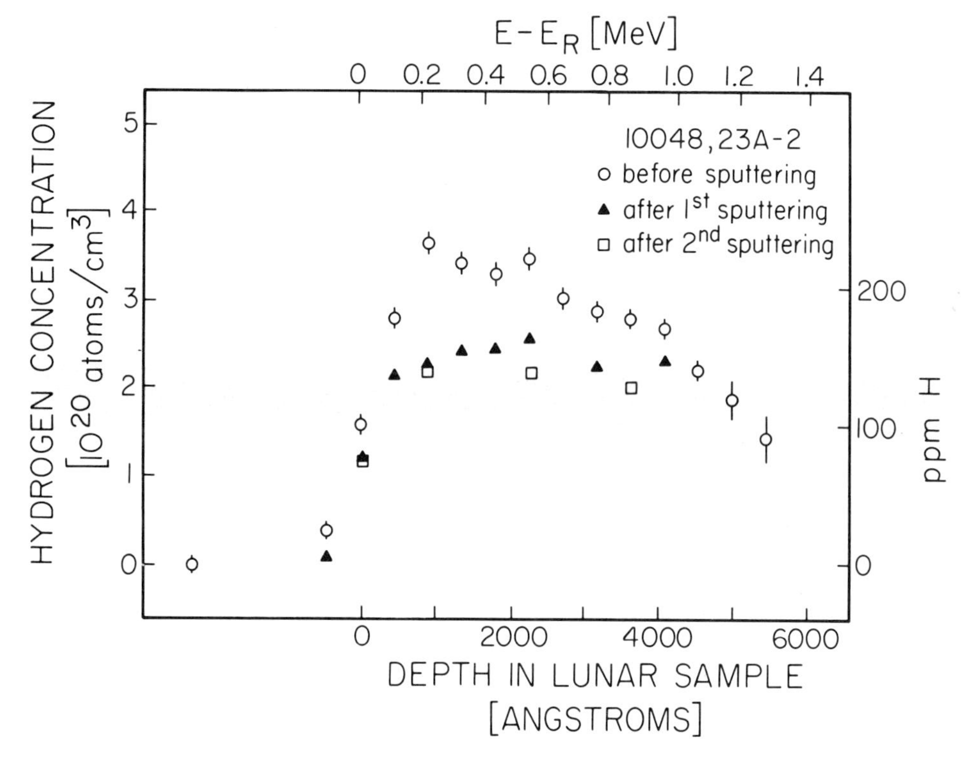

Fig. 3. The depth distribution of hydrogen in sample 10048,23 before and after it was subjected to the sputter-desorption process. Only about 20% of the total hydrogen was removed by the first F bombardment; a subsequent sputtering of equal dose caused no further change in the hydrogen content or distribution.

does not complicate our interpretation of the differences between sputtered and unsputtered spectra in terms of surface roughness. Very similar difference spectra were observed for all of the soil breccia samples; consequently, we have used a standard distorted spectral shape (shown as the solid curve in Fig. 4a) for all the soil breccia samples.

There is no direct experimental way to evaluate the spectral distortion for a volume component; however, if we make the assumption that the scale of surface roughness is large compared to 2.5 μ (our maximum depth of measurement) then the proton spectrum from each depth interval will suffer the same distortion as that from the surface, and a distorted volume component can be calculated using the measured distortion of the adsorbed CO/CO_2 layer. This calculation is described in Appendix A. Figure 4b compares the calculated volume component spectral shape for a rough lunar sample surface with the measured spectrum from a smooth calcite sample.

Figure 5 shows a typical fit to an actual lunar sample spectrum by the distorted surface and volume spectral shapes. A reduced χ^2 value near unity is an

Table 1. Measured surface and volume carbon concentrations.

Sample	Surface C concentration in $10^{15}/cm^2$ Before sputtering	After sputtering	F-dose in $10^{16}/cm^2$	Volume C concentration (ppm)(e)
60009,2066	3.7 ± 0.4(a)	2.8 ± 0.3	3.0	≤10
60010,1040	5.7 ± 0.5	2.1 ± 0.3	5.5	68 ± 22
60010,1042	3.4 ± 0.3	3.2 ± 0.3	3.0	36 ± 5
60010,1044	6.0 ± 0.5	2.9 ± 0.5	4.0	50 ± 14
60010,1060	3.8 ± 0.3	n.m.	—	78 ± 22
70019,17F	15 ± 1.3	6.5 ± 0.5	3.0	176 ± 30
70019,17F3	19 ± 1.6	n.m.	—	165 ± 30
70019,17B	19 ± 1.6	7.8 ± 0.7	3.0	158 ± 30,535(f)
10048,23A-2	8.7 ± 0.9	4.0 ± 0.4	1.8	151 ± 21
10048,23A-3	6.7 ± 0.7	3.1 ± 0.4	2.5	155 ± 25
10068,23A-1	9.7 ± 1.0	4.8 ± 0.4	6.0	181 ± 28
10068,23A-2	10.8 ± 1.0	5.6 ± 0.5	2.5	194 ± 16
15059,28	4.2 ± 0.4(b)	4.6 ± 0.4(b)	3.0	≤10
15299,191A	5.9 ± 0.5	4.1 ± 0.4	2.8	30 ± 9
15299,191B	3.9 ± 0.4	1.5 ± 0.2	3.5	42 ± 4
79135,59A	7.2 ± 0.6	3.7 ± 0.4	3.2	105 ± 16
79135,59B	5.7 ± 0.5	2.7 ± 0.3	3.0	149 ± 12
65315,7	2.5 ± 0.6(d)	≤0.1	1.5	≤12
terrestr. quartz	1.6 ± 0.2(c)	≤0.4	3.0	0
terrestr. quartz	1.2 ± 0.2(c)	≤0.5	1.0	0
terrestr. quartz	3.0 ± 0.3(c)	n.m.	—	0

(a)glass surface; approximately 50% dust covered.
(b)glass surface of unpitted lunar bottom.
(c)before handling: 0.6–1.0 × $10^{15}/cm^2$.
(d)interior surface of unexposed anorthosite.
(e)errors are standard deviation of all measurements, both before and after sputtering.
(f)after sputtering.

indication of a good fit (Bevington, 1969), and in general the χ^2 values are lower for the distorted spectra than for the undistorted spectra used previously. However, the data points are systematically above the calculated best-fit curve in the region of 1–2 microns depth. Conceivably there are small systematic errors remaining in our adopted reference surface and volume spectra, but it is also possible that there is some carbon at these intermediate depths (perhaps small C-rich grains) and that there may be considerable deviation of the true C depth distribution from our simple 2-component model.

We previously noted that sputter cleaning produced large variations in the volume component. We now understand the reason for the observed variation in that we were under-estimating the effects of surface roughness. With the distorted spectral shapes we now obtain good reproducibility of the volume component. An exception is the sample 70019,17B. This sample showed a large decrease in surface C and a very large increase in volume C upon sputtering (Table 1), suggestive of redistribution. The after-sputtering spectrum shows an

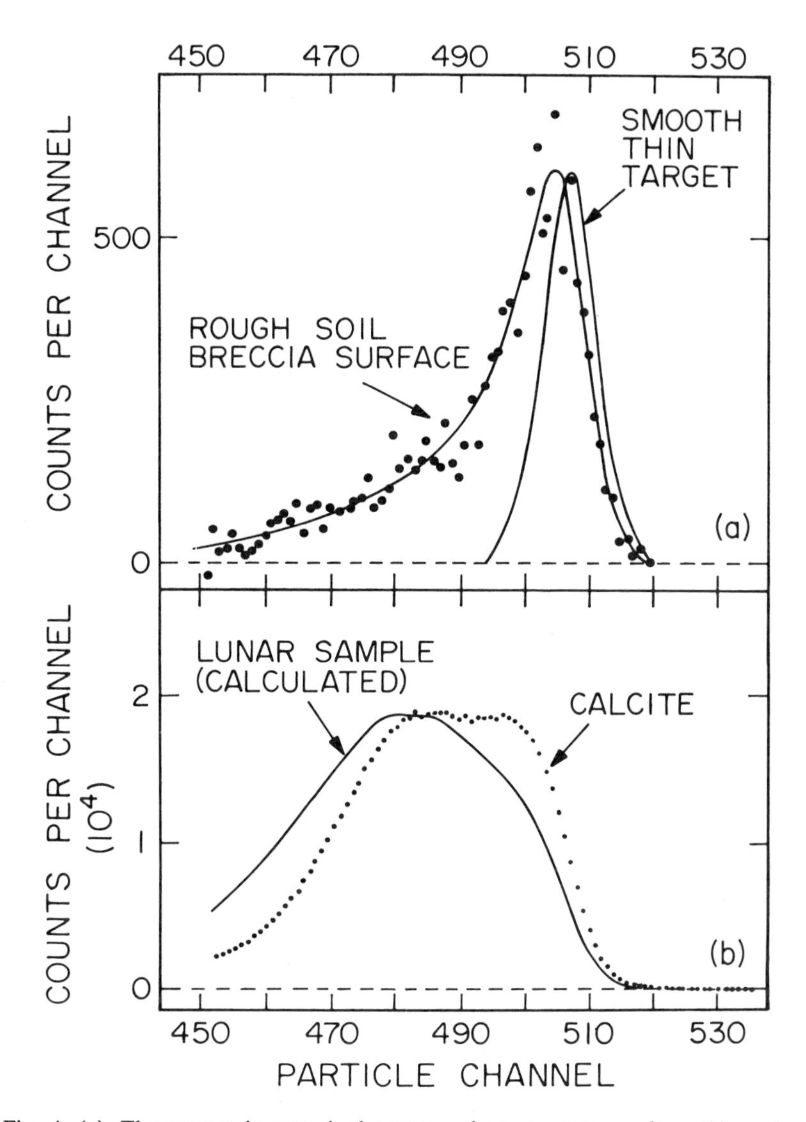

Fig. 4. (a). The symmetric curve is the measured proton spectrum for a thin carbon layer on a smooth surface of a carbon-free material. The data points (through which an asymmetric curve has been drawn) correspond to the difference spectrum from two measurements in which different amounts of CO/CO_2 were deliberately adsorbed on the sample. The distortion from the symmetric curve is caused by the extreme surface roughness of the breccia. (b). The data points shown are for a proton spectrum from a smooth calcite sample (uniform depth distribution of carbon). The solid curve is the calculated spectrum for a uniform depth distribution of carbon that has been distorted by the surface roughness effect shown in (a). (Refer to Appendix A.)

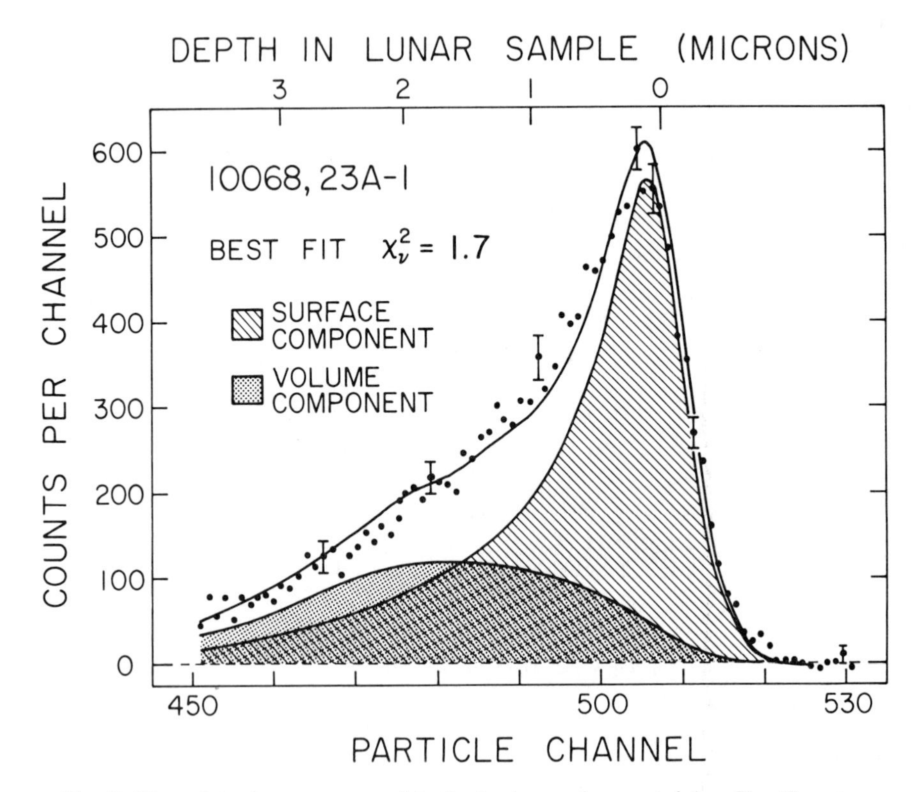

Fig. 5. The points shown correspond to the background-corrected (see Fig. 1) proton spectrum from carbon for sample 10068,23. The two cross-hatched regions show how this spectrum has been decomposed into surface and uniform volume components, each of which has the distorted shape that is a consequence of surface roughness. The solid curve through the data is the sum of the two contributions and has $\chi^2 = 1.7$.

unusual shape suggestive of a C profile that increases inward. This may imply an unreasonably large amount of redistribution. Alternatively, the 70019,17B data can be explained by a small difference in the area analyzed with the deuteron beam before and after sputtering, such that a sub-surface C-rich grain or clast was irradiated in the post-sputtering analysis.

In Table 1 we report surface C concentrations in 10^{15} atoms/cm^2 after sputter-cleaning and volume C concentrations in ppm. The errors given for the surface C in Table 1 are one standard deviation obtained from the regression analysis. They include counting statistics and uncertainties in background correction. For the Apollo 16 samples our surface C concentrations are a factor of roughly 4 higher than those reported previously because most of the tailing of the surface peak was assigned to the volume component. Similarly our volume components are lower by up to a factor of 2 compared to those given by Filleux *et al.* (1977).

As a test of the use of the distorted spectral shapes for most of the samples, we analyzed a smooth glass sample from the unpitted lunar bottom of 15059. Microscopic examination after irradiation showed a very smooth, although curved, surface. There were large cavities (collapsed bubbles?) at the margin of the sample but the C analyses were confined to the smooth central part. As expected, the difference spectra, before and after sputtering, showed only slight distortion from the line shape of the thin C film (Fig. 4a). Most importantly, the measured surface C concentration agrees with those found in the soil breccias. This agreement may be the result of a combination of contamination and coincidence; however, to the extent that such a coincidence is unlikely, the agreement supports the reliability of the surface concentrations obtained with the distorted spectral shapes. Moreover, it appears that the bottom of 15059 has had some solar wind exposure. No impact pits greater than 50 microns are present on ~0.2 cm^2 of our 15059 sample. Using impact pit production rates given by Hörz *et al.* (1977), this is compatible with our observed surface C; however, further study of this sample is warranted.

Our 'volume' component in Table 1 corresponds to an average over ~2.5 μ depth and is not necessarily representative for greater depths. Contributions from surface carbon on the 'back sides' of 1–3 μ grains would mimic high volume concentrations. Inward diffusion of surface C beyond about 0.5 μ would also result in higher volume C. A peculiarity of the two component model is that an over-estimation of the volume C will result in an under-estimation of surface C (and vice-versa) and might introduce a certain bias in the ratio of C_s/C_v. Also, our measured 'volume' C concentration should not agree with the measured bulk C for a given soil breccia, although comparison with available data in the literature might lead to such a suggestion; for example, the bulk carbon value reported by Petrowski *et al.* (1974) for 70019 is 142 ppm.

In an attempt to check whether there are systematic differences between soils and soil breccias, we analyzed a less than 125 μ size fraction of 14259. The bulk C value reported by Moore *et al.* (1972) is 160 ppm. The resulting C depth profile is shown in Fig. 6. The shape of the profile is very different from those obtained on soil breccias and cannot be fit by a simple surface + volume distribution (the best fit is the solid curve in Fig. 6 corresponding to $6 \times 10^{15}/cm^2$ surface concentration). The volume component, if present at all, is very small ($\leq$20 ppm), and there is an excess of C in the 1–2 μ range compared to the soil breccias. The amount of surface carbon may be very low also, because the spectrum shown on Fig. 6 still has contributions from adsorbed CO/CO_2. No change was observed upon sputtering, probably due to electrostatic stirring during bombardment. It may be that most of the C in 14259 resides as a volume component in micron-sized secondary particles. The presence of C in small grains would be needed to explain the excess carbon observed in the 1–2 μ range of our 14259 soil spectrum. However, we cannot completely rule out fine particulate contamination during sizing. Based on the assumption that saturation equilibrium prevails, Jull and Pillinger (1977) also propose that C and N must reside on the surface of grains smaller than about three microns. However, it is not clear that the assumption of saturation is reasonable. The absence of a 'volume' C

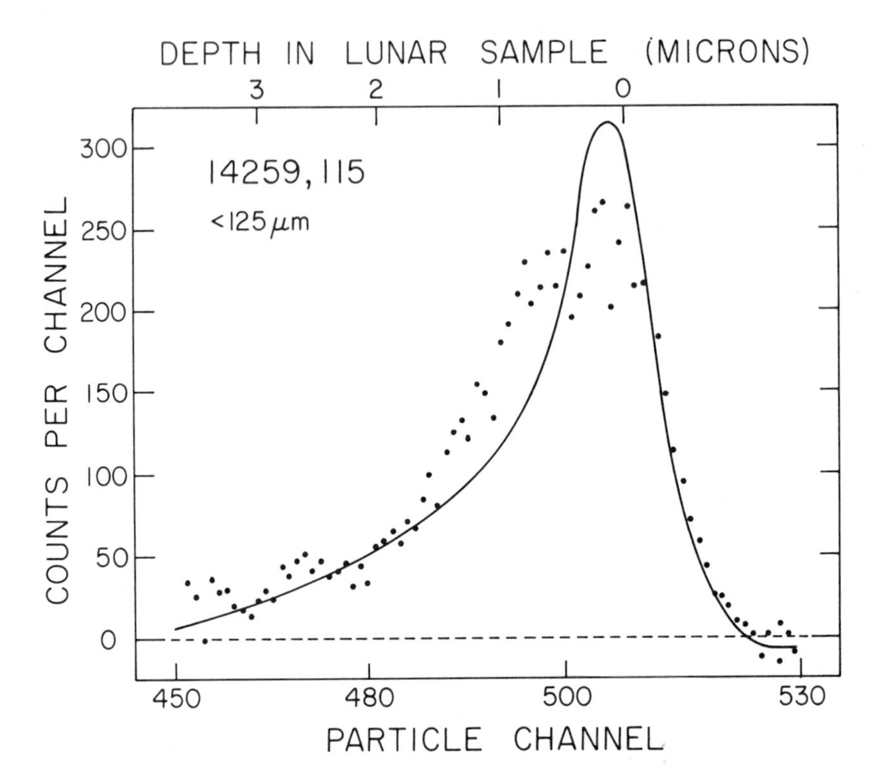

Fig. 6. The proton spectrum from carbon in the soil sample, 14259,115. The solid curve is the best fit possible in terms of the surface plus uniform volume components described in the text and shown in detail in Fig. 5. The best fit corresponds to a surface concentration of 6×10^{15} C atoms/cm^2 and a volume concentration less than 20 ppm. The relatively poor fit indicates that the actual distribution of carbon in the soil grains is more complicated than provided for by the model.

component is very difficult to understand. It is conceivable that our sieved fraction (dry sieving) has a very low percentage of composite particles (agglutinates, microbreccias). Bogard *et al.* (1978) report that trapped solar wind gases are associated with the glass phases for agglutinates, and higher abundances of glassy agglutinates have been reported for larger grain sizes (DesMarais *et al.*, 1973, 1975). These observations would be consistent with a low volume C for our fine fractions of 14259. On the other hand Etique *et al.* (1978) present evidence that agglutinates in smaller grain size fractions contain higher trapped gas concentrations. Either the observations of DesMarais *et al.* are not generally valid or the behavior of C or rare gases is very different. Alternatively, it is conceivable that the 'volume' C we observe in soil breccias may have been formed in the brecciation event, but there is no other strong evidence for significant chemical/mineralogical differences between soils and soil breccias.

Kerridge (pers. comm.) has suggested that our results may be high because of CO/CO_2 adsorbed in pores or other highly re-entrant locations not accessible to the F beam. This is suggested by the large (factor of 5) increase in specific surface area for soils obtained by gas adsorption measurements [see, for example, Holmes and Gammage (1975)] compared to calculated areas from grain size distributions (see, e.g., Signer *et al.*, 1977). However, it cannot be firmly concluded that the difference in these two specific surface areas can be ascribed to a high degree of porosity in lunar soils. The calculated specific surface areas will be too low because of (1) the assumption of spherical grains and (2) biases against small grains in the measured grain size distributions. Also, the surface area inferred from gas adsorption measurements can be high if lunar soils adsorb more than one monolayer of gas. Further, our results are not quantitatively consistent with the Kerridge suggestion in that in the great majority of cases over half of the surface C was removed by F bombardment, whereas only ~20% removal would be predicted if only the "geometrical" (as opposed to the true) surface were accessible to the F beam. Finally, unless the "walls" of these hypothetical re-entrant cavities were very thin ($<0.5\ \mu$), any CO/CO_2 trapped in them would most likely be included in the "volume" component, if analyzed at all. Summarizing, it is quite possible that there is adsorbed terrestrial CO/CO_2 which is inaccessible to our F sputtering beam, but it is also highly likely to be inaccessible to our analyzing d beam, consequently we believe that our measurements do provide relatively accurate values for average lunar surface C concentrations.

Discussion

Figure 7 is a graphical summary of our measured surface and volume C concentrations for the soil breccia samples and for the glass surfaces of 60009 and 15059. The majority of our measured surface C values lie within the range of $2\text{–}5 \times 10^{15}/cm^2$. The fact that we observe a very limited range of surface C concentrations for a wide range of volume C suggests that our surface concentrations represent a steady state value. If so, then the time scale to achieve a steady state surface C concentration is short compared to the time scale to form volume-correlated carbon and also short compared to the mean lifetime of grains at the lunar surface. The $10^3\text{–}10^4$ yr. average surface residence time estimated by Poupeau *et al.* (1975) appears compatible with our measured surface C concentrations. The approximately constant surface C is striking, considering that the analyzed breccia fragments represent a fairly wide range in degree of maturity. Our visual observations of sample maturity correlate reasonably well with the measured volume C concentration in that 70019 and the Apollo 11 soil breccias were typical regolith samples but that 79135 and 15299 had a much larger proportion of mineral and lithic clasts. However, even for mature soils, it is unreasonable to assume that all grains will have a steady state surface C concentration. Also, smaller grains (which have most of the surface area) may have systematically lower surface residence times (M. Maurette, pers. comm.). Our measurements represent an average over many grains, preferentially applying to the smaller grains which account for most of the surface area. The presence of variable proportions of unexposed grains undoubtedly contributes to

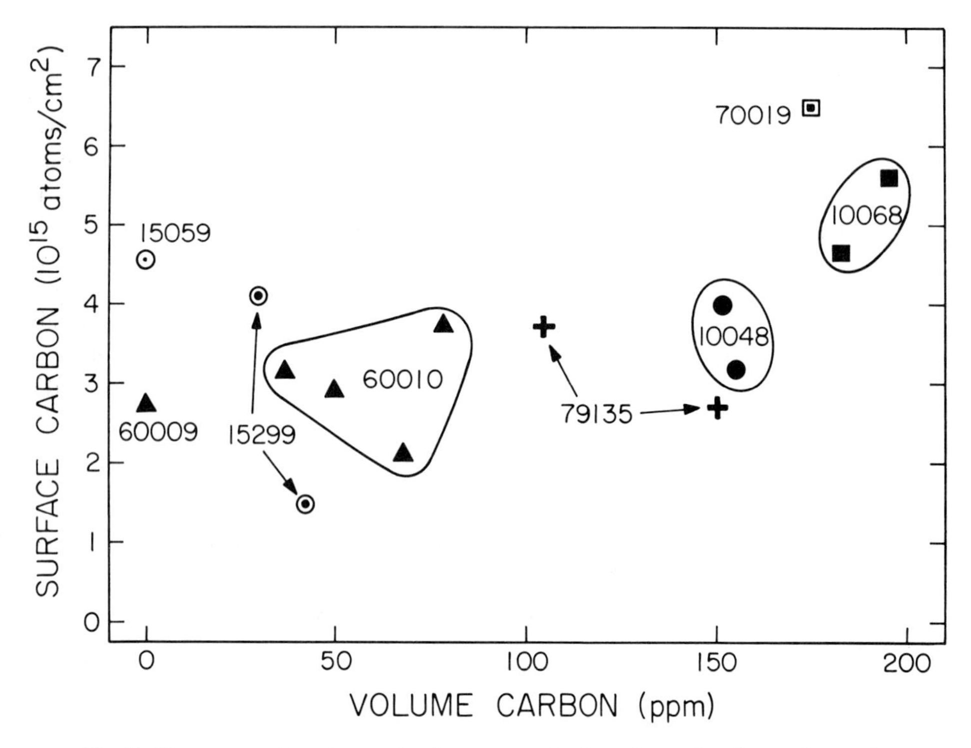

Fig. 7. The measured surface and volume concentrations for carbon for each sample are compared. It is clear that there is no strong correlation between surface concentration and volume concentration for these samples.

the observed variation in surface C. For example, our 10048 surfaces were estimated to have more mineral clasts than the 10068 surfaces analyzed, which is consistent with a higher surface C for 10068. From this point of view the best estimate of the steady state level in solar wind exposed material would be 5–8 $\times$ 10^{15}/cm^2.

Alternatively, the steady state level of surface C could also be composition dependent and mare soils could have higher steady state concentrations, e.g., due to the presence of ilmenite or glass. For example, Bibring *et al.* (1975) concluded that sputtering saturation concentrations should be highest for ilmenite. Our measurements indicate high surface C concentrations for ilmenite-rich samples such as 10068 or 70019, although 10048, which is also ilmenite-rich, does not appear significantly different from the Apollo 16 or 15 breccia samples. Another possibility is that glasses may have systematically higher steady state surface C contents. This is suggested by the high surface C content observed for the 15059 glass and the glass-rich 70019 sample, although the relatively low surface C for the primarily-glass 60009 would require a special explanation. This alternative may be unreasonable if glasses have the highest sputtering rates as proposed by

Bibring *et al.* (1975). Glasses also show the highest amounts of rare gas diffusion loss (see, for example, Müller *et al.*, 1976); however, it may be that C is more readily chemically bound in glass than in mineral grains, unlike rare gases, leading to a higher steady state value if sputtering is not important.

A third explanation for the observed range in surface C concentrations is that none of the results represent a steady state value. The soils from 60009 and 60010 are regarded as immature (McKay *et al.*, 1976) and the accompanying soil breccias may likewise have had a relatively brief exposure. Similarly the 79135 and 15299 samples can be regarded as primarily rock breccias. Moreover, it is conceivable that there is a weak correlation between the surface and volume C on Fig. 7, particularly if the 15059 point is ignored. However, we feel that this explanation is somewhat contrived and that our measured surface concentrations are good approximations to the steady state lunar surface values, perhaps being slightly low (by no more than a factor of 2) due to the presence of unexposed grains on our analyzed surfaces.

An important question is the distribution of carbon between surface and volume components in lunar soils. We assume for the following discussion that our analyzed soil breccias are representative for the soils from which they are derived. This should at least be true for the 10048, 10068 and 70019 samples. In order to derive the relative amounts of surface and volume C, we use 800 cm^2/g (Signer *et al.*, 1977) as the specific surface area for mature lunar soils. Their value is much lower than the BET surface area (see, for example, Holmes and Gammage, 1975). Our surface C concentrations and the BET specific surface area are incompatible in that, when combined, they indicate much more total carbon than is actually observed in a bulk analysis. Using 800 cm^2/g, the range of observed surface C for our soil breccias contributes about 35 to 85 ppm to the total soil. Comparison with bulk C concentrations in the literature indicates that about $^1/_3$ to $^1/_2$ of the total carbon is surface-correlated. This may make C different from solar wind rare gases which, as shown by Signer *et al.* (1977), are concentrated, in microbreccias and agglutinates for the 150–200 micron grain size fraction. Alternatively, even within agglutinates, a significant portion of the carbon and solar wind rare gases may be located on surfaces of individual mineral grains and glasses which are buried inside a single agglutinate particle, i.e., the rare gases within agglutinates may still be surface correlated. A third possibility is that the specific surface area adopted above may be too high.

An unreasonably low specific surface area would be required to make our percentage surface C compatible with that inferred by Epstein and Taylor (1975) from partial fluorination studies. This may represent a basic incompatibility in our results and those of Epstein and Taylor. However, recalling that our surface C represents an average down to a depth of about 0.5 micron, it is also possible that the partial fluorination treatment affects a much shallower depth and that most of our "surface" C actually lies below 0.1 micron.

Chang and Lennon (1975) infer saturation fluences of $\sim 10^{18}$ C ion/cm^2 for implantation of 1 KeV/nucleon C^+ ions into pre-pyrolyzed (1100°C) lunar soil samples. The limitations in comparing the laboratory data to lunar soils are carefully discussed by Chang and Lennon; however, the large difference between the $10^{18}/cm^2$ saturation levels and our observed surface concentrations of 5–8 $\times$ $10^{15}/cm^2$ probably illustrates the importance of loss mechanisms for implanted solar wind C. Alternatively, but less likely, this large difference might be used to argue that our measured surface concentrations are not steady state values.

The "total" carbon contents (surface $\times$ 800 cm^2/g + volume) correspond to 90 ppm for the Apollo 16 double drive tube samples and 210 to 280 ppm for the Apollo 11 samples. Values above 200 ppm seem somewhat high, which may either cast doubt on the 800 cm^2/g used or may indicate that our "volume" component is sampling a depth/size range (0–2.5 μ) with higher than average C content compared to the interiors of all grains in the sample.

Surface C concentrations can also be derived from grain size studies (DesMarais *et al.*, 1975; Chang and Lennon, 1975). Assuming spherical particles DesMarais *et al.* compute surface C concentrations of about 1 $\times$ 10^{16} atoms C/cm^2 which, although close, still appear somewhat higher than our revised values. Increased specific surface areas would be obtained for nonspherical particles which could result in lower surface C concentrations and may account for the differences between our results and those of DesMarais *et al.* Although there is considerable scatter in the grain size data analyzed (from Moore *et al.*, 1973) Chang and Lennon estimate surface C concentrations of 5 $\times$ $10^{15}/cm^2$, in good agreement with our values, but here a large correction for nonspherical grains would lead to lower surface C concentrations than our results. The values for surface C reported by Gardiner *et al.* (1977) for hydrolysable carbon are lower by a factor of 2–3 than our values, but might have been expected to have been substantially lower (factor of 10–15; Kerridge *et al.*, 1978) since our results apply to all C species. Correction factors of up to 4 for nonspherical grains, as assumed by Gardiner *et al.* (1977), would tend to remove the discrepancy, but this seems like a rather large correction. It should be emphasized that the surface concentrations obtained from grain size analyses are not directly measured and are model dependent. Introducing a grain size dependent volume concentration for solar wind implanted species (Bogard, 1977; Schultz *et al.*, 1977) could also change the surface C concentrations determined from grain size studies. In summary, our directly measured surface C concentrations are in fair agreement with those from C analyses of size fractions.

Our surface C concentrations are only upper limits to the amount of directly-implanted solar wind carbon; however, they permit this source to be important in the C balance for a lunar soil. On the other hand they do not rule out contributions to the surface carbon from meteoritic or lunar sources. In addition, some of the measured C could be due to glassy accretionary particles (Blandford *et al.*, 1974). If the accretionary particles are formed by impacts in soils and carbon compounds have not outgassed, these particles represent a means of recycling volume C back onto surfaces. However, if the rock surface glass for 15059 and 60009 is typical of accretionary particle glass (clearly a tentative assumption) then the accretionary particles appear to have been outgassed of C during their formation. Thus, the accretionary particles would not directly affect the surface C concentration but they would bury old C-containing surfaces and thus indirectly increase the steady state value for surface C when averaged over a 0.5 micron depth.

Acknowledgments—The initial experiments in this project were carried out by R. Goldberg and R. Ollerhead. We acknowledge profitable conversations with R. Weller and constructive criticism of the manuscript by J. Kerridge. One of us (C. F.) wishes to acknowledge partial support from The Swiss National Science Foundation. This research was supported by NASA Grant NGR-05-002-333 and by NSF Grant PHY76-83685.

REFERENCES

Bevington P. R. (1969) *Data Reduction and Error Analysis for the Physical Sciences.* McGraw Hill, N.Y. 336 pp.

Bibring J. P., Borg J., Burlingame A. L., Langevin Y., Maurette M. and Vassent B. (1975) Solar wind and solar flare maturation of the lunar regolith. *Proc. Lunar Sci. Conf. 6th*, p. 3471–3493.

Blandford G. E., Fruland R. M., McKay D. S. and Morrison D. A. (1974) Lunar surface phenomena: Solar flare track gradients, mircocraters and accretionary particles. *Proc. Lunar Sci. Conf. 5th*, p. 2501–2526.

Bogard D. D. (1977) Effects of soil maturation on grain size dependence of trapped solar gases. *Proc. Lunar Sci. Conf. 8th*, p. 3705–3718.

Bogard D. D., Morris R. V., McKay D. S., Fuhrman R. and Hirsch W. C. (1978) Relative abundances of fine-grained metal and noble gas in magnetic separates of crushed agglutinate particles (abstract). In *Lunar and Planetary Science IX*, p. 114–116. Lunar and Planetary Institute, Houston.

Chang S. and Lennon K. (1975) Implantation of carbon and nitrogen ions into lunar fines: Trapping efficiencies and saturation concentrations. *Proc. Lunar Sci. Conf. 6th*, p. 2179–2188.

DesMarais D. T. (1978) Carbon isotopes, nitrogen and sulfur in lunar rocks (abstract). In *Lunar and Planetary Science IX*, p. 247–249. Lunar and Planetary Institute, Houston.

DesMarais D. T., Basu A., Hayes J. M. and Meinschein W. G. (1975) Evolution of carbon isotopes, agglutinates, and the lunar regolith. *Proc. Lunar Sci. Conf. 6th*, p. 2353–2373.

DesMarais D. T., Hayes J. M. and Meinschein W. G. (1973) The distribution in lunar soil of carbon released by pyrolysis. *Proc. Lunar Sci. Conf. 4th*, p. 1543–1558.

Epstein S. and Taylor H. P. Jr. (1973) The isotopic composition and concentration of water, hydrogen, and carbon in some Apollo 15 and 16 soils and in the Apollo 17 orange soil. *Proc. Lunar Sci. Conf. 4th*, p. 1559–1575.

Epstein S. and Taylor H. P. Jr. (1975) Investigation of the C, H, O, and Si isotope and concentration relationships on the grain surfaces of a variety of lunar soils and in some Apollo 15 and 16 core samples. *Proc. Lunar Sci. Conf. 6th*, p. 1771–1798.

Etique Ph., Baur H., Derksen U., Funk H., Horn P., Signer P. and Wieler R. (1978) Light noble gases in agglutinates: A record of their evolution? (abstract). In *Lunar and Planetary Science IX*, p. 297–299. Lunar and Planetary Institute, Houston.

Filleux C., Tombrello T. A. and Burnett D. S. (1977) Direct measurement of surface carbon concentrations. *Proc. Lunar Sci. Conf. 8th*, p. 3755–3772.

Gardiner L. R., Woodcock M. R. and Pillinger C. T. (1977) Carbon chemistry and magnetic properties of bulk and agglutinate size fractions from soil 15601. *Proc. Lunar Sci. Conf. 8th*, p. 2817–2839.

Goldberg R. H., Tombrello T. A. and Burnett D. S. (1976) Fluorine as a constituent in lunar magmatic gases. *Proc. Lunar Sci. Conf. 7th*, p. 1597–1613.

Holmes H. F. and Gammage R. B. (1975) Interaction of gases with lunar materials; revised results for Apollo 11. *Proc. Lunar Sci. Conf. 6th*, p. 3343–3350.

Hörz F., Morrison D. A., Gault D. F., Oberbeck V. R., Quaide W. L., Vedder J. F., Brownlee D. E. and Hartung J. B. (1977) The micrometeorite complex and evolution of the lunar regolith. In *Proc. of the Soviet-American Conf. on the Cosmo-Chemistry of the Moon and Planets*, p. 605–636. NASA SP-370. Washington, D.C.

Jull A. J. T. and Pillinger C. T. (1977) Effects of sputtering on solar wind element accumulation. *Proc. Lunar Sci. Conf. 7th*, p. 3817–3833.

Kerridge J. F., Kaplan I. R. and Petrowski C. (1978) Carbon isotope systematics in the Apollo 16 regolith. In *Lunar and Planetary Science IX*, p. 618–620. Lunar and Planetary Institute, Houston.

Leich D. A., Goldberg R. H., Burnett D. S. and Tombrello T. A. (1974) Hydrogen and fluorine in the surface of lunar samples. *Proc. Lunar Sci. Conf. 5th*, p. 1869–1884.

Leich D. A., Tombrello T. A. and Burnett D. S. (1973) The depth distribution of hydrogen and fluorine on lunar samples. *Proc. Lunar Sci. Conf. 4th*, p. 1597–1612.

McKay D. S., Morris R. V., Dungan M. A., Fruland R. M. and Fuhrman R. (1976) Comparative studies of grain size separates of 60009. *Proc. Lunar Sci. Conf. 7th*, p. 295–313.

Moore C. B., Lewis C. F., Cripe J., Delles F. M., Kelly W. R. and Gibson E. K. Jr. (1972) Total carbon, nitrogen and sulfur in Apollo 14 lunar samples. *Proc. Lunar Sci. Conf. 3rd*, p. 2051–2058.

Müller H. W., Jordan J., Kalbitzer S., Kiko J. and Kirsten T. (1976) Rare gas ion probe analysis of helium profiles in individual lunar soil particles. *Proc. Lunar Sci. Conf. 7th*, p. 937–951.

Petrowski C., Kerridge J. F. and Kaplan I. R. (1974) Light element geochemistry of the Apollo 17 site. *Proc. Lunar Sci. Conf. 5th*, p. 1939–1948.

Poupeau G., Walker R. M., Zinner E. and Morrison E. A. (1975) Surface exposure history of individual crystals in the lunar regolith. *Proc. Lunar Sci. Conf. 6th*, p. 3433–3448.

Schultz L., Weber H. W., Spettel B., Hintenberger H. and Wänke H. (1977) Noble gas and element distribution in agglutinate and bulk grain size fractions of soil 15601. *Proc. Lunar Sci. Conf. 8th*, p. 2799–2815.

Signer P., Baur H., Derksen V., Etique P., Funk H., Horn P. and Wieler R. (1977) Helium, neon and argon records of lunar soil evolution. *Proc. Lunar Sci. Conf. 8th*, p. 3657–3683.

Zinner E., Walker R. M., Chaumont J. and Dran J. C. (1976) Ion probe analysis of artificially implanted ions in terrestrial samples and surface enhanced ions in lunar samples 76215,77. *Proc. Lunar Sci. Conf. 7th*, p. 953–984.

APPENDIX A. SURFACE ROUGHNESS DISTORTION FOR THE PROTON SPECTRUM FROM A UNIFORM C DISTRIBUTION

We assume that the scale of the irregularly shaped surface is large compared to 2.5 μ, the maximum depth of our measurement. Consequently, protons from any depth interval between 0 and 2.5 μ should suffer the same energy dispersion as those from the surface.

We first consider the number of protons N(E) with energy E reaching the detector. For a *smooth target* N(E) can be calculated as follows: The number of protons of energy E produced at a depth x is

$$dN(E,x) = \text{const. } \sigma(E_d(x)) \cdot P(E,x)\, dx \tag{1}$$

where σ is the cross section of the $^{12}C(d,p_o)$ reaction, $E_d(x)$ the deuteron energy at the reaction site, and P(E,x) is the probability that a proton produced at depth x will have energy E at the detector. Since the largest sources of dispersion are the Al foil over the detector and the detector itself, we can write

$$P(E,x) = \text{const. } \exp\left\{-2.772 \left(\frac{E-\bar{E}}{\mu}\right)^2\right\}, \tag{2}$$

where $\bar{E}$ is the mean proton energy from depth x:

$$\bar{E} = E_p(E_d) - X \cdot dE/dx_p \cdot \sec(160°) - \Delta E_{Al}$$

where $E_p(E_d)$ is the proton energy corresponding to the mean deuteron energy at depth x, $|dE/dx|_p$ is the proton energy loss rate (assumed constant) and ΔE_{Al} is the proton energy lose in the Al foil (also assumed constant). In equation (2) μ is taken to be a constant (51 KeV) from the measured width of the proton distribution from a thin C film. We have thus neglected dispersion of the incoming deuteron and out-going proton due to energy loss straggling in the sample. The total number of counts for protons of energy E is

$$N(E) \cong \text{const.} \cdot \int_O^{\bar{E}_o} \sigma\,(E_d)\,\exp\left\{-2.772\,(\frac{E-\bar{E}}{\mu})^2\right\} d\bar{E} \qquad (3)$$

where the integral over E is equivalent to a depth integral. E_o is the average proton energy for a surface C atom. Strictly speaking, the lower limit of the integral is a function of E, but because μ (51 KeV) is much smaller than the range of values of E (300 KeV), the only significant contributions to the integral occur when E is close to E, and it is sufficient to replace the exact lower limit by O. The proton energy spectral shape calculated from equation (3) agrees well with our measured calcite spectral shape, giving confidence in the overall method of calculation. The calculation of the equivalent spectral shape for a rough surface involved replacing the gaussian probability distribution P(E,x) in equation (2) and (3) with the experimental spectral shape shown in Fig. 4a, with E now representing the most probable energy of the distorted spectral shape. Numerical integration of the equivalent of equation (3) produced the distorted volume spectral shape shown in Fig. 4b.

Proc. Lunar Planet. Sci. Conf. 9th (1978), p. 1619–1627.
Printed in the United States of America

Nitrogen isotope systematics of two Apollo 12 soils

RICHARD H. BECKER*

Enrico Fermi Institute, University of Chicago
Chicago, Illinois 60637

ROBERT N. CLAYTON

Enrico Fermi Institute and Departments of
Chemistry and of the Geophysical Sciences
University of Chicago, Chicago, Illinois 60637

Abstract—Soils 12023 and 12037 were analyzed for nitrogen by step-wise heating. Helium contents, $^{40}Ar/^{36}Ar$ ratios and spallation-^{15}N ages were determined in addition to nitrogen contents and isotopic compositions. The results for 12023 show it to be a typical lunar soil, similar to many Apollo 16 and 17 soils in its nitrogen content and $\delta^{15}N$ value, with a complex history. Soil 12037 appears to have had a simple, two-stage surface exposure history; one exposure occurring recently for about 10–20 m.y. and the other between about 3 and 4 $\times$ 10^9 yr ago. The $\delta^{15}N$ value of nitrogen implanted in the earlier exposure was $-125‰$. This nitrogen may reside in the minor components of 12037 such as the KREEP. If it should reside in the basaltic component of 12037, however, then the variation in the $\delta^{15}N$ of implanted nitrogen with time probably has been more complex than previously considered.

INTRODUCTION

Although it has been demonstrated that the $\delta^{15}N$ value of nitrogen implanted in the lunar regolith has varied with time, ranging from about $-105‰$ to $+120‰$, the cause of this variation is still a matter of conjecture (Kerridge, 1975; Becker and Clayton, 1975, 1977; Becker *et al.*, 1976; Kerridge *et al.*, 1977). In an attempt to obtain additional information which might bear on the question of the cause of the variation in $\delta^{15}N$, and working on the assumption that samples which are somehow unusual or atypical would be most informative, we chose to analyze two Apollo 12 soils which had been reported to have some notable properties. These soils were 12023 and 12037.

Sample 12023 was supposed to consist of essentially 100% KREEP (Morris *et al.*, 1977). We now know that this is not the case, and that the sample thought by Morris *et al.* (1977) to be 12023 had been mislabeled (Morris, pers. comm.). Our data for 12023 will thus be used as a reference with which to compare 12037,

*Present address:
Division of Geological and Planetary Sciences
California Institute of Technology
Pasadena, California 91125

since such monitors of solar wind exposure as I_s/FeO (Morris, 1978) and helium, nitrogen and carbon contents as well as nitrogen isotopic composition (Kerridge *et al.*, 1978) all indicate that 12023 is representative of the typical submature soils to be found at the Apollo 12 site. Soil 12037 was chosen for analysis because it was found by Kerridge *et al.* (1978) to have the lowest $\delta^{15}N$ value of any bulk soil yet analyzed. This makes it significant from the point of view of establishing the time span over which the change in $\delta^{15}N$ has occurred. Also, any other unusual properties which can be found for this soil might relate to the mechanism which caused the variation in $\delta^{15}N$.

Bulk nitrogen data have already been reported by Kerridge *et al.* (1978) for the two soils we analyzed, along with carbon, sulfur and helium data. The advantage of our analytical method, involving step-wise heating of the soils, is its information on the relative amounts of different generations of nitrogen and the cosmic-ray exposure age. Approximate $^{40}Ar/^{36}Ar$ ratios were also obtained. This is a parameter which may well have some relevance to the time of surface exposure of a soil (Yaniv and Heymann, 1972), and thus should be related to some extent to $\delta^{15}N$.

ANALYTICAL PROCEDURE

The procedures used have been described by Becker and Clayton (1975). Overnight outgassing of the two samples to remove adsorbed terrestrial N_2 was done at 25°C. The $^{40}Ar/^{36}Ar$ ratios for the bulk samples were obtained by taking a weighted average of the ratios for the individual temperature fractions. The amounts of argon in each fraction were determined from the sum of the peak heights at masses 36, 38 and 40 and the peak height at mass 28 in each fraction. Since there are uncertainties in the total amount of gas in each fraction, in the Ar/N_2 ratio in each fraction, and in the individual $^{40}Ar/^{36}Ar$ ratios because of discrimination in the mass spectrometer, the $^{40}Ar/^{36}Ar$ ratios of the bulk samples should not be considered to be known to better than about ±20%.

RESULTS

The nitrogen isotopic results for the step-wise heating of 12023,92 and 12037,22 are shown in Figs. 1 and 2, respectively. Included in the figures are the bulk values for the nitrogen, obtained by summing the nitrogen yields for the different temperature fractions and weighting the $\delta^{15}N$ values for the different fractions by their yields. Our values are in good agreement with those of Kerridge *et al.* (1978) for these two soils. The dashed line in Fig. 1 represents the weighted average of the two fractions above 1000°C, and is included to allow direct comparison with Fig. 2.

In Table 1 are shown the helium contents, $^{40}Ar/^{36}Ar$ ratios, spallation-^{15}N contents and nitrogen spallation ages obtained for these two soils. The spallation-^{15}N contents were determined as described by Clayton and Becker (1977) and Becker and Clayton (1977). The ^{15}N ages were calculated assuming the soils have had a near-surface cosmic-ray exposure history, using the ^{15}N production rate found for rocks (Becker *et al.*, 1976). This assumption is probably not valid for 12037, since it appears that soils with low $\delta^{15}N$ values have spent much of their histories buried at some depth in the regolith after having had an initial

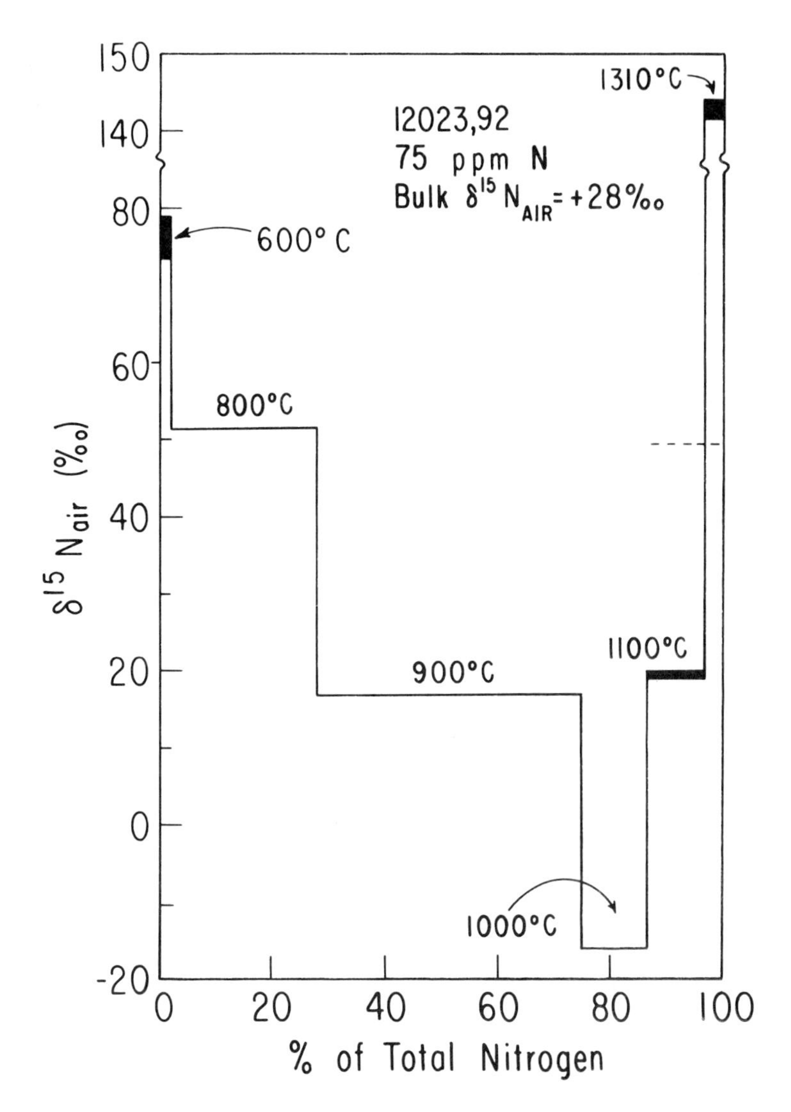

Fig. 1. Isotopic compositions of nitrogen fractions released from soil 12023,92 during step-wise heating, as a function of the amount of nitrogen released. Total nitrogen yield is 75 ppm. Temperatures are approximate. Uncertainty in $\delta^{15}N$ is indicated by thickness of horizontal line. Dashed line indicates average $\delta^{15}N$ of the two highest temperature fractions.

surface exposure to solar wind (Becker and Clayton, 1977).

The helium contents in Table 1 are in good agreement with the values of 9.56 and 7.12 cm^3/g obtained by Kerridge *et al.* (1978) for 12023 and 12037, respectively. There do not appear to be any $^{40}Ar/^{36}Ar$ data or noble gas spallation ages for these soils with which to compare our results.

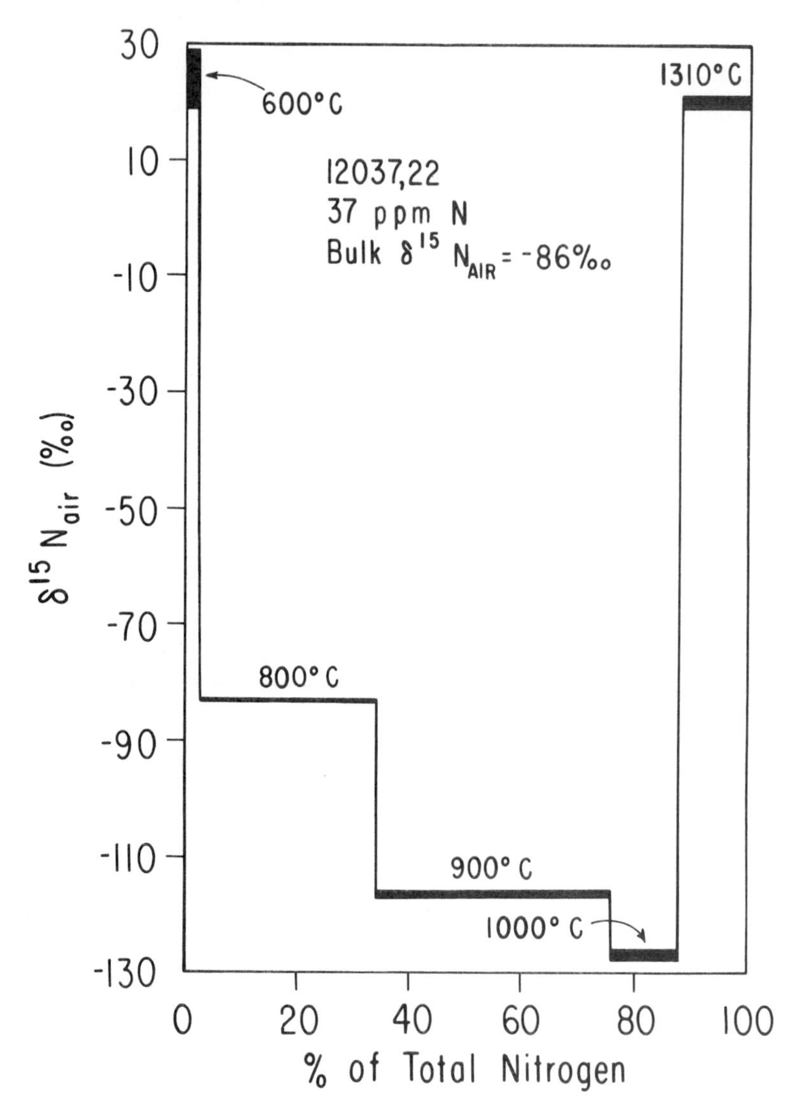

Fig. 2. Isotopic composition of nitrogen fractions released from soil 12037,22 during step-wise heating. Total nitrogen yield is 37 ppm. Value of −127‰ in the 1000° cut is the lowest $^{15}N/^{14}N$ ratio yet seen in nature.

Table 1. Helium, argon and spallation nitrogen data for soils 12023,92 and 12037,22.

Sample	Wt. (gm)	He content (cm³ STP/g)	$^{40}Ar/^{36}Ar$*	Spallation-^{15}N (ng/g)	^{15}N age (m.y.)
12023,92	0.973	0.0984	0.5	2.48 ± 0.29	690 ± 80
12037,22	0.858	0.0696	2.2	2.57 ± 0.27	715 ± 75

*Uncertainty of about ±20%. See analytical procedures.

Discussion

The step-wise heating profile obtained for sample 12023 (Fig. 1) is typical of those seen in several other soils. One component, released below about 900°C, is relatively heavy isotopically, while a second component, isotopically light, comes off above 900°C. As is usually the case, these two components are not well separated, so that the 900°C fraction, containing a significant fraction of the total nitrogen, has an isotopic composition lying approximately halfway between the high and low temperature components. A third component, which is again isotopically heavy, is found in the highest-temperature fractions where melting occurs. The total nitrogen and bulk $\delta^{15}N$ values are also typical of those seen in many Apollo 16 and 17 samples (Kerridge *et al.*, 1975; Becker and Clayton, 1975).

Our explanation for the profile seen in Fig. 1, supported by partial stripping experiments (Becker *et al.*, 1976), has been as follows: that the 800°C component (ignoring the 600°C fraction which is very small and may be contaminated by the analytical blank) represents nitrogen currently on grain surfaces, and consists of nitrogen from the most recent surface exposure of the soil plus nitrogen implanted in the grains during surface exposures in the more distant past, while the 1000°C component comes from former grain surfaces now buried in agglutinates. These buried surfaces contain a higher ratio of old relative to recent nitrogen at any given time than do the exposed surfaces, due to the recycling nature of the agglutination process (Basu, 1977), and thus have a lower $^{15}N/^{14}N$ ratio than the exposed surfaces, due to the secular increase in the $\delta^{15}N$ of implanted nitrogen. The heavy component seen on melting consists of ^{15}N-rich spallation nitrogen mixed with the light component, and it undoubtedly also contains some indigenous lunar nitrogen in amounts on the order of 1 ppm or less and with a $\delta^{15}N$ value of about +10‰ (Becker *et al.*, 1976).

A sample of fresh regolith exposed only once in the past would be expected to yield a flat profile, except in the melt fraction. Later exposures, either a single one in the period just before collection on the moon or multiple ones over a long period of lunar history, would cause the flat pattern to evolve into the type seen in Fig. 1, as would the admixture of soils having had different histories. A soil such as 12023, with roughly equal amounts of heavy and light components after correction for spallation-^{15}N, could thus be considered to have had just two surface exposures, of approximately equal duration, at very different times. The more likely situation, however, is that 12023 consists of grains exposed at many different times which average out to yield a typical isotopic profile. Choosing between the two histories does not appear possible for 12023 on the basis of the nitrogen data alone.

The situation for 12037 (Fig. 2) is somewhat different. As expected from the bulk $\delta^{15}N$ value (Kerridge *et al.*, 1978), both the low and high-temperature implanted components in 12037 are very light, the lightest yet seen for lunar samples. (The $\delta^{15}N$ value of about −125‰ extends the total range of $\delta^{15}N$ in lunar samples, excluding spallation-^{15}N influenced values, to 250‰.) The relatively small difference in $\delta^{15}N$ between the 800°C and 1000°C fractions, comparable to that seen in 61221 (Becker *et al.*, 1976) and 74240 (Becker and Clayton, 1977) but much less than in other soils, the rather good separation between components, as indicated by the similarity in $\delta^{15}N$ of the 900°C and 1000°C fractions, and the low $\delta^{15}N$ value of −83‰ in the 800°C fraction all suggest a relatively simple two-stage surface history for this soil.

This latter point can be seen as follows. The amount of nitrogen with $\delta^{15}N$ of +125‰, the apparent value of recently implanted nitrogen (Becker *et al.*, 1976), present in the sample can be no more than 10% of the total, or about 3.5 ppm, if the initial $\delta^{15}N$ were −125‰. Even if the total N content is a factor of 2 too low (see below), the amount of nitrogen from the most recent exposure is limited to about 7 ppm. Since implantation of this amount of nitrogen would require only a few million years at the present solar wind flux, and since the Bench Crater event with which 12037 is associated apparently has an age on the order of 20 m.y. from the two-stage exposure model for rocks (Burnett *et al.*, 1975), there does not seem to be much room in the history of 12037 for anything but a very old and a very recent surface exposure.

Given that 12037 consists mainly of basaltic fragments (Marvin *et al.*, 1971), and that the Apollo 12 basalts were formed about 3.25×10^9 yr ago (Nyquist *et al.*, 1977), the old surface exposure would appear to have taken place at, or more recently than, 3.25×10^9 yr ago. However, previous measurements indicated that implantation of nitrogen with $\delta^{15}N$ of −80‰ occurred prior to 3.6×10^9 yr ago (Becker and Clayton, 1977). The value of −80‰, determined on soil 74240, is considered to be a good measure of the implanted component because of the excellent agreement in isotopic composition between the 900° and 1000° fractions, implying that neither was seriously contaminated by a heavier component. Thus the observations on 74240 and 12037 are in conflict with the assumption that the $\delta^{15}N$ of the implanted component has increased monotonically with time, and suggests that perhaps the $\delta^{15}N$ of implanted nitrogen has varied in a more complex way. There is, however, a complication with regard to 12037 which may obviate the need for such a conclusion.

Soil 12037 has several characteristics which are atypical of Apollo 12 soils. It contains the highest proportion of basaltic fragments (54%) (Marvin *et al.*, 1971), and it has been suggested that some portion of this basaltic material may have been derived, after collection on the moon, by abrasion of the friable basalt 12036 which was carried in the same sample bag (Marvin *et al.*, 1971; McKay *et al.*, 1971). In terms of its bulk chemistry, 12037 is more basaltic and less KREEPy than other Apollo 12 soils (Wänke *et al.*, 1971), and Wänke (in Wasson and Baedecker, 1972) considers this to be due to contamination of 12037 by 12036. The content of meteoritic siderophile elements in 12036 is half that in 12070, a mature Apollo 12 soil, but comparable to that in the immature soils 12032 and 12033 (Laul *et al.*, 1971), which again could be the result of dilution of 12037 by rock fragments from 12036. It should be noted, however, that if such dilution took place, the rock fragments must have had a grain-size distribution similar to an immature soil, since the size distribution in 12037 does not show a marked enrichment in very coarse grains relative to other Apollo 12 soils (King *et al.*, 1971). In light of the above indications that some of the basaltic component in 12037 may have been derived from 12036 during sample return, it is possible that the major carrier of nitrogen is not the abundant basaltic fragments, but a minor component which is considerably older, such as KREEP. Wasson and Baedecker (1972) argue that some of the KREEP in Apollo 12 soils comes from material underlying the basalts at Apollo 12, and this material could have had a surface exposure history prior to the formation of the basalts.

Contamination of soil 12037 by material from 12036 would have two further consequences with regard to our data. As the rock contains little or no solar wind, such contamination would serve to dilute the solar wind gases in the soil, perhaps by as much as a factor of two if most of the basalt fragments were derived from the rock, and this dilution would make the sample look less mature than it actually was. It would also lower the spallation age, since the rock has a much lower cosmic ray exposure than the soil (Burnett *et al.*, 1975). The latter effect could account for the fact that 12023 and 12037 have similar exposure ages when the previous tendency was for samples with low $\delta^{15}N$ values to have higher spallation ages (Kerridge, 1975).

There remain two problems, however, associated with the view that the nitrogen in 12037 is present in an older component such as KREEP, both associated with argon. The $^{40}Ar/^{36}Ar$ ratio for 12037, while significantly larger than that for 12023 as expected, is lower than the ratio seen in several other soils with low $\delta^{15}N$, such as 61221 and 74240. Correction for ^{40}K-decay will yield a still lower value for the trapped argon component. Since the highest trapped $^{40}Ar/^{36}Ar$ ratio and the lowest $\delta^{15}N$ value do not occur in the same soil, these two apparent indicators of the antiquity of surface exposure must be decoupled in some way. One such way would, of course, be to have $\delta^{15}N$ vary in some complex way with time.

The second problem involving argon has to do with the observation that the K-Ar ages of KREEP in Apollo 12 soils 12032 and 12033 have been reset about 800 m.y. ago (Eberhardt *et al.*, 1973; Alexander *et al.*, 1977). An event which extensively outgassed argon might be expected to have had some effect on nitrogen as well. If the nitrogen in 12037 was in KREEP, and if all the KREEP was involved in the outgassing event, we might have expected to see some mass fractionation effects in nitrogen leading to ^{15}N enrichment, the opposite to what is seen. However, if KREEP in 12037 is the principal nitrogen carrier, it cannot be the same KREEP as is found in 12032 and 12033, since the latter soils do not have the low $\delta^{15}N$ found in 12037 (Kerridge *et al.*, 1978). Thus the thermal histories of 12032 and 12033 do not necessarily apply to 12037.

It is clear that the location of the nitrogen in 12037 can be determined by selecting samples of KREEP and basalts and analyzing them separately. It is not yet clear, however, that the associated problems will be as easy to solve.

Conclusion

Analysis of soil 12023 by step-wise heating yields little information on its surface exposure history other than that it was probably complex and that a significant portion of it was fairly recent. This is not surprising, in view of its relatively high maturity, at least for the Apollo 12 site, and its associated high cosmic-ray age. Sample 12037 on the other hand appears to have had a simple, two-stage surface exposure history, with the most recent surface residence having lasted on the order of 10–20 m.y. or less. The earlier exposure may have occurred at about 3.25×10^9 yr ago, in which case the variation of the $\delta^{15}N$ value of implanted nitrogen in lunar soils with time has been something other than monotonic, with a decrease from the −80‰ seen in 74240 (Becker and Clayton, 1977) to the −125‰ seen in 12037 preceding an increase of 250‰. Alternatively the nitrogen in 12037 may have been brought in by grains of other materials than the local mare basalts, such as KREEP for example, the grains of which were exposed on the surface as long ago as perhaps 4×10^9 yr. This requires that these grains did not undergo degassing such as that seen in 12032 and 12033 to any significant degree. It also requires that the lunar wind at that time had a lower ^{40}Ar content than it had at the time such samples as 61221 and 74240 received their trapped argon components.

Acknowledgment—This research was supported by NASA grant 14-001-169.

REFERENCES

Alexander E. C., Jr., Coscio M. R., Jr., Dragon J. C., Pepin R. O. and Saito K. (1977) K/Ar dating of lunar soils III: Comparison of ^{39}Ar-^{40}Ar and conventional techniques; 12032 and the age of Copernicus. *Proc. Lunar Sci. Conf. 8th*, p. 2725–2740.

Basu A. (1977) Steady state, exposure age, and growth of agglutinates in lunar soils. *Proc. Lunar Sci. Conf. 8th*, p. 3617–3632.

Becker R. H. and Clayton R. N. (1975) Nitrogen abundances and isotopic compositions in lunar samples. *Proc. Lunar Sci. Conf. 6th*, p. 2131–2149.

Becker R. H. and Clayton R. N. (1977) Nitrogen isotopes in lunar soils as a measure of cosmic-ray exposure and regolith history. *Proc. Lunar Sci. Conf. 8th*, p. 3685–3704.

Becker R. H., Clayton R. N. and Mayeda T. K. (1976) Characterization of lunar nitrogen components. *Proc. Lunar Sci. Conf. 7th*, p. 441–458.

Burnett D. S., Drozd R. J., Morgan C. J. and Podosek F. A. (1975) Exposure histories of Bench Crater rocks. *Proc. Lunar Sci. Conf. 6th*, p. 2219–2240.

Clayton R. N. and Becker R. H. (1977) Nitrogen spallation ages of lunar soil samples (abstract). In *Lunar Science VIII*, p. 190–192. The Lunar Science Institute, Houston.

Eberhardt P., Geiss J., Grögler N. and Stettler A. (1973) How old is the crater Copernicus? *The Moon* **8**, 104–114.

Kerridge J. F. (1975) Solar nitrogen: Evidence for a secular increase in the ratio of nitrogen-15 to nitrogen-14. *Science* **188**, 162–164.

Kerridge J. F., Kaplan I. R. and Petrowski C. (1975) Nitrogen in the lunar regolith: Solar origin and effects (abstract). In *Lunar Science VI*, p. 469–471. The Lunar Science Institute, Houston.

Kerridge J. F., Kaplan I. R., Lingenfelter R. E. and Boynton W. V. (1977) Solar wind nitrogen: Mechanism for isotopic evolution. *Proc. Lunar Sci. Conf. 8th*, p. 3773–3789.

Kerridge J. F., Kaplan I. R., Kung C. C., Winter D. A., Friedman D. L. and DesMarais D. J. (1978) Light element geochemistry of the Apollo 12 site. *Geochim. Cosmochim. Acta* **42**, 391–402.

King E. A., Butler J. C. and Carman M. F., Jr. (1971) The lunar regolith as sampled by Apollo 11 and Apollo 12; Grain size analyses, modal analyses and origins of particles. *Proc. Lunar Sci. Conf. 2nd*, p. 737–746.

Laul J. C., Morgan J. W., Ganapathy R. and Anders E. (1971) Meteoritic material in lunar samples: Characterization from trace elements. *Proc. Lunar Sci. Conf. 2nd*, p. 1139–1158.
McKay D. S., Morrison D. A., Clanton U. S., Ladle G. H. and Lindsay J. F. (1971) Apollo 12 soil and breccia. *Proc. Lunar Sci. Conf. 2nd*, p. 755–773.
Marvin U. B., Wood J. A., Taylor G. J., Reid J. B., Jr., Powell B. N., Dickey J. S., Jr. and Bower J. F. (1971) Relative proportions and probable sources of rock fragments in the Apollo 12 soil samples. *Proc. Lunar Sci. Conf. 2nd*, p. 679–699.
Morris R. V. (1978) The maturity of lunar soils: Concepts and more values of I_s/FeO (abstract). In *Lunar and Planetary Science IX*, p. 760–762. Lunar and Planetary Institute, Houston.
Morris R. V., Warner J. L. and McKay D. S. (1977) Nearly pure Apollo 12 KREEP: Soil sample 12023. *Proc. Lunar Sci. Conf. 8th*, p. 2449–2458.
Nyquist L. E., Bansal B. M., Wooden J. L. and Wiesmann H. (1977) Sr-isotopic constraints on the petrogenesis of Apollo 12 mare basalts. *Proc. Lunar Sci. Conf. 8th*, p. 1383–1415.
Wänke H., Wlotzka F., Baddenhausen H., Balacescu A., Spettel B., Teschke F., Jagoutz E., Kruse H., Quijano-Rico M. and Rieder R. (1971) Apollo 12 samples: Chemical composition and its relation to sample locations and exposure ages, the two component origin of the various soil samples and studies on lunar metallic particles. *Proc. Lunar Sci. Conf. 2nd*, p. 1187–1208.
Wasson J. T. and Baedecker P. A. (1972) Provenance of Apollo 12 KREEP. *Proc. Lunar Sci. Conf. 3rd*, p. 1315–1326.
Yaniv A. and Heymann D. (1972) Atmospheric Ar^{40} in lunar fines. *Proc. Lunar Sci. Conf. 3rd*, p. 1967–1980.

Proc. Lunar Planet. Sci. Conf. 9th (1978), p. 1629–1645.
Printed in the United States of America

Solar cosmic ray produced neon and xenon isotopes and particle tracks in feldspars from lunar fines 14148 and 24087

N. B. BHAI, K. GOPALAN, J. N. GOSWAMI, M. N. RAO, and T. R. VENKATESAN

Physical Research Laboratory
Navrangpura, Ahmedabad 380009, India.

Abstract—Feldspars from lunar fines 14148 and 24087 were separated into different grain size fractions and were selectively etched to remove the surficial solar wind components in order to determine the less-abundant solar and galactic cosmic ray (SCR and GCR) spallation noble gas components. The etched feldspar separates were heated stepwise and the resulting gas fractions were analysed mass-spectrometrically to further resolve the SCR and GCR components. The SCR-^{21}Ne in multicomponent neon mixtures in the case of lunar feldspars is estimated. Significant amounts of SCR-^{132}Xe are observed in 14148 feldspars which resulted from high abundances of the target elements (Ba, REE) and high dose of solar flare irradiation as revealed by the track studies. The amounts of SCR-^{132}Xe observed in 24087 feldspars are smaller as they are highly depleted in target elements Ba, even though the percentage of solar flare irradiated grains in this soil is as high as 95%. The solar flare exposure ages calculated by ^{21}Ne and ^{132}Xe methods for 14148 and 24087 soils agree within the limits of experimental error with surface exposure ages based on the fossil track data.

INTRODUCTION

Galactic and solar cosmic rays (GCR and SCR) interact with the lunar surface materials, producing spallation noble gas components having characteristic isotopic compositions. In favourable cases, it is possible to use the characteristic spallation Ne and Xe spectra to distinguish SCR effects from those due to GCR in lunar fines (Rao *et al.*, 1971; Gopalan *et al.*, 1977). The cumulative production of noble gas isotopes due to SCR and GCR spallation processes allows study of temporal changes characterising the dynamic processes in the evolution of the lunar regolith on centimeter to meter scales over long time periods.

A considerable amount of work has been carried out to delineate the GCR effects in lunar soils, but only a few experiments have been tried in noble gas studies to understand the interactions of SCR with lunar surface samples. One of the major impediments to studies of SCR produced noble gases in lunar surface samples is the large amount of solar wind gases present in the lunar fines. To overcome this difficulty we have developed a technique which involves selective etching of grain size separates to remove the major amount of surficial solar wind gases and subsequent step-wise heating of the etched samples to quantitatively resolve the volume correlated SCR and GCR components (Gopalan *et al.*, 1977). Using this technique we have previously demonstrated that the SCR spallation Xe spectrum is different from that due to GCR spallation and calculated the solar flare exposure ages for lunar fines 14163 and 10084 using the SCR produced ^{132}Xe.

In this paper we describe techniques to isolate both Xe and Ne components produced by SCR-spallation in mineral grain size separates from two lunar fines: 14148 from Apollo 14 and 24087 from Luna 24 site. The results on feldspar separates are reported here and the work on pyroxene fractions is in progress.

EXPERIMENTAL PROCEDURES

Feldspar and pyroxene mineral fractions were separated from about 100 mg of Apollo 14 soil 14148 and 70 mg of Luna 24 soil 24087 by heavy liquids (bromoform and methylene iodide). All the chemical operations were carried out in a clean room. The fractions of the mineral separates with density <2.9 g/cc are designated as feldspar and those with density between 2.9 and 3.3 g/cc are designated as pyroxene. These mineral separates were further divided into several size fractions by dry sieving: 40–90 micron (A), 90–200 micron (B), 40–200 micron (C), 200–1000 micron (D). The purity of the feldspar mineral separates was tested by randomly selecting grains from the processed separates and analysing them in a scanning electron microscope (Cambridge S-4/10) equipped with a Kevex energy dispersive X-ray analyser. Ca, Al, Na and Si signals were commonly found, but no Mg signal was detected.

The etching procedures were similar to those described by Gopalan *et al.* (1977). Based on atomic absorption spectrometry of the supernate resulting from the etching of feldspar grains, we estimate that ~1–2 microns of the surface layer of each grain, on the average, were removed by etching. No Mg could be detected in the supernate, indicating high purity of the feldspar separates. Mass-spectrometric and fossil track techniques used are similar to those described earlier by Gopalan *et al.* (1977) and Gopalan and Rao (1976).

RESULTS AND DISCUSSION

Particle track irradiation records

To estimate the percentage of solar flare irradiated grains in each size fraction, particle track analysis was carried out in the unetched feldspar size separates of the two soil samples 14148 and 24087. The result of this analysis is shown in Fig. 1. Grains having a track density gradient and/or high central track densities ($\gtrsim 10^8 cm^{-2}$) are identified as solar flare irradiated grains. Such track records are produced only when the grains reside within the top millimeter of the lunar regolith, and thus serve as a good indicator for identifying solar flare irradiated grains. It can be easily seen from Fig. 1 that both the samples 14148 and 24087 are highly mature soils with about 80% and 90% of the crystals having received solar flare irradiation, respectively. Also, we do not see any significant differences in the track records of the two grain size fractions in the two cases. Thus the track data indicate that the feldspar size separates of both these soil samples have received a high dose of solar flare irradiation during their near surface exposure in the lunar regolith.

Noble gas elemental and isotopic abundances

A comparison of the elemental abundances measured in the different size

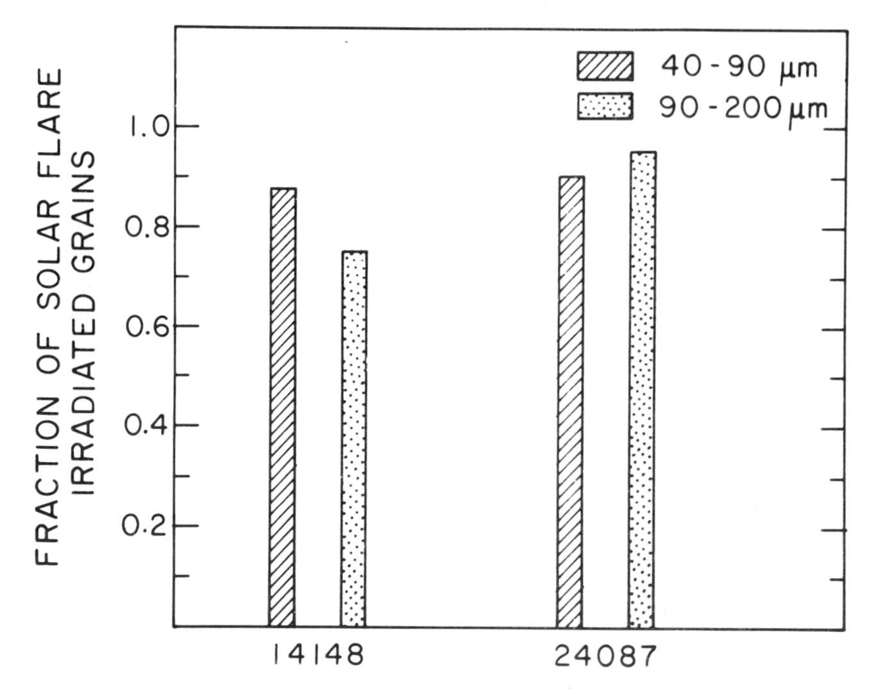

Fig. 1. Fraction of solar flare irradiated feldspar grains for different size separates of lunar soils 14148 and 24087.

fractions of the etched feldspars with those measured by Bogard and Nyquist (1972) in the unetched 14148 bulk fines as well as with the unetched 24087 bulk fines (measured by us) indicate that a major portion of solar wind noble gases ($\gtrsim$60%) was removed by etching from the samples under study.

Neon: The presence of a solar cosmic ray produced Ne component in lunar soils remained ambiguous in spite of repeated attempts made by Walton *et al.* (1974), Frick *et al.* (1975), and Gopalan *et al.* (1977). The Ne isotopic data for 14148 and 24087 etched feldspars obtained in the present work are given in Tables 1 and 2. In Fig. 2 we have plotted the Ne data from step-wise heating experiments on the etched 14148 and 24087 feldspars on a three-isotope diagram. Ideally, a mixture of solar wind and GCR components in the case of high Ca-feldspars is required to move along the mixing line A. Any deviation from this line is an indication of the presence of an additional component—namely SCR spallation. This holds good for a pure component under the assumption that proper blank corrections account for the atmospheric component and the blank-corrected Ne data could be described as a three component system. Any addition of a SCR-produced Ne component with isotopic ratios of $20/22 = 2.0$ and $21/22 = 0.2$, as determined for anorthite by Walton *et al.* (1974), to a point on

Table 1. Concentration and isotopic composition of neon in the two size fractions of etched feldspar separates of 14148 soil*.

Size fraction (wt. in mg)	Temperature °C	^{22}Ne 10^{-8} cc STP/g	$^{20}Ne/^{22}Ne$	$^{21}Ne/^{22}Ne$
A	600	308 ± 31	12.41 ± 0.12	0.0356 ± 0.0005
40–90 micron	1200	918 ± 92	11.98 ± 0.12	0.0641 ± 0.0006
(6.86)	1600	37 ± 4	6.38 ± 0.07	0.4085 ± 0.0041
	Total	1263 ± 97	11.92 ± 0.18	0.0672 ± 0.0042
B	600	418 ± 42	12.45 ± 0.12	0.0391 ± 0.0004
90–200 micron	1200	669 ± 67	11.73 ± 0.12	0.0700 ± 0.0007
(7.65)	1600	20 ± 2	8.17 ± 0.08	0.2807 ± 0.0030
	Total	1107 ± 79	11.94 ± 0.19	0.0621 ± 0.0031

^{22}Ne Blanks: (10^{-8} cc STP): 0.016(600°C); 0.016(1200°C); 0.031(1600°C)

*Errors quoted in this and subsequent tables are $\pm 1\sigma$.

Table 2. Concentration and isotopic composition of neon in the two size fractions of etched feldspar separates of Luna 24 soil 24087.

Size fraction (wt. in mg)	Temperature °C	^{22}Ne 10^{-8} cc STP/g	$^{20}Ne/^{22}Ne$	$^{21}Ne/^{22}Ne$
C	600	351 ± 35	9.972 ± 0.099	0.0790 ± 0.0008
40–200	1200	664 ± 66	10.53 ± 0.11	0.0839 ± 0.0008
micron	1600	12.3 ± 1.5	7.712 ± 0.077	0.4044 ± 0.0041
(3.34)	Total	1027.3 ± 74.7	10.31 ± 0.17	0.0861 ± 0.0043
D	600	231 ± 23	10.11 ± 0.11	0.1371 ± 0.0014
200–1000	1200	284 ± 28	10.15 ± 0.10	0.1098 ± 0.0011
micron	1600	4.2 ± 0.6	5.15 ± 0.06	0.5650 ± 0.0057
(6.16)	Total	519 ± 36.2	10.094 ± 0.016	0.1256 ± 0.0060

^{22}Ne Blanks: (10^{-8} cc STP) 0.028 (600°C); 0.031(1200°C); 0.053(1600°C)

the SW-GCR mixing line results in its movement downwards as shown in Fig. 2 (e.g., the 1600°C point for 14148-B).

Frick *et al.* (1975) have analysed different size fractions of feldspar and pyroxene separates from lunar soil 15421 to determine their Ne isotopic compositions. The diamond shaped symbols in Fig. 2 represent the feldspar data of Frick *et al.* (1975). As these feldspars are from a soil sample, some SCR contribution may be expected in them. Using GCR spallation Ne isotopic composition in meteoritic felspar (20/22 = 0.9, 21/22 = 0.8) as a guide, they have suggested possible presence of SCR-^{21}Ne in their samples. We have used the

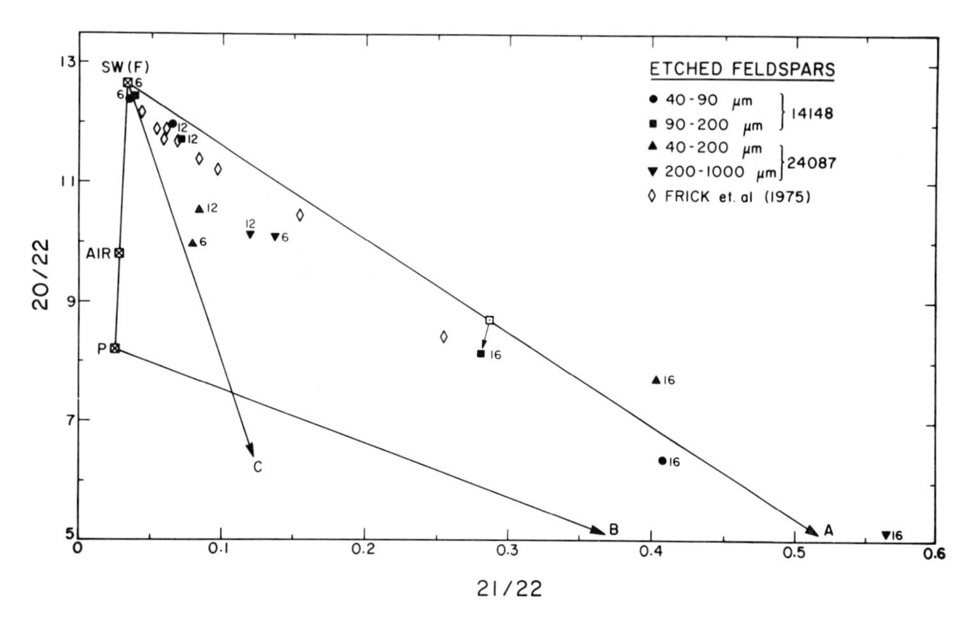

Fig. 2. Neon three isotope correlation diagram for etched feldspar separates of lunar soils 14148 and 24087. P represents planetary neon composition while SW(F) represents the solar wind neon composition as deduced from the 15421 feldspar data of Frick *et al.* (1975). The lines A and C join SW(F) to the GCR and SCR neon compositions respectively: 20/22 = 2; 21/22 = 0.72 (GCR), and 20/22 = 2.0; 21/22 = 0.2 (SCR). Line B joins P to the GCR composition. For the sake of clarity, the error bars are not shown here, but given in the tables.

above mentioned values in fixing the SW-GCR and SW-SCR mixing lines in the 3-isotope diagram for Ne.

For decomposing the Ne-mixtures into SCR(s), GCR(g) and solar wind (t) components, we use the following compositions for the respective end members and apply the 3-isotope Ne-systematics:

$$(20/22)_s = 1, 2 \ \& \ 3; \quad (21/22)_s = 0.2$$
$$(20/22)_g = 0.9; \quad (21/22)_g = 0.8$$
$$(20/22)_t = 12.65; \quad (21/22)_t = 0.032$$

Using the blank corrected $(21/22)_m$ and $(20/22)_m$ ratios and the above compositions we have calculated the fractional amount of SCR, GCR and solar wind Ne present in each temperature fraction of the 14148 and 24087 etched feldspar samples.

The ^{21}Ne excesses attributed to SCR production, as estimated in various temperature fractions of 14148 and 24087 etched feldspars, are given in Table 3. The values for $(20/22)_s = 3$ correspond to a shielding depth of 0 g/cm^2 for the samples, while those for $(20/22)_s = 1$ correspond to a shielding depth of ~7 g/cm^2. An intermediate shielding depth is given by $(20/22)_s = 2$. As the

Table 3. ^{21}Ne excesses due to SCR, estimated in various temperature fractions of 14148 and 24087 etched feldspars for three assumed values of $(20/22)_s$.

Sample/size fraction	Temperature °C	SCR ^{21}Ne (10^{-8} cc STP/g)		
		$(20/22)_s = 1$	$(20/22)_s = 2$	$(20/22)_s = 3$
14148	600	1.25 ± 0.37	1.41 ± 0.44	1.61 ± 0.54
40–90	1200	3.62 ± 1.11	4.06 ± 1.31	4.64 ± 1.59
micron	1600	0.42 ± 0.05	0.47 ± 0.06	0.53 ± 0.07
(A)	Total	5.29 ± 1.17	5.94 ± 1.38	6.78 ± 1.68
14148	600	0.84 ± 0.48	0.95 ± 0.58	1.08 ± 0.70
90–200	1200	4.99 ± 0.91	5.61 ± 1.07	6.40 ± 1.28
micron	1600	0.30 ± 0.03	0.33 ± 0.04	0.38 ± 0.04
(B)	Total	6.13 ± 1.03	6.89 ± 1.22	7.86 ± 1.46
24087	600	15.15 ± 1.55	17.02 ± 1.74	19.42 ± 2.00
40–200	1200	19.39 ± 2.05	21.79 ± 2.32	24.87 ± 2.67
micron	1600	—	—	—
(C)	Total	34.54 ± 2.57	38.81 ± 2.90	44.29 ± 3.34
24087	600	4.74 ± 0.53	5.33 ± 0.60	10.51 ± 1.11
200–1000	1200	8.19 ± 0.85	9.21 ± 0.96	6.08 ± 0.70
micron	1600	—	—	—
(D)	Total	12.93 ± 1.00	14.54 ± 1.13	16.59 ± 1.31

feldspars analysed are essentially Mg-free, a value of $(21/22)_s = 0.2$ is used. As shown in Table 3, significant amounts of SCR-^{21}Ne seem to be present in all the temperature fractions of 14148(A) and (B) samples and the 600°C and 1200°C temperature fractions of 24087(C) and (D) samples. These values are valid for $(20/22)_t = 12.65$, $(21/22)_t = 0.032$ as obtained by Frick *et al.* (1975). But if we use a value of 12.5 for $(20/22)_t$, the SCR-^{21}Ne contents of 14148 samples become marginal. However, even with this low value, considerable amounts of SCR-^{21}Ne can be observed in the 24087(C) and (D) feldspar samples. This problem is currently under further study in our laboratory. The SCR-^{21}Ne contents in different etched feldspar separates are used for calculating the solar flare exposure ages of these soils, as discussed later.

Xenon: The elemental and isotopic abundances of Xe in different size fractions of etched feldspar separates of 14148 and 24087 samples are given in Tables 4 and 5. The Xe isotopic ratios given in these tables are plotted in several 3-isotope correlation diagrams (Figs. 3–6). The analytical methods used for isolation of SCR-produced ^{129}Xe, ^{131}Xe and ^{132}Xe are similar to those discussed earlier for the case of 14163 and 10084 (Gopalan *et al.*, 1977). The basic principle is the same for Xe as for Ne: SCR contributions are identified by departure of measured compositions from the mixing line connecting postulated trapped and GCR compositions. The use of ^{128}Xe for normalization instead of ^{130}Xe does not alter the conclusions reached here.

Table 4. Concentration and isotopic composition of xenon for the two size fractions of etched feldspar separates of 14148 soil.

Size fraction	Temperature °C	^{132}Xe 10^{-12}cc STP/g	$^{124}Xe/^{132}Xe$	$^{126}Xe/^{132}Xe$	$^{128}Xe/^{132}Xe$	$^{129}Xe/^{132}Xe$	$^{130}Xe/^{132}Xe$	$^{131}Xe/^{132}Xe$	$^{134}Xe/^{132}Xe$	$^{136}Xe/^{132}Xe$
A 40–90 micron	600	816 ± 82	0.0514	0.0388	0.1068	1.091	0.1402	0.887	0.3975	0.2991
			±0.0013	±0.0010	±0.0021	±0.022	±0.0035	±0.018	±0.0080	±0.0060
	1200	4692 ± 470	0.0476	0.0806	0.1897	1.113	0.2198	1.184	0.3520	0.2802
			±0.0012	±0.0020	±0.0038	±0.022	±0.0055	±0.024	±0.0070	±0.0056
	1600	1538 ± 704	0.0697	0.1278	0.2482	1.170	0.2453	1.334	0.3440	0.2853
			±0.0017	±0.0032	±0.0050	±0.023	±0.0061	±0.027	±0.0069	±0.0057
	Total	7046 ± 850	0.0523	0.0861	0.1925	1.123	0.2161	1.182	0.355	0.284
			±0.0025	±0.0039	±0.0066	±0.039	±0.0089	±0.040	±0.013	±0.010
B 90–200 micron	600	469 ± 47	0.0237	0.0314	0.1091	1.183	0.1609	0.903	0.4334	0.2835
			±0.0006	±0.0008	±0.0022	±0.024	±0.0032	±0.018	±0.0087	±0.0057
	1200	5593 ± 560	0.0348	0.0583	0.1611	1.102	0.2065	1.084	0.3387	0.2635
			±0.0009	±0.0015	±0.0032	±0.022	±0.0041	±0.022	±0.0068	±0.0053
	1600*	3221 ± 1610	0.0293	0.0552	0.1454	1.086	0.1857	1.028	0.3639	0.2399
			±0.0007	±0.0014	±0.0031	±0.022	±0.0037	±0.021	±0.0073	±0.0048
	Total	9283 ± 1705	0.0323	0.0469	0.1533	1.101	0.1970	1.055	0.352	0.2563
			±0.0013	±0.0022	±0.0050	±0.039	±0.0064	±0.035	±0.013	±0.0091

^{132}Xe Blanks: (10^{-12}cc STP) 0.47(600°C); 0.34(1200°C); 2.26(1600°C)

*Air contamination is suspected. The applied corrections (using $^{136}Xe_{air} = 0.66\ ^{136}Xe_m$) may not be adequate.

Table 5. Concentration and isotopic composition of xenon in two size fractions of etched feldspar separates of Luna 24 soil 24087.

Size fraction	Temperature °C	^{132}Xe 10^{-12}cc STP/g	$^{124}Xe/^{132}Xe$	$^{126}Xe/^{132}Xe$	$^{128}Xe/^{132}Xe$	$^{129}Xe/^{132}Xe$	$^{130}Xe/^{132}Xe$	$^{131}Xe/^{132}Xe$	$^{134}Xe/^{132}Xe$	$^{136}Xe/^{132}Xe$
C 40–200 micron	600	482 ± 48	–	–	0.0911	1.074	0.1801	0.891	0.3969	0.2733
					±0.0020	±0.022	±0.0045	±0.018	±0.0079	±0.0055
	1200	3061 ± 306	0.0080	0.0095	0.0873	1.034	0.1642	0.781	0.3761	0.3112
			±0.0001	±0.0002	±0.0017	±0.021	±0.0040	±0.016	±0.0074	±0.0062
	1600	976 ± 98	0.0179	0.0246	0.1101	1.082	0.1788	0.905	0.3715	0.3109
			±0.0004	±0.0006	±0.0022	±0.022	±0.0045	±0.018	±0.0074	±0.0062
	Total	4519 ± 325	0.0093	0.0117	0.0926	1.049	0.1690	0.819	0.378	0.3071
			±0.0004	±0.0006	±0.0034	±0.022	±0.0075	±0.030	±0.013	±0.0104
D 200–1000 micron	600	162 ± 16	–	–	0.0964	1.042	0.1848	0.911	0.3564	0.3093
					±0.0019	±0.021	±0.0037	±0.018	±0.0071	±0.0062
	1200	1540 ± 154	0.0153	0.0235	0.1104	1.057	0.1801	0.902	0.3560	0.2950
			±0.0004	±0.0006	±0.0022	±0.021	±0.0036	±0.018	±0.0071	±0.0059
	1600	342 ± 34	0.0234	0.0386	0.1386	1.140	0.2050	0.982	0.3601	0.3397
			±0.0006	±0.0010	±0.0028	±0.023	±0.0041	±0.019	±0.0072	±0.0068
	Total	2044 ± 159	0.0154	0.0242	0.1139	1.070	0.1846	0.916	0.357	0.304
			±0.0007	±0.0012	±0.0040	±0.037	±0.0066	±0.032	±0.012	±0.011

^{132}Xe Blanks (10^{-12}cc STP) 0.76(600°C); 0.84(1200°C); 1.53(1600°C)

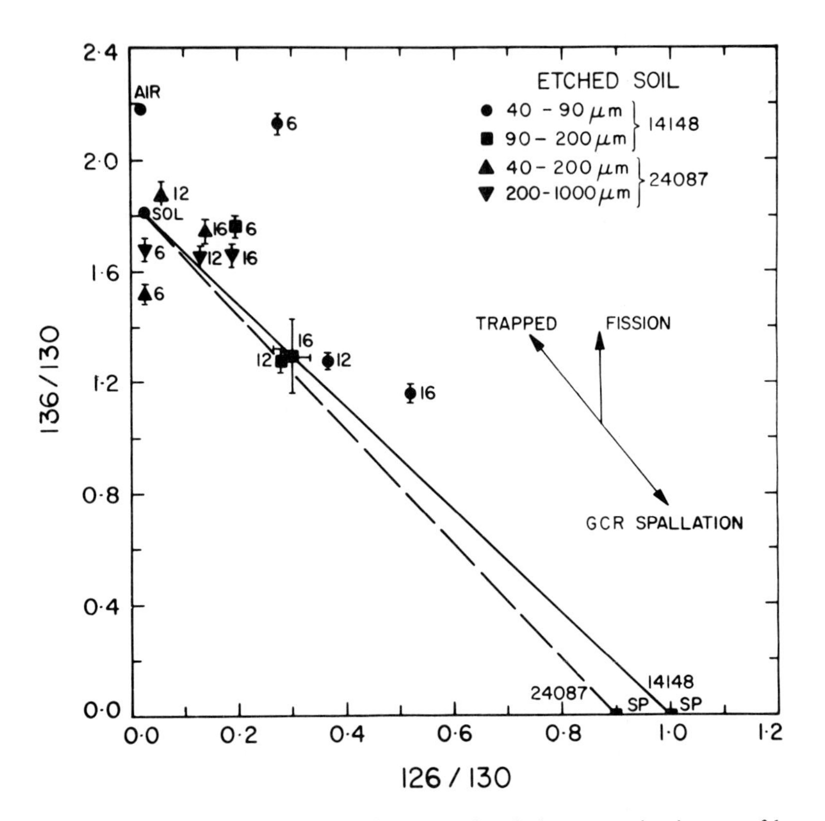

Fig. 3. Three isotope xenon diagram for computing fission correction in case of lunar feldspars from 14148 and 24087 etched samples, assuming $(136/130)_{Sp} = 0$. The choice of the trapped and spallation components is discussed in the text. The GCR spallation end points for these soils are taken from Gopalan *et al.* (1977). Numerals close to the data points are release temperatures in hundreds of degrees centigrade. Deviations from the mixing line in the direction of excess ^{136}Xe are used to calculate the amount of $(^{136}Xe)_f$.

The ^{132}Xe and ^{129}Xe excesses due to SCR-production, as estimated in various temperature fractions of 14148 and 24087 etched feldspars, are shown in Table 6. The SCR-^{132}Xe and ^{129}Xe contents seem to be significant in the 14148(A) and (B) samples, whereas they are marginal in the 24087(C) and (D) samples. The low SCR-contents in the case of 24087 samples seem to be due to the low target element (Ba) abundance (~40 ppm in Luna 24 soil samples: Ma and Schmitt, 1977), even though the samples contain large percentages of solar flare irradiated grains (~90%). In the case of 14148 samples, the Ba content is about 750–800 ppm (Lindstrom *et al.*, 1972).

If we consider a GCR spallation ratio of 1.1 for 126/130 instead of 1.0 as used in Figs. 4 and 5, the SCR-excesses estimated for 24087 samples vanish. However, in both the 14148(A) and (B) samples the 600°C temperature fractions clearly

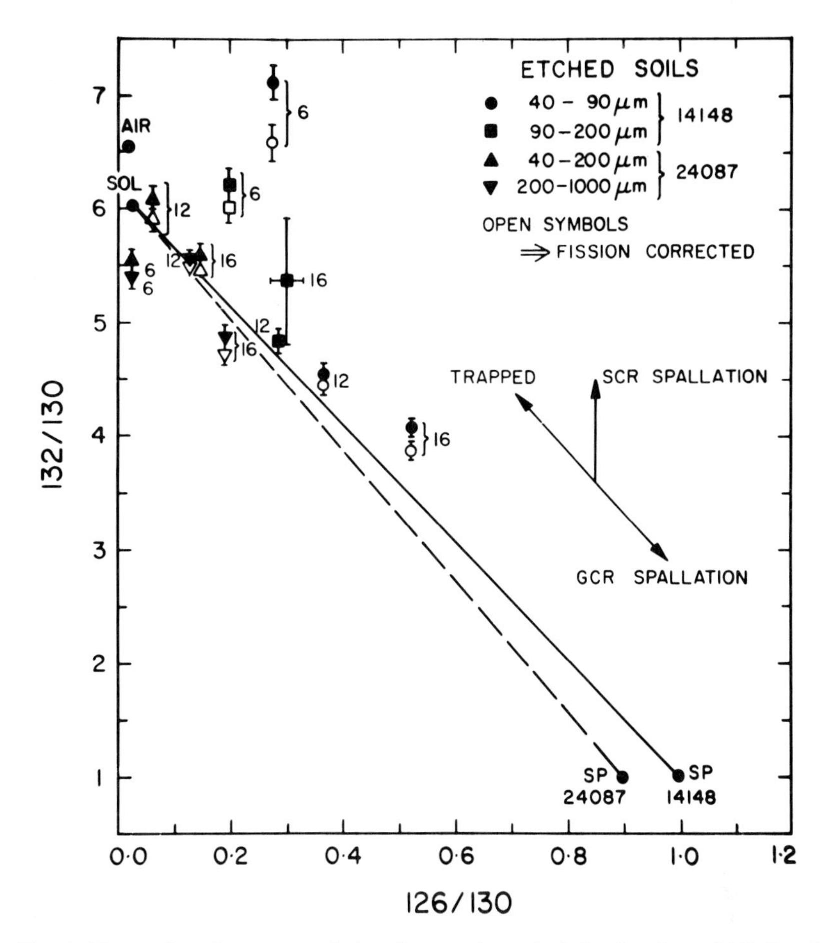

Fig. 4. Xenon three isotope correlation diagram for etched size fractions of 14148 and 24087 feldspars. The isotopic ratios of 132/130 after fission correction are indicated by open symbols. Deviations from the line in the direction of ^{132}Xe are used to compute the concentration of ^{132}Xe due to SCR-production. Other nomenclature is same as in Fig. 3.

show ^{132}Xe excesses. Also, in the case of 1600°C temperature fraction of 14148(A), a SCR-^{132}Xe content of 70×10^{-12} cc STP/g is detectable. But the 1200°C temperature fraction of 14148(A) and the 1200°C and 1600°C temperature fractions 14148(B) are zero. The total amounts of SCR-produced ^{132}Xe in 14148(A) and (B) samples could be considered to be similar within the limits of experimental error.

As can be seen from Figs. 4 and 5 the gas release pattern of SCR-^{129}Xe is similar to that of SCR-^{132}Xe in different size fractions of both of these etched samples. The SCR-^{129}Xe contents in 14148 samples and in 24087 samples are given in Table 6. The earlier study of etched 14163 and 10084 soil samples (Gopalan *et al.*, 1977) also showed similar SCR-^{129}Xe release patterns.

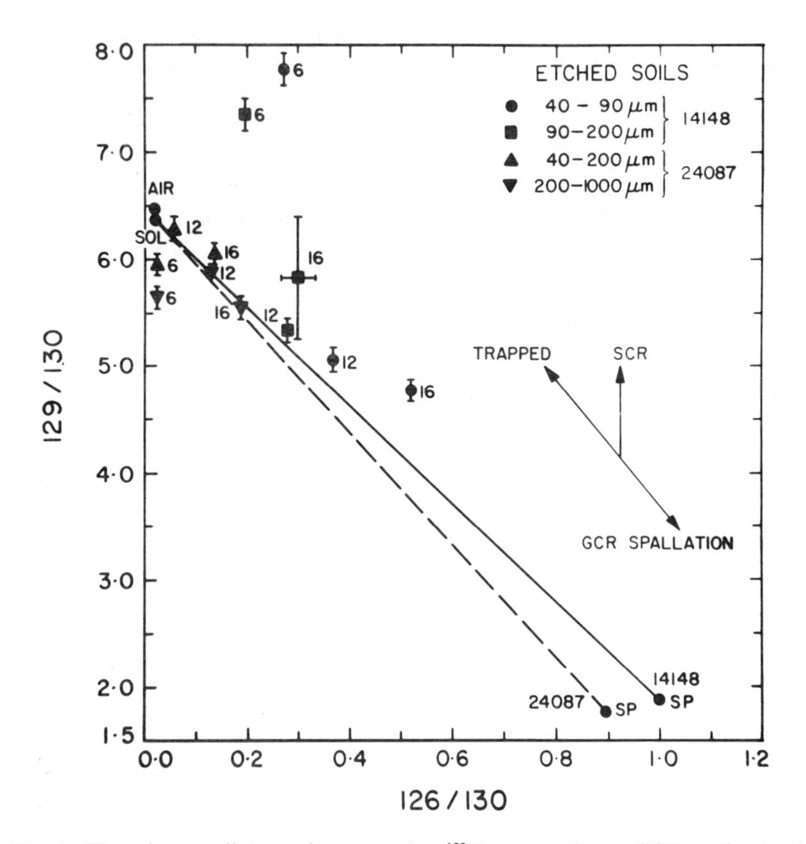

Fig. 5. Three isotope diagram for computing ^{129}Xe excess due to SCR-production in feldspar separates from 14148 and 24087. This diagram shows that SCR-^{129}Xe excesses exist irrespective of whether trapped gas is represented by solar or air composition. Note the similarity in the release patterns of ^{132}Xe and ^{129}Xe excesses in Figs. 4 and 5.

The estimation of SCR-produced ^{131}Xe is difficult (Gopalan *et al.*, 1977) because there is no direct way of finding the amount of $^{131}Xe_n$ due to (n,γ) capture on ^{130}Ba. Furthermore, $^{131}Xe_{sp}$ yields are not accurately known. In Fig. 6, the total $^{131}Xe_{sp}$ + $^{131}Xe_n$ is treated as a single component, using a value of (131/130) = 5.5 (Gopalan *et al.*, 1977). The ^{131}Xe excesses above the SW-"GCR + n" mixing line are attributed to an SCR-contribution. The weighted average SCR-^{131}Xe of 14148 is 355 × 10^{-12}cc STP/g. In 24087-C and -D the SCR-^{131}Xe is difficult to estimate because the experimental points lie close to the mixing line. The relatively small ^{131}Xe excesses for 24087 samples are due to the low Ba-content of these samples and may also be due in part to the difficulty in fixing the 131/130 ratio to represent the "GCR + n" component.

We have plotted 126/130 versus 128/130 (figure not shown here) for all the temperature fractions of 14148 and 24087 samples studied here. We considered GCR spallation composition as one end member and SUCOR (Podosek *et al.*,

Table 6. The ^{132}Xe and ^{129}Xe excesses due to SCR as estimated in various temperature fractions of etched 14148 and 24087 feldspars.

Sample/size fraction	Temperature °C	$^{132}Xe_s$	$^{129}Xe_s$
		(10^{-12}cc STP/g)	
	600	212 ± 22	276 ± 20
14148	1200	198 ± 198*	267 ± 175*
40–90 micron	1600	152 ± 98	253 ± 93
(A)	Total	562 ± 222	796 ± 199
	600	67 ± 12	134 ± 14
14148	1200	165 ± 165*	185 ± 185*
90–200 micron	1600	454 ± 337*	455 ± 347*
(B)	Total	686 ± 375	774 ± 393
	600	–	–
24087	1200	72 ± 40*	43 ± 40*
40–200 micron	1600	26 ± 19*	50 ± 22*
(C)	Total	98 ± 44	93 ± 46
	600	–	–
24087	1200	25 ± 25*	18 ± 18*
200–1000	1600	–	6 ± 6*
micron	Total	25 ± 25	24 ± 24
(D)			

*If we use $(126/130)_{sp} = 1.1$, these excesses are negligible or zero.

1971) and air compositions as other end members. Within the experimental error, the data points fall on the line joining SUCOR to the spallation end point. This indicates that atmospheric contamination is not a serious problem in the temperature fractions studied here.

SCR spallation xenon spectrum

The weighted average SCR-129 and -131 excesses, normalized to 132 ≡ 1.00, for both A and B size fractions of 14148 feldspars are shown in Fig. 7. In this figure, the SCR-spallation Xe spectra, obtained earlier from the studies on 14163 and 10084 etched soils (Gopalan *et al.*, 1977) are also included. For comparison, the low energy Xe spallation spectra, obtained from 38 MeV and 50 MeV proton irradiation of Ba targets by Kaiser (1977) and Hussain (pers. comm., 1976) respectively are also shown in this figure.

All the SCR-^{129}Xe yields normalized to ^{132}Xe fall between the limits set by 38 MeV and 50 MeV proton irradiations on Ba, indicating that these lunar soils were irradiated by solar cosmic rays of average energy in this range. The ^{131}Xe

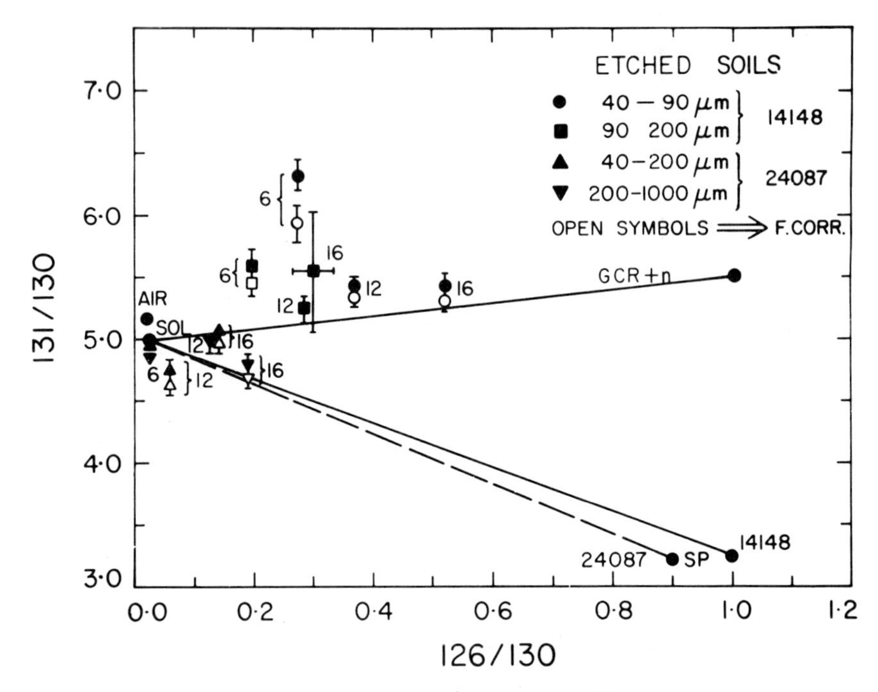

Fig. 6. Three isotope correlation diagram for calculating ^{131}Xe excess due to SCR-production in feldspar separates from 14148 and 24087. The compositions of the end members in this diagram are discussed in the text. For comparison, we show pure GCR spallation as represented by 'SP' (Marti *et al.*, 1973). Deviations from the "GCR + n" line in the direction of ^{131}Xe excess are used to find out the amount of SCR-^{131}Xe.

yields of 14148 are in good agreement with our earlier results on 14163 and 10084 samples. The ^{131}Xe yields, due to SCR-spallation in all the samples discussed here, are much lower than the ^{131}Xe yields of Kaiser (1977) and Hussain (pers. comm., 1976). This may be partly due to the over-correction for $^{131}Xe_{(GCR + n)}$, but such large differences are difficult to account for by these corrections only. Detailed study of low energy proton irradiations on Ba can hopefully shed light on these discrepancies.

SCR xenon and neon production rates

We have analysed several excitation functions for (p,α), (p,αn), (p,2p2n), (p,2p3n) and similar (p,xpyn) reactions where 3 to 5 neucleon evaporation is involved. Most of these excitation functions were determined experimentally (or partly extrapolated) by bombarding medium to heavy elements with protons in the energy interval 10–100 MeV (Sachdev *et al.*, 1967; Kantello and Hogan, 1976; Keller *et al.*, 1973). The general shape of the excitation functions for the (p,α) and (p,αn) reactions shows that the curve rises steeply from the threshold to

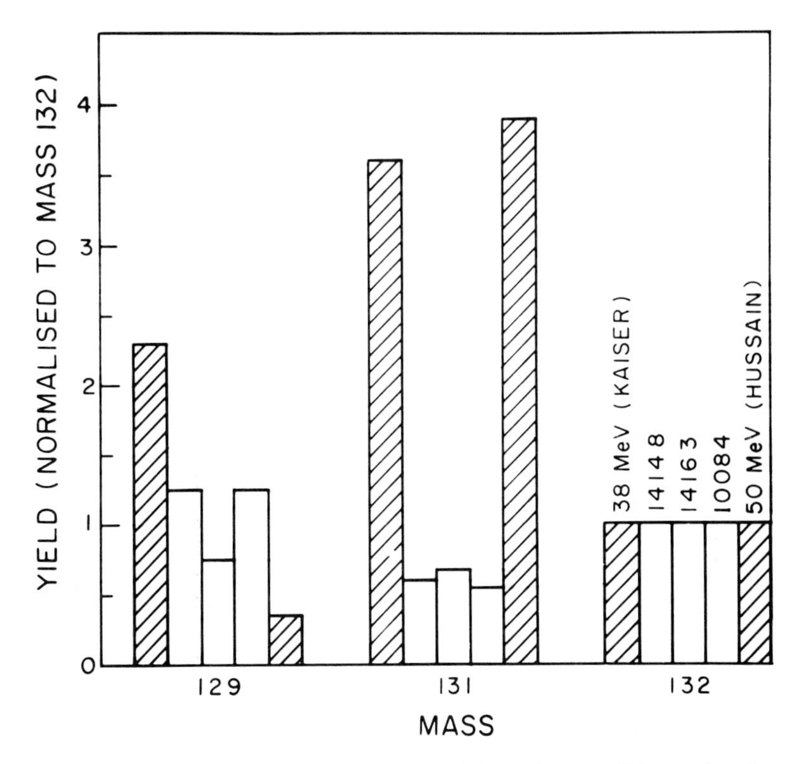

Fig. 7. Comparison of the spallation spectra as deduced from the SCR-produced xenon isotopic excesses (weighted average) observed in the etched feldspar separates from 14148 along with the earlier results for 10084 and 14163. The results obtained from 38 MeV and 50 MeV proton irradiation on Ba (Kaiser, 1977; Hussain, pers. comm., 1976) are also shown for comparison. The low yields for ^{131}Xe are probably due to over-correction of "(GCR + neutron capture)" component.

a proton energy of about 20 to 30 MeV, where the cross-section is maximum. Later it passes through a minimum at about 40–50 MeV and then starts increasing as a function of energy up to about 80 MeV, after which it gradually levels off. This behaviour is typical for several low energy reactions which involve the emission of an α particle and a proton and/or a neutron. For some of these reactions on Sr and Y, as well as on Rb, Zr and Nb isotopes (Sachdev *et al.*, 1967), it is found that σ_{max} is about 30–40 mb and occurs at about 20 MeV. For (p,α2n) reactions, the thresholds are higher and σ_{max} occurs at about 30–35 MeV and has a value of about 50 mb. After 50 MeV, other competing reactions are significant and the cross-sections reach a steady value of about 40 mb.

The cross-sections measured by Kaiser (1977) for several proton-induced reactions on Ba leading to Xe production at various energies forms one of the important sets of information regarding Xe spallation systematics available to date. However, his data for the 75 and 38 MeV irradiations suffer from "leaching" effects and "air contamination" and "monitor" problems respectively.

Work is presently in progress to redo some of these irradiations, particularly at low energies. In view of the problems associated with the direct cross-section measurements, we have estimated an average cross-section of about 20 mb for the production of ^{132}Xe from natural Ba by bombardment of protons in the energy interval of 15 to 75 MeV, based on the above systematics of several low energy proton induced reactions in medium heavy elements in the neighbourhood of Ba. Using a solar flare proton spectra given by $J_{E\geq 10\ MeV} = 100$ protons/cm^2.sec(4 π) and $R_o = 100$ MV, as suggested by Reedy and Arnold (1972), and Lal (1972), we calculate an average ^{132}Xe production rate of 5×10^{-15}cc STP/10^{-6}g Ba per m.y. for lunar soil with 2 π irradiation geometry.

As we are dealing with feldspar mineral separates from 14148 and 24087 soil samples, we must use production rates of SCR-^{21}Ne pertaining to this mineral. Considering the measured cross-sections by Walton *et al.* (1976), we use a production rate for SCR-^{21}Ne of $(0.3 \pm 0.1) \times 10^{-8}$cc STP/g sample per m.y. for lunar feldspars (Reedy, pers. comm., 1976).

"Solar flare" exposure ages of lunar fines 14148 and 24087

Arrhenius *et al.* (1971) have proposed a model to obtain integrated surface exposure ages of a soil layer in the lunar regolith based on the observed track records in individual grains in the layer. A determination of a solar flare exposure age of a lunar soil layer from track data alone is not possible because the total fluence of solar cosmic rays received by a soil sample, as well as the average track production rate for a soil layer in the top few millimeters of the lunar regolith, cannot be meaningfully estimated. However, the surface exposure ages of a soil layer based on the quartile track density method (Arrhenius *et al.*, 1971) could be considered as an upper limit to the solar flare exposure age. The solar flare exposure age of a soil sample based on the observed content of the solar cosmic ray produced Ne and Xe isotopes can be calculated meaningfully provided the production rates for these isotopes are correctly established. The advantage with the "average" production rate for SCR-^{21}Ne and ^{132}Xe, in the first centimeter or so in the lunar regolith, is that it is not so depth-sensitive as in the case of tracks.

For the 14148(A) and (B) as well as 24087(C) and (D) samples, the SCR-^{21}Ne and ^{132}Xe are given in Tables 3 and 6 respectively. Using the production rates discussed above, the solar flare exposure ages, based on SCR-^{21}Ne and SCR-^{132}Xe [calculated using $(126/130)_{sp} = 1.1$] are given in Table 7. For comparison, the particle track based surface exposure ages of these lunar soils are also given. The exposure ages of soil 14148 determined from SCR-^{21}Ne show good agreement with those determined from SCR-^{132}Xe, within experimental errors. These ages reasonably agree, in turn, with the track based surface exposure ages which could be treated as an upper limit. With $(126/130)_{sp} = 1$, the exposure ages for soil 14148 are much higher than the track based exposure age. A firm conclusion regarding this discrepancy can be reached

Table 7. Solar flare exposure ages of lunar soils.

Sample	Solar flare exposure age* (m.y.)		Surface exposure age (m.y.) from track data†
	^{132}Xe**	^{21}Ne	
14148-A	53 ± 16	18 ± 7	20
14148-B	15 ± 5	22 ± 9	
24087-C	—	128 ± 43	100
24087-D	—	48 ± 16	

*See text for production rate.
**Based on $(126/130)_{sp}$ = 1.1.
†Based on Arrhenius *et al.* (1971) model.

only after better experimental data on xenon spallation systematics and production cross-sections of interest become available.

Summary

We have used the method of selective etching and stepwise heating of mineral separates from lunar soils to isolate SCR-produced noble gas components and to deduce solar flare exposure ages of lunar soils. On this basis we have been able to obtain the following results:

1) SCR-produced ^{21}Ne and ^{132}Xe can be identified in lunar feldspars exposed to solar flare irradiation for time periods >10 m.y.
2) The solar flare exposure ages of lunar soils have been calculated by using the production rate of SCR-produced ^{21}Ne and ^{132}Xe based on known solar flare proton energy spectra and known reaction cross-sections in case of Ne and estimated cross-sections in case of xenon. The ^{21}Ne and ^{132}Xe exposure ages are in agreement with each other (within the limits of experimental error) and, in turn, agree with particle track based surface exposure ages.

We would like to point out that this method has the following features:

(i) Being a stable isotope method, it can be used for calculating the integrated solar flare exposure ages of lunar samples in the time scale >5 m.y., in which the conventional methods using ^{26}Al and ^{53}Mn isotopes are inapplicable due to saturation effects.
(ii) In the time interval 10–50 m.y., this method is complementary to the particle track method for determining surface exposure ages of lunar soils. For higher ages, this is the only method, we believe, that can be used to determine the integrated near surface (0–1 cm) exposure ages of lunar soils.

(iii) A combined study of both the GCR- and SCR- produced noble gas isotopes and particle tracks provides us with a unique way of studying the dynamics of the lunar regolith in two extreme depth scales, 0–1 cm and submeter to meter, over long periods of time.

Acknowledgments—We thank Prof. D. Lal and Prof. K. Marti for valuable discussions regarding this work. The assistance rendered by Mr. J. R. Trivedi, P. Sharma and J. T. Padia during this work is sincerely appreciated. We thank Prof. F. Podosek and Prof. Ch. Hohenberg and the three reviewers of this paper for many valuable comments and suggestions regarding this work. We are grateful to NASA and the Soviet Academy of Sciences for making the precious Apollo and Luna samples available for our studies.

REFERENCES

Arrhenius G., Liang S., MacDougall D., Wilkening L., Bhandari N., Bhat S., Lal D., Rajagopalan G., Tamhane A. S. and Venkatavardan V. S. (1971) The exposure history of Apollo 12 regolith. *Proc. Lunar Sci. Conf. 2nd*, p. 2583–2589.

Bogard D. D. and Nyquist L. E. (1972) Noble gas studies on regolith materials from Apollo 14 and 15. *Proc. Lunar Sci. Conf. 3rd*, p. 1797–1819.

Frick U., Baur H., Ducati H., Funk H., Phinney D. and Signer P. (1975) On the origin of helium, neon, and argon isotopes in sieved mineral separates from an Apollo 15 soil. *Proc. Lunar Sci. Conf. 6th*, p. 2097–2129.

Gopalan K., Goswami J. N., Rao M. N., Suthar K. M. and Venkatesan T. R. (1977) Solar cosmic ray produced noble gases and tracks in lunar fines 10084 and 14163. *Proc. Lunar Sci. Conf. 8th*, p. 793–811.

Gopalan K. and Rao M. N. (1976) Rare gases in Bansur, Udaipur and Madhipura chondrites. *Meteoritics* **11**, 131–136.

Kaiser W. A. (1977) The excitation functions of Ba (p,X) ^{M}Xe (M = 124–136) in the energy range 38–600 MeV: The use of "cosmogenic" xenon for estimating "burial" depths and "real" exposure ages. *Phil. Trans. Roy. Soc. London A.* **285**, 337–362.

Kantelo M. V. and Hogan J. J. (1976) Charged particle emission in reactions of ^{90}Zr with 10–86 MeV protons. *Phys. Rev. C.* **14**, 64–75.

Keller K. A., Lange J., Munzel H. and Pfenning G. (1973) Q-values and excitation functions of nuclear reactions. *Landolt-Bornstein* (New Series) **5**. Springer-Verlag, Berlin.

Lal D. (1972) Hard rock cosmic ray archaeology. *Space Sci. Rev.* **14**, 3–102.

Lindstrom M. M., Duncan A. R., Fruchter J. S., McKay S. M., Stoeser J. W., Goles G. G. and Lindstrom D. J. (1972) Compositional characteristics of some Apollo 14 clastic materials. *Proc. Lunar Sci. Conf. 3rd*, p. 1201–1214.

Ma M.-S. and Schmitt R. A. (1977) Luna 24 soils: A chemical study (abstract). In *Papers Presented to the Conference on Luna 24.* p. 98–101. The Lunar Science Institute, Houston.

Marti K., Lightner B. D. and Osborn T. W. (1971) Krypton and xenon in some lunar samples and the age of North Ray Crater. *Proc. Lunar Sci. Conf. 4th*, p. 2037–2048.

Podosek F. A., Huneke J. C., Burnett D. S., Wasserburg G. J. (1971) Isotopic composition of xenon and krypton in the lunar soil and in the solar wind. *Earth Planet. Sci. Lett.* **10**, 199–216.

Rao M. N., Gopalan K., Venkatravardan V. S. and Wilkening L. (1971) Solar flare effects in lunar xenon. *Nature Phys. Sci.* **233**, 114–117.

Reedy R. C. and Arnold J. R. (1972) Interaction of solar and galactic cosmic ray particles with the moon. *J. Geophys. Res.* **77**, 537–555.

Sachdev D. R., Porile N. T. and Yaffe L. (1967) Reactions of ^{88}Sr with protons of energies 7–85 MeV. *Can. J. Chem.* **45**, 1149–1160.

Walton J. R., Heymann D., Jordan J. L. and Yaniv A. (1974) Evidence for solar cosmic ray proton-produced neon in fines 67701 from the rim of North Ray Crater. *Proc. Lunar Sci. Conf. 5th*, p. 2045–2060.

Walton J. R., Heymann D., Yaniv A., Edgerley D. and Rowe M. W. (1976) Cross sections for He and Ne isotopes in natural Mg, Al and Si, He isotopes in CaF_2, Ar isotopes in natural Ca, and radionuclides in natural Al, Si, Ti, Cr, and stainless steel induced by 12 to 45 MeV protons. *J. Geophys. Res.* **81**, 5689–5699.

Proc. Lunar Planet. Sci. Conf. 9th (1978), p. 1647–1654.
Printed in the United States of America

Carbon-14 in lunar soil and in meteorites*

E. L. FIREMAN

Center for Astrophysics
Harvard College Observatory and Smithsonian Astrophysical Observatory
Cambridge, Massachusetts 02138

Abstract—^{14}C is measured in grain-size fractions of lunar soil, 10084, and in the meteorites Bruderheim, Allan Hills #5, Allan Hills #6, and Allan Hills #8. The Allan Hills meteorites are from an Antarctic site where a large number of meteorites have been found. The measurements for the lunar soil fractions provide information on the cosmic-ray and solar-wind bombardment of the moon. The measurements for Bruderheim, a recent fall, serve as a comparison for the studies with the lunar material and Antarctic meteorites. ^{14}C is released at temperatures below melting for the small soil grains ($<74\ \mu$) but not from the large soil grains ($>74\ \mu$) or from the meteorites. The below-melting ^{14}C contents increase with decreasing grain size in a manner similar to the solar-wind rare-gas contents. The ^{14}C released above melting temperatures is independent of the grain size and the activity is half that measured in the Bruderheim meteorite. The lunar ^{14}C results are consistent with the idea that the below-melting ^{14}C is solar-wind implanted and the above-melting ^{14}C is produced by cosmic-ray spallations. No ^{14}C is observed in Allan Hills #5, Allan Hills #6, or in Allan Hills #8; the ^{14}C limits correspond to fall times of more than 25,000 yrs. ago.

1. INTRODUCTION

Several studies indicate that there is implanted ^{14}C in lunar surface material. (1) Begemann *et al.* (1972) measured excess ^{14}C on the top of rock 12053; they found 72 dpm kg^{-1} in the top 0.5 cm compared to 33 and 30 dpm kg^{-1} at lower depths. The depth profile of ^{14}C in 12053 could not be interpreted as galactic and solar cosmic-ray spallations unless solar cosmic rays were much more intense during the past 10,000 yrs. than during the past 30 yrs. (2) Fireman *et al.* (1977) found that ^{14}C was released in 600–1000°C heatings of lunar surface soils but not in similar heatings of subsurface soils. The carbon contents of lunar soils are large compared to those in lunar rocks. Chang *et al.* (1972, 1973), Gibson and Moore (1973), and Simoneit *et al.* (1973) showed that the carbon compounds in lunar soils are released principally between 500 and 1200°C and interpreted both the amounts and the temperature-release patterns to result from carbon implantation. There are two processes that can account for the carbon: solar-wind implantation and carbon condensation from meteor impacts. The amount of ^{14}C condensed on lunar soil from the impact vapors is negligible. Fireman *et al.* (1977) therefore interpreted the 600–1000°C ^{14}C from soils as solar-wind implanted. (3) Skim soil (0 to ~1 cm depth) had more 600–1000°C ^{14}C than scooped soil (0 to ~5 cm depth) according to Fireman *et al.* (1977) and the

*Center for Astrophysics. Preprint Series No. 953.

excess in the skim soil was approximately equal to the excess ^{14}C on the top of rock 12053.

It is desirable to supplement these ^{14}C studies with a study of the grain-size dependence of the ^{14}C in lunar soil. Spallation-produced ^{14}C should be independent of grain size, whereas solar-wind-implanted ^{14}C is probably not. In soil 10084, Eberhardt *et al.* (1970) found that the solar-wind-implanted rare-gas isotope contents increased as the grain size decreased. Since soil 10084 is a surface soil, the first contingency material collected on the moon, and since 10084 soil had been thoroughly studied by physical, chemical, and isotopic methods, it was an obvious choice for a ^{14}C grain-size study. King *et al.* (1971) sieved 14.3 g of 10084 into <10, 10–30, 30–37, 37–74, 74–125, and >125 μ fractions, and ~1 g samples of these size fractions were available for our grain-size ^{14}C experiment. The measurement of ^{14}C in 1-g samples of lunar material requires long counting time (~several weeks) with small (~1 cm^3) low-level counters.

We also measured ^{14}C in meteorites with the same technique used for the lunar samples for comparison of the current technique with previous work. The chondrite, Bruderheim, is a good meteorite to use for comparison purposes: its fall date, March 4, 1960, is known; and its ^{14}C content has been measured to be 63 ± 5 and 55.8 ± 3.0 dpm kg^{-1} by Goel and Kohman (1962) and by Suess and Wänke (1962), respectively. It is important to measure ^{14}C in additional meteorites. An interesting group of meteorites whose ^{14}C contents have not been measured are the recently found Antarctic meteorites. W. A. Cassidy kindly gave us ~10-g samples of the Allan Hills #5, #6, #7, and #8 meteorites (Cassidy *et al.*, 1977). L. Rancetelli *et al.* (pers. comm., 1977) found contemporary amounts of ^{26}Al radioactivity in Allan Hills #5, #6, and #7. The ^{14}C content of these meteorites is important information for the timely topic of the Antarctic meteorites and the history of the Antarctic ice sheet.

2. Experimental Procedure

The procedures for carbon extraction, conversion to CO_2, and CO_2 purification have not changed (Fireman *et al.*, 1976a,b). The carbon compounds are extracted from a lunar soil sample in a molybdenum crucible, by raising the temperature with resistance heating and then holding it constant for 4 hrs. while the gases are removed with an automatic toepler pump. The temperatures used for the lunar samples and for the Allan Hills #5 meteorite were 400, 600, 800, and 1000°C. Bruderheim, Allan Hills #6, and Allan Hills #8 were melted after a 550°C extraction. The sample is heated above its melting point by induction heating and the gases are removed. The melting is repeated at a higher (>1600°C) temperature and the gases are removed again. The collected gas is passed over CuO at 650°C, which converts the carbon compounds to CO_2. The CO_2 is condensed in a trap at −196°C, recovered from the trap at −78°C, and measured in a standard volume. To purify the CO_2 from radon, the CO_2 is converted to CO over zinc at 300°C. The gases that condense from CO at −196°C are removed. The purified CO is then reconverted to CO_2 over CuO. On the basis of stoichiometry, no carbon is lost in the zinc purification. Occasionally, ^{14}C was observed in the remelted sample (Fireman *et al.*, 1977), which indicated that not all the ^{14}C was extracted. An additional remelting and ^{14}C extraction have been performed on some of these samples. The melt extractions are now done at a higher temperature than previously used.

For counting the ^{14}C, a small, low-level proportional gas counter of the Davis type is used (Fireman *et al.*, 1976a). The counters range in size from 0.7 to 5.0 cm^3 volume and are filled with CO_2 and argon. The CO_2 pressure in the counter is kept under 1 atm; argon is added until the gas pressure is 1.5 atm. The energy resolution of the counters is 20% for an external ^{55}Fe source (5.8 keV) and is independent of counter size. The counter backgrounds with 0.5 atm of CO_2 from a petroleum source are approximately 3, 6, and 10 counts day^{-1} above 6.8-keV energy for the 0.7, 2.5, and 5.0 cm^3 counters, respectively. The counting efficiencies for ^{14}C under these conditions are 25, 33, and 35%, respectively. The counter backgrounds and efficiencies rise slightly with CO_2 pressure above 0.5 atm. When the CO_2 pressure exceeds 1.0 atm, the operating voltage is in the neighborhood of 2500 v and voltage breakdown pulses occur.

3. RESULTS

Table 1 gives the ^{14}C temperature-release patterns from the grain-size separates of soil 10084. There is little or no ^{14}C in the 400°C extractions. In the 600°C extractions, there is a small amount of ^{14}C in the less than 10 μ fraction but none in the larger grains. There is a considerable amount of ^{14}C in the 800°C extractions from the small grains but little from the large grains. The 800°C ^{14}C contents increase approximately linearly with the reciprocal of the average grain diameter. The 400°, 600°, and 800°C data are plotted in Fig. 1. In the 1000°C extractions, there is also ^{14}C from the smaller grain sizes and none from the 74–125 μ size fraction; however, the ^{14}C does not decrease monotonically with increasing grain size. The largest amount, 18 dpm kg^{-1}, is released from the 10–30 μ fraction. The sum of the activities in the 600°, 800°, and 1000°C extractions for the various size fractions are given in Table 1. The below-melting ^{14}C contents decrease with increasing grain size. Eberhardt *et al.* (1970) found a

Table 1. ^{14}C temperature-release patterns.

Sample	10084.91	10084.1519	10084.939	10084.938	10084.937
Weight (g)	1.26	1.01	1.28	1.86	1.00
Grain Size (μ)	<10	10–30	30–37	37–74	74–125
Temp. °C	^{14}C dpm kg^{-1}	^{14}C dpm kg^{-1}	^{14}C dpm kg^{-1}	^{14}C dpm kg^{-1}	^{14}C dpm kg^{-1}
400	2.1 ± 0.9	2.3 ± 1.4	0 ± 0.8	0.6 ± 0.8	2.2 ± 1.2
600	4.5 ± 0.9	0 ± 1.1	0.5 ± 0.8	−0.4 ± 0.8	2.4 ± 1.2
800	22.5 ± 2.0	8.0 ± 1.2	4.2 ± 1.3	1.6 ± 0.8	1.5 ± 1.2
1000	10.0 ± 0.9	18.0 ± 1.8	3.6 ± 1.0	4.0 ± 1.0	−0.2 ± 1.2
Sum (600–1000)	37.0 ± 2.7	26.0 ± 2.5	8.3 ± 1.7	5.2 ± 2.0	3.7 ± 2.1
Melts	29.8 ± 2.0	21.3 ± 2.1	8.0 ± 1.5	32.4 ± 3.0	30.1 ± 3.0
Remelt (1st)	−2.0 ± 2.0	3.6 ± 2.0	14.0 ± 1.0	−1.0 ± 1.0	−3.0 ± 2.5
Remelt (2nd)	—	0.5 ± 1.5	3.1 ± 1.5	—	—
(Melt and Remelts)	29.8 ± 3.0	24.9 ± 3.3	25.1 ± 3.0	32.4 ± 3.0	30.1 ± 3.0
Total	66.8 ± 4.1	50.9 ± 4.2	33.4 ± 3.5	37.6 ± 3.7	33.8 ± 3.8

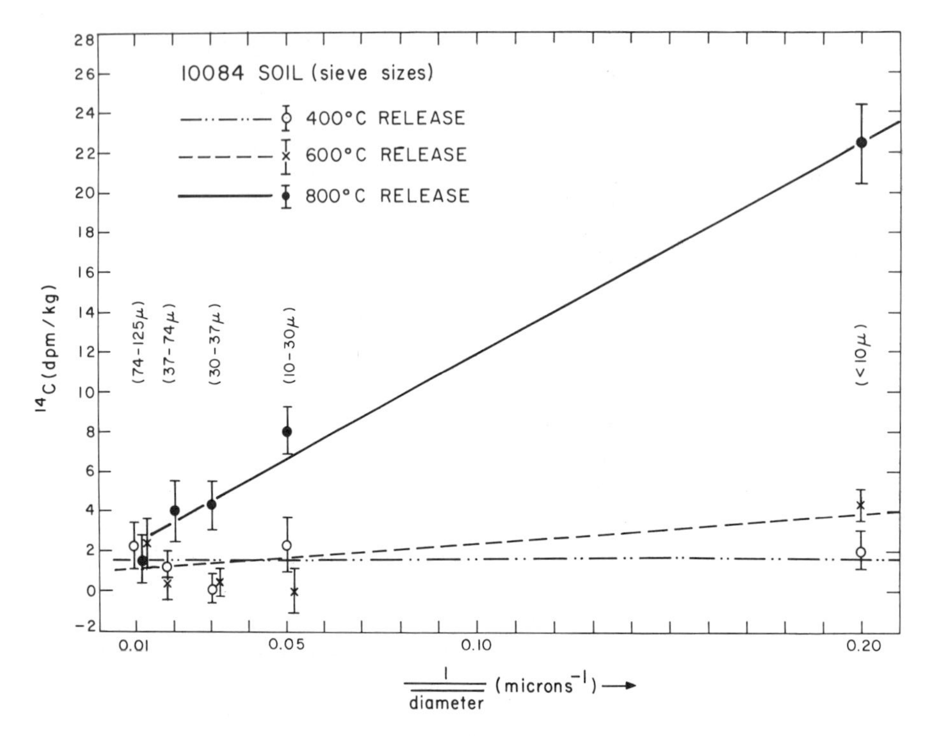

Fig. 1. ^{14}C releases at 400°, 600°, and 800°C from size fractions of 10084 soil.

similar dependence of the solar-wind rare-gas concentrations on grain size for 10084 soil.

The amounts of ^{14}C released above the melting temperature, on the other hand, are independent of the grain size: 29.8 ± 3.0, 32.4 ± 3.0, and 30.1 ± 3.0 dpm kg^{-1} were observed in the melts of the <10, 37–74, and 74–125 μ size fractions. Since no activities were observed in the remelts of these fractions, the ^{14}C appears to be totally extracted. The melts of the 10–30 and 30–37 μ fractions contained less activity than the other grain-size fractions; however, these melt extractions were not complete as evidenced by ^{14}C in remelts. The sum of the activities in the melts and remelts from the 10–30 and 30–37 μ fractions were, within the errors, the same as in the other size fractions. We attribute the melt plus remelt ^{14}C to spallation caused by galactic and solar cosmic rays. On the basis of the ^{22}Na and ^{26}Al contents of bulk soil 10084, Yokoyama *et al.* (1975) determined that the 10084 material came from the top 5 cm of the moon. For lunar material at this depth, Reedy and Arnold (1972) calculate a ^{14}C production rate of 19 dpm kg^{-1} from galactic cosmic rays and 6 dpm kg^{-1} from solar cosmic rays with R_0 = 100 MV and a flux of 100 protons cm^{-2} sec^{-1} above 10 MeV, in essential accordance with the melt plus remelt results, which range from 24.9 to 32.4 dpm kg^{-1}.

The sum of the ^{14}C contents in the 600–1000°C extractions, the activities in the melt plus remelts, and the total ^{14}C contents are plotted versus grain size in

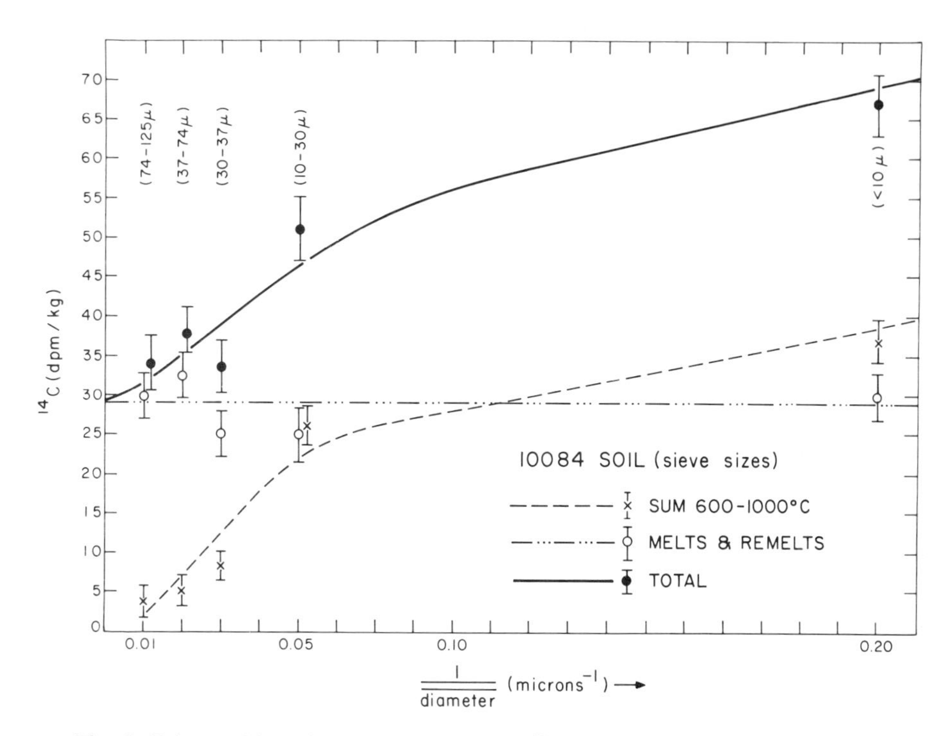

Fig. 2. Below melting, above melting, and total ^{14}C from size fractions of 10084 soil.

Fig. 2. If the weight percentage of bulk 10084 (King *et al.*, 1971) for each grain-size fraction is multiplied by its ^{14}C content, the average ^{14}C content of the analyzed material, 43 ± 8 dpm kg^{-1} is obtained. This average value for the (<125 μ) 10084 fractions which represent 74% of the bulk, agrees with the 39 ± 5 dpm kg^{-1} value in 10084 bulk measured by Wänke *et al.* (1970). The highest and lowest ^{14}C contents of bulk soil are 57.5 ± 5.0 dpm kg^{-1} and 21.5 ± 3.5 dpm kg^{-1} measured by Fireman *et al.* (1977) from skim surface, 73221, and trench bottom, 73261.

Table 2 gives the amounts of CO_2 and ^{14}C activities in the meteorites: Bruderheim, Allan Hills #5, Allan Hills #6, and Allan Hills #8. Bruderheim is an L-6 chondrite that fell on May 4, 1960. For the ^{14}C measurements in Bruderheim, we used a 10.06-g sample from a fragment that had 58 ± 3 dpm kg^{-1} of ^{26}Al (Fireman, 1967). This ^{26}Al activity is approximately the amount expected in a recently fallen L-6 chondrite that had not been heavily shielded. We carried out carbon extractions at 500°C, then at temperatures of ~1600°C and >1600°C for a melt and a remelt extraction. In the CO_2 from the 500°C extraction, there was less than 3 ct day^{-1} of activity; in the CO_2 from the melt, 270 ± 10 ct day^{-1} were observed. In the remelt CO_2, less than 2 ct day^{-1} were observed. These count rates correspond to 57 ± 3 dpm kg^{-1} of ^{14}C in the melt and less than 0.7 and 0.5 dpm kg^{-1} in the 500°C and remelt extractions. This

Table 2. Amounts of CO_2 and ^{14}C in meteorites.

Meteorite (type)	Wt. (g)	Extraction Temperature	CO_2 (cm^3 STP)	^{14}C (dpm kg^{-1})
		500°C	2.05	<0.7
Bruderheim (L6)	10.06	Melt	1.02	57 ± 3
		Remelt	0.12	<0.5
		400°C	0.46	–
		600°C	2.20	–
Allan Hills #5 (Eu)	9.99	800°C	0.31	<1
		Melt	0.46	<0.5
		Remelt	0.72	<0.5
		350°C	7.25	–
Allan Hills #6 (H6)	9.05	550°C	6.23	–
		Melt	1.02	<0.7
		Remelt	3.7	<1
		300°C	7.3	–
Allan Hills #8 (H6)	10.08	550°C	4.03	–
		Melt	4.04	<1
		Remelt	0.32	<0.7
		550°C	–	–
		800°C	1.10	–
Allan Hills #7 (L6)	11.00	1000°C	0.29	–
		Melt	0.16	<0.5
		Remelt	1.31	–

Bruderheim value is in accord with the 63 ± 5 and 55.8 ± 3.0 dpm kg^{-1} values for Bruderheim by Goel and Kohman (1962) and by Suess and Wänke (1962).

The classifications of the Allan Hills meteorites given in Table 2 were obtained from E. Olsen (pers. comm., 1977). For Allan Hills #5 and #6, there are time-of-fall estimates of less than 100,000 and less than 200,000 yrs. based on the measurements of their ^{26}Al activities by the Battelle group (L. Rancetelli, pers. comm., 1977). Because of the long half-life of ^{26}Al (7.4×10^5 yrs.) and uncertainties arising from the dependence of the ^{26}Al production rate on the chemical composition and shielding, contemporary falls can not be distinguished from falls that occurred 100,000–200,000 yrs. ago by ^{26}Al. Whether or not cosmic-ray-produced radioactivities with half-lives shorter than ^{26}Al are present is therefore of interest. For Allan Hills #5, we observed no activities in the CO_2 from the 800°C extraction, the 1600°C melt, or the >1600°C remelt. From the count rate limit of these three extractions (<3 ct day^{-1}), an upper limit of 2 dpm kg^{-1} is assigned for the total ^{14}C content of Allan Hills #5. By comparing the <2 dpm kg^{-1} value with the 57 ± 3 dpm kg^{-1} for Bruderheim, it can be concluded that the ^{14}C present in Allan Hills #5 at the time of fall has decayed by at least 5 half-lives and that its time of fall occurred more than 25,000 yrs. ago.

In Allan Hills #6 and #8, no activities were observed in the melt and remelt extractions. For these meteorites upper limits of 2 dpm kg^{-1} are also obtained for ^{14}C. On the basis of a comparison with the ^{14}C in Bruderheim, lower limits of 25,000 yrs. are assigned for their terrestrial ages. Several hundred meteorites from approximately fifty different falls have been recovered from the Allan Hills site (W. A. Cassidy, pers. comm., 1978). If the current ^{26}Al and ^{14}C data are representative, most of these meteorites fall between 25,000 and 200,000 years ago. The distribution of fall times for the Allan Hills meteorites, which can be obtained from their radioactivity contents, is a new source of information on the rate of meteorite falls and the history of the Antarctic ice sheet.

Acknowledgments—I wish to thank Prof. W. A. Cassidy for the Allan Hills meteorite samples and J. DeFelice and J. D'Amico for their assistance with the measurements. This research was supported in part by grant NGR 09-015-145 from the National Aeronautics and Space Administration.

REFERENCES

Begemann F., Born W., Palme W., Vilcsek E. and Wänke H. (1972) Cosmic-ray produced radioisotopes in Apollo 12 and 14 samples. *Proc. Lunar Sci. Conf. 3rd*, p. 1693–1702.

Cassidy W. A., Olsen E. and Yanai K. (1977) Antarctica: A deep-freeze storehouse for meteorites. *Science* **198**, 727–731.

Chang S., Mack R., Gibson E. K. Jr. and Moore G. W. (1973) Simulated solar wind implantation of carbon and nitrogen ions into terrestrial and lunar fines. *Proc. Lunar Sci. Conf. 4th*, p. 1509–1522.

Chang S., Smith J., Sakai H., Petrowski C., Kuenvolden K. A. and Kaplan I. R. (1972) Carbon, nitrogen, and sulfur released during pyrolysis of bulk Apollo 15 fines. In *The Apollo 15 Lunar Samples* (J. W. Chamberlain and C. Watkins, eds.). p. 291–293. The Lunar Science Institute, Houston.

Eberhardt P., Geiss J., Graf H., Grögler N., Krähenbühl U., Schwaller H., Schwarzmuller J. and Stettler A. (1970) Trapped solar wind noble gases, exposure age and K/Ar age in Apollo 11 lunar fine material. *Proc. Apollo 11 Lunar Sci. Conf.*, p. 1037–1070.

Fireman E. L. (1967) Radioactivities in meteorites and cosmic ray variations. *Geochim Cosmochim. Acta* **31**, 1691–1700.

Fireman E. L., DeFelice J. and D'Amico J. (1976a) The abundance of ^{3}H and ^{14}C in the solar wind. *Earth Planet. Sci. Lett.* **32**, 85–190.

Fireman E. L., DeFelice J. and D'Amico J. (1976b) Solar wind ^{3}H and ^{14}C abundances and solar surface processes. *Proc. Lunar Sci. Conf. 7th*, p. 525–531.

Fireman E. L., DeFelice J. and D'Amico J. (1977) ^{14}C in lunar soil: Temperature-release and grain-size dependence. *Proc. Lunar Sci. Conf. 8th*, p. 3749–3754.

Gibson E. K. Jr. and Moore G. W. (1973) Carbon and sulfur distributions and abundances in lunar fines. *Proc. Lunar Sci. Conf. 4th*, p. 1577–1586.

Goel P. S. and Kohman T. (1962) Cosmogenic carbon-14 in meteorites and terrestrial ages of "finds" and craters. *Science* **136**, 875–876.

King E. A. Jr., Butler J. C. and Carman M. F. Jr. (1971) The lunar regolith as sampled by Apollo 11 and Apollo 12: Grain size analysis, model analyses, and origins of particles. *Proc. Lunar Sci. Conf. 2nd*, p. 737–746.

Reedy R. C. and Arnold J. R. (1972) Interaction of solar and galactic cosmic rays with the Moon. *J. Geophys. Res.* **77**, 537–555.

Simoneit B. R., Christiansen P. C. and Burlingame A. L. (1973) Volatile element chemistry of selected lunar, meteoritic, and terrestrial samples. *Proc. Lunar Sci. Conf. 4th*, p. 1635–1650.

Suess H. E. and Wänke H. (1962) Radiocarbon content and terrestrial age of twelve stony meteorites and one iron meteorite. *Geochim. Cosmochim. Acta* **26**, 475–480.

Wänke H., Begemann F., Vilcsek E., Rieder R., Teschke F., Born W., Quijano-Rico M. and Wlotzka F. (1970) Major and trace elements and cosmic-ray produced radioisotopes in lunar samples. *Science* **167**, 523–525.

Yokoyama Y., Reyss J. and Guichard F. (1975) ^{22}Na-^{26}Al studies of lunar regolith. *Proc. Lunar Sci. Conf. 6th*, p. 1823–1843.

Proc. Lunar Planet. Sci. Conf. 9th (1978), p. 1655–1665.
Printed in the United States of America

Distribution properties of implanted rare gases in individual olivine crystals from the lunar regolith

J. KIKO, T. KIRSTEN and D. RIES

Max-Planck-Institut für Kernphysik, Heidelberg, Germany

Abstract—The characteristic double humped ^{4}He profiles observed by Gas Ion Probe (GIP) analysis in individual olivine crystals from mare soil 71501 are reflected in bimodal gas release patterns of $^{20,22}Ne$ and $^{36,38}Ar$ obtained by stepwise heating experiments performed on the same crystals. The two reservoirs correspond to a saturated highly radiation-damaged 300 Å surface layer with strongly fractionated implanted gases and a less damaged zone underneath which is populated by range straggling of incident solar wind ions and, to a lesser extent, by ions from the high energetic tail of the Maxwellian velocity distribution of the solar wind. The absence of parentless ^{40}Ar in the deeper region excludes the possibility that this zone is populated by solid state diffusion. For the first time, parentless ^{40}Ar has been physically separated from ^{36}Ar. The transient K-coating hypothesis is not feasible.

INTRODUCTION

In lunar soil particles, the record of the interaction of solar wind and other heavy ions incident on the exposed skin of the lunar regolith reflects the complexity of the relevant processes. Apart from the flux, the primary composition, and the energy spectrum of the incident corpuscular radiations, the present concentration profiles of implanted species are governed by mass-dependent trapping probabilities during implantation and by diffusive redistribution which, in turn, is controlled by the specific mineral properties, by radiation damage, and by the trajectory of the catcher particle within the stirred lunar regolith. In view of this complexity, it is not even sufficient to analyze single particles, but second generation experiments require additional microresolution. Suitable techniques for this purpose are GIP (Gas Ion Probe) analysis (Müller *et al.*, 1976; Kiko *et al.*, 1976; 1977; Kiko *et al.* in paper submitted to *Int. J. Mass Spectrum. Ion Phys.*, 1978), and linear or stepwise heating of single soil particles (Ducati *et al.*, 1973; Deubner and Kirsten, 1976). GIP analyses of olivine crystals from mare soil 71501 (Kiko *et al.*, 1977) have revealed a peculiar behavior of olivines which consistently displayed a double humped ^{4}He profile with maximum concentrations: 1) at the very surface and 2) at a depth of ~420 Å (Fig. 1).

The question then arises whether the second maximum has been generated by the He-migration from the outer 300 Å layer into which solar wind He implantation has originally occurred, or whether it is produced by direct implantation into the comparatively less radiation-damaged lattice underneath the heavily damaged outer ~300 Å layer. The latter alternative could apply due to range straggling, to a temporarily anomalously high mean solar wind energy,

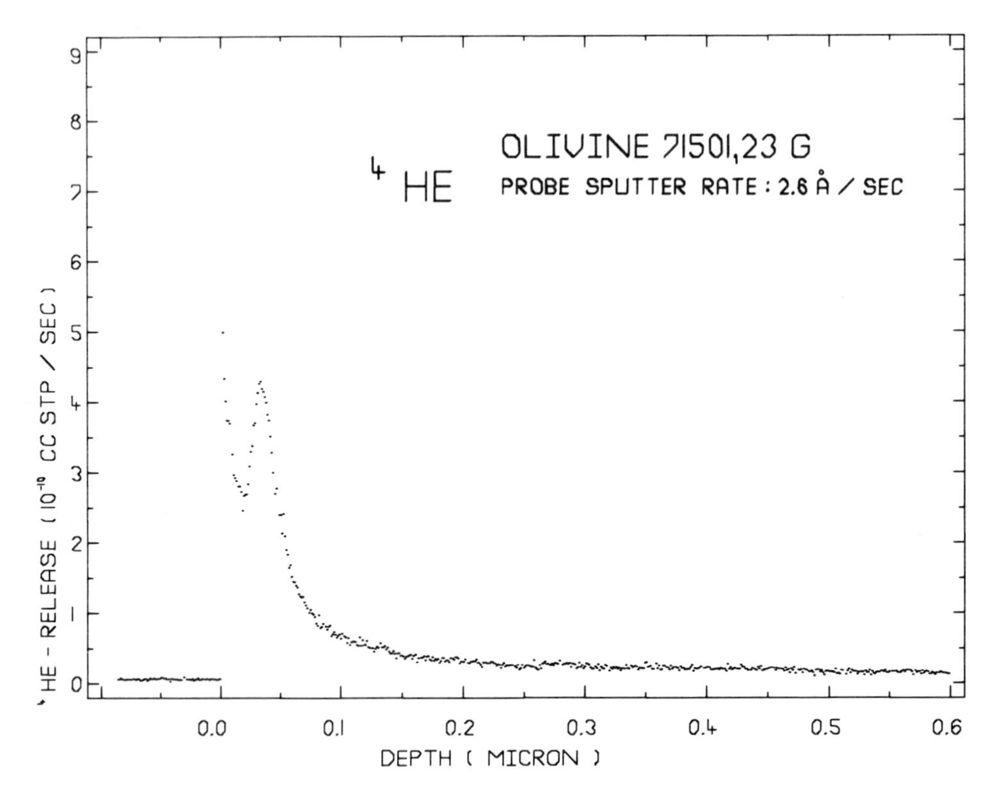

Fig. 1. ^{4}He-depth profile in olivine 71501,23 G measured with the GIP by Kiko *et al.*, 1977. The analyzed surface area was 3×10^{-4}cm^2. This double humped profile typically has been found in olivines.

or to a direct implantation of ions from the high energetic Maxwellian tail of the solar wind velocity spectrum.

If the low diffusion constants for non-amorphous olivine have, in fact, prevented the exchange of fractionated gas from the heavily damaged outer 300 Å with the much less damaged underlying interior, this would be of considerable practical importance for the analysis of unfractionated ancient solar wind abundances. A critical test for this proposition can be provided by thermal gas release patterns of the same crystals after GIP analysis and, in particular, by the behavior of surface-correlated ^{40}Ar since this isotope is incorporated with much lower energy than solar wind proper (Manka and Michel, 1971; Heymann and Yaniv, 1970), and no direct implantation at depths greater than 300 Å could occur unless the magnetic field at the moon and/or the solar wind velocity in the past have been considerably higher than at present (Yaniv and Heymann, 1972). This is also the case if a transient potassium coating is the source of parentless ^{40}Ar (Baur *et al.*, 1972). We have therefore decided to perform mass spectrometric stepwise heating experiments on the same olivine crystals for which GIP profiles had been measured and the results shall be reported in this paper.

EXPERIMENTS

After the GIP analysis of the handpicked olivine crystals, most of their surface was still intact since the lateral crystal dimensions ranged from 300–700 μm, and the GIP-ion beam affected only a 200 μm diameter circular area on one site of each crystal. The crystals were removed from the indium substrate, cleaned, wrapped in Al-foil, and loaded into the extraction line. Sample weights ranged

Table 1. Olivine 71501,23 C (0.27 mg).

Extraction Temperature (°C)	Gas Release per Step (% of total)					Isotope Ratios	
	^{4}He	^{20}Ne	^{36}Ar	^{40}Ar	^{20}Ne/^{22}Ne	^{36}Ar/^{38}Ar	^{40}Ar/^{36}Ar
200	–	0.1	–	–	–	–	–
400	1.0	2.0	–	12.7	7.7[b)]	–	>80
600	69.1	46.8	21.8	51.0	14.3	5.2	2.8[a)]
800	29.2	27.4	23.5	36.3	13.0	5.5	1.8[a)]
900	0.4	7.3	6.9	–	9.2	5.6[a)]	<1
1200	0.3	16.0	46.3	–	11.8	5.3	<0.5
1500	–	.2[b)]	0.9[b)]	–	–	4.7[b)]	–
1800	–	.3[b)]	0.7[c)]	–	0.95[b)]	–	–
Total (10^{-4}cc STP/g)	52.0 ±2.6	0.49 ±.03	0.016 ±.002	0.019 ±.003	12.3 ±1.4	5.3 ±.2	1.2 ±.2
Olivine 71501,23 E (0.53 mg).							
200	–	–	–	–	–	–	–
400	1.2	1.0	0.4	10.1	13.4	3.1[c)]	28[c)]
600	49.0	24.5	5.0	43.7	12.2	4.2[a)]	10
800	48.1	37.8	30.3	18.7	12.6	5.27	0.7
900	1.0	13.1	13.7	6.8[b)]	11.4	5.4	0.6[b)]
1200	0.6	23.1	47.9	8.0[b)]	11.8	5.5	0.2[c)]
1500	0.1	0.6	2.7	12.8[c)]	2.0[a)]	4.1[a)]	6[c)]
Total (10^{-4}cc STP/g)	50.3 ±2.3	1.09 ±.05	0.051 ±.005	0.058 ±.010	11.8 ±.3	5.3 ±.2	1.2 ±.3

from 0.12 to 0.90 mg. After pre-degassing for 3 hours at 100°C, the samples were subsequently heated in a Mo-crucible for 10 minutes each at the temperatures as stated in Table 1. The temperature was controlled by a Pt-Rh-Pt thermocouple with a reproducibility of ~20°C. The released gases were immediately subjected to mass spectrometric analysis with a dry ice/acetone-cooled cold trap serving as the only purification device. Blank values are given in Table 2.

RESULTS

The released quantities of He, Ne, and Ar isotopes are given for all extraction temperatures in Table 1. A typical release pattern is shown in Fig. 2. ^{20}Ne, ^{22}Ne, ^{36}Ar, and ^{38}Ar are released in two clearly resolved fractions with maximum release temperatures around 600°C and 1200°C, respectively in all samples

Table 1. Olivine 71501,23 B (0.59 mg). *(cont'd.)*

Extraction Temperature (°C)	Gas Release per Step (% of total)					Isotope Ratios	
	^{4}He	^{20}Ne	^{36}Ar	^{40}Ar	^{20}Ne/^{22}Ne	^{36}Ar/^{38}Ar	^{40}Ar/^{36}Ar
200	0.1[a]	0.1[a]	–	–	–	–	–
400	2.9	0.8	–	–	12.6[a]	–	–
500	22.7	5.3	–	–	14.3	–	–
600	34.8	31.9	–	–	13.2	–	–
700	35.0	17.7	(30)[x]	(60)[x]	12.4	–	(2.6)[x]
800	2.7	4.2	2.8	10.0	13.0	6.0[a]	1.0
1200	1.4	37.9	62.5	24.4[a]	13.1	5.7	0.1[b]
1500	0.3	2.0	4.6	5.6[b]	10.0	5.2	0.3[b]
1800	–	0.1[a]	0.1[b]	–	0.8[b]	1.3[b]	–
Total (10^{-4}cc STP/g)	66.8 ±3.5	0.49 ±.03	0.016 ±.002	0.021 ±.004	12.8 ±.6	5.3 ±.2	1.3 ±.3

Olivine 71501,23 G (0.12 mg).

Extraction Temperature (°C)	^{4}He	^{20}Ne	^{36}Ar	^{40}Ar	^{20}Ne/^{22}Ne	^{36}Ar/^{38}Ar	^{40}Ar/^{36}Ar
200	–	–	–	–	–	–	–
300	0.2	0.1[b]	0.2[b]	5[c]	–	2.5[c]	12[c]
400	0.9	0.8[a]	0.3[a]	8[c]	6.6[c]	2.0[c]	16[c]
450	5.1	1.1	0.3[a]	13.3[c]	20[c]	2.3[c]	22[c]
500	16.5	2.8	0.7[a]	15.1[b]	12.2[b]	1.4[c]	11[b]
550	19.7	6.1	1.4[a]	12.6[b]	13.4[a]	4.7[b]	4.6[b]
600	26.3	26.2	4.2	5.0[c]	12.7	5.3[a]	0.6[c]
650	18.0	21.8	11.1	10.0[c]	13.0	5.5[a]	0.5[c]
700	11.3	16.1	19.4	17.2[b]	11.6[a]	5.5[a]	0.4[b]
750	1.1	8.5	6.3	–	12.0[a]	5.8[b]	<0.4
800	0.4	5.0	7.4	–	12.7[a]	5.1[b]	<0.4
900	0.3	3.9	10.8	1.4[c]	10.0[a]	5.4[a]	0.1[c]
1150	0.1	4.3	19.0	–	10.2[a]	5.1[a]	<0.07
1400	0.2	3.0	16.9	12[c]	7.5[b]	4.5[a]	0.4[c]
1600	–	0.3[a]	2.0[a]	–	5.7[c]	5.8[c]	–
Total (10^{-4}cc STP/g)	131 ±14	3.34 ±.40	0.14 ±.02	0.07 ±.05	12.0 ±1.8	5.0 ±.9	0.51 +45 −30

except sample H (Table 1). The ^{20}Ne/^{22}Ne and ^{36}Ar/^{38}Ar ratios remain roughly constant between 600°C and 1200°C extraction temperature (Table 1). Above 1200°C, spallogenic contributions reduce these ratios. ^{4}He is mainly released between 400°C and 800°C, and we could not resolve two components even though a special attempt was made by analyzing sample G in 50°C rather than 200°C temperature increments.

The ^{40}Ar release pattern is quite distinct from that of ^{36}Ar (^{38}Ar). Apart from radiogenic ^{40}Ar released at the highest temperatures, most of the ^{40}Ar released below 800°C, and there is no ^{40}Ar counterpart to the high temperature release of

Table 1. Olivine 71501,23 A (0.90 mg). *(cont'd.)*

Extraction Temperature (°C)	Gas Release per Step (% of total) ^{4}He	^{20}Ne	^{36}Ar	^{40}Ar	^{20}Ne/^{22}Ne	Isotope Ratios ^{36}Ar/^{38}Ar	^{40}Ar/^{36}Ar
200	0.1	–	–	–	–	–	–
400	0.3	0.7	0.2[a)]	4.6	7.2[b)]	2.6[c)]	34
600	54.0	20.6	4.5	15.8	11.9	5.45	4.2
800	36.1	42.0	46.8	41.0	11.7	5.30	1.0
900	3.4	18.9	12.3	8.6	11.9	5.5	0.8
1000	1.6	10.5	14.6	6.1	10.3	5.30	0.5
1200	1.6	5.5	19.0	6.4[b)]	6.3	5.33	0.4[b)]
1400	2.7	1.4	2.4	13.3[b)]	1.16	2.01	6.7[b)]
1600	0.3	0.2[a)]	0.2[a)]	4.2[c)]	.95[a)]	1.21[b)]	22[c)]
Total (10^{-4}cc STP/g)	17.9 ±.9	0.58 ±.03	0.064 ±.006	0.076 ±.012	9.65 ±.25	5.10 ±.15	1.2 ±.2
Olivine 71501,23 H (0.24 mg).							
400	2.0	1.5	0.3[b)]	3.3[c)]	9.8[b)]	3[c)]	3[c)]
800	96.7	79.1	50.7	53.5[a)]	12.6	5.5	0.37[a)]
1200	1.2	16.6	31.4	29.1[b)]	11.8	5.4	0.33[b)]
1300	0.1	2.2	12.9	14.2[c)]	11.1[b)]	5.1	0.39[c)]
1600	–	0.7	4.7	–	3.2[b)]	5.4[b)]	–
Total (10^{-4}cc STP/g)	101 ±10	1.85 ±.20	0.078 ±.016	0.028 ±.015	12.1 ±.6	5.4 ±.3	0.35 ±.20

Fractional gas release from single olivine crystals previously analyzed by the Gas Ion Probe technique. Relative errors of gas amounts for the individual temperature fractions including blank corrections are generally below 10% (see also Table 2). Larger errors due to small gas quantities or larger relative blank corrections are marked as follows: a (10–20%), b (20–50%), and c (>50%).

[x)] Ar for fractions up to 700°C could not be reliably determined for this sample. For normalization of the remaining fractions, the contribution of the undetermined fractions is assumed to be 30% for ^{36}Ar and 60% for ^{40}Ar as judged from the behavior of the other samples.

^{36}Ar. This is reflected in ^{40}Ar/^{36}Ar ratios decreasing with increasing temperature as illustrated for two samples in Figures 3a and 3b. Again, sample H is an exception.

Discussion

The bimodal gas release patterns observed for Ne and ^{36,38}Ar indicate the existence of two distinct reservoirs. This implies either two distinct activation energies or the reservoirs must be spaced far apart. The latter alternative is unlikely in view of the narrow He distribution profile. Consequently, the GIP He profiles enable us to infer the locations of the two reservoirs as the outer 200–300

Table 2.

°C	^{4}He	^{20}Ne	^{22}Ne	^{36}Ar	^{38}Ar	^{40}Ar	
≤1000	<4	.2	<.1	<.02	<.01	<.5	($\pm$1.5
1200	<4	.4	<.1	.02	<.01	1.0	−0.8)
1400	<10	.4	<.1	.02	.01	2.4	
1600	<30	.4	<.1	.04	.02	7.5	
1800	<30	.6	<.1	.11	.03	30	

Mean values of blank runs for different extraction temperatures in 10^{-10}ccSTP.

Å layer for the low retentivity phase and the layer below the second He-maximum at ~400 Å depth for the high retentivity phase. The difference in activation energies is due to the intense radiation damage of the outer 300 Å as a result of the high H- and ^{4}He doses implanted into this layer by the bulk of solar wind ions with mean energies of ~0.9 keV/nucleon. In comparison, the radiation damage of the underlying zone is much less intense since it is mainly caused by solar flare particles incident with much lower fluxes. For the highly mobile ^{4}He, the resulting difference in the activation energy of the two reservoirs is apparently not sufficient to be resolved in the stepwise heating experiment.

In order to establish the significance of the rare gases now located deeper than ~400 Å and degassed in the second release maximum, two feasible mechanisms must be considered for the population of this zone.

1) Inward diffusion from the outer, heavily damaged saturated reservoir in steady state equilibrium into the less damaged underlying material which is characterized by a lower diffusion constant.
2) Direct implantation of ions with higher than average solar wind energy.

A distinction between these two mechanisms can be made by comparing the behavior of ^{40}Ar and ^{36}Ar, if ^{40}Ar is not part of the solar wind but is reimplanted from the latent lunar atmosphere with typical energies of ~0.1 keV/nucleon as at present; whereas, the mean energy of solar wind ^{36}Ar is higher by approximately one order of magnitude (Heymann and Yaniv, 1970; Manka and Michel, 1971). Consequently, even considering range straggling or the high energetic Maxwellian tail of the velocity distribution, there is no way to directly implant ^{40}Ar into depths of ~500 Å, quite contrary to the case of ^{36}Ar. A similar argument could be made if ^{40}Ar is due to the decay of a transient K-coating (Baur *et al.*, 1972; Signer *et al.*, 1977). In the other case, if the 500 Å zone is populated by diffusion from the outer amorphous zone, the ^{40}Ar/^{36}Ar ratio in both reservoirs should be nearly identical. The data clearly favor an interpretation in terms of population of the 500 Å zone by direct implantation since they demonstrate for the first time a clear physical separation of surface-correlated ^{36}Ar and ^{40}Ar (Figs. 2–3b). Parentless ^{40}Ar is absent in the high temperature regime in which the second

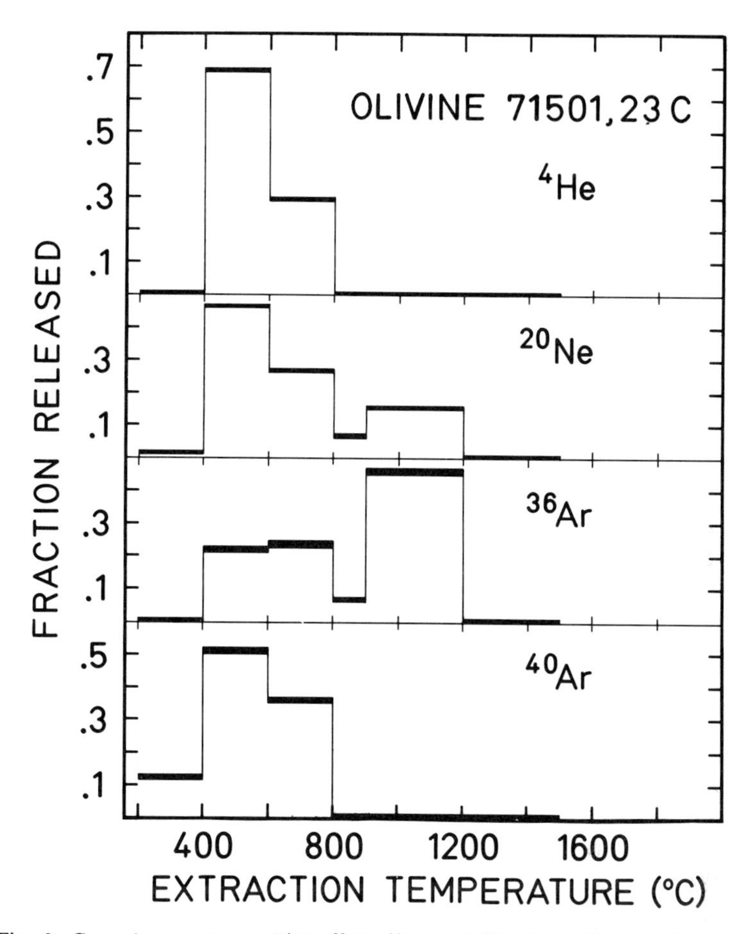

Fig. 2. Gas release pattern of ^{4}He, ^{20}Ne, ^{36}Ar, and ^{40}Ar from olivine 71501,23 C.

release maximum for Ne and ^{36,38}Ar occurs and, accordingly, the ^{40}Ar/^{36}Ar ratios decrease with increasing temperature.

With direct implantation identified as the responsible mechanism, we attempt to explain the higher than average 420 Å-range of the responsible ions as a result of a temporarily higher mean energy of the solar wind. However, this possibility is unlikely since one would expect relatively unfractionated elemental abundance ratios in the deeper layer. For instance, the ^{20}Ne/^{36}Ar ratio in the 1000–1200°C release fractions should be ~40 instead of ~10 as observed. The measured ratio may be lowered by the laboratory degassing procedure in that a portion of ^{20}Ne from the deeper region is degassed already below 1000°C, but we estimate from the diffusion characteristics that this could at most account for a 30% fractionation.

In contradiction, if range straggling is responsible, a mass fractionation is to be expected since straggling is mass dependent.

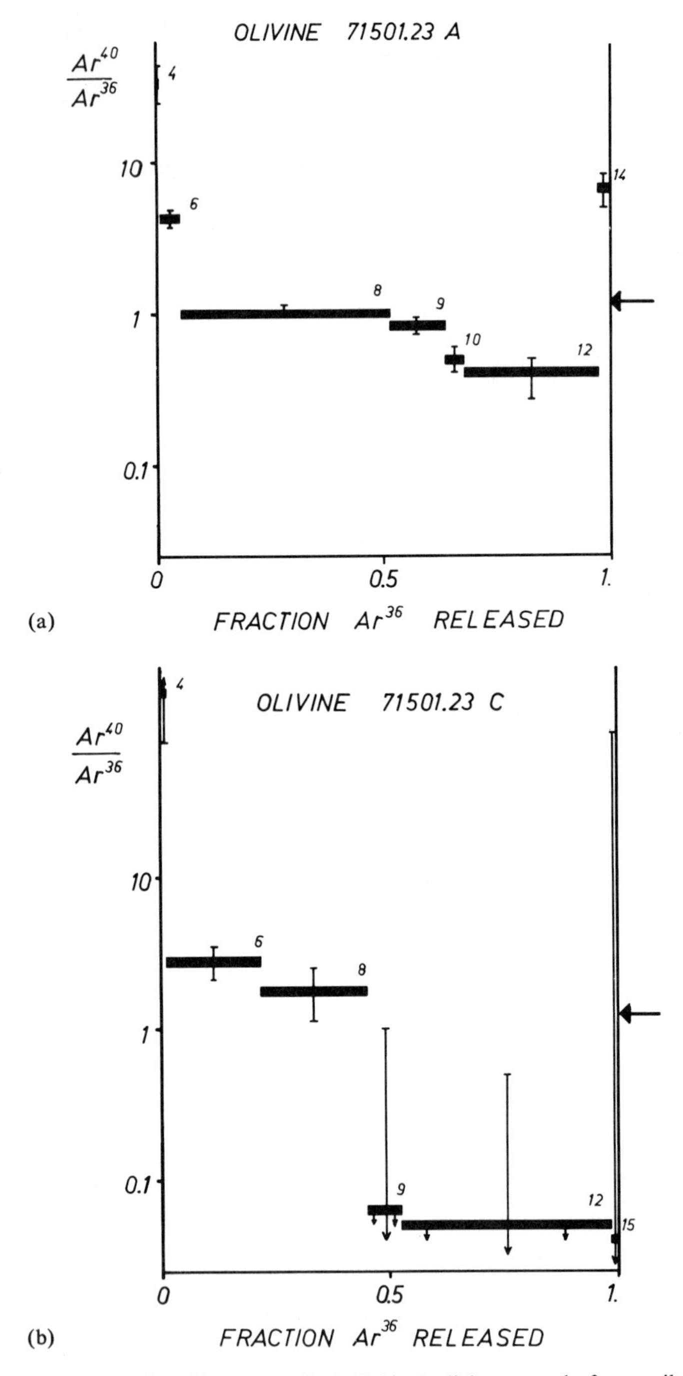

Figs. 3a and 3b. $^{40}Ar/^{36}Ar$ ratios in individual olivine crystals from soil 71501 vs. fraction of ^{36}Ar released during stepwise heating. Extraction temperature in units of 100°C are indicated. The arrow at the right hand ordinate marks the mean $^{40}Ar/^{36}Ar$ ratio. 3 σ errors are indicated. They include blank uncertainties. The upper limit is without any blank correction for ^{40}Ar. Small arrows at the bars imply zero values after blank correction.

Figure 4 shows the expected distribution of ^{4}He, ^{20}Ne, and ^{36}Ar after implantation of identical doses of ions with an energy of 1 keV/nucleon into a silicate mineral integrated over all directions of incidence which apply under lunar conditions. The profiles are calculated according to the LSS-theory (Lindhard *et al.*, 1963) and making use of data submitted to *Earth Planet. Sci. Lett.* by Agrawal and Tamhane (1978).

Obviously, the fraction of an isotope implanted deeper than a given depth is strongly mass dependent, and the elemental ratios within the deeper reservoir depend sensitively on the depths at which the second reservoir begins in which no significant diffusion has occurred under lunar conditions.

In Table 3 we compare the measured elemental ratios ^{4}He/^{20}Ne and ^{20}Ne/^{36}Ar with unfractionated solar wind ratios and with the calculated values considering range straggling under lunar conditions. The measured ratios in column 3 are the mean values from samples B, C, and E which allow an unequivocal partitioning between the two reservoirs. The fraction of ^{4}He belonging to the deeper reservoir was taken from the GIP results and the ^{20}Ne and ^{36}Ar portions from the stepwise heating experiments. Range straggling alone does not fit the measured values at a depth of 420 Å (column 4) which is the mean depth at which the second maximum of ^{4}He was found with the GIP. However, if one takes into account a small contribution of solar wind implanted from the high energetic Maxwellian

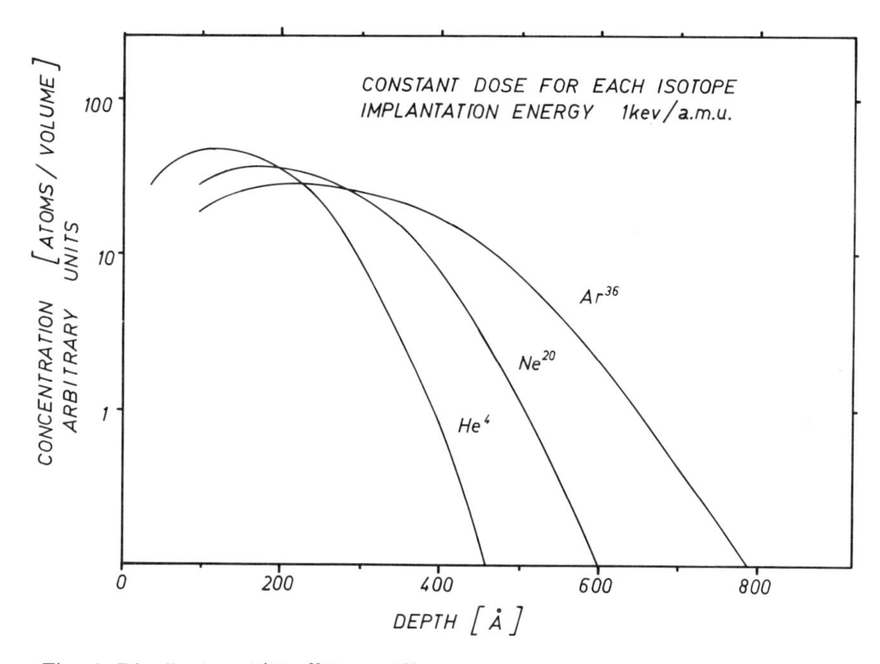

Fig. 4. Distribution of ^{4}He, ^{20}Ne, and ^{36}Ar implanted into a silicate mineral calculated according to the theory of Lindhard, *et al.* (1963) and adjusted for lunar incidence conditions by Agrawal and Tamhane in data submitted to *Earth Planet. Sci. Lett.* (1978).

Table 3.

	Unfractionated Solar Wind	Measured High Temperature Fraction In Olivine	LSS-Calculation For Reservoir Below 420 Å	LSS-Calculation For Reservoir Below 420 Å Plus 13% Unfractionated Solar Wind
$^{4}He/^{20}Ne$	540[1)]	95	17	85
$^{20}Ne/^{36}Ar$	43[2)]	12	7	12

Elemental ratios of implanted gases.
First column: Unfractionated solar wind.
Second column: Measured ratios inferred from the mean of olivine crystals 71501,23, B, C, E (see text) for the high temperature fraction. For He, the required partitioning is based on the measured GIP-He profiles.
Third column: Calculated ratios for reservoir below 420 Å (as inferred from the GIP-He profile) according to LSS theory including range straggling.
Fourth column: Calculated ratios as in column 3 with a 13% contribution of unfractionated solar wind implanted due to the high energetic Maxwellian tail of the solar wind velocity spectrum.
[1)]Geiss *et al.* (1972).
[2)]Kirsten *et al.* (1972).

tail of the solar wind velocity spectrum, the calculated ratios would shift in favor of the lighter elements. If 13% of the gases now found below 420 Å are due to unfractionated solar wind, then the calculated ratios would fit the experimental data (Table 3).

We conclude that range straggling is the dominant process generating the deeper reservoir. If this process can be treated quantitatively, the better retention properties of the material underneath the amorphous outer layers of solar wind exposed olivine crystals may be utilized for solar wind analyses and to avoid the complications caused by saturation and amorphization of the outer zone.

A final remark concerns the origin of parentless ^{40}Ar. If the given interpretation is correct, it follows that parentless ^{40}Ar is confined to the outer 300 Å and has not been redistributed into the interior. This rules out the transient potassium coating hypothesis of Baur *et al.* (1972) since the generation of the measured ^{40}Ar quantities within a shell of only 300 Å thickness during 4 b.y. would require unreasonable K-concentrations in the order of 0.5 g/g.

References

Baur H., Frick U., Funk H., Schultz L. and Signer P. (1972) Thermal release of helium, neon, and argon from lunar fines and minerals. *Proc. Lunar Sci. Conf. 3rd*, p. 1947–1966.

Deubner J. and Kirsten T. (1976) Rare gas retention under lunar conditions: A systematical simulation study (abstract). In *Lunar Science VII*, p. 199–201. The Lunar Science Institute, Houston.

Ducati H., Kalbitzer S., Kiko J., Kirsten T. and Müller H. W. (1973) Rare gas diffusion studies in individual lunar soil particles and in artificially implanted glasses. *The Moon* **8**, 210–227.

Geiss J., Buehler F., Cerutti H., Eberhardt P. and Filleux Ch. (1972) Solar wind composition experiment. *Apollo 16 Prelim. Sci. Rep.* NASA SP-315, p. 14-1–14-10.

Heymann D. and Yaniv A. (1970) Ar^{40} anomaly in lunar samples from Apollo 11. *Proc. Apollo 11 Lunar Sci. Conf.*, p. 1261–1267.

Kiko J., Büchler K., Jordan J., Kalbitzer S., Kirsten T., Müller H. W., Plieninger T. and Warhaut M. (1976) A comparison of rare gas concentration profiles in lunar and artificially implanted ilmenites measured with the gas ion probe. *Meteoritics* **11**, 309–312.

Kiko J., Kirsten T. and Warhaut M. (1977) He and Ne depth profiles in olivine from lunar soil 71501,23. *Meteoritics* **12**, 274–275.

Kirsten T., Deubner J., Horn P., Kaneoka I., Kiko J., Schaeffer A. O. and Thio S. K. (1972) The rare gas record of Apollo 14 and 15 samples. *Proc. Lunar Sci. Conf. 3rd*, p. 1865–1889.

Linhard J., Scharff M. and Schiøtt H. E. (1963) Range concepts and heavy ion ranges (Notes on Atomic Collisions II). *Klg. Danske Videnskab. Selskab., Mat.-Fys. Medd.* **33**, 1–42.

Manka R. H. and Michel F. C. (1971) Lunar atmosphere as a source of lunar elements. *Proc. Lunar Sci. Conf. 2nd*, p. 1717–1728.

Müller H. W., Jordan J., Kalbitzer S., Kiko J. and Kirsten T. (1976) Rare gas ion probe analysis of helium profiles in individual lunar soil particles. *Proc. Lunar Sci. Conf. 7th*, p. 937–951.

Signer P., Baur H., Etique P., Frick U. and Funk H. (1977) On the question of the ^{40}Ar excess in lunar soils. *Philos. Trans. Roy. Soc. London* **A285**, 385–390.

Yaniv A. and Heymann D. (1972) Atmospheric ^{40}Ar in lunar fines. *Proc. Lunar Sci. Conf. 3rd*, p. 1967–1980.

Proc. Lunar Planet. Sci. Conf. 9th (1978), p. 1667–1686.
Printed in the United States of America

Surface concentrations of Mg, Ti, Fe and surface features in individual plagioclase crystals from lunar soil samples

E. ZINNER[1,2], S. DUST[1,2], J. CHAUMONT[3] and J. C. DRAN[3]

[1]The Lunar and Planetary Institute, Houston, Texas 77058.
[2]McDonnell Center for the Space Sciences,
Washington University, St. Louis, Missouri 63130.
[3]Laboratoire René Bernas, University of Paris, 91406 Orsay, France.

Abstract—Surface enhancements of Mg, Ti and Fe were measured with an ARL-IMMA and a CAMECA IMS-300 ion microprobe in 100–200 mesh plagioclase grains from eight different lunar soils. Enhancements of Mg, Ti and Fe were found in almost all grains. The solar wind Fe/Mg ratio is not very different from the Fe/Mg measured for the bulk lunar soils which makes it difficult to use Fe/Mg to differentiate between implanted solar wind effects and other causes of surface enhancements such as glass splashes or vapor deposits. The Ti/Mg ratio which differs by a factor of more than 10 between bulk soils and solar wind is potentially more useful in separating various effects. The Ti/Mg ratios of surface enhancements vary widely but many grains have values which are much smaller than that of the average agglutinate or bulk soil.

The Fe/Mg and Ti/Mg ratios are not sufficient to decide unambiguously on the source of surface enhancements. Surface concentrations were also compared with other surface features such as microcraters and glass splashes. The presence of Mg, Ti and Fe on uncratered grains, which apparently have not been exposed to the solar wind, and comparison of surface concentrations with the percentage of glass coverage and with the measured depth of Mg enhancements indicate that glass splashes (pancakes) are the most likely sources of elemental enhancements on the surfaces of many lunar soil grains. Vapor deposits appear to contribute at best only minor amounts. In a few cratered grains, where the relatively low glass coverage cannot account for the measured surface concentrations and in cases where the Ti/Mg ratio is low, the solar wind remains the most probable source of surface Mg and Fe.

INTRODUCTION

In the last few years we have combined ion probe measurements of elemental surface concentrations in individual lunar crystals with measurements of solar flare tracks and microcraters in an effort to compare the solar wind and solar flare particle flux with the flux of interplanetary dust (Poupeau *et al.*, 1975; Morrison and Zinner, 1975 and 1977; Zinner *et al.*, 1976 and 1977). Crystals exposed to free space on the moon accumulate microcraters (~0.1 μ diameter) and solar wind implanted elements, both of which affect the crystal surface on approximately the same depth scale of a few hundred Å. This fact allows a *direct* comparison of the two associated fluxes which is not always possible for microcraters and solar flare tracks, since tracks can be accumulated under some dust shielding or the microcrater record can be eradicated by erosion processes while the solar flare tracks at lower depth will still be preserved (Poupeau *et al.*, 1975; Comstock, 1978; Poupeau and Johnson, 1978). Only three techniques allow direct measurements of concentration depth profiles with a sufficiently high depth resolution to identify the location of solar wind implanted species: nuclear depth profiling (Goldberg *et al.*, 1975, 1976; Filleux *et al.*, 1977, 1978); ESCA analysis combined with ion sputtering (Housley and Grant, 1977; Grant and Housley, 1978); and the rare gas ion probe (Müller *et al.*, 1976; Kiko *et al.*, 1978). With the ion probe it is possible to measure concentration depth profiles of selected elements with a depth resolution of approximately 30Å in areas of $\sim$40 $\mu \times$ 40 μ (Zinner *et al.*, 1976), surpassing the other methods in spatial resolution. This allows us in principle to measure amounts and depth distributions of solar wind implanted ions by comparing these two quantities with those of artificially implanted calibration isotopes ("marker ions").

Using an ion microprobe we measured surface enhancements of Mg and Fe in two separate plagioclase crystals of rock 76215 (Zinner *et al.*, 1976 and 1977) and interpreted enhancements as due to solar wind implanted ions by eliminating alternative possible sources (glass splashes, vapor deposits). This elimination was based on: (1) the low Ti/Mg ratio of the enhancements, inconsistent with the much higher ratio expected for vapor deposits and accreta (Baron *et al.*, 1977; Gold *et al.*, 1977; Zook, 1975); and (2) the small density of glassy disks on the crystal surface (Blanford *et al.* 1974) which therefore could not account for the measured enhancements. Furthermore, the solar wind exposure age derived from the Mg and Fe concentrations agrees well with the exposure age obtained from solar flare track and microcrater densities (Morrison and Zinner, 1977).

We also reported preliminary ion probe measurements of Mg, Ti and Fe surface concentrations in plagioclase grains selected from soil 67601 (Zinner *et al.*, 1977) and concluded on the basis of the relatively low Ti/Mg ratio of the enchancements in many grains that the solar wind was the most likely source. This interpretation, however, required a solar wind erosion rate much smaller than the presently assumed value (Bibring *et al.*, 1977; McDonnell, 1977) for at least certain orientations of crystal surfaces on the moon. Microcrater densities and their ratios relative to surface concentrations were variable but we showed (Zinner *et al.* 1977) that a wide spread in this ratio is expected if the individual grains experienced only a few surface exposure episodes. However, the 67601 grains analyzed in the ion probe were not selected according to uniform criteria and the analyzed areas were not completely documented for all grains. For these reasons the conclusions were necessarily limited. The expected wide variation of the ratio of solar wind implanted ion concentration to microcrater density also made it necessary to analyze a large number of grains in order to obtain sufficient statistics for a comparison of the two quantities. Here we report the analyses from seven additional soil samples which were subjected to uniform selection criteria. As before, most of the ion probe measurements were performed with the ARL IMMA (ion microprobe mass analyzer) at the Johnson Space Center but were supplemented by measurements with a CAMECA IMS 300 at the University of Paris (laboratory of Prof. C. Allègre).

Sample description and experimental procedure

Plagioclase grains from the 100–200 mesh sieve fraction were cleaned and mounted in indium as described by Zinner *et al.* (1977). Table 1 lists all the mounts whose analyses are reported here. Included is sample 67601,9003 for which data were already reported last year (Zinner *et al.*, 1977) but for which we give additional analyses of these data. Some of the selected soil samples (67601, 64421, and both 60007 core samples) had been examined for solar flare tracks and microcraters by Poupeau *et al.* (1975) and Poupeau and Johnson (1978). Other soils were chosen on the basis of the track density data reported by Crozaz and Dust (1977), Crozaz (1978) and Borg (pers. comm.). The mounted grains were implanted with ^{25}Mg ($2 \times 10^{15}/cm^2$ at 1 keV/nuc), ^{47}Ti($10^{14}/cm^2$ at 1 keV/nuc) and ^{57}Fe($5 \times 10^{14}/cm^2$ at .877 keV/nuc) except for mounts 64421,9005 and 60007,9029 which were implanted with ^{26}Mg ($3 \times 10^{15}/cm^2$ at 2 keV/nuc), ^{53}Cr ($2 \times 10^{14}/cm^2$ at 0.94 keV/nuc) and ^{47}Ti and ^{57}Fe as above. After implantation and coating with 50–100 Å of gold, a low magnification (50–100X) SEM polaroid mosaic was made of each mount and from this mosaic grains were selected with smooth surfaces which did not show large amounts of microcrud (glassy accreta). These superior grains were further documented using stereoscopic micrographs taken at magnifications of 250–700X as positive images on 35 mm film by inverting the SEM display signal on the CRT screen. These micrographs were used in a stereoscopic slide viewer to further select grains according to the above criteria with the added requirement that the inclination of the grain surface relative to the indium surface be small. On each of the grains surviving this final selection a ~40 $\mu \times 60\ \mu$ region was chosen with a low density of visible glass splashes and photographed at high magnification (7000–10,000X) in the SEM in a series of 35 mm photographs taken with the signal inverted as for the stereopair pictures. Surface features such as glassy disks and microcraters were documented by projecting the film onto a screen.

Ion probe depth profiles were measured in the ARL IMMA as described by Zinner *et al.* (1976) for grains of all mounts except 60007,9029. The number of grains analyzed is given in Table 1. The experimental conditions were the same as given by Zinner *et al.* (1977) except that the NO_2^- beam was rastered over an area of ~40 $\mu \times 60\ \mu$. This area was the same as the area documented at high magnification within the accuracy of the alignment that could be achieved in the optical microscope of the IMMA. A certain misalignment could be tolerated since the electronic aperture limited the region from which secondary ions were counted to only the central 25–35% of the rastered area. Depth profiles were measured at m/e = 24, 25, 26, 27, 28, 40, 47, 48, 54, 56, 57, and 197; for some selected crystals, m/e = 49 was included.

Mount 60007,9029 was analyzed with the CAMECA IMS 300 ion microscope (Morrison and Slodzian, 1975) in order to compare the Ti and Fe measurements obtained with the ARL IMMA, for which the contributions of molecular species, especially in the case of the Fe isotopes, had to be

Table 1. List of all soil samples from which individual grains were studied at high magnification in the SEM and analyzed in the ion probe.

Sample No.	Parent No.	Depth In Core (cm)	Grains Mounted	Grains Analyzed In Ion Probe
67601,9003	20	Surface Soil	64	16
60007,9027	51	9.5	85	14
60007,9029	79	2.5	78	11
64421,9005	5	Surface Soil	87	9
15002,562	97	175	61	2
15006,209	14	38	81	2
24077,9002	85	*	64	1
24210,9002	58	*	59	6

*See Bogard and Hirsch (1977).

subtracted in order to obtain elemental surface concentrations (Zinner *et al.*, 1977). Grains on this mount were selected for analysis from single low magnification (300–400X) photographs and depth profiles were measured at m/e = 24, 25, 26, 27, 28, 40, 47, 48, 54, 56 and 57 using an on-line computer system interfaced with the IMS 300. This system allows scans across the various peaks while integrating the counts. The instrument was operated at low mass resolution with energy filtering (Shimizu *et al.*, 1977; Zinner *et al.*, 1977; Shimizu *et al.*, 1978) to suppress molecular secondary ions. A 14.5 kV unseparated O^- beam of 1.0–2.0 nA (spot size ~25 μ) was rastered over an area of approximately 40 μ × 50 μ.

In order to study the spatial distribution of different elements on the grain surfaces, elemental images of Mg, Al, Si and Ca were obtained for a few grains with the ARL IMMA with a 0.4 nA beam of 15 kV O^- ions rastered over the area to be imaged, the secondary ion mass spectrometer tuned to m/e = 24, 27, 28 and 40 respectively, and the secondary ion signal modulating the brightness on a CRT.

After ion probe analysis the grains were viewed in the SEM to determine whether the sputtered areas coincided with the preselected regions. Some of the samples were recoated and high magnification micrographs taken of a few grains in order to compare features on the grain surfaces before and after sputtering.

RESULTS AND DISCUSSION

(a) Measurement of elemental concentrations

Surface concentrations of Mg, Ti and Fe were measured by integrating the areas under the depth profiles and by comparing these with those of artificially implanted marker isotopes whose concentrations were known within better than 10% (Zinner *et al.*, 1974; Zinner and Walker, 1975). Surface enhancements (Table 2) are present in almost all measured grains.

For all three Mg isotopes interferences from molecular and doubly charged ions in the ARL IMMA are less than 1–2% (Zinner *et al.*, 1976 and 1977). M/e = 24 and 26 signals are proportional to ^{24}Mg and ^{26}Mg isotopic abundances throughout the profiles. This is not the case for Fe, for which the background of Al_2^+ at m/e = 54 and CaO^+ at m/e = 56 is substantial. We used the profiles at m/e = 54 for the determination of surface concentrations by subtracting the background at large depth consisting of Al_2^+ and of structural $^{54}Fe^+$ since the Al_2^+ signal (Zinner *et al.*, 1977) as well as the $^{54}Fe^+$ signal from a uniform Fe concentration (Zinner, 1978) were shown to exhibit no surface enhancement. In the ARL IMMA we do not have any way of knowing what the relative contributions of the structural $^{54}Fe^+$ and Al_2^+ are but, except for a larger uncertainty on the surface Fe concentration because of a higher level of background, detailed knowledge of the composition of a *constant* background signal is not necessary for the measurement of the Fe surface enhancement. With the CAMECA IMS 300 with high mass resolution we can separate the structural $^{54}Fe^+$ signal from the Al_2^+ background and have shown these to be approximately equal (Zinner *et al.*, 1977). This time we measured 60007,9029 with the CAMECA IMS 300 employing energy filtering. Figure 1 shows depth profiles at m/e = 24 and 25 as well as at m/e = 54 and 56 which were measured in this way. Since the m/e = 54 to 56 ratio is that of $^{54}Fe/^{56}Fe$ we know that molecular ions are sufficiently suppressed. The long (in comparison to ARL profiles) tails are caused by atomic mixing under the primary beam (Zinner *et al.*, 1976; Ishitani *et al.*, 1974 and 1975) and by edge effects since no electronic aperture was used. In order to integrate the areas under the depth profile curves, the measured points of the decreasing tails were fitted to an exponential function of the form $Ae^{-Bt} + C$ where t is the time of sputtering.

A comparison of the ARL and CAMECA measurements of Mg and Fe surface concentrations is shown in Fig. 2 as Fe versus Mg concentrations for two Apollo 16 core samples analyzed in the two different instruments. Absolute values of surface enhancements measured in the CAMECA are higher than those measured in the ARL probe by up to a factor of five. We think the reason for this fact is that because of the poor viewing system of the IMS 300, individual grains were placed under the incident primary ion beam not by visual alignment but the alignment was made by maximizing the secondary Al^+ signal. As a consequence of this and of the 45° primary beam incidence, many grains were not only sputtered on the top surface where the marker ions had been implanted but also on side faces as could be seen in the SEM photographs taken after ion probe analysis. This misalignment should affect the absolute values measured by our marker ion method but not elemental ratios. Fe/Mg ratios measured in the ARL probe are somewhat higher than those measured in the CAMECA ion microscope. This difference might reflect a systematic error due to Al_2^+ interference in the ARL IMMA or it might result from the fact that natural isotopes were also sputtered from faces where no artificial isotopes were implanted. A third explanation is that the difference is a consequence of possible different bulk concentrations of the two samples taken from different locations in the Apollo 16 deep drill core (see Table 1). The slight difference is not significant in light of the fact that the Fe/Mg ratio is not a strong indicator for the possible sources (solar wind implanted ions, vapor deposits, thin glass splashes) of the surface enhancements. In Fig. 2 the two lines correspond to Cameron's solar system ratio (Cameron, 1973) and the average bulk soil composition of 60007,114 (Nava and Philpotts, 1973) which is the closest analyzed bulk sample to our analyzed grains. Since measured solar wind ratios are known to agree with the Cameron values only within a factor of two (Geiss, 1973) the uncertainty in the solar wind Fe/Mg ratio is larger than the difference between the Cameron value and the average soil composition.

The Ti/Mg ratio should provide a much better way of differentiating between solar wind and lunar sources of surface enhancements because the Ti/Mg ratio is much higher on the moon than it is expected to be in the sun. In the past we have considered our Ti measurements in the ARL IMMA to be only upper limits since, although we could subtract the $^{48}Ca^+$ background at m/e = 48, we could not exclude the presence of molecular interferences such as Mg_2^+. In the CAMECA microscope such interferences are suppressed and the values derived from measurements at m/e = 48 represent Ti concentrations rather than upper limits. Figure 3 compares the Ti and Mg measurements in both instruments. Again, absolute Ti values in 60007,9029 are higher than in 60007,9027 as are Mg and Fe

Table 2. Surface concentrations of Mg, Fe and Ti, percentage of glass coverage and microcrater densities in individual soil grains. Grains from sample 60007,9029 were not documented at high magnification in the SEM. A blank space in the microcrater density column of the other samples indicates that no microcraters were seen on this grain but that the low quality of the SEM micrographs makes it impossible to claim the absence of any microcraters with certainty.

Sample	Grain #	Surface Concentration (Atoms/cm^2)			Glass Coverage (%)	Microcrater Density (#/cm^2)
		Mg($\times 10^{15}$)	Fe($\times 10^{15}$)	Ti($\times 10^{13}$)		
67601,9003	1	7.1	2.8	8.2	13.0	
	2	6.8	1.5	14.	not measured	
	3	0.95	3.9	4.0	12.4	0
	4	1.15	11.4	3.3	5.8	0
	5	5.76	4.0	12.6	25.6	
	6	1.30	2.56	4.4	19.9	0
	7	0.61	0.76	<2.8	5.5	8.1×10^7
	8	0.80	0.85	1.7	1.	0.9×10^7
	9	0.34	0.45	<1.8	0.8	0
	10	2.25	1.76	4.5	21.4	0
	11	2.02	1.73	1.9	4.1	3.1×10^7
	12	1.87	1.80	2.8	2.7	1.8×10^6
	13	0.61	0.76	<1.0	4.2	0
	14	2.06	1.74	3.3	21.7	3.6×10^6
	15	0.46	0.73	4.8	8.2	
	16	1.45	1.25	0.57	22.7	5.1×10^7
60007,9027	21	1.7	2.3	4.5	17.1	
	22	7.0	3.52	20.7	47.2	
	23	3.0	4.2	7.8	9.3	
	24	0.54	0.44	<1.84	13.2	
	25	1.31	1.03	4.3	1.7	
	26	1.3	0.85	4.1	10.6	
	27	0.47	0.50	<0.92	4.0	
	28	~0.08	0.12	≤3.2	1.9	
	29	≤0.06	≤0.35	≤2.6	0.4	0
	30	13.3	5.4	24.8	88.8	
	31	0	<0.12	0	3.9	0
	32	0.18	0.39	0.6	2.8	0
	33a	1.07	0.74	1.5	6.9	0
	33b	1.9	1.2	5.1	14.2	0
60007,9029	41	1.7	0.7	50.		
	42	6.0	3.15	33.		

values for the probable reason mentioned above. In contrast to the Fe/Mg ratio the Ti/Mg obtained from the CAMECA measurements of 60007,9029 is slightly higher than that measured in 60007,9027 with the ARL IMMA. After the measurement of the Apollo 16 core grains we measured profiles also at $m/e = 49$ in addition to $m/e = 48$ in several grains from the Apollo 15 and Luna 24 core with the ARL machine. Unlike in the case of crystal 76215,91 from a rock surface (Zinner *et al.*, 1977) where the profile at $m/e = 49$ and that at $m/e = 48$ with $^{48}Ca^+$ subtracted did not agree with the Ti isotopic abundances (which led us to infer the presence of molecular interferences), these more recent measurements give good agreement. From this agreement, the comparison of the ARL and CAMECA measurements (Fig. 3), and on the basis of laboratory experiments which show that the contribution of Mg_2^+ is small, we tend to believe

Table 2. *(cont'd.)*

Sample	Grain #	Surface Concentration (Atoms/cm^2) Mg($\times 10^{15}$)	Fe($\times 10^{15}$)	Ti($\times 10^{13}$)	Glass Coverage (%)	Microcrater Density (#/cm^2)
60007,9029	43	17.0	14.5	560.		
	44	5.3	3.15	14.5		
	45	2.4	4.6	62.		
	46	13.3	4.5	69.		
	47	3.3	1.2	21.		
	48	7.7	8.8	51.		
	49	6.2	4.0	40.		
	50	10.0	4.3	40.		
	51	2.4	1.1	10.		
64421,9005	61	$\leq$0.02	$\leq$0.10	<0.5	1.6	0
	62	0.57	0.42	<1.0	7.1	
	63	0.92	0.78	3.7	13.1	
	64a	1.97	1.66	13.8	27.1	
	64b	6.1	4.4	22.8	29.8	
	65	0.57	0.42	2.3	13.3	0
	66	0.57	0.66	<0.8	3.3	
	67	$\lesssim$0.11	$\lesssim$0.07	<0.5	7.2	0
	68	0.66	0.43	1.5	6.4	
15002,562	71	1.13	0.64	31.0	6.2	$\lesssim 3 \times 10^4$
	72	2.12	1.28	20.8	5.1	1.2×10^7
15006,209	76	3.23	1.88	24.3	19.0	0
	77	0	0	0	1.7	0
24077,9002	81	3.17	3.01	20.6	14.4	1.0×10^7
24210,9002	91	<0.04	<0.1	0	2.8	0
	92	<0.03	<0.1	<1.3	4.2	0
	93	0.87	0.65	15.5	18.2	$\lesssim 6 \times 10^4$
	94	0.79	1.02	5.1	2.8	$\lesssim 9 \times 10^4$
	95	1.43	1.41	20.9	21.1	0
	96	5.75	5.77	23.6	16.2	8.0×10^5

that most of the ARL values also represent actual Ti concentrations within the large errors (30–50%) of these measurements. In any case, conclusions involving Ti/Mg ratios are based only on their *upper limit* thus do not depend on the detailed understanding of the interference problem.

The Mg, Ti and Fe surface concentrations measured in all samples are combined in Fig. 4 where Ti/Mg ratios are plotted against Fe/Mg ratios and where we have also added our previous results on 67601 (Zinner *et al.*, 1977). Again, these ratios are compared to the Cameron values and the ratios of average soil and agglutinate compositions. For the samples 64421 and 67601 elemental concentrations of agglutinates have been measured (Rhodes *et al.*, 1975) although they do not differ very much from bulk soil values. The points in Fig. 4 scatter widely and all the grains show Ti/Mg ratios which are higher than the Cameron value. However, most grains still have a lower Ti/Mg ratio than the average agglutinate or soil. The Ti/Mg average of all cratered grains (which are denoted by numbers) is lower than the average of all uncratered ones. This could indicate that for grains exposed on the surface there is a solar wind contribution present. This obviously could not be the case for unexposed (uncratered) grains.

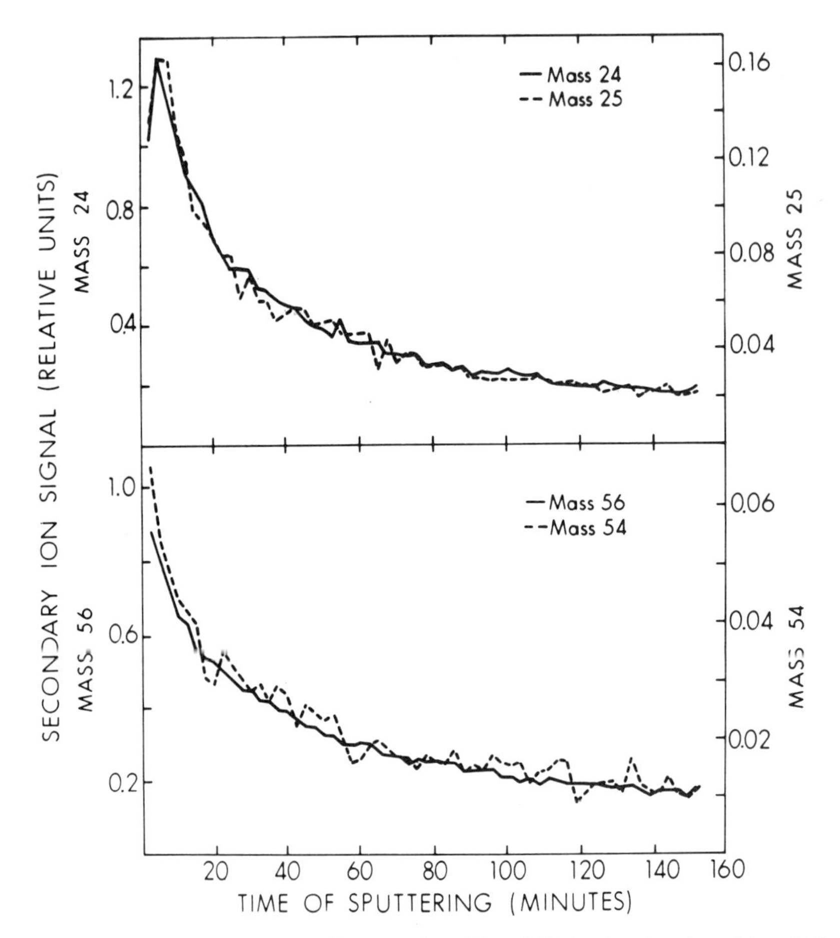

Fig. 1. Concentration depth profiles at m/e = 24 and 25 (top) and m/e = 54 and 56 (bottom) measured with the CAMECA IMS 300 in a plagioclase grain from soil sample 60007,9029. The vertical scales of both graphs are proportional to the normal isotopic abundances of ^{24}Mg and ^{25}Mg and ^{54}Fe and ^{56}Fe respectively.

However, on the basis of elemental ratios alone, it is not possible to identify the source of the surface enhancements unambiguously.

(b) Surface features

Listed in Table 2 are densities of microcraters for the cases when present and surface coverage by glassy disks as observed in the analyzed areas. Whereas glassy disks (pancakes) are present on the surface of practically all grains, only a limited number of grains exhibit microcraters. For some grains the quality of the high magnification SEM pictures was not sufficient to exclude the presence of microcraters of $\lesssim 0.1\ \mu$ size with certainty even if none were detected by the film

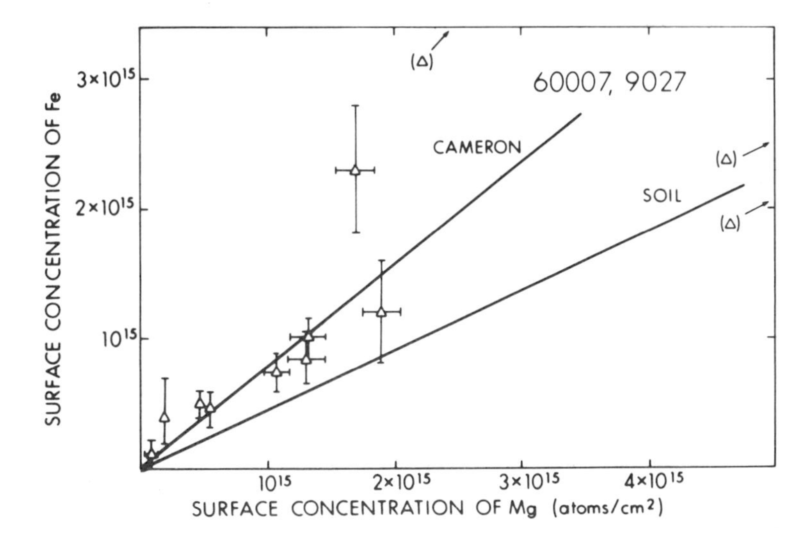

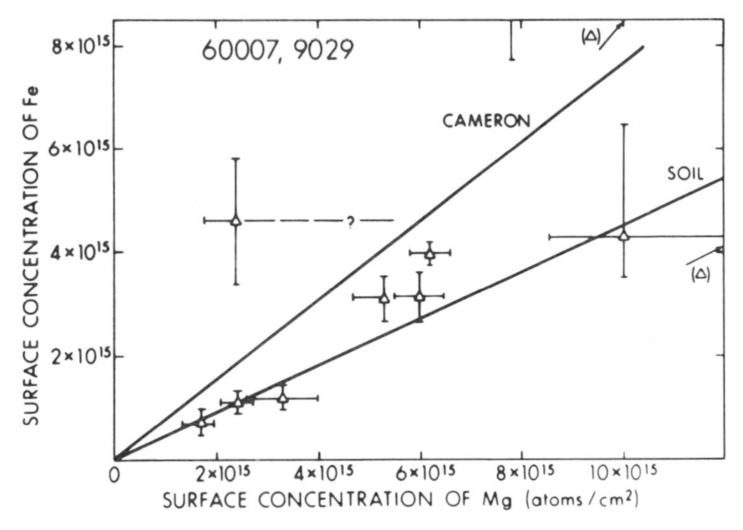

Fig. 2. Surface concentrations of Fe versus surface concentrations of Mg measured in plagioclase grains of two Apollo 16 core samples. Measurements of 60007,9027 were made with the ARL IMMA, those of 60007,9029 with the CAMECA IMS 300. The scales are different in the two graphs to account for the fact that surface enhancements measured with the CAMECA are larger than those measured with the ARL machine. The lines labeled CAMERON and SOIL represent the elemental abundance ratios of the solar system (Cameron, 1973) and the average composition of soil material from the Apollo 16 deep drill core. The first ratio is assumed for the solar wind composition, the second is expected if glass splashes and/or vapor deposits are responsible for the surface enhancements. The soil line is taken from a bulk soil analysis of 60007,114 (Nava and Philpotts, 1973).

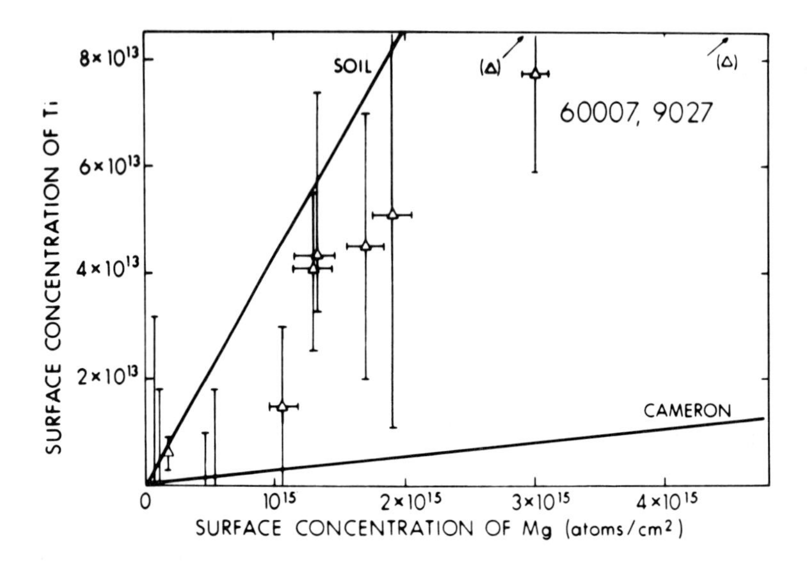

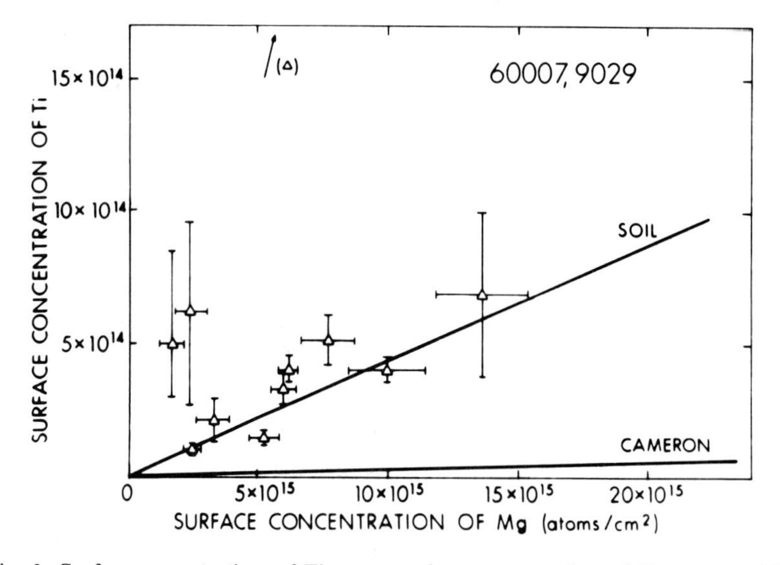

Fig. 3. Surface concentrations of Ti versus surface concentrations of Mg measured in the same grains as in Fig. 2. Vertical bars without central points represent uncertainties of the possible presence of Ti in cases where no surface enhancements were seen in the m/e = 48 depth profile. The lines labeled CAMERON and SOIL have the same meaning as in Fig. 2.

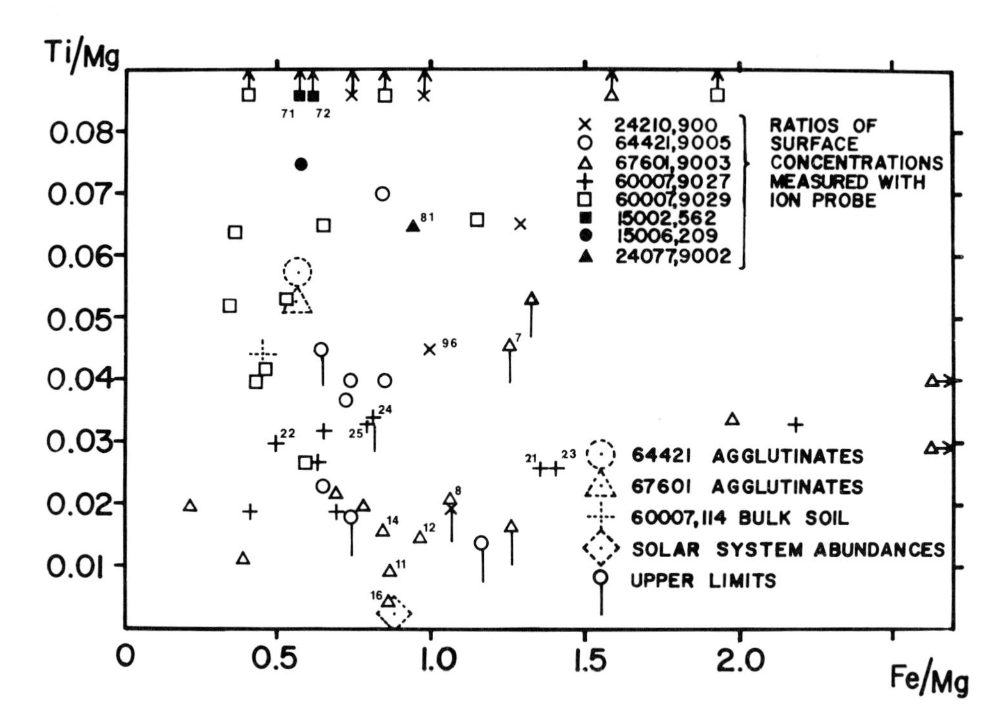

Fig. 4. Scatter plot of Ti/Mg versus Fe/Mg ratios for all samples measured. No error bars are drawn for simplification but are approximately of the same magnitude as in Figs. 2 and 3, i.e., 15–20% for the Fe/Mg ratios and 30–50% for the Ti/Mg ratios. "Upper limits" reflect the experimental uncertainties in cases where no actual surface enhancements were seen at m/e = 48. Numbered points represent grains with microcraters.

scanner. The much higher percentage of cratered grains on the 67601 mount is explained by the different selection criteria applied to this sample: grains with microcraters which were not necessarily flat were also analyzed. The presence of Mg and Fe surface enhancements on grains without any apparent exposure to free space as would be indicated by the presence of microcraters (Morrison and Zinner, 1977; McDonnell, 1977) is evidence for sources other than the solar wind such as glassy pancakes, vapor deposits from meteorite impacts, and coatings produced by the sputtering ejecta from the solar wind ions incident on nearby grains. Also, it has been remarked that Fe in lunar plagioclase crystals could be highly zoned. A zoning pattern which would result in surface enhancements on a depth scale of a few hundred Å but over an area of ~40 $\mu \times 40$ μ would be possible only if the analyzed surface of a given grain is part of the original surface of a naturally grown plagioclase crystal. However, from the SEM micrographs it is clear that an average lunar grain in the size range of 100–200 μ results from a break-up of larger crystals. We thus practically always analyze interior crystal surfaces.

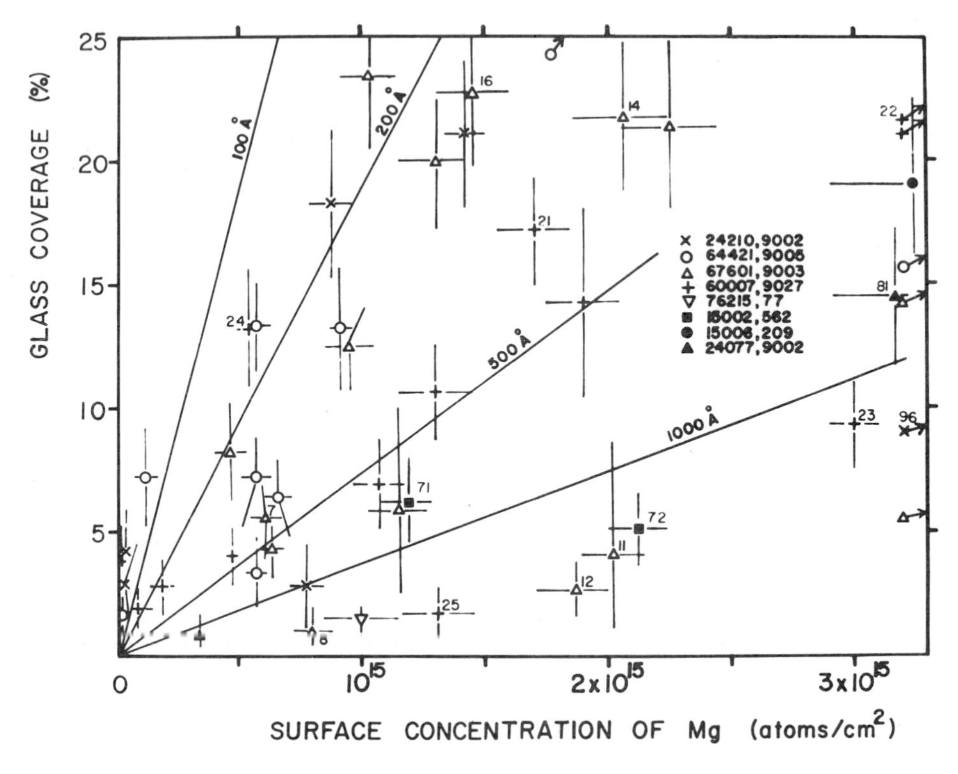

Fig. 5. Percentage of glass coverage of soil grain surfaces versus Mg surface concentration. Included are also measurements on 76215,77 (Zinner *et al.*, 1976 and 1977). The radial lines from the origin are lines of constant thickness for an assumed uniform glass coating of 64421 agglutinate composition which would account for the measured Mg concentrations. Numbered points represent cratered grains. Crystal 76215,77 from a rock is also cratered, although not numbered.

In order to estimate the contribution of the pancakes we plotted in Fig. 5 the percentage of coverage by glass splashes on the surface of grains against the Mg surface concentration. If we assume a Mg concentration of the glass we can calculate how thick the pancakes would have to be on the average in order to account for the observed Mg concentrations. We assumed the 64421 agglutinate composition (Rhodes *et al.*, 1975) to obtain the thicknesses for the lines in Fig. 5. Most derived "glass thicknesses" (we use quotation marks to emphasize the model dependent nature of this quantity) range from ~150Å to a little more than 500Å which is the right order of magnitude for pancake thicknesses based on SEM pictures. This thickness range also agrees roughly with the depth of the Mg enhancements measured from the ion probe depth profiles. This depth can be measured with some accuracy only for the ARL runs, in which the use of a low energy (9kV) NO_2^- beam in combination with an electronic aperture yields good depth resolution (Zinner *et al.*, 1976).

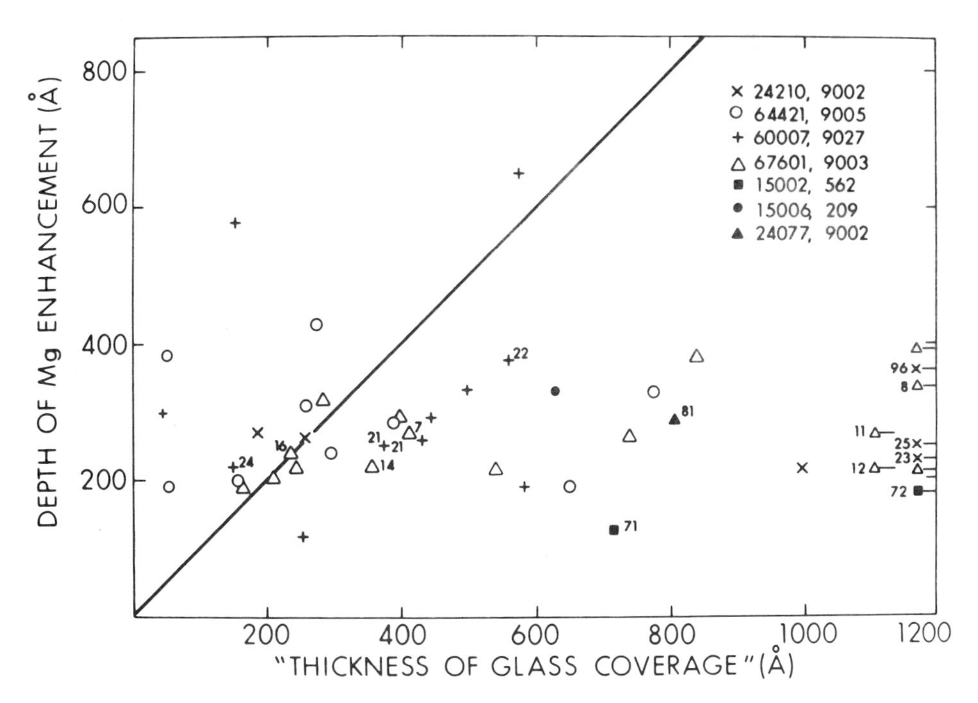

Fig. 6. "Thickness of pancakes" calculated from glass coverage and Mg surface concentrations in Fig. 5 versus the half-maximum depth of the Mg surface enhancement in lunar soil grains. Grains for which these two thicknesses are equal would be plotted on the 45° line. Of the grains with "glass thicknesses" of more than 1200 Å which fall outside the plot and are indicated by symbols with short horizontal lines at the right edge of the figure, all except two are cratered.

Figure 6 shows the "thickness of pancakes" calculated from Fig. 5 plotted against the half-maximum depth of the Mg surface enhancement. This depth was obtained from comparison with the maximum of the Mg marker ion profiles which was calculated to be at 324 Å for 1 keV/nuc and at 620 Å for 2 keV/nuc Mg ions from the LSS (Lindhard *et al.*, 1963) and WSS (Winterbon *et al.*, 1970) theories. There exists no simple correlation between these two quantities. "Glass thicknesses" are on the average a little larger than the Mg fall-off depths but show a much wider scatter. If our belief that glass splashes are the source of the surface enhancements is correct, then the wider variation in the "glass thickness" reflects the variation of the Mg concentration of the glass whereas the more uniform fall-off depth reflects a more uniform physical thickness of the pancakes.

Let us now consider the alternative possible source for the surface enhancements on uncratered grains: namely vapor deposits from meteorite impacts (Zook, 1975) and redeposition of solar wind sputtered material (Cassidy and Hapke, 1975; Baron *et al.*, 1977; Gold *et al.*, 1977; Switkowski *et al.*, 1977). Since we expect these deposits to be distributed uniformly on surfaces and not to

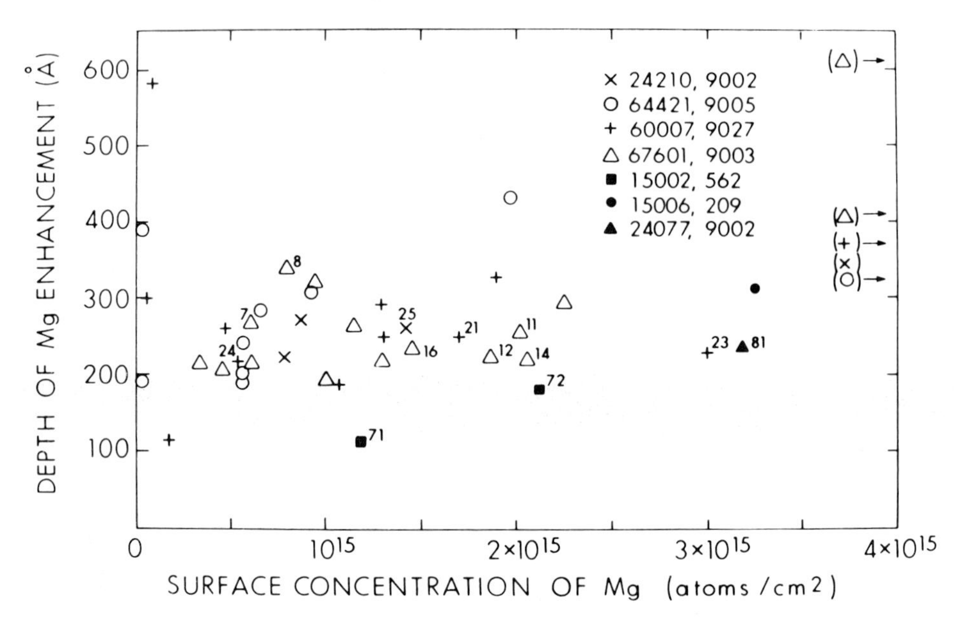

Fig. 7. Half-maximum depth of Mg depth profiles versus Mg surface concentration. For vapor deposits as the source of surface Mg one expects a linear correlation between these two quantities.

be concentrated like the pancakes, the expected thicknesses of these coatings would be much less than 100 Å. This can be seen from Fig. 5 where 100% coverage with a 100 Å thick coating would result in an average surface Mg concentration of $>2.5 \times 10^{15}$ atoms/cm^2 whereas most grains show smaller surface concentrations. Second, since vapor deposits would be the result of averaging over a much larger area from which the vapor originates and would represent a larger number of individual impact events than pancakes, we would expect the deposits to have a more uniform Mg concentration than the pancakes. A variation in the Mg surface concentration would then correspond to a variation of the deposit thickness and we would expect some linear correlation between these two quantities. This is not borne out by the data shown in Fig. 7 where the depth of the Mg enhancements as measured in the ion probe versus the Mg surface concentrations is shown. We conclude that the thickness is apparently quite independent of the surface concentration. This lack of a correlation as well as the measured depth values themselves provide evidence that pancakes, rather than vapor deposits are the most likely source of elemental surface enhancements for most grains. For sputter deposits it is conceivable that because of a particular geometry the source of deposits is only a few grains. In such special cases larger variations of Mg concentrations are possible but we still would expect a correlation (be it with more scatter) between Mg concentration and thickness of coating.

Preliminary attempts have been made to verify this conclusion. Scanning ion images were obtained in the ARL ion microprobe but poor spatial resolution compared to the average size of the pancakes as well as the fact that the secondary ion image is dominated by topographic effects of rough crystal surfaces limit interpretation. Nonetheless, in a few cases we succeeded in obtaining images which showed that the distribution of the major elements Al, Si and Ca was uniform indicating a homogeneous plagioclase surface, but with a different Mg distribution. Figure 8 compares the SEM image of a plagioclase grain from sample 15002,562 with the Mg^+ and Al^+ image obtained in the ion probe. A Scanning Auger Microscope (PEI) which has a high enough spatial resolution so that pancakes can be probed with the Auger spectrometer was also utilized. Charging problems and machine availability limited the amount of information we could obtain from this investigation but given the right sample preparation such an instrument should be capable of comparing the Mg and Fe concentration of clean areas on grains with areas covered by glass splashes. The sensitivity of an Auger spectrometer is high enough to detect Fe and Mg at the concentration level of a few percent expected in glassy splashes. Unfortunately, sensitivity is too low to detect elements implanted from the solar wind in glass-free areas.

Additional evidence concerning the thickness of pancakes is provided by comparing grain surfaces as they appear in the SEM before and after sputtering with the ion beam. Neither glassy disks nor their outlines could be recognized after sputtering (removal of ~1500 Å), and previously existing pancakes seem to have been completely removed. It should be possible to obtain direct measurements of pancake thicknesses in a Scanning Auger Microscope in combination with ion etching.

So far we have discussed the nature of the elemental surface enhancements in uncratered grains which presumably have never been exposed to free space. While almost all grains show glass splashes only a much smaller fraction has microcraters. One explanation is that the re-entrant nature of the lunar regolith makes it apparently more likely for grains to "see" other grains, leading to the acquisition of glassy surface splashes, than to "see" the sun and the sky directly. But it is also possible that a grain acquires all its pancakes from a single larger impact event. For cratered grains with glass splashes we expect these splashes to contribute to the surface enhancements in the same way as for uncratered grains. However, if we look at the distribution of cratered grains on the scatter plots of Figs. 5 and 6 we notice that it differs from the distribution of uncratered grains. Only about 50% of the cratered grains plot in the same locations as uncratered ones, while the other 50% fall into quite distinct locations (beyond the 1000Å "glass thickness" line in Fig. 5 and off the plot to the right in Fig. 6). If glass is to account for the Mg enhancement in this second group, glass thicknesses would be required which are much larger than the Mg depth measured in the ion probe even if we assume pure olivine or pyroxene composition for the glassy pancakes. Furthermore, even if Mg enhancements are due to glass pancakes in *all* grains it would not be clear why the glass composition of *cratered* grains should be

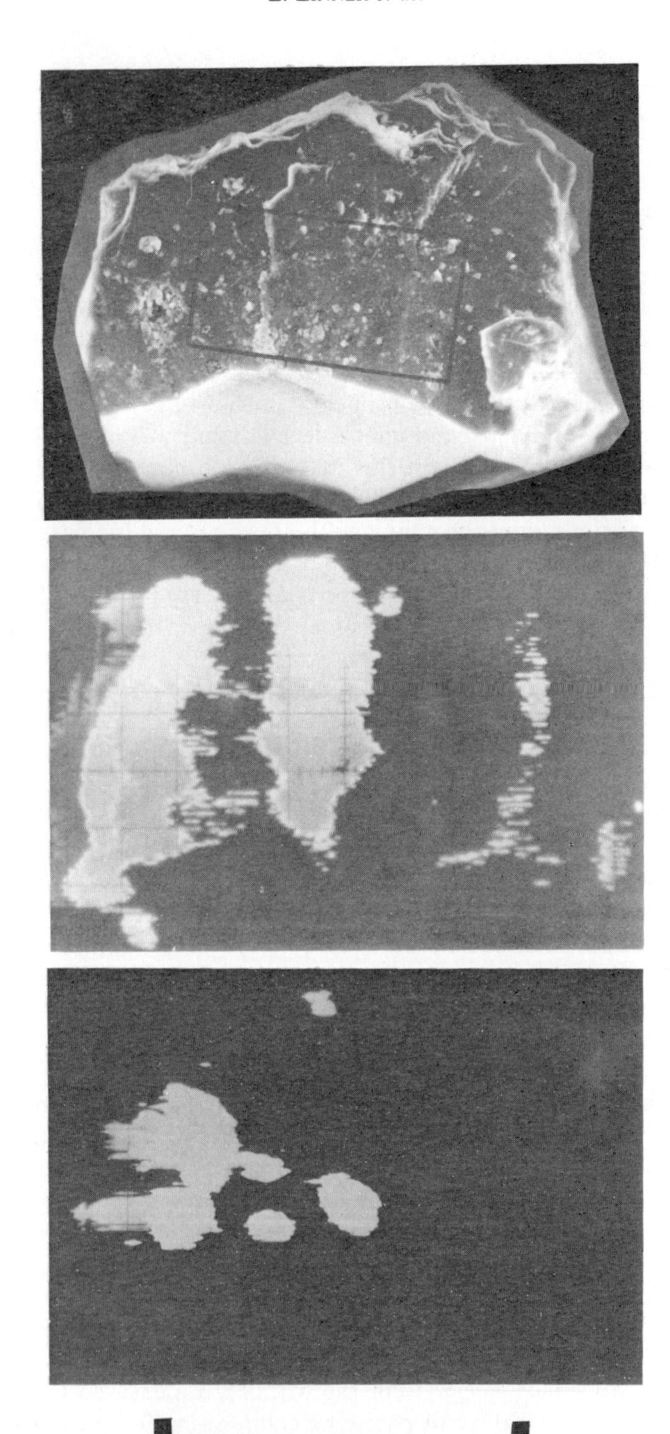

different from that of *uncratered* ones. Alternatively, we can interpret the high surface enhancements relative to the glass coverage as the addition of Mg from a different source. Also in Fig. 4 cratered grains plot closer to the Cameron solar system value than uncratered grains. The Ti/Mg ratio of the Apollo 15 soils falls outside the range of Fig. 4 but both cratered grains from the Apollo 15 core still have a much smaller Ti/Mg ratio than the soil value. The Ti/Mg ratio of the Apollo 17 soils is even higher but for 76215 the measured surface concentration ratios are between a factor of 3 and 10 below this value (Zinner *et al.*, 1976 and 1977). In this crystal from a rock we interpreted the surface Mg and Fe as being due to solar wind implanted ions. In Fig. 5, 76215,77 falls also well beyond the 1000Å "glass thickness" line. According to our argument above that this requires the addition of Mg from a source different from pancakes we thus still interpret the 76215 enhancements as being of solar wind origin. We also conclude for the few soil grains with low glass coverage and/or low Ti/Mg ratios that the solar wind is still the most likely source of a large part of their Mg and Fe surface enhancements.

We have etched several of these grains for solar flare tracks. In no case could we find any steep gradients at the exposed surface. Track densities were much higher than we would have expected if tracks were produced only during exposure to free space for a solar wind exposure time derived from the surface Mg. This indicates as before (Poupeau *et al.*, 1975) that most tracks were accumulated under some shielding.

Unfortunately, some glassy pancakes are present in all cratered grains with large surface enhancements, and since we do not know how to subtract their contribution, there is considerable uncertainty in the solar wind contribution. Even if we allow a glass contribution, Mg surface concentrations attributable to the solar wind range up to $1.5 - 2.0 \times 10^{15}$ atoms/cm^2. Previously we (Zinner *et al.*, 1977) and others (Filleux *et al.*, 1977; Jull and Pillinger, 1977) discussed how the solar wind sputtering affects the accumulation of solar wind implanted elements. The maximum surface concentrations given above interpreted as solar wind implanted elements would still require a very low solar wind erosion rate. Recently, Filleux *et al.* (1978) have revised their earlier measurements of C concentrations (Filleux *et al.*, 1977) and added new measurements which would also imply very low erosion rates even if still being larger than ours. The solar wind sputter rate remains an extremely complicated question (Kerridge *et al.*, 1978; Carey and McDonnell, 1978) and, as Filleux *et al.* (1978) already pointed out, until it is solved in a satisfactory way any discussion of solar wind equilibrium concentrations takes place on shaky ground.

The fact that after searching through many hundreds of relatively large soil

---

Fig. 8. A plagioclase grain from 15002,562 imaged in the SEM (top), in the ARL ion probe at m/e = 27 (middle) and m/e = 24 (bottom). The scale is the same for all three pictures; the scale bar is 100 microns. The Mg and Al distributions are not the same for this grain. High Mg areas can be associated with glassy accreta on the surface.

grains and after analyzing dozens of these we found only a handful of them where we believe we can observe solar wind implanted ions points to the difficulty in measuring solar wind species with the ion probe: we have to compete with seemingly ever-present glass coatings on the grain surfaces. Clean grains which have been exposed to free space (i.e., those which show microcraters, but which are essentially free of glassy pancakes) are extremely rare. The plagioclase crystals from 76215 are exceptions. This rock has the shortest exposure age of all returned Apollo samples and has not gone through the complex gardening process of the average soil grain. An additional problem is the fact that the thickness of glass pancakes is approximately the same as the implantation depth of solar wind ions. If we want to look for elements not indigenous to the moon for which glass splashes would not represent a background problem, we are frustrated by the low ion probe sensitivity for these elements (N, noble gases) or by problems of terrestrial contamination (C).

The measurement of depth profiles in the clean areas between the pancakes would require a new instrument which is a combination of an ion probe with an SEM, in which the pancakes could be seen during analysis and in which the ion beam could be focused into a finer spot and rastered over an area of several microns on a side. Basic designs of instruments incorporating these features exist (Liebl, 1971 and 1972) but their practical realization lies in the future.

Acknowledgments—We thank R. M. Walker for his active interest in this effort and acknowledge his many helpful suggestions. This work was made possible through the Lunar and Planetary Institute and the Planetary and Earth Science Division of the Johnson Space Center, which provided us with the experimental use of the ARL ion microprobe. At the JSC D. Phinney, D. Anderson, C. Meyer, and T. Rohrer helped us in many ways for which we are most grateful. We thank C. Allègre at the University of Paris who granted us access to the CAMECA IMS 300 as well as M. P. Semet and N. Shimizu who supported us through the experimental runs on the CAMECA instrument. F. Lalu was instrumental for the ion implantations and P. Maurice was responsible for low magnification SEM photographs of one of the grain mounts. We acknowledge the extensive efforts of P. Swan who not only prepared and mounted the samples but who also is responsible for almost all the SEM documentation on which the present work is based. Careful reviews by D. Burnett and I. Steele have improved this manuscript. We thank B. Wilcox and C. White for manuscript preparation. This work was in part supported by NASA grant No. NGL 26-008-065. A portion of this work was performed at the Lunar and Planetary Institute, which is operated by the Universities Space Research Association under contract No. NSR-09-051-001 with the National Aeronautics and Space Administration. This paper constitutes Lunar and Planetary Institute Contribution No. 338.

REFERENCES

Baron R. L., Bilson E., Gold T., Colton R. J., Hapke B. and Steggert M. A. (1977) The surface composition of lunar soil grains: a comparison of the results of Auger and X-ray photoelectron (ESCA) spectroscopy. *Earth Planet. Sci. Lett.* **37**, 263–272.

Bibring J. P., Chaumont J., Dran J. C., Lalu F., Langevin Y., Maurette M. and Vassent B. (1977) Solar wind erosion of lunar dust grains. A progress report (abstract). In *Lunar Science VIII*, p. 106–108. The Lunar Science Institute, Houston.

Blanford G. E., Fruland R. M., McKay D. S. and Morrison D. A. (1974) Lunar surface phenomena: solar flare track gradients, microcraters, and accretionary particles. *Proc. Lunar Sci. Conf. 5th*, p. 2501–2526.

Bogard D. D. and Hirsch W. C. (1977) Noble gases in Luna 24 core soils (abstract). In *Papers Presented to the Conference on Luna 24*, p. 44–47. The Lunar Science Institute, Houston.

Cameron A. G. W. (1973) Abundances of the elements in the solar system. *Space Sci. Rev.* **15**, 121–146.

Carey W. C. and McDonnell J. A. M. (1978) The role of diffusion and heterogeneous target characteristics in the sputter mechanism: Monte Carlo simulations and laboratory He^+ ion sputter measurements (abstract). In *Lunar and Planetary Science IX*, p. 152–154. Lunar and Planetary Institute, Houston.

Cassidy W. and Hapke B. (1975) Effects of darkening processes on surfaces of airless bodies. *Icarus* **25**, 371–383.

Comstock G. M. (1978) How dusty are lunar rocks? Evidence from track profiles (abstract). In *Lunar and Planetary Science IX*, p. 189–191. Lunar and Planetary Institute, Houston.

Crozaz G. (1978) Nuclear particle tracks and the regolith at the Luna 24 site. In *Mare Crisium: The View From Luna 24* (R. B. Merrill and J. J. Papike, eds.), p. 165–169. Pergamon, N.Y.

Crozaz G. and Dust S. (1977) Irradiation history of lunar cores and the development of the regolith. *Proc. Lunar Sci. Conf. 8th*, p. 3001–3016.

Filleux C., Spear R., Tombrello T. A. and Burnett D. S. (1978) Carbon depth distributions for soil breccias (abstract). In *Lunar and Planetary Science IX*, p. 317–319. Lunar and Planetary Institute, Houston.

Filleux C., Tombrello T. A. and Burnett D. S. (1977) Direct measurement of surface carbon concentrations. *Proc. Lunar Sci. Conf. 8th*, p. 3755–3772.

Geiss J. (1973) Solar wind composition and implications about the history of the solar system. Cosmic Ray Conference, 13th International. *Invited lectures; rapporteur papers* **5**, 3375–3398. Dept. Phys. and Astron., Univ. Denver.

Gold T., Bilson E. and Baron R. L. (1977) The search for the cause of the low albedo on the moon. *J. Geophys. Res.* **82**, 4899–4908.

Goldberg R. H., Burnett D. S. and Tombrello T. A. (1975) Fluorine surface films on lunar samples: evidence for both lunar and terrestrial origins. *Proc. Lunar Sci. Conf. 6th*, p. 2189–2200.

Goldberg R. H., Tombrello T. A. and Burnett D. S. (1976) Fluorine as a constituent in lunar magmatic gases. *Proc. Lunar Sci. Conf. 7th*, p. 1597–1613.

Grant R. W. and Housley R. M. (1978) An XPS study of Apollo 17 orange and black glasses (abstract). In *Lunar and Planetary Science IX*, p. 405–407. Lunar and Planetary Institute, Houston.

Housley R. M. and Grant R. W. (1977) An XPS (ESCA) study of lunar surface alteration profiles. *Proc. Lunar Sci. Conf. 8th*, p. 3885–3899.

Ishitani T., Shimizu R. and Tamura H. (1974) Knock-on effects in surface analysis. Proc. 2nd Int. Conf. on Solid Surfaces, 1974. *Japanese J. Appl. Phys. Suppl. 2, Pt. 2*, p. 591–594.

Ishitani T., Shimizu R. and Tamura H. (1975) Atomic mixing in ion probe microanalysis. *Appl. Phys.* **6**, 277–279.

Jull A. J. T. and Pillinger C. T. (1977) Effects of sputtering on solar wind element accumulation. *Proc. Lunar Sci. Conf. 8th*, p. 3817–3833.

Kerridge J. F., Kaplan I. R. and Petrowski C. (1978) Carbon isotope systematics in the Apollo 16 regolith (abstract). In *Lunar and Planetary Science IX*, p. 618–620. Lunar and Planetary Institute, Houston.

Kiko J., Kirsten T. and Ries D. (1978) Peculiarities of solar wind in lunar olivines (abstract). In *Lunar and Planetary Science IX*, p. 621–623. Lunar and Planetary Institute, Houston.

Liebl H. (1971) Design of a combined ion and electron microprobe apparatus. *Int. J. Mass Spectrom. Ion Phys.* **6**, 401–412.

Liebl H. (1972) A coaxial combined electrostatic objective and anode lens for microprobe mass analysers. *Vacuum* **22**, 619–621.

Lindhard J., Scharff M. and Schiott H. E. (1963) Range concepts and heavy ion ranges (notes on atomic collisons II). *Mat. Fys. Medd. Dan. Videnskab. Selskab* **33** (14), 1–42.

McDonnell J. A. M. (1977) Accretionary particle studies on Apollo 12054,58: *in-situ* lunar surface microparticle flux rate and solar wind sputter rate defined. *Proc. Lunar Sci. Conf. 8th*, p. 3835–3857.

Morrison D. A. and Zinner E. (1975) Studies of solar flares and impact craters in partially protected crystals. *Proc. Lunar Sci. Conf. 6th*, p. 3373–3390.

Morrison D. A. and Zinner E. (1977) 12054 and 76215: new measurements of interplanetary dust and solar flare fluxes. *Proc. Lunar Sci. Conf. 8th*, p. 841–863.

Morrison G. H. and Slodzian G. (1975) Ion microscopy. *Anal. Chem.* **47**, 932A–943A.

Müller H. W., Jordan J., Kalbitzer S., Kiko J. and Kirsten T. (1976) Rare gas ion probe analysis of helium profiles in individual lunar soil particles. *Proc. Lunar Sci. Conf. 7th*, p. 937–951.

Nava D. F. and Philpotts J. A. (1973) A lunar differentiation model in light of new chemical data on Luna 20 and Apollo 16 soils. *Geochim, Cosmochim. Acta* **37**, 963–973.

Poupeau G., Michel-Lévy M. C., Mandeville J. C., Johnson J. and Romary Ph. (1978) Microcrater and solar-flare tracks maturation of the lunar regolith. In *Mare Crisium: The View From Luna 24* (R. B. Merrill and J. J. Papike, eds.), p. 137–155. Pergamon, N.Y.

Poupeau G., Walker R. M., Zinner E. and Morrison D. A. (1975) Surface exposure history of individual crystals in the lunar regolith. *Proc. Lunar Sci. Conf. 6th*, p. 3433–3448.

Rhodes J. M., Adams J. B., Blanchard D. P., Charette M. P., Rodgers K. V., Jacobs J. W., Brannon J. C. and Haskin L. A. (1975) Chemistry of agglutinate fractions in lunar soils. *Proc. Lunar Sci. Conf. 6th*, p. 2291–2307.

Shimizu N., Semet M. P. and Allègre C. J. (1978) Geochemical applications of quantitative ion microprobe analysis. *Geochim. Cosmochim. Acta* **42**. In press.

Shimizu N., Semet M. P., Lorin J. C. and Allègre C. J. (1977) The energy filtering and geochemical applications of quantitative ion microprobe analysis (abstract V41). *EOS (Trans. Amer. Geophys. Union)* **58**, 521.

Switkowski Z. E., Haff P. K., Tombrello T. A. and Burnett D. S. (1977) Mass fractionation at the lunar surface by solar wind sputtering. *J. Geophys. Res.* **82**, 3797–3804.

Winterbon K. B., Sigmund P. and Sanders J. B. (1970) Spatial distributions of energy deposited by atomic particles in elastic collisons. *Mat. Fys. Medd. Dan. Videnskab. Selskab* **37** (14), 1–73.

Zinner E. (1978) The effect of preferential sputtering on ion microprobe depth profiles in minerals. In *13th Annual Conf. of the Microbeam Analysis Society*, Paper #32.

Zinner E. and Walker R. M. (1975) Ion-probe studies of artifically implanted ions in lunar samples. *Proc. Lunar Sci. Conf. 6th*, p. 3601–3617.

Zinner E., Walker R. M., Borg J. and Maurette M. (1974) Apollo 17 lunar surface cosmic ray experiment—measurement of heavy solar wind particles. *Proc. Lunar Sci. Conf. 5th*, p. 2975–2989.

Zinner E., Walker R. M., Chaumont J. and Dran J. C. (1976) Ion probe analysis of artifically implanted ions in terrestrial samples and surface enhanced ions in lunar sample 76215,77. *Proc. Lunar Sci. Conf. 7th*, p. 953–984.

Zinner E., Walker R. M. Chaumont J. and Dran J. C. (1977) Ion probe surface concentration measurements of Mg and Fe and microcraters in crystals from lunar rock and soil samples. *Proc. Lunar Sci. Conf. 8th*, p. 3859–3883.

Zook H.A. (1975) The state of meteoritic material on the moon. *Proc. Lunar Sci. 6th*, p. 1653–1672.

Proc. Lunar Planet. Sci. Conf. 9th (1978), p. 1687–1709.
Printed in the United States of America

Sputtering: Its relationship to isotopic fractionation on the lunar surface

J. F. KERRIDGE and I. R. KAPLAN

Institute of Geophysics and Planetary Physics
University of California, Los Angeles, California 90024

Abstract—Silicon, oxygen, sulfur and, possibly, potassium show systematic enrichment of heavier isotopes with increasing soil maturity, probably as a result of solar-wind sputtering, but such a pattern is not exhibited by solar wind-implanted species. Apparently sputter-erosion is not penetrating to their implantation depth, about 200 Å. This suggests that sputtering on the moon is being impeded by deposition of vapor condensate following meteoritic impact.

Although neither secular variation in the isotopic composition at the source of regolith carbon, probably the solar wind, nor isotopic fractionation of carbon after implantation in the regolith can be ruled out, carbon isotope systematics currently yield no evidence in support of either process.

INTRODUCTION

It was recognized early that irradiation of the lunar surface by solar wind with energies around 1 KeV per nucleon could sputter material from the surfaces of grains and produce physical and chemical changes in the lunar regolith (Wehner *et al.*, 1963a,b; Hapke, 1966). More recently, interest has revived in this process as a possible mechanism for: (a) producing observed changes in lunar albedo (Hapke, 1973; Gold *et al.*, 1974); (b) reducing indigenous ferrous iron to superparamagnetic metal grains (Yin *et al.*, 1972; Hapke *et al.*, 1975); (c) enriching grain surfaces in heavy elements (Hapke *et al.*, 1975; Pillinger *et al.*, 1976a); and (d) causing isotopic fractionations for some elements in soil samples (Epstein and Taylor, 1971; Pillinger *et al.*, 1976b; Switkowski *et al.*, 1977; Haff *et al.*, 1977). We focus here upon the last of these topics, i.e., the concept that preferential loss of light species from the moon by means of sputtering is capable of explaining observed enrichments in the heavy isotopes of some elements in the lunar surface.

An important detail is that sputtering in the lunar environment is a complex process which may be divided into several stages, each capable of producing its own characteristic fractionation, and that different models have emphasized different stages. Thus, Switkowski *et al.* (1977) considered gravitational sorting of sputtered atoms in ballistic trajectories from the lunar surface; Haff *et al.* (1977) expanded on this idea to include mass-dependent recoil within the sputtered surface; whereas Cassidy and Hapke (1975) and Paruso *et al.* (1978) emphasized preferential sticking of heavy species during redeposition of sputtered material.

Insofar as isotopic fractionation is concerned, the most detailed theoretical treatment of the lunar sputtering process to date is that by Switkowski *et al.* (1977) and Haff *et al.* (1977). They considered establishment of an equilibrium layer on grain surfaces from which atoms are removed by sputtering and to which atoms are added both by diffusion from underlying material and by reimplantation of previously sputtered material which remained gravitationally bound to the moon. A net loss of material from the moon resulted from the fraction of atoms which were sputtered with greater than lunar escape velocity. A general equation was derived for the equilibrium surficial enrichment of the heavier of two components, such as two isotopes of the same element, e.g., ^{34}S and ^{32}S, or two elements, e.g., silicon and oxygen. This expression included a term which depended upon the surface binding energy for the species in question. It was pointed out that, although such binding energies are typically a few eV for normal oxides, the heavily radiation-damaged lunar surface would probably be characterized by somewhat lower values, leading to greater calculated fractionation. A value of 1 eV appeared to be reasonable and gave an acceptable match with observed fractionations, which are typically on the order of a few percent for surficial material. In order to test their model, Switkowski *et al.* (1977) and Haff *et al.* (1977) employed results of analyses for $\delta^{18}O$, $\delta^{30}Si$, $\delta^{34}S$ and Si/O. These data will be discussed in more detail later in this paper; the match between theory and observation is good and justifies the conclusion that "mass fractionation by solar wind is an important phenomenon on the lunar surface" (Haff *et al.*, 1977). We shall take as a working hypothesis that solar wind sputtering is in fact responsible for the fractionations given above and we shall consider the implications of this for other isotopic systems.

Before proceeding, however, it is necessary to point out that other possible mechanisms for producing these isotopic fractionations may exist. One possibility is that, instead of a physical interaction between the solar wind and indigenous atoms, chemical reaction between solar wind hydrogen and elements such as silicon, oxygen and sulfur could produce volatile compounds, SiH_4, H_2O, H_2S, which could be readily lost by diffusion. Faster diffusion of compounds containing the lighter isotopes would then enrich the residue in the heavier isotopes (Kaplan *et al.*, 1970). Similarly, diffusive loss of atomic species at the very high temperatures reached during meteorite impact might result in preferential loss of light isotopes (Clayton *et al.*, 1974). Finally, for relatively volatile elements such as sulfur, diffusion at lunar daytime temperatures may be significant.

Intuitively, it seems unlikely that diffusion of atomic species is capable of explaining the observations. The observed isotopic fractionations for silicon, oxygen and sulfur are comparable, suggesting that diffusive loss of each was also comparable, whereas they differ greatly in volatility, and, at any temperature, they would have been depleted to very different degrees. Examples of fractionation probably produced by diffusive loss will be considered later in connection with the noble gases and will be shown to differ significantly from the silicon, oxygen and sulfur systematics. Diffusive loss of products of reaction with solar wind hydrogen cannot be so easily dismissed as a fractionation mechanism. Fortunately, an observational test may be applied to this hypothesis, as fractionation should in that case be restricted to elements which form volatile hydrides. The key elements for this test are calcium and potassium, although results to date have not wholly resolved this question, as will be discussed later.

A complication may arise in assessing the role of volatility in establishing isotopic fractionation patterns as the behavior of an element during ion sputtering may closely resemble its behavior during thermal volatilization (Housley and Grant, 1977).

In what follows, we review the evidence for isotopic variations among lunar samples, except for nuclear (spallogenic or radiogenic) effects. The systematics of those fractionations believed to be due to sputtering are investigated with the objective of establishing the "signature" of the sputtering process. Similar systematics are then sought among other elements in order either to characterize other fractionation processes or to identify the cause of unknown fractionations. An immediate goal is the interpretation of isotopic variations in regolith carbon.

A concept which will feature prominently in this discussion is that of regolith maturity. The approach adopted here is to regard maturation as a continuous process leading to a homogeneous soil, each component of which has had a similar history of surface exposure. This is a conscious over-simplification; in reality virtually all soils are heterogeneous, consisting of components characterized by different exposure histories. For example, sub-mature soils, such as those from Station 13, Apollo 16, are invariably mixtures of mature, "country" regolith with relatively fresh ejecta from recent craters, e.g., North Ray Crater (Heymann *et al.*, 1975). However, although attempts have been made to model heterogeneous maturation quantitatively, either through the evolution of grain size parameters (e.g., Mendell and McKay, 1975; Lindsay, 1975), or by identification of major components of characterizable maturity (e.g., Heymann *et al.*, 1975; Heymann, 1978; Pillinger *et al.*, 1977), those models do not yet have sufficient predictive capability to enable them to be generally applied to interpretation of analytical data such as the carbon isotope systematics considered here.

Observed Isotopic Variations

Silicon and oxygen

Because isotopic analyses of silicon and oxygen are frequently conducted on the same aliquots and because the two elements show very similar effects, it is convenient to consider them together. Relative to their parental rocks, bulk soils are slightly enriched in the heavier isotopes of these elements, but these effects are greatly magnified if the silicon and oxygen are removed progressively from soil samples by partial fluorination (Epstein and Taylor, 1971; Taylor and Epstein, 1973). The same authors also showed that these enrichments correlated with soil maturity, as indicated by contents of solar wind hydrogen and noble gases, and that enrichment in $^{30}Si/^{28}Si$ was twice as large as that in $^{29}Si/^{28}Si$, demonstrating that the effect was produced by a mass-dependent process. Clayton *et al.* (1974) showed similarly that enrichments in $^{18}O/^{16}O$ were twice those in $^{17}O/^{16}O$. Maximum enrichments of about 25‰ in $\delta^{30}Si$ and 50‰ in $\delta^{18}O$ were observed in the first cut of material removed by fluorination, the enrichments decreasing as more material was stripped. This was initially interpreted as a purely surficial effect, although subsequently it was shown that partial fluorination attacked agglutinates, i.e., formerly exposed surfaces, as well as existing surfaces (Pillinger *et al.*, 1976b). Because the depth resolution of the partial fluorination technique is poorly controlled, the maximum measured enrichments represent lower limits to the values characteristic of the actual surfaces.

Sulfur

Soils are enriched, relative to rocks, in $^{34}S/^{32}S$ and this enrichment is surface correlated (Rees and Thode, 1974) and also correlated with soil maturity as measured by contents of agglutinates or solar wind nitrogen (Kerridge *et al.*, 1975a). This last relationship is illustrated in Fig. 1 and shows that bulk enrichments can reach 12‰ relative to rock values, although enrichments of about 17‰ may be observed in grain size fractions (Rees and Thode, 1974), leading to a lower limit of about 20‰ for actual surface enrichments. That the enrichment in $^{34}S/^{32}S$ resulted from a mass-dependent process was demonstrated by Rees and Thode (1972), who showed that enrichments in $^{36}S/^{32}S$ and $^{33}S/^{32}S$ were twice and half, respectively, those in $^{34}S/^{32}S$.

Thus isotopic variations in sulfur closely resemble those in silicon and oxygen, both qualitatively and quantitatively, even though sulfur is, compared with the other elements, much less abundant, significantly more volatile and of partly (up to about 30%) meteoritic origin (Kerridge *et al.*, 1975a). As mentioned earlier, the observed fractionations in these elements match well the predictions of models of sputter-fractionation (Switkowski *et al.*, 1977; Haff *et al.*, 1977). It should be noted that the proportion of total sulfur which is surficial, and therefore vulnerable to sputter-fractionation, is much greater than for silicon and oxygen. This does not preclude direct comparison of predictions and observations because the relevant experimental data pertain only to the surficial fraction in each case. However, it is clear that some process, presumably vaporization and recondensation, is responsible for transfer of some initially volume-correlated, indigenous sulfur onto surfaces of grains, and it is possible that this process could be accompanied by some isotopic fractionation. Therefore the observed surficial enrichment in ^{34}S may represent the sum of two fractionation processes, which could cause a discrepancy between the observed isotopic composition of surficial sulfur and that predicted by a model involving only sputter fractionation.

Because sputtering efficiency is a function of atomic mass and surface binding energy and does not depend upon chemical reactivity, fractionations should not vary greatly for the common rock-forming elements, provided that they are exposed on grain surfaces. Consequently, the presence or absence of mass-

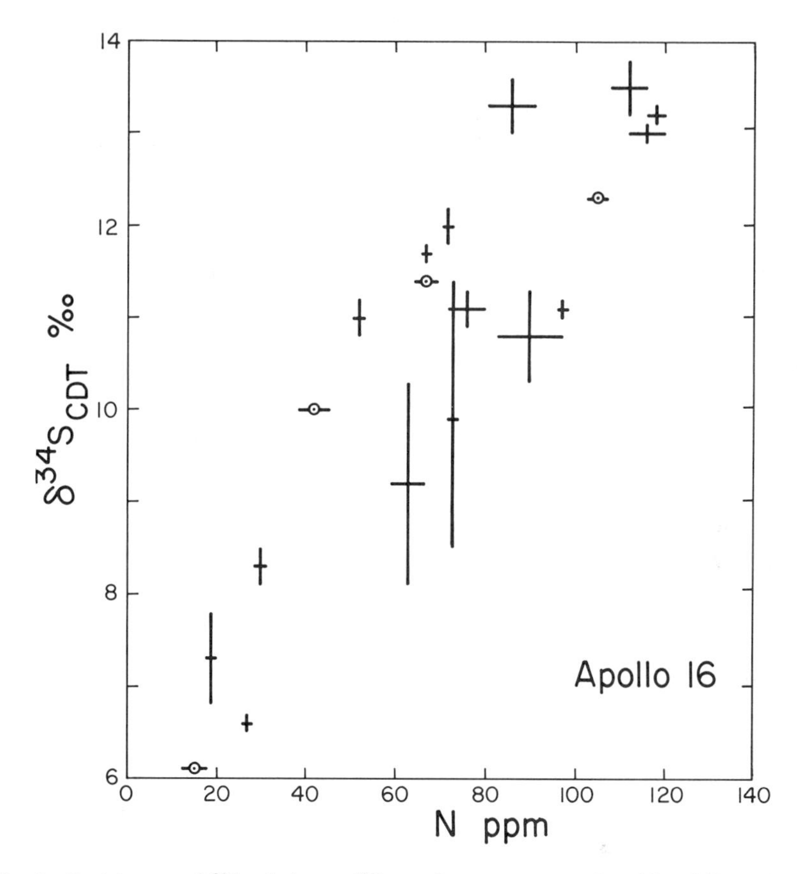

Fig. 1. Enrichment of ^{34}S relative to ^{32}S correlates very strongly with soil maturity, measured by content of solar wind-implanted nitrogen. Rocks at the Apollo 16 site are characterized by $\delta^{34}S$ values around +1.8‰ relative to Canyon Diablo troilite, so that maximum observed enrichments in bulk samples are about 12‰. From Kerridge *et al.* (1975a).

dependent isotopic variations in elements such as calcium and potassium, which are comparable in mass to silicon and sulfur but which do not form volatile hydrides, can provide a useful test of the sputter-fractionation hypothesis. Results of such analyses are therefore considered next.

Calcium and potassium

In order to extract the calcium associated with isotopically fractionated silicon and oxygen, Russell *et al.* (1977) used aqueous leaching of material previously activated by partial fluorination following the technique of Epstein and Taylor (1971). Enrichments of about 3‰ were found for $^{44}Ca/^{40}Ca$ which, though significantly above their limit of detectability, were an order of magnitude smaller than the analogous effects in silicon and oxygen. Although this result may represent a problem for the sputter hypothesis, a number of other possible explanations cannot be excluded at this time. These include the possibility that the leaching procedure does not in fact extract the most heavily fractionated calcium or, alternatively, that it is diluting fractionated calcium with normal material dissolved from easily soluble phases (Russell *et al.*, 1977). That sputtering can, in fact, produce fractionations of about 10‰ in $^{44}Ca/^{40}Ca$ was demonstrated by Griffith *et al.* (1978) in simulation experiments using terrestrial fluorite and plagioclase. These experiments reproduced only the processes occurring within the sputtered zone and did not simulate sorting within the lunar gravitational field or possible fractionation during redeposition. Consequently, significantly larger fractionations, possibly some tens per mill, might be expected for sputtering under lunar conditions. The experiments of Griffith *et al.* (1978) therefore support the idea that sputtering can generate mass fractionations to the degree observed for silicon, oxygen and sulfur, but fail to explain why such an effect is not observed for calcium.

Because isotopic analysis of potassium is effectively restricted to the two isotopes ^{39}K and ^{41}K, corrections for instrumental and chemical fractionations are more difficult than for calcium and resulting data have significantly greater uncertainty. Nonetheless, Barnes *et al.* (1973), Garner *et al.* (1975) and Church *et al.* (1976) have reported enrichments of up to about 13‰ in $^{41}K/^{39}K$ for some bulk soils and respect to rocks. There is a tendency for this enrichment to increase with soil maturity. Thus, the potassium isotope systematics resemble those of silicon, oxygen and sulfur, although the magnitude of the experimental uncertainties prevents the data from being used as a rigorous test of any fractionation model. It is worth noting that there is evidence that some potassium is surface correlated in analysis of grain size fractions (Evensen *et al.*, 1973), so that, like sulfur, it may be subject to fractionation during mobilization onto grain surfaces as well as after emplacement on surfaces.

Noble gases

The foregoing elements have had in common the fact that they are predominantly indigenous in origin, an exception being sulfur which is partly meteoritic. The remainder of our discussion deals with solar wind-implanted elements, as most of those analyzed to date exhibit isotopic variations which may stem from fractionation on the lunar surface.

It is well known that the light noble gases of solar wind origin are depleted in lunar soils relative to either the heavier gases or more retentive species such as carbon and nitrogen. It is generally agreed that this depletion resulted from diffusive loss of gas (Huneke, 1973; Signer *et al.*, 1977), probably from the heavily radiation-damaged margins of soil grains (Ducati *et al.*, 1973). Mineral-specific retention factors have also been shown to be related to activation energies for diffusion, rather than to differences in sputter rate (Signer *et al.*, 1977). It is

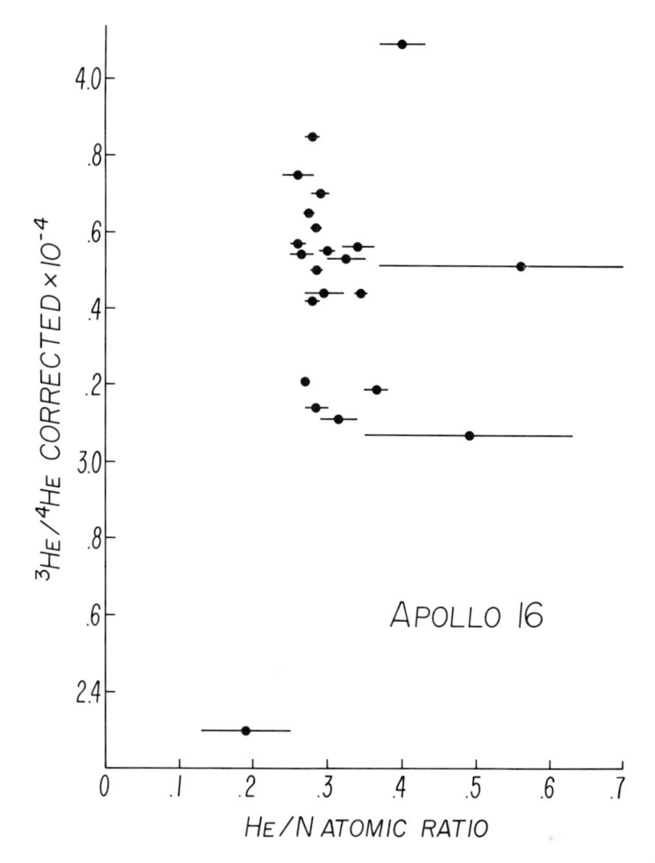

Fig. 2. The $^3He/^4He$ ratio, corrected for spallation, does not appear to be related to degree of helium depletion, as measured by the He/N ratio. Helium isotopic data from Bogard and Nyquist (1973), Eberhardt *et al.* (1976), Heymann *et al.* (1975), Hintenberger and Weber (1973) and Kirsten *et al.* (1973).

therefore of interest to compare the systematics believed to result from thermally driven diffusion with those described earlier for sputter fractionation.

We have shown previously that the He/N ratio decreases systematically with increasing maturity (see Fig. 2 in Kerridge *et al.*, 1976). It might therefore be supposed that this depletion of helium would be accompanied by a decrease in the $^3He/^4He$ ratio, although the observed variation in He/N represents a small increment compared with the overall depletion relative to the solar ratio. Because most isotopic fractionation probably occurs during the earliest stages of irradiation (Huneke, 1973), it is possible that any maturation-dependent fractionation could be obscured. In fact, Fig. 2 shows that there is no systematic relationship between $^3He/^4He$ and He/N ratio. It is possible that secular changes in isotopic composition of solar wind helium may be masking a lunar surface fractionation pattern. However, Geiss (1973) has pointed out that, though such a secular

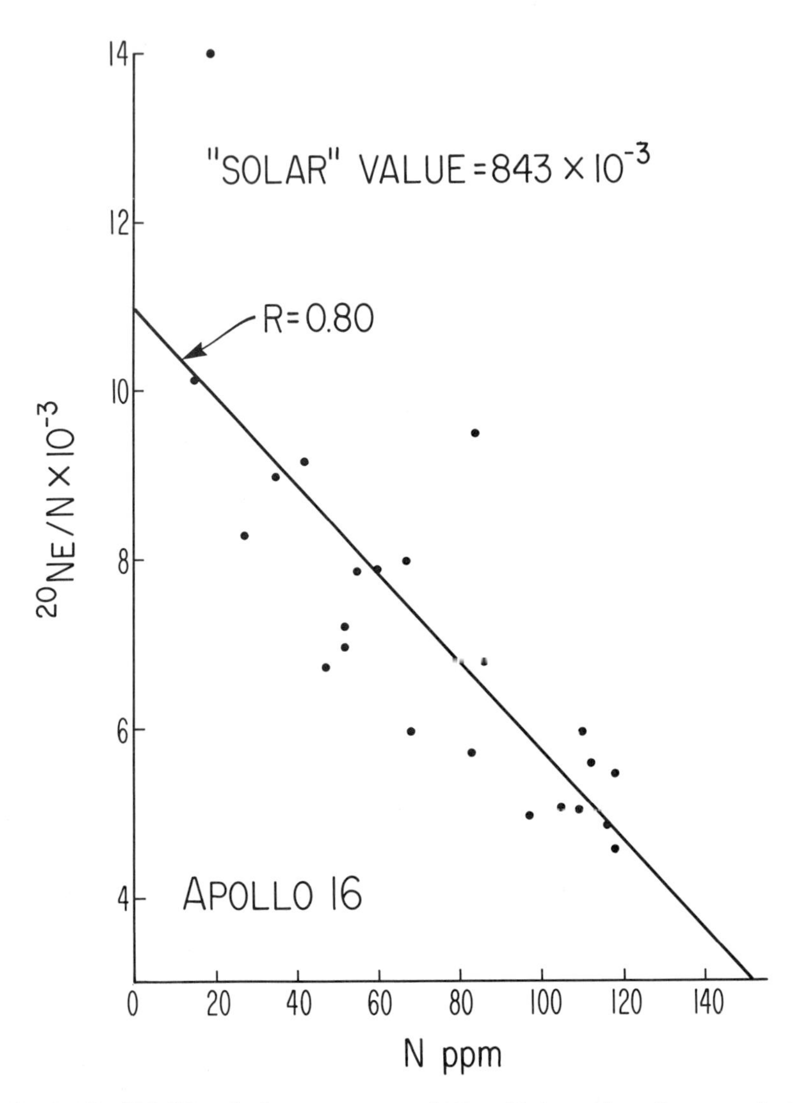

Fig. 3. The $^{20}Ne/N$ ratio decreases systematically with increasing nitrogen content, indicating progressive depletion of neon with increasing soil maturity. Neon data from sources cited in Fig. 2. "Solar" value for $^{20}Ne/N$ from Cameron (1973).

increase in $^{3}He/^{4}He$ may be inferred from comparison of soil analyses with old breccias, soils themselves do not show a systematic change with age. It is also possible that short term variations in the $^{3}He/^{4}He$ ratio of the solar wind could be masking a fractionation trend (Geiss, pers. comm.).

Similarly, the Ne/N ratio systematically decreases with maturity (Fig. 3) indicating progressive depletion of neon possibly accompanied by mass fractionation of the neon isotopes. Although a comparison of the isotopic composition of

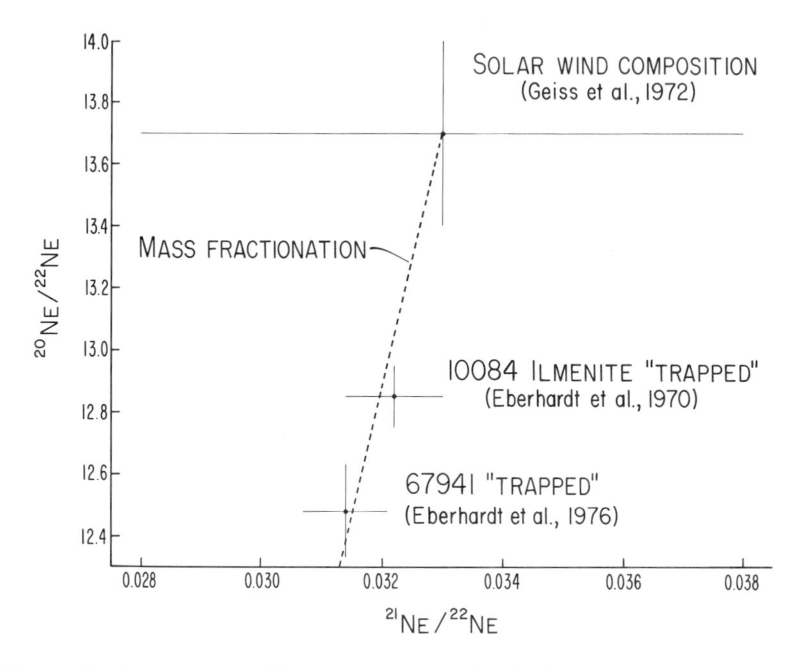

Fig. 4. The isotopic compositions of neon trapped in both a retentive sample, 10084 ilmenite, and a "leaky" sample, plagioclase-rich 67941, lie on a mass-fractionation line through the measured solar wind composition, although observational uncertainties are considerable.

solar wind neon (Geiss *et al.*, 1972) with that of neon trapped in a retentive mineral, ilmenite (Eberhardt *et al.*, 1970), and with neon trapped in a poorly retentive, plagioclase-rich, sample, 67941, (Eberhardt *et al.*, 1976) reveals a trend which is consistent with mass-fractionation (Fig. 4), experimental uncertainties do not permit a unique determination of the trend line. The isotopic compositions plotted in Fig. 4 represent the true trapped components from which spallogenic neon has been subtracted by means of the Eberhardt technique (Eberhardt *et al.*, 1970). Such precise determinations of the trapped composition are available for only a few samples: too few for a reliable test of a possible dependence of this composition upon maturity. If bulk analyses are used in their place, the quantity of data increases but no consistent pattern of decreasing $^{20}Ne/^{22}Ne$ with increasing maturity emerges (Fig. 5); in fact, a suggestion of the opposite trend may be discerned. Again, it is possible that a real fractionation pattern may be masked by either secular variations or experimental uncertainty. Note the size of some of the error bars in Fig. 5 which reflect inter-laboratory discrepancies, and also the fact that observed variations in $^{20}Ne/^{22}Ne$ span a range of about 100‰, a factor of five greater than observed for, say, $^{34}S/^{32}S$.

None of the light noble gases, therefore, yield the kind of systematic dependence of isotopic composition upon maturity which characterizes sulfur, silicon and oxygen. The closest approach to such a dependence may be found in

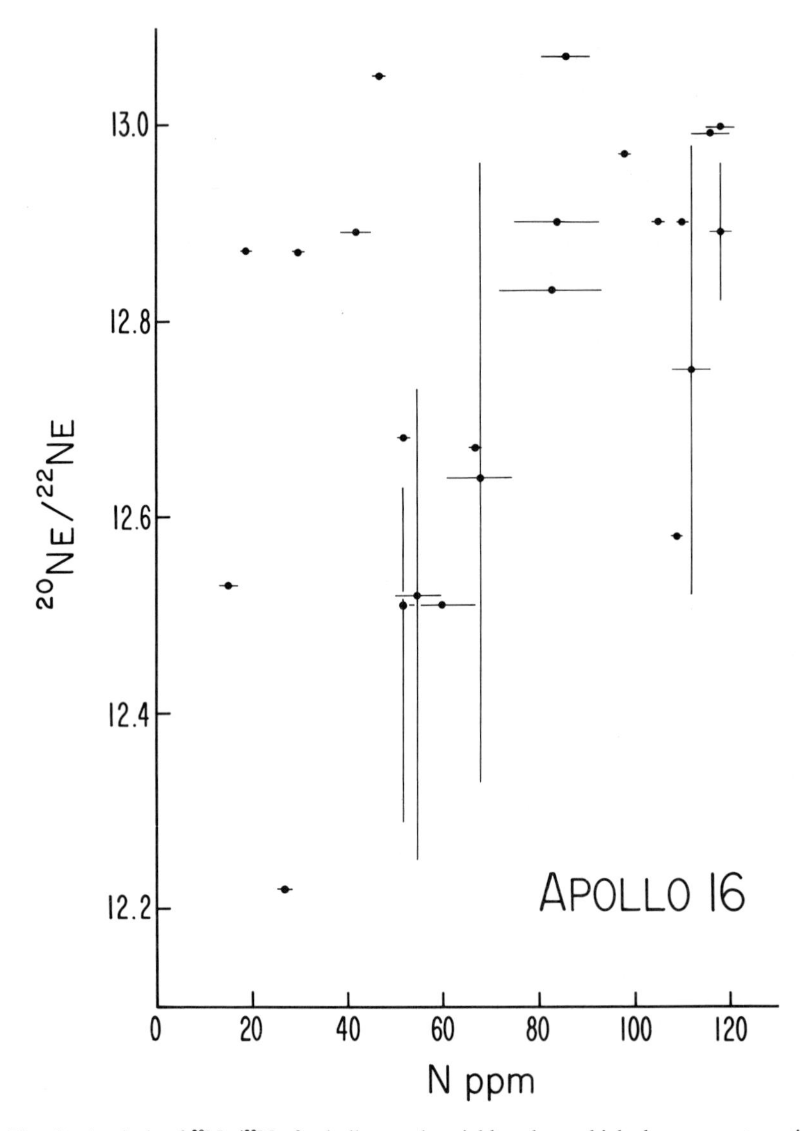

Fig. 5. Analysis of $^{20}Ne/^{22}Ne$ for bulk samples yields values which show no systematic dependence upon nitrogen content, i.e., soil maturity. Error bars for neon data reflect inter-laboratory discrepancies. Neon isotopic data from sources given in Fig. 2.

the relationship between the $^{20}Ne/^{36}Ar$ ratio for bulk samples, and maturity, in the form of nitrogen content (Fig. 6). This reveals a statistically real trend for the ratio to decrease with increasing maturity, but the correlation coefficient is only −0.67 compared with 0.89 for the relationship between $\delta^{34}S$ and nitrogen content (Fig. 1). If, indeed, depletion of the light noble gases is caused by diffusive loss, as seems likely, this process does not apparently progress uniformly with sample

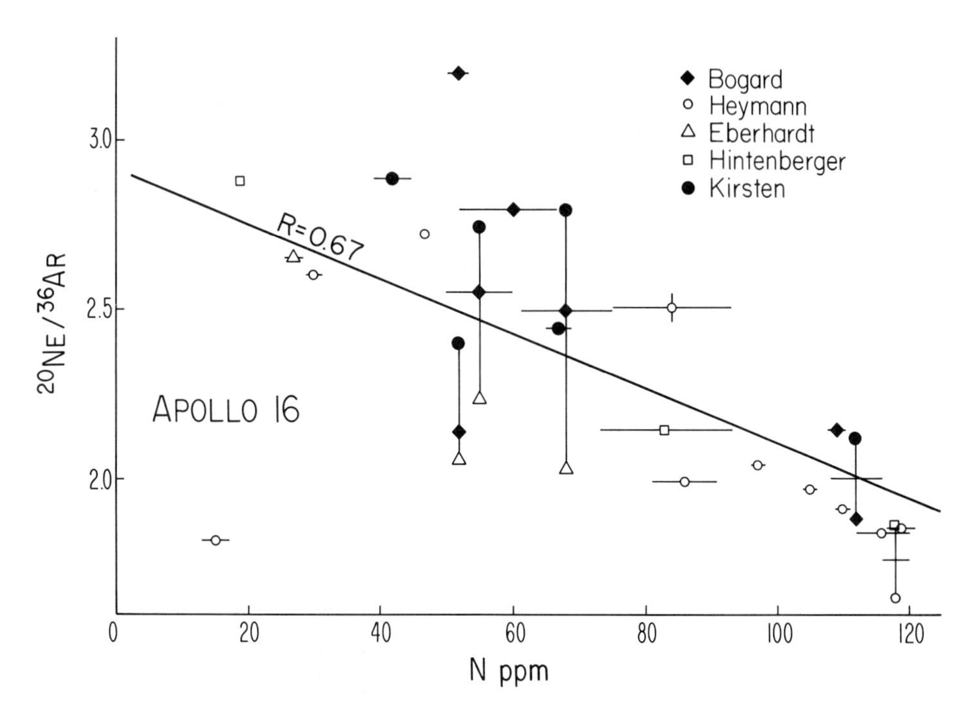

Fig. 6. The $^{20}Ne/^{36}Ar$ ratio decreases systematically with increasing soil maturity, although scatter in the data is significant. Noble gas data from sources given in Fig. 2.

maturation but appears to proceed more randomly. This suggests that the isotopically selective loss of gas from a sample is dominated by a small number of heating events, presumably involving unusually large meteorite impacts.

Mass fractionation trends have been reported for the isotopes of trapped krypton and xenon (Bogard *et al.*, 1974); however, it is not clear whether this fractionation occurred after implantation in the lunar surface or at some earlier stage. Insufficient data exist at present to enable these effects to be investigated by the methods employed above.

Nitrogen

Although variations of about 25% have been reported for the $^{15}N/^{14}N$ ratio of nitrogen apparently implanted in lunar soils by the solar wind, it is generally agreed that these variations represent a secular change in the source composition and do not reflect post-implantation fractionation (Kerridge, 1975; Becker and Clayton, 1975; Becker *et al.*, 1976; Kerridge *et al.*, 1977). The secular trend (see Fig. 2 in Kerridge *et al.*, 1977) consists of an increase in $^{15}N/^{14}N$ by at least 25% over an uncertain time interval, probably about 1 to 4 Gy. The correlation between $^{15}N/^{14}N$ ratio and time of implantation is good but shows some scatter

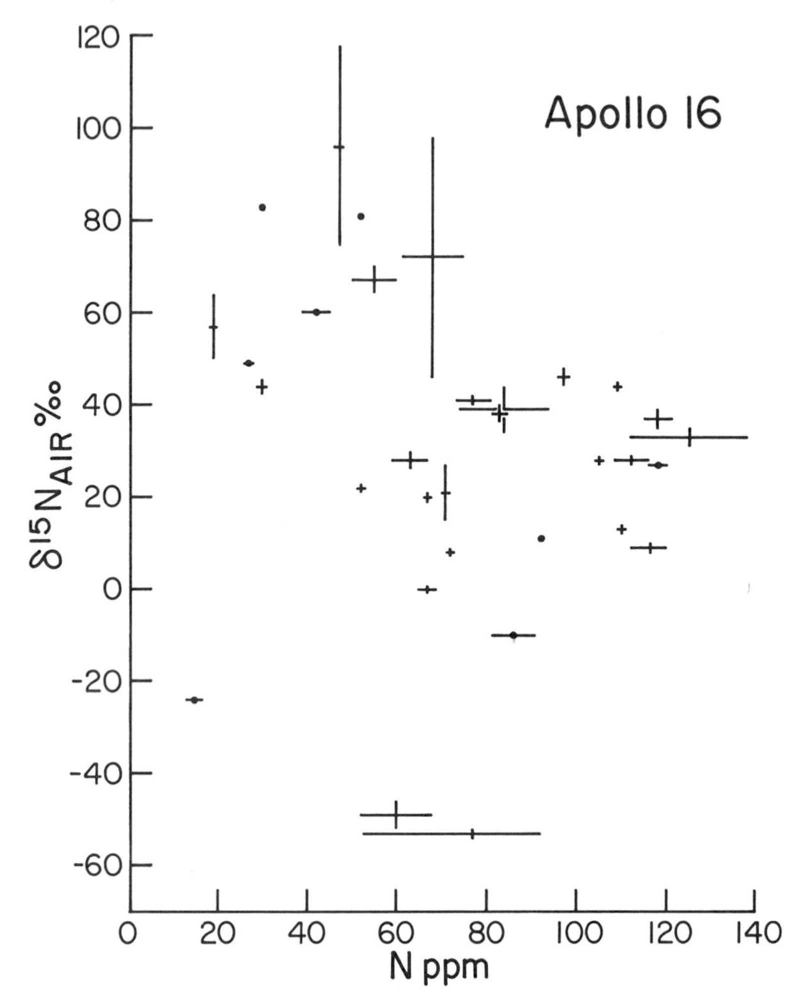

Fig. 7. No systematic variation of $\delta^{15}N$ with soil maturity, given by nitrogen content, may be discerned. This conclusion also holds true for subsets of data chosen to minimize the effects of the secular variation in $\delta^{15}N$, see text.

and the possibility must be considered that this scatter reflects subsequent fractionation. However, Fig. 7 shows a lack of any systematic relationship between $\delta^{15}N$ and maturity, as indicated by nitrogen content. This lack of a trend persists even if subsets of data are selected within which the secular trend is minimized. This is done by taking samples within a narrow range of implantation age and searching for a systematic relationship between maturity and deviation from a linear dependence of $^{15}N/^{14}N$ upon implantation age. There is no evidence at all that the $^{15}N/^{14}N$ ratio of implanted material has changed after implantation. This observation constitutes a serious constraint on any theory involving the processing of solar wind-implanted material.

Carbon

Although it is universally agreed that most carbon in regolith samples is of extralunar origin, it is still not clear whether the solar wind is the sole important source of carbon or whether other sources such as meteorites are also significant. So far as actual abundances are concerned, the solar wind flux is adequate to supply all the extralunar carbon observed in soils. However, additional sources cannot be ruled out, particularly if substantial loss of carbon occurs, either during or following implantation. Jull and Pillinger (1977) have argued that, if solar wind nitrogen is 100% retained during agglutination, observed C/N ratios in soils would allow solar wind carbon in the regolith to be depleted by a factor of two relative to solar abundance. The retentiveness of carbon is not known and although it is likely that carbon could enter into solid solution in the metallic iron known to be present in soils (Jull *et al.*, 1976), this could account for less than 10% of the total carbon.

Variations of about 30‰ have been reported for the $^{13}C/^{12}C$ ratio in lunar soils and these have generally been attributed, at least in part, to selective loss of ^{12}C from the lunar surface (Kaplan *et al.*, 1970; Epstein and Taylor, 1972; Kerridge *et al.*, 1974; DesMarais *et al.*, 1975). Other possible explanations for the observed isotopic variations include addition of extra components, such as meteoritic carbon, and variation in the isotopic composition at the source of one or more of the components. A relatively simple model involving addition of only solar wind to indigenous carbon and subsequent mass fractionation induced by proton stripping provided a fair match with observed data, although it appeared that the solar wind flux was barely able to supply sufficient material (Kerridge *et al.*, 1974). Further failings of this model are considered later. A qualitative model interpreting observed variations in $^{13}C/^{12}C$ in terms of a combination of sputter fractionation and diffusive loss during agglutination has been proposed by Pillinger *et al.* (1976b). Switkowski *et al.* (1977) suggested that sputter-fractionation would lead to an approximately 10‰ increase in $^{13}C/^{12}C$ in surficial carbon.

There is some evidence that the concentration of implanted carbon on the surfaces of grains reaches an equilibrium, quasi-saturation value, attributable to concurrent implantation and sputter-erosion (DesMarais *et al.*, 1975; Filleux *et al.*, 1977, 1978), although total carbon contents could exceed the equilibrium value as a result of cycling of implanted surficial carbon into agglutinates (DesMarais *et al.*, 1973; Kerridge *et al.*, 1974). By analysis of grain size fractions, DesMarais *et al.*, (1975) concluded that, although the finest grains were relatively depleted in ^{13}C, surficial carbon was characterized by $\delta^{13}C$ values which were 6 to 14‰ heavier than bulk analyses; but Epstein and Taylor (1975) inferred from mass balance calculations following analysis of partially fluorinated samples that a small carbon component, enriched by 100 to 200‰ in ^{13}C, resided on grain surfaces. However, subsequent investigation of partially fluorinated residues (Pillinger *et al.*, 1976b) showed that they were effectively devoid of agglutinates and heavily depleted in hydrolysable carbon, known to be greatly concentrated in agglutinates (Cadogan *et al.*, 1973). In view of the fact that agglutinates contain a major fraction of the total carbon in a sample, being enriched by factors of 2 to 4 over non-agglutinate material (DesMarais *et al.*, 1975) it follows that partial fluorination must remove more carbon than inferred from the mass balance calculations; consequently, the isotopic composition of the stripped carbon cannot be as extreme as appeared at first. This conclusion is supported by results of Becker *et al.* (1976) who showed that similar partial fluorination experiments removed about 25% of the nitrogen and it is likely, though not certain, that carbon and nitrogen are similarly sited. We conclude that the evidence for a surficial component of "superheavy" carbon is not convincing at this time.

We have recently presented carbon abundance and isotopic data for a suite of 31 soils from the Apollo 16 landing site (Kerridge *et al.*, 1978; see also Kerridge *et al.*, 1975b). We have argued elsewhere (Kerridge *et al.*, 1974) that the effectively uniform major element composition and useful range of crater ages make this the optimum Apollo landing site for the study of regolith maturation. The soils analyzed were chosen to span as wide a range of maturity as possible, using parameters such as nitrogen content, FMR intensity (I_s/FeO) and agglutinate content. Nitrogen abundances and δ^{15}N values and helium contents were determined simultaneously with the carbon data. Figure 8 illustrates the close relationship between abundances of carbon and nitrogen, with results from individual replicates plotted separately. The raw data correlate with a coefficient of 0.92. Because of the omnipresent danger of terrestrial contamination in analysis of such small quantities of carbon, we chose to try to estimate the incidence of such contamination. Experience has shown that contamination increases the C/N ratio and decreases the δ^{13}C value, tending towards a value of −30‰ relative to the PDB standard. Consequently we adopted a trial and error procedure whereby, for each analysis characterized by a high C/N ratio, the amount of "excess" carbon, compared with aliquots of comparable nitrogen content, was related to its measured δ^{13}C value, again compared with the same neighboring aliquots. Twelve analyses were identified in which the "excess" carbon was calculated to have a δ^{13}C value of about −30‰, thus most plausible being due to contamination. These analyses are identified in Fig. 8 by open symbols; the remaining data, solid symbols, we term "preferred" data. We emphasize that these data were not selected merely to improve the apparent fit to the trend line in Fig. 8, but were chosen on the basis of an independent, objective criterion, namely isotopic composition. In view of the prevalence of terrestrial contamination in such analyses we believe that some such critical approach is desirable; however, this view was challenged at the Conference as "throwing the baby out with the bath water," so until this issue is settled it is perhaps desirable to present, side by side, interpretations based both upon total data and upon selected data. Thus, although our conclusions here will be illustrated using "preferred" data, we have determined that none of them would be affected by employing total data in their stead.

As noted above, regolith samples display a significant range of δ^{13}C values and it is of interest to see if these data fit into any of the fractionation patterns

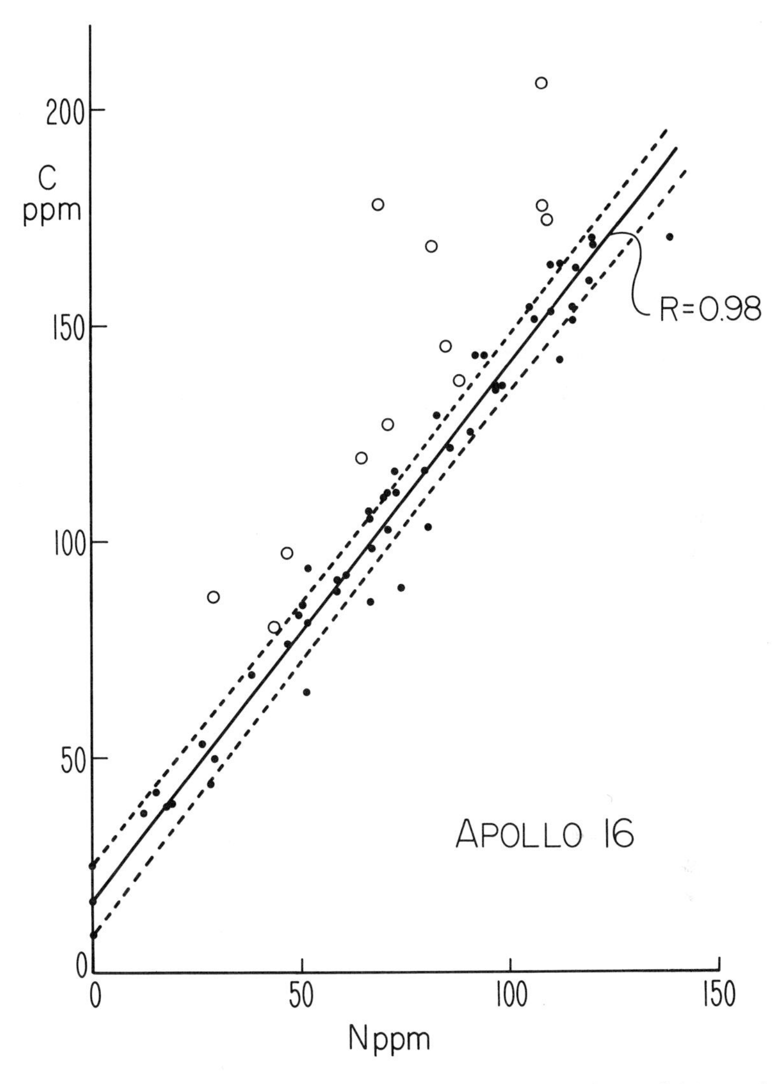

Fig. 8. Carbon and nitrogen abundances for 62 replicate analyses of 31 samples correlate strongly. Open symbols represent data believed to have been affected by terrestrial contamination (see text). Solid symbols, "preferred" data, yield the least squares regression line illustrated with 1σ uncertainty limits. From Kerridge *et al.* (1978).

described earlier. In Fig. 9 we plot $\delta^{13}C$ against carbon content. Because of the extremely strong association between carbon and nitrogen contents (correlation coefficient equal to 0.98 for the preferred data used here), carbon content may be regarded as a reliable measure of maturity, analogous to nitrogen content. Figure 9 also permits comparison with the predictions of our model (Fig. 1 in Kerridge

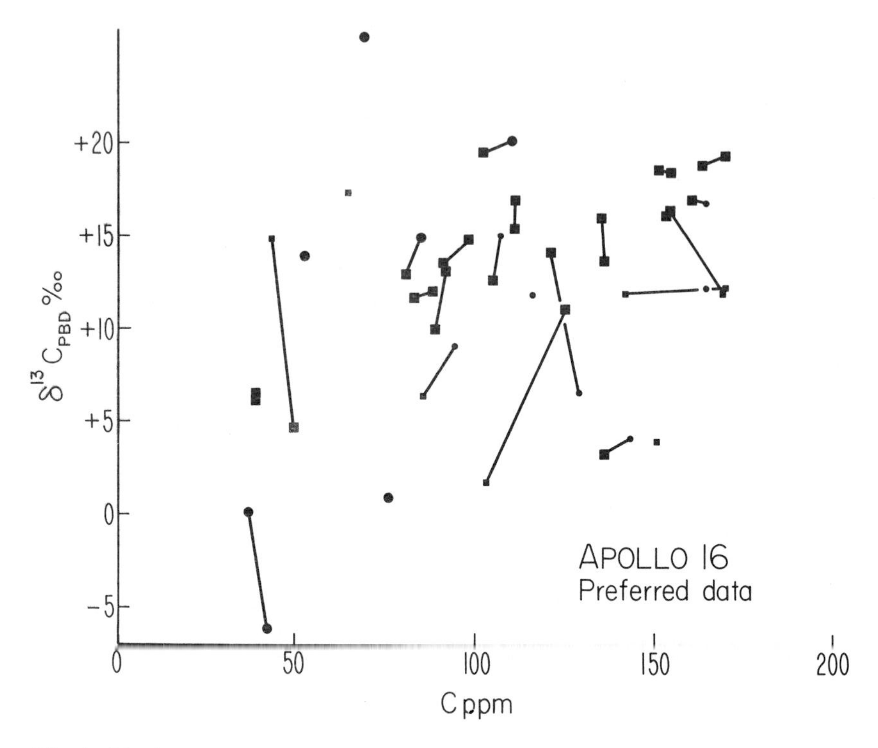

Fig. 9. No clear systematic relationship may be discerned between $\delta^{13}C$ and C content for soils, using "preferred" data. This conclusion also applies if all data are used, or if only those data falling within 1σ of the trend line in Fig. 8 are used (large symbols here). Square symbols represent points on the carbon-poor side of the trend line in Fig. 8; round symbols, carbon-rich. Tie lines join replicate analyses. From Kerridge *et al.* (1978).

et al., 1974) and shows, in fact, that the data do not support that model, which will not be considered further here. More generally, the data show no convincing evidence for a systematic dependence of $\delta^{13}C$ value upon maturity, of the kind found for sulfur, oxygen and silicon. This suggests strongly that, if carbon is being fractionated on the lunar surface, it is by a different process from that responsible for those other elements. As the evidence is in favor of sputter-fractionation for sulfur, silicon and oxygen, it follows that sputtering is unlikely to be the cause of observed variations in $^{13}C/^{12}C$ in the regolith.

The observed association between carbon and nitrogen abundances is so close that a common origin seems highly likely, although independent, exposure-related origins could conceivably mimic such an association. In view of the secular change in isotopic composition of the source of the nitrogen, described above, it is worth searching for evidence of a corresponding secular change in $\delta^{13}C$. Figure 10 shows, however, that no clear relationship exists between $\delta^{15}N$ and $\delta^{13}C$. A hint of a possible positive trend may be discerned, and during

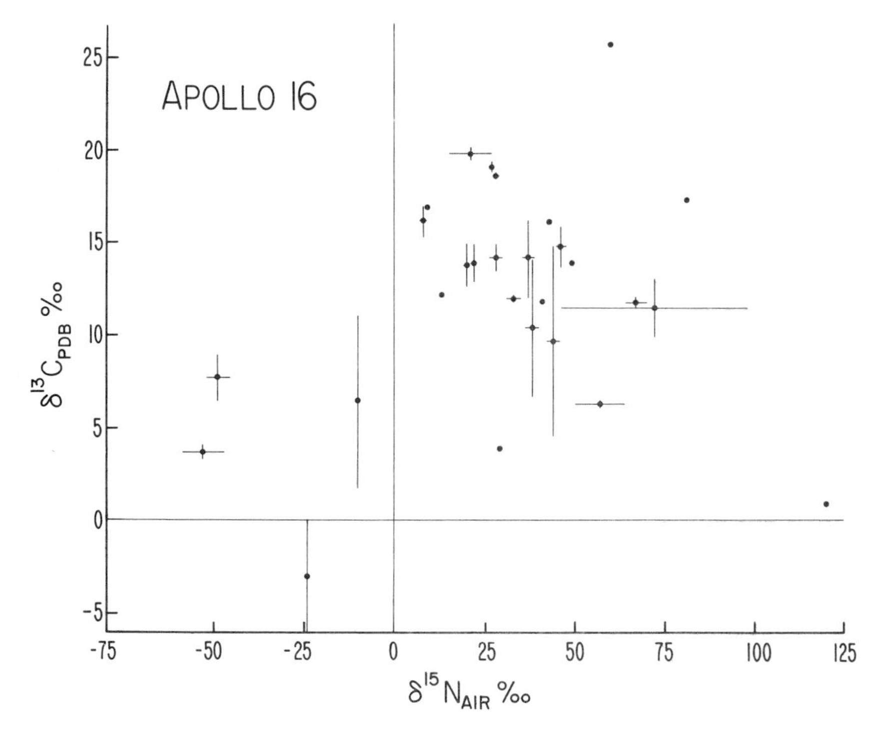

Fig. 10. Although a hint of a positive trend may be discerned in this plot of $\delta^{13}C$ against $\delta^{15}N$, using "preferred" data, it does not show convincing evidence for an actual secular variation in $\delta^{13}C$.

discussion at the Conference, R. H. Becker suggested that combined data for all landing sites revealed significant evidence for such a trend. We are unconvinced by these trends and therefore do not believe that there is currently any evidence for a secular change in $\delta^{13}C$ at the source(s) of the carbon. The possibility that the observed systematics are due to a combination of both secular change and exposure-related fractionation seems unlikely because outlying points in either Fig. 9 or Fig. 10 are not generally characterized by extreme values on the other plot. Nonetheless, it is probably premature to deny the possibility that a secular change may have occurred in the $^{13}C/^{12}C$ ratio of the solar wind.

Although it is intuitively unlikely that the behavior of carbon should resemble that of the noble gases, it is worth comparing the carbon data with the fractionation pattern characteristic of the light noble gases. As noted above, the parameter which seems most faithfully to be reflecting the mass fractionation undergone by noble gases on the lunar surface is the $^{20}Ne/^{36}Ar$ ratio, so in Fig. 11 we plot this ratio against $\delta^{13}C$ value. Again, no trend is discernible, suggesting that diffusive loss is unlikely to be the cause of the isotopic variations in carbon.

We are forced to conclude at this time that there is no evidence that carbon is

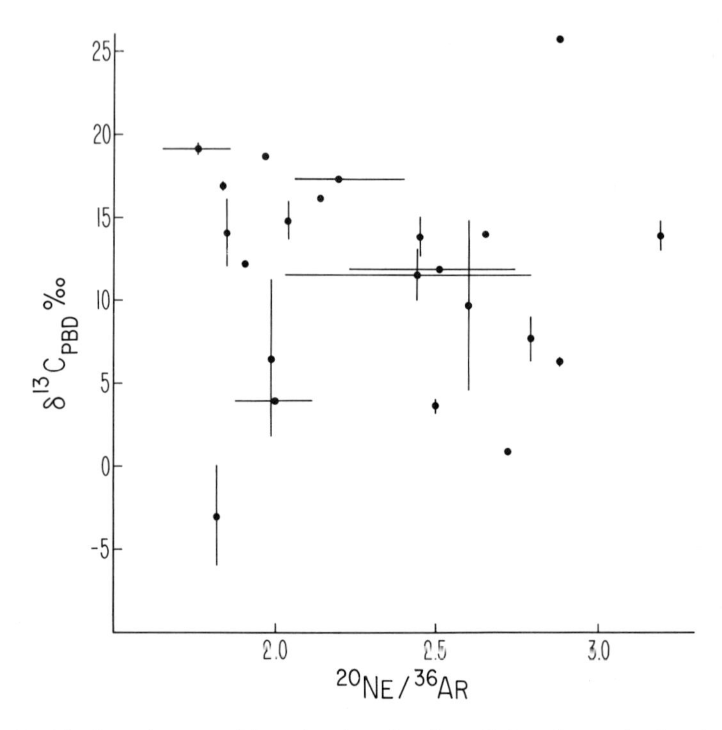

Fig. 11. The isotopic composition of carbon, "preferred" data, is not clearly related to the depletion of trapped neon relative to trapped argon, believed to be due to diffusive loss following solar wind implantation in soil. Noble gas data from sources cited in Fig. 2.

being isotopically fractionated on the lunar surface. If such fractionation is, in fact, occurring, but masked by other effects, then it is highly unlikely that it is caused by solar wind sputtering. At present it is not even clear whether there is any net loss of solar wind carbon from the moon during or following implantation, as the observed C/N ratio, 1.39 from Fig. 8, lies within the uncertainty limits for the solar spectroscopic value, though lower than the nominal value (Cameron, 1973).

It should be emphasized that scenarios involving multiple, independent processes can readily be constructed to "explain" *qualitatively* the carbon abundance and isotopic data (e.g., Pillinger *et al.*, 1976b). However, we believe that a true, quantitative model is more likely to be achieved via the instructive failings of oversimplified but *quantitative* models, than by attempting to insert numbers into a qualitative scenario.

In general, the striking differences in isotope systematics between the silicon-oxygen-sulfur suite, on the one hand, and the solar wind-implanted elements, on the other, suggest strongly that different fractionation mechanisms are active in the two cases. If, as seems likely, solar wind sputtering is responsible for the effects in the former elements, it is therefore unlikely to be the cause of fractionations among the solar wind elements. Even if sputtering is not fractionating silicon, oxygen and sulfur, it is unlikely to be affecting the solar wind elements. This follows because a major fraction of the sulfur is deposited upon grain surfaces and therefore lies directly above the species implanted by the solar wind. It is thus implausible that sputtering could excavate solar wind elements without disturbing, i.e., fractionating the surficial sulfur. Our conclusions about the fractionation of solar wind-implanted species agree with independent reasoning by Huneke (1973), Ducati *et al.* (1973) and Signer *et al.* (1977) which has identified diffusive loss as the likely cause of mass fractionation among trapped noble gases of solar wind origin.

It appears, therefore, that sputtering is not affecting those elements implanted by the solar wind in the outer 200 Å or so of lunar soil grains. Because there is no reason for indigenous atoms to be sputtered preferentially to implanted ones, it follows that the front of sputter erosion is not penetrating in general to solar wind-implantation depths. The lowest value estimated for the solar wind sputtering rate is about 0.02 Å/y (McDonnell, 1977) so that a depth of about 200 Å would be reached in less than 10^4 y. Either this time is greater than the lifetime of a grain against burial or agglutination or the effective sputtering rate on the lunar surface is lower than that found by simulation. The former explanation is unlikely because of the observed surface concentrations of solar wind elements, e.g., carbon (Filleux *et al.*, 1978), which require exposure time for individual grains at least a factor of two greater than the upper limit given above.

Consequently we conclude that the lunar surface sputtering rate is being effectively slowed down below the laboratory simulation values by the operation of some other process. That process is probably the deposition on grain surfaces of material vaporized during meteorite impact. It is well known that mature lunar soils contain at least 1% by weight of meteoritic material (Anders *et al.*, 1973) and, although much of this material is probably incorporated into agglutinates, a proportion is likely to be deposited directly onto grain surfaces. This was suggested by Boynton and Wasson (1977) in view of distributions of volatile and siderophile elements in grain size fractions. It is important to note, in addition, that in the vapor from which such meteoritic elements condense, target atoms would predominate over those from the projectile. Consequently, the concentration of vapor-deposited material on grain surfaces is considerably greater than estimated from concentrations of meteoritic elements alone. It does not appear practical at present to quantify this process but a very rough estimate may be made, taking 1% as the content of meteoritic material in a soil, assuming that 90% of this goes into agglutinates and 10% into a vapor deposit within which it is diluted by about 10 times its weight of indigenous, target material. This leads to a surface deposit about 200 Å thick, a figure which is comparable to the solar wind implantation depth. It is also comparable to the amount of material which must be lost from the system in order to generate the observed oxygen isotope fractionation, assuming Rayleigh fractionation. Clayton *et al.* (1974) pointed out that enrichment of bulk soils in ^{18}O by about 0.5‰ required loss of about 1% of total oxygen. Taking 800 cm^2/g as the surface area of a mature lunar soil, this corresponds to loss of a surface layer about 200 Å in thickness. The dimensions of this surface layer correspond closely to those measured by Dran *et al.* (1970) for the amorphous rims of micrometer-sized soil grains. However, it is not suggested that vapor deposition is producing these rims because it has been demonstrated that they result from solar wind-irradiation damage (e.g., Bibring *et al.*, 1973). Nonetheless, it is likely that a surficial vapor deposit would be promptly metamictized by the solar wind irradiation, to become effectively indistinguishable from amorphous indigenous material.

Conclusions

1. Isotopic fractionations exhibited by solar wind-implanted elements in lunar soils are of variable magnitude and do not show the systematic dependence upon maturity characteristic of fractionations observed in silicon, oxygen and sulfur.
2. It seems likely that fractionations in silicon, oxygen and sulfur are caused by solar wind sputtering but, whether or not this is the case, sputtering does not appear to be responsible for fractionating the solar wind elements.
3. The active sputtered zone is therefore probably not reaching to the depth within grains where most solar wind-implanted species reside.
4. It follows that the rate of erosion by solar wind sputtering is probably being impeded by deposition from vapor produced during meteorite impact and it is possible that these two processes are close to being in dynamic balance on the lunar surface.
5. Carbon isotope systematics give no evidence for either secular variation in isotopic composition of solar wind carbon or fractionation of carbon isotopes on the lunar surface. However, neither possibility can be excluded at this time.
6. Diffusive loss of light noble gases may be dominated by a small number of major degassing events, presumably large impacts.

Acknowledgments—This paper has benefited from a helpful review by R. H. Becker. Financial support by NASA through grant NGR 05-007-289 is gratefully acknowledged. Publication #1832, Institute of Geophysics and Planetary Physics, University of California, Los Angeles, California 90024.

References

Anders E., Ganapathy R., Krähenbühl U. and Morgan J. W. (1973) Meteoritic material on the moon. *The Moon* **8**, 3–24.

Barnes I. L., Garner E. L., Gramlich J. W., Machlan L. A., Moody J. R., Moore L. J., Murphy T. J. and Shields W. R. (1973) Isotopic abundance ratios and concentrations of selected elements in some Apollo 15 and 16 samples. *Proc. Lunar Sci. Conf. 4th*, p. 1197–1207.

Becker R. H. and Clayton R. N. (1975) Nitrogen abundances and isotopic compositions in lunar samples. *Proc. Lunar Sci. Conf. 6th*, p. 2131–2149.

Becker R. H., Clayton R. N. and Mayeda T. K. (1976) Characterization of lunar nitrogen components. *Proc. Lunar Sci. Conf. 7th*, p. 441–458.

Bibring J. P., Chaumont J., Comstock G., Maurette M. and Meunier R. (1973) Solar wind and lunar wind microscopic effects in the lunar regolith (abstract). In *Lunar Science IV*, p. 72–74. The Lunar Science Institute, Houston.

Bogard D. D., Hirsch W. C. and Nyquist L. E. (1974) Noble gases in Apollo 17 fines: Mass fractionation effects in trapped Xe and Kr. *Proc. Lunar Sci. Conf. 5th*, p. 1975–2003.

Bogard D. D. and Nyquist L. E. (1973) $^{40}Ar/^{36}Ar$ variations in Apollo 15-16 regolith. *Proc. Lunar Sci. Conf. 4th*, p. 1975–1985.

Boynton W. V. and Wasson J. T. (1977) Distribution of 28 elements in size fractions of lunar mare and highland soils. *Geochim. Cosmochim. Acta* **41**, 1073–1082.

Cadogan P. H., Eglinton G., Gowar A. P., Jull A. J. T., Maxwell J. R. and Pillinger C. T. (1973) Location of methane and carbide in Apollo 11 and 16 lunar fines. *Proc. Lunar Sci. Conf. 4th*, p. 1493–1508.

Cameron A. G. W. (1973) Abundances of the elements in the solar system. *Space Sci. Rev.* **15**, 121–146.

Cassidy W. A. and Hapke B. (1975) Effects of darkening processes on surfaces of airless bodies. *Icarus* **25**, 371–383.

Church S. E., Tilton G. R., Wright J. E. and Lee-Hu C.-N. (1976) Volatile element depletion and $^{39}K/^{41}K$ fractionation in lunar soils. *Proc. Lunar Sci. Conf. 7th*, p. 423–439.

Clayton R. N., Mayeda T. K. and Hurd J. M. (1974) Loss of oxygen, silicon, sulfur and potassium from the lunar regolith. *Proc. Lunar Sci. Conf. 5th*, p. 1801–1809.

DesMarais D. J., Basu A., Hayes J. M. and Meinschein W. G. (1975) Evolution of carbon isotopes, agglutinates and the lunar regolith. *Proc. Lunar Sci. Conf. 6th*, p. 2353–2373.

DesMarais D. J., Hayes J. M. and Meinschein W. G. (1973) The distribution in lunar soil of carbon released by pyrolysis. *Proc. Lunar Sci. Conf. 4th*, p. 1543–1558.

Dran J. C., Durrieu L., Jouret C. and Maurette M. (1970) Habit and texture studies of lunar and meteoritic materials with a 1 MeV electron microscope. *Earth Planet. Sci. Lett.* **9**, 391–400.

Ducati H., Kalbitzer S., Kiko J., Kirsten T. and Müller H. W. (1973) Rare gas diffusion studies in individual lunar soil particles and in artificially implanted glass. *The Moon* **8**, 210–227.

Eberhardt P., Eugster O., Geiss J., Grögler N., Guggisberg S. and Mörgeli M. (1976) Noble gases in the Apollo 16 special soils from the East-West split and the permanently shadowed area. *Proc. Lunar Sci. Conf. 7th*, p. 561–585.

Eberhardt P., Geiss J., Graf H., Grögler N., Krähenbühl U., Schwaller H., Schwarzmüller J. and Stettler A. (1970) Trapped solar wind noble gases, exposure age and K/Ar age in Apollo 11 lunar fine material. *Proc. Apollo 11 Lunar Sci. Conf.*, p. 1037–1070.

Epstein S. and Taylor H. P. (1971) O^{18}/O^{16}, Si^{30}/Si^{28}, D/H, and C^{13}/C^{12} ratios in lunar samples. *Proc. Lunar Sci. Conf. 2nd*, p. 1421–1441.

Epstein S. and Taylor H. P. (1972) O^{18}/O^{16}, Si^{30}/Si^{28}, C^{13}/C^{12}, and D/H studies of Apollo 14 and 15 samples. *Proc. Lunar Sci. Conf. 3rd*, p. 1429–1454.

Epstein S. and Taylor H. P. (1975) Investigation of the carbon, hydrogen, oxygen, and silicon isotope and concentration relationships on the grain surfaces of a variety of lunar soils and in some Apollo 15 and 16 core samples. *Proc. Lunar Sci. Conf. 6th*, p. 1771–1798.

Evensen N. M., Murthy V. R. and Coscio M. R. (1973) Rb-Sr ages of some mare basalts and the isotopic and trace element systematics in lunar fines. *Proc. Lunar Sci. Conf. 4th*, p. 1707–1724.

Filleux C., Spear R., Tombrello T. A. and Burnett D. S. (1978) Carbon depth distributions for soil breccias (abstract). In *Lunar and Planetary Science IX*, p. 317–319. Lunar and Planetary Institute, Houston.

Filleux C., Tombrello T. A. and Burnett D. S. (1977) Direct measurement of surface carbon concentrations. *Proc. Lunar Sci. Conf. 8th*, p. 3755–3772.

Garner E. L., Machlan L. A. and Barnes I. L. (1975) The isotopic composition of lithium, potassium, and rubidium in some Apollo 11, 12, 14, 15, and 16 samples. *Proc. Lunar Sci. Conf. 6th*, p. 1845–1855.

Geiss J. (1973) Solar wind composition and implications about the history of the solar system. In *Proc. 13th Intern. Cosmic Ray Conf.* **5**, 3375–3398. Univ. Denver, Denver, Colorado.

Geiss J., Buehler F., Cerutti H., Eberhardt P. and Filleux C. (1972) Solar wind composition experiment. *Apollo 16 Prelim. Sci. Rep.* NASA SP-315, p. 14-1 to 14-10.

Gold T., Bilson E. and Baron R. L. (1974) Observation of iron-rich coating in lunar grains and a relation to low albedo. *Proc. Lunar Sci. Conf. 5th*, p. 2413–2422.

Griffith J. E., Papanastassiou D. A., Russell W. A., Tombrello T. A. and Weller R. A. (1978) Simulation experiments and solar wind sputtering (abstract). In *Lunar and Planetary Science IX*, p. 419–421. Lunar and Planetary Institute, Houston.

Haff P. K., Switkowski Z. E., Burnett D. S. and Tombrello T. A. (1977) Gravitational and recoil contributions to surface mass fractionation by solar-wind sputtering. *Proc. Lunar Sci. Conf. 8th*, p. 3807–3815.

Hapke B. (1966) Optical properties of the Moon's surface. In *The Nature of the Lunar Surface*, p. 141–154. Johns Hopkins Press, Baltimore.

Hapke B. (1973) Darkening of silicate rock powders by solar wind sputtering. *The Moon* **7**, 342–355.

Hapke B., Cassidy W. and Wells E. (1975) Effects of vapor-phase deposition processes on the optical, chemical, and magnetic properties of the lunar regolith. *The Moon* **13**, 339–353.

Heymann D. (1978) Light and dark soils at Apollo 16 revisited (abstract). In *Lunar and Planetary Science IX*, p. 506–508. Lunar and Planetary Institute, Houston.

Heymann D., Walton J. R., Jordan J. L., Lakatos S. and Yaniv A. (1975) Light and dark soils at the Apollo 16 landing site. *The Moon* **13**, 81–110.

Hintenberger H. and Weber H. W. (1973) Trapped rare gases in lunar fines and breccias. *Proc. Lunar Sci. Conf. 4th*, p. 2003–2019.

Housley R. M. and Grant R. W. (1977) An XPS (ESCA) study of lunar surface alteration profiles. *Proc. Lunar Sci. Conf. 8th*, p. 3885–3899.

Huneke J. C. (1973) Diffusive fractionation of surface implanted gases. *Earth Planet. Sci. Lett.* **21**, 35–44.

Jull A. J. T., Eglinton G., Pillinger C. T., Biggar G. M. and Batts B. D. (1976) The identity of lunar hydrolysable carbon. *Nature* **262**, 566–567.

Jull A. J. T. and Pillinger C. T. (1977) Effects of sputtering on solar wind element accumulation. *Proc. Lunar Sci. Conf. 8th*, p. 3817–3833.

Kaplan I. R., Smith J. W. and Ruth E. (1970) Carbon and sulfur concentration and isotopic composition in Apollo 11 lunar samples. *Proc. Apollo 11 Lunar Sci. Conf.*, p. 1317–1329.

Kerridge J. F. (1975) Solar nitrogen: Evidence for a secular increase in the ratio of nitrogen-15 to nitrogen-14. *Science* **188**, 162–164.

Kerridge J. F., Kaplan I. R. and Lesley F. D. (1974) Accumulation and isotopic evolution of carbon on the lunar surface. *Proc. Lunar Sci. Conf. 5th*, p. 1855–1868.

Kerridge J. F., Kaplan I. R., Lingenfelter R. E. and Boynton W. V. (1977) Solar wind nitrogen: Mechanisms for isotopic evolution. *Proc. Lunar Sci. Conf. 8th*, p. 3773–3789.

Kerridge J. F., Kaplan I. R. and Petrowski C. (1975a) Evidence for meteoritic sulfur in the lunar regolith. *Proc. Lunar Sci. Conf. 6th*, p. 2151–2162.

Kerridge J. F., Kaplan I. R. and Petrowski C. (1976) Retention of solar wind noble gases in the regolith (abstract). In *Lunar Science VII*, p. 446–447. The Lunar Science Institute, Houston.

Kerridge J. F., Kaplan I. R. and Petrowski C. (1978) Carbon isotope systematics in the Apollo 16 regolith (abstract). In *Lunar and Planetary Science IX*, p. 618–620. Lunar and Planetary Institute, Houston.

Kerridge J. F., Kaplan I. R., Petrowski C. and Chang S. (1975b) Light element geochemistry of the Apollo 16 site. *Geochim. Cosmochim. Acta* **39**, 137–162.

Kirsten T., Horn P. and Kiko J. (1973) ^{39}Ar-^{40}Ar dating and rare gas analysis of Apollo 16 rocks and soils. *Proc. Lunar Sci. Conf. 4th*, p. 1757–1784.

Lindsay J. F. (1975) A steady state model for the lunar soil. *Bull. Geol. Soc. Amer.* **86**, 1661–1670.

McDonnell J. A. M. (1977) Accretionary particle studies on Apollo 12054,58: *In-situ* lunar surface microparticle flux rate and solar wind sputter rate defined. *Proc. Lunar Sci. Conf. 8th*, p. 3835–3857.

Mendell W. W. and McKay D. S. (1975) A lunar soil evolution model. *The Moon* **13**, 285–292.

Paruso D., Cassidy W. A. and Hapke B. (1978) An experimental investigation of fractionation by sputter deposition (abstract). In *Lunar and Planetary Science IX*, p. 868–869. Lunar and Planetary Institute, Houston.

Pillinger C. T., Eglinton G., Gowar A. P., Jull A. J. T. and Maxwell J. R. (1977) The exposure history of the Apollo 16 site: An assessment based on methane and hydrolysable carbon. In *The Soviet-American Conference on Cosmochemistry of the Moon and Planets*, p. 541–551. NASA SP-370. Washington, D.C.

Pillinger C. T., Gardiner L. R. and Jull A. J. T. (1976a) Preferential sputtering as a method of producing metallic iron, inducing major element fractionation and trace element enrichment. *Earth Planet. Sci. Lett.* **33**, 289–299.

Pillinger C. T., Jull A. J. T., Eglinton G. and Epstein S. (1976b) A model for the isotopic evolution of light elements in lunar soil (abstract). In *Lunar Science VII*, p. 703–705. The Lunar Science Institute, Houston.

Rees C. E. and Thode H. G. (1972) Sulphur concentrations and isotope ratios in lunar samples. *Proc. Lunar Sci. Conf. 3rd*, p. 1479–1485.

Rees C. E. and Thode H. G. (1974) Sulfur concentrations and isotope ratios in Apollo 16 and 17 samples. *Proc. Lunar Sci. Conf. 5th*, p. 1963–1973.

Russell W. A., Papanastassiou D. A., Tombrello T. A. and Epstein S. (1977) Ca isotope fractionation on the moon. *Proc. Lunar Sci. Conf. 8th*, p. 3791–3805.

Signer P., Baur H., Derksen U., Etique P., Funk H., Horn P. and Weiler R. (1977) Helium, neon, and argon records of lunar soil evolution. *Proc. Lunar Sci. Conf. 8th*, p. 3657–3683.

Switkowski Z. E., Haff P. K., Tombrello T. A. and Burnett D. S. (1977) Mass fractionation of the lunar surface by solar wind sputtering. *J. Geophys. Res.* **82**, 3797–3804.

Taylor H. P. and Epstein S. (1973) $0^{18}/0^{16}$ and Si^{30}/Si^{28} studies of some Apollo 15, 16, and 17 samples. *Proc. Lunar Sci. Conf. 4th*, p. 1657–1679.

Wehner G. K., KenKnight C. E. and Rosenberg D. L. (1963a) Sputtering rates under solar-wind bombardment. *Planet. Space Sci.* **11**, 885–895.

Wehner G. K., KenKnight C. E. and Rosenberg D. L. (1963b) Modification of the lunar surface by solar-wind bombardment. *Planet. Space Sci.* **11**, 1257–1261.

Yin L. I., Ghose S. and Adler I. (1972) Investigations of possible solar wind darkening of the lunar surface by photoelectron spectroscopy. *J. Geophys. Res.* **77**, 1360–1367.

Proc. Lunar Planet. Sci. Conf. 9th (1978), p. 1711–1724.
Printed in the United States of America

An experimental investigation of fractionation by sputter deposition

D. M. PARUSO*, W. A. CASSIDY and B. W. HAPKE,

Department of Earth and Planetary Sciences,
University of Pittsburgh, Pittsburgh, Pennsylvania 15260

Abstract—Fractionation during sputter deposition has been shown to be important when the target material is a fine, loose powder such as lunar soil. Only atoms which are sputtered in backward directions will escape directly from the powder without first striking an adjacent grain surface. Results of this work indicate that the backward-sputtered component is enriched in elements of low atomic weight, while the forward-sputtered component shows enrichment in heavy atoms. This effect may account for a previously-reported positive relationship between atomic weight and enrichment in the film, because those analyses were taken from an area of the film where forward-sputtered atoms were deposited. Some backward-sputtered atoms and all forward-sputtered atoms will strike a nearby soil grain surface upon ejection. This fraction will be slightly enriched in heavy atoms, and individuals will not escape from the powder unless they are able to desorb from that surface. This study shows that elements of low binding energy in the oxide are more likely to desorb from the first adsorption surface. However, these atoms tend to accumulate on shielded surfaces in the soil, resulting in a net enrichment of elements with volatile oxides in finer grain-size fractions of the soil, where sputter-deposited coatings are more volumetrically significant.

INTRODUCTION

Three fractionation mechanisms associated with solar wind sputtering of lunar soil have been subjects of recent interest: (1) Formation of a fractionated transition layer: Preferential sputtering away of higher-yield species in a multi-component target produces a transition layer ~100 A thick (Switkowski *et al.*, 1977) at the target surface. Species are depleted in this layer according to their relative sputtering yields (Wehner, 1961; Hapke, 1965; Pillinger *et al.*, 1976). Once this layer is established, however, an equilibrium condition is attained, and sputtering proceeds in a stoichiometric manner that reflects the bulk composition of the target material rather than the composition of the fractionated transition layer (Tarng and Wehner, 1971; Coburn and Kay, 1975; Werner and Warmoltz, 1976). (2) Fractionation by preferential escape from the moon of lighter sputtered atoms: the velocity distribution of sputtered atoms is such that most are able to escape lunar gravity. The fraction which falls back to the surface is enriched in heavy atoms. The enriched material becomes part of the transition layer mentioned above, and the end result is a layer slightly enriched in heavy

*Present address: Westinghouse Research and Development Center, Pittsburgh, PA 15235.

atoms (Cassidy and Hapke, 1975; Swtikowski *et al.*, 1977). This model assumes that all sputtered atoms with sufficient velocity escape from the lunar surface. (3) Fractionation upon deposition of sputtered species: The complex micromorphology of a fine powder, such as lunar soil, dictates that most sputtered atoms will strike adjacent grain surfaces. There the sputtered atoms will become part of a sputter deposit on the undersides of soil grains, rather than directly escaping from the soil (Cassidy and Hapke, 1975), and will be shielded from resputtering. This was confirmed by weight loss experiments which show that 90% of sputtered atoms are retained in a powdered target (Hapke, 1977). Details of this work are described in Hapke and Cassidy (1978), where the importance of sputter deposition is emphasized through a calculation showing that the mass of sputter-deposited material is 100 times that of the transition layer. Note that this figure is a lower limit, allowing for gardening processes at the lunar surface. Experimental studies have shown that the composition of a sputter-deposited film may differ significantly from that of the target material (Cassidy and Hapke, 1975).

Clearly, any fractionation during sputter deposition must be included in a realistic model of fractionation by solar wind sputtering. Two concerns exist: (1) Which species are most likely to be ejected in directions such that they would escape from a rough surface without striking an adjacent grain? (2) Of those atoms which do strike adjacent grains, which are most likely to desorb and have a second chance to escape?

The second question was investigated by Cassidy and Hapke (1975) in an experiment where they bombarded a terrestrial basalt glass with 2-keV hydrogen ions, and sputtered atoms were given only one chance to strike a Mo foil substrate. Electron microprobe analyses of the original target and sputter-deposited film indicated a linear relationship between atomic weight and the tendency of a sputtered atom to stick to the "first-bounce" surface. The alkali elements were exceptions; they showed extreme depletions in the film. These authors suggested that mass-dependent "first-bounce" fractionation could contribute to reported enrichments in lunar soil of heavy trace elements (Keays *et al.*, 1970; Ganapathy *et al.*, 1970; Laul *et al.*, 1971) and depletions of light isotopes of several elements (Epstein and Taylor, 1971; Grossman *et al.*, 1974; Clayton *et al.*, 1974).

EXPERIMENTAL

In a further experimental investigation of first-bounce fractionation, artificial glass targets composed of elements varying widely in atomic weight were prepared, and each was irradiated in the open configuration shown in Fig. 1. With a current density of .33mA/cm^2, 2-keV hydrogen ions struck the sample at an angle of incidence of 45°. The target was prevented from charging by low-energy electron flooding. A pressure of 1×10^{-5} torr was maintained by use of a 500 l/sec ion pump. The background gas during irradiation was H_2 leaking from the ion source. With the ion gun off, the base pressure was $\sim 1 \times 10^{-9}$ torr. Sputtered atoms were caught on a molybdenum foil substrate, as shown, and films up to 10μm thick were accumulated. This required sputtering times ranging from 75 to 205 hrs. A carbon shield prevented sputtering of the deposited film by the ion beam.

Table 1. Composition of starting mixtures from which glass samples were made.

Glass No. 3		Glass No. 4		Glass No. 5	
Element	Weight percent of oxide	Element	Weight percent of oxide	Element	Weight percent of oxide
Na	5.968	Na	4.431	Na	10.571
Al	1.228	Al	2.831	Al	2.174
Si	43.394	Si	53.380	Si	56.365
Ti	1.923	Ca	3.114	Ca	2.391
Sr	3.553	Y	6.269	Y	4.814
In	3.341	Ba	8.514	Ba	6.538
La	3.922	Gd	10.063	Nd	7.174
Sm	4.197	Pb	12.392	Gd	7.728
Pb	26.865			Hf	2.244
Bi	5.609				

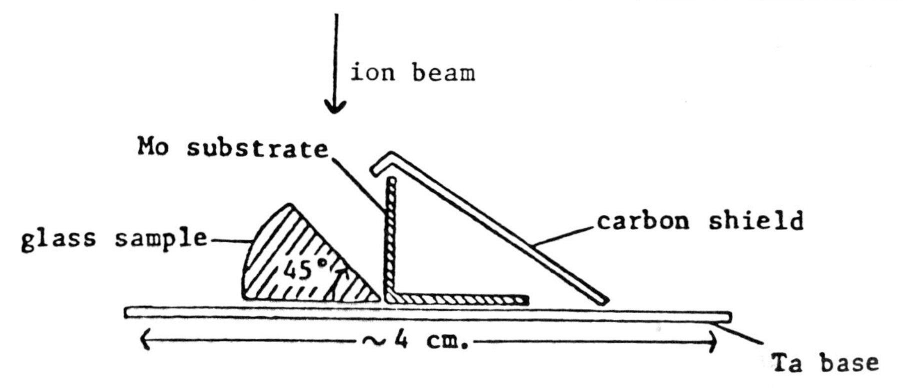

Fig. 1. Geometry, open configuration. Molybdenum was chosen as the substrate material to be comparable to the substrate used by Cassidy and Hapke (1975). Carbon was used as the shield material because, if carbon were introduced into the film, it would not interfere with subsequent electron microprobe analyses. Tantalum was chosen for the base, being a component of neither the glass nor the substrate. This was considered important because we routinely checked for adequate thickness of deposited films by analyzing for Mo. If material from the base had been sputtered onto the thin film, it at least would not appear that the film was too thin for analysis.

The exact temperature of the target is unknown. An upper limit may be estimated by assuming that the irradiated surface loses heat only by radiation; this gives about 310°C. However, thermal conduction will cause the effective radiating surface to exceed the bombarded surface by a factor of about 5, so that the average temperature of the target will be about 120°C. This temperature is comparable to daytime lunar surface values. Taking the thermal conductivity of the target glass to be ~.003 cal/cm.-sec.-°C (Clark, 1966) indicates that the temperature gradient across the sample is ~50°C/cm. This calculation ignores conductive losses to the support surfaces. Thus, we conservatively estimate that the maximum temperature of the target is below 200°C. At this temperature the vapor pressure of PbO, one of the most volatile oxides in our glasses, is $\sim 1 \times 10^{-19}$bar. (Samsonov, 1973). The time necessary to evaporate a monolayer of this oxide is 4000 years. Thus, evaporation is negligible, and all transport from target to substrate is by true sputtering.

In order to gain information on multiple-bounce effects, one of the glasses was irradiated in the configuration shown in Fig. 2. The ion beam entered the box through an opening in the ceiling and struck the target at an angle of incidence of 45°. Sputtered atoms which did not stick to the first surface they encountered could desorb and be accommodated elsewhere. Deposits on the outer substrate and the central section of the front wall are unique because they represent areas shielded from direct impingement by sputtered atoms; any atoms found there must have desorbed from at least one previous site. These may then be thought of as "multiple-bounce" areas of deposition. For some multiple-bounce surfaces, an irradiation time of 784 hrs. was required to build up deposits of sufficient thickness for analysis, about 8μm.

Analyses of the sputter-deposited films and unsputtered target glasses were done using the electron microprobe at the MIT Dept. of Earth and Planetary Sciences. The directly sputtered surface was not analyzed because the depth of the transition layer is insignificant compared to the electron beam penetration depth. Elemental concentrations in the films were taken as the average of ten or more evenly-spaced data points. Five points on the target glass were averaged to give the parent composition. The ratios (atoms of an element in the film)/(atoms of that element in parent glass) were calculated and are shown as film/parent ratios in the results presented here.

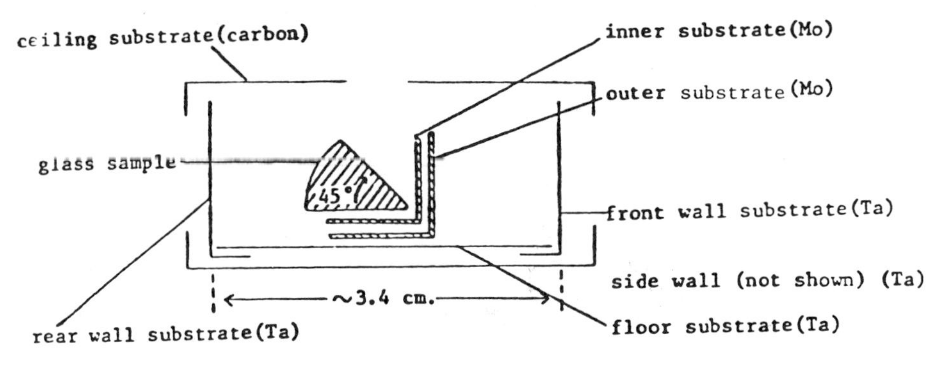

Fig. 2. Geometry, closed box configuration. Substrate materials were chosen for the reasons outlined in Fig. 1.

RESULTS AND DISCUSSION

In Figs. 3, 4 and 5, where Na is the only alkali present, it is depleted. These three figures also indicate that the mass-dependent trend observed by Cassidy and Hapke (1975) is somewhat evident up to atomic weight 60 (the limit in their experiments), but either does not hold or is masked by other factors in higher mass ranges. Low ratios for Pb and Bi in Figs. 3 and 4 suggest that heavy atoms might have low yields when sputtered by 2-keV H^+ ions. Pillinger *et al.* (1976) suggested that heavy atoms would not be sputtered at all. However, results from the closed configuration experiment do not support this assertion, as will be discussed later.

Widely differing ratios for yttrium in Figs. 4 and 5 and for aluminum in Figs. 3, 4, and 5 suggest that the sticking coefficient of an atom may depend in part on the elements which it must compete with for accommodation at the substrate.

In an effort to uncover other factors influencing sticking coefficient, film/parent ratios from Figs. 3–5 were plotted versus binding energy of the metal atoms

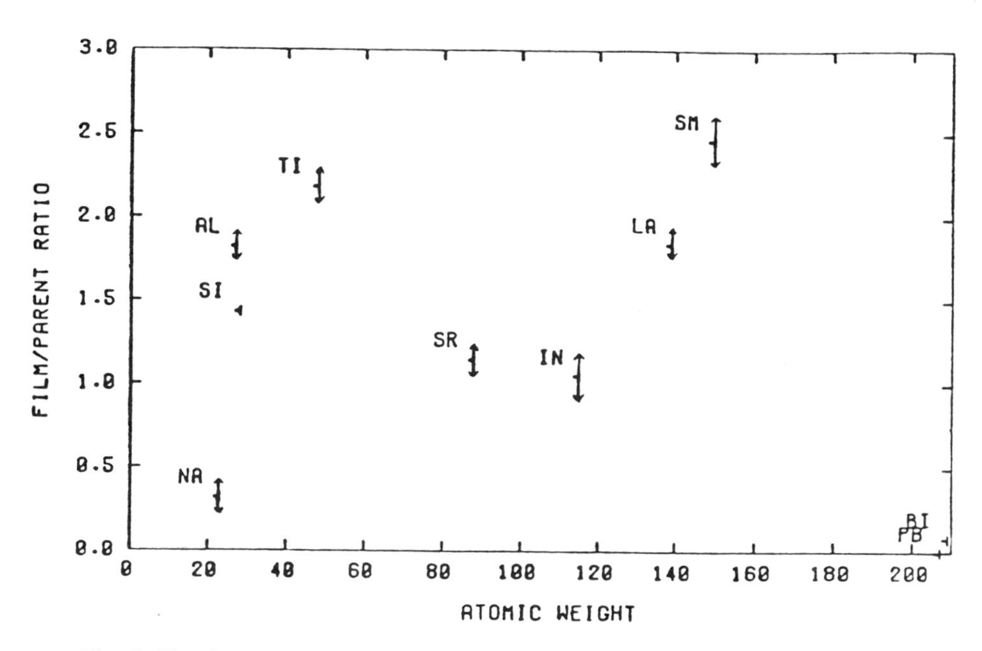

Fig. 3. First-bounce sputter deposition fractionation. Glass No. 3, open configuration.

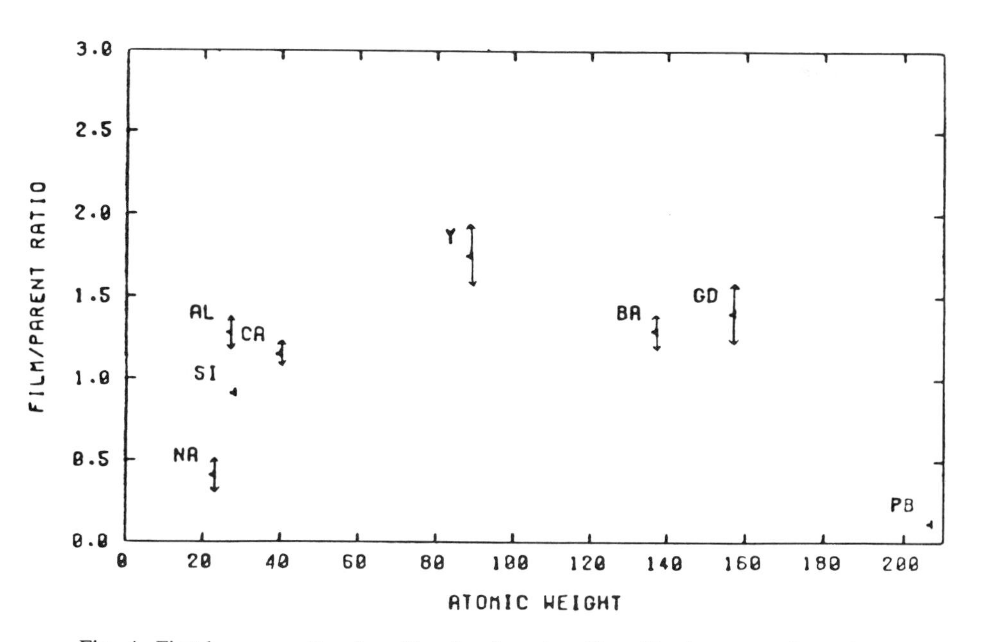

Fig. 4. First-bounce sputter deposition fractionation. Glass No. 4, open configuration.

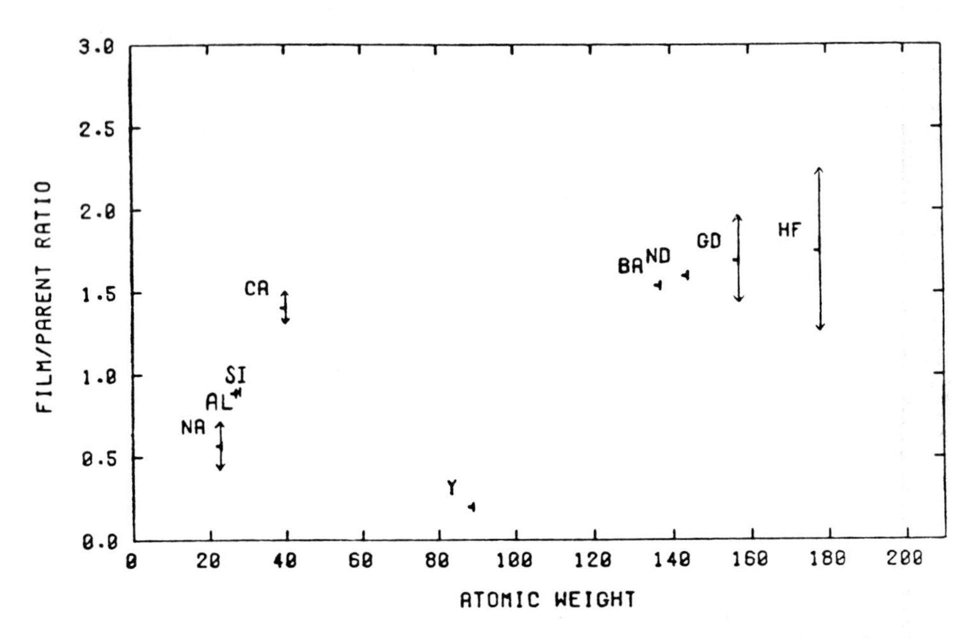

Fig. 5. First-bounce sputter deposition fractionation. Glass No. 5, open configuration.

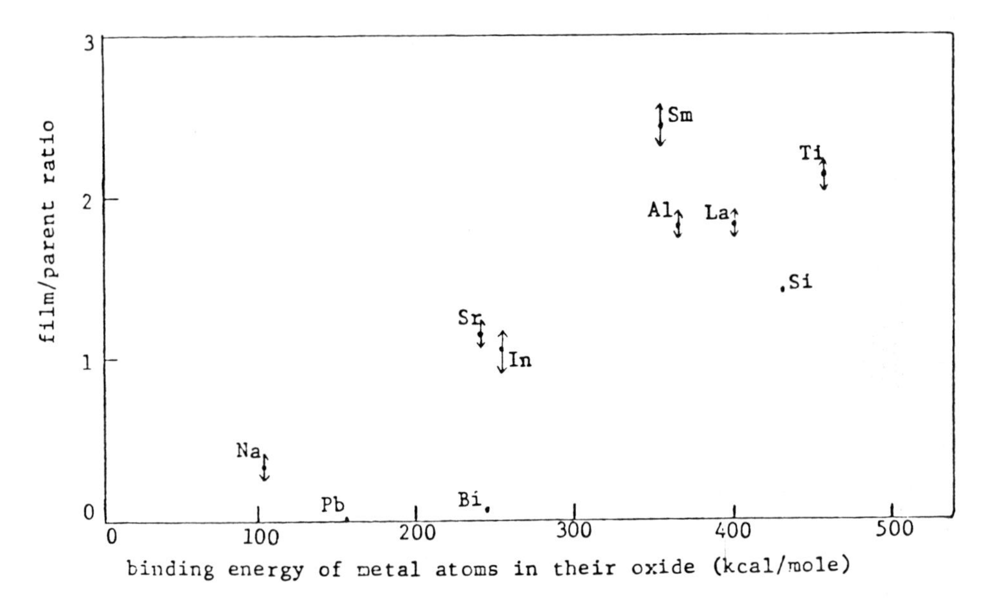

Fig. 6. Glass No. 3 sputtered in open configuration. Elements of high binding energy tend to stick to the first-bounce surface.

in their oxide. These plots are shown in Figs. 6, 7 and 8. These figures reveal a positive relationship between ability to stick to a first-bounce surface and binding energy in the oxide.

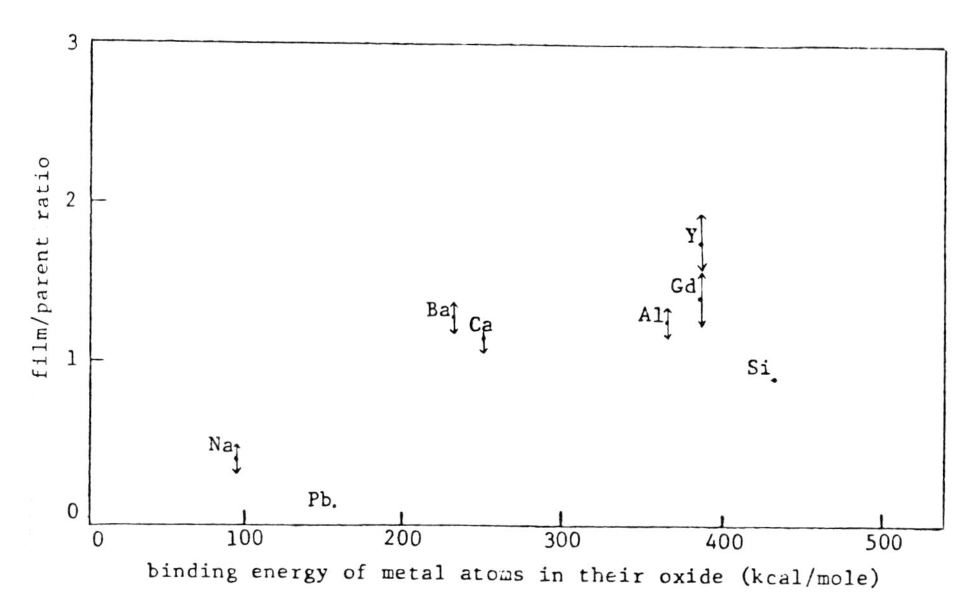

Fig. 7. Glass No. 4 sputtered in open configuration. Elements of high binding energy tend to stick to the first-bounce surface.

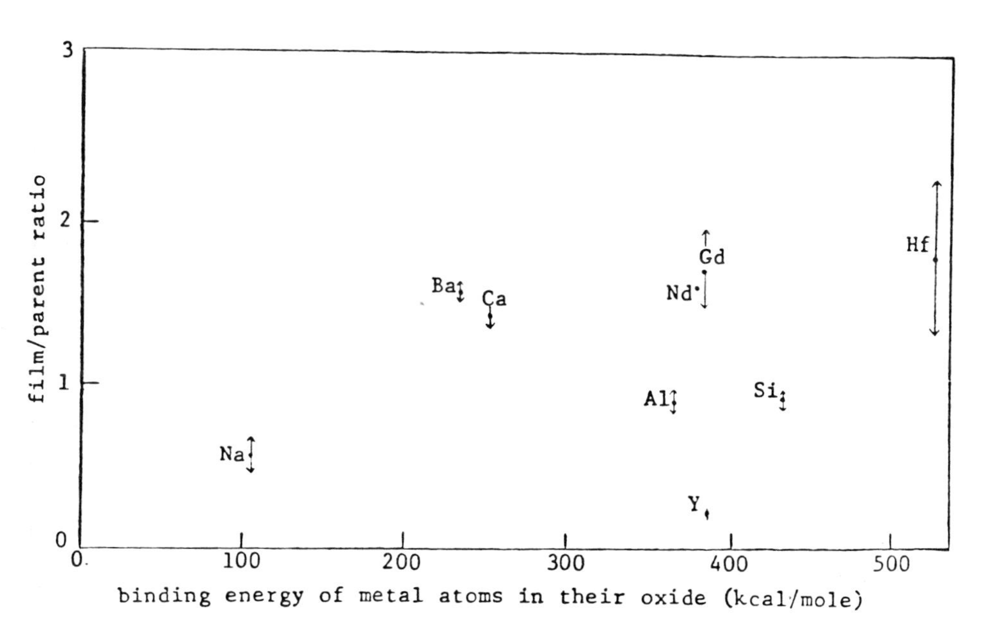

Fig. 8. Glass No. 5 sputtered in open configuration. Elements of high binding energy tend to stick to the first-bounce surface. Y exhibits anomalous behavior in this experiment.

The error bars shown in Figs. 3–8 reflect variations in the concentration of each element in a ten-point scan from the top to bottom of the substrate. Figures 10 and 11 show the gradients in composition along the substrates. As is consistent

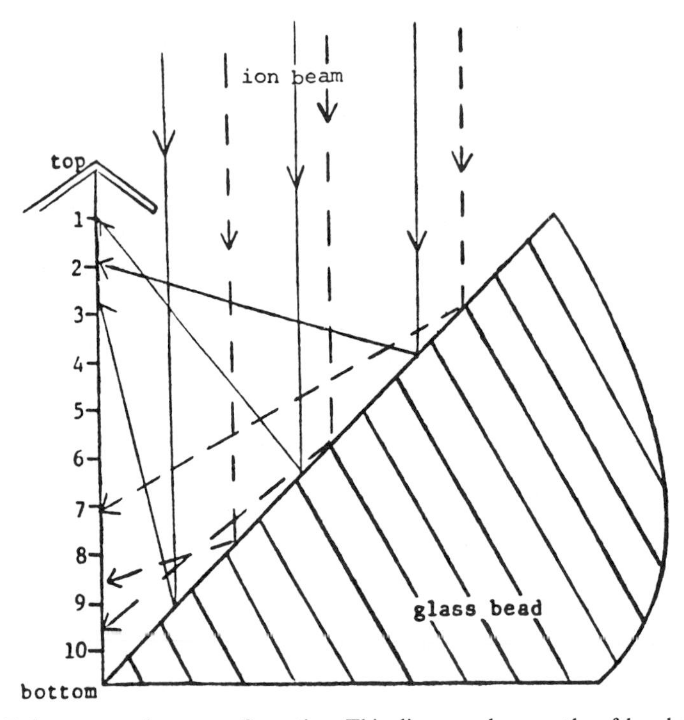

Fig. 9. Enlargement of open configuration. This diagram shows paths of bombarding ions and atoms ejected in both forward (dotted lines) and backward (solid lines) directions. Note that the bottom area of the substrate receives mostly forward-scattered atoms, while the top receives backward-scattered atoms.

with the "billiard ball" or "hard-sphere" model of sputtering, atoms are more easily ejected in forward directions from an inclined target, requiring less change in the direction of momentum. Forward ejection may be accomplished by a short cascade, or even by a single glancing blow. Backward ejection, however, requires a longer cascade to accomplish a reversal of momentum direction, and energy is lost in each collision. It is not surprising that heavy atoms show their highest film/parent ratios in areas near the bottom of the substrate, where forward-scattered atoms are received (Fig. 9). This leaves the top of the film enriched in light elements, as shown in Figs. 10 and 11. This effect has also been observed by Wehner (1977), who reported that backward-scattered species from both Cu and Mo targets bombarded at normal incidence were enriched in light isotopes. The mass effect reported by Cassidy and Hapke (1975) was based on measurements taken near the bottom of the substrate where the film was thickest. Therefore it is likely that their reported enrichment of high-mass elements results from mass-related directional sputtering.

The application of this directional effect at the lunar surface would mean that lighter elements and isotopes would be favored to be ejected in backward directions, escaping directly through the openings which admit bombarding ions,

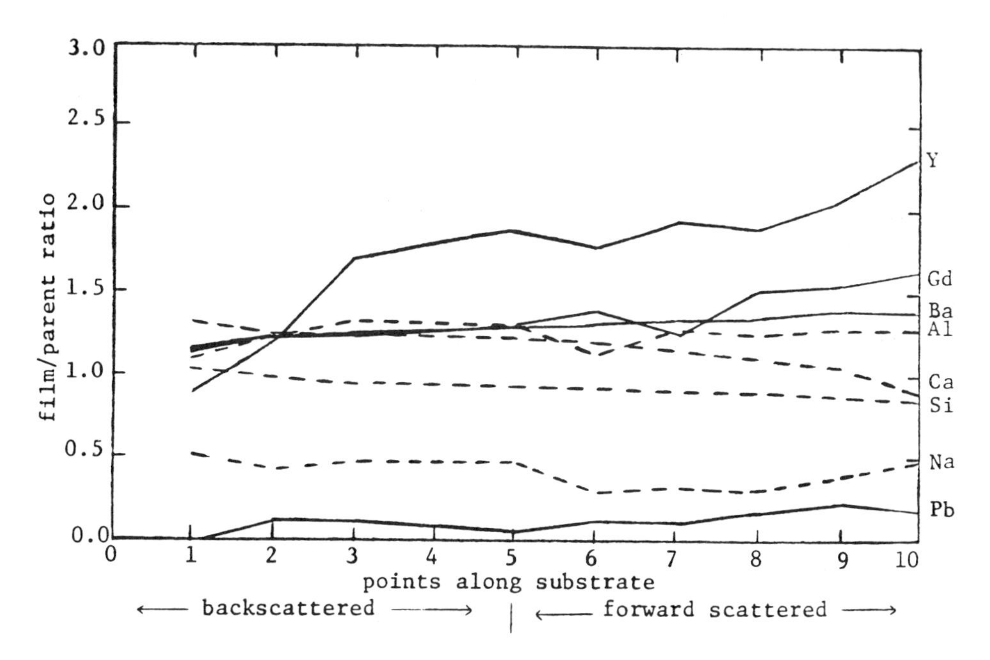

Fig. 10. Variation in sputter deposition fractionation with ejection angle. Glass No. 4, open configuration.

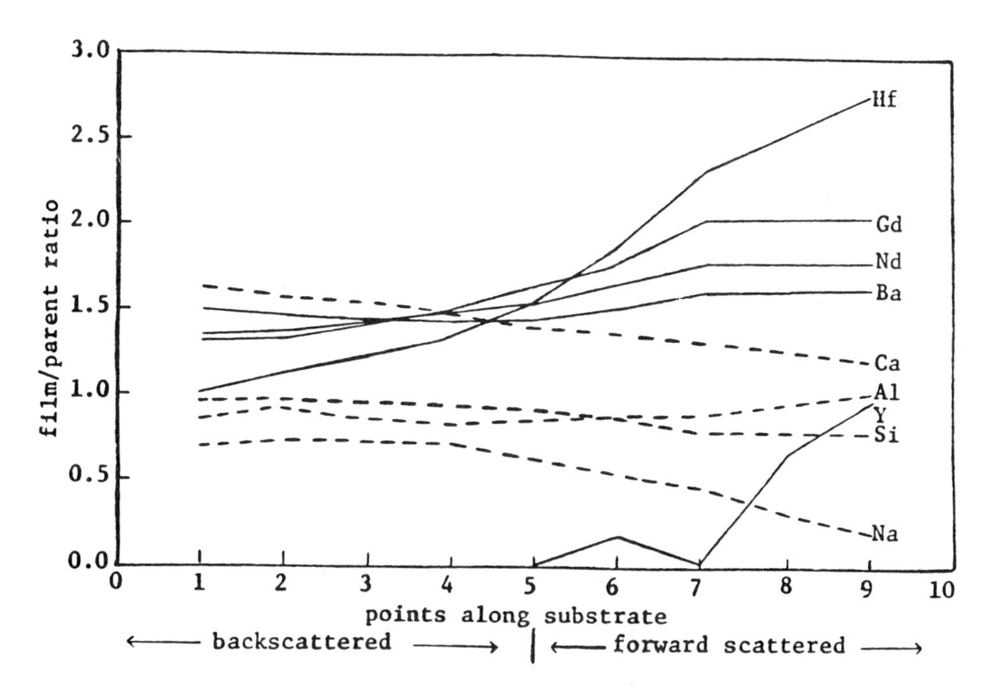

Fig. 11. Variation in sputter deposition fractionation with ejection angle. Glass No. 5, open configuration.

without first striking an adjacent grain surface. They would be preferentially lost regardless of sticking coefficients. Conversely, heavy elements and isotopes would tend to be forward-sputtered deeper into the soil and be preferentially retained. This process would contribute to the reported enrichments of heavy elements and isotopes. Escape of oxygen, a light element, may help to explain the presence of reduced forms of Fe and possibly Ti, Si and Al found on the surfaces of fine grains of mature lunar soil (Vinogradov *et al.*, 1972; Housley *et al.*, 1974; Hapke *et al.*, 1975; Pillinger *et al.*, 1976; and Dikov *et al.*, 1978). Reduced metal was predicted by Hapke (1966) on the basis of laboratory experiments. The case of oxygen is particularly interesting, but, unfortunately, we were not able to analyze for this element using the electron microprobe. In addition to the depletion of oxygen by directional sputtering, further fractional loss may occur upon deposition as a result of a low sticking coefficient (Becker, 1957; Eisinger and Law, 1959; Hapke, 1965).

It has been shown above that elements with refractory oxides tend to stick to the first-bounce surface, but the fate of an atom which desorbs is an additional consideration. Analyses of the films deposited on multiple-bounce substrates of the closed configuration showed that elements of low binding energy in the oxide are concentrated on such shielded areas, while those of high binding energy are strongly depleted, having already been accommodated at a previous substrate (Figs. 12, 13 and 14). This result is emphasized by a comparison of the

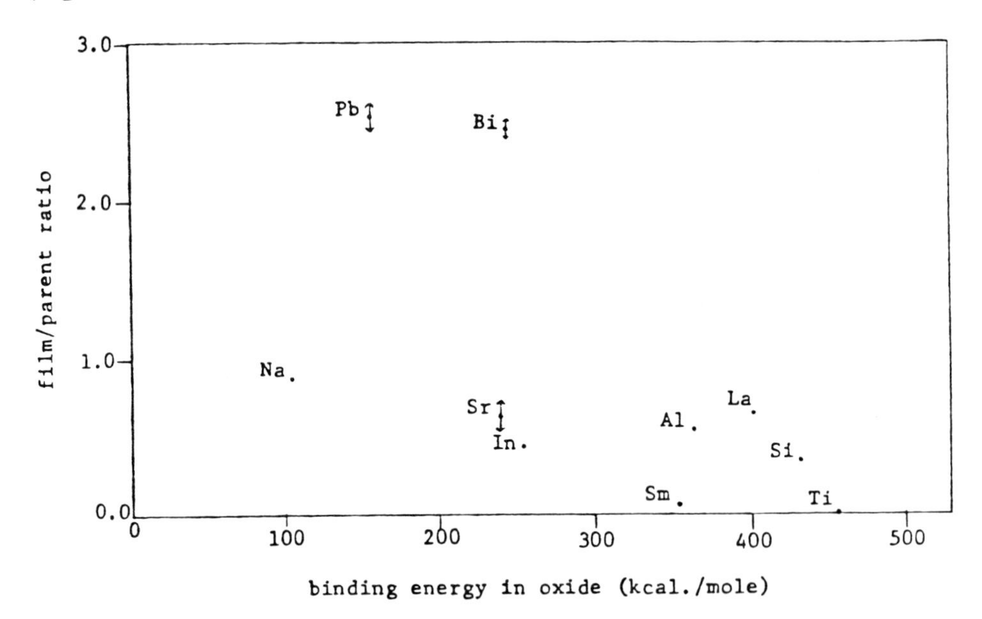

Fig. 12. Sputter deposition fractionation of Glass No. 3 elements at a multiple-bounce surface, the outer substrate of the closed configuration. Note that elements of high binding energies are depleted here, having already been accommodated at another surface. (See Fig. 4.) Pb and Bi are indeed sputtered by protons of solar wind energies.

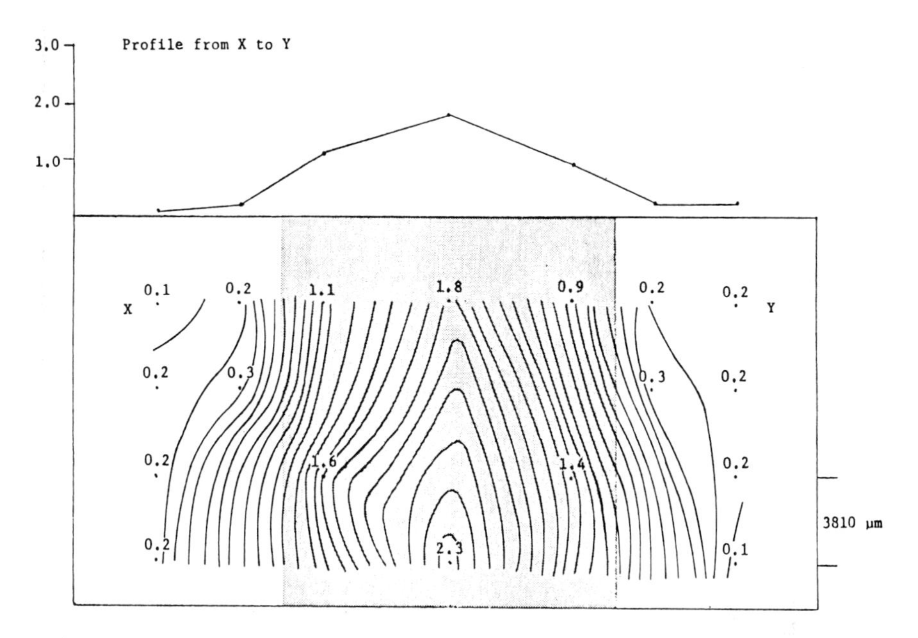

Fig. 13. Contour map of film/parent ratios for Pb in Glass No. 3 on the front wall substrate, closed configuration. Shading indicates the section shielded from direct impingement by sputtered atoms. Pb, an element of low binding energy in its oxide, is enriched in this multiple-bounce area.

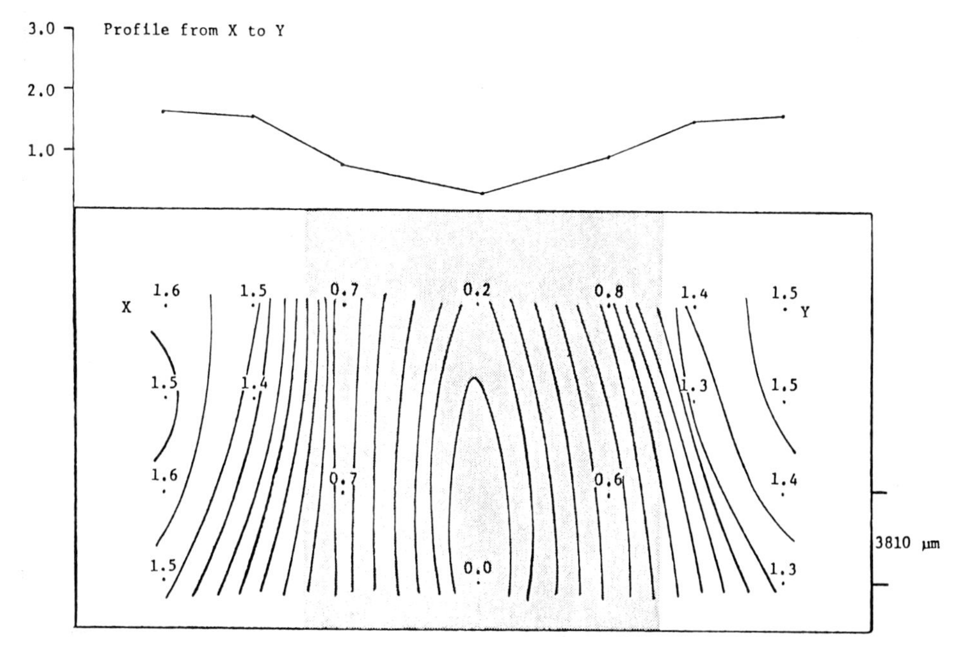

Fig. 14. Contour map of film/parent ratios for Ti in Glass No. 3 on the front wall substrate, closed configuration. Note the depletion of Ti in the shaded, multiple-bounce area, indicating that it is accommodated before reaching such an area.

enrichment ratios of the first-bounce substrate (Fig. 6) with multiple-bounce ratios (Fig. 12). The high enrichments of Pb and Bi in the multiple-bounce film show that heavy elements are efficiently sputtered by hydrogen. Their depletions in open configuration films were caused primarily by low sticking coefficients rather than low yields. These results suggest that elements with volatile oxides might be concentrated in shielded, multiple-bounce areas in the lunar soil. Silver (1970), Allen *et al.*, (1973, 1974), Jovanovic and Reed (1973), Reed *et al.* (1977), Boynton and Wasson (1977), and Blanchard and Budahn (1978) have found progressive bulk enrichments of volatile elements such as Na, Pb, Cd, Bi, Zn, Tl, and In in lunar soil fractions of progressively decreasing grain size. Enrichment factors up to 20 have been reported. Sputter-deposited coatings would be more volumetrically significant in smaller grains. Coatings which have spalled off would also be part of the fine fraction.

CONCLUSIONS

Further insight into fractionation in the sputtering of a fine powder has been gained through this experimental investigation. We have demonstrated that the following are likely to occur at the lunar surface: (1) Atoms of low atomic weight are most likely to be sputtered in backward directions such that they might escape from the powder surface without striking an adjacent grain. This would also happen to lighter isotopes of various elements. Heavy elements would be forward-sputtered and be preferentially retained in the soil. (2) Binding energy of an element in its oxide is one factor influencing the sticking coefficient of a sputtered atom. Elements of high binding energy will have a greater tendency to stick to the first bounce surface, while atoms of low binding energy will be likely to desorb. Easily desorbed atoms, such as Pb and Bi, tend to accumulate on shielded surfaces rather than being lost. Because they are on the surfaces rather than the interiors of grains, and because they are volatile, they will tend to remain mobile in the soil and be less likely to be incorporated into agglutinates during small impacts.

Acknowledgments—We thank Dr. E. N. Wells for his assistance. This work was supported by a grant from the National Aeronautics and Space Administration, Lunar and Planetary Programs Office.

REFERENCES

Allen R. O. Jr., Jovanovic S., and Reed G. W. Jr. (1973) Geochemistry of primordial Pb, Bi, and Zn in Apollo 15 samples. *Proc. Lunar Sci. Conf. 4th*, p. 1169–1175.

Allen R. O. Jr., Jovanovic S. and Reed G. W. Jr. (1974) A study of ^{204}Pb partition in lunar samples using terrestrial and meteoritic analogues. *Proc. Lunar Sci. Conf. 5th*, p. 1617–1623.

Becker J. A. (1957) Study of surfaces using new tools. In *Solid State Physics 7* (F. Seitz and D. Turnbull eds.). p. 379–424. Acad. Press, N.Y.

Blanchard D. P. and Budahn J. R. (1978) Compositional variations among size fractions in a homogeneous, unmatured regolith: Drive core 74001/2 (abstract). In *Lunar and Planetary Science IX*, p. 100–101. Lunar and Planetary Institute, Houston.

Boynton W. V. and Wasson J. T. (1977) Distribution of 28 elements in size fractions of lunar mare and highland soils. *Geochim. Cosmochim. Acta* **41**, 1073–1082.

Cassidy W. A. and Hapke B. W. (1975) Effects of darkening processes on surfaces of airless bodies. *Icarus* **25**, 371–383.

Clayton R. N., Mayeda T. K., and Hurd J. M. (1974) Loss of O, Si, S and K from the lunar regolith (abstract). In *Lunar Science V*, p. 129–131. The Lunar Science Institute, Houston.

Clark S. P. Jr. (1966) Thermal conductivity. In *Handbook of Physical Constants* (S. P. Clark, Jr., ed.), p. 459–482. Geol. Soc. Amer. Mem. 97, New York.

Coburn J. W. and Kay E. (1975) Composition profiling—A comparison of surface analysis techniques vs. methods involving the detection of sputtered species. *J. Vac. Sci. Tech.* **12**, 403.

Dikov Y. P., Bogatikov O. A., Barsukov J. L., Florensky C. P., Ivanov A. V., Nemoshkalenko V. V., and Alyoshin V. G. (1978) Reduced forms of elements in the surface parts of regolith particles: ESCA studies (abstract). In *Lunar and Planetary Science IX*, p. 250–252. Lunar and Planetary Institute, Houston.

Eisinger J. and Law J. T. (1959) Adsorption of oxygen on silicon. *J. Chem. Phys.* **30**, 410–412.

Epstein S. and Taylor H. P. Jr. (1971) O^{18}/O^{16}, Si^{30}/Si^{28}, D/H, and C^{13}/C^{12} ratios in lunar samples. *Proc. Lunar Sci. Conf. 2nd*, p. 1421–1441.

Ganapathy R., Keays R., Laul J. C., and Anders E. (1970) Trace elements in Apollo 11 lunar rocks: Implications for meteorite influx and origin of moon. *Proc. Apollo 11 Lunar Sci. Conf.*, p. 1117–1142.

Grossman L., Clayton R. N., and Mayeda T. K. (1974) Oxygen isotopic compositions of lunar soils and Allende inclusions and the origin of the moon (abstract). In *Lunar Science V*, p. 298–300. The Lunar Science Institute, Houston.

Hapke B. (1965) Effects of a simulated solar wind on the photometric properties of rocks and powders. *Ann. N. Y. Acad. Sci.* **123**, 711–721.

Hapke B. (1966) Optical properties of the moon's surface. In *The Nature of the Lunar Surface* (W. Hess, D. Menzel and J. O'Keefe, eds.). p. 141–154. Johns Hopkins Univ. Press, Baltimore.

Hapke B. W. (1977) The albedo of lunar soil (abstract). In *Lunar Science VIII*, p. 398–400. The Lunar Science Institute, Houston.

Hapke B. W. and Cassidy W. A. (1978) Is the moon really as smooth as a billiard ball? Remarks concerning recent models of sputter-fractionation on the lunar surface. *Geophys. Res. Lett.* In press.

Hapke B., Cassidy W., and Wells E. (1975) Effects of vapor-phase deposition processes on the optical, chemical, and magnetic properties of the lunar regolith. *The Moon* **13**, 339–353.

Housley R. M., Cirlin E., Paton N. E., and Goldberg I. B. (1974) Solar wind and micrometeorite alteration of the lunar regolith. *Proc. Lunar Sci. Conf. 5th*, p. 2623–2642.

Jovanovic S. and Reed G. W. Jr. (1973) Volatile trace elements and the characterization of the Cayley Formation and the primitive lunar crust. *Proc. Lunar Sci. Conf. 4th*, p. 1313–1324.

Keays R. E., Ganapathy R., Laul J. C., Anders E., Herzog G. F., and Jeffery P. M. (1970) Trace elements and radioactivity in lunar rocks: Implications for meteorite infall, solar wind flux, and formation conditions of moon. *Science* **167**, 490–493.

Laul J. D., Morgan J. W., Ganapathy R., and Anders E. (1971) Meteoritic material in lunar samples: Characterization from trace elements. *Proc. Lunar Sci. Conf. 2nd*, p. 1139–1158.

Pillinger C. T., Gardiner L. R., and Jull A. J. T. (1976) Preferential sputtering as a method of producing metallic iron, inducing element fractionation and trace element enrichment. *Earth Planet. Sci. Lett.* **33**, 289–299.

Reed G. W. Jr., Allen R. O. Jr., and Jovanovic S. (1977) Volatile metal deposits on lunar soils—relation to volcanism. *Proc. Lunar Sci. Conf. 8th*, p. 3917–3930.

Samsonov G. V. (1973) *The Oxide Handbook*, IFI/Plenum, N.Y.

Silver L. T. (1970) Uranium-thorium-lead isotopes in some Tranquility base samples and their implications for lunar history. *Proc. Apollo 11 Lunar Sci. Conf.*, p. 1533–1574.

Switkowski Z., Haff P., Tombrello T., and Burnett D. (1977) Mass fractionation of the lunar surface by solar wind sputtering. *J. Geophys. Res.* **82**, 3897–3904.

Tarng M. L. and Wehner G. K. (1971) Alloy sputtering studies with *in situ* auger electron spectroscopy. *J. Appl. Phys.* **42**, 2449–2452.

Vinogradov A. P., Nefedov V. I., Urusov V. S., and Zhavoronokov N. M. (1972) ESCA-investigation of lunar regolith from the Seas of Fertility and Tranquility. *Proc. Lunar Sci. Conf. 3rd*, p. 1421–1427.

Wehner G. K. (1961) Sputtering effects on the moon's surface. *Amer. Rocket Soc. J.* **31**, 438–439.

Wehner G. K. (1977) Isotope enrichment in sputter deposits. *Appl. Phys. Lett.* **30**, 185–187.

Werner H. W. and Warmoltz N. (1976) The influence of selective sputtering on surface composition. *Surface Science* **57**, 706–714.

Proc. Lunar Planet. Sci. Conf. 9th (1978), p. 1725–1744.
Printed in the United States of America

Monte Carlo sputter simultations and laboratory ion sputter measurements of lunar surfaces

W. C. CAREY and J. A. M. MCDONNELL

Space Science Laboratories, The University of Kent, Canterbury, England

Abstract—Results from a Monte Carlo computer simultation of the growth of submicron-sized accreta population distributions are presented, which characterize the development of such a population and its approach to solar wind sputter equilibrium. The accreta influx used was derived from the Apollo consortium sample 12054. The first laboratory measurement of sputter on a lunar sample due to a variable incidence ion beam representing the lunar cycle is reported. It is shown that sputter induced diffusion plays a major role in the development of microscale surface roughness; the sputter model verifies that atoms ejected from an amorphous or polycrystalline lunar material form a cosinal angular distribution. Redeposition of sputtered material confirms our earlier model result that microscale surface roughness is sufficient to retain some 70% of ejected atoms.

INTRODUCTION

Decoding the temporal aspects of micrometeoroid impact data gained from lunar material exposed to multivarious processes (regolith gardening, sputter degradation, cosmic ray track production, etc.), which occur on the lunar surface, remains a difficult task due to competitive processes being highly interactive. The particular process investigated in this paper is the growth of submicron-sized accreta populations and the approach of such a population to sputter equilibrium. These particle populations can influence greatly the information deduced from returned lunar rocks on micrometeorite crater counts and cosmic ray track densities. Indeed, Morrison *et al.* (1973) have shown that the areal density of accreta particles exceeds microcrater populations typically by one and a half orders of magnitude. We attempt to utilize the characteristics of such an accreta population distribution to enable the deduction of limits to certain important parameters, such as the sputter rate, the original accreta influx and the exposure age from accreta population measurements on exposed lunar material. In the pursuit of this aim, we have investigated the following aspects of lunar surface dynamic processes:

(a) The growth and development of accreta population distributions using a Monte Carlo (MC) computer simulation model, both with and without the effects of sputtering by solar wind ions.
(b) The behaviour of accreta populations under solar wind equivalent sputtering in the laboratory.

(c) An MC simulation study of the dynamic processes occurring on exposed lunar material, namely differential sputtering and ion-impact induced atomic diffusion.
(d) The angular distribution of atoms ejected from exposed amorphous or polycrystalline lunar material using an MC computer simulation.

MONTE CARLO ACCRETA SIMULATION

As the lunar surface is continuously exposed to meteoritic bombardment, a considerable source of material in the form of comminution products (this term includes both high and low velocity material ejected upon primary impact of a micrometeoroid) is made available for secondary impact on the lunar regolith. This impact-induced source of particles together with the probable contribution from electrostatically levitated particles provides an accreta influx which accumulates concurrently with the primary impact crater population distribution. Earlier work by Morrison *et al.* (1973) characterized the physical appearance of these particles into three general categories:

(a) angular, fragmented particles which predominate within the 1–10 μm size regime,
(b) below 1 μm, a larger number of almost precisely spherical particles are seen,
(c) low velocity glass splashes or "pancakes" are in evidence over both these size ranges and are observed down to below 0.1 μm.

The larger ($\sim$10–100 μm) size of accreta particle, which is loosely bound, can prevent the accumulation of cosmic ray tracks on the masked area (Morrison and Zinner, 1975), and recently, Zook *et al.* (1978) have predicted evidence from studies of consortium sample 12054 that this loosely bound dust can also prevent the growth of submicron impact pits or low velocity accreta pancakes. In an attempt to gain insight into the characterisation of accreta particle population distribution, we developed a Monte Carlo model to simulate the growth of an accreta population. As has been mentioned by previous authors of MC models (Oberbeck *et al.*, 1973; Hartung *et al.*, 1973) a compromise has to be reached between the size of the test surface area and the number of accreta particle "impacts" due to computer storage limitation. To maintain computer usage within reasonable limits, this study selected a test surface area consisting of a grid of 50 $\times$ 50 unit cells. By varying the cell size, different regions of accreta particle diameters could be investigated.

A total of 2,500 impact coordinates are available which were chosen by a random number generator to give spatial randomness. The model uses as input a differential accreta particle influx as a function of accreta diameter, which is converted in the program to a probability distribution as a function of particle size. The selection of an accreta particle influx for this study was chosen from the measured accreta production population distribution of consortium sample 12054

(McDonnell, 1977), representing the *total* number of low velocity particles impacting a lunar rock. After a particle impacts the test surface area, a record is kept of its impact site and diameter. As we are concerned mainly with areal densities (i.e., projected areal coverage of test surface), our model particles are considered cylindrical with a height-to-diameter of unity, representing the projection of spherical particles on the surface. On the size scale involved in our simulation the most important parameters affecting the lifetime of an accreta particle are sputter degradation and obscuration of one particle by another; both of these events are included in our simulation. The particles arrive onto the test surface at equal time intervals. During this interval any particle resident on the surface is subject to sputter erosion if it is in an "exposed state". The criterion for determining when a particle is considered to be shielded from sputter, i.e., in a "non-exposed state", is when its centre is covered by a particle of diameter equal to or greater than its own diameter. For the case of a large particle centre being obscured by a smaller particle, the covered particle is still available for sputter degradation. This degradation takes the form of a reduction in diameter of all particles which are in an "exposed state" at the completion of each cycle in the main program. In this way we can observe the build-up of accreta population distributions on exposed lunar surfaces taking into account both sputter erosion and single particle obscuration. Figures 1a and 1b show an example of the output obtained from the program.

Figure 1a shows the extent to which the test surface area is covered by incoming particles after an exposure equivalent to 5×10^4 years, i.e., ⅓ the lifetime of sample 12054. Compare this with the areal coverage and population distribution shown in Fig. 1b, which corresponds to an identical exposure time but in this simulation sputter degradation has been included representing a sputter rate of 0.031 A.yr^{-1} (McDonnell, 1977). The rationale behind this value for the sputter rate is given in the conclusion at the end of the paper. In this second simulation, nearly all of the particles in the smallest size range have been destroyed. This, together with a diminishing of the larger particles, results in a significant reduction in the projected areal coverage and a radical modification to the particle size distribution.

Results of MC

The results in this paper correspond to several computer runs simulating 10^3 impacts onto the test surface with a printout of the current status of all model variables every 10^2 impacts.

Cumulative accreta populations

Figure 2 shows a plot of the measured cumulative accreta production curve together with the MC model equivalent, the slope of the curves in each case being 1.74 in the particle diameter region ≤ 1.0 μm. The first sputtered curve (a) in Fig. 2 represents the cumulative population distribution after an exposure equal

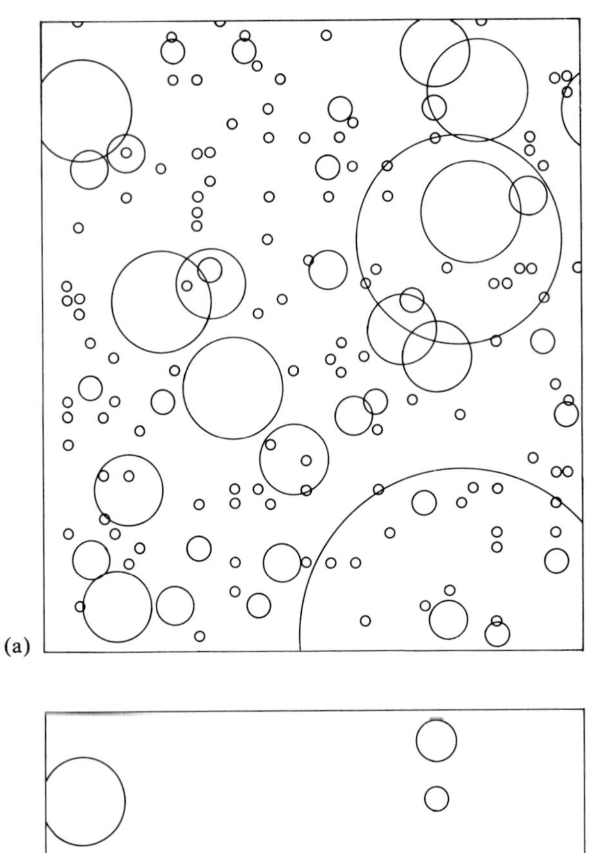

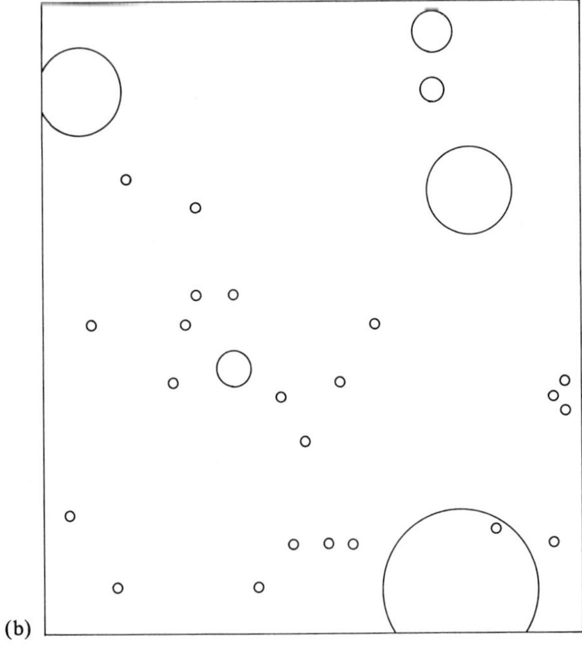

Fig. 1. Monte Carlo model accreta population distribution after an exposure equivalent to 5×10^4 years for an initially clean surface, with an exposure geometry equivalent to Apollo 12054. (a) Shows the areal coverage achieved if the sample surface had been shielded from sputter. (b) Shows the resultant distribution if the sample surface was exposed to a solar wind sputter rate of 0.031 Å yr^{-1} for a similar exposure of 5×10^4 years. Both figures represent a test surface 0.5 $\mu m \times 0.5$ μm in area.

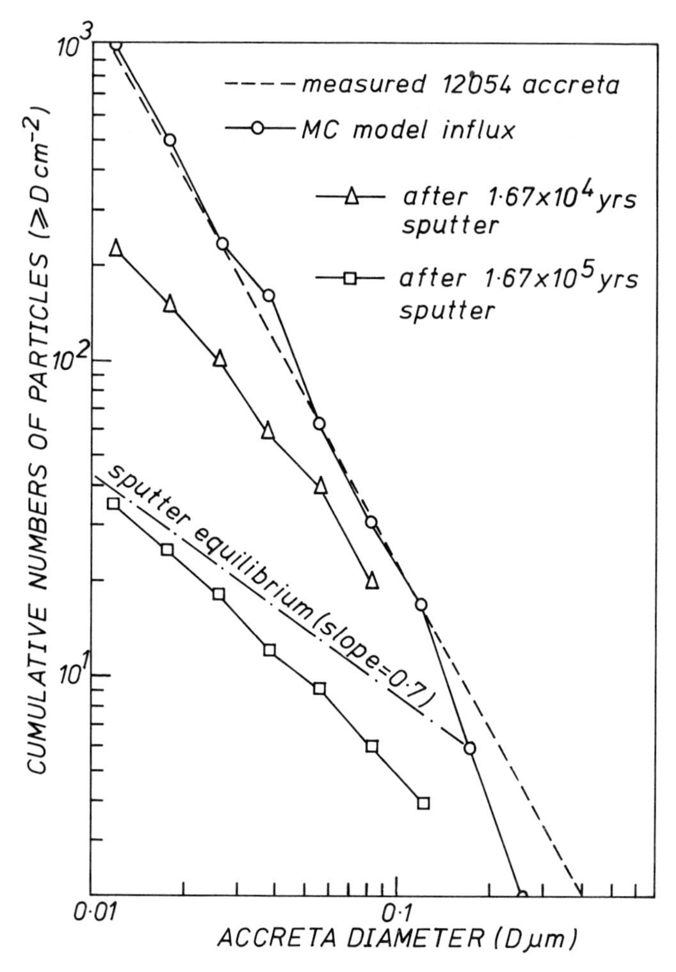

Fig. 2. Measured accreta production obtained from Apollo sample 12054 together with its Monte Carlo model equivalent. The slope of these curves in the diameter range shown is 1.74. (a) Represents the predicted behavior of the population after an exposure of 1.67×10^4 years at a sputter rate of 0.03 Å yr^{-1}. (b) The predicted behaviour after an exposure of 1.8×10^5 years at the same sputter rate. The slope of the theoretical equilibrium curve is drawn in to highlight the approach to equilibrium of the sputtered distributions with increasing exposure age.

to 10% of the sample lifetime, in this case 12054, whose age has been measured at 1.75×10^5 years (Morrison and Zinner, 1977). The second sputtered curve (b) in Fig. 2 corresponds to that obtained when the test surface has been exposed for a period almost equal to the sample lifetime. From this model simulation, we can observe the gradual build-up of an accreta particle population distribution with time together with the approach to sputter equilibrium.

We now present an analytical derivation to show that the slope of an accreta particle population distribution in sputter equilibrium is n-1, where n is the slope of the accreta production curve. The number of particles in equilibrium at a particular diameter d is given by

$$n_{eq}(d) = n_i(d)t(d), \tag{1}$$

where $n_i(d)$ is the number of incident particles at diameter d, and t(d) is a time dependent function proportional to d/R, because $t(d) = \text{const.} \frac{d}{\alpha R}$ (const. $\simeq 1$), and where R represents the sputter rate. The cumulative influx is given by

$$\Phi = Ad^{-\alpha}, \tag{2}$$

where A and α are constants.
Differentiating (2) to obtain the differential influx we find

$$\psi = -A\alpha d^{-\alpha-1}. \tag{3}$$

Substituting for $n_i(d)$ into (1) from Eq. (3) we obtain

$$n_{eq}(d) = -A\alpha d^{-\alpha} \cdot \frac{1}{R}.$$

Thus, the cumulative number of particles ≥ a particular diameter d is given by

$$N_{eq}(\geq d) = -\frac{A\alpha}{R} \cdot \frac{d^{1-\alpha}}{(1-\alpha)}. \tag{4}$$

The slope of the second sputtered distribution at the small-sized end of the curve is ~0.7, thus agreeing with theory which predicts the slope of an equilibrium distribution to be n-1, where the primary influx slope equals n. We have been able to simulate the behaviour of accreta population distributions on 12054 with respect to their development under sputter and approach to equilibrium using our MC model. Here, then, we have been able to reproduce a measured accreta particle distribution on 12054 using our MC model together with a known particle influx and sputter rate. By using different accreta influx distributions and sputter rate, it is possible to match the MC model sputtered curves to any other particular measured accreta population distributions, allowing the exposure age of the rock surface on which the accreta resides to be determined.

Projected areal coverage

The model output for areal coverage as a function of time is shown in Fig. 3. Where no sputter degradation of the particle occurs, after an exposure corresponding to 5×10^4 years (i.e., ⅓ of the sample age) approximately half (51.2%) of the test surface has been covered by accreta particles. After an exposure corresponding to 1.7×10^5 years (i.e., equal to the sample age), this percentage areal coverage has risen to 91.2%. This picture of the development of test surfaces under MC particle influx has been observed in a previous simulation by Hörz *et al.* (1974) in a study of micrometeoroid abrasion of lunar rocks. They

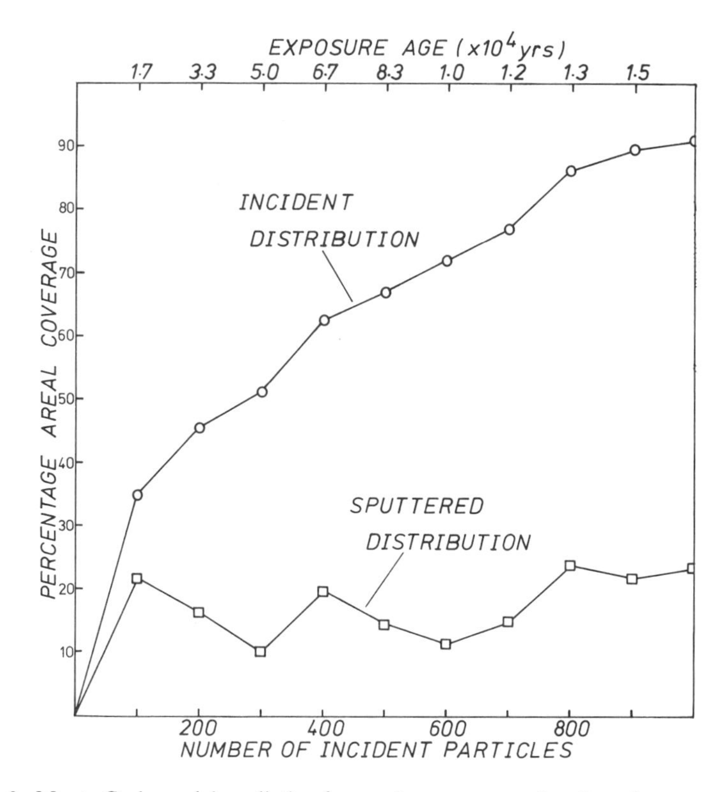

Fig. 3. Monte Carlo model prediction for areal coverage as a function of exposure age. On a sputter shielded sample, ~50% of the surface is rapidly covered in a time equivalent to only ⅓ of the sample age. After an exposure *equal* to the sample age, this figure has risen to ~ 90%. The sputtered distribution of particles though predicts a much slower growth in the areal coverage with the main contribution coming increasingly from large particles as the exposure time lengthens.

also found that 50% of their test surface was cratered in a very short time compared to that taken to crater the complete surface. If it takes one unit in time to crater 50% of the area, then it took 19 units to crater the whole surface.

In our model we only used an influx of 10^3 particles, which resulted in an areal coverage of 91.2%, presumably, if we extended this to say 10^4 particles, we would obtain ~100% areal coverage and arrive at a figure similar to that of Hörz *et al.* (1974) for the time taken to achieve complete surface coverage.

If we now look at the curve in Fig. 3 which represents the percentage areal coverage due to the sputtered particle distribution population, the situation is quite different. The percentage value oscillates due to the impact of randomly sized particles onto the surface together with the gradual degradation of exposed particles due to sputter. The interesting point to note here is that almost as much of the surface is covered after only 10^2 incident particles as there is after 10^3 incident particles, and the percentage areal coverage never rises above 25%; with a larger number of incident particles, this percentage would be expected to increase due to a greater number of large particles impacting the surface. From recent work by Ashworth (1978) using an analytical approach, it is suggested that 100% of the area is covered in 1.75×10^5 years for sample 12054 neglecting the effects of overlap, which is in very close agreement with our results, which indicate ~91% is covered after an exposure of 1.67×10^5 years.

Accreta particle lifetimes

A previous analytical approach to accreta lifetimes by Ashworth (1978) is confirmed by our model here, with the sputter lifetime being shorter than the obscuration lifetime for particle diameters ~0.01 μm and the reverse being the case for particle diameters >0.1 μm.

The sputter lifetimes in both models are identical with a ~0.01 μm diameter particle having a sputter lifetime of 4×10^3 years. The ratio of observation lifetime to sputter lifetime for a particle of diameter 0.01 μm is ~16 in our model and ~5 in that of Ashworth. For a particle of diameter 0.1 μm, the ratio of obscuration lifetime to sputter lifetime in our model is ~0.8 and ~0.6 in that of Ashworth. However, it should be noted that lifetime values for any particular diameter can vary from a factor of two times to nine times the sputter lifetime depending on the size of the covering particle and whether it itself is covered by a subsequent impact. Even in our relatively simplified model then, the actual history of any individual particle can be extremely complex. However, we can predict the behaviour of the population as a whole in some considerable detail using our MC approach to simulation modelling.

EXPERIMENTAL SPUTTER OF 12054 ACCRETA

The target used in our laboratory simulation was lunar sample 15205,9003, whose surface exposure age has been measured at 8×10^4 years by Hartung and Storzer (1974) using solar flare track dating techniques. The sample was mounted on a microscope stub with its glass splash coated surface facing upwards and coated in a sputter deposition unit with ~300 Å of gold.

In order to be able to study different situations of sputter shielding and redeposition, photomosaics of the following three types of area in each of our selected regions were constructed:

(a) crater pits which are heavily shielded from ions incident at near grazing incidence;
(b) spall zones being usually at some shallow angle to the surface are shielded to a much lesser degree from grazing incidence ions;
(c) flat surfaces for which the only shielding is due to atomic roughness and any accreta particles which may be resident on the area.

The first experiment run consisted of a total exposure of 3 hours to normally incident N_2^+ ions at an extraction energy of 4.0 KeV. Assuming an average solar wind velocity of ~400 kms^{-1}, the energy of the constituent nitrogen ions of the solar wind is calculated to be 11.48 KeV. The ion source used at Canterbury limits at 10 KeV, but from experimental measurements by Almen and Bruce (1961a, 1961b) and Rol *et al.* (1960) on the sputter yield of nitrogen ions on copper, McDonnell and Ashworth (1972) have shown the sputter yield to be approximately equal at energies of 4.0 KeV and 11.48 KeV. Thus, by using N_2^+ ions at an energy of 4.0 KeV, we can simulate solar wind sputter in the laboratory. During this experimental run the average beam current density was ~40 μA cm^{-2}, with crater E (cf., Fig. 5) shielded from the ion beam.

Photographs of a spallation zone near the centre of the incident ion beam both before and after sputter are shown in Fig. 4a and 4b, respectively. In the pre-sputter case, Fig. 4a, we can suggest from the fact that the most angular accreta particles in the picture are those resident in the void (bottom centre of the photo) that this population distribution has already experienced a certain amount of sputter degradation while exposed on the lunar surface. However, after the three hour laboratory simulation, features of heavy sputtering can be seen (Fig. 4b). Etching of areas on the substrate directly adjacent to accreta particles has occurred due to incident ions bouncing off the sides of the particles at extreme grazing incidence angles. Also where particles have been completely destroyed (from the figure nearly all particles ≤0.4 μm have disappeared), this etching results in a depression being left in the substrate. These particles could not have been removed from the surface by electrostatic forces as the sample surface was prevented from charging up by a neutralising beam of electrons.

The cumulative accreta population distribution has been modified extensively by the action of laboratory simulated normal incidence sputter. This normal incidence experiment can yield information on general sputter features, but we know from our CANSP sputter model that using a time-averaged flux of ions onto a target surface results in a completely different picture of the surface behaviour on the atomic level.

Our second sputter simulation experiment was conducted on the previously shielded area around crater E (cf., Fig. 5) on sample 15205,9003.

On this occasion, however, the angle of incidence of the ion beam was variable to mimic more closely the actual conditions on the lunar surface where possible incident directions cover a range 0–2 π radians. We chose 15 incident directions equally spaced covering the complete range of 180°, i.e., each direction was ~12.86° apart. In order to avoid any propagation of sputter-induced features, the incident directions were chosen in a random sequence, and the sample was exposed for 8 minutes at each angle giving a total exposure time of 2 hours. The current density averaged out at ~32 μA cm^{-2}. A comparison of a glass splash impact site both before and after sputter is shown in Fig. 6a and 6b. In this case, the sputter features are less pronounced than in the normal incidence case, especially that of etching the substrate around accreta particles, and the smaller particles show a definite reduction in diameter.

By measuring the size distribution of the particles in Fig. 6a, we obtained a cumulative population distribution which was exposed to sputter degradation on the lunar surface prior to retrieval by the Apollo astronauts. This distribution is labelled curve (a) in Fig. 7. We then took the post-laboratory sputter size distribution measurements and obtained another population distribution which represents exposure to sputter erosion on the lunar surface *and* in the laboratory (labelled curve (b) in Fig. 7). The behaviour is identical to that predicted in our MC model.

Differential Sputtering and Diffusion

In a previous paper (Carey and McDonnell, 1976), we described a Monte Carlo sputter simulation model, CANSP, which has now been extended to include, during the sputter process, the possibility of a surface atom gaining enough energy to enable it to diffuse to a new lattice site.

For an average solar wind velocity of 400 kms^{-1}, the sputter yield of constituent ion species of the solar wind have been derived from experimentally measured sputter yields (McDonnell and Ashworth, 1972). The major constituents of the solar wind are H^+ and He^+, both of which have a sputter yield less than unity. For He^+, approximately three incident ions result in one atom being sputtered, and for H^+ about 24 incident ions result in a single target atom being ejected. This implies that even though a collision cascade sequence may not result

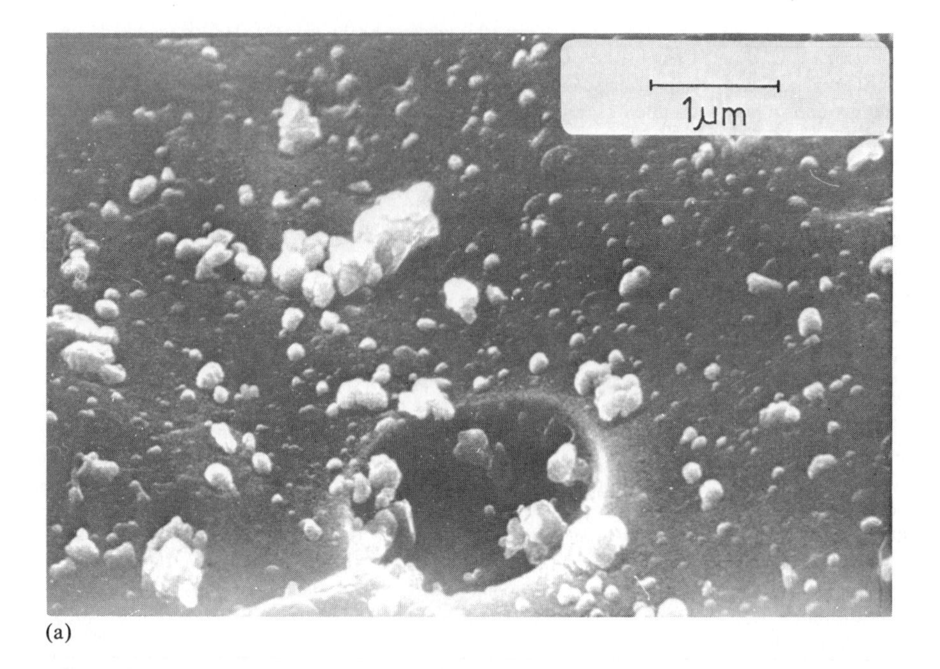

(a)

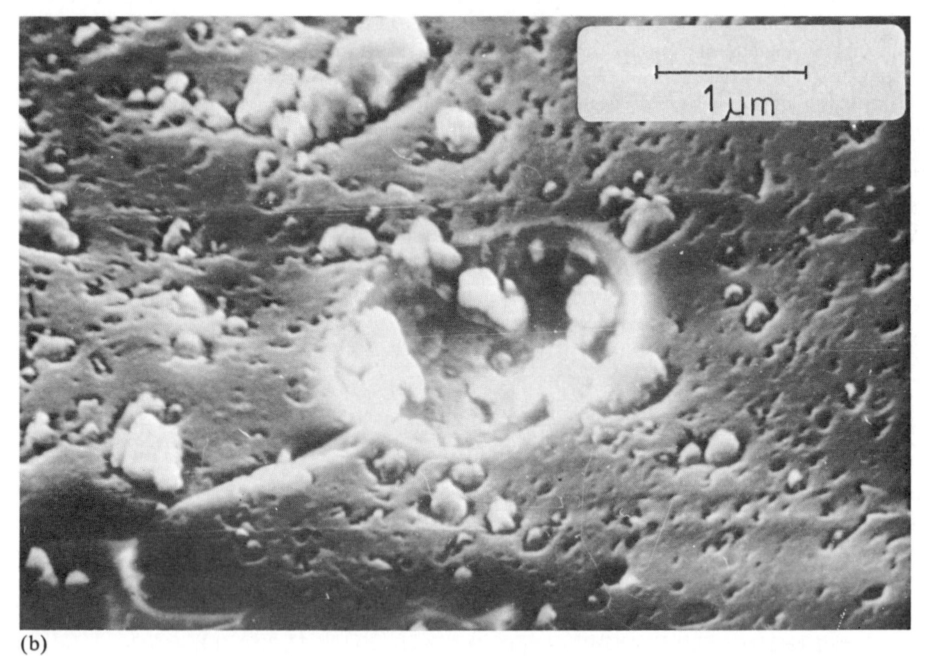

(b)

Fig. 4. Shown above are photographs of a spallation zone accreta particle population near the centre of the incident ion beam. The beam current density used was ~37 μA cm^{-2} for a total exposure time of 3 hours to normally incident N_2^+ ions of energy 4.0 KeV. (a) Region *before* laboratory sputter simulation and (b) shows the same area after sputter simulation.

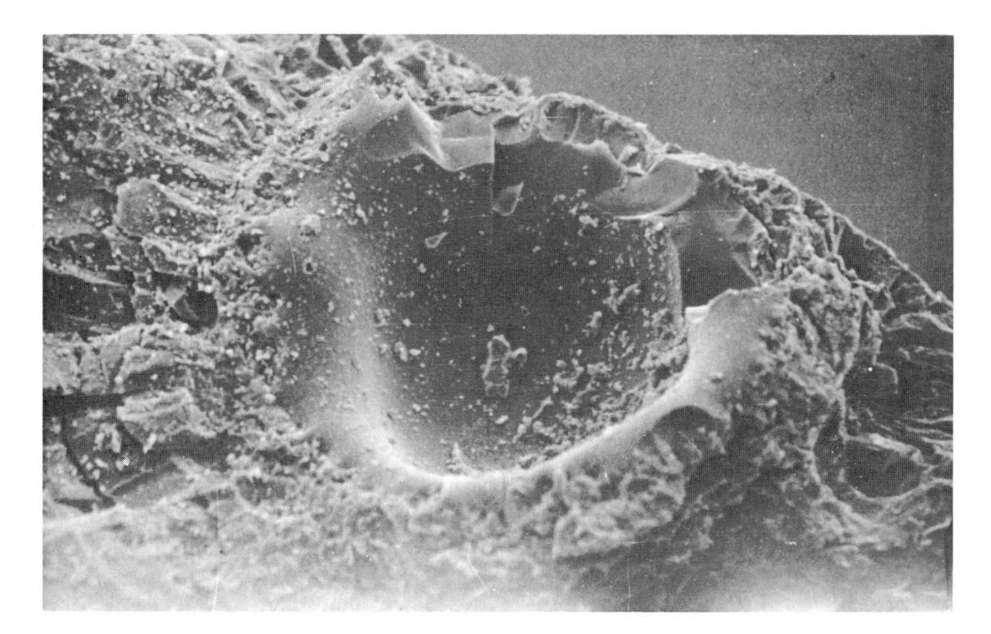

Fig. 5. Crater E on Apollo lunar sample 15205,9003 which is a hypervelocity impact site of pit diameter ~16 μm. The glass splash area shown in Fig. 6 is situated on the spall zone to the left of the lefthand side crater rim. The variable incidence angle of the ion beam was conducted in a plane roughly north to south as one views the crater.

in the ejection of an atom from a surface, sufficient energy may be imparted to the primary-impact atom to enable it to diffuse through the target material.

The proportion of atoms not sputtered upon impact, but which may diffuse to neighbouring sites, is difficult to determine experimentally in an amorphous material, the picture being complicated by the presence of atoms of differing sputter ratio within lunar target material itself (Haff *et al.*, 1977); the magnitude of this impact-induced diffusion could have a significant influence in the dynamical processes occurring in the outer 200–300 A damage layer. We can, however, conclude that atoms of high sputter yield will be preferentially diffused on a local atomic scale, suggesting that individual areas on an exposed lunar sample may differ greatly from the measured bulk diffusion value. Impact induced differentiation of target atoms would *add* to the effect of sputter induced fractionation of elements (Housley and Grant, 1975) and isotopes (Taylor and Epstein, 1973; Clayton *et al.*, 1974) in a target material. In order to ascertain the influence of sputter induced diffusion and surface microscale topography, we arbitrarily selected that 90% of struck target atoms which are *not* sputtered would diffuse to a new lattice site. The results of the CANSP model for an initially flat target surface are compared in Table 1.

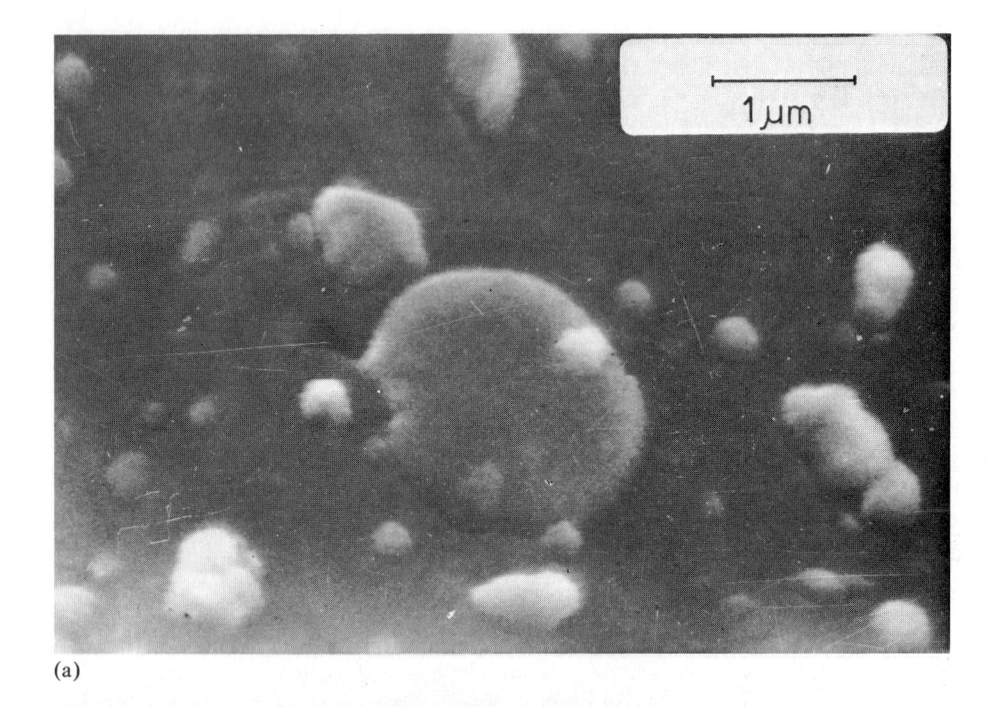

(a)

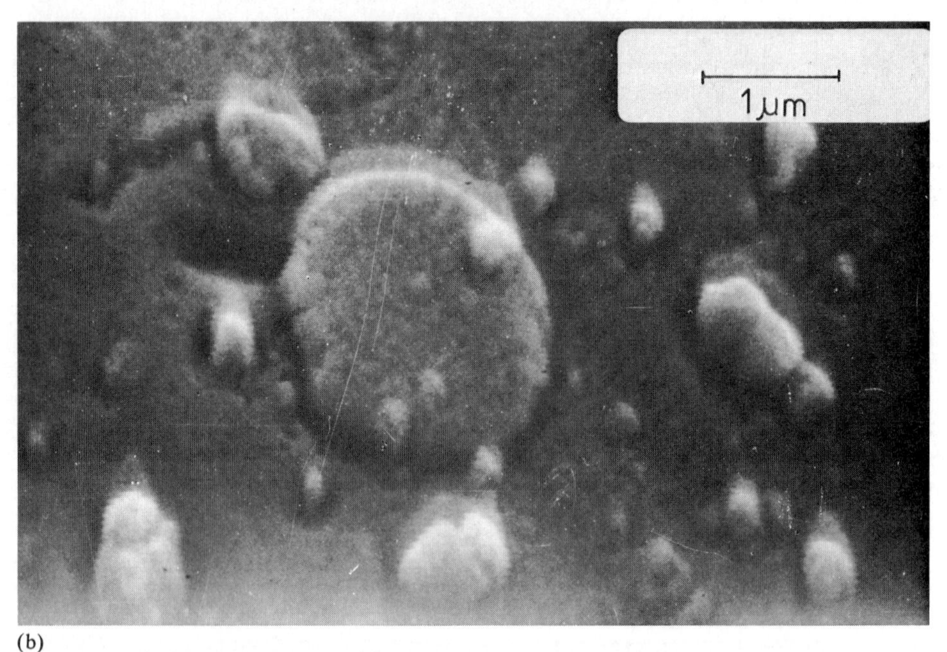

(b)

Fig. 6. A comparison is made here between the accreta population on a selected area of the spall zone of crater E (cf., Fig. 5). (a) Shows the distribution prior to sputtering by a time-averaged incident ion beam. (b) Shows the same area after a 2 hour sputter at a beam current density of $\sim 32\ \mu A\ cm^{-2}$. Area of photograph = $20\ \mu m^2$.

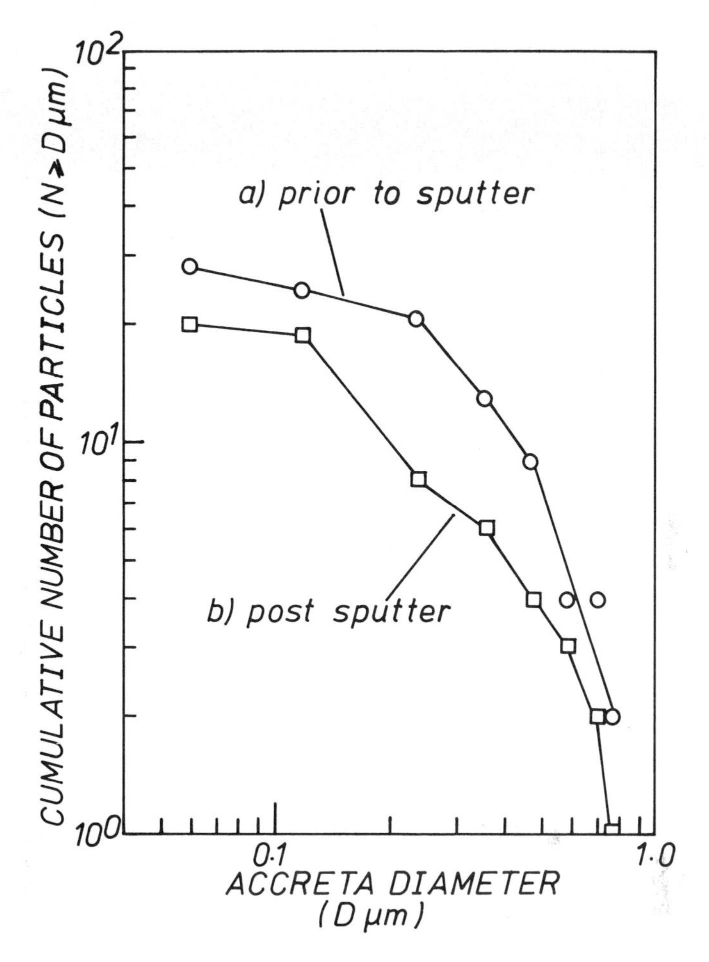

Fig. 7. (a) This measured accreta particle population distribution represents that which has been sputter degraded on the lunar surface only prior to collection (cf., Fig. 6a). (b) Represents the measured distribution after laboratory sputtering by 4.0 KeV N_2^+ ions.

The term sputter ratio here refers to the bulk sputter yield for the complete target surface. From this table it can be seen that the bulk sputter ratio for the surface in the case of a normally incident (NI) ion beam is 0.19. In the time average (TA) case, this value has increased by about 20% to 0.23, because of less ejected material being recaptured due to a lesser degree of surface roughness developing. On including the possibility of diffusion into the model, the bulk sputter ratio is increased to 0.32 in the normal incidence case, i.e., an increase of ~70%. This is not easily interpreted as the resultant topology has a greater surface roughness than the no diffusion situation and should therefore recapture a significant proportion of ejected atoms, resulting in a *decrease* of the bulk

Table 1. Normal incidence (NI) and time-average (TA) sputter ratios for various MC models are listed. The term sputter ratio here refers to the bulk sputter yield for the complete target surface.

	NI	TA
–Diffusion –Differential Sputter	0.19	0.23
+ Diffusion	0.32	0.36
+ Differential Sputter + Diffusion	0.32	0.37
+ Differential Sputter	0.23	0.26

sputter ratio. We suggest that the surface develops a larger number of 'spicules' when diffusion is allowed, resulting in a greater number of atoms being sputtered off the tops of these spicules reducing the recapture probability. During bombardment, however, these spicules quickly develop again due to the sputter induced diffusion, so a smoothing of the surface is not observed.

This increase in the bulk sputter yield due to diffusion also is predicted for the TA case, where an increase in the bulk sputter ratio is found to be ~60%, to give a value 0.36. However, the surface roughness developed during ion bombardment is almost negligible, so the increase here in bulk sputter ratio can be explained wholly by a significant reduction in the recapture probability of ejected atoms due to a decrease in the surface roughness of the target.

In a recent publication, Jull and Pillinger (1978) suggested a process of preferential sputtering as a method of producing metallic iron [see also, Gold *et al.* (1975) and Yin *et al.* (1975)]. Housley and Grant (1975) proposed the sputter mechanism to be responsible for major element fractionation as well as a number of trace element enrichments. To investigate the effects of an ion-bombarded surface experiencing differential sputter, the test surface of our CANSP model consisting of 101 atoms was modified to include nine atoms, regularly spaced, which possessed a sputter yield six times the bulk sputter yield. This model simulates 'light' atom impurities in an otherwise homogeneous surface. Table 1 shows that for both the NI and TA situations, a slight increase in the bulk sputter yield follows from an inclusion of differential sputter alone, i.e., no diffusion is occurring. Also, the resultant surface microtopography does not develop any significant spicules where one would expect etching about the lattice sites of the light impurity atoms. When we include the possibility of both diffusion together with differential sputtering, the effect on both the bulk sputter yield and final physical profile of the surface is identical to the diffusion case alone.

The behaviour of the surface is summarized in Fig. 8 for the situation where light atom inclusions are present in the surface. Graph (a) indicates no significant deviation from the mean surface profile height for the light atom positions (circled data point) implying that diffusion of atoms on the surface is so great, any effects of etching about light atom inclusions is masked. In graph (b), where no diffusion is occurring on the target surface, etching about the light atom inclusions is definitely indicated.

By decreasing the sputter yield of our inclusion atoms to $^1/_6$ the normal surface yield, we can investigate the behaviour of heavy atom inclusions. We would expect this to produce needle-like features with the heavy atoms sitting on top, i.e., a preferential etching of the normal surface atoms. As Fig. 8 clearly shows, this is exactly what occurs, although it is more readily indicated in the situation where diffusion is absent [(b) in Fig. 8] than in the situation where diffusion is present [(a) in Fig. 8]. Thus, we must conclude that on the *atomic level*, at least, impact induced diffusion can mask the effects of heterogeneity, i.e., differential sputter.

Angular Distribution of Sputtered Atoms

An important parameter in the study of lunar surface dynamic processes is that of the functional form of the angular distribution of atoms sputtered from exposed lunar material. Almost two decades ago, Wehner and Rosenberg (1960) measured the angular distribution of sputtered atoms from polycrystalline targets by collecting deposits on glass strips and determining the density distributions photometrically. Their results indicated that ejection was "under cosine", i.e., more material ejected in non-normal directions than in the pure cosine case. These results though are for Hg^+ ions incident at energies in the range 0.1–1 KeV on various metals and, hence, may not be analogous to light ion sputtering (namely H^+ and He^+), the major mechanism of sputtering on the lunar surface. Numerous experiments to investigate this phenomena have been undertaken (cf., Carter and Colligon, 1968) with all the results being either slightly "under" or "over cosinal." In theories based on the random cascade model (Thompson, 1968; Sigmund, 1969), the angular distribution of sputtered atoms is predicted to obey a cosine law for amorphous materials with respect to the local surface normal regardless of the incident direction of the ion beam.

In this paper we confirm these earlier results using our CANSP sputter model which produced a cosinal ejection characteristic for atoms sputtered from polycrystalline or amorphous materials by the impact of light ions (H^+ or He^+) at solar wind energies, i.e., $\cos^n\theta$, where θ is the angle of ejection relative to the surface normal and n represents the cosinal index. The intrinsic angular distribution used in the model is a uniformly random one, and we assume a sticking coefficient of unity for ejected atoms intersecting the surface. Thus, the cosinal ejection characteristic, which results in our simulation, arises *solely* from surface roughness. A recent theory on the sputter mechanism has been suggested (Pollitt *et al.*, 1978) which places even greater emphasis on surface microtopology

"LIGHT" ATOM INCLUSIONS

SPUTTER YIELD OF LIGHT ATOMS 6x NORMAL SURFACE ATOMS

a) time-average with diffusion

b) time-average without diffusion

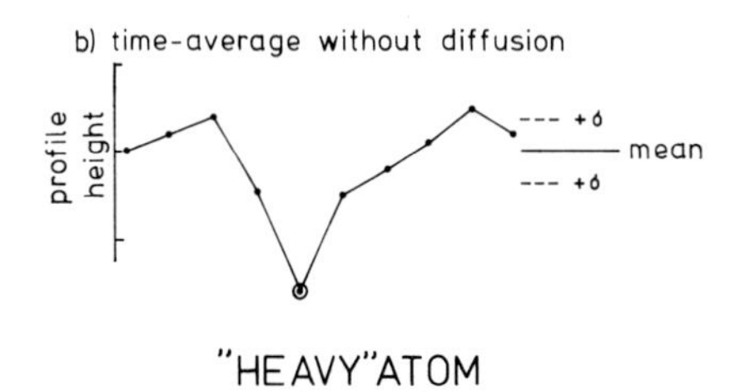

"HEAVY" ATOM INCLUSIONS

SPUTTER YIELD OF HEAVY ATOMS 1/6x NORMAL SURFACE ATOMS

a) time-average with diffusion

b) time-average without diffusion

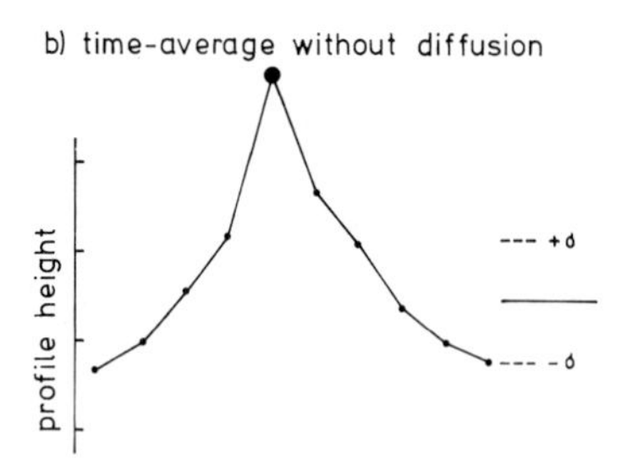

Fig. 8. Here we give the results of our MC model in the influence of 'light' and 'heavy' atom inclusions upon the development of surface roughness. (a) Shows that for a time-average flux (i.e., that pertaining on the lunar surface) the effect of sputter-induced diffusion is more efficient at increasing surface roughness than etching due to differential sputtering of target atoms. (b) Gives a similar result for the case of heavy atom inclusions.

Table 2. Variation of cosinal index n as a function of angle of incidence to a flat target surface. A decrease in the index towards grazing incidence angles is predicted by the MC model.

Angle of Incidence to flat surface, I°	Cosinal Index n of sputter function $S(\theta) = \cos^n\theta$, $\theta = \lvert I-90 \rvert$
30°	0.54
60°	0.94
90° (N.I.)	1.32
120°	0.81
150°	0.42

than the previously most accepted theory of Sigmund (1969). This new theory suggests that the bombarding species electronically excite the surface atoms of the target, resulting in a reduction of the binding energy of the atoms to the surface. When the binding energy becomes zero the atoms will be ejected from the surface. The majority of the atoms or molecules leaving the surface are predicted to have neutral charge. The results of our MC model are consistent with this new theory which predicts that the surface microstructure will play a major role in the angular distribution of sputtered atoms.

The value of the cosinal index for various ion-beam incident directions is shown in Table 2.

These results imply that in laboratory experiments of sputter simulation great care must be taken in arriving at sputter rate measurements as the cosinal index is indeed affected by the direction of the incident ion beam. This change in the index with incident ion direction can be explained by the influence of surface roughness on the ejected target atoms. As an initially flat target surface develops, microscale roughening under uni-directional sputter erosion, the percentage of the projected surface in effective shadow increases when θ approaches grazing incidence. This results in the majority of sputtered atoms being ejected from the "high spots" of the target surface, which implies a greater number of atoms will be ejected at near grazing incidence than in the case of normal incidence sputter. Hence, we observed a "flattening" of the ejection characteristic towards high values of θ. Our results also show that for laboratory sputter experiments, where the angle of incidence of the ion beam is normal to the surface, the cosinal index $n \sim 4/3$, i.e., the angular distribution is "over-cosine."

Summary

A Monte Carlo computer model has been produced which is able to reproduce the behaviour of measured accreta particle population distributions under sputter erosion by solar wind ions on exposed lunar surfaces. With increasing exposure age, the cumulative plot of the sputtered distribution reproduces the theoretical prediction for the value of the slope. By using measured accreta influx distributions together with a well-defined sputter rate, the exposure age of the rock surface on which the accreta resides may be determined.

Our model also indicates that the exposure history, even for particles well-attached to rock surfaces, can be extremely complex due to overlapping. The first experimental results of time-averaged sputter upon accreta particle populations reproduce exactly the behaviour predicted in our MC model.

From our CANSP sputter model, we conclude that on the *atomic level* at least, impact induced diffusion can mask effects of heterogeneity, i.e., mass fractionation by differential sputtering can be affected by impact induced diffusion. The CANSP model also showed that by using an intrinsic *uniform* angular distribution for sputtered atoms, a *cosinal* ejection characteristic results purely by the action of surface roughness.

The results from our model also confirm the validity of a 'low' sputter rate for material exposed on the lunar surface, as opposed to the 'high' values suggested by some other groups. Before the advent of the space age, Wehner *et al.* (1965) predicted a figure of ~1 Å yr^{-1} for the solar wind sputter rate. Early work by McDonnell and Ashworth (1971) suggested a much lower rate of ~0.021 Å yr^{-1} and later McDonnell and Flavill (1974), from He^+ ion bombardment of lunar rock, measured a rate of 0.043 Å yr^{-1} from direct measurements of the sputter rates of micron-sized grains. Zinner and Walker (1975) estimated a figure of ~0.025 Å yr^{-1} from ion microprobe data and, more recently, McDonnell (1977), by observing accretionary particle populations on sample 12054,58, obtained the first direct evaluation of the *in situ* solar wind sputter rate of lunar grains as being ~0.031 Å yr^{-1}. This discrepancy between various groups, however, now seems to have been finally resolved. Maurette (pers. comm.) has shown that in laboratory sputter of lunar grains the *initial* sputter rate is very high (~0.25 Å yr^{-1}), but after an equivalent exposure time of ~20,000 years, the sputtering rate gets *much smaller* and approaches a figure of ~0.025 Å yr^{-1}. In all models of lunar surface dynamic processes which include the solar wind sputter rate as a parameter, a figure of 0.02–0.03 Å yr^{-1} is suggested for the sputter rate from the most reliable laboratory and *in situ* measurements to date.

Acknowledgments—We thank Professor R. C. Jennison for the support of the University of Kent and the Science Research Council for financial support. We appreciate the invaluable assistance of the Computing Laboratory staff (especially Mrs. Z. Kemp) and Mrs. J. Elmes and Miss D. Paine for manuscript preparation.

REFERENCES

Ashworth D. G. (1978) The temporal development of microcrater accreta populations on lunar rocks subjected to meteoroid and solar wind bombardment—I. To be submitted to *Planet. Space Sci.*

Almen O. and Bruce G. (1961a) Collection and sputtering experiments with noble gas ions. *Nucl. Instrum. Methods* **11**, 257–278.

Almen O. and Bruce G. (1961b) Sputtering experiments in the high energy region. *Nucl. Instrum. Methods* **11**, 279–289.

Bibring J. P., Borg J., Burlingame A. L., Langevin Y., Maurette M. and Vassent B. (1975) Solar wind and solar flare maturation of the lunar regolith. *Proc. Lunar Sci. Conf. 6th*, p. 3471–3493.

Carey W. C. and McDonnell J. A. M. (1976) Lunar surface sputter erosion: A Monte Carlo approach to microcrater erosion and sputter redeposition. *Proc. Lunar Sci. Conf. 7th*, p. 913–926.

Carter G. and Colligon J. S. (1968) *Ion Bombardment of Solids*. Heinemann Educational Books, London. 446 pp.

Clayton R. N., Mayeda K. T. and Hurd J. M. (1974) Loss of oxygen, silicon, sulphur and potassium from the lunar regolith. *Proc. Lunar Sci. Conf. 5th*, p. 1801–1809.

Gold T., Bilson E. and Baron R. L. (1975) Auger analysis of the lunar soil: Study of processes which change the surface chemistry and albedo. *Proc. Lunar Sci. Conf. 6th*, p. 3285–3303.

Haff P. K., Switkowski Z. E., Burnett D. S. and Tombrello T. A. (1977) Gravitational and recoil contributions to surface mass fractionation by solar wind sputtering. *Proc. Lunar Sci. Conf. 8th*, p. 3807–3815.

Hartung J. B., Hörz F., Aitken F. K., Gault D. E. and Brownlee D. E. (1973) The development of microcrater populations on lunar rocks. *Proc. Lunar Sci. Conf. 4th*, p. 3213–3234.

Hartung J. B. and Storzer D. (1974) Lunar microcraters and their solar flare track record. *Proc. Lunar Sci. Conf. 5th*, p. 2527–2541.

Hörz F., Schneider E. and Hill R. E. (1974) Micrometeoroid abrasion of lunar rocks: A Monte Carlo simulation. *Proc. Lunar Sci. Conf. 5th*, p. 2397–2412.

Housley R. M. and Grant R. W. (1975) ESCA studies of lunar surface chemistry. *Proc. Lunar Sci. Conf. 6th*, p. 3269–3275.

Jull A. J. T. and Pillinger C. T. (1978) The carbon content of finely-divided lunar iron (abstract). In *Lunar and Planetary Science IX*, p. 609–611. Lunar and Planetary Institute, Houston.

McDonnell J. A. M. (1977) Accretionary particle studies on Apollo 12054,58: *In situ* lunar surface microparticle flux rate and solar wind sputter rate defined. *Proc. Lunar Sci. Conf. 8th*, p. 3835–3857.

McDonnell J. A. M. and Ashworth D. G. (1972) Erosion phenomena on the lunar surface and meteorites. In *Space Research XII*, p. 333–347. Akademie Verlag, Berlin.

Morrison D. A., McKay D. S., Fruland R. M. and Moore H. J. (1973) Microcraters on Apollo 15 and 16 rocks. *Proc. Lunar Sci. Conf. 4th*, p. 3235–3253.

Morrison D. A. and Zinner E. (1975) Studies of solar flares and impact craters in partially protected crystals. *Proc. Lunar Sci. Conf. 6th*, p. 3373–3390.

Morrison D. A. and Zinner E. (1977) 12054 and 76215: New measurements of interplanetary dust and solar flare fluxes. *Proc. Lunar Sci. Conf. 8th*, p. 841–863.

Oberbeck V. R., Quaide W. L., Mahan M. and Paulson J. (1973) Monte Carlo calculations of lunar regolith thickness distributions. *Icarus* **19**, 87–107.

Pollitt K. R., Robb J. C. and Thomas D. W. (1978) Mechanism of sputtering of solid surfaces by ion-impact. *Nature* **272**, 436–437.

Rol P. K., Fluit J. M. and Kistemaker J. (1960) Sputtering of copper by bombardment with ions. *Physica* **26**, 1000–1008.

Sigmund P. (1969) Theory of sputtering I. Sputtering yield of amorphous and polycrystalline targets. *Phys. Rev.* **184**, 383–416.

Taylor H. P. and Epstein S. (1973) O^{18}/O^{16} and Si^{30}/Si^{28} studies of some Apollo 15, 16 and 17 samples. *Proc. Lunar Sci. Conf. 4th*, p. 1657–1679.

Thompson M. W. (1968) II. The energy spectrum of ejected atoms during the high energy sputtering of gold. *Philos. Mag.* **18**, 377–414.

Wehner G. K., Kenknight C. and Rosenberg D. L. (1965) Sputtering rates under solar wind bombardment. *Planet. Space Sci.* **11**, 885–895.

Wehner G. K. and Rosenberg D. (1960) Angular distribution of sputtered material. *J. Appl. Phys.* **31**, 177–179.

Yin L., Tsang T. and Adler I. (1975) ESCA studies on solar wind reduction mechanisms. *Proc. Lunar Sci. Conf. 6th*, p. 3277–3284.

Zinner E. and Walker R. M. (1975) Ion-probe studies of artificially implanted ions in lunar samples. *Proc. Lunar Sci. Conf. 6th*, p. 3601–3617.

Zook H., Hartung J. B. and Hauser E. (1978) Loosely bound dust, impact pits and accreta on lunar rock 12054,54 (abstract). In *Lunar and Planetary Science IX*, p. 1300–1302. Lunar and Planetary Institute, Houston.

Proc. Lunar Planet. Sci. Conf. 9th (1978), p. 1745–1747.
Printed in the United States of America

Transport of terrestrial atmospheric gases to the moon

H. K. HILLS and J. W. FREEMAN

Rice University, Houston, Texas 77001

Abstract—The SIDE instruments on the lunar surface have identified fluxes of ions which may be different from the solar wind both in elemental and in isotopic abundances. At present, only O^+ is (tentatively) identified, but the mechanism operates to produce ions at the moon which have their origin in the earth's atmosphere. Consequently, the "solar wind" component of the surface-correlated gases is effectively a combination of terrestrial atmospheric ions and the actual solar wind. The results differ from solar wind results if the terrestrial ion abundances strongly differ from those of the solar wind.

The solar wind interaction with a planetary body depends on the strength and orientation of the planetary magnetic field. In the case of the earth, the magnetic field is sufficient to stand off the solar wind, and a complex magnetosphere structure is established, extending downstream to distances far beyond the moon. The whole moon is shielded from solar wind impact from any direction while the moon is inside the magnetosphere. Since the moon keeps one face toward the earth, the solar wind can never hit the surface vertically near the center of this side of the moon. Thus, solar wind-protected lunar material is available in suitably shadowed spots (see, for example, Hartung *et al.*, 1977). However, the Suprathermal Ion Detector Experiments (SIDEs) have shown that even while the moon is thus shielded from the solar wind, there is a significant flux of protons and heavier ions, apparently of terrestrial origin, striking the moon (Hardy *et al.*, 1977). As a result, the lunar near-side is bombarded not only by solar wind, but also part of the time by a "planetary wind" which may be of much different ion composition, both in elemental abundances and in isotopic abundances (Hills and Freeman, 1978). The present lunar far-side material is being bombarded by solar wind, but not by the planetary wind, although there may be an ion circulation pattern with a return flow (earthward in the tail) of the planetary wind which does strike the lunar far-side (Freeman *et al.*, 1977).

Hills and Freeman (1978) point out that the ratio of heavy ions to protons in the planetary wind is about 10 to 50 times that reported by Bame (1972) for the solar wind oxygen, although the proton flux is about 10^{-2} that of the solar wind. Consider now the average flux of oxygen ions impacting the lunar surface, due to the combined effects of the solar wind and planetary wind. The moon is in the solar wind outside the earth's magnetospheric tail approximately 70% of the time. The earth-facing side of the moon is shielded from the solar wind 50% of the time and exposed to it approximately 20% of the time. It is in the magnetosheath about 20% of the time and in the inner part of the tail about

10% of the time. A solar wind flux of 10^7 protons/cm^2 sec and 4×10^3 oxygen ions/cm^2 sec, together with a planetary wind flux of 10^5 protons/cm^2 sec and 10^3 oxygen ions/cm^2 sec, will result in an average flux of $.2 \times 10^7 + .1 \times 10^5 \approx 2 \times 10^6$ protons/cm^2 sec, and an oxygen flux of $.2 \times 4 \times 10^3 + .1 \times 10^3 = .8 \times 10^3 + .1 \times 10^3$ ions/cm^2 sec. Thus, there is a 1 part in 8 enrichment in the average oxygen abundance, compared to solar wind alone. This calculation ignores the fluxes encountered the remaining 20% of the time, while the moon is in the magnetosheath. Fluxes encountered there probably have a mixture of solar wind and terrestrial sources, but the composition is not established and we will not consider them further here, except to note that they represent yet another effect which bears on the surface-correlated gases in the lunar material.

The sample calculation above showed a relatively small effect due to the observed oxygen ions in the magnetospheric tail. However, if the terrestrial elemental or isotopic abundances display large differences from those of the solar wind, then even the small fraction of the time they are observed is sufficient to produce an average ion flux at the moon which is much different from solar wind composition. For instance, if the solar wind ratio of Ar^{40}/Ar^{36} is the same as in meteorites, $\sim 10^{-3}$, then the planetary wind ions, with terrestrial ratio of ~ 300, produce a relatively large effect, with the combination yielding an average ratio of approximately 6×10^{-3}, again ignoring the magnetosheath, and assuming the same ratio of argon to protons in both fluxes. This is an arbitrary assumption, since no non-lunar mass peaks higher than oxygen have been seen in the SIDE data. It should be noted that most such peaks would be at energies above the upper limit of the SIDE.

This planetary effect is not expected to account for the lunar Ar^{40}/Ar^{36} ratios (Eberhardt *et al.*, 1976; Heymann *et al.*, 1975), but we re-iterate that this effect can make the effective "solar wind" fluxes different in composition from the actual solar wind. Hills and Freeman (1978) have speculated that this effect may be operative, for example, in the results of Epstein and Taylor (1972), Kerridge *et al.* (1977), and Becker and Clayton (1975). The far side of the moon should be relatively little affected by this planetary wind, since the far side is never shielded from the solar wind by the magnetosphere and is shielded from the direct planetary wind by the solid moon.

The transport of atmospheric gas to the moon as discussed herein is not limited to the earth. It should occur, in any case, where a planet's satellite passes through its magnetospheric tail.

Acknowledgments—This work was supported by the Atmospheric Research Section of the National Science Foundation and by NASA grant NSG-7157. We thank R. R. Hodges for pointing out an error in the average flux computation.

REFERENCES

Bame S. J. (1972) Spacecraft observations of the solar wind composition. In *Solar Wind* (C. P. Sonett, P. J. Coleman, Jr., and J. M. Wilcox, eds.), p. 535. NASA SP-308. Washington, D.C.

Becker R. H. and Clayton R. N. (1975) Nitrogen abundances and isotopic compositions in lunar samples. *Proc. Lunar Sci. Conf. 6th*, p. 2131–2149.

Eberhardt P., Eugster O., Geiss J., Grögler N., Guggisberg S., and Mörgeli M. (1976) Noble gases in the Apollo 16 special soils from the East-West split and the permanently shadowed area. *Proc. Lunar Sci. Conf. 7th*, p. 563–585.

Epstein S. and Taylor H. P. (1972) O^{18}/O^{16}, Si^{30}/Si^{28}, C^{13}/C^{12}, and D/H studies of Apollo 14 and 15 samples. *Proc. Lunar Sci. Conf. 3rd*, p. 1429–1454.

Freeman J. W., Hills H. K., Hill T. W., Reiff P. H., and Hardy D. A. (1977) Heavy ion circulation in the earth's magnetosphere. *Geophys. Res. Lett.* **4**, 195-197.

Hardy D. A., Freeman J. W., and Hills H. K. (1977) Double-peaked ion spectra in the lobe plasma: Evidence for massive ions? *J. Geophys. Res.* **82**, 5529–5540.

Hartung J. B., Plieninger T., Müller H. W., and Schaeffer O. A. (1977) Helium, neon, and argon on sunlit and shaded surfaces of lunar rock 12054 by laser probe mass spectrometry (abstract). In *Lunar Science VIII*, p. 409–411. The Lunar Science Institute, Houston.

Heymann D., Walton J. R., Jordan J. L., Lakatos S., and Yaniv A. (1975) Light and dark soils at the Apollo 16 landing site. *The Moon* **13**, 81-110.

Hills H. K. and Freeman J. W. (1978) Transport of terrestrial atmospheric gases to the moon (abstract). In *Lunar and Planetary Science IX*, p. 518–520. Lunar and Planetary Institute, Houston.

Kerridge J. F., Kaplan I. R., Lingenfelter R. E., and Boynton W. V. (1977) Solar wind nitrogen: Mechanisms for isotopic evolution (abstract). In *Lunar Science VIII*, p. 537–539. The Lunar Science Institute, Houston.

Proc. Lunar Planet. Sci. Conf. 9th (1978), p. 1749–1764.
Printed in the United States of America

Gravitational and radiative effects on the escape of helium from the moon

DR. R. RICHARD HODGES, JR.

University of Texas at Dallas, Richardson, Texas 75080

Abstract—On the moon, and probably on Mercury and other similar regolith-covered bodies with tenuous atmosphere, the dominant gas is ^{4}He. It arises as the radiogenic product of the decay of uranium and thorium within any planet, but its major source appears to be the α particle flux of the solar wind. The moon intercepts solar wind helium at an average rate of 1.1×10^{24} atom/sec, and loses it at the same rate. Some helium may escape directly as the result of the process of solar wind soil bombardment which may release previously trapped helium at superthermal speeds. Atmospheric models have been calculated with the total helium influx as source. Subsequent comparison of model and measured helium concentrations indicates that the fraction of helium escaping via the atmosphere may range from 20% to 100% of the solar wind influx. Of the escaping atmosphere, most of the helium (about 93%) becomes trapped in earth orbit, while about 5% gets trapped in satellite orbits about the moon. Owing to a 6 month lifetime for helium in solar radiation, the satellite atoms form a lunar corona that exceeds the lunar atmosphere in total abundance by a factor of 4 to 5.

INTRODUCTION

The theory of escape of light gases, notably helium and hydrogen, from the moon has undergone notable changes in recent years. Initially it was thought that Jeans' escape mechanism of thermal evaporation should dominate, but analytical application of this idea proved to be inaccurate for several reasons. One difficulty was the problem of determining gas concentrations in lunar daytime, where the surface is hottest and escape most likely. Elementary exospheric transport theory suggests that concentration varies as the $-5/2$ power of temperature over the surface (Hodges and Johnson, 1968), but a large lateral flow to supply daytime escape should result in an even larger night to day concentration ratio than the $T^{-5/2}$ result.

Uncertainties in analytical techniques led to the adoption of a Monte Carlo atmospheric simulator by Hodges (1973) and an integral method by Hartle and Thomas (1974) to determine the behavior of helium and hydrogen on the moon. These atmosphere models were able to establish that the Apollo 17 mass spectrometer measurements of neutral helium were approximately in accord with thermal escape of the entire solar wind influx of α particles to the lunar surface. An extremely low level of atmospheric atomic hydrogen on the moon established by the orbital UV spectrometer on Apollo 17 (Fastie *et al.*, 1973) was interpreted by Hodges and by Hartle and Thomas to imply the escape of the bulk of the solar wind proton influx as molecular hydrogen. Upper bounds for H_2 at the Apollo 17 site are commensurate with this explanation (Hodges *et al.*, 1974) but do not rule out other nonthermal escape processes.

Further study of the helium escape problem showed photoionization to be a small but nonnegligible loss process (Hodges *et al.*, 1974). A less obvious, but far more important loss mechanism arises from the perturbations of atmospheric atom trajectories due to solar and terrestrial gravitational potentials and to the noninertial motion of the moon itself through space. Kopal (1966) notes that a particle with speed of 97.8% of the escape speed can pass through the inner Lagrangian L1 point and then be captured by earth. This speed was adopted by Hodges (1975) as an approximate effective escape velocity. Owing to the large sensitivity of the Maxwellian velocity distribution to small changes in helium atomic speed near the lunar escape velocity, this slight decrease in the effective escape speed reduced the model helium lifetime by about a factor of 2.

The large change in the helium atmosphere model brought about by the adoption of an effective escape speed raised the obvious question of the propriety of assuming a blanket reduction of the escape speed to account for gravitational perturbations. In reality, the energy needed to reach the Lagrangian L1 point is not clearly related to the escape problem. What is important is the fact that a slight perturbation of a particle's velocity may be sufficient to raise the perilune of its trajectory above the lunar surface, trapping the atom in a satellite orbit about the moon, where it eventually becomes an ion due to charge exchange with a solar wind ion or to photoionization.

A preliminary Monte Carlo simulation of the behavior of lunar atmospheric helium in the moon-earth-sun gravitational system was reported by Hodges (1977a). Trajectories of sub-escape particles were computed by numerical solution of the equation of motion in the moving, moon-centered coordinate system. A salient feature of this model was the formation of a corona in which the abundance of trapped satellite helium atoms exceeds the total content of the entire bound lunar atmosphere.

This paper is an extension of previous work on the helium escape problem. Mainly, it is a necessary refinement of the calculations reported in Hodges (1977a). A significant change in the orbit calculations has been made to include the particle acceleration due to solar radiation pressure, particularly that due to resonant scatter at 584A. In addition, the present calculations provide a comparison of two velocity distribution functions for particles leaving the lunar surface; a modified form of the Maxwellian distribution used in Hodges (1973) and subsequent models; and the Maxwellian flux distribution suggested by Smith *et al.* (1978), which is roughly analogous to the distribution incorporated in Hodges (1977a). For either distribution, the directional probability is proportional to the cosine of the angle from local vertical to the velocity vector, giving an approximate representation of the obvious fact that a horizontally travelling particle will very likely strike a crater wall or other orographic feature long before it traverses a significant lateral distance, regardless of its speed.

Method of atmosphere simulation

Owing to the lack of interaction among neutral atoms on the moon, it is convenient to construct atmospheric models by averaging the behaviors of a large number of atoms. A Monte Carlo technique reported by Hodges (1973) does this by tracing the random paths of individual particles over the planet, from creation to escape or annihilation, recording locations of impacts with the lunar surface. The calculation method used here is similar to the above, but it incorporates numerous improvements which add realism to the simulator.

An important feature of lunar helium is the time dependence of its source. Rapid variations in the atmospheric abundance of helium, detected by the mass spectrometer at the Apollo 17 landing site, and an apparent correlation of these variations with solar activity, suggests that the neutral helium source depends on the solar wind momentum flux (Hodges and Hoffman, 1974). The dominant source of helium must be solar wind α particles, which impact the moon at energies the order of 4000 eV and, hence, are implanted in soil grains. A small radiogenic source also may be present (Hodges *et al.*, 1973), but its average amplitude is unlikely to exceed 10% of the solar wind supply (Hodges, 1975). More important, the radiogenic helium is probably vented impulsively, along with ^{40}Ar. The short lifetime of helium (a few days) and the long time between ventings of radiogenic gases (2-4 months; cf., Hodges, 1977b) makes this source of little interest in the description of the persistent atmosphere of the moon.

A recurring conjecture in the literature is the idea that the absence of solar wind in the geomagnetic tail should produce a time modulation of the source of neutral helium and other gases of solar wind origin (cf., Hodges *et al*, 1973). The only experimental support for this hypothesis is the aforementioned conclusion of Hodges and Hoffman (1974) that release of implanted helium is correlated with the solar wind influx. In the calculations reported here it has been assumed that the source of neutral helium is constant when the moon is not in the geomagnetic tail and zero when it is in the tail. The width of the geomagnetic tail is approximated as 50° of lunar phase centered on full moon. It is assumed that the average solar wind flux of α particles is 4.5% of the proton flux, or about 1.3×10^7 cm^{-2} sec^{-1} (Johnson *et al.*, 1972). As a result, the average source of new helium atoms is

$$\pi R^2 \times \frac{360\text{-}50}{360} \times 1.3 \times 10^7 = 1.1 \times 10^{24}, \qquad (1)$$

where R is the radius of the moon. This source is augmented by those atmospheric helium ions which impact the moon, become reimplanted in soil grains, and subsequently are recycled to the atmosphere, a process described by Manka and Michel (1970). By analogy with ^{40}Ar results (Hodges, 1977b), the recycled component is assumed to be 10% of the bound helium ionization rate.

Points of origin of new helium atoms are confined to the daytime hemisphere where the solar wind impacts. In the calculations, these points are chosen from a random deviate of the cosine of the moon-centered angle to the subsolar point. In addition, the phase of the moon, ϕ , at the creation of a new atom is chosen randomly in the range $|\phi| < 155°$, i.e., outside the geomagnetic tail.

The average source of helium must balance the average rate of loss of atmosphere. In a Monte Carlo atmosphere simulation it is useful to express this continuity relation as

$$\psi = 4\pi R^2 K\, N_L, \tag{2}$$

where ψ is the total source (i.e., the sum of expression 1 and the recycled component), N_L is the number of simulated atoms lost from the atmosphere, and K is a constant. Equation 2 is the Monte Carlo analogue of the integral

$$\psi = R^2 \int_0^{2\pi} d\lambda \int_0^{\pi/2} d\theta \sin\theta \int d^3\mathbf{v}\, f(\mathbf{v},\lambda,\theta)\, v_{r,L}, \tag{3}$$

where $f(\mathbf{v}, \lambda, \theta)$ is the velocity distribution function of upgoing atoms at longitude λ and colatitude θ on the lunar surface. Again, the velocity $v_{r,L}$ limits the integral to a special class of particles: those which are lost from the atmosphere due to escape, entrapment in satellite orbits, or ionization. The constant K is seen as the proportionality which converts a Monte Carlo sum of events to a flux.

Time variation of helium due to periodic passage of the moon through the geomagnetic tail is monitored by dividing the synodic lunation (29.53d) into 16 equal parts, each equivalent to 1.85 d. Each atmospheric parameter calculated in the simulation is accumulated in 16 individual registers, one for each phase zone.

An interesting, time varying parameter is the total abundance of helium atoms of the bound lunar atmosphere. This is also a difficult quantity to compute. Owing to the extreme altitudes reached by ordinary thermal atoms and the non-Maxwellian nature of the velocity distribution function, the barometric approximation is of doubtful accuracy. An alternative is to calculate the bound abundance as

$$N_B = R^2 \int_0^{2\pi} d\lambda \int_0^{\pi} d\theta \sin\theta \int d^3\mathbf{v}\, f(\mathbf{v})\, v_{r,B} t(\mathbf{v}), \tag{4}$$

where t is the time of flight of a particle with initial velocity $\mathbf{v}$, and the integration is restricted to particles in trajectories which return to the planet. While Eq. 4 is not analytically tractable, it can be evaluated approximately by use of the Monte Carlo analog

$$N_B = 16 \times 4\pi R^2 K \sum t. \tag{5}$$

The constant K is that found by evaluating Eq. 1, and the factor 16 is needed to account for the dilution of the statistical weight of each trajectory by the accumulation of this sum in 16 phase zones. The time of flight, t, of an atom which passes from one phase zone to another is divided accordingly in its contributions to N_B. For short trajectories, the flight time is arbitrarily divided equally between zones of origin and impact; but for longer, high energy orbits which cross several phase zones in flight, the contribution to each zone is based on actual time spent there.

One of the goals of this study is to determine the total number of helium atoms trapped in satellite orbits about the moon. These orbits are populated by trajectory perturbations due to the sun and earth and are depopulated by ionization. Also classified with the trapped component are the small number of atoms which become ionized while their trajectories are being traced. By analogy with Eq. 5 the number of trapped atoms, averaged over a lunation, is

$$N_T = 4\pi R^2 K \sum t_i \ , \tag{6}$$

where t_i is time of flight of gravitationally trapped atoms which become ionized while in the vicinity of the moon. The ionization time for helium at the moon is estimated to be 1.7×10^7 sec. At the start of each trajectory, a uniform random deviate u, in the range 0 to 1, is used to determine t_i via the equation

$$t_i = -1.7 \times 10^7 \ln(u). \tag{7}$$

If the time the atom spends in sunlight exceeds t_i, the atom is considered ionized.

The most controversial phenomenological parameter of the lunar atmosphere is concentration, which must be defined in terms of its detection. The only quantity that has been measured *in situ* on the moon is the downward flux of helium, ϕ_d, detected by the mass spectrometer at the Apollo 17 site (Hoffman *et al.*, 1973). These data have been transformed to an effective concentration by adaptation of the Maxwellian relationship

$$n = 4\ \phi_d \ / <v>, \tag{8}$$

where $<v>$ is the mean atomic speed corresponding to local surface temperature. This definition necessarily neglects the non-Maxwellian nature of the downcoming flux. It also leaves out the upgoing flux of escaping new atoms, a negligible quantity for helium but an important one for hydrogen (Hartle and Thomas, 1974). However, it is useful to maintain this definition in order that the simulator results can be compared wih the Apollo 17 data.

Since concentration varies greatly over the moon and with time in response to the variation of the source, it is useful to subdivide the lunar surface into 134 equal areas and to calculate average concentrations for each of these areas as was done by Hodges (1973). This arrangement is reproduced in each of the 16 phase zones discussed earlier, so that 16 separate global distributions of helium are produced. The concentration for one of these areas is given by

$$n = 16 \times 134 \times \frac{4K\ N_i}{<v>}, \tag{9}$$

where N_i is the number of times an atom has impacted in the area in question, and $<v>$ is the mean atomic speed corresponding to the surface temperature at the point of impact.

The surface temperature model is an analytical approximation adopted from earlier work (Hodges, 1973). Local radiative equilibrium is assumed in daytime, so that temperature is given by

$$T = 384(\cos \chi)^{1/4}, \tag{10}$$

where χ is solar zenith angle. This expression is applied only for $T > 140K$. At night the temperature is given by the empirical formula

$$T = 60 +]\left[22 + 24e^{.07466(90^\circ - \lambda)} + 25e^{.00456(90^\circ - \lambda)}\right] (\sin\theta)^{1/4}, \tag{11}$$

where θ is colatitude, and λ is longitude measured from the subsolar meridian. In the daytime region near the terminator where $T > 140$ K, a linear extrapolation between these expressions is applied.

Velocity distribution function

In a Monte Carlo simulation of an exosphere each penetration of the exobase by a downcoming atom must be followed by the generation of a new upgoing atom with velocity chosen from some random distribution. Atoms emerging from a Maxwellian thermosphere, such as on earth, should have a Maxwell-Boltzman flux distribution (cf., Brinkmann, 1970 and Smith *et al.*, 1978). However, on the Moon, Mercury, and other similar bodies the velocity of an atom leaving the surface depends on its interaction with solid rocks. It was assumed by Hodges (1973) that this should result in a Maxwellian velocity distribution, while Smith *et al.* (1978) have adopted a flux distribution. Either of these distributions is at best an approximation of questionable validity, according to Shemansky and Broadfoot (1977), because a realistic surface accommodation process requires that the velocity of a departing atom reflect a memory of its impact velocity. The absence of high energy downcoming atoms in an exosphere should be expected to cause the probability of high energy, escaping atoms to decrease more rapidly than the tail of either of the aforementioned Maxwellian distributions.

An obvious ramification of any mechanism which decreases the likelihood of escape is that it must also increase lifetime. Hodges and Hoffman (1974) have presented experimental data from the Apollo 17 mass spectrometer which shows noticeable increases and decreases in the helium abundance on the moon which have time scales of the order of several days. Hodges (1977b) has argued that this data fixes the average helium lifetime in the range of 2–5 days. Atmospheric models based on the Maxwellian velocity distribution give lifetimes near the low end of this range, suggesting that accommodation effects may at most contribute a factor of 2 decrease in the overall probability of escape and, hence, that there is no serious depletion of the high energy tail of the velocity distribution. This can be reconciled with the arguments of Shemansky and Broadfoot (1977) by postulating that each impact of an atmospheric atom with the lunar surface is in fact a sequence of several collisions with the surfaces of adjacent soil grains, so that an adequate high energy population of atoms is produced. Lacking a rigorous velocity distribution function for atoms leaving the lunar regolith, it is necessary to employ reasonable approximations. An atom that leaves the surface of a soil grain may enter a long range ballistic trajectory, but it seems more likely that it should strike another grain, and the likelihood of a subsequent collision must increase with increasing zenith angle of the velocity vector. On a larger scale, an atom leaving the lunar surface has greatest probability of striking a nearby boulder, a crater wall, or another orographic feature if its velocity is near the horizontal. This behavior suggests a directional distribution that vanishes at the horizon as the cosine of local zenith angle of the velocity vector.

Helium atoms interacting with lunar soil must attain speeds characteristic of the vibrational temperature of the atoms of the soil grains. This has prompted the assumption (Hodges, 1973) that the distribution of speeds of lunar exospheric helium must resemble the Maxwell-Boltzmann (M-B) function. The directional considerations mentioned above suggest that modification of the M-B distribution by the cosine of local zenith angle is a reasonable approximation of the true velocity distribution. Denoted hereafter by M-B-C, this distribution is given by

$$f_{\text{M-B-C}} = 4\left(\frac{\alpha}{\pi}\right)^{3/2} e^{-\alpha v^2} \cos\chi, \qquad (12)$$

where χ is the local zenith angle of the velocity vector. The α parameter is

$$\alpha = m/2kT, \qquad (13)$$

where m is the atomic mass of helium, k is Boltzmann's constant and T is the soil temperature.

The Maxwell-Boltzmann flux distribution (denoted M-B-F), given by

$$f_{\text{M-B-F}} = 2\frac{\alpha^2}{\pi} v e^{-v^2} \cos\chi, \qquad (14)$$

describes the velocities of atoms crossing a plane in a Maxwellian gas. Brinkmann (1970) has noted that this distribution must be used for Monte Carlo simulations of the escape process in the terrestrial thermosphere-exosphere transition region. Smith *et al.* (1978) have noted that the M-B-F distribution is the statistical analogue needed for Monte Carlo duplication of most analytical treatments of exospheric lateral transport. While there is no clear argument that the M-B-F distribution should be physically related to the process of helium atoms bouncing off the rocks of a regolith, its bias toward higher energies provides a test of the effect of total thermalization of the lunar exospheric gas.

Trajectory calculations

Trajectories and times of flight are calculated by several methods, depending on initial kinetic energy. Atoms with energies less than 80% of that required for lunar escape travel in essentially elliptic orbits. These are treated in the manner outlined in Hodges (1973). Atoms with energies greater than 20% of escape generally originate in daytime, have rather high apolunes, and spend most of their flight time in sunlight. Hence, the time in the solar environment is approximated as total time of flight, which is calculated for an elliptic trajectory. Flight times of slower atoms (i.e., with energies less than 20% of escape) are computed using a parabolic trajectory approximation. The time that a slow particle spends in the solar environment is computed as the part of the parabolic

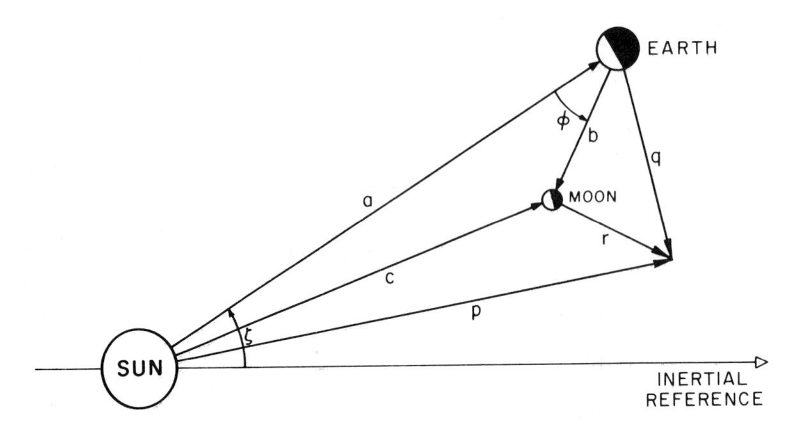

Fig. 1. Vector relationships of a point in the sun-earth-moon system.

flight time spent above the antisunward, cylindrical extension of the lunar terminator.

As initial energy increases above 80% of the escape energy the influences of the terrestrial and solar gravitational fields, the noninertial motion of the moon, and solar photon scatter on atom trajectories become noticeable. Owing to the relatively short life and flight times of these particles it is reasonable to make several simplifying assumptions, including neglect of the tilt of the lunar orbit plane with respect to the ecliptic and neglect of the eccentricities of the orbit of the moon about earth and of earth about the sun. In addition, it is convenient to neglect the mass of the moon with respect to that of earth in determining the lunar orbit. Vector relationships of the sun, earth, moon, and a lunar atomsphere atom are shown in Fig. 1. The sun is treated as an inertial reference point and the sun-earth line at the time the particle leaves the lunar surface is established as a fixed directional reference. The angular movement of earth about the sun is

$$\zeta = \Omega\, t, \tag{15}$$

where t is elapsed time from the origin of the trajectory. The assumed circular orbit of earth implies

$$\Omega = \left(\frac{GM\odot}{a^3}\right)^{1/2}. \tag{16}$$

Phase of the moon, ϕ, is assumed to increase approximately linearly with time, so that

$$\phi = \omega t + \phi_o, \tag{17}$$

where ω is the rate of synodic rotation of the moon, and ϕ_o is the initial phase angle.

The equation of motion of the moon relative to the sun is

$$\frac{d^2\mathbf{c}}{dt^2} = -\mathbf{b}\frac{GM_\oplus}{b^3} - \mathbf{c}\frac{GM_\odot}{c^3}. \tag{18}$$

With respect to earth this transforms to

$$\frac{d^2\mathbf{b}}{dt^2} = -\mathbf{b}\left(\frac{GM_\oplus}{b^3} + \Omega^2\right) + \mathbf{c}\Omega^2\left(1 - \frac{a^3}{c^3}\right). \tag{19}$$

If the last term of 19 is neglected for the moment, the scalar part of the first term is identifiable as the square of the sidereal orbit rate of the moon:

$$\omega + \Omega = \left(\frac{GM_\oplus}{b^3} + \Omega^2\right)^{1/2}. \tag{20}$$

The second term of 19 represents the semi-synodic perturbation of the lunar orbit.

An atom of the lunar atmosphere at point p moves according to the equation

$$\frac{d^2\mathbf{p}}{dt^2} = -\mathbf{r}\frac{GM_☾}{r^3} - \mathbf{q}\frac{GM_\oplus}{q^3} - \mathbf{p}\frac{GM_\odot}{p^3} + \mathbf{S}, \tag{21}$$

where **S** is the acceleration due to the momentum transferred to the atom by resonant scatter of solar photons. Using the vector relationships shown in Fig. 1 and Eqs. 19 and 21, the moon referenced equation of motion of the particle is

$$\begin{aligned}\frac{d^2\mathbf{r}}{dt^2} &= -\mathbf{r}\left(\frac{GM_☾}{r^3} + \frac{GM_\oplus}{q^3} + \frac{GM_\odot}{p^3}\right)\\ &-\mathbf{b}\left(\frac{GM_\oplus}{q^3} - \frac{GM_\oplus}{b^3}\right)\\ &-\mathbf{c}\,\Omega^2\left(\frac{a^3}{p^3} - \frac{a^3}{c^3}\right)\\ &+\mathbf{S}.\end{aligned} \tag{22}$$

At this point it is convenient to make some approximations which amount to neglecting terms of order $(r/a)^2$ and $(b/a)^2$. In addition, it is helpful to convert the coefficients involving earth and sun parameters to terms involving orbit frequencies. The result is the approximation

$$\begin{aligned}\frac{d^2\mathbf{r}}{dt^2} &\cong -\mathbf{r}\left(\frac{GM_☾}{r^3} + \epsilon^2\frac{b^3}{q^3} + \Omega^2\right)\\ &-\mathbf{b}\,\epsilon^2\left(\frac{b^3}{q^3} - 1\right)\ 3\mathbf{a}\Omega^2\frac{\mathbf{a}\cdot\mathbf{b}}{a^2} + \mathbf{S},\end{aligned} \tag{23}$$

where

$$\epsilon^2 = \frac{GM_\oplus}{b^3} = \omega(\omega + 2\Omega) \tag{24}$$

is determined from Eq. 20.

The trajectory of a lunar atmospheric atom is traced by formulating Eq. 23 as a difference equation. Initial conditions of velocity and position on the lunar surface are converted to give the first two points on the trajectory and each subsequent point is computed on the basis of the previous two points. The time interval between points is reset every 50 points to

$$\Delta t = 0.03 \sqrt{R_☾ \Big/ \left|\frac{d^2\mathbf{r}}{dt^2}\right|} \ . \tag{25}$$

This is an empirical result which gives reasonably accurate trajectories while conserving computation time.

The acceleration due to photon scatter, S, is treated as an impulsive phenomenon. The dominant source of photon momentum is assumed to be the 584A line for which Meier and Weller (1972) suggest a photon scatter rate of 3.1×10^{-5} sec^{-1}. To account for the weaker nearby resonant lines of neutral helium the effective scatter rate has been increased 10% to 3.4×10^{-5} sec^{-1}, making the effective time constant for photon scatter 2.9×10^4 sec. Initially, and after each photon scatter event a logarithmic deviate of this time is computed, and when the flight time from the last scatter event exceeds this deviate, an impulsive momentum increment is added to the atom. This increment is the vector sum of the incident photon momentum (directed antisunward) and an equal amplitude random vector. The former accounts for the absorbed photon, the latter for the recoil of the atom due to emission of a photon in an arbitrary direction. The average velocity change per photon scatter event is 170.3 cm/sec, corresponding to the momentum per helium atomic mass of a 584A photon.

The trajectory calculation is terminated by one of several occurrences. If the particle impacts the moon, it is treated as any other returning particle, and the sums needed to evaluate concentration (expression 9), bound atmosphere abundance (expression 5), and accumulated lifetime are updated. Those atoms which become ionized contribute to the total loss flux (expression 1) as do those with sub-escape initial speeds which leave the lunar vicinity and become trapped in earth orbit. An atom which completes two successive orbits of the moon without impact is deemed to be trapped in a satellite orbit and contributes to the number of these particles via expression 6.

MODEL ATMOSPHERE RESULTS

Summaries of bulk model atmosphere parameters for the two velocity distribution functions are presented in Table 1. An ovbious ramification of the excess high energy population of the M-B-F distribution (Eq. 14) is that it produces a considerably lower residence time for bound atmospheric helium atoms than does the M-B-C distribution (Eq. 12). As a result, the average bound atmospheric abundance of helium predicted by the M-B-F distribution is only 60% that predicted by the M-B-C distribution.

Figure 2 shows the variation of the bound helium abundance as a function of lunar phase. The solid histogram represents M-B-C distribution data while the dashed lines are for the M-B-F distribution. The continuous curve overlaying each histogram consists of an exponential decay with rate equal to the inverse of the residence time in the geomagnetic tail and an exponential recovery fitted to the remaining data. Statistical uncertainties of the data do not permit assessment of the accuracy of an exponential decay in the geomagnetic tail, but there is qualitative assurance in the greater relative depth of the M-B-F data (a 70% decrease versus 53% for the M-B-C data).

Effects of passage of the moon through the geomagnetic tail are also apparent in the global distribution of helium at the lunar surface. Figure 3 shows night and day surface concentration corresponding to averages over 9% of the lunar sphere centered at the antisolar and subsolar points, respectively. Again, the solid histogram represents M-B-C data and the dashed graph shows M-B-F results, and as before, the longer residence time for the former model produces lesser geomagnetic tail decreases than the former. At night the average concentration for the M-B-C model is 2.4 times that for the M-B-F distribution, while in daytime this ratio is about 2.2. The corresponding total abundance ratio for these models is only 1.6 (cf., Table 1). The excess of the concentration ratio is mainly due to the greater population of low speed atoms in the M-B-C distribution, which produces a nonbarometric height distribution with excess concentration near the surface (Smith *et al.*, 1978). Average night to day concentration ratios for the M-B-C and M-B-F models are 24:1 and 20:1, respectively.

The only practical assessment of the accuracies of these models is to compare them with the mass spectrometer helium measurements made at the Apollo 17 site in 1973. The continuous curve over the central, nighttime, portion of Fig. 4 is a 9 lunation average of the Apollo 17 data, which has been converted to effective concentration by use of Eq. 8. The histograms show model concentration results. The M-B-F data (dashed lines) are in reasonable agreement with the measurements; whereas, the M-B-C results generally exceed the measurements by about a factor of 2.5. These results differ little from previously published models. In Hodges (1977a), it was noted that close agreement was obtained using a computation method which approximates the M-B-F distribution. Excess model concentrations found from earlier M-B distribution calculations were attributed to a nonthermal mechanism for release of trapped solar wind helium from lunar soil by Hodges and Hoffman (1974).

While conditions in the bound atmosphere have roughly a factor of 2 uncertainty attributable to the velocity distribution of atoms leaving the lunar surface, the fate of escaping atoms is more certain. In Table 1 it can be noted that to an accuracy of ±10%, the total abundance of helium in trapped satellite orbits about the moon is 1.2×10^{30} atoms.

In the loss rate data of Table 1, it can be noted that both models are based on a total supply (and loss) rate of 1.1×10^{24} helium atoms/sec, which corresponds to a generally accepted solar wind α particle influx rate of 4.5% of the proton flux to the moon (Johnson *et al.*, 1972). Neutral helium atoms are lost from the moon by

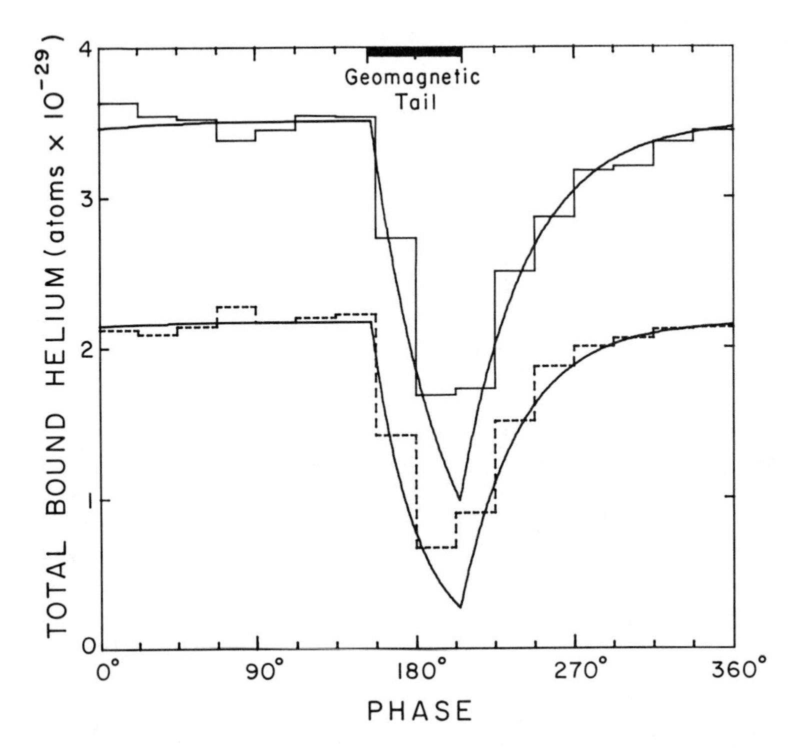

Fig. 2. Total abundance of bound helium for the M-B-C and M-B-F distribution models (solid and dashed line histograms, respectively) as functions of lunar phase. Each of the superimposed curves represents an exponential decay within the geomagneitc tail with time constant equal to the appropriate model residence time and exponential recovery fitted to the model data.

several processes, but almost all are bound to the earth-moon system until ionized. Bound atoms are ionized and accelerated from the moon by induced fields of the solar wind at a rate of roughly 2×10^{22} atoms/sec, accounting for only 1.8% of the helium loss. About 5% of the helium loss (i.e., 5×10^{22} atom/sec) is to the lunar corona, which contains 4–6 times as much lunar helium as the bound atmosphere. However, the bulk of the lunar helium (93%) resides in earth orbit as part of the terrestrial helium corona. It is interesting to note that the transfer of neutral helium from the moon to the terrestrial corona, at the indicated rate of 10^{24} atoms/sec, is three times as great as the total rate of helium escape from earth (Kockarts and Nicolet, 1962).

Relative effects of resonant photon scatter and the influence of the sun-earth-moon gravitational system on the perturbations of helium atom trajectories have been estimated by computation of a model lunar helium atmosphere in which photon scatter was ignored. This model used a M-B-F velocity distribution for atoms leaving the lunar surface. The major effect caused by neglect of photon

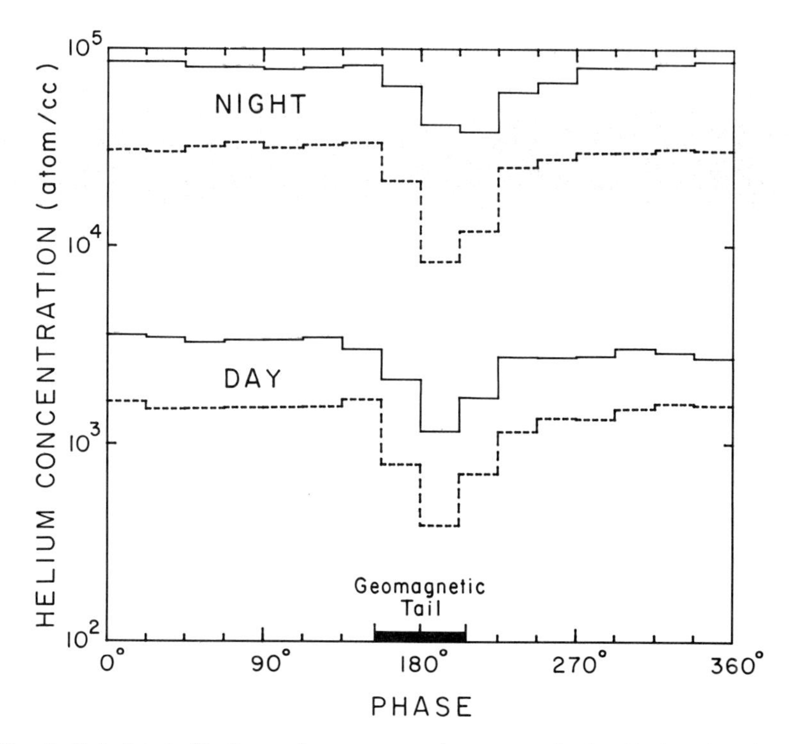

Fig. 3. Calculated effective surface concentrations near subsolar (day) and antisolar (night) points. Solid lines represent M-B-C velocity distribution results while dashed lines are for the M-B-F distribution.

scatter was the reduction of the rate of trapping of helium atoms in satellite orbits about the moon. The resulting lunar corona helium abundance was 6.8×10^{29} atoms versus 1.1×10^{30} atoms for the M-B-F model with photon scatter.

An interesting aspect of the helium loss process is that, for atoms with initial speeds below the lunar surface escape velocity, escape from the moon to earth orbit is slightly more likely than is trapping in satellite orbit about the moon. Surprisingly, this fraction is nearly constant, regardless of velocity distribution or whether photon scatter is included. In total, the sub-escape atoms account for about 15% of the helium loss, which includes roughly 9% of the helium escaping to earth orbit, as well as, the entire supply of helium atoms to the lunar corona.

Conclusions

Perturbations of the trajectories of helium atoms of the gravitationally bound lunar atmosphere due to the sun-earth-moon gravitational system and to resonant scatter of solar photons act to decrease the average residence time for helium on

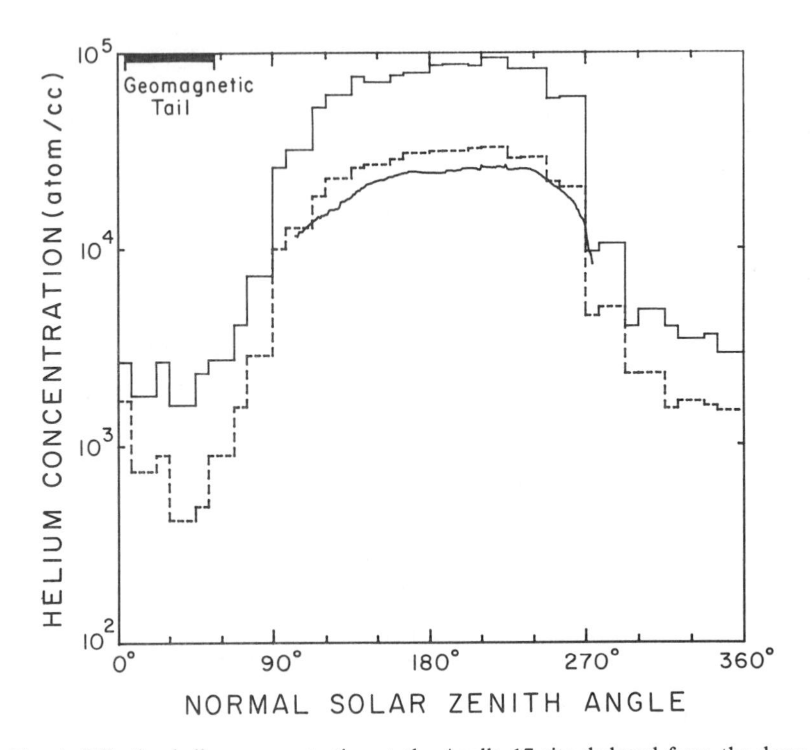

Fig. 4. Effective helium concentration at the Apollo 17 site deduced from the downward flux. Solid curve through the nighttime portion of graph is the 9 lunation average of the helium measurements from the Apollo 17 mass spectrometer. Solid histogram gives calculated value for the M-B-C velocity distribution calculations while dashed results are for the M-B-F distribution.

the moon. For an atom with sub-escape initial speed, the perturbations may act to increase flight time by temporarily raising perilune above the lunar surface and, hence, increasing the probability of photoionization. A permanent increase in the height of perilune can result in trapping of the atom in a lunar satellite orbit until it becomes ionized. Alternatively, the atom may travel outside the range of gravitational influence of the moon and become part of the terrestrial helium corona.

Numerically, the loss of sub-escape helium atoms from the moon accounts for about 15% of the total helium escape rate. This fraction seems noteworthy, but it loses significance when compared with bulk atmosphere uncertainties on the order of a factor of 3 associated with the unknown nature of the velocity distribution function for helium atoms leaving the lunar surface. In the preceding calculations, use has been made of two hypothetical velocity distributions, the Maxwell-Boltzmann distribution with vertical referenced cosine directional probability (M-B-C) and the Maxwellian flux distribution (M-B-F). Owing to the generally higher energies associated with the latter distribution, it resulted in a

Table 1. Model Atmosphere Parameters for Lunar Helium.

parameters	Velocity Distribution Function M-B-C	M-B-F
No. of atoms:		
bound atmosphere	3.1×10^{29}	1.9×10^{29}
lunar corona	13.3×10^{29}	10.9×10^{29}
terrestrial corona	170×10^{29}	174×10^{29}
Loss rates (atoms/sec)		
ionization	2.3×10^{22}	1.6×10^{22}
lunar corona	5.7×10^{22}	4.9×10^{22}
terrestrial corona	102×10^{22}	104×10^{22}
Mean residence time for bound atoms (sec)	2.79×10^{5}	1.69×10^{5}

much shorter helium residence time and comparably lower surface flux and total abundance than the M-B-C distribution model.

The M-B-F model is in reasonable agreement with the helium data from the mass spectrometer at the Apollo 17 site, while the M-B-C distribution predicts concentrations almost 3 times the measured values. However, this fact alone does not prove anything. Hodges and Hoffman (1974) have noted that the erosion process which releases trapped solar wind helium from soil grains may produce a significant flux of superthermal atoms or ions capable of direct escape from the moon. Therefore, the bound atmosphere need only account for a small fraction of the helium escape rate, implying that a velocity distribution with lower average speed than the M-B-F distribution would be suitable. Thus, the M-B-C distribution could account for about ⅓ of the helium loss, decreasing the corresponding abundance, concentration, and loss rate data of Figs. 2, 3 and 4 and Table 1 by a factor of 3. A still lower energy distribution would require more direct, erosional loss, a lower atmospheric escape rate, and smaller atmospheric bulk parameters. There is a lower limit to the atmospheric escape rate due to the necessity of a helium residence time of about 5 days or less (cf., the time fluctuations of helium on the moon reported by Hodges and Hoffman, 1974). Since the rather subtle differences between the M-B-F and M-B-C distributions caused the greatly varied residence times of 2 d and 3.2 d, respectively, any velocity distribution which results in an acceptable residence time cannot be grossly different from those used here. Therefore, it seems reasonably safe to assert that the abundance and loss rate parameters for the M-B-F model in Table 1 are upper bounds, and that the M-B-C model values exceed lower bounds by not more than a factor of 5.

Acknowledgments—It is a pleasure to acknowledge useful discussions with D. E. Shemansky and G. R. Smith. This research was supported by NASA grant NSG-7034.

REFERENCES

Brinkmann R. T. (1970) Departures from Jeans' escape rate for H and He in the earth's atmosphere. *Planet Space Sci.* **18**, 449–477.

Fastie W. G., Feldman P. D., Henry R. C., Moos H. W., Barth C. A., Thomas G. E. and Donahue T. M. (1973) A search for far ultraviolet emissions from the lunar atmosphere. *Science* **182**, 710–711.

Hartle R. E. and Thomas G. E. (1974) Neutral and ion exosphere models for lunar hydrogen and helium. *J. Geophys. Res.* **79**, 1519–1525.

Hodges R. R. (1973) Helium and hydrogen in the lunar atmosphere. *J. Geophys. Res.* **78**, 8055–8064.

Hodges R. R. (1975) Formation of the lunar atmosphere. *The Moon* **14**, 139–157.

Hodges R. R. (1977a) Formation of the lunar helium corona and atmosphere. *Proc. Lunar Sci. Conf. 8th*, p. 537–549.

Hodges R. R. (1977b) Release of radiogenic gases from the moon. *Phys. Earth Planet. Inter.* **14**, 282–288.

Hodges R. R. and Hoffman J. H. (1974) Measurements of solar wind helium in the lunar atmosphere. *Geophys. Res. Lett.* **1**, 69–71.

Hodges R. R., Hoffman J. H. and Johnson F. S. (1974) The lunar atmosphere. *Icarus* **21**, 415–426.

Hodges R. R., Hoffman J. H., Johnson F. S. and Evans D. E. (1973) Composition and dynamics of lunar atmosphere. *Proc. Lunar Sci. Conf. 4th*, p. 2855–2864.

Hodges R. R. and Johnson F. S. (1968) Lateral transport in planetary exospheres. *J. Geophys. Res.* **73**, 7307–7317.

Hoffman J. H., Hodges R. R., Johnson F. S. and Evans D. E. (1973) Lunar atmospheric composition results from Apollo 17. *Proc. Lunar Sci. Conf. 4th*, p. 2865–2875.

Johnson F. S., Carrol J. M. and Evans D. E. (1972) Lunar atmospheric measurements. *Proc. Lunar Sci. Conf. 3rd*, p. 2231–2242.

Kockarts G. and Nicolet M. (1962) Le probleme aeronomique de l'helium et de l'hydrogene neutres. *Ann. Geophys.* **18**, 269–290.

Kopal Z. (1966) Dynamics of the earth-moon system. In *An Introduction to the Study of the Moon.* p. 52–64. Gordon and Breach, N.Y.

Manka R. H. and Michel F. C. (1970) Lunar atmosphere as a source of argon-40 and other lunar surface elements. *Science* **169**, 278–280.

Meier R. R. and Weller C. S. (1972) EUV resonance radiation from helium atoms and ions in the geocorona. *J. Geophys. Res.* **77**, 1190–1204.

Shemansky D. E. and Broadfoot A. L. (1977) Interaction of the surfaces of the Moon and Mercury with their exospheric atmospheres. *Rev. Geophys. Space Phys.* **15**, 491–497.

Smith G. R., Shemansky D. E., Broadfoot A. L. and Wallace L (1978) Monte Carlo modeling of exospheric bodies: Mercury. *J. Geophys. Res.* **83**, 3783–3790.

Proc. Lunar Planet. Sci. Conf. 9th (1978), p. 1765–1786.
Printed in the United States of America

Plausible depositional histories for the Apollo 15, 16 and 17 drill core tubes

Y. LANGEVIN and M. MAURETTE

Laboratoire René-Bernas du Centre de Spectrométrie Nucléaire, et de Spectrométrie de Masse, 91406 ORSAY (FRANCE)

Abstract—A Monte-Carlo soil mixing code (ORSOLMIX) has been applied to randomly generate 200 distinct depositional histories in the lunar regolith, which were then used to compute the depth concentration profiles of four cosmic ray induced isotopes (^{21}Ne, ^{126}Xe, ^{131}Xe, ^{158}Gd) up to depths of 600 $g.cm^{-2}$ in the lunar regolith (GCRi profiles). The major aim of this work was to tackle the difficult problem of solving a set of integral equations from which the depositional histories of the Apollo 15, 16 and 17 core tubes could be inferred from the corresponding experimental GCRi profiles so far available for these core tubes. For this purpose, a "graphical" method for finding the solution of these equations was developed, which relies on the analysis of a great number (~2000) of features (sudden raises, plateaus) imprinted on the 200 sets of theoretical GCRi profiles. The major conclusions of these investigations are: (1) The depositional histories of the Apollo 15, 16 and 17 core tubes can be interpreted as resulting from 2, 5 and 6 major deposition episodes, respectively, that involved changes in the elevation of the lunar surface larger than 50 $g.cm^{-2}$, and the corresponding total deposition time scales are 800 ± 200 m.y. (Apollo 15), 1000 ± 250 m.y. (Apollo 16) and 1500 ± 350 m.y. (Apollo 17); (2) The major uncertainty in these evaluations arises from the graphical method itself, and results from the requirement that this method, when applied to the theoretical sets of GCRi profiles, should account for ~98% of the 200 corresponding depositional histories; (3) In the theoretical depositional histories, fast deposition episodes, resulting in the accumulation of ~400 $g.cm^{-2}$ of regolith in less than 200 m.y. (bottom section of Apollo 15) are surprisingly frequently found (35 out of 200). In addition, the rapid succession of several 100 $g.cm^2$ craters and/or layers during a single theoretical depositional history (Apollo 16 and 17) is very common (120 out of 200). On the other hand, a long period (~500 m.y.) during which the elevation of the surface varied within only ~100 $g.cm^{-2}$ (top section of Apollo 15) is very unlikely (4 out of 200).

I. INTRODUCTION

As a result of the meteoritic gardening, lunar dust grains experience a complex depth motion in the lunar regolith, during which they accumulate a variety of depth and time dependent effects resulting from their exposure to solar wind ions, and solar and galactic cosmic rays. A wealth of data concerning such effects have been reported over the past few years, which were then used for very different purposes. In particular, the depth distribution of solar flare tracks (SFT) in lunar core tubes, as well as that of galactic cosmic rays induced isotopes (GCRi) have already been exploited to get information concerning the depositional history of lunar core tubes, by using two very different types of soil-mixing models, which can be quoted as Monte-Carlo models (MC) and "parameter fitting" models (PF) intended to empirically fit one type of experimental data.

Unfortunately, instead of giving concordant conclusions, these different models generated new unsolved problems. For example, in the case of Apollo 15, the thermal neutron fluence profile has been interpreted as resulting either from a rapid deposition followed by a prolonged *in situ* irradiation (Russ *et al.*, 1972) or from an extensive mass loss (Fireman, 1974). Furthermore, the proponents of one of the PF models (Curtis and Wasserburg, 1975a) concluded that meteoritic mixing cannot account for the fluence values measured in lunar cores, whereas we argued the opposite using a Monte-Carlo code intended to describe the meteoritic mixing of the lunar regolith (Langevin and Maurette, 1976). This code (ORSOLMIX) was already applied in order to tentatively describe the accumulation of solar wind effects (Bibring *et al.*, 1975; Borg *et al.*, 1976), solar flare tracks (Bibring *et al.*, 1975) and galactic cosmic rays isotopes (Langevin and Maurette, 1976) in the lunar regolith, by using a given set of "meteoritic" input data concerning the meteoritic mass spectrum and crater shape parameters, chosen four years ago. In particular, we have already been able to generate a great number of depth profiles for two GCRi isotopes (^{21}Ne and ^{158}Gd) which match those measured in the deep drill core tubes returned to the earth by the Apollo 15, 16 and 17 missions.

The major purpose of the present work was to correlate the GCRi profiles so far measured in the Apollo core tubes to the depositional histories of these cores. We start in randomly generating 200 ORSOLMIX depositional histories for a test core 600 g.cm^{-2} deep, which appear quite distinct. For each one of these histories, we compute four different GCRi profiles using formulae which involve the relevant production function, for ^{21}Ne, ^{126}Xe, ^{131}Xe and the neutron fluence as inferred from ^{158}Gd concentrations. We then tackle the difficult mathematical problem of solving the complex set of integral equations derived from these formulae, to obtain the depositional history of a core from the corresponding set of GCRi profiles. In fact, these equations can only be solved unambiguously for histories which are both continuous and monotonous functions of time, whereas the real depositional histories resulting from meteoritic mixing involve discrete changes in elevation (thick ejecta blankets and deep craters), which generate random-walk like histories. We have thus developed a graphical method intended to find plausible solutions of this set of integral equations, compatible with a meteoritic mixing dominated evolution, by analysing the characteristics of a great number (~2000) of features imprinted on the 200 sets of theoretical profiles, as well as their relationship with the nature and magnitude of the major events in the corresponding depositional histories. We can thus identify and characterize each major event from its associated features in the set of GCRi profiles, using relatively simple formulae. The depositional history is then inferred from the succession of all major events. The major uncertainties attached to this "graphical" method result from the high "confidence level" that we required by fixing at 98% the proportion of theoretical depositional histories which should correctly be inferred by applying the graphical method to the 200 sets of theoretical GCRi profiles. To achieve such a high confidence level, we had to allow an uncertainty of ±10% on the corresponding depositional time scale. After checking that the experimental sets of GCRi profiles so far available for the Apollo 15, 16 and 17 core tubes do show the general characteristics expected from the theoretical sets, we finally applied the graphical method to tentatively decipher the depositional history of these three cores. We finally point out the present limitations of our method and briefly discuss its specific advantages with respect to those already reported.

II. Results of the Monte-Carlo Computations

II.1—Outline of the Monte-Carlo computations

In these computations which have already been described (Duraud *et al.*, 1975; Langevin and Maurette, 1976), ORSOLMIX gives the depth variation $d(t,r,d_0)$ of a test particle with a radius r (which can be varied from 1 μm up to 5 cm) as a function of both the time t and the initial depth d_0 of the particle in the lunar regolith. With respect to other Monte-Carlo codes, ORSOLMIX has a unique mode of operation which follows the d(t) trajectories backward in time (thus stripping of strata and filling up craters) in generating $d(-t,r,d_f)$ trajectories, where d_f is the final "present day" depth of the test particle. Thus d_f replaces the completely unknown "initial" d_0 value, and this considerably restricts the number of d(-t) trajectories for the test particles. A great number of trajectories are randomly generated by the code, which can be coupled with a variety of production rates to describe the effects accumulated by the test particle as a result of its direct exposure to solar wind nuclei, and solar and galactic cosmic rays.

One of the interesting results of our soil mixing computations has been to show (Langevin and Maurette, 1976) that the d(-t) trajectories can be divided in two distinct parts, identified as the "plateau" and "burial" history. In fact, if we attribute all stratigraphic units in the core to separate layering events, the test particle has been cycled at least once through an ejecta blanket a few centimeters thick, which corresponds to its "last" present day parent stratum, now found at a depth d_f below the lunar surface. When ORSOLMIX is run backward in time from the d_f position, this present day parent stratum was at one time the top stratum of the regolith. The "plateau" history represents the whole dynamic evolution of the test particle between its incorporation to the regolith and its last deposition on the top surface of the lunar regolith. The "burial" history then corresponds to the gradual burial of the test particle from this last top surface exposure down to its present day depth, d_f. In lunar core tubes, the burial history is expected to generate GCRi profiles that should reflect the deposition history of the constituent grains of the core, and that will be superimposed on pre-irradiation levels acquired during the plateau history of the same grains.

We showed recently (Langevin and Maurette, 1976) that the plateau contribution could be represented as a first approximation by the following mean pre-irradiation levels[1]: Fluence, 1.85 $n.cm^{-2}$; ^{21}Ne, 50.10^{-8}ccSTP/g; ^{126}Xe,

[1]These values are slightly lower than those reported last year, due to adjustments in the GCRi production rates.

45.10^{-8} ccSTP/g(Ba); ^{131}Xe, 240.10^{-8}ccSTP/g(Ba). This feature can be qualitatively understood from the "group" behaviour of the constituent particles of a given soil sample during their dynamical evolution in the regolith. Indeed, with the exception of the most immature soils, the d(-t) trajectory of a given test particle during its plateau history is mostly uncorrelated to that of its future neighbors in the core, whereas in the burial history, the relative positions of all the constituent particles of the core get "frozen" during the progressive burial of their parent strata. Consequently, mean values can be defined for the GCRi pre-irradiations levels, whereas the burial history of the grains at a given point of the lunar surface results in very unique GCRi profiles.

We considered four distinct GCRi profiles, F, N, X26 and X31, which refer to the neutron fluence as inferred from ^{158}Gd, and the ^{21}Ne, ^{126}Xe and ^{131}Xe cosmogenic concentrations, respectively. For this purpose, we have used the neutron flux dF/dt that was directly measured in the lunar regolith by Woolum *et al.*, (1975) that we normalized to the chemical composition of the Apollo 12 landing site (Fig. 1,a) as well as the ^{21}Ne, ^{126}Xe and ^{131}Xe production rates derived from the work of Reedy and Arnold (1972) (Fig. 1,b) and Reedy (1976) (Fig. 1,c and 1,d).

In order to obtain the four GCRi profiles corresponding to a given burial history, we take a standard 600 g.cm^{-2} core tube with a mare type chemical

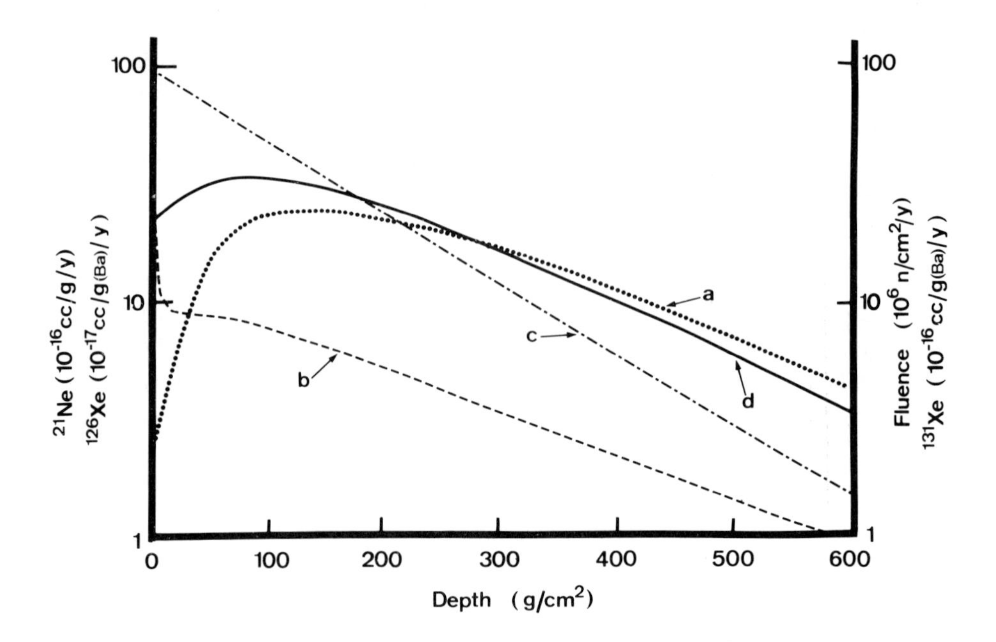

Fig. 1. Production rates of cosmogenic nuclides as a function of depth in lunar core tubes, as normalized to a mare type chemical composition (Apollo 12): a. Thermal neutron flux (Woolum *et al.*, 1975); b. Neon 21 (Reedy and Arnold, 1972); c. Xenon 126; d. Xenon 131 (Reedy, 1976).

composition, along which we consider 40 points separated by intervals of 15 $g.cm^{-2}$, which outline the position of the constituent strata of the core. The Monte-Carlo routine is then run backward in time for the bottom point, now at a final depth $D_c = 600\ g.cm^{-2}$, thus generating its burial trajectory $d_c(-t)$. All the overlaying points follow this trajectory at constant depth intervals, and for a point with a present day final depth D_s, the burial trajectory is: $d_s(-t) = d_c(-t) + D_s - D_c$. Consequently the trajectory of the bottom point $d_c(-t)$ describes *completely* the depositional history of a core. Then the GCRi are accumulated at each point according to its depth. Whenever one of these 40 points "reaches" the surface in the backward mode, the burial history is ended at a time $-T(D_s)$, which is the shortest time t such as $d_c(-t) = D_c - D_s$, and the "constant" preirradiation level is just added to the accumulated GCRi concentrations. The time $-T(D_s)$ which represents the date of deposition of stratum s in the core, will be quoted as the "age" of this stratum. In summary, if $C_i(D_s)$ is the profile of isotope i along the core, $\Psi_i(d)$ the associated production function, and M_i the corresponding mean pre-irradiation level, we have:

$$C_i\ (D_s) = M_i + \int_{-T(D_s)}^{O} \Psi_i(d_c(t) + D_s - D_c)\ dt \quad (i = F, N, X26, X31)$$

If we want now to tackle the reciprocal problem of determining the depositional history of a given core tube (i.e., the trajectory $d_c(-t)$ and the ages $T(D_s)$ from the corresponding GCRi profiles these straightforward formulae become a complex set of integral equations relating the four profiles $C_i(D_s)$ to the trajectory $d_c(-t)$. These equations can only be solved unambiguously for functions $d_c(-t)$ which are both continuous and monotonous functions of time, whereas the real depositional history at a given site is dominated by discrete deposition events (thick ejecta blankets, deep craters) and consequently the real trajectories $d_c(-t)$ present discontinuities, and have a "random-walk" character. In the next chapter, we propose a "graphical" method based on the analysis of both the 200 ORSOLMIX depositional histories and their associated sets of theoretical profiles, which is intended to derive such "real" solutions compatible with a meteoritic mixing dominated evolution. It can also be shown that for such likely solutions, the analysis of one GCRi profile alone cannot provide an accurate determination. Thus several isotopes with productions functions as different as possible have to be considered to reduce uncertainties in the determination of $d_c(-t)$ and $T(D_s)$.

II.2—Major features of the theoretical GCRi profiles, and the "graphical" method to decipher the depositional history of a core

Any one of the 200 ORSOLMIX depositional histories results essentially from the contribution of several major deposition episodes, which trigger a change $\delta H > 40\ g.cm^{-2}$ in the elevation of the lunar surface, which can be conveniently described as: cratering events (CE); fast deposition episodes (FDE) corresponding to sedimentation rates in excess of 100 g.cm^{-2}/100 m.y., and which can be due either to a single thick ejecta blanket or to a rapid succession of thinner layer; slow deposition episodes (SDE) which correspond to the steady deposition of thin layers ($\delta H < 40$ g.cm^{-2}). In our computation, a given depositional history is characterized by the following important parameters: the time of occurence, t_j, and the duration, δt_j of each major deposition episode; the corresponding δH_j value; the elevation, H_j of the lunar surface, evaluated with respect to the present day level H_0 just after the cessation of the deposition episode; the total deposition time scale Σ of the 600 g.cm^{-2} core tube.

When the F, N, X26 and X31 theoretical profiles are presented on a dimensionless form by using as unit the corresponding pre-irradiation levels M_i, they show a striking similarity in shape. In particular, the most common feature appearing at the same depth in the four profiles of a given set simultaneously (Fig. 2) is an upward discontinuity ("jump") which can be followed by different types of profile variations such as a steady rise, a plateau, a slow decrease, etc. . . . In addition, the F, N and X31 profiles nearly match one another, with each sharp "peak" in the N profile corresponding to a "shoulder" and a "square step" with a similar amplitude in the F and X31 profiles respectively. On the

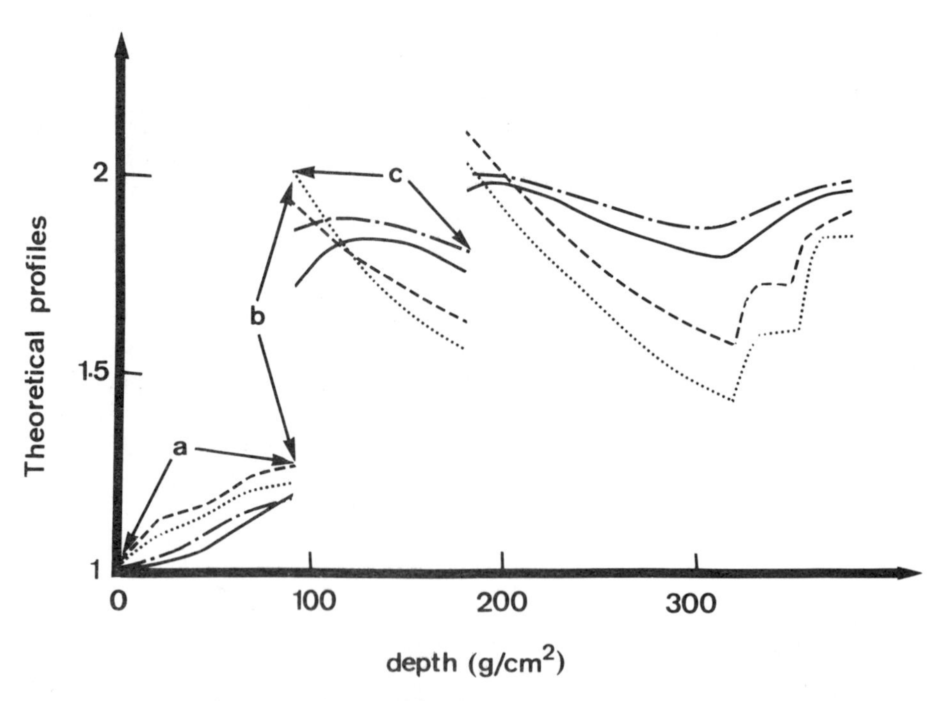

Fig. 2. Set of four theoretical GCRI profiles: Fluence (solid line); Neon 21 (dashed line); Xenon 126 (dotted line); Xenon 131 (dots dashes). The three main types of features are delineated by the arrows: a. "steady rise" region; b. "jump" region; c. "leveling or decrease" region.

other hand, the X26 profile, which looks very much like the N profile, drops progressively below the three others at depths ~200 $g.cm^{-2}$.

From the analysis of a great variety of features simultaneously imprinted on the four constituent profiles of the 200 ORSOLMIX sets, we have derived a graphical method intended to trace back the corresponding depositional history from the GCRi profiles. In this method, we start from the present day surface, and the first feature appearing simultaneously on all the GCRi profiles is analyzed. The nature of the corresponding deposition episode (CE, FDE, or SDE) is then simply inferred from the shape of this first feature, whereas its characteristics δt_1 and δH_1 are evaluated from simple formulae which relate them to the amplitude Δ_{1i} (i = F, N, X26, X31) of the corresponding variations observed on the GCRi profiles. The contribution of this episode to the GCRi profiles is then subtracted. A similar analysis is next applied to the adjacent deeper GCRi feature, which yields a new deposition episode, and this analytical procedure is repeated all the way down to the bottom point of the core tube. Finally, from this analysis we derive the total deposition time scale Σ of the core, as well as constraints on both the trajectory $d_c(-t)$ and the "sedimentation profile" $T(D_s)$ giving the ages of the constituent strata of the core as a function of their present day depth. This information is very useful to trace back the past activity of various lunar surface processes, by applying the "lunar skin" sampling technique described in our companion paper (Dartyge *et al.*, 1978) to extract grains in lunar core tubes exposed at a well-defined epoch at the surface.

More specifically, the characteristics of the various deposition episodes are determined as follows:

1. Cratering episode. A crater is reflected by a jump in all GCRi profiles simultaneously (Fig. 2,b) which is observed at a depth position corresponding to the bottom of the crater. This feature can be qualitatively understood by noting that the material which subsequently fills up the crater has only been exposed once in the high GCRi production zone (0 → 1 m) whereas the region below the bottom of the crater has been exposed twice in the same zone, thus resulting in an upward discontinuity in *all* GCRi profiles. For about 80% of the cratering events, the δt_j and δH_j values can be derived with an accuracy of ±10% from the following set of equations for episode j

$$\delta t_j \sim \Delta_{ij}/\langle\Psi_i(\delta H_j)\rangle \text{ with } \langle\Psi_i\delta H_j)\rangle = \int_0^{\delta H_j} \Psi_i(u)\, du \quad (i = F, N, X26, X31)$$

(For a crater, the δt_j value corresponds to the time required for the deposition of the material excavated subsequently by the crater). Such "normal" types of craters can be identified by checking whether or not the δt_j equations are simultaneously verified for all four isotopes by a single value of δH_j. Most of the residual cratering events (~20%) for which this condition is not fulfilled can then be correctly described by the following "delayed" cratering scenario: the now excavated region was earlier deposited at a constant rate, during a time interval δt_{j1}; then, before the formation of the crater, no major change in the elevation H of the lunar surface occurred over a time interval δt_{j2}. In this scenario, the values of δt_{j1}, δt_{j2} and δH_j can be evaluated from the following set of equations:

$$\Delta_{ij} \sim \delta t_{j1} <\Psi_i(\delta H_j)> + \delta t_{j2} \Psi_i(\delta H_j)$$

Finally, about 2% of the cratering episodes cannot be described by either one of these two scenarios with an uncertainty of 10% on δt.

2. Fast deposition episode. The GCRi signature of a FDE appears as both a progressive decrease in the N and X26 profiles, and a levelling of the F and X31 profiles (Fig. 2,c), and these features are observed over a core tube section characterised by a length equal to the δH_j(FDE) value. However, on the basis of the GCRi profiles alone, it is not possible to decide whether the FDE is due to a single thick ejecta blanket or to the rapid succession of thinner ones. Fortunately, for the Apollo core tubes returned to the earth, a variety of stratification indices can be used to infer the single or multi-layer structure of the FDE.

3. Slow deposition episodes. This type of episode produces a smooth rise in all four GCRi profiles simultaneously, which is observed along a core tube section of length δH_j(SDE). Then, $\delta t_j \sim \Delta_{ij}/<\Psi_i(\delta H_j)>$ where Δ_{ij} is now the increase in the profile i from the top to the bottom of the corresponding core section (Fig. 2,a).

We have verified the validity of this method by applying it to the 200 sets of theoretical GCRi profiles, and comparing the results to the 200 corresponding depositional histories first inferred from the Monte-Carlo computations. The major problems deal with the analysis of the cratering episodes, which introduce discontinuities in the sedimentation profile $T(D_s)$ in excavating all strata deposited during an unknown interval δt. In order to describe correctly 98% of the cratering episodes by our two scenarios, we had to admit an uncertainty of ~10% on the δt values, which is inherent to the graphical method in being related to the requirement of achieving a high "confidence level" in the derived depositional histories. The remaining 2% of the cratering events include very unlikely sequences such as the rapid succession of 2 very large events nearly cancelling each other or a prolonged burial of the whole core well below the production region of the GCRi, followed by a cratering event which brings it back to the surface. The accuracy of the graphical method when applied to the Apollo core tubes will be further limited by experimental uncertainties in the determinations of GCRi profiles, as well as by the poor depth resolution of the measurements (see below).

III. Depositional Histories of the Apollo 15, 16 and 17 Drill Core Tubes

III.1—Chemical normalisation of the experimental GCRi profiles

The experimental GCRi profiles so far available for the 15, 16 and 17 cores have to be corrected for the influence of the local chemical composition, which can greatly vary either with the depth in a given core tube or from one Apollo landing site to the other. Chemical composition variations can indeed affect the GCRi depth dependent production rates in two distinct ways: first these production rates are directly proportional to the concentrations (expressed in weight per cent) of several target nuclei present in the sample under investigation. Second, the average chemical composition of the overlaying material can alter both the intensity and the energy spectrum of the nuclear active particles (protons, neutrons) reaching the sample. We have normalised the experimental F, N, X26 and X31 profiles in the following ways:

1. Neutron fluence. As the F profiles are directly determined from Gd isotopic ratios they are independent of the sample composition. On the other hand neutron absorbing elements present in the surrounding material can drastically reduce the thermal neutron flux reaching the sample. As the concentrations of these elements (rare earths, etc. . .) are highly variable from one site to the other, the thermal neutron production rates were certainly different in the three Apollo core tubes. As a first approximation we used the depth independent normalization factors proposed by Russ (1974), which are about 1.5, 2.90 and 1 for the 15, 16 and 17-cores, respectively.

2. Neon 21. Cosmogenic ^{21}Ne is a high energy spallation product (Reedy and Arnold, 1972), mostly produced by the flux of $\gtrsim$100 MeV secondary neutrons, which is roughly independent of the composition of the overlaying material. In addition, solar cosmic rays contribute to the production of ^{21}Ne at very shallow depths.

   The production of this isotope will drastically depend on the chemical abundance of three major target elements (Mg, Al, Si) in the samples. In all lunar soil samples the Si concentration is roughly constant (~20%). On the other hand, the Al and Mg concentrations are variable. Namely, in the "Mare" type 15 and 17-cores, the Al and Mg concentrations, which are fairly constant from the top to the bottom sections of the cores, are about 7% and 6%, respectively. But in the "highland" type 16-core, the measurements so far reported (Nava, 1973; Nava *et al.*, 1976) show that the Al and Mg concentrations both depart from a mare type composition and also vary along the 16-core (these features are most likely due to a variable admixing of Al-rich and Mg-poor anorthosite). Indeed, for the top and bottom sections of the 16-core, the Al and Mg concentrations range from about 16% to 14% (Al) and 2.5% to 4% (Mg). As the relative contribution of Mg, Al and Si to the production of ^{21}Ne are still difficult to assess, we simply normalized the neon profile to the surface value, in considering that the ^{21}Ne production rate was linearly anticorrelated with the anorthosite content of the sample.

3. Xenon 126. As a high energy spallation product, ^{126}Xe is relatively insensitive to change in the chemical composition of the overlaying material. In addition Ba is the major parent nuclei of this isotope and as suggested by Pepin *et al.* (1974) we simply performed the chemical normalization by expressing the ^{126}Xe concentration per g of Ba.
4. Xenon 131. As this element also partially results from the high energy spallation of Ba, we evaluated the ^{131}Xe concentration in ccSTP/g (Ba). But in addition, at depths ~200 g. cm^{-2}, ^{131}Xe is generated via neutron capture reactions (Reedy, 1976), and consequently the ^{131}Xe concentration should be corrected for the presence of neutron absorbing "poisonous" elements in the surrounding material. However we did not make this last correction, which is difficult to perform, as many elements have to be considered.

Finally we did not take into consideration the possible loss of the cosmogenic rare gases via diffusion processes. In order to minimize this effect, which should be most marked for ^{21}Ne, we intend to replace the N profile by the ^{38}Ar profile, which is expected to be much less affected by diffusive losses. In addition ^{38}Ar has the definite advantage of being easily corrected for the chemical composition of the sample, as it is mainly produced from ^{40}Ca. However, this isotope is somewhat difficult to sort out from the solar wind component.

III. 2—General features of the dimensionless experimental GCRi profiles, and comparison with the theoretical profiles

The four experimental GCRi profiles are available for the A15 core (Fig. 3), from ~60 g.cm^{-2} down to the bottom point of the core [N, X26, X31: Nava *et al.* (1976); F: Russ *et al.* (1972)]. But as illustrated in Figs. 4 and 5, only 3 profiles have been so far reported for the 16-core [N, X26 : Bogard *et al.* (1973); F : Russ (1973)] and the 17-core [N, X26 : Pepin *et al.* (1975); F : Curtis and Wasserburg (1975b)]. In addition we note that the depth resolution of the experimental profiles (distance of 2 successive samples), which ranges from about 40 g.cm^{-2} (Curtis and Wasserburg (1975b) to 20 g.cm^{-2} (Bogard *et al.* (1973), is larger than that (~15 g.cm^{-2}) used in our computations.

With the exception of the very low levels reached by all four profiles in the bottom section of the 15-core, the sets of "dimensionless" GCRi profiles so far available for the A15, A16 and A17 cores show *all the general features of the theoretical ones*, when the appropriate chemical normalization has been performed. The anomaly observed for the bottom section of the 15-core is easily explained by noting that this section is very immature, and consequently it is likely that the plateau contributions had not built up to the mean pre-irradiation

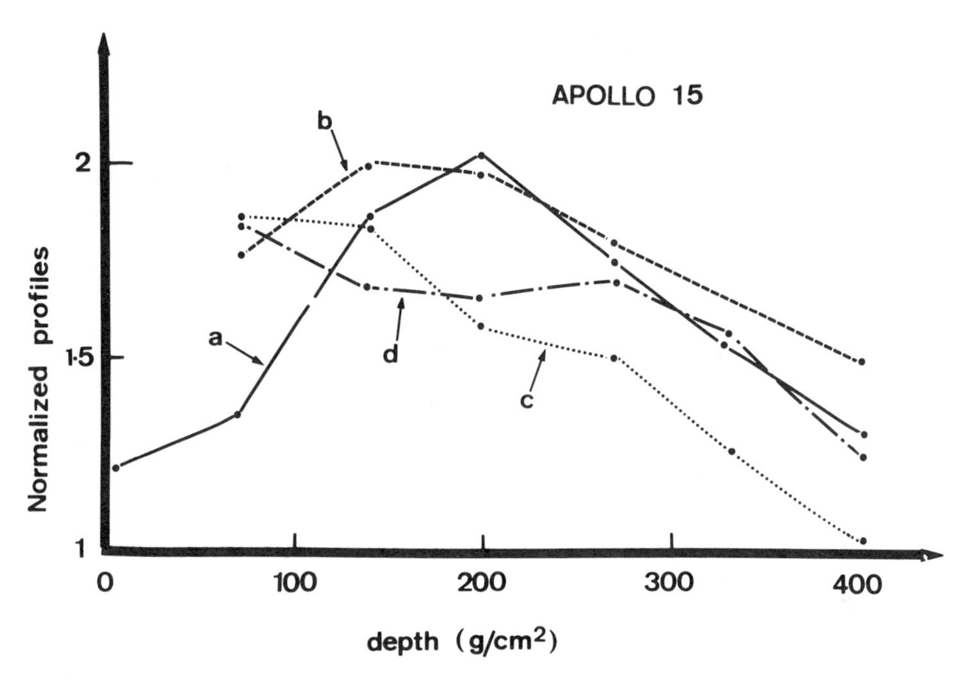

Fig. 3. Normalized cosmogenic profiles in the Apollo 15 core: a. Fluence; b. Neon 21; c. Xenon 126; d. Xenon 131.

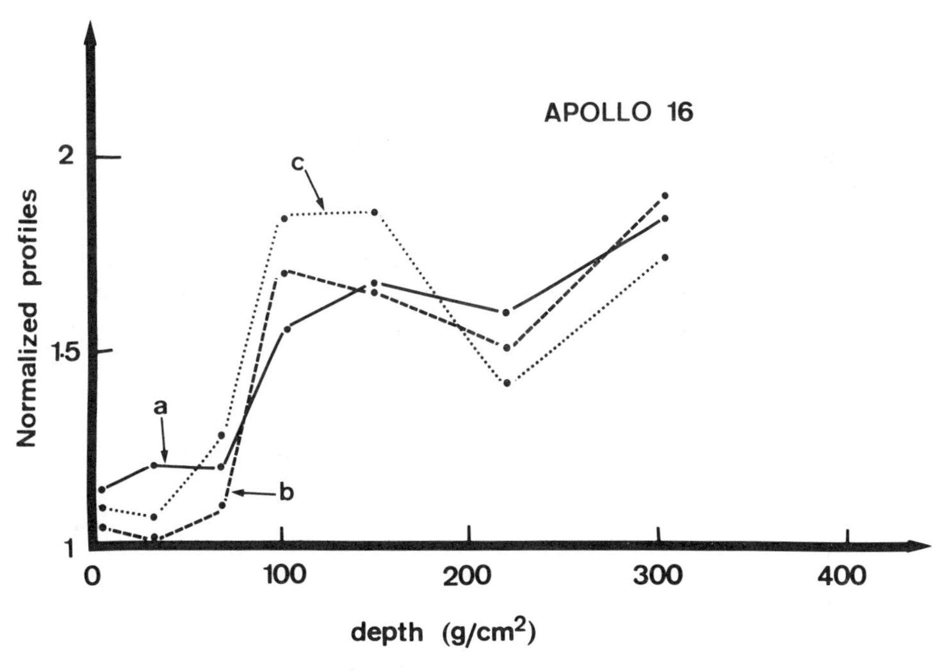

Fig. 4. Normalized cosmogenic profiles in the Apollo 16 core: a. Fluence; b. Neon 21; c. Xenon 126.

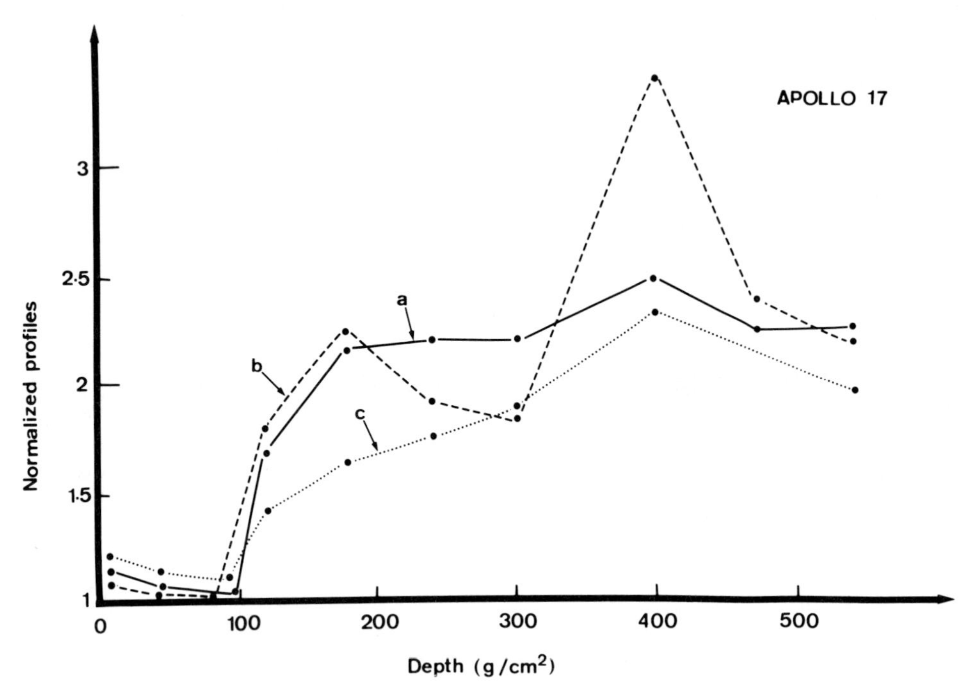

Fig. 5. Normalized cosmogenic profiles in the Apollo 17 core: a. Fluence; b. Neon 21; c. Xenon 126.

levels. One marked jump is prominent in the profiles of the 16-core and 17-core, at depths of ~85 g.cm^{-2} and 100 g.cm^{-2} respectively. As suggested by the N profiles [Pepin *et al.* (1974); Pepin *et al.* (1975)] other jumps might also be present at other positions in the 17-core (depth ~370 g.cm^{-2}) and 15-core (depth ~600 g.cm^{-2}).

We consider that this qualitative agreement between the expected and observed GCRi features when associated with the remark that the Apollo core tubes have been drilled at stable depositional areas justifies the use of the graphical method as outlined in section II.2.

III.3.—The graphical depositional histories of the Apollo 15, 16 and 17 core tubes

Apollo 15 (Σ_{15} ~ 800 ± 200 m.y.) Two major deposition episodes can be inferred from the profiles (F, N, X26, X31) available:

1. A cratering event in the 0–70 g.cm^{-2} top section. The Δ_N, Δ_{X26} and Δ_{X31} values are best interpreted by assuming that during the last 500 m.y. the

elevation of the lunar surface was varying between +20 $g.cm^{-2}$ and −60 $g.cm^{-2}$, as a result of a succession of small ($\delta H_j \lesssim 40$ $g.cm^{-2}$) craters and layers.

2. A fast deposition episode, from 70 to 400 $g.cm^{-2}$, in order to explain the very low Δ_F, Δ_N, Δ_{X26} and Δ_{X31} values observed in this section. In addition, if the existence of two additional peaks in the N and X26 profiles is confirmed by further work at 110 $g.cm^{-2}$ and 230 $g.cm^{-2}$, then these features would strongly suggest that two shallow craters have delayed this fast deposition episode by ~200 m.y. In this last scenario, the δt_2 value of the 70–400 $g.cm^{-2}$ bottom section is ~300 m.y., and the total deposition time scale, Σ_{15}, of the 15-core is thus, Σ_{15} = 800 ± 200 m.y. (Fig. 6).

Apollo 16 (Σ_{16} = 1000 ± 250 m.y.) In this core tube, for which only the F, N and X26 profiles are available, the existence of a void at a depth ~90 $g.cm^{-2}$, as well as severe chemical normalization problems, greatly complicate the graphical interpretation of the data so far available which requires five distinct episodes:

1. A fast deposition episode ($\delta t_1 \sim 150$ m.y.) for the 0–80 $g.cm^{-2}$ top section.
2. A cratering event at a depth of 90 $g.cm^{-2}$, if the void observed below this level reflects a genuine shift of material. Then for this event, $\delta H_2 \sim 100$ $g.cm^{-2}$, $\delta t_2 \sim 400$ m.y., and the shocked material at the bottom of the crater could have triggered the formation of the void.
3. Either a steady deposition episode or a small crater could as well interpret the GCRi features observed in the 100–160 $g.cm^{-2}$ section ($\delta H_3 = 60$ $g.cm^{-2}$; $\delta t_3 \sim 150$ m.y.).
4. A fast deposition episode ($\delta t_4 \lesssim 100$ m.y.) in the 160–240 $g.cm^{-2}$ sections.
5. A cratering event ($\delta H_5 \sim$ 100–200 $g.cm^{-2}$; $\delta t_5 \sim 300$ m.y.) somewhat between 240 and 310 $g.cm^{-2}$ (Fig. 7).

Apollo 17 (Σ_{17} = 1500 ± 350 m.y.) The three GCRi profiles so far available (F, N, X26) possibly reflect six distinct deposition episodes:

1. A fast deposition episode ($\delta t_1 \lesssim 200$ m.y.) in the 0–100 $g.cm^{-2}$ top section.
2. A cratering event in the 100–120 $g.cm^{-2}$ section. The very small Δ_{X26} value as well as the nearly equal value of Δ_F and Δ_N suggest the occurence of a large crater ($\delta H_2 \gtrsim 150$ $g.cm^{-2}$; $\delta t_2 \sim 500$ m.y.).
3. A cratering event in the 120–180 $g.cm^{-2}$ section ($\delta H_3 \gtrsim 100$ $g.cm^{-2}$; $\delta t_3 \sim 400$ m.y.).
4. A fast deposition episode in the 180–320 $g.cm^{-2}$ section ($\delta t_4 \sim 200$ m.y.).
5. A cratering event in the 320–400 $g.cm^{-2}$ section. The 400 $g.cm^{-2}$ point represents a puzzle in the profiles and consequently the δt_5 value ($\gtrsim$200

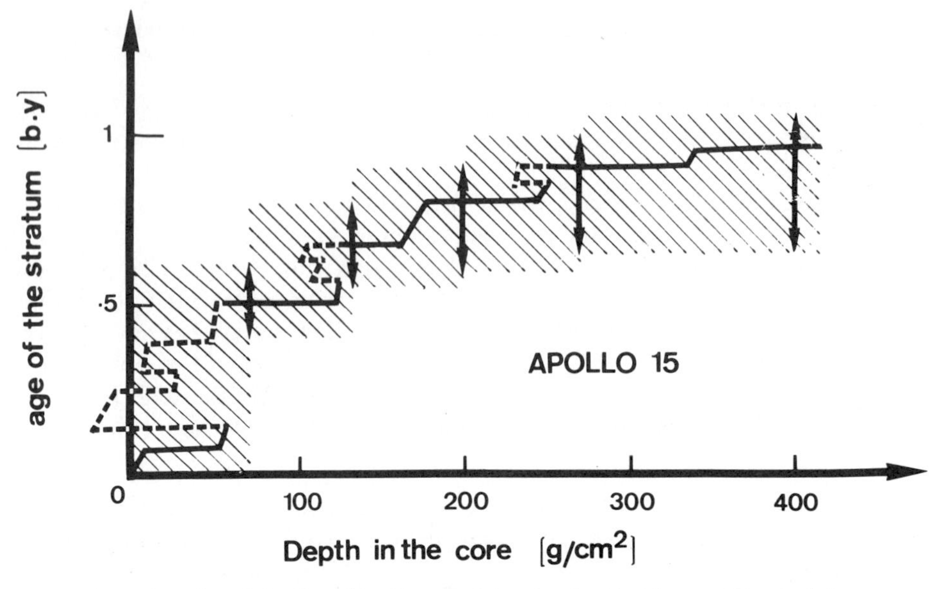

Fig. 6. Results of the "graphical" method for the Apollo 15 core. The shaded area represents the constraints derived from this method on the "age" of core strata (time elapsed since their deposition) as a function of their depth in the core. The curve represented by the solid and dotted lines is a Monte-Carlo generated possible fit to these constraints. The solid line corresponds to the deposition of material now in the core, whereas the material deposited during the dotted-line parts of this history has subsequently been excavated.

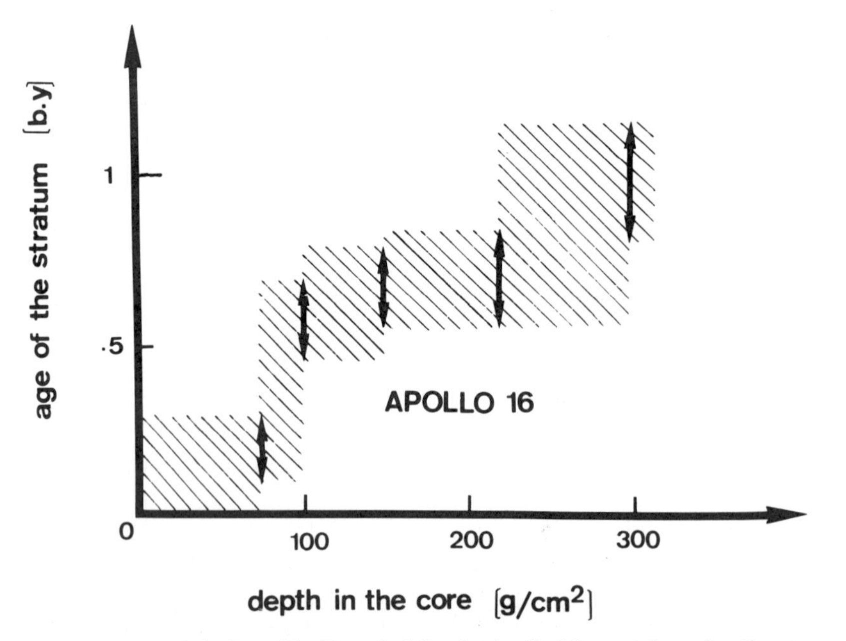

Fig. 7. Results of the "graphical" method for the Apollo 16 core tube, using the same symbolism as in Fig. 6.

m.y.) is poorly determined. Indeed at this point the experimental profiles give $\Delta_N > \Delta_{X26} > \Delta_F$, whereas we expect either $\Delta_F > \Delta_N > \Delta_{X26}$ for a large crater, or $\Delta_{X26} > \Delta_N > \Delta_F$ for a rapid succession of small craters.

6. A fast deposition episode at depths $\gtrsim$400 g.cm^{-2} ($\delta t_6 \sim 200$ m.y.) (Fig. 8).

For each one of the cores, these scenarios give constraints on the variation of the age of a constituent stratum as a function of depth in the core. These variations, which have been reported in Figs. 6, 7 and 8 for the Apollo 15, 16 and 17 cores, reflect a "plausible" depositional history of the core, that necessarily belongs to the "normal" set of the Monte-Carlo solutions (see section II.2).

The error bars reported in Figs. 6, 7 and 8, that reflect the uncertainties attached to each one of these graphical histories, have been evaluated by only considering errors in the experimental data as well as the "model" dependant error of ±10% that results from our choice of a confidence level of 98% for applying the graphical method. Consequently we did neglect in this first approximation attempt to decipher the depositional histories of the Apollo core tubes other uncertainties also attached to our model, that are more conveniently discussed in IV.2, and that mostly deal with the input parameters injected in the ORSOLMIX code and the assumed constancy of the pre-irradiation levels. In

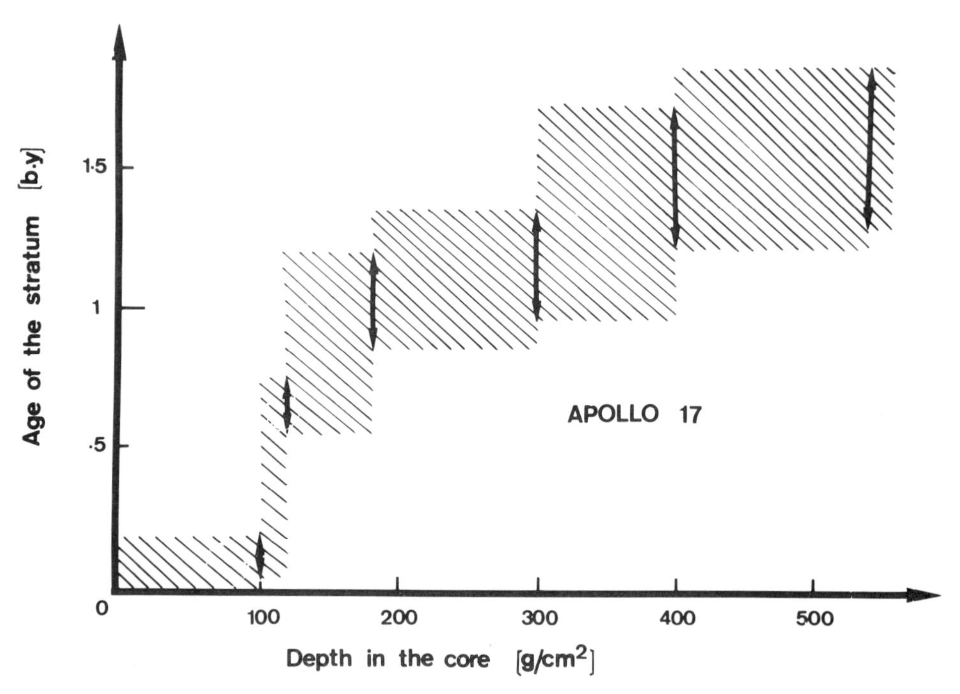

Fig. 8. Results of the "graphical" method for the Apollo 17 core tube, using the same symbolism as in Fig. 7.

addition we assumed that the void observed at a depth of ~90 g.cm^{-2} in the 16-core represents a shift of material but not a loss of soil during the handling of this core.

About the experimental errors, we note that the measurements of the cosmogenic concentrations of ^{21}Ne, ^{26}Xe, ^{131}Xe are now obtained with an accuracy of 2% and 10% respectively, whereas the neutron fluence are determined with an accuracy of better than 2%. Consequently in addition to the relatively poor accuracy in the Xenon measurement most of the experimental errors result from the following causes: only three GCRi profiles have been reported for the 16 and 17 cores. Furthermore, the depth resolution (distance between two successive sampling points) of the measurements (~40 g.cm^{-2}) is not sufficient for several regions of the cores for discriminating between several possible deposition episodes. In addition, Ne could be lost as a result of diffusion processes. Finally, the dependance of both the thermal neutron flux and the ^{126}Xe and ^{131}Xe production rates on the chemical composition of the soil column overlaying the sample is not sufficiently known.

IV. DISCUSSION

IV.1—Comparison of the depositional histories of the Apollo core tubes with the theoretical ones

This comparison reveals the following features:

1. The overall deposition rate of the three Apollo core tubes (300 to 400 g.cm^{-2}/b.y.) agree well with the mean ORSOLMIX value (335 g.cm^{-2}/b.y.). The single disturbing feature is the relative narrow spread in the derived values with respect to that observed for the theoretical ones, which vary from 200 to 1000 g.cm^{-2}/b.y. It could thus be argued that the selection of flat and smooth surfaces for the drilling of the Apollo core tubes has introduced a statistical bias.
2. The complex and unique scenarios proposed for each one of the Apollo core tubes we investigated are not likely to be exactly reproduced by any one of the 200 ORSOLMIX deposition histories. However, the following conclusions can be made: (i) fast deposition episodes resulting in the accumulation of 400 g.cm^{-2} of regolith in less than 200 m.y. (bottom section of the 15-core) are surprisingly frequently found in the theoretical histories (35 out of 200). In most of the theoretical histories, such FDE result from the rapid succession of small (δH < 40 g.cm^{-2}) or medium ($40 \lesssim \delta H \lesssim 100$ g.cm^{-2}) sized layers, and not from the deposition of a single thick layer; (ii) the succession of several ~100 g.cm^{-2} craters and layers during a single depositional history (such as in the 16 and 17 scenarios) is very common (120 out of 200); (iii) on the other hand, a long period (~500 m.y.) during which the surface elevation varies within ~100 g.cm^{-2} is very unlikely (4 out of 200). In fact, such a peculiar situation can

only result from a dynamic equilibrium between several small ($\delta H < 40$ g.cm^{-2}) craters and layers.

Finally we were able to find theoretical histories matching the constraints derived for the depositional histories of the 15, 16 and 17-cores by using the following artifices: for the 16-core, we simply shifted the t and H origins of one of the 200 theoretical histories (solid line in Fig. 6); for the 15-core and the 17-core, we joined parts of two distinct theoretical histories at a depth of respectively 100 and 200 g.cm^{-2}. However, we note that these fits should not be considered as *the* depositional histories of the Apollo core tubes, but only as particular solutions of the set of integral equations involving the experimental cosmogenic profiles, which are just as likely as any other history included within the error bands in Figs. 6, 7 and 8.

IV.2—Limitations of the present studies

Several additional uncertainties discussed below, and concerning the validity of the major assumptions underlaying the ORSOLMIX computations, could possibly contribute in limiting the accuracy of the graphical method:

1. "Constancy" of the plateau contributions. These contributions were added as constant pre-irradiation levels to the GCRi concentrations accumulated during the burial history of the grains. This hypothesis, which was used as a first approximation in the present work, is certainly not correct at all for immature lunar soils (i.e., the bottom section of the 15-core). This point is further strengthened by the spread in the GCRi concentrations measured for surface soil samples, that only reflect the plateau contributions (Langevin and Maurette, 1976), and which vary within ±30% of the corresponding mean values. We consider such as spread as indicating different degree of "intrinsic" GCRi maturity. We are currently trying to analyse possible correlation between these GCRi maturities and other surface maturation indices such as solar flare track densities, which are only accumulated when the grains reside in the most superficial layers of the regolith (depths $\lesssim$ 1 cm). A preliminary result of this study indicates that such a correlation is indeed quite loose, as a change of 10 in the indices of surface exposure only triggers a variation $\lesssim$3 in the GCRi concentrations. Although loose, such types of correlations should soon yield a "second order" approximation for the plateau contributions, which would in turn greatly help in improving the accuracy of the deposition histories and the meaningfulness of the rare gas model ages so far used in lunar science.
2. Production functions of the GCRi. The thermal neutron flux alone has been directly measured in the lunar regolith as a function of depth (Woolum *et al.*, 1975). However, theoretical estimations of spallation production functions such as that of ^{21}Ne, ^{126}Xe and part of ^{131}Xe are

rather reliable, as these types of cross-sections vary relatively smoothly as a function of the atomic number of the product nuclei and the energy of the particle. We thus consider that these production functions should be accurate within ±10%.

3. Meteoritic input parameters (meteorite mass spectrum and crater shape parameters). Last year, we already pointed out (Langevin and Maurette, 1976) that we could only account simultaneously for the very distinct effects due to solar wind ions, solar flare and galactic cosmic rays in the lunar regolith by using a unique set of meteoritic input parameters, which is compatible with the recent work of Neukum and König (1976). Consequently, as such a fit in the theoretical estimations of solar wind, solar flare and galactic cosmic ray effects can hardly be accidental, we concluded that our choice of meteoritic input parameters was essentially a valid one.

IV.3—Comparison with other deposition models

Parameter fitting models (PFM) have already been used to infer the deposition history of the Apollo core tubes:

1. *Apollo 15.* By fitting the F profile, Russ *et al.* (1972) suggest a steady deposition period during about 400 m.y., which would be followed by a 450 m.y. *in situ* exposure of the accumulated layers. From rare gas measurements, Pepin *et al.* (1974) and Reedy (1976) infer a similar scenario, with a total deposition time scale Σ of 900 and 500 m.y. respectively.
2. *Apollo 16.* From the N profile, Russ (1973) proposes a "two-slabs" scenario, in which two slabs were successively deposited 400 m.y. and 900 m.y. ago.
3. *Apollo 17:* From the N profile, Curtis and Wasserburg (1975b) infer a very short (~200 m.y.) "three-slabs" deposition model. On the basis of rare gas studies, Dragon *et al.* (1975) also propose a similar scenario, which extends on a much longer time scale of ~900 m.y.

Our graphical deposition histories thus yield deposition time scales ($\Sigma_{15} \sim 800 \pm 200$ my; $\Sigma_{16} \sim 1000 \pm 250$ my; $\Sigma_{17} \sim 1500 \pm 350$ my) which are in rough agreement with the values derived by PF models from *rare gas measurements*. They do not agree with the predictions of PF models based on fluence measurements, except in the case of the A-15 core. In this case, the agreement is somewhat fortuitous as our pre-irradiation levels and our production rates are respectively higher and lower than those previously used. On the basis of the deposition time scale alone, it could be concluded that our MC model yields no significantly new result.

However, the deposition time scales are far from telling the whole deposition history of a lunar core. Indeed, there are basic major differences in our interpretation of the deposition histories leading to these time scales, which are:

1. The four "constant" pre-irradiation levels that we apply to all the core tubes are no longer completely arbitrary parameters: they directly result from our choice of production rates and meteoritic mixing parameters, the latter having already been checked by predicting the accumulation of other depth dependent effects observed in the regolith (solar flare tracks, solar wind amorphous coatings, . . .). Furthermore, they are in good agreement with the values observed in surface lunar soils.
2. In PF models, deposition is considered either as proceeding at a constant rate and/or via thick "slabs", or stopping for long periods of time. *In addition, in these models excavating events are not considered.* Consequently, discontinuities in the profiles are attributed either to different pre-irradiation levels or long *"in-situ"* exposures. In striking contrast, in our model, such discontinuities are attributed to medium and large sized cratering events ($\delta H_c \gtrsim 100$ g.cm^{-2}), and this interpretation looks more compatible with the general absence of very thick ($\gtrsim 100$ g.cm^{-2}) stratigraphic units in the Apollo core tubes (the only exception being the top section of the 17-core). The validity of our interpretation could be checked by searching for shock features in grains extracted from the 100–150 g.cm^{-2} section of the 17 core, and possibly, from the 80–100 g.cm^{-2} section of the 16-core.

In addition, all the previous PF models try to infer the depositional history of lunar core tubes from the consideration of a given GCRi profile only.We already pointed out that the complex set of integral equations that relates the GCRi profiles to the corresponding deposition history cannot be solved by considering one GCRi profile only, as the number of plausible solutions is not sufficiently constrained. Consequently, our general analytical procedure that considers simultaneously all CGRi profiles so far available is likely to give the "most" plausible depositional history of the Apollo 15, 16 and 17 core tubes.

We also note that the MC model independently proposed by Arnold (1975) yields similar *mean* total deposition times for lunar core tubes when the same meteoritic mixing input parameters are injected in both codes. However, average values are of little help in deciphering the unique deposition history at a given point of the lunar surface and our model is the first meteoritic mixing model giving a detailed plausible scenario for the unique and distinct depositional history of the lunar core tubes so far returned to the earth.

Finally, we discuss below the contributions of solar flare track studies in deciphering the depositional history of the Apollo core tubes, as this controversial issue was raised by one of the referees. In our view the most striking advantage of track studies is to delineate clear boundaries between the constituent strata of a given core, by plotting the solar flare track densities registered in individual grains as a function of depth in the core. Such type of work when conducted with a depth resolution of $\sim$ 1 mm is very useful in revealing a layering which cannot be detected by other methods (Goswami *et al.*, 1976). On the other hand, we believe that previous studies (e.g., Arrhenius *et al.*, 1971) dealing with the use of track data for determining the mean sedimentation rates of lunar core tubes are meaningless for the two following reasons:

1. As a result of the very short ranges (≲1 mm) of solar flare ions in the lunar regolith, solar flare tracks are only registered during the plateau history of the grains. As most of the constituent strata of lunar core tube are mature, their constituent grains have been exposed more than once in the solar flare cosmic rays during their plateau history. Consequently, the track densities registered in the grains reflect a complex history of "pre-irradiation" which cannot be related at all to the last top surface exposure of the present day parent stratum of the grains on the lunar surface, τ_s, from which a meaningful total deposition time scale could be obtained just by summing up the τ_s values of all strata.
2. Any method based on track data cannot account at all for the effects of large craters that considerably delay the formation of the core by excavating a great number of strata. In fact, as indicated in one of our previous paper (Duraud *et al.*, 1975), up to 60% of the strata deposited on the lunar surface should have been on the average "recycled" by excavation.

CONCLUSIONS

The Monte-Carlo model developed at Orsay has already accounted with some success for the accumulation of the very distinct depth dependent effects produced in the lunar regolith by solar wind ions, solar cosmic rays, and galactic cosmic rays. In addition some of these results are consistant with those derived independently by Arnold (1975) by means of another complex, Monte-Carlo code which operates in a completely different way when the same set of "meteoritic" input data are injected in both codes. Consequently, we consider that uncertainties specifically related to our choice of meteoritic input parameters should not be important. In addition, we conclude that the meteoritic mass spectrum as well as the three distinct types of nuclear particle fluxes considered in our work should have been fairly constant (within a factor of 2) over a time scale of several billion years (this does not preclude cyclic variations on a shorter time scale).

However, major uncertainties are still attached to our deposition histories, that are mostly related to the GCRi production rates, to our postulated constancy of the pre-irradiation levels, and to the relatively poor depth resolution of the experimental measurements.

It is interesting to note that changes in the GCRi production rates would simply multiply all time constants by a common amount, without changing the nature of the deposition episodes. In order to reduce the uncertainties attached to the pre-irradiation levels, it will be necessary to investigate possible correlation between such levels, which can be assimilated to CGRi "maturity indices," and several other indices of surface exposure maturity (i.e., track densities, glassy agglutinate content, and so forth). Finally the GCRi profile measurements should be conducted on a depth scale of about 10 $g.cm^{-2}$ at several critical positions in the Apollo core tubes (100–150 $g.cm^{-2}$ section and 400 $g.cm^{-2}$ region in the 17-core; 0–80 $g.cm^{-2}$ section in the 15-core; regions close to the void observed at ~90 $g.cm^{-2}$ in the 16-core).

In spite of these limitations, the depositional histories inferred in the present work (see Figs. 6, 7 and 8) are potentially important for tracing back the past activity of the ancient solar wind and solar flare and galactic cosmic rays, as well as that of other lunar surface processes, by applying our so-called "lunar-skin" sampling technique (Borg *et al.*, 1976; see also our companion paper, Dartyge *et al.*, this volume).

It would be quite interesting if the lunar core tubes which possibly will be returned by future Luna missions were drilled at the bottom of a relatively deep crater ($\delta H > 200$ $g.cm^{-2}$). Indeed in this case our model predicts much higher near surface values for the GCRi concentrations, and the bottom section of such a core tube should sample material deposited more than 2 b.y. ago on the top surface of the regolith.

Acknowledgments—One of us (Y. L.) thanks J. R. Arnold and P. Russ, III for very stimulating comments and relevant data, and G. J. Wasserburg for useful discussion. One of us (M. M.) is grateful to R. Klapisch and D. Lal, for their support and interest.

REFERENCES

Arnold J. R. (1975) A Monte-Carlo model for the gardening of the lunar regolith. *The Moon* **13**, 159–172.

Arrhenius G., Liang S., McDougall C., Wilkening L., Bhandari N., Bhat S., Lal D., Rajagopalan G., Tamhane A. S. and Venkatavaradan V. S. (1971) The exposure history of the Apollo 12 regolith. *Proc. Lunar Sci. Conf. 2nd*, p. 2583–2598.

Bibring J. P., Borg J., Burlingame A. L., Langevin Y., Maurette M. and Vassent B. (1975) Solar wind and solar flare maturation of the lunar regolith. *Proc. Lunar Sci. Conf. 6th*, p. 3471–3493.

Bogard D. D., Nyquist L. E., Hirsch W. C. and Moore D. R. (1973) Trapped solar and cosmogenic noble gases abundances in Apollo 15 and 16 deep drill samples. *Earth Planet. Sci. Lett.* **21**, 52–69.

Borg J., Comstock G. M., Langevin Y., Maurette M., Jouffrey B. and Jouret C. (1976) A Monte-Carlo model for the exposure history of lunar dust grains in the ancient solar wind. *Earth Planet. Sci. Lett.* **29**, 161–174.

Curtis D. B. and Wasserburg G. J. (1975a) Neutron fluence in lunar soils (abstract). In *Lunar Science VI*, p. 172–174. The Lunar Science Institute, Houston.

Curtis D. B. and Wasserburg G. J. (1975b) Apollo 17 neutron stratigraphy. *The Moon* **13**, 185–227.

Dartyge E., Duraud J. P., Langevin Y. and Maurette M. (1978) A new method for investigating the past activity of ancient solar flare cosmic rays over a time scale of a few billion years. *Proc. Lunar Planet. Sci. Conf. 9th*. This volume.

Dragon J. C., Johnson N. L., Pepin R. O., Bates A., Coscio M. R. and Murthy V. R. (1975) The Apollo 17 deep drill core: A possible depositional model (abstract). In *Lunar Science VI*, p. 196–198. The Lunar Science Institute, Houston.

Duraud J. P., Langevin Y., Maurette M., Comstock G. M. and Burlingame A. M. (1975) The simulated depth history of dust grains in the lunar regolith. *Proc. Lunar Sci. Conf. 6th*, p. 2397–2415.

Fireman E. L. (1974) Regolith history from Cosmic Ray produced nuclides. *Proc. Lunar Sci. Conf. 5th*, p. 2075–2092.

Goswami J. N., Braddy D. and Price P. B. (1976) Microstratigraphy of the lunar regolith and the compaction ages for lunar breccias. *Proc. Lunar Sci. Conf. 7th*, p. 55–74.

Langevin Y. and Maurette M. (1976) A Monte-Carlo simulation of galactic cosmic rays effects in the lunar regolith. *Proc. Lunar Sci. Conf. 7th*, p. 75–91.

Nava D. F. (1973) Chemical composition of core-soil samples from the Apollo 15, Apollo 16 and Luna 20 lunar sites (abstract). In *Lunar Science IV*, p. 555–557. The Lunar Science Institute, Houston.

Nava D. R., Lindstrom M. M., Schumann P. J., Linstrom D. J. and Philpotts J. A. (1976) Chemical compositions of fines from the Apollo 16 deep drill core (abstract). In *Lunar Science VII*, p. 613–615. The Lunar Science Institute, Houston.

Neukum G. and König B. (1976) Dating of individual lunar craters. *Proc. Lunar Sci. Conf. 7th*, p. 2867–2881.

Pepin R. O., Basford J. R., Dragon J. C., Coscio M. R. and Murthy V. R. (1974) Rare gases and trace elements in Apollo 15 drill core fines. *Proc. Lunar Sci. Conf. 5th*, p. 2149–2184.

Pepin R. O., Dragon J. C., Johnson N. L., Bates A., Coscio M. R. and Murthy V. R. (1975) Rare gases and Ca, Sr, Ba, in Apollo 17 drill core fines. *Proc. Lunar Sci. Conf. 6th*, p. 2027–2055.

Reedy R. C. (1976) Spallation rare gas production calculations and studies of lunar surfaces processes (abstract). In *Lunar Science VII*, p. 721–723. The Lunar Science Institute, Houston.

Reedy R.C. and Arnold J. R. (1972) Interaction of solar and galactic cosmic rays with the moon. *J. Geophys. Res.* **77**, 537–555.

Russ P., III (1973) Apollo 16 neutron stratigraphy. *Earth and Planet. Sci. Lett.* **19**, 275–289.

Russ P., III (1974) Neutron stratigraphy in the lunar regolith. Ph.D. thesis, California Institute of Technology, Pasadena.

Russ P., III, Burnett D. S. and Wasserburg G. J. (1972) Lunar neutron stratigraphy. *Earth and Planet. Sci. Lett.* **15**, 172–186.

Woolum D. S., Burnett D. S., Furst M. and Weiss J. R. (1975) Measurement of the lunar neutron density profile. *The Moon* **12**, 231–250.

Proc. Lunar Planet. Sci. Conf. 9th (1978), p. 1787–1800.
Printed in the United States of America

Regolith irradiation stratigraphy at the Apollo 16 and 17 landing sites

GHISLAINE CROZAZ

Department of Earth and Planetary Sciences and
McDonnell Center for the Space Sciences, Washington University, St. Louis, Missouri 63130

Abstract—Additional fossil track measurements in the Apollo 17 deep drill stem, as well as detailed track studies in section 3 of the Apollo 16 deep drill core are reported. Although the upper part of the Apollo 17 core seems to have accreted rapidly, no evidence for a rapid accretion of the lower part, as postulated by some authors, is found. Despite the apparent inhomogeneity of section 60003, its track record is unexpectedly homogeneous; all levels are heavily irradiated and emplacement of big slabs of material is not favored.

INTRODUCTION

The deep drill cores which are representative of the top 2 to 3 m of the lunar surface are the best samples available to study the long-term evolution of the regolith and they continue to be studied by a variety of techniques.

We have pursued here our attempts to establish the chronologies and modes of deposition of lunar cores using the nuclear particle track method. Additional measurements in the Apollo 17 deep drill stem as well as detailed studies of section 3 of the Apollo 16 deep drill stem are reported. The latter is part of a consortium study also including noble gases (D. Heymann and his group) and ferromagnetic resonance (R. Housley and colleagues) investigations.

Tracks in lunar soils are produced by both solar flare and galactic heavy cosmic ray nuclei with respective penetration depths in lunar soils of a few millimeters and a few tens of centimeters. Because of the complexity of the regolith processes, it is common in a given layer to find grains with track densities spanning a range of about three orders of magnitude. These grains have obviously suffered very different irradiation exposures, only part of which occurred *in situ* (i.e., at the sampling site). In special cases, which fortunately are not too uncommon, it is possible to distinguish with certainty between tracks acquired *in situ* and those acquired at a previous location (pre-irradiation).

The track record can be unambiguously deciphered when a layer a few centimeters thick, containing grains initially devoid of tracks at the time of deposition, remains undisturbed, except for its topmost part, until it gets buried by another depositional layer. Minimum track densities in grains from such a layer exhibit a characteristic smooth decrease with depth, a consequence of the rapid attenuation of galactic heavy cosmic ray nuclei. Such layers are a common feature in lunar cores. However, in order to recognize them, it is necessary to carry long and painstaking analyses on samples separated by only a few millimeters using either individual grains or impregnated core sections. Such layers have been recognized, for example by Crozaz and Dust (1977) in the double drive tube collected near the Apollo 16 landing site. The most spectacular example of an undisturbed layer is the 60 cm coarse-grained layer occurring near the top of the Apollo 17 deep drill stem (Crozaz and Plachy, 1976).

Some authors (Price *et al.*, 1975; Goswami *et al.*, 1976, 1977), studying impregnated sections, have observed abrupt changes in the track record over distances less than 1 mm and have argued that they may correspond to discrete depositional events. However, in this case, the observations do not lead to a unique and unambiguous interpretation. The observed microlayers, indeed, may be caused by discrete depositional events but they could equally be explained as ancient surfaces of thicker layers intensely reworked by the micrometeorite bombardment before burial by the next layers or even as the result of the deposition of an inhomogeneously mixed thicker layer (a few centimeters or more) of mature material (whose grains contained tracks, in similar amounts as presently found, before deposition at the sampling site). Essentially, the question is whether the tracks were acquired *in situ* or not. In some cases, as for the top of the Apollo 17 deep drill stem (see below) independent evidence may provide the necessary additional constraint which helps to decide where the irradiation took place.

Of the various track indices that have proven useful in defining the irradiation state of a lunar soil, we have used the track density distribution as an index of mixing and reworking. All the time constraints on the layering chronology, that have been derived from analyses carried out in our laboratory, have been reached using the minimum track density method in which a large number of crystals are examined and only those with the lowest densities are used to calculate a model age. This has to be kept in mind when comparing results obtained in different laboratories. Some groups prefer to rely on the quartile track density (density above which 75% of the grains are found). We have avoided using it as it is sensitive to the amount of irradiation that the material received prior to final emplacement and can thus hardly be used to establish exposure times between the deposition of successive layers.

EXPERIMENTAL

Samples from the Apollo 17 deep drill averaged about 50 mg in weight each whereas samples from section 3 of the Apollo 16 deep drill only averaged about 1 mg (the same samples from 60003 were analyzed for ferromagnetic resonance by Housley and his group). Only crystals larger than 150 μm were selected in 70007, 70006 and 70004 whereas all suitable crystals, regardless of their sizes, were picked from the 60003 samples. Individual feldspar grains were mounted and etched as described by Crozaz and Dust (1977). As noted by these authors, feldspars are the preferred track detectors as their spallation track production rate is relatively low and their iron track registration is relatively homogeneous. In view of the large number of 60003 samples, track densities in these samples were determined using only an optical microscope. Two series of observations were made after total etching times of one and two hours in a NaOH solution (6N).

Nuclear particle tracks in 70007, 70006 and 70004 and the deposition of the Apollo 17 deep drill core

Previous work (Crozaz and Plachy, 1976) on the top two sections (70009 and 70008) of the Apollo 17 deep drill has shown that the following depositional model for the topmost part of this core is most plausible: emplacement of the 60 cm coarse-grained layer, now located between ~20 and 80 cm, some 100 m.y. ago associated either with Camelot Crater of the Central Cluster Craters. At that time the present top of the coarse-grained layer was capped by some 25 cm of material which were recently (~2 m.y.) excavated. The resulting depression has been partially (~18 cm) and gradually filled. This interpretation takes into account track and thermoluminescence data (Crozaz and Plachy, 1976) as well as the ^{22}Na and ^{26}Al observations of Fruchter *et al.* (1976). The track densities in the coarse-grained material are consistently low and decrease with depth as expected for an undisturbed layer whereas track densities in the top fine-grained 19 cm of the core range over three orders of magnitude, typical of most lunar soils. In this case, because of the ^{26}Al data which indicated that the top 18 cm were emplaced during the last 2 m.y., the bulk of the tracks were acquired in a prior location. It it had not been for the constraint provided by the ^{26}Al data, it would have been much more difficult on the basis of the tracks alone to distinguish between a mode of slow deposition where thin layers are successively emplaced and the mode of rapid deposition of largely pre-irradiated material which is preferred here.

Isotopic measurements by Curtis and Wasserburg (1977) were conducted on the entire length of the Apollo 17 deep drill and led these authors to suggest that the entire core accumulated rapidly during the past 200 m.y. Two sources of materials were postulated, each with distinctly different average predepositional neutron fluences; a lightly irradiated source for the shallow section of the core and a heavily irradiated source for the deep section. Whether or not this interpretation, which is highly model dependent, is correct is a question which may be answered when track studies are conducted in the deeper part of the core. In this context it is interesting to note that Pepin *et al.* (1975) using rare gas and isotopic measurements concluded that their data could be explained by models involving long-term continuous accretion as well as models of very recent rapid accumulation. They favored long-term continuous accretion as they could not find geological evidence for the deposition of the entire regolith section during the short time interval advocated by Curtis and Wasserburg (1977).

The top three sections of the core (70009, 70008 and 70007) have been extensively investigated by petrologists and geochemists. Ehmann and Ali (1977), Taylor *et al.* (1977) and Vaniman and Papike (1977) all observed an increase of the highland component with depth in core section 70007. This evidence as well as the observations reported above led Taylor *et al.* (1977) to propose a two-stage model in which the lower part of the core (70007 part enriched in highland material and below) was deposited when the Central Cluster Craters formed whereas the upper part was deposited when the 700 m diameter Camelot Crater located ~700 m west of the sampling site was excavated. The abundant non-mare lithic fragments in the lower part of 70007 (and presumably below) are interpreted as fragments of the projectiles that made the Central Cluster Craters. If the Central Cluster Craters were produced by secondary impacts from the distant crater Tycho, as argued by Arvidson *et al.* (1976), it may be possible to infer the nature of the Tycho ejecta. Combined petrological (Taylor), chemical (Schmitt) and track (Crozaz) investigations of lithic fragments are currently in progress in order to test this hypothesis.

In this work, track densities in feldspars from three levels in 70007, one in 70006 and five in 70004 have been measured. The results are shown in Figs. 1 and 2.

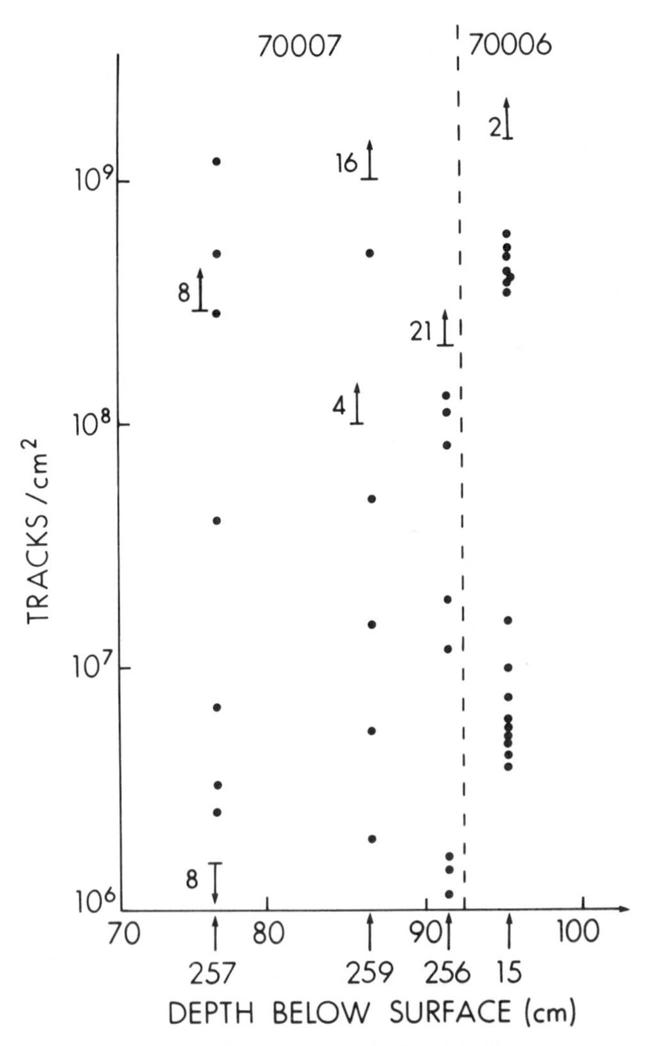

Fig. 1. Track densities in individual feldspar grains of sections 70007 and 70006 of the Apollo 17 deep drill core. Lower numbers are sample numbers.

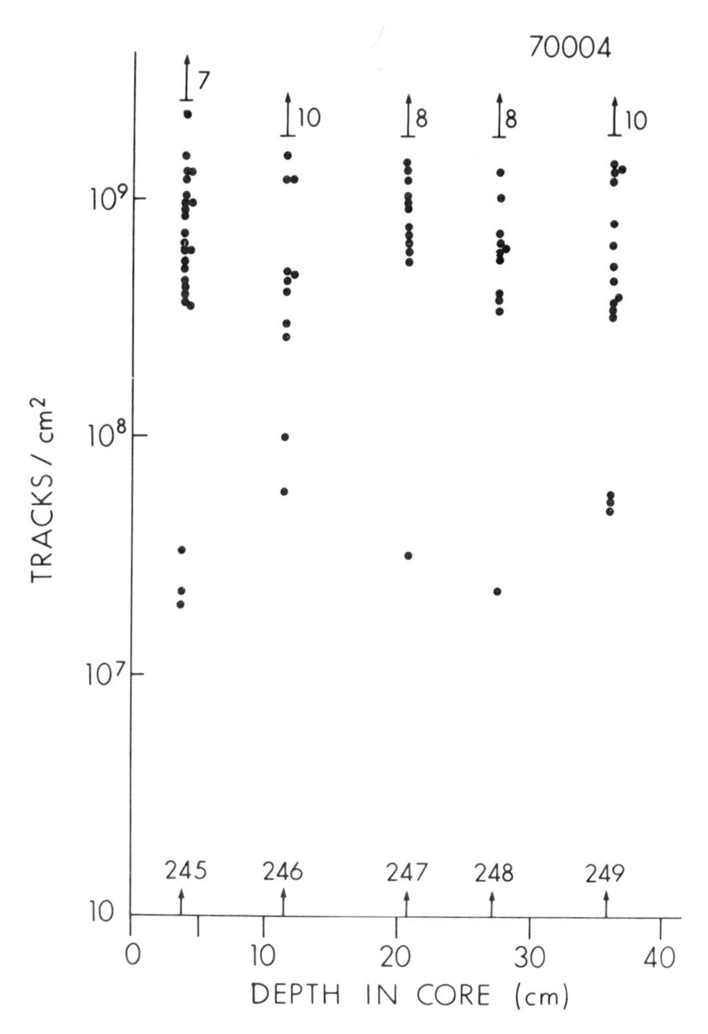

Fig. 2. Track densities in individual feldspar grains of section 70004 of the Apollo 17 deep drill core. Numbers above the abcissa are sample numbers.

In section 70007, only sample 257 belongs to the unique coarse-grained layer mentioned above. And, indeed, from the track viewpoint this sample was deposited at the site at the same time as the rest of the coarse-grained layer as evidenced by the fact that eight crystals (~⅓ of total) are devoid of tracks, i.e., have extremely low track densities in agreement with the ones expected at this depth if sample 257 is part of the same undisturbed layer recognized earlier in 70008 (25 to 63 cm below the lunar surface) and the bottom of 70009 (20 to 25 cm below the surface) (Crozaz and Plachy, 1976). However, compared to the 70008 samples where less than 5% of the feldspars had track densities in excess of $10^6 \cdot cm^{-2}$, sample 257 contains over 60% of feldspars with track densities in excess of this value. Presumably, the fraction of pre-irradiated grains increases as the lower boundary of the coarse-grained layer is approached. Samples 70007,259 and 70007,256 located at the bottom of section 7 have track densities ranging from $10^6 \cdot cm^{-2}$ to more than $10^9 \cdot cm^2$ with over 80% of the crystals being heavily irradiated (track densities in excess of $10^8 \cdot cm^{-2}$). If the low track density crystals in these two samples were devoid of tracks at the time of their deposition, their emplacement must have preceded the deposition of the coarse-grained layer by a few millions of years.

The material at the top of section 6 (70006,15) was apparently deposited at still an earlier time. The reason for this assertion is the strong clustering of track densities observed below $10^7 \cdot cm^{-2}$. The track density distribution is definitely bimodal. No grains are observed with track densities between 2×10^7 and $3 \times 10^8 \cdot cm^{-2}$. The soil comprising 70007,15, at the time of its deposition, most probably included a heavily irradiated component as well as a fresh component devoid of tracks. The low track densities observed today in 70007,15 can be interpreted as having been acquired after deposition of this part of the core but before the deposition of the overlying coarse-grained layer. This interpretation thus favors, at least for this part of the core, a model where layers are successively deposited. Time intervals of only a few m.y. between successive deposition are sufficient to explain the track data and fall within the constraints of the preferred model of Curtis and Wasserburg (1977).

The material in section 70004 is uniformly very rich in tracks (less than 10% of the crystals have less than 10^8 tracks$\cdot cm^{-2}$). Even the crystal with the lowest track density (in sample 70004,245) exhibits a track gradient indicative of exposure to solar flares. Essentially all the crystals analyzed in section 70004 were once exposed at the very lunar surface. If indeed the whole core was deposited during the last 200 m.y., the soil in 70004 was irradiated at another location than the sampling site.

On the basis of these data alone, it is not possible to exclude the possibility that the lower part (below the coarse-grained layer) of the Apollo 17 deep drill stem was deposited in a single event. However, if it were, it should be possible to find crystals (however few they are) which exhibit the same track density dependence as a function of depth as the coarse-grained layer. None have been found yet. The resolution of this problem will have to await further detailed measurements. However, it is worth mentioning that the apparent maturity of the soils below the coarse-grained layer based on track distributions is also reflected in the I_s/FeO measurements of Morris *et al.* (1978).

The track record and the mode of deposition of section 60003

Section 60003 is part of the 224 cm Apollo 16 deep drill core collected ~100 m southwest of the landing site. The deep drill core consists of six sections (60002–60007) which Vaniman *et al.* (1976) have divided into four major stratigraphic units, A, B, C and D based on model abundances of clasts sizes and compositions. Unit D consists only of the top 6 cm (upper part of 60007) and is followed by unit C which extends down to 55 cm (lower part of 60007 and entire section 60006). Unit B covers the depth interval 55–190 cm (section 5—which has been disturbed and has lost material—60004, 60003 and the upper 6 cm of 60002) and unit A (190–224 cm) only includes the rest of section 60002.

Russ (1973), studying Gd and Sm isotopes, established that the low energy neutron fluence increases monotonically with depth throughout the core and concluded that accretion was the dominant process at this location. However, his data could not distinguish between two very different modes of accretion. The data may be fit by a model of continuous accretion (up to the present) of pre-irradiated material at a rate of ~0.5 cm/m.y. as well as models involving as few as two slabs of material in which the first slab could have been deposited as early as 10^9yrs ago. Track studies have tended to support the former model whereas ferromagnetic resonance studies have supported the latter.

Track studies, particularly in impregnated sections, are time consuming and have been conducted only for a small number of locations along the core. The ones that are relevant to the problem of regolith deposition are briefly summarized below.

The first detailed study of a 3 cm impregnated section from the upper part of 60003 was carried out by Price *et al.* (1975) who claimed that they could recognize both micro (≤1 mm) and mini (~several millimeters) layers. A similar study by Goswami *et al.* (1976) in a 3 cm section from 60002 led to the same conclusions. The authors believed that they could at least recognize one discrete 6 mm-depositional event. A detailed study of 12 impregnated sections crossing several of the stratigraphic boundaries of Duke and Nagle (1976) was undertaken by Blanford and Morrison (1976). This work has been recently extended by Blanford and Wood (1978). Their data show that two and possibly three of the contacts have time significance, indicating again that a model of successive deposition of rather small layers may be more appropriate than models which postulate the deposition of large slabs of material. Crozaz and Dust (1977) analyzing only six locations in core section 60007 concluded that core 60007 consists largely of pre-irradiated material most likely derived from a number of local events. Fleischer *et al.* (1974) had previously examined this core and had reached the same conclusions, including an accretion rate of ~0.3 cm/m.y.

In contrast with the conclusions derived from track studies, Gose and Morris (1977) believe that the entire core section below 13 cm was deposited in a single impact event and subjected to meteoritic gardening for about 450 m.y. They see abrupt changes in I_s/FeO (FMR signal normalized to the total iron content) at 13 cm (close to the contact C-D) and 190 cm (at the boundary A-B). They interpret contact C-D as a fossil surface and contact A-B as a stratigraphic boundary without time significance (in contradiction with the interpretation of Blanford and Morrison, 1976).

Core section 60003 which has been extensively studied here is characterized by the presence of intricately laminated and marbled regions of light and dark lithologies. When it was opened, it broke into two parts, one of which was processed according to usual procedures. Ferromagnetic resonance studies on this part have been reported by Housley *et al.* (1976) and Morris and Gose (1976) and they indicate the same trend: most of the soils appear mature with the exception of the soils at each end (above 8 cm and below 28 cm in the section) which are submature.

The second part was carefully dissected in order to obtain as pure aliquots as possible of the various lithologies. Ninety-five small samples were collected of which 57 were studied by the track method.

The stratigraphic description of section 60003 (Duke and Nagle, 1977) as well as the location of the 95 special samples are shown in Fig. 3. As indicated on the figure, a number of units are marbled (units 27, 25, 23, 21, 17 and 15 in order of increasing depth). Track density distributions obtained in 57 of the special samples are presented in Figs. 4, 5 and 6. The number of track density distributions presented here has been reduced in order to simplify the presentation. Track densities, in all massive units and in a limited number of other cases, have been grouped according to depth. This treatment is justified because, in a massive layer, there is no lateral variation of the track densities. A general observation pertaining to the whole length of the section is that all samples are heavily irradiated. These are all typical mature lunar soils with track densities ranging from 10^6 to $10^9 \cdot cm^{-2}$; densities above $10^8 \cdot cm^{-2}$ could not be resolved as an optical microscope was used. Three massive zones respectively located between 2.2 and 4.2 cm (unit 28), 17.7 and 18.1 cm (unit 22) and 37.5–38.5 cm (unit 14) depth are compared in Fig. 4. There is no indication of a smooth decrease of the minimum track density with depth within a unit or between units. Thus, no evidence for the presence of cm layers is found.

The careful analysis of three laminated zones in units 27, 23 and 15 also fails to reveal any systematic pattern. All samples collected in the first laminated zone (unit 27) have been analyzed and results (in order of increasing depth) are shown in Fig. 5. At each depth, contrasting light and dark lithologies were encountered but no obvious difference in the track density distributions can be perceived. The same conclusion applies to the two other laminated zones investigated (unit 23 between 14.1 and 16.5 cm and unit 15 between 34.1 and 34.8 cm—see Fig. 6). The irradiation level of massive and laminated zones is indistinguishable. In Fig. 6, adjacent light and dark lithologies have also been compared. Again, no systematic variation is apparent, thus, despite the apparent visual inhomogeneity of this section, its track record is unexpectedly homogeneous.

As usual when a series of heavily irradiated soils is encountered, one is faced with a variety of alternatives to explain the data. The two extreme alternatives include: 1) the deposition at the sampling site of largely pre-irradiated material which could have accumulated rapidly (layers a few centimeters thick or larger); and 2) slow accumulation (mm layers) of material which may have been fresh (no pre-irradiation) or heavily irradiated at the time of last deposition.

We have seen above that Gose and Morris (1977) favor a depositional model in which the whole Apollo 16 deep drill core below 13 cm (including 60003) was deposited in a single impact event (or possibly in a series of closely spaced events, i.e., separated by less than a few millions of years) and was subjected to meteoritic gardening for ~450 m.y. If this were the case and if the soils, at the

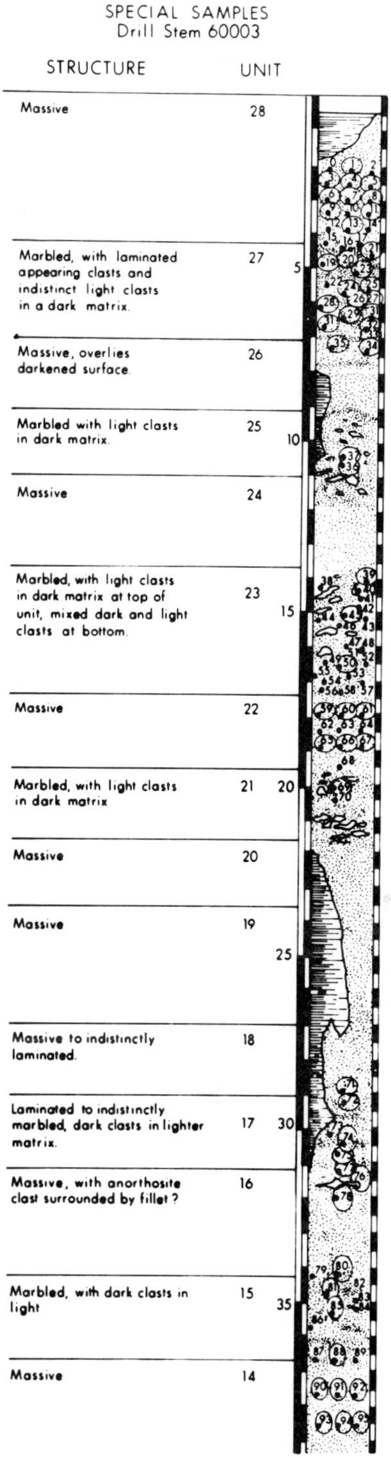

SPECIAL SAMPLES
Drill Stem 60003

STRUCTURE	UNIT
Massive	28
Marbled, with laminated appearing clasts and indistinct light clasts in a dark matrix.	27
Massive, overlies darkened surface.	26
Marbled with light clasts in dark matrix.	25
Massive	24
Marbled, with light clasts in dark matrix at top of unit, mixed dark and light clasts at bottom.	23
Massive	22
Marbled, with light clasts in dark matrix	21
Massive	20
Massive	19
Massive to indistinctly laminated.	18
Laminated to indistinctly marbled, dark clasts in lighter matrix.	17
Massive, with anorthosite clast surrounded by fillet ?	16
Marbled, with dark clasts in light	15
Massive	14

Fig. 3. Stratigraphic description (Duke and Nagle, 1977) and location of the special samples in section 60003 of the Apollo 16 deep drill core.

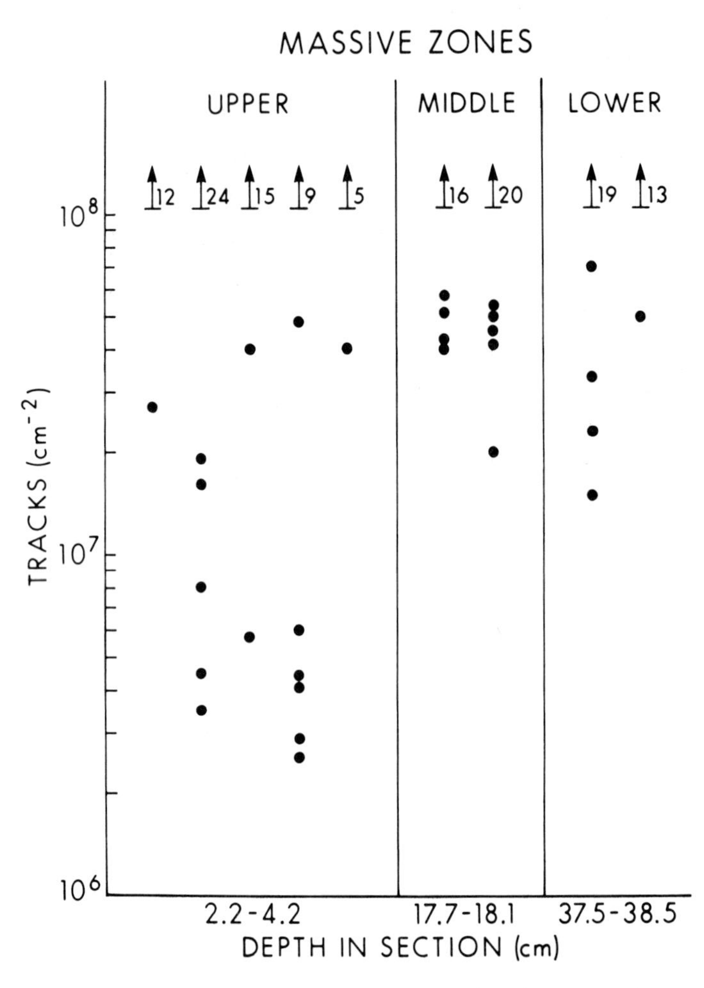

Fig. 4. Track densities in individual feldspar grains of section 60003 of the Apollo 16 deep drill core.

time of deposition contained unirradiated feldspar crystals, we should expect to see very low density crystals and the track density dependence characteristic of an undisturbed layer (as it is estimated that 450 m.y. are not sufficient to carefully garden the material down to a depth of ~2 meters). Low density crystals have carefully been looked for and were not found. This result is best interpreted by a model in which small layers (millimeters thick) slowly accumulated. Even cm layers are not apparent in 60003. The soils of different lithologies have similar track records and thus comparable histories. However, on the basis of tracks alone, it is not possible to determine if they acquired most of their

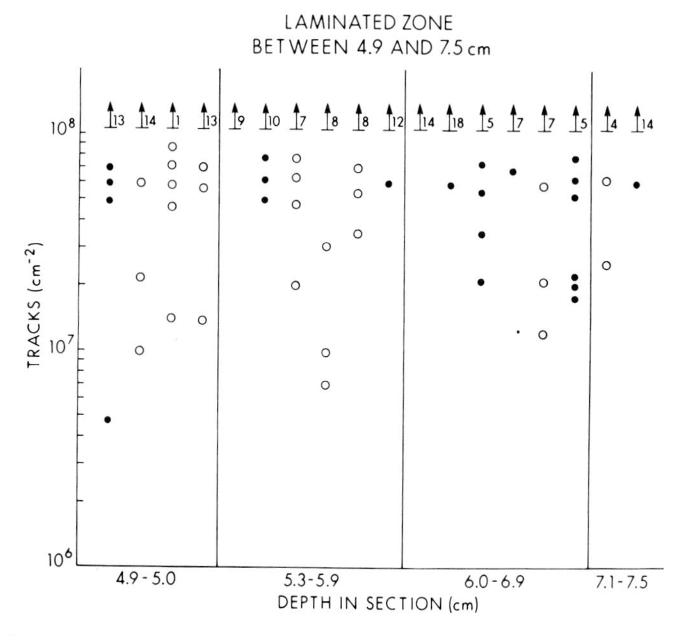

Fig. 5. Same as for Fig. 4. Open and closed points refer respectively to light and dark lithologies.

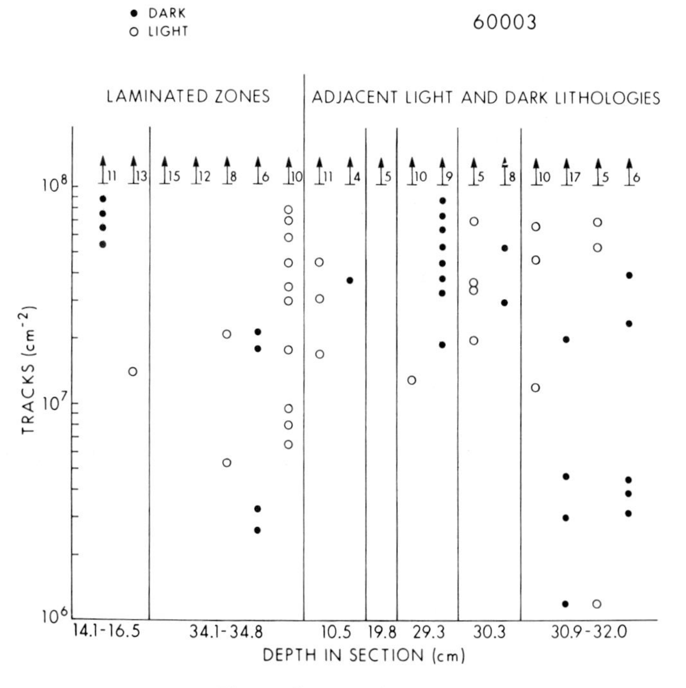

Fig. 6. Same as for Fig. 5.

irradiation *in situ* or if they were already typical mature soils when they were last emplaced.

In addition, Ray *et al.* (1978) have recently reported detailed noble gas measurements on some of the 60003 special samples (laminated regions in units 27 and 23). They feel that their data agree better with a model of slow deposition by "dribs and drabs."

This model of slow accumulation has been challenged by Gose and Morris (1977) who contend that mm to cm thick layers do not have time stratigraphic significance but rather reflect the random inhomogeneities of an ejecta deposit due to mixing. In support of this suggestion, they mention that cores taken in the ejecta blanket of the Ries Crater (Hörz *et al.*, 1977) show many sharp contacts and stratigraphic layering. However, it is not obvious that conclusions derived from a terrestrial impact which penetrated a variety of sedimentary layers are directly applicable to the surface of the moon.

If indeed the whole Apollo 16 core below 13 cm was deposited in one event, one might expect to see evidence for a lateral extension of this ejecta blanket. Failure by Gose and Morris (1977) to see any correlation between the Apollo 16 deep drill material and the soils from the double drive tube 60010/60009 collected only 50 meters away would appear to weaken the case for their interpretation. More important, cm-layers corresponding to distinct depositional events have been recognized in 60010/60009 (Crozaz and Dust, 1977).

Obviously there is a basic difference in philosophy on the part of people who advocate slow accumulation (mainly researchers working with tracks) and emplacement of big slabs (essentially Morris and Gose). Slow accretion (mm to cm layers) should be common compared to occasional large deposition just because the frequency of impacts decreases steeply as the size of the impacting body (and its ejecta blanket) increases. This is precisely what trackologists deduce from their studies: they have seen repeated evidence for the deposition of small layers (mm to cm) and occasional evidence for the deposition of a much larger layer (for example the coarse-grained layer in the Apollo 17 deep drill stem). It is not clear to this author that the ferromagnetic resonance data cannot be interpreted in the same way. The data also show variations on a millimeter to centimeter scale which could reflect the slight maturity variations of successive thin layers of local origin.

Ferromagnetic data, which are generally given every 5 mm in a core are usually averaged over much larger depth intervals in order to look for trends or correlations with the stratigraphy defined by petrologists. Such a correlation is often found and often provides the basis for favoring big layers depositional models. However, it should not be forgotten that petrologists usually do not attribute a time significance to the layers they define as they are not able to register the arrival of small layers of material which presumably is of a local origin and of very similar composition to the material previously exposed at the surface. In conclusion, except for the Apollo 17 deep drill coarse-grained layer, no evidence for the accumulation of large slabs of material is found in this work.

Acknowledgments—The technical assistance of Lou Ross is deeply appreciated. I thank Barbara Wilcox for manuscript preparation. This work was supported by the National Aeronautics and Space Administration under contract NGL 26-008-065.

REFERENCES

Arvidson R., Drozd R., Guinness E., Hohenberg C., Morgan C., Morrison R. and Oberbeck V. (1976) Cosmic ray exposure ages of Apollo 17 samples and the age of Tycho. *Proc. Lunar Sci. Conf. 7th*, p. 2817–2832.

Blanford G. E. and Morrison D. A. (1976) Stratigraphy in Apollo 16 drill section 60002. *Proc. Lunar Sci. Conf. 7th*, p. 141–154.

Blanford G. E. and Wood G. C. (1978) Irradiation stratigraphy in Apollo 16 cores (abstract). In *Lunar and Planetary Science IX*, p. 106–108. Lunar and Planetary Institute, Houston.

Crozaz G. and Plachy A. L. (1976) Origin of the Apollo 17 deep drill coarse-grained layer. *Proc. Lunar Sci. Conf. 7th*, p. 123–131.

Crozaz G. and Dust S. (1977) Irradiation history of lunar cores and the development of the regolith. *Proc. Lunar Sci. Conf. 8th*, p. 3001-3016.

Curtis D. B. and Wasserburg G. J. (1977) Stratigraphic processes in the lunar regolith—additional insight from neutron fluence measurements on bulk soils and lithic fragments from the deep drill cores. *Proc. Lunar Sci. Conf. 8th*, p. 3575–3593.

Duke M. B. and Nagle J. S. (1976) *Lunar Core Catalog and Supplements*. NASA JSC 09252.

Duke M. B. and Nagle J. S. (1977) *Lunar Core Catalog and Supplements*. NASA JSC 09252.

Ehmann W. D. and Ali M. Z. (1977) Chemical stratigraphy of the Apollo 17 deep drill cores 70009-70007. *Proc. Lunar Sci. Conf. 8th*, p. 3223–3241.

Fleischer R. L., Hart H. R., Jr. and Giard W. B. (1974) Surface history of lunar soil and soil solumns. *Geochim. Cosmochim Acta* **38**, 365–380.

Fruchter J. S., Rancitelli L. A. and Perkins R. W. (1976) Recent and long-term mixing of the lunar regolith based on ^{22}Na and ^{26}Al measurements in Apollo 15, 16 and 17 deep drill stems and drive tubes. *Proc. Lunar Sci. Conf. 7th*, p. 27–39.

Gose W. A. and Morris R. V. (1977) Depositional history of the Apollo 16 deep drill core. *Proc. Lunar Sci. Conf. 8th*, p. 2909–2928.

Goswami J. N., Braddy D. and Price P. B. (1976) Microstratigraphy of the lunar regolith and compaction ages of lunar breccias. *Proc. Lunar Sci. Conf. 7th*, p. 55–74.

Goswami J. N., Borg J., Langevin Y., Maurette M. and Price P. B. (1977) Microstratification in Apollo 15 and 16 core tubes: implications to regolith dynamics (abstract). In *Lunar Science VIII*, p. 365–367. The Lunar Science Institute, Houston.

Hörz F., Gall H. and Oberbeck V. R. (1977) Shallow drilling in the "Bunte breccia" impact deposits, Ries Crater, Germany (abstract). In *Lunar Science VIII*, p. 457–459. The Lunar Science Institute, Houston.

Housley R. M., Cirlin E. H., Goldberg I. B. and Crowe H. (1976) Ferromagnetic resonance studies of lunar core stratigraphy. *Proc. Lunar Sci. Conf. 7th*, p. 13–26.

Morris R. V. and Gose W. A. (1976) Ferromagnetic resonance and magnetic studies of cores 60009/60010 and 60003: compositional and surface exposure stratigraphy. *Proc. Lunar Sci. Conf. 7th*, p. 1–11.

Morris R. V., Gose W. A. and Lauer H. V., Jr. (1978) Maturity and FeO profiles for the Apollo 17 deep drill core (abstract). In *Lunar and Planetary Science IX*, p. 766–768. Lunar and Planetary Institute, Houston.

Pepin R. O., Dragon J. C., Johnson N. L., Bates A., Coscio M. R., Jr. and Murthy V. R. (1975) Rare gases and Ca, Sr, and Ba in Apollo 17 drill-core fines. *Proc. Lunar Sci. Conf. 6th*, p. 2027–2055.

Price P. B., Hutcheon I. D., Braddy D. and MacDougall D. (1975) Track studies bearing on solar-system regoliths. *Proc. Lunar Sci. Conf. 6th*, p. 3449–3469.

Ray J., Dziczkaniec M., Walker A. and Heymann D. (1978) Stratigraphy in Apollo drill core segment 60003 (abstract). In *Lunar and Planetary Science IX*, p. 937–939. Lunar and Planetary Institute, Houston.

Russ P. G. III (1973) Apollo 16 neutron stratigraphy. *Earth Planet Sci. Lett.* **19**, 275–289.
Taylor G. J., Keil K. and Warner R. D. (1977) Petrology of Apollo 17 deep drill core—I: depositional history based on modal analyses of 70009, 70008, and 70007. *Proc. Lunar Sci. Conf. 8th*, p. 3195–3222.
Vaniman D. T. and Papike J. J. (1977) The Apollo 17 drill core: characterization of the mineral and lithic component (sections 70007, 70008, 70009). *Proc. Lunar Sci. Conf. 8th*, p. 3123–3159.
Vaniman D. T., Lellis S. F., Papike J. J. and Cameron K. L. (1976) The Apollo 16 drill core: modal petrology and characterization of the mineral and lithic component. *Proc. Lunar Sci. Conf. 7th*, p. 199–239.

Proc. Lunar Planet. Sci. Conf. 9th (1978), p. 1801–1811.
Printed in the United States of America

In situ reworking (gardening) of the lunar surface: Evidence from the Apollo cores

RICHARD V. MORRIS

Code SN7, NASA Johnson Space Center, Houston, Texas 77058

Abstract—The *in situ* reworking (gardening) of the lunar surface by impacting projectiles creates an *in situ* reworking zone extending horizontally over the entire regolith surface and extending vertically from the surface to a depth which varies from place-to-place on the moon. On the basis of available evidence, the "high-maturity" zones observed at the top of the lunar cores have resulted from the *in situ* reworking of the present-day lunar surface. The temporal variation of the *in situ* reworking depth was investigated using depths inferred from maturity (I_s/FeO) and ^{26}Al profiles of Apollo cores. The observed temporal variation of the *in situ* reworking depth is described by the equation

$$D_R = 2.2\ (t)^{0.45}$$

where D_R is in units of centimeters and time is in units of million years.

INTRODUCTION

During the past few years, maturity profiles (i.e., maturity as a function of depth) have been determined quantitatively for the Apollo 15, 16, and 17 deep drill cores and the 60009/10 and 74001/2 drive tube cores using the FMR maturity index I_s/FeO (Heiken *et al.*, 1976; Gose and Morris, 1977; Morris *et al.*, 1978a,b; Morris and Gose, 1976, 1977). This maturity index is discussed in detail by Morris (1976, 1978). For the purposes of this paper, it is necessary to emphasize that the fine-grained metal, upon which I_s/FeO is based, is produced predominantly in the upper 1 mm or so of the lunar regolith (e.g., Housley *et al.*, 1973). Thus, variations of I_s/FeO with depth are a consequence of processes such as meteoric impact which transport and mix soil and are not due to variations in the production of fine-grained metal with depth. One feature common to all of the above cores (although not very distinct in the Apollo 16 deep drill core) is a decrease in maturity from the lunar surface to a depth that varies from core to core. These "high-maturity" zones are attributed to *in situ* reworking of the present-day lunar surface, and the reasons for this conclusion are reviewed below.

Probabilistic cratering models have been developed (e.g., Gault *et al.*, 1974; Arnold, 1975) to model the lunar *in situ* reworking (gardening) process. Since the relative frequency of impact of meteorites decreases systematically with increasing meteoritic mass, these models predict a mixing zone (i.e., the *in situ* reworking or gardening zone) extending from the surface to some depth. The effect of the reworking is to increase the average maturity of the soil throughout the entire thickness of the *in situ* reworking zone. This is evident because the upper 1 mm of soil in which the fine-grained metal is produced by micrometeorites, is "constantly" worked into the other soil in the *in situ* reworking zone by larger meteorites. The above observation was used by Heiken *et al.* (1976) and Morris and Gose (1976) to suggest that the high-maturity zones observed at the top of the Apollo 15 deep drill core and the 60009/10 core could be due to *in situ* reworking. The recent finding of high-maturity zones at the top of the Apollo 17 deep drill core and the 74001/2 core (Morris *et al.*, 1978a, b) strengthens the conclusion that the high-maturity zones at the top of the cores are a consequence of *in situ* reworking of the present-day lunar surface. That is, the occurrence of these zones in cores from three Apollo landing sites strongly suggests that they are due to the moon-wide *in situ* reworking process occurring subsequent to final core emplacement and are not features present at the depositional time of the individual cores.

Two of the cores discussed above, 60009/10 and 74001/2, have been the subject of multidisciplinary studies (including petrography, grain-size analysis, FMR, noble gases, tracks, and chemistry) of the same samples selected at various locations along the cores (60009/10; McKay *et al.*, 1976, 1977; Bogard and Hirsch, 1976, 1977; Blanford *et al.*, 1977; Blanchard and Brannon, 1977; and 74001/2: McKay *et al.*, 1978; Bogard and Hirsch, 1978; Morris *et al.*, 1978b). A multidisciplinary approach generates many different types of data and consequently strongly constrains the number of viable interpretations. One result of these studies is that the high-maturity zones of the top of 60009/10 and 74001/2 are best interpreted as *in situ* reworking zones resulting from *in situ* reworking of the present-day lunar surface.

In summary, the *in situ* reworking interpretation for the origin of the high-maturity zones at the top of the cores seems well supported. This paper first presents a synthesis of the manifestations of the *in situ* reworking process in the Apollo cores. The synthesis is strongly based on the results of and the insights gained in the above studies. The remainder of the paper focuses on (1) the observed temporal variation of the *in situ* reworking depth and (2) a comparison of the observed temporal variation to the corresponding predictions of probabilistic cratering models.

THE *IN SITU* REWORKING PROCESS

In situ reworking is a moon-wide process which can be envisioned as composed of two parts operating concurrently (McKay *et al.*, 1977). One part of the process consists of micrometeorite reworking of the upper 1 mm or so of the regolith. This reworking forms agglutinates and the fine-grained metal on which the index I_s/FeO is based. (Fine-grained metal is contained within agglutinitic glass and does not exist as discrete soil particles.) The concentrations of solar wind gases (including N and the trapped noble gases), the densities of solar flare tracks, and other surface-correlated components increase in the upper 1 mm of regolith during micrometeorite reworking. The other part of the process consists of mixing of this uppermost few millimeters into underlying soil by somewhat larger impacts. The net result of this two-part process is a "percolation" or "churning" of fine-grained metal, agglutinates, and other surface-correlated components vertically into the regolith, increasing the fine-grained metal, agglutinate, etc. content of the underlying soil. The vertical extent of the percolation is the *in situ* reworking depth, and this depth defines the lower limit of the *in situ* reworking zone, the upper limit being the very surface. Statistically, the *in situ* reworking depth will increase with the time *in situ* reworking has acted on a surface. One might expect that maturity should decrease from the surface through the reworking zone because of the systematic decrease in the relative frequency of larger impacts. However, this is not necessarily the case. The shape of the maturity profile in an *in situ* reworking zone depends on the shape of the maturity profile before *in situ* reworking and the relative timing of the larger impacts which perform the mixing. This position is born out observationally; the maturity profiles of the *in situ* reworking zones have variable shapes, but are on the average more mature than the inferred pre-existing maturity (c.f., next section).

It is readily apparent that a boundary in the depth profile of any parameter that can be measured for soils (e.g., composition) will tend to be smeared out when acted upon by the *in situ* reworking process. This homogenization occurs both by mechanical mixing of the soil particles and by the melting together of particles, such as in the formation of agglutinates by micrometeorite impact. An important subset of the above parameters are the stable nuclides which are produced by interactions with cosmic rays and consequently whose production rates are a function of depth in the regolith (e.g., cosmogenic ^{21}Ne and neutron-capture produced isotopes of the rare earths). Because soil is stirred and mixed in the *in situ* reworking zone, the observed profiles of these parameters are not expected to reflect their production profiles in the *in situ* reworking zone. That is, the depth of any soil component in the *in situ* reworking zone is time-dependent and, consequently, the observed profiles of the stable nuclides should not reflect their respective production profiles. Similar to the maturity profiles, it is not possible to exactly predict the difference between the observed and the production profiles of the stable nuclides in the *in situ* reworking zones because the differences depend on the pre-irradiation history and on the relative timing of the mixing events. For volatile species such as cosmogenic ^{21}Ne, there is the added complication of their loss from the *in situ* reworking zone during heating associated with such events as agglutination (e.g., Bogard, 1977).

Since meteoritic impact ejects material horizontally, lateral mixing in addition to vertical mixing of soil is inherent to the *in situ* reworking process. Consequently, a soil column of initially uniform composition (as an example) can develop an *in situ* reworking zone having a somewhat different composition due to the lateral transport of "foreign" material. Obviously, the extent of this "contamination" is related to the location of the sampling site relative to the lateral boundary of the initially uniform soil unit and to the period of time the soil has been subject to *in situ* reworking. This lateral mixing of soil during *in situ* reworking has clearly introduced the "foreign" material into the *in situ* reworking zones of

74001/2 and the Apollo 17 deep drill core. The zone of 74001/2 has mare basalt fragments, and the underlying soil does not (McKay *et al.*, 1978); the zone of the Apollo 17 deep drill core has a somewhat higher highland component than immediately underlying soil (Taylor *et al.*, 1977; Vaniman and Papike, 1977; Ehmann and Ali, 1977; Morris *et al.*, 1978b).

The same impacts which produce vertical and horizontal mixing within the *in situ* reworking zones also cause a temporal variation in the position of the lunar surface. This follows because impacts can lower the position of the surface when soil is excavated by cratering and can raise the position of the surface when crater ejecta is deposited. Since many impacts are involved in *in situ* reworking, the position of the lunar surface will presumably tend to fluctuate around an average value. An obvious consequence of this temporal variation is that the observed concentrations of cosmogenic ^{21}Ne, neutron-capture produced isotopes of the rare earth, etc. will reflect production over a range of shielding depths. For soil beneath the *in situ* reworking zone (i.e., where mixing does not occur), the difference between the maximum and minimum shielding depths is probably on the order of the *in situ* reworking depth. Thus, even if a thick slab of soil having no or constant pre-irradiation is emplaced and subsequently irradiated by cosmic rays and subjected to *in situ* reworking, the observed profile of the neutron fluence, for example, is expected to deviate from its production profile because of both shielding changes and, within the *in situ* reworking zone, mixing (as discussed earlier). Mixing will produce much greater deviations, and its relative influence on a given production profile depends on the ratio of the *in situ* reworking depth to the total depth over which the production profile extends.

TEMPORAL VARIATION OF THE *IN SITU* REWORKING DEPTH

Time/depth data from the Apollo cores

The maturity profiles of the near surface regions of the Apollo 15 and 17 deep drill cores and the 60009/10 and 74001/2 drive tube cores are shown in Fig. 1. The *in situ* reworking depths (D_R) are denoted by arrows and are compiled in Table 1. The dashed lines are extrapolations of the maturity of soil underlying the *in situ* reworking zones and denote the inferred maturity of each core at the time of core emplacement, i.e., before *in situ* reworking raised the maturity to the presently-observed levels.

For the time *in situ* reworking has acted on each core, the time of emplacement of the section of core containing the *in situ* reworking zone and immediately underlying soil is required. The times adopted are compiled in Table 1, and a discussion of their selection follows. Based on neutron fluence profiles, Curtis and Wasserburg (1977) argue that the Apollo 15 deep drill core has experienced static irradiation for ~450 m.y. These authors did not consider the effects of *in situ* reworking on the neutron fluence profile, so this age may need some revision. However, as can be seen later, an uncertainty of ~100 m.y. will not significantly alter the conclusions of this paper. The ~125 m.y. cosmogenic ^{21}Ne age of

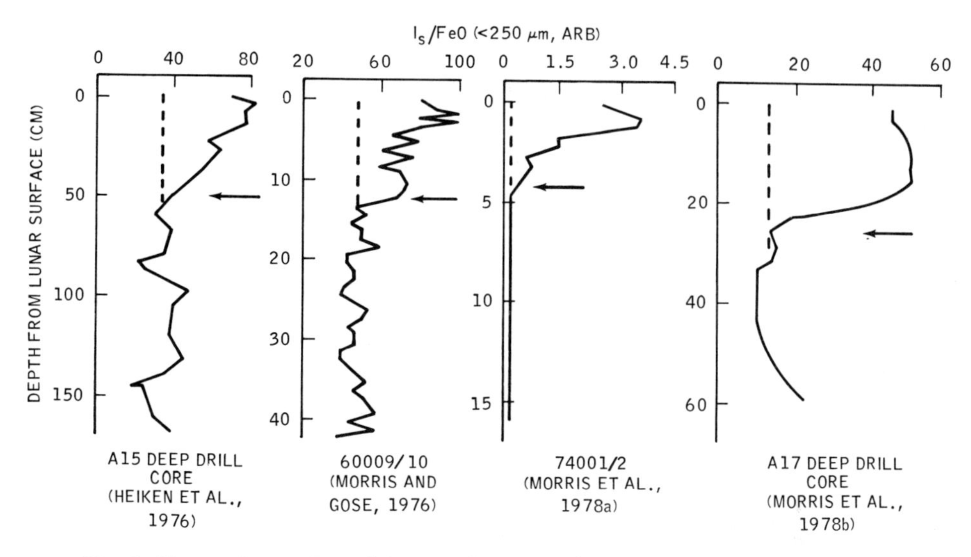

Fig. 1. Near surface regions of the maturity profiles for the Apollo 15 and 17 deep drill cores and 60009/10 and 74001/2 drive tube cores. The maturity is in terms of the FMR maturity index I_s/FeO. The dashed lines represent the inferred maturity at the time of emplacement. The arrows denote the *in situ* reworking depth D_R.

Bogard and Hirsch (1977) is used for the 60009/10 drive tube core; this age was calculated from data obtained from plagioclase separates from several samples of soil from 60009/10. For the Apollo 17 deep drill core, the ~110 m.y. ^{81}Kr-Kr age reported by Drozd *et al.* (1977) for Apollo 17 rocks collected near Camelot Crater was adopted. This age presumes that at least the upper part of the Apollo 17 deep drill was emplaced contemporaneously with the Camelot (or Central Cluster) event, as has been argued by many workers (e.g., Taylor *et al.*, 1977; Heiken and McKay, 1974; Crozaz and Dust, 1977). The 110 m.y. age is also comparable to the ages based on neutron fluence profiles of the Apollo 17 deep drill core. For the 74001/2 drive tube core, selection of the "correct" age is somewhat more critical than for the other cores discussed above because the *in situ* reworking depth is quite small. An age of 14 ± 4 m.y. is adopted and

Table 1. *In situ* reworking depths and times for maturity profiles from Apollo cores.

Core	Depth, D_R (cm)	Reference	Time m.y.	Reference
A15 DDC†	50 ± 10	Heiken *et al.* (1976)	450 ± 100	Curtis and Wasserburg (1977)
60009/10	12.5 ± 1	Morris and Gose (1976)	125	Bogard and Hirsch (1977)
A17 DDC	25 ± 5	Morris *et al.* (1978b)	110	Drozd *et al.* (1977)
74001/2	5 ± 1	Morris *et al.* (1978a)	14 ± 4	Bogard and Hirsch (1978); Crozaz (1978)

†DDC = Deep Drill Core.

overlaps the emplacement ages inferred by Bogard and Hirsch (1978) from cosmogenic ^{21}Ne data and by Crozaz (1978) from particle track data.

The *in situ* reworking depth/time data inferred from the maturity profiles are plotted in Fig. 2. Note that the trend of the four data points is the expected increase in *in situ* reworking depth with time. In fact, the trend is surprisingly systematic. A large degree of scatter could have resulted because meteoritic impact is a random process and consequently each value of time is characterized not by a unique value of D_R but by a distribution of values of D_R.

Mixing depths on a shorter time scale than the available maturity-derived data have been inferred from ^{26}Al and ^{53}Mn profiles. Fruchter *et al.* (1977) argue that they generally observe in their ^{26}Al profiles mixing on the order of 2 to 3 cm in depth in the time frame 10^5 to 10^6 years. This mixing depth is indicated in Fig. 2 at a time of 0.73 m.y., the half-life of ^{26}Al. It is appropriate to consider this mixing depth as an *in situ* reworking depth because the time scale is established

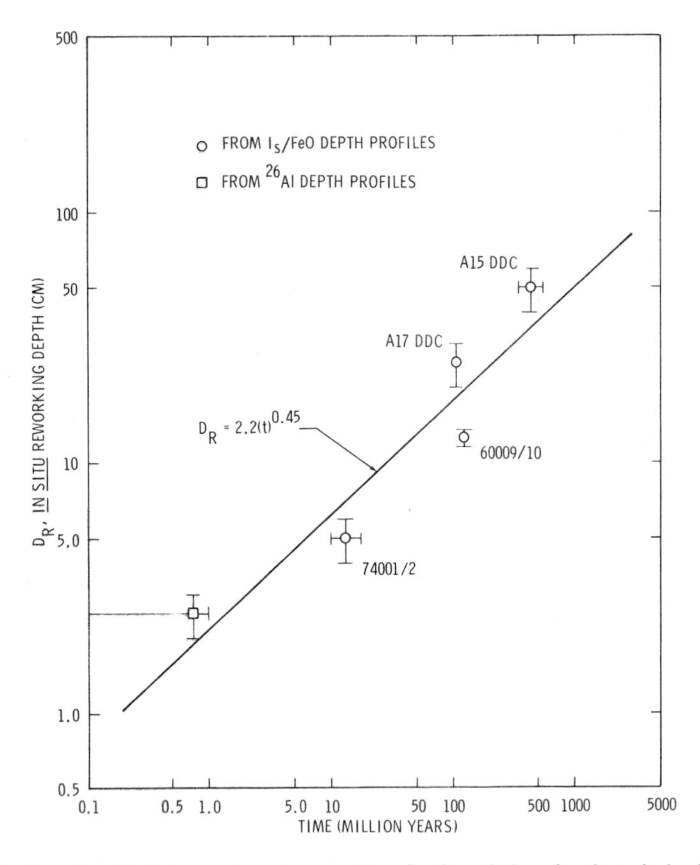

Fig. 2. Relationship between *in situ* reworking depth and time for data derived from I_s/FeO and ^{26}Al depth profiles in Apollo cores. The solid line was derived from a linear regression analysis of the data; the correlation coefficient is 0.94. The solid line is regarded to represent the 50% probability *in situ* reworking depth D_R after a time t.

by the half-life of ^{26}Al. It is readily apparent that the time/depth data from ^{26}Al and maturity profiles form a systematic trend. Nishiizumi *et al.* (1976) have inferred *in situ* reworking depths from ^{53}Mn profiles (half-life for ^{53}Mn is ~3.7 m.y.) and, according to the authors, the time scale for this mixing is within the last 10 m.y. The ^{53}Mn-derived data are generally compatible with the data in Fig. 2 but were not plotted because the time estimates are too broad.

When the data plotted in Fig. 2 is subjected to linear regression analysis, the following equation is obtained:

$$D_R = 2.2\ (t)^{0.45}. \tag{1}$$

In Eq. 1, the *in situ* reworking depth (D_R) is in units of centimeters and time is in units of million years. Eq. 1 holds only over the ~1.0 to 500 m.y. range of the data. The form of Eq. 1 implies that a unique relationship exists between D_R and time. As indicated earlier, in actuality a distribution of values of D_R exist for each value of time so that D_R is predictable only at a given probability level. Thus, it is desirable to attach a probability level to the values of D_R calculated from Eq. 1. (The statistics associated with the linear regression method are not applicable because the method assumes that the deviations from the best fit are entirely due to experimental error.) If it is presumed, as seems likely, that the five data points in Fig. 2 include values of D_R from both greater than and less than 50% probability levels, then the values of D_R calculated from Eq. 1 are most reasonably regarded as the *in situ* reworking depth at a 50% probability level. A direct, rigorous treatment in which values of D_R for various probability levels are determined awaits a much larger data base.

In summary, *in situ* reworking is an important lunar surface process because of the significant depths to which *in situ* reworking penetrates on a fairly short time scale. The chemical and physical manifestations of *in situ* reworking, as discussed earlier, should be considered in the interpretation of the depositional and evolutional history of the lunar regolith. As a specific example, consider whether one would expect to observe in the regolith production profiles of ^{21}Ne and other stable nuclides produced by solar cosmic rays. Since these production profiles are confined to the upper few centimeters of regolith and since the data in Fig. 2 indicate *in situ* reworking (and thus mixing) to a depth of ~2 cm in a million years, the probability is quite low of observing one of these production profiles for surfaces older than about 1 m.y.

Comparison to predictions of probabilistic cratering models

Probabilistic cratering models, including those of Gault *et al.* (1974) and Arnold (1975), have been developed to predict at a given probability level the *in situ* reworking depth as a function of time from input assumptions concerning the mass frequency of meteorites, the absolute flux thereof, and cratering mechanics. The predictions of the Gault *et al.* (1974; Fig. 9) and Arnold (1975) models at a 50% probability level are compared to the experimental data in Fig. 3. The 50% probability level was chosen for the comparison because of the presumption in the

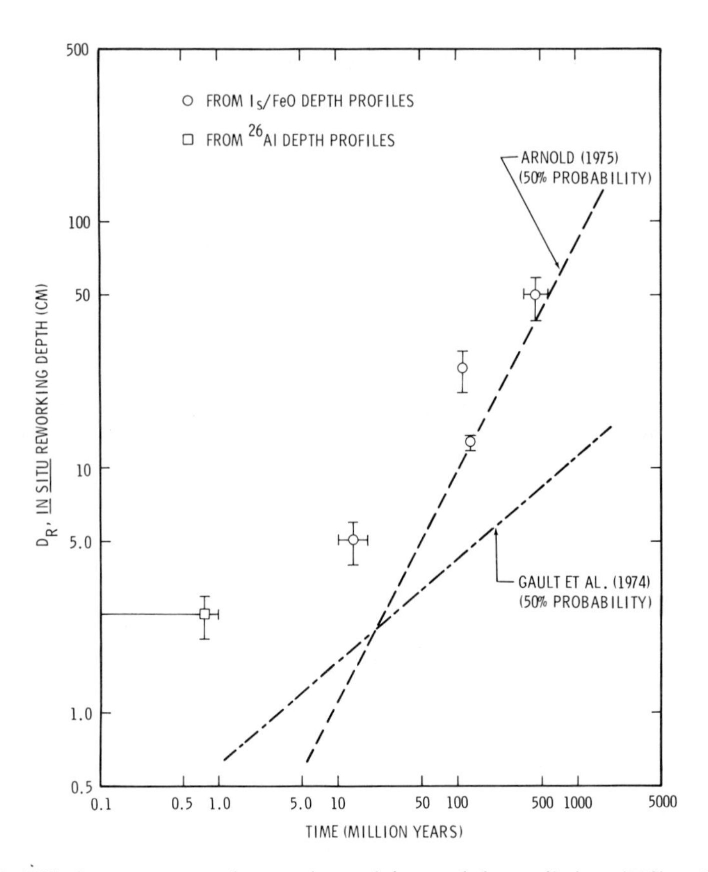

Fig. 3. This figure compares the experimental data and the predictions (50% probability) of the probabilistic cratering models of Gault *et al.* (1974; Fig. 9) and Arnold (1975).

last section that it is the most likely situation that the experimental data includes values of D_R from greater than and less than 50% probability levels. In any event, the predictions of the two models are comparable with each other only at the same probability level. All predictions referred to below are at the 50% probability level.

Fig. 3 shows there is good agreement between the experimental data and the predictions of Arnold (1975) for model times greater than about 10 m.y. For times less than about 10 m.y., both the Gault *et al.* (1974) and the Arnold (1975) models predict too shallow *in situ* reworking depths. Do these results mean that the input parameters used in the Arnold (1975) model are more representative of the actual situation than those of the Gault *et al.* (1974) model for times greater than about 10 m.y.? Not necessarily. The Arnold (1975) model uses for the projectiles larger than 1.0 gm the mass distribution given by Neukum and Dietzel (1971), and that mass distribution tends to maximize the calculated *in situ* reworking depth. In addition, neither model incorporates the contribution of secondary projectiles (crater ejecta) to *in situ* reworking. Thus, at this point, it does not seem possible for model times greater than ~10 m.y. to distinguish whether (1) the difference between the Gault *et al.* (1974) model and the observed data is due to ignoring the effect of secondary projectiles, or (2) the input parameters of the Arnold (1975) model are such that they compensate for the effect of secondary projectiles, or (3) the effect of secondary projectiles is small and input parameters of the Arnold (1975) model are more representative of the actual situation, or (4) some combination of the above.

SUMMARY AND CONCLUSIONS

(1) The *in situ* reworking (gardening) of the lunar surface by impacting projectiles forms in the regolith an *in situ* reworking zone extending horizontally over the entire regolith surface and extending vertically from the surface to a depth which varies from place-to-place on the moon. On the basis of available evidence, the "high-maturity" zones at the top of lunar cores are the *in situ* reworking zones that have resulted from the *in situ* reworking of the present-day lunar surface. Within the *in situ* reworking zone, both vertical and (very local) horizontal mixing of the soil occur as important inherent characteristics of *in situ* reworking. A temporal variation in the vertical distance between the lunar surface and any arbitrary point below the *in situ* reworking zone also occurs.

(2) The observed temporal variation of the *in situ* reworking depth is described by the equation $D_R = 2.2\ (t)^{0.45}$, where D_R is in units of centimeters and time is in units of million years.

(3) The experimental *in situ* reworking depth/time data and the corresponding predictions (50% probability) of the probabilistic cratering models of Gault *et al.* (1974) and Arnold (1975) were compared. The Arnold (1975) model is in reasonable agreement with the experimental results for times longer than about 10–20 m.y., but it predicts too shallow *in situ* reworking depths for shorter times. The Gault *et al.* (1974) model predicts too shallow *in situ* reworking depths at all times.

(4) The empirically-derived temporal dependence of the *in situ* reworking depth demonstrates that *in situ* reworking is an important lunar surface process because significant depths are reworked and mixed in a relatively short time. Thus, one must be cognizant of the extent and properties of *in situ* reworking zones when formulating models for the history of lunar cores.

Acknowledgments—This paper has benefited from discussions with Drs. F. Hörz and D. McKay and the reviews of Drs. D. Morrison and D. Woolum. C. Hardy and B. Allison are thanked for typing the manuscript.

REFERENCES

Arnold J. R. (1975) Monte Carlo simulation of turnover processes in the lunar regolith. *Proc. Lunar Sci. Conf. 6th*, p. 2375–2395.

Blanchard D. P. and Brannon J. C. (1977) Effects on composition of maturation in a well documented, isochemical suite of soils from drive tube 60009/10 (abstract). In *Lunar Science VIII*, p. 121–123. The Lunar Science Institute, Houston.

Blanford G. E., McKay D. S. and Wood G. C. (1977) Particle track densities in double drive tube 60009/10. *Proc. Lunar Sci. Conf. 8th*, p. 3017–3025.

Bogard D. D. (1977) Effects of soil maturation on grain size-dependence of trapped solar gases. *Proc. Lunar Sci. Conf. 8th*, p. 3705–3718.

Bogard D. D. and Hirsch W. C. (1976) Noble gases in 60009–60010 drive tube samples: Trapped gases and irradiation history. *Proc. Lunar Sci. Conf. 7th*, p. 259–279.

Bogard D. D. and Hirsch W. C. (1977) Noble gas evidence for the depositional and irradiational history of 60010-60009 core soils. *Proc. Lunar Sci. Conf. 8th*, p. 2983–2999.

Bogard D. D. and Hirsch W. C. (1978) Depositional and irradiational history and noble gas contents of orange-black droplets in the 74002/1 core from Shorty Crater. *Proc. Lunar Planet. Sci. Conf. 9th*. This volume.

Crozaz G. (1978) History of regolith deposition at the Apollo 17 and Luna 24 landing sites and at Shorty Crater (abstract). In *Lunar and Planetary Science IX*, p. 203–205. Lunar and Planetary Institute, Houston.

Crozaz G. and Dust S. (1977) Irradiation history of lunar cores and the development of the regolith. *Proc. Lunar Sci. Conf. 8th*, p. 3001–3016.

Curtis D. B. and Wasserburg G. J. (1977) Stratigraphic processes in the lunar regolith—additional insights from neutron fluence measurements on bulk soils and lithic fragments from the deep drill cores. *Proc. Lunar Sci. Conf. 8th*, p. 3575–3593.

Drozd R. J., Hohenberg C. M., Morgan C. J., Podosek F. A. and Wroge M. L. (1977) Cosmic-ray exposure history of Taurus-Littrow. *Proc. Lunar Sci. Conf. 8th*, p. 3027–3043.

Ehmann W. D. and Ali M. Z. (1977) Chemical stratigraphy of the Apollo 17 deep drill cores 70009-70007. *Proc. Lunar Sci. Conf. 8th*, p. 3223–3241.

Fruchter J. S., Rancitelli L. A., Laul J. C. and Perkins R. W. (1977) Lunar regolith dynamics based on analysis of the cosmogenic radionuclides ^{22}Na, ^{26}Al, and ^{53}Mn. *Proc. Lunar Sci. Conf. 8th*, p. 3595–3605.

Gault D. E., Hörz F., Brownlee D. E. and Hartung J. B. (1974) Mixing of the lunar regolith. *Proc. Lunar Sci. Conf. 5th*, p. 2365–2386.

Gose W. A. and Morris R. V. (1977) Depositional history of the Apollo 16 deep drill core. *Proc. Lunar Sci. Conf. 8th*, p. 2909–2928.

Heiken G. H. and McKay D. S. (1974) Petrography of Apollo 17 soils. *Proc. Lunar Sci. Conf. 5th*, p. 843–860.

Heiken G. H., Morris R. V., McKay D. S. and Fruland R. M. (1976) Petrographic and ferromagnetic resonance studies of the Apollo 15 deep drill core. *Proc. Lunar Sci. Conf. 7th*, p. 93–111.

Housley R. M., Grant R. W. and Paton N. E. (1973) Origin and characteristics of excess Fe metal in lunar glass welded aggregates. *Proc. Lunar Sci. Conf. 4th*, p. 2737–2749.

McKay D. S., Dungan M. A., Morris R. V. and Fruland R. M. (1977) Grain size, petrographic, and FMR studies of the double core 60009/10: A study of soil evolution. *Proc. Lunar Sci. Conf. 8th*, p. 2929–2952.

McKay D. S., Heiken G. H. and Waits G. H. (1978) Core 74001/2: Grain size and petrology as a key to the rate or *in-situ* reworking and lateral transport on the lunar surface. *Proc. Lunar Planet. Sci. Conf. 9th*. This volume.

McKay D. S., Morris R. V., Dungan M. A., Fruland R. M. and Fuhrman R. (1976) Comparative studies of grain size separates of 60009. *Proc. Lunar Sci. Conf. 7th*, p. 295–313.

Morris R. V. (1976) Surface exposure indices of lunar soils: A comparative FMR study. Proc. Lunar Sci. Conf. 7th, p. 315–335.

Morris R. V. (1978) The surface exposure (maturity) of lunar soils: Some concepts and I_s/FeO compilation. *Proc. Lunar Planet. Sci. Conf. 9th.* This volume.

Morris R. V. and Gose W. A. (1976) Ferromagnetic resonance studies of cores 60009/60010 and 60003: Compositional and surface-exposure stratigraphy. *Proc. Lunar Sci. Conf. 7th*, p. 1–11.

Morris R. V. and Gose W. A. (1977) Depositional history of core section 74001: Depth profiles of maturity, FeO, and metal. *Proc. Lunar Sci. Conf. 8th*, p. 3113–3122.

Morris R. V., Gose W. A. and Lauer H. V. (1978a) Depositional and surface exposure history of the Shorty Crater core 74001/2: FMR and magnetic studies. *Proc. Lunar Planet. Sci. Conf. 9th.* This volume.

Morris R. V., Gose W. A. and Lauer H. V. (1978b) Maturity and FeO profiles for the Apollo 17 deep drill core (abstract). In *Lunar and Planetary Science IX*, p. 766–768. Lunar and Planetary Institute, Houston.

Neukum G. and Dietzel H. (1971) On the development of the crater population on the Moon with time under meteoroid and solar wind bombardment. *Earth Planet. Sci. Lett.* **12**, 59–66.

Nishiizumi K., Imamura M., Honda M., Russ G. P. III, Kohl C. P. and Arnold J. R. (1976) ^{53}Mn in the Apollo 15 and 16 drill stems: Evidence for surface mixing. *Proc. Lunar Sci. Conf. 7th*, p. 41–54.

Taylor G. J., Keil K. and Warner R. D. (1977) Petrology of the Apollo 17 deep drill core—I: Depositional history based on modal analyses of 70009, 70008, and 70007. *Proc. Lunar Sci. Conf. 8th*, p. 3195–3222.

Vaniman D. T. and Papike J. J. (1977) The Apollo 17 drill core: Modal petrology and glass chemistry (sections 70007, 70008, 70009). *Proc. Lunar Sci. Conf. 8th*, p. 3161–3193.

Proc. Lunar Planet. Sci. Conf. 9th (1978), p. 1813–1826.
Printed in the United States of America

The Apollo 16 drive tube 60009/60010. Part I: Modal petrology

S. B. SIMON,* J. J. PAPIKE and D. T. VANIMAN

Planetary Regolith Studies, Department of Earth and Space Sciences
State University of New York, Stony Brook, New York 11794

Abstract—Detailed petrographic examination of 22 thin sections from the impregnated portion of the Apollo 16 double drive tube supports the findings of previous workers. A well-gardened unit exists at the top of 60010, indicated petrographically by an increase in agglutinates near the lunar surface. There is a coarse-grained anorthositic layer near the bottom of 60009. Few, if any, of the modal variations are stratigraphically correlatable with the deep drill core 60002-7, which was recovered 50 m southwest of drive tube 60009/60010.

INTRODUCTION

The drive tube 60009/60010 sampled ~60 cm of regolith and was undisturbed during sampling and subsequent handling according to the ^{22}Na profiles obtained by Fruchter *et al.* (1977). We have examined a continuous string of 22 polished thin sections from the drive tube (12 from 60009; 10 from 60010). Our study is divided into two parts: Part I (reported here) concerns the modal petrology; Part II (Vaniman *et al.*, this volume) concerns chemical characterization of the monomineralic, lithic, and glass component, and the total major element partitioning among the regolith components.

Based on petrography, layers and trends can be identified which were previously recognized by workers using different analytical methods such as INAA, FMR, and track studies. The mature (as defined in glossary of the Proceedings of the Fifth Lunar Science Conference, 1974) unit at the top of 60010 is indicated by decreasing monomineralic component and increasing agglutinate percentage toward the lunar surface. McKay *et al.* (1977) also noted this increase in agglutinates in the top ~12 cm of 60010. Morris and Gose (1976) recognized the unit as a layer of well gardened, mature soil. Nagle (1978) suggested that the material in this unit was deposited more slowly than that in the rest of the core, resulting in the buildup of agglutinates. We believe that the arguments summarized by Morris (1978) for significant *in situ* reworking of the upper 12 cm of drive tube 60010 are sound and see no reason to call upon slow deposition of the upper portion of the core to account for the agglutinate buildup. The immature anorthositic layer near the bottom of 60009 is characterized

*Present address: Department of Geology, University of Massachusetts, Amherst, Massachusetts 01002.

petrographically by a high content of plagioclase single crystals, few rock fragments, and a low glass content. It is characterized chemically by high Ca and Al contents and low Fe, Mg, Sc, and Co (Ali and Ehmann, 1977, and Vaniman *et al.*, 1978).

ANALYTICAL METHODS

1000 grid points on each half of 20 polished thin sections (P.T.S.), 1000 on P.T.S. 60009,6030 and 3000 on P.T.S. 60009,6029 were identified using a Zeiss photomicroscope equipped with both transmitted and reflected light optics. Spacing between points was 0.05 mm. The 44,000 optical identifications are the data on which this report is based.

The use of reflected light allows accurate determination of matrix percentages because grain boundaries can be seen more clearly than in transmitted light. Reflected light is also an aid in distinguishing agglutinates from dark matrix breccias (DMB's) on the basis of vesicular texture and content of metal spherules. The modal data are presented in Figs. 1–5 and in the appendix (Tables A-1 and A-2).

MODAL CLASSIFICATION

The classification system used here is very similar to that which was used previously for the Apollo 16 deep drill core (Vaniman *et al.*, 1976). The individual modal components are the same and will not be described in detail here. Some of the groupings have been changed, however. Recrystallized moritic breccias (RNB's) and pyroxene poikilitic rocks (POIK's) were separated petrographically on the basis of texture, but were plotted together because they are chemically equivalent. Pyroxene and olivine could not always be separated petrographically and are, therefore, grouped together. The definitions of the clast sizes are the same as in Vaniman *et al.* (1976): matrix, <0.02 mm; small clasts, 0.02–0.2 mm; large clasts, 0.2–2.0 mm; and extra large clasts (not included in percentage calculations), >2 mm. Glasses are grouped by color, using the classification scheme of Naney *et al.* (1976) and Vaniman *et al.* (1976).

Clast size abundance

The total modal abundances of the different clast sizes are shown in Fig. 1. The stratigraphic units are those of Duke and Nagle (1974). The coarse-grained unit near the bottom of 60009 (P.T.S. 6029 and 6028) has significantly less matrix and more large clasts; the large clasts in this immature unit are predominantly feldspar. Conversely, the mature upper part of 60010 shows a higher matrix percentage and a low amount of large clasts with very few large feldspar grains. Overall, the material in the drive tube is coarser grained than that in the deep drill core. The drill core averages 66% matrix (Vaniman *et al.*, 1976), while the drive tube averages 57% matrix and has more material in the small clast category. The senior author has calibrated this comparison by re-analyzing thin sections from the A-16 drill core used in our previous studies (Vaniman *et al.*, 1976).

At this point we should elaborate on what is meant by matrix in our

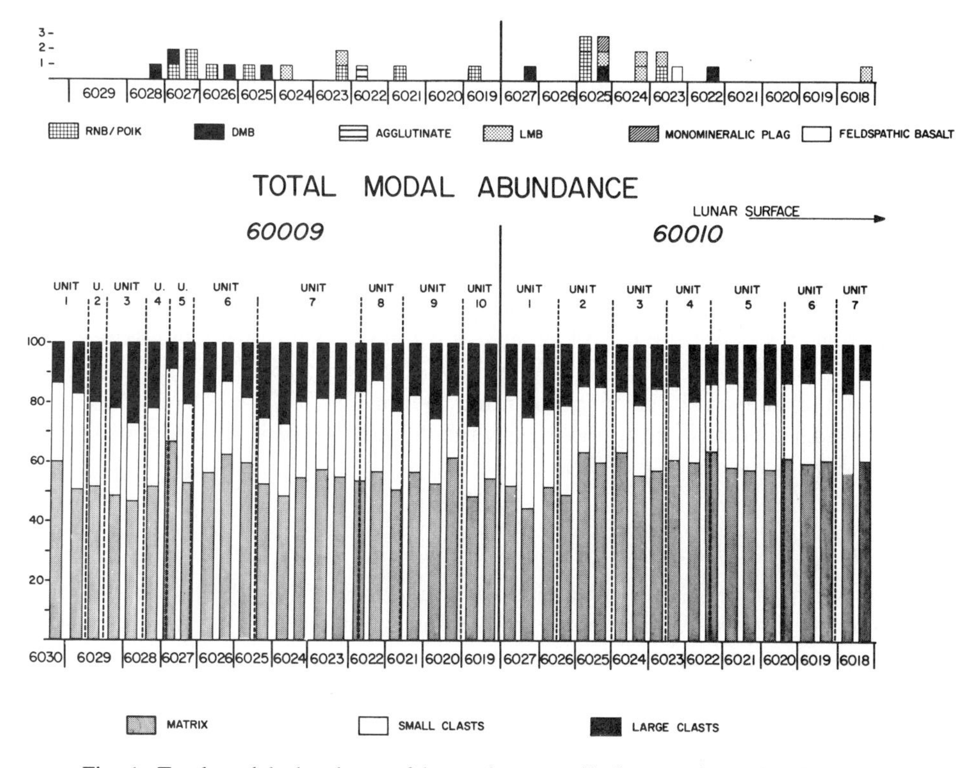

Fig. 1. Total modal abundance of large clasts, small clasts, and matrix. Fragments >2.0 mm were not included in modal analyses but are shown as individual "extra large" clasts. Vertical scale for modal abundance is in percent; vertical scale for extra large clasts is in number of clasts.

tabulations. As described above this includes everything in the plane of the thin section (as observed in reflected light) that is <20 μm in size. Thus, in our usage matrix includes not only particles in the <20 μm size range but also the pore space that is now filled with epoxy. By comparing our modal data of the epoxy impregnated thin sections with the data on weight % of the various size fractions (McKay *et al.*, 1976, 1977), we can make estimates of porosity of the drive tube *after extrusion.* Our average matrix estimate for the <2 mm size fraction is 54.96%. We can compare this to the average wt.% of the <20 μm fraction of McKay *et al.* (1976, 1977) which equals 22.58% (this value has been adjusted to take into account the 1–2 mm size fraction which was not sieved by McKay's group). By comparing the 22.58% value of McKay *et al.* with our estimate of 54.96% matrix, we calculate a mean porosity for the <20 μm fraction of 58.9% and a porosity of 32.4% for the <2 mm size fraction. The 32.4% value for the drive tube porosity is low by about 10% relative to those porosity values reported in a review of lunar soil petrology by Heiken (1975). We believe that this reduction in porosity is due to the core extrusion which reduced the length of the drive tube by ~13%. Thus, there is no incompatibility in the data of McKay *et al.* (1976, 1977) and ours. The difference is that McKay *et al.* are reporting wt.% of particles in the <20 μm fraction whereas we are reporting volume % of the <20 μm particles plus pore space in our matrix category.

Glass data

Glass distribution throughout the drive tube is summarized in Fig. 2. The figure clearly shows that immature soil such as that near the bottom of 60009 and in Stratigraphic Unit 7 of 60009 contains little glass. A more mature soil, such as that in stratigraphic unit 8, has more glass. The most abundant glass types in the drive tube are clear and yellow. Orange, green, brown, and gray glasses are also present in varying but minor quantities. The Apollo 16 drill core contains, on the average, more glass than the drive tube, and has a higher brown glass content (Naney *et al.*, 1976).

Large clasts (0.2–2 mm)

Modal petrography of the large clasts is illustrated in Fig. 3. Rock types that are chemically equivalent have been grouped together. Immature units have high abundances of large clasts, and these coarse-grained layers are enriched in monomineralic fragments. Increasing maturity of the soil towards the top of

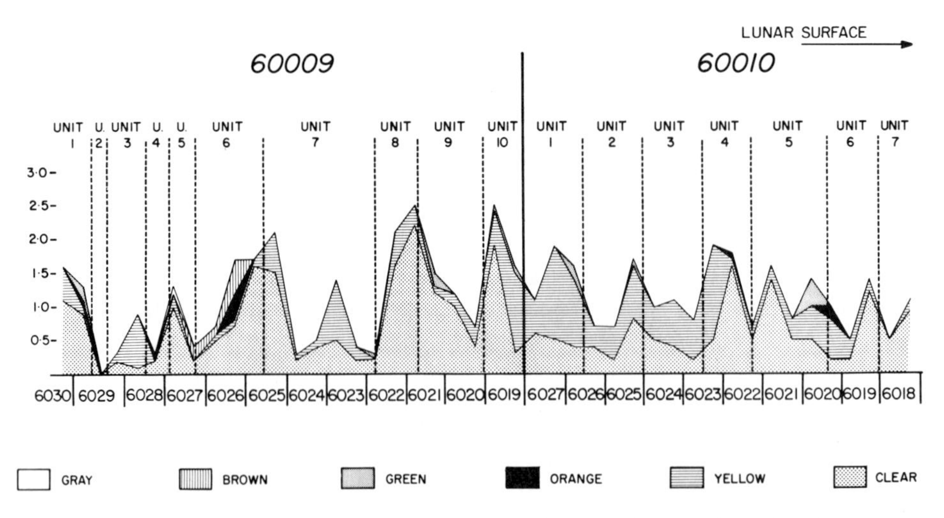

Fig. 2. Stratigraphic distribution of total large and small clast glass abundances. Vertical scale is percentage of total soil.

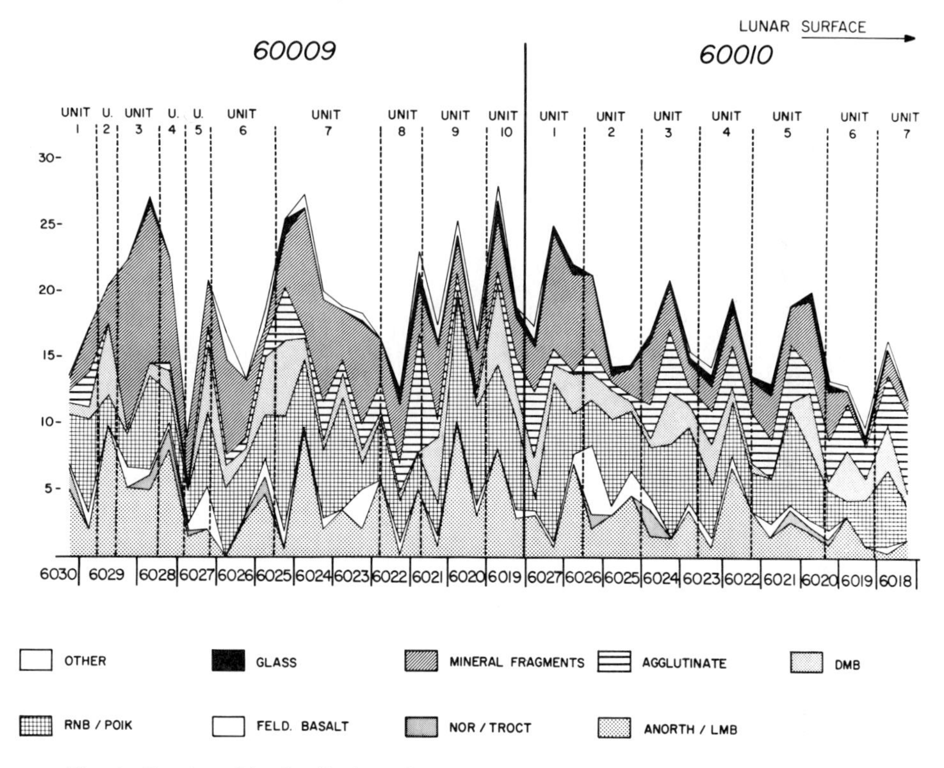

Fig. 3. Stratigraphic distribution of large clast lithic, mineral, glass, and fused soil fragments. Vertical scale is percentage of total soil.

60010 is shown by a decreasing abundance of large clasts and increase in agglutinates. Stratigraphic Unit 8 in 60009 also has a relatively high percentage of agglutinates. This was also recognized by Nagle (1978) who interpreted it as an indication of a much longer surface exposure time than the rest of 60009 (on the order of 7×10^6 yr). This agglutinate buildup may reveal a buried lunar surface. With the exception of the immature feldspathic layer at the base of 60009, the large clasts are predominantly lithic fragments in the POIK/RNB and LMB/anorthosite suites.

Small clasts (0.02–0.2 mm)

Small clast lithologies are summarized in Fig. 4. This size range is dominated by monomineralic fragments except at the top of 60010 where agglutinates become more, and mineral fragments less abundant. As can be seen by comparing Figs. 3 and 4, rock fragments tend to occur as large clasts whereas mineral and glass fragments are more common as small clasts. Grain size in the

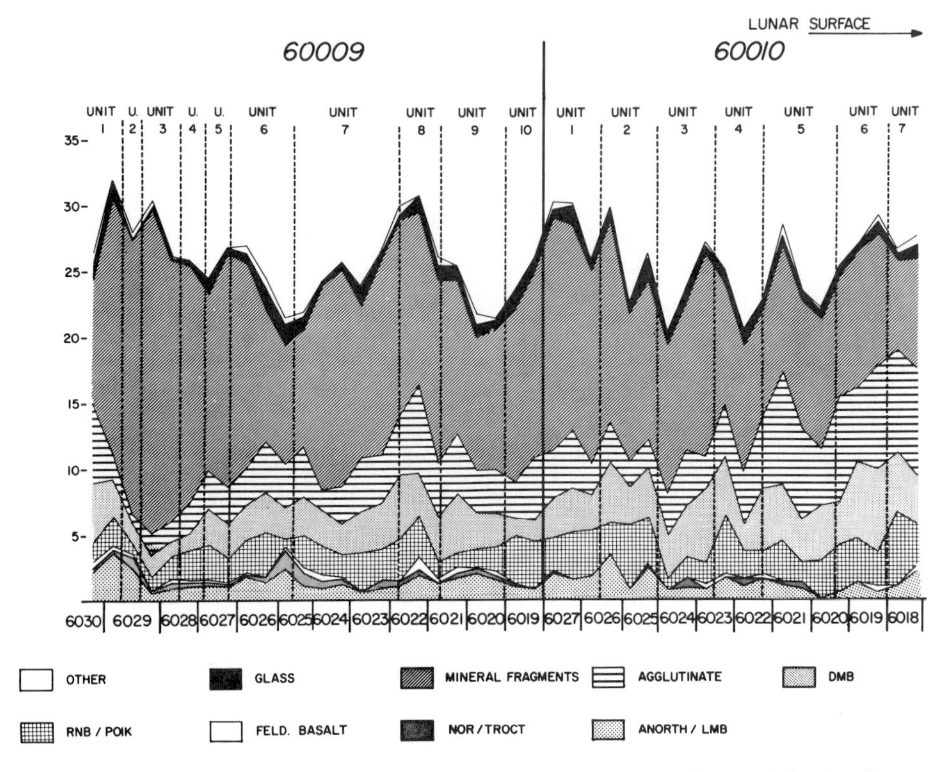

Fig. 4. Stratigraphic distribution of small clast lithic, mineral, glass, and fused soil fragments. Vertical scale is percentage of total soil.

source rocks is such that the <0.2 mm fraction is dominantly monomineralic. The immature layers, which have not undergone much mixing, have a high monomineralic content. The general findings here are similar to those from the drill core, where plagioclase fragments dominate this clast size range.

Monomineralic component

The diagram summarizing the composition of the monomineralic component (Fig. 5) largely reflects the abundance of plagioclase derived from ANT rocks. Plagioclase crystals comprise more than one third of the anorthositic layer near the bottom of 60009, and then drop to about 10% at the surface. The drill core does not exhibit such a pronounced decrease in plagioclase near the surface as is present here (Vaniman *et al.*, 1976). This, along with the lack of an agglutinate buildup in the drill core may indicate that the regolith at the site of the drive tube has been reworked *in situ* more than the surface soils at the drill core site. Neither the drill core nor the drive tube show any significant variations in pyroxene-olivine content; both contain ~1–2% throughout.

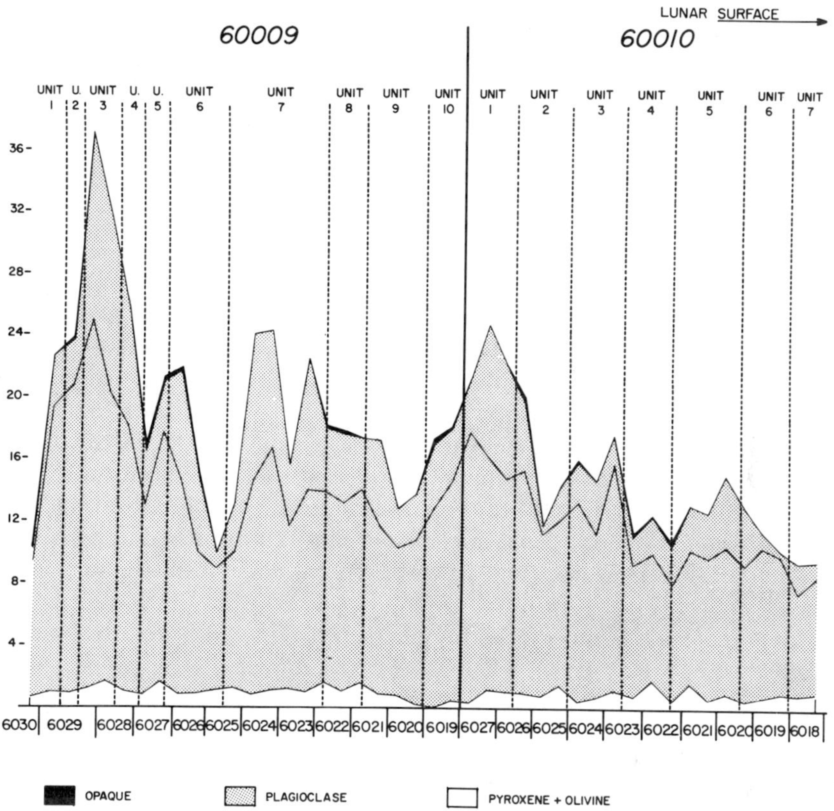

Fig. 5. Stratigraphic distribution of total large and small clast monomineralic component. Lower section of stippled area is plagioclase small clasts; upper part is plagioclase large clasts. Vertical scale is percentage of total soil.

Lithic fragments

The lithic fragments in the drive tube are highland types with the exception of a few mare basalt fragments. Poikilitic rocks are most abundant with slightly fewer coarse-grained anorthositic rocks. Note however, that the immature unit at the base of 60009 has more anorthositic fragments than POIK and RNB fragments. Feldspathic basalts, norites and troctolites are much less abundant (<1% total). This is different from the drill core, where the suite of ANT rocks and light-matrix breccias is three times as abundant as the POIK suite (Vaniman *et al.*, 1976).

Comparative modal petrology: 60009/60010 and the drill core 60001-60007

As a first approximation, the stratigraphic succession in double drive-tube

60009/60010 and the Apollo 16 drill core (60001-60007) are comparable: both consist of coarse-grained, immature basal units overlain by more mature units of finer grain-size (see also Morris, 1978). However, the detailed modal differences between these two soil sections are more striking than their stratigraphic similarities. The modal differences are summarized in Table 1.

What is the nature of the difference between coarse-grained, immature soil and fine-grained, mature soil? In both the double drive tube and the drill core, the coarse-grained soils are marked by an abundance of ANT and LMB lithic fragments, but only in the double drive tube does the coarse-grained unit also have significantly more monomineralic feldspar. It is important to note that the coarse-grained layer in 60009 is significantly more feldspathic and less mafic than the upper part of the double-drive tube; this difference was emphasized in the INAA of Ali and Ehmann (1976) and by Vaniman *et al.* (1978). However, there is essentially no chemical difference between the coarse-grained unit at the base of the deep drill core and the overlying units (Nava *et al.*, 1976) though Gose and Morris (1977) note a soil with lower FeO much higher in the drill core. Re-excavation and mixing of a single soil composition has been an important process at the drill core site, but the soil column at the double drive tube site has preserved an immature soil beneath a mixed and well-gardened upper unit that contains material from at least two different sources which can still be recognized (McKay *et al.*, 1977)

There are more RNB and POIK fragments in the double drive tube than in the deep drill core. McKay *et al.* (1977) have recognized POIK as one of these primary components in the 60009/60010 soils. Moreover, although the double drive tube is more feldspathic than the deep drill core, most of the feldspar occurs as small rather than large clasts (unlike the drill core: compare small- vs. large-clast percentages in Table A-1 and A-2).

Table 1. Summary Modal Comparisons: 60009/60010 and Drill Core 60001–60007 *(0.02–2.0 mm size fraction).*

	60009/60010		60001–60007	
	Coarse-grained Base	Upper Soil	Coarse-grained Base	Upper Soil
DMB and Agglutinates	5.8	12.1	5.5	4.6
RNB and POIK	6.1	8.2	3.8	2.4
ANT and LMB	8.6	4.9	16.6	6.2
Feld. Basalt	0.9	0.6	0.7	0.5
Plagioclase	27.0	17.9	10.5	10.1
Pyroxene and Olivine	1.2	1.0	1.8	1.1
Glasses	0.5	1.2	3.3	2.2
Large Clasts	21.9	17.8	37.5	21.0
Small Clasts	28.4	25.5	8.3	9.1
Matrix	49.7	56.7	54.2	69.9

Morris and Gose (1976) noted an increase of I_s/FeO in the upper 12.5 cm of 60010 and suggested *in situ* reworking to this depth; in a later study of the Apollo 16 drill core, Gose and Morris (1977) found an incresae in I_s/FeO at 13 cm depth, representing a fossil soil surface, overlain by a later deposition of 13 cm of soil with less surface exposure. The period of gardening which has affected the upper 12.5 cm of 60010 may be about as old as the *in situ* residence of the less mature upper 13 cm of the drill core. The ^{38}Ar data of Bogard and Hirsch (1976) give a maximum age of 125 m.y. for the reworking of 60010, and the neutron fluence data of Russ (1973) suggest an age of 50–100 m.y. for the upper 13 cm of the Apollo 16 drill core.

The least mature components of the double drive tube are at the base of 60009 and have ~6% DMB and agglutinates (Table 1). McKay *et al.*, (1977), Bogard and Hirsch (1977), and Blanford *et al.*, (1977) recognize two primary components in the 60010 soils and note an increase of ~2.5 × in the abundance of agglutinates and "brown matrix breccias" as the other maturity indices of these soils (e.g., I_s/FeO) increase. From our data we have noted a 2 × increase in the abundance of agglutinates and DMB (from 6% to 12%) towards the lunar surface in the upper 13 cm of 60010. Therefore, we suggest that an increase of ~6% DMB and agglutinates has occurred in the maturation of 60010 over the last 50-100 million years.

In our previous study of the deep drill core (Vaniman *et al.*, 1976) we remarked on a characteristic two-pyroxene high Ca-low Ca pair in the monomineralic component of one restricted unit above the coarse-grained layer. No lithic correlatives could be found in the drill core, but the same two-pyroxene component is found in the double drive tube with abundant fragments of the POIK source rock (see the companion paper by Vaniman *et al.*, 1978, this volume, for further discussion).

Summary and Conclusions

According to McKay *et al.* (1974), the two main processes in the evolution of lunar regolith are (1) major-impact mixing of components and (2) reworking by micrometeorite impacts. The data reported here, in agreement with the general model of Blanchard *et al.* (1976) and the specific interpretation of 60009/60010 FMR data by Morris and Gose (1976) indicate that in 60009 the mixing of an immature anorthositic soil with a mature soil is the dominant effect above the basal immature unit. At the top of 60010 the reworking and maturation process has greater effect. Ali and Ehmann (1977) reached a similar conclusion. A coarser average grain size and lower glass content in the drive tube may indicate that it is less mature than the drill core.

Acknowledgments—The authors would like to acknowledge Claude Mercier for her skillful drafting of the diagrams, and Joan Coffey and Pauline Papike for aid in the preparation of the manuscript. This work was supported by NASA Grant #NSG-9044 which we gratefully acknowledge.

REFERENCES

Ali M. Z. and Ehmann W. D. (1976) Chemical characterization of lunar core 60009. *Proc. Lunar Sci. Conf. 7th,* p. 241–258.

Ali M. Z. and Ehmann W. D. (1977) Chemical characterization of lunar core 60010. *Proc. Lunar Sci. Conf. 8th,* p. 2967–2981.

Blanchard D. P., Jacobs J. W., Brannon J. C. and Brown R. W. (1976) Drive tube 60009: A chemical study of magnetic separates of size fractions from five strata. *Proc. Lunar Sci. Conf. 7th,* p. 281–294.

Blanford G. E., McKay D. S. and Wood G. C. (1977) Particle track densities in double drive tube 60009/10. *Proc. Lunar Sci. Conf. 8th,* p. 3017-3025.

Bogard D. D. and Hirsch W. C. (1976) Noble gases in 60009-60010 drive tube samples: Trapped gases and irradiation history. *Proc. Lunar Sci. Conf. 7th,* p. 259-279.

Bogard D. D. and Hirsch W. C. (1977) Noble gas evidence for the depositional and irradiational history of 60010-60009 core soils. *Proc. Lunar Sci. Conf. 8th,* p. 2983–2999.

Duke M. B. and Nagle J. S. (1974) Lunar Core Catalog, 1977 Supplement. JSC 90252. NASA Johnson Space Center, Houston.

Fruchter J. S., Rancitelli L. A., Laul J. C. and Perkins R. W. (1977) Lunar regolith dynamics based on analysis of the cosmogenic radionuclides ^{22}Na, ^{26}Al, and ^{53}Mn. *Proc. Lunar Sci. Conf. 8th,* p. 3595–3605.

Gose W. A. and Morris R. V. (1977) Depositional history of the Apollo 16 deep drill core. *Proc. Lunar Sci. Conf. 8th,* p. 2909–2928.

Heiken G. (1975) Petrology of lunar soils. *Rev. Geophys. Space Phys.* **13**, 567-587.

McKay D. S., Dungan M. A., Morris R. V. and Fruland R. M. (1977) Grain size, petrographic, and FMR studies of the double core 60009/10: A study of soil evolution. *Proc. Lunar Sci. Conf. 8th,* p. 2929–2952.

McKay D. S., Fruland R. M. and Heiken G. H. (1974) Grain size and the evolution of lunar soils. *Proc. Lunar Sci. Conf. 5th,* p. 887-906.

McKay D. S., Morris R. V., Dungan M. A., Fruland R. M. and Fuhrman R. (1976) Comparative studies of grain size separates of 60009. *Proc. Lunar Sci. Conf. 7th,* p. 295–313.

Morris R. V. (1978) *In situ* reworking (gardening) of the lunar surface: evidence from the Apollo cores (abstract). In *Lunar and Planetary Science IX,* p. 757–759. Lunar and Planetary Institute, Houston.

Morris R. V. and Gose W. A. (1976) Feromagnetic resonance and magnetic studies of cores 60009/60010 and 60003: Compositional and surface exposure stratigraphy. *Proc. Lunar Sci. Conf. 7th,* p. 111.

Nagle J. S. (1978) The authigenic component in lunar cores (abstract). In *Lunar and Planetary Science IX,* p. 790–792. Lunar and Planetary Science Institute, Houston.

Naney M. T., Crowl D. M. and Papike J. J. (1976) The Apollo 16 drill core: Statistical analysis of glass chemistry and the characterization of a high alumina-silica poor (HASP) glass. *Proc. Lunar Sci. Conf. 7th,* p. 155–184.

Nava D. F., Lindstrom M. M., Schumann P. J., Lindstrom D. J. and Philpotts J. A. (1976) The remarkable chemical uniformity of Apollo 16 layered deep drill core section 60002. *Proc. Lunar Sci. Conf. 7th,* p. 133–139.

Russ G. P., III (1973) Apollo 16 neutron stratigraphy. *Earth Planet. Sci. Lett.* **17**, 275–289.

Vaniman D. T., Lellis S. F., Papike J. J. and Cameron K. L. (1976) The Apollo 16 drill core: Modal petrology and characterization of the mineral and lithic component. *Proc. Lunar Sci. Conf. 7th,* p. 199-239.

Vaniman D. T., Papike J. J. and Schweitzer E. L. (1978) Apollo 16 drive tube 60009/60010. Part II. Petrology and major element partitioning among the regolith components. *Proc. Lunar Planet. Sci. Conf. 9th.* This volume.

Table A 1. Apollo 16 Drive Tube 60009: Modal Data.

		6030	6029			6028		6027		6026		6025		6024		6023		6022		6021		6020		6019	
			B	M	T	B	T	B	T	B	T	B	T	B	T	B	T	B	T	B	T	B	T	B	T
	Lithic Fragments																								
	Mare Component	—	—	—	—	—	—	—	—	—	—	—	—	—	—	—	0.3	—	—	—	0.7	—	0.7	—	—
	Highland Component																								
	Anorthosites	0.6	1.3	—	0.7	0.7	4.8	1.5	1.8	—	0.5	0.8	—	1.0	1.0	—	0.9	0.3	0.2	1.8	0.3	2.3	—	0.7	1.6
	Norite/Troctolite	1.6	—	—	—	1.1	1.5	0.4	—	0.1	0.4	1.3	—	0.3	—	—	—	—	—	—	—	0.2	—	—	—
	LMB	4.4	0.8	9.8	4.5	4.3	3.2	—	0.3	—	2.2	3.8	0.6	8.4	1.1	3.4	1.2	5.4	—	3.2	0.5	7.7	3.1	7.3	1.3
	Feldspathic Basalt	0.2	1.1	—	1.5	0.3	0.5	0.5	3.1	—	—	1.4	1.2	—	0.7	—	2.5	—	0.8	—	—	—	—	—	0.5
	RNB	1.4	0.1	0.7	—	2.1	—	—	0.3	0.7	0.9	0.4	2.3	—	0.6	4.7	—	0.3	1.3	—	0.3	4.4	—	1.2	—
	POIK	2.4	7.1	1.7	2.6	5.3	2.3	2.7	5.3	4.4	3.4	3.0	6.6	5.2	4.7	3.9	2.1	4.4	1.9	2.8	2.2	5.0	7.5	5.3	7.0
Large	Fused Soil Component																								
Clasts	DMB	0.8	0.9	5.3	0.2	0.9	1.7	—	5.1	1.6	0.7	4.3	5.6	1.6	0.6	1.8	1.0	0.5	0.6	—	5.0	0.4	0.7	6.0	1.9
	Agglutinate	1.1	2.6	—	0.2	—	0.7	0.4	1.4	0.8	0.4	1.3	4.0	0.4	3.0	1.0	2.0	2.2	2.2	9.0	1.3	1.4	0.6	0.9	2.8
0.2–2.0 mm	Mineral Fragments																								
	Olivine/Pyroxene	—	0.3	—	0.4	0.3	—	—	—	—	—	—	—	—	—	—	—	—	—	—	—	—	—	—	—
	Plagioclase	0.9	3.0	3.0	12.2	11.5	7.8	3.5	3.5	7.2	4.8	1.0	4.1	9.5	7.7	3.9	7.7	3.2	4.5	3.4	5.6	2.6	3.0	4.2	3.4
	Opaques	—	—	—	—	—	—	—	—	—	—	—	—	—	—	—	—	—	—	—	—	—	—	—	—
	Glass Fragments																								
	Orange	—	—	—	—	—	—	—	—	—	—	—	—	—	—	—	—	—	—	—	—	—	—	—	—
	Yellow	—	—	—	—	0.7	—	—	—	—	—	—	—	—	—	—	—	—	—	—	—	0.2	—	—	0.5
	Green	—	—	—	—	—	—	—	—	—	—	—	—	—	—	—	—	—	—	—	—	—	—	—	—
	Brown	—	—	—	—	—	—	—	—	—	—	—	—	—	—	—	—	—	—	—	—	—	—	—	—
	Gray	—	—	—	—	—	—	—	—	—	—	—	—	—	—	—	—	—	—	—	—	—	—	—	—
	Clear	0.2	—	—	—	—	—	—	—	—	—	—	1.1	—	—	—	0.1	—	1.0	1.3	0.3	—	—	1.3	—
	Miscellaneous																								
	Devitrified Glass	—	—	—	—	—	—	—	—	2.1	—	1.4	—	0.9	0.2	0.2	0.6	—	—	1.5	1.3	1.1	1.5	1.1	—
	Other	—	—	—	—	—	—	—	—	—	—	—	—	—	0.4	—	—	—	—	—	—	—	—	—	—

Table A 1. *(cont'd.)*

		6030	6029			6028		6027		6026		6025		6024		6023		6022		6021		6020		6019	
			B	M	T	B	T	B	T	B	T	B	T	B	T	B	T	B	T	B	T	B	T	B	T
	Lithic Fragments																								
	Mare Component	–	–	–	–	–	–	–	–	–	–	–	–	–	–	–	–	–	–	–	–	–	–	–	–
	Highland Component																								
	Anorthosites	1.1	1.9	0.3	–	0.2	0.5	0.5	0.6	1.5	1.1	1.8	0.6	0.6	0.8	0.6	0.2	0.5	1.3	1.0	1.5	1.3	1.0	0.4	0.3
	Norite/Troctolite	0.3	0.2	0.9	0.1	0.4	0.3	0.3	0.2	0.2	0.4	1.4	1.0	0.6	0.4	–	0.7	0.3	0.6	0.1	0.4	0.4	0.3	0.1	–
	LMB	1.1	1.7	2.1	0.6	0.7	0.5	0.6	0.4	0.3	0.3	0.6	0.5	0.2	0.4	–	0.6	0.6	0.4	0.3	0.2	0.7	0.5	0.6	0.5
	Feldspathic Basalt	0.4	0.3	0.1	–	0.3	0.2	0.1	0.1	0.1	–	0.1	0.2	0.5	0.1	–	–	–	0.7	0.2	0.3	–	0.3	–	–
	RNB	0.4	0.7	0.3	–	0.5	–	0.1	0.3	0.4	1.0	–	0.2	0.4	0.1	0.3	–	0.2	0.1	–	–	0.1	0.1	–	–
	POIK	1.2	1.5	0.6	1.2	1.3	2.3	2.6	1.7	2.3	2.4	0.8	2.4	1.8	1.7	2.7	2.3	3.1	3.1	1.3	1.3	1.4	1.9	3.8	3.7
Small	Fused Soil Component																								
Clasts	DMB	4.4	2.9	1.7	1.6	1.1	1.3	2.8	2.5	2.5	3.0	2.4	2.9	2.7	2.3	3.1	3.5	4.8	3.4	3.4	4.4	2.8	2.5	1.3	1.6
	Agglutinate	5.9	2.0	0.6	1.7	1.5	2.3	2.8	2.8	2.7	3.9	3.3	3.9	2.6	2.8	4.1	3.7	4.6	6.7	4.1	4.6	3.2	3.3	2.8	4.7
0.02–0.2 mm	Mineral Fragments																								
	Olivine/Pyroxene	0.7	1.0	0.9	0.8	1.4	1.0	0.8	1.6	0.9	0.9	1.1	1.3	0.9	1.1	1.2	1.0	1.6	1.1	1.6	0.9	0.8	0.3	–	0.5
	Plagioclase	8.7	18.3	19.8	23.7	18.5	17.1	12.2	16.1	14.7	9.1	7.9	7.7	13.7	15.6	10.5	14.0	13.3	12.1	12.4	10.8	9.5	10.5	12.9	14.1
	Opaques	0.1	–	0.1	–	–	–	0.5	0.1	0.1	0.1	–	–	–	–	–	–	0.1	0.1	–	–	–	–	0.3	0.1
	Glass Fragments																								
	Orange	–	0.1	–	–	–	0.1	–	–	–	0.4	–	–	–	–	–	–	–	–	–	–	–	–	–	–
	Yellow	0.5	0.1	–	0.1	–	–	0.2	–	0.1	0.1	0.1	0.6	0.1	0.1	0.9	0.2	–	0.5	0.3	0.1	–	0.3	0.5	0.7
	Green	–	0.2	–	–	–	–	–	–	–	–	–	–	–	–	–	–	0.2	–	–	0.2	–	–	0.1	0.1
	Brown	–	–	–	–	–	–	–	0.2	0.1	0.5	–	–	–	–	–	–	–	–	–	–	–	–	–	–
	Gray	–	–	–	–	–	–	0.1	–	–	–	–	–	–	–	–	–	–	–	–	–	–	–	–	–
	Clear	0.9	0.9	–	0.2	0.1	0.2	1.0	0.2	0.5	0.7	1.6	0.4	0.2	0.4	0.5	0.2	0.1	0.6	0.9	0.9	1.0	0.4	0.6	0.3
	Miscellaneous																								
	Devitrified Glass	0.7	0.1	0.4	0.3	–	–	–	–	0.6	0.7	0.5	0.1	–	–	0.1	0.2	0.7	–	0.6	–	0.6	0.1	0.3	–
	Other	–	–	–	–	0.1	–	–	–	–	–	–	0.1	–	–	–	–	–	–	–	–	–	–	–	–
Totals	Large Clasts	13.6	17.2	20.5	22.3	27.2	22.5	9.0	20.8	14.8	13.3	18.7	25.5	27.3	20.0	18.9	18.4	16.3	12.5	23.0	17.5	25.3	17.1	28.0	19.5
	Small Clasts	26.4	31.9	27.8	30.3	26.1	25.8	24.6	26.8	27.0	24.6	21.6	21.9	24.3	25.8	24.0	26.6	30.1	30.8	26.2	25.6	21.8	21.5	23.7	26.1
	Matrix	60.0	50.9	51.7	47.4	46.7	51.7	66.4	52.4	58.2	62.1	59.7	52.6	48.4	54.2	57.1	55.0	53.6	56.7	50.8	56.9	52.9	61.4	48.3	54.4

Table A 2. Apollo 16 Drive Tube 60010: Modal Data.

		6027		6026		6025		6024		6023		6022		6021		6020		6019		6018	
		B	T	B	T	B	T	B	T	B	T	B	T	B	T	B	T	B	T	B	T
	Lithic Fragments																				
	Mare Component	0.2	–	0.1	3.1	0.5	–	–	–	–	0.6	–	–	–	–	0.4	–	–	–	–	–
	Highland Component																				
	Anorthosites	1.2	0.4	0.4	0.7	2.0	1.9	1.3	0.4	1.5	0.6	0.9	1.2	0.4	2.5	0.5	0.2	0.9	0.6	–	–
	Norite/Troctolite	–	–	–	1.1	–	–	2.0	–	–	–	–	–	–	1.0	0.5	0.4	–	–	–	–
	LMB	2.0	1.8	6.6	1.4	1.2	2.8	0.3	0.9	2.0	0.1	5.7	1.1	1.2	0.2	1.4	0.8	2.2	0.3	0.4	1.4
	Feldspathic Basalt	–	0.2	0.5	2.0	–	1.7	0.8	–	0.5	–	0.9	–	1.0	0.1	–	0.7	–	–	0.5	–
	RNB	–	1.0	–	0.4	2.3	1.6	–	–	–	0.4	0.4	–	0.5	–	–	–	0.3	0.8	–	–
	POIK	1.0	9.8	3.2	3.1	4.3	3.0	3.9	7.2	5.7	3.7	3.7	4.0	3.0	7.4	5.7	3.0	0.9	2.7	5.8	2.6
Large	Fused Soil Component																				
	DMB	3.0	1.3	3.1	2.0	2.1	0.6	0.6	3.9	1.8	3.1	1.2	0.7	–	0.5	3.9	0.7	3.8	1.6	3.1	0.8
Clasts	Agglutinate	5.3	1.2	0.1	1.9	0.7	0.5	2.6	4.7	1.3	2.5	3.1	3.8	2.8	4.4	1.9	3.1	3.6	2.4	4.0	6.1
0.2–2.0 mm	Mineral Fragments																				
	Olivine/Pyroxene	–	0.4	–	–	–	–	–	–	–	–	0.3	–	0.5	–	0.5	–	–	–	–	–
	Plagioclase	3.4	8.5	7.4	5.5	0.5	2.1	4.7	3.4	1.9	1.9	2.4	2.5	2.9	2.9	4.6	3.8	1.0	0.3	2.0	1.1
	Opaques	–	–	–	–	–	–	–	–	–	–	–	–	–	–	–	–	–	–	–	–
	Glass Fragments																				
	Orange	–	–	–	–	–	–	–	–	–	–	–	–	–	–	–	–	–	–	–	–
	Yellow	–	0.4	0.4	–	–	–	0.1	0.2	0.2	0.8	–	–	–	–	0.4	0.4	–	–	–	–
	Green	–	–	–	–	–	–	–	–	–	–	–	–	–	–	0.3	–	–	–	–	–
	Brown	–	–	–	–	–	–	–	–	–	–	–	–	–	–	–	–	–	–	–	–
	Gray	–	–	–	–	–	–	–	–	–	–	–	–	–	–	–	–	–	–	–	–
	Clear	0.4	–	0.3	–	–	–	–	0.1	–	–	0.7	–	0.8	–	–	–	–	0.4	–	–
	Miscellaneous																				
	Devitrified Glass	0.9	–	–	0.1	0.7	0.2	–	–	0.5	0.6	–	0.3	–	–	–	0.2	0.3	0.7	0.6	–
	Other	–	–	–	–	–	–	–	–	–	–	–	–	–	–	–	–	–	–	–	–

Table A 2. *(cont'd.)*

		6027		6026		6025		6024		6023		6022		6021		6020		6019		6018	
		B	T	B	T	B	T	B	T	B	T	B	T	B	T	B	T	B	T	B	T
	Lithic Fragments																				
	Mare Component	—	—	—	—	—	—	—	—	—	—	—	—	—	—	—	—	—	—	—	0.1
	Highland Component																				
	Anorthosites	1.4	1.6	0.9	1.8	0.6	2.1	0.8	0.7	0.8	1.8	1.0	1.7	1.0	0.7	0.1	0.6	0.7	0.4	0.7	1.5
	Norite/Troctolite	0.1	—	—	—	—	0.2	—	0.8	—	—	0.5	—	1.2	0.5	—	—	—	—	—	—
	LMB	0.7	1.2	0.9	1.7	0.2	0.4	—	0.2	0.2	—	0.1	—	0.2	0.2	—	—	0.7	0.3	0.4	0.7
	Feldspathic Basalt	—	—	—	—	—	—	—	—	0.2	0.1	—	0.2	—	—	—	—	—	0.3	—	—
	RNB	—	—	—	—	0.5	0.2	—	—	—	—	0.1	—	—	—	—	—	—	0.1	—	0.1
	POIK	2.7	2.4	3.6	2.4	4.5	3.4	0.9	1.6	1.6	4.5	2.1	2.0	3.1	1.5	2.9	3.6	3.3	2.6	5.5	3.4
	Fused Soil Component																				
	DMB	2.8	3.2	2.5	4.5	2.8	3.7	3.3	3.9	5.5	3.7	2.0	4.5	4.2	3.2	4.2	3.2	5.8	6.2	4.6	3.6
Small	Agglutinate	3.7	4.5	2.4	3.1	1.9	2.2	3.2	4.2	2.5	4.7	3.9	5.6	8.6	7.0	4.3	8.1	5.7	8.0	7.3	8.2
Clasts	Mineral Fragments																				
	Olivine/Pyroxene	0.4	0.8	1.1	1.0	0.8	1.5	0.5	0.7	1.2	0.8	1.5	0.5	1.1	0.6	0.5	0.5	0.7	1.0	0.9	1.0
0.02–0.2 mm	Plagioclase	17.4	15.0	13.7	14.3	10.5	10.7	10.7	10.6	14.5	8.4	8.2	7.5	8.6	9.1	9.4	8.7	9.7	8.9	6.5	7.4
	Opaques	—	—	—	0.3	—	—	0.1	—	—	0.2	—	0.2	—	—	—	0.1	0.1	0.1	—	0.1
	Glass Fragments																				
	Orange	—	—	—	—	—	—	—	—	—	—	0.1	—	—	—	—	0.2	—	—	—	—
	Yellow	0.5	1.0	0.6	0.3	0.5	0.8	0.4	0.5	0.4	0.6	0.1	0.2	0.2	0.3	0.1	0.2	0.3	0.2	—	0.2
	Green	—	—	0.2	—	—	0.1	—	—	—	—	—	—	—	—	0.1	—	—	—	—	—
	Brown	—	—	—	—	0.1	—	—	—	—	—	—	—	—	—	—	—	—	—	—	—
	Gray	—	—	—	—	—	—	—	—	—	—	—	—	—	—	—	—	—	—	—	—
	Clear	0.2	0.5	0.1	0.4	0.2	0.8	0.5	0.3	0.2	0.5	0.9	0.5	0.6	0.5	0.5	0.2	0.2	0.8	0.5	0.9
	Miscellaneous																				
	Devitrified Glass	0.2	0.1	—	0.2	—	0.4	—	0.2	0.2	—	0.1	—	0.9	—	0.1	0.1	—	0.5	0.4	0.6
	Other	0.3	—	—	—	—	—	—	—	—	—	—	—	—	—	—	—	—	—	—	—
Totals	Large Clasts	17.4	25.0	22.1	21.3	14.3	14.4	16.3	20.8	15.4	14.3	19.3	13.6	13.1	19.0	20.1	13.3	13.0	9.8	16.4	12.0
	Small Clasts	30.4	30.3	26.0	30.0	22.6	26.5	20.4	23.7	27.3	25.3	20.6	22.9	28.7	23.6	22.2	25.5	27.2	29.4	26.8	27.8
	Matrix	52.2	44.7	51.9	48.7	63.1	59.1	63.3	55.5	57.3	60.4	60.1	53.5	58.2	57.4	57.7	61.2	59.8	60.8	56.8	60.2

Proc. Lunar Planet. Sci. Conf. 9th (1978), p. 1827–1860.
Printed in the United States of America

The Apollo 16 drive tube 60009/60010. Part II: Petrology and major element partitioning among the regolith components

D. T. VANIMAN, J. J. PAPIKE and E. L. SCHWEITZER

Planetary Regolith Studies, Department of Earth and Space Sciences,
State University of New York, Stony Brook, New York 11794

Abstract—Compositions of pyroxene, olivine and plagioclase preserve a record of the lithic components which contributed to the >20 μm soils in double drive tube 60009/60010. Variation of the mineral component with depth does not necessarily indicate stratification; in most cases this variation is preserved simply because of incomplete mixing. The preservation of mineral compositional clusters permits identification of specific poikilitic and ANT/LMB rock types which may be otherwise poorly preserved. The dominant poikilitic source rock in 60009/60010 is represented by both lithic fragments and a compositional cluster of pyroxenes at several levels of the double drive tube, but this same cluster of pyroxene compositions was previously found in the Apollo 16 deep drill core without a lithic correlative. The mineral data provide one means of correlating source lithologies between the A-16 deep drill core and 60009/60010.

Pyroxene, olivine and glass components must be considered in addition to lithic fragments and feldspar in order to characterize the mixing processes which formed the 60009/60010 soils. In addition to previously recognized feldspar, POIK and crystalline ANT components of the mixture there are important contributions of pyroxenes that can be related to specific crystalline ANT or POIK sources, olivines from a specific ANT source and a variety of other sources, feldspars attributable to either melt rock or ANT sources, and glasses of anorthositic gabbro composition which reflect the dominant highland melt rock composition. The importance of considering a large number of lithic, mineral and glass components is brought out when one calculates a bulk soil chemistry directly from modal data.

In this paper we use the soil components we actually observe petrographically, their modal proportions, and their chemistry as determined by microprobe analyses to calculate the major-element soil chemistry by a modal recombination technique. By using this approach we can reliably calculate the chemistry of the >0.02 mm size fraction and, more significantly, derive the total major element partitioning among the regolith components.

Glasses in the double drive tube have a compositional range similar to glasses in the Apollo 16 drill core; the Apollo 16 glasses do not include mare-highland mixtures with evidence of silica volatilization as found at the Apollo 17 site. Exotic components (e.g., mare rocks/minerals/glasses) form ≤0.1% of the >20 μm soil.

INTRODUCTION

This paper summarizes petrographic and electron microprobe data from 22 polished thin sections that represent the total length of double drive tube 60009/60010 and discusses a new application of the method of modal recombination to estimate bulk chemistry. A companion paper by Simon *et al.* (1978) describes the detailed modal data from 60009/60010. Figure 1 shows the layers obtained in various studies of 60009/60010: X-ray study before core extrusion

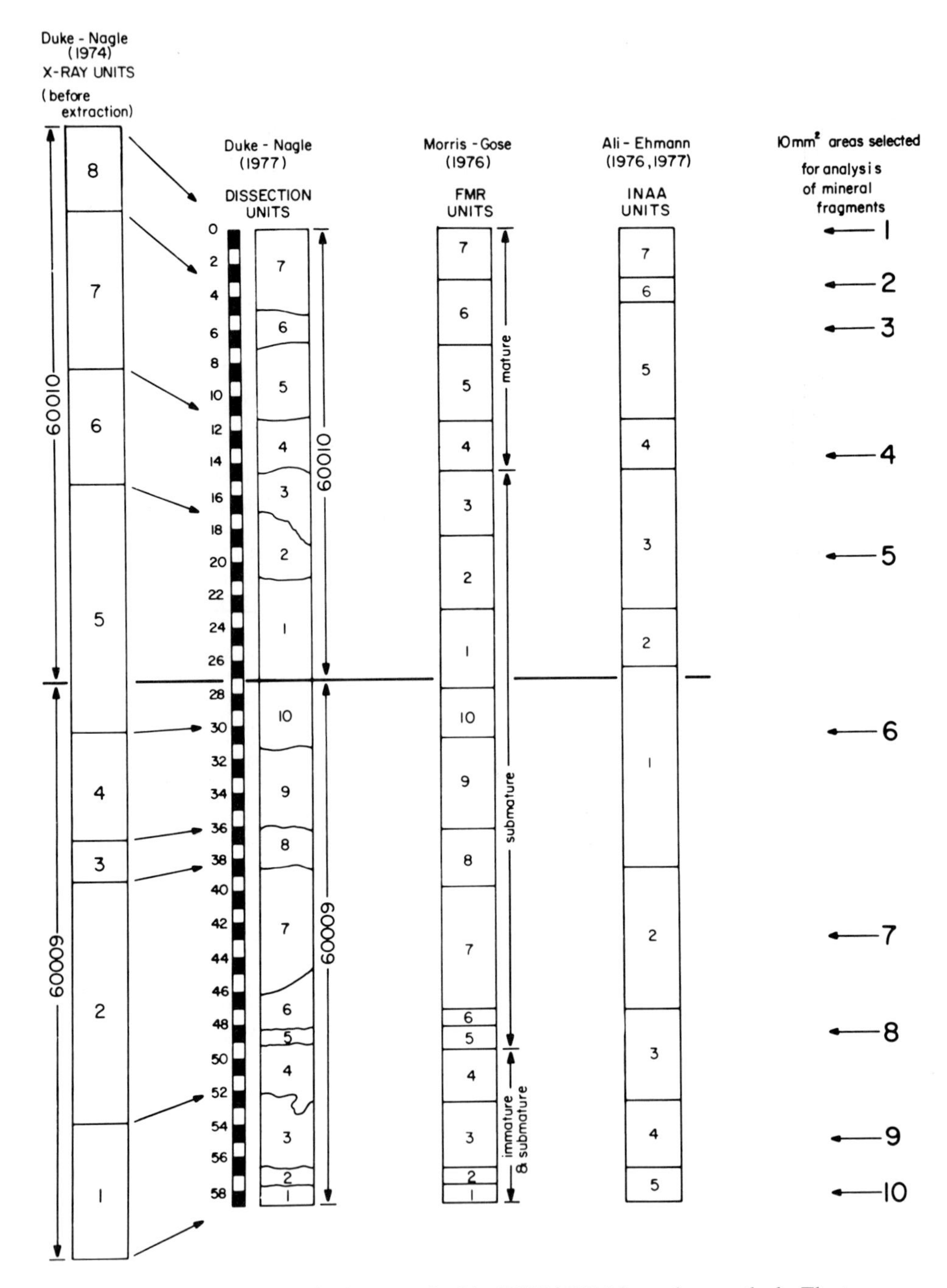

Fig. 1. A comparison of units recognized in 60009/60010 by various methods. The ten levels represented to the right are the depths of the ten areas described in the text and in subsequent diagrams.

and dissection studies following extrusion (Duke and Nagle, 1974, 1977; note ~13% shortening of the core due to the process of extrusion); chemical studies (Ali and Ehmann, 1976, 1977) and study of I_s/FeO maturity (Morris and Gose, 1976). At the extreme right of the figure are ten arrows that indicate levels of the double drive tube where we collected compositional data on minerals and glasses (lithic fragments were analyzed throughout the double drive tube). These ten levels were chosen to fall within layers of the core sample which were recognized by one or more analysts.

The double drive tube 60009/60010 (69 cm in length) was taken on the rim of a shallow 50–60 cm crater 60 m SW of the landing module at station 10 (LSPET, 1973). This station was also the site of the deep drill-core sample (60001–60007), which was collected ~60 m SW of 60009/60010. Although these two soil sections are close together, it is not possible to correlate strata over this distance (Gose and Morris, 1977). The combined studies of neutron fluence (Russ, 1973), petrographic and grain-size analysis (McKay *et al.*, 1976, 1977) and the maturity index I_s/FeO (Morris and Gose, 1976; Gose and Morris, 1977) show that both the lower drill core (below 13 cm) and 60009/60010 were emplaced in single events or in a rapid series of discrete events. Other data require that these two emplacement events are not related. The neutron fluence data of Russ (1973) and rare gas data of Bogard and Hirsch (1975) show that the lower drill core (below 13 cm) was emplaced ~450 m.y. ago, but the ^{38}Ar data of Bogard and Hirsch (1976) show a maximum emplacement age of 125 m.y. for 60009/60010. Neutron fluence data (Russ, 1973) suggest that the upper 13 cm of the deep drill core were deposited ~50–100 m.y. ago; since this is within the age estimate for deposition of 60009/60010, could the upper drill core and the double drive tube represent a single stratum? The petrographic dissimilarities rule out this possibility (ref. McKay *et al.*, 1976, 1977, and Simon *et al.*, 1978). Although the deep drill core and 60009/60010 were collected only 60 meters apart, they represent two separate emplacements and share no stratigraphic horizons except the present lunar surface.

Although the deep drill core and 60009/60010 cannot be related stratigraphically, it is possible that specific rock and mineral compositions can be correlated between these two cores. It is reasonable to assume that the same source regions contributed to closely-spaced ejecta deposits even though the emplacements may be widely separated in time. It is also possible that unique soils can be traced from the deep drill core to 60009/60010 on the basis of soil composition rather than stratigraphy. Gose and Morris (1977) suggest that a mature FeO-poor soil at ~13–16 cm depth in the drill core may be similar to the immature FeO-poor soil at ~53–57 cm (post-extrusion depth) in 60009/60010 (data of Ali and Ehmann, 1976). Even though these unique soils might be correlated, the FeO-poor layer in the drill core must be far older (~450 m.y.) than the FeO-poor layer in 60009 (<125 m.y.; Gose and Morris, 1977). We have approached the problem of correlation by comparing the lithic and mineral compositions of the drill core and 60009/60010. Using this method, we can recognize and correlate a specific poikilitic (POIK) melt-rock from pyroxene data in the drill core and from pyroxene and lithic data in the double drive tube. This "POIK" lithology is described in greater detail in this paper.

The POIK melt-rock type is a major contributor of lithic and mineral fragments in 60009/60010. McKay *et al.* (1976, 1977) recognize two mixing trends involving three major components in the double drive tube: the lower soils (60009) can be described as mixtures of plagioclase fragments and an end-member of 4:1 ratio of POIK:metamorphosed breccia (recrystallized ANT), and the upper soils (60010) are mixtures of the same plagioclase-fragment composition with an end-member of 1:1 ratio of POIK:metamorphosed breccia. These three categories of feldspar, POIK and LMB/ANT describe the three most abundant lithic components which we have found in 60009/60010 (Simon *et al.*, 1978), though we have also found that clear glass (anorthositic gabbro composition), pyroxene and olivine are primary soil components which must be taken into consideration in describing the total variation in >20 μm soils of the double drive tube. The importance of these other components is brought out when a bulk soil composition is calculated from modal data.

Many analysts have remarked on the presence of a coarse-grained, feldspar-rich (anorthitic) layer at the base of 60009 (~53–57 cm depth after extrusion) overlain by finer-grained soils (Duke and Nagle, 1974; McKay *et al.*, 1976; Blanchard *et al.*, 1976; Ali and Ehmann, 1976; Blanford *et al.*, 1977; Morris, 1978). The coarse-grained soil is the plagioclase-rich component in the mixing models of McKay *et al.* (1976, 1977), and is a markedly immature soil (Morris, 1978) with a more anorthitic (34% Al_2O_3) and FeO-poor (1.7% FeO) composition compared to the average soil (28% Al_2O_3, 5% FeO) of the upper drill core (Blanchard *et al.*, 1976; Ali and Ehmann, 1976, 1977). The anorthitic soil at the base of 60009 has been compared to rock 60025 by Blanchard *et al.*, (1976). This composition is rare, for the soils from most other Apollo 16 stations have Al_2O_3 contents in the range of 27–29% and FeO contents in the range of 4–6% (LSPET, 1973). The data of Blanchard *et al.* (1976) and Ali and Ehmann (1976, 1977) also show that the coarse, anorthitic soils in 60009 have a unique rare-earth element signature. They have Sm/chondritic ratios of ~10× with a positive Eu anomaly, whereas the more "normal" soils higher in the core have Sm/chondritic ratios of ~30× with a negative Eu anomaly (Fig. 2). In the nomenclature of Reid *et al.* (1972) the major- and trace-element composition of the coarse, anorthitic soil is that of *gabbroic anorthosite*, whereas the composition of more typical Apollo 16 soil and the upper 60009/60010 core is that of *anorthositic gabbro*. In this paper we use modal recombination to arrive at a bulk soil chemistry and to illustrate the change from the coarse anorthitic layer to the more fine-grained soils higher in the core as a modal exchange (disregarding agglutinates and other maturity indices) of POIK, pyroxene, olivine and clear glass for the coarse feldspar and ANT components of the anorthitic layer.

Figure 1 shows that some of the units in 60009/60010 are recognized by a variety of analytical methods (visual discrimination, FMR, chemical analysis). Verification of a core layer boundary by more than one method inspires greater confidence that the layer has some stratigraphic significance. For example, the

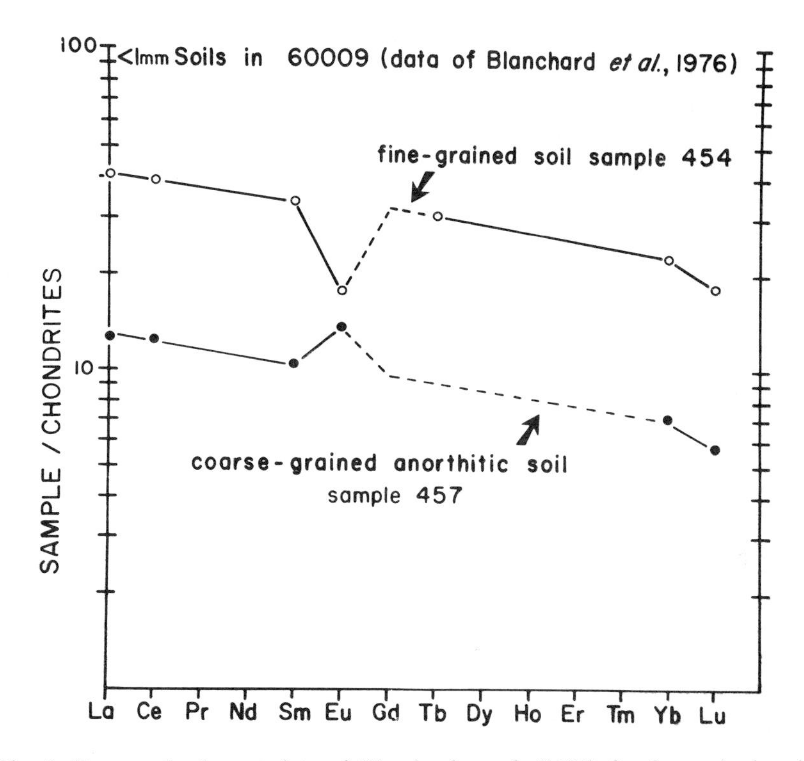

Fig. 2. Rare earth element data of Blanchard *et al.* (1976) for fine-grained and coarse-grained anorthitic soils of 60009. The fine-grained soils are more typical of Apollo 16 (LSPET, 1973).

12.5 cm-deep mature surface in 60010 (FMR) corresponds with a dissection boundary; and the top of the coarse, basal anorthitic layer in 60009 (53 cm depth) corresponds with chemical, FMR and dissection units. We mention this point because much of the discussion that follows is based on the distinction of mineral compositions from 10 separate levels of 60009/60010. We do not suggest that all 10 of these levels represent stratigraphic units. Some of the differences represent fine-scale concentrations of distinctive lithic and mineral components which have been poorly mixed. These concentrations do not necessarily have stratigraphic significance, but they provide information about the nature of lithic sources which can be correlated between 60009/60010 and the Apollo 16 drill core. Moreover, these variations in modal concentration can be used to model chemical variations between soil levels while accounting for the total distribution of major elements between soil components.

An important goal of lunar regolith studies is to determine the soil components that have mixed to form the observed soil depth/composition profiles in lunar cores and the compositions of surface soils from a variety of lunar localities. A common approach to this problem is to obtain a chemical analysis, assume reasonable "end-member components" and use mathematical fitting procedures to model the data (see, for example, Blanchard *et al.*, 1976; Koretev, 1976). This approach has given valuable insights into the nature of the lunar regolith and is especially useful when dealing with trace elements and/or fine-grained soils (<0.02 mm). We have approached the problem of chemical mixing in lunar soils using a different approach in order to account for the major element distributions among the soil components in the >0.02 mm fraction. We emphasize major elements in this size range because our chemical analyses are obtained with the electron microprobe, and our modal data is only for the >0.02 mm fraction for reasons mentioned above.

METHODS

Electron microprobe analysis

Electron microprobe analyses of glasses and minerals were obtained on an ARL EMX-Sm automated electron microprobe. Data were corrected by the procedures of Bence and Albee (1968) and Albee and Ray (1970). Apollo 17 orange glass (74220, 3a) was used as an internal standard during analysis of glass beads. In order to obtain an unbiased sample of the major mineral components at various depths, all >20 μm fragments of pyroxene, olivine and feldspar were analyzed in areas of ~10 mm^2 at each of the ten levels in Fig. 1. Multiple analyses were made of pyroxene grains to account for compositional zoning. The precision of the microprobe data is 1–2% relative for each element. Mineral structural formulae were calculated before the oxide data were rounded off.

Lithic fragments and glasses were analyzed in 14 thin sections. The total data base from microprobe analysis is 230 pyroxene, 893 feldspar, and 179 olivine analyses in monomineralic fragments, 403 mineral and glass analyses in lithic fragments, and 130 analyses of glass beads. Glasses were analyzed for Si, Al, Ti, Fe, Mn, Mg, Ca, Na, K and Cr; pyroxenes were analyzed for Si, Al, Ti, Fe, Mn, Mg, Ca, Na, and Cr; olivines were analyzed for Si, Al, Ti, Fe, Mn, Mg, Ca, and Cr; feldspars were analyzed for Si, Al, Fe, Mg, Ca, Na, and K; and oxide minerals in lithic fragments were analyzed for Si, Fe, Mg, Ti, Mn, Cr, and Al.

Modal recombination

In our regolith mixing models we use the soil components we actually observe, their modal proportions, and their chemistry as determined by microprobe analysis. We can calculate the total bulk chemistry of the soil by combining the average chemical composition of each soil component according to its modal weight proportion (note that the tables in Papike *et al.*, 1978, are different from the tables presented here for two reasons: (1) we incorrectly used volume modal proportions rather than weight modal proportions to estimate bulk chemistry and (2) we have refined our estimates of the average compositions of the modal components listed in table 1.) Modal weight proportions are obtained by using density to normalize the volume proportions obtained in point counting. The densities and oxide weight compositions of average lithic, mineral and glass components are listed in Table 1. Chemical data for each of the mineral and glass components are the average compositions which we have determined from ~1500 microprobe analyses. The compositions of the mare basalt and feldspathic basalt components were determined by modal recombination of microprobe data from lithic fragments actually in the double drive tube. Attempted modal recombinations proved that this method could not be used for ANT and RNB/POIK lithologies; for these two lithic components we have used data from the literature (see Table 1).

We can check our caluculations by comparing the calculated chemistry with the chemistry determined by direct analytical methods (e.g., XRF or INAA). Obviously the chemical analysis must

Table 1. Chemistries of Regolith Components Used in Mixing Models.

Regolith Component*	SiO_2	Al_2O_3	"Cr_2O_3"**	"TiO_2"	MgO	"FeO"	CaO	Na_2O	K_2O	Σ	Density
1. Mare Basalt	46.32	8.33	0.24	7.22	8.08	19.70	10.25	0.32	0.04	100.50	3.33
2. ANT = Anorthosite + Norite + Troctolite + LMB	44.80	31.54	0.04	0.09	2.42	3.41	18.09	0.26	0.01	100.66	2.82
3. Feldspathic Basalt	45.5	26.4	0.14	0.56	6.9	4.8	15.3	0.47	0.10	100.17	2.91
4. RNB + POIK	45.8	19.6	0.20	1.11	11.2	8.9	12.0	0.46	0.25	99.52	3.04
5. Olivine	37.70	0.10	—	0.01	36.16	25.75	0.17	—	—	99.89	3.68
6. Pyroxene	51.95	1.58	0.42	0.86	18.96	15.26	10.36	—	—	99.39	3.39
7. Plagioclase	44.49	35.01	—	—	—	0.16	19.47	0.43	0.03	99.59	2.76
8. Opaques	—	—	0.08	10.68	0.44	88.70	—	—	—	99.90	5.16
9. Orange glass	41.64	10.63	0.45	8.43	9.29	17.87	10.68	0.28	0.03	99.30	3.04
10. Yellow glass	46.76	16.20	0.26	3.07	10.02	11.61	10.84	0.53	0.27	99.56	2.84
11. Green glass	44.14	8.14	0.44	0.45	16.06	21.67	8.61	0.22	0.08	99.81	3.04
12. Clear glass	44.70	25.79	0.13	0.56	7.47	5.95	14.82	0.26	0.12	99.80	2.79

*Sources of analyses: 1. Modal recombination, this study; 2. Apollo 16 rock 67075 (LSPET, 1973); 3. Modal recombination, this study; 4. Average of Apollo 16 rocks 60315, 61156, 62235, 64567, 64815 (Simonds *et al.*, 1973); 5-8. Average mineral analyses, this study; 9-12. Based on data from this study and Naney *et al.*, 1976.

**Quotation marks indicate that multiple valance states are present.

be obtained from the same size fraction (>0.02 mm) that we are modelling. This type of complete chemical data (majors including silicon, minors and trace elements) for different grain size separates in lunar cores has been obtained for the first time by Laul (1978) on the Apollo 17 drill core. Agreement between Laul's chemical data and our "modal recombination" chemical data for the >0.02 mm size fraction of the Apollo 17 drill core is quite good and adds credibility to the approach used in this paper, where a direct check was not available.

THE LITHIC COMPONENTS IN 60009/60010

The lithic categories we use are defined in a previous paper on the Apollo 16 deep drill core (ref. Vaniman *et al.*, 1976). These categories can be extended to lithic types in the double drive tube 60009/60010 with some modification. Although the general mineralogy and bulk compositions of rock fragments are similar in the deep drill core and in 60009/60010, there are more large lithic

Table 2. Lithic Fragments > 2 mm in 60009/60010.

Lithic Category	No. of Fragments
Agglutinate	1
DMB	7
RNB/POIK	12
LMB	6
Feldspathic Basalt	1

fragments in the double drive tube. We have found 27 >2 mm lithic fragments in the 59 cm length of 60009/60010, compared to only 7 >2 mm lithic fragments in the 225 cm-long deep drill core. A summary of the >2 mm lithic fragment data for 60009–60010 is given in Table 2. Several lithic types are illustrated in Fig. 3.

Fused soil (agglutinate and dark matrix breccia)

Our resolution in the distinction of agglutinates from dark matrix breccias (DMB) has improved with the introduction of a rapid-switching transmission-reflection Zeiss photomicroscope into our point counting procedure. Agglutinates and DMB's are distinguished in transmitted light by a bonding matrix of opaque or brown glass. Both agglutinates and DMB's may contain clasts of lithic, mineral and angular to spherical glass fragments (Delano *et al.*, 1973; Agrell *et al.*, 1970). In reflected light, agglutinates can be distinguished from DMB's by complex flow schlieren in the glass matrix, by abundant vesicles, and by flow-oriented trails of immiscible Fe droplets which may have troilite rims (Agrell *et al.*, 1970). Delano *et al.* (1973) point out that the Apollo 16 DMB's are similar in composition to the bulk soils. Other studies have suggested that

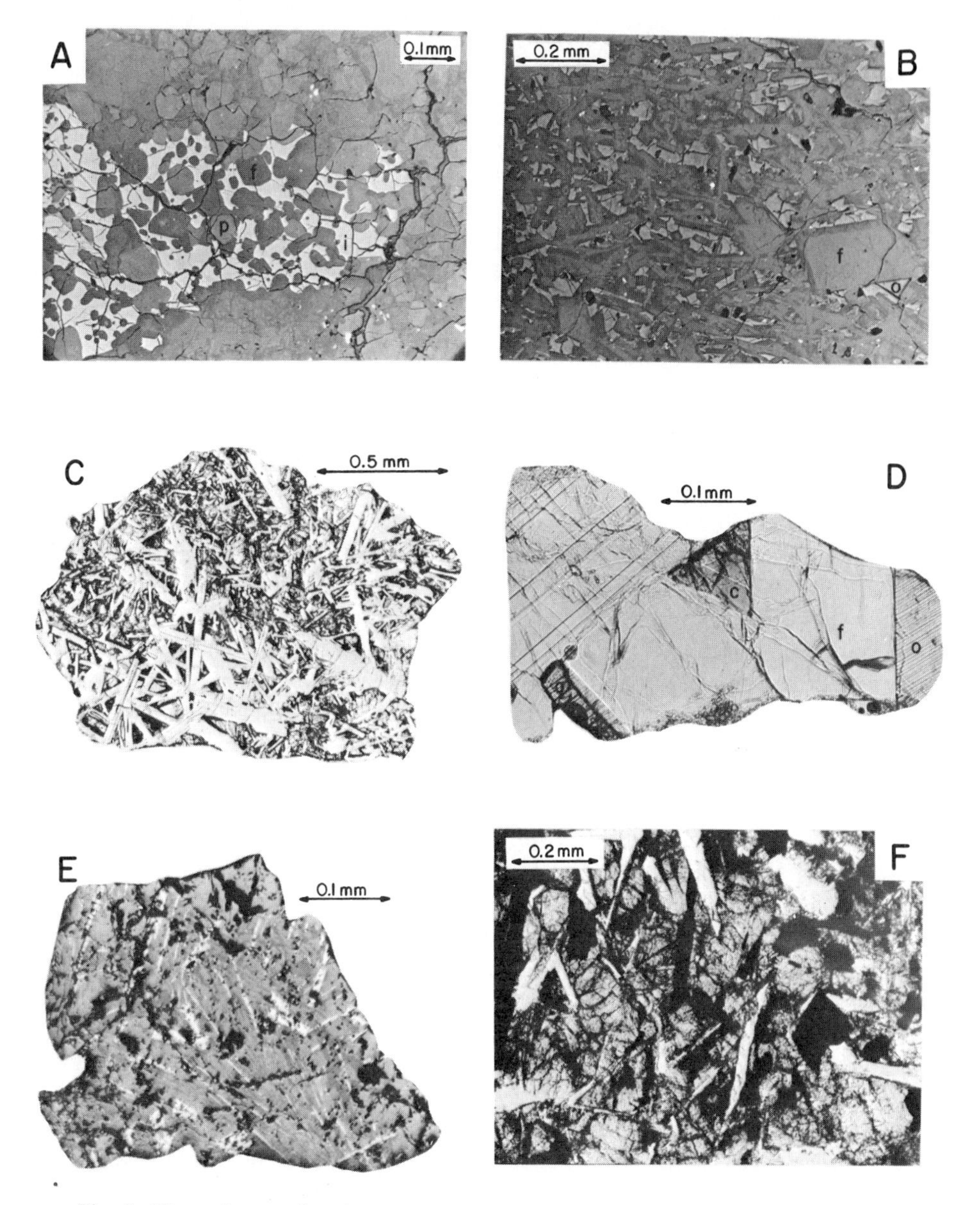

Fig. 3. Photomicrographs of rocks in 60009/60010. (A) POIK rock 21, described in the text and listed in Table 3. This fragment is representative of the lithology from which the two-pyroxene cluster (Table 2) is derived. Symbols: i = poikilitic ilmenite; p = pyroxene; and f = feldspar. (B) Troctolitic/feldspathic basalt 66, described in the text and listed in Table 3. Symbols: f = blocky feldspar; o = intergranular olivine. (C) Feldspathic basalt 70, described in the text and listed in Table 3. This rock has laths of feldspar with intergranular pyroxene. (D) Cumulate-textured two-pyroxene ANT rock. This lithology and LMB represent the sources of pyroxenes described in Table 4. Symbols: o = orthopyroxene; c = clinopyroxene; f = feldspar. (E) Mare fragment "A", similar to Apollo 17 olivine porphyritic ilmenite basalts. (F) Mare fragment "B", similar to Apollo 12 ilmenite basalts.

Photographs A, B and E were taken in reflected light; all others were taken in plane-polarized light.

agglutinates also represent bulk soil compositions (Hu and Taylor, 1977), though there is some question about this assumption (Rhodes *et al.*, 1975). For our purposes, the most important feature of these DMB/agglutinate types is their accretionary origin. Because of their fused-soil origin, the petrology of this component is not treated in detail in this paper. We are concerned primarily with those lithic types which provide single-crystal clasts to the soil mixture through comminution.

Poikilitic rocks (POIK) and recrystallized noritic breccias (RNB)

Poikilitic rocks (POIK) contain poikilitic grains (or oikocrysts) of low-Ca pyroxene ± Mg – augite ± olivine ± ilmenite ± armalcolite ± ulvöspinel. The included mineral (or chadacryst) is plagioclase. Plagioclase chadacrysts may be euhedral and are small (~0.01 to 0.1 mm). The poikilitic phases may exceed 0.5 mm size. Pyroxene zonation varies between POIK samples: one particularly large sample (~5 mm) has strongly zoned pigeonite (Wo_3En_{80} to $Wo_{10}En_{54}$) and augite (av. $Wo_{35}En_{46}$) oikocrysts (Fig. 4); ilmenite is also an oikocryst phase. Similar pyroxene zonation was found in POIK fragments of the deep drill core (Vaniman *et al.*, 1976). However, this POIK type is rare in 60009/60010 where most POIK fragments (as well as RNB fragments) have unzoned orthopyroxene + Mg – augite oikocrysts (olivine may be intergranular or poikilitic and may be zoned). The composition of these equilibrated pyroxenes may vary somewhat between POIK lithic fragments, but the most abundant two-pyroxene composition is $Wo_{4.5}En_{70.5}$ + $Wo_{40}En_{47}$. This two-pyroxene composition is important, for it occurs among pyroxene clasts at the 150–170 cm depth level of the deep drill core even though that core does not contain any lithic fragments with pyroxenes of that composition. The same cluster of two-pyroxene composition occurs among mineral clasts at several levels in 60009/60010 (discussion below). The source of this two-pyroxene cluster in the deep drill core was previously unknown, and had been attributed to either a mafic ANT cumulate rock or to a POIK/RNB source (Vaniman *et al.*, 1976). The discovery of POIK rocks in 60009/60010 with the appropriate mineralogy resolves this question. Low-Ca and high-Ca pyroxenes from this POIK type are listed in Table 3 along with representatives of the two-pyroxene compositional clusters in 60009/60010 and the deep drill core.

The recrystallized noritic breccias (RNB's) do not differ significantly from the unzoned POIKS. The RNB texture is generally finer-grained and sub-poikilitic or recrystallized. However, single samples with a gradation between POIK, RNB and feldspathic basalt texture occur in both the drill core and in 60009/60010. The range of Ca-Mg-Fe pyroxene variation and Mg-Fe olivine variation in these rock types is shown in Fig. 4, and the Ab-An-Or variation is shown in Fig. 5. Note that the range of feldspar zonation in POIK and RNB extends from $An_{97}Ab_3$ to $An_{82}Ab_{16}Or_2$, a compositional range that occurs in POIKS with both zoned and unzoned pyroxenes (RNB's generally have feldspar of more restricted composition, $An_{96}Ab_4$ to $An_{94}Ab_6$).

Table 3. Two-Pyroxene Cluster: In POIK Lithic Fragments of 60009–60010 and in Mineral Fragments of 60009–60010 and Drill Core Segment 60003.

	POIK Lithic		Mineral Fragments 60010		Mineral Fragments 60003	
	1	2	3	4	5	6
SiO_2	53.2	50.4	53.2	51.0	53.2	50.9
Al_2O_3	1.34	2.80	1.73	2.71	1.44	2.50
TiO_2	1.03	2.13	0.92	2.16	1.08	2.01
FeO	16.4	8.6	16.5	8.5	16.5	8.7
MnO	0.19	0.17	0.15	0.06	–	–
MgO	25.4	16.0	25.7	16.5	25.6	16.7
CaO	2.25	18.9	2.15	18.8	2.21	18.5
Cr_2O_3	0.34	0.68	0.33	0.68	0.40	0.77
Σ	100.15	99.68	100.68	100.41	100.43	100.08
Si	1.936	1.876	1.926	1.880	1.931	1.883
Al^{IV}	0.058	0.123	0.074	0.118	0.062	0.109
$\Sigma_{tet.}$	1.994	1.999	2.000	1.998	1.992	1.992
Al^{VI}	0.000	0.000	0.000	0.000	0.000	0.000
Ti	0.028	0.060	0.025	0.060	0.029	0.056
Fe	0.499	0.268	0.498	0.264	0.499	0.270
Mn	0.006	0.005	0.004	0.002	–	–
Mg	1.376	0.886	1.386	0.906	1.382	0.921
Ca	0.087	0.753	0.083	0.742	0.085	0.734
Cr	0.010	0.020	0.009	0.020	0.011	0.023
$\Sigma_{oct.}$	2.006	1.992	2.005	1.994	2.006	2.004
$\Sigma_{cations}$	4.000	3.991	4.005	3.992	3.998	3.996
Wo	4.4	39.4	4.2	38.8	4.3	38.1
En	69.9	46.3	70.3	47.3	70.4	47.9
Fs	25.7	14.3	25.5	13.9	24.3	14.0

All analyses normalized to 6 oxygens. Columns 1 and 2: Pyroxenes of POIK lithic fragment 21 (60010,6025); Columns 3 and 4: Pyroxene fragments of level 3, 60010, 6 cm depth; Columns 5 and 6: Pyroxene fragments of 60003 at 170 cm depth in the Apollo 16 drill core.

Feldspathic basalts

Feldspathic basalts are rare in 60009/60010 (<2% of all lithic fragments), as in the Apollo 16 deep drill core (Vaniman *et al.*, 1976). The feldspathic basalts in 60009/60010 are fine-grained (0.1–0.2 mm) intersertal to intergranular rocks with abundant laths of early-formed plagioclase. Delano *et al.* (1973) distinguished two types of feldspathic basalt in Apollo 16 2–4 mm surface soils: (1) skeletal, lathy to variolitic basalts with the same composition as the soil (~anorthositic gabbro, 27–28% Al_2O_3) and (2) low-Al basalts (22–24% Al_2O_3) with a variety of textures. Vaniman *et al.*, (1976) found that feldspathic basalts in the drill core have the more aluminous composition, similar to anorthositic gabbro. The data of both groups (Delano *et al.*, 1973, and Vaniman *et al.*, 1976)

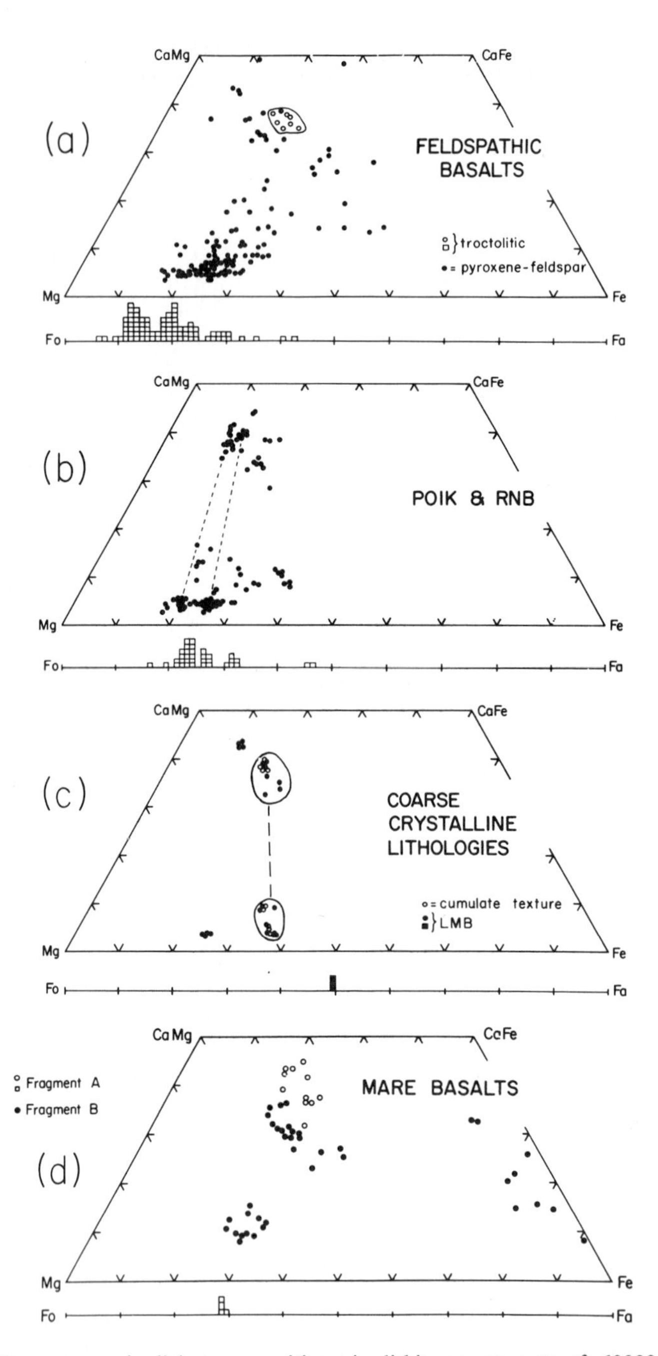

Fig. 4. Pyroxene and olivine compositions in lithic components of 60009/60010. Pyroxene tie-lines in (b) connect common two-pyroxene compositions of two POIK rock types, and in (c) connect a two-pyroxene composition found in both LMB and cumulate textured ANT lithologies.

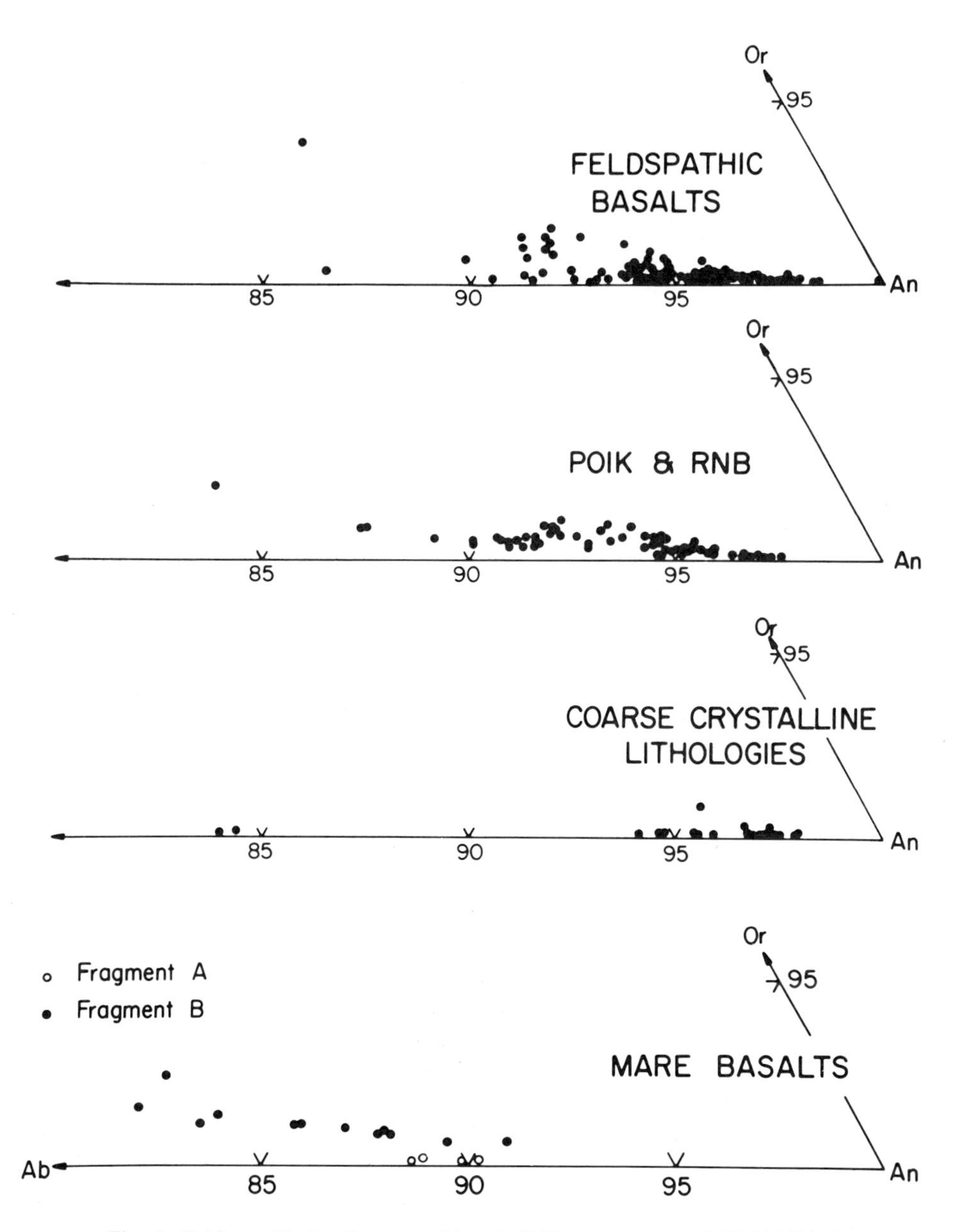

Fig. 5. Feldspar Ab-An-Or compositions in lithic components of 60009/60010.

Table 4. Analyses of Three > 1.5 mm Melt Rock Fragments Determined by Combined Modal and Microprobe Analysis.

	Basalt 70	Basalt 61	Troctolitic Basalt 66
SiO_2	45.4	46.4	43.0
TiO_2	1.07	0.20	0.17
Al_2O_3	26.5	27.1	24.2
FeO	5.4	3.4	6.4
MnO	0.06	0.05	0.06
MgO	5.2	6.2	12.4
CaO	16.0	15.0	13.8
Na_2O	0.35	0.65	0.36
K_2O	0.03	0.20	0.03
Cr_2O_3	0.13	0.08	0.20
Σ	100.1	99.3	100.6
Mg/(Mg + Fe)	0.63	0.76	0.78
CaO/Al_2O_3	0.60	0.55	0.57
Qz	21.7	27.3	6.8
An	52.4	52.4	49.8
Ol	20.7	19.9	41.6
Wo	5.2	0.4	1.8

Analytical method described in Vaniman and Papike (1977a); the four-component norm is determined following the method of Walker *et al.* (1972), modified by T.L. Grove. The normative components Qz-An-Ol are plotted in Fig. 6.

were obtained by uncorrected broad-beam analysis and may be in error by >10% (Albee *et al.*, 1977).

In order to get a better estimate of feldspathic basalt compositions, we selected three fragments of exceptional size (1.5 mm to 5 mm) from 60009/60010 and used a modal recombination method (standard microprobe analysis plus modal analysis). The results are plotted in Fig. 6 and listed in Table 4. Two of the basalt fragments (numbers 61 and 70) fall within the field of "anorthositic gabbro" (Fig. 6) which Naney *et al.* (1977) and Vaniman and Papike (1978) outlined for melt rocks plotted in the pseudoternary projection. A third feldspathic basalt (number 66) is troctolitic. The poikilitic melt rocks are less aluminous, and plot near the olivine-plagioclase cotectic (Simonds *et al.*, 1973) in the field of "low-K Fra Mauro" (LKFM) compositions. The fields of anorthositic gabbro and LKFM plotted in Fig. 6 are taken from the data of Naney *et al.* (1977).

There are sharp mineralogical as well as chemical differences between the basalts in the anorthositic gabbro field and the troctolitic basalts. The troctolitic basalts of 60009/60010 are low in silica (<44% SiO_2) and contain feldspar and olivine, (Fo_{89} to Fo_{57}) + Mg, Al spinel ± minor high-Ca augite; whereas the more siliceous (>44% SiO_2) feldspathic basalts contain feldspar + pyroxene (zoned from Wo_4En_{79} pigeonite to Fe-rich augite) + Mg, Al spinel ± olivine (~Fo_{70}) ± ilmenite. Feldspathic basalts of both types may be represented by feldspathic vitrophyres with feldspar + glass ± olivine.

Figure 4 shows the range of Ca-Mg-Fe pyroxene composition in lithic fragments of 60009/60010. The data we have obtained for all primary rock types show that only feldspathic basalts and mare basalts have pyroxenes with Fe/(Fe + Mg) >0.45. Light matrix breccias may also contain such iron-rich pyroxenes (Delano *et al.*, 1973), but LMB is not a primary rock type; its Fe-rich pyroxenes must be inherited from some primary rock type. Delano *et al* (1973) and Vaniman *et al.* (1976) did not find any major primary rock type in the Apollo 16 soils with such Fe-rich pyroxenes. However, feldspathic basalts with such Fe-rich pyroxenes in the assemblage feldspar + pyroxene + ilmenite occur in 60009/60010 (e.g., fragment 70, Fig. 3c) and may be the source of Fe-rich pyroxenes which Delano *et al.* (1973) have observed in LMB's and other workers (Meyer and McCallister, 1974, 1976; Meyer H. O. A. *et al.*, 1975; and Vaniman *et al.*, 1976 and this paper) have seen as mineral fragments in the Apollo 16 core samples.

Anorthosites, norites and troctolites

Pristine or metamorphosed highland cumulate rocks (ANT) are poorly represented as lithic fragments in 60009/60010 (<1.5% of all >20 μm fragments). Most anorthosites (>90% modal plagioclase) and norites and troctolites (70–90% modal plagioclase) have been metamorphosed or shocked and recrystallized. We have included as anorthosites some rare shock-melted rocks which are composed totally of anorthitic plagioclase in variolitic clusters with minor amounts of glass. Undoubtedly, coarse-grained ANT rocks are the source of most of the large (≥1 mm) unshocked feldspar crystals which are found throughout 60009/60010 and concentrated in the coarse-grained layer at ~53–57 cm depth.

Since most ANT lithic fragments have been shocked or metamorphosed, there are few examples of the coarse-grained cumulate rocks that must be the sources of the large, unshocked plagioclase crystals. However, there is one unmetamorphosed fragment of noritic rock with cumulate texture (Fig. 3d) in soil at the depth of level 5 (Fig. 1). This rock has blocky feldspar crystals >0.5 mm with intercumulus grains (~0.2 mm) of orthopyroxene and augite. The pyroxene compositions are shown in Fig. 4. Table 5 lists the compositions of similar low-Ca and high-Ca pyroxenes which are found in the soil of level 7 (Fig. 1) in 60009. It is likely that the association of pyroxene grains found at this level represents an anorthositic or noritic source rock.

Light matrix breccia (LMB)

Light matrix breccias (LMB's) are abundant (~6%) in the coarse layer at the base of 60009 (ref. Simon *et al.*, 1978) and common (~3%) higher in the double drive tube. LMB fragments are polymict feldspathic breccias which are highly shocked, poorly recrystallized and may have patchy fracture fillings of yellow to clear glass. Delano *et al.* (1973) found that LMB's of the 2–4 mm Apollo 16

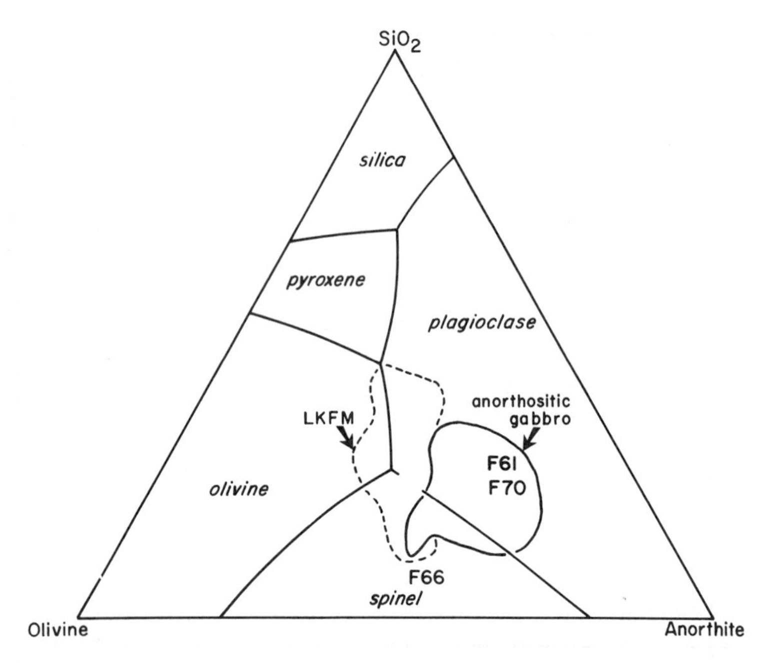

Fig. 6. Pseudoternary diagram projected from clinopyroxene onto the plane olivine-anorthite-SiO_2 (after Walker *et al.*, 1972). Data for modal/microprobe calculations of POIK (P) and feldspathic basalt (F) lithic fragment compositions are listed in Table 3. The fields for anorthositic gabbro and low-K Fra Mauro melt rock compositions are taken from the combined data of Naney *et al.* (1977) and Vaniman and Papike (1978).

surface soils contained Fe-rich pyroxenes [Fe/(Fe + Mg) >0.45] that did not occur in any of the common primary lithologies. The double drill core 60009/60010 contains feldspathic basalts that may represent the source of these pyroxene grains (ref. discussion above). We note however that LMB's are not a homogeneous clan: none of the LMB's in 60009/60010 have such Fe-rich pyroxenes (Fig. 3). Instead, they have pyroxenes that reflect equilibrated ANT sources. Some of the 60009/60010 LMB's contain pyroxenes that are very similar to those of the cumulate-textured rock described in the preceeding section (see Fig. 4).

Exotic lithic fragments: Mare basalts

Nine fragments of mare basalt have been found in 60009/60010, but only two are large enough to permit an accurate petrographic description. Mare fragment "A" (Fig. 3e) is a high-Ti basalt with reddish high-Ca augites ($TiO_2 \sim 7\%$, $Al_2O_3 \sim 5.5\%$, $Cr_2O_3 \sim 0.7\%$), olivine of Fo_{72}, plagioclase, ilmenite and glass. Mare fragment "B" (Fig. 3f) is a low-Ti basalt (~4.5% TiO_2) with pyroxene + plagioclase + ilmenite + glass + FeS; the pyroxene and plagioclase

Table 5. End members of low-Ca to high-Ca pyroxene trend and representative olivine (Fo_{62-63} cluster) of area 7 (60009,6024).

	1	2	3
SiO_2	53.3	49.9	36.7
Al_2O_3	0.55	2.88	0.14
TiO_2	0.46	2.22	0.00
FeO	22.0	8.5	32.6
MnO	0.31	0.09	0.33
MgO	22.3	16.1	30.9
CaO	1.18	18.7	0.24
Cr_2O_3	0.07	0.77	0.00
Σ	100.17	99.16	100.96
Si	1.979	1.867	0.995
Al^{IV}	0.024	0.128	0.004
$\Sigma_{tet.}$	2.000	1.995	0.999
Al^{VI}	0.003	0.000	0.000
Ti	0.013	0.062	0.000
Fe	0.684	0.265	0.738
Mn	0.010	0.003	0.007
Mg	1.237	0.899	1.250
Ca	0.047	0.748	0.006
Cr	0.002	0.023	0.000
$\Sigma_{oct.}$	1.996	2.000	2.001
$\Sigma_{cations}$	3.996	3.995	3.000
Wo	2.4	39.1	–
En(Fo)	62.5	46.9	62.6
Fs(Fa)	35.1	14.0	37.4

Columns 1 and 2: Pyroxenes normalized to six oxygens; Column 3: Olivine normalized to four oxygens.

Table 6. Model Data for Mare Basalt Fragments.

	Fragment A	Fragment B
pyroxene	40.6	58.4
plagioclase	25.9	26.7
olivine	5.8	–
ilmenite	16.2	8.1
glass	11.5	6.7
FeS	–	0.1

compositions in fragment "B" with the absence of olivine are similar to the Apollo 12 ilmenite basalts (ref. Papike *et al.*, 1976, and Dungan and Brown, 1977). Modal analyses of these mare basalts are listed in Table 6 and the silicate mineral chemistries are shown in Figs. 4 and 5.

Delano (1975) described 7 mare fragments from Apollo 16 and found that most fragments were similar to Apollo 17 high-Ti basalts or Luna 16 high alumina basalts. Fragment "A" in 60009/60010 is very similar to Apollo 17 olivine porphyritic ilmenite basalts (ref. Papike *et al.*, 1974), but fragment "B" is neither a high-Ti nor an aluminous basalt. The presence of a third, different mare basalt type emphasizes the heterogeneity of the mare basalt component at Apollo 16.

MINERAL FRAGMENTS IN 60009/60010

Approximately 120 to 170 mineral fragments have been analyzed at each of the ten levels shown in Fig. 1. These data are summarized in plots of Ab-An-Or (Fig. 7, 8) and Ca-Fe-Mg (pyroxene) or Fe-Mg (olivine, Fig. 9). The feldspar mineral data mark the coarse-grained layer at 53 to 57 cm in 60009/60010 (Fig. 1) which was recognized by core dissection (Duke and Nagle, 1974, unit 60009-3), by low maturity of the index I_s/FeO (Morris and Gose, 1976, unit 60009-3), and by INAA analysis (Ali and Ehmann, 1976, unit 60009-4). This coarse-grained layer has also been recognized as an immature feldspathic end member in petrographic mixing models of McKay *et al.* (1976, 1977) for the 60009/60010 soils. The recognition of the same stratigraphic unit by four independent methods lends credibility to this stratigraphy. We add two types of supporting evidence for this stratigraphy: (1) modal data (Simon *et al.*, 1978) that indicate a feldspar + LMB-rich and matrix-poor layer at this depth, and (2) chemical data for mineral fragments that indicate a homogeneous population of ANT-type plagioclase in this level.

Figure 8 shows that there is a cluster of plagioclase composition between An_{96} and An_{97} at all levels of 60009/60010. Feldspars from all lithic fragments except the mare basalts can be found in this range, but only the coarse-crystalline ANT lithologies are effectively restricted to anorthitic ($>An_{94}$) plagioclase composition (Fig. 5). Area 9 in Figs. 7 and 8 corresponds to the coarse-grained feldspathic unit in 60009, and the feldspars in this unit have an exceptionally restricted range of composition (An_{95}–An_{98}). This ANT-type composition with the absence of more sodic melt rock (POIK, feldspathic basalt) plagioclase indicates an ANT source for the feldspars that distinguish the coarse-grained unit.

This common composition centered an An_{96}–An_{97} is found at all levels of 60009/60010, though more sodic plagioclase is intermixed above (in levels 1–8) and in level 10 below the coarse-grained unit. The same cluster of feldspar composition occurs in the upper 190 cm of the Apollo 16 deep drill core (units B–D of Vaniman *et al.*, 1976). Below 190 cm in the drill core, the feldspar fragments have been shocked and recrystallized and do not cluster between An_{96}

and An_{97} in composition. In texture and in composition, the feldspars throughout 60009/60010 are similar to the unshocked feldspar fragments found in the upper 190 cm of the nearby drill core.

Pyroxene and olivine compositions vary between the ten levels of Fig. 1, but cannot by themselves define a series of stratigraphic units. Instead the concentrations of specific pyroxene compositions at various levels provide an indication of the nature of the source rocks. Levels 3 and 5 (Fig. 1) have concentrations of the two-pyroxene composition $Wo_{4.5}En_{70.5}$ and $Wo_{40}En_{47}$, a composition which represents the common equilibrated POIK lithology of 60009/60010. One would expect such concentrations of mineral fragments that reflect the common lithic fragments of the soil. Area 7, however, has a trend of pyroxene compositions that connect high-Ca and low-Ca members (Table 5) with a tie-line orientation that suggests an ANT source which equilibrated to low temperature (Dixon and Papike, 1975). Moreover, there is an olivine cluster in area 7 which has an Fe/(Fe + Mg) ratio (0.38) similar to that of the low-Ca member of the pyroxene trend. A comparison with pyroxene data for lithic fragments (Fig. 4) suggests that the pyroxene trend in area 7 is from a noritic/troctolotic cumulate rock, or from such a rock which has been worked into an LMB and then into the soil. The olivine cluster at Fe/(Fe + Mg) = 0.38 is not found in Fig. 4, but this absence may simply be due to an inadequate number of lithic samples.

Glass Data

Homogeneous glass beads of the double tube have a compositional range similar to glass beads of the Apollo 16 drill core (Naney *et al.*, 1976). Figure 10a shows wt.% CaO/wt.% Al_2O_3 (an indication of plagioclase content) plotted against wt.% SiO_2 for glasses of 60009/60010, with fields that show the range of compositions in the Apollo 16 and Apollo 17 drill cores. There is a field (stippled) of intermediate mare/highland glass compositions which are common at the Apollo 17 site, but are not found in the Apollo 16 soils. These glasses are more silica-poor than any common highland rock type (minimum $SiO_2 \sim 40\%$) and probably result from the combined processes of (1) impact mixing of mare ($CaO/Al_2O_3 \sim 1.0$) and highland ($CaO/Al_2O_3 \sim 0.55$) rock types, a process which would occur on a large scale near mare-highlands contacts (e.g., Apollo 17) but would not contribute abundant glass fragments to a distant highland site (Apollo 16), and (2) impact volatilization and loss of silica. Ivanov and Florensky (1975) discussed the evidence for silica loss through volatilization, and Naney *et al.* (1976) described a volatilized high-alumina, silica poor (HASP) glass in the Apollo 16 drill core. The original definition of HASP glass by Naney *et al.* (1976) describes a restricted composition with $CaO/Al_2O_3 \simeq 0.55$ and SiO_2 <35 wt.%. Hartung *et al.* (1978) found that HASP glass occurs in the linings of some microcraters and grades into glasses that lack silica depletion deeper in the microcrater walls. Vaniman and Papike (1977b) found glasses of HASP composition at Apollo 17, but noted the expanded field of less aluminous glasses which appear to be silica depleted—the stippled field in Fig. 10a.

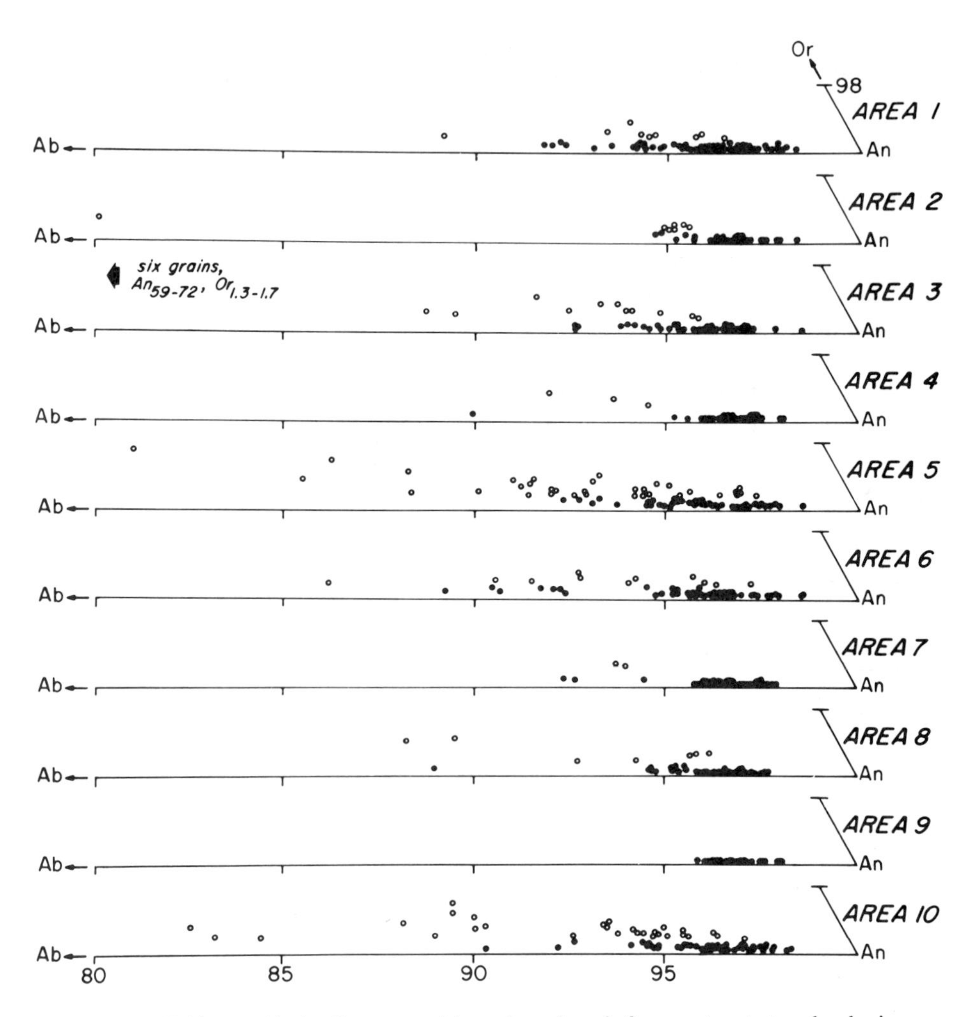

Fig. 7. Feldspar Ab-An-Or compositions in mineral fragments at ten levels in 60009/60010. The ten areas indicated (each about 10 mm^2) were selected at the ten levels indicated in Fig. 1; within each area all single-crystal fragments of feldspar larger than 0.02 mm were analyzed.

The two processes of (1) mixing mare and highland sources and (2) volatilization of SiO_2 can be displayed by a comparison of the sources for glass compositions and the actual range of glass compositions. Figure 10b shows common mare basalt compositions and highland rock types plotted on the same axes used for glass data in Fig. 10a; soils from a highland site (Apollo 16), a high-Ti mare site (Apollo 17), a low-Ti mare site (Apollo 15) and a VLT mare site (Luna 24) are also shown. The heavy arrows which trend toward low silica values cover a field which has no lithic or soil representatives; these arrows indicate silica volatilization from impact melting of lithic or soil sources. The

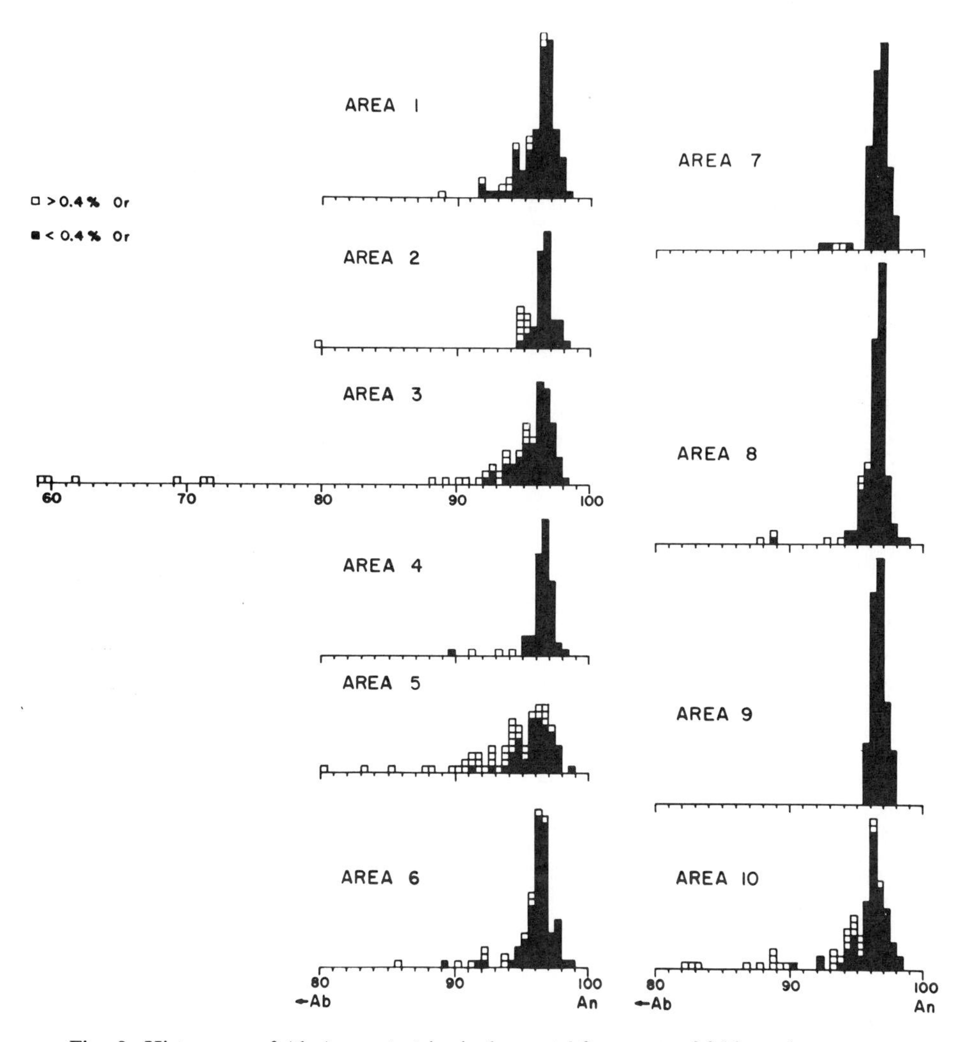

Fig. 8. Histograms of Ab-An content in single-crystal fragments of feldspar (areas are defined in Figs. 1 and 7). Note the common concentration of feldspar fragments with compositions between An_{96} and An_{97} throughout 60009/60010. Potassic feldspars (greater than 0.4 mole % Or) common only to melt rock lithologies are concentrated in areas 3, 5 and 10, and completely absent from the coarse-grained anorthitic soils in area 9.

glass mixtures of mare and highland compositions may result from one-stage (mixture of lithic sources during impact) or multistage (impact melting of a soil mixture) processes.

Figure 10b shows that most of the mare glass types are not affected by silica volatilization. This observation suggests that a process other than impact has operated in the formation of many mare glasses. Lack of silica volatilization in

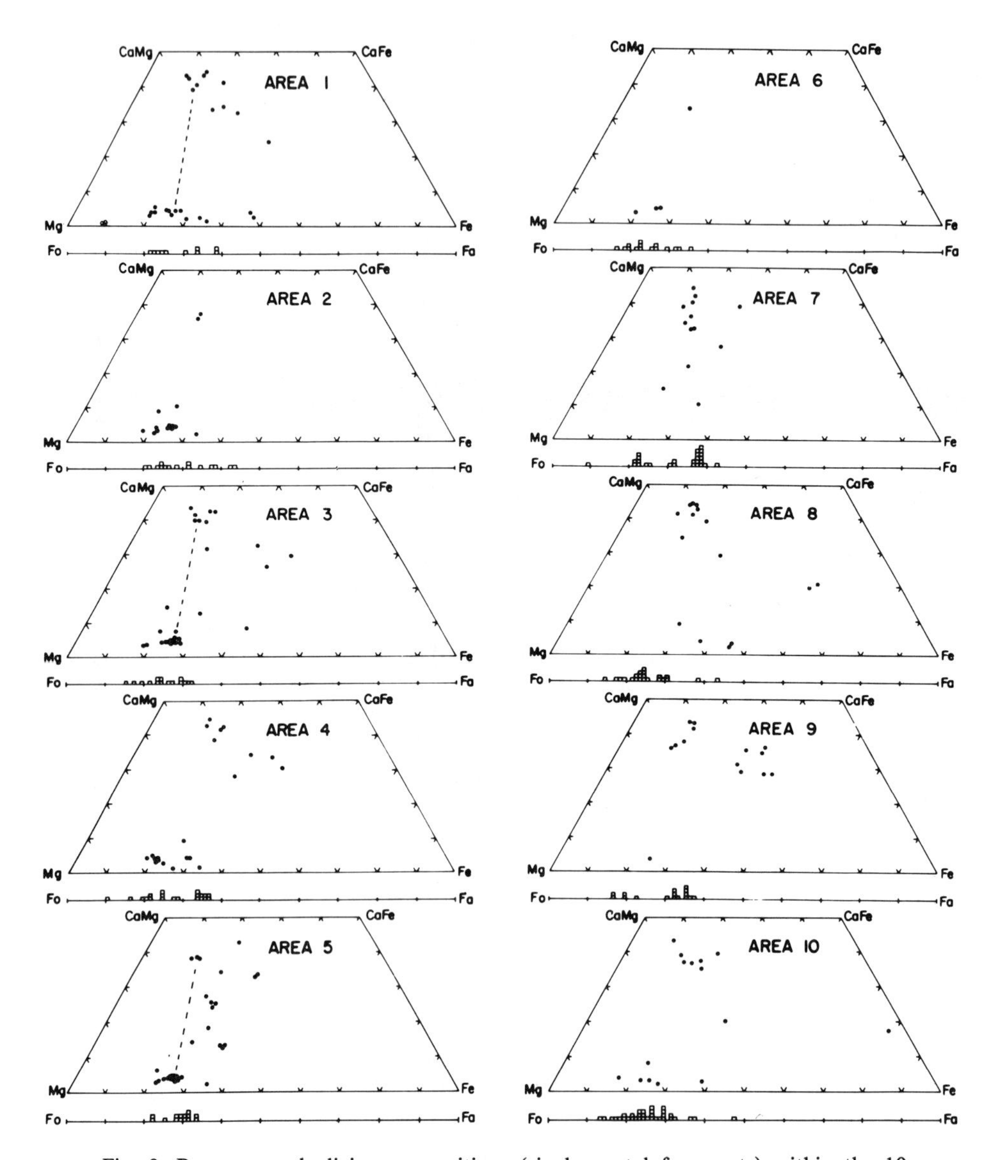

Fig. 9. Pyroxene and olivine compositions (single-crystal fragments) within the 10 areas defined in Figs. 1 and 7. Areas 1, 3 and 5 have clusters of two-pyroxene composition (dashed lines) similar to the equilibrated two-pyroxene compositions of a common POIK lithology in 60009/60010 (Figs. 4 and 3a). Area 7 has a concentration of pyroxene and olivine that suggests an ANT cumulate or LMB source.

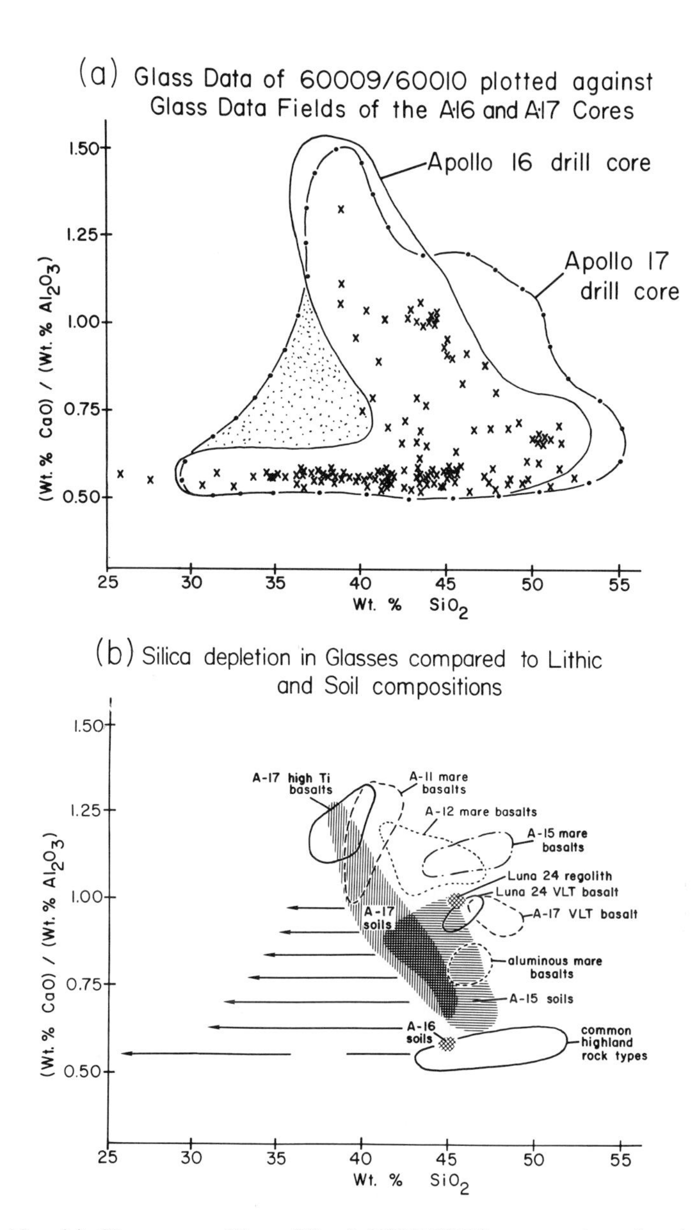

Fig. 10. (a) Glass compositions (X) of 60009/60010 compared to the glass data fields of the Apollo 16 and Apollo 17 drill cores. Note that there is a stippled field of silica-poor glasses with intermediate CaO/Al_2O_3 ratios (i.e., intermediate between highland and mare compositions) in the Apollo 17 soils but not in the Apollo 16 soils. (b) Data from mare rock types, highland rock types, and soils plotted with the same axes and at the same scale as Fig. 10a. Note that the arrows pointing in the direction of silica loss (volatilization) cover an area of common glass compositions in the Apollo 16 and/or Apollo 17 soils.

mare glasses is further evidence that mare glass spheres (e.g., Apollo 15 green glass and Apollo 17 orange glass) were derived by fire-fountaining rather than by impact (ref. sphere surface-layer studies by Chou *et al.*, 1974, 1975; Meyer C. *et al.*, 1975; Butler, 1978).

The recognition of volatilized silica-poor compositions among highland glasses may be useful in selecting those glasses which represent highland rock types. The feldspathic highland rock types (Fig. 10b) fall in a narrow compositional range of 43% $<SiO_2<$53% and 0.50 $<(CaO/Al_2O_3)<$0.65. We are currently employing restrictions such as this to select highland glasses which are most likely to represent highland rock types (Vaniman and Papike, 1978).

Expansion of lithic, mineral and glass data by modal recombination: Calculated partitioning of soil chemistry among soil components

We summarize above the nature of the lithic, mineral and glass components in 60009/60010. As in most petrologic studies, rock types are used as a basis of classification. Thus certain concentrations of lithic and mineral fragments are seen as clues to the nature of the parent rocks, and glass compositions are related to known rock types or to mixtures and volatilized derivatives of known rock types. This detailed characterization of each component provides a basis for correlation between soil cores by using characteristic lithic or mineral compositions. However, there is another quite different application of these data on the composition of soil components which involves recombination with soil modal data. This method allows one to calculate the partitioning of soil chemistry among soil components.

The modal data for 60009/60010 (Simon *et al.*, 1978) are summarized (selectively) in Fig. 11. Several things are evident in these diagrams: (1) there is a significant increase in plagioclase with depth, (2) pyroxene and olivine are rather constant with depth except for a low concentration at ~28 cm, (3) RNB + POIK show a decrease with depth while ANT + plagioclase increase, (4) clear glass and yellow glass are in relatively low concentrations and are highly variable but there is a build-up in yellow glass in the 13 to 30 cm depth interval. Another feature of the modal data that is worth noting is the increase in the fused soil component (DMB + agglutinate) near the lunar surface. This effect has been noted by McKay *et al.* (1977) and Morris (1978). Morris argues that much of this effect is due to *in situ* reworking (gardening) of the upper 13 cm of the drive tube and that the mixing is efficient vertically but not laterally. These observations coupled with recent studies of agglutinates (e.g., Hu and Taylor, 1977) which show lack of chemical fractionation during the agglutinate formation process indicate that many of the agglutinates formed in the near vicinity and represent an average soil composition for a portion of the local regolith. Note, however, that Taylor *et al.* (1978) argue that agglutinates in the Apollo 17 drill core are recorders of fossil soil compositions that could have been transported significant distances laterally. Although the problems of soil maturation and agglutinate formation are important, we have chosen to remove the

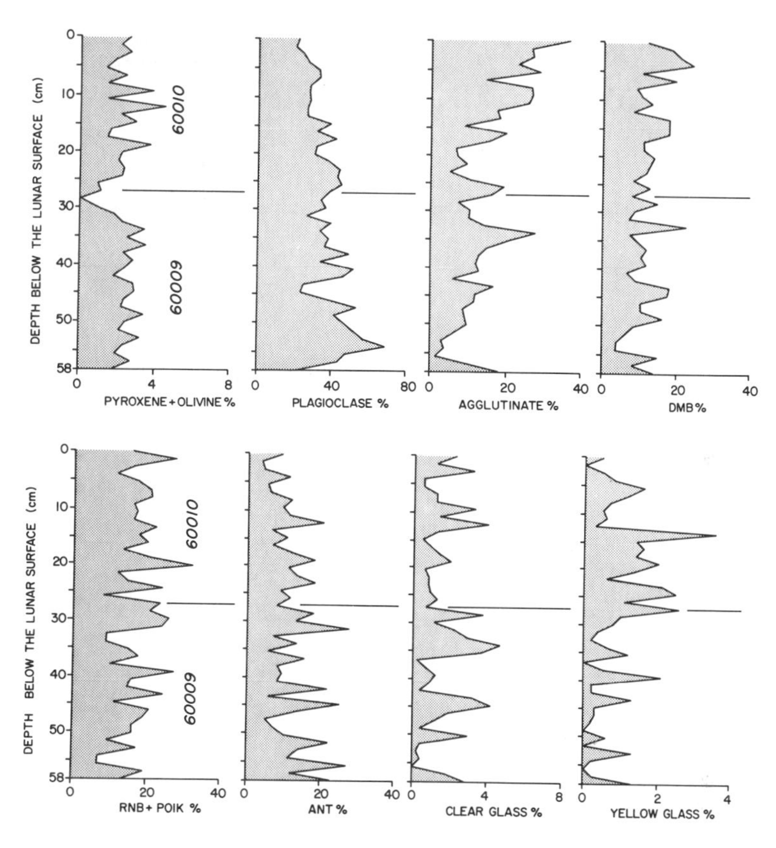

Fig. 11. Apollo 16 drive tube 60009/60010 modal data. Note horizontal scales vary from diagram to diagram.

fused soil component (agglutinate + DMB) from the modal data and renormalize to a "fused soil free" mode. We feel justified in doing this for our chemical modelling of >0.02 mm soil in drive tube 60009/60010 because it is quite likely that the agglutinates (at least where they are abundant in upper part of the core) formed locally (Morris, 1978) and are quite close to the average soil composition. We note that our calculated soil compositions correspond closely to the INAA data of Ali and Ehmann (1976, 1977).

Disregarding DMB's and agglutinates, our best estimates for the chemical compositions of the 12 soil components needed to model the chemistry of 60009/60010 are given in Table 1. Densities for each component are either obtained from tables of mineral data, calculated from modal composition for rocks (or taken from the literature, in the case of ANT and RNB/POIK), or taken from the literature for glasses (Anderson *et al.*, 1970 and Warren *et al.*,

Table 7. Representative Chemical Distribution Matrix for 1 of 44 Core Depth Levels Studied (P.T.S. 60009,6029M, depth $\simeq$ 57 cm, dissection unit 2).

Regolith Components	Modal Data (wt. %) 0.02-2 mm size fraction	SiO_2	Al_2O_3	"Cr_2O_3"*	"TiO_2"	MgO	"FeO"	CaO	Na_2O	K_2O
1. Mare basalt	0.0	—	—	—	—	—	—	—	—	—
2. ANT	32.5	14.58	10.28	0.01	0.03	0.79	1.12	5.88	0.09	0.00
3. Feld. basalt	0.3	0.12	0.07	0.00	0.00	0.02	0.01	0.04	0.00	0.00
4. RNB + POIK	8.9	4.05	1.74	0.02	0.10	0.99	0.79	1.06	0.04	0.03
5. Olivine	1.1	0.42	0.00	0.00	0.00	0.41	0.29	0.00	0.00	0.00
6. Pyroxene	1.3	0.66	0.02	0.01	0.01	0.24	0.19	0.13	0.00	0.00
7. Plagioclase	55.5	24.68	19.42	0.00	0.00	0.00	0.09	10.80	0.24	0.02
8. Opaques	0.5	0.00	0.00	0.00	0.05	0.00	0.40	0.00	0.00	0.00
9. Orange glass	0.0	—	—	—	—	—	—	—	—	—
10. Yellow glass	0.0	—	—	—	—	—	—	—	—	—
11. Green glass	0.0	—	—	—	—	—	—	—	—	—
12. Clear glass	0.0	—	—	—	—	—	—	—	—	—
Σ	100.1	44.51	31.53	0.04	0.19	2.45	2.89	17.91	0.37	0.05
	Sum = 99.94	CaO/Al_2O_3 = 0.57				Fe/(Fe + Mg) atomic = 0.40				

*Quotation marks indicate that multiple valance states are present.

Table 8. Representative Chemical Distribution Matrix for 1 of 44 Core Depth Levels Studied (P.T.S. 60010,6022B, Depth $\simeq$ 12 cm, dissection unit 4).

Regolith Components	Modal Data (wt. %) 0.02-2 mm size fraction	SiO_2	Al_2O_3	"Cr_2O_3"*	"TiO_2"	MgO	"FeO"	CaO	Na_2O	K_2O
1. Mare basalt	0.0	—	—	—	—	—	—	—	—	—
2. ANT	27.3	12.25	8.63	0.01	0.03	0.66	0.94	4.95	0.07	0.00
3. Feld. basalt	3.1	1.41	0.82	0.00	0.02	0.21	0.15	0.47	0.01	0.00
4. RNB + POIK	22.6	10.36	4.43	0.04	0.25	2.53	2.01	2.72	0.11	0.05
5. Olivine	3.0	1.14	0.00	0.00	0.00	1.09	0.78	0.01	0.00	0.00
6. Pyroxene	3.4	1.76	0.05	0.01	0.03	0.64	0.52	0.35	0.00	0.00
7. Plagioclase	34.6	15.38	12.10	0.00	0.00	0.00	0.06	6.73	0.15	0.01
8. Opaques	0.0	—	—	—	—	—	—	—	—	—
9. Orange glass	0.4	0.15	0.04	0.00	0.03	0.03	0.06	0.04	0.00	0.00
10. Yellow glass	0.3	0.16	0.05	0.00	0.01	0.03	0.04	0.04	0.00	0.00
11. Green glass	0.0	—	—	—	—	—	—	—	—	—
12. Clear glass	5.3	2.36	1.36	0.01	0.03	0.39	0.31	0.78	0.01	0.01
Σ	100.0	44.97	27.48	0.07	0.40	5.58	4.87	16.09	0.35	0.07
	Sum = 99.88	CaO/Al_2O_3 = 0.59				Fe/(Fe + Mg) atomic = 0.33				

*Quotation marks indicate that multiple valance states are present.

Table 9. Some Representative "Calculated" Major Element Analyses for the Core Dissection Units *(0.02–2 mm size fraction).*

P.T.S.	Dissection Unit	SiO_2	Al_2O_3	"Cr_2O_3"*	"TiO_2"	MgO	FeO	CaO	Na_2O	K_2O	Σ
60010,6018T	7	44.64	26.84	0.09	0.57	5.75	5.70	15.69	0.38	0.10	99.76
" 6019B	6	44.53	28.90	0.07	0.43	4.24	4.53	16.65	0.38	0.08	99.81
" 6021T	5	45.04	27.91	0.09	0.45	5.17	4.45	16.15	0.40	0.11	99.77
" 6022B	4	44.97	27.48	0.07	0.40	5.58	4.87	16.09	0.35	0.07	99.88
" 6024T	3	45.05	27.91	0.09	0.48	5.13	4.37	16.14	0.41	0.11	99.69
" 6025B	2	45.19	26.08	0.11	0.70	6.26	5.53	15.34	0.41	0.12	99.74
" 6027T	1	45.03	28.14	0.08	0.47	4.94	4.24	16.26	0.41	0.10	99.67
60009,6019T	10	44.83	27.97	0.09	0.56	4.89	4.70	16.14	0.42	0.11	99.71
" 6020B	9	45.03	28.03	0.08	0.41	5.09	4.57	16.24	0.37	0.09	99.91
" 6022T	8	44.68	28.44	0.08	0.47	4.67	4.47	16.41	0.40	0.09	99.71
" 6023T	7	44.84	30.31	0.05	0.32	3.39	3.02	17.35	0.40	0.07	99.75
" 6024T	7	44.85	30.00	0.06	0.28	3.71	3.15	17.16	0.40	0.08	99.69
" 6026B	6	44.63	29.58	0.06	0.37	3.87	3.72	16.92	0.41	0.08	99.64
" 6027T	5	44.72	28.56	0.07	0.39	4.72	4.30	16.48	0.40	0.08	99.72
" 6028T	4	44.75	31.20	0.04	0.20	2.80	2.68	17.77	0.38	0.05	99.87
" 6028B	3	44.87	29.80	0.06	0.30	3.81	3.34	17.09	0.40	0.07	99.74
" 6029M	2	44.51	31.53	0.04	0.19	2.45	2.89	17.91	0.37	0.05	99.94
" 6030	1	44.67	28.53	0.08	0.44	4.55	4.77	16.48	0.36	0.08	99.96

*Quotation marks indicate that multiple valance states are present.

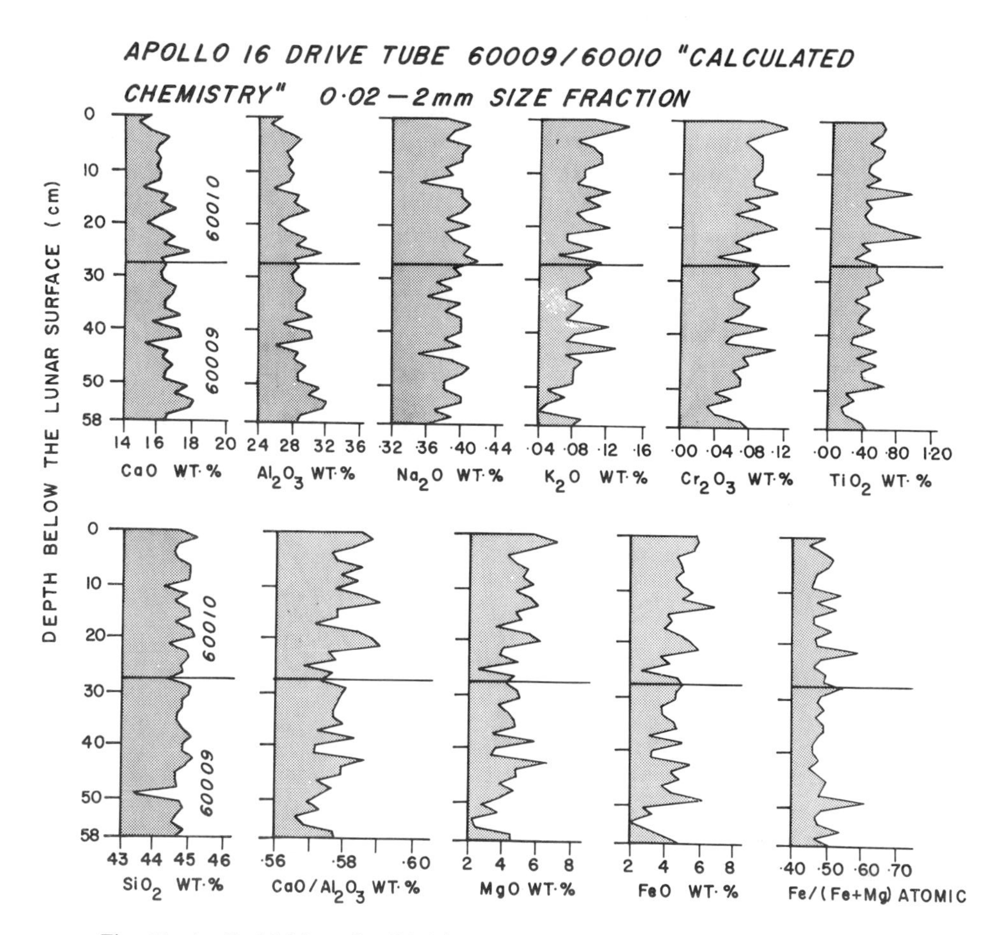

Fig. 12. Apollo 16 drive tube 60009/60010 "calculated" major element chemistry. See text for discussion.

1972). The densities in Table 1 were used to convert modal volume percentages to modal weight percentages. The chemistries of the 12 most important regolith components were combined with the modal weight data to yield "calculated" chemical analyses. Tables 7 and 8 are examples of how we calculate the bulk chemistry by our modal recombination technique. (The Cal. Tech. group led by A. Albee has been doing similar recombinations in an automated mode for some time with their electron microprobe.) Using our computer program, we generated 44 such matrices to model the >0.02 mm soil fraction in drive tube 60009/60010 as a function of depth and grain size. Representative "calculated" chemical analyses for all core dissection units (Duke and Nagle, 1974) are presented in Table 9.

Figure 12 is a stratigraphic summary of the 44 calculated soil compositions. This figure shows that the CaO and Al_2O_3 contents decrease dramatically toward the lunar surface, reflecting the decrease in modal plagioclase. The two peaks in TiO_2 that show up in Fig. 12 reflect two >1 mm high Ti mare basalts and show the dangers in analyzing small soil samples that contain relatively large clasts of exotic chemistry. When our calculated analyses (0.02–2 mm size fraction) are compared to directly measured bulk major element chemistry for the core (Ali and Ehmann, 1976, 1977) the agreement is remarkably good.

There are many ways in which data thus generated can be used. We illustrate one example here. We compared Al_2O_3-rich layer 60009, 6029M (~57 cm depth) with Al_2O_3-poor layer 60010, 6022B (~12 cm depth) to reconcile the chemical differences with the modal differences. A balanced "chemical exchange" can be written in wt.% as:

$$\frac{60009,\ 6029\ \text{M}}{4.0\ Al_2O_3 + 1.8\ CaO} \rightleftarrows \frac{60010,\ 6022\ \text{B}}{0.5\ SiO_2 + 0.2\ TiO_2 + 3.1\ MgO + 2.0\ FeO}\ .$$

In other words, in moving from the lower core level to the upper, these are the chemical changes that occur. We can also define the "modal exchange" in modal wt.% as:

5.2 ANT + 20.8 Plag + 0.5 Opaque ⇄ 2.8 Feld. Basalt + 13.7 (RNB + POIK) + 1.9 Olivine + 2.1 Pyroxene + 0.4 Orange Glass + 0.3 Yellow Glass + 5.3 Clear Glass.

The modal exchange and chemical exchange data may be combined into an "exchange matrix" (Table 10). This matrix illustrates the mixing that would have to take place in converting the chemistry and modal petrology of core level 60009, 6029M (~57 cm depth) to that of core level 60010, 6022B (~12 cm depth). In Table 10 one can readily find the portions of the matrix which involve the largest chemical changes. As a first approximation it is the loss of plagioclase and gain of RNB/POIK + olivine + pyroxene that accounts for the loss of Al_2O_3 and gain in MgO + FeO in going from the lower to the upper core level.

CONCLUSIONS

The lithic components in the Apollo 16 double drive tube 60009/60010 are predominantly highland breccias and melt rocks. The principal mineral component in these soils is An_{96}-An_{97} plagioclase derived from cumulate or brecciated-cumulate rocks. Concentrations of this coarse-grained anorthitic plagioclase are most prominent at 53–57 cm depth in the double drive tube; concentrations of pyroxene and olivine from the same anorthitic cumulate rock or a light-matrix breccia equivalent occur at higher levels. The presence of a recurring two-pyroxene cluster ($Wo_{4.5}En_{70.5}Fs_{25}$ + $Wo_{39}En_{47}Fs_{14}$) above 53 cm depth reflects the contribution of mineral fragments from a two-pyroxene + feldspar + ilmenite ± armalcolite ± ulvöspinel POIK lithology. The same cluster of pyroxene compositions was found at 150–170 cm depth in the Apollo 16 drill core (Vaniman *et al.*, 1976). Above 53 cm in the double drive tube and below 57 cm

Table 10. "Exchange Matrix" Comparing Core Level (60009,6029M, depth ~ 57 cm) to Core Level (60010,6022B, depth ~ 12 cm).

Regolith Component	SiO_2	Al_2O_3	TiO_2	MgO	FeO	CaO	Σmodal (wt. %)
ANT	−2.3	−1.6	0	−0.1	−0.2	−0.9	−5.1
Feld. Basalt	+1.3	+0.7	0	+0.2	+0.1	+0.4	+2.7
RNB + POIK	+6.3	+2.7	+0.2	+1.5	+1.2	+1.6	+13.5
Olivine	+0.7	0	0	+0.7	+0.5	0	+1.9
Pyroxene	+1.1	0	0	+0.4	+0.3	+0.2	+2.0
Plagioclase	−9.2	−7.2	0	0	0	−4.0	−20.4
Opaques	0	0	−0.1	0	−0.4	0	−0.5
Orange glass	+0.2	0	0	0	+0.1	0	+0.3
Yellow glass	+0.2	+0.1	0	0	0	0	+0.3
Clear glass	+2.4	+1.4	0	+0.4	+0.3	+0.8	+5.3
Σoxide (wt. %)	+0.7	−3.9	+0.1	+3.1	+1.9	−1.9	ΣΣ = 0

there is a sharp increase in the concentration of sodic plagioclase, reflecting the increased contribution of a highland melt rock (POIK) component. A variety of analytical methods (dissection, INAA, and I_s/FeO) indicate that the coarse feldspathic layer at 53–57 cm is a distinct stratigraphic unit. The variations in composition of the mineral component above and below the coarse-grained unit help to identify the source rocks mixed into the 60009/60010 soil, but do not necessarily define a more detailed stratigraphy. The pyroxene component with Fe/(Fe + Mg) >0.45, for example, represents feldspathic basalt sources that are dispersed throughout the core. A few of these more Fe-rich pyroxenes may be derived from mare basalts, but the mare component forms $\ll$0.1% of the >20 μm components. The diversity of the few mare lithic fragments which are large enough to characterize suggests a number of distant source areas (high-Ti, low-Ti and aluminous mare sources). Glasses in 60009/60010 include the dominant anorthositic gabbro composition previously observed at Apollo 16; glass data from both the Apollo 16 drill core and 60009/60010 differ from glasses at Apollo 17 by lacking a prominent range of mare-highland mixtures which show evidence of silica loss through volatilization.

Modal recombination is an effective way of modelling the major element chemistry of the >0.02 mm soil fraction. It has major advantages over the usual procedure of mathematical fitting procedures of *assumed* components to a bulk chemical analysis. The modal recombination technique combines *real* soil components that are actually present in the soil under investigation to give a calculated bulk chemistry. However, the primary advantage in the modal recombination technique is not that the chemistry can be calculated but, because of the method used, the total major element partitioning among the regolith components is defined.

Acknowledgments—This paper was improved by the reviews of W. Quaide, D. Blanchard, D. McKay, J. M. Rhodes and K. Keil.

We acknowledge with thanks the help given by K. Baldwin in preparing the computer software used and in obtaining the plots in Figs. 11 and 12. This research was supported by NASA Grant #NSG-9044 for which we are extremely grateful.

REFERENCES

Agrell S. O., Scoon J. H., Muir J. D., Long J. B. P., McConnell J. D. C. and Peckett A. (1970) Observations on the chemistry, mineralogy and petrology of some Apollo 11 lunar samples. *Proc. Apollo 11 Lunar Sci. Conf.*, p. 93–128.

Albee A. L., Quick J. E. and Chodos A. A. (1977) Source and magnitude of errors in "broad-beam analysis" (DBA) with the electron microprobe (abstract). In *Lunar Science VIII*, p. 709. The Lunar Science Institute, Houston.

Albee A. L. and Ray L. (1970) Correction factors for electron probe microanalysis of silicates, oxides, carbonates, phosphates and sulfates. *Anal. Chem.* **42**, 1408–1414.

Ali M. Z. and Ehmann W. D. (1976) Chemical characterization of lunar core 60009. *Proc. Lunar Sci. Conf. 7th*, p. 241–258.

Ali M. Z. and Ehmann W. D. (1977) Chemical charactization of lunar core 60010. *Proc. Lunar Sci. Conf. 8th*, p. 2967–2981.

Anderson O. L., Scholz C., Soga N., Warren N. and Schreiber E. (1970) Elastic properties of a micro-breccia, igneous rock and lunar fines from Apollo 11 mission. *Proc. Apollo 11 Lunar Sci. Conf.*, p. 1959–1973.

Bence A. E. and Albee A. L. (1968) Empirical correction factors for electron microanalysis of silicates and oxides. *J. Geol.* **76**, 382–403.

Blanchard D. P., Jacobs J. W., Brannon J. C. and Brown R. W. (1976) Drive tube 60009: A chemical study of magnetic separates of size fractions from five strata. *Proc. Lunar Sci. Conf. 7th*, p. 281–294.

Blanford G. E., McKay D. S. and Wood G. C. (1977) Particle track densities in double drive tube 60009/10. *Proc. Lunar Sci. Conf. 8th*, p. 3017–3025.

Bogard D. D. and Hirsch W. C. (1975) Noble gas studies on grain size separates of Apollo 15 and 16 deep drill cores. *Proc. Lunar Sci. Conf. 6th*, p. 2057–2083.

Bogard D. D. and Hirsch W. C. (1976) Noble gases in 60009-60010 drive tube samples: Trapped gases and irradiation history. *Proc. Lunar Sci. Conf. 7th*, p. 259–279.

Butler P. Jr. (1978) Recognition of lunar glass droplets produced directly from endogenous melts: The evidence from S-Zn coatings (abstract). In *Lunar and Planetary Science IX*, p. 143–145. Lunar and Planetary Institute, Houston.

Chou C.-L., Baedecker P. A., Bild R. W. and Wasson J. T. (1974) Volatile-element systematics and green glass in Apollo 15 lunar soils *Proc. Lunar Sci. Conf. 5th*, p. 1645–1657.

Chou C.-L., Boynton W. V., Sundberg L. L. and Wasson J. T. (1975) Volatiles on the surface of Apollo 15 green glass and trace-element distributions among Apollo 15 soils. *Proc. Lunar Sci. Conf. 6th*, p. 1701–1727.

Delano J. W. (1975) Petrology of the Apollo 16 mare component: Mare Nectaris. *Proc. Lunar Sci. Conf. 6th*, p. 15–47.

Delano J. W., Bence A. E., Papike J. J. and Cameron K. L. (1973) Petrology of the 2–4 mm soil fraction from the Descartes Region of the moon and stratigraphic implications. *Proc. Lunar Sci. Conf. 4th*, p. 537–551.

Dixon J. R. and Papike J. J. (1975) Petrology of anorthosites from the Descartes region of the moon: Apollo 16. *Proc. Lunar Sci. Conf. 6th*, p. 263–291.

Duke M. B. and Nagle J. S. (1974) Lunar Core Catalog and Supplements. JSC 09252. NASA Johnson Space Center, Houston. 242 pp.

Duke M. B. and Nagle J. S. (1977) Lunar Core Catalog and Supplements. JSC 09252, p. 16–58 to 16–78. NASA Johnson Space Center, Houston.

Dungan M. A. and Brown R. W. (1977) The petrology of the Apollo 12 ilmenite basalt suite. *Proc. Lunar Sci. Conf. 8th*, p. 1339–1381.

Gose W. A. and Morris R. V. (1977) Depositional history of the Apollo 16 deep drill core. *Proc. Lunar Sci. Conf. 8th*, p. 2909–2928.

Hartung J. B., El Goresy A. and Nagel K. (1978) Chemical composition variations in lunar microcrater pit glasses (abstract). In *Lunar and Planetary Science IX*, p. 462–464. Lunar and Planetary Institute, Houston.

Hu H. and Taylor L. A. (1977) Lack of chemical fractionation in major and minor elements during agglutinate formation. *Proc. Lunar Sci. Conf. 8th*, p. 3645–3656.

Ivanov A. V. and Florensky K. P. (1975) The role of vaporization processes in lunar rock formation. *Proc. Lunar Sci. Conf. 6th*, p. 1341–1350.

Koretev R. L. (1976) Geochemistry of grain-size fractions of soils from the Taurus-Littrow Valley floor. *Proc. Lunar Sci. Conf. 7th*, p. 695–726.

Laul J. C. (1978) Chemical study of size fractions of Apollo 17 deep drill cores 70009–70006 (abstract). In *Lunar and Planetary Science IX*, p. 637–639. Lunar and Planetary Institute, Houston.

LSPET (Lunar Sample Preliminary Examination Team) (1973) The Apollo 16 lunar samples: Petrographic and chemical description. *Science* **179**, 23–34.

McKay D. S., Dungan M. A., Morris R. V. and Fruland R. M. (1977) Grain size, petrographic and FMR studies of the double core 60009/10: A study of soil evolution. *Proc. Lunar Sci. Conf. 8th*, p. 2929–2952.

McKay D. S., Morris R. V., Dungan M. A., Fruland R. M. and Fuhrman R. (1976) Comparative studies of grain size separates of 60009. *Proc. Lunar Sci. Conf. 7th*, p. 295–313.

Meyer C., McKay D. S., Anderson D. H. and Butler P. (1975) The source of sublimates on the Apollo 15 green and Apollo 17 orange glass samples. *Proc. Lunar Sci. Conf. 6th*, p. 1673–1699.

Meyer H. O. A. and McCallister R. H. (1974) Apollo 16: Core 60004-preliminary study of <1 mm fines. *Proc. Lunar Sci. Conf. 5th*, p. 907–916.

Meyer H. O. A. and McCallister R. H. (1976) Mineral, lithic and glass clasts <1 mm size in Apollo 16 core section 60003. *Proc. Lunar Sci. Conf. 7th*, p. 185–198.

Meyer H. O. A., McCallister R. H. and Tsai H. M. (1975) Mineralogy and petrology of <1 mm fines from Apollo 16 core sections 60002 and 60004. *Proc. Lunar Sci. Conf. 6th*, p. 595–614.

Morris R. V. (1978) *In situ* reworking (gardening) of the lunar surface: Evidence from the Apollo cores (abstract). In *Lunar and Planetary Science IX*, p. 757–759. Lunar and Planetary Institute, Houston.

Morris R. V. and Gose W. A. (1976) Ferromagnetic resonance and magnetic studies of cores 60009/60010 and 60003: Compositional and surface-exposure stratigraphy. *Proc. Lunar Sci. Conf. 7th*, p. 1–11.

Naney M. T., Crowl D. M. and Papike J. J. (1976) The Apollo 16 drill core: Statistical analysis of glass chemistry and the characterization of a high alumina-silica poor (HASP) glass. *Proc. Lunar Sci. Conf. 7th*, p. 155–184.

Naney M. T., Papike J. J. and Vaniman D. T. (1977) Lunar highland melt rocks: Major element chemistry and comparisons with lunar highland glass groups (abstract). In *Lunar Science VIII*, p. 717–719. The Lunar Science Institute, Houston.

Papike J. J., Bence A. E. and Lindsley D. H. (1974) Mare basalts from the Taurus-Littrow region of the moon. *Proc. Lunar Sci. Conf. 5th*, p. 471–504.

Papike J. J., Hodges F. N., Bence A. E., Cameron M. and Rhodes J. M. (1976) Mare basalts: Crystal chemistry, mineralogy and petrology. *Rev. Geophys. Space Phys.* **4**, 475–540.

Papike J. J., Vaniman D. T., Schweitzer E. L. and Baldwin K. (1978) Apollo 16 drive tube 60009/60010. Part III: Total major element partitioning among the regolith components or chemical mixing models with petrologic credibility (abstract). In *Lunar and Planetary Science IX*, p. 862–864. Lunar and Planetary Institute, Houston.

Reid A. M., Warner J., Ridley W. I. and Brown R. W. (1972) Major element compositions of glasses in three Apollo 15 soils. *Meteoritics* **7**, 395–415.

Rhodes J. M., Adams J. B., Blanchard D. P., Charette M. P., Rodgers K. V., Brannon J. C. and Haskin L. A. (1975) Chemistry of agglutinate fractions in lunar soils. *Proc. Lunar Sci. Conf. 6th*, p. 2291–2307.

Russ G. P. III (1973) Apollo 16 neutron stratigraphy. *Earth Planet. Sci Lett.* **17**, 275–289.
Simon S. B., Papike J. J. and Vaniman D. T. (1978) The Apollo 16 drive tube 60009/60010. Part I: Modal petrology. *Proc. Lunar Planet. Sci. Conf. 9th.* This volume.
Simonds C. H., Warner J. L. and Phinney W. C. (1973) Petrology of Apollo 16 poikilitic rocks. *Proc. Lunar Sci. Conf. 4th*, p. 613–632.
Taylor G. J., Wentworth S., Warner R. D. and Keil K. (1978) Petrology of Apollo 17 deep drill core. II. Agglutinates as recorders of fossil soil compositions (abstract). In *Lunar and Planetary Science IX*, p. 1146–1148. Lunar and Planetary Institute, Houston.
Vaniman D. T., Lellis S. F., Papike J. J. and Cameron K. L. (1976) The Apollo 16 drill core: Modal petrology and characterization of the mineral and lithic component. *Proc. Lunar Sci. Conf. 7th*, p. 199–239.
Vaniman D. T. and Papike J. J. (1977a) Very low Ti (VLT) basalt: A new mare rock type from the Apollo 17 drill core. *Proc. Lunar Sci. Conf. 8th*, p. 1443–1471.
Vaniman D. T. and Papike J. J. (1977b) The Apollo 17 drill core: Modal petrology and glass chemistry (sections 70007, 70008, 70009). *Proc. Lunar Sci. Conf. 8th*, p. 3161–3193.
Vaniman D. T. and Papike J. J. (1978) The lunar highland melt-rock suite. *Geophys. Res. Lett.* In press.
Walker D., Longhi J. and Hays J. F. (1972) Experimental petrology and origin of Fra Mauro rocks and soil. *Proc. Lunar Sci. Conf. 3rd*, p. 797–817.
Warren N., Anderson O. L. and Soga N. (1972) Applications to lunar geophysical models of the velocity-density properties of lunar rocks, glasses and artificial lunar glasses. *Proc. Lunar Sci. Conf. 3rd*, p. 2587–2598.

Proc. Lunar Planet. Sci. Conf. 9th (1978), p. 1861–1874.
Printed in the United States of America

Apollo 16 deep drill: A review of the morphological characteristics of oxyhydrates on *Rusty* particle 60002, 108, determined by SEM

STEPHEN E. HAGGERTY

Department of Geology, University of Massachusetts, Amherst, Massachusetts 01003

Abstract—The Apollo 16 deep drill (60001-60007) recovered a regolith core of approximately 2 meters in length. The lowermost portion of the core contains a 3 mg metallic Fe-Ni spherule (60002, 108) and a soil clod (60002, 178) that are indurated with iron oxyhydroxides. These oxidized particles are present in a distinctive horizon dated at 1 + 0.5 b.y., and this layer is very different petrologically and in terms of exposure history from the rest of the core. Lunar oxyhydrates generally are considered to have formed by the interaction of meteoritic lawrencite ($FeCl_2$) and H_20 to yield akaganéite ($\beta FeO.OH$). Although the lunar highlands are recognized as having a volatile-rich component, and although the occurrence of oxyhydrates is widespread in highlands materials, an origin by terrestrial water vapor contamination for the existence of akaganéite, nevertheless, has been favored by a number of, but not all, previous investigators.

This SEM study demonstrates that the surface encrustation on metal particle 60002, 108 constitutes a complex and stratified mineralogy of at least five morphologically distinct layers, with akaganéite commonly being the most distant from the metal substrate. The metal particle occurs at 200 cm from the lunar surface, and although akaganéite has been demonstrated experimentally to be unstable under lunar surface conditions, ~140°C and 10^{-7} torr, there is evidence that the spherule was incorporated into the regolith soon after deposition. Among several other factors, which argue against terrestrial contamination and are in support of a probable lunar origin by reaction of water at very low partial pressures, is the fact that residual lawrencite was still present after a three-year storage period in the Curatorial Facility. Lawrencite is extremely hygroscopic and would not have survived under normal terrestrial conditions. It is concluded that there is ample justification for reopening the debate that small concentrations of indigenous lunar water may still be present. However, the mineralogy and exposure history of the particle may be unique, but this must be established by detailed and systematic examination of other akaganéite occurrences in lunar samples by SEM and related techniques.

INTRODUCTION

The Apollo 16 deep drill penetrated and recovered a regolith profile of approximately 2.2 meters from station 10, the ALSEP (Apollo Lunar Surface Experiments Package) site, located 105 m southwest of the Lunar Module landing site in the Descartes area in close proximity to Spook, Gator and Eden Valley craters (Duke and Nagle, 1974). The core is 2.25 cm in diameter, is in seven sections, and was returned in three units: the drill bit (60001), the lowermost section (60002, 60003, and 60004), and the uppermost section (60005, 60006, and 60007).

The core has been divided into 46 stratigraphic units (Duke and Nagle, 1974) and 4 major lithologic units (Vaniman *et al*, 1976). Unit A, at the base of the core (190–224 cm) is distinguished by the abundance of large lithic clasts and abundant yellow glass; Unit B (55–190 cm) is characterized by fewer lithic clasts and an abundance of green glass; Unit C (6–55 cm) contains dominantly large plagioclase fragments and lacks orange glass; and Unit D (0–6 cm) has fewer mafic mineral fragments and fewer glasses than the remainder of the core (Vaniman *et al.*, 1976). Each of these units may be indicative of separate cratering events as demonstrated by Russ (1973), who suggests that the core does not represent a single-stage depositional model. Based on $^{4}He/^{20}Ne$ and $^{40}Ar/^{36}Ar$, this is confirmed in part by Bogard and Hirsh (1975), who have shown that Unit A has an apparent age of 1 ± 0.5 b.y. and that this basal section was very probably an ancient regolith surface (i.e., between Unit A and Unit B) derived from deeper fragmental material at the Apollo 16 site. In a review by Meyer and McCallister (1977) on the Apollo 16 deep drill, these authors note, in addition, that: (1) accumulation of the regolith has been irregular in deposition, but whether such deposition was by a small limited number of events or by a combination of fast and steady-state depositions coupled with cratering events is still uncertain; (2) track density studies indicate both mini (mm) and micro (<1 mm) stratigraphy in the core; (3) mineralogical and chemical analyses suggest that the major portion of the core consists of highlands material with mare components being restricted to glasses; and (4) almost all sections of the core show some degree of mixing and small scale regolith gardening.

A unique feature of section 60002 (the lowermost section of the drill core), but one which is relatively common among the Apollo 16 rocks (Taylor *et al*, 1973), is the occurrence of oxidized metal particles which were retrieved from the base of the section as a soil clod (60002, 178) and at the center of the section as a discrete metal sphere (60002, 108); the apparent depths below the lunar surface for these particles are at a maximum of 216 cm and 200 cm respectively, and these are in the 1 ± 0.5 b.y. stratigraphic horizon of Unit A.

The major component of the oxyhydrate is the mineral akaganéite (βFeO.OH) which is associated with K and Ca chlorine-rich complexes (Carter, 1975). It has been convincingly argued by Taylor *et al.* (1973, 1974a, 1974b) and by Taylor and Burton (1976) that metal oxidation in other lunar highland samples results from the interaction of meteoritic lawrencite ($FeCl_2$) with water. However, the following questions have been posed: Is the chlorine necessarily of meteoritic origin? Is the water indigenous to the moon? Did the reaction take place during the return flight, or did the oxyhydrate crystallize in the Curatorial Facility? Although the cores were not returned in vacuum containers, each section was tightly plugged after the drilling operation was completed, and processing of the dissected core took place in cabinets under positive nitrogen pressure. It cannot be unequivocally demonstrated that the reaction did not take place under terrestrial conditions, but it can be argued that the oxidized particles in the deep drill string offer one of the best opportunities of resolving the problem of lunar versus terrestrial oxidation. A significant facet of the argument relates to the important experiments by Taylor and Burton (1976) who have shown that akaganéite is unstable under high vacuum conditions at lunar surface temperatures of ~140°C and the fact that lawrencite was still present when it was examined initially by Carter (1975) three years after the core was collected. The present report is a Scanning Electron Microscopy (SEM) study of the morphological features of 60002, 108. This study, which demonstrates that the oxyhydrate is not entirely akaganéite but a complex mixture of minerals encrusting the metallic Fe-Ni spherule, supports the observations of Carter (1975). The new observations record a series of morphologies which previously have not been presented, and these show that the oxide-oxyhydrate assemblage results in a coherently stratified sequence from the metal substrate to the outermost surface of the particle.

Scanning Electron Microscopy

The metal spherule weighs 3 mg and is 1.2 × 0.8 × 0.6 mm in dimension. The particle as received in this laboratory was gold-coated for Carter's (1975) previous SEM study, and although he described the surface texture as being highly irregular and the encrustation as being yellow, orange, and red, neither of these features were obvious on binocular examination. This is an unfortunate factor because the variety of textures, crystal morphologies, and types of encrustations observed in the present study has prevented a color contrast and textural correlation.

Carter (1975) noted that the particle had been bombarded by low velocity micrometeorites, and craters of the order of 10–50 microns are apparent in some regions of the metal spherule as illustrated in Fig. 1. He classified the particle into four basic types of material: (1) silicious materials that include silicate splashes rich in Al and Ca, in association with fractured mixtures of plagioclase, pyroxene, and olivine; (2) metallic Ni-Fe; (3) a yellowish granular material; and (4) relatively smooth, brilliant deep ruby-red mounds. The metallic substrate is an Fe-Ni alloy; the yellowish material has a Ni content approximately one half that of the metallic spherule and rare splashed glass drapes around this material; the ruby-red mounds are of two forms, and these are superimposed on the yellow material, having high Fe, Ni and Cl, and high Ca, Cl, and K (with minor Na and S), respectively. Based on the associated silicates, Carter (1975) suggests that the metal particle is derived from either an igneous rock or a metamorphosed breccia.

The present SEM study substantiates Carter's (1975) observations; but, in exfoliated or cracked regions of the encrustation, a series of additional morphologies in texturally distinctive horizons are apparent from the outermost surface of the metal particle to the outermost rind of the oxyhydrated encrustation. Following the descriptive morphologies of earlier studies and the scanning electron microbeam analyses by Carter (1975), it is assumed that the bifurcating crystals shown in Fig. 2 are akaganéite. Crystalline akaganéite forms mounds (Fig. 2a) that are typically present as contorted prisms (Fig. 2b), and these crystalline mounds constitute the outermost layer of the reaction rind. Surrounding these mounds and at depths below the mounds are four additional morphologies which are classified as follows: (1) spray rosettes of almost cylindrical

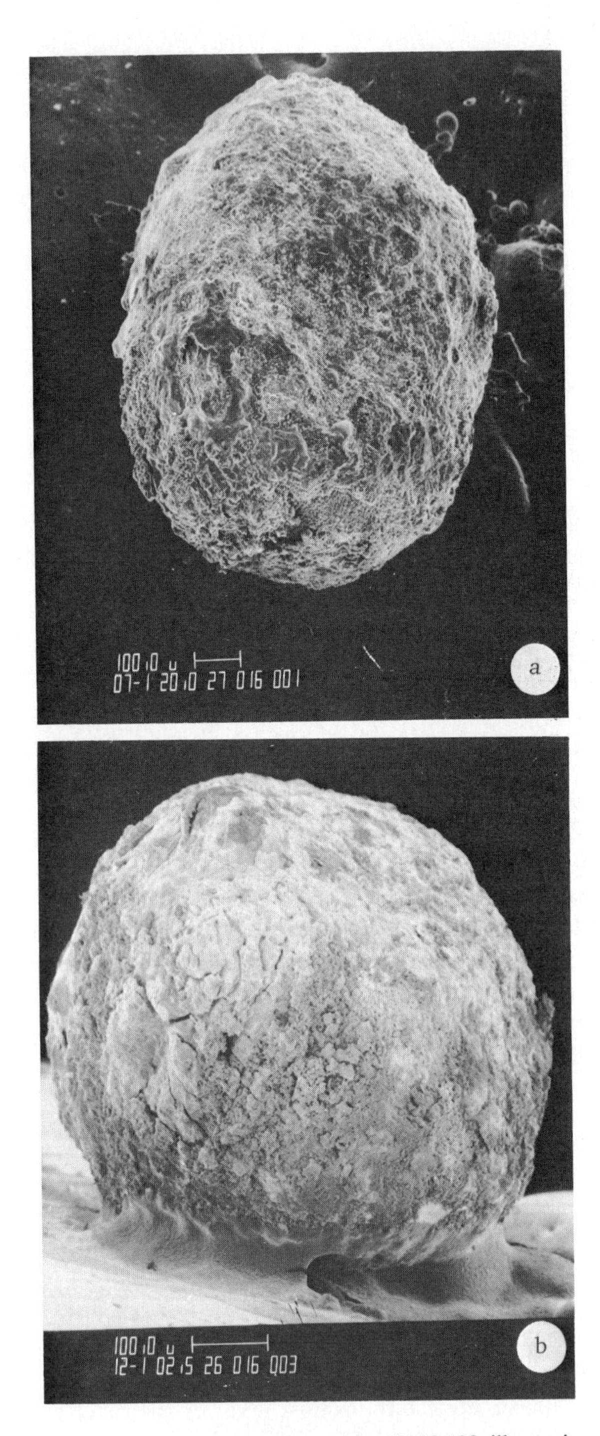

Fig. 1. (a,b) SEM photomicrographs of particle 60002,108 illustrating the surface oxidation encrustation.

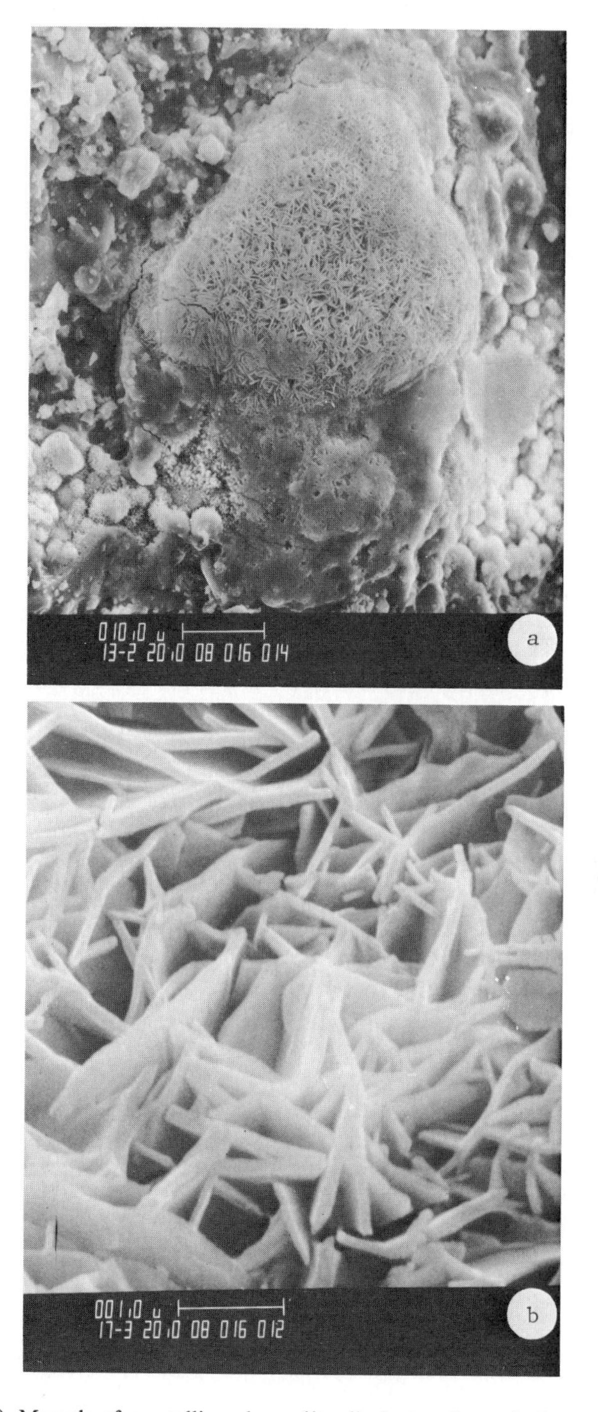

Fig. 2. (a,b) Mounds of crystalline akaganéite displaying the typical morphology of interlaced and bifurcating aggregates.

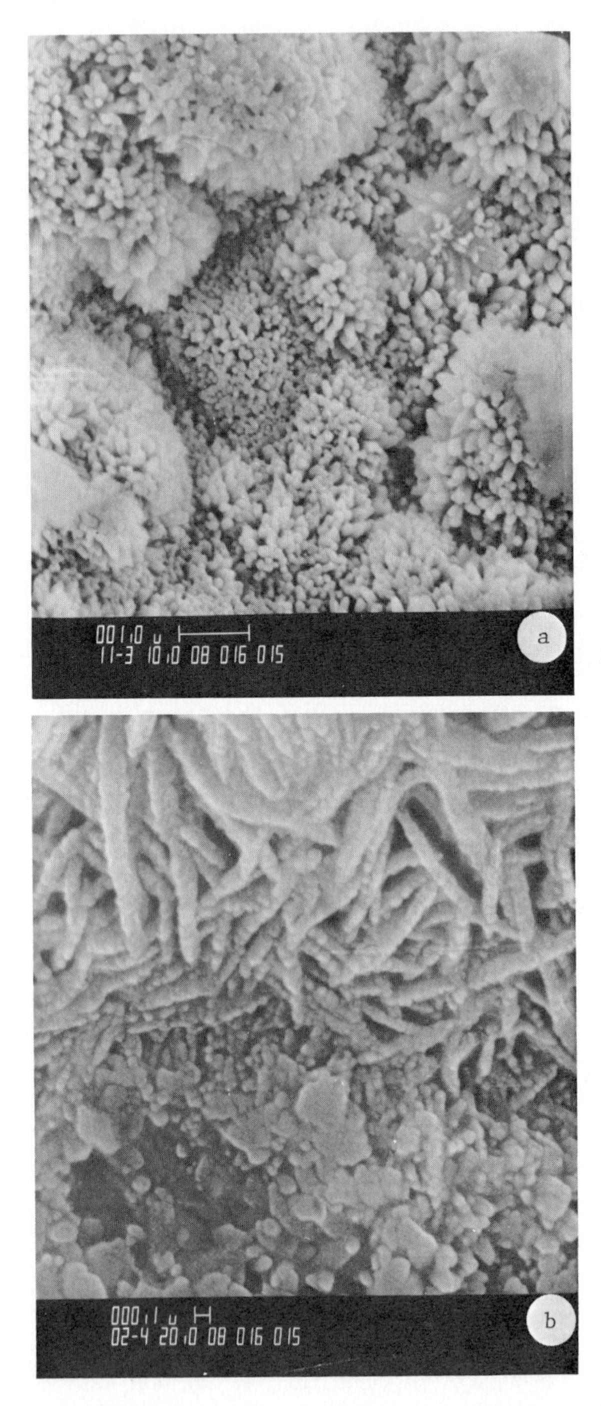

Fig. 3. (a,b) Spray rosettes of an unidentified mineral which is almost cylindrical in habit and has bulbous protrusions at right angles to the length of the crystal.

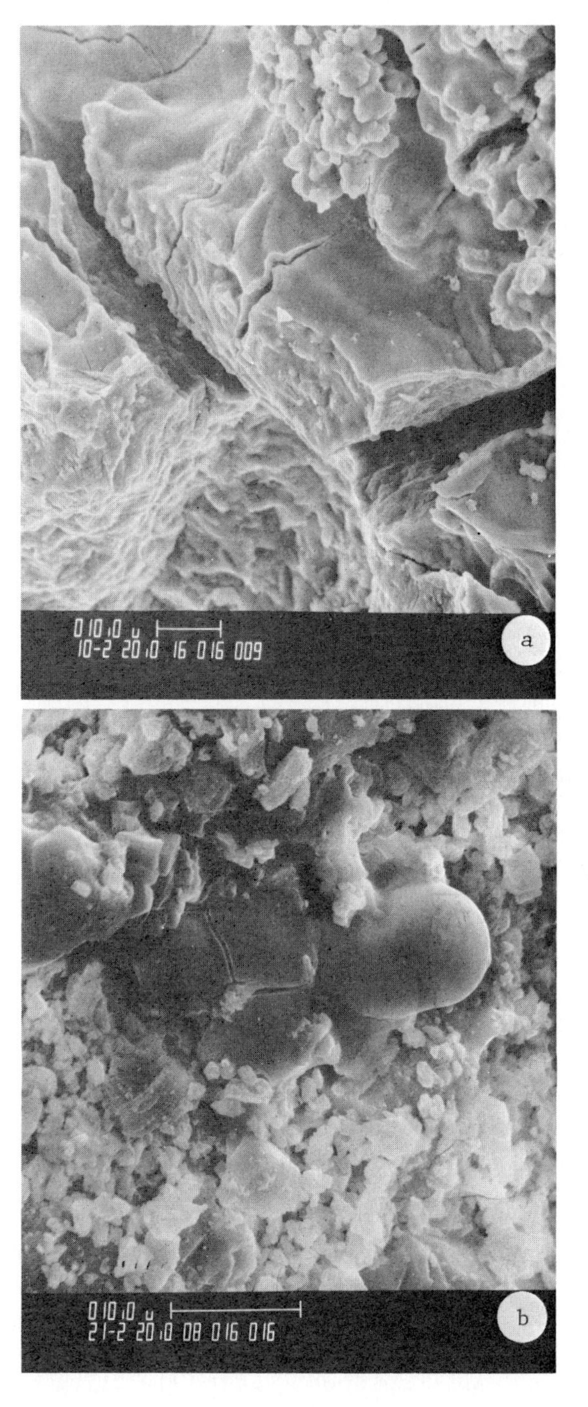

Fig. 4. (a,b) Gelatinous material which may be either weakly stratified (Fig. 4a) or is alternatively present as irregularly shaped domes (Fig. 4b).

Fig. 5. Curved and flatly bladed honeycombed textured crystals of an unidentified mineral.

crystals which are lined with bulbous buds at right angles to the length of the crystal (Fig. 3); (2) a stratified layer which is gelatinous in appearance and is free of crystals at the 70 A resolution limit (Fig. 4); (3) curved and flatly bladed honeycombed textured crystals (Fig. 5); and (4) immediately adjacent to the Fe-Ni spherule substrate (Fig. 6a) and surrounding splashed glass droplets (Fig. 6b) is a convoluted, dimpled, and *con-carne* textured layer which Carter (1975) has previously described as yellow, wrinkled and shatter-coned.

Qualitative scanning electron microbeam analyses of each of the five textural forms show that Fe is a major component with minor, but varying, proportions of Ni and Cl. The metal substrate showed a strong Fe and Ni signal, and although Co was not sought in the analysis, it is assumed that the particle is probably of lunar origin. A significant observation by Carter (1975) that is confirmed by this study is that the crystalline form of akaganéite illustrated in Fig. 2 is not in direct contact with the metallic Fe-Ni particle. This observation suggests that the

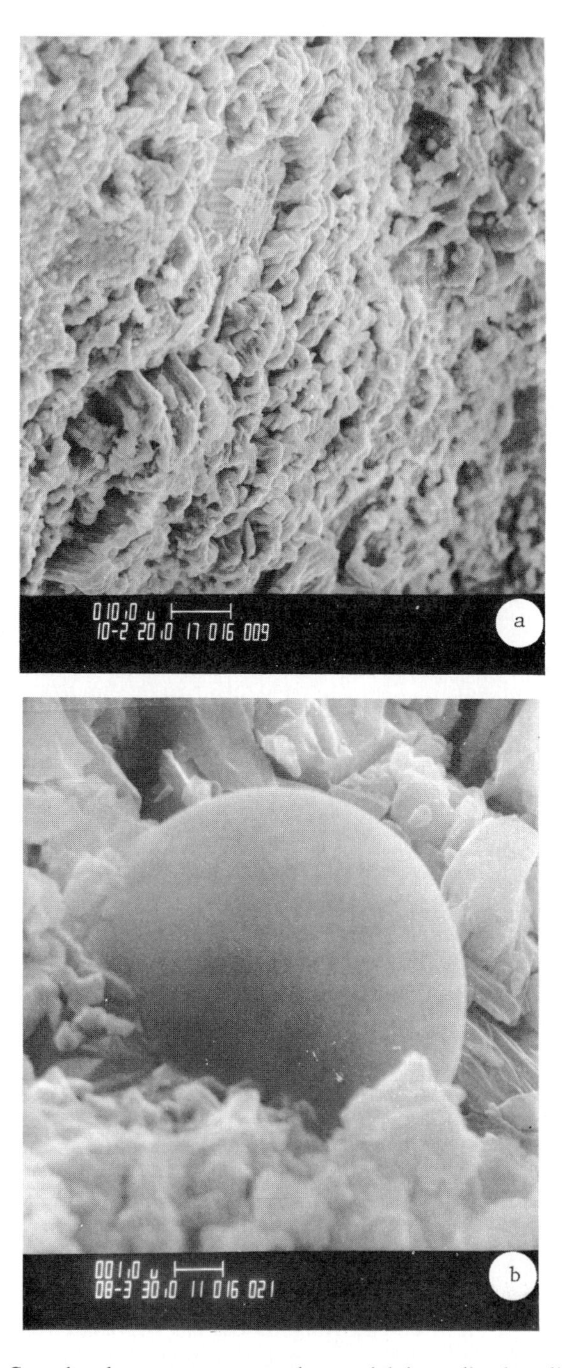

Fig. 6. (a,b) Convoluted *con-carne* textured material immediately adjacent to the Fe-Ni substrate. A portion of the metal surface is shown in the upper left-hand part of Fig. 6a. This material also surrounds splashed glass droplets, and the glass is presumably in direct contact with the underlying metal.

formation of akaganéite, which is clearly the dominant surface material that has previously been identified by Taylor and coworkers (1973, 1974a, 1974b) using X-ray techniques, is at least a two-stage or perhaps even a multiple-stage process which ultimately results in βFeO.OH. An alternative explanation is that the cracked surface of the encrustation through to the substrate is the result of partial dehydration and shrinkage and that the variations in color may, therefore, be due to anhydrous oxides such as hematite (αFe_2O_3) or maghemite (γFe_2O_3).

DISCUSSION

Several factors, in addition to those suggested by Carter (1975), argue for reopening the question of small quantities of lunar water based on the oxidized particles in the Apollo 16 deep drill:

(1) Although the experiments by Taylor and Burton (1976) have demonstrated that the mineral akaganéite is unstable on the lunar surface at temperatures of ~140°C and pressures of 10^{-7} torr, it is important to note the spherule has a low exposure crater density and that the particle is at the interface of two micro-stratigraphic horizons in Unit A. This suggests that the particle may have been buried rather soon after it had been deposited. It would appear, therefore, that in order for akaganéite to be effectively stabilized, the reaction must have taken place at some depth, if the oxyhydrated assemblage is indeed a byproduct of lawrencite ($FeCl_2$) and lunar water or, alternatively, of indigenous chlorine + water.

(2) The variety of distinctive and stratified mineral morphologies that are identified in this study and by Carter (1975) suggest that the oxyhydrate on particle 60002, 108 is not entirely akaganéite but a complex mixture of microcrystalline phases which appear to be well ordered with respect to the Fe-Ni metal substrate. This again argues that the stability experiments by Taylor and Burton (1976) may not necessarily apply to the entire *rusty* assemblage. It is not certain, of course, that all other lunar samples, which contain akaganéite, are as complex as the Apollo 16 deep drill particle, nor is it certain that all *rust* is formed by the same process. Epstein and Taylor's (1974) oxygen isotope measurements have shown that at least some akaganéite-bearing samples from the Descartes site are contaminated by terrestrial water. However, it is clear that a re-examination of the comparative mineralogies of these samples is necessary before an all-encompassing model is proposed.

(3) Although the drill string was not returned in a vacuum container from the moon, the *rusty* particle is at an apparent depth of 2.00 meters from the surface and is present in a densely compacted horizon. Teflon plugs were removed from the ends of the drill string, and dissection of the core

tubes took place in dry nitrogen cabinets in the Curatorial Facility. Particles 60002, 108 (the metal sphere) and 60002, 178 (the soil clod) were observed to be oxidized with a surrounding red halo at the instant that these areas were uncovered during the spatula dissection (Nagle, pers. comm., 1972) and shortly thereafter by Haggerty. It would appear that these particles were reasonably well protected from surface terrestrial oxidation; but it can not be demonstrated that the core was sufficiently compacted to inhibit the percolation of water vapor into the interior of the core.

(4) There are a large number of metal particles in the drill core and an abundance of metal spheres in glasses and agglutinates; yet, the horizons at 200 cm and at 216 cm are the only layers in which the metals are obviously oxidized. Lawrencite or a chlorine-rich component appears to be a necessary association to trigger the oxidation reaction, but it is also possible that oxyhydration only took place in specific soil horizons because these were relatively enriched in very small concentrations of water vapor. A significant aspect of the distribution of oxidized metal particles in the drill core is that these are restricted to the lowermost 1 ± 0.5 b.y. old horizon, a stratigraphic layer which has been shown to be radically different from the remainder of the core (e.g., Vaniman *et al.*, 1976; Bogard and Hirsch, 1975; Jordan and Heymann, 1976).

(5) Undoubtedly the strongest point, which argues against the terrestrial contamination of particle 60002, 108 by water vapor in the Curatorial nitrogen cabinets or by water vapor in the sample boxes of the Apollo 16 spacecraft, is that lawrencite was identified by Carter (1975) three years after the core was returned from the moon. Lawrencite has been demonstrated to be highly reactive in the presence of water vapor, and in view of the large concentrations that are potentially available during space flight or laboratory storage, it is anticipated that it would have readily reacted to completion. On the other hand, if oxidation did occur under extremely low partial pressures of H_2O in the lunar regolith, some lawrencite may still have been preserved in association with akaganéite.

(6) The stratified sequence of yellow, wrinkled and *con-carne* textured material below that of akaganéite suggests that this oxidized material was subsequently hydrated and, as noted by Carter (1975), is partially covered by splash glass, suggesting at the very minimum that some degree of oxidation must have taken place on the lunar surface or during regolith formation. An alternative explanation is that the variety of morphologies are due to either the oxyhydrates αFeO.OH (goethite), γFeO.OH (lepidocrocite), or the oxides αFe_2O_3 (hematite), γFe_2O_3 (maghemite) and that it is one of these phases which subsequently interacted with chlorine to form the present assemblage. In all cases, the lower thermal stability limits are in excess of 140°C.

(7) There is now a mounting body of evidence (e.g., Cadogan *et al.*, 1973; Epstein and Taylor, 1973; Gibson and Moore, 1973; Jovanovic and Reed,

1973; El Goresy *et al.*, 1973) to show that rocks and soils from the lunar highlands have an extremely large volatile component. Although there is still considerable debate on whether these volatiles are indigenous to the moon or were introduced by meteoritic contamination, there is no evidence which is either contradictory or which demonstrates unequivocally that the moon was always *totally* devoid of small concentrations of water vapor. It would indeed by surprising, if the models of planetary condensation are accepted, to suppose that the moon formed in a *totally* anhydrous environment. Lawrencite is highly reactive and only small concentrations of H_2O are required to induce the reaction of iron to akaganéite. Two proposals have been suggested for the formation of minute concentrations of lunar water: (1) by the reduction of iron-bearing silicates or oxides through a process that involves the remobilization of solar-wind implanted hydrogen (Carter, 1975) and (2) the incorporation of small amounts of carbon into the lunar regolith to form metallic iron and water (Carter and McKay, 1972).

SUMMARY AND CONCLUSIONS

There is a high incidence of oxyhydrates on metallic iron particles in samples from the ancient lunar highlands. Rocks and the regolith in the highlands are enriched in volatiles in comparison to those of the mare and, whether this difference is due to a fundamental property of the evolution of the moon or whether it is merely a function of surface meteoritic bombardment and mixing, is not known. It has been argued that the surface oxidation of metal particles by the reaction of lawrencite ($FeCl_2$) + H_2O to yield the iron oxyhydroxide, akaganéite ($\beta FeO.OH$), results from the terrestrial contamination of this assemblage by water vapor (e.g., Epstein and Taylor, 1973).

An oxidized metal particle from the Apollo 16 Decartes deep drill core, which was uncovered at a depth of 200 cm, is shown to have a complexly textured mineralogy which overlies the metallic Fe-Ni substrate. The only positively identified phase is akaganéite; this mineral, along with four distinctively shaped minerals of contrasting morphologies, contains iron as a major constituent, with varying proportions of Ni and chlorine.

Although it cannot be conclusively demonstrated that the oxidation of this metal particle did not take place under terrestrial conditions, the following factors are strong arguments for reopening the debate for the presence of small concentrations of water at low partial pressures on the moon: (1) the particle was probably buried soon after its incorporation into the lunar regolith and, this event would stabilize the formation of akaganéite, which is unstable at the lunar surface; (2) this study has demonstrated that it is not akaganéite alone that contributes to the oxyhydrated encrustation but that it includes other hydroxides and possibly also hematite and maghemite; (3) the particle was recovered at 200 cm from the lunar surface in a densely compacted horizon, and it was observed to be oxidized on dissection; (4) only two horizons of the drill string contain oxidized particles and these are restricted to the lowermost 1 ± 0.5 b.y. old Unit A layer; and (5) the most pertinent evidence against terrestrial oxidation is that residual lawrencite continued to persist even after a storage period of three years in the Curatorial Facility—an unexpectedly long preservation time, given the abundance of water vapor and the intensely reactive nature of $FeCl_2$ in the presence of H_2O.

This particle may well be very different from other oxidized metal particles that are widely distributed in lunar highland materials, but this supposition needs to be demonstrated by careful and systematic SEM examination.

Acknowledgments—This research was supported by NASA under grant NGR-22-010-89, which is gratefully acknowledged. The careful dissecting skills of Stu Nagle and the prompt action of the Lunar Sample Analysis Planning Team, under the duress of a proposal put forward by Peter M. Bell and Stephen E. Haggerty to remove these particular samples, should serve as a clear endorsement to NASA that the lunar sample collection is in good hands. The ETEC Corporation and their personnel are acknowledged for their help in SEM photography. The manuscript was critically and constructively reviewed by James L. Carter, Sudhir Mehta, and Roger H. Hewins. To all I express my appreciation.

References

Bogard D. D. and Hirsch W. L. (1975) Noble gas studies on grain size separates of Apollo 15 and 16 deep drill cores. *Proc. Lunar Sci. Conf. 6th*, p. 2057–2083.

Cadogan P. H., Eglinton G., Gowar A. P., Jull A. J. T., Maxwell J. R. and Pillinger C. T. (1973) Location of methane and carbide in Apollo 11 and 16 lunar fines. *Proc. Lunar Sci. Conf. 4th*, p. 1493–1508.

Carter J. L. (1975) Surface morphology and chemistry of rusty particle 60002, 108. *Proc. Lunar Sci. Conf. 6th*, p. 711–718.

Carter J. L. and McKay D. S. (1972) Metallic mounds produced by reduction of material of simulated lunar composition and implications on the origin of metallic mounds on lunar glasses. *Proc. Lunar Sci. Conf. 3rd*, p. 953–970.

Duke M. B. and Nagle J. S. (1974) Lunar core catalog and Supplements. JSC 09252. NASA Johnson Space Center, Houston.

El Goresy A., Ramdohr P., Pavićević M., Medenbach O., Müller O. and Gentner W. (1973) Zinc, lead, chlorine and FeO.OH-bearing assemblages in the Apollo 16 sample 60095: Origin by impact of a comet or a carbonaceous chondrite. *Earth Planet. Sci. Lett.* **18**, 411–419.

Epstein S. and Taylor H. P. (1973) The isotopic composition and concentration of water, hydrogen, and carbon in some Apollo 15 and 16 soils and in the Apollo 17 orange soil. *Proc. Lunar Sci. Conf. 4th*, p. 1559–1575.

Epstein S. and Taylor H. P. (1974) D/H and $^{18}O/^{16}O$ ratios of H_2O in the rusty breccia 66095 and the origin of "lunar water". *Proc. Lunar Sci. Conf. 5th*, p. 1839–1854.

Gibson E. K. Jr. and Moore G. W. (1973) Carbon and sulfur distributions and abundances in lunar fines. *Proc. Lunar Sci. Conf. 4th*, p. 1577–1586.

Jordan J. L. and Heymann D. (1976) Inert gas stratigraphy of sections 60002 and 60004 of the Apollo 16 deep drill core (abstract). In *Lunar Science VII*, p. 434–436. The Lunar Science Institute, Houston.

Jovanovic S. and Reed G. W. (1973) Volatile trace elements and the characterization of the Cayley formation and the primitive lunar crust. *Proc. Lunar Sci. Conf. 4th*, p. 1313–1324.

Meyer H. O. A. and McCallister R. H. (1977) The Apollo 16 deep drill core. *Proc. Lunar Sci. Conf. 8th*, p. 2889–2907.

Russ G. P. III (1973) Apollo 16 neutron stratigraphy. *Earth Planet. Sci. Lett.* **19**, 275–289.

Taylor L. A. and Burton J. C. (1976) Experiments on the stability of βFeO.OH on the surface of the moon. *Meteoritics* **11**, 225–230.
Taylor L. A., Mao H. K. and Bell P. M. (1973) "Rust" in the Apollo 16 rocks. *Proc. Lunar Sci. Conf. 4th*, p. 829–839.
Taylor L. A., Mao H. K. and Bell P. M. (1974a) βFeO.OH, akaganéite, in lunar rocks. *Proc. Lunar Sci. Conf. 5th*, p. 743–748.
Taylor L. A., Mao H. K. and Bell P. M. (1974b) Identification of the hydrated iron oxide mineral akaganéite in Apollo 16 lunar rocks. *Geology* **1**, 429–432.
Vaniman D. T., Lellis S. F., Papike J. J. and Cameron K. L. (1976) The Apollo 16 deep drill core: Modal petrology and characterization of the mineral and lithic component. *Proc. Lunar Sci. Conf. 7th*, p. 199–239.

Proc. Lunar Planet. Sci. Conf. 9th (1978), p. 1875–1884.
Printed in the United States of America

Irradiation stratigraphy in the Apollo 16 deep drill section 60002

GEORGE E. BLANFORD and GORDON C. WOOD

University of Houston at Clear Lake City, Houston, Texas 77058

Abstract—Particle track density frequency distributions, abundance of track rich grains and minimum track densities are reported for the upper 20 cm of the 60002 section of the Apollo 16 deep drill core. The principal stratigraphic feature is a boundary approximately 7 cm from the top of the section. Experimental evidence does not conclusively determine whether this contact is an ancient regolith surface or is simply a depositional boundary. If it is an ancient surface, it has a model exposure age of 3–7 $\times 10^6$ y and a reworking depth of ~0.5 cm. However, because track density frequency distributions indicate the mixing of soils of different maturities, we favor interpreting this contact as a depositional boundary. There may be a second depositional boundary approximately 19 cm below the top of 60002.

INTRODUCTION

We report measurements of particle track densities in soil grains from 185–205 cm in the Apollo 16 deep drill which is the upper 20 cm of core section 60002. In particular we show the relation between solar flare and galactic cosmic ray irradiated soil grains with other indicators of stratigraphy such as color and texture (Duke and Nagle, 1976), trapped rare gases (Bogard and Hirsch, 1975; Jordan and Heymann, 1976), ferromagnetic resonance surface exposure index I_s/FeO (Gose and Morris, 1977), and in a general way with chemistry (Nava *et al.*, 1976) and petrology (Vaniman *et al.*, 1976). This work differs from that of Blanford and Morrison (1976) in that we have continuous sampling which eliminates stratigraphic ambiguities which they experienced.

Particle track density data are commonly expressed as frequency distributions, the relative amount of track rich grains and mimimum, quartile or median track densities. When measured as a function of depth, these data are especially useful in revealing stratigraphy in a lunar core.

Track density frequency distributions can often reveal lunar soil evolutionary paths. Because track production on the moon is a decreasing power function of depth (Walker and Yuhas, 1973; Hutcheon *et al.*, 1974; Blanford *et al.*, 1975), grains that are within millimeters of the lunar surface acquire large track densities rapidly, whereas grains buried more than a centimeter aquire tracks much more slowly. Since the track production rate at ~1 cm is ~1 $\times 10^6$ cm^{-2} $(10^6\ y)^{-1}$, we can conclude that soil grains with track densities less than 10^8 cm^{-2} most probably have never been exposed at the immediate lunar surface. Such soils are immature. Because the track production rate at ~100 μm is ~3 $\times 10^9$ cm^{-2} $(10^6\ y)^{-1}$, grains with track densities $\geq 1 \times 10^9$ cm^{-2} have significant

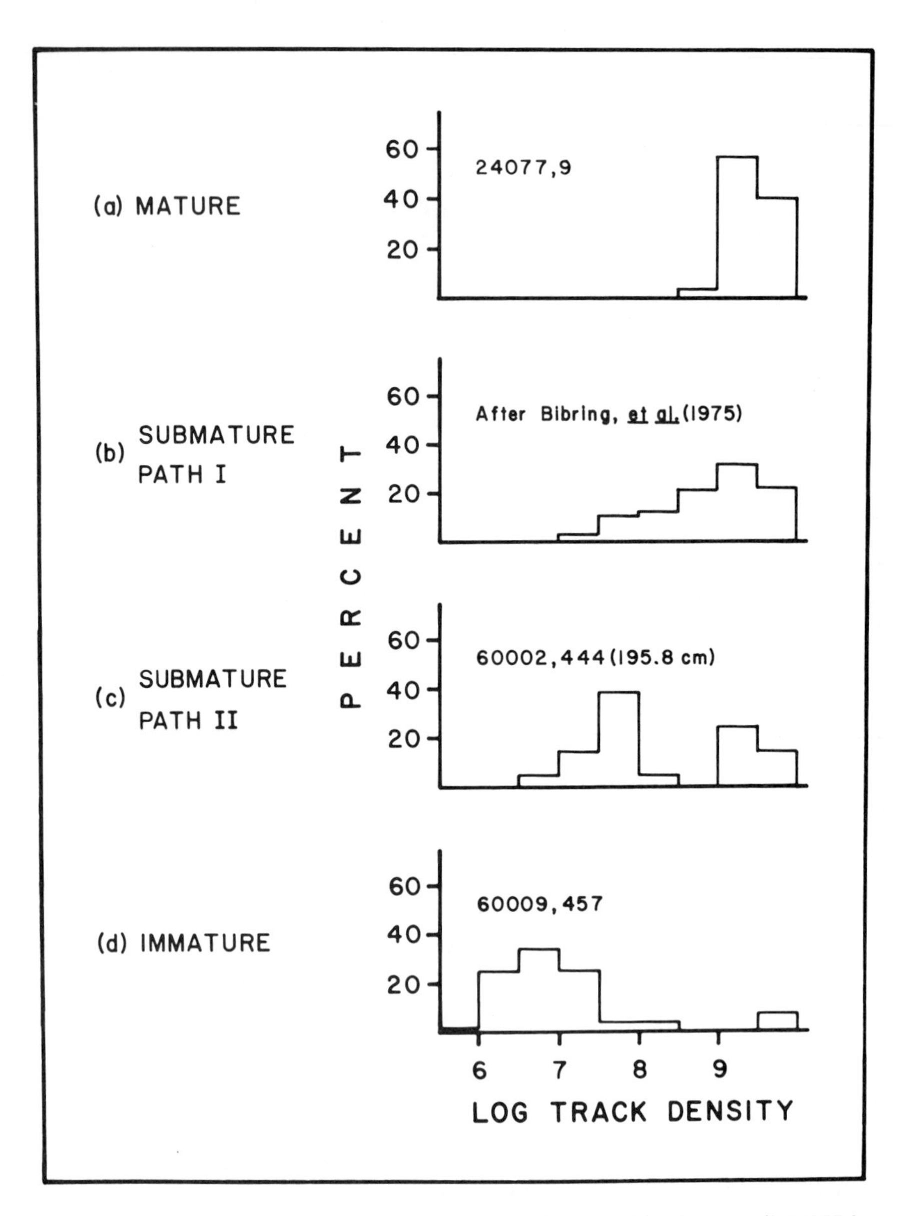

Fig. 1. Examples of track density frequency distributions. (a) Mature soil 24077,9 (Blanford and Wood, 1978b). Since track densities are nearly all greater than 10^9 cm^{-2}, almost all of these grains have been exposed within millimeters of the lunar surface. (b) Submature soil from the work of Bibring *et al.* (1975), who show that this distribution agrees with those predicted from Monte Carlo simulations of meteoroid reworking (Duraud *et al.*, 1975). McKay *et al.* (1974) refer to maturation by reworking as evolutionary path I. (c) Submature soil 60002,444 measured at a depth of 195.8 cm in the Apollo 16 drill string (this work). This soil is a mixture of an immature and a mature component. McKay *et al.* (1974) refer to maturation by mixing as evolutionary path II. (d) Immature soil 60009,457 (Blanford *et al.*, 1977). Since track densities are nearly all less than 10^8 cm^{-2}, these grains have most probably never been exposed within a few millimeters of the lunar surface.

exposure at the immediate lunar surface, and the corresponding soils should have high values of other surface exposure indices such as agglutinate contents and I_s/FeO. Examples of track density frequency distributions in lunar soils are shown in Fig. 1 a,d. A soil that has nearly all of its grains with track densities that are greater than 10^9 cm^{-2} is considered mature, and a soil that has nearly all of its grains with track densities less than 10^8 cm^{-2} is considered immature. McKay *et al.* (1974) point out that soils advance from immature to submature by two evolutionary paths or a combination of them. Path I occurs by *in situ* reworking from meteoroid bombardment; whereas, path II occurs by a mixing of immature with very mature material. Figure 1b shows a submature soil that follows evolutionary path I; Bibring *et al.* (1975) show that this track density distribution fits the distribution calculated by Monte Carlo simulations of soil reworking (Duraud *et al.*, 1975). Figure 1c shows a soil in which no transition exists between the immature and mature components and, therefore, this distribution represents mixed or evolutionary path II soil.

As indicated above, track-rich grain abundance, when this represents the solar-flare irradiated component of the soil, is an index of maturity for the soil. As a maturity index it has some limitations. Blanford *et al.* (1977) have shown that it saturates with respect to I_s/FeO. Also it is often associated only with the plagioclase component of the soil, as in this work, and Etique *et al.* (1978) have shown that this bias in track density measurements can result in poor correlation between track-rich grain abundance and trapped solar-wind rare gas abundance. However, because Apollo 16 soils are rich in plagioclase, we expect that the abundance of track-rich grains will correlate well with other maturity indices.

Minimum track densities (ρ_{min}) are especially useful in lunar cores because they can be related to a model from which an exposure age can be calculated (Crozaz *et al.*, 1970). If fresh, unirradiated grains are incorporated into a soil, and that soil remains undisturbed at some depth below the lunar surface, the mimimum density grains will acquire tracks that follow the track production curve (Walker and Yuhas, 1973; Blanford *et al.*, 1975). Therefore, if we find minimum density grains in a core that follow the track production curve, we can determine the interface which was the ancient exposed surface and calculate its exposure age.

In this paper, we report measurements of track density distributions, the abundance of track-rich grains and mimimum track densities in the 185–205 cm portion of the Apollo 16 deep drill core. We discuss these measurements with regard to other measurements in the core and relate them to the stratigraphy of the core.

Experimental Methods

We work with polished thick sections prepared by the curatorial staff from impregnated core materials (Nagle and Duke, 1974). After photographing the sections, we measure them with a magnifying scale. We estimate saw-cuts to be 0.7 mm wide, a value that we found to be consistent over the whole core lengths of core sections 60009/10. A factor of two error in our saw-cut estimation would lead to a several millimeter systematic error in depth assignments at the deeper end of our sampling length; we use the Duke and Nagle (1976) maximum depths in this report. We initially etch samples for 1.5 h in boiling 1 N NaOH (~118° C). After this time the samples peal off the silica slides, and we remount them on epoxy bases. First stage etching is completed in an additional 4.5 h (total of 6 h), which enlarges tracks to ~60 nm average diameter. We measure track densities from scanning electron micrographs of plagioclase-bearing grains taken in a 1 mm wide band at selected depths. To insure internal comparability and to avoid a bias toward large grains, we limit the grain-size selection from ~40 μm to ~120 μm. Although we restrict our measurements to the mineral plagioclase, we do not restrict our measurements to clean, monomineralic fragments. We attempt to measure every neighboring grain that fits our size criteria, selecting a third of the grains each at the middle and at both ends of the 1 mm wide bands. Because we restrict our track density measurements

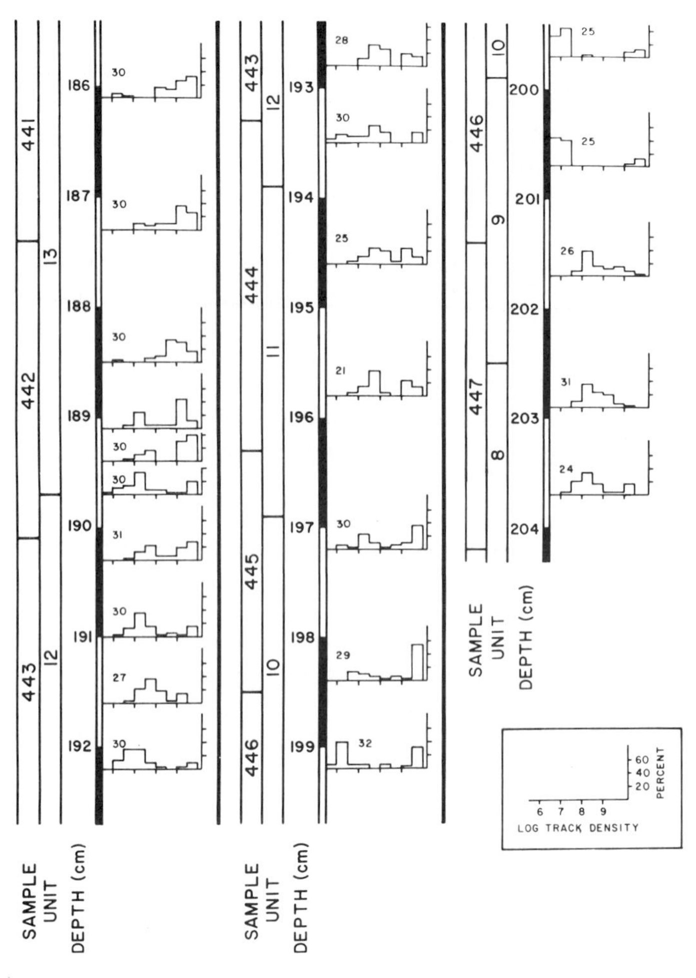

Fig. 2. Histograms of track density frequency distributions in the upper half of 60002 from the Apollo 16 deep drill core. Sample numbers refer to daughter numbers assigned to the impregnated core sections we measured, unit numbers refer to dissection units of Duke and Nagle (1976), and depths are the maximum depths given by Duke and Nagle (1976).

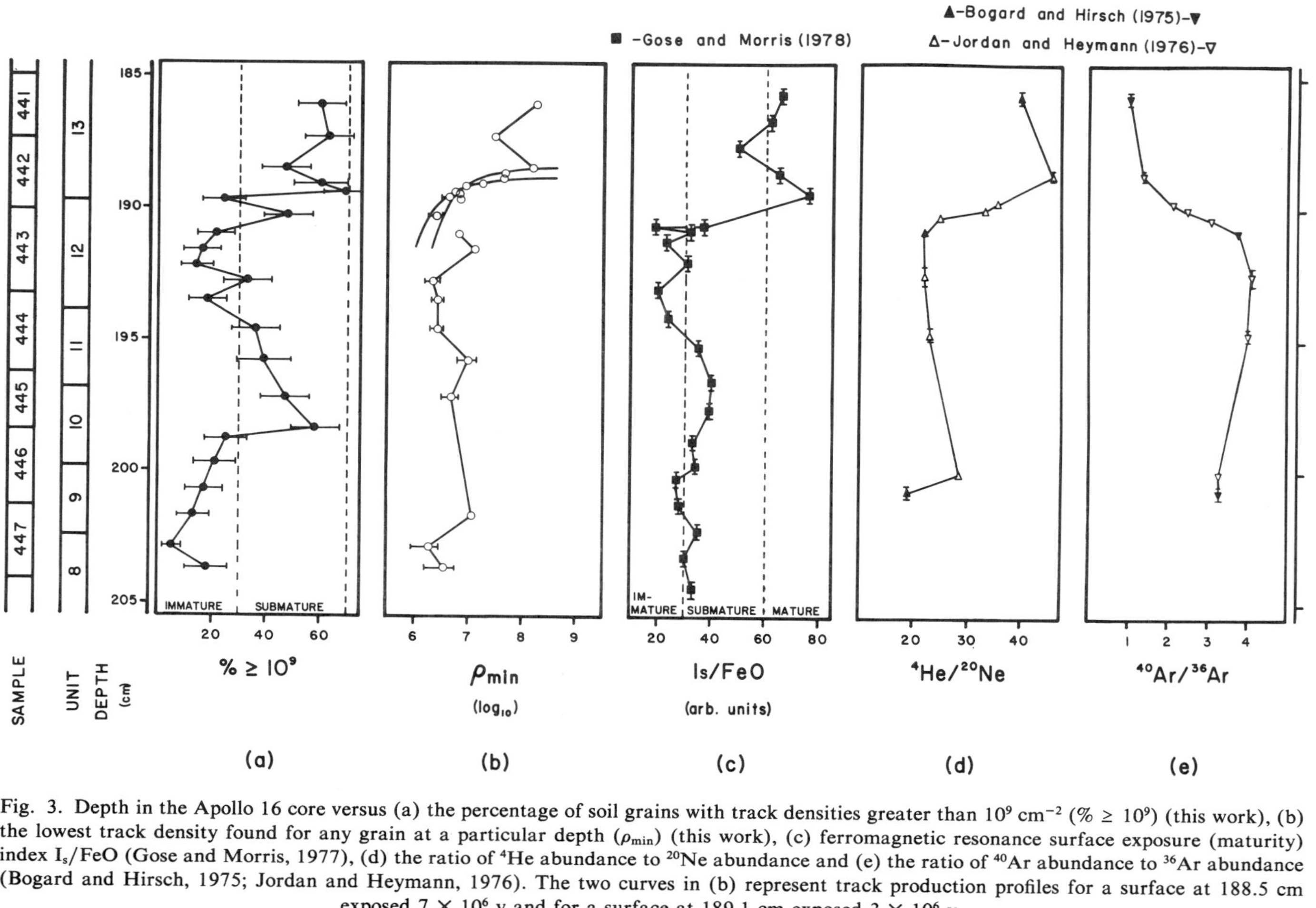

Fig. 3. Depth in the Apollo 16 core versus (a) the percentage of soil grains with track densities greater than 10^9 cm^{-2} (% $\geq 10^9$) (this work), (b) the lowest track density found for any grain at a particular depth (ρ_{min}) (this work), (c) ferromagnetic resonance surface exposure (maturity) index I_s/FeO (Gose and Morris, 1977), (d) the ratio of ^{4}He abundance to ^{20}Ne abundance and (e) the ratio of ^{40}Ar abundance to ^{36}Ar abundance (Bogard and Hirsch, 1975; Jordan and Heymann, 1976). The two curves in (b) represent track production profiles for a surface at 188.5 cm exposed 7×10^6 y and for a surface at 189.1 cm exposed 3×10^6 y.

to plagioclase, we cannot claim that our sampling is completely free of bias, but we have made a determined effort to minimize biases which lead to discrepancies with other measures of soil maturity (e.g., Etique *et al.*, 1978). From the initial measurements, we prepare histograms of track density frequency distributions at each depth (Fig. 2) and determine the percentage of soil grains with track densities $\geq 10^9$ cm^{-2} ($\% \geq 10^9$) (Fig. 3a). In Fig. 3a, we place boundaries classifying soils as immature and submature by using comparisons of earlier track density data to other maturity classifications (Blanford and Wood, 1978b). These boundaries may not always correlate exactly with maturity classifications by other methods, but they offer a starting point for comparison, After we obtain the frequency distributions and $\% \geq 10^9$, we etch the sections for an additional 12 h (total of 18 h). We determine minimum density grains at each level by remeasuring the 5 lowest density grains. With larger tracks we can use lower magnification micrographs and the statistical significance of the measurements greatly improve. We show results of depth versus track density for mimimum density grains in Fig. 3b.

RESULTS AND DISCUSSION

Units 8 and 9: Three track density distributions from 201.7 cm to the bottom of our sampling interval are similar to each other and are distinct from track density distributions above this point (Fig. 2). The distributions, $\% \geq 10^9$, and ρ_{min} (Fig. 3a,b) indicate that the soil is immature, but the track density distributions appear to indicate a small admixture of submature, path I soil. Gose and Morris (1977) state that this material is submature, but it is clear from their graph (Fig. 3c) that it is near their immature-submature boundary.

Units 9, 10, 11, and 12: Track density distributions from 192.2–200.8 cm are all bimodal and indicate submature, path II soils (Fig. 2). The mature soil component increases slowly with depth from 192.2–197.2 cm and then decreases (Fig. 3a). Track-rich grains decrease abruptly between 197.2 and 198.4 cm and then decrease more gradually with depth. Ferromagnetic resonance intensity also shows a similar increasing and decreasing trend within this interval, but I_s/FeO does not decrease sharply (Fig. 3c). The sharp decrease of $\% \geq 10^9$ near the unit 9–10 boundary may have influenced Blanford and Morrison (1976) to conclude that this was an ancient lunar surface, but we conclude from the path II type track density distributions that variations in $\% \geq 10^9$ from 192.2–200.8 cm indicate smoothly varying degrees of heterogeneous mixing.

Units 12 and 13: At 188.5 cm and above, track density distributions indicate submature, path I soils (Fig. 2) and I_s/FeO oscillates about the submature-mature boundary (Fig. 3c). Track-rich grain abundances, mimimum track densities, I_s/FeO, and ratios of rare gas isotypes exhibit sharp changes between 188.5–191.0 cm (Fig. 3). Minimum track densities fit a track production curve with an ancient surface between 188.5–189.1 cm and an exposure age from $3\text{–}7 \times 10^6$ y (Fig. 3b). On the other hand, track density distributions over this interval indicate path II mixing rather than *in situ* reworking. Blanford and Morrison (1976) have concluded that the unit 12–13 boundary is an ancient lunar surface, whereas Gose and Morris (1977) have concluded that it is not an ancient surface but simply a depositional boundary.

Blanford and Morrison (1976) argued that the 12–13 boundary is an ancient lunar surface because minimum density grains fit a track production curve, because % $\geq 10^9$ decreases in a way that is consistent with a reworking zone and because track density distributions varied across the boundary zone in a manner that is consistent with *in situ* reworking. It is clear from our work that Blanford and Morrison (1976) did not correctly identify minimum density grains, which led them to overestimate the supposed exposure age of the contact. However, the measurements we are reporting also fit a track production curve (Fig. 3b), and the abundance of track-rich grains also vary in the boundary zone in a manner consistent with either a 0.5 cm (188.5–189 cm) or 2 cm (188.5–190.8 cm) reworking zone (Fig. 3a). If these data are to be used to support an ancient regolith surface, then we must conclude that there is 0.5 cm reworking depth for otherwise we would not expect the data for miminum density grains to fit a track production curve because it would be obscured by mixing. However, we should be skeptical that minimum density grains do fit a track production curve with depth. The data fit the curve over a depth during which track densities are changing rapidly. If the average level of preirradiation was less than $\sim 2 \times 10^6$ cm^{-2}, the production curve should fit the data at depths greater than 190.2 cm. Because this is not true, the agreement between data and the production curve has a strong chance of being fortuitous. Although a 0.5 cm reworking depth is consistent with Monte Carlo calculations of a meteoroid reworking depth for a surface that is exposed $3–7 \times 10^6$ y (Arnold, 1975), it is not consistent with data obtained from cosmic ray induced ^{22}Na and ^{26}Al contents (Fruchter *et al.*, 1977), which would predict a mixing depth of several centimeters for a surface exposed $\sim 5 \times 10^6$ y. Finally, although the variation of track density distributions across the boundary zone observed by Blanford and Morrison (1976) follows a trend that one would expect for *in situ* reworking (Fig. 1b), the data that we present show that the boundary zone is a mixing zone (Fig. 2, 189.1–189.7 cm). We do not understand the reason for this discrepancy, but, because experimental techniques employed in this work were slightly more refined, we believe the boundary zone is more likely to be a mixing zone.

Gose and Morris (1977) argued that the unit 12–13 boundary represents a depositional stratigraphic boundary because I_s/FeO changes too rapidly to be associated with a reworking zone. Unfortunately, no one knows how an I_s/FeO depth profile evolves with time. Gose and Morris (1977) show an I_s/FeO profile development from a soil with an initially constant value for I_s/FeO (Fig. 4a). Two simplistic models of profile development can be constructed consistent with their diagram and with the fact that Morris (1978) shows that I_s/FeO determined reworking depths vary as a power function of time. The first model has the increase in I_s/FeO (ΔI_s/FeO) occur as a linearly increasing function of depth (Fig. 4b). In the second model, I_s/FeO increases rapidly at first, and then the reworking process causes ΔI_s/FeO to become almost constant (Fig. 4c). With the exception of lunar core 74002, all lunar core data (three cores) agree with the model with $\Delta I_s/\text{FeO} = \sim 45 \pm 10$ even though core ages are different up to an order of magnitude (Morris, 1978). For 60002, $\Delta I_s/\text{FeO} = 45 \pm 10$ for the unit

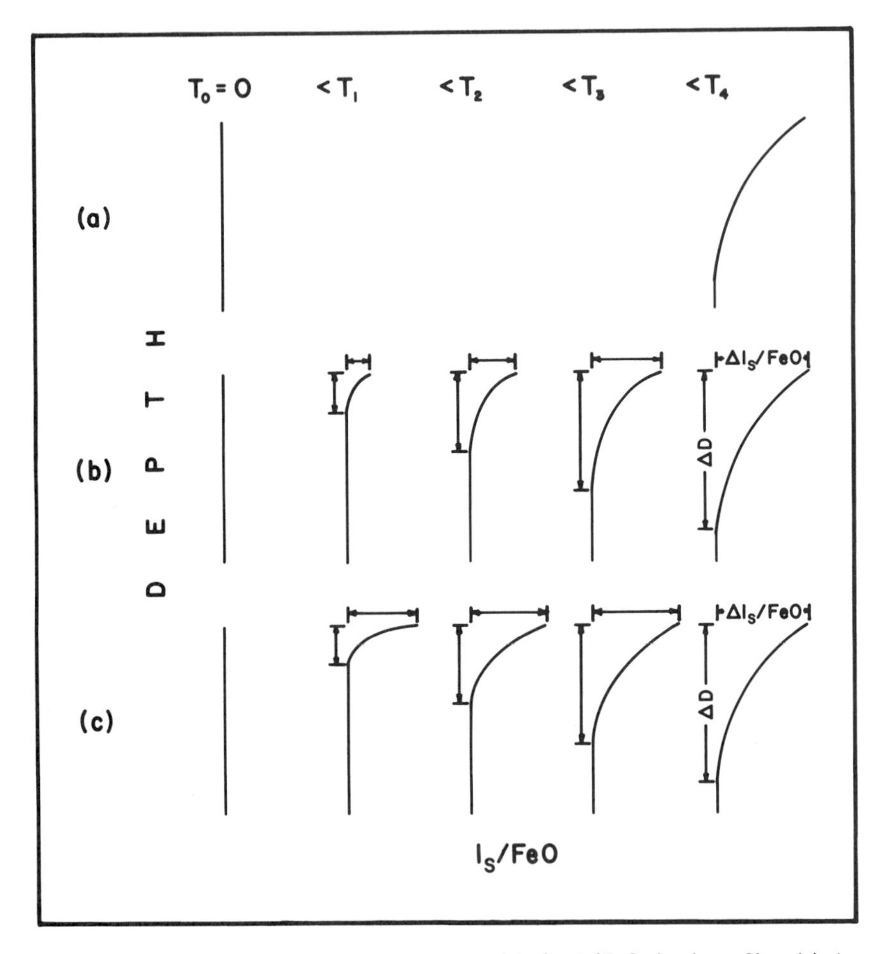

Fig. 4. Schematic diagrams of evolutionary models for I_s/FeO depth profiles. (a) A diagram given by Gose and Morris (1977). (b) A model in which $\Delta I_s/FeO$ increases linearly with depth. (c) A model in which $\Delta I_s/FeO$ increases rapidly at first and then increases very slowly or almost remains constant.

12–13 boundary (Fig. 3c) (Gose and Morris, 1977). We are not claiming that the second model approximates the true evolutionary development of an I_s/FeO depth profile. Because the depth profile for 74002 does not fit the second model and because its age is similar to the track model age for the unit 12–13 boundary, such a simple model is not valid. Nevertheless, it is our opinion that Gose and Morris (1977) do not have sufficiently strong experimental evidence to state that the sharpness of the unit 12–13 boundary requires that it be a depositional boundary and not the surface of an ancient regolith.

Solar wind rare gas abundances (Bogard and Hirsch, 1975; Jordan and Heymann, 1976) (Fig. 3a,b), chemistry (Nava *et al.*, 1976) and modal petrography (Vaniman *et al.*, 1976) all clearly indicate that the unit 12–13 boundary is a distinct stratigraphic boundary. Neutron fluence data is consistent with a stratigraphic boundary between units 12 and 13 whether is is time related (a three slab model) or not (a two slab model) (Russ, 1973). Solar wind rare gas abundances, chemistry and modal petrography also indicate that the material below this boundary is unique Apollo 16 material which has been sampled in only one surface soil, 61220 (Jordan and Heymann, 1976). In other words, many techniques point to the uniqueness of the material below the unit 12–13 boundary, and therefore, it can be either a depositional contact or an ancient, short-lived lunar surface.

We would conclude that the unit 12–13 boundary probably is not an ancient regolith surface. We base this reversal of opinion (Blanford and Morrison, 1976; Blanford and Wood, 1978a) on the fact that track density distributions are of path II rather than path I character.

Summary

Our experimental results and many others indicate that the main feature of the upper part of section 60002 in the Apollo 16 deep drill core is a single stratigraphic boundary between units 12 and 13. The experimental evidence is not conclusive whether this contact is an ancient regolith surface or is simply a depositional boundary. If it is an ancient surface, it has a model exposure age of $3 - 7 \times 10^6$ y and a reworking depth ~0.5 cm. Our current assessment of the data leads us to believe that it is a depositional boundary. There also may exist another small depositional boundary about 2 cm thick between units 8 and 9.

Acknowledgments—We appreciate a constructive review and useful discussions with R. V. Morris. We wish to acknowledge and thank J. Blanford, P. Grover, G. Hart and J. Ragland for their help in experimental work and data reduction. We thank T. Fernandez, L. Prichard and J. Ragland for their help in manuscript preparation. This research was sponsored in part by NASA grant NSG 9043.

References

Arnold J. R. (1975) Monte Carlo simulation of turnover processes in the lunar regolith. *Proc. Lunar Sci. Conf. 6th*, p. 2375–2395.

Bibring J. P., Borg J., Burlingame A. L., Langevin Y., Maurette M. and Vassent B. (1975) Solar-wind and solar-flare maturation of the lunar regolith. *Proc. Lunar Sci. Conf. 6th*, p. 3471–3493.

Blanford G. E., Fruland R. M. and Morrison D. A. (1975) Long-term differential energy spectrum for solar-flare iron-group particles. *Proc. Lunar Sci. Conf. 6th*, p. 3557–3576.

Blanford G. E., McKay D. S. and Wood G. C. (1977) Particle track densities in double drive tube 60009/10. *Proc. Lunar Sci. Conf. 8th*, p. 3017–3025.

Blanford G. E. and Morrison D. A. (1976) Stratigraphy in Apollo 16 drill section 60002. *Proc. Lunar Sci. Conf. 7th*, p. 141–154.

Blanford G. E. and Wood G. C. (1978a) Irradiation stratigraphy in Apollo 16 cores (abstract). In *Lunar and Planetary Science IX*, p. 106–108. Lunar and Planetary Institute, Houston.

Blanford G. E. and Wood G. C. (1978b) Particle track densities in the Luna 24 core. In *Mare Crisium: The View from Luna 24* (R. B. Merrill and J. J. Papike, eds.). p. 157-163. Pergamon, N.Y.

Bogard D. D. and Hirsch W. C. (1975) Noble gas studies on grain size separates of Apollo 15 and 16 drill cores. *Proc. Lunar Sci. Conf. 6th*, p. 2057–2083.

Crozaz G., Haack U., Hair M., Maurette M., Walker R. M. and Woolum D. (1970) Nuclear track studies of ancient solar radiations and dynamic lunar surface processes. *Proc. Apollo 11 Lunar Sci. Conf.*, p. 2051–2080.

Duke M. B. and Nagle J. S. (1976) Lunar Core Catalog and Supplements. NASA Publication JSC 09252.

Duraud J. P., Langevin Y., Maurette M., Comstock A. and Burlingame A. L. (1975) The simulated depth history of dust grains in the lunar regolith. *Proc. Lunar Sci. Conf. 6th*, p. 2397–2415.

Etique Ph., Funk H., Poupeau G., Romary Ph., Signer P. and Wieler R. (1978) Solar-wind gas and solar flare track correlations in lunar soils; revisited (abstract). In *Lunar and Planetary Science IX*, p. 300–302. Lunar and Planetary Institute, Houston.

Fruchter J. S., Rancitelli L. A., Laul J. C. and Perkins R. W. (1977) Lunar regolith dynamics based on an analysis of the cosmogenic radionuclides ^{22}Na, ^{26}Al, and ^{53}Mn. *Proc. Lunar Sci. Conf. 8th*, p. 3595–3605.

Gose W. A. and Morris R. V. (1977) Depositional history of the Apollo 16 deep drill core. *Proc. Lunar Sci. Conf. 8th*, p. 2909–2928.

Hutcheon I. D., Macdougall D. and Price P. B. (1974) Improved determination of the long term average Fe spectrum from 1 to 460 Mev/a.m.u. *Proc. Lunar Sci. Conf. 5th*, p. 2561–2576.

Jordan J. L. and Heymann D. (1976) Inert gas stratigraphy of sections 60002 and 60004 of the Apollo 16 deep drill core (abstract). In *Lunar Science VII*, p. 434–436. The Lunar Science Institute, Houston.

McKay D. S., Fruland R. M. and Heiken G. H. (1974) Grain size and the evolution of lunar soils. *Proc. Lunar Sci. Conf. 5th*, p. 887–906.

Morris R. V. (1978) *In situ* reworking (gardening) of the lunar surface: Evidence from the Apollo cores. Proc. Lunar Planet. Sci. Conf. This volume.

Nagle J. S. and Duke M. B. (1974) Stabilization of lunar core samples. *Proc. Lunar Sci. Conf. 5th*, p. 967–973.

Nava D. F., Lindstrom M. M., Schuhmann P. J., Lindstrom D. J. and Philpotts J. A. (1976) The remarkable chemical uniformity of Apollo 16 layered deep drill core section 60002. *Proc. Lunar Sci. Conf. 7th*, p. 133–139.

Russ G. P. III (1973) Apollo 16 neutron stratigraphy. *Earth Planet. Sci. Lett.* **17**, 275–289.

Vaniman D. T., Lellis S. F., Papike J. J. and Cameron K. L. (1976) The Apollo 16 drill core: Modal petrology and characterization of the mineral and lithic component. *Proc. Lunar Sci. Conf. 7th*, p. 199–239.

Walker R. and Yuhas D. (1973) Cosmic ray production rates in lunar materials. *Proc. Lunar Sci. Conf. 4th*, p. 2379–2389.

Proc. Lunar Planet. Sci. Conf. 9th (1978), p. 1885–1912.
Printed in the United States of America

Inert gas measurements in the Apollo-16 drill core and an evaluation of the stratigraphy and depositional history of this core

D. HEYMANN, J. L. JORDAN, A. WALKER, M. DZICZKANIEC,
J. RAY and R. PALMA

Department of Geology; Department of Space Physics and Astronomy
Rice University, Houston, Texas 77001

Abstract—Inert gas measurements on a large number of <1 mm fines from barrels 60003 and 60002 of the Apollo 16 drill core are discussed in the context of a survey of the entire core. We show or confirm that the mixing of three major soil components, called α, β, and γ, is a major Leitmotiv of the core, and that these soil components have close relatives among the soils now at the surface of the regolith at Apollo 16 and in the double drive tube 60009/60010. Most soils in the core are therefore locally (within 5–10 km) derived. We discuss three categories of boundaries between dissection units (DU's): *1* major, well-established stratigraphic breaks which are buried surfaces (DU 12/DU 13 and DU 43/DU 44); *2* stratigraphic breaks which may also be time-stratigraphic boundaries (DU 9/DU 10; DU 5/DU 6; DU 27/DU 28); and *3* stratigraphic breaks for which supporting evidence is weak, and which need more study (DU 17/DU 18; DU 24/DU 25; DU 13/DU 14; DU 31/DU 32; DU 45/DU 46; DU 34/DU 35; and DU 3/DU 4). We suggest the following depositional history: *1* MPU-A (Modal Petrologic Unit) represents deposition, probably inside a secondary crater, in a boulder-strewn environment (the boulders were predominantly light matrix breccia); *2* when the boulders were essentially comminuted, a sub-unit called MPU-BO3 was deposited by further filling in of the crater from its own walls; *3* the sub-unit called MPU-BO4 may then have been emplaced as a single soil slab ranging from 13 to more than 45 cm thickness; *4* MPU-C represents again deposition in a boulder-rich environment (the boulders were plagioclase-rich): *5* MPU-D represents principally β-rich soils "pushed" onto the core site by the North Ray Crater event or by the South Ray Crater event, or by both (MPU-D may therefore represent two discrete events; the break seems to occur at DU 45/DU 46). Erosion may have occurred but cannot be firmly documented.

INTRODUCTION

The Apollo 16 drill core was collected at station 10 approximately 100 meters southwest of the Lunar Module, and about 40 meters southwest of the double drive tube 60009/60010. Details are given in report NASA SP-315, "Apollo 16 Preliminary Science Report." Upon arrival at JSC, the six core barrels and the drill bit were disconnected, X-radiographs were taken, and samples were dug out from the drill bit and from the tops or bases of barrels 60006, 60004, 60003, and 60002 for early allocations. Details are given in the Lunar Core Catalog and its supplements edited by Duke and Nagle (1976). The X-radiographs revealed that barrel 60005, the third from the top, was only partly filled. This has resulted in considerable uncertainty about true lunar depth, discussed in detail by Duke and Nagle (1976). Figure 1 shows their conclusions for three plausible models.

The X-radiographs established 46 X-ray units along the entire core. The subsequent dissections of the individual core barrels, details of which are given in

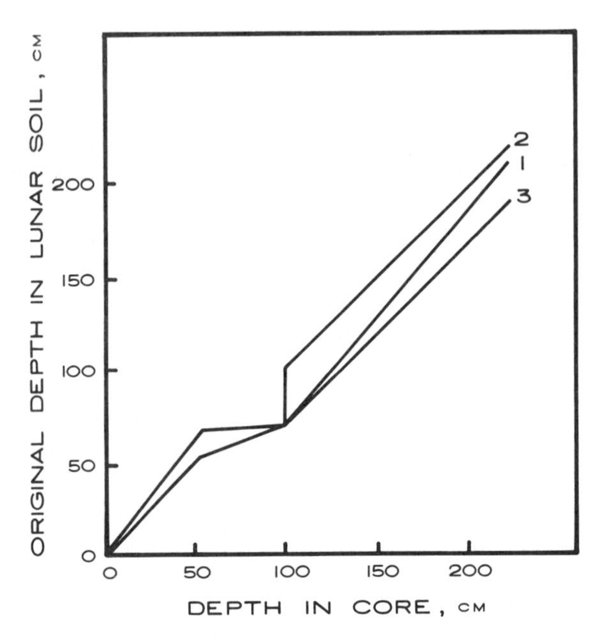

Fig. 1. Relationship between depth in drill core and deduced original lunar depth for three models: 1. original core recovery was roughly 80% with a void of 1.5 barrels in 60005–60007; subsequent migration of sample upwards into 60005–60007; 2. original core recovery was roughly 100%; upon separation of 60005 and 60004 on the moon, sample fell out of the bottom of 60005; 3. original core recovery 100%, but missing 29 cm fell out at the bottom (60001) of the drill stem (Duke and Nagle, 1976).

Duke and Nagle (1976), established 46 dissection units (abbreviated hereafter as DU) in the entire core. Initially, the DU's were given roman numerals in each individual barrel, e.g., 60002, I through 60002, XII; 60003, I through 60003, XV. When all the barrels were processed, the DU numbering system was changed to a sequential system from 1 through 46 (see Fig. 2). In the present paper, X-ray units will not be used, but the sequential numbering system of DU's will be. Vaniman *et al.* (1976) have reported four modal petrologic units (called MPU's hereafter) A through D on the basis of petrologic studies of thin sections. These will be used in the present paper.

The following types of samples were generated during core processing:

1.) unsieved bulk soils.
2.) <1 mm fines.
3.) >1 mm fines, "rocks", or other special lithologies.
4.) chemically clean samples (the soils immediately adjacent to the inner walls of the cylindrical core barrels turned out to be contaminated with terrestrial lead and other heavy elements; the clean samples were taken from near the central axis of the core).

BARRELS	DISSECTION UNITS	PETROLOGIC UNITS	DU BREAKS	LUNAR DEPTH,cm Min.	Max.
60007	42-46	D	43/44	13	13
		C		22.5	22.5
60006	39-41			55.0	55.0
60005	38	C		69.9	104.2
60004	28-37	B		108.3	142.6
60003	14-28	B		147.7	182.0
60002	2-13		12/13	155.4	189.7
		A		183.7	218.0
60001	1			189.7	224.0

Fig. 2. Relationships between core barrels (and drill bit) 60001 through 60007, dissection units (Duke and Nagle, 1976), modal petrologic units (Vaniman *et al.*, 1976), and lunar depth. We have placed the boundary MPU-C/MPU-D at 13 cm lunar depth (see text).

5.) samples collected under "red light" for thermoluminescence studies.
6.) "Peels" (see Duke and Nagle, 1976).
7.) Epoxy-impregnated soils from all barrels except 60005 (none from the drill bit).
8.) Epoxy-impregnated "scabs" and "platelets," which had fallen off from a core segment immediately after the opening of the retaining metal barrel.

These samples have been studied by a great variety of techniques and with widely different objectives. Because the six core barrels were processed over a span of several years, the resulting scientific papers are scattered over a number

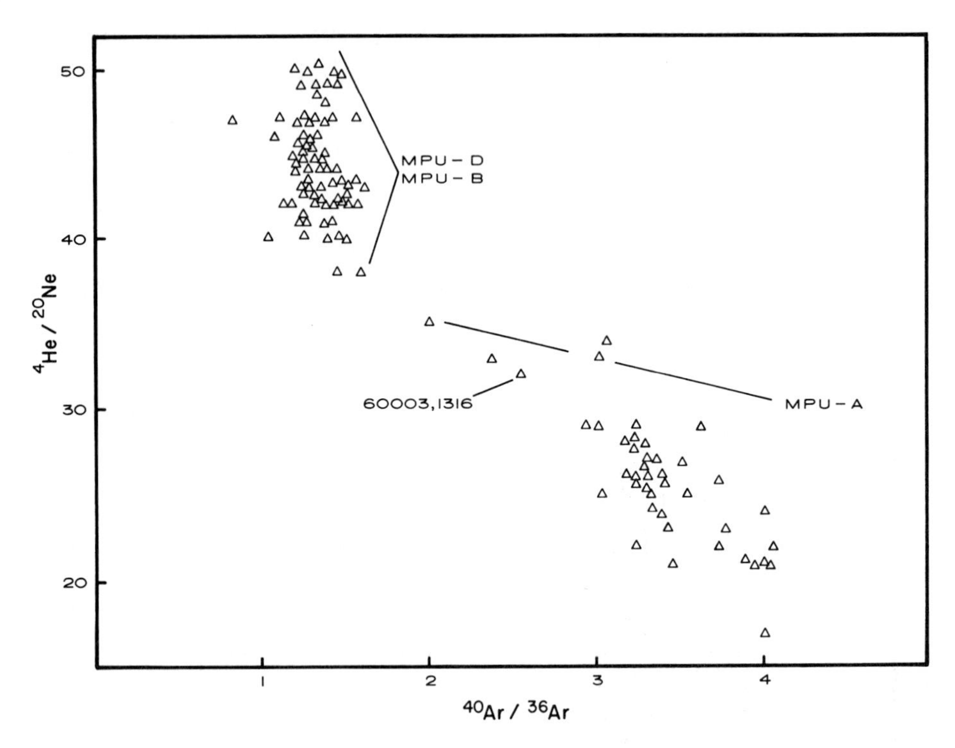

Fig. 3. $^4He/^{20}Ne$ vs. $^{40}Ar/^{36}Ar$ of the drill core soils.

of Proceedings of Lunar Science Conferences and other scientific journals. In this paper we present previously unpublished inert gas data obtained in our laboratory on samples from core barrels 60003 and 60002, but in the framework of an attempt to review the following issues:

1.) What is the nature and provenance of the soils in the drill core and what is their relationship to soils collected at the surface of the regolith at Apollo 16?
2.) Which are the most salient stratigraphic features of the drill core?
3.) What is the depositional history of this core?

Our objectives are similar to those of the review by Meyer and McCallister (1977), but the present paper intends to be a long rather than a brief review. The readers are referred to the very useful Fig. 3 of Meyer and McCallister's paper, where the locations of all allocations from this core are shown.

RELATIONSHIPS OF CORE AND SURFACE SOILS

In Table 1 we present all known results of inert gas measurement in the Apollo

16 drill core. With respect to previously unpublished results, our experimental techniques and data reduction have been described elsewhere (Heymann and Yaniv, 1970). Other inert gas data have been reported in the following papers and abstracts: Bogard and Nyquist (1973); Bogard *et al.* (1973); Bogard and Hirsch (1975); Jordan and Heymann (1976); Heymann *et al.* (1977); and Ray *et al.* (1978).

Heymann *et al.* (1975), in their discussion of light and dark surface soils at the Apollo 16 site found that the inert gas ratios $^{4}He/^{20}Ne$ and $^{40}Ar/^{36}Ar$ are useful diagnostic tools for first-order, gross classifications of soil types. Figure 3 shows a plot of $^{4}He/^{20}Ne$ vs. $^{40}Ar/^{36}Ar$ of *all* drill core samples, while Fig. 4 shows the same plot for all surface samples. $^{4}He/^{20}Ne$ are directly measured, unmassaged ratios. A number of $^{40}Ar/^{36}Ar$ ratios, specifically those from Bogard and Hirsch (1975), were obtained by the ordinate-intercept method (Yaniv and Heymann, 1970). Other ratios are corrected for estimated radiogenic ^{40}Ar. Such corrections seldom exceed 5% of the total ^{40}Ar measured in a sample.

The remarkable similarity of the two distributions reveals more eloquently than words that the drill core contains, with one major exception, the variety of soil types at the surface and in trenches at Apollo 16 today. The major exception is soils from the North Ray Crater ejecta blanket (stations 11 and 13). Judging

Table 1. He, Ne, and Ar contents; elemental and isotopic ratios; $(^{40}Ar/^{36}Ar)_{CORR.}$; and $^{21}Ne_{SP.}$ of Apollo drill core samples measured in our laboratory.

Core Segment	Sample Parent Number	Dissection Interval, cm from top	^{4}He $10^{-3}cm^{3}g^{-1}$	^{20}Ne $10^{-3}cm^{3}g^{-1}$	^{36}Ar $10^{-4}cm^{3}g^{-1}$	$^{4}He/^{3}He$	$^{20}Ne/^{22}Ne$	$^{21}Ne/^{22}Ne$	$^{40}Ar/^{36}Ar$
60004	,334	4.0–4.5	60.2	1.30	5.53	2350	13.0	0.037	1.14
	,316	8.0–8.5	67.3	1.12	4.24	2260	13.0	0.035	1.13
	,287	14.0–16.0	62.1	1.23	4.75	2610	12.9	0.038	1.26
	,260	21.0–21.6	42.6	0.99	3.26	2240	13.2	0.040	1.57
	,211	31.5–32.0	51.6	1.12	4.53	2170	12.4	0.040	1.20
	,186	38.0–38.5	50.6	1.18	4.72	2300	13.1	0.040	1.32
60003	,204	4.0–4.3	41.9	0.89	4.05	2310	13.4	0.039	1.49
"Normal	,202	4.3–4.7	47.7	0.94	4.24	2460	13.5	0.039	1.48
Dissec-	,180	4.7–4.9	48.2	0.97	4.34	2390	12.9	0.040	1.52
tion"	,176	5.1–5.6	40.7	0.90	4.06	2370	12.9	0.038	1.44
	,174	5.6–5.8	48.6	0.99	4.48	2370	12.5	0.036	1.31
	,170	6.0–6.4	38.6	0.81	3.57	2300	12.6	0.038	1.36
	,138	13.5–14.1	50.0	1.01	4.66	2320	13.0	0.035	1.32
	,134	14.3–14.5	41.9	0.86	3.52	2530	12.8	0.037	1.39
	,132	14.5–14.7	42.6	0.93	4.19	2310	12.9	0.037	1.40
	,130	14.7–15.0	40.8	0.90	4.26	2280	12.8	0.037	1.31
	,122	16.5–16.7	36.7	0.71	3.57	2230	12.4	0.035	1.33
	,120	16.7–17.0	45.9	0.91	4.24	2380	12.8	0.038	1.32
	,118	17.0–17.3	47.1	1.01	4.75	2660	12.7	0.038	1.37
	,101	19.5–20.1	30.2	0.64	2.91	2360	12.5	0.040	1.32
	,100	20.1–20.8	43.2	0.94	4.21	2350	12.8	0.039	1.36
	, 37	34.0–34.3	46.1	1.07	5.46	2400	12.8	0.037	1.46
	, 35	34.3–34.5	51.2	1.05	4.24	2460	12.5	0.038	1.49
	, 34	34.5–34.7	48.8	1.02	3.86	2350	12.9	0.039	1.50

Table 1. *(cont'd.)* He, Ne, and Ar contents; elemental and isotopic ratios; $(^{40}Ar/^{36}Ar)_{CORR.}$; and $^{21}Ne_{SP.}$ of Apollo Drill Core Samples measured in our laboratory.

Core Segment	Sample Parent Number	DU	$(^{40}Ar/^{36}Ar)_{CORR.}$	$^4He/^{20}Ne$	$^{21}Ne_{SP.}$ $10^{-8}cm^3g^{-1}$
60004	,334	37	1.10	46	57
	,316	36	1.08	60	28
	,287	34	1.22	51	65
	,260	33	1.52	43	65
	,211	29	1.16	46	78
	,186	28	1.28	43	75
				Avge:	61
60003 "Normal Dissection"	,204	27	1.44	47	47
	,202	27	1.44	51	50
	,180	27	1.48	49	64
	,176	27	1.39	45	45
	,174	27	1.26	49	34
	,170	27	1.31	47	40
	,138	23	1.27	47	28
	,134	23	1.33	49	38
	,132	23	1.35	46	39
	,130	23	1.27	45	41
	,122	23	1.28	52	19
	,120	23	1.27	50	47
	,118	23	1.33	47	50
	,101	21	1.25	47	45
	,100	21	1.32	46	54
	, 37	15	1.42	43	47
	, 35	15	1.44	47	52
	, 34	15	1.45	48	60
				Avge:	44

Core Segment	Sample Parent Number	Dissection Interval, cm from top	^4He $10^{-3}cm^3g^{-1}$	^{20}Ne $10^{-3}cm^3g^{-1}$	^{36}Ar $10^{-4}cm^3g^{-1}$	$^4He/^3He$	$^{20}Ne/^{22}Ne$	$^{21}Ne/^{22}Ne$	$^{40}Ar/^{36}Ar$
60003 "Special Dissection"	.1300	2.2	42.4	0.993	3.77	2393	12.6	0.0370	1.58
	.1304	2.7	39.5	0.947	3.44	2410	12.6	0.0376	1.55
	.1308	3.2	45.1	1.08	4.04	2357	12.4	0.0388	1.50
	.1310	3.7	38.5	0.914	3.43	2351	11.9	0.0384	1.59
	.1313	4.2	33.2	0.788	2.90	2378	12.5	0.0391	1.61
	.1315	4.6	42.5	1.06	4.50	2253	12.2	0.0364	1.50
	.1316	4.5	48.3	1.50	5.17	2370	12.7	0.0406	2.59
	.1317	4.5	47.2	1.11	4.58	2379	12.4	0.0353	1.43
	.1318	4.7	48.4	1.13	4.53	2313	12.8	0.0380	1.55
	.1319	5.0	47.1	1.13	4.48	2308	13.3	0.0389	1.62
	.1320	5.0	46.0	1.15	4.71	2269	13.3	0.0394	1.52
	.1321	5.0	63.1	1.51	6.18	2295	13.1	0.0387	1.43
	.1322	5.7	56.0	1.29	5.51	2351	13.2	0.0384	1.48
	.1323	5.3	29.9	0.779	3.10	2152	12.8	0.0398	1.48
	.1324	5.8	63.1	1.58	6.24	2263	12.8	0.0368	1.46
	.1325	5.7	28.5	0.705	2.92	2221	12.8	0.0372	1.36
	.1326	6.0	56.8	1.37	5.65	2176	12.9	0.0375	1.24
	.1327	5.9	21.5	0.533	2.27	2282	12.7	0.0419	1.59
	.1328	6.1	81.2	1.96	8.00	2334	12.8	0.0383	1.49
	.1329	6.5	50.1	1.18	4.92	2316	12.8	0.0373	1.20
	.1330	6.4	35.2	0.838	3.54	2312	12.8	0.0394	1.53
	.1331	6.8	45.9	1.12	4.54	2257	12.9	0.0396	1.34
	.1332	6.7	49.3	1.20	5.03	2300	12.8	0.0388	1.31

Core Segment	Sample Parent Number	Dissection Interval, cm from top	^4He $10^{-3}cm^3g^{-1}$	^{20}Ne $10^{-3}cm^3g^{-1}$	^{36}Ar $10^{-4}cm^3g^{-1}$	$^4He/^3He$	$^{20}Ne/^{22}Ne$	$^{21}Ne/^{22}Ne$	$^{40}Ar/^{36}Ar$
60003	.1333	7.0	43.4	1.04	4.38	2266	12.8	0.0387	1.37
"Special	.1334	7.5	38.2	0.910	3.77	2217	12.9	0.0442	1.44
Dissec-	.1335	7.5	57.9	1.36	5.42	2318	12.8	0.0375	1.30
tion"	.1338	14.1	35.6	0.791	3.10	2245	12.5	0.0386	1.43
	.1339	13.9	28.6	0.668	2.90	2261	12.6	0.0412	1.32
	.1340	14.2	34.9	0.819	3.40	2356	12.4	0.0387	1.71
	.1341	14.3	42.7	0.951	3.87	2333	12.3	0.0353	1.38
	.1342	14.5	41.3	0.914	3.83	2264	12.5	0.0364	1.30
	.1343	14.7	49.9	1.12	4.48	2318	12.2	0.0365	1.29
	.1344	14.8	50.9	1.15	4.69	2319	12.3	0.0368	1.32
	.1345	14.8	24.9	0.650	3.26	2303	12.4	0.0388	1.68
	.1346	15.0	29.5	0.658	2.66	2294	12.5	0.0367	1.36
	.1347	16.0	29.6	0.671	2.87	2339	12.5	0.0383	1.29
	.1348	16.1	39.6	0.894	3.85	2267	12.4	0.0367	1.44
	.1349	16.5	62.0	1.38	5.76	2327	12.3	0.0362	1.38
	.1350	16.7	42.5	0.981	4.05	2368	12.4	0.0367	1.43
	.1351	16.8	47.8	1.10	4.61	2340	12.4	0.0361	1.34
	.1352	17.0	41.1	0.974	4.01	2315	12.5	0.0355	1.38
	.1353	17.2	41.2	0.953	4.04	2319	12.5	0.0373	1.32
	.1354	17.2	45.4	1.01	4.05	2354	12.4	0.0369	1.37
	.1355	17.1	42.1	1.02	4.07	2283	12.6	0.0371	1.32
	.1356	17.4	49.1	1.20	4.95	2300	12.7	0.0380	1.31
	.1357	17.5	47.5	1.04	4.20	2336	12.3	0.0369	1.32
	.1359	17.7	27.4	0.615	2.40	2282	11.9	0.0379	1.35
	.1363	18.3	42.3	0.958	4.01	2428	12.5	0.0371	1.29
	.1367	19.1	50.5	1.15	4.21	2368	12.4	0.0379	1.34
	.1387	36.5	44.1	1.07	3.77	2410	12.5	0.0367	1.49
	.1391	37.5	49.1	1.14	4.04	2395	12.2	0.0354	1.42
	.1395	38.5	50.6	1.16	4.10	2383	12.3	0.0358	1.43

Core Segment	Sample Parent Number	DU	$(^{40}Ar/^{36}Ar)_{CORR.}$	$^4He/^{20}Ne$	$^{21}Ne_{SP}$
60003	.1300	28	1.52	42.7	39.7
"Special	.1304	28	1.49	41.8	41.7
Dissection"	.1308	28	1.45	41.6	59.0
	.1310	28	1.53	42.1	48.2
	.1313	28	1.54	42.1	44.4
	.1315	27	1.46	40.1	37.1
	.1316	27	2.55	32.3	102.0
	.1317	27	1.38	42.5	28.1
	.1318	27	1.50	42.9	52.6
	.1319	27	1.58	41.7	58.1
	.1320	27	1.48	40.0	64.2
	.1321	27	1.40	41.7	77.1
	.1322	27	1.44	43.3	62.7
	.1323	27	1.42	38.4	47.3
	.1324	27	1.43	39.8	58.6
	.1325	27	1.29	40.4	28.1
	.1326	27	1.21	41.5	57.0
	.1327	27	1.50	40.3	41.8
	.1328	27	1.46	41.5	95.2
	.1329	27	1.16	42.4	47.7
	.1330	27	1.47	42.0	48.0
	.1331	27	1.29	41.1	65.8
	.1332	27	1.27	41.2	63.1

Table 1. *(cont'd.)* He, Ne, and Ar contents; elemental and isotopic ratios; $(^{40}Ar/^{36}Ar)_{CORR.}$; and $^{21}Ne_{SP.}$ of Apollo Drill Core Samples measured in our laboratory.

Core Segment	Sample Parent Number	DU	$(^{40}Ar/^{36}Ar)_{CORR.}$	$^{4}He/^{20}Ne$	$^{21}Ne_{SP}$
60003	.1333	27	1.32	41.6	54.2
"Special	.1334	26	1.38	41.9	87.0
Dissection"	.1335	26	1.26	42.4	57.9
	.1338	23	1.37	45.0	41.5
	.1339	23	1.26	42.9	49.3
	.1340	23	1.65	42.6	43.9
	.1341	23	1.33	44.9	24.6
	.1342	23	1.25	45.2	31.3
	.1343	23	1.24	44.6	39.8
	.1344	23	1.28	44.5	43.1
	.1345	23	1.62	38.4	35.7
	.1346	23	1.28	44.9	23.9
	.1347	23	1.22	44.2	33.5
	.1348	23	1.39	44.3	32.9
	.1349	23	1.35	44.9	45.4
	.1350	23	1.38	43.3	36.5
	.1351	23	1.30	43.3	35.0
	.1352	23	1.33	42.2	26.3
	.1353	23	1.27	43.2	40.0
	.1354	23	1.32	45.1	38.8
	.1355	23	1.27	41.3	40.1
	.1356		1.27	41.0	55.6
	.1357		1.27	45.8	40.5
	.1359	22	1.27	44.5	29.9
	.1363	22	1.23	44.1	38.3
	.1367	22	1.30	44.1	53.6
	.1387	14	1.44	41.2	39.3
	.1391	14	1.37	43.2	30.0
	.1395	14	1.38	43.6	34.1
				Avge:	72

Core Segment	Sample Parent Number	Dissection Interval, cm from top	^{4}He $10^{-3}cm^{3}g^{-1}$	^{20}Ne $10^{-3}cm^{3}g^{-1}$	^{36}Ar $10^{-4}cm^{3}g^{-1}$	$^{4}He/^{3}He$	$^{20}Ne/^{22}Ne$	$^{21}Ne/^{22}Ne$	$^{40}Ar/^{36}Ar$
60002	, 45	6.0–6.4	49.3	1.21	4.55	2400	12.7	0.037	1.42
	, 51	6.4–6.7	36.2	0.78	3.56	2330	12.8	0.036	1.64
	, 51	6.4–6.7	39.2	0.82	3.60	2360	13.0	0.038	1.40
	, 53	6.7–7.2	33.3	0.75	3.25	2250	12.6	0.037	1.63
	, 53	6.7–7.2	33.4	0.60	2.71	2710	12.8	0.043	1.56
	, 56	7.2–7.4	44.2	1.25	4.26	2380	12.7	0.039	2.08
	, 58	7.4–7.7	48.6	1.46	4.69	2450	12.8	0.040	2.46
	, 61	7.7–8.0	31.7	1.28	3.51	1960	12.8	0.041	3.10
	, 69	8.5–9.1	22.6	1.01	3.08	2130	11.9	0.038	3.83
	, 73	9.1–9.4	22.2	1.05	2.96	2120	11.9	0.039	3.97
	, 75	9.4–9.7	6.8	0.29	1.05	1670	12.3	0.054	4.20
	, 75	9.4–9.7	11.5	0.50	1.75	2010	13.1	0.050	3.90
	, 78	9.7–10.4	23.9	1.09	3.54	2480	12.3	0.040	4.14
	, 80	10.4–10.9	12.3	0.71	2.68	2410	11.9	0.041	4.08
	, 82	10.9–11.4	19.2	0.92	2.74	2080	11.9	0.041	4.08
	, 84	11.4–12.0	18.0	0.88	2.82	2040	12.2	0.040	3.54
	, 87	12.0–12.5	37.8	1.83	4.76	2260	12.5	0.038	4.01
	, 89	12.5–13.2	27.6	1.33	3.51	2110	12.2	0.039	4.10
	, 92	13.2–13.8	30.5	1.17	3.65	2060	12.2	0.040	3.37
	, 94	13.8–14.4	23.0	1.06	3.64	2120	12.6	0.039	3.31

Core Segment	Sample Parent Number	Dissection Interval, cm from top	^{4}He $10^{-3}cm^{3}g^{-1}$	^{20}Ne $10^{-3}cm^{3}g^{-1}$	^{36}Ar $10^{-4}cm^{3}g^{-1}$	$^{4}He/^{3}He$	$^{20}Ne/^{22}Ne$	$^{21}Ne/^{22}Ne$	$^{40}Ar/^{36}Ar$
60002	,106	17.2–17.4	27.0	0.95	2.93	2300	12.4	0.040	3.32
	,109	17.4–17.6	31.9	1.12	3.36	2210	12.0	0.044	3.32
	,110	17.6–18.0	19.6	0.57	2.25	2260	12.4	0.042	3.18
	,110	17.6–18.0	18.8	0.66	2.34	2090	12.7	0.043	3.12
	,115	18.5–19.0	21.3	0.75	2.35	2060	12.0	0.039	3.39
	,117	19.0–19.5	26.0	0.97	3.11	2140	12.1	0.040	3.45
	,119	19.5–20.0	20.5	0.78	2.41	2100	12.0	0.038	3.49
	,121	20.0–20.5	22.4	0.84	2.77	2120	12.1	0.040	3.60
	,140	25.0–25.2	21.7	0.75	2.56	2050	12.1	0.038	3.03
	,142	25.2–25.6	26.8	1.02	3.26	2110	12.1	0.038	3.31
	,144	25.6–26.0	22.9	0.96	2.98	2140	12.0	0.039	3.43
	,149	26.5–27.0	15.8	0.57	1.89	2060	12.0	0.041	3.31
	,151	27.0–27.6	23.6	0.81	2.70	2110	12.1	0.040	3.09
	,155	28.1–28.5	24.9	0.91	2.92	2060	12.1	0.040	3.37
	,157	28.5–29.0	24.8	1.08	3.37	2130	12.0	0.038	3.48
	,159	29.0–29.5	26.4	0.99	3.13	2120	12.1	0.039	3.38
	,161	29.5–30.0	21.3	0.81	2.62	2090	12.2	0.040	3.28
	,165	30.5–31.1	22.1	0.93	3.07	2130	11.8	0.038	3.48
	,167	31.1–31.8	24.0	0.95	3.19	2140	12.0	0.038	3.61
	,169	31.8–32.2	21.3	0.86	2.88	2090	12.0	0.040	3.42
	,174	32.9–33.3	25.5	1.02	3.38	2090	12.0	0.038	3.41
	,176	33.3–33.9	27.0	1.04	3.46	2090	12.1	0.038	3.30

Core Segment	Sample Parent Number	DU	$(^{40}Ar/^{36}Ar)_{CORR.}$	$^{4}He/^{20}Ne$	$^{21}Ne_{SP.}$ $10^{-8}cm^{3}g^{-1}$
60002	, 45	13	1.38	41	55
	, 51	13	1.59	46	27
	, 51	13	1.35	48	38
	, 53	13	1.57	44	30
	, 53	13	1.49	56	56
	, 56	12	2.03	35	73
	, 58	12	2.39	33	94
	, 61	12	3.05	25	92
	, 69	12	3.77	22	51
	, 73	12	3.90	21	58
	, 75	12	4.03	24	55
	, 75	12	3.79	23	72
	, 78	12	4.08	22	76
	, 80	12	4.01	17	54
	, 82	12	4.01	21	74
	, 84	11	3.47	21	60
	, 87	11	3.97	21	101
	, 89	11	4.04	21	72
	, 92	11	3.32	26	82
	, 94	11	3.26	22	61
	,106	10	3.25	28	61
	,109	9	3.26	29	122
	,110	9	3.09	34	48
	,110	9	3.04	33	52
	,115	9	3.30	28	44
	,117	9	3.38	27	62
	,119	9	3.40	26	39
	,121	8	3.53	27	54
	,140	5	2.95	29	36
	,142	5	3.25	26	50

Table 1. *(cont'd.)* He, Ne and Ar contents; elemental and isotopic ratios; $(^{40}Ar/^{36}Ar)_{CORR.}$; and $^{21}Ne_{SP.}$ of Apollo Drill Core Samples measured in our laboratory.

Core Segment	Sample Parent Number	DU	$(^{40}Ar/^{36}Ar)_{CORR.}$	$^{4}He/^{20}Ne$	$^{21}Ne_{SP.}$ $10^{-8}cm^{3}g^{-1}$
60002	,144	5	3.36	24	53
	,149	5	3.20	28	43
	,151	5	3.02	29	51
	,155	4	3.31	27	62
	,157	3	3.42	23	58
	,159	3	3.31	27	60
	,161	3	3.21	26	53
	,165	3	3.41	24	53
	,167	3	3.55	25	50
	,169	3	3.36	25	57
	,174	2	3.35	25	49
	,176	2	3.24	26	47
				Avge:	59

Core Segment	Sample Parent Number	Dissection Interval, cm from top.	DU	^{4}He	$^{4}He/^{20}Ne$	$^{40}Ar/^{36}Ar$	Reference
60007	, 83	1.5–2.0	46	38.2	47	0.84	1,2
	, 6	21.0–21.5	45	24.0	47	1.13	1,2
60006	, 3	7.0–7.5*	39	36.3	50	1.25	1,2
	, 3	7.0–7.5*	39	25.3	43	1.3	1,2
	, 63	18.5–19.0	40	–	58**	1.06	3
60004	, 7	0.5–1.0	37	43.4	44	1.42	1,2
	, 3	0.0–0.7*	28	51.5	49	1.42	1,2
60003	, 3	0.0–0.5*	14	45.8	50	1.36	1,2
	, 3	0.0–0.5*	14	32.5	43	1.50	1,2
60002	, 31	3.0–3.5	13	–	40**	1.05	3
	, 65	8.3–8.5	12	–	26**	3.74	3
	,112	18.0–18.5	9	–	27**	3.26	3
	,146	26.0–26.5	5	–	29**	3.64	3
	,171	32.2–32.9	3	–	26**	3.42	3
60001	, 3	5.0–5.5		23.3	31	2.94	1,2

Notes: *From base of segment
**Estimated from <20 μm fraction.

References: 1) Bogard and Nyquist (1973)
2) Bogard *et al.* (1973)
3) Bogard and Hirsch (1975). These authors have reported data on size fractions.

from the geologic description of the landing site (Muehlberger *et al.*, 1972), one might expect ejecta from South Ray Crater at the core site, possibly overlying earlier ejecta from the older North Ray Crater. Although tangible rocks ejected by the two cratering events have been firmly identified, no firm identification of <1 mm fines from either ejecta blanket in the drill core has come to the fore. Nishiizumi *et al.* (1976) have suggested that the top 6 cm of the core either

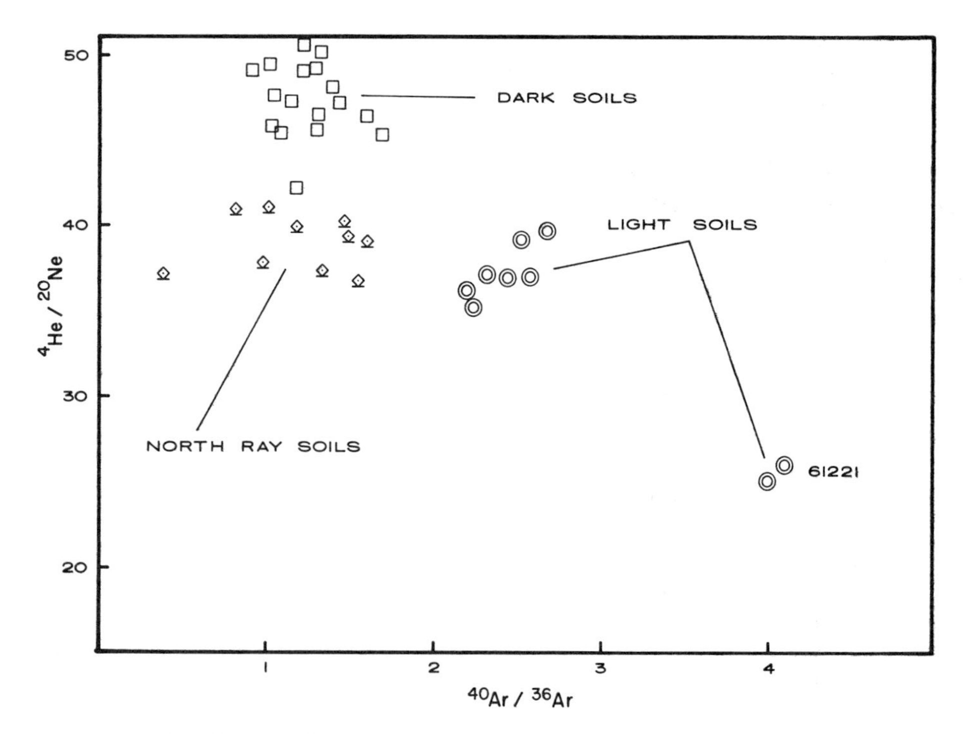

Fig. 4. $^4He/^{20}Ne$ vs. $^{40}Ar/^{36}Ar$ of Apollo 16 surface soils. Note the position of the trench soil 61221.

represents South Ray Crater material or is a small infilled crater from local soils.

The distribution of core samples in Fig. 3 is not random. *All* but one sample with $^{40}Ar/^{36}Ar > 2.0$ in the southeastern portion of the trend come from MPU-A in barrel 60002. The exception is 60003,1316.* The range $1.5 < {}^{40}Ar/^{36}Ar < 2.0$ is populated with a number of samples from MPU-A (60002) and MPU-B (60003, 60004), but because a much greater number of samples from 60003 has

*Special dissection samples were obtained from ninety-two distinct spots of barrel 60003. Each of the 92 parent samples was immediately split into four daughter samples. Thus, four sets of daughter samples were generated. The first set was numbered 60003,1200 through 60003,1292; the second set 60003,1300 through 60003, 1392, etc. Therefore, samples 60003,1200; 60003,1300; 60003,1400; and 60003,1500 are, in principle, identical in their lithologies and chemical composition, etc. However, because the weights of the daughter samples are small, i.e., about 10 milligram or less, and because it is difficult to homogenize soil samples without causing significant disturbance of agglutinate particles, one may expect differences between the four daughter samples generated from a single parent to occur occasionally.

been analyzed than from 60002, or 60004, the relative proportions in this interval are not meaningful. All of the remaining core samples, *but none from MPU-A*, fall in the northwestern portion of the trend, i.e., at $^{40}Ar/^{36}Ar < 1.5$.

It was on the basis of such observations that Bogard and Hirsch (1975) and Jordan and Heymann (1976) concluded that the portion of the core now known as MPU-A is, indeed, distinctly different from the remainder of the core.
This conclusion is now amply supported by the following observations:

1.) MPU-A is one of the two "light" (i.e., more highly reflecting) portions of the core (Duke and Nagle, 1976), the other being MPU-C (we shall see, however, that there are fundamental petrologic differences between the two "light" MPU's).
2.) MPU-A is spectacularly laminated and marbled (Duke and Nagle, 1975, 1976). Only certain portions of MPU-B in barrel 60003 are similar to MPU-A in this respect.
3.) The <250 μm size-fractions from soils in MPU-A show much lower I_s/Fe values (<35) than those from soils in MPU's B, C, and D (Gose and Morris, 1977).
4.) Soils in MPU-A show the largest neutron dose ($\psi = 10^{17}$ n/cm^2; alas only one measurement in MPU-A is available) of all soils in the core (Russ, 1973).
5.) Track measurements (Behrmann *et al.*, 1973; Price *et al.*, 1973) and measurements of cosmogenic gases (Bogard *et al.*, 1973; this work) confirm the heavy exposure of some soils in MPU-A to cosmic rays and their secondaries ($^{21}Ne_c$ in Table 1).
6.) Soils in MPU-A are relatively coarse-grained (Meyer and McCallister, 1974; Meyer *et al.*, 1975; Meyer and McCallister, 1976; Heiken *et al.*, 1973).
7.) Some soils in MPU-A are relatively poor in trapped solar wind gases (Bogard and Nyquist, 1973; Bogard *et al.*, 1973; Bogard and Hirsch, 1975; Jordan and Heymann, 1976; Heymann *et al.*, 1977; this work).

Returning to the nature and provenance of the soils in the drill core, there is ample evidence (see next sections) that the soils at many positions in the core have been pre-irradiated and pre-matured to different degrees elsewhere, *prior* to their deposition in the regolith section sampled by the core. Figure 5a–f illustrate another Leitmotiv, namely that of mixing of relatively few, distinct soil components. Whether the mixing occurred before or during deposition cannot be concluded from these figures. Heymann *et al.* (1975); Bogard and Hirsch (1975); Blanchard *et al.* (1976); and McKay *et al.* (1974) have used the terms "soil component" or "soil group" in an operational sense. A "component" or "group" may itself be a mixture from different sources, but each of these mixtures is present so abundantly or ubiquitously in the landing site that it is useful as an "end-member" for mixing models.

The variables of Fig. 5a–f are large clast (>200 μm) content and monomineralic plagioclase content of large clasts taken from Vaniman *et al.* (1976). Our analysis has shown that these variables bring out the nature and mixing of three major soil components in the core much better than any other set of variables from the petrologic (thin section) study of Vaniman *et al.* Although standard deviations are not given by the authors, one can deduce from the number of points they have analyzed that standard deviations in Figs. 5 a-f are seldom greater than 10% of the stated values, and are more typically less than 5% of the stated values. One assumes that the standard deviation is equal to the square root of the number of points measured.

Fig. 5f shows that all data points (none are available for barrel 60005) fall roughly in a triangle whose corners are indicated by α, β, and γ. Soils near the α corner come from MPU-A only; these are "light", immature soils, with I_s/FeO 30 or less; FeO ~ 5%, and $^{40}Ar/^{36}Ar$ 2 or greater. Fig. 5a shows the location of all data from MPU-A. The closest relatives of α-soils are surface soils from stations 1 and 2, especially 61220 (see Heymann *et al.*, 1975). Soils near the β-corner occur most abundantly in MPU-D (Fig. 5e) and in much of barrel 60004 of MPU-B (Fig. 5c). This soil type is, however, also present to some extent in MPU's A and C (Figs. 5a and 5d) as well as in barrel 60003 of MPU-B (Fig. 5b). The archetype soil is dark, mature, agglutinate-rich, $I_s/FeO > 80$; FeO is 7% or greater, $^{40}Ar/^{36}Ar$ less than 1.5. The closest relatives are the ubiquitous dark surface soils on Stone Mountain and at station 10 (Heymann *et al.*, 1975). Soils near the γ-corner occur most abundantly in MPU-C (Fig. 5d). These are light soils, rich in monomineralic plagioclase, mature-to-submature, i.e., $30 < I_s/FeO < 60$ in general. FeO is probably less than 3%. Gose and Morris (1977) have shown that γ-rich soils have been extensively reworked (i.e., matured and mixed) such that the most mature and contaminated versions are now located near the top of MPU-C. The γ-rich soil horizon of MPU-C is almost certainly the same as the plagioclase-rich horizon reported by Blanchard *et al.* (1976) and McKay *et al.* (1976) in 60009, the lower half of the double drive tube collected only 40 meters away from the drill core. The plagioclase contents of soils from 60009 range up to 70% (compare with Fig. 5d) and the plag-rich soils in 60009 are clearly less mature than those available to Vaniman *et al.* (1976) as indicated by I_s/FeO for 60009 soils (Morris and Gose, 1976). There is no puzzle here. If thin sections from 60005 had been available, the distribution of data points in Fig. 5d would almost certainly have extended to larger values of large clast content and larger values of plagioclase content, i.e., into the area where the plag-rich soils from 60009 are located.

Geochemically and petrologically the closest relatives of γ among the surface soils are soils from stations 11 and 13 on the ejecta blanket of North Ray Crater (see also Nagle, 1977), but MPU-C has definitely not been deposited by the North Ray Crater event.

Figures 5b and 5c show that MPU-B is, petrologically, not a single unit, but that its lower portion, in barrel 60003, is distinctly different from its upper portion in barrel 60004. The upper portion is richer in β-component. Whether

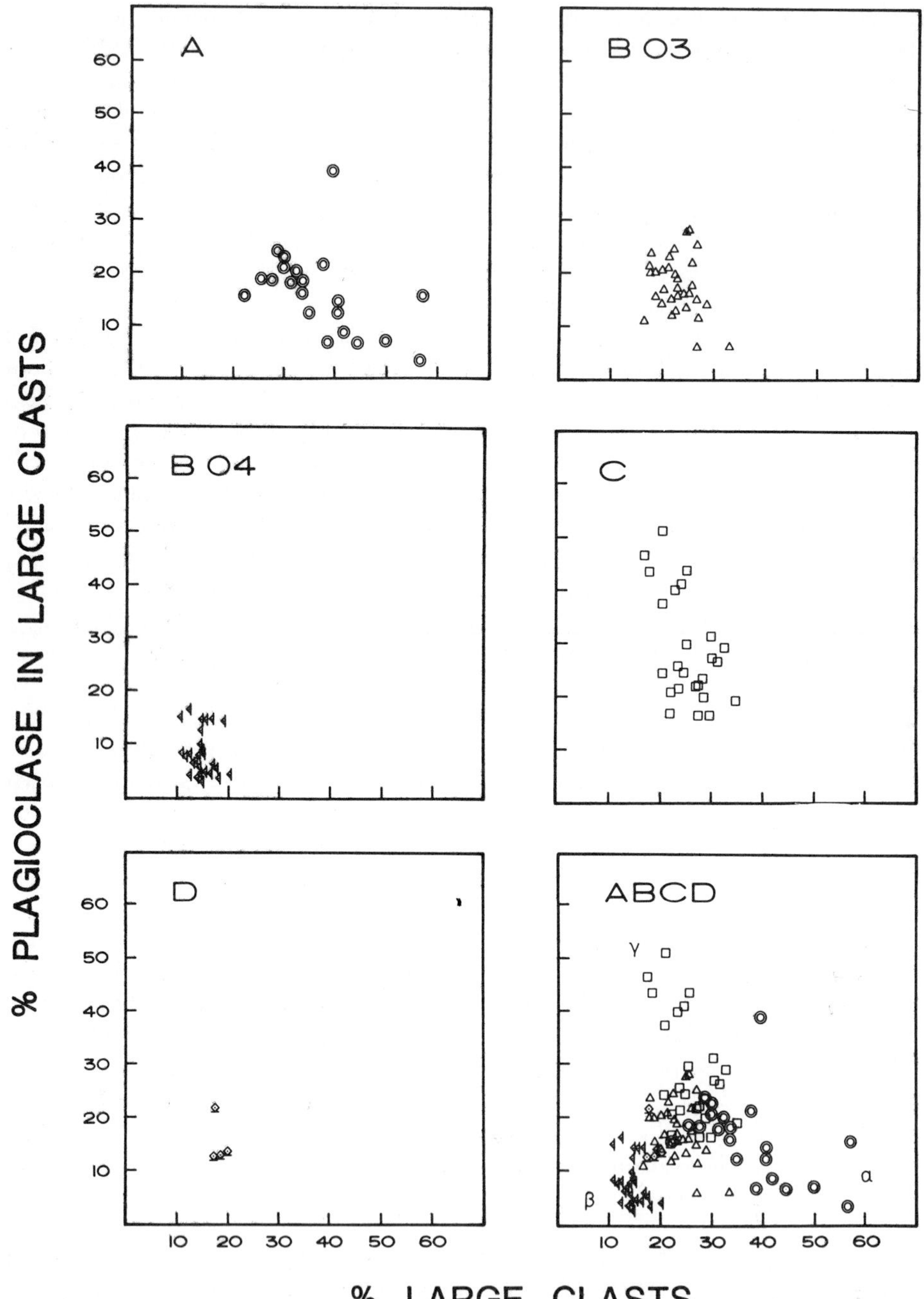
A
B O3
B O4
C
D
ABCD
γ
β
α
60
50
40
30
20
10
% PLAGIOCLASE IN LARGE CLASTS
% LARGE CLASTS

there is a sharp or only a gradual change between the two petrologic subunits BO3 and BO4 is difficult to say. There is a known, distinct break between DU's 27 and 28 at about 4 cm from the top of barrel 60003 (Duke and Nagle, 1976; Heymann *et al.*, 1977; Ray *et al.*, 1978), but it is unclear whether this is the dividing boundary in the core for the samples in Figs. 5b and 5c (see next sections).

The results of Vaniman *et al.* (1976) can now be used to characterize the three soil types more fully. The most prominent lithology of the large clasts of the α-soil is light matrix breccia. The mean LMB content of the samples in Fig. 5a is 26%. The largest value of 65% is seen roughly 10 cm below the MPU A-B boundary. LMB is not absent in MPU's BO3 and BO4, but is seldom more than 20% of the large clasts. In MPU-C, LMB is even less abundant with a mean of 14% in the large clasts. When one considers non-agglutinitic glass (i.e., total glass abundance in the core; Vaniman *et al.*, 1976) one observes that MPU's A and BO3 are on the whole the most glass-rich. One might conclude from this that α-soils are glass-rich, which is consistent with the report by Heiken *et al.* (1973) that 61221 is among the most glass-rich soils of the Apollo site. There are strong hints that the yellow glass (Naney *et al.*, 1976) is associated with the α-soils. Judging from the type sample, thin section 60002, 382-1 (Vaniman *et al.*, 1976), which contains 65% LMB, 16% monomineralic plagioclase, but only 6% dark matrix breccia, one concludes that the archetype α-soil is indeed, near-pure LMB.

The type section for γ-soil is 60006, 388-1 (Vaniman *et al.*, 1976) which contains 51% monomineralic plagioclase in the large clasts. In the small clasts of this thin section (20–200 μm), the plagioclase content is 72%. We have seen that the plagioclase content in the related 60009 soils runs as high as 70%. Undoubtedly the source rocks for γ are very plagioclase-rich.

There exists no type section for β soil. This soil type is very agglutinate-rich (40–50% in MPU-D according to Heiken *et al.*, 1973) such that much of the parent lithologies which have gone into β soil have been obliterated into agglutinitic glass, hence can no longer be recognized. Generally speaking, this soil is rich in dark matrix breccia, recrystallized noritic breccia, and norite and troctolite which make up the bulk of the non-agglutinitic lithologies.

Several chemical studies have been made on this core (Philpotts *et al.*, 1973; Nava and Philpotts, 1973; Nava *et al.*, 1976; and Ehmann *et al.*, 1977). These

---

Fig. 5. Top from left to right: 5a. Large clast content vs. % monomineralic plagioclase for all thin section studies by Vaniman *et al.* (1976). MPU-A only. Soils in this modal unit are obviously α-rich. The dominant lithology is light matrix breccia. 5b. Same variables as 5a; MPU-B03 only. 5c. Same variables as 5a; MPU-B04 only. Bottom from left to right: 5d. Same variables as 5a; MPU-C only. Soils in this modal unit are γ-rich. The most dominant lithology is monomineralic plagioclase. 5e. Same variables as 5a; MPU-D only. 5f. Same variables as 5a. Composite of 5a through 5e. The immature plagioclase-rich soils from 60009 (McKay *et al.*, 1976) would be located towards the northeast.

show that the lower portions of the core are generally enriched in mafic constituents, whereas the upper portions are richer in plagioclase. This observation is one of the cornerstones of Nagle's (1977) hypothesis that the soils of MPU-A have come from the west, specifically the Eden Valley Crater, and that the soils of MPU-C have come from the north, i.e., Palmetto and/or Gator Craters.

The study by Nava *et al.* (1976) reveals an interesting chemical "break" in MPU-A, which has thus far been overlooked. These authors have reported that the soils in barrel 60002 are remarkably uniform in their major and minor element chemistry despite the obvious layering in this barrel. The authors failed to find a clear-cut chemical "break" at the MPU-A/MPU-B boundary, but they have analyzed only 1 sample above this boundary. However, above 16 cm depth in 60002, the average FeO content is 5.55%, whereas it is 5.15% below 18 cm. The compositional change is quite abrupt. At about 17 cm depth there is the contact DU-9/DU-10 and when one considers the dissection descriptions (Duke and Nagle, 1976), one finds that the soils in MPU-A *above* this contact are generally sandy, whereas those below the contact are generally gravelly.

In summary, the three major soil types of the Apollo 16 drill core and their known or suspected parent rock-types occur abundantly within less than 10 km and probably less than 5 km of the core site. The possible presence of mare-type glasses (Meyer and Tsai, 1975; Naney *et al.*, 1976) signals the occurrence of exotic components, but it is not clear at the present time whence this matter has come.

STRATIGRAPHY AND STRUCTURE OF THE CORE

The stratigraphy proposed by Vaniman *et al.* (1976), i.e., four modal petrologic units A–D from base to top is an excellent first-order approximation, but we have already seen that MPU-B can be grossly subdivided into BO3 and BO4, and that MPU-A may consist of two major zones, divided at the boundaries of DU's 9 and 10.

By far the most detailed stratigraphy is the one that has emerged from the dissection observations (Duke and Nagle, 1976), which have generated 46 major dissection units and their numerous sub-units. Let us examine which DU boundaries have been confirmed by other observations, how strong the confirmations are, and what this may mean.

Category 1: Very strongly confirmed DU boundaries.

There are two of these: DU 12/DU 13 which corresponds to the boundary of MPU's A and B and DU 43/DU 44 which corresponds to the boundary between MPU's C and D. We have already seen much of the evidence for DU 12/DU 13. At this boundary there is a sharp break in reflectivity and modal petrology. The change in major and minor element chemistry, however, is subtle (Nava *et al.*, 1976); only slight increases in MgO (6.7 to 7.0) and FeO (5.9 to 6.7) are noteworthy. Jordan and Heymann (1976) noted a distinct break in $^{40}Ar/^{36}Ar$ values. I_s/FeO changes from an average of 27 ± 7 in DU 12 to 65 ± 8 in DU 13 (Gose and Morris, 1977). The principal issue concerning this boundary is whether it represents a buried fossil regolith surface or whether it does not. Jordan and Heymann (1976) noted that, while $^{40}Ar/^{36}Ar$ and $^{4}He/^{20}Ne$ (for revised data see Table 1) show a distinct break at the boundary, these ratios also display a systematic trend downward in DU 12. The complete set of data for DU 12 in Table 1 fully confirms the earlier conclusion. In the context of the mixing model of the preceding section this can be interpreted in terms of mixing of α-rich soil of MPU-A with the more β- and γ-rich soils of DU 13. The trend observed by Jordan and Heymann has the earmark of reworking at a regolith surface which is now buried. Blanford and Morrison (1976) and Blanford and Wood (1978), on the strength of track density measurements in DU's 12 and 13, have arrived at the same conclusion and have estimated exposure times of this buried surface between 15 and 25 $\times 10^6$ yr. Gose and Morris (1977), on the strength of I_s/FeO data, have concluded that MPU's A and B were deposited simultaneously, or essentially simultaneously, thus denying that this surface could have been reworked for as long as the period claimed by Blanford and Morrison.

Since inert gas measurements and I_s/FeO measurements have been done on aliquots from the same parent samples, they lend themselves more readily for comparison. Figure 6 presents $^{40}Ar/^{36}Ar$ vs. I_s/FeO for a selection of core and surface soils (see also Heymann *et al.* 1978). The slanted curve has been constructed by the on-paper mixing of two end-members: *1* an average of the dark surface soils 60601, 64421, 66041, 66081, 68501, and 69941; and *2* the light trench soil 61221. This is essentially one of the main trends of Figs. 1 and 2 when one ignores the North Ray Crater soils. A regolith surface à la Blanford and Morrison exists today at station 1 where a trench has been sampled by the astronauts. The soil from the trench top is 61241, that from the bottom is 61221. Soil 61181 has been collected from the very surface, *also on the rim of Plum Crater*, only 4–5 meters from 61241. The sequence 61221–61241–61181 shows the same $^{40}Ar/^{36}Ar$ trend which Jordan and Heymann (1976) have noted in DU 12. Soils 61241 and 61181 fall above the mixing trend of Fig. 6 but this is exactly what one would expect for reworking at the surface, i.e., α-rich soils on the rim of Plum crater have not only become "contaminated" with the ubiquitous dark β-rich soils of the landing area, but the immature α-component has become matured by exposure in the active regolith zone at the top.

If one now compares the trench to MPU-A, one finds the essentially horizontal trend of Fig. 6, which is mixing of α-rich with β, γ-rich soils, but no trace of maturation as seen at the top of the trench! Apparently the Gose-Morris (1977) interpretation is correct. However, Heymann *et al.* (1978) have pointed out that Morris and Gose (1976) have not measured enough samples from DU 12 near the MPU A-B boundary (this is not the fault of the investigators, but is due to the so-called interim storage of lunar samples). Thus, in our opinion, the Blanford-Morrison (1976) interpretation must be preferred at this time.

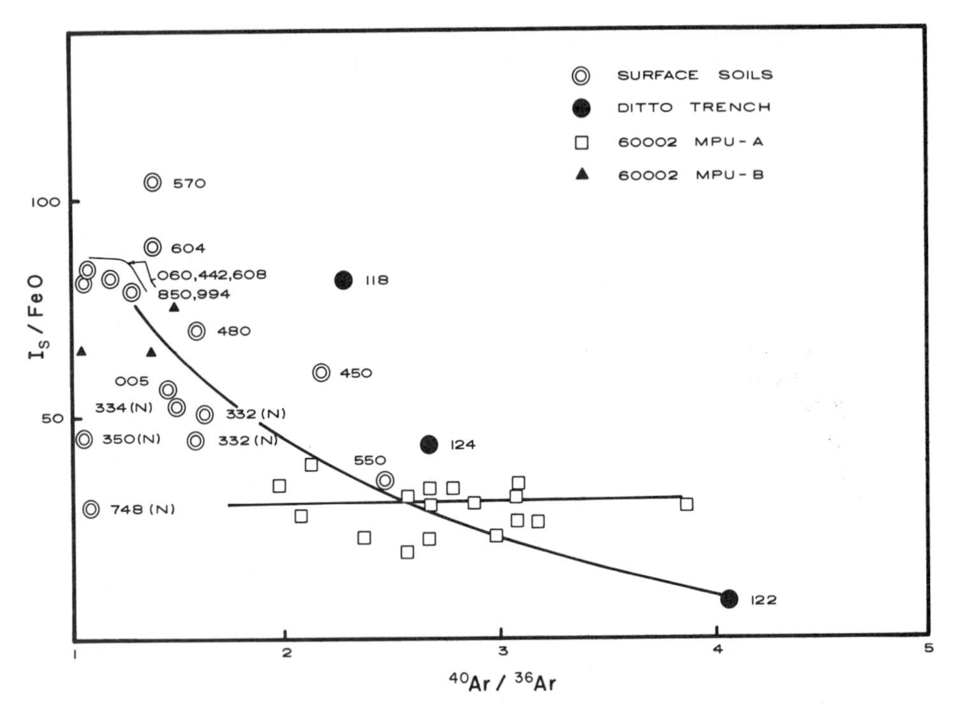

Fig. 6. I_s/FeO vs. $^{40}Ar/^{36}Ar$ for a number of surface soils and drill core soils. Magnetic data from Morris (1976) and Gose and Morris (1977). The slanted curve is a calculated mixing line of two end-members: *1* average of 60601, 64421, 66041, 66081, 68501, and 69941; and *2* 61221. The North Ray Crater soils (N) fall to the left of the mixing line (see Heymann *et al.*, 1975). Soils 61221, 61241, and 61181 represent a reworking sequence of the trench at station 1. No such sequence is seen for soils from MPU-A (A); these represent mixing only along the horizontal trend. However, the control of magnetic data near the MPU-A/MPU-B boundary is insufficient to rule out significant reworking of the top of MPU-A (Blanford and Morrison, 1976; Gose and Morris, 1977).

Much less supporting evidence is available for the confirmation of the DU 43/DU 44 boundary. Nevertheless, this, or alternatively DU 42/DU 43, can be accepted as strongly confirmed boundaries on the basis of the petrologic and magnetic observations. Samples 60007,38 and 34 (Gose and Morris, 1977) straddle the DU 43/DU 44 boundary very well, and the abrupt and sustained change of FeO from >5% to <4% as well as the change of I_sFeO from 82 to 128 at the DU 43/DU 44 boundary mark it as the probable break between MPU's C and D. The Gose-Morris (1977) interpretation of this boundary as a buried surface which has been reworked for very long times (~450×10^6 yr) is undisputed.

Category 2: moderately strongly confirmed DU boundaries.

Gose and Morris (1977) have argued that MPU-A was deposited as a single slab by a single event. This does not preclude the existence of genuine DU's in MPU-A, because there is no *a priori* reason that a single slab must be unstratified. However, there are indications that their interpretation may not be correct. We have already seen that the chemistry of MPU-A changes at the DU 9/DU 10 boundary. Bogard and Hirsch (1975) have argued that this boundary is a buried surface. Blanford and Morrison (1976) have supported this view and have a suggested an exposure time of about 10×10^6 yr. Inert gases have been measured in samples both above and below the boundary (see Table 1). These measurements confirm a possible break at DU 9/DU 10; above this boundary, $^{40}Ar/^{36}Ar$ ratios are generally greater than 4.0 (DU 12), while below it these ratios are all less than 3.6. Regrettably, the magnetic control at this boundary is insufficient, because I_s/FeO has been measured in only one (parent 110) of the three samples (parents 109, 110, and 112) for which inert gas data are available. Our verdict is that DU 9/DU 10 is certainly a genuine compositional boundary, that it may be a buried surface, but that the evidence for the latter is not yet compelling.

Blanford and Morrison (1976) have suggested a third buried surface at the boundary DU 5/DU 6 in MPU-A with a short exposure time of only a few million years.

We now come to the enigmatic boundary DU 27/DU 28 in barrel 60003 and in MPU-B, which may well be the boundary between the subunits BO3 and BO4. This boundary must be discussed in the context of the entire section of the core contained between it and the strongly confirmed boundary DU 12/DU 13. The inert gas data present the following gross picture. At the base of this section (DU 13), the soils appear to be relatively β-rich as $^{40}Ar/^{36}Ar$ is about 1.4. Inert gas data in 60003, both from normal and special dissection procedures (Table 1) show a distinct gross trend. The $^{40}Ar/^{36}Ar$ ratio is about 1.4–1.5 in DU's 13 and 15, is about 1.3–1.4 in DU's 21, 22, and 23, then increases upward through DU's 26 and 27 to values between roughly 1.5 and 1.6, *which continue into DU 28*! Grossly one observes higher-lower-higher, but the results from the special dissection indicate that there is significant variance of $^{40}Ar/^{36}Ar$ ratios in DU's 23, 26, and 27 among samples which were taken only a few millimeters from one another in different, or what appeared to be different lithologies.

Unpublished magnetic measurements of R. N. Housley (essentially Fe^o contents) on *all* 92 special dissection samples in 60003, summarized in Table 2, show a somewhat different picture. The average Fe^o contents decrease systematically upward in the sequence DU's 14–15–22–23–27–28. In each of the three laminated units 15, 23, and 27, the mean Fe^o content of the light lithologies is less than that of the dark lithologies. The I_s/FeO magnetic data reported by Gose and Morris (1977) on normal dissection samples show a trend which corresponds more closely to that of the inert gas trend, namely submature soils in DU's 14 and 16, mature soils in DU's 18 through 26, and submature soils in DU's 27 and 28, *again without any distinct break at DU 27/28*!

Visually, the DU 27/DU 28 boundary was quite striking to all (including one of the authors of this paper) who have seen the core at JSC prior to dissection. However, careful studies of photos (especially S-76-26367 and S-76-26366) as well as the results of our inert gas studies (Heymann *et al.*, 1977) and Housley's magnetic measurements show that this may well have been an optical illusion due to unknown structural alterations during core taking, transport, and processing. Unit 28 has been classified by Duke and Nagle (1976) as uniform and massive, whereas unit 27 has been classified as marbled and laminated. The inert gas and magnetic data from the special dissection samples show that unit 28 is far from uniform. Photo S-76-26367 shows mm-sized dark clasts and distinct light-dark lamination in DU 28, but the lamination in DU 28 is more chaotic than in DU 27, where a horizontal direction is more preferred. In addition, there are distinct light clasts discernible in DU 28. All of this raises again the old, but serious issue whether much, if not all of the marbling and lamination in DU 27 is man-made, i.e., that the rotary coring device as it penetrated into the regolith encountered, here and there, sizeable or smaller, but relatively friable light or dark soil clasts which it crushed up and smeared along the periphery of the core proper. Our verdict is that DU 27/DU 28 is, indeed, an optical boundary. It is not a compositional boundary.

This verdict calls again for the question whether there is a recognizable DU boundary for the two modal petrologic subunits BO3 and BO4. One possibility is that there is no recognizable DU boundary but a gradual change from BO3 to BO4. The only hint for a genuine boundary comes from the work of Gose and Morris (1977) who report that DU 28, which straddles the physical contact of barrels 60003 and 60004 is submature in 60003, but mature in 60004. Our search for the BO3/BO4 boundary has been frustrated, but for once it might be that this boundary is located near the break of two core barrels.

Category 3: poorly confirmed and unconfirmed DU boundaries.

Virtually the only measurements which have been made in small steps (about 0.5 cm) along the entire core are magnetic measurements. Housley (pers. comm.) has noted from his study of the special dissection samples of 60003 that there is frequently a distinct "jump" in Fe^{o} at reported DU boundaries of this core segment. Morris and Gose (1976) report a distinct increase of I_s/FeO at the boundary DU 16/DU 17, and continued increase of this maturity indicator upward into DU 17. No other measurements have been done on this part of the core. The magnetic data may be interpreted as soil mixing in the transitional DU 17, but one cannot exclude the possibility that DU 17/DU 18 is a buried regolith surface. The data of Morris and Gose show a sharp drop of I_s/FeO at DU 25/DU 26. No other data are available here. Taken at face value, the observation can be interpreted to mean that a mature regolith surface was buried here, and elsewhere in the core when a mature unit is *overlain* by submature or immature soils, by a relatively thick deposit which contained submature or immature soils *at its base*.

The crucial question is where the soils in DU 25 became mature. If the maturation occurred when DU 25 was, locally, at the top of the regolith, then DU 25/DU 26 is a genuine buried surface. However, if the maturation occurred elsewhere and at a time when DU 25 was not yet in place, then DU 25/DU 26 is not a buried regolith surface. Because dark, mature soils with I_s/FeO between 70 and 90 are unbiquitous in the landing area today and presumably were so in the past, magnetic data at boundaries such as DU 25/DU 26 cannot be interpreted unambiguously. Judging from Gose and Morris (1977), the following DU boundaries are in this category: DU 13/DU 14, and DU 31/DU 32.

The placement of the MPU-C/MPU-D boundary at DU 43/DU 44, i.e., at about 13 cm lunar depth is not the original one of Vaniman *et al.* (1976), who placed the boundary at 6 cm depth. The latter places the MPU-C/MPU-D boundary in the middle of what appears to be the rather massive and uniform unit 45. Duke and Nagle (1976) have reported four sub-units in DU 45, but none of the boundaries between these subunits have the earmarks of a genuine stratigraphic boundary. Considering the boundary DU 44/DU 45 at about 10.5 cm lunar depth, there is a distinct change in I_s/FeO (92 to 57) and FeO (5.2 to 8.2%) at this boundary (Gose and Morris, 1977), but similar abrupt changes are seen in two places inside DU 45. However, there is a distinct albeit subtle difference between I_s/FeO and FeO content in DU 46, the upper 2 cm of the core, and the upper half of DU 45. If MPU-D is sub-stratified, the major break occurs, in our opinion, at DU 45/DU 46 at 2 cm lunar depth.

Track studies have direct bearings on the stratigraphy of cores, but in the case of the Apollo 16 drill core, the results of track studies are contradictory. However, all authors agree that the core cannot be a single slab of previously unirradiated matter. Behrmann *et al.* (1973) have reported a very heavily irradiated (by solar cosmic rays) sample at ~117 cm maximum lunar depth. This observation has prompted us to review the inert gas data from barrel 60004, however few in number these are (Table 1). Except for sample 60004,7 (Bogard *et al.*, 1973) all samples above DU 34 contain more than 60×10^{-8} cm^3 STP/g of solar-wind derived ^{4}He; those in or below DU 34 contain less than 52×10^{-8} cm^3 STP/g of ^{4}He. Whether this observation is significant is difficult to say with only eight samples analyzed, but it is possible that the boundary DU 34/DU 35 represents some kind of break in the otherwise uniform barrel 60004.

Fleischer *et al.* (1974) have studied samples from what appears to be DU 42, hence from near the top of MPU-C, but their study does not seem to pertain to the issue of the location of the MPU-C/MPU-D boundary.

Price *et al.* (1975) have reported track studies of a 3 cm long impregnated section which straddles the boundary DU 27/DU 28 mentioned above. Their study sheds some more light on the nature of this enigmatic boundary. Figure 7 is a faithful sketch, made from photo S-76-26366. One must keep in mind that the section of Price *et al.* comes from the other half of the core, roughly ½ cm away from the sketched surface. Price *et al.* report that the layer 3.62–3.72 cm depth (from top of barrel 60003) is heavily irradiated. This layer comes from the light but massive DU 28 above the DU 28/DU 27 boundary, roughly at the positions

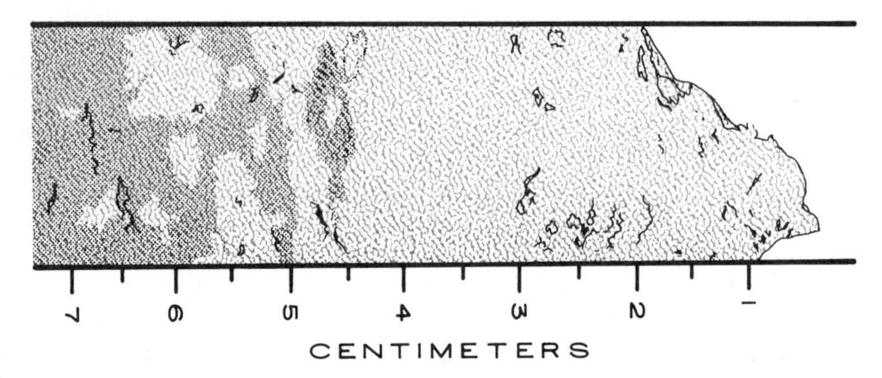

Fig. 7. Sketch of the DU 27/DU 28 boundary in barrel 60003 from photo S-76-26366.

of special dissection samples 60003,1309, 1310, and 1311. The layer 3.7–4.1 cm shows gradual decrease in level of irradiation with depth. This layer is still above the DU 28/DU 27 boundary, roughly at the positions of special dissection samples 60003,1312, 1313, and 1314. At 4.48–4.55 there is an extremely heavily irradiated layer; from 4.5–4.8 cm the layer is only very modestly irradiated. From here to 6.25 cm there is only weak irradiation with a local maximum at 5.3–5.4 cm depth. On the whole, the results of Price *et al.* seem to say that DU 27/DU 28 is, indeed, a genuine boundary of some sort, and in detail, their results conform roughly with the light-dark lamination near this boundary. It appears that strongly irradiated and weakly irradiated soils have become "scrambled" at this boundary, but we have noted earlier, it is unclear whether the scrambling occurred on the moon or during or after the taking of the core.

Goswami *et al.* (1976) as well as Price *et al.* (1975) report micro- (<1 mm) and mini-layers (several mm thick) and the former think that DU 4 has been deposited as a discrete event.

SUMMARY

1.) Boundaries DU 12/DU 13 and DU 43/DU 44 are major stratigraphic breaks. Both must be considered buried fossil surfaces.
2.) Boundaries DU 9/DU 10; DU 5/DU 6; and DU 27/DU 28 are stratigraphic breaks. The first may be a buried surface. The nature of the second and third are unclear.
3.) The following boundaries may be stratigraphic breaks, but evidence is weak: DU 17/DU 18; DU 24/DU 25; DU 13/DU 14; DU 31/DU 32; DU 45/DU 46; DU 34/DU 35; DU 3/DU 4. Whether any of these represent time-stratigraphic boundaries (buried surfaces) is unknown at this time.

Depositional History of the Core

The time-stratigraphic history of this core as evidenced by its neutron stratigraphy, cosmogenic gases and tracks is complex. Accordingly, the models and theories of different investigators are not in agreement. One anchor point is the neutron stratigraphy of the core reported by Russ (1973). Neutron doses (fluences) were deduced for seven samples along the entire core. For a static slab irradiation, one expects a ψ (fluence) profile which first increases with depth, reaches a maximum, then decreases again. In real cores serious complications arise from the fact that core soils may have been pre-irradiated elsewhere before deposition at their present locale, that cores may have been accreted in a variety of ways (ranging from dribs-and-drabs to the other extreme which is deposition of the entire core by a single event), and that erosional surface are liable to occur in a core of 2 meters length. Hence, interpretations are model-dependent.

Russ reported that his data are consistent with:

1.) Continuous accretion (dribs-and-drabs) at a mean rate of 7×10^{-7} g/cm^2 yr with soils now contained in 60004 more strongly pre-irradiated than the rest. The total accretion time of the core would then be $\sim 500 \times 10^6$ yr, but if one allows the degree of pre-irradiation to vary freely, the core could have been deposited in its entirety the day before the astronauts landed.
2.) A three slab model with *i* the upper ~ 100 g/cm^2 deposited $\sim 50 \times 10^6$ yr ago, *ii* slab 60004 about 10^8 years older, *iii* a lower third slab, heavily pre-irradiated.

Gose and Morris (1977) have modified the triple-slab model to a double-slab model as follows:

1.) Deposition of MPU's -A, B, and C *simultaneously* about 500×10^6 yr ago, with MPU-A more heavily pre-irradiated than MPU's B and C.
2.) *In situ* reworking of MPU-C to considerable depth, but relative quiescence during the following 450×10^6 yr.
3.) Deposition of MPU-D as a slab about 50×10^6 yr ago.

Bogard *et al.* (1973), from cosmogenic gases measured at only seven points arrive at models similar to those of Russ, with the proviso that the first 0–62 cm of the core (essentially MPU's C and D) come from North Ray Crater.

Let us try to sort out these ideas. There are several problems connected with a 60 cm thick deposit from North Ray Crater. For one thing, no such deposit is seen at stations 1 and 2 which are roughly as distant from North Ray Crater as the core. Second, the areas of stations 1, 2 and 10 are, if anything, on or near rays from South Ray Crater (Muehlberger *et al.*, 1972), but Bogard *et al.* (1973) report that they see no previously unirradiated matter ejected by South Ray Crater in the upper 60 cm of the core. Third, the evidence presented by Gose and Morris (1977) for extensive reworking of the top of MPU-C during a much longer period of time than 50×10^6 yr is so strong that it is implausible that any solids below the Du 43/DU 44 boundary at 13 cm depth come from the North Ray Crater event.

This raises the question whether MPU-D, the upper 13 cm of the core, is due either to North Ray or to South Ray or to both. Petrologically MPU-D shows some affinity with the soils of the North Ray Crater ejecta blanket, but this could mean that MPU-D might have been emplaced by North Ray and/or South Ray secondaries which "pushed" significant amounts of the ubiquitous dark β-soils either from Stone Mountain or Smoky Mountain into the valley. The apparent $^{21}Ne_c$ radiation ages at the top of the core are closer to 100×10^6 yr than to 50×10^6 yr, the accepted age of North Ray Crater. One would expect some dilution of the dark soils with unirradiated matter from the North Ray and/or South Ray ejecta, which would reduce the apparent $^{21}Ne_c$ of the mix relative to that of the unmixed dark β-rich soils at Stone Mountain. Our conclusion is that MPU-D represents the emplacement, not necessarily by North Ray Crater only, of dark surface soils at the core site proper during the last 50×10^6 yr, probably as ejecta from North Ray and/or South Ray-induced secondaries. Now that the thickness has been reduced from the order of one meter to the order of ten centimeters, it is no longer so embarrassing that so little evidence is seen at station 1. In fact, the trench at station 1 does show a "patina" of a few cm (soil 61241) of dark soils whose presence here may well be due to the same causes which emplaced MPU-D at station 10. It is possible that DU's 44 and 45 were so emplaced by the North Ray Crater event, and that DU 46 was so emplaced by the South Ray Crater event.

Let us now consider the base of the core, i.e., MPU-A. If one accepts, as we do, the basic interpretation of Gose and Morris (1977) *for MPU-C*, i.e., reworking for 450×10^6 yr, then MPU-A has been buried for at least 500×10^6 yr. We cannot accept the single slab deposition of MPU-A as long as the nature of the boundary DU 9/DU 10 is unresolved.

The question arises whether MPU-A is the top of a very ancient regolith or merely a "wad" or uncommon soils deposited at the landing site on top of more common β-rich soils now at the surface. If the core had been 10 meters instead of 2 meters long, this question might have been settled. Now if MPU-A represents the top of a very ancient regolith one wonders what the landing area looked like before DU 13 was emplaced. Stone Mountain and Smoky Mountain were undoubtedly present then, but North Ray and South Ray craters were not. It is unclear whether Palmetto and Gator craters were in place, even the very subdued Eden Valley crater may not have been in place. Is it conceivable that α-rich soils dominated the surface in the valley between the two mountains? If that was so, why do we not find much, if any *matured* samples of α-rich soils in the cores or at the surface (61221!)? Must there not have been thick deposits of mature α-rich soils?

There are several possible answers to this enigma. One is that an unusually strong base surge of unknown origin has swept away much of the mature α-rich soils, such that we are witnessing a major erosional event. Another possibility is that we see both at Plum crater (station 1) and at the site of the drill core overturned rims of craters à la 74001/74002 (Bogard and Hirsch, 1978). Since the rims are topographic highs, they are less liable to become covered with mature soils nearby. In this interpretation, the mature α-rich soils at the core site are located in the unsampled regolith beneath the base of the core.

If one accepts uniformitarianism, at least since 3×10^9 yr ago, then MPU-A is not the top of the ancient regolith, but represents the deposition of what may be a scoop from an ancient regolith on top of either β-rich or α-rich soils located beneath the base of the core. Jordan and Heymann (1976) have suggested that MPU-A is the result of a ray-deposit in the valley. This suggestion differs from the idea of Nagle (1977) only insofar as Jordan and Heymann thought in terms of rays from a distant crater which covered much of the valley and perhaps portions of Stone Mountain where α-rich soils have been found, whereas Nagle thought in terms of a more local ray deposit from Eden Valley. The idea of a ray deposit is at once fetching and challenging. Models of regolith gardening and evolution (cf. Arnold, 1975) usually consider the throwing hither-and-yon of soils in what is essentially a three-dimensional sandbox with some gravel thrown in for good measure. The evolution of a regolith beneath a rock-strewn field has not been considered in detail, but the remarkable structuring of MPU-A may well be the result of regolith evolution in such an environment. Rocks now at the lunar surface nearly always have fillets. Fillet soils tend to be relatively coarse-grained and may be somewhat shielded against solar wind implantation. On the other hand, as the experience of the "permanently shadowed" soil at station 13 has shown, non-fillet soils are thrown from time-to-time under and up against boulders, covering their proper fillets.

The local formation of regolith soils via the stage of boulder fillets has another interesting aspect. Once the boulders are completely disaggregated, the source for soils derived from them has dried up. It is intriguing to note that the supply of fresh α-rich soils appears to have dried up above DU 12, although, not unexpectedly, α-rich ghosts appear higher up in the stratigraphic column, e.g., sample 60003,1316.

One may thus reconcile the Gose-Morris (1977) and Blandford-Morrison (1976) models as follows. MPU-A was, indeed, initiated by a single event, was not deposited as a single slab of soil, but in the form of some soil and many boulders. Conceivably the rock (or rocks) which contributed to MPU-A found themselves inside a secondary crater. The regolith in which the secondary had formed was α-poor, but relatively β-rich. Some γ-containing soils may have been in the crater walls also. MPU-A grew in thickness by "filleting", interrupted from time-to-time by the deposition of soils from the crater wall, either by slumping or by small-crater-forming impacts. Thus one can understand the report of Behrmann *et al.* (1973) that the soils in drill bit 60001 at the base of MPU-A contain both "young" and "old" soil components. One can understand the occurrence of "buried" surfaces. What may have happened at DU 9/DU 10 is that DU 9 represents a deposit on top of a fillet DU 8.

Within the context of this model, at least MPU-BO3 probably represents the

Table 2. Summary of Fe° measurements on special dissection samples from 60003 by R. N. Housley. Average Fe° in weight %.

Zone	Depth from top of barrel	DU	Samples: Parent Numbers	Fe° (Average) Light	Dark	Total
Upper massive	2–4	28	1200–1214	–	–	0.285
Upper laminated	4–7	27	1215–1235	0.280	0.334	0.307
Middle laminated	14–17	23	1238–1258	0.309	0.343	0.326
Middle massive	17.7–18.1	22	1259–1267	–	–	0.345
Lower laminated	34.1–35.8	15	1279–1286	0.326	0.353	0.340
Lower massive	36.5–38.5	14	1287–1295	–	–	0.389

continued infill of the crater from its own walls. However, MPU-BO4 is so different that we are inclined to accept a single deposit of a soil slab ranging in thickness from a minimum of 13 cm (if DU 34/DU 35 is the MPU-BO3/MPU-BO4 boundary) to more than 45 cm if DU 27/DU 28 is the boundary. It seems possible that the scrambled nature of much of the core in barrel 60003 could have been generated during the emplacement of MPU-BO4.

The model of Gose-Morris (1977) for MPU-C is entirely acceptable, with the provision that, as in the case of MPU-A, the regolith evolved in a rather rock-strewn environment, i.e., MPU-C need not have been deposited as a single slab of soil.

Acknowledgments—We wish to thank Drs. J. S. Nagle, M. B. Duke, R. V. Morris, D. S. McKay, and D. A. Morrison for stimulating discussions about this core. Supported by NASA-NGL-44-006-127.

REFERENCES

Arnold J. R. (1975) Monte-Carlo simulation of turnover processes in the lunar regolith. *Proc. Lunar Sci. Conf. 6th*, p. 2375–2395.

Behrmann C., Crozaz G., Drozd R., Hohenberg C., Ralston C., Walker R. and Yuhas D. (1973) Cosmic-ray exposure history of North Ray and South Ray material. *Proc. Lunar Sci. Conf. 4th*, p. 1957–1974.

Blanchard D. P., Jacobs J. W., Brannon, J. C. and Brown R. W. (1976) Drive tube 60009; A chemical study of magnetic separates of size fractions from five strata, 1976. *Proc. Lunar Sci. Conf. 7th*, p. 281–294.

Blanford G. E. and Morrison D. A. (1976) Stratigraphy in Apollo 16 drill section 60002. *Proc. Lunar Sci. Conf. 7th*, p. 141–154.

Blanford G. E. and Wood G. C. (1978) Irradiation stratigraphy in Apollo 16 cores (abstract). In *Lunar and Planetary Science IX*, p. 106–108. Lunar and Planetary Institute, Houston.

Bogard D. D. and Hirsch W. L. (1975) Noble gas studies on grain size separates of Apollo 15 and 16 deep drill cores. *Proc. Lunar Sci. Conf. 6th*, p. 2057–2083.

Bogard D. D. and Hirsch W. L. (1978) Noble gas contents and irradiation history of orange-black glass in the 74001–74002 core (abstract). In *Lunar and Planetary Science IX*, p. 111–113. Lunar and Planetary Institute, Houston.

Bogard D. D. and Nyquist L. E. (1973) $^{40}Ar/^{36}Ar$ variations in Apollo 15 and 16 regolith. *Proc. Lunar Sci. Conf. 4th*, p. 1975–1985.

Bogard D. D., Nyquist L. E., Hirsch W. C. and Moore D. R. (1973) Trapped solar and cosmogenic noble gas abundances in Apollo 15 and 16 deep drill samples. *Earth Planet. Sci. Lett.* **21**, 52–69.

Duke M. B and Nagle J. S. (1975) Stratification in the lunar regolith-A preliminary view. *The Moon* **13**, 143–158.

Duke M. B. and Nagle J. S. (1976) Lunar Core Catalog, and Supplements. JSC 09252. NASA Johnson Space Center, Houston.

Ehmann W. D., Ali M. D. and Hossain T. I. M. (1977) Chemical characterization of lunar core 60003 (abstract). In *Lunar Science VIII*, p. 275–277. The Lunar Science Institute, Houston.

Fleischer R. L., Hart H. R. and Giard W. R. (1974) Surface history of lunar soils and soil columns. *Geochim. Cosmochim. Acta* **38**, 365–380.

Gose W. A. and Morris R. V. (1977) Depositional history of the Apollo 16 drill core. *Proc. Lunar Sci. Conf. 8th*, p. 2909–2928.

Goswami J. N., Braddy D. and Price P. B. (1976) Microstratigraphy of the lunar regolith and compaction ages of lunar breccias. *Proc. Lunar Sci. Conf. 7th*, p. 55–74.

Heiken G. H., McKay S. and Fruland R. M. (1973) Apollo 16 soils: Grain size analysis and petrography. *Proc. Lunar Sci. Conf. 4th*, p. 251–266.

Heymann D., Ray J., Dziczkaniec M. and Palma R. (1978) Mixing of major soil components in the Apollo 16 drill core (abstract). In *Lunar and Planetary Science IX*, p. 512–514. Lunar and Planetary Institute, Houston.

Heymann D., Ray J., Walker A., Dziczkaniec M. and Palme R. (1977) Inert gas stratigraphy of the lower half of the Apollo 16 drill core (abstract). In *Lunar Science VIII*, p. 441–443. The Lunar Science Institute, Houston.

Heymann D., Walton J. R., Jordan J. L., Lakatos S, and Yaniv A. (1975) Light and dark soils at the Apollo 16 landing site. *The Moon* **13**, 81–110.

Heymann D. and Yaniv A. (1970) Inert gases in the fines from the Sea of Tranquility. *Proc. Apollo II Lunar Sci. Conf.*, p. 1247–1260.

Jordan J. L. and Heymann D. (1976) Inert gas stratigraphy of Sections 60002 and 60004 of the Apollo-16 deep drill core (abstract). In *Lunar Science VII*, p. 434–436. The Lunar Science Institute, Houston.

McKay D. S., Fruland R. M. and Heiken G. H. (1974) Grain size and evolution of lunar soils. *Proc. Lunar Sci. Conf. 5th*, p. 887–906.

McKay D. S., Morris R. V., Dungan M. A., Fruland R. M. and Fuhrman R. (1976) Comparative studies of grain size separates in 60009. *Proc. Lunar Sci. Conf. 7th*, p. 295–314.

Meyer H. O. A. and McCallister R. H. (1974) Apollo 16: Core 60004-Preliminary study of <1 mm fines. *Proc. Lunar Sci. Conf. 5th*, p. 907–916.

Meyer H. O. A. and McCallister R. H. (1976) Mineral, lithic, and glass clasts <1 mm size in Apollo 16 core section 60003. *Proc. Lunar Sci. Conf. 7th*, p. 185–198.

Meyer H. O. A. and McCallister R. H. (1977) The Apollo 16 drill core. *Proc. Lunar Sci. Conf. 8th*, p. 2889–2908.

Meyer H. O. A., McCallister R. H. and Tsai H. M. (1975) Mineralogy and petrology of <1 mm fines from Apollo 16 core sections 60002 and 60004. *Proc. Lunar Sci. Conf. 7th*, p. 595–614.

Meyer H. O. A. and Tsai H. M. (1975) Lunar glass compositions: Apollo 16 core sections 60002 and 60004. *Earth Planet. Sci. Lett.* **28**, 234–240.

Morris R. V. (1976) Surface exposure indices of lunar soils: A comparative FMR study. *Proc. Lunar Sci. Conf. 7th*, p. 315–336.

Morris R. V. and Gose W. A. (1976) Ferromagnetic resonance and magnetic studies of cores 60009/60010 and 60003: Compositional and surface-exposure stratigraphy. *Proc. Lunar Sci. Conf. 7th*, p. 1–11.

Muehlberger W. R., Batson R. M., Boudette E. L., Duke C. M., Eggleton R. E., Elston D. P., England A. W., Freeman V. L., Hait M. H., Hall T. A., Head J. W., Hodges C. A., Holt H. E., Jackson E. D., Jordan J. A., Larson K. B., Milton D. J., Reed V. S., Rennilson J. J., Schaber G. G., Schafer J. P., Silver L. T., Stuart-Alexander D., Sutton R. L., Swann G. A., Tyner R. L., Ulrich G. E., Wilshire H. G., Wolfe E. W. and Young J. W. (1972) Preliminary investigation of the Apollo 16 landing site. *Apollo 16 Prelim. Sci. Rep.* NASA SP-315, p. 6–1 to 6–81.

Nagle J. S. (1977) Possible sources of immature soil at the Apollo 16 ALSEP site (abstract). In *Lunar Science VIII*, p. 709–711. The Lunar Science Institute, Houston.

Naney M. T., Crowl D. M. and Papike J. J. (1976) The Apollo 16 drill core: Statistical analysis of glass chemistry and the characterization of a high alumina-silica poor (HASP) glass. *Proc. Lunar Sci. Conf. 7th*, p. 155–184.

Nava D. F., Lindstrom M. M., Schuhmann P. J., Lindstrom D. J. and Philpotts J. A. (1976) The remarkable chemical uniformity of Apollo 16 layered deep drill core section 60002. *Proc. Lunar Sci. Conf. 7th*, p. 133–139.

Nava D. F. and Philpotts J. A. (1973) A lunar differentiation model in light of new chemical data on Luna 20 and Apollo 16 soils. *Geochim. Cosmochim. Acta* **37**, 963–973.

Nishiizumi K., Imamura M., Honda M., Russ G. P. III, Kohl C. P. and Arnold J. R. (1976) ^{53}Mn in the Apollo 15 and 16 drill stems: Evidence for surface mixing. *Proc. Lunar Sci. Conf. 7th*, p. 41–54.

Philpotts J. A., Schuhmann S., Kouns C. W., Lum R. K. L., Bickel A. L. and Schnetzler C. C. (1973) Apollo 16 returned lunar samples: Lithophile trace element abundances. *Proc. Lunar Sci. Conf. 4th*, p. 1427–1436.

Price P. B., Chan J. H., Hutcheon I. D., Macdougall D., Rajan R. S., Shirk E. K. and Sullivan J. D. (1973) Low-energy heavy ions in the solar system. *Proc. Lunar Sci. Conf. 4th*, p. 2347–2361.

Price P. B., Hutcheon I. D. and Braddy D. (1975) Track studies bearing on solar-system regoliths. *Proc. Lunar Sci. Conf. 6th*, p. 3449–3469.

Ray J., Dziczkaniec M., Walker A. and Heymann D. (1978) Stratigraphy in Apollo drill core segment 60003 (abstract). In *Lunar and Planetary Science IX*, p. 937–939. Lunar and Planetary Institute, Houston.

Russ G. P. III (1973) Apollo 16 neutron stratigraphy. *Earth Planet. Sci. Lett.* **19**, 275–289.

Vaniman D. T., Lellis S. F., Papike J. J. and Cameron K. L. (1976) The Apollo 16 drill core: Modal petrology and characterization of the mineral and lithic component. *Proc. Lunar Sci. Conf. 7th*, p. 199–239.

Yaniv A. and Heymann D. (1970) Ar^{40} anomaly in lunar samples from Apollo 11. *Proc. Apollo 11 Lunar Sci. Conf.*, p. 1261–1268.

Proc. Lunar Planet. Sci. Conf. 9th (1978), p. 1913–1932.
Printed in the United States of America

Core 74001/2: Grain size and petrology as a key to the rate of *in-situ* reworking and lateral transport on the lunar surface

DAVID S. MCKAY[1], GRANT H. HEIKEN[2], and GEORGANN WAITS[3]

[1]NASA Johnson Space Center, Houston, Texas 77058
[2]Los Alamos Scientific Laboratory, Los Alamos, New Mexico 87544
[3]Lockheed Electronics Co., Inc., Houston, Texas 77058

Abstract—We have studied a suite of samples from the double drive tube 74001/2. Material from this core mainly consists of orange and black droplets interpreted to be volcanic pyroclastic ejecta. Grain size analysis indicates that this material is very homogeneous in its grain size properties. It is also the finest and best sorted suite of soil samples in the lunar collection having a mean grain size of 40 μm and a mean standard deviation of 1.75ϕ.

The upper 5.5 cm of this core has apparently undergone *in situ* reworking by meteorites over a period of about 10 m.y. This reworked zone contains "exotic" grains including basalt, mineral fragments, vitric breccias, and agglutinates. One type of agglutinate is unique and has been made primarily from orange and black glass droplets melted and welded together by micrometeorite impacts. Other agglutinates are made mainly from basaltic fragments and minerals.

The amount of "exotic" material added to the core combined with an estimate of the location of the source areas for the "exotic" material allows us to estimate that a maximum of about 0.36 gm per cm^2 of regolith surface is added from a radius of about 1 meter in 10 m.y. Furthermore, no more than about 0.01 gm per cm^2 of regolith surface is added from a radius of about 100 meters in 10 m.y.

INTRODUCTION

It has been known since the time of the Apollo 17 mission that the Shorty Crater core or drive tube 74001/2 contained a unique sample. Early analysis of material from this drive tube confirmed that it consisted primarily of orange glass droplets and their partly crystallized black equivalents (McKay and Heiken, 1973; Heiken and McKay, 1974; Heiken *et al.*, 1974). These droplets were interpreted by most investigators to be volcanic pyroclastics. Recently the double drive tube 74001/2 was opened and dissected by Nagle (1978) who described a variety of units based on subtle difference in color, texture, and grain size. We have now studied, in some detail, selected samples from this core. The samples were chosen to be representative of major units described by Nagle and were also chosen to sample closely the upper 5 cm which had been shown by the FMR data of Morris *et al.* (1978) to have been modified by micrometeorite reworking. Samples which we studied and their depths in the core are given in Table 1.

In order to extract the maximum amount of information from our suite of samples we adopted an interdisciplinary approach similar to that used on core samples 60009/10 (McKay *et al.*, 1977) and on the Luna 24 core samples (McKay *et al.*, 1978). Samples were first split into two fractions and the larger fraction was then wet sieved into 8 grain size fractions (1000–500 μm; 500–250 μm; 250–150 μm; 90–150 μm; 75–90 μm; 45–75 μm; 20–45 μm; and <20 μm). In addition, R. Fuhrman, in our laboratory, analyzed the <20 μm grain size distribution with a Coulter Counter. After sieving was completed, splits were distributed to 5 groups for a variety of investigations. Blanchard and Budahn (1978) present major and trace element chemistry; Morris (1978) presents ferromagnetic (FMR) and vibrating sample magnetometer (VSM) data; and Bogard and Hirsch (1978) present rare gas data. Additionally, Clanton *et al.* (1978) present scanning electron microscope (SEM) studies of grain surfaces, and Heiken and McKay (1978) discuss the petrography, textures, olivine morphology, and olivine compositions. G. Blanford is currently investigating solar and cosmic ray tracks in the samples.

In this paper we present the grain size data of the entire suite of samples and the petrography of the upper 5 cm which has been reworked by micrometeorites.

Grain Size Parameters

Grain size data from the sieve analyses are presented in Table 1 and summary statistics are presented in Table 2. Variation with depth of mean grain size and standard deviation are shown in Fig. 1. Several important features should be noted from Table 2 and Fig. 1. First, the 74001/2 samples as a group are the finest grained of any soils returned from the moon. The *mean* mean grain size for the 14 analyzed samples is 4.63ϕ (40 μm). This can be compared to a *mean* mean grain size of 3.66ϕ (79 μm) for 38 surface and trench Apollo 17 soils (McKay *et al.*, 1974). It can also be compared to the 11 analyzed soils from the 60009/10 core which have a *mean* mean grain size of 3.23ϕ (107 μm) (McKay *et al.*, 1976; McKay *et al.*, 1977). None of the approximately 100 normal soils analyzed by our group is finer grained than 40 μm, the mean of the 74001/2 samples. Similarly the standard deviation or sorting parameter σ_I is the lowest (best sorted) for any group of lunar soils. The mean for the 14 analyzed soils is 1.75ϕ. This can be compared to the mean σ_I of 2.49 for 38 normal surface and trench soils from Apollo 17. For the 11 analyzed samples from 60009/10 the mean sorting parameter σ_I is 2.83ϕ, also considerably greater than that for 74001/2. All of these values are for the subcentimeter grain size data.

Not only are thc samples in 74001/2 finer grained and better sorted than any other set of lunar soils, they are also the most homogeneous in terms of grain size parameters. The variation (one sigma) from the mean grain size for 74001/2 is only 3.9% compared to 13.1% for the 38 Apollo 17 soils and 12.1% for the 11 analyzed 60009/10 samples. Similarly the variation (one sigma) of the sorting parameter σ_I is only 6.9% for 74001/2 but is 16.9% for the 38 normal Apollo 17 soils and is 12.7% for the 11 soils from 60009/10.

In spite of the pronounced homogeneity of the grain size characteristics of these samples, real variations exist and are most apparent in Fig. 1. A general trend of decreasing mean grain size goes from the bottom of the core to the top.

Table 1. Sieve Weights from 74001 and 74002 (all weights in milligrams).

	74002							
Sample	175	176	177	178	179	180	181	182
Parent	2	89	86	80	74	60	45	30
Depth (cm)	0–0.1	0.5–1	1.5–2	3–3.5	5–5.5	11.5–12	18–18.5	25.5–26
500μm–1mm	1.64	4.15	2.29	0.94	0.90	1.74	0.29	2.75
250–500μm	2.56	16.43	20.49	18.47	18.45	18.01	18.65	17.95
150–250μm	5.00	28.80	33.38	37.03	36.87	37.10	21.48	22.58
90–150μm	9.28	47.84	52.09	56.34	55.98	50.97	38.72	48.80
75–90μm	4.79	21.99	20.34	21.82	19.01	22.73	24.00	23.53
45–75μm	14.25	65.24	62.57	64.30	65.38	71.36	90.98	84.08
20–45μm	19.65	91.35	89.97	88.22	100.69	100.16	108.30	104.79
<20μm	28.47	139.33	134.25	124.49	133.07	120.63	123.56	120.18
Total	85.64	415.13	415.38	411.61	430.35	422.70	425.98	424.66

	74001					
Sample	98	107	113	119	125	2
Parent	21	36	50	63	78	2
Depth (cm)	32–32.5	37–37.5	44–44.5	50.5–51	57.5–58	66.8–67.7
500μm–1mm	4.35	5.04	7.26	7.83	8.09	1.69
250–500μm	24.59	22.30	27.19	24.11	28.46	18.13
150–250μm	26.55	31.08	34.30	34.47	30.78	18.16
90–150μm	53.27	57.79	57.90	58.51	53.09	27.63
75–90μm	25.58	30.65	28.41	24.28	21.51	11.66
45–75μm	98.00	90.85	79.08	79.30	68.23	36.87
20–45μm	105.53	121.36	81.52	83.85	95.84	58.09
<20μm	113.89	105.29	85.66	106.17	100.50	65.19
Total	451.76	464.36	401.32	418.52	406.50	237.42

Table 2. Statistical grain size parameters based on subcentimeter data.

				Mean grain size (M_z)		Standard Deviation (σ_I)
Core	Sample	Parent	Depth (cm)	ϕ	μm	ϕ
74002	175	2	0–0.1	4.87	34	1.83
74002	176	89	0.5–1	4.86	34	1.91
74002	177	86	1.5–2	4.80	36	1.91
74002	178	80	3–3.5	4.74	38	1.87
74002	179	74	5–5.5	4.73	38	1.79
74002	180	60	11.5–12	4.65	40	1.75
74002	181	45	18–18.5	4.79	36	1.61
74002	182	30	25.5–26	4.71	38	1.66
74001	98	21	32–32.5	4.54	43	1.66
74001	107	36	37–37.5	4.48	45	1.54
74001	113	50	44–44.5	4.27	52	1.73
74001	119	63	50.5–51	4.45	46	1.88
74001	125	78	57.5–58	4.48	45	1.75
74001	2	2	66.8–67.7	4.47	45	1.63

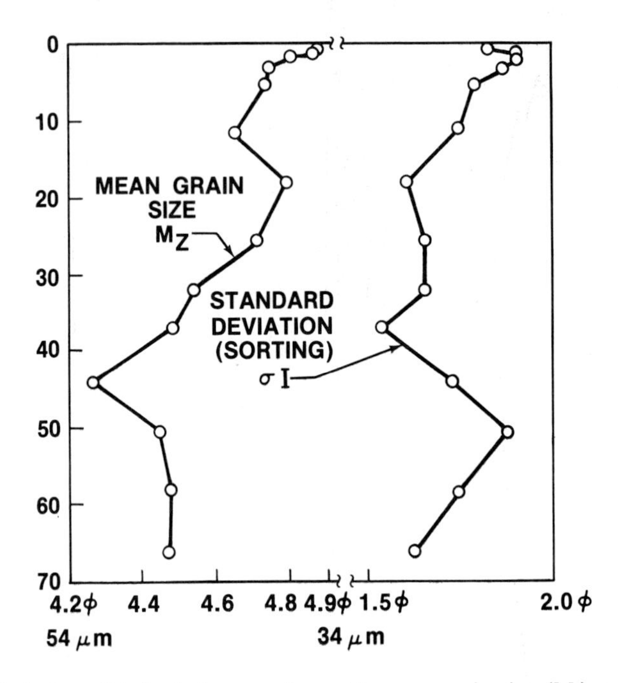

Fig. 1. Variation with depth in core of graphic mean grain size (M_z) and inclusive standard deviation (σ_I).

Below 5 cm, this trend may be related to the higher percentage of compound droplets in the lower part of the core compared to the upper part (Heiken and McKay, 1978). The formation of compound droplets would tend to create coarser material, as smaller droplets stuck to larger droplets during fire fountaining. However, at intermediate depths in the core, the mean grain size shows no correlation to the percentage of compound droplets, so the relationship near the top and bottom may be fortuitous. The mean grain size is apparently not related to the degree of breakage (Heiken and McKay, 1978); if anything, the coarser samples tend to have a slightly higher ratio of broken to unbroken droplets.

For other groups of normal lunar soils, the mean grain size Mz correlates inversely with the inclusive standard deviation σ_I (e.g., McKay *et al.*, 1978). However, for the 74001/2 samples no correlation exists between these parameters (Fig. 2). This is the first set of lunar soils for which such a correlation does not exist—a fact which further emphasizes the unusual nature of these samples.

Samples in the upper 5 cm reworked zone are somewhat finer grained and better sorted than the underlying samples. We originally thought that comminution effects of reworking had dominated the effects of the addition of "exotic" coarser grained material during *in situ* reworking (McKay *et al.*, 1978). Nagle (1977) states that glass particles are much more fragmented in the upper part of the core. However, a petrographic survey of our own polished grain mounts

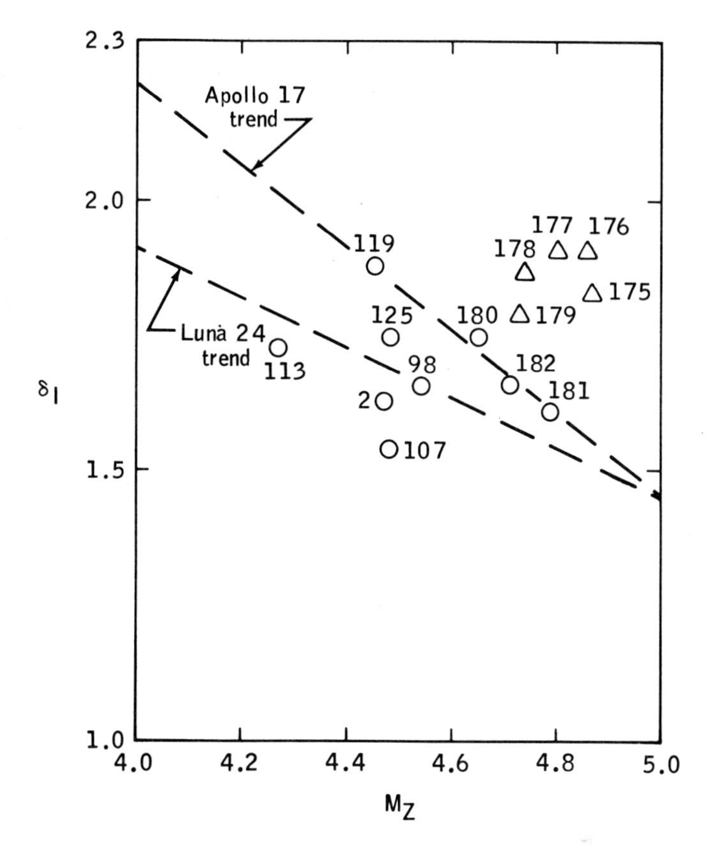

Fig. 2. Plot of graphic mean grain size (M_z) against inclusive standard deviation (σ_I). No clear correlation exists between these parameters. For brevity only the daughter numbers are shown; the full sample numbers and depths are given in Table 1. The triangular symbols indicate samples from the upper 5.5 cm reworked zone (see text). For comparison, the correlation least squares line for other Apollo 17 samples and for Luna 24 samples are shown.

(Table 3) shows little significant difference between the sample at 1 cm (74002,176) known to be most affected by meteorite reworking (Morris, 1978; Bogard and Hirsch, 1978 and petrographic data in this paper), the sample at 5 cm (74002,179), which is least affected by meteorite reworking, and a sample at 11 cm (74002,180), which has apparently not been affected at all by reworking. Our data on droplet breakage are presented in Table 3. The ratio of broken to unbroken (+ chipped) droplets for these 3 samples is essentially the same as that found throughout most of the core and is actually less than that found near the bottom of the core (Heiken and McKay, 1978). We are forced to conclude that most of the droplet breakage is related to an earlier history, probably the pyroclastic eruption, and that little is due to meteorite comminution in the upper 5 cm. The reason for the somewhat finer grain size in the upper 5 cm remains an

Table 3. Degree of droplet breakage (vol. %).

Sample	Grain Size	Whole Droplets	Chipped Droplets (>90% whole)	Broken Droplets	Number Counted
74002,176	250–500μm	40.4	18.1	41.5	98
"	150–250μm	23.0	9.0	68.0	300
"	90–150μm	32.0	10.0	58.0	300
"	75– 90μm	26.7	5.3	68.0	300
"	45– 75μm	11.7	3.7	84.7	300
"	20– 45μm	13.0	4.7	82.7	300
"	Weighted mean	19.2	6.5	74.3	
74002,179	250–500μm	44.4	11.8	43.8	144
"	150–25–μm	22.3	8.7	69.0	300
"	90–150μm	21.7	6.7	71.7	300
"	75– 90μm	14.0	4.3	81.7	300
"	45– 75μm	15.3	5.7	79.0	300
"	20– 45μm	10.0	2.0	88.0	300
"	Weighted mean	16.0	12.7	71.4	
74002,180	250–500μm	37.7	22.7	39.6	154
"	150–250μm	30	11.7	58.3	300
"	90–150μm	26.3	8.7	65.0	300
"	75– 90μm	18.3	9.3	72.3	300
"	45– 75μm	16.7	5.0	78.3	300
"	20– 45μm	13.3	3.3	83.3	300
"	Weighted mean	20.2	7.3	72.5	

unresolved problem. Perhaps it was an initial characteristic of this region of the core which predated the *in situ* reworking. Alternatively it may result from effects of reworking other than comminution such as a preferential addition of very fine grained material to the reworked zone. Our grain size data (Table 1) show that the <20 μm fraction of the reworked zone contains on the order of 10% (relative) more material than that found in lower regions of the core. Korotev (1976) suggests that fine grained material may be more mobile at the Apollo 17 site compared to coarser material. Laul (1977) noted a pronounced enrichment in meteoritic component in the <20 μm fractions of more mature soils in the Apollo 17 deep drill core. The data of Blanchard and Budahn (1978) do show an average enrichment in Ni in the <20 μm size fraction compared to bulk composition for the upper 5 cm reworked zone. Perhaps these samples contain a fine-grained meteoritic addition. However, this addition is not apparent in the metal data of Morris *et al.* (1978). The question must await a more detailed analysis of chemical data on many elements.

Grain Size Histograms

Figure 3 shows typical grain size histograms for samples from 74001/2. The histograms are generally very similar, but subtle differences exist. The deepest sample shown (74002,182) displays the sharpest peak and is the most symmetrical and best sorted. This sample is from the region of the core in which the degree of crystallization of glass droplets is increasing rapidly with depth and which also contains a higher proportion of compound droplets relative to samples higher in the core (Heiken and McKay, 1978). For the sample at 11.5–12 cm (74002,180) the proportion of uncrystallized glass is significantly higher, by about a factor of 10 over the 25 cm sample. The 11.5–12 cm sample also has a lower proportion of compound droplets. These changes in petrographic texture may be reflected in the changes in the grain size histogram which is less peaked and more poorly sorted. The sample at 0.5–1 cm is in the *in situ* reworking zone (see following section). While finer grained, it is notably broader and less well sorted. Of particular interest is the coarse grained tail not found on the other histograms. As discussed in the next section, this coarse grained tail consists entirely of "exotic" particles which are not orange and black glass droplets. We propose that they have been mixed into the upper 5 cm of the core by *in situ* reworking. Nagle (1977) also included a considerable fraction of orange and gray clods in the material coarser than 1 mm. In our allocated samples most of this clod material was disaggregated by sieving and normal sample transfer. Therefore we have chosen not to include it in the 1 mm to 1 cm fractions of our grain size calculations but have limited this larger size category to the basalts and agglutinates described by Nagle (1977). Consequently, all of this coarser material is "exotic" to the orange and black glass.

In-Situ Reworking of the Upper 5 Cm and the Presence of "Exotic" Particles

Several lines of evidence indicate that the upper ~5 cm has been subjected to meteorite reworking and gardening. The FMR profile of Morris *et al.* (1978) clearly shows a break from a background value of 0.2 units for most of the core to systematically higher values beginning at about 4.5 cm and extending upward to

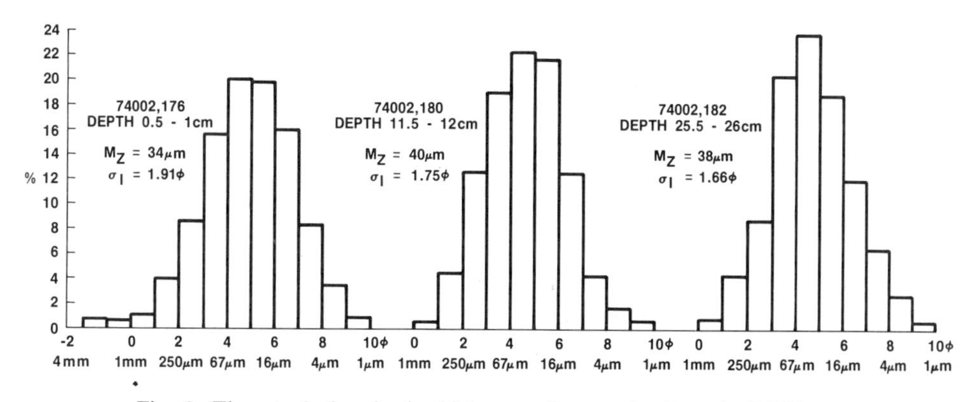

Fig. 3. Three typical grain size histograms for samples from the 74002 core.

the surface. Bogard and Hirsch (1978) show an increase in rare gas which most obviously begins at the 3.0–3.5 cm sample (74002,178) and increases toward the surfaces. The beginning of increase may be detectable in the 5.0–5.5 cm sample (74002,179), particularly for ^{22}Ne. Crozaz observes solar flare tracks in the upper ~5 cm but not below this depth (oral presentation, Ninth Lunar and Planetary Science Conference). Blanchard and Budahn (1978) see the first significant increase in Ni in the 5.0–5.5 cm sample.

Our own petrographic data (Table 4) for the upper 5 samples extending from the surface to 5.5 cm show the presence of "exotic" particles. We call these particles "exotic" because they are neither the orange nor the black glass which makes up nearly all of the core and are thus exotic to the core. However, as discussed in the next section, these "exotic" particles may come from very local sources. These particles have clearly been added to this zone by meteorite reworking. The particles consist of basalt fragments and their mineral constituents, agglutinates, vitric breccias, highland breccias, ropy glasses, and miscellaneous glass particles. The total percentage of these "exotic" particles is shown in Fig. 4 for two grain sizes. The "exotic" particles increase from nearly zero at 5–5.5 cm to a maximum at the surface for the 250–500 μm fraction and a maximum at 0.5–1 cm for the 90–150 μm fraction. The 90–150 μm fraction data are based on a larger number of grains (about 700 grains per sample compared to less than 200 grains per sample for the 250–500 μm) and is therefore statistically more reliable. It is of interest to note that the 0.5–1.0 cm sample which shows the highest percentage of exotic grains is also highest in rare gases (Bogard and Hirsch, 1978) and has the highest value of Is/FeO (Morris *et al.*, 1978). This sample appears to have been subjected to slightly more reworking than the sample at the very surface, perhaps because of a recent small cratering event which overturned the upper ~0.5 cm. At the lower end of the reworking zone, our petrographic data detected only a trace of "exotic" material (maximum of 0.6% in 20–45 μm fraction). We thus consider the 5.0–5.5 cm sample to be the very bottom of the reworked zone. Below this sample no "exotic" grains were detected in the approximately 18,000 grains described, all of which were orange and black (crystallized) glass. It is important to emphasize that not a single agglutinate was found below the 5 cm reworked zone or even in the 5.0–5.5 cm sample at the base of the reworked zone in any of the petrographically identified particles larger than 20 μm. Clearly, none of the samples which we studied below 5.5 cm had any petrographically detectable surface exposure at any time in their past history.

The presence of "exotic" particles by itself does not prove that *in situ* reworking has occurred (see McKay *et al.*, 1977 and Morris, 1978 for a definition and discussion of *in situ* reworking). The exotic material may represent fallback of ejecta from Shorty Crater or ejecta from a single large nearby impact which was mixed with the uppermost few centimeters, or even local slumping. However, the rather systematic upward increase of the exotic component (Fig. 4) is not likely to occur, for example, from simple fall-back of Shorty Crater ejecta.

Table 4. Particle Types and Relative Amounts in the Upper 5.5 cm of 74002. All values in percent.

Sample	74002,175			
Parent	74002,2			
Depth From Lunar Surface	0–0.1 cm			
Grain Size	500–1000 μm	250–500 μm	150–250 μm	90–150 μm
Orange and Black Glass	–	72	75.8	86.7
Basalt	100	13	10.0	2.4
Agglutinates–mainly Basaltic	–	2	4.6	3.7
Agglutinates–mainly Orange & Black Glass	–	11	3.4	1.4
Vitric Breccia	–	–	1.7	0.6
Orange and Black Glass Clods	–	–	1.0	1.4
Highland Breccia	–	2	1.0	0.7
Pyroxene-Olivine	–	–	1.2	1.6
Plagioclase	–	–	0.5	0.4
Ilmenite	–	–	–	–
Ropy Glass	–	–	0.5	0.6
Misc. Glass	–	–	0.2	–
Total Percent	100	100	100	100
Total Grains Counted	2	46	410	700

Sample	74002,176						
Parent	74002,89						
Depth From Lunar Surface	0.5–1.0 cm						
Grain Size	500–1000 μm	250–500 μm	150–250 μm	90–150 μm	75–90–150 μm	45–75 μm	20–45 μm
Orange and Black Glass	–	71	75.4	80.9	84.9	84.8	88.6
Basalt	57	13	7.8	5.9	4.5	6.2	3.5
Agglutinates–mainly Basaltic	–	2	4.1	4.6	3.7	1.4	0.8
Agglutinates–mainly Orange & Black Glass	–	4	2.5	1.6	2.1	0.4	0.4
Vitric Breccia	28	3	3.4	2.4	0.7	0.6	0.1
Orange and Black Glass Clods	–	4	2.1	1.1	–	0.6	0.1
Highland Breccia	–	1	1.4	0.6	0.1	0.1	0.1
Pyroxene-Olivine	–	–	2.1	0.9	2.5	4.3	4.2
Plagioclase	–	–	0.5	0.6	0.6	0.7	0.3
Ilmenite	–	–	0.2	–	0.1	–	0.4
Ropy Glass	14	2	0.2	0.9	0.4	0.4	0.6
Misc. Glass	–	1	0.2	0.1	0.3	0.4	0.7
Total Percent	100	100	100	100	100	100	100
Total Grains Counted	7	160	435	700	709	723	708

Table 4. *(cont'd.)*

Sample Parent	74002,177 74002,86			
Depth From Lunar Surface	1.5–2.0 cm			
Grain Size	500–1000 μm	250–500 μm	150–250 μm	90–150 μm
Orange and Black Glass	—	69	86.0	90.0
Basalt	43	11	4.4	3.6
Agglutinates—mainly Basaltic	—	5	3.0	1.8
Agglutinates—mainly Orange & Black Glass	29	4	2.8	1.4
Vitric Breccia	14	5	0.9	1.2
Orange and Black Glass Clods	—	4	0.7	1.1
Highland Breccia	—	—	0.5	0.2
Pyroxene-Olivine	—	1	0.5	1.4
Plagioclase	—	—	0.2	0.1
Ilmenite	—	—	—	—
Ropy Glass	14	1	0.5	0.1
Misc. Glass	—	—	0.2	0.3
Total Percent	100	100	100	100
Total Grains Counted	7	185	430	659

Sample Parent	74002,178 74002,80			
Depth From Lunar Surface	3.0–3.5 cm			
Grain Size	500–1000 μm	250–500 μm	150–250 μm	90–150 μm
Orange and Black Glass	100	94	97.5	97.2
Basalt	—	2	0.9	0.9
Agglutinates—mainly Basaltic	—	—	0.2	0.2
Agglutinates—mainly Orange & Black Glass	—	1	0.2	0.5
Vitric Breccia	—	—	—	0.2
Orange and Black Glass Clods	—	1	0.2	0.3
Highland Breccia	—	—	—	—
Pyroxene-Olivine	—	—	0.5	0.6
Plagioclase	—	—	—	0.2
Ilmenite	—	—	0.2	—
Ropy Glass	—	2	0.2	—
Misc. Glass	—	—	—	—
Total Percent	100	100	100	100
Total Grains Counted	3	168	433	643

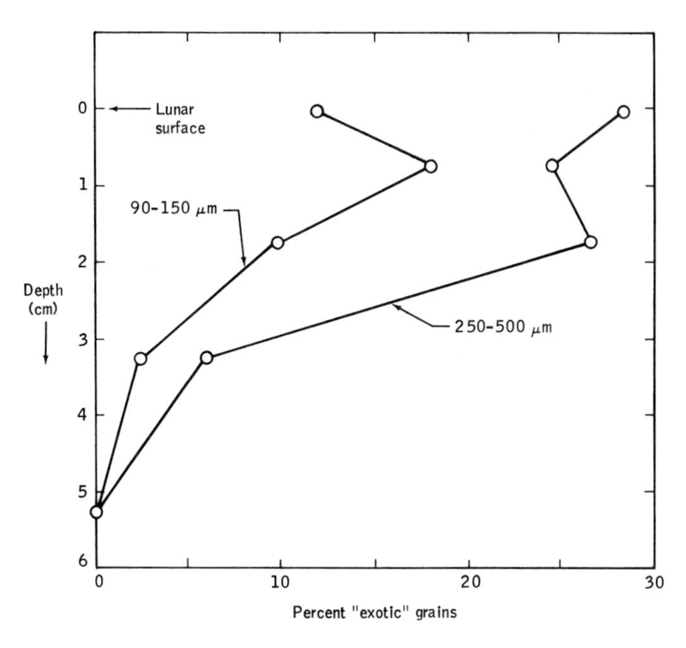

Fig. 4. Variation of the "exotic" component with depth for two grain size fractions. "Exotic" component includes all material which is not the orange and black glass droplets which make up the bulk of the core.

Table 4. *(cont'd.)*

Sample Parent	74002,179 74002,74						
Depth From Lunar Surface	5.0–5.5 cm						
Grain Size	500– 1000 μm	250– 500 μm	150– 250 μm	90– 150 μm	75– 90 μm	45– 75 μm	20– 45 μm
Orange and Black Glass	100	100	99.7	99.9	99.6	99.8	99.4
Basalt	–	–	–	–	–	0.06	0.2
Agglutinates–mainly Basaltic	–	–	–	–	–	–	–
Agglutinates–mainly Orange & Black Glass	–	–	–	–	–	–	–
Vitric Breccia	–	–	–	–	–	–	–
Orange and Black Glass Clods	–	–	–	–	–	–	–
Highland Breccia	–	–	–	0.1	0.1	0.06	0.2
Pyroxene-Olivine	–	–	0.3	–	0.2	0.06	–
Plagioclase	–	–	–	–	0.1		–
Ilmenite	–	–	–	–	–	–	0.2
Ropy Glass	–	–	–	–	–	–	–
Misc. Glass	–	–	–	–	–	–	–
Total Percent	100	100	100	100	100	100	100
Total Grains Counted	2	125	781	819	1000	1750	1000

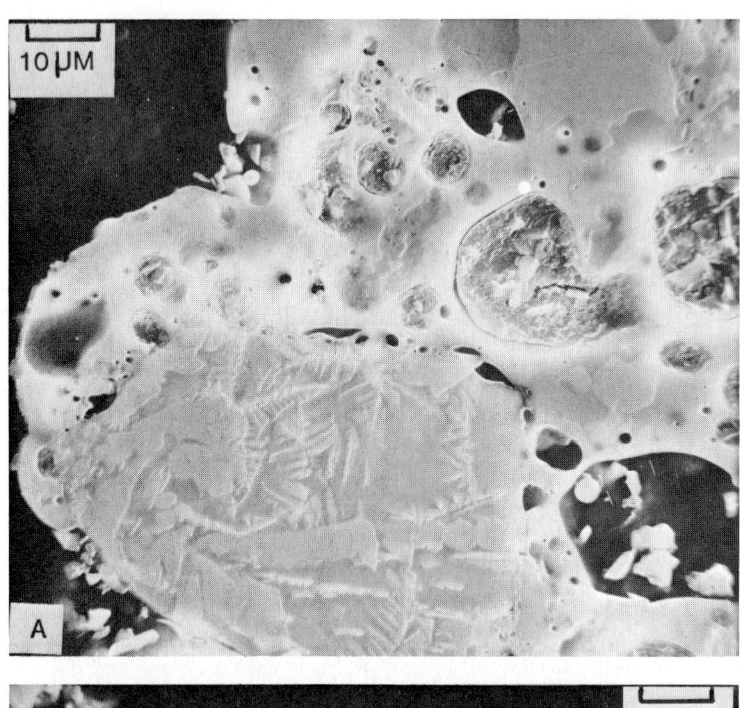

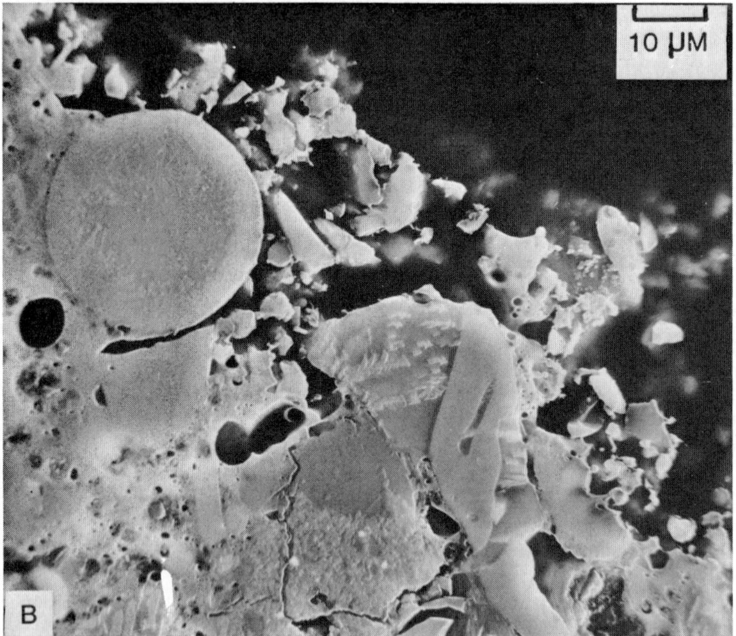

Fig. 5. Scanning electron microscope photographs of polished and ion etched agglutinates from 74002,176 (150–250 μm fraction). These agglutinates are made primarily from orange and black glass droplets which have been melted and welded together. Most of the identifiable clasts are either orange or black droplet fragments.

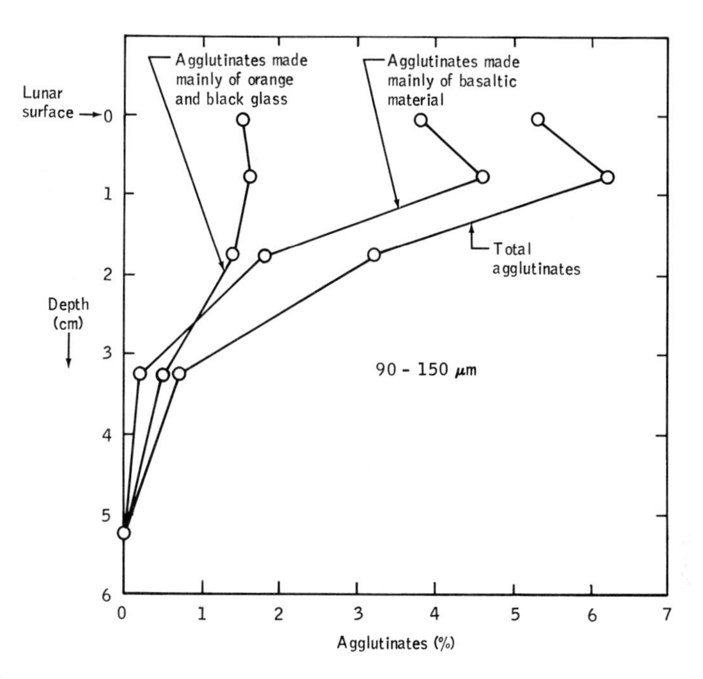

Fig. 6. Variation of agglutinate content with depth.

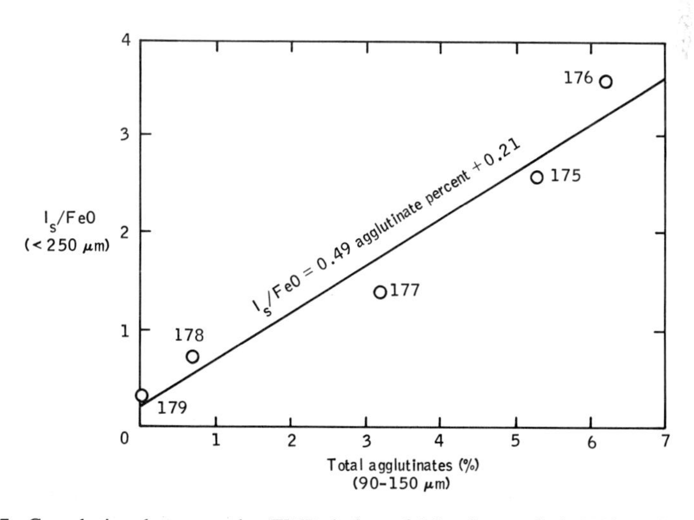

Fig. 7. Correlation between the FMR index of Morris *et al.* (1978) and our total agglutinate content in the 90–150 μm size fraction. Daughter numbers are shown; full sample numbers and depth are given in Table 1.

One of the strongest arguments for *in situ* reworking consists of the presence in the upper 5 cm of agglutinates which are formed mainly from orange and black glass. We have identified many of these agglutinates on the basis of their clast population by optical and SEM petrography. Figure 5 illustrates two agglutinates which come from the sample that has been most reworked (74002,176). These agglutinates were clearly formed by micrometeorite impact into a soil rich in orange and black glass. Such agglutinates could not be produced by Shorty Crater fallback, by simple mixing of a normal basaltic soil caused by a nearby impact, or by slumping of normal soil. They could only be produced by micrometeorite impact into orange and black glass.

More normal agglutinates are also present which contain clasts mainly of pyroxene and plagioclase, presumably derived from local basalt. The variation of both types of agglutinates with depth for the 90–150 μm size fraction is shown in Fig. 6. Both types reach maximum abundance in sample 74002,176 at 0.5–1 cm, the sample which is also richest in total "exotic" particles, I_s/FeO and rare gases. Apparently normal agglutinates and orange-black glass agglutinates have similar fine grained metal contents and rare gas contents. Figure 7 shows the correlation between total agglutinates and the FMR index I_s/FeO (Morris *et al.*, 1978). The correlation is good. Similarly total agglutinates correlate well with rare gases; a plot of total agglutinates against ^{36}Ar from Bogard and Hirsch (1978) is shown in Fig. 8. These close correlations are particularly significant considering the low abundance of agglutinates, only about 6% at a maximum. These correlations show the close correspondence between several independent maturity indices for these core samples and strengthen the arguments for reworking of the upper zone of the core.

Source of the "Exotic" Component

Figure 9 shows that the abundance of "exotic" component varies with grain size. The data in Table 4 allow us to calculate this variation of total "exotic" material as a function of grain size. For this calculation we use sample 74002,176 because it has the largest proportion of "exotic" particles and therefore is the most statistically reliable. Figure 9 plots the midpoints (in phi units) of each grain size fraction against the total % "exotic" grains. The graph shows that "exotic" material becomes much less abundant at finer grain sizes. The relationship seems to be approximately linear on the log phi plot and a least squares equation is shown, although this curve should not really be extrapolated beyond the limits of the data.

We can combine the "exotic" grain size variation in Fig. 9 with the overall grain size distribution for sample 74002,176 to arrive at a calculated grain size distribution for the exotic component. This calculated grain size distribution is shown in Fig. 10 over the size range from 4 mm to 16 μm. For comparison the nearby gray soil 74241 is also shown. The calculated mean grain size of "exotic" component (subcentimeter) is 83 μm or considerably coarser than the host soil 74002,176, which is 34 μm. The calculated "exotic" distribution also shows a slight resemblance to 74241 in the sense that they are both rather coarse grained and both show a coarse tail which is clearly bimodal for 74241 and marginally bimodal for the "exotic" material.

We can also compare the petrographic composition of the "exotic" material with that of 74241 (Table 5). The orange and black glass droplets are removed from both sets of data and they are normalized to 100%. Some clear resemblances are apparent. Both materials contain basalt as the major component. Basalt plus basalt derivatives including vitric breccias, pyroxene, plagioclase, and basaltic agglutinates make up about 70% of both materials. The biggest difference is in the abundance of agglutinates, which is higher in the "exotic" component and in the abundance of ropy glasses, which is higher in 74241. Note that both have only a small fraction of highland breccia fragments. In spite of the differences we conclude that the "exotic" material is similar enough to 74241 that a significant fraction of it could be derived from this soil or its closely related neighbor 74261. Figure 11 is a sketch map of the area around Shorty Crater from which core 74001/2 was taken. The map was adapted from ALGIT (1973) and AFGIT (1975). Also shown is the trench from which 74220 was sampled. At either end of the trench are the gray soils, 74240 away from the crater and 74260 on the inner rim. The contacts between the gray and orange soils are inferred to be steeply dipping and roughly parallel to the rim crest. A large basalt boulder is located within about 2 meters from the core site and basalt fragments associated with the boulder are within a meter of the core site. Gray soils 74240 and 74260

Table 5. Composition of "Exotic" Component Compared with Nearby Sample 74240.

	74002,176 90–150μm	74240,6* 90–150μm
Basalt	30.6	30
Vitric Breccia	12.7	14.9
Pyroxene	4.5	11.3
Plagioclase	3.0	4.6
Olivine, Ilmenite	–	1.3
Basaltic Agglutinates	20.9	8
Subtotal Basaltic	71.7	70.1
Orange Clods	6.0	–
Orange Glass Agglutinates	5.2	–
Mixed Agglutinates	6.0	–
Highland Breccia	3.0	2.6
Ropy Glass	4.5	14.3
Misc. Glass	0.7	8.2
(Percent of Total Sample)	(19)	(96)

*Heiken and McKay (1974).

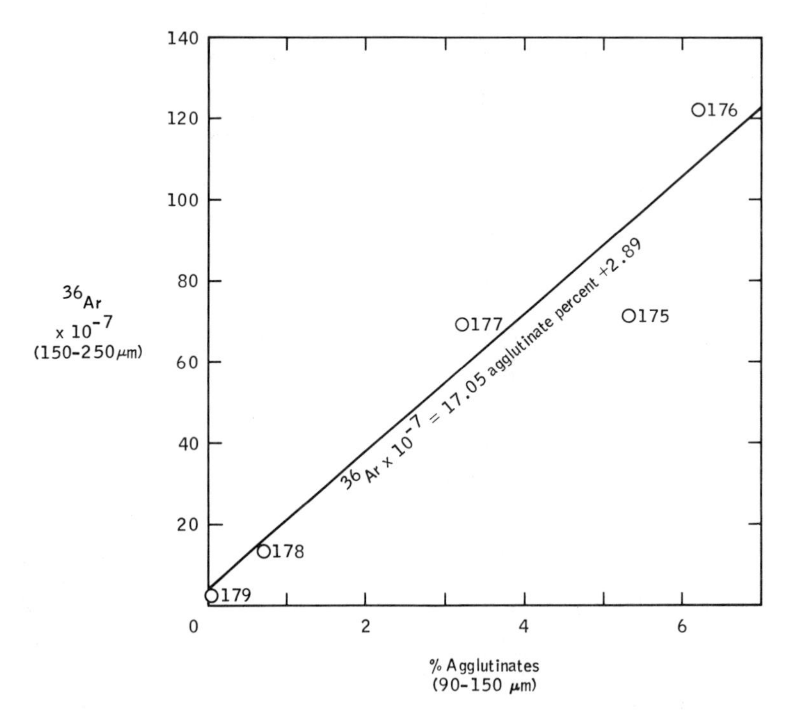

Fig. 8. Variation of ^{36}Ar in the 150–250 μm fraction with agglutinate content in the 90–150 μm fraction. Daughter numbers are shown; full sample numbers and depths are given in Table 1.

are as close as about 40 cm if the inferred contacts are accurate. Consequently, one does not have to go far to find the material in the exotic component; it is virtually all located within about a meter of the core site. A mixture of gray soil and fresh basalt could easily be reworked into the regolith surface by *in situ* reworking. The lateral movements necessary are minimal and are well within the concepts of *in situ* reworking. The 1 cm gray layer which appears to cover the area in the region of the core site and the trench (ALGIT, 1973) appears to be a natural consequence of *in situ* reworking in which the underlying material is being slowly matured and gardened, with some lateral movement and homogenization.

Rate of Lateral Movement

This site can provide us some information on the rate of lateral transportation on the lunar surface. Bogard and Hirsch (1978) estimate that the core 74001/2 has been in its present position for about 10 m.y. and during that time we conclude that *in situ* reworking has been going on and has gardened to a depth of about 5 cm. We can calculate the total amount of "exotic" material that has been added to the area of the core in 10 m.y. from the data of Table 4. For example,

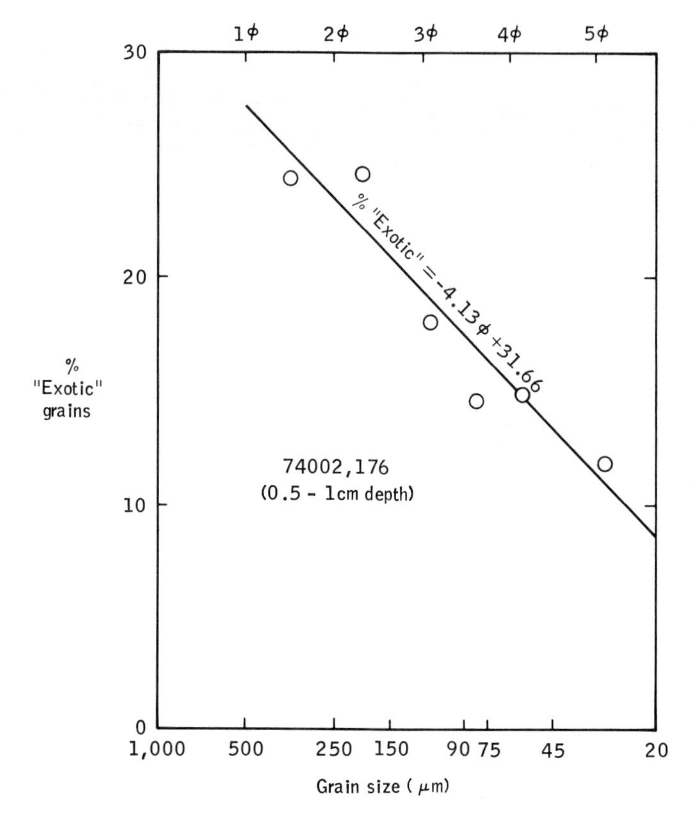

Fig. 9. Variation of percent "exotic" material with grain size for the sample containing the most "exotic" component.

making the assumption that volume percent in each size fraction is approximately equal to weight percent we calculate that for sample 74002,176, 15% of this sample is "exotic" material. Using the variation of the 90–150 μm as a guide to the relative variation of "exotic" with depth we calculate that the reworked zone from 0 to 5.5 cm contains an average of 3.33% exotic material. A column 5.5 cm deep and 1 cm^2 contains 11 gms of orange and black glass using the average density of 74002 of 2.0. This column then contains 0.36 gms of gardened in exotic material. As shown in Fig. 11, sources for this material can be found within a meter, some as close as 40 cm. Using a one meter average distance, we conclude that 1 cm^2 of lunar surface will be covered by 0.36 gm from a radius of 1 meter in 10 m.y. The absorbed material is gardened to a depth of about 5 cm during this same time interval.

Shorty Crater penetrated the light mantle and is surrounded by it at an average distance of about 100 m. As estimated from photos in AFGIT (1973) the light mantle approaches as close as 70 m. Yet the "exotic" component contains an extremely small proportion of highland breccias characteristic of the light

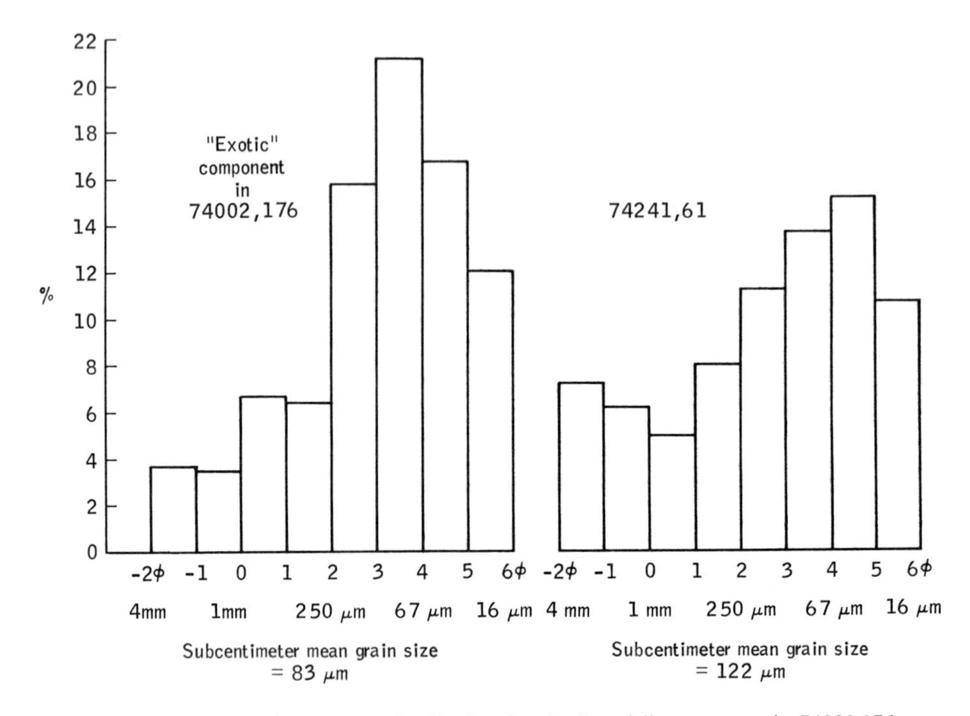

Fig. 10. Calculated grain size distribution for the "exotic" component in 74002,176. This distribution was calculated by combining the data in Fig. 11 with the original weight data in Table 1. For the fractions coarser than 1 mm, data from Nagle (1978) were used.

mantle. The sample with the largest proportion of "exotic" material contains on the order of 0.1% highland breccia. A 1 cm^2 column 5.5 cm deep would contain only 0.01 gm of highland breccia. Thus, from a radius of about 100 m, only 10 mg of material per cm^2 can be expected to migrate in 10 m.y. This is an upper limit; virtually all of the highland component could come from very local sources such as 74240.

Perhaps these calculations can be scaled up to larger areas. Hörz (1978) suggests that lateral transport of highland material is overwhelmed by vertical excavation of highland material through thin basalt flows in mare basins. Our calculations could be used in testing this hypothesis. Clearly, our calculations have many assumptions and assume very simple scenarios. However, the 74001/2 core is unique and may be our only chance to really study under controlled conditions such processes as lateral transport, *in situ* reworking, the physical and chemical effects of micrometeorite reworking, and other surface processes. The unique advantage of 74001/2 is that we know many of the starting conditions such as the chemistry, the grain size, and the petrology. We know the time span that the material has been subjected to micrometeorite reworking. We know the distances to other material of significantly different petrology and chemistry. We

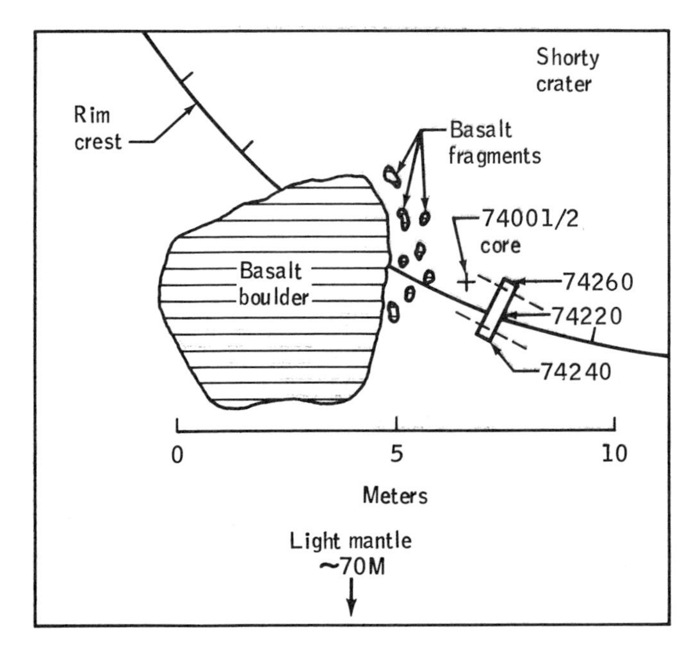

Fig. 11. Sketch map of the coring site for 74001/2 adapted from ALGIT (1973) and AFGIT (1975).

know that the starting material had essentially no surface exposure. These samples constitute in many ways the best controlled experiment in regolith dynamics that we are likely to see from the lunar samples. Detailed analysis of data from this lunar experiment should give us the key to many of the basic questions of regolith evolution.

Summary and Conclusions

1. The 74001/2 core consists primarily of orange and black glass droplets interpreted to be pyroclastic ejecta.
2. The core contains the finest grained and best sorted material in the lunar collection, having a mean grain size of 40 μm and a mean sorting coefficient of 1.75ø.
3. Above 5.5 cm the core material has undergone *in situ* reworking. The main effect of this reworking has been to add "exotic" particles, probably from local sources within a meter, and to create agglutinates from the orange and black glass droplets.
4. Making assumptions about the location of the source area for the "exotic" component it is possible to estimate that a maximum of 0.36 gm/cm^2/10 m.y. has been gardened into the regolith surface from a source radius of about 1 meter and a maximum of 0.01 gm/cm^2/10 m.y. has been gardened

into the regolith surface from a source area at a radius of about 100 meters.

Acknowledgments—We thank Richard Fuhrman who performed the grain size analysis on the <20 μm fraction using a Coulter Counter. Useful discussions with Richard Morris, Donald Bogard, Douglas Blanchard, and J. Stewart Nagle are also gratefully acknowledged.

REFERENCES

AFGIT (Apollo Field Geology Investigation Team) (1975) Documentation and environments of the Apollo 17 sampels. U.S.G.S. Interagency report: *Astrogeology 73*, Part I, p. 112–115.

ALGIT (Apollo Lunar Geology Investigation Team) (1973) Preliminary geology analysis of the Apollo 17 site. U.S.G.S. Interagency Report: *Astrogeology 72*, p. 18, 92–94.

Blanchard D. P. and Budahn J. R. (1978) Chemistry of orange/black soils from core 74001/2. *Proc. Lunar Planet. Sci. Conf. 9th.* This volume.

Bogard D. D. and Hirsch W. C. (1978) Depositional and irradiational history and noble gas contents of orange-black droplets in the 74002/1 core from Shorty Crater. *Proc. Lunar Planet. Sci. Conf. 9th.* This volume.

Clanton U. S., McKay D. S., Waits G. and Fuhrman R. (1978) Sublimate morphology on 74001 and 74002 orange and black glassy droplets. *Proc. Lunar Sci. Conf. 9th.* This volume.

Heiken G. H. and McKay D. S. (1974) Petrography of Apollo 17 soils. *Proc. Lunar Sci. Conf. 5th*, p. 843–860.

Heiken G. H., McKay D. S. and Brown R. W. (1974) Lunar deposits of possible pyroclastic origin. *Geochim. Cosmochim. Acta* **38**, 1703–1718.

Heiken G. and McKay D. (1978) Petrology of a sequence of pyroclastic rocks from the valley of Taurus-Littrow (Apollo 17 landing site). *Proc. Lunar Sci. Conf. 9th.* This volume.

Hörz F. (1978) How thick are lunar mare basalts? (abstract) In *Lunar and Planetary Science IX*, p. 540–542. Lunar and Planetary Institute, Houston.

Korotev R. L. (1976) Geochemistry of grain-size fractions of soils from the Taurus-Littrow valley floor. *Proc. Lunar Sci. Conf. 7th*, p. 695–726.

Laul J. C. (1978) Chemical study of size fractions of Apollo 17 deep drill cores 70009–70006 (abstract). In *Lunar and Planetary Science IX*, p. 637–639. Lunar and Planetary Institute, Houston.

McKay D. S. and Heiken G. H. (1973) Petrography and scanning electron microscope study of Apollo 17 orange and black glass (abstract). *EOS* (Trans, Amer. Geophys. Union) **54**, 599–600.

McKay D. S., Fruland R. M. and Heiken G. H. (1974) Grain size and the evolution of lunar soils. *Proc. Lunar Sci. Conf. 5th*, p. 887–906.

McKay D. S., Morris R. V., Dungan M. A., Fruland R. M. and Fuhrman R. (1976) Comparative studies of grain size separates of 60009. *Proc. Lunar Sci. Conf. 7th*, p. 295–313.

McKay D. S., Dungan M. A., Morris R. V. and Fruland R. M. (1977) *Proc. Lunar Sci. Conf. 8th*, p. 2929–2952.

McKay D. S., Basu A. and Waits G. (1978a) Grain size and the evolution of Luna 24 soils. In *Mare Crisium: The View from Luna 24* (R. B. Merrill and J. J. Papike, eds.), p. 125–136. Pergamon Press, N.Y.

McKay D. and Waits G. (1978b) Grain size distribution of samples from core 74001 and 74002 (abstract). In *Lunar and Planetary Science IX.* p. 723–725. Lunar and Planetary Institute, Houston.

Morris R. V. (1978) *In situ* reworking (gardening) of the lunar surface: Evidence from the Apollo cores. *Proc. Lunar Sci. Conf. 9th.* This volume.

Morris R. V., Gose W. A. and Lauer H. V. Jr. (1978) Depositional and surface exposure history of the Shorty Crater core 74001/2: FMR and magnetic studies. *Proc. Lunar Sci. Conf. 9th.* This volume.

Nagle J. S. (1978) Drive tubes 74002/74001: Dissection and description. *Lunar core catalog*, NASA Johnson Space Center, Houston. Supplement VIII. 5. i. 49 pp.

Proc. Lunar Planet. Sci. Conf. 9th (1978), p. 1933–1943.
Printed in the United States of America

Petrology of a sequence of pyroclastic rocks from the valley of Taurus-Littrow (Apollo 17 landing site)

GRANT HEIKEN

Los Alamos Scientific Laboratory, Los Alamos, New Mexico 87545

DAVID MCKAY

NASA Johnson Space Center, Houston, Texas 77058

Abstract—Samples collected from the rim of Shorty Crater, located in the valley of Taurus-Littrow (Apollo 17 landing site), include the orange glass droplets that have been the subject of a great deal of research and some controversy as to their origin. We have studied 13 samples from core 74001/2, collected at Shorty Crater; these samples represent most of the major stratigraphic units described by Nagle (1978). Below 5.5 cm, the samples consist entirely of whole and broken orange glass droplets and their partly to completely crystalline-black equivalents. We believe these droplets are pyroclastic ejecta from a lunar volcano. The crystalline droplets contain olivine and ilmenite as major phases. Minor phases include pyroxene, Cr-spinel, and metallic Fe. Four different properties of these droplets suggest that a wide range of cooling rates is represented. These properties are the grain shape, degree of crystallization, olivine shape or texture, and olivine composition. Many droplets contain vesicles indicating that a gas phase was involved in the eruption. Above a depth of 5.5 cm the core sequence has undergone *in situ* reworking by micrometeorites and contains "exotic" fragments including basalt and agglutinates. We conclude that the sequence below 5.5 cm, to a depth of 70 cm, represents 3.5 b.y. old volcanic pyroclastic ejecta which was deposited in a relatively short time period, was buried, and was subsequently brought to the lunar surface by the Shorty Crater impact where it was subjected to minor *in situ* reworking.

INTRODUCTION

One of the major goals of the Apollo 17 mission was to sample the low-albedo "dark mantle" deposits that rim Mare Serenitatis. These had been interpreted as young pyroclastic deposits draping mare lava flows (Scott *et al.*, 1972; El Baz, 1972). At the west end of the valley of Taurus Littrow, the crew of Apollo 17 sampled material around Shorty Crater, located on the "light mantle," a landslide deposit from the adjacent highland. Shorty Crater is a 120 m diameter, fresh crater of impact origin. On its rim the crew described several unconsolidated orange bands. A trench was dug across one of the bands and a 68 cm long core tube collected adjacent to the trench. The purpose of this paper is to describe the petrology of this core.

Before the core was opened and described there were many studies of samples scooped from the crater rim (orange clastic rock 74220) and of a small sample removed from the base of the core (74001) before it was extruded and dissected. These samples consist of fine-grained (median grain size of 40 μm) ultramafic droplets that are remarkably homogeneous and contain a trace of olivine phenocrysts (Fo_{80-81}). The glass droplets are orange and crystallized equivalents are black. The petrographic and chemical character of the samples is summarized by Heiken *et al.* (1974). Most investigators agree that the droplets are part of a pyroclastic deposit and were formed by lava fountaining along the rim of Mare Serenitatis. Other hypotheses presented for the genesis of these droplets are presented in Heiken *et al.* (1974).

Heiken and McKay (1974, 1977) proposed that the orange band sampled at Shorty Crater was part of a deposit originally situated below thin basalt flows and regolith developed on those flows. The Shorty Crater impact penetrated this sequence and deposited, intact, a cohesive block of the ultramafic pyroclastic deposit within the "rim flap" of the crater. This has given us the opportunity to study a short but continuous section of ancient tephra.

The core, consisting of two sections (lower-74001, upper-74002) was recently opened, described, and dissected by Nagle (1978). He has described a continuous section of ultramafic droplets, varying from mostly black in 74001 to mostly orange near the top of 74002. He has also subdivided the cores into nine major depositional units and numerous subunits. These units reflect changes in grain size, sorting and grain shape, not composition. Only the uppermost 5 cm has been

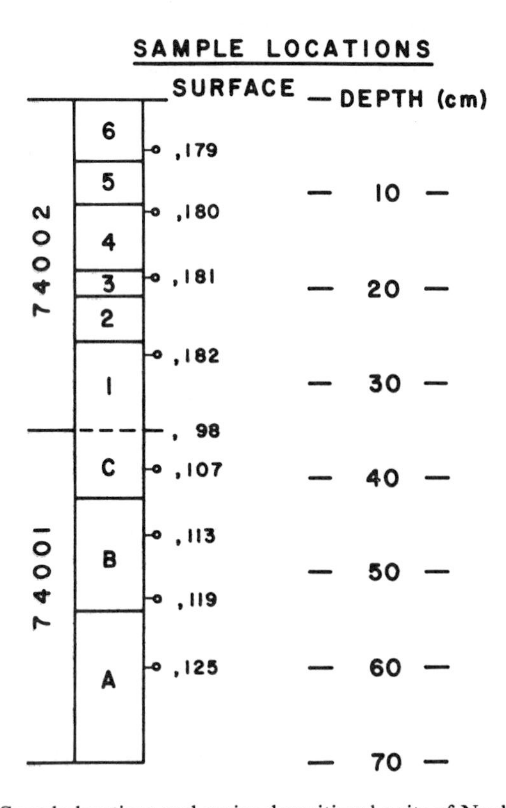

Fig. 1. Sample locations and major depositional units of Nagle (1978).

mixed into the regolith by meteorite reworking and "exotic" particles of basalt, agglutinates, etc., are present (McKay *et al.* 1978; Morris *et al.* 1978). The remainder of this paper is concerned with the section below 5.5 cm.

We studied samples from each major stratigraphic unit of Nagle (1978) (Fig. 1). Each was sieved (McKay *et al.* 1978) and the sieve fractions made into polished thin sections (>20 μm for 74001 and >45 μm for 74002).

For each droplet, we characterized the grain shape, degree of crystallinity, breakage, vesicularity, predominant olivine shapes, and phenocryst content. Approximately 18,000 droplets or droplet fragments were described. If a droplet was crystalline, the predominant olivine shape was identified, using a modified version of Donaldson's (1976) classification, used by Heiken and McKay (1977) to describe sample 74001.

Grain shapes

Droplets were characterized as spheres, ovoids, or compound forms. As has been described previously (McKay and Heiken, 1973; Heiken *et al.* 1974), the spheres and ovoids are smooth-skinned forms, covered only by a thin deposit of micromounds. Compound forms consist of a central droplet with smaller droplets on the surface that had impacted into the droplet surface while they were still partly molten. Nearly all of the coarser-grained particles (>150 μm) are composed of compound forms. Near the base of the core, there are 41.5% compound droplets (Table 1, Fig. 2a). The ratio of compound to simple droplets changes systematically from the bottom of the core to the surface, where surface sample 74220 consists of mostly simple spheres; 70 to 78% of the droplets in all samples are broken (Fig. 2b). For comparison, 60 to 78% of basaltic tephra from the 1959–1960 eruption of Kilauea Iki, Hawaii were broken during the eruption (Heiken *et al.* 1974).

Vesicles

A main objection often made to dispute a volcanic origin for these fine-grained droplets concerns the lack of vesicles. Vesicles have been described in these samples before, although they are rare (Heiken *et al.* 1974). Within this core every sample contains some droplets with vesicles that always occur in glassy or partly crystalline droplets; i.e., those droplets that are rapidly chilled after their formation. Vesicles consist of spherical bubbles close to droplet centers, although some have reached grain surfaces (Fig. 3). Up to 6.0 volume % of the droplets contain single vesicles (Fig. 2d). Vesicle abundance varies randomly throughout the core.

Phenocrysts

Olivine phenocrysts are present in all units in concentrations of 0.3 to 2.7% of all particles from each sample. Nearly all are subequant, subhedral grains, about

Table 1.

Sample No.	Unit No.	Depth (cm)	Droplet Shapes (Vol. %)			Degree of Crystallization (Vol. %)			Predominant Olivine Shapes in all Crystalline or Partly Crystalline Droplets (Vol. %)			
			Sphere	Ovoid	Compound	Glass Droplets	Partly Crystalline Droplets	Crystalline Droplets	Subequant, skeletal	Acicular, skeletal	Thin, tabular, parallel	Dendritic
74002,179	6	5.0	47.0	42.1	10.9	46.3	29.7	24.0	4.1	13.1	53.8	29.1
74002,180	4	11.5	61.1	28.6	10.3	49.3	22.5	28.2	11.8	12.0	43.2	33.0
74002,181	3	18.0	45.8	36.7	17.5	15.0	18.8	66.2	28.6	8.3	21.2	41.9
74002,182	1	25.5	41.7	38.3	20.0	5.9	11.6	82.5	45.8	5.1	13.4	35.8
74001,98	C-13	32.0	70.1	26.3	3.6	3.6	10.7	85.7	41.1	9.5	17.8	31.7
74001,107	C-12	37.0	74.4	23.2	2.3	4.2	11.3	84.5	31.9	10.6	22.4	35.1
74001,113	B-19	44.0	53.2	34.1	12.6	1.8	8.6	89.7	38.3	8.3	18.0	35.4
74001,119	B-6	50.5	40.2	36.2	23.7	1.2	8.4	90.5	44.3	7.2	17.1	31.4
74001,125	A-3	57.5	35.1	23.4	41.5	2.1	4.6	93.3	49.6	9.6	9.3	24.5

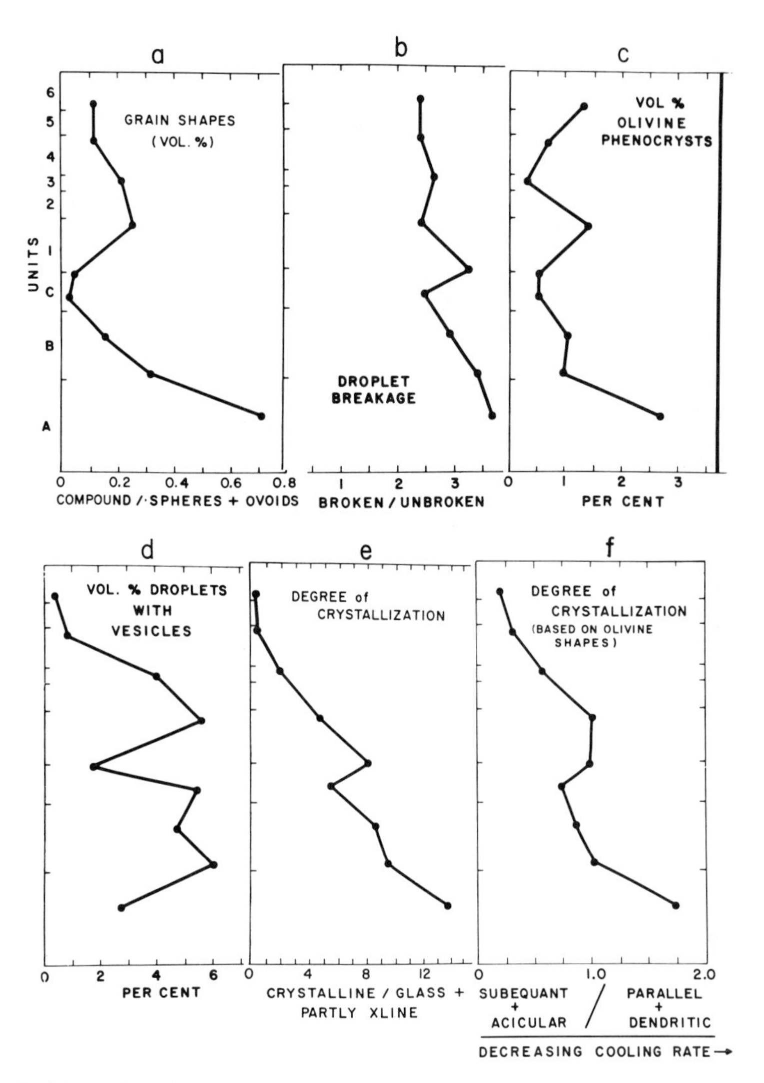

Fig. 2. (a) Ratios of grain shapes. (b) Ratio of broken to unbroken droplets. (c) Volume percent olivine phenocrysts. (d) Volume percent droplets with vesicles. (e) Ratio of crystalline to glass plus partly crystalline droplets. This ratio is roughly equivalent to the ratio of "black" to orange glass. (f) Degree of crystallization; ratio of subequant and acicular to parallel and dendritic olivine shapes.

400 μm in diameter. Most have compositions of Fo_{81}. The percentage of droplets with phenocrysts decreases significantly in going upward from unit A to unit B, then varies randomly throughout the core (Fig. 2c).

Crystallization textures as evidence for variation in cooling history of the droplets

Only part of the droplets are glassy; many exhibit crystalline quench textures.

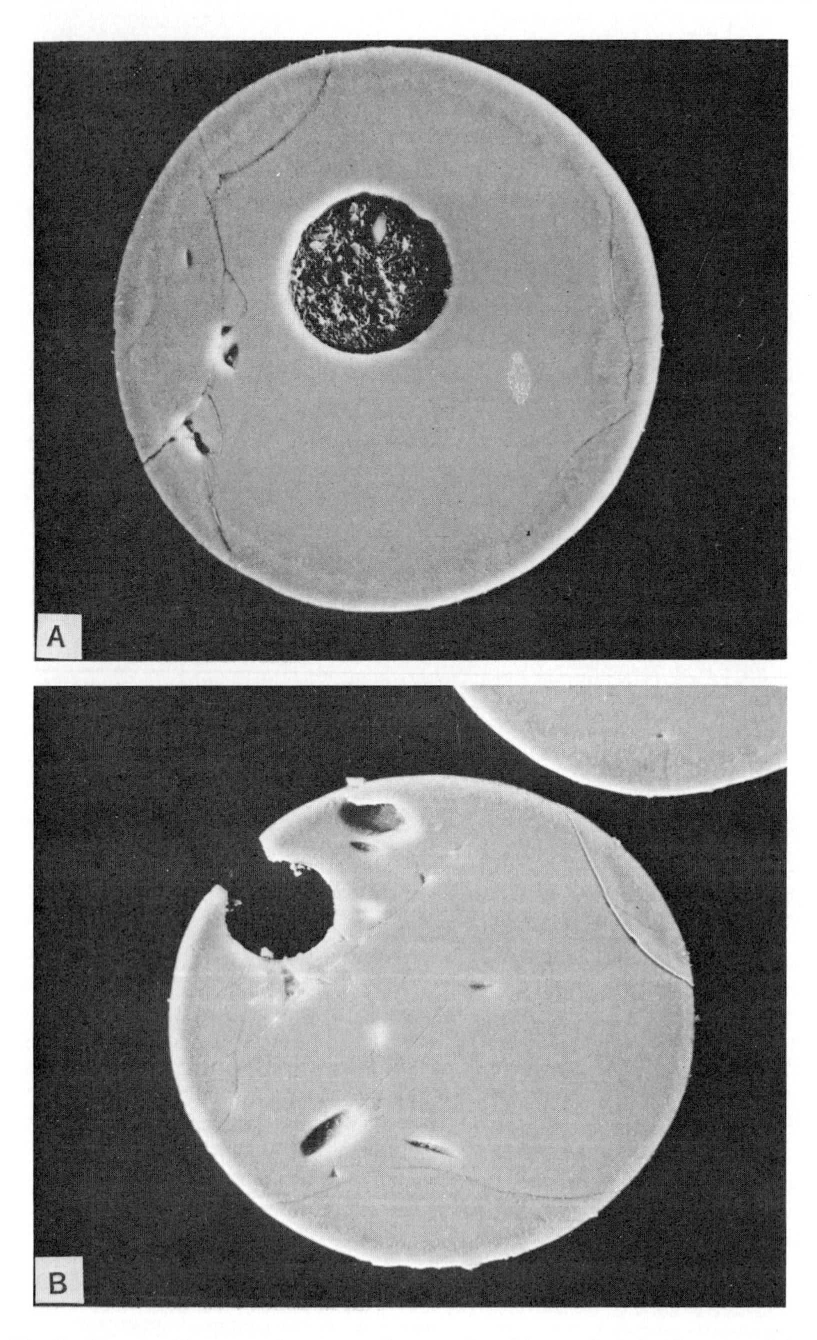

Fig. 3. Scanning electron microscope photographs of ion etched droplets containing vesicles. A. Vesicle near center of the droplet. Note the finely crystallized outer margin of the droplet. The remainder of the droplet is glassy. Diameter of droplet is 140 μm. B. Droplet with vesicle near outer rim. The vesicle has partly broken through to the droplet surface. Longest dimension of droplet is 130 μm.

Crystalline or partly crystalline forms contain olivine and ilmenite as major phases, minor pyroxene and chrome spinel, and traces of metallic iron.

Near the base of the core, 93% of the droplets are crystalline; near the top, only 24% are crystalline (Table 1, Fig. 2e). The variation in the degree of quench crystallization from base to top indicates that there was a systematic change in the cooling rate of erupted droplets. This trend is similar to that of grain shapes; where compound forms dominate most of the droplets are crystalline.

Donaldson (1976) has determined that changes in olivine shapes correspond to a change in cooling rate and the degree of supercooling. Olivine quench textures have been used to study the petrologic history of lunar lavas at several different lunar landing sites (Lofgren *et al.* 1974; Usselman *et al.* 1975). Heiken and McKay (1977) used a modified version of Donaldson's 1976 classification to describe samples 74001 (base of core) and 74220 (surface) and to infer cooling rates for the droplets and their eruptive history.

We have used the same approach to study the 74001/2 core samples. A progression from most rapidly to most slowly cooled droplets is indicated by the following sequence of forms: (1) branching, dendritic, (2) thin, tabular, parallel growth, (3) acicular skeletons, and (4) subequant skeletons. Photomicrographs of typical olivine shapes within these droplets are illustrated in the paper by Heiken and McKay (1977). The ratio of subequant plus acicular to parallel plus dendritic olivines (Table 1, Fig. 2f) is used here as another indicator of the cooling rate. Using this indicator, the variation from base to top of the core is parallel to the variation of cooling rate as indicated by the ratio of crystalline to glassy droplets (Fig. 2e).

Donaldson *et al.* (1975) noted that there is a decrease in the Mg content of olivines from the same melt, with increased cooling rates. This effect was also noted by Heiken and McKay (1977) for olivines within ultramafic droplets from the Apollo 17 landing site. Nearly all of the phenocrysts from the 74001/2 core had compositions of about Fo_{81}, whereas variation between olivine shapes ranged from Fo_{78} for subequant, skeletal forms to Fo_{62} for dendrites (Fig. 4).

Discussion

The orange unit sampled at Shorty Crater consists of ultramafic glass droplets and their crystalline or partly crystalline equivalents. Only the upper 5 cm of this deposit has been exposed to micrometeorite bombardment that generated regolith at the lunar surface (McKay *et al.* 1978; Morris, 1978), for a time period on the order of 10 m.y. (Bogard and Hirsch, 1978). An analysis of solar-wind derived gases and cosmic-ray produced ^{21}Ne and ^{38}Ar by Bogard and Hirsch (1978) indicates that the deposit was exposed at the lunar surface for a very short period of time, about 3.6 b.y. ago, then buried until the Shorty Crater event brought a section of the pyroclastic unit to the lunar surface. Their data support the interpretation that the section is upside-down, confirming an earlier opinion of Heiken and McKay (1974, 1977) that the orange band at Shorty Crater was part of an upside-down rim flap.

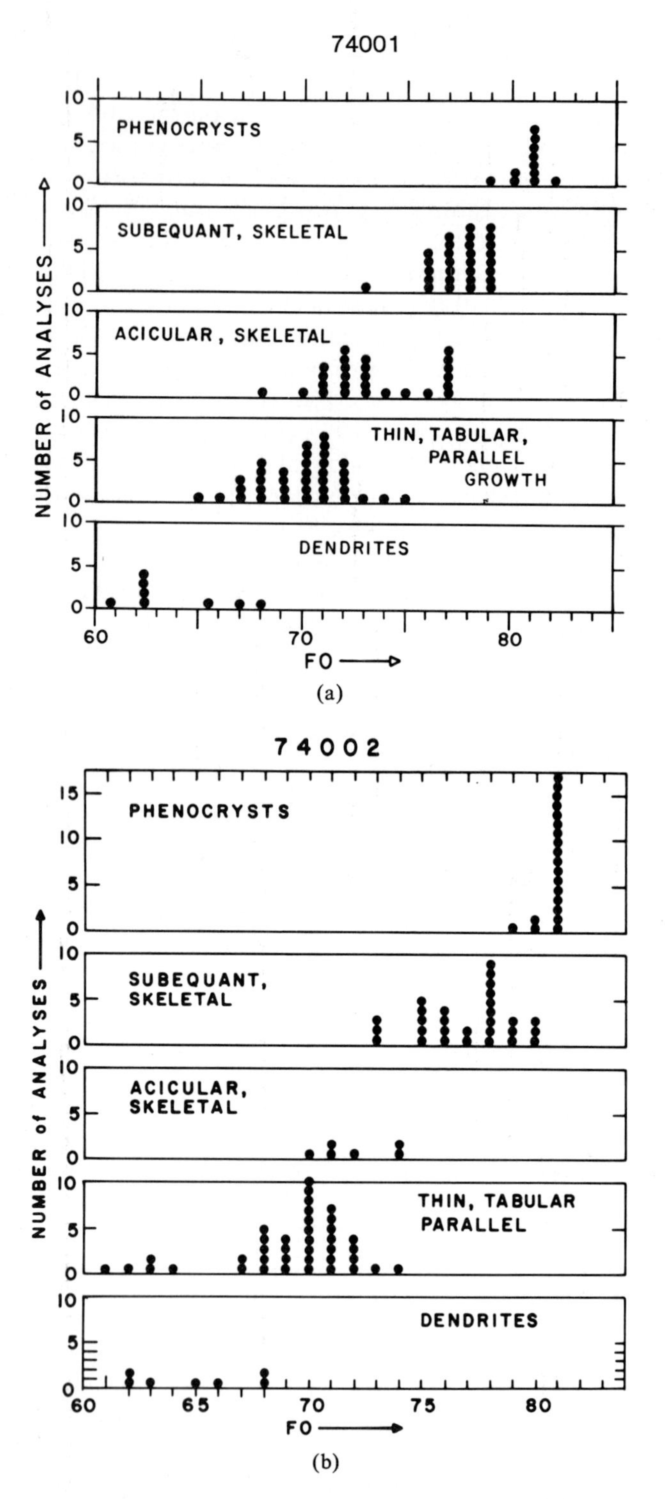

Fig. 4. Olivine compositions as a function of olivine shapes. a. 74001 b. 74002.

The section is chemically uniform throughout (Blanchard and Budahn, 1978). Olivine phenocryst compositions are also uniform throughout the section. The morphology and petrographic character of all droplets studied throughout the section remain the same as described in earlier studies of samples 74220 and 74001 (base of core) (McKay and Heiken, 1973; Heiken *et al.* 1974; Butler and Meyer, 1976). Studies of micromounds on grain surfaces (Butler, 1978; Clanton *et al.* 1978; Grant and Housley, 1978) indicate that they are similar to those seen earlier on sample 74220; they are uniform throughout and are best explained as sublimates deposited from gas phases driving a lava fountain.

The homogeneous nature of the droplets and the presence of phenocrysts and vesicles all support the hypothesis that these droplets were formed in a lava fountain at the lunar surface. However, the question of the composition of the gas driving the eruption remains unsolved.

Within the 74001/2 core "cooling indicators" such as degree of quench crystallization and the shape of olivine crystals indicate that there were systematic changes in the dispersion or strength of a lava fountain as proposed by Heiken and McKay (1977). If the section is upside-down, as proposed by Bogard and Hirsch (1978), then the section sampled is part of a deposit formed during a change from widely dispersed spray that produced simple, rapidly chilled droplets to a narrow jet where compound, crystalline droplets were "recycled" through the fountain several times.

Time span of deposition

The chemical homogeneity of the sequence in 74001/74002 (Blanchard and Budahn, 1978) suggests that this sequence came from the same source and was deposited in a relatively short time interval. With the exception of the upper 5.5 cm which has been reworked, the chemical homogeneity of the pyroclastic core sequence approaches the homogeneity of samples known to be from a single volcanic eruption (e.g., Laidley and McKay, 1971). Terrestrial ejecta known to be from the same volcano but separated by times on the order of years to hundreds of years usually have chemical variations which far exceed that of 74001/74002 (e.g., Aramaki, 1963; Tsuya, 1968). If these terrestrial analogies are valid, the chemical homogeneity of 74001/74002 argues against multiple eruptions separated by years, let alone millions of years. The close similarity of grain-size parameters among the samples of 74001/74002 also supports the idea that the samples originated from the same event (McKay *et al.*, 1978).

A deposit such as 74001/74002 could be produced in hours or days by a terrestrial volcano and we know of no reason why the 74001/74002 sequence could not have been produced in a similarly short time span. A total lack of micrometeorite-produced agglutinates in the sequence below 5.5 cm argues against any significant exposure of these regions of the core material at the lunar surface. Furthermore, Morris *et al.* (1978) see no indication of intermediate fossil surfaces, although the time resolution of the technique is limited to the order of 1 m.y. Crozaz (oral presentation, 9th Lunar and Planetary Science

Conference) did not see any horizon below 5.5 cm which contained solar flare tracks. Bogard and Hirsch (1978) detect trapped solar wind in all of the core samples which increases in abundance at finer grain sizes. However, they suggest that these gases were not directly implanted on grain surfaces by the solar wind and propose that these gases were acquired during the eruption of the pyroclastic ejecta 3.5 b.y. ago by contamination with regolithic derived solar wind gases and small regolith particles. Bogard and Hirsch conclude that the Shorty Crater event brought the 74001/74002 sequence to the surface about 10 m.y. ago and it has been subjected to *in situ* reworking since that time.

In summary, we believe that the 74001/74002 samples represent an ultramafic pyroclastic deposit brought to the lunar surface by the Shorty Crater impact. This deposit, which may be considerably thicker than the 68 cm represented in the core, is part of an eruptive sequence of a lava fountain that was active within the valley of Taurus-Littrow 3.5 b.y. ago. The sequence sampled by 74001/74002 was likely deposited in a relatively short time, perhaps hours to weeks, and likely does not represent a time span of thousands to millions of years. The sequence shows a range of cooling times which can be explained by different phases of a single lava fountain eruption or which may represent pulses of varying magnitude from a single vent over a somewhat longer time period.

Acknowledgments—David Mann made the excellent polished thin sections used in this study and Ron Gooley provided considerable help with analysis of the olivines. The paper was improved by comments from J. Smyth, J. Eichelberger, K. Cameron, H. O. A. Meyer, and an anonymous reviewer. This research was funded by a grant from NASA.

REFERENCES

Aramaki S. (1963) Geology of Asama Volcano. *Journal of the Faculty of Science*, Section II, **XIV**, 229–443. Univ. Tokyo, Japan.

Blanchard D. P. and Budahn J. R. (1978) Compositional variations among size fractions in a homogeneous, unmatured regolith: drive core 74001/2 (abstract). In *Lunar and Planetary Science IX*, p. 100–102. Lunar and Planetary Institute, Houston.

Bogard D. D. and Hirsch W. C. (1978) Noble gas contents and irradiation history of orange-black glass in the 74001-74002 core (abstract). In *Lunar and Planetary Science IX*, p. 111–113. Lunar and Planetary Institute, Houston.

Butler P. Jr. (1978) Recognition of lunar glass droplets produced directly from endogenous melts: The evidence from S-Zn coatings (abstract). In *Lunar and Planetary Science IX*, p. 143–145. Lunar and Planetary Institute, Houston.

Butler P. Jr. and Meyer C. Jr. (1976) Sulfur prevails in coatings on glass droplets: Apollo 15 green and brown glasses and Apollo 17 orange and black (devitrified) glasses. *Proc. Lunar Sci. Conf. 7th*, p. 1561–1581.

Clanton U., McKay D., Waits G. and Fuhrman R. (1978) Unusual surface features on volcanic droplets from 74001 and 74002 (abstract). In *Lunar and Planetary Science IX*, p. 172–174. Lunar and Planetary Institute, Houston.

Donaldson C. H. (1976) An experimental investigation on olivine morphology. *Contrib. Mineral. Petrol.* **57**, 187–213.

Donaldson C. H., Usselman T. M., Williams R. J. and Lofgren G. E. (1975) Experimental modeling of the cooling history of Apollo 12 olivine basalts. *Proc. Lunar Sci. Conf. 6th*, p. 843–869.

El Baz F. (1972) The cinder field of the Taurus Mountains. *The Apollo 15 Prelim. Sci. Rep.* NASA SP-289, p. 25–66 to p. 25–71.

Grant R. W. and Housley R. M. (1978) An XPS study of Apollo 17 orange and black glasses (abstract). In *Lunar and Planetary Science IX*, p. 405–407. Lunar and Planetary Institute, Houston.

Heiken G. H. and McKay D. S. (1974) Petrography of Apollo 17 soils. *Proc. Lunar Sci. Conf. 5th*, p. 843–860.

Heiken G. H. and McKay D. S. (1977) A model for eruption behavior of a volcanic vent in eastern Mare Serenitatis. *Proc. Lunar Sci. Conf. 8th*, p. 3243–3255.

Heiken G. H., McKay D. S. and Brown R. W. (1974) Lunar deposits of possible pyroclastic origin. *Geochim. Cosmochim. Acta*, **38**, p. 1703–1718.

Laidley R. A. and McKay D. S. (1971) Geochemical Examination of Obsidians from Newberry Caldera, Oregon. *Contrib. Mineral. Petrol.* **30**, 336–342.

Lofgren G. E., Donaldson C. H., Williams R. J., Mullins O. and Usselman T. M. (1974) Experimentally reproduced textures and mineral chemistry of Apollo 15 quartz-normative basalts. *Proc. Lunar Sci. Conf. 5th*, p. 549–568.

McKay D. S. and Heiken G. H. (1973) Petrography and scanning electron microscope study of Apollo 17 orange and black glass (abstract). *EOS Trans. Amer. Geophys. Union* **54**, 599–600.

McKay D. S., Heiken G. H. and Waits G. A. (1978) Core 74001/2: Grain size and petrology as a key to the rate of *in situ* reworking and lateral transport on the lunar surface. *Proc. Lunar Planet. Sci. Conf. 9th.* This volume.

Morris R. V., Gose W. A. and Louer H. V. Jr. (1978) Depositional and reworking history of core 74001/2 (abstract). In *Lunar and Planetary Science IX*, p. 763–765. Lunar and Planetary Institute, Houston.

Nagle J. S. (1978) Drive Tubes 74002/74001, Dissection and Description. Lunar Sample Curator's Doc. VII-5. NASA Johnson Space Center, Houston. 49 pp.

Scott D. H., Lucchitta B. K. and Carr M. H. (1972) Geologic maps of the Taurus-Littrow region of the moon. U.S. Geol. Survey Map I-800.

Tsuya H. (1968) Geology of Volcano Mt. Fuji. Abstract of the explanatory text of the geologic map (1:50,000). Geologic Survey of Japan.

Usselman T. M., Lofgren G. E., Donaldson C. H. and Williams R. J. (1975) Experimentally reproduced textures and mineral chemistries of high-titanium mare basalts. *Proc. Lunar Sci. Conf. 6th*, p. 997–1020.

APPENDIX

Analytical procedures

All olivines and pyroxenes were analyzed using a model Camebax Cameca electron microprobe, with matrix correction procedures of Bence-Albee. Analyses of dendritic olivines were made by analyzing the "stalks" at their thickest point. All petrographic data were collected using polished thin sections.

Proc. Lunar Planet. Sci. Conf. 9th (1978), p. 1945–1957.
Printed in the United States of America

Sublimate morphology on 74001 and 74002 orange and black glassy droplets

U. S. CLANTON[1], D. S. MCKAY[1], G. WAITS[2] and R. FUHRMAN[2]

[1]NASA Johnson Space Center, Houton, Texas 77058
[2]Lockheed Electronics Co., Inc., Houston, Texas 77058

Abstract—SEM-EDX analysis of the surface morphology of the 74001 and 74002 orange and black glassy droplets has revealed the presence of ubiquitous micromound coatings. Assuming that the diameter of an individual micromound is an indication of total coating thickness, then the thickness of this volatile and metal-rich layer ranges from less than 20Å to over 300Å in thickness.

On rare droplets the micromound coating is partially covered by a sparse population of crystals and masses that are essentially NaCl in composition. Additionally, an irregular mass on the surface of one black droplet may contain free sulfur. Amoeboid masses of iron, some with incipient crystal faces, are also present on a few droplet surfaces. The above features, i.e., halite crystals, sulfur-rich masses and amoeboid iron, are rare; the more common occurrence of metals and volatiles is in the micromound coating.

Based on our observations of several hundred spheres from both Apollo and Luna missions, the classic micromound coating is unique to the Apollo 15 green and Apollo 17 orange and black droplets. Although some investigators have proposed various impact sequences to produce these unique droplets, the volcanic fire fountain origin is favored.

INTRODUCTION

Glassy droplets have been observed in regolith samples from all of the Apollo and Luna missions. However, the Apollo 15 green and Apollo 17 orange and black glasses appear to be unique because of the high concentration of droplets, narrow grain size, and total lack of shock debris contamination. The petrographic and chemical character of these unusual samples have been summarized by Heiken *et al.* (1974). Four origins for the droplets have been proposed: (1) volcanic fire fountain; (2) meteorite impact on solid lunar material; (3) meteorite impact into liquid lava; (4) vapor condensation. Roedder and Weiblen (1973a, 1973b) discuss the limitations of each hypothesis and argue that meteorite impact into liquid lava best explains the unusual character of these samples. Most investigators (e.g., Carter *et al.*, 1973; Reid *et al.*, 1973; Heiken *et al.*, 1974; Meyer *et al.*, 1975), however, agree that the droplets are part of a pyroclastic deposit that was formed by lava fountaining. Indeed the large amounts of volatile/sublimate observed (e.g., Tatsumoto *et al.*, 1973; Chou *et al.*, 1974, 1975; Morgan *et al.*, 1974) on the Apollo 15 green and Apollo 17 orange and black glassy droplets cannot be easily explained except by a volcanic eruption with a high volatile to ejecta ratio. The presence of a volatile phase is demonstrated by relatively abundant vesicles in the droplets in 74001/2 (Heiken and McKay, 1978).

Numerous authors (e.g., Meyer *et al.*, 1975; Butler and Meyer (1976); Wasson *et al.*, 1976; Reed *et al.*, 1977; Cirlin and Housley, 1977) have analyzed surface coatings on these and the related droplets from 74220. The most complete survey of the earlier literature regarding sublimate coatings and volatile constituents is in the paper by Meyer *et al.* (1975). Ion microprobe analysis by Meyer *et al.* (1975) found the Apollo 17 droplet coatings enriched in Zn, Ga, Pb, Cu, Tl, S, F and Cl. Butler and Meyer (1976) found that the elements in the A-17 coating were Zn, K, Cl, Cu, Ni and P in order of decreasing abundance. Wasson *et al.* (1976) determined that six elements were strongly concentrated on the surface of the A-17 orange droplets. In order of decreasing abundance the enrichments were In, Cd, Zn, Ga, Ge, and Au. Chou *et al.* (1975) determined by leaching and etching experiments that Zn, Ge, Cd, In and Au were concentrated in the surficial deposits on the A-15 green glass droplets.

Grant *et al.* (1974), based on studies using Auger spectroscopy (AES), reported a sulfur-rich coating on 74220 grains that was approximately 30Å thick. They suggest that this vapor deposited coating is of micrometeorite origin. This work is in close agreement with the earlier work by Carter *et al.* (1973) regarding the enrichment of one or more of the elements Na, S, or K.

Reed *et al.* (1977) and Allen *et al.* (1977) conducted a series of parallel leaching and volatilization experiments in order to resolve the question of the chemical nature of the ^{204}Pb, Zn, Bi, and Tl deposits on grain surfaces. They reject the suggestions of Allen *et al.* (1977), Meyer *et al.* (1975), and Butler and Meyer (1976) that the volatiles may be chlorides or sulfides and suggest instead that the difference in volatilities and solubilities is best explained by differences in the metal ion migration into the surface of the droplet. Pb and Bi are weakly chemsorbed, can be volatilized at 600°C and dissolved with weak HNO_3. Zn and Tl with strong lithophile affinities are strongly chemsorbed, are retained at 600°C but can be dissolved with weak HNO_3.

Butler and Meyer (1976) and Wasson *et al.* (1976) reject the idea that free sulfur could occur on the surface of the droplets and suggest that sulfide minerals (e.g., ZnS and FeS) may be present. Additionally, Wasson *et al.* (1976) suggest that Cl may be present as NaCl; the presence of $PbCl_2$ and possibly other chlorides has been suggested by Cirlin and Housley (1977). KCl crystals on Apollo 15 green glass spheres were reported by Carusi *et al.* (1972a, 1972b); Clanton *et al.* (1978) provide details of NaCl crystals on Apollo 17 glassy droplets.

ANALYSIS

We report here on studies by Scanning Electron Microscope (SEM) and Energy Dispersive X-Ray Analysis (EDX) of the surfaces of some of the orange and black glassy droplets. Specifically the intent of the research was to characterize the surface morphology of the droplets and to attempt to correlate morphology with chemical composition.

Special care was taken to prevent contamination of the droplets during sieving, picking and mounting. All of these operations were performed in laminar flow clean benches. Ultrapure liquid freon was used in the sieving procedure and in the ultrasonic cleaning of the particles. Individual droplets were hand picked with a vacuum needle from the 90 to 150 μm size fraction, or in a few cases the 45 to 90 μm size fraction, for SEM analysis. The selected grains were mounted with PVA after the technique outlined by Clanton and Ladle (1975) on spectrographic grade carbon planchettes and then sputter-coated with AuPd.

Micromound coating

Micromound coatings were first described on Apollo 15 green glass by McKay *et al.* (1973) who interpreted the coating as a vapor deposit. Additionally, this paper illustrates numerous examples of particle and micromound abrasion by both fluid and solid material. The morphology of the surface features of the Apollo 15 green and Apollo 17 orange/black droplets are very similar. The McKay *et al.* (1973) and Heiken *et al.* (1974) papers should be consulted for additional morphology of these droplets.

A micromound coating is present on all of the 74001–74002 orange and black glass droplets that have been examined. Figure 1 illustrates one type of abrasion where the still plastic micromounds have been scraped and smeared over the glassy substrate. Figure 2 provides high resolution details of a typical micromound surface. Individual micromounds range in size from about 20Å to over 200Å in diameter. In some areas either the micromounds did not form or individual mounds are below the resolution limit of the SEM. Additionally, some of the larger micromounds have developed hexagonal outlines (McKay *et al.*, 1973). If changes in the diameter of the micromound in two dimensions reflects a comparable change in the third dimension, then the thickness of the micromound

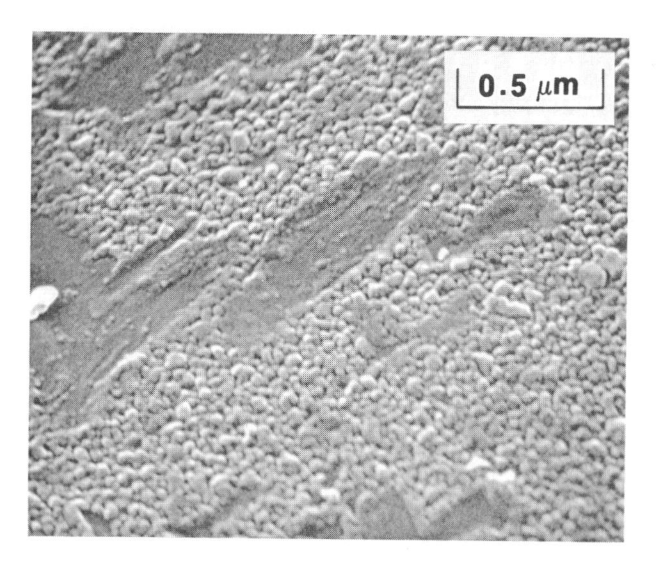

Fig. 1. SEM micrograph of an abraded micromound surface, sample 74001,98. Some of the abrasion on the sphere surface occurred while the micromounds were soft and plastic and could be scraped and smeared.

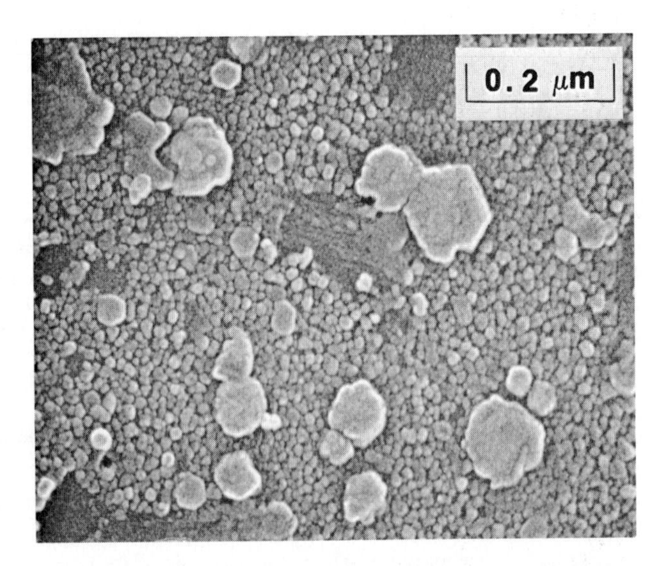

Fig. 2. SEM micrograph of a well developed micromound surface on an Apollo 17 orange glass sphere. Individual micromounds smaller than 25Å in diameter have been observed; the median size is about 150Å. Some of the larger aggregates have outlines suggestive of incipient crystal faces. Barren areas are produced by the failure of the micromounds to form or by the existence of micromounds smaller than the SEM resolution. NASA Photograph S-77-26179.

coating ranges from less than 20Å to over 300Å across a typical surface. Not only does the density and size of the micromounds vary across the surface of a single sphere but it also varies from droplet to droplet. As density increases, the micromounds interlock and coalesce, eventually forming a continuous coating of increasing thickness. In places this thicker coating has partially vesiculated (Fig. 3) as a volatile constituent has evolved or has cracked (Fig. 4A) from shrinkage and cooling or from desiccation.

EDX spectra of the cracked coating show S, Na, and K (Fig. 4B) in addition to peaks from the silicate substrate. The small peak at 8.6 KEV suggests a minor contribution from Zn. The Zn L line overlaps the peak identified as Na on the EDX spectrum. However, the ratio of Zn L to Zn K lines [1.6 to 1, see Butler and Meyer (1976) for details] establishes a maximum contribution of Zn L to the Na K peak. Consequently, while the low Zn K peak at 8.6 KEV clearly indicates the presence of Zn, the peak at ~1.1 KEV is higher than can be attributed to the Zn L line, the major contributor must be the Na K line.

Sodium chloride crystals

The SEM laboratory at JSC has thousands of photographs from over 400 droplets from not only the Apollo missions but also the Luna missions. Only three examples of droplets with a coating containing chloride crystals have been noted.

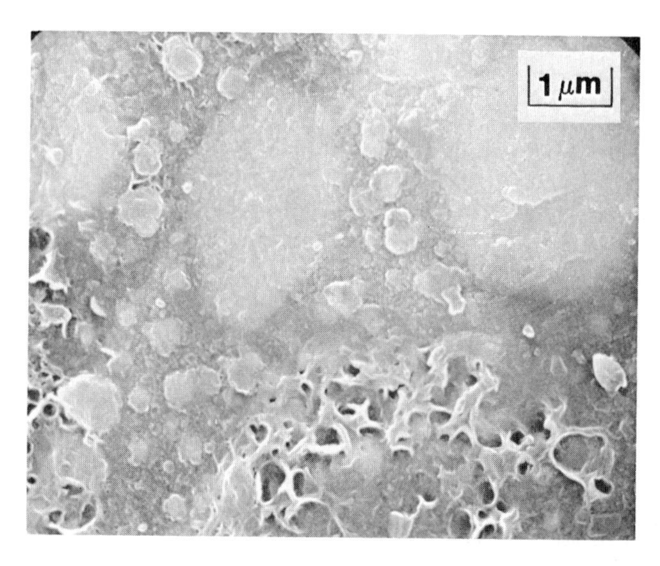

Fig. 3. SEM micrograph of a rare vesiculated surface on an Apollo 17 black glass sphere. The large diamond shaped white "ghosts" are sphere crystals that occur near the surface of the droplet but are below the micromound coating. The area adjoining the vesiculated area has a well developed micromound coating. NASA Photograph S-78-26193.

Of these examples only one is sufficiently unique for us to suggest that the NaCl crystals are clearly lunar in origin.

Sample 74002,179 appears to demonstrate that in this case the chloride crystals must be lunar in origin. Some thirty five droplets were hand picked from this sample for SEM study utilizing the procedures outlined above. Twenty-six grains were orange glass droplets, nine were black glassy droplets. Of the thirty-five droplets, only one black sphere has a sparse population of chloride crystals. Additionally this sphere has been fractured and a fragment has spalled from the surface. There are no halite crystals in the fracture or in the spall area. NaCl crystals occur on the original droplet surface and are most abundant in irregular surface depressions.

The halite crystals (Figs. 5 and 6A) are euhedral and average from 1000 to 3000Å in length along crystallographic axes; the crystals display several habits including the cube, octahedron, and dodecahedron. EDX analysis (Fig. 6B) shows Na and Cl peaks in addition to peaks from the glassy silicate substrate. The substrate contributes to the X-ray spectrum because the electron beam excites a volume that is larger than the submicron chloride crystal.

This droplet also has a larger irregular mass of NaCl (Fig 7A). Although displaying a suggestion of crystal faces, this mound resembles some of the splashes of glass that have been observed on all droplets. The spectra, however, are primarily NaCl (Fig. 7B) with only minor peaks from the glassy silicate substrate. The chlorine rich mound also contains partially imbedded inclusions of

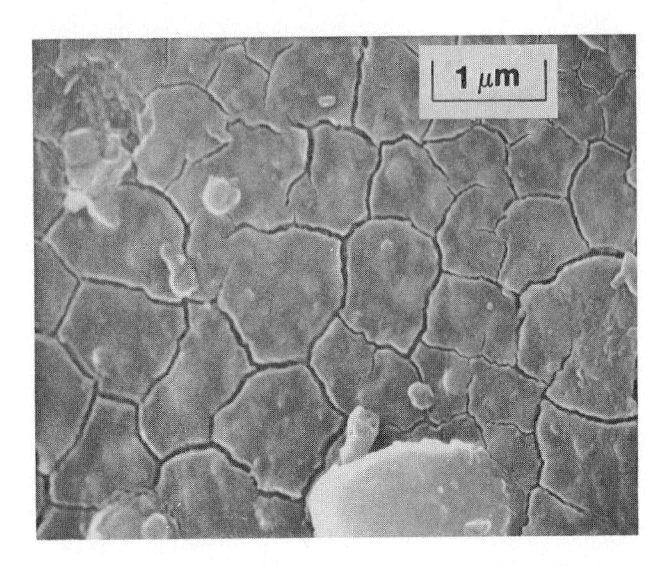

Fig. 4a. SEM micrograph of a crazed surface on an Apollo 17 black glass sphere. These rare features, resembling "mud cracks", may represent a surface where the micromounds have coalesced and the surface cracked during the cooling or desiccation of the molten droplet. NASA Photograph S-78-26188.

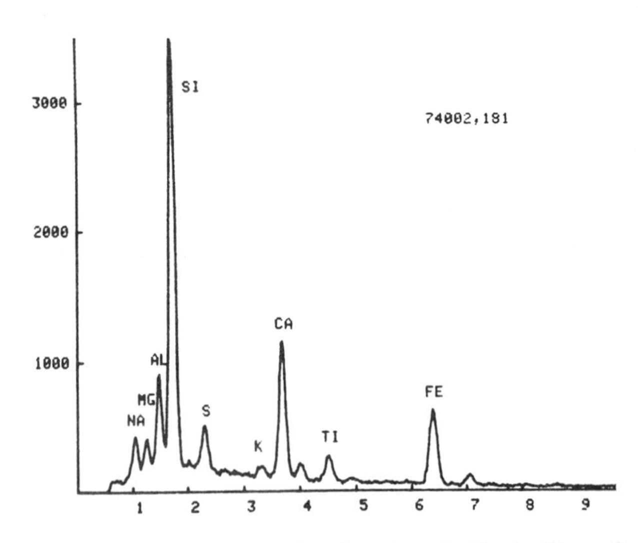

Fig. 4b. EDX spectrum from the crazed surface shown in Fig. 4a. The surface contains Na, S, K and perhaps a small percentage of Zn based on the presence of a low Zn Ka peak at 8.6 KEV. The other elements represent the substrate that has also been excited by the EDX analysis (Mg, Al, Si, Ca, Ti, and Fe).

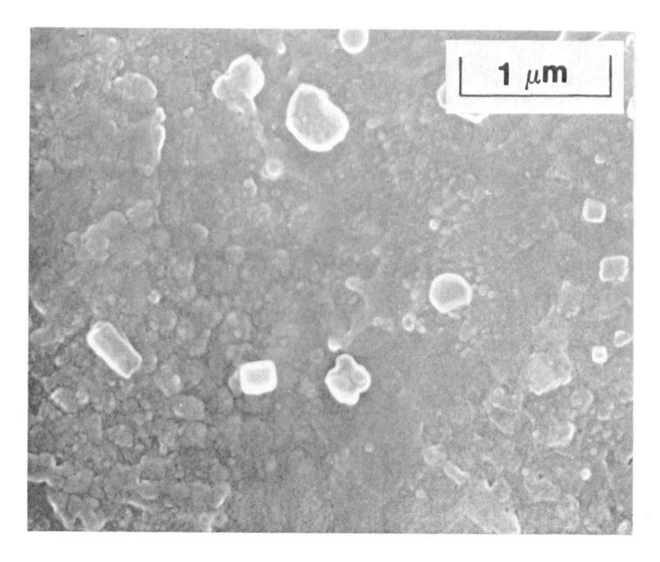

Fig. 5. SEM micrograph of euhedral NaCl crystals on a black glassy droplet from 74002. This low magnification view illustrates the relative abundance of halite crystals on the droplet surface. NASA Photograph S-78-26250.

silicate glass. The contact with both the micromound substrate surface and the glass inclusions is sharp and the irregular mass clearly coats the micromound surface.

Sulfur

A sulfur-rich feature (Fig. 8A) has also been observed on the 74001,179 black glass droplet that contains abundant NaCl crystals. EDX spectra of this mound (Fig. 8B) show a strong S peak with lesser peaks from Na, and K in addition to the substrate. The small peak at 8.6 KEV indicates that some Zn is also present, but again, Na is a major component of the combined Na + Zn peak at ~1.1 KEV. The morphology of the particle suggests that more than one phase may be present; based on EDX, sulfides of Na, K, Fe and Zn may be important constitutents of the mass. However, the large size of the S peak compared with the Na, K, Fe and Zn peaks leaves open the possibility that elemental sulfur may also be present. A comparison of EDX spectra from the mound and the adjacent substrate indicates that the mound is not primarily FeS; most of the Fe peak originates in the silica substrate. This sulfur-rich mound may represent a condensed phase of some of the volatiles present in the plume of a fire fountain or may represent a splash of a once-molten phase.

Metallic Fe

Figure 9 illustrates a form of metallic iron that has been observed on some of

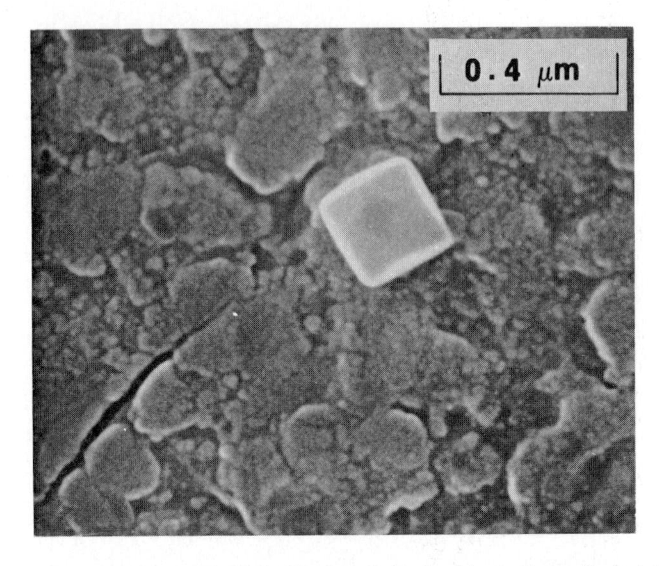

Fig. 6a. SEM micrograph of a NaCl crystal. Both cube and octahedral faces are present on the crystal. The euhedral crystal sits on an iron-rich micromound surface. NASA Photograph S-78-26267.

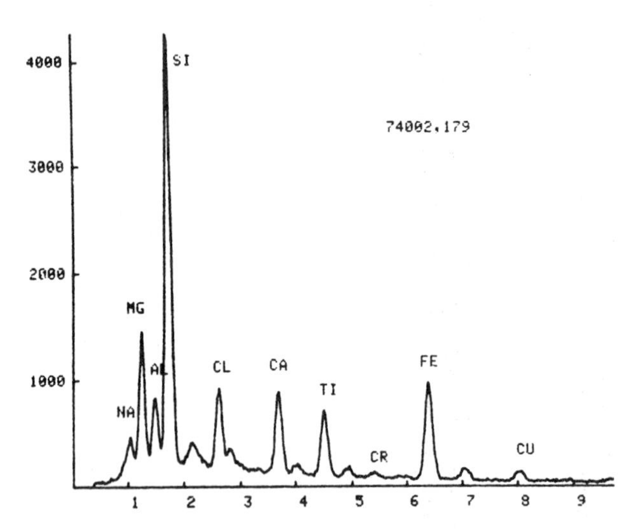

Fig. 6b. EDX spectrum from the crystal and substrate shown in Fig. 6a. The electron beam penetrates the submicron crystal and excites both substrate and crystal to produce the composite spectrum. The NaCl peaks are not present when only the micromound surface is analyzed by EDX.

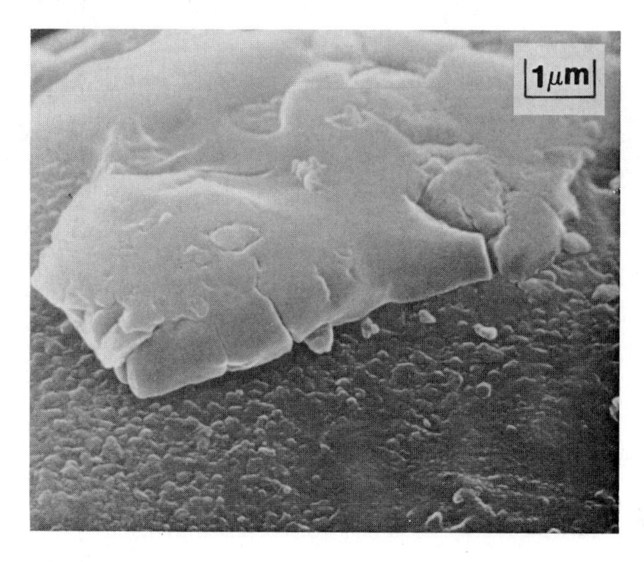

Fig. 7a. SEM micrograph; the large smooth mass with some suggestion of a regular outline or crystal faces is enriched in NaCl. The NaCl mass rests upon a well developed micromound surface. This geometric relationship suggests that the Cl-rich mass formed after the micromound coating had fully devleloped. NASA Photograph S-78-26252.

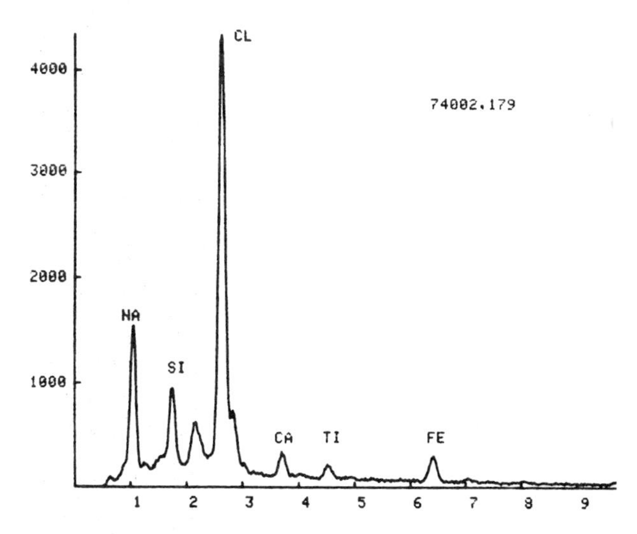

Fig. 7b. EDX spectrum from the smooth Cl-rich mass in the center of Fig. 7a. The major chemistry of the mass is NaCl although some contribution from the substrate is evidenced by the Si, Ca, Ti and Fe peaks in the spectrum. The Na and Cl peaks are not present when only the micromound surface is analyzed by EDX.

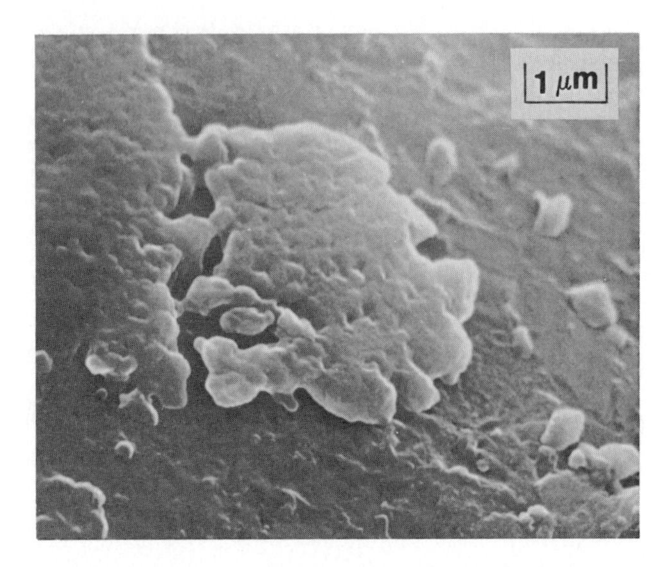

Fig. 8a. SEM micrograph of a sulfur-rich feature that was observed on the droplet shown in Figs. 1, 2 and 4. The morphology of the particle, rough and scaley vs. smooth and globular, suggests that more than one phase may be present. NASA Photograph S-78-26253.

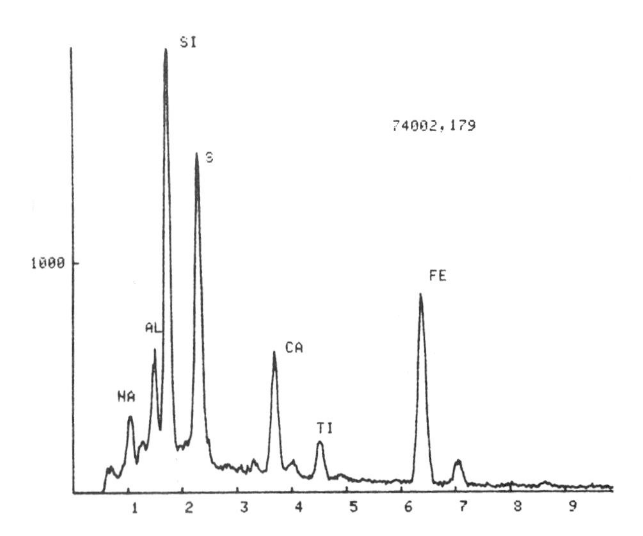

Fig. 8b. EDX spectrum from the feature shown in Fig. 8a. Based on a comparison of EDX spectra from the S-rich mass and the substrate, sulfides of Fe, Na, K and Zn appear to be important constituents in the S-rich mass. The intensity of the S peak compared to the weaker Fe, Na, K, and Zn peaks suggests the possibility that elemental sulfur may be present. The mass is not primarily FeS; most of the Fe originates in the substrate. Because of the small sample size, a composite EDX spectra from both silica substrate and coating is recorded.

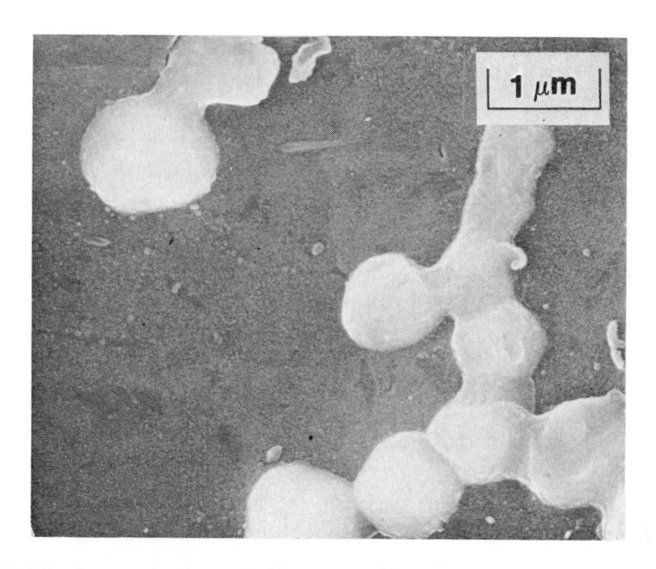

Fig. 9. SEM micrograph of amoeboid masses of iron on the surface of a black glassy sphere. A close inspection of some of the iron particles shows a partial development of incipient crystal faces. The iron occurs on a surface with well developed micromounds. NASA Photograph S-78-26172.

the black glassy droplets. A somewhat similar feature was reported by Carter and Padovani (1973) from a silicate sphere from Apollo 16. Based on EDX, the amoeboid masses are relatively pure Fe; Ni, P, Co, and S, if present, are below the detection limit, i.e., less than 0.5%. A close inspection of the iron masses shows a partial development of incipient crystal faces; the iron mass extends well above the micromound surface. If one accepts the volcanic origin of the glassy spheres, then the iron may be a vapor condensate or a splash of once-molten material. However, the iron may have been reduced from the silicate melt in a process somewhat similar to the hydrogen reduction experiments of Carter and McKay (1972). Metallic iron is chemically incompatible with elemental sulfur; the two should combine to form iron sulfide. However, if these phases were deposited on the droplets during a regime of rapidly decreasing temperature, the sulfur may have been deposited at a temperature much lower than that associated with the iron and may have been quenched before significant reaction with metallic iron could occur. Thus, the presence of both metallic iron and elemental sulfur are not incompatible if, as is likely, the condensates represent extreme non-equilibrium conditions.

Conclusions

1. Rare examples of halite crystals and irregular masses of NaCl occur on glassy droplets from Apollo 17 that have been interpreted as volcanic tephra.

2. A sulfur-rich feature was found on one black glass droplet. Although the evidence is circumstantial, the feature appears to contain elemental sulfur.
3. Metallic masses, some with crystal faces, are present on some of the droplets. The presence of both metallic iron and elemental sulfur are not necessarily incompatible considering the non-equilibrium conditions of very rapid cooling.
4. The micromound coating contains the bulk of the metal and volatile constituents and forms a discontinuous surface on droplets. Barren areas occur where the micromounds did not form and where abrasion has removed the coating. Micromounds range in size from about 20Å to over 300Å in diameter. If the diameter in two dimensions can be extended into the third (depth), such changes in "thickness" would explain the varying amount but constant ratio of volatile constituents observed by other workers.
5. Based on our observations of several hundred glassy spherules from all Apollo lunar missions, the classic micromound surfaces occur only on the Apollo 15 green and Apollo 17 orange and black droplets. Additionally none of the more than 200 Luna 24 spherules that we have examined displayed micromounds. Based on all available data, we consider that this unique micromound coating is directly related to the origin of the Apollo 15 green and Apollo 17 orange and black droplets. It is our consensus that a volcanic fire fountain origin best fits the extensive observational and experimental data that have accumulated. The micromound surface is the unique product of this volcanic origin and may be used to determine if glassy droplets from other places on the moon have had a similar origin.

REFERENCES

Allen R. O., Jr., Jovanovic S. and Reed G. W., Jr. (1977) Volatile metals-mode of transport (abstract). In *Lunar Science VIII*, p. 22–24. The Lunar Science Institute, Houston.

Butler P., Jr. and Meyer C., Jr. (1976) Sulfur prevails in coatings on glass droplets: Apollo 15 green and brown glasses and Apollo 17 orange and black (devitrified) glasses. *Proc. Lunar Sci. Conf. 7th*, p. 1561–1581.

Carter J. L. and McKay D. S. (1972) Metallic mounds produced by reduction of material of simulated lunar composition and implications on the origin of metallic mounds on lunar glasses. *Proc. Lunar Sci. Conf. 3rd*, p. 953–970.

Carter J. L. and Padovani E. (1973) Genetic implications of some unusual particles in Apollo 16 less than 1 mm fines 68841,11 and 69941,13. *Proc. Lunar Sci. Conf. 4th*, p. 323–331.

Carter J. L., Taylor H. C. J. and Padovani E. (1973) Morphology and chemistry of particles from Apollo 17 soils 74220, 74241 and 75081. *EOS (Trans. Amer. Geophys. Union)* **54**, 582–584.

Carusi A., Cavaretta G., Cinotti F., Civitelli G., Coradini A., Fulchignoni M. and Funiciello R. (1972a) The source area of Apollo 15 "green glass." In *The Apollo 15 Lunar Samples* (J. W. Chamberlain and C. Watkins, eds.), p. 5–9. The Lunar Science Institute, Houston.

Carusi A., Cavaretta G., Cinotti F., Civitelli G., Coradini A., Funiciello R., Fulchignoni M. and Taddevcci A. (1972b) Lunar glasses as an index of the impacted sites lithology: The source area of Apollo 15 "green glasses." *Geol. Romana* **11**, 137–151.

Chou C.-L., Baedecker P. A., Bild R. W. and Wasson J. T. (1974) Volatile-element systematics and green glass in Apollo 15 lunar soils. *Proc. Lunar Sci. Conf. 5th*, p. 1645–1657.

Chou C.-L., Boynton W. V., Sundberg L. L. and Wasson J. T. (1975) Volatiles on the surface of Apollo 15 green glass and trace-element distribution among Apollo 15 soils. *Proc. Lunar Sci. Conf. 6th*, p. 1701–1727.

Cirlin E. H. and Housley R. M. (1977) A flameless atomic absorption study of the volatile trace metal lead in lunar samples. *Proc. Lunar Sci. Conf. 8th*, p. 3931–3940.

Clanton U. S. and Ladle G. H. (1975) Polyvinyl acetate-methyl alcohol solution as a scanning electron microscope particle mounting medium. *Amer. Mineral.* **60**, 327.

Clanton U. S., McKay D. S., Waits G. and Fuhrman R. (1978) Unusual surface features on volcanic droplets from 74001 and 74002 (abstract). In *Lunar and Planetary Science IX*, p. 172–174. Lunar and Planetary Institute, Houston.

Grant R. W., Housley R. M., Szalkowski F. J. and Marcus H. L. (1974) Auger electron spectroscopy of lunar samples. *Proc. Lunar Sci. Conf. 5th*, p. 2423–2439.

Heiken G. H. and McKay D. S. (1978) Petrology of a sequence of pyroclastic rocks from the Valley of Taurus-Littrow (Apollo 17 landing site). *Proc. Lunar Planet. Sci. Conf. 9th*. This volume.

Heiken G. H., McKay D. S. and Brown R. W. (1974) Lunar deposits of possible pyroclastic origin. *Geochim. Cosmochim. Acta* **38**, 1703–1718.

McKay D. S., Clanton U. S. and Ladle G. (1973) Scanning electron microscope study of Apollo 15 green glass. *Proc. Lunar Sci. Conf. 4th*, p. 225–238.

Meyer C., Jr., McKay D. S., Anderson D. H. and Butler P., Jr. (1975) The source of sublimates on the Apollo 15 green and Apollo 17 orange glass samples. *Proc. Lunar Sci. Conf. 6th*, p. 1673–1699.

Morgan J. W., Ganapathy R., Higuchi H., Krahenbuhl U. and Anders E. (1974) Lunar basins; tentative characterization of projectiles from the meteoritic elements in Apollo 17 boulders. *Proc. Lunar Sci. Conf. 5th*, p. 1703–1736.

Reed G. W. Jr., Allen R. O. Jr. and Jovanovic S. (1977) Volatile metal deposits on lunar soils—relation to volcanism. *Proc. Lunar Sci. Conf. 8th*, p. 3917–3930.

Reid A. M., Lofgren G. E., Heiken G. H., Brown R. W. and Moreland G. (1973) Apollo 17 orange glass, Apollo 15 green glass and Hawaiian lava fountain glass (abstract). *EOS (Trans. Amer. Geophys. Union)* **54**, 606–607.

Roedder E. and Weiblen P. W. (1973a) Origin of orange glass spheres in Apollo 17 sample 74220 (abstract). *EOS (Trans. Amer. Geophys. Union)* **54**, 612–613.

Roedder E. and Weiblen P. W. (1973b) Apollo 17 "orange soil" and meteorite impact on liquid lava. *Nature* **244**, 210–212.

Tatsumoto M., Nunes P. D., Knight R. J., Hedge C. E. and Unruth D. M. (1973) U-Th-Pb, Rb-Sr and K measurements of two Apollo 17 samples (abstract). *EOS (Trans. Amer. Geophys. Union)* **54**, 614–615.

Wasson J. T., Boynton W. V., Kallemeyn G. W., Sundberg L. L. and Wai C. M. (1976) Volatile compounds released during lunar lava fountaining. *Proc. Lunar Sci. Conf. 7th*, p. 1583–1595.

Proc. Lunar Planet. Sci. Conf. 9th (1978), p. 1959–1967.
Printed in the United States of America

Agglutinates as recorders of fossil soil compositions

G. J. TAYLOR, S. WENTWORTH, R. D. WARNER and K. KEIL

Department of Geology and Institute of Meteoritics
University of New Mexico, Albuquerque, New Mexico 87131

Abstract—The regolith sampled by the upper three drill stems (70009–70007) of the Apollo 17 drill core is a mixture of mature regolith, mare basalt and orange glass excavated when Camelot and the Central Cluster craters were formed, and possibly fragments of the projectiles that formed these craters. Similarly, soils from South Massif are mixtures of an older regolith and the materials that slid downhill during the landslide. Since the average composition of glass in agglutinates appears to have the composition of the soil in which it is formed, we measured the compositions of the agglutinates in polished sections from the Apollo 17 drill core and in sections of soils from North and South Massifs in an attempt to deduce the nature of the pre-Camelot, pre-Central Cluster regolith in the Taurus-Littrow valley. Except for samples from North Massif, average agglutinate compositions in the soils studied differed from the compositions of their host soils. Least squares mixing calculations indicate that the pre-Central Cluster regolith was composed of local materials from the valley floor and the massifs. We also show that mixing calculations on bulk soil compositions can be misleading. Soil 70181, for example, ostensibly contains 55% mare basalts and equal amounts (~15%) of orange glass, ANT, and noritic rock. However, when agglutinates (representing mature regolith) are added as a component to the mix, the results indicate that soil 70181 actually is a mixture of 60% mature regolith, 30% mare basalts, 9% ANT, and 1% norite. We conclude that chemical mixing models must take into account the petrologic characteristics of each soil.

INTRODUCTION

Lunar soils are littered with dark-colored, irregularly-shaped, vesicular, glass-bonded aggregates of rock, glass, and mineral fragments. These aggregates are called agglutinates. In polished thin sections, agglutinates have unmistakable petrographic characteristics: they contain fragments of practically everything found in lunar soils, and are bound by a highly vesicular, usually brownish glass that has a swirly appearance due to the presence of numerous, typically submicron-sized, flow-aligned droplets of metallic iron. Excellent photomicrographs of agglutinates appear in papers by Gibbons *et al.* (1976), Basu (1976), and Hu and Taylor (1977).

An interesting property of agglutinates is that the *glass* in them tends to have the composition of the soil in which the agglutinates are formed (Marvin *et al.*, 1971; Taylor *et al.*, 1972; Gibbons *et al.*, 1976; Hu and Taylor, 1977). Although minor differences between average agglutinate glass compositions and local soils are common, they may in some cases be pronounced. However, significant chemical fractionation, as suggested by Adams *et al.* (1975) and Rhodes *et al.* (1975), clearly does not take place during the formation of agglutinate glass.

Taylor *et al.* (1977) showed that modal analyses readily divide the top three stems of the Apollo 17 drill core into three major layers, which we designate the upper zone (0–19 cm depth), the coarse layer (24 to 56 cm), and the lower zone (65–94 cm). In the same paper, we devised a model that describes the depositional history of the core. The model's essential ingredient is that the identifiable layers in the core are mixtures of mature regolith and mare basalt freshly excavated by the impacts that formed the Central Cluster craters and Camelot. We also proposed, along with Delano (1977), that the core contains fragments of the projectiles responsible for forming the Central Cluster craters and Camelot, and that these fragments came from Tycho; see also Wolfe *et al.* (1975) and Arvidson *et al.* (1976).

If the layers in the core contain varying proportions of excavated bedrock and mature regolith, the agglutinates in the core ought to have the composition of the regolith that covered the Apollo 17 site in the vicinity of Central Cluster and Camelot *before* those craters were blasted out. With that idea in mind, we measured the compositions of agglutinates in our thin sections of the top three stems (70009–70007) of the Apollo 17 deep drill core. We also analyzed agglutinates in soils (90–150 μm size fraction) from North and South Massifs. The Massif soils were supplied to us by Dave McKay and include sections 76241,24A, 76261,26A, 76281,6, 76321,10, and 76501,1 from North Massif and 72150,2A, 72321,7A, 72441,7A, 72501,1A, 72501,29A, 72701,1A, and 72701,29A from South Massif. Our results and conclusions are presented in this paper.

EXPERIMENTAL PROCEDURES

In each of 35 polished thin sections we made electron microprobe analyses of all agglutinates that contained suitably large ($>\sim$20 μm) areas of the swirly glass. Because such areas are rare, most agglutinates were not suitable for analysis. We used a beam size of $\sim$10 μm and usually analyzed one spot per agglutinate. When we did more than one analysis on a given particle, we averaged them before averaging the whole set. The microprobe data were corrected first for background and drift and then for differential matrix effects by the method of Bence and Albee (1968) and Albee and Ray (1970).

RESULTS

Average agglutinate compositions for specific depth intervals in the drill core and for North and South Massifs appear in Table 1. Except for the uppermost 5 cm, the intervals correspond to the three major layers we defined previously (Taylor *et al.*, 1977). The top 5 cm correspond approximately to 70181, the core reference soil, which was sampled by a scoop that dug into the lunar surface to a depth of $\sim$5 cm. For comparison, we also list bulk compositions for soils corresponding to those in which the agglutinates occur.

To assess the compositional similarity of the agglutinates and their host soils, we determined the 95% confidence limits for the means of each oxide in the agglutinate analysis (Table 1). These limits were calculated from standard

Table 1. Compositions (wt.%) of and calculated lithologic abundances in agglutinates, compared to host bulk soils.

	0-5 cm			0-19 cm			24-56 cm			65-94 cm			North Massif			South Massif		
	1		2	3		4	5		6	7		8	9		10	11		12
N	36			182			45			62			24			19		
SiO_2	41.2	(1.1)	40.9	41.9	(0.4)	40.7	42.4	(0.7)	40.3	42.1	(0.4)	42.0	43.7	(0.7)	43.3	44.0	(1.0)	44.9
TiO_2	7.5	(1.1)	8.3	6.9	(0.4)	8.5	5.4	(0.6)	8.6	6.0	(0.4)	6.0	3.3	(0.5)	3.6	3.6	(1.1)	1.4
Al_2O_3	11.9	(1.0)	12.3	13.2	(0.6)	11.8	14.0	(1.1)	10.9	14.1	(0.6)	13.2	18.7	(1.5)	17.7	17.7	(1.6)	20.9
Cr_2O_3	0.37	(.05)	0.45	0.36	(.03)	0.41	0.43	(.04)	0.45	0.40	(.02)	0.43	0.26	(.03)	0.29	0.28	(.03)	0.22
FeO	16.6	(1.0)	16.5	15.5	(0.5)	17.2	15.2	(1.1)	18.1	14.5	(0.5)	16.6	10.9	(1.0)	11.2	11.6	(1.5)	8.5
MnO	0.23	(.03)	0.22	0.18	(.01)	0.22	0.22	(.03)	0.23	0.19	(.02)	0.21	0.14	(.02)	0.16	0.12	(.02)	0.12
MgO	10.9	(0.3)	9.8	10.3	(0.3)	10.2	11.0	(0.6)	9.8	10.5	(0.2)	10.5	10.5	(0.5)	10.5	10.6	(0.4)	9.9
CaO	10.6	(0.2)	11.0	11.0	(0.2)	10.6	11.0	(0.5)	10.9	11.0	(0.2)	10.9	12.3	(0.4)	12.0	11.5	(0.4)	12.9
Na_2O	0.28	(.03)	0.36	0.29	(.02)	0.40	0.38	(.03)	0.42	0.35	(.03)	0.41	0.32	(.04)	0.43	0.37	(.04)	0.44
K_2O	0.09	(.01)	0.08	0.09	(.01)	0.09	0.08	(.01)	0.09	0.12	(.01)	0.11	0.13	(.02)	0.11	0.13	(.01)	0.16
P_2O_5	0.06	(.01)	0.06	0.09	(.01)		0.07	(.01)		0.06	(.01)		0.08	(.01)	0.08	0.12	(.02)	0.14
Total	99.73		99.97	99.81		100.12	100.18		99.79	99.32		100.36	100.33		99.37	100.02		99.58
A17 Mare Basalt	38.8		54.8	41.9		50.5	19.6		53.0	31.4		24.5	10.1		14.2	13.8		5.1
Orange Glass	28.0		15.4	16.4		23.7	32.4		27.5	20.5		36.9	15.7		14.6	13.0		2.1
ANT	9.9		14.3	12.6		13.4	20.3		14.7	18.2		22.5	42.6		35.4	29.0		46.8
Norite	23.3		15.4	29.1		12.5	27.7		4.8	29.9		16.0	31.6		35.8	44.2		46.1
% ANT in Nonmare	29.8		48.1	30.2		51.7	42.3		75.4	37.8		58.4	57.4		49.7	39.6		50.4
PMD	1.7		1.6	2.2		1.3	3.5		3.8	1.7		4.9	0.7		1.2	1.7		1.2

(1) Agglutinates in upper 5 cm of core. (2) Core reference soil 70181; average of analyses by Rhodes *et al* (1974) and Rose *et al* (1974). (3) Agglutinates in top 19 cm of core. (4) Average of bulk analyses of 70009,432 and 434 and calculated bulk of 70009,433 (Laul *et al.*, 1978). (5) Agglutinates in coarse-grained layer of core. (6) Average of bulk analysis of 70008,456 and calculated bulk analyses of 70008,455, 457, and 458 (Laul *et al*, 1978). (7) Agglutinates in 65 to 94 cm depth interval in core. (8) Average of bulk analysis of 70007,167 and calculated bulks of 70007, 168 and 169. (9) Agglutinates in North Massif soils. (10) Average of Sta 6 and 7 soils (Rhodes *et al.*, 1974). (11) Agglutinates in soils from South Massif. (12) Average of Sta 2 and 2A soils (Rhodes *et al*, 1974).

N is the number of analyses.

Numbers in parentheses are 95% confidence limits; computed from standard deviations and Student's t-distribution.

PMD is percent mean difference between observed and calculated composition (see text).

deviations and Student's t-distribution. The results show that agglutinates in the 0 to 19 cm and 24–56 cm depth intervals in the drill core and those in South Massif soils are distinctly different from the bulk soils in which they occur. In all three cases, the concentrations of TiO_2, Al_2O_3, FeO, MgO, and Na_2O are significantly different in the agglutinates compared to their host soils. Conversely, compositions of agglutinates in North Massif soils are not significantly different from the compositions of the soils from North Massif. The other two cases, the top 5 cm and the 65 to 94 cm interval in the core, are ambiguous. Only two elements in each are significantly different (MgO and CaO in the upper 5 cm and Al_2O_3 and FeO in the 65 to 94 cm interval).

DISCUSSION

Do agglutinates have soil compositions?

The glass that holds agglutinates together obviously forms by impact-melting of the lunar regolith. It also seems to have the same composition as its host soil, if that soil is mature and not mixed with another soil of a different composition or mixed with freshly-excavated rock debris. Reasonable agreement between average agglutinate composition and host soil were found for at least some soils from Apollo 12 (Marvin *et al.* 1971), Apollo 14 (Taylor *et al.*, 1972). Apollo 16 (Gibbons *et al.* 1976; Hu and Taylor, 1977), and Apollo 17 (Hu and Taylor, 1977). It is tempting, therefore, to assume that an average agglutinate composition represents the composition of a mature regolith. However, a few uncertainties make such an assumption risky. One problem is that the composition of lunar soils varies with size fraction, manifested mostly clearly by a pronounced change

Table 2. Compositions (wt.%) of lithologic components used in mixing calculations.

	A	B	C	D
SiO_2	38.9	38.6	46.2	44.5
TiO_2	12.0	8.8	1.50	0.39
Al_2O_3	9.0	6.3	18.0	26.0
Cr_2O_3	0.41	0.75	0.20	0.06
FeO	19.3	22.0	9.2	5.8
MnO	0.25	0.30	0.12	–
MgO	8.2	14.4	12.3	8.0
CaO	10.7	7.7	11.2	14.9
Na_2O	0.39	0.36	0.65	0.25
K_2O	0.06	0.09	0.26	–
P_2O_5	–	0.04	0.28	–

(A) Apollo 17 mare basalt (literature complication; 89 analyses); (B) Orange glass (Rhodes *et al.*, 1974); (C) Noritic breccia (Rhodes *et al.*, 1974); (D) ANT (Taylor, 1975).

at ~20 μm (Laul *et al.*, 1978). Consequently, if agglutinates are made preferentially from the smaller size fractions, their compositions will not represent that of the bulk soils. On the other hand, there are no data that indicate the sizes of the soil particles that melted to form agglutinate glass. Another uncertainty is that the <20 μm fraction might be included in the glass as schlerien (D. Vaniman, pers. comm.). The apparent glass compositions, therefore, might be biased toward the <20 μm composition, which is enriched in KREEP in Apollo 17 soils (Laul *et al.*, 1978). Unfortunately this problem has not been addressed and we cannot assess its importance. A third problem is the larger scatter in agglutinate analyses (Table 1). Why should a set of analyses that scatter so much have a mean composition that corresponds to a bulk soil? We do not know the answer, but we note that in many cases, average agglutinates seem to have nearly the same compositions as the mature soils in which they are found. Consequently, in spite of the uncertainties we will assume in the following sections that agglutinates have the composition of the soil in which they are formed, but not necessarily the compositions of the soils in which they are found.

Nature of the pre-Tycho regolith

Apollo 17 bulk soil compositions must have been affected by the formation of Camelot and the Central Cluster craters, ~100 m.y. ago (Arvidson *et al.*, 1976). Our thesis is that we can use the average agglutinate compositions to deduce the nature of the regolith before secondary projectiles from Tycho impacted the Taurus-Littrow Valley. As discussed in the next section, we cannot use bulk soil compositions alone because the soils are mixtures of mature regolith and another component (freshly excavated mare basalt in the core, the land slide at South Massif).

We can guess that the regolith would be dominated by the four main lithologies present among the Apollo 17 samples: high-Ti mare basalt, orange glass, ANT, and norite (Apollo 17 KREEP). We made least-squares mixing calculations (Wright and Doherty, 1970) to compute the abundances of these four main lithologies. Compositions of the components used appear in Table 2. Due to their low abundances and resulting large analytical uncertainties, we did not use MnO, K_2O, or P_2O_5 in the calculations. As a measure of the quality of the fit, we computed the percent mean (absolute) difference between the observed and calculated compositions (PMD) excluding Cr_2O_3 and Na_2O because their low concentrations yield large percentage differences (e.g., Cr_2O_3 is usually ~0.4 wt.%; a difference of only 0.02 wt.% is 5%). Since the percent mean analytical uncertainty in the microprobe measurements of the six remaining elements is ~3.0%, we consider PMD values smaller than 3.0% to be indicative of a reasonable fit to the data.

The results of the mixing calculations appear in Table 1. Except for the 24–56 cm interval in the core (the coarse-grained layer), the PMD values are all acceptably small. The high quality of the fits indicate that the components and their compositions adequately describe the lithologies present at Apollo 17. It also suggests that no significant elemental fractionation took place during the formation of the agglutinate glasses.

Assuming the agglutinates have faithfully recorded fossil soil compositions, the data suggest that the pre-Tycho mare soils in the core were all broadly similar (except for the coarse-grained layer). They contain more mare basalt than orange glass, 33 to 48 wt.% nonmare component, and the amount of ANT in the nonmare component ranges from 30 to 38%. The variations in the amount of nonmare component might reflect slight variations in the maturities of the soils in which the agglutinates formed. As a mare soil matures, the amount of nonmare material in it (in this case from North and South Massifs) ought to increase because of horizontal movement across the lunar surface. The narrow range in the composition of the nonmare component, 30–38% ANT, suggests a single source, presumably the local massifs. However, North and South Massifs appear to contain different percentages of ANT in the fossil soils now present on their flanks. North Massif contains 57% ANT, and South Massif 40%. It appears, therefore, that South Massif is the more likely source for the nonmare materials at the drill core site.

The average agglutinate in the coarse-grained layer (24–56 cm depth in the core) has a somewhat unusual composition, containing more orange glass than high-Ti mare basalt. This is manifested in the bulk composition (Table 1) by higher Cr_2O_3 and MgO and lower TiO_2 than agglutinates in the other layers. If the average agglutinate composition in the coarse layer really reflects the petrologic make-up of a fossil soil, then that soil had more orange glass than mare basalt in it. On the other hand, perhaps the agglutinates in the coarse layer preferentially incorporated the orange-glass component into their glasses. It is not clear however, why such an enrichment would have occurred in one group of agglutinates and not in the others.

What do mixing calculations on bulk soil compositions mean?

The soil in the Apollo 17 drill core represents a mixture of mature regolith, basalt recently excavated from Camelot and Central Cluster craters, and fragments of the projectiles that formed those craters. Similarly, the South Massif soils are mixtures of mature regolith and the materials that slid down during the landslide. Obviously, mixing calculations on bulk soil compositions are not necessarily meaningful. In this section we try to elucidate this point by comparing the results of mixing calculations using the four lithologic components with those using these four plus a fossil-regolith component (whose composition is given by the average agglutinate present). The results are given in Table 3. (We did not perform the exercise for North Massif soils because the agglutinates in it do not differ significantly in composition from the bulk soil, implying that the North Massif soils are not mixtures of mature soils and freshly excavated rock.)

Mixing calculations ignoring a mature regolith component in soil 70181 (the reference soil for the drill core) suggest that it is composed of 55% mare basalt

Table 3. Proportions (wt.%) of lithologic components in bulk soils.[1]

	70181		0–19 cm		24–56 cm		65–94 cm		South M.	
	A[2]	B[3]	A	B	A	B	A	B	A	B
A17 Mare Basalt	54.8	30.2	50.3	30.4	53.0	54.1	24.5	0.0	5.1	0.9
Orange Glass	15.4	0.0	23.7	14.2	27.5	0.0	36.9	0.0	2.1	0.0
ANT	14.3	9.3	13.4	5.7	14.7	0.0	22.5	0.0	46.8	40.4
Norite	15.4	0.9	12.5	0.0	4.8	0.0	16.0	0.0	46.1	34.3
Fossil Regolith[4]	–	60.5	–	49.7	–	45.9	–	100	–	24.4
% ANT in Nonmare	48.1	91.2	51.7	100	75.4	–	58.4	–	50.4	54.0
PMD[5]	1.6	0.6	1.3	0.5	3.8	2.8	4.9	3.4	1.2	1.0

[1]Bulk soil compositions from Table 1

[2]Calculation A does not include a fossil-regolith component.

[3]Calculation B includes a fossil-regolith component.

[4]Represented by agglutinate compositions (Table 1).

[5]Precent mean difference between observed and calculated compositions (see text).

and roughly equal amounts of orange glass, ANT, and norite. The calculation is right, but it is incomplete. In reality, the top 5 cm of the drill core (which corresponds to 70181) is a mixture of a mature regolith (itself a mixture of agglutinates, mare basalts, orange glass, and nonmare rock fragments), recently added mare basalt (from Camelot and Central Cluster craters) and possibly nonmare rock (from the projectiles that blasted out the craters). We assume that the mature regolith has the composition given by the agglutinates in the 0-5 cm interval (Table 1). Adding this component into the mixing calculations yields a significantly different result than above (Table 3): Soil 70181 appears to be a mixture of 60% mature regolith, 30% mare basalt, and 9% ANT.

A similar situation prevails for the 0–19 cm interval in the drill core (Table 3). The mixing calculation not using agglutinates yields a mix of ~50% mare basalt, 24% orange glass, and roughly equal amounts (~13%) of ANT and norite. The mixing calculation that includes the fossil regolith component, in contrast, contains almost 50% regolith, 30% basalt, 14% orange glass, 5% ANT and no norite. Note that both of these soils in the upper part of the core contain a preponderance of ANT over norite. We suggest that this nonmare component originated as fragments of the projectiles that formed the Central Cluster craters, and that, therefore, these projectiles were predominantly ANT.

The same comparison for the coarse-grained layer (24–56 cm) is also illuminating. A mixing calculation that ignores the fossil-regolith component suggests that the soil is made of 53% mare basalt, 27% orange glass, and 20% nonmare lithologies. In contrast, the calculation that includes the fossil regolith (which is clearly present) suggests that the coarse-grained layer is a mixture of 46% mature regolith and 54% freshly excavated mare basalt. The calculations on the 65–94 cm interval suggest that it is 100% fossil regolith, but the fit is poor (PMD >3.0%).

South Massif soils are a mixture of a mature regolith and the material (possibly including agglutinates) that flowed downhill during the landslide. Assuming that the agglutinate compositions reflect a pre-landslide soil, a mixing calculation indicates that the South Massif soils consist of 24% mature regolith, 40% ANT, and 34% norite. The percentage of ANT in the landslide debris is 54%.

The obvious conclusion from these calculations is that mixing models based solely on bulk chemical composition, while certainly useful, are not as informative as those that combine bulk compositional with petrologic information. Papike *et al.* (1978) express the same view, but take a different approach to the problem. They calculate bulk soil compositions by combining modal analyses and measured chemical compositions of petrographically identifiable components. Our approach complements their's and the combination should advance our understanding of lunar soils and their evolution.

Suggestions for future work

The thesis of this paper is that agglutinates can be used to determine the nature of a fossil regolith present in a soil that is a mixture of that fossil regolith and other materials (basalts, nonmare rocks, or a compositionally distinct mature regolith). The idea rests on the somewhat shaky foundation that average agglutinates represent the compositions of bulk soils and are not affected by the obvious variations in soil compositions with grain size (Korotev, 1976; Laul *et al.*, 1978). We suggest, therefore, that experiments bearing on the mechanism of agglutinate formation be performed, perhaps using materials that have profound compositional differences between size fractions. Until such experiments are done, we cannot be certain that agglutinates have the compositions of the soils in which they are formed.

Assuming experiments verify the foundation on which our hypothesis rests, a useful next step would be to separate agglutinates from soils and determine the concentrations of major, minor, and trace elements in them and in their host soils. Such information would then allow for more elaborate and accurate mixing models.

Acknowledgments—We thank Dave McKay and Dave Vaniman for useful and illuminating reviews, and Ted Bornhorst for helpful discussions. This work was supported by NASA Grant NGL 32-004-063.

REFERENCES

Adams J. B., Charette M. P. and Rhodes J. M. (1975) Chemical fractionation of the lunar regolith by impact melting. *Science* **190**, 380–381.

Albee A. L. and Ray L. (1970) Correction factors for electron probe microanalyses of silicates, oxides, carbonates, phosphates and sulfates. *Anal. Chem.* **42**, 1408–1414.

Arvidson R., Drozd R., Guinness E., Hohenberg C. and Morgan D. (1976) Cosmic ray exposure ages of Apollo 17 samples and the age of Tycho. *Proc. Lunar Sci. Conf. 7th*, p. 2817–2832.

Basu A. (1976) An example of a thermally metamorphosed agglutinate. *Meteoritics* **11**, 207–216.

Bence A. E. and Albee A. L. (1968) Empirical correction factors for the electron microanalysis of silicates and oxides. *J. Geol.* **76**, 382–403.

Delano J. W. (1977) The highlands component in the Apollo 17 soils (abstract). In *Lunar Science VIII*, p. 236–238. The Lunar Science Institute, Houston.

Gibbons R. V., Hörz F. and Schaal R. B. (1976) The chemistry of some individual lunar soil agglutinates. *Proc. Lunar Sci. Conf. 7th*, p. 405–421.

Hu H.-N. and Taylor L. A. (1977) Lack of chemical fractionation in major and minor elements during agglutinate formation. *Proc. Lunar Sci. Conf. 8th*, p. 3645–3656.

Korotev R. L. (1976) Geochemistry of grain-size fractions of soils from the Taurus-Littrow valley floor. *Proc. Lunar Sci. Conf. 7th*, p. 695–726.

Laul J. C., Vaniman D. T., Papike J. J. and Simon S. (1978) Chemistry and petrology of size fractions from Apollo 17 deep drill cores 70009–70006. *Proc. Lunar Planet. Sci. Conf. 9th*. This volume.

Marvin U. B., Wood J. A., Taylor G. J., Reid J. B., Jr., Powell B. N., Dickey J. S., Jr. and Bower J. F. (1971) Relative proportions and probable sources of rock fragments in the Apollo 12 soil samples. *Proc. Lunar Sci. Conf. 2nd*, p. 679–699.

Papike J. J., Vaniman D. T., Schweitzer E. L. and Baldwin K. (1978) Apollo 16 drive tube 60009/60010. Part III: Total major element partitioning among the regolith components or chemical mixing models with petrologic credibility (abstract). In *Lunar and Planetary Science IX*, p. 862–864. Lunar and Planetary Institute, Houston.

Rhodes J. M., Adams J. B., Blanchard D. P., Charette M. P., Rodgers K. V., Jacobs J. W., Brannon J. and Haskin L. A. (1975) Chemistry of agglutinate fractions in lunar soils. *Proc. Lunar Sci. Conf. 6th*, p. 2291–2307.

Rhodes J. M., Rodgers K. V., Shih C., Bansal B. M., Nyquist L. E., Wiesmann H. and Hubbard N. J. (1974) The relationships between geology and soil chemistry at the Apollo 17 landing site. *Proc. Lunar Sci. Conf. 5th*, p. 1097–1117.

Rose H. J., Jr., Cuttitta F., Berman S., Brown F. W., Carron M. K., Christian R. P., Dwornik E. J. and Greenland L. P. (1974) Chemical composition of rocks and soils at Taurus-Littrow. *Proc. Lunar Sci. Conf. 5th*, p. 1119–1133.

Taylor G. J., Keil K. and Warner R. D. (1977). Petrology of Apollo 17 deep drill core-I: Depositional history based on modal analyses of 70009, 70008, and 70007. *Proc. Lunar Sci. Conf. 8th*, p. 3195–3222.

Taylor G. J., Marvin U. B., Reid J. B., Jr. and Wood J. A. (1972) Noritic fragments in the Apollo 14 and 12 soils and the origin of Oceanus Procellarum. *Proc. Lunar Sci. Conf. 3rd*, p. 995–1014.

Taylor S. R. (1975) *Lunar Science: A Post-Apollo View*. Pergamon Press, N.Y. 372 pp.

Wolfe E. W., Lucchitta B. K., Reed V. S., Ulrich G. E. and Sanchez A. G. (1975) Geology of the Taurus-Littrow valley floor. *Proc. Lunar Sci. Conf. 6th*, p. 2463–2482.

Wright T. L. and Doherty P. C. (1970) A linear programming and least squares computer method for solving petrologic mixing problems. *Bull. Geol. Soc. Amer.* **81**, 1995–2008.

Proc. Lunar Planet. Sci. Conf. 9th (1978), p. 1969–1980.
Printed in the United States of America

Chemistry of orange/black soils from core 74001/2

Douglas P. Blanchard

NASA Johnson Space Center, Houston, Texas 77058

James R. Budahn

Lockheed Electronics Co., Inc., Houston, Texas 77058

Abstract—We have cooperated with the core study group at JSC and University of Houston at Clear Lake City in the study of double core 74001/2. We have analyzed bulk soils and the 90–150 μm and <20 μm size fractions of 13 soils from the core for both major and trace elements by instrumental neutron activation analysis.

The orange and black glassy soils in the core 74001/2 are homogeneous in composition and due to their unusual compositional characteristics probably have derived from a single source with the same composition.

The top of the core has been reworked to a depth of about 5 cm. The reworked layer is marked by compositional differences in the first 3 cm which reflect the addition (at least 10%) of materials including mare basalt and a CaO-rich component. The I_s/FeO profile clearly indicates reworking depth of about 5 cm. It is likely that at least some of the ferro-magnetically active component (including agglutinitic glass) has been "inherited" with the added material.

Introduction

In a strict sense, the 74001/2 core soils are not lunar regolith; they have not been derived by comminution and homogenization of bedrock by meteorite bombardment as have most lunar soils. Because these soils are unique, they are of special interest for the study of soil evolution and surface processes.

The 74001/2 orange/black glass soils have the finest mean grain size of any lunar soils and are among the best sorted materials returned from the moon (McKay and Waits, 1978). The textures of orange and black droplets which make up the soil range from completely glassy to totally crystalline (Heiken *et al.*, 1978). Except for the near surface soils from 74002, there is virtually no evidence for surface exposure (Morris *et al.*, 1978). If the slightly higher I_s/FeO values for the near surface are indicative of *in situ* surface reworking, 74001/2 represents an opportunity to observe the effects of incipient maturation in a suite of soils with a simple history and homogeneous composition.

We are reporting the results of chemical analyses of bulk soils and size fractions. These samples have been shared with others in the working group at JSC and the University of Houston at Clear Lake City, and the results of their work are also reported in this volume (McKay *et al.*, 1978; Bogard and Hirsch, 1978; and Morris, *et al.*, 1978).

SAMPLES, METHODS, AND UNCERTAINTIES

We have analyzed samples from 13 depths in the double core 74001/2. Three of the samples (74001,175; 176; 177) are from very near the lunar surface (<2.0 cm); the remaining 10 samples were taken at more or less even intervals down the length of the core. Each of the samples was sieved into its grain size fractions in the JSC-SEM laboratory (McKay *et al.*, 1978). We analyzed an aliquant of each bulk, unsieved soil and aliquants of the 90–150 μm and <20 fractions of each soil with a few omissions.

Samples were first analyzed for FeO, Na_2O, Cr_2O_3, and trace elements that have long half-lives using our usual techniques (Jacobs *et al.*, 1977). The samples of bulk soils and selected <20 μm fractions were repackaged and analyzed for Al_2O_3, TiO_2, MgO, CaO, MnO, and V by short irradiation-rapid counting techniques similar to those used by Schmitt *et al.* (1970).

Results of analyses for major elements are reported in Table 1. Table 2 contains the trace element analyses of the bulk soils and size fractions plus the FeO, Na_2O, and Cr_2O_3 values for all the size fractions. The values for USGS standard rock BCR-1 as determined in this laboratory against primary standards are also listed in Tables 1 and 2. BCR-1 was used as a flux monitor for most elements in both the long and short irradiations; DTS-1 was used as a flux monitor for MgO in the short activation and for Cr_2O_3 and Ni in the long activation. Pure, synthetic grossularite was used for determining the Al contribution to Mg by the (n,p) reaction and as a flux monitor for Al and Ca in the short irradiation.

Uncertainties for the analytical data are estimated by several methods. Because we can not control the absolute amount of flux to each sample during the short irradiations to better than ±2%, we

Table 1. Major Element Compositions of 74001/2 Samples.

	Depth Below Surface (cm)	TiO_2	Al_2O_3	FeO	MnO	MgO	CaO	Na_2O	K_2O	Cr_2O_3
74002,										
172 Bulk	0–0.1	9.0	6.7	22.3	0.27	14	9.1	0.45		0.74
<20		8.6	7.2	22.1	0.26	13	8.1	0.52	0.13	0.73
176 Bulk	0.5–1.0	8.6	6.9	22.4	0.26	14	8.4	0.45		0.74
<20		8.5	7.1	21.6	0.26	13	7.4	0.50		0.73
177 Bulk	1.5–2.0	8.7	6.5	22.8	0.26	14	8.3	0.46		0.75
<20		8.6	6.9	22.8	0.26	13	8.4	0.50	0.10	0.70
178 Bulk	3.0–3.5	8.9	5.9	23.8	0.27	14	7.7	0.42		0.76
179 Bulk	5.0–5.5	9.1	5.8	23.6	0.27	14	7.6	0.43		0.77
180 Bulk	11.5–12.0	8.9	5.8	23.8	0.27	15	7.9	0.42		0.78
181 Bulk	18.0–18.5	8.8	5.8	23.8	0.26	15	7.2	0.43		0.79
<20		9.0	6.1	23.1	0.27	13	7.6	0.43		0.78
182 Bulk	25.5–26.0	8.9	5.6	24.1	0.27	15	7.2	0.43		0.78
74001,										
98 Bulk	32.5–33.0	8.7	5.9	22.9	0.28	16	7.3	0.42		0.76
107 Bulk	37.0–37.5	–								
<20		9.0	6.0	21.7	0.27	14	7.2	0.43		0.72
113 Bulk	44.0–44.5	9.1	6.2	23.5	0.27	15	7.9	0.40		0.74
<20		8.7	6.0	23.1	0.27	14	7.7	0.45		0.74
119 Bulk	50.5–51.0	8.7	5.7	23.8	0.28	15	7.1	0.40		0.75
124 Bulk	57.5–58.0	9.3	6.0	23.8	0.28	16	8.2	0.39		0.75
<20		9.1	5.8	23.1	0.27	14	7.2	0.40		0.74
BCR-1		2.21*	13.6	12.2*	0.176*	3.45	7.0*	3.26*	1.71*	
DTS-1						49.8*				0.669*
% uncertainty		4.0%	2.0%	2.0%	2.0%	6.5%	4.0%	2.0%	10%	4.0%

*These values used as flux monitor values.

Table 2. (Part 1) Bulk Soils from 74001/2.

Bulk	Depth Below Surface (cm)	Sc	V	Co	Ni	Ba	Sr	La	Ce	Nd	Sm	Eu	Tb	Yb	Lu	Hf	Ta	Th	Zn
74002,																			
175	0–0.1	50.2	112	58.1	110	–	250	6.3	23	21	7.42	1.84	1.7	4.6	0.68	6.0	1.3	0.6	210
176	0.5–1.0	48.7	112	60.2	120	–	260	6.7	24	20	7.55	1.85	1.6	4.8	0.67	6.0	1.3	0.6	220
177	1.5–2.0	47.6	116	61.2	55	–	280	6.2	25	17	7.21	1.86	1.7	4.2	0.63	5.6	1.4	0.3	190
178	3.0–3.5	48.1	110	64.7	–	–	290	6.1	22	20	6.95	1.92	1.7	4.1	0.59	6.4	1.1	0.4	200
179	5.0–5.5	48.1	113	63.6	–	–		6.3	21	18	7.21	1.83	1.6	4.2	0.61	5.7	1.2	0.4	210
180	11.5–12.0	48.0	109	65.1	50	–	280	6.1	21	17	7.04	1.82	1.6	4.1	0.60	5.9	1.1	0.4	200
181	18.0–18.5	47.3	113	65.2	65	–	250	6.0	21	17	6.95	1.80	1.6	4.2	0.59	5.8	1.0	0.4	180
182	25.5–26.0	48.0	112	67.2	115	–	220	6.0	20	18	7.14	1.86	1.6	4.1	0.58	6.2	1.1	0.3	190
74001,																			
98	32.5–33.0	47.7	114	65.0	–	(250)	360	5.6	19.4		6.60	1.93	1.54	4.3	0.56	6.8	1.3	0.5	170
107	37.0–37.5																		
113	44.0–44.5	47.0	121	66.7	52	140	–	5.6	20.9		6.72	2.01	1.37	4.2	0.60	7.0	1.2	0.3	–
119	50.5–51.0	47.9	108	67.6	–	–	306	5.6	19.5		6.6	1.89	1.49	4.3	0.59	7.0	1.4	0.5	190
124	57.5–58.0	47.7	125	66.3	–	–	–	5.79	19.5		6.74	1.86	1.50	4.2	0.59	5.6	1.1	0.5	180
BCR-1		31.6	412	36.0	2390(DTS1)	675	330	25.2	54.2		6.8	1.97	1.15	3.48	0.526	5.2	0.9	6.0	120
% uncertainty		3.0%	2.5%	4.0%	10%	10%	10%	2.5%	3.0%	7.0%	3.0%	4.0%	9.0%	6.0%	6.0%	7.0%	20.0%	8.0%	10–30%

Table 2. (Part 2) 90–150 μm fractions from 74001/2.

90–150 μm	Depth Below Surface (cm)	Sc	V	Co	Ni*	Ba	Sr	La	Ce	Nd	Sm	Eu	Tb	Yb	Lu	Hf	Ta	Th	FeO	Na_2O	Cr_2O_3	Zn
74002,																						
175	0–0.1	49.0		57.0		—	240	6.0	21	23	6.97	1.74	1.6	4.3	0.63	5.8	1.3	0.5	22.2	0.40	0.74	110
176	0.5–1.0	51.0		59.0		—	260	6.1	22	20	7.06	1.83	1.6	4.5	0.68	6.0	1.1	0.5	22.6	0.41	0.74	130
177	1.5–2.0	48.8		60.8		60	295	6.3	23	16	7.21	1.87	1.7	4.2	0.64	6.0	1.2	0.5	23.1	0.39	0.71	100
178	3.0–3.5	48.7		67.0		—	210	6.1	22	19	7.11	1.90	1.6	4.1	0.60	6.0	0.9	0.4	23.9	0.39	0.77	105
179	5.0–5.5	47.2		64.2		75	250	5.8	21	20	6.87	1.83	1.6	3.9	0.59	5.6	1.0	0.4	23.2	0.38	0.76	120
180	11.5–12.0																					
181	18.0–18.5	48.0		66.8		—	210	5.8	23	20	7.03	1.81	1.7	4.1	0.60	5.5	0.6	0.2	24.2	0.42	0.78	130
182	25.5–26.0																					
74001,																						
98	32.5–33.0	47.0		65.9		300	322	5.6	19.2		6.69	1.85	1.4	4.2	0.55	6.6	1.1	0.5	23.8	0.39	0.78	100
107	37.0–37.5	46.7		68.2		—	231	5.8	20.9		6.91	1.89	1.6	4.2	0.58	6.6	1.2	0.5	24.7	0.40	0.83	100
113	44.0–44.5	46.2		66.2		135	307	5.5	20.9		6.58	1.94	1.4	4.2	0.55	6.4	1.2	—	23.4	0.39	0.75	100
119	50.5–51.0	46.9		67.7		—	—	5.5	20.5		6.52	1.84	1.3	4.2	0.56	6.1	1.3	—	239	0.38	0.75	130
124	57.5–58.0	47.1		66.4		(150)	317	5.6	18.9		6.64	1.86	1.47	4.1	0.60	5.9	1.2	0.25	24.0	0.37	0.76	140

*Ni not reported due to contamination—see text.

Table 2. (Part 3) <20 μm fractions from 74001/2.

20 μm	Depth Below Surface (cm)	Sc	V	Co	Ni*	Ba	Sr	La	Ce	Nd	Sm	Eu	Tb	Yb	Lu	Hf	Ta	Th	FeO	Na_2O	Cr_2O_3	Zn
74002,																						
175	0–0.1	47.3	111	59.5		–	255	7.0	25	25	7.54	1.95	1.8	4.5	0.66	6.2	1.1	0.5	22.1	0.52	0.73	320
176	0.5–1.0	46.0	107	56.3		–	(385)	6.9	25	22	7.28	1.86	1.7	4.4	0.65	6.1	1.2	0.3	21.6	0.50	0.73	360
177	1.5–2.0	47.6	108	60.2		–	(500)	6.8	25	17	7.51	1.94	1.6	4.5	0.65	6.0	1.2	0.3	22.8	0.50	0.70	350
178	3.0–3.5	47.5		64.0		130	–	6.5	22	21	7.33	1.90	1.6	4.4	0.64	6.3	1.2	0.4	23.1	0.52	0.75	325
179	5.0–5.5	47.5		63.3		–	(410)	6.5	22	18	7.38	1.80	1.7	4.3	0.63	5.8	1.0	0.4	22.7	0.52	0.75	350
180	11.5–12.0																					
181	18.0–18.5	47.7	114	62.4		–	215	6.2	21	18	7.11	1.85	1.7	4.3	0.61	5.6	1.0	0.4	23.1	0.43	0.78	315
182	25.5–26.0																					
74001,																						
98	32.5–33.0	46.9	–	61.8		(290)	312	6.4	22		6.90	2.02	1.7	4.4	0.60	6.6	1.3	0.5	23.0	0.45	0.78	270
107	37.0–37.5	46.7	118	60.0		–	215	5.8	20		6.62	1.87	1.6	4.2	0.57	6.6	1.3	–	21.7	0.44	0.72	270
113	44.0–44.5	48.3	115	64.0		(108)	271	5.6	21		6.7	2.01	1.5	4.3	0.55	6.9	1.6	0.7	22.1	0.45	0.74	290
119	50.5–51.0	47.7	–	63.2		–	335	6.1	22.2		6.8	1.97	1.3	4.1	0.58	6.4	1.4	–	23.1	0.44	0.74	290
124	57.5–58.0	47.6	119	62.1		–	–	5.9	21.2		6.9	1.90	1.5	4.4	0.61	6.6	1.4	0.5	23.1	0.40	0.74	190

*Ni not reported due to contamination—see text.

estimate the uncertainties for the elements determined by those irradiations as ±2% *or* as the uncertainties associates with radioassay (including uncertainties propagated by interelement interference corrections), whichever is greater. Uncertainties for elements by the long irradiation are estimated based on observed dispersion among replicate analyses of the same rock powder over a period of years (see Jacobs *et al.*, 1977, for details).

RESULTS

The average composition of 74001/2 bulk soils closely resembles analyses of samples of orange soil 74220 (Rhodes *et al.*, 1974; Korotev, 1976) (Table 3). If the sample standard deviations (±s) are used as criteria, only Na_2O, TiO_2, and FeO are different by more than 1s.

Except for the uppermost three samples, the major and trace element abundances of 74001/2 soils are not distinguishable within analytical uncertainties. For purposes of comparison, concentrations in the samples have been normalized to those in sample 74002,181 and plotted in Figs. 1 and 2. The choice of 74001,181 was arbitrary; any sample from a depth greater than 3 cm could have been used. We do not regard the seemingly systematic trend in La as significant. There are no corresponding changes in La/Sm ratios or other trace or major element concentrations. This suggests that the La variations are random and that there are no changes in components or compositions among the lower 10 samples.

In the upper three soils, all sampled within 3 cm of the lunar surface, there are systematic and significant differences in composition. Relative to the rest of the core, these samples are clearly depleted in FeO, MgO, and Co and enriched in Al_2O_3, CaO, and perhaps La and Sc. Some of these differences would not be analytically significant except that they occur together, thus reducing the probability that the variations are random. Some of the compositional variations in the uppermost soils are even more pronounced when the average compositions of the <20 μm fractions are compared (Table 3). Normalizing again to 74002,181, the ratios plotted on Fig. 2 show that the <20 μm fractions are slightly less depleted than the bulk soils in both heavy and light REE, but especially the light REE.

Ni values are reported only for the bulk soil analyses. There is clear evidence that the size fractions have been seriously contaminated by the process of sieving with metal grid sieves. The contamination is most readily apparent in the finest size fractions where concentrations of more than 1000 ppm were detected. Since the <20 μm fraction normally comprise 25–35% of the bulk soil, the calculated contribution from these fractions would greatly exceed the abundances determined directly in the original bulk soil samples. This can only be explained by contamination during the sieving process.

DISCUSSION

Although the soils from 74001/2 have different proportions of orange and black glass, they are homogeneous in composition; therefore, there appear to be

Table 3. Compositions of orange glass 74220 and average soils from 74001/2.

	74220		74001/2									
					0–2 cm				2–68 cm			
	Rhodes et al. (1974) Bulk	Korotev (1976) 90–150 μm	Bulk Soil Average		Bulk Average		<20 μm Average		Bulk Average		<20 μm Average	
TiO_2	8.81		8.9	±.2	8.8	±.2	8.6	±.1	8.9	±.2	9.0	±.2
Al_2O_3	6.32		6.1	±.4	6.7	±.2	7.1	±.2	5.8	±.2	6.0	±.1
FeO	22.04	23.0	23.4	±.6	22.5	±.3	22.2	±.6	23.7	±.3	23.1	±.3
MnO	0.30		0.27	±.01	0.26	±.01	0.27	±0	0.27	±.01	0.27	±0.0
MgO	14.44		15	±1	14	±1	13	±1	15	±1	14	±1
CaO	7.68		7.8	±.6	8.6	±.4	8.0	±.5	7.6	±.4	7.4	±.3
Na_2O	0.36	0.37	0.43	±.02	0.45	±.01	0.51	±.01	0.42	±.02	0.45	±.05
Cr_2O_3	0.75	0.71	0.76	±.02	0.74	±.01	0.72	±.02	0.76	±.02	0.76	±.01
Sc		47.1	48	±1	49	±1	47	±1	48	±1	48	±1
Co		61.2	64	±3	60	±2	59	±2	66	±1	63	±1
Ni	89	105	80	±30								
La		6.23	6.0	±.3	6.4	±.3	6.9	±.1	5.9	±.3	6.1	±.3
Ce		22.4	21	±2	24	±1	25	±0	21	±1	21	±1
Sm		7.0	7.0	±.3	7.4	±.2	7.4	±.1	6.9	±.2	7.0	±.3
Eu		1.78	1.87	±.06	1.85	±.01	1.92	±.05	1.88	±.07	1.92	±.08
Tb		1.7	1.6	±.1	1.7	±.1	1.7	±.1	1.6	±.1	1.6	±.1
Yb		4.4	4.3	±.2	4.5	±.3	4.5	±.1	4.2	±.1	4.3	±.1
Lu		0.63	0.61	±.04	0.66	±.03	0.65	±.01	0.59	±.01	0.60	±.03
Hf		5.8	6.2	±.5	5.9	±.2	6.1	±.1	6.3	±.6	6.4	±.4
Ta		1.2	1.2	±.1	1.3	±.1	1.2	±.1	1.2	±.1	1.3	±.2
Th		0.4	0.4	±.1	0.5	±.1	0.4	±.1	0.4	±.1	0.5	±.1

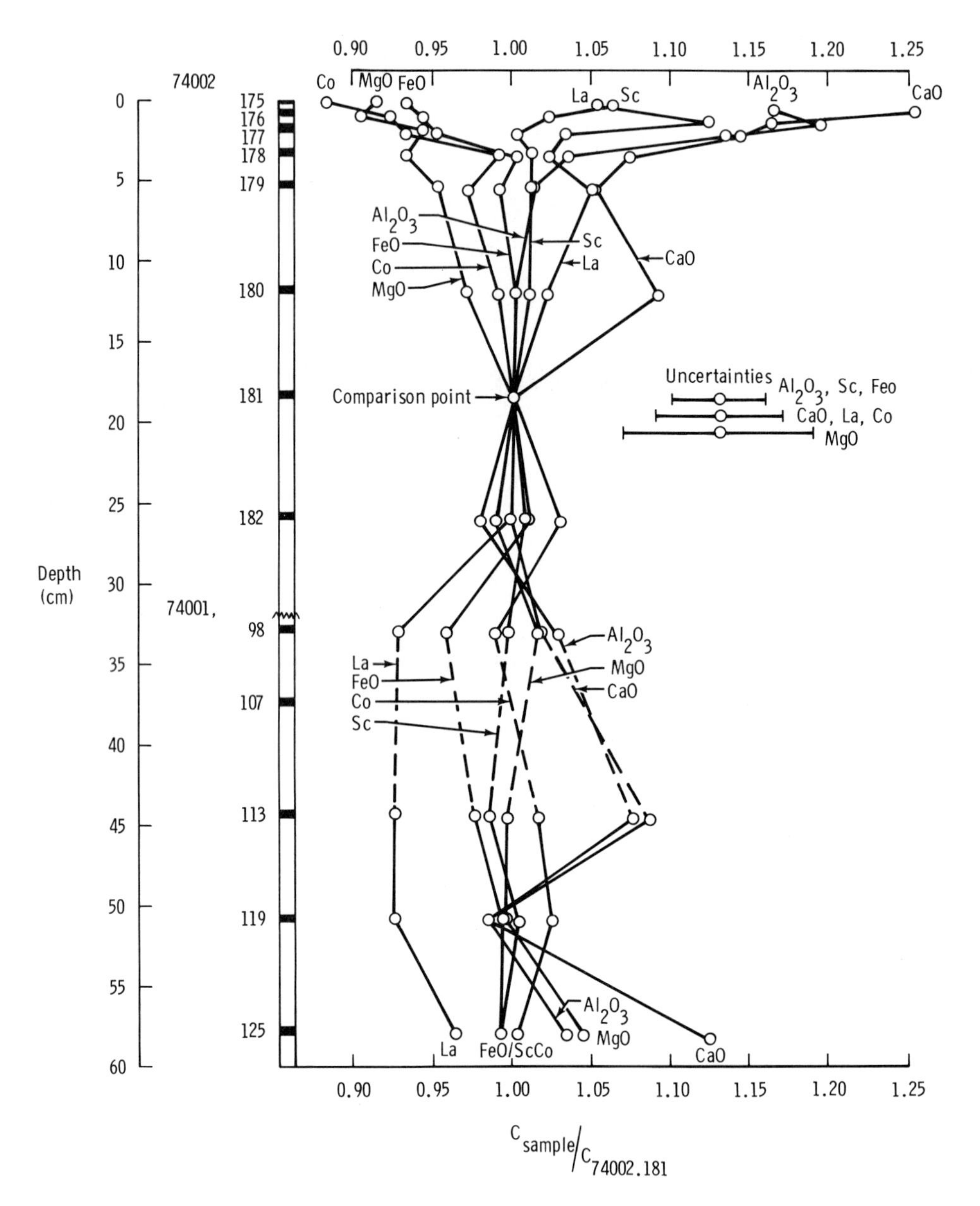

Fig. 1. Profiles of elemental distributions in 74001/2 bulk soils vs. depth. Profiles are arbitrarily normalized to 74002,181 bulk soil. Note the change in composition near the lunar surface (0–2 cm).

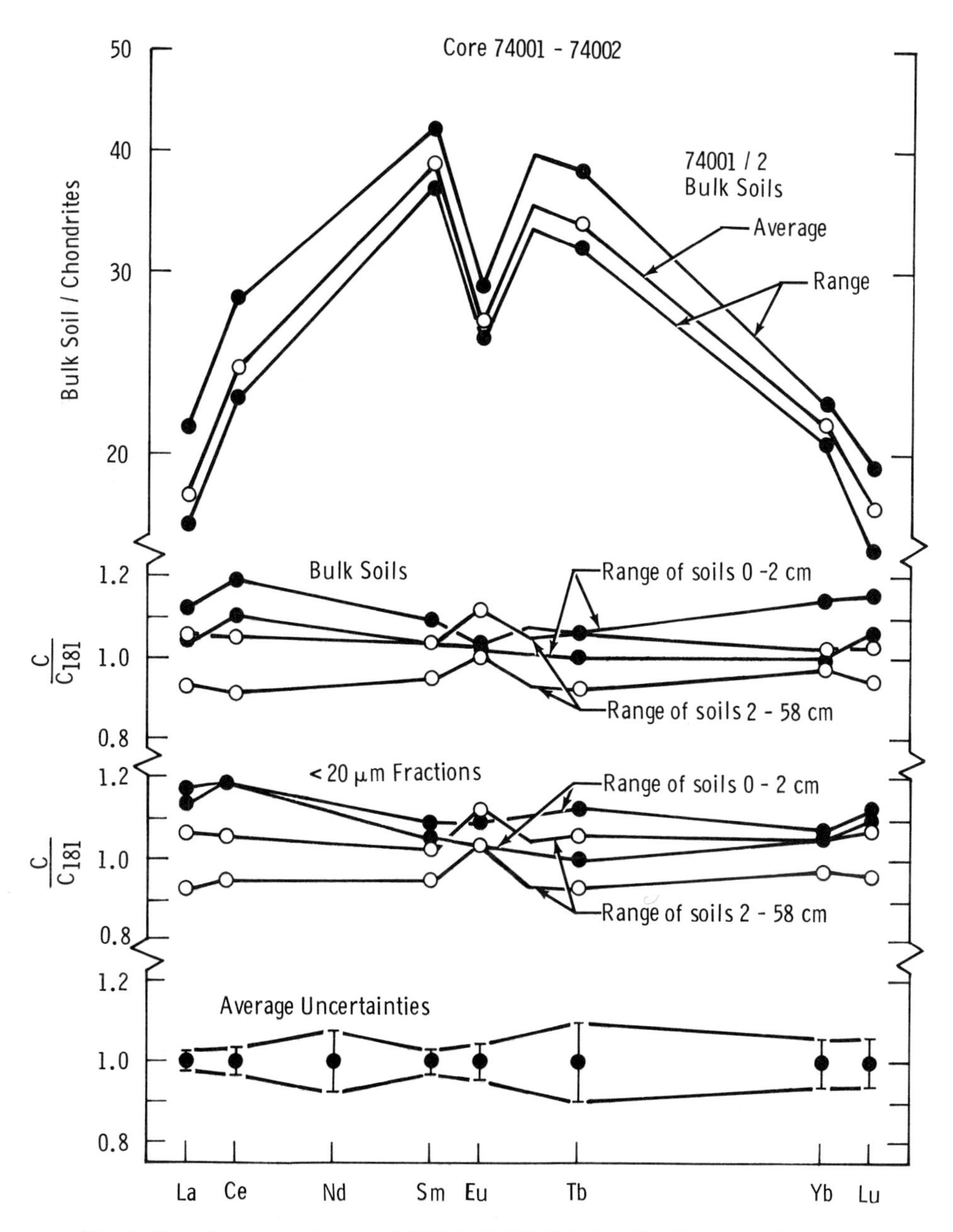

Fig. 2. Top: Average and range of REE in 74001/2 bulk soils. Upper middle: Ranges of bulk soils from surface samples (0–2 cm) and deeper samples (2–68 cm), both normalized to 74002,181. Lower middle: Ranges of <20 μm fractions from surface samples and deeper samples, both normalized to 74002,181. Bottom: typical uncertainties.

no compositional differences between the orange and black components. This inference agrees with the conclusions of Heiken and McKay (1974) that the two glasses differ in cooling rate history and degree of crystallinity rather than composition.

Because the compositions of 74001/2 and 74220 are strikingly similar and so unusual, especially the REE distributions, that these soils are probably derived from a single source. Heiken and McKay (1974) have suggested that the orange/black soil was deposited as ejecta from a fire fountain. As argued previously, there is evidence for admixture of other materials to the top 3 samples in the core; thus, the lower part of the core probably represent the composition of the original material, perhaps better than 74220.

The uppermost soils appear to be mixtures of the orange/black soil with less mafic, more REE-rich material with somewhat less severely fractionated REE. We have used the unusually high Co concentrations in the orange-black soil as an index to estimate extent of mixing. Assuming the extreme case of no Co in the admixed material, a lower limit of about 10% added material is present in the surface soils. If Co is present in the added material, the estimate becomes larger. Assuming a lower limit of 10% admixed material, we then estimate the extreme limits of composition of the admixed material. The composition calculated in this way has about 12% FeO, 4% MgO, 17.6% CaO, and 14.8% Al_2O_3 and has a REE pattern that is relatively depleted in the light and the heavy REE (La = 33X chondrites, Sm = 66X, Lu = 38X). Again, this is an extreme composition based on 10% addition; greater percentages of added material move the calculated composition toward the observed composition of the bulk soil. The calculated I_s/FeO is about 40, not unreasonable for a mare soil from the Taurus-Littrow valley floor. The REE pattern of the added component suggests that it was clearly mare basaltic in nature; however, the low FeO and MgO and relatively high Al_2O_3 and CaO are not consistent with addition of only mare basalt.

The $<20\ \mu m$ fractions in the surface soils are less mafic and more REE-rich than the bulk compositions, suggesting that the $<20\ \mu m$ fractions contain a larger proportion of the admixed material than does the bulk soils, i.e., the admixed material enters as very fine material. An exception is CaO which is, if anything, slightly depleted in the $<20\ \mu m$ fraction relative to the bulk. Apparently whatever carries CaO into the soil is too coarse to the included preferentially in the <20 fraction.

Overall, the compositional evidence then suggests that at least 10% of the surface soils is not orange-black glass. The components of this 10% includes a fine-grained, mare component plus a coarser-grained CaO-rich component. This, in part at least, agrees with the observation of mare basalt fragments in the surface soils (Heiken *et al.*, 1978).

The higher values of Is/FeO (Morris *et al.*, 1978) probably result, at least in part, from "inherited" surface exposure from one or more of the added components described above. A reworking depth of about 5 cm is defined by I_s/FeO data. Compositional evidence for mixing extends to 3 cm. Clearly the surface has been reworked; however, it is not clear that the agglutinates and ferro-magnetic iron particles contained in the surface soils have been formed in this soil exclusively.

If, as suggested by Bogard and Hirsch (1978), the soil layer sampled by 74001/2 had a previous history of shielded exposure and redeposition, these events have not detectably altered the composition of this core. The surface reworking most likely post-dates any redeposition.

SUMMARY

The orange and black glassy soils in the core 74001/2 are homogeneous in composition and due to their unusual compositional characteristics probably have derived from a single source with the same composition.

Compositional evidence suggests that the top of the core has been reworked to a depth of at least 3 cm. The reworked layer is marked by compositional differences which reflect the addition of at least 10% of materials including both mare basalt and a CaO-rich component. The I_s/FeO profile (Morris *et al.*, 1978) indicates a reworking depth of about 5 cm; however, it is likely that at least some of the ferro-magnetically active component (including agglutinitic glass) has been "inherited" with the added material.

Acknowledgments—We thank Bob Browning and the crew of the Texas A&M reactor for irradiating the samples; G. Waits for carefully preparing the size separates, and C. Hardy and B. Allison for help in preparing this manuscript. The manuscript has benefited from the thoughtful reviews of J. C. Laul and M.-S. Ma.

REFERENCES

Blanchard D. P., Brannon J. C., Aaboe E. and Budahn J. R. (1978) Major trace element chemistry of Luna 24 samples from Mare Crisium. In *Mare Crisium: The View from Luna 24* (R. B. Merrill and J. J. Papike, eds.), p. 613–630. Pergamon, N.Y.

Bogard D. D. and Hirsch W. C. (1978) Depositional and irradiational history and noble gas contents of orange-black droplets in the 74002/1 core from Shorty Crater. *Proc. Lunar Planet. Sci. Conf. 9th*. This volume.

Heiken G. and McKay D. S. (1974) Petrology of Apollo 17 soils. *Proc. Lunar Sci. Conf. 5th*, p. 843–860.

Heiken G., McKay D. S. and Gooley R. (1978) Petrology of a sequence of pyroclastic rocks from the valley of Taurus-Littrow (Apollo 17 landing site) (abstract). In *Lunar and Planetary Science IX*, p. 491–493. Lunar and Planetary Institute, Houston.

Jacobs J. W., Korotev R. L., Blanchard D. P. and Haskin L. A. (1977) A well tested procedure for instrumental neutron activation analysis of silicate rock and minerals. *J. Radioanal. Chem.* **40**, 93–114.

Korotev R. L. (1976) Geochemistry of grain size fractions of soils from the Taurus-Littrow valley floor. *Proc. Lunar Sci. Conf. 7th*, p. 695–726.

McKay D. S., Heiken G. and Waits G. (1978) Grain size and petrology as a key to the rate of *in situ* reworking and lateral transport on the lunar surface. *Proc. Lunar Planet. Sci. Conf. 9th*. This volume.

McKay D. S. and Waits G. (1978) Grain size distribution of samples from core 74001 and 74002. (abstract). In *Lunar and Planetary Science IX*, p. 723–725. Lunar and Planetary Institute, Houston.

Morris R. V., Gose W. A. and Lauer H. V., Jr. (1978) Depositional and surface exposure history of the Shorty Crater core 74001/2: FMR and magnetic studies. *Proc. Lunar Planet. Sci. Conf. 9th.* This volume.

Rhodes J. M., Rodgers K. V., Shih C., Bansal B. M., Nyquist L. E., Wiesmann H. and Hubbard N. J. (1974) The relationship between geology and soil chemistry at the Apollo 17 landing site. *Proc. Lunar Sci. Conf. 5th*, p. 1097–1117.

Schmitt R. A., Linn T. A. Jr. and Wakita H. (1970) The determination of fourteen common elements in rocks via sequential instrumental activation analysis. *Radiochim. Acta* **13**, 200–212.

Proc. Lunar Planet. Sci. Conf. 9th (1978), p. 1981–2000.
Printed in the United States of America

Depositional and irradiational history and noble gas contents of orange-black droplets in the 74002/1 core from Shorty Crater

D. D. BOGARD and W. C. HIRSCH*

NASA Johnson Space Center, Houston, Texas 77058
(*Also Northrop Services, Inc., Houston, Texas)

Abstract—The isotopic concentrations of the noble gases have been measured in grain size separates of 14 soils from different depths in the 74002–74001 drive tube taken on the rim of Shorty Crater. The orange-black droplets in this core were probably formed from pyroclastic eruptions $\sim 3.6 \times 10^9$ yr. ago, excavated by Shorty Crater ~10 m.y. ago, and except for the uppermost ~5 cm, contain extremely low values of indices of surface exposure. All soils contain trapped solar wind gases in concentrations $\sim 10^{-2}$–10^{-3} that of typical lunar soils. This gas may be either adsorbed onto droplet surfaces or contained in a mature soil contaminant which may occur in the core in variable concentrations of ~0.5%. Isotopic compositions of these trapped gases are similar to that in typical lunar soils and are $^4He/^3He \simeq 2785$, $^{20}Ne/^{22}Ne \simeq 12.6$, $^{22}Ne/^{21}Ne \simeq 32$, $^{36}Ar/^{38}Ar \simeq 5.3$, and $^{40}Ar/^{36}Ar \simeq 3.8$. Concentrations of cosmogenic ^{21}Ne and ^{38}Ar generally show a steady increase with increasing subsurface depth in the core for samples below 5 cm depth. These data are inconsistent with the entire cosmic ray irradiation of the core having occurred in its recovered orientation and with any reasonable model of slow core accumulation. We suggest that the core has received a two-stage irradiation by cosmic rays, and that the core stratigraphy was inverted between stages. The latest irradiation stage began when the droplets were excavated and inverted by Shorty Crater, which we infer to have occurred ~10 m.y. ago. The first irradiation stage possibly occurred for ~20 m.y. immediately after the pyroclastic deposition of the droplets ~3600 m.y. ago. Droplets in the lowermost ~10 cm of the core show appreciably higher concentrations of cosmogenic gases and apparently experienced a somewhat different irradiation history. Droplets in the uppermost ~5 cm of the core show evidence of recent *in situ* reworking and contain correlated, excess concentrations of cosmogenic Ne and Ar. The excess cosmogenic gases in these near-surface soils have $[^{21}Ne/^{38}Ar]_c \simeq 0.65$, which is considerably different from the values of 2.0–2.4 shown by core soils of >5 cm depth. Two possible explanations for this difference are that near-surface soils contain a few percent of a pre-irradiated, highland component (ropy glasses?), or that near-surface soils have an appreciable cosmogenic component produced by solar flare irradiation over the past ~10 m.y.

INTRODUCTION

The double drive tube 74001-74002 is an ~67 cm deep section of the lunar regolith taken on the rim of Shorty Crater. The core contains orange and black droplets which have been proposed to be pyroclastic deposits of volcanic eruptions (Heiken *et al.*, 1974; Heiken and McKay, 1978) which occurred on the surface of the moon $\sim 3.6 \times 10^9$ yrs ago, the measured K-Ar age of the material (Husain and Schaeffer, 1973; Alexander *et al.*, 1978). The droplets are primarily composed of glass, olivine, and ilmenite, and the black droplets are apparently the crystallized equivalent of the orange droplets (Heiken *et al.*, 1974; Heiken and McKay, 1978). Agglutinates and fine-grained metal, which are indices of surface maturity, are virtually absent from the core except for small amounts in the uppermost ~5 cm, which presumably result from recent, *in situ* reworking by meteorites (Nagle, 1977; McKay *et al.*, 1978; Morris *et al.*, 1978). The orange-black droplet soils show little variation in chemical composition (Blanchard and Budahn, 1978) and in distribution of grain sizes (McKay *et al.*, 1978). Therefore, it appears that the orange-black droplets were excavated from appreciable depth by the relatively young Shorty Crater event (Muehlberger *et al.*, 1973), were deposited on the lunar surface as a single unit, and except for the uppermost ~5 cm have not since been contaminated with other lunar soils or exposed at the lunar surface. These conclusions suggest that any solar wind gases found in the orange-black material would result from surface exposure $\sim 3.6 \times 10^9$ yrs ago and might differ in composition from more recent solar wind. Furthermore, the above conclusions suggest that the 74001-74002 core ought to have had a simple irradiation history and to offer an ideal example to examine the effects of surface exposure on a chemically-homogeneous material with a relatively simple history. Unfortunately, as we shall show in this paper, the 74001-74002 core appears to have had a complex history which makes its exposure history difficult to determine.

EXPERIMENTAL

Fourteen samples from different depths in the 74001-74002 core were each sieved into 8 grain size fractions (McKay *et al.*, 1978). We measured the isotopic abundances of the noble gases by standard mass spectrometric techniques in the <20 μm sizes of all 14 samples, in the 150–250 μm sizes of 13 samples, and in a few additional grain sizes. Noble gas data are given in Table 1. These same grain size separates were also studied by Heiken and McKay (1978) for petrology as a function of grain size, by Morris *et al.* (1978) for FMR and magnetic properties, and by Blanchard and Budahn (1978) for selected chemical compositions.

TRAPPED SOLAR GASES

Concentrations of the noble gases in these core soils increase with decreasing grain size by approximately the amounts to be expected if the gas is located on grain surfaces. Gas concentrations are lower by a factor of 10^2–10^3, however, compared to an analogous grain size of typical lunar soils. The noble gases present consist of one component with a composition similar to that of the solar wind and another component produced by nuclear interactions of cosmic rays. Figure 1 is an isotope correlation diagram for Ne which demonstrates that Ne in 150–250 μm size fractions of these soils consists of a cosmogenic component and a trapped solar wind component. Ne in <20 μm grain sizes is almost entirely solar wind in origin, with relatively small amounts of cosmogenic Ne. There is no evidence in these soils of a significant component of Ne which isotopically differs from Ne in the recent solar wind, as might be expected for any indigenous lunar gas. Because it seems highly probable that this solar gas was trapped in the

Table 1. Isotopic concentrations (units of cm^3 STP/g) and ratios of noble gases in grain size separates of soils from the 74002-74001 drive tube. Subsurface soil depths are also indicated. Absolute uncertainties in concentrations are estimated as +5–10% for He, Ne, and Ar and ±10–15% for Kr and Xe. Relative uncertainties given for isotopic ratios are one sigma of the mean of multiple measurements of individual ratios plus one-half the applied blank corrections. Absolute isotopic ratios have an additional uncertainty of ±0.1%/mass unit in applied discrimination corrections. Blank corrections for <20 μm grain sizes were: $^4He \lesssim 1\%$, $^{22}Ne \lesssim 2\%$, $^{36}Ar \leq 1\%$, and $^{40}Ar \lesssim 2\%$, except for soil 74001,113 which had somewhat higher blank corrections. Blank corrections for 150-250 μm grain sizes were generally: $^4He \leq 15\%$, $^{22}Ne \leq 25\%$, $^{36}Ar \lesssim 26\%$, and $^{40}Ar \leq 6\%$. Soil 74001,113 again showed somewhat higher blank corrections and was the only exception to the above limits for He and Ar blanks.

Sample	Wt. (mg)	^3He $\times 10^{-7}$	^4He $\times 10^{-4}$	^{22}Ne $\times 10^{-7}$	$^{20}Ne/^{22}Ne$	$^{22}Ne/^{21}Ne$	^{36}Ar $\times 10^{-7}$	$^{36}Ar/^{38}Ar$	$^{40}Ar/^{36}Ar$	^{84}Kr $\times 10^{-9}$	^{132}Xe $\times 10^{-9}$
74002,175 (<0.1 cm)											
45-75 μm	7.12	23.7	60.7	58.4	12.59 ± .04	24.23 ± .41	71.0	5.128 ± .005	7.03 ± .10	5.69	10.7
<20 μm	7.25	125.	336.	347.	12.62 ± .04	30.73 ± .18	434.	5.238 ± .005	5.52 ± .02	25.7	39.2
74002,176 (0.5–1.0 cm)											
150–250 μm	13.26	26.2	64.5	70.7	12.59 ± .03	24.55 ± .32	122	5.147 ± .005	4.96 ± .03	8.32	13.6
<20 μm	10.81	155.	420.	412.	12.70 ± .03	31.27 ± .16	558	5.233 ± .005	4.53 ± .01	31.3	46.6
74002,177 (1.5–2.0 cm)											
150–250 μm	12.31	19.5	47.9	43.8	12.53 ± .14	23.04 ± .24	68.7	5.117 ± .005	6.45 ± .05	5.82	11.2
<20 μm	10.93	97.9	287.	284.	12.62 ± .03	30.79 ± .28	391.	5.250 ± .015	5.46 ± .02	21.2	32.5
74002,178 (3.0–3.5 cm)											
150-250 μm	13.53	4.08	7.42	5.72	11.88 ± .07	12.37 ± .36	13.2	4.965 ± .009	23.5 ± .4	2.85	8.74
<20 μm	11.47	45.7	124.	139.	12.55 ± .03	29.27 ± .34	176	5.317 ± .015	7.55 ± .05	9.43	15.0
74002,179 (5.0–5.5 cm)											
150–250 μm	14.20	1.89	1.85	1.71	9.28 ± .38	4.37 ± .46	2.12	3.907 ± .009	133. ± 8	0.86	2.71
<20 μm	11.42	20.4	56.1	81.4	12.39 ± .04	28.27 ± .75	97.2	5.286 ± .010	11.3 ± .2	5.35	10.7
20–45 μm											
Split B	2.41	3.32	4.87	7.30	11.36 ± .13	11.0 ± 1.0	8.8	4.65 ± .04	48.2 ± .7	5.91	15.2
Split D	5.54	5.89	12.4	13.4	11.57 ± .16	17.1 ± .4	12.5	4.99 ± .02	32.7 ± .3	1.60	2.57
74002,180 (12 cm)											
150–250 μm	9.70	1.72	1.43	1.30	7.76 ± .74	3.29 ± .65	1.65	3.62 ± .13	169. ± 16	1.13	3.79
<20 μm	13.78	13.8	36.7	60.5	12.31 ± .03	27.79 ± .45	70.6	5.17 ± .05	15.6 ± .2	6.02	15.1

Table 1. *(cont'd.)*

Sample	Wt.(mg)	^{3}He $\times 10^{-6}$	^{4}He $\times 10^{-2}$	^{22}Ne $\times 10^{-7}$	$^{20}Ne/^{22}Ne$	$^{22}Ne/^{21}Ne$	^{36}Ar $\times 10^{-7}$	$^{36}Ar/^{38}Ar$	$^{40}Ar/^{36}Ar$	^{84}Kr $\times 10^{-9}$	^{132}Xe $\times 10^{-9}$
74002,181 (18 cm)											
150–250 μm	9.32	1.56	1.26	0.94	6.41 ± 2.25	2.46 ± .64	0.99	3.06 ± .20	276 ± 22	1.13	4.17
<20 μm	12.18	8.06	19.4	39.6	12.25 ± .03	24.44 ± .10	49.5	5.146 ± .005	16.5 ± .1	2.75	5.66
74002,182 (26 cm)											
150–250 μm	10.15	1.51	1.10	0.65	3.16 ± 1.95	1.59 ± .56	0.63	2.33 ± .20	434 ± 50	0.48	1.50
<20 μm	11.63	4.44	9.52	17.6	12.02 ± .04	19.15 ± .18	27.3	5.292 ± .036	20.6 ± .2	1.75	3.77
74001,98 (32 cm)											
150–250 μm	13.12	2.15	2.06	1.17	7.95 ± .29	2.65 ± .26	1.61	3.35 ± .04	206 ± 5	5.49	1.89
<20 μm	10.68	21.9	52.7	47.1	12.71 ± .15	24.84 ± .42	99.3	5.185 ± .005	5.95 ± .07	12.0	2.88
74001,107 (37 cm)											
150–250 μm	12.95	1.91	1.79	1.17	7.73 ± .26	2.62 ± .19	2.00	3.60 ± .07	169 ± 7	6.00	2.16
<20 μm	10.74	4.08	7.87	11.1	11.90 ± .03	15.16 ± .26	22.9	5.14 ± .15	22.3 ± .5	3.91	1.21
74001,113 (44 cm)											
150–250 μm	10.25	1.19	0.98	0.65	6.31 ± 1.90	2.03 ± 2.5	0.62	3.57 ± .28	238 ± 46	1.88	0.63
90–150 μm	10.12	1.98	1.39	1.63	9.34 ± .24	3.56 ± .35	2.87	4.07 ± .41	95. ± 3	3.33	1.12
<20 μm	8.66	4.59	9.02	11.1	11.97 ± .07	13.46 ± .75	25.2	5.10 ± .03	21.3 ± .4	2.98	0.85
74001,119 (50 cm)											
150–250 μm	12.58	1.61	1.28	1.26	7.69 ± .30	2.66 ± .10	1.94	3.53 ± .06	183 ± 7	7.20	2.59
<20 μm	8.00	3.59	6.78	9.54	11.86 ± .18	14.24 ± .39	30.0	5.15 ± .03	21.6 ± .5	7.89	3.48
74001,125 (57 cm)											
150–250 μm	12.31	1.63	0.87	0.94	6.26 ± .67	1.99 ± .25	1.44	3.14 ± .09	175 ± 6	8.97	4.15
90–150 μm	10.83	2.60	0.88	0.84	1.68 ± 1.23	1.18 ± .18	0.65	1.59 ± .34	417 ± 43	2.06	1.02
20–45 μm	7.31	2.81	1.70	1.51	6.50 ± .65	2.05 ± .20	2.40	3.14 ± .06	131 ± 2	2.53	1.26
<20 μm	10.00	5.15	8.41	9.28	11.75 ± .03	9.83 ± .24	23.0	4.93 ± .03	18.6 ± .4	5.28	2.60
74001,2 (67 cm)											
150–250 μm	12.60	2.69	1.34	1.55	5.57 ± .51	1.84 ± .14	2.08	2.88 ± .06	132 ± 3	0.81	0.202
45–75 μm	12.50	13.2	2.14	1.81	6.94 ± .38	2.18 ± .18	2.47	3.14 ± .07	124 ± 3	1.51	0.42
<20 μm	9.74	4.15	5.95	6.88	11.24 ± .06	7.48 ± .23	14.8	4.77 ± .02	25.8 ± .3	2.32	0.57

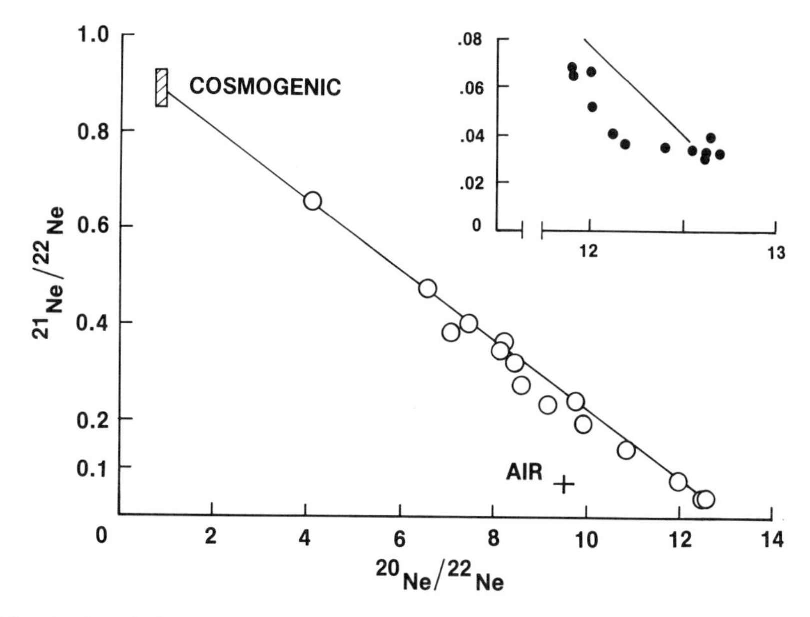

Fig. 1. Correlation diagram for Ne isotopes in 74002–74001 core soils. Open symbols represent 150–250 μm grain sizes; closed symbols in the Figure inset represent <20 μm grain sizes. The mixing line shown connects the typical composition of Ne produced by cosmic rays (upper left) with trapped solar wind Ne (lower right).

orange-black droplets at the time of their formation, it is worthwhile to discuss the isotopic composition of this gas in some detail.

The ordinate-intercept technique (Eberhardt *et al.*, 1970) is the standard way to determine the isotopic composition of the trapped component and the abundance of the cosmogenic component in a suite of grain size separates of lunar soils. For most of our core soils, however, we only analyzed two grain sizes for each soil. To determine the composition of the trapped component, therefore, we combined data for the <20 μm grain sizes for all soils below the upper ~5 cm reworking zone and treated these in a common ordinate-intercept plot. Such a common treatment seems valid for the following reasons. To use an ordinate-intercept analysis for a suite of samples the trapped component must be surface correlated and have the same isotopic composition among samples, and the abundances of the cosmogenic component must be the same among samples. Because soils below 5 cm depth in the core had a common origin, appear to contain no agglutinate or soil contaminants, and are extremely similar in mineralogy and major element composition, they are likely to contain trapped gases with nearly identical composition. Abundances of cosmogenic gases in core soils below the 5 cm reworking zone are also similar, as we shall show later. Two exceptions to this statement are the two deepest soils (74001,2 and ,125) which contain higher concentrations of cosmogenic gases, and which we have commonly excluded from our ordinate-intercept analyses. Data for all grain sizes of soils in the upper ~5 cm reworking zone of the core were analyzed in a separate ordinate-intercept plot in case the composition of trapped gas in these soils differed from that in the rest of the core. The results of these ordinate-intercept analyses are described below.

Trapped $^4He/^3He$ in <20 μm grain sizes for soils below 5 cm depth (omitting 74001,125 and ,2 for the reason given above) was 2785 ± 105, whereas this value for soils both above and below the 5 cm level was 2790 ± 60. (Uncertainties are one sigma of the least squares intercept values). A correction of 1.65×10^{-4} cm^3/g, which assumes an age of 3.6×10^9 yr. (Alexander *et al.*, 1978) and a U concentration of 155 ppb (Hutcheon *et al.*, 1974), was applied to all ^4He concentrations to correct for radiogenic ^4He. The $^4He/^3He$ ratios derived, however, are not very sensitive to this correction, as applying no correction would produce a negligible difference in the derived ratios. The $^4He/^3He$ ratios derived are the same within uncertainties and are similar to $^4He/^3He$ ratios which have been measured on other lunar soils, i.e., generally 2300–2900. Much of this observed variation in $^4He/^3He$ in soils has been attributed to isotope fractionation during gas loss. Values of $^4He/^3He$ of 1860–2450 were measured in Al foils exposed on the lunar surface during several of the Apollo missions (Geiss, *et al.*, 1972), and indicate that the value of $^4He/^3He$ in the solar wind is variable over relatively short time periods. Thus, long-term variability of $^4He/^3He$ in the solar wind may also have contributed to the observed variation of $^4He/^3He$ in lunar soils. The observation that $^4He/^3He$ in 74001/2 soils are similar to $^4He/^3He$ in many other lunar soils suggests that no major change in this ratio in the solar wind occurred between 3600 m.y. ago and the past ~200 m.y. Several meteorites which contain solar gases possibly implanted as long as 4500 m.y. ago also show $^4He/^3He$ of ~3000 (Heymann 1971).

Trapped $^{20}Ne/^{22}Ne$ derived from an ordinate-intercept analysis of all grain sizes of soils of <5 cm depth is 12.65 ± .03. Trapped $^{20}Ne/^{22}Ne$ derived for <20 μm size fractions of all soils of >5 cm depth is 12.58 ± .09, but if data for soils 74001,098 and ,2 are omitted, the $^{20}Ne/^{22}Ne$ becomes 12.41 ± .04. Trapped $^{22}Ne/^{21}Ne$ of near-surface soils is 32.5 ± .6, whereas <20 μm size fractions of soils from the rest of the core (excluding soils 74001,125 and ,2) give a value of 32.0 ± .4. These derived isotopic ratios for trapped Ne are similar to each other and to values commonly obtained on lunar soils.

Trapped $^{36}Ar/^{38}Ar$ obtained by ordinate-intercept analysis of all grain size separates of near-surface soils 74002,175–177 is 5.26 ± .02 (relative to an atmospheric value of 5.32). Values of $^{36}Ar/^{38}Ar$ in <20 μm grain sizes of soils below 3.5 cm depth in the core show considerable scatter on an ordinate-intercept plot. For three soils (74002,178, 179, and 182) the measured $^{36}Ar/^{38}Ar$ are higher (5.32, 5.29, and 5.29, respectively) than the 5.26 value derived for near-surface soils. Although the trapped $^{36}Ar/^{38}Ar$ ratio for those soils below the reworking zone is not well defined, the data are consistent with a value of 5.32, which is similar to values of $^{36}Ar/^{38}Ar$ observed in many lunar soils. Values of $^{40}Ar/^{36}Ar$ show some scatter and a nonlinear trend on an ordinate-intercept plot.

Data for <20 μm grain sizes from seven soils 74002,176 through ,181, however, are rather linear and give a trapped $^{40}Ar/^{36}Ar$ value of 3.8 ± .7. This value is higher than that observed in most lunar soils, but is similar to values of ~4 which were observed in soils with high cosmic ray exposure but low surface maturity in the 60002 drill core section (Bogard and Hirsch, 1975).

The isotopic composition of Xe was measured in many, but not all of the grain size fractions analyzed. When plotted on three isotope correlation diagrams, data for the most abundant isotopes of Xe tend to fall between the atmospheric composition and the composition typically measured in lunar soils. This suggests that our samples contained adsorbed atmospheric Xe in amounts greater than our extraction blanks, which is perhaps not surprising considering the small amounts of Xe measured in these samples and the fact that our samples were not preheated. This gas of presumed atmospheric origin makes it impossible to determine the isotopic composition of trapped Xe in these soils.

To summarize, the isotopic composition of solar-derived He, Ne and Ar trapped in the 74001-74002 core soils does not appear to differ substantially from solar gases found in the majority of lunar soils. Because several lines of evidence indicate that trapped gases in these soils were acquired $\sim 3.6 \times 10^9$ yrs ago, this observation suggests that the isotopic composition of noble gases in the solar wind $\sim 3.6 \times 10^9$ yrs ago did not differ substantially from that over the past $\sim 10^8$ yrs. However, because of the uncertain effects of isotopic fractionation after gas implantation, differences of a few percent cannot be ruled out. The observation that the $^{40}Ar/^{36}Ar$ ratio is higher in these droplets compared to typical lunar soils can be explained by a decrease with time in the flux of radiogenic ^{40}Ar from the lunar interior (e.g., Heymann *et al.*, 1975) and does not reflect solar wind composition.

The relative concentrations of trapped solar gases for <20 μm grain sizes as a function of core depth are shown in Fig. 2. The higher [^{36}Ar] for soils in the upper ~5 cm of the core is consistent with the presence of agglutinates, non-orange-black materials, and values of I_s/FeO of ~2–4, all of which have been attributed to *in situ* reworking by meteorites since core deposition (McKay *et al.*, 1978; Morris *et al.*, 1978). Below 5 cm depth both ^{36}Ar concentrations and $^{20}Ne/^{36}Ar$ ratios show a steady decrease with increasing depth. The ratio of orange droplets to black droplets for >20 μm grain sizes in the core shows a nearly identical trend as a function of depth, and the total fraction of orange glass (uncrystallized) decreases by a factor of ~20 from top to bottom (Heiken and McKay, 1978). Depth profiles of concentrations of volatile Cd and Pb (Cirlin and Housley, 1978) and of C (Gibson and Andrawes, 1978) similar to that for ^{36}Ar have also been observed in the core. It appears that these concentration profiles are somehow related to the ratio of orange to black droplets, and several possible mechanisms can be considered.

First, the [^{36}Ar] and $^{20}Ne/^{36}Ar$ may reflect different retention efficiencies of solar wind Ne and Ar, as the Ne/Ar ratio is known to be sensitive to the host phase such as glass or ilmenite. However, this explanation will not account for the observed profiles in Cd and Pb, which are not solar in origin, and it cannot

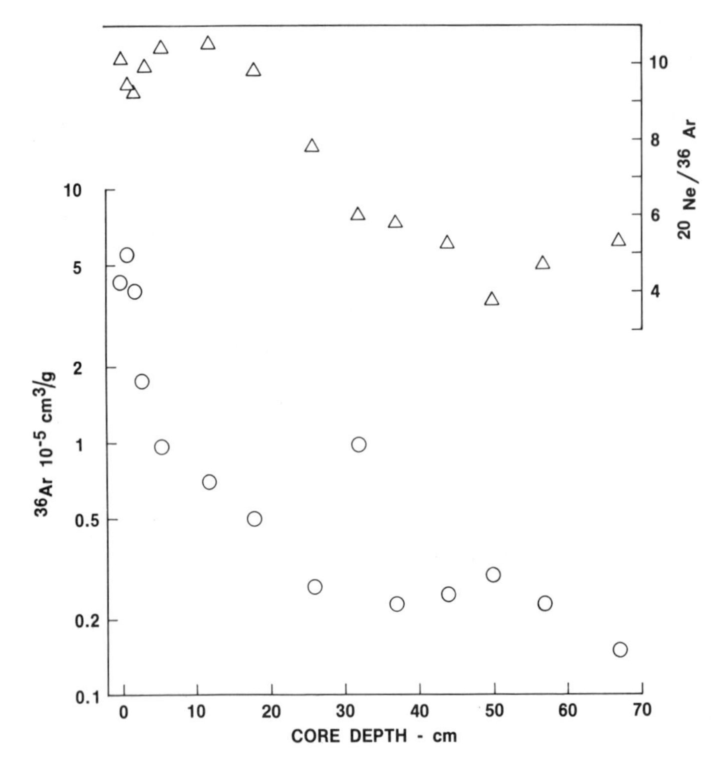

Fig. 2. Concentrations of ^{36}Ar and values of $^{20}Ne/^{36}Ar$ as a function of subsurface depth for <20 μm size fractions of soils in the 74002–74001 core.

readily explain why [^{36}Ar] and Ne/Ar are lower in ilmenite + olivine compared to glass (Baur *et al.*, 1972). To test this explanation the 20–45 μm fraction of 74002,179 was repeatedly passed through a Franz magnetic separator using different current settings. The presence of metal in the black droplets makes these much more magnetic than the orange droplets. Four splits were obtained showing decreasing magnetic susceptibility from A to D, and with most of the yield in split C. We analyzed splits B and D (Table 1) which had values of Fe° of 0.01 and 0.16, respectively (Morris *et al.*, 1978). If the trends in Fig. 2 are caused by different efficiencies of gas retention between orange and black phases, the ^{36}Ar concentration and $^{20}Ne/^{36}Ar$ ratio of split B (largely orange droplets) ought to be considerably higher than those of split D (largely black droplets). Just the opposite is observed. Concentrations of He, Ne, and Ar are higher in split D over split B by ~ a factor of two and $^{20}Ne/^{36}Ar$ is ~30% higher in split D. This experiment then indicates that the profiles of Fig. 2 are not caused directly by the orange/black ratios, but may be related to core depth.

A second explanation of the observed profiles of [^{36}Ar] and $^{20}Ne/^{36}Ar$ is that they reflect fractionated gas loss due to different post-depositional temperatures experienced by core soils of different depths. Cirlin and Housley (1978) suggest that the Cd and Pb profiles may have been produced after the core materials were deposited hot, either in the original pyroclastic deposition or as a result of ejection by Shorty Crater. The volatile Cd and Pb were then driven from deeper material and redeposited on cooler material at shallow depth. It seems difficult to explain the profiles for solar wind Ne and Ar by this mechanism, however, because of their greater volativity. Furthermore, the higher [^{36}Ar] for soil 74001,98 (Fig. 2), which correlates with a higher I_s value for the <20 μm fraction of this soil (Morris *et al.*, 1978), should not exist if the [^{36}Ar] profile was determined by a temperature gradient along the core.

A third possible explanation is that both noble gases and volatile metals were adsorbed onto surfaces of semi-molten droplets from the gas and vapor phase associated with the pyroclastic eruption. This implies that the trapped solar gases were not directly implanted by the solar wind into the droplets, but were adsorbed onto surfaces of droplets, either directly from the gas cloud or as very fine grained regolith particles suspended by the eruption. There is good reason to believe that solar gases contained in these core soils were not directly implanted by the solar wind. Eberhardt (1974) has estimated that it would require $\sim 1.5 \times 10^6$ yr to implant 3×10^{-4} cm^3/g of solar ^{36}Ar into a 0.6 cm deep layer of typical lunar soil. If we scale this estimate to the observed ^{36}Ar content and total depth of the 74001-74002 core, it would require on the order of 10^6 yr of solar wind irradiation of a continuously accumulating core to account for the total solar ^{36}Ar observed. In sharp contrast, Heiken and McKay (1978) argue by analogy with terrestrial pyroclastic eruptions that the total eruption time of the orange-black droplets most likely was orders of magnitude less than 10^6 yr, and probably on the order of days to months.

Still another possible explanation of the variation in [^{36}Ar] with core depth is that the solar gases observed reside in a soil contaminant which was mobilized by the escaping gases of the pyroclastic eruption and which was deposited along with the orange-black glass $\sim 3.6 \times 10^9$ yr ago. This contaminant presumably would be typical lunar soil, which would be required to occur in the core in abundances of $\sim$0.5–1%. The profile of [^{36}Ar] with core depth could then reflect the relative amount of soil contaminant which would have been added to a rapidly accumulating core. Neither petrographic studies nor I_s/FeO measurements (Heiken and McKay, 1978; Morris *et al.*, 1978) can exclude the possible presence of up to $\sim$0.5% of a relatively mature soil in the core. This explanation, however, requires that the similarity in the depth profiles of concentrations of ^{36}Ar and Cd and Pb are fortuitous, as the volatile metals are undoubtedly associated with the orange-black droplets. The solar gases in this possible soil contaminant would still be ancient, however.

Cosmogenic Gases:

Concentrations of cosmogenic ^{21}Ne and ^{38}Ar and the cosmogenic ratios ^{21}Ne/^{38}Ar and ^{22}Ne/^{21}Ne for the coarser grain size fractions of soils in the

Table 2. Calculated concentrations (10^{-8} cm^3 STP/g) of cosmogenic Ne and Ar in grain size separates of soils from the 74002–74001 drive tube. Relative uncertainties in concentrations are estimated as ±2.6% for ^{21}Ne and ±3.7% for ^{38}Ar (see text). These concentrations have been calculated from the non-blank-corrected data by the technique of Bogard and Cressy (1973), which is relatively insensitive to the magnitude of the blank correction.

Sample	Grain Size μm	^{21}Ne	^{38}Ar	$^{21}Ne/^{38}Ar$	$^{22}Ne/^{21}Ne$
74002,175	45– 75	6.08	3.96	1.54	–
74002,176	150–250	6.94	5.81	1.19	–
74002,177	150–250	5.51	4.16	1.32	–
74002,178	150–250	3.85	2.03	1.90	–
74002,179	150–250	3.49	1.64	2.13	1.34 ± .22
Split B	20– 45	4.17	–		–
Split D	20– 45	3.55	–		–
74002,180	150–250	3.65	1.66	2.20	1.4 ± .3
74002,181	150–250	3.68	1.57	2.34	1.36 ± .08
74002,182	150–250	3.99	1.73	2.31	1.28 ± .12
74001,098	150–250	4.13	2.03	2.03	1.17 ± .07
74001,107	150–250	4.21	2.04	2.06	1.17 ± .10
74001,113	90–150	4.11	1.89	2.17	1.14 ± .09
74001,119	150–250	4.48	2.11	2.12	1.10 ± .09
74001,125	150–250	4.53	2.14	2.12	1.13 ± .06
	90–150	7.07	3.24	2.18	1.12 ± .04
	20– 45	7.04	3.57		
	< 20	6.68	3.86		
74001,2	150–250	8.16	3.74	2.18	1.13 ± .03
	45– 75	7.96	3.71		
	< 20	7.22	3.65		

74001–74002 core are given in Table 2. These data are calculated from the non-blank corrected values according to the technique of Bogard and Cressy (1973) and the trapped ratios derived from ordinate-intercept analyses (discussed earlier). Except for the upper three soils, calculated abundances of these cosmogenic gases are not sensitive to uncertainties in the composition of the trapped component. For most soils an uncertainty of ±1 in trapped $^{22}Ne/^{21}Ne$ produces an uncertainty of ~±0.4% in $[^{21}Ne]_c$, and a variation of 5.32 to 5.26 in trapped $^{36}Ar/^{38}Ar$ produces a total uncertainty in $[^{38}Ar]_c$ of ~2%. For the purpose of discussing the depositional and irradiational history of this core, we estimate relative uncertainties of these data from two times the standard deviation of the mean of 29 sets of instrument sensitivity calibrations made during the time of analyses of these core soils as ±2.6% for $[^{21}Ne]_c$ and ±3.7% for $[^{38}Ar]_c$. Uncertainties of $[^{21}Ne]_c$ and $[^{38}Ar]_c$ in soils 74002,175–177 are greater due to the much larger trapped gas contents, and are probably ~±8% for $[^{21}Ne]_c$ and $[^{38}Ar]_c$. Uncertainties given for $[^{22}Ne/^{21}Ne]_c$ are estimated as 2σ of the measured ratios (Table 1). Because of the relatively large gas contents in near surface soils, the $[^{22}Ne/^{21}Ne]_c$ could not be accurately calculated and are not given in Table 2. Because of the large uncertainty in isotopic composition of trapped Xe the uncertainties in calculated cosmogenic Xe are large. We attempted to use the measured $^{129}Xe/^{136}Xe$ ratio to deduce the relative amounts of atmosphere and solar Xe in 150–250 μm grain sizes and to calculate concentrations of cosmogenic ^{128}Xe from these values. The $[^{128}Xe]_c$ so derived show a trend which is roughly similar to the $[^{21}Ne]_c$ and $[^{38}Ar]_c$ as a function of core depth, but there is appreciable scatter in the $[^{128}Xe]_c$ data.

Concentrations of cosmogenic ^{21}Ne and ^{38}Ar show very similar trends as a function of subsurface depth in the 74001–74002 core (Figs. 3 and 4, respectively). The solid line (Fig. 3) represents the approximate production rate of ^{21}Ne from cosmic rays as a function of depth (Reedy and Arnold, 1972; Reedy, 1976), and has been arbitrarily normalized to the datum of 74001,107 (37 cm). The discontinuity in the model curve at 32 cm occurs because the average density of soils in drive tube 74002 vary from ~1.75 g/cm^3 at the surface of ~2.35 g/cm^3 at the bottom, whereas soils in drive tube 74001 (32–67 cm) have essentially the same density of ~2.35 g/cm^3 (Nagle, 1977; and pers. comm., 1978). We have assumed average densities of 2.0 g/cm^3 for 74002 soils and 2.35 g/cm^3 for the 74001 soils, and have normalized the ^{21}Ne production rate curve separately for each drive tube to the 37 cm datum. The production rate curve for cosmogenic ^{38}Ar (Reedy and Arnold, 1972) would have a shape quite similar to that shown for ^{21}Ne.

Concentrations of ^{21}Ne and ^{38}Ar as a function of depth can be considered in three categories. Soils in the upper ~5 cm reworking zone show higher $[^{21}Ne]_c$ and $[^{38}Ar]_c$ compared to most other soils, and these higher concentrations correlate very well with surface exposure indices such as agglutinates (Heiken and McKay, 1978) and I_s/FeO (Morris *et al.*, 1978). Soils in the upper ~5 cm contain ~20% (90–150 μm fraction) of components other than orange-black glass (McKay *et al.*, 1978), and these components may have contributed extra $[^{21}Ne]_c$ and $[^{38}Ar]_c$ from pre-irradiation. Soils between ~5 cm and ~57 cm depth show a nearly smooth increase in $[^{21}Ne]_c$ and $[^{38}Ar]_c$ with increasing depth. The much less precise $[^{3}He]_c$ data show a roughly similar trend, but with considerable scatter. These soils, as far as can be determined, are pure orange-black droplets and show no evidence of *in situ* reworking (Morris *et al.*, 1978; Heiken and McKay, 1978). All grain sizes of the deepest soil analyzed (74001,2) and <150 μm grain sizes of the second deepest soil (74001,125) show much higher $[^{21}Ne]_c$ and $[^{38}Ar]_c$. These higher gas concentrations cannot be attributed to soil maturation, because, unlike soils from the upper ~5 cm of the core, 74001,2 and ,125 contain no appreciable evidence of extraneous components or surface maturation (Morris *et al.*, 1978; Heiken and McKay, 1978). Below we discuss these data in detail, beginning with soils of >5 cm depth.

Irradiation history and the age of Shorty Crater: The data (Figs. 3 and 4) for soils of 5–57 cm depth are clearly inconsistent with a simple irradiation history of the core in its recovered stratigraphy (solid curve). The data are also inconsistent

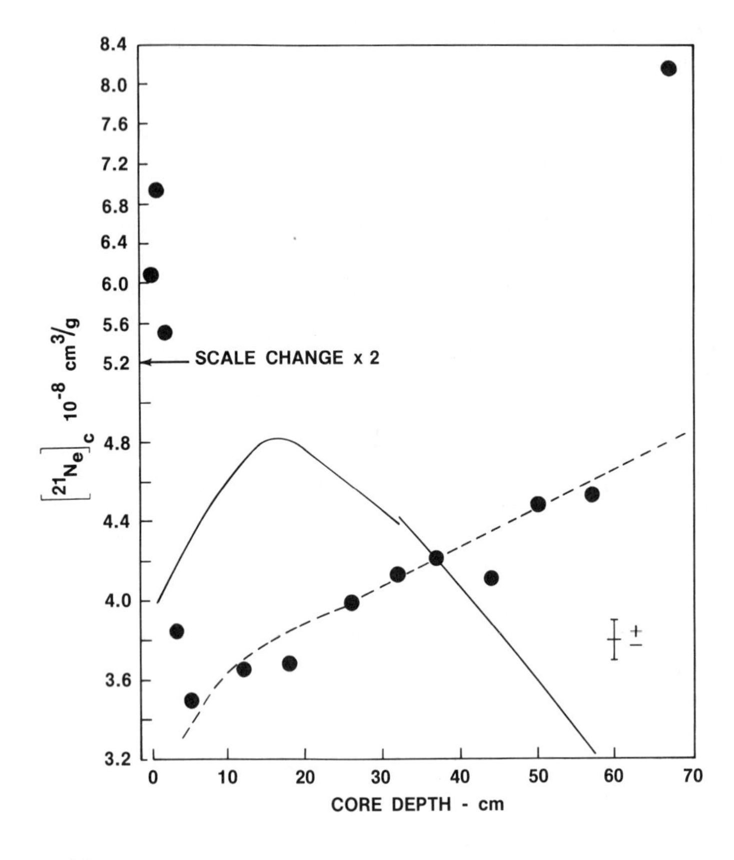
8.4
8.0
7.6
7.2
6.8
6.4
6.0
5.6
5.2
4.8
4.4
4.0
3.6
3.2
SCALE CHANGE x 2
$[^{21}Ne]_c$ 10^{-8} cm^3/g
0
10
20
30
40
50
60
70
CORE DEPTH - cm

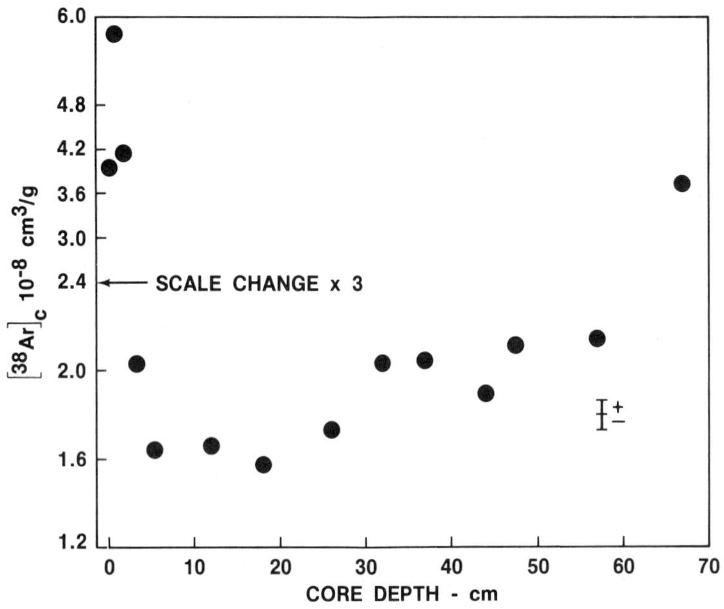
6.0
4.8
4.2
3.6
3.0
2.4
2.0
1.6
1.2
SCALE CHANGE x 3
$[^{38}Ar]_c$ 10^{-8} cm^3/g
0
10
20
30
40
50
60
70
CORE DEPTH - cm

with any reasonable model of slow accumulation of the core. The one accumulation model which might reproduce the data trend requires unreasonable depositional mechanisms for this core. This model is that soils with the same pre-irradiation level of $\sim 3.4 \times 10^{-8}$ cm^3/g of ^{21}Ne must have existed and must have uniformly accumulated over the past ~8 m.y. without acquiring any additional components or being reworked at the lunar surface. Because simple *in situ* and core accumulation models can be ruled out, more complex, multi-stage irradiation models must be considered. The core data can be explained by a two-stage irradiation by cosmic rays if the stratigraphy of the core was inverted between these irradiation stages. An inversion of stratigraphy is the simplest way to explain the increase in cosmogenic gas concentrations with increasing sample depth. The latest irradiation stage probably began when the orange-black droplets were excavated from considerable depth (several meters) by Shorty Crater, which also inverted the core stratigraphy. Such an inversion is an expected consequence of crater ejecta located on crater rims (Shoemaker, 1960). In this second irradiation stage $[^{21}Ne]_c$ and $[^{38}Ar]_c$ would have accumulated in the core as a function of depth according to the production curves (solid line in Fig. 3). Unfortunately the ^{22}Ne/^{21}Ne data, which can be an indication of irradiation depth, is defined too imprecisely to constrain the depth of this irradiation. The first irradiation stage possibly occurred immediately after the pyroclastic deposition of this material ~3600 m.y. ago, and possibly was terminated abruptly when additional volcanic material or crater ejecta was laid on top of the orange-black droplets. In this envisioned stage of irradiation $[^{21}Ne]_c$ and $[^{38}Ar]_c$ would have accumulated according to the production curve, which for depths $\gtrsim 20$ cm is a smooth function of depth, and subsequently would have been inverted.

The dashed line in Fig. 3 is a best fit model of a two-stage irradiation with an intermediate inversion of stratigraphy. This model assumes that 60% of the total cosmogenic ^{21}Ne produced in the core by cosmic rays occurred in the early stage

---

Fig. 3. Concentrations of cosmogenic ^{21}Ne as a function of depth for 150–250 μm grain sizes of soils in the 74002–74001 core. Soils of $\lesssim 5$ cm depth fall within the micrometeorite reworking zone and contain excess $[^{21}Ne]_c$. The solid line is the approximate production profile (Reedy and Arnold, 1972), which has been arbitrarily normalized to the datum at 37 cm depth, and which has been calculated by assuming average soil densities of 2.0 and 2.35 g/cm^2 for the 74002 (upper) and 74001 (lower) drive tubes, respectively. The dotted line is a best-fit irradiation model characterized by 60% irradiation in the first stage, 40% irradiation in the second stage, and an inversion of core stratigraphy between stages. The estimated relative uncertainty in $[^{21}Ne]_c$ is also shown.

Fig. 4. Concentrations of cosmogenic ^{38}Ar as a function of depth for 150–250 μm grain sizes of soils in the 74002–74001 core. Soils of $\lesssim 5$ cm depth fall within the micrometeorite reworking zone and contain excess $[^{38}Ar]_c$. The estimated relative uncertainty in $[^{38}Ar]_c$ for soils of >5 cm depth is indicated.

at depths of ~100–250 g/cm^2, and that 40% of the total cosmogenic ^{21}Ne was produced in the late, post-Shorty stage with the core in its recovered stratigraphy. Changes in the proportions of this two-stage irradiation (e.g., 50%–50% or 70%–30%) produces noticeably poorer agreements between the calculated curve and the data. From the total $[^{21}Ne]_c$ and $[^{38}Ar]_c$ in the core soils and estimated production rates of ~0.18×10^{-8} cm^3/g and ~0.076×10^{-8} cm^3/g, respectively, for depths of ~0–40 cm (Bogard and Cressy, 1973), we estimate the duration of the second-stage, post-Shorty irradiation as ~9 m.y. from ^{21}Ne and ~10 m.y. from ^{38}Ar (± a few m.y.). The duration of the first stage of the irradiation would be ~20 m.y., if the irradiation depth were ~150 g/cm^2 and production rates were ~2/3 of those above (Reedy and Arnold, 1972). Actually, the duration and shielding depth of the first irradiation stage are not defined by this model, but are required to be >14 m.y. and >50 g/cm^2, respectively. Greater depths would permit relatively greater irradiation times for the first stage.

Soils in the bottom ~10 cm of the 74001 drive tube show much higher $[^{21}Ne]_c$ and $[^{38}Ar]_c$, which represents a major discontinuity in the trend established by other soils of >5 cm depth. Apparently 74001,125 falls in the transition zone as <150 μm grain sizes show the higher $[^{21}Ne]_c$ and $[^{38}Ar]_c$ concentrations, whereas <150 μm grain sizes of 74001,125 and both 150–250 μm and <20 μm grain sizes of 74001,119 do not show these high concentrations. As discussed earlier, these soils show no evidence of *in situ* reworking, and it is highly unlikely that they contain any contaminant which has contributed significant $[^{21}Ne]_c$ and $[^{38}Ar]_c$. One possible explanation of this discontinuity in the concentration profiles is that in the first irradiation stage soil 74001,2 was much nearer the surface than the rest of the core soils and accumulated greater amounts of ^{21}Ne and ^{38}Ar. In the excavation by Shorty Crater the material between 74001,2 and ,119 was presumably removed, creating the discontinuity. On the other hand, 74001,2 may also represent a pyroclastic eruption which preceeded the main eruption by several millions of years, and the soils were later brought together. We consider this second explanation less likely. We caution, therefore, that soils in the bottom of the 74001 drive tube have apparently experienced a considerably different irradiation history from the other core soils, and that these soils should be considered separately in any model of the depositional and irradiational history of the core.

The calculations presented above for the two-stage irradiation model of the 74001–74002 core imply an age for Shorty Crater of ~10 m.y. Two laboratories (Kirsten *et al.*, 1973; Hutcheon *et al.*, 1974) have reported particle track ages of ~10 m.y. and ~11 m.y., respectively, for orange glass in soil 74220, which was collected from a trench less than a meter from the core. Measurement of cosmic ray-produced tracks in soils in the 74002 drive tube have led Crozaz (1978) to conclude that these soils were deposited in one event some 10 m.y. ago. These ages are considerably different from an earlier, inferred age of Shorty Crater of ~30 m.y., which was based on particle tracks (Fleischer *et al.*, 1974) and cosmogenic noble gases (Kirsten *et al.*, 1973; Hintenberger *et al.*, 1974) determined in orange glass from trench soil 74220. The $[^{21}Ne]_c$ and $[^{38}Ar]_c$ reported for 74220 are similar to our data for 5–57 cm depths in the 74001–74002 core. The 74220 data are also consistent with our two-stage irradiation model described above, and with a Shorty Crater age of ~10 m.y. The age of ~30 m.y. inferred earlier from noble gases in 74220 assumed that cosmic ray irradiation was initiated by Shorty Crater, and did not consider the consequence of a two-stage irradiation of the orange-black droplets.

Eugster *et al.* (1978) have suggested a somewhat different irradiation model for the 74001–74002 core, which is based on their derived age for Shorty Crater of ~19 m.y. (Eugster *et al.*, 1977) and on noble gas analyses of four 74001 drive tube soils. Production profiles of various cosmogenic noble gases as a function of shielding depth were compared with experimental data for these core soils, and the conditions of the best mutual agreement were taken to derive their irradiation model. Eugster *et al.* (1978) conclude that soils in the 74001 drive tube have had a multi-stage irradiation, without inversion of stratigraphy by Shorty Crater, but which was characterized by various levels in the drive tube existing at the lunar surface for periods as long as several millions of years.

We believe three problems may exist in the core irradiation model derived by Eugster *et al.* (1978). First, the existence of fossil lunar surfaces in the 74001–74002 core for periods longer than ~0.5 m.y. is inconsistent with the extremely low I_s/FeO and agglutinate maturity indices for soils of >5 cm depth and with the presumed rapid accumulation of the droplets. Second, the major trends which these authors note in their data occur across the discontinuity which apparently exists in the core at a depth of ~60 cm. We cautioned earlier against directly using data from these deepest soils in deriving an irradiation model for the rest of the core, as these deepest soils have apparently experienced a different irradiation history from the rest of the core. Additional analyses by the Bern group of soils between 5 and 55 cm depths would resolve this objection.

The third problem which we believe may exist in the irradiation model derived by Eugster *et al.* (1978) for the 74001–74002 core is in their age of 19 m.y. for Shorty Crater. This age was derived by assuming that basalt rocks 74275 and 74255 were initially excavated by Shorty Crater, that a 2.8 m.y. track age for 74275 also applied to 74255, and that this age represented the second stage of a two-stage irradiation with major shielding changes between stages (Eugster *et al.*, 1977). As was the case for 74001 soils, production profiles of various cosmogenic noble gases as a function of shielding depth were compared with experimental data for these rocks, and the conditions of the best mutual agreement were taken to deduce the crater age and the fact that both basalt rocks were shielded by 100–140 g/cm^2 (~40–70 cm) until ~2.8 m.y. ago. The early stage of the irradiation was deduced to be ~16 m.y. for 74255 and ~40 m.y. for 74275. The 19 m.y. age of Shorty Crater was derived by adding 16 m.y. for 74255 to the 2.8 m.y. late irradiation stage. Rock 74275 was presumed to have had an even earlier pre-irradiation, i.e., a three-stage irradiation. However, rock 74275 was recovered less than a meter away from the 74001–74002 core, and the 2.8 m.y. track age for this rock cannot reflect a local shielding change of ~140 g/cm^2, for this would readily be observed in the core. Therefore, rock 74275 must have been thrown to its recovered position 2.8 m.y. ago by an unknown cratering event. As the track age for rock 74255 (which was chipped from a boulder a few meters from the core) has not been reported, it may or may not be the same as that for 74275. It is not clear whether the data reported by Eugster *et al.* (1977) for rock 74255 is consistent with a two-stage irradiation model in which Shorty Crater has an age of ~10 m.y. The event which produced major shielding changes in 74275 and possibly in 74255 also is not readily identifiable.

THE EXOTIC COMPONENT IN NEAR-SURFACE SOILS:

Soils of <5 cm depth in the core apparently contain higher trapped [^{36}Ar] and cosmogenic [^{21}Ne] and [^{38}Ar] because of additional components which have been added during *in situ* reworking in the past ~10 m.y. McKay *et al.* (1978) have found several percent basalt fragments, ropy glasses, and breccias and agglutinates not made from the orange-black droplets in these near-surface soils. The ratio of cosmogenic ^{21}Ne to cosmogenic ^{38}Ar for near-surface soils is distinctly different from other core soils, which enables us to place strong constraints on the nature of the "exotic" component. On a plot of [^{21}Ne/^{38}Ar]$_c$ versus [^{36}Ar] (Fig. 5), all core soils below 5 cm depth cluster in the lower right with [^{21}Ne/^{38}Ar]$_c$ values of 2.0–2.4. This group includes soil 74001,2, which is further evidence that excess cosmogenic gas in this soil is not contained in an "exotic" component. In contrast, near surface soils 74002,175–177, and to a lesser extent, 178, show much lower values of [^{21}Ne/^{38}Ar]$_c$. Trench soil 74220 also shows a slightly lower [^{21}Ne/^{38}Ar]$_c$ (Hintenberger *et al.*, 1974), which suggests that it too contains a small excess of cosmogenic gases. Values of [^{21}Ne/^{38}Ar]$_c$ calculated for orange-black droplets and for local basalts from chemical compositions according to the technique of Bogard and Cressy (1973) are indicated as arrows in Fig 5. The calculated ratio for orange-black droplets falls at the upper range of the measured data, and the calculated value for Apollo 17 subfloor basalts falls within the range of measured data for the Apollo 17 deep drill core (Pepin *et al.*, 1975). With the likely assumption that [^{21}Ne/^{38}Ar]$_c$ in near-surface soils is determined by a mixture of orange-black droplets and an exotic component, the calculated [^{21}Ne/^{38}Ar]$_c$ of the exotic component in soils 74002,175–177 is ~0.65. This value is much lower than either the calculated or the observed ratios for mare basalts, but it is quite consistent with values of [^{21}Ne/^{38}Ar]$_c$ which have been observed for highland materials such as those from Apollo 16. Ropy glasses, which comprise ~15% of the gray soils at the surface adjacent to the 74001–74002 core, have a highland-like composition (Fruland *et al.*, 1977) and may be ray material produced from a large highland impact (Heiken and McKay, 1974). These ropy glasses occur in near-surface soils of the 74002 drive tube in amounts of ~1–2% (McKay, pers. comm.), and may have contributed the excess [^{21}Ne]$_c$ and [^{38}Ar]$_c$ to these soils.

A second possible explanation of the [^{21}Ne/^{38}Ar]$_c$ value of 0.65 in the near-surface soils from the core is that the excess cosmogenic gas has not been

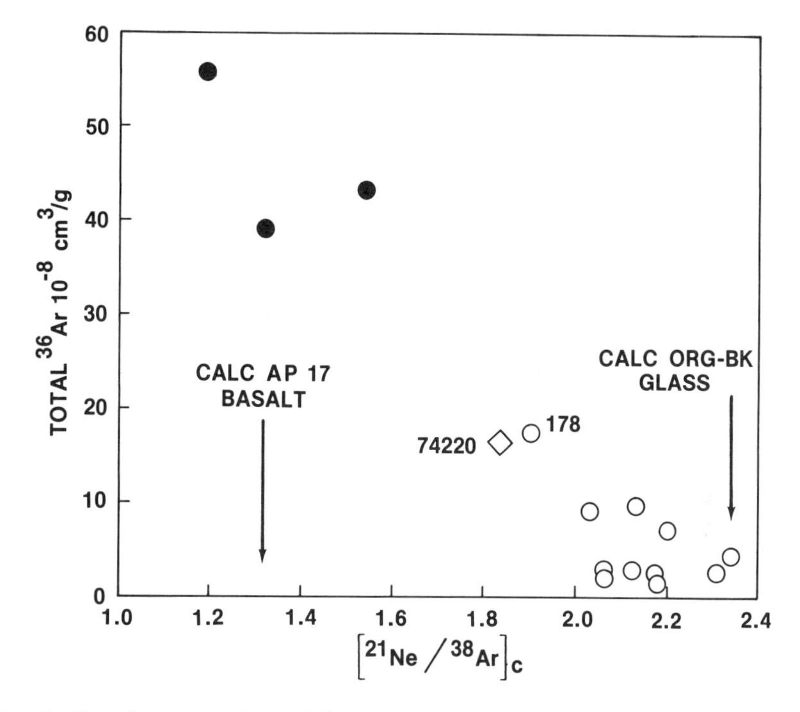

Fig. 5. Plot of concentrations of ^{36}Ar in <20 μm grain sizes against the ratios of cosmogenic ^{21}Ne to cosmogenic ^{38}Ar in 150–250 μm grain sizes of soils from the 74002–74001 core. Filled symbols represent soils of 0–2 cm depths. Soil 74002,178 (3–3.5 cm) and trench soil 74220 (Hintenberger *et al.*, 1974) are also indicated. Arrows represent values of $[^{21}Ne/^{38}Ar]_c$ for the orange-black droplets and for Apollo 17 subfloor basalts which have been calculated by the technique of Bogard and Cressy (1973).

contributed by an exotic soil component, but has been produced in these soils from nuclear reactions induced by irradiation with solar flares over the past ~10 m.y. The near-surface soils in the 74002 drive tube are particularly well suited for detecting solar flare effects because of their low concentrations of trapped ^{36}Ar and cosmogenic gases compared to most soils, and because a large fraction of their cosmic ray irradiation has apparently occurred at the surface. Walton *et al.* (1978) have calculated the expected ratio $[^{21}Ne/^{38}Ar]_c$ produced in soil with a chemical composition that of the orange glass from irradiation by solar flares and high energy, galactic cosmic rays to be ~0.6–0.7 and ~1.3–1.7, respectively. Their calculations predict the $[^{21}Ne/^{38}Ar]_c$ produced in the 74002 drive tube would be ~0.67 at depths of <1 cm and would increase to ~1.1 at a depth of ~5 cm. The value of $[^{21}Ne/^{38}Ar]_c$ predicted from solar flares is essentially identical to the observed ratio of the excess component in near-surface soils. Therefore, solar flares may well be the source of the excess $[^{21}Ne]_c$ and $[^{38}Ar]_c$ seen in these soils. If so, the production rates of ^{21}Ne and ^{38}Ar in these soils from solar flare irradiation over ~10 m.y. are ~2×10^{-9} and ~3×10^{-9} cm^3/g-m.y., respectively, and are similar to the production rates predicted by Walton *et al.* (1978), e.g., $^{38}Ar = 1.8 \times 10^{-9}$ cm^3/g-m.y. Concentrations of ^{21}Ne and ^{38}Ar produced in these soils by solar flares are expected to smoothly decrease by an order of magnitude from the surface to a depth of ~2 cm (Reedy and Arnold, 1972). Excess $[^{21}Ne]_c$ and $[^{38}Ar]_c$ do decrease with increasing depths below 0.5 cm and correlate closely with values of the I_s/FeO surface maturity index (Morris *et al.*, 1978) and with agglutinate content (McKay *et al.*, 1978). This suggests that the depth profiles of excess $[^{21}Ne]_c$ and $[^{38}Ar]_c$ are determined mainly by *in situ* soil reworking whereby the excess gases are produced in the uppermost few millimeters and reworked to greater depths.

The fact that Walton *et al.*'s predicted $[^{21}Ne/^{38}Ar]_c$ from galactic cosmic rays is much lower than either the observed values or the calculated values of Bogard and Cressy (1973) need not effect the conclusion that excess ^{21}Ne and ^{38}Ar in these near-surface soils could have been produced by solar flares. Walton *et al.* (1978) calculated $[^{21}Ne/^{38}Ar]_c$ for solar flare production directly from cross section data. On the other hand, calculated $[^{21}Ne/^{38}Ar]_c$ for galactic production was based on the observed depth profiles of $[^{21}Ne]_c$ and $[^{38}Ar]_c$ in the Apollo 15 deep drill core. Not only is the $[^{21}Ne]_c$ profile in this core poorly understood (Pepin *et al.*, 1974), but there apparently has been significant loss of $[^{21}Ne]_c$ from samples of <80 cm depth (Bogard and Hirsch, 1975).

Acknowledgments—D. S. McKay supplied the grain-size separates. We have benefited from discussions with D. McKay, R. Morris, G. Heiken, and D. Blanchard.

REFERENCES

Alexander E. C. Jr., Coscio M. R. Jr., Dragon J. C. and Saito K. (1978) ^{40}Ar-^{39}Ar studies of glasses from lunar soils (abstract). In *Lunar and Planetary Science IX*, p. 7–10. Lunar and Planetary Institute, Houston.

Baur H., Frick U., Funk H., Schultz L. and Signer P. (1972) Thermal release of He, Ne, and Ar from lunar fines and minerals. *Proc. Lunar Sci. Conf. 3rd*, p. 1947–1966.

Blanchard D. P. and Budahn J. R. (1978) Compositional variations among size fractions in a homogeneous, unmatured regolith: Drive core 74001/2. *Proc. Lunar and Planet. Sci. Conf. 9th.* This volume.

Bogard D. D. and Cressy P. J. Jr. (1973) Spallation production of ^{3}He, ^{21}Ne, and ^{38}Ar from target elements in the Bruderheim chondrite. *Geochim. Cosmochim. Acta* **37**, 527–546.

Bogard D. D. and Hirsch W. W. (1975) Noble gas studies on grain size separates of Apollo 15 and 16 deep drill cores. *Proc. Lunar Sci. Conf. 6th*, p. 2057–2083.

Cirlin E. H. and Housley R. M. (1978) Flameless atomic absorption studies of volatile trace metals in Apollo 17 samples (abstract). In *Lunar and Planetary Science IX*, p. 169–171. Lunar and Planetary Institute, Houston.

Crozaz G. (1978) History of regolith deposition at the Apollo 17 and Luna 24 landing sites and at Shorty Crater (abstract). In *Lunar and Planetary Science IX*, p. 203–205. Lunar and Planetary Institute, Houston.

Eberhardt P. (1974) The solar wind as deduced from lunar samples. Paper presented at the Third Solar Wind Conference, Asilomar, March, 1974.

Eberhardt P., Geiss J., Graf H., Grögler N., Krähenbühl U., Schwaller H., Schwarzmuller J. and Stettler A. (1970) Trapped solar wind noble gases, exposure age and K/Ar-age in Apollo 11 fine material. *Proc. Apollo 11 Lunar Sci. Conf.*, p. 1037–1070.

Eugster O., Eberhardt P., Geiss J., and Grögler N. (1978) The solar wind and cosmic-ray exposure history of soil from drive tube 74001 (abstract). In *Lunar and Planetary Science IX*, p. 306–308. Lunar and Planetary Institute, Houston.

Eugster O., Eberhardt P., Geiss J., Grögler N., Jungck M. and Margeli M. (1977) The cosmic-ray exposure history of Shorty Crater samples; the age of Shorty Crater. *Proc. Lunar Sci. Conf. 8th*, p. 3059–3082.

Fleisher R. L., Hart H. R. and Giard W. R. (1974) Surface history of lunar soil and soil columns. *Geochim. Cosmochim. Acta* **38**, 365–380.

Fruland R. M., Morris R. V., McKay D. S. and Clanton U. S. (1977) Apollo 17 ropy glasses. *Proc. Lunar Sci. Conf. 8th*, p. 3095–3111.

Geiss J., Buehler F., Cerutti H., Eberhardt P. and Filleux C. (1972) Solar wind composition experiment. NASA SP-315, p. 14–1 to 14–10.

Gibson E. K. Jr. and Andrawes F. (1978) Carbon and sulfur abundances in 74001/74002 drive tubes from Shorty Crater (abstract). In *Lunar and Planetary Science IX*, p. 377–379. Lunar and Planetary Institute, Houston.

Heiken G. and McKay D. S. (1974) Petrography of Apollo 17 soils. *Proc. Lunar Sci. Conf. 5th*, p. 843–860.

Heiken G. and McKay D. S. (1978) Petrology of a sequence of pyroclastic rocks from the valley of Taurus-Littrow (Apollo 17 landing site). *Proc. Lunar Planet. Sci. Conf. 9th*. This volume.

Heymann D. (1971) The inert gases. In *Handbook of Elemental Abundances in Meteorites* (B. H. Mason, ed.), p. 29–66. Gordon and Breach, N.Y.

Heymann D., Walton J. R. and Jordan L. (1975) Light and dark soils at the Apollo 16 landing site. *The Moon* **13**, 81–110.

Hintenberger H., Weber H. W. and Schultz L. (1974) Solar, spallogenic, and radiogenic rare gases in Apollo 17 soils and breccias. *Proc. Lunar Sci. Conf. 5th*, p. 2005–2022.

Husain L. and Schaeffer O. A. (1973) Lunar volcanism: Age of the glass in the Apollo 17 orange soil. *Science* **180**, 1358–1360.

Hutcheon I. D., MacDougall D. and Stevenson J. (1974) Apollo 17 particle track studies: Surface residence times and fission track ages for orange glass and large boulders. *Proc. Lunar Sci. Conf. 5th*, p. 2597–2608.

Kirsten T., Horn P., Heymann D., Hubner W. and Storzer D. (1973) Apollo 17 crystalline rocks and soils: Rare gases, ion tracks, and ages (abstract). *EOS (Trans. Amer. Geophys. Union)* **54**, 595–597.

McKay D. S., Heiken G. and Waits G. (1978) Core 74001/2; Grain size and petrology as a key to the rate of *in-situ* reworking and lateral transport on the lunar surface. *Proc. Lunar Planet. Sci. Conf. 9th*. This volume.

Morris R. V., Gose W. A. and Lauer H. V. (1978) Depositional and surface exposure history of the Shorty Crater core 74001/2: FMR and magnetic studies. *Proc. Lunar Planet. Sci. Conf. 9th*. This volume.

Muehlberger W., Batson R., Cernan E., Freeman V., Hait M., Holt H., Howard K., Jackson E., Larson K., Reed V., Rennilson J., Schmitt H., Scott D., Sutton R., Stuart D., Swann G., Trask N., Ulrich G., Wilshire H. and Wolfe E. (1973) Preliminary geologic investigation of the Apollo 17 landing site. NASA SP-330, p. 6-1 to 6-91.

Nagle J. S. (1977) Drive tubes 74002/74001. *Supplement to the Lunar Core Catalog*, p. V11.5.1–V11.5.49. JSC 09252 NASA Johnson Space Center, Houston.

Pepin R. O., Basford J. R., Dragon J. C., Coscio M. R. and Murthy V. R. (1974) Rare gases and trace elements in Apollo 15 drill core fines: Depositional chronologies and K-Ar ages, and production rates of spallation-produced ^{3}He, ^{21}Ne, and ^{38}Ar versus depth. *Proc. Lunar Sci. Conf. 5th*, p. 2149–2184.

Pepin R. O., Dragon J. C., Johnson N. L., Bates A., Coscio M. R. and Murthy V. R. (1975) Rare gases and Ca, Sr, and Ba in Apollo 17 drill core fines. *Proc. Lunar Sci. Conf. 6th*, p. 2027–2055.

Reedy R. C. (1976) Spallation rare gas production calculations and studies of lunar surface processes (abstract). In *Lunar Science VII*, p. 721–723. The Lunar Science Institute, Houston.

Reedy R. C. and Arnold J. R. (1972) Interaction of solar and galactic cosmic ray particles with the moon. *J. Geophys. Res.* 77, 537–555.

Shoemaker E. M. (1960) Penetration mechanics of high velocity meteorites, illustrated by Meteor Crater, Arizona. *Internat. Geol. Conf.*, 21st session, pt. 18, Copenhagen, p. 418–434.

Walton J. R., Heymann D. and Yaniv A. (1978) The search for solar cosmic ray proton-produced neon-21 and argon-38 in lunar soils (abstract). In *Lunar and Planetary Science IX*, p. 1199–1201. Lunar and Planetary Institute, Houston.

Proc. Lunar Planet. Sci. Conf. 9th (1978), p. 2001–2009.
Printed in the United States of America

Regolith depositional history at Shorty Crater

GHISLAINE CROZAZ

Department of Earth and Planetary Sciences and McDonnell Center for the Space Sciences
Washington University, St. Louis, Missouri 63130

Abstract—Nuclear particle track measurements in the 68 cm double drive tube 74002–74001 indicate that the whole core was deposited in one event some 10 m.y. ago. Significant reworking of the soil only occurred down to a few cm from the lunar surface since this event. Complementary investigations in this core by other groups are discussed. Most of the evidence available leads to a two stage model in which the orange and black soils collected at Shorty Crater were first irradiated for ~25 m.y. at some depth and then were deposited only a few m.y. ago as an overturned ejecta blanket.

INTRODUCTION

The bright orange soils at Shorty Crater have continued to attract the attention of many investigators even though the early hope that they might represent recent lunar volcanism was soon dashed when the Apollo 17 samples were returned to earth. The orange soils were formed some 3.6 b.y. ago, shortly after the extrusion of the Apollo 17 basalts (3.65–3.8 b.y.) (Huneke and Wasserburg, 1978). They are composed essentially of orange and black glass droplets of similar composition whose origin is assumed to be pyroclastic (Heiken *et al.*, 1974). The black glass is the partially crystallized equivalent of the orange glass. Heiken and McKay (1977), studying the olivine textures within the black glass droplets, concluded that the black glass experienced cooling rates two or three orders of magnitude slower than the orange glass. The same authors proposed models for the sequential formation of black and orange droplets as a spray from a lava fountain.

Shorty Crater is a 120 m diameter impact crater, and the orange soil is assumed to be Shorty ejecta (Wolfe *et al.*, 1975). The first orange soils studied (74220) were collected at a depth of 5–7.5 cm on the southwest rim of the crater. Their analysis, as well as the study of rocks from the rim of Shorty Crater, has led to widely varying exposure ages, as well as various assumptions as to the time of formation of the crater and of the emplacement of the unique orange soils.

Arvidson *et al.* (1975) tentatively dated Shorty Crater at ~30 m.y. on the basis of cosmogenic ^{38}Ar results obtained by Huneke *et al.* (1973) and Eberhardt *et al.* (1974). Huneke *et al.* reported a 30 m.y. exposure age for the orange soil, 74220; whereas, Eberhardt *et al.* measured a 25 m.y. exposure for rock 74275. However, track studies by Goswami and Lal (1974) on the same rock, 74275, were only consistent with a surface exposure age of ~3 m.y. Thus, it appeared that while this rock spent some 25 m.y. at a depth of ~1 meter, where cosmogenic isotopes accumulated, it is only recently (~3 m.y. ago) that the rock was brought to its sampling location where the track record accumulated.

Additional cosmic ray exposure ages (based on the contents of cosmogenic noble gases) were reported by Eugster *et al.* (1977) for samples from the rim of Shorty Crater. Three rocks and a variety of soils, including the orange and black soils, were investigated. A somewhat complex picture emerged: two of the rocks (74255 and 74275) experienced a change of shielding conditions about 3 m.y. ago. In addition, the boulder from which sample 74255 was selected was excavated from the bedrock (i.e., from a shielded position) about 19 m.y. ago. The orange and black soils have cosmic-ray exposure ages of 30–50 m.y.; whereas, considerably larger exposure times (~200 m.y.) are found for rock 74235 and the remaining soils. Eugster *et al.* (1977) concluded that the Shorty impact occurred less than 30 m.y. ago and probably 19 m.y. ago when the large boulder was excavated. Thus, there seemed to be considerable uncertainty as to the time of formation of Shorty Crater and of emplacement of the orange soils.

The orange soil, 74220, was also investigated using the nuclear particle track method. Results, again, were widely variable with exposure ages varying from ~10 m.y. (Storzer *et al.*, 1973; Hutcheon *et al.*, 1974) to 20–35 m.y. (Fleischer *et al.*, 1974). Although Storzer *et al.* (1973) and Hutcheon *et al.* (1974) reached similar conclusions as to the exposure age, their data differed substantially.

The recent availability of samples from the 68 cm double drive tube (74002–74001), collected near the trench from which the orange soil, 74220, was removed, provided a unique opportunity to re-examine the depositional history of the glassy soils. The 68 cm soil column (Nagle, 1978) is essentially made of relatively homogeneous layers of orange and black glass; the former is predominant above 25 cm; whereas, the latter is found in largest concentrations below 25 cm. Minor concentrations of olivine and pyroxene are present at all levels; whereas, soil clasts and agglutinates are recognized only in the near-surface samples. A number of stratigraphic units, which may or may not have a time significance, have been recognized by Nagle (1978). Potentially, track studies, as a function of depth, can determine the time of deposition and the mode of emplacement (one or multiple events) of the soils. If these soils have had a simple history, they are essential to our understanding of the rate of regolith reworking at the lunar surface.

EXPERIMENTAL

Six samples from the upper half (74002) and three samples from the lower half (74001) of the double drive tube were analyzed. Each weighed ~250 mgms. This unusually large amount of material for a track analysis was required in order to study the accessory phases present along with the glass. Sample numbers and depths from the top of the drive tube are as follows: 74002,2078 (2.5–3 cm); 74002,2091 (7–7.5 cm); 74002,2093 (12.5–13 cm); 74002,2103 (17.5–18 cm); 74002,2112 (22.5–23 cm); 74002,2121 (28.5–29 cm); 74001,2073 (35–35.5 cm); 74001,2074 (44–44.5 cm); 74001,2075 (56.5–57 cm). Among the minor phases present, olivines were found to be best suited for measurements. Individual olivine grains and glass droplets and fragments >50 μm in size were mounted in epoxy resin, polished and etched to reveal nuclear particle tracks. A fluoboric etchant

(Macdougall, 1971) was used for the glass (etching time: 6 minutes); whereas, olivines were etched for two to four hours in the olivine etchant described by Kirshnaswami *et al.* (1971). Most track densities were determined by using an optical microscope. Track densities in excess of 3×10^7 cm^{-2} (requiring the use of a steroscan electron microscope) were found only in the sample originally located closest to the surface.

Track Densities in the 74002–74001 Core

Track densities in the olivines vary systematically with the depth in the core (see Fig. 1 and below). In Fig. 1, measured track densities in olivines have been corrected for comparison with track production rates in feldspars. Track densities in olivines were multiplied by a factor of two to account for the different registration efficiencies of olivines and feldspars (Bhandari *et al.*, 1972). Although this correction factor has been systematically used by other authors, it should be noted that the true value may be somewhat different in these olivines; the absolute age is directly dependent on this correction factor. For estimating the depth, a density of 2 g cm^{-3} was used. Also taken into account was a loss of 2 cm at the top of the core as reported by Fruchter in his presentation to the last Lunar and Planetary Science Conference.

The solid curves represent expected track densities in feldspar grains irradiated in an undisturbed soil column for 10 m.y. (the two curves represent densities for two extreme orientations of the grains with respect to the surface) (Walker and Yuhas, 1973). The track densities reported are averages, usually for about 20 grains, at each depth except for the uppermost sample located at ~3 cm (5 cm from the original surface if the 2 missing cm referred to above are taken into consideration). Indeed, sample 74002,2078 is the only one in which crystals with track densities significantly exceeding the values reported in Fig. 1 are found. Over 80% of the crystals at this level have track densities in excess of 10^8 cm^{-2}. The reported density at this level is the average for the low track density grains (i.e., grains with less than 7×10^6 tracks cm^{-2}—no grain with a track density comprised between 7×10^6 and 10^8 cm^{-2} was observed). In view of the very low track densities measured at depth, only results for the 4 uppermost samples have been plotted in Fig. 1. Except for the top sample in 74002, extreme track densities in individual grains of a given level in this section only vary by a factor of ~2, as expected for an undisturbed layer. At larger depths, the track densities follow the decreasing trend indicated by the upper samples, but the total number of tracks observed is so small as to make track densities statistically meaningless. As an example, only one track was observed in about 100 olivine grains from the three levels in the lower half of the drive tube.

These results show that the core material was deposited in one event some 10 m.y. ago. The majority of the core lay undisturbed since deposition with significant soil reworking strictly limited to the top few cm. The precise age of emplacement, however, is critically dependent on both the correction factor applied to olivines and the amount of missing surficial material.

Measurements in glass were comparatively more difficult but are in agreement with the above interpretation. As observed by Heiken *et al.* (1978), glass droplets

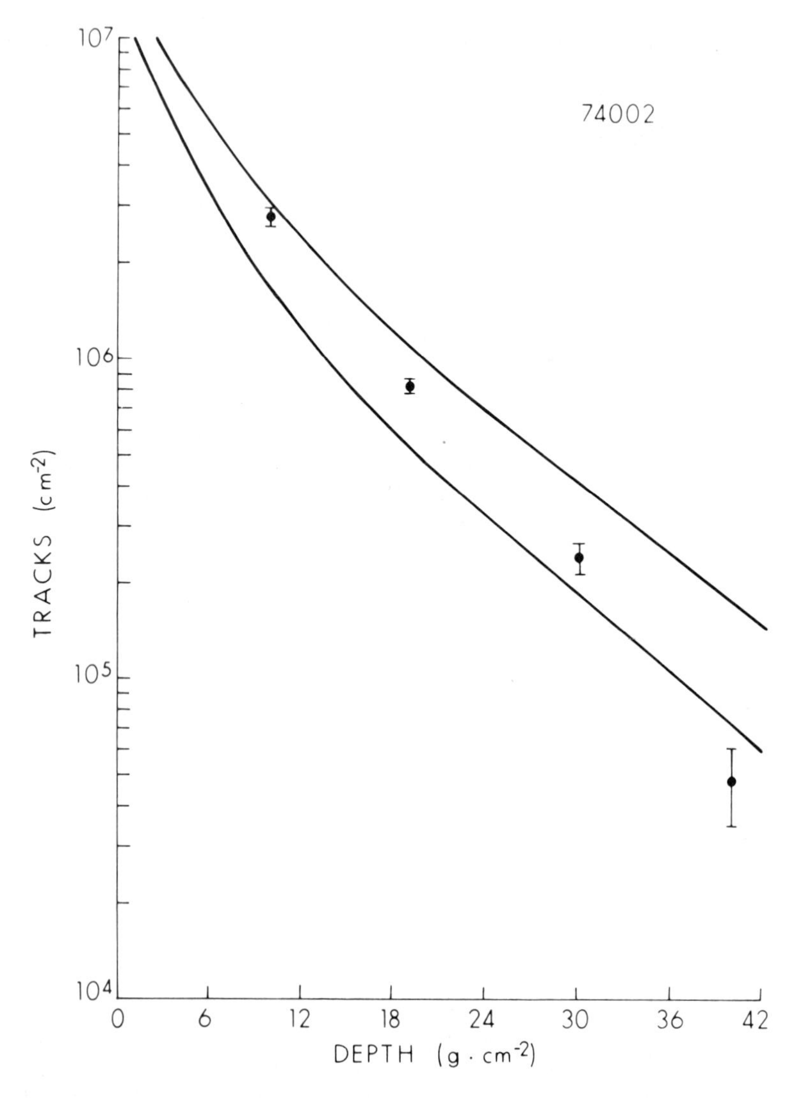

Fig. 1. Corrected track densities in olivines from the double drive tube 74002-74001 collected at Shorty Crater. The two smooth curves correspond to the track production as a function of depth from the lunar surface for two extreme orientations of the grains with respect to the surface and an exposure time of 10 m.y. The error bars are $\pm 1\ \sigma$ estimated from the square root of the total number of tracks in all the grains included in the present analysis, at each level.

contain abundant vesicles, which often resemble tracks and make low track density measurements quite uncertain. These glasses contain ~120 ppb of uranium (Fleisher and Hart, 1974), which over the lifetime of the glass have produced fission tracks in amounts similar to the cosmic ray tracks present. In addition, tracks in the glasses may have been annealed to an unknown extent. Because of these facts, no attempt was made to obtain quantitative data in the glass.

However, cosmic ray track densities in glass are in agreement with the ones measured in olivines, which reinforces the conclusion that these soils were emplaced some 10 m.y. ago and have been undisturbed since and also indicates that, prior to emplacement, the glasses were shielded from heavy cosmic rays, probably at depths exceeding about one meter. The presence in the glasses of a background of very shallow tracks, which could be due to spallation recoils induced by cosmic ray proton bombardment (Fleischer *et al.*, 1970), is probably indicative of a former residence at a deeper location shielded from heavy cosmic rays but accessible to cosmic ray protons where, presumably, some of the reported cosmogenic noble gases were accumulated.

Thus, the track exposure age measures a true age since final emplacement; whereas, cosmogenic isotopes measure integrated effects over the time of exposure of the samples within the first few meters of the lunar surface. The apparent discrepancy of the present conclusion, with the ones derived by Fleischer *et al.* (1974) from track measurements, is readily explained if one considers that these authors based their 20–35 m.y. maximum exposure age on difficult observations in glasses from a sample whose depth was not well known (5–7.5 cm) and that they used median rather than minimum track densities in their calculations. Actually, using minimum track densities led them to derive a time of ~7 m.y. "since the last disturbance".

Regolith Deposition At Shorty Crater: The Global View

In this section, I would like to discuss the composite picture of recent regolith evolution at Shorty Crater that emerges when results reported at this conference by other authors and using other methods are considered. The complementary investigations of direct relevance to the present subject include: 1) petrological studies (Heiken *et al.*, 1978; Nagle, 1978); 2) determinations of the grain size distributions (McKay and Waits, 1978); 3) ferromagnetic resonance analyses (Morris and Gose, 1977; Morris *et al.*, 1978); as well as, 4) noble gas investigations (Bogard and Hirsch, 1978; Eugster *et al.*, 1978).

The first three types of studies, in various ways, all support our conclusion that the core material was emplaced in one event which occurred relatively recently.

Although Nagle (1978), in his preliminary analysis of the double drive tube, recognized a number of stratigraphic units on the basis of compositional differences, there is no reason to believe that these units were deposited at distinct times in the past. If they had been, we should expect to find surface maturation products, such as agglutinates, in the middle of the core, indicating the location of past surfaces. Except for the uppermost 5 cm, there are no agglutinates or rock fragments present within this core (Heiken *et al.*, 1978). The only surface ever exposed and subjected to reworking is the present one. Thus, the material in the core was deposited in one event.

McKay and Waits (1978) also find agglutinates only in the top 5 cm. Although their grain size distributions along the whole core are very similar, the mean grain size tends to decrease from the bottom to the top of the core. These core samples include the best sorted lunar material ever analyzed. The top few cms are finest grained, possibly a consequence of comminution accompanying *in situ* reworking.

The ferromagnetic resonance studies of Morris and Gose (1977) and Morris *et al.* (1978) also strongly imply that the 74002/1 core was emplaced as one slab and that only the upper 5 cms were reworked since deposition. Using the FMR maturity index I_s/FeO, these authors found that all the soils in the drive tube are immature, the soils below 5 cm being exceptionally immature (they exhibit the lowest I_s/FeO ratio so far measured in lunar soils). Below 5 cm from the surface, I_s/FeO not only is extremely low but also constant; whereas, above 5 cm there is a sharp increase in maturity attributed to *in situ* reworking. Below 5 cm, none of the soils were ever exposed at the surface for an appreciable time since their formation 3.6 b.y. ago, and they have not been mixed with a mature soil (Morris *et al.*, 1978).

Thus, the ferromagnetic results are consistent with a model in which the soils were deposited very rapidly, probably in a single episode or several closely spaced episodes of lunar fire fountaining, some 3.7 b.y. ago. The soils were then, and for most of their existence, shielded (no production of tracks). It is only relatively recently that they were excavated. The noble gas observations that now need to be considered will show that the excavation event was not straightforward or unique.

Two rare gas groups (Bogard and Hirsch, 1978; Eugster *et al.*, 1978), making similar measurements, reached conflicting conclusions. Bogard and Hirsch (1978) fully endorse the above model which is challenged by Eugster *et al.* (1978 and oral presentation at the conference). Both groups observe solar wind derived noble gases present in low concentrations in all soils analyzed. These solar gases, most likely, were implanted ~3.6 b.y. ago when the orange and black soils were formed. Bogard (oral presentation) tends to favor the injection of solar wind in the glasses from the degassing of mature soils during the episode of fire fountaining, which he believes was of short duration; whereas, Eugster *et al.* (1978) conclude that the solar gases were acquired while the glasses or some minor associated component were residing at the lunar surface. As mentioned above, the absence in the soils of surface maturity indicators tends to disfavor the latter hypothesis.

The above authors agree that the cosmogenic data are inconsistent with a simple, single stage irradiation but differ in their interpretations. Eugster and his colleagues, assuming a late stage irradiation duration of 19 m.y. (their favored age for Shorty Crater corresponding to the emplacement of the boulder from which sample 74255 was collected), conclude that, at the time of formation of the glassy soils, successive layers were deposited. Times between the deposition of the successive layers range from 4 to 15 m.y. If this were indeed the case, we should expect to see tracks produced by this early irradiation. None are found.

In contrast, Bogard and Hirsch (1978) offered an interpretation which is compatible with all the other types of measurements (if not with the rare gas observations of the Bern group). They first observed that the cosmic ray produced ^{21}Ne and ^{38}Ar are present in concentrations consistent with exposure times of ~25 m.y. In first approximation, the depth profiles of both ^{21}Ne and ^{38}Ar are identical and the mirror images of what is expected for an undisturbed layer. This led Bogard and Hirsch to postulate a two-stage irradiation model in which the samples were first irradiated for ~25 m.y. at some depth and then were deposited only a few million years ago as an overturned ejecta blanket. This assumes that the core stratigraphy was preserved but overturned during the last event.

In summary, it thus appears that during the last few millions of years the orange and black soils had a rather simple and undisturbed history. Only the top few centimeters were reworked in agreement with the predictions of the Monte Carlo simulations. Prior to their final emplacement, these soils, probably overturned, may have resided at depths exceeding about one meter for up to 30 m.y. However, they spent most of the time since their formation at still larger depths, where they were completely removed from the influence of cosmic rays. The nature of the two successive excavating events, which occurred during the last 30 m.y. and whose effects have been described here, is still an enigma.

The accumulation of large slabs (>5 cm) of material largely devoid of tracks, which have remained undisturbed since deposition, is a rare occurrence, and in this respect, the soils collected at Shorty Crater are unique. The only other known example of a layer exceeding 5 cm in thickness and showing the characteristic track profile of an undisturbed layer is the 60 cm coarse-grained layer in the topmost part of the Apollo 17 deep drill. At the contact between the coarse-grained layer and the above material, there is a region of about 5 cm which shows strong analogies with the top of the core at Shorty Crater: agglutinates are present, as well as grains with track densities ranging from $\sim 10^6$ to more than 10^8 cm^{-2}. Below this region, track densities and agglutinate contents decrease dramatically. However, it is not clear that this comparison is valid, if one considers that the regolith at the Apollo 17 landing site and at Shorty Crater had very different depositional histories. The top 19 cm of fine-grained material which overlies the coarse-grained layer in the deep drill most probably accumulated during the last 2 m.y. (Crozaz and Plachy, 1976); whereas, the glassy soils at Shorty Crater remained undisturbed and uncovered since deposition about 10 m.y. ago.

Acknowledgments—The assistance of Lou Ross in all phases of the experimental work is gratefully acknowledged. This work was supported by the National Aeronautics and Space Administration under contract NGL-26-008-065.

References

Arvidson R., Crozaz G., Drozd R. J., Hohenberg C. M. and Morgan C. J. (1975) Cosmic ray exposure ages of features and events at the Apollo landing sites. *The Moon* **13**, 259–276.

Bhandari N., Goswami J. N., Lal D., MacDougall D. and Tamhane A. S. (1972) A study of the vestigial records of the cosmic rays in lunar rocks using a thick section technique. *Proc. Indian Acad. Sci.*, Sect. A, **76**, 27–50.

Bogard D. D. and Hirsch W. C. (1978) Noble gas contents and irradiation history of orange-black glass in the 74001–74002 core (abstract). In *Lunar and Planetary Science IX*, p. 111–113. Lunar and Planetary Institute, Houston.

Crozaz G. and Plachy A. L. (1976) Origin of the Apollo 17 deep drill coarse-grained layer. *Proc. Lunar Sci. Conf. 7th*, p. 123–131.

Eberhardt P., Eugster O., Geiss J., Graf H., Grögler N., Guggisberg S., Jungck M., Maurer P., Morgeli M. and Stettler A. (1974) Solar wind and cosmic radiation history of Taurus Littrow Regolith (abstract). In *Lunar Science V*, p. 197–199. The Lunar Science Institute, Houston.

Eugster O., Eberhardt P., Geiss J. and Grögler N. (1978) The solar wind and cosmic-ray exposure history of soil from drive tube 74001, an unmixed regolith (abstract). In *Lunar and Planetary Science IX*, p. 306–308. Lunar and Planetary Institute, Houston.

Eugster O., Eberhardt P., Geiss J., Grögler N., Jungck M. and Morgeli M. (1977) The cosmic-ray exposure history of Shorty crater samples; the age of Shorty crater. *Proc. Lunar Sci. Conf. 8th*, p. 3059–3082.

Fleischer R. L., Haines E. L., Hart H. R. Jr., Woods R. T. and Comstock G. M. (1970) The particle track record of the Sea of Tranquility. *Proc. Apollo 11 Lunar Sci. Conf.*, p. 2103–2120.

Fleischer R. L. and Hart H. R. Jr. (1974) Uniformity of the Uranium content of lunar green and orange glasses. *Proc. Lunar Sci. Conf. 5th*, p. 2251–2255.

Fleischer R. L., Hart H. R. Jr. and Giard W. B. (1974) Surface history of lunar soil and soil columns. *Geochim. Cosmochim. Acta* **38**, 365–380.

Goswami J. N. and Lal D. (1974) Cosmic ray irradiation pattern at the Apollo 17 site: Implications to lunar regolith dynamics. *Proc. Lunar Sci. Conf. 5th*, p. 2643–2662.

Heiken G. and McKay D. S. (1977) A model for eruption behavior of a volcanic vent in eastern Mare Serenitatis. *Proc. Lunar Sci. Conf. 8th*, p. 3243–3255.

Heiken G. H., McKay D. S. and Brown R. W. (1974) Lunar deposits of possible pyroclastic origin. *Geochim. Cosmochim. Acta* **38**, 1703–1718.

Heiken G., McKay D. S. and Gooley R. (1978) Petrology of a sequence of pyroclastic rocks from the valley of Taurus-Littrow (Apollo 17 landing site) (abstract). In *Lunar and Planetary Science IX*, p. 491–493. Lunar and Planetary Institute, Houston.

Huneke J. C., Jessberger E. K., Podosek F. A. and Wasserburg G. J. (1973) $^{40}Ar/^{39}Ar$ measurements in Apollo 16 and 17 samples and the chronology of metamorphic and volcanic activity in the Taurus-Littrow region. *Proc. Lunar Sci. Conf. 4th*, p. 1725–1756.

Huneke J. C. and Wasserburg G. J. (1978) ^{40}Ar-^{39}Ar ages of single orange glass balls and highland breccia phenocrysts (abstract). In *Lunar and Planetary Science IX*, p. 567–569. Lunar and Planetary Institute, Houston.

Hutcheon J. D., MacDougall D. and Stevenson J. (1974) Apollo 17 particle track studies: Surface residence times and fission track ages for orange glass and large boulders. *Proc. Lunar Sci. Conf. 5th*, p. 2597–2608.

Krishnaswami S., Lal D., Prabhu N. and Tamhane A. S. (1971) Olivines: Revelation of tracks of charged particles. *Science* **174**, 287–291.

MacDougall D. (1971) Fission track dating of volcanic glass shards in marine sediments. *Earth Planet. Sci. Lett.* **10**, 403–406.

McKay D. and Waits G. (1978) Grain size distribution of samples from core 74001 and 74002 (abstract). In *Lunar and Planetary Science IX*, p. 723–725. Lunar and Planetary Institute, Houston.

Morris R. V. and Gose W. A. (1977) Depositional history of core section 74001: Depth profiles of maturity, FeO and metal. *Proc. Lunar Sci. Conf. 8th*, p. 3113–3122.

Morris R. V., Gose W. A. and Lauer H. V. Jr. (1978) Depositional and reworking history of core 74001/2 (abstract). In *Lunar and Planetary Science IX*, p. 763–765. Lunar and Planetary Institute, Houston.

Nagle J. S. (1978) Drives tubes 74002/74001, dissection and description. Report to the Lunar Sample Curator. Lunar Curatorial Facility, Johnson Space Center, Houston.
Storzer D., Poupeau G. and Krätschmer W. (1973) Track-exposure and formation ages of some lunar samples. *Proc. Lunar Sci. Conf. 4th*, p. 2363–2377.
Walker R. and Yuhas D. (1973) Cosmic ray track production rates in lunar materials. *Proc. Lunar Sci. Conf. 4th*, p. 2379–2389.
Wolfe E. W., Lucchitta B. K., Reed V. S., Ulrich G. E. and Sanchez A. G. (1975) Geology of the Taurus-Littrow valley floor. *Proc. Lunar Sci. Conf. 6th*, p. 2463–2482.

Proc. Lunar Planet. Sci. Conf. 9th (1978), p. 2011–2017.
Printed in the United States of America

Sulfur abundances in the 74001/74002 drive tube core from Shorty Crater, Apollo 17

EVERETT K. GIBSON, JR.

SN7, Geochemistry Branch, NASA Johnson Space Center, Houston, Texas 77058

FIKRY F. ANDRAWES

Lockheed Electronics Co., Inc., 16811 El Camino Real, Houston, Texas 77058

Abstract—Sulfur abundances have been measured for seven samples of the 74001/74002 drive tube core from Shorty Crater at Apollo 17 landing site. Total sulfur abundances ranged from 420 to 750 μgS/g and a mean of 548 μgS/g. Sulfur abundances were genreally lower for samples from the bottom section of the double drive tube as compared to samples from the upper section. Heating to 1000°C under vacuum removed around 83 ± 12% of the total sulfur present in the samples. The majority of the removed sulfur is believed to be the surface related component.

INTRODUCTION

The soils in drive tube core 74001/74002 collected at Shorty Crater during Apollo 17 consist mainly of droplets of ultramafic composition consisting of homogeneous orange glass or their partially crystallized equivalent, black glass droplets (Heiken *et al.*, 1974). There have been a variety of explanations for the origins of these glasses; hypotheses include volcanic and impact processes and an impact of a meteorite into a lava lake (models summarized in Heiken *et al.*, 1974). Presently, most investigators prefer that soils 74001/74002 and the related 74220 sample are of volcanic origin.

Gibson and Moore (1973a) noted variations in the carbon and sulfur abundances for orange soil 74220. Carbon concentrations ranged from 3.6 to 100 μgC/g. At the present time, sulfur abundances (10 analyses reported) range from 212 to 820 μgS/g for 74220 (Lunar Sample Curator's Data Base, 1978). Detailed studies of the surfaces of the glassy particles have shown that sulfur-bearing vapors had been condensed or sublimed onto the surfaces of the glass droplets (Meyer *et al.*, 1975; Butler and Meyer, 1976; Clanton *et al.*, 1978). Clanton *et al.* (1978) using SEM-EDX methods noted the apparent presence of elemental sulfur on the surfaces of several spheres from 74001/74002. The surface condensates of sulfur along with other volatile trace elements suggest that 74220 and 74001/74002 were derived from a pyroclastic event and that the surface coatings represent volcanic volatiles (Heiken and McKay, 1977).

In this paper we report sulfur abundances in seven samples from 74001/74002, the orange soil 74220, and the Apollo 15 green glass 15427. A portion of each sample has been heated under vacuum in order to remove the majority of the surface sulfur. Analysis of the heat-treated samples provides information on the original sulfur contents of the samples prior to the addition of encrustations of the volatiles on the surfaces or the addition of extralunar elements (Gibson and Moore, 1973b).

EXPERIMENTAL

Total sulfur measurements have been made using the procedures of Gibson and Moore (1973b). A LECO IR-32 total sulfur analyzer was used in which samples were combusted in oxygen at 1600°C, and the resulting SO_2 was detected using an infrared Luft Cell detector. Detection limit of the analyzer was 1 μgS. Accuracy of the sulfur analysis was checked against standard steels and reference rock samples. Multiple analysis of the various recognized sulfur standards throughout calibration and operation of the analyzer indicates the precision and accuracy of our analytical techniques. For example, five replicate analyses of N.B.S. standard steel 55e (certified sulfur content of 110 μgS/g) gave 113, 112, 107, 105, 110 μgS/g. Replicate analysis of N.B.S. standard steel 101f (certified sulfur content of 80 μgS/g) gave values of 81, 80, 81, 84, and 79 μgS/g. Analysis of British Certified Standard BCS260/4 (reported sulfur concentration of 40 μgS/g) gave analytical values of 40, 41, 40, 40, and 40 μgS/g. Ten analyses of the U.S.G.S. reference basalt BCR-1 resulted in a mean sulfur abundance of 470 μgS/g, a spread from 457–481 μgS/g and a standard deviation from the mean value of 8. Previous workers have reported sulfur concentrations for the reference basalt BCR-1 to be 464 ± 10 μgS/g (Cripe, 1973). The experimentally obtained sulfur values were well within ±5% of the previously reported values.

Portions of all samples were heated in platinum crucibles to 1000°C for 24 hours under vacuum (10^{-6} torr, $P_{O_2} < 10^{-10}$ atms.) using the methods of Gibson and Hubbard (1972) to remove surface sulfur. After the volatilization experiments were completed, the sample residues were analyzed for their total sulfur contents.

DISCUSSION OF RESULTS

The experimentally determined sulfur concentrations for the samples studied are given in Table 1. Duplicate analyses were made for the unheated samples and only a single analysis of the heated samples, because of the limited amounts of samples. The analytical uncertainties given in Table 1 were calculated from the analysis of the instrument blanks, reference standards, and replicates of the samples studied. These reflect the standard deviations within the reported data. Differences between the duplicate analysis of the same sample reflect sample inhomogeneity.

Total sulfur abundances for the 74001/74002 samples ranged from 420 to 750 μgS/g, with a mean sulfur value of 549 μgS/g. Lower sulfur abundances were noted for samples from 74001 than for samples from 74002 (with one exception 74002,1104; 1106; 1108). The sulfur contents are less (with one exception) than the abundances observed for other soils at the Apollo 17 site. Gibson and Moore (1974) noted that sulfur concentrations ranged from 630 to 1130 μgS/g and had a mean of 950 μgS/g for 10 Apollo 17 soils. Heating the 74001/74002 soils under vacuum at 1000°C for 24 hours reduced the sulfur abundances (Table 1) between 71 and 94 percent (mean of 83 percent for 7 samples) of their original concentrations. Residual sulfur contents ranged from 30 to 150 μgS/g. Orange soil 74220 and green glass 15427 contained initially 820 and 330 μgS/g

Table 1. Total sulfur abundances, μgS/g.

		Unheated Samples			Heated Samples	
Sample	Depth, cm	Sample wt. mg.	Sulfur content	Mean	Sample wt. mg.	Sulfur Content
74220,20	4–7 cm	59.7	820	820 ± 20	72.5	80 ± 10
		97.4	830			
74002,1104,1106,1108	5.5–7.5	69.5	420	420 ± 20	90.2	120 ± 10
		56.0	410			
74002,1093,1095,1097	13.5–15.5	70.0	645	680 ± 40	88.0	110 ± 10
		76.3	720			
74002,1084,1086,1088	18–20	68.0	560	590 ± 30	79.0	150 ± 10
		63.3	615			
74002,1071,1073,1075	26.5–28.5	56.4	710	750 ± 50	92.1	65 ± 10
		86.9	795			
74001,1076	37	65.8	555	510 ± 40	93.6	30 ± 5
		75.9	470			
74001,1080	44	70.8	450	440 ± 20	84.6	70 ± 10
		72.3	440			
74001,1089	57.5	75.7	460	450 ± 20	83.2	70 ± 10
		68.9	440			
15427,44	Surface	70.2	325	330 ± 20	75.9	225 ± 20
		75.6	330			

respectively, and upon heating the sulfur concentrations were reduced to 80 and 225 μgS/g respectively.

Temperatures selected for the volatilization of the surface sulfur from the soils studied were chosen from data in the literature and other investigators' experimental studies. Gibson and Moore (1973b) heated mature lunar soils under vaccum to 1100°C and found that 85 to 94% of the total sulfur present in the samples was lost; whereas heating to 950°C under vacuum removed only 50 to 71 percent of the sulfur present in the samples. These experiments indicated a significant amount of the sulfur present in typical lunar soils would be lost by heating under vacuum in the temperature interval 950–1100°C.

Samples of 74220, 74001, and 74002 are not typical lunar soils (Heiken *et al.*, 1974) and the loss of sulfur during heating may be different from typical mature lunar soils. As noted previously by Clayton *et al.* (1974) up to 20 to 30 percent of the original abundances of sulfur in the lunar regolith may have been lost by a wide variety of processes (e.g., vaporization by micrometeorite impact, ion sputtering, thermal volatilization within ejecta blankets, etc.). The volatilization processes may redistribute some of the species within the lunar regolith. These deposited mobile species are mixed with other components in soils. In the case of sulfur, the deposited vapors (e.g., sulfides) are added to soils already bearing sulfides. Soils associated with Shorty Crater have different histories from typical lunar soils (Heiken *et al.*, 1974). The lunar pyroclastic event (Heiken and

McKay, 1977) associated with the formation of 74220, 74001 and 74002 is reflected in the nature of the volatiles associated with these soils. Their surfaces have been coated with condensates from the volcanic event (Meyer *et al.*, 1975; Butler and Meyer, 1976; Clanton *et al.*, 1978). Gas release studies on Apollo 17 basalts (Gibson, unpublished data) indicate that sulfur loss (e.g., SO_2 and H_2S) does not occur until temperatures approaching 1100°C (reaction products of troilite and the silicate matrix) whereas gas release studies of normal lunar soils indicates sulfur loss at temperatures below 1000°C. We feel that the sulfur loss below 1000°C is related to the more volatile forms of sulfur which are surface related. This sulfur may represent sulfur which has been previously deposited by vapor transport processes.

In order to understand the behavior of sulfur on the surfaces of orange and black glasses, samples of 74220 were heated under the same experimental conditions used in our previous studies and examined by electron microprobe and scanning electron microscope equipped with EDX analyzers (SEM-EDX). Fifty orange particles were studied for their surface metal and sulfide abundances. It was found by P. Butler (pers. comm. 1978), using electron microprobe and SEM-EDX, that the vacuum heating at 900°C did not significantly change the sulfur abundances present on the samples which had been heated to 900°C under vacuum for the 50 particles studied. However, it was found that the heating had significantly mobilized the zinc, which was depleted in the heated samples as compared to unheated orange samples. Thus, it appears that heating to 900°C under vacuum did not significantly remove a major fraction of the surface sulfur. However, heating to 1000°C under vaccum removes an average of 83% of the sulfur present in the sample (Table 1). The majority of the sulfur loss occurring during heating under vacuum to 1000°C undoubtedly results from the volatilization of the sulfur present on the surfaces of the glass particles.

The varied distribution of sulfur among the samples of 74220 appear to be reflected in the reported sulfur abundances. Examination of the Lunar Sample Curator's Data Summary (1978) for all sulfur analysis for 74220 indicates that 10 analyses have been made and the reported values vary by a factor of four, from a low sulfur abundance of 212 μgS/g obtained by acid hydrolysis (Chang *et al.*, 1974) to a value of 820 μgS/g reported in this work. A summary of the reported sulfur values for 74220 is given in Table 2. It is evident from examination of the data in Table 2 that the lower analytical values were obtained by techniques employing acid hydrolysis for the release of the sulfur present in the sample. Methods which use combustion of the samples in oxygen, typically yield the higher analytical values. The analyses reported by LSPET (1973) and Duncan *et al.* (1974) employ X-ray fluorescence procedures and the sulfur values are expected to be lower than the combustion methods because sulfur is lost during fusion of the sample prior to analysis (J. M. Rhodes, pers. comm., 1977). Differences in analytical techniques have been recognized previously (Kerridge *et al.*, 1975, 1978; Gibson *et al.*, 1975; Moore and Cripe, 1977). The spread in the analytical data for 74220 must also reflect the variable sulfur content of the sample. Detailed petrographic and mineralogical studies of 74220, 74001, and

Table 2. Previously reported sulfur abundances for 74220.

Sulfur Abundance $\mu gS/g$	Method	Reference
212	Acid Hydrolysis	Chang *et al.* (1974)
420	Acid Hydrolysis	Thode and Rees (1976)
442	Acid Hydrolysis	Chang *et al.* (1974)
550	Combustion	Gibson and Moore (1973a)
564	Acid Hydrolysis	Chang *et al.* (1974)
600	Electron Microprobe	Mao *et al.* (1973)
700	XRF	LSPET (1973)
730	XRF	Duncan *et al.* (1974)
750	Combustion	Gibson and Moore (1973a)
820	Combustion	Gibson and Andrawes (this work)

74002 samples indicates widely varying abundances of sulfur-containing phases associated with the surfaces of these materials. Clanton *et al.* (1978) reported the apparent presence of elemental sulfur while other workers have noted metal sulfides (Meyer *et al.*, 1975; Butler and Meyer, 1976). Sample inhomogeneity undoubtedly also must play a role in the variations noted for the reported sulfur abundances along with differences between analytical methods used for determining sulfur in the materials.

Sulfur abundances for the samples from 74001/74002 and 74220 are only one-quarter to one-half the mean sulfur abundances for the Apollo 17 mare basalts (mean value of 1880 $\mu gS/g$) (Gibson, 1977). It has previously been shown (Gibson *et al.*, 1976, 1977) that the sulfur abundances in lunar basalts is directly related to their TiO_2 contents. Because the orange and black glasses have mafic compositions similar to the Apollo 17 basalts (Heiken *et al.*, 1974), we can approximate the sulfur contents of their source materials based upon the fact that they contain approximately 8.8 to 9.0% TiO_2. Using data in Gibson *et al.* (1976, 1977) it is found that the orange soil compositions would contain around 1600 ± 200 $\mu gS/g$. The question on what was the sulfur contents of the interior glass of 74220 was addressed by Butler and Meyer (1976) and in this work. Butler and Meyer calculated an estimated value of 187 $\mu gS/g$ for the interior sulfur content of the glass associated with 74220. The heating results from this work found between 30 and 150 $\mu gS/g$ retained by the glass samples after volatilization at 1000°C. Thus, it appears that the sulfur contents of the interior 74220 and corresponding 74001/74002 material is somewhere on the order of 30 to 187 $\mu gS/g$. The "interior" sulfur abundances are less than 1/10th the expected values (1600 ± 200 $\mu gS/g$—based upon the S correlation with TiO_2 in mafic lunar compositions). The low sulfur abundances for the orange and black glasses, excluding the surface component, requires that these glasses have been very effectively outgassed of their sulfur or their sources were depleted in sulfur initially. However, the abundances of surface sulfur implies that the sulfur was condensed or sublimed onto the droplets' surfaces at the time of formation from a volatile phase rich in sulfur-bearing species (Meyer *et al.*, 1975; Butler and Meyer, 1976; Clanton *et al.*, 1978). Naughton *et al.* (1972) noted that S_2 would be the most abundant gaseous species in equilibrium with lunar basaltic liquid, nearly 40 mole %. The puzzle now exists as to why the orange and black glasses' interior sulfur contents are so low (30 to 187 μgS/g) as compared to the Apollo 17 mare basalts which have related chemistries. At the present time, we do not have an explanation for this depletion of "interior" sulfur in the materials of this mafic composition (e.g., orange and black glasses). Future experimental work will be required before the ultimate answer is arrived at.

CONCLUSIONS

The study of the sulfur abundances in samples from 74001/74002 and soils 74220 and 15427 has produced the following conclusions:

1. Total sulfur abundances for 74001/74002 ranged from 420 to 750 μgS/g and had a mean sulfur value of 549 μgS/g. Lower sulfur abundances were found at greater depth within the double drive tubes.
2. Between 71 and 94% (mean of 83% for 7 samples) of the total sulfur was lost by heating the samples under vacuum at 1000°C. The majority of the sulfur removed from the samples was believed to be the major function of the surface condensed on the surfaces of the glass droplets.
3. The residual sulfur in the glass (sulfur remaining after heating to 1000°C) is extremely low (30 to 150 μgS/g) in abundance for lunar materials of similar bulk composition. The source materials for 74001, 74002 and 74220 were extremely depleted in sulfur as compared to the lunar basalts found at the Apollo 17 site, yet the surfaces of the glass droplets are enriched in sulfur phases.

Acknowledgments—The authors acknowledge discussions with D. D. Bogard, R. V. Morris, R. Brett, R. Housley, and D. S. McKay. Dr. Pat Butler kindly provided his unpublished data on sulfur abundances on heated samples of the orange soil. The assistance of Prof. Thomas Cobleigh in the early stages of this work is recognized.

REFERENCES

Butler P., Jr. and Meyer C., Jr. (1976) Sulfur prevails in coating on glass droplets: Apollo 15 green and brown glasses and Apollo 17 orange and black (devitrified) glasses. *Proc. Lunar Sci. Conf. 7th*, p. 1561–1581.

Chang S., Lennon K., and Gibson E. K., Jr. (1974) Abundances of C, N, H, He, and S in Apollo 17 soils from Stations 3 and 4: Implications for solar wind exposure ages and regolith evolution. *Proc. Lunar Sci. Conf. 5th*, p. 1785–1800.

Clanton U., McKay D. S., Waits G., and Fuhrman R. (1978) Unusual surface features on volcanic droplets from 74001 and 74002 (abstract). In *Lunar and Planetary Science IX*, p. 172–174. Lunar and Planetary Institute, Houston.

Clayton R. N., Mayeda T. K. and Hurd J. M. (1974) Loss of oxygen, silicon, sulfur, and potassium from the lunar regolith. *Proc. Lunar Sci. Conf. 5th*, p. 1801–1809.

Cripe J. D. (1973) The total sulfur content of lunar samples and terrestrial basalts. Master's thesis, Arizona State University, Tempe.

Duncan A. R., Erlank A. J., Willis J. P., Sher M. K. and Ahrens L. H. (1974) Trace element evidence for a two-stage origin of some titaniferous mare basalts. *Proc. Lunar Sci. Conf. 5th*, p. 1147–1157.

Gibson E. K., Jr. (1977) Volatile elements, carbon, nitrogen, sulfur, sodium, potassium, and rubidium in the lunar regolith. *Phys. Chem. Earth* **10**, 57–62.

Gibson E. K. Jr., Brett R. and Andrawes F. (1977) Sulfur in lunar mare basalts as a function of bulk composition. *Proc. Lunar Sci. Conf. 8th*, p. 1417–1428.

Gibson E. K., Jr., Chang S., Lennon K., Moore G. W. and Pearce G. W. (1975) Sulfur abundances and distributions in mare basalts and their source magmas. *Proc. Lunar Sci. Conf. 6th*, p. 1287–1301.

Gibson E. K., Jr. and Hubbard N. J. (1972) Thermal volatilization studies on lunar samples. *Proc. Lunar Sci. Conf. 3rd*, p. 2003–2014.

Gibson E. K. Jr. and Moore C. B. (1973a) Variable carbon contents of lunar soil 74220. *Earth Planet. Sci. Lett.* **20**, 404–408.

Gibson E. K. Jr. and Moore G. W. (1973b) Carbon and sulfur distributions and abundances in lunar fines. *Proc. Lunar Sci. Conf. 4th*, p. 1577–1586.

Gibson E. K. Jr. and Moore G. W. (1974) Sulfur abundances and distributions in the valley of Taurus-Littrow. *Proc. Lunar Sci. Conf. 5th*, p. 1823–1837.

Gibson E. K. Jr., Usselman T. M. and Morris R. V. (1976) Sulfur in the Apollo 17 basalts and their source regions. *Proc. Lunar Sci. Conf. 7th*, p. 1491–1428.

Heiken G. and McKay D. S. (1977) A model for eruption behavior of a volcanic vent in eastern Mare Serenitatis. *Proc. Lunar Sci. Conf. 8th*, p. 3243–3255.

Heiken G. H., McKay D. S. and Brown R. W. (1974) Lunar deposits of possible pyroclastic origin. *Geochim. Cosmochim. Acta* **38**, 1703–1718.

Kerridge J. F., Kaplan I. R., Kung C. C., Winter D. A., Friedman D. L. and DesMarais D. J. (1978) *Geochim. Cosmochim. Acta* **42**, 391–402.

Kerridge J. F., Kaplan I. R., Petrowski C. and Chang S. (1975) Light element geochemistry of the Apollo 16 site. *Geochim. Cosmochim. Acta* **39**, 137–162.

Lunar Sample Curator's Data Base (1978) Curator's Office, NASA Johnson Space Center, Houston.

LSPET (Lunar Sample Preliminary Examination Team) (1973) Apollo 17 lunar samples: Chemical and petrographic description. *Science* **182**, 659–690.

Mao H. K., Virgo D. and Bell P. M. (1973) Sample 74220: Analysis of the Apollo 17 orange soil from Shorty Crater. *EOS (Trans. Amer. Geophys. Union)* **54**, 598–601.

Meyer C., Jr., McKay D. S., Anderson D. H. and Butler P., Jr. (1975) The source of sublimates on the Apollo 15 green and Apollo 17 orange glass samples. *Proc. Lunar Sci. Conf. 6th*, p. 1673–1699.

Moore C. B. and Cripe J. D. (1977) The distribution of sulfur in lunar rocks and its relationship to carbon content. *The Moon* **16**, 295–310.

Naughton J. J., Hammond D. A., Margolis S. V., and Muenow D. W. (1972) The nature and effect of the volatile cloud produced by volcanic and impact events on the Moon as derived from a terrestrial volcanic model. *Proc. Lunar Sci. Conf. 3rd*, p. 2015–2024.

Thode H. G. and Rees C. E. (1976) Sulpher isotopes in grain size fractions of lunar soils. *Proc. Lunar Sci. Conf. 7th*, p. 459–468.

Proc. Lunar Planet. Sci. Conf. 9th (1978), p. 2019–2032.
Printed in the United States of America

Lunar surface processes and cosmic ray histories over the past several million years

J. S. FRUCHTER, L. A. RANCITELLI, J. C. EVANS, and R. W. PERKINS

Battelle Northwest, Richland, Washington 99352

Abstract—Measurements of the ^{26}Al and ^{53}Mn in interior portions of lunar rocks have shown that lunar surface processes which move a significant fraction of kilogram size rocks on the lunar surface occur on time scales of a few million years. These measurements, together with noble gas age dating have made it possible to define the history for nine rock samples selected from whole rock counting data because of anomalously low ^{26}Al relative to ^{22}Na. Six of the rocks from the Apollo 15 and 16 missions showed evidence of movement during the past five million years. Of these six, only two are of an age consistent with their origin from the South Ray Crater Event. In addition, our measurements of ^{22}Na and ^{26}Al in Apollo 17 double drive tube 74001-74002 suggest that one to two cm of soil is missing from the top of this core tube. Even with this loss, at least two cm of gardening is indicated in the top portion of 74002.

INTRODUCTION

Measurements of the dynamic mixing processes in the lunar regolith are of particular importance as we attempt to understand the lunar surface processes on the moon and other planets devoid of atmosphere in the solar system. Determinations of cosmogenic radionuclides in lunar surface materials are providing useful information on recent exposure history, mixing rates of the lunar regolith, erosion rates of lunar rocks, and variations in intensity, energy spectrum, and incident angle of solar and galactic flux. The half lives of ^{53}Mn (3.7 m.y.) and ^{26}Al (0.72 m.y.) permit the surface history of lunar surface samples to be determined on a time scale through 10 m.y. These radionuclide measurements complement noble gas determinations because they are not subject to long-term memory effects, and they provide information on variations in shielding which may affect the accuracy of noble gas of other types of surface or near surface exposure and age estimates. In this study we are reporting two types of measurements. The first deals with the use of ^{26}Al and ^{22}Na (2.6 yr) in defining natural vs. man-induced mixing that has occurred in the upper portion of the lunar cores that were collected both by the deep drilling and drive tube operations. This work is a continuation of our previous work which showed evidence of both kinds of mixing in certain cores. The second portion of our study involved the use of ^{53}Mn and ^{26}Al concentrations to determine the surface exposure histories of individual rocks over the past 10 m.y. This work on the surface exposure ages of lunar rocks was undertaken because previous analyses of the whole rock ^{22}Na-^{26}Al data had indicated a substantial proportion (20–30%) of the rocks were undersaturated even with respect to the 0.72 m.y. isotope ^{26}Al (Yokoyama *et al.*, 1974; Keith and

Clark, 1974). Although it is difficult to rigorously apply published soil mixing models (Gault, 1974; Arnold, 1975) to the movement of large rocks, these numbers are clearly in excess of the percentage of large rocks which would be expected to show substantial movement during the past two million years. We have, therefore, analyzed sections of lunar rocks from below the solar cosmic ray penetration depth for ^{26}Al and ^{53}Mn in order to better define the surface histories of some of these samples.

SAMPLES AND EXPERIMENTAL METHODS

The core samples employed in this study were subsamples of 0.5 cm long core sections which were dissected at the Lunar Curatorial Facility at JSC. From one to six of these subsamples with total weights of from 0.5 to 10 g from consecutive positions in the core were counted in a high sensitivity coincidence gamma ray spectrometer using techniques described in a previous paper (Rancitelli *et al.*, 1974). For defining the lunar surface exposure history of the individual rocks, ^{26}Al and ^{53}Mn measurements were made on documented specimens from the interiors of the rocks. Most of the rocks were obtained from interior regions with at least 2 cm of shielding on all sides in order to minimize solar proton effects. Because of the small size of some of the rocks it was only possible to maintain 1.5 cm of shielding in some cases. At 2 cm depth, a small contribution to cosmogenic radionuclide production by solar proton effects remains but this can be corrected for in the calculations. In each case, a 5 g specimen was nondestructively counted for ^{26}Al using techniques described in earlier papers (Perkins *et al.*, 1970; Rancitelli *et al.*, 1974). Manganese-53 was determined radiochemically on a 0.5 g specimen adjacent to the 5 g sample as described previously (Fruchter *et al.*, 1977; Fruchter, 1971; Imamura *et al.*, 1974).

RESULTS

The observed ^{26}Al and ^{53}Mn concentrations in nine lunar rocks from Apollo 15, 16, and 17 are presented in Table 1. Also shown in Table 1 is the percentage of the calculated saturation value which each concentration represents, assuming a 2π exposure of the rocks to galactic radiation on the surface of the moon.

Table 1. ^{26}Al and ^{53}Mn Concentrations in Interior Samples of Lunar Rocks and Percent of Saturation.

	^{26}Al dpm/kg Samples	% ^{26}Al Saturation (Calculated)	^{53}Mn dpm/kg Fe	% ^{53}Mn Saturation (Calculated)
12002,OP-6	64 ± 16*	100%	285 ± 18*	~100%
15205,98	26.8 ± 3.5	50%	173 ± 16	58%
64435,31	74.5 ± 2.4	72%	85 ± 15	28%
66095,161	57.8 ± 2.5	64%	71 ± 10	24%
67095,64	67.9 ± 3.4	91%	261 ± 26	87%
67935,18	51.0 ± 2.9	75%	156 ± 20	50%
68815,182	63 ± 2.4*	85%	71 ± 6*	27%
69955,7	70 ± 3.0	100%†	148 ± 20	60%†
70019,48	51.9 ± 3.4	86%	245 ± 25	82%

*From our previously reported measurement (Fruchter *et al.*, 1977).
†At an average depth of 0.35 meter.

Table 2. Calculated ^{26}Al and ^{53}Mn Apparent Surface Residence Ages Compared with Noble Gas Ages.

	^{26}Al Age (m.y.)	^{53}Mn Age (m.y.)	Noble Gas Age* (m.y.)
12002,OP-3	>3	>15	94 (Kr)
15205,98	0.7 ± 0.1	4.5 ± 0.5	169 (Kr)
64435,31	1.4 ± 0.3	1.7 ± 0.3	2 (Ar)
66095,161	0.9 ± 0.2	1.4 ± 0.3	1 (Ne)
67095,64	>2.5	>12	50 (Kr)
67935,18	0.5 ± 0.2	3.8 ± 0.9	50 (Kr) (67955)
68815,182	2.1 ± 0.3	1.7 ± 0.2	2 (Kr)
69955,7	>3	5 ± 1	4 (Kr)
70019,48	2.2 ± 1.0	9 ± 2	–

*References for noble gas ages given in text.

Most of the rock samples were selected for this study because previous whole rock measurements had indicated that they were undersaturated with respect to ^{26}Al (Rancitelli *et al.*, 1972; Yokoyama *et al.*, 1974). In Table 2 we show apparent surface exposure ages of the lunar rocks based on ^{53}Mn and ^{26}Al and compare these with previously reported noble gas ages where available. The relationships between the three ages are shown in Fig. 1. In those cases where the three ages are in agreement, the irradiation history is relatively simple and the ages can be taken as the true surface exposure ages. Where the three ages are not substantially in agreement, a more complicated irradiation history must have occurred. Exposure histories which seem to be in accord with the observed radionuclides and noble gases are presented in the following section.

Table 3. ^{26}Al and ^{22}Na in the Apollo 17 Double Drive Tube 74001-74002.

	Depth Range cm	Average Depth g/cm^2	^{22}Na dpm/kg	^{26}Al dpm/kg
74002,2,4,6,8,10	0–0.5	0.3	125 ± 18	78 ± 3
74002,1067,2070	0.5–1.0	1.33	142 ± 16	85 ± 9
74002,1066,2068	1.0–1.5	2.35	135 ± 25	75 ± 7
74002,1065,2066	1.5–2.0	3.37	110 ± 10	71 ± 3
74002,1063,2062	2.5–3.0	5.41	103 ± 12	54 ± 3
74002,1061,1062, 2059,2061	3.0–1.0	6.73	74 ± 9	48 ± 3
74001,36–41	36.5–39.5	77.7	38 ± 8	33 ± 2
74001,56–61			34 ± 6	33 ± 2
74001,78–83	57.0–60.0	125	29 ± 8	33 ± 2

*Average Densities: 74002:2.04 g/cm^3
74001:2.79 g/cm^3

The observed ^{22}Na and ^{26}Al concentrations in sections of the Apollo 17 double drive tube 74001-74002 are tabulated in Table 3 and compared with calculated values in Fig. 2. The galactic portion of the calculated curves are normalized to experimental results for the deeper portions of the core tube while the solar proton portion is calculated using the model of Reedy and Arnold (1972) for the rigidities and fluxes shown. The upper portion (0–1 cm) of the calculated solar proton curve has also been verified experimentally by nondestructive measurements of Apollo 17 skim soil samples reported by our group in the Apollo 17 Preliminary Science Report (1973).

DISCUSSION

Several of our recent papers (Fruchter *et al.*, 1976, 1977) and others (Nishiizumi, 1976) have employed the observed depth profiles of cosmogenic nuclides to determine mixing rates of the lunar regolith. These measurements

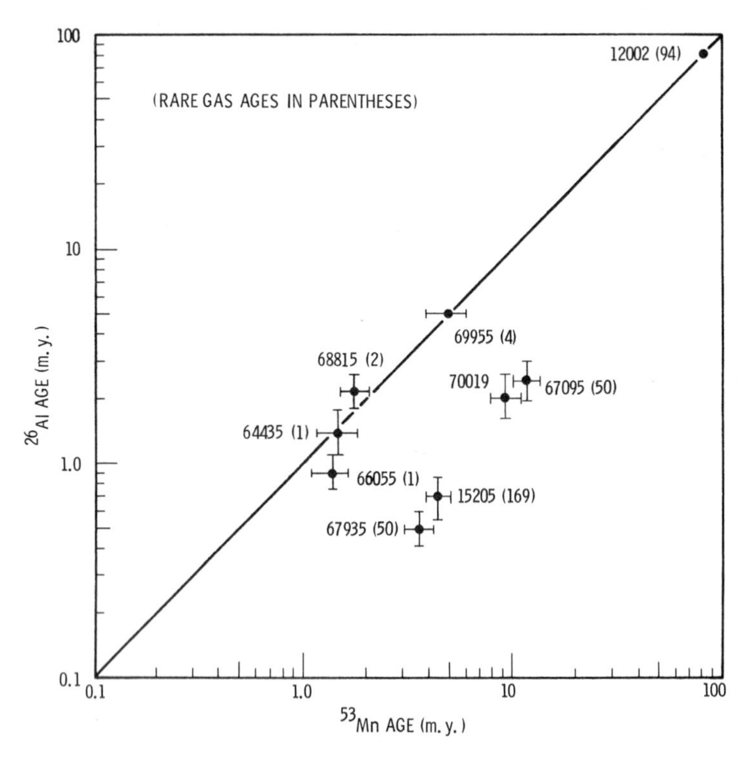

Fig. 1. Calculated ^{26}Al ages and calculated ^{53}Mn ages for nine lunar rocks. Noble gas ages for each rock are shown in parentheses.

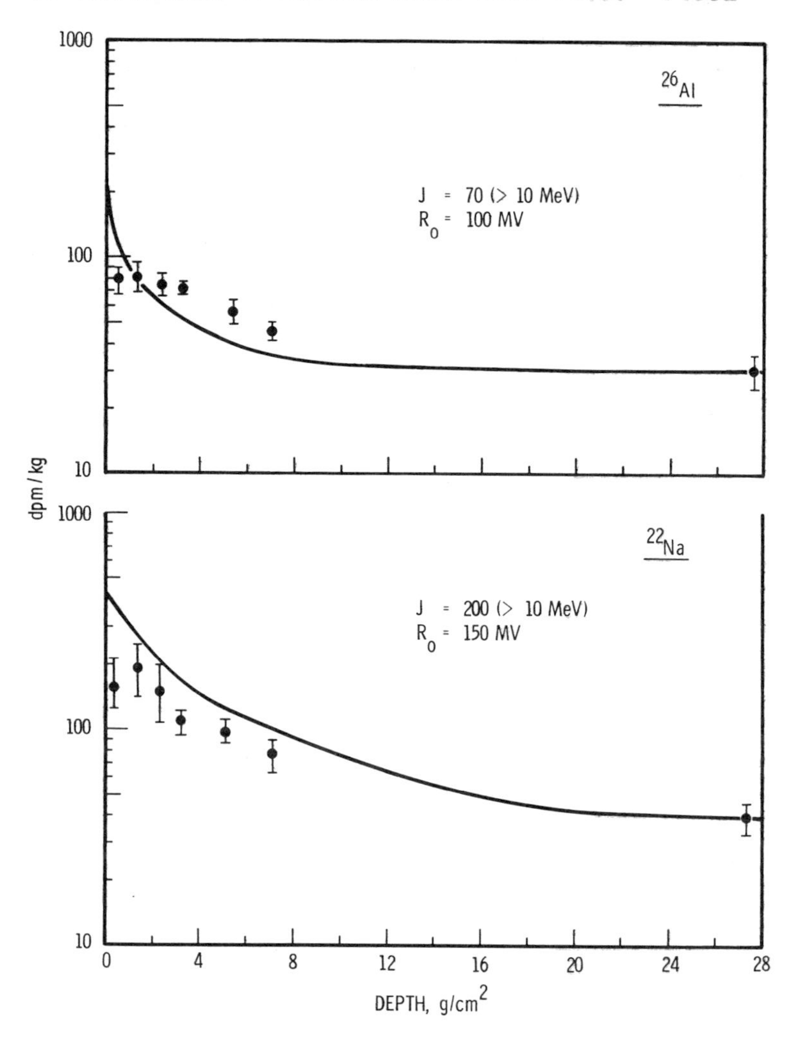

Fig. 2. Comparison of calculated and measured ^{22}Na and ^{26}Al profiles in the top portion of the Apollo 17 double drive tube 74001-74002.

complement stable isotope studies of lunar cores in that they extend the measurements to much shorter time periods (<1 yr to 10 m.y.). To ensure that the observed mixing is due to natural lunar surface processes and not a man-induced disturbance of the core, we also measured ^{22}Na (2.6 yr). In this paper we have employed these cosmogenic radionuclides in a study of the Apollo 17 double drive tube 74001-74002.

In a second phase of the study we are measuring the radionuclides ^{26}Al and ^{53}Mn on individual rocks in order to gain some understanding of the time scale and extent of movement of larger objects on the lunar surface. The goal is to eventually achieve a unified understanding of the rates of movement of both soils and the various sizes of rocks on the lunar surface.

The lunar rock samples for this study were chosen from those which previous nondestructive analyses had shown to be low in ^{26}Al compared to ^{22}Na. This ^{26}Al undersaturation, therefore, may be indicative of a short lunar surface exposure or a recent disturbance on the lunar surface. To date, a total of over 150 lunar rocks have been measured nondestructively for a number of cosmogenic nuclides including ^{22}Na and ^{26}Al. Rocks with low values of ^{26}Al were noted by several groups involved in nondestructive counting. These data were summarized by Yokoyama *et al.* (1974) who also took into account size dependence and chemical composition where possible. Their analysis showed over one-third of the rocks measured to be either definitely or possibly undersaturated on a two million year time scale. Although it is difficult to apply published soil models (Gault, 1974; Arnold, 1975) to the movement of rocks and boulders in a precise manner, it is clear that a much smaller number than one-third would be expected to show under-saturation for ^{26}Al. In fact, the number of rocks with surface exposure ages of less than 2 m.y. would be expected to be in the range of a few percent. Therefore, an additional goal of this study is to resolve this dilemma. For this work we have selected two rock specimens with apparent normal ^{26}Al content and seven that showed evidence of undersaturation.

Calculation of expected ^{22}Na, ^{26}Al, and ^{53}Mn values

Expected ^{22}Na, ^{26}Al, and ^{53}Mn concentrations for both the core and rock samples were calculated using the model of Reedy and Arnold (1972). This model determines both depth and composition dependence. The absolute values were normalized to experimental data available from careful depth studies in various rocks (Finkel *et al.*, 1971; Wahlen *et al.*, 1972; Russ *et al.*, 1978). The corrections to the model from normalization were generally less than 20%. The expected concentrations of ^{22}Na in the top sections of core tube 74002 were obtained from calculations which include the contribution from the large solar flare of August, 1972. These values were in harmony with an absolute value for the ^{22}Na content of surface soil which was obtained from a previous nondestructive measurement of an Apollo 17 one centimeter thick skim soil samples (Apollo 17 Preliminary Science Report, 1973). Values for ^{26}Al were calculated for the core in a similar manner. Galactic production values for ^{26}Al and ^{53}Mn were calculated for each rock on the basis of their chemical compositions from the literature. These calculations were made for appropriate average depths which ranged from four to eight grams per square centimeter and a relatively small but significant correction was made for solar proton effects. The calculations also assume that the average cosmic ray flux has remained essentially constant during the past 6 million years.

Individual rock histories

Several lunar rocks have relatively simple surface histories as indicated by the agreement in their ^{26}Al, ^{53}Mn, and noble gas ages (see Table 2). Rocks in this category include 12002, 64435, 66095, 67095, and 68815. Rock 12002 shows saturation in both ^{26}Al and ^{53}Mn contents, which is in accord with its 94 m.y. surface exposure estimated from Kr measurement (Marti and Lugmair, 1971). The data for 12002 are plotted in Fig. 1 without error bars, since only minimum exposure ages can be deduced from the ^{26}Al and ^{53}Mn contents for this rock. Rock 64435 is clearly undersaturated in both ^{53}Mn and ^{26}Al. The surface exposure age for ^{53}Mn and ^{26}Al are in fair, if not good, agreement with the 1 m.y. Ar age reported by Bogard *et al.* (1973), but disagree with a more precise age of 0.7 ± .3 reported subsequently by Bogard and Gibson (1975). In any event, it appears that this rock was ejected to the surface around 1 m.y. ago, making it clearly younger than the South Ray Crater Event. Rock 66095 is substantially undersaturated as indicated by the apparent ages from both ^{26}Al and ^{53}Mn measurements and is again in reasonable agreement with the Ne age of 1 m.y. reported by Heymann and Hubner (1974). Based on the agreement among these three ages, this rock is also clearly younger than the South Ray Crater Event and, therefore, must have been ejected to the surface in a separate and more recent event. Data for Sample 67095 show a possible shielding effect of about 10% for both radionuclides. However, because of uncertainties in the method and the compositional heterogeneity of this rock, no firm conclusions can be drawn on this point. Thus, it is likely that this rock has lain in its present position for the 50 m.y. period indicated by its Kr age. Rock Sample 68815 was reported to have been taken from a prominent top position of a boulder that was apparently ejected from the South Ray Crater Event. The ^{26}Al and ^{53}Mn ages are in good agreement with the reported 2 m.y. Kr age for this rock (Drozd *et al.*, 1974). The ^{26}Al and ^{53}Mn ages were erroneously reported as somewhat younger in our Lunar and Planetary Science IX abstract (Fruchter *et al.*, 1978) due to a mix-up in sample depth documentation. The shallower than actual depth led us to calculate a somewhat higher expected saturation value for this sample. The calculated ages all now point to a simple exposure history for the rock having been ejected to the lunar surface from a depth of greater than one meter by the South Ray Crater Event approximately 2 m.y. ago.

Sample 69955 appears from sample documentation to have had a relatively simple history. This sample was taken from the bottom of a 0.5 m boulder at an average depth of approximately 0.35 m which was suggested to have come from South Ray Crater. The sample position is shown in Fig. 3. The ^{26}Al is saturated for that depth (see Table 2). The ^{53}Mn is about 60% of saturation, which is consistent with a surface exposure age of about 5 ± 1 m.y. This age is in good agreement with the 4.2 m.y. Kr age reported by Drozd *et al.* (1974) showing that

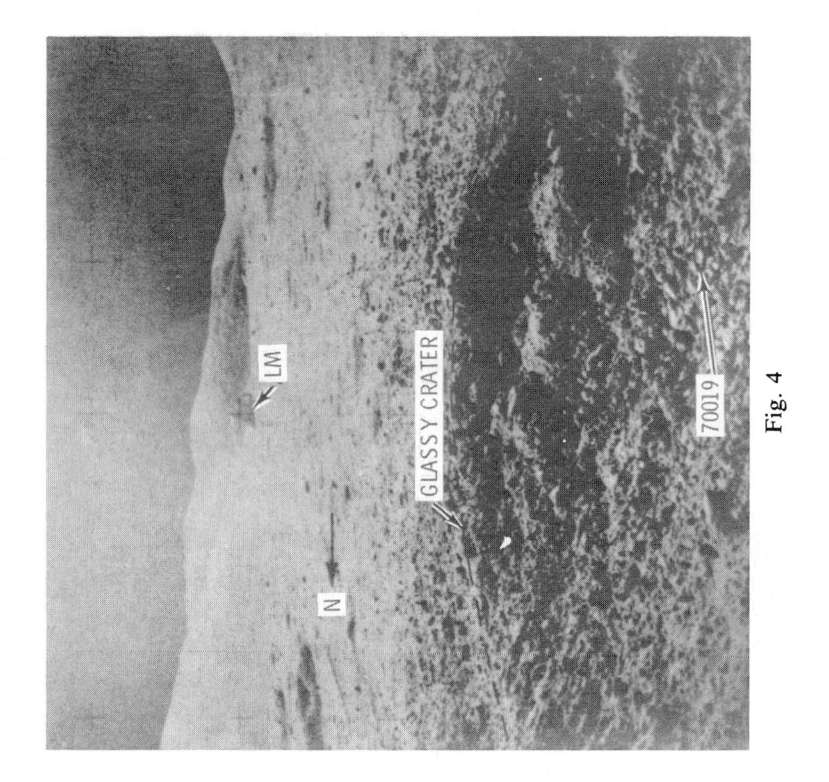

Fig. 4

Fig. 3

Figs. 3–6. Lunar surface photographs of four of the rocks for which surface histories were determined in this study.

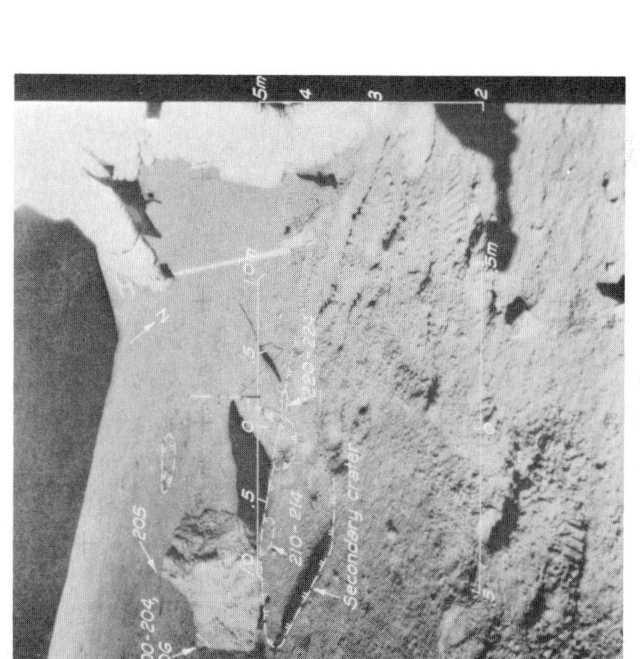

Fig. 5

Fig. 6

this rock clearly predates the South Ray Crater Event by at least 2 m.y. It is highly unlikely that even a shallow burial at the pre-South Ray Crater site for a considerable length of time followed by ejection 2 m.y. ago could have led to the good agreement between the apparent ^{53}Mn and Kr ages.

Rock Sample 70019 is about 85% saturated in both ^{26}Al and ^{53}Mn. Figure 4 shows the 3 m crater in which 70019 resided on the lunar surface. The 85% saturation is approximately what one would predict for the partially shielded location of this rock at the bottom of this crater. Thus, the rock shows a simple exposure history for a time period which is long compared to the half life of ^{53}Mn. We expect that when a noble gas measurement is made that it will confirm a long surface exposure age.

The exposure histories of 15205 and 67935 are somewhat more complex. 15205 was taken from the top of a 1 m boulder as shown in Fig. 5. From the ^{53}Mn and ^{26}Al measurements which show both isotopes to be about 50% saturated, we conclude that 15205 was shielded at a depth of approximately 1 m for a time period which was long compared to the half life of ^{53}Mn. The reported 169 m.y. Kr age suggests that the rock was on or near the surface for a long period of time, especially since the 1 m of shielding would lower the Kr production and, thus, underestimate the true surface age. Thus, it appears certain that after a surface exposure age of about 200 m.y. that the geometry changed to its present configuration less than 1 m.y. ago. This is required since the ^{26}Al is considerably undersaturated. An exposure history that is consistent with the three exposure ages—^{26}Al, 0.7 m.y.; ^{53}Mn, 4.5 m.y.; and Kr, 169 m.y.—requires that the boulder from which this sample was taken remained buried just below the surface for over 200 m.y. and was then ejected by a small cratering event to its present position, and has remained there for less than 100,000 yr. The ^{26}Al surface age is consistent with the 70,000 to 80,000 surface track and pit ages reported by Bhandari (1977) and Hartung and Storzer (1974) for this rock.

Rock 67935 was taken from a "shatter cone" on an approximately 2 m diameter boulder, pictured in Fig. 6. In this sample the ^{26}Al is at 75% of the saturation value for the lunar surface but is essentially saturated for its partially shielded position on the side of this large boulder. The ^{53}Mn, however, is present at about 50% of lunar surface saturation value. The lack of saturation of ^{53}Mn, together with the essentially complete saturation of ^{26}Al, required that the shatter cone was formed about 2 m.y. ago, possibly during the South Ray Crater Event. A Kr age of 50 m.y. was obtained for Rock 67955, which was sampled immediately adjacent to 67936 in the boulder and this suggests a possible long exposure age for Rock 67935 also.

Apollo 17 double drive tube 74001-74002

Double drive tube 74001-74002 is the orange-black glass core sample taken from the rim of Shorty Crater. The unusual chemical composition and density of this core, plus the fact that it is partially shielded by a near-by 5 m diameter boulder, make the interpretation of this core somewhat more difficult than usual. The position of the core relative to surrounding features is shown in Fig. 7. The composition and shielding have been taken into account in constructing the theoretical curves plotted in Fig. 2. The high concentrations of ^{22}Na in the tops of the Apollo 17 cores because of the large August, 1972 solar flare were in harmony with the ^{22}Na values observed in the 1 cm skin soil samples reported by our group and others in the Apollo 17 Preliminary Science Report (1973). On comparing the shape and concentration in the ^{22}Na depth profile for the core, it is evident that the top 1–2 cm of this core tube was lost during sampling or in subsequent handling. In the remaining upper portion of the core, the ^{26}Al measurements show the effects of gardening in the top few centimeters over a 1 m.y. time scale. The effect of this mixing over a depth of 2–3 cm has been observed on all six cores which have been measured to date for ^{26}Al, indicating that such gardening is a common feature on the lunar surface. The reported possible inversion of the material that made up this core 7 m.y. ago as shown by noble gas measurements (Bogard and Hirsch, 1978) would have no measureable effect on the ^{26}Al results, but should be observable when ^{53}Mn measurements are made.

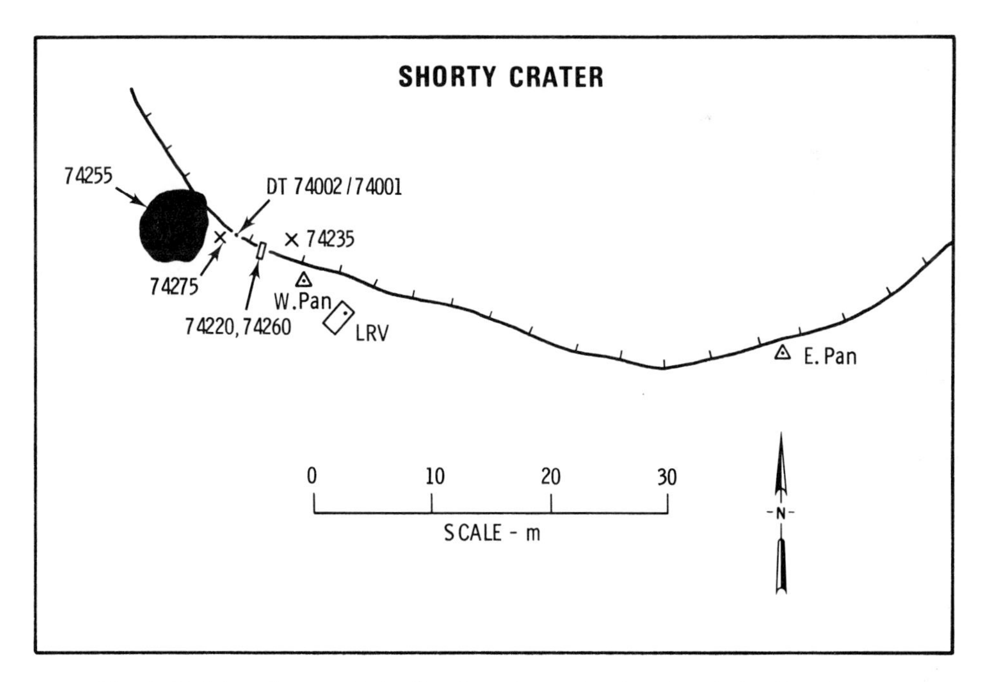

Fig. 7. Plainimetric map of Apollo 17 station 4 showing the relationship of double drive tube 74001-74002 to surrounding features.

SUMMARY AND CONCLUSIONS

A major objective in our present work has been to conduct precise measurements of ^{53}Mn and ^{26}Al on interior portions of lunar rocks which would permit us to confirm or otherwise explain the apparent short lunar surface exposure ages based on ^{22}Na-^{26}Al measurements. Our second objective was to identify material and/or man-induced mixing which had occurred in the lunar surface at the site sampled by the Apollo 17 double drive tube 74001-74002. From these studies we conclude:

(1) One to two centimeters of lunar soil are apparently missing from the top of double drive tube 74001-74002.
(2) Even with the loss of the top 1 or 2 cm of soil, the ^{26}Al profile indicates at least 2 or 3 cm of gardening in the top portion of 74002. A similar amount of mixing has also been observed in all five of the drill stems and drive tubes previously analyzed for ^{26}Al. Therefore, a few centimeters of gardening over a million year time scale appear to be a common feature of the lunar surface.
(3) Of the nine rocks selected for study, three showed no evidence of recent exposure changes; four, in fact, were undersaturated because of a short lunar surface exposure, and two had a more complex history.
(4) Of the six rocks with recent exposure changes, two have an age which can be related to the South Ray Crater Event.
(5) The rocks examined to date in our study were selected from some of the more obvious examples of undersaturation. The four cases found so far which are not related to South Ray Crater show that frequent activity of a sufficient magnitude to move rocks does take place on the lunar surface on a time scale of a few million years. The South Ray Crater Event is excluded from this consideration because there is an obvious bias towards this event in the samples of the Apollo 16 site.

A primary objective of our future work includes a documentation of lunar surface exposure history for the lunar rocks which appear to be undersaturated based on ^{26}Al content. This documentation can be done in a reasonably absolute way where we have interior samples of the rocks for ^{53}Mn and ^{26}Al analysis and where noble gas and track measurements are made. This information will allow the development of more realistic models of lunar surface dynamics.

Acknowledgments—We wish to thank R. M. Campbell, D. R. Edwards, C. L. Nelson, J. G. Pratt, and particularly J. H. Reeves of this laboratory for their aid in standards preparation and in data acquisition; and to E. S. Getchell for typing the manuscript. The unique and sensitive instrumentation which made this work possible was developed during the past decade under sponsorship of the United States Atomic Energy Commission Division of Biomedical and Environmental Research, now part of the United States Department of Energy. This paper is based on work supported by the National Aeronautics and Space Administration, Johnson Space Center, Houston, Texas, under Contract NAS-9-11712.

REFERENCES

Arnold J. R. (1975) Monte Carlo simulation of turnover processes in the lunar regolith. *Proc. Lunar Sci. Conf. 6th*, p. 2375–2395.

Apollo 17 Prelimin. Sci. Rep. (1973) NASA SP-330. 710 pp.

Bhandari N. (1977) Solar flare exposure ages of lunar rocks and boulders based on ^{26}Al. *Proc. Lunar Sci. Conf. 8th*, p. 3606–3615.

Bogard D. D. and Gibson E. K. (1975) Volatile gases in breccia 68115 (abstract). *Lunar Science VI*, p. 63–65. The Lunar Science Institute, Houston.

Bogard D. D., Nyquist L. E., Hirsch W. C. and Moore D. R. (1973) Trapped solar and cosmogenic noble gas abundances in Apollo 15 and 16 drill samples. *Earth Planet. Sci. Lett.* **21**, 52–69.

Bogard D. D. and Hirsch W. C. (1978) Noble gas contents and irradiation history of orange-black glass in the 74001-74002 core (abstract). In *Lunar and Planetary Science IX*, p. 111–113. Lunar and Planetary Institute, Houston.

Drozd R. J., Hohenburg C. M., Morgan C. J. and Ralston C. (1974) Cosmic-ray exposure history at the Apollo 16 and other lunar sites: Lunar surface dynamics. *Geochim. Cosmochim. Acta* **38**, 1625–1642.

Finkel R. C., Arnold J. R., Imamura I., Reedy R. C., Fruchter J. S., Loosli H. H., Evans J. C., Delany A. C. and Shedlovsky J. P. (1971) Depth variation of cosmogenic nuclides in a lunar surface rock and lunar soil. *Proc. Lunar Sci. Conf. 2nd*, p. 1773–1789.

Fruchter J. S. (1971) Measurement of ^{53}Mn and ^{54}Mn. Profiles in lunar rocks and meteorites. Ph.D. Thesis, Univ. of California, San Diego. 161 pp.

Fruchter J. S., Rancitelli L. A. and Perkins R. W. (1976) Recent and long-term mixing of the lunar regolith based on ^{22}Na and ^{26}Al measurements in Apollo 15, 16 and 17 deep drill stems and drive tubes. *Proc. Lunar Sci. Conf. 7th*, p. 27–39.

Fruchter J. S., Rancitelli L. A., Laul J. C. and Perkins R. W. (1977) Lunar regolith dynamics based on analysis of the cosmogenic radionuclides ^{22}Na, ^{26}Al, and ^{53}Mn. *Proc. Lunar Sci. Conf. 8th*, p. 3595–3605.

Fruchter J. S., Evans J. C., Rancitelli L. A. and Perkins R. W. (1978) Lunar surface processes and cosmic ray histories over the past several million years (abstract). In *Lunar and Planetary Science IX*, p. 350–352. Lunar and Planetary Institute, Houston.

Gault D. E., Hörz F., Brownlee D. E. and Hartung J. B. (1974) Mixing of the lunar regolith. *Proc. Lunar Sci. Conf. 5th*, p. 2365–2386.

Hartung J. B. and Storzer D. (1974) Lunar microcraters and their solar flare records. *Proc. Lunar Sci. Conf. 5th*, p. 2527–2571.

Heymann D. and Hubner W. (1974) Origin of inert gases in "Rusty Rock" 66095. *Earth Planet. Sci. Lett.* **22**, 423–426.

Imamura M., Nishiizumi K., Honda M., Finkel R. C., Arnold J. R. and Kohl C. P. (1974). Depth profiles of ^{53}Mn in lunar rocks and soils. *Proc. Lunar Sci. Conf. 5th*, p. 2093–2103.

Keith J. E. and Clark R. S. (1974) The saturation activity of ^{26}Al in lunar samples as a function of chemical composition and the exposure ages of some lunar samples. *Proc. Lunar Sci. Conf. 5th*, p. 2105–2119.

Marti K. and Lugmair G. W. (1971) Kr^{81}-Kr and K-Ar^{40} ages, cosmic-ray spallation products, and neutron effects in lunar samples from Oceanus Procellarum. *Proc. Lunar Sci. Conf. 2nd*, p. 1591–1605.

Nishiizumi K., Imamura M., Honda M., Russ G. P. III, Kohl C. P. and Arnold J. R. (1976) ^{53}Mn in the Apollo 15 and 16 drill stems: Evidence for surface mixing. *Proc. Lunar Sci. Conf. 7th*, p. 41–54.

Perkins R. W., Rancitelli L. A., Cooper J. A., Kaye J. H. and Wogman N. A. (1970) Cosmogenic and primordial radionuclide measurements in Apollo 11 lunar samples by nondestructive analysis. *Proc. Apollo 11 Lunar Sci. Conf.*, p. 1455–1469.

Rancitelli L. A., Perkins R. W., Felix W. D. and Wogman N. A. (1972) Lunar surface processes and cosmic ray characterization from Apollo 12–15 analysis. *Proc. Lunar Sci. Conf. 3rd*, p. 1681–1691.

Rancitelli L. A., Perkins R. W., Felix W. D. and Wogman N. A. (1974) Solar flare and lunar surface process characterization at the Apollo 17 site. *Proc. Lunar Sci. Conf. 5th*, p. 2185–2203.
Reedy R. C. and Arnold J. R. (1972) Interaction of solar and galactic cosmic ray particles with the moon. *J. Geophys. Res.* **77**, 537.
Russ G. P. III, Kohl C. P., Murrell M. T. and Arnold J. R. (1978) An experiment on the constancy of the SCR flux over the past two million years (abstract). In *Lunar and Planetary Science IX*, p. 982–984. Lunar and Planetary Institute, Houston.
Wahlen M., Honda M., Imamura M., Fruchter J. S., Finkel R. C., Kohl C. P., Arnold J. R. and Reedy R. C. (1972) Cosmogenic nuclides in football-sized rocks. *Proc. Lunar Sci. Conf. 3rd*, p. 1719–1732.
Yokoyama Y., Reyss J. L. and Guichard F. (1974) ^{22}Na-^{26}Al chronology of lunar surface processes. *Proc. Lunar Sci. Conf. 5th*, p. 2231–2247.

Proc. Lunar Planet. Sci. Conf. 9th (1978), p. 2033–2048.
Printed in the United States of America

Depositional and surface exposure history of the Shorty Crater core 74001/2: FMR and magnetic studies

RICHARD V. MORRIS

Code SN7, NASA Johnson Space Center, Houston, Texas 77058

WULF A. GOSE

Geophysics Laboratory, University of Texas at Austin, Marine Science Inst.
Galveston, Texas 77550

HOWARD V. LAUER, JR.

Lockheed Electronics Co., Inc., Houston, Texas 77058

Abstract—The values of the FMR surface exposure (maturity) index I_s/FeO and concentrations of FeO and metallic iron were determined for almost every 0.5 cm interval of soil in core section 74002. These data, together with those of Morris and Gose (1977), provide depth profiles of the above parameters for the entire 74001/2 drive tube core which sampled a pyroclastic deposit on the rim of Shorty Crater. Results of FMR and magnetic studies of grain size separates of 13 soils selected from various depths in the core are also reported. The values of I_s/FeO systematically decrease from the lunar surface to a depth of 4.5 cm; the range of values is 3.6 to 0.2 units. From 4.5 cm to the bottom of the core, I_s/FeO is essentially constant at 0.2 units. All of the soil in the core is thus very immature. The relatively higher maturity zone at the top of the core is attributed to *in situ* reworking (gardening) to a depth of 4.5 cm since emplacement of the core in one event (most likely Shorty Crater) which occurred 10–15 m.y. ago. No evidence was found for fossil (buried) surfaces anywhere in the core. The FeO content of the core is very constant. The values of Fe^{0}_{VSM} (essentially total metal) range between 0.06 and 0.18 wt.%. The origin of this variation is not completely understood, but it in part reflects changes in the relative proportions of orange and black droplets.

INTRODUCTION

The drive tube core 74001/2 was driven about 67 cm into the regolith during the Apollo 17 mission on the rim of Shorty Crater. This core is apparently unique among the returned cores because it is virtually entirely composed of orange and black droplets (e.g., Heiken *et al.*, 1974; Heiken and McKay, 1978). The black droplets are the partially or wholly crystallized equivalent of the orange droplets (e.g., Heiken *et al.*, 1974); the formation age of the droplets is about 3.6 billion years (e.g., Huneke *et al.*, 1973; Huneke and Wasserburg, 1978; Alexander *et al.*, 1978). Most investigators have adopted the view that the orange and black droplets are volcanic in origin and were formed by lunar fire fountaining of low viscosity magmas (e.g., Heiken *et al.*, 1974; Heiken and McKay, 1978; Housley, 1978).

In this paper we report depth profiles for FeO, the FMR surface exposure (maturity) index I_s/FeO, and total metal for the upper core section 74002. These profiles were determined from FMR and static magnetic measurements of bulk soil from almost every 0.5 cm dissection interval of 74002 (75 samples). Corresponding profiles for the lower core section 74001 are given by Morris and Gose (1977). In addition to the depth profiles, we report the results of FMR and static magnetic analyses of grain size separates of 13 samples selected from various depths along the core. Collectively, these data provide information about the surface exposure history of the 74001/2 core and the distribution and origin of metallic phases in the orange and black droplets.

EXPERIMENTAL

The experimental procedures are the same as those of Morris and Gose (1976) except for a modification of the procedure to determine FeO and an additional capability to measure metallic iron concentrations with the FMR spectrometer.

The procedure for the FeO determination was modified by a slight change in the equation to convert the experimentally-measured paramagnetic susceptibility (χ_p) to a FeO concentration. We currently use the following equation:

$$\mathrm{FeO} = (4.97 \pm 0.11)\,\chi_p + (0.41 \pm 0.31). \quad (1)$$

In this equation, χ_p is in units of 10^{-5} emu/g and FeO is in units of wt.%. Applying this equation to our previous work produces a negligible change for the Apollo 16 data (Morris and Gose, 1976; Gose and Morris, 1977), but does lower the FeO values for 74001 (Morris and Gose, 1977) by 5% (relative). Equation 1 was calculated from a linear least squares analysis of a plot of chemically-determined values of FeO versus χ_p for 80 lunar soils. The correlation coefficient was 0.98 and the standard deviations permit calculation of the accuracy of our magnetically-determined values of FeO. The accuracy is a function of FeO content and varies monotonically from about ±0.4 wt.% for soils having 5 wt.% FeO to about 0.9 wt.% for soils having 20 wt.% FeO at the one standard deviation level. The accuracy is less than the precision of the measurement of the values of χ_p. Since the paramagnetic susceptibilities of fayalite (Fe_2SiO_4), pyroxene ($FeSiO_3$) and ilmenite are 14.2, 13.4, and 0.18×10^{-5} emu/(g FeO) respectively (Strangway, 1970), it seems likely that the relatively lower accuracy as compared to precision is due to small variations in the average value of χ_p for FeO for individual lunar soils. That is, Equation 1 assumes χ_p for FeO is the same for all lunar soils while in actuality χ_p for FeO is somewhat variable. And finally, Equation 1 is, strictly speaking, applicable only to values of χ_p determined in the manner described by Morris and Gosė (1976). Namely, values of χ_p calculated from a least squares fit of the following equation using as data the magnetization curve between 12.0 and 15.0 KGauss:

$$J = \chi_p H + J_s\left(1.0 - \frac{a}{H}\right) \quad (2)$$

In this equation, J, J_s, and H are the magnetization, saturation magnetization, and magnetic field strength, respectively. The parameter a is an empirical constant whose value was set equal to 0.5 KGauss after Pearce and Simonds (1974). Slightly different values for the constants in Equation 1 will result if the value of *a* is changed and/or the magnetization curve is measured over a different interval. We have found, however, that the value of J_s (from which Fe^0_{VSM} is calculated) is not very sensitive to the above changes in experimental and calculational procedures.

The procedure to obtain metallic iron concentrations from measurements made with a FMR spectrometer is discussed by Morris (unpublished manuscript). The FMR-determined metal concentrations reported in this study are denoted by Fe^0_{330} and represent the concentration of metal in, according to Housley *et al.* (1976), the diameter range between ~40 and ~330 A. The magnetically-determined metal concentrations are denoted by Fe^0_{VSM} and represent the concentration of metal in, according to Pearch and Simonds (1974), the diameter range greater than ~20–40 A. For semantic convenience, Fe^0_{330} and Fe^0_{VSM} are referred to as the concentrations of fine-grained metal and total metal, respectively. Fe°_{330} is roughly the absolute concentration equivalent of I_s, the relative concentration of fine-grained metal.

RESULTS AND DISCUSSION

The FMR and magnetically-derived data for the bulk samples of soil are compiled in Appendix A. The corresponding data for the grain-size separates of 13 soils are compiled in Appendix B. The average values of FeO, I_s/FeO, and Fe°_{VSM} for each of the stratigraphic units of Duke and Nagle (1974) for 74002 are given in Table 1. A similar table for 74001 is given by Morris and Gose (1977). The depth profiles of FeO, I_s/FeO, and Fe°_{VSM} for 74002 are shown in Fig. 1. The ranges for those parameters for 74001 are shown at the bottom of the figure.

The FeO profile

Figure 1 shows that the FeO concentration in the 74001/2 core is very uniform; the average concentration of FeO and its standard deviation are 21.0 ± 0.5 wt.%. These values of FeO are systematically about 10% (relative) lower than the values reported from INAA analysis of 12 bulk soils from 74001/2 by Blanchard and Budahn (1978). This difference is at about the two standard deviation level of the accuracy of our magnetically-determined values and we presume that the difference between magnetically and INAA determined values of FeO is due to the difference between the actual (and unknown) value of χ_p for FeO for the orange and black droplets and the value of χ_p for FeO used in Equation 1, the latter being an average χ_p for FeO for lunar soils.

The I_s/FeO (surface exposure) profile

As shown in Fig. 1, the values of the FMR surface exposure (maturity) index range between 0.1 and 0.3 units over the entire 67 cm length of 74001/2 except

Table 1. Average values of FeO, I_s/FeO, and Fe°_{VSM} for the stratigraphic study of 74002 as given by Duke and Nagle (1974). The uncertainties are standard deviations of the parameters for each stratigraphic unit. Depth intervals are from the top of 74002.

Unit	Depth Interval (cm)	FeO (wt. %)	I_s/FeO (Arb) 250 m	Fe°_{VSM} (wt. %)	Comment
6	0.0–5.5	20.0 ± .3	1.8 ± 1.2	0.11 ± .01	Immature
5	5.5–10.0	20.0 ± .4	0.2	0.10 ± .01	Immature
4	10.0–17.0	20.2 ± .6	0.2	0.11 ± .03	Immature
3	17.0–20.0	20.3 ± .3	0.3	1.16 ± .01	Immature
2	20.0–25.0	20.6 ± .2	0.2	0.17 ± .01	Immature
1	25.0–32.0	20.6 ± .4	0.2	0.17 ± .01	Immature

*Maturity classification after Morris (1976).

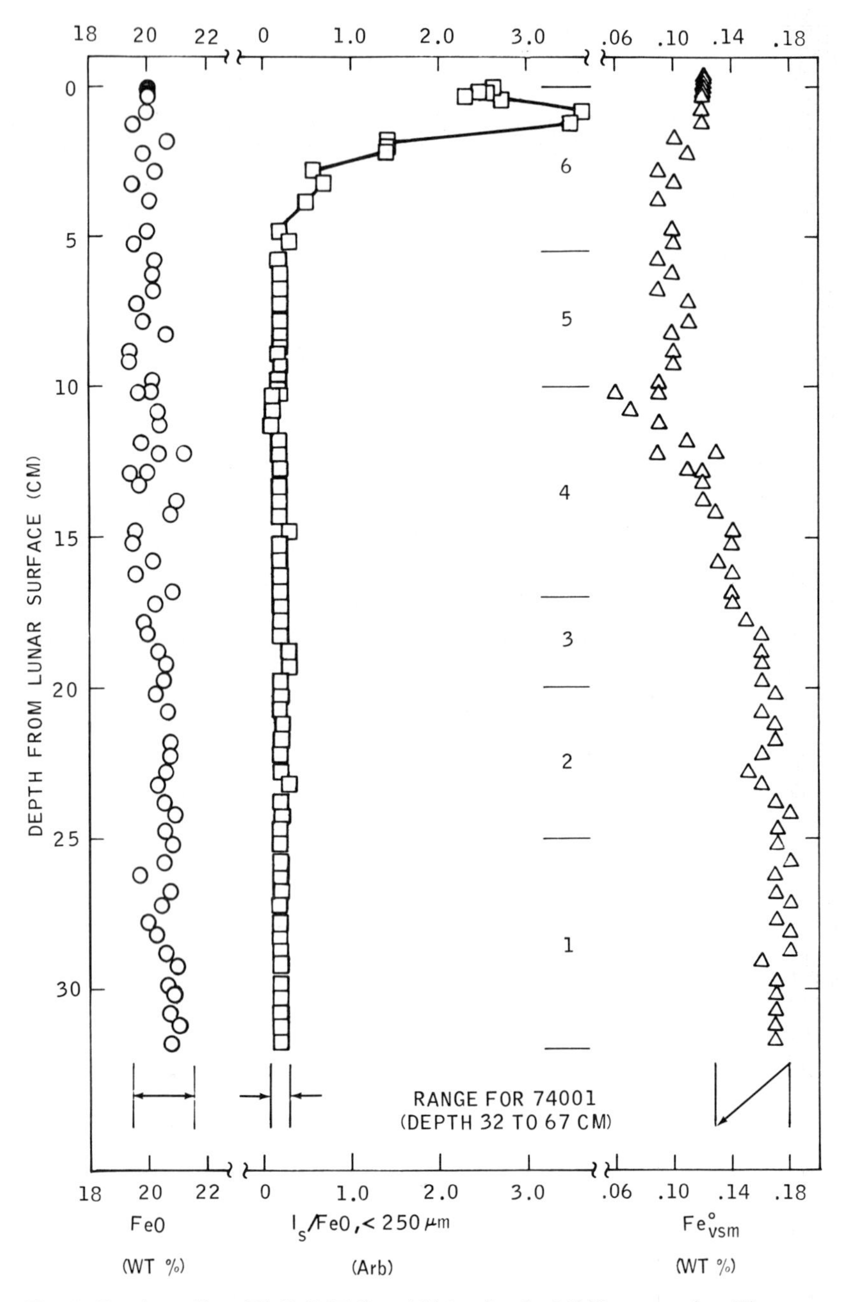

Fig. 1. Depth profiles of FeO, I_s/FeO and Fe^o_{VSM} for the 74002 core section. The range of the above parameters for the lower core section 74001 are shown at the bottom of the figure (after Morris and Gose, 1977). The six stratigraphic units were taken from Duke and Nagle (1974). Since the immature range for the FMR surface exposure (maturity) index is 0.0 to 29.0 units (Morris, 1976, 1978), all of the soils in 74001/2 are very immature.

for the zone extending from the lunar surface to a depth of 4.5 cm. The values of I_s/FeO vary systematically in this zone and reach a maximum value of 3.6 units. Since soils having values of I_s/FeO from 0.0 through 29.0 units are immature (Morris, 1976, 1978a), all of the soil in the 74001/2 core is very immature. In fact, the 0.1 and 0.3 values of I_s/FeO for most of 74001/2 are the lowest thus far measured for lunar soil, except for the value of 0.3 units for the green glass soil 15426 (Morris, 1978a). Thus, relative to other lunar soils that have been sampled and analyzed for their values of I_s/FeO, the 74001/2 soil below 4.5 cm has seen the least amount of surface exposure.

The discussion now focuses on the region of relatively higher values of I_s/FeO in the upper 4.5 cm of 74001/2. Based on our previous studies of other cores (Heiken *et al.*, 1976; Morris and Gose, 1976; McKay *et al.*, 1977), we believe the shape of I_s/FeO profile alone strongly suggests that this "higher-maturity" region is the *in situ* reworking (gardening) zone for the 74001/2 core. That is, we envision that the upper 4.5 cm of 74001/2 also had I_s/FeO ~0.2 units when it was implaced at some time in the past. Between that time and the present, *in situ* reworking progressed to a depth of 4.5 cm and consequently produced the observed I_s/FeO profile. Morris (1978b) has shown this 4.5 cm depth is part of a systematic increase in *in situ* reworking depth with time for data obtained from Apollo cores. The *in situ* reworking interpretation is also supported by or consistent with the petrographic and grain-size studies of McKay *et al.* (1978), the noble gas study of Bogard and Hirsch (1978), the particle track data of Crozaz (1978), and, as discussed next, the dependence of I_s on soil particle grain size.

Figure 2 is a plot of I_s (relative concentration of fine-grained metal) versus soil-particle grain size for 13 soils from 74001/2. Open symbols denote soils from the *in situ* reworking zone (i.e., the upper 4.5 cm). The closed symbols denote the average values for the 9 soils from depths greater than 4.5 cm; the uncertainty represents the range of the observed values. The values of I_s/FeO ($<250\ \mu m$) are also noted on the figure. The general trend of the data is for the grain-size dependence of I_s to become increasingly bowl-shaped with increasing maturity. This trend is compatible with the post-emplacement *in situ* reworking but is not compatible with a mixing mechanism occurring at the time of emplacement of the core. As discussed by Morris (1977), pre-existing fine-grained metal is transported from smaller to larger soil-particle diameters through such constructional events as agglutination. The net effect of this process is that, until a steady-state is reached, the slope of a log-log plot of I_s versus soil particle diameter increases with increasing maturity, which is the trend observed in Fig. 2. If the trend in Fig. 2 were due only to the dilution of previously matured soil with variable amounts of orange and black droplet soil, the arguments given by Morris (1977) indicate no trend or the opposite trend would have been observed because the orange and black droplet soils have a subcentimeter mean grain size which is about the same as or smaller than that of other Apollo 17 soils (McKay *et al.*, 1974, 1978).

The negative of the slope (n = -slope for grain size data $\leq 250\ \mu m$) for the soil

74002,80 (I_s/FeO = 0.7 units) in Fig. 2 is 0.78. Compared to corresponding data for other lunar soils (Morris, 1977), that value falls in a group which has the maximum observed values of n. These values of n were inferred by Morris (1977) to most likely reflect the production of fine-grained metal in the regolith as a function of soil-particle diameter. However, with increasing maturity, the values of n for the 74001/2 soils decrease much more rapidly than observed by Morris (1977). We believe this is due to the difference between the grain size distribution of the 74001/2 soils and that of the bedrock-derived soils upon which the trend of Morris (1977) is based. The data of McKay *et al.* (1978) indicate that the 74001/2 orange and black droplet soils have about 14% of their mass between 150 μm and 1 mm and essentially no mass between 1 mm and 1 cm. In marked contrast, Apollo 17 soils derived predominantly from bedrock typically

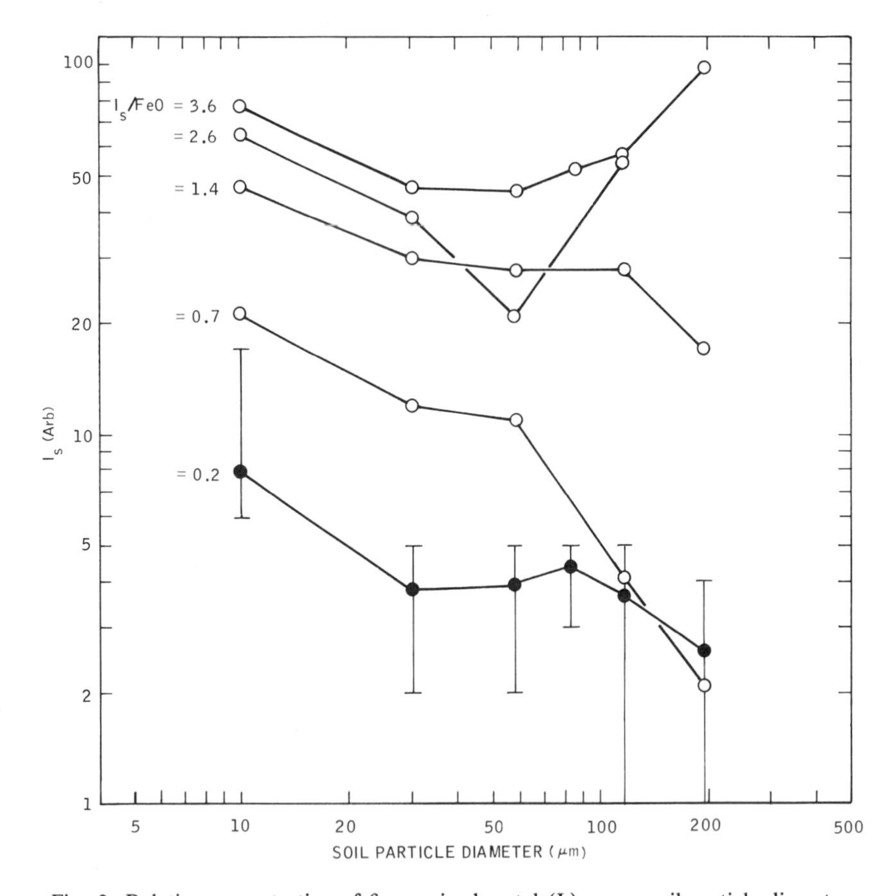

Fig. 2. Relative concentration of fine-grained metal (I_s) versus soil-particle diameter. The open symbols denote soils from the *in situ* reworking zone (i.e., from the upper 4.5 cm). The closed symbols denote average values for 9 soils from depths below 4.5 cm; the uncertainties are the ranges of the observed values. The values of I_s/FeO are for bulk soil (<250 μm).

have 30% of their mass between 150 μm and 1 cm and 10% of their mass between 1 mm and 1 cm (D. S. McKay, pers. comm.). Thus, for the 74001/2 soils there is not as large a reservoir of material to replenish material in the larger sieve fractions as material is removed by destructive processes such as comminution. Thus, the concentration of constructive particles such as agglutinates (and the fine-grained metal contained therein) will increase relatively faster in the larger sieve fractions of the 74001/2 soils as compared to the bedrock-derived soils, which in turn gives the more rapid rate of decrease for n for the 74001/2 soils. These results cause a slight modification of the interpretation given to the dependence of n on I_s/FeO as discussed by Morris (1977). Namely, data points for soils initially formed with a grain size distribution similar to the orange and black droplet soils and falling off the general trend do not necessarily indicate a Path II evolution (McKay *et al.*, 1974).

We now discuss the region of the I_s/FeO profile below 4.5 cm and specifically consider arguments for fossil surfaces. We argued earlier (Morris and Gose, 1977) that a surface exposure time on the order of 0.3 m.y. will produce a value of $I_s/FeO \sim 0.2$ units for a 0.5 cm interval of soil. Thus, we conclude no point in the core below 4.5 cm was at the lunar surface for a period of time longer than about 0.3 m.y. Therefore, these data do not support the presence of 10 m.y. surfaces in 74001 as required by the model of Eugster *et al.* (1978). Integrating 0.3 m.y. per 0.5 cm over the entire 67 cm length of 74001/2 gives ~40 m.y. as an upper estimate for the accumulation time of the orange and black droplet soil in 74001/2. We emphasize that the 40 m.y. accumulation time is an upper limit because it presumes that values of I_s/FeO are entirely due to surface exposure during accumulation of the orange and black droplet soils. If, as was also pointed out by Morris and Gose (1977), the observed values of I_s/FeO are instead due in part or wholly to admixtures during accumulation of very small amounts of a matured soil or if, at this very low level of values of I_s/FeO, some other mechanism contributes significantly to the total concentration of fine-grained metal, accumulation times even as short as minutes are entirely consistent with the I_s/FeO profile. Our data are, therefore, entirely compatible with the hours to weeks time frame for the accumulation of 74001/2 proposed by Heiken and McKay (1978) by analogy to terrestrial fire fountaining.

Bogard and Hirsch (1978) have followed a similar line of reasoning for solar-wind derived ^{36}Ar. Instead of a maximum accumulation time of ~40 m.y., however, they calculate a maximum accumulation time of ~1 m.y. We take the difference as evidence that the values of I_s/FeO below 4.5 cm in part reflect contamination by a previously matured soil and in part an origin other than surface exposure, several of which are considered by Morris and Gose (1977). A rough upper limit for the concentration of the "contaminant" soil can be calculated by assuming its I_s/FeO value equals 60 units and the 0.2 value for the 74001/2 soils below 4.5 cm is entirely due to the contaminant. The calculation yields 0.3 wt.%. At this low level, and particularly if it is concentrated in the <20 μm sieve fraction, it is not surprising that Heiken and McKay (1978) do not report the presence of "contaminant soil" particles in their petrographic analyses of sieve fractions >20 μm for soils below 4.5 cm.

In summary, the evidence seems quite conclusive that *in situ* reworking has produced the relatively high maturity zone that is the upper 4.5 cm of the 74001/2 core. We found no compelling evidence for fossil (buried) surfaces, and thus 74001/2 was most likely emplaced at its present location by one event.

The Fe^{o}_{VSM} profile

Figure 1 shows there are no sharp contacts in the Fe^{o}_{VSM} profile as it varies between 0.06 and 0.18 wt.%. The lowest values of Fe^{o}_{VSM} occur in the upper 11 cm and the highest values occur in the region between about 20 and 30 cm. We now consider origins for the metal in the 74001/2 core.

Profiles of Fe^{o}_{330} (metal between ~40 and ~330 Å in diameter), Fe^{o}_{VSM} (metal ≳20–40 Å in diameter), the volume percentage of crystalline droplets, and the sum of the volume percentages of crystalline and partly crystalline droplets are shown in Fig. 3. The degree of crystallization data was obtained from Heiken and McKay (1978), and they point out that crystalline droplets are roughly synonymous with black droplets. Comparison of the absolute magnitudes of Fe^{o}_{330} and Fe^{o}_{VSM} (<250 μm data) shows that Fe^{o}_{330} constitutes less than 2% and between 6 and 17% of Fe^{o}_{VSM} respectively for soils deeper than and shallower than 4.5 cm. Thus, except for soils in the upper 4.5 cm, the contribution of metal from surface-exposure-induced reduction of FeO is small. Actually, the 2% value for this mechanism is an upper limit because we argued in the last section that the fine-grained metal probably is not attributable predominantly to surface exposure in these exceptionally low maturity soils below 4.5 cm.

Heiken and McKay (1977, 1978) conclude that differential cooling rates of initially molten droplets in the plume of the fire fountain produce the range in the degree of crystallization of the droplets. The noncrystalline droplets are the orange glass and cooled the quickest. The completely crystalline droplets are the black droplets and cooled the slowest. Droplets having both glass and crystalline phases and thus reflecting an intermediate cooling rate are also found in appreciable quantities. We suggested previously (Morris and Gose, 1977) that the higher concentration of metal in the mostly black droplet soils of 74001 (Ave. = 0.15 wt.%) as compared to the orange soil (0.07 wt.%) was due to the crystallization of metal along with the silicate and oxide phases to form the black droplets. The covariance in 74002 of the profiles of Fe^{o}_{VSM}, the percentage of crystalline droplets, and the percentage of crystalline plus partly crystalline droplets supports this suggestion. However, the metal and degree of crystallization profiles contravary below about 30 cm. Perhaps the conditions (e.g., oxygen fugacity) for crystallization varied sufficiently during fire fountaining such that a range of metal concentrations are possible during crystallization. And finally, there is apparently no available value of Fe^{o}_{VSM} for pure orange glass. The petrographic analyses of Heiken and McKay (1974) suggest 74220 is at best 92% orange glass (in the 90–150 μm sieve fraction). Thus, there may be some significant contribution to Fe^{o}_{VSM} which is not genetically related to crystalliza-

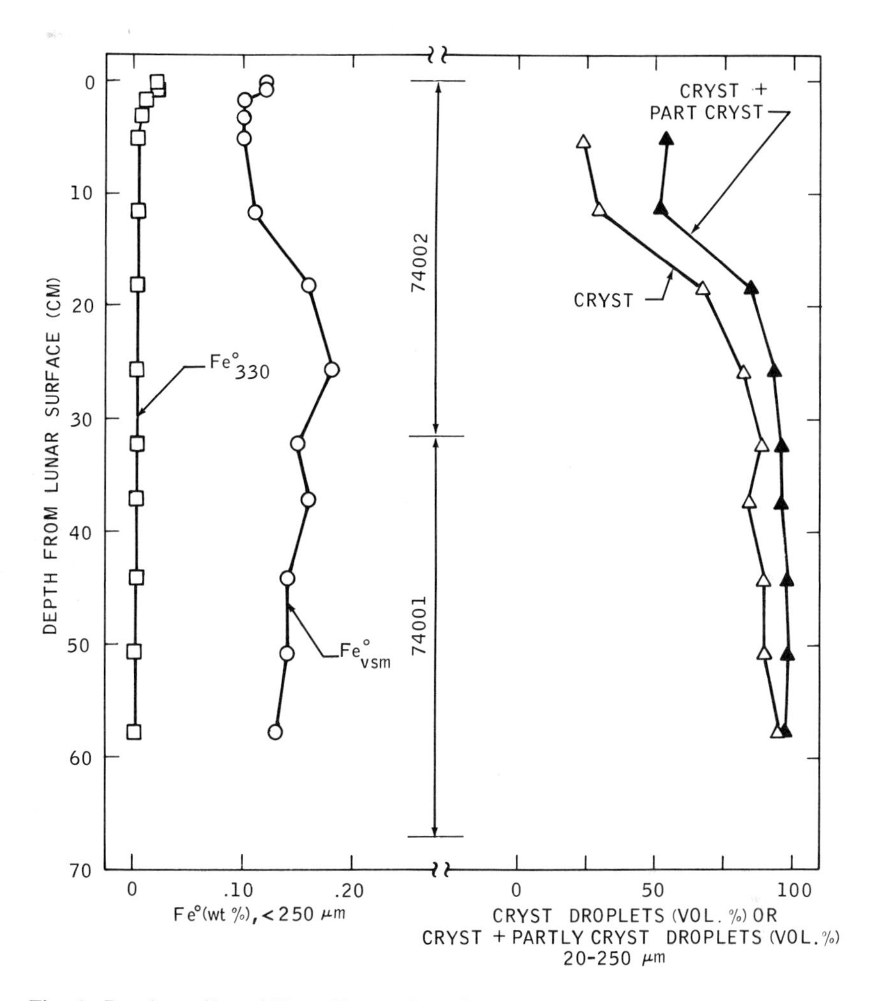

Fig. 3. Depth profiles of Fe°_{330}, Fe°_{VSM}, the volume percentage of crystalline droplets, and the volume percentage of the sum of crystalline and partly crystalline droplets. The degree of crystallization data was taken from Heiken and McKay (1978). The values of Fe°_{330} and Fe°_{VSM} were obtained from measurements on the <250 μm sieve fraction of soil. The degree of crystallization data was obtained on the 20–250 μm sieve fraction of soil.

tion and which may dominate the depth variations in the Fe°_{VSM} profile below about 30 cm where the degree of crystallization changes slowly.

Summary and Exposure History

After considering the results of this and other studies, we propose the five stage model shown schematically in Fig. 4 for the exposure history of the 74001/2 core. Each stage is discussed next in more detail.

EXPOSURE HISTORY OF THE 74001/2 CORE

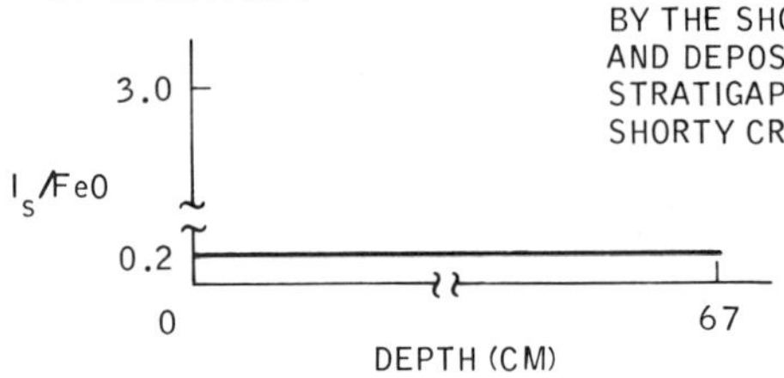

STAGE	TIME	I_s/FeO	COMMEMTS
1. ACCUMULATION	3.6 B.Y. AGO	I_s/FeO~0.2	74001/2 SOILS DEPOSITED BY LUNAR FIRE FOUNTAINING IN HOURS TO WEEKS.
2. SHALLOW BURIAL	FIRST 20M.Y. AFTER ACCUMULATION	I_s/FeO~0.2	74001/2 SOILS BURIED DEEP ENOUGH TO ISOLATE FROM IN SITU REWORKING BUT SHALLOW ENOUGH FOR COSMOGENIC GASES.
3. DEEP BURIAL	FROM THE END OF STAGE 2 UNTIL~10-15 M.Y. AGO	I_s/FeO~0.2	74001/2 SOILS BURIED DEEP ENOUGH TO ISOLATE FROM BOTH IN SITU REWORKING AND PRODUCTION OF COSMOGENIC GASES.
4. EXCAVATION	~10-15 M.Y. AGO		74001/2 SOILS EXCAVATED BY THE SHORTY CRATER IMPACT AND DEPOSITED WITH INVERTED STRATIGAPHY ON THE RIM OF SHORTY CRATER.
5. IN SITU REWORKING	~10-15 M.Y. AGO UNTIL PRESENT		IN SITU REWORKING PROGRESSED TO A DEPTH OF ~5 CM SINCE EMPLACEMENT.

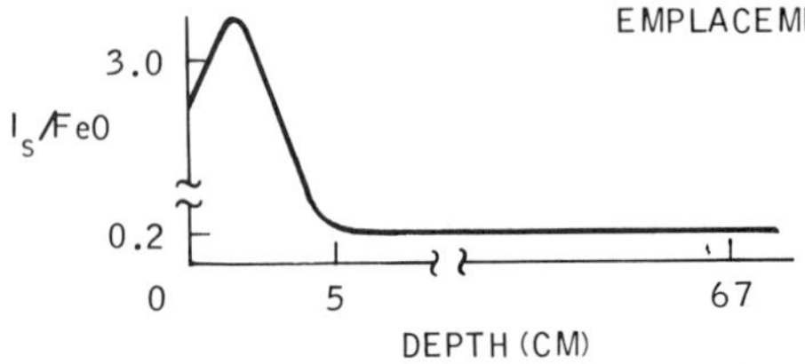

Fig. 4. A five stage model for the exposure history of the 74001/2 core. A more detailed discussion is given in the text.

Stage 1: Accumulation. The black and orange droplet soils were most likely formed by lunar fire fountaining and accumulated as a pyroclastic deposit in a period of hours to months (Heiken and McKay, 1978). The 74001/2 soil as accumulated had I_s/FeO ~0.2 units which, in part, reflects contamination by a small amount of soil (≲0.3%) having a previous history of surface exposure. This contaminant may also be the origin of the small amounts of trapped noble gases now observed in the 74001/2 core below 4.5 cm (Bogard and Hirsch, 1978). According to dating studies of the orange and black droplets (e.g., Huneke *et al.*,

1973; Huneke and Wasserburg, 1978; Alexander *et al.*, 1978), the fire fountaining occurred about 3.6 b.y. ago.

Stage 2: Shallow burial. Bogard and Hirsch (1978) suggest that the soil in 74001/2 could have been exposed to cosmic rays for the first ~20 m.y. after accumulation at a shielding depth of ~25 cm. Burial is also required from the I_s/FeO data so that the soil in the core does not experience surface exposure due to *in situ* reworking. The above shielding depth is sufficient to meet this constraint since the *in situ* reworking depth/time data of Morris (1978) indicate a reworking depth on the order of 9 cm in 20 m.y.

Stage 3: Deep burial. This is the longest stage, comprising virtually all of the 3.6 b.y. age of the soil. Deep burial, which could have begun around 20 m.y. after accumulation according to Bogard and Hirsch (1978), is required to isolate the 74001/2 soil from *in situ* reworking for 3.6 b.y. and to shield the soil from galactic cosmic rays so that high concentrations of cosmogenic gases are not produced (e.g., Huneke *et al.*, 1973; Heiken *et al.*, 1974). A burial depth of at least 2.5 meters is indicated from the *in situ* reworking depth/time data of Morris (1978); an even greater depth is probably required to meet the constraints for production of cosmogenic gases. Heiken *et al.* (1974) suggest an ejecta blanket or a basalt flow may have been the overburden.

According to Bogard and Hirsch (1978), the ~20 m.y. time and ~25 cm shielding depth are only one possible set of times and shielding depths that satisfy the cosmogenic gas data; all that is required is that the time be $\gtrsim$14 m.y. and the shielding depth be $\gtrsim$25 cm within the constraint that longer times require appropriately deeper shielding depths. Thus, the ~20 m.y. time for Stage 2 should not be regarded as firm. In addition, it is also possible that a period of deep burial occurred between Stages 1 and 2. That is, the period of shallow burial could have occurred at any time in the period between accumulation and excavation. We adopted the scenario as given because it is the simplest and because the fire fountaining provides a convenient mechanism for shallow burial.

Stage 4: Excavation. This stage is the excavation of the soil sampled by the 74001/2 core from depth, presumably by the impact which formed Shorty Crater, and its emplacement on the rim of Shorty Crater. Bogard and Hirsch (1978) conclude from the cosmogenic ^{21}Ne and ^{38}Ar profiles of the core that the stratigraphy of the core is inverted from what it was during the irradiation of Stage 2. Ages based on particle tracks (Crozaz, 1978) and cosmogenic ^{21}Ne and ^{38}Ar (Bogard and Hirsch, 1978) data put the emplacement of the core 10–15 m.y. ago.

Stage 5: In situ reworking. During the past 10–15 m.y. the soil at the top of the core was subjected to *in situ* reworking (gardening), and the entire core was irradiated by galactic cosmic rays (e.g., Bogard and Hirsch, 1978). The I_s/FeO profile shows that during the above time *in situ* reworking progressed to a depth of 4.5 cm.

Acknowledgments—This paper benefited from discussions with Drs. D. Bogard and D. McKay and from the reviews of G. Crozaz and J. Laul. C. Hardy and B. Allison are thanked for typing the manuscript. This work was supported in part by NASA grant NSG-9021. University of Texas Marine Science Institute Contribution No. 295.

REFERENCES

Alexander E. C. Jr., Coscio M. R. Jr., Dragon J. C. and Saito K. (1978) ^{40}Ar-^{39}Ar studies of glasses from lunar soils (abstract). In *Lunar and Planetary Science IX*, p. 7–10. Lunar and Planetary Institute, Houston.

Blanchard D. P. and Budahn J. R. (1978) Compositional variations among size fractions in a homogeneous, unmatured regolith: Drive core 74001/2. *Proc. Lunar Planet. Sci. Conf. 9th*. This volume.

Bogard D. D. and Hirsch W. C. (1978) Depositional and irradiational history and noble gas contents of orange-black droplets in the 74002/1 core from Shorty Crater. *Proc. Lunar Planet. Sci. Conf. 9th*. This volume.

Crozaz G. (1978) History of regolith deposition at the Apollo 17 and Luna 24 landing sites and at Shorty Crater (abstract). In *Lunar and Planetary Science IX*, p. 203–205. Lunar and Planetary Institute, Houston.

Duke M. B. and Nagle J. S. (1974) *Lunar Core Catalog and Supplements*, JSC 09252.

Eugster O., Eberhardt P., Geiss J. and Grögler N. (1978) The solar wind and cosmic ray exposure history of soil from drive tube 74001 (abstract). In *Lunar and Planetary Science IX*, p. 306–308. Lunar and Planetary Institute, Houston.

Gose W. A. and Morris R. V. (1977) Depositional history of the Apollo 16 deep drill core. *Proc. Lunar Sci. Conf. 8th*, p. 2909–2928.

Heiken G. and McKay D. S. (1974) Petrography of Apollo 17 soils. *Proc. Lunar Sci. Conf. 5th*, p. 843–860.

Heiken G. and McKay D. S. (1977) A model for eruption behavior of a volcanic vent in eastern Mare Serenitatis. *Proc. Lunar Sci. Conf. 8th*, p. 3243–3255.

Heiken G. and McKay D. S. (1978) Petrology and origin of a sequence of pyroclastic rocks from the valley of Taurus-Littrow (Apollo 17 landing site). *Proc. Lunar Planet. Sci. Conf. 9th*. This volume.

Heiken G. H., McKay D. S. and Brown R. W. (1974) Lunar deposits of possible pyroclastic origin. *Geochim. Cosmochim. Acta* **38**, 1703–1718.

Heiken G. H., Morris R. V., McKay D. S. and Fruland R. M. (1976) Petrographic and ferromagnetic resonance studies of the Apollo 15 deep drill core. *Proc. Lunar Sci. Conf. 7th*, p. 93–111.

Housley R. M. (1978) Modeling lunar volcanic eruptions (abstract). In *Lunar and Planetary Science IX*, p. 549–551. Lunar and Planetary Institute, Houston.

Housley R. M., Cirlin E. H., Goldberg I. B. and Crowe H. (1976) Ferromagnetic resonance studies of lunar core stratigraphy. *Proc. Lunar Sci. Conf. 6th*, p. 13–26.

Huneke J. C., Jessberger E. K., Podosek F. A. and Wasserburg G. J. (1973) ^{40}Ar/^{39}Ar measurements in the Apollo 16 and 17 samples and the chronology of metamorphic and volcanic activity in the Taurus-Littrow region. *Proc. Lunar Sci. Conf. 4th*, p. 1725–1756.

Huneke J. C. and Wasserburg G. J. (1978) ^{40}Ar-^{39}Ar ages of single orange glass balls and highland breccia phenocrysts (abstract). In *Lunar and Planetary Science IX*, p. 567–569. Lunar and Planetary Institute, Houston.

McKay D. S., Dungan M. A., Morris R. V. and Fruland R. M. (1977) Grain size, petrographic, and FMR studies of the double core 60009/10: A study of soil evolution. *Proc. Lunar Sci. Conf. 8th*, p. 2929–2952.

McKay D. S., Fruland R. M. and Heiken G. H. (1974) Grain size and the evolution of lunar soils. *Proc. Lunar Sci. Conf. 5th*, p. 887–906.

McKay D. S., Heiken G. and Waits G. (1978) Grain size and petrology as a key to the rate of *in-situ* reworking and lateral transport on the lunar surface. *Proc. Lunar Sci. Conf. 9th*. This volume.

Morris R. V. (1976) Surface exposure indices of lunar soils: A comparative FMR study. *Proc. Lunar Sci. Conf. 7th*, p. 315–335.

Morris R. V. (1977) Origin and evolution of the grain-size dependence of the concentration of fine-grained metal in lunar soils: The maturation of lunar soils to a steady state stage. *Proc. Lunar Sci. Conf. 8th*, p. 3719–3747.

Morris R. V. (1978a) The surface exposure (maturity) of lunar soils: Some concepts and I_s/FeO compilation. *Proc. Lunar Planet. Sci. Conf. 9th*. This volume.

Morris R. V. (1978b) *In situ* reworking of the lunar surface: Evidence from the Apollo cores. *Proc. Lunar Planet. Sci. Conf. 9th*. This volume.

Morris R. V. and Gose W. A. (1976) Ferromagnetic resonance and magnetic studies of cores 60009/10 and 60003: Compositional and surface-exposure stratigraphy. *Proc. Lunar Sci. Conf. 7th*, p. 1–11.

Morris R. V. and Gose W. A. (1977) Depositional history of core section 74001: Depth profiles of maturity, FeO, and metal. *Proc. Lunar Sci. Conf. 8th*, p. 3113–3122.

Pearce G. W. and Simonds C. H. (1974) Magnetic properties of Apollo 16 samples and implications for their mode of formation. *J. Geophys. Res.* **79**, 2953–2959.

Strangway D. W. (1970) *History of the Earth's Magnetic Field*. McGraw-Hill, N.Y. 168 pp.

Appendix A

The table given below contains the FMR and magnetics data for the 74002 soils. The soils are identified by their parent numbers. The stratigraphic units were obtained from Duke and Nagle (1974).

Parent Number	Depth (cm)	Unit	ΔH (Gauss) <250 μm	FeO (wt. %)	I_s/FeO (Arb) <250 μm	Fe°_{VSM} (wt. %)
2	0.05	6	750	20.0	2.6	0.12
4	0.15	6	720	20.1	2.5	0.12
6	0.25	6	745	20.0	2.5	0.12
8	0.35	6	745	20.2	2.3	0.12
10	0.45	6	730	20.0	2.7	0.12
89	0.8	6	750	19.8	3.6	0.12
87	1.2	6	750	19.5	3.5	0.12
86	1.8	6	730	20.7	1.4	0.10
84	2.2	6	730	19.8	1.4	0.11
82	2.8	6	720	20.3	0.6	0.09
80	3.2	6	730	19.5	0.7	0.10
78	3.8	6	760	20.1	0.5	0.09
76	4.8	6	700	20.0	0.2	0.10
74	5.2	6	740	19.6	0.3	0.10
73	5.8	5	740	20.3	0.2	0.09
72	6.2	5	760	20.2	0.2	0.10
71	6.8	5	730	20.2	0.2	0.09
70	7.2	5	770	19.7	0.2	0.11
69	7.8	5	770	19.9	0.2	0.11
68	8.2	5	770	20.7	0.2	0.10
67	8.8	5	790	19.4	0.2	0.10
66	9.2	5	780	19.4	0.2	0.10
65	9.8	5	780	20.2	0.2	0.09
64	10.2	4	750	19.7	0.1	0.06
63	10.2	4	775	20.1	0.2	0.09
62	10.8	4	750	20.4	0.1	0.07

Appendix A *(cont'd.)*

Parent Number	Depth (cm)	Unit	ΔH (Gauss) <250 μm	FeO (wt. %)	I_s/FeO (Arb) <250 μm	Fe°_{VSM} (wt. %)
61	11.2	4	750	20.4	0.1	0.09
60	11.8	4	750	19.8	0.2	0.11
59	12.2	4	790	21.3	0.2	0.13
58	12.2	4	750	20.4	0.2	0.09
57	12.8	4	800	19.4	0.2	0.12
56	12.8	4	780	20.0	0.2	0.11
55	13.2	4	800	19.7	0.2	0.12
54	13.8	4	780	21.0	0.2	0.12
53	14.2	4	800	20.8	0.2	0.13
52	14.8	4	785	19.6	0.3	0.14
51	15.2	4	810	19.5	0.2	0.14
50	15.8	4	795	20.2	0.2	0.13
49	16.2	4	795	19.6	0.2	0.14
48	16.8	4	792	20.9	0.2	0.14
47	17.2	3	800	20.3	0.2	0.14
46	17.8	3	810	19.9	0.2	0.15
45	18.2	3	810	20.0	0.2	0.16
44	18.8	3	830	20.4	0.3	0.16
43	19.2	3	830	20.7	0.3	0.16
42	19.8	3	815	20.6	0.2	0.16
41	20.2	2	800	20.3	0.2	0.17
40	20.8	2	830	20.7	0.2	0.16
39	21.2	2	830	20.5	0.2	0.17
38	21.8	2	850	20.8	0.2	0.17
37	22.2	2	835	20.8	0.2	0.16
36	22.8	2	835	20.7	0.2	0.15
35	23.2	2	840	20.4	0.3	0.16
34	23.8	2	850	20.6	0.2	0.17
33	24.2	2	825	21.0	0.2	0.18
32	24.8	2	820	20.6	0.2	0.17
31	25.2	1	840	20.9	0.2	0.17
30	25.8	1	840	20.6	0.2	0.18
29	26.2	1	840	19.7	0.2	0.17
28	26.8	1	850	20.8	0.2	0.17
27	27.2	1	840	20.5	0.2	0.18
26	27.8	1	830	20.0	0.2	0.17
25	28.2	1	825	20.3	0.2	0.18
24	28.8	1	820	20.6	0.2	0.18
23	29.2	1	825	21.0	0.2	0.16
22	29.8	1	815	20.7	0.2	0.17
21	30.2	1	820	20.9	0.2	0.17
20	30.8	1	800	20.8	0.2	0.17
19	31.2	1	805	21.1	0.2	0.17
18	31.8	1	805	20.8	0.2	0.17

APPENDIX B

The table given below contains the FMR and magnetics data for size separates of 13 soils which were selected from various depths in the 74001/2 core. The subsample number is the parent number.

Sample and Depth	Sieve Fraction (μm)	ΔH (Gauss)	I_s (Arb)	I_s/FeO (Arb)	Fe°_{FMR} (wt. %)	Fe°_{VSM} (wt. %)
74002,2	<20	745	64			0.16
0.0 to	20–45	745	39			0.11
0.1 cm	45–75	771	21			0.09
	90–150	767	54			
	<250	750	53	2.6	0.02	0.12
74002,89	<20	721	77			0.19
0.5 to	20–45	708	47			0.12
1.0 cm	45–75	760	46			0.09
	75–90	764	52			0.13
	90–150	763	56			0.07
	150–250	790	100			0.07
	<250	750	72	3.6	0.02	0.12
74002,86	<20	721	48			0.17
1.5 to	20–45	744	30			0.10
2.0 cm	45–75	767	28			0.08
	90–150	767	28			0.05
	150–250	665	17			0.12
	<250	730	29	1.4	0.01	0.10
74002,80	<20	721	21			0.17
3.0 to	20–45	755	12			0.09
3.5 cm	45–75	765	11			0.07
	90–150	730	4			0.05
	150–250	780	2			0.04
	<250	730	13	0.7	0.006	0.10
74002,74	<20	715	9			0.17
5.0 to	20–45	685	3			0.11
5.5 cm	45–75	765	2			0.06
	90–150	825	1			0.04
	150–250	910	1			0.03
	<250	740	5	0.3	0.003	0.10
74002,60	<20	740	7			0.16
11.5 to	20–45	850	3			0.11
12.0 cm	45–75	900	3			0.08
	90–150	895	2			0.04
	150–250	900	2			0.03
	<250	750	4	0.2	0.002	0.11
74002,45	<20	795	7			0.20
18.0 to	20–45	850	5			0.16
18.5 cm	45–75	850	4			0.14
	90–150	900	5			0.10
	150–250	840	2			0.08
	<250	810	5	0.2	0.003	0.16

Appendix B *(cont'd.)*

Sample and Depth	Sieve Fraction (μm)	ΔH (Gauss)	I_s (Arb)	I_s/FeO (Arb)	Fe°_{FMR} (wt. %)	Fe°_{VSM} (wt. %)
74002,30	<20	855	6			0.23
25.5 to	20–45	840	4			0.19
26.0 cm	45–75	820	4			0.16
	90–150	905	5			0.14
	150–250	880	3			0.12
	<250	840	4	0.2	0.002	0.18
74001,21	<20	735	17			0.23
32.0 to	20–45	800	2			0.19
32.5 cm	45–75	840	4			0.17
	75–90	820	4			
	90–150	835	4			0.16
	150–250	845	4			0.11
	<250	815	4	0.2	0.002	0.15
74001,36	<20	795	6			0.20
37.0 to	20–45	830	4			0.18
37.5 cm	45–75	845	4			0.17
	75–90	830	5			
	90–150	840	4			0.15
	150–250	855	3			0.11
	<250	840	4	0.2	0.002	0.16
74001,50	<20	840	7			0.19
44.0 to	20–45	845	5			0.17
44.5 cm	45–75	850	5			0.15
	75–90	845	5			
	90–150	830	5			0.14
	150–250	845	3			0.13
	<250	850	5	0.2	0.003	0.14
74001,63	<20	840	6			0.20
50.5 to	20–45	840	4			0.17
51.0 cm	45–75	835	5			0.15
	75–90	860	5			
	90–150	850	4			0.12
	150–250	855	3			0.11
	<250	840	5	0.2	0.002	0.14
74001,78	<20	802	7			0.19
57.5 to	20–45	815	4			0.16
58.0 cm	45–75	850	4			0.14
	75–90	820	3			
	90–150	800	3			0.12
	150–250	830	2			0.10
	<250	820	4	0.2	0.002	0.13

Proc. Lunar Planet. Sci. Conf. 9th (1978), p. 2049–2063.
Printed in the United States of America

Studies of volatiles in Apollo 17 samples and their implication to vapor transport processes

E. H. CIRLIN, R. M. HOUSLEY and R. W. GRANT

Rockwell International Science Center, Thousand Oaks, California 91360

Abstract—We report further developments in our study of vapor transport processes through analysis of Pb, Cd, Zn and S in orange and black droplets from drive tube sample 74002/74001 and nearby surface samples 74220 and 74241, using flameless atomic absorption (FLAA), SEM, X-ray photoemission spectroscopy (XPS), and grain size separation techniques. Our FLAA and XPS studies on grain size separates imply that the volatile trace metals are enriched in smaller size grain fractions which are mostly broken droplet pieces, thus indicating that part of the volatiles were deposited after the droplets were broken. The concentrations of Pb and Cd show continuous decreases as core tube depth increases. This result was unexpected since the droplets were collected at the rim of Shorty Crater where it is unlikely that any original stratigraphy would be well preserved. Assuming that most of the droplet breakage occurred during the Shorty Crater event we can infer that the volatiles were redistributed by the cratering event.

We also report our preliminary studies of Zn on individual regolith grains. Thermal release profiles on black droplets show that most of the Zn is surface Zn, whereas no surface Zn was observed on the mineral phases of basaltic rock 75035. In North Ray Crater samples 67701 and 67712, we detected no Zn in pure anorthite grains, high temperature Zn in breccia and anorthite grains with metal inclusions and a high content of surface Zn on agglutinates.

INTRODUCTION

Three vapor transport processes are considered of major importance in the distribution of volatiles on the moon. Early in its history, the moon underwent a period of extensive volcanic activity. The Apollo 15 green glass and Apollo 17 orange and black glasses are considered to be of volcanic origin and surface coatings on them provide strong evidence of vapor transport associated with this volcanic activity. Another important vapor transport process on the moon occurs in hot ejecta blankets produced by major meteoritic impacts. The existence of sizable vapor grown crystals gives evidence to support the importance of this type of vapor transport process. Solar wind sputtering and micrometeorite bombardment also contribute to the distribution of volatiles in the lunar regolith. The near instantaneous redistribution of volatiles during meteoritic impacts themselves is another possible process which has not yet been widely considered by researchers and will be discussed further in the following sections.

Last year we reported the development of a flameless atmoic absorption (FLAA) spectroscopy technique for analyzing volatile trace metals in lunar and meteoritic samples. In our preliminary studies, we have shown that the FLAA technique can detect concentrations of less than one nanogram Pb in submilligram size samples and can distinguish $PbCl_2$ from other chemical forms of Pb.

We can also distinguish surface Pb from interior Pb through their thermal release profiles. By using this technique, we obtained the Pb and $PbCl_2$ results on the lower drive tube sample 74001 reported last year (Cirlin and Housley, 1977). In this paper, we report further developments of our FLAA technique for analyzing the volatile trace metals Pb, Cd and Zn in lunar samples. We also report SEM, FLAA and XPS studies on orange and black glass drive tube samples 74002/74001 and nearby surface samples 74220 and 74241 and discuss the implications of the volatile trace metal distributions in the orange and black glass droplets. In addition, we report survey results of Zn analyses in black glass droplets and individual mineral grains from the basaltic rock sample 75035 and in various types of grains from the North Ray Crater regolith samples 67701 and 67712.

EXPERIMENTAL

The instruments and primary experimental techniques used in our FLAA studies are similar to those described previously (Cirlin and Housley, 1977). We will therefore only discuss further developments and new calibration procedures.

Pb analysis

Under our experimental conditions $PbCl_2$ volatilizes as an undissociated molecule and thus gives no atomic absorption signal. In our earlier studies in order to obtain the total Pb including $PbCl_2$, the lunar samples were treated with 5 μl of concentrated HF directly in the atomizer. Because PbF_2 does dissociate to free atoms, we then inferred that the difference in concentration between untreated and HF treated samples gave the $PbCl_2$ concentration (Cirlin and Housley 1977). Although this method was a good way of obtaining the total Pb and $PbCl_2$ concentrations, unfortunately thermal release profiles were drastically changed due to dissolution of silicates by the HF. For this reason we have developed a new procedure using dilute HNO_3 instead of HF so that the silicates are not dissolved and the thermal release profiles are preserved.

First it was necessary to determine the optimum concentration of HNO_3 to convert $PbCl_2$ present in lunar orange and black droplets to $Pb(NO_3)_2$. The concentration of HNO_3 must also be dilute enough so that it does not dissolve the silicate phases in the samples. Known concentrations of $Pb(NO_3)_2$ were first converted to $PbCl_2$ by adding 5 μl of 5% HCl to $Pb(NO_3)_2$ directly in the atomizer (5 μl of 5% HCl is the optimum concentration of HCl to use when making $PbCl_2$ so that no chloroplumbates or chloroplumbites are formed). After the $PbCl_2$ was formed, various concentrations of HNO_3 were added to learn when the $PbCl_2$ is completely converted back to $Pb(NO_3)_2$.

It was found that 5 μl of 0.1% HNO_3 is sufficient to convert about 5 nanogram $PbCl_2$ to $Pb(NO_3)_2$. In order to make sure that all the $PbCl_2$ is converted to $Pb(NO_3)_2$ we used 5 μl of 1% HNO_3.

Figure 1 shows the thermal release profiles of a 74002 sample before and after HNO_3 treatment. It can be seen that the surface lead (the first peak) is considerably increased by the HNO_3 treatment while the rest of the profile remains unchanged. Although surface and interior Pb are clearly resolved (Fig. 1) the absorbance does not go to zero between the surface peak and the rest of the profile. For purposes of tabulation and discussion we somewhat arbitrarily define surface Pb to correspond to the area before the first minimum which is about 970 to 1000°C in Fig. 1, and interior Pb to correspond to the area after it. The total Pb and $PbCl_2$ concentrations obtained by dilute HNO_3 treatment, and those obtained by HF treatment are in very good agreement (see Table 1), thus our current data are directly comparable to data on 74001 and 74220 (Cirlin and Housley, 1977) which were obtained by HF treatment. For convenient comparison, data on 74001 and 74220 are included in Table 1.

In the future we plan to use the NH_4NO_3 treatment described in the next section for our Pb analysis. The current measurements were completed before we adopted it.

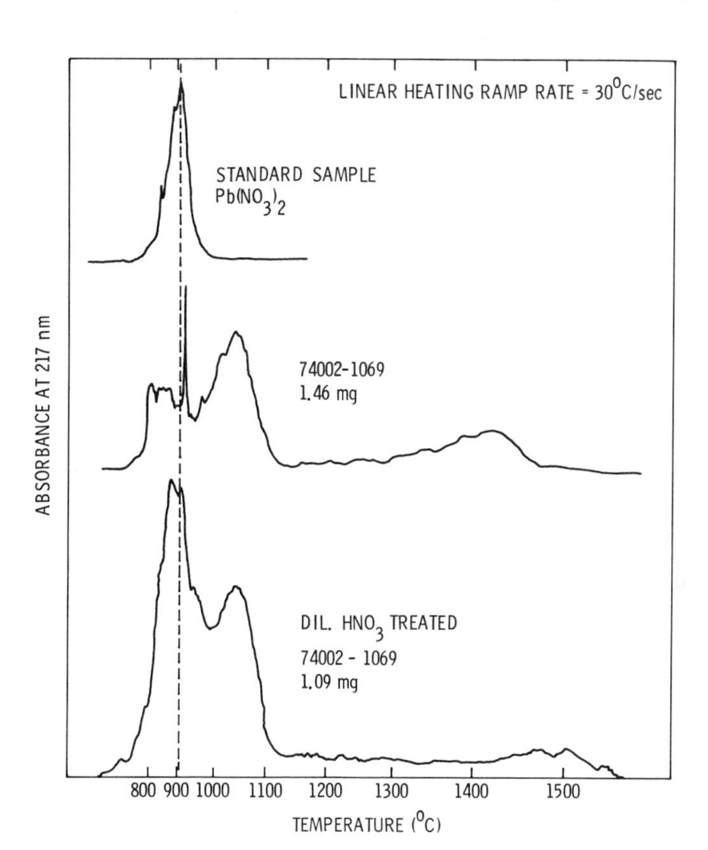

Fig. 1. Thermal release profiles of Pb from sample 74002,1069 before and after HNO_3 treatment. Note increase in strength of low temperature surface peak because of conversion of $PbCl_2$ to $Pb(NO_3)_2$.

Cd analysis

A freshly prepared dilute solution of CdI_2 (2.89×10^{-7} mg Cd) was used as a standard calibration sample. Both standard and unknown samples were treated with 5 μl of 0.2% NH_4NO_3 to obtain the total Cd concentrations. This treatment converts the cations to nitrates or oxides while the resulting NH_4 halides volatize at low temperatures, eliminating any potential molecular interference problems (Manning and Slavin, 1978) as does the HNO_3 treatment. We used the 228.8 nm atomic line from a standard hollow cathode lamp. The spectral bandwidth was set at 0.5 nm.

Zn analysis

Our laboratory experiments showed that $ZnCl_2$ is extremely hygroscopic, deliquesces in moist air and becomes hydrolized to $Zn(OH)_2$ or ZnO, thus giving an atomic absorption signal without any chemical treatment. The bromide and iodide are expected to behave similarly. It is, therefore, very unlikely that any of the Zn is still present as $ZnCl_2$ in returned lunar samples even if it once was since most of the $ZnCl_2$ would probably have hydrolyzed. For this reason, no chemical treatment was used for our Zn analyses. A 213.9 nm atomic line from a standard hollow cathode lamp was used. The spectral bandwidth was set at 0.5 nm. Only qualitative analysis of Zn on individual grains will be presented in this paper, therefore the calibration procedure will not be discussed.

Grain size separation

Five grain size separates (75–1000 μm, 45–75 μm, 30–45 μm, 20–30 μm, and <20 μm) were made on sample 74001,105 (from a depth of 39–39.5 cm) by dry sieving on a laminar flow clean bench. The sieves (Perforated Products, Inc.) used had electroformed pure nickel screens in stainless steel frames. It is clear from microscopic examination that some fine material adheres to grains in all the coarser fractions.

X-ray photoemission spectroscopy (XPS)

The instruments and experimental techniques used are similar to those described previously (Housley and Grant, 1977). The 75–100 μm sample was cemented into a recessed platen using colloidal graphite suspended in propanol. One 20 μm sample was prepared in the normal way by sedimentation from ethanol and one was packed dry into a recessed platen.

Results and Discussion

Orange and black glass 74002/74001 background

A 68 cm long double drive tube sample 74002/74001 was collected at the southwest rim of Shorty Crater in Taurus-Littrow valley. This core tube contains unusually homogenous orange and black droplets. It is generally assumed that these glasses are of volcanic origin and were formed during lava fountaining about 3.5 b.y. ago (Heiken and McKay, 1974). A cosmic ray track density study of the upper tube 74002 indicates that this material was deposited in one or a series of closely spaced events and was later excavated by the Shorty Crater impact some 10 m.y. ago (Crozaz, 1978). Several workers have proposed that these orange and black droplets formed during different lava fountains based on petrographic studies (Heiken and McKay, 1977) and on leaching and volatilization experiments (Reed *et al.*, 1977). *In situ* reworking of the upper 5 cm is also apparent from track density studies and maturity index I_s/FeO measurements (Crozaz, 1978; Morris and Gose, 1978), however, no buried surface with an age as great as 500,000 years is possible.

Available geologic evidence suggests that it is very unlikely that either the original stratigraphy or an inverted stratigraphy would be well preserved at the particular place on the rim of Shorty Crater where the samples were collected. The following observations (Nagle, 1978) indicate that the Shorty Crater impact did not cause a simple inversion of the original stratigraphy, but instead, a much more random type mixing during excavation. The impact that formed Shorty Crater excavated some meter-sized blocks. Many patchy areas of orange, black and grey colored regolith are evident in photographs of the rim. Since this is a very young crater, these cannot be due to secondary degradation of ejecta blankets. It is also reported that one large boulder near the rim is microdiabase in texture while other rock fragments collected nearby are fine grained. Although orange and black fines in the double drive tube have been referred to as horizontally stratified, this cannot really be ascertained with only one core, and vertical stratified areas were also seen in the crater trench. Although laboratory experiments showed well preserved inverted stratigraphy for impacts into sand targets (Gault, *et al.*, 1974), no experimental work was reported regarding impacts into inhomogeneous targets such as a soft regolith layer underlain by bedrock.

SEM study

Our SEM study of <20 μm material from 74001,105 (Fig. 2) shows that most of the fragements appear to be broken pieces rather than whole droplets. Since our other studies show surface enrichments and surface coatings even in this fine size fraction it becomes important to consider when the breakage is most likely to have occurred. The most likely possibilities seem to be either during the initial fountaining and deposition on the lunar surface or during the recent Shorty Crater impact.

The range of the mean grain size of the droplets is 34–52 μm (McKay and Waits, 1978) indicating that the landing velocity would have to have been very high for the droplets to have been broken in such a quantity. This suggests that the droplets were probably broken by the Shorty Crater impact, although no experimental study of breakage versus impact velocity has been made to definitively test the probability of such a degree of breakage during the initial deposition.

FLAA results for Pb and Cd

Figure 3 (see Table 3 also) shows Pb and Cd analyses on four dry sieved grain size separates (45–75 μm, 30–45 μm, 20–30 μm, and <20 μm) from 74001,105. This figure shows that the Pb and Cd concentrations increase as the grain size decreases in spite of the abundance of broken pieces in small size fractions. This result strongly suggests that these volatile trace metals were deposited or redeposited after the droplets were broken. Depth profile data obtained from unsieved samples are tabulated for Pb in Table 1 and Cd in Table 2.

Figure 4 shows that the total Pb and $PbCl_2$ concentration in both orange and black glasses decreases continuously as core tube depth increases. Figure 5 shows that the total Cd concentration in both upper and lower core tube samples also decreases continuously as a function of depth. Continuity in concentration is expected if volatile trace metals were redistributed by the Shorty Crater impact. Since we do not feel that original stratigraphy was preserved during the impact, we feel it would be very improbable otherwise.

We feel that it is probable that the droplets were dominantly broken by the Shorty Crater impact and that the volatiles were redistributed during this cratering event. The presence of significant solar wind rare gas in these droplets observed by Bogard and Hirsch (1978) could also be explained by the same model if we assume the inert gases were trapped on the droplets from the gas released during the Shorty Crater impact along with the volatile trace metals.

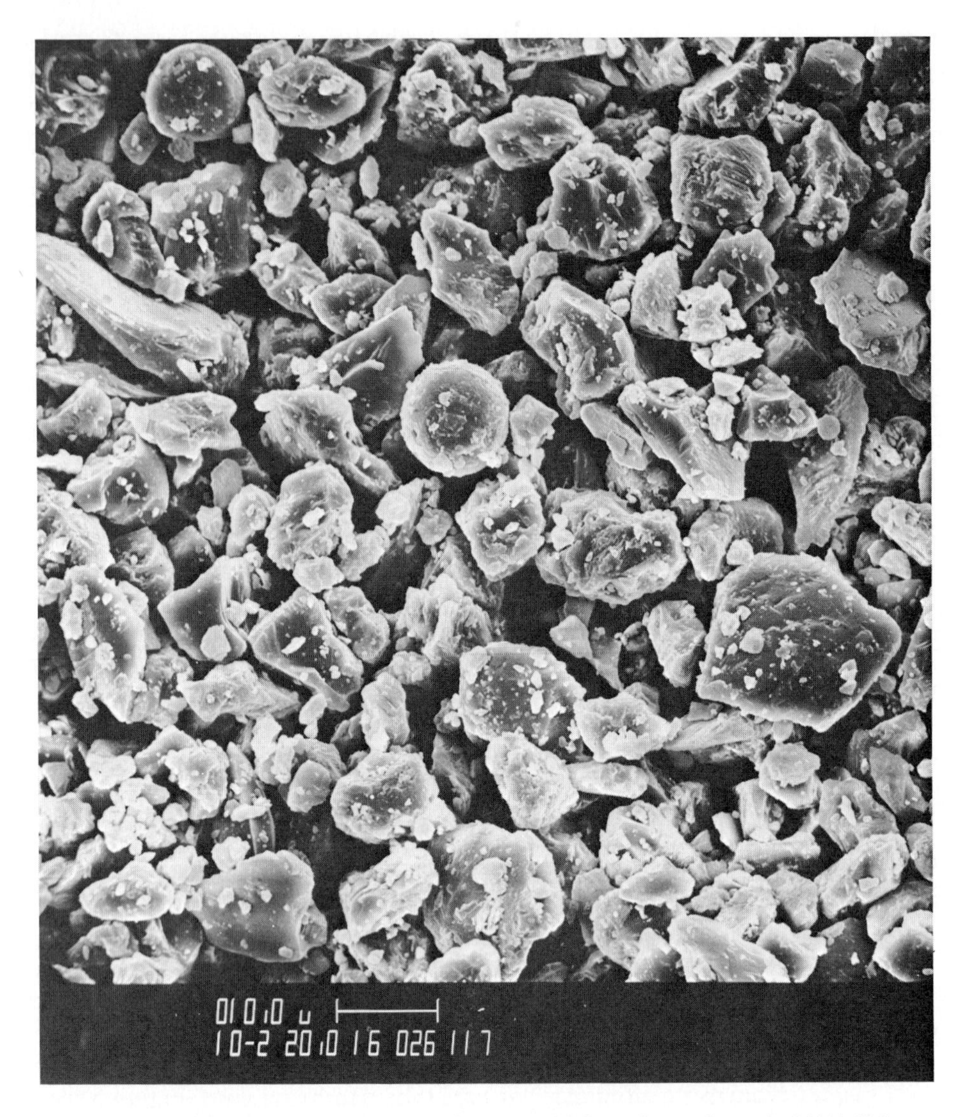

Fig. 2. Scanning electron micrograph of <20 μm XPS specimen of sample 74001,105 showing that most of the small grains are sharply angular broken pieces.

This picture has the desirable features that the solid grains could be cold and that the deposition could take place fast enough to physically trap inert gases.

XPS study of other volatiles

Enrichment of volatiles on the surface of these droplets led several researchers to believe that coatings such as S and halogens are the condensed portion of the

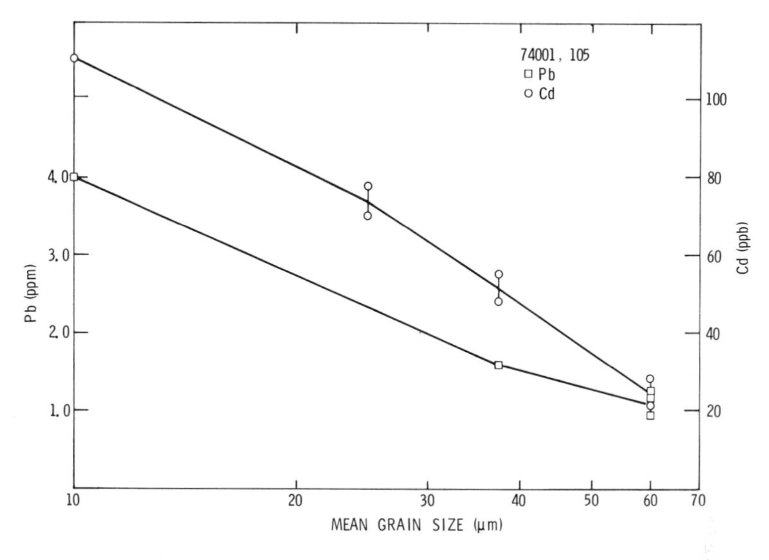

Fig. 3. Pb and Cd concentrations versus grain size showing surface correlation. Binocular microscopic examination showed that our dry sieving was not complete and that very fine materials still adhered to larger grains, thus Pb and Cd concentrations of larger size separates might be lower than what we observed.

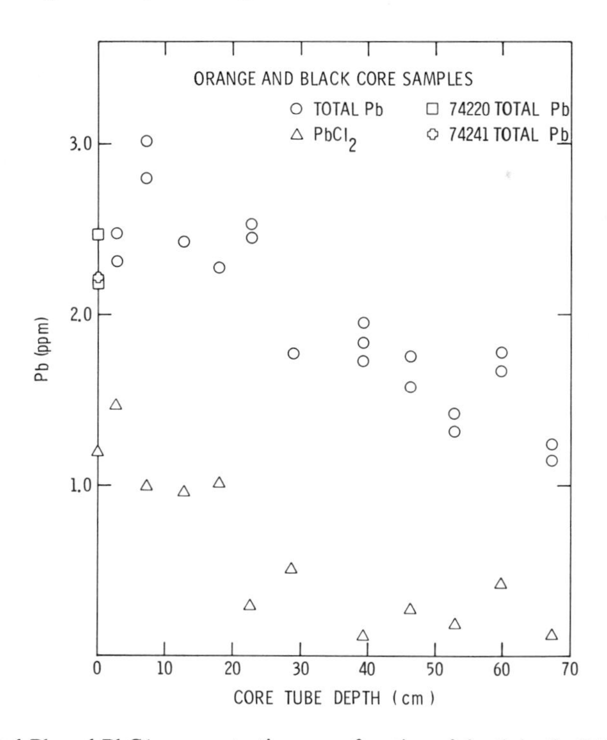

Fig. 4. Total Pb and $PbCl_2$ concentrations as a function of depth in the 74001/74002 double drive tube. Note smooth decrease from the surface downward.

Table 1. 74002/74001 drive tube Pb results.

Sample Fines (<1 mm)	Depth (cm)	Wt.(mg)	Total Pb (ppm)	$PbCl_2$(ppm)	Total Surface Pb (ppm)	Total Interior Pb (ppm)
74220,545	surface	1.038	2.21 (HF)	1.20 (HF)		
		0.977	2.19 (HF)			
		0.660	2.47 (HF)			
74241,155	surface	1.152				0.61
		1.129	2.21 (HNO_3)	0.060 (HNO_3)	1.62	0.59
74002,1069	2.5–3.0	1.458				0.60
		1.090	2.31 (HNO_3)	1.32 (HNO_3)		0.79
		0.952	2.48 (HF)	1.49 (HF)	1.52	
74002,1080	7.0–7.5	1.228				0.80
		0.843	3.01 (HNO_3)	1.10 (HNO_3)		0.85
		1.016	2.80 (HF)	0.89 (HF)	2.16	
74002,1083	12.5–13.0	1.215				0.52
		1.298				0.47
		0.942	2.43 (HNO_3)	0.96 (HNO_3)	1.52	0.91
74002,1092	17.5–18.0	0.903				0.48
		1.312	2.28 (HNO_3)	1.02 (HNO_3)	1.65	0.63
74002,1101	22.5–23.0	1.266				0.37
		1.143				0.39
		1.084	2.53 (HNO_3)	0.30 (HNO_3)	1.83	0.69
		0.981	2.45 (HNO_3)		1.79	0.67
74002,1113	28.5–29.0	0.955				0.34
		1.035	1.78 (HNO_3)	0.51 (HNO_3)	1.46	0.31
74001,105	39.0–39.5	0.434	1.95 (HF)	0.12 (HF)		
		1.026	1.84 (HF)			
		1.155	1.73			
74001,111	46.0–46.5	1.003	1.76 (HF)	0.27 (HF)		
		1.087	1.58 (HF)			
74001,117	52.5–53.0	0.580	1.42 (HF)	0.19 (HF)		
		0.843	1.32 (HF)			
74001,123	59.5–60.0	0.786	1.67 (HF)	0.42 (HF)		
		0.988	1.78 (HF)			
74001,129	67.0–67.5	0.915	1.15 (HF)	0.12 (HF)		
		0.687	1.24 (HF)			

gas that drove the lava fountain and that the coatings were formed during the lava fountaining when the droplets had just solidified and were still in flight.

Chou *et al.* (1975) and Meyer *et al.* (1975) hypothesized that halogens were in the vapor of the lava fountains because of the increased volatility of metals as chlorides and because some Cl was found on the surface of the glass. Our study shows that some of the Pb in orange and black glasses is indeed in the form of $PbCl_2$.

Constant ratios of Zn and S on individual droplets were found in SEM-EDAX studies of orange and black glass spheres by Butler and Meyer (1976). This led these workers to believe that the coatings have not since suffered significant gain or loss of Zn or S.

We performed XPS studies of material from two size fractions, <20 μm and

Table 2. Cd concentration in surface and core tube samples.

Fines (<1mm)	Depth (cm)	wt. (mg)	Concentration (ppb)
74241,155	surface	1.626	208
		1.538	185
74220,545 <45 μm	surface	1.294	401
		0.797	404
74002,1069	2.5–3.0	0.955	352
		1.464	341
74002,1080	7.0–7.5	1.427	181
		2.305	194
74002,1092	17.5–18.0	2.054	160
		0.762	176
74002,1101	22.5–23.0	1.047	125
		2.027	115
74001,105	39.0–39.5	1.146	48
		1.450	62
74001,117	52.5–53.0	1.387	20
		2.058	18
74001,129	67.0–67.5	1.595	12
		2.483	11
		2.577	12

75–1000 μm of 74001,105 from a depth of 39–39.5 cm. We could detect no significant differences between spectra of a sample of <20 μm material prepared dry and those of one prepared by our usual sedimentation procedure. Therefore, sputter profiles were determined on the latter.

Initial searches in the range from 0–1200 eV binding energy revealed strong surface enrichments in Na and Zn as previously reported by Cadenhead (1975) on both size fractions. The only other element found to be comparably enriched was S. Surface F was detected at a level roughly 10% that of Zn. Surface K and Cl were not observed at a sensitivity that would have detected a level about 10% that of S.

Spectra in the 0–200 eV binding energy range obtained for both size fractions before sputtering or baking are compared to a similar spectrum of ZnS in Fig. 6. By comparing the intensity of the Zn 3s line with those of the Si lines or the Al 2s line, it is apparent that the <20 μm material has about 0.7 as much surface Zn per unit area as the 75–1000 μm material.

Examination under a binocular microscope revealed that more than 50% of the area measured for the 75–1000 μm material consisted of unbroken exterior surfaces. Scanning electron microscope examination of the <20 μm material seems to show less than about 10% exterior surfaces (Fig. 2) although distin-

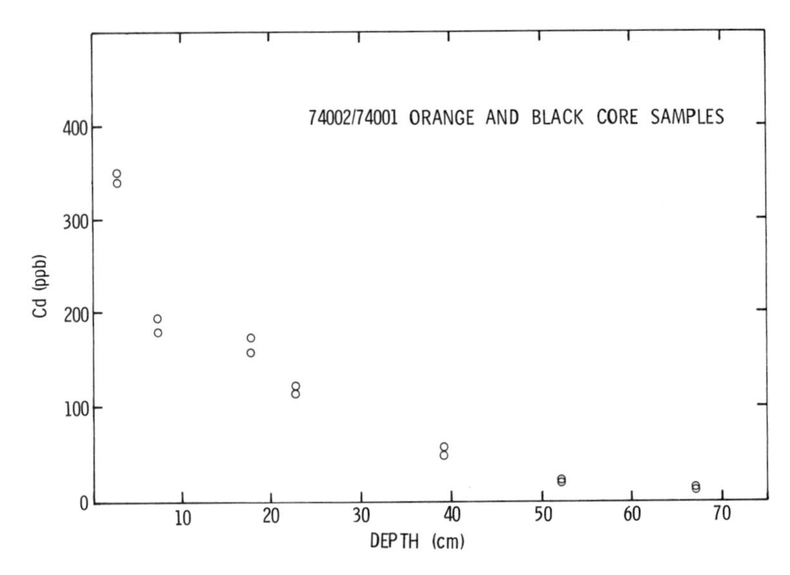

Fig. 5. Total Cd versus depth profile. Note similarity to total Pb in Fig. 6.

guishing exterior surfaces from fracture surfaces in many cases requires some subjective judgement.

The relatively small reduction in surface Zn concentration, in the face of a large reduction in the fraction of exterior surface material exposed, strongly argues that Zn has been redistributed after the grains were broken. Although not conclusive this supports the similar inference drawn from the FLAA size fraction data for Pb and Cd.

Table 3. Total Pb and Cd concentrations in size fractions of black glass fines 74001,105.

Size Fraction	Pb		Cd	
	Wt. (mg)	Concentration (ppm)	Wt. (mg)	Concentration (ppb)
45–75 μm	0.547	1.11	2.159	28
	0.362	1.22	1.448	21
	0.424	0.96		
30–45 μm	0.796	1.58	1.227	55
			1.494	48
20–30 μm	—	—	1.352	70
			2.198	77
<20 μm	0.677	4.00	1.702	106
bulk (<1 mm)	0.434	1.95	1.146	48
	1.026	1.84	1.450	62
	1.155	1.73		

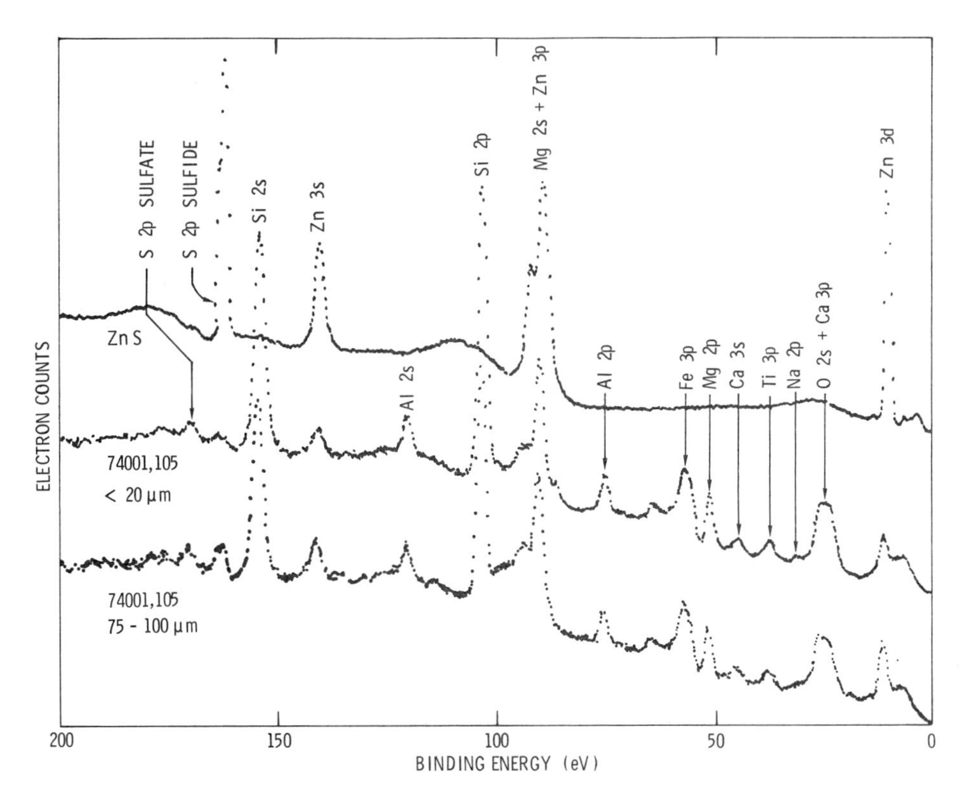

Fig. 6. XPS spectra comparing surface Zn and S on <20 μm and 75–1000 μm size fractions of sample 74001,105 with ZnS reference sample. Note oxidized S and deficiency of S with respect to Zn.

It can also be seen that a substantial fraction of the S in the lunar samples is oxidized and appears in the peaks labeled "sulfate" in Fig. 6. Even if we assume that all the S is associated with Zn we find from a comparison of S and Zn peak intensities that about 40% of the surface Zn in the 75–1000 μm sample and about 65% of the surface Zn in the <20 μm sample must be in another chemical form. From the chemical shift it is not in the form of metal and no other anions were recognized in sufficient abundances. Therefore, we feel that most of this remaining Zn is in the form of oxide, hydroxide, or silicate.

The 75–1000 μm sample was baked at 300°C for one hour prior to sputtering. As would be suggested by the mass spectrometric observations of Simoneit *et al.* (1976) this produced a small but noticeable increase in the ratio of sulfide to sulfate.

Profiles of the raw Zn line intensities versus depth obtained by sputtering these samples are shown in Fig. 7. The nominal sputtering rate for a flat surface was about 1 Å per minute, although this sputtering gun has not been calibrated on this type of sample. The interpretation of the tails of the profiles is complicated

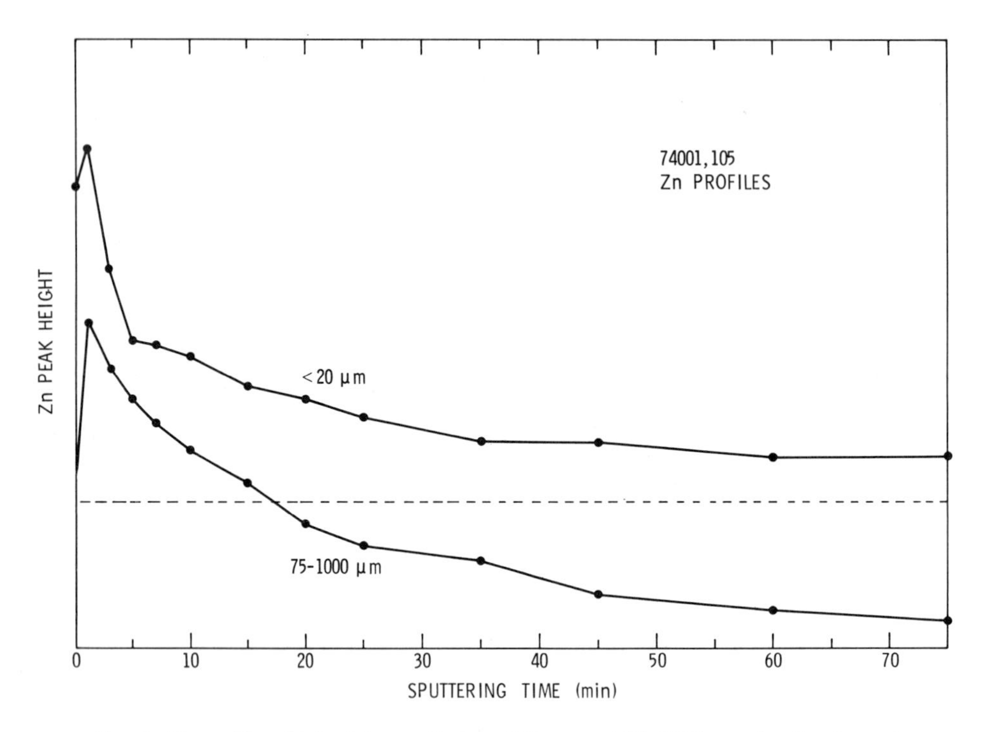

Fig. 7. Zn profiles obtained by sputtering <20 μm and 75–1000 μm size fractions of sample 74001,105. The 20 μm data are offset upward as indicated by the dotted baseline. Sputtering rate is estimated to be about 1 Å per minute.

by the rough nature of the surfaces. The initial rise is due to the removal of C contamination.

Despite these limitations the profiles show that the bulk of the surface Zn is within about 5 Å of the surface for the <20 μm material and within about 20 Å for the 75–1000 μm material.

It appears that these results can be reconciled with the SEM-EDAX studies of Butler and Meyer (1976) by assuming that they primarily saw much thicker coatings of small aerial extent.

In preliminary simulation experiments with ZnS and Fe_7S_8 (pyrrhotite) we have failed to detect oxidized S in surfaces that have been partially oxidized by exposure to air. Instead the oxygen seems to be present as oxide or hydroxide and carbonate. This leads us to suspect that the sulfate in the surfaces of the lunar samples is associated with the Na.

Our currently favored hypothesis is that the Na and most of the Zn were transported and deposited as metals and that these deposits were partially converted to sulfides on the moon by reaction with S_2. A fraction of the Zn may have been transported as $ZnCl_2$. On return of the samples to earth any remaining Na metal would have quickly formed NaOH. $ZnCl_2$ which is very hygroscopic would have reacted with this NaOH to produce NaCl and any sulfate generated would have reacted to produce Na_2SO_4. Possibly the $PbCl_2$ we observe was formed by a similar process.

Zn analysis survey results

Studying the volatile trace metals in various types of individual grains can provide additional information about vapor transport processes and the evolutionary processes of the grains.

Taking advantage of the high sensitivity of FLAA for Zn (Culver, 1975) (absolute sensitivity of Zn is 0.1×10^{-12} g) and the relative abundance of Zn in lunar materials, it is possible to analyze for Zn in individual submillimeter grains.

We report preliminary Zn analyses on 74001 black glass droplets, individual

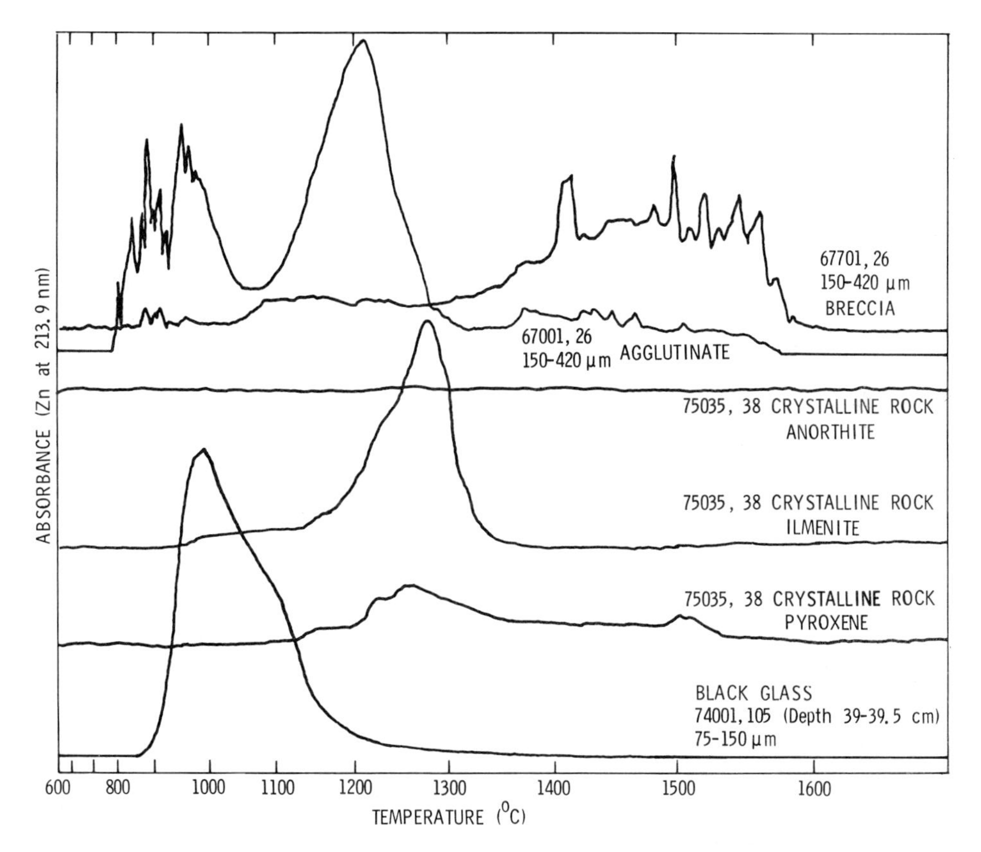

Fig. 8. This figure shows a typical Zn thermal release profile of a single black droplet from sample 74001, thermal release profiles of Zn in individual pyroxene and ilmenite mineral grains from basaltic rock sample 75035 showing no surface Zn, absence of Zn in anorthite grain, and thermal release profiles of typical breccia and agglutinate grains from North Ray Crater sample 67701.

mineral grains from basaltic rock 75035 and various types of grains from North Ray Crater samples 67701 and 67712.

The bottom curve in Fig. 8 shows a typical thermal release profile of a black crystalline droplet from sample 74001. It shows that most of the Zn is present as a surface coating, although the ratio of the surface to the interior Zn (the shoulder at higher temperature) varies considerably from one droplet to another.

In studying individual mineral grains obtained by gentle crushing of basaltic rock sample 75035, we found that, unlike the volcanic droplets, no surface Zn was present on pyroxene and ilmenite grains (the fourth and fifth curves in Fig. 8). In the case of anorthite grains, neither surface Zn nor interior Zn was observed. These results suggest that it may be possible to learn how much of the Zn has been outgassed during the formation of these droplets by comparing the interior Zn in basaltic rocks with the interior Zn in the droplets.

It would be interesting to learn how Zn is distributed and transported during the evolutionary processes of individual grains in relatively young highland samples such as those from North Ray Crater. We learned that the concentration of Zn is extremely low in pure anorthite grains from regolith samples 67701 and 67712. Anorthite grains with metal inclusions and breccia samples, however, exhibited a high temperature Zn release as shown in Fig. 8 suggesting that the source of the high temperature Zn in these samples is the same.

Agglutinates from sample 67701 showed a considerable amount of both surface and interior Zn (last curve in Fig. 8). The high concentrations of surface Zn were probably due to the fact that the agglutinates are an agglomeration of fine material whose surface area is high.

Study of volatile trace metals in individual grains seems to open up a whole new area of research which may further extend our understanding of vapor transport mechanism and lunar regolith evolutionary processes.

Summary and Conclusion

We report concentration versus depth profiles for Pb and Cd based on a series of equally spaced samples spanning the length of the 74001/74002 Shorty Crater double drive tube. We report Pb and Cd concentration versus size data for a set of dry sieved fractions from 74001,105 corresponding to a depth interval of 39–39.5 cm. We present XPS data on surface enrichments observed particularly for Na, Zn, and S on <20 μm and 75–1000 μm fractions of 74001,105. We finally present survey data showing Zn thermal release profiles from a variety of submillimeter lunar grains.

The size fraction data and the XPS data in conjunction with SEM observations strongly suggest that the volatiles were deposited or redistributed after the grains were broken, although the evidence does not seem to be completely conclusive. If we accept this and the argument suggesting that the dominant breakage occurred during the Shorty Crater event, then it follows that the volatiles must have been redistributed during the Shorty Crater event.

Alternatively if we accept the improbability of preserving original stratigraphy during the Shorty Crater event then the continuity of the Pb and Cd depth profiles provides a strong argument for the redistribution of volatiles during the Shorty Crater event.

Acknowledgments—This manuscript has benefited from thoughtful reviews by P. Butler, D. Blanchard and R. Korotev. The work was supported by NASA Contract NAS9-11539.

References

Bogard D. D. and Hirsch W. C. (1978) Noble gas contents and irradiation history of orange-black glass in the 74001–74002 core (abstract). In *Lunar and Planetary Science IX*, p. 111–113. Lunar and Planetary Institute, Houston.

Butler P., Jr. and Meyer C., Jr. (1976) Sulfur prevails in coatings on glass droplets: Apollo 15 green and brown glasses and Apollo 17 orange and black (devitrified) glasses. *Proc. Lunar Sci. Conf. 7th*, p. 1561–1581.

Cadenhead D. A. (1975) An Auger-ESCA study of Taurus-Littrow orange soil. In *Papers presented to the Conference on Origins of Mare Basalts and Their Implications for Lunar Evolution*, p. 20–24. The Lunar Science Institute, Houston.

Chou C. L., Boynton W. V., Sundberg L. L. and Wasson J. T. (1975) Volatiles on the surface of Apollo 15 green glass and trace-element distribution among Apollo 15 soils. *Proc. Lunar Sci. Conf. 6th*, p. 1701–1727.

Cirlin E. H. and Housley R. M. (1977) A flameless atomic absorption study of the volatile trace metal lead in lunar samples. *Proc. Lunar Sci. Conf. 8th*, p. 3931–3940.

Crozaz G. (1978) History of regolith deposition at the Apollo 17 and Luna 24 landing sites at Shorty Crater (abstract). In *Lunar and Planetary Science IX*, p. 203–205. Lunar and Planetary Institute, Houston.

Culver B. R. (1975) *Analytical methods for carbon rod atomizer*. Varian Techtron Pty. Ltd., Springvale, Vic., Australia.

Gault D. E., Hörz F., Brownlee D. E. and Hartung J. B. (1974) Mixing of the lunar regolith. *Proc. Lunar Sci. Conf. 5th*, p. 2365–2386.

Heiken G. and McKay D. S. (1974) Petrography of Apollo 17 soils. *Proc. Lunar Sci. Conf. 5th*, p. 843–860.

Heiken G. and McKay D. S. (1977) A model for eruption behavior of a volcanic vent in eastern Mare Serenitatis. *Proc. Lunar Sci. Conf. 8th*, p. 3243–3255.

Housley R. M. and Grant R. W. (1977) An XPS (ESCA) study of lunar surface alteration profiles. *Proc. Lunar Sci. Conf. 8th*, p. 3885–3599.

Manning D. C. and Slavin W. (1978) Determination of Pb in a chloride matrix. *Atomic Absorption Newsletter* **172**, 43–46.

McKay D. and Waits G. (1978) Grain size distribution of samples from core 74001 and 74002 (abstract). In *Lunar and Planetary Science IX*, p. 723–725. Lunar and Planetary Institute, Houston.

Meyer C., Jr., McKay D. S., Anderson D. H. and Butler P., Jr. (1975) The source of sublimates on the Apollo 15 green and Apollo orange glass samples. *Proc. Lunar Sci. Conf. 6th*, p. 1673–1699.

Morris R. V. and Gose W. A. (1978) Depositional and reworking history of core 74001/2 (abstract). In *Lunar and Planetary Science IX*, p. 763–765. Lunar and Planetary Institute, Houston.

Nagle J. S. (1978) Drive tubes 74002/74001: Dissection and description. Lunar Sample Curator's Doc. VII-5. NASA Johnson Space Center, Houston. 49 pp.

Reed G. W., Jr., Allen R. O., Jr. and Javonovic S. (1977) Volatile metal deposits on lunar soils—relation to volcanism. *Proc. Lunar Sci. Conf. 8th*, p. 3917–3930.

Simoneit B. R., Neil J. M., Wszolek P. C. and Burlingame A. L. (1976) Hydration and loss of low temperature volatiles from the Apollo 17 orange and black soils. In *Lunar Science VII*, p. 815–816. The Lunar Science Institute, Houston.

Proc. Lunar Planet. Sci. Conf. 9th (1978), p. 2065–2097.
Printed in the United States of America

Chemistry and petrology of size fractions of Apollo 17 deep drill core 70009–70006

J. C. LAUL

Earth and Planetary Chemistry Section
Physical Sciences Department, Battelle-Northwest
Richland, Washington 99352

D. T. VANIMAN, J. J. PAPIKE and S. SIMON

Planetary Regolith Studies, Department of Earth and Space Sciences
State University of New York, Stony Brook, New York 11794

Abstract—We have used INAA methods to analyze 34 major, minor and trace elements in 48 bulk soils and size fractions (90–1000 μm, 20–90 μm and <20 μm) of the Apollo 17 deep drill core sections 70009–70006 (upper 130 cm). Modal data were also obtained for the >20 μm size fraction. The cores are highly heterogenous chemically and petrographically. The >90 μm and 20–90 μm coarse fractions are chemically identical but different from the <20 μm fine fraction. The coarse fractions are highly enriched in mare material (high Ti basalt + orange glass) in sections 70009 and 70008; the 28, 38 and 47 cm layers of coarse core 70008 are almost entirely mare material. Below 58 cm depth, mare material decreases and highland material increases with increasing depth. The <20 μm fine fraction comprises about 15–25 wt.% of the bulk soil and is considerably enriched in highland material. Except for the coarse section 70008, highland material in the <20 μm fraction increases with increasing depth. The dominant source of highland material is KREEPy and mafic instead of anorthositic. Relative to the coarse fractions, the fine fractions contain ~10% more low-K KREEPy material. Low-K KREEP can be obtained by lateral transport of fines from the local massifs (Korotev, 1976). Enrichment of KREEP in the Apollo 17 fines follows similar observations at other sites (Apollo 11 soil 10084 and Apollo 15 soil 15100) and points to enhanced transport of fines on a moonwide basis, with the constraint that local sources predominate at each site.

INTRODUCTION

The regolith at the Apollo 17 site is quite complex due to the interface of mare and highland materials at the Taurus-Littrow site. In order to characterize and understand the formation of the regolith, we have focused our attention on the Apollo 17 deep drill core. The Apollo 17 drill core (sections 70009–70001) represents the deepest soil column (295 cm) returned from the moon. This core was collected at the ALSEP site, about one crater diameter SE of the 400 m Camelot crater and NW of the "Central Cluster" craters, and is believed to contain ejecta from these craters (LSPET, 1973; Duke and Nagle, 1974; Taylor *et al.*, 1977a). The neutron fluence study of Curtis and Wasserburg (1975) suggests that the entire drill core was deposited less than 200 m.y. ago, a conclusion which places an upper age limit on nearby contributing craters. From track studies in 70009 and 70008 Crozaz and Plachy (1976) suggest that the upper 80 cm were emplaced some 100 m.y. ago, and the upper 25 cm was recently (~2 m.y.) excavated by a small impact and was then gradually filled by heavily irradiated surface soil.

The entire core has three major stratigraphic units: an upper, coarse-grained interval (107 cm) dominated by mare basalt fragments; a middle, fine-grained interval (56 cm) consisting mainly of highland material; and a lower interval (132 cm) containing a variety of breccias and crystalline fragments (LSPET, 1973). These major stratigraphic units can be subdivided on the basis of X-radiography or core dissection observations (Duke and Nagle, 1974). Ehmann and Ali (1977) analyzed 26 bulk samples from 70009–70007 for major and trace elements by INAA, and they reported ten chemical units that were distinct from the 12 X-ray units and 19 stratigraphic units of Duke and Nagle (1974). Based on petrographic observations, Vaniman and Papike (1977a,b) and Taylor *et al.*, (1977a) suggested three major zones in 70009–70007: an upper zone (0 to 21 cm) characterized by high agglutinate content; a middle zone (22 to 66 cm) characterized by low agglutinate content and high abundance of coarse mare basalt fragments, and a lower zone (67 to 100 cm) distinguished by medium agglutinate content and a large quantity of non-mare lithic fragments. These three zones are consistent with breaks in the FMR maturity index (Morris *et al.*, 1978). Section 70009 is submature, 70008 is immature, and 70007 is submature. Vaniman and Papike (1977a) characterized >20 μm lithic and mineral fragments in 70009–70007 and suggested that the anorthositic component predominates over the noritic (KREEPy) component in the upper drill core (70009, 70008), but noritic compositions become more abundant in 70007 where the mare component decreases and the highlands component increases.

Korotev (1976) analyzed major and trace elements in size fractions of 90–150 μm and <20 μm from 70008 and seven surface soils and noted that the 90–150 μm fraction is enriched in mare material (basalt plus orange glass), whereas the <20 μm fraction is more feldspathic in composition and enriched in highland material and some orange glass. Korotev (1976) investigated in detail the processes responsible for the observed chemical differences in the two size fractions, and showed that the differences can be explained by mixing distinct components. Chemical differences can also result between different size fractions of comminuted lunar basalt (Haskin and Korotev, 1977). However, Korotev (1976) showed that the later process is a second-order effect in the compositional differences between size fractions of the Apollo 17 soils. He used a multielement mixing model with five components and reported ~30% highland material in 1:1 proportions of noritic breccia (KREEP) and anorthositic gabbro in <20 μm fraction of 70008. We studied three size fractions (90–1000 μm, 20–90 μm and <20 μm) from core sections 70009–70006 and our findings in the <20 μm fraction of 70008, using Th as an indicator for KREEP, are consistent with Korotev's observations. We point out that the present work on 70009–70006 is part of our ongoing study of the 70009–70001 core sections. Thus, detailed quantitative treatment of the data by a multielement mixing model will await the completion of our studies of core sections 70005–70001. In this paper we focus on general features and characterization of the <20 μm fractions. Since the <20 μm soil fraction is different from the coarser soils, its chemistry is exceedingly important in understanding lateral and vertical regolith mixing. In addition, the chemical information from the >20 μm fraction is important to test the approach of chemical mixing by modal recombination techniques as demonstrated for drive tube 60009/60010 by Papike *et al.* (1978) and Vaniman *et al.* (1978).

In our chemical-petrographic consortium effort, we selected 14 soils (<1 mm size fraction) from about 10 cm depth intervals in core sections 70009–70006. The sample number, parent number, and corresponding depth intervals are shown in Fig. 1. Each soil was sieved into three size fractions (90–1000 μm, 20–90 μm and <20 μm). These soil fractions were studied by instrumental neutron activation (INAA) at Battelle-Northwest (BNW), whereas petrologic studies of >90 μm and 20–90 μm fractions were done at Stony Brook. A complete string of polished thin sections that represent the full length of core sections 70009–70006 has been examined by the Stony Brook group. The detailed mineralogy, petrology and stratigraphy of sections 70009–70007 are already published (Vaniman and Papike, 1977a,b). The modal data for new grain mounts from depths corresponding to the BNW chemical study on sections 70009–70006 are discussed here.

METHODS

Sieving

About 80 mg of each bulk soil (<1 mm fraction) were sieved into three size fractions (90–1000 μm, 20–90 μm and <20 μm) in the clean facility of Dr. D. S. McKay at JSC using the procedure of McKay *et al.* (1974). Intermediate sieve sizes of 45 μm and 150 μm were used to facilitate sieving operations. Ultrapure Freon TF (DuPont) was used throughout the sieving operations. The material loss in sieving operations was typically 10%, largely from the <20 μm fraction. A small aliquot from each 90–1000 μm and 20–90 μm size fraction was reserved for petrographic study at Stony Brook, while the remaining material (along with the <20 μm fraction) was used for INAA analysis at Battelle-Northwest. To check the mass-balance of sieved fractions, several small aliquots of bulk soil were saved prior to sieving.

Figure 2 displays the percentage weight recoveries in >90 μm, 20–90 μm and <20 μm size fractions of each soil. The weight of the <20 μm size fraction varied from 15–25% and was generally ~20% of the bulk soil weight. The >90 μm and 20–90 μm fractions were about equal in weight proportion in core sections 70009, 70007 and 70006, but the >90 μm fraction was significantly greater in 70008 reflecting the dominance of a coarse grained layer.

Instrumental neutron activation analysis (INAA)

Soils were analyzed by sequential INAA using a high efficiency 130 cc Ge(Li) detector (25%, FWHM 1.8 KeV for 1332 Kev γ of ^{60}Co), a 2048 channel analyzer, and employing the unique BNW coincidence and non-coincidence Ge(Li)-NaI(Ti) counting system. The coincidence-noncoincidence approach enables us to improve the sensitivities of various elements and obtain additional elements such as Ho, Ga and Zn that are not accurately determined in normal Ge(Li) counting. This approach is particularly useful when a small size sample is involved. The details of our INAA procedure and the systematics of coincidence-noncoincidence counting are described in Laul and Rancitelli (1977) and Laul *et al.* (1978). Silicon was determined using the 14 MeV neutron activation facility of Dow Chemical at Midland, Michigan, whereas the remaining 33 elements were determined after

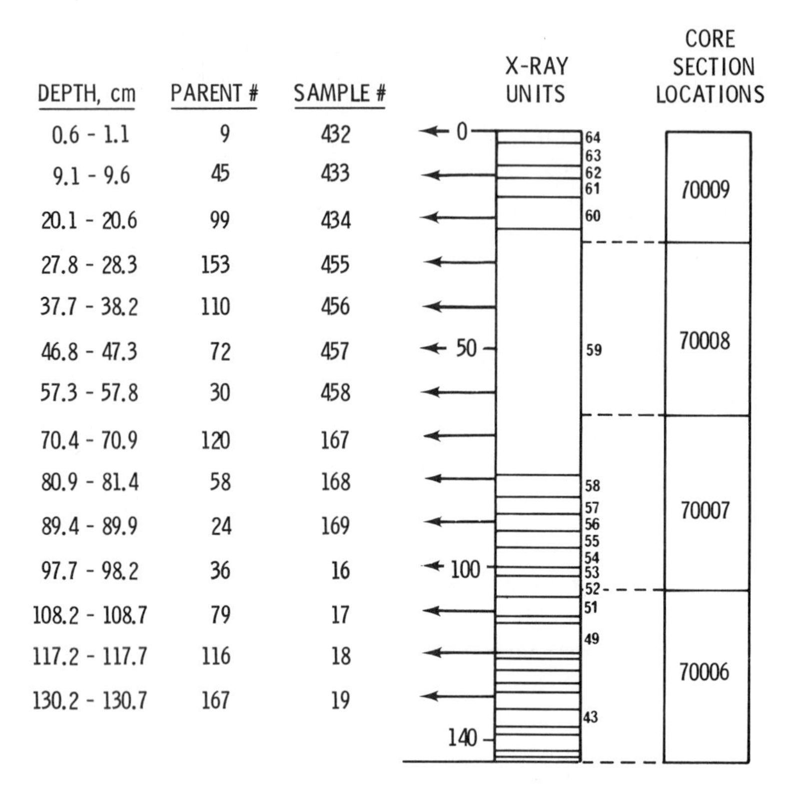

Fig. 1. X-ray stratigraphy and core section locations, 70009–70006. Locations are shown for sample numbers used for chemical-petrographic studies.

irradiation at the Oregon State University (OSU) TRIGA reactor. In each irradiation, two Apollo 17 soils, 73121 and 75081, and the USGS standards BCR-1 (duplicate), GSP-1, and PCC-1, and Allende standard (USNM) were included as internal standards to check for systematic errors and to minimize geometry problem in the counting procedure.

Modal and microprobe analyses

The procedure used here for modal and microprobe analyses is the same as previously reported by Vaniman and Papike (1977b). Modal petrography was based on optical characterization of polished thin sections. Microprobe analyses were obtained on an ARL-EMX SM automated microprobe with one semi-fixed (for silicon) and three positionable spectrometers. Data were reduced by applying the matrix correction procedure of Bence and Albee (1968).

PETROGRAPHY AND MINERALOGY

We have subdivided the drill core lithic fragments into three major categories: mare basalt, fused soil, and highland rocks. The fused soils (agglutinates and dark matrix breccias) include mixtures of highland and mare rock types; Taylor

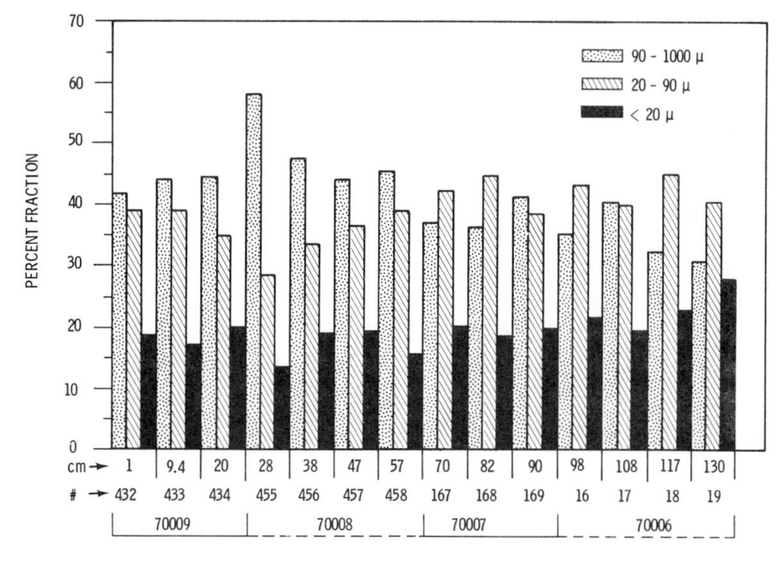

Fig. 2. Percent weight recoveries in 90–1000 μm, 20–90 μm and <20 μm size fractions of soils in 70009–70006.

et al. (1978) indicate ~30–50% highland material in agglutinate glasses of the A-17 drill core. In the discussion below we will deal with the highland and mare rocks, but not with the fused soil components.

Mare rock types

Ilmenite basalts. The ilmenite basalt category includes a variety of textures and a wide range in mineral chemistry. At least two magma types are represented:

1) The commonest high-Ti mare basalt is the Apollo 17 olivine porphyritic to plagioclase poikilitic series. This magma series spans a range in texture and mineral chemistry from vitrophyres with olivine (Fo_{77-65}) + armalcolite + CrAl-spinel, to plagioclase poikilitic rocks with pyroxene + plagioclase + ilmenite + cristobalite. Olivines may be zoned from Fo_{77} to Fo_{60}; plagioclase compositions vary from $An_{88}Or_0$ to $An_{75}Or_4$. Poorly zoned high-Ca augites in olivine porphyritic vitrophyres have ~9 wt.% Al_2O_3 and ~6 wt.% TiO_2; pyroxene atomic Al/Ti ratios >2 reflect some Al in octahedral as well as tetrahedral coordination. The more slowly cooled, coarse plagioclase-poikilitic basalts of this series have pyroxenes zoned to more Ca-poor augites with final development of Fe-rich rims. In the slowly cooled rocks the pyroxene Al/Ti ratio is fixed at 2 (essentially all Al in tetrahedral coordination).
2) A much rarer ilmenite basalt in the Apollo 17 core is diabasic/ophitic in texture. Pyroxene compositions are similar to those in the slowly

cooled ilmenite basalt described above, but olivines may be more Fe-rich (Fo_{65-50}). The greatest difference in mineralogy between the diabasic/ophitic basalt and the other Apollo 17 high-Ti basalts is in feldspar composition; the diabasic/ophitic basalt has plagioclase more Ca-rich (An_{90-95}).

Very low Ti (VLT) basalt. This is a new mare basalt type first recognized in the Apollo 17 drill core by Vaniman and Papike (1977c) and Taylor *et al.* (1977b). Although VLT basalt is a minor core component (~0.1% of all lithic fragments), this lithic type is important in our discussion because the Apollo 17 drill core is the "type locality" for VLT basalt. Variants of this rock type have since been found at Mare Crisium (ref. *Papers Presented to the Conference on Luna 24*, 1977). The mineral chemistry of Apollo 17 VLT basalt is characterized by early crystallization of CrAl spinel rather than ilmenite, by calcic plagioclase (An_{93-85}), by olivine cores of high Mg content (~Fo_{75}), and pyroxene zonation from low-Ca pigeonite (Wo_7En_{20}) to Fe-rich augite rims. One of the most distinctive features of VLT mineralogy is an extremely high Al:Ti ratio (~10) in initial stages of pyroxene crystallization (Vaniman and Papike, 1977c).

HIGHLAND ROCK TYPES

The initial survey by LSPET (1973) emphasized two highland rock types, an anorthositic gabbro with plagioclase cumulate texture (e.g., 77017) and a KREEPy (low-K Fra Mauro) noritic rock type (e.g., 77135). The KREEPy rock type is represented by textural variants including poikilitic rocks (POIK), recrystallized noritic breccia (RNB) and feldspathic basalt (Bence *et al.*, 1974).

Recrystallized noritic breccia (RNB) and poikilitic rocks (POIK)

The RNB/POIK rocks have two pyroxenes + feldspar ± olivine ± ilmenite, Fe and FeS. Textures and mineralogy range from POIK melt rocks with lath-shaped feldspar chadacrysts in augite + pigeonite oikocrysts to RNB textures (annealed grain boundaries) with orthopyroxene + diopsidic augite. Feldspars range in composition from $An_{97}Or_0$ to $An_{88}Or_2$. Pyroxenes have Al/Ti ratios ≥2, and olivine compositions range from Fo_{90} to Fo_{70}.

Feldspathic basalts

Feldspathic basalts are rare in the drill core (Table 1). The major-element and Ti-Al-Cr compositions of pyroxene, olivine and plagioclase in the feldspathic basalts are similar to the mineral compositions of RNB and POIK rocks. Pyroxene rims in feldspathic basalts may be more Fe-rich [Fe/(Fe + Mg) >0.45] than pyroxene in the RNB and POIK rocks.

Table 1. Simplified Lithic Modal Data, 70009–70006.

	modal data unit 1 X-ray units 60–64	modal data unit 2 X-ray unit 59	modal data unit 3 X-ray units 40–58
	(a) Modal percentages of 0.02–2.0 mm lithic fragments in total soil.		
Ilmenite Basalts	8.0	12.0	5.0
Other Mare Basalts	<0.1	<0.1	<0.1
ANT	0.3	0.4	0.4
Feldspathic Basalt	<<0.1	<<0.1	<<0.1
POIK/RNB	0.5	0.5	1.2
Anorthositic Gabbro	0.3	0.4	0.2
Light Matrix Breccia	0.2	0.6	0.4
Mare: highland ratio	5.7	6.3	2.5
	(b) Number of lithic fragments > 2.0 mm.		
Mare Basalts	5	21	13
POIK	–	–	1

Anorthositic gabbro

The term "anorthositic gabbro" is used for rocks with the texture and mineral composition of 77017 (LSPET, 1973; McCallum *et al.*, 1974). Anorthositic gabbros have large (~0.25–0.5 mm) blocky feldspar grains ($An_{98}Or_0$ to $An_{94}Or_{0.5}$) enclosing beads or "necklaces" of olivine and surrounded by a mortar-like matrix of poikilitic Ca-poor pyroxenes and Ca-rich pyroxenes. The pyroxene Ca-Fe-Mg compositions overlap with the pyroxene compositions of the high-Ti mare basalts

Other highland lithologies

Other highland lithologies in the drill core include small fragments of ANT cumulate rocks, light matrix breccias (LMB) with ~5–30% glassy matrix, and one fragment of recrystallized peridotite. These rock types have feldspar compositions of generally restricted Ca-Na zonation ($An_{90-95}Or_{<0.5}$); but crystal rims and small grains of feldspar in the glassy LMB matrices may be more Na-rich. Pyroxenes of the ANT lithologies may be more Mg-rich [(Mg/Mg + Fe) >0.85] than pyroxenes of other highland and mare rock types (Fig. 5). Olivines of ANT and LMB lithic fragments may be as magnesian as Fo_{90}.

Bence *et al.* (1974) describe surface-soil samples of a spinel cataclasite which is not present as lithic fragments in the drill core. Spinel cataclasite is characterized by Al-spinel + olivine (Fo_{90}) + Al-rich enstatite. The aluminous pyroxene has a high Al/Ti ratio (~10), similar to the Al/Ti ratio in VLT mare basalts (VLT pyroxenes are, however, less Mg-rich: Spinel cataclasite enstatites may range from $En_{79}Wo_{<1}$ to $En_{90}Wo_{<1}$.)

MODAL DATA

Lithic modal data from polished thin sections are summarized in Table 1. The data reflect the dominant high-Ti mare basalt component in the drill core. There is an abundance of high-Ti mare fragments (ilmenite basalts) in the coarse layer (unit 2). However, there is an increase in POIK and RNB (KREEPy) melt rocks in unit 3 corresponding with a decrease in mare basalt fragments. This transition from upper layers rich in mare basalt to lower layers with an increased highland component is also seen in the distribution of extra-large (>2 mm) lithic fragments.

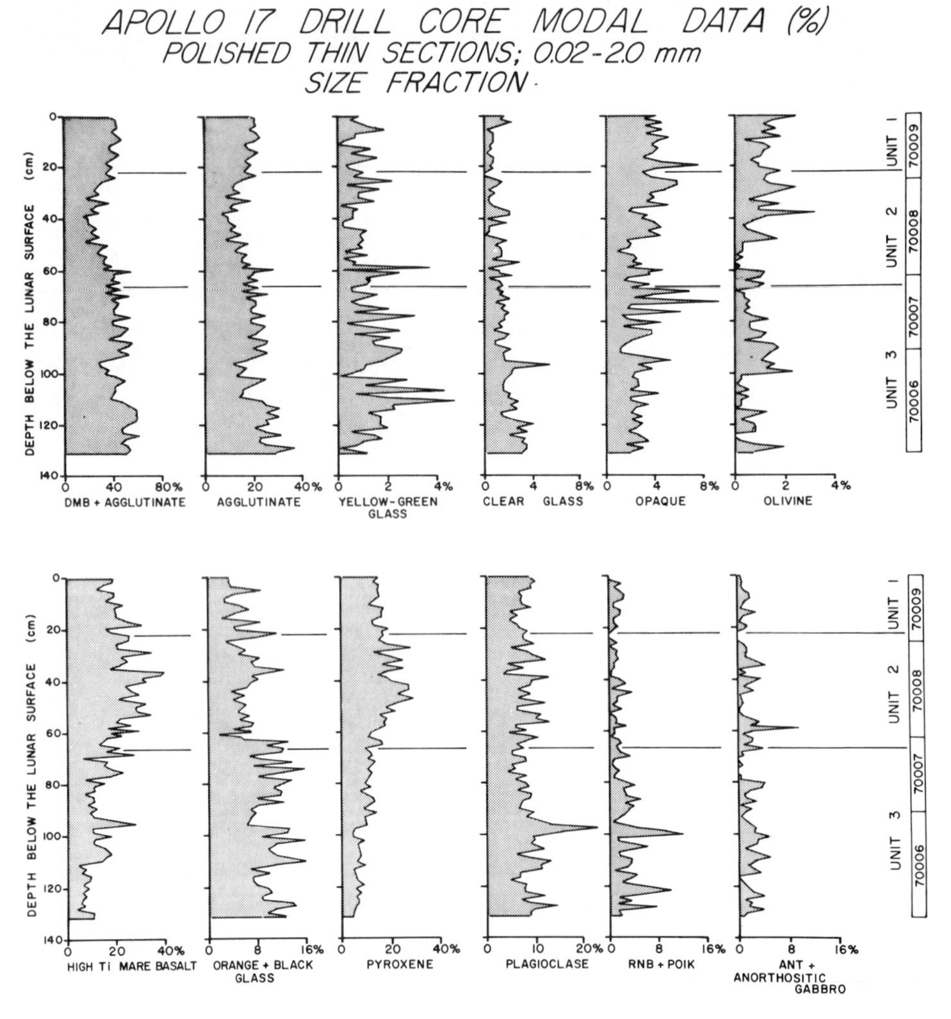

Fig. 3. Stratigraphic distribution of lithic, mineral and glass components in core segments 70009–70006. Data are from polished thin sections; 0.02 to 2.0 mm clast data is normalized to 100%. Note that the horizontal scale varies between columns.

The data from continuous strings of polished thin sections are plotted in Fig. 3. This type of data array can be compared directly with the INAA data displayed as a function of depth (ref. Fig. 6, 7, 8 and 11 in following sections of this paper). Particularly notable is the increase of DMB and agglutinate fragments above and below unit 2 (the coarse-grained, immature mare unit). Unit 2 is marked by an abundance of pyroxene fragments (pyroxene composes ~50% of the common A-17 high-Ti mare basalts) as well as an abundance of high-Ti mare basalt lithic fragments. We note that the orange/black glasses (a high-Ti mare composition) increase in unit 3, below the predominantly mare soils. The increase of high-Ti mare glasses below the high-Ti mare unit may reflect either (1) the inversion of a stratigraphy in which orange/black glass overlay a crystalline mare unit, or (2) mixture of crystalline mare soil with a highland + high-Ti glass soil in unit 3. The presence of an inverted flap in the upper A-17 drill core has been proposed from studies of site geology (LSPET, 1973), core dissection (Duke and Nagle, 1974), track and thermoluminescence data (Crozaz and Plachy, 1976), petrographic data (Vaniman and Papike, 1977b, and Taylor *et al*, 1977a) and I_s/FeO data (Morris *et al.*, 1978). Further studies of the deeper drill core will resolve the question of whether the orange/black glass increase in unit 3 was delivered to the site with the overturned flap or mixed into the overturned flap as part of an underlying highland + high-Ti glass soil.

There is a pronounced increase of highland material with depth in unit 3. Figure 3 shows that in unit 3 highland components such as colorless glass, plagioclase and RNB + POIK increase with depth. Morris *et al.* (1978) show a continuous increase in I_s/FeO with depth in 70007 (upper unit 3) which reflects an inhomogeneous but monotonically increasing mixture of mature highland soils with a corresponding decrease in immature mare soils.

Modal data were also obtained from polished grain mounts. The grain mounts were made from 20–90 μm and >90 μm soil splits taken from each of the 14 soil allocations which were sieved and analyzed by INAA. Data from the polished grain mounts are summarized in Table 2. These data can be correlated directly with the 20–90 μm and >90 μm soil analyses reported below. There is some advantage in having grain mounts from each soil analyzed, but we have found in this study that the grain mounts are less reliable for modal analysis because section thickness is irregular, making the identification of colored grains difficult in plane polarized light and identification of anisotropic grains difficult in cross-polarized light. One of the most critical problems is the identification of glass color types in abnormally thin sections (<30 μm thick) where orange and yellow glasses cannot be distinguished. For this reason, much of the glass data in Table 2 has been grouped into a broader category: (orange/black + yellow/green).

Stratigraphy of Mineral Compositions

Some features of major- and minor-element mineral chemistry can be used to distinguish between possible mare and highland sources. Figures 4–5 show

Table 2. Model data from 0.02–0.09 mm and 0.09–1.0 mm grain mounts (each size range normalized to 100%).

	0.09–1.0 mm Grain Mounts	70009,432 70009,442	70009,433 70009,443	70009,434	70008,455 70008,467	70008,456 70008,468	70008,457 70008,469	70008,458 70008,470	70007,167 70008,414	70007,168 70007,415	70007,169 70007,416	70006,16 70006,301	70006,17 70006,302	70006,18 70006,303	70006,19 70006,304
	Depth	8.5 mm	93.5 mm	203.5 mm	280.5 mm	379.5 mm	470.5 mm	575.5 mm	706.5 mm	811.5 mm	896.5 mm	989.5 mm	1084.5 mm	1174.5 mm	1304.5 mm
	MARE BASALTS	19.75	12.40	15.44	26.41	41.58	28.06	20.35	16.17	18.13	11.05	14.56	9.52	4.20	15.47
	Highland Component														
	Anorthosite	–	–	0.19	–	–	–	1.10	2.45	2.93	3.40	–	1.83	1.68	2.38
	Norite/Troctolite	0.79	–	–	–	–	–	2.20	0.49	–	0.85	–	–	0.84	–
	LMB	0.79	0.62	0.57	–	1.89	2.76	1.10	1.96	2.93	5.10	5.82	2.19	3.36	2.38
	Feldspathic Basalt	–	–	0.19	–	1.89	–	–	–	0.59	0.85	–	1.10	1.68	2.38
	RNB/POIK	1.58	0.62	2.05	1.39	–	0.92	1.10	1.96	3.52	3.40	0.97	1.83	0.84	–
0.09–1.0 mm size fraction	FUSED SOIL COMPONENT														
	DMB	6.32	7.44	2.61	1.39	–	1.38	12.10	13.72	9.94	5.95	22.33	5.49	10.92	3.57
	Agglutinate	37.92	35.34	26.97	16.68	9.45	9.66	14.85	19.60	28.07	41.65	18.45	35.90	49.58	51.19
	MINERAL FRAGMENTS														
	Olivine	–	5.58	0.19	1.39	1.89	2.76	1.10	0.49	–	–	–	1.10	–	–
	Pyroxene	11.06	17.36	28.09	23.63	20.79	25.30	12.10	9.80	10.54	6.80	11.65	13.92	9.24	2.38
	Plagioclase	7.11	8.68	10.04	12.51	5.67	10.58	11.00	6.86	5.26	5.95	7.76	6.95	4.20	8.33
	Opaques	3.95	4.34	3.17	1.39	7.56	7.36	4.95	2.45	1.18	1.70	0.97	3.30	–	–
	GLASS FRAGMENTS														
	Orange/black, Yellow/green	9.48	6.20	9.86	15.29	9.45	10.58	14.85	23.03	12.88	11.05	13.59	13.55	8.40	8.33
	Brown	–	1.24	–	–	–	–	–	–	–	–	–	0.37	–	–
	Gray	–	–	–	–	–	–	–	–	–	–	–	–	–	–
	Colorless	0.79	0.62	–	–	–	0.46	1.65	–	1.77	–	1.94	1.83	2.52	2.38
	OTHER COMPONENTS	–	–	0.38	–	–	0.46	1.65	0.98	2.34	1.70	1.94	1.11	2.52	1.19

Table 2. *(cont'd).*

	0.02–0.09 mm Grain Mounts	70009,436	70008,459	70008,460	70008,462	70008,464	70007,260	70007,261	70007,262	70006,183	70006,184	70006,185	70006,186
	Depth	93.5 mm	280.5 mm	379.5 mm	470.5 mm	575.5 mm	706.5 mm	811.5 mm	896.5 mm	989.5 mm	1084.5 mm	1174.5 mm	1304.5 mm
0.02–0.09 mm size fraction	MARE BASALTS	8.0	12.3	10.2	8.7	19.5	16.1	13.2	9.0	11.0	6.7	5.1	3.3
	Highland Component												
	Anorthosite	0.1	0.3	—	0.6	0.7	0.9	1.2	1.3	1.1	1.1	0.4	0.5
	Norite/Troctolite	—	—	—	—	—	0.1	—	—	—	—	0.2	—
	LMB	0.9	0.2	0.2	0.5	1.1	1.2	0.6	1.3	1.9	0.7	0.3	—
	Feldspathic Basalt	—	—	—	—	0.1	—	—	—	—	—	0.1	—
	RNB/POIK	1.5	0.6	0.5	1.0	0.2	0.9	0.6	0.6	0.4	1.0	0.5	0.1
	FUSED SOIL COMPONENT												
	DMB	13.9	7.8	5.3	7.2	9.7	15.6	9.3	9.1	14.2	13.8	23.7	24.2
	Agglutinate	18.8	8.1	11.4	12.2	12.7	14.5	11.7	19.3	16.1	22.1	14.8	23.0
	MINERAL FRAGMENTS												
	Olivine	0.1	0.1	0.3	0.5	0.2	0.3	0.6	0.4	0.1	1.0	1.3	1.5
	Pyroxene	17.1	16.2	26.3	24.3	25.0	13.3	18.2	17.7	21.0	14.1	22.6	19.2
	Plagioclase	19.4	23.7	15.1	13.4	7.0	13.0	12.0	15.1	12.0	20.4	10.5	10.4
	Opaques	7.3	9.9	9.7	9.8	5.8	5.1	6.9	4.2	5.2	4.2	2.8	3.5
	GLASS FRAGMENTS												
	Orange/black, Yellow/green	9.8	17.4	18.2	19.6	9.7	11.1	12.1	14.5	12.4	9.1	13.1	6.5
	Brown	0.4	0.6	0.1	—	0.7	1.2	2.8	1.3	1.4	0.6	1.4	2.0
	Gray	0.2	0.1	0.1	—	2.0	0.2	0.3	0.1	—	0.1	—	0.2
	Colorless	2.1	2.3	2.1	1.5	4.7	6.3	9.6	4.7	2.9	4.4	2.8	5.5
	OTHER COMPONENTS	0.4	0.4	0.5	0.7	0.9	0.2	0.9	0.8	0.3	0.7	—	0.2

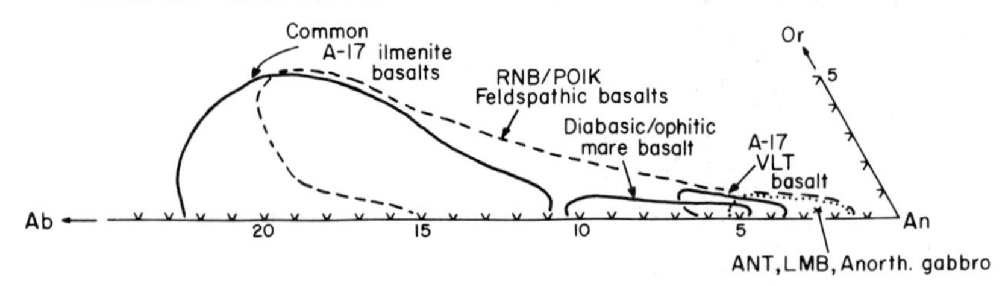

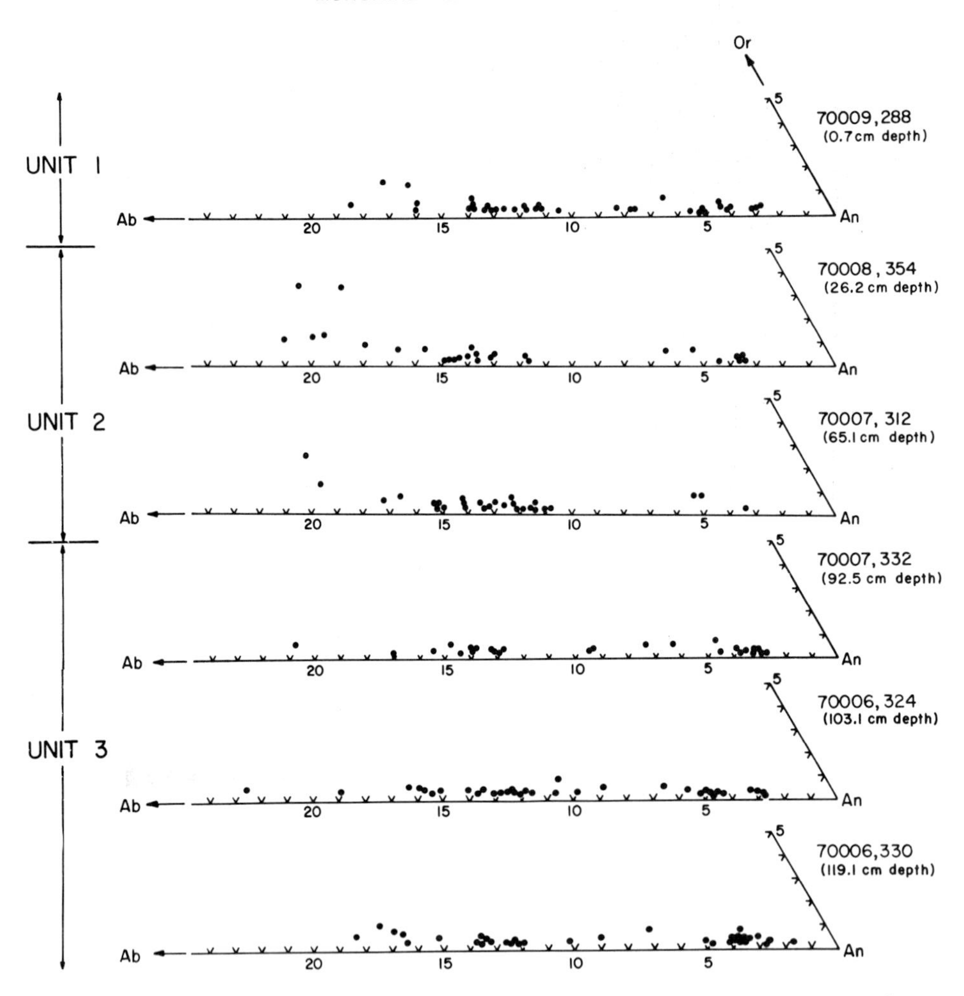

Fig. 4. Feldspar compositional data. The uppermost diagram shows the ranges of feldspar composition in Apollo 17 lithic types; lower six diagrams show the stratigraphy of feldspar mineral fragment compositions in six levels of 70009–70006.

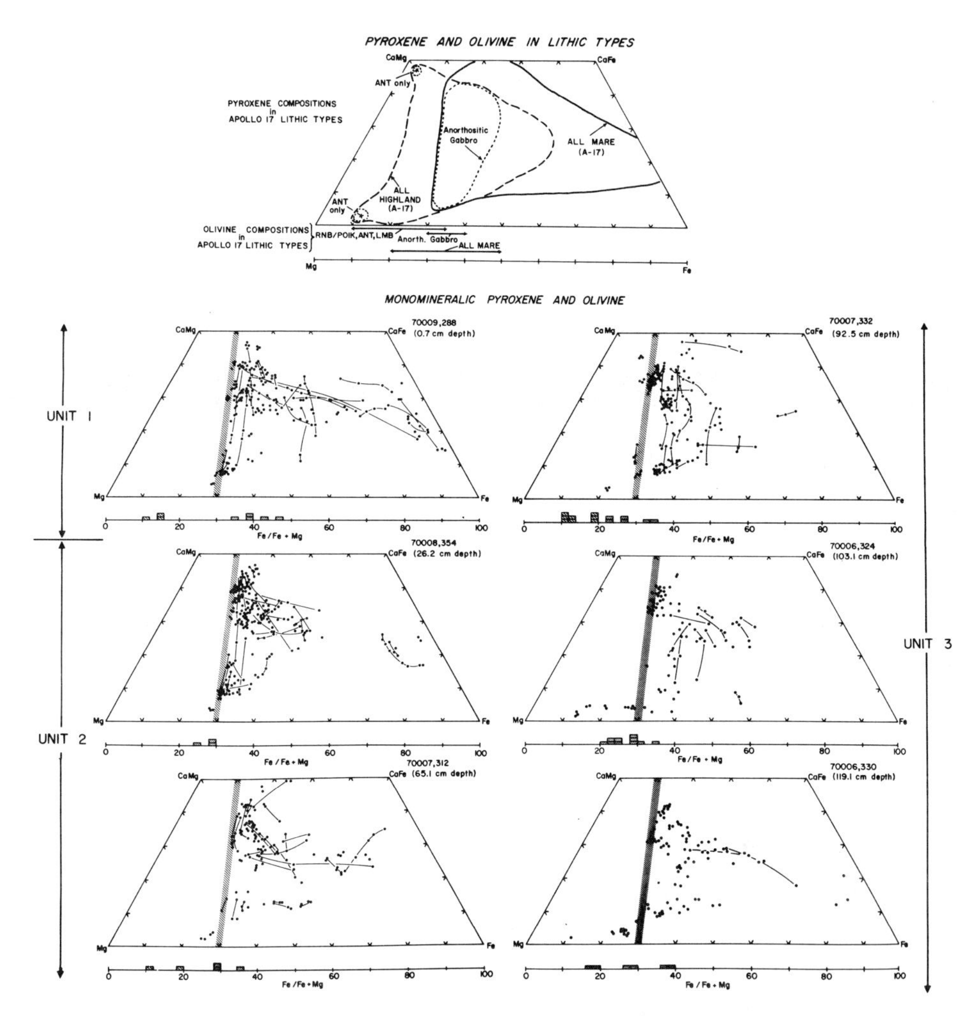

Fig. 5. Pyroxene and olivine compositional data. The uppermost diagram shows the ranges of pyroxene and olivine composition in Apollo 17 lithic types. Note that low-Ca pyroxenes more magnesian than En_{70} come from highland rather than mare sources, and olivines more magnesian than Fo_{80} come from highland sources. The lower six diagrams show the stratigraphy of pyroxene and olivine mineral fragments in six levels of 70009–70006.

variation in mineral composition (feldspar Ab-An-Or and pyroxene Ca-Mg-Fe) at six levels within core segments 70006–70009. At each of the six levels an area of $\sim$10 mm^2 was selected; within each of these areas all feldspar, pyroxene and olivine grains were analyzed. Several analyses were made within each pyroxene grain to account for zonation. At the top of each figure (4–5) is a summary diagram showing the mineral composition fields of different rock types. These two figures point out several important conclusions regarding source-rock petrology:

1) The common high-Ti Apollo 17 basalts have feldspars less calcic than An_{88}. Concentrations of these relatively Ca-poor feldspars are seen in unit 2 (Fig. 4). Other mare basalts (VLT and diabasic/ophitic) have feldspars that are more calcic, but these other mare rocks are rare (Table 1). Therefore, it is reasonable to assume that most feldspar grains more calcic than An_{88} are from highland source rocks. Such calcic feldspars are concentrated in units 1 and 3.
2) Olivines more Mg-rich than Fo_{80} and low-Ca pyroxenes more Mg-rich than En_{70} are derived from highland sources (Fig. 5). There is a marked increase of Fo_{80-90} olivine and low-Ca pyroxene with Mg $>En_{70}$ in unit 3, below the coarse-grained unit. This corresponds with the well-documented increase of highland components below the coarse unit (Duke and Nagle, 1974). However, the nature of the highland component is not the same in all levels of unit 3. Note that the increased highland component is represented by Mg-rich olivine in 70007, 332 but Mg-rich pyroxene is dominant in 70006. These highland mineral types could represent one or several rock types (ANT, LMB, POIK/RNB, feldspathic basalt or spinel cataclasite). However, the evidence for an increased KREEPy component in all soils of 70006 (discussion of INAA data below) as well as the modal data indicating an increase of KREEP-bearing POIK in 70006 suggest that POIK rocks account for a significant amount of the increase in highland component.

CHEMICAL SYSTEMATICS

Results

The data for 34 major-, minor- and trace-elements determined by INAA in 48 bulk soils and size fractions (each 4–30 mg) are presented in Tables 3 to 6 along with controls 73121, 75081 and (USGS) BCR-1. Tables 3 and 4 list the major- and minor-element abundances whereas Tables 5 and 6 contain trace-element abundances. In cases where the actual bulk samples were analyzed, these are compared with the calculated bulk values to test the mass balance. Overall, the agreement between the actual and calculated bulk soil compositions is remarkably good for both major- and trace-elements. The only exception we note is in sample 70008,456 (38 cm depth) where the mass balance is poor for trace elements. The calculated bulk LIL (Ba, Sr, REE, Hf, Th, etc.) and K_2O abundances are ~30% higher than the actual bulk analysis (Table 3). Since K correlates strongly with LIL trace elements in "KREEPy" rocks and soils, this imbalance is most likely due to variability between soil samples.

For SiO_2, FeO, Sc, Co, Hf, Ta and Zn, we used (USGS) BCR-1 values as an internal standard. These values are listed in Tables 4 and 6 and the agreement with the recommended values of Flanagan (1973) is excellent. For Th, we used GSP-1 = 105 ppm. For Au, Ir and Ni, we used the Allende standard (USNM) analyzed by E. Anders' group (Takahashi *et al.*, 1978). Their mean values for six

Table 3. Major and minor abundances (%) via INAA in bulk and size fractions of 70009 and 70008 cores.

Sample (Depth)	Size (μm)	Weight Fraction	Weight (mg)	SiO_2	TiO_2	Al_2O_3	FeO	MgO	CaO	Na_2O	K_2O	MnO	Cr_2O_3	Total
70009,432	Bulk	—	16.8	40.4	8.3	12.1	17.1	10.7	10.8	0.39	0.085	0.224	0.410	100.5
Parent 9	Calculated			40.3	8.2	12.3	17.1	10.6	11.1	0.40	0.094	0.220	0.417	100.7
0.6–1.1 cm	>90	0.418	21.0	39.8	8.2	12.1	16.9	10.2	11.8	0.40	0.086	0.226	0.407	100.1
	20–90	0.390	20.1	40.8	8.5	11.7	17.4	11.4	10.4	0.40	0.090	0.223	0.430	101.3
	<20	0.192	10.4	40.5	7.4	14.2	16.9	9.6	11.1	0.41	0.12	0.191	0.413	100.8
70009,433	>90	0.441	30.0	40.8	8.9	10.7	17.4	9.9	10.8	0.38	0.092	0.234	0.427	99.6
Parent 45	20–90	0.390	26.8	40.8	8.6	12.2	17.5	10.6	11.3	0.39	0.086	0.222	0.417	102.1
9.1–9.6 cm	<20	0.169	12.5	42.2	7.8	14.3	15.5	8.4	10.7	0.42	0.12	0.183	0.400	100.0
70009,434	Bulk	—	3.65	40.7	8.5	11.5	17.5	10.0	10.0	0.41	0.085	0.224	0.420	99.4
Parent 99	Calculated			40.8	8.6	11.5	17.4	10.3	10.1	0.41	0.091	0.227	0.407	99.8
20.1–20.6 cm	>90	0.445	28.0	40.4	8.7	10.9	18.0	10.2	10.4	0.41	0.081	0.230	0.430	99.7
	20–90	0.354	21.7	40.4	8.9	11.1	17.6	11.0	9.8	0.40	0.094	0.238	0.400	99.9
	<20	0.201	12.8	42.3	7.8	13.4	15.5	9.4	10.0	0.43	0.11	0.202	0.370	99.5
70008,455	>90	0.579	41.0	39.9	9.8	10.0	18.3	9.0	10.9	0.40	0.078	0.240	0.400	99.0
Parent 153	20–90	0.286	20.3	40.1	9.0	10.4	19.2	10.3	10.6	0.40	0.089	0.247	0.460	100.8
27.8–28.3 cm	<20	0.135	9.08	41.1	7.6	12.7	16.8	9.6	10.9	0.52	0.13	0.201	0.470	100.0
70008,456	Bulk	—	5.89	39.5	9.1	11.1	18.3	9.6	11.2	0.43	0.068	0.230	0.460	100.0
Parent 110	Calculated			40.2	9.0	10.9	18.4	9.6	11.0	0.43	0.089	0.232	0.450	100.2
37.7–38.2 cm	>90	0.474	28.7	39.9	9.6	10.4	18.7	9.3	10.8	0.40	0.078	0.240	0.440	99.9
	20–90	0.336	20.3	39.9	8.9	10.5	19.2	10.3	11.1	0.43	0.082	0.240	0.470	101.3
	<20	0.190	12.5	41.7	7.4	12.6	16.2	9.0	11.1	0.49	0.13	0.199	0.420	99.2
70008,457	>90	0.439	29.4	40.6	8.5	11.0	17.5	9.0	11.5	0.40	0.12	0.238	0.410	99.3
Parent 72	20–90	0.367	23.8	40.5	8.6	10.6	19.0	10.7	10.9	0.42	0.095	0.233	0.450	101.5
46.8–47.3 cm	<20	0.194	13.2	41.4	7.3	12.7	17.3	9.1	10.1	0.50	0.14	0.205	0.455	99.2
70008,458	>90	0.453	23.5	40.6	7.7	10.9	18.0	10.2	11.2	0.41	0.085	0.232	0.470	99.8
Parent 30	20–90	0.390	21.4	40.8	7.5	11.4	18.2	10.8	10.0	0.41	0.094	0.236	0.450	99.9
57.3–57.8 cm	<20	0.157	7.37	42.5	7.0	13.5	16.5	9.7	11.1	0.50	0.13	0.200	0.470	101.6

Table 4. Major and minor abundances (%) via INAA in bulk and size fractions of 70007 and 70006 cores.

Sample (Depth)	Size (μm)	Weight Fraction	Weight (mg)	SiO_2	TiO_2	Al_2O_3	FeO	MgO	CaO	Na_2O	K_2O	MnO	Cr_2O_3	Total
70007,167	Bulk	—	4.03	41.8	6.8	12.3	17.9	11.0	10.0	0.42	0.11	0.225	0.480	101.0
Parent 120	Calculated			41.6	6.7	12.1	17.7	10.9	10.5	0.42	0.10	0.222	0.463	100.7
70.4–70.9 cm	>90	0.372	20.5	41.7	7.0	11.6	18.2	10.7	10.4	0.41	0.085	0.226	0.470	100.8
	20–90	0.428	25.6	41.5	6.7	11.6	17.7	11.3	10.3	0.40	0.10	0.227	0.450	100.3
	<20	0.200	11.1	41.6	6.2	14.1	16.8	10.3	10.9	0.48	0.15	0.206	0.480	101.2
70007,168	>90	0.366	22.0	41.7	6.0	13.9	15.9	10.1	11.3	0.39	0.087	0.206	0.440	100.0
Parent 58	20–90	0.450	27.0	41.8	5.8	12.5	17.2	11.0	11.0	0.39	0.099	0.217	0.420	100.4
80.9–81.4 cm	<20	0.184	10.6	41.7	5.8	15.0	15.7	9.4	11.7	0.46	0.13	0.188	0.440	100.5
70007,169	>90	0.415	23.1	42.5	5.0	14.2	15.3	9.9	12.2	0.44	0.12	0.192	0.345	100.2
Parent 24	20–90	0.386	21.9	42.3	5.7	12.7	15.9	11.0	11.1	0.38	0.10	0.199	0.380	99.8
89.4–89.9 cm	<20	0.199	11.9	42.5	5.3	15.3	14.8	9.1	11.4	0.46	0.13	0.183	0.405	99.6
70006,16	>90	0.351	23.4	42.2	6.0	13.4	16.0	10.0	11.0	0.41	0.10	0.205	0.405	99.7
Parent 36	20–90	0.434	28.4	42.5	5.7	12.8	16.0	10.7	10.9	0.39	0.10	0.212	0.380	99.7
97.7–98.2 cm	<20	0.215	14.0	43.1	5.4	15.2	14.4	9.8	11.5	0.45	0.14	0.188	0.390	100.6
70006,17	Bulk	—	5.38	42.1	6.1	13.0	16.3	10.1	10.5	0.43	0.11	0.200	0.450	99.3
Parent 79	Calculated			42.4	6.1	12.7	16.2	10.3	10.9	0.43	0.11	0.210	0.430	99.8
108.2–108.7 cm	>90	0.407	20.2	42.3	6.3	12.1	16.5	10.1	10.8	0.43	0.10	0.213	0.410	99.3
	20–90	0.400	20.5	42.1	6.0	12.3	16.5	10.6	10.9	0.41	0.10	0.218	0.450	99.6
	<20	0.193	9.48	43.3	5.7	14.7	15.0	9.9	11.0	0.46	0.16	0.188	0.430	100.8
70006,18	>90	0.324	19.7	42.0	5.6	13.7	15.8	10.3	11.7	0.39	0.096	0.207	0.430	100.2
Parent 116	20–90	0.451	29.9	42.7	5.1	13.6	15.6	10.5	11.6	0.41	0.10	0.202	0.370	100.2
117.2–117.7 cm	<20	0.225	13.9	42.8	5.4	15.8	14.2	10.0	11.5	0.42	0.14	0.185	0.390	100.8
70006,19	Bulk	—	9.31	42.6	5.5	14.0	15.6	10.0	11.7	0.42	0.12	0.200	0.415	100.5
Parent 167	Calculated			42.5	5.6	14.3	15.6	10.2	11.7	0.41	0.12	0.198	0.413	101.0
130.2–130.7 cm	>90	0.309	17.8	42.6	5.8	13.8	15.8	9.8	11.9	0.42	0.095	0.205	0.430	100.8
	20–90	0.407	24.9	42.2	5.5	13.7	16.2	10.9	11.7	0.40	0.10	0.203	0.410	101.3
	<20	0.284	16.5	42.8	5.4	15.7	14.6	9.8	11.5	0.42	0.16	0.183	0.400	101.0
Controls														
73121,17				45.4	1.4	20.6	8.5	10.0	13.1	0.39	0.14	0.110	0.210	99.9
75081,21				40.4	9.1	11.1	17.3	9.6	10.9	0.44	0.080	0.230	0.430	99.6
BCR-1				≡54.5	2.2	13.6	≡12.2	3.0	6.95	3.20	1.70	0.180	0.0025	—

Estimated errors based on counting statistics are: ±0.5–3% for SiO_2, TiO_2, Al_2O_3, FeO, MnO, Na_2O and Cr_2O_3; ±5% for MgO, CaO and K_2O.

Table 5. Trace abundances via INAA in bulk and size fractions of 70009 and 70008 cores. (All values in ppm unless noted.)

Sample (Depth)	Size (μm)	Weight Fraction	Sc	V	Co	Ba	Sr	La	Ce	Nd	Sm	Eu	Tb	Dy	Ho	Yb	Lu	Hf	Ta	Th	U	Ga	Zn	Au (ppb)	Ni*	Ir (ppb)
70009,432	Bulk	—	56.6	100	32.3	120	210	7.90	28	23	8.06	1.76	1.9	11.4	2.9	7.11	1.07	6.60	1.20	0.95	0.23	6.3	44	3	150	<10
Parent 9	Calculated		57.6	95	35.5	120	200	8.08	29	23	8.10	1.82	1.9	11.9	2.8	7.11	1.06	6.50	1.23	1.08	0.26	6.6	46	—	—	—
0.6–1.1 cm	>90	0.418	61.4	100	28.8	130	190	7.27	27	22	7.71	1.80	1.9	11.7	2.8	7.40	1.10	6.42	1.13	0.96	0.20	5.4	32	3	120	<10
	20–90	0.390	58.9	90	39.7	100	190	7.53	28	22	7.97	1.70	2.0	12.1	2.8	7.00	1.06	6.52	1.20	0.98	0.23	6.2	40	3	200	12
	<20	0.192	46.8	90	41.7	120	250	11.0	36	26	9.20	2.10	2.0	12.1	2.7	6.70	0.99	6.47	1.50	1.55	0.45	9.8	88	20	580	25
70009,433	>90	0.441	67.0	90	31.4	110	180	7.67	26	22	8.10	1.78	2.0	12.4	2.8	6.89	1.00	7.21	1.35	0.94	0.24	6.4	30	2	140	<10
Parent 45	20–90	0.390	58.7	90	32.5	110	180	7.84	27	24	8.00	1.84	2.0	11.3	2.7	6.71	1.00	6.33	1.28	1.05	0.25	5.6	38	3	240	12
9.1–9.6 cm	<20	0.169	43.0	90	38.2	130	200	11.5	39	26	8.66	2.00	2.1	11.8	2.5	6.10	0.90	6.70	1.35	1.90	0.55	9.5	80	16	600	30
70009,434	Bulk	—	60.0	100	34.9	110	180	8.30	29	25	8.80	1.90	2.1	12.5	2.8	7.40	1.10	6.60	1.35	0.94	—	—	40	2	100	<10
Parent 99	Calculated		58.7	95	34.0	105	185	8.58	30	24	8.80	1.86	2.1	12.3	2.9	7.12	1.06	6.40	1.29	0.97	0.28	6.8	49	—	—	—
20.1–20.6 cm	>90	0.445	65.0	100	35.5	95	180	7.60	30	24	9.34	1.93	2.3	13.2	3.0	8.00	1.17	6.84	1.40	0.94	0.24	5.4	35	3	220	12
	20–90	0.354	60.6	90	32.1	110	190	9.00	30	24	8.32	1.72	2.0	12.0	2.9	6.65	1.00	5.81	1.15	0.83	0.29	6.8	40	2	140	<10
	<20	0.201	41.6	90	34.1	120	190	10.0	31	26	8.44	1.95	1.9	11.0	2.7	6.01	0.91	6.43	1.30	1.30	0.34	10.0	95	(30)	440	30
70008,455	>90	0.579	72.1	100	22.3	100	160	6.54	25	24	8.50	1.75	2.2	12.6	2.6	8.21	1.16	6.50	1.44	0.40	—	5.2	30	—	60	<10
Parent 153	20-90	0.286	62.5	100	35.4	110	150	7.28	27	25	8.80	1.91	2.1	12.3	2.9	7.61	1.10	6.54	1.30	0.60	—	6.3	45	3	220	<15
27.8–28.3 cm	<20	0.135	44.5	95	36.6	150	200	9.48	31	26	8.49	1.97	2.0	10.6	2.6	6.32	0.93	6.10	1.37	1.10	0.4	14.0	150	(70)	600	40
70008,456	Bulk	—	63.6	100	28.4	80	150	5.40	20	19	6.30	1.70	1.8	10.0	2.5	6.64	0.95	5.56	1.30	0.60	—	9.0	40	2	110	<10
Parent 110	Calculated		59.2	97	31.8	120	190	7.57	26	24	8.42	1.88	2.1	12.4	2.8	7.50	1.06	6.72	1.24	0.78	—	9.5	67	—	—	—
37.7–38.2 cm	>90	0.474	64.6	100	27.4	120	210	6.81	25	23	8.10	1.77	2.2	13.0	2.7	7.70	1.10	6.74	1.20	0.60	—	9.0	50	3	200	<15
	20–90	0.336	60.8	90	35.5	110	170	7.50	28	25	8.78	1.90	2.2	12.3	3.0	7.80	1.10	7.12	1.30	0.80	0.3	6.6	50	3	350	<15
	<20	0.190	43.0	100	36.1	140	180	9.60	28	26	8.59	2.05	1.9	11.3	2.6	6.50	0.93	6.00	1.25	1.20	0.4	16.0	140	(100)	410	25
70008,457	>90	0.439	68.8	100	23.5	120	150	6.96	27	23	8.80	1.85	2.2	14.0	2.8	8.00	1.10	7.10	1.40	0.50	—	6.4	40	2	150	<10
Parent 72	20–90	0.367	61.5	90	35.8	110	140	7.70	28	24	8.50	1.90	2.1	12.0	2.7	7.52	1.06	6.75	1.40	0.70	—	6.4	60	—	710	30
46.8–47.3 cm	<20	0.194	47.9	85	38.5	110	200	10.5	36	29	9.46	2.10	2.1	12.0	2.6	6.46	0.97	6.58	1.40	1.40	0.4	15.0	160	17	1100	50
70008,458	>90	0.453	65.0	110	32.4	90	160	7.67	27	23	8.17	1.72	2.1	12.1	3.0	7.48	1.05	6.58	1.35	0.85	—	6.4	40	2	150	<10
Parent 30	20–90	0.390	54.9	90	38.5	110	170	7.70	29	22	7.62	1.76	1.9	11.0	2.7	7.03	0.98	6.58	1.10	1.10	0.3	7.8	60	3	150	<10
57.3–57.8 cm	<20	0.157	42.9	85	40.7	130	210	10.1	31	24	7.82	1.90	2.0	10.4	2.4	5.85	0.87	6.40	1.28	1.30	0.4	12.0	130	5	290	15

Table 6. Trace abundances via INAA in bulk and size fractions of 70007 and 70006 cores. (All values in ppm unless noted).

Sample (Depth)	Size (μm)	Weight Fraction	Sc	V	Co	Ba	Sr	La	Ce	Nd	Sm	Eu	Tb	Dy	Ho	Yb	Lu	Hf	Ta	Th	U	Ga	Zn	Au (ppb)	Ni*	Ir (ppb)
70007,167	Bulk	—	50.0	95	37.5	100	180	8.60	26	20	7.20	1.72	1.9	10.3	2.4	6.10	0.90	6.10	1.10	1.10	0.30	9.0	60	2.0	150	<10
Parent 120	Calculated		52.0	97	38.3	110	180	9.24	29	22	7.39	1.71	2.0	11.0	2.7	6.36	0.92	6.12	1.10	1.25	0.40	9.1	70	—	—	—
70.4–70.9 cm	>90	0.372	60.0	95	35.0	90	180	9.30	30	24	7.70	1.80	2.2	12.0	2.9	7.20	1.05	6.70	1.11	1.30	0.40	8.5	40	3.0	120	<10
	20–90	0.428	50.1	100	38.9	120	180	8.57	27	20	7.10	1.60	1.8	10.5	2.6	6.05	0.85	5.60	1.00	1.10	0.40	7.8	65	3.0	180	<10
	<20	0.200	41.2	95	43.4	120	170	10.6	31	23	7.44	1.78	1.9	10.3	2.4	5.50	0.81	6.20	1.10	1.50	0.40	13.0	140	7.0	400	20
70007,168	>90	0.366	49.3	85	31.8	100	170	6.65	23	18	6.30	1.55	1.6	9.6	2.4	6.00	0.84	5.45	1.20	0.95	—	5.5	40	2.0	150	<10
Parent 58	20–90	0.450	50.5	85	39.0	100	160	8.46	26	21	7.00	1.60	1.7	10.4	2.6	6.10	0.85	6.00	1.00	1.05	0.40	6.7	60	3.0	210	<15
80.9–81.4 cm	<20	0.184	39.7	85	41.0	110	210	10.7	31	23	7.25	1.70	2.0	10.1	2.4	5.70	0.84	6.10	1.20	1.60	0.48	12.0	100	7.2	600	26
70007,169	>90	0.415	46.4	80	49.0	150	160	10.7	36	28	8.20	1.60	1.9	11.4	2.8	7.00	1.00	6.60	1.15	1.60	0.40	6.7	40	6.0	430	20
Parent 24	20–90	0.386	47.0	80	36.7	120	170	8.50	27	20	6.71	1.48	1.6	9.9	2.3	5.90	0.83	5.70	1.15	1.25	0.35	7.5	50	3.0	270	<15
89.4–89.9 cm	<20	0.199	37.6	90	41.0	160	190	11.4	33	24	7.33	1.70	1.7	9.7	2.5	5.60	0.85	5.90	1.14	1.75	0.50	10.0	90	14.0	710	50
70006,16	>90	0.351	53.0	80	37.5	130	180	9.00	28	20	6.92	1.60	1.6	9.7	2.4	6.10	0.85	5.60	1.00	1.20	0.40	7.1	45	3.0	290	15
Parent 36	20–90	0.434	47.9	85	35.4	110	160	9.15	28	21	7.20	1.48	1.7	10.6	2.4	6.36	0.92	5.70	0.95	1.30	0.40	6.0	45	3.0	250	<15
97.7–98.2 cm	<20	0.215	36.7	80	39.7	190	170	12.3	38	26	8.38	1.70	1.8	11.0	2.4	5.91	0.90	6.11	1.00	2.05	0.50	10.0	80	(70)	550	25
70006,17	Bulk	—	50.0	90	36.9	120	190	9.36	28	23	7.15	1.70	1.9	10.0	2.4	6.10	0.90	6.00	1.05	1.35	0.40	6.6	40	3.0	220	<15
Parent 79	Calculated		50.2	95	36.2	120	180	9.50	29	23	7.38	1.63	1.8	10.5	2.5	6.25	0.93	5.80	1.02	1.40	0.40	6.4	46	—	—	—
108.2–108.7 cm	>90	0.407	54.9	100	31.9	130	160	8.18	28	22	7.22	1.60	1.7	10.2	2.5	6.38	0.95	5.79	1.10	1.25	0.40	5.0	35	3.0	170	<10
	20–90	0.400	50.4	90	38.6	100	190	9.52	29	23	7.30	1.60	1.9	10.8	2.6	6.25	0.92	5.64	0.95	1.30	0.40	6.0	40	4.0	640	15
	<20	0.193	39.8	90	40.5	130	190	12.3	34	25	7.90	1.77	2.0	10.6	2.5	6.00	0.89	6.02	1.00	1.80	0.50	10.0	80	18.0	970	30
70006,18	>90	0.324	51.9	100	38.5	130	150	8.50	26	20	6.43	1.56	1.7	9.4	2.3	5.54	0.83	5.45	0.92	1.30	0.40	6.0	40	3.0	250	<15
Parent 116	20–90	0.451	48.1	95	36.0	120	160	8.86	27	21	6.76	1.51	1.6	10.0	2.4	5.99	0.87	5.52	1.00	1.30	0.40	6.3	45	3.0	250	<15
117.2–117.7 cm	<20	0.225	37.7	80	39.0	130	160	11.7	34	23	7.37	1.66	1.9	11.4	2.6	5.90	0.84	5.80	1.10	1.90	0.55	8.0	60	9.5	420	25
70006,19	Bulk	—	47.0	90	35.7	130	160	11.0	33	24	7.60	1.58	2.0	10.3	2.6	6.29	0.94	5.73	1.00	1.50	0.50	6.1	40	7.0	260	17
Parent 167	Calculated		45.5	90	37.2	120	150	10.1	30	23	7.10	1.55	1.8	10.4	2.4	6.06	0.87	5.82	0.97	1.50	0.40	6.5	48	—	—	—
130.2–130.7 cm	>90	0.309	48.1	90	32.8	120	120	10.0	27	24	7.00	1.46	2.0	11.0	2.5	6.17	0.91	5.50	0.85	1.40	0.40	6.0	35	3.0	240	<15
	20–90	0.407	49.2	95	37.8	120	140	9.40	29	22	7.17	1.55	1.6	9.8	2.5	6.36	0.88	6.01	0.92	1.30	0.40	5.2	40	3.0	270	<15
Controls	<20	0.284	37.5	80	41.2	130	190	11.2	33	23	7.00	1.64	1.9	10.6	2.2	5.51	0.81	5.90	1.16	1.90	0.50	9.0	75	7.7	520	30
73121,17			18.3	50	31.0	170	160	15.1	38	26	7.20	1.30	1.5	9.2	2.2	5.40	0.76	5.50	0.73	2.90	0.85	6.0	20	4.0	330	16
75081,21			67.0	100	30.0	100	160	7.70	28	25	8.40	1.80	2.1	13.0	3.0	7.40	0.98	7.40	1.40	0.90	0.25	6.4	35	—	110	<10
BCR-1			≡32.0	410	≡36.0	670	330	25.5	54	30	6.70	2.00	1.0	6.3	1.3	3.40	0.53	≡4.70	≡0.80	6.00	1.70	24.0	≡130	—	—	—

Estimated Errors Based on Counting statistics are: ±0.5–3% for Sc, Co, La, Sm, Eu, Yb and Lu; ±5% for V, Tb, Dy, Hf and Ta; ±5–10% for Ba, Ce, Nd, Ho, Th, Ga, Zn and Ni; ±10–20% for Sr, U, Au, and Ir. The upper limits are based on 3σ of the peak area. The values in parenthesis are suspect of contamination.

*The Ni, Au and Ir values are highly suspect of contamination from the Rh-plated Ni sieves.

replicates are Au = 0.140 ppm; Ir = 0.780 ppm; and Ni = 14200 ppm, respectively. Unfortunately, our Ni and possible Au and Ir values especially in the <20 μm fractions are highly suspect because of possible Ni contamination from abrasion of the sieves. Au and Ir may also be carried as impurities with Rh and Ni from the sieves. Thus, any interpretation of the meteoritic component using these siderophile elements should be made with caution.

Typical errors in our INAA procedure based on counting statistics using BCR-1 and other controls are: ±0.5–3% for SiO_2, TiO_2, Al_2O_3, FeO, MnO, Na_2O, Cr_2O_3, Sc, Co, La, Sm, Eu, Yb and Lu; ±5% for MgO, CaO, K_2O, V, Tb, Dy, Hf and Ta; ± 5–10% for Ba, Ce, Nd, Ho, Th, Sr, Ga, Zn and Ni; 10–20% for U, Au and Ir respectively. The analyses of controls 73121, 75081 and BCR-1 agree very well with the several replicate analyses reported by others (Laul and Schmitt, 1973; Korotev *et al.*, 1976; Baedecker *et al.*, 1974; Wänke *et al.*, 1974; *Proc. Apollo 11 Lunar Sci. Conf.*, 1970; *Proc. Lunar Sci. Conf. 2nd* through *8th*, 1971, 1972, 1973, 1974, 1975, 1976, 1977.

The major- and trace-element abundances of the >90 μm and 20–90 μm fractions are identical (Tables 3–6), and we have averaged these two coarse fractions for comparison to the <20 μm fine fraction whose chemistry is quite different.

Major element characteristics

Among the major elements, FeO and particularly TiO_2 are strong indicators of mare material, whereas Al_2O_3 is a strong indicator of highland material. Thus, we have plotted TiO_2 and Al_2O_3 for bulk, coarse and fine fractions as a function of depth in Figs. 6 and 7. For consistency, we have plotted calculated bulk values in all figures. For Figs. 6 and 7, we make the following observations.

As TiO_2 content decreases from the top of section 70009 to the bottom of 70006, the Al_2O_3 content correspondingly increases, suggesting that decrease in the mare content is coupled with an increase in the highland component. However, the <20 μm fine fraction is consistently more feldspathic in composition at all depths; i.e., it is high in Al_2O_3, CaO, K_2O, Na_2O, light REE, and low in TiO_2, FeO, MnO, Cr_2O_3, Sc and heavy REE relative to the >90 μm and 20–90 μm fractions. The fine (<20 μm) fraction is considerably enriched in highland material, consistent with the observations of Korotev (1976). The bulk soil chemistry is governed by the coarse fractions because of their greater weight. The highland contribution in the <20 μm fraction generally increases from 45 to 81 cm depth, and below this depth the highland contribution stays about constant. The TiO_2 content in the <20 μm and coarse fractions tends to merge at ~4.5–5.5% TiO_2 below 90 cm depth, suggesting that the coarse and bulk fractions like the fine fractions are dominated by highland instead of mare material. This is also supported by the large-ion lithophile (K, REE and Th) trace-element data discussed below.

The X-ray Unit 59, which includes core 70008, is a coarse-grained interval which extends from 25 to 80 cm depth (Duke and Nagle, 1974). We analyzed

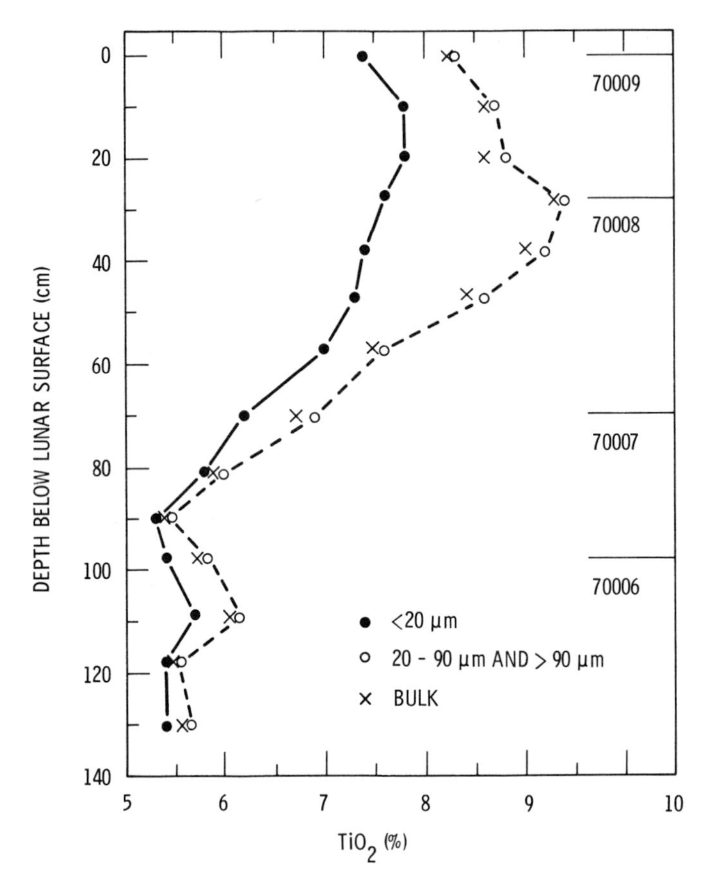

Fig. 6. Depth profiles of TiO_2 in bulk, coarse (>90 μm and 20–90 μm) and <20 μm fine fractions of 70009–70006.

five samples at ~10 cm depth intervals in 70008. The TiO_2 contents in samples at 28, 38 and 47 cm depth intervals are quite high (8.3–9.4%), in the range of typical high Ti mare basalt. Vaniman and Papike (1977a) and Taylor *et al.* (1977a) found that this unit consists almost entirely of mare material (basalt + orange glass). Our modal data (Table 2) and LIL patterns (discussed later) are consistent with this interpretation.

Figure 8 shows the depth profile of FeO and Sc in bulk soils and in size fractions. In the coarse fractions (>90 μm and 20–90 μm), FeO and Sc tend to follow TiO_2 content as indicators of the ilmenite basalt component. When the TiO_2 content is high (7–9%) the FeO (18–19%), and Sc (60–70 ppm) are also high, suggesting that much of the Sc is in the minerals of the ilmenite basalt. The distribution coefficients (D) of Sc in pyroxene and ilmenite of ilmenite basalt are 3.3 and 1.8 respectively (Haskin and Korotev, 1977). Because of the higher proportion of ilmenite basalt in coarse soil fractions, the FeO/Sc ratios deviate significantly from the lunar FeO/Sc ratio of 5000 ± 500 (Laul and Schmitt,

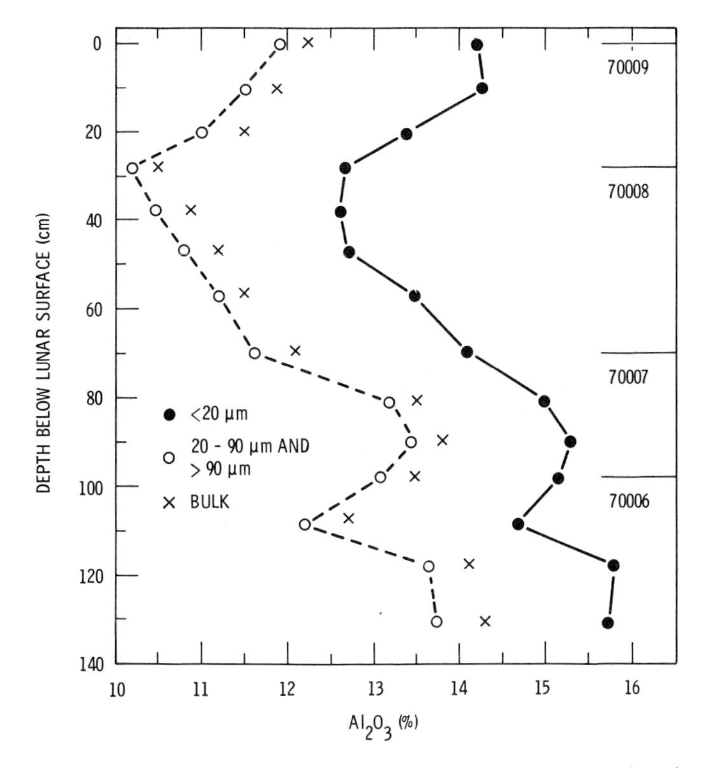

Fig. 7. Depth profiles in Al_2O_3 in bulk, coarse (>90 μm and 20–90 μm) and <20 μm fine fractions of 70009–70006.

1973; Wänke *et al.*, 1974; Laul *et al.*, 1978). When TiO_2 is low (5.5%) as in core segment 70006, FeO (~16%) and Sc (~50 ppm) are also low. In the cases of low TiO_2 content (1–2%), the FeO/Sc ratio follows the lunar ratio. In general, the coarse fractions contain considerable amounts of mare basalt and core section 70008 contains the highest mare contribution. The fine fractions contain consistently less mare basalt (less TiO_2, FeO, Sc, etc.) relative to the coarse fractions. This again emphasizes the dominance of highland material in the fine fractions.

The strong correlation between FeO and MnO with the FeO/MnO ratio of 80 ± 5 has been well established in a wide variety of lunar rocks and soils (Laul *et al.*, 1972, 1974, 1978; Miller *et al.*, 1974; Wänke *et al.*, 1974). Mn^{++}(0.80 Å) can readily replace Fe^{++}(0.74 Å) in oxides and Mg-Fe silicates (pyroxene, olivine, etc.). Ehmann and Ali (1977), in their analyses of 26 bulk samples from 70009–70007 reported a range of 62–85 in the FeO/MnO ratios. This prompted them to suggest a third component with a much lower FeO/MnO ratio in order to account for their generally low FeO/MnO ratios. However, in the samples analyzed by us (Tables 3 and 4), the FeO/MnO ratios are remarkably constant at 80 ± 5 in bulk soils and all size fractions. Thus, we find no evidence of fractionation between FeO and MnO.

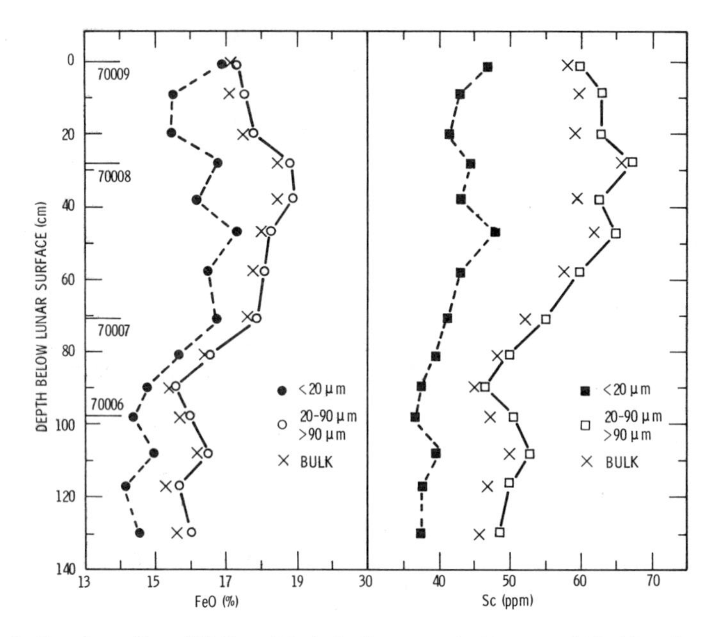

Fig. 8. Depth profiles of FeO and Sc in bulk, coarse (>90 μm and 20–90 μm) and <20 μm fine fractions of 70009–70006. When TiO_2 is high (7–9%, see Fig. 6), FeO and Sc are high; when TiO_2 is low (~5%), FeO and Sc are low, suggesting high Sc is related partly to the high ilmenite content.

The ratio CaO/Al_2O_3 reflects the relative proportion of plagioclase. In bulk soils and size fractions this ratio is constant above and below the coarse-grained layer (unit 2). The coarse layer (unit 2: 70008) has the higher CaO/Al_2O_3 suggesting a significant contribution of mare material (pyroxene + ilmenite + orange glass), in both fine and coarse fractions. The coarse fractions at 28, 38 and 47 cm depths with CaO/Al_2O_3 ratios of 1.05 approach high-Ti mare basalt (CaO/Al_2O_3 = 1.1 to 1.2) and thus appear to be derived almost totally from mare basaltic material. Our qualitative interpretation is consistent with our LIL trace data and with the petrographic studies on the monomineralic component (0.02–2 mm) by Vaniman and Papike (1977b), which suggest 61% pyroxene and 26% plagioclase in the coarse section of 70008, and about 50% pyroxene and 30% plagioclase in sections 70009 and 70007. The Fe/Fe + Mg ratios, which reflect the pyroxene/olivine proportions, do not vary significantly both in the coarse and fine fractions of the entire core 70009–70006, indicating that the relative proportions of pyroxene and olivine are about constant. This is also supported from the modal data in coarse fractions (Table 2).

Large-ion lithophile patterns (K, REE and Th)

The chondritic normalized K, REE and Th (LIL) patterns in bulk soils and >90 μm, 20–90 μm and <20 μm size fractions are displayed in Fig. 9. The LIL patterns of intermediate depths essentially follow the same general trend. Solid circles shown for some of the <20 μm fractions are representative analyses plotted to show the well defined patterns. In other samples, only lines are drawn in to avoid crowding. These patterns reveal the following observations:

1) The steep positive slopes of light REE or La/Sm ratios in the coarse fractions (>90 μm and 20–90 μm) are typical of mare material, whereas the enhanced La/Sm ratio in the fine fraction <20 μm indicates a significant contribution of highland material, consistent with the observation of Korotev (1976).
2) With increasing depth (except in coarse core 70008) the light REE slopes (La/Sm ratio) in the coarse fractions tend to flatten with an increase in highland-type material. The <20 μm fractions also become more enriched in highland material with increasing depth.
3) The absolute concentration of La-Sm is in general higher in the fine than in the coarse fraction, though both fractions tend to merge at or below depths of 118 cm.
4) Both coarse and fine fractions exhibit a significant negative Eu anomaly. The absolute concentration of Eu is consistently higher in the fine fraction, although the Sm/Eu ratio of 1.65 is about the same in both coarse and fine fractions.
5) The K concentration in the fine fraction is consistently higher relative to the coarse fractions. In both fine and coarse fractions K increases with depth in the core, but the relative differences between the two fractions do not change significantly.
6) In the coarse fractions, Th inflection is downward (characteristic of mare material) from the core top to 47 cm depth. Below this depth, it levels off (58 cm depth) and then shows an upward trend (characteristic of highland material). In the fine fractions, Th inflection is always upward (except for layers at 28 and 38 cm where it levels off).

The observations (1) and (2) confirm our earlier conclusion based on stratigraphy of TiO_2 and Al_2O_3 content; i.e., the fine fraction in each core segment is considerably enriched in highland material whose contribution, except in 70008 core, generally increases with depth. The coarse fractions at the top (70009) core are also rich in the mare material and become gradually enriched in highland material with depth. The coarse and fine fractions tend to merge towards the bottom core 70006 (Fig. 9).

Is the highland component anorthositic (anorthosite-anorthositic gabbro) or KREEPy? Korotev (1976) used a 5-component (mare basalt, orange glass, KREEP, anorthositic gabbro and meteoritic) mixing model to show that both low-K KREEP and anorthositic gabbro are required as highland components to satisfy mass balance in the 90–150 μm and <20 μm fractions. As mentioned earlier, we plan to follow a similar multielement mixing model after complete study of the remaining core (70005–70001). However, using our present data on

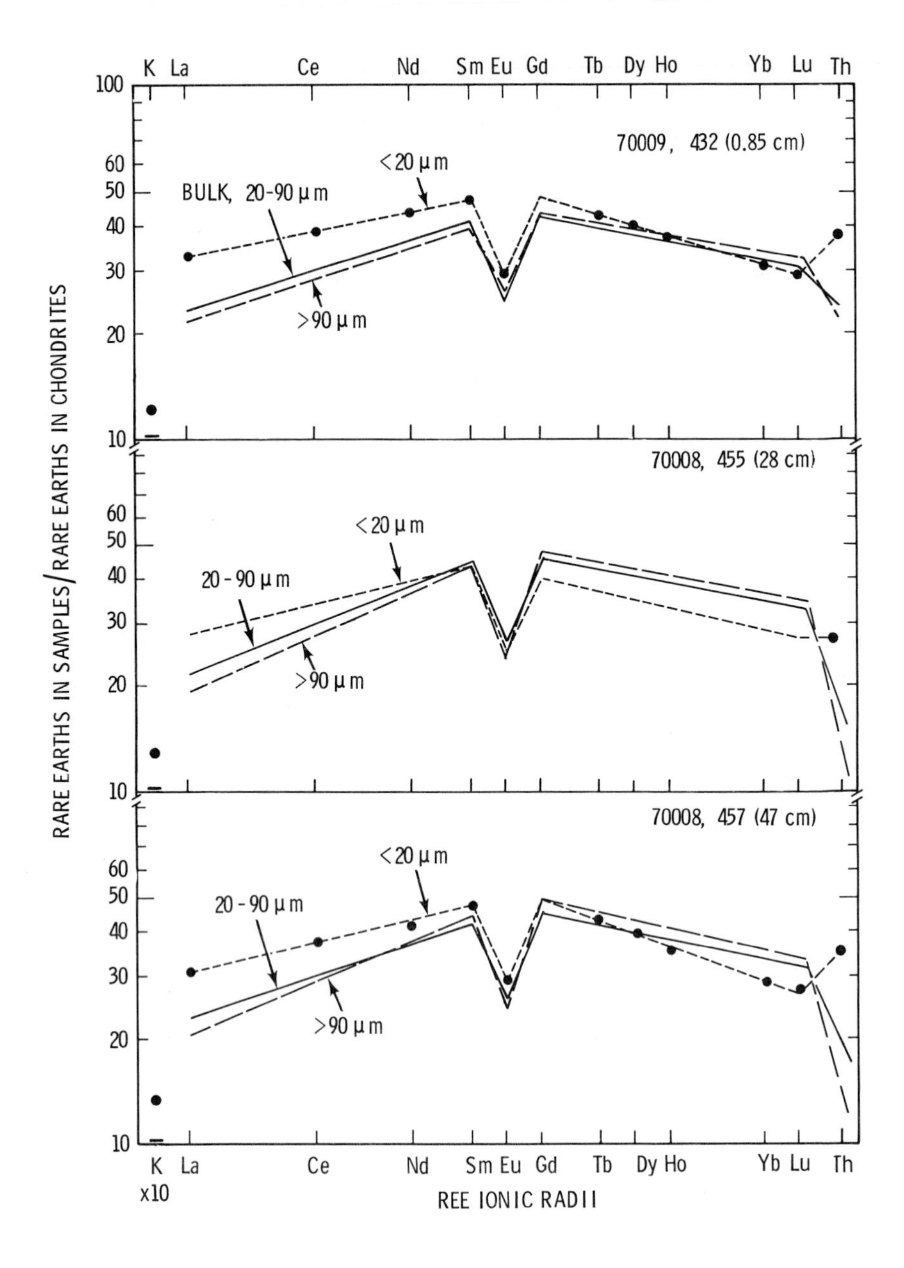
K La Ce Nd Sm Eu Gd Tb Dy Ho Yb Lu Th
100
60
50
40
30
20
10
70009, 432 (0.85 cm)
<20 μm
BULK, 20-90 μm
>90 μm
70008, 455 (28 cm)
<20 μm
20-90 μm
>90 μm
70008, 457 (47 cm)
<20 μm
20-90 μm
>90 μm
RARE EARTHS IN SAMPLES/RARE EARTHS IN CHONDRITES
K x10
REE IONIC RADII

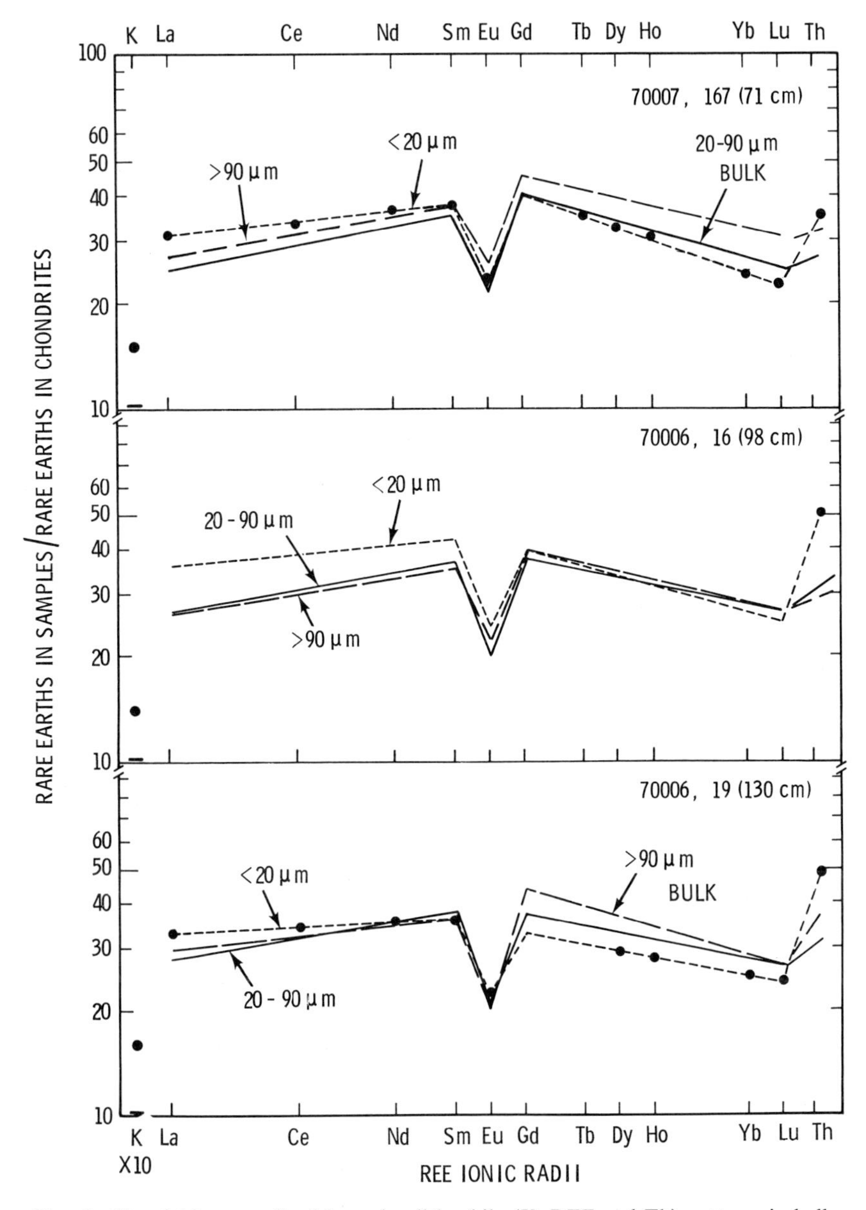

Fig. 9. Chondritic normalized large-ion lithophile (K, REE and Th) patterns in bulk, >90 μm, 20–90 μm and <20 μm size fractions at six depth levels. Solid circles are the experimental points to show well-defined patterns. In other samples, only lines are drawn in to avoid crowding.

La, Sm, Eu and Th patterns we can characterize the dominant highland source and estimate the KREEP contribution in the <20 μm fractions (Table 7). Our conclusions are consistent with Korotev (1976); e.g., our mean value of ~16% low-K KREEP in the <20 μm fraction of 70008 core agrees well with the 16% low-K KREEP estimate obtained by Korotev (1976).

Figure 10 shows the LIL (K, REE and Th) patterns and their ranges in anorthosites 60015,63 and 15415 (97% plagioclase), anorthositic gabbros 76230, 77017 and 78155 (72% plagioclase), mare basalts 70017, 75055 and 70135, med-K KREEP 76055, 72435, 76315, 77135 and 72275, and high-K KREEP (LSPET, 1973; *Proc. Lunar Sci. Conf. 5th* through *8th*, 1974, 1975, 1976, 1977.

These patterns show that relative Th abundance is a strong indicator of KREEP and can easily distinguish KREEPy from mare and anorthositic material. The La/Sm ratios (light REE) can distinguish mare from KREEPy and anorthositic materials. Thus REE concentrations and Sm/Eu ratios play an important role in distinguishing among these components. With these guidelines we can reexamine the LIL patterns (Fig. 9) more closely, particularly in light of Th.

In the top of the core (70009,432), Th in the coarse fraction is depleted relative to Lu (inflection downward) characteristic of mare material, whereas the fine fraction has upward Th inflection characteristic of KREEPy material. The same trend exists in layers at 10 and 20 cm in 70009. In the 28 cm layer of the coarse unit (70008) Th in the coarse fractions is almost characteristic of mare material

Table 7. Low-K KREEP* contribution in <20 μm fine fractions of 70009–70006 cores.

Sample	Depth (cm)	In excess of coarse fraction: Based on Th (%)	In excess of coarse fraction: Based on La (%)	Total value[a] Based on Th (%)
70009,432	0.85	11	11	22
70009,433	9.4	17	12	28
70009,434	20.4	8.0	7.0	17
70008,455	28.0	11	8.4	13
70008,456	38.0	12	8.0	15
70008,457	47.0	15	10	19
70008,458	57.5	6.2	7.8	17
70007,167	70.6	5.8	5.4	21
70007,168	81.0	11	10	23
70007,169	89.6	6.3	4.1	26
70006,16	98.0	14	11	30
70006,17	108	12	11	27
70006,18	117	12	9.9	29
70006,19	130	10	6.0	29

*Based on Th = 130× (chondrites); La = 90×.
[a]Mare indigenous value is taken as 10×.

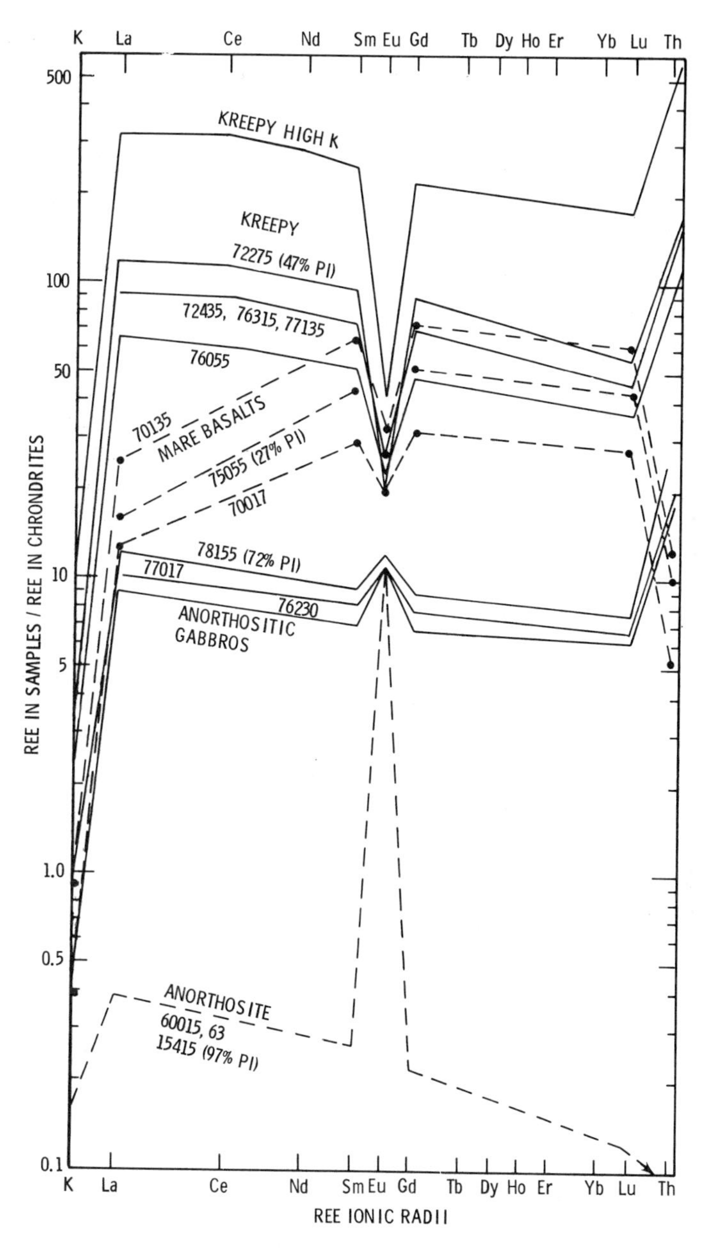

Fig. 10. Chondritic normalized LIL (K, REE and Th) patterns and their ranges in anorthosites 60015,63 and 15415, anorthositic gabbros 76230, 77017 and 78155, mare basalts 70017, 75055 and 70135, med-K KREEP 76055, 72435, 76315, 77135 and 72275, and high-K KREEP. Sources for data are from LSPET (1973) and *Proc. Lunar Sci. Conf. 5th* through *8th*, 1974, 1975, 1976, 1977.

because of its steep downward inflection, but the fine fraction shows virtually no inflection of Th suggesting low contribution of KREEPy material. The same trends are found at 38 cm depth (70008,456). At 47 cm and 58 cm depths, Th in the fine fractions shows slight upward trends while it still has a downward inflection in the coarse fractions. At 70 cm depth (70007,167) Th in both the coarse and fine fractions shows upward trends. These upward trends increase somewhat with depth and level off towards the base of section 70006. Such smooth transitional changes in Th inflections from the top to the bottom of the core suggest that the highland source in these cores (and especially in the fine fraction) is KREEPy instead of anorthositic. A pure anorthositic source can be ruled out because the absolute REE (except Eu) and Th concentrations are far too low to cause any observable change between coarse to fine fractions. If anorthositic gabbro were the sole source it would have to be present in great abundance (≥50%) to affect the magnitude of chemical differences we observe between the coarse and fine fractions. However, this would also affect the Sm/Eu ratios in the fine relative to the coarse fraction contrary to our observation (4): the Sm/Eu ratio is about constant at 1.65 in both the coarse and fine fractions. Based on Eu constraints we estimate about 10% anorthositic gabbroic material in the fine relative to the coarse fraction. This is in reasonable agreement with the estimates of Korotev (1976).

The noritic breccias collected at the Apollo 17 site are mostly low-K KREEP (LSPET, 1973). The low-K KREEP is very similar to low-K Fra Mauro basalts abundant in the highlands (Apollo 14, 15 and 16, and Luna 20 sites). In order to estimate the KREEP contribution in the fine fraction, we use La and Th as indicators for KREEP. Since low-K KREEP shows a range in LIL abundances (Fig. 10), we have simply taken the mean value of La (90X chondritic) and Th (130X) in KREEPy rocks to represent low-K KREEP. The mean Th in mare basalts is taken as 10X (Fig. 10).

We have calculated KREEPy contribution in the fines in two ways: 1) by subtracting the coarse contribution (>90 μm and 20–90 μm) using La and Th as indicators; and 2) by subtracting the mare basalt contribution (Th = 10X chondritic). The latter method yields the total KREEP contribution in the fine fraction, since Th in mare material (basalt + orange glass) is relatively small. The results are tabulated in Table 7. The total low-K KREEP varies from 13–30% in the fine fraction. It is low in the coarse unit core 70008, and increases below 57 cm depth and levels off to 29% at the base of the 70006. The average total KREEP contribution is ~22% in the fine soil fraction. Moreover, the fine fractions contain on the average ~11% (6–17%) more low-K KREEPy material than the coarse soils.

Transport of fine fraction

The highland enrichment in the fine fraction raises other questions. Is the transport of fine fraction a local phenomenon or does it extend from distant localities? Is transport mainly lateral or vertical? Korotev (1976) in his meticulous grain size study of Apollo 17 soils suggested the north and south massifs as the local source for "KREEPy" components added to the valley floor soil as fine-grained material. The soils and boulders from the massifs are feldspathic in composition, and their LIL abundances lie in the range of medium-K and low-K KREEP respectively (LSPET, 1973; *Proc. Lunar Sci. Conf. 4th* through *6th*, 1973, 1974, 1975). The argument for a source area in the massifs is supported by the fact that the KREEP contribution in the coarse (90–150 μm) and fine (<20 μm) fractions from the south massif soil 72150 (Korotev, 1976) show the expected pattern of low-K KREEP enrichment relative to the valley soils.

Korotev (1976) reported a similar highland enrichment in the <20 μm fraction from Apollo 11 soil 10084. Boynton and Wasson (1977) noticed a similar effect in their 20–7 μm and <7 μm fractions of Apollo 15 soil 15100. Their work, coupled with our systematic study of Apollo 17 soils, tends to suggest that the fine fraction from mare soils is in general enriched in highlands (KREEPy) material.

In the case of the Apollo 17 site, the massifs are at higher elevation than the mare valley and have been constantly bombarded by meteorites and micrometeorites with the result that highland material tends to be finer than mare materials on the valley floor. At the Apollo 17 site the elevated massifs provide favorable sites for lateral and downward mixing of highland materials into the dominating mare regolith in the valley of Taurus Littrow. However, the lack of large-scale geochemical gradients at highland/mare contacts argues against efficient lateral transport of highland material across mare areas (Hörz, 1978; Hubbard and Villas, 1977). Cratering calculations of Arvidson *et al.* (1975), and Quaide and Oberbeck (1975) also argue against any significant lateral transport over distances >>10 Km. It is perhaps possible that the scaling laws do not extend to fine fractions. Nevertheless, our arguments for lateral transport of a few Km (~10 Km) is consistent within the constraints of cratering and orbital X-ray calculations.

Depth profile of Zn

Figure 11 shows the depth profile of Zn in <20 μm fine fractions of 70009–70006. For typical size distribution, we have included two histograms; one at the surface and the other at 108 cm depth. Zn is unusually high and its depth profile is quite variable. We have ruled out the possibility of Zn contamination because the calculated mass balance agrees well with bulk soil analyses (Tables 5 and 6). We attribute such high Zn abundances to indigenous mare volatile concentrations (Zn, Cd, Se, etc.).

The source for volatiles appears to be high-Ti mare glasses. The drill core contains a significant amount of high-Ti orange and black glasses (~10–15% of the >20 μm fraction; Vaniman and Papike, 1977b; Taylor *et al.*, 1977a,b). Morgan *et al.* (1974) reported 140 ppm of Zn in orange glass and 45 ppm of Zn in dark glass. About 30% contribution of orange glass will yield 42 ppm Zn, which matches our Zn values in bulk soils. The orange glasses are considered to

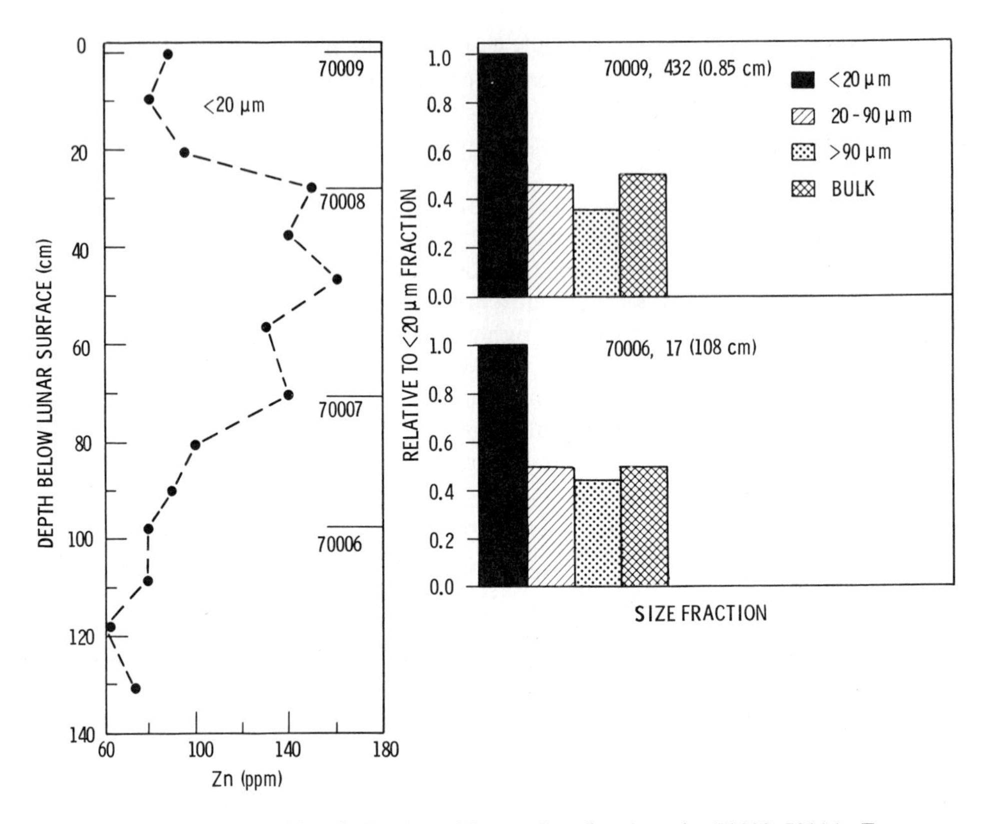

Fig. 11. Depth profile of Zn in <20 μm fine fractions in 70009–70006. Two histograms, one at the surface and the other at 108.5 cm depth show typical distribution of Zn among size fractions.

have been formed as a result of volcanic vent activity (Meyer *et al.*, 1975; Heiken and McKay, 1977). The released volatiles seem to reside on the surface of glass droplets and thus are surface correlated (Boynton and Wasson, 1977; Krähenbühl *et al.*, 1977). Because of the large surface concentration and small grain size, Zn enrichment would be greatly increased in the fine fraction, consistent with our data. The Zn in the >90 μm and 20–90 μm fractions are about the same, suggesting about equal proportions of glasses in both coarse fractions in all segments of the core. This is also supported by our modal data (Table 2). The bulk and coarse fractions contain only 30–50% of the Zn concentration found in the fine fraction. Among the fine fractions, the coarse unit 70008 has much higher Zn content relative to its neighboring cores, implying a high contribution of orange glass in the coarse layer. This is consistent with the petrographic observations of Vaniman and Papike (1977b) and Taylor *et al.* (1977a).

Except for the top 20 cm where Zn is uniform because of recent reworking of surface soil (Crozaz and Plachy, 1976), the depth profile of Zn below 20 cm is the same pattern displayed by FeO and Sc and it is a profile complementary to pattern displayed by Al_2O_3. These features suggest that the original core layering has not been disturbed since the time of emplacement from Camelot Crater and "Central Cluster" craters (LSPET, 1973). It is difficult to establish the exact number of chemical units present since we analyzed relatively few samples at ~10 cm depth intervals.

CONCLUSIONS

This paper is essentially a progress report and presentation of data. Our final conclusions must await completion of our study, including core sections 70005 through 70001. Preliminary conclusions are listed below.

1. The chemistry of the >90 μm and 20–90 μm coarse fractions is identical but quite different from the <20 μm fine fraction.
2. The upper 50 cm of the drill core is highly enriched in mare material (basalt + orange glass). The coarse fractions at 28, 38 and 47 cm depth appear to consist almost totally of mare material. Below the 58 cm depth all size fractions show a gradual decrease of mare material and an increase of highland material. At ~90 cm depth these changes level out.
3. The dominant source of highland material is KREEPy instead of anorthositic. Relative to the coarse fractions, the fine fraction contains about 10% more of this low-K KREEPy material. This KREEPy excess can be best explained by lateral (~10 km) movement of fines from local massifs (consistent with the conclusion of Korotev, 1976).
4. Highland enrichment in the fine fractions of Apollo 17 soils, coupled with similar observations in Apollo 11 soil 10084 (Korotev, 1976) and Apollo 15 soil 15100 (Boynton and Wasson, 1977), suggests that this enrichment may perhaps be moonwide. However, the contributing sources are likely to be local at each site.
5. Indigenous volatiles such as Zn are quite high in all size fractions. The source of volatiles is the high-Ti orange and black glasses which are abundant in the drill core.

Acknowledgments—J. C. Laul is highly indebted to W. C. Richey of Battelle Northwest Laboratory for his assistance in the INAA experiments. He thanks R. A. Schmitt and M.-S. Ma of Oregon State University for the use of their counting equipment for short irradiations. We benefitted from the critical reviews of L. A. Haskin and G. J. Taylor. The assistance of the reactor crew at the OSU TRIGA reactor is appreciated. Si was measured using Dow Chemical's facility at Midland, Michigan, and the assistance of M. Kocsis in Si determinations is appreciated. This work was supported by NASA Grants NAS 9-15357 to J. C. Laul and NSG-9044 to J. J. Papike, which we gratefully acknowledge.

REFERENCES

Arvidson R., Drozd R. J., Hohenberg C. M., Morgan C. J. and Poupeau G. (1975) Horizontal transport of the regolith, modification of features, and erosion rates on the lunar surface. *The Moon* **13**, 67–79.

Baedecker P. A., Chou C. L., Sundberg L. L. and Wasson J. T. (1974) Volatile and siderophilic trace elements in the soils and rocks of Taurus-Littrow. *Proc. Lunar Sci. Conf. 5th*, p. 1625–1643.

Bence A. E. and Albee A. L. (1968) Empirical correction factors for electron microanalysis of silicates and oxides. *J. Geol.* **76**, 382–403.

Bence A. E., Delano J. W., Papike J. J. and Cameron K. L. (1974) Petrology of the highlands massifs at Taurus-Littrow: An analysis of the 2–4 mm soil fraction. *Proc. Lunar Sci. Conf. 5th*, p. 785–827.

Boynton W. V. and Wasson J. T. (1977) Distribution of 28 elements in size fractions of lunar mare and highlands soils. *Geochim. Cosmochim. Acta* **41**, 1073–1082.

Crozaz G. and Plachy A. L. (1976) Origin of the Apollo 17 deep drill coarse-grained layer. *Proc. Lunar Sci. Conf. 7th*, p. 123–131.

Curtis D. and Wasserburg G. J. (1975) Apollo 17 neutron stratigraphy sedimentation and mixing in the lunar regolith. *The Moon* **13**, 185–227.

Duke M. B. and Nagle J. S. (1974) Lunar Core Catalog and Supplements, NASA SP 09252. 242 pp.

Ehmann W. D. and Ali M. Z. (1977) Chemical stratigraphy of the Apollo 17 deep drill cores 70009–70007. *Proc. Lunar Sci. Conf. 8th*, p. 3223–3241.

Flanagan F. J. (1973) 1972 values for international geochemical reference sample. *Geochim. Cosmochim. Acta* **37**, 1189–1200.

Haskin L. A. and Korotev R. L. (1977) Test of a model for trace element partition during closed-system solidification of a silicate liquid. *Geochim. Cosmochim. Acta* **41**, 921–939.

Heiken G. and McKay D. S. (1977). A model for eruption behavior of a volcanic vent in eastern Mare Serenitatis. *Proc. Lunar Sci. Conf. 8th*, p. 3243–3255.

Hörz F. (1978) How thick are lunar mare basalts? (abstract). In *Lunar and Planetary Science IX*, p. 540–542. Lunar and Planetary Institute.

Hubbard N. J. and Vilas F. (1977) From Serenity to Langemak: A regional chemical setting for Crisium. In *Papers Presented to the Conference on Luna 24*, p. 85–88. The Lunar Science Institute, Houston.

Korotev, R. L. (1976) Geochemistry of grain-size fraction of soils from the Taurus-Littrow Valley floor. *Proc. Lunar Sci. Conf. 7th*, p. 695–726.

Krähenbühl U., Grütter A., Von Gunten H. R., Meyer G., Wegmuller F. and Wyttenbach A. (1977) Volatile and non-volatile elements in grain-size fraction of Apollo 17 soils 75081, 72461, and 72501. *Proc. Lunar Sci. Conf. 8th*, p. 3901–3916.

Laul J. C., Hill D. W. and Schmitt R. A. (1974) Chemical studies of Apollo 16 and 17 samples. *Proc. Lunar Sci. Conf. 5th*, p. 1047–1066.

Laul J. C. and Rancitelli L. A. (1977) Multielement analysis by sequential instrumental and radiochemical neutron activation. *J. Radioanal, Chem.* **38**, 461–475.

Laul J. C. and Schmitt R. A. (1973) Chemical composition of Apollo 15, 16 and 17 samples. *Proc. Lunar Sci. Conf. 4th*, p. 1349–1367.

Laul J. C., Vaniman D. T. and Papike J. J. (1978) Chemistry, mineralogy and petrology of seven >1 mm fragments from Mare Crisium. In *Mare Crisium: The View from Luna 24* (R. B. Merrill and J. J. Papike, Eds.), p. 537–568. Pergamon, N.Y.

Laul J. C., Wakita H., Schowalter D. L., Boynton W. V. and Schmitt R. A. (1972) Bulk, rare earth and other trace elements in Apollo 14 and 15 and Luna 16 samples. *Proc. Lunar Sci. Conf. 3rd*, p. 1181–1200.

LSPET (Lunar Sample Preliminary Examination Team) (1973) *Apollo 17 Prelim. Sci. Rep.* NASA SP-330, p. 1–37.

McKay D. S., Fruland R. M. and Heiken G. H. (1974) Grain size and evolution of lunar soils. *Proc. Lunar Sci. Conf. 5th*, p. 887–906.

McCallum J. S., Mathez E. A., Okamura F. P. and Gose S. (1974) Petrology and crystal chemistry of poikilitic anorthositic gabbro 77017. *Proc. Lunar Sci. Conf. 5th*, p. 287–302.

Meyer C. Jr., McKay D. S., Anderson D. H. and Butler P. Jr. (1975). The source of sublimates on the Apollo 15 green and Apollo 17 orange glass samples. *Proc. Lunar Sci. Conf. 6th*, p. 1673–1699.

Miller M. D., Pacer R. A., Ma M.-S., Hawke B. R., Lookhart G. L. and Ehmann W. D. (1974) Compositional studies of the lunar regolith at the Apollo 17 site. *Proc. Lunar Sci. Conf. 5th*, p. 1079–1086.

Morgan J. W., Ganapathy R., Higuchi H., Krähenbühl U. and Anders E. (1974) Lunar basins: Tentative characterization of projectiles from meteoritic elements in Apollo 17 boulders. *Proc. Lunar Sci. Conf. 5th*, p. 1703–1736.

Morris R. V., Gose W. A. and Lauer H. V. Jr. (1978) Maturity and FeO profiles for the Apollo 17 deep drill core (abstract). *Lunar and Planetary Science IX*, p. 766–768. Lunar and Planetary Institute, Houston.

Papers Presented to the Conference on Luna 24 (1977). The Lunar Science Institute, Houston. 213 pp.

Papike J. J., Vaniman D. T., Schweitzer E. L. and Baldwin K. (1978) Apollo 16 drive tube 60009/60010, Part III. Total major element partitioning among the regolith components or chemical mixing models with petrologic credibility (abstract). *Lunar and Planetary Science IX*, p. 862–864. Lunar and Planetary Institute, Houston.

Proc. Apollo 11 Lunar Sci. Conf. (A. A. Levinson, ed.) (1970) Pergamon, N.Y. 2492 pp.

Proc. Lunar Sci. Conf. 2nd (A. A. Levinson, ed.) (1971) Pergamon, N.Y. 2818 pp.

Proc. Lunar Sci. Conf. 3rd (E. A. King Jr., ed.) (1972) Pergamon, N.Y. 3263 pp.

Proc. Lunar Sci. Conf. 4th (W. A. Gose, ed.) (1973) Pergamon, N.Y. 3290 pp.

Proc. Lunar Sci. Conf. 5th (W. A. Gose, ed.) (1974) Pergamon, N.Y. 3134 pp.

Proc. Lunar Sci. Conf. 6th (R. B. Merrill, ed.) (1975) Pergamon, N.Y. 3637 pp.

Proc. Lunar Sci. Conf. 7th (R. B. Merrill, ed.) (1976) Pergamon, N.Y. 3651 pp.

Proc. Lunar Sci. Conf. 8th (R. B. Merrill, ed.) (1977) Pergamon, N.Y. 3965 pp.

Quaide W. L. and Oberbeck V. R. (1975) Development of the mare regolith: Some model considerations. *The Moon* **13**, 27–55.

Takahashi H., Janssens M. J., Morgan J. W. and Anders E. (1978) Further studies of trace elements in C_3 chondrites. *Geochim. Cosmochim. Acta* **42**. In press.

Taylor G. J., Wentworth S., Warner R. D. and Keil K. (1978) Petrology of Apollo 17 deep drill core-II: Agglutinates as recorders of fossil soil compositions (abstract). *Lunar and Planetary Science IX*, p. 1146–1148. Lunar and Planetary Institute, Houston.

Taylor G. J., Keil K. K. and Warner R. D. (1977a) Petrology of Apollo 17 deep drill core: Depositional history based on modal analyses of 70009, 70008, and 70007. *Proc. Lunar Sci. Conf. 8th*, p. 3195–3222.

Taylor G. J., Keil K. and Warner R. D. (1977b) Very low-Ti mare basalts. *Geophys. Res. Lett.* **4**, 207–210.

Vaniman D. T. and Papike J. J. (1977a) The Apollo 17 drill core: Characterization of the mineral and lithic component (Sections 70007, 70008, 70009). *Proc. Lunar Sci. Conf. 8th*, p. 3123–3159.

Vaniman D. T. and Papike J. J. (1977b) The Apollo 17 drill core: Modal petrology and glass chemistry (Section 70007, 70008, 70009). *Proc. Lunar Sci. Conf. 8th*, p. 3161–3193.

Vaniman D. T. and Papike J. J. (1977c) Very low Ti (VLT) basalts: A new mare rock type from the Apollo 17 drill core. *Proc. Lunar Sci. Conf. 8th*, p. 1443–1471.

Vaniman D. T., Papike J. J. and Schweitzer E. (1978) Apollo 16 drive tube 60009/60010. Part II: Petrology and major element partitioning among the regolith components. *Proc. Lunar Planet. Sci. Conf. 9th*. This volume.

Wänke H., Palme H., Baddenhausen H., Dreibus G., Jagoutz E., Kruse H., Spettel B., Teschke F. and Thacker R. (1974) Chemistry of Apollo 16 and 17 samples: Bulk composition late-stage accumulation and early differentiation of the moon. *Proc. Lunar Sci. Conf. 5th*, p. 1307–1355.

Proc. Lunar Planet. Sci. Conf. 9th (1978), p. 2099–2109.
Printed in the United States of America

Natural radioactivity of regolith in Mare Crisium

YU. A. SURKOV and G. A. FEDOSEYEV
V. I. Vernadsky Institute of Geochemistry and Analytical Chemistry
U.S.S.R. Academy of Sciences, Moscow V-334, U.S.S.R.

Abstract—Using data from the Mare Crisium region of the moon recently explored by Luna 24, the reliability of forecasting the composition of surface rocks by remote-control methods is estimated. To determine more accurately the radioactivity distribution pattern across the surface of Mare Crisium, a comparison was made between the orbital and laboratory evaluations of the radioelement content in the regolith of this region. The data on the distribution of radioelements in the soil column brought from Mare Crisium were analyzed within the framework of Th-U and K-U systematics of lunar rocks in the light of the possible regolith formation processes at the Luna 24 landing site.

INTRODUCTION

Comprehensive orbital studies of the lunar surface have shown that, amidst various formations of the eastern part of the moon's visible side, the region of Mare Crisium is of special interest. Its round basin (about 500 km in diameter, with an area inside the crest of the surrounding rim of about 200,000 km^2) with a low bottom is characterized by a considerable gravitational anomaly. Although the overall radioactivity level in the moon's eastern maria including Mare Crisium is lower than in western ones; nevertheless, it exceeds, as a rule, the radioactivity level of surrounding highland regions (Trombka *et al.*, 1977). The delivery of a soil column from the southeastern part of Mare Crisium to earth and its analyses have made it possible to understand the nature of the observed radioactivity level contrasts in these regions.

According to the stratigraphic model of the formation of mare basins, the excavation of the Mare Crisium basin apparently took place about 4,000 million years ago (Tatsumoto *et al.*, 1977) during giant impact events which were accompanied by intensive processes of melting and crystallization of the lunar crust. In the opinion of many researchers, processes linked with the chemical separation of rocks enriched, in particular, by potassium, rare earth elements and phosphorus (KREEP), occurred during this period. Undoubtedly the origin and distribution of the rocks of this type on the moon figure prominently in the general genesis of lunar rocks.

One of the approaches to the solution of these problems is linked with studies into the regional features in the distribution of chemical elements, including radioactive ones (K, U, Th) by which KREEP-rocks are enriched. From this point of view it is of interest to analyze the data of X-ray-fluorescent and γ-ray radiometric orbital measurements conducted over the surface of Mare Crisium and the areas adjacent to it and to compare these data with the results of laboratory analyses of the lunar soil sample brought by Luna 24. Such a comparison will permit the estimation of the reliability of forecasting the type of surface rocks by the maps of elemental composition (and including in particular, the distribution of natural radioactivity levels) for lunar regions covered by the trajectories of orbital stations' flights. The data on the distribution of natural radioelements in regolith of the newly-sampled Mare Crisium region on the moon should also be taken into account while analyzing the K-U systematics of lunar rocks from all sampled regions and in the assessment of the average K, U and Th content for the moon on the whole. If we take into consideration that these values are among the "boundary conditions" for models of the moon's thermal history, then, in the light of the recently revised experimental data on the lunar heat flow (Langseth *et al.*, 1976), it is of interest to compare the estimates of the overall contents of radioelements obtained by various methods.

MARE CRISIUM FROM THE CIRCUMLUNAR ORBIT: ANALYSIS OF X-RAY-FLUORESCENT AND RADIOMETRIC MEASUREMENTS

Some conclusions about the composition of the surface rocks of Mare Crisium, on the territory where the Luna 24 was landed, can be drawn from the known orbital data on X-ray-fluorescent and gamma-spectrometric analysis (Trombka *et al.*, 1977). Table 1 lists the average data on the content of a number of elements and their relationships in the surface layer of regolith in Mare Crisium. For the sake of comparison, data are also given from the elemental mapping of other adjacent eastern maria (Mare Serenitatis, Mare Tranquillitatis and Mare Fecunditatis) and highland areas adjacent to Mare Crisium.

Rather large variations in the chemical element content between individual regions recorded in orbital measurements unfortunately do not allow us to draw unambiguous conclusions about the composition of regolith-forming rocks. A clear-cut difference is recorded only between the rocks of highland and mare regions. However, in Mare Crisium there is a trend towards a rise in the values of Al/Si ratios and towards a decrease in Mg/Si ratios as compared with adjacent

Table 1.

Region	Al/Si	Mg/Si	Fe %	Ti %	K %	U*ppm	K/U
Mare Crisium	0.39 ± 0.11	0.18 ± 0.02	10.5	—	—	—	—
Mare Serenitatis	0.29 ± 0.10	0.26 ± 0.07	10.7	2.6	0.17	0.64	2660
Mare Tranquillitatis	0.35 ± 0.08	0.22 ± 0.04	12.1	2.9	0.11	0.53	2080
Mare Fecunditatis	0.36 ± 0.09	0.23 ± 0.05	10.1	2.2	0.125	0.52	2400
Highland south-west of Mare Crisium	0.51 ± 0.10	0.23 ± 0.05	9.3	0.8	0.12	0.39	3080
Highland south-east of Mare Crisium	0.54 ± 0.09	0.22 ± 0.03	8.6	—	0.10	0.36	2780

*Th/U derived from Th data based on a ratio of about 3.6.

maria, which gives some grounds to suppose that the composition of surface rocks is not quite typical of mare basalts.

For the approximate determination (from the data of elemental mapping) of the type of rocks prevailing in this or that region let us make use of the classification suggested by Gleadow *et al.*, (1974) and Lovering and Wark (1975). According to this classification, lunar surface rocks, depending on the chemical composition, can be subdivided into five basic groups (Table 2) while mare basalts (group II) are subdivided into two subgroups (with the high and the low titanium content). A rather sharply pronounced correlation of the content of natural radioelements with the above-mentioned groups of rocks has been noted by Lovering and Wark (1975). It finds its expression in an increase of the K and U content in the rocks with the transfer from group I to group II, from group II to group III, etc. For the convenience of comparisons of the tabulated materials, the data given in Table 2 are modified by the authors with due account of the specific features of orbital measurements (the values of Al and Mg ratios to silicon and the concentrations of Fe, Ti, K and U are taken in weight per cent).

Passing over to analysis of the data referring to the surface rocks of Mare Crisium, we should take into consideration two additional facts: first, orbital photography in this region has been conducted only over its southern and southwestern part; and, second, until recently we have had at our disposal an incomplete set of the numerical data for this region as compared to other areas (Trombka *et al.*, 1977) (see Table 1).

The comparison of the data available in Tables 1 and 2 leads one to the conclusion that the surface rocks of Mare Crisium can belong to the rocks of only two of the groups considered—group II or group IV, and by the Al/Si and Mg/Si ratios they seem to come even closer to group IV of highland basalts enriched by KREEP-components. Although on the one hand an analysis of the distribution of the radioactivity on the flight paths of Apollo 15 and 16 performed by Trombka *et al.* (1977) characterized the region of Mare Crisium as

Table 2.

	Group I	Group II		Group III	Group IV	Group V
Elements	Anorthosites and gabbro-anorthosites	Mare basalts		Highland basalts		
		Rich in Ti	Poor in Ti	Poor in KREEP	Rich in KREEP	Monzonites and granites
Al/Si	0.75 ± 0.09	0.29 ± 0.03	0.28 ± 0.07	0.55 ± 0.095	0.39 ± 0.09	0.29 ± 0.07
Mg/Si	0.14 ± 0.07	0.27 ± 0.06	0.25 ± 0.065	0.27 ± 0.09	0.21 ± 0.08	0.07 ± 0.055
Fe %	2.8 ± 1.4	14.4 ± 1.1	14.2 ± 2.5	5.4 ± 1.5	6.4 ± 2.1	2.2 ± 1.3
Ti %	0.14 ± 0.15	8.0 ± 1.1	1.3 ± 0.7	0.50 ± 0.28	1.0 ± 0.48	0.36 ± 0.12
K %	0.046 ± 0.30	0.060 ± 0.026	0.062 ± 0.037	0.20 ± 0.14	0.89 ± 0.50	4.78 ± 0.66
$U.10^{-4}$%	0.130 ± 0.136	0.131 ± 0.050	0.22 ± 0.15	1.19 ± 0.77	3.5 ± 1.3	8.0 ± 3.2
K/U	$(3.5 \pm 4.2)10^3$	$(4.4 \pm 2.7)10^3$	$(2.8 \pm 2.5)10^3$	$(1.7 \pm 1.6)10^3$	$(2.5 \pm 1.7)10^3$	$(6.0 \pm 2.5)10^3$

to how most contrast one in comparison with surrounding highlands, on the other hand, Metzger *et al.* (1977) draw the unexpected conclusion that this region has the lowest thorium content among the large eastern maria.

In the light of these facts the interest increases in the data on a direct radiometric analysis of lunar soil from Mare Crisium.

RADIOACTIVE ELEMENTS IN THE LUNAR SOIL COLUMN FROM MARE CRISIUM

At present the data on the natural radioelement content in the Luna 24 lunar soil column have been obtained by two methods: nondestructive gamma-spectrometry with the use of a semiconductor detector and a neutron-activation analysis.

In the first case (Surkov *et al.*, 1977b) the assessment of the overall level of the natural radioelement content at the probe's landing site was made. With this aim in view the average soil sample was measured. It weighed 5 g and actually consisted of ten individual specimens weighing about 0.5 g and selected throughout the depth of the column. The following values were obtained for the K, Th and U contents: 0.035 ± 0.01 (weight %), 1.35 ± 0.35 ppm and 0.4 ± 0.1 ppm, respectively. It should be noted that data given by Surkov *et al.* (1977b) for thorium coincide with the corresponding "orbital" value of 1.4 ppm recently obtained by Metzger *et al.* (1977) for regolith of Mare Crisium by processing a statistically low amount of orbital gamma-spectra information, which earlier had not been considered. Apparently the level of the radioactive element content at the Luna 24 landing site is much lower than the radioactivity level (0.89 ± 0.50 weight % for potassium and 3.5 ± 1.3 ppm for uranium; see Table 2) of highland basalts enriched by KREEP. Consequently, in establishing the main type of the surface rocks in Mare Crisium, the rocks of group IV can be excluded from the analysis. If we add to the above-mentioned analysis the data of ground-based optical-instrumental measurements of lunar spectral reflection which provided an approximate assessment of the overall TiO_2 content in mature mare soils (for Mare Crisium this value is 2 ± 0.5 weight %) (Charette *et al.*, 1974), we can make the final choice in the example of interest to us between two subgroups of group II rocks in favour of Ti-depleted basalts (Table 2).

It is worth noting the resulting conclusion to the effect that the surface rocks of Mare Crisium are represented mainly by mare basalts with the low titanium content which coincides with data of direct petrographic and chemical analyses of Luna 24 samples generally determined as alumina mare basalts with low Ti and alkali content (Barsukov *et al.*, 1977; Tarasov *et al.* 1977).

Let us now consider the data from a neutron-activation analysis (Surkov *et al.*, 1977a), (Barsukov *et al.*, 1977) (Fig. 1 a, b) for the content of Th, U and K in the fine-grained fraction of regolith (dark circles) and in individual fragments of magmatic rocks (circles with points) from selectively taken specimens. Figure 1 compares these data and also Th/U and K/U ratios with the depth distributions of radioelements in soil columns selected by Apollo 15 and Apollo 17 expeditions (Fruchter *et al.*, 1975).

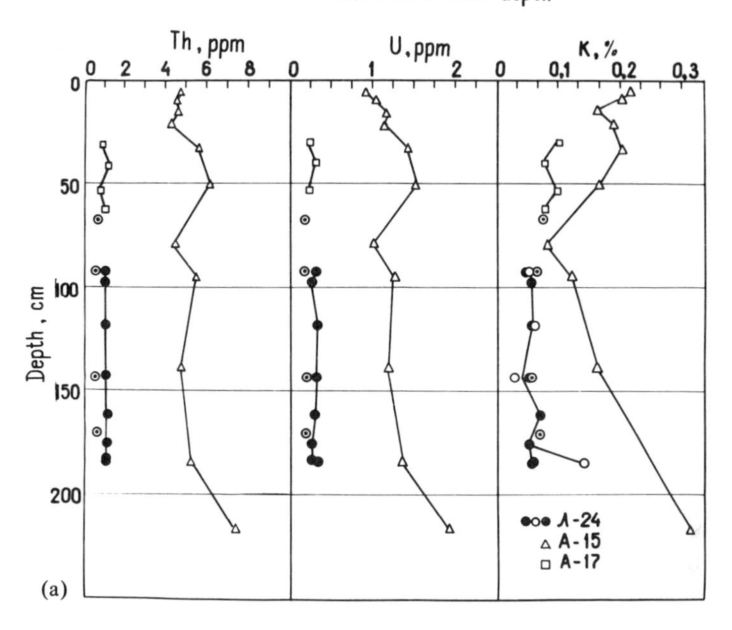

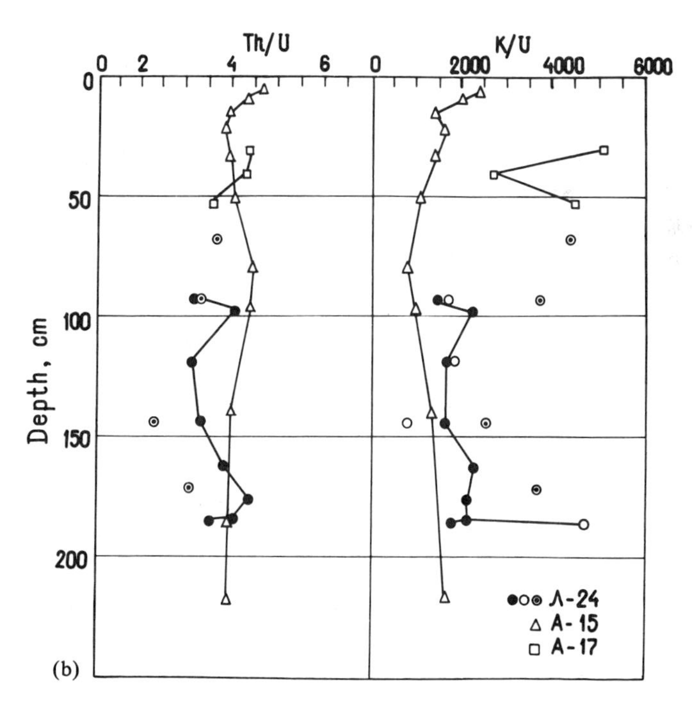

Fig. 1. (a) The distribution of radioelements (the Th, U and K content, ppm) and (b) variations of their ratios (Th/U and K/U by the depth of the Luna 24, Apollo 15 and Apollo 17 soil columns. (The data for the content of K are denoted by unhatched circles is determined by the method of flame photometry).

As is evident from the figure, the entire Luna 24 regolith column is characterized by the rather stable U and Th content (at the level of 0.3 and 1.0 ppm, respectively) with the noticeable variation of potassium contents (0.025–0.141%) by the depth (in particular, the high content of K_2O-0.17% at 184.4 cm mark). One cannot escape noticing the similarity in the character of the distribution in depth and the level of the Th, U and K content in the Luna 24 and Apollo 17 regolith columns in contrast to the Apollo 15 column where four-fold variations in potassium concentrations with depth are recorded (0.077–0.312%) and approximately two-fold variations of uranium (0.90–1.89 ppm) and thorium (4.33–7.35 ppm). While at least some soil layers from the Apollo 15 column are supposed to be enriched by the KREEP-component due to the deposition of ejecta from the craters Aristilles or Autolycus, in the analyzed specimens of the Luna 24 and Apollo 17 columns which have considerably less radioelement content, the content of KREEP is, in all probability, very low. Nevertheless, as evident from the figures, a slight, but regular excess of the Th and U contents in the fine-grained fraction of Luna 24 regolith, as compared to their content in the specimens of magmatic rocks, is due to admixtures of KREEP-components. In this respect the comparison of magmatic rocks and fine-grained fractions of regolith by the potassium content in them is less meaningful due to considerable losses of alkaline metals from mare regolith in the evaporation processes under the effect of exogenic factors (Ivanov, 1977).

The rather monotonous course of the radioelement concentration distribution curves, especially in the upper part of the Luna 24 column (approximately up to the 150 cm mark from the surface), and appreciable variations in the potassium content in deeper layers coincide with the conclusions from the stratigraphic description of the column (Rode *et al.*, 1977) about the homogeneous character of the soil matter in its upper zones and the sharply pronounced stratified nature of its lower horizons. K. P. Florensky *et al.* (1977) revealed the possible causes of such a structure of regolith. In particular, a geological-morphological analysis of the Luna 24 landing site (40 km from the ledge of highland mountains and 18 km from the crater Fahrenheit formed 0.5 to 1.10^9 years ago) points to the possibility that material from the adjacent highland and deep parts of the lava filling of the mare got into the regolith column. At the same time, petrographic studies of the Luna 24 sample (Tarasov *et al.*, 1977) show the extremely limited quantity of highland matter in the soil column. This fact, along with the above-mentioned considerations that KREEP-components are represented in Luna 24 soil sample only as admixtures, are decisive for understanding the pattern of the radioelement distribution in this region—the absence of noticeable contrasts with the overall rise in the radioactivity level over the mare region as compared to the highland area. Such an idea agrees with the general ratios of the content of natural radioelements in the series of rocks (Table 2) when the transfer takes place from anorthosites ($K \sim 0.046\%$, $U \sim 0.13 \cdot 10^{-4}\%$) to mare basalts with low titanium content ($K \sim 0.06\%$, $U \sim 0.22 \cdot 10^{-4}\%$).

Th-U and K-U Systematics of Lunar Specimens

An analysis of the natural radioelement content in lunar rocks is usually conducted using the ratios of K to U and Th to U.

The Th-U ratio is used much more rarely since both U and Th display, as a rule, rather low fractionation in lunar materials without deviating considerably from the Th/U range of ratios of 3.6–3.8. Most meteorites and terrestrial rocks are characterized by these values (Surkov and Fedoseyev, 1977). However, lately, while discussing the problems of lunar petrogenesis, researchers have begun to give much more attention to Th-U fractionation.

In collections of the last Apollo expeditions lunar anorthosite rocks with low Th/U ratios in the range of 1–3 were detected by Laul and Fruchter (1976). It has been found out that all the basalts from Apollo 17 collections also have low Th/U ratios (in the 2.1–3.5 range). As it has been shown by subsequent more careful studies, in Apollo 11 basalts poor in potassium the values of Th/U ratios are approximately 10% lower than in basalts enriched by potassium of the same series of specimens. Further studies led to the conclusion that Th will fractionate with respect to U in the crystallization process due to the difference in their separation coefficients, primarily for such an important rock-forming mineral as clinopyroxene (Silver, 1977).

As is evident from the data of Surkov *et al.* (1977a) and Barsukov *et al.* (1977) three of four measured fragments of the magmatic rocks from the Luna 24 column are also characterized by reduced Th/U values up to the minimum value of 2.3 in case of the basaltic fragment selected from a depth of 143 cm. It has been noted by Silver (1977) that in minerals, Th-U fractionation is, as a rule, masked when the KREEP-component is noticeably represented in the rock. However, at low levels of the natural radioelement content, as can also be seen in Luna 24 rocks, Th-U fractionation becomes especially sharply pronounced. No doubt special attention should be given to this fact in further lunar studies, particularly in the light of the low radioactivity level over highland areas on the moon's other side obtained as a result of orbital gamma-measurements. At any rate, the Th/U ratio of 4.0 assumed by some authors, for instance Tera *et al.* (1974), as the average value for the moon, is specifically in the light of the latest Apollo 17 and Luna 24 data, certainly overestimated.

The K-U systematics of rocks more wide-spread in geochemical practice has found its application to lunar studies in elucidating the problems linked with the thermal history and the chemical fractionation of elements in the course of the differentiation of lunar matter (by the ratio of its low-temperature and high-temperature fractions). In particular, with the aid of this systematics attempts are being made to evaluate the average potassium and uranium content and the corresponding K/U ratio for the moon on the whole (Surkov and Fedoseyev, 1977; Toksöz and Johnston, 1977).

Figure 2 gives data in K-U coordinates for lunar specimens from all nine explored areas, adding the results of analysis of the fragments of rocks from the Luna 24 column. A detailed representation of correlation lines plotted on the basis of regression analysis of various selections characterizing specimens from various regions of the moon can be found in Surkov *et al.* (1978). There it has been pointed out that the region of the intersection of the analyzed correlation lines characterizes the matter of the moon's upper mantle by the K and U content if we suppose the complementary character of the composition of lunar rocks corresponding to them.

In Fig. 2, the addition of the link representing the specimens of primary Luna 24 rocks (specimens of basalt 24067.4–0.12 with K/U of 4350; dolerite 24092, 4–001 with K/U of 3690; gabbro 24190, 1–6.2 with K/U 3670; the basalt specimen 24143. 4–006.1 with K/U of 2550 stands somewhat aloof), supplements the third chain of rocks displaying increased K/U ratios. It is worth noting that the coordinates of the intersection point of two extreme correlation lines (I and III), which most differ from each other by their slopes, correspond to the values of about 0.05 ppm for uranium and about 140 ppm for potassium (for a K/U ratio of 2800). This result agrees rather well with the revised data (0.043 ± 0.006 ppm for uranium) proceeding from the results of direct measure-

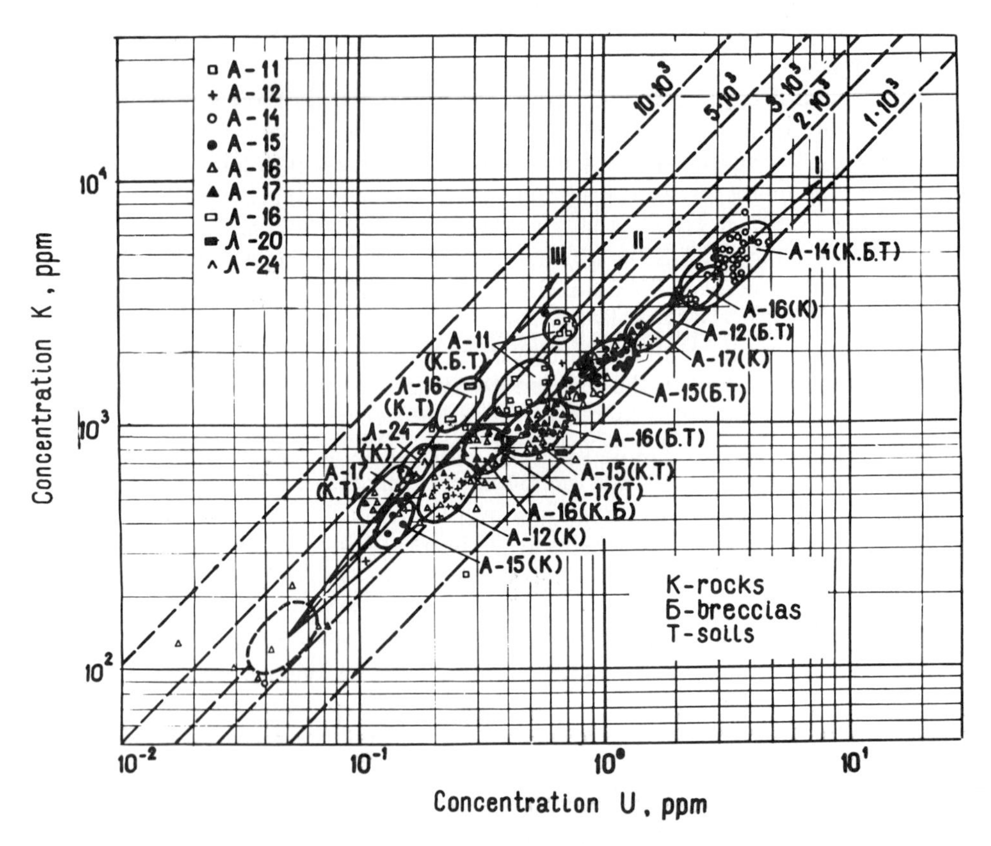

Fig. 2. The potassium-uranium systematics of lunar samples. (K-crystalline rocks, B-breccias, T-fine regolith fractions).

ments of the heat flow on the lunar surface (Langseth *et al.*, 1976).

In this connection it is interesting to compare once again the values of the average values of U and K characterizing the interior of the earth and the moon. In one of our papers dealing with a gamma-spectrometric analysis of lunar soil from the Luna 16 column (Surkov *et al.*, 1972), we have concluded that there is a similarity between average K/U ratios (about $2.8 \cdot 10^3$) for the earth and the moon on the whole, evidence for the kinship of their primary non-differentiated matter.

However, the calculated values of the overall uranium and potassium content on these neighbouring celestial bodies seemed to differ considerably. That is why hypotheses were put forward to the effect that the lunar body could form in an independent circumsolar orbit with its subsequent capture by the earth. In further estimates of the average potassium and uranium content on the moon obtained by various methods, as new information came in, there were trends towards reducing the supposed level of these average contents, and the latest data for K of about 100 ppm (Taylor, 1977) to 140 ppm and U of about 0.020 ppm (Hubbard and Minear, 1976) to 0.050 ppm are considerably lower than the initial data (K of about 250 ppm and U of about 86 ppm).

At present we can state the coincidence of K/U ratios and, as is evident from Table 3, a very close similarity for the values of the average bulk potassium and uranium content in the earth and the moon.

In its turn, this seems to testify quite convincingly in our opinion to the genetic kinship of the matter which formed these two planetary bodies in the course of accretion. This parental matter was equally enriched (by three or four times) by refractory elements and depleted (about four-fold) as far as volatile elements are concerned with respect to the matter of carbonaceous chondrites (C1).

Table 3.

	U, ppm	K, ppm
Earth	0.031 to 0.035[1]	90 to 170[2]
Moon	0.037 to 0.050[3]	100 to 140[4]
Chondrites Cl	0.011[5]	510[5]

[1]The average U content is obtained on the basis of the data on the thermal current at the earth's surface (MacDonald, 1964; Tera *et al.*, 1974).

[2]The limits of the K content are assessed for K^{40} on the basis of the data on the Ar^{40} content in the earth's atmosphere (Hurley, 1968; Larimer, 1971).

[3]The average content is obtained on the basis of the data on the thermal current at the moon's surface (Langseth, 1976) and from K-U systematics of lunar rocks (the present paper).

[4]The average K content is determined from the K/U ratio (Taylor, 1977) and from the K-U systematics of lunar samples (this paper).

[5]Data from Surkov and Fedoseyev (1977).

REFERENCES

Barsukov V. L., Dmitriev L. V., Tarasov L. S., Kolesov G. M., Shevaleevsky I. D., Ramendik G. I. and Garanin A. V. (1977) Geochemical and petrochemical peculiarities of regolith and lithic clasts from Mare Crisium: preliminary data. *Geokhimia* **10**, 1477–1487.

Charette M. P., McCord T. B. and Pieters C. (1974) Application of remote spectral reflectance measurements to lunar geology classification and determination of titanium content of lunar soils. *J. Geophys. Res.* **79**, 1605–1613.

Florensky K. P., Basilevsky A. T. and Pronin A. A. (1977) Geologic setting of the vicinity of Luna 24 landing site, southeast of Mare Crisium. *Geokhimia* **10**, 1449–1464.

Fruchter J. A., Rancitelli L. A. and Perkins R. W. (1975) Primordial radionuclide variations in the Apollo 15 and 17 deep core samples and in Apollo 17 igneous rocks and breccias. *Proc. Lunar Sci. Conf 6th*, p. 1399–1406.

Gleadow A. J. W., Le Maitre R. W., Sewell D. K. B. and Lovering J. F. (1974) Chemical discrimination of petrographically defined clast groups in Apollo 14 and 15 lunar breccias. *Chem. Geol.* **14**, 39–61.

Hubbard N. J. and Minear J. W. (1976) Petrogenesis in a modestly endowed moon. *Proc. Lunar Sci. Conf. 7th*, p. 3421–3435.

Hurley P. M. (1968) Absolute abundance and distribution of Rb, K and Sr in the Earth. *Geochim. Cosmochim. Acta* **32**, 273–284.

Ivanov A. V. (1977) On the intensity of sodium vaporization from the mare regolith (abstract). In *Lunar Science VIII*, p. 496–498. The Lunar Science Institute, Houston.

Langseth M. G., Keihm S. J. and Peters K. (1976) The revised lunar heat flow values. *Proc. Lunar Sci. Conf. 7th*, p. 3143–3171.

Larimer J. W. (1971) Composition of the earth: Chondritic or achondritic. *Geochim. Cosmochim. Acta* **35**, 769–786.

Laul J. C. and Fruchter J. S. (1976) Thorium, uranium and potassium systematics in Apollo 17 basalts. *Proc. Lunar Sci. Conf. 7th*, p. 1545–1559.

Lovering J. F. and Wark D. A. (1975) The lunar crust-chemically defined rock groups and their potassium-uranium fractionation. *Proc. Lunar Sci. Conf. 6th*, p. 1203–1217.

Mac Donald G. J. F. (1964) Dependence of the surface heat flow in the radioactivity of the earth. *J. Geophys. Res.* **69**, 2933–2946.

Metzger A. E., Haines E. L., Parker E. R. and Radocinsky R. G. (1977) Thorium concentrations in the lunar surface. I: Regional values and crystal content. *Proc. Lunar Sci. Conf. 8th*. p. 949–999.

Rode O. D., Ivanov A. V., Tarasov L. S. and Korina M. I. (1977) General lithologic-morphologic characteristics of Luna 24 regolith core. *Geokhimia* **10**, 1465–1476.

Silver L. T. (1977) Implication of major mineral fractionation of thorium and uranium in some lunar materials (abstract). In *Lunar Science VIII*, p. 809–811. The Lunar Science Institute, Houston.

Surkov Yu. A., Kolesov G. M. and Ivanov I. N. (1977a) Neutron-activation analysis of lunar soil returned by Luna 24 from Mare Crisium. *Proc. 4th All-Union Conf. on Neutron-Activation Analysis, Tbilisi*, p. 75–76.

Surkov Yu. A., Moskalyova L. P., Kharyukova V. P., Manvelyan O. S., Dudin A. D. and Trofimov V. I. (1977b) Preliminary analysis of radioactivity of lunar soil returned by Luna 24 probe from Mare Crisium. *Geokhimia* **10**, 1510–1515.

Surkov Yu. A. and Fedoseyev G. A. (1977) Radioactivity of the moon, planets and meteorites. In *The Soviet-American Conference on the Cosmochemistry of the Moon and Planets*, p. 201–218. NASA SP-370. Washington, D.C.

Surkov Yu. A., Fedoseyev G. A. and Sobornov O. P. (1972) Gamma-spectrometric analysis of lunar samples from Luna 16. *Kosm, Issled.* **10**, 938–942.

Surkov Yu. A., Fedoseyev G. A. and Sobornov O. P. (1978) Investigation of the radioactivity of lunar soil returned from region of crater Apollonium-C. Soil from the moon's highland. Nauka, Moscow. In press.

Tarasov L. S., Nazarov M. A., Shevaleevsky I. D., Kudryashova A. F., Gaverdovskaya A. S. and Korina M. I. (1977) Rock types and mineral chemistry of lunar soil from Mare Crisium. *Geokhimia* **10**, 1488–1509.

Tatsumoto M., Nunes P. D. and Unruh D. M. (1977) Early history of the moon: implications of U-Th-Pb and Rb-Sr systematics. In *The Soviet-American Conference on Cosmochemistry of the Moon and Planets*, p. 507–523. NASA SP-370. Washington, D.C.

Taylor S. R. (1977) Geochemical constraints on the composition of the moon (abstract). In *Lunar Science VIII*, p. 855–857. The Lunar Science Institute, Houston.

Tera F., Papanastassiou D. A. and Wasserburg G. J. (1974) Isotopic evidence for a terminal lunar cataclism. *Earth Planet. Sci. Lett.* **22**, 1–22.

Toksöz M. N. and Johnston D. H. (1977) The evolution of the moon and the terrestrial planets. In *The Soviet-American Conference on Cosmochemistry of the Moon and Planets*, p. 295–327. NASA SP-370. Washington, D.C.

Trombka J. I., Arnold J. R., Adler I., Metzger A. E. and Reedy R. C. (1977) Lunar elemental analysis obtained from the Apollo gamma-ray and x-ray remote sensing experiment. In *The Soviet-American Conference on Cosmochemistry of the Moon and Planets*, p. 153–182. NASA SP-370. Washington, D.C.

Proc. Lunar Planet. Sci. Conf. 9th (1978), p. 2111–2124.
Printed in the United States of America

Some features of the main element conditions in surface layers of the regolith particles of the Luna automatic stations samples: X-ray photoelectronic spectroscopy studies

YU. P. DIKOV, O. A. BOGATIKOV

Institute of Ore Deposits Geology, Petrography, Mineralogy and Geochemistry
U.S.S.R. Academy of Sciences, Moscow, U.S.S.R.

V. L. BARSUKOV, K. P. FLORENSKY, A. V. IVANOV

V. I. Vernadsky Institute of Geochemistry and Analytical Chemistry
U.S.S.R. Academy of Sciences, Moscow, U.S.S.R.

V. V. NEMOSHKALENKO, V. G. ALYOSHIN

Institute of Metallophysics, Ukrainian S.S.R. Academy of Sciences, Kiev, U.S.S.R.

M. G. CHUDINOV

State Institute of Nitrogen Industry, Moscow, U.S.S.R.

Abstract—The reduced form of elements and structural-chemical parameters of surface layers of particles from the fine fraction of regolith samples of Luna missions have been studied by X-ray photoelectron spectroscopy using the step-wise etching by argon ions. The presence of iron, titanium, silicon, and aluminum reduced to elemental forms has been revealed on the surface of particles in all samples. The content of these reduced forms decreases toward the interior of particles. From structural-chemical parameters of the particles studied, one can deduce that their surfaces are covered by amorphous films of different thicknesses. The revealed modifications of surface characteristics of regolith particles is mainly due to the *in situ* action of solar wind hydrogen.

INTRODUCTION

Surface layers of the lunar regolith are subjected to the long-term influence of the exogenous agents—micrometeoritic bombardment and solar and galactic irradiation. The action of these agents can result in alteration of the chemical composition and structural condition of the matter. The uppermost surficial parts of regolith particles would be subjected to the most radical alteration.

At present, X-ray photoelectron spectroscopy is one of the most effective methods for analyzing the electron structure of solid bodies and the chemical state of atoms in them (Siegbahn *et al.*, 1971; Nemoshkalenko and Alyoshin, 1976). Using this method it is possible to determine the charge state, type of chemical bond and chemical composition of the matter under study. This method allows the study of thin surface layers of matter because the effective depth of electron release is about 20–30 Å.

The first X-ray photoelectronic spectroscopy studies of regolith samples from the lunar mare regions (Vinogradov *et al.*, 1972a,b; Nefedov *et al.*, 1972) have revealed that, compared with the bulk of matter in the regolith particles, the surface layers of the particles are characterized by a number of peculiarities such as the presence of fine-dispersed metallic iron and a notable chemical distinction of these layers—silicon enrichment and calcium depletion. These results were confirmed by other investigators (Housley and Grant, 1975, 1976, 1977; Baron *et al.*, 1977).

Changes in concentrations of some other elements at the surface of some samples has also been revealed. Etching of the regolith particles surfaces by argon atoms has allowed Housley and Grant (1977) to determine the character of concentration changes of some elements and of reduced iron with depth. The concentration changes have been traced to the depth of hundreds of angstroms.

Having analyzed a fine fraction of regolith samples by X-ray photoelectron spectroscopy, Dikov *et al.*, (1977a,b,c) have also determined the presence of elemental state titanium, silicon and aluminum, at the surface of the regolith particles in addition to metallic iron, and have traced the character of their concentration changes with depth using surface etching by argon ions.

The present paper deals with the continuing study of the lunar regolith particles surfaces by use of X-ray photoelectronic spectroscopy.

Research Technique

The study was carried out with an X-ray electron spectrometer VIEE-15. The K-magnesium radiation with a photon energy of 1253.6 ev was used for the excitation of photoelectrons. An electron bond energy for narrow lines was determined with an accuracy of ±0.1 ev; the resolution was not worse than 1.2 ev. The spectrometer vacuum was $3 \cdot 10^{-7}$ torr. The I_s-carbon line was used for spectrum calibration; the value of 285.0 ev was taken as the electron bond energy.

A coating of undisturbed sample was applied to the fine-ribbed aluminum cylinder so that the sample completely covered the surface of the cylinder. The effectiveness of this method of coating was checked by the possibility of "lighting" of the backing through the cover of samples that did not contain aluminum. The coatings of some samples were applied to an adhesive tape.

The etching of the sample surfaces was carried out by argon ions with an energy of 0.7 ev, current 6 ma, argon pressure in the sample preparation camera 0.14 torr. A glass specimen electrochemically modified by copper ions was used for an etching depth calibration. It was found that the rate of successive etching of the outer layer of the glass up to the appearance of a copper signal was 170 Å/min under the etching conditions stated above. The etching rate of powdery specimens of lepidocrocite with adsorbed layer of Co, Ni and Mn of a fixed thickness was 150–180 Å/min. On the basis of these data, the average rate of the etching was taken as 170 Å/min.

Results

It is known that the intensity reworking of various grain size fractions of regolith under the influence of exogenous processes increases with decreasing size. That is why we have selected for study the fine fraction of the regolith delivered by the Soviet automatic stations of the Luna series. For Luna 16 and Luna 20 we examined the upper sections of the cores (samples 1603-1 and 2001-1 respectively, the <83 μm fraction). The Luna 24 core characteristic of notable stratification was studied in more detail: the <74 μm fractions of four base samples (24092, 4-1; 24118-4-1; 24143, 4-1; 24184, 4-1), were examined as well as the fine dust collected from the core-barrel tape in the upper part of the column at the depth interval 21–47 cm (sample 24021,27). The sample weights were 30–50 mg.

X-ray and photoelectron spectra of inner-shell electrons of the main elements—Si, Al, Fe, Ti, Ca, Mg, Na, K and O were studied in all samples. The 2p and 3p-levels in iron, the 2p and 2s-levels in titanium, the 1s-level in sodium and oxygen, and the 2p-level in the rest of the elements were examined.

At the present stage of study of lunar regolith samples by X-ray photoelectron spectroscopy, most attention has been paid to investigation of reduced forms of elements and structural condition of silicate matrix in the surface layers of regolith particles.

Reduced Forms of Elements

Rather clearly resolved peaks (mainly for iron) or well reproducible shoulders of various intensities are revealed on the low-energy side of the main spectral lines corresponding to an oxidized state of an element in examining the 2p-electrons lines of Fe, Ti, Si and Al in all regolith samples (Fig. 1). A computer decomposition of the original spectra has revealed the existence of additional peaks with bond energies very close to those found for corresponding elements in non-oxidized form. On this basis we identify the additional peaks as belonging to reduced forms of corresponding elements.

An evaluation of the relative content of the reduced forms was carried out using the ratio between the areas under lines of reduced and oxidized forms of elements (Table 1). It is significant that the relative content of the reduced forms $\frac{(Me^{\circ})}{Me_{total}}$ as a rule corresponds to the sequence: Fe, Ti, Si, Al, which in general corresponds to the increase of the Me-O bond energy. At the same time the relative contents of reduced forms of elements at surfaces of regolith particles of the various samples fluctuate within considerable limits: Fe- from 3 to 16%, Ti- from 1 to 10%, Si- from 2 to 5%. The content of Al° was not evaluated because of the nearness of energy positions of the Al° and Al^{3+} lines. The Me°/Me_{total} ratios cited by us earlier (Dikov *et al.*, 1978) were calculated only on the basis of formal conditions of spectra separation in the Lorenz approximation without considera-tion of the distorting influence of the high energy part of the spectrum. In the present paper the reduced forms contents were estimated using a correction for the above distortions. Peculiarities of the low-energy slopes of the lines of various terrestrial silicates and lunar samples not comprising reduced forms were also taken into account.

An investigation of the changes in reduced element contents with depth by sample etching with argon ions (Fig. 2) has revealed that samples are significant-ly distinguished by this feature; moreover, a depth of disappearance of the reduced forms depends directly on their quantity at the surfaces of the regolith particles. The thickness of the layer containing the reduced forms can be quite

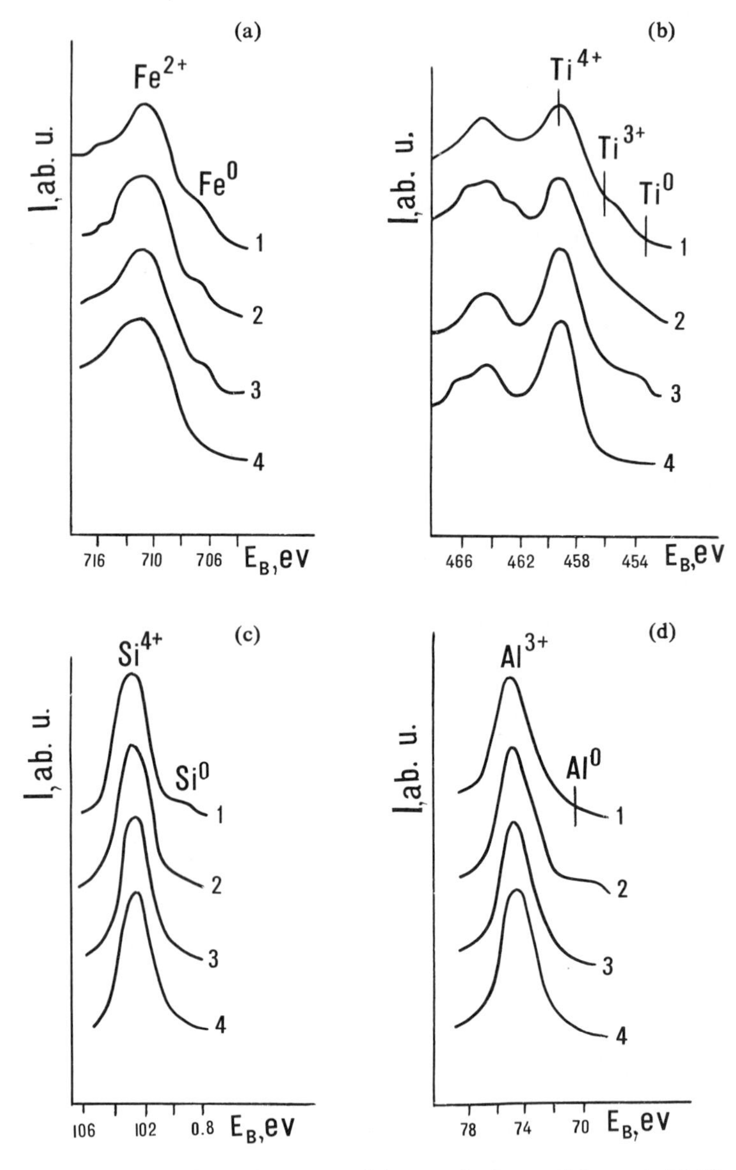

Fig. 1. X-ray photoelectronic spectra of elements in the regolith sample 24092, 4-I. (a) Spectrum of the Fe 2p3/2 electrons; (b) Spectrum of the Ti 2p electrons; (c) Spectrum of the Si 2p electrum; (d) Spectrum of the Al 2p electrons; I—at particles surface; 2—at depth of I70 Å; 3—at depth of 340 Å; 4—at depth of 5IO Å.

Table 1. Contents of reduced forms of elements (in numerator—% of a total content of the element) and depth of disappearance of reduced forms (in denominator—in angströms) in a fine fraction of the regolith samples.

Sample	Fe	Ti	Si
1603-1	9/2500	6/1000	4/1000
2001-1	7/1000	–	3/700
24021.27	16	10	5
24092.4-1	11/600	8/500	4/350
24118.4-1	10/400–	6/400	4/200
24143.4-1	6/400	–	3/200
24184.4-1	3/200	–	2/150

large; the presence of Fe° in sample 1603-1 is detected at a depth of more than 1500 Å.

It should be mentioned that the maximum content of reduced forms of the elements is observed not at the very surface layer but at some depth generally different for different samples.

Structural Condition of Silicate Matrix

Besides the constant presence of the reduced forms at the surface of particles, attention is directed to the comparatively narrow range of inner-shell electron bond energies in all the major elements in oxidized state (Si, Al, Mg, Ti, Ca and Fe). Thus if the position of the 2p- level of Si in lunar minerals could change from 102.0 ev (in case of olivine) to 103.6 ev (in case of cristobalite) (Alyoshin *et al.*, 1975), in the fine fraction of the regolith this position is within the range of 102.8 ± 0.2 ev. The same applies to the lines of the rest of the elements including oxygen in which the ls-electron bond energy in the regolith samples is fixed at 531.9 ± 0.1 ev. At the same time it is known (Alyoshin *et al.*, 1977) that in crystal silicates this energy increases with the increase of the SiO_4 tetrahedron

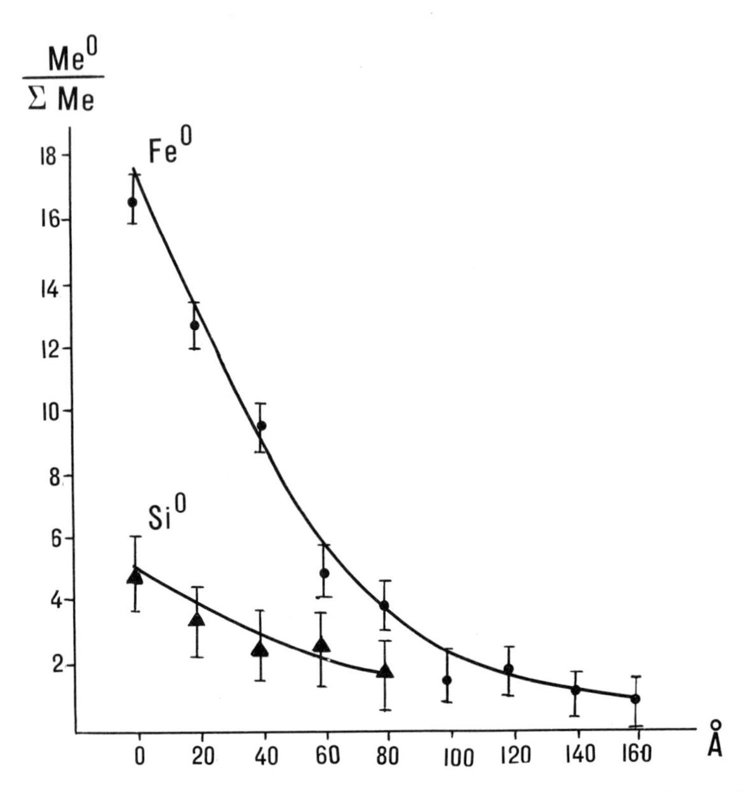

Fig. 2. A relative content change of Fe and Si reduced forms along regolith particles depth of the sample 24021,27.

formation from 531.3 ev (nesosilicates) to 532.9 ev (quartz). The 2 p electrons of Al for which bond energy is 74.9 ± 0.1 ev and the 2p electrons of Ca with bond energy of 348.2 ± 0.2 ev (2p $\frac{3}{2}$) and 351.7 ± 0.2 ev (2p $\frac{1}{2}$) behave analogously.

Such uniformity in the energy positions of the X-ray photoelectron spectrum line for various elements suggests that the more or less homogeneous state of charge characteristics in the outer layers of the regolith particles is most probably related to phase homogeneity of these layers. The values of bond energy of the Si observed for various regolith samples, Al, Ca, Mg 2p-electrons and the 1s electrons of 0 are extremely close to that of silicate glasses of basic composition (Alyoshin *et al.*, 1975); this it is possible to suppose that the phase homogeneity of the outer layers of the regolith particles is caused by amorphouzation of initial rock-forming silicates under influence of exogenous agents.

A somewhat special position among the rock-forming elements belongs to titanium. Earlier having used a sufficiently large number of titanium-bearing oxides and silicates (Nemoshkalenko *et al.*, 1977) it has been established that the bond energy of the Ti 2 $p_{3/2}$ electrons in oxides where Ti has coordination number 6 does not exceed the 459.2–459-6 ev limits and in silicates with the same

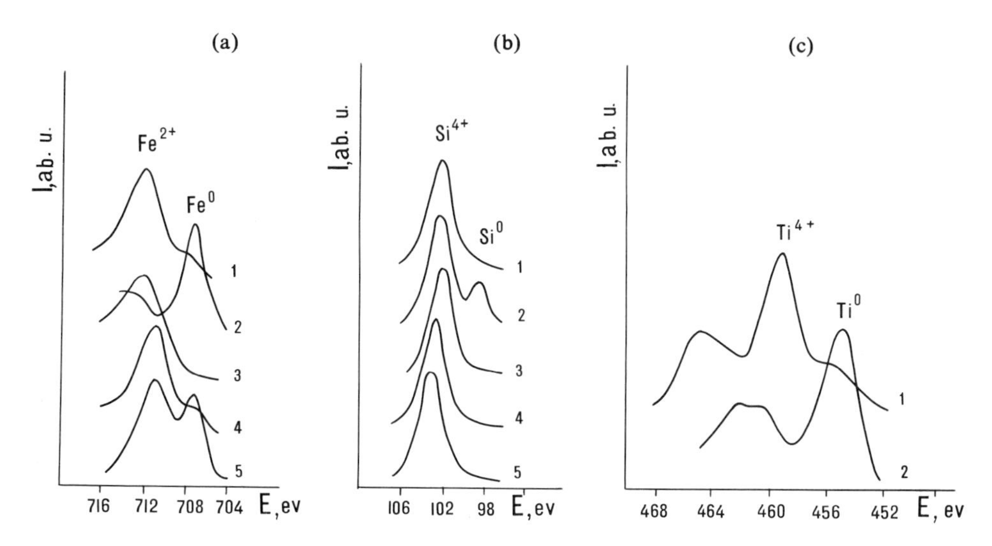

Fig. 3. Change of the X-ray photoelectronic spectra of elements in the regolith sample I608-I and in the earth's anorthosite sample under heating in hydrogen atmosphere. (a) Spectrum of the Fe 2p3/2 electrons; (b) Spectrum of the Si 2p electrons; (c) Spectrum of the Ti 2p electrons; I—the regolith in initial condition; 2—the regolith after heating in hydrogen atmosphere under 500°C; 3—the anorthosite in initial condition; 4—the anorthosite after heating in hydrogen atmosphere under 500°C; 5—the same, under 650°C.

titanium coordination of Ti, the bond energy of the 2 $p_{3/2}$ electrons rises to 459.6–460.0 ev (in oxides) and to 460.0–461.0 ev (in silicates). Study of various regolith samples shows that the bond energy of the Ti 2 $p_{3/2}$ electrons changes in them from 459.4 to 460.0 ev suggesting that in the outer layers of the regolith particles there is strong interaction of titanium-oxygen groups with silicon-oxygen ones; moreover, that tetrahedral groups in TiO_4^{4-} as well as SiO_4^{4-} tetrahedrons may be functioning as anions.

It is also necessary to note the constant and considerable broadening of the Ti 2p electrons baseline on the low-energy side that gives evidence of a large number of intermediate forms of Ti oxidation.

In studying the matter of particles as a function of depth by means of ion etching, individualization of all lines positions is observed; moreover, this individualization is, as a rule, related to the moment of disappearance of reduced forms. Among the most distinct examples of such individualization is the gradual decrease of the bond energy of the 2p level of Si in various samples changing from 103.0 to 102.6–102.2 ev, i.e., from the values characteristic of glasses to the values characteristic of inosilicates and nesosilicates and high-calcium alumosilicates. Simultaneously, a decrease of the bond energy of the 2 $p_{3/2}$ level of Ti is also observed, which, for example, in the sample 24184 reaches 458.6 ev and in all the rest of the cases comes down to 459.2–458.9 ev, i.e., in inner layers of the regolith particles, titanium is characterized by octahedral coordination and is distributed both in silicate and in oxide phases. The 2p electrons of aluminum also show tendency of displacement towards the lower bond energies that generally gives evidence of its tetrahedral rather than octahedral coordination.

An interesting regularity is also noted in the behaviour of the 2p electrons of Ca: while in the outer layers of regolith particles the value of spin-doublet splitting of the Ca (2 $p_{1/2}$–2 $p_{3/2}$) level is in the interval of 3.5–3.8 ev which is characteristic of Ca-bearing silicate glasses and ionic compounds of this element (such as $CaCl_2$, CaF_2, Ca aluminates). In the process of ion etching one can observe in some cases a decrease of the Ca 2 $p_{1/2}$–2 $p_{3/2}$ level to 3.3 ev which has been noted only in cases of crystal silicates with strongly distorted polyhedron (for example, in wollastonite, Ca-seydozerite, lovenite, etc.).

In other words, the regolith fractions are generally two-layer structures: in centers of grains are relics of crystal phases and in their peripheries are products of its amorphouzation and partial reduction.

Change of Surface Characteristics Under Heating

The peculiarities in structural-chemical parameters of surface layers and distribution in these layers of the reduced forms confirm the generally accepted viewpoint that these peculiarities are related to exogenous agents. It was expedient to study the influence of hydrogen on reducing processes in lunar matter itself because one of the leading agents for this reduction is considered to be cosmic radiation, the main component of which are the protons with an energy of 1–10 Kev (Zeller, 1976). Experiments with heating samples of lunar regolith and a terrestrial magmatic rock were conducted under various conditions: in a hydrogen atmosphere and in vacuum.

The undisturbed regolith sample 1608-1 was heated in the camera of the VIEE-15 spectrometer in an atmosphere of hydrogen purified by diffusion through a palladium capillary. The sample was exposed for 20 min at a temperature of about 500 °C and hydrogen pressure about 900 torr; and then, without contact with the atmosphere, it was moved into the electron spectrometer.

After an examination of the electron lines corresponding to inner-shell 2p-electrons of Fe, Ti and Si, the sample was exposed to air for 3 min, 1 hour, 18 hours, 82 hours and 10 days. After each air exposure an examination of the electron spectra was conducted.

As a consequence of the heating in the hydrogen atmosphere, the iron, titanium, and silicon in the lunar regolith were rapidly reduced; the line of corresponding oxide for iron practically disappeared, the quantity of reduced titanium exceeded that of oxidized titanium, the content of reduced, elemental silicon also considerably increased (Fig. 3).

The exposure of a lunar regolith sample in air results in a gradual decrease of intensities of the lines corresponding to the reduced forms of the elements and in an increase of intensities of the lines corresponding to the oxides. Silicon is somewhat resistant to the oxidizing process while iron and titanium, even after

3-min. exposure of the regolith sample in air, are somewhat oxidized. Longer exposure times lead to an intense oxidation of titanium and to a somewhat smaller degree iron, while silicon is oxidized appreciably more slowly. However, even after 10-days exposure of the sample in air, the portion of Ti° and Fe° found is significant and comparable with the original quantity of these forms found in the lunar regolith matter prior to the hydrogen treatment.

At the same time, the heating in a hydrogen atmosphere of a ground, <60 μm sample of the terrestrial magmatic rock (gabbro-anorthosite) at the same temperature as the lunar regolith sample—500°C (and even under a higher temperature—650°C) and otherwise identical conditions results in considerably weaker effects on shape of the electron lines under study. The reduction processes in this case effect only iron; however, the quantity of reduced iron is less than that of oxidized iron. The reduced forms of titanium and silicon were not observed.

At the same time one should note one circumstance related to position of the Si 2p-electron line: exposure of the gabbro-anorthosite sample to the hydrogen atmosphere results in considerable—up to 1.0 ev (from 102.5 to 103.5 ev)—shift of the line to the higher bond energy. At the same time the corresponding line in the lunar regolith after its hydrogen treatment does not undergo any displacement; however, in the oxidation process it shifts to the lower bond energy (from 102.9 ev to 102.2 ev).

For a vacuum heating experiment the sample 2004-1 was used. The experiment was carried out according to the following scheme: the sample was heated in succession for 20 min. under temperatures of 50°, 150°, 250°, 350°, 450°, 550° and 580°C and at each stage of heating the Fe, Si, Al, Ti, Ca and O 2p-electron lines were studied without exposure of the sample to air. Simultaneously the gabbro-anorthosite sample was being studied according to the same scheme.

The results of this experiment are shown in Fig. 4 and in the Table 3.

The results distinguish two types of regolith changes with heating in nonhy-

Table 2. Position of core electrons of Te, Si, Al, Ca and O in regolith samples 2004-1 during vacuum heating (ev).

T°C	Fe $2p_{3/2}$	Ca $2p_{1/2}$	Ca $2p_{3/2}$	Si 2p	Al 2p	O Is
Room	711.4	351.8	348.4	102.7	74.8	532.0
50°C	710.8	352.0	348.5	10.27	74.8	532.0
150°C	711.0	352.0	348.5	102.8	74.6	532.0
250°C	710.4	352.0	348.6	102.8	74.8	532.0
350°C	710.4	352.0	348.6	102.7	74.9	532.0
450°C	711.4	352.9	349.4	103.2	75.4	532.4
550°C	—	352.9	349.4	103.4	75.4	532.4
580°C	711.2	352.9	349.2	103.4	75.5	532.4
600°C(+H_2)	711.2	352.9	349.3	103.4	75.5	532.4

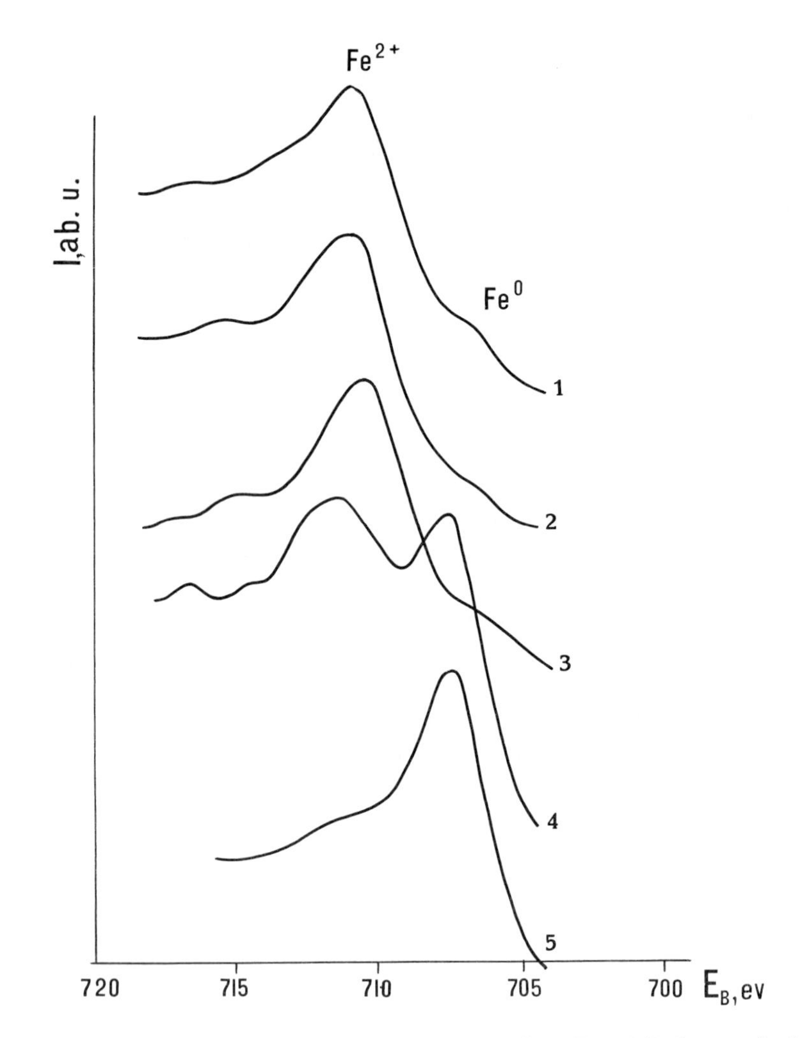

Fig. 4. Change of the X-ray photoelectronic spectra of the Fe 2p3/2 electrons in the regolith sample 2004-I under heating in non-hydrogen (2–4) and hydrogen (5) atmospheres. I—in initial condition; 2—after heating up to 50°C; 3—after heating up to 350°C; 4—after heating up to 450°C; 5—after heating up to 600°C in hydrogen atmosphere.

drogen atmosphere: 1) The low-temperature type (in interval of the 50–350°C) with the characteristics of moderate increase in reduced forms of Fe and Si, with the energy of the main peaks corresponding to the oxidized state of the elements, remains invariable throughout the temperature interval; and 2) the high-temperature type (450° and more) with the characteristics of a sharp increase in reduced Fe accompanied by disappearance of reduced Si. Along with the Si^{0}

oxidation there is a sharp displacement of the peak related to the oxidized state of Si towards higher energy which is close to those of pure oxides and hydroxides. However calibration with carbon at 450°C is not completely certain because of considerable broadening of the 1 s-line of carbon. Therefore for this and higher temperatures, the calibration was carried out using the 2 $p_{3/2}$ line for which the value of 707.0 ev obtained by carbon calibration at room temperature was accepted. As the layer under study is at the very surface of the regolith particles, its charging could possibly not reflect the charging of deeper particles layers. Therefore it is possible that the coherent increases of bond energies of the Ca, Si, Al 2 p-electrons and the 1 s-electron of O are related to such an effect.

A further heating of regolith transformed at high-temperature in hydrogen atmosphere does not lead to any essential changes and has the effect of only some increase of the portion of reduced iron. In other words the regolith after high-temperature roasting losses its potential for reduction and begins to behave in this respect analogously to terrestrial magmatic rocks.

Conclusions

The results of the X-ray photoelectron spectroscopy study of the fine fraction of the lunar regolith delivered by the Luna automatic stations indicate that in all samples notable changes of a number of spectrum parameters of the outer surface zone of the regolith particles are observed as compared to the parameters of deeper layers. The parameters in particular indicate the presence at the surfaces of reduced forms of a number of elements and an amorphous state of surface layers of the matter.

It should also be noted that the intensity of changes in the parameters at the surfaces of the regolith particles correlate rather well with many other lunar regolith parameters related to processes of its reworking by exogenous agents, in particular, content of agglutinates, rare gases and some others. The more objective criterion for the reduced forms is not their absolute concentration at the surface but the width of the zone where they are detected. The experiment has determined some decrease in the concentration of the reduced forms at the very surface. This decrease varies for individual samples and is obviously related to oxidation in the earth's conditions.

An investigation of structural peculiarities of the oxidation states of main elements gives evidence of formation at the surfaces of regolith particles of an amorphousized (metamictized) layer where the reduced forms of elements are located. It is reputed that the amorphouzation and the reduction are various stages of one process of destruction of crystalline silicate matter.

The first includes process and phenomena related to the constant meteorite and micrometeorite gardening of the lunar surface: melting, evaporation and condensation of the lunar rock and meteorite bodies and related dissociation of elements in an explosion cloud. Wide development of evaporation processes on the moon is rather well known and is revealed both in chemical and isotopic studies. Both the highly volatile elements such as alkalis and the less volatile petrogenic elements including iron and silicon (Clayton *et al.*, 1974; Dowty *et al.*, 1973; Ivanov and Florensky, 1975) are subjects of evaporative loss. Examples are also known of the presence on surfaces of particles of condensation films enriched in the same elements (Kurat and Keil, 1972; Ivanov *et al.*, 1975). The scale of the phenomena observed in many cases (impoverishment of coarse particles on the whole in the considered elements and relatively thick films of condensation origin) undoubtedly gives evidence that their cause was the process of meteorite bombardment.

The second complex of the processes is related to a constant flux of cosmic radiation influencing the lunar surface. The result of these processes is implantation of the protons and the heavier particles of the solar wind into the lunar matter accompanied by destruction of a mineral crystal lattice and ablation of lunar grain surfaces.

The intensity of manifestation of the meteorite bombardment and cosmic radiation processes is a function of a surface exposure age and both the complexes of the processes could to a certain degree contribute in the effects observed. However, the process of matter condensation out of a vapour cloud formed by a meteorite explosion repeated many times should result in formation of a surface film characteristic of rather irregular change of composition. On the other hand an effect of implantation of solar wind ions should to the utmost degree be displayed at a grain outer surface and gradually decrease with a depth.

Monotonic decreases in concentrations of reduced forms of various elements with depth observed under investigation of regolith particles gives apparent evidence that the element reduction in surface layers of the particles is principally related to a cosmic radiation influence. Analogous conclusions were drawn by the other investigators studying finely-dispersed iron in lunar samples.

Mechanics of the reduction process requires further research. However, the character of the distribution with depth of reduced forms and the results of the experiments in heating of regolith samples in hydrogen atmospheres and in vacuum certainly indicate that the reduction of elements takes place under influence of implanted hydrogen ions *in situ* and not in the process of evaporation under influence of ion bombardment and the resulting concentration of matter on the moon's surface as it has been suggested by Pillinger *et al.* (1976). The principal influence of implanted hydrogen on element reduction in the lunar regolith has also been pointed out by Huguenin (Zeller, 1976; Huguenin comment). However, it is necessary to caution that our conclusions are based on statistical features of experiments which measure the average response of a large number of fine grains for each sample.

We believe that the reduction processes on the moon act on not only Fe°, Ti°, Si° and Al° but also on the other electropositive elements. Lack of reliable experimental evidence for the existence of reduced forms of the other elements is probably explained by the high rates of their oxidation under terrestrial conditions.

REFERENCES

Alyoshin V. G., Dikov Yu. P., Nemoshkalenko V. V., Salova T. P., Senkevich A. I., Taldenko Yu. D., Chichagov A. V. and Epelbaum M. B. (1975) Some results of X-ray electron investigation of silicates. *Metallophysica* **60**, 39–44. SSR Acad. Sci., Kiev. (In Russian.)

Alyoshin V. G., Chichagov A. V. and Epelbaum M. B. (1977) The determination of basicity of minerals based on X-ray spectra. Preprint of Inst. Metallophysics, Ukrain. SSR Acad. Sci., Kiev. (In Russian).

Baron R. L., Bilson E., Gold T., Colton R. J., Hapke B. and Steggert M. A. (1977) A comparison of ESCA-XPS and Auger electron spectroscopy for the study of the surface chemical composition of lunar soil samples (abstract). In *Lunar Science VIII*, p. 56–58. The Lunar Science Institute, Houston.

Clayton R. N., Mayeda T. K. and Hurd J. M. (1974) Loss of oxygen, silicon, sulfur, and potassium from the lunar regolith. *Proc. Lunar Sci. Conf., 5th*, p. 1801–1809.

Dikov Yu. P., Nemoshkalenko V. V., Alyoshin V. G., Ivanov A. V. and Bogatikov O. A. (1977a) Reduced titanium in lunar regolith. *Dokl. Acad. Nauk. SSSR.* **234**, 176–179. (In Russian.)

Dikov Yu. P., Bogatikov O. A., Alyoshin V. G., Nemoshkalenko V. V., Barsukov V. L. and Ivanov A. V. (1977b) Reduced silicon in lunar regolith. *Dokl. Acad. Nauk. SSSR* **235**, 1410–1412. (In Russian.)

Dikov Yu. P., Bogatikov O. A., Nemoshkalenko V. V., Alyoshin V. G., Barsukov V. L., Florensky K. P. and Ivanov A. V. (1977c) Peculiarities of state of rock-forming elements in surface layers of Luna 24 regolith particles. *Geochimia* **10**, 1524–1533. (In Russian.)

Dikov Yu. P., Bogatikov O. A., Barsukov V. L., Florensku K. P., Ivanov A. I., Nemoshkalenko V. V. and Alyoshin V. G. (1978) Reduced forms of elements in the surface parts of regolith particles: ESCA studies (abstract). In *Lunar and and Planetary Science IX*, p. 250–252. Lunar and Planetary Institute, Houston.

Dowty E., Keil K. and Prinz M. (1973) Major-element vapor fractionation on the lunar surface: an unusual lithic fragment from the Luna 20 fines. *Earth Planet. Sci. Lett.* **21**, 91–96.

Housley R. M. and Grant R. W. (1975) ESCA studies of lunar surface chemistry. *Proc. Lunar Sci. Conf. 6th*, p. 3269–3275.

Housley R. M. and Grant R. W. (1976) ESCA studies of the surface chemistry of lunar fines. *Proc. Lunar Sci. Conf. 7th*, p. 881–889.

Housley R. M. and Grant R. W. (1977) An XPS (ESCA) study of lunar surface alteration profiles. *Proc. Lunar Sci. Conf. 8th*, p. 3885–3899.

Ivanov A. V. and Florensky K. P. (1975) The role of vaporization processes in lunar rock formation. *Proc. Lunar Sci. Conf. 6th*, p. 1341–1350.

Ivanov A. V., Florensky K. P., Nazarov M. A. and Shevaleevsky I. D. (1975) A manifestation of the processes of vaporization and condensation during the formation of lunar regolith particles. *Dokl. Akad. Nauk. SSSR* **221**, 202–250. (In Russian.)

Kurat G. and Keil K. (1972) Effects of vaporization and condensation on Apollo 11 glass spherules: implications for cooling rates. *Earth Planet. Sci. Lett.* **14**, 7–13.

Nefedov V. I., Urusov V. S. and Zhavoronkov N. M. (1972) Difference of surface and bulk main elements concentration of the lunar regolith particles. *Dokl. Acad. Nauk. SSSR* **207**, 698–701. (In Russian.)

Nemoshkalenko V. V. and Alyoshin V. G. (1976) Electron spectroscopy of crystals. *Naukova Dumka Publ.*, Kiev. (In Russian.)

Nemoshkalenko V. V., Dikov Yu. P., Alyoshin V. G., Sheludchenko V. M. and Senkevich A. I. (1977) Investigation of the charge of titanium in titanates and titanosilicates. *Dokl. Akad. Nauk. Ukrain. SSSR* **6**, 545–548. (In Russian.)

Pillinger C. T., Gardiner L. R. and Jull A. J. T. (1976) Preferential sputtering as a method of producing metallic iron, inducing major element fractionation and trace element enrichment. *Earth Planet. Sci. Lett.* **33**, 289–299.

Siegbahn K., Nordling C., Johansson G., Hedman J., Heden P. F., Hamrin K., Gelius U., Bergmark T., Wermel L. O., Maune R. and Baer Y. (1971) *ESCA applied to free molecules*. North-Holland Publ., Amsterdam-London.

Vinogradov A. P., Nefedoc V. I., Urusov V. S. and Zhavoronkov N. M. (1972a) X-ray electron investigations of metallic iron in lunar regolith. *Acad. Nauk SSSR* **207**, 433–436. (In Russian.)

Vinogradov A. P., Nefedov V. I., Urusov V. S. and Zhavoronkov N. M. (1972b) ESCA investigation of lunar regolith from the Seas of Fertility and Tranquility. *Proc. Lunar Sci. Conf. 3rd*, p. 1421–1427.

Zeller E. J. (1976) High energy protons—an early source of regolith water. *Proc. Colloquium on Water in Planetary Regoliths*. Hanover, New Hampshire, p. 31–33.

Proc. Lunar Planet. Sci. Conf. 9th (1978), p. 2125–2135.
Printed in the United States of America

Track studies in four samples of Luna 24 core

L. L. KASHKAROV, L. I. GENAEVA and A. K. LAVRUKHINA
V. I. Vernadsky Institute of Geochemistry & Analytical Chemistry
USSR Academy of Sciences Moscow, USSR

Abstract—A study of tracks formed by nuclei of the iron group of cosmic rays in mineral grains (olivine, pyroxene, feldspar) which were isolated from four levels of the Luna 24 core (basic samples 24092,4-4; 24118,4-4; 24143,4-4; 24184,4-4) has been made. In all samples grains were detected as having been subject to irradiation by solar cosmic rays on the lunar surface under a $\lesssim$1 mm thick layer of matter. Of the total of ~200 crystals studied of the fraction 200–370 μm, none were found to have a track density less than $\sim 10^6$ cm^{-2}. The non-uniformity of the soil radiation characteristics from the top of the core to the deeper layers has been established. The most highly irradiated soils proved to be the samples 24092 and 24118. The material of sample 24143 has received a lower dose of integral irradiation. In sample 24184 the presence of strongly irradiated matter together with a noticeable amount of a weakly irradiated component (represented mainly by feldspars) points to the bimodal character of this sample. The data obtained confirm the idea that a process of successive deposition of finely crushed soil layers plays an important role in the lunar regolith formation.

Cosmic ray fossil tracks have been studied in feldspar, olivine and pyroxene crystals (fraction 200–370 μm) from the Luna 24 core. Tracks produced by solar flare and galactic cosmic rays were observed. The crystals examined came from lunar fines samples 24092,4-4; 24118,4-4; 24143,4-4 and 24184,4-4*). Each sample represents a bulk probe on four cm of core. The first two samples belong to zone II which is composed of homogeneous regolith; sample 24143 was taken from zone III; and sample 24184 from zone IV which is the deepest part of the core (Barsukov *et al.*, 1977). A total of 204 crystals were studied. The depth interval for each sample and the number of studied crystals are given in Table 1.

The observation and measurement of the track density (ρ) was carried out on the polished inner surfaces of the silicate grains with the aid of an optical microscope at an ~1000-fold magnification. For polishing, etching and examination convenience the crystals were polymerized into transparent tablets of epoxide resin. The revelation of tracks in various minerals was accomplished by known methods of chemical etching. Feldspars were etched for 6–10 minutes in a boiling alkali solution (3g NaOH: 8g H_2O). A more concentrated solution (3g NaOH: 4g H_2O) was used to reveal tracks in pyroxenes with the etching time in this case being necessarily increased up to 20–80 minutes (Lal *et al.*, 1968). Olivine was etched for 2–10 hours in a WN solution at pH 8 described by Krishnaswami *et al.* (1971).

*More abbreviated designation of samples are given below in text.

Table 1.

Sample	Depth from surface (cm)	Number of crystals analyzed	Track density (10^7 cm^{-2})			Fraction of grains exposed at the surface
			Average	Quartile	Median	
24092,4-4	91–95	47 Olivine (6) Pyroxene (15) Feldspar (26)	3,7	3,2	4,1	0,26
24118,4-4	117–121	54 Olivine (9) Pyroxene (16) Feldspar (29)	3,6	3,0	3,7	0,22
24143,4-4	142–146	45 Olivine (6) Pyroxene (15) Feldspar (24)	3,0	1,8	3,2	0,13
24184,4-4	183–187	58 Olivine (16) Pyroxene (12) Feldspar (30)	1,9	1,0	3,0	0,34

Reported track densities have been increased relative to the raw data by dividing by 1.0, 0.7 and 0.5 for feldspar, pyroxene and olivine respectively in order to allow for differing etching and registration efficiencies of the detectors (Fleischer *et al.*, 1974; Goswami and Lal, 1974).

Various examples of tracks observed in the samples of the Luna 24 soil core are given in Fig. 1. The presence of a track density gradient (Fig. 1a), i.e., of the abrupt decrease of track density when passing from the surface into the interior of individual crystals, indicates that the given grains were subjected to irradiation by low-energy nuclei of the VH-group of solar cosmic rays shielded by a layer of matter not more than a few tenths of microns thick. The uniform density of tracks exceeding 10^8 cm^{-2} (Fig. 1b) also indicates that the grains had accumulated tracks mainly from solar cosmic rays during regolith exposure beneath a few mm of shielding. An example of tracks from nuclei of the VH-group of galactic cosmic rays being characterized by a relatively low density ($\lesssim 10^7$cm^{-2}) is given in Fig. 1c.

In the statistical analysis of the data we used the following parameters applied elsewhere (Arrhenius *et al.*, 1971; Fleischer *et al.*, 1974; Goswami *et al.*, 1976). $\bar{\rho}$ is the mean arithmetical track density in crystals having $\rho < 6 \cdot 10^7$ cm^{-2}, the maximum value usually resolvable by optical microscopy. ρ_q is the track density at which an inequality, $\rho < \rho_q$, is maintained for 25% of all investigated grains; ρ_{med} is the track density at which the inequality; $\rho < \rho_{med}$ is maintained for 50% of all investigated grains of each sample. The use of one of these parameters may

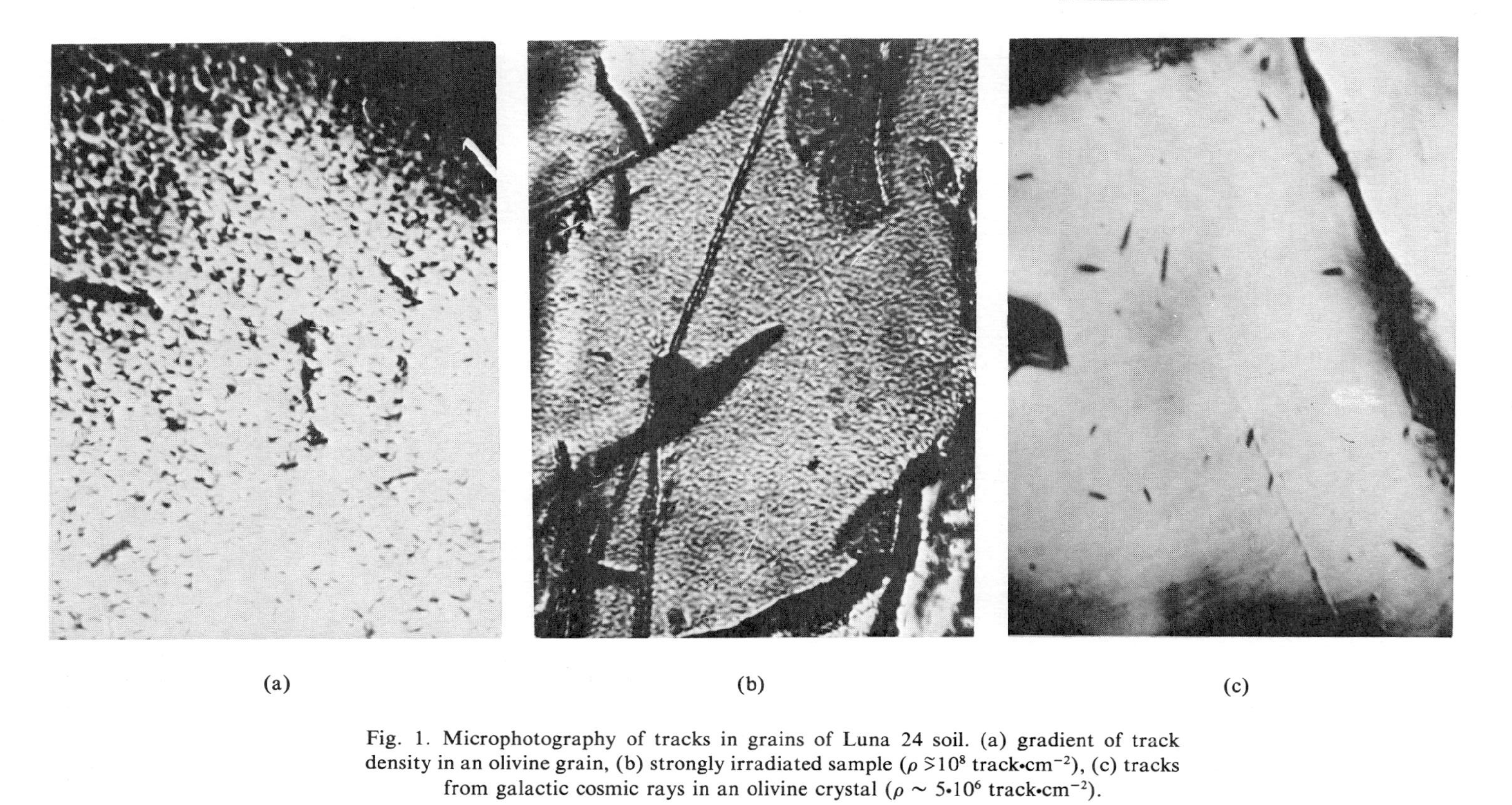

Fig. 1. Microphotography of tracks in grains of Luna 24 soil. (a) gradient of track density in an olivine grain, (b) strongly irradiated sample ($\rho \gtrsim 10^8$ track·cm^{-2}), (c) tracks from galactic cosmic rays in an olivine crystal ($\rho \sim 5 \cdot 10^6$ track·cm^{-2}).

prove to be insufficient, only in combination do they lead to a more definite estimation of the integral dose received by a soil sample at a given depth of the core. The parameter N_H/N characterizes the fraction of grains with high density ($\rho > 6 \cdot 10^7$ cm^{-2} in our case) or with a track density gradient. Since these grains belong to a material which was irradiated in the superficial soil layer ranging from a few microns to a few mm thick, the parameter N_H/N is an indicator of near-surface irradiation.

Results of track measurements obtained for specimens of four different samples of Luna 24 soil core are given in Table 1. In Fig. 2 the distributions of a number of mineral grains are represented in the form of a histogram depending on their observed track densities.

Samples 24092 and 24118 substantially differ from the deeper samples 24143 and 24184. The main difference is that 96% of the crystals in sample 24092 and 95% in sample 24118 show track densities exceeding $2 \cdot 10^7$ cm^{-2}. Only ~5% of all studied grains in these samples have received lower irradiation doses. Grain distributions according to track density in these samples have sharply pronounced maxima falling in the interval of $\rho = (2\text{–}6) \cdot 10^7$ cm^{-2}. A substantial portion of the crystals (15% and 11% for 24092 and 24118, respectively) in these samples show track densities exceeding $6 \cdot 10^7$ cm^{-2}. As can be seen in Table 1, all track parameters ρ, ρ_q, ρ_{med} and N_H/H for these two samples are similar, although for sample 24092 the values of N_H/N and ρ_{med} exceed analogous values for the sample 24118, which may indicate a lower degree of irradiation in the lower level.

In the specimen taken from greater depth (sample 24143) there is a noticeable increase in the proportion of less irradiated grains. The distribution histogram according to track density of crystals from this specimen (Fig. 2c) is characterized by a greater area in the region of $\rho < 2 \cdot 10^7$ cm^{-2}. The relative number of grains with $\rho > 2 \cdot 10^7$ cm^{-2} is about 70% and the values $\bar{\rho}$, ρ_q and N_H/N are considerably lower than for samples 24092 and 24118 (see Table 1).

The lowest values of the parameters $\bar{\rho}$, ρ_q, ρ_{med} are found in sample 24184. In this specimen more than 50% of the crystals have track densities of $\rho < 2 \cdot 10^7$ cm^{-2}. However, the fraction of strongly irradiated grains in this sample gave the highest value, equal to 34%, for the parameter N_H/N of the four samples investigated. It is of interest to note that in specimen 24184 the least irradiated crystals proved to be feldspars. In Fig. 3 three separate histograms of grain distributions are given for monomineralic fractions (feldspar, pyroxene and olivine) according to track density. The olivine crystals were the most irradiated ($\rho_{med} = 4.7 \cdot 10^7$ cm^{-2}), whereas feldspar crystals had the least irradiation dose ($\rho_{med} = 1.3 \cdot 10^7$ cm^{-2}). Pyroxenes occupy an intermediate position ($\rho_{med} = 3.2 \cdot 10^7$ cm^{-2}). For specimens of the upper levels (24092, 24118, 24143) we have not noticed a substantial difference in the distribution of crystals of various minerals according to the degree of irradiation.

Crystals having a track density gradient were revealed in all four samples of the Luna 24 soil core. The proportion of crystals in which the gradient of track density was revealed is equal to 13, 13, 4 and 10% for the samples 24092, 24118,

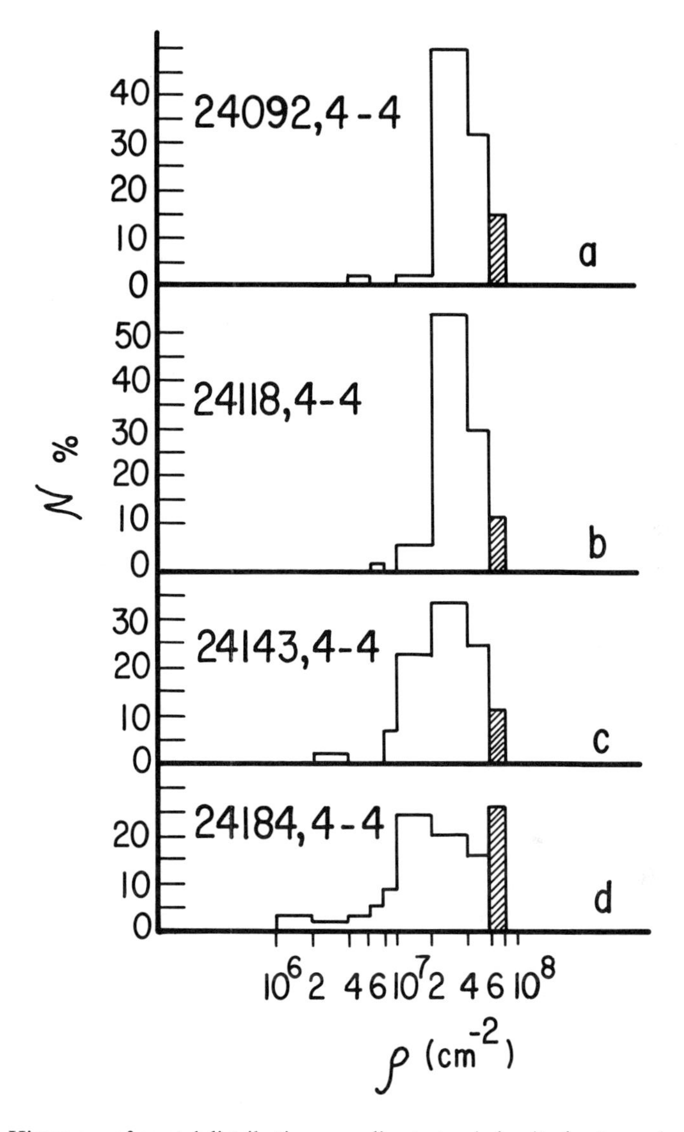

Fig. 2. Histogram of crystal distribution according to track density for 4 samples of the Luna 24 core. Each distribution normalized to 100 grains. Data for all minerals are plotted together with the account of track recording efficiency. Crystals having a density of $>6\cdot10^7$ track$\cdot$cm^{-2} are marked by a hatched field.

24143 and 24184 respectively. However, these figures may prove to be too low because in crystals with high track densities the gradient of track density may not be revealed with the aid of an optical microscope; in addition, the crystals studied in this paper represented a relatively coarse-grained fraction (200–370 μm). Examples of track density variation with respect to depth in separate crystals are

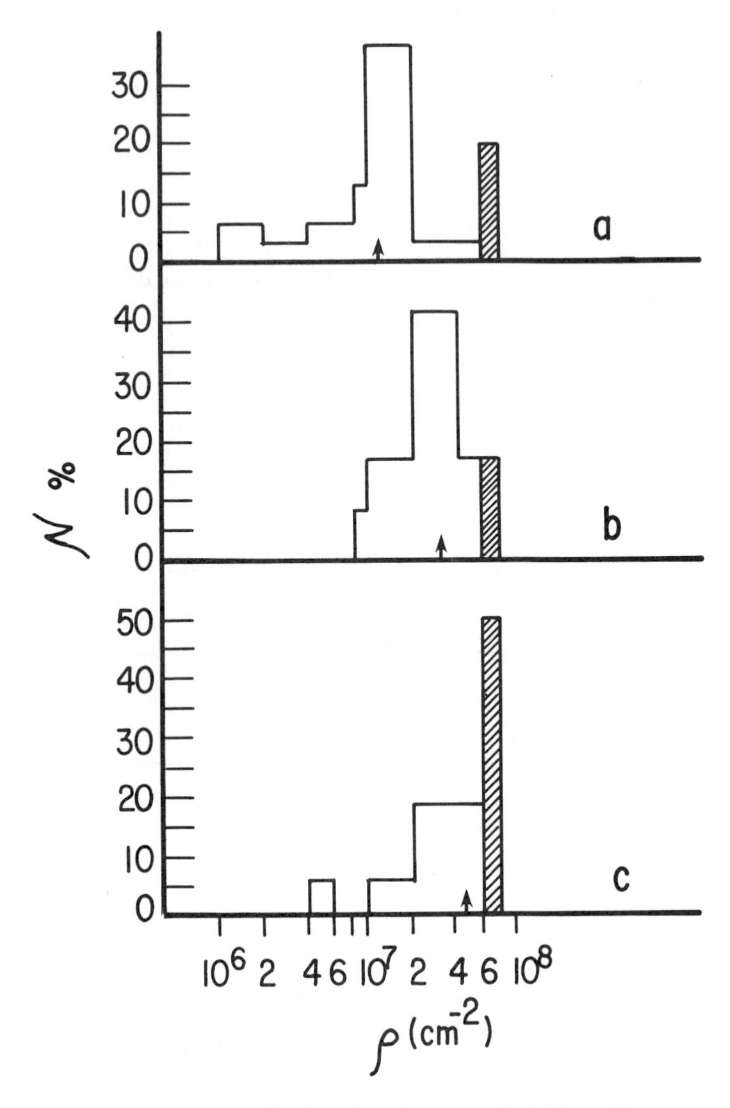

Fig. 3. Histogram of grain distribution of various minerals: feldspar (a), pyroxene (b), and olivine (c) according to track density in sample 24184 with account of track recording efficiency. The ρ_{med} values are marked by arrows. The hatched field corresponds to crystals with $\rho > 6 \cdot 10^7$ cm^{-2}.

shown in Fig. 4. Comparison of the values of inclinations and character of these curves with data obtained earlier for crystals from soil cores of Luna 16 and Luna 20 (Kashkarov *et al.*, 1975, 1976) does not show any substantial difference. All curves have a characteristic break corresponding to the depth of X $\lesssim$20 μm from the surface of the crystal. The maximum values of the exponent entering

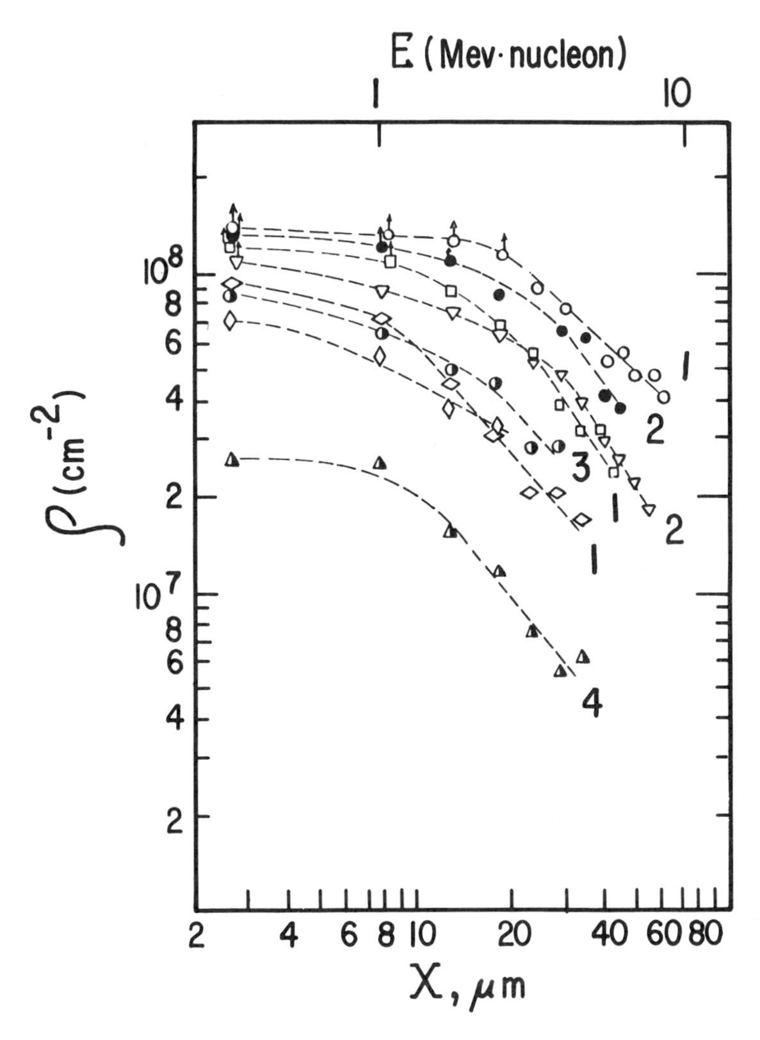

Fig. 4. Gradients of track density according to the depth of separate grains of olivine in samples 24092-(1); 24118-(2); 24143-(3) and plagioclase in sample 24184-(4).

the equation $d\rho/dX = \text{const}\cdot X^{-\alpha}$, as in the case of the Luna 16 and Luna 20 cores, are equal to ~1–1.5.

Discussion of the Results

The grains that we had available for analysis were located within 4-centimeter sections of the core, but the location of these grains within these sections is not known to us. Therefore the data obtained only characterize sample properties averaged over 4 cm.

Examining the results obtained for the samples 24092 and 24118 one may see that, in spite of the great similarity of track radiation characteristics for these specimens, they can not be attributed to a single large outburst layer of an earlier unirradiated soil possibly representing a material from the crater Fahrenheit (Florensky *et al.*, 1977). If we assume that a layer at least 30 cm thick (distance between the levels of samples 24092 and 24118) of an earlier unirradiated soil was deposited, the crystals below sample 24118 could not accumulate such high track densities, even if they were present at the given site during the entire history of the moon (Arrhenius *et al.*, 1971). This is associated with the fact that the efficiency of regolith mixing under the influence of meteoritic bombardment quickly decreases with depth (Gault *et al.*, 1974). If a thick layer of previously unirradiated material had been deposited in a single event, one would expect to find a characteristic decrease of the track density with depth. For samples 24092 and 24118 such an effect has not been observed. It may be assumed that the regolith accumulated gradually in thin layers (e.g., during the slow crumbling of the matter of the surrounding hills), thus allowing sufficient time to receive a substantial dose of radiation. During this transportation small particles of the soil roll on the surface, are irradiated, and accumulate at the given site. Such a mechanism was suggested by Goswami & Lal (1974) as an explanation of the substantial radiation dose in a larger layer of soil; the possibility of the presence within the 30-centimeter layer of material with a different radiation history was not excluded.

Lower values of the track parameters $\bar{\rho}$, ρ_q, ρ_{med} and N_H/N for sample 24143 are evidence that the integral dose of irradiation by solar cosmic rays was smaller in comparison with the upper samples 24092 and 24118.

In sample 24184 the presence of at least two (or more) layers is possible, each of which has its own radiation history of irradiation. Examples of detection by the track method of very thin (~1 mm) layers of strongly irradiated matter have been described earlier (Price *et al.*, 1975; Goswami, *et al.*, 1976). For example, Price *et al.*, (1975) have revealed five different microlayers less than 1 mm thick on the 3 cm long section of the Apollo 16 core. The average dose of irradiation of each of these layers exceeds the degree of irradiation of the materials above and below the given layer. It may be assumed that sample 24184 consists of two components differing in degree of irradiation: (1) the sublayer enriched in plagioclase which had a relatively short exposure to irradiation in the superficial layer; and (2) the sublayer enriched in olivine which had longer exposure time to irradiation in the upper layer of lunar soil. Other assumptions can also be made which agree with the data. If this whole 4 cm layer represents a homogeneous mixture of strongly and weakly irradiated grains, it could have come from some adjacent part of the lunar surface in a meteorite impact into a soil which had a layered structure. After transportation of the material of this soil to a new place, a homogeneous layer could be formed which would consist of a mixture of grains having different degrees of irradiation. We should not exclude the possibility of contamination of sample 24184 by mineral grains from the neighboring sample 24182, which contains a very small portion of grains exposed to reworking on the surface (Ryder *et al.*, 1977). The authors of this paper also note the bimodal distribution of grains according to size. In this context it may be assumed that the detection of a considerable amount of weakly irradiated crystals in sample 24184 is conditioned by the presence of material differing in its irradiation history from average regolith. However, a definitive affirmation of the presence at this level of layers with different radiation histories requires a more detailed study of the given 4-centimeter section of the core.

The variation of track parameters obtained from the top of the Luna 24 core to the bottom agrees with the data of X-ray investigations, as well as with the measurement of magnetic properties and the composition of the core material (Barsukov *et al.*, 1977; Rode *et al.*, 1977; Ivanov *et al.*, 1978).

Of the samples we examined, 24092 and 24118 were subject to the greatest exogenous action. They are enriched in finely dispersed iron (Ivanov *et al.*, 1978), have a low coefficient of reflection, and do not contain large fragments. These data, as well as the high amount of agglutinates (Rode *et al.*, 1977) point to the high degrees of maturity of these samples, which agree with the large values of the parameter N_H/N and with the average value of track density for the samples. At the same time it should be noted that for sample 24184 we obtained a large value for the parameter N_H/N (see Table 1), which points out the significance of the strongly irradiated component of this sample in the observed distribution of grains according to track density.

In Fig. 5 the values of the track parameters $\bar{\rho}$, ρ_q, ρ_{med} and N_H/N are compared with the content of finely dispersed iron (Ivanov *et al.*, 1978) and agglutinates (Rode *et al.*, 1977) observed in four basic samples of the Luna 24 core. The samples 24143 and 24184, in which there are crystals with low track density, are characterized by a heightened amount of coarse-grained material and a lower content of agglutinates and metallic iron, respectively. These signs indicate that these samples contain a considerable admixture of grains which have not been subjected to protracted exogenous reworking on the lunar surface.

In analysing grain tracks in the Luna 24 core, Blanford and Wood (1977) note that the lower samples, 24182 and 24210, have a significant percentage of grains whose tracks are cosmogenic. They conclude that the soil at such levels is a mixture of mature and immature components. This corresponds to our results for sample 24184.

Track densities reported by Blanford and Wood (1977) and Goswami and Lal (1978) are generally higher than obtained in this study. This could be due either to the instrumental effect (the use of an optical microscope as opposed to an electron microscope) or to a grain size effect.

The results obtained for two grain-size fractions, 40–90 μm and 100–250 μm (Goswami and Lal, 1978), indicate that parameters ρ_{min}, ρ_q and N_H/N for coarse fractions are lower than for thin fractions. In our case (grain-size fraction 200–370 μm) these track parameters may be lower. Chaillou *et al.* (1978) found that ~90% of the feldspars from the size fraction 200–370 μm show only galactic cosmic ray tracks. This agrees with our results.

Some minor differences in figures may be explained by the fact that our

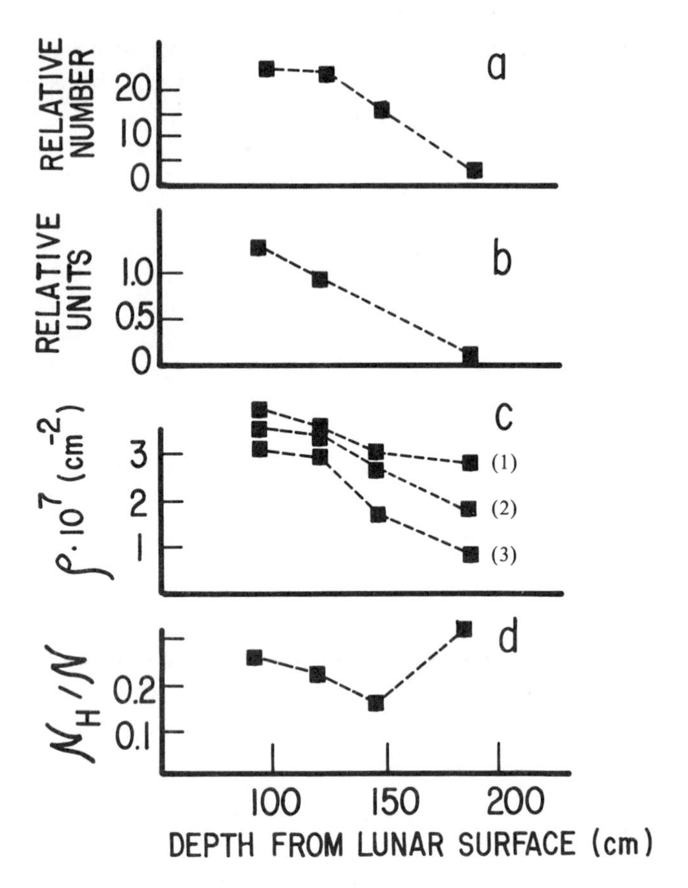

Fig. 5. Variation of various characteristics for basic samples of the Luna 24 soil core. (a) Content of agglutinates according to data (Rode *et al.*, 1977); (b) content of finely dispersed iron (Ivanov *et al.*, 1978); (c) track parameters ρ_{med}-(1); $\bar{\rho}$-(2); ρ_q-(3); and (d) parameter N_H/N.

sample 24184,4 represents the 4 cm core interval while Chaillou *et al.*'s sample 24184,1 represents the 1 cm core interval.

In Conclusion

The track characteristics obtained for four samples of the Luna 24 core point to the following most essential peculiarities: a) there is a noticeable change of radiation characteristics of Luna 24 samples from the top to the interior; b) there is an absence of weakly irradiated ($\rho \sim 10^5$ cm^{-2}) crystals in the Luna 24 regolith in comparison with the Luna 20 regolith; c) there is a bimodal track density distribution in sample 24184, with feldspar crystals being the least irradiated.

REFERENCES

Arrhenius G., Liag S., Macdougall D., Wilkening L., Bhandari N., Bhat S., Lal D., Rajagopalan G., Tamhane A. S. and Venkatavaradan V. S. (1971) The exposure history of the Apollo 12 regolith. *Proc. Lunar Sci. Conf. 2nd*, p. 2583–2598.

Barsukov V. L., Ivanov A. V., Nazarov M. A., Rode O. D., Stakheev Yu. I., Tarasov L. S., Tobelko K. I. and Florensky C. P. (1977) Preliminary description of the regolith core from Mare Crisium (abstract). In *Lunar Science VIII*, p. 67–70. The Lunar Science Institute, Houston.

Blanford G. E. and Wood G. C. (1977) Particle track densities in the Luna 24 core (abstract). In *Papers Presented to the Conference on Luna 24*, p. 41–43. The Lunar Science Institute, Houston.

Chaillou D., Pellas P., Levy M. C. and Storzer D. (1978) Cosmic ray track record and maturity of the Luna 24 regolith. In *Mare Crisium: The View From Luna 24* (R. B. Merrill and J. J. Papike, eds.), p 171–177. Pergamon, New York.

Fleischer R. L., Hart R., Jr. and Giard W. R. (1974) Surface history of lunar soil and lunar columns. *Geochim. Cosmochim. Acta* **38**, 365–380.

Florensky C. P., Basilevsky A. T. and Burba G. A. (1977) Geologo-morphological analysis on landing place of Luna 24 station. *Dokl. Akad. Nauk. SSSR* **233**, 936–939.

Gault D. E., Hörz F., Brownlee D. G. and Hartung J. B. (1974) Mixing of the lunar regolith. *Proc. Lunar Sci. Conf. 5th*, p. 2365–2386.

Goswami J. N., Braddy D. and Price P. B. (1976) Microstratigraphy of the lunar regolith and compaction ages of lunar breccias. *Proc. Lunar Sci. Conf. 7th*, p. 55–57.

Goswami J. N. and Lal D. (1974) Cosmic ray irradiation pattern at the Apollo 17 site: Implication to regolith dynamics. *Proc. Lunar Sci. Conf. 5th*, p. 2643–2662.

Goswami J. N. and Lal D. (1978) Particle tracks and microcraters in Luna 24 drill core soil samples (abstract). In *Lunar and Planetry Science IX*, p. 400–402. Lunar and Planetary Institute, Houston.

Ivanov A. V., Gorshkov E. S., Zukov V. V., Marov I. N. and Urusov V. S. (1978) Distribution of metallic iron in regolith of Luna 24 core. *Geokhimia*. In press.

Kashkarov L. L., Genaeva L. I. and Lavrukhina A. K. (1975) Track studies of radiation history of matter returned by Soviet automatic stations Luna 16 and Luna 20. In *Cosmochimia of Moon and Planet*, p. 593–601, Nauka, Moscow.

Kashkarov L. L., Genaeva L. I. and Lavrukhina A. K. (1976) Track investigation of energy spectrum of nuclei iron group solar cosmic rays in lunar soil returned by automatic stations Luna 16 and Luna 20. *Izvestia Akad. Nauk SSSR*, seria fiz. **40**, 535–538.

Krishnaswami S., Lal D., Prabhu N. and Tamhane A. S. (1971) Olivine: revelation of tracks of charged particles. *Science* **174**, 287–291.

Lal D., Muralli A. V., Rajan R. S., Tamhane A. S. Lorin J. C. and Pellas P. (1968) Technique for proper revelation and viewing etch-tracks in meteoritic and terrestrial minerals. *Earth Planet. Sci. Lett.* **5**, 111–119.

Price P. B., Hutcheon I. D., Braddy D. and Macdougall D. (1975) Track studies bearing on solar-system regoliths. *Proc. Lunar Sci. Conf. 6th*, p. 3449–3469.

Rode O. D., Ivanov A. V., Tarasov L. S. and Korina M. I. (1977) General lithologic-morphologic characteristics of Luna 24 regolith core. *Geokhimia* **10**, 1465–1476.

Ryder G., McSween H. Y. and Marvin U. (1977) Basalts from Mare Crisium. *The Moon* **17**, 263–287.

Proc. Lunar Planet. Sci. Conf. 9th (1978), p. 2137–2147.
Printed in the United States of America

Luna 24: Mineral chemistry of 90–150 micron clasts

HENRY O. A. MEYER
JIANN-YANG HWANG
ROBERT H. MCCALLISTER

Department of Geosciences, Purdue University
West Lafayette, Indiana 47907

Abstract—Seven grain mounts with clasts in the size range 0.09–0.15 mm from the Luna 24 Core have been examined to determine the modal abundances of various clasts and mineral compositions. With respect to the distribution of clasts, the only significant variation is found in samples from the 24174 horizon. At this level the greatest number of mafic mineral clasts is found. Consequently, the inference is that the sample represents a less reworked horizon or one which received a greater average influx of gabbroic and/or basaltic ejecta.

Analyses of the mineral clasts indicate that both clinopyroxene and olivine have a wide range of composition within the same sample. In addition, the pyroxene is commonly strongly zoned. Even with these compositional variations, most of the mafic minerals fall within a range which indicates they were derived from the comminution of clasts of very low titanium basalts and/or gabbros. Also, the possibility exists that a small number of mineral clasts may have been derived from a Mg-rich gabbro. The relatively high iron content of the analyzed plagioclase suggests that no Highland component is present.

Most of the regolith at the Luna 24 site is probably a mixture of locally derived mare basalt and ejecta from Fahrenheit crater.

INTRODUCTION

The Soviet unmanned Luna 24 mission returned from its landing site in Mare Crisium with an approximately 160 cm core of lunar regolith. The total slope depth cored was about 225 cm at an angle about 30° off vertical, but the upper 50 cm of the returned core sleeve was mostly devoid of material (Barsukov *et al.*, 1977; Florensky, *et al.*, 1977).

The landing site was in the south-eastern section of Mare Crisium about 40 km inside the encircling Highland rim. The structure of the mare is fairly complex with a major topographic break occurring between a depressed inner mare and a higher outer mare shelf (Butler and Morrison, 1977). The Luna 24 module landed on the surface of the inner mare region about 15 km from the margin of the 6 km diameter Fahrenheit crater. It is believed by Butler and Morrison (1977) that the regolith sampled at the Luna 24 site may consist of material derived from at least four different sources. The major portion of material is probably representative of local mare basalt mixed with ejecta from Fahrenheit crater. Some older, darker mantling material may have reached the site as well as fragments of Highland rocks (Butler and Morrison, 1977). A small possibility exists according to these same authors that some material from a larger crater,

Giordano Bruno, about 1320 km distant may be present at the Luna 24 site.

Barsukov (1977) believes that the ejecta blanket from Fahrenheit crater may be in the region of 50 to 100 cm thick at the Luna 24 site. An increase in magnetic susceptibility at around 133 cm depth in the core may mark the lower limit of this ejecta (Barsukov, 1977).

Preliminary analysis of the core (Florensky *et al.*, 1977; Tarasov *et al.*, 1977) indicated a high proportion of gabbroic fragments followed in abundance by mare basalt fragments and relatively rare Highlands material. More recently, Bence *et al.* (1977) and Albee *et al.* (1977) have described the gabbroic rocks, whereas Ryder *et al.* (1977), Vaniman and Papike (1977a) and Taylor *et al.* (1977) have discussed the nature and origin of basaltic material. All have noted the occurrence of a low-Ti mare basalt, somewhat akin to the VLT Apollo 17 basalt of Vaniman and Papike (1977b).

This report considers the mineralogy, composition and source relations of monomineralic clasts in the size range 0.09 to 0.15 mm. Identification of the various phases was accomplished using standard petrographic and microprobe techniques on polished grain mounts. Sample numbers, their position in the core length and relative abundances of pyroxene, olivine, plagioclase, opaques (spinel, ilmenite and Fe-metal) and other components (breccias, lithic clasts, glasses, agglutinates, etc.) are listed in Table 1 and shown in Fig. 1.

ABUNDANCES OF MINERAL CLASTS

The predominant mineral clasts in the seven grain mounts studied are pyroxene, olivine and plagioclase. Opaque phases, spinel and ilmenite, are rare and generally amount to less than 1 modal % of the total clast population. These opaque phases in the Luna 24 samples have been discussed by Haggerty (1977) and Friel and Goldstein (1977).

In the 0.09 to 0.15 mm clast size it is apparent (Fig. 1 and Table 1) that the

Table 1. Relative abundances of mineral clasts in 0.09 to 0.15 mm fraction.

	Percentage of various clasts					
Sample Number	Olivine	Pyroxene	Plagioclase	Opaque Phases	Others*	Total grains
24077,48	11	29	15	1	44	438
24109,43	13	34	9	1	43	314
24149,52	15	27	11	2	45	344
24174,40 } 24174,41 }	15	47	9	1	28	463
24182,40	11	41	13	1	34	326
24210,46	13	39	10	2	35	214

*Includes rock and breccia clasts, agglutinates and glasses.

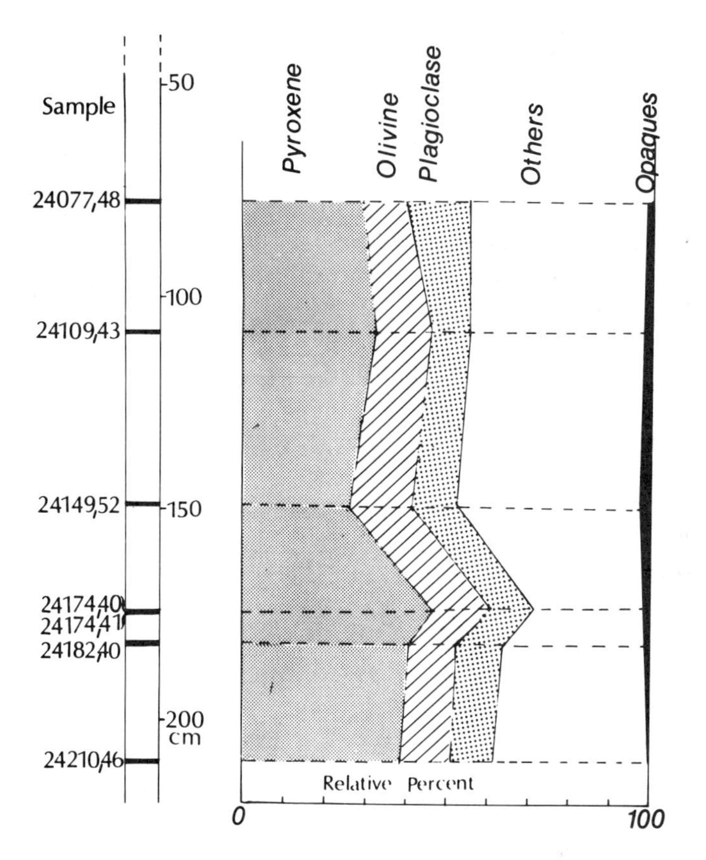

Fig. 1. Position of samples studied within the core length and relative abundances of various components in the samples (0.09 to 0.15 mm size range).

plagioclase is relatively constant in modal percentage, ranging between 9 and 15%. However, a modal study by Simonds *et al.* (1977) indicates a larger range in abundances for the same size range (i.e., 9 to 20 volume %). Possibly the different proportions are an indication of the error in determining modal values from different grain amounts based on relatively small total samples.

The mafic minerals, olivine and pyroxene, tend to be slightly more abundant at greater depth in the core (Fig. 1). Interestingly, McKay *et al.* (1977) noted a decrease in agglutinate content with depth and Nagle (1977) observed a greater proportion of crystalline fragments in the samples 24174 and 24182 than in the samples above. The major discontinuity in the overall trend occurs at the level of sample 24174. This horizon appears to contain the greatest proportion of mafic constituents. The phenomenon is probably real since it was observed in two separate grain mounts (24174,40 and 24174,41) examined from this horizon. McKay *et al.* (1977) observed that this sample (24174) had the smallest mean grain size (M_z) of all the Luna 24 samples. Furthermore, these authors also noted an increase in abundance of pyroxene (and plagioclase) with decreasing particle size.

Basu *et al.* (1977) have examined the variation of clasts in the <0.250 mm fraction of the Luna 24 samples, and their results are substantially in agreement with those reported in this study. Small differences in relative abundances of the monomineralic clasts between both studies can be explained satisfactorily on the basis of the limited grain size ranges.

Causes of variation in modal abundances for lunar samples are complex. Such factors as source, comminution, agglutinate formation and overall maturity of

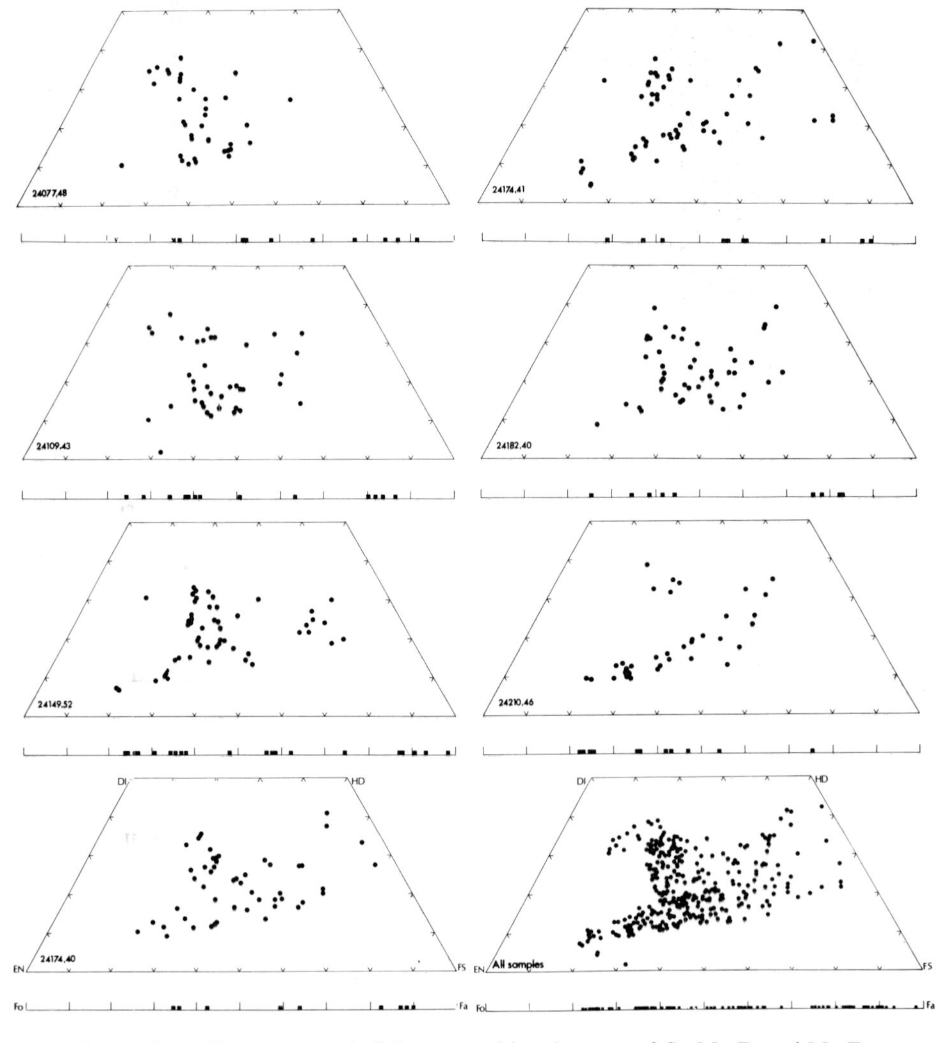

Fig. 2. Luna 24 pyroxene and olivine compositions in terms of Ca-Mg-Fe and Mg-Fe, respectively.

sample need to be addressed. Undoubtedly, the whole sample (i.e., total grain size and population) must be considered in this context since it is noted (Nagle and Walton, 1977) that there are significant differences in the major components (or clast types) with variation in grain size. Unfortunately, bulk soil compositions are of little help since the bulk compositions from all horizons are fairly similar (Blanchard *et al.*, 1977). Nevertheless, within the constraints of the 0.09 to 0.15 mm grain size data one can suggest that mixing or reworking was not as extensive in the horizon represented by 24174 as in other samples, or that this horizon received a larger component of gabbroic or basaltic material. Interestingly, the mafic minerals analyzed in this sample have a high percentage of extreme Fe-rich types (Fig. 2).

Mineral Chemistry

Representative clasts of pyroxene, olivine, plagioclase and opaque minerals were analyzed in all grain mounts using an automated MAC (ETEC) electron microprobe. Standards were either glasses or analyzed minerals and the data reduction was done by methods similar to Bence and Albee (1968) and Albee and Ray (1970).

Pyroxene

The range of pyroxene compositions is shown for each sample in Fig. 2 and representative analyses are presented in Table 2. In general, the pyroxenes from all horizons display similar chemical trends and are virtually identical with those of the VLT basalts and gabbros described by others (e.g., Vaniman and Papike,

Table 2. Representative analyses of pyroxenes in Luna 24 (90–150 micron fraction).

	24077,48		24109,43		24149,52		24174,40		24182,40		24210,46	
SiO_2	52.0	52.6	50.4	49.8	45.1	51.1	48.6	52.7	50.3	49.1	53.1	50.4
TiO_2	0.29	0.24	0.20	0.99	0.86	0.22	1.25	0.15	1.09	1.19	0.19	0.79
Al_2O_3	0.75	1.56	1.09	1.46	0.97	1.27	1.43	1.36	1.09	1.16	1.24	2.27
CR_2O_3	0.23	0.61	0.57	0.10	<0.01	0.57	0.19	0.68	0.06	0.07	0.64	0.69
FeO	24.9	13.7	24.4	28.8	37.8	16.5	25.3	15.6	28.6	27.4	17.7	13.8
MnO	0.41	0.24	0.41	0.43	0.61	0.35	0.41	0.27	0.43	0.36	0.42	0.27
MgO	15.0	15.5	16.3	7.17	5.20	17.0	9.57	22.3	8.09	4.01	21.4	14.2
CaO	6.52	15.7	6.34	12.4	8.88	11.9	12.7	6.35	11.1	17.2	5.35	15.9
Na_2O	0.18	0.04	0.23	0.05	<0.01	0.05	0.16	0.23	0.13	<0.01	0.01	0.11
K_2O	<0.01	0.01	0.02	<0.01	<0.01	<0.01	<0.01	<0.01	<0.01	<0.01	<0.01	<0.01
	100.6	100.2	100.0	101.2	99.4	99.0	99.6	99.6	100.9	100.5	100.1	98.4
Wo	14	33	13	27	19	25	28	13	25	39	11	34
En	45	45	47	22	16	50	29	63	25	13	61	43
Fs	41	22	40	50	65	25	43	24	50	48	28	23

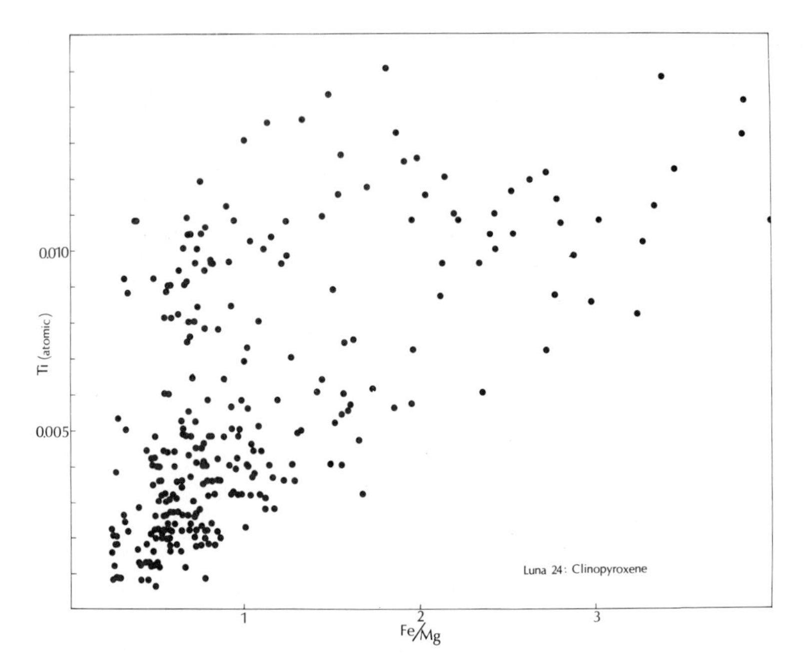

Fig. 3. Atomic Ti versus atomic Fe/Mg for all pyroxenes analyzed in the 0.09 to 0.15 mm fraction of Luna 24.

1977a; Bence *et al.*, 1977; Albee *et al.*, 1977; Papp *et al.*, 1977).

Several important points may be noted with regard to the distribution of pyroxene compositions within the quadrilateral. One is the large compositional range of the pyroxene from $Wo_7En_{75}Fs_{18}$ or $Wo_{37}En_{50}Fs_{13}$ to $Wo_{42}En_2Fs_{50}$. Zoning is extremely common and within one grain in 24210 the composition varies from $Wo_{35}En_{39}Fs_{26}$ to $Wo_{35}En_{16}Fs_{49}$. These compositional ranges and zoning trends are similar to those noted by Albee *et al.* (1977). Only one Ca-poor pyroxene was found (24109,43). Although the trend to iron-enrichment is obvious in pyroxenes from all horizons, it appears that it is most marked in the 24174 samples (Fig. 2).

From petrographic observations of the 0.09 to 0.15 mm grains, two types of pyroxene are clearly evident on the basis of color. One type is pale pink and has TiO_2 contents about or in excess of 1 wt.%. The other is usually colorless and has relatively low TiO_2 contents ($\ll 1$ wt.%) (Table 2). However, this distinction may be an artifact of size since Roedder and Weiblen (1977) noted that coarse pyroxene grains often showed a clear colorless zone and a later pale pinkish one. In Fig. 3, all pyroxene data are plotted in terms of atomic Ti versus atomic Fe/Mg. Although the majority of pyroxenes have relatively low Ti contents and Fe/Mg values, there is evidence of a second group characterized by higher Ti contents and considerably larger range in Fe/Mg. A similar distribution occurs for Ti versus Ca/Mg and it is likely that this partly represents the overall trend of

Fe and Ti enrichment noted by Grove and Bence (1977).

Although some pyroxenes do fall within the field of high-Mg gabbros as defined by Bence *et al.* (1977), the evidence for origin from such a source is not equivocal. Also, in terms of Ca-Fe-Mg there is some overlap of these Luna 24 Mg-rich pyroxenes with the field of Highland pyroxenes (e.g., Meyer and McCallister, 1976; Vaniman *et al.*, 1976). In general, on the basis of pyroxene chemistry it is logical to consider that most, if not all, pyroxene clasts have been derived from the comminution of rocks similar to those described as very low titanium basalts or gabbros. These rock types have been sampled directly in the coarser fragments of the Luna 24 material (e.g., Ryder *et al.*, 1977; Albee *et al.*, 1977).

Olivine

Olivine is less abundant than pyroxene in the samples studied but the clast population displays a considerable range in Mg/(Mg + Fe) (Fig. 2; Table 3). For example, in sample 24077,48 olivine clasts vary from composition of Fo_{78} to Fo_8. For the most part the clustering in the olivine compositions as noted by Papp *et al.* (1977), and Basu *et al.* (1977) is not entirely evident. This is probably due to an insufficient number of olivine grains having been analyzed.

It is reasonable to suggest that the fayalite-rich olivine is genetically related to the ferrobasalts or gabbros described by others. The more magnesian olivine is also likely to be associated with the same rock types, but is probably indicative of earlier, higher temperature crystallization. Because most of the Mg-rich olivine (Fo_{78}) is comparable to Highlands olivine in composition, its provenance is uncertain. Basu *et al.* (1977) consider that the Mg-rich olivine in Luna 24 samples may be derived from a hitherto unsampled coarse cumulate rock. This

Table 3. Representative analyses of olivine in Luna 24 (90-150 micron fraction).

	24077,48	24109,43	24149,52	24174,40	24182,40	24210,46
SiO_2	32.4	36.1	31.0	36.7	32.9	38.1
TiO_2	0.06	0.05	0.07	0.07	0.03	0.03
Al_2O_3	0.04	<0.01	0.02	0.05	<0.01	<0.01
Cr_2O_3	0.05	0.13	0.01	0.11	0.03	0.07
FeO	57.2	30.5	56.4	30.2	56.4	23.7
MnO	0.59	0.31	0.59	0.30	0.50	0.18
MgO	9.45	32.1	10.8	31.9	9.84	38.4
CaO	0.55	0.29	0.32	0.23	0.47	0.29
Na_2O	0.15	0.12	<0.01	0.07	<0.01	<0.01
K_2O	<0.01	<0.01	<0.01	<0.01	<0.01	<0.01
	100.5	99.6	99.2	99.6	100.2	100.8
NiO less than 0.01 in all samples						
Fo	23	65	25	65	24	74

rock type may be equivalent to a high Mg-gabbro whose existence is inferred from the compositions of some pyroxene grains (Bence *et al.*, 1977).

Plagioclase

Plagioclase is the second most abundant mineral clast. The plagioclase is calcic with a relatively restricted range between An_{74} and An_{98}, and very little orthoclase (Fig. 4, Table 4). Compared to plagioclase in Highland rocks, the Fe contents of the clasts analyzed are considerably higher, and several grains showed marked zoning with respect to Fe-content. It is likely that most of the feldspar clasts examined in this limited size range have a mare origin.

Opaque phases

Individual clasts of opaque phases are relatively rare in most sections. However, several grains of ilmenite as well as Cr-spinel inclusions in clinopyroxene and plagioclase were analyzed (Table 5). All of the analyzed opaque minerals fall within the compositional region mentioned by Haggerty (1977), although no ulvöspinel was observed.

DISCUSSION

The results of this study of the mineralogy and mineral chemistry of seven samples of 0.09–0.150 mm size fraction from Luna 24 are generally similar to that reported elsewhere (e.g., Papp *et al.*, 1977; Basu *et al.*, 1977). Also, the majority of the mineral clasts have compositions equivalent to analyzed mineral constituents of the basaltic and gabbroic rocks that are present in the larger fragments.

Table 4. Representative analyses of plagioclase in Luna 24 (90-150 micron fraction).

	24077,48	24109,43	24149,52	24174,40	24182,40	24210,46
SiO_2	48.6	44.1	43.7	45.8	47.5	50.8
TiO_2	0.04	<0.01	0.02	0.02	0.02	0.03
Al_2O_3	32.9	34.4	36.0	34.1	32.9	30.6
Cr_2O_3	0.02	<0.01	<0.01	0.04	<0.01	<0.01
FeO	0.37	0.32	0.13	0.44	0.48	0.68
MnO	<0.01	0.01	0.02	<0.01	0.01	<0.09
MgO	0.27	0.20	0.07	0.15	0.20	0.04
CaO	18.0	18.9	18.3	18.5	18.7	14.4
Na_2O	1.54	0.52	1.25	0.91	0.83	2.83
K_2O	0.04	0.01	0.04	0.02	0.01	0.35
	101.8	98.5	99.5	100.0	100.7	99.8
An	87	95	89	92	93	74

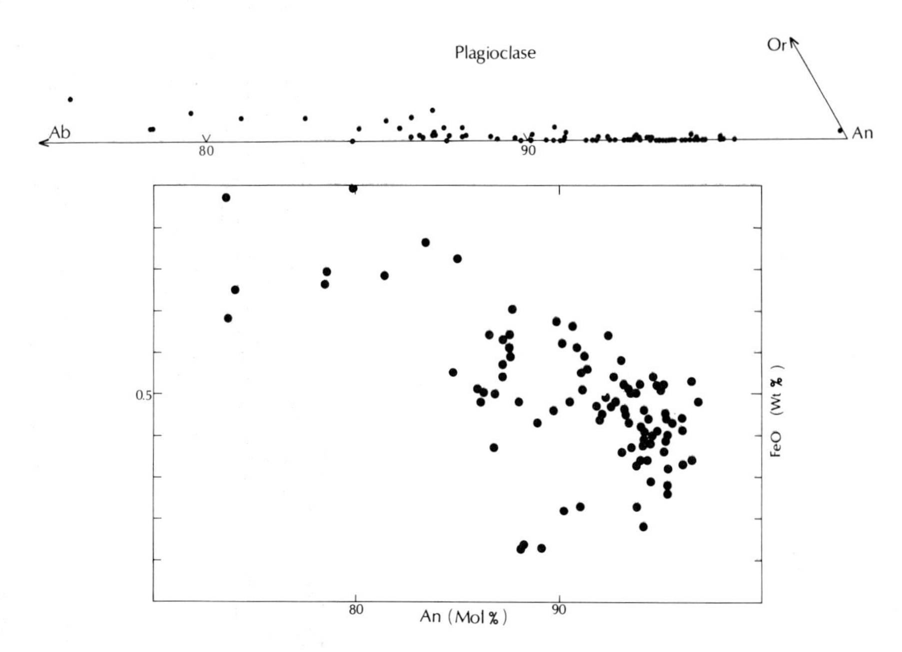

Fig. 4. Feldspar grains in Luna 24 samples (0.09 to 0.15 mm fraction).

The ubiquitous monomineralic nature of almost all the clasts in the size range studied plus the lack of fine grained lithic clasts strongly suggests that the parent rocks from which the fragments were derived were relatively coarse grained. Furthermore, the apparent similarity of all compositional ranges of the mineral clasts also suggests that if there are grain size differences in the parent rocks (i.e., basalt versus gabbro), both rocks are compositionally and mineralogically

Table 5. Representative analyses of opaque minerals in Luna 24 (90–150 micron fraction).

		Spinel		
	Ilmenite	in CPX	in Plag	Free
SiO_2	0.04	0.23	0.21	0.05
TiO_2	51.7	2.87	1.36	1.04
Al_2O_3	0.18	17.0	21.1	13.9
Cr_2O_3	0.09	40.4	41.8	53.3
FeO	45.8	35.3	0.34	22.5
MnO	0.22	0.45	28.9	0.34
MgO	0.23	3.39	6.26	7.90
CaO	<0.01	0.17	0.17	0.01
	98.3	99.8	100.1	99.0

similar. Haggerty (1977), on the basis of opaque minerals, believed that both gabbro and basalt were closely related in space.

Unfortunately, in terms of defining provenance it appears that the regolith at the Luna 24 landing site is almost totally of locally derived mare material. Possibly, the gabbroic material may represent the more deeply excavated material from Fahrenheit Crater; whereas, the basalt is from near surface lava flows. This idea of course necessitates that most of the regolith at the landing site is a mix of Fahrenheit and local ejecta. The evidence for Highlands material based on mineral clasts alone in the samples studied is extremely tenuous. Bence and Grove (1977) and Maxwell and El-Baz (1977) believe such fragments do occur. Furthermore, Simonds *et al.* (1977) and Basu *et al.* (1977) have commented on the presence of Highlands glass in other samples of Luna 24. No clasts are sufficiently exotic as to warrant considering them as representatives of dark mantling material or ejecta from Giordano Bruno (Butler and Morrison, 1977). However, Ryder *et al.* (1977) have suggested that olivine vitrophyre may represent the dark mantling material 20 km east of Luna 24.

Acknowledgments—This study was supported by NASA grant NGR 15-005-157. We appreciate the help of Dr. N. Z. Boctor in obtaining some analyses, as well as the opportunity provided us by NASA to study these samples. The initial manuscript benefitted greatly from the critical comments of Drs. Dymek and Neilson, and an anonymous reviewer.

REFERENCES

Albee A. L., Chodos A. A. and Dymek R. F. (1977) Petrography of Luna 24 sample 24170 (abstract). In *Papers Presented to the Conference on Luna 24*, p. 5–7. The Lunar Science Institute, Houston.

Albee A. and Ray L. (1970) Correction factors for electron probe microanalysis of silicates, oxides, carbonates, phosphates and sulphates. *Anal. Chem.* **42**, 1408–1414.

Barsukov V. L. (1977) Preliminary data for the regolith core brought to earth by the automatic lunar station Luna 24. *Proc. Lunar Sci. Conf. 8th*, p. 3303–3318.

Barsukov V. L., Tarasov L. S., Dmitriev L. V., Kolesov G. M., Shevaleevsky I. D. and Garanim A. V. (1977) The geochemical and petrochemical features of regolith and rocks from Mare Crisium (preliminary data). *Proc. Lunar Sci. Conf. 8th*, p. 3319–3332.

Basu A., McKay D. S. and Fruland R. (1977) Petrography, mineralogy and source rocks of Luna 24 drill core soils (abstract). In *Papers Presented to the Conference on Luna 24*, p. 18–21. The Lunar Science Institute, Houston.

Bence A. E. and Albee A. (1968) Empirical correction factors for the electron microanalysis of silicates and oxides. *J. Geol.* **76**, 382–403.

Bence A. E. and Grove T. L. (1977). The Highland component in the Luna 24 core (abstract). In *Papers Presented to the Conference on Luna 24*, p. 22–24. The Lunar Science Institute, Houston.

Bence A. E., Grove T. L. and Scambos T. (1977) Gabbros from Mare Crisium: An analysis of the Luna 24 soil. *Geophys. Res. Lett.* **4**, 493–496.

Blanchard D. P., Haskin L. A., Brannon J. C. and Aaboe E. (1977) Chemistry of soils and particles from Luna 24 (abstract). In *Papers Presented to the Conference on Luna 24*, p. 37–40. The Lunar Science Institute, Houston.

Butler P. and Morrison D. A. (1977) Geology of the Luna 24 landing site. *Proc. Lunar Sci. Conf. 8th*, p. 3281–3301.

Florensky C. P., Basilevsky A. T., Iranov A. V., Pronin A. A. and Rode D. D. (1977) Luna 24: Geologic setting of landing site and characteristics of sample core (preliminary data). *Proc. Lunar Sci. Conf. 8th*, p. 3257–3279.

Friel J. J. and Goldstein J. I. (1977) Metallic phases in the Luna 24 soil samples. *Geophys. Res. Lett.* **4**, 481–483.

Grove T. L. and Bence A. E. (1977) Petrogenesis of gabbros from Mare Crisium (abstract). In *Papers Presented to the Conference on Luna 24*, p. 64–67. The Lunar Science Institute, Houston.

Haggerty S. E. (1977) Luna 24: Opaque mineral chemistry of gabbroic and basaltic fragments from Mare Crisium. *Geophys. Res. Lett.* **4**, 489–492.

Maxwell T. A. and El-Baz F. (1977) Sources of Highland material in Mare Crisium regolith (abstract). In *Papers Presented to the Conference on Luna 24*, p. 110-113. The Lunar Science Institute, Houston.

McKay D. S., Basu A., Waits G., Clanton U., Fuhrman R. and Fruland R. (1977) Grain size and evolution of Luna 24 soils (abstract). In *Papers Presented to the Conference on Luna 24*, p. 114–117. The Lunar Science Institute, Houston.

Meyer H.O.A. and McCallister R. H. (1976) Mineral, lithic and glass clasts <1 mm size in Apollo 16 core section 60003. *Proc. Lunar Sci. Conf. 7th*, p. 185–198.

Nagle J. S. (1977) Luna 24, Clast population (abstract). In *Papers Presented to the Conference on Luna 24*, p. 129–131. The Lunar Science Institute, Houston.

Nagle J. S. and Walton W. J. A. (1977) Luna 24: Catalog and preliminary description. NASA, Johnson Space Center, Houston. 85 pp.

Papp H., Steele I. M. and Smith J. V. (1977) Luna 24 90–150 micron fraction: Implications for sampling of planetary and asteroidal regoliths (abstract). In *Papers Presented to the Conference on Luna 24*, p. 143–146. The Lunar Science Institute, Houston.

Roedder E. and Weiblen P. W. (1977). Petrology of melt inclusions in Luna 24 samples (abstract). In *Papers Presented to the Conference on Luna 24*, p. 152–155. The Lunar Science Institute, Houston.

Ryder G., McSween H. Y. and Marvin U. B. (1977) Basalts from Mare Crisium. *The Moon* **17**, 263–288.

Simonds C. H., Warner J. L., McGee P. E., Cochran A. and Brown R. W. (1977) Luna 24: Petrochemical potpourri (abstract). In *Papers Presented to the Conference on Luna 24*, p. 170–173. The Lunar Science Institute, Houston.

Tarasov L. S., Nazarov M. A., Sheraloevsky I. D., Kudryashova A. F., Gaverdovskaya A. S. and Korina M. I. (1977) Mineralogy and petrology of lunar rocks from Mare Crisium. *Proc. Lunar Sci. Conf. 8th*, p. 3333–3356.

Taylor G. J., Warner R., Wentworth S., Keil K. and Sayeed U. (1977) Petrology and chemistry of Luna 24 mare basalts and basaltic glasses (abstract). In *Papers Presented to the Conference on Luna 24*, p. 189–192. The Lunar Science Institute, Houston.

Vaniman D. T., Lellis S. F., Papike J. J. and Cameron K. L. (1976) The Apollo 16 drill core. Modal petrology and characterization of the mineral and lithic component. *Proc. Lunar Sci. Conf. 7th*, p. 199–239.

Vaniman D. T. and Papike J. J. (1977a). Ferrobasalts from Mare Crisium: Luna 24. *Geophys. Res. Lett.* **4**, 497–500.

Vaniman D. T. and Papike J. J. (1977b) Very low Ti (VLT) basalts: A new mare rock type from the Apollo 17 drill core. *Proc. Lunar Sci. Conf. 8th*, p. 1443-1471.

Proc. Lunar Planet. Sci. Conf. 9th (1978), p. 2149–2165.
Printed in the United States of America

Progress towards the direct measurement of $^{13}C/^{12}C$ ratios for hydrolysable carbon in lunar soil by static mass spectrometry

L. R. GARDINER, A. J. T. JULL and C. T. PILLINGER

Planetary Sciences Unit, Department of Mineralogy and Petrology
University of Cambridge, England

Abstract—At the present time, no valid measurement exists for the $^{13}C/^{12}C$ isotopic composition of hydrolysable carbon in lunar soils. An attempt is being made directly to measure carbon isotopes from CD_4 released by deuterated acid dissolution studies using static mass spectrometry. This paper describes and evaluates the method. Two major problems exist in terms of using the method to obtain a measure of the isotopic composition of C_{hyd} relative to PDB. The first involves fractionation of carbon isotopes due to partial deuterium labelling. Limited experimental evidence suggests that a correction factor established assuming that CD_4 and CD_3H have the same carbon isotopic composition is insufficient. Therefore, ^{13}C may be preferentially concentrated in the CD_3H and consequently $^{13}C_{CD_4}$ underestimates isotopic enrichment of C_{hyd} relative to PDB. The second problem involves yield of species, other than deuteromethane, from acid dissolution. Higher molecular weight deuterocarbons are known to be formed and carbon may be precipitated, or true carbide unreacted. At present, there is no evidence to indicate how carbon isotopes would be fractionated. Some possible solutions to the problems are offered. Production of species other than methane invalidates estimates of the isotopic composition of C_{hyd} based on measurements performed on carbon dioxide obtained by oxidation of CH_4 released from C_{hyd} using protonated acids.

Although no absolute measure of $^{13}C/^{12}C$ ratios of C_{hyd} is possible, relative values of $^{13}C_{CD_4}$ may provide useful information. We tentatively identify two trends for particle separates from 12023. Agglutinates may be enriched in ^{13}C relative to less magnetic material, and bulk soil. Fine fractions of bulk material may be depleted in ^{13}C relative to coarser grained material.

INTRODUCTION

Studies of carbon isotope systematics for the lunar soil have in the main been directed towards measurement of the $^{13}C/^{12}C$ ratio of carbon exhaustively extracted from samples by high temperature pyrolysis or combustion (see *inter alia* Epstein and Taylor, 1973; Chang *et al.*, 1974; Kerridge *et al.*, 1975, 1978 Kaplan *et al.*, 1976; DesMarais *et al.*, 1975). The rationale for studying total carbon is easy to follow: any change observed in the isotopic ratio of the sample relative to a standard is real (assuming no contamination) and not the result of fractionation due to the extraction procedure. All the investigators agree that the bulk carbon in lunar soils is enriched by as much as +20‰ relative to PDB; carbon isotope fractionation appears to be a function of maturity. A variety of mechanisms have been proposed to account for the enrichment including thermal diffusion, hydrogen stripping and sputtering. It is impossible to rule out that the solar wind itself is enriched in ^{13}C.

One problem with total extraction is that it homogenises the carbon present in the sample so that the isotope ratio measured is the weighted average of the ratios of the individual phases. A study of different particle types (e.g., DesMarais *et al.*, 1975) does not necessarily help solve this problem. In the case of the lunar soil, much potentially useful information could be lost since three major phases (>10%, indigenous lunar carbon, hydrolysable carbon (C_{hyd}) and carbon implanted into lunar silicates) and one important minor phase (1–5%, trapped hydrocarbons, particularly methane) are present. Attempts to measure the carbon isotopic ratio of the individual phases, with the exception of indigenous lunar carbon which can be determined from lunar igneous rocks (e.g., DesMarais, 1978), are comparatively rare. Stepwise pyrolysis has been employed to only a limited extent (e.g., Chang *et al.*, 1974); one measurement suggests that trapped CH_4 has a $\delta^{13}C_{PDB}$ of 30–45‰. The remaining data are not so easily interpreted since carbon released as CO and CO_2 over the temperature range 400–1170°C, although all enriched in ^{13}C relative to PDB, unlike nitrogen (Becker and Clayton, 1975) shows no easily recognisable major peaks in the isotope distribution.

Epstein and Taylor (1975) examined the residues of soil samples which had been treated with fluorine for unextracted carbon. Their data suggested that although only a minor portion (15%) of the total carbon was removed, the decrease in $^{13}C/^{12}C$ ratio implied that the fraction lost was enriched relative to PDB in ^{13}C by as much as +100 to +200‰. Three of the fluorine treated residues were subsequently examined by the DC1 acid dissolution and magnetic methods and a limited modal analysis carried out (Pillinger *et al.*, 1976, and unpublished data). The results obtained were tantalisingly ambiguous and unfortunately further samples were unavailable. The two samples which would be considered as mature or submature soils, 10084 and 14149 respectively, were substantially depleted in both C_{hyd} and CH_4, and one of them, 10084, had a greatly reduced abundance of finely divided iron. No particles recognisable as glassy agglutinates were found in the treated material. It is possible to equate the missing portion of the total carbon with C_{hyd} and CH_4, and postulate that C_{hyd} is considerably enriched in ^{13}C.

Other efforts to measure the isotopic composition of C_{hyd} (Chang *et al.*, 1971; Chang *et al.*, 1974; Kerridge *et al.*, 1975) have involved the release of C_{hyd} as CH_4 by 6N H_2SO_4 dissolution, together with the trapped methane. Conversion of the total gas to CO_2 is followed by isotopic abundance determination by standard techniques. Estimates of the isotopic enrichment relative to PDB for total carbon released as CH_4 are between +12 and +20‰; the C_{hyd} fraction must be isotopically lighter if it is assumed that trapped methane, which constitutes *ca.* 25–30% of the fraction, is *ca.* +40‰ relative to PDB (Kerridge, pers. comm.).

Both procedures for determining the isotopic ratio of hydrolysable carbon can be criticised. Results from the method based on the analysis of fluorine treated residues are extremely tenuous. No isotopic data were obtained from C_{hyd}; the conclusions reached were supposition which relied on a measurement made by difference on samples which could have been contaminated by any of the handling steps, and no overall mass balance was obtained. Even the method used by Chang *et al.* (1971; 1974) and Kerridge *et al.* (1975) was not necessarily immune from contamination as the following example could illustrate. Chang *et al.* (1971) have measured the $\delta^{13}C$ of hydrocarbons released from lunar soil 12023 as either 21 $\mu g/g$ or 26 $\mu g/g$ of C as CO_2 with an isotope ratio of +14 and +5‰ respectively. It was suggested (Chang *et al.*, 1971) that the value having the higher total carbon and the lower $\delta^{13}C$ value, reflected contamination introduced during the measurement of the very small amounts of gas involved. Thus, even the higher measurement for $\delta^{13}C$ represents a minimum value for hydrolysable carbon, since contamination may still be present (Chang *et al.*, 1971). Our measurements of CD_4 and CH_4 released by DC1 from 12023 (Cadogan *et al.*, 1972; Pillinger *et al.*, 1978) suggest that contamination could account for *ca.* 8 $\mu g/g$ and 13 $\mu g/g$ of the 21 and 26 $\mu g/g$ respectively of C measured as CO_2 by Chang *et al.* (1971). The solution of simultaneous equations indicates that the hydrolysable carbon, and the trapped CH_4 are, on average, enriched relative to PDB in ^{13}C (+43‰), whereas the contamination is light ($\delta^{13}C_{PDB}$ = −33‰) and typical of terrestrial hydrocarbons.

It appears that we know very little about the carbon isotopic composition of hydrolysable carbon. With this knowledge, the isotopic composition of three of the four carbon components in lunar soil would be available. It would then be possible to make valuable inferences about the $^{13}C/^{12}C$ ratio of the fourth component, the solar wind carbon implanted into lunar silicates. At the moment, the implanted carbon cannot be measured directly because it cannot be isolated from other forms of the element. The isotopic composition of carbon in C_{hyd} could also be an important factor in deciding between mechanisms postulated for the production of finely divided iron in lunar soils by exposure induced reduction. It would almost certainly reflect diffusion losses of carbon during formation of agglutinates and may indicate the mode of incorporation of carbon into the metal.

In view of the importance we attach to the measurement, we have attempted to develop a method for the direct determination of the $^{13}C/^{12}C$ ratio of C_{hyd} from CD_4. Such a procedure should be relatively immune from problems of contamination. The technique used involves static mass spectrometry (Gardiner, 1978; Gardiner *et al.*, 1977; Pillinger *et al.*, 1978) and has considerable sensitivity (5-10ng of carbon is a typical sample size). We shall discuss in this paper the operation of the method, its precision, and some problems (and possible solutions) which are associated with its application. Preliminary data on fractions from 12023 will also be reported.

THE STATIC MASS SPECTROMETRIC METHOD OF CARBON ISOTOPE DETERMINATION

The mass spectrometer is a V. G. Micromass 603 incorporating a Nier-type electron impact source operated under static conditions with an analogue electron multiplier as the detector. Source parameters include operation of a LaB_6 (Re) filament at reduced electron emission (trap current 4 μA) and electron energy 60eV. The filament temperature is calculated to be *ca.* 700°C. Ion accelerating voltage is fixed at 4 Kv and repeller voltage constant at +1.0V with respect to the source. A collector slit of 1.0 mm together with the flight tube geometry (90° sector, 6.3 cm radius) gives a resolving power of *ca.* 44 (10% valley).

Gas handling

Gases released from samples by acid dissolution are expanded from the reaction vessel (cooled to −196°C) into the inlet manifold valved off from the ion source. The manifold contains a liquid nitrogen cooled U tube that holds back D_2O and C_2–C_3 hydrocarbon vapours which are potential interferences at masses 19, 20 and 21. A Zr-Al getter (SORB-AC, SAES getters) operated at room temperature allows the selective removal of H_2, D_2, CO, N_2 and CO_2 without interference to CD_4. Some additional CH_4 is also generated from the getter during this procedure. A molecular sieve filled trap (−196°C) quantitatively traps down CD_4, CH_4 and Ar and allows the separation of He and Ne which are analysed separately as an untrapped fraction. The CD_4 is then quantitatively desorbed off the sieve and expanded into the statically operated ion source. The cold SORB-AC getter is operated continually during isotope measurements to stabilize the background pressure in the spectrometer during static operation. Source pressure changes over the measurement period are relatively small and are attributed to background rise of Ar^{40} (10^{-9} cc STP/minute). Impurity gases associated with CD_4 are consequently restricted to CH_4 and Ar.

Isotope measurements

Carbon isotopic analyses are performed by measurement of ions due to $^{12}CD_4$ and $^{13}CD_4$ (m/e 20 and m/e 21). Data are collected by repetitive scanning (usually every 30 seconds) to cover the mass range m/e 19.5 to 21.5. The gain of the amplifier is increased a factor of 100 between the major and minor ions. In the case of 15601 soil separates, appropriate values for the baseline and the flat peak top were measured initially with a pen recorder. Data could also be collected by digitisation of the amplifier signal using a digital voltmeter, with paper-tape output. At least twenty data points were taken at the top of the peak and five complete scans were collected. The CD_4/CD_3H ratios were obtained separately. A standard was run approximately 40 minutes before the sample gas and again 5 minutes after the sample gas had been pumped away. Amplifier range settings were the same for sample and standard in order to eliminate recorder bias.

By the time 12023 studies commenced, some considerable instrumental improvements had been made. Data collection and reduction is now accomplished by an on-line DEClab 11/03 computer system which includes a 12-bit A/D converter. A VT-55 graphics display unit provides a real-time spectrum and amplifier gain is controlled by software. Routinely, data from 15–30 consecutive scans are recorded, each scan averages typically 40 measurements of the baseline before m/e 20 and before and after m/e 21. Peak heights are averaged over the full extent (*ca.* 60 points) of the flat region at the top of each peak to improve the precision of the ion current measurement. A subroutine checks that peak positions have not changed relative to the first scan. The running sequence has now been slightly altered so that after the sample has been analysed, a standard spike is introduced and the 21/20 ratio of the mixture is measured. After pump-out, a fresh standard is analysed immediately. CD_4/CD_3H ratios for the samples are measured from a comparison of the m/e 20 and m/e 19 peaks from six consecutive scans. Ratios are calculated from the baseline-corrected ion intensity ratio m/e 21/20 for each scan. Additionally, m/e 21 from any one scan is compared to m/e 20 from the previous scan. Isotopic enrichments are calculated according to the formula:

$$\delta^{13}C_{CD_4} = \left[\frac{(21/20)_{sample} - (21/20)_{std}}{(21/20)_{std}}\right] \times 1000$$

In order to compare the observed carbon isotopic enrichments with literature values for lunar carbon, six aliquots of our working CD_4 standard were sent to Indiana University for combustion and conventional measurement as CO_2. The average ^{13}C value relative to PDB was −28.9‰.

System Evaluation

Ion beam stability

The choice of a low work function LaB_6 filament reduces many of the problems that would otherwise affect the study of active gases under static conditions. The emitter characteristics of this material provide a means of maximising electron emission (and therefore sensitivity) for the lowest operating temperature.

Unfortunately for the analysis of carbon species, low wattage emitters are not sufficient in themselves to ensure a stable ion beam. Other effects interfere even at the reduced operating temperatures. The filament is able thermally to decompose certain gases by a process involving thermal cracking. Our filament is therefore operated at reduced electron emission (trap current 4 μA) to minimise these pumping effects. Despite this, the behaviour of CO_2 in the static mass spectrometer is found to be very unstable and obviously would be unsuitable for isotopic studies since substantial isotopic fractionation effects accompany filament pumping and adsorption.

On the other hand, the behaviour of the isotopic methanes (CH_4 and CD_4) exhibit remarkably high thermodynamic stability in the source and relatively stable ion currents are achieved under the operating conditions employed. Thus, the half life of CD_4, assuming an exponential decay, is found to be *ca.* 11 hours. Gas residence time in the source is approximately fifteen minutes and therefore during an analysis <2% of the deuterocarbon is degraded. Although a small difference is found in the removal rate of the ions at masses 20 and 21 during very long term experiments, implying that the 20/21 ratio is time dependent, no detectable change in this ratio, within error limits, is observed for gas residing in the source over a period of 2 hours. Hence for the time scale and conditions of these experiments, gas degradation does not contribute significantly to isotope fractionation. Further studies are in progress to accurately measure the fractionation factor and establish models that may explain the isotope kinetics of CD_4 in the source.

One peculiarity of the system is that sensitivity increases during the course of operation; concurrently, response to the ^{12}C isotope is increased so that standards appear to vary in $^{13}C/^{12}C$ during a day's operation. This effect is most marked during the hour after the first standard of the day is introduced. We believe that the enhanced gain is in some way connected with the activation of the conversion dynode of the electron multiplier. It is not a fractionation in the working standard since after overnight pumping, a similar series of absolute peak heights and isotopic ratios for CD_4 may be obtained the following day. To avoid major problems, a warm-up standard is left in the instrument for an hour before the normal operating sequence of standard-sample-standard is commenced.

Interferences

Mass spectrometer background and blank contributions at masses 20 and 21 are less than 0.1% of typical sample peaks and can be considered to be negligible. No memory effects above background are associated with CD_4 presumably because the relative stability and non-polar nature of the molecule discourage adsorption on the walls of the vacuum chamber. Pump-out time is <20 seconds, but the time between introducing samples and standards is considerably longer, *ca.* 3 minutes. Interference from ^{20}Ne and ^{21}Ne released simultaneously from lunar samples by the acid treatment is avoided by cryogenic separation of CD_4 on 5 A molecular sieve. Removal of neon is made by successively expanding the gas from the inlet manifold to the spectrometer source (the sieve is not directly pumped). A mass spectrometric analysis of both the neon 20/22 ratio and the size of the mass 21 peak demonstrates conclusively that 99.9% of the CD_4 is retained by the trap. Interferences from $^{40}Ar^{++}$ may likewise be ignored. Under the source conditions employed, $^{40}Ar^{++}$ is *ca.* 0.24% of $^{40}Ar^{+}$ and during a typical analysis, ^{40}Ar from background and lunar gas is $<10\%$ of the CD_4.

A significant interference during sample analysis arises from partial labelling of the methanes released by the deuterated acid dissolution method. Thus, a variable amount of CD_3H is always formed along with CD_4 e.g., CD_4/CD_3H ratios for the samples analysed fall within the range 2.5–10 with the majority *ca.* 3–6 (Fig. 1). The CD_4/CD_3H ratio of the working standard, determined over a period of 6 months, is relatively constant at 18.8 ± 0.1 (1σ). As $^{12}CD_3H$ occurs at m/e 19 there must be an isotopic contribution to m/e 20 from $^{13}CD_3H$. The $^{13}C/^{12}C$ ratio derived from the measured 21/20 ratio is therefore artificially low. If it were possible to assume that CD_3H has the same $^{13}C/^{12}C$ abundance as CD_4,

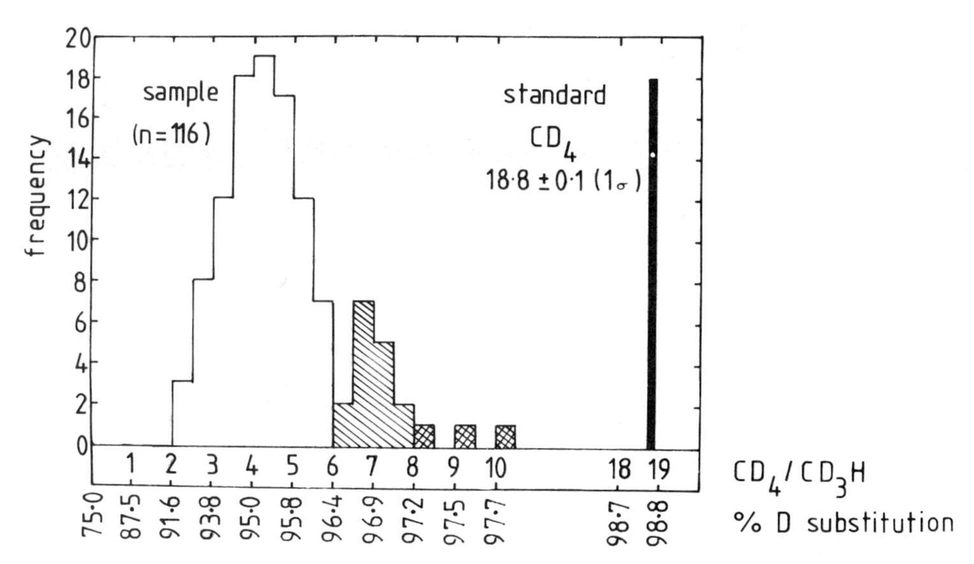

Fig. 1. CD_4/CD_3H ratio variability and equivalent deuterium substitution in atom % D. Normal acid (□) freshly opened new acid (▧) freshly opened acid on single >250 μm grains (▩).

a relatively simple correction factor could be applied to the data. This would take the form:

$$^{13}R = \frac{R_1R_2 - \sqrt{(R_1R_2)^2 - 4R_1R_2}}{2R_2} \tag{1}$$

where R_1 = mass 20/mass 19; R_2 = mass 20/mass 21; ^{13}R = "true" $^{13}CD_4/^{12}CD_4$ ratio.

Both R_1 and R_2 can be measured directly.

Abundance sensitivity

Abundance sensitivity is a measure of the tailing between two masses M and M + ΔM, and in the case of CD_4 is defined as the ratio of the number of m/e 20 ions relative to the number of m/e 20 ions contributing to m/e 21. Multiplier output ion currents for a typical CD_4 standard are 5.78×10^{-7} amp (m/e 20) and 6.03×10^{-9} amp (m/e 21). The contribution of the m/e 20 tail to m/e 21 is found to be 0.20% of m/e 21 corresponding to 1.21×10^{-11} amp. Abundance sensitivity is therefore calculated to be 4.8×10^4 at an instrument resolution of 44. This figure is found to vary between *ca.* 3–5 × 10^4 for sample CD_4 gas (*ca.* 10–20 ng C as CD_4). Nevertheless, no abundance sensitivity correction factor is required since allowance is made during data collection by baseline subtraction for both mass 20 and 21 peaks.

Precision

Table 1 includes data obtained from 10 acid dissolution experiments and illustrates the variation in precision found for the absolute 20/21 ratio and $\delta^{13}C_{CD_4}$. Precision is expressed in terms of the standard error of the mean ratio (quoted as 95% confidence limits) found by averaging n scans. An average reproducibility of 1.3‰ and 2.2‰ is demonstrated for the absolute 20/21 ratio of standard and sample respectively, on gas samples of *ca.* 10 ng C. The different precisions for samples and standards are not understood. The presence of CH_4 and argon with the CD_4 sample gas during the analysis may exert a destabilising influence on the isotopic measurement. Detailed studies are planned to characterise the magnitude of this effect by controlled spiking of CD_4 gas with variable amounts of impurity gases. In this regard, it is significant that measurements on the 21/20 ratio after adding a CD_4 spike to the sample invariably decreased the 21/20 ratio compared to the expected value by *ca.* 2–3‰. We suspect a pressure

Table 1. Precision of isotope measurements on CD_4.

Run	Standard				Sample				$\delta^{13}C_{CD_4}$	
	$\bar{x}$ 20/21	n scans	S_x	$CV_{\bar{x}}$ ‰	$\bar{x}$ 20/21	n scans	S_x	$CV_{\bar{x}}$ ‰	‰	$\pm 2\sigma$
47	95.883	22	0.305	1.4	89.669	14	0.224	1.4	+70	1.8
	96.016	14	0.215	1.3						1.7
48	96.016	14	0.215	1.3	89.943	16	0.475	3.1	+68	2.7
	96.154	16	0.242	1.4						2.8
49	96.154	16	0.242	1.4	90.945	8	0.306	2.8	+56	2.6
	96.059	14	0.098	0.6						2.3
50	96.016	8	0.211	1.8	90.733	13	0.339	2.2	+60	2.5
	96.303	12	0.108	0.7						2.0
51	96.059	14	0.098	0.6	90.809	13	0.296	2.0	+59	1.8
	96.231	7	0.158	1.5						2.1
52	95.929	16	0.204	1.1	89.520	13	0.383	2.6	+72	2.4
	96.054	9	0.163	1.3						2.1
53	95.681	12	0.230	1.5	89.518	13	0.421	2.9	+70	2.8
	95.802	18	0.146	0.8						2.5
54	95.439	11	0.172	1.2	89.136	22	0.376	1.9	+73	2.0
	95.929	16	0.204	1.1						1.9
55	96.054	9	0.163	1.3	89.611	18	0.264	1.5	+71	1.7
	96.016	8	0.211	1.8						2.0
56	95.781	10	0.258	1.9	89.717	22	0.346	1.7	+69	2.2
	96.123	18	0.200	1.0						1.8

Notes: $\bar{x}$ is the mean 20/21 ratio determined from n scans with an estimated standard deviation S_x. $CV_{\bar{x}}$ defines the coefficient of variation of the mean ratio and is expressed in terms of the 95% confidence limits ($CL = \pm tS_x/\sqrt{n}$) for the mean ratio by 1000 $CL/\bar{x}$ ‰.

Errors for $\delta^{13}C$ are compounded from the estimated standard error of the mean ($S_{\bar{x}}$) for sample and standard.

effect that will require careful investigation and compensation by using a variable volume source for future higher precision measurements.

A mean error of 2.4‰ in $\delta^{13}C_{CD_4}$ is associated with the static mass spectrometric determination of CD_4 released by lunar samples. This error is governed by the precision of the sample rather than the standard gas measurement. An investigation of the internal statistics with long runs and reproducibility of the mean for different runs of the same standard will enable values to be set on the limiting precision. Using a model developed by Peterson and Hayes (Peterson, 1976), the theoretical precision of the analog electron multiplier system should enable sample requirements to be less than 1 ngC as CD_4 for a 2 σ precision in $\delta^{13}C_{CD_4}$ of 1‰. Further work is required to see how close this figure can be approached in practice. However, practical multiplier noise levels may well be a major factor limiting further improvements and a system based on Faraday cup collection/voltage to frequency (rather than direct A/D) conversion may offer better possibilities for securing higher precision providing some sensitivity is sacrificed.

Isotopic Fractionation During Acid Dissolution

For the isotopic enrichments measured by static mass spectrometric analysis of CD_4 to be of any value, it is essential that the isotopic composition of CD_4 accurately or reproducibly reflects the composition of the material from which the gas is released. To use the example quoted by Von Unruh and Hayes (1976), if a degradative reaction has a kinetic isotope effect such that ^{12}C reacts 1.03 times more rapidly than ^{13}C, then the isotopic composition of the total product will be within 1.0‰ of that of the starting material, if the reaction is 99% complete. In the case of the CD_4 method the problem is slightly different. Competitive reactions such as partial deuterium labelling and generation of a range of deuterocarbons from carbon present as C_{hyd} could mean that the carbon isotope composition of CD_4 differs from that of C_{hyd}.

Partial labelling effects

The deuterated acid dissolution method yields mainly CD_4 but, in addition, a variable proportion of CD_3H is also generated. From the CD_4/CD_3H ratio it is possible to calculate the atom % D substitution in methane produced by the reaction. Figure 1 shows that the methane released by acid dissolution of the lunar samples is between 92 and 98 atom % D labelled with the majority around 94 to 96 atom % D. Both CD_4 and CD_3H are mixtures of ^{12}C and ^{13}C, not necessarily in the same ratio, since four possible competitive reactions exist: ^{12}C with H, ^{12}C with D, ^{13}C with H, and ^{13}C with D. Without knowing any details of the reaction mechanisms involved, it is difficult to decide whether any difference exists between relative reaction rates. We are therefore unable to speculate as to whether CD_3H or CD_4 is enriched in the heavy isotope of carbon, or whether $^{13}C/^{12}C$ ratios derived from the 21/20 ratios are over or under estimates of the true isotopic composition of C_{hyd}. Kinetic isotope effects between H and D can be very large implying that considerable fractionation could be involved.

With appropriate data it should be possible to calculate the magnitude of any fractionation effect. However, the obvious answer to the problem would be to reduce partial labelling. Some data we have gathered suggests that improvement beyond *ca.* 98 atom % D, the theoretical maximum from a 99.5 atom % D acid, is unlikely to be possible and 98 atom % D may only be obtained under special circumstances. The acid is added to the reaction vessels with clean, dry unused glass pipettes. The reaction vessels are pretreated with the identical acid before use. A quantity of acid, *ca.* 3 times the amount used in the experiment, is placed in the vessel which is then sealed and left in an oven at *ca.* 140°C for several days. Complete exchange of at least any labile protonated species in the glass should have taken place. A study of the CD_4/CD_3H ratios for over a hundred experiments suggests that the higher values occur for experiments performed with freshly opened batches of acid, and coarse size fractions compared to fine material (Fig. 1). This latter observation is very pronounced in the case of >250 μm grains which gave 97.3 to 97.8 atom % D, the highest CD_4/CD_3H values found during any sample analysis. It appears that a source of protons could be the surface of the lunar grains, even though specimens for analysis are evacuated to better than 10^{-6} torr during sample preparation. Sources of H could include terrestrial water, lunar water, hydroxyl groups, solar wind implanted hydrogen, etc. Stepwise acid dissolution studies on three samples of 15601 appear to confirm this conclusion. In each case (Table 2) the higher temperature fraction has a higher CD_4/CD_3H ratio.

Another possible answer to the partial labelling problem might be a systematic study of the variation of isotope ratio with the CD_4/CD_3H ratio. An appropriate systematic investigation has not yet been carried out, but some relevant information is available from samples where duplicate analyses have been performed

Table 2. Stepwise[a] acid dissolution of 15601 light agglutinates.

Grain size μm		C_{hyd} μgC/g	CD_4/CD_3H
150–250[(b)]	20°C	15.5	4.5
	100°C	14.2	6.2
53–75[(b)]	20°C	15.2	4.3
	100°C	13.8	6.0
40–53[(c)]	20°C	23.1	3.9
	100°C	11.1	5.6

[(a)] 1st fraction released at 20°C overnight. 2nd fraction from same sample released at 100°C.
[(b)] Highly magnetic, $\rho < 2.96$ g/cm^3.
[(c)] As for b but magnetically repurified.

Table 3. Variation of CD_4/CD_3H ratio with $\delta^{13}C_{CD_4}$ for duplicate runs of 12023 separates.

Run	Grain Size μm	CD_4/CD_3H	$\delta^{13}C_{CD_4}$ ‰	$\delta^{13}C^*_{CD_4}$ ‰
2		3.2	+70	+73
23	150–250	4.1	+67	+69
30		4.4	+61	+63
24	75–106	4.8	+64	+66
45		7.8	+68	+68
26	53–75	4.8	+67	+69
32		5.1	+65	+67
7	30–40	4.9	+56	+58
28		5.0	+64	+65
48	20–30	7.4	+68	+69
58		4.9	+65	+67
33		5.6	+55	+56
46	"smoke"	7.2	+61	+62
60		5.3	+56	+58
9	150–250	4.1	+68	+71
31	$\rho < 2.96$, M3.5†	4.5	+68	+70
14	53–75	4.3	+69	+71
65	$\rho < 2.96$, M3.5	5.5	+65	+67
13	53–75	3.8	+65	+67
66	$\rho < 2.96$, M3.0	6.5	+67	+68
47	53–75	7.6	+70	+71
70	$\rho < 2.96$, M2.5	4.2	+66	+69

*Corrected for partial isotope labelling effects using equation (1) with R_1 (standard) = 18.8.

†M distance moved to magnetic separator (Pillinger *et al.*, 1978).

(Table 3). In eight out of ten cases, it is seen that an increase in the CD_4/CD_3H ratio leads to an increase in the enrichment relative to the working standard. If it is assumed that both CD_4 and CD_3H have the same isotopic composition, then a correction can be made using the formula given earlier for the effect of the CD_3H interference. Even after the correction factor has been considered, there is no change in the pattern. This suggests that CD_3H may have a higher $^{13}C/^{12}C$ ratio than CD_4. Thus all the enrichments measured for C_{hyd} could be artificially low by a factor of up to 3‰.

Yield of deuterocarbons from C_{hyd}

In addition to the problems of partially labelled species, a potentially more serious problem may be that C_{hyd} must afford C_2 and even higher deuterocarbon species in addition to CD_4 (Abell *et al.*, 1970; Chang *et al*, 1970; Holland *et al.*, 1972). A proportion of the carbon could be precipitated as elemental carbon during the reaction (Krapf *et al.*, 1972). It should be pointed out that this problem has previously been overlooked during attempts to measure the isotopic composition of C_{hyd} from the oxidised product of the total methane released by acid dissolution (Chang *et al.*, 1971; Chang *et al.*, 1974; Kerridge *et al*, 1975).

The problem is similar to, although more complex than, the measurement of oxygen isotopes from carbonates. Acid release of CO_2 from CO_3^{--} can give only two oxygen atoms and thus some fractionation must be involved in the reaction. Difficulties are avoided by standardising the reaction conditions as far as possible, and quoting isotope enrichments relative to a standard carbonate (PDB) which is treated in the same way as the sample. As far as possible, we have attempted to standardise conditions during the DC1 acid treatment and yields of CD_4 appear to be reproducible. Unfortunately, most investigators using the technique have considered CD_4, which is by far the most abundant deuterocarbon released from the lunar samples, to be representative of C_{hyd}. The little information that is available suggests that CD_4/C_2D_6 and CD_4/C_2D_4 ratios are variable (Abell *et al.*, 1971; Wszolek and Burlingame, 1973; Wszolek *et al.*, 1973; Jull *et al.*, 1976). However, Holland *et al.*, (1972) were able to demonstrate that C_2D_6 abundance correlated with various solar wind exposure parameters. There is no evidence, either for or against, of carbon species other than hydrocarbons being products of the acid dissolution. Because of the very high sensitivity available to the static mass spectrometer, it would be feasible to attempt to measure the $^{13}C/^{12}C$ isotopic composition of C_2D_6 from the relative abundance of the m/e 36 and m/e 37 peaks. Partial isotope labelling may present a difficulty in interpreting the result. If the reaction conditions and yield are reproducible, the problem then becomes: what is a suitable standard sample of carbon in iron? It would be simple to measure the isotopic ratio of CD_4 released from the standard iron, or even an iron meteorite for which the $^{13}C/^{12}C$ ratio had been determined in a conventional manner. An absolute scale would then be established. This approach, however, could be misleading, since iron in the lunar soil is so finely disseminated (predominantly <130 Å) that acid attack on such small grains might be *via* an entirely different mechanism.

ISOTOPIC MEASUREMENTS ON CD_4 RELEASED FROM 12023 SOIL SEPARATES

Notwithstanding the concerns expressed above, CD_4 isotopic data have been gathered for *ca.* fifty 12023 soil separates. All the samples analysed show a substantial enrichment relative to both our working standard, and PDB. Although there is considerable scatter in the data, we tentatively identify two trends: (i) There is an enrichment in ^{13}C for the most magnetic samples relative to less magnetic material and bulk sieved fractions (Table 4). Other data obtained by us (Pillinger *et al.*, 1978) for the same suite of samples suggest that progressively more magnetic grains of $\rho < 2.96$ g/cm^3 may represent an evolu-

Table 4. $\delta^{13}C$ measured from $^{13}CD_4/^{12}CD_4$ (m/e 21/20) of sample compared to a standard gas.

		$\delta^{13}C_{CD_4}$ ‰	$\delta^{13}C^*_{CD_4}$ ‰	$[\delta^{13}C^*_{CD_4}]^{(a)}_{PDB}$ ‰
Bulk sieved fractions	(12)	$+62.2 \pm 4.0$	$+63.9 \pm 4.2$	33
Magnetic separates M1.5–3.0 mm†:	(22)	$+66.3 \pm 2.9$	$+68.4 \pm 2.8$	38
Magnetic separates M $\geq$ 3.5 mm:	(15)	$+69.2 \pm 2.9$	$+71.0 \pm 2.5$	40

*Corrected for partial isotope labelling effects using equation (1) with R_1 (standard) = 18.8.
†M distance moved to magnetic separator (Pillinger *et al.*, 1978).
[(a)]With standard $CD_4 = -28.9‰$ PDB units and using equation given by Craig (1957).

tion sequence for agglutinates. Hence, we would interpret the trend in ^{13}C enrichment to be a measure of carbon loss and fractionation during the recycling of material into agglutinates. Previously, DesMarais *et al.* (1975) have observed that the total carbon in agglutinates is enriched in ^{13}C by +5 to +12‰ relative to non-agglutinates. There is clear evidence that volatile species, such as trapped hydrocarbons, are lost relative to C_{hyd} during agglutination (Cadogan *et al.* 1973a,b). Billetop *et al.* (1976) have also suggested that solar wind carbon implanted into the silicate matrix may have been substantially reduced relative to C_{hyd} by agglutination processes. Gibson and Moore (1973) have inferred from pyrolysis data that implanted solar wind carbon is released before hydrolysable carbon. Thus, the CO temperature profile obtained by Bibring *et al.* (1974) from agglutinates could be interpreted to demonstrate that a large proportion of the total carbon in these particles is in the form of C_{hyd}. Hence, the absolute values of up to +17‰ obtained by DesMarais *et al.* (1975) could be reflecting the C_{hyd} isotopic composition. (ii) DesMarais *et al.* (1975) reported that $\delta^{13}C$ values for total carbon from sieve fractions (14240, 72501) maximise at 40–50 μm, and then decrease, whereas the absolute amounts of carbon continue to increase uniformly for much finer samples. A similar trend, although less clear, may be observed for the $^{13}CD_4$ isotopic enrichments and absolute CD_4 abundances from 12023 (Table 5) i.e., isotopic enrichment in C_{hyd} could be reflecting the behaviour of total carbon. Similar observations have been made by Becker and Clayton (1975) from nitrogen data. So far, no completely satisfactory explanation has been presented to explain the observed trends. DesMarais *et al.* (1975) proposed that an additional unidentified carbon component is present in fine fractions. Pillinger *et al.* (1976) have suggested that fresh, isotopically light, relatively unfractionated solar carbon added to particle surfaces could be outweighing isotopically heavier volume components in fine material. More recent data (Pillinger *et al.*, 1978) suggest there are inconsistencies in this latter argument.

SUMMARY

Clearly, the difficulties regarding partial isotopic labelling and yield of CD_4

Table 5. C_{hyd} and $\delta^{13}C_{CD}$ variation with bulk grain size for 12023.

Grain size μm	CD_4 μgC/g	$\delta^{13}C_{CD_4}$ ‰	$\delta^{13}C^{(*)}_{CD_4}$ ‰	$[\delta^{13}C^{(*)}_{CD_4}]^{(a)}$ ‰
250–500	5.0	+61	+63	+32
150–250	5.9	+66	+68	+37
106–150	7.3	+60	+63	+32
75–106	7.7	+66	+67	+36
53–75	8.1	+66	+68	+37
40–53	9.2	+68	+70	+39
30–40	10.3	+64	+65	+34
20–30	11.6	+67	+68	+37
15–20	14.5	+57	+58	+27
10–15	16.9	+60	+62	+31
< 10	22.3	+59	+60	+29
"smoke"	29.6	+57	+58	+27

(*) Corrected for partial isotope labelling effects using equation (1) with R_1 (standard) = 18.8.

(a) With standard CD_4 = −28.9‰ PDB units and using equation given by Craig (1957).

preclude any speculation, at the present time, regarding the true isotopic composition of hydrolysable carbon in finely divided iron metal. If partial deuterium labelling does cause a fractionation, it may mean that the CD_4 carbon isotope method is slightly underestimating the ^{13}C enrichment. No information is available concerning the problem of deuterocarbon yield. However, if this problem prevents the CD_4 method from providing information concerning the true isotopic composition of C_{hyd}, by the same reasoning it infers that previous investigations (Chang *et al.*, 1971, 1974; Kerridge *et al.*, 1975) based on the oxidation of CH_4 are equally in error. Further investigations and more ingenuity are necessary to tackle the various problems we have discussed.

None of the enrichments we have observed approach values suggested by Epstein and Taylor (1975) for the ultra-heavy carbon phases postulated on the basis of partial fluorination studies. Substantial fractionation in favour of ^{12}C would have to be involved during the DCl treatment if C_{hyd} were to be identifiable with ultra-heavy carbon as suggested by Pillinger *et al.* (1976).

Although definite answers concerning the absolute variation of $^{13}C/^{12}C$ ratio in C_{hyd} are not yet forthcoming, relative changes between fractions isolated from a single lunar soil may be distinguished. Hence, useful information with respect to agglutinate evolution and the formation and growth of metallic iron in the lunar soil may yet be obtained. Particularly intriguing seems to be the paradoxical relationships between particle size on the one hand, and light element concentrations and isotopic abundance on the other.

Development of the CD_4 carbon isotopic method has provided some insight into a few of the problems which may arise when we attempt to expand the usefulness of the technique. The extreme sensitivity of static mass spectrometry to CD_4 and the small sample requirements (detection limit 5×10^{-14} gC) gained over conventional differential (dynamic) isotope methods has encouraged us to investigate the possibility of converting other forms of carbon to CD_4 for static isotope measurement.

Finally it must be emphasized that in describing a novel technique for carbon isotopic analysis, we attach prime importance to the design of experiments that validate the integrity of the method and allow the data obtained to be interconvertable with those from conventional isotope ratio measurements. A comparison of the isotope data determined on two CD_4 samples of different isotopic composition by the static method with results from the analysis of the same gases analysed by conventional means (dynamically as CD_4 or *via* combustion to CO_2) should go a long way in addressing this question. Other experiments including conversion of CO_2 of known isotopic composition to CD_4 for static measurement and the investigation of CD_4 released from a lunar sample by a conventional dynamically operated mass spectrometer will also require our attention.

Acknowledgments—This work is supported by the Science Research Council (grant no. SGD-00772) and the Nuffield Foundation. LRG thanks the Nuffield Foundation for a fellowship; AJTJ is grateful to the National Research Council of Canada for a fellowship. We would like to thank Dr. J. M. Hayes and Mr. E. Vogler of Indiana University, for isotopic analysis of our CD_4 standard gases, and many useful comments. We appreciate the constructive criticism of Dr. A. Fallick, and thanks are also due to Mrs. D. Fabian for help with the separations and manuscript preparation. We are grateful to Drs. J. M. Hayes and D. J. DesMarais for their reviews of this manuscript.

REFERENCES

Abell P. I., Eglinton G., Maxwell J. R., Pillinger C. T. and Hayes J. M. (1970) Indigenous lunar methane and ethane. *Nature* **226**, 251–252.

Abell P. I., Cadogan P. H., Eglinton G., Maxwell J. R. and Pillinger C. T. (1971) Survey of lunar carbon compounds. I. The presence of indigenous gases and hydrolysable carbon compounds in Apollo 11 and Apollo 12 samples. *Proc. Lunar Sci. Conf. 2nd*, p. 1843–1863.

Becker R. H. and Clayton R. N. (1975) Nitrogen abundances and isotopic compositions in lunar samples. *Proc. Lunar Sci. Conf. 6th*, p. 2131–2149.

Becker R. H., Clayton R. N. and Mayeda T. K. (1976) Characterization of lunar nitrogen components. *Proc. Lunar Sci. Conf. 7th*, p. 441–458.

Bibring J. P., Burlingame A. L., Langevin Y., Maurette M. and Wszolek P. C. (1974) Simulation of lunar carbon chemistry: II. Lunar winds contribution. *Proc. Lunar Sci. Conf. 5th*, p. 1763–1784.

Billetop M. C. J., Eglinton G., Gardiner L. R., Jull A. J. T. and Pillinger C. T. (1976) Pyrolysis acid dissolution studies of lunar carbon chemistry (abstract). In *Lunar Science VII*, p. 61–63. The Lunar Science Institute, Houston.

Cadogan P. H., Eglinton G., Firth J. N. M., Maxwell J. R., Mays B. J. and Pillinger C. T. (1972) Survey of lunar carbon compounds II. The carbon chemistry of Apollo 11, 12, 14 and 15 samples. *Proc. Lunar Sci. Conf. 3rd*, p. 2069–2090.

Cadogan P. H., Eglinton G., Gowar A. P., Jull A. J. T., Maxwell J. R. and Pillinger C. T. (1973b) Location of methane and carbide in Apollo 11 and 16 fines. *Proc. Lunar Sci. Conf. 4th*, p. 1493–1508.

Cadogan P. H., Eglinton G., Maxwell J. R. and Pillinger C. T. (1973a) Distribution of methane and carbide in Apollo 11 fines. *Nature Phys. Sci.* **241**, 81–82.

Chang S., Smith J. W., Kaplan I., Lawless J., Kvenvolden K. A. and Ponnamperuma C. (1970) Carbon compounds in lunar fines from Mare Tranquillitatis IV: Evidence for oxides and carbides. *Proc. Apollo 11 Lunar Sci. Conf.*, p. 1857–1869.

Chang S., Kvenvolden K. A., Lawless J., Ponnamperuma C. and Kaplan I. R. (1971) Carbon, carbides and methane in an Apollo 12 sample. *Science* **171**, 474–477.

Chang S., Lawless J., Romiez R., Kaplan I. R., Petrowski C., Sakai H. and Smith J. W. (1974) Carbon, nitrogen and sulphur in lunar fines 15012 and 15013: Abundances, distribution and isotopic compositions. *Geochim. Cosmochim. Acta* **38**, 853–872.

Craig H. (1957) Isotopic standards for carbon and oxygen and correction factors for mass spectrometric analysis of carbon dioxide. *Geochim. Cosmochim. Acta* **12**, 133–149.

DesMarais D. J., Basu A., Hayes J. M. and Meinschein W. G. (1975) Evolution of carbon isotopes, agglutinates, and the lunar regolith. *Proc. Lunar Sci. Conf. 6th*, p. 2353–2374.

DesMarais D. J. (1978) Carbon isotopes, nitrogen and sulfur in lunar rocks (abstract). In *Lunar and Planetary Science IX*, p. 247–249. Lunar and Planetary Institute, Houston.

Epstein S. and Taylor H. P. Jr. (1973) The isotopic composition and concentration of water, hydrogen and carbon in some Apollo 15 and 16 soils and in the Apollo 17 orange soil. *Proc. Lunar Sci. Conf. 4th*, p. 1559–1575.

Epstein S. and Taylor H. P. Jr. (1975) Investigation of the carbon, hydrogen, oxygen and silicon isotope and concentration relationships on the grain surfaces of a variety of lunar soils and in some Apollo 15 and 16 core samples. *Proc. Lunar Sci. Conf. 6th*, p. 1771–1798.

Gardiner L. R., Woodcock M. R., Pillinger C. T. and Stephenson A. (1977) Carbon chemistry and magnetic properties of bulk and agglutinate size fractions from soil 15601. *Proc. Lunar Sci. Conf. 8th*, p. 2817–2839.

Gardiner L. R. (1978) Static mass spectrometry of active gases: Applications to lunar science. Ph.D. thesis, University of Bristol.

Gibson E. K. Jr. and Moore G. W. (1973) Carbon and sulfur distributions and abundances in lunar fines. *Proc. Lunar Sci. Conf. 4th*, p. 1577–1586.

Holland P. T., Simoneit B. R., Wszolek P. C. and Burlingame A. L. (1972) Study of carbon compounds in Apollo 12 and 14 samples. *Space Life Sci.* **3**, 551–561.

Jull A. J. T., Eglinton G., Pillinger C. T., Batts B. D. and Biggar G. M. (1976) The identity of lunar hydrolysable carbon. *Nature* **262**, 566–567.

Kaplan I. R., Kerridge J. F. and Petrowski C. (1976) Light element geochemistry of the Apollo 15 site. *Proc. Lunar Sci. Conf. 7th*, p. 481–492.

Kerridge J. F., Kaplan I. R., Petrowski C. and Chang S. (1975) Light element geochemistry of the Apollo 16 site. *Geochim. Cosmochim. Acta* **39**, 137–162.

Kerridge J. F., Kaplan I. R. and Petrowski C. (1978) Carbon isotope systematics in the Apollo 16 regolith (abstract). In *Lunar and Planetary Science IX*, p. 618–620. Lunar and Planetary Institute, Houston.

Krapf G., Lutz J. L., Melnick L. M. and Bandi W. R. (1972) A DTA-EGA study of the chemical isolation of Fe_3C, amorphous carbon and graphite from steel and cast iron. *Thermochim. Acta* **4**, 257–271.

Peterson D. W. (1976) Signal to noise ratios of mass spectrometer detection systems. M.Sc. thesis, University of Indiana.

Pillinger C. T., Eglinton G., Jull A. J. T. and Epstein S. (1976) A model for carbon isotopic evolution of the lunar regolith (abstract). In *Lunar Science VII*, p. 703–705. The Lunar Science Institute, Houston.

Pillinger C. T., Jull A. J. T., Woodcock M. R. and Stephenson A. (1978) Maturation of the lunar regolith: Some implications from magnetic measurements and hydrolysable carbon data on bulk soils, and particle separates from 12023 and 15601. *Proc. Lunar and Planet. Sci. Conf. 9th*. This volume.

Von Unruh G. E. and Hayes J. M. (1976) Intramolecular distribution of carbon isotopes in fatty acids. *Proc. 2nd Int. Conf. on Stable Isotopes* (Klein and Klein, eds.), p. 561–568. U.S. ERDA, Conf. 751027.

Wszolek P. C. and Burlingame A. L. (1973) Carbon chemistry of the Apollo 15 and 16 deep drill cores. *Proc. Lunar Sci. Conf. 4th*, p. 1681–1692.

Wszolek P. C., Simoneit B. R. and Burlingame A. L. (1973) Studies of magnetic fines and volatile-rich soils: Possible meteoritic and volcanic contributions to lunar carbon and light element chemistry. *Proc. Lunar Sci. Conf. 4th*, p. 1693–1706.

Proc. Lunar Planet. Sci. Conf. 9th (1978), p. 2167–2193.
Printed in the United States of America

Maturation of the lunar regolith: Some implications from magnetic measurements and hydrolysable carbon data on bulk soils and particle separates from 12023 and 15601

C. T. PILLINGER, A. J. T. JULL and M. R. WOODCOCK

Planetary Sciences Unit, Department of Mineralogy and Petrology
University of Cambridge, England

A. STEPHENSON

Institute of Lunar and Planetary Science, School of Physics
The University, Newcastle upon Tyne, England

Abstract—Soil 12023 has been extensively separated according to size, density and magnetic properties. The relative abundance of the various fractions obtained indicate an increasing portion of more complex agglutinates in finer size fractions. A large number of fractions have been measured for magnetic susceptibility (χ), isothermal remanent magnetisation (IRM), hydrolysable carbon (C_{hyd}), and solar wind rare gases. The abundance of all species in size fractions show the expected increasing relationship with decreasing particle size. It can be shown that χ and IRM are similar to I_s (when respectively normalised to FeO content), and C_{hyd} similar to nitrogen, as maturity indices. Treatment of a suite of agglutinates from 15601 with a reagent ($CuCl_2 \cdot 2KCl$) which should selectively remove surface metal, has led us to question the existence of very surface components of iron metal and C_{hyd} as less than 2% of the iron metal was accessible to the reagent. In order to satisfy apparent relationships between the finely divided iron and its associated carbon, and particle size, a model which involves a volume component which increases in fine fractions is invoked in preference to surface and volume related species. The analyses of 12023 agglutinate size fractions tend to support this model.

Agglutinate/bulk soil concentration ratios have been established for the C_{hyd} and ^{20}Ne data. They show the preferential retention of C_{hyd} over neon and suggest that progressively, more magnetic agglutinates may be considered as more evolved or mature. The concentration ratios also support the increasing volume component model, and may be used in a variety of ways to provide an index of maturity.

A particularly interesting feature of the magnetic studies is that for material separated at the same magnetic separation distance, grains in the size range 53–75 μm have a magnetic susceptibility which is 20–25% more than grains 250–500 μm in size. We attribute this difference to a variation in the iron particle size distribution in the <130 Å range. If different soils are affected to a different extent by fluctuations in iron particle size, the scatter in a plot of χ *vs.* C_{hyd} may be explained. Thus finely divided iron in the lunar soil could have a relatively narrow carbon concentration range.

INTRODUCTION

The object of this paper is to provide information relevant to the better understanding of lunar surface maturation. In this context we adhere to the definition of maturity given by Morris (1978). Thus our current interest centres on (i) effects which occur as a direct result of unshielded exposure to solar wind bombardment, e.g., the collection of solar wind elements and sputtering, and (ii) modifications superimposed on ion irradiated materials during the formation of agglutinates and other complex particles by micrometeorite impact, in the upper 1 mm of the regolith.

It seems increasingly apparent that the abundance of finely divided iron metal (Fe°, <300 Å in diameter) in a lunar soil as measured by the ferromagnetic resonance parameter I_s, (e.g., Morris, 1976) when normalised to total FeO content, is one of the best maturity indices currently available. Accordingly our investigations have embodied a more searching enquiry into the distribution of superparamagnetic (<130 Å in diameter, Fe^{o}_{sp}) and single domain iron metal (130–300 Å, Fe^{o}_{sd}) and hydrolysable carbon (C_{hyd}). Here we report magnetic susceptibility data (χ) as a measure of Fe^{o}_{sp} and the isothermal remanent magnetisation (IRM) acquired in a field of 2.6 KOe as a measure of Fe^{o}_{sd} for a suite of separates obtained from soil 12023 by sieving, density and magnetic separations. Selected fractions from the same suite of samples have also been treated with DCl to release light rare gases, and CD_4 as a measure of C_{hyd}. Carbon isotope data obtained from the CD_4, and major element chemistry of agglutinate separates are considered elsewhere in papers by Gardiner *et al.* (1978) and Woodcock and Pillinger (1978) respectively. The current studies may be considered as an extension of our earlier work on 15601 (Gardiner *et al.*, 1977) and the pathfinder investigations of 10086 and 60501 (Cadogan *et al.*, 1973a,b). The 15601 data indicated that certain of the correlations reported previously by Pillinger *et al.* (1974) for CD_4 and finely divided iron in bulk soils could profitably be reexamined. Thus, magnetic susceptibility measurements have been performed for a series of 55 bulk soil samples.

In Table 1 we have compiled all the data so far obtained for the various fractions separated from 12023. Values for magnetic susceptibility and IRM are available for almost all samples. The CD_4, He, Ne and FeO data are quoted for a representative number. Most fractions below 53 μm in size remain to be studied. The relevant data for bulk soils are given in Table 2.

EXPERIMENTAL

Analytical procedures

Methods for the determination of magnetic susceptibility have been reported previously (Stephenson and de Sa, 1970); and IRM (Stephenson, 1970); however, these techniques have recently been modified to allow measurements to be made on samples <1 mg. Procedures for the measurement of CD_4 and light rare gases, on the same sample, by static mass spectrometry have been outlined by Gardiner *et al.* (1977). A complete account of the methods involved is in preparation. Acid dissolution experiments are routinely performed on *ca.* 0.3 mg samples. Unfortunately it was impossible to obtain data on trapped CH_4 with samples this small because of variable blank contributions. Agglutinate size fractions from 15601 (*ca.* 1 mg) were treated with $CuCl_2{\cdot}2KCl$ in 2 ml aqueous solution and the solution analysed by atomic absorption. Exposed metallic iron should be dissolved very rapidly by the copper reagent due to the replacement reaction $Fe^{o} + Cu^{2+} \rightarrow Fe^{2+} + Cu^{o}$ (Smerko and Flinchbaugh, 1968; Krapf *et al.*, 1972). Standards (designed to mimic samples) were prepared by hand shaking 0–5 mgs of iron metal together with 2 g powdered tachylite basaltic glass, Snake River, Idaho in $CuCl_2.2KCl$ solution. Standards were diluted to give a working range of 0–1 ppm Fe and a straight line calibration plot was obtained.

Separation of 12023

Fourteen size fractions (>1 mm, 500 μm–1 mm, 250–500, 150–250, 106–150, 75–106, 53–75, 40–53, 30–40, 20–30, 15–20, 10–15, <10 μm and "smoke") were obtained by wet sieving 2 grams of

Table 1. Magnetic susceptibility (χ), isothermal remanent magnetisation (IRM), hydrolysable carbon (CD_4) and light rare gases (He, Ne) for 12023 soil separates.

Sample description	Magnetic susceptibility (χ) 10^{-3}G cm^3 g^{-1} Oe^{-1}	IRM 10^{-3}G cm^3 g^{-1}	CD_4 μgC/g	He arbitrary units $\times 10^{-2}$	Ne	FeO wt %
bulk soil	2.72	–	12.5	–	–	16.22
size fractions (μm)						
1000–500	1.65	61	–	–	–	–
500–250	1.21	49	5.0	35	127	–
250–150	1.79	65	5.9	74	247	15.27
150–106	1.94	59	7.3	68	207	16.58*
106–75	2.59	67	7.7	95	310	16.16*
75–53	2.25	65	8.1	108	391	16.57*
53–40	2.43	75	9.2*	107*	398*	15.95*
40–30	2.61	77	10.3*	158*	510*	16.04*
30–20	3.02	93	11.6	220	654	15.86
20–15	3.16	108	14.5	387	994	15.35
15–10	3.13	113	16.9	527	1303	14.33
<10	3.10	139	22.3	937	1934	–
Smoke	3.94	173	29.6*	1018*	2247*	12.97*
500–250 μm ρ < 2.96						
M3.5	5.52	207	–	–	–	–
M3.0	4.48	163	–	–	–	–
M2.5	3.78	124	–	–	–	–
M2.0	2.88	109	–	–	–	–
M1.5	1.82	91	–	–	–	–
ρ = 2.96–3.3						
M2.0	2.54	131	–	–	–	–
M1.5	1.68	89	–	–	–	–
M1.0	0.64	36	–	–	–	–
250–150 μm ρ < 2.96						
M3.5	6.28	153	23.9*	145*	497*	16.81
M3.0	5.01	115	21.9	127	484	–
M2.5	4.28	132	–	–	–	–
M2.0	3.11	107	19.8	114	534	–
M1.5	1.76	81	14.1	85	411	–
MT	0.61	28	4.6	41	164	–

Table 1. *(cont'd.)*.

Sample description	Magnetic susceptibility (χ) 10^{-3}G cm^3 g^{-1} Oe^{-1}	IRM 10^{-3}G cm^3 g^{-1}	CD_4 μgC/g	He arbitrary units $\times 10^{-2}$	Ne	FeO wt %
ρ = 2.96–3.3						
M2.0	3.57	126	10.2	150	482	–
M1.5	2.05	92	9.9	110	369	–
M1.0	1.11	47	2.9	68	214	–
MT	0.35	20	1.8	34	128	–
$\rho > 3.3$	0.30	7.3	1.0	34	133	–
150–106 μm $\rho < 2.96$						
M3.5	6.51	155	25.7	154	499	15.83
M3.0	5.45	127	23.3	124	454	15.83
M2.5	4.01	116	21.3	–	456	16.42
M2.0	3.06	105	16.9	–	422	15.38
M1.5	1.92	73	10.2	65	274	13.57
ρ = 2.96–3.3						
M2.0	4.02	135	–	–	–	17.27
M1.5	2.28	87	–	–	–	18.04
M1.0	1.32	52	–	–	–	15.52
106–75 μm $\rho < 2.96$						
M4.0	7.68	176	26.4	134	525	–
M3.5	6.96	157	25.2	147	494	16.02
M3.0	5.92	127	21.5	139	479	15.66
M2.5	4.74	109	17.8	109	421	15.24
M2.0	3.69	114	16.1	99	387	14.52
M1.5	2.62	90	–	–	–	12.60
ρ = 2.96–3.3						
M2.5	4.82	142	–	–	–	–
M2.0	3.42	109	–	–	–	–
M1.5	2.63	88	–	–	–	–
M1.0	1.66	52	–	–	–	–
75–53 μm $\rho < 2.96$						
M4.0	8.80	163	24.8	149*	550*	15.98
M3.5	7.60	159	23.3*	154*	540*	15.12
M3.0	6.28	146	21.8*	146*	568*	15.13
M2.5	5.17	115	17.6*	135*	461*	14.44
M2.0	4.07	100	15.7	116	394	14.08
M1.5	2.53	85	11.8	103	323	12.59
MT	–	–	5.1	62	184	–

Table 1. *(cont'd.).*

Sample description	Magnetic susceptibility (χ) 10^{-3}G cm^3 g^{-1} Oe^{-1}	IRM 10^{-3}G cm^3 g^{-1}	CD_4 μgC/g	He arbitrary units $\times 10^{-2}$	Ne	FeO wt %
ρ = 2.96–3.3						
M2.5	5.14	141	15.7	156	580	18.70
M2.0	3.88	111	11.9	128	471	18.17
M1.5	2.60	77	8.9	115	421	17.14
MT	—	—	4.1	81	269	15.38
40–53 μm ρ < 2.96						
M4.0	8-80	180	24.3*	167*	625*	14.74
M3.5	7-92	152	21.6*	156*	578*	—
M3.0	6.19	148	—	—	—	—
M2.5	4.97	117	—	—	—	—
M2.0	3.61	100	—	—	—	—
M1.5	2.55	74	—	—	—	—
M1.0	1.90	55	—	—	—	—
30–40 μm ρ < 2.96						
M4.5	8.84	190	—	—	—	—
M4.0	7.96	173	—	—	—	—
M3.5	7.47	172	—	—	—	—
M3.0	6.17	149	—	—	—	—
M2.5	5.97	136	—	—	—	—
M2.0	3.76	105	—	—	—	—
M1.5	3.66	91	—	—	—	—
M1.0	1.93	59	—	—	—	—

Footnotes

CD_4 measurements by static mass spectrometry.

Maximum errors: magnetic measurements $\chi \pm 0.02$–0.06×10^{-3}G cm^3 g^{-1} Oe^{-1}

IRM $\pm 2 \times 10^{-3}$G cm^3 g^{-1}

mass spectrometric data CD_4, He and Ne $\pm$ 5%

XRF data FeO $\pm$ 0.3 wt %

— = data not available, * = replicate analyses, density — ρ: g/cm^3.

12023,7 using methanol (spectroscopic grade) as the liquid phase. The sieving was carried out in two stages, down to 30 μm and then below. The smoke sized fraction was the portion of the sample which was in suspension when the methanol was decanted off the <30 μm fraction after the first sieving. Since the size fractions cannot be weighed until after recovery, and the recovery procedure results in progressively increasing losses of the finer material, the sieving operation is not capable of providing useful size distribution data. Microscopic and electron microscopic examination, however, confirm that the size fractions are uniform in size.

All the size fractions below 500 μm and above 30 μm been density separated using tetrabromoethane (ρ = 2.96 g/cm^3) and diiodomethane (ρ = 3.3 g/cm^3) according to our previously reported procedures (Gardiner *et al.*, 1977). The low (ρ<2.96 g/cm^3) and medium (ρ = 2.96 − 3.3 g/cm^3)

Table 2. A comparison of CD_4, magnetic susceptibility and FMR data on bulk soils.

Sample	CD_4 (μgC/g)	χ*	I_s (arbitrary units)	Sample	CD_4 (μgC/g)	χ*	I_s (arbitrary units)
10086	18.6	2.44	1232	60501	5.4	2.34	440
12001	12.0	2.44	943	61141	7.2	1.11	297
12023	12.5	2.72	954	61161	7.4	1.10	443
12032	1.7	0.56	181	61501	4.6	1.30	296
12033	0.84	0.29	65	63320	3.1	0.68	221
12037	3.0	1.09	363	63340	4.2	0.67	243
12042	15.0	2.70	1024	63500	5.7	0.84	216
14003	11.0	2.32	686	64421	8.8	1.80	415
14141	1.1	0.35	58	64501	3.6	1.01	317
14148	11.6	2.29	770	66040	10.6	1.66	540
14149	8.7	1.67	530	66081	10.8	1.69	496
14156	13.0	2.33	686	67701	3.4	0.61	164
14163	7.7	1.95	593	67941	2.4	0.38	122
14240	2.6	1.62	478	67960	2.2	0.50	90
14298	13.0	2.74	934	68121	12.2	1.25	329
14421	10.2	3.03	—	68501	5.7	1.37	451
14422	8.5	1.96	—	69921	9.5	1.60	504
15031	16.4	2.17	1020	69941	9.7	1.73	485
15041	23.5	3.27	1344	69961	15.7	1.83	524
15091	12.5	2.10	863	71501	15.1	1.45	641
15231	14.8	1.89	829	72141	21.1	2.32	1094
15240	8.3	1.40	553	72501	10.5	1.68	705
15261	14.6	2.23	932	72701	8.3	1.63	537
15291	8.5	1.86	731	73221	6.0	0.97	383
15431	9.5	1.17	467	73281	5.1	0.94	299
15471	11.5	1.63	558	75080	10.4	1.54	684
15501	14.7	1.99	850	76501	7.3	1.51	597
15601	8.2	1.33	557				

I_s data from Morris (1976) and Morris (1978).
*unit 10^{-3}G cm^3 g^{-1} Oe^{-1}.
Errors for CD_4, ±8%, χ, ±0.02 to 0.04 × 10^{-3}G cm^{3}g^{-1} Oe^{-1}.
CD_4 data (by gas chromatography) from Pillinger *et al.*, 1974; Pillinger *et al.*, 1977; Cadogan *et al.*, 1972 and previously unpublished data.

density fractions were further magnetically separated using our magnetic separator (Cadogan *et al.*, 1973a), which has been modified to incorporate a 0.01 mm micrometer. Fractions were collected by decreasing (0.5 mm increments) the distance between the sample and a soft iron rod located at the pole of the magnet. The most magnetic material was collected at least 4.5 mm from the soft iron rod and is referred to as M4.5. A fraction of intermediate magnetic properties, e.g., collected on decreasing the magnet distance from 4 to 3.5 mm is referred to as M3.5. Samples collected by touching the sample are called MT, fractions not collected are NM. The fields produced by the magnetic separator vary from *ca.* 250 Oe at 4 mm to *ca.* 500 Oe at 1.5 mm.

During our magnetic separations of 15601, it was discovered there was a tendency for less or non-magnetic material, particularly in the fine size ranges, to be collected along with the highly magnetic fractions (Gardiner *et al.*, 1977). Thus, we found it was necessary to repurify some of the fractions of 15601. In view of the contention regarding magnetic separations (Hu and Taylor, 1977; McKay *et al.*, 1977), we have included information regarding our current techniques. Although these require patience, the results of the current invetigation suggest that they afford "clean" samples of considerable usefulness. The greatest enemy of a good magnetic separation is the lazy operator. Thus, difficulties arise when the distance between the sample and the magnet becomes foreshortened by the operator: (i) attempting to process too much material at once, i.e., the layer in the separating vessel is not approximately monoparticulate; (ii) not frequently removing the particles collected on the tip of the soft iron rod so that a layer of finite thickness develops; or (iii) moving the separator sidewise thereby causing a current in the liquid which levitates grains and allows them to be collected. The first error may also result in grains which should have been collected at a given separation distance being buried and thus prevented from jumping to the soft iron rod. Our method, which is intended to minimise some of the above problems, is as follows: An initial separation of the total sample under investigation (*ca.* 50 mg) is carried out with each of the separated fractions transferred to a separate collection vessel. Each fraction, now less than 10 mg, is reseparated starting with a separation distance 0.5 mm greater than the nominal distance. Any material which is collected is combined with the appropriate fraction. The separator is lowered 0.5 mm and the major portion of the sample is collected. Any material remaining is added to the next lowest fraction. The procedure is repeated three times for each fraction.

The density and magnetic separation procedures for a single size fraction are carried out continuously; samples are not reweighed until the final magnetic fractions are stored. Although many handling steps are involved, the use of specially designed collection vessels helps to minimise sample losses. Once the magnetic separations have been carried out, magnetic fractions can be manipulated using a magnet which facilitates almost complete recovery. The overall sample losses for a complete density-magnetic separation are typically only *ca.* 5% by weight for coarse fractions (105–250 μm) to 10% by weight for fine material (30–40 μm). Thus, we believe that the modal analysis data afforded by density and magnetic separations are meaningful within a particular size fraction.

Modal analysis

The distribution of grains having particular density and magnetic properties is plotted in histogram form in Fig. 1, which was prepared assuming that no losses occurred during density-magnetic separation. A number of observations may be made concerning the data: (i) The abundance of particles within a certain density range is higher in the $\rho<2.96$ fraction than the ρ = 2.96–3.3 fraction. The amount in the $\rho>3.3$ fraction is the lowest (data not shown). The proportion of each size range having a density $\rho<2.96$ g/cm^3 seems approximately constant. (The abundances for the 40–53 and 30–40 μm fractions may be artificially low since the 2.96–3.3 g/cm^3 material was not magnetically separated and some sample losses would have been inevitable). (ii) It is very apparent that the light fraction accommodates grains substantially more magnetic than the denser fraction. Because the separation was carried out in a liquid, there is a variation of collection distance with density. We calculate that this is unlikely to exceed a few percent and is unlikely to account for the observed variations in χ. Although the >3.3 g/cm^3 fraction was not magnetically separated, χ and CD_4 data suggest that it is unlikely that material of this density included a substantial portion of magnetic particles (see Table 1). (iii) The magnetic fractions of <2.96 g/cm^3 material show a clear bimodal distribution. Bearing in mind that the MT fraction includes all grains which move between 0 and 1.5 mm to the soft iron rod, one maximum of the distribution for all size ranges except 30–40 μm is probably the NM fraction. The amount of non-magnetic material in the 30–40 μm size range is severely depleted. We interpret this to mean that with high surface area particles, the abundance of splashed glass and welded detritus becomes sufficiently large to allow non-magnetic plagioclase grains to be collected on contact with the soft iron rod. The second maximum in the distribution is variable but seems to increase to higher values of M in finer size fractions. There is no readily discernable bimodal distribution for the 2.96–3.3 g/cm^3 material. The abundance of non-magnetic material in the 2.96–3.3 density range is low. It is possible that ferromagnetic minerals are to some extent collected (Hu and Taylor, 1977) on contact with the soft iron rod.

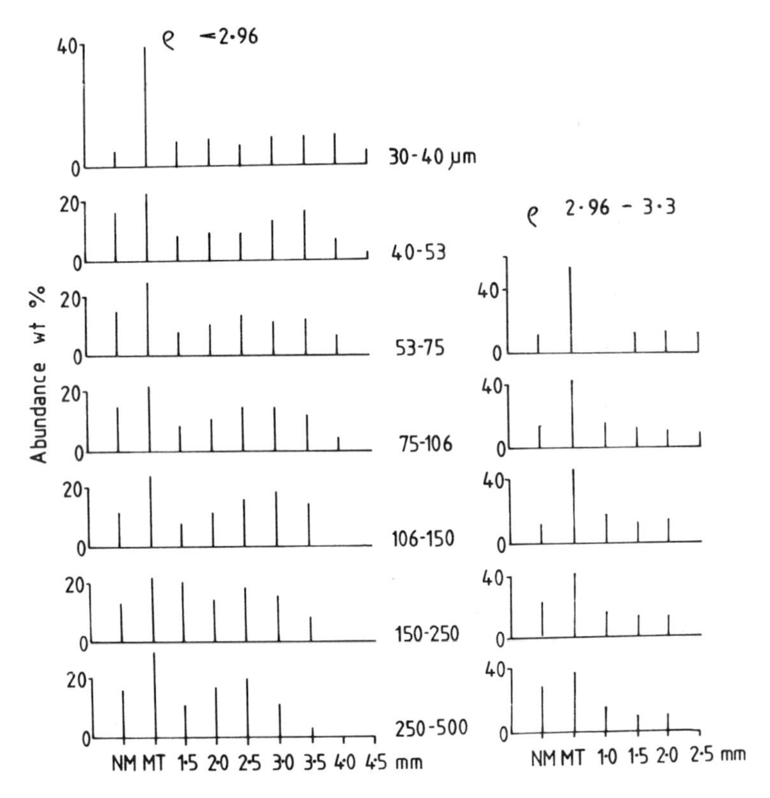

Fig. 1. Distribution of magnetic material in the <2.96 and 2.96–3.3 g/cm³ density separates from size fractions of soil 12023.

Sample description

Binocular microscope examination of $\rho<2.96$, magnetic material suggests that all samples above M 2.0 are homogeneous fractions of dark cindery agglutinates. Spherules and glassy shards start to appear at M1.5 and are common in the MT fractions. Non-magnetic material is essentially crystalline plagioclase. Dark cindery grains are the most abundant constitutents of M1.5, M2.0 and M2.5 fractions of the 2.96–3.3 g/cm³ material. The non-magnetic material at this higher density appears to be mainly plagioclase intergrown with other minerals such as pyroxene. Scanning electron microscopy on the 75–106 μm fraction confirms the above impressions and reveals a variety of surface textures for the magnetic particles; some areas having smooth bubbly surfaces, others almost completely covered with detritus. Grains from M2.0 to M3.5 ($\rho<2.96$) and those from M1.5 to M2.5 ($\rho = 2.96 - 3.3$) are indistinguishable by their exterior appearance. It is very clear that these samples are not variable mixtures of two end members, highly complex reworked grains on the one part and relatively unaltered rock or mineral fragments on the other. Less magnetic fractions contain a large proportion of grains that are superficially indistinguishable from material which makes up the more magnetic fractions. It should be pointed out that scanning electron microscopy (SEM) shows that even the non-magnetic crystalline grains have considerable amounts of splashed glass and detritus adhering to their surfaces. Petrographic observations in transmitted and reflected light substantiate the other investigations, thus, material less magnetic than M1.5 is a mixture. Grains identified as complex in the $\rho<2.96$, magnetic fractions are glassy, vesicular and incorporate an abundance of included fine crystalline fragments. Reflected light shows a larger proportion of highly reflecting, possibly very fine metal grains. The complex grains in M3.5 appear to contain more glass, more metal and be more vesicular than those in M2.0. The complex grains in the 2.96–3.3 g/cm^3 fraction are more densely packed with crystalline material. A statistical survey is required, but it seems likely that the conclusion we reached previously (Gardiner *et al.*, 1977,) is applicable. Highly magnetic low density fragments are considered as agglutinates, whereas magnetic material 2.96–3.3 g/cm^3, although possibly containing some agglutinates, could also have a large microbreccia population.

It should be emphasised that differences in magnetic properties which cause fractions to be collected at different magnetic separation distances are located within the complex grains themselves. They are not the result of weak attractive forces between strongly and less strongly magnetic grains. This means that a sample collected at an intermediate M value, e.g., 2.0 is not just a contaminated sample; the material which has reduced the magnetic properties is physically part of the M2.0 agglutinate. Thus, trends for other data which we shall subsequently discuss in this paper, and elsewhere (Gardiner *et al.*, 1978; Woodcock and Pillinger, 1978), are a result of internal variations in the composition of agglutinates and not inadequate separations.

Both sample description and modal analysis data for 12023 are in keeping with that might be expected if this soil is in a continuous state of evolution according to the models of Basu (1977) and Morris (1977). The sample contains sources of fresh minerals from which agglutinate and microbreccia formation could be initiated. The progressively more magnetic agglutinates and microbreccias correspond to increasingly more mature particles within the sample, and would appear to be evidence for recycling of material.

Analysis of Grain Size Material

The data from bulk sieved fractions for χ, IRM, CD_4, He and Ne (Table 1) may all be utilised according to the conventional techniques to demonstrate the apparent existence of surface and volume correlated components. Thus, all four parameters plot successfully against $1/r$ and show a linear increase for decreasing grain size.

An alternative method of assessing surface and volume components from a suite of size separates is the log c *versus* log d plot, (where c is the concentration of a species and d the average grain diameter) as suggested by Eberhardt *et al.* (1972), and used extensively by Bogard (1977) and Morris (1977). From the fit of data to the resulting curve, a negative parameter n, which is a measure of the retention of a species as a volume related component, may be obtained. In Table 3, values of n for 12023 are compared with values obtained previously for soil 15601, and the Luna 24 core samples. Values of n for our He and Ne data are in good agreement with the data presented by Bogard (1977) for 13 soils having $I_s/FeO>45$ units. The n values of 0.21 and 0.18 for χ and IRM respectively are comparable to the 0.18 defined by Morris (1977) on the basis of I_s/FeO data as a steady state condition, where fine grained metal production is counterbalanced by constructive and destructive processes in the soil. Magnetic susceptibility and IRM seem to be as good a measure of fine grained metal as I_s. Values of n for CD_4 suggest that C_{hyd} is better retained than all the rare gases, and probably as good a measure of solar wind exposure as nitrogen (Gardiner *et al.*, 1977). Interpolating between the n values for χ and CD_4 for 12023 and corresponding values for I_s/FeO and N (Morris, 1977; Bogard, 1977) for other soils indicates that 12023 is just mature. This is in agreement with the value of 60.0 for the I_s/FeO maturity index (Morris, 1978).

Table 3. Values of the exposure parameter n($C \propto d^{-n}$) for χ, IRM and CD_4 for 12023, 15601† and Luna 24† soils.

Soil	Values of n				
	χ	IRM	CD_4	He	Ne
12023	0.21	0.18	0.35	0.73	0.64
15601	0.40	0.47	0.32	–	–
24090	0.25	0.18	0.22	–	0.81
24125	0.47	0.33	0.52	–	1.17
24196	0.50	0.44	0.57	–	1.03
13 soils*	–	–	N_2 = 0.33	0.68	0.66

*Average of 13 soils with $I_s > 45$ (Bogard, 1977).
†Gardiner *et al.,* 1977; Pillinger *et al.,* 1978.

How significant are surface related components?

A number of investigators (Zinner *et al*., 1977; Filleux *et al*., 1977; Müller *et al*., 1976) have now presented evidence from depth profiling experiments for the existence of surface concentrations of solar wind derived elements. However, the major treatments of surface and volume related components are conjecture, based on c *vs*. 1/r or log c *vs*. log d models. In order to obtain surface and volume related components from a c *vs*. 1/r plot one has to assume the volume related component is constant for all sized particles and equal to the y intercept. Such a notation is probably valid as long as single crystalline grains are involved, as in perhaps the study of grains from disaggregated gas rich meteorites, or mineral separates. However, in bulk lunar soils there is no guarantee that the volume correlated component is constant since the abundance of complex reworked grains or even the complexity within the grains could probably increase with decreasing grain size. Chang *et al.* (1976) on the basis of the data have suggested that volume components increase with grain size. Morris (1977) has argued extensively that I_s is entirely due to volume correlated metal and that changes in the value of n result from variations in the concentration of the volume correlated species. Bogard (1977) demonstrates how the log c vs. log d model behaves when either a constant volume related component or a volume related component which increases by a factor of two for an order of magnitude decrease in size is added to a surface related component. Increasing the volume related component in finer fractions shifts the value of n towards unity. Unfortunately, although the model shows what will happen, there is no way of distinguishing an increasing volume related component from a surface component. The factor of two chosen by Bogard (1977) was based on the studies performed on the 60009 core by McKay *et al.* (1976), who showed that the abundance of the magnetic fraction from the <20 μm material was about twice that from the 90–150 μm size range. Later work by the same group (McKay *et al*., 1977) suggests that the weight of magnetic material as determined by their method is not a good maturity index.

The problem of whether or not volume related components change is unresolved. However, CD_4 and χ measurements may be able to provide some relevant insight. Our studies on 15601 (Gardiner *et al.*, 1977) gave 1/r plots which demonstrate that both Fe^{o}_{sp} and C_{hyd} increase with decreasing size of agglutinates (all the material ρ<2.96 g/cm^3 collected at >2.0 mm from the magnetic separator). Although this could be interpreted (Gardiner *et al.*, 1977) as co-existence of surface and volume related components, we have some new evidence to suggest that the proportion of Fe^{o} which exists at the very surface of these agglutinate grains is small compared to the amount in the interior. Table 4 gives the results obtained by atomic absorption performed on $CuCl_2$•2KCl solutions which were used to treat agglutinates from 15601. For samples of 15601 which were hand shaken several times and allowed to stand overnight, the four finest fractions gave values less than the blank, and three others suggest that less than 2% by weight of the Fe^{o}_{sp} in the sample was taken into solution. The 150–250 μm fraction yielded a solution which indicated that about 1μg of Fe^{o} metal had been dissolved. This was the only positive result obtained, and it is not inconceivable that a single iron meteorite fragment could have been responsible. Subsequent retreatment of all the samples by ultrasonic agitation in fresh solution suggests that 1.7 to 4% by weight of the iron metal in the sample becomes accessible to the reagent. The dissolved iron could have been attacked by the copper solution penetrating the micropore structure, but it is equally likely that the ultrasonic treatment exposed metal in the interior by partial disaggregation of agglutinates. In this respect, it is noteworthy that coarser grains release more iron metal than the high surface area finer particles. Although further work is necessary on the agglutinate residues, we are able to draw some tentative conclusions: (i) most (possibly >98%) of the iron metal in 15601 is not present as a very surface component even in 10–20 μm sized grains; (ii) the increased

Table 4. 15601 $CuCl_2$/2KCl Experiment.

Grain Size μm	untreated*	% Fe Metal Detected by Atomic Absorption	
		treated for 12 hrs.	20 mins. sonication
>250	1.12	0.020	0.041
150–250	1.49	0.144	0.060
106–150	1.74	0.027	0.058
75–106	1.88	0.033	0.043
53–75	2.06	−0.005	0.050
40–53	2.23	−0.002	0.057
30–40	2.60	−0.001	0.076
20–30	2.30	—	—
10–20	1.85	−0.003	0.033

*% Fe metal calculated from magnetic susceptibility.
Error in one measurement ± 0.005.

amounts of χ and CD_4 from fine 15601 agglutinate fractions indicate the increasing volume related component in terms of a greater complexity within the agglutinates themselves; (iii) metal or C_{hyd} produced on the very surfaces of grains by preferential sputtering (Pillinger *et al.*, 1976), sputter deposition (Hapke and Cassidy, 1977), or any other mechanism during exposure could have been destroyed by oxidation in the terrestrial atmosphere (Pillinger *et al.*, 1976). The detection of iron in the zero valent state by ESCA (Housley and Grant, 1977) does not necessarily imply that a large proportion of the metallic iron in the lunar soil is on the surfaces of grains.

Results from the current study of 12023 also provide information relevant to the study of surface and volume related components. Our separations of agglutinates from 12023 were designed to select precisely defined magnetic material from each size fraction. We expected that any indication of increasing amounts of Fe^{o}_{sp} or C_{hyd} being associated with finer agglutinate fractions would have to come from material balance calculations. These will be performed once sufficient data are available. Preliminary data (Pillinger and Jull, 1978) from DCl acid dissolution on the M3.5 fractions of 12023 suggested that the amounts of C_{hyd} and ^{20}Ne for these materials of different sizes were the same within the experimental error. We speculated that for this to be so, any or all of the following rationales could be appropriate: (i) The volume related component of both species in the M 3.5 particles far outweighs the surface component; (ii) the two components exist in variable proportions with finer particles having a reduced volume related component to counterbalance a surface component which increases as size decreases; or (iii) the magnetic separation is able to concentrate material with a constant ratio of surface to volume component of solar wind derived species. We anticipated that χ data would confirm the above observations. However, surprisingly, χ increases as particle size decreases for material between 50–500 μm collected at a constant magnet separation distance. The χ data are of considerable significance and are discussed in detail in a later section. Further C_{hyd} data have, however, tended to confirm our original impression. A plot of CD_4 *vs*. M values (Fig. 2) suggests that the C_{hyd} concentration for grains of different size (but collected at constant magnetic separation distance) is constant within the limits of experimental error. Additional data indicate that the concentration of neon increases slightly in the finer grain size materials (see Fig. 3 and discussion in next section). Of the three rationales outlined above it would appear that the explanation which involves volume related components much larger than surface components for C_{hyd} + ^{20}Ne is the most reasonable. The slight increase in ^{20}Ne in finer fractions could be the real surface component of the rare gas. Grains are separated during the magnetic separation according to their increased volume related component of free metal, which reflects an increased maturity within the grain (i.e., more exposure and cycling of the constituent particles.)

Before continuing we should emphasize that we are not attempting to deny that freshly implanted surface components of solar wind species such as rare gases, N and C exist at particle surfaces. These must exist now and have been

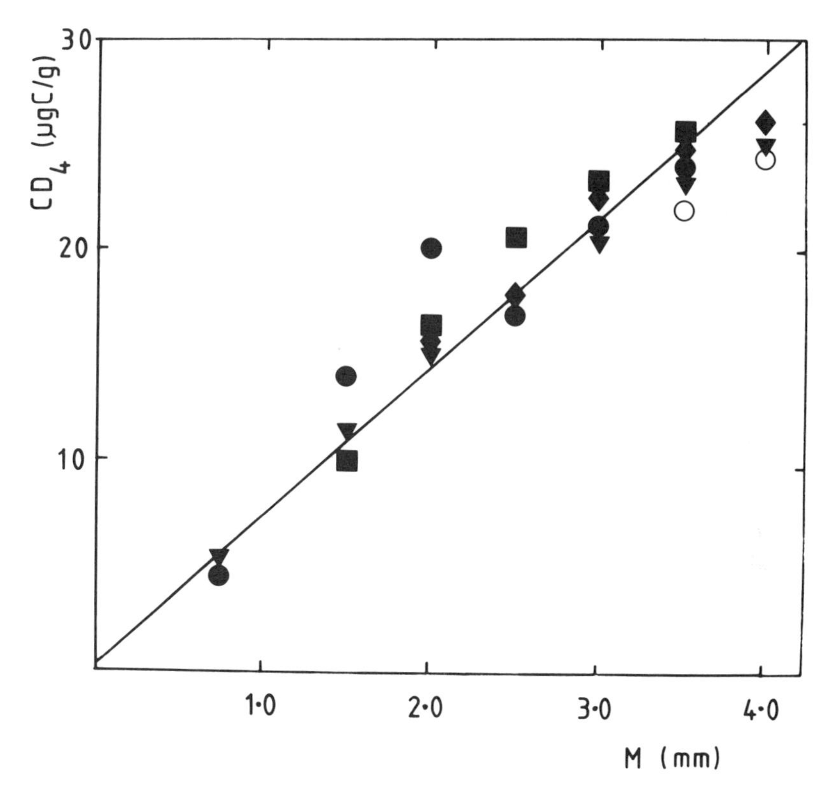

Fig. 2. CD_4 *vs.* magnetic separation (M). Different symbols refer to different particle sizes (μm): 150–250(●), 106–150(■), 75–106(◆), 53–75 (▼), and 40–53(○). Correlation coefficient 0.95.

recycled into agglutinates during maturation (e.g., Jull and Pillinger, 1977). Likewise, surface components of Fe° and C_{hyd} formed by any mechanism could exist whilst grains are on the lunar surface; they too have become incorporated into agglutinates and preserved. The data we have presented above suggest that there is only a very small surface component of Fe° and C_{hyd} on the two samples we have studied and yet both parameters give the apparent "surface correlation" with grain diameter. Therefore we suggest that the χ and CD_4 measurements for size separates are best explained in terms of a very small surface component and a volume related component which increases with decreasing particle size. The rare gases, N and C, would have a larger surface component due to freshly implanted material, but could still have a variable volume component. Our model only differs from that of Bogard (1977) in terms of the relative magnitude of surface and volume related components. Bogard's model suggests a surface component greater than the volume in fine fractions (*ca.* 20 μm). Our model suggests a much larger volume component.

An idea of the relative magnitudes of surface and volume related components for rare gases may be obtained from the data of Signer *et al.* (1977) on minerals and agglutinates from the 150–200 μm size fraction of numerous soils. The ^{20}Ne and ^{36}Ar content of the agglutinates is usually 10–20 times that of the mineral grains. If there is no change in the volume component in going from 200 to 20 μm, the surface and volume component would become about equal; if the volume component increased then obviously it would be the larger. Earlier we suggested that C_{hyd} could be considered as almost entirely volume correlated; therefore, it is possible to obtain an indication of the extent to which the volume component is increasing from the CD_4 *vs.* 1/r plot. The data for 12023 suggest a factor of 2 to 3 for an order of magnitude decrease in particle size.

Good evidence for the increasing volume component model may be forthcoming from the absolute measurement of surface carbon concentrations. Filleux *et al.* (1978) report that typical values for the surface component of carbon measured in depth profile by a ^{12}C (d,p) nuclear reaction is 2 to 7×10^{15} atoms/cm^2. These authors point out that this is less than estimates of 10^{16} atoms/cm^2 made using the slopes of carbon *vs.* 1/r plots published by DesMarais *et al.* (1975) and discussed by Chang and Lennon (1975). Erroneously high values for surface concentrations would be forthcoming if a major portion of the observed slope was due to an increasing volume related component. A calculation based on a sputter-equilibrium model (Jull and Pillinger, 1977) and using a sputter rate of 0.031 A/yr. (McDonnell, 1977) suggest a maximum surface concentration (1.7 to 3×10^{15} atoms/cm^2) of carbon which is very close to that measured by Filleux *et al.* (1977).

Theoretical considerations

Having accepted that an increasing volume component model is able to satisfy grain size separate data, it is relatively easy to demonstrate how the lunar soil might reach this condition. Consider an absolutely fresh unexposed, well-mixed sample of powered rock fragments which starts to accumulate solar wind elements and iron on its surface. A micrometeorite impact occurs and a series of agglutinates is formed. A 100 μm agglutinate incorporates material <100 μm, for arguments sake of average particle size 50 μm. A 10 μm agglutinate can only incorporate material <10 μm, again for the sake of discussion average size 5 μm. Assuming the losses of volatiles accompanying the agglutination process are about equal in proportion to the amounts present on the surfaces of the constituent grains, then the volume correlated component in the 10 μm agglutinate per unit weight will be ten times that of the 100 μm particle. Thus the concentration of solar wind species in the agglutinate is proportional to 1/r. Repeated agglutination processes will decrease the 10:1 ratio as small agglutinates are brought into the large ones, and as large ones are comminuted and then reagglutinated. All the time the volume component will be increasing relative to any possible surface component. A constant volume component cannot be reached until infinite maturity when the soil would be completely homogenised and give a horizontal line on a 1/r plot. Indeed, infinite maturity may never be obtained because a steady state (Morris, 1977; Basu, 1977) may be achieved first when constructional effects are balanced by the input and mixing of fresh material and comminution. If the probability of agglutination is greater in finer fractions, then constructional processes would be counteracted by an increase in the absolute abundance of agglutinates in these fractions compared to coarser fractions. The slope of the 1/r plot would also increase. Measurements of the content of petrographic agglutinates suggest that the later process probably does not contribute more than a factor of two (McKay *et al.*, 1976). Our modal analysis data for 12023 (Fig. 1) suggest that the absolute abundance of agglutinates does not vary much between 500 and 30 μm. Increasing maturity would be indicated in any soil following the increased volume component model by an increase in the intercept and decreasing slope for the 1/r plot, or a decreasing value of n on a log c *vs.* log d plot. The addition of fresh unexposed ejecta would give rise to slight variations in intercept and n.

The model equation for constant surface and volume related components according to Eberhardt (1972) and Bogard (1977) is

$$C \propto C_s d^{-n} + C_v \quad (1)$$

where C_s is the surface and C_v the volume related component. If we add a volume component which depends on grain size we get (similar to Bogard, 1977)

$$C \propto C_s d^{-n} + C_v a^{-\log d} \quad (2)$$

where a is the factor by which the volume component increases for an order of magnitude decrease in grain diameter. As $x^{\log y} = y^{\log x}$, equation (2) can be written

$$C \propto C_s d^{-n} + C_v d^{-\log a} \quad (3)$$

substituting n' for log a

$$C \propto C_s d^{-n} + C_v d^{-n'} \quad (4)$$

Thus, a volume component dependent on $d^{-n'}$ is indistinguishable from a surface component plus a volume component on a log/log plot. If $C_s << C_v$ values of n' are obtained by plotting log c against log d and n' is an indication of soil maturity and a is the measure of how rapidly the volume component is increasing with decreasing size. At $n' = O$, a is 1 or infinite maturity; at $n' = 1$, a becomes 10, a very immature soil.

Magnetic Separate to Bulk Soil Comparisons

To support the arguments that n decreases with increasing volume correlation (increasing maturity), Bogard (1977) compared the concentration of rare gases found in grain size separates with those determined for the corresponding magnetic separates and hand-picked agglutinates. Similarly, concentration ratios (Signer *et al.*, 1977) have been compared for rare gases in agglutinates and minerals. The data which have been obtained on 12023 may also be manipulated to give concentration ratios (Table 5 shows values for <2.96 g/cm^3 separates).

Table 5. Concentration ratios* for magnetic separates from $\rho < 2.96$ g/cm^3 fractions of size separates of 12023.

Size range	M4.0				M3.5				M3.0			
(μm)	CD_4	χ	^{20}Ne	^{4}He	CD_4	χ	^{20}Ne	^{4}He	CD_4	χ	^{20}Ne	^{4}He
250–500	← not separated →				–	4.6	–	–	–	3.7	–	–
150–250	← not separated →				4.1	3.5	2.0	3.7	3.7	2.8	2.0	1.7
106–150	← not separated →				3.5	3.4	2.4	2.3	3.2	2.8	2.2	1.8
75–106	3.4	2.9	1.7	1.4	3.3	2.7	1.6	1.5	2.8	2.3	1.5	1.5
53–75	3.1	3.6	1.4	1.4	2.9	3.1	1.4	1.4	2.7	2.6	1.5	1.4
40–53	2.6	3.7	1.6	1.6	2.4	3.0	1.5	1.5	–	2.5	–	–
Size range	**M2.5**				**M2.0**				**M1.5**			
(μm)	CD_4	χ	^{20}Ne	^{4}He	CD_4	χ	^{20}Ne	^{4}He	CD_4	χ	^{20}Ne	^{4}He
250–500	–	3.1	–	–	–	2.4	–	–	–	1.5	–	–
150–250	–	2.4	–	–	3.3	1.7	2.2	1.5	2.4	1.0	1.7	1.1
106–150	2.9	2.1	2.2	–	2.3	1.6	2.0	–	1.4	1.0	1.3	1.0
75–106	2.3	1.8	1.4	1.2	2.1	1.4	1.0	–	–	1.0	–	1.3
53–75	2.2	2.1	1.2	1.3	1.9	1.7	1.0	1.1	1.5	1.0	0.8	1.0
40–53	–	2.0	–	–	–	1.5	–	–	–	–	–	–

*concentration ratio = $[\text{parameter}]_{agg}/[\text{parameter}]_{bulk}$ (Bogard, Signer, 1977).

The conclusions which we draw may be compared and contrasted to those of Bogard (1977).

(i) All the magnetic fractions down to M2.0 are enriched in ^{20}Ne, C_{hyd} and Fe^{o}_{sp} relative to the bulk sieved material.

(ii) The enrichments for Fe^{o}_{sp} and C_{hyd} decrease substantially in all cases with decreasing magnet separation distance. However, the enrichments for ^{20}Ne are more constant over the range M4.0 to M2.0 for the <2.96 g/cm^3 fraction. This conclusion may be illustrated by plotting CD_4 vs. ^{20}Ne for bulk sieved fractions and separates. Figure 3 shows both the raw data, and CD_4 normalised to FeO content. The plots show the fractionation of the rare gas relative to the involatile hydrolysable carbon, demonstrating the retentivity of the carbon associated with iron. Slightly different slopes for the different sized particles probably reflect the change in the real surface component of the rare gas as discussed in a previous section. Note the markedly decreased slopes for separates having $\rho = 2.96 - 3.3$ g/cm^3 suggesting that neon is less fractionated. Either microbrecciation is a less severe process than agglutination (Cadogan *et al.*, 1973b), or the melting temperatures of glasses of higher FeO content are reduced (Basu, 1977; Pillinger *et al.*, 1977) and consequently cause less volatile loss. We interpret the plots in Fig. 3 to indicate an agglutinate evolution sequence. The most magnetic grains have been through the greatest number of cycles and have the largest volume related components. Well retained species have volume related components which evolve rapidly. From the modal analysis data we know that there is a tendency for the more magnetic fraction to increase in finer grain sizes. A greater population of more evolved agglutinates in fine grains sizes is in keeping with our increasing volume component model.

(iii) The concentration ratios for ^{20}Ne and CD_4 decrease substantially in all cases in going from coarse to fine fractions. However, the ratios for χ are much nearer constant (see below) assuming that the value for 75–106 μm bulk fraction is abnormally high. Bogard (1977) observed a similar trend for rare gases and suggested it might be due in part to difficulties in obtaining pure magnetic separates for fine material. In

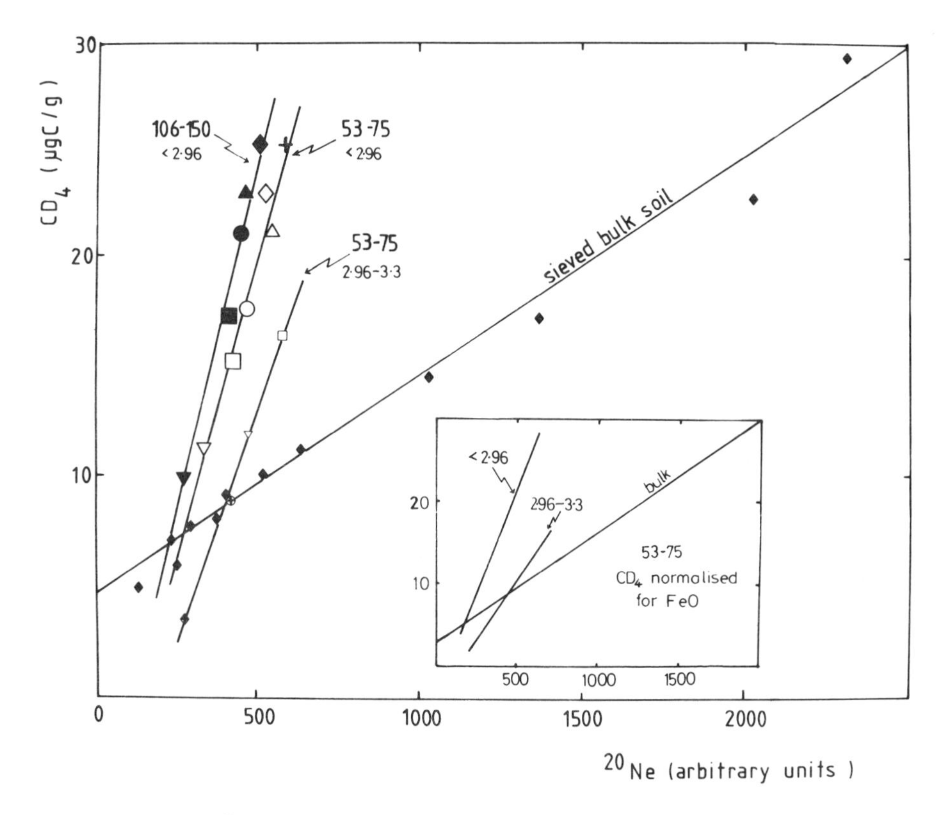

Fig. 3. CD_4 *vs.* ^{20}Ne for bulk sieved fractions (◆) and magnetic separates of ρ <2.96 and ρ = 2.96–3.3 g/cm^3 density material, 106–150 and 53–75 μm in size. Different symbols refer to different magnetic separation distances: M4.0(+), M3.5(◇), M3.0(△), M2.5(○), M2.0(□), M1.5(▽), M1.0(⊕), and MT(◈). Inset, data for CD_4 for bulk sieved fractions and 53–75 μm magnetic separates normalised to a constant FeO content of 16.22 wt.%.

view of the care taken during our magnetic separations we believe the trend for ^{20}Ne and CD_4 is real. The data could be explained by the increasing proportion of more complex agglutinate material in finer fractions, as indicated by the modal analysis. The most magnetic agglutinates have reached almost a steady state, since they do not appear to be produced in ever increasing complexity, as indicated by the small increase in CD_4 released by M4.0 compared to M3.5 particles. However, modal analysis shows that the finer fractions contain a greater proportion of agglutinates of increased complexity (i.e., the peak in the distribution of magnetic grains moves to high M values). The concentration ratio for the M4.0 and M3.5 fraction to the bulk soil will decrease for finer fractions. Again such an observation might be expected in terms of the increasing volume component model. Note that the absolute abundance of particles recognised as agglutinates need not increase, only their complexity.

(iv) The absolute values of the enrichments observed for ^{20}Ne in 150–250 μm and 106–150 μm fractions are in the range 2.0–2.4. For magnetic separates and hand picked agglutinates from similar size fractions of 15271 and 60010,1076, Bogard (1977) reported a range of 1.57–1.96. High concentration ratios could be an indication of immaturity (Bogard, 1977). However, in this case, since 12023 in mature, they reflect the greater purity of our magnetic separates. Samples used for Bogard's study must have contained magnetic material of different densities (concentration ratios for ^{20}Ne in M2.5, 2.0 and 1.5 material from 12023, ρ 2.96 − 3.3 g/cm^3, are 1.1 to 1.9).

(v) The absolute values for χ and CD_4 concentration ratios are high relative to those for ^{20}Ne from the same fractions, e.g., 4.1 and 3.5 respectively for χ and CD_4 from 150–250 μm, M3.5, compared to 2.0 for ^{20}Ne. The high values again reflect the increased volume related components for Fe^{o}_{sp} and C_{hyd} produced by retention during agglutination. The magnitude of the concentration ratio for χ and CD_4 could become a valuable maturity index. Data for immature/submature soil 15601 (Gardiner *et al.*, 1977) suggest the $CD_{4\,agg}/CD_{4\,bulk}$ and χ_{agg}/χ_{bulk} vary from 3.8 to 4.2 and 3.4 to 4.7 respectively for fractions from 53 to 250 μm. These values are high compared to 12023. Because these values are averages for all fractions of M $\geq$2.0 mm, due to the less sophisticated separation for 15601, the most magnetic particles presumably have even higher concentration ratios. If CD_4 concentration ratios for all particles M $\geq$2.0 are averaged for 12023, values of 2.5–2.9 are obtained in the size range 53–250 μm. Thus very high concentration ratios for Fe^{o}_{sp} and C_{hyd} reflect immaturity (cf. arguments by Bogard, 1977, based on Ar, Kr and Xe data). Ratios for χ and CD_4 from the soils 24090, 24125 and 24196 are 3.4, 5.4 and 7.9 and 2.7, 6.6

and 9.1 respectively. The lower layers of the Luna 24 core are either very immature or contain agglutinates which were not derived from *in situ* reworking (Pillinger *et al.*, 1978). Data for the double core tube 60009/10 (McKay *et al.*, 1977) show that I_s concentration ratios increase for low absolute values of I_s.

(vi) Concentration ratios of χ and CD_4 for the most magnetic grains over the least magnetic may exceed values of 20 whereas for He and Ne *ca.* 4 is a maximum. (Data not shown in Table 4, but compare M3.5 for the <2.96 g/cm^3 fraction of 150–250 μm with $\rho > 3.3$ in the same size range, Table 1).

As we have suggested above, concentration ratios might be a useful maturity index. The comparison of individual fractions, however, could lead to errors, for example, due to a discrepancy in an experimental measurement or because a sample analysed was impure or not carefully separated. An alternative parameter which could provide the same measure of maturity, but which can be determined more precisely, is the value of r when the extrapolation of the best fit line for bulk soil size fractions (on a 1/r plot) reaches the volume related component of separated agglutinates. Because of averaging, the agglutinate/bulk soil intercept could be precisely found. Data for 15601 (Gardiner *et al.*, 1977) give a value for r of 5.5 μm considering agglutinates to be all particles $M \geq 2.0$. The data for 12023 give $r = 7.3$ μm; the higher value reflects the greater maturity.

The Carbon Content of Finely Divided Iron

The composition of fine grained iron in lunar soils is likely to be of great importance in elucidating its origins (Gardiner *et al.*, 1977). The abundance of C_{hyd} provides one possible compositional parameter since CD_4 is thought to be the measure of carbon present in solid solution in α-Fe (Jull *et al.*, 1976). Carbon studies were the first to show clearly that the metal must have been formed by an exposure related reduction of Fe^{II} indigenous to lunar silicates (Pillinger *et al.*, 1974). The exact nature of the reduction process remains the subject of considerable debate.

On the basis of a plot of CD_4 *vs* χ for 29 separate fractions of 15601 we were able to suggest that α-iron in this soil has a relatively constant carbon concentraion *ca.* 0.16 ± 0.4 wt.% (Gardiner *et al.*, 1977). Earlier, Pillinger *et al.* (1974), had reported a correlation between CD_4 and I_s measured by three different investigators on 27 different bulk soils. Unfortunately, 15601 was not included among the samples studied, and as I_s values have not been absolutely calibrated, an estimate for the carbon content for the iron in these bulk soils could not be obtained. However, if both sets of data are considered together they suggest the concentration of carbon in iron might be relatively constant for a wide range of lunar soils of very different composition and maturity. Such an observation would place severe constraints on mechanisms which may be invoked for the origin of fine grained lunar iron, particularly as the estimated carbon concentration for iron from 15601 is above the equilibrium solid solubility predicted by the Fe/C phase diagram. Since the implications are considerable, it is important to provide rigorous proof of any limitations to the carbon in iron concentration.

The method which we are employing to measure the carbon in iron concentration is to establish the C_{hyd}/iron metal ratio from a plot of CD_4 *vs.* χ. If this is to be a viable proposition, it will be necessary to demonstrate that procedures employed for determining CD_4 and χ provide a consistent estimate of C_{hyd} and Fe^o respectively. At the present time, there is no evidence that DCl dissolution gives a variable yield of deuterocarbons with different aliquots of the sample provided the acid strength is kept constant (Cadogan *et al.*, 1972). We have attempted to prepare samples for dissolution in DCl as reproducibly as possible. However, without knowing the nature of the chemical reaction involved in the deuterolysis, it is impossible to be absolutely certain about its consistency with different samples. On the other hand, in the following discussion we shall present evidence to the effect that measurements of finely-divided iron, particularly magnetic susceptibility, may be dependent on the size distribution of the metal.

In an earlier section we noted the rather surprising result that χ increases as particle size decreases for grains collected at a constant magnetic separation distance. This result is best illustrated by a plot of χ *vs.* the distance of magnetic separation M (Fig. 4) which shows a regular increase in susceptibility over an order of magnitude change in grain size from 53–500 μm. Magnetic susceptibility seems to maximise at 53 μm since further decrease in particle size does not result in further increase in χ. The 53–75 μm fraction has a χ value 20% more than material 250–500 μm in size. Thus, it appears to be easier to collect coarse grains than fine grains. Assuming that the magnetic separation is governed by the abundance of free metal in the sample, one explanation currently acceptable for the data in Fig. 4 is that the distribution of finely divided iron grains within large agglutinates (250–500 μm) is different from small agglutinates (53–75 μm). This can be explained if the magnetisation curves of the different fractions are non-linear and of different shape. Although all the grains which are collected at a particular separation distance must have the same induced magnetisation M′ in the field H′ produced by the magnet at that distance, grains with different non linear magnetisation curves passing through the point M′H′ would have a different initial susceptability. Other rationalisations of Fig. 4 which may be dismissed include: (i) The observed results are not some arbitrary change in the magnetic separator (i.e., decrease in field strength), since the size fractions were not magnetically separated in order of decreasing size; (ii) both measures (χ and M) of the magnetic properties of the separates are per unit weight, density changes cannot account for more than a 6% variation in χ; (iii) as small particles show increased magnetic susceptibility, arguments concerning inadequacies of the magnetic separation for fine grains are not applicable (Gardiner *et al.*, 1977; Bogard, 1977); (iv) normalisation of χ to constant FeO content increases the differences between coarse and fine grains therefore chemical effects may be

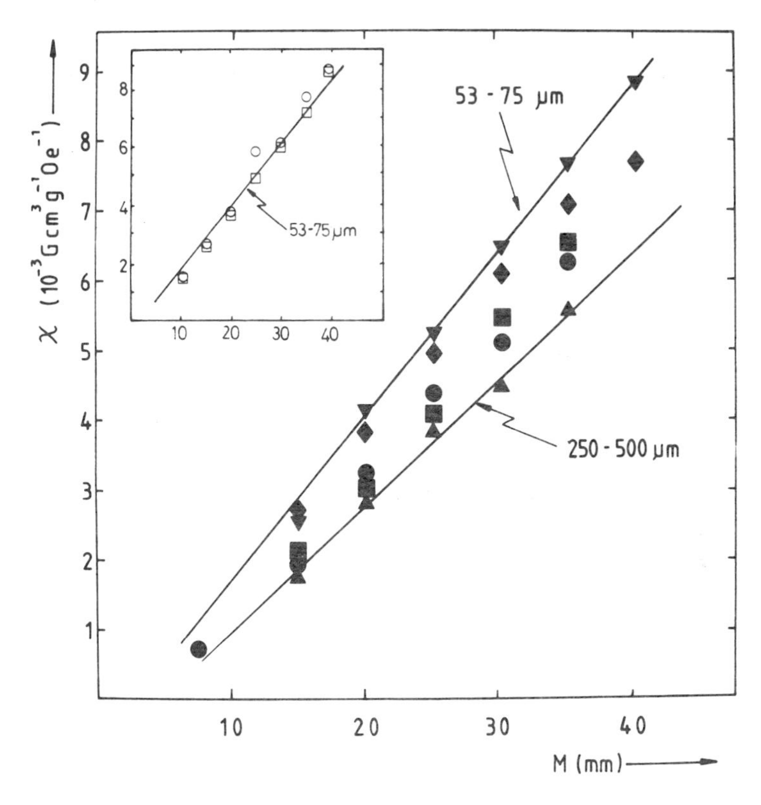

Fig. 4. Magnetic susceptibility (χ) *vs.* magnetic separation distance (M). Different symbols refer to different particle sizes (μm): 250–500 (▲), 150–250 (●), 106–150(■), 75–106 (◆), 53–75 (▼) and inset 40–53 (□), and 30–40 (O).

ruled out; and (v) a near surface distribution of metal within the grains should favour the collection of fine grains relative to coarse ones.

To examine further the hypothesis that the distribution of finely divided iron grains changes as a function of particle size we have measured IRM for the 12023 separates. This parameter measures iron grains in the size range of about 130–300 A. Values of IRM/χ (Table 1) decrease in the finer fractions but a plot of IRM *vs.* M (Fig. 5) fails (note the different order of symbols for any M value) to reveal any systematic change in IRM with particle size in contrast to χ versus M (Fig. 4). It is likely that the IRM is constant within the experimental errors for fractions collected at any given magnetic separator distance, independent of grain size since IRM and M plot with a correlation coefficient of 0.93. Thus, the size distribution change might be confined to the <130 Å range. Fe^{o}_{sp} metal occurs in an average size range which gives a greater magnetic susceptibility for 53–75 μm particles.

Whatever the cause of the systematic variation of χ with grain size, this effect seems capable of explaining our χ *vs.* CD_4 plot (Jull *et al.*, 1978) for bulk soils. In Fig. 6, we show data for 55 bulk soils including 12023 and 15601. Although there is a considerable degree of scatter, a general trend is evident; two thirds of the samples define a confidence interval for the line shown. The remaining soils plot outside this range. Superimposed on Fig. 6 are the best fit lines for χ *vs.* CD_4 plots of separates from 12023 and 15601, and three Luna 24 soils (Gardiner *et al.*, 1977; Pillinger *et al.*, 1978). The data points for 12023 and 15601 are with the 2σ errors of the respective lines for separates. The slopes of the χ *vs.* CD_4 lines for the separates increase as the average value of the ratio IRM/χ decreases. Our revised (compared to Jull *et al.*, 1978) interpretation of Fig. 6 is as follows: The scatter on the graph is predominantly a real effect; experimental errors are insufficient to account for the observed departure from a linear correlation. Data

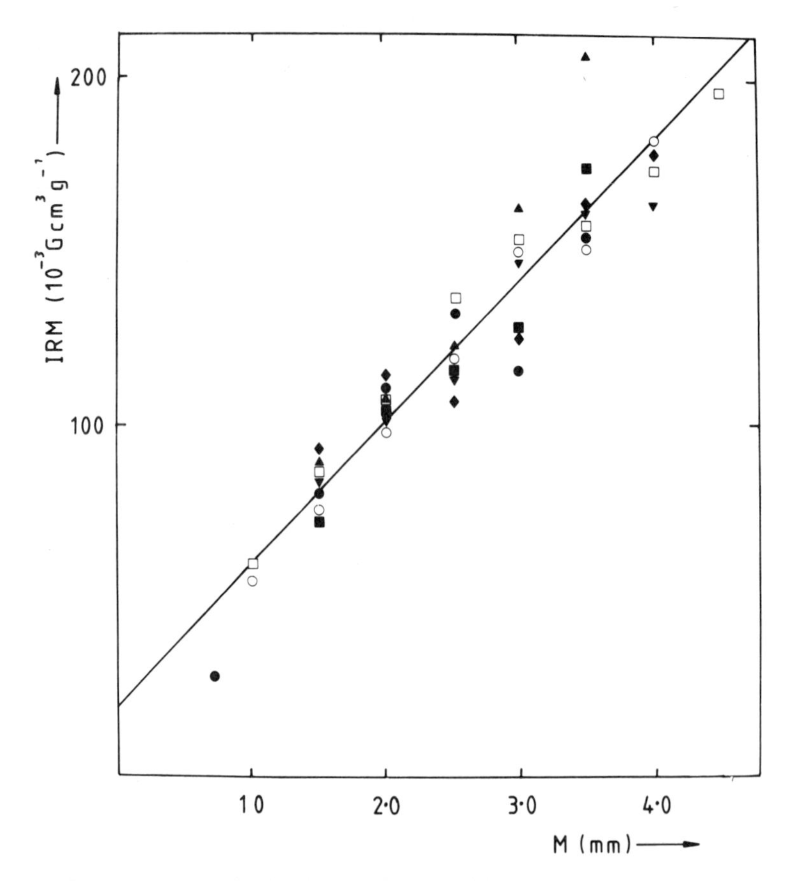

Fig. 5. Isothermal remanent magnetisation (IRM) *vs.* magnetic separation (M). Different symbols refer to different particle sizes (μm): 250–500(▲), 150–250(●), 106–150(■), 75–106(◆), 53–75(▼), 40–53(□), and 30–40(○). Correlation coefficient 0.93.

for particle separates suggest that individual relationships exist between χ and CD_4 for different materials, such that a series of lines of the form

$$\chi = k\,u\,(C_{hyd}) + b$$

may be possible, where k is a constant, but u is a variable depending on the sample, and b is a constant, related to the abundance of magnetic minerals. The factor u may be dependent in some way on the variation of χ inferred from Fig. 5 and not due to changes of carbon/iron ratio. If the change in magnetic susceptibility reflects the size distribution of metal grains, u becomes an important parameter, which may provide information about (i) the reduction mechanism producing Fe°, (ii) the growth of iron grains, and (iii) the nature of the process by which carbon is incorporated into the iron. The carbon content of iron in most lunar soils could be constrained between comparatively narrow limits. In Fig. 7 we show a plot of I_s (data from Morris) vs. CD_4. For all the bulk soils studied there is a correlation coefficient of $R = 0.87$ and only three samples fall outside the 95% confidence interval for the remaining 50 samples which define a line with $R = 0.89$. The three anomalous results may be due to experimental errors. A plot of χ vs. I_s gives a correlation intermediate between those observed for χ vs. CD_4 and I_s vs. CD_4. Variations in the grain size of iron

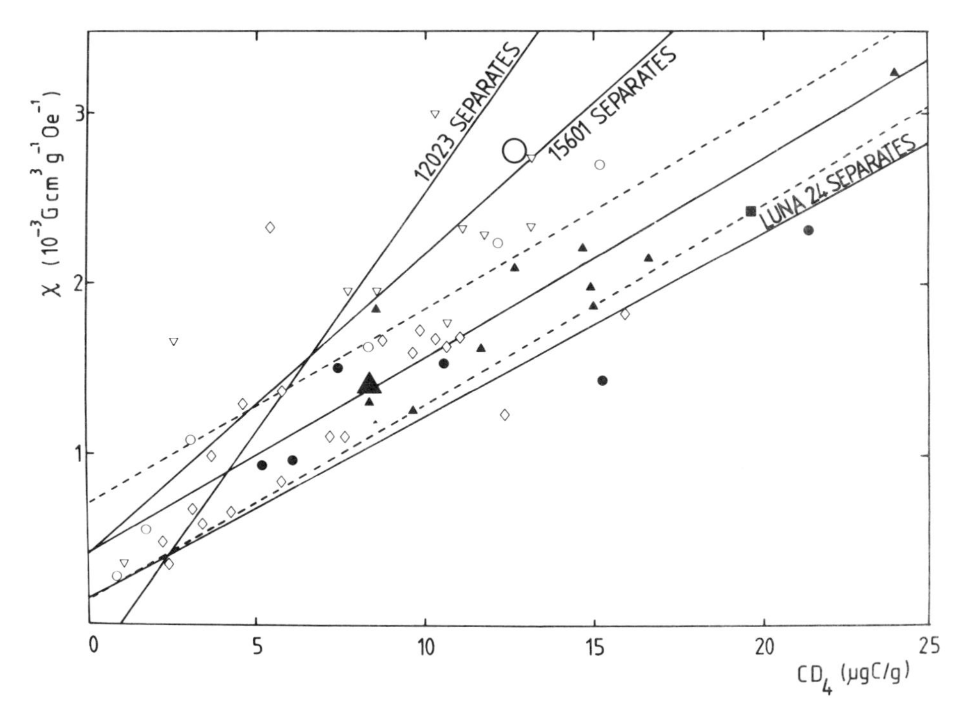

Fig. 6. Magnetic susceptibility (χ) *vs.* CD_4, for bulk soils and particle separates. Different symbols refer to samples from different missions: A11(■), A12(○), A14(▽), A15(▲), A16(◇), and A17(●). Large symbols are for 12023 and 15601.

might be reflected in the I_s values to a smaller extent than χ as I_s measures iron <300 Å (Gose and Morris, 1977).

CONCLUSIONS

The finely divided iron metal thought to provide an excellent measure of lunar soil maturity is widely dispersed throughout the particles which compose the sample. Provided stringent precautions are observed, magnetic separation methods are capable of giving a variety of relatively homogeneous fractions for subsequent analysis by more precise magnetic techniques and other methods. Although magnetic separations have received much criticism, they provide the only relatively simple means of obtaining fractions having defined physical properties. Handpicking of agglutinates is more subjective and there is no guarantee that a pure sample will be collected. Handpicking of very small grains is not practical. Our separations show that there are grains in 12023 which have densities less than 2.96 g/cm^3 and in the range 2.96–3.3 g/cm^3, which are collected at between 2.0 and 4.0 mm by the magnetic separator, and are indistinguishable from each other on the basis of exterior appearance. Magnetic measurements, C_{hyd} and rare gas data for these grains suggest they have substantially different maturities. In view of the apparent evolution sequence afforded by magnetic separates of light density material, all fractions would

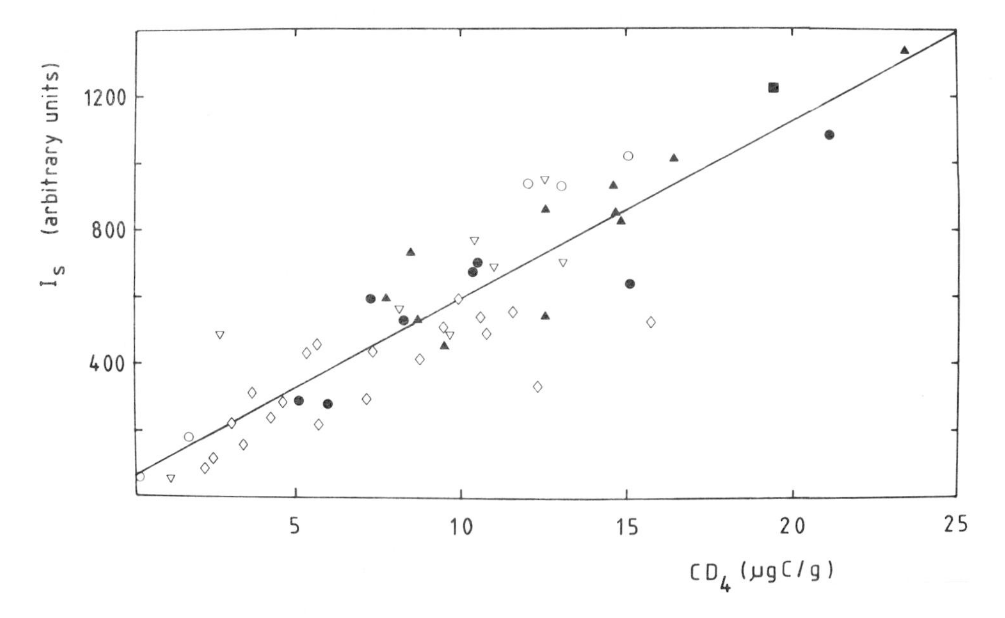

Fig. 7. I_s *vs.* CD_4 for 53 bulk soils. Different symbols refer to different missions: A11(■), A12(○), A14(▽), A15(▲), A16(◇), and A17(●). Correlation coefficient 0.87.

benefit from an unbiased and less cursory petrographic description than we have hitherto been able to provide.

From the study of bulk soil size fractions and a comparison between these and magnetic separates it can be demonstrated that χ (notwithstanding variations within metal grain size) and CD_4 are both good maturity indices when normalised to FeO content. The IRM results suggest that this parameter may be superior to χ as a measure of metallic iron concentration. The n values for χ and IRM, are similar to I_s and CD_4 comparable to nitrogen. Hydrolysable carbon could provide a better measure of exposure than total carbon. It is well retained in volume-related components and its measurement is less likely to be affected by contamination or the existence of indigenous material. A number of other related parameters, such as agglutinate/bulk concentration ratios, or agglutinate/bulk intercept values, may provide useful exposure indices as these measurements essentially provide an indication of agglutinate abundance. We have provided some preliminary rare gas data; however, more detailed measurement will be carried out by the European Consortium (Schultz, Mainz, and Signer, Zurich).

Our observation that the very surface component of Fe° on 15601 is minor, requires further study especially as ESCA studies on surfaces show the presence of reduced iron (Housley and Grant, 1977; Dikov *et al.*, 1978). The implications drawn with respect to the increasing volume component model for lunar soil requires evaluation. However, a number of other pieces of evidence, including agglutinate/bulk concentration ratios, support the conclusion that c vs. 1/r and log c *vs.* log d plots can be interpreted on the basis of an increased volume related component due to increasing complexity within agglutinates in fine fractions. The trend identified for magnetic susceptibility changes in agglutinates of different size fractions may explain the scatter in the χ *vs.* CD_4 plots. However, if the changes in χ can be explained in terms of a change in the grain size distribution of iron, χ may provide additional information concerning maturity. In order to demonstrate conclusively that C_{hyd} and IRM remain constant or follow divergent trends from magnetic susceptibility in agglutinates of different particle diameter, even more carefully controlled magnetic separations may be advantageous and it should be taken into consideration that magnetic field strength decreases as M^2. In the future, it may be necessary to study single grains. Certainly material separated from disaggregated agglutinates (Etique *et al.*, 1978; Bogard *et al.*, 1978) needs to be investigated. The number of fractions requiring analysis increases rapidly; however, the scientific returns in terms of our understanding of maturational effects may eventually compensate.

Acknowledgments—This work is supported by grants (SGD 00772, SGD 00963) from the Science Research Council. MRW acknowledges the receipt of a research studentship from the SRC, and AJTJ is grateful for a fellowship from the National Research Council of Canada. We also thank Mrs. D. Fabian for help with the sample separations, and preparation of the manuscript.

References

Basu A. (1977) Steady state, exposure age and growth of agglutinates in lunar soils. *Proc. Lunar Sci. Conf. 8th*, p. 3617–3632.

Bogard D. D. (1977) Effects of soil maturation on grain size dependence of trapped solar gases. *Proc. Lunar Sci. Conf. 8th*, p. 3705–3718.
Bogard D. D., Morris R. V., McKay D. S., Fuhrman R. and Hirsch W. C. (1978) Relative abundances of fine grained metal and noble gases in magnetic separates of crushed agglutinate particles (abstract). In *Lunar and Planetary Science IX*, p. 114–116. Lunar and Planetary Institute, Houston.
Cadogan P. H., Eglinton G., Firth J. N. M., Maxwell J. R., Mays B. J. and Pillinger C. T. (1972) Survey of lunar carbon compounds: II. The carbon chemistry of Apollo 11, 12, 14 and 15 samples. *Proc. Lunar Sci. Conf. 3rd*, p. 2069–2090.
Cadogan P. H., Eglinton G., Gowar A. P., Jull A. J. T., Maxwell J. R. and Pillinger C. T. (1973a) Location of methane and carbide in Apollo 11 and 16 fines. *Proc. Lunar Sci. Conf. 4th*, p. 1493–1508.
Cadogan P. H., Eglinton G., Maxwell J. R. and Pillinger C. T. (1973b) Distribution of methane and carbide in Apollo 11 fines. *Nature Phys. Sci.* **241**, 81–82.
Chang S. and Lennon K. (1975) Implantation of carbon and nitrogen ions into lunar fines: Trapping efficiencies and saturation effects. *Proc. Lunar Sci. Conf. 6th*, p. 2179–2188.
DesMarais D. J., Basu A., Hayes J. M. and Meinschein W. G. (1975) Evolution of carbon isotopes, agglutinates and the lunar regolith. *Proc. Lunar Sci. Conf. 6th*, p. 2353–2373.
Dikov Yu P., Bogatikov O. A., Barsukov V. L., Florensky C. P., Ivanov A. V., Nimoshkalenko V. V. and Alyoshin V. G. (1978) Reduced forms of elements in the surface parts of regolith particles: ESCA studies (abstract). In *Lunar and Planetary Science IX*, p. 250–252. Lunar and Planetary Institute, Houston.
Eberhardt P., Geiss J., Graf H., Grögler N., Mendia D., Mörgeli M., Schwaller H., Stettler A., Krähenbühl U. and von Gunten H. R. (1972) Trapped solar wind noble gases in Apollo 12 lunar fines 12001 and Apollo 11 breccia 10046. *Proc. Apollo 11 Lunar Sci. Conf.*, p. 1821–1856.
Etique P., Baur H., Derksen U., Funk H., Horn P., Signer P. and Wieler R. (1978) Light noble gases in agglutinates: A record of their evolution? (abstract). In *Lunar and Planetary Science IX*, p. 297–299. Lunar and Planetary Institute, Houston.
Filleux C., Spear R., Tombrello T. A. and Burnett D. S. (1978) Carbon depth distribtuion for soil breccias (abstract). In *Lunar and Planetary Science IX*, p. 317–319. Lunar and Planetary Institute, Houston.
Gardiner L. R., Jull A. J. T. and Pillinger C. T. (1978) Progress towards the direct measurement of $^{13}C/^{12}C$ ratios for hydrolysable carbon in lunar soil by static mass spectrometry. *Proc. Lunar Sci. Conf. 9th*. This volume.
Gardiner L. R., Woodcock M. R., Pillinger C. T. and Stephenson A. (1977) Carbon chemistry and magnetic properties of bulk and agglutinate size fractions from soil 15601. *Proc. Lunar Sci. Conf. 8th*, p. 2817–2839.
Gose W. A. and Morris R. V. (1977) Depositional history of the Apollo 16 deep drill core. *Proc. Lunar Sci. Conf. 8th*, p. 2909–2928.
Hapke B. and Cassidy W. A. (1978) Is the Moon really as smooth as a billiard ball?; comments on some recent models of sputter fractionation. *Geophys. Res. Lett.* **5**, 297–300.
Housley R. M. and Grant R. W. (1977) An XPS (ESCA) study of lunar surface alteration profiles. *Proc. Lunar Sci. Conf. 8th*, p. 3885–3899.
Hu H.-N. and Taylor L. A. (1977) Lack of chemical fractionation in major and minor elements during agglutinate formation. *Proc. Lunar Sci. Conf. 8th*, p. 3645–3656.
Jull A. J. T., Eglinton G., Pillinger C. T., Batts B. D. and Biggar G. M. (1976) The identity of lunar hydrolysable carbon. *Nature* **262**, 566–567.
Jull A. J. T. and Pillinger C. T. (1977) Effects of sputtering on solar wind element accumulation. *Proc. Lunar Sci. Conf. 8th*, p. 3817–3833.
Jull A. J. T., Pillinger C. T. and Stephenson A. (1978) The carbon content of finely divided lunar iron (abstract). In *Lunar and Planetary Science IX*, p. 609–611. Lunar and Planetary Institute, Houston.

Krapf G., Lutz J. L., Melnick L. M. and Bandi W. R. (1972) A DTA-EGA study of the chemical isolation of Fe_3C, amorphous carbon and graphite from steel and cast iron. *Thermochim. Acta* **4**, 257–271.

McDonnell J. A. M. (1977) Accretionary particle studies on Apollo 12054, 58: *In-situ* lunar surface microparticle flux rate and solar wind sputter rate defined. *Proc. Lunar Sci. Conf. 8th*, p. 3835–3857.

McKay D. S., Dungan M. A., Morris R. V. and Fruland R. M. (1977) Grain size, petrographic and FMR studies of the double core 60009/10: A study of soil evolution. *Proc. Lunar Sci. Conf. 8th*, p. 2929–2952.

McKay D. S., Morris R. V., Dungan M. A., Fruland R. M. and Fuhrman R. (1976) Comparative studies of grain size separates of 60009. *Proc. Lunar Sci. Conf. 7th*, p. 295–313.

Morris R. V. (1976) Surface exposure indices of lunar soils: A comparative FMR study. *Proc. Lunar Sci. Conf. 7th*, p. 315–336.

Morris R. V. (1977) Origin and evolution of the grain size dependence of the concentration of fine grained metal in lunar soils: The maturation of lunar soils to a steady state stage. *Proc. Lunar Sci. Conf. 8th*, p. 3719–3747.

Morris R. V. (1978) The surface exposure (maturity) of lunar soils: Some concepts and I_s/FeO compilation. *Proc. Lunar Sci. Conf. 9th*. This volume.

Müller H. W., Jordan J., Kalbitzer S., Kiko J. and Kirsten T. (1976) Rare gas ion probe analysis of helium profiles in individual lunar soil particles. *Lunar Sci. Conf. 7th*, p. 937–951.

Pillinger C. T., Davis P. R., Eglinton G., Gowar A. P., Jull A. J. T., Maxwell J. R., Housley R. M. and Cirlin E. H. (1974) The association between carbide and finely-divided metallic iron in lunar fines. *Proc. Lunar Sci. Conf. 5th*, p. 1949–1961.

Pillinger C. T., Eglinton G., Gowar A. P., Jull A. J. T. and Maxwell J. R. (1977) The exposure history of the Apollo 16 site: an assessment based on methane and hydrolysable carbon. *The Soviet American Conference on the Cosmochemistry of the Moon and Planets*, p. 541–551. NASA SP-370. Washington, D.C.

Pillinger C. T., Gardiner L. R. and Jull A. J. T. (1976) Preferential sputtering as a method of producing metallic iron, inducing major element fractionation and trace element enrichment. *Earth Planet. Sci. Lett.* **33**, 289–299.

Pillinger C. T., Gardiner L. R., Jull A. J. T., Woodcock M. R. and Stephenson A. (1978) Magnetic properties and carbon chemistry studies pertinent to the evolution of the regolith at the Luna 24 site. In *Mare Crisium: The View from Luna 24* (R. B. Merrill and J. J. Papike, eds.), p. 217–228. Pergamon, N. Y.

Pillinger C. T. and Jull A. J. T. (1978) Carbon chemistry studies of agglutinate separates from soil 12023 (abstract). In *Lunar and Planetary Science IX*, p. 910–912. Lunar and Planetary Institute, Houston.

Signer P., Baur H., Derksen U., Etique P., Funk H., Horn P. and Wieler R. (1977) Helium, neon and argon records of lunar soil evolution. *Proc. Lunar Sci. Conf. 8th*, p. 3657–3683.

Smerko R. G. and Flinchbaugh D. A. (1968) Recent progress in the chemical extraction of non-metallic inclusions in steel. Techniques and Applications. *J. Metals* **20**, No. 7, 43–51.

Stephenson A. (1970) Single domain grain distributions II. The distribution of single domain iron grains in Apollo 11 lunar dust. *Phys. Earth Planet. Inter.* **4**, 361–369.

Stephenson A. and de Sa A. (1970) A simple method for the measurement of the temperature variation of the initial magnetic susceptibility between 77 and 1000°K. *J. Sci. Instrum.* **3**, 59–61.

Woodcock M. R. and Pillinger C. T. (1978) Major element chemistry of agglutinate size fractions. *Proc. Lunar Sci. Conf. 9th*. This volume.

Zinner E., Walker R. M., Chaumont J. and Dran J. C. (1977) Ion probe surface concentration measurements of Mg and Fe and microcraters in crystals from lunar rock and soil samples. *Proc. Lunar Sci. Conf. 8th*, p. 3859–3883.

Proc. Lunar Planet. Sci. Conf. 9th (1978), p. 2195–2214.
Printed in the United States of America

Major element chemistry of agglutinate size fractions

M. R. WOODCOCK and C. T. PILLINGER

Planetary Sciences Unit, Department of Mineralogy and Petrology
University of Cambridge, England

Abstract—Low density, highly magnetic material, considered to represent "pure" agglutinates, has been isolated from sieved fractions of a number of soils. Care has been taken to eliminate interference from magnetic non-agglutinates. Partial major chemical analysis demonstrates that agglutinates from 60501 and 68121 are enriched in FeO. However, more detailed studies on 15601 and 12023 show that agglutinates from these soils are depleted in ferromagnesium elements and enriched in feldspathic elements. The chemical changes observed appear to be most pronounced in finest fractions. Two samples from the Luna 24 core behave similarly to 15601 and 12023. For magnetic separates from 12023, it can be shown that progressively more evolved agglutinates are enriched in total iron. Possible mechanisms for the observed chemical changes include mixing of mature and immature components, the addition of non-native material and fractionation by sputtering.

INTRODUCTION

In recent years there has been a considerable interest in the major and trace element chemistry of agglutinates. Activity in this area was precipitated by Rhodes *et al.* (1975) who argued from elemental analyses of magnetic separates that agglutinates are enriched, relative to bulk soils, in ferromagnesian and lithophile elements and depleted in species normally associated with plagioclase. These observations were explained in terms of a multistage partial melting model, which caused selective incorporation of mesostasis and mafic material into agglutinitic glass. Additional data (Rhodes *et al.*, 1976), obtained on a suite of Apollo 14 samples, exhibited the same trend.

The concept of partial melting to give glasses enriched in ferromagnesian elements has been questioned by Taylor and his co-workers. A painstaking study of agglutinitic glass by electron microprobe analysis has failed to reveal any fractionation of major or minor elements during the agglutination process (Hu and Taylor, 1977). Thus, agglutinate glasses may result from a "whole soil" impact melt (Gibbons *et al.*, 1976). The Taylor group (Via and Taylor, 1976; Hu and Taylor, 1977) attribute the results obtained by Rhodes *et al.* (1975, 1976) to a bias introduced during the magnetic separation procedure. A petrographic study of magnetic separates, prepared according to the method of Rhodes *et al.* (1975), suggests that "agglutinate" fractions contain magnetic non-agglutinates–glass free microbreccias with 30–60 μm droplets of FeNi and pyroxene and olivine fragments. Conversely, the residue, after removal of "agglutinates", contains non-magnetic agglutinates composed of plagioclase and feldspathic glasses.

Our group is currently involved in a detailed investigation of the distribution of finely divided iron metal and hydrolysable carbon (C_{hyd}) phases within lunar soil (Gardiner *et al.*, 1977; Pillinger *et al.*, 1978a) To this end, we have carried out extensive separations on two soils, 15601 and 12023, using sieving and density and magnetic properties. It is now well established (Pillinger *et al.*, 1974; Morris, 1976) that both finely divided metal and associated C_{hyd} produced in the lunar soil by surface exposure is governed by the bulk ferrous iron content of the silicates and the length of surface exposure. The understanding of the exposure induced Fe^{o} and C_{hyd} survey depends, to some extent, on a knowledge of the major element chemistry of the fractions under study, therefore, the enrichment/non-enrichment controversy is of some importance. We have developed a technique to allow analysis of $1.O^{mg}$ quantities of our separates. Although we can confirm some observations made by both Rhodes *et al.* and the Taylor group, the majority of our data conflicts with ideas previously presented. For most of the samples investigated here, mature agglutinates seem to be depleted in ferromagnesian elements and enriched in feldspathics. In order to explain the observations, an alternative model may be invoked. Our studies differ substantially from previous work in that we consider "end members" (i.e., pure agglutinates) obtained from a narrow range of particle sizes; therefore, the data obtained should reflect more accurately the true composition of agglutinates. A study of very fine material is also included.

EXPERIMENTAL

Sample separation procedures for 15601, 12023 and Luna 24 are discussed in detail by Gardiner *et al.*, (1977), Pillinger *et al.* (1978a) and Pillinger *et al.* (1978b). In brief, the samples were sieved into a large number of narrow size ranges. Each size range was density separated with two heavy liquids, at 2.96 g/cm^3 and 3.3 g/cm^3. The low and intermediate density separates were subdivided according to progressive changes in magnetic susceptibility. For 15601 and the Luna 24 soils, three magnetic fractions were obtained, high (particles moving $\geq$2 mm to the magnetic separator), medium and non-magnetic. In the case of 12023, fractions were collected at 0.5 mm intervals from 4.0 mm to zero (designated M4.0, M3.5, M3.0, etc.). Apollo 16 soils, 60501, 66040 and 68121 were processed in a manner exactly analogous to the 12023 soil, except that a density fraction $<$2.6 g/cm^3 was used for the magnetic separations.

The microanalytical procedure used (X.R.F. measurement of ion exchange resin loaded paper) was an adaptation of the method of Campbell *et al.* (1966). Sample aliquots for major element (iron, magnesium, titanium, aluminum and calcium) analyses *ca.* 1 mg were digested in HF/HNO_3 in P.T.F.E. beakers. Solutions were made up to 40 ml with doubly distilled water and adjusted to pH 2. Each solution was passed (seven times) through discs of Reeve-Angel SA-2 cation exchange resin paper. Dried discs were analysed using a Siemens, focusing crystal, X-ray spectrometer, in a manner analogous to pressed pellets. Quantitation was obtained by recourse to discs prepared from standard solutions and international geochemical standards. Table 1 gives recommended and obtained oxide values for seven geochemical standards, when calibrated against the standard solutions only. As an example, Fig. 1 shows the calibration graph for iron where standard solutions and international geochemical standards were used. For an estimate of the maximum error in one analysis, the points furthest from the line of the calibration graphs were used. Values of $\pm$0.3% for Fe0 and Mg0, $\pm$0.1% for $Ti0_2$, $\pm$0.5% for $A1_20_3$ and $\pm$0.7% for Ca0 were obtained. Matrix effects were negligible at the concentrations involved because, by weight, approximately 99% of the target material was the disc.

Results obtained by XRF for Fe0, Mg0 and Ca0 by the above method compare favourably with

Table 1. Recommended (R) and Obtained (O) values for International Geochemical Standards.

		BCR1	BR	G2	JB1	W1	AGV1	JG1
FeO	R	12.2	11.7	2.5	8.1	10.0	6.1	2.0
	O	12.6	11.8	2.7	8.1	9.7	6.3	2.1
MgO	R	3.3	13.3	0.5	7.7	6.6	1.5	0.7
	O	3.6	13.2	0.5	7.5	6.5	1.4	0.6
TiO_2	R	2.2	2.6	0.5	1.3	1.1	1.1	0.3
	O	2.3	2.6	0.5	1.3	1.1	1.2	0.3
Al_2O_3	R	13.6	10.2	15.4	14.5	14.8	17.0	14.2
	O	13.7	10.1	15.9	14.5	14.3	16.8	14.0
CaO	R	6.9	13.7	2.0	9.2	11.0	5.0	2.2
	O	6.8	13.2	2.6	9.6	11.4	4.6	1.6

atomic absorption analysis performed directly on solutions. Our data for Fe0 are in good agreement with values obtained by the Wänke group on selected samples from 15601. The bulk Luna 24 analyses are comparable to those presented by other workers (see Pillinger *et al.*, 1978b).

Magnetic Separates

Detailed descriptions of most of the fractions which have been used in this study have been provided elsewhere (Gardiner *et al.*, 1977; Pillinger *et al.*, 1978a,b). However, in view of the scepticism concerning magnetic separations by the Taylor group (Via and Taylor, 1976; Hu and Taylor, 1977) some comments here are appropriate. The density separation at 2.96 g/cm^3 (2.6 g/cm^3, Apollo 16) discriminates between vesicular glassy material (low density, highly magnetic ≡ agglutinates) and compact particles without glass (intermediate density, highly magnetic ≡ microbreccias). The interpretation that $\rho < 2.96$ (2.6), M $\geq$2.0 particles are predominantly agglutinates is based on binocular, petrographic and electron microscopic examination. This conclusion is strongly supported by the observation that low density, highly magnetic particles have the highest abundance of fine grained metal as measured by magnetic susceptibility (Fig. 2) and release by far the largest concentrations of carbon species and rare gases (Gardiner *et al.*, 1977; Pillinger *et al.*, 1978a,b; Schultz *et al.*, 1977; Schultz *et al.*, 1978). Highly magnetic particles with densities in the range 2.96–3.3 g/cm^3 are essentially indistinguishable from the less dense fraction on the basis of external appearance. However, they can have a factor of two less metal (Fig. 2) than the most magnetic grains of $\rho < 2.96$ material and have lower concentrations of carbon compounds and rare gases. Polished thin section studies suggest the presence of complex grains with a vesicular glassy matrix which could be considered as agglutinates and more densely packed particles which are microbreccias. Interference to the chemical analysis from ferromagne-

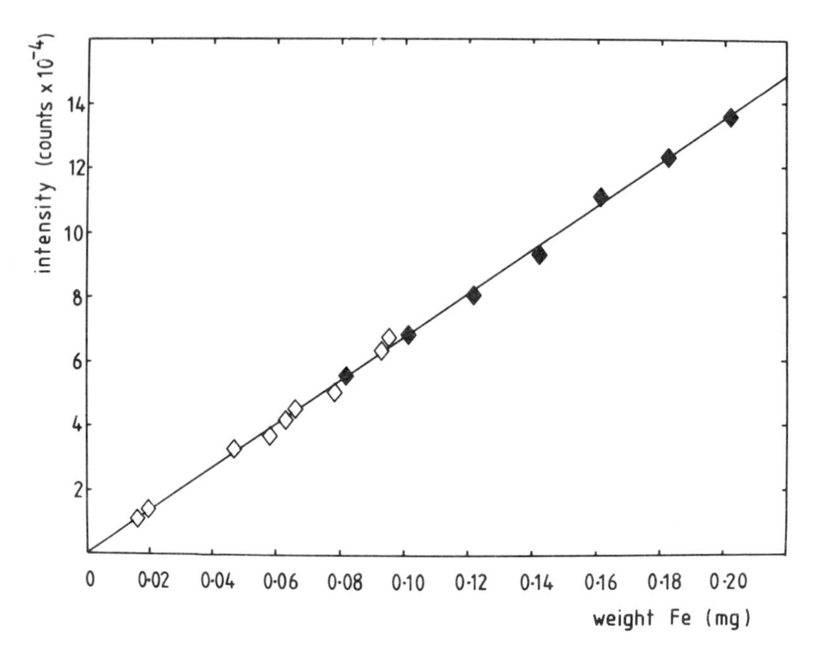

Fig. 1. XRF intensity (counts $\times 10^{-4}$) vs. weight of iron (mg). Different symbols refer to: standard solutions (◆) and international geochemical standards (◇).

sian, mineral and metal grains has been eliminated since these particles are in the fraction $\rho > 3.3$. Some pyroxene and ferromagnesian minerals intergrown with plagioclase have densities $\rho = 2.96 - 3.3$, but these occur in non- or low-magnetic susceptibility fractions.

Plagioclase is restricted to the non- and least-magnetic fraction of $\rho < 2.96$. Density fractions $\rho > 3.3$ contain little or no magnetic material as indicated by attempts to obtain appropriate fractions; the magnetic susceptibility of a $\rho > 3.3$ fraction of 12023 was very small (Pillinger *et al.*, 1978a)

Results

Total Fe, expressed as FeO, MgO, TiO_2, Al_2O_3 and CaO for all the fractions investigated, are compiled in Tables 2, 3 and 4. Additionally, the 15601 and 12023 analyses are plotted as a function of particle size in Figs. 3 and 4. In Fig. 5, the major element results for 12023 magnetic separates are plotted against magnetic separation distance. A number of observations may be made concerning the data:

(i) The elemental composition of the bulk soil fractions of 12023 suggest that this sample is fairly typical of the Apollo 12 site. It had previously been reported (Morris *et al.*, 1977) that 12023 was almost pure KREEP.

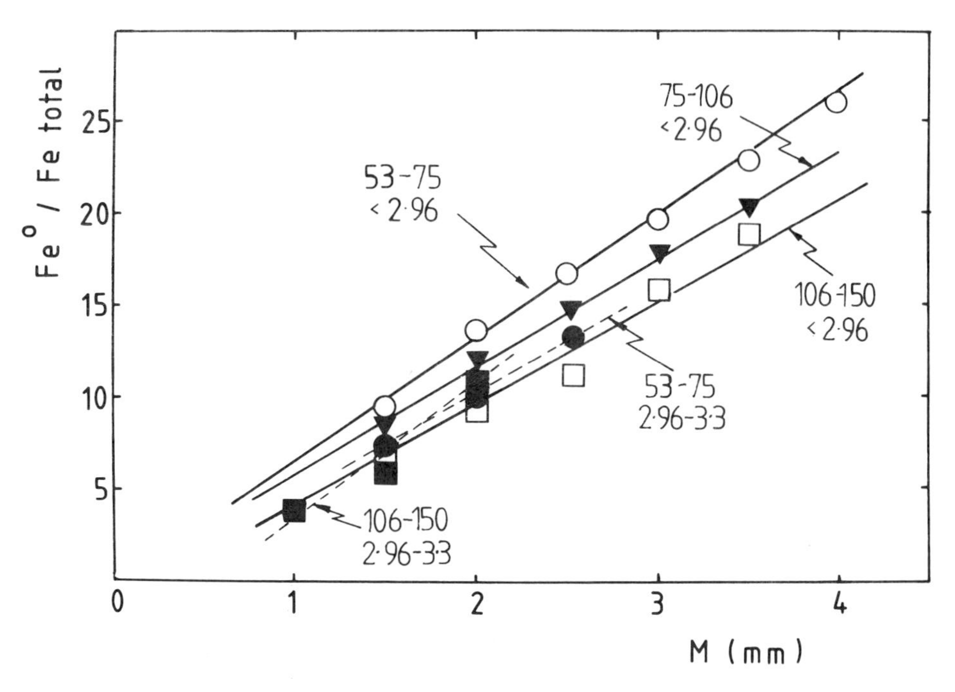

Fig. 2. Finely divided iron metal in 12023 as measured by magnetic susceptibility as a percentage of total iron vs. magnetic separation distance. The data demonstrate that material less than 2.96 g/cm^3 contains by far the greatest proportion of iron metal and should be considered as agglutinates. For the most mature particles in 12023, 20–25% of the total iron is present as metal.

This conclusion has now been withdrawn; the sample analysed by Morris *et al.* (1977) derived from a specimen returned to the curatorial facility by another PI and was mistakenly labelled (Morris, pers. comm.).

(ii) For both the soils measured in detail, 15601 and 12023, there is a tendency for fine fractions (<30 μm) of bulk material to be depleted relative to coarse fractions in FeO and MgO but enriched in Al_2O_3 and CaO. The trend for 12023 is substantiated by the analysis of a smoke sized fraction (very fine material, *ca.* 7 μm in diameter; Pillinger *et al.*, 1978a). Increased abundance of certain major, minor and trace elements indicative of anorthosite and KREEP have been observed in the finest fractions of numerous mare soils (see *inter alia* Finkelman, 1973; Evensen *et al.*, 1974; Basford, 1974; Krahenbuhl *et al.*, 1977; Korotev, 1976).

(iii) Agglutinates ($\rho < 2.96$, M $\geq$ 2.0 mm) for both 15601 and 12023 are predominantly depleted in FeO and MgO and enriched in Al_2O_3 and CaO relative to the bulk soil. For 12023, the depletion of FeO is more marked for agglutinates with decreased magnetic susceptibility (compare M2.0 with M3.5). When the weighted average chemical composi-

Table 2. Bulk Chemical Data Soil 15601.

Sample (μm)		FeO %	MgO %	TiO_2 %	Al_2O_3 %	CaO %
Bulk*		20.1	11.3	1.7	11.1	9.9
	B	20.1	12.4	1.3	8.1	6.6
>250	LA	17.3	9.5	1.7	12.9	9.5
	HA	16.4	12.0	1.5	10.3	8.6
	B	–	–	–	–	–
150–250	LA	18.0	10.8	1.9	12.7	9.3
	HA	16.6	10.1	1.5	10.3	9.2
	B	20.2	11.5	1.6	9.8	7.1
106–150	LA	16.4	9.1	1.5	14.4	9.9
	HA	15.4	10.2	1.4	10.8	10.4
	B	20.7	12.1	1.4	9.9	7.6
75–106	LA	15.9	8.7	1.5	13.2	9.0
	HA	19.1	10.2	1.4	9.9	10.0
	B	21.6	13.1	1.4	9.4	7.6
53–75	LA	14.3	9.2	1.5	13.6	9.3
	HA	18.4	8.0	1.3	11.2	10.3
	B	20.7	13.8	1.4	9.0	8.9
40–53	LA	14.0	8.5	1.4	15.7	10.5
	HA	17.7	10.2	1.5	10.8	10.7
	B	20.9	12.6	1.8	10.9	7.7
30–40	LA	12.9	7.9	1.3	15.3	11.4
	HA	16.7	10.1	1.3	12.0	10.1
	B	20.0	12.2	1.9	12.2	8.7
20–30	LA	–	–	–	–	–
	HA	19.6	8.7	1.5	8.3	9.6
	B	18.6	12.2	1.9	11.9	9.8
10–20	LA	10.7	7.3	1.2	18.0	11.7
	HA	18.4	10.7	1.6	9.6	8.7
	B	17.5	11.4	1.7	15.5	11.4
<10	LA	–	–	–	–	–
	HA	–	–	–	–	–

*Average of three determinations.

B = Bulk soils. LA = Light agglutinates, $p < 2.96$ g/cm^3, M ≥ 2.

HA = Heavy agglutinates, microbeccias, $p = 2.96$–3.30 g/cm^3, M ≥ 2.

Errors FeO $\pm$ 0.3 MgO $\pm$ 0.3 TiO_2 $\pm$ 0.1 Al_2O_3 $\pm$ 0.5 CaO $\pm$ 0.7%.

Table 3. Bulk Chemical Data Soil 12023.

Sample	FeO %	MgO %	TiO_2 %	Al_2O_3 %	CaO %
Bulk*	16.2	11.2	2.7	13.7	10.9
Sieve Fractions (μm)					
150–250	15.3	10.2	2.0	14.0	14.0
106–150†	16.6	11.1	3.1	13.1	11.1
75–106†	16.2	10.8	2.5	13.2	10.9
53–75 †	16.6	11.0	2.3	13.2	10.9
40–53 †	15.9	11.2	2.4	13.1	11.4
30–40 †	16.0	10.4	2.6	13.5	10.6
20–30	15.9	10.0	2.9	13.0	11.6
15–20	15.3	10.3	3.4	13.7	11.8
10–15	14.3	9.6	3.1	15.0	11.1
Smoke†	13.1	7.3	2.9	16.8	11.5
106–150 (μm)					
$p < 2.96$ (g/cm^3)					
M = 3.5	15.8	10.6	2.6	15.3	10.9
3.0	15.8	11.2	2.5	15.4	10.9
2.5	16.4	12.4	2.7	14.1	11.3
2.0	15.4	10.9	2.3	14.9	9.0
1.5	13.6	10.9	2.3	18.2	9.1
M T	9.9	8.1	1.7	19.6	10.4
N M	4.9	5.5	0.7	27.1	12.7
≥2‡	15.9	11.3	2.5	14.9	10.7
$p = 2.96$–3.3 (g/cm^3)					
M = 2.0	17.3	11.2	2.3	13.8	10.2
1.5	18.0	11.0	2.5	10.9	11.1
1.0	17.5	10.1	2.6	13.8	10.1
M T	14.5	9.1	3.0	14.0	11.1
N M	13.5	8.7	2.5	14.1	11.4
$p > 3.3$ g/cm^3)	23.4	14.9	3.0	3.3	6.7
75–106 (μm)					
$p < 2.96$ (g/cm^3)					
M = 3.5	16.0	9.6	3.0	14.8	11.3
3.0	15.7	9.9	3.0	14.9	11.9
2.5	15.2	9.6	2.6	14.2	11.9
2.0	14.5	10.0	2.8	14.4	11.5
1.5	12.6	10.1	2.4	17.4	12.2
≥ 2‡	15.4	9.7	2.8	14.6	11.7

Table 3. *(cont'd.)*

Sample	FeO %	MgO %	TiO_2 %	Al_2O_3 %	CaO %
53–75 (μm)					
$p < 2.96$ (g/cm^3)					
M = 4.0	16.0	9.4	2.5	16.0	11.4
3.5	15.5	10.6	2.8	14.7	11.3
3.0	15.1	10.5	2.9	14.7	10.6
2.5	14.4	10.6	2.6	16.4	11.8
2.0	14.1	9.9	2.7	16.6	11.9
1.5	12.6	8.4	2.2	18.1	12.4
M T	9.0	6.8	1.7	22.1	13.8
N M	4.6	3.6	1.0	27.7	17.3
≥2‡	14.9	10.3	2.7	15.6	11.4
$p = 2.96$–3.3 (g/cm^3)					
M = 2.5	18.7	10.8	3.1	12.0	11.5
2.0	18.2	11.0	3.0	10.4	11.1
1.5	17.1	11.2	2.7	9.9	10.5
M T	15.4	9.3	2.9	13.5	12.6
N M	14.0	8.2	2.5	15.6	10.5
$p > 3.3$ (g/cm^3)	25.9	14.3	4.2	5.3	6.4
40–53 (μm)					
$p < 2.96$ (g/cm^3)					
M = 3.5	14.7	10.1	2.8	15.5	11.9

Errors—See Table 1. M = distance in mm moved by sample to soft iron magnet pole piece.
*Average of three determinations.
†Average of two determinations.
‡Weighted average.

tion for all separates having $\rho < 2.96$ M ≥ 2.0 (Table 2) for 12023 is compared to the analytical data for 15601 $\rho < 2.96$, M ≥ 2.0 (Table 1), it is clear that agglutinates from the Apollo 15 soil show the greatest change from the bulk material. Soil 15601 is immature to submature (Gardiner *et al.*, 1977; Morris, 1976); whereas, 12023 is just mature (Pillinger *et al.*, 1978a). The two agglutinate fractions of Luna 24 show depletions in FeO and MgO and an enrichment in Al_2O_3 relative to bulk material of the same size. The Luna 24 core sample 24090 and 24125 are considered to be submature and submature/immature, respectively. Agglutinates from 60501 and 68121 are enriched in FeO relative to the bulk. The fraction from 68121 is also enriched in MgO and depleted in Al_2O_3. Other elements in these samples are similar to the bulk composition. Agglutinates from 66040 are almost identical in chemistry to the bulk sample, except possibly for CaO. The proportion of material in the $\rho < 2.6$, M ≥ 2.5, 106–150 μm fractions from the three Al6 soils is in

Table 4. Bulk Chemical Data for Apollo 16 and Luna 24 Soils.

Sample	FeO %	MgO %	TiO_2 %	Al_2O_3 %	CaO %
24090					
106–150 (μm)					
Bulk	18.0	10.7	0.6	11.8	14.3
Agglutinate*	16.6	9.2	0.9	14.0	12.8
24125					
106–150 (μm)					
Bulk	18.9	9.0	0.6	11.5	11.9
Agglutinate	17.1	8.7	1.0	13.6	11.4
60501					
106–150 (μm)					
Bulk	4.5	5.5	0.6	28.2	17.7
Agglutinate	5.5	5.6	0.6	28.9	17.7
66040					
106–150 (μm)					
Bulk	5.8	6.0	0.6	26.7	16.7
Agglutinate	5.5	6.1	0.6	26.9	17.5
68121					
106–150 (μm)					
Bulk	5.2	6.0	0.6	27.6	17.5
Agglutinate	5.9	6.5	0.6	26.5	17.5

Errors See Table 1.
*$p < 2.96$ g/cm^3 M $\geq$ 2–5 mm.

the order 66040 > 68121 > 60501, suggesting a maturity in that order. Other maturity indices for different size fractions of the same soils give a different order of maturity (Kerridge *et al.*, 1978; Morris *et al.*, 1977; Pillinger *et al.*, 1977).

(iv) Fine agglutinates are depleted in FeO, and enriched in Al_2O_3 and CaO relative to coarse agglutinates for both 15601 and 12023. Additionally, fine agglutinates in 15601 are depleted in MgO. A similar conclusion was reached for bulk sieved fractions [see (ii) above]; however, in the case of agglutinates, the effect becomes apparent at much larger grain sizes. It is likely that the two observations are connected, as an agglutinate must be made up of material finer than itself. Agglutinates could have a chemistry which reflects the contribution made to fine fractions by non-native components.

(v) Highly magnetic M $\geq$ 2.0 material in the density range 2.96–3.3 (heavy agglutinates and microbreccias) in 15601 follow similar trends to $\rho < 2.96$ agglutinates and are depleted in FeO, MgO and possibly TiO_2 and are enriched in Al_2O_3 and CaO (except for the very finest sieve

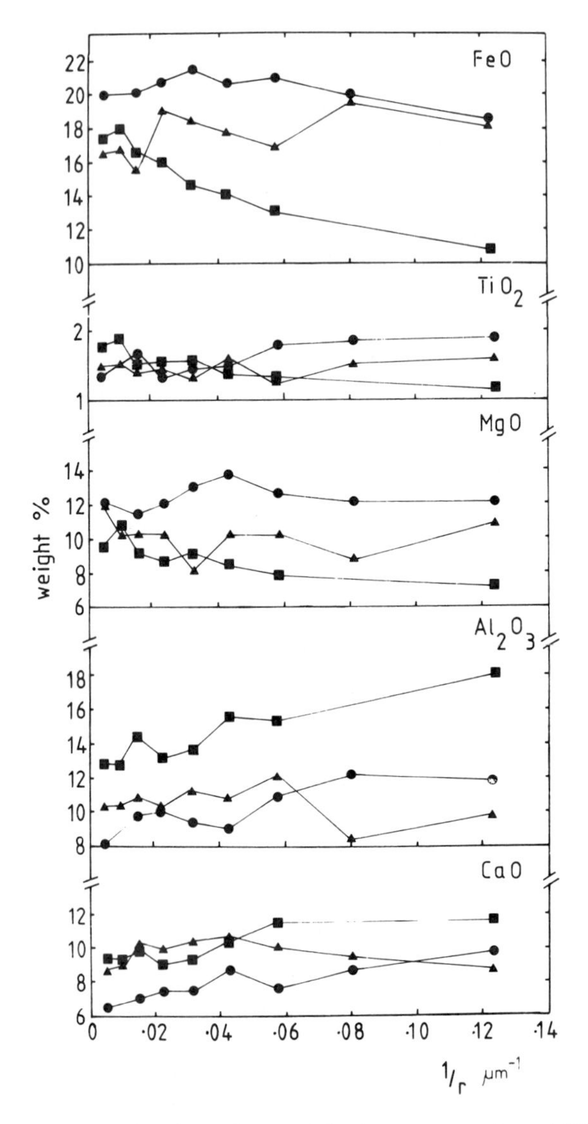

Fig. 3. Major element abundance for particle size separates from 15601. Different symbols refer to: bulk soil fractions (●) $\rho < 2.96$ g/cm^3, magnetic separation distance M $\geq$ 2.0 mm (■) and ρ 2.96–3.3 g/cm^3 M $\geq$ 2.0 mm (▲).

fractions) relative to bulk samples. However, the enrichments and depletions observed are less. In the case of 12023, the very few samples analysed suggest that the microbreccias and heavy agglutinates are enriched in FeO relative to bulk samples. The behaviour of MgO, TiO_2, Al_2O_3 and CaO is unclear.

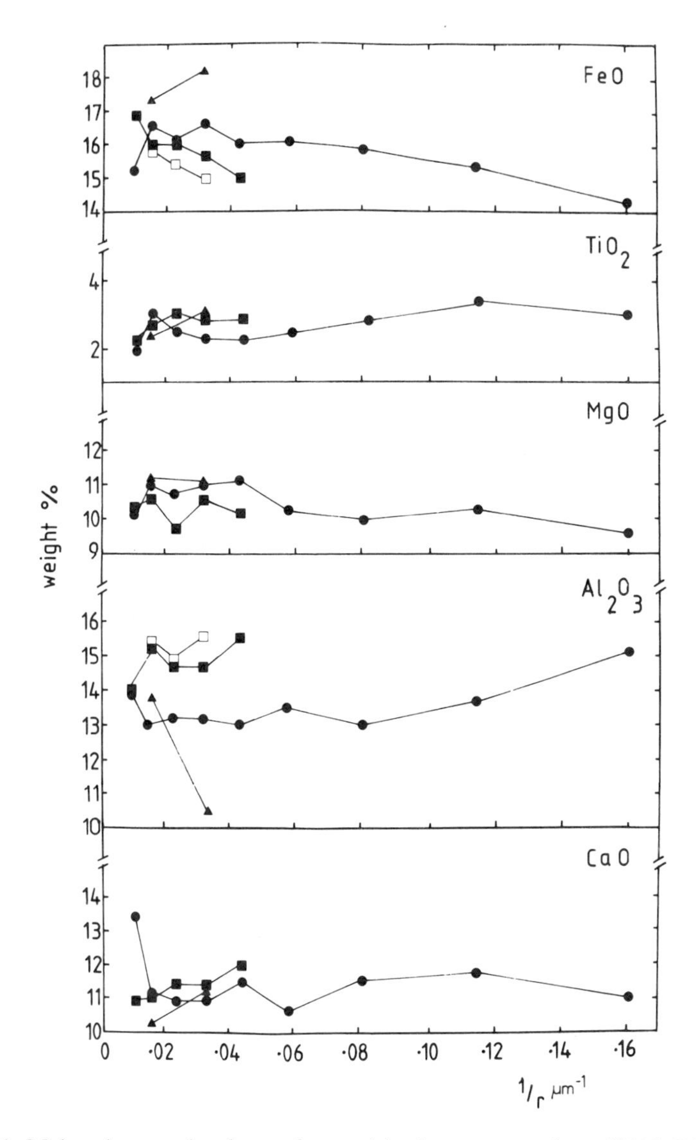

Fig. 4. Major element abundances for particle size separates from 12023. Different symbols refer to: bulk soil fractions (●) $\rho < 2.96$ g/cm, magnetic separation distance M 3.5 mm (■), $\rho < 2.96$ g/cm^3 M ≥ 2.0, weighted average (□) and ρ 2.96–3.3 g/cm^3 M ≥ 2.0 (▲).

Discussion

Apollo 16 soils

Taylor and his co-workers (Via and Taylor, 1976; Hu and Taylor, 1977) have demonstrated that although agglutinitic glass shows a wide range of composi-

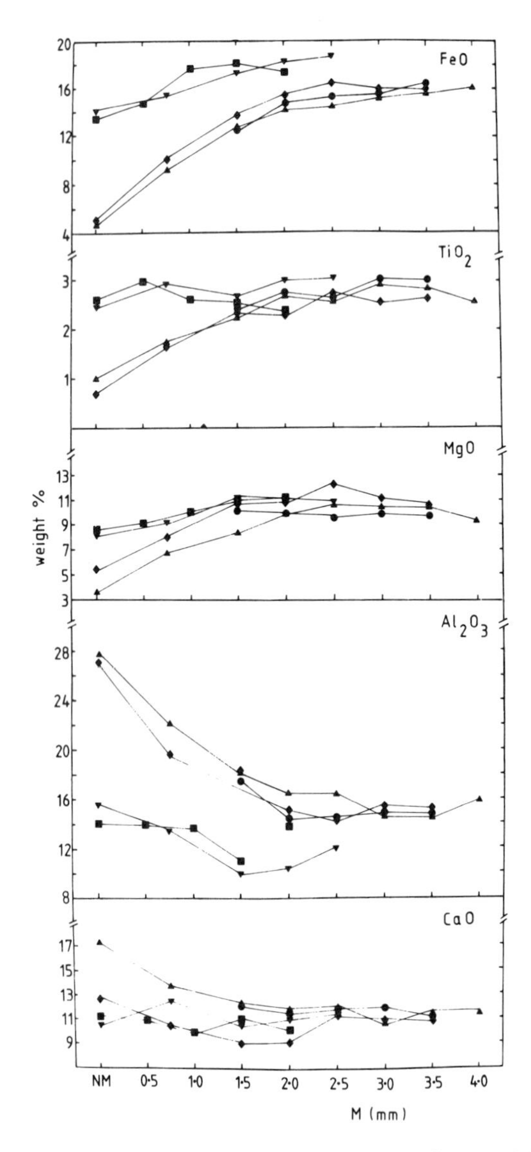

Fig. 5. Major element abundances for magnetic separates of 12023. Different symbols refer to different particle sizes (μm) and density fractions: $\rho < 2.96$ g/cm^3, 106–150 (◆) 75–106 (●) and 53–75 (▲); ρ 2.96–3.3, 106–150 (■) and 53–75 (▲).

tions, its average chemistry is not substantially changed relative to the soil from which it was isolated. Hu and Taylor (1977) point out that the slight enrichments, which they see for FeO in agglutinitic glass from 64421 and 61241, are less than the enrichments observed by Rhodes *et al.* (1975) for the total magnetic separate. The Taylor group argues that since the glass only makes up a fraction (*ca.* 20–25%) of the magnetic separates, it cannot be responsible for the major chemical changes observed by Rhodes *et al.* (1975). The alternative, which is suggested, is that the magnetic separates contain <5% of magnetic non-agglutinates (i.e., microbreccias and ferromagnetic minerals). The results of our detailed study on 15601 and 12023 appear to indicate that the magnetic and density separation procedures employed by us remove magnetic non-agglutinates from the light agglutinate fractions. Thus, we have tested the Taylor alternative for the Apollo 16 soils by studying "pure" agglutinates. Of the three Apollo 16 soils investigated in this study, the two least mature have agglutinates which are enriched in FeO relative to the bulk soil. For the third, the FeO content of agglutinates and bulk soil are the same within the experimental error. Rhodes *et al.* (1976) previously pointed out that the least mature Apollo 16 soils show the greatest enrichments in ferromagnesian elements and the two samples studied by Hu and Taylor (1977) have FeO enrichments and Al_2O_3 depletions which are in inverse maturity order. All the above results for Apollo 16 samples might be explicable in terms of a simple mixing model for the site. Carbon chemistry measurements, considered in conjunction with published data on bulk chemistry, rare gases, primordial and cosmogenic radionuclides, and agglutinate abundances, have led to an interpretation of the Apollo 16 site in terms of three major components: mature Cayley Plains material and immature North and South Ray crater ejecta (Pillinger *et al.*, 1977). It was invisaged that the two immature materials were low in FeO (*ca.* 4.2% North Ray and 5.2% South Ray) compared to the mature Cayley Plains soil (*ca.* 5.7 to 6.0%). Similar conclusions were reached by Heymann *et al.* (1975; 1978). If all the soils at the Apollo 16 site were mixtures of the mature soil and one (or both) of the immature components, then the bulk chemistry would be an average of the end members. However, agglutinates isolated from the resultant mix would most likely reflect the chemistry of the mature component. The greatest apparent enrichments for ferromagnesian elements in agglutinates over bulk soil would be for the samples containing the least agglutinates (i.e., the least mature).

15601, 12023 and Luna 24

For the two samples we have studied in detail (15601 and 12023), and for the Luna 24 soils, we have identified exactly opposite trends to those observed by Rhodes *et al.* (1975), Hu and Taylor (1977), Stroube and Ehmann (1978), and ourselves for Apollo 16 soils. Here agglutinates appear to be depleted in ferromagnesian elements and enriched in feldspathics. The data for 15601 has been confirmed by Schultz *et al.* (1977) who, in addition, demonstrated that agglutinates in this soil are enriched in KREEP. On the basis of the previous discussion on Apollo 16 soils, it would be tempting to suggest that 15601, 12023 and Luna 24 were mixtures of a mature component depleted in ferromagnesian elements and an immature component which was enriched. A simple mixing model of this sort could be operative to some extent. However, in samples where the agglutinate population is approximately constant for the various size ranges

(Pillinger *et al.*, 1978a), it is difficult to explain why the finest agglutinates are most depleted in FeO and correspondingly enriched in Al_2O_3 on the basis of simple mixing. Here, we present a model involving mixing of a different kind, which could account for the observed elemental abundances in agglutinates of any size.

Agglutination, as a maturational process, is generally considered to be active in the upper 1–2 mm of the lunar regolith (Basu, 1977; Housley *et al.*, 1973; Morris, 1978). The active layer is replenished by continual stirring in of material from below (Mendell and McKay, 1975; Basu, 1977), perhaps from as deep as 50 cm (Morris, 1978) or episodically from greater depths, and involving lateral transportation. The chemical analysis of Apollo 16 agglutinates would fit in with this generally applicable model. However, to explain the 15601 and 12023 data, we suggest an additional source of replenishment is the gradual in-fall of fine grained fragments (Evensen *et al.*, 1973, 1974). KREEP and anorthositic components are undoubtedly present at both the Apollo 15 (Wänke *et al.*, 1972; Evensen *et al.*, 1973) and the Apollo 12 (Papanastassiou and Wasserburg, 1971; Murthy *et al.*, 1971) sites. Meteoritic impacts in the highlands or other areas could lead to a continuous input of very fine grained particles of, for example, anorthosite and/or KREEP composition at the very surface of an evolving mare regolith. Since the input occurs right at the active margin for agglutinate formation, there is a high probability that non-native material would be incorporated into agglutinates. The input may be small but still sufficient to affect the overall chemistry of agglutinates relative to local composition. Small agglutinates being formed directly during micrometeorite impact at the very surface or from the smallest droplets of splashed glass, will tend to incorporate a greater proportion of the very surface non-native material. Hence, the chemistry of fine grained agglutinates will show the largest change relative to bulk material. As a soil reaches increased maturity, the contribution from non-native components will be counterbalanced by the accepted reworking in of locally derived material from below the active layer. Hence, the differences between agglutinates and bulk material would be greatest for immature soils as seen for 15601 compared to 12023 and to some extent Luna 24. In Fig. 6, we show plots of FeO versus Al_2O_3 for 15601 and 12023. For both samples, the agglutinates show depletion of FeO and enrichment in Al_2O_3 relative to the bulk sieved fractions. The greater degree of overlap in the case of 12023 is possibly due to the higher maturity, and greater reworking of this sample. We should point out that the effect of the additional component could work in the reverse direction in sites where the local rock is highland and a mare or other iron rich component is added. An input of a mare type material at the regolith interface could lead to agglutinates enriched (compared to bulk samples) in ferromagnesian elements. For 12023, we have observed an increased TiO_2 abundance in fine fractions of bulk sieved material and agglutinates. A possible explanation might be a non-native component rich in titanium.

An alternative source to impact generated infall for fine grained material depleted in FeO but enriched in Al_2O_3, for incorporation into fine agglutinates,

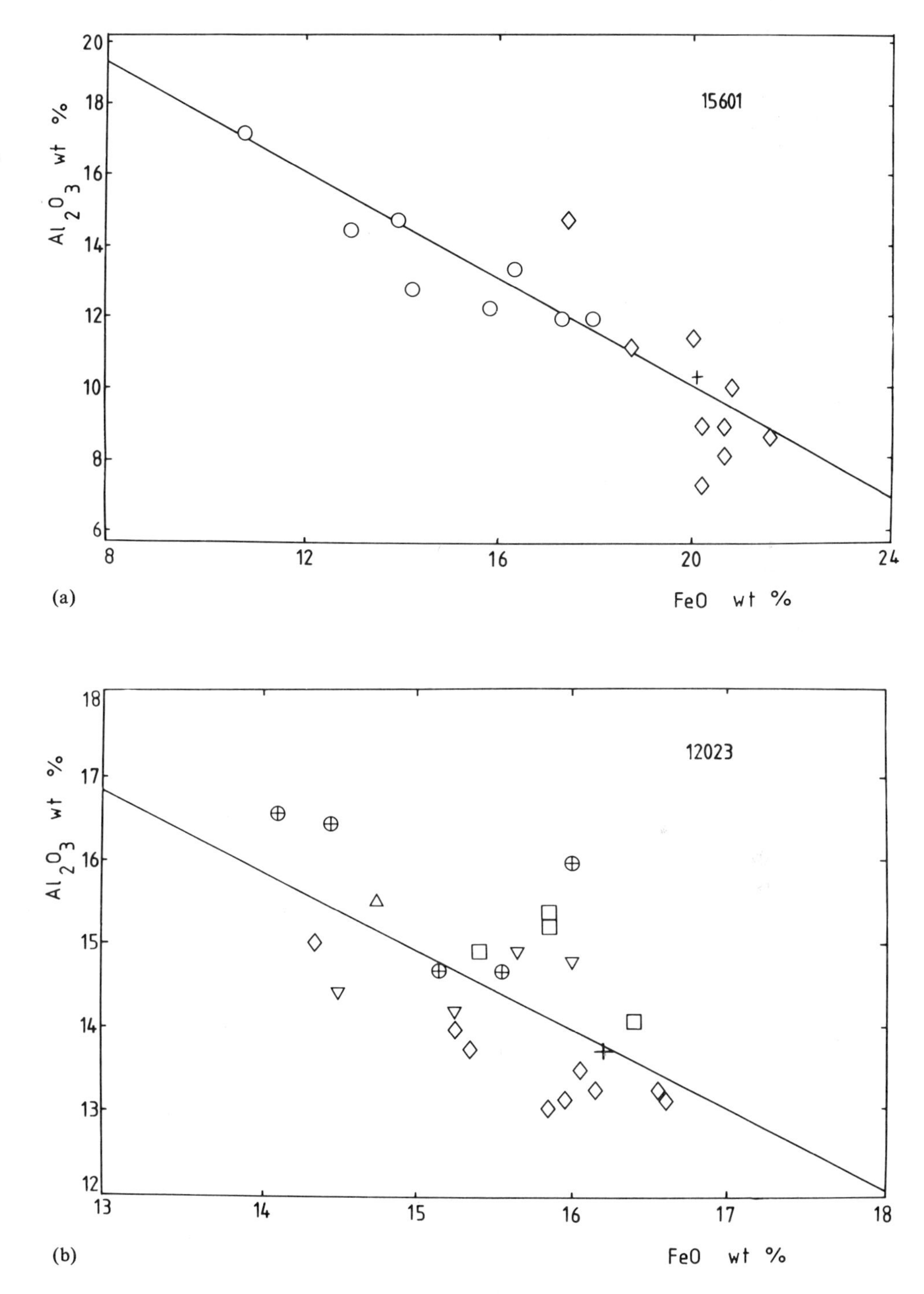

Fig. 6. FeO vs. Al_2O_3 for 15601 and 12023 bulk sieved fractions and agglutinates ($\rho < 2.96$ g/cm^3 magnetic separation distance $M \geq 2.0$ mm). Different symbols refer to: bulk soil (+) bulk sieved fractions (◇) 15601 agglutinates (○); 12023 agglutinates 106–150 (□) 75–106 (▽) 53–75 (⊕) and 40–53 (△).

might be differential comminution of rocks (Korotev, 1976; Haskin and Korotev, 1977). With this source, however, a mechanism for separation of the coarse and fine grains prior to agglutination is required. The infall mechanism specifically contributes fine grained material at the active margin for agglutinate formation.

Elemental fractionation in agglutinates

So far we have purposely avoided using the term fractionation to describe the enrichments or depletions of various elements in agglutinates. Fractionation implies a change in chemical composition produced by distillation, melting, crystallisation, or some other physical process; whereas, we have been considering only mixing events. However, if real fractionation processes such as preferential sputtering (Pillinger *et al.*, 1976) or preferential deposition of sputtered material (Hapke and Cassidy, 1978) are actively changing the relative elemental abundances, at grain surfaces, then these changes could be carried over into agglutinates formed from exposed fine grained material. Agglutination will work as a concentrating effect since the true exposed surface in an agglutinate is the sum of the surfaces of the constituent particles. Thus, chemical fractionations could become more apparent in agglutinates just as the abundances of accumulated solar wind elements are enhanced. Pillinger *et al.* (1976) have presented arguments to demonstrate how the chemistry of agglutinates might change relative to bulk chemistry, if sputtering effects are responsible for the reduction of Fe^{II} in silicates to metallic iron. The mechanisms implied may be causing enrichments in FeO and TiO_2 and depletion of Al_2O_3 and CaO in agglutinates relative to the original material, irrespective of the source of that material.

From Fig. 2, it can be seen that for fine agglutinates (53–75 μm) from 12023, *ca.* 20–25% of the total iron is present as iron metal. A careful appraisal of the bulk chemical analyses suggests that there is a slight increase in FeO content for M4.0 fractions compared to M2.0 material (Fig. 5). Total iron data from all fractions $M \geq 2.0$ have been replotted (with greatly increased scale) in Fig. 7 against metallic iron content of the samples as measured by magnetic susceptibility (data from Pillinger *et al.*, 1978a). A clear linear trend is apparent for the agglutinates from the 53–75 μm and 75–106 μm fractions; the trend could also exist for the 106–150 μm material, but the M2.5 fraction is a considerable distance off the line, and its analysis may be in error. We suggest the increase in total iron content observed for the most magnetic agglutinates of a particular size fraction is due to the increased metallic iron abundance with the grains studied. Since, on the basis of CD_4, χ and rare gas measurements, the agglutinates in M2.0 to M4.0 fractions of a given size range may be considered as a suite of progressively more complex particles (Pillinger *et al.*, 1978a), the iron enrichment could be attributed to a fractionation process associated with exposure, e.g., sputtering. At present, there is insufficient data to distinguish any recognisable trends for elements other than iron in progressively more evolved agglutinates.

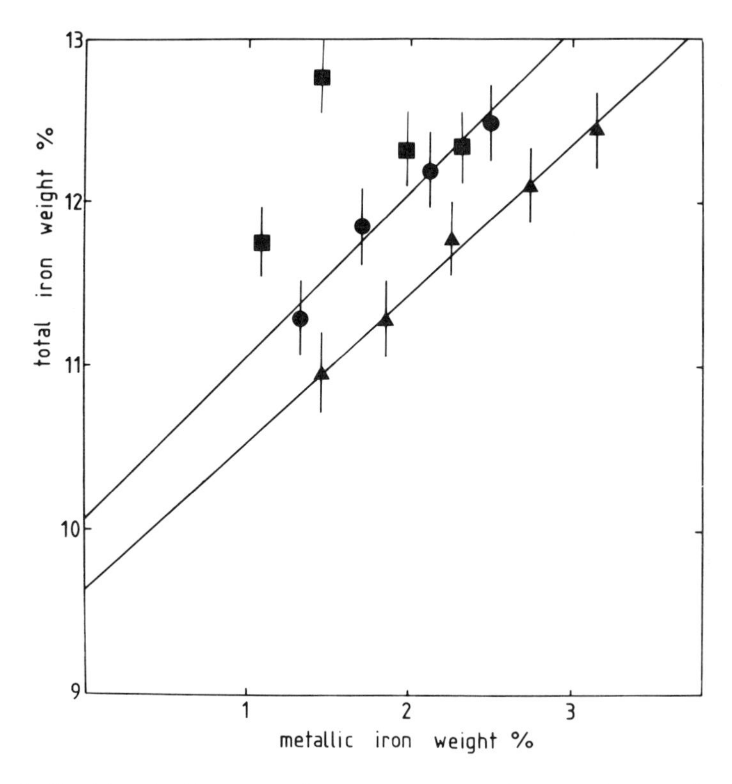

Fig. 7. Total iron (as Fe) vs. metallic iron as measured by magnetic susceptibility for agglutinates ($\rho < 2.96$ g/cm^3 M ≥ 2.0) from 12023. Different symbols refer to different particle size (μm) 106–150 (■) 75–106 (●) and 53–75 (▲).

CONCLUSIONS

We have attempted, as far as possible, to provide "clean" separates of end members for study. Our techniques, although still operator-dependent (Pillinger *et al.*, 1978a), are an improvement on separation methods (e.g., Rhodes *et al.*, 1975), which only split a bulk sample into two fractions (magnetic and non-magnetic concentrates) with no regard to particle size or density. We are confident that the differences observed between our fractions arise from differences occurring within the grains which make up the fraction and not from inadequate separation of two or more components. We suggest that physical (i.e., density and magnetic) separations of agglutinates may be preferable to handpicking, since the eye is incapable of resolving the subtle differences, which are detected by density and magnetic properties and substantiated by precise analyses involving carbon chemistry, rare gas and magnetic measurements. Electron microprobe studies should be extended to include agglutinates from finer size fractions and comparisons ought to be made between the composition of glass in very fine material and the chemistry of the soil. The wide spread of elemental abundances observed by Hu and Taylor (1977) might be reduced if more precisely separated material was to be used for electron microprobe study.

Finally, we conclude that chemical differences do exist between agglutinates and the soils from which they are isolated. The observed differences may be caused by any or all of the following mechanisms:

(i) The mixing of mature and immature components;
(ii) the addition of fine grained non-native material of high mobility to the very surface of the layer where agglutinate formation is active;
(iii) the intervention of processes associated with sputtering at the surfaces of grains incorporated into agglutinates.

Processes (i) and (ii) may result in the enrichment or depletion of ferromagnesian elements according to the chemistry of the immature or non-native component. Process (iii) would be most important for FeO enrichment but could be involved in the increasing abundance of TiO_2 and depletion of Al_2O_3 and CaO. Clearly the chemistry of agglutinates is not as simply rationalised as previously imagined by either Rhodes *et al.* (1975), the Taylor group (Hu and Taylor, 1977) or ourselves (Pillinger *et al.*, 1976).

Further investigations of major, minor and trace elements in agglutinates might be expected to establish the relative importance of the various mechanisms. A full understanding of the processes involved will be required if agglutinates are to be used as recorders of fossil soil compositions (Taylor *et al.*, 1978).

Acknowledgments—This work is supported by the Science Research Council (SGD 00772, SGD 00963). MRW thanks the SRC for a research training award. We are grateful to Mrs. D. Fabian for sample separation and help with the preparation of the manuscript.

REFERENCES

Basford J. (1974) K-Ar analysis of Apollo 11 fines 10084. *Proc. Lunar Sci. Conf. 5th*, p. 1375–1388.

Basu A. (1977) Steady state, exposure age, and growth of agglutinates in lunar soils. *Proc. Lunar Sci. Conf. 8th*, p. 3617–3632.

Campbell W. J., Spano E. F. and Green T. F. (1966) Micro and trace analysis by a combination of ion exchange resin loaded papers and X-ray spectrography. *Anal. Chem.*, **38**, 987–996.

Evensen N. M., Murthy V. R. and Coscio M. R. Jr. (1973) Rb-Sr ages of some mare basalts and the isotopic and trace element systematics in lunar fines. *Proc. Lunar Sci. Conf. 4th*, p. 1707–1724.

Evensen N. M., Murthy V. R. and Coscio M. R. Jr. (1974) Provenance of KREEP and the exotic component: Elemental and isotopic studies of grain size fractions in lunar soils. *Proc. Lunar Sci. Conf. 5th*, p. 1401–1417.

Finkelman R. B. (1973) Analysis of the ultrafine fraction of the Apollo 14 regolith. *Proc. Lunar Sci. Conf. 4th*, p. 179–189.

Gardiner L. R., Woodcock M. R., Pillinger C. T. and Stephenson A. (1977) Carbon chemistry and magnetic properties of bulk and agglutinate size fractions from soil 15601. *Proc. Lunar Sci. Conf. 8th*, p. 2817–2839.

Gibbons R. V. and Hörz F. (1976) The chemistry of some individual lunar soil agglutinates. *Proc. Lunar Sci. Conf. 7th*, p. 405–422.

Hapke B. and Cassidy W. A. (1978) Is the Moon really as smooth as a billiard ball?: Comments on some recent models of sputter fractionation. *Geophys. Res. Lett.* In press.

Haskin L. A. and Korotev R. L. (1977) Test of a model for trace element partition during closed system solidification of a silicate liquid. *Geochim. Cosmochim. Acta*, **41**, 921–939.

Heymann D. (1978) Light and dark soils at Apollo 16 revisited (abstract). In *Lunar and Planetary Science IX*, p. 506–508. Lunar and Planetary Institute, Houston.

Heymann D., Walton J. R., Jordan J. L., Lakatos S. and Yaniv A. (1975) Light and dark soils at the Apollo 16 landing site. *The Moon*, **13**, 81–92.

Housley R. M., Grant R. W. and Paton N. E. (1973) Origin and characteristics of excess Fe metal in lunar glass welded aggregates. *Proc. Lunar Sci. Conf. 4th*, p. 2737–2749.

Hu H.-N. and Taylor L. A. (1977) Lack of chemical fractionation in major and minor elements during agglutinate formation. *Proc. Lunar Sci. Conf. 8th*, p. 3645–3656.

Kerridge J. F., Kaplan I. R. and Petrowski C. (1978) Carbon isotope systematics in the Apollo 16 regolith (abstract). In *Lunar and Planetary Science IX*, p. 618–620. Lunar and Planetary Institute, Houston.

Korotev R. L. (1976) Geochemistry of grain size fractions of soils from the Taurus-Littrow valley floor. *Proc. Lunar Sci. Conf. 7th*, p. 695–726.

Krähenbühl U., Grütter A., von Gunten H. R., Meyer G., Wegmüller F. and Wyttenbach A. (1977) Volatile and non-volatile elements in grain-size fractions of Apollo 17 soils 75081, 72461, 72501. *Proc. Lunar Sci. Conf. 8th*, p. 3901–3919.

Mendell W. W. and McKay D. S. (1975) A lunar soil evolution model. *The Moon* **13**, 285–292.

Morris R. V. (1976) Surface exposure indices of lunar soils: A comparative FMR study. *Proc. Lunar Sci. Conf. 7th*, p. 315–336.

Morris R. V. (1978) *In situ* reworking (gardening) of the lunar surface: Evidence from the Apollo cores (abstract). In *Lunar and Planetary Science IX*, p. 757–759. Lunar and Planetary Institute, Houston.

Morris R. V., Warner L. and McKay D. S. (1977) Nearly pure Apollo 12 KREEP: Soil sample 12023. *Proc. Lunar Sci. Conf. 8th*, p. 2449–2458.

Murthy V. R., Evensen N. M., Jahr Bar-ming and Coscio M. R. Jr. (1971) Rb-Sr ages and elemental abundances of K, Rb, Sr and Ba in samples from the Ocean of Storms. *Geochim. Cosmochim. Acta* **35**, 1139–1153.

Papanastassiou D. A. and Wasserburg G. J. (1971) Lunar chronology and evolution from Rb-Sr studies of Apollo 11 and 12 samples. *Earth Planet. Sci. Lett.* **11**, 37–62.

Pillinger C. T., Davis P. R., Eglinton G., Gowar A. P., Jull A. J. T., Maxwell J. R., Housley R. M. and Cirlin E. H. (1974) The association between carbide and finely-divided metallic iron in lunar fines. *Proc. Lunar Sci. Conf. 5th*, p. 1949–1961.

Pillinger C. T. Eglinton G., Gowar A. P., Jull A. J. T. and Maxwell J. R. (1977) The exposure history of the Apollo 16 site: An assessment based on methane and hydrolysable carbon. In *The Soviet-American Conference on Cosmochemistry of the Moon and Planets*, p. 541–551. NASA SP-370. Washington, D.C.

Pillinger C. T., Gardiner L. R. and Jull A. J. T. (1976) Preferential sputtering as a method of producing metallic iron, inducing major element fractionation and trace element enrichment. *Earth Planet. Sci. Lett.* **33**, 289–299.

Pillinger C. T., Gardiner L. R., Jull A. J. T., Woodcock M. R. and Stephenson A. (1978b) Magnetic properties and carbon chemistry studies pertinent to the evolution of the regolith at the Luna 24 site. In *Mare Crisium: The View From Luna 24.* (R. B. Merrill and J. J. Papike, eds.) p. 217–228. Pergamon, N.Y.

Pillinger C. T., Jull A. J. T., Woodcock M. R. and Stephenson A. (1978a) Maturation of the lunar regolith: Some implications from magnetic measurements and hydrolysable carbon data on bulk soils, and particle separates from 12023 and 15601. *Proc. Lunar Planet. Sci. Conf. 9th.* This volume.

Rhodes J. M., Adams J. B., Blanchard D. P., Charette M. P., Rodgers K. V., Jacobs J. W., Brannon J. C. and Haskin L. A. (1975) Chemistry of agglutinate fractions in lunar soils. *Proc. Lunar Sci. Conf. 6th*, p. 2291–2307.

Rhodes J. M., Blanchard D. P., Adams J. B., Charette M. P., Brannon J. C. and Rodgers K. V. (1976) The chemistry of agglutinate fractions in lunar soils (abstract). In *Lunar Science VII*, p. 733–735. The Lunar Science Institute, Houston.

Schultz L., Weber H. W., Spettel B., Hintenberger H. and Wänke H. (1977) Noble gas and element distribution in agglutinate in bulk grain size fractions of soil 15601. *Proc. Lunar Sci. Conf. 8th*, p. 2799–2815.

Schultz L., Weber H. W., Spettel B., Hintenberger H. and Wänke H. (1978) Noble gas and element distribution in agglutinates of different densities (abstract). In *Lunar and Planetary Science IX*, p. 1021–1023. Lunar and Planetary Institute, Houston.

Stroube W. B. Jr. and Ehman W. D. (1978) An examination of chemical fractionation during agglutinate formation (abstract). In *Lunar and Planetary Science IX*, p. 1125–1127. Lunar and Planetary Institute, Houston.

Taylor G. S., Wentworth S., Warner R. D. and Keil K. C. (1978). Petrology of Apollo 17 deep drill core. Agglutinates as recorders of fossil soil compositions (abstract). In *Lunar and Planetary Science IX*, p. 1146–1148. Lunar and Planetary Institute, Houston.

Via W. N. and Taylor L. A. (1976) Chemical aspects of agglutinate formation: Relationships between agglutinate composition and the composition of bulk soil. *Proc. Lunar Sci. Conf. 7th*, p. 393–403.

Wänke H., Baddenhausen H., Balacescu A., Teschke F., Spettel B., Dreibus G., Palme H. Quijano-Rico M., Kouse H., Wlotzka F. and Begemann F. (1972) Multi-element analysis of lunar samples and some implications of the results. *Proc. Lunar Sci. Conf. 4th*, p. 1251–1268.

Proc. Lunar Planet. Sci. Conf. 9th (1978), p. 2215–2220.
Printed in the United States of America

Primordial Pb, radiogenic Pb and lunar soil maturation

GEORGE W. REED, Jr. and STANKA JOVANOVIC

Chemistry Division, Argonne National Laboratory
Argonne, Illinois 60439

Abstract—^{204}Pb is directly correlated with the Fe° measured by FMR. A similar correlation has been noted for hydrolyzable carbon (Pillinger *et al.*, 1974). An enrichment of these elements appears to have occurred during soil maturation.

In contrast to ^{204}Pb, radiogenic Pb is reported to be lost during soil maturation (Church *et al.*, 1976a). Radiogenic Pb is present in mineral grains and may be lost by solar wind sputtering (or volatilization) and not resupplied. ^{204}Pb coating grain surfaces acts as a reservoir to provide the ^{204}Pb being extracted in the Fe° formation process. Venting or some other volatile release mechanism may replenish the surface-related ^{204}Pb.

INTRODUCTION

Three phenomena have been studied that relate trace elements to soil maturation processes. Most extensively investigated are solar wind implanted elements, in particular the rare gases, hydrogen, carbon and nitrogen. In these cases the amounts found are taken to be measures of the exposure histories of the samples. The interpretation is not necessarily straightforward, however, since not only solar wind, hence particle irradiation damage and sputtering effects, but also heating due to micrometeorite impacts enter. These may cause gas diffusion and element volatilization resulting in a dynamic equilibrium between trace element implantation and loss.

This phenomenon has been studied and the reader is referred to the work of Morris (1976) in which an attempt is made to explain and summarize much of the solar wind related results in terms of a maturity index based on reduced Fe measured by ferromagnetic resonance (FMR) techniques. This Fe° results, presumably, from solar-wind hydrogen implantation and micrometeorite heating. The FMR intensities (I_s) normalized by the FeO content of the soils (I_s/FeO) is designated a maturity index. Nitrogen and ^{36}Ar are found to correlate reasonably well with this index. The lighter rare gases tend to deviate, possibly because of their greater mobility.

Two other phenomena related to soil maturation will be the topic of this paper. They are loss from soils of non-solar wind trace elements and incorporation of such trace elements into soils. The trace element we address specifically is Pb: radiogenic Pb is reported (Church *et al.*, 1976a) to have been lost from mature soils and especially agglutinates which are glass bonded aggregates resulting from soil comminution and micrometeorite impact melting. ^{204}Pb, in contrast, is reported to increase in soils as the amount of FMR iron increases (Allen *et al.*, 1974).

EXPERIMENTAL BACKGROUND

^{204}Pb is usually measured in <1 mm soil fractions but also occasionally on bulk soils and size splits. Fast neutron irradiation produces 52-hr ^{203}Pb via ^{204}Pb (n,2n) reaction. The procedure is free from contamination or blank corrections. However, the low ^{204}Pb concentrations, $\sim 10^{-9}$ g/g, do not permit accuracies of better than about 10%. On the average the data are good to 15–20% and are in good agreement with results determined mass spectrometrically; see comparison in Allen *et al.* (1972). Irradiated samples are leached in pH 5–6 hot HNO_3 solution containing Pb and other heavy element carriers (~10 mg each) to remove and determine any soluble Pb (Pb_l). The remaining Pb is defined as residual Pb_r. This is an important step since from 10–50% of the ^{204}Pb may be of this labile type which also has been observed as 600°C volatile Pb [see Reed *et al.* (1977) and also Silver (1972) for earlier references].

The Fe^o measurements are made on aliquants taken randomly from the 100–300 mg samples used for the Pb measurement. The ferromagnetic Fe is measured by ESR with the assistance of J. Norris (ANL) by a procedure similar to that described by Tsay *et al.* (1973). The absolute ferromagnetic Fe^o concentrations in Fig. 1 could contain a systematic uncertainty. The same procedure was used for each sample, hence the relative numbers are reliable.

DISCUSSION

We have presented data showing a correlation of residual (non-surface) primordial Pb, via ^{204}Pb, with the FMR iron (=Fe^o in this paper) measured by ESR (Allen *et al.*, 1974). Figure 1 is an updated plot of the data. Most of the results are from our work, a few points based on literature Fe^o measurements are included (see Allen *et al.*, 1974). Immature soils such as fillet 67460, the two trench bottom soils and 74220 (not plotted) tend to plot near $Fe^o = 0$. Soil breccias 14047, 14049 and 15505, also not plotted, fall close to the trend line. Samples with FeO ranging from 4–17 wt.% establish the trend.

A similar direct correlation between hydrolyzable carbon (C_{hyd}) and $I(\Delta H)^2$, a measure of FMR Fe^o, has been reported by Pillinger *et al.* (1974) and Jull *et al.* (1978).

We will attempt to understand the direct vs. the FeO normalized types of correlations. The result of comminution processes is the production of fine grained feldspathic and mafic material. During soil maturation, some of this material is converted into the glass which encompasses and binds lithic and mineral detritus together to form agglutinates (Adams and McCord, 1973). The non-magnetic feldspathic material acts as a diluent to the Fe^o present (Gardiner *et al.*, 1977). The role of such a diluent becomes clear when a solar wind implanted species is considered.

Solar wind ^{36}Ar is implanted in mafic and in feldspathic glass and detritus alike. The Fe^o is proportional to both the length of surface exposure and to the amount of FeO available for reduction. Normalization to FeO leaves only the surface exposure dependence; hence the $^{36}Ar - I_s/FeO$ (or Fe^o/FeO) correlation. CH_4 is another such volatile component and correlates with ^{36}Ar and, therefore, is associated with all constituents which have been exposed to solar wind.

The hydrolyzable carbon (C_{hyd}) component measured as CD_4 (above) may be present as an iron carbide (Pillinger *et al.*, 1974) which means that it has not been merely implanted but has chemically reacted with metallic Fe.

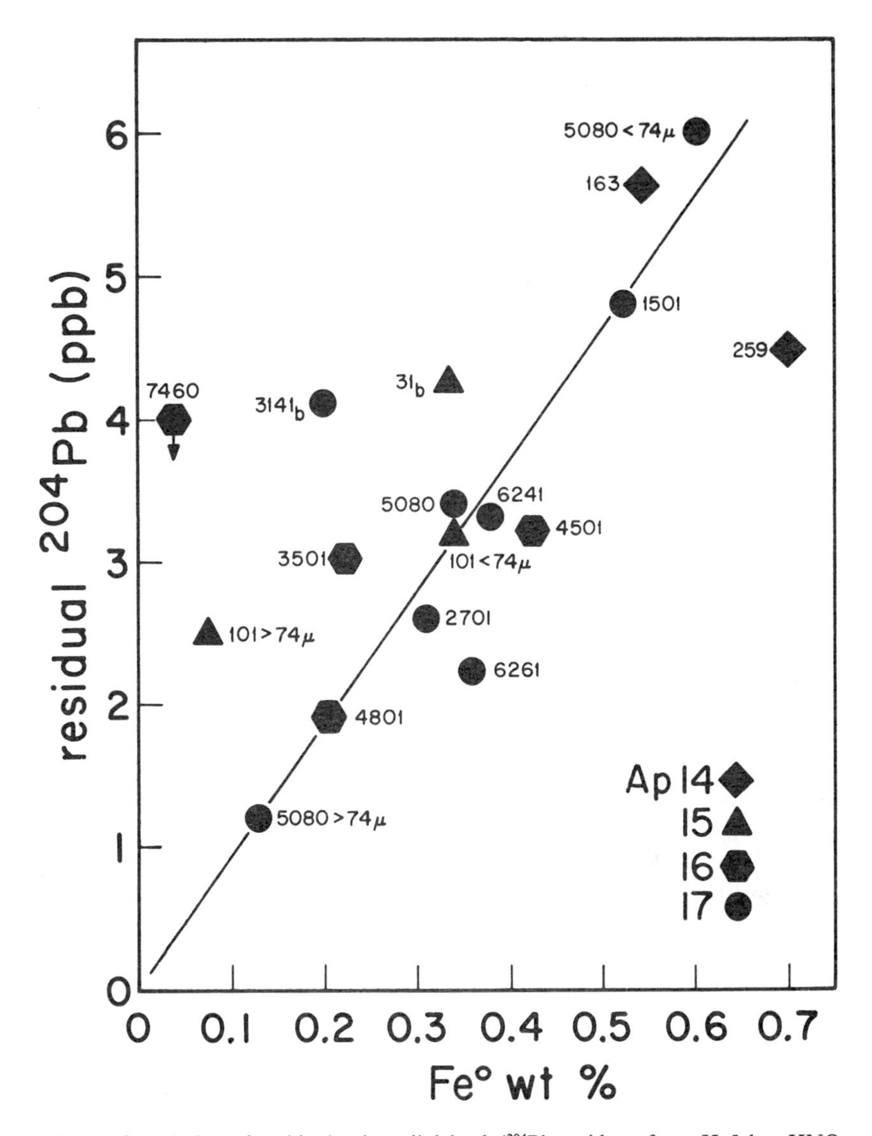

Fig. 1. Correlation of residual primordial lead (^{204}Pb residue after pH 5 hot HNO_3 leach) and FMR metal Fe° determined by ESR at our lab in lunar soils. A few data points are based on literature Fe° (Allen *et al.*, 1974); b = trench bottom soils. Immature soils tend to fall toward Fe° = O; these include 67460, trench bottom sample 73141, 15101 > 74 μ and 74220 (not plotted). Soil breccias 14047, 14049 and 15505 (also not plotted) fall close to the trend line.

There is also a two component aspect to ^{204}Pb, but in a very different sense. The Pb is not solar wind implanted. The more labile component is surficial. The residual component may be chemically associated with a Fe° phase since it correlates with Fe°. Thus the ^{204}Pb and CD_4 cases are similar.

A source of ^{204}Pb is needed. The ^{204}Pb on surfaces ranges from 10–50% of the total. This has been attributed to depositional processes (Silver, 1972; Allen *et al.*, 1974) which we have attempted to relate to volcanic venting. We suggested that these deposits provided the ^{204}Pb that became associated with Fe°. Pillinger *et al.* (1976) have suggested that preferential sputtering should be considered along with volatile transport. This sputtering mechanism also requires implantation. It could be postulated that knock-on processes occur between solar wind hydrogen and He and surface Pb. This should not be a very efficient method for uniform momentum transfer. Another possibility is acceleration in the lunar atmosphere (Manka and Michel, 1971) but the concentrations of Pb, even if venting occurs, so accelerated should be extremely low.

We proposed that agglutinate formation by micrometeorite impact was the mechanism for incorporation of surface deposited ^{204}Pb and its association with the Fe° phase (Allen *et al.*, 1974). On the basis of a meteorite analogy, it was proposed that this phase could be an Fe-FeS eutectic. S should also have been a surface deposit. If Pb has a solid solubility in Fe of ~0.001 wt.% (Shunk, 1965), then S may not be necessary.

Considerations of a reservoir for Pb and sputtering losses (Pillinger *et al.*, 1976) may explain the difference between the trend we note for ^{204}Pb and that for radiogenic Pb during soil maturation. Whereas, we observe that ^{204}Pb is being incorporated into maturing soils, Church *et al.* (1976a) note the loss of radiogenic Pb and ascribe this to volatile loss during agglutinate formation. The sources of the ^{204}Pb being incorporated are surface deposits. The main reservoir for radiogenic Pb is probably the mineral grains themselves (Silver, 1972) as a result of U and Th decay. This Pb is implanted by recoil during radioactive decay. It may be lost by the sputtering process without being replaced as is necessary for the preferential sputtering of Pillinger *et al.* (1976).

A test of this possibility may lie in the Pb model age trends. In separated agglutinate fractions, Church *et al.* (1976b) report larger model ages from ^{207}Pb/^{235}U than from ^{208}Pb/^{232}Th and ^{206}Pb/^{238}U. The ^{207}Pb/^{235}U model ages for agglutinates are nearer the bulk soil ages. These observations may be explained as follows. Of the radiogenic Pb isotopes, a larger fraction of ^{207}Pb than ^{206}Pb and ^{208}Pb would be formed during early lunar history. This early radiogenic Pb along with primordial Pb could have become a reservoir of parentless Pb during this very active period of lunar evolution. This is the Pb that could later be deposited on soil grain surfaces. Compared to the Pb produced later, this parentless Pb would be relatively enriched in ^{207}Pb. If this parentless Pb was incorporated into agglutinates, the contribution of ^{207}Pb would partially compensate for that being lost by the sputtering mechanism discussed above. The contributions of ^{206}Pb and ^{208}Pb from this source would be much smaller.

In summary, we outline an explanation of the different behavior of ^{204}Pb (parentless Pb) and radiogenic Pb during lunar soil maturation processes as follows:

Radiogenic Pb	*^{204}Pb (parentless Pb)*
Generated in or recoiled into crystals by radioactive decay.	Present as surface deposits; volcanic origin?
Loss from grains by solar wind sputtering and/or micrometeorite impact heating.	Incorporated into agglutinate glass by extraction process related to FMR metal formation.
^{207}Pb is exceptional. The relatively greater amounts (*vis à vis* ^{206,208}Pb) would be reflected in the parentless Pb reservoir.	Since amount of ^{204}Pb present varies directly with the amount of this metal, an adequate reservoir was always available.

Acknowledgments—Fruitful discussions with J. R. Norris are appreciated. The paper has been greatly benefited by critical reviews by R. V. Morris and D. A. Papanastassiou. Special thanks are extended to L. Koehl for manuscript preparation. Work supported by the National Aeronautic and Space Administration.

REFERENCES

Adams J. B. and McCord T. B. (1973) Vitrification darkening in the lunar highlands and identification of Descartes material at the Apollo 16 site. *Proc. Lunar Sci. Conf. 4th*, p. 163–177.

Allen R. O. Jr., Jovanovic S. and Reed G. W. Jr. (1972) ^{204}Pb in Apollo 14 samples and inferences regarding primordial Pb lunar geochemistry. *Proc. Lunar Sci. Conf 3rd*, p. 1645–1650.

Allen R. O. Jr., Jovanovic S. and Reed G. W. Jr. (1974) A study of ^{204}Pb partition in lunar samples using terrestrial and meteoritic analogues. *Proc. Lunar Sci. Conf. 5th*, p. 1617–1623.

Church S. A., Tilton G. R. and Chen J. H. (1976b) Lead isotopic studies of lunar soils: Their bearing on the time scale of agglutinate formation. *Proc. Lunar Sci. Conf. 7th*, p. 351–371.

Church S. A., Tilton G. R., Wright J. E. and Lee-Hu C.-N. (1976a) Volatile element depletion and ^{39}K/^{41}K fractionation in lunar soils. *Proc. Lunar Sci. Conf. 7th*, p. 423–439.

Gardiner L. R., Woodcock M. R., Pillinger C. T. and Stephenson H. (1977) Carbon chemistry and magnetic properties of bulk and agglutinate size fractions from soil 15601. *Proc. Lunar Sci. Conf. 8th*, p. 2817–2839.

Jull A. J. T., Pillinger C. T. and Stephenson A. (1978) The carbon content of finely-divided lunar iron (abstract). In *Lunar and Planetary Science IX*, p. 609–611. Lunar and Planetary Institute, Houston.

Manka R. H. and Michel F. C. (1971) Lunar atmosphere as a source of lunar surface elements. *Proc. Lunar Sci. Conf. 2nd*, p. 1717–1728.

Morris R. V. (1976) Surface exposure indices of lunar soils: A comparative FMR study. *Proc. Lunar Sci. Conf. 7th*, p. 315–335.

Pillinger C. T., Davis P. R., Eglinton G., Gowar A. P., Jull A. J. T., Maxwell J. R., Housley R. M. and Cirlin E. H. (1974) The association between carbon and finely divided metallic iron in lunar fines. *Proc. Lunar Sci. Conf. 5th*, p. 1949–1961.

Pillinger C. T., Gardiner L. R. and Jull A. J. T. (1976) Preferential sputtering as a method of producing metallic iron, inducing major element fractionation and trace element enrichment. *Earth Planet. Sci. Lett.* **33**, 289–299

Reed G. W. R., Allen R. O. Jr. and Jovanovic S. (1977) Volatile metal deposits on lunar soils—relation to volcanism. *Proc. Lunar Sci. Conf. 8th*, p. 3917-3930.

Shunk F. A. (1965) *Constitution of Binary Alloys*, 2nd Suppl. McGraw-Hill, N.Y. 339 pp.
Silver L. T. (1972) Lead volatilization and volatile transfer processes on the moon (abstract). In *Lunar Science III*, p. 617–619. The Lunar Science Institute, Houston.
Tsay F. D., Manatt S. L., Live D. H. and Chan S. I. (1973) Metallic Fe phases in Apollo 16 fines: Their origin and characteristics as revealed by electron spin resonance studies. *Proc. Lunar Sci. Conf. 4th*, p. 2751–2761.

Proc. Lunar Planet. Sci. Conf. 9th (1978), p. 2221–2232.
Printed in the United States of America

Noble gas and element distribution in agglutinate grain size separates of different density

L. SCHULTZ, H. W. WEBER, B. SPETTEL, H. HINTENBERGER and H. WÄNKE
Max-Planck-Institut für Chemie, 65 Mainz, West Germany

Abstract—The concentration and isotopic composition of He, Ne, and Ar as well as the concentration of ^{84}Kr and ^{132}Xe have been measured in grain size fractions of heavy agglutinates ($2.96 < \zeta < 3.3$ g/cm^3) separated from soil 15601. These results are compared with measurements on bulk and light agglutinate size fractions ($\zeta \leq 2.96$ g/cm^3) reported previously by Schultz *et al.* (1977a).

Trapped solar gases in agglutinates are interpreted as the sum of a volume-correlated and a surface-correlated component. The concentration of volume-correlated gases is similar in light and heavy agglutinates. The surface-correlated component, however, shows differences which can be explained by the smaller surface area per gram of denser grains and the incorporation of minerals with higher retentivity for He and Ne. From elemental ratios of the trapped gases, it is concluded that the agglutination process does not fractionate the noble gases with the possible exception of helium. The chemical differences observed between grain size fractions of light agglutinates are not found in heavy agglutinates. This shows that a separation according to the density of the agglutinates produces a chemical fractionation due to the inhomogeneity of these composite particles. Therefore, the observed large chemical differences among grain size fractions of agglutinates is interpreted as an artifact introduced by the separation technique.

INTRODUCTION

The interaction of interplanetary dust with the surface of atmosphere-free planetary bodies results in a sedimentary process not observable on the Earth's surface: Micro-meteorite impacts create new particles, the so called agglutinates. Agglutinates are fragile aggregates of crystalline grains and lithic fragments welded together by impact-produced glasses (Duke *et al.*, 1970; McKay *et al.*, 1970). The agglutinate content of lunar soils is directly related to the duration of their surface exposure and can be taken as a measure of soil maturity (e.g., McKay *et al.*, 1972; Charette and Adams, 1975; Morris, 1976, 1977; Bogard, 1977).

The dominant noble gas component of lunar soils originates from trapped ions of the solar wind. As a consequence of this implantation mechanism the solar gases are surface-correlated. However, for composite particles like breccias or agglutinates, in addition to the surface-correlated component, a volume-correlated component is expected due to the incorporation of smaller particles with earlier exposure to the solar wind. Therefore, the concentration of solar noble gases in agglutinates is higher than that in the bulk sample of equal grain size from the same soil (Bogard and Nyquist, 1973; Bogard *et al.*, 1973; Hübner *et al.*, 1975; Signer *et al.*, 1977). In a recent paper (Schultz *et al.*, 1977a), we showed that the grain size dependency of the solar noble gas concentration in agglutinates can be described by the equation:

$$C_y(d) = S_y \left(\frac{d}{d_o}\right)^{-n_y} + V_y, \tag{1}$$

with trapped solar wind concentration C_y of an isotope y in a grain size fraction with average diameter d; d_o is an arbitrary reference grain size with a surface-correlated gas concentration S_y. An ideal surface correlation produces a straight line with a slope $n_y = 1$ in a log concentration versus log grain-size diagram. The addition of a grain-size independent volume-correlated gas concentration V_y transforms this straight line to an upward concaved curve. If a straight line is fitted to an array of data points, on such a curve n_y-values less than unity are computed.

Schultz *et al.* (1977a,b) reported also large chemical fractionations between different magnetically separated grain-size fractions of agglutinates. In the finer fractions, feldspathic and incompatible elements were enriched, but the mafic elements were depleted compared to coarse grain sizes and bulk analyses.

In this paper we present further measurements of agglutinates separated from soil 15601. The aim was to shed light on the following questions:

—Are the chemical differences between grain size fractions due to chemical fractionation in the agglutination process or the result of a bias of the separation technique?

—Does a fractionation of trapped solar noble gases occur during agglutination?

—Do agglutinates have longer regolith residence times compared to mineral fragments?

A preliminary report of this study has been given by Schultz *et al.* (1978a).

EXPERIMENTAL PROCEDURES AND RESULTS

The agglutinitic samples were prepared by C. T. Pillinger and his coworkers in Cambridge, England. By wet sieving in acetone, 9 grain-size fractions of the bulk soil 15601 were obtained. Using heavy liquids, each of these fractions was divided into a light ($\zeta \leq 2.96$ g/cm^3) and a heavy fraction ($2.96 < \zeta < 3.3$ g/cm^3). A magnetic separation yielded highly magnetic particles from each grain-size and density fraction. Highly magnetic particles with density $\leq$2.96 g/cm^3 are called *light agglutinates*, whereas highly magnetic particles with densities between 2.96 and 3.3 g/cm^3 are called *heavy agglutinates*. A complete description of the separation procedure and a detailed sample characterization have been given by Gardiner *et al.* (1977).

The experimental procedure of the mass spectrometric noble gas analyses has been described in earlier publications (Hintenberger *et al.*, 1970; Hintenberger and Weber, 1973). Before and after the measurements of agglutinates, two samples of the Bruderheim-Berkeley-Standard have been analysed. Their results are given in Table 1. These last two measurements yield somewhat higher He and Ne concentrations compared to the mean of the previous analyses. This might be the result of a calibration with a gas mixture of "solar" elemental composition. A summary of all available analyses of the Bruderheim-Standard is given by Schultz and Kruse (1978b).

The reproducibilities in the concentrations of an individual sample are believed to be smaller than 5% for He, Ne, and Ar; for Kr and Xe smaller than 10%. This is based on estimates of all possible experimental uncertainties. However, the small sample weights available for this study may introduce an additional error due to sample inhomogeneities, especially for coarse grain-size fractions.

Table 1. Noble gas concentrations in the Bruderheim Standard Bru-7-1 determined before and after the analyses of lunar agglutinates. Except ^{3}He and ^{21}Ne all other isotopes are affected by memory effects due to the relative large trapped gas concentration in previously measured lunar samples.

	^{3}He	^{4}He	^{21}Ne	^{38}Ar	^{40}Ar	^{84}Kr	^{132}Xe
			in 10^{-8} cm^3STP/g				
Oct. 20, 1976	51.3	643	10.7	1.77	1200	0.014	0.014
Dec. 12, 1977	51.2	570	10.7	1.47	1090	0.0095	0.013
Mean of 12 measurements between 1969 and 1977	47.7 ± 1.9	511 ± 30	9.65 $\pm .71$	1.43 $\pm .08$	1200 ± 100	0.013 $\pm .006$	0.013 $\pm .005$

The noble gas results obtained on heavy agglutinates are given in Table 2. A graphic presentation of the data is given in Figs. 1 and 2. These figures also include the results obtained on light agglutinates (Schultz *et al.*, 1977a). The curves shown are fits to the light agglutinate data points according to eq. (1). In this case, n_y was taken to be unity.

The chemical analyses of 19 elements were carried out by instrumental neutron activation techniques as described by Wänke *et al.* (1973). The uncertainties in the concentrations of these elements are given in the last column. The nature of these errors has been discussed by Wänke *et al.* (1977).

The results of the chemical analyses on 5 grain-size fractions of heavy agglutinates are given in Table 3.

Table 2. Concentrations (in cm^3STP/g) in heavy agglutinates (density between 2.96 and 3.3 g/cm^3) separated from soil 15601. The weight of the samples analysed varied between 0.62 and 4.4 mg.

	Range of grain size (μm)								
Isotope	10–20	20–30	30–40	40–53	53–77	77–106	106–150	150–250	250–1000
^{333}He (10^{-5})	3.44	2.37	1.89	1.61	1.13	1.05	1.02	1.26	1.10
^{334}He (10^{-2})	9.03	6.16	4.64	4.01	2.74	2.58	2.45	3.18	2.89
^{320}Ne (10^{-5})	281	196	144	128	92.7	82.8	82.4	95.4	93.0
^{321}Ne (10^{-5})	.777	.561	.425	.391	.292	.254	.263	.299	.278
^{322}Ne (10^{-5})	23.1	15.8	11.6	10.2	7.37	6.77	6.53	7.73	7.48
^{336}Ar (10^{-5})	49.6	35.6	31.7	24.8	20.9	19.8	19.5	20.6	20.7
^{338}Ar (10^{-5})	9.35	6.75	6.02	4.67	3.98	3.73	3.68	3.93	3.99
^{340}Ar (10^{-5})	37.3	28.6	25.6	20.7	17.5	17.3	19.9	21.2	20.0
^{384}Kr (10^{-8})	30.5	23.8	21.3	17.0	13.7	12.7	13.1	13.6	13.0
^{132}Xe (10^{-8})	5.17	3.81	3.36	2.57	1.94	2.13	1.61	2.18	1.86

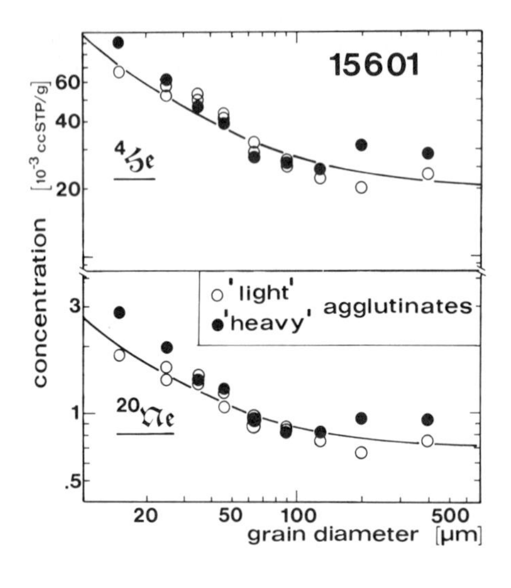

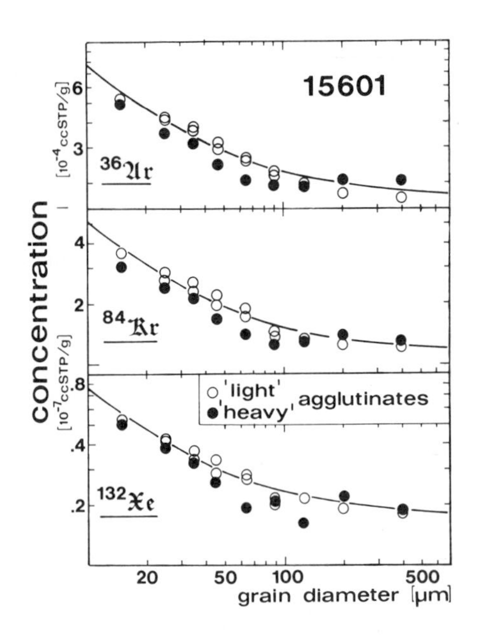

Fig. 1. Trapped ^{4}He and ^{20}Ne in grain-size fractions of heavy and light agglutinates of soil 15601. Curves are fitted [according to eq. (1) with n = 1] to the light agglutinate data assuming a volume-correlated component and a grain-size dependent, surface-correlated component.

Fig. 2. Trapped ^{36}Ar, ^{84}Kr, and ^{132}Xe concentrations in grain-size fractions of light and heavy agglutinates of soil 15601. Curves are fitted [according to eq. (1)] to the light agglutinate data assuming a volume-correlated component and a grain-size dependent, surface-correlated component.

DISCUSSION

Noble gas pattern of agglutinates

In our paper on light agglutinates (Schultz *et al.*, 1977a), we have used eq. (1) with n = 1 for calculating a volume-correlated and a surface-correlated trapped component, V_y and S_y, respectively.

The dependence of the solar gas concentration on the grain size of regolithic particles, which were irradiated uniformly and with equal probability at the lunar surface, is expected to follow the relationship

$$C \sim d^{-1} \tag{2}$$

(C = gas concentration in cm^3STP/g), d = effective grain diameter; e.g., Eberhardt *et al.*, 1970). However, by fitting curves to experimentally determined gas concentrations in bulk soil or mineral grain size suites, n-values ranging from ~0.5 to ~1.2 are obtained. In addition, several grain size suites display no linear relationship in a log C versus log d plot. Several suggestions have been made to

Table 3. Ca, Fe (in wt %) and several trace elements (in ppm) in *heavy* agglutinates from soil 15601.

Element	Grain size (μm) 150–250	106–150	53–77	40–53	20–30	mean	Estimated precision of a single analysis (%)
Ca	6.9	8.25	–	6.7	7.0	7.2 ± 0.7	10
Fe	15.7	16.0	15.8	14.8	14.8	15.42 ± 0.58	3
Na	2420	2280	2220	2480	2340	2348 ± 104	2
K	1010	970	945	1060	940	985 ± 50	7
Sc	39.8	41.7	42	39.5	39.2	40.4 ± 1.3	2
Cr	3410	3890	3570	3520	3470	3572 ± 187	2
Mn	1980	2050	2000	1930	1910	1974 ± 56	2
Co	83.5	67.9	59.3	59.2	56.3	65.2 ± 11.1	3
La	14.9	14.2	14.0	15.7	16.9	15.1 ± 1.2	3
Ce	45	37	35	46	45	41.6 ± 5.2	10
Sm	7.2	7.2	7.0	7.6	8.0	7.4 ± 0.4	3
Eu	1.06	0.99	0.98	1.09	1.09	1.04 ± .05	5
Tb	1.6	1.8	1.9	1.8	2.0	1.82 ± .15	15
Dy	10.5	10.3	9.7	10.8	11.4	10.5 ± .6	5
Ho	2.1	2.0	2.1	2.2	2.3	2.1 ± .1	10
Yb	5.7	5.5	5.2	5.7	6.1	5.6 ± .3	5
Lu	0.77	0.69	0.78	0.79	0.90	0.79 ± .08	10
Hf	5.2	6.0	6.2	6.1	6.5	6.0 ± 0.5	10
Th	1.7	2.3	1.8	1.8	2.4	2.0 ± 0.3	15

explain this deviation from the prediction (e.g., Eberhardt *et al.*, 1972; Criswell, 1975; Frick *et al.*, 1975; Bogard, 1977; Boynton *et al.*, 1977). No general agreement has been reached, but a n-value of 1 for surface-correlated solar gases in lunar soil particles seems to describe the measurements best. Furthermore, best fits to our data on light agglutinates according to eq. (1) with variable n-values show that the highest correlation coefficients are found for n-values between 0.85 and 1.10. This variation in n produces changes in the calculated concentrations of volume- or surface-correlated gas components which are within the given limits of error for those calculated with $n = 1$. Therefore, we will use $n = 1$ also for the surface correlation of trapped gases in the heavy agglutinates.

The procedure of separating the volume-correlated and surface-correlated components by using different grain-size fractions according to eq. (1) assumes that the volume-correlated component is the same in all grain size fractions as well as a grain-size independent surface concentration measured in $ccSTP/cm^2$. This assumption of a grain-size independent volume-correlated component is not true, especially for small grains (Etique *et al.*, 1978). For grains with diameters less than 10 μm, more than about 50% of their volume is accessible to surface-correlated gases because the penetration depth (including secondary effects like diffusion) of solar wind ions is on the order of 1 μm (e.g., Müller *et al.*, 1976). However, the volume correlated component is determined mainly from the larger grain sizes, and this component contributes to a much lesser content to the total gas amounts in the small grain size fractions.

Table 4 gives a comparison of surface (S_y) and volume (V_y) correlated

Table 4. Surface-(S_y) and volume-(V_y) correlated concentration in light agglutinates of 15601 (given in cm^3STP/g) calculated with eq. (1) using n = 1. S_y is the surface-correlated concentration of the reference grain-size (100 μm).

	light agglutinates		heavy agglutinates	
Element	S_y	V_y	S_y	V_y
^{4}He (10^{-4})	106	130	108	170
	±13	±20	±9	±30
^{20}Ne (10^{-4})	2.46	5.5	3.34	5.6
	±.25	±.4	.23	±.7
^{36}Ar (10^{-4})	0.72	1.44	0.51	1.53
	±.06	±.13	±.03	±.10
^{84}Kr (10^{-8})	4.1	10.4	3.0	11.0
	±.2	±.5	±.3	.8
^{132}Xe (10^{-8})	0.68	1.52	0.57	1.45
	±.04	±.09	.06	.18

concentrations in light and heavy agglutinates. Within the limits of error, the volume-correlated component is equal in both kinds of agglutinates. The surface-correlated Ar, Kr, and Xe, however, show differences in concentration between light and heavy agglutinates, which may be explained, at least partly, by the effect that the concentration of surface-correlated gases (in ccSTP/g) is dependent on the density of the grains investigated. Lighter particles have per unit weight more surface area. Consequently, the concentrations of surface-correlated noble gases of light agglutinates should be higher than those in heavy agglutinates. This is observed for Ar, Kr, and Xe. Furthermore, small size fractions of light agglutinates contain more feldspars compared to the bulk sample (Schultz *et al.*, 1977a). This mineral has a small retentivity for He and Ne (Frick *et al.*, 1975), and the surface-correlated solar gases in light agglutinates are more strongly effected by diffusive losses. Thus, the higher concentration of Ne in heavy agglutinates could be explained by this secondary effect.

Table 5 gives elemental ratios for the surface- and volume-correlated gas components of light and heavy agglutinates. The new values for heavy agglutinates corroborate our earlier conclusion that the agglutination process does not fractionate the heavy noble gases Ar, Kr, and Xe because the elemental ratios between these elements for surface- and volume-correlated gases are equal within the limits of error. This is in agreement with conclusions by Alexander *et al.* (1978) that the agglutination process leaves no discernable record in the K/Ar systematics of the agglutinates. Differences between the light noble gas ratios of volume- and surface-correlated gases are not solely attributable to a fractionation process during agglutination. The variations in those ratios are in the observed range of different regolithic minerals. Therefore, these effects can be partly the result of noble gas fractionations like saturation or diffusion prior to agglutination.

Table 5. Elemental ratios of surface-and volume-correlated rare gas components in light and heavy agglutinates.

	Volume-correlated heavy agglutinates	Volume-correlated light agglutinates	Surface-correlated heavy agglutinates	Surface-correlated light agglutinates
$^{4}He/^{20}Ne$	30 ±10	24 ±5	32 ±6	43 ±9
$^{20}Ne/^{36}Ar$	3.7 ±.7	3.8 ±.6	6.5 ±.9	3.4 ±.6
$^{36}Ar/^{84}Kr$	1400 ±200	1400 ±200	1700 ±300	1750 ±250
$^{36}Ar/^{132}Xe$	10500 ±2000	9500 ±1500	9000 ±1500	10600 ±1300
$^{84}Kr/^{132}Xe$	7.6 ±1.5	6.8 ±.8	5.3 ±1.2	6.1 ±.7

Table 6. Isotopic composition, spallogenic isotopes and $^{40}Ar_r$-component which is not accompanied by ^{36}Ar calculated from isotope correlation plots in bulk, light and heavy agglutinates from soil 15601. Concentrations are 10^{-8}cc STP/g.

	Agglutinates light	Agglutinates heavy	Bulk
Spallogenic			
$^{3}He_s$	—	80 ±30	180 ±10
$^{21}Ne_s$	46 ±6	44 ±8	60 ±4
$^{38}Ar_s$	—	50 ±30	30 ±20
Radiogenic			
$^{40}Ar_r$	2000 ±900	6300 ±1400	2000 ±300
Trapped			
$(^{4}He/^{3}He)_t$	2550 ±40	2670 ±70	2800 ±20
$(^{20}Ne/^{22}Ne)_t$	12.6 ±.1	12.11 ±.09	12.6 ±.1
$(^{22}Ne/^{21}Ne)_t$	32.2 ±1.0	31.6 ±.6	32.5 ±.4
$(^{36}Ar/^{38}Ar)_t$	5.31 ±.02	5.32 ±.03	5.35 ±.03
$(^{40}Ar/^{36}Ar)_t$	0.84 ±.04	0.62 ±.05	0.76 ±.02

Table 6 compares spallogenic gas concentrations and trapped isotopic ratios as calculated from isotope correlation plots.

The high concentrations of trapped gases introduce large uncertainties in the calculated spallogenic gas concentrations. However, the spallogenic ^{3}He and ^{21}Ne in bulk samples is higher compared to agglutinates. This implies that some loss of spallogenic ^{3}He and ^{21}Ne occurs during the formation of agglutinates, also. The large amounts of trapped solar gases do not allow the exact determination of spallogenic ^{38}Ar which is less affected by diffusive losses. However, exposure ages of bulk soil and agglutinates calculated from the ^{38}Ar are equal within a factor of two. Thus, it is not possible to test theories of agglutinate evolution (e.g., Mendell and McKay, 1975; J. F. Lindsay, 1975; Basu, 1977) using these exposure ages.

Distribution of other elements

Figure 3 shows the concentration of some of the non-gaseous elements measured in the different grain-size fractions of the heavy agglutinates, normalized to the concentration in the 150–250 μm fraction. There is no obvious

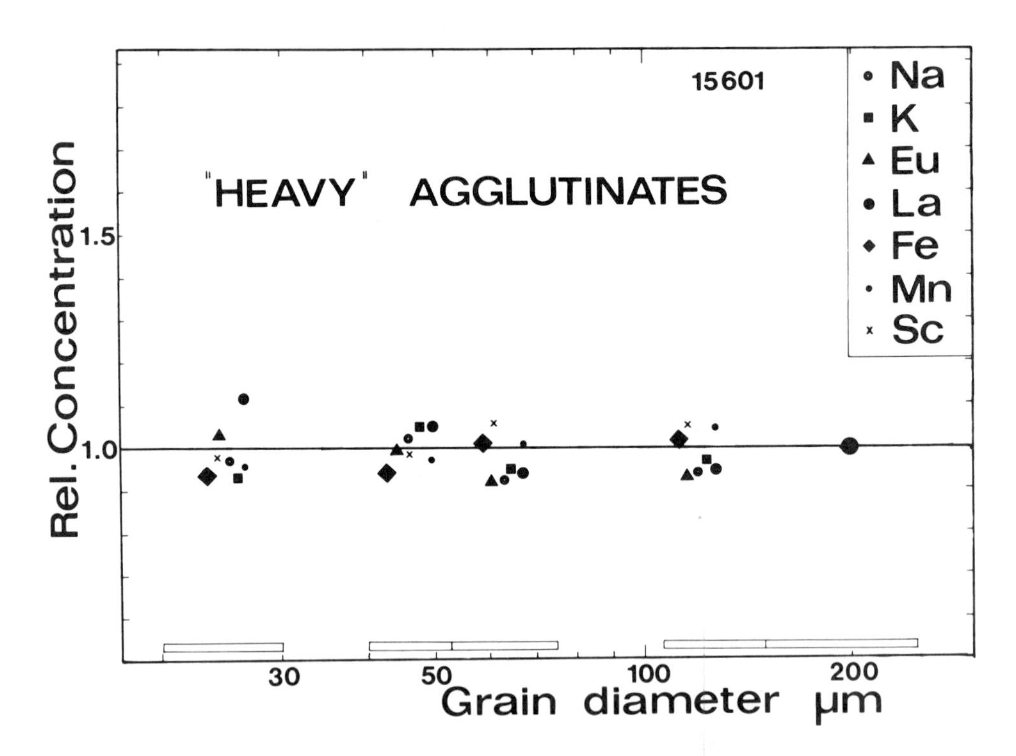

Fig. 3. Elemental concentrations in size fractions of heavy agglutinates. The data are normalized to the 150–250 μm size fraction. A comparison with Fig. 4 shows that chemical differences observed in light agglutinates are not present in the grain-size suite of heavy agglutinates.

dependence of the concentration on grain-size, neither for these elements nor for the remaining ones in Table 3.

Table 3 also contains mean values for the concentrations in all five grain-size fractions. A comparison of this value with the estimated precision of a single measurement shows that only the concentrations of Co, La and Cr are outside the 2σ–range of the uncertainties of the measurement. However, the variation for Co, Cr and La is due to a higher concentration in only one fraction. Therefore, no systematic trend with grain-size is observed. This is in contradiction to the results obtained for the light agglutinates (Schultz *et al.*, 1977a) which for comparison are displayed in Fig. 4. Thus, large variations in the chemical composition with grain size are *not* a common feature in agglutinates from soil 15601.

Rhodes *et al.* (1975) explained their experiments on the chemical composition of agglutinates by a multistage partial melting during the process of agglutination and a selective incorporation of mesostasis and mafic material. The enrichment or depletion of certain elements in grain-size suites of agglutinates is, according to Woodcock and Pillinger (1978), caused by mixing between exotic material and soils of different maturity as well as by sputtering by the solar corpuscular irradiation. Via and Taylor (1976) and Hu and Taylor (1977) have

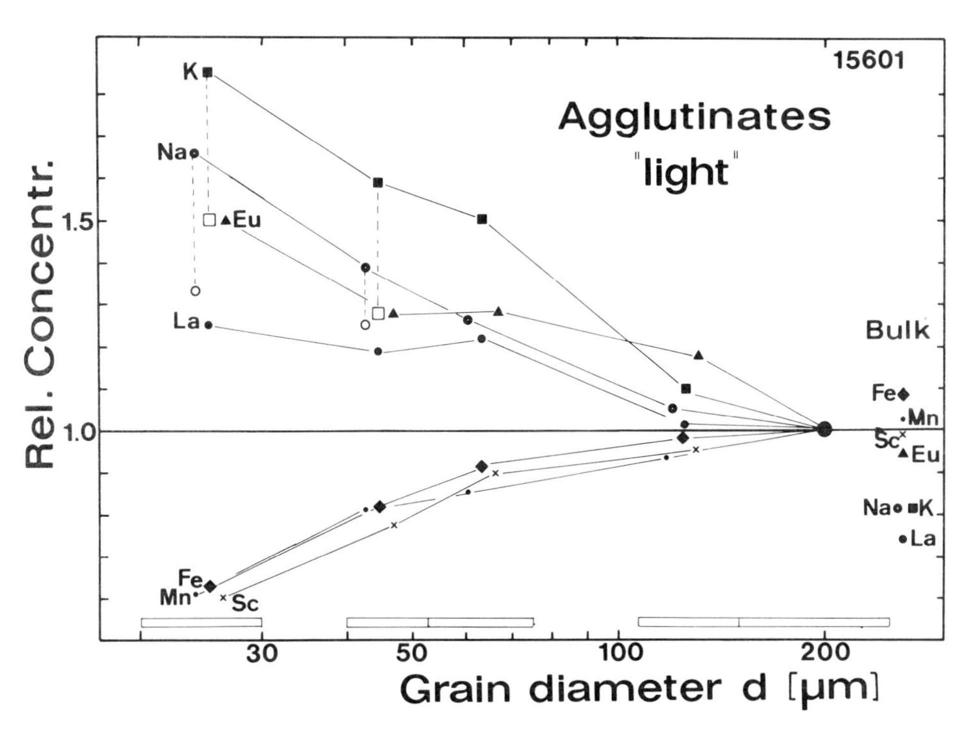

Fig. 4. Elemental concentrations in size fractions of light agglutinates. The data are normalized to the 150–250 μm size fraction. Bars on the abscissa correspond to grain size ranges. These data are taken from Schultz *et al.* (1977a).

concluded from electron microprobe studies on agglutinitic glass that no chemical fractionation cccurs for major and minor elements during agglutinate formation but that the enrichment or depletion of elements for the agglutinates of a soil may be entirely the result of a bias introduced by the magnetic separation technique. This view also could explain the chemical fractionations in our light agglutinates, if one assumes that small mineral grains become magnetic by splashes of agglutinitic glass. The chemical properties of those magnetic particles are determined by the chemistry of the mineral. For large agglutinates this effect disappears because these particles have incorporated many individual grains and, their chemical composition is similar to the bulk soil. Our observations are compatible with the assumption that chemical fractionation observed in small grain-size fractions of agglutinates are presumably the result of a mineral fractionation, because the smaller grains may become magnetic without loosing their original chemical characterisation.

Acknowledgments—This paper is part of the investigation of soil 15601 by the European Consortium. We appreciate the cooperation of Dr. C. T. Pillinger and his co-workers in discussions and the preparation of the grain-size separates. We thank Mrs. R. Löhr and Mr. P. Deibele for their assistance in the laboratory and Mrs. S. Link for her help in manuscript preparation. We have benefited from constructive criticism by Prof. F. Begemann, Mainz, Dr. O. Eugster, Bern, and Prof. P. Signer, Zürich.

REFERENCES

Alexander E. C., Coscio M. R., Dragon J. C. and Saito K. (1978) ^{40}Ar-^{39}Ar studies of glasses from lunar soils (abstract). In *Lunar and Planetary Science IX*, p. 7–9. Lunar and Planetary Institute, Houston.

Basu A. (1977) Steady state, exposure age and growth of agglutinates in lunar soils. *Proc. Lunar Sci. Conf. 8th*, p. 3617–3632.

Bogard D. D. (1977) Effects of soil maturation on grain size-dependence of trapped solar gases. *Proc. Lunar Sci. Conf. 8th*, p. 3705–3718.

Bogard D. D. and Nyquist L. E. (1973) $^{40}Ar/^{36}Ar$ variations in Apollo 15 and 16 regolith. *Proc. Lunar Sci. Conf. 4th*, p. 1975–1985.

Bogard D. D., Nyquist L. E., Hirsch W. C. and Moore D. R. (1973) Trapped solar and cosmogenic noble gas abundances in Apollo 15 and 16 deep drill samples. *Earth Planet. Sci. Lett.* **21**, 52–69.

Boynton W. V., McEwing C. E. and Wasson J. T. (1977) On the amounts of volatile metals and rare gases actually on the surface of lunar soil grains (abstract). In *Lunar Science VIII*, p. 139–141. The Lunar Science Institute, Houston.

Charette M. P. and Adams J. B. (1975) Agglutinates as indicators of lunar soil maturity: The rare gas evidence at Apollo 16. *Proc. Lunar Sci. Conf. 6th*, p. 2281–2289.

Criswell D. R. (1975) The Rosiwal Principle and the regolithic distribution of solar-wind elements. *Proc. Lunar Sci. Conf. 6th*, p. 1967–1987.

Duke M. B., Woo C. C., Sellers G. A., Bird M. L. and Finkelman R. B. (1970) Genesis of lunar soil at Tranquility base. *Proc. Apollo 11 Lunar Sci. Conf.*, p. 347–362.

Eberhardt P., Geiss J., Graf H., Grögler, N. Krähenbühl U., Schwaller H., Schwarzmüller J. and Stettler A. (1970) Trapped solar wind noble gases, exposure age and K/Ar-age in Apollo 11 lunar fine material. *Proc. Apollo 11 Lunar Sci. Conf.*, p. 1037–1070.

Eberhardt P., Geiss J., Graf H., Grögler N., Mendia M. D., Mörgeli M., Schwaller H., Stettler A., Krähenbühl U. and von Gunten H. R. (1972) Trapped solar wind noble gases in Apollo 12 lunar fines 12001 and Apollo 11 breccia 10046. *Proc. Lunar Sci. Conf. 3rd*, p. 1821–1856.

Etique P., Baur H., Derksen U., Funk H., Horn P., Signer, P. and Wieler R. (1978) Helium, neon, and argon in 61501 agglutinates: Implications to gas studies on complex samples. *Proc. Lunar and Planet. Sci. Conf. 9th.* This volume.

Frick U., Baur H., Ducati H., Funk H., Phinney D. and Signer P. (1975) On the origin of helium, neon, and argon isotopes in sieved mineral separates from an Apollo 15 soil. *Proc. Lunar Sci. Conf. 6th*, p. 2097–2129.

Gardiner L. R., Woodcock M. R., Pillinger C. T. and Stephenson A. (1977) Carbon chemistry and magnetic properties of bulk and agglutinate size fractions from soil 15601. *Proc. Lunar Sci. Conf. 8th*, p. 2817–2839.

Hintenberger H. and Weber H. W. (1973) Trapped rare gases in lunar fines and breccias. *Proc. Lunar Sci. Conf. 4th*, p. 2003–2019.

Hintenberger H., Weber H. W., Voshage H., Wänke H., Begemann F. and Wlotzka F. (1970) Concentrations and isotopic abundances of the rare gases, hydrogen and nitrogen in Apollo 11 Lunar matter. *Proc. Apollo 11 Lunar Sci. Conf.*, p. 1269–1281.

Hu H.-N and Taylor L. A. (1977) Lack of chemical fractionation in major and minor elements during agglutinate formation. *Proc. Lunar Sci. Conf. 8th*, p. 3645–3656.

Hübner W., Kirsten T. and Kiko J. (1975) Rare gases in Apollo 17 soils with emphasis on analysis of size and mineral fractions of soil 74241. *Proc. Lunar Sci. Conf. 6th*, p. 2009–2026.

Lindsay J. F. (1975) A steady-state model for the lunar soil. *Bull. Geol. Soc. Amer.* **86**, p. 1661–1670.

McKay D. S., Greenwood W. R. and Morrison D. A. (1970) Origin of small lunar particles and breccia from the Apollo 11 site. *Proc. Apollo 11 Lunar Sci. Conf.*, p. 673–694.

McKay D. S., Heiken G. H., Taylor R. M., Clanton U. S., Morrison D. A. and Ladle G. H. (1972) Apollo 14 soils: Size distribution and particle types. *Proc. Lunar Sci. Conf. 3rd*, p. 983–994.

Mendell W. W. and McKay D. S. (1975) A lunar soil evolution model. *The Moon* **13**, p. 285–292.

Morris R. V. (1976) Surface exposure indices of lunar soils: A comparative FMR study. *Proc. Lunar Sci. Conf. 7th*, p. 315–335.

Morris R. V. (1977) Origin and evolution of the grain-size dependence of the concentration of fine-grained metal in lunar soils: The maturation of lunar soils to a steady-state stage. *Proc. Lunar Sci. Conf. 8th*, p. 3719–3747.

Müller H. W., Jordan J., Kalbitzer S., Kiko J. and Kirsten T. (1976) Rare gas ion probe analysis of helium profiles in individual lunar soil particles. *Proc. Lunar Sci. Conf. 7th*, p. 937–951.

Rhodes J. M., Blanchard D. P., Adams J. B., Charette M. P., Rodgers K. V., Jakobs J. W., Brannon J. L. and Haskin L. A. (1975) Chemistry of agglutinate fractions in lunar soils. *Proc. Lunar Sci. Conf. 6th*, p. 2291–2307.

Schultz L. and Kruse H. (1978b) Light Noble Gases in Stony Meteorites—A Compilation. *Nuclear Track Detection* **2**, 65–103.

Schultz L., Weber H. W., Spettel B., Hintenberger H. and Wänke H. (1977a) Noble gas and element distribution in agglutinate and bulk grain size fractions of soil 15601. *Proc. Lunar Sci. Conf. 8th*, p. 2799–2815.

Schultz L., Weber H. W., Spettel B., Hintenberger H. and Wänke H. (1977b) Agglutinates: Noble gas and element distribution in grain size fractions of the Apollo 15 soil 15601 (abstract). In *Lunar Science VIII*, p. 852–854. The Lunar Science Institute, Houston.

Schultz L., Weber H. W., Spettel B., Hintenberger H. and Wänke H. (1978a) Noble gas and element distribution in agglutinates of different densities (abstract). In *Lunar and Planetary Science IX*, p. 1021–1023. The Lunar and Planetary Institute, Houston.

Signer P., Baur H., Derksen U., Etique P., Funk H., Horn P. and Wieler R. (1977) Helium, Neon, and Argon records of lunar soil evolution. *Proc. Lunar Sci. Conf. 8th*, p. 3657–3683.

Via W. N. and Taylor L. A. (1976) Chemical aspects of agglutinate formation: Relationships between agglutinate composition and the composition of the bulk soil. *Proc. Lunar Sci. Conf. 7th*, p. 393–403.

Wänke H., Baddenhausen H., Dreibus G., Jagoutz E., Kruse H., Palme H., Spettel B and Teschke F. (1973) Multielement analyses of Apollo 15, 16, and 17 samples and the bulk composition of the moon. *Proc. Lunar Sci. Conf. 4th*, p. 1461–1481.

Wänke H., Kruse H., Palme H. and Spettel B. (1977) Instrumental neutron activation analysis of lunar samples and the identification of primary matter in the lunar highlands. *J. Radioanal. Chem.* **38**, p. 363–378.

Woodcock M. R. and Pillinger C. T. (1978) Major element chemistry of agglutinate size fractions. *Proc. Lunar and Planet. Sci. Conf. 9th*. This volume.

Proc. Lunar Planet. Sci. Conf. 9th (1978), p. 2233–2267.
Printed in the United States of America

Helium, neon, and argon in 61501 agglutinates: Implications to gas studies on complex samples

PHILIPPE ETIQUE, UWE DERKSEN, HERBERT FUNK, PETER HORN,
PETER SIGNER and RAINER WIELER

Swiss Federal Institute of Technology
CH-8092 Zurich, Sonneggstrasse 5, Switzerland

Abstract—Samples with agglutinitic affinity from soil 61501 were measured for He, Ne, and Ar concentrations and isotopic ratios. We differentiate between magnetically separated agglutinitic material and handpicked "dendritic" and "scoriaceous" agglutinates. Agglutinitic material can be viewed, with respect to its gas contents, as a mixture of the two types of agglutinates.

A grain size suite of magnetically prepared agglutinitic material, consisting of nine size fractions between 25 and 500 μm, was also investigated. Some aliquots were etched to remove surficially trapped gases. The results show that both surface and volume correlated gases in agglutinitic material are not grain size independent. For neon, data evaluation by means of three-isotope correlation and ordinate intercept plots and interpretation in terms of a two-component system, i.e., "trapped" and "spallogenic", fail. Theoretical aspects of a three-component model are discussed. It is demonstrated that superposition of three gas components can lead to a linear array of data points in ordinate intercept and three-isotope correlation plots. The final conclusion is that the conventional application of three-isotope correlation and ordinate intercept plots to deduce isotopic ratios and concentrations of rare gas components in agglutinitic material (and bulk soils) becomes questionable.

In an agglutinate disintegration experiment, acid-etched agglutinates of the scoriaceous type were separated into their main constituents, which were analyzed for He, Ne, and Ar. The gases in the agglutinates are not homogeneously distributed. They are found in different concentrations and with varying fractionation in the glassy and in mineral-rich fractions.

1. INTRODUCTION

The investigation of noble gases in lunar soil constitutents has increased considerably the understanding of the lunar regolith. The distinction between mineral or primary particles, and constructional or secondary particles, like agglutinates, has been fruitful in dealing with the regolith evolution (Signer *et al.*, 1977a; Basu and Meinschein, 1976; Morris, 1976). It was found that agglutinates which are usually very abundant in lunar soils contain the highest concentrations of trapped solar gases. Therefore, noble gas characteristics of bulk soils are governed by those in agglutinates.

For bulk soils the correlation between gas concentrations and grain size has been viewed as superposition of a grain size independent volume component, consisting of solar and spallogenic gas, and a surface component, comprising solar gas only, with concentrations proportional to the grain surface area (cf. Eberhardt *et al.*, 1972). Actual evidence of a volume distributed component of trapped gases exists from etching experiments (e.g., Hintenberger *et al.*, 1970; Kirsten *et al.*, 1970; Hübner *et al.*, 1975; Leich *et al.*, 1975; Gopalan *et al.*, 1977) but, so far, no systematic investigation to ascertain its independence of grain size has been reported.

Without specific assumptions about the geometrical distribution of noble gases, their isotopic ratios also have been interpreted in terms of a two-component system by means of the ordinate intercept and three-isotope correlation plots. In general, a linear array of data points in both types of plots has been taken as justification for this treatment.

On the average, agglutinates have acquired their trapped solar gases over longer periods of time than minerals. However, the record of the ancient solar irradiation has been obscured by the continuous reworking of agglutinates taking place at the lunar surface (Bogard and Nyquist, 1972; Basu, 1977). Obviously, agglutinates are not homogeneous and a detailed investigation of the distribution of noble gases within agglutinate grains is called for (Basu and Meinschein, 1976; Schultz *et al.*, 1977; Signer *et al.*, 1977a).

In view of the difficulties encountered in applying a definition of "agglutinates" accepted and understood by everyone, we analyzed He, Ne, and Ar in different samples of agglutinitic affinity. Results were obtained for grain size fractions of magnetically separated "agglutinitic material", which can be compared to results for handpicked "agglutinates" previously studied by Signer *et al.* (1977a). The second part of our investigation was aimed at determining volume and surface correlated components. The approach used was to directly verify the existence of these components by their separation through etching experiments. The third part of our study was aimed at an understanding of the high gas concentrations and at an analysis of the gas distribution in agglutinates. For this purpose several subsamples from coarse-grained agglutinates were obtained by etching, crushing, and magnetic and density separation procedures.

2. SAMPLE PREPARATIONS AND SAMPLE DEFINITIONS

Samples analyzed were prepared from submature/mature highland soil 61501 (Morris, 1978), whereby splits with specific numbers ,16 and ,37 had been combined in our laboratory and subsequently were handled together.

2.1 Comparison of agglutinitic material with agglutinates and grain size suite of agglutinitic material separates

Figure 1 shows the genealogy of the samples analyzed. Although the total of the samples indicated in this figure has been prepared only in the 150–200 μm size fraction, the scheme also gives the preparational steps for the samples in the other grain size fractions. These are 300–500 μm for agglutinate comparison; 300–500 μm, 200–300 μm, 125–150 μm, 100–125 μm, 80–100 μm, 64–80 μm, 42–64 μm, and 25–42 μm for agglutinitic material.

Those samples for which analytical results are given in this paper are underlined. The ones which had been reported by Signer *et al.* (1977a) are indicated by an asterisk. The other sample types given are present in the soil, but

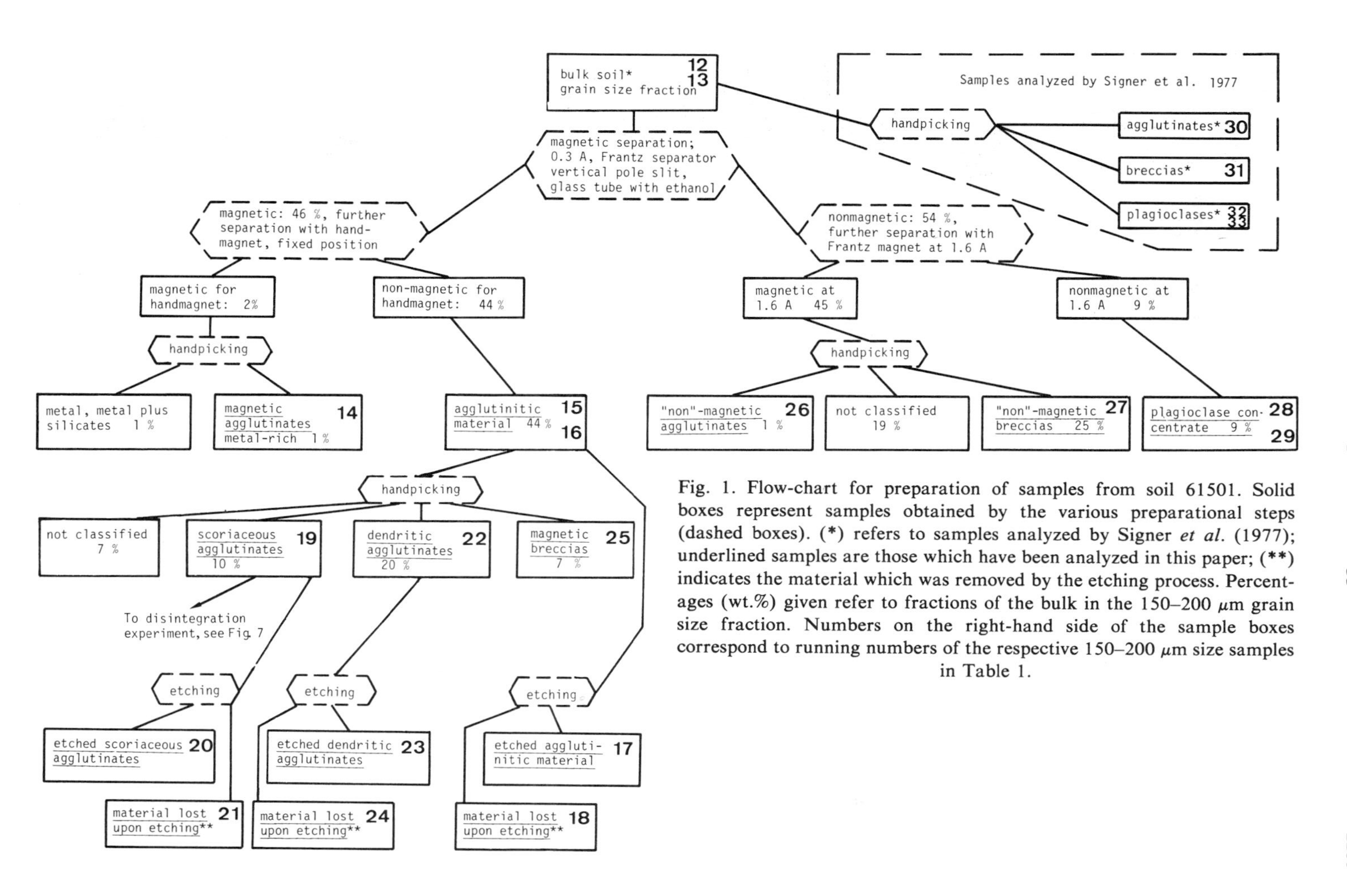

Fig. 1. Flow-chart for preparation of samples from soil 61501. Solid boxes represent samples obtained by the various preparational steps (dashed boxes). (*) refers to samples analyzed by Signer *et al.* (1977); underlined samples are those which have been analyzed in this paper; (**) indicates the material which was removed by the etching process. Percentages (wt.%) given refer to fractions of the bulk in the 150–200 μm grain size fraction. Numbers on the right-hand side of the sample boxes correspond to running numbers of the respective 150–200 μm size samples in Table 1.

are regarded as ill-defined. These samples are the hybrid samples labelled "not classified", and metal-silicate aggregates.

Magnetic separation Size fractions were obtained by wet sieving on nylon sieves, using ethanol as the washing agent. Subsequent magnetic separation was carried out in a 12 mm i.D. glass tube attached to the pole slit of a Frantz magnetic separator, run in the first separation with a magnet current of 0.3 A. The glass tube was in vertical position with a 5° forward tilt and was filled with ethanol. The well suspended samples were then repeatedly fed into the glass tube and were allowed to sediment through the magnetic field. This technique is comparable to that applied by Adams and McCord (1973). The reproducibility of the method proved to be better than 3%. We note that for the specific samples ,16 and ,37 of soil 61501 (<1000 μm) we so obtained a recovery of 52% by weight of magnetic particles.

This figure compares well with 56% separated by Rhodes *et al.* (1975) out of the <250 μm fraction of the same soil. For the grain size ranges 300–500 μm and 200–300 μm we had a yield of magnetic particles of 37% and 39%, respectively, whereas for the smaller grain sizes the yield was 50 ± 4%. This indicates that there is no significant change in the concentration of magnetic particles in this soil from 200 μm down to 25 μm. The relatively low abundance of magnetic particles in the coarser sieve fractions might be the result of the rapid settling of the large particles at our experimental conditions.

From the magnetic fractions, the strongly magnetic particles, such as iron or iron-silicate aggregates as well as magnetic agglutinates were removed by a hand-magnet. The latter were separated from these fractions by handpicking. Throughout all grain size ranges these strongly magnetic fractions comprise about 1–2% of the respective bulk soil fractions. The magnetic material which is not magnetic for the handmagnet we call *agglutinitic material*, as opposed to agglutinitic glass, a term previously coined for analogous samples by Adams and McCord (1973), as it is by far not only glass these particles consist of.

Crystalline and glassy phases in agglutinitic material We determined the proportions of glassy to crystalline phases in the agglutinitic material photometrically. Standardized (single layer, dense packing of particles) grain mounts of agglutinitic samples from the different size fractions were prepared by crushing and sieving to sizes <36 μm. These mounts were scanned point by point using a microscope photometer, with the polarizers of the microscope in crossed position. Thus, the relative contents of anisotropic *and* transparent phases versus glass contents could be measured.

By referring to a calibration curve, prepared by mixing pure terrestrial bytownite and a mafic impact glass, we arrive at concentrations of crystalline material in our agglutinitic samples of 58% for the 200–300 μm sample, increasing to 65% for the 25–42 μm sample. As we have an analytical uncertainty of ±5%, the concentrations of crystalline non-opaques (mostly plagioclases) in the agglutinitic material are essentially constant over the grain size range analyzed.

In the agglutinitic material one finds the different agglutinates as described further below, and minerals and composite particles with relatively large quantities of magnetic glass adhering. In addition, the agglutinitic material contains *magnetic breccias*. These are the more magnetic particles of those specimens named breccias in Signer *et al.* (1977a). We estimate their concentration in the agglutinitic material to be below 10%.

The agglutinitic material comprises also agglutinates. Among the agglutinates we distinguish between *scoriaceous agglutinates* and *dendritic agglutinates*. The former are the well known "textbook" agglutinates, whereas the latter we named for their branching forms (Fig. 2a and 2b). The concentration of crystalline phases in a handpicked sample of agglutinates (scoriaceous plus dendritic agglutinates) has been determined photometrically (see above). It is 37% for the 200–300 μm grain size fraction and, as expected, significantly lower than in the respective agglutinitic material. The scoriaceous agglutinates are optically similar to the particles handpicked out of the bulk grain size fractions in Signer *et al.* (1977a), which then had been named, less specifically, agglutinates (see Fig. 1).

The dendritic and scoriaceous subgroups of the agglutinates represent extremes of a continuous series and can be found in any soil and at any grain size, at least down to 25 μm. In the 200–300 μm grain size fraction, for example, the proportion of scoriaceous to dendritic agglutinates is 1:2.

The differences in shape between dendritic and scoriaceous agglutinates (Fig. 2a and 2b) lead to a weight-per-grain difference, whereby the dendritic particles have weights which are on the average only 70% of the weights of the scoriaceous agglutinates in the same size class.

In both types of agglutinates, a deficiency of mineral and rock fragments smaller than about 10 μm is observed in thin sections, as if the impact glasses had been formed at the expense of the smaller grains. Also no difference with regard to void space or bubble content can be detected (Fig. 3). It amounts to about 30% on the average, leading to an overall density of about 2 g/cm^3 for the agglutinates. This figure has been calculated independently from the number of grains per unit weight assuming spherical grains.

Besides the morphology, scoriaceous and dendritic agglutinates differ in the degree of dust cover, which is higher in the dendritic agglutinates as shown in Fig. 2c and 2d.

Among the handpicked samples from the non-magnetic concentrates which have been analyzed in this paper are *non-magnetic agglutinates*. These are agglutinates structurally, but are non-magnetic and of light color. Some are in essence plagioclase glass. They are regarded as melt products of one and only one impact on anorthositic rocks or on a single plagioclase grain only. The nonmagnetic agglutinate concentration is negligible ($\leq$1%).

Furthermore, we handpicked *non-magnetic breccias* from the nonmagnetic fraction. Structurally they compare to the magnetic breccias mentioned before. They represent the non-magnetic type of those breccias analyzed by Signer *et al.* (1977a).

The least magnetic fraction of the magnetic separation with the Frantz magnet is a plagioclase concentrate. It amounts to about 9% of the 150–200 μm bulk soil fraction and consists almost entirely of plagioclases, to a minor amount of maskelynite or plagioclase glass. The surfaces of the grains in this fraction are slightly contaminated by glass-splashes. The purity of this concentrate is almost comparable to the purity of handpicked samples of plagioclase (Signer *et al.*, 1977a).

Etching of samples From some grain size fractions of the agglutinitic material as well as from scoriaceous and dendritic agglutinates in the grain size ranges 300–500 μm and 150–200 μm, aliquots

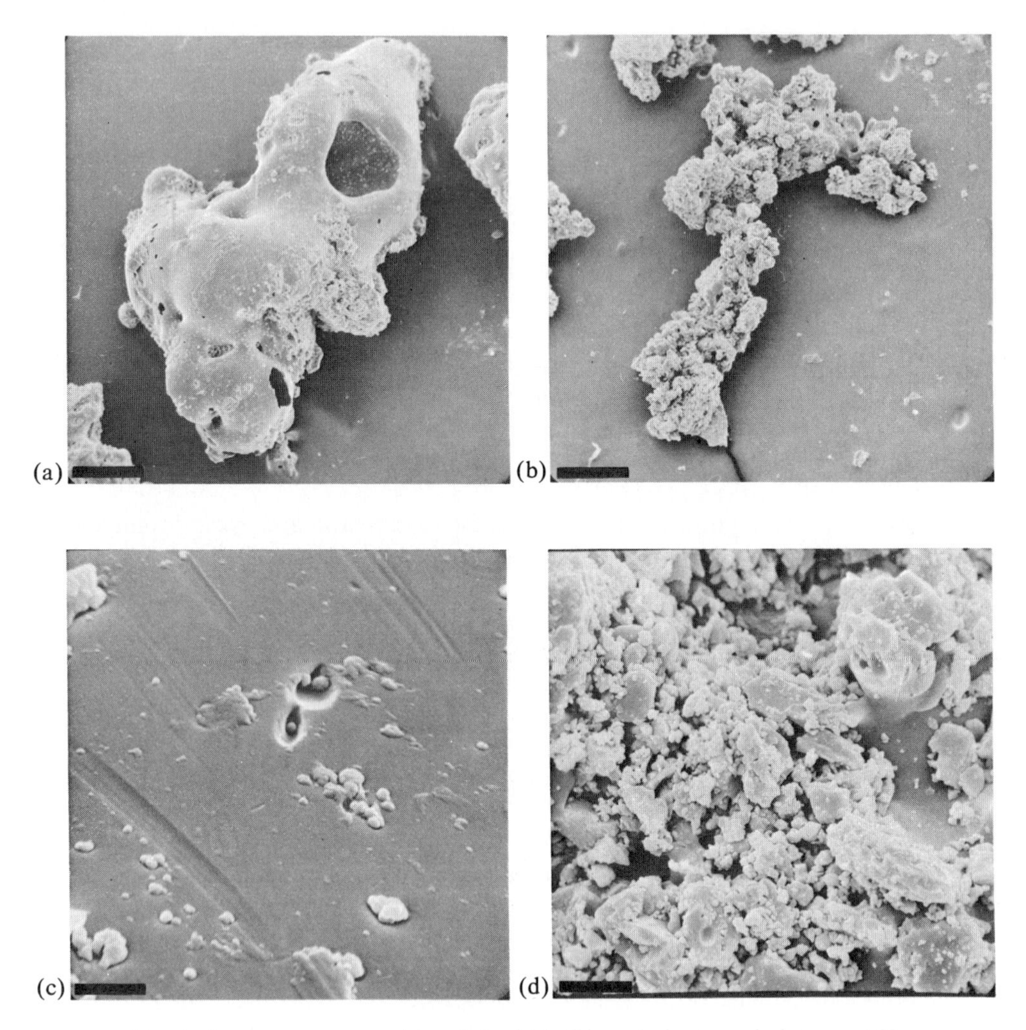

Fig. 2. Typical scoriaceous (a) and dendritic (b) agglutinate; scale bars are 100 μm and 200 μm, respectively. (c) Surface of a scoriaceous agglutinate with little mineral debris adhering; scale bar is 1.3 μm; (d) large abundance of minerals (mostly plagioclases) on a debris-rich area of a dendritic agglutinate; scale bar is 5.5 μm.

were etched. Etching was performed at room temperature with a 1:1:2 mixture of HF (conc.), H_2SO_4 (conc.), and H_2O, for 10 sec. Weight losses upon etching were reproducible within 20%, as judged from comparison of duplicate etching experiments. The thickness of the layers removed from the scoriaceous and dendritic agglutinates were nominally 5 μm and 7μm, respectively. Figures 3a–3d are SEM-images of the etched agglutinates and show many voids and bubbles. They show up so clearly only after etching, but are not produced by the etching. It can also be seen that the mineral particles, mostly plagioclases, have slightly lower etching rates than the matrix glass.

With respect to the weights entering the determinations of gas concentrations in the etched samples it should be noted that SEM-pictures (Fig. 3d) revealed that upon etching an insoluble material remained, which had not been noticed previously. Apparently, this artificial substance, presumably fluorides, survived the subsequent washings only partially.

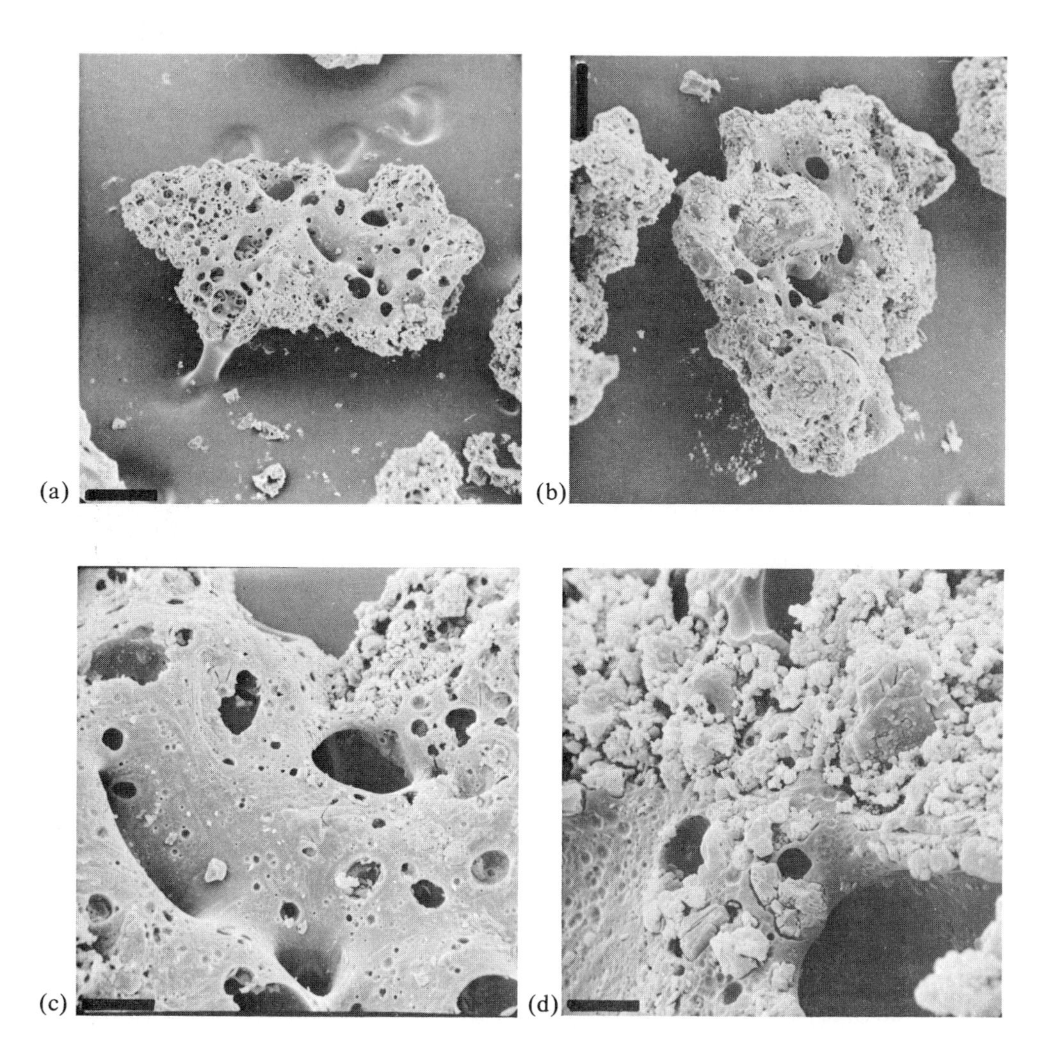

Fig. 3. (a) Etched scoriaceous agglutinate with typical abundance (30 vol.%) of bubbles, and (b) etched dendritic agglutinate; scale bars are 100 μm. (c) Enlarged view of scoriaceous agglutinate; scale bar is 50 μm; (d) enlarged view of dendritic agglutinate; light particles are predominantly plagioclases, bright white patches are possibly fluorides formed upon etching; scale bar is 10 μm.

2.2 Agglutinate disintegration experiment

In the experiment designed to locate the gases within the agglutinates the scoriaceous agglutinates were used. This type of agglutinate was investigated because it was much easier to handpick than the fragile dendritic one. The flow-chart showing the separation and treatment of this sample and the subsamples derived from it is given below (see Section 4.3, Fig. 7).

To remove the solar gases from the outside of the grains, so that gases measured would be relatively enriched in volume distributed gases, the 300–500 μm fraction was etched. The etched samples were then gently crushed and sieved to 46–110 μm. Subsequent separations of subsamples were performed magnetically, by handpicking and by density separation.

Density separation The residual "R" from the handpicked magnetic fraction (see Fig. 7) has been further separated by the density gradient method. To set up the density gradient, a glass tube with 25 mm i.D. was filled with bromoform and after the sample was added, the tube was further filled with acetone and when the tube was slightly shaken, a density gradient formed. By inserting a marker glass of density 2.6 g/cm^3, we had a reference density layer, relative to which the respective density fractions were removed with an injection needle after the tube had been centrifuged to obtain complete separation. The samples were then cleaned and dried.

3. EXPERIMENTAL PROCEDURE AND RESULTS

The experimental procedure of the mass spectrometric noble gas analysis has been described in Signer *et al.* (1977a).

The He, Ne, and Ar concentrations and isotopic abundances are given in Table 1. Results on ^{40}Ar are not given; they will be discussed elsewhere (Signer *et al.*, manuscript in preparation). The samples are listed in three groups according to the three experiments reported here. In a fourth group, results of replicate analyses on <25 μm bulk samples are assembled. These analyses, interspersed over the time during which the measurements reported here were carried out, serve as a stringent check of the mass spectrometric procedure. From these data one notes that isotope ratios are measured with a precision better than 0.5% and concentrations have a precision around 2%. The precision of aliquot analyses of other samples, identified by labels "AL 1, AL 2, ..." in column REMARK, is affected due to the inevitable heterogeneity of coarse grained separates where on the order of only 30–100 grains are used for an analysis.

The accuracy of the analytical result on untreated samples (TRMT: "UE") is believed to be about 10% for concentrations and $^4He/^3He$ ratios and about 2% for the other isotope ratios. The uncertainty of gas concentrations detected in etched samples (TRMT: "ER") is larger. This is, on the one hand, due to the possible presence of insoluble fluorides which may cause the weight to be high, and, on the other hand, due to the possibility that substance may have lost gas without having been dissolved. These additional uncertainties are estimated to be less than 20% (from comparison of spallogenic Ar) and do not affect isotope ratios.

Gas concentrations of samples with a nominal sample weight of 1000 mg are no actual measurements. These concentrations refer to gas lost either during etching ("EL") or crushing and sieving ("CL", sample numbers 68, 69 in Table 1). They were computed as difference between the gas amounts in the untreated sample and in the etched (or crushed and sieved) samples by assuming that the gases lost were homogeneously distributed over the matter dissolved upon etching (or lost in crushing and sieving). The loss upon etching of the corresponding sample is given by the percentage numbers in the column REMARK. In cases where several aliquots of the untreated parent samples were analyzed, the average of the gas concentrations was used for the "EL" computation. As an example for the accuracy of the "EL" gases, we have given two values for the gases lost during etching of the 300–500 μm scoriaceous agglutinates. The sample listed as number 7 in Table 1 is based on the average of the "UE" samples number 2 and 3 and the "ER" sample number 5. The "EL" sample 8 is computed from "UE" sample number 4 and the "ER" sample number 6.

4. DISCUSSION

4.1 Gas concentrations and element abundance ratios

4.1.1 Comparison of different types of samples, etched and unetched, in the same size class. Table 2 gives the gas concentrations in the various samples analyzed in the 150–200 μm grain size range normalized to the gas concentrations in the bulk of the same grain size range. Also given are the fractions of the measured ^{36}Ar concentrations, weighted by the abundances of the respective sample types in the bulk soil.

Besides allowing a comparison of the individual samples, the main objective of this table is to show that we have not missed any soil constituent which contributes significantly to the gases in the bulk soil. Those soil phases which constitute 10% by weight or more of the bulk soil have high relative gas concentrations, ranging from a factor of 1 to 2.8. The other constituents have gas concentrations as low as 12% of the bulk soil. The sum of ^{36}Ar concentrations in the individual samples constituting the agglutinitic material is 83% of the bulk gas content, compared to actually measured 92% for the agglutinitic material. The difference is in part due to the fact that the "unclassified material" in the agglutinitic material was not analyzed; furthermore, sampling reproducibility is about 15% only in the 150–200 μm grain size range. Similarly the ^{36}Ar content of the bulk soil calculated from the ^{36}Ar content in the soil constituents is 126% of the actually measured bulk soil.

From the fact that the agglutinitic material which amounts to 44% of the mass of the bulk soil comprises about 90% of the ^{36}Ar of the bulk sample, we find our rationale for analyzing the agglutinitic material in all grain sizes.

The gas concentrations and elemental abundance data for the main samples with agglutinitic affinity in the grain size ranges 150–200 μm and 300–500 μm are shown in Fig. 4. Also plotted are the values for the gases remaining after etching and those calculated for the gases removed in the etching process. Dendritic agglutinates have slightly higher gas concentrations than scoriaceous agglutinates. This can be attributed to the larger surface area per unit weight of the dendritic agglutinates due to their more complex shapes. The dendritic agglutinates also show lower $^{4}He/^{36}Ar$ and $^{20}Ne/^{36}Ar$ ratios. An explanation for this is that the dendritic agglutinates are more densely covered with small mineral fragments than the scoriaceous ones (Figs. 2c and 2d). In soil 61501, these grains consist mostly of plagioclase, which, for small grain sizes, is also strongly depleted in ^{4}He relative to ^{36}Ar (Signer *et al.*, 1978). The depletion of He and Ne on dendritic agglutinate surfaces may also be seen from the element ratios of the gases lost upon etching. For large grain sizes, the thickness of the removed layer is small relative to grain diameter. Therefore, gases removed by etching are a good approximation to the gases residing actually near the very surfaces of the agglutinates.

Analysis of etched samples reveals two remarkable features: (a) Etched dendritic samples have higher $^{4}He/^{36}Ar$ and $^{20}Ne/^{36}Ar$ ratios than the etched scoriaceous samples. (b) Etched dendritic samples have higher $^{4}He/^{36}Ar$ and $^{20}Ne/^{36}Ar$ ratios than the unetched samples, while for scoriaceous agglutinates the opposite is true. This will be further discussed in Section 4.1.2.

The data points of the agglutinitic material in the 150–200 μm range fall close to the point representing the agglutinates (Signer *et al.*, 1977a) in the same grain size range. They plot also close to or between the respective points for scoriaceous and dendritic agglutinates. This we take as justification for analyzing a grain size suite of magnetic separates instead of handpicked agglutinates. Handpicking is certainly difficult or even impossible to perform for small grain sizes.

4.1.2 Grain size suite of agglutinitic material and some etched aliquots thereof. He, Ne, and Ar concentrations for the grain size suite are shown in Fig. 5. The gas concentrations show the well known increase with decreasing grain size. For the coarsest grain size fraction (300–500 μm) we observe a drop-off in the gas concentrations for the agglutinitic sample (bars labelled *a*). In this grain size range gas concentrations are considerably lower than for both, scoriaceous and dendritic agglutinates. This we explain by the relatively poor sampling reproducibility due to the small number of grains in coarse grain size fractions (about 30 grains in the 300–500 μm fraction).

The fact that ^{4}He and ^{20}Ne concentrations increase with decreasing grain size by about a factor of 2 more than ^{36}Ar leads to a dependence of the element ratios on grain size (see Fig. 6). The cause for this feature will be discussed later in this section.

As in most grain size suite studies, the curves fitted to the data points in Fig. 5 show a flattening with increasing grain size for all three elements. This is usually taken as indication for the presence of a volume correlated gas component (e.g., Eberhardt *et al.*, 1972; Schultz *et al.*, 1977, 1978). To test this assumption, we have analyzed some etched aliquots of the agglutinitic material grain size suite. The gas concentrations in the etched samples are displayed in Fig. 5 by dashed bars. Data for etched dendritic and scoriaceous samples of the larger grain sizes are included. The dashed bars in Fig. 5 show that concentrations of the volume distributed gases vary with grain size: ^{4}He concentrations in etched samples increase with decreasing grain size, whereas ^{36}Ar concentrations are positively correlated with grain size while ^{20}Ne shows an intermediate pattern. Results of

---

Table 1. ^{4}He, ^{20}Ne, and ^{36}Ar concentrations in $10^{-8}cm^{3}STP/g$ and isotope ratios of samples with agglutinitic affinity of soil 61501. Sample size in μm, sample weight in mg. Abbreviations in column "TYPE": SCOR, scoriaceous agglutinates; DEND, dendritic agglutinates; TINI, agglutinitic samples; AGAT, agglutinates as defined in Signer *et al.* (1977a) (already published there, * in column REMARK); AGGL, agglutinates; BREC, breccias; PLAG, plagioclases; MGN, magnetic; NMG, nonmagnetic; CON, concentrate; HPK, handpicked. In part 2 and 3 of the table the abbreviations are the same as in Fig. 7. Roman numerals refer to experiment I (Etique *et al.*, 1978) and II (this paper). Abbreviations in column TRMT: UE = unetched sample; ER = residual after etching; EL = loss upon etching. Entry "1000" in column WEIGHT indicates a fictive sample consisting of material removed by etching. In column REMARK duplicate samples are marked AL, percentage numbers give weight loss upon etching. Samples listed repeatedly are indicated. The numbers in column ZH # give our laboratory code for the respective samples.

Table 1, Part I. Comparison of Types of Agglutinates.

NR	Size, Type, TRMT.	Weight	4HE	4/3	20NE	20/22	22/21	36AR	36/38	Remark	ZH #	NR
1	300–500, TINI, UE	2.91	581000	2077	14960	11.97	18.42	10960	5.208		02,79	1
2	300–500, SCOR, UE	2.97	695000	2310	19650	12.05	21.61	14260	5.253	AL 1	03,57	2
3	300–500, SCOR, UE	2.68	683000	2089	19800	12.10	21.73	14310	5.251	AL 2	03,60	3
4	300–500, SCOR, UE	1.94	669000	2308	20150	12.15	24.31	14350	5.275	AL 3	03,83	4
5	300–500, SCOR, ER	3.70	290700	2342	13450	11.72	17.01	11690	5.208	AL 1, 10%	03,55	5
6	300–500, SCOR, ER	2.21	289300	2639	12760	11.88	20.06	11080	5.231	AL 2, 11%	03,84	6
7	300–500, SCOR, EL	1000.	4122000	2113	73800	12.68	40.31	36700	5.381	"AL 1"	04,34	7
8	300-500, SCOR, EL	1000.	3593000	2141	77000	12.52	34.06	39250	5.338	"AL 2"	04,37	8
9	300–500, DEND, UE	2.25	848000	2585	34250	12.16	26.28	26790	5.153		03,39	9
10	300–500, DEND, ER	2.93	506000	2775	20480	12.01	23.01	15220	5.248	16%	03,58	10
11	300–500, DEND, EL	1000.	2644000	2419	106500	12.32	30.89	87500	5.069		04,39	11
12	150-200, BULK	2.31	720000	2212	18470	12.28	19.57	10680	5.201	AL 1, *	01, 5	12
13	150-200, BULK	4.43	912000	2243	23070	12.13	19.97	12390	5.183	AL 2, *	01,56	13
14	150–200, AGGL, MGN	.70	1312000	2459	41970	12.17	26.40	30430	5.305		03,63	14
15	150–200, TINI, UE	3.09	1181000	2312	34650	12.13	23.10	22040	5.275	AL 1	03, 1	15
16	150–200, TINI, UE	1.00	1345000	2359	38720	12.15	23.40	25290	5.267	AL 2	03,70	16
17	150–200, TINI, ER	1.49	526000	2197	13720	11.64	16.23	8640	5.105	14%	03,77	17
18	150–200, TINI, EL	1000.	5490000	2406	170300	12.40	30.02	110000	5.357		04,40	18
19	150–200, SCOR, UE	.62	1323000	2306	33530	12.19	24.45	23650	5.277		03,71	19
20	150–200, SCOR, ER	1.63	372000	2287	11950	11.55	15.30	9190	5.147	12%	03,75	20
21	150–200, SCOR, EL	1000.	8300000	2313	191800	12.50	34.71	129700	5.349		04,32	21
22	150–200, DEND, UE	.29	1394000	2504	45830	12.25	26.36	32580	5.293		03,72	22
23	150–200, DEND, ER	.65	581000	2396	13810	11.52	16.83	7180	5.068	19%	03,79	23
24	150–200, DEND, EL	1000.	4860000	2562	182400	12.51	32.92	140900	5.344		04,33	24
25	150–200, BREC, MGN	1.16	781000	2162	19080	11.81	16.56	9550	5.127		03,62	25
26	150–200, AGGL, NMG	1.69	652000	1947	12990	11.90	14.81	4469	5.031		03,61	26
27	150–200, BREC, NMG	1.20	851000	2092	24800	11.87	19.22	12620	5.185		03,59	27
28	150–200, PLAG, CON	2.21	98600	2203	2419	11.41	11.33	2071	4.741	AL 1	03,11	28
29	150–200, PLAG, CON	2.93	95500	2229	2809	11.38	12.21	2105	4.702	AL 2	03,74	29
30	150–200, AGAT, HPK	1.24	1280000	2369	34000	12.36	24.68	25630	5.292	*	01,61	30
31	150–200, BREC, HPK	1.47	659000	2007	13530	11.97	14.38	5250	5.014	*	01,41	31
32	150–200, PLAG, HPK	3.80	105100	2256	2735	11.58	12.28	1859	4.542	AL 1, *	01,40	32
33	150–200, PLAG, HPK	1.96	98400	2149	2322	11.70	12.33	1815	4.705	AL 2, *	01,45	33

Table 1, Part II. Grain-Size Suite of Agglutinitic Material.

NR	Size, Type, TRMT.	Weight	4HE	4/3	20NE	20/22	22/21	36AR	36/38	Remark	ZH #	NR
34	25– 42, TINI, UE	.78	4157000	2426	104300	12.18	27.57	46650	5.306	AL 1	03,45	34
35	25– 42, TINI, UE	.31	3856000	2441	102300	12.21	27.77	48070	5.302	AL 2	03,68	35
36	25– 42, TINI, ER	.46	1489000	2423	16830	11.26	16.70	1628	4.557	54%	03,78	36
37	25– 42, TINI, EL	1000.	6270000	2431	177800	12.27	29.43	85800	5.319		04,42	37
38	42– 64, TINI, UE	.76	2103000	2366	58400	12.19	24.88	33950	5.290	AL 1	03, 6	38
39	42– 64, TINI, UE	.56	2618000	2375	70500	12.18	26.40	36700	5.294	AL 2	03,44	39
40	64– 80, TINI, UE	1.33	1971000	2346	54000	12.15	25.07	30320	5.288		03,43	40
41	80–100, TINI, UE	2.47	1651000	2316	47220	12.12	24.67	28380	5.286	AL 1	03, 5	41
42	80–100, TINI, UE	1.26	1707000	2330	48860	12.14	24.70	28980	5.283	AL 2	03,47	42
43	80–100, TINI, UE	.81	1830000	2332	50400	12.20	24.35	27680	5.275	AL 3	03,69	43
44	80–100, TINI, ER	1.11	493100	1864	7180	10.88	10.20	1852	4.488	37%	03,76	44
45	80–100, TINI, EL	1000.	3751000	2458	118200	12.29	29.45	73700	5.324		04,41	45
46	100–125, TINI, UE	1.89	1532000	2328	44350	12.13	24.51	28130	5.290		03, 8	46
47	125-150, TINI, UE	3.21	1349000	2341	37920	12.16	23.78	23370	5.278		03, 3	47
48	150–200, TINI, UE	3.09	1181000	2312	34650	12.13	23.10	22040	5.275	Same as 15	03, 1	48
49	150–200, TINI, UE	1.00	1345000	2359	38720	12.15	23.40	25290	5.267	Same as 16	03,70	49
50	150–200, TINI, ER	1.49	526000	2197	13720	11.64	16.23	8640	5.105	Same as 17	03,77	50
51	150–200, TINI, EL	1000.	5490000	2406	170300	12.40	30.02	110000	5.357	Same as 18	04,40	51
52	200–300, TINI, UE	2.98	1117000	2231	33330	12.12	23.04	21140	5.255	AL 1	02,43	52
53	200–300, TINI, UE	4.71	1120000	2293	32230	12.15	22.95	20610	5.254	AL 2	02,80	53
54	300–500, TINI, UE	2.91	581000	2077	14960	11.97	18.42	10960	5.208	Same as 1	02,79	54

Table 1, Part III. Agglutinate Disintegration.

NR	Size, Type, TRMT.		Weight	4HE	4/3	20NE	20/22	22/21	36AR	36/38	Remark	ZH #	NR
Starting Matter, 300–500 μm													
55	300–500, S	I	2.97	695000	2310	19650	12.05	21.61	14260	5.253	Same as 2	03,57	55
56	300–500, S	I	2.68	683000	2089	19800	12.10	21.73	14310	5.251	Same as 3	03,60	56
57	300–500, S	II	1.94	669000	2308	20150	12.15	24.31	14350	5.275	Same as 4	03,83	57
Residual After Etching													
58	300–500, E	I	3.70	290700	2342	13450	11.72	17.01	11690	5.208	Same as 5	03,55	58
59	300–500, E	II	2.21	289300	2639	12760	11.88	20.06	11080	5.231	Same as 6	03,84	59
Loss on Etching													
60	300–500, EL	I	1000.	4122000	2113	73800	12.68	40.31	36700	5.381	Same as 7	04,34	60
61	300–500, EL	II	1000.	3593000	2141	77000	12.52	34.06	39250	5.338	Same as 8	04,37	61
Crushed and Sieved													
<46 μm Bulk													
62	0–46, F	I	1.74	298700	2286	10930	11.67	17.67	9620	5.201		03,54	62
63	0–46, F	II	1.05	235600	2204	9970	11.59	15.16	9450	5.181	AL 1	03,85	63
64	0–46, F	II	1.78	236100	2221	9580	11.53	14.92	9110	5.181	AL 2	03,90	64
46–110 μm Bulk													
65	46–110, C	I	1.44	128900	2109	8010	11.59	15.29	8590	5.200		03,53	65
66	46–110, C	II	.85	219400	2258	10730	11.68	15.66	10410	5.189	AL 1	03,86	66
67	46–110, C	II	1.24	239300	2281	11240	11.61	15.52	10890	5.195	AL 2	05, 2	67
Loss on Crushing													
68	CL	I	1000.	579000	2459	24480	11.80	18.11	18250	5.214		04,35	68
69	CL	II	1000.	447900	3448	18170	12.31	39.85	12940	5.319	27%	04,38	69

Table 1. Magnetic Separation.

NR	Size, Type, TRMT.		Weight	4HE	4/3	20NE	20/22	22/21	36AR	36/38	Remark	ZH #	HR
Nonmagnetic Fraction													
70	46-110, NM	I	.38	52500	915	1469	10.30	5.39	836	4.318		03,51	70
71	46-110, NMD	II	.43	79100	1223	2652	9.19	4.57	2642	4.731	Dark	03,91	71
72	46-110, NML	II	1.67	88200	1246	3413	10.38	7.29	2589	4.790	Light	03,92	72
Magnetic Glass Rich													
73	46-110, G	I	.33	42260	1787	4196	11.53	9.80	5460	5.145		03,52	73
74	46-110, G	II	.61	125400	2007	7700	11.30	13.06	8910	5.177		03,88	74
Magnetic Mineral Rich													
75	46-110, M	I	.27	149900	2018	8490	11.98	16.34	7660	5.217		03,50	75
76	46-110, M	II	.44	369100	2633	17550	12.04	21.23	16030	5.245		03,89	76
Magnetic Residual, Unseparated													
77	46-110, R	I	.42	182100	2161	8460	11.88	15.52	8260	5.189		03,49	77
78	46-110, R	II	1.09	265500	2423	14150	11.73	18.11	14260	5.230	AL 1	03,93	78
79	46-110, R	II	1.39	264300	2439	13000	11.74	18.43	12760	5.231	AL 2	05, 3	79
Magnetic Residual, Separated													
Density >2.6													
80	46-110, R >	II	.63	229300	1773	9160	11.35	13.57	7870	5.150		03,94	80
Density = 2.6													
81	46-110, R =	II	.47	288800	2670	14310	11.88	19.34	14380	5.235		05, 1	81
Density <2.6													
82	46-110, R <	II	.83	271700	2793	16500	11.81	20.81	17020	5.247	AL 1	03,99	82
83	46-110, R <	II	.42	268000	2831	16400	11.88	21.03	17130	5.255	AL 2	05, 4	83

Table 1, Part IV. Replica Runs, <25 Micron Bulk Soils.

NR	Size, Type, TRMT.	Weight	4HE	4/3	20NE	20/22	22/21	36AR	36/38	Remark	ZH #	NR
65501,8,2	<25 μm											
84	<25, Bulk	.49	6330000	2556	159900	12.42	28.91	0	.000		02,63	84
85	<25, Bulk	.57	6590000	2544	161100	12.36	28.97	59800	5.336		02,65	85
86	<25, Bulk	.26	6630000	2549	165100	12.39	29.26	60600	5.339		02,68	86
87	<25, Bulk	.46	6530000	2531	162300	12.33	29.09	59300	5.331		02,69	87
88	<25, Bulk	.50	6410000	2537	159600	12.40	28.96	59100	5.352		03, 7	88
89	<25, Bulk	.36	6750000	2525	166200	12.35	28.88	60800	5.340		03,46	89
61501,16+37	<25 μm											
90	<25, Bulk	.99	9310000	2421	224000	12.40	30.46	87400	5.344		02,66	90
91	<25, Bulk	.78	8760000	2440	208900	12.37	30.57	82600	5.341		02,67	91
92	<25, Bulk	.46	8730000	2467	205500	12.41	30.60	80700	5.342		02,70	92
93	<25, Bulk	.40	8680000	2479	202500	12.43	30.48	82200	5.351		03,20	93
94	<25, Bulk	.33	8560000	2492	201000	12.37	30.48	82300	5.355		03,23	94

Table 2. ^{4}He, ^{20}Ne, and ^{36}Ar concentrations in the different sample types (unetched samples only) of 150–200 μm grain size normalized to the gas concentrations in the 150–200 μm bulk soil. Column 2 gives the modal abundances of the respective samples in the soil. In column 4 the normalized ^{36}Ar concentrations weighted by modal abundances are listed. Figures in the last column refer to sample numbers in Table 1.

Sample type from soil 61501; grain size 150–200 μm	Sample, wt-fraction of bulk	Noble gases in samples, normalized to bulk soil			^{36}Ar (fraction of bulk)	Sample number (cf. Table 1)
		^{4}He	^{20}Ne	^{36}Ar		
Bulk*	1.0	1.0	1.0	1.0	1.0	12; 13
Agglutinitic material	0.44	1.6	1.8	2.1	0.92	15; 16
Scoriaceous agglutinates	0.10	1.6	1.6	2.1	0.21	19
Scoriaceous* agglutinates	(0.10)	1.6	1.6	2.2	(0.22)	30
Dendritic agglutinates	0.20	1.7	2.2	2.8	0.56	22
Magnetic agglutinates	.01	1.6	2.0	2.6	0.03	14
Magnetic breccias	0.07	0.96	0.92	0.83	0.06	25
Metal	.01	n.d.	n.d.	n.d.	n.d.	
Not classified	0.07	n.d.	n.d.	n.d.	n.d.	
Agglutinitic material	0.44[a]				0.83[a]	
Non-magnetic breccias	0.25	1.0	1.2	1.1	0.28	27
Non-magnetic agglutinates	0.01	0.8	0.6	0.4	0.004	26
Breccias*	(0.32)	0.8	0.6	0.5	(0.16)	31
Plagioclase concentrate	0.09	0.12	0.13	0.18	0.016	28
Plagioclase*	(0.09)	0.12	0.12	0.16	(0.014)	32; 33
Not classified	0.19	n.d.	n.d.	n.d.	n.d.	
Bulk	1.0[a]				1.26[a]	

*Samples analyzed by Signer *et al.* (1977a).
Figures in brackets do not enter balance calculations.
n.d.: not determined.
[a]Sum of figures in the respective sub-columns.

etching experiments performed on agglutinates in one grain size range (Hübner *et al.*, 1975) and on bulk soil grain size fractions (Kirsten *et al.*, 1970; Gopalan *et al.*, 1977) are in qualitative agreement with our experiments.

The element abundance ratios are given in Fig. 6 as a function of grain size. $^{4}He/^{36}Ar$ and $^{20}Ne/^{36}Ar$ ratios increase with decreasing grain size both for the

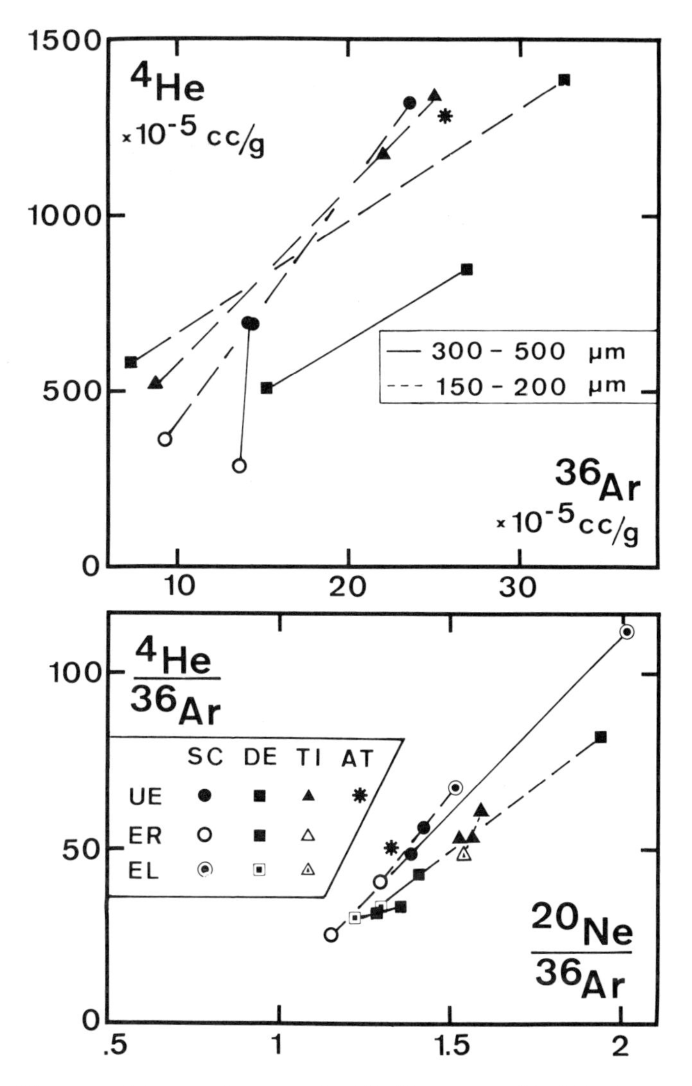

Fig. 4. ^4He versus ^{36}Ar and element ratios $^4He/^{36}Ar$ versus $^{20}Ne/^{36}Ar$ for scoriaceous (SC) and dendritic (DE) agglutinates in the 300–500 μm and 150–200 μm grain size ranges as well as for agglutinitic material (TI) and "agglutinates" (AT) as defined in Signer *et al.* (1977a) in the 150–200 μm range of soil 61501. "UE" unetched samples; "ER" residues after etching; "EL" loss on etching (computed).

unetched and the etched samples, whereby the increase is very pronounced for the volume distributed gases remaining after etching, namely a factor of 50 for $^4He/^{36}Ar$ and a factor of 10 for $^{20}Ne/^{36}Ar$, respectively.

Before we interpret these results we give a glossary of terms which we will apply: a) We distinguish between gas *components* and *secondary components*: Noble gas derived from a reservoir with specific element and isotope abundances is called a component. If the elemental and isotopic patterns of such

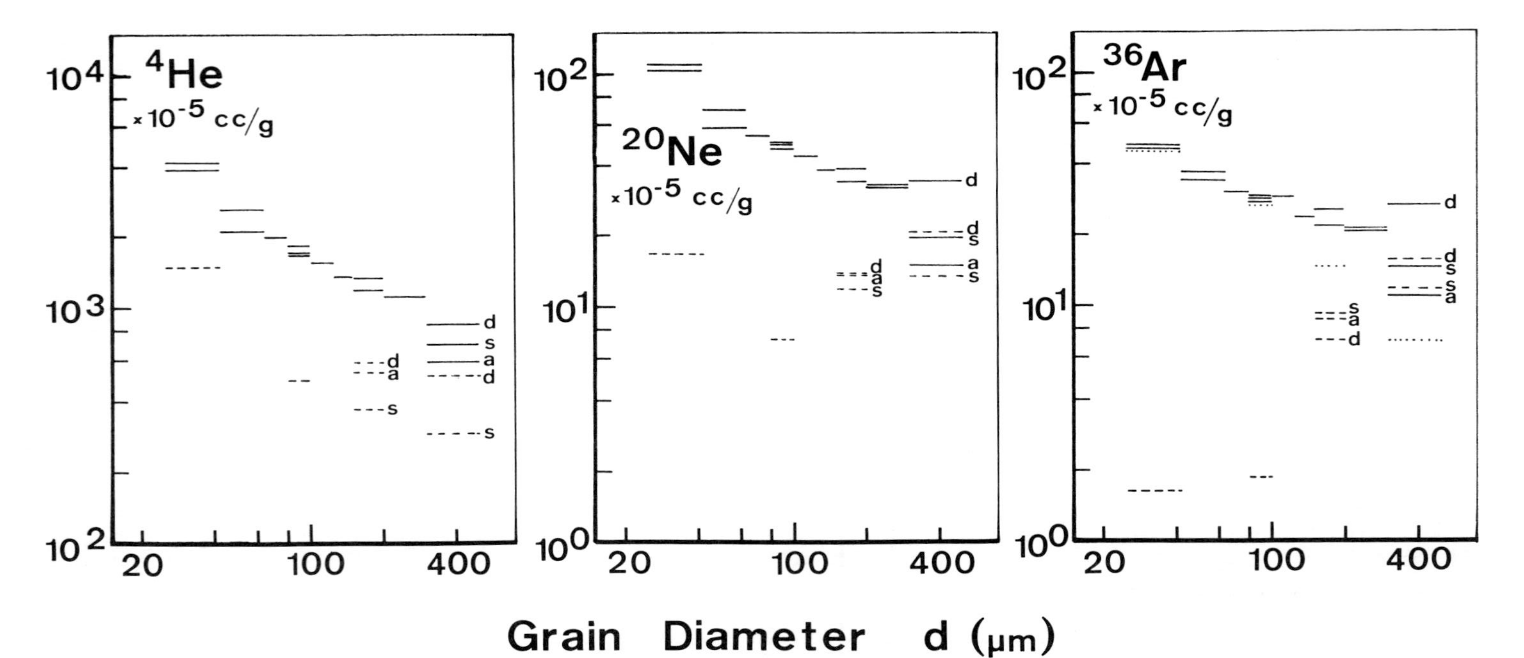

Fig. 5. Gas concentrations of ^{4}He, ^{20}Ne, and ^{36}Ar versus grain diameter in logarithmic scales for the suite of agglutinitic material (solid bars) and the residues after etching for 4 grain size fractions (dashed bars). Whenever data of scoriaceous and dendritic agglutinates are included, they are marked by "s" and "d", respectively. Agglutinitic samples are marked by "a". In the ^{36}Ar diagram, the dotted bars stand for the gas lost upon etching.

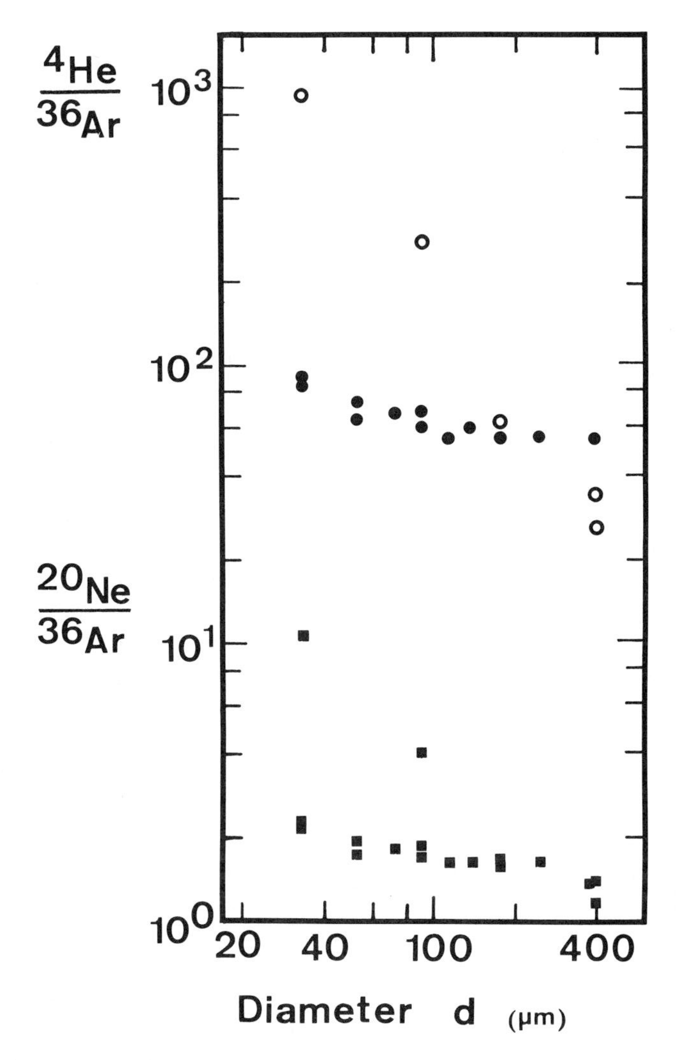

Fig. 6. $^4He/^{36}Ar$ and $^{20}Ne/^{36}Ar$ ratios of agglutinitic material versus grain diameter *d* (solid symbols: unetched samples, open symbols: residues after etching).

a component are modified by any processes, such that these patterns are distinct, we speak of a secondary component (Smith *et al.*, 1978). b) Noble gases of solar origin may be located in different loci of lunar soil grains with different element and isotope abundance ratios. We distinguish between: i) *Trapped gas*: this is solar gas retained in grain surfaces. As it may be distinctly fractionated, trapped gas may be a secondary component (Huneke, 1973). ii) *Migrated gas*: this was trapped gas which has migrated into the grain volume, whereby He and Ne become enriched relative to Ar. Migrated gas is detectable only in etched samples. iii) *Redistributed gas*: this was trapped or migrated gas that has been redistributed by agglutination into the grain volume. In Table 3 we list these gas types. c) If the concentration of a gas component or secondary component is proportional to the grain volume or the grain surface we speak of *volume correlated* or *surface correlated* gas, respectively. The terms *volume distributed* and *surface distributed* are used to express the actual site of a type of gas. Migrated gas for example is volume distributed, but may be surface correlated, because it originates from the trapped gas.

With this terminology we try to interpret the observations from Figs. 5 and 6. Signer *et al.* (1978) demonstrate that in a plagioclase suite of soil 61501, the light gases He and Ne, but not Ar, migrate from the surface into the volume of the grains. The same effect has been observed by Leich *et al.* (1975) on etched ilmenites, and our data indicate it to be true for agglutinates also. We thus regard the ^{36}Ar data in the etched agglutinates as a good approximation of the sum of redistributed and spallogenic ^{36}Ar. The computed ^{36}Ar removed by etching (Fig. 5) is, on the other hand, a good approximation for the trapped ^{36}Ar. This, however, does not hold for He and Ne. Because of the high mobility of these gases, a considerable amount of migrated gas is superimposed on the redistributed gas. The determination of the concentrations of redistributed He and Ne is therefore not readily possible. It may be achieved by etching aliquots to different depths. For neon the data suggest, however, that the concentrations of the redistributed gas increase with grain size as they do for Ar. Only in the 25–42 μm fraction migrated gas dominates. For He migrated gas masks the redistributed gas completely.

In section 4.1.1 we noted that $^{4}He/^{36}Ar$ and $^{20}Ne/^{36}Ar$ ratios are higher in etched dendritic agglutinates than in etched scoriaceous agglutinates which means that migrated gas is more prominent in dendritic samples. We mention two reasons: a) dendritic particles have higher surface-to-volume ratios than scoriaceous particles; b) in the small plagioclase grains adhering to the dendritic particles, migration is a prevailing feature, because these minerals do not contain any redistributed gas (Signer *et al.*, 1978). Consequently, we believe that the redistributed He and Ne is better represented by the concentrations of these gases in etched scoriaceous than in etched dendritic agglutinates. This goes together with the expectation that He and Ne are depleted in the redistributed gases and with the observation from Fig. 4 that etched scoriaceous agglutinates indeed have lower $^{4}He/^{36}Ar$ and $^{20}Ne/^{36}Ar$ ratios than their unetched aliquots.

Noble gas concentrations in grain size suites have been generally interpreted as superposition of a surface correlated and a volume correlated secondary component (e.g., Schultz *et al.*, 1977, 1978). The latter is usually determined as the ordinate intercept of the best fit straight line in a c versus 1/d plot (c = gas concentrations in cm^3STP/g, d = grain diameter). However, we note from Fig. 5 that the volume distributed ^{36}Ar concentrations increase strongly with grain size, i.e., they are far from being volume correlated. Therefore, for particles such as agglutinates, the above determination becomes meaningless, at least for Ar and most likely for He and Ne, too.

Pillinger *et al.* (1978) find in an agglutinate grain size suite of soil 12033 concentrations of Fe^0_{sp} (i.e., small grained superparamagnetic iron reduced by solar wind hydrogen) increasing with decreasing grain size, in contrast to the concentrations of redistributed ^{36}Ar in our suite of agglutinitic material. Fe^0_{sp} is thought to measure the degree of reworking (Housley *et al.*, 1973; Morris, 1976),

Table 3.

Gas type	Progenitor gas	Processes involved	Distribution of gas within grain	Fractionation in comparison to progenitor gas
Trapped gas	Solar wind, solar flare	Implantation, diffusion, sputtering metamictization	Surface layer only	Light gases depleted
Migrated gas	Trapped	Diffusion into undamaged host	Concentration–gradient from surface to center	Light gases enriched
Redistributed gas	Trapped and migrated	Agglutination, mechanical mixing	Throughout grain, but varying concentrations in different phases	Light gases depleted
Spallogenic, radiogenic gas	–	Spallation, radioactive decay	Throughout grain, surface layer depleted	If diffusion is significant, light gases depleted

just as the redistributed ^{36}Ar should, so the two findings contradict each other. One may, however, argue that in small grains, due to diffusion, hydrogen is effectively distributed in a way that reduction of iron takes place more effectively.

Although the gas concentrations in agglutinates from 10 soils are 10–60 times higher than in plagioclases of the same grain size range, Signer *et al.* (1977a) expected the concentrations of trapped ^{36}Ar to be the same for the two types of particles, and they concluded that most of the gas in agglutinates should be volume distributed. Our etching experiments now show that the concentration of trapped ^{36}Ar in 61501 agglutinitic samples is, for equally sized grains, in fact about 5 times higher than in plagioclases of the same soil, if measured in cm^3STP/g. Thereby, the concentrations of trapped ^{36}Ar in the agglutinitic samples have been calculated as $c_{tr} = c_{UE} - c_{ER}$, where c_{UE} and c_{ER} are the concentrations in the unetched samples and their etched residues, respectively (see dotted bars in Fig. 5). In plagioclases, there is essentially only trapped Ar (Signer *et al.*, 1978). Towards resolution of this discrepancy of a factor of 5 we mention the following: a) Agglutinitic material has a lower density than plagioclase (2 and 2.8 g/cm^3, respectively). Thus, the trapped gas concentrations of agglutinates, expressed in cm^3STP/cm^2 surface area, are only about 3.5 times those of plagioclase. b) Agglutinitic material is densely covered by very small minerals with high gas concentrations. c) Small cavities and vugs connected to the grain surface may have been filled by fine-grained soil, not removed by the washing of the samples in ethanol.

The best fit straight line for the trapped ^{36}Ar concentrations (dotted bars in Fig. 5) in the log c versus log d plot has a slope n = −0.8. This is distinctly different from the commonly expected value of −1 (Bogard, 1977; Schultz *et al.*, 1977, 1978, but see also discussions in Criswell, 1975; Becker, 1977; Criswell and Basu, 1978).

4.1.3 Agglutinate disintegration experiment. The goal of the experiment outlined in Fig. 7 was to investigate the gas concentrations as well as element and isotope abundances in the different constituents of agglutinates. We concentrate here on results of a refined version of the disintegration experiment reported by Etique *et al.*, (1978). The results of the two experiments are distinguished in Table 1, part 3, by roman numerals I and II in the "TRMT" column. Generally, our previous results are confirmed; however, in some samples gas concentrations differ as much as by a factor of 3. The differences tend to be higher when more experimental steps are involved for the preparation of the respective samples. It should also be said that in the first experiment some inadequacies in preparing the samples may have occurred.

The results of the present experiment are given in Fig. 8. Confirming the earlier results (Etique *et al.*, 1978) the glass-rich fraction (G) contains less noble gases than the mineral-rich fraction (M) and the residue (R). At the same time gases in the glass-rich fraction are more fractionated. Density separation of the residue (R) yielded samples with densities smaller (R<), equal to (R=), or larger (R>) than 2.6 g/cm³, respectively. The elemental patterns of sample R<, consisting mainly of vuggy glass, and of sample R>, enriched in minerals, resemble those of the glass-rich and the mineral-rich fractions, respectively. The non-magnetic dark (NM_D) and light (NM_L) separates contain even smaller gas concentrations than the glass-rich phase, but show rather different element fractionation. The light portion, comprising 80% of the whole non-magnetic fraction, shows the smallest fractionation among all agglutinate phases which is similar to that of plagioclase. This concurs with the finding that a large portion of this separate is plagioclase, maskelynite and plagioclase glass. Note that the non-magnetic agglutinates and the NM_L sample (sample numbers 26 and 72 in Table 1) have similar $^4He/^{36}Ar$ and $^{20}Ne/^{36}Ar$ ratios. We regard this material as "first generation" agglutinates. Upon repeated agglutination, i.e., reworking, gas concentrations will increase and the agglutinates will become increasingly magnetic (Morris, 1976). Thereby He and Ne will be depleted relative to Ar.

The point labelled "CL" in Fig. 8 stands for the calculated gas loss during the crushing process. In agreement with Funkhouser *et al.* (1971) we believe this to be gas released from voids within the agglutinates. The gas in the voids is enriched, in comparison to its "parent sample" (E), in mobile He and Ne relative to Ar. In addition, the $^{20}Ne/^{22}Ne$ ratio is higher in "CL" than in "E."

Our previous conclusion (Etique *et al.*, 1978) that, within the resolution of our separation, all phases of scoriaceous agglutinates contain light noble gases has been confirmed. Their degree of element fractionation correlates with the presumed intensity of reworking. The gases within the volume of the agglutinates

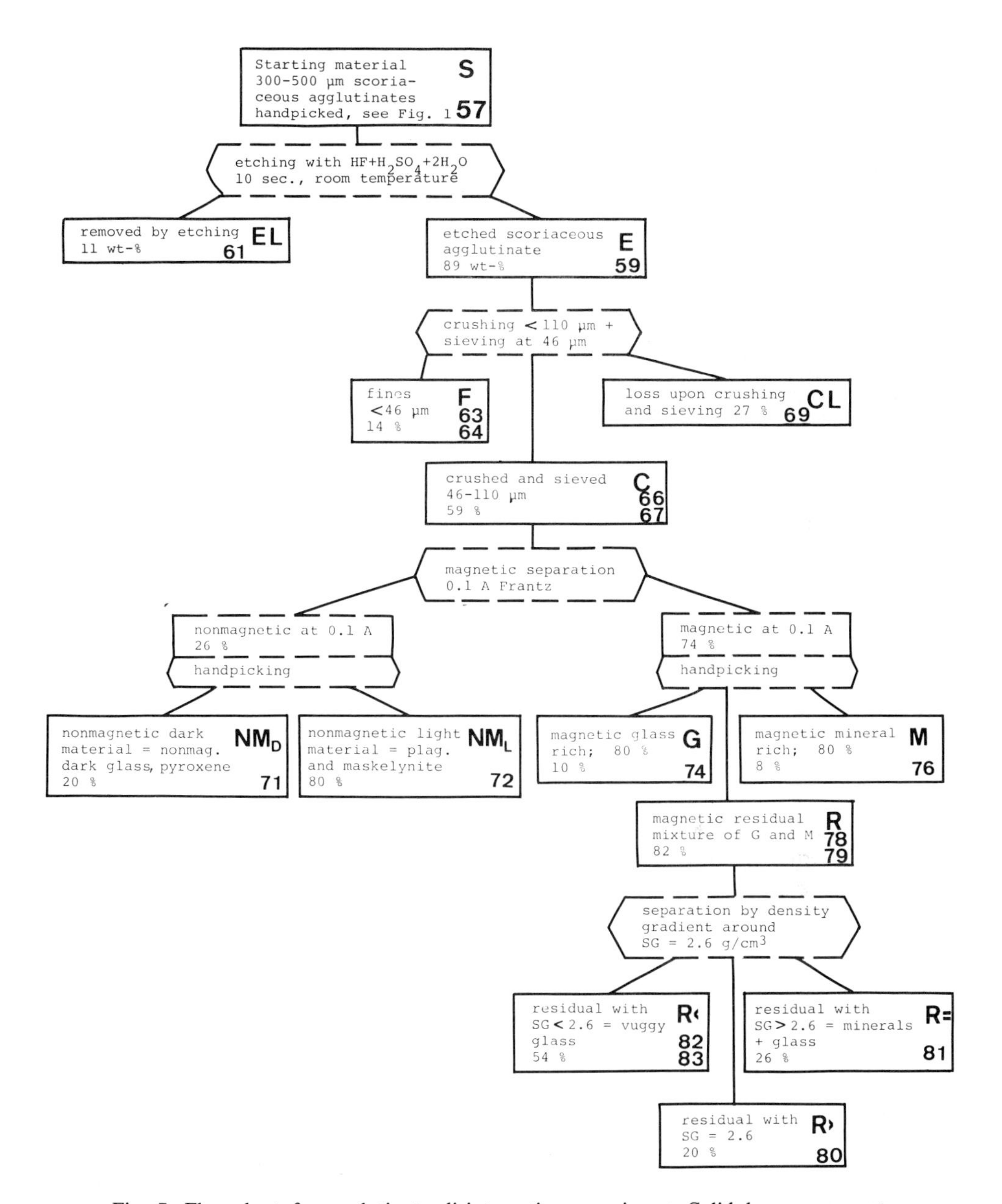

Fig. 7. Flow-chart for agglutinate disintegration experiment. Solid boxes represent samples, material removed by etching and material losses, respectively. Dashed boxes refer to operational steps. Letters in bold print refer to respective letters in Fig. 8. Percentages given are wt.% and correspond to the related parent sample. Numbers in the lower right corner of sample boxes are running numbers from Table 1.

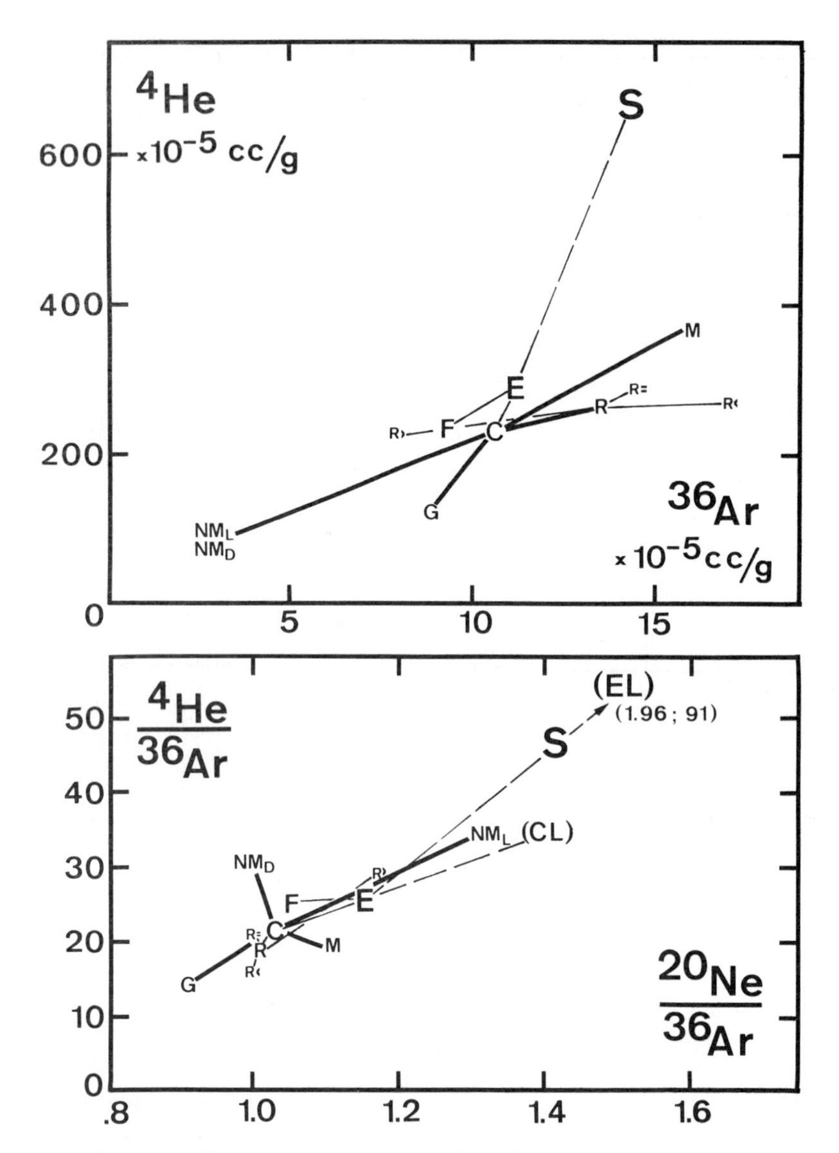

Fig. 8. ^{4}He versus ^{36}Ar and ^{4}He/^{36}Ar versus ^{20}Ne/^{36}Ar in a 300–500 μm scoriaceous agglutinate sample and all the separates derived therefrom in the disintegration experiment. Symbols as in Fig. 7. Increasing number of steps for the preparation of the samples is indicated by decreasing size of symbol. Symbols in parenthesis represent calculated values.

are far from being distributed homogeneously. Gas concentrations as well as element and isotope ratios vary substantially. Qualitatively our results are in agreement with those of Bogard *et al.* (1978). The quantitative differences can be explained by the fact that these authors did not etch their samples before crushing and separating.

Unfortunately we have not been able to identify an agglutinate constituent the gas content of which reflects the agglutinate history better than does the gas content of the whole agglutinate.

4.2 Isotope abundances

The discussion of the elemental patterns for the light noble gases in various types of agglutinate samples has led us to introduce the concepts of migrated and redistributed gases. In this section, we will examine the consequences therefrom in interpreting isotope abundances.

4.2.1 Neon. In the $^{20}Ne/^{22}Ne$ vs. $^{21}Ne/^{22}Ne$ correlation diagram shown in Fig. 9 the data points for the grain size suite of agglutinitic material define a "best fit line" labeled *a*. The point for 300–500 μm fraction (labelled 9) has been omitted (see section 2.). The line *a* is considerably flatter than the best fit lines (labelled *b* and *c*) for a plagioclase suite (Frick *et al.*, 1975) and a mare bulk soil (Pepin *et al.*, 1970). Line *e* connects data points for etched samples with points which represent isotope abundances of the gas in the material lost upon etching. We note that the $^{20}Ne/^{22}Ne$ ratios in the etched samples are systematically lower than those of the agglutinitic material and do not lie on the line *a*. Correspondingly, the calculated data points representing the neon isotopic ratios in the surficial material removed by etching are above line *a*. These latter data points have considerable uncertainties (see section 3). With an assumed spallogenic $^{20}Ne/^{22}Ne$ ratio of 0.9, the line *a* gives a spallogenic $^{21}Ne/^{22}Ne$ ratio well above the range 0.6–0.9 commonly observed (cf. Frick *et al.*, 1975; Schultz *et al.*, 1977, 1978).

Figure 10 presents the same data in the $^{20}Ne/^{22}Ne$ vs. $1/^{22}Ne$ ordinate intercept plot. In this plot the importance of spallogenic Ne is reduced because ^{21}Ne—with its prominent spallogenic portion—does not occur. Again the line *a* fitted to the data points of the untreated samples (without the 300–500 μm separate) does not coincide with the line *e* defined by the data points from "etched residuals" and "etch loss" samples. The large deviations of the data points of the "etched residual" from the line *a* are not due to experimental uncertainties. If one extrapolates line *a* to a $^{20}Ne/^{22}Ne$ ratio of 0.9, one obtains a concentration of spallogenic ^{21}Ne of $(24 \pm 5) \times 10^{-8} cm^3/g$. This value is much smaller than the $(45 \pm 7) \times 10^{-8} cm^3/g$ which, for agglutinitic material, is the average spallogenic gas concentration computed with the assumptions and the formalism given by Signer *et al.* (1977a).

The data for the untreated samples from the grain size suite, as can be seen in Figs. 9 and 10, show patterns which conform with the conventional two-component model (Eberhardt *et al.*, 1970). However, the neon revealed by our etching experiment manifests the presence of at least a third component. Therefore, besides the spallogenic neon, we consider trapped and redistributed

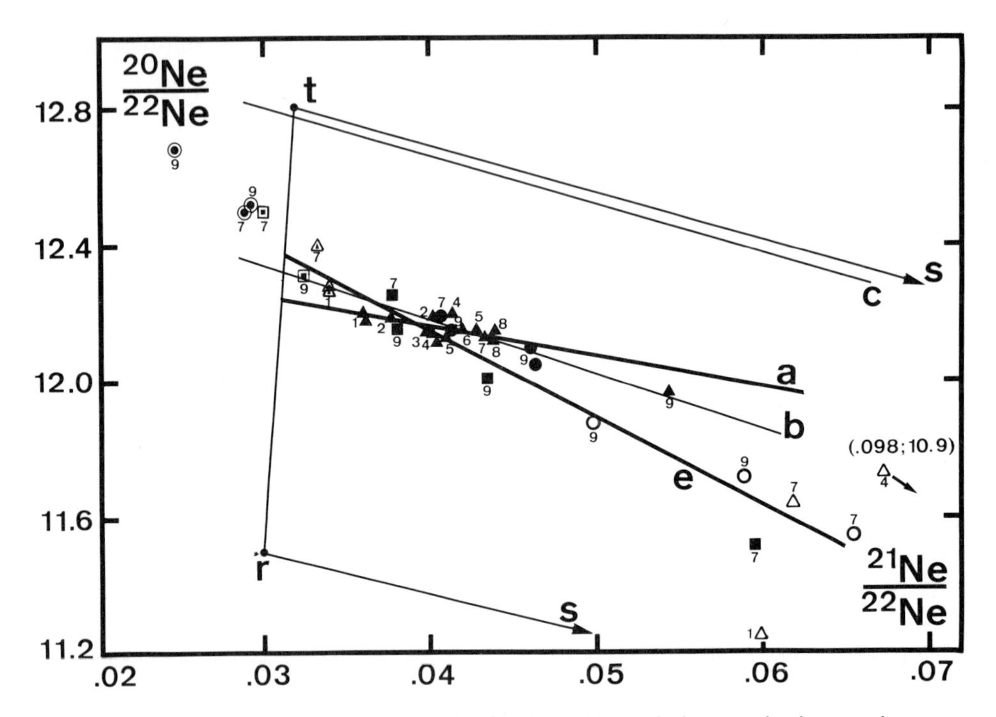

Fig. 9. Ne three-isotope correlation plot for the 61501 agglutinate grain size samples and some etched aliquots. Symbols as in Fig. 4. Numbers at the symbols indicate grain size (in μm): (1) 25–42, (2) 42–64, (3) 64–80, (4) 80–100, (5) 100–125, (6) 125–150, (7) 150–200, (8) 200–300, (9) 300–500. *a*: best fit line for data from unetched samples (data point of 300–500 μm separate omitted, see text). *e*: best fit line for data from etched samples and for computed values corresponding to the gas lost upon etching. *b*: line given by Frick *et al.* (1975) for 15421 plagioclase suite. *c*: "Rock line" given by Pepin *et al.* (1970). *t*, *s*, and *r* denote the points representing trapped, spallogenic, and redistributed pure gas components. Numerical values for points *t*, *r*, and *s*: $(^{20}Ne/^{22}Ne)_t = 12.8$ (Eberhardt *et al.*, 1972), $(^{20}Ne/^{22}Ne)_r = 11.5$ (chosen as maximum value so that all data points lie in the "mixing area"). $(^{21}Ne/^{22}Ne)_{r,t}$ values are chosen such that their fractionation per mass unit relative to solar wind composition is the same as for the respective $^{20}Ne/^{22}Ne$ ratios. Spallogenic component: $^{20}Ne/^{22}Ne = 0.9$, $^{21}Ne/^{22}Ne = 0.9$.

neon as secondary components derived from solar gas. Migrated gas is not taken into account for the moment.

In a three-isotope correlation plot, superposition of three gas components leads to a straight line, only if special conditions govern the mixing. Since the data points from the untreated agglutinitic material show a well pronounced linear array, we will investigate the mixing conditions prevailing.

For convenience we do not distinguish in this paragraph between component and secondary component and we make the following assumptions about the three components: (1) The spallogenic component (*s*) is volume correlated in all grain size fractions; (2) the surface distributed trapped component (*t*) has constant isotope ratios but variable concentrations; and (3) the volume distributed

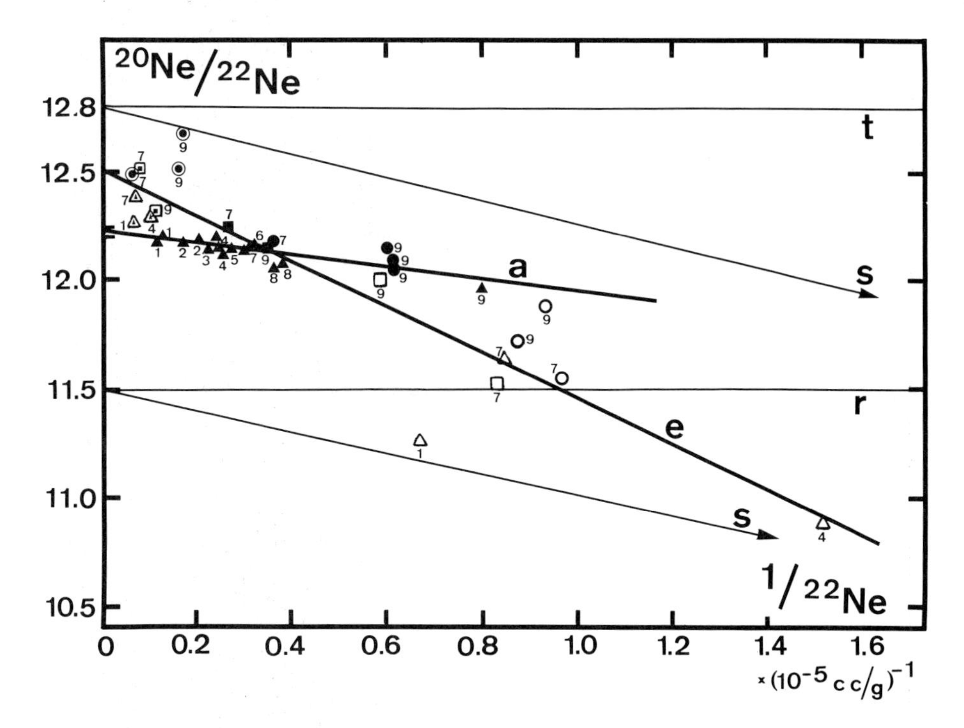

Fig. 10. Ordinate intercept plot $^{20}Ne/^{22}Ne$ versus $1/^{22}Ne$. Symbols as in Fig. 9. *a*: best fit line for the 61501 agglutinate grain size suite (data point of 300–500 μm separate omitted, see text). *e*: best fit line for data from etched samples and for computed values corresponding to the gas lost upon etching. *t* and *r* represent the $^{20}Ne/^{22}Ne$ ratios of the trapped and redistributed gas components; two straight lines point to "spallation component" *s*. *t*, *r*, and *s* are given according to assumptions in caption of Fig. 9.

redistributed component (*r*) has constant isotope ratios, different from those of the component (*t*), and again variable concentrations. For the ordinate intercept plot A/B vs. 1/B, the geometrical construction of a data point is demonstrated in Fig. 11*. The point t_2s represents a mixture of the two components t and s in concentrations corresponding to points t_2 and *s*. This point is given by the intersection of the two straight lines, *l* and *m*.

$$\frac{A}{B} = \frac{B_s}{B}\left[\left(\frac{A}{B}\right)_s - \left(\frac{A}{B}\right)_t\right] + \left(\frac{A}{B}\right)_t \qquad (l)$$

$$\frac{A}{B} = \frac{B_{t2}}{B}\left[\left(\frac{A}{B}\right)_t - \left(\frac{A}{B}\right)_s\right] + \left(\frac{A}{B}\right)_s \qquad (m)$$

To this *t*- and *s*-superposition (t_2s) we admix now gas from the *r*-component with the concentration corresponding to r_2 and obtain point t_2sr_2. Similarly, a mixture of the three components in amounts corresponding to t_1, *s*, and r_2 gives point (t_1sr_2). Any mixture of the three components *r*, *s*, *t* with abscissa values between t_1 to t_2 and r_1 to r_2, respectively, leads to a point in the shaded field. Mixtures

*Letters t, r, and s are used as indices to denote components and as point labels.

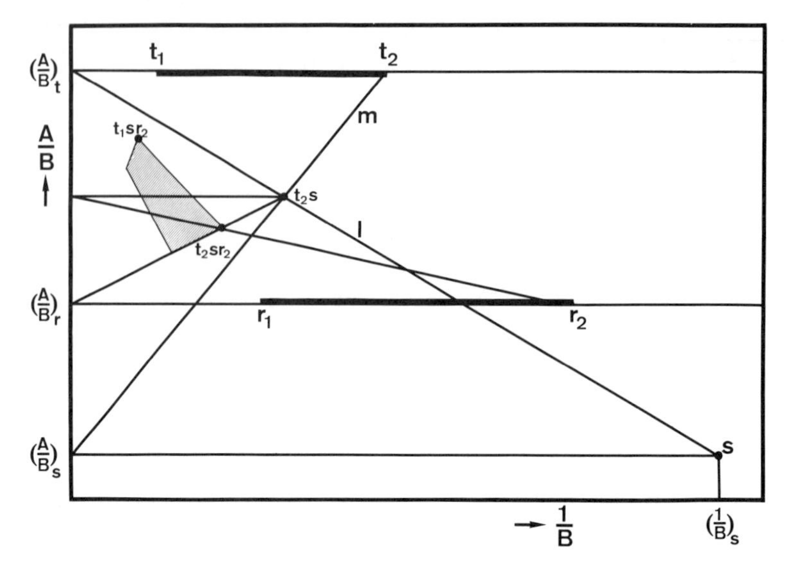

Fig. 11. A schematic ordinate intercept plot A/B versus 1/B. *t*: trapped gas; *s*: spallogenic gas; *r*: redistributed gas. $(A/B)_t$: ratio of the trapped component; $(A/B)_r$: ratio of the redistributed component, and $(A/B)_s$: ratio of the spallogenic component. These isotope ratios are indicated by horizontal lines. The inverse concentrations of isotope B ranges between t_1 and t_2 for the trapped and r_1 and r_2 for the redistributed component. $(1/B)_s$ is the inverse concentration of the spallogenic portion of isotope B. Straight lines *l* and *m* are the loci of data for mixtures of two components, one with constant and one with varying concentrations. Line *l* has a constant spallogenic component $(1/B)_s$ and variable trapped gas, line *m* a constant trapped (t_2) and variable spallogenic gas. Point t_2s represents a mixture of trapped gas with the inverse concentration t_2 and spallogenic gas (point *s*). Point t_2sr_2 represents a mixture of trapped (t_2), spallogenic (*s*), and redistributed (r_2) gases. Variations in trapped (t_1 to t_2) and redistributed (r_1 to r_2) gas together with spallogenic gas (*s*) lead to data points within the hatched area.

with, from sample to sample smoothly varying concentrations of components *t* and *r*, lead to arrays of points as shown in Fig. 12. We have selected three examples of mixing variations as indicated on the righthand side of Fig. 12, where the logarithm of concentrations of isotope B is plotted as a function of a parameter *p*, for example the logarithm of the grain size. In mixing mode 1, the component *t* decreases with *p* while the component *r* remains constant. In mode 2, *t* decreases and *r* increases with *p*. Finally, mode 3 shows both *t* and *r* decreasing with *p*. The three modes are not unrealistic choices: on the basis of Fig. 5, if we decompose the gases into trapped gas and redistributed gas, we obtain curves resembling those of the modes given. The lefthand side of Fig. 12 shows the data point arrays resulting from the three mixing modes. This part of Fig. 12 is an enlargement of the shaded area in Fig. 11. Mode 1 gives a straight line, extrapolating to ratio $(A/B)_t$. Mode 2 results in a curved line. Finally, mode 3 gives again a straight line which, however, if extrapolated, gives a meaningless spallogenic abundance and a meaningless ordinate intercept.

In applying this model to our data, we choose the following numerical values (Figs. 9 and 10): a) The spallogenic component (*s*) has a ratio $^{20}Ne/^{22}Ne = ^{21}Ne/^{22}Ne = 0.9$ and a constant ^{22}Ne concentration of $45 \times 10^{-8} cm^3 STP/g$. This value is an average spallogenic concentration from all

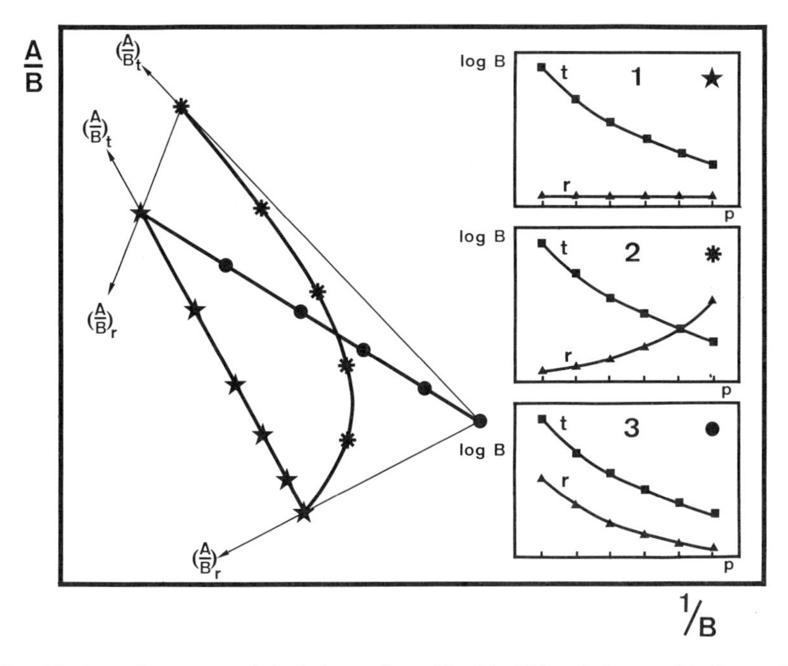

Fig. 12. An enlargement of shaded area from Fig. 11. Abbreviations and indices as in Fig. 11. On the right-hand side three examples for the variation of trapped and redistributed gas of isotope B as a function of a parameter *p* representing e.g., grain size are given. Data points resulting from the different mixing modes are plotted on the left-hand side with the symbols as indicated.

our samples and was calculated according to the assumptions and the formalism given in Signer *et al.* (1977a). b) The trapped neon (*t*), a secondary component, has $^{20}Ne/^{22}Ne = 12.8$. This value was obtained on the one hand for grain size suites of ilmenites (Eberhardt *et al.*, 1972) and on the other hand by extrapolation of the line connecting the data points of the 300–500 μm, etched and unetched scoriaceous agglutinates. For the largest grain size fraction the redistributed gas can be closely approximated by the gas remaining after etching. Here the contribution of migrated gas is the smallest. Consequently the computed gas amounts lost upon etching represent the best approximation to the trapped gas. In Fig. 9 the line connecting *t* and *r* goes through the solar wind composition point (Geiss *et al.*, 1972) and the mass fractionation for $^{20}Ne/^{22}Ne$ is twice as large as that for $^{21}Ne/^{22}Ne = 11.5$. This value was chosen so that all the data points lie in the mixing area.

If the layer removed by etching would contain only all the surficially distributed trapped Ne, all etched samples should fall onto the line connecting *r* and *s*. Consequently, the data points for neon lost upon etching should fall on the line connecting *t* and *s* in Figs. 9 and 10. Obviously, this is not the case. Furthermore, data points plotting to the left of the line connecting points *t* and *r* cannot be explained with the three-component model. We expect that the actual situation is more complex than assumed in the model because (1) migrated Ne is not accounted for, and (2) redistributed Ne as well as spallogenic Ne may not be distinct components.

The model considerations illustrate a very important point: Superposition of three components with certain, smoothly varying mixing ratios can lead to linear arrays of data points, which simulate a two-component system. In such cases, ordinate intercepts and extrapolation to the concentration of the spallogenic component are misleading. Values for trapped and spallogenic Ne for bulk soils deduced by three-isotope and ordinate intercept plots are suspect to be affected by the same artifact. Also the spallogenic neon in individual samples of complex genesis cannot be deduced if only isotopic values for trapped and spallogenic components are assumed.

4.2.2 Argon. From Figs. 9 and 10 we see that the $^{20}Ne/^{22}Ne$ ratios in etched samples plot considerably below the straight line fitted to the data of the untreated agglutinates. For the $^{36}Ar/^{38}Ar$ ratios the analogous effect is expected to be less pronounced, because: a) the relative mass difference between ^{36}Ar and ^{38}Ar is only half that between ^{20}Ne and ^{22}Ne; b) during agglutination argon becomes less depleted than neon; and c) diffusive migration is less pronounced for Ar than for Ne. Therefore, even if migrated gas would be present, it would hardly be detectable, and the isotopic ratios of trapped and redistributed argon are most likely very similar.

In Fig. 13 we present the Ar data in a $^{36}Ar/^{38}Ar$ versus $1/^{38}Ar$ ordinate intercept plot. Line *a* is the best fit for the data points of the grain size suite of agglutinitic material and intersects the ordinate at the value of 5.34. Line *e* is fitted to the points for etched and unetched samples, excluding the points for the 300–500 μm dendritic agglutinates. These latter points define a line with a positive slope and an ordinate intercept of 5.03, features we have no explanation for. Lines *a* and *e* have similar slopes and ordinate intercepts. This implies that we cannot distinguish between the trapped and the redistributed components and it is not clear whether the interpretation of the Ar isotopic data requires a superposition of 3 components. Therefore, we regard the value of 5.34 as the proper $^{36}Ar/^{38}Ar$ ratio of trapped argon. With a spallogenic $^{36}Ar/^{38}Ar$ ratio of 0.63 we compute a ^{38}Ar concentration of $(80 \pm 10) \times 10^{-8} cm^3 STP/g$. The exposure age computed with the production rates and chemical composition given by Signer *et al.* (1977a) is 540 ± 60 m.y. The data points for etched and unetched scoriaceous agglutinates of the largest grain size range (300–500 μm) define a line which extrapolates to an ordinate intercept value above 5.4. A high value for the trapped $^{36}Ar/^{38}Ar$ was also observed for feldspar separates in soil 15421 by Frick *et al.* (1975).

4.2.3 Helium. Interpretation of the helium isotope data is complicated due to diffusional loss of spallogenic ^3He and trapped ^4He and inward migration of trapped ^4He. Furthermore, reworking of the agglutinates affects the ^4He more

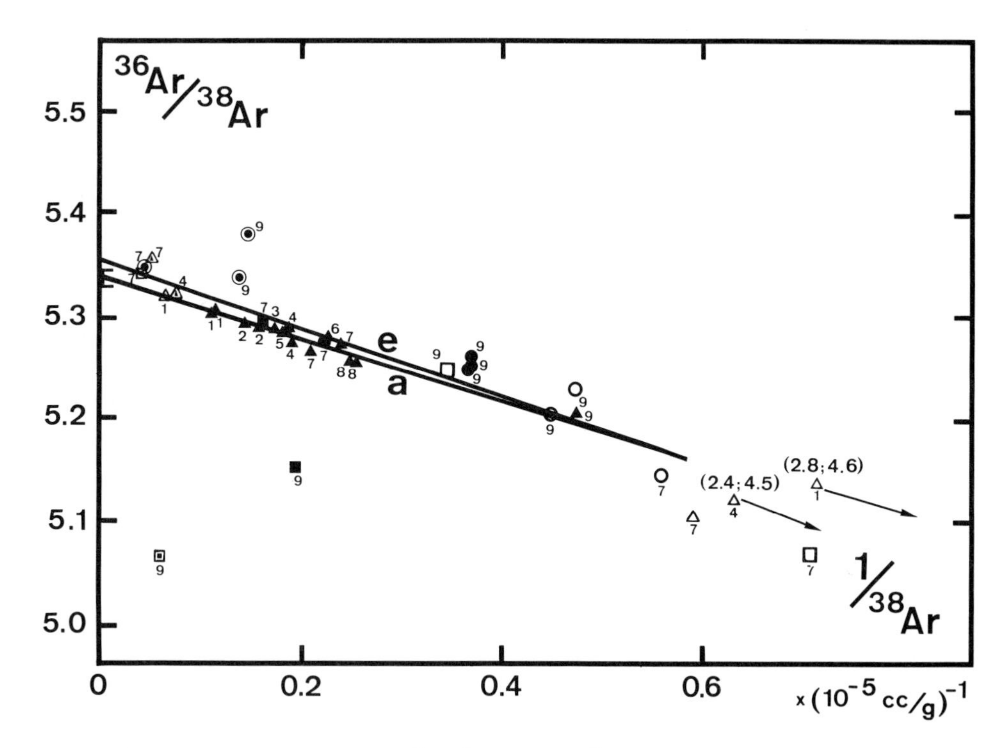

Fig. 13. Ordinate intercept plot $^{36}Ar/^{38}Ar$ versus $1/^{38}Ar$. Symbols as in Fig. 9. Straight line *a* is a best-fit for the 61501 agglutinate grain size suite. Line *e* is a best-fit for data from etched samples and for computed values corresponding to the gas lost upon etching (data for dendritic agglutinates of the grain size range 300–500 μm are omitted).

than Ne and Ar. In feldspars only about 1% of solar ^{4}He is retained and the corresponding $^{4}He/^{36}Ar$ ratios range around 60 (Frick *et al.*, 1975). Since we obtained for our agglutinate samples quite similar values for these ratios, we expect that loss of He from our samples is very high, too. Furthermore, spallogenic ^{3}He may amount to a substantial portion of the total ^{3}He and thus any small differences in diffusion loss of He has large effects on measured $^{4}He/^{3}He$ ratios. Consequently, the large and unsystematic scatter in a $^{3}He/^{4}He$ versus $1/^{3}He$ plot (not given) is not surprising.

5. Summary and Conclusions

In the various *magnetic materials of agglutinitic affinity from soil 61501* He, Ne, and Ar concentrations *are only slightly different*. These variations may be attributed to different degrees of surface coverage by fine grained soil debris which has gas concentrations, element abundances and isotopic compositions similar to those in plagioclase. Furthermore, different surface-to-volume ratios can lead to different contributions by diffused gases. In contrast, the *non-*

magnetic agglutinate particles have low gas concentrations reflecting a lower degree of reworking.

The He, Ne, and Ar distribution within the agglutinates is quite heterogeneous with respect to concentrations, element abundances and isotopic compositions. The mineral-rich phase has high gas concentrations and a $^{4}He/^{36}Ar$ ratio which is larger than that of the glass-rich phase. In the mineral-rich phase, the $^{4}He/^{36}Ar$ ratio is about half that of a plagioclase concentrate. In the glass-rich phase the $^{4}He/^{36}Ar$ ratio is only about a fourth of the plagioclase value. This shows that the depletion of ^{4}He with respect to ^{36}Ar in the different agglutinate phases correlates to the intensity of the processes forming the respective agglutinate fragments. Different degree of reworking of agglutinitic material may again be seen from a comparison of gas concentrations and element abundances in light, non-magnetic, and dark, magnetic agglutinate fragments.

A grain size suite of agglutinitic material and etched samples thereof reveal a strong grain size dependence of the volume distributed He and Ar not removed upon etching. The concentrations of the volume distributed Ar decrease with decreasing grain size whereas those of He increase with decreasing grain size. The Ne concentrations detected after etching are similar in large and small grains but are lower in medium sized grains. The different behaviour of He, Ne, and Ar is explained by diffusion of surficially distributed trapped He and Ne into the grain volume. The *interpretation of the gases* in the grain size suite of agglutinitic material *as the sum of grain size independent volume and surface components fails* certainly for Ar, but presumably also for He and Ne. Due to the eminence of agglutinitic material as gas carrier in bulk soils, the above interpretation applied to bulk soil grain size suites also becomes dubious. At present, the significance of the grain size dependent surface component and volume component is not clear; possibly, they are important tell-tales for the soil evolution and regolith dynamics.

The *Ne isotope data* obtained on the grain size suite of agglutinitic material *give linear arrays* of data points in three-isotope and ordinate intercept plots. However, the isotopic compositions of Ne retained in the samples after etching plot considerably below the straight lines fitted to the data of the unetched samples. This is indicative for the superposition of more than two types of neon, all being isotopically different. We show that in a three-component model, mixtures of varying amounts of two of the three components can indeed result in linear data arrays. In such cases, *however*, *ordinate intercept values escape a simple interpretation* and the deduction of isotopic composition of "trapped" Ne and concentration of spallogenic Ne in agglutinitic material—as well as in bulk soils—is not possible.

The *Ar isotopic data* in agglutinates *do not require superposition of three types of Ar*. This is presumably due to the smaller mobility of Ar in comparison with Ne. An interpretation of the *He isotope abundances* in terms of two or three component models *is not possible* because only minor portions of solar and spallogenic He are retained in highland soils.

In agglutinates interpretation of noble gases and presumably other extraneous elements cannot be safely accomplished by simple two-component models. If an understanding of extraneous elements in agglutinates and other complex soil constituents or even in bulk soils is to be achieved, analytical methods capable of spatial resolution have to be applied. On the other hand, it may well be that not the abundant and gas-rich agglutinates but the comparatively rare and gas-poor mineral grains, owing to their simple history, are the *stepping stone* to the understanding of solar gases in the regolith.

Acknowledgments—We thank Dr. R. Schmid for his helpful suggestions for the photometric analyses, and E. Schärli for the skillful preparation of thin-sections. Our colleague H. Baur contributed to this work by forcing us to clarify some of our ideas. We gratefully acknowledge P. Waegli for providing the SEM pictures. U. Frick and J. Hartung provided very helpful and careful reviews. We are grateful to A. Reber for typing this manuscript. We thank NASA and its curatorial staff for generously supplying lunar samples. Financial support by Swiss National Science Foundation Grant No. 2.489-0.75 is appreciated.

REFERENCES

Adams J. B. and McCord T. B. (1973) Vitrification darkening in the lunar highlands and identification of Descartes material of the Apollo site. *Proc. Lunar Sci. Conf. 4th*, p. 163–177.

Basu A. (1977) Biography of an agglutinate. *Meteoritics* **12**, 174–175.

Basu A. and Meinschein W. G. (1976) Agglutinates and carbon accumulation in Apollo 17 lunar soils. *Proc. Lunar Sci. Conf. 7th*, p. 337–349.

Becker R. H. (1977) Does application of the Rosiwal Principle to lunar soils require that concentrations of solar wind-implanted species be grain-size independent? *Earth Planet. Sci. Lett.* **34**, 136–140.

Bogard D. D. (1977) Effects of soil maturation on grain size-dependence of trapped solar gases. *Proc. Lunar Sci. Conf. 8th*, p. 3705–3718.

Bogard D. D., Morris R. V., McKay D. S., Fuhrmann R. and Hirsch W. C. (1978) Relative abundances of fine-grained metal and noble gases in magnetic separates of crushed agglutinate particles (abstract). In *Lunar and Planetary Science IX*, p. 114–116. Lunar and Planetary Institute, Houston.

Bogard D. D. and Nyquist L. E. (1972) Noble gas studies on regolith materials from Apollo 14 and 15. *Proc. Lunar Sci. Conf. 3rd*, p. 1797–1819.

Bogard D. D. and Nyquist L. E. (1973) $^{40}Ar/^{36}Ar$ variations in Apollo 15 and 16 regolith. *Proc. Lunar Sci. Conf. 4th*, p. 1975–1985.

Criswell D. R. (1975) The Rosiwal Principle and the regolithic distributions of solar-wind elements. *Proc. Lunar Sci. Conf. 6th*, p. 1967–1987.

Criswell D. R. and Basu A. (1978) Rosiwal Principle and surface exposure of lunar soil grains (abstract). In *Lunar and Planetary Science IX*, p. 197–199. Lunar and Planetary Institute, Houston.

Eberhardt P., Eugster O. and Marti K. (1965) A redetermination of the isotopic composition of atmospheric neon. *Z. Naturforsch.* **20a**, 623–624.

Eberhardt P., Geiss J., Graf H., Grögler N., Krähenbühl U., Schwaller H., Schwarzmüller J. and Stettler A. (1970) Trapped solar wind noble gases, exposure ages and K/Ar-age in Apollo 11 lunar fine material. *Proc. Apollo 11 Lunar Sci. Conf.*, p. 1037–1070.

Eberhardt P., Geiss J., Graf H., Grögler N., Mendia M. D., Mörgeli M., Schwaller H., Stettler A., Krähenbühl U. and von Gunten H. R. (1972) Trapped solar wind noble gases in Apollo 12 lunar fines 12001 and Apollo 11 breccia 10046. *Proc. Lunar Sci. Conf. 3rd*, p. 1821–1856.

Etique Ph., Baur H., Derksen U., Funk H., Horn P., Signer P. and Wieler R. (1978) Light noble gases in agglutinates: A record of their evolution? (abstract). In *Lunar and Planetary Science IX*, p. 297–299. Lunar and Planetary Institute, Houston.

Frick U., Baur H., Ducati H., Funk H., Phinney D. and Signer P. (1975) On the origin of helium, neon, and argon isotopes in sieved mineral separates from an Apollo 15 soil. *Proc. Lunar Sci. Conf. 6th*, p. 2097–2129.

Funkhouser J., Jessberger E., Müller O. and Zähringer J. (1971) Active and inert gases in Apollo 12 and Apollo 11 samples released by crushing at room temperature and by heating at low temperatures. *Proc. Lunar Sci. Conf. 2nd*, p. 1381–1396.

Geiss J., Bühler F., Cerutti H., Eberhardt P. and Filleux Ch. (1972) Solar wind composition experiment. In *Apollo 16 Preliminary Science Report*, p. 14-1 to 14-10. NASA SP-315, Washington, D.C.

Gopalan K., Goswami J. N., Rao M. N., Suthar K. M. and Venkatesan T. R. (1977) Solar cosmic ray produced noble gases and tracks in lunar fines 10084 and 14163. *Proc. Lunar Sci. Conf. 8th*, p. 793–811.

Hintenberger H., Weber H. W., Voshage H., Wänke H., Begemann F. and Wlotzka F. (1970) Concentration and isotopic abundances of the rare gases, hydrogen and nitrogen in Apollo 11 lunar matter. *Proc. Apollo 11 Lunar Sci. Conf.*, p. 1269–1282.

Housley R. M., Grant R. W. and Paton N. E. (1973) Origin and characteristics of excess Fe metal in lunar glass welded aggregates. *Proc. Lunar Sci. Conf. 4th*, p. 2737–2749.

Hübner W., Kirsten T. and Kiko J. (1975) Rare gases in Apollo 17 soils with emphasis on analysis of size and mineral fractions of soil 74241. *Proc. Lunar Sci. Conf. 6th*, p. 2009–2026.

Huneke J. C. (1973) Diffusive fractionation of surface implanted gases. *Earth Planet. Sci. Lett.* **21**, 35–44.

Kirsten T., Müller O., Steinbrunn F. and Zähringer J. (1970) Study of distribution and variations of rare gases in lunar material by a microprobe technique. *Proc. Apollo 11 Lunar Sci. Conf.*, p. 1331–1343.

Leich D. A., Niemeyer S., Rajan R. S. and Srinivasan B. (1975) Rare gases in etched 10084 ilmenite: A search for trapped solar-flare rare gases. *Proc. Lunar Sci. Conf. 6th*, p. 2085–2096.

Mendell W. W. and McKay D. S. (1975) A lunar soil evolution model. *The Moon* **13**, 285–292.

Morris R. V. (1976) Surface exposure indices of lunar soils: A comparative FMR study. *Proc. Lunar Sci. Conf. 7th*, p. 315–335.

Morris R. V. (1978) The maturity of lunar soils: concepts and more values of Is/FeO (abstract). In *Lunar and Planetary Science IX*, p. 760–762. Lunar and Planetary Institute, Houston.

Nier A. O. (1950) A redetermination of the relative abundances of the isotopes of carbon, nitrogen, oxygen, argon, and potassium. *Phys. Rev.* **77**, 789–793.

Nyquist L. E., Funk H., Schultz L. and Signer P. (1973) He, Ne, and Ar in chondritic NiFe as irradiation hardness sensors. *Geochim. Cosmochim. Acta* **37**, 1655–1685.

Pepin R. O., Nyquist L. E., Phinney D. and Black D. C. (1970) Rare gases in Apollo 11 lunar material. *Proc. Apollo 11 Lunar Sci. Conf.* p. 1435–1454.

Pillinger C. T., Jull A. J. T., Woodcock M. R. and Stephenson A. (1978) Maturation of the lunar regolith: Some implications from magnetic measurements and hydrolysable carbon data on bulk soils and particle separates from 12023 and 15601. *Proc. Lunar Planet. Sci. Conf. 9th.* This volume.

Rhodes J. M., Adams J. B., Blanchard D. P., Charette M. P., Rodgers K. V., Jacobs J. W., Brannon J. C. and Haskin L. A. (1975) Chemistry of agglutinate fractions in lunar soils. *Proc. Lunar Sci. Conf. 6th*, p. 2291–2307.

Schultz L., Weber H. W., Spettel B., Hintenberger H. and Wänke H. (1977) Noble gas and element distribution in agglutinate and bulk grain size fractions of soil 15601. *Proc. Lunar Sci. Conf. 8th*, p. 2799–2815.

Schultz L., Weber H. W., Spettel B., Hintenberger H. and Wänke H. (1978) Noble gas and element distribution in agglutinates of different densities (abstract). In *Lunar and Planetary Science IX*, p. 1021–1023. Lunar and Planetary Institute, Houston.

Signer P., Baur H., Derksen U., Etique Ph., Funk H., Horn P. and Wieler R. (1977a) Helium, neon, and argon records of lunar soil evolution. *Proc. Lunar Sci. Conf. 8th*, p. 3657–3684.

Signer P., Baur H., Derksen U., Etique Ph., Funk H., Horn P. and Wieler R. (1977b) Admixtures of fresh material, agglutination, and "reworking" as reflected in the noble gas record of lunar soil constituents. *Meteoritics* **12**, 362–363.

Smith S. P., Huneke J. C. and Wasserburg G. J. (1978) Neon in gas-rich samples of the carbonaceous chondrites Mokoia, Murchison, and Cold Bokkeveld. *Earth Planet. Sci. Lett.* **39**, 1–13.

Proc. Lunar Planet. Sci. Conf. 9th (1978), p. 2269–2286.
Printed in the United States of America

Trace and major elements in grain-size fractions of separated minerals and agglutinates of soil 70160 and implications on their origin

H. R. VON GUNTEN[1,2], U. KRÄHENBÜHL[1], A. GRÜTTER[2], D. JOST[1], G. MEYER[1], K. PENG[2], F. WEGMÜLLER[1], and A. WYTTENBACH[2]

[1]Anorganisch-chemisches Institut, Universität Bern
CH-3012 Bern, Switzerland
[2]Eidgenössisches Institut für Reaktorforschung
CH-5303 Würenlingen, Switzerland

Abstract—Apollo 17 soil 70160 was separated into 7 grain-size fractions with mean diameters between 3.6 and 450 μm. The fractions >7 μm were further partitioned into mineral-rich and agglutinate-rich portions using a magnetic separator. Five major and more than 20 volatile and non-volatile trace elements were measured in each fraction using INAA and RNAA techniques.

The concentrations of the volatile trace elements in the agglutinate-rich fraction are always higher (up to a factor of 10) and are predominantly volume correlated. The grain-size dependency of the volatiles in the mineral-rich fraction is much more pronounced. In some cases a surface correlation is approached.

Major elements and non-volatile trace elements are about equally distributed among the agglutinates and minerals and do not show a systematic trend in their size dependency. Variations in the concentrations may be due to changes in the mineralogical composition of different size fractions.

The enrichments of siderophilic and chalcophilic trace elements in the regolith are due to meteoritic (Au, Co, Ge, Te) and volcanic (Cd, Zn) contributions. Mare basalts represent the main source for the lithophilic elements in the regolith.

INTRODUCTION

Considerable variations in the concentrations of volatile elements are found on the moon among different landing sites and various sample types: lunar soils are generally enriched in volatile elements, breccias and igneous rocks are generally depleted. The higher concentrations of volatiles in the regolith soil suggest that the enrichment took place after the formation of the lunar crust. The present work is a continuation of an investigation which is performed to elucidate further on processes and mechanisms leading to the observed distribution of volatile trace elements on the moon.

In our previous work (Giovanoli *et al.* 1977; Krähenbühl *et al.* 1977 and 1977a; von Gunten *et al.* 1978) and in the investigations of the group of J. T. Wasson (Boynton *et al.* 1976 and 1976a; Wasson *et al.* 1976, Boynton and Wasson 1977) it was demonstrated that volatile trace elements (e.g., Au, Cd, Ga, Ge, Hg, In, Sb, Te, and Zn) are enriched in the finer fractions of lunar soils. This suggests a surface relation of the concentration of these elements and hence their transport through the lunar "atmosphere" with a subsequent deposition on surfaces. Non-volatile trace elements do not show these grain-size dependent changes of their concentrations.

Krähenbühl *et al.* (1977a) found deviations from a pure surface relation of the concentrations of volatile elements and, therefore, postulated that agglutinates may play an important role for the distribution of these elements among different size fractions of lunar soils. A mature soil with a relatively high content of agglutinates seemed most suitable for the investigation of this postulate. Sample 70160 was selected since it fulfilled these criteria quite well and since this surface-sample can further be used as reference for the drill core samples 70004, 70007, and 70009 which will be examined later on in our work.

Sieve fractions of sample 70160 were, therefore, separated into mineral-rich and agglutinate-rich portions using a magnetic separation. From the distribution of different elements among these enriched fractions of the regolith, possible mechanisms and sources are proposed which could have produced the observed enrichments of volatile elements in lunar soils.

EXPERIMENTAL

Samples

Two samples, 70160 and 70161, were used in this investigation. The fillet sample 70160 was recovered about 12 m beside the landing site of the LM. The sample consists of unsieved fines of the mare type, but contains also highland material. It belongs to the class of mature soils (McKay *et al.*, 1974; and Morris 1976). An unsieved sample was used in order to minimize the contamination of trace elements during the sieving operation. The mineralogical composition of 70160 was determined by Heiken and McKay (1974) for the size fraction 90–150 μm: agglutinates 34%, basalt 15%, breccias 7%, plagioclase 9%, pyroxene 22%, ilmenite 5%, black glass 5%, orange glass 2%, and other glasses 1%.

Sample 70161 is a sieved (<1 mm) portion of 70160. The two samples are expected to be chemically equivalent for the major elements, with the possible exception of the size fraction >149 μm. 70161 may further be somewhat more homogeneous due to the sieving operation.

Separation into grain-size fractions, minerals, and agglutinates

Seven size fractions were obtained by wet sieving in acetone using polypropylene and nylon sieves of 7, 15, 30, 45, 74, and 149 μm mesh-width. The mean diameters of the fractions thus obtained were 3.6, 11, 22, 36, 60, 110, and 450 μm. The sieving procedure is described in detail by Meyer *et al.* (1978).

The size fractions >7 μm were further separated applying a magnetic field. The samples were sedimented in acetone in an inclined glass tube and passed between the modified poles of a magnet from a Varian magnetic balance. The agglutinate-rich particles were held back by the magnet, whereas the mineral-rich fraction sedimented. The quality of the separations was checked microscopically and was comparable to that obtained in conventional "Frantz" separators (current ~0.3 A, dia of glass tube 1.5 cm, compared size range 60–200 μm) (Horn, pers. comm., 1977). Good reproducibilities in the separations were obtained in tests with samples 75080 and 75081. Such a separation does not lead to pure fractions of agglutinates and minerals (e.g., Via and Taylor 1976; Hu and Taylor 1977). The size fraction $\leq$7 μm was not subjected to the magnetic treatment since its rate of sedimentation was too low.

Determination of the main elements and of the mineralogical composition

An aliquot (~100 mg) of sample 70161 was separated into the 7 grain-size fractions and was subjected to the magnetic treatment. The 14 samples were irradiated with thermal neutrons and were analyzed for the major elements Al, Ca, Mn, Na, Mg, and Ti. The detailed procedure is given by Krähenbühl *et al.* (1977a). Powder diffraction patterns were obtained with a 114.6 mm Gandolfi and a Guinier-de Wolff camera using FeK_α radiation.

Determination of the trace elements

412 mg of sample 70160 were consumed for the determination of the trace elements. The sieving of the sample and the magnetic separations were performed after irradiation with thermal neutrons. The risk of contamination was thus reduced to a minimum. The irradiation and the analytical procedures were carried out according to Krähenbühl *et al.* (1977a). In addition to our earlier work Au was also determined.

Results

Major elements and non-volatile trace elements

The results for the partition of the major elements Al, Ca, Cr, Fe, Mg, Mn, Na, and Ti among the agglutinate-rich and the mineral-rich fractions of 70161 are shown in Table 1. Table 2 presents the results for the partition of the rare earth elements Ce, Eu, La, Sm, Tb, and Yb between the mineral-rich and the agglutinate-rich fractions of sample 70160. Also given are the weight distributions in the sieve fractions and the portion of agglutinate-rich and mineral-rich material in each fraction.

The major elements and the rare earths are in general about equally distributed (within 2 σ errors) among the agglutinate-rich and the mineral-rich fractions. However, considerable variations in the concentrations of most of the elements of up to ~30% are observed in the grain-size distributions of both fractions. Even larger deviations from this distribution pattern exist for La at 450 μm, for Ce at 11 μm, and for Al and Ti. The latter two elements increase and decrease by a factor of about 1.5, respectively, from larger to finer grain-sizes. All these variations are very probably due to changes in the mineralogical composition. Indeed, similar changes as for Al and Ti are observed from x-ray powder diffraction patterns for the ratio of plagioclase/ilmenite.

The distribution pattern for the trace elements Co, Eu, Hf, Sc, Ta, and V in the agglutinate-rich and in the mineral-rich size fractions are shown in Fig. 1. Hf, Sc, Ta, and V (Eu as an example from Table 2) are again about equally partitioned (within 2 σ errors) between agglutinate-rich and mineral-rich fractions. The variations in the concentrations of the investigated elements between different grain-sizes are of the order of 30%. A significant deviation from this behaviour is observed for Co. Its concentration in the agglutinate-rich fraction is more than twice as high as in the mineral-rich fraction with the exception of the point at 110 μm. Irregularities at ~110 μm are also observed for Ce, La, and Te (see Table 2 and Fig. 2).

Table 1. Partition of major elements between agglutinate-rich (A) and mineral-rich (M) fractions of grain-size separates of sample 70161. Errors (1 σ) are 3-7%.

Mean grain-size	Fe%		Al%		Ca%		Mg%		Ti%		Na%		Mn%		Cr%	
μm	M	A	M	A	M	A	M	A	M	A	M	A	M	A	M	A
450	10.7	13.9	4.90	5.85	7.65	6.35	4.30	4.60	5.95	6.50	0.26	0.30	0.18	0.18	0.39	0.48
110	12.4	12.3	4.85	6.20	6.90	7.30	4.70	4.65	5.85	5.45	0.29	0.28	0.19	0.17	0.41	0.41
60	12.3	12.7	5.05	7.05	7.10	6.95	4.50	4.55	5.85	4.40	0.25	0.29	0.18	0.16	0.39	0.42
36	12.8	12.9	5.95	7.00	7.30	7.30	5.00	4.70	5.85	4.10	0.28	0.28	0.20	0.15	0.41	0.41
22	11.4	13.6	6.40	7.00	6.85	7.80	4.25	4.40	4.85	5.00	0.30	0.28	0.16	0.15	0.39	0.43
11	10.3	13.6	7.40	6.70	6.90	5.75	4.40	3.85	3.60	4.65	0.35	0.29	0.14	0.13	0.38	0.44
3.6	9.9		8.40		7.05		4.00		3.80		0.33		0.12		0.35	
Mean concentration	11.78	13.21	5.37	6.56	7.21	6.86	4.53	4.47	5.66	5.15	0.27	0.29	0.18	0.16	0.40	0.43

Table 2. Weight distribution of grain-size fractions of sample 70160 and results obtained by INAA of rare earth elements in mineral-rich (M) and agglutinate-rich (A) fractions. Errors (1 σ) are $\simeq$ 10% for La, Yb, Ce, and Sm, 15% for Tb, and 20% for Eu.

Mean grain-size	Weight fractions			La(ppm)		Ce(ppm)		Sm(ppm)		Eu(ppm)		Tb(ppm)		Yb(ppm)	
μm	%[a]	M%	A%	M	A	M	A	M	A	M	A	M	A	M	A
450	19.1	71	29	4.8	11	32	48	6.9	12.	1.8	2.6	1.2	1.5	7.8	10
110	15.2	50	50	4.9	4.9	27	26	7.2	7.1	1.8	2.2	1.2	1.1	7.9	7.7
60	23.6	64	36	6.2	9.1	38	37	7.1	8.8	1.7	1.5	1.1	1.3	7.3	7.3
36	16.8	51	49	6.4	8.8	33	35	7.7	8.5	2.4	1.8	1.2	1.2	7.3	7.7
22	11.9	40	60	7.2	9.2	30	35	7.8	8.8	1.8	1.8	1.0	1.1	6.9	7.6
11	8.4	30	70	9.5	9.9	19	38	9.1	9.0	2.5	2.1	1.3	1.2	6.8	7.6
3.6	5.0			12		74		9.5		2.2		1.1		6.5	
Mean concentration				6.0	8.6	33	36	7.3	8.8	1.9	1.9	1.2	1.2	7.4	7.8

[a] Sample weight 412 mg = 100%.

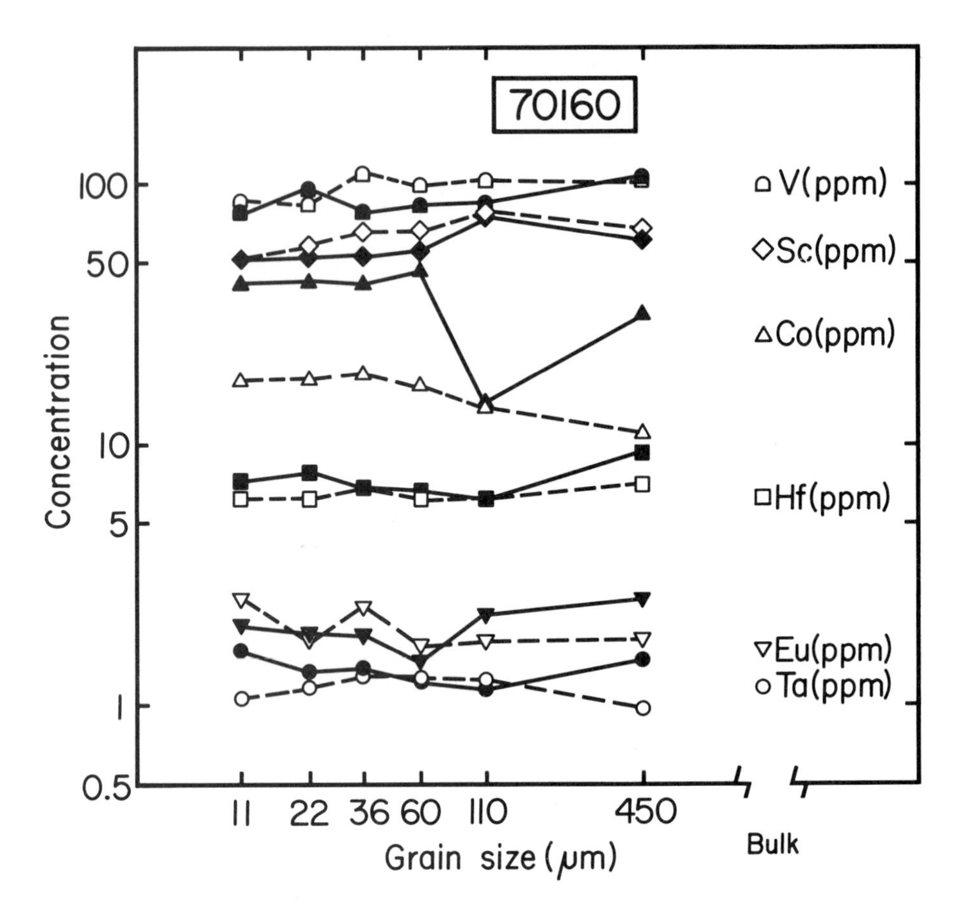

Fig. 1. INAA results for the partition of the trace elements Co, Eu, Hf, Sc, Ta, and V in grain-size fractions of 70160 between an agglutinate-rich (full lines and symbols) and a mineral-rich fraction (dotted lines, open symbols). Note the significant difference for Co between the two fractions (exception at 110 μm). Errors (1 σ) are of the order of 5–15% (20% for Eu).

Volatile trace elements

Figure 2 demonstrates the grain-size distribution of the concentration of the volatile trace elements Au, Cd, Ge, Hg, In, Te, and Zn in sample 70160. The data in Fig. 2 were obtained by weighted averaging the results from the mineral-rich and agglutinate-rich fractions of each element. The distribution patterns (log concentration *vs.* log grain-size) for these volatile trace elements are comparable to the results for other samples (e.g., Krähenbühl *et al.* 1977a), except for In and Hg which are higher. Since we have observed large irregularities for these two elements in earlier investigations, their concentrations should be considered with care and will not be used further.

The partition of the elements Au, Cd, Ge, Te, U, and Zn between the

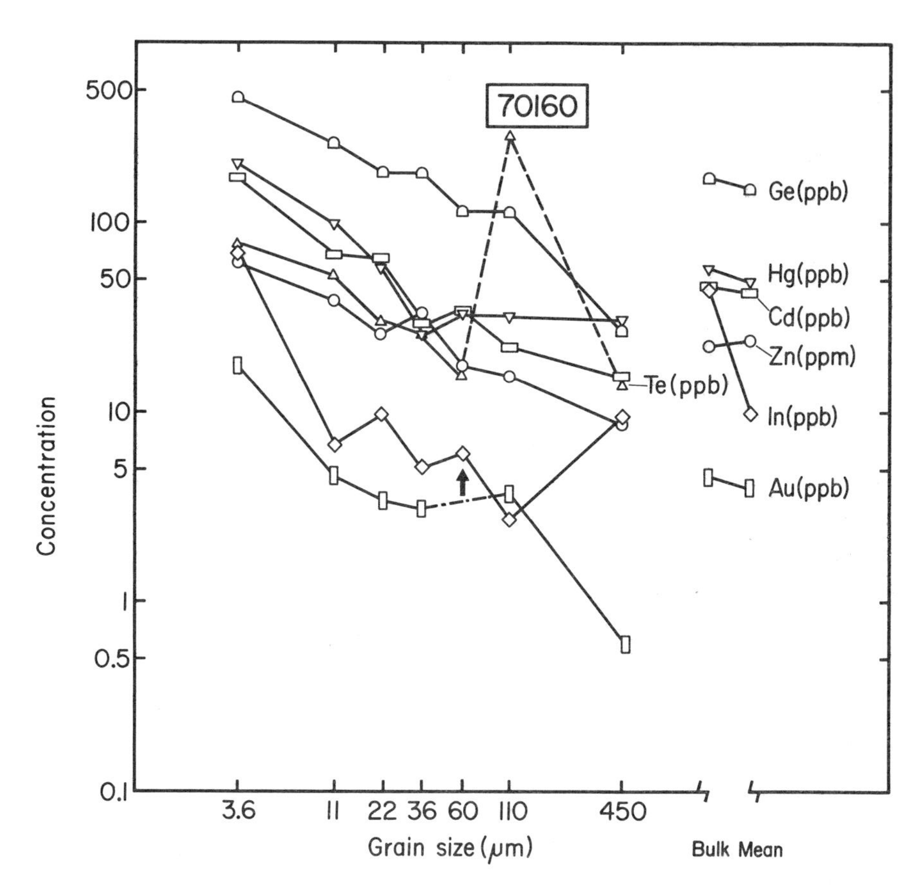

Fig. 2. Distribution of volatile trace elements in grain-size separated samples of 70160. The plots are weighted averages of the results for the mineral-rich and the agglutinate-rich fractions obtained by magnetic separation of the size fractions. See text for discussions of the high values of Te (110 μm) and Au (60 μm). The "bulk"-value resulted from small aliquots of the unseparated samples. The "mean"-value is the weighted average of the individual size fractions. The mean for Au does not include the high value. Errors (1 σ) are of the order of 15–25%.

agglutinate-rich and the mineral-rich fractions is shown in Table 3. The weight distribution of these fractions can be seen in Table 2. These data are plotted in Fig. 3 for the siderophilic elements Au and Ge and for the chalcophilic elements Cd, Te, and Zn. The concentrations of the volatile trace elements are always higher in the agglutinate-rich fraction. Enrichment factors of up to ten are observed between the mineral-rich and agglutinate-rich part.

Much larger fluctuations exist in the concentrations of the investigated elements in the mineral-rich fraction. This could indicate an influence of the observed changes in the mineralogical composition and/or variations in the amount of unseparated material (admixtures of agglutinates).

Table 3. Trace element concentrations determined by RNAA in agglutinate-rich (A) and mineral-rich (M) size fractions of sample 70160. The mean values for Au and Te do not include the high results in parentheses.

Mean grain-size μm	Au (ppb)		Cd (ppb)		Ge (ppb)		Te (ppb)		U (ppb)		Zn (ppm)	
	M	A	M	A	M	A	M	A	M	A	M	A
450	0.25 ± 0.02	1.45 ± 0.15	5.1 ± 0.4	41 ± 3	5.2 ± 0.5	80 ± 7	13 ± 3	17 ± 3	135 ± 8	490 ± 30	3.1 ± 0.2	22.2 ± 1.4
110	0.95 ± 0.10	6.5 ± 0.6	7.9 ± 0.7	36 ± 3	13 ± 1	205 ± 18	20 ± 4	(542)	205 ± 13	295 ± 18	4.5 ± 0.3	26.3 ± 1.6
60	(2730)	4.9 ± 0.5	29 ± 3	42 ± 4	28 ± 3	230 ± 21	4.2 ± 0.8	36 ± 7	150 ± 9	355 ± 22	8.6 ± 0.5	33.6 ± 2.1
36	1.4 ± 0.1	4.8 ± 0.5	13 ± 1	46 ± 4	93 ± 8	280 ± 25	21 ± 4	32 ± 6	145 ± 9	310 ± 20	33.2 ± 2.0	34.1 ± 2.1
22	1.1 ± 0.1	5.0 ± 0.5	35 ± 3	86 ± 7	32 ± 3	280 ± 25	11 ± 2	43 ± 8	305 ± 19	300 ± 19	12.2 ± 0.8	35.3 ± 2.2
11	1.8 ± 0.2	5.8 ± 0.6	96 ± 8	56 ± 5	76 ± 7	345 ± 31	31 ± 6	63 ± 12	315 ± 20	415 ± 26	22.6 ± 1.4	46.0 ± 2.8
3.6	18 ± 2		173 ± 14		455 ± 41		75 ± 15		415 ± 26		61.0 ± 3.7	
Mean concentration	1.03	4.96	21.8	51.3	35.6	245	13.3	36	177	346	12.3	33.2
Mean all fractions	3.96		43		150		34		257		24.0	
Bulk	—		46 ± 4		173 ± 15		52 ± 10		248 ± 15		22.4 ± 1.4	
Bulk 70161	—		38 ± 4		224 ± 25		—		—		32 ± 3	
BCR-1 this work	0.39 ± 0.20		174 ± 20		1290 ± 140		3.6 ± 0.8		1280 ± 80		120 ± 8	

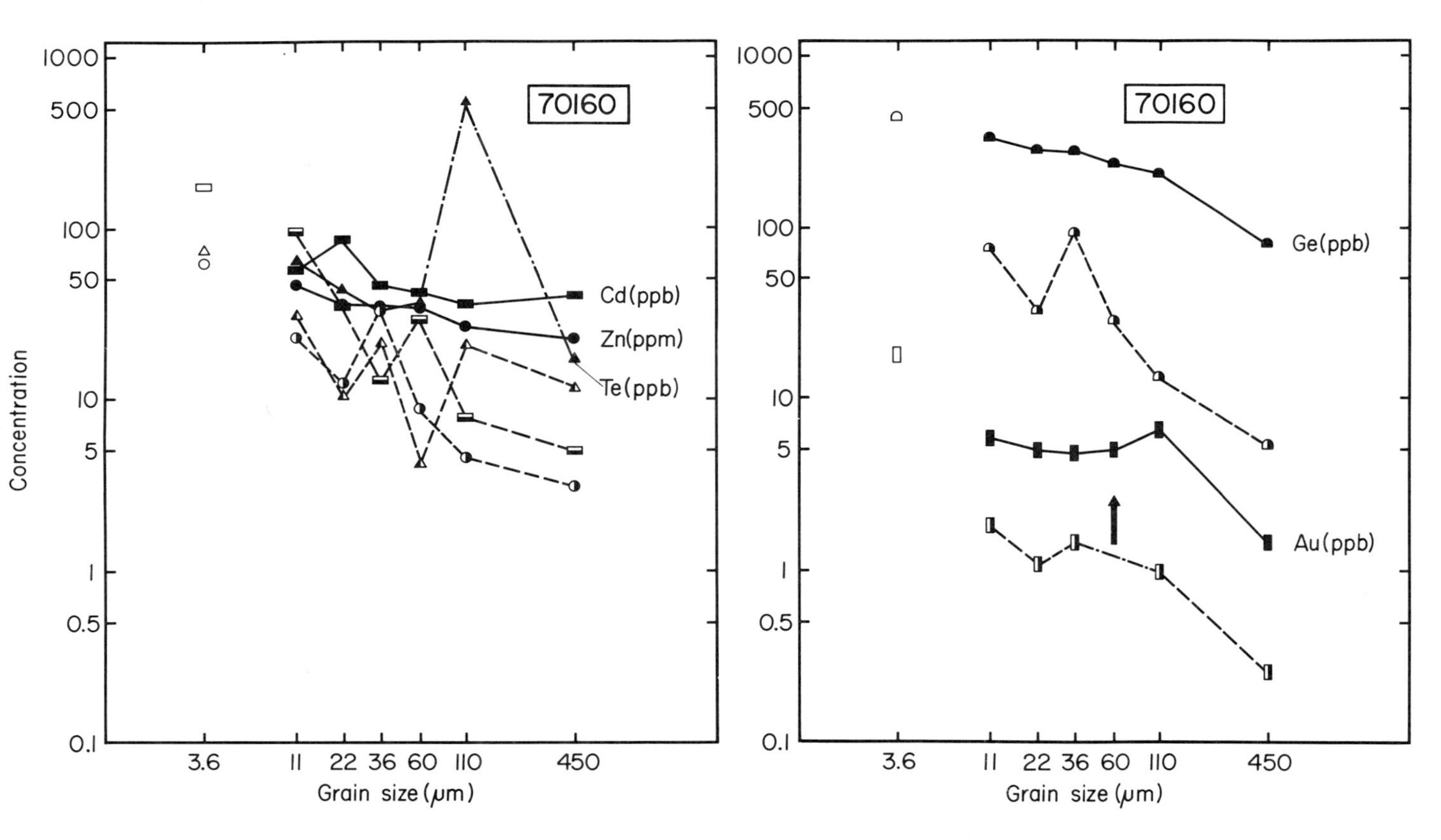

Fig. 3. RNAA results on the partition of the siderophilic trace elements Au and Ge (right-hand picture), and the chalcophilic trace elements Cd, Te, and Zn between agglutinate-rich (full lines and symbols) and mineral-rich (dotted lines, half-filled symbols) size fractions of 70160. The points at 3.6 μm (open symbols) represent the unseparated size fraction ≤ 7 μm. 2700 ppb Au were found in the 60 μm mineral-rich fraction. Errors ($1\ \sigma$) are of the order of 10–15%.

DISCUSSION

Distribution of major elements and non-volatile trace elements

Our results indicating an almost equal partition of major elements and non-volatile trace elements among the mineral-rich and agglutinate-rich fractions (except Co) are in disagreement with measurements of Rhodes *et al.* (1975) on Apollo 11 and Apollo 17 soils. These authors found generally higher concentrations in the agglutinate-rich fraction for many of the elements. The observed variations in their work in the distribution of the measured elements between several samples were, however, quite large. The disagreement between the two investigations could result from a less clean separation in our work: the handling of the highly radioactive material and the very small samples involved made the separation a very difficult task.

Distribution of volatile trace elements

The grain-size correlation of the concentration of the volatile trace elements is much more pronounced in the mineral-rich fraction. If linear least-squares fits are performed with the data points, slopes of ~ -0.6 are obtained for most of the elements. Linear fits are quite arbitrary for the data for some of the elements. The fits for two reasonably justified cases are shown in Fig. 4: slopes of -0.6 and -0.7 are obtained for Zn and Ge, respectively. This suggests a surface correlation of volatile trace elements in the mineral-rich fraction. The deviations from slopes of -1 which would result for a pure surface correlation may still originate from small amounts of agglutinates which were not separated by our technique. Other reasons for deviations from pure surface relations were discussed by Krähenbühl *et al.* (1977a).

The observed slopes of ~ -0.6 to -0.7 for most of the investigated volatile elements in this fraction are close to values reported for solar wind implanted noble gases in separated minerals (Eberhardt *et al.* 1970 and Eugster *et al.* 1975). Solar wind implanted gases ought to be on the surfaces of the minerals. The partition of the volatile trace elements between mineral-rich and agglutinate-rich fractions is also similar to that of these noble gases. It is, therefore, likely that in lunar soils comparable processes have led to the distributions of volatile elements as well as of solar wind implanted noble gases, even though these two classes of elements originated from different sources.

The grain-size/concentration correlations in the agglutinate-rich fractions are much less expressed and almost do not exist for some of the volatile trace elements investigated. The linear least-squares fits (Fig. 4) for Zn and Ge show slopes of -0.2 and -0.4, respectively, thus indicating that volatile trace elements in the agglutinates are predominantly volume distributed.

The slopes for the volatile trace elements in the agglutinate-rich fractions (Figs. 3 and 4) are in agreement with slopes for solar wind noble gases of ~ -0.2 reported by Signer *et al.* (1977) for separated agglutinates. These authors and

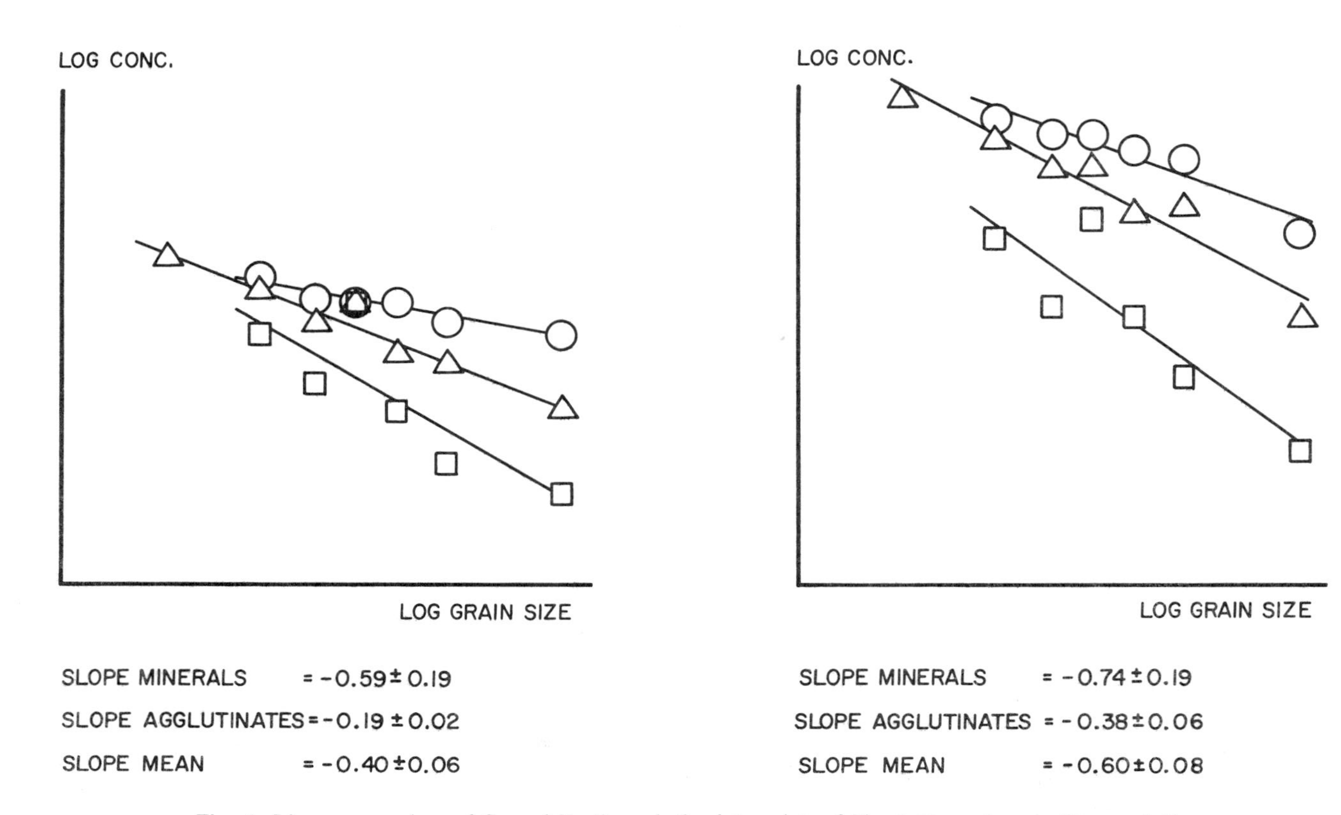

Fig. 4. Linear regressions of Ge and Zn through the data points of Fig. 3. □ = minerals, ○ = agglutinates, △ = mean, weighted averages of the agglutinate-rich and mineral-rich fractions. The data of 3.6 μm were only used for the mean. Experimental errors on both axes were neglected for the regressions.

Signer *et al.* (1977a) suggested that in the process of agglutination, surface related noble gases become volume distributed. A similar mechanism could help to produce the distributions found in our work for volatile trace elements. Upon the impacts of micrometeorites fine grains of regolith are melted together to form agglutinates. Volatile elements on the surfaces of the grains and meteoritic compounds are partly incorporated into the formed agglutinates and partly redistributed among new surfaces. Several reworking sequences with a supply of volatile trace elements from sources other than the regolith are needed to lead to the observed concentrations in the agglutinate-rich fraction. 10–10^2 turnovers of the regolith (Basu, 1977) combined with the proposed reworking mechanism could produce the measured enrichments.

Figure 4 also presents linear fits of the weighted averages of the agglutinate-rich and mineral-rich fractions. The slopes of -0.4 and -0.6 for this averaged fraction are compatible with results for grain-size separated (sieved) samples in our earlier work (e.g., Krähenbühl *et al.*, 1977a), thus demonstrating that agglutinates should have been separated also in these earlier investigations.

Sources for the enrichment of trace elements in the regolith

In Table 4 the results for the volatile and for some nonvolatile trace elements determined in this work are compared with several possible sources for the supply of these materials. The coarsest mineral-rich fraction (>149 μm) in Table 4 represents the sample with the highest volume-to-surface ratio. Mean values for the agglutinate-rich and for the mineral-rich samples were calculated for comparison with the suggested sources for the enrichment of the trace elements. Mechanisms are proposed in the following discussion which could lead to the observed concentrations of the investigated elements.

VOLATILE TRACE ELEMENTS

Gold: The concentrations are much higher in all fractions of sample 70160 than in mare basalt. The excess can reasonably be explained by adding 1.75% of Cl-equivalent material. This meteoritic admixture was calculated for Apollo 17 samples by Baedecker *et al.* (1974) from Au, Ge, Ir, and Ni data. The main amount of Au is directly incorporated into the agglutinates upon the impact of the micrometeorite. The pronounced surface correlation in the mineral-rich fraction indicates volatile Au-species. Partial volatilization of the added meteoritic material leads to a concentration much above mare basalt even in the >149 μm fraction. A very high value for Au (2700 ppb) was found in the 60 μm mineral-rich fraction. Since sample 70160 was recovered at the landing site of the LM, and since many of the installed apparatuses were gold-coated, a contamination seems likely. Indeed, one gold flake of $\sim$60 μm dia and 1 μm thickness would produce the observed effect.

Germanium and Tellurium: The excess Ge and Te can easily be obtained by contributions from meteoritic impacts. In fact this contribution (again based on

Table 4. Comparison of concentrations of trace elements in sample 70160 (this work) with concentrations in several lunar materials and in two types of meteorites. "Coarsest minerals" indicate concentrations found in the >149 μm mineral-rich fraction.

Element	Concentrations in 70160			Concentrations in lunar materials				Concentrations in meteorites	
	Mean agglutinate-rich fractions	Mean mineral-rich fractions	Coarsest minerals (>149 μm)	Mare basalt	KREEP	Anorthosite	Orange glass	C1	Irons
Au ppb	5	1	0.3	0.02[a]	0.05[c]	0.007[k]	1[a]	150[h]	1'400[i]
Cd ppb	51	22	5	1[a]	18[c]	0.57[e]	320[a]	639[h]	50[i]
Ge ppb	245	36	6	2[a]	65[c]	1.2[e]	250[a]	31'000[h]	30'000[i]
Te ppb	36	13	13	2[a]	48[c]	2.1[e]	62[a]	3'000[h]	50[i]
Zn ppm	33	12	3	2[a]	3[c]	0.26[e]	230[a]	300[h]	22[i]
Co ppm	37	16	11	19[b]	27[d]	0.6[f]	63[g]	510[i]	5'000[i]
La ppm	9	6	5	6[b]	93[d]	0.12[f]	7[g]	0.3[i]	10^{-4}[i]
Sc ppm	58	66	67	80[b]	28[d]	0.4[f]	47[g]	5[i]	
U ppb	346	177	136	135[a]	4'900[c]	<0.6[f]	115[a]	9[h]	<0.1[i]

[a]Morgan *et al.* (1974), [b]Rhodes *et al.* (1976), [c]Morgan *et al.* (1973), [d]Schonfeld and Meyer (1972), [e]Ganapathy *et al.* (1976), [f]Wänke *et al.* (1974), [g]Miller *et al.* (1974), [h]Krähenbühl *et al.* (1973), [i]Mason (1971), [k]Ganapathy *et al.* (1974).

1.75% for Cl-equivalent material) should lead to concentrations which are about twice as high as the observed values. Less primitive chondrites contain less Ge and Te than Cl material and would lead about to the measured enrichments. Part of Ge and Te volatilizes upon the impacts; the rest is directly incorporated into the agglutinate melt. The diurnal temperatures on the lunar surface would probably be sufficient to mobilize very volatile Ge- and Te- compounds (e.g., GeS or TeH_2) and condense them in cooler areas of the moon. This process would help to reduce the excess concentrations of the two elements if the meteoritic input consists indeed of Cl material. No obvious reason can be given for the irregularity in the concentration of Te in the 110 μm agglutinate-rich fraction.

Cadmium and Zinc: The enrichments for these two elements in 70160 can only partly (to about 25%) be due to meteoritic inputs. Admixtures from lunar surface material, i.e., orange and other glasses, (Heiken and McKay 1974) could produce the observed concentrations in the mineral-rich fractions if the glass remains predominantly in this fraction in our separation procedure, an assumption which seems somewhat unrealistic. However, we were not able to investigate this fact, since we needed to dissolve the samples soon after the separation due to the decaying radioactivity. The observed concentrations in the agglutinate-rich fractions cannot be obtained by admixtures of glass, especially not if the assumption of an enrichment of glass in the mineral-rich fraction during the separation should be correct. Therefore, we propose, in agreement with Reed *et al.* (1977), that part of the Cd and Zn originated from the interior of the moon and was added during volcanic activities. This proposition is supported by high amounts of Cd found on surfaces of glasses (Chou *et al.* 1975 and Cirlin and Housley 1978) which were probably condensed during magmatic fountaining. Cirlin and Housley (1977) report, furthermore, metal Pb and $PbCl_2$ on surfaces. $PbCl_2$ (cotunnite) is a characteristic compound of fumarolic activities on earth.

NON-VOLATILE TRACE ELEMENTS

Cobalt: The higher concentration of Co in the agglutinate-rich fraction is in qualitative agreement with results of Blanchard *et al.*, (1976) and Rhodes *et al.* (1975) and suggests an addition of Co during the process of agglutination. The enrichment cannot be explained by admixtures from the other fraction and also not by addition of orange or other glasses since their concentrations in this sample are too low (Heiken and McKay, 1974). The observed enrichment is, however, compatible with the amount of Co added to the lunar surface by meteoritic impacts, i.e., 1.5–2% of Cl-equivalent material in the mare formations (Baedecker *et al.* 1974).

The low values of Co in the mineral-rich fractions of Table 4 compared to the contents in mare basalt can be explained by an admixture of anorthositic material.

Lanthanum, Scandium, and Uranium: Sc shows in all fractions concentrations corresponding about to those of mare basalt, taking in account anorthositic contributions. The high value for U in the agglutinates could be produced by a small (~4%) admixture of KREEP. This addition would also explain the increase for La between the coarsest mineral-rich fraction and the agglutinate-rich portion. The other elements in Table 4 are not much influenced by small additions of KREEP. Rhodes *et al.* (1975) proposed a model which involves preferential melting and incorporation of material which is rich in lithophile elements, e.g., K, REE, Th, and U. This model could help to understand the reason for the selective enrichment of U in the agglutinate-rich fraction, since agglutinates undergo a melting process during their formation.

The previous discussion of the individual trace elements proposed three different sources for the observed concentrations in the mineral-rich and agglutinate-rich fractions of lunar soil: a) admixtures of meteoritic material lead to the higher values of the siderophilic elements Au, Ge, and Co. The supply for the excess chalcophilic Te is also from this source. Meteoritic sources for these elements were already proposed by Laul *et al.* (1971) and by Morgan *et al.* (1977). b) The enrichment of the chalcophilic elements Cd and Zn in soil 70160 probably results mainly from volcanic activities. c) The concentrations of the lithophilic trace elements (e.g., La, Sc, U) are comparable to those in mare basalt which is probably the most significant source for the formation of the mare soils. Similar concentrations of the major elements in the soil and in the mare basalt support this hypothesis. However, it should be borne in mind that only the composition of the last mixing episode of the soil under investigation is observed. The composition of the soil in which the agglutinates were formed does not necessarily need to be identical with that of the soil in which they are found.

Chemical form of volatile trace elements

Cd, Te, and Zn are volatile enough to escape in elemental form. Ge may be volatilized as GeS. The formation of Ge-halides seems less probable due to the small concentrations of the elements involved and due to the fact that several halogen atoms are needed to form a molecule. Halides may, however, play an important role in the volatilization of that part of the chalcophilic elements which originates from volcanic activities. The importance of halogen compounds was discussed by Meyer *et al.* (1975) and by Chou *et al.* (1975). Halides are indeed found on surfaces of orange glass and other samples (Allen *et al.* 1975; Cirlin and Housley 1977).

The volatilization of Au is more difficult to understand. Most of the Au compounds are not easily formed and decompose at quite low temperatures. In order to avoid these problems Chou *et al.* (1975) suggested complexes (e.g., AuCl•HCl) which could form under lunar conditions and which are probably volatile. The formation of these proposed compounds seems unlikely since presently only small amounts of halogens are available and would react preferentially with less noble elements. However, if the concentrations of halogens have been much higher in the past the formation of halogen compounds cannot completely be excluded. Nevertheless, we favor a direct volatilization of Au atoms and a subsequent deposition on cooler surfaces. Upon an impact of a micrometeorite its gold would be incorporated into the molten material. The vapor pressures of Au at the melting temperatures of the soils would suffice to produce the observed surface concentrations.

CONCLUSIONS

i) Elements are redistributed during agglutinate formation, and may change from surface-correlated to volume-correlated or vice versa.

ii) Solar wind noble gases and volatile trace elements are similarly partitioned between mineral-rich and agglutinate-rich fractions of lunar soil. Comparable processes may be responsible for their volatilization and distribution.

iii) Volatile trace elements are enriched in the agglutinate-rich fraction. Major elements and non-volatile trace elements are about equally distributed among mineral-rich and agglutinate-rich fractions of the regolith. Co is an exception among the investigated non-volatile trace elements and is enriched in the agglutinate-rich fraction.

iv) The concentrations of volatile trace elements in the mineral-rich and agglutinate-rich fractions, respectively, are predominantly surface- and volume-correlated.

v) The enrichment of the siderophile elements Au, Ge, and Co and of the chalcophile Te in the regolith of Apollo 17 is due to meteoritic admixtures. The meteoritic material leading to the enrichments is probably less primitive than Cl material.

The concentrations of lithophile elements (e.g., La., Sc, U) in Apollo 17 soils are comparable to those found in mare basalt. Admixtures of small amounts of KREEP and anorthosite could be responsible for deviations from these concentrations.

The enrichment of the chalcophile elements Cd and Zn is partly produced by volcanic contributions from the interior of the moon.

Acknowledgments—The assistance of Miss E. Rössler and Mr. R. Keil and valuable discussions with Profs. J. Geiss, N. Grögler, and P. Signer and Drs. P. Baertschi and P. Horn are highly appreciated. The authors thank Miss R. Colombo and Miss E. Jenny for the preparation of the manuscript and Mr. W. Pinter for the drawing of the figures. The helpful comments of the reviewers Drs. D. P. Blanchard, E. H. Cirlin, and C. Meyer are acknowledged. Part of the work was supported by the Swiss National Science Foundation.

REFERENCES

Allen R. O., Jovanovic S. and Reed G. W. (1975) Agglutinates: Role in element and isotope chemistry and inferences regarding volatile-rich rock 66095 and glass 74220. *Proc. Lunar Sci. Conf. 6th*, p. 2271–2279.

Baedecker P. A., Chou C.-L., Sundberg L. L. and Wasson J. T. (1974) Volatile and siderophile trace elements in the soils and rocks of Taurus-Littrow. *Proc. Lunar Sci. Conf. 5th*, p. 1625–1643.

Basu A. (1977) Steady state, exposure age, and growth of agglutinates in lunar soils. *Proc. Lunar Sci. Conf. 8th*, p. 3617–3632.

Blanchard D. P., Jacobs J. W., Brannon J. C. and Brown R. W. (1976) Drive tube 60009: A chemical study of magnetic separates of size fractions from five strata. *Proc. Lunar Sci. Conf. 7th*, p. 281–294.

Boynton W. V., Chou C.-L., Bild R. W., Baedecker P. A. and Wasson J. T. (1976) Element distribution in size fractions of Apollo 16 soils: Evidence for element mobility during regolith processes. *Earth Planet. Sci. Lett.* **29**, 21–33.

Boynton W. V., Robinson K. L., Warren P. H. and Wasson J. T. (1976a) Differences in elemental distributions in the size fractions of mare- and highlands-type soils (abstract). In *Lunar Science VII*, p. 88–90. The Lunar Science Institute, Houston.

Boynton W. V. and Wasson J. T. (1977) Distribution of 28 elements in size fractions of lunar mare and highlands soils. *Geochim. Cosmochim. Acta* **41**, 1073–1082.

Chou C.-L., Boynton W. V., Sundberg L. L. and Wasson J. T. (1975) Volatiles on the surface of Apollo-15 green glass and trace-element distributions among Apollo-15 soils. *Proc. Lunar Sci. Conf. 6th*, p. 1701–1727.

Cirlin E. H. and Housley R. M. (1977) A flameless atomic absorption study of the volatile trace metal lead in lunar samples. *Proc. Lunar Sci. Conf. 8th*, p. 3931–3940.

Cirlin E. H. and Housley R. M. (1978) Flameless atomic absorption studies of volatile trace metals in Apollo 17 samples (abstract). In *Lunar and Planetary Science IX*, p. 169–171. Lunar and Planetary Institute, Houston.

Eberhardt P., Geiss J., Graf H., Grögler N., Krähenbühl U., Schwaller H., Schwarzmüller J. and Stettler A. (1970) Trapped solar wind noble gases, exposure age and K/Ar-age in Apollo 11 lunar fine material. *Proc. Apollo 11 Lunar Sci. Conf.*, p. 1037–1070.

Eugster O., Eberhardt P., Geiss J., Grögler N., Jungck M. and Mörgeli M. (1975) Solar-wind-trapped and cosmic-ray-produced noble gases in Luna 20 soil. *Proc. Lunar Sci. Conf. 6th*, p. 1989–2007.

Ganapathy R., Morgan J. W., Higuchi H., Anders E. and Anderson A. T. (1974) Meteoritic and volatile elements in Apollo 16 rocks and in separated phases from 14306. *Proc. Lunar Sci. Conf. 5th*, p. 1659–1683.

Ganapathy R., Morgan J. W., Krähenbühl U. and Anders E. (1973) Ancient meteoritic components in lunar highland rocks: Clues from trace elements in Apollo 15 and 16 samples. *Proc. Lunar Sci. Conf. 4th*, p. 1239–1261.

Giovanoli R., von Gunten H. R., Krähenbühl U., Meyer G., Wegmüller F., Grütter A. and Wyttenbach A. (1977) Volatile and non-volatile elements in grain-size separated samples of Apollo 17 lunar soils. *Helv. Phys. Acta* **50**, 190–192.

Heiken G. and McKay D. S. (1974) Petrography of Apollo 17 soils. *Proc. Lunar Sci. Conf. 5th*, p. 843–860.

Hu H.-N. and Taylor L. A. (1977) Lack of chemical fractionation in major and minor elements during agglutinate formation. *Proc. Lunar Sci. Conf. 8th*, p. 3645–3656.

Krähenbühl U., Grütter A., von Gunten H. R., Meyer G., Wegmüller F. and Wyttenbach A. (1977) Distribution of volatile and non-volatile elements in grain-size fractions of Apollo 17 lunar soils (abstract). In *Lunar Science VIII*, p. 561–563. The Lunar Science Institute, Houston.

Krähenbühl U., Grütter A., von Gunten H. R., Meyer G., Wegmüller F. and Wyttenbach A. (1977a) Volatile and non-volatile elements in grain-size fractions of Apollo 17 soils 75081, 72461 and 72501. *Proc. Lunar Sci. Conf. 8th*, p. 3901–3916.

Krähenbühl U., Morgan J. W., Ganapathy R. and Anders E. (1973) Abundance of 17 trace elements in carbonaceous chondrites. *Geochim. Cosmochim. Acta* **37**, 1353–1370.

Laul J. C., Morgan J. W., Ganapathy R. and Anders E. (1971) Meteoritic material in lunar samples: Characterization from trace elements. *Proc. Lunar Sci. Conf. 2nd*, p. 1139–1158.

Mason B. (1971) Handbook of elemental abundances in meteorites. Gordon and Breach, N. Y. 555 pp.

McKay D. S., Fruland R. M. and Heiken G. H. (1974) Grain-size and the evolution of lunar soils. *Proc. Lunar Sci. Conf. 5th*, p. 887–906.

Meyer G., Grütter A., von Gunten H. R., Krähenbühl U. and Rössler E. (1978) Apparatus for semi-automatic grain-size separations of small samples. *Rev. Sci. Instrum.* **49**, 326–328.

Meyer C., McKay D. S., Anderson D. H. and Butler P. (1975) The source of sublimates on the Apollo 15 green and Apollo 17 orange glass samples. *Proc. Lunar Sci. Conf. 6th*, p. 1673–1699.

Miller M. D., Pacer R. A., Ma M. S., Hawke B. R., Lookhart G. L. and Ehmann W. D. (1974) Compositional studies of the lunar regolith at the Apollo 17 site. *Proc. Lunar Sci. Conf. 5th*, p. 1079–1086.

Morgan J. W., Ganapathy R., Higuchi H. and Anders E. (1977) Meteoritic material on the moon. In *Soviet-American Conference on Cosmochemistry of the Moon and Planets*, p. 659–689. NASA SP-370. Washington, D.C.

Morgan J. W., Ganapathy R., Higuchi H., Krähenbühl U. and Anders E. (1974) Lunar basins: Tentative characterization of projectiles, from meteoritic elements in Apollo 17 boulders. *Proc. Lunar Sci. Conf. 5th*, p. 1703–1736.

Morgan J. W., Krähenbühl U., Ganapathy R. and Anders E. (1973) Trace element abundances and petrology of separates from Apollo 15 soils. *Proc. Lunar Sci. Conf. 4th*, p. 1379–1398.

Morris R. V. (1976) Surface exposure indices of lunar soils: A comparative FMR study. *Proc. Lunar Sci. Conf. 7th*, p. 315–335.

Reed G. W., Allen R. O. and Jovanovic S. (1977) Volatile metal deposits on lunar soils—relation to volcanism. *Proc. Lunar Sci. Conf. 8th*, p. 3917–3930.

Rhodes J. M., Adams J. B., Blanchard D. P., Charette M. P., Rodgers K. V., Jacobs J. W., Brannon J. C. and Haskin L. A. (1975) Chemistry of agglutinate fractions in lunar soils. *Proc. Lunar Sci. Conf. 6th*, p. 2291–2307.

Rhodes J. M., Hubbard N. J., Wiesmann H., Rodgers K. V., Brannon J. C. and Bansal B. M. (1976) Chemistry, classification, and petrogenesis of Apollo 17 mare basalts. *Proc. Lunar Sci. Conf. 7th*, p. 1467–1489.

Schonfeld E. and Meyer C. (1972) The abundance of components of the lunar soils by a least-squares mixing model and the formation age of KREEP. *Proc. Lunar Sci. Conf. 3rd*, p. 1397–1420.

Signer P., Baur H., Derksen U., Etique Ph., Funk H., Horn P. and Wieler R. (1977) Light noble gas records on lunar soil evolution (abstract). In *Lunar Science VIII*, p. 868–870. The Lunar Science Institute, Houston.

Signer P., Baur H., Derksen U., Etique Ph., Funk H., Horn P. and Wieler R. (1977a) Helium, neon, and argon records of lunar soil evolution. *Proc. Lunar Sci. Conf. 8th*, p. 3657–3683.

Via W. N. and Taylor L. A. (1976) Chemical aspects of agglutinate formation: Relationships between agglutinate composition and the composition of the bulk soil. *Proc. Lunar Sci. Conf. 7th*, p. 393–403.

von Gunten H. R., Krähenbühl U., Meyer G., Wegmüller F., Grütter A., Peng K. and Wyttenbach A. (1978) On the partition of volatile trace elements between minerals and agglutinates in grain-size fractions of Apollo 17 soils (abstract). In *Lunar and Planetary Science IX*, p. 436–438. Lunar and Planetary Institute, Houston.

Wänke H., Palme H., Baddenhausen H., Dreibus G., Jagoutz E., Kruse H., Spettel B., Teschke F. and Thacker R. (1974) Chemistry of Apollo 16 and 17 samples: Bulk composition, late stage accumulation and early differentiation of the moon. *Proc. Lunar Sci. Conf. 5th*, p. 1307–1335.

Wasson J. T., Boynton W. V., Kallemeyn G. W., Sundberg L. L. and Wai C. M. (1976) Volatile compounds released during lunar lava-fountaining. *Proc. Lunar Sci. Conf. 7th*, p. 1583–1595.

Proc. Lunar Planet. Sci. Conf. 9th (1978), p. 2287–2297.
Printed in the United States of America

The surface exposure (maturity) of lunar soils: Some concepts and I_s/FeO compilation

RICHARD V. MORRIS

Code SN7, NASA Johnson Space Center, Houston, Texas 77058

Abstract—Surface exposure (or, equivalently, maturity) indices of lunar soils are a measure of residence time of soil in the upper one millimeter of the regolith. Several concepts concerning the use of and terminology associated with maturity indices are discussed. For reasons which include its generally-applicable nature and large data base, the FMR surface exposure (maturity) index I_s/FeO is particularly suitable and useful. A compilation of values of I_s/FeO for 164 Apollo surface and trench soils and six Luna 24 core soils is given.

INTRODUCTION

Since the moon is a planetary body without an atmosphere, its surface is exposed to unattenuated bombardment by micrometeorites, the solar wind, and galactic cosmic rays. The micrometeorites and solar wind interact with the upper one millimeter or so of the lunar regolith (i.e., virtually only the surface), in part producing fine-grained metal (from reduction of Fe^{2+} in the silicate and oxide phases of soil), agglutinate particles, and solarflare particle tracks and implanting solar wind gases (including N, C, and the trapped rare gases). In contrast, the galactic cosmic rays interact with the upper few meters of regolith, in part producing the cosmogenic rare gases and galactic particle tracks. It has been the practice of this and some other laboratories to use the term maturity when discussing the degree of surface exposure of lunar soils, the surface being defined as the upper one millimeter of regolith. Still other laboratories, however, use the same term when discussing the degree of exposure of lunar soils in the upper few meters. Obviously, conceptual difficulties have developed when discussing the exposure history of lunar soils. The problem is removed if, as is already the general practice, the term maturity is associated only with the surface exposure of lunar soils. The term age can be associated with exposure in the upper few meters of the regolith. Thus, a soil located at 20 cm depth is aging but not maturing.

The remainder of this paper is divided into two parts. In the first part, some concepts regarding surface exposure (maturity) indices are discussed. It is argued that the FMR surface exposure index, I_s/FeO, is particularly suitable as a general, quantitative index of surface exposure (maturity). The surface exposure index I_s/FeO is the ratio of the value of the intensity ($I_s = (\Delta H)^2 A$, where ΔH = linewidth and A = amplitude) in arbitrary units of the FMR resonance at $g \sim 2.1$ to the value of the FeO concentration (Morris, 1976). The FMR resonance at $g \sim 2.1$ in lunar soils is due to non-interacting fine-grained metal particles in, according to Housley *et al.* (1976), the diameter range from about 40 to 330 Å. It is necessary to normalize I_s to the FeO content of the soil in order to obtain a surface exposure index because I_s (i.e., the amount of fine-grained metal) is proportional to both the length of surface exposure and the amount of FeO available for reduction (e.g., Morris, 1976). The second part of the paper is a compilation of values of I_s/FeO for 164 Apollo surface and trench soils and six Luna 24 core soils.

CONCEPTS

Generally-applicable indices of surface exposure

In lunar soil studies, it sometimes becomes appropriate to correlate the values of experimentally or theoretically-derived parameters with the values of a surface exposure index representative of the bulk soil, i.e., to construct maturity correlations. Obviously, a generally-applicable index of surface exposure is the most suitable for this purpose. Generally-applicable means that there is no available evidence that the numerical values of a particular surface exposure index saturate or unduly reflect variations other than in the length of surface exposure. In an earlier paper (Morris, 1976), the FMR maturity index I_s/FeO was shown to be a generally-applicable index of surface exposure, and two other often-used indices, agglutinates and mean grain size, were shown not to be generally-applicable indices. The number percentage of petrographic agglutinates both saturates and is somewhat dependent on bulk soil composition. Mean grain size based on size data less than 1 mm is an especially insensitive surface exposure index (also see Morris, 1977). Mean grain size based on size data less than 1 cm is much more sensitive, but systematic offsets are present between different landing sites. Solar-flare track densities were not considered by Morris (1976), but they are not generally-applicable because the density of tracks saturates (e.g., Blanford *et al.*, 1977).

The preceding discussion should not be taken to imply that I_s/FeO can be used to the exclusion of other surface exposure indices. Agglutinate contents, solar-flare track densities, etc. have properties that are not inherent to I_s/FeO. Two examples illustrate this. The percentage of petrographically-determined agglutinates in a particular size range can be used to normalize the modal analysis of that size range to an "agglutinate free" basis (e.g., McKay *et al.*, 1977); this procedure allows modal analyses to be compared without the effect of maturation. Solar-flare track densities, because they are measured on individual mineral grains, can sometimes distinguish whether a soil has components having different maturities (e.g., Blanford *et al.*, 1977; Crozaz and Dust, 1977). It should also be pointed out that if an index is found with respect to which I_s/FeO saturates, then that index would be the appropriate index for maturity correlations.

In addition to its generally-applicable property, there are three additional factors that make I_s/FeO a particularly useful and suitable surface exposure (maturity) index for maturity-correlation plots. (1) The data base is large. The compilation at the end of this paper contains values of I_s/FeO for 164 surface and trench soils collected during the Apollo missions and 6 soils collected during the Luna 24 mission. The compilation thus contains greater than 96% of all surface and trench soils in the Apollo sample collection. Depth profiles of I_s/FeO are also available for five Apollo cores (Apollo 15 deep drill core, Heiken *et al.*, 1976; Apollo 16 deep drill core, Gose and Morris, 1977; 60009/10 core, Morris and Gose, 1976; Apollo 17 deep drill core, Morris *et al.*, 1978a; 74001/2 core, Morris and Gose, 1977, and Morris *et al.*, 1978b). (2) The values of I_s/FeO are based on measurements of a substantial size fraction of soil (<250 μm). (3) The values of I_s/FeO were all determined in one laboratory so that inter-laboratory biases are not possible.

Immature, submature, and mature

It is a semantic convenience when discussing the relative length of surface exposure of lunar soils to have them classified into immature, submature, and mature groups. This terminology was apparently first applied to lunar soils by McKay *et al.* (1974), and the classification is accomplished in the following manner. The observed range of values of a generally-applicable index of surface exposure is arbitrarily, but sensibly, divided into three parts. The part having the lowest values of the index, and thus reflecting relatively the shortest duration of surface exposure, is designated the immature range for the index. The part having the intermediate values, and thus reflecting relatively an intermediate duration of surface exposure, is designated the submature range. And similarly, the part having the highest values, and thus reflecting relatively the longest duration of surface exposure, is designated the mature range. For I_s/FeO, the immature range is from 0.0 through 29.0 units, the submature range is from 30.0 through 59.0 units, and the mature range is greater than or equal to 60.0 units.

Unfortunately, the classification immature, submature, and mature loses some of its usefulness because consistent ranges have not been maintained among the indices. This condition appears to have arisen largely because some of the classifications were based on surface exposure indices which subsequent studies have shown to be not generally-applicable. In any event, the correlations of various surface exposure indices given by Morris (1976) and/or the compilation given later in this paper can be used to define, in as much as is possible, consistent immature, submature, and mature ranges. As an example, the above ranges for I_s/FeO define the following equivalent ranges for petrographically-determined agglutinates: soils having about 0 to 28% agglutinates are immature; soils having about 28 to 45% agglutinates and <10.0 wt.% FeO or about 28 to 55% agglutinates and ≥10.0 wt.% FeO are submature; and soils having ~45% agglutinates and <10.0 wt.% FeO or ~55% agglutinates and ≥10.0 wt.% FeO can also be mature. The complex definition for the classification and lack of resolution at high agglutinate contents is due to the aforementioned compositional and saturation effects associated with agglutinates.

Regolith depth and maximum I_s/FeO

Earlier in this paper it was pointed out that saturation (i.e., after a certain time of surface exposure a surface exposure index reaches a maximum value and becomes insensitive to additional periods of surface exposure) limits the usefulness of a surface exposure index. Examination of the I_s/FeO compilation shows that the numerical values for I_s/FeO range from near 0.0 to over 100.0 units. Does the maximum value of I_s/FeO reflect a saturation effect or does it reflect a maximum surface exposure that is an inherent property of lunar soils? Available evidence indicates the latter is the case. The data of Morris (1978, unpublished material) shows that on the average only about 2% of the total FeO available for reduction has been reduced to fine-grained metal through surface exposure. Thus, for any soil examined to date, neither does I_s/FeO approach its absolute upper limit (the complete reduction of FeO to fine-grained metal) nor does the concentration of FeO available for reduction change appreciably with maturity. Nevertheless, I_s/FeO could still saturate before it reaches its upper limit, say by a process which coalesces fine-grained metal so that it is no longer observed in the FMR experiment. However, I_s/FeO and nitrogen do not saturate with respect to each other (Morris, 1976), indicating both saturate at the same rate with respect to a third and as yet unidentified index or, more likely, neither saturates. The conclusion that neither saturates seems reasonable on phenomonological grounds, but cannot be rigorously justified, because the processes associated with the derivations of fine-grained metal and nitrogen are so different, the former being reduction of FeO and the latter being direct implantation by the solar wind. In summary, there does not seem to be any available evidence that the maximum value of I_s/FeO for lunar soils is due to saturation of the index. What, then, limits the observed values of I_s/FeO? An argument adapted from one given by McKay *et al.* (1974) provides an answer.

According to the calculations of Quaide and Oberbeck (1975), the ratio of the volume of bedrock to the volume of regolith excavated per unit time by impacts decreases as the thickness of the regolith increases. Since crushed bedrock necessarily has no surface exposure, the maximum values of I_s/FeO should thus be determined ultimately by regolith thickness, and thin regoliths should have relatively lower maximum values of I_s/FeO than thicker ones. This appears to be the case, although there are the associated problems with applying the statistical treatment of Quaide and Oberbeck (1975) to relatively few data points. The Apollo 12 and 16 landing sites have, respectively, among the thinnest and thickest regoliths sampled during the Apollo missions (e.g., Quaide and Oberbeck, 1975). From the compilation, the maximum values of I_s/FeO for the Apollo 12 and 16 landing sites are, respectively, about 60 and 100 units.

The above argument does not endorse the general use of maximum values of I_s/FeO over a particular area on the moon to estimate regolith depths. A recent large impact excavating bedrock will leave a low maturity imprint on the surrounding area independent of the regolith thickness. Thus, the limited Luna 24 I_s/FeO data (maximum I_s/FeO = 39.0 units) should not be used to infer that the regolith at the Luna 24 site is thin, although it could be.

In summary, there is no available evidence that the maximum values of I_s/FeO observed in the lunar regolith are due to the saturation of the index. It seems most likely that maximum observed values of I_s/FeO reflect the input of unexposed material, which is related to the thickness of the regolith.

I_s/FeO Compilation

Table 1 is a compilation of values of I_s/FeO (<250 μm) for 164 Apollo surface

Table 1. Compilation of values of the FMR maturity index I_s/FeO for Apollo (excepting cores) and Luna 24 soils. Also given for each soil is its concentration of FeO and value of the FMR linewidth (ΔH). Soils having values of I_s/FeO (<250 μm) from 0.0 through 29.0 units are immature; from 30.0 through 59.0 units, submature; and greater than or equal to 60.0 units, mature.

No.	Soil	FeO (wt. %)	ΔH (Gauss)	I_s/FeO (<250 μm, Arb.)	Comment	No.
			Apollo 11			
1.	10010	15.1*	790	75.0	Mature	1.
2.	10011	14.6*	794	69.0	Mature	2.
3.	10084	15.8	790	78.0	Mature	3.
			Apollo 12			
4.	12001	16.8	752	56.0	Submature	4.
5.	12003	15.4	737	57.0	Submature	5.
6.	12023	16.0	737	60.0	Mature	6.
7.	12024	14.6*	705	30.0	Submature	7.
8.	12030	14.3	768	14.0	Immature	8.
9.	12032	15.1	760	12.0	Immature	9.
10.	12033	14.2	764	4.6	Immature	10.
11.	12037	17.3	742	21.0	Immature	11.
12.	12041	14.2	742	63.0	Mature	12.
13.	12042	16.8	747	61.0	Mature	13.
14.	12044	15.7	750	57.0	Submature	14.
15.	12057	16.6	739	40.0	Submature	15.
16.	12060	16.9	745	24.0	Immature	16.
17.	12070	16.5	725	47.0	Submature	17.
			Apollo 14			
18.	14003	10.4	600	66.0	Mature	18.
19.	14141	10.2	560	5.7	Immature	19.
20.	14148	10.4	600	74.0	Mature	20.
21.	14149	10.0	600	53.0	Submature	21.
22.	14156	10.4	600	68.0	Mature	22.
23.	14161	10.2	599	48.0	Submature	23.
24.	14163	10.4	595	57.0	Submature	24.
25.	14240†	10.4	582	46.0	Submature	25.
26.	14259	10.5	597	85.0	Mature	26.
27.	14260	10.0	590	72.0	Mature	27.

Table 1. *(cont'd.)*

No.	Soil	FeO (wt. %)	ΔH (Gauss)	I_s/Fe0 (<250 μm, Arb.)	Comment	No.
			Apollo 15			
28.	15012	12.4	660	66.0	Mature	28.
29.	15013	15.0	717	77.0	Mature	29.
30.	15021	15.0	730	70.0	Mature	30.
31.	15031	15.0	717	68.0	Mature	31.
32.	15041	14.3	731	94.0	Mature	32.
33.	15071†	16.4	715	52.0	Submature	33.
34.	15081	15.4	715	68.0	Mature	34.
35.	15091†	11.6	670	74.0	Mature	35.
36.	15101†	11.6	670	70.0	Mature	36.
37.	15201	11.9*	663	68.0	Mature	37.
38.	15211	11.7	685	60.0	Mature	38.
39.	15221†	11.5	665	63.0	Mature	39.
40.	15231†	11.6	663	71.0	Mature	40.
41.	15241†	12.3	658	45.0	Submature	41.
42.	15251	12.0	676	75.0	Mature	42.
43.	15261	12.1	678	77.0	Mature	43.
44.	15271	12.2	664	63.0	Mature	44.
45.	15291	11.6	666	63.0	Mature	45.
46.	15301	15.5	701	48.0	Submature	46.
47.	15401	18.3	665	5.6	Immature	47.
48.	15411†	13.2	680	43.0	Submature	48.
49.	15426	19.7	560	0.3	Immature	49.
50.	15431†	11.9	762	39.0	Submature	50.
51.	15471†	16.4	715	34.0	Submature	51.
52.	15501†	16.6	738	51.0	Submature	52.
53.	15531†	19.2	755	27.0	Immature	53.
54.	15601	19.2	746	29.0	Immature	54.
			Apollo 16			
55.	60051	4.5	557	57.0	Submature	55.
56.	60501†	5.5	554	80.0	Mature	56.
57.	60601	5.5	560	85.0	Mature	57.
58.	61141	5.3	570	56.0	Submature	58.
59.	61161	5.4	575	82.0	Mature	59.
60.	61181	5.5	575	82.0	Mature	60.

Table 1. *(cont'd.)*

No.	Soil	FeO (wt. %)	ΔH (Gauss)	I_s/FeO (<250 μm, Arb)	Comment	No.
61.	61221	4.9	568	9.2	Immature	61.
62.	61241	5.4	560	47.0	Submature	62.
63.	61281†	5.4	556	69.0	Mature	63.
64.	61501	5.6	563	53.0	Submature	64.
65.	62231	5.1*	594	91.0	Mature	65.
66.	62241†	5.2	583	100.0	Mature	66.
67.	62281	5.5	590	76.0	Mature	57.
68.	63321	4.7	545	47.0	Submature	68.
69.	63341	4.5	550	54.0	Submature	69.
70.	63501	4.7	555	46.0	Submature	70.
71.	64421	5.0	551	83.0	Mature	71.
72.	64501	5.2	555	61.0	Mature	72.
73.	64801	5.2	530	71.0	Mature	73.
74.	64811	5.6	555	54.0	Submature	74.
75.	65501	6.0	570	38.0	Submature	75.
76.	65511	6.0*	572	55.0	Submature	76.
77.	65701	5.7	565	106.0	Mature	77.
78.	65901	5.8	570	99.0	Mature	78.
79.	66031	5.5	573	102.0	Mature	79.
80.	66041	6.0	562	90.0	Mature	80.
81.	66081	6.2	560	80.0	Mature	81.
82.	67010	4.2*	555	26.0	Immature	82.
83.	67461	4.3	550	25.0	Immature	83.
84.	67481	4.2	545	31.0	Submature	84.
85.	67511	4.2*	555	8.8	Immature	85.
86.	67601	4.0	552	45.0	Submature	86.
87.	67701	4.2	550	39.0	Submature	87.
88.	67711	3.0	550	2.8	Immature	88.
89.	67941	4.2*	540	29.0	Immature	89.
90.	67960	4.6*	545	20.0	Immature	90.
91.	68121†	5.4	544	61.0	Mature	91.
92.	68501	5.3	558	85.0	Mature	92.
93.	68821†	5.2	555	84.0	Mature	93.
94.	68841	5.6	570	70.0	Mature	94.

Table 1. *(cont'd.)*

No.	Soil	FeO (wt. %)	ΔH (Gauss)	I_s/FeO (<250 μm, Arb.)	Comment	No.
95.	69921	5.6	555	90.0	Mature	95.
96.	69941	5.7	562	85.0	Mature	96.
97.	69961	5.7	572	92.0	Mature	97.
			Apollo 17			
98.	70011	16.0	795	54.0	Submature	98.
99.	70161	17.1	792	46.0	Submature	99.
100.	70181	16.4	790	47.0	Submature	100.
101.	70251	16.6*	790	43.0	Submature	101.
102.	70271	16.2*	790	56.0	Submature	102.
103.	70311	17.5*	795	39.0	Submature	103.
104.	70321	16.5*	785	42.0	Submature	104.
105.	71041	17.7	772	29.0	Immature	105.
106.	71061	17.8	775	14.0	Immature	106.
107.	71131	18.2*	783	33.0	Submature	107.
108.	71151	18.0*	795	34.0	Submature	108.
109.	71501	18.3	797	35.0	Submature	109.
110.	72131	17.2*	790	60.0	Mature	110.
111.	72141	13.5	750	81.0	Mature	111.
112.	72150	14.5	765	82.0	Mature	112.
113.	72161	14.9	765	87.0	Mature	113.
114.	72221	9.6*	652	58.0	Submature	114.
115.	72241	9.1*	660	64.0	Mature	115.
116.	72261	9.6*	657	59.0	Submature	116.
117.	72321	8.7	665	73.0	Mature	117.
118.	72431	9.8*	658	63.0	Mature	118.
119.	72441	8.7	656	68.0	Mature	119.
120.	72461	8.6	655	71.0	Mature	120.
121.	72501	8.7	660	81.0	Mature	121.
122.	72701	8.8	657	61.0	Mature	122.
123.	73121	8.5	670	78.0	Mature	123.
124.	73131	6.8*	635	16.0	Immature	124.
125.	73141	8.1	665	48.0	Submature	125.
126.	73151	9.3*	655	68.0	Mature	126.
127.	73211	9.4*	673	39.0	Submature	127.

Table 1. *(cont'd.)*

No.	Soil	FeO (wt. %)	ΔH (Gauss)	I_s/FeO (<250 μm, Arb.)	Comment	No.
128.	73221	8.9	675	43.0	Submature	128.
129.	73241	8.8	680	18.0	Immature	129.
130.	73261	8.9	680	45.0	Submature	130.
131.	73281	8.9	680	34.0	Submature	131.
132.	74111	10.2*	705	31.0	Submature	132.
133.	74121	10.0	700	88.0	Mature	133.
134.	74220	22.0	730	1.0	Immature	134.
135.	74241	14.9	685	5.1	Immature	135.
136.	74261	15.3	660	5.0	Immature	136.
137.	75061	18.0	790	33.0	Submature	137.
138.	75081	17.1	782	40.0	Submature	138.
139.	75111	16.0*	781	54.0	Submature	139.
140.	75121	16.0	787	67.0	Mature	140.
141.	76031	11.7*	720	64.0	Mature	141.
142.	76121	15.2*	770	71.0	Mature	142.
143.	76131	12.3*	737	70.0	Mature	143.
144.	76221	10.9*	720	66.0	Mature	144.
145.	76240	10.9	735	56.0	Submature	145.
146.	76261	10.9	720	58.0	Submature	146.
147.	76281	11.3	720	45.0	Submature	147.
148.	76321	9.8	720	93.0	Mature	148.
149.	76501	10.3	718	58.0	Submature	149.
150.	77511	12.3*	728	80.0	Mature	150.
151.	77531	11.7	735	79.0	Mature	151.
152.	78121	14.2*	740	68.0	Mature	152.
153.	78221	11.7	736	93.0	Mature	153.
154.	78231	13.1*	735	81.0	Mature	154.
155.	78421	12.0*	740	92.0	Mature	155.
156.	78441	12.4	740	77.0	Mature	156.
157.	78461†	12.2	732	83.0	Mature	157.
158.	78481	13.1	740	82.0	Mature	158.
159.	78501	13.2	727	36.0	Submature	159.
160.	79121	16.5	780	57.0	Submature	160.
161.	79221	15.4	795	81.0	Mature	161.

Table 1. *(cont'd.)*

No.	Soil	FeO (wt. %)	ΔH (Gauss)	I_s/Fe0 (<250 μm, Arb)	Comment	No.
162.	79241	15.6	788	51.0	Submature	162.
163.	79261	15.0*	785	43.0	Submature	163.
164.	79511	15.3	785	61.0	Mature	164.
			Luna 24			
165.	24077	19.9	787	39.0	Submature	165.
166.	24109	20.6	783	31.0	Submature	166.
167.	24149	20.3	798	21.0	Immature	167.
168.	24174	20.9	790	27.0	Immature	168.
169.	24182	20.2	793	19.0	Immature	169.
170.	24210	21.1	790	19.0	Immature	170.

*Determined magnetically. The other FeO values were determined chemically and obtained from the Curatorial Data Base (J. Warner, compiler).

†These samples were reissued by the RSPL. The pedigree of these samples is thus not as secure as it is for the other samples.

and trench soils and six Luna 24 core soils. The compilation includes the data of Morris (1976) and new data. The experimental technique for the measurement of I_s (relative concentration of fine-grained, i.e., ~40 to 330 Å, metal) is given in the above paper. Except as noted in the compilation, the values of FeO determined by chemical analysis were used to compute I_s/FeO. The other FeO values were determined magnetically as chemically-determined values were apparently not available. Since the accuracy of the magnetic method is not quite as good as that of chemical methods, some of the values of I_s/FeO in the compilation may require updating as chemical data becomes available.

SUMMARY

(1) Surface exposure, or maturity, indices of lunar soils are a measure of residence time of soil in the upper one millimeter of the regolith. Confusion in terminology will be avoided if the term maturity is associated only with the surface exposure of lunar soils, the surface being defined as the upper one millimeter of soil. The terms immature, submature, and mature represent a semiquantitative maturity grouping denoting, respectively, successively longer periods of surface exposure.

(2) For reasons which include a generally-applicable nature and a large data base, the FMR surface exposure (maturity) index I_s/FeO is particularly suitable and useful. A compilation of values of I_s/FeO for 164 Apollo surface and trench soils and six Luna 24 core soils is given.

References

Blanford G. E., McKay D. S. and Wood G. C. (1977) Particle track densities in double drive tube 60009/10. *Proc. Lunar Sci. Conf. 8th*, p. 3017–3025.

Crozaz G. and Dust S. (1977) Irradiation history of lunar cores and the development of the regolith. *Proc. Lunar Sci. Conf. 8th*, p. 3001–3016.

Gose W. A. and Morris R. V. (1977) Depositional history of the Apollo 16 deep drill core. *Proc. Lunar Sci. Conf. 8th*, p. 2909–2928.

Heiken G. H., Morris R. V., McKay D. S. and Fruland R. M. (1976) Petrographic and ferromagnetic resonance studies of the Apollo 15 deep drill core. *Proc. Lunar Sci. Conf. 7th*, p. 93–111.

Housley R. M., Cirlin E. H., Goldberg I. B. and Crowe H. (1976) Ferromagnetic resonance studies of lunar core stratigraphy. *Proc. Lunar Sci. Conf. 7th*, p. 13–26.

McKay D. S., Dungan M. A., Morris R. V. and Fruland R. M. (1977) Grain size, petrographic, and FMR studies of the double core 60009/10: A study of soil evolution. *Proc. Lunar Sci. Conf. 8th*, p. 2929–2952.

McKay D. S., Fruland R. M. and Heiken G. H. (1974) Grain size and the evolution of lunar soils. *Proc. Lunar Sci. Conf. 5th*, p. 887–906.

Morris R. V. (1976) Surface exposure indices of lunar soils: A comparative FMR study. *Proc. Lunar Sci. Conf. 7th*, p. 315–335.

Morris R. V. (1977) Origin and evolution of the grain-size dependence of the concentration of fine-grained metal in lunar soils: The maturation of lunar soils to a steady-state stage. *Proc. Lunar Sci. Conf. 8th*, p. 3719–3747.

Morris R. V. and Gose W. A. (1976) Ferromagnetic resonance studies of cores 60009/60010 and 60003: Compositional and surface-exposure stratigraphy. *Proc. Lunar Sci. Conf. 7th*, p. 1–11.

Morris R. V. and Gose W. A. (1977) Depositional history of core section 74001: Depth profiles of maturity, FeO, and metal. *Proc. Lunar Sci. Conf. 8th*, p. 3113–3122.

Morris R. V., Gose W. A. and Lauer H. V. (1978a) Maturity and FeO profiles for the Apollo 17 deep drill core (abstract). In *Lunar and Planetary Science IX*, p. 766–768. Lunar and Planetary Institute, Houston.

Morris R. V., Gose W. A. and Lauer H. V. (1978b) Depositional and surface exposure history of the Shorty Crater core 74001/2: FMR and magnetic studies. *Proc. Lunar Planet. Sci. Conf. 9th*. This volume.

Quaide W. and Oberbeck V. (1975) Development of the mare regolith: Some model considerations. *The Moon* **13**, 27–55.

Proc. Lunar Planet. Sci. Conf. 9th (1978), p. 2299–2310.
Printed in the United States of America

Evidence for the constancy of the solar cosmic ray flux over the past ten million years: ^{53}Mn and ^{26}Al measurements

C. P. KOHL, M. T. MURRELL, G. P. RUSS III* and J. R. ARNOLD

Department of Chemistry
University of California, San Diego, La Jolla, California 92093

Abstract—^{53}Mn and ^{26}Al activities have been measured in fourteen samples carefully ground from the upper 1.5 cm of rock 68815 to discern whether the magnitude and shape of the solar cosmic ray flux in the last 2 m.y. has differed from the recent SCR flux and that averaged over the last ~10 m.y. Our results yield no evidence that the SCR flux has varied among these time intervals. Assuming the appropriate erosion rates, the measured profiles of rock 68815 (with a 2 m.y. surface exposure) and of rocks 12002 and 14321 (saturated with respect to ^{53}Mn and ^{26}Al) can all be fit with the SCR parameters of R_o = 100 MV and J = 70 protons • cm^{-2} • sec^{-1} (4 π, E > 10 MeV). The measured profiles of 68815 and 14310 are incompatible with the short surface exposure age (<3 m.y.) proposed for 14310 on the basis of track studies. Activity vs. depth profiles for three different faces of 68815 show surface activities nearly independent of sample inclination.

INTRODUCTION

The history of the charged particle flux emitted by the sun is of inherent and practical interest. It is well known that the flux of solar cosmic rays (SCR) varies throughout the 11-year solar cycle with the bulk being emitted in flares clustered near the solar maximum. Measurements of the SCR flux have only been made for the last two decades. Studies of the constancy of this flux over longer time periods rely on inferences from sunspot counts and the record of the nuclear interactions left by these particles in extraterrestrial materials, particularly lunar samples. Eddy (1976) has recently reviewed the sunspot evidence and concluded that between 1645 and 1715 there were essentially no sunspots. This implies a cessation of the SCR flux as well in this period. Eddy also notes that this interval was abnormally cold in Europe, lending support to the idea of climate being correlated with flares and presumably the SCR flux.

Previous work in this laboratory (SHRELLDALFF, 1970; Finkel *et al.*, 1971; Wahlen *et al.*, 1972; Imamura *et al.*, 1974) has shown that depth profiles of SCR produced radionuclides measured in the upper 2 cm of lunar rocks are useful probes for determining the average energy spectrum and number of SCR particles over a time scale of a few half-lives. By studying nuclides of various half-lives, this technique can be extended from a few months (^{56}Co) to 10^7 years

*Present Address: Hawaii Institute of Geophysics, Dept. of Chemistry, University of Hawaii at Manoa, Honolulu, Hawaii 96822

(^{53}Mn). In addition to learning the sun's past, a knowledge of the flux history enables one to interpret the SCR data measured in lunar cores in terms of depositional history and regolith mixing rates (Nishiizumi *et al.*, 1976; Nishiizumi *et al.*, 1978).

The suite of Apollo 16 rocks determined by ^{81}Kr-Kr to have been exposed on the lunar surface for only 2.0 ± 0.1 m.y. and attributed to South Ray Crater (Behrmann *et al.*, 1973) provide a unique opportunity to study the SCR flux in the last 2 m.y. Although ^{26}Al ($t_{1/2}$ = 0.72 m.y.) (Samworth *et al.*, 1972) has been measured in 12002 and 14321 (Finkel *et al.*, 1971; Wahlen *et al.*, 1972) and provides some information about this recent time interval, the interpretation of the data is somewhat uncertain in that nuclides of different half-lives were compared to establish the SCR history. Such comparisons are subject to uncertainties in the excitation functions and to the absence of a strict cut-off time. Since the South Ray Crater samples have a well-defined 2 m.y. exposure history, it is possible to compare the effective production rates of various nuclides to each other during this discrete time interval. A comparison can also be made between long-lived nuclides produced in this 2 m.y. period and those produced during the longer, unbounded interval sampled by the other rocks. ^{53}Mn ($t_{1/2}$ = 3.7 m.y.), ^{10}Be ($t_{1/2}$ = 1.5 m.y.), ^{26}Al ($t_{1/2}$ = 0.72 m.y.) and ^{36}Cl ($t_{1/2}$ = 0.3 m.y.) have all been measured in lunar rocks, but of these, only ^{53}Mn and ^{26}Al are of current practical interest for detailed depth profile studies. We report here the results of such a study for a member of the South Ray suite: 68815.

Two rocks, 68815 and 60025, were chosen from this group for a preliminary study to confirm whether or not their exposure to cosmic rays was indeed only 2 m.y. These preliminary samples were cut from depths ≥3 cm (~10 g/cm^2) in order to sample only galactic cosmic ray (GCR) produced activity which decreases more slowly with depth than the SCR contribution. A high GCR component would have indicated a longer or more complicated exposure; for example, the rock might have been exposed below the surface for a time or rolled over during exposure on the surface. Such a history would complicate the interpretation of the SCR profiles. ^{53}Mn and ^{26}Al measurements were consistent with a simple 2 m.y. exposure for both rocks (Kohl *et al.*, 1977). An independent ^{53}Mn analysis of an aliquot of our 68815 sample was in good agreement with our data (Fruchter *et al.*, 1977). A non-destructive ^{26}Al measurement (Yokoyama *et al.*, 1974) on 68815 was also compatible with a 2 m.y. exposure on the lunar surface. Rock 68815 was selected for our detailed SCR study because of its higher Fe concentration (the principal target for ^{53}Mn production), larger size, good documentation and the availability of data from rare gas and track studies.

Rock 68815 is a breccia composed of a variety of basaltic and anorthositic clasts (Brown *et al.*, 1973) welded together by a glassy matrix. It was collected by breaking it from the top of a meter-high boulder. The lunar orientation of the rock was reconstructed at the LRL by comparison with documentation photographs (Apollo 16 Preliminary Science Report, 1972). The exposure history of this rock has been studied by rare gas and track techniques. The ^{81}Kr-Kr age of

2.04 ± .08 m.y. (Drozd *et al.*, 1974) is in good agreement with other rocks in the suite and is supported by ^{38}Ar data measured by the same workers. No neutron effects were observed in the Kr data for this rock, indicating that any subsurface exposure occurred at depths in excess of 7 m (Drozd *et al.*, 1974; Behrmann *et al.*, 1973). Pepin *et al.* (1974) interpret the relatively high $^{38}Ar/^{21}Ne$ to imply an exposure at a depth of ~7 m for ~70 m.y. Such an exposure would not affect the radionuclide activities. If the 2 m.y. age is adopted, particle track data (Walker and Yuhas, 1973) show no evidence (within 20%) for a change in the spectrum or flux of VH nuclei in the last 2 m.y. over that observed for solar cycle 19. A follow-up study by Dust and Crozaz (1977) on a column cut from our sample was in agreement with this interpretation; also they find no evidence for large scale surface chipping, at least above their column. Our piece of 68815 was selected for radionuclide measurements on the basis of its location near the top of the rock and because it appeared free of abrasion caused by breaking the parent rock from the boulder.

EXPERIMENTAL

Sample 234 of rock 68815 was cut at the LRL from an area near the top of 68815 when viewed in its lunar orientation (Fig. 1). Its mass was 55.08 g. After the sample arrived, we cut a 5 × 5 × 20 mm column from one corner and transferred this piece to the Walker group at Washington University for track studies (Dust and Crozaz, 1977). The remaining piece, with surface area of ~30 cm^2 was mounted on a plexiglas plate with Caulk orthodontic resin. The plate was mounted at the approximate angle the sample experienced on the moon, and sunlight was simulated with a spot light. This illumination was used to choose three faces for separate processing. It was expected that the three faces would have different cosmogenic activities because of their different exposure angles to the solar cosmic rays. Face A had an average inclination of ~45° to the vertical and was tilted toward the west on the moon, face B had an inclination of ~37° and faced east, and face C was the most horizontal with an angle of ~12° and a southwest exposure.

The plexiglas plate on which the sample was mounted could be reproducibly positioned inside a plastic dish which was used to collect the material as it was ground from the surface. The dish was in turn mounted on an X-Y stage. In the plane of the stage, the rock could be positioned to within 0.1 mm. Height contours of the rock were measured with a micrometer probe mounted normal to the X-Y plane. Maps of the height of the surface before and after grinding allowed determination of the thickness of the layer removed. Surface heights could be measured to ±0.01 mm and reproduced to ~0.05 mm. For steeply sloped faces, the micrometer probe could be angled in the X-Z plane. Control points on the plexiglas plate, dish and rock were routinely measured to correct for variations in rock positioning. Layers of nominal thickness 0–0.5, 0.5–1.0, 1.0–2.0 and 2.0–4.0 mm were separately ground and collected from the three faces. The 1 σ uncertainty in the thickness of each layer ground was ~30%. Such sampling allowed us to search for anisotropy as well as secular variations in the SCR flux. For the 0.5 mm layers, the rock surface was mapped on a square grid with the measurement points separated by $\sqrt{2}$ mm. For the 1 and 2 mm thick layers a 2 mm grid was used. A single 4–8 mm layer was ground from parts of faces A and C (2 mm map grid). Also a 10–15 mm layer was ground from near the bottom of the rock from faces A and C.

At the start of each grinding session, the mounted rock was weighed, positioned and mapped. Grinding was done with a dental drill using diamond cutting tools. The diamonds were mounted in Ni with steel shafts. Isopropanol was dripped over the area to wet the powder and prevent loss of dust. A teflon tube connected to the house vacuum through a suction flask was used to control the volume of alcohol in the dish. This aspirator was also useful for drying the work area and for partial collection of the sample. A 0.5 μ millipore filter and second flask were placed between the collection flask and the vacuum source as a safety feature. After grinding, the bulk of the ground sample was collected and

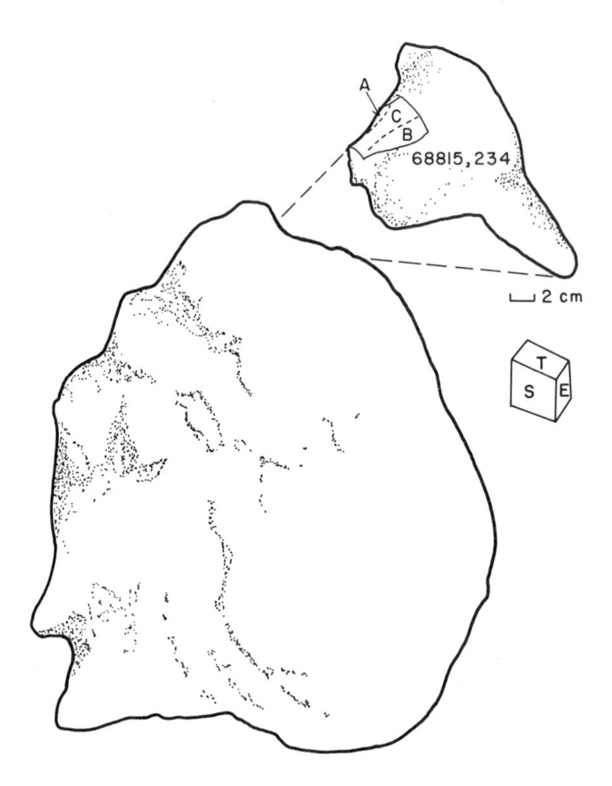

Fig. 1. This sketch shows roughly the relation between 68815, 234 and its parent boulder on the moon. The scale marker and lunar orientation cube relate only to the outline of 68815 and not to the boulder.

then the rock, dish and millipore paper were cleaned ultrasonically to collect the remaining powder. This material was combined with that in the collection flask, centrifuged, dried and weighed. The rock was weighed both before and after each layer was ground in order to monitor recovery of the sample. Recovery yields were in all cases nearly 100%. The dental tool was also weighed before and after grinding. Weight losses from the grinding tools ranged up to a few mg but are not significant in that the sample weights were 0.7 to 3.1 g. Addition of Fe from abrasion of the cutting tool shafts was considered to be unimportant. No other elemental contamination was of concern for our analyses. The grinding and mapping operation was carried out in a desk top plastic tent.

The samples were sufficiently fine for dissolution and were subjected to no further grinding. Aliquots of the powder were given to Professor Kurt Marti for rare gas work; others were taken for chemical analysis. The remaining material was dissolved using an HF-HNO_3 mixture in a teflon beaker. The HF was removed by fuming twice with $HClO_4$; the sample was taken to dryness, dissolved in 9 M HCl and separated into various elemental fractions using the ion exchange procedures used in our earlier work (SHRELLDALFF, 1970). A sample of standard terrestrial rock JB-1 was processed in the same manner and was used as a counting blank.

The Mn fraction from the main chemistry was analyzed for ^{53}Mn using chemical purifications and irradiation procedures which have been described by Imamura *et al.* (1973). The samples were irradiated in isotope tray #2 of the Argonne CP-5 reactor for 113 days at a flux of 3.1×10^{12} neutrons/cm$^2 \cdot$ sec. The amplification factor obtained in this irradiation was 8.27×10^3 dpm^{54}Mn/dpm^{53}Mn.

^{54}Mn was counted using a 45 cc Ge(Li) detector which has been described by Nishiizumi *et al.* (1976). The background count rate under the peak was 0.018 cpm (52,000 minutes counting time) which constitutes a correction of <2% for the samples. The average efficiency was approximately 4.7%.

The Al fraction from the main chemistry cation column was purified as described by Finkel (1972). An unidentified pure β activity was found in some samples; so several additional steps described by SHRELLDALFF (1970) were added to the chemistry. However, these did not appreciably lower the contamination. The JB sample was also found to contain this activity, and since it gave the same background in our β–γ counting system as an Al_2O_3 sample which did not have this additional β activity, the contamination was considered not to be a problem. The samples were ignited as usual at 1100°C before mounting, but careful examination over the first few hours after removal from the furnace showed a weight gain of ~3% due to hydration. To minimize this difficulty, the samples were cooled in a desiccator, weighed as soon as cool and weighed at intervals until a stable weight was achieved; a correction was then applied to the material weighed into the sample holders. The sample holders were redesigned slightly to maximize the counting efficiency by placing the sample over the most active area of the β counter.

^{26}Al was counted using a β–γ coincidence system which has been described in detail by SHRELLDALFF (1970). The efficiency for the 511 keV $\beta+$ annihilation peak is approximately 9.5% considering only the positron emission of the standard. The background under the peak averages 0.004 counts per minute. Sample count rates varied from 0.01 to 0.03 cpm.

The chemical analyses were performed on aliquots of the ground samples using essentially the same dissolution technique as for the main samples. Mn, Fe and Al compositions were determined using the standard additions methods for atomic absorption spectrometry using a Varian Techtron AA-6.

RESULTS

The data for the neutron activation of 68815 along with the terrestrial reference samples and ^{53}Mn, Fe and stable Mn standards are presented in Table 1. The ^{54}Fe(n,p)^{54}Mn correction was less than 0.4% for all samples of 68815. The ^{55}Mn(n,2n)^{54}Mn correction ranged from 15% to 37%.

Because Fe is the only significant target for the production of ^{53}Mn in lunar samples, the effects of variations in chemical composition can be normalized by reporting the activities as ^{53}Mn dpm/kg Fe. For a given set of operating conditions for the atomic absorption spectrometer, the Fe and Mn concentrations can be determined with 1 σ precisions of 3% and 2% respectively.

Our chemical analysis of Al, Fe and Mn in 68815 can be found in Table 1. The compositon of 68815 appears to be reasonably homogenous with depth. There seem to be some small but consistent variations in composition among the faces. For example face A has higher Fe and Mn at all depths than faces B or C. A weighted average of the Al, Fe and Mn concentrations in the three faces and at all depths gives values of 14.54%, 3.60% and 441 ppm, respectively. These weighted averages have been used in the theoretical production rate calculations made using the model of Reedy and Arnold (1972) and are in good agreement with the values reported by the Apollo 16 Preliminary Examination Team (1973) and Wänke *et al.* (1974). The Fe to Mn ratios reported here for terrestrial standards JG-1 and JB-1 are in reasonable agreement with our previous work (Imamura *et al.*, 1974; Nishiizumi *et al.*, 1976).

The results for ^{53}Mn and ^{26}Al for the three faces of rock 68815 as a function of depth are given in Table 2. The ^{26}Al activities were calculated using a

Table 1.

68815 depth g/cm^2 (FACE)	Sample description			Sample counted after irradiation		^{54}Mn (cpm/mg Mn)[(2)]	^{54}Mn (cpm/mg Mn) after (n,2n) correction[(3)]	^{53}Mn (dpm/kg Fe)[(4)]
	Al (%)	Fe (%)	Mn (ppm)	Mn (μg)	Net ^{54}Mn (cpm)[(1)]			
0–0.14 (A)	13.96	4.28	547	181	1.710 ± .032	9.42 ± .26	7.78 ± .26	254 ± 13
0–0.14 (B)	14.32	3.43	430	166	1.600 ± .030	9.61 ± .26	7.97 ± .26	256 ± 13
0–0.14 (C)	14.12	4.16	486	149	1.572 ± .027	10.52 ± .28	8.88 ± .28	265 ± 13
0.14–0.28 (A)	14.39	4.15	517	158	1.287 ± .028	8.14 ± .24	6.50 ± .24	207 ± 11
0.14–0.28 (B)	14.89	3.46	433	176	1.590 ± .032	9.02 ± .26	7.38 ± .26	236 ± 12
0.14–0.28 (C)	14.05	3.95	480	193	1.752 ± .027	9.06 ± .23	7.42 ± .23	231 ± 12
0.28–0.56 (A)	14.03	4.13	524	210	1.444 ± .030	6.85 ± .20	5.21 ± .20	169 ± 9
0.28–0.56 (B)	14.86	3.74	449	180	1.373 ± .027	7.61 ± .21	5.97 ± .21	183 ± 10
0.28–0.56 (C)	14.45	3.86	456	245	1.994 ± .037	8.13 ± .22	6.49 ± .22	196 ± 10
0.56–1.12 (A)	14.10	3.72	466	210	1.284 ± .028	6.10 ± .18	4.46 ± .18	143 ± 8
0.56–1.12 (B)	14.66	3.62	433	210	1.370 ± .024	6.52 ± .17	4.88 ± .17	149 ± 8
0.56–1.12 (C)	14.81	3.23	394	159	1.053 ± .017	6.61 ± .17	4.97 ± .17	155 ± 8
1.12–2.24	16.02	2.44	323	188	1.069 ± .024	5.68 ± .17	4.04 ± .17	137 ± 8
2.8–4.2	14.02	3.32	379	229	1.032 ± .023	4.49 ± .13	2.85 ± .13	83 ± 5

[(1)] Weighted averages. All activities have been corrected to July 5, 1977. The uncertainties are 1 σ counting statistics.

[(2)] A correction of 0.0073 ± .0001 cpm ^{54}Mn/μg Fe for the ^{54}Fe(n,p)^{54}Mn contribution is included. This correction is $<0.4\%$ for this irradiation. A 2% uncertainty for the Mn yield has been added quadratically to the counting error. The uncertainty in the (n,p) correction is negligible.

[(3)] For this irradiation the correction for the ^{55}Mn(n,2n)^{54}Mn contribution was 1.64 ± .02 cpm^{54}Mn/mg Mn. The total uncertainty was calculated by quadratically adding the uncertainty in the (n,2n) correction to that for the ^{54}Mn(cpm/mg Mn).

[(4)] For this irradiation, the value for cpm^{54}Mn/dpm ^{53}Mn is 391 ± 6. The following uncertainties have been added quadratically: 2% for Mn concentration, 3% for Fe concentration, 1.5% for cpm ^{54}Mn/dpm ^{53}Mn, and the percentage uncertainty from the previous column. The 5% uncertainty in the original standardization of the ^{53}Mn standard has not been included.

Table 2.

depth (g/cm^2)[a]	FACE C (12°)		FACE B (37°)		FACE A (45°)	
	^{53}Mn[b]	^{26}Al[c]	^{53}Mn[b]	^{26}Al[c]	^{53}Mn[b]	^{26}Al[c]
0–0.14	265 ± 13	334 ± 30	256 ± 13	337 ± 30	254 ± 13	290 ± 26
0.14–0.28	231 ± 12	263 ± 25	236 ± 12	287 ± 27	207 ± 11	274 ± 26
0.28–0.56	196 ± 10	207 ± 21	183 ± 10	224 ± 19	169 ± 9	203 ± 18
0.56–1.12	155 ± 8	193 ± 17	149 ± 8	169 ± 19	143 ± 8	161 ± 13
1.12–2.24[d]	137 ± 8	147 ± 18				
2.8–4.2[d]	83 ± 5	96 ± 12				

[a]density of 68815 was taken to be 2.8 g/cm^3.
[b]dpm/kg Fe.
[c]dpm/kg rock.
[d]The two deepest samples were ground from areas on both faces C and A.

self-absorption factor based on a half-thickness for Al_2O_3 of 50.7 mg/cm^2; this factor was ~0.8. The uncertainties associated with the ^{26}Al values are: a 1 σ counting error, 5% for the efficiency, 5% for the chemical yield and 3% for the self-absorption factor. These uncertainties were quadratically added to give the final uncertainty.

The ^{26}Al values presented here are substantially lower than the numbers we reported in our abstract (Russ *et al.*, 1978). This change is due primarily to the difference in geometry between the sample holders used to count the samples from 68815 and the sample holder we had used for the ^{26}Al standard. The standard which was used to determine the activities which were reported by Russ *et al.* (1978) was mounted on a holder of an older style which did not cover the active area of the β counter as well as the newer design. When a NBS ^{26}Al standard was deposited on the new style holders and compared to the activity of the old standard, the new holder design was found to have increased the efficiency of the β-γ counter by ~20%. Based on additional counting, the new ^{26}Al standard mounted in the appropriate holder, and the ^{26}Al branching ratio for positron emission of 82.1% (Samworth *et al.*, 1972), the ^{26}Al activities for 68815 have been lowered by roughly 20%; as will be seen below, this change eliminates the need for a two-stage model SCR flux which was previously required to explain the ^{53}Mn and ^{26}Al results.

Discussion

Rock 68815, with a 2 m.y. surface exposure, provides for the first time an opportunity to determine the effective energy spectrum and magnitude of the SCR flux over a closed time interval. In contrast, our previous work with rocks 12002 and 14321, which have had long surface exposures compared to the half-lives measured, has looked at the SCR flux in a time interval which was essentially open-ended. The improved grinding procedure used for 68815 provides better depth resolution than in our previous work, and the detailed maps that were made during grinding will allow for the first time a calculation of the expected ^{53}Mn and ^{26}Al activities on a point to point basis. Therefore, the actual geometry of the sample can be taken into account.

The ^{53}Mn results for 12002, 14321 and 68815 are plotted in Fig. 2. The ^{26}Al results for these rocks are plotted in Fig. 3. The data for 12002 are from Finkel *et al.* (1971), and the data for 14321 are from Wahlen *et al.* (1972). A density of

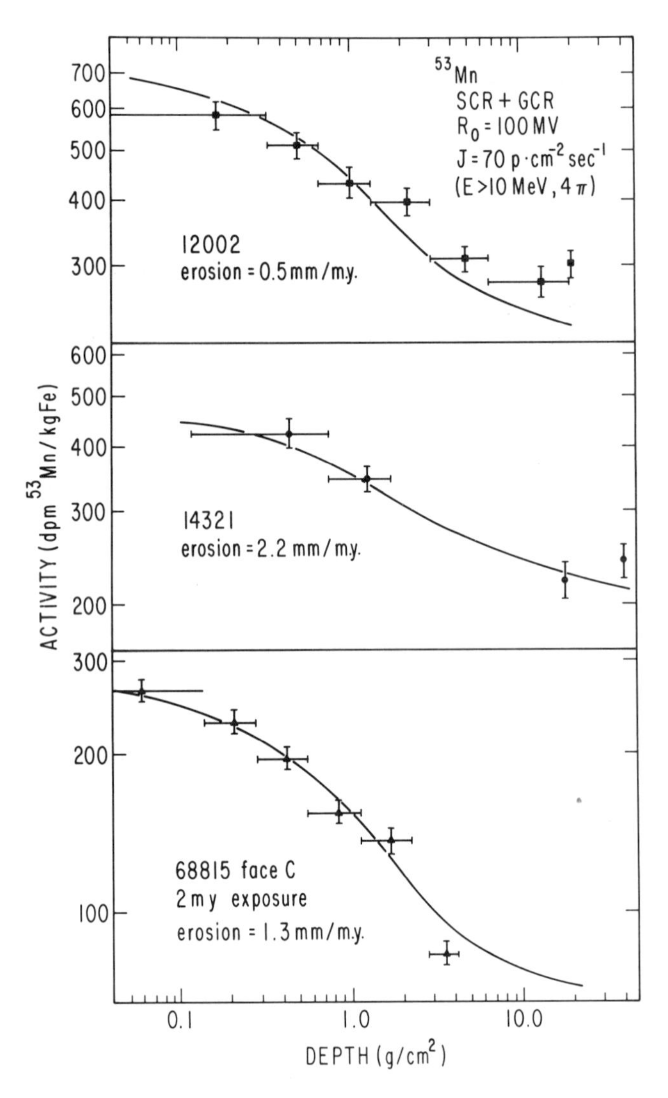

Fig. 2. Measured ^{53}Mn activity depth profile in rocks 12002, 14321 and 68815. For an explanation of the curves see the text.

3.3 g/cm^3 was assumed for 12002, 2.65 g/cm^3 for 14321 and 2.8 g/cm^3 for 68815. The curves shown are the expected profiles calculated according to the model of Reedy and Arnold. Each curve is the sum of the more penetrating GCR component and the lower energy SCR component. Based on earlier work with lunar cores in the 20 to 100 g/cm^2 region, the GCR contribution to the ^{53}Mn activity has been raised by 40% (Kohl *et al.*, 1977), and the ^{26}Al GCR component has been lowered by 13% (Kohl, 1975) relative to the values originally calculated by Reedy and Arnold. The SCR component was calculated

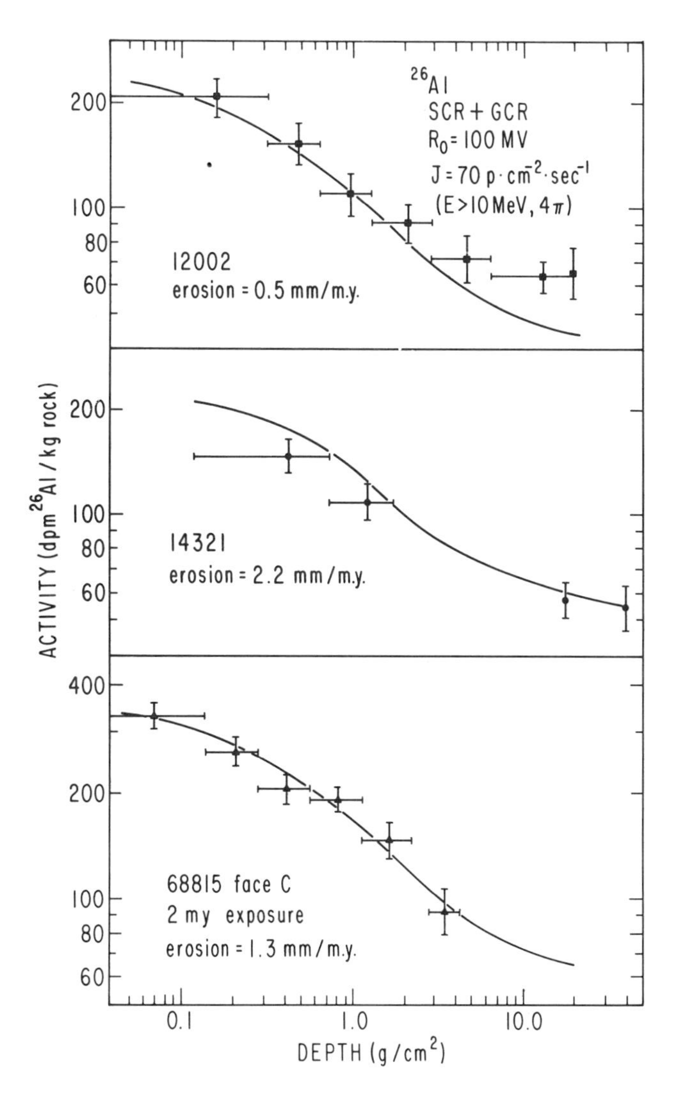

Fig. 3. Measured ^{26}Al activity depth profile in rocks 12002, 14321 and 68815. For an explanation of the curves see the text.

using the shape parameter R_o = 100 MV and a 4 π integral flux of 70 proton • cm^{-2} • sec^{-1} above 10 MeV. These SCR parameters provide a good fit to both the ^{53}Mn and ^{26}Al results for all three rocks and in addition fit our previous work with lunar cores (Nishiizumi *et al.*, 1976; Kohl *et al.*, 1977; Nishiizumi *et al.*, 1978). Except for the normalization of the GCR component, all theoretical curves have been calculated according to the model of Reedy and Arnold; there was no change made in any of the cross sections. For rock 68815, the theoretical curves are based on a 2 m.y. surface exposure; rocks 12002 and 14321 are considered saturated in both activities. A correction for the 12° slope of face C of 68815 has not been applied to its predicted curves, since a similar correction to 12002 and 14321, which also have faces which are not totally horizontal, is difficult to estimate. For the theoretical Reedy and Arnold curves shown in Figs. 2 and 3, the rocks were assumed to be hemispherical with a radius of 23 g/cm^2. The effects of the assumed radius of the rock are small for near surface regions. With increasing depth, smaller radii tend to increase the expected activity and thus flatten the predicted profile. The assumption of a radius of 23 g/cm^2 appears to underestimate the production rate in 12002, the smallest of the three rocks, at the greater depths. The fit for 14321 and 68815 at the deeper points is better. It should be noted that the ^{26}Al activities observed, and also the profile predicted for 68815, are higher than for either 12002 or 14321. This is due to an increased $^{27}Al(p,pn)^{26}Al$ production in 68815 because of its higher Al content.

Each curve in Figs. 2 and 3 has been calculated with an appropriate erosion rate. As the erosion rate is an indication of the hardness of the rock, the higher erosion for 14321 is in agreement with the fragile nature found in cutting the sample. The erosion rate of 1.3 mm/m.y. for 68815 is in good agreement with value of 1 mm/m.y. obtained by Walker and Yuhas (1973) from their track work. Blanford *et al.* (1975) calculated an erosion rate for this rock, based on track counts, of 2.5 to 3 mm/m.y. but noted that it could be local or an artifact.

Figures 2 and 3 demonstrate that rocks 12002, 14321 and 68815 can all be fit with the same set of SCR parameters for both ^{26}Al ($t_{1/2}$ = 0.72 m.y.) and ^{53}Mn ($t_{1/2}$ = 3.7 m.y.). Since the depth dependence for the production of ^{53}Mn and ^{26}Al is similar, the SCR parameters and erosion rates are not uniquely determined. For example, with an appropriate change in the erosion rates, the data for the three rocks can be fit with R_o = 150 and J = 45. The most important feature of the data is that all three rocks can be fit with the same set of SCR parameters for both nuclides. ^{26}Al for all three rocks samples almost the same effective time interval (~2 m.y.), since in 68815 it is 86% saturated; however, ^{53}Mn has been produced during quite a different time interval for 68815 (the past 2 m.y.), than for 12002 and 14321 (the past 10 m.y.). ^{53}Mn is only 31% saturated in 68815 as reflected by the lower measured activities. The agreement of SCR parameters found for the three rocks strongly indicates that this flux has been constant within the past 10 m.y. Of course small scale or short time variations could have occurred without being reflected in the data, since the activities represent an accumulation over time.

Imamura *et al.* (1974) reported measurements of ^{53}Mn in rock 14310 which can be fit with essentially the same SCR parameters and erosion rate as those for 12002, indicating saturated ^{53}Mn and a long recent surface exposure for 14310 in contrast to track reports (Yuhas *et al.*, 1972; Crozaz *et al.*, 1972). The present 2 m.y. data on 68815 reinforce our earlier conclusion of saturated ^{53}Mn in 14310.

In Table 2, it is apparent that both the ^{53}Mn and ^{26}Al values for faces B and C are essentially identical in spite of the large differences in the slope of the faces. The inclination of the face influences the amount of sky or SCR flux to which it is exposed. Face B is expected to have an activity 15% less than face C for both nuclides, assuming that the average flux is isotropic. And although the data for face A are somewhat lower than for the other faces, they are not as low as expected for a slope of 45°. The reason for the near independence of surface activities on sample inclination is not well understood. If erosion is strongly related to the amount of sky to which each face is exposed, decreased erosion could effectively cancel the decrease in SCR flux. The relatively high values for the angled faces might also be due to an anisotropy in the SCR flux, or to an effect caused by the surface roughness of the rock, or it may be due to the rather complex geometry of the rock. Additional discussion will follow when we have calculated the expected activities on a point to point basis using measurements made during grinding.

Acknowledgments—The authors wish to thank Patrick Butler for his help in the cutting of 68815, Robert Sutton for providing us with the rock's proper lunar orientation, Norman Fong for assisting with the grinding procedure and Philip Davis for his help with computer model studies. Florence Kirchner furnished essential support to the work. This work was supported by NASA Grant NAS NGL 05-009-148.

References

Apollo 16 Preliminary Examination Team (1973) The Apollo 16 lunar samples: A petrologic and chemical description. *Science* **179**, 23–34.

Apollo 16 Prelim. Sci. Rep. (1972) NASA SP-315, p. 6–80.

Behrmann C., Crozaz G., Drozd R., Hohenberg C., Ralston C., Walker R. and Yuhas D. (1973) Cosmic-ray exposure history of North Ray and South Ray material. *Proc. Lunar Sci. Conf. 4th*, p. 1957–1974.

Blanford G. E., Fruland R. M. and Morrison D. A. (1975) Long-term differential energy spectrum for solar-flare iron-group particles. *Proc. Lunar Sci. Conf. 6th*, p. 3557–3576.

Brown G. M., Peckett A., Phillips R. and Emeleus C. H. (1973) Mineral-chemical variations in the Apollo 16 magnesio-feldspathic highland rocks. *Proc. Lunar Sci. Conf. 4th*, p. 505–518.

Crozaz G., Drozd R., Hohenberg C. M., Hoyt H. P. Jr., Ragan D., Walker R. M. and Yuhas D. (1972) Solar flare and galactic cosmic ray studies of Apollo 14 and 15 samples. *Proc. Lunar Sci. Conf. 3rd*, p. 2917–2931.

Drozd R. J., Hohenberg C. M., Morgan C. J. and Ralston C. E. (1974) Cosmic-ray exposure history at the Apollo 16 and other lunar sites: Lunar surface dynamics. *Geochim. Cosmochim. Acta* **38**, 1625–1642.

Dust S. and Crozaz G. (1977) 68815 Revisited. *Proc. Lunar Sci. Conf. 8th*, p. 2315–2319.

Eddy J. A. (1976) The Maunder minimum. *Science* **192**, 1189–1201.

Finkel R. C. (1972) Depth profiles of galactic cosmic ray produced radionuclides in lunar samples. Ph.D. thesis, University of California, San Diego.

Finkel R. C., Arnold J. R., Imamura M., Reedy R. C., Fruchter J. S., Loosli H. H., Evans J. C. and Delany A. C. (1971) Depth variations of cosmogenic nuclides in a lunar surface rock and lunar soil. *Proc. Lunar Sci. Conf. 2nd*, p. 1773–1789.

Fruchter J. S., Rancitelli L. A., Laul J. C. and Perkins R. W. (1977) Lunar regolith dynamics based on analysis of the cosmogenic radionuclides ^{22}Na, ^{26}Al and ^{53}Mn. *Proc. Lunar Sci. Conf. 8th*, p. 3595–3605.

Imamura M., Finkel R. C. and Wahlen M. (1973) Depth profile of ^{53}Mn in the lunar surface. *Earth Planet. Sci. Lett.* **20**, 107–112.

Imamura M., Nishiizumi K., Honda M., Finkel R. C., Arnold J. R. and Kohl C. P. (1974) Depth profiles of ^{53}Mn in lunar rocks and soils. *Proc. Lunar Sci. Conf. 5th*, p. 2093–2103.

Kohl C. P. (1975) Galactic cosmic ray produced radioactivity in lunar and chondritic materials. Ph.D. thesis, University of California, San Diego.

Kohl C. P., Russ III G. P., Arnold J. R., Nishiizumi K., Imamura M. and Honda M. (1977) ^{53}Mn in lunar cores: Evidence for the time scale of surface gardening (abstract). In *Lunar Science VIII*, p. 552–554. The Lunar Science Institute, Houston.

Nishiizumi K., Imamura M., Honda M., Russ III G. P., Kohl C. P. and Arnold J. R. (1976) ^{53}Mn in the Apollo 15 and 16 drill stems: Evidence for surface mixing. *Proc. Lunar Sci. Conf. 7th*, p. 41–54.

Nishiizumi K., Imamura M., Honda M., Murrell M. T. and Arnold J. R. (1978) A study of gardening in the lunar regolith using ^{53}Mn (abstract). In *Lunar and Planetary Science IX*, p. 811–813. Lunar and Planetary Science Institute, Houston.

Pepin R. O., Basford J. R., Dragon J. C., Coscio M. R. Jr. and Murthy V. R. (1974) Rare gases and trace elements in Apollo 15 drill core fines: Depositional chronologies and K-Ar ages, and production rates of spallation-produced ^{3}He, ^{21}Ne and ^{38}Ar versus depth. *Proc. Lunar Sci. Conf. 5th*, p. 2149–2184.

Reedy R. C. and Arnold J. R. (1972) Interaction of solar and galactic cosmic-ray particles with the moon. *J. Geophys. Res.* **77**, 537–555.

Russ G. P. III, Kohl C. P., Murrell M. T. and Arnold J. R. (1978) An experiment on the constancy of the SCR flux over the past two million years (abstract). In *Lunar and Planetary Science IX*, p. 982–984. Lunar and Planetary Science Institute, Houston.

Samworth E. A., Warburton E. K. and Engelbertink G. A. P. (1972) Beta decay of the ^{26}Al ground state. *Phys. Rev.* **C 5**, 138–142.

SHRELLDALFF (1970) Pattern of bombardment-produced radionuclides in rock 10017 and in lunar soil. *Proc. Apollo 11 Lunar Sci. Conf.*, p. 1503–1532.

Wahlen M., Honda M., Imamura M., Fruchter J. S., Finkel R. C., Kohl C. P., Arnold J. R. and Reedy R. C. (1972) Cosmogenic nuclides in football-sized rocks. *Proc. Lunar Sci. Conf. 3rd*, p. 1719–1732.

Walker R. and Yuhas D. (1973) Cosmic ray track production rates in lunar materials. *Proc. Lunar Sci. Conf. 4th*, p. 2379–2389.

Wänke H., Palme H., Baddenhausen H., Dreibus G., Jagoutz E., Kruse H., Spettel B., Teschke F. and Thacker R. (1974) Chemistry of the Apollo 16 and 17 samples: Bulk composition, late stage accumulation and early differentiation of the moon. *Proc. Lunar Sci. Conf. 5th*, p. 1307–1335.

Yokoyama Y., Reyss J. L. and Guichard F. (1974) ^{22}Na-^{26}Al chronology of lunar surface processes. *Proc. Lunar Sci. Conf. 5th*, p. 2231–2247.

Yuhas D. E., Walker R. M., Reeves H., Poupeau G., Pellas P., Lorin J. C., Chetrit G. C., Berdot J. L., Price P. B., Hutcheon I. D., Hart H. R. Jr., Fleischer R. L., Comstock G. M., Lal D., Goswami J. N. and Bhandari N. (1972) Track consortium report on rock 14310. *Proc. Lunar Sci. Conf. 3rd*, p. 2941–2947.

Proc. Lunar Planet. Sci. Conf. 9th (1978), p. 2311–2344.
Printed in the United States of America

Comparisons between observed and predicted cosmogenic noble gases in lunar samples

C. M. HOHENBERG[1,2], K. MARTI[4], F. A. PODOSEK[1,3],
R. C. REEDY[5] and J. R. SHIRCK[1,2]

[1]McDonnell Center for the Space Sciences, Washington University
St. Louis, Missouri 63130
[2]Department of Physics, Washington University
St. Louis, Missouri 63130
[3]Department of Earth and Planetary Sciences, Washington University
St. Louis, Missouri 63130
[4]Department of Chemistry, University of California
San Diego, California 92093
[5]Los Alamos Scientific Laboratory
Los Alamos, New Mexico 87545

Abstract—Comparisons are made between cosmogenic production rates and isotopic ratios predicted on the basis of nuclear systematics with those observed in well-documented lunar samples. Although some significant differences are found, the agreement in general is surprisingly good. The predicted production rates for ^{38}Ar and ^{126}Xe agree reasonably well with observation but the predicted rate for ^{21}Ne appears to be too high; in most cases the predicted rate for ^{83}Kr agrees to within 50%, although some larger differences are apparent. There are some prominent cases in which predicted isotopic compositions differ from observation: ^{78}Kr/^{83}Kr trends in the wrong sense with the Zr/Sr ratio; ^{124}Xe/^{126}Xe is too large by about 13%; ^{128}Xe/^{126}Xe is too small by about 6%; ^{130}Xe/^{126}Xe is too low by 30–40%.

INTRODUCTION

Noble gases found in extraterrestrial materials are superpositions of several different components, where a "component" is defined operationally as an individual constituent or source of gas with a unique and well-defined isotopic composition. In general terms, noble gas components can be divided into two classes: "trapped" components which are incorporated from external sources, and "*in situ*" components which are generated by nuclear processes within the sample. Three major sources dominate the inventory of noble gases in extraterrestrial samples: a trapped component, normally gas of the "solar" or "planetary" types (Signer and Suess, 1963), a cosmogenic component from cosmic ray induced nuclear reactions, and components from various radioactive decay processes. Successful spectral decomposition—determination of the amounts and/or compositions of various superposed components—is usually an important part of noble gas studies and, in practice, often the most difficult to accomplish. In this paper we focus on the cosmogenic component.

Cosmogenic gases are produced in significant quantities in samples within about two meters from a surface exposed to cosmic rays, and their abundances are the principal means by which such exposure can be studied. Cosmogenic gases have thus figured importantly in studies of meteorite orbital and parent body histories, in the dating of lunar craters, and in more general constraints on the dynamics of the lunar regolith. Whether or not the cosmogenic gases are of intrinsic interest, they must frequently be resolved and be removed ("corrected for") in order to allow an accurate assessment of the other components present.

It would be convenient if there were "a" cosmogenic component, a single component of unique composition, so that the problem of cosmogenic noble gases could be reduced simply to determination of the amount present in a given sample. Unfortunately, this simplification is inadequate for most applications. The production of a given nuclide depends on the abundances of multiple target nuclides and the flux and spectrum of cosmic ray primary and secondary protons and neutrons, which are in turn dependent on shielding depth and the bulk composition of the target material. Variations in relative production rates are nearly as important as variations in absolute rates. Although surface exposure ages and first order corrections for cosmogenic gases can sometimes be made for samples with single stage exposure histories (buried at depth until excavation to final surface residence in one step), it has become increasingly apparent that such simple surface exposure histories are rare among lunar samples, most of which have experienced complex exposure histories during which these parameters may have changed substantially. In special circumstances chronologically meaningful parameters and spectral decompositions can, in principle, be extracted through multi-stage models for cosmic ray exposure but so far with only limited success, due largely to inadequate knowledge of the appropriate production parameters.

Two approaches to a quantitative evaluation of cosmogenic noble gases have been used, which may be figuratively denoted as *a priori* and *a posteriori* approaches. The *a posteriori* or empirical approach consists of attempts to identify cosmogenic gases in favorable samples, specifically in cases where correction for non-cosmogenic gases can be made with the least possible ambiguity. Where cosmogenic nuclides can be identified clearly, it may be possible to establish, strictly on an empirical basis, generalizations about both absolute and relative production ratios, in the expectation that these generalizations may usefully be extrapolated to cases where most cosmogenic contributions are not so easily identified. Examples of treatments guided essentially by this viewpoint are given by Marti *et al.* (1966), Bochsler *et al.* (1969), Podosek and Huneke (1971), and Bogard and Cressy (1973).

The *a priori* approach consists of predicting the production of cosmogenic nuclides on the basis of nuclear properties and the physics of cosmic ray interactions. The principle strength of this approach for the noble gas isotopes is that it is potentially capable of supplying parameters that are observationally elusive but necessary for adequate treatment of specific data sets. This treatment has been developed most extensively by Reedy (1976) (also see Rudstam, 1966; Silberberg and Tsao, 1973a,b; Reedy and Arnold, 1972), and has been quite successful in the prediction of lunar radionuclide abundances (Reedy and Arnold, 1972).

In practice, many investigations concerned with cosmogenic gases combine these two approaches in varying degree. Recent examples are given by Pepin *et al.* (1974), Burnett *et al.* (1975), Drozd *et al.* (1977), Niemeyer (1977), Bernatowicz *et al.* (1977), Eugster *et al.* (1977, 1978), Eugster (1978), and Reedy *et al.* (1978). These investigations are limited applications of the Reedy (1976) systematics to specific problems, generally attempts to untangle "complex" exposure histories: accumulation of cosmogenic gases in more than one shielding configuration.

In this paper we attempt a comprehensive comparison of the *a priori* and *a posteriori* approaches. Specifically, we will evaluate cosmogenic noble gas components in a number of lunar rocks for which target chemistry and the duration and circumstances of cosmic ray exposure can best be determined and compare these data with *a priori* predictions. Our goal is to systematically identify areas in which the predictions do or do not agree with observation, thus identifying cases in which prediction can be used with the greatest confidence and hopefully indicating the means whereby modified calculational procedures or input parameters can improve the predictive model. We will confine the scope of this paper to consideration of lunar samples with simple exposure histories and the appropriate 2π geometry; a summary of preliminary results from a similar consideration of light cosmogenic noble gases in meteorites (with a more complicated geometry) is presented by Reedy *et al.* (1978).

Theoretical Systematics

In this section we describe theoretical calculations leading to predictions for the production of individual cosmogenic nuclides. There are two basic steps in these calculations. The first is an estimation of the flux and spectrum of the particles which induce the nuclear reactions of interest. The calculations considered here all assume 2π exposure geometry, i.e., energetic particles incident on a plane surface of infinitely thick target material. The second is determination of the rate at which specific cosmogenic nuclides are produced, i.e., for each target the integral over energy of the inducing particle flux times the reaction cross section. Three components of reaction-inducing particles are considered separately: solar cosmic rays (SCR), energetic galactic cosmic ray (GCR) particles, and very low energy neutrons, generated as secondaries of the GCR, which interact only through capture reactions.

Solar cosmic rays (SCR)

Solar (flare) protons constitute a relatively low energy component which attenuates quickly with depth. Because of the high flux, however, SCR are a major component in the first few g/cm^2 of shielding, especially for the light noble gases. On the basis of activities of ^{53}Mn and ^{26}Al in lunar samples, the average flux of solar protons incident on the moon over the last several million years can be described as an omnidirectional flux of 70 protons/cm^2-sec above 10 MeV with an exponential-rigidity spectral shape [cf., Reedy and Arnold (1972) with R_o = 100 MV (Nishiizumi *et al.*, 1976; Kohl *et al.*, 1978)]. For this incident flux the models of Reedy and Arnold (1972) were used to calculate proton spectrum and flux inside target material. The proton reaction cross sections of Walton *et al.* (1976) were used to calculate the production of Ne and Ar isotopes and the proton reaction cross sections of Kaiser (1977) were used for the production of Xe isotopes from Ba. There are no experimental cross sections for the production of Kr isotopes by protons with energies below 100 MeV. Excitation functions for proton reactions with Sr (Sachdev *et al.*, 1967) were used to estimate the cross sections for the production of Kr isotopes.

Energetic GCR particles

The production of noble gases by GCR particles with energies above about 1 MeV was calculated using the model and the lunar GCR fluxes of Reedy and Arnold (1972). For lunar radionuclides, this model has been very successful in calculating the shape of the production rate as a function of depth and reasonably good in calculating the absolute production rates (cf., Nishiizumi *et al.*, 1976). For most noble gas isotopes, a large fraction of the production is induced by GCR particles with energies above several hundred MeV. Experimental cross sections for the production of noble gases by high-energy protons have been reported by Goebel *et al.* (1964) for Ne, Stoenner *et al.* (1970) for Ar, and Funk *et al.* (1967) for Kr from Sr and Xe from Ba [the Xe cross sections agreeing well with those of Kaiser (1977)]. Xenon isotope ratios were measured by Hohenberg and Rowe (1970) for reactions of 730-MeV protons with Cs, Ce, Nd, and Dy. The experimental data were used to normalize spallation cross sections calculated with the formulae of Rudstam (1966) for energies at which there were no experimental cross sections.

For the production of Xe from rare earth elements, the normalizations determined for proton reactions with Ba were used to determine the absolute magnitudes of the cross sections. Since most of the GCR particles with energies below 100 MeV are secondary neutrons, the cross sections which should be used for such energies are those for neutron-induced reactions (cf. Reedy and Arnold, 1972). Thus the cross sections used at low energy for the GCR production of ^{20}Ne from Al, of ^{21}Ne and ^{22}Ne from Mg, Al and Si, and of ^{38}Ar from Ca were modified from available proton cross sections.

Because of the lack of relevant experimental data none of the Kr or Xe production cross sections were so modified, and thus certain neutron-induced reactions, e.g., $^{130}Ba(n,p)^{130}Cs$ (which decays to ^{130}Xe), were not included in these production-rate calculations.

Slow neutron capture

Neutrons with energies below about 0.5 MeV cannot induce spallation reactions (i.e., reactions which emit nucleons other than the incident particle from the excited nucleus), but scatter from nuclei until they escape from the lunar surface or are slowed down until they are captured by (n,γ) reactions. We consider here neutron-capture reactions with ^{130}Ba, ^{79}Br, and ^{81}Br producing ^{131}Xe, ^{80}Kr, and ^{82}Kr, respectively. Lingenfelter *et al.* (1972) (hereafter denoted LCH) calculated the fluxes of low-energy neutrons as a function of depth in the moon and, using these fluxes, determined production rates for many neutron-capture products. The results of Apollo 17 Lunar Neutron Probe Experiment indicated that the actual lunar neutron densities are 0.8 of the LCH-calculated ones (Woolum *et al.*, 1975). For the production rates of ^{80}Kr and ^{82}Kr from neutron capture in Br we thus take those of LCH times 0.8. For ^{130}Ba(n,γ), LCH used an average cross section of 212 barns from 1 eV to 1 keV. The resonance parameters for this reaction were measured by Berman and Browne (1973). Using these resonance parameters and the LCH lunar-neutron fluxes (E. H. Canfield, pers. comm., 1973), the effective cross section from 1 eV to 1 keV was found to be about 0.14 of that assumed by LCH. Since most of the production of ^{131}Ba is made by neutrons with energies between 30 and 100 eV, ^{131}Ba production rates were calculated using an effective cross section of 163 barns and 0.8 times the LCH fluxes for this narrow energy group. These rates calculated using the energy interval 30–100 eV were about 0.16 and 0.14 of the LCH rates at the lunar surface and at depths of about 400 g/cm^2, respectively.

Cosmogenic production rates

Production rates for individual cosmogenic nuclides at various shielding depths, calculated by the procedures outlined above, are presented in Table 1. Production rates are given in units of atoms per minute per kilogram of target element (1 atom/min/kg = 1.96×10^{-11} cm^{3}STP/g-m.y.), appropriate for 2 π exposure geometry. The tabulated figures include both direct production and indirect production (radioactive species which decay to noble gas isotopes). GCR and SCR (at 1 AU) spallation rates are shown separately. Secondary neutron capture reactions are shown separately under the Br entry for Kr and under the second Ba entry for Xe. The La production for Xe includes contributions from other rare earth elements. The target elements included in Table 1 collectively account for essentially all cosmogenic production in natural silicates. The general trend for the depth dependence of Kr and Xe isotope production by GCR particles is illustrated in Figs. 1 and 2. In general terms, the data in Table 1 are similar to those used by Reedy (1976), and which have been privately circulated by R. C. Reedy. They are somewhat different, however, because of the use of recently available cross section data for Ne and because of the explicit consideration of low energy neutron capture.

The Table 1 data are calculated using excitation functions based on experimental cross sections or pure target production ratios and the Reedy and Arnold (1972) particle fluxes and spectra. In the sense important in this paper, however, it is important to note that these data are strictly "theoretical," in that they contain no input parameters which are selected or adjusted to match noble gas

Table 1a. Predicted Cosmogenic[a] Production Rates for Neon (atoms/min/kg).

	Na[d]	Mg	Al	Si	Ca	Fe
			Surface			
20	–	239 + 4780	90.4 + 692	88.4 + 287	10.2 + .227	1.73 + .021
21	128 + –	232 + 1550	119 + 259	73.3 + 88.7	18.2 + .758	2.14 + .028
22	309 + 3830	296 + 1470	142 + 743	105 + 222	15.5 + .888	2.01 + .025
			1 g/cm² Shielding			
20	–	248 + 976	91.4 + 140	89.1 + 114	10.1 + .171	1.70 + .016
21	130 + –	240 + 350	120 + 130	74.0 + 41.6	18.0 + .562	2.10 + .021
22	317 + 937	302 + 504	145 + 302	105 + 114	15.4 + .651	1.98 + .019
			2 g/cm² Shielding			
20	–	257 + 503	92.6 + 74.3	89.8 + 70.3	9.95 + .141	1.68 + .013
21	132 + –	249 + 204	121 + 88.1	74.6 + 28.3	17.8 + .459	2.07 + .017
22	326 + 518	308 + 311	147 + 187	106 + 78.1	15.3 + .529	1.94 + .015
			5 g/cm² Shielding			
20	–	287 + 161	96.3 + 26.0	92.3 + 28.0	9.60 + .092	1.60 + .009
21	140 + –	278 + 80.6	126 + 40.5	76.9 + 13.5	17.3 + .293	1.96 + .011
22	353 + 183	328 + 127	156 + 75.4	109 + 36.3	14.9 + .334	1.85 + .010
			10 g/cm² Shielding			
20	–	334 + 53.7	102 + 9.77	96.2 + 10.9	9.09 + .054	1.49 + .005
21	151 + –	323 + 31.8	132 + 17.4	80.1 + 6.17	16.5 + .170	1.82 + .006
22	397 + 65.8	359 + 50.2	168 + 29.3	113 + 15.7	14.4 + .191	1.72 + .006
			20 g/cm² Shielding			
20	–	404 + 13.9	109 + 2.94	101 + 3.24	8.28 + .025	1.34 + .002
21	167 + –	390 + 9.60	140 + 5.65	83.9 + 2.14	15.2 + .075	1.61 + .003
22	459 + 18.2	402 + 15.0	185 + 8.60	118 + 5.11	13.5 + .084	1.54 + .003
			40 g/cm² Shielding			
20	–	468	111	100	6.83	1.07
21	176	452	140	82.9	12.7	1.29
22	510	430	195	116	11.5	1.24
			65 g/cm² Shielding			
20	–	480	104	91.9	5.38	.837
21	169	462	129	75.7	10.2	.995
22	508	417	186	105	9.28	.960
			100 g/cm² Shielding			
20	–	435	87.6	76.2	3.86	.593
21	145	418	107	62.4	7.39	.701
22	451	362	159	86.3	6.81	.678
			150 g/cm² Shielding			
20	–	368	68.2	58.2	2.45	.370
21	116	353	81.5	47.2	4.76	.435
22	372	292	126	64.7	4.44	.422
			225 g/cm² Shielding			
20	–	281	47.5	39.6	1.28	.190
21	82.6	269	54.8	31.6	2.54	.222
22	227	211	89.9	42.8	2.41	.216

Table 1a. *(cont'd.)*.

	Na[d]	Mg	Al	Si	Ca	Fe
			500 g/cm² Shielding			
20	–	52.6	7.85	6.32	.125	.018
21	14.0	50.1	8.46	4.90	.260	.021
22	50.1	36.7	15.1	6.42	.253	.020

Table 1b. Predicted Cosmogenic[a] Production Rates for Argon (atoms/min/kg).

	K	Ca	Ti	Fe		K	Ca	Ti	Fe
		Surface					*40 g/cm² Shielding*		
36	300 + 4300	215 + 2400	64.8 + 3.93	10.4 + .550	36	568	393	70.0	7.29
37	173 + 713	288 + 913	26.2 + 2.80	6.66 + .322	37	221	628	25.5	4.80
38	313 + 5230	313 + 2510	60.4 + 31.3	14.8 + .575	38	580	576	56.5	10.9
39	296 + 71.3	37.9 + 18.3	37.7 + 29.1	7.83 + 1.38	39	910	99.5	39.1	5.77
40	5.42 + 56.7	2.46 + 35.9	10.5 + 20.8	3.86 + .256	40	10.8	4.60	11.6	2.82
		1 g/cm² Shielding					*65 g/cm² Shielding*		
36	311 + 575	222 + 553	65.4 + 2.79	10.3 + .411	36	579	397	61.8	5.81
37	175 + 282	301 + 301	26.3 + 2.03	6.60 + .242	37	209	653	22.2	3.85
38	324 + 972	323 + 846	60.5 + 19.1	14.7 + .442	38	589	584	48.9	8.83
39	318 + 29.6	40.1 + 8.03	37.9 + 16.0	7.77 + .957	39	984	106	34.8	4.65
40	5.62 + 15.0	2.54 + 9.84	10.6 + 8.92	3.82 + .188	40	11.1	4.66	10.5	2.28
		2 g/cm² Shielding					*100 g/cm² Shielding*		
36	322 + 291	229 + 302	66.0 + 2.22	10.2 + .336	36	523	357	49.2	4.21
37	178 + 175	315 + 188	26.4 + 1.64	6.54 + .198	37	177	600	17.5	2.80
38	335 + 509	334 + 489	60.6 + 14.2	14.6 + .368	38	531	525	38.2	6.47
39	342 + 18.8	42.5 + 5.22	38.1 + 11.3	7.71 + .753	39	928	98.9	27.9	3.41
40	5.85 + 8.34	2.63 + 5.49	10.6 + 5.86	3.79 + .152	40	10.1	4.19	8.63	1.68
		5 g/cm² Shielding					*150 g/cm² Shielding*		
36	357 + 95.3	253 + 106	67.8 + 1.35	9.87 + .216	36	440	298	35.3	2.70
37	186 + 71.6	358 + 78.4	26.7 + 1.03	6.37 + .127	37	139	513	12.4	1.80
38	370 + 170	369 + 177	61.0 + 7.61	14.2 + .244	38	445	439	27.0	4.20
39	416 + 7.86	50.1 + 2.29	38.9 + 5.75	7.53 + .448	39	816	86.0	20.3	2.21
40	6.55 + 2.92	2.92 + 1.94	10.9 + 2.69	3.70 + .096	40	8.55	3.50	6.46	1.10
		10 g/cm² Shielding					*225 g/cm² Shielding*		
36	412 + 33.5	290 + 38.1	70.3 + .750	9.43 + .125	36	335	225	21.6	1.43
37	198 + 28.3	425 + 31.4	27.1 + .585	6.12 + .075	37	97.6	398	7.57	.962
38	426 + 59.9	424 + 62.8	61.4 + 3.74	13.7 + .146	38	337	332	16.5	2.27
39	534 + 3.18	62.1 + .961	39.9 + 2.70	7.26 + .241	39	651	67.8	12.9	1.20
40	7.66 + 1.03	3.36 + .694	11.4 + 1.19	3.56 + .054	40	6.55	2.65	4.26	.599
		20 g/cm² Shielding					*500 g/cm² Shielding*		
36	494 + 9.27	345 + 10.5	72.6 + .314	8.70 + .056	36	62.5	41.2	2.73	.143
37	215 + 8.50	528 + 9.52	27.3 + .253	5.69 + .033	37	16.2	75.9	.971	.097
38	507 + 16.4	505 + 16.9	61.1 + 1.38	12.9 + .068	38	62.4	61.1	2.15	.233
39	716 + .975	80.4 + .304	40.9 + .959	6.78 + .098	39	129	13.2	1.79	.123
40	9.30 + .278	4.02 + .189	11.8 + .402	3.32 + .023	40	1.23	.485	.636	.063

Table 1c. Predicted Cosmogenic[a] Production Rates for Krypton (atoms/min/kg).

	Rb	Sr	Y	Zr	Br[b]		Rb	Sr	Y	Zr	Br[b]
			Surface					*40 g/cm^2 Shielding*			
78	27.1 + 64.6	20.4 + 17.1	17.8 + 15.0	11.7 + 5.80	0	78	33.6	22.6	19.8	10.1	0
80	38.9 + 157	47.8 + 64.6	39.3 + 50.4	38.9 + 34.0	5710	80	51.7	55.1	44.1	40.6	29500
81	34.1 + 126	53.5 + 95.0	42.3 + 74.6	39.2 + 38.4	0	81	42.8	64.7	49.9	41.2	0
82	53.3 + 255	63.2 + 156	49.2 + 117	46.9 + 61.3	2470	82	69.3	79.7	60.9	50.6	12800
83	101 + 1260	67.7 + 201	58.8 + 144	58.7 + 94.0	0	83	178	88.6	74.7	65.2	0
84	132 + 1570	46.3 + 204	47.0 + 108	20.1 + 34.2	0	84	220	64.5	57.2	22.6	0
86	7.68 + 29.0	.911 + 1.60	.005 + .004	.242 + .662	0	86	8.93	.921	.006	.318	0
		1 g/cm^2 Shielding						*65 g/cm^2 Shielding*			
78	27.4 + 35.8	20.6 + 11.2	18.0 + 9.77	11.7 + 3.60	0	78	31.4	20.4	17.8	8.65	0
80	39.5 + 71.3	48.3 + 38.2	39.6 + 29.1	39.1 + 21.0	6400	80	49.1	50.2	40.0	36.2	42400
81	34.5 + 52.4	54.1 + 53.0	42.7 + 39.5	39.4 + 22.7	0	81	40.2	59.6	45.8	36.8	0
82	54.0 + 88.2	64.1 + 78.6	49.9 + 55.8	47.3 + 33.6	2770	82	65.7	74.3	56.5	45.6	18400
83	104 + 395	68.8 + 94.8	59.6 + 74.2	59.2 + 47.5	0	83	179	83.2	69.8	59.1	0
84	136 + 397	47.1 + 81.4	47.6 + 51.9	20.3 + 17.9	0	84	219	61.6	52.9	20.6	0
86	7.75 + 9.84	.914 + .700	.005 + .003	.246 + .338	0	86	8.25	.820	.005	.299	0
		2 g/cm^2 Shielding						*100 g/cm^2 Shielding*			
78	27.8 + 25.0	20.8 + 8.58	18.1 + 7.44	11.6 + 2.67	0	78	26.4	16.5	14.4	6.71	0
80	40.2 + 46.2	48.8 + 27.7	40.0 + 21.0	39.4 + 15.6	7089	80	41.9	41.2	32.7	29.1	53300
81	35.0 + 33.6	54.8 + 37.6	43.2 + 27.7	39.7 + 16.6	0	81	33.9	49.4	37.8	29.6	0
82	54.8 + 54.7	65.0 + 53.5	50.6 + 38.1	47.6 + 23.8	3069	82	55.8	62.2	47.1	37.1	23100
83	107 + 223	69.9 + 63.1	60.5 + 50.8	59.7 + 33.0	0	83	160	70.1	58.5	48.3	0
84	140 + 222	48.1 + 52.2	48.2 + 35.0	20.5 + 12.3	0	84	194	52.6	43.9	16.9	0
86	7.83 + 5.93	.918 + .451	.005 + .002	.250 + .232	0	86	6.87	.660	.004	.253	0
		5 g/cm^2 Shielding						*150 g/cm^2 Shielding*			
78	29.0 + 11.9	21.3 + 4.76	18.7 + 4.12	11.6 + 1.41	0	78	20.4	12.2	10.6	4.71	0
80	42.2 + 19.8	50.5 + 14.1	41.2 + 10.7	40.2 + 8.24	9180	80	33.1	30.9	24.4	21.3	52900
81	36.5 + 14.4	57.1 + 18.5	44.8 + 13.5	40.5 + 8.61	0	81	26.5	37.5	28.6	21.8	0
82	57.5 + 22.7	68.1 + 24.8	52.8 + 18.0	48.7 + 11.8	3970	82	43.9	47.8	36.1	27.5	22900
83	117 + 78.7	73.6 + 28.7	63.4 + 23.8	61.4 + 15.9	0	83	133	54.3	45.0	36.1	0
84	152 + 80.5	51.0 + 22.7	50.2 + 16.3	21.1 + 5.81	0	84	160	41.5	33.5	12.6	0
86	8.08 + 2.36	.931 + .195	.005 + .001	.263 + .109	0	86	5.27	.487	.003	.196	0
		10 g/cm^2 Shielding						*225 g/cm^2 Shielding*			
78	30.7 + 5.14	22.3 + 2.36	19.4 + 2.05	11.5 + .674	0	78	14.1	7.83	6.83	2.87	0
80	45.4 + 8.01	52.9 + 6.53	43.0 + 4.96	41.3 + 3.94	12650	80	23.3	20.4	16.0	13.6	44500
81	38.8 + 5.87	60.4 + 8.33	47.2 + 6.11	41.7 + 4.07	0	81	18.4	25.1	19.1	14.0	0
82	61.6 + 9.16	72.6 + 10.8	56.1 + 7.93	50.4 + 5.40	5480	82	30.9	32.5	24.4	17.9	19300
83	134 + 27.7	79.0 + 12.3	67.7 + 10.4	63.8 + 7.15	0	83	99.7	37.3	30.7	23.8	0
84	171 + 29.6	55.5 + 9.45	53.2 + 7.13	22.0 + 2.56	0	84	119	29.2	22.6	8.37	0
86	8.47 + .934	.950 + .081	.006 + .001	.282 + .048	0	86	3.61	.316	.002	.135	0
		20 g/cm^2 Shielding						*500 g/cm^2 Shielding*			
78	33.0 + 1.62	23.2 + .859	20.2 + .750	11.2 + .239	0	78	2.26	1.10	.955	.377	0
80	49.6 + 2.42	55.7 + 2.24	44.9 + 1.71	42.3 + 1.39	18100	80	3.89	3.01	2.34	1.92	6160
81	41.8 + 1.80	64.4 + 2.79	50.0 + 2.05	42.8 + 1.43	0	81	3.01	3.82	2.88	1.99	0
82	67.0 + 2.80	78.2 + 3.51	60.1 + 2.61	52.1 + 1.83	7840	82	5.12	5.09	3.79	2.63	2670
83	157 + 7.43	86.0 + 4.00	73.2 + 3.38	66.5 + 2.40	0	83	18.2	5.92	4.81	3.53	0
84	198 + 8.37	61.4 + 2.96	56.8 + 2.35	23.0 + .844	0	84	21.4	4.80	3.48	1.26	0
86	8.94 + .286	.966 + .026	.006 + .000	.308 + .015	0	86	.579	.047	.000	.022	0

observations in natural samples and were, in fact, calculated without regard to such data. We feel that these are the best such calculations presently available, but this should not be interpreted to mean the best calculations which could be made in this way. These calculations are clearly incomplete in that many of the relevant cross sections have not been measured and must be estimated, and in that no account is taken of a few minor target elements, of rare earth element compositional variation, or of the dependence of exciting particle flux and spectrum on total chemical composition of the target material. As stated, an important objective of this paper is identification of areas in which these theoretical calculations may most significantly be improved.

Observations of Cosmogenic Gases

As an observational data base with which to compare the theoretical systematics of the previous section, we have calculated cosmogenic noble gas contributions in a number of lunar rocks. In order to make the most meaningful comparisons with the predictions, samples were selected according to the following criteria:

(i) Relevant noble gas and target element data must be available.

(ii) In order to conform to the requirement for 2π irradiation exposure samples must be small (in comparison with the cosmic ray attenuation scale).

(iii) Lunar surface exposure times must be relatively brief, about 50 million years or less, so that shielding changes during exposure, by impact fragmentation or turnover, are unlikely to be important, and the effects of surface erosion can be explicitly taken into account.

(iv) For shielding depth during exposure to be known, adequate lunar surface photodocumentation must be available. Determinations of shielding depths for individual samples were made from the surface documentation and from cutting diagrams.

(v) Absolute surface exposure ages must be known. This criterion leads to potential circularity, since most exposure ages are based on cosmogenic noble gases and other possible approaches (impact pits, heavy ion tracks) are calibrated by gas ages. Circularity can be mostly avoided by use of ^{81}Kr-^{83}Kr ages, and all absolute ages used in this paper will be ^{81}Kr ages. This method (Marti, 1967) determines an age from the ratio of (radioactive) cosmogenic ^{81}Kr to (stable) cosmogenic ^{83}Kr, and does not depend on target element concentration, absolute production rates, or depth (or other geometrical effects) of shielding, as long as these parameters remain constant. The only circularity is that the method requires an assumption about the relative production rates of ^{81}Kr and ^{83}Kr; we can anticipate that uncertainty in these relative rates will be small, so that circularity will be a second-order effect and will be unimportant for our purposes. The limited target irradiation experimental data available are in reasonable agreement with adopted production ratios.

These criteria can be satisfied for a number of rocks from the ejecta of Cone

Table 1d. Predicted Cosmogenic[a] Production Rates for Xenon (atoms/min/kg).

	Ba	La[c]	Ba[b]		Ba	La[c]	Ba[b]
	Surface				*40 g/cm² Shielding*		
124	21.0 + 2.47	177 + 5.67	0	124	18.2	125	0
126	33.9 + 8.35	210 + 11.6	0	126	31.6	155	0
128	46.3 + 24.0	241 + 23.3	0	128	47.8	188	0
129	50.0 + 44.6	256 + 34.7	0	129	54.7	206	0
130	24.8 + 12.3	36.6 + 7.90	0	130	24.4	31.8	0
131	73.5 + 128	411 + 160	10.5	131	91.2	372	68.5
132	18.8 + 11.3	19.3 + 16.6	0	132	18.6	19.5	0
134	1.32 + .314	1.63 + 1.60	0	134	1.15	1.85	0
136	.164 + .082	.008 + .008	0	136	.150	.009	0
	1 g/cm² Shielding				*65 g/cm² Shielding*		
124	20.9 + 1.89	175 + 4.37	0	124	15.3	99.8	0
126	33.9 + 6.07	209 + 8.72	0	126	27.2	126	0
128	46.6 + 16.5	240 + 17.1	0	128	42.3	155	0
129	50.4 + 26.4	254 + 24.8	0	129	49.2	171	0
130	24.9 + 7.99	36.5 + 5.55	0	130	21.3	26.9	0
131	74.4 + 75.0	411 + 102	12.3	131	84.8	318	92
132	18.8 + 7.00	19.4 + 9.59	0	132	16.3	17.2	0
134	1.32 + .213	1.65 + .990	0	134	.972	1.67	0
136	.164 + .050	.009 + .005	0	136	.129	.008	0
	2 g/cm² Shielding				*100 g/cm² Shielding*		
124	20.9 + 1.56	173 + 3.64	0	124	11.6	72.4	0
126	34.0 + 4.89	207 + 7.14	0	126	21.1	92.0	0
128	46.8 + 12.9	238 + 13.8	0	128	33.7	115	0
129	50.8 + 20.0	253 + 19.8	0	129	39.8	129	0
130	25.0 + 6.16	36.5 + 4.39	0	130	16.8	20.5	0
131	75.4 + 54.9	411 + 77.1	14.0	131	70.8	247	111
132	18.9 + 5.31	19.5 + 6.90	0	132	12.9	13.7	0
134	1.32 + .167	1.67 + .740	0	134	.744	1.36	0
136	.164 + .038	.009 + .003	0	136	.100	.006	0

	5 g/cm² Shielding				*150 g/cm² Shielding*		
124	20.9 + 1.02	168 + 2.41	0	124	7.87	46.6	0
126	34.3 + 3.02	202 + 4.57	0	126	14.7	60.0	0
128	47.8 + 7.56	235 + 8.61	0	128	24.3	76.4	0
129	52.2 + 11.1	250 + 12.1	0	129	29.3	86.4	0
130	25.3 + 3.53	36.4 + 2.62	0	130	12.0	14.1	0
131	78.8 + 28.4	412 + 42.7	19.2	131	54.3	173	114
132	19.2 + 2.97	19.8 + 3.51	0	132	9.28	9.93	0
134	1.31 + .099	1.72 + .400	0	134	.512	1.01	0
136	.165 + .020	.009 + .002	0	136	.071	.005	0
	10 g/cm² Shielding				*225 g/cm² Shielding*		
124	20.8 + .598	161 + 1.44	0	124	4.50	24.8	0
126	34.5 + 1.67	195 + 2.64	0	126	8.71	32.6	0
128	49.1 + 3.95	229 + 4.83	0	128	15.1	42.6	0
129	54.2 + 5.57	245 + 6.64	0	129	18.8	49.0	0
130	25.7 + 1.82	36.2 + 1.41	0	130	7.40	8.22	0
131	83.7 + 13.2	412 + 21.5	27.4	131	36.7	104	85.8
132	19.5 + 1.49	20.2 + 1.64	0	132	5.78	6.26	0
134	1.31 + .053	1.80 + .195	0	134	.300	.649	0
136	.165 + .010	.009 + .001	0	136	.043	.003	0
	20 g/cm² Shielding				*500 g/cm² Shielding*		
124	20.3 + .265	149 + .660	0	124	.508	2.50	0
126	34.4 + .688	182 + 1.16	0	126	1.06	3.41	0
128	50.2 + 1.53	217 + 2.05	0	128	1.99	4.68	0
129	56.2 + 2.05	235 + 2.74	0	129	2.62	5.55	0
130	26.0 + .693	35.4 + .564	0	130	.967	.978	0
131	89.9 + 4.48	407 + 8.13	42.2	131	5.64	13.1	18.9
132	19.7 + .555	20.6 + .578	0	132	.771	.866	0
134	1.27 + .021	1.88 + .071	0	134	.036	.092	0
136	.163 + .004	.009 + .000	0	136	.006	.000	0

[a] Where two entries are given the first represents GCR spallation and the second SCR spallation.
[b] Slow neutron capture contribution.
[c] Includes contributions from other rare earth elements, grouped as Ce + Pr, Nd, Sm + Eu + Gd, and Tb + Dy + Ho + Er, assumed present in ratios (to La) 3, 1.7, 1 and 1.2, respectively.
[d] Dashes indicate cases where production is expected to be negligible and no explicit calculations were made.

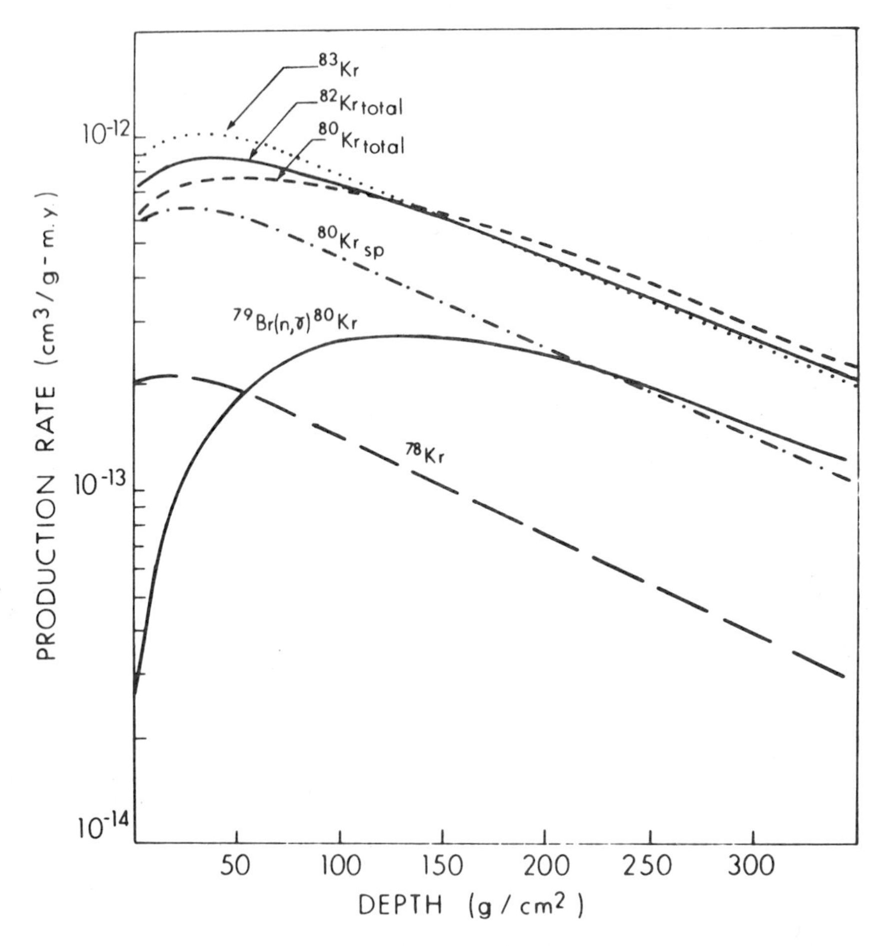

Fig. 1. Predicted cosmogenic production rates of Kr isotopes as a function of depth in the moon. For cosmogenic ^{80}Kr the contributions from spallation reactions (in which one or more nucleons leave the target nuclide) and from the ^{79}Br (n, γ) reaction are shown separately. The production rates are based on data in Table 1, except that the SCR contributions (important only in the uppermost few g/cm^2) are omitted, and are appropriate for the chemical composition of soil at Apollo 15 deep drill core (concentrations in ppm: Rb = 6.6, Sr = 132, Y = 120, Zr = 460).

crater at Apollo 14 and North Ray Crater and South Ray Crater at Apollo 16, whose ages we will take to be 25, 50 and 2 million years, respectively (cf., Arvidson *et al.*, 1975). The ^{81}Kr-^{83}Kr ages for these samples cluster quite strongly so that interlopers or samples which have suffered shielding changes would not be recognized as members of these groups.

In the samples considered here, as in general, resolution of cosmogenic components is frequently not easy. Details of the calculations for each gas are given below. It is worth explicit note that for a few isotopes, ^{21}Ne and ^{126}Xe (and ^{124}Xe), identification of cosmogenic contributions is usually very easy, since they

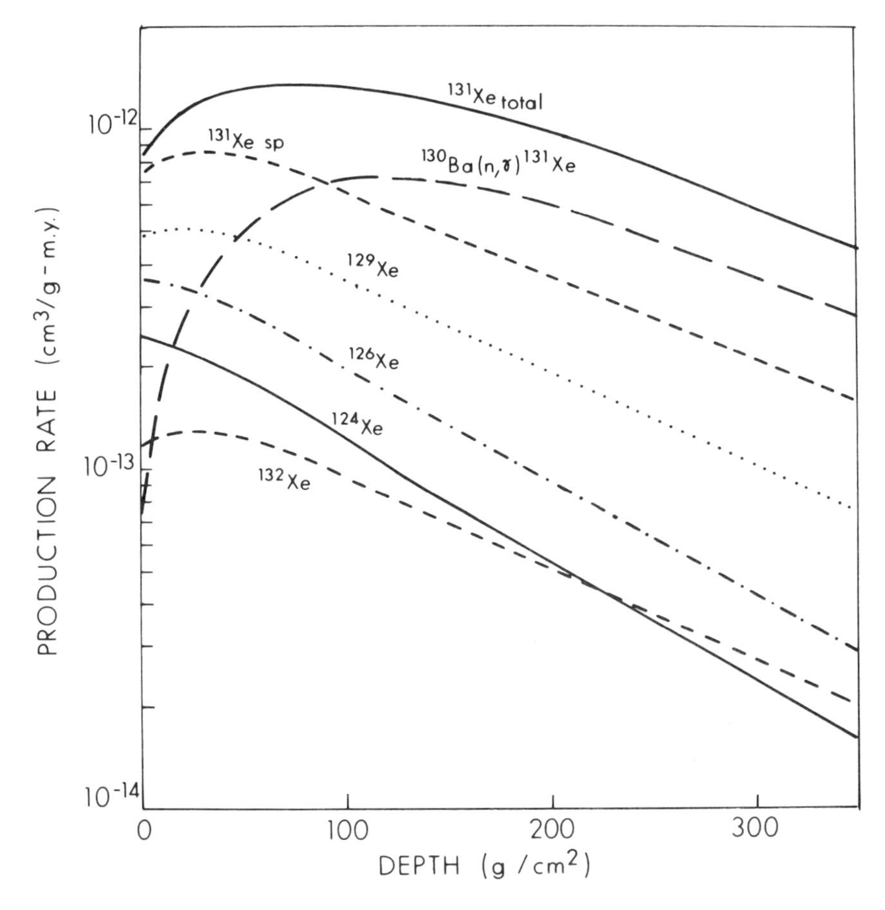

Fig. 2. Predicted cosmogenic production rates of Xe isotopes as a function of depth. Remarks are the same as for Fig. 1 (Ba = 320, La = 30).

are very underabundant in trapped gas and in any situation where cosmogenic contributions are at all significant at other isotopes they are dominant at ^{21}Ne, ^{124}Xe and ^{126}Xe. We will adopt ^{21}Ne and ^{126}Xe as normalization isotopes, i.e., compute their absolute abundances and calculate ratios of other isotopes relative to them. For Ar we will use ^{38}Ar as normalization, since it is the most prominent cosmogenic Ar isotope, qualitatively in the same fashion as ^{21}Ne and ^{126}Xe although to a much smaller degree. For Kr we will use ^{83}Kr as normalization; it is not the most prominent cosmogenic Kr isotope but ^{78}Kr is often subject to large experimental uncertainty and ^{80}Kr and ^{82}Kr often vary irregularly because of Br neutron capture contributions.

Ne was decomposed by *assuming* the observed ^{21}Ne/^{20}Ne ratio represents a superposition of a cosmogenic component with ^{21}Ne/^{20}Ne = 1 and a trapped component with ^{21}Ne/^{20}Ne = 0.00276, and subtracting trapped ^{22}Ne assuming ^{22}Ne/^{20}Ne = 0.0781 (the trapped composition is that determined in soil 10084 by Marti *et al.*, 1970). This procedure gives a reliable evaluation of cosmogenic ^{21}Ne, which in all cases dominates total ^{21}Ne, and a good evaluation of cosmogenic $^{22}Ne/^{21}Ne$; it also amounts to abandoning the effort to determine cosmogenic ^{20}Ne. Similarly, cosmogenic ^{38}Ar is calculated assuming observed $^{38}Ar/^{36}Ar$ results from superposition of a cosmogenic component, with $^{36}Ar/^{38}Ar = 0.65$, and a trapped component, with $^{36}Ar/^{38}Ar = 5.35$. The procedure makes no attempt to independently determine cosmogenic $^{36}Ar/^{38}Ar$ and entirely abandons ^{40}Ar, which is always completely dominated by the radiogenic contribution from decay of ^{40}K.

For Kr, we first make a correction for ^{238}U spontaneous fission contributions assuming an accumulation time of 3.9 AE (1 AE = 10^9 yr); this correction is very small for the samples considered here. The remaining Kr is assumed to be a superposition of cosmogenic Kr and trapped Kr with the BEOC-12 composition determined by Eberhardt *et al.* (1972). Cosmogenic ^{86}Kr is estimated from the (fission-corrected) $^{86}Kr/^{83}Kr$ ratio assuming cosmogenic $^{86}Kr/^{83}Kr = 0.015 \pm 0.015$. Cosmogenic contributions at other Kr isotopes are obtained by subtraction of BEOC-12 composition based on the trapped ^{86}Kr.

Xe is traditionally the most complicated noble gas to resolve into independent components. Cosmogenic contributions are frequently important at all but the two heaviest isotopes ^{134}Xe and ^{136}Xe, and fission contributions to the four heaviest isotopes are also frequently important, so that there is no isotope which reliably serves as an accurate measure of trapped Xe. For the lunar highland samples considered here there is an additional problem with variability of the trapped Xe composition (cf. Bernatowicz *et al.*, 1978). In principle fission corrections could be made for Xe in the same manner as for Kr, but for the samples under consideration this is not very satisfactory since the corrections are quite significant and uncertainties arising in U variability, the possibility of ^{235}U neutron-induced fission, and the possibility of ^{244}Pu fission (both *in situ* and trapped) result in substantial uncertainty in the cosmogenic contributions to most isotopes.

In spite of these difficulties it is still possible to make good estimates of cosmogenic contributions to the three lightest isotopes, ^{124}Xe, ^{126}Xe and ^{128}Xe, since these three isotopes are prominent in the cosmogenic component, not very abundant in trapped Xe, and are not produced in any radioactive decay processes. A common graphical presentation is a three isotope correlation diagram in which abscissa and ordinate are $^{126}Xe/^{130}Xe$ and $^{124}Xe/^{130}Xe$, respectively. On such a diagram, mixing of two components—trapped and cosmogenic Xe—produces a linear correlation whose slope can be determined by a least squares fit. The slope is a function of the two compositions and the defining equation can be rearranged to yield

$$
\begin{aligned}
(^{124}Xe/^{126}Xe)_c &= S + \frac{(^{124}Xe/^{130}Xe)_t - S(^{126}Xe/^{130}Xe)_t}{(^{126}Xe/^{130}Xe)_c} \\
&= S + \frac{(0.0289) - (0.0263)\ S}{(^{126}Xe/^{130}Xe)_c}
\end{aligned}
\qquad (1)
$$

where subscript c designates cosmogenic, subscript t designates trapped, and S is the slope of the correlation line. We assume the trapped composition to be the SUCOR composition of Podosek *et al.* (1971); the slope S can be determined by a least squares fit to the data or, with the same assumption for trapped composition, from a single datum (e.g., total melt rather than stepwise heating analysis). A rigorous solution for $(^{124}Xe/^{126}Xe)_c$ is not possible without knowledge of $(^{126}Xe/^{130}Xe)_c$, but in practice we can take $(^{126}Xe/^{130}Xe)_c = 0.9 \pm 0.2$ to cover the range of observation. Since the second term is small (less than 3% of the first) we can compute $(^{124}Xe/^{126}Xe)_c$ with an uncertainty dictated by the input measurements (carried in the uncertainty in S) and not by ambiguity in decomposition. Similarly and with not much more uncertainty we can use the same approach to calculate $(^{128}Xe/^{126}Xe)_c$. This approach is not suitable for the other Xe isotopes, however, because $(^{i}Xe/^{130}Xe)_t$ is much larger so that the uncertainty in $(^{126}Xe/^{130}Xe)_c$ is much more crucial, and because there are in general additional contributions besides the cosmogenic and trapped components.

Table 2 lists the abundances of cosmogenic ^{21}Ne, ^{38}Ar, ^{83}Kr and ^{126}Xe, and the relative abundances of other cosmogenic isotopes, as calculated by these techniques in a suite of samples selected by the criteria previously described. Table 2 also lists the comparable contributions predicted from the data presented in Table 1. This tabulation is the principal basis for comparison of the *a priori* predictive systematics with actual observation on the basis of the target chemistry (target element concentrations assumed uncertain by 20%) and the shielding depths listed. The predicted production rates and predicted cosmogenic compositions include adjustments for shielding changes by surface erosion at the rate of 0.3 g/cm^2/m.y. (Crozaz *et al.*, 1972; Yuhas, 1974). Because of SCR contributions the erosional shielding changes are important for the shallow samples. Wahlen *et al.* (1972) have observed rather large variations in erosion rates and we consequently tabulate production rates not corrected for erosion for comparison.

As described above, the calculations represented in Table 2 are limited in that for several isotopes the cosmogenic contributions in the samples considered can be observed only poorly or not at all. At a price, we can expand the basis for comparison by considering additional lunar samples in which cosmogenic contributions are relatively much more enhanced, so that cosmogenic isotopes are more accurately resolvable. The price is that for these additional samples we cannot make reasonable estimates of the shielding depth during irradiation. Indeed, since such samples generally have had substantially longer exposures than the samples considered in Table 2, it is almost certain that they have experienced significant shielding changes during their exposure histories, and no generally applicable and testable techniques are available for deciphering such histories. Also, since the ^{81}Kr-^{83}Kr exposure age calculation necessarily assumes exposure at a single shielding depth, we cannot make determinations of total exposure time.

In Table 3 we list cosmogenic noble gas compositions for several such samples in which more cosmogenic isotopes can be identified with greater precision. As noted, however, for these cases the irradiation history must be considered unknown and we can make no meaningful comparisons of absolute production rates and the observed compositions can be compared only with predicted compositions (from data in Table 1) for a range of shielding depths. The smaller the observed $^{86}Kr/^{78}Kr$ and $^{136}Xe/^{130}Xe$ measured ratios the more the spectra are dominated by the cosmogenic component and the greater the accuracy of the "observed" cosmogenic component. We have therefore tabulated these ratios in Table 3 as a gauge for the quality of the derived ("observed") cosmogenic components.

In the same spirit we include in Table 2d the predictions for cosmogenic Xe composition derived from the regularities observed by Podosek and Huneke (1971). This consists of a set of linear relationships among cosmogenic Xe ratios observed in lunar and meteoritic samples (also cf. Kaiser, 1977) such that the $(^{124}Xe/^{126}Xe)_c$ ratio, which by Eq. (1) can usually be determined without sensitive dependence on specific models, can be used to predict the other cosmogenic ratios. In Table 2d the listed $(^{124}Xe/^{126}Xe)_c$ and the first listing of $(^{128}Xe/^{126}Xe)_c$ are computed directly from the data as described in connection with Eq. (1); the second listing for $(^{128}Xe/^{126}Xe)_c$ and the listings for the heavier Xe isotopes are computed *not* from the measured data from these rocks but from the observed $(^{124}Xe/^{126}Xe)_c$ by application of the correlation parameters given by Podosek and Huneke (1971).

This approach for determining "observed" cosmogenic composition for heavy Xe isotopes is clearly not very definitive. Xe cosmogenic compositions depend on

Table 2a. Comparison of Observed and Predicted Cosmogenic Ne.[z]

Sample Shielding Depth Exposure Age	Target Chemistry (%)	$^{21}Ne_c$ (observed) / $^{21}Ne_c$ (predicted) ($\times 10^{-9}$ cm^{3}STP/g-m.y.)	Observed Cosmogenic Composition	Predicted Cosmogenic Composition
68815,113	0.395 Na			
(w)	3.85 Mg	0.85 ± .14	(20 ≡ 0.750)	0.750 ± .020
	14.2 Al		21 = 0.781 ± .028	0.769 ± .010
18 g/cm^2	21.8 Si		22 ≡ 1.000	1.000
2 m.y.	10.8 Ca	1.11 ± .22	(y)	
	3.87 Fe	(1.11)		
	(a)			
14321,92	0.525 Na			
FM 3D	6.40 Mg	0.8 ± .2	20 = –	0.815 ± .020
(p)	8.1 Al		21 = 0.80 ± .03	0.800 ± .016
	21.9 Si		22 ≡ 1.000	1.000
18 g/cm^2	7.9 Ca	1.14 ± .23		
25 m.y.	10.5 Fe	(1.14)		
	(b)			

Table 2b. Comparison of Observed and Predicted Cosmogenic Ar[(z)]

Sample Shielding Depth Exposure Age	Target Chemistry (%)	$^{38}Ar_c$ (observed) / $^{38}Ar_c$ (predicted) ($\times 10^{-9}$ cm^{3}STP/g-m.y.)	Predicted Cosmogenic Composition
67015,14	0.15 K	1.1 ± .1	36 = 0.666 – 0.668
(p)	11.0 Ca		38 ≡ 1.000
	0.29 Ti		40 = 0.011 – 0.011
1–2 g/cm^2	2.83 Fe	1.25 – 1.21 ± .24	
51.1 m.y.	(c)	(2.5 – 1.8)	
67075,8	0.014K	1.2 ± .1	36 = 0.669 – 0.683
(p)	12.2 Ca		38 ≡ 1.000
	0.60 Ti		40 = 0.012 – 0.011
2–10 g/cm^2	3.07 Fe	1.35 – 1.24 ± .25	
50.2 m.y.	(c)	(2.0 – 1.2)	
62255,17	0.069 K	2.1 ± .3	36 = 0.664 – 0.650
(p)	11.0 Ca		38 ≡ 1.000
	0.29 Ti		40 = 0.012 – 0.012
1–3 g/cm^2	2.83 Fe	2.16 – 1.49 ± .43	
1.9 m.y.	(d)	(2.5 – 1.6)	
14321,91	0.31 K		
FM 1,2	7.8 Ca	1.26 ± .13	36 = 0.666 ± .00
(p)	1.4 Ti		38 ≡ 1.000
	10.9 Fe	1.14 ± .23	40 = 0.020 ± .003
1 g/cm^2	(b)	(1.91)	
23.8 m.y.			
14321,92	0.48 K	0.76 ± .08	36 = 0.705 ± .006
FM 3D	7.9 Ca		38 ≡ 1.000
(p)	1.4 Ti		40 = 0.019 ± .003
18 g/cm^2	9.5 Fe	0.90 ± .18	
25 m.y.	(b)	(0.89)	
14321,123	0.40 K		36 = 0.705 ± .006
FM 5	7.8 Ca	1.05 ± .11	38 ≡ 1.000
(p)	1.4 Ti		40 = 0.017 ± .002
40 g/cm^2	10.5 Fe	0.96 ± .19	
27.2 m.y.	(b)	(0.96)	

Table 2c. Comparison of Observed and Predicted Cosmogenic Kr.[z]

Sample Shielding Depth Exposure Age	Target Chemistry (ppm)	$^{83}Kr_c$ (measured) / $^{83}Kr_c$ (predicted) ($\times 10^{-12} cm^3 STP/g$-m.y.)	Observed Cosmogenic Spectrum	Predicted Cosmogenic Spectrum	$^{81}Kr_c$ (measured) / $^{81}Kr_c$ (predicted) ($\times 10^{-13} cm^3 STP/g$)
14066,21,2.01	6.19 Rb		78 = 0.234 ± .002	0.211 ± .0076	
(r)	180 Sr	3.58 ± .20	80 = 0.530 ± .004	0.670 ± .0089	6.96 ± .42
	115 Y		81 = .0070 ± .0001	.0074 ± .0001	
10 g/cm^2	550 Zr		82 = 0.776 ± .008	0.833 ± .008	2.60 ± .53
27.8 m.y.	(0.168 Br)	1.27 ± .26	83 ≡ 1.000	1.000	
	(e,f,g,h)	(1.28)	84 = 0.240 ± .010	0.511 ± .026	
14066,31,1	26 Rb		78 = 0.231 ± .004	0.202 ± .006	
(r)	150 Sr	3.55 ± .20	80 = 0.526 ± .007	0.645 ± .006	6.86 ± .41
	200 Y		81 = .0070 ± .0003	.0072 ± .0001	
10 g/cm^2	950 Zr	1.95 ± .39	82 = 0.767 ± .014	0.803 ± .006	3.89 ± .78
27.7 m.y.	(.168 Br)	(1.97)	83 ≡ 1.000	1.000	
	(e,h,i)		84 = 0.250 ± .036	0.500 ± .027	
14306,26L	32 Rb		78 = 0.264 ± .015	0.200 ± .006	
(t)	230 Sr	2.31 ± .35	80 = 0.556 ± .036	0.652 ± .007	4.27 ± .64
	148 Y		81 = .0073 ± .0007	.0080 ± .0001	
7 g/cm^2	1150 Zr	2.35 ± .47	82 = 0.854 ± .052	0.808 ± .007	4.8 ± 1.0
25.4 m.y.	(.270 Br)	(2.48)	83 ≡ 1.000	1.000	
	(h,j)		84 = 0.318 ± .074	0.492 ± .026	
14306,26D	14 Rb		78 = 0.239 ± .008	0.201 ± .006	
(t)	195 Sr	4.21 ± .63	80 = 0.577 ± .021	0.653 ± .006	7.6 ± 1.1
	297 Y		81 = .0077 ± .0004	.0088 ± .0001	
7 g/cm^2	1370 Zr	2.76 ± .55	82 = 0.790 ± .027	0.812 ± .006	5.7 ± 1.1
23.4 m.y.	(.270 Br)	(2.91)	83 ≡ 1.000	1.000	
	(h,j)		84 = 0.270 ± .038	0.478 ± .024	
14171,2	26 Rb		78 = 0.236 ± .002	0.184 ± .004	
(t)	150 Sr	2.75 ± .41	80 = 0.534 ± .004	0.605 ± .007	5.00 ± .75
	200 Y		81 = .0074 ± .0002	.0070 ± .0001	
1 g/cm^2	950 Zr	2.31 ± .46	82 = 0.787 ± .007	0.784 ± .007	5.6 ± 1.1
24.5 m.y.	(0.39 Br)	(3.24)	83 ≡ 1.000	1.000	
	(k)		84 = 0.303 ± .005	0.515 ± .003	
67095,9	6.0 Rb		78 = 0.208 ± .002	0.219 ± .007	
(t)	150 Sr	0.78 ± .12	80 = 0.519 ± .002	0.677 ± .012	1.45 ± .22
	61 Y		81 = .0037 ± .0001	.0044 ± .0001	
5 g/cm^2	250 Zr	0.74 ± .14	82 = 0.781 ± .001	0.848 ± .010	1.63 ± .33
50.2 m.y.	(0.140 Br)	(0.80)	83 ≡ 1.000	1.000	
	(l,m,e)		84 = 0.491 ± .005	0.561 ± .026	
68815,113	8.8 Rb		78 = 0.211 ± .007	0.214 ± .007	
(t)	160 Sr	1.17 ± .18	80 = 0.587 ± .020	0.905 ± .067	2.33 ± .35
	64.4 Y		81 = 0.100 ± .004	0.101 ± .001	
18 g/cm^2	331 Zr	0.86 ± .17	82 = 0.825 ± .031	0.938 ± .031	1.73 ± .35
2 m.y.	(0.720 Br)	(0.86)	83 ≡ 1.000	1.000	
	(a,n)		84 = 0.18 ± .05	0.546 ± .027	
67015,14	3.0 Rb		78 = 0.218 ± .003	0.211 – 0.216	
(p)	150 Sr	1.21 ± .18	80 = 0.566 ± .010	0.710 – 0.730	2.19 ± .35
	87 Y		81 = .0038 ± .0006	0.0036 – 0.0038	
1-2 g/cm^2	260 Zr	0.85 ± .17	82 = 0.784 ± .009	0.867 – 0.874	2.07 – 2.35 ± 20%
51.1 m.y.	(0.4 Br)	(1.07 – 1.28)	83 ≡ 1.000	1.000	
	(p,m)		84 = 0.357 ± .015	0.568 – 0.564	

Table 2c. *(cont'd.)*.

Sample Shielding Depth Exposure Age	Target Chemistry (ppm)	$^{83}Kr_c$ (measured) / $^{83}Kr_c$ (predicted) ($\times 10^{-12}cm^3STP/g$-m.y.)	Observed Cosmogenic Spectrum	Predicted Cosmogenic Spectrum	$^{81}Kr_c$ (measured) / $^{81}Kr_c$ (predicted) ($\times 10^{-13}cm^3STP/g$)
67075,8	0.67 Rb		78 = 0.150 ± .006	0.249 – 0.264	
(p)	127 Sr	0.53 ± .08	80 = 0.508 ± .011	0.837 – 0.988	1.09 ± .28
	1.4 Y		81 = .0037 ± .0011	0.0045 – 0.0055	
2-10 g/cm^2	7.6 Zr	0.24 – 0.26 ± .05	82 = 0.751 ± .016	0.988 – 1.049	0.5–0.7 ± 20%
50.2 m.y.	(0.265 Br)	(0.24 – 0.35)	83 ≡ 1.000	1.000	
	(c,p,q)		84 = 0.529 ± .065	0.714 – 0.703	
14321,92	15.7 Rb		78 = 0.250 ± .001	0.202 ± .006	
FM 3D	188 Sr	1.96 ± .29	80 = 0.576 ± .002	0.683 ± .014	5.02 ± .80
(p)	216 Y		81 = 0.010 ± .002	.0080 ± .0001	
18 g/cm^2	900 Zr	1.91 ± .38	82 = 0.812 ± .003	0.827 ± .009	
	(0.317 Br)	(1.93)	83 = 1.000	1.000	3.83 ± .77
25 m.y.	(p,x,o,h)		84 = 0.348 ± .006	0.506 ± .026	

at least two independent parameters, shielding depth and Ba/La ratio (more if the rare earth elements fractionate), so that variations even in principle cannot be represented by a single parameter such as $(^{124}Xe/^{126}Xe)_c$. There is also no fundamental reason why either dependence should be linear. Thus, these anticipated heavy isotope cosmogenic compositions cannot be considered rigorously valid. For our purposes their only important virtue is that they are entirely empirical, and are at least one step better than use of a constant average composition or compositional range for the heavy isotopes. A comparison of the "observed" and predicted cosmogenic contributions to the heavy Xe isotopes is rather important since these isotopes, especially ^{130}Xe, are critical to the problem of spectral resolution of Xe.

Discussion

In Tables 2 and 3 we provide a data base by which predictions of cosmogenic noble gas production rates (Table 1) can be compared with observation. For some isotopes the agreement between the two is quite satisfactory, for others considerably less so. This comparison cannot be considered either complete or definitive, however. As already noted, the predictive calculations are themselves incomplete, omitting some minor targets and in many cases relying on estimates of unmeasured reaction parameters. In addition, the input parameters, shielding and target element abundances are in many cases not ideally determined. The "observations," to varying degree, are also not free from difficulty: target element variability, imperfect data, and the problems associated with resolution of cosmogenic components. Also, assessment of the predicted production rates is complicated by the fact that our observations are made in multi-element targets having variable composition and shielding, not individual target elements or samples with controlled composition and shielding. We will nevertheless be able to make some generalizations from these comparisons, but in view of the limitations described above these generalizations must be considered provisional.

Neon

Table 2a lists only two samples for comparison. The resolution of cosmogenic ^{21}Ne is not difficult and the principal targets are major elements, so that the observed cosmogenic Ne should not be subject to a large uncertainty. In both cases the predicted production is about 25–30% higher than the observed. A preliminary comparison with meteoritic production rates (Reedy *et al.*, 1978) suggests the opposite trend, but there are considerable uncertainties in the particle fluxes used for meteorites. Since the predictions of ^{21}Ne and ^{22}Ne are largely independent it is surprising that the cosmogenic ^{22}Ne/^{21}Ne ratio is predicted so much better, within a few percent (Table 2; Reedy *et al.*, 1978). Cosmogenic Ne can occasionally be demonstrated to have been partially depleted by diffusive loss, and this may be a factor. Since both ^{21}Ne and ^{22}Ne are relatively "soft" products (much of their production comes from lower energy secondary particles) it may also be that the calculations systematically overestimate their production while not seriously erring in their ratio.

Argon

Table 2b lists six samples for comparison. The principal target for ^{38}Ar production is the major element Ca, so that uncertainties due to sampling heterogeneity should not be severe. In all six cases there is satisfactory agreement between predicted and observed production rates of ^{38}Ar. It is unfortunately not possible to independently determine the cosmogenic ^{36}Ar/^{38}Ar ratio in these samples, but the predicted ^{36}Ar/^{38}Ar ratios are quite compatible with ranges found in other samples in which cosmogenic Ar is more dominant. It is noteworthy that four of these cases represent low shielding, well within the range where SCR effects are important, and the predicted average production rates are quite markedly affected by the adjustments for surface erosion (failure to account for surface erosion would produce apparent discrepancies of about a factor of two).

Krypton

For Kr we can make independent evaluations of the absolute abundances of two species in Table 2c, the normalization isotope ^{83}Kr and the radioactive cosmogenic isotope ^{81}Kr. Since the abundance of ^{81}Kr reflects production only during the last mean life (3×10^5 yr), it should provide a more direct test of the theoretical systematics since it will be independent of any earlier shielding changes; the corresponding disadvantage is that its low abundance leads to larger experimental uncertainty in its measurement. For near-surface samples the

Table 2d. Comparison of Observed and Predicted Xe.(z)

Sample Shielding Depth Exposure Age	Target Chemistry (ppm)	$^{126}Xe_c$ (measured) / $^{126}Xe_c$ (predicted) ($\times 10^{-12} cm^3$/g-m.y.)	Observed Cosmogenic Spectrum	Predicted Cosmogenic Spectrum
14066,21,2.01	800 Ba		124 = 0.577 ± .020	0.657 ± .015
(r)			126 ≡ 1.000	1.000
	61 La	1.16 ± .08	128 = 1.479 ± .054	1.387 ± .019
10 g/cm^2	(e)		129 = 1.42 ± .25	1.546 ± .024
			130 = 0.864 ± .050	0.595 ± .037
		0.79 ± .16	131 = 2.75 ± .22	3.17 ± .16
27.8 m.y.		(0.80)	132 = 0.54 ± .25	0.443 ± .030
			134 = —	0.030 ± .002
14066,31,1	900 Ba		124 = 0.571 ± .014	0.664 ± .016
(r)			126 ≡ 1.000 ±	1.000
	79 La	1.13 ± .07	128 = 1.462 ± .049	1.378 ± .019
10 g/cm^2	(e)		129 = 1.48 ± .41	1.534 ± .026
			130 = 0.898 ± .066	0.578 ± .039
		0.93 ± .19	131 = 2.87 ± .33	3.13 ± .15
27.8 m.y.		(0.94)	132 = 0.68 ± .41	0.429 ± .032
			134 = —	0.029 ± .002
14306,26L	1300 Ba		124 = 0.559 ± .008	0.634 ± .011
(s)			126 ≡ 1.000	1.000
	55 La	1.16 ± .11	128 = 1.59 ± .13	1.417 ± .014
7 g/cm^2	(j)		1.64 ± .12	
			129 = 1.70 ± .30	1.590 ± .019
			130 = 1.09 ± .07	0.656 ± .028
24.4 m.y.		1.13 ± .23	131 = 4.60 ± .60	3.23 ± .16
		(1.15)	132 = 0.96 ± .18	0.494 ± .023
			134 = 0.07 ± .06	0.032 ± .001
14306,26D	1350 Ba		124 = 0.544 ± .008	0.661 ± .016
(s)			126 ≡ 1.000	1.000
		1.18 ± .12	128 = 1.55 ± .17	1.382 ± .019
7 g/cm^2	110 La		1.55 ± .12	
	(j)		129 = 1.70 ± .30	1.543 ± .026
			130 = 1.12 ± .10	0.587 ± .038
24.4 m.y.		1.38 ± .28	131 = 4.60 ± .70	3.08 ± .13
		(1.41)	132 = 1.01 ± .23	0.438 ± .032
			134 = 0.07 ± .06	0.029 ± .002
14171,2	920 Ba		124 = 0.579 ± .005	0.660 ± .017
(s)			126 ≡ 1.000	1.000
	79 La	1.07 ± .11	128 = 1.49 ± .15	1.405 ± .023
1 g/cm^2	(k)		1.53 ± .12	
			129 = 1.7 ± .3	1.605 ± .034
			130 = 0.91 ± .09	0.596 ± .041
24.5 m.y.		1.00 ± .20	131 = 4.00 ± .70	3.155 ± .092
		(1.06)	132 = 0.61 ± .20	0.452 ± .034
			134 = —	0.029 ± .002

Table 2d. *(cont'd.)*.

Sample Shielding Depth Exposure Age	Target Chemistry (ppm)	$^{126}Xe_c$ (measured) / $^{126}Xe_c$ (predicted) ($\times 10^{-12}cm^3/g$-m.y.)	Observed Cosmogenic Spectrum	Predicted Cosmogenic Spectrum
67095,9	220 Ba		124 = 0.591 ± .009	0.668 ± .016
(s)			126 ≡ 1.000	1.000
	21 La	0.287 ± .029	128 = 1.43 ± .17	1.375 ± .020
5 g/cm²	(l,m,n)		1.56 ± .12	
			129 = 1.7 ± .3	1.534 ± .027
			130 = 0.87 ± .11	0.569 ± .040
50.2 m.y.		0.235 ± .047	131 = 3.90 ± .70	3.11 ± .14
		(0.245)	132 = 0.53 ± .3	0.423 ± .033
			134 = —	0.028 ± .002
14321,91	520 Ba		124 = 0.569 ± .007	0.664 ± .017
FM 1,2			126 ≡ 1.000	1.000
(u)	48 La	0.65 ± .10	128 = 1.53 ± .11	1.399 ± .023
	(o)		1.43 ± .12	
1 g/cm²			129 = 1.70 ± .31	1.598 ± .034
			130 = 0.99 ± .07	0.586 ± .042
		0.58 ± .12	131 = 4.26 ± .60	3.133 ± .092
25 m.y.		(0.61)	132 = 0.76 ± .13	0.444 ± .034
			134 = —	0.029 ± .002
14321,92	628 Ba		124 = 0.586 ± .005	0.671 ± .017
FM 3L			126 ≡ 1.000 ±	1.000
(u)	70.6 La	0.67 ± .10	128 = 1.47 ± .15	1.373 ± .020
			1.49 ± .12	
18 g/cm²	(o)		129 = 1.70 ± .40	1.529 ± .027
			130 = 0.87 ± .11	0.553 ± .041
25 m.y.		0.68 ± .14	131 = 3.89 ± .73	3.33 ± .20
		(0.69)	132 = 0.53 ± .27	0.407 ± .034
			134 = —	0.027 ± .002
14321,92	750 Ba		124 = 0.583 ± .006	0.660 ± .016
FM 3D			126 ≡ 1.000	1.000
(u)	69 La	1.24 ± .19	128 = 1.46 ± .17	1.386 ± .019
			1.52 ± .12	
18 g/cm²	(o)		129 = 1.70 ± .40	1.546 ± .025
			130 = 0.84 ± .11	0.578 ± .039
25 m.y.		0.75 ± .15	131 = 3.82 ± .70	3.40 ± .22
		(0.77)	132 = 0.48 ± .28	0.428 ± .032
			134 = —	0.028 ± .002

Table 2d. *(cont'd.).*

Sample Shielding Depth Exposure Age	Target Chemistry (ppm)	$^{126}Xe_c$ (measured) / $^{126}Xe_c$ (predicted) ($\times 10^{-12}cm^3/g$-m.y.)	Observed Cosmogenic Spectrum	Predicted Cosmogenic Spectrum
14321,123	730 Ba		124 = 0.580 ± .007	0.645 ± .015
FM 5	66 La		126 ≡ 1.000	1.000
(u)	(o)	0.97 ± .13	128 = 1.50 ± .13	1.424 ± .020
			1.53 ± .12	
40 g/cm^2			129 = 1.70 ± .35	1.614 ± .027
			130 = 0.929 ± .08	0.600 ± .038
			131 = 4.07 ± .64	4.35 ± .41
25 m.y.		0.64 ± .13	132 = 0.642 ± .17	0.448 ± .031
		(0.65)	134 = —	0.029 ± .002
67015,14	220 Ba		124 = 0.593 ± .016	0.671 ± .017
(v)		0.22 ± .03	126 ≡ 1.000	1.000
	22 La		128 = 1.46 ± .20	1.376 ± .022
2 g/cm^2	(p)		1.54 ± .12	
			129 = 1.70 ± .40	1.543 ± .030
			130 = 0.86 ± .12	0.565 ± .042
50 m.y.		0.24 ± .05	131 = 3.85 ± .96	3.06 ± .12
		(0.26)	132 = 0.50 ± .34	0.421 ± .034
			134 = —	0.028 ± .002
67075,8	8.85 Ba		124 = 0.61 ± .09	0.633 ± .012
(v)		$\lesssim$0.006	126 ≡ 1.000	1.000
	0.39 La		128 = 1.61 ± .25	1.421 ± .015
2–8 g/cm^2	(p)		129 = —	1.598 ± .019
			130 = —	0.653 ± .020
		0.0076 – 0.0079	131 = —	3.333 ± .17
		± .0016	132 = —	0.494 ± .023
49 m.y.		(0.0079 – 0.0084)	134 = —	0.032 ± .001

(a) Wänke *et al.,* 1974.
(b) Wahlen *et al.,* 1972.
(c) Wänke *et al.,* 1975.
(d) Chemical abundances not published; assumed to be same as 67015 which is petrographically similar (Wilshire *et al.,* 1973).
(e) Laul *et al.,* 1972.
(f) Mark *et al.,* 1973.
(g) Y not measured, assumed from measured Zr (Ref. e) and Zr/Y in 14066, 31 (Rose *et al.,* 1972).
(h) Br not measured, assumed from measured K (Ref. e) and average Br/K in 14321, 14063, 14310 and 14047 (Morgan *et al.,* 1972; Lindstrom *et al.,* 1972; Laul *et al.,* 1972; Philpotts *et al.,* 1972; Rose *et al.,* 1972).
(i) Rose *et al.,* 1972.
(j) Taylor *et al.,* 1972.
(k) Chemical abundances not published; assumed to be same as 14066, 31 which is petrographically similar (Simonds *et al.,* 1977).
(l) Laul *et al.,* 1974.
(m) Rb, Sr, Y not measured, assumed from measured K, Ca, and Zr and average Rb/K, Sr/Ca and

Y/Zr ratios in Apollo 16 breccias (LSPET, 1972). Br is assumed and falls within range of values for most Apollo 16 rocks (Jovanovic and Reed, 1973; Krähenbühl *et al.*, 1973).

(n) Jovanovic and Reed, 1973.

(o) Ba, La, Rb values are estimated from measured K values and average Ba/K, La/K, and Rb/K ratios in Apollo 14 breccias.

(p) Unpublished data (La Jolla).

(q) Y not measured, estimated from measured La, Zr (Ref. p) and an average of La/Y = 0.35 ± .1 and Y/Zr = 0.26 ± .07 in Apollo 16 rocks. (Haskin *et al.*, 1973).

(r) Srinivasan, 1974.

(s) Morgan, 1975.

(t) Drozd *et al.*, 1974.

(u) Marti, *et al.*, 1973a.

(v) Marti, *et al.*, 1973b.

(w) Unpublished data (St. Louis).

(x) Y not measured, estimated from measured Zr (Ref. p) and average Y/Zr ratio in 14321,222 (Strasheim *et al.*, 1972).

(y) The "observed" cosmogenic $^{21}Ne/^{22}Ne$ was obtained by defining the $(^{20}Ne/^{22}Ne)_c$ but there is only a weak dependency on the specific value assumed.

(z) A 20% uncertainty in elemental abundances and a surface erosion rate of $0.3 g/cm^2$–m.y. are assumed in these calculations. Predicted cosmogenic production rates without correction for erosion are shown in parentheses.

Table 3a. Comparisons of Observed and Predicted Cosmogenic Krypton Composition in Selected Lunar Rocks.

Sample $^{86}Kr/^{78}Kr$	Target Chemistry (ppm)	Observed Cosmogenic Spectrum	Predicted Cosmogenic Spectrum 10 g/cm^2 – 500 g/cm^2
12021,61	1.2 Rb	78 = 0.175 ± .002	0.233 – 0.162
(a)	100 Sr	80 = 0.507 ± .005	0.647 – 0.599
	50 Y	82 = 0.762 ± .004	0.848 – 0.837
	115 Zr	83 ≡ 1.000	≡ 1.000
.09	(.018 Br)	84 = 0.499 ± .044	0.588 – 0.654
	(b,c)		
15475,135	0.73 Rb	78 = 0.167 ± .001	0.238 – 0.165
(l)	117 Sr	80 = 0.497 ± .004	0.794 – 1.760
	29 Y	82 = 0.756 ± .005	0.925 – 1.355
	89 Zr	83 ≡ 1.000	≡ 1.000
0.13	(.235 Br)	84 = 0.548 ± .005	0.603 – 0.682
	(d,e)		
10044,20	5.64 Rb	78 = 0.183 ± .005	0.218 – 0.151
(k)	167 Sr	80 = 0.486 ± .010	0.676 – 0.889
	167 Y	82 = 0.787 ± .005	0.839 – 0.938
	460 Zr	83 ≡ 1.000	≡ 1.000
1.07(m)	(.216 Br)	84 = 0.402 ± .005	0.539 – 0.592
	(f,g,h)		

Table 3b. Comparison of Observed and Predicted Cosmogenic Xenon Composition in Selected Lunar Rocks.

Sample	Target Chemistry (ppm)	Observed Cosmogenic Spectrum	Predicted Cosmogenic Spectrum 10 g/cm² – 500 g/cm²
$^{136}Xe/^{130}Xe$			
10044,20	290 Ba	124 = 0.62 ± .02	0.67 – 0.54
(k)	25.7 La	126 ≡ 1.000	≡ 1.000
	(i)	128 = 1.47 ± .08	1.37 – 1.76
		129 = 1.55 ± .06	1.53 – 2.28
		130 = 0.90 ± .02	0.57 – 0.77
0.44		131 = 3.87 ± .12	3.03 – 18.87
0.18(m)		132 = 0.70 ± .03	0.42 – 0.62
		134 = 0.04 ± .01	0.029 – 0.032
12021,61	80 Ba	124 = 0.584 ± .006	0.66 – 0.53
(a)	6.6 La	126 ≡ 1.000	≡ 1.000
	(b)	128 = 1.55 ± .02	1.38 – 1.77
		129 = 1.82 ± .02	1.53 – 2.29
		130 = 1.020 ± .012	0.58 – 0.78
		131 = 5.19 ± .06	3.05 – 19.10
0.06		132 = 1.06 ± .03	0.43 – 0.63
		134 = 0.097 ± .018	0.029 – 0.033
15475,135	61.2 Ba	124 = 0.56 ± .003	0.67 – 0.54
(l)	5.76 La	126 ≡ 1.000	≡ 1.000
	(d)	128 = 1.550 ± .011	1.37 – 1.76
		129 = 1.844 ± .015	1.52 – 2.28
		130 = 1.079 ± .006	0.57 – 0.77
		131 = 5.71 ± .04	3.01 – 18.66
0.046		132 = 1.075 ± .009	0.42 – 0.62
		134 = 0.086 ± .002	0.03 – 0.03

(a) Lugmair and Marti, 1971.
(b) Morrison *et al.,* 1971.
(c) Anders *et al.,* 1971.
(d) Rhodes and Hubbard, 1973.
(e) Br not measured, assumed to be the same as in 15065 (Ganapathy *et al.,* 1973) which petrographically similar (Ref. d).
(f) Philpotts and Schnetzler, 1970.
(g) Y not measured, estimated from measured Zr (Goles *et al.,* 1970) and average Zr/Y ratio in Apollo 11 igneous rocks (Compston *et al.,* 1970).
(h) Ganapathy *et al.,* 1970.
(i) Goles *et al.,* 1970.
(j) La not measured, assumed from measured Ce (Ref. a) and average La/Ce ratio in Apollo 12 rocks (Haskin *et al.,* 1971).
(k) Hohenberg, *et al.,* 1970.
(l) Morgan, 1975.
(m) Most cosmogenic fraction.

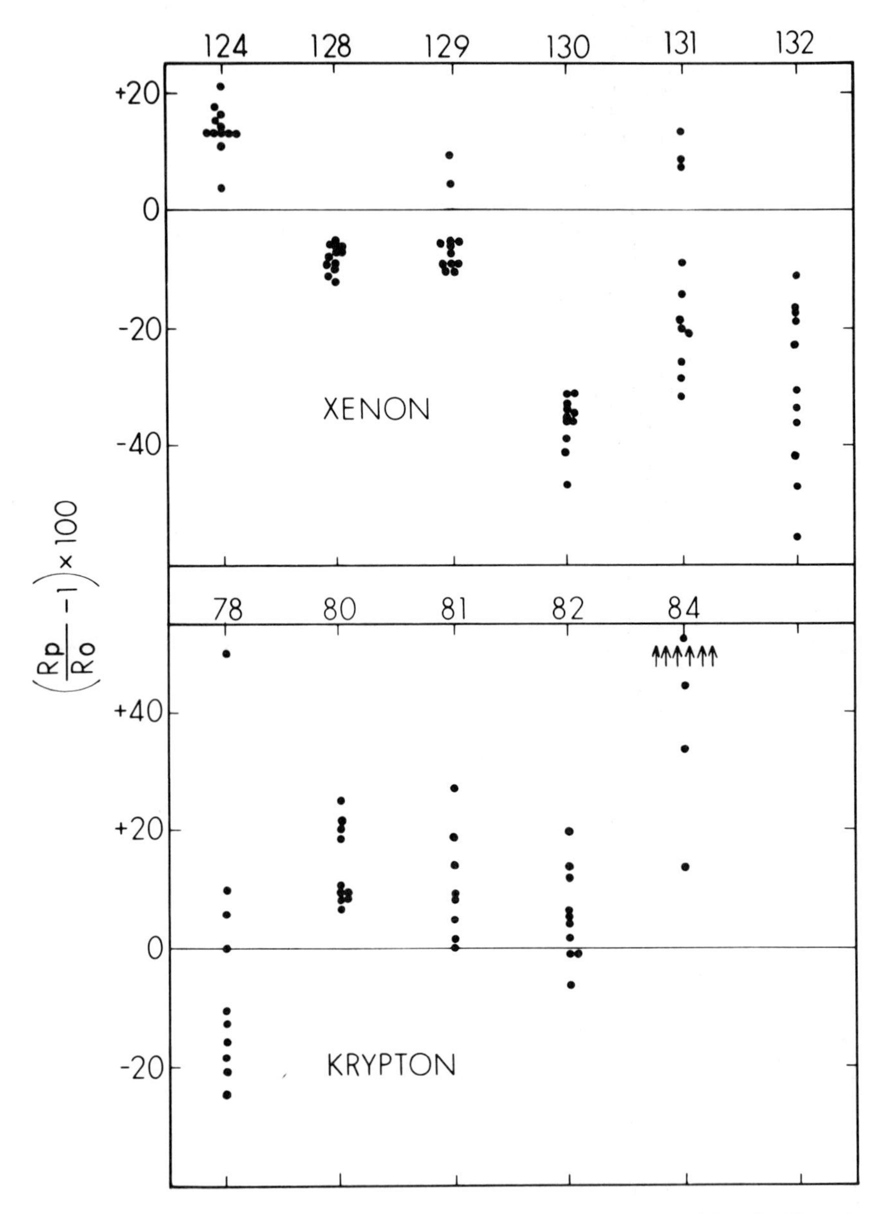

Fig. 3. A comparison of predicted and "observed" cosmogenic compositions for Xe and Kr; all data from Table 2c and 2d. R is the ratio of a cosmogenic isotope to ^{126}Xe or ^{83}Kr, R_p is the predicted ratio (from data in Table 1) and R_o is the "observed" ratio (for heavy Xe isotopes R_o is a ratio anticipated on the basis of empirical correlations—see text). For ^{130}Xe, for example, the figure indicates that the predicted production of cosmogenic ^{130}Xe is consistently about 30–40% lower than is observed. Slow neutron effects are not included for ^{80}Kr and ^{82}Kr.

cosmogenic $^{81}Kr/^{83}Kr$ ratio is much more sensitive to erosion, in particular to statistical variations in the erosion rate, than are the ratios of stable cosmogenic nuclides. Correspondingly one might expect to find less accord among the shallow samples. Considering both ^{83}Kr and ^{81}Kr in Table 2c, we note that in most cases there is satisfactory agreement between theoretical prediction and observation, but that in others there is significant disagreement, up to a factor of two, with the predicted abundances systematically too low.

For Kr there are an additional four isotopes in which cosmogenic contributions can be resolved, and which accordingly allow comparisons of isotopic ratios with considerably greater precision. Fig. 3 shows some systematic trends. ^{78}Kr is generally predicted too low, although there is a large amount of scatter. Since ^{78}Kr is a prominent spallation product with a low abundance in the trapped component, the resolution of $^{78}Kr_c$ is not a serious problem. Thus departures from observation are viewed as deficiencies with the predictive model and the degree of scatter is suggestive of modeling problems more complex than simple scaling. Predicted relative ^{80}Kr yields are higher than the observed data, a surprising result since Fig. 3 does not include contributions from the Br (n,γ) reactions. This suggests that either the predictive systematics substantially overestimates the ^{80}Kr or that very little contribution from $^{79}Br(n,\gamma)$ has in fact occurred, or both.

Finally, if one compares Table 3a with Table 2c, comparatively greater discrepancies are noted at ^{78}Kr and ^{84}Kr among the samples rich in cosmogenic gas. Cosmogenic compositions can be deduced with greater confidence for these latter samples but, since the depth history is unknown, comparisons are made with predicted compositions at depths ranging from 10 to 500 g/cm^2. Even so, the observed $(^{78}Kr/^{83}Kr)_c$ and $(^{84}Kr/^{83}Kr)_c$ values seldom fall within the predicted range. Apparent discrepancies will occur at both $^{78}Kr_c$ and $^{84}Kr_c$ if it is the normalization isotope $^{83}Kr_c$ that is in error. Since these samples all have had complex exposure histories it is probably true that, in general, significant cosmogenic Kr accumulated under rather heavily shielded conditions. Enhanced production of $^{83}Kr_c$ under heavily shielded conditions has previously been noted (Drozd *et al.*, 1974) and we suggest that production of ^{83}Kr by reactions such as $^{86}Sr(n,\alpha)$ ^{83}Kr may be underestimated. This is clearly an area where pure target measurements can improve the predictive model.

The problem at ^{78}Kr can be delineated by consideration of sample chemistry. Marti *et al.* (1973b) and Lugmair and Marti (1977) observe a convincing correlation between the $(^{78}Kr/^{83}Kr)_c$ ratio and the Zr/Sr ratio under constant exposure conditions for both lunar and meteoritic samples. Such a correlation can also be seen in Table 2c. 14321,92,FM3D has a Zr/Sr ratio of 4.8 and an observed $(^{78}Kr/^{83}Kr)_c$ ratio of 0.25; the corresponding numbers for 67075,8 are 0.06 for the Zr/Sr ratio, and 0.15, for the $(^{78}Kr/^{83}Kr)_c$ ratio. This pattern is clearly consistent with the trend observed by Marti *et al.* (1973b) and Lugmair and Marti (1977) but disagrees with the predictive model in its present form which suggests a trend in the *opposite* direction (Table 2c). Until appropriate corrections can be applied to the model it seems ill-advised to rely upon the predicted $(^{78}Kr/^{83}Kr)_c$ ratios to deduce any critical parameters in complex exposure modeling of the type suggested by Eugster (1978).

As pointed out above, an inaccuracy in the model probably also occurs at ^{80}Kr. Specific and meaningful comparison for this isotope is uncertain for two reasons. First, Br is a volatile trace element that is highly variable in lunar samples (variations more than an order of magnitude have been observed in individual samples); we have no assurance that the Br concentrations tabulated are applicable to these particular samples. Second, the major Br component is often leachable (Jovanovic and Reed, 1973) and may not in fact contribute to the cosmogenic Kr inventory at all. Low temperature release or quantitative loss of Kr and Xe from (n,γ) reactions on I and Br have previously been noted in neutron irradiated samples (Hohenberg, 1968; Alexander *et al.*, 1972). Furthermore, in contrast to predictions for many of the samples in Table 2c, it has been pointed out (cf. Marti *et al.*, 1973b) that the Br (n,γ) reactions are normally not very important contributors to $^{80}Kr_c$ and $^{82}Kr_c$ in the observed cosmogenic spectra for most lunar rocks. To summarize, we suggest that the $Br(n,\gamma)$ products are probably not quantitatively retained, and in addition, that the model probably overestimates other cross sections for the production of ^{80}Kr, but this conclusion is only provisional; the problem clearly deserves further study.

More generally, we can identify Kr as the gas for which there is the most serious disagreement between theoretical prediction and observation. In terms of the absolute gas concentrations it is not clear whether this is a failure of the theory or difficulty in identification of observed abundances. Discrepancies in isotopic ratios point out the need for improvements in the theoretical systematics, however. That Kr should be the "worst case" is not surprising. Of the gases considered here Kr is most liable to significant experimental uncertainties (in terms of absolute abundances and interferences in mass spectrometry) and its targets are trace elements and thus more liable to sampling heterogeneity problems. In the theoretical systematics Kr is also the case least solidly founded on experimental cross section measurements.

Xenon

Table 2d contains a large number of cases in which comparisons between prediction and observation may be made. Xe, like Kr, has several prominent cosmogenic isotopes whose ratios may be observed with less ambiguity than their absolute abundances. Comparison between observed and predicted ratios is illustrated in Fig. 3.

The agreement between predicted and observed absolute amounts of cosmogenic ^{126}Xe is reasonably good, mostly within about 20%. We can detect no significant correlation between the disagreement and either the shielding depth or the ratio of Ba to La. Since the predicted levels are not consistently high or low, we are inclined to suspect that the scatter is as likely due to the observations, chemical data, and documentation as it is due to the systematics. Such scatter could not be attributable to the difficulty in resolving cosmogenic ^{126}Xe, but might reasonably be due to a typical 10% experimental uncertainty in measurement of absolute gas amounts or to variability in the abundances of the trace element targets. The case of ^{126}Xe is probably the clearest instance in which we can recommend that the production rates in Table 1 are a fair representation of reality.

Two other cosmogenic isotopes, ^{124}Xe and ^{128}Xe, are directly determinable from raw data with little systematic uncertainty. As seen in Fig. 3, the differences between prediction and observation in these cases are fairly consistent and coherent, with the predicted ^{124}Xe/^{126}Xe ratio characteristically about 13% too high, and the predicted ^{128}Xe/^{126}Xe ratio about 6% too low. These variations are small enough that appeal to the absolute production rates cannot determine which of ^{124}Xe, ^{126}Xe or ^{128}Xe is predicted too high or too low.

The comparison between prediction and observation for the heavier Xe isotopes shows considerably greater scatter (Fig. 3). This is almost certainly due to scatter in the "observations" rather than in the predictions, since cosmogenic contributions to these isotopes are much more difficult to identify. It must also be remembered that the heavy isotope "observations" (Table 2) on which Fig. 3 is based follow not from the data for these isotopes but from the observed ^{124}Xe/^{126}Xe ratio and application of assumed linear correlation among these cosmogenic ratios, a procedure subject to the substantial qualifications expressed earlier. Nevertheless, trends are apparent relative to ^{126}Xe: the ^{129}Xe prediction is about 5% to 10% too low and the ^{130}Xe and ^{132}Xe predictions are both consistently about 30% to 40% too low. Independently of the assumptions made concerning heavy isotopes in Table 2d, the same conclusion follows from appeal to the comparisons in Table 3b, where the heavy isotope data result from direct calculation rather than assumption of linear correlation. These trends also suggest that the deficiency in predicted ^{130}Xe and ^{132}Xe may result from failure to explicitly consider the moderate energy neutron spallation reactions, specifically ^{130}Ba (n, p) and ^{132}Ba (n, p). Kaiser (1977) reached the same conclusion from comparison of similar calculations with lunar Xe measurements. At any rate it is clear that the Table 1 data should not be used for prediction of cosmogenic ^{130}Xe and ^{132}Xe. More cross section data are needed.

^{131}Xe again must be discussed as a special case. For the samples in Table 2, shielding depths are sufficiently shallow that ^{131}Xe production by spallation is usually more important than production by neutron capture. In this regime the predictions and observations, on the average, suggest an underproduction in the predicted ratios, although there is considerable scatter (see also Fig. 3) and the same reservations as expressed above. The range of predicted cosmogenic ratios due to the assumed 20% uncertainty element is larger for ^{131}Xe and only two samples of one rock (14306 L and D) have cosmogenic ^{131}Xe/^{126}Xe ratios that lie outside of this range. Target element chemistry for these particular samples was obtained from 14306,45 ("dark" and "white"), a different parent specimen, and we have no assurance that this lithology is representative of the samples analysed. In addition to the discrepancy at ^{131}Xe, 14306,26D appears to have too much

$^{83}Kr_c$ when compared with the predicted (Table 2c), suggestive perhaps of too large assumed Ba and Sr abundances. The possibility of a pre-irradiation at depth for 14306, which would provide an alternative explanation, cannot be excluded.

At greater depths, ^{131}Xe production is greatly enhanced by neutron capture. The substantially higher $^{131}Xe/^{126}Xe$ ratios corresponding to greater shielding are well known experimentally, but we have no basis for a quantitative comparison, and in the absence of natural samples irradiated at known depths in excess of about 100 g/cm^2 a rigorous comparison is unlikely to be forthcoming. Indeed, the $^{131}Xe/^{126}Xe$ ratio is the most sensitive depth indicator available, and depth estimates are usually based on theoretical calculations like those in Table 1. Since the parameters for the ^{130}Ba (n,γ) reaction are among those best founded in experiment, however, we expect that the predicted depth dependence of $^{131}Xe/^{126}Xe$ should be among the most realiable of the predictions.

SUMMARY AND CONCLUSIONS

To restate our objective and approach, we have attempted to compare theory and observation of the rates of production of cosmogenic noble gas isotopes in lunar rocks in order to delineate areas of agreement and disagreement, and to identify areas where further work is most appropriate. The results are neither complete nor totally definitive, and it is clear that this work must be considered a first attempt. The theoretical approach is incomplete, since many important excitation functions are assumed rather than measured. In spite of the wealth of noble gas analyses available in the literature, it has proved surprisingly difficult to establish a firm library of observed cosmogenic noble gas production rates in samples of well-defined age, chemistry and shielding conditions.

Overall, the most important generalization we can make is that in spite of these limitations and qualifications the agreement between observation and theoretical production is better than might perhaps have been expected. The agreement is, in nearly all cases, better than a factor of two, and in many cases is good to the 10% or 20% level, approaching experimental uncertainties. At this level, however, there are also several clear cases where the theoretical systematics fails to match observation; several instances have been mentioned, the most important of which are absolute production rates of ^{83}Kr and, to a lesser extent, ^{21}Ne and relative production ratios for $^{78}Kr/^{83}Kr$, $^{81}Kr/^{83}Kr$ and $^{130}Xe/^{126}Xe$. There is the clear need to examine the applicability of bulk Br chemistry for Kr production by neutron capture. There are also some important instances where there are insufficient data for rigorous comparison, e.g., $^{20}Ne/^{21}Ne$, $^{36}Ar/^{38}Ar$ and relative yields of heavy isotopes. In many cases, evident discrepancies are sufficiently large that the theoretical systematics can be identified as unsuitable for some of the detailed applications in which they have been used.

While the general applicability of the theoretical systematics seem well established, it is equally evident that significant improvements can be made. Additional proton and neutron cross section measurements for critical target elements and at various energies are needed. More precision measurements are needed on selected samples with well-documented shielding conditions and aliquots to determine target abundances. After available beam energies have been exploited for pure target cross section measurements, some of the more elusive parameters, such as specific excitation functions, may have to be adjusted in order to get optimum agreement with a controlled data base.

Acknowledgments—We gratefully acknowledge the assistance of B. Wilcox and C. White in preparation of this manuscript, and helpful discussions with M. Kennedy, T. Bernatowicz and B. Hudson. This work was supported in part by NASA grants NSG-07-016 and NSG-05-009-170 and NASA Work Order W-14,084.

REFERENCES

Alexander E. C., Davis P. K. Jr. and Reynolds J. H. (1972) Rare-gas analyses on neutron irradiated Apollo 12 samples. *Proc. Lunar Sci. Conf. 3rd*, p. 1787–1795.

Anders E., Ganapathy R., Keays R. R., Laul J. C. and Morgan J. W. (1971) Volatile and siderophile elements in lunar rocks: Comparison with terrestrial and meteoritic basalts. *Proc. Lunar Sci. Conf. 2nd*, p. 1021–1036.

Arvidson R., Crozaz G., Drozd R. J., Hohenberg C. M. and Morgan C. J. (1975) Cosmic-ray exposure ages of features and events at the Apollo landing sites. *The Moon* **13**, 259–276.

Bernatowicz T. J., Hohenberg C. M., Hudson B., Kennedy B. M. and Podosek F. A. (1978) Excess fission xenon at Apollo 16. *Proc. Lunar Sci. Conf. 9th*. This volume.

Bernatowicz T., Drozd R. J., Hohenberg C. M., Lugmair G., Morgan C. J. and Podosek F. A. (1977) The regolith history of 14307. *Proc. Lunar Sci. Conf. 8th*, p. 2763–2783.

Berman B. L. and Browne J. C. (1973) Microscopic $^{130}Ba(n,\gamma)$ cross section and the origin of ^{131}Xe on the moon. *Phys. Rev.* **C 7**, 2522–2531.

Bogard D. and Cressy P. (1973) Spallation production of ^{3}He, ^{21}Ne, and ^{38}Ar from target elements in the Brüderheim chondrite. *Geochim. Cosmochim. Acta* **37**, 527–546.

Bochsler P., Eberhardt P., Geiss J. and Grögler N. (1969) Rare gas measurements in separate mineral phases of the Otis and Elenovka chondrites. *Meteorite Research* (P. M. Millman, ed.) p. 857–873. D. Reidel, Dordrecht.

Burnett D. S., Drozd R. J., Morgan C. J. and Podosek F. A. (1975) Exposure histories of Bench Crater rocks. *Proc. Lunar Sci. Conf. 6th.*, p. 2219–2240.

Compston W., Chappell B. W., Arriens P. A. and Vernou M. J. (1970) The chemistry and age of Apollo 11 lunar material. *Proc. Apollo 11 Lunar Sci. Conf.*, p. 1007–1027.

Crozaz G., Drozd R., Hohenberg C., Hoyt H., Ragan D., Walker R. and Yuhas D. (1972) Solar flare and galactic cosmic-ray studies of Apollo 14 and 15 samples. *Proc. Lunar Sci. Conf. 3rd*, p. 2475–2499.

Drozd R., Hohenberg C., Morgan C. and Ralston C. (1974) Cosmic-ray exposure history at the Apollo 16 and other lunar sites: lunar surface dynamics. *Geochim. Cosmochim. Acta* **38**, 1625–1642.

Drozd R. J., Hohenberg C. M., Morgan C. J., Podosek F. A. and Wroge M. L. (1977) Cosmic-ray exposure history at Taurus-Littrow. *Proc. Lunar Sci. Conf. 8th*, p. 3027–3043.

Eberhardt P., Geiss J., Graf H., Grögler N., Mendia M. D., Mörgeli M., Schwaller H. and Stettler A. (1972) Trapped solar wind noble gases in Apollo 12 lunar fines 12001 and Apollo 11 breccia 10046. *Proc. Lunar Sci. Conf. 3rd*, p. 1821–1856.

Eugster O. (1978) A refined method for the calculation of residence times and shielding depths for two-stage irradiation models and the determination of the depth dependency of cosmogenic $^{131}Xe/^{126}Xe$ and $^{83}Kr/^{78}Kr$ (abstract). In *Lunar and Planetary Science IX*, p. 303–305. Lunar and Planetary Institute, Houston.

Eugster O., Eberhardt P., Geiss J., Grögler N., Jungck M. and Mörgeli M. (1977) The cosmic-ray exposure history of Shorty Crater samples; the age of Shorty Crater. *Proc. Lunar Sci. Conf. 8th*, p. 3059–3082.

Eugster O., Eberhardt P., Geiss J. and Grögler N. (1978) The solar wind and cosmic-ray exposure history of soil from drive tube 74001, and unmixed lunar regolith (abstract). In *Lunar and Planetary Science IX*, p. 306–308. Lunar and Planetary Institute, Houston.

Funk H., Podosek F. and Rowe M. W. (1967) Spallation yields of krypton and xenon from irradiation of strontium and barium with 730 MeV protons. *Earth Planet. Sci. Lett.* **3**, 192–196.

Ganapathy R., Keays R. R., Laul J. C. and Anders E. (1970) Trace elements in Apollo 11 lunar rocks: implications for meteorite influx and origin of moon. *Proc. Apollo 11 Lunar Sci. Conf.*, p. 1117–1142.

Ganapathy R., Morgan J. W., Krähenbühl U. and Anders E. (1973) Ancient meteoritic components in lunar highland rocks: clues from trace elements in Apollo 15 and 16 samples. *Proc. Lunar Sci. Conf 4th*, p. 1239–1261.

Goebel K., Schultes H. and Zaehringer J. (1964) Production cross-sections of tritium and rare gases in various target elements. CERN report CERN-64-12 78 pp.

Goles G. G., Randle K., Osawa M., Lindstrom D. J., Jerome D. Y., Steinborn T. L., Beyer R. L., Martin M. R. and McKay S. M. (1970) Interpretations and speculations on elemental abundances in lunar samples. *Proc. Apollo 11 Lunar Sci. Conf.*, p. 1177–1194.

Haskin L. A., Helmke P. A., Allen R. O., Anderson M. R., Korotev R. L. and Zweifel K. A. (1971) Rare-earth elements in Apollo 12 lunar materials. *Proc. Lunar Sci. Conf. 2nd*, p. 1307–1317.

Haskin L. A., Helmke P. A., Blanchard D. P., Jacobs J. W. and Telander K. (1973) Major and trace element abundances in samples from the lunar highlands. *Proc. Lunar Sci. Conf. 4th*, p. 1275–1296.

Hohenberg C. M. (1968) Studies in I-Xe and Pu-Xe Dating of Pile-Irradiated Achondrites, Ph. D. thesis, Univ. California, Berkeley.

Hohenberg C. M., Davis P. K., Kaiser W. A., Lewis R. S. and Reynolds J. H. (1970) Trapped and cosmogenic rare gases from stepwise heating of Apollo 11 samples. *Proc. Apollo 11 Lunar Sci. Conf.*, p. 1283–1309.

Hohenberg C. M. and Rowe M. W. (1970) Spallation yields of xenon from irradiation of Cs, Ce, Nd, Dy, and a rare earth mixture with 730-MeV protons. *J. Geophys. Res.* **75**, 4205–4209.

Jovanovic S. and Reed G. W., Jr. (1973) Volatile trace elements and the characterization of the Cayley formation and the primitive lunar crust. *Proc. Lunar Sci. Conf. 4th*, p. 1313–1324.

Kaiser W. A. (1977) The excitation functions of Ba(p,x)^{m}Xe (m = 124–136) in the energy range 38–600 MeV; the use of 'cosmogenic' xenon for estimating 'burial' depths and 'real' exposure ages. *Phil. Trans. Roy. Soc. London* **A. 285,** 337–362.

Kohl C. P., Murell M. T., Russ G. P. III and Arnold J. R. (1978) Evidence for the constancy of the solar cosmic ray flux over the past 10 million years: ^{53}Mn and ^{26}Al measurements. *Proc. Lunar Sci. Conf. 9th.* This volume.

Krähenbühl U., Ganapathy R., Morgan J. W. and Anders E. (1973) Volatile elements in Apollo 16 samples: implications for highland volcanism and accretion history of the moon. *Proc. Lunar Sci. Conf. 4th*, p. 1325–1348.

Laul J. C., Wakita H., Showalter D. L., Boynton W. V. and Schmitt R. A. (1972) Bulk, rare earth and other trace elements in Apollo 14 and 15 and Luna 16 samples. *Proc. Lunar Sci. Conf. 3rd*, p. 1181–1200.

Laul J. C., Hill D. W. and Schmitt R. A. (1974) Chemical studies of Apollo 16 and 17 samples. *Proc. Lunar Sci. Conf. 5th*, p. 1047–1066.

Lindstrom M. M., Duncan A. R., Fruchter J. S., McKay S. M., Stoeser J. W., Goles G. G. and Lindstrom D. J. (1972) Compositional characteristics of some Apollo 14 clastic materials. *Proc. Lunar Sci. Conf. 3rd*, p. 1201–1214.

Lingenfelter R. E., Canfield E. H. and Hampel V. E. (1972) The lunar neutron flux revisited. *Earth Planet. Sci. Lett.* **16**, 355–369.

Lugmair G. W. and Marti K. (1971) Neutron capture effects in lunar gadolinium and the irradiation histories of some lunar rocks. *Earth Planet. Sci. Lett.* **13**, 32–42.

Lugmair G. W. and Marti K. (1977) Sm-Nd-Pu Timepieces in the Angra Dos Reis Meteorite. *Earth Planet. Sci. Lett.* **35**, 273–284.

LSPET (Lunar Sample Preliminary Examination Team) (1972) Preliminary examination of lunar samples. In *Apollo 16 Prelim. Sci. Rep.*, NASA SP-315.

Mark R. K., Cliff R. A., Lee-Hu C. and Wetherill G. W. (1973) Rb-Sr studies of lunar breccias and soils. *Proc. Lunar Sci. Conf. 4th*, p. 1785–1795.

Marti K. (1967) Mass-spectrometric detection of cosmic ray produced ^{81}Kr in meteorites and the possibility of Kr-Kr dating. *Phys. Rev. Lett.* **18**, 264–266.

Marti K., Eberhardt P. and Geiss J. (1966) Spallation. fission and neutron capture anomalies in meteoritic krypton and xenon. *Z. Naturforsch.* **21a**, 398–413.

Marti K., Lightner B. D. and Lugmair G. W. (1973a) On ^{244}Pu in lunar rocks from Fra Mauro and implications regarding their origin. *The Moon* **8**, 241–250.

Marti K., Lightner B. D. and Osborn T. (1973b) Krypton and xenon in some lunar samples and the age of North Ray crater. *Proc. Lunar Sci. Conf. 4th*, p. 2037–2048.

Marti K. and Lugmair G. W. (1971) Kr^{81}-Kr and K-Ar^{40} ages, cosmic-ray spallation products, and neutron effects in lunar sample from Oceanus Procellarum. *Proc. Lunar Sci. Conf. 2nd*, p. 1591–1605.

Marti K., Lugmair G. W. and Urey H. C. (1970) Solar wind gases, cosmic ray spallation products, and the irradiation history. *Science* **167**, 548–550.

Morgan C. J. (1975) Exposure age dating of lunar features: lunar heavy rare gases. Ph.D. thesis, Washington Univ., St. Louis, Missouri.

Morgan J. W., Laul J. C., Krähenbühl U., Ganapathy R. and Anders E. (1972) Major impacts on the moon: characterization from trace elements in Apollo 12 and 14 samples. *Proc. Lunar Sci. Conf. 3rd*, p. 1377–1395.

Morrison G. H., Gerard J. T., Potter N. M., Grangadharam E. V., Rothenberg A. M. and Burdo R. A. (1971) Elemental abundances of lunar soil and rocks from Apollo 12. *Proc. Lunar Sci. Conf. 2nd*, p. 1169–1185.

Niemeyer S. (1977) Exposure histories of lunar rocks 71135 and 71569. *Proc. Lunar Sci. Conf. 8th*, p. 3083–3093.

Nishiizumi K., Imamura M., Honda M., Russ G. P., Kohl C. P. and Arnold J. R. (1976) ^{53}Mn in the Apollo 15 and 16 drill stems: evidence for surface mixing. *Proc. Lunar Sci. Conf. 7th*, p. 41–54.

Pepin R. O., Basford J. R., Dragon J. C., Coscio M. R. and Murthy V. R. (1974) Rare gas and trace elements in Apollo 15 drill core fines: depositional chronologies and K-Ar ages, and production rates of spallation-produced ^{3}He, ^{21}Ne, and ^{38}Ar versus depth. *Proc. Lunar Sci. Conf. 5th*, p. 2149–2184.

Philpotts J. A. and Schnetzler C. C. (1970) Potassium, rubidium, strontium, barium and rare-earth concentrations in lunar rocks and separated phases. *Science* **167**, 493–495.

Philpotts J. A., Schnetzler C. C., Nava D. F., Bottino M. L., Fullagar P. D., Thomas H. H., Schuhmann S. and Kouns C. W. (1972) Apollo 14: some geochemical aspects. *Proc. Lunar Sci. Conf. 3rd*, p. 1293–1305.

Podosek F. A. and Huneke J. C. (1971) Isotopic composition of ^{244}Pu fission xenon in meteorites: reevaluation using lunar spallation xenon systematics. *Earth Planet. Sci. Lett.* **12**, 73–82.

Podosek F. A., Huneke J. C., Burnett D. S. and Wasserburg G. J. (1971) Isotopic composition of xenon and krypton in the lunar soil and in the solar wind. *Earth Planet. Sci. Lett.* **10**, 199–216.

Reedy R. C. (1976) Spallation rare gas production calculations and studies of lunar surface processes (abstract). In *Lunar Science VII*, p. 721–723. The Lunar Science Institute, Houston

Reedy R. C. and Arnold J. R. (1972) Interaction of solar and galactic cosmic-ray particles with the moon. *J. Geophys. Res.* **77**, 537–555.

Reedy R. C., Herzog G. F. and Jessberger E. K. (1978) Depth variations of spallogenic nuclides in meteorites (abstract). In *Lunar and Planetary Science IX*, p. 940–942. Lunar and Planetary Science Institute, Houston.

Rhodes J. M. and Hubbard N. J. (1973) Chemistry, classification, and petrogenesis of Apollo 15 mare basalts. *Proc. Lunar Sci. Conf. 4th*, p. 1127–1148.

Rose H. J. Jr., Cuttitta F., Annell C. S., Carron M. K., Christian R. P., Dwornik E. J., Greenland L. P. and Ligon D. T. Jr. (1972) Compositional data for twenty-one Fra Mauro lunar materials. *Proc. Lunar Sci. Conf. 3rd*, p. 1215–1229.

Rudstam G. (1966) Systematics of spallation yields. *Z. Naturforsch.* **21a**, 1027–1041.

Sachdev D. R., Porile N. T. and Yaffe L. (1967) Reactions of ^{88}Sr with protons of energies 7 to 85 MeV. *Can. J. Chem.* **45**, 1149–1160.

Signer P. and Suess H. E. (1963) Rare gases in the sun, in the atmosphere, and in meteorites. In *Earth Sci. and Meteorites* (J. Geiss and E. D. Goldberg, eds.), p. 241–272. Wiley, N.Y.

Silberberg R. and Tsao C. H. (1973a) Partial cross-sections in high-energy nuclear reactions, and astrophysical applications. I Targets with $Z \leq 28$. *Astrophys. J. Suppl.* **25**, 315–334.

Silberberg R. and Tsao C. H. (1973b) Partial cross-sections in high-energy nuclear reactions, and astrophysical applications. II Targets heavier than nickel. *Astrophys. J. Suppl.* **25**, 335–368.

Simonds C. H., Phinney W. C., Warner J. L., McGee P. E., Geeslin J., Brown R. W. and Rhodes J. M. (1977) Apollo 14 revisited, or breccias aren't so bad after all. *Proc. Lunar Sci. Conf. 8th*, p. 1869–1893.

Srinivasan B. (1974) Lunar breccia 14066: ^{81}Kr-^{83}Kr exposure age, evidence for fissiogenic xenon from ^{244}Pu, and rate of production of spallogenic ^{126}Xe. *Proc. Lunar Sci. Conf. 5th*, p. 2033–2044.

Stoenner R. W., Lyman W. J. and Davis R. Jr. (1970) Cosmic-ray production of rare-gas radioactivities and tritium in lunar material. *Proc. Apollo 11 Lunar Sci. Conf.*, p. 1583–1594.

Strasheim A., Jackson P. F. S., Coetzee J. H. J., Strelow F. W. E., Wybenga F. T., Gricius A. J., Kokot M. L. and Scott R. H. (1972) Analysis of lunar samples 14163, 14259, and 14321 with isotopic data for ^{7}Li/^{6}Li. *Proc. Lunar Sci. Conf. 3rd*, p. 1337–1342.

Taylor S. R., Kaye M., Muir P., Nance W., Rudowski R. and Ware N. (1972) Composition of the lunar uplands: chemistry of Apollo 14 samples from Fra Mauro. *Proc. Lunar Sci. Conf. 3rd*, p. 1231–1249.

Wahlen M., Honda M., Imamura M., Fruchter J. S., Finkel R. C., Kohl C. P., Arnold J. R. and Reedy R. C. (1972) Cosmogenic nuclides in football-sized rocks. *Proc. Lunar Sci. Conf. 3rd*, p. 1719–1732.

Walton J. R., Heymann D., Yaniv A., Edgerley D. and Rowe M. W. (1976) Cross sections for He and Ne isotopes in natural Mg, Al, and Si, He isotopes in CaF_2, Ar isotopes in natural Ca, and radionuclides in natural Al, Si, Ti, Cr, and stainless steel induced by 12- to 45-MeV protons. *J. Geophys. Res.* **81**, 5689–5699.

Wänke H., Palme H., Baddenhausen H., Dreibus G., Jagoutz E., Kruse H., Spettel B., Teschke F. and Thacker R. (1974) Chemistry of Apollo 16 and 17 samples: bulk composition, late stage accumulation and early differentiation of the moon. *Proc. Lunar Sci. Conf. 5th*, p. 1307–1335.

Wänke H., Palme H., Baddenhausen H., Dreibus G., Jagoutz E., Kruse H., Palme C., Spettel B., Teschke F. and Thacker R. (1975) New data on the chemistry of lunar samples: primary matter in the lunar highlands and the bulk composition of the moon. *Proc. Lunar Sci. Conf. 6th*, p. 1313–1340.

Wilshire H. G., Stuart-Alexander D. E. and Jackson E. D. (1973) Apollo 16 rocks: petrology and classification. *J. Geophys. Res.* **78**, 2379–2392.

Woolum D. S., Burnett D. S., Furst M. and Weiss J. R. (1975) Measurement of the lunar neutron density profile. *The Moon* **12**, 231–250.

Yuhas D. (1974) The particle track record in lunar silicates: Long term behavior of solar and galactic VH nuclei and lunar surface dynamics. Ph.D. thesis, Washington Univ., St. Louis, Missouri.

Proc. Lunar Planet. Sci. Conf. 9th (1978), p. 2345–2362.
Printed in the United States of America

^{40}Ar-^{39}Ar microanalysis of single 74220 glass balls and 72435 breccia clasts

J. C. HUNEKE

The Lunatic Asylum, Division of Geological and Planetary Sciences*
California Institute of Technology, Pasadena, California 91125

Abstract—We report ^{40}Ar-^{39}Ar age measurements on single orange glass balls from the Apollo 17 soil 74220 and individual clasts from the Apollo 17 highland breccia 72435. The measurements required the use of newly established microanalytical techniques to obtain high quality analyses on ~0.5 mg particles with only a few hundred ppm K. An age of 3.60 ± 0.04 AE is determined for the orange glass. No corrections for a trapped ^{40}Ar component were required. The glass forming event occurred at the very end of or after the extrusion of the mare basalts at the Apollo 17 site. An extremely well defined age plateau at 3.86 ± 0.04 AE was determined for a 72435 plagioclase clast with attached matrix. A second large plagioclase crystal yielded significantly older ages over the last 60% of Ar release at high temperatures and is a relict clast incompletely degassed at the time of breccia formation. 72435 also contains plagioclase clasts with primitive Sr and a 4.55 AE old dunite clast. The Ar results provide additional evidence for the association of chemically unequilibrated, relict clasts with both primitive Sr and older K/Ar ages.

INTRODUCTION

^{40}Ar-^{39}Ar gas retention ages have been measured on individual balls from the Apollo 17 orange soil (74220) and plagioclase clasts from the Apollo 17 breccia 72435. Each of the particles weighed less than a milligram and required the use of new microanalytical techniques to obtain high quality ^{40}Ar-^{39}Ar thermal release age spectra. The new techniques are one to two orders of magnitude more sensitive than previous methods. Individual large glass balls, in which nonradiogenic ^{40}Ar contributions are minimized, were measured in an effort to clarify differences of up to 250 m.y. in ^{40}Ar-^{39}Ar ages so far reported for the 74220 glass. Single plagioclase crystals from 72435 were studied in search of material with K/Ar ages older than 3.9 AE which correlate with primitive Sr found in relict plagioclase clasts in this breccia.

K/Ar ages reported here are calculated using newly recommended ^{40}K decay constants and K isotopic composition (Steiger and Jaeger, 1977). K/Ar ages previously reported have been altered to be consistent with the present results. The effect of the change is to reduce K/Ar ages which were older than 2.0 AE and to increase ages which were younger than 2.0 AE. For guidance we note that ages of 2.5 AE, 3.0 AE, 3.5 AE, 4.0 AE and 4.5 AE calculated with decay constants and K compositions previously in general use are reduced by 0.015 AE,

*Contribution No. 3071 (265)

0.025 AE, 0.045 AE, 0.060 AE and 0.080 AE, respectively, when recalculated using the newly recommended values. In particular, ages near the time of the terminal lunar cataclysm about 4.0 AE ago are reduced by a large fraction of the 100–200 m.y. time interval within which most nearside basins were formed. K/Ar ages about 4.5 AE are reduced by about 80 m.y., which is large compared to the mean lives of ^{244}Pu, ^{129}I and ^{26}Al, which are extremely relevant to discussions of early solar system history. The shift at 4.5 AE is large compared to the resolution in time obtainable by all major absolute chronometers. Precise ^{40}Ar-^{39}Ar ages are generally accurate to about 0.04 AE. K/Ar ages above 3.4 AE are reduced by more than this amount.

Analytical system for ^{40}Ar-^{39}Ar dating of micro-samples

There has been a clear need to obtain reliable ages on very small particles. Potential applications of microanalytic dating techniques are the analysis of single soil particles and breccia clasts and the analysis of rare samples with little mass available for study. Microanalytic K/Ar dating by Ar extraction with focussed laser beam heating has been under development for several years (Megrue, 1973; Schaeffer *et al.*, 1977 and references cited). The laser extraction method lacks the advantages of ^{40}Ar-^{39}Ar dating by stepwise thermal release for evaluating natural ^{40}Ar losses and for the identification of nonradiogenic ^{40}Ar components. We have recently established microtechniques to obtain detailed ^{40}Ar-^{39}Ar thermal release age patterns on samples with as little as 100 ppm K and weighing less than a milligram (10^{-7} μg K and 3×10^{-9}ccSTP^{40}Ar).

In the simplest design of the extraction assembly, used for the analyses reported here, the bare sample is placed on a Re ribbon and surrounded with a second piece of Re ribbon which retains the fragment in place and serves as a radiation shield (Fig. 1). The filament assembly is mounted to W leads and inserted into the center of a glass extraction vessel, well away from the vessel components to avoid degassing during extraction. A shutter is included in the extraction vessel to prevent buildup of volatilized metal and silicate on a portion of the vessel walls. Prior to sample loading, the filament is degassed and calibrated for temperature using an optical pyrometer and assuming a spectral emissivity of 0.3. Temperatures below 900°C are extrapolated from current vs. temperature measurements at higher temperatures. The open structure of this simple filament assembly does not allow for accurate control of sample temperature. An indication of the temperature gradients in the sample region is provided by the temperature of the outside of the Re cover, which can be several hundred degrees cooler than a filament temperature at ~1600°C. Some of the finer details in the ^{40}Ar-^{39}Ar age spectra may be obscured as different portions of the sample degas simultaneously at slightly different temperatures, but the plateau age should be unaffected. Filament currents of less than 1.6 A were sufficient to degas and melt the 74220 samples, while up to 2.5 A were required for the 72435 fragments. Filament currents above 2 A can begin to desorb air Ar from the W leads, but neither 72435 sample shows evidence for an extraction blank larger than normal. Reactions with the silicate melt did not cause rapid filament deterioration and the Re filament and shield remained flexible after extensive high temperature operation.

Ar was extracted for 40 min. at successively higher filament temperatures. Extracted gas was briefly exposed to SAES (Zr-Al) getters heated to ~400°C, prior to admitting Ar to the mass spectrometer for isotopic measurement. The Ar ion beam was measured using an electron multiplier at a gain of $\sim 10^4$, with a sensitivity of 4.0×10^{-9}ccSTP/V and discrimination of 0.05% per mass unit favoring lighter masses.

The Ar blank of the system was monitored throughout the experiments and was $\sim 2 \times 10^{-11}$ccSTP air Ar. Heating of an empty filament assembly subjected to the sample loading procedures showed no increase in the blank over the cold system blank. Mass spectrometer interferences were present at mass 39 at the level of 5×10^{-13}ccSTP Ar equivalent and at masses 35–38 due to Cl at the level of 2×10^{-11}ccSTP Ar equivalent, with $HCl^+/Cl^+ \approx 0.05$. The low Ar extraction blanks and mass spectrometer interferences allowed good ^{40}Ar-^{39}Ar age spectra to be measured on 0.5–1 mg of a

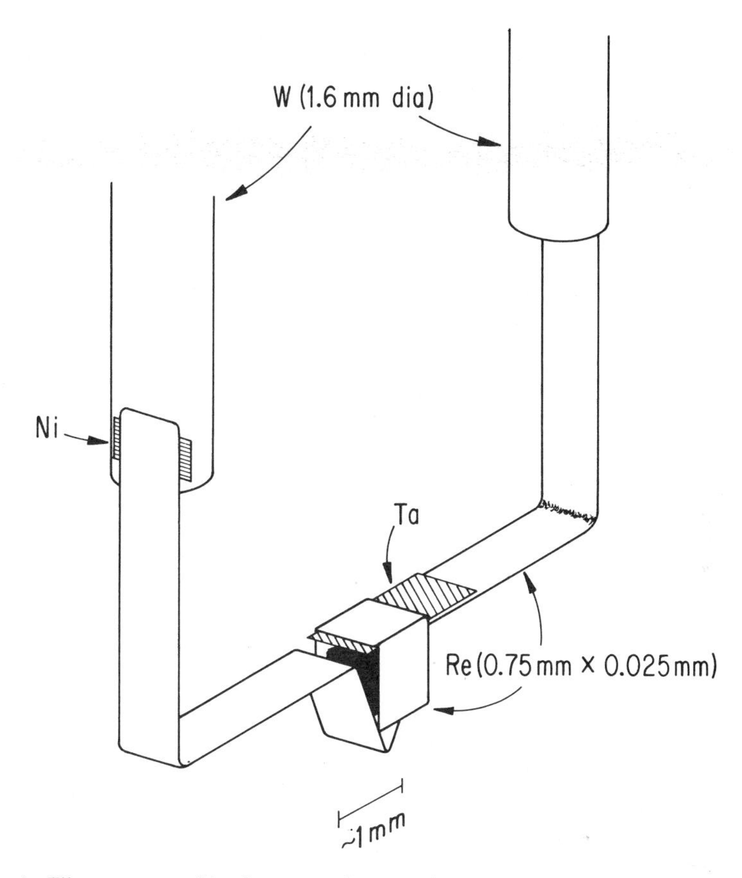

Fig. 1. Filament assembly for extracting Ar from sub mg-sized fragments, with minimal Ar contamination from the extraction system. The sample is placed directly on a resistance-heated Re filament supported from W leads. Additional Re is added to retain the sample on the filament and provide radiation shielding. Ta and Ni are intermediates for spot welding. The assembly requires only 2 A to attain filament temperatures of ~1600°C.

sample with only 100 ppm K. Higher K contents enable age measurements on correspondingly smaller samples.

The sensitivity of the ^{40}Ar-^{39}Ar microanalytical techniques described here is comparable to that of the focussed laser extraction method for K/Ar dating. Detailed thermal release ^{40}Ar-^{39}Ar age spectra can be obtained on about (500 μm)^3 of material containing several hundred ppm K. Control on the microvolume analyzed is achieved by selecting and removing the desired grains from the host rock. The measurement of a K/Ar age by Ar extraction with a focussed laser beam requires the extraction of Ar from about (500 μm)^2 × 50 μm of material. The ~(50 μm)^3 volumes from which Ar is extracted are optically selected and melted *in situ*, maintaining the petrographical context of the volume analyzed. This is the primary advantage of the laser extraction method. The resulting K/Ar age suffers the normal ambiguities in interpretation related to possible loss of ^{40}Ar and/or presence of nonradiogenic ^{40}Ar. Dating by stepwise thermal release of Ar from separated small fragments provides some possibility of resolving such ambiguities on the same microscopic scale as the laser extraction method.

We present ^{40}Ar-^{39}Ar measurements on single glass particles from the 74220 soil and then measurements on the plagioclase clasts from the 72435 breccia. These results were first presented at the Ninth Lunar and Planetary Science Conference (Huneke and Wasserburg, 1978) together with ^{40}Ar-^{39}Ar microanalyses on two Luna 24 basalt fragments (Wasserburg, *et al.*, 1978).

Neutron irradiation

The particles were neutron irradiated at the G.E. Test Reactor at Pleasanton, CA. All five 74220 particles were irradiated in a single Al packet. Likewise, the two 72435 crystals were part of a group of six irradiated in a single packet. The relative fluence variations in the irradiation cannister were monitored by Ni flux wires placed between samples; the variations between adjacent monitor positions were $\leq 2\%$. The conversion coefficients from this irradiation (LAV 6) determined from hornblend (hb3gr) and CaF_2 monitors are

$$C_{39}(K) = 6.48 \times 10^{-4} \text{ccSTP/g}$$
$$C_{37}(Ca) = 7.75 \times 10^{-6} \text{ccSTP/g (1 April 1977).}$$

Ar interferences from neutron reactions on Ca were $(^{36}Ar/^{37}Ar/^{38}Ar/^{39}Ar)_{Ca} = 0.0121 \pm 0.0002/1/0.004$ (cf., Turner *et al.*, 1971)/0.0277 ± 0.0001. Ar interferences from neutron reactions on K are assumed to be $(^{38}Ar/^{39}Ar/^{40}Ar)^* = 0.01/1/0.01$ (Turner *et al.*, 1971).

74220 orange glass

The orange glass from the Taurus-Littrow region is a unique sampling of a distinctive style of lunar volcanic activity. Of all lunar glasses, the evidence is strongest for the orange glass that it is the result of fire-fountaining of a basaltic magma onto the lunar surface (cf., Heiken and McKay, 1977). A precise determination of the age of the glass is required to establish in what way the fire fountain event may be related to the extrusion of the lunar mare basalts. Numerous attempts to measure the ^{40}Ar-^{39}Ar gas retention age of the glass have been reported. Efforts so far have resulted in ages significantly different from each other primarily as a result of the large corrections required to account for nonradiogenic ^{40}Ar contributions. The proper correction for trapped ^{40}Ar contributions is very difficult to establish. The variation in K/Ar ages reflects the multiplicity of trapped Ar compositions and the fractionation of trapped Ar components from each other during differential thermal release. The problem remains to determine a precise ^{40}Ar-^{39}Ar age for the orange glass despite the presence of nonradiogenic ^{40}Ar in 74220 soil particles. The approach of the present study is to measure single large balls in which trapped ^{40}Ar is a minor component and to identify the trapped ^{40}Ar present as well as possible by differential thermal release.

Five glass particles ranging in size from 0.15 mg to 0.8 mg were neutron irradiated and the Ar from each analyzed using the microanalytic extraction system described above. All particles were devitrified orange glass. The largest particle was prolate with an axial ratio of 2:1; the remaining four were relatively spherical. The surfaces of all five were rough from incomplete assimilation of late accreting glass droplets. The Ar amounts and isotopic compositions from

Table 1. Ar amounts and isotopic composition in five neutron-irradiated glass balls from the Apollo 17 soil 74220.[f]

Filament Temp °C[h]	^{40}Ar[a] 10^{-10}ccSTP	^{36}Ar[b]/^{40}Ar $\times 10^{-5}$	^{37}Ar[c]/^{40}Ar $\times 10^{-4}$	^{38}Ar[d]/^{40}Ar $\times 10^{-5}$	$^{39}Ar^{*}$[e]/^{40}Ar $\times 10^{-4}$	Age AE
A (0.6 mg, 470 ppmK)						
800	0.1	2400	78	3500	106	4.10
		±2500	± 91	±2800	± 94	±1.45
875	0.3	2030	100	1900	108	4.06
		± 800	± 47	± 720	± 44	± .65
950	0.7	1960	59	1530	115	3.96
		± 340	± 16	± 260	± 20	± .28
1025	3.5	842	87	527	119	3.915
		± 44	± 4	± 23	± 5	± .072
1100	5.2	129	114	164	131	3.756
		± 17	± 3	± 9	± 4	± .043
1150	12	133	179	171	154	3.506
		± 8	± 3	± 4	± 3	± .024
1025r[g]	0.2	610	220	580	177	3.3
		± 740	±170	± 460	±130	±1.1
1100r	0.2	250	380	160	148	3.44
		± 560	±240	± 180	±100	± .96
1150r	4.6	96	201	176	148	3.574
		± 20	± 5	± 11	± 4	± .042
1200	67	54	192	163	146	3.595
		± 10	± 1	± 1	± 1	± .008
1200r	15	56	200	174	146	3.599
		± 8	± 3	± 5	± 3	± .027
1250	12	56	204	175	145	3.598
		± 8	± 3	± 4	± 2	± .021
1300	2.0	124	199	2130	145	3.601
		± 47	± 12	± 110	± 8	± .009
Total	123	110	186	228	146	3.602
		± 9	± 2	± 16	± 1	± .009

the stepwise thermal release analysis of the five single particles are given in Table 1.

The weights of the particles and the K contents calculated from the total Ar released are also given in Table 1. The weight is estimated from ^{37}Ar using $C_{37}(Ca)$ and a Ca content of 5.1 wt.% (Mao *et al.*, 1973). The K concentrations are calculated from ^{39}Ar using the inferred weight and $C_{39}(K)$. The K concentrations range from 430 ppm K to 480 ppm K and agree reasonably well with other measurements on large spheres (Alexander *et al.*, 1978).

The stable Ar isotopes are partitioned between trapped Ar_t, cosmic ray produced Ar_s and radiogenic $^{40}Ar^{*}$ using the compositions $(^{36}Ar/^{38}Ar/^{40}Ar)_t = 1/0.187/0$ (see below) and $(^{36}Ar/^{38}Ar/^{40}Ar)_s = 0.6/1/0.15 \pm 0.15$. If

Table 1. *(cont'd.)*

Filament Temp °C	$^{40}Ar^{(a)}$ $\times 10^{-10}$ccSTP	$^{36}Ar^{(b)}/^{40}Ar$ $\times 10^{-5}$	$^{37}Ar^{(c)}/^{40}Ar$ $\times 10^{-4}$	$^{38}Ar^{(d)}/^{40}Ar$ $\times 10^{-5}$	$^{39}Ar^{*(e)}/^{40}Ar$ $\times 10^{-4}$	Age AE
B (0.7 mg, 480 ppmK)						
700	0.5	620	≤194	1540	17	7.2
		±840		±260	± 16	±1.7
900	2.3	1620	≤55	124	83	4.50
		±220		± 6	± 5	± .10
1010	6.3	539	102	389	116	3.950
		± 77	± 17	± 16	± 3	± .031
1080	5.7	106	110	170	134	3.729
		± 84	± 19	± 16	± 3	± .030
1180	5.9	128	171	179	149	3.563
		± 84	± 19	± 16	± 3	± .027
1270	113	56	191	164	144	3.611
		± 4	± 1	± 1	± 1	± .009
1350	22	47	191	163	142	3.636
		± 20	± 5	± 5	± 1	± .015
1410	0.6	≤707	≤160	170	115	3.98
				±140	± 19	± .27
Total	156	104	180	194	141	3.639
		± 10	± 2	± 3	± 1	± .008
C (0.25 mg, 450 ppmK)						
880	0.3	2110	64	2170	99	4.21
		±810	± 61	±770	± 44	± .73
910	1.0	990	81	110	116	3.95
		±130	± 15	± 12	± 11	± .16
1105	3.2	177	114	180	131	3.765
		± 33	± 5	± 23	± 4	± .048
1120	28	63	196	158	146	3.592
		± 4	± 1	± 4	± 1	± .009
1220	8.3	78	208	179	146	3.597
		± 13	± 3	± 10	± 2	± .019
1250	4.2	122	212	186	147	3.578
		± 26	± 5	± 19	± 3	± .036
1270	2.5	197	212	209	146	3.590
		± 45	± 9	± 31	± 6	± .058
1320	2.9	178	211	200	144	3.614
		± 38	± 8	± 29	± 5	± .050
1320r	0.6	240	219	194	151	3.54
		±170	± 36	±115	± 22	± .22
Total	51	120	193	198	144	3.610
		± 8	± 1	± 7	± 1	± .009

a) ^{40}Ar corrected for blanks and contributions from neutron reactions on K.

b) ^{36}Ar corrected for spectrometer background, blank and contributions from neutron reactions on Ca.

c) ^{37}Ar corrected for spectrometer background, radioactive decay between 1st of April 1977 and time of analysis, and neutron fluence relative to the monitor (1.024).

d) ^{38}Ar corrected for spectrometer background, blank and contributions from neutron reactions on K and Ca.

Table 1. *(cont'd.)*

Filament Temp °C	$^{40}Ar^{(a)}$ $\times 10^{10}$ccSTP	$^{36}Ar^{(b)}/^{40}Ar$ $\times 10^{-5}$	$^{37}Ar^{(c)}/^{40}Ar$ $\times 10^{-4}$	$^{38}Ar^{(d)}/^{40}Ar$ $\times 10^{-5}$	$^{39}Ar^{*(e)}/^{40}Ar$ $\times 10^{-4}$	Age AE
D (0.3 mg, 430 ppmK)						
880	3.3	2017	42	879	95	4.273
		± 91	± 24	± 40	± 4	±.061
960	2.7	298	87	215	124	3.850
		± 81	± 29	± 36	± 5	±.068
1000	3.0	226	165	191	132	3.746
		± 76	± 28	± 33	± 5	±.060
1060	5.6	312	185	238	151	3.542
		± 42	± 15	± 20	± 3	±.031
1090	30	90	211	181	143	3.624
		± 8	± 3	± 4	± 2	±.015
1120	11	94	232	190	144	3.617
		± 24	± 9	± 11	± 2	±.021
1150	1.6	269	175	250	137	3.69
		±158	± 57	± 70	±10	±.11
1200	3.7	67	217	164	139	3.667
		± 59	± 22	± 25	±40	±.045
1290	0.5	356	306	253	92	4.34
		±409	±155	±178	±20	±.36
Total	61	238	195	229	139	3.662
		± 14	± 4	± 6	± 1	±.011
E (0.15 mg, 470 ppmK)						
800	0.4	670	140	2440	82	4.51
		±830	±350	±650	±26	±.53
1000	1.1	500	153	660	111	4.03
		±260	± 93	±100	±10	±.15
1060	1.1	0	190	210	116	3.96
		±270	±100	±110	±11	±.15
1130	2.3	312	130	120	142	3.639
		±111	± 41	± 37	± 6	±.065
1200	12	26	185	158	146	3.597
		± 22	± 8	± 8	± 2	±.017
1260	11	27	198	170	146	3.594
		± 24	± 9	± 9	± 2	±.017
1380	6.7	17	198	169	148	3.573
		± 41	± 15	± 14	± 3	±.029
Total	34	64	186	203	143	3.620
		± 21	± 8	± 11	± 1	±.013

e) $^{39}Ar^*$ corrected for spectrometer background, radioactive decay between 1st of April 1977 and time of analysis, neutron fluence relative to the monitor (1.024) and contributions from neutron reactions on Ca.

f) Errors include standard deviation in the extrapolation of the ratios to the inlet time into the spectrometer, errors in mass spectrometer and irradiation interferences, and a conservative error of ±50% in the amount of blank subtraction. The uncertainty in measured ^{40}Ar amount is 5%.

g) "r" denotes a second extraction at the indicated temperature.

h) Temperature measured by optical pyrometer (assuming spectral emissivity 0.3) or extrapolated from high temperature measurements versus filament current. A spectral emissivity of 0.4 would decrease 1000°C to 980°C, 1300°C to 1270°C, 1600°C to 1555°C.

$^{36}Ar/^{38}Ar < 0.6$, the presence of ^{38}Ar produced during the reactor irradiation of the sample by $^{37}Cl(n,\gamma)$ is inferred. All particles contained large amounts of $^{38}Ar_{Cl}$. The effect of $^{38}Ar_{Cl}$ on partitioning is to increase the inferred amount of cosmic ray produced Ar_s at the expense of trapped Ar_t.

The ^{36}Ar in the large $^{39}Ar^*$ release fractions from three spheroids (A, B, and C) can be accounted entirely as $^{36}Ar_s$. A ^{21}Ne cosmic ray exposure age of 27 m.y. has been measured for the 74220 soil by Hintenberger *et al.* (1974). For a 27 m.y. cosmic ray exposure, a ratio of $^{36}Ar_s/^{37}Ar_{Ca} \approx 0.039$ is estimated for the LAV 6 neutron-irradiated samples using Fe = 17.2% and Ca = 5.1% for the orange glass (Mao *et al.*, 1973), $P_{36}(Ca) = 0.8 \times 10^{-8}$ccSTP$^{36}Ar_s$/g Ca/m.y. and $P_{36}(Ca)/P_{36}(Fe) = 10$ for the production of $^{36}Ar_s$ from Ca and Fe. For samples A, B and C, the large Ar releases providing the most precise measurements yielded $^{36}Ar/^{37}Ar = 0.030$ in reasonably good agreement with the estimated value. The minimum $^{36}Ar/^{37}Ar$ of sphere D is significantly larger than for A, B and C, indicating $^{36}Ar_t$ is present in all releases from D.

Since the $^{38}Ar_{Cl}$ present prevents the decomposition of ^{36}Ar and ^{38}Ar between trapped and cosmic ray produced Ar by usual procedures, the composition of trapped Ar in the low temperature releases is calculated using $^{36}Ar_s/^{37}Ar = 0.030$ to subtract $^{36}Ar_s$ from the total ^{36}Ar. The calculated values of $^{36}Ar_t/^{40}Ar$ are plotted in Fig. 2 versus $^{39}Ar^*/^{40}Ar$ for the three spheres with the best resolution in the low temperature release. In this graph, pure K-derived Ar* plots on the abcissa and pure trapped Ar_t plots on the ordinate. Mixtures of Ar* and Ar_t lie along a line joining the two compositions. The existence of a linear data array would indicate that there is a well-defined Ar_t component as well as a constant $^{39}Ar^*/^{40}Ar^*$ ratio (or age).

The data of Huneke *et al.* (1973) shown for comparison in Fig. 2 illustrate the complexity of orange glass Ar analyses. At least two trapped Ar components with different thermal retentivities were indicated. Ar released initially was dominated by Ar_t very enriched in $^{40}Ar_t$. $^{36}Ar/^{40}Ar$ then increased as relatively larger amounts of solar wind $^{36}Ar_t$ were released at successively higher temperatures. Trapped Ar was less dominant in the intermediate and high temperature fractions, but the Ar isotopic variations were still complex. Huneke *et al.* (1973) could at best identify a co-linear sequence of three intermediate temperature release fractions which contained ~50% of the $^{39}Ar^*$ and lay along a line intersecting the abcissa at a composition corresponding to an age of 3.49 ± 0.06 AE. No Ar release fraction was free of significant amounts of $^{40}Ar_t$. Possible effects of ^{39}Ar recoil redistribution between phases were not considered.

The single glass ball analyses suffer from similar but much less severe problems. In comparison, all Ar releases from the single balls contain relatively less Ar_t than the most radiogenic releases from the bulk sample, and the last ~80% of Ar released from four of the balls apparently contains *no* Ar_t. These are the only analyses of 74220 glass to date for which this is the case, and in this respect they provide the least ambiguous age information. All of the spheres contain trapped ^{36}Ar in the low temperature Ar releases and the isotopic variation patterns are again complex. Lunar parentless $^{40}Ar_t$ is apparently much

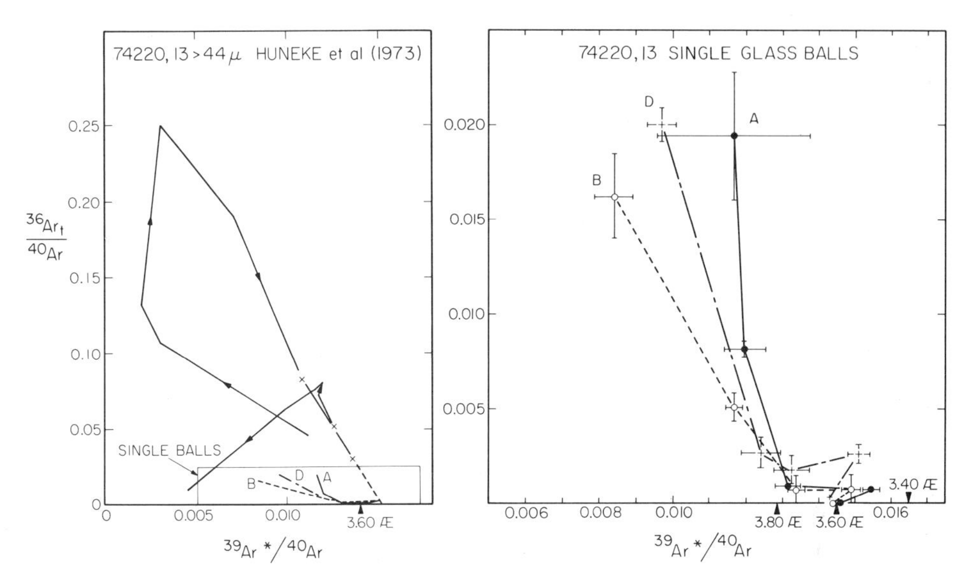

Fig. 2. Correlation of $^{36}Ar_t/^{40}Ar$ with $^{39}Ar^*/^{40}Ar$ for Ar released from single balls (right) and a >44 μ sized fraction (left) from the orange soil 74220. Trapped Ar_t compositions plot on the ordinate and K-derived Ar* compositions plot on the abscissa. A linear data array indicates an admixture of a single well-defined Ar_t component to Ar* of constant composition (and age). The >44 μ sample measurements show the difficulty of correcting for the large amounts of $^{40}Ar_t$ present to determine a K/Ar age for the glass. The large variation in composition of the low temperature releases results from the differential release of at least two Ar_t components. No Ar release was free of $^{40}Ar_t$. An age of 3.49 AE was inferred from the extrapolation of the line through only the three intermediate temperature releases (x) containing ~50% of the Ar*. The isotopic variation patterns for the single balls decrease smoothly to the abcissa with a single large fluctuation parallel to the abcissa. The fluctuation is ascribed to $^{39}Ar^*$ recoil redistribution among phases of the devitrified glass. Comparison in the left figure shows the Ar_t in the single large balls is richer in $^{40}Ar_t$ but that the contribution of Ar_t is much smaller. The Ar_t composition is different for different balls and varies with Ar release from a single ball. The last 80%–90% of Ar released has essentially no $^{36}Ar_t$ and yields an age of 3.60 ± 0.04 AE.

more abundant relative to solar wind $^{36}Ar_t$ than in any other 74220 glass sample measured. The trapped Ar composition is not the same for different balls nor does it appear that the composition is constant for Ar_t released from a single sphere. The Ar isotopic variation patterns are generally the same, and the three analyses illustrated show a sharp fluctuation in $^{39}Ar/^{40}Ar$ just prior to the large Ar release on melting. The fluctuations occur very close to and parallel with the abcissa, indicating they are due to variations in Ar* and not in Ar_t. The balls are devitrified glass and are mineralogically heterogeneous on a very fine scale. The sharp fluctuations probably reflect recoil redistribution of $^{39}Ar^*$ among phases of different K content and Ar retentivity. Extreme $^{39}Ar^*$ recoil redistribution effects were noted by Huneke and Smith (1976) in devitrified K-rich glass from the

Bhola achondrite. The effect on the remainder of the isotopic variation pattern of the process yielding the observed fluctuations in the orange glass data is unknown. The fluctuations lead to apparent age variations of ±0.1 AE in the same 10–20% of the Ar released affected by trapped Ar. The presence of trapped Ar and the complexity of the isotopic variation patterns prevent an assessment of low temperature ^{40}Ar losses in the lunar environment.

Apparent ages

The apparent ages for all five spheres are shown in Fig. 3 as a function of $^{39}Ar^*$ released. There are no corrections to ^{40}Ar for any $^{40}Ar_t$ since a consistent correction based on $^{36}Ar_t$ could not be inferred. In all cases the apparent ages begin initially high due to $^{40}Ar_t$ contributions and/or $^{39}Ar^*$ recoil effects and decrease rapidly to an apparent age plateau extending over the last 80–90% of $^{39}Ar^*$ release at high temperatures. In three instances the age spectrum passes through a minimum just prior to achieving the age plateau. The largest Ar releases (yielding 40–70% of the total Ar) were observed to coincide with the melting of the glass in two instances, and this is presumably true for all of the particles. After melting Ar is degassed from a homogenized reservoir, and the uniformity of composition does not constitute an age plateau in the normal sense. The designation is retained for convenience. The small variations in composition after melting serve to demonstrate the good reproducibility of the microanalytical techniques. In no case has the apparent age reached a constant value prior to the largest release which would indicate a resolution of the complexities present in the low temperature releases (for example complete separation of excess ^{40}Ar from radiogenic ^{40}Ar or complete degassing of regions or sites affected by low temperature ^{40}Ar losses). To this extent the interpretation of the plateau ages is ambiguous.

The plateau ages of the single balls containing no $^{36}Ar_t$ scatter very little about the average age of 3.60 AE. This is the strongest argument that the K/Ar ages of 3.60 ± 0.04 AE (including errors of ±1.5% in $^{40}Ar/K$ of hb3gr and ±0.5% in relative fluences) represent the solidification age of the glass. Both samples which did not exhibit minima in the age spectra just prior to the age plateau yield plateau ages lower by 0.01 AE, which suggests that whatever the cause of the minima, the effect on the age plateau is minor.

The presence of Ar_t with large $(^{40}Ar/^{36}Ar)_t$, for example Ar_t very enriched in ^{40}Ar dissolved in the magma and incompletely degassed before solidification of the glass, would not be identified by ^{36}Ar excesses. The radii of the balls differ by only 50% and their degassing characteristics are quite similar. Small amounts of magmatic ^{40}Ar either incompletely degassed or trapped during fire fountaining would probably be similar and difficult to resolve.

The question of $^{40}Ar^*$ losses is also not resolved, but such losses are assumed negligible since the Ar^* is released at relatively high extraction temperatures.

Discussion of orange glass results

The age for the orange glass event determined in this study is 3.60 ± 0.04 AE.

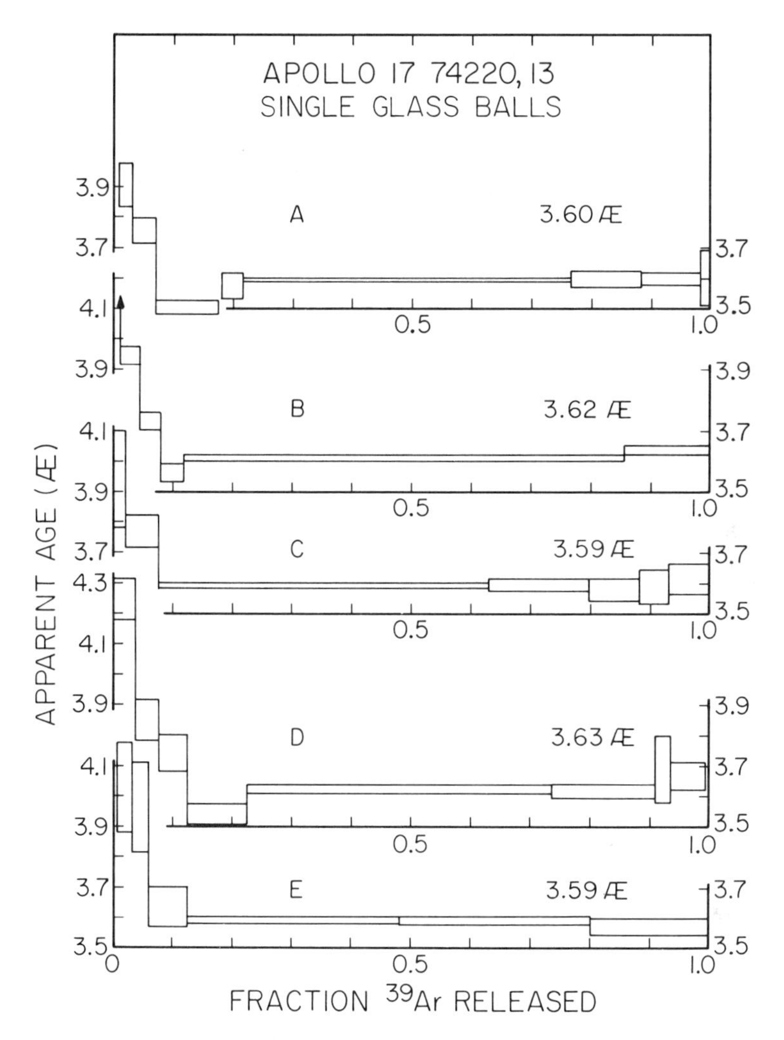

Fig. 3. ^{40}Ar-^{39}Ar age vs. $^{39}Ar^*$ release for five individually analyzed glass balls from the orange soil 74220. The ages are initially high due to contributions from $^{40}Ar_t$ and decrease rapidly to constant ages over the last 80–90% of Ar release corresponding to Ar released from melts. Three age spectra exhibit age minima just prior to melting which are ascribed to $^{39}Ar^*$ recoil effects. Only ball D retains detectable $^{36}Ar_t$ in the melt. The age plateaux of the other four balls are essentially identical and define an age for the orange glass event of 3.60 ± 0.04 AE.

A comparable orange glass age of 3.66 ± 0.06 AE was reported by Husain and Schaeffer (1973) from ^{40}Ar-^{39}Ar measurements on ≥150 μ sized fraction from 74220, and a range of ages from 3.55 AE to 3.66 AE was reported by Eberhardt *et al.* (1973) from the ^{40}Ar-^{39}Ar analysis of four sized fractions. A marginally lower age of 3.49 ± 0.06 AE was inferred by Huneke *et al.* (1973) from Ar released at intermediate temperatures from a >44 μ sized fraction. (The reported ages have been decreased by 0.05 AE for comparison to ages calculated using newly recommended decay constants and K isotopic composition.) Most recently Alexander *et al.* (1978) inferred an age of 3.66 ± 0.06 AE from total Ar extractions on sized separates. In all cases but the present one, significant corrections for nonradiogenic ^{40}Ar were required and were deduced from systematic variations in Ar composition assuming a single, well-defined Ar_t component. The data from all of the ^{40}Ar-^{39}Ar studies of the 74220 glass conclusively show this assumption is invalid. The ages presented here are less subject to the whims of the Ouija board manipulations necessary for $^{40}Ar_t$ corrections and are to be preferred. No previous study considered the effects of $^{39}Ar^*$ recoil redistribution among phases of the devitrified glass, which is the likely cause of significant apparent age fluctuations in the present measurements.

The ^{39}Ar-^{40}Ar age of the single large balls is significantly older than the precise ^{207}Pb-^{206}Pb age of 3.48 ± 0.03 AE reported by Tera and Wasserburg (1976) for reasons which are unclear. The difference may be due to undetected $^{40}Ar_t$ (unaccompanied by $^{36}Ar_t$) or to complications in the Pb-Pb systematics or perhaps reflects a more complex history for the orange glass. The differences between ^{40}Ar-^{39}Ar ages presumably arise from the difficulty of correcting for $^{40}Ar_t$ and for effects of $^{39}Ar^*$ recoil, but this is not definite. The suggestion of a discrepancy between Pb-U and Pb-Pb measurements (Tera and Wasserburg, 1976) has also not been resolved.

The K/Ar age of 3.60 ± 0.04 AE of the orange glass is younger than reliably determined ages of mare basalts at the Apollo 17 site, which range over 3.63–3.84 AE from Rb—Sr measurements (summary by Nyquist, 1977) and 3.65–3.79 AE from ^{40}Ar-^{39}Ar measurements (summary by Turner, 1977). We conclude that the lava fountain event resulting in the orange glass droplets occurred at least at the very end of and probably distinctly after mare basalt volcanic activity at this site. An age as young as 3.48 AE, as indicated by the Pb measurements, yields a large hiatus in volcanic activity of 200 million years and would suggest the two types of volcanic activity were unrelated.

72435 breccia clasts

Several lunar highland brecciae have been shown to contain chemically and isotopically unequilibrated plagioclase clasts. Sr in these relict clasts is isotopically more primitive than Sr in the enclosing breccia matrix. Ar measurements on plagioclase mineral separates from one such highland rock (65015) yielded Ar age spectra increasing from an age plateau at 3.90 AE to an apparent age of ~4.4 AE in the high temperature Ar release (Jessberger, *et al.*, 1974). The older ages correlated with the fraction of plagioclase with primitive Sr in the separates. Some relict clasts appear to represent lithic fragments from the ancient lunar crust.

Huneke *et al.* (1977) reported Ar measurements on several other breccia with primitive Sr and Pb and relict petrographic features which suggested survival of parent material from the era prior to the end of the lunar cataclysm at 3.9 AE. In particular, they studied the Apollo 17 breccia 72435 which contained a large dunite clast yielding a Rb-Sr age of 4.5 AE and primitive initial Sr, as well as plagioclase clasts with primitive Sr (Papanastassiou and Wasserburg, 1975). Although some clasts were demonstrably ancient, Ar measurements on a 1.5 mg sample comprised of four large plagioclase crystals failed to show evidence for K/Ar ages older than 3.9 AE. The ^{40}Ar-^{39}Ar age spectrum of the 72435 plagioclase was relatively constant at 3.87±0.07 AE, the date of formation of the breccia. Huneke *et al.* (1977) concluded that either Ar, but not Sr, was more easily redistributed during breccia formation or primitive Sr does not necessarily indicate older ages for the clasts. Ar contributions from an older but K-poor clast may have been masked by Ar from younger crystals formed or thoroughly degassed at 3.87 AE. This possibility is minimized in the present study by the measurement of individual breccia fragments.

Two neutron irradiated single fragments from 72435 were analyzed using the microanalytic methods. One fragment (1) appeared to be a single large plagioclase crystal. The second fragment (2) was comprised of ~25% of matrix material welded to a large plagioclase crystal. The latter crystal was the only one of the six which retained matrix and was separated out for individual analysis because of the possibility the remaining matrix material would dominate the Ar release patterns. The Ar amounts and isotopic compositions are given in Table 2. Weights and K contents calculated as for the 74220 particles are also given.

Partitioning of the stable Ar isotopes between Ar_t, Ar_s and Ar^* was also as described for 74220 with the exception that composition of Ar_t was assumed to be $(^{40}Ar/^{36}Ar)_t = 1$ and the corresponding amounts of $^{40}Ar_t$ subtracted from the total ^{40}Ar. Typically $^{36}Ar/^{38}Ar < 0.6$, and significant amounts of $^{38}Ar_{Cl}$ are inferred.

The apparent ages are plotted in Fig. 4 as a function of ^{39}Ar released. K/Ca calculated from $^{39}Ar/^{37}Ar$ is also shown. The age spectrum of fragment 2 is essentially constant throughout the Ar release and yields an extremely well defined age of 3.86±0.04 AE. The age spectrum is similar to that measured previously on the group of plagioclase clasts (Huneke *et al.*, 1977) and the ages are identical. The K content of fragment 2 is very high and the K/Ca ratio varies by an order of magnitude. K/Ca generally diminishes with increasing $^{39}Ar^*$ release but is always much greater than the ratio observed in the plagioclase fragment, indicating that the Ar release (and age) of fragment 2 is dominated by the matrix material attached to the plagioclase crystal. The age of 3.86 AE is the same as the Rb-Sr age of 3.85±0.18 (2σ) AE measured on a lithic clast in the 72435 breccia (Papanastassiou and Wasserburg, 1975) and presumably reflects the time of breccia formation. A well-defined ^{40}Ar-^{39}Ar age of 3.87 AE has been obtained on a felsite glass in breccia 73215 which petrographic evidence indicates melted and degassed during breccia formation (Jessberger *et al.*, 1977; James and Hammarstrom, 1977) in the Serenitatis basin-forming event.

The age spectrum of fragment 1 is constant over the first 40% of $^{39}Ar^*$

Table 2. Ar amounts and isotopic composition in two neutron-irradiated fragments of plagioclase plus matrix and plagioclase from the Apollo 17 breccia 72435.[f]

Filament Temp °C	$^{40}Ar^{(a)}$ 10^{-10}ccSTP	$\frac{^{36}Ar^{(b)}}{^{40}Ar}$ $\times 10^{-5}$	$\frac{^{37}Ar^{(c)}}{^{40}Ar}$ $\times 10^{-4}$	$\frac{^{38}Ar^{(d)}}{^{40}Ar}$ $\times 10^{-5}$	$\frac{^{39}Ar^{*(e)}}{^{40}Ar}$ $\times 10^{-4}$	Age AE
Fragment 1 (0.6 mg, 190 ppmK)						
870	0.3	1021	238	415	130	3.75
		±472	±117	±265	± 52	± .63
920	0.3	738	195	623	144	3.61
		±394	±103	±336	± 55	± .59
970	0.3	1088	156	404	114	3.97
		±518	± 95	±295	± 44	± .63
1010	0.6	726	287	426	125	3.83
		±177	± 59	±136	± 22	± .28
1050	1.0	648	383	701	137	3.68
		±130	± 53	±120	± 17	± .19
1080	1.0	650	416	809	140	3.66
		±122	± 54	±122	± 16	± .19
1110	1.0	653	534	881	119	3.92
		±120	± 63	±129	± 14	± .19
1180	1.3	693	534	955	124	3.84
		± 99	± 50	±107	± 12	± .15
1180r	0.5	564	559	846	118	3.92
		±215	±122	±238	± 25	± .34
1240	1.8	483	618	924	128	3.796
		± 54	± 37	± 69	± 8	± .095
1330	0.5	852	138	2016	173	3.33
		±184	± 27	±401	± 34	± .30
1390	15.9	629	912	1789	123	3.861
		± 10	± 7	± 17	± 2	± .020
1450	7.4	707	952	1381	113	3.996
		± 19	± 15	± 35	± 2	± .028
1500	16.3	793	1135	1654	112	4.004
		± 11	± 8	± 20	± 1	± .014
1550	1.7	894	1131	1676	109	4.049
		±109	±106	±165	± 11	± .156
1610	4.7	834	1105	1606	110	4.042
		± 42	± 29	± 49	± 3	± .049
1700	5.3	897	1187	1682	110	4.037
		± 30	± 27	± 52	± 3	± .041
Total	59.7	735	967	1542	117	3.930
		± 10	± 8	± 15	± 1	± .015

released at an age similar to the well-defined age plateau of fragment 2. The apparent ages then increase dramatically to a higher age plateau at 4.04 AE for the remaining 60% of the Ar release. K/Ca is initially high but decreases rapidly to a low and relatively constant value over the last 80% of the Ar release, as

Table 2. *(cont'd.)*

Filament Temp °C	$^{40}Ar^{(a)}$ 10^{-10}ccSTP	$^{36}Ar^{(b)}/^{40}Ar$ $\times 10^{-5}$	$^{37}Ar^{(c)}/^{40}Ar$ $\times 10^{-4}$	$^{38}Ar^{(d)}/^{40}Ar$ $\times 10^{-5}$	$^{39}Ar^{*(e)}/^{40}Ar$ $\times 10^{-4}$	Age AE
Fragment 2 (0.4 mg, 1200 ppmK)						
770	1.1	103	97	329	116	3.95
		± 148	± 75	± 46	± 14	± .19
930	13.6	55	24	76	115	3.964
		± 14	± 7	± 4	± 1	± .019
1020	18.5	68	55	102	122	3.872
		± 10	± 5	± 3	± 1	± .016
1070	12.8	71	67	123	122	3.875
		± 14	± 7	± 5	± 1	± .021
1110	10.9	84	78	143	121	3.888
		± 17	± 9	± 4	± 2	± .023
1140	7.8	104	80	163	121	3.894
		± 24	± 12	± 7	± 2	± .030
1170	6.0	131	80	205	122	3.872
		± 31	± 16	± 8	± 3	± .039
1200	6.8	140	105	261	122	3.875
		± 29	± 15	± 8	± 3	± .034
1230	32.5	185	240	723	125	3.836
		± 6	± 4	± 8	± 1	± .014
1260	18.9	178	203	375	125	3.832
		± 11	± 6	± 6	± 1	± .018
1290	51.4	109	132	469	126	3.824
		± 4	± 2	± 4	± 1	± .011
1310	3.6	239	236	399	119	3.916
		± 50	± 27	± 19	± 5	± .064
1340	5.3	216	247	385	119	3.910
		± 34	± 19	± 14	± 3	± .041
1440	11.2	214	271	419	124	3.848
		± 17	± 9	± 7	± 2	± .023
1640	24.0	207	277	411	123	3.861
		± 8	± 5	± 4	± 1	± .014
1870	20.8	206	262	397	123	3.866
		± 10	± 5	± 5	± 1	± .015
Total	245	143	166	374	123.1	3.859
		± 3	± 2	± 3	± 0.4	± .005

[(a-f)] See footnotes to Table 1. Neutron fluence of 72435 samples relative to the monitor is 1.005.

expected of the plagioclase. The older age plateau at high Ar release temperatures shows fragment 1 is indeed a relict plagioclast incompletely degassed at 3.86 AE, the time of breccia formation. The age of 4.04 AE is a lower limit for the time of formation of the clast.

Breccia 72435 is the third lunar highlands breccia in which the occurrence of relict clasts with primitive Sr has been shown to correlate with clast K/Ar ages

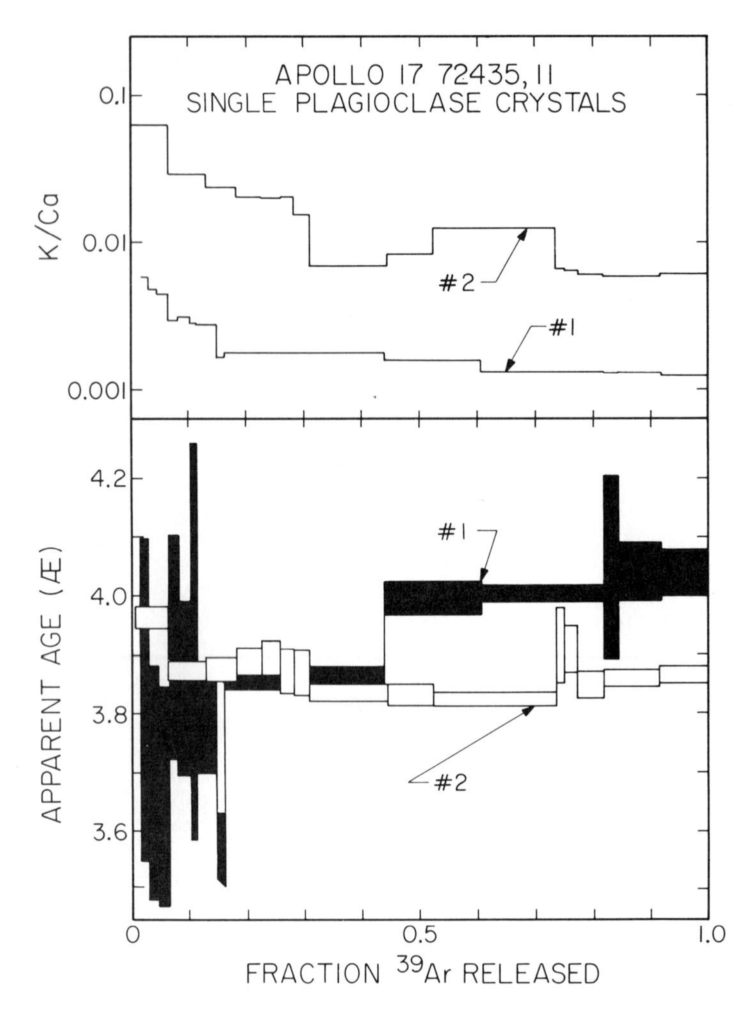

Fig. 4. ^{40}Ar-^{39}Ar ages and K/Ca vs. $^{39}Ar^*$ release for two fragments from the highland breccia 72435, which contains unequilibrated plagioclasts with primitive Sr and a dunite clast dated at 4.55 AE. Fragment #2, with ~25% attached matrix, yields a well-defined age plateau at 3.86 ± 0.04 AE. K/Ca for #2 is larger and much more variable than for #1. K/Ca and the apparent age spectra of #2 appear to be dominated by the attached matrix. The ages for #1 are initially comparable to #2 but then increase to 4.04 AE over the last ~50% of $^{39}Ar^*$ release. Crystal #1 is a relict clast formed earlier than 4.04 AE and incompletely degassed at 3.86 AE.

significantly older than the brecciation age. In only the case of the breccia 65015, however, were the K/Ar ages of the clasts truly ancient. The very limited sampling from 72435 has yielded a lower limit at only 4.0 AE. Thorough sampling of the 73215 breccia has yielded numerous older ages, but the ages do not exceed 4.2 AE. The 73215 breccia is not as strongly metamorphosed as

72435, and the plagioclase clasts presumably were not as thoroughly degassed. However, the present results confirm the existence of older relict clasts, which in the case of the dunite clast was shown to be an ancient crustal material by Rb/Sr measurements. The weakly metamorphosed highlands breccia 67435 contains plagioclase clasts yielding K/Ar ages of 4.34 AE (Dominik and Jessberger, 1978). From the point of view of K/Ar dating, the question of whether some clasts with primitive Sr formed as recently as 4.0 AE from Rb-depleted source material must remain open since a K/Ar age is at best a gas retention age. Nevertheless, in instances where relict clasts with primitive Sr are present, close examination reveals clasts with K/Ar ages older than the breccia formation time. This and the presence of older K/Ar ages in breccias of lower metamorphic grade would suggest the relict clasts with primitive Sr are indeed ancient crustal material extensively reworked by impacts and partly degassed of $^{40}Ar^{*}$ but still isotopically unequilibrated.

Acknowledgments—This research was supported by NASA grant NGL 05-002-188. Contribution No. 3071 of the Division of Geological and Planetary Sciences, California Institute of Technology.

REFERENCES

Alexander E. C. Jr., Coscio M. R. Jr., Dragon J. C. and Saito K. (1978) ^{40}Ar-^{39}Ar studies of glasses from lunar soils (abstract). In *Lunar and Planetary Science IX*, p. 7–9. Lunar and Planetary Institute, Houston.

Dominik B. and Jessberger E. K. (1978) Early lunar differentiation: 4.42-AE-old plagioclase clasts in Apollo 16 breccia 67435. *Earth Planet. Sci. Lett.* **38**, 407–416.

Eberhardt P., Geiss J., Grögler N., Maurer P. and Stettler A. (1973) ^{39}Ar-^{40}Ar ages of lunar material (abstract) *Meteoritics* **8**, 360–361.

Heiken G. and McKay D. S. (1977) A model for eruption behavior of a volcanic event in eastern Mare Serenitatis. *Proc. Lunar Sci. Conf. 8th*, p. 3243–3255.

Hintenberger H., Weber H. W. and Schultz L. (1974) Solar, spallogenic and radiogenic rare gases in Apollo 17 soils and breccias. *Proc. Lunar Sci. Conf. 5th*, p. 2005–2022.

Huneke J. C., Jessberger E. K., Podosek F. A. and Wasserburg G. J. (1973) $^{40}Ar/^{39}Ar$ measurements in Apollo 16 and 17 samples and the chronology of metamorphic and volcanic activity in the Taurus-Littrow region. *Proc. Lunar Sci. Conf. 4th*, p. 1725–1756.

Huneke J. C. and Smith S. P. (1976) The realities of recoil: ^{39}Ar recoil out of small grains and anomalous age patterns in ^{39}Ar-^{40}Ar dating. *Proc. Lunar Sci. Conf. 7th*, p. 1987–2008.

Huneke J. C., Radicati di Brozolo F. and Wasserburg G. J. (1977) ^{40}Ar-^{39}Ar measurements on lunar highlands rocks with primitive $^{87}Sr/^{86}Sr$ (abstract). In *Lunar Science VIII*, p. 481–483. The Lunar Science Institute, Houston.

Huneke J. C. and Wasserburg G. J. (1978) ^{40}Ar-^{39}Ar ages of single orange glass balls and highland breccia phenocrysts (abstract). In *Lunar and Planetary Science IX*, p. 567–569. Lunar and Planetary Institute, Houston.

Husain L. and Schaeffer O. A. (1973) Lunar volcanism: age of the glass in the Apollo 17 orange soil. *Science* **180**, 1358–1360.

James O. B. and Hammarstrom J. G. (1977) Petrology of four clasts from consortium breccia 73215. *Proc. Lunar Sci. Conf. 8th*, p. 2459–2494.

Jessberger E. K., Kirsten T. and Staudacher Th. (1977) One rock and many ages—Further K-Ar data on consortium breccia 73215. *Proc. Lunar Sci. Conf. 8th*, p. 2567–2580.

Jessberger E. K., Huneke J. C., Podosek F. A. and Wasserburg G. J. (1974) High resolution Ar analysis of neutron-irradiated Apollo 16 rocks and separated minerals. *Proc. Lunar Sci. Conf. 5th*, p. 1419–1449.

Mao H. K., Virgo D. and Bell P. M. (1973) Analytical and experimental study of iron and titanium in orange glass from Apollo 17 soil sample 74220. *Proc. Lunar Sci. Conf. 4th*, p. 397–412.

Megrue G. (1973) Spatial distribution of $^{40}Ar/^{39}Ar$ ages in lunar breccia 14301. *J. Geophys. Res.* **78**, 3216–3221.

Nyquist L. E. (1977) Lunar Rb-Sr chronology. *Phys. Chem. Earth* **10**, 103–142.

Papanastassiou D. A. and Wasserburg G. J. (1975) Rb-Sr study of a lunar dunite and evidence for early lunar differentiates. *Proc. Lunar Sci. Conf. 6th*, p. 1467–1489.

Schaeffer O. A., Müller H. W. and Grove T. L. (1977) Laser ^{39}Ar-^{40}Ar study of Apollo 17 basalts. Proc. Lunar Sci. Conf. 8th, p. 1489–1499.

Steiger R. H. and Jaeger E. (1977) Subcommission on geochronology: convention on the use of decay constants in geo- and cosmochronology. *Earth Planet. Sci. Lett.* **36**, 359–362.

Tera F. and Wasserburg G. J. (1976) Lunar ball games and other sports (abstract). In *Lunar Science VII*, p. 858–860. The Lunar Science Institute, Houston.

Turner G., Huneke J. C., Podosek F. A. and Wasserburg G. J. (1971) ^{40}Ar-^{39}Ar ages and cosmic ray exposure ages of Apollo 14 samples. *Earth Planet. Sci. Lett.* **12**, 19–35.

Turner G. (1977) Potassium-argon chronology of the moon. *Phys. Chem. Earth* **10**, 145–195.

Wasserburg G. J., Radicati di Brozolo F., Papanastassiou D. A., McCulloch M. T., Huneke J. C., Dymek R. F., DePaolo D. J., Chodos A. A. and Albee A. L. (1978) Petrology, chemistry, age and irradiation history of Luna 24 samples. In *Mare Crisium: The View from Luna 24* (R. B. Merrill and J. J. Papike, eds.), p. 657–678.

Proc. Lunar Planet. Sci. Conf. 9th (1978), p. 2363–2373.
Printed in the United States of America

^{39}Ar-^{40}Ar and petrologic study of Luna 24 samples 24077,13 and 24077,63

O. A. SCHAEFFER, A. E. BENCE, G. EICHHORN,
J. J. PAPIKE and D. T. VANIMAN

Department of Earth and Space Sciences, State University of New York at Stony Brook
Stony Brook, New York 11794

Abstract—The ferrogabbro 24077,13, with pyroxene and plagioclase major and minor element chemical ranges comparable to the dominant ferrogabbro component in the Luna 24 core, has a ^{39}Ar-^{40}Ar temperature release age of $3.33 \pm 0.21 \times 10^9$ yr. The metaferrobasalt 24077,63 is very fine-grained with a texture and mineralogy consistent with contact metamorphism between superimposed basalt flows or within a single flow. The ^{39}Ar-^{40}Ar plateau age of 24077,63 is $3.26 \pm .04 \times 10^9$ yr. The basalt flows in Mare Crisium are about the same age as those of Mare Imbrium and younger than those of Mare Tranquillitatis and Mare Serenitatis.

INTRODUCTION

The Luna 24 Soviet space probe in August 1976 returned a 160 cm core of the lunar regolith from the surface of the south east edge of Mare Crisium. The landing site lies 18 km east of the 6 km diameter Fahrenheit crater and 40 km west of the highland terrain of the edge of Mare Crisium. The sampling site contains mare material from the surface, deep seated mare material excavated by the Fahrenheit cratering event, terra material from the nearby basin rim, material from local features, and material from distant craters which have associated rays crossing the landing site. Local mare material, ejecta from Fahrenheit crater, should dominate the Luna 24 core sample (Florensky *et al.*, 1977; Butler and Morrison, 1977).

The regolith sampled by the core is uniform in composition and is mainly gabbroic. All fragments are low in titanium as well as alkalies and high in aluminum (Barsukov *et al.*, 1977; Tarasov *et al.*, 1977).

Mare Crisium is the eastern most mare basin and allows a comparison of the rock types and times of volcanic activity in the eastern surface of the moon to the other mare surfaces of the moon. We report the results of petrographic studies and ^{39}Ar-^{40}Ar ages for two samples: 24077,13, a ferrogabbro, and 24077,63, a recrystallized basalt.

Petrography and phase chemistry of ferrogabbro 24077,13

Ferrogabbro 24077,13, a 3.5 mg fragment, Fig. 1, is medium-grained (maximum grain size ~0.25 mm) and consists largely of subhedral calcic plagioclase (An_{88}-An_{84}) and iron-rich, subcalcic clinopyroxene (Wo_{33} En_{34} Fs_{33}) trending

Fig. 1. Ferrogabbro 24077,13, a 3.5 mg sample, consisting of calcic plagioclase and iron-rich clinopyroxene. The sample is about 1 × 2 mm in size.

Table 1. 24077,13 Mineral Chemistry.

	Pyroxene					Plagioclase		Ilmenite
SiO_2	49.0	48.6	46.7	45.7		46.8	47.7	0.00
Al_2O_3	1.92	1.59	1.10	0.99		33.3	32.7	0.62
TiO_2	1.25	1.34	0.87	1.04		—	—	52.7
FeO	20.3	24.8	35.7	40.4		0.63	1.00	46.2
MnO	0.37	0.34	0.51	0.60		—	—	0.58
MgO	11.1	8.00	4.15	1.74		0.04	0.00	0.14
CaO	15.1	15.2	10.1	8.81		17.7	17.4	—
Na_2O	0.00	0.00	0.00	0.00		1.36	1.62	—
K_2O	—	—	—	—		0.05	0.16	—
Cr_2O_3	0.43	0.26	0.09	0.06		—	—	0.30
$^{\Sigma}$Oxides	99.5	100.2	99.2	99.4		99.8	100.5	100.5
Si	1.912	1.924	1.945	1.943		2.158	2.189	0.000
Al^{IV}	0.088	0.076	0.055	0.057		1.813	1.769	0.018
Al^{VI}	0.000	0.000	0.000	0.000				
Ti	0.037	0.040	0.027	0.033		—	—	0.990
Fe	0.663	0.821	1.242	1.437		0.024	0.039	0.965
Mn	0.012	0.011	0.018	0.022		—	—	0.012
Mg	0.645	0.472	0.258	0.110		0.003	0.000	0.005
Ca	0.631	0.645	0.452	0.401		0.875	0.854	—
Na	0.000	0.000	0.000	0.000		0.122	0.144	—
K	—	—	—	—		0.003	0.009	—
Cr	0.013	0.008	0.003	0.002		—	—	0.006
$^{\Sigma}$Cations	4.001	3.995	3.999	3.998		4.997	5.004	1.998
Wo	32.6	33.3	13.1	20.6	Ab	12.2	14.3	
En	33.3	24.3	22.9	5.7	An	87.5	84.8	
Fs	34.2	42.4	64.0	73.7	Or	0.3	0.9	

towards iron-enriched compositions ($\sim Wo_{20}\ En_5\ Fs_{75}$). Accessory phases include ilmenite and troilite. Representative analyses of the plagioclase, clinopyroxene, and ilmenite are presented in Table 1. Olivine, though not present in the material separated for petrographic examination, is a likely additional phase. Soil sample 24170, which consists largely (~98%) of ferrogabbro components (single grains and lithic fragments), contains approximately 10% olivine (Fo_{51}-Fo_8, Bence *et al.*, 1977).

Clinopyroxene in ferrogabbro 24077,13 has generally comparable major element crystallization trends as the pyroxene in sample 24170, Fig. 2. However, there are significant differences in the minor element trends. The most magnesian (and hence early crystallizing) calcic clinopyroxene in 24077,13 has significantly higher Ti and lower Cr contents than the magnesian, calcic clinopyroxene from 24170, Fig. 3. Furthermore, the 24077,13 pyroxene does not plot on any of the Ti-Al or Cr-Al crystallization trends determined for the ferrogabbro in sample 24170. These differences suggest either slight differences

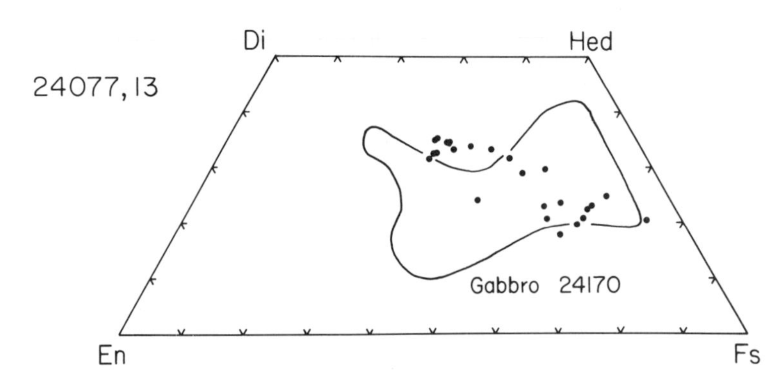

Fig. 2. Pyroxenes from 24077,13.

in the crystallization history of the two samples or slight differences in the composition of the crystallizing magma.

Plagioclase in ferrogabbro 24077,13 has a restricted compositional range (An_{88}-An_{84}) which corresponds to the most sodic plagioclase observed in the gabbro layer (24170). This suggests that, if both were derived from a common parental magma, 24077,13 represents a slightly more differentiated fraction of that magma. Alternatively, multiple, low Ti magmas may have contributed to the gabbroic component at the Luna 24 site.

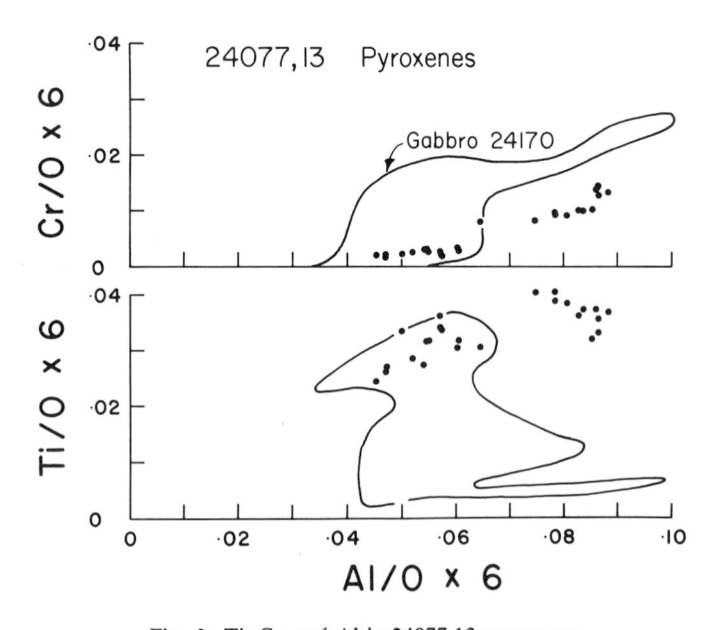

Fig. 3. Ti, Cr, and Al in 24077,13 pyroxenes.

Petrography of recrystallized basalt 24077,63

The Luna 24 ferrobasalt composition is found in several textural variants: brown glasses, ferrogabbros, subophitic ferrobasalts, and recrystallized (or "meta") ferrobasalts. Fragment 24077,63 is a large, 9 mg, recrystallized ferrobasalt which has been analyzed by INAA (Ma *et al.*, 1977), by microprobe analysis of polished mounts (Laul *et al.*, 1978), and has been dated in this study. The recrystallized ferrobasalt is a fine-grained (~0.01 mm) feldspar, pyroxene, olivine, ulvöspinel, and FeS rock; its texture is equigranular hornfelsic. Mineral compositons in the recrystallized ferrobasalt have been re-equilibrated, in contrast to the unequilibrated mineral zonation trends in the subophitic ferrobasalt (ref., pyroxene compositions, Figs. 2 and 4). Pyroxenes of the recrystallized type are intergrown (or exsolved) on a fine scale with compositions ranging from $Wo_{33}\,En_{28}\,Fs_{39}$ (Al_2O_3 ~1.2%; TiO_2 ~0.8%; Cr_2O_3 ~0.2%) to $Wo_{13}\,En_{32}\,Fs_{55}$ (Al_2O_3 ~0.7%; TiO_2 ~0.5%; Cr_2O_3 ~0.1%). Olivines are tightly restricted to the composition $Mg_{26}\,Fe_{73}\,Mn_1$. Plagioclase compositions are within the range $An_{93\pm2}\,Ab_{7\pm2}\,Or_{\sim0}$. The chromian spinels and silica which appear in the subophitic ferrobasalts do not occur in the recrystallized ferrobasalts.

Ryder *et al.* (1978) use the term "metabasalt" for this recrystallized rock type, implying that the texture and mineralogy result from contact metamorphism between superimposed basalt flows. McSween *et al.* (1977) suggest an alternative origin by foundering of basalt crusts within a flow. Grove and Vaniman (1978) find that equilibrium experiments on the Luna 24 ferrobasalt composition result in mineral compositions similar to those found in the ferrobasalt; the trend of hign-Ca to low-Ca pyroxenes at a fixed crystal-melt equilibration temperature may result from nucleation and growth of both high-Ca and low-Ca pyroxene domains. An equilibration temperature of ~1035°C may be calculated for 24077,63 from the experimental data. Grove and Vaniman (1978) calculate that

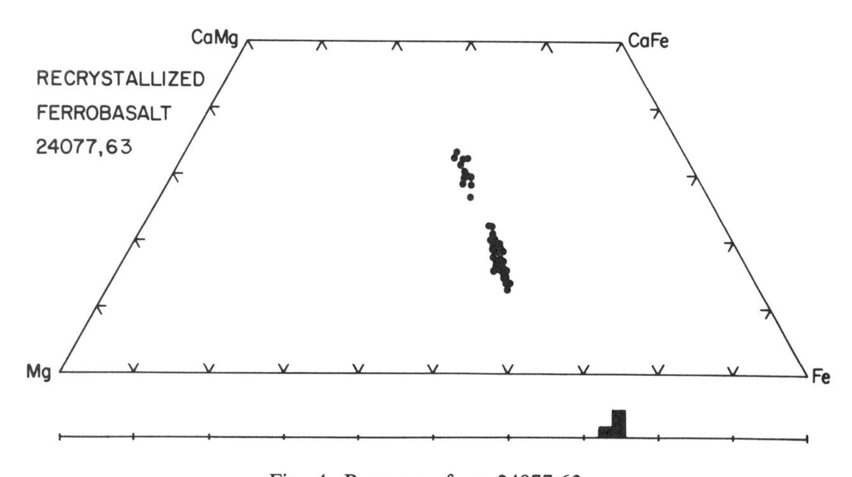

Fig. 4. Pyroxenes from 24077,63.

Table 2. 24077,63 Mineral Chemistry.

	Pyroxene			Olivine		Plagioclase	Ulvö-spinel
SiO_2	50.4	50.0		32.6		45.1	0.24
Al_2O_3	1.21	0.84		0.25		35.1	1.79
TiO_2	0.77	0.64		0.14		–	30.0
FeO	21.7	31.2		56.6		0.99	58.8
MnO	0.27	0.36		0.54		–	–
MgO	10.5	11.1		11.4		0.00	0.88
CaO	15.0	6.85		0.31		18.9	–
Na_2O	0.00	0.00		–		0.66	–
K_2O	–	–		–		0.03	–
Cr_2O_3	0.19	0.17		0.09		–	8.76
ΣOxides	100.0	101.1		101.9		100.8	100.5
Si	1.958	1.962		0.995		2.072	0.008
Al^{IV}	0.042	0.038		0.005		1.903	–
Al^{VI}	0.014	0.001		–		–	0.077
Ti	0.023	0.019		0.003		–	0.824
Fe	0.704	1.026		1.444		0.038	1.793
Mn	0.009	0.012		0.014		–	–
Mg	0.608	0.647		0.519		0.000	0.048
Ca	0.626	0.288		0.010		0.932	–
Na	0.000	0.000		–		0.059	–
K	–	–		–		0.002	–
Cr	0.006	0.005		0.002		–	0.252
ΣCations	3.990	3.998		2.992		5.006	3.002
Wo	32.1	14.6		–	Ab	6.0	–
En	31.2	32.8	Fo	26.3	An	93.8	–
Fs	36.7	52.6	Fa	73.7	Or	0.2	–

the contact temperature for a ferrobasalt flow is between 700–800°C, and therefore favor the hypothesis of foundered flow crust recrystallized at higher sub-liquidus temperatures in the flow interior.

We found that the recrystallized ferrobasalt fragment 24077,63 is essentially isochemical with the best estimated composition for subophitic ferrobasalts (Ma *et al.*, 1977; Laul *et al.*, 1978). However, we note that McSween *et al.* (1977) describe recrystallized ferrobasalt variants which are depleted in TiO_2 and Al_2O_3. If portions of a largely solidified chill margin were foundered and recrystallized in a low interior, the residual liquid within the flow crust may be drawn off into the surrounding magma. The residual liquids lost from the recrystallizing ferrobasalt will be enriched in Al_2O_3 and TiO_2, and this process may account for the observed compositional variation among recrystallized ferrobasalts (Stolper and Grove, pers. comm.). The chemical analysis appears in Table 2.

Table 3. Isotopic argon results for ^{39}Ar-^{40}Ar thermal release study.[a]

Temp (°C)	$^{36}Ar/^{38}Ar$[b]	$^{38}Ar/^{37}Ar$	$^{39}Ar/^{37}Ar$	$^{40}Ar^*/^{39}Ar^*$	$^{39}Ar \times 10^{-8}$ ccSTP/g	Exp. age ($\times 10^6$ yr)	Age ($\times 10^9$ yr)
24077,13 (3.5 mg)							
600	5.919 ± 4.2	.17722 ± 4.16	.0486 ± 5.5	106.0 ± 5.2	.68 ± 5.1		4.447 ± 1.6
700	4.870 ± 8.5	.04795 ± 8.3	.0448 ± 6.2	37.8 ± 31	1.45 ± 6.1	107 ± 94	2.843 ± 15
800	4.948 ± 1.8	.04032 ± 1.1	.0324 ± 2.0	25.9 ± 51	3.48 ± 1.9	76 ± 23	2.324 ± 29
Lost fractions			.0981 ± 2.1		29.19 ± 1.5		
1100	1.749 ± 41	.00821 ± 18	.0105 ± 5.5	49.4 ± 9.7	.93 ± 5.4	138 ± 32	3.237 ± 4.0
1125	2.037 ± 16	.01069 ± 16	.0084 ± 54	51.4 ± 68	1.18 ± 54	165 ± 26	3.296 ± 27
1150	1.659 ± 15	.00880 ± 14	.0044 ± 46	77.1 ± 55	1.54 ± 46	151 ± 20	3.928 ± 18
1200	1.735 ± 25	.01147 ± 22	.0132 ± 5.4	98.0 ± 6.3	.89 ± 5.3	193 ± 33	4.317 ± 2.0
1250	1.714 ± 12	.01027 ± 8.7	.0016 ± 9.2	330.2 ± 12	.61 ± 9.2	174 ± 13	6.389 ± 2.5
1300	1.410 ± 52	.00528 ± 28	.0043 ± 11	241.5 ± 12	.50 ± 11	97 ± 42	5.844 ± 2.8
1400	.830 ± 34	.01730 ± 6.6	.0109 ± 7.3	512.9 ± 6.4	.43 ± 6.3	364 ± 10	7.165 ± 1.2
24077,63 (8.0 mg)[c]							
500	10.868 ± 42	.00661 ± 31	.0402 ± 16	85.1 ± 16	.23 ± 16		4.138 ± 5.3
600	5.128 ± 17	.02322 ± 15	.0101 ± 12	44.8 ± 12	.28 ± 12	25 ± 409	3.137 ± 5.4
700	5.001 ± 5.5	.04207 ± 5.1	.0132 ± 5.4	38.7 ± 5.6	1.27 ± 5.4	71 ± 83	2.924 ± 2.6
800	4.501 ± 1.6	.06058 ± 1.5	.0061 ± 7.8	54.5 ± 8.2	1.45 ± 7.8	247 ± 10	3.434 ± 3.2
900	2.973 ± 3.1	.02702 ± 2.4	.0060 ± 1.9	48.6 ± 1.9	2.81 ± 1.8	309 ± 6	3.26 ± .8
1000	1.873 ± 2.5	.02239 ± 2.3	.0057 ± 1.7	47.7 ± 1.7	3.65 ± 1.6	374 ± 4.4	3.231 ± .7
1050	2.244 ± 8.0	.02345 ± 8.0	.0049 ± 3.4	51.1 ± 3.3	1.67 ± 3.3	350 ± 14	3.334 ± 1.3
1100	3.435 ± 2.1	.03581 ± 1.2	.0031 ± 10	48.1 ± 10	.39 ± 10	330 ± 5.2	3.244 ± 4.3
1200	1.884 ± 1.1	.02353 ± .5	.0009 ± 4.9	52.3 ± 5.3	.68 ± 4.9	392 ± 2.8	3.370 ± 2.1
1400	.945 ± 1.2	.01788 ± 1.1	.0005 ± 7.5	27.4 ± 7.7	.59 ± 7.5	379 ± 2.9	2.442 ± 4.2

[a] All errors are in %; values are corrected for blanks; ^{39}Ar, ^{38}Ar, and ^{36}Ar are corrected for n-induced contributions from Ca; ^{38}Ar is also corrected for n-induced contributions from K.
[b] For the trapped component, $^{40}Ar/^{36}Ar$ is taken as 5.8 ± 2.2.
[c] For the trapped component, $^{40}Ar/^{36}Ar$ is taken as 0.

^{39}Ar-^{40}Ar temperature release study

The experimental procedure for the temperature release age determination and the method of calculation, using the new decay constants and isotopic ratios for the K-Ar system, was described in detail by Schaeffer and Schaeffer (1977). The values for the $^{40}Ar/^{39}Ar$ ratios in the monitors were 33.2 ± 0.1 for 24077,13 and 32.18 ± 0.04 for 24077,63.

The Ar isotopic data for the two samples are given in Table 3. Unfortunately, in obtaining the middle temperature fractions of sample 24077,13, the getter container sprang a small leak and these fractions contained excessive amounts of atmospheric ^{40}Ar. Only the values for ^{39}Ar and ^{37}Ar could be obtained.

The ferrogabbro 24077,13 gives a thermal release of argon which shows the presence of solar wind in all fractions. The data are therefore presented as a three isotope plot, Fig. 5. Due to the loss of the most important middle temperature fractions, the remaining data points represent only about 30% of the total argon and show considerable scatter. The age obtained from the best fit line through the low and high temperature fractions is 3.33 ± 0.21 G.y.

The metaferrobasalt 24077,63 also contains solar wind. Fig. 6 and Fig. 7 show the three isotope plot and the temperature release age spectrum respectively. The points in the three isotope plot fall very close to an isochron corresponding to an

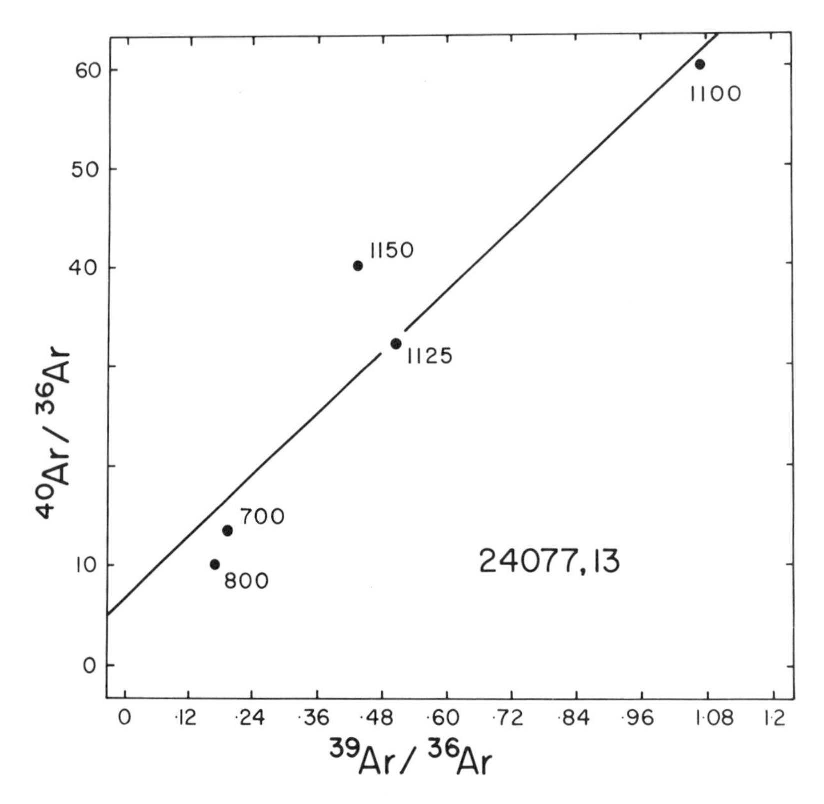

Fig. 5. $^{40}Ar/^{36}Ar_{sw}$ vs. $^{39}Ar/^{36}Ar_{sw}$ for 24077,13.

age of 3.24 ± 0.06 G.y. The $^{40}Ar/^{36}Ar$ intercept of zero indicates a negligible amount of trapped ^{40}Ar. The ^{39}Ar-^{40}Ar plateau age for the 900–1100°C fractions with 63% of the ^{39}Ar is 3.26 ± 0.04 G.y. This is the best age for this sample and is listed in Table 4 together with some other ages obtained from this regolith.

The ^{39}Ar-^{40}Ar ages for the metamorphosed gabbro 24077,63, the ferrogabbro 24077,13, and the leucocratic microgabbro 24170 (Wasserburg *et al.*, 1978) are all the same within error limits. The two point Sm-Nd isochron for the leucocratic microgabbro (Wasserburg *et al.*, 1978) agrees with the ^{39}Ar-^{40}Ar age of the same rock. These ages for the 24077 and 24170 layers are among the youngest found on the moon. On the other hand, the fine-grained lithic fragment, L24A, gives a ^{39}Ar-^{40}Ar age of 3.65 ± 0.12 (Stettler and Albarede, 1978) and is older. As the petrologic description of this sample is incomplete, it is not possible to decide how this sample is related to the basaltic flows in Mare Crisium. As a result, the question of the time duration of volcanism in Mare Crisium is an open question.

Cosmic ray exposure ages have also been determined for the two samples. The exposure age for 24077,13 for the fractions released between 1100°C and

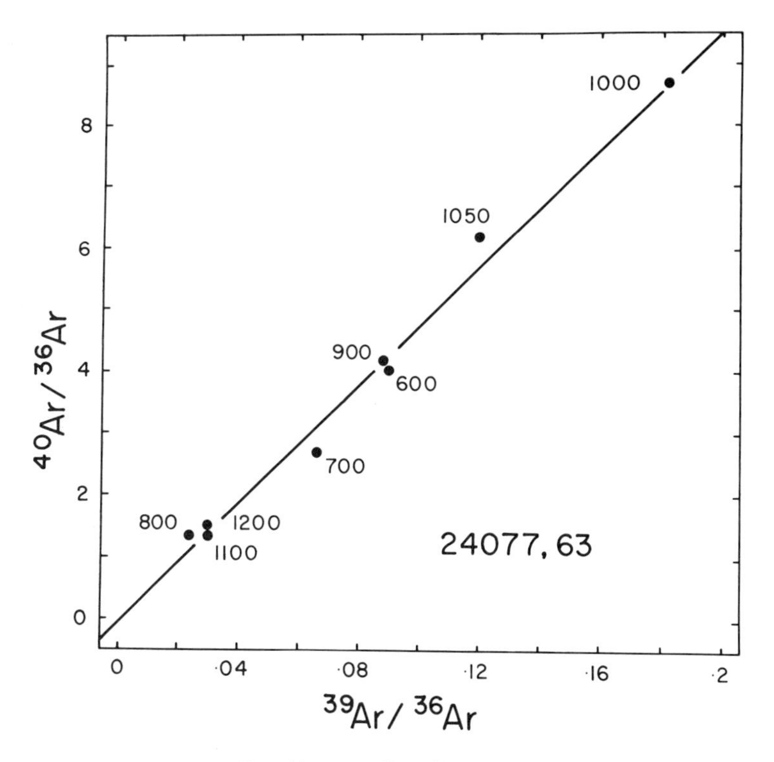

Fig. 6. $^{40}Ar/^{36}Ar_{sw}$ vs. $^{39}Ar/^{36}Ar_{sw}$ for 24077,63.

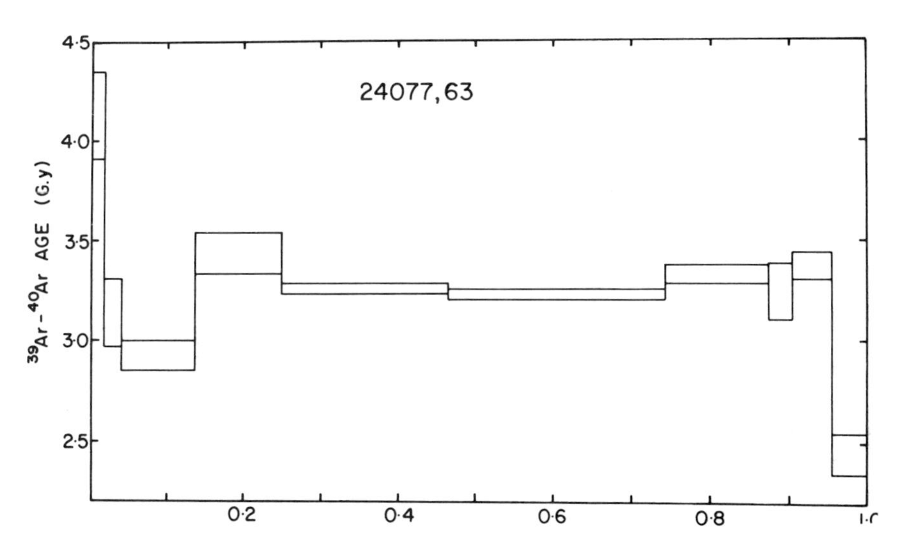

Fig. 7. ^{39}Ar-^{40}Ar age, 10^9 yr. vs. fractional ^{39}Ar released for 24077,63.

Table 4. Ages from Mare Crisium, Luna 24 Samples.

Sample No.	Rock Type	Mineral Age 10^9 years	Exposure Age 10^6 years	Method and References
24077,63	metamorphosed gabbro	3.26 ± .04	350 ± 20	^{39}Ar–^{40}Ar (this work)
24077,13	ferrogabbro	3.33 ± .21	170 ± 40	^{39}Ar–^{40}Ar (this work)
L24A	fine-grained lithic fragment	3.65 ± .12	660	^{39}Ar–^{40}Ar (Stettler and Albarede, 1978)
24170	leucocratic micro gabbro	3.30 ± .04	165	^{39}Ar–^{40}Ar (Wasserburg *et al.*, 1978)
24170	leucocratic micro gabbro	3.30 ± .05	—	Sm-Nd (Wasserburg *et al.*, 1978)

1250°C averages to 170 ± 40 m.y. Sample 24077,63 on the other hand, gives an exposure age of 350 ± 20 m.y. for the fractions released between 800°C and 1200°C. These ages are typical for those found in soils from other missions. There is no strong evidence that either of these ages corresponds to a cratering event.

The K/Ca ratios for sample 24077,13 are around 0.02 for the low temperature fractions and between 0.002 and 0.007 for the high temperature fractions. For sample 24077,63 the K/Ca ratio was 0.022 for the 500°C fraction and between 0.001 and 0.007 for the other fractions. The fractions released between 800°C and 1050°C showed a constant K/Ca ratio of 0.003.

CONCLUSION

Mare flows were emplaced in Mare Crisium about 3.3×10^9 years ago. From the relatively small number of ages obtained at this time, it is difficult to make any conclusions about the time span of the volcanism at Mare Crisium. The chronological data presented here and the additional chemical data of other workers indicate that the fine-grained ferrobasalts, metaferrobasalts, and gabbro formed at about the same time. This is consistent with thermal annealing of the metaferrobasalts by superimposed basaltic flows (Ryder *et al.*, 1977). These ages are close to the ages from the Imbrium basin which range from 3.1–3.4×10^9 years (Husain, 1974), and they are younger than the basalt flows in Mare Tranquillitatis, which were emplaced 3.5–3.9×10^9 years ago (summarized by Stettler *et al.*, 1974) and in Mare Serenitatis which were emplaced 3.7–3.9×10^9 years ago (Schaeffer *et al.*, 1976). Evidently, relatively young flows also exist on the extreme eastern side of the mare region of the moon.

Acknowledgments—We should like to thank L. Husain, D. A. Papanastassiou, T. Grove, and an unidentified reviewer for comments which improved the manuscript. In addition we should like to acknowledge the skillful work of T. Ludkewycz in obtaining the argon isotopic data. The work was supported by NASA Grants NGL-33-015-174, NGL-33-015-130, and NSG 9044.

REFERENCES

Barsukov V. L., Tarasov L. S., Dmitriev L. V., Lokesov G. M., Shevaleevsky I. D. and Garanin A. V. (1977) The geochemical and petrochemical features of regolith and rocks from Mare Crisium (preliminary data). *Proc. Lunar Sci. Conf. 8th*, p. 3319–3332.

Bence A. E., Grove T. L. and Scambos T. (1977) Gabbros from Mare Crisium: An analysis of the Luna 24 soil. *Geophys. Res. Lett.* **4**, 493–496.
Butler P. Jr. and Morrison D. A. (1977) Geology of the Luna-24 landing site. *Proc. Lunar Sci. Conf. 8th*, p. 3281–3301.
Florensky C. P., Basilevsky A. T., Ivanov A. V., Pronin A. A. and Rode O. D. (1977) Luna 24: Geological setting of the landing site and characteristics of the sample core (preliminary data). *Proc. Lunar Sci. Conf. 8th*, p. 3257–3279.
Grove T. L. and Vaniman D. T. (1978) Experimental petrology of very low Ti (VLT) basalts. In *Mare Crisium: The View from Luna 24* (R. B. Merrill and J. J. Papike, eds.), p. 445–471. Pergamon, N.Y.
Husain L. (1974) ^{40}Ar-^{39}Ar chronology and cosmic ray exposure ages of Apollo 15 samples. *J. Geophys. Res.* **79**, 2588–2606.
Laul J. C., Vaniman D. T. and Papike J. J. (1978) Chemistry, mineralogy, and petrology of seven >1 mm fragments from Mare Crisium. In *Mare Crisium: The View from Luna 24* (R. B. Merrill and J. J. Papike, eds.), p. 537–568. Pergamon, N.Y.
Ma M.-S., Schmitt R. A., Taylor G. J., Warner R. D., Lange D. E. and Keil K. (1977) Chemical and petrographic study of 18 Luna 24 lithic fragments (abstract). In *Papers Presented to the Conference on Luna 24*, p. 102–105. The Lunar Science Institute, Houston.
McSween H. Y. Jr., Taylor L. A. and Clark J. C. (1977) Luna 24 metabasalts: Possible evidence for assimilation of highland materials? (abstract). In *Papers Presented to the Conference on Luna 24*, p. 118–120. The Lunar Science Institute, Houston.
Ryder G. and Marvin U. B. (1978) On the origin of Luna 24 basalts and soils. In *Mare Crisium: The View from Luna 24* (R. B. Merrill and J. J. Papike, eds.), p. 339–355. Pergamon, N.Y.
Schaeffer G. A. and Schaeffer O. A. (1977) ^{39}Ar-^{40}Ar ages of lunar rocks. *Proc. Lunar Sci. Conf. 8th*, p. 2253–2300.
Schaeffer O. A., Husain L. and Schaeffer G. A. (1976) The duration of volcanism in Mare Serenitatis. *Earth Planet. Sci. Lett.* **31**, 358–368.
Stettler A. and Albarede F. (1978) Ar^{39}-Ar^{40} systematics of two mm-sized rock fragments from Mare Crisium. *Earth Planet. Sci. Lett.* **38**, 401–406.
Stettler A., Eberhardt P., Geiss J., Grögler N. and Maurer P. (1974) On the duration of lava flow activity in Mare Tranquillitatis. *Proc. Lunar Sci. Conf. 5th*, p. 1557–1570.
Tarasov L. S., Nazarov M. A., Shevaleevsky I. D., Kudryashova A. F., Gaverdovskaya A. S. and Korina M. I. (1977) Mineralogy and petrography of lunar rocks from Mare Crisium (preliminary data). *Proc. Lunar Sci. Conf. 8th*, p. 3333–3356.
Wasserburg G. J., di Brozolo F. R., Papanastassiou D. A., McCulloch M. T., Huneke J. C., Dymek R. F., DePaolo D. J., Chodos A. A. and Albee A. L. (1978) Petrology, chemistry, age and irradiation history of Luna 24 samples. In *Mare Crisium: The View from Luna 24* (R. B. Merrill and J. J. Papike, eds.), p. 657–678. Pergamon, N.Y.

Proc. Lunar Planet. Sci. Conf. 9th (1978), p. 2375–2398.
Printed in the United States of America

A new method for investigating the past activity of ancient solar flare cosmic rays over a time scale of a few billion years

E. DARTYGE

Laboratoire de Physique du Solide—91400 Orsay-Campus
Orsay, France

J. P. DURAUD

Service de Chimie-Physique du CEA—91190 Gif-sur-Yvette
Orsay, France

Y. LANGEVIN and M. MAURETTE

Laboratoire René Bernas—91406 Orsay-Campus
Orsay, France

Abstract—Monte-Carlo computations describing the accumulation of solar flare tracks in lunar dust grains first indicate that *igneous* lunar coarse fines with a size of ~1 mm are potentially very useful probes to trace back the past activity of the ancient solar flare cosmic rays (SFCR) over a very long time scale (≳1 b.y.) never explored before. For this purpose these privileged grains should be extracted at very specific depth positions in lunar core tubes by using thick (≳1 mm) impregnated core tube sections. Then coarse fines showing high densities (≳$10^9 cm^{-2}$) of tracks due to iron-group nuclei (VH-nuclei) as well as anisotropic track gradients should be selected for track studies intended to determine the energy spectrum and the chemical composition of VVH-nuclei heavier than the VH-nuclei ($20 \le Z \le 30$) at well known epochs in the distant past. However, the usual track method cannot be used meaningfully for exploiting such a fossil track record in lunar dust grains. Consequently a new method was developed, which is based on new concepts dealing with the formation, the thermal annealing and the chemical etching of tracks in minerals.

In this method a critical annealing temperature is used to selectively erase the fossil tracks due to VH-nuclei in the grains with a rejection factor better than 10^{-4}, thus leaving a very clean record of "residual" VVH-tracks with shortened lengths. So far this new method was just tried on feldspar grains hand-picked in several surface soil samples from the Apollo 16 mission as no adequate impregnated core section is yet available. The preliminary results of this study are: 1. Two marked peaks repeatedly appear at ~3 μm and 17 μm in the length distribution of the partially annealed VVH-tracks in all the grains when the degree of thermal annealing of the VH-tracks is adjusted by requiring that their residual lengths be ≲100 Å. 2. As the charge resolution of the method is of about two units between iron and krypton, the first peak in this track length distribution is confidently attributed to elements heavier than Ga. Consequently the ratios of the track densities measured in the grains before and after the critical thermal annealing yield measurements of the VVH ($Z > 31$)/VH ($20 \lesssim Z \le 30$) ratios. For feldspar grains with a size of 100 μm, which are thought to record the average properties of the SFCR over a time scale of 1 b.y., these VVH/VH ratios are narrowly clustered around a value of 5×10^{-4}, which is compatible with universal abundances. 3. On the other hand, the charge resolution of the method beyond krypton is still not sufficiently well-defined to clearly identify the group of VVH-nuclei with a low abundance, which are responsible for the second peak at 17 μm observed in the length distribution of the VVH-tracks, which is tentatively attributed to nuclei of the platinum and lead group ($Z \gtrsim 75$).

INTRODUCTION

One of the most exciting prospects of fossil track studies in lunar minerals deals with the measurements of both the energy spectrum and the chemical abundance of various groups of VVH-nuclei with atomic number $Z > 30$, in the ancient solar flare cosmic rays (SFCR), at *known epochs* in the distant past, by using as a reference the VH-nuclei ($20 \lesssim Z \leq 30$), that are much more abundant. Indeed these characteristics are only very poorly known in the contemporary SFCR, and their knowledge should help in fixing new constraints on the processes responsible for the emission of SFCR. In addition, the past activity of energetic solar flares can possibly be traced back from such measurements, over a time scale of a few billion years never explored before, by using the VVH-nuclei as probes of solar flare activity.

These exciting objectives have not been achieved as a result of two major difficulties: first, no method was as yet available to select lunar dust grains with a simple and tractable exposure history in the ancient SFCR; furthermore, the usual track method, based on track length measurements with an optical microscope, that was previously applied by most authors, cannot give meaningful determinations of the so-called VVH/VH ratio in the SFCR (this ratio weights the relative abundance of various groups of VVH-nuclei with respect to that of the VH-nuclei). The major purpose of the present paper is to tackle these two severe difficulties.

In the next section we summarize a few of the results of complex statistical Monte-Carlo computations, intended to both trace back the depth trajectories of lunar dust grains in maria-type regoliths and to predict in detail the accumulation of solar flare tracks in lunar dust grains. We show how this code gives important clues concerning the grain-size dependent exposure history of lunar grains in the ancient SFCR, and pinpoints the privileged role of 1 mm-coarse fines, extracted from impregnated core tube section with a thickness $\gtrsim$1 mm, showing a very high density of SFCR-tracks as well as anisotropic track gradients.

Then we discuss the problem of determining the chemical composition of VVH-nuclei in the ancient SFCR from etched track measurements. We first point out why the usual track method (Fleischer *et al.*, 1975), which is based on track length criteria, has so far not yielded good measurements of VVH/VH ratios in the ancient solar flare and galactic cosmic rays, although over the last 10 years various authors reported many measurements of such ratios in the galactic cosmic rays (GCR). We then present a new method intended to infer such ratios from lunar dust grains with very high track densities ($\gtrsim 10^{-9}cm^{-2}$); such grains are the only ones with a tractable exposure history in the ancient SFCR. However, their high track densities prevent the use of the usual track method, as the individual lengths of the tracks cannot be measured with the optical microscope. This new method, which relies on the determination of the microscopic structure of partially annealed latent tracks by small angle X-ray scattering techniques, is finally applied to lunar dust grains in surface soil samples from the Apollo 16 mission in order to demonstrate its feasibility.

Selection of Lunar Dust Grains With a Simple "One-Stage" Exposure History in the Ancient Solar Flare Cosmic-Rays

Over the past few years we have developed a Monte-Carlo statistical code (Bibring *et al.*, 1975; Duraud *et al.*, 1975; Borg *et al.*, 1976; Langevin and Maurette, 1978), mostly intended to account for the accumulation of the very distinct effects due to solar wind ions and solar flare and galactic cosmic rays in the lunar regolith, over a time scale of 4 b.y. and up to depths of 5 m. The results of these "ORSOLMIX" computations which are relevant to the present work are:

1. Lunar dust grains were cycled through a number, N_b, of ejecta blankets before having been deposited in their present day parent stratum, now found at a final depth, d_f, in the lunar regolith. On the average, N_b varies from about 1 to 10, going from immature to mature mare-type soils. We defined this very active period of ejecta blanket cycling in the grain history, which lasted for ~1 b.y., as the "plateau" history.
2. During this plateau history, most of the grains were deposited within at least one lunar "skin", i.e., the most superficial 5 mm-thick layer of the lunar regolith. This peculiar layer is very frequently turned over to depths of up to 5 mm by micrometeoroids with a mass $\lesssim 10^{-5}$ g. Consequently, the constituent grains of the lunar skin have a high probability of being exposed on the top surface of the lunar regolith during the lifetime, $\tau_B \sim 20$ m.y., of their parent stratum vs. being buried by an ejecta blanket. In particular, the 100 μm-sized grains in this layer have a probability $\gtrsim 90\%$ of being exposed more than 5 times on the lunar surface and thus of being uniformly irradiated on all "their faces" in the ancient SFCR. More specifically, it can be shown that each cycling of the 100 μm-grains through a lunar skin both increases the central track density, ρ_c, by approximately $10^9 cm^{-2}$, and produces a very uniform track gradient characterized by a ρ_5/ρ_c ratio of about 2, where ρ_5 is the track density measured at a depth of about 5 μm in the grains*.
3. All grains with a size of ~100 μm, and showing ρ_5 values $\gtrsim 10^9 cm^{-2}$ as well as uniform track gradients, can potentially be used as SFCR-track detectors because the contribution of GCR to the track density is negligible. In contrast, a high proportion of the grains showing ρ_5 values

*G. Comstock recently suggested (1977) that our computer treatment is oversimplistic and can only generate a similar ρ_c-distribution in all lunar soil samples. We note that our program traces back the motion of lunar dust grains on both a depth and a time scale (5 meters and 4 b.y., respectively), which are much larger than those used by Comstock (0.5 cm and 1 m.y., respectively). In this way we have not had to fix arbitrarily, as Comstock did, the values of important parameters such as the number of "close surface exposures". The second statement is also inaccurate. In fact, in Bibring *et al.* (1975) we report (Fig. 5), *as an example* of the ρ_c-distributions, those expected both for an immature soil and a random mixture of many lunar soils, which can then be relevantly compared to the corresponding experimental data.

smaller than $10^8 cm^{-2}$ has never approached the lunar surface at depths $\lesssim 5$ mm; consequently the track record of some of these grains could be mostly due to GCR. Such grains should exhibit neither a uniform track gradient nor impact pits with a size of 0.1 μm.

4. The integrated residence time of lunar dust grains in the 10 cm-thick superficial layer of the lunar regolith, where they are subjected to the lunar thermal cycle, is about 100 m.y.

Consequently high-ρ_c grains, with sizes of $\sim$100 μm and uniform track gradients, should be selected for SFCR-track studies. However, they have an ill-defined exposure history in the ancient SFCR, characterized by successive exposure steps of about 300,000 yr in integrated duration, which randomly occurred any time during their plateau history. Such grains could only be useful in determining the characteristics of ancient SFCR as averaged over a period of $\sim$1 b.y. We note that in mature lunar soils about half of the grains with high-ρ_c values ($\gtrsim 10^9 cm^{-2}$) do indeed show anisotropic track gradients. As argued elsewhere (Bibring *et al.* 1975), this feature, which is not predicted by our code, easily can be attributed to grains that were released from the surface of coarser grains by micrometeoroid chipping. Such grains, that have to be rejected as probes of regolith mixing, do not have a tractable exposure history in the ancient SFCR.

These severe limitations only hold for grains with a size $\sim$100 μm but not for those in the 0.5–1 mm size range. We will define these grains as the 1 mm-coarse fines (1 mm-CF). This is related to a very useful "rejection" effect due to the high abundance of micrometeoroids with a size of 20 to 50 μm which will act very selectively upon a direct hit by crushing grains with a size 10 to 20 times as large (such as the 1 mm-CF) into smaller fragments. In contrast, upon a direct hit by the same micrometeoroids, smaller grains will be completely destroyed and larger coarse fines will be sandblasted, with a resulting erosion rate of about 1 mm/m.y. Like the 100 μm-grains, the 1 mm-CF have a high probability of being exposed several times on the top surface of the lunar regolith *as long as they reside within a lunar skin.*

Thus we should extract igneous 1 mm-CF in the fossil lunar skins, which can possibly be found in the constituent strata of lunar core tubes. The grains of such igneous coarse fines could have been exposed only once in the ancient SFCR, i.e., when their parent stratum now found at a depth, d_f, in lunar core tubes was at one time, T_i, the top stratum of the lunar regolith. We can further select the 1 mm-CF relevant to SFCR-track studies by applying the following criteria deduced from the ORSOLMIX computations, which insure that the coarse fines really have been irradiated as individual grains in the SFCR: the maximum values of the ρ_5 and ρ_5/ρ_c parameters should be larger than $5 \times 10^8 cm^{-2}$ and 5, respectively. Since the values of T_i can be inferred from the depth profiles of GCR-induced isotopes in the same core tubes (Langevin and Maurette, 1978) we can hopefully use SFCR-track records in 1 mm-CF to deduce the characteristics of the ancient SFCR by applying the thermal annealing method described in the next section (such peculiar coarse fines would also be very useful probes to study the past activity of other processes active on the lunar surface at the same time, such as the fluxes of micrometeoroids and accretionary particles).

However, the application of the lunar skin sampling technique to the 1 mm-CF requires the delineation of clear boundaries between the constituent strata of lunar core tubes with an accuracy of ~1 mm, in order to identify the position of the fossil lunar skins from which the 1 mm-CF should then be extracted. This difficult task can be done by studying impregnated core sections with a thickness ≳1 mm with the track technique developed by Goswami and Lal (1975). Unfortunately, such thick impregnated cores are not available as yet. Consequently in this preliminary work we extracted feldspar grains with sizes ranging from 100 μm up to 1 mm in several surface soil samples from the Apollo 16 mission, with the intent to demonstrate the feasibility of the new method described in the following section, and to exploit the SFCR-track record in lunar dust grain in a meaningful way.

By measuring the values of the ρ_5 and ρ_5/ρ_c parameters for the solar flare VH-tracks registered in the grains, we selected eight feldspar grains with sizes of 100 μm as well as one feldspar-rich 1 mm-CF that showed a VH-track record compatible with those expected for grains which were irradiated as individual regolith grains and which have not suffered a complex history of fracturing, agglutination, etc., during such an irradiation. From the Monte-Carlo computations outlined above we infer that the 100 μm-grains should reflect the characteristics of the ancient SFCR *as averaged over a time scale of ~1 b.y.* comparable to the duration of their plateau history. In contrast, the 1 mm-CF selected in this work should have been exposed only once for a much shorter time (≲1 m.y.) in the ancient SFCR. The date of this one-stage irradiation cannot be evaluated because the depth of sampling of the coarse fine in the top layer of the Apollo 16 regolith was not determined.

Determination of VVH/VH Ratios in the Ancient Solar Flare Cosmic Rays

A need for a change

The usual track method for determining VVH/VH ratios (Fleischer *et al.*, 1975) is based on measurements of track length distribution with an optical microscope, as well as on the definition of the "etchable range", ΔL(Z), of an ion along which an etched track can be formed. This very important parameter is evaluated from the computation of a set of J(Z,E) curves, where J is the primary rate of ionization, by assuming that the tracks are formed by a threshold track formation mechanism, only operating when J exceeds a critical value, J_c. The etchable range ΔL(Z) is used to identify various groups of heavy nuclei in the track length distribution by assimilating the "maximum" etchable length of the tracks, as generally revealed with the TINT* etching technique (Lal *et al.*, 1969); $\Delta L(Z)$ is also used to relate the density of the selected tracks, $\rho(\Delta L)$, to the chemical abundance, $A(Z)$, of the track-producing nuclei, as $A(Z) \propto \rho(\Delta L(Z))/\Delta L(Z)$.

This method has been applied by many authors over the last 10 years to measure VVH/VH ratios in the ancient galactic cosmic rays. On the other hand, only three determinations of this ratio in the ancient SFCR have so far been reported in the literature. The results are quite contradictory, as Bhandari *et al.* (1973) and Bhandari and Padia (1974) in two successive papers report values of 10^{-3} and $\gtrsim 10^{-2}$, whereas Bull and Durrani (1975) give a value of 3×10^{-3}. However, in view of our recent work we believe that all these previous determinations, although certainly useful as a first approximation, cannot be considered sufficiently reliable. We will further discuss this point elsewhere (Dartyge *et al.* 1978). We simply note here that the usual track method has the two following major limitations, with the first one being mostly relevant to lunar studies:

1. Individual track lengths can only be measured with an optical microscope in grains showing VH-track density $\lesssim 5 \times 10^7$ cm^{-2}. On the moon, grains indeed exposed in the ancient SFCR always show higher track densities. (For this reason we believe that the VVH/VH ratio estimated by Bull and Durrani (1975) for lunar dust grains showing both ρ_c values $\lesssim 10^8$ cm^{-2} and no track gradient, are relevant to galactic cosmic rays but not to SFCR.) In addition, this density of VH-tracks already seems too low, with respect to the low abundance of VVH-nuclei that can be inferred from universal abundance curves (Trimble, 1975), to allow the finding of a sufficient number of VVH-tracks in a single grain.
2. In the usual track method the ΔL(VVH) values, which should be applied to the fossil tracks, cannot be evaluated with reasonable accuracy. On the one hand, all the fossil tracks observed in the most reliable track detectors (feldspar and to a lesser extent olivine) have already suffered some natural thermal annealing as a result of the lunar thermal cycle. Such an annealing was found to be equivalent to an annealing at 450°C for 2 hours in the laboratory, which both shortens artificial Fe and Kr-tracks and broadens their length distributions. Consequently, the ΔL(VVH) values inferred from the $J_c(Z)$ concept should be corrected for this effect; this is not possible within the framework of the mechanisms of track formation and track annealing so far reported in the literature. We also note that the "non-annealed" $\Delta L(Z)$ cut-off value of about 20 μm generally used by previous authors to separate VH-tracks from the VVH-tracks, which all

*In the TINT (TRACK-IN-TRACK) technique, long "host" tracks intersecting the external surface of the grain give the etchant access within the crystal, allowing the tracks which in turn intersect the host tracks to be etched.

are supposed to have lengths ≳20 μm, may not be correct al all. In fact, an important proportion of the fossil VVH-tracks have lengths smaller than 20 μm as a result of the natural broadening of the fossil track length distribution; thus these tracks are lost for the VVH count. Finally, as argued elsewhere (Dartyge *et al.*, 1976, 1978) the J_c-concept used to derive the non-annealed ΔL(Z) values does not satisfactorily account for the registration of nuclear particle tracks in minerals.

Recently, a variant of this method in which the etching rate of the tracks, V(Z,E), is considered as a function of J(Z,E) (Price *et al.*, 1973), has been applied (Bull *et al.*, 1978; Goswami and Lal, 1975) to the determination of VVH/VH ratios in both the solar flare and the galactic cosmic rays. In particular, Goswami and Lal (1975) reported a value of the VVH/VH ratio in the ancient SFCR of about 5×10^{-2} for energies ≲5 MeV/amu. We demonstrate elsewhere (Dartyge *et al.*, 1978) that the concentration of the type of radiation damage defects along a track that fixes the value of the etching rate is not directly related to J(Z,E). We thus believe that this "improved" method is no more reliable *as yet* than the usual ones.

The difficulty in determining an accurate ΔL(VVH) scaling function as well as the high track densities expected in grains irradiated in the ancient SFCR hamper the reliable application of the usual track method. Consequently, we have developed a method based on new concepts and intended to exploit the SFCR-track record in lunar dust grains in a meaningful way. For this purpose we investigated over the last two years the complex processes responsible for the formation, the thermal annealing, and the chemical etching of latent tracks. This work, which will be presented in detail elsewhere (Dartyge *et al.*, 1978, submitted to *J. Appl. Phys.*), is only qualitatively outlined in the next sections.

The "gap-model" for the chemical etching of tracks

Our investigations were conducted by combining small angle X-ray scattering analysis of the structure of the latent tracks with the measurements of their etching rates, as a function of the atomic number and energy of the track producing ions, both before and after thermal annealing. The tracks were produced by irradiating flat targets (mica, labradorite, olivine) with parallel beams of artificially accelerated ions (Dartyge and Lambert, 1974; Dartyge *et al.*, 1976, 1977a, 1977b, 1978).

The small angle X-ray scattering analysis reveals that the microscopic structure of the latent tracks consists of large clusters of small defects that we will quote as extended defects (ExD), which are linked by much smaller "point" defects (PtD) (Fig. 1). This microscopic structure of the latent tracks can be characterized by the linear densities (number per μm), N(Z,E) and n(Z,E,) of the ExD and PtD, respectively, which are directly inferred from the X-ray analysis. Upon heating the PtD first disappear in the low temperature range, while both the size, and the linear density, N(Z,E), of the ExD keep the

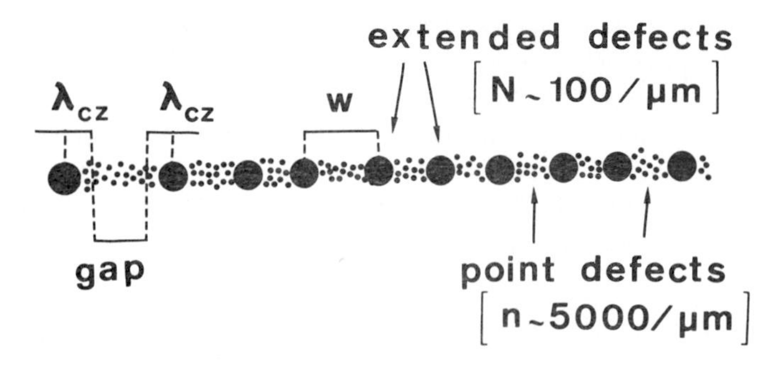

Fig. 1. Schematic representation of the microscopic structure of a latent track due to 1 MeV/amu Fe-ions in labradorite. This structure was inferred from the small angle X-ray scattering analysis of single crystals of labradorite, exposed to a high flux ($\gtrsim 10^{11}$ cm^{-2}) of artificially accelerated ions.

non-annealed values. When the ExD start to anneal at higher temperature, two of their characteristics (size and maximum concentration of defects within an ExD) are markedly altered, but their linear density is not affected up to the highest annealing temperature.

The microscopic structure of the latent tracks is similar (same value of N(Z,E)) in the two minerals so far ranked in previous studies (Fleischer *et al.*, 1975) as the most (mica) and the least (olivine) "sensitive" minerals for the registration of etchable tracks. This important finding is not compatible with a threshold mechanism that would operate at a lower level in mica. We show elsewhere (Dartyge *et al.*, 1978) that the sensitivity for etched track registration of these two minerals is indeed correlated to marked differences in the etching rate of their crystallographic planes and not to a greater efficiency of the radiation damage mechanism in mica.

In our "gap" model for the chemical etching of tracks a latent track consists of a microscopic row of tiny etchable zones showing two very different levels of chemical reactivity both associated with a marked increase in the free energy of the crystal along the defect-rich zones:

1. The ExD showing a high concentration of lattice defects can be assimilated to "core-zones" with an effective diameter, λ_{cz}, along which the etching rate, V_{cz}, exceeds the normal value, V_0, as measured for the non-irradiated crystal along the track direction. This last part of the conclusion is particularly well supported by our electron microscope observations of xenon tracks in mica, etched in a very dilute solution of HF, which were intended to reconfirm much earlier observations (Price and Walker, 1962). These observations show that the etching rate of the radiation-damaged core of xenon-tracks, where the ExD are known to overlap from the X-ray analysis, are $\gtrsim$1000 times higher than the rate of subsequent enlargement of the tiny primary etch canal resulting from the very fast etching of the ExD-core of the tracks.

2. As a result of either random fluctuations and/or small values of the linear density of the ExD, and/or the shrinking of the size of the ExD upon annealing, the distance, w, between the centers of two successive ExD can exceed λ_{cz}. In this case a "gap-zone" of length w-λ_{cz}, that is much less reactive than the core zones, appears. This zone can only be etched at a rate V_g, much smaller than V_{cz}, but still higher than V_0.

Theoretical prediction of track length distributions

From the gap-model we have derived a method which allows for the first time the computation of detailed track length distributions, both before and after thermal annealing. This method is applied below for the simple etching geometry where the tracks intersect the external surface of the mineral. We first divide a given latent track in successive elementary segments, i, with a length $dl_i = 1\ \mu m$, and we identify the position of these track segments with respect to the external surface of the mineral by using as length coordinate the residual range, $R_i(Z, E_i)$, of the track-producing ion in the mineral. We further assume that the linear density $N(R_i)$ and $n(R_i)$ of the ExD and PtD, respectively, are constant along any one of these track segments, and we use the experimental plots determined from the X-ray analysis to infer the variations of these linear densities (number per μm) of defects with both Z and R_i. The Monte-Carlo computation is started by distributing at random on the first track segment (i = 1), which is exposed to the chemical reagent, the number, $N(R_1)$, of ExD available. The computer code then follows the progression of the etchant along the track by evaluating that part, $\delta_1\tau$ of the standard etching time (τ_0) allocated for etching the whole track, which is consumed in etching out this first track segment. The probability that a given interval, w, between successive ExD exceed λ_{cz} is $\exp|-\lambda_{cz}.N(R_1)|$. Consequently, the number of gap zones, N_{g1}, is distributed randomly around a mean value $N(R_1).\exp|-\lambda_{cz}.N(R_1)|$. Then the integrated length of gap zones, W_{g1}, along this first segment is also randomly distributed around a mean value, $\exp|-\lambda_{cz}.N(R_1)|$.

$$\text{We thus obtain } \delta_1\tau = \frac{W_{g1}}{V_g} + \frac{1\text{-}W_{g1}}{V_{cz}}$$

We then etch out the next track segment dl_2. This procedure is repeated until $\sum_{i=1}^{n} \delta_i\tau = \tau_0$. The ordinate R_n of the track segment at which etching stops then fixes the etched length of a model track, $L_0 = (R_n - R_1) - \tau_0.V_0/\sin\theta$, where V_0 and θ are the perpendicular etching rate, respectively, of the external surface used as an observation plane and the angle between the track direction and this plane.

With these computations, that depend on 5 basic parameters (the 3 etching rates V_0, V_g, V_{cz}, the linear density N(R), and the effective diameter λ_{cz} of the ExD), a number of model tracks similar to the real tracks used to plot the experimental track length distribution can be generated. In addition, the computations can be extended to any annealing temperature when the temperature dependence of these basic parameters is known. Consequently a very detailed and quite meaningful comparison between the theoretical predictions and their experimental counterparts can be made. Indeed the values of these basic parameters are determined from methods which do not rely on any curve-fitting technique, in which the values of the parameters would be adjusted in order to obtain a good fit between the experimental and theoretical distributions. As this difficult part of our work is discussed more appropriately elsewhere (Dartyge *et al.*, 1978), we only point out here that the values of N(R) are directly inferred from the small angle X-ray scattering analysis of latent tracks, without referring to any etched track measurement.

Validity of the gap model

We have performed many comparisons between experimental and theoretical track length distributions, as we consider that the various features imprinted on the experimental distributions constitute the most severe challenge to be met by any theory dealing with the registration of etched tracks in minerals. We illustrate this work in Fig. 2 by reporting the variation of the track length distribution of 7 MeV/amu Fe-ions in labradorite as a function of the etching time, τ. It can be seen that the experimental histogram (Fig. 2a), which is narrowly peaked at a value L $\sim$ 6 μm, gradually broadens in shifting to greater lengths when τ increases from 40 up to 160 minutes. For a further increase in τ, the histogram switches back to a narrowly peaked distribution which is spread over a range of L values of about 6 μm, very similar to that initially observed for $\tau \sim$ 40 minutes.

Before the present work no model was available to predict etched track length distribution, and the unique set of histograms reported in Fig. 2a could not be understood. Now our Monte-Carlo computations generate a theoretical histogram (Fig. 2b) which fits very well the experimental one and give the following clues about the effects of increasing the etching time on track length distribution. The key to understanding these striking variations of the track length distributions due to 7 MeV/amu Fe-ions in labradorite is the variation of the linear density, N(R), of the ExD with the residual range, R, of the ion, as well as the dependence of the bulk etching rate of the track on N(R). In fact the N(R) curve is very unique* in showing both a maximum peaked at R $\sim$ 8 μm and steeply falling edges at both high and low values of R. In our model, the integrated length of the gap-zones, which are much more difficult to etch out than the core-zones, varies as exp $[-\lambda_{cz}N(R)]$. Thus when the etching of the tracks starts

*In particular, the N(R) curves as inferred from X-ray analysis are very distinct from the J(R) curves extensively used by previous authors in showing both a maximum shifted to much smaller values of R and a very steep decrease at high values of R. This observation strongly suggests that the track formation mechanism is not directly correlated to a J_c-threshold concept.

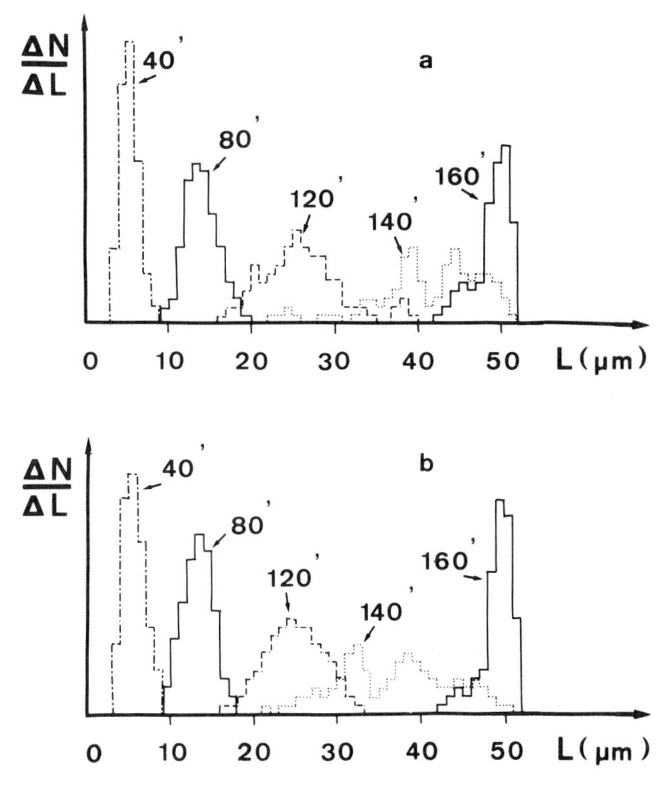

Fig. 2. Variatons of the track length distribution of 7 MeV/amu Fe-ions in labradorite with the etching time in minutes. The experimental histogram (curve a) is very well fitted by the theoretical one (curve b), which was obtained by requiring that the Monte-Carlo computer code generate a total number of tracks similar to those used to plot the experimental histogram. These curves cast serious doubts upon the validity of the TINT-etching technique.

at high values of R (small value of τ), the etchant penetrates through track segments showing a low value of N(R) and, consequently, a high integrated length of gap-zones. The bulk etching rate, V(R), of the tracks is relatively small and there is a dispersion in the V(R) values, which mostly reflects the random fluctuations in the position of the ExD from one track to the next, and which is responsible for the spread of about 4 μm in the etched track lengths. Such a spread can be interpreted as a delay in the etching of the shortest tracks with respect to the longest tracks. This initial delay in the track etching rates is propagated almost unchanged up to a value of R where the etchant reaches ExD-rich track segments which are etched out at a much faster rate than the gap-rich segments. Then, as the length of the longest tracks suddenly increases by as much as 20 μm while that of the short tracks keeps a small value, the spread in the track lengths jumps to ~15 μm when $\tau \sim$ 140 minutes. For a further increase in the etching time, the shortest etched tracks reach the ExD-rich segments and their length is considerably increased, while the longest tracks are blocked in the gap-rich zones at the end of the range. Thus the spread of the track length distribution is switched back to a much smaller value ~5 μm.

Such very good agreement between theoretical and experimental track length distributions gives us confidence in applying our new concepts, presented in the next section, dealing with the registration and the thermal annealing of etched tracks in minerals to the measurements of VVH/VH ratios in the ancient SFCR. In addition, this work gives very useful clues about using fossil tracks registered in natural minerals in a meaningful way. For example, the results reported in Fig. 2 clearly demonstrate that the TINT etching technique widely used by previous authors is inadequate for measuring VVH/VH ratios in minerals such as feldspar and olivine, which are characterized by relatively isotropic etching rates. In fact, in such minerals, the single peak expected for the track length distribution of a given ion continuously shifts toward longer length upon etching (Fig. 2) as new gap-zones along the tracks can be etched out. As the diameter of the long host tracks used in the TINT technique also enlarges upon etching, fresh tracks continuously intersect these host tracks. Although the value of the standard etching time, τ_0, is fixed, the *effective* etching time of the tracks used to plot the length distribution ranges anywhere between 0 and τ_0, and the distribution cannot reflect the chemical composition of VVH-nuclei. This severe criticism does not hold for the TINCLE technique, in which internal cleavage and/or fracture surface are used to inject the etchant into the tracks. Unfortunately, as such internal surfaces are very infrequent in lunar dust grains, we could not apply this interesting technique. For these reasons we decided to mostly rely in the present work on the simple etching geometry where the tracks intersect the external surface of the mineral exposed to the chemical etchant. However, the Monte-Carlo computations outlined above can be easily extended to the TINT and TINCLE etching geometries when necessary, just by propagating etchant *both* ways along the tracks, from the point where the tracks intersect either host tracks (TINT technique) or internal surfaces (TINCLE technique).

Outline of the new method

From the concepts discussed in the previous sections dealing with both the registration and the thermal annealing of tracks in minerals, we propose the following new method for determining VVH/VH ratios, which is applicable to fossil tracks at high densities registered in lunar *feldspar* grains, and observed on a polished surface of the grains.

For artificial ions ranging from iron up to xenon there is a sharply defined annealing temperature, $T_c(Z)$, beyond which the ratio $\rho(T_c)/\rho_o$ of the track densities, measured with a scanning electron microscope (SEM) both before and after annealing, drops below 10^{-4}. In fact for all ions so far observed (Fe, Cu, Kr), such a spectacular decrease in the $\rho(T)$ values occurs over a very narrow range of temperature of ≲20°C (Fig. 3).

This sharp drop cannot be described within the framework of the usual track

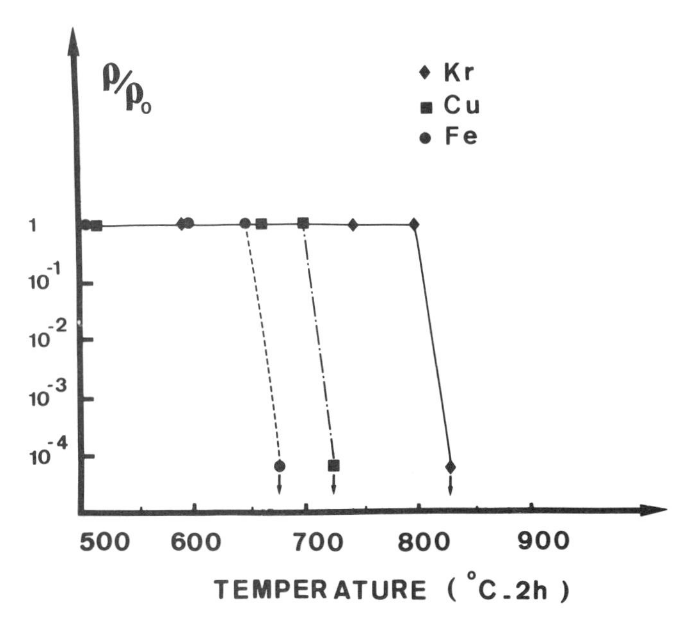

Fig. 3. Experimental variation of the track density with the annealing temperature in labradorite for Fe, Cu and Kr ions, with an energy of ~1 MeV/amu. The sharp variation of $\rho(T)$ occurs over a very narrow temperature range ($\lesssim 20°C$) for both types of ions, and there is a 150°C difference between the annealing temperature of Fe and Kr tracks. These curves, plotted for artificially accelerated ions, are very similar in shape to those reported earlier by Bull and Durrani (1975) for fossil VH-tracks in lunar pyroxene grains.

annealing model (Fleischer *et al.*, 1975), in which a latent track constituted of many defects in assimilated to a single defect characterized by a single activation energy, and which anneals following an Arrhenius-type equation. On the other hand, these observations can at least qualitatively be understood from our model: each one of the ExD that delineates the latent tracks due to a given type of ion is a cluster of many atomic defects (>100 per ExD, when $Z \gtrsim 26$). In addition there are a great number ($\gtrsim$100 per micron of range) of ExD along a given track. Consequently, as soon as one of the ExD "disappears" along a given track during the high temperature annealing, all the constituent ExD of all tracks will "statistically" reach the same degree of thermal annealing and the whole track population will disappear over a very narrow range of temperatures.

There is a drastic change in the atomic structure of the latent tracks between Fe-tracks where the ExD are always separated by gap-zones, and Kr-tracks where the "core-zones" start to overlap within an energy range $0.5 \lesssim E \lesssim 6$ MeV/amu. Consequently, upon thermal annealing, the etching rate of Fe-tracks will be much more severely affected than that of Kr-tracks and the $T_c(Z)$ values measured for Fe and Kr-tracks do indeed differ by as much as 150°C (Fig. 3), while the $T_c(Z)$ value observed for Cu-tracks is 50°C higher than $T_c(Fe)$. This set of $T_c(Z)$ curves allows the following conclusions: 1. As the upper end member of the VH-group is zinc, which is next to copper in the periodic table, we conclude that by a slight increase ~30°C of the $T_c(Cu)$ values of copper, it should be possible to eliminate all VH-tracks from the population of SFCR-tracks registered in lunar dust grains (T_c-method). 2. The differences observed in the T_c values reported for Fe, Cu and Kr clearly demonstrate that the charge resolution of the T_c-method should be of about 2 charge units between Fe and Kr.

However, the charge resolution of this method clearly worsens beyond Kr as soon as the ExD-rich parts of the tracks get saturated with "core-zones". This saturation is reflected by plateau values in the *etching rates* of Kr and Xe tracks for $E \gtrsim 0.5$ MeV/amu, which are very similar for both types of ions, with the exception that the plateau extends to much higher energy (so far inaccessible to experiment) for xenon. Upon heating, these plateaus shrink in extent and there is no marked difference between the etching rates of Kr and Xe tracks. Consequently the T_c-method cannot as yet be used beyond Kr in order to identify various subgroups of VVH nuclei. We should thus rely on track length measurements with the optical microscope, but we already pointed out that this technique, as applied to non-annealed tracks, has so far not yielded meaningful measurements of VVH/VH ratios.

Fortunately our work again helps in improving this method. We define the etchable range of an ion as that part of the ion range which is delineated by two critical energies, E_{min}-E_{max}, beyond which the linear density of the ExD is not sufficiently high to trigger a preferential chemical corrosion of the tracks, still observable with the SEM as shallow etch pits with depths $\gtrsim$1000 A. Before thermal annealing it is not possible either to experimentally measure the value of $\Delta_o L(VVH)$, as the corresponding values of E_{max} fall far beyond the maximum energy ($\lesssim$10 MeV/amu) accessible with heavy ion accelerator, or to theoretically compute these values from a track formation mechanism which has yet to be found. Our thermal annealing runs indicate that the $\Delta_c L(VVH)$ values applicable to the VVH tracks, exposed to the high temperature annealing intended to reject the VH-tracks, can now be measured from artificial heavy ion irradiations because the shortening of the tracks is sufficiently pronounced to shift the corresponding E_{max} value within the reach of heavy ion accelerators.

These $\Delta_c L(VVH)$ values can be obtained with good accuracy by using the etching method first proposed by Price *et al.* (1973). In this method the external surface of the target is irradiated with a perpendicular beam of monoenergetic ions characterized by a residual range R_0, then the observational surface to be etched is cut by polishing the irradiated surface at an angle θ. All values of the residual ion range from R_0 to zero are imprinted on the surface of such a polished wedge. After a long etching time, intended to reveal tracks as etch pits for the SEM observation, the extent of the area of the wedge covered with etch pits then directly yields the value of $\Delta_c L(Z)$. With this simple technique we have obtained values of $\Delta_c L(Kr) \sim 12$ μm when the annealing temperature was adjusted to a value $T_c(Zn) \sim T_c(Cu) + 30°C$. For such a temperature, and consequently for such values of $\Delta_c L(Kr)$, both the Fe and Cu tracks are no longer detectable with the SEM but can still be observed as very shallow etch pits with depths $\lesssim 100$ Å, using the Nomarsky phase contrast attachment of the optical microscope (No-OM). The same wedge technique, which can be applied to the non-annealed Fe-tracks, yields a $\Delta L_0(Fe)$ value of about 50 μm.

If the results of the simulation experiments could be directly applicable to the fossil tracks registered in lunar dust grains, our T_c-method should then be run as follows: 1. the track densities are measured both before (ρ_0) and after ($\rho(T_c)$) the critical annealing temperature, $T_c(VH) \sim T_c(Zn) = T_c(Cu) + 30°C$, which eliminates all the tracks due to VH-nuclei up to zinc from the fossil track length distribution. 2. As we further know the values of $\Delta_0 L(VH) \sim \Delta_0 L(Fe) \sim 50$ μm, and $\Delta_c L(VVH) \sim \Delta_c L(Kr) \sim 12$ μm* we can thus directly infer the chemical abundance of the VVH-nuclei with respect to that of the VH-nuclei from the $\rho(T_c)/\rho_0$ ratio.

Unfortunately, such a direct extrapolation of the laboratory experiments is not possible because it is well known that the fossil Fe-tracks that constitute up to 90% of the tracks observed before the T_c-annealing, are much more stable with respect to both a thermal and an ionization annealing than the fresh Fe-tracks used for calibration purpose. In addition, the length of these tracks has already been shortened by natural annealing processes triggered by the thermal cycle active in the lunar regolith (Plieninger *et al.*, 1972; Burnett *et al.*, 1972; Borg *et al.*, 1973; Price *et al.*, 1973; Bastin *et al.*, 1974; Maurette and Price, 1975). Consequently we had to modify the T_c-method as follows: 1. the $T_c(Fe)$ value of about 720°C applicable to the *fossil* Fe-tracks, which is ~70°C higher than that evaluated for "fresh" Fe-tracks, was directly measured in lunar feldspars loaded with SFCR-tracks by measuring the temperature at which the fossil track density drops below 10% of the initial value. 2. We then assumed that the $T_c(Z)$ values applicable to heavier SFCR-nuclei were all shifted from the same amount (~70°C) to higher temperatures, thus keeping their relative separation inferred from the simulation experiments and reported in Fig. 3. With this hypothesis the effective $T_c(VH)$ value applicable to the fossil tracks should be of about 800°C. 3. The distinct chemical composition of individual feldspar grains on the moon could possibly affect their $T_c(VH)$ values, so we slightly adjusted the values of $T_c(VH)$ around 800°C in each one of the grains in order to get the same degree of "Nomarskization" of the VH-tracks in all the grains. With this adjustment we can also safely assume that the effective value of $\Delta_c L(VVH) \sim \Delta_c L(Kr)$ to be applied to lunar dust grains is similar to the value of about 12 μm inferred from the thermal annealing runs of artificial Kr tracks, when a similar degree of

*This last approximation on the value of $\Delta_c L(VVH) \sim \Delta_c L(Kr)$ is based on the observation that the universal abundance of VVH elements up to Zr is much higher (×10) than that of the heavier nuclei. Consequently, these elements are mostly responsible for the VVH-tracks and their etchable ranges are roughly similar to that of Kr.

"Nomarskization" of the fresh Fe-Cu tracks has been achieved. 4. At this point the only unknown still to be fixed is the effective value of $\Delta L(VH) \sim \Delta L(Fe)$ for lunar dust grains, which is definitively smaller than the value of 50 μm determined for non-annealed artificial Fe-tracks. Values of $\Delta L(VH) \sim$ 10–15 μm were already reported by several authors (Plieninger *et al.*, 1972; Price *et al.*, 1971) from measurements of VH-track length distribution in grains extracted at a depth of a few hundred microns in lunar igneous rocks. Such grains, which contain relatively low densities of SFCR tracks, have been exposed at most a few m.y. in the lunar thermal cycle because the high erosion rate (~1 mm/m.y.) of lunar rocks by micrometeoroids would have released them in the lunar regolith. Consequently this value of $\Delta L(VH) \sim$ 15 μm is certainly an upper limit, because the residence time of individual lunar dust grains in the lunar thermal cycle, which is active up to depth ~10 cm in the lunar regolith, is ~100 m.y. (cf., first section of this paper). Fortunately we can fix a lower limit of $\Delta L(VH)$ of ~5 μm by noting that the VH-track etch pits observed with the SEM before the artificial annealing are not "flat"-bottomed; from our simulation experiments this implies that $\Delta L(VH) \gtrsim 5$ μm. Consequently from now on we will adopt a value of ~10 μm for the etchable range of fossil VH-tracks in lunar dust grains.

Preliminary results and discussion

100 μm-sized feldspar grains as well as igneous 1 mm-CF enriched in feldspar were extracted from several surface soil samples from the Apollo 16 mission, and then polished and etched in a boiling solution of NaOH (3g NaOH, 4g H_2O) for a standard etching time, τ_o ~2 minutes. The fossil track densities were measured as a function of depth into all the grains in order to obtain the track gradient of the VH-tracks. We then selected the SFCR-irradiated grains (eight 100 μm-sized grains and one 1 mm-CF) by using the ORSOLMIX criteria quoted above to eliminate 100 μm-grains either showing ρ_c values $<10^9 cm^{-2}$ or anisotropic track gradients, as well as 1 mm-CF with no well-defined gradients ($\rho_5/\rho_c<5$). Next, the grains were annealed at a temperature $T_c(VH) = 800°C$ for two hours and then repolished and etched for a longer time, τ_I~15 minutes. At this stage the degree of "Nomarskization" of the VH-tracks was measured with the No-OM in order to insure that no residual VH-tracks with lengths ≳1000 Å could be counted as VVH-tracks with the SEM. Finally the residual track length distributions were measured with both the SEM and the optical microscope.

Our most interesting result is the observation of the following four distinct populations of solar flare tracks in *all the grains* so far investigated:

1. Population I corresponds to the non-annealed fossil tracks observed with the SEM standard etching time, τ_o (Fig. 4a).
2. After annealing at $T_c(VH)$, three additional track populations with very different length distributions easily can be separated after an etching time τ_I~7 τ_o. The constituent tracks of population II appear as very shallow etch pits with depths ≲100 A, *easily* observable only with the No-OM in *all*

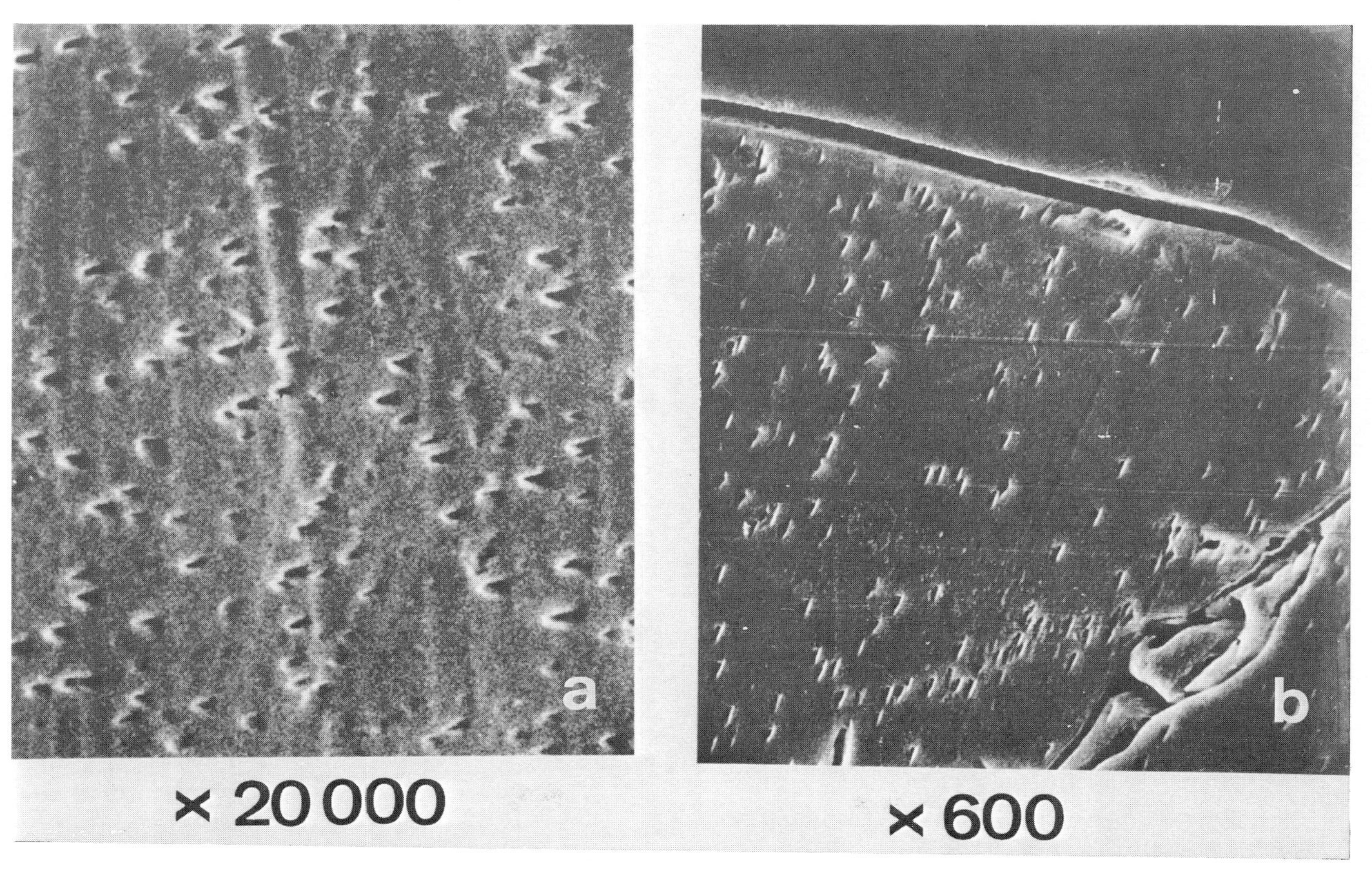

Fig. 4. Scanning electron microscope observations of SFCR-tracks in a lunar feldspar grain, both before (micrograph a) and after (micrograph b) a thermal annealing of 2 hours at 800°C.

Fig. 5. Nomarsky phase contrast observation of a lunar feldspar grain that was etched after an annealing of 2 hours at 800°C. The very shallow etch pits with depths $\lesssim$100 Å, which are no longer observable with the SEM, represent VH-tracks almost completely annealed.

grains (Fig. 5), and characterized by a ρ_{II}/ρ_I ratio of ~1/20. On the other hand, populations III and IV correspond to tracks with much greater lengths ($\gtrsim$1 μm) (Fig. 4b), which are grouped within the "3 μm" and "17 μm" peaks of the residual track length distribution, respectively (Fig. 6). At the center of all grains, the ρ_{III}/ρ_I and ρ_{IV}/ρ_I ratios are ~1/2000 and ~1/80,000, respectively. In addition we conclude that the ρ_{III}/ρ_I ratio is constant from edge to edge in the 1 mm-CF, within the experimental errors (Fig. 7), which are mostly due to the requirements of using a relatively large area of ~(100 μm)^2 to measure the relatively low density ($\lesssim 10^6$cm^{-2}) of VVH-tracks.

From both universal abundance tables (Trimble, 1975) and study of iron-group elements in the contemporary SFCR (Crawford *et al.*, 1975), it is inferred that Fe-tracks represent about 90% of the tracks found in population I, and consequently $\rho_I \sim \rho_{Fe}$. Next we have to identify the atomic number of the ions responsible for the tracks found in populations II, III and IV by relying on the two distinct gaps observed in the track length distribution both in the 0.1–1 μm and 6–9 μm length intervals.

As the tracks in population II are only observable with the No-OM after the T_c(VH)-annealing, we can safely assume that they represent the partially annealed tracks of the end members (zinc, copper and possibly nickel) of the VH-nuclei group. Then the ρ_{II}/ρ_I ratio would yield an abundance ratio for these nuclei relative to iron of ~1/20. This ratio is compatible with universal abundances, assuming that the tracks due to Ni ions are still visible with the No-OM after the T_c(VH)-annealing. Indeed only the universal abundance of Ni (~1/20), which is much higher than that expected for Cu (1/1500) or Zn (1/700), can match the ratio inferred from track studies.

The first gap in the experimental track length distribution of the partially annealed VVH-tracks is observed between a length (~100 A) corresponding to the depths of the very shallow etch pits detected with the No-OM and the minimum length (~1 μm) of the residual tracks detected with the SEM, which well exceeds the limit of depth resolution of this instrument (~1000 A). We believe that Zn tracks have been eliminated from the SEM track count after the T_c(VH)-annealing, and that the charge resolution of the method is of about 2 charge units (see previous section). Then this first gap should necessarily correspond to a severe depletion of nuclei adjacent to Zn in the periodic table, which should be compatible with universal abundances, since such very marked depletions *over a charge range of two units at most* have no reason to be completely suppressed during the emission of SFCR. As ^{31}Ga does indeed fulfill these two requirements we conclude that the T_c-method as applied in our work selects VVH-nuclei with atomic number $\geq$32. Consequently we can directly infer the VVH(Z $\geq$32)/VH ratios from the experimental values of $(\rho_{III} + \rho_{IV})/\rho_o \sim \rho_{III}/\rho_I$, by using the appropriate etchable ranges determined in the previous section: ΔL(VH) ~ 10 μm and Δ_cL(VVH) ~ Δ_cL(Kr) ~ 12 μm.

In the eight 100 μm-sized feldspar grains we have thus determined the

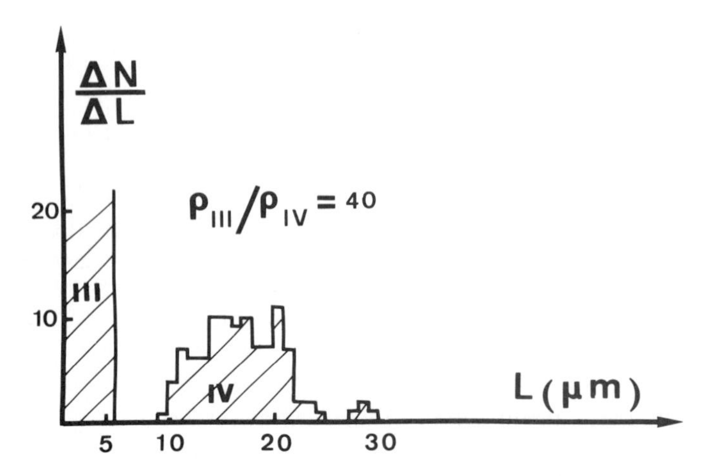

Fig. 6. Track length distribution of VVH-tracks in the 1 mm-CF, as observed after rejection of the VH-tracks by a thermal annealing. The two peaks in the track distribution, which are separated by a gap of ~3 μm, most likely correspond to distinct groups of ions with atomic numbers ranging from 32 to at least 54 and ≳75, respectively.

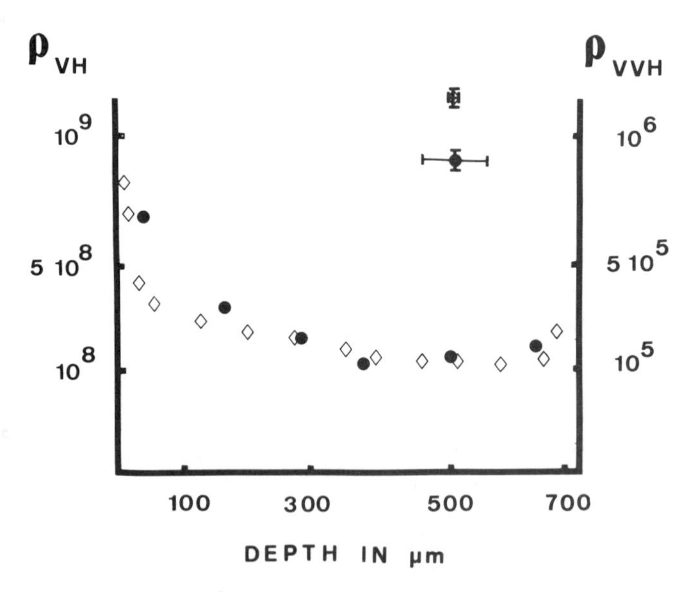

Fig. 7. Variation of the track density with depth in the 1 mm-CF (track gradient). The open squares and the black dots correspond to VH-tracks and VVH-tracks observed before and after the thermal annealing, respectively. The striking similarity of these two track gradients implies that the VVH/VH ratio in the ancient SFCR is constant from about 5 to 30 MeV/amu.

following values of the VVH($Z \geq 32$)/VH ratios as expressed in units of 10^{-4}: 4.8, 5, 4, 4.5, 5.9, 3.3, 5.6, 6.3. There is a marked clustering of these ratios at a value of 5×10^{-4} that should then represent an average value of the VVH/VH ratio over a very long time scale of 1 b.y., that corresponds to the plateau history of the 100 μm-grains. In contrast, the only value of this ratio so far obtained for a 1 mm-CF with a single exposure in the SFCR of about 1 m.y. in duration is about 2 times higher than this long-term value. It could be that the short-term (1 m.y.) value of the VVH/VH ratio differs from the long-term (1 b.y.) average value. In addition, the constancy of the ρ_{III}/ρ_{I} ratio through the 1 mm-CF (Fig. 7) implies a similarity in the energy spectrum of the VH and VVH nuclei from about 5 to 30 MeV/amu (this upper limit corresponds to the center of the 1 mm-CF). These results would then confirm and extend the earlier ones obtained by Crawford *et al.* (1975) on the contemporary SFCR, for nuclei up to iron and for a maximum energy of 20 MeV/amu.

The exact identification of the subgroups of VVH-nuclei responsible for both the two distinct peaks and the gap observed in Fig. 6 is a very difficult task at the present time because of the lack of appropriate irradiation with ions heavier than xenon. First we know from the No-OM observations that when the residual lengths of artificial Fe-tracks are $\lesssim 100$ Å, then the $\Delta L_c(Kr)$ value determined with the wedge technique for Kr is 12 μm. As the fossil VH-tracks reach the same residual length of $\lesssim 100$ Å after the $T_c(VH)$ annealing we can assume that the $\Delta_c L(Kr)$ value associated with Kr ions in the ancient SFCR, is about 12 μm. These $\Delta_c L(Z)$ values are very useful for relating fossil track densities to the chemical abundance of the corresponding track-producing nuclei. However, for tracks *with low etching rates* such as partially annealed VVH-tracks, $\Delta_c L(Z)$ does not correspond to the maximum etched length of the tracks as measured with the optical microscope on the external surface of the grains, because there is a balance between the low etching rate of the tracks and that of the observational plane. In fact, we have verified that for artificial Kr tracks partially annealed at $T_c(VH)$ the maximum length, L_{max}, of the tracks can be approximated by $L_{max} \sim \Delta_c L(Kr)/2 \sim 7$ μm, and this conclusion agrees well with the previous (Blanford *et al.*, 1975) one inferred from the study of fossil VH tracks.

This result implies that VVH-nuclei up to krypton should give tracks within the first peak in the track length distribution. But we know that after the $T_c(VH)$-annealing the etching rates of Kr and Xe tracks are very similar, as a result of the overlapping of the ExD along the radiation-damaged core of the tracks. Consequently we believe that after the standard etching time, tracks of all elements up to at least xenon should be included in the first peak of the VVH-track length distribution.

It is not possible to understand either the gap between 6 and 9 μm or the second peak in the VVH-track length distribution, or whether this conclusion of "similar etching rates" holds true for elements much heavier than xenon. We thus feel that there is possibly a critical atomic number, Z_c, still to be identified, beyond which this basic conclusion fails. For $Z > Z_c$ the very severe overlapping of ExD would trigger the formation of a new type of atomic structure of the latent tracks, which would be much more stable against heating than that of lighter ions. Consequently the etching rates as well as the resulting track lengths observed after the standard etching time would suddenly jump to higher values for these much heavier ions. In assuming that marked gaps in universal abundances should be reflected in the chemical composition of SFCR-nuclei, and using the rule of thumb which states that $\Delta L(Z) \sim 2\ L_{max}$, we tentatively attribute the second peak in Fig. 7 to elements in the so-called platinum and lead-group of nuclei ($Z \gtrsim 75$).

If these conclusions are verified by studying artificial tracks due to ions heavier than xenon, then the charge resolution of the T_c-method for heavy ions beyond Kr would be much better than we presently think, as the VVH-tracks responsible for the first peak appearing in Fig. 6 should be preferentially erased by an annealing conducted at temperatures slightly higher than T_c(VH).

CONCLUSIONS AND FUTURE PROSPECTS

Although much more work is needed to definitively ascertain the validity of these conclusions, we believe that the thermal annealing method when handled within the framework of a detailed annealing and etching model for the tracks constitutes an interesting new approach for exploiting the fossil track record in extraterrestrial materials. In particular, the present results already strongly suggest that there is no marked enrichment of VVH-nuclei heavier than gallium in the ancient solar flare cosmic rays, over an energy range of about 5 to 30 MeV/amu.

There are bright prospects for this new method within the framework of the lunar skin sampling technique (Borg *et al.*, 1976; Langevin and Maurette, 1978). This technique can be applied to select both 1 μm-grains and 1 mm-CF, which have been exposed at the same epoch to the ancient solar wind and solar flare cosmic rays. The thickness of the amorphous coating of solar wind radiation-damaged material on the 1 μm-grains can be used to determine the energy of the ancient solar wind, whereas the 1 mm-CF will serve as SFCR-detectors. Consequently, the prospect of tracing back the past activity of the sun over a very long time scale never explored before has a greater potential now that we can look at two different types of solar "radiations".

It would also be very interesting to extract larger feldspar-rich coarse-fines with a size of about 5 mm in both lunar soil samples and gas-rich meteorites. In such grains the tracks observed at depths $\lesssim$1 mm are mainly due to SFCR, whereas those registered at greater depths represent the contribution of ancient galactic cosmic rays. With lunar coarse fines we could meaningfully compare VVH/VH ratios in the ancient SFCR and GCR from tracks registered at about the same epoch in a single coarse fine at a known distance from the sun. Then, by comparing the track record in lunar and meteoritic coarse fines, we possibly could probe the ancient cosmic rays at very different epochs in the past and/or at different distances to the sun.

REFERENCES

Bastin G., Comstock G. M., Dran J. C., Duraud J. P., Maurette M. and Thibaud C. (1974) Lunar soil maturation, part III: Short-term and long-term aging of radiation damage features in the regolith (abstract). In *Lunar Science V*, p. 44–47. The Lunar Science Institute, Houston.

Bhandari N., Goswami J. N., Lal D. and Thamhane A. S. (1973) Long-term fluxes of heavy cosmic-ray nuclei based on observations in meteorites and lunar samples. *Ap. J.* **185**, 975–983.

Bhandari N. and Padia J. T. (1974) Secular variations in the abundances of heavy nuclei in cosmic-rays. *Science* **185**, 1043–1045.

Bibring J. P., Borg J., Burlingame A. M., Langevin Y., Maurette M. and Vassent B. (1975) Solar-wind and solar-flare maturation of the lunar regolith. *Proc. Lunar Sci. Conf. 6th*, p. 3471–3493.

Blanford G. E., Fruland R. M. and Morrison D. A. (1975) Long-term differential energy spectrum for solar flare iron group particles. *Proc. Lunar Sci. Conf. 6th*, p. 3557–3576.

Borg J., Comstock G. M., Langevin Y., Maurette M., Jouffrey B. and Jouret C. (1976) A Monte-Carlo model for the exposure history of lunar dust grains in the ancient solar wind. *Earth Planet. Sci. Lett*, **29**, 161–174.

Borg J., Dran J. C., Comstock G., Maurette M., Vassent B., Duraud J. P. and Legressus C. (1973) Nuclear particle track studies in the lunar regolith: Some new trends and speculations (abstract). In *Lunar Science IV*, p. 82–34. The Lunar Science Institute, Houston.

Bull R. K. and Durrani S. A. (1975) Annealing and etching studies of fossil and fresh tracks in lunar and analogous crystals. *Proc. Lunar Sci. Conf. 6th*, p. 3619–3637.

Bull R. K., Green P. F. and Durrani S. A. (1978) The energy spectrum of the VVH-cosmic ray averaged over the last 19 Myr (abstract). In *Lunar and Planetary Science IX*, p. 131–133. Lunar and Planetary Institute, Houston.

Burnett D., Hohenberg C. M., Maurette M., Monnin M., Walker R. and Woolum D. (1972) Solar cosmic ray, solar wind, solar flare and neutron albedo measurements. *Apollo 16 Prelim. Sci. Rep.* NASA SP-315, p. 15-19 to 15-32.

Comstock G. M. (1977) On deciphering the particle track record of lunar regolith history. *J. Geophy. Res.* **10**, 357–367.

Crawford H. J., Price P. B., Cartwright B. G. and Sullivan J. D. (1975) Solar flare particles: energy-dependent composition and relationship to solar composition. *Ap. J.* **195**, 213–221.

Dartyge E., Dran J. C., Duraud J. P., Langevin Y. and Maurette M. (1977b) Thermal annealing of nuclear particle tracks and the chemical composition of very heavy cosmic rays (abstract). In *Lunar Science VIII*, p. 218. The Lunar Science Institute, Houston.

Dartyge E., Duraud J. P. and Langevin Y. (1977a) Thermal annealing of iron tracks in muscovite, labradorite and olivine. *Radiat. Eff.* **34**, 77–79.

Dartyge E. and Lambert M (1974) Formation de défauts dans les échantillons de mica muscovite irradiés par des ions de grande énergie. *Radiat. Eff.* **21**, 71–79.

Dartyge E., Lambert M. and Maurette M. (1976) Structure et enregistrement des traces latentes d'ions argon et fer dans l'olivine et le mica muscovite. *J. Phys.* (France) **37**, 137–141.

Duraud J. P., Langevin Y., Maurette M., Comstock G. and Burlingame A. L. (1975) The simulated depth history of dust grains in the lunar regolith. *Proc. Lunar Sci. Conf. 6th*, p. 2397–2415.

Fleischer R. L., Price P. B. and Walker R. M. (1975) *Nuclear tracks in solids*. University of California Press, Berkeley. 605 pp.

Goswami J. N. and Lal D. (1975) Compositional studies of solar-iron group and heavier nuclei of kinetic energy below 10 MeV/nucleon. *Proc. Lunar Sci. Conf. 6th*, p. 3541–3555.

Lal D., Rajan R. S. and Tamhane A. S. (1969) Chemical composition of nuclei of Z >22 in cosmic rays using minerals as detectors. *Nature* **221**, 33–37.

Langevin Y. and Maurette M. (1978) Plausible depositional histories for the Apollo 15, Apollo 16 and Apollo 17 deep drill cores. *Proc. Lunar Planet. Sci. Conf. 9th*. This volume.

Maurette M. and Price P. B. (1975) Electron microscopy of irradiation effects in space. *Science* **187**, 121–129.

Plieninger T., Krätschmer W. and Gentner W. (1972) Charge assignment to cosmic ray heavy ion tracks in lunar pyroxene. *Proc. Lunar Sci. Conf. 3rd*, p. 2933–2939.

Price P. B., Lal D., Tamhane A. S and Perelygin V. P. (1973) Characteristics of tracks of ions of $14 \leq Z \leq 36$ in common rock silicates. *Earth Planet. Sci. Lett.* **19**, 377–395.

Price P. B., Rajan R. S. and Shirk E. K. (1971) Ultra-heavy cosmic rays in the moon. *Proc. Lunar Sci. Conf. 2nd*, p. 2621–2627.

Price P. B. and Walker R. M. (1962) Chemical etching of charged particle tracks. *J. Appl. Phys.* **33**, 3407–3412.

Trimble V. (1975) The origin and abundances of the chemical elements. *Rev. Mod. Phys.* **47**, 877–968.

Proc. Lunar Planet. Sci. Conf. 9th (1978), p. 2399–2414.
Printed in the United States of America

On the absence of effective modulation of galactic cosmic rays in the solar system during the ice-age

A. K. LAVRUKHINA and G. K. USTINOVA

Vernadsky Institute of Geochemistry and Analytical Chemistry
USSR Academy of Sciences, Moscow, USSR

Abstract—From an analysis of ^{53}Mn experimental data it has been shown that over a period of 2–6 million years ago the lunar surface and chondrites were exposed to galactic cosmic rays with the least modulated spectrum close to the spectrum recorded near the earth in April–May, 1965 or in August–December, 1976. From this it follows that (1) the intensity of unmodulated galactic cosmic rays was constant over the last ~6 million years, and (2) the solar system was free of the effective modulation of galactic cosmic rays over a period of 2–6 million years ago, which is coincident with the last global ice-age on the earth.

In recent years the sun spot activity as a cause of temporal variations of galactic cosmic rays in the solar system has been considered to be of primary importance. In addition to the comprehensively studied 11-year variations, there are a variety of alterations of the galactic cosmic-ray intensity, including long-time variations which correspond to the cyclic changes of solar activity. On the other hand, the solar activity variations are reflected in numerous processes on the earth. There exists a hypothesis (Shklovsky, 1975) that the last global ice-age of the earth was due to an appreciable decrease in solar luminosity [and in its temperature (Dilke and Gough, 1972)] because of relatively fast sudden mixing of the matter in the solar core several million years ago (Fowler, 1972). Fowler's idea was due to the problem of neutrino fluxes from the sun. In detail the inter-relations between solar activity, ice-ages and cosmic rays have been discussed by Kocharov, (1975). Geological data imply that the global glaciation takes place every 200–300 million years with a duration of about 10 million years. At the present time the earth is in an inter-glacial period, characterized by separate glaciation phases [e.g., there were nine glaciation phases over $6 \cdot 10^5$ years of the Pleistocene period (Zeuner, 1959)]. Due to variation of the luminosity, the modern temperature of the earth must be, indeed, 10–15° lower than in "normal" state. But the last major ice ages appear to pertain to a time of the order of 10 millions years. On the other hand, it is easy to explain the negative results of solar neutrino experiment (Davis and Evans, 1974) if the last mixing in the solar interior took place less than 10 million years ago. So Fowler's suggestion can be made to resolve the solar neutrino discrepancy provided we assume that the sun is now in close of a reviving phase after having been mixed ~10 million years ago. Then it is natural to suppose that the modulation mechanism of galactic cosmic rays, in particular 11-year variations, has to be broken with reducing solar luminosity and, apparently, its spot activity ~10 million years ago (Kocharov, 1975). This effect can be determined from the formation of cosmogenic radionuclide ^{53}Mn ($T_{1/2} = 3.7 \cdot 10^6$ years) in cosmic bodies. It has been established (Lavrukhina and Ustinova 1971, 1972) that long-lived radionuclides ($T_{1/2} \gg 11$ years) up to ^{26}Al ($T_{1/2} = 7.4 \cdot 10^5$ years) on the moon and in the chondrites with orbit aphelia $q' \lesssim 2$ AU are produced at an average intensity of 0.24 particles $cm^{-2}s^{-1}sr^{-1}$ ($R > 0.5$ GV, 1962 spectrum (Vernov *et al.*, 1974)) per solar cycle. In the chondrites with large orbits ($q' \gtrsim 4$ AU) they are formed at a maximum galactic cosmic-ray intensity of 0.35 particles $cm^{-2}s^{-1}sr^{-1}$ ($R > 0.5$ GV, April–May, 1965 or August–December, 1976 spectrum) registered in stratosphere in the minima of 19 and 20 solar cycles and close to the unmodulated interstellar intensity (Vernov *et al.*, 1974). In the chondrites with intermediate orbits they are formed in part at the average intensity and in part at the maximum intensity, e.g., in the case of the Lost City chondrite with the known orbit ($q' = 2.35$ AU) theoretical values of ^{26}Al radioactivity are in agreement with experimental data if over half of its orbit period the chondrite was exposed to the average spectrum of galactic cosmic rays and over the other half to the spectrum close to the unmodulated one. If the modulation mechanism was operative over the last 5–6 million years (an accumulation period of ~70% nowaday observed ^{53}Mn radioactivity in cosmic bodies with radiation ages of $\gtrsim 10^7$ years is $5.6 \cdot 10^6$ years) and the galactic cosmic-ray intensity was in conformity with the modern one, the above features should be reflected in ^{53}Mn variations in cosmic bodies. If the modulation mechanism was not operative, ^{53}Mn on the moon and in chondrites with different orbits had to be produced by the unmodulated spectrum of galactic cosmic rays. Also if the galactic cosmic-ray intensity was higher than the modern one several million years ago (for example, because of nearby supernova explosions), the radioactivity of ^{53}Mn should be above theoretical values not only on the moon but also in the chondrites with $q' > 2$ AU.

In order to settle this question, we have analyzed experimental data from a measurement of ^{53}Mn in the Apollo 15 lunar soil core from a depth of 60–416 g/cm^2 (Imamura *et al.*, 1973) and in some chondrites. As follows from the last reference, the measured depth profile of ^{53}Mn radioactivity in lunar soil core lies above the theoretical curve calculated by the method of Reedy and Arnold (1972) for average galactic cosmic ray intensity according to neutron monitor data by ~25%. The same result has been obtained by Rancitelli *et al.* (1975) who have drawn also a conclusion about temporal variations of galactic cosmic rays over the last 8 million years. In this work the calculation of the ^{53}Mn production rates in the lunar surface layer and in the chondrites of different sizes has been performed by the previously developed method of the description of nuclear cascade processes in straight-forward approximation (Lavrukhina and Ustinova, 1967; Lavrukhina *et al.*, 1969) for the average cosmic-ray spectra (1962 and April–May, 1965). However, because of extremely low threshold of the excitation function of ^{53}Mn from iron ($\lesssim 8$ MeV for production by neutrons, Fig. 1), the necessity has arisen for taking into account the contribution of the third generation of nuclear-active particles in the lunar or chondrite matter to the radionuclide production. By solving the differential equation, which describes the production of the third generation particles by the flux of secondaries and their absorption due to nuclear interactions we have obtained the depth distributions of the third generation cascade particles and evaporation neutrons in the lunar surface layer and in chondrites of different sizes. Thus in the case of 2π-incidence on the moon, for example, the fluxes of nuclear-active particles of galactic cosmic rays are of the form:

a) the primaries:

$$I_p(d) = 2\pi I_0[\exp(-\mu_a d) + \mu_a d \mathrm{Ei}(-\mu_a d)] \tag{1}$$

b) the secondary neutrons and showers (ionisation losses are absent or small):

$$I_s(d) = 2\pi I_0 \bar{S} \frac{\mu_p}{\mu_s - \mu_a} [\exp(-\mu_a d) + \mu_a d \ \mathrm{Ei}(-\mu_a d) - \\ -\exp(-\mu_s d) - \mu_s d \ \mathrm{Ei}(-\mu_s d)] \tag{2}$$

c) the secondary charged particles (π-mesons and protons):

$$I_s^{\pi}(d) = 2.2\pi I_0 \bar{S} \frac{\mu_p}{a} \{J(d) \exp[-0.2(\mu_s - \mu_a)/a] - \\ \frac{a}{\mu_s - \mu_a} [\exp(-\mu_a d) + \mu_a d \ \mathrm{Ei}(-\mu_a d) - \exp(-\mu_s d) - \mu_s d \ \mathrm{Ei}(-\mu_s d)]\},$$

where

$$J(d) = \int_0^1 \exp(-\mu_s d/x) \ J(x) dx; \\ J(x) = \int_{0.2}^{ad/x + 0.2} \exp(\mu_s - \mu_a) \ u/a \ u^{-0.4} du; \tag{3}$$

$$I_s^p(d) = 2\pi \ I_o\bar{S}; \frac{\mu_p}{a} \{Y_1(d) \frac{\mu_s - \mu_a}{a} \exp[-0.2(\mu_s - \mu_a)/a] - 0.1 Y_2(d) + \\ + (0.31 + \frac{0.07a}{\mu_s - \mu_a}) [\exp(-\mu_s d) + \mu_s d \ \mathrm{Ei}(-\mu_s d)] - \\ \frac{0.07a}{\mu_s - \mu_a} [\exp(-\mu_a d) + \mu_a d \ \mathrm{Ei}(-\mu_a d)]\}, \text{ where}$$

$$Y_1(d) = \int_0^1 \exp(-\mu_s d/x) \ Y(x) dx, \\ Y(x) = \int_{0.2}^{ad/x + 0.2} \exp[(\mu_s - \mu_a) u/a] \ u^{-0.7} du, \\ Y_2(d) = \int_0^1 (ad/x + 0.2)^{-0.7} \exp(-\mu_a d/x) \ dx \tag{4}$$

d) the third generation cascade particles and evaporation neutrons:

$$I_t^{casc}(d) = 2\pi I_o \bar{S}_t^{casc} \bar{S}_s' \frac{\mu_p}{\mu_s - \mu_a} \{ \frac{\mu_s}{\mu_s - \mu_a} [\exp(-\mu_a d) + \mu_a d\ Ei(-\mu_a d) - \exp(-\mu_s d) - \mu_s d\ Ei(-\mu_s d)] + \mu_s d\ Ei(-\mu_s d)\}; \quad (5)$$

$$I_t^n(d) = 2\pi I_o \bar{S}_t^n \bar{S}_s' \frac{\mu_p}{\mu_s - \mu_a} \{ \frac{\mu_s}{\mu_n - \mu_a} [\exp(-\mu_a d) + \mu_a d\ Ei(-\mu_a d)] - \frac{\mu_s}{\mu_n - \mu_s} [\exp(-\mu_s d) + \mu_s d\ Ei(-\mu_s d)] - [\frac{\mu_s}{\mu_n - \mu_a} - \frac{\mu_s}{\mu_n - \mu_s}] \cdot [\exp(-\mu_n d) + \mu_n d\ Ei(-\mu_n d)]\}. \quad (6)$$

In Eqs. (1)–(6) d is a depth from the surface; I_o—is an average intensity of incident galactic cosmic rays; $\mu_a = N\rho\bar{\sigma}_a/A$ is an effective absorption coefficient of primary particles (N—Avogadro number; ρ—density of matter; $\bar{\sigma}_a$—effective absorption cross section of primaries; A—mass number); $\mu_{p,s,t} = N\rho\bar{\sigma}_{p,s,t}/A$ are interaction coefficients of primaries (p), secondaries (s) and the third generation particles (t) ($\bar{\sigma}_{p,s,t}$—their cross sections for interactions). For compound material

Table 1. Material and particle parameters.

Parameter \ Matter*)	Iron	Chondrite /a/	Apollo-11 rock /b/	Apollo-15 soil /b/
ρ, g/cm^3	7.8	3.6	~3.1	~1.7
$\bar{A}$	56	30.5	28.5	27.9
σ_a, mbarn	573	330	305	300
σ_p, mbarn	691	445	420	415
μ_a, cm^{-1}	0.048	0.023	0.020	0.011
μ_p, cm^{-1}	0.058	0.032	0.028	0.015
μ_s, cm^{-1} (showers, E > 1000 MeV)	0.052	0.028	0.025	0.014
μ_s, cm^{-1} (p, π, n, 100 ≲ E < 1000 MeV)	0.058	0.032	0.028	0.015
μ_n, cm^{-1} (n, 1 ≲ E < 100 MeV)	0.125	0.071	0.062	0.034

*) Chemical composition from: /a/—Yavnel and Dyakonova (1958); /b/—Wänke (1974).

the matter parameters must be averaged according to the chemical composition. For some materials values of the parameters are given in Table 1. Ionization coefficients "a" are tabulated by Williamson *et al.* (1966). The multiplicities ($\bar{S}$) for production of secondary particles in primary interactions have been obtained by weighting their production functions according to the galactic cosmic ray average spectra in 1962 and April-May 1965 (Lavrukhina, 1972). Also the multiplicities ($\bar{S}_t$) for the production of the third generation particles in secondary interactions have been obtained by weighting their production functions according to the spectra of secondaries. The production multiplicities in

Table 2. Average production multiplicities of nuclear-active particles.*

Type of particles	Energy range, MeV	$\bar{S}$, mean number per interaction act: 1962 spectrum, Fe	1962 spectrum, Al	April–May 1965 spectrum, Fe	April–May 1965 spectrum, Al
the second generation shower particles (p,π,n)	>1000	0.6	0.6	0.42	0.42
cascade protons	200–1000	0.4	0.4	0.4	0.4
cascade mesons	150–1000	0.6	0.6	0.42	0.42
cascade neutrons	100–1000	1.0	1.0	1.0	1.0
total number of secondary particles ($\bar{S}'_s$)	100	2.6	2.6	2.24	2.24
evaporation and cascade neutrons	1–100	6.8	5.2	6.46	5.0
the third generation cascade particles	100	0.4	0.4	0.37	0.37
evaporation and cascade neutrons	1–100	3.82	2.92	3.72	2.88

*Averaging (statistical weighting) according to the spectra of nuclear active particles has been carried out by the formula

$$\bar{S} = \frac{\int_{E_1}^{E_2} S(E)F(E)dE}{\int_{E_1}^{E_2} F(E)dE}, \text{ where}$$

S(E) is production function of nuclear-active particles, and $\int_{E_1}^{E_2} F(E)dE$ is an integral spectrum of primary or secondary cosmic radiation in energy range of interest.

aluminium have been used in the case of lunar and chondrite matter (Table 2). The production rate $H_i(d)$ of i-radionuclide (which is equal to its disintegration rate under condition of secular equilibrium) is a sum of its production rates by all types of nuclear-active particles:

$$H_i(d) = \frac{N}{A}\left[\bar{\sigma}_i^p\, I_p(d) + \sum_s \bar{\sigma}_i^s\, I_s(d) + \sum_t \bar{\sigma}_i^t\, I_t(d) \ldots\right], \quad (7)$$

where $\bar{\sigma}_i^{p,s,t}$ are average production cross sections of i-radionuclide by all components of primary, secondary and the third generation cosmic rays in energy range of interest (where the particles are nuclear-active). Because of ionization losses the contribution from galactic cosmic-ray particles having rigidities below the cut-off value R = 0.5 GV is negligible except for locations within a very few

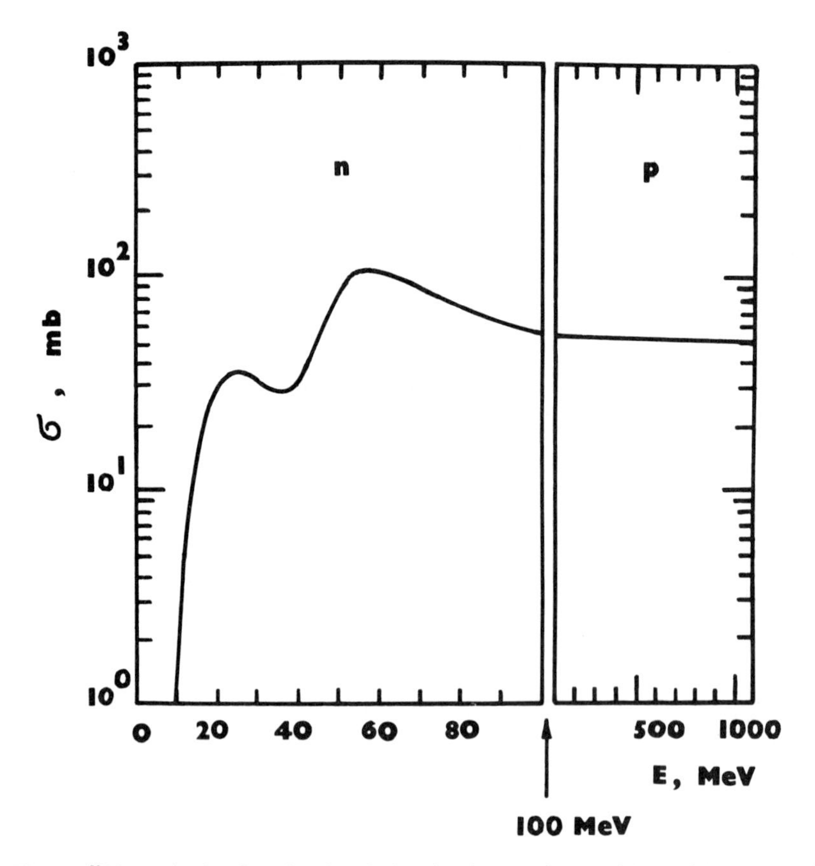

Fig. 1. ^{53}Mn excitation function for the bombardment of natural iron with protons of energy above 100 MeV and with neutrons below 100 MeV (which should be used for estimation of ^{53}Mn radioactivity produced by *galactic* cosmic rays at a depth of ≳20 g/cm² in cosmic bodies).

cm of the surface of a meteoroid which is removed by ablation when it falls through the atmosphere. Also in the upper few cm of the lunar surface the effect of solar protons dominates but at the depths $\gtrsim 6$ cm it is negligible (Lavrukhina and Ustinova, 1971). So we suppose that in the Apollo 15 lunar soil core from a depth of 60–416 g/cm^2, and in most cases of meteorite samples, radionuclides have been produced by galactic cosmic rays of $E > 100$ MeV and by nuclear-active secondaries. Due to ionization losses, the secondary charged particles (protons and mesons) as well are nuclear-active if $E > 100$ MeV. But the production by secondary neutrons is considerable at $E > 1$ MeV for most radionuclides. Further, it is assumed that production cross sections of radionuclides by all nuclear-active particles are the same as for protons above 100 MeV and below this energy the radionuclides are formed by secondary neutrons only. (Of course, if one investigates the effects of solar protons ($E < 400$ MeV), the excitation functions for protons up to the reaction thresholds should be considered). The necessary cross sections can be taken from accelerator experiments, from reaction systematics or can be derived by the formerly developed approach of Lavrukhina and Ustinova (1971). For example, in Fig. 1 ^{53}Mn excitation function for the bombardment of natural iron with protons of energy above 100 MeV and with neutrons below 100 MeV is presented. Above 100 MeV the ^{53}Mn cumulative cross sections have been determined from the experimental ^{54}Mn production cross sections from iron (Rayudu, 1964, 1968) by the formula (Lavrukhina and Ustinova, 1971):

$$\sigma(^{53}Mn_{25}) = \exp [P \cdot (53 - 54) - R_{53} \mid 25 - S \cdot 53 + T \cdot (53)^2 \mid^{3/2} + R_{54} \mid 25 - S \cdot 54 + T \cdot (54)^2 \mid^{3/2}]\sigma(^{54}Mn_{25})^{\text{experim.}} \quad (8)$$

(P,R,S,T—Rudstam coefficients) accompanied by addition of experimental ^{53}Fe production cross sections from iron (Lavrukhina *et al.*, 1963). Below 100 MeV the ^{53}Mn cumulative cross sections of the reactions

$$^{56}Fe_{26}\,(n,p3n + n,4n)^{53}Mn_{25};\ ^{54}Fe_{26}(n,pn + n,2n)^{53}Mn_{25}$$

have been determined from analogous peripherical reactions with protons (see, c.g., Lavrukhina, 1972). The upper limit of errors of the calculated ^{53}Mn cross sections (20–25%) is due to experimental errors of the measured cross sections used (^{54}Mn and ^{53}Fe production cross sections from iron, etc.). The ^{53}Mn excitation function obtained is similar to that of Reedy and Arnold (1972), but above 100 MeV the cross sections are ~30% higher than those in the reference. The average cross sections of the ^{53}Mn production (Table 3) have been obtained in conformity with content of iron, manganese and nickel in the lunar and chondrite matter by weighting the ^{53}Mn excitation functions from these elements according to the spectra of primary and secondary cosmic radiation.

The analytical method described above has been verified by the calculations (and comparisons with measurements) of radioactivities of short-lived radionuclides in the lunar samples identified for which the irradiation conditions were well determined (Lavrukhina and Ustinova, 1971; Vinogradov *et al.*, 1974). In

Table 3. Average cross sections of the ^{53}Mn production in some materials (mbarns).

	Primaries** E ≳ 100 MeV		Secondaries and third generation particles**				
Matter*	1962 spectrum	1965 spectrum	showers E> 1000 MeV	p E = 200–1000 MeV	π E = 150–1000 MeV	n E = 100–1000 MeV	n E = 1–100 MeV
Lunar soil 15001–15006[a]	4.4	4.5	4.3	4.5	4.6	4.8	1.4
The Bruderheim chondrite[b]	12.5	12.6	12.2	12.6	12.8	13.5	3.9
Iron	52.9	53.0	51.5	52.3	53.3	56.4	17.2

*Chemical composition from: [a]—Imamura *et al.* (1973).
[b]—Baadsgaard *et al.* (1961).

**Averaging (statistical weighting) of the cross sections according to the spectra of nuclear-active particles has been carried out by the formula

$$\bar{\sigma}_i = \frac{\int_{E_1}^{E_2} \bar{\sigma}_i(E)\, F(E)dE}{\int_{E_1}^{E_2} F(E)dE}, \text{ where}$$

$\bar{\sigma}_i(E)$ is excitation function of i-radionuclide, and $\int_{E_1}^{E_2} F(E)dE$ is an integral spectrum of primary or secondary cosmic radiation in energy range of interest.

addition our calculations of radioactivities in meteorites and lunar samples are in quantitative agreement with ones fulfilled by other methods, e.g., Monte Carlo method of Armstrong and Alsmiller (1970) or Barashenkov *et al.* (1972) (see the comparison in reference of Lavrukhina *et al.*, 1973), and thick target method of Kohman and Bender (1967) (see Fig. 2 for ^{53}Mn in iron meteorites; the average galactic cosmic-ray spectrum given in the last reference has been used in the calculations). There is also close agreement with the method of Reedy and Arnold (1972) in the calculations of radioactivities of radionuclides which production cross sections are well known. Besides that to determine the range of validity of our method, a modelling nuclear reactions in isotropically irradiated thick targets has been fulfilled (Lavrukhina *et al.*, 1973). While revolving about two mutually perpendicular axes, an iron sphere, having a radius of 10 cm, was irradiated by a 660 MeV proton beam of Dubna cynchrocyclotron. Cross section of the proton beam was expanded to cover the sphere. As a result of such rotation, the surface of the sphere was irradiated isotropically. The activity of ^{24}Na in thin aluminium plates and ^{52}Mn, ^{48}V, ^{44m}Sc, ^{47}Sc and ^{47}Ca in iron plates placed at various depths along the diameter of the sphere was measured. In Fig. 3

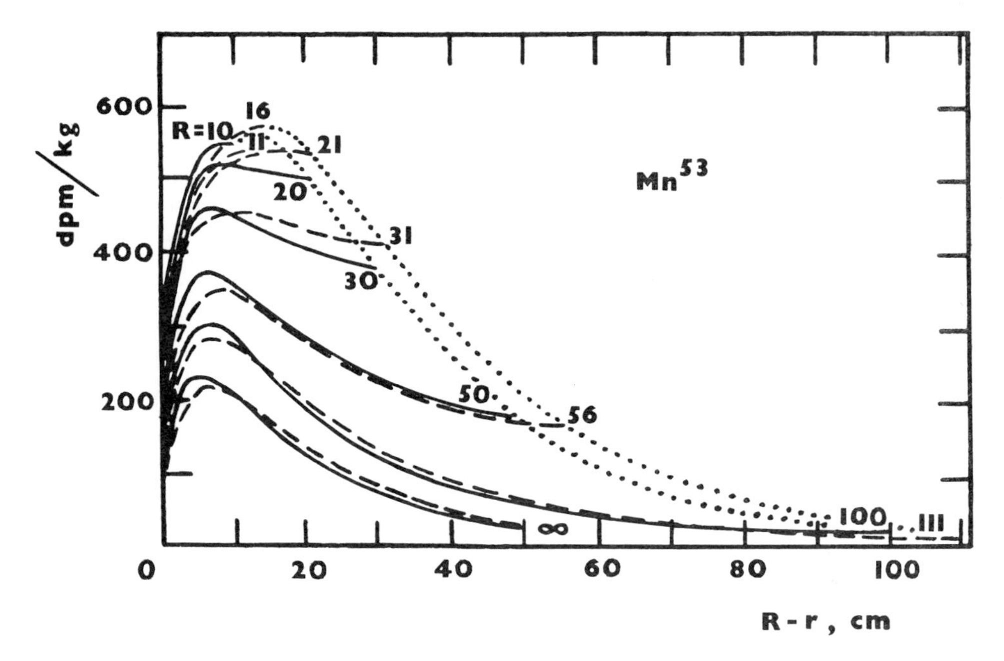

Fig. 2. Comparison of ^{53}Mn depth distributions in iron meteorites with different radii (R) calculated by the analytical method (this work; solid curves) and thick target method (Khoman and Bender, 1967; dashed curves). Dotted curves show ^{53}Mn alterations in the centres of meteorites; r is the distance from the centre.

the experimental results are compared with calculated curves obtained by the analytical method. It is seen that at a depth of $\gtrsim 2$ cm below the surface the activity calculated by the analytical method closely (quantitatively) agrees with the experimental results.

The depth distributions of ^{53}Mn in the Apollo 15 lunar soil core so obtained are presented in Fig. 4. It is seen that curve 1, describing the ^{53}Mn distribution produced by the modern mean modulated flux of galactic cosmic rays, lies below the measured profile of ^{53}Mn by 25–50% (depending upon the depth) in agreement with Imamura *et al.* (1973) and Rancitelli *et al.* (1975), and curve 2 for the maximum flux of galactic cosmic rays recorded near by the earth over the period of the least modulation in April–May, 1965 fits experimental data. According to Eq. (7) the errors of the radioactivity calculations are due to errors in the production cross sections because these errors exceed those of the intensity measurements by an order of magnitude (Vernov *et al.*, 1974). The upper limit of errors of ^{53}Mn cross sections at different energies determined by Eq. (8) is 20–25%. It is the uppermost limit of the errors of the ^{53}Mn radioactivity calculations by the analytical method in cosmic bodies because the errors of the statistical weighted cross sections used by the method (see footnote 2 under Table 3) should be far less than ones of separate values. And besides, as follows from Fig. 3, the analytical method fits experimental data within a standard deviation

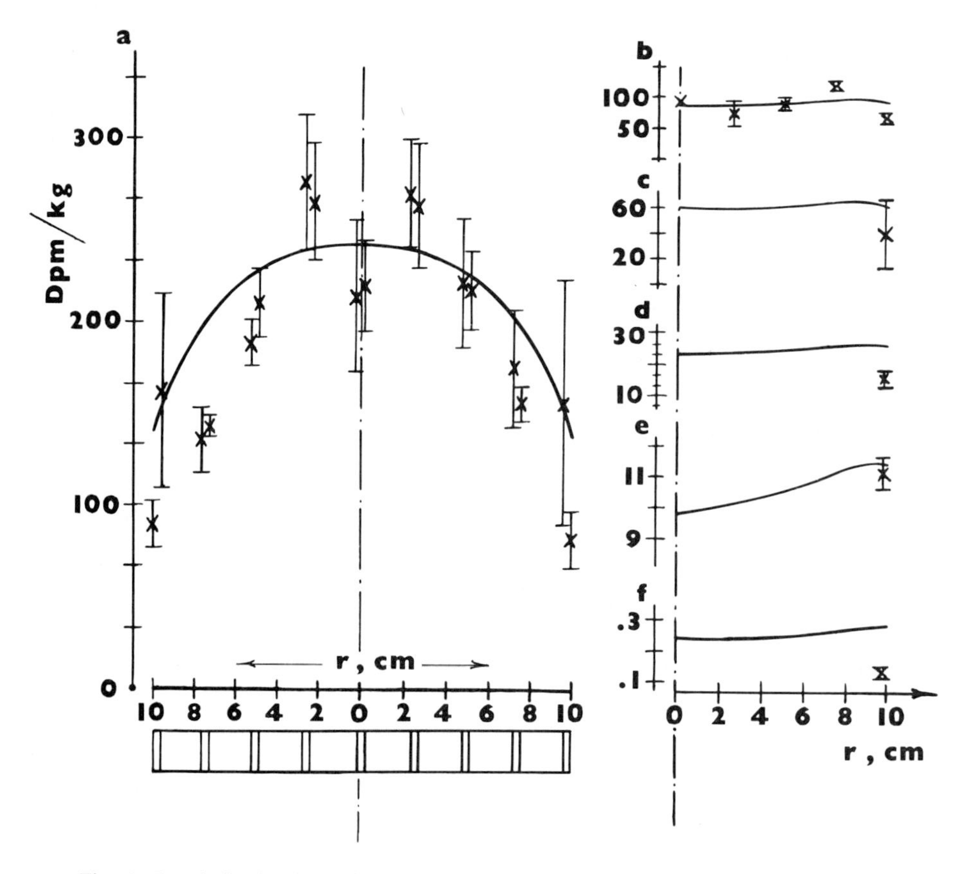

Fig. 3. Depth distributions of radionuclides within isotropically irradiated thick target (iron sphere of R = 10 cm; E_p = 660 MeV; I_0 = 1 proton·cm^{-2}s^{-1}sr^{-1}). Points are experiment; curves are calculation by the analytical method. (a) is ^{24}Na in aluminium plates; indicated errors are standard deviations of five independent experiments; under the graph a scheme of disposition of the aluminium plates along the diameter of the iron sphere is shown; (b), (c), (d), (e) and (f) are ^{52}Mn, ^{48}V, ^{44m}Sc, ^{47}Sc and ^{47}Ca in iron plates, successively; the errors are average deviations of two measurements at symmetrical points of the diameter; r is the distance from the centre.

at a depth of ≳15 g/cm^2 (≳2 cm of iron) below the surface. One can see the same in Fig. 4: curve 2 fits experimental data within a standard deviation at all depths, and the difference between curves 1 and 2 well exceeds a range of calculation errors, totaling ≳100% at a depth of ≳400 g/cm^2. Therefore a valuable conclusion follows from the data of Fig. 4: over the most part of time of accumulation of the observed ^{53}Mn radioactivity (5–6 m.y.) the moon was irradiated by the flux of galactic cosmic rays, corresponding to the least modulated modern flux. According to the ^{26}Al data in the Luna 16 and Luna 20 specimens, it has been found (Vernov and Lavrukhina, 1975; see also Rancitelli *et al.*, 1975) that already a million years ago the moon was bombarded by the mean modulated flux of

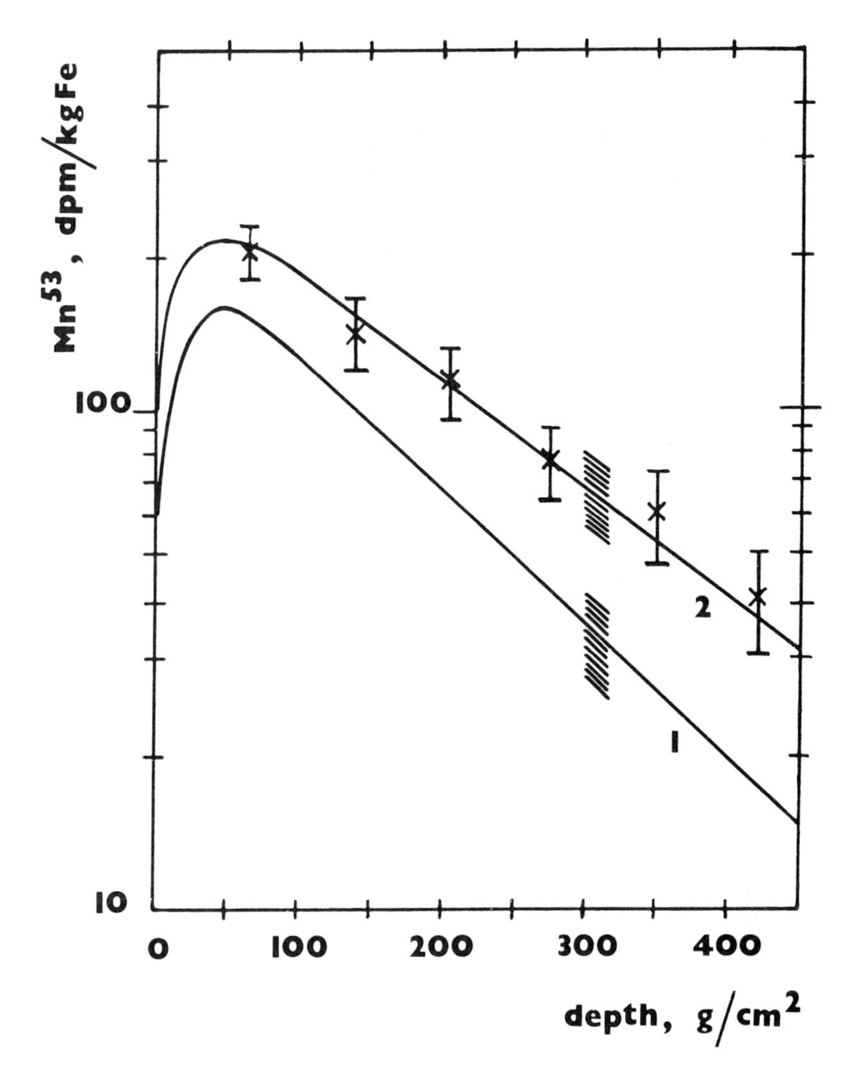

Fig. 4. ^{53}Mn depth distribution in the Apollo 15 lunar soil core 15001–15006. The experimental data are from Imamura *et al.* (1973). Curve 1 is the calculated ^{53}Mn depth profile in the surface layer of the moon using the average spectrum of galactic cosmic rays in 1962 (average intensity per solar cycle). Curve 2 is the same as curve 1 but based on the average galactic cosmic-ray spectrum of April–May, 1965 (maximum intensity in the 19 solar cycle). Calculation errors grow with depth but do not exceed ~20% marked by dashes.

galactic cosmic rays close to the modern average flux. Thus the time of the moon irradiation by the least modulated flux probably covered a period of 2–6 million years ago which coincides with the expected period of reduced solar activity and global ice-age on the earth. It is also in agreement with the results obtained by using experimental data on content of ^{10}Be and ^{26}Al in sea sediment cores (Kocharov *et al.*, 1975). According to the work the intensity of comsic rays was roughly constant or varied slowly for the last 1.5 million years preceded by a significant increase between 2.0 and 3.5 million years ago.

It still remains to prove that the measured radioactivity of ^{53}Mn in the lunar soil was not due to higher intensity of cosmic rays in the Galaxy at that period of time. For this purpose we have analysed the radioactivities of cosmogenic radionuclides in some chondrites with exposure ages $T_{exp} \gtrsim 10^7$ years (Table 4). [Unfortunately, there are no ^{53}Mn data for two chondrites of known orbits: Pribram ($q' = 4.05$ AU) and Lost City ($q' = 2.35$ AU)]. To estimate pre-atmospheric sizes of the chondrites, we have used the ratio of the ^{60}Co radioactivity in bulk meteorite to ^{39}Ar in metallic phase (Fig. 5). This ratio should be recognized as an extremely sensitive and universal indicator of chondrite sizes as its value is practically independent of temporal and space variations of cosmic rays (see the solid and dashed curves in Fig. 5) and varies with chondrite sizes by four orders of magnitude. For the chondrites under consideration R is equal to ~20–30 cm (Table 4, column 6). The eventual calculation of ^{26}Al radioactivity in the chondrites of the indicated sizes enables us to evaluate the orbital extent of each chondrite by the recently developed original method (Lavrukhina and Ustinova, 1972); the aphelion range is $q' \sim 2$–3 AU (Table 4, column 8). Next the production rates of ^{53}Mn in the chondrites have

Table 4. Chondrite characteristics and radioactivity of cosmogenic radionuclides

Chondrite	$T^{x)}_{exp}$, 10^6 year	Fe,$^{x)}$ %	$^{60}Co_{bulk}$, dpm/kg	$^{39}Ar_{Fe+Ni}$, dpm/kg	R, cm	^{26}Al, dpm/kg	q′, AU	^{53}Mn, dpm/kg: Experiment data	Calculation: 1965 spectrum	Calculation: 1962 spectrum
1	2	3	4	5	6	7	8	9	10	11
St. Severin	11	20.2	9.1 ± 2.0 /a/	23.2 ± 1.5 /b/	~30	53.7 ± 8 /c/	2.04	70 ± $10^{xx)}$ /d/	76 ± $8^{xx)}$	58 ± $6^{xx)}$
Peace River	34	22.3	2.1 ± 0.4 /e/	20.3 ± 0.8 /b/	20–30	56.5 ± 3 /f/	2.28	85 ± 8 /g/	93 ± 9	71 ± 7
Ehole	10	27.4	4,8 ± 1.2 /h/	–	20–30 xxx)	57.5 ± 6 /i/	2.38	110 ± 20 /h/	114 ± 12	87 ± 9
Bruderheim	21	22.5	8.5 ± 1.1 /j/	24.9 ± 1.2 /b/	25–30	60 ± 6 /k/	2.41	87 ± 15 /l/	97 ± 10	73 ± 8

x)—from many references in literature; xx)—at ~5 cm from preatmospheric surface; xxx)—by ^{60}Co radioactivity only; /a/—Bibron *et al.* (1974); Tobailem *et al.* (1967); Cressy (1970); /b/—Begemann and Vilcsek (1969); /c/—Bibron *et al.* (1974); Tobailem *et al.* (1967); Marti *et al.* (1969); Fireman (1967); Spannagel and Sonntag (1967); Herpers *et al.* (1969); /d/—Bibron *et al.* (1974); /e/—Shedlovsky *et al.* (1967); /f/—Fireman (1967); Shedlovsky *et al.* (1967); /g/—Shedlovsky *et al.* (1967); Herr *et al.* (1971); Cressy (1964); /h/—Honda and Arnold (1964); /i/—Honda and Arnold (1964); Rowe and Clark (1971); /j/—Honda*et al.* (1961) Fireman (1966); /k/—Fireman (1967); Spannagel and Sonntag (1967); Rowe and Clark (1971); Honda *et al.* (1961); Biswas *et al.* (1963); /l/—Honda and Arnold (1964); Honda *et al.* (1961).

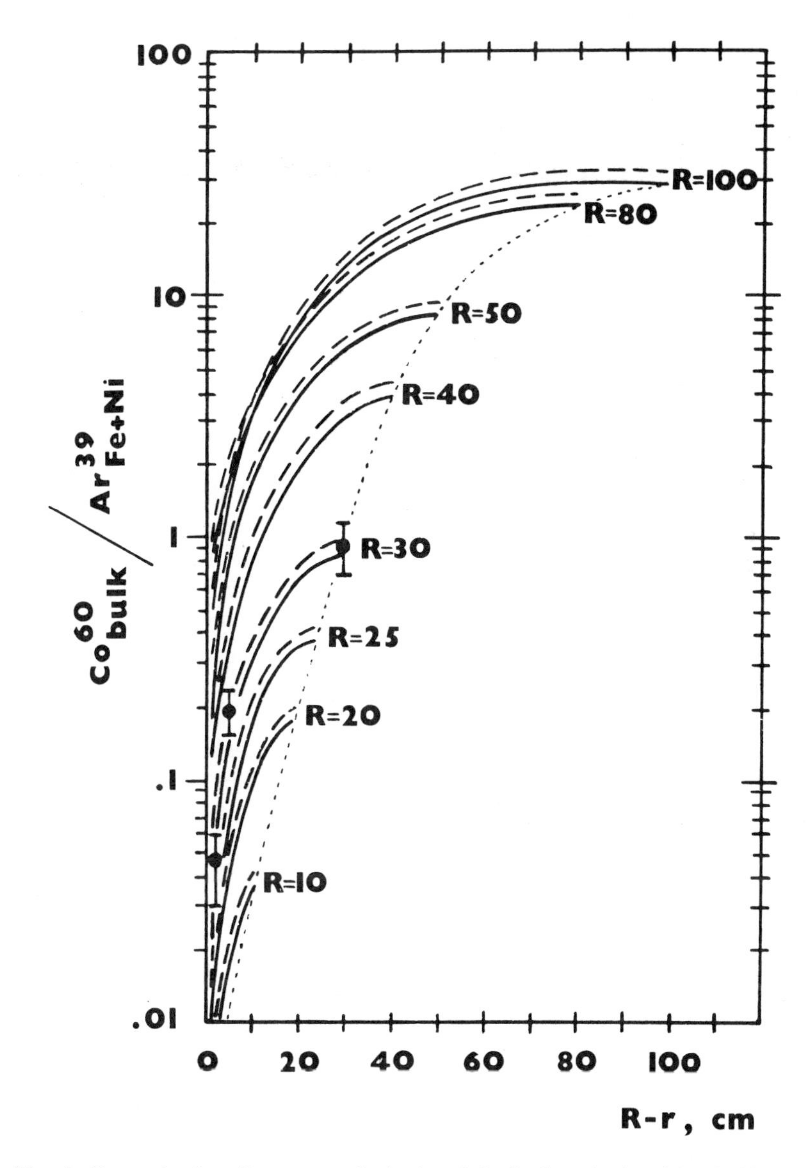

Fig. 5. Determination of pre-atmospheric size of the St. Severin chondrite. Solid curves are the calculation of $^{60}Co_{bulk}/^{39}Ar_{Fe+Ni}$ in the chondrites with different radii R (r is the distance from the center) using the average galactic cosmic-ray 1962 spectrum; dashed curves are the same for the spectrum of April–May, 1965; dotted curve is the $^{60}Co_{bulk}/^{39}Ar_{Fe+Ni}$ variation at the center of chondrites of different sizes; cobalt content in the chondrite is 430 ppm (Bibron *et al.*, 1974). Experimental data (dpm/kg): $^{39}Ar_{Fe+Ni}$—23.2 ± 1.5 (Begemann and Vilcsek, 1969); $^{60}Co_{bulk}$—1.1 ± 0.3 (Bibron *et al.*, 1974) (at ~2 cm from the pre-atmospheric surface); 4.5 ± 0.6 (Bibron *et al.*, 1974) (at ~5 cm from the pre-atmospheric surface); 21 ± 4 (Tobailem *et al.*, 1967) (for the central part of the meteorite); the sample depth is identified by heavy ion track density (Cantelaube *et al.*, 1969).

been calculated taking into account their iron content and using the mean galactic cosmic ray spectra of 1962 and April–May, 1965 (Vernov *et al.*, 1974). Comparison between these results and the ^{53}Mn experimental data (Table 4, column 9–11) indicates that over the production period of the observed present radioactivity of ^{53}Mn in the chondrites, they were exposed to galactic cosmic rays with intensity no larger than the maximum recorded in the earth stratosphere in April–May of 1965 (Vernov *et al.*, 1974), i.e., the intensity of unmodulated galactic cosmic rays in interstellar space was constant over the last 5–6 m.y.

Thus, high ^{53}Mn radioactivity in the lunar soil is just due to the absence of the effective modulation of galactic cosmic rays in the solar system over a period of 2–6 million years ago because of decrease in solar activity. This is the experimental proof of the hypotheses stated at the beginning of our paper, in particular, Fowler's idea of cardinal processes on the sun over that period, as well as accuracy of dating the global ice-age on the earth.

In conclusion it is interesting to note some other opportunities to investigate the problems discussed above. In addition to an analysis of ^{53}Mn radioactivity in long exposure age meteorites it is desirable that the same be carried out in short ($\lesssim$2 m.y.) exposure age meteorites (Bhandari, pers. comm., 1978). In the first case ^{53}Mn was produced by the least modulated spectrum of galactic cosmic rays irrespective or orbit sizes of the meteorities but in the second case ^{53}Mn as ^{26}Al and other long-lived radionuclides should reflect spatial variations of galactic cosmic rays along the meteorite orbits due to the solar modulation mechanism being operative over the last $\lesssim$2 million years. Unfortunately, at present, all the necessary data (see Table 4) to accomplish the complex analysis of the radioactivities do not exist for any of those meteorites.

Acknowledgments—We are very grateful to Prof. N. Bhandari for some valuable comments.

REFERENCES

Armstrong T. W. and Alsmiller R. G. (1970) Calculation of cosmogenic radionuclides in the moon and comparison with Apollo measurements. Preprint ORNL-TM-3267.

Baadsgaard H., Campell F. A., Folinsbee R. E. and Cumming G. L. (1961) The Bruderheim meteorite. *J. Geophys. Res.*, **66**, 3574–3577.

Barashenkov V. S., Sobolevsky N. M. and Toneev V. D. (1972) Accumulation of cosmogenic isotopes in iron meteorites, Preprint JINR P2-6225, Dubna.

Begemann F. and Vilcsek E. (1969) Clorine-36 and argon-39 production rates in the metal of stones and stony-iron meteorites. In *Meteorite Research* (P. M. Millman, ed.) p. 355–362. D. Reidel, Dordrecht.

Bibron R., Leger C., Tobailem J., Yokoyama Y., Mabuchi H. and Bailard N. (1974) Radionuclides produits par le rayonnement cosmique dans la météorite Saint-Séverin. *Geochim. Cosmochim. Acta* **38**, 197–205.

Biswas M. M., Mäyer-Böricke C. and Gentner W. (1963) Cosmic-ray produced ^{22}Na and ^{26}Al activities in chondrites. In *Earth Science and Meteoritics* (J. Geiss and E. D. Goldberg, eds.), p. 207–218. North-Holland, Amsterdam.

Cantelaube Y., Pellas P., Nordemann D. and Tobailem J. R. (1969) Reconstitution de la météorite Saint-Séverin dans l'espace. In *Meteorite Research* (P. M. Millman, ed.), p. 705–713. D. Reidel, Dordrecht.

Cressy P. J. (1964) Cosmogenic radionuclides in stony meteorites. USAEC Report NYC-8924. CIT, Pittsburgh.

Cressy P. J. (1970) Multiparameter analysis of gamma radiation from the Barwell, St. Séverin and Tatlith meteorites. *Geochim. Cosmochim. Acta* **34**, 771–779.

Davis R. and Evans J. C. (1974) Report on the Brookhaven solar neutrino experiment. *Proc. 6th Leningrad Intern. Seminar on Cosmophysics,* (G. E. Kocharov and V. A. Dergachev, eds.), p. 91–110. LINP Press, Leningrad.

Dilke F. W. W. and Gough D. O. (1972) The solar spoon. *Nature* **240**, 262–264; 293–294.

Fireman E. L. (1966) Neutron exposure ages of meteorites. *Z. Naturforschg.*, **21a**, 1138–1146.

Fireman E. L. (1967) Radioactivities in meteorites and cosmic-ray variations. *Geochim. Cosmochim. Acta* **31**, 1691–1710.

Fowler W. A. (1972) What cooks with solar neutrinos? *Nature* **238**, 24–26.

Herpers U., Herr W. and Wölfe R. (1969) Evaluation of ^{53}Mn by (n, γ)-activation, ^{26}Al and special trace elements in meteorites by γ-γ-coincidence techniques. In *Meteorite Research* (P. M. Millman, ed.), p. 387–396. D. Reidel, Dordrecht.

Herr W., Herpers U. and Wölfe R. (1971) Spallogenic ^{53}Mn ($T \sim 2 \cdot 10^6$ y) in lunar surface material by neutron activation. *Proc. Lunar Sci. Conf. 2nd*, p. 1797–1802.

Honda M. and Arnold J. R. (1964) Effects of cosmic rays on meteorites. *Science* **143**, 203–212.

Honda M., Umemoto S. and Arnold J. R. (1961) Radioactive species produced by cosmic rays in Bruderheim and other stone meteorites. *J. Geophys. Res.* **66**, 3541–3546.

Imamura M., Finkel R. C. and Wahlen M. (1973) Depth profile of ^{53}Mn in the lunar surface. *Earth Planet. Sci. Lett.* **20**, 107–112.

Kocharov G. E. (1975) Nuclear astrophysics of the sun. *Proc. 14th Int. Cosm. Ray Conf.* **11**, 3521–3549. Munich.

Kocharov G. E., Dergachev V. A. and Gordeichik N. I. (1975) The rate of production of cosmogenic isotopes in the past and the solar activity. *Proc. 14th Int. Cosm. Ray Conf.* **3**, 1009–1012, Munich.

Kohman T. P. and Bender M. L. (1967) Nuclide production by cosmic rays in meteorites and on the moon. USAEC Report No. NYO-844-72, Pittsburgh.

Lavrukhina A. K. (1972) *Nuclear Reactions in Cosmic Bodies.* Nauka, Moscow.

Lavrukhina A. K. and Ustinova G. K. (1967) On distribution of the intensity of nuclear-active component of cosmic rays in meteorites. *Astronomichesky Journal* **44**, 1081–1086.

Lavrukhina A. K. and Ustinova G. K. (1971) Solar proton medium flux constancy over a million years. *Nature* **232**, 462–463.

Lavrukhina A. K. and Ustinova G. K. (1972) Cosmogenic radionuclides in stones and meteorite orbits. *Earth Planet. Sci. Lett.* **15**, 347–360.

Lavrukhina A. K., Revina L. D., Malyshev V. V. and Satarova L. M. (1963) Spallation of Fe nuclei with 150 Mev protons. *Soviet Phys. JETP* **17**, 960–963.

Lavrukhina A. K., Ustinova G. K. Ibraev T. A. and Kuznetsova R. I. (1969) Cosmic-ray produced radioactivity in the moon and meteorites and origin of meteorites. In *Meteorite Research* (P. M. Millman, ed.), p. 227–245. D. Reidel, Dordrecht.

Lavrukhina A. K., Ustinova G. K., Malyshev V. V. and Satarova L. M. (1973) Modelling nuclear reactions in isotropically irradiated thick targets. *Atomnaya Energiya*, **34**, 23–28.

Marti M., Shedlovsky J. P., Lindstrom R. M., Arnold J. R. and Bhandary N. G. (1969) Cosmic-ray produced radionuclides and rare gases near the surface of Saint Séverin meteorite. In *Meteorite Research* (P. M. Millman, ed.), p. 246–266. D. Reidel, Dordrecht.

Rancitelli L. A., Fruchter J. S., Felix W. D., Perkins R. W. and Wogman N. A. (1975) Cosmogenic isotope production in Apollo deep-core samples. *Proc. Lunar Sci. Conf. 6th*, p. 1891–1900.

Rayudu G. F. (1964) Formation cross sections of various radionuclides from Ni, Fe, Si, Mg, O and C for protons of energies between 130 and 400 Mev. *Can. J. Chem.*, **42**, 1149–1154.

Rayudu G. F. (1968) Formation cross sections of various radionuclides from Ni, Fe, Si, Mg, O and C for protons of energies between 0.5 and 2.0 Gev. *J. Inorg. Nucl. Chem.* **30**, 2311–2315.

Reedy R. C. and Arnold J. R. (1972) Interaction of solar and galactic cosmic ray particles with the moon. *J. Geophys. Res.* **77**, 537–555.

Row M. W. and Clark R. S. (1971) Estimation of error in the determination of ^{26}Al in stone meteorites by indirect γ-ray spectrometry. *Geochim. Cosmochim. Acta* **35**, 727–730.
Shedlovsky J. P., Cressy P. J. and Kohman T. P. (1967) Cosmogenic radioactivities in the Peace River and Harleton chondrites. *J. Geophys. Res.* **72**, 5051–5058.
Shklovsky I. S. (1975) *The stars.* Nauka, Moscow.
Spannagel C. and Sonntag C. (1967) Cosmic-ray produced activities in chondrites. In *Radioactive Dating and Methods of Low-Level Counting,* p. 231–238. IAEA, Vienna.
Tobailem J., David E. and Nordemann D. (1967) Radioactivité induité par le rayonnement cosmique dans la météorite Saint Séverin. In *Radioactive Dating and Methods of Low-Level Counting,* p. 207–213. IAEA, Vienna.
Vernov S. N. and Lavrukhina A. K. (1975) Primary cosmic radiation on the surface of the moon. In *Cosmochemistry of the Moon and Planets* (A. P. Vinogradov, ed.), p. 542–546. Nauka, Moscow.
Vernov S. N., Charakhchyan A. N., Stojkov Yu. I. and Charakhchyan T. N. (1974) 11-year cycle of galactic cosmic rays in the interplanetary space. Preprint No. 107. Phys. Inst. Acad. Sci. USSR, Moscow.
Vinogradov A. P., Lavrukhina A. K., Gorin V. D. and Ustinova G. K. (1974) Cosmogenic ^{26}Al and ^{22}Na in Luna 16 drill soil core. In *Lunar Soil from Mare Fecunditatis* (A. P. Vinogradov, ed.), p. 410–416, Nauka, Moscow.
Wänke H. (1974) Chemistry of the moon. *Fortschritte der chemischen Forschung,* **44**, 115–154.
Williamson C. F., Boujot J. P. and Picard J. (1966) Tables of range and stopping power of chemical elements for charged particles of energy 0.05 to 500 MeV. Rapport CEA-R3042, Centre d'etudes nucleaires de Saclay.
Yavnel' A. A. and Dyakonova M. I. (1958) The chemical compositions of meteorites. *Meteoritica,* **15**, 136–146.
Zeuner F. E. (1959) *The Pleistocene Period.* Hutchinson Sci. and Techn., London.

Proc. Lunar Planet. Sci. Conf. 9th (1978), p. 2415–2431.
Printed in the United States of America

Studies of the charge and energy spectra of the ancient VVH cosmic rays

R. K. BULL, P. F. GREEN and S. A. DURRANI

Department of Physics, University of Birmingham
Birmingham B15 2TT, England

Abstract—Measurements of track etch rates as a function of residual range have been used to determine the charge of heavy track-forming particles incident on meteorite crystals. By using a lower limit etch rate of 30 μm.hr^{-1} to differentiate VVH ($Z > 30$) galactic cosmic ray tracks from VH ($20 \leq Z \leq 28$) tracks, the VVH track density ρ_{VVH} at a number of depths within the Shalka meteorite has been examined. The resultant ρ_{VVH} versus depth profile is consistent with the assumption of an energy spectrum for the VVH cosmic rays having the same index as that which is believed to describe the ancient VH spectrum.

Preliminary data on the ancient cosmic-ray charge spectrum between $Z = 31$ and 47 show no gross differences from contemporary charge spectrum results.

INTRODUCTION

Extraterrestrial mineral crystals contain a record of their irradiation by heavy cosmic ray particles extending back many millions of years.

The charge and energy spectra of the heavy cosmic rays provide information on the production, acceleration and propagation mechanisms for these particles. Cosmic ray studies by the fossil track technique allow the possibility of time variations in charge and/or energy spectra to be explored. Also, because of the long exposure times for many meteoritic and lunar rocks, these entities may provide a unique means of detecting very rare cosmic ray species.

A number of determinations of the ancient iron-group galactic-cosmic-ray energy spectrum have been made by measuring variations in track density as a function of depth within meteorites (Cantelaube *et al.*, 1967; Lal *et al.*, 1969; Maurette *et al.*, 1969) and lunar rocks (Bhandari *et al.*, 1971; Walker and Yuhas, 1973; Hutcheon *et al.*, 1974). In the region above about 500 MeV/n results have been obtained which are consistent with an energy spectrum of the form:

$$\frac{dN}{dE} = (A + BE)^{-\gamma}$$

The index γ lies in the range 1.9 to 2.5, in agreement with most measurements on the contemporary VH flux.

Evaluation of the VVH energy spectrum is a more difficult problem in view of the very low abundance of these ultra-heavy particles. There is, however, some evidence that there are no gross differences between the energy spectra of ancient galactic VH and VVH nuclei (Bhandari and Padia, 1974).

The identification of different charge groups on the basis of fossil track measurements has proved more difficult. Most attempts so far have utilized the total etchable range concept for particle identification; and while a certain degree of success has been achieved (Plieninger *et al.*, 1972, 1973) for the VH group, extension of this technique to the VVH group suffers from some problems. Total etchable ranges for ions of $Z \gg 30$ will be hundreds of microns, and long etch times are necessary in order to fully reveal such tracks. Also, Price *et al.* (1973) have shown that track etch rates do not fall sharply to the bulk etch velocity at some critical ionization level: the validity of the total etchable range concept is thus questionable.

Price *et al.* (1968b) suggested that track etch rate measurements may provide a useful means of charge identification of cosmic ray tracks in minerals.

Recently Krätschmer and Gentner (1976) have developed a method of track etch rate measurement to deduce VH group abundances. In this paper we describe a similar method in which variations in track etch rate as a function of residual range are used to separate VVH ($Z > 30$) from VH ($20 \leq Z \leq 28$) tracks and to resolve some details of the charge spectrum within the VVH group.

THE METEORITE SAMPLES

Most of the measurements described below were made on tracks in hypersthene crystals from the Shalka meteorite. In earlier papers (Bull and Durrani, 1976; 1977) the pre-atmospheric radius (assuming a spherical shape, zero space erosion) was deduced to be 20 ± 2 cm from VH track densities, and pre-atmospheric depths for each of the sampling locations were calculated. Also, from a comparison of Fe ion track-in-track and track-in-cleavage lengths with fossil track length distributions it was suggested that the degree of fossil track shortening was relatively small for these samples.

A number of cosmogenic rare-gas studies have been conducted on Shalka (Eberhardt *et al.*, 1965; Megrue, 1968; Müller and Zähringer, 1969). Exposure ages around 20 M. yr are obtained in all cases. Concordancy between ^3He, ^{21}Ne and ^{38}Ar ages suggest little diffusive loss of ^3He.

In addition, hypersthene crystals were removed from a fusion-crusted fragment of Patwar. These show VH track densities of 1.0 (± 0.07) $\times 10^7$ tracks cm^{-2} from both optical and scanning electron microscope (SEM) measurement. Price *et al.* (1967) made a detailed study of tracks in Patwar, and on the basis of their results a track density of 10^7 tracks cm^{-2} would correspond to a pre-atmospheric sample depth of 3.6 cm (for meteorite radius $R_o = 16.5$ cm, no erosion) or 2.7 cm ($R_o = 14$ cm, erosion rate 0.1 cm per M. yr). However, Price *et al.* (1967) used an assumed exposure age of 50 M. yr for Patwar. Using the measured age of 63 M. yr (Eberhardt and Stettler, 1970), a pre-atmospheric depth (zero-erosion case) of about 4 cm is obtained.

The low uranium content in Shalka hypersthene crystals ($\ll 1$ ppb) and anisotropic angular distributions of fossil tracks (Bull, 1975) suggest that fission tracks do not present an important contribution to the total track density. In Patwar hypersthene, also, the fission track contribution is probably negligible.

Shalka and Patwar provide a range of pre-atmospheric sampling depths from 4 to 13 cm, and VVH track density measurements at these depths should allow constraints to be placed on the VVH energy spectrum at around 1 GeV/n.

EXPERIMENTAL PROCEDURES

Crystals were mounted in epoxy, ground and polished. Etching of both fossil and calibration ion tracks was performed in 60% (by weight) NaOH solution, boiling under reflux. Samples were cleaned in an ultrasonic bath with a dilute solution of Dowfax surfactant. Etching was carried out in 30 minute stages, the etchant being renewed after each such interval.

The tracks were measured with an optical microscope having a total magnification of 600×. A micrometer eyepiece attachment and calibrated fine focus drum were used. The estimated precision achieved was ±0.5 μm in the object plane and ±1.0 μm normal to this surface.

The V_T-R method

This method consists, essentially, in measuring track etch rate as a function of residual range for both fossil and calibration tracks. This is achieved by measuring the tracks after a number of etching stages, the tracks being re-located by a mapping technique.

The initial cone length L yields an etch rate $V_T = L/t$, which applies to a mean residual range R given by $R_t - L/2$ where R_t is the total length of the fully etched track.

Fossil tracks formed by 'stopping' particles yield a track etch rate V_T and a residual range R. In such cases an estimate of the particle charge may be made. For 'non-stopping' particles (i.e., those for which etching proceeds towards the high energy end of the track) the etch rate measurement allows a lower limit to be placed on the charge.

Calibration was carried out by irradiating polished Shalka and Patwar crystals with Ca, Ti, Fe, Ni, Cu, Kr, Xe and U ions at various energies between 1.04 and 9.6 MeV/n. In some cases, in order to examine etch rates over a wider spread of residual ranges, irradiated samples were ground and polished at a small angle to the original surface prior to etching (this is the L − R plot method of Price *et al.*, 1973).

Some of the raw calibration data for Shalka are shown in Fig. 1. In order to compute V_T versus R profiles for any ion, it is necessary to relate V_T to some fundamental track-formation parameter J such that V_T is a monotonically increasing function of J. Following Price *et al.* (1973), we use the primary ionization rate given by

$$J = \text{const} \times \frac{Z^2_{\text{eff}}}{\beta^2} \ln \frac{\beta^2}{(1-\beta^2)} + K - \beta^2 \quad (1)$$

where β is the particle velocity relative to that of light, K is a constant and Z_{eff} is the effective charge given by

$$Z_{\text{eff}} = Z\,[1 - \exp(-130\,\beta/Z^{2/3})] \quad (2)$$

where Z is the atomic number of the particle.

The constants in Eq. (1) are calculable only for a stopping medium of hydrogen. We adopt here the usual practice of calculating J in arbitrary units and using K as a fitting parameter.

Price *et al.* (1968a), have selected a K value of 9.7 on the basis of comparisons between etchable and calculated ranges for low energy I, Br and Fe ions incident on hypersthene. In this analysis we have plotted V_T against $J(\beta,K)$ for a range of

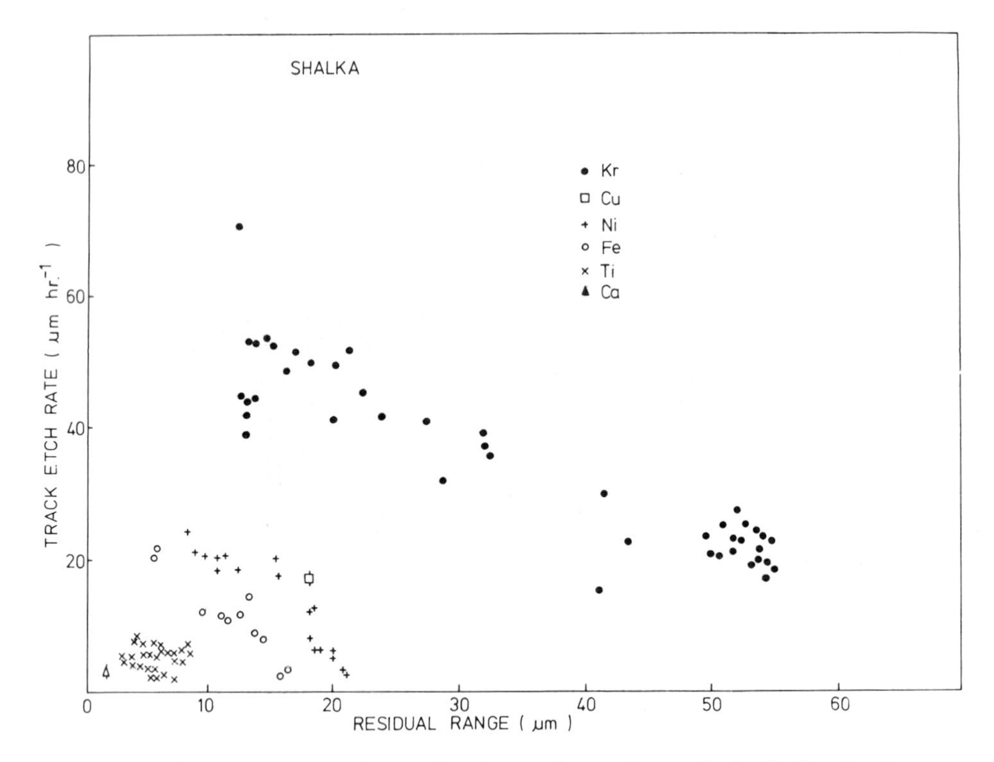

Fig. 1. Some raw track-etch-rate data for accelerator ion tracks in Shalka. Tracks measured after a 30 min etch, to yield initial length L. Longer etch times reveal range R_t. Track etch rate = 2 L μm hr^{-1}. This is plotted at an average residual range given by $R = R_t - L/2$. For Ca and Cu, data points represent an average over many tracks from a monoenergetic irradiation.

K values between 8.5 and 10.5, using Eqs. (1) and (2) and the range-energy code of Henke and Benton (1967). In each case a third order polynomial

$$V_T = \sum_{i=0}^{3} a_i J^i$$

was fitted to the data. The K value for which the smallest least squared deviation was obtained was selected; this was found to be 8.9. Figure 2 shows V_T versus J for this K value. The experimental track etch-rate data used here represent average values for each range interval rather than raw data.

It became apparent during the calibration experiments that etch rates obtained for Ca and Ti did not fit in with those obtained for other ions, being considerably higher than those pertaining to high-energy portions of Fe and Ni tracks where the same J values hold. Also, in the primary ionization region represented by U ions, a saturation or at least a considerable flattening-off in the V_T versus J curve

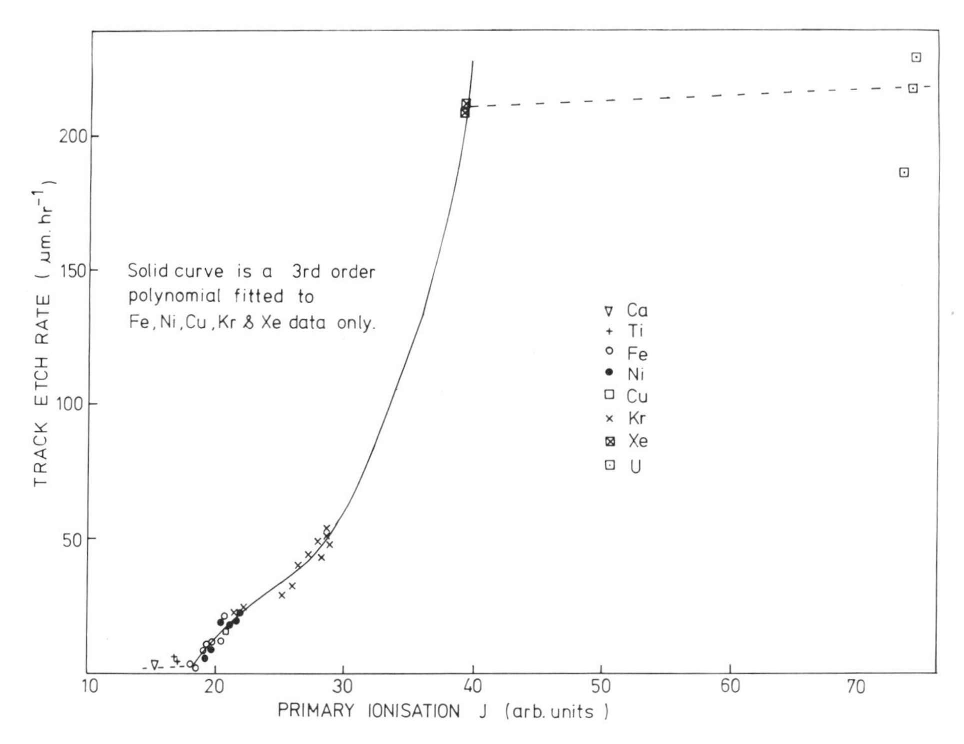

Fig. 2. Track etch rate versus primary ionisation for K = 8.9. The solid line represents a third order polynomial fitted to the Fe, Ni, Cu, Kr and Xe data.

is evident. For the purpose of selecting K and curve fitting, only the Fe, Ni, Cu, Kr and Xe data were used.

The response curve is used to compute V_T versus R profiles for a number of ions. From Fig. 3 it may be seen that the fit to the experimental data is quite good. Because of the anomalous behaviour of Ca and Ti ion tracks and the provisional nature of the U data, V_T versus R profiles for $Z < 24$ and $Z > 54$ are not likely to be reliable. However, this does not affect the results discussed below.

It is clear from Fig. 3 that selection of fossil tracks with etch rates >30 μm hr^{-1} effectively excludes all VH group tracks. More details of our calibration procedures will be published separately (Green *et al.*, 1978).

Annealing

Thermal annealing of fossil tracks during ~20 M. yr could affect their track etch rates and thus necessitate modification of the $V_T - R$ curves computed to fit calibration data. We have not made a correction to allow for annealing for the following reasons:

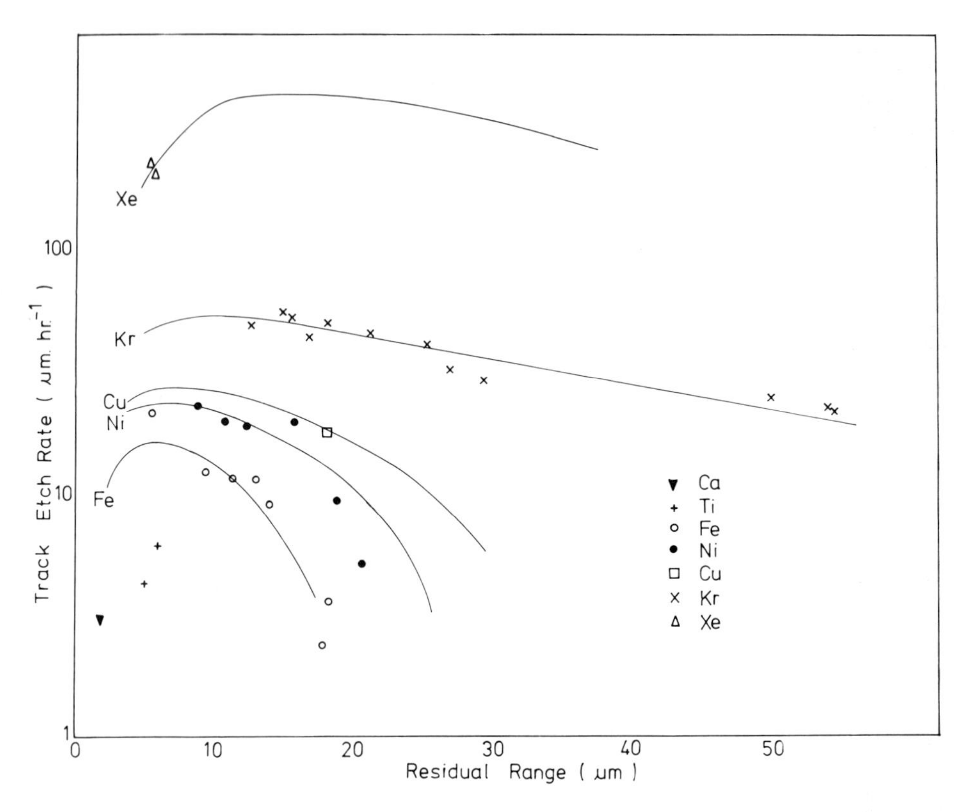

Fig. 3. Track etch rate versus residual range profiles calculated using the calibration curve of Fig. 2 and the range-energy code of Henke and Benton (1967). The data points represent average values derived from the raw track-etch-rate data.

1. The most prominent peaks in broad total-track-length distributions for fossil VH tracks in Shalka lie at ~12–15 μm compared to ~14–17 μm for fresh Fe ion tracks (Bull and Durrani, 1976; 1977). Thus, annealing effects on VH tracks appear to be small. Figure 4 shows V_T versus R data for all fossil tracks >1 μm long (and therefore predominantly VH group tracks) appearing in a careful scan over a small area of a crystal from Shalka. Whilst the spread in data points is large, the calibration Fe profile (solid line) passes through the most densely populated region of the graph and the most straightforward interpretation is that little annealing of Fe group tracks has taken place.

It is possible to argue (Dartyge *et al.*, 1977) that the entire spread in total track lengths is primarily produced by variable degrees of annealing acting on Fe tracks. This reasoning, applied to the Shalka length distributions, would imply shortening by >70% for some Fe tracks. Although the cosmic-ray abundance of elements just below Fe may not be very high,

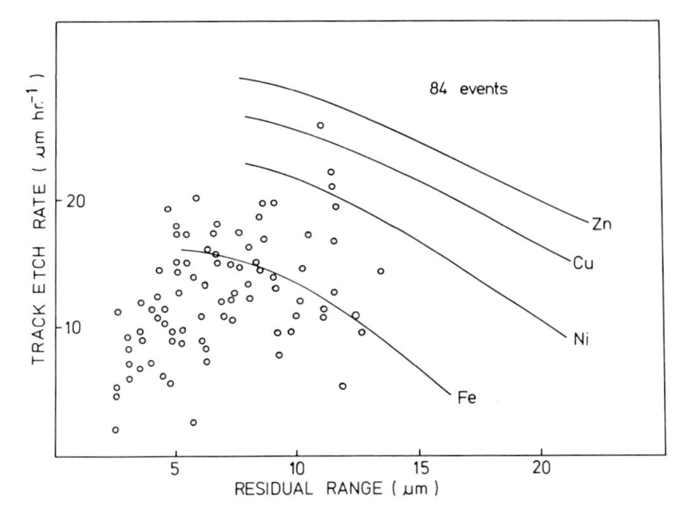

Fig. 4. V_T versus R data for 84 fossil tracks measured in a scan of a small area of a crystal from Shalka, location 8. The solid lines are calculated V_T versus R profiles.

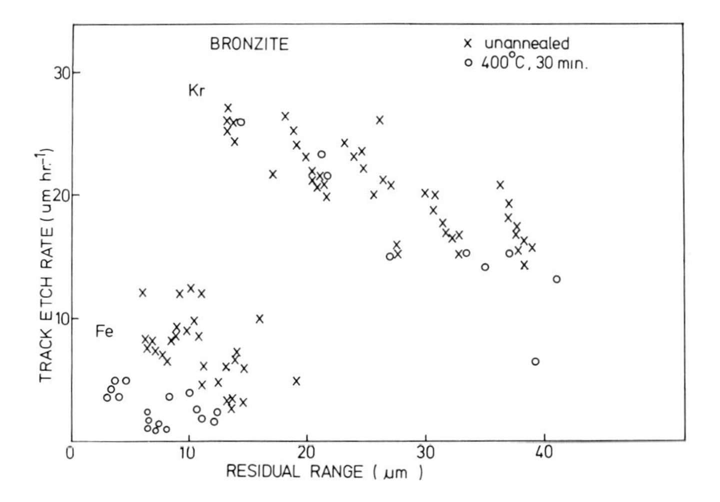

Fig. 5. The effects of annealing on the etch rates of Kr and Fe tracks in a terrestrial bronzite. The annealing step has a much larger effect on the Fe tracks than on the Kr tracks.

fragmentation of Fe nuclei during their passage through ~6–10 cm of meteoritic material is sufficient to explain the observed number of short tracks corresponding to nuclei just below iron in charge (Bull and Durrani, 1976). Whilst we cannot rule out the possibility of a contribution to this track length region from annealed Fe tracks, this effect does not seem to be a large one.

2. The rate of the annealing process seems to depend on the degree of track damage present initially, heavy ion tracks being more resistant than those formed by lighter ions (Maurette, 1970).

   Figure 5 shows L − R plot data for Kr and Fe ion tracks in bronzite, both unannealed and annealed for 30 minutes at 400°C. This annealing step is sufficient to reduce Fe ion track etch rates by ~50%, whereas Kr ion track etch rates are almost unaltered except at the highest residual ranges. Thus, the presence of small degrees of fading in VH group tracks need not imply any significant change in the etching properties of VVH tracks particularly if, as is the case in this work, attention is focussed on the more highly damaged (and therefore high etch rate) regions of these tracks.

   The two lines of evidence given above, when taken together, provide plausible but by no means conclusive evidence against significant etch-rate reductions along VVH tracks. The nature of the annealing process which takes place at low temperatures over very long time scales may be quite different from that observed in laboratory experiments. Clearly this is an area requiring further study.

   Similar arguments to those outlined above can be applied to the Patwar samples, and it may also be noted that Krätschmer and Gentner (1976) found no necessity for annealing corrections in their work on tracks in Patwar feldspars.

Charge resolution

The spread in initial cone lengths for monoenergetic ions incident on Shalka hypersthene defines a charge resolution of ~±1 charge unit. However, there is a small variation in the mean of the track-etch-rate distributions (possibly due to variations of response with crystal plane) from one crystal to another and, since it is not possible to perform a calibration experiment on every crystal from which fossil track data are extracted, the use of an average calibration curve leads to a worsening of the charge resolution in practical cases. Also, since the dip angle of fossil tracks is not known independently, calculation of the total track lengths relies on the relatively inaccurate measurement of the depth of the tip of the etch-cone. An estimate of the true charge resolution of ~±2 units is probably reasonable. Note from Fig. 4 that few data points for VH tracks (predominantly Fe) lie beyond the Ni profile, thus tending to support this estimate, although uncertainties about the annealing question prevent any firm conclusions on the charge resolution being made from this plot.

Scanning

The crystal sections were etched for 30 minutes and then scanned at a magnification of 600×. In order to obviate the necessity for measuring the total length of large numbers of VH tracks, a projected-length threshold of 13.4 μm was imposed (corresponding to 100 micrometer units). This allowed rapid scanning, although some steeply inclined VVH tracks with etch rates above the 30 μm hr^{-1} limit are lost by this method. This fraction is, however, readily calculable. Of the tracks selected for accurate measurement only those with total lengths >15 μm were assigned to the VVH ($Z > 30$) group and relocated after longer etching times.

Results and Discussion

VVH energy spectrum

The density, ρ_{VVH}, of tracks with initial etched lengths >15 μm, is shown in Fig. 6 and Table 1 as a function of pre-atmospheric sample depth. Recent calibration work indicates that track etch rates in Patwar hypersthene are ~50% higher than in Shalka hypersthene and the acceptance threshold for assignment to the VVH group has been adjusted accordingly for this meteorite. The Patwar point has been normalized to an exposure age of 20 M.yr.

Also shown on Fig. 6 are curves representing calculated ρ_{VVH} versus depth profiles for two different values of the in-space ratio of VVH to VH ions. These are adapted from the work of Fleischer *et al.* (1967) and are calculated for Rb ions assuming a VH type of energy spectrum (index $\gamma = 2.2$) and a fragmentation parameter, for VVH to VH nuclei, of 0.25.

In order to compare these calculated profiles with experimental data it is necessary to consider the value of the etchable range ΔR appropriate to these VVH ions. In the procedure described above, only tracks which etched out to a cone length of >15 μm in 30 minutes were assigned to the VVH group. If the sample surface intersects a latent damage trail at a distance R_t from the low energy end of the trail, such that $R_t \leq 15$ μm, then the etched track formed at this point cannot exceed a length of 15 μm; this defines R_{tMIN}. At large ranges the damage trail will be intersected at a point where the local V_T value is too low for a track of length >15 μm to result, thus at R_{tMAX}, $V_T (R_{tMAX}) = 30$ μm hr^{-1} (for the standard etch time of 30 minutes). The length of trajectory over which an identifiable VVH track results is given by $\Delta R = R_{tMAX} - R_{tMIN}$. $V_T - R_t$ curves for each ion are readily obtained from the $V_T - R$ profiles such as those shown in Fig. 3 (note that $R_t = R + L/2$) and ΔR values may be calculated. The interpretation of this parameter ΔR has been discussed in some detail for the case of slowing-down VH group solar-flare particles by Blanford *et al.* (1975). This analysis is adequate where all tracks formed when the surface intersects the latent damage trail between R_{tMAX} and R_{tMIN} are counted. However the use of the projected length limit S, introduced to facilitate rapid scanning, imposes restrictions on the number of tracks accepted. For tracks of length L, only those with dip angles δ, such that $\sin \delta < S/L$, will be counted. If an isotropic angular distribution of tracks is assumed (a reasonably good approximation since we are averaging over a large number of crystals), the fraction f of tracks of length L

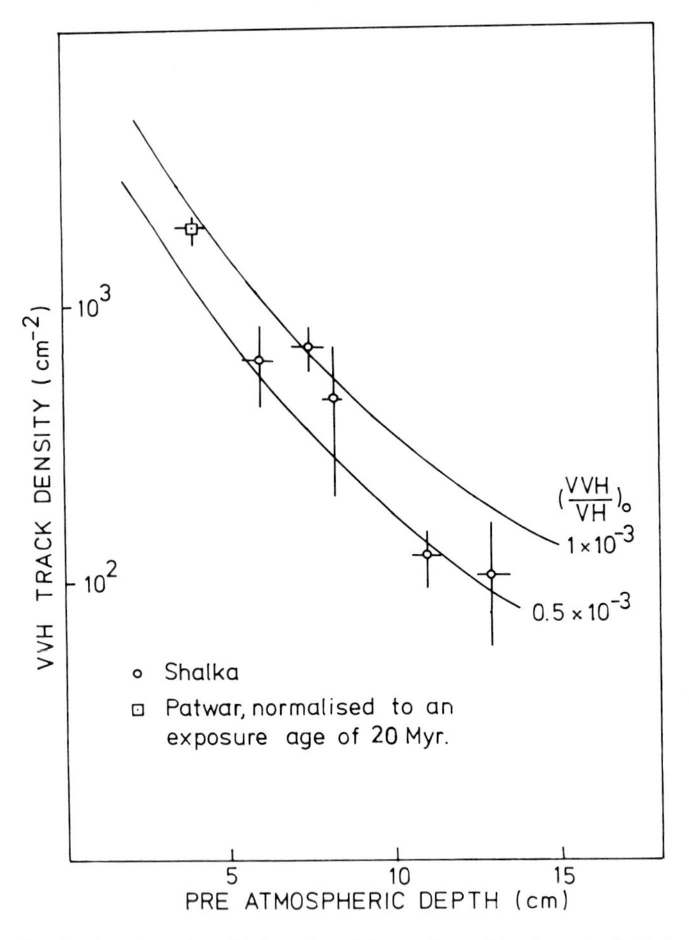

Fig. 6. The density of tracks with lengths >15 μm after a 30 min etch, (>23.5 μm for Patwar) as a function of pre-atmospheric depth. The solid lines are calculated track density-depth profiles adapted from Fleischer *et al.* (1967), for two different values of the in-space ratio $(VVH/VH)_o$.

counted is given by $f = 1 - \sin \delta_{lim}$ where $\delta_{lim} = S/L$. For any ion a numerical integration of this f(L) along the $V_T - R_t$ profile between R_{tMIN} and R_{tMAX} must be carried out. An effective etchable range ΔR_{eff} is thus obtained from

$$\Delta R_{eff} = \int_{R_{tMIN}}^{R_{tMAX}} f \cdot dR$$

(note that if $f = 1$, we recover the simple expression $\Delta R = R_{tMAX} - R_{tMIN}$). Ideally ΔR_{eff} would be computed for each VVH species and weighted with the abundance of that species in order to obtain some mean value $\overline{\Delta R}_{eff}$ appropriate to the VVH group. As a first approximation, however, we have taken a rather

Table 1. Track density data.

Sample	Pre-atmospheric depth* (cm)	No. of VVH tracks measured**	VVH track density*** (cm^{-2})	VH track density (10^6 cm^{-2})†	ρ_{VVH}/ρ_{VH}
Patwar	4	67	5600 (±700)	10 (±1)	5.6 (±0.9) × 10^{-4}
Shalka 1	6	8	600 (±200)	1.20 (±0.1)	5.0 (±1.7) × 10^{-4}
Shalka 3	8.2	3	500 (±300)	0.75 (±0.15)	6.7 (±4.2) × 10^{-4}
Shalka 4	11	22	130 (±30)	0.61 (±0.10)	2.1 (±0.6) × 10^{-4}
Shalka 5	13	4	100 (±50)	0.32 (±0.07)	3.1 (±1.7) × 10^{-4}
Shalka 8	7.5	32	700 (±100)	1.09 (±0.10)	6.4 (±1.1) × 10^{-4}

*Based on VH group track densities.
**No. of tracks >15 μm long after 30 min etch time for Shalka.
No. of tracks >23.5 μm long after 30 min etch time for Patwar.
***1-σ errors calculated from counting statistics.
†VH track densities after 30 minutes of etching. Slightly higher values obtained after 1–2 hours were used in the sample-depth calculations (see Bull and Durrani, 1976).

arbitrary 'typical' VVH ion as being represented by Rb ($Z = 37$). This ion was used for the calculation of ρ_{VVH} versus depth profiles by Fleischer *et al.*, (1967) who took ΔR for Rb to be 35 μm. Using our calibration data we obtain $\Delta R_{eff} = 5.5$ μm for a projected length threshold of 13.4 μm and an etch rate limit of 30 μm $hr.^{-1}$. The curves of Fleischer *et al.* (1967) are therefore scaled down by a factor of 5.5/35. Fleischer *et al.* (1967) gave depth profiles for the extreme values of the sample orientation angle β of 0 and $\pi/2$ (β is the angle between the normal to the sample surface and the radius vector of the meteorite). We have chosen a profile intermediate between these two extremes.

The calculated curves apply to a meteorite radius of 20 cm and, since the Patwar point should be referred to a meteorite radius of 16.5 cm (if erosion is neglected; Price *et al.*, 1967), the comparison here is not strictly valid. However at shallow depths the theoretical profiles for different meteorite radii tend to merge into each other.

The data follow the trend of the calculated curves fairly well and lend support to the assumption that VVH and VH spectra are roughly parallel in the energy region (~1 GeV/n) pertaining to these depths. The best estimate of the ratio (VVH/VH) from our results is 8(±2) × 10^{-4}, where VVH refers to $Z > 30$.

In view of our estimated charge resolution of ~ ±2 charge units, some overspill from the abundant element Zn ($Z = 30$) into the $Z > 30$ region is likely. However, it should be noted that the calculated V_T profile for Zn approaches the etch rate cut-off of 30 μm hr^{-1} only over a very limited region (see Fig. 8) and the number of Zn tracks which can spill over into the $Z > 30$ group is quite small. Analysis of the initial (30 minute etch time) cone lengths shows no peak near to the 15 μm cut-off which could be associated with a large Zn overspill.

The energy spectrum of contemporary VVH ions has recently been the subject of some controversy. Work by Shirk *et al.* (1973) suggested a very steep energy spectrum for nuclei of $Z \geq 60$ in the region of 1 GeV/n, whereas Fowler *et al.* (1977) favour a normal VH type spectrum for these ultraheavy nuclei. The data presented here for ancient VVH cosmic rays do not seem to support a very steep spectrum although, as will be shown in the next section, the bulk of the fossil tracks measured here were formed by nuclei of $30 < Z < 47$ and so the results are not directly relevant to this question. It is of interest that Binns *et al.* (1973) have found a spectrum for $Z \geq 32$ particles intermediate in steepness between the VH spectrum and the very steep form of Shirk *et al.* (1973).

Bhandari and Padia (1974) found no evidence for large changes in the VVH/VH ratio with energy in the ancient cosmic ray flux.

It should be emphasized that the results described here do not conclusively rule out differences between the VVH and VH spectra. Further work in gaining improved statistics for these ancient VVH events is important. Also more realistic calculated ρ_{VVH} profiles taking account of erosion (neglected in this analysis) and of deviations in meteorite shape from an assumed spherical form are essential.

VVH charge spectrum

Figure 7 shows the results of track length measurements as a function of etch time for 4 VVH tracks in Shalka hypersthene. On the basis of such data, 'non-stopping' tracks (convex growth curve) are distinguished from 'stopping' tracks (steep rise followed by an abrupt end to track growth). For the latter events a point may be plotted on a $V_T - R$ diagram and an estimate of the charge made. Figure 8 shows such a diagram on which 22 data points are plotted along with calculated profiles for even-charge nuclei from Fe to Sn and for Xe. The data points are for Shalka tracks only.

The varying probability of detecting tracks due to ions of different charge must be allowed for. These corrections are taken into account by means of the ΔR_{eff} values discussed earlier. The meaning of R_{tMIN} is different to that used earlier. For a track to yield a point on the $V_T - R$ diagram of Fig. 8, it is necessary not merely that it exceed 15 μm in length after a 30 minute etching stage but also that it be not fully etched after this etching step, otherwise no track-etch-rate data will be obtained. This imposes the condition that $R_{tMIN} = L(R_{tMIN})$, and this lower limit will be different for every ion. The ΔR_{eff} values rise quite steeply with increasing Z. That this will be so is clear from inspection of Fig. 8, where it may be seen, for example, that a Ge ion has an etch rate in excess of the 30 μm hr^{-1} limit over a much smaller length of trajectory than Xe.

The 22 data points shown on Fig. 8 are converted to a charge abundance spectrum as follows. For each charge the number of VVH tracks is normalized to

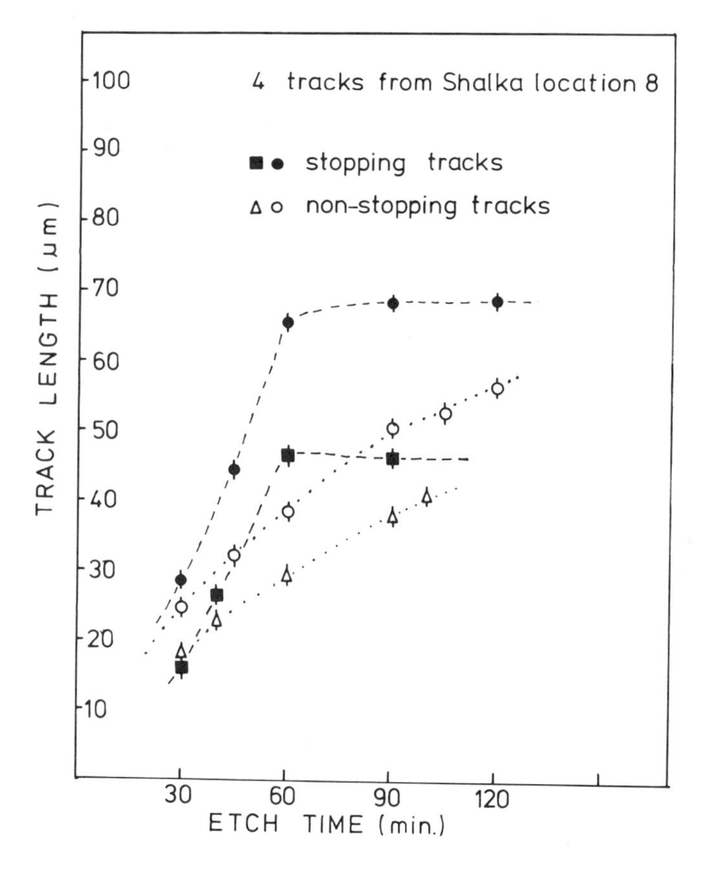

Fig. 7. Track length versus etch time growth curves for some VVH tracks in Shalka. Those for which a plateau is reached are taken to represent tracks formed by particles which were stopped in the sample.

a ΔR_{eff} value of unity. The number of VH tracks counted in the sample area over which the VVH tracks were obtained was 186,500. Assuming that all VH tracks >1 μm long were counted, a ΔR_{eff} value of 10 μm is obtained for the VH group ions. The number of VH tracks is also normalized to a ΔR_{eff} value of unity. The ratio of the normalized densities at each charge to the normalized VH group density is computed and multiplied by 10^6 (to normalize the abundances to VH $\equiv 10^6$). In addition, multiplication of the VVH densities by a factor of 2 is required to allow for the fact that only tracks due to slowing-down particles can appear on a $V_T - R$ plot, whereas no such restriction applies to the counting of VH tracks. Finally, an estimated ~30% of the VVH tracks are lost due to intersection of the growing track with a crystal edge or a failure to reach the end-point of the particle trajectory.

The relative abundances of the different charges after allowance for the corrections indicated above are collected into intervals 4 charge units wide and

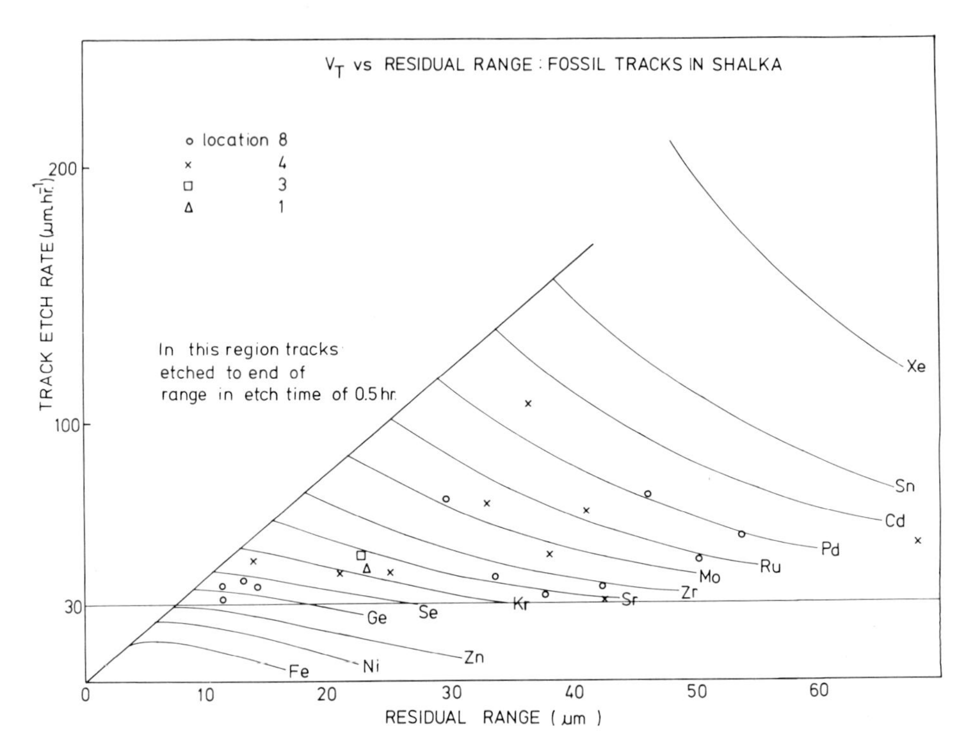

Fig. 8. V_T versus residual range plotted for 22 'stopping' VVH tracks in Shalka. Also shown are calculated $V_T - R$ profiles for even Z nuclei from Fe to Sn and for Xe.

plotted in Fig. 9 along with the solar-system abundances of Cameron (1973) and the contemporary cosmic-ray data of Blanford *et al.* (1973). Also shown are the actual numbers of VVH tracks measured in each interval. One track was assigned to Z = 31 and does not appear on Fig. 9.

These results are not corrected for the passage of the cosmic rays through ~6 − 10 cm of meteoritic material and are as yet only preliminary. Correction for this effect will tend to enhance the VVH to VH ratios particularly for the higher charge groups.

No definite conclusions can be drawn from this charge spectrum as yet, although no gross deviations from the solar system abundances or contemporary cosmic ray abundances are evident.

CONCLUSIONS

Track etch rate measurements may provide a useful means of deciphering the past record of the galactic cosmic rays. It is highly desirable to utilize meteorite and lunar rock samples with simple exposure histories and with demonstrably small environmental annealing effects. The method may not be useful in samples where very high track densities are present (e.g., solar-flare irradiated grains),

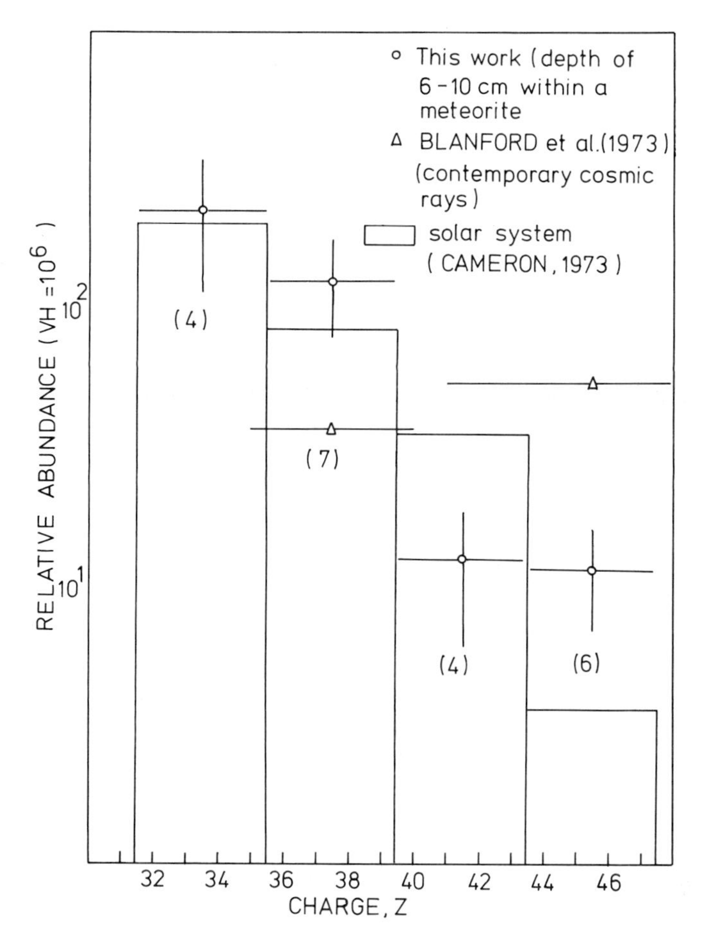

Fig. 9. Relative abundances of ancient cosmic rays of $Z \geq 32$ based on the 'stopping' track data of Fig. 8 and corrected for varying effective ranges and scanning efficiencies for the different charge groups. The histogram is taken from the solar-system abundance compilation of Cameron (1973). Also shown for comparison are some contemporary cosmic-ray data corrected to the top of the atmosphere from Blanford *et al.* (1973). Horizontal error bars represent the charge interval over which the data are averaged. For the contemporary cosmic-ray point at $Z = 46$ this interval extends to $Z = 50$. Figures in brackets represent the numbers of VVH tracks counted in each charge interval.

since it becomes difficult to make measurements on individual tracks. The differential annealing technique of Dartyge *et al.* (1978) may provide a more useful approach in such instances.

Acknowledgments—We wish to thank the respective staffs of the accelerators at which our samples were irradiated, namely the LINAC at the University of Manchester, the UNILAC at GSI, Darmstadt, and the Heavy Ion accelerator at Dubna. The Shalka samples were provided from fragment BM 33761 by Dr. R. Hutchison of the British Museum (Natural History) and the Patwar crystals by Professor D. Lal. We acknowledge the financial support of the Science Research Council.

REFERENCES

Bhandari N., Bhat S., Lal D., Rajagopalan G., Tamhane A. S. and Venkatavaradan V. S. (1971) High resolution time averaged (millions of years) energy spectrum and chemical composition of iron-group cosmic ray nuclei at 1 A.U. based on fossil tracks in Apollo samples. *Proc. Lunar Sci. Conf. 2nd*, p. 2611–2619.

Bhandari N. and Padia J. T. (1974) On the variation of cosmic ray composition in the past. *Proc. Lunar Sci. Conf. 5th*, p. 2577–2589.

Binns W. R., Fernandez J. I., Israel M. H., Klarmann J., Maehl R. C. and Mewaldt R. A. (1973) Chemical composition of cosmic rays with $Z \geq 30$ and $E \geq 325$ MeV/N. In *Proc. 13th Intern. Cosmic Ray Conf.* **1**, 260–264. Univ. of Denver, Denver, Colorado.

Blanford G. E., Friedlander M. W., Klarmann J., Pomeroy S. S., Walker R. M., Wefel J. P., Fowler P. H., Kidd J. M., Kobetich E. J., Moses R. T. and Thorne R. T. (1973) Observation of cosmic ray particles with $Z > 35$. *Phys. Rev.* **D8**, 1707–1722.

Blanford G. E., Fruland R. M. and Morrison D. A. (1975) Long-term differential energy spectrum for solar-flare iron-group particles. *Proc. Lunar Sci. Conf. 6th*, p. 3557–3576.

Bull R. K. (1975) Studies of charged particle tracks in terrestrial and extra-terrestrial crystals. Ph.D. thesis, Univ. Birmingham, England.

Bull R. K. and Durrani S. A. (1976) Cosmic ray tracks in the Shalka meteorite. *Earth Planet. Sci. Lett.* **32**, 35–39.

Bull R. K. and Durrani S. A. (1977) Studies of fresh and fossil tracks in meteoritic hypersthene. *Nuclear Track Detection* **1**, 75–80.

Cameron A. G. W. (1973) Abundances of the elements in the solar system. *Space Sci. Rev.* **15**, 121–146.

Cantelaube Y., Maurette M. and Pellas P. (1967) Traces d'ions lourds dans les minéraux de la chondrite de Saint Séverin. *Symposium of Radioactive Dating and Methods of Low-Level Counting, Monaco*, p. 215–229. International Atomic Energy Agency, Vienna.

Dartyge E., Dran J. C., Duraud J. P., Langevin Y. and Maurette M. (1977) Thermal annealing of nuclear particle tracks in minerals and the composition of the very heavy cosmic rays (abstract). In *Lunar Science VIII*, p. 218–220. Lunar Science Institute, Houston.

Dartyge E., Duraud J. P., Langevin Y. and Maurette M. (1978) A new method for investigating the past activity of ancient solar flare cosmic rays. Part II: A thermal annealing method for exploiting high densities of tracks in lunar minerals (abstract). In *Lunar and Planetary Science IX*, p. 218–220. The Lunar and Planetary Institute, Houston.

Eberhardt P., Eugster O., Geiss J. and Marti K. (1965) Rare gas measurements in 30 stone meteorities. *Z. Naturforsch.* **21a**, 214–426.

Fleischer R. L., Price P. B., Walker R. M. and Maurette M. (1967) Origins of fossil charged-particle tracks in meteorites. *J. Geophys. Res.* **72**, 331–353.

Fowler P. H., Alexandre C., Clapham V. M., Henshaw D. L., O'Ceallaigh C., O'Sullivan D. and Thompson A. (1977) High resolution study of nucleonic cosmic rays with $Z \geq 34$. *Nucl. Instrum. Methods* **147**, 195–199.

Green P. F., Bull R. K. and Durrani S. A. (1978) Particle identification from track-etch rates in minerals. *Nucl. Instrum. Methods*. In press.

Henke R. P. and Benton E. V. (1967) A computer code for the computation of heavy-ion range energy relationship in any stopping material. U. S. Naval Radiological Defense Laboratory Report TR-67-122.

Hutcheon I. D., Macdougall D. and Price P. B. (1974) Improved determination of the long-term average Fe spectrum from 1 to 460 MeV/a.m.u. *Proc. Lunar Sci. Conf. 5th*, p. 2561–2576.

Krätschmer W. and Gentner W. (1976) The long-term average of the galactic cosmic-ray iron group composition studied by the track method. *Proc. Lunar Sci. Conf. 7th*, p. 501–511.

Lal D., Lorin J. C., Pellas P., Rajan R. S. and Tamhane A. S. (1969) On the energy spectrum of iron-group nuclei as deduced from fossil-track studies in meteoritic minerals. In *Meteorite Research* (P. Millman, ed.) p. 275–285. D. Reidel, Dordrecht, Holland.

Maurette M. (1970) Some annealing characteristics of heavy ion tracks in silicate minerals. *Radiat. Eff.* **5**, 15–19.

Maurette M., Thro P., Walker R. and Webbink R. (1969) Fossil tracks in meteorites and the chemical abundance and energy spectrum of extremely heavy cosmic rays. In *Meteorite Research* (P. Millman, ed.) p. 286–315. D. Reidel, Dordrecht, Holland.

Megrue G. (1968) Rare gas chronology of hypersthene achondrites and pallasites. *J. Geophys. Res.* **73**, 2023–2033.

Müller H. W. and Zähringer J. (1969) Rare gases in stony meteorites. In *Meteorite Research* (P. Millman, ed.) p. 845–856. D. Reidel, Dordrecht, Holland.

Plieninger T., Krätschmer W. and Gentner W. (1972) Charge assignment to cosmic ray heavy ion tracks in lunar pyroxenes. *Proc. Lunar Sci. Conf. 3rd*, p. 2933–2939.

Plieninger T., Krätschmer W. and Gentner W. (1973) Indications for time variations in the galactic cosmic ray composition derived from track studies on lunar samples. *Proc. Lunar Sci. Conf. 4th*, p. 2337–2346.

Price P. B., Fleischer R. L. and Moak C. D. (1968a) Identification of very heavy cosmic ray tracks in meteorites. *Phys. Rev.* **167**, 277–282.

Price P. B., Lal D., Tamhane A. S. and Perelygin V. P. (1973) Characteristics of tracks of ions of $14 < Z < 36$ in common rock silicates. *Earth Planet. Sci. Lett.* **19**, 377–395.

Price P. B., Rajan R. S. and Tamhane A. S. (1967) On the pre-atmospheric size and maximum space erosion rate of the Patwar stony-iron meteorite. *J. Geophys. Res.* **72**, 1377–1388.

Price P. B., Rajan R. S. and Tamhane A. S. (1968b) The abundance of nuclei heavier than iron in the cosmic radiation in the geological past. *Astrophys. J.* **151**, L109–L116.

Shirk E. K., Price P. B., Kobetich E. J., Osborne W. Z., Pinsky L. S., Eandi R. D. and Rushing R. B. (1973) Charge and energy spectra of trans-iron cosmic rays. *Phys. Rev.* **D7**, 3220–3232.

Walker R. and Yuhas D. (1973) Cosmic ray production rates in lunar materials. *Proc. Lunar Sci. Conf. 4th*, p. 2379–2389.

Proc. Lunar Planet. Sci. Conf. 9th (1978), p. 2433–2450.
Printed in the United States of America

Nature of the gases released from lunar rocks and soils upon crushing

EVERETT K. GIBSON, JR.

SN7, Geochemistry Branch, NASA Johnson Space Center, Houston, Texas 77058

FIKRY F. ANDRAWES

Lockheed Electronics Co., Inc., 1830 NASA Road 1, Houston, Texas 77058

Abstract—An analytical system has been developed and operated for the qualitative and quantitative analysis of inorganic gases released from lunar basalts, breccias, and soils by crushing. The sample is crushed in a flowing stream of purified helium with a 1 cm diameter piston apparatus and the gases released are analyzed with a dual-column gas chromatograph equipped with He-ionization detectors. Studies have been carried out to demonstrate that the newly developed crushing technique eliminates previously identified problems associated with adsorption of the released gases by the crushed sample.

Nine lunar basalts and three breccia samples were analyzed in our study. Nitrogen was released during crushing of four basalts at concentration levels of 1.5 to 2.5 ngN/g but not detected in five basalts (maximum concentration <1.5 ngN/g). The upper concentration levels for other gases released upon crushing nine basalts were: H_2-0.12 ng/g; CH_4-0.15 ng/g; 0_2-1.5 ng/g; Ar-1.8 ng/g; CO-1.5 ng/g and CO_2-0.12 ng/g. Three breccia samples had nitrogen concentrations ranging from 0.6 to 7.2 ngN/g. Absence of carbon-containing gases in the rocks is surprising because gas phase equilibrium calculations suggest that significant amounts of CO, CH_4, along with H_2, should have been present at the time of formation of lunar basalts.

Seventeen lunar soils were analyzed (N range <1 to 818 ngN/g) and a direct correlation between the amount of nitrogen released upon crushing with Morris' (1976) lunar soil maturity indicator (I_s/FeO) was found. Only one to two percent of the total nitrogen present in lunar soils is releasable upon crushing. The finest grain size fractions of the soils (<20 microns) contain the least amount of nitrogen releasable upon crushing.

INTRODUCTION

At the time when the lunar mare basalts were formed, no samples of the volcanic gases were collected by analysts. However, efforts have been made to identify the composition and proportions of the gases associated with lunar volcanism. Most of the studies to date have been from a theoretical point of view or result from petrologic conclusions based on the mineral chemistry of lunar basalts. To date, few studies have attempted to measure the composition and proportions of the active gases which may presently be trapped in vugs, vesicles, and melt inclusions within lunar basalts (Funkhouser *et al.*, 1971). We wish to report the results of the analysis of the active gases (e.g., H_2, CH_4, N_2, CO, CO_2, Ar, O_2) released during crushing of lunar mare basalts, highland rocks, and lunar soils.

The general chemical nature of the gases of lunar basalts had been anticipated on the basis of rock mineralogy and bulk chemistry. The mineral assemblage Fe-FeS-$FeTiO_3$ and the absence of ferric iron and hydrated minerals have shown that lunar basalts crystallized under conditions of low P_{O_2}, P_{S_2}, and P_{H_2O} (Wellman, 1970; Sato *et al.*, 1973; Sato, 1976). Ringwood (1970) and Gerlach (1974) noted that for lunar basalts H_2O and CO_2 would be unstable relative to H_2 and CO as a result of the low P_{O_2} which existed at the time of formation of the lunar basalts. Studies of Gerlach (1974), Sato *et al.* (1973) and Sato (1976) suggested that the lunar basalt gases had C/H ratios less than 1.0 and C/S ratios greater than 10.0, and that they were essentially hydrogen-rich C-O-H-S-N gases. Compared to terrestrial volcanic gases they were probably much poorer in oxygen and sulfur and much richer in hydrogen and nitrogen (Gerlach, 1974).

Gas inclusions in lunar rocks have been identified previously (Roedder and Weiblen, 1970) which indicate that gases were present in the lunar magmas and a portion may still be retained within the samples. A portion of these gases is believed to be trapped in the silicate melt inclusions and as occasional gas-rich inclusions or vesicles. Small amounts of noncondensable gases were reported by Roedder and Weiblen (1970) in some bubbles within glass samples from the lunar soils, though many bubbles might have contained gases below the limits of detection used in their study. Crushing experiments of Funkhouser *et al.* (1971) indicated that the gases within inclusions included H_2, N_2, and CH_4 along with the rare gases. Crushing studies and analysis of the noble gases in an Apollo 11 soil breccia have been reported by Heymann and Yaniv (1971). They did not analyze any of the other inorganic gases in the sample. The noble gases measured by mass spectrometry by Heymann and Yaniv (1971) and Funkhouser *et al.* (1971) were identified to be of solar wind origins by their isotopic compositions. The potential for production of some nitrogen (e.g., ^{15}N) in lunar rocks by spallation processes has been considered by Becker *et al.* (1976). The distribution and production rates for nitrogen in lunar basalts resulting from spallation processes has been studied but at the present time the amounts of nitrogen released upon crushing lunar basalts from spallation processes are below the detection limits. In addition, some lunar samples such as the Apollo 17 orange soil may contain surface sublimates condensed from the gas phase during volcanic emanations (Heiken *et al.*, 1974; Butler and Meyer, 1976).

Because of the apparent evidence and additional possibilites for active gases being trapped or held within selected lunar materials, we have analyzed a variety of samples in hope of identifying the nature of the gas present at the time of formation of both basalts and soils. A sample crusher-gas chromatographic system was developed which would permit the analysis of trace quantities of the active gases. The system was designed and operated in a manner which eliminated previously identified problems (Barker and Torkelson, 1975; Piperov and Penchev, 1973) associated with adsorption of the released gas by the fresh sample surfaces generated during the crushing operation. The system is described in greater detail in Andrawes and Gibson (1978a,b).

Experimental

Samples were crushed using a heat-treated tool steel crusher shown in Fig. 1. Samples were crushed at room temperature under a flowing helium gas stream (purity He = 99.999%) using a hydraulically-operated press. The crusher was constructed and operated in a manner which minimized any metal to metal contact which might introduce contaminant gases (e.g., H_2 and CH_4). The crusher was constructed so that a series of "dead-volumes" of pure carrier gas were held along the crusher's piston. The purified carrier gas filled the "dead volumes" between the viton O-rings of the piston as it was pushed into the crusher. The He flow during degassing and crushing operations was 2 cc/sec. The total volume of space within the crushing chamber depends upon the size of the sample being crushed. The maximum interior volume of the crusher with a 400 mg rock chip is less

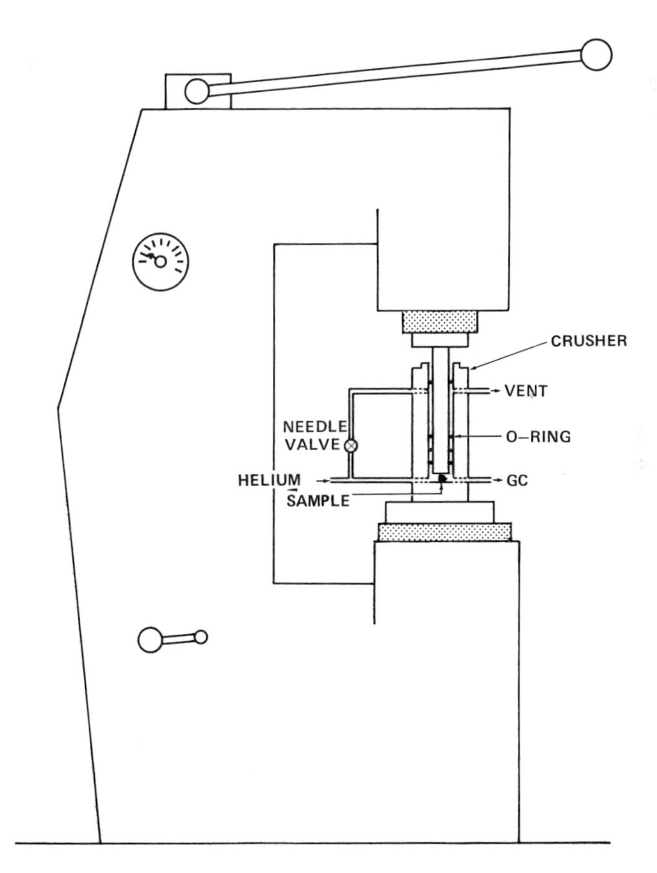

Fig. 1. Crusher used for sample studies. The diameter of the crushing piston is 1 cm. Note the O-ring baffles located along the piston which serve as storage spaces for the purified carrier gas and which prevent introduction of contaminants into the sample area.

than 0.5 cc and without a sample less than 0.1 cc. The gases released from the samples upon crushing are instantly removed from the crushing chamber by a longitudinal diffusion process. Thus, the gases released from the samples are not permitted to remain within the chamber for potential adsorption by the fresh surfaces generated during crushing. This procedure eliminates problems identified in previous investigations (Barker and Torkelson, 1975; Barker and Sommer, 1973).

The sample crusher was constructed with an exterior resistance heater capable of heating the sample and crushing chamber to temperatures of 400°C in order to remove any adsorbed gases. Samples were heated for 24 hours at 150°C under a flowing helium (99.9999% purity) stream, cooled to room temperature, and analyzed as comparisons with samples which had not been heated at 150°C. No differences, within the analytical uncertainty, were noted between heated and unheated samples during analysis of their gases after crushing. The temperatures generated in samples during crushing are not high enough to remove any contaminant terrestrially-adsorbed gases, as noted by the study of very immature lunar soils which did not release atmospheric gases (e.g., N_2, O_2, and Ar) upon crushing. Thus, the analytical values generated for atmospheric gases are values for gases held within the samples and not contaminants or surface-adsorbed gases. Sample sizes ranged from 90 to 400 mg for basalt and breccia chips to soil samples of 15 to 50 mg, but as small as 2 mg for hand-picked agglutinates. The diameter of the piston inside the crusher was 1 cm. The hydraulic press was capable of applying a force of 55 tons to the crushing area. The maximum pressure which could be applied to the crushing area was 6.96×10^2 kN/cm^2. However, initial crushing pressures resulting from the 1 ton applied force were 12.7 kN/cm^2. The maximum applied force of 25 tons used to crush the samples in this study resulted in a pressure of 3.17×10^2 kN/cm^2.

The crushing procedure used was stepwise so that the gases released at each interval (e.g., pressures recorded on hydraulic press gauge) could be analyzed. The sum of the released gases comprises the total amount of gas released for each sample. The gases released upon crushing were split 50/50 and fed directly into two columns (Molecular Sieve 5A, 80/100 mesh, 2 and 6 meters long and 2.1 mm I.D., and Porapak Q, 80/100 mesh, 3 meters long and 2.1 mm I.D.) of a gas chromatograph where they were separated and analyzed using two helium ionization detectors. A Varian 1700 gas chromatograph was used in this work. The molecular sieve columns were used to separate H_2, Ar, O_2, N_2, CH_4, and CO. Porapak Q columns were used to separate CO_2 and light hydrocarbons. Molecular sieve columns were conditioned at 350°C with helium flow of 60 ml/min. for 48 hours and the Porapak Q column was conditioned at 200°C, with helium flow of 60 ml/min. for 48 hours. Operation of the helium ionization detectors was according to the recently established procedures of Andrawes and Gibson (1978a,b).

For calibration, a glass exponential dilution flask was used. The calibration gas (certified gases of purity >99.9999%) was introduced into the exponential dilution flask by a gas sampling valve to provide accurate calibration. Helium used in the exponential dilution flask was the same as that going to the analytical column and through the crusher. A Spectra Physics SP-4000 data system was used to integrate peak areas. Chromatograms were obtained on a strip-chart recorder at 0.5 millivolt full scale. All valves and tubing were made of stainless steel. The schematic of the analytical system is given in Fig. 2. The detection limits for the analytical system are: H_2-0.04 ng/g; N_2-0.6 ng/g; CH_4-0.05 ng/g; O_2-0.4 ng/g; Ar-0.7 ng/g; CO-0.6 ng/g; CO_2-0.05 ng/g and C_2H_6-0.07 ng/g.

The detection limits for the helium ionization detector used in our work are significantly different from those reported by Funkhouser *et al.* (1971). It is not unusual for gas chromatograph's detectors to differ slightly in their response. However, the detection limits reported by Funkhouser *et al.* (1971) for some of the inorganic gases are obviously incorrect. They reported identical detection limits (Table 1, Funkhouser *et al.*, 1971) for H_2, CO_2, N_2, CH_4, and C_2H_2. These detection limits are in error with the long established theory of the helium ionization detectors (Hartman and Dimick, 1966). The theory of the helium ionization detector is based on the ionization of gases upon collision with metastable helium. Gases with different ionization potentials will produce different detector responses. For example, the helium ionization detector's response for CO_2 is twenty-five times greater than for H_2 (Hartman and Dimick, 1966). Our recent work (Andrawes and Gibson, 1978a) on the helium ionization detector has shown that the response of the detector would vary significantly upon the addition of small concentrations of additive gases to the helium carrier gas. Without accurately knowing the response of the detector, values reported previously are meaningless. Discussions with

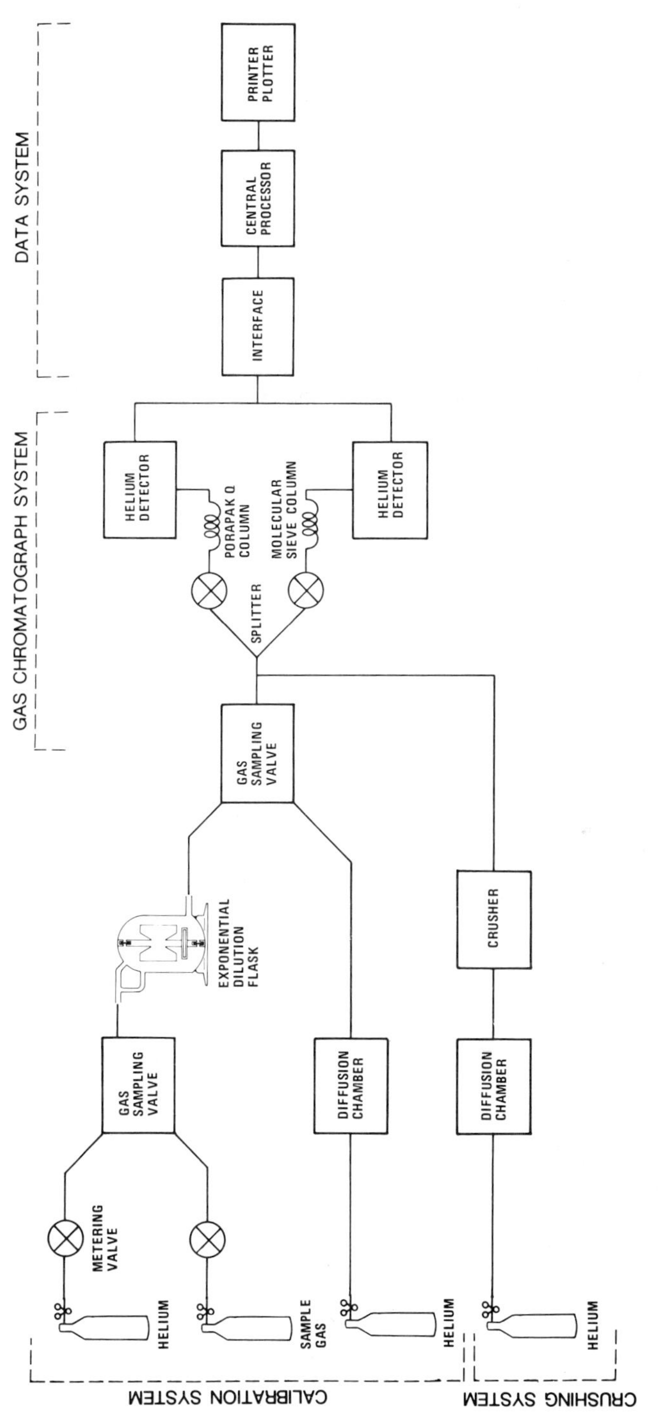

Fig. 2. Schematic of the crushing system. The system is composed of four separate units: (a) calibration system, (b) crushing system, (c) gas chromatograph system, and (d) data system.

Funkhouser (pers. comm., 1978) indicate that the analytical values reported in their work may not be "true" values but only "qualitative" for gases released from the samples and that the mode of operation of their crusher may have introduced additional CH_4 and H_2 into their samples. Thus, we cannot fully accept the accuracy of the work reported by Funkhouser *et al.* (1971).

It is well known that when rocks and minerals are crushed, many new, clean surfaces are generated, and these have the potential for adsorbing some of the volatiles released by the crushing under certain experimental conditions (Barker and Sommer, 1973; Barker and Torkelson, 1975; Roedder, 1972; Piperov and Penchev, 1973). In the case of lunar samples, their adsorption properties have been studied for some gases (Cadenhead *et al.*, 1972; Grossman *et al.*, 1972). Previous gas adsorption studies have not included all possible inorganic gases which might be released from lunar samples upon crushing. In fact, H_2O, CO_2, H_2 and N_2 have been studied in some detail (Cadenhead *et al.*, 1972; Grossman *et al.*, 1972) and the conditions under which these studies have been made are of only limited use for comparison with our studies because under our experimental conditions the gases are immediately removed from the samples upon crushing by the flowing He stream. Thus, the released gases are not permitted to remain in the crushing chamber to adsorb on the sample as in previous experiments of Barker and Torkelson (1975) and Piperov and Penchev (1973).

In order to determine whether the fresh surfaces generated during crushing adsorbed any gases which might be released from the samples, the following experiments were carried out. (1) The sample crusher (without a sample in place for crushing) was placed in line between the gas sampling valve and the chromatographic column. (2) A known amount (100 μl) of standard gas mixture containing 9 ppm H_2, 9.5 ppm O_2, 17 ppm N_2, 25 ppm CO, 4.7 ppm CO_2, 4.7 ppm CH_4, 4.8 ppm C_2H_2, 4.3 ppm C_2H_4 and 4.7 ppm C_2H_6 was injected to the system passing through the crusher. The gas was analyzed after passing through the crusher. The peak areas were integrated using the data system. (3) The lunar basalt sample 70215 (150 mg) was loaded into the crusher, degassed for 24 hours in a flowing stream of helium. The sample was crushed in the flowing stream of helium carrier gas. (4) The standard gas mixture was introduced into the sample chamber and passed over the newly crushed sample. The standard gas mixture was analyzed after passing over the newly crushed sample. The peak areas were determined using the data system. A comparison of the reference gas mixture analysis before crushing and post-crushing the sample indicated that no loss of the reference gas mixture occurred by adsorption on the newly generated surfaces of the crushed sample. If the sample adsorbed any of the gases in the reference gas sample, the analysis of the gases (post-crushing) would have indicated an absence or decrease in peak areas of the adsorbed gas as compared to the analysis carried out with the empty chamber. The previously identified problems of adsorption of CO_2 by new surfaces (Barker and Torkelson, 1975; Piperov and Penchev, 1973) have apparently been eliminated by the design of our crushing apparatus. The rapid removal of released gases by the purified helium carrier gas reduces the opportunity for adsorption on the newly generated surfaces.

Reproducibility of the analytical system can be seen from four samples of lunar soil 75111. Crushing separate samples ranging in size from 29 to 48 mg produced the following nitrogen results: 113, 108, 147 and 102 ngN/g and a mean of 117 and a range of $\pm$10–15% of the value present. The analytical differences are quite good considering the sample's heterogeneity.

The efficiency of crushing has been checked by crushing various terrestrial materials (e.g., phenocrysts from basalts, minerals with gas inclusions, etc.) along with lunar soils (Andrawes and Gibson, 1978b). Samples of lunar soils were crushed, removed from the crusher, disaggregated, and crushed a second and third time. A single crush (applied force of 25 tons) of the lunar soil removed 92 percent of the total releasable gases. The second crush removed the additional 8 percent of the releasable gases. Additional crushes failed to release any more detectable gases.

The efficiency of the crusher for releasing volatiles was checked against known quartz samples which had been studied by Kvenvolden and Roedder (1971). These results have been reported elsewhere (Andrawes and Gibson, 1978b).

To date, various types of geological samples (ranging from submarine basalt glasses with their associated phenocrysts to volcanic glasses with fluid inclusions) have been analyzed with the crusher-gas chromatographic system and reported elsewhere (Andrawes and Gibson, 1978b). The ability of the analytical system to provide both qualitative and quantitative data on a variety of samples has been demonstrated. Despite the fact that CO_2 and CO were not observed in the crushing of lunar samples, these gases have been analyzed in numerous terrestrial materials (Andrawes and Gibson, 1978b).

EXPERIMENTAL RESULTS

Lunar Rocks:

Nine basalts and three highland rocks were crushed and the released gases analyzed. The analytical results are given in Table 1. The only gas detected in the basalts was nitrogen (found in four of the nine samples). Concentration levels resulting from crushing were from 1.5 to 2.5 ngN/g. The upper concentration levels for other possible gases released upon crushing nine basalts were: H_2-0.12 ng/g; CH_4-0.15 ng/g; O_2-1.5 ng/g; Ar-1.8 ng/g; CO-1.5 ng/g, and CO_2-0.12 ng/g. Müller (1972) performed crushing studies on 15556 followed by analysis of the gases using the methods of Funkhouser *et al.* (1971). He also failed to find N_2 (detection limit 1 ngN/g) but did find trace amounts of CH_4 ($<10^{-7}$ cc). As noted in the Experimental section of this paper, the analyses carried out using Funkhouser *et al.* (1971) procedure are probably incorrect.

Table 1. Lunar rocks studied.

BASALTS		
No Active Gases	Active Gases Present	Lunar Highland Rocks
15065	15058–2.5 ngN/g	64435 (gray portion) 7.2 ngN/g
15556	15499–2.0 ngN/g	64435 (white portion) 0.6 ngN/g
74275	70215–1.5 ngN/g	67016 1.2 ngN/g
75035	74255–1.5 ngN/g	
78505		

The three lunar highland samples crushed contained trapped nitrogen levels ranging from 0.6 to 7.2 ngN/g (Table 1). The gray and white breccia 64435 released 0.6 ngN/g in the white anorthositic gabbro and 7.2 ngN/g in the gray-matrix portion. The gray-matrix breccia portion of 64435 also released methane in trace amounts (approaching the blank values) and in amounts similar to those observed for some immature lunar soils. The highland melt rock 67016 released 1.2 ngN/g.

The absence of any hydrogen and carbon gases released during crushing was unexpected because previous gas phase equilibrium calculations suggested the presence of significant quantities of H_2 and CO (Gerlach, 1974; Funkhouser *et al.*, 1971; Wellman, 1970; Barker, 1974) and which may be retained within melt inclusions, vesicles, etc. of lunar basalts (Roedder and Weiblen, 1970). Diffusion of hydrogen in silicate materials is well established (Jost, 1960) and the absence of hydrogen in the basalts is not surprising because sufficient time has passed

(3.2 to 3.6 b.y.) for the hydrogen to have escaped. In fact, all silicate materials have a finite permeability.

The origin of the nitrogen in the lunar rocks may be from multiple sources: (1) trapped indigenous lunar or solar wind gas similar to that observed by Roedder and Weiblen (1970) and Funkhouser *et al.* (1971), or (2) nitrogen which may have been introduced into the samples during processing under nitrogen (the gas used in the Curatorial Processing Laboratories). At the present time it is difficult to distinguish between the two sources because of the small quantity of gas present. Methane and hydrogen were not found in any of the basalts during crushing; however, trace amounts of methane and hydrogen could be produced during the crushing operation by the action of the steel piston and the crusher base plate unless precautions are taken. In fact at crushing pressures above 3.16×10^2 kN/cm^2 (hydraulic press force of 25 tons), hydrogen and methane are released from the heat-treated tool steel crusher simply by pressing the piston against the base plate. The design of our crusher minimized contact between the piston and the base plate. The absence of methane or hydrogen above the system blank levels in our work implies that the large quantities of CH_4 and H_2 observed by Funkhouser *et al.* (1971) may have been an artifact of their crushing technique and this has been verified by discussions with Funkhouser (pers. comm., 1978).

Crushing studies carried out on nine lunar basalts have failed to identify any of the trapped gases other than nitrogen. From the observations of Roedder and Weiblen (1970) on the presence of melt inclusions, the next obvious place to search for trapped gases in lunar basalts is the phenocrysts within the basalts which have melt inclusions. Numerous Apollo 12 olivine basalts contain melt inclusions within their large olivine crystals. Analysis of a "concentrate" of these crystals may provide the missing inorganic gases associated with the evolution of the lunar basalts. Studies carried out to date have been on bulk basalt samples and the "gas-rich" phenocrysts. Some terrestrial basalts typically contain phenocrysts with gases trapped within their melt inclusions (Andrawes and Gibson, 1978b) which comprise less than 10 percent of the basalt and must be concentrated in order to enhance the probabilities of identifying trapped gases.

Lunar Soils:

The lunar regolith soils have complex histories which are related to the processes operating in the lunar environment. Previous work by Carrier *et al.* (1972) on the strength and compressibility of returned lunar soils has shown that active gases may be released from lunar soils when they are subjected to shear. The composition of the released gases was not studied in detail by Carrier *et al.* (1972). Carrier (pers. comm., 1978) noted that pressure loads as small as 6.9×10^{-4} kN/cm^2 (1 psi) on lunar soils were sufficient to result in gas release (as noted by their mass spectrometer traces) from the samples studied in shear tests. We have studied the nature of the inorganic gases released from seventeen lunar soils during crushing. Nitrogen was the major gas phase seen during the crushing operations. Trace amounts of argon (5 to 15 ng Ar/g) were also noted, but will not be reported in this paper. Argon separations were made using a 6 meter molecular sieve column to distinguish between O_2 and Ar. A summary of the released nitrogen data is given in Table 2.

Table 2. Lunar Soils—Crushing Data.

Sample	ngN/g	I_s/FeO*
10084	140, 160	78
14003	173	66
14163	95	57
15271	170	63
15427	<1	0.3
15601	34, 35	29
61221	17	9.2
69961	190	92
70011	491, 818, 626	54
74220	<2.5	1.0
75111	117	58
75121	102	68
76240	8	56
76260	21	58
76280	5.6	45
76501	318	58
78501	77	36

*After Morris (1976).

The gas release profiles produced by crushing two lunar soils are given in Figs. 3 and 4. Chromatograms given for 64221 soil (Fig. 3) indicate the nature of the gases released with increasing pressure placed on the sample. Release of nitrogen begins to occur from the soil when the hydraulic press registers a force of 1 ton (pressure on piston = 12.7 kN/cm^2) applied to the sample. Maximum nitrogen and methane evolution occur between 1 and 20 tons applied force (pressure equivalent to 12.7 to 253 kN/cm^2). At pressures above 316 kN/cm^2 (applied force of 25 tons on the crushing piston) evolution of methane occurs from the outgassing of the crusher (note the large CH_4 peaks at the 40 and 55 ton applied force chromatograms given in Fig. 3). The total individual (e.g., N_2) gas released from the sample is the sum of the gases released at each of the crushing stages below the 25 ton mark. A release profile for immature lunar soil 61221 is given in Fig. 4. The levels of gas released from the samples are approaching the limits of detection. Only with the interfaced chromatographic control system can significant differences between background and gases released from samples be determined.

Nitrogen releases measured changed from less than 1 ngN/g for the Apollo 15 green glass (15427) to a mean of 646 ngN/g for soil 70011 (which was collected

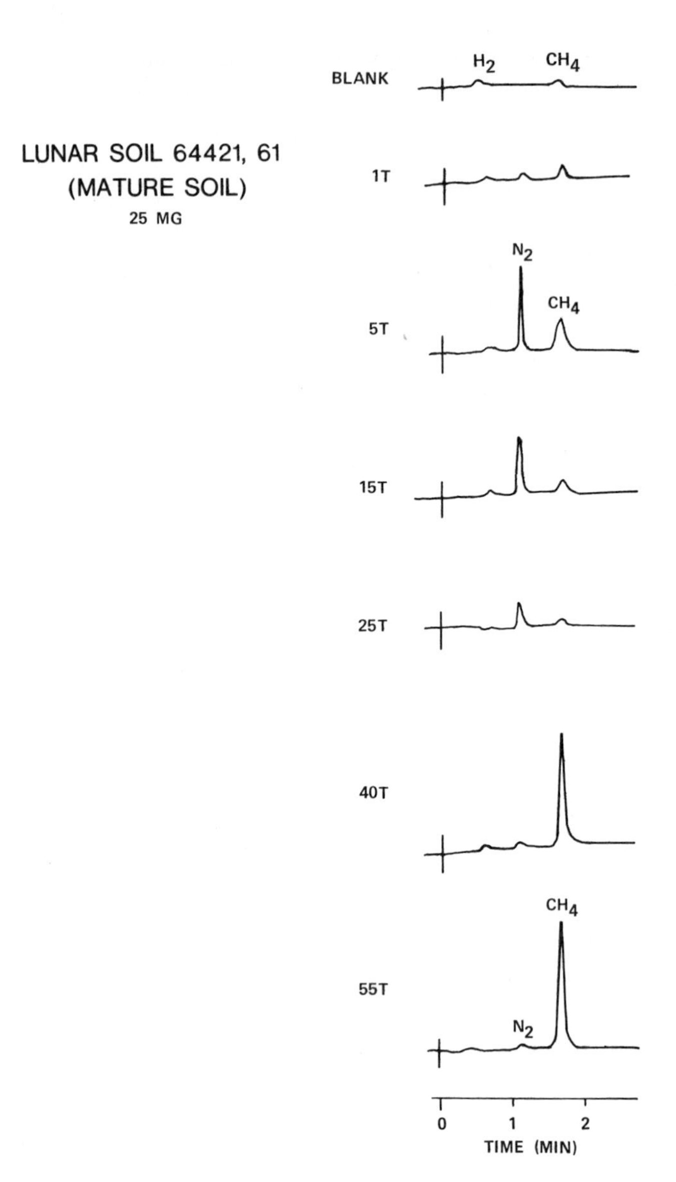

Fig. 3. Gas released upon crushing mature lunar soil 64421. Sample size 25 mg of less than 1 mm fines. The top chromatogram shows the system blank. With the application of one ton (1T) force (pressure on samples = 12.7 kN/cm^2) to the samples, nitrogen begins to be released from the sample. Nitrogen and methane release is maximized around 5 tons of applied force (pressure of 63.5 kN/cm^2 on the samples). At pressures greater than 317.5 kN/cm^2 (applied force of 25 tons), methane evolution occurs from outgassing of the piston and base plate. Gas evolution for geologic samples is based only on gases released at pressures below 317.5 kN/cm^2 (applied force of 25 tons).

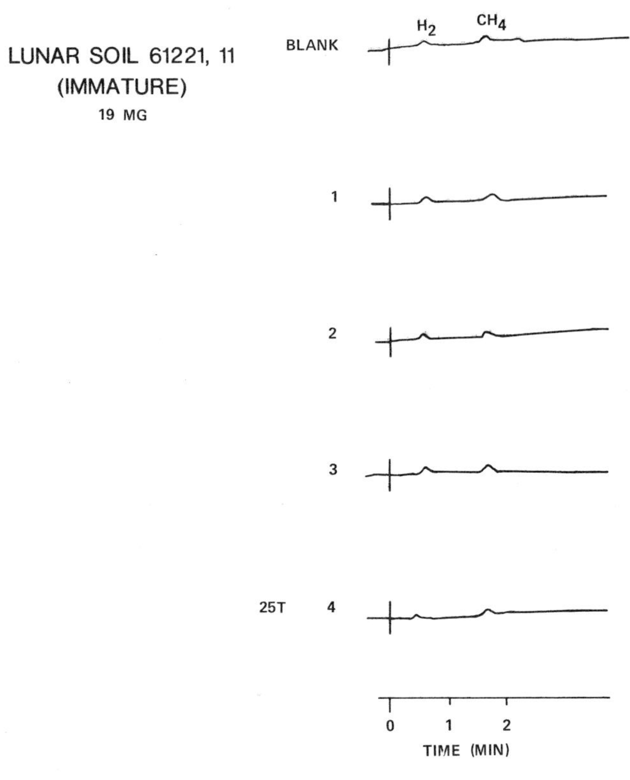

Fig. 4. Gas released upon crushing immature lunar soil 61221. Sample size 19 mg of less than 1 mm lunar fines. The absence of releasable gases from this sample is contrasted with the quantities of gases released from the mature soil 64421 shown in Fig. 3.

beneath the exhaust port of the Lunar Module). Typical lunar soils contained between 50 and 200 ngN/g releasable upon crushing. Comparison of the analytical data for N released upon crushing with the lunar soil maturity indicator I_s/FeO of Morris (1976) shown in Fig. 5 indicates that a strong correlation exists for most soils. The correlation coefficient is +0.95 for twelve soils. Five soils do not correlate and must be considered separately. Soil 76501 collected from the rim of a fresh crater and having a typical I_s/FeO value for a mature soil (Morris, 1976) had a factor of three more releasable nitrogen than other soils of similar maturity. Becker and Clayton (1975) analyzed 76501 for its nitrogen and isotopic composition and did not find the soil unusually enriched in total nitrogen. The enrichment in nitrogen releasable by crushing for 76501, as compared to other lunar soils of similar maturity, is not understood at the present time. Soil 70011 was collected directly beneath the Lunar Module exhaust port and was expected to be contaminated with the hydrazine exhaust products. Becker and Clayton (1977) have reported the total nitrogen and isotopic

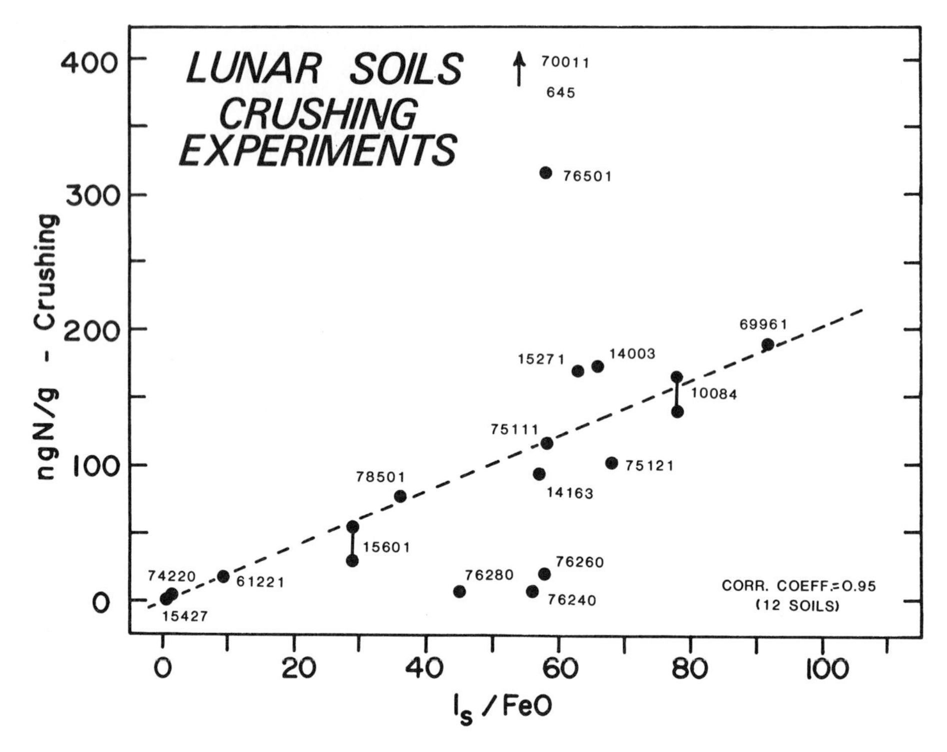

Fig. 5. Relationship between amount of nitrogen released upon crushing and the maturity of the soils as measured by I_s/FeO by Morris (1976). Correlation coefficient for 12 soils (excluding 70011, 76501, 76240, 76260 and 76280) is +0.95. See text for discussion of excluded samples.

compositions for 70011 to be similar to lunar soils of similar maturity. It was noted in our studies that three separate samples of 70011 which were crushed produced a mean nitrogen abundance of 645 ngN/g. This nitrogen value was a factor of 5 to 6 higher than other soils of similar maturities. At the present time, no grain size information is available for soil 70011, therefore its petrographic characteristics are unknown. Thus, the large amount of nitrogen released from this soil upon crushing is a puzzle. The collection site for this soil makes one suspicious of contamination with exhaust products.

The three Apollo 17, Station 6 soils 76240, 76260, and 76280 represent a very unusual situation. The collection location for these three soils was between two of the large boulders at Station 6. In fact, the smaller downslope boulder (5 × 4 × 3 meters in size) was originally a piece of the larger Station 6 boulder. The two boulders have been matched photographically and petrographically by Phinney *et al.* (1974) in their consortium study of samples from the Station 6 boulders. Their studies have shown that the smaller downslope boulder separated from the larger Station 6 boulder, landing beside the larger boulder and slid one to two meters downslope to its present position. The collection site for soils 76240, 76260 and 76280 was at the site where the smaller boulder landed and slid downslope. Thus, the sampling site has been disturbed by both the pressure applied by the boulder falling on the collection site and the shear forces operating on the soils when the boulder slid to its present location. Carrier *et al.* (1972) reported that when a vertical stress (ranging from 2.3×10^{-4} to 6.8×10^{-3} kN/cm^2) was placed on lunar soil it underwent compression and released gases. The compression was believed to result from the crushing of fragile particles (e.g., agglutinates) within the soil and filling of void space. Carrier *et al.* (1972) and Carrier (pers. comm., 1978) report that the shear forces operating on the soils when the boulder slid downslope were the most effective in breaking fragile particles and releasing gases.

The pressures (excluding shear effects) exerted by the Station 6 boulder landing were around 3.1×10^{-3} kN/cm^2. This pressure was similar to the values of 6.7×10^{-3} to 2.3×10^{-4} kN/cm^2 observed by Carrier *et al.* (1972) for gas release from lunar soils when they were vertically stressed. Thus, it appears that the pressures exerted by the boulder when it landed were sufficient for some gas release from soils. The maximum pressures applied to the soils crushed in our experiments were several orders of magnitude greater than the stresses applied to the soils at the Station 6 site when the boulder landed on the three soils collected. However, in our work gas release occurs in the soils studied at the lowest applied stresses (12.7 kN/cm^2) (Fig. 3). The nitrogen values obtained for the three soils 76240, 76260, and 76280 were lower (Fig. 5) than other soils of similar I_s/FeO maturities. The previously applied vertical and shear stresses which have been applied to these three soils by the boulder may account for the lower amounts of nitrogen released by crushing of these soils in our studies. The applied stresses are similar to those observed previously (Carrier *et al.*, 1972) which release gases from samples. As discussed later in this section, the fragile agglutinates which contain the greatest fraction of the releasable gas could be broken by the stresses applied during the movement of the boulder.

The fraction of nitrogen released during crushing represents only a small portion of the total nitrogen present in the lunar soils. A comparison of total nitrogen abundances determined by different investigators with the nitrogen released upon crushing is shown in Fig. 6. It can be seen that, for ten soils which have had total nitrogen abundances measured, a direct correlation exists between the two nitrogen fractions. The correlation coefficient is +0.94 (excluding soils 70011 and 76501 as discussed earlier). The correlation shown in Fig. 6 indicates that the crushing releases only approximately 1 to 2 percent of the total nitrogen present in the soils.

Previous studies (Goel *et al.*, 1975) have shown that the total nitrogen abundances increase with increasing surface area (decreasing grain size). The vast majority of the nitrogen present in lunar soils has been shown to be extra-lunar and has predominantly a solar wind origin.

In order to determine the phases containing the nitrogen released upon crushing, three soils of differing maturity were sieved and hand picked. The bulk

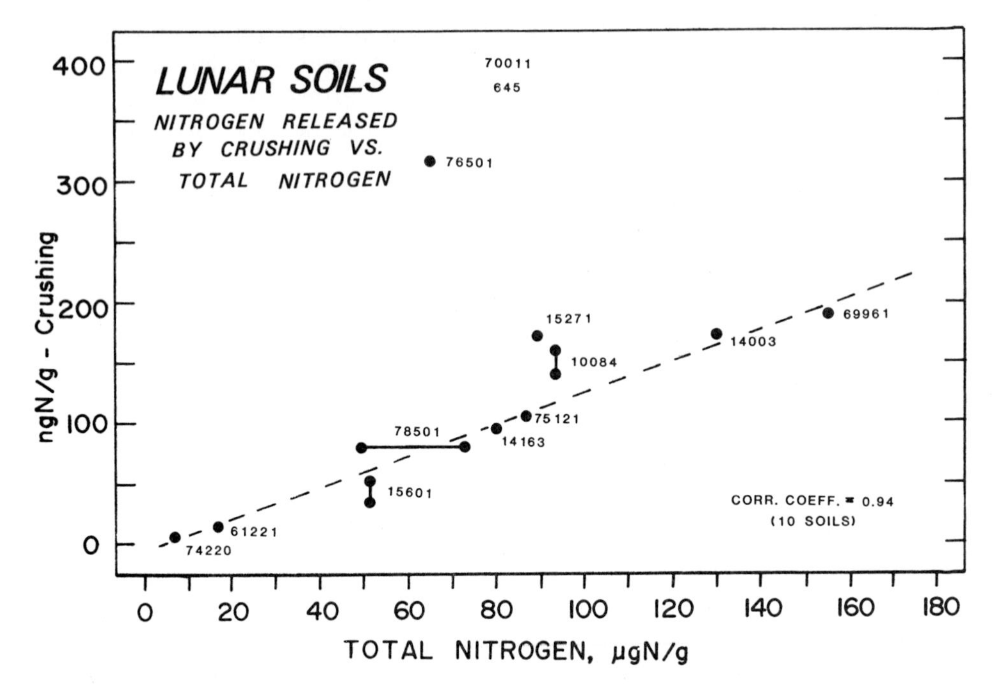

Fig. 6. Relationship between amount of nitrogen released upon crushing and total nitrogen abundances measured by other investigators. Correlation coefficient is +0.94 for ten soils (excluding 70011 and 76501 as described previously). It appears that only 1 to 2 percent of the total nitrogen present in a lunar soil can be released from the sample crushing. Total nitrogen abundance data is from: Chang *et al.* (1974); Becker *et al.* (1976); Goel *et al.* (1975); Kaplan *et al.* (1976); Kerridge *et al.* (1975); Moore and Lewis (1976); Müller (1972, 1973, 1974); and Petrowski *et al.* (1974).

soils (<1 mm fines) were sieved into two fractions of greater and less than 90 microns. Agglutinates were hand picked from the >90 micron fraction. A <20 micron fraction was prepared from a separate sample of the same bulk soil. The bulk soils along with the four separates (>90 micron agglutinates, >90 micron agglutinate-poor fraction, <90 micron bulk fraction, and <20 micron fraction) were crushed. The analytical results are given in Fig. 7.

It is obvious from the analytical data presented in Fig. 7 that the majority (greater than 70, 84 and 85 percent) of the nitrogen released upon crushing is found in the agglutinate fraction. The agglutinate-poor >90 micron fraction is significantly depleted in nitrogen released upon crushing along with the <90 micron fractions. Analysis of the <20 micron fractions indicated that less than 3 percent of the nitrogen released upon crushing is found in this fraction which contains the largest solar wind component of the total nitrogen. An inverse relationship appears to exist for nitrogen; samples with the greatest surface area contain the solar wind nitrogen whereas the nitrogen fraction released upon crushing is greatest in the fraction with the least surface area. The lack of

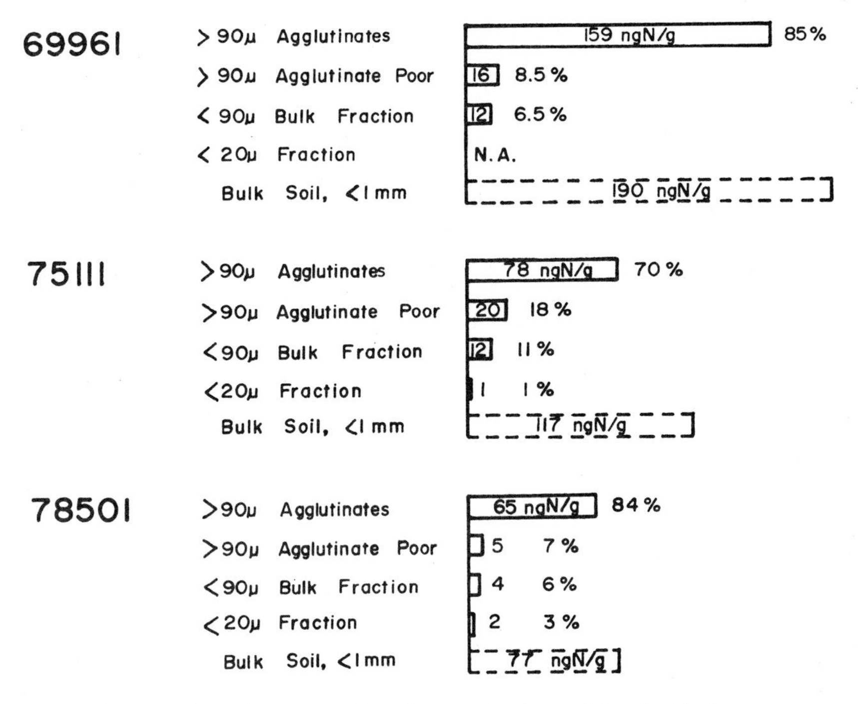

Fig. 7. Nitrogen released upon crushing from grain-size and hand-picked separates. Note that the majority of the nitrogen released upon crushing is found in the agglutinate fraction. In addition, the lowest amount of nitrogen released upon crushing is found in the finest grain size separates.

releasable gas in the <20 μ size fraction is consistent with the observations of Piperov and Penchev (1973). They noted that mineral fragments in the 20 μ size range are as small as the vacuoles containing the trapped gases and such grains are free of trapped gases because the particles could not contain vacuoles greater than the particles themselves.

The nitrogen released during crushing of lunar soils originates during the maturation process occurring on the lunar surface. It has been shown that the nitrogen resides in the glass or agglutinate fraction of the soil. During the formation of a lunar agglutinate, solar wind implanted nitrogen is released (at temperatures between 700–1000°C) and becomes trapped within the glassy constructional particles. Increased maturation of the lunar regolith produces more glass and results in the entrapment of greater amounts of nitrogen.

Conclusions

The study of gases released by crushing lunar rocks and soils has produced the following conclusions:

Lunar rocks:

1. No H_2, CO, CO_2, CH_4, O_2 and Ar were found in nine lunar basalts which were crushed. Gas release limits for gases released upon crushing must be below the following levels: H_2-0.12 ng/g; CO-1.5 ng/g; CO_2-0.12 ng/g; CH_4-0.15 ng/g; O_2-1.5 ng/g; and Ar-1.8 ng/g.
2. Nitrogen was present in four basalts (15058, 15499, 70215 and 74255) at release levels of less than 2.5 ngN/g. Five basalts (15065, 15556, 74275, 75035 and 78505) did not contain any gases detected upon crushing.
3. The source(s) of the nitrogen measured is difficult to assess because: (a) samples were processed in N_2 atmosphere by the curatorial operations; (b) solar wind nitrogen cannot be completely ruled out; and (c) nitrogen is difficult to model in gas phase equilibria studies because it serves as a "buffer" or dilutant.
4. The active gases which formed the vugs and vesicles in lunar basalts have either: (a) leaked from the samples studied, or (b) been hydrogen-rich (per Gerlach, 1974) and the hydrogen has leaked or diffused from the samples after solidification.

Lunar soils:

1. Typical mature lunar fines contain between 50 and 200 ngN/g nitrogen which is released upon crushing.
2. Nitrogen released by crushing amounts to only 1 to 2 percent of the total nitrogen present in the soils.
3. Nitrogen released by crushing is not proportional to surface-correlated nitrogen (e.g., solar wind-derived nitrogen) as shown by the less than 20 micron fraction being depleted in nitrogen released upon crushing.
4. Nitrogen released by crushing is associated with agglutinates or constructional particles.
5. Methane and argon are also released upon crushing and are enriched in the agglutinate fractions.
6. The origin of the gases measured in soils is related to the surface processes; e.g., micrometeorite impacting soil, heat-released volatiles present in target material, glass produced upon melting traps some of the volatiles within the agglutinates and glass particles. Crushing releases this trapped component.

Acknowledgments—The authors wish to acknowledge the assistance of Oscar Mullins in construction of the crusher and oven assemblies. The patience of Richard Morris in hand-picking pure agglutinates from the sieve fractions is appreciated. Discussions with Drs. David S. McKay, Richard V. Morris, S. Chang and D. D. Bogard are recognized as assisting with the improvement of this project. Discussions with Colin Barker and John Funkhouser concerning their past crushing experiments were informative to the authors.

References

Andrawes F. F. and Gibson E. K. Jr. (1978a) The effect of gaseous additives on the response of the helium ionization detector. *Anal. Chem.* **50**, 1146–1151.

Andrawes F. F. and Gibson E. K. Jr. (1978b) An improved method for the qualitative and quantitative analysis of gases released from geological samples upon crushing. *Amer. Mineral.* In press.

Barker C. (1974) Composition of the gases associated with the magmas that produced rocks 15016 and 15065. *Proc. Lunar Sci. Conf. 5th*, p. 1737–1746.

Barker C. and Sommer M. A. (1973) Mass spectrometric analysis of the volatiles released by heating or crushing rocks. *American Society of Testing Materials Publication STP-539*, p. 56–70.

Barker C. and Torkelson B. E. (1975) Gas adsorption on crushed quartz and basalt. *Geochim. Cosmochim. Acta* **39**, 212–218.

Becker R. H. and Clayton R. N. (1975) Nitrogen abundances and isotopic compositions in lunar samples. *Proc. Lunar Sci. Conf. 6th*, p. 2131–2149.

Becker R. H. and Clayton R. N. (1977) Nitrogen isotopes in lunar soils as a measure of cosmic-ray exposure and regolith history. *Proc. Lunar Sci. Conf. 8th*, p. 3685–3704.

Becker R. H., Clayton R. N. and Mayeda T. K. (1976) Characterization of lunar nitrogen components. *Proc. Lunar Sci. Conf. 7th*, p. 441–458.

Butler P. Jr. and Meyer C. Jr. (1976) Sulfur prevails in coatings on glass droplets: Apollo 15 green and brown glasses and Apollo 17 orange and black (devitrified) glasses. *Proc. Lunar Sci. Conf. 7th*, p. 1561–1581.

Cadenhead D. A., Wagner N. J., Jones B. R. and Stetter J. R. (1972) Some surface characteristics and gas interactions of Apollo 14 fines and rock fragments. *Proc. Lunar Sci. Conf. 3rd*, p. 2243–2257.

Carrier W. D. III, Bromwell L. G. and Martin R. T. (1972) Strength and compressibility of returned lunar soil. *Proc. Lunar Sci. Conf. 3rd.*, p. 3223–3234.

Chang S., Lennon K. and Gibson E. K. Jr. (1974) Abundances of C, N, H, He, and S in Apollo 17 soils from Stations 3 and 4: Implications for solar wind exposure ages and regolith evolution. *Proc. Lunar Sci. Conf. 5th*, p. 1785–1800.

Funkhouser J., Jessberger E., Müller O. and Zähringer J. (1971) Active and inert gases in Apollo 12 and Apollo 11 samples released by crushing at room temperature and by heating at low temperatures. *Proc. Lunar Sci. Conf. 2nd*, p. 1381–1396.

Gerlach T. M. (1974) the C-O-H-S gas system and its application to terrestrial and extraterrestrial volcanism. Ph.D. dissertation, Univ. Arizona, Tucson.

Goel P. S., Shukla P. N., Kothari B. K. and Garg A. N. (1975) Total nitrogen in lunar soils, breccias and rocks. *Geochim. Cosmochim. Acta* **39**, 1347–1352.

Grossman J. J., Mukherjee N. R. and Ryan J. A. (1972) Microphysical, microchemical and adhesive properties of lunar materials—III. Gas interaction with lunar material. *Proc. Lunar Sci. Conf. 3rd*, p. 2243–2257.

Hartman C. H. and Dimick K. P. (1966) Helium detector for permanent gases. *J. Gas Chromatog.* **4**, 163–167.

Heiken G. H., McKay D. S. and Brown R. W. (1974) Lunar deposits of possible pyroclastic origin. *Geochim. Cosmochim. Acta* **38**, 1703–1718.

Heymann D. and Yaniv A. (1971) Breccia 10065: Release of inert gases by vacuum crushing at room temperature. *Proc. Lunar Sci. Conf. 2nd*, p. 1681–1692.

Jost W. (1960) *Diffusion in Solids, Liquids, Gases*. Academic Press, N.Y. 558 pp.

Kaplan I. R., Kerridge J. F. and Petrowski C. (1976) Light element geochemistry of the Apollo 15 site. *Proc. Lunar Sci. Conf. 7th*, p. 481–492.

Kerridge J. F., Kaplan I. R., Petrowski C., and Chang S. (1975) Light element geochemistry of the Apollo 16 site. *Geochim. Cosmochim. Acta* **39**, 137–162.

Kvenvolden K. and Roedder E. (1971) Fluid inclusions in quartz crystals from South-West Africa. *Geochim. Cosmochim. Acta* **35**, 1209–1229.

Moore C. B. and Lewis C. F. (1976) Total nitrogen contents of Apollo 15, 16 and 17 lunar rocks and breccias (abstract). In *Lunar Science VII*, p. 571–573. The Lunar Science Institute, Houston.

Morris R. V. (1976) Surface exposure indices of lunar soils: A comparative FMR study. *Proc. Lunar Sci. Conf. 7th*, p. 315–335.

Müller O. (1972) Chemically bound nitrogen abundances in lunar samples, and active gases released by heating at lower temperatures (250 to 500°C). *Proc. Lunar Sci. Conf. 3rd*, p. 2059–2068.

Müller O. (1973) Chemically bound nitrogen contents of Apollo 16 and Apollo 15 lunar fines. *Proc. Lunar Sci. Conf. 4th*, p. 1625–1634.

Müller O. (1974) Solar wind nitrogen and indigenous nitrogen in Apollo 17 lunar samples. *Proc. Lunar Sci. Conf. 5th*, p. 1907–1918.

Petrowski C., Kerridge J. F. and Kaplan I. R. (1974) Light element geochemistry of the Apollo 17 site. *Proc. Lunar Sci. Conf. 5th*, p. 1939–1948.

Phinney W. C., Anders E., Bogard D. D., Butler P., Gibson E. K. Jr., Gose W., Heiken G., Hohenberg C., Nyquist L., Pearce W., Rhodes J. M., Silver L. T., Simonds C., Strangway D., Turner G., Walker R., Warner J. L. and Yuhas D. (1974) Progress report: Apollo 17, Station 6 Boulder Consortium (abstract). In *Lunar Science V, Supplement A*, p. 7–13. The Lunar Science Institute, Houston.

Piperov N. B. and Penchev N. P. (1973) A study on gas inclusions in minerals. Analysis of the gases from micro-inclusions in allanite. *Geochim. Cosmochim. Acta* **37**, 2075–2097.

Ringwood A. E. (1970) Petrogenesis of Apollo 11 basalts and implications for lunar origin. *J. Geophys. Res.* **75**, 6453–6479.

Roedder E. (1972) Composition of fluid inclusions. *U.S. Geol. Survey Prof. Paper 440-JJ*, JJ-1-JJ-163.

Roedder E. and Weiblen P. W. (1970) Lunar petrology of silicate melt inclusions, Apollo 11 rocks. *Proc. Apollo 11 Lunar Sci. Conf.*, p. 801–823.

Sato M. (1976) Oxygen fugacity and other thermochemical parameters of Apollo 17 high-Ti basalts and their implications on the reduction mechanism. *Proc. Lunar Sci. Conf. 7th*, p. 1323–1344.

Sato M., Hickling N. L. and McLane J. E. (1973) Oxygen fugacity values of Apollo 12, 14 and 15 lunar samples and reduced state of lunar magmas. *Proc. Lunar Sci. Conf. 4th*, p. 1061–1079.

Wellman T. R. (1970) Gaseous species in equilibrium with the Apollo 11 holocrystalline rocks during their crystallization. *Nature* **225**, 716–717.

Proc. Lunar Planet. Sci. Conf. 9th (1978), p. 2451–2467.
Printed in the United States of America

Carbon, nitrogen and sulfur in Apollo 15, 16 and 17 rocks

DAVID J. DES MARAIS

NASA-Ames Research Center, Moffett Field, California 94035

Abstract—Seven lunar rocks were examined for their carbon, nitrogen and sulfur contents and their carbon isotopic compositions. Samples were combusted sequentially at three temperatures to resolve any terrestrial contamination from indigenous lunar volatiles.

The sulfur abundances agree within experimental uncertainty with previous combustion data from the UCLA laboratory, but are lower than sulfur measurements obtained by x-ray fluorescence and infrared. Recombustion of the samples at 1250°C yielded negligible additional volatiles.

Nitrogen abundances in these samples are less than 1.0 $\mu g/g$, and average 0.4 $\mu g/g$. Two rocks, 15058 and 66095, yielded less than 0.1 $\mu g/g$.

The samples contain two carbon components. One component is released by combustion at 420°C; the other is released when the rocks melt. Different allocation splits of the same sample yielded different amounts of carbon at low temperatures, whereas all splits yielded identical high temperature abundances. These observations indicate that, prior to analysis, all samples contain variable amounts of terrestrial carbon contamination. The carbon observed at high temperatures in the basalts reflects indigenous lunar abundances which range from 2.5 to 6 $\mu g/g$. Two Apollo 16 rocks with shocked textures contain more than 10 $\mu g/g$ carbon, which may reflect either the presence of an ancient meteoritic component or chemically-bound atmospheric carbon dioxide.

Carbon in these rocks has $\delta^{13}C$ values of typically −19 to −26 permil. An equation is derived which predicts the abundance of spallation-produced ^{13}C in the samples. In five of the rocks, the estimated spallation-induced shifts in $\delta^{13}C$ values are less than 2 permil.

Assuming carbon and nitrogen abundances in the lunar crust of 4.0 and 0.4 $\mu g/g$ respectively, the moon is substantially depleted in these two elements relative to the earth, C and H meteorites and cosmic abundances.

INTRODUCTION

The complex geochemistry of carbon and nitrogen tends to complicate theoretical models which attempt to explain elemental abundances in the lunar crust. The ability of these two elements to form solid solutions with iron allows them to accompany iron to at least some extent during planetary differentiation and possibly planetary accretion. Carbon can condense at moderate nebular temperatures as amorphous coatings on silicate grains (Bauman *et al.*, 1973; Bunch and Chang, 1978). Carbon and nitrogen form highly volatile species which condense at very low nebular temperatures and which may be lost from the moon during magmatic activity. Given these considerations, carbon may accrete over a wide span of nebular temperatures and both carbon and nitrogen probably experience substantial redistribution during planetary evolution.

The earth and moon occupy markedly different positions in the observable spectrum of volatile element abundances in the terrestrial planets (see e.g., Gast *et al.*, 1970; Anders and Owen, 1977). Relative to terrestrial basalts, lunar basalts are markedly enriched in refractory elements, slightly enriched in iron, and depleted by one to two orders of magnitude in the volatile elements (Ganapathy *et al.* 1970). The markedly different bulk compositions of these two planets offer an opportunity to study the geochemical behavior of carbon and nitrogen in a wide spectrum of sample compositions. For example, because carbon and nitrogen are considered to be volatiles, their lunar crustal abundances, relative to earth, could parallel the abundances observed for the other volatile elements. On the other hand, the distinctive chemical behavior of carbon and nitrogen could cause these two elements to behave differently than the other volatiles during planetary accretion and differentiation.

The present work re-examines carbon, nitrogen and sulfur abundances in lunar rocks. The suite of rocks analyzed include an Apollo 15 quartz normative basalt and an olivine normative basalt, an Apollo 16 cataclastic anorthosite, a VHA basalt, a poikilitic KREEP-rich rock, and two Apollo 17 basalts.

EXPERIMENTAL

The light element analyses are performed using procedures optimized for the very low abundances in lunar rocks. To minimize contamination, the samples are processed in a glovebox and handled using small aluminum foil trays which have been extracted with pure solvents. A typical, 1.5 to 2 g sample is split into two unequal portions for replicate analyses. A portion is then ground to approximately 40 mesh size using a sapphire mortar and pestle which has been combusted overnight in a 560°C oven. The carbon and nitrogen contamination accompanying the sample handling, crushing and weighing operations was assessed using clean fused quartz chips previously fired in oxygen at 1220°C. As Table 1 shows, typically 0.2 μg carbon per gram of chips and no nitrogen and sulfur are added to the quartz sample during these procedures. Prior to loading a crushed sample into the combustion boat, the boat is cleaned for 2 hours in a 1220°C oven in an atmosphere of Matheson UHP Grade oxygen. A blank combustion is then performed. The boat is removed and reinserted into the oven, and, after a 2 hour evacuation at 150°C, the boat is recombusted in 40 torr oxygen at 1220°C for one hour.

As Table 1 shows, the typical combustion blanks are 0.09 μg carbon, 0.06 μg nitrogen and 1.5 μg sulfur. The surprisingly high sulfur value derives from residual sulfur remaining in the line from the previous lunar rock combustion, which is generally performed less than 24 hours prior to the blank analysis.

The combustion line design (Fig. 1) generally resembles that described by Sakai *et al.*, (1977),

Table 1. Procedural blanks, minimum sample requirements and typical sample sizes encountered in rock analyses.

	C, μg	N, μg	S, μg
Blanks			
Crushing, etc.	0.2	0	0
Combustion	0.09	0.06	1.5
Minimum Samples			
Manometer	0.01	0.03	0.03
Isotopes	0.5	1.4	1.6
Typical Samples	0.5–10	0–1	5–600

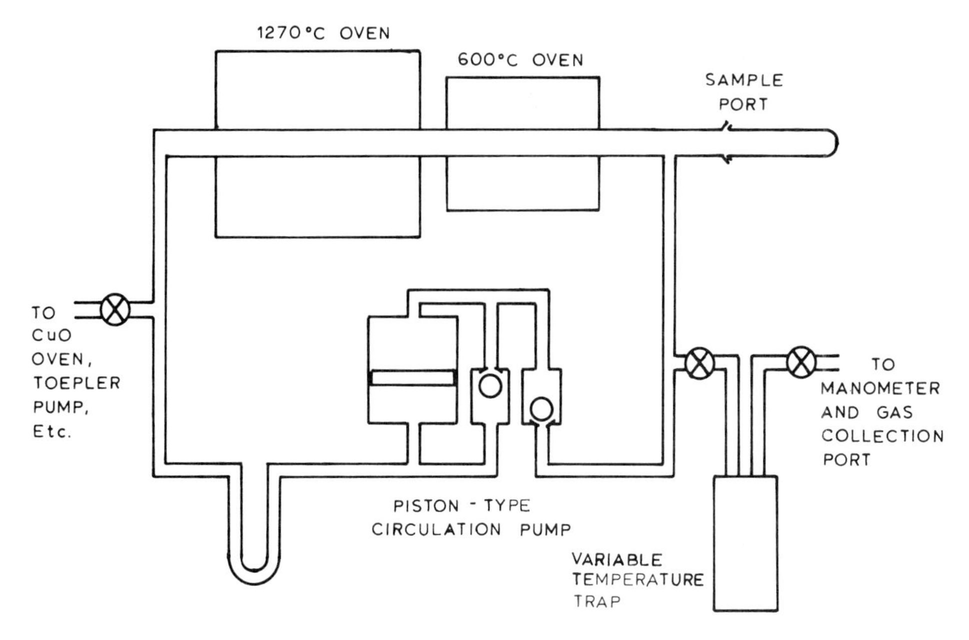

Fig. 1. Sample combustion section of lunar analysis line. The "600°C" oven is cooled to provide the 150°C, 420°C and 500°C temperatures required by the analysis. The section of line not shown is discussed by Sakai *et al.*, (1977).

except for a number of innovations needed to assess terrestrial contamination in the rock samples, maintain minimal blank levels, and handle the small amounts of carbon dioxide and nitrogen obtained from the samples. The sample combustion loop contains a high temperature oven (operable to 1270°C), a low temperature oven (operable to 600°C), a simple U-shaped cold trap, and a pyrex glass circulating pump. This pump consists of a piston and two valves (Fig. 1) which are operated by an external set of motor-driven magnets. When the piston moves down, the left valve is open and the right valve is closed, permitting the gas to move from one end of the piston chamber to the other. When the piston moves up, the left valve is closed and the right valve is open, and the gas is pumped around the combustion loop. The combustion loop is made of pyrex glass everywhere except for the portion made of mullite which passes through the two ovens. The sample and quartz boat are admitted to the combustion loop via a glass ball joint fitted with a viton o-ring. The boat is held in a quartz boat holder to which an encapsulated magnet is attached. This magnet permits the use of an external magnet to move the sample into the ovens. After it is loaded, the sample is moved into the low temperature oven, which is maintained at 150°C, and the entire combustion loop is evacuated overnight.

The analysis procedure includes three sequential combustions of a rock sample at the temperatures 420°C, 500°C and 1220°C (or 1270°C, depending upon the melting point of the rock). The high temperature oven is always maintained above 1200°C during the entire analysis, thus ensuring the combustion of gases liberated from the sample when it is at 420°C or 500°C in the low temperature oven. The three step combustion procedure removes terrestrial contaminants before it melts the rock and liberates lunar volatiles. Control experiments show that both a strand of cotton cellulose and the carbon from a fingerprint volatilize at 420°C in an oxygen atmosphere. The 500°C combustion constitutes a check of this contaminant removal. The 1220°C (or 1270°) combustion then extracts uncontaminated carbon and nitrogen from the rock sample. Incidentally, a sequential combustion of the quartz chips processed through the rock grinding procedure (see above) typically yielded 0.2 μg/g carbon at 420°C and essentially no carbon, nitrogen or sulfur in excess of line blank levels at the higher temperatures. For each of the three combustions, oxygen is transferred to the loop from a 900°C CuO oven using a toepler pump until the combustion loop pressure attains 40 mm Hg. During the 420°C, 500°C and 1220°C combustion intervals, which are 40 min., 40 min. and 90 min., respectively, the oxygen is circulated continuously using the piston pump. In the last 20 min. of each interval, the U-trap in the combustion loop is maintained at −196°C to trap CO_2, SO_2 and H_2O. After each combustion step, the rock is cooled to below 150°C. The O_2 is readsorbed into a 500°C CuO oven and the noncondensible N_2 is pumped via a toepler pump into a manometer for measurement and subsequent collection. To shorten the analysis time, the N_2 from the 420°C combustion is not measured. The condensible gases are transferred from the combustion loop to a variable temperature trap and purified as described by Des Marais (1978). The CO_2 and SO_2 are released from the trap by 15 min. distillations at −150°C and −90°C respectively. Only trace amounts of water were recovered from the combustions.

The manometers for measuring both the N_2 and the CO_2 and SO_2 are constructed of 1.5 mm diameter precision bore tubing. The mercury can be raised to control both the volume and pressure of the measurement. As Table 1 shows, the minimum measurable amounts are well below the combustion blank levels. Levels of detectability, established by the reproducibility of the blank values, are 0.15 μg for carbon, 0.07 μg for nitrogen and 2 μg for sulfur. Calibration of the manometers using known volumes of gas determined their accuracy to be better than ±4%. Samples were transferred to the isotope mass spectrometer in 6 mm pyrex tubes which were subsequently opened using the device described by Des Marais and Hayes (1976).

Carbon isotope values were obtained using a Nuclide 6–60 RMS mass spectrometer fitted with a special inlet system which enables small (down to 0.05 μmole) samples to be analyzed to a precision typically exceeding 1 permil. The sample flows from a 2 mm I.D. pyrex tube into the mass spectrometer via a viscous metal leak. The sample pressure in the pyrex tube is maintained by a vertical column of mercury which is raised by a motor driven bellows assembly. In this fashion, the sample pressure in the instrument is held constant even though over 90 percent of the sample can be utilized for the isotope measurement.

RESULTS AND DISCUSSION

Sulfur

The sulfur abundances reported in Table 2 are in agreement with the values obtained by the UCLA laboratory using combustion. X-ray fluorescence analyses by Duncan *et al.* (1973) and Hubbard *et al.* (1973) yielded systematically higher values. Agreement between combustion, hydrolysis and x-ray fluorescence methods is generally mediocre, and it is not yet clear which method provides more accurate results. Petrowski *et al.* (1974) discuss these methods and conclude that combustion and hydrolysis are the two most precise techniques. They suggest that the agreement between these two techniques may be affected by the chemical form of the sulfur in the samples. Despite these reservations, the general agreement between sulfur combustion analyses by UCLA and this laboratory indicates that the analytical yields were identical in the two laboratories. This fact should be remembered during the discussion of the carbon results later in this paper.

The release of sulfur increases monotonically with combustion temperature. This behavior suggests that sulfur is being released from within the rock sample and that its extraction is more efficient at higher temperatures, especially after the rock has melted. A terrestrial contaminant such as elemental or organic sulfur would be situated on sample surfaces and therefore oxidize or volatilize readily at 420°C. If an appreciable contamination were present, one might expect a bimodal sulfur release pattern with temperature, with maxima occurring at 420°C and >1200°C and a minimum occurring at 500°C. The absence of such a bimodal pattern, together with the absence of a probable source of terrestrial sulfur contamination during handling, suggests that the released sulfur is lunar in origin.

The chemical nature of rock sulfur, carbon and nitrogen probably controls their relative rates of volatilization. Given that minimal amounts of sulfur contamination are present, the oxygen at 420°C volatilizes from one to seven percent of the indigenous sulfur from the rock (see Table 2).

Nitrogen

Lunar rocks apparently contain very little nitrogen, typically less than 1 μg/g, as previously observed by the Chicago and UCLA laboratories. The nitrogen yields from the 500°C combustion analyses of 15058, 65015, 66095 and 75035 were equal to the combustion system blank. This result verifies that the samples were free of terrestrial nitrogen before the high temperature combustions. It should be noted here that Niemeyer and Leich (1976) observed the evolution of adsorbed terrestrial xenon from rock 60015 at temperatures exceeding 900°C. Because they were measuring very small amounts of xenon (approximately 10^{-10} ccSTP, compared to the present study's nitrogen blank level of 5×10^{-5} ccSTP), and because xenon has a substantially lower vapor pressure than nitrogen at room temperatures, the amounts of terrestrial nitrogen retained by the rock samples above 500°C is relatively insignificant. The nitrogen yields at high temperature range from 0.1 μg/g to 0.7 μg/g and indicate that varying amounts of nitrogen do exist in excess of spallation-produced ^{15}N (see Becker and Clayton, 1975). It should be noted that the typical nitrogen detection limit for these samples was 0.1 μg/g. Because the analyzed rock samples weighed less than one gram, the amount of nitrogen obtained was insufficient for a precise determination of $\delta^{15}N$ values. Approximate $^{15}N/^{14}N$ measurements on the samples from rocks 65015, 70215 and 75035 did reveal a several percent ^{15}N enrichment relative to terrestrial atmospheric nitrogen. This observation is consistent with the discovery of spallation-produced ^{15}N in lunar rocks by Becker and Clayton (1975).

Carbon

The carbon abundances shown in Table 3 are substantially lower than the literature values. The total carbon released from a given sample in the present study averages only 40 percent of the corresponding literature values. The lower values reported here do not reflect a relatively incomplete extraction of carbon from the sample. First, essentially identical combustion methods and carbon

Table 2. Nitrogen and Sulfur Abundance Data.

Sample Number	Weight analyzed, g	T, °C	N, μg/g	S, μg/g
15058,207	0.802	420	n.m.[a]	0
		500	<0.1	25
		1270	<0.1	646
		Total	<0.1	671
		Literature	–	400[b],960[c],700[d]
15555,479	0.809	420	n.m.	10.4
		500	n.m.	2.5
		1220	<0.8	lost
		Total	<0.8	–
		Literature	<2[e]	855[c],650[e],500[d]
61016,323	0.603	370	n.m.	0.3
		500	n.m.	2.0
		1260	<1.0	lost
		Total	<1.0	–
		Literature	0[f]	395[f],320[f],100[g], 510[h],518[i]
65015,174	0.600	420	n.m.	5.7
		500	<0.1	39
		1270	0.7	597
		Total	0.7	642
		Literature	0[f]	600[f],790[h] 820[j],1300[k]
66095,198	0.740	420	n.m.	6
		500	<0.1	104
		900	<0.1	454
		1250	<0.1	192
		Total	<0.1	756
		Literature	0[f]	775[f],920[h], 1200[g],1400[k]
70215,110	0.400	440	n.m.	173
		440	n.m.	50
		1260	<0.9	1190
		Total	<0.9	1410
		Literature	<8.0[l],3.0[m], 88[n]	1689[o],1700[p],1581[i] 2210[q],2040[b]

dioxide recovery procedures were used both in the literature analyses and the present work. Second, a recombustion of samples 15555 and 66095 at 1250°C yielded no additional carbon and nitrogen. Given this evidence, the literature data apparently suffer from very substantial terrestrial contamination.

The data in Table 3 reveal that two carbon components are in the samples. The

Table 2. *(cont'd.)*

Sample Number	Weight analyzed, g	T, °C	N, μg/g	S, μg/g
70215,113	0.494	420	n.m.	95
		500	n.m.	123
		1260	<1.0	1180
		Total	<1.0	1400
75035,57	0.812	420	n.m.	13.6
		500	<0.1	88
		1220	0.4	1750
		Total	0.4	1850
		Literature	85[n]	1849[o],3140[b],2770[q]
75035,57	0.664	420	n.m.	58
		500	<0.1	169
		1220	0.3	1530
		Total	0.3	1760

[a]Not measured.
[b]Moore (1975).
[c]Gibson *et al.* (1975).
[d]Rhodes and Hubbard (1973).
[e]Kaplan *et al.* (1976).
[f]Kerridge *et al.* (1975).
[g]LSPET (1972).
[h]Duncan *et al.* (1973).
[i]Rees and Thode (1974).
[j]Hughes *et al.* (1973).
[k]Hubbard *et al.* (1973).
[l]Müller (1974).
[m]Müller *et al.* (1976).
[n]Moore and Lewis (1976).
[o]Petrowski *et al.* (1974).
[p]Rhodes *et al.* (1974).
[q]Gibson and Moore (1974).

low and high temperature extractions yielded the most carbon, whereas in all but two cases, the 500°C extractions yielded very little carbon. The two samples which yielded appreciable carbon at 500°C, namely 61016 and 66095, are discussed later. This bimodal release pattern contrasts with the sulfur release pattern discussed above, which increases monotonically with temperature. Although the sulfur released at the lowest temperature is probably indigenous, it constitutes only 1 to 7 percent of the total sulfur obtained from the samples. By contrast, the low temperature carbon components are highly variable and often constitute 50 percent or more of the total carbon yield. The cause of this behavior becomes apparent if the replicate analyses of rocks 70215 and 75035 are examined.

Figure 2 depicts the results for the allocation splits of 70215 and 75035. Two analyses of 70215 were performed upon two separate splits (numbers 110 and 113) whereas both analyses of 75035 were performed on the same split (number 57). Note that the high but unequal low temperature carbon yields from the two different splits of 70215 contrast with the lower and more uniform yields from 75035. These data indicate that the different splits received varying amounts of terrestrial contamination. Note that, despite this nonuniform contamination, the

Table 3. Carbon abundance and isotope data.

Sample Number	Weight Analyzed, g	T, °C	C, μg/g	$\delta^{13}C_{PDB}$
15058,207	0.802	420	2.2	−33.1 ± 0.2
		500	0.6	−27.1 ± 1
		1270	2.5	−3.3 ± 0.2
		Total	5.3	
		Literature	27[a],10[a]	–
15555,479	0.809	420	2.1	−32.6 ± 0.3
		500	<0.1	–
		1220	5.6	−26.0 ± 0.1
		Total	7.7	−27.8
		Literature	12[b]	−14[b]
61016,323	0.603	370	26	−29.3 ± 0.1
		500	6.7	−30.8 ± 0.1
		1260	8.3	−33.4 ± 0.2
		Total	41.1	−30.8
		Literature	35[c],28[c]	−35.7[c],−32.8[c]
65015,174	0.600	420	3.7	
		500	0.9	
		1270	2.1	
		Total	6.7	
		Literature	9[c],45[c]	−16.7[c]
66095,198	0.740	420	19.4	−22.7 ± 0.1
		500	4.3	−13.1 ± 0.2
		900	5.2	−15.2 ± 0.2
		1250	1.1	−19.4 ± 1
		Total	30.0	
		Literature	38[c],88[c]	−24.2[c],−23.5[c]
70215,110	0.400	440	14.2	−24.7 ± 0.1
		440	0.5	–
		1260	5.4	−21.0 ± 0.2
		Total	20.1	−23.7
		Literature	36[d],31[e]	−39.3[d]
70215,113	0.494	420	7.0	−26.2 ± 0.2
		500	0.7	−25.6 ± 0.5
		1260	4.6	−19.3 ± 0.1
		Total	12.3	−23.5

replicate carbon yields at >1200°C are in good agreement. This agreement would be expected if the terrestrial contamination had been removed from the samples. The notion that the low temperature carbon fractions represent contamination receives additional support from the carbon isotope data. The $\delta^{13}C$ values

Table 3. *(cont'd.)*

Sample Number	Weight Analyzed, g	T, °C	C, μg/g	$\delta^{13}C_{PDB}$
75035,57	0.812 g	420	2.4	−30.1 ± 0.1
		500	0.5	−24.0 ± 1.0
		1220	4.8	−25.3 ± 0.2
		Total	7.7	−26.7
		Literature	23[d],64[e]	−28.5[d]
75035,57	0.664 g	420	3.0	−27.0 ± 0.1
		500	0.7	−21.5 ± 0.2
		1220	4.9	−21.6 ± 0.2
		Total	8.6	−23.5

[a] Moore *et al.* (1973).
[b] Kaplan *et al.* (1976).
[c] Kerridge *et al.* (1975).
[d] Petrowski *et al.* (1974).
[e] Moore (1975).

of these fractions lie within the range −24 to −33 permil versus PDB. These values are commonly obtained from cloth fibers, plastics and hydrocarbons.

Given the presence of this contamination, any estimate of indigenous carbon abundances in lunar rocks is obviously compromised by the volatilization of indigenous carbon at combustion temperatures below 500°C. The generally low carbon yields at 500°C (except for 61016 and 66095, which are discussed below) suggest that a no larger percentage of lunar carbon is liberated at low temperatures than the percentage of lunar sulfur liberated at low temperatures (compare Tables 2 and 3). The carbon abundances measured at >1200°C therefore constitute at least 80 to 90 percent of the original lunar carbon component.

The carbon yields from anorthositic breccias 61016 and 66095 merit additional comment (see Table 3). Both samples released more carbon than the other rocks, especially during the 500°C extractions. Breccia 66095 released most of its carbon by 900°C. The substantial yield of carbon between 400°C and 500°C is attributed by Gibson and Moore (1975) to carbonate formed as secondary alteration product. They propose that these breccias represent metamorphosed regolith, whose solar wind-derived carbon has been chemically transformed by metamorphism into a carbonate phase. Two observations in the present study do not support this explanation. The $\delta^{13}C$ values of this "carbonate" carbon are much lower than the values typically observed in the Apollo 16 regolith (see e.g., Kerridge *et al.*, 1975). Second, little or no nitrogen is found in these samples. Because nitrogen is roughly two thirds as abundant in the regolith as carbon (see e.g., Kerridge *et al.*, 1975) and because nitrogen is generally better retained during soil maturation processes than carbon (Des Marais *et al.*, 1974), one

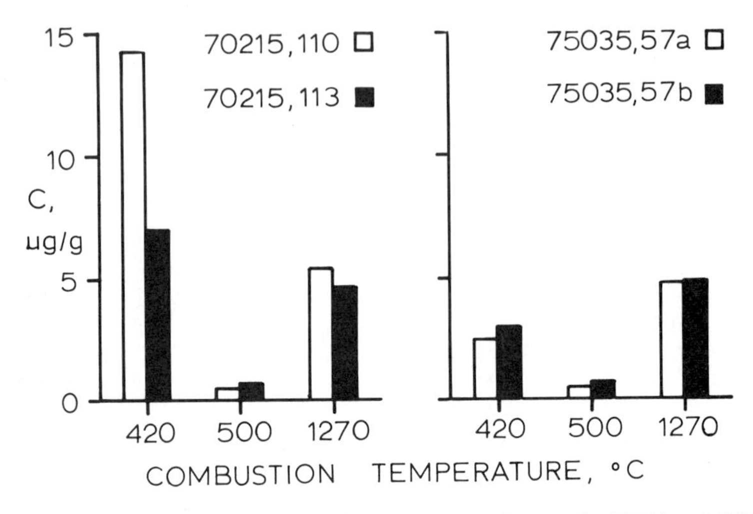

Fig. 2. Carbon yield versus combustion temperature for samples 70215 and 75035.

would expect to find appreciable nitrogen components in these two rocks if the lunar regolith were the source of the carbon.

Other possible sources of carbon include meteorites and some form of contamination which must be rather more refractory than lint or hydrocarbons. Some evidence supports a meteoritic origin for the carbon. El Goresy *et al.*, (1971) observed cohenite (Fe_3C) and metal phases of meteoritic composition in 66095. Deines and Wickman (1975) measured $\delta^{13}C$ values between −18 and −19 permil for cohenite in iron meteorites. These data compare closely to the value of −19.9 permil obtained for the total carbon released from the combustion of 66095. A small amount of isotopically light terrestrial carbon contamination in the 420°C fraction could easily produce the 1 permil difference found between the combustion isotopic value and the values reported by Deines and Wickman. Meteoritic cohenite is one possible source of the carbon released during the combustion of 66095. Carbon released at higher temperatures could derive directly from the combustion of cohenite. Perhaps the carbon released at lower temperatures derives from cohenite carbon which had been chemically altered prior to analysis, perhaps by the metamorphism of 66095 (see Gibson and Moore, 1975).

Some of the Apollo 16 breccias apparently have chemical and mineralogical peculiarities (El Goresy *et al.*, 1973) which make them especially susceptible to contamination from the terrestrial atmosphere. For example, 60015 and 60025 appear to have trapped terrestrial xenon, which is released only above 900°C (Leich and Niemeyer, 1975); and the water in 66095 also appears to be terrestrial (Epstein and Taylor, 1974). The carbon released at low temperatures from 66095 could derive from atmospheric carbon dioxide which had reacted with ferrous iron in a lawrencite grain to give $FeCO_3$ (R. N. Clayton, pers. comm., 1978). This carbonation reaction should be accompanied by a kinetic isotope effect which yields carbonate that is isotopically lighter than terrestrial atmospheric carbon dioxide, which is −7 permil versus PDB. In fact, $CaCO_3$ with $\delta^{13}C$ ranging from −19 to −25 permil has been made by allowing CaO to absorb atmospheric carbon dioxide (R. N. Clayton, pers. comm., 1978). Atmospheric carbon dioxide is certainly another possible source for the "excess" carbon in 61016 and 66095.

Carbon isotopes

The $\delta^{13}C$ values of four rocks cluster in the range −19 to −26 permil. This range compares closely with values obtained from reduced carbon in terrestrial igneous rocks (−24 to −28 permil; Hoefs, 1973) and the total carbon in H, L, C4 and achondrite meteorites (−23 to −31 permil; Boato, 1954; Belsky and Kaplan, 1970). All of these occurrences share additional common features, namely the carbon is in a reduced form, and the host phases, whether they be rock or meteoritic, experienced relatively high temperatures of formation.

Porphyritic basalt 15058 contains only 2.5 μg of carbon with a $\delta^{13}C$ value of −3.3 permil. The rather high ^{13}C abundance in 15058 warrants an assessment of spallation-produced ^{13}C in these rocks. Armstrong and Alsmiller (1971) calculated spallation production rates for a model composition of lunar basalt. Using Fig. 2 of their paper, production rates for ^{13}C, ^{15}N, and ^{21}Ne are approximately 0.55, 1.7 and 0.14 nuclei/cm^2-sec respectively. By measuring $\delta^{15}N$ values on rocks with known spallation exposure ages, Becker *et al.* (1976) obtained a ^{15}N spallation production rate of $3.6 \pm 0.8 \times 10^{-6}$ μg ^{15}N/g sample/my. From the data of Armstrong and Alsmiller (1971), the ^{13}C spallation production rate is expected to be 0.55/1.7 or 0.32 times the ^{15}N rate. Using Becker *et al.*'s nitrogen value, a ^{13}C production rate of 1.0×10^{-6} μg ^{13}C/g sample/my is obtained.

The increase in $\delta^{13}C$ as a function of time due to spallation is given by the equation

$$\delta^{13}C_{PDB}(t) = \delta^{13}C_{PDB}(0) + [1000\ P(^{12}C/^{13}C)_{PDB}]\ (t/[C]), \qquad (1)$$

where $\delta^{13}C_{PDB}(0)$ is the $\delta^{13}C$ value of the non-spallation carbon, P is the production rate in μg^{13}C/g sample/my, t is the spallation age in m.y. and [C] is the total carbon content of the rock in μg/g. Using a value 88.9 for $(^{12}C/^{13}C)$ and the value 1.0×10^{-6} μg ^{13}C/g sample/my for P, Equation (1) can be rewritten as

$$\delta^{13}C_{PDB}(t) = \delta^{13}C_{PDB}(0) + 8.9 \times 10^{-2} my^{-1}\ \mu g\ g^{-1}\ (t/[C]). \qquad (2)$$

This equation is analogous to the equation for the ^{15}N spallation effect reported by Becker *et al.* (1976):

$$\delta^{15}N_{AIR}(t) = \delta^{15}N_{AIR}(0) + 1.0\ my^{-1}\ \mu g\ g^{-1}\ (t/[N]). \qquad (3)$$

where [N] is the total nitrogen concentration in μg/g. A comparison of these two equations reveals that ^{13}C spallation production effects are an order of magnitude smaller than ^{15}N spallation effects. In addition, the ^{13}C spallation effect is attenuated further by the markedly higher carbon contents observed in rocks. Using Equation (2), available spallation ages, and carbon data from the present study, the expected $\delta^{13}C$ shifts due to spallation can be estimated (Table 4). The [C] values used are from the high temperature combustion fractions only. Becker *et al.* (1976) found that spallation nitrogen was released in soils only at high temperatures. Because the rock matrix should protect spallation carbon from oxidation, it is assumed that the carbon is released only at high temperatures when the rock melts. As Table 4 shows, the predicted $\delta^{13}C$ shift for these rocks is less than 2 permil. If we assume for the moment that the high ^{13}C content of rock 15058 derives from spallation, Equation (2) permits a spallation age to be calculated. Assuming a $\delta^{13}C_{PDB}$ (o) value of −20 permil, the calculated spallation age of 15058 is 470 m.y. Unfortunately the validity of this calculation cannot be checked, because no independent determination of spallation age exists for 15058 at this time.

Light element relationships with iron

Although carbon and nitrogen are soluble in iron, their abundances do not correlate with Fe^{o} abundances in these rocks (Table 5). One possible exception is the meteoritic cohenite in 66095 (discussed above). Jull *et al.* (1976) observed that acid solutions can hydrolyze carbon which is dissolved in α-iron. The very minor abundance of hydrolyzable carbon in lunar rocks (Cadogan *et al.*, 1972) suggests that the iron in these rocks contains little carbon. Gibson *et al.*, (1975) suggest that desulfuration of the Apollo 15 and 17 basaltic melts produced the metallic iron found in lunar basalts. Both the apparent absence of hydrolyzable

Table 4. Predicted spallation ^{13}C contributions in five lunar rocks.

Rock	Spallation age, my	Calculated $\Delta\delta^{13}C^{(a)}$	Measured $\delta^{13}C_{PDB}(t)^{(b)}$	Calculated $\delta^{13}C(0)$
	17			
15555	81 ± 7[c]	1.2	−26.0	−27.2
61016	≤7[d]	0.07	−33.4	−33.5
66095	1.1[e]	0.01	−15.9	−15.9
70215	100[f]	1.7	−20.1	−21.8
75035	71.7 ± 1.8[g]	1.2	−23.5	−24.7

[a] $\Delta\delta^{13}C = \delta^{13}C(t) - \delta^{13}C_{PDB}(0)$.
[b] This work. High temperature combustion fractions only.
[c] Behrmann *et al.* (1972).
[d] Stettler *et al.* (1973).
[e] Heymann *et al.* (1973) in Ganapathy *et al.* (1974).
[f] Kirsten and Horn (1974).
[g] Crozaz *et al.* (1974).

carbon and the low total carbon abundances in the rocks indicate that very minor amounts of carbon were partitioned into the iron during the desulfuration of these magmas.

Sato (1976) suggests that the loss of carbon constituted the dominant mechanism for the reduction of lunar basaltic magmas and the production of Fe^o. The carbon mechanism would proceed by reactions such as the following:

$$C + FeO = Fe^o + CO \quad (4)$$

$$C + 2FeO = 2Fe^o + CO_2. \quad (5)$$

Initial carbon concentrations of at least 30 to 100 μg/g are required to produce the observed Fe^o, and the CO and CO_2 are lost from the basaltic magmas during the reduction process. According to Sato, the little carbon that is left unreacted in the basalt would be locked in the metallic globules trapped in the early silicates. Although this explanation is attractive in theory, both the absence of a correlation between carbon and iron and the very low and uniform carbon abundances do not support the contention that the lunar carbon inventory was sufficient to assume a major role in the reduction of lunar basalts. A more definitive evaluation of this proposed carbon mechanism must await additional planned rock analyses.

Carbon and nitrogen inventories on the moon and elsewhere

Because rocks from three different Apollo missions show remarkably consistent carbon and nitrogen contents, let us assume for this discussion that these

Table 5. Relationship between carbon, nitrogen, sulfur, Fe° and FeO in lunar rocks.

Sample	C, μg/g	N, μg/g	S, μg/g	Fe°, wt.%	FeO, wt.%
15058	2.5	<0.1	670	0.071[a,b,c]	19.97[e]
15555	5.6	—	—	0.070[a,b,c]	22.47[e]
61016	14.5+	—	—	0.75[d]	5.12[f]
65015	2.1	0.7	640	0.96[d]	8.59[f]
66095	9.0+	0.1	760	1.05[b,d]	7.46[f]
70215	5.0	—	1410	0.111[b,c]	19.22[g]
75035	4.9	0.4	1810	0.061[c]	18.8[h]

[a]Nagata *et al.* (1972).
[b]Huffman *et al.* (1974).
[c]Gibson *et al.* (1975).
[d]Kerridge *et al.* (1975).
[e]Rhodes and Hubbard (1973).
[f]Duncan *et al.* (1973).
[g]Rose *et al.* (1974).
[h]Laul *et al.* (1974).

values accurately represent lunar crustal abundances. Samples 61016 and 66095 are excluded because they may contain "exotic" carbon. The proposed carbon and nitrogen crustal abundances are therefore 4.0 μg/g and 0.4 μg/g respectively. Table 6 lists these two values together with estimated abundances for terrestrial basalts, C1 and H meteorites and the solar system. The last two columns contain lunar crustal abundances relative to abundances in these other bodies. Comparisons both between the values in Table 6 and between Table 6 and abundance data for other elements offer a number of interesting observations.

a. The carbon to nitrogen abundance ratios in the lunar crust and in other solid bodies are uniformly greater than the solar system ratio of 3.2. These larger ratios possibly reflect a preferential incorporation of carbon, relative to nitrogen, during the condensation of these bodies.

b. The lunar crust is more depleted in carbon and nitrogen than any other occurrence listed. The moon is depleted in all volatile elements, relative to these other occurrences (Ganapathy *et al.*, 1970).

c. There is reason to believe that differences in elemental abundances between terrestrial and lunar basalts reflect differences in the bulk compositions of the two planets (Ganapathy and Anders, 1974). If metallic element abundances in both lunar and terrestrial basalts are normalized to their respective bulk planetary abundances, the two resulting abundance patterns are similar (see Anders, 1977). The similar patterns imply similar geochemical behavior of the earth and moon during planetary differentiation and basalt formation. Relative to terrestrial basalts, lunar basalts are typically depleted in volatile metals by one to two orders of magnitude (Ganapathy *et al.*, 1970). Lunar basalts are also depleted in carbon and nitrogen by one to two orders of magnitude, as shown on Table 6. Therefore, the differences in basaltic carbon, nitrogen and volatile metal

Table 6. Carbon and nitrogen abundances in lunar rocks and elsewhere.

Occurrence	Typical Abundance					Abundance Ratio[a] (Moon/elsewhere)	
	μg/g sample		atoms/10^6 atoms Si				
	C	N	C	N	C/N[a]	C	N
Lunar rocks	4.0[b]	0.4[b]	42	3.6	12	1.00	1.00
Unaltered terrestrial mafic rocks	200[c]	3[d]	1.9×10^3	24	79	2.3×10^{-2}	1.5×10^{-1}
H meteorites	1.19×10^3[e]	8[f]	1.46×10^4	84	174	2.9×10^{-3}	4.3×10^{-2}
Cl meteorites	3.6×10^4[e]	1.48×10^3[g]	7.7×10^5	2.8×10^4	28	5.5×10^{-5}	1.3×10^{-4}
Solar system	–	–	1.18×10^7[h]	3.74×10^6[h]	3.2	3.6×10^{-6}	9.6×10^{-7}

[a]Ratios calculated using units atoms/10^6 atoms Si.
[b]This work.
[c]Hoefs (1965), Fuex and Baker (1973), Pineau *et al.* (1976).
[d]Becker and Clayton (1977).
[e]Otting and Zähringer (1967).
[f]Kung and Clayton (1978).
[g]Injerd and Kaplan (1974).
[h]Cameron (1973).

abundances that exist between the earth and moon may reflect differences in the planetary inventories which originated during the accretion of the earth-moon system.

Acknowledgments—I thank E. K. Gibson and the curatorial staff at Johnson Space Center for their efforts in supplying these samples. I also appreciate the helpful criticism of the manuscript by R. N. Clayton and R. H. Becker. This work was supported by a NASA grant.

References

Anders E. (1977) Chemical compositions of the moon, earth, and eucrite parent body. *Phil. Trans. Roy. Soc. London* **A285**, 23–40.

Anders E. and Owen T. (1977) Mars and Earth: Origin and abundance of volatiles. *Science* **198**, 453–465.

Armstrong T. W. and Alsmiller R. G. Jr. (1971) Calculation of cosmogenic radionuclides in the moon and comparison with apollo measurements. *Proc. Lunar Sci. Conf. 2nd*, p. 1729–1745.

Bauman A. J., Devaney J. R. and Bollin E. M. (1973) Allende meteorite carbonaceous phase: Intractable nature and scanning electron morphology. *Nature* **241**, 264–267.

Becker R. H. and Clayton R. N. (1975) Nitrogen abundances and isotopic compositions in lunar samples. *Proc. Lunar Sci. Conf. 6th*, p. 2131–2149.

Becker R. H. and Clayton R. N. (1977) Nitrogen isotopes in igneous rocks (abstract). *EOS* (*Trans. Amer. Geophys. Union*) **58** (6), 536.

Becker R. H., Clayton R. N. and Mayeda T. K. (1976) Characterization of lunar nitrogen components. *Proc. Lunar Sci. Conf. 7th*, p. 441–458.

Behrmann C., Crozaz G., Drozd R., Hohenberg C. M., Ralston C., Walker R. M. and Yuhas D. (1972) Rare gas and particle track studies of Apollo 15 samples. In *The Apollo 15 Lunar Samples* (J. W. Chamberlain and C. Watkins, eds.), p. 329–332. The Lunar Science Institute, Houston.

Belsky T. and Kaplan I. R. (1970) Light hydrocarbon gases, ^{13}C, and the origin of organic matter in carbonaceous chondrites. *Geochim. Cosmochim. Acta* **34**, 257–278.

Boato G. (1954) The isotopic composition of hydrogen and carbon in the carbonaceous chondrites. *Geochim. Cosmochim. Acta* **6**, 209–220.

Bunch T. E. and Chang S. (1978) Carbonaceous chondrite (CM) phyllosilicates: Condensation or alteration origin? (abstract). In *Lunar and Planetary Science IX*, p. 134–136. Lunar and Planetary Institute, Houston.

Cadogan P. H., Eglinton G., Firth J. N. M., Maxwell J. R., Mays B. J. and Pillinger C. T. (1972) Survey of lunar carbon compounds: II. The carbon chemistry of Apollo 11, 12, 14 and 15 samples. *Proc. Lunar Sci. Conf. 3rd*, p. 2069–2090.

Cameron A. G. W. (1973) Abundances of the elements in the solar system. *Space Sci. Rev.* **15**, 121–146.

Crozaz G., Drozd R., Hohenberg C., Morgan C., Ralston C., Walker R. and Yuhas D. (1974) Lunar surface dynamics: Some general conclusions and new results from Apollo 16 and 17. *Proc. Lunar Sci. Conf. 5th*, p. 2475–2499.

Deines P. and Wickman F. E. (1975) A contribution to the stable carbon isotope geochemistry of iron meteorites. *Geochim. Cosmochim. Acta* **39**, 547–557.

Des Marais D. J. (1978) A simple variable temperature cryogenic trap for the separation of gas mixtures. *Anal. Chem.* **50**, 1405–1406.

Des Marais D. J. and Hayes J. M. (1976) Tube cracker for opening glass-sealed ampoules under vacuum. *Anal. Chem.* **48**, 1651–1652.

Des Marais D. J., Hayes J. M. and Meinschein W. G. (1974) Retention of solar wind-implanted elements in lunar soils (abstract). In *Lunar Science V*, p. 168–170. The Lunar Science Institute, Houston.

Duncan A. R., Erlank A. J., Willis J. P and Ahrens L. H. (1973) Composition and interrelationships of some Apollo 16 samples. *Proc. Lunar Sci. Conf. 4th*, p. 1097–1113.

El Goresy A., Ramdohr P. and Medenbach O. (1973) Lunar samples from Descartes site: Opaque mineralogy and geochemistry. *Proc. Lunar Sci. Conf. 4th*, p. 733–750.

Epstein S. and Taylor H. P. Jr. (1974) D/H and $^{18}O/^{16}O$ ratios of H_2O in the "rusty" breccia 66095 and the origin of "lunar water". *Proc. Lunar Sci. Conf. 5th*, p. 1839–1854.

Fuex A. N. and Baker D. R. (1973) Stable carbon isotopes in selected granitic, mafic, and ultramafic igneous rocks. *Geochim. Cosmochim. Acta* **37**, 2509–2521.

Ganapathy R. and Anders E. (1974) Bulk compositions of the moon and earth, estimated from meteorites. *Proc. Lunar Sci. Conf. 5th*, p. 1181–1206.

Ganapathy R., Keays R. R., Laul J. C. and Anders E. (1970) Trace elements in Apollo 11 lunar rocks: Implications for meteorite influx and origin of the moon. *Proc. Apollo 11 Lunar Sci. Conf.*, p. 1117–1142.

Ganapathy R., Morgan J. W., Higuchi H., Anders E. and Anderson A. T. (1974) Meteoritic and volatile elements in Apollo 16 rocks and in separated phases from 14306. *Proc. Lunar Sci. Conf. 5th*, p. 1659–1683.

Gast P. W., Hubbard N. J. and Wiesmann H. (1970) Chemical composition and petrogenesis of basalts from Tranquility Base. *Proc. Apollo 11 Lunar Sci. Conf.*, p. 1143–1163.

Gibson E. K. Jr., Chang S., Lennon K., Moore G. W. and Pearce G. W. (1975) Sulfur abundances and distributions in mare basalts and their source magmas. *Proc. Lunar Sci. Conf. 6th*, p. 1287–1301.

Gibson E. K. Jr. and Moore G. W. (1974) Sulfur abundances and distributions in the valley of Taurus-Littrow. *Proc. Lunar Sci. Conf. 5th*, p. 1823–1837.

Gibson E. K. Jr. and Moore G. W. (1975) Breccias and crystalline rocks from Apollo 16 which contain carbonate-like phases (abstract). In *Lunar Science VI*, p. 287–289. The Lunar Science Institute, Houston.

Hoefs J. (1965) Ein beitrag zur geochemie des kohlenstoffs in magmatischen und metamorphen gesteinen. *Geochim. Cosmochim. Acta* **29**, 399–428.

Hoefs J. (1973) Ein beitrag zur isotopengeochemie des kohlenstoffs in magmatischen gesteinen. *Contrib. Mineral. Petrol.* **41**, 277–300.

Hubbard N. J., Rhodes J. M., Gast P. W., Bansal B. M., Shih C.-Y., Wiesman H. and Nyquist L. E. (1973) Lunar rock types: The role of plagioclase in non-mare and highland rock types. *Proc. Lunar Sci. Conf. 4th*, p. 1297–1312.

Huffman G. P., Schwerer F. C., Fisher R. M. and Nagata T. (1974) Iron distributions and metallic-ferrous ratios for Apollo lunar samples: Mossbauer and magnetic analyses. *Proc. Lunar Sci. Conf. 5th*, p. 2779–2794.

Hughes T. C., Keays R. R. and Lovering J. F. (1973) Siderophile and volatile trace elements in Apollo 14, 15 and 16 rocks and fines: Evidence for extralunar component and Ti-, Au-, and Ag-enriched rocks in the ancient lunar crust (abstract). In *Lunar Science IV*, p. 400–402. The Lunar Science Institute, Houston.

Injerd W. G. and Kaplan I. R. (1974) Nitrogen isotope distribution in meteorites (abstract). *Meteoritics* **9**, 352-353.

Jull A. J. T., Eglinton G., Pillinger C. T., Biggar G. M. and Batts B. D. (1976) The identity of lunar hydrolysable carbon. *Nature* **262**, 566–567.

Kaplan I. R., Kerridge J. F. and Petrowski C. (1976) Light element geochemistry of the Apollo 15 site. *Proc. Lunar Sci. Conf. 7th*, p. 481–492.

Kerridge J. F., Kaplan I. R., Petrowski C. and Chang S. (1975) Light element geochemistry of the Apollo 16 site. *Geochim. Cosmochim. Acta* **39**, 137–162.

Kirsten T. and Horn P. (1974) Chronology of the Taurus-Littrow region III: Ages of mare basalts and highland breccias and some remarks about the interpretation of lunar highland rock ages. *Proc. Lunar Sci. Conf. 5th*, p. 1451–1475.

Kung C.-C. and Clayton R. N. (1978) Nitrogen abundances and isotopic compositions in stony meteorites. *Earth Planet. Sci. Lett.* **38**, 421–435.

Laul J. C., Hill D. W. and Schmitt R. A. (1974) Chemical studies of Apollo 16 and 17 samples. *Proc. Lunar Sci Conf. 5th*, p. 1047–1066.

Leich D. A. and Niemeyer S. (1975) Trapped xenon in lunar anorthositic breccia 60015. *Proc. Lunar Sci. Conf. 6th*, p. 1953–1965.

LSPET (Lunar Sample Preliminary Examination Team) (1972) Preliminary examination of lunar samples. *Apollo 16 Preliminary Science Report*. NASA SP–315. Washington, D.C.

Moore C. B. (1975) A check list of total carbon and sulfur analyses. Contr. 94, Center for Meteorite Studies, Tempe.

Moore C. B. and Lewis C. F. (1976) Total nitrogen contents of Apollo 15, 16 and 17 lunar rocks and breccias (abstract). In *Lunar Science VII*, p. 571–573. The Lunar Science Institute, Houston.

Moore C. B., Lewis C. F. and Gibson E. K. Jr. (1973) Total carbon contents of Apollo 15 and 16 lunar samples. *Proc. Lunar Sci. Conf. 4th*, p. 1613–1623.

Müller O. (1974) Solar wind nitrogen and indigenous nitrogen in Apollo 17 lunar samples. *Proc. Lunar Sci. Conf. 5th*, p. 1907–1918.

Müller O., Grallath E. and Tolg G. (1976) Nitrogen in lunar igneous rocks. *Proc. Lunar Sci. Conf. 7th*, p. 1615–1622.

Nagata T., Fisher R. M., Schwerer F. C., Fuller M. D. and Dunn J. R. (1972) Summary of rock magnetism of Apollo 15 lunar materials. In *The Apollo 15 Lunar Samples* (J. W. Chamberlain and C. Watkins, eds.), p. 442–445. The Lunar Science Institute, Houston.

Niemeyer S. and Leich D. A. (1976) Atmospheric rare gases in lunar rock 60015. *Proc. Lunar Sci. Conf. 7th*, p. 587–597.

Otting W. and Zähringer J. (1967) Total carbon content and primordial rare gases in chondrites. *Geochim. Cosmochim. Acta* **31**, 1949–1960.

Petrowski C., Kerridge J. F. and Kaplan I. R. (1974) Light element geochemistry of the Apollo 17 site. *Proc. Lunar Sci. Conf. 5th*, p. 1939–1948.

Pineau F., Javoy M. and Bottinga Y. (1976) $^{13}C/^{12}C$ ratios of rocks and inclusions in popping rocks of the mid-Atlantic ridge and their bearing on the problem of isotopic composition of deep-seated carbon. *Earth Planet. Sci. Lett.* **29**, 413–421.

Rees C. E. and Thode H. G. (1974) Sulfur concentrations and isotope ratios in Apollo 16 and 17 samples. *Proc. Lunar Sci. Conf. 5th*, p. 1963–1973.

Rhodes J. M. and Hubbard N. J. (1973) Chemistry, classification, and petrogenesis of Apollo 15 mare basalts. *Proc. Lunar Sci. Conf. 4th*, p. 1127–1148.

Rhodes J. M., Rodgers K. V., Shih C., Bansal B. M., Nyquist L. E., Wiesmann H. and Hubbard N. J. (1974) The relationships between geology and soil chemistry at the Apollo 17 landing site. *Proc. Lunar Sci. Conf. 5th*, p. 1097–1117.

Rose H. J., Cuttitta F., Berman S., Brown F. W., Carron M. K., Christian R. P., Dwornik E. J. and Greenland L. P. (1974) Chemical composition of rocks and soils at Taurus-Littrow. *Proc. Lunar Sci. Conf. 5th*, p. 1119–1133.

Sakai H., Smith J. W., Kaplan I. R. and Petrowski C. (1977) Microdeterminations of C, N, S, H. He, metallic Fe, $\delta^{13}C$, $\delta^{15}N$ and $\delta^{34}S$ in geologic samples. *Geochem. J.* **10**, 85–96.

Sato M. (1976) Oxygen fugacity and other thermochemical parameters of Apollo 17 high-Ti basalts and their implications on the reduction mechanism. *Proc. Lunar Sci. Conf. 7th*, p. 1323–1344.

Stettler A., Eberhardt P., Geiss J., Grögler N. and Maurer P. (1973) ^{39}Ar-^{38}Ar ages and ^{37}Ar-^{38}Ar exposure ages of lunar rocks. *Proc. Lunar Sci. Conf. 4th*, p. 1865–1888.

Proc. Lunar Planet. Sci. Conf. 9th (1978), p. 2469–2484.
Printed in the United States of America

Dust, impact pits, and accreta on lunar rock 12054

HERBERT A. ZOOK

NASA Johnson Space Center, Houston, Texas 77058

Abstract—The large variation in the areal density of meteoroid impact pits observed over the face of lunar rock chip 12054,54 was previously attributed to variable amounts of shielding of this chip by loosely bound dust. However, an examination of the directionality of flow of the disk and splash accreta on this chip now suggests that the regions with low areal densities of pits and disk accreta were geometrically shielded by a formerly overhanging part of the rock. The "overhang" is postulated to have broken off during the return of this rock to earth. This new interpretation is contrary to that which was presented at the 9th Lunar and Planetary Science Conference. There remains, nevertheless, considerable evidence for shielding of rock and grain surfaces on the lunar surface by loosely bound dust. The areal extent and depth of deposition of dust is still poorly understood, however. Populations of disk (or pancake) accreta and "solid" accreta that show evidence of melting are shown to be correlated with impact pit densities. The population of solid angular accreta with exteriors that exhibit little evidence of melting shows no correlation with impact pit densities.

INTRODUCTION

We had earlier reported observations on lunar rock 12054 (Zook *et al.*, 1978) that demonstrated that the areal densities of submicron impact pits and accreta varied greatly from one place to another on chip 12054,54. The variations were far too great to be accounted for by differential shielding due to different rock geometry in one place compared to another. For this reason the variations were attributed at that time to loosely bound dust which was presumed to have rapidly accreted into a mild local topographic low on this chip shortly after rock emplacement on the lunar surface.

This hypothesis is no longer considered correct, however. As is shown in the next section, it is now thought very probable that a presently missing overhang was attached to the crystalline outcrop above the region of no pits and geometrically shielded this region from accumulating impact pits and disk accreta. It is postulated that the overhang was broken off during rock collection or during return to earth.

As reviewed in a later section, however, loosely bound dust remains an important, if enigmatic, element of the various processes occurring at the surfaces of lunar rocks. Dust can shield rock surfaces from the accumulation of solar wind implanted atoms, meteoroid impact generated vapor deposits, solar flare tracks, meteoroid impact pits, and various kinds of accreta.

An explanation for the variability of pits and accreta

Lunar rock 12054 and its location on the lunar surface is described in Hartung

Fig. 1. One of a series of photographs (originals in color) taken of lunar rock 12054 during the preliminary examination at the NASA Johnson Space Center. In the lunar surface orientation, the camera would be facing south and depicting the west end of the north "face" of this rock. The arrow shows the local lunar vertical. The location of chip 12054,54 is shown outlined in white. The dotted line shows the original chip size before some of the fragile glass surface was accidently knocked off. The region between the white pointers is the "clean" region of no pits or disks. The overhang is postulated to have been attached to the outcrop just above the clean region. NASA Photograph S-70–23000.

et al. (1978). The location of chip 12054,54 on this rock is outlined in Fig. 1. This photograph was taken during the preliminary examination at the Johnson Space Center. With the rock positioned as it would be on the lunar surface this is a view looking south at the west end of the north face of lunar rock 12054.

Photomosaics made at a nominal magnification of $10^4\times$ at three locations several millimeters apart on chip 12054,54 gave the following areal densities for pit diameters larger than 0.1 μm: $2 \times 10^7/\text{cm}^2$, $2 \times 10^6/\text{cm}^2$, and $\leq 10^5/\text{cm}^2$. For pits larger than 0.3 μm the corresponding areal densities were: $3.5 \times 10^6/\text{cm}^2$, $2 \times 10^5/\text{cm}^2$, and $\leq 10^5/\text{cm}^2$. At $\sim 10^4\times$ no difficulty is expected in detecting all pits larger than 0.3 μm. Hence, no selection effect problems are expected at this diameter. The photomosaic with the highest areal density of 0.1 μm and 0.3 μm pits was that reported earlier by Morrison and Zinner (1977). A Scanning Electron Microscope (SEM) photomosaic of the area of intermediate density of

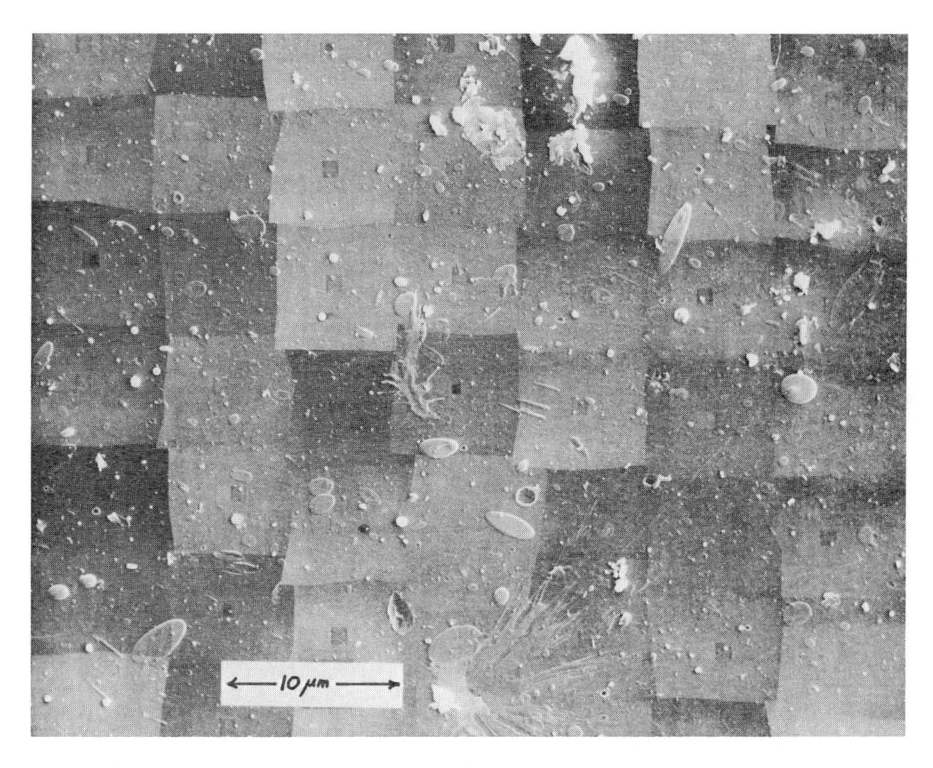

Fig. 2. A photomosaic taken of a region where the areal density of pits and accreta is intermediate between high and low values on chip 12054,54. The original magnification was 12,600×. The data from this mosaic is later referred to as "MDS" (Medium Density Scan). The scale bar is 10 μm long.

pits is shown in Fig. 2. Also seen in this photomosaic (original magnification = 12,600×) are various kinds of accreta, including splash and disk accreta.

Actually no pits or accretionary disks of any size whatever were noted on the one mosaic ($2.1 \times 10^{-5} cm^2$ in area), although a minor amount of particulate accreta gave assurance that resolution and focusing had not deteriorated. The location of this mosaic on chip 12054,54 is between the pointers shown in Fig. 1 and is in a mild local topographic low. It is indented approximately 1 mm into the face of the rock relative to the crystalline outcrop above it.

A SEM photomosaic taken at 100× of chip 12054,54 is shown in Fig. 3. The cube shows the directions of the north face and the top of the rock in relation to this chip. The region of no pits or accretionary disks is in the upper left portion of this figure and extends about 3 mm in an east-west direction by about 1 mm in an up-down direction.

As mentioned in the introduction, the presently observed rock topography

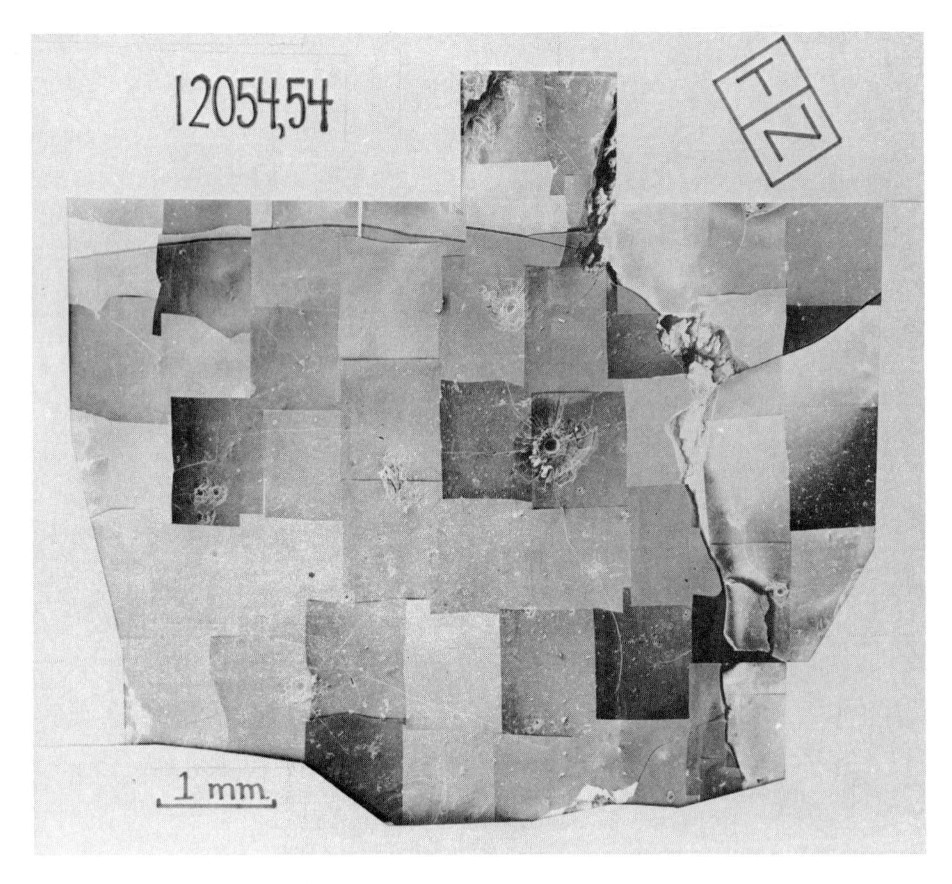

Fig. 3. An SEM photomosaic of chip 12054,54 obtained at an original magnification of 100×. The vertical and north directions are indicated by T and N in the inset cube at upper right. Scale is inset at lower left. Courtesy of E. Hauser.

could not have geometrically shielded the region of no pits nor the region of intermediate areal pit density (just below the region of no pits in Fig. 3) nearly enough to cause the very low observed areal densities of pits and disk accreta. On the other hand, it is also no longer believed, as in Zook *et al.* (1978), that loosely bound dust quickly accreted onto this area and shielded it from impact pits and disk accreta. It is now thought very probable that material formerly attached to the crystalline outcrop above the region of no pits formed an overhang over this region and geometrically shielded it from accumulating impact pits and disk accreta. The reasons for this conclusion are as follows:

1. A strip of photos was taken (at a magnification of $\sim 10^3 \times$) with one part of the strip in the region of no disk accreta (or pits) and the other end in the

region of disk accreta and pits. It was observed that the disk accreta became very "directional" in the transition region. That is, the disks were often elongated and the direction of elongation was very strongly oriented along the most direct line between the pitted and the nonpitted (and no disk accreta) region. Also the direction of flow of the accreta appeared to be from the pitted to the nonpitted region. This elongated disk behavior can be understood if there was an overhang protecting this area while the rock was on the moon but which was broken off during the return to earth. Hence, the overhang was not observed after return to earth. It is difficult to explain the directional disk effect using only loosely bound dust as shielding.

The directionality of flow of the disk and splash accreta can be seen, to some extent, in Fig. 2. None of the disk and splash accreta appear to have arrived from the right in this figure. The direction to the right in this figure is toward the top of the rock and projects under the postulated overhang.

2. An "overhang" of glass was indeed observed on the west end of the rock. This overhang was chipped off before sending the west end of the rock to the alternate lunar sample storage site (near San Antonio, Texas). The west end overhang was quite fragile and took little effort to remove. It revealed an original, fresh, unpitted, and unfractured glass surface beneath it. This experience proves that fragile overhangs that are easily removed did exist on this rock.
3. The fact that there were no disk accreta of any size on the one region indicates that this region's unshielded space exposure was probably less than 10^{-3} times either the time or the solid angle of the heavily pitted region. The calculated surface exposure age of this rock ranges from $\sim 3 \times 10^4$ yr (Zook, 1978) as deduced from the areal density of meteoroid craters to $\sim 2 \times 10^5$ yr (Morrison and Zinner, 1977) as deduced from solar flare track analyses. If there were no overhang restricting the solid angle it would follow that loosely bound dust would have had to thoroughly have covered this small region in times less than 30 to 200 years after rock emplacement. This would mean either that the dust was put there with the event that caused rock emplacement or that the flux of dust on the lunar surface is very high. As no dust was observed to be welded to the unpitted region, unlike the dust welded onto the rock-soil boundary (Hartung *et al.*, 1978), it is concluded that the emplacement event did not put dust in this region. To entirely cover this region by the normal flux of lunar dust in less than 30 (or 200) years, would suggest not only that the flux is improbably high but that it would gather on the upper side of this chip underneath the crystalline outcrop. Geometric shielding by an overhang that reduces the solid angle by 10^3 removes the requirement for such a flux. Reason 1 is felt to be the conclusive one. Reasons 2 and 3 provide additional evidence that make the hypothesis of geometric shielding by a fragile overhang not an unreasonable one. The question of loosely bound dust will be returned to later.

Size distribution variations

Photomosaics at a nominal magnification of $10^4\times$ of parts of the spall zones of several of the larger impact craters on 12054,54 were prepared and scanned for submicron impact pits and accreta. One purpose was to observe possible changes in the size distributions of impact pits from one location to another. Because each crater and its associated spall zone is formed at a different time in the past, possible changes in the pit size distribution at different times in the past might thus also be observed. The spall zones are, in general, easier to scan than are the insides of the large pits because they usually present relatively large, flat, and shallow surfaces.

The distributions obtained are shown in Fig. 4. Numbers were arbitrarily assigned to the craters on chip 12054,54 for convenient reference. Hence "C-6

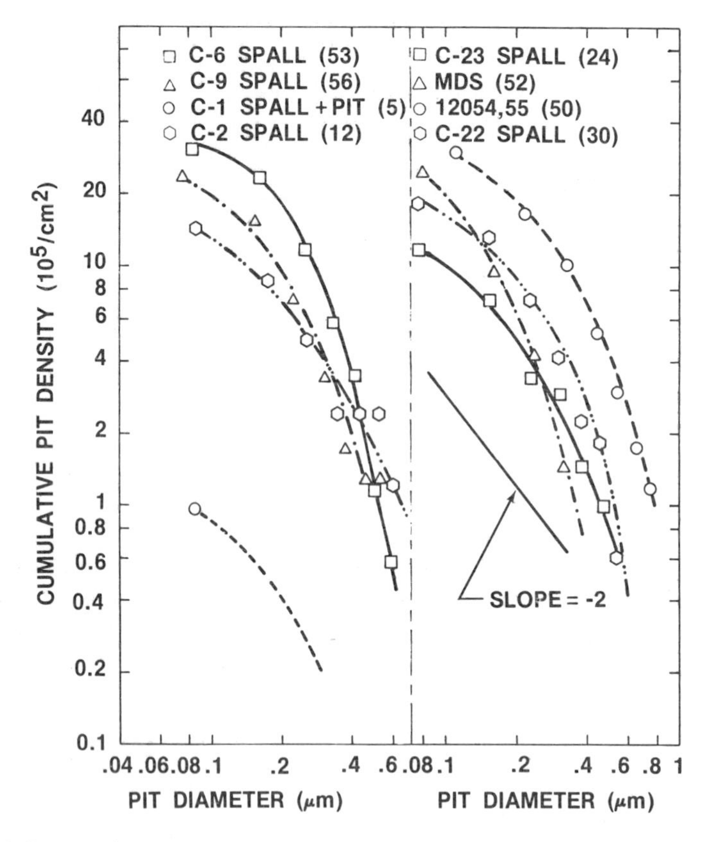

Fig. 4. Impact pit areal densities on different regions (mostly crater spall zones) of 12054,54 versus pit diameter. Numbers in parentheses are numbers of pits cumulatively counted down to the smallest pit plotted in each case. MDS is an area of intermediate pit density on the original glass surface (= medium density scan). Also shown are results for one spot on chip 12054,55. Line of slope = − 2 shown for reference. NASA S-78–10842.

spall" refers to the spall zone on crater number 6. The numbers in parentheses in this figure are the number of pits actually counted on the referenced mosaic. The photomosaic magnifications ranged from 11,700× to 13,200× except for the mosaic taken on chip 12054,55 (adjacent to 12054,54). This latter photomosaic was earlier prepared by D. Morrison (and the results presented in Morrison and Zinner, 1977). Approximately half of it is used in the present analysis. It was taken at a magnification of 9000×. The data labeled MDS (= medium density scan) was taken from the SEM photomosaic shown in Fig. 2. This region is located on chip 12054,54 just below the region totally shielded by the postulated overhang. Hence it is presumed to be partly shielded. Figures 4 and 5 differ, in part, from those earlier presented (Zook *et al.*, 1978) because the actual magnifications are used here rather than the nominal 10^4× previously used.

There do appear to be some variations between the shapes of the different distributions observed on the different areas scanned. However, the statistics are such that there is little assurance that the variances can be attributed to true changes in the meteoroid flux from one place to another. One exception may be the size distribution shown for MDS. This distribution has the steepest average slope of all the distributions shown; i.e., it is weighted heavily toward small pits.

Interestingly, the HEOS data (Hoffman *et al.*, 1975) would lead one to deduce that the size spectrum should, indeed, be heavily weighted toward small pits, as observed, in this location. The HEOS data demonstrate that particles smaller than about 10^{-12}g are divided into two main groups. Those meteoroids that are smaller than about 10^{-14}g tend to arrive at the meteoroid sensor from the ecliptic north and ecliptic south directions while most meteoroids with masses between 10^{-14}g and 10^{-12}g arrive at the sensor from the satellite heliocentric apex direction. Hence, one could reasonably expect from the HEOS data that a surface shielded from "apex" particles (those arriving preferentially from the apex direction) might show a very steep submicron pit population relative to surfaces that are exposed to apex particles. Apex particles probably are responsible for most impact pits with diameters between 0.8 μm and 6 μm. The normal to the face of chip 12054,54 makes an angle relative to the north lunar axis (and thus approximately to the ecliptic N-S axis) of about 35°. Therefore it should normally receive a substantial number of apex particles when the moon rotates this face into the heliocentric apex direction. However, the postulated overhang undoubtedly would protect the MDS region from many of the apex particles—thus weighting the sub-micron pit distribution preferentially toward very small ones—as observed. The statistics are not conclusive, however, and other distributions with fairly steep slopes are also observed.

Accreta and crater density correlations

Areas of different pit areal densities were also examined for areal densities of various kinds of accreta. Specifically, it was examined how areal densities of "disk" accreta with diameters ≥1.0 μm, and "solid" accreta with diameters

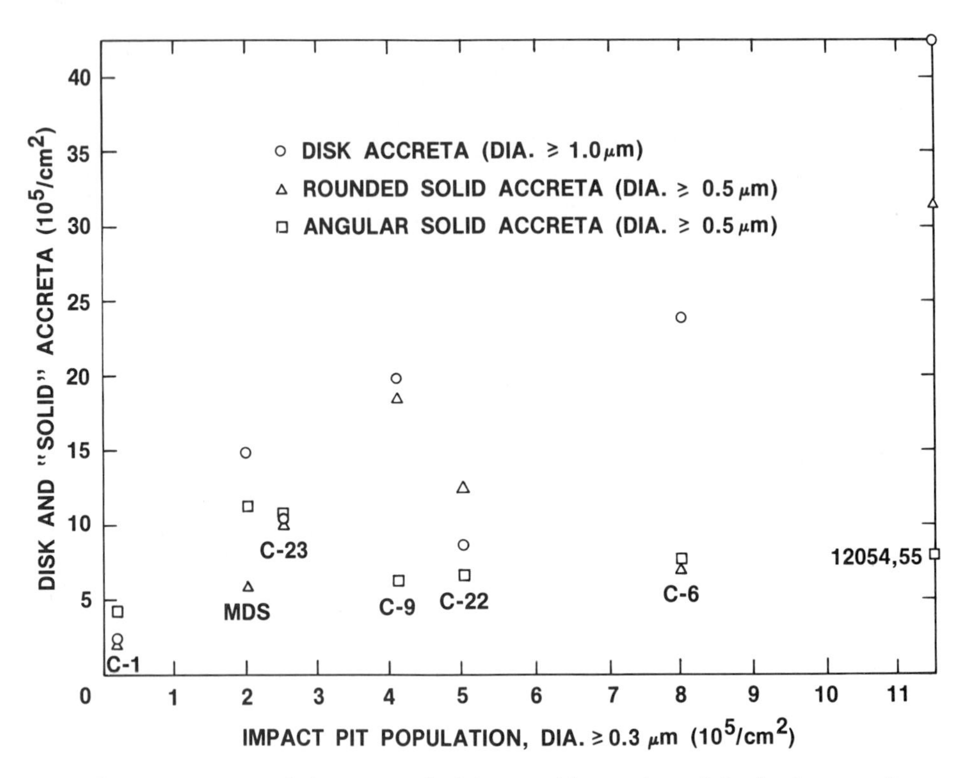

Fig. 5. Accreta populations versus the 0.3 μm and larger pit areal density for most of the regions shown in Fig. 4. NASA S-78–10841.

≥ 0.5 μm correlated with the areal densities of impact pits with diameters larger than 0.3 μm. These correlations are shown in Fig. 5. Such correlations have been earlier suggested (Morrison *et al.*, 1973; Hartung and Comstock, 1978). The areal density of pits ≥ 0.3 μm on C-1 is estimated from the extrapolated line in Fig. 4. No pits larger than 0.1 μm diameter were observed on C-1. The solid accreta referred to herein are equivalent to the glassy, crystalline, or fragmental accreta of McDonnell (1977) that have apparently suffered little from impact deformation during accretion. The solid accreta were divided into two separate groups; the "rounded" (often globular) and the "angular" accreta in order to look for possible different "sticking" coefficients of these two groups. Examples of disk, rounded, and angular accreta are shown in Fig. 6. The division between "rounded" and "angular" was somewhat arbitrary and was not always sharp. Angular fragments were defined to be those whose edges were relatively sharp and which showed little or no evidence of melting (or remelting) at the edges as might occur for grains from a highly shocked soil (Gibbons *et al.*, 1975).

A fit of a straight line by the method of least squares was made for each accreta population versus pit areal density. The corresponding correlation

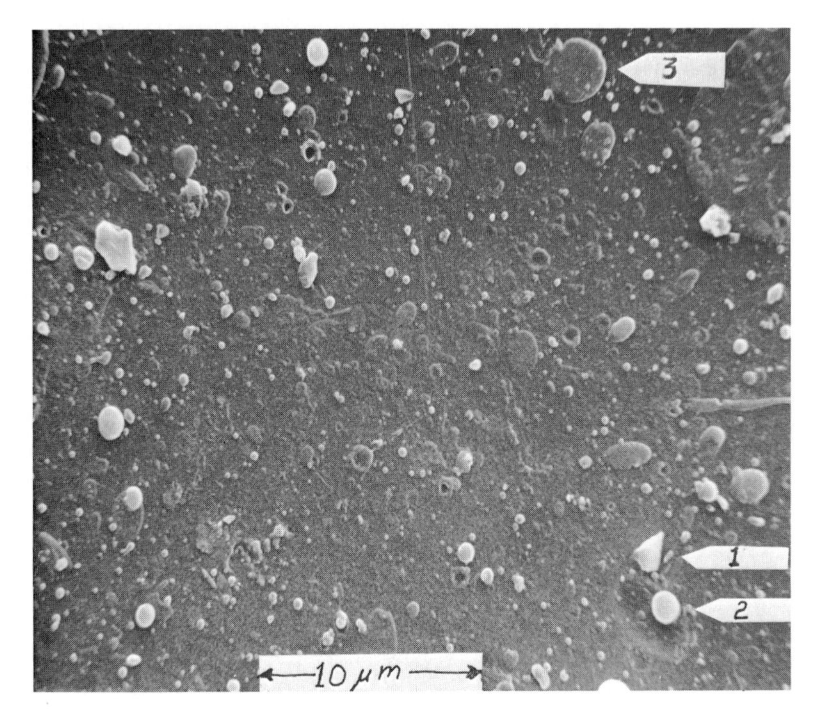

Fig. 6. A region of high areal densities of pits and accreta. The arrows point to examples of (1) angular accreta, (2) rounded accreta, and (3) disk accreta. Many rounded accreta are angularly shaped but rounded at edges and corners.

coefficients obtained from these fits were as follows: $r = 0.91$ for disk accreta versus pits; $r = 0.78$ for rounded accreta versus pits and $r \approx 0$ for angular accreta versus pits. Hence, there is an excellent correlation of disk accreta with pits (less than 1% probability of obtaining such a correlation by random chance), a good correlation of rounded accreta with pits, and no apparent correlation at all of angular accreta with pits. These variations generally agree with visual impressions of the photomosaics.

The impact pits are presumably made by meteoroids from interplanetary flight and the accreta doubtless derive from the lunar surface. Hence, the correlations mean that if the relative exposure geometries of the different surfaces are not too dissimilar (relative to both space exposure and exposure to accreta sources), pit populations and accreta populations increase together with increasing time. In the case of MDS where exposure time is presumably the same as for chip 12054,55, the indication is that the postulated overhang shielded this area similarly from impact pits and accreta.

The lack of correlation of the angular accreta with pits indicates either that these accreta do not build up on these surfaces with time or that they are so loosely bonded that they are removed by the later processing of this sample, leaving only some equilibrium distribution of angular accreta. The processing includes handling during sample return, sawing and chipping to remove this chip from the parent rock, ultrasonic cleaning in liquid freon, removal of the first gold coating on this chip (as it was too thick) in a solution of potassium iodide, further ultrasonic cleaning, and re-coating with gold. At any rate, it is clear that caution is advised when using all kinds of solid accreta as an exposure clock as suggested by McDonnell (1977).

McDonnell (1977) also suggests that the primary agent for bonding solid accreta to rock surfaces is solar wind sputtering which would form bonding fillets with time. There is little certainty, however, that solar wind sputtering is, in fact, the responsible agent for the observed accreta rounding and bonding. Although there is a fairly sharp break in morphological shapes as one proceeds from disk or "pancake" accreta to glassy and rounded solid accreta, there is, indeed, a moderate population of accreta with intermediate morphologies. Theoretically, if low viscosity glass droplets are created, it would seem reasonable that those droplets which arrive from greater distances should have cooled to high viscosity droplets. Hence they would tend not to flatten out into disks so easily. Also, one expects that impacts into soil would not only produce glass droplets but would also produce highly heated, but not completely melted, crystalline fragments (e.g., see Gibbons *et al.*, 1975). Their corners and edges might well be melted, for example, and thus provide a mechanism for sticking them to any surface they should happen to impact. Although sputtering and vapor deposition may play a role, the observations of accreta here reported appear entirely consistent with bonding due to originally melted surfaces on the accreta. Further experimental work is required to settle this question.

Composition of a large glass spatter

There is one additional type of accreta which has been occasionally reported by others (e.g., see Fig. 1 in Fechtig *et al.*, 1977) and that consists of what appear to be large "splashes" of molten glass. One of these, about a millimeter in diameter, is shown in Fig. 7 and is the only such feature on chip 12054,54. No other glass splashes within a factor of 10 of the diameter of this one were observed on this chip. The glass splash was chemically analyzed by energy-dispersive X-ray analysis at five points. The resulting average major element composition, as oxides, is 6% MgO, 11% Al_2O_3, 41% SiO_2, 11% CaO, 4% TiO_2, and 16% FeO. The percentages do not total to 100 because three of the analyses were performed on small globules of high relief in the molten-appearing splash. The composition of the background glass covering 12054 is (from an average of 2 points) 8% MgO, 7% Al_2O_3, 44% SiO_2, 11% CaO, 7% TiO_2, and 23% FeO. See also Schaal and Hörz (1977) for chemical analyses of 12054. It is concluded from this analysis that the molten glass that gave rise to this splash derived not from meteoritic material but from lunar surface material. The percentages of calcium and aluminum in the splash material are, for example, higher than is normal for bulk meteorites.

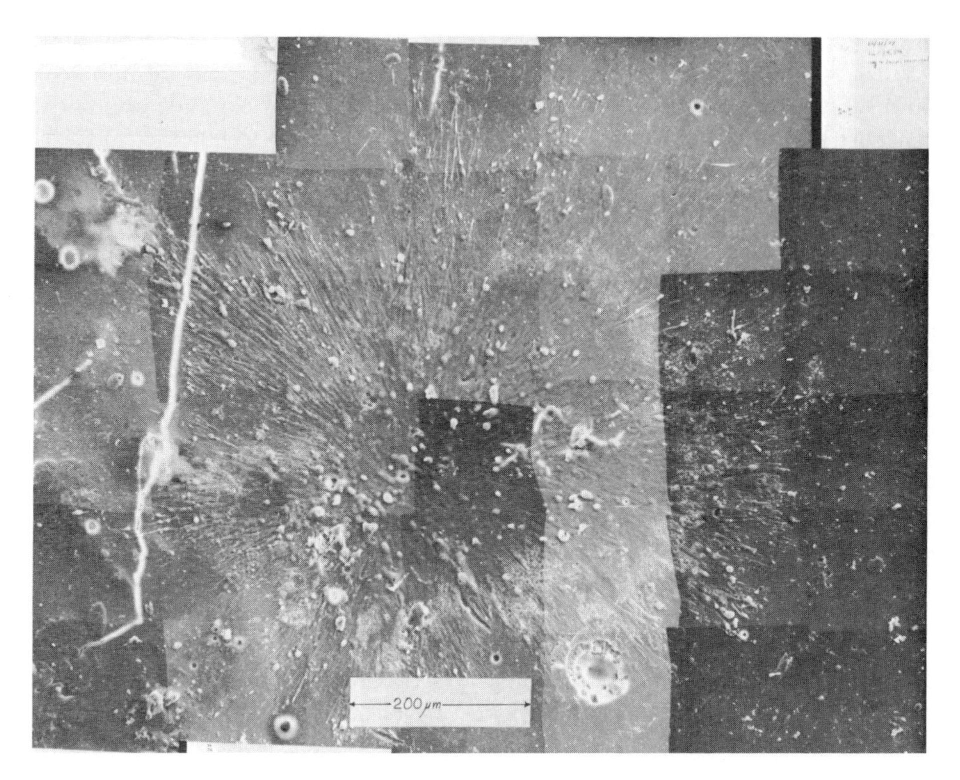

Fig. 7. A large, approximately 1 mm diameter, "molten appearing" splash on chip 12054,54. Chemistry is indicative of origin from lunar surface. The original SEM photomosaic was prepared at ~700×. NASA photograph S-78–23277.

Evidence for the importance of dust

Although the large variations in pit and accreta areal densities that were observed on chip 12054,54 are not now thought to be due to variable obscuration by loosely bound dust, this does not mean that obscuration by such dust is unimportant. For example, the fine-grained portion of the dust that the astronauts stirred up during their traverses over the lunar surface proved to be quite "sticky" and appeared to adhere to practically any surface with which it came in contact. Thus, the suits worn by the astronauts, the lunar rover, and even the lens of the lunar TV camera (which often required cleaning) all became coated with varying amounts of dust during extra-vehicular activity by the astronauts.

Similarly, when the lunar rocks were taken out of the plastic sampling bags in which they were stored for the return journey from the moon, they were usually found to be largely coated with dust (D. Morrison, P. Butler, and F. Hörz; pers. comm.). Likewise, the interiors of the sampling bags were also coated with dust. Some of the dust could be blown off by a jet of dry nitrogen gas. Much of it, however, was too sticky to be removed with this technique and was removed with a brush in order to permit effective preliminary examinations of the rocks. However, these observations only tell us that the dust is sticky, not whether the transport rate of dust on the lunar surface is high enough to quickly cause rock surface obscuration.

Observations of certain rock surface phenomena by different investigators occasionally led to contradictory results. Morrison and Zinner (1975), for example, noted that if one ratios the areal density of 0.1 μm and larger diameter impact pits on a rock surface to the areal density of solar flare tracks at a depth of 100 μm below the same lunar rock surface, the value is approximately unity. This ratio, however, was about 50 times higher than that obtained by Hutcheon (1975). Morrison and Zinner suggested that the most likely explanation for this discrepancy was that the surfaces examined by Hutcheon were shielded by a thin layer of dust for most of their exposure periods. Such shielding would permit a track buildup while inhibiting the formation of microcraters.

Poupeau *et al.* (1975) measured solar flare tracks, microcraters, and glass accreta on a large number of crystals from several mature soils. In a rather elegant analysis, they noted that many crystals have high track densities and/or surface track gradients showing near surface exposure but do not have impact pits. They suggested, quite reasonably, that the results were compatible with a thin, fine dust coating on most of the grains during their near-surface exposure.

Hartung *et al.* (1975) measured the areal densities of submicron pits (diameters larger than 0.7 μm) inside of larger pits (pit diameters between 100 μm and 260 μm) on lunar rock 12002. They observed that the number of large impact pits with very low areal densities of submicron impact pits was highly enhanced over the number of large pits with high areal densities. If the fluxes of the particles causing the submicron, as well as those causing larger pits, were both constant with time, there should be a uniform distribution of areal density populations between high and low values. They favored the idea at that time that equilibrium superposition of large impact craters was the likely cause for very few craters having high areal densities of submicron pits. They were, however, well aware that obscuration by lunar dust was a possibility.

Later, Hartung *et al.* (1977) and Hartung and Comstock (1978) examined the distribution of the areal densities of submicron impact pits (diameters $\geq$0.7 μm) and accretionary disks (diameters $\geq$1.4 μm) inside large impact pits (diameters $\geq$100 μm) on lunar rock 60015. They observed that the population of accretionary disks was distributed such that the number of large impact craters having very low areal densities of accretionary disks was highly enhanced over those having higher areal densities. The submicron impact pits were similarly distributed. As rock 60015 was clearly not saturated with impact craters and was therefore in production, impact crater superposition was not a possible explanation. They also noted that the mass range of meteoroids that generates the large impact pits is most probably the same meteoroid mass range that generates the accretionary disks. Therefore, it was not possible for the rate of production of large impact craters and the rate of production of the molten glass droplets responsible for accretionary disks to vary with respect to one another. They noted that one way out of this difficulty was to suppose that dust was intermittently coating portions of lunar rocks and shielding them from accumulating accretionary disks and submicron impact pits.

Later work by Hauser *et al.* (1978) on submicron impact pit and disk accreta distributions inside of larger craters appears to confirm this earlier work. We also note, interestingly, that the accreta of McDonnell (1977) are similarly distributed. Also, Hartung *et al.* (1978) note that, because submicron pits and disk accreta are absent on chip 12054,59, it probably was coated with dust while on the lunar surface.

These observations, taken together, rather powerfully implicate lunar dust as an important agent in the obscuration of lunar surface features against the action of other surface or near-surface processes. In addition, there is both near-lunar photometric (McCoy, 1976) and lunar surface (Berg *et al.*, 1976) experimental evidence for both high and low altitude electrostatic transport of lunar dust. In fact, the Lunar Ejecta and Meteoroid (LEAM) data of Berg *et al.* (1976) indicate that electrostatic transport of dust probably dominates the transport of the micron to submicron particles. In spite of these observations, however, we remain quite unsure as to both the rate and the extent of surface obscuration by dust. The fact that the astronauts observed many glass covered rocks whose surfaces were yet quite shiny suggests that at least some rock surfaces remain essentially free of loosely bound dust throughout their surface exposure periods.

It was hoped that the lunar surface close-up stereoscopic photography (Gold, 1969; Gold *et al.*, 1970; Gold, 1971) could shed some light on this subject. Numerous impact pits are observed in these close-up photographs taken of the lunar surface. Some of the pits are observed to have dust in them and some appear to be quite shiny and free of dust. However, these photographs were taken near the lunar module. It is reasonable to presume that, during its descent and landing, the lunar module rocket nozzle scoured the dust out of some of the pits or, alternatively, blew dust into some of the pits. Hence, these data also do not give a firm solution to the problem of loosely bound dust.

Discussion

The extent and the rapidity with which rock surfaces are covered by dust while residing on the lunar surface remains an important problem. Until this phenomenon is better understood, considerable caution must be used in attempting to analyze other surface phenomena. For example, the track-within-pit data of Hartung and Storzer (1974) was interpreted by them as possibly due to a recently (during the last 10^4 yr) increasing rate of meteoritic bombardment. Zook *et al.* (1977) later explored in more detail the suggestion that their observed distribution of track densities within the pits might also be due to variable solar flare activity. The latter explanation was perferred therein over a variable meteoritic impact rate. However, it is also possible that loosely bound lunar dust gathered in the pits in such a way as to cause the observed track density

Fig. 8. Another photograph of the series of 12054 taken during the preliminary examination. In the lunar orientation, the camera would now be facing north and showing the south "face" of this rock. Fine grained lunar dust can be observed in most of the local topographic lows in this view. The dusty effect was observed over most of the rock. NASA photograph S-70–22988.

distribution. Recent work (Hartung and Comstock, 1978) lends more credence to that idea. There is even a question (Comstock and Hartung, 1977) as to whether or not the observed tracks were truly due to solar flare Fe particles.

The question of cause for the track-within-pit distribution observed by Hartung and Storzer cannot yet be firmly settled as several alternate explanations are still viable. A somewhat different problem, however, is the question of why the meteoroid impact exposure ages derived by Zook (1978) for several rocks (including 12054) are much younger than are the solar flare track ages reported for these same rocks. The comparisons are made in Zook (1978). The track ages for these rocks were typically obtained from track profiles at depths in excess of 100 μm and the meteoroid impact exposure ages were obtained from the areal density of impact pits with diamters larger than 20 μm. Therefore, a rather thick surface coating of dust, indeed, would be required to significantly suppress either pit or track densities below the densities that would be obtained on an unshielded surface. A thick dust coating could well give the appearance of younger pit ages than track ages. More research is needed, however, to establish whether or not such coverings of dust are, in fact, significantly suppressing the creation of 20 μm impact pits. It is certainly not now appropriate to assume that such suppression is occurring.

Potentially useful work might yet be done on lunar rock 12054. Figure 8 is a photograph of a portion of the south side of this rock that was taken during preliminary examination. Loosely bound dust can be seen in many of the local topographic lows (it is especially easy to see in the original color photographs). Not yet known is whether this dust was on the rock while it rested on the lunar surface or whether it was acquired in the sample bag during return to earth. There should be few problems with "vanished overhangs" on this side of the rock, permitting more confident analysis of any results that might, in the future, be obtained.

One final note is that none of the various pit size distributions that were obtained on the different surfaces of this chip have nearly as flat a slope as that obtained by Brownlee *et al.* (1973) on rock 15286. The reason for this difference is not yet understood. However, as noted in the analysis of the very steep slope of the MDS region of 12054,54, changes in slope can indicate that certain parts of the impacting meteoroid population can be shielded against, thus leading to slope changes. It is important to be alert to even rather subtle changes.

Acknowledgments—It is a pleasure to acknowledge the usefulness of many discussions with Jack Hartung who, as 12054 consortium leader, provided much of the motivation for this research. I also appreciate the help of Emilie Hauser who, along with Don Morrison, established the orientation of chip 12054,54 on lunar rock 12054. Don was also helpful in alerting me to many pertinent observations already made by him and by others. I thank Dave McKay for carrying out the energy-dispersive X-ray analysis reported herein and for supervising the maintenance of the SEM—thus making it possible to obtain the high resolution photomosaics required for this study. Finally, I appreciate the manuscript preparation by Elinor Stockton.

REFERENCES

Berg O. E., Wolf H. and Rhee J. (1976) Lunar soil movement registered by the Apollo 17 cosmic dust experiment. In *Interplanetary Dust and Zodiacal Light* (H. Elsässer and H. Fechtig, eds.), p. 233–237. Springer-Verlag, N.Y.

Brownlee D. E., Hörz F., Vedder J. F., Gault D. E. and Hartung J. B. (1973) Some physical parameters of micrometeoroids. *Proc. Lunar Sci. Conf. 4th*, p. 3197–3212.

Comstock G. M. and Hartung J. B. (1977) Found: Solar flare iron-group tracks in microcrater glass (abstract). In *Lunar Science VIII*, p. 205–206. The Lunar Science Institute, Houston.

Fechtig H., Nagel K., Stähle V., Grögler N., Schneider E. and Neukum G. (1977) Impact phenomena on an Apollo 12 sample. *Proc. Lunar Sci. Conf. 8th*, p. 889–899.

Gibbons R. V., Morris R. V., Hörz F. and Thompson T. D. (1975) Petrographic and ferromagnetic resonance studies of experimentally shocked regolith analogs. *Proc. Lunar Sci. Conf. 6th*, p. 3143–3171.

Gold T. (1969) Apollo 11 observations of a remarkable glazing phenomenon on the lunar surface. *Apollo 11 Prelim. Sci. Rep.* NASA SP-214, p. 193–197. (See also the stereoscopic views p. 187–193).

Gold T. (1971) Lunar surface closeup stereoscopic photography. *Apollo 14 Prelim. Sci. Rep.* NASA SP-272, p. 239–244.

Gold T. Pearce F. and Jones R. (1970) Lunar surface closeup stereoscopic photography. *Apollo 12 Prelim. Sci. Rep.* NASA SP-235, p. 183–188.

Hartung J. B., Breig J. J. and Comstock G. M. (1977) Microcrater studies on 60015 do not support time variation of meteoroid flux (abstract). In *Lunar Science VIII*, p. 406–408. The Lunar Science Institute, Houston.

Hartung J. B. and Comstock G. M. (1978) New lunar microcrater evidence against a time varying meteoroid flux. *Space Research.* In press.

Hartung J. B., Hauser E. E., Hörz F., Morrison D. A., Schonfeld E., Zook H. A., Mandeville J.-C., McDonnell J. A. M., Schaal R.B. and Zinner E. (1978) Lunar surface processes: Report of the 12054 consortium. *Proc. Lunar Planet. Sci. Conf. 9th.* This volume.

Hartung J. B., Hodges F., Hörz F., and Storzer D. (1975) Microcrater investigations on lunar rock 12002. *Proc. Lunar Sci. Conf. 6th*, p. 3351–3371.

Hartung J. B. and Storzer D. (1974) Lunar microcraters and their solar flare track record. *Proc. Lunar Sci. Conf. 5th*, p. 2527–2541.

Hauser E. E., Hartung J. B., Morrison D. A. and Zook H. A. (1978) Impact craters and accretionary disks on lunar rock 12054 (abstract). In *Lunar and Planetary Science IX*, p. 471–473. Lunar and Planetery Institute, Houston.

Hoffman H.-J., Fechtig H., Grün E. and Kissel J. (1975) Temporal fluctuations and anisotropy of the micrometeorid flux in the earth-moon system measured by HEOS 2. *Planet. Space Sci.* **23**, 985–991.

Hutcheon I. D. (1975) Micrometeorites and solar flare particles in and out of the ecliptic. *J. Geophys. Res.* **80**, 4471–4483.

McCoy J. E. (1976) Photometric studies of light scattering above the lunar terminator from Apollo solar corona photography. *Proc. Lunar Sci. Conf. 7th*, p. 1087–1112.

McDonnell J. A. M. (1977) Accretionary particle studies on Apollo 12054,58: *In-situ* lunar surface microparticle flux rate and solar wind sputter rate defined. *Proc. Lunar Sci. Conf., 8th*, p. 3835–3857.

Morrison D. A., McKay D. S., Fruland R. M. and Moore H. J. (1973) Microcraters on Apollo 15 and 16 rocks. *Proc. Lunar Sci. Conf., 4th*, p. 3235–3253.

Morrison D. A. and Zinner E. (1975) Studies of solar flares and impact craters in partially protected crystals. *Proc. Lunar Sci. Conf., 6th*, p. 3373–3390.

Morrison D. A. and Zinner E. (1977) 12054 and 76215: New measurements of interplanetary dust and solar flare fluxes. *Proc. Lunar Sci. Conf., 8th*, p. 841–863.

Poupeau G., Walker R. M., Zinner E. and Morrison D. A. (1975) Surface exposure history of individual crystals in the lunar regolith. *Proc. Lunar Sci. Conf. 6th*, p. 3433–3448.

Schaal R. B. and Hörz F. (1977) Shock metamorphism of lunar and terrestrial basalts. *Proc. Lunar Sci. Conf. 8th*, p. 1697–1729.

Zook H. A., Hartung J. B. and Storzer D. (1977) Solar flare activity: Evidence for large scale changes in the past. *Icarus* **32**, 106–126.

Zook H. A. (1978) Temporal and spatial variations of the interplanetary dust flux. In *COSPAR: Space Research.* Vol. XVIII (M. J. Rycroft and A. C. Strickland, eds.), p. 411–422. Akademie-Verlag, Berlin.

Zook H., Hartung J. and Hauser E. (1978) Loosely bound dust, impact pits, and accreta on lunar rock 12054,54 (abstract). In *Lunar and Planetary Science IX*, p. 1300–1302. Lunar and Planetary Institute, Houston.

Proc. Lunar Planet. Sci. Conf. 9th (1978), p. 2485–2493.
Printed in the United States of America

Chemical investigations of impact features on sample 12001,520

K. NAGEL, A. EL GORESY
Max-Planck-Institut für Kernphysik
Heidelberg, West Germany

N. GRÖGLER
Physikalisches Institut der Universität
Bern, Switzerland

Abstract—Investigations on FeNi spherules partially embedded in the glass surface and as inclusions in polished sections of sample 12001,520 demonstrate that they have a complex structure consisting of a FeNi-schreibersite core surrounded by a troilite shell. Since large metal spherules have thick troilite shells and small particles have thinner shells, the Ni to Fe ratios of analyzed spherules in SEM depend on the spherule diameter. The formation of the crumbled edges in some particles sticking in the glass surface are the result of the brittleness of troilite shells. Quantitative electron microprobe analyses revealed differences in the chemical composition of the glass and the underlying breccia of sample 12001,520. These results negate a genetic relationship between glass and breccia. The biggest glass splash on the surface of sample 12001,520 could be genetically related to a neighbouring crater on the same sample.

INTRODUCTION

Earlier surface investigations of lunar samples exposed to the interplanetary medium were focussed on measuring dust fluxes and exposure ages via crater statistics (e.g., Fechtig *et al.*, 1974; Hartung and Storzer, 1974). Recently interest developed to detect projectile residues in lunar microcraters (Brownlee *et al.*, 1975; Schaal *et al.*, 1976; Nagel *et al.*, 1975, 1976). Chemical composition of lunar samples put severe restrictions to the interpretation of these measurements. The results of suitable samples (anorthositic rocks) showed that two different kinds of projectile residues are present (Nagel *et al.*, 1975):

1) Iron-silicate components mixed with the glass-linings of the craters; and
2) Iron-nickel spherules sticking in the glass-linings.

Chemical analyses of the iron-nickel residues revealed that the Ni to Fe ratios are a function of spherule diameter (Nagel *et al.*, 1976). The origin of the dependence of these ratios is, however, disputed. Two alternative explanations have been proposed:

1) Condensation under extremely reducing conditions (Nagel *et al.*, 1975);
2) A highly heterogeneous mixture of melts and possibly vaporized species of

both target and projectile existing under extremely high pressures gives rise to various degrees of fractionation (Gibbons *et al.*, 1976).

However, both alternatives did not satisfactorily explain this consistent dependence. Therefore, additional investigations have been performed to get further information. We analysed metallic spherules partially embedded in the glass surface of sample 12001,520 as well as metallic spherules completely inclosed in the glass matrix of the same sample using a Si(Li) solid state detector attached to a scanning electron microscope. The chemical composition of the glass splash surrounding the largest crater of sample 12001,520 and contaminations of two neighbouring craters have been measured applying a Si(Li) detector.

Quantitative electron microprobe measurements of the glass and the underlying breccia of a polished section of sample 12001,520 were also carried out.

MEASUREMENTS AND RESULTS

On the surface of sample 12001,520 many metallic spherules have been detected. These spherules have been found in various degrees of penetration in the glass. This variation ranges from just sitting on the surface to complete inclusion in the glass. Some of these particles are of special interest. Figure 1 displays one spherule along with X-ray spectra of two points located in the micrograph. The crumbled edge of the spherule, point 1, is distinctly enriched in sulphur as compared with the area of point 2 on this particle, as shown by their respective analyses.

A polished cross section of a small piece of sample 12001,520 showed metallic inclusions far below the target surface. Subsequent investigations using a scanning electron microscope revealed complex mineral compositions. The metal spherules are multiphase objects with distinct structures (Fig. 2). The particles are usually composed of a FeNi core eutectically intergrown with schreibersite. Both phases are covered by an almost complete shell of troilite which may also contain small inclusions of schreibersite. These features are well documented in the X-ray scanning micrographs (Fig. 2). Similar zoning features were also observed during the investigations of small particles in the glass-linings of a crater on sample 65315,68. The spherules in this sample (Fig. 3) also consist of FeNi cores and troilite shells. Very similar dendritic spherules in lunar soil have already been described by Goldstein *et al.* (1972). According to Blau *et al.* (1973), the origin of these dendritic spheroids may be ascribed to the shock melting of sulfide-metal-phosphide areas within a meteoritic projectile when it hits its target with crater forming energy. If this assumption is correct then the multiphase spherules encountered are not representative for the bulk composition of the impacting projectile.

The multiphase texture of the spherules offers a good explanation for the dependence of FeNi ratios on spherule size. Because small spherules have thinner troilite shells than larger ones, the size becomes an important factor in the penetration of the electron beam through the troilite shell to the FeNi rich core. In large particles, the excited area covers mainly the troilite shell which is poor in

Ni. The smaller the spherule the deeper the penetration of the beam to the FeNi rich core. This means Ni increase with decreasing spherule diameter. Similar results have been found in earlier studies, measuring the NiFe ratios of residues as a function of their diameter in some lunar microcraters (Nagel *et al.*, 1975).

We performed low velocity (2–3 km/sec) impact cratering experiments in order to simulate the crumbling (Fig. 1) and partially damaging (Fig. 2) effects observed in the outer shells of spherules. These simulation experiments demonstrated that the edge of the particle is squeezed and exhibits a spherical lip (Fig. 4). Contrary to the homogeneous artificial iron particle, the outer shell of the multiphase spherules, consisting of a brittle mineral, crumbled during penetration of the particle into the glass of sample 12001,520. All particles incorporated in the glass matrix as well as those sitting on the glass surface have very similar structures. Probably all these particles have been formed in a very early stage of the impact process.

For quantitative measurements an automated computer-controlled ARL-SEMQ instrument was used with 15 kV accelerating voltage, 0.015 μ-amp beam current, and wavelength dispersive spectrometers. The excited volume by the electron beam bombardment during the analyses was about 2–3 μm^3. ZAF corrections were carried out for every single bulk analysis in an on-line computer. Table 1 gives the average values for 26 quantitative point analyses of the glass matrix. These results are confirmed by similar microprobe measurements of G. M. Brown, A. Peckett and B. Beddoe-Stephens, University of Durham (unpublished data), on another section of our sample. The investigations revealed a very homogeneous glass composition and showed that the elemental compositions of the breccia and the overlying glass are too dissimilar to justify the glass formation by melting of the breccia during an impact process. The Al content of the breccia is higher than that of the glass, and the higher K-content in the glass negates a genetic relationship between breccia and glass.

The largest splash surrounding the largest crater on 12001,520 (A in Fig. 5) showed a genetic relationship between crater and splash formation whereas all other splashes on this sample did not. The splash has a radial structure emanating from a center (Fig. 5). The Si(Li) measurements at SEM showed that some particles found near the center were strongly enriched in Fe and Cr or Fe and Ti. However, the splash showed a composition very similar to the undisturbed glass substrate. These particles cannot have originated from the glass matrix as evidenced by the significant contrast in chemical compositions (Table 1). In the two smaller craters (B, C in Fig. 6) Fe, Cr and Fe, Ti remnants were detected, whereas in the large (A in Fig. 5 and 6) no remnants were found. Therefore, crater A was probably formed subsequent to the splash and a portion of the splash was disturbed during the formation of crater A. The fused crater glass in crater C contains melted CrFe and TiFe rich particles, while such particles in crater B are on the surface of the glass. For these reasons we suggest that the splash originated during the formation of crater C by a Ti, Cr, Fe rich projectile. However, these qualitative measurements are insufficient evidence to determine whether the projectile was of lunar or extralunar origin.

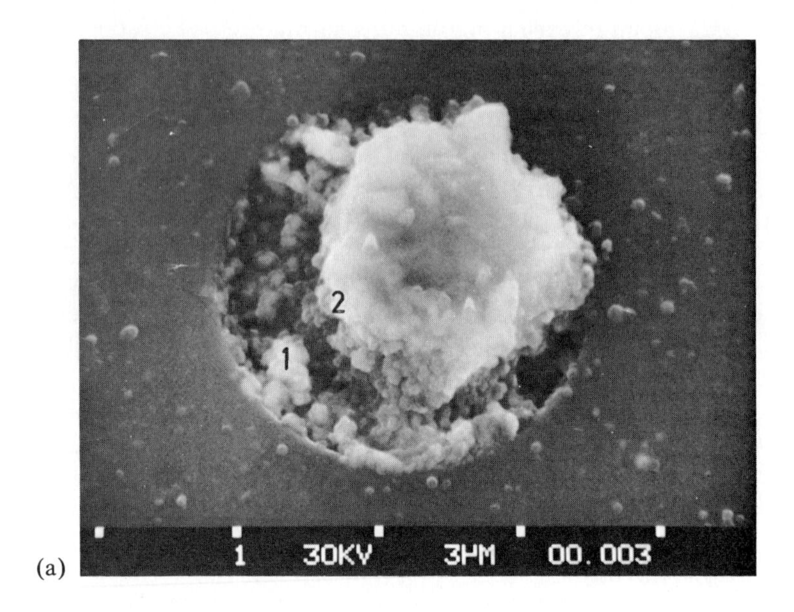

(a)

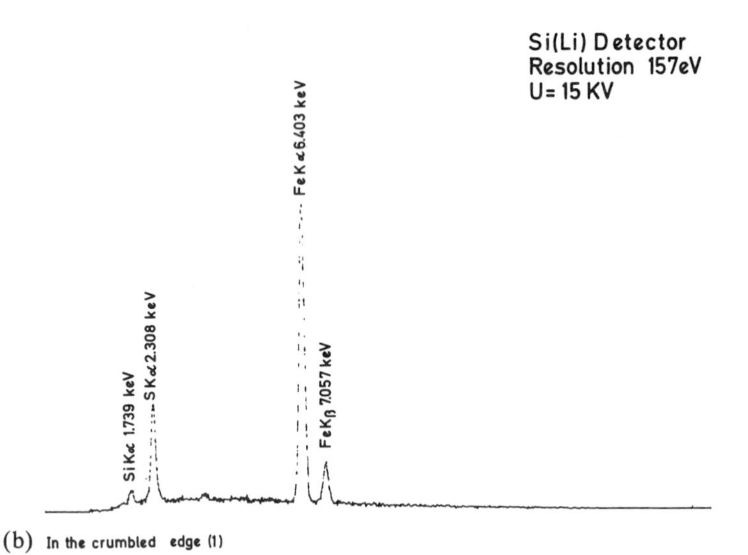

(b) In the crumbled edge (1)

Fig. 1. (a) Scanning electron micrograph of a spherule partially embedded in the surface of sample 12001,520 (Scale: 3 μm from bar to bar). (b) X-ray spectrum of point 1. (c) X-ray spectrum of point 2.

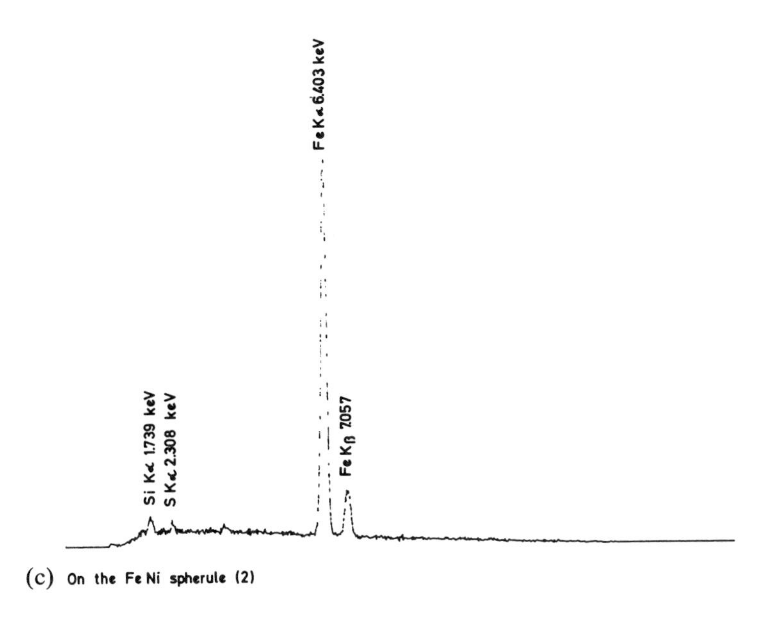

(c) On the Fe Ni spherule (2)

Conclusions

Investigations of metallic inclusions in the glass matrices of sample 12001,520 have shown that meteoritic particles changed during the impact processes. In many cases multiphase objects consisting of a FeNi core eutectically intergrown with schreibersite and covered by an almost complete shell of troilite have been produced. The origin of this zoning in these objects is not yet unequivocally explained. Further investigations are in progress to obtain more detailed information about chemical classification of the projectiles.

Chemical analyses show no evidence that the metallic inclusions have been partially dissolved in the surrounding glass. As concluded in earlier simulation experiments (Nagel *et al.*, 1976), oxygen fugacity of the surrounding atmosphere plays an important role in elemental dissolution in glass. Due to the very low fO_2 in the lunar environment Fe and Ni present in the projectiles will not be oxidized and dissolved in the impactite glasses.

The dissimilarity of chemistries of breccia and glass is clearly inconsistent with a genetic relationship. The glass was most probably formed during an indepen-

Table 1. Chemical composition of the glass substrate of sample 12001, 520 (average of 26 measurements).

Ti	Ca	Mn	Fe	Cr	Si	Al	Na	Mg	Total
2.74	10.3	0.24	14.3	0.30	46.3	14.4	0.5	11.2	100.3

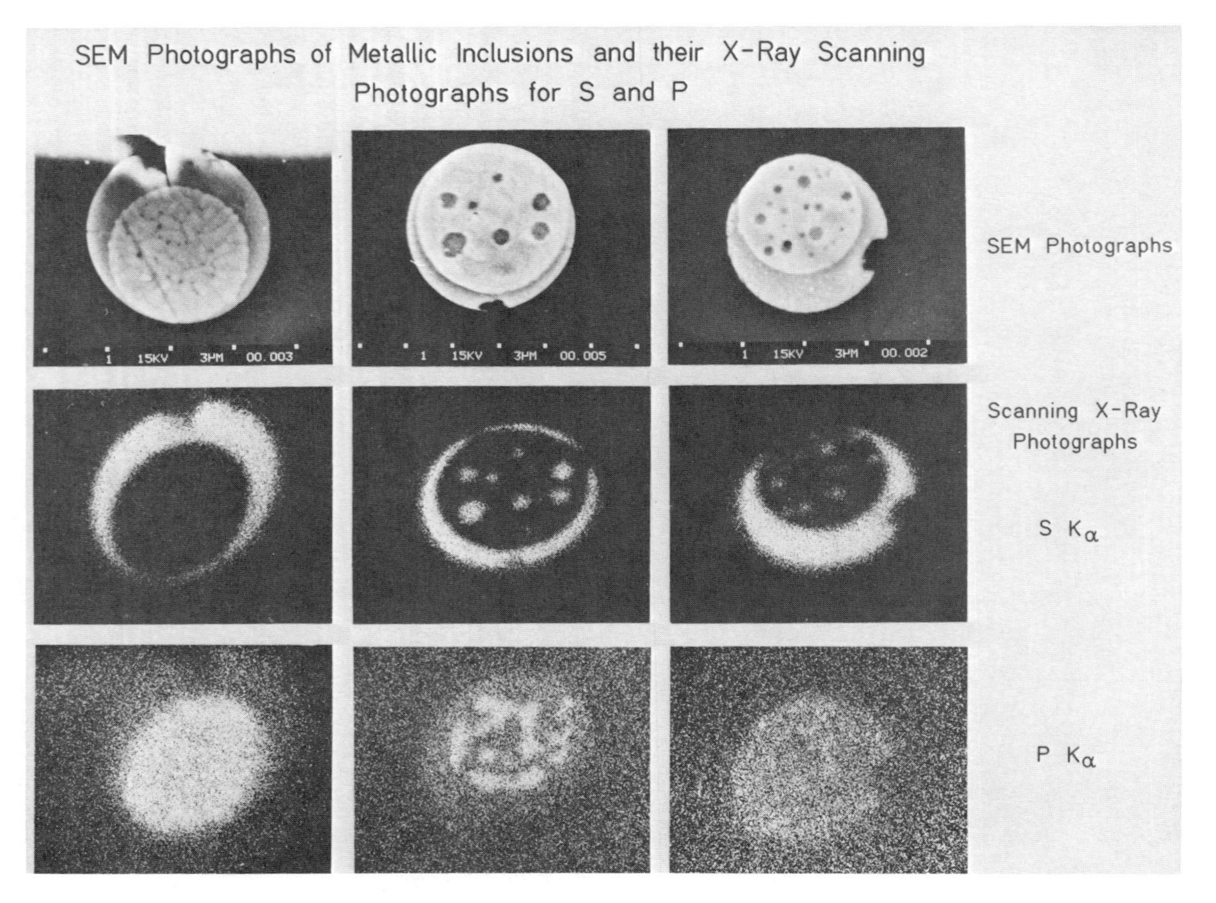

Fig. 2. SEM micrographs of metallic inclusions of sample 12001,520 and their X-ray scanning micrographs for sulfur and phosphorus.

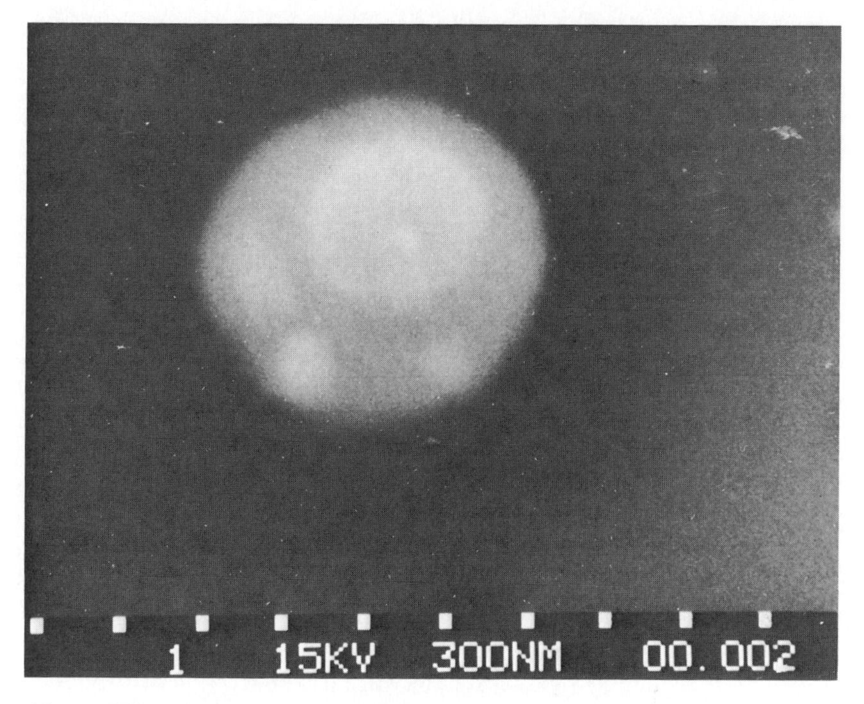

Fig. 3. SEM micrograph of a multiphase spherule completely inclosed in the glass-linings of a crater on 65315,68 (Scale: .3 μm from bar to bar).

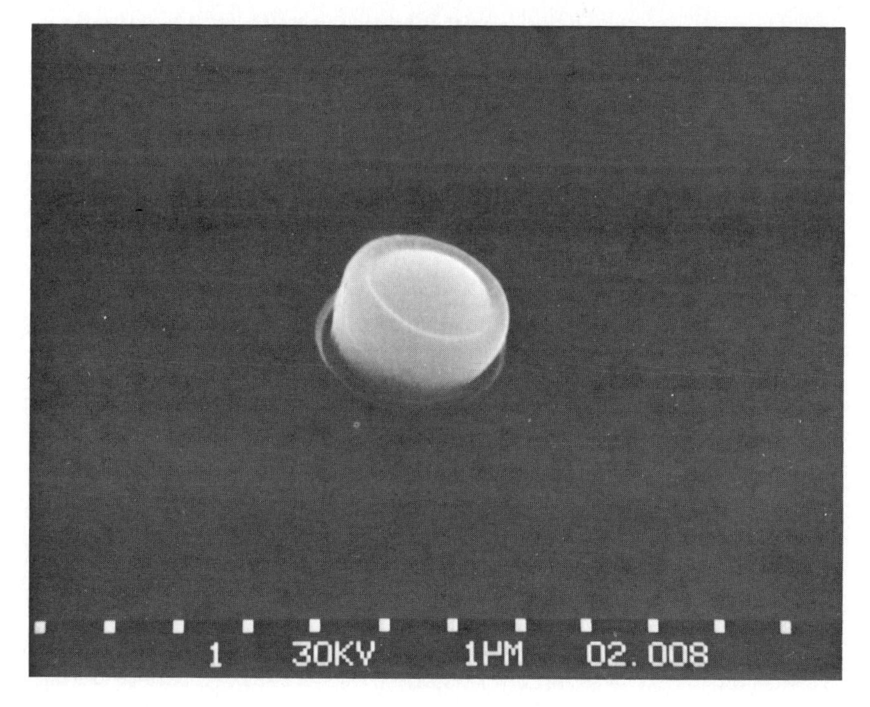

Fig. 4. Impact crater simulation experiment: a homogeneous iron projectile (velocity range: 2–3 km/sec) is embedded in a glass target.

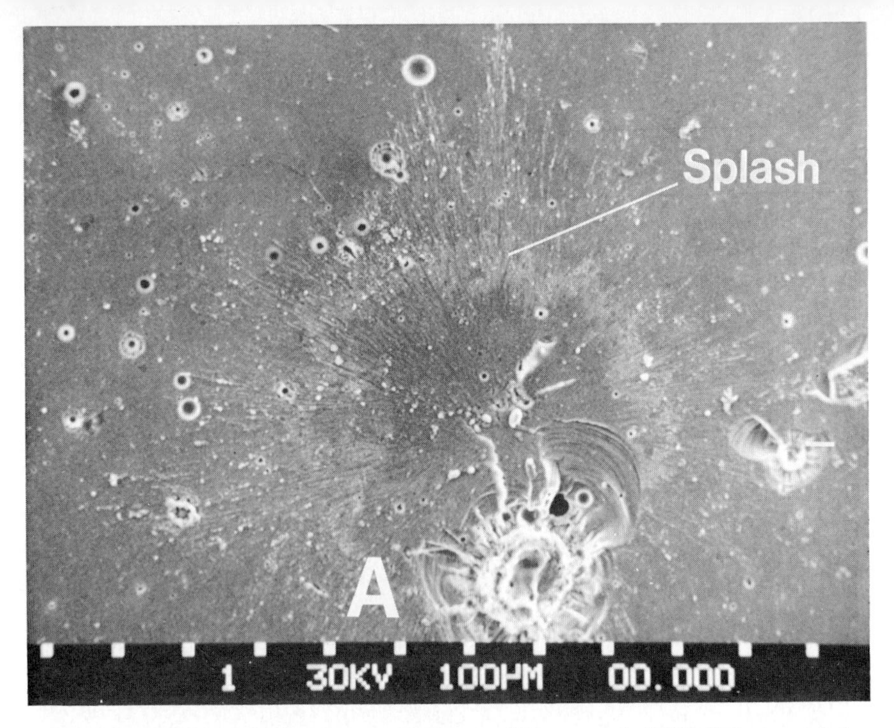

Fig. 5. Magnified view of the largest microcrater (A) on sample 12001,520 surrounded by a glass splash with radial structure.

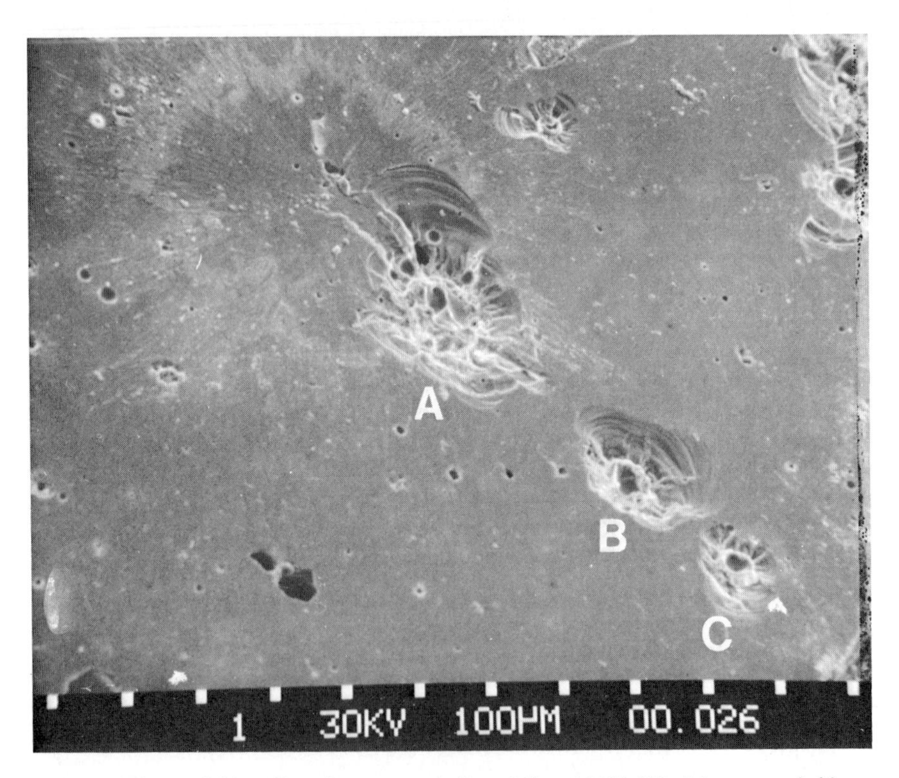

Fig. 6. Three neighbouring microcraters A, B, and C on 12001,520; A is surrounded by a glass splash.

dent impact event. The presence of multiphase spherules in all parts of the glass indicates that the projectile was an iron-nickel meteorite. Released impact energy affected the transport of silicate impact melts and projectile residues into the breccia. Subsequently, more residues of both types were deposited on 12001,520. Therefore, we must consider the possibility that part of the glass splashes on the sample surface could have been derived from impact which formed the glass. The larger splash with radial structures, however, is associated with a later event, because:

a) The homogeneous glass shows no enrichments in Fe, Cr or Fe, Ti; and
b) Particles showing these enrichments were only detected in the center of this splash and in two neighbouring craters.

These craters were presumably formed during the sample exposure time of about 7.10^5 years.

Acknowledgments—Part of this work was supported by the Deutsche-Forschungsgemeinschaft and by the Swiss National Science Foundation NF 2.886.77. We thank Prof. Geiss for his co-operation and for providing the sample. The authors are indebted to Prof. G. M. Brown, Dr. A. Peckett and Dr. B. Beddoe-Stephens for permission to use unpublished data.

REFERENCES

Blau P., Goldstein J. I. and Axon H. J. (1973). An investigation of the Canyon Diablo metallic spheroids and their relationship to the breakup of the Canyon Diablo meteorite. *J. Geophys. Res.* **78**, 363–374.

Brownlee D. E., Hörz F., Hartung J. B. and Gault D. E. (1975). Density, chemistry, and size distribution of interplanetary dust. *Proc. Lunar Sci. Conf. 6th*, p. 3409–3416.

Fechtig H., Hartung J. B., Nagel K., Neukum G. and Storzer D. (1974). Lunar microcrater studies, derived meteoroid fluxes and comparison with satellite-borne experiments. *Proc. Lunar Sci. Conf. 5th*, p. 2463–2474.

Gibbons R. V., Hörz F., Thompson T. D. and Brownlee D. E. (1976). Metal spherules in Wabar, Monturaqui and Heubury impactites. *Proc. Lunar Sci. Conf. 7th*, p. 863–880.

Goldstein J. I., Axon H. J. and Yen F. (1972). Metallic particles in the Apollo 14 lunar soil. *Proc. Lunar Sci. Conf. 3rd*, p. 1037–1064.

Hartung J. B. and Storzer D. (1974). Lunar microcraters and their solar flare track record. *Proc. Lunar Sci. Conf. 5th*, p. 2527–2541.

Nagel K., Neukum G., Eichhorn G., Fechtig H., Müller O. and Schneider E. (1975). Dependencies of microcrater formation on impact parameters. *Proc. Lunar Sci. Conf. 6th*, p. 3417–3432.

Nagel K., Neukum G., Fechtig H. and Gentner W. (1976). Density and composition of interplanetary dust particles. *Earth Planet. Sci. Lett.*, **30**, 234–240.

Schaal R. B., Hörz F. and Gibbons R. V. (1976). Shock metamorphic effects in lunar microcraters. *Proc. Lunar Sci. Conf. 7th*, p. 1039–1054.

Proc. Lunar Planet. Sci. Conf. 9th (1978), p. 2495–2506.
Printed in the United States of America

Chemical composition variations in microcrater pit glasses from lunar anorthosite, 65315

J. B. HARTUNG*, K. NAGEL and A. EL GORESY

Max-Planck-Institut für Kernphsik, Heidelberg, Federal Republic of Germany

Abstract—Glasses lining pits formed by meteoroid impacts onto an almost monomineralic lunar anorthosite were analyzed using standard electron microprobe techniques. Concentrations of all detectable elements were measured at ~2 μm-sized spots located along profiles on polished surfaces perpendicular to the originally exposed surface of the pit. Concentration profiles in two black glass pits show dramatic increases in FeO and MgO in the outermost 5 μm of the glass. These enrichments may be derived from the impacting meteoroid. Profiles in the black portion of a pit containing both clear and black glass indicate that lunar soil material was melted and incorporated into the pit glass during the impact. Profiles in one pit containing clear glass yielded no evidence of either a meteoroidal or a lunar soil component in the impact glass. Slight near-surface enrichments of FeO and MgO were observed in another clear glass pit. At topographically elevated sites in both clear glass pits, SiO_2 and Na_2O were depleted and Al_2O_3 and CaO were enriched in the outer few μm of the glass. This effect may be due to selective volatilization of SiO_2 and Na_2O or to selective condensation of Al_2O_3 and CaO.

INTRODUCTION and BACKGROUND

Glass-lined pits with diameters ranging from $<10^{-5}$ cm to 1 cm occur on essentially all lunar surfaces exposed to space. The origin of these pits by meteoroid impact is well established. Several investigators have analyzed the chemical composition of the glass linings of these pits to obtain information about the composition of the impacting meteoroids. Results have been negative or, at best, very uncertain in most cases. A possible reason for the lack of evidence for meteoroid material in glass-lined pits is that meteoroids may vaporize upon hitting the lunar surface (Gault *et al.*, 1972; O'Keefe and Ahrens, 1977). However, binocular microscope observations of impact pits indicate that the color of some pit glasses is not consistent with their being derived *only* from melting of host rock minerals. In one study roughly 10% of the pits with black or dark glass linings occur on light or clear-melting minerals (Hörz *et al.*, 1971).

Material on the surface of impact pits in a metal grain was analyzed by Fredriksson *et al.* (1970) and found to have a composition similar to that of lunar soil. (See cover of *Science* **167**, No. 3918). They concluded "that the impacting particles were not meteoritic but lunar material. . . ." Morrison *et al.* (1973) showed, however, that essentially all exposed lunar surfaces are coated on the μm

*Present Address: Department of Earth and Space Sciences, State University of New York at Stony Brook, N.Y. 11794/USA

scale with accretionary material probably derived from small impacts nearby. Thus, it is likely that Fredriksson *et al.* (1970) analyzed such a coating, which was not related to the impacting object.

Previous workers (Chao *et al.*, 1970; Bloch *et al.*, 1971; Brownlee *et al.*, 1975; Nagel *et al.*, 1975, 1976; Hartung *et al.*, 1975) have analyzed impact pit glasses and found metal-rich spherules or mounds in or on the glass. The origin of these spherules may be meteoritic, indigenous metal-rich grains, or metallic material produced by the reduction of oxides through the action of implanted solar wind hydrogen and subsequent impact melting (Housley *et al.*, 1971). Quantitative analyses are required to associate positively metallic spherules in impact pits with metal-bearing meteoroids.

Identification of nonmetallic meteoroid material in pit glasses requires quantitative analysis of both the impact glass and host material. Mehl (1974) reported electron microprobe results for a 2 mm-diameter pit which was sectioned vertically. Unfortunately, the host material was not homogeneous on a millimeter scale, so the results are difficult to interpret conclusively. Brownlee *et al.* (1975) analyzed three vertical sections of pits in homogeneous plagioclase feldspar. For these analyses, a defocussed (100 μm-diameter) electron beam was used and a comparison of the composition of the pit glasses and the underlying host material was made. For two of the pits, slight enhancements of Fe and Mg were observed and attributed either to the impacting meteoroid or to Fe-Mg-rich lunar material on the pre-impact surface or in the volume of material melted by the impact. Schaal *et al.* (1976) analyzed vertical sections of pits in a variety of host materials, including three in almost-pure plagioclase. Maps of the sections show the analysis sites nearest the exposed surface to be a few tens of micrometers below the surface. Despite directed efforts, Schaal *et al.* (1976) found "no evidence for elemental fractionation via selective volatilization" and "no evidence for meteoritic contamination" of impact pit glasses. Using semi-quantitative energy dispersive X-ray analysis techniques on exposed impact glass surfaces, Nagel *et al.* (1976) concluded "stony-iron meteorite residual material is homogeneously mixed with the glass linings", especially those which are dark in color.

In summary, the work done to date by investigators using carefully selected and prepared samples and quantitative methods has led to apparently contradictory results. Questions remain. Is silicate meteoroidal material identifiable in impact pit glasses? Is selective volatilization an important process during hypervelocity impacts?

EXPERIMENT DESCRIPTION

The rock selected for this study was 65315, an almost monomineralic anorthosite containing plagioclase (An_{97}) and rare grains of pyroxene and olivine, which are not larger than 20 μm along a side in our sections. Wänke *et al.* (1974) measured the concentrations of the elements listed in Table 1 using both instrumental neutron activation and X-ray fluorescence analysis for all elements except Na. The extremely low values for Fe and Mg make 65315 an ideal potential detector of meteoroidal material in impact pit glasses.

Sixteen pits, 0.5 to 2.4 mm in diameter, on about 2 cm^2 of rock surface were studied optically and with a scanning electron microscope equipped with an energy dispersive X-ray analyzer. Clear or transparent glass lined ten pits. In three of these the glass was darker away from the center of the pit. Black glass, usually with a more uneven surface and containing more vesicles, lined four pits. The surface of one of these was dotted everywhere with mounds up to 2 μm in diameter and rich in Fe, Ni, and S. Two pits were irregular, containing both transparent and black glass.

Five pits, representative of the different types of pits observed optically, were selected for detailed examination: two with clear glass, two with black glass (one of these with Fe-Ni-S mounds), and one with both clear and black glass. Vertical polished sections through each pit were prepared. Numerous qualitative analyses at a variety of locations on the polished surfaces were made using energy dispersive X-ray analysis techniques to find interesting and representative areas deserving more detailed study.

Wavelength dispersive electron microprobe analysis was used to obtain quantitative elemental compositions. An automated computer-controlled ARL-EMX instrument was used with an accelerating voltage of 15 kV and beam current of 0.001 μamp. Stability of the beam was maximized by cooling the objective lens with liquid N_2 and executing a complete analysis in a short time, about six minutes, without moving the sample. Na was measured first to minimize possible losses due to vaporization under the beam. In addition, repeated analysis at the same spot indicated no significant Na loss occurred during the analysis.

Concentrations of detectable elements were measured at spots located along profiles perpendicular to the exposed surface of the pit. Figure 1 is a scanning electron micrograph taken at an oblique angle and shows that essentially no rounding, on the micrometer scale, occurred due to polishing at the intersection of the polished and originally exposed surfaces.

Also visible in Fig. 1 are several areas about 2 μm in diameter which were affected by the microprobe electron beam during the analyses. Such areas are clearly related to analysis profiles based on scanning electron microscope observations of the areas analyzed both before and after microprobe measurements. This shows that we can expect the lateral resolution associated with the incident electron beam to be about 2 μm. The resolution for the analysis profiles is limited by the size of source volume of X-rays which may be only slightly more than 2 μm in diameter, based on the measured distances over which distinct changes in composition occur and the distance required for analysis totals to decrease from nearly 100% to below 50% at the edge of the sample.

Several profiles were measured on each impact pit. No particular criteria were used to select the

Table 1. Major and Minor Elements in 65315.

Element	Concentrations, wt. %	
	Wänke *et al.* (1974)	This Work
O	46.5 ± 0.5	46.0*
Si	20.7 ± 0.2	21.0 ± 0.5**
Al	18.45 ± 0.37	19.3 ± 0.3
Ca	13.62 ± 0.54	14.3 ± 0.4
Na	0.22 ± 0.01	0.27 ± 0.06
Fe	0.24 ± 0.02	0.11 ± 0.03
Mg	0.15 ± 0.01	0.08 ± 0.02
Ti	0.0072 ± 0.0004	—
Total	99.9	101.1

*Calculated value.

**Uncertainties correspond approximately to one standard deviation of the data.

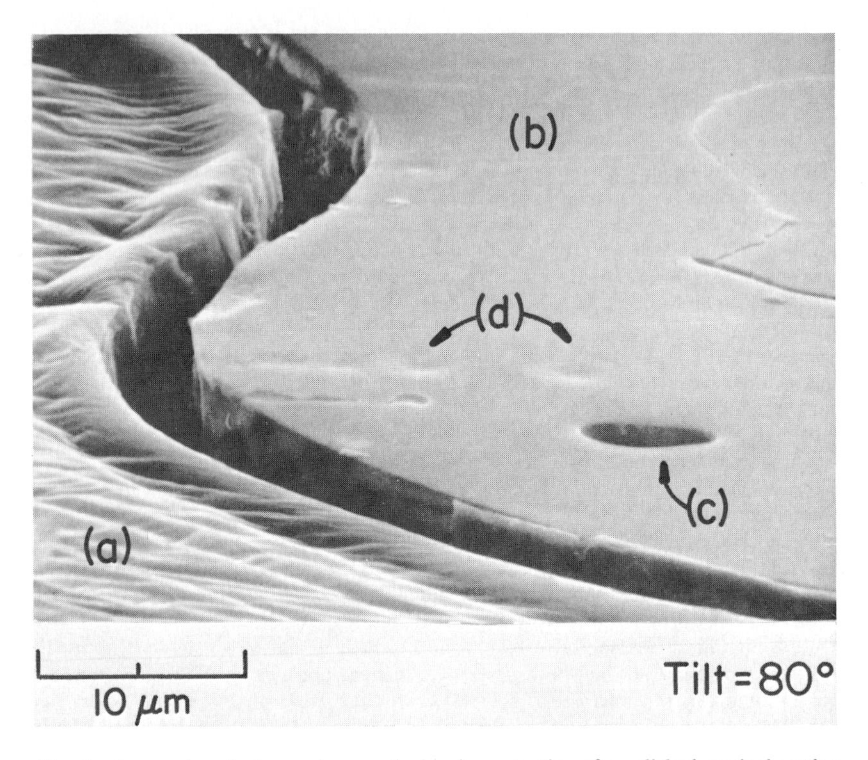

Fig. 1. A scanning electron micrograph showing a portion of a polished vertical section through a microcrater pit. The imbedding epoxy (a) is separated from the polished section of the pit (b) leaving the deep valley. The viewing angle is highly inclined (10° above the polished surface), as is indicated by the sectioned vesicle (c), which is circular in plan view. The row of diffuse spots (d) identify the analysis sites for one element concentration profile.

lateral location of the profiles. Profile depths extended only as far as was necessary, that is, to a depth such that consecutive analyses yielded compositions equal to one another and to that of the host rock. Our analyses showed that no significant variations occurred below this depth, which was usually between about 10 and 30 μm. An average composition for the host rock based on numerous spot analyses at depths greater than 10 μm, but still in impact melted material, is also shown in Table 1. The composition we obtained differs from that of Wänke *et al.* (1974) only in that they found higher concentrations of Fe and Mg in the host rock. This may be explained by the fact that our spots probably do not include any contribution from the rare pyroxene and olivine grains present in the rock.

RESULTS and INTERPRETATIONS

Data will be presented for each pit separately. Only one profile out of several taken for each pit will be presented. A particular profile was selected for presentation because it effectively illustrates important trends in the data. That profile is illustrative in the sense that the other profiles measured are usually more irregular or less impressive but show characteristics similar to the one

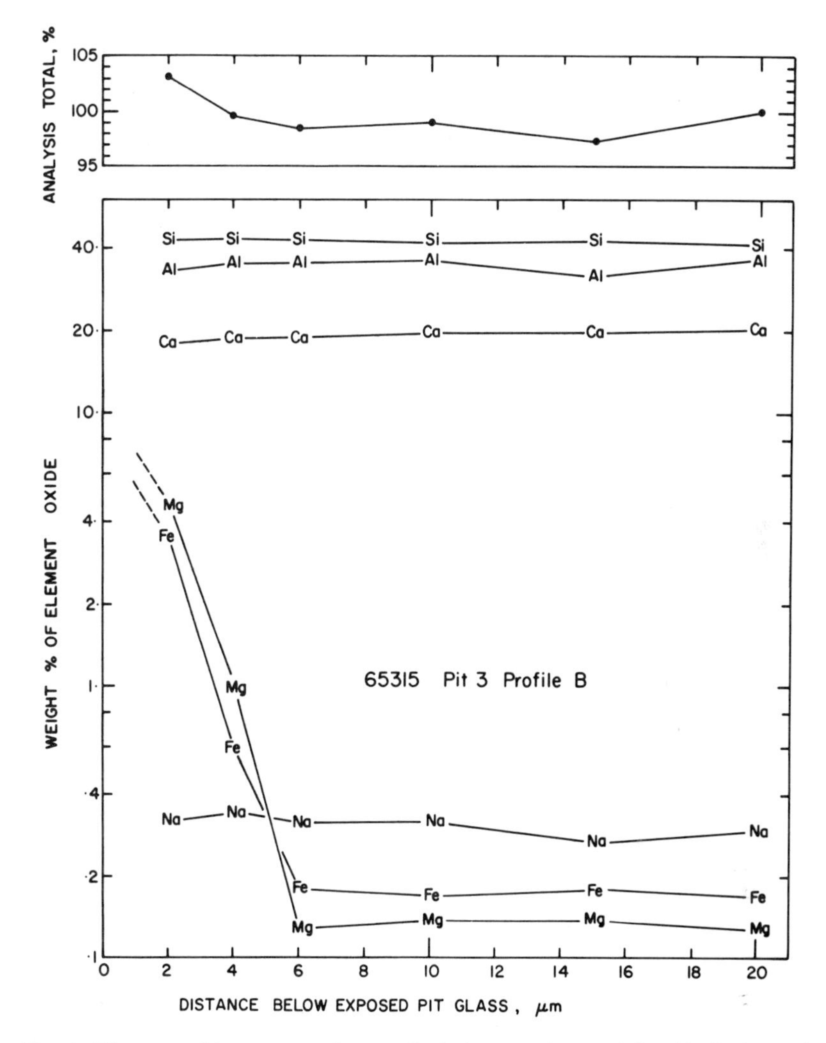

Fig. 2. Element oxide concentration profile in impact pit containing black glass. The sharp increases in FeO and MgO may be due to a contribution of meteoroidal material contained in the glass.

presented. Different profiles for the same pit usually showed composition changes at different depths and of different amounts.

The relative uncertainty associated with a single concentration measurement is indicated approximately by those given in Table 1. We estimate the depth uncertainty to be less than about 2 μm.

An element oxide concentration profile into the pit with black glass is shown in Fig. 2. At depths shallower than 6 μm below the originally exposed surface, the concentrations of FeO and MgO increase dramatically (note logarithmic scale) with decreasing depth below the surface. Al_2O_3 and CaO are correspondingly reduced, while SiO_2 remains unchanged. An attempted analysis at a depth shallower than 2 μm was not acceptable (the analysis total was much less than 100%), but the concentrations of FeO and MgO corresponding to an assumed total of 100% reached about 6% and 8%, respectively. We interpret these data as indicating the presence of meteoroid material in the outer few micrometers of pit glass although contamination from lunar soil on the rock at the impact site cannot be ruled out. Also, from these data we suggest that the meteoroid or soil material was an Fe-Mg silicate with a MgO/FeO ratio of about 1.5.

A concentration profile in the black glass containing Fe-Ni-S mounds is shown in Fig. 3. The profile is similar to that of Fig. 2, except that in this case the S content approaches 1 weight% of the metal at some near-surface spots. Energy dispersive X-ray analysis of the Fe-Ni-S mounds or spherules indicates that at least two phases are present, an Fe-Ni-rich phase and a surrounding S-rich phase. More detailed results of work on these metal and sulfide mounds will be reported separately. Apparently, the meteoroid which formed this pit contained Fe, Ni, and S, presumably in metal and/or sulfide phases, and possibly also a Mg silicate phase.

Profiles in one of the pits containing clear glass provided essentially no evidence of meteoroid material in the glass. In the other clear glass pit, small (factor-of-2) enrichments of FeO and MgO were often, but not always, present in about the outermost 10 μm of the pit glass. The FeO and MgO profiles shown in Fig. 4 are typical in this regard. We suggest, but with less certainty than for the previous case, that the impacting meteoroid contained FeO and MgO.

A striking pattern was observed in four of thirteen profiles measured in the clear glass pits. This pattern for one profile is shown in Fig. 4. SiO_2 and Na_2O are depleted near the surface, while Al_2O_3 and CaO are correspondingly enriched. We interpret these trends as being due to preferential loss of SiO_2 and Na_2O or retention of Al_2O_3 and CaO occurring during the meteoroid impact event and arising because of corresponding differences in the vaporization or condensation temperatures of these oxides. It is not immediately clear whether preferential SiO_2 and Na_2O volatilization occurred or whether preferential Al_2O_3 and CaO condensation occurred from an oxide vapor "cloud". The fact that all four profiles showing the effect are located at topographic "highs" within the pit favors the latter possibility. The CaO, Al_2O_3, and SiO_2 concentrations that we measured for the fractionated near-surface glass are nearly the same as those established for high-alumina silica-poor (HASP) glass spherules found in Apollo 16 (Naney *et al.*, 1976) and 17 (Vaniman and Papike, 1977) core soils, thus supporting a local impact origin for these and similar glasses.

Data for the clear glass in the pit containing about half clear and half black glass provided no evidence for impacting meteoroid material. A typical profile in the black glass in this pit is shown in Fig. 5. The glass is enriched in FeO, MgO, and TiO_2 throughout the outermost 10 μm; CaO and Al_2O_3 are correspondingly depleted. Based on our previous interpretations, one could argue that these data indicate the impact of a Fe-Mg-Ti-bearing silicate. However, this is totally

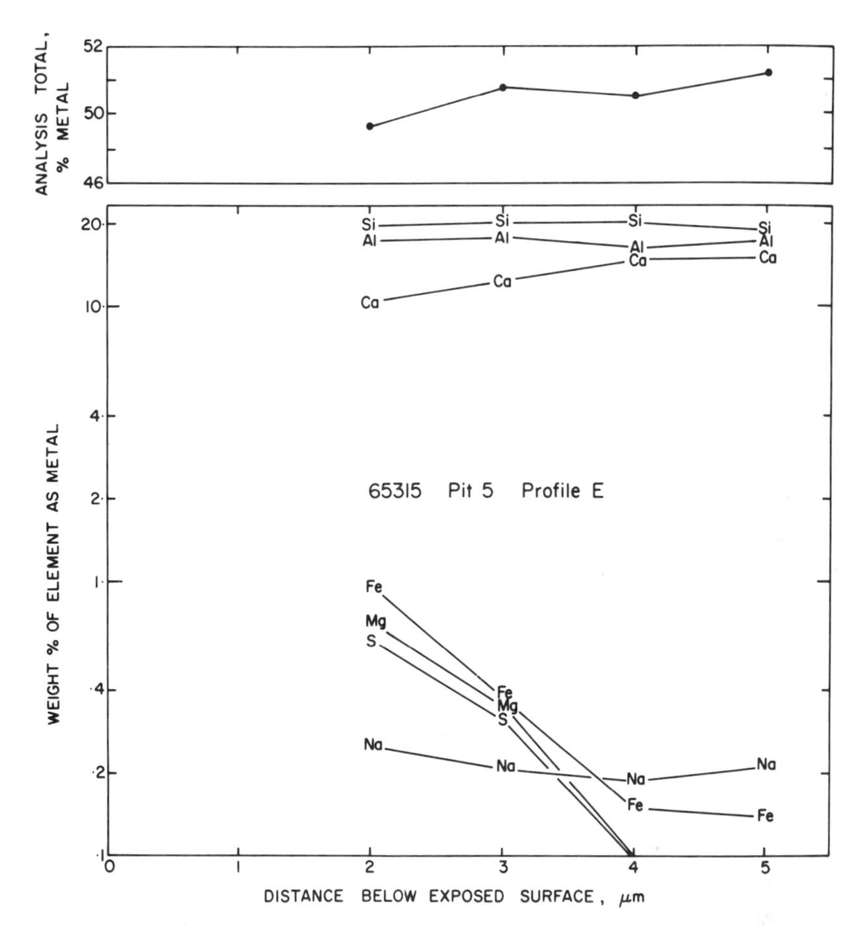

Fig. 3. Element concentration profile in the glass of the pit containing black glass and Fe-Ni-S-bearing mounds and spherules. Near-surface increases in Fe, Mg, and S are probably due to material derived from the impacting meteoroid and incorporated into the glass during the impact. Analysis totals are near 50% because we did not assume all metals occurred as oxides, so the weight of oxygen is not included in the total.

inconsistent with the fact that the other half of the pit contains essentially no evidence of contamination by the impacting meteoroid. Furthermore, the thickness of the Fe-Mg-Ti-rich zone is greater than that observed in other pits; and high TiO_2 concentrations, which reach several percent for another profile, seem more consistent with a lunar rather than a meteoroidal origin. The extent of the black glass, 500 μm laterally and 10 μm in depth, seems too large to be caused by an Fe-Mg-rich grain present in the rock at the time of the impact. We conclude, therefore, that the black glass in this pit was derived from lunar soil which was adhering to the rock surface at the time of the impact and was melted and incorporated into the pit glass by the impact.

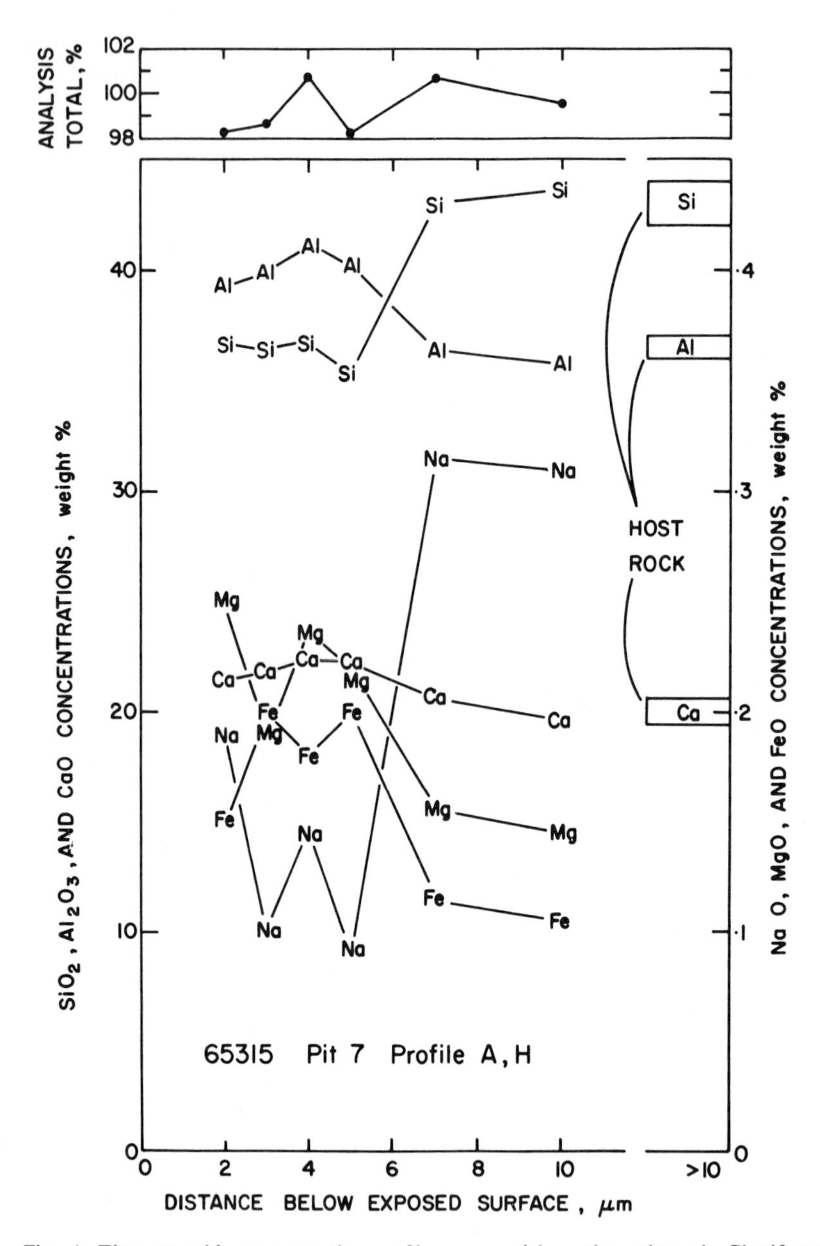

Fig. 4. Element oxide concentration profile measured in a clear glass pit. Significant near-surface depletion of Na_2O and SiO_2 and corresponding increases in Al_2O_3 and CaO may be due to either selective volatilization or selective condensation occurring immediately after the impact.

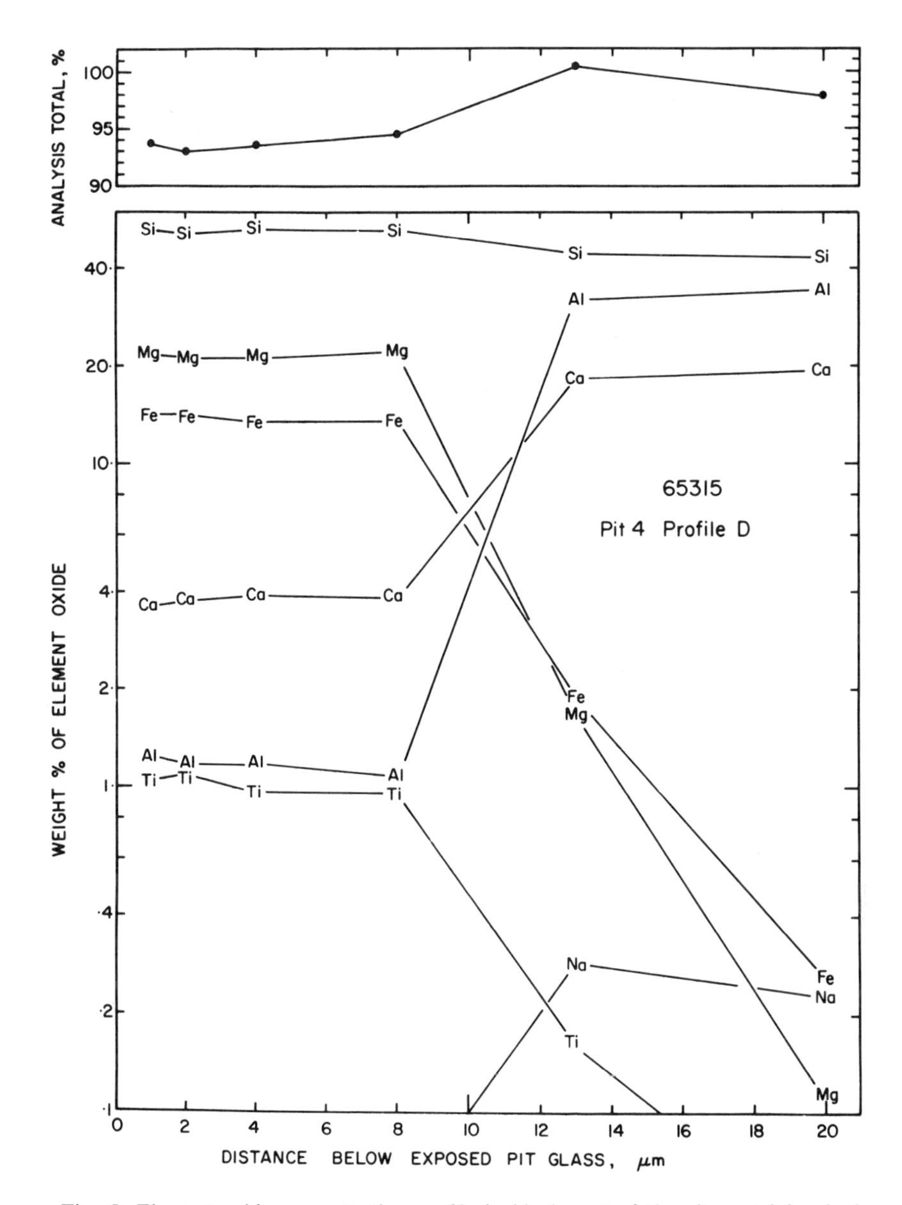

Fig. 5. Element oxide concentration profile in black part of the pit containing both clear and black glass. The near-surface Fe-Mg-Ti enrichment observed may be due to lunar soil which adhered to the rock surface before the impact and was incorporated into the pit glass during the impact.

DISCUSSION

One contribution of this paper is to reconcile apparently divergent results and interpretations which exist in the literature and are summarized in our introduction. Workers using quantitative electron microprobe techniques restricted their analysis sites to depths usually greater than 10 μm and found little or no variation from host-rock compositions. We have shown that essentially all evidence for the presence of meteoroid or lunar soil material and selective volatilization or condensation is to be found within the outermost 10 μm of the pit glasses. Workers using semi-quantitative energy dispersive X-ray analysis techniques did sample the outer 1 or 2 μm and found evidence for meteoroid material. We improved upon this result by obtaining quantitative results and establishing with greater accuracy the depth range of significant chemical variations.

We have confirmed that pit glass color is closely related to the composition of the glass and have shown that detailed analyses of different types of pits yield data requiring different interpretations. The chemical composition variation within each of the five pits that we analyzed was significantly different.

We have provided further evidence that meteoroid composition, pre-impact lunar soil contamination, and selective volatilization or condensation may cause pit glass chemistry variations. We expect that impact energy is also a factor which could lead to variations in chemical composition. For example, impacts delivering higher total energies will form larger craters and produce correspondingly larger masses of melt and vaporized material. Significantly thicker zones affected by selective volatilization or condensation and by meteoroid contamination may be expected to develop. Because the pits we studied were all similar in size, we could not study the effects of impacts having different total energies. We do, however, have the possibility of observing the effects of impacts having different specific energies, that is, different energies per unit mass of impacting object, which corresponds to impacts of objects with similar masses but different velocities. From theoretical considerations we would expect higher specific energy impacts to produce more melt and vapor from the host rock compared to that derived from the impacting object (O'Keefe and Ahrens, 1977). Therefore, we tentatively conclude that clear glass pits which contain little or no meteorid material were probably formed by higher specific energy, i.e., higher velocity, impacts than those pits containing larger amounts of meteorid material. Clearly, a much larger number of pits must be analyzed to establish the relative importance of the different variables which control pit glass composition differences.

We have interpreted near-surface enhancements of certain elements in two cases as a result of meteoroid material and in one case as a result of pre-impact lunar soil contamination of the host rock surface. The possibility that all enhancements were caused by lunar soil contamination cannot be ruled out absolutely on chemical grounds. We suggest, however, that the restricted thickness of the zones of enrichment (<10 μm) and the lateral continuity of these zones are more consistent with the idea of a meteoroid source for the enrichments.

Perhaps the most noteworthy outcome of this study has been the association of a surprising degree of chemical fractionation with a single millimeter-sized impact event. The mechanism of this fractionation deserves more attention. Previous workers have considered the possible importance of selective volatilization. We can visualize Na_2O and SiO_2 molecules preferentially leaving a hot plagioclase-composition melt. We suggest, however, that the process of selective condensation also should be considered as a fractionation mechanism. In this view, CaO and Al_2O_3 molecules in a plagioclase-composition vapor would preferentially condense on nearby surfaces. In support of this suggestion, we have noted that those profiles showing the highest degree of fractionation are located at topographic "highs" within the impact pit. This suggestion would be strengthened if it could be shown that the outermost few micrometers of material, including that derived from the impacting meteoroid, was all vapor deposited. This may be the case because at moderate impact velocities most of the meteoroid may be expected to vaporize (Gault *et al.*, 1972; O'Keefe and Ahrens, 1977). No vapor deposited material has been observed in spall zones surrounding impact pits, but this could be explained if the spallation process were slow compared to the vapor-deposition process. Interestingly, Hörz *et al.* (1971), using an optical microscope, observed a "concentric arrangement of thin film coating outside the spall zone on one extremely large pit crater on rock 12051". They interpreted this observation in terms of condensation of silicate vapors produced by the pit-forming impact. Available data do not permit firm conclusions, but selective condensation of certain element oxides and vapor deposition of meteoroid and host rock material on nearby surfaces may be processes which operate normally at the lunar microsurface.

Acknowledgments—We thank J. Janicke and W. Horn for laboratory assistance and M. O'Dowd, C. Mercier, and S. de Nicolas for assistance in preparing the manuscript. The work of JBH was supported by NASA grant NSG-9013.

REFERENCES

Bloch M. R., Fechtig H., Gentner W., Neukum G. and Schneider E. (1971) Meteorite impact craters, crater simulations, and the meteoroid flux in the early solar system. *Proc. Lunar Sci. Conf. 2nd*, p. 2639–2652.

Brownlee D. E., Hörz F., Hartung J. B. and Gault D. E. (1975) Density, chemistry, and size distribution of interplanetary dust. *Proc. Lunar Sci. Conf. 6th*, p. 3409–3416.

Chao E. C. T., Borman J. A., Minkin J. A., James O. B. and Desborough G. A. (1970) Lunar glasses of impact origin: Physical and chemical characteristics and geologic implications. *J. Geophys. Res.* **75**, 7445–7479.

Fredriksson K., Nelen J. and Melson W. G. (1970) Petrography and origin of lunar breccias and glasses. *Proc. Apollo 11 Lunar Sci. Conf.*, p. 419–432.

Gault D. E., Hörz F. and Hartung J. B. (1972) Effects of microcratering on the lunar surface. *Proc. Lunar Sci. Conf. 3rd*, p. 2713–2734.

Hartung J. B., Hodges F., Hörz F. and Storzer D. (1975) Microcrater investigations on lunar rock 12002. *Proc. Lunar Sci. Conf. 6th*, p. 3351–3371.

Hörz F., Hartung J. B. and Gault D. E. (1971) Micrometeorite craters on lunar rock surfaces. *J. Geophys. Res.* **76**, 5570–5798.

Housley R. M., Grant R. W., Muir A. H. Jr., Blander M. and Abdel-Gawad M. (1971) Mossbauer studies of Apollo 12 samples. *Proc. Lunar Sci. Conf. 2nd*, p. 2125–2136.
Mehl A. (1974) Investigation on the chemical modification in an impact crater on a lunar sample. In *Space Research XIV*, p. 741–743. Akademie-Verlag, Berlin.
Morrison D. A., McKay D. S. and Fruland R. M. (1973) Microcraters on Apollo 15 and 16 rocks. *Proc. Lunar Sci. Conf. 4th*, p. 3235–3253.
Nagel K., Neukum G., Eichhorn G., Fechtig H., Müller O. and Schneider E. (1975) Dependencies of microcrater formation on impact parameters. *Proc. Lunar Sci. Conf. 6th*, p. 3417–3432.
Nagel K., Neukum G., Fechtig H. and Gentner W. (1976) Density and composition of interplanetary dust particles. *Earth Planet. Sci. Lett.* **30**, 234–240.
Naney M. T., Crowl D. M. and Papike J. J. (1976) The Apollo 16 drill core: Statistical analysis of glass chemistry and the characterization of a high alumina-silica poor (HASP) glass. *Proc. Lunar Sci. Conf. 7th*, p. 155–184.
O'Keefe J. D. and Ahrens T. J. (1977) Impact-induced energy partitioning, melting, and vaporization on terrestrial planets. *Proc. Lunar Sci. Conf. 8th*, p. 3357–3374.
Schaal R. B., Hörz F. and Gibbons R. V. (1976) Shock metamorphic effects in lunar microcraters. *Proc. Lunar Sci. Conf. 7th*, p. 1039–1054.
Vaniman D. T. and Papike J. J. (1977) The Apollo 17 drill core: Modal petrology and glass chemistry (sections 70007, 70008, 70009). *Proc. Lunar Sci. Conf. 8th*, p. 3161–3193.
Wänke H., Palme H., Baddenhausen H., Dreibus G., Jagoutz E., Kruse H., Spettel B., Teschke F. and Thacker R. (1974) Chemistry of Apollo 16 and 17 samples: Bulk composition, late stage accumulation and early differentiation of the moon. *Proc. Lunar Sci. Conf. 5th*, p. 1307–1335.

Proc. Lunar Planet. Sci. Conf. 9th (1978), p. 2507–2537.
Printed in the United States of America

Lunar surface processes: Report of the 12054 consortium

J. B. HARTUNG and E. E. HAUSER

Department of Earth and Space Sciences, State University of New York at Stony Brook
Stony Brook, New York 11794

F. HORZ, D. A. MORRISON, E. SCHONFELD and H. A. ZOOK

NASA Johnson Space Center, Houston, Texas 77058

J.-C. MANDEVILLE

ONERA/CERT/DERTS, 2 Avenue Edouard Belin, B.P. 4025-31055
Toulouse Cedex, France

J. A. M. MCDONNELL

Space Science Laboratory, University of Kent, Canterbury, England

R. B. SCHAAL

Lockheed Electronics Company, Inc., Houston, Texas 77058

E. ZINNER

McDonnell Center for the Space Sciences, Washington University
St. Louis, Missouri 63130

Abstract—We have studied a variety of lunar surface phenomena using a well-characterized, glass-coated, ilmenite basalt, 12054, which had a simple surface residence history. Cosmogenic ^{3}He, ^{26}Al, and cosmic ray track data indicate, however, a complex, near-surface, regolith residence history. Petrographic and electron microprobe studies show that the rock was thoroughly shocked to a pressure of about 300 kilobars and that the glass was derived from a basalt composition melt and not from lunar soil. An ^{39}Ar-^{40}Ar stepwise temperature release pattern apparently was affected by the shocked state of the sample.

Optical and scanning electron microscope crater counts on surfaces selected to give the highest crater densities yielded a standard size frequency distribution for microcraters with central pits from 0.1 to 1000 μm in diameter. Analysis of accreta on a variety of surfaces has provided a production size frequency distribution for accreta. Solar wind ^{36}Ar concentrations measured on host and microcrater spall zone surfaces are both lower by about a factor of ten than would be expected if it were retained efficiently. ^{36}Ar loss by solar wind sputtering may cause such an equilibrium level on exposed surfaces. Solar flare track data indicate a surface residence time of $(1.75 \pm 0.5) \times 10^5$ yr, based on the track production rate of Blanford *et al.* (1975). Surface-related ^{26}Al γ-ray and solar wind sputter degradation data support this result. Based on one estimate of the present-day microcrater production rate, too few microcraters are present on 12054 for this exposure time.

The accumulation of loose dust or unbonded accreta on the μm scale has been established as an important process. This process may explain, among other things, the disproportionately large numbers of apparently "young" impact pits, and, at least in part, the wide range in the shapes of microcrater production size frequency distributions, especially for smaller-sized craters.

INTRODUCTION

Processes affecting lunar rock surfaces interact strongly with one another. For example, areal densities of microcraters and depth profiles of solar flare track densities are affected by meteoroid impact and sputter erosion and surface dust and accreta accumulation. Therefore, our primary objective was to study the effects and relative importance of as many processes as possible using samples from the same well-documented rock. Our second objective was to determine the rates of these processes, both relative to one another and absolutely, if possible. Surface processes related to the following effects were studied.

1. Microcraters.
2. Solar flare and comsic ray tracks.
3. Cosmogenic ^{26}Al.
4. Solar wind sputtering.
5. Accreta or accretionary material.
6. Solar wind implanted noble gases.
7. Loose dust accumulation.

The potential of rock 12054 for surface studies was recognized immediately because:

1. Most of the surface exposed to space at the time of collection was covered by a smooth glass coating.
2. This surface was essentially in production with respect to crater superposition, and the amount of superposition present could be evaluated optically.
3. Exposed glass surfaces faced distinctly different directions and were large, several cm^2.
4. No microcraters were on the unexposed portion of the rock, thus suggesting, but not proving, a simple, stationary, exposure history.
5. The precise orientation of the rock at the time of its collection is known based on lunar surface photography.
6. Preliminary gamma-ray analysis indicated ^{26}Al radioactivity had not reached equilibrium, thus increasing the likelihood of a short, simple, surface residence history.
7. A cm-deep crack, one or two mm wide and lined with glass, cuts across one exposed face of the sample, thus providing the possibility of limited exposure detector surfaces within the crack.

ASTRONAUT AND CURATORIAL ACTIVITIES

Conrad: "Al, look at these rocks. They look a little bit different. Let's grab some. Look at the glass in the bottom of that one. They look like granites, don't they?"

Bean: "They do. They look just like granites. Here's a beauty over . . . here's a beauty. Right here. That is a nice rock. Right around here. Let's get this one for sure. Won't fit in the bag, but it sure is different. It seems to have some . . ."

Conrad: "Got a big glass splotch on it."

Bean: "Yes, that's a good one. That's a real good rock. Get some pictures." (Bailey and Ulrich, 1975).

At approximately 6:04 GMT on November 20, 1969, the Apollo 12 crew photographed a fist-sized glass-coated lunar rock that became rock 12054. This sample was retrieved from near a fresh-appearing four-meter-diameter crater (Shoemaker *et al.*, 1970) near the edge of the much larger Surveyor crater. The sampling location was about 200 meters directly south of the lunar module. See, for example, Fig. 10–1 in Shoemaker *et al.* (1970) for the sampling location and Fig. 1–31 in Wade (1970) for a view of the lunar module from near the sampling location.

The lunar coordinates of the Apollo 12 landing site were 23.4°W longitude and 3.2°S latitude (Stephenson, 1970). The lunar sub-solar point at the time of photographing sample 12054 was located (from the American Ephemeris and Nautical Almanac) at 49.82°E longitude and 1.45°S latitude. The corresponding elevation and azimuth angles of the sun at the Apollo 12 landing site are then calculated to be 16.9° above the horizon and 0.4° south of due east.

Before any sawing was done, rock 12054 was placed in a large glass and stainless steel nitrogen flushed cabinet at JSC (Johnson Space Center) with a lamp simulating the position and angular size of the sun. The rock was then oriented so that the shadows seen in the lunar surface documentation photographs were reproduced. Figure 1a is a "cross-sun" photo taken by the Apollo 12 crew and Fig. 1b is the corresponding photo taken at JSC. Likewise Fig. 2a is the "down-sun" photo taken on the moon and Fig. 2b is its JSC counterpart. We feel confident that we have reestablished the original lunar orientation of the rock to within an accuracy of ±5° about all rotatation axes. We concur with the general orientation established earlier by R. L. Sutton (Shoemaker *et al.*, 1970, Fig. 10–46).

Additional photos were then taken of the rock as it was rotated about a vertical axis to document the slopes of some of the faces of the rock. The slopes were measured by holding a straight-edge parallel to the face in profile and measuring the angle relative to the horizontal (the lower edge of the cabinet) with a protractor. Angles were not determined from the photographs. Because the rock is somewhat rounded and irregular in shape, one must estimate the slopes of surfaces not shown in profile. Figure 3 depicts the rock as viewed from the north-west and from somewhat above. Inclination angles with respect to the local vertical of two northeast faces are indicated. Because the normal to the face inclined 37° points down to the lunar surface, the inclination for this face should be considered to be negative 37°. Figure 4 shows the west end of the rock as viewed looking toward the east. The rock exhibits a triangular profile in this view with a ridge-line at the top. The ridge-line trends almost exactly in an east-west

(a)

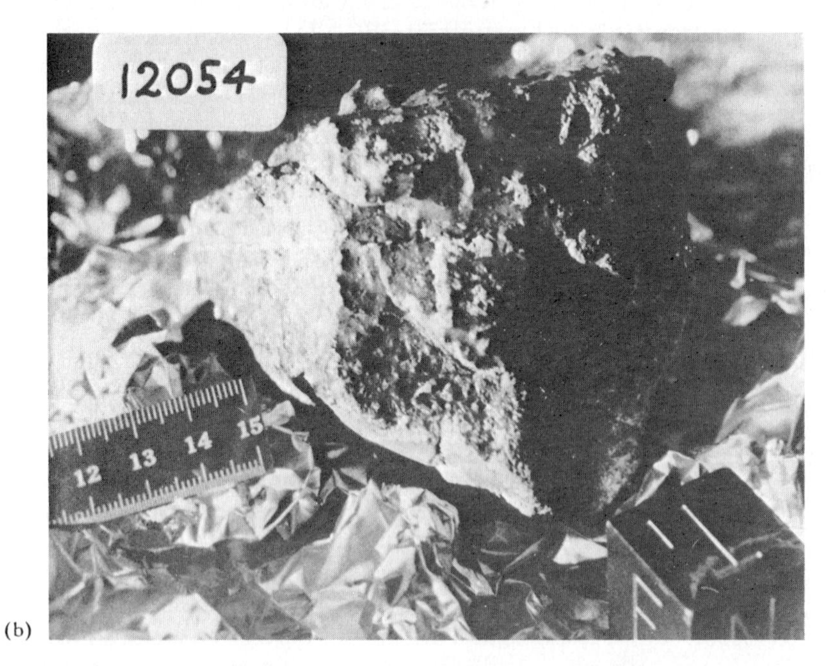

(b)

Fig. 1 (a) "Cross-sun" photo of 12054 taken, before sampling, by the Apollo 12 crew on the lunar surface (AS12-49-7315). (b) "Cross-sun" photo of 12054 taken in the lunar sample curatorial laboratory (S-75-33592).

(a)

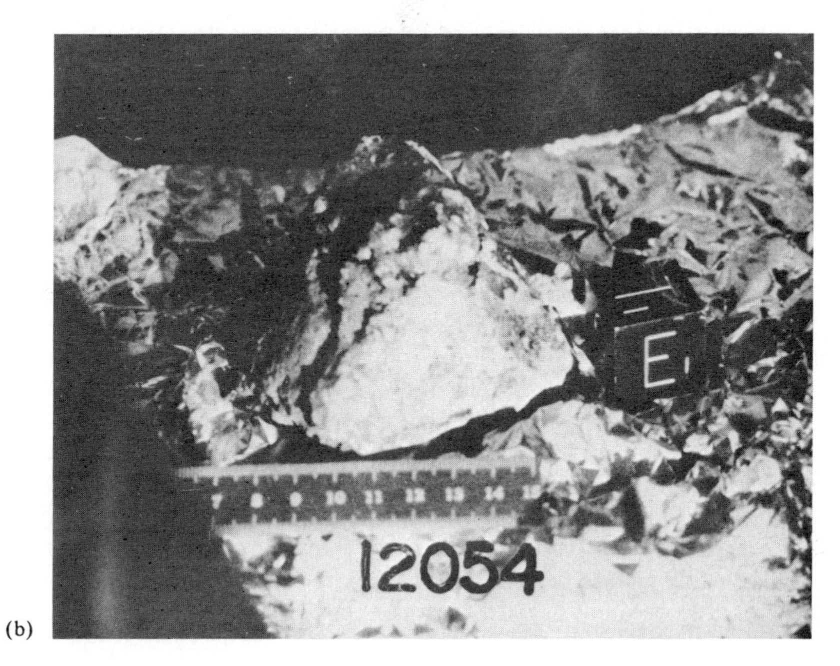

(b)

Fig. 2. (a) "Down-sun" lunar surface photo of 12054 (AS12-49-7313). (b) "Down-sun" photo in laboratory (S-75-33593).

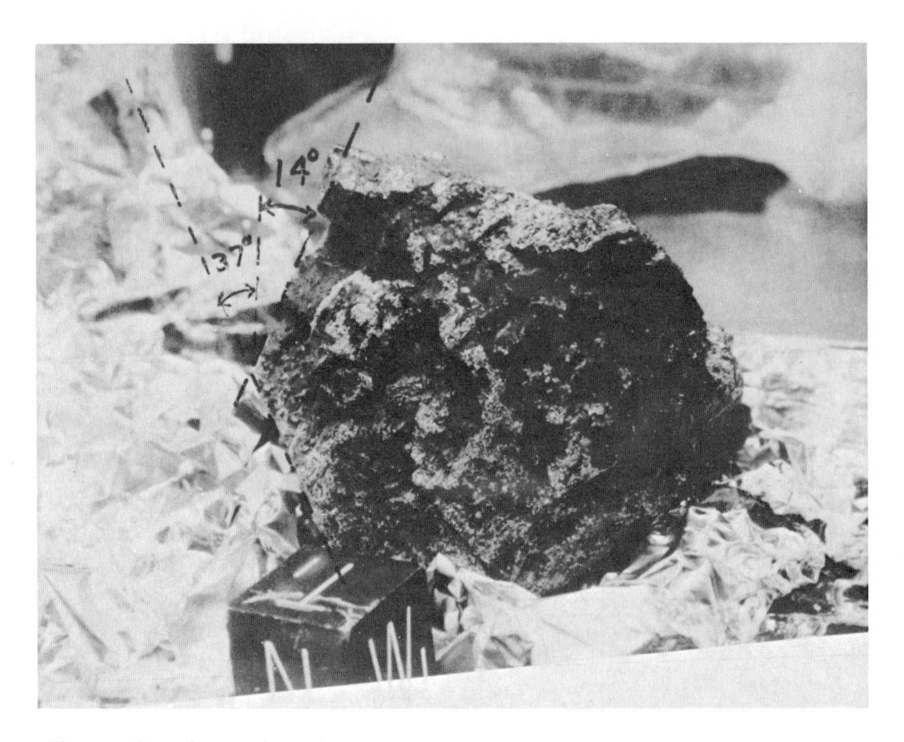

Fig. 3. View of north face of 12054 as seen from the northwest. The block indicates true directions. Glass covered portions of the rock appear dark in the photograph.

direction. Various inclination angles are shown to aid in estimating the exposure geometry of the different parts of the north (left side) and south (right side) faces of the rock. We assume that the actual horizon, as viewed from the location of the rock, is equivalent to the astronomical or ideal horizontal calculated for this site. Hence, the exposure geometry for a particular sub-sample of 12054 may be estimated once the orientation of the sub-sample with respect to the local horizontal is established. Figure 5 shows the south face of 12054. In this view, angles with respect to the local horizontal are indicated for the ridge-line, the trend of the large crack, and two downward-facing surfaces.

Cutting and chipping to obtain appropriate sub-samples of 12054 was accomplished in three stages. First, representative samples were chipped from the west end of the rock. Then, a one-cm-thick slab was cut roughly perpendicular to the east-west axis of the rock using the band saw in the lunar sample curatorial facility. This cutting caused considerable flaking off of the thin glass coating, but a smaller wire saw could not be used without exposing the entire rock to air in the lunar thin section laboratory. After the slab was cut, about 5 cm of the east end and 3 cm of the west end remained. The west end was sent to permanent storage at a location away from the Johnson Space Center. The east end remains available for future allocations. Finally, a variety of sub-samples was chipped or

Fig. 4. View of the west face of 12054 showing parts of the north and south faces in profile.

trimmed with the wire saw. Figure 6 shows the locations of most of the sub-samples taken so far from 12054. More complete and detailed information is available from the lunar sample curator.

Whole-Rock Description and Data

Stereoscopic microscope observations

The exterior surfaces of 12054 were examined thoroughly using a stereoscopic microscope. Microcraters were found to be absent on the face of the rock known to be "down" or unexposed at the time of its collection. In addition, a "soil line" separated the cratered and uncratered regions of the rock. Using criteria established earlier (Hörz and Hartung, 1971) we identified independently the exposed and unexposed portions of 12054, which were consistent with those derived from the lunar surface photography. This information is illustrated in Fig. 7 and serves as a basis for the conclusion that the surface residence history of 12054 was simple, i.e., no tumbling occurred while the rock was exposed at the surface.

A large portion of the buried crystalline surface displays striations which are

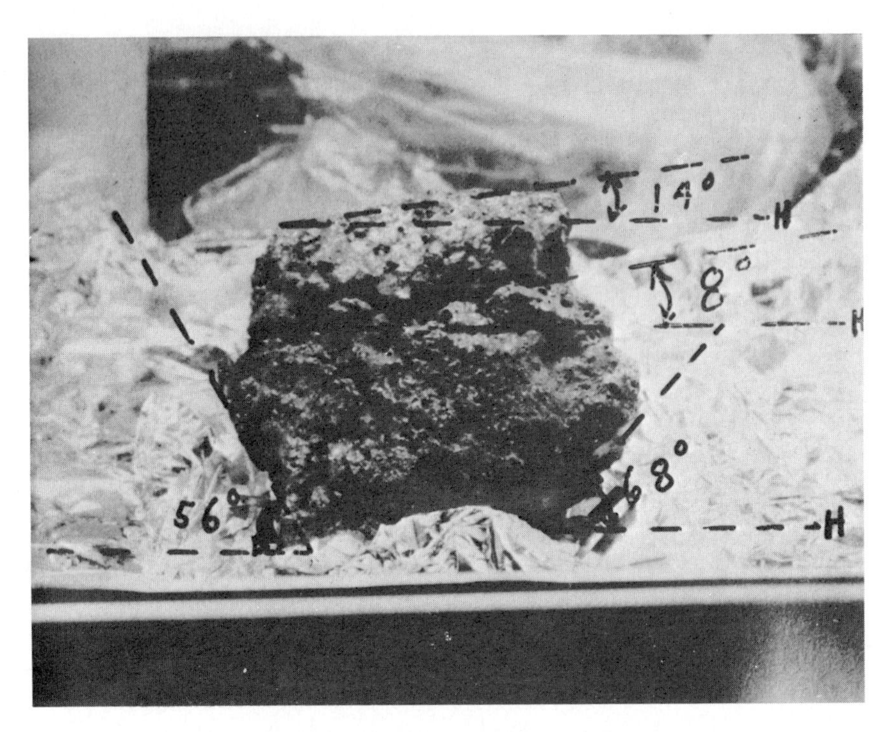

Fig. 5. View of the south face of 12054.

illustrated in Fig. 8. They are similar to the "slickensides" reported by Morrison *et al.* (1972) for some Apollo 14 rocks. A grayish powder up to .5 mm thick is compacted to the rock surface and displays pronounced striations. The fine scale gouges and striations are entirely confined to this compacted layer of rock flour. Their vertical relief is typically a few tens of μm, and they can be several cm long. The subparallel sets strike generally in one direction; however, on cm scales they can be divergent and fan-like. Where exposed, the underlying feldspar mineral grains display a milky, highly microfractured character indicative of shock, as discussed in more detail below. This striated surface is not a bonafide slickenside typical for terrestrial tectonic environments. The surface is neither polished nor competent. Furthermore, the features are not shattercones as was suggested during preliminary examination of the sample in the Lunar Receiving Laboratory. In some respects the striations may resemble those reported from the Ries ejecta by Chao (1976). The major difference between the two types of striations is that those described by Chao (1976) are grooves in competent host rock, while those we observed are confined to a compacted "dust" layer. We can say with certainty only that the striations are unique features and that they are caused by relative movement of two rocks.

Glass, which appears dark in the photographs of Fig. 7, coats approximately 70% of the rock. Surface features visible optically on the glass include loose dust,

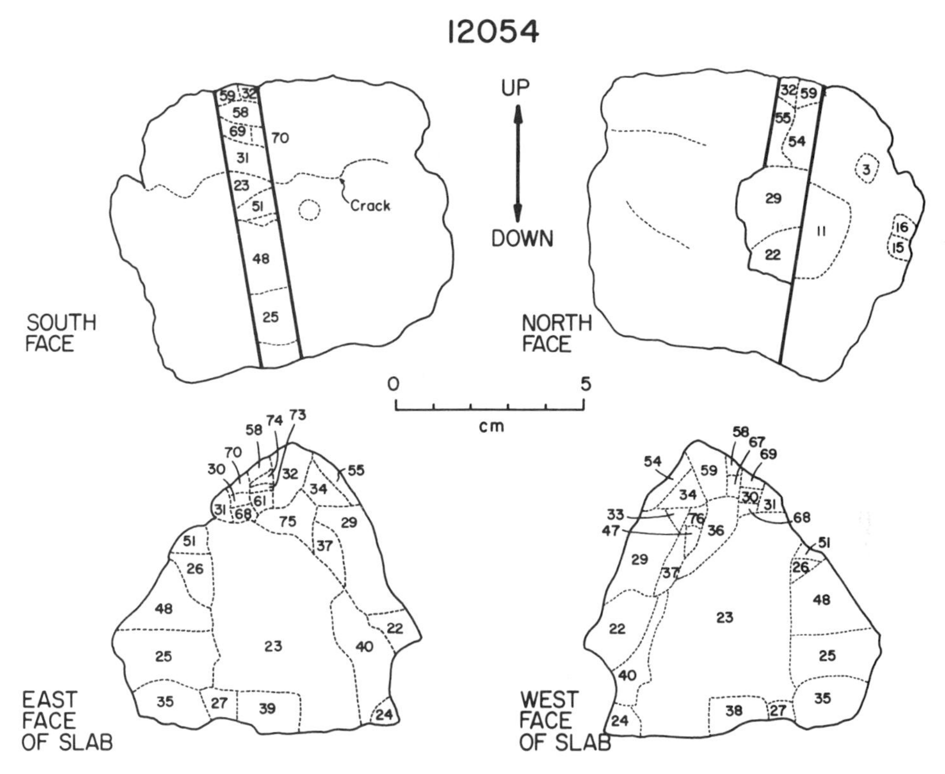

Fig. 6. Schematic drawing showing the location and lunar surface orientation of most of the sub-samples of 12054.

welded particles, and patinas (Blanford *et al.*, 1974). The glass surface shown in Fig. 9a is dark and "shiny." Microcraters of all sizes and minor amounts of dust adhering to the crystalline substrate within the spall zones of relatively large microcraters are visible. The area shown in Fig. 9a is the upper-east corner of the north face of 12054. Figure 9b shows a portion of the nearly vertical east face, which is coated with glass that is significantly duller in appearance. Local topographic highs are still relatively free of dust, which apparently has accumulated preferentially in depressions. Spall zones of intermediate-sized craters are often almost completely filled with dust. Figure 9c shows an area on the south face that is heavily coated with dust. The visible glass surfaces have a "patina" described by Blanford *et al.* (1974). In some places the dust obscures entirely the underlying glass substrate. The surface appearance is dull, including those areas that contain only small amounts of dust. In some cases such relatively dust free areas are dull because they have a sand-blasted, highly pitted, sugary surface texture attesting possibly to some microscopic mass wasting process. Finally, Fig. 9d displays an area close to the soil line. The surface material is a mixture of glass melt, fine-grained dust, and moderately-sized minerals and lithic clasts.

There is a clear relationship between the amount of dust adhering to a surface and the location of the surface. More dust occurs at locations near the soil line. There is also evidence that dust accumulated preferentially in local depressions. While most of the dust deposition may have occurred continually or sporadically during the surface exposure time of rock 12054, the frothy, dust- and-clast-laden glass zones at the soil line indicate incorporation of regolith material into the melt splash when 12054 landed on the regolith surface while the glass splash was still hot. Essentially all exposed glass surfaces have a patina which apparently is caused by the accretion of submicroscopic, μm-sized grains and melt droplets. This causes once shiny surfaces to take on a slightly duller appearance.

Mineralogy and petrography

The Apollo 12 mare basalts may be grouped into 4 major categories: olivine,

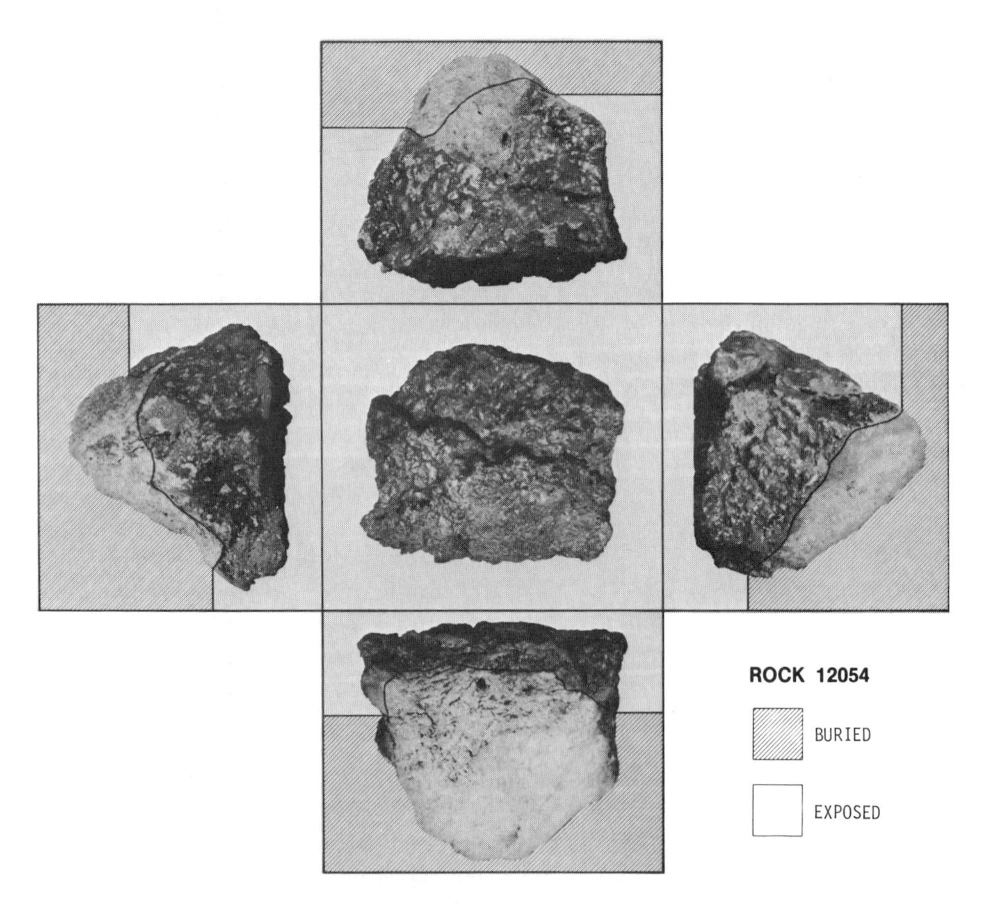

Fig. 7. Orthogonal views of rock 12054 indicating its approximate lunar surface orientation and "exposed" and "buried" surfaces.

Fig. 8. Striations on the buried crystalline surface. Note the darker albedo of the finely crushed, compacted surface layer to which the striations are confined (S-70-23018).

ilmenite, pigeonite, and feldspathic basalts (Rhodes *et al.*, 1977). Rock 12054 is a member of the ilmenite basalt suite. Detailed petrographic descriptions of the Apollo 12 ilmenite basalts have been given by Dungan and Brown (1977). Rock 12054 is a medium-grained basalt of equigranular-ophitic texture which may be derived from a parental magma approximated by vitrophyr 12008 via olivine fractionation. The reader is referred to the above papers for detailed geochemical data and petrogenetic reasoning.

A detailed petrographic study of thin section TS 12054,79, which was made from sub-sample 12054,25, was performed by Schaal and Hörz (1977). The modal composition is: 29% plagioclase, 62% clinopyroxene, 11% ilmenite and traces of troilite and kamacite. Detailed mineral chemistry is documented by

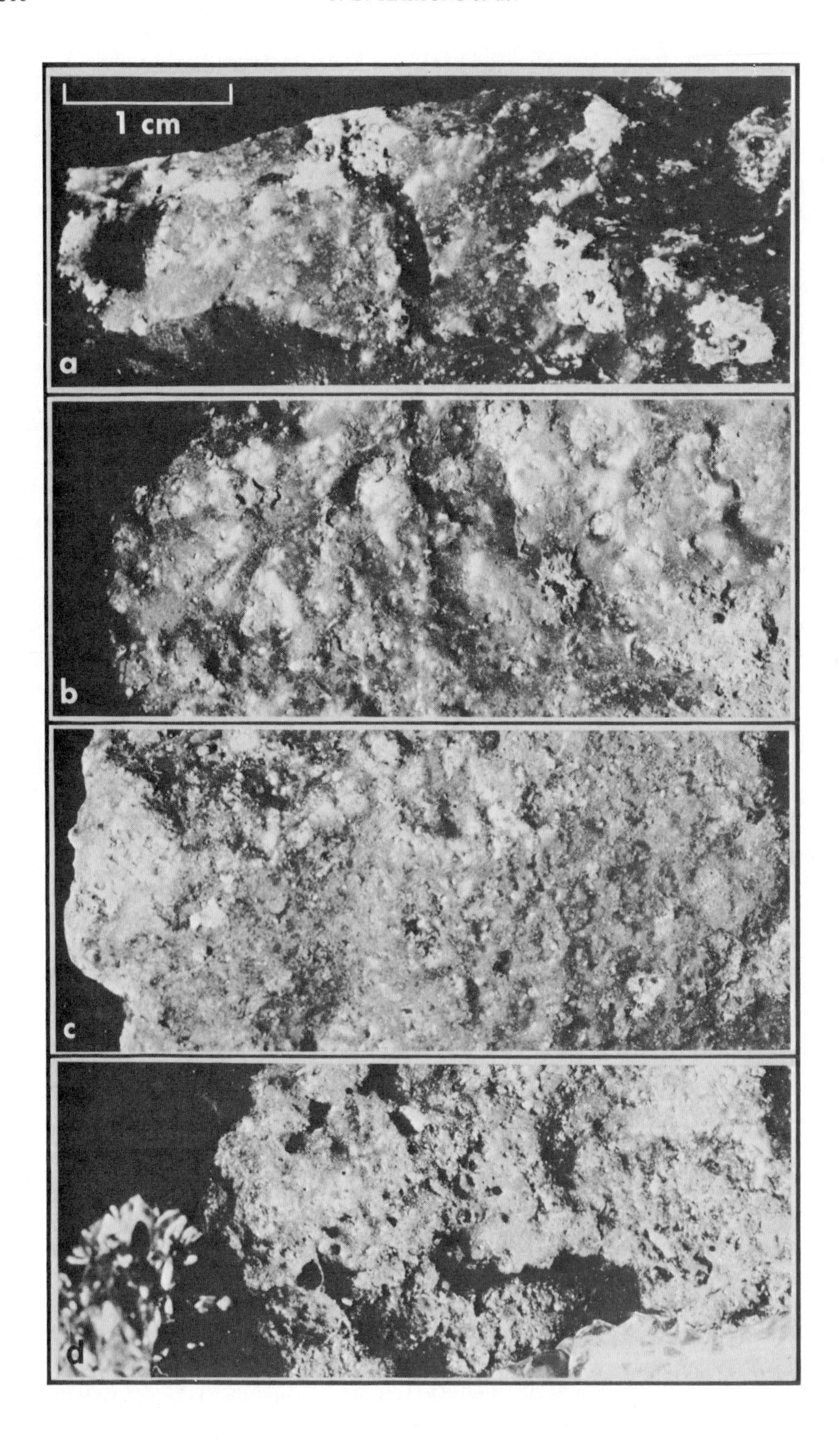
1 cm
a
b
c
d

Schaal and Hörz (1977). The rock displays pervasive weak to moderate shock effects, such as diaplectic feldspars and diaplectic feldspar glasses (maskelynite) as well as fine shock lamellae in feldspars. Pyroxenes are intensely fractured or granulated and have suffered a marked decrease in birefringence. Ilmenite grains are also fractured, and many display bent twin lamellae less than 15 μm wide. The assemblage of these shock features suggests that shock pressures did not exceed 450 kb and most likely were between 250–300 kb.

The glass coating rock 12054 was studied petrographically in cross-section, and its composition was determined using an electron microprobe. Below the glass there is a thermal aureole approximately 1 mm deep, which is accompanied by more severe shock effects than are present in the interior of the rock. Feldspar grains are partially or totally melted and show signs of incipient flow (Fig. 10). Edges of most pyroxene grains in contact with the glass are rounded by edge melting and form a veneer and individual schlieren of yellow green (5 GY 7/2) to pale green (5 GL 7/2) glasses less than .1 mm thick and up to .4 mm in length. Compositions of these glasses are nearly identical to pyroxenes in the host rock (Schaal and Hörz, 1977, Table 4). Reddish-brown (10R 4/6) glass patches and schlieren indicate enrichment in ilmenite-derived material. The predominant color of the glass is dark yellowish orange (10 YR 6/6) to light brown (5 YR 5/6). Chemical analysis of this material indicates that it is a "whole-rock" melt, a mixture of all minerals in the basalt, although in varying proportions.

These glass analyses are compared to those for individual 12054 mineral species and to the bulk compositions of 12054, Apollo 12 pigeonite basalts, Apollo 12 ilmenite basalts and Apollo 12 basaltic soils in Fig. 11. Considerable scatter is caused by the presence of different proportions of the various minerals in each microprobe analysis, i.e., within individual schlieren. However, the bulk of the brown glasses clusters relatively tightly around the average for ilmenite basalts (Papike *et al.*, 1976); it does not exactly match the bulk composition of 12054 (Rhodes *et al.*, 1977). The cluster differs significantly from Apollo 12 bulk soil analyses of Frondel *et al.* (1971). We conclude, therefore, that the glass on 12054 is an impact-generated ilmenite basalt melt. On the basis of presently available data, an origin from impact melting of nearby soils can be excluded.

Relative chronology of rock 12054

Based on visual inspection of the rock surfaces and several types of measured ages we know rock 12054 has undergone a complex history. Though no unique history can be established at present, a variety of scenarios can be developed. This variety serves to illustrate the complexity of lunar regolith dynamics.

Fig. 9. Surface characteristics of the glass coating 12054. (a) Essentially dust free glass surfaces from the northeast corner of rock 12054 (S-70-22999). (b) Moderately dust coated surface on the vertical east face of 12054. Note abundance of dust adhering in the local depressions, especially within the spall zones of moderately-sized impact craters (S-70-22994). (c) Heavily dust coated and patinated surface on the south face (S-70-22992). (d) Frothy glass surface with abundant mineral and lithic clasts typical for glasses near the soil line (S-70-22990).

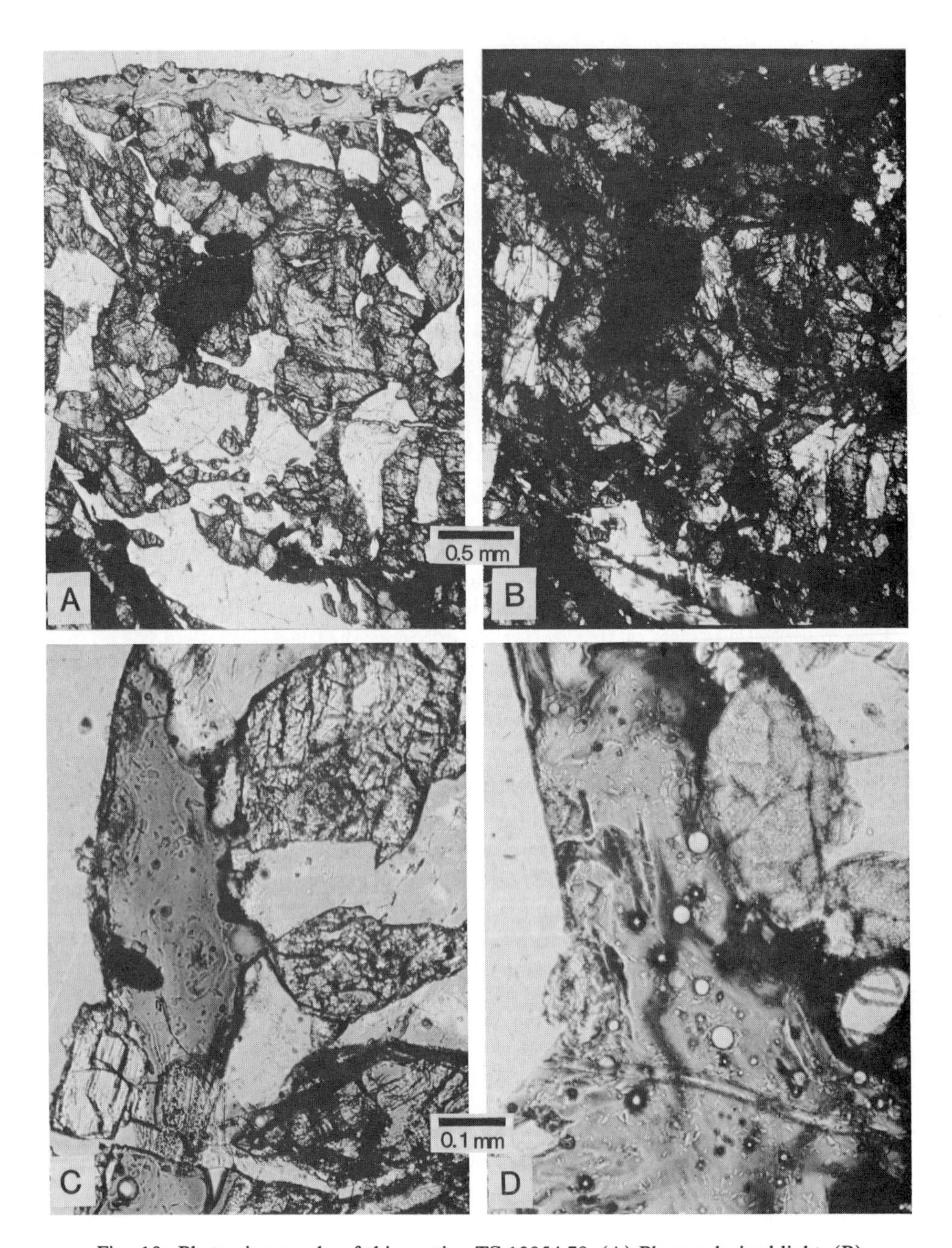

Fig. 10. Photomicrographs of thin section TS 12054,79. (A) Plane polarized light. (B) Cross-polarized light. Note the abundance of isotropic maskelynite. (C) Cross-section through contact zone between glass coating and the crystalline host. (D) Enlargement of glass coating.

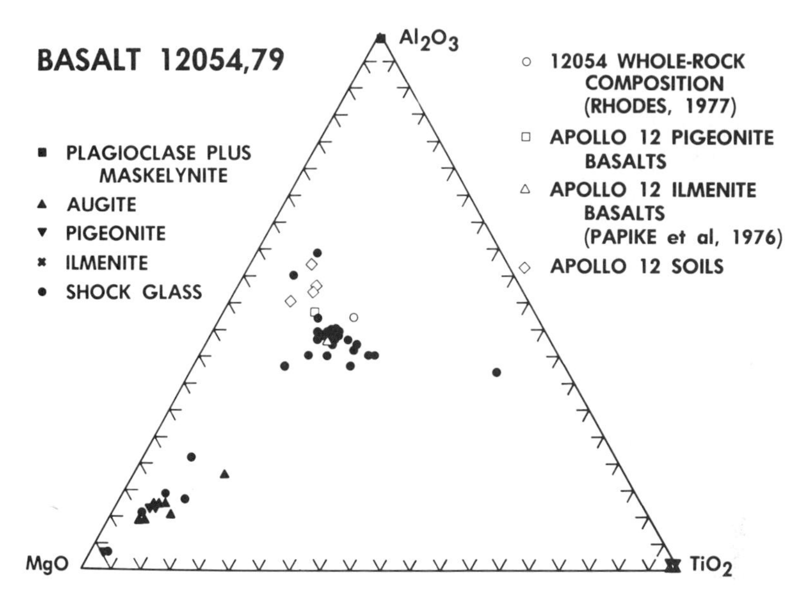

Fig. 11. Ternary diagram showing composition variation of shock glasses coating 12054 and the compositions of the minerals in the crystalline host material. Glass compositions and the whole rock composition more closely resemble compositions of Apollo 12 pigeonite and ilmenite basalts than the composition of the local soils.

Rock 12054 probably crystallized about 3.3×10^9 yr ago along with other Apollo 12 ilmenite basalts which have yielded reliable ages. Preliminary ^{39}Ar-^{40}Ar analysis of a shocked crystalline sample, 12054,74, resulted in a temperature release pattern which could not be interpreted readily in terms of either a crystallization age or a shock event time. It may be that the shock event was the cause of the unsatisfactory temperature release pattern.

We tried to obtain glass material suitable for ^{39}Ar-^{40}Ar analysis. Because of solar wind gases present on exposed surfaces, the extreme thinness of the glass on available shielded samples, and thc difficulty in removing contaminating, unmelted, host-rock grains from the glass, dating of the glass was not pursued.

Cosmogenic ^3He measurements (Bogard *et al.*, 1971) indicate a near surface (depth $\lesssim 1$ m) irradiation of at least 1.8×10^8 yr. A much longer previous residence time at greater depths, perhaps as part of the basaltic substrate, is possible and even likely.

^{26}Al γ-ray activity corresponding to the top of the rock was 53 ± 10 dpm/kg, a value which is below the equilibrium activity expected for ilmenite basalts exposed at the surface for long times. This indicates the surface residence of 12054 must have been short compared to the half-life of ^{26}Al, 0.74×10^6 yr. The activity measured for the bottom of 12054 was only 10% lower and still within the 2σ uncertainty range of the top measurement. This activity was higher than expected if the rock had a simple, one-stage exposure history, i.e., excavated from depth and placed on the lunar surface in the orientation in which it was found. Therefore, a more complex, multistage exposure history for 12054 at depths less than a few tens of cm is required. Depending upon the model for exposure history assumed, times for the surface residence of 12054 between 0.1 and 0.3 $\times$ 10^6 yr were obtained based on ^{26}Al measurements.

Galactic cosmic ray track densities were measured near the top and bottom of 12054 and were found to be about 2 $\times$ 10^7 tracks/cm^2 near both faces. This apparent lack of track gradient is also consistent with a period of irradiation below but near the surface. On the basis of these results we conclude 12054 must have experienced a complex history in the lunar regolith. Because of this, exposure ages based on effects of galactic cosmic rays, which occur up to tens of cm below the surface, could not be used as independent measures of surface residence time, thus eliminating the possibility of comparing these ages with other measures of surface exposure, such as solar flare track or microcrater densities. Plans to measure depth profiles of cosmogenic, short-lived, radioactive nuclides were abandoned due to the high activities expected at depth in the rock.

Ages referring to crystallization, regolith residence and surface residence provide an interpretative framework for the history of 12054, which may be still more complex than indicated above. The time of the shock event, the glass formation event, and the formation of the "slickensided" surface have not been established. These might represent three additional cratering events. If the worst case of three independent events is assumed, the shock event would be first, followed by the slickenside and glass formation events. A shock level of >250 kb would probably have destroyed both of these delicate surface features. Since the glass coating appears to wet the slickenside (Fig. 8), the glass forming event was last. All these observations indicate relative sequences only, and it is also entirely plausible that all of the above features were generated during the course of one single impact event.

The clast and regolith laden glass zones apparently associated with the present soil line of rock 12054 and the "basaltic" composition of the glass coating itself, however, suggest that the glass-forming event and the excavation event which deposited 12054 on the lunar surface may have been the same event. Such an event may have been an impact of sufficient size to penetrate a basaltic bedrock. This possibility seems, however, to be at odds with a complex regolith history outlined earlier.

SURFACE PROCESSES

Microcrater populations

Crater counts using a stereoscopic microscope have been made on all major faces of 12054. Although some of the exposed crystalline surfaces contained microcraters, detailed counts were made only on glass surfaces to assure a common time for the onset of exposure and to provide a standard host surface on which the craters may be seen easily. Data obtained are presented in Fig. 12. The data for the south, east, and north faces are essentially identical, when appropriately corrected for their respective solid angles of exposure. The northwest corner face contains fewer microcraters than the other surfaces. This apparent discrepancy may be caused by: (a) inferior crater statistics (only 51 craters were observed); (b) relatively rough surface texture of the glass splash, making quantitative crater counting rather difficult; (c) micrometeoroid dynamics which cause disproportionately fewer impacts on west faces; or (d) some kind of temporary shielding which may have protected this particular face during part of the surface residence time of 12054.

Microcraters were counted using scanning electron microscope (SEM) techniques on samples from the northwest (12054,3), north (12054,54 and 12054,55) (Morrison and Zinner, 1977; Hauser, 1978; Zook, 1978b) and south (12054,58) (Mandeville, 1977, 1978) faces. An additional sample (12054,59), selected from near the top of 12054, showed a complete absence of small craters (diameter ≤10

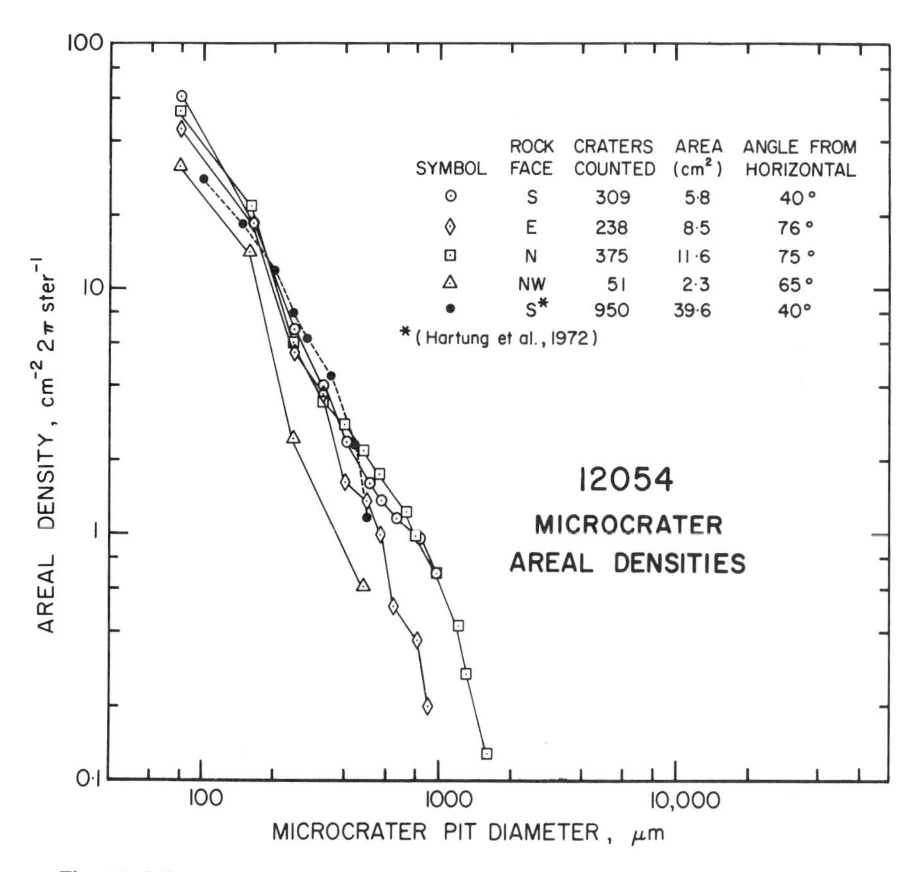

Fig. 12. Microcrater cumulative size distributions measured on 12054 using optical microscopy.

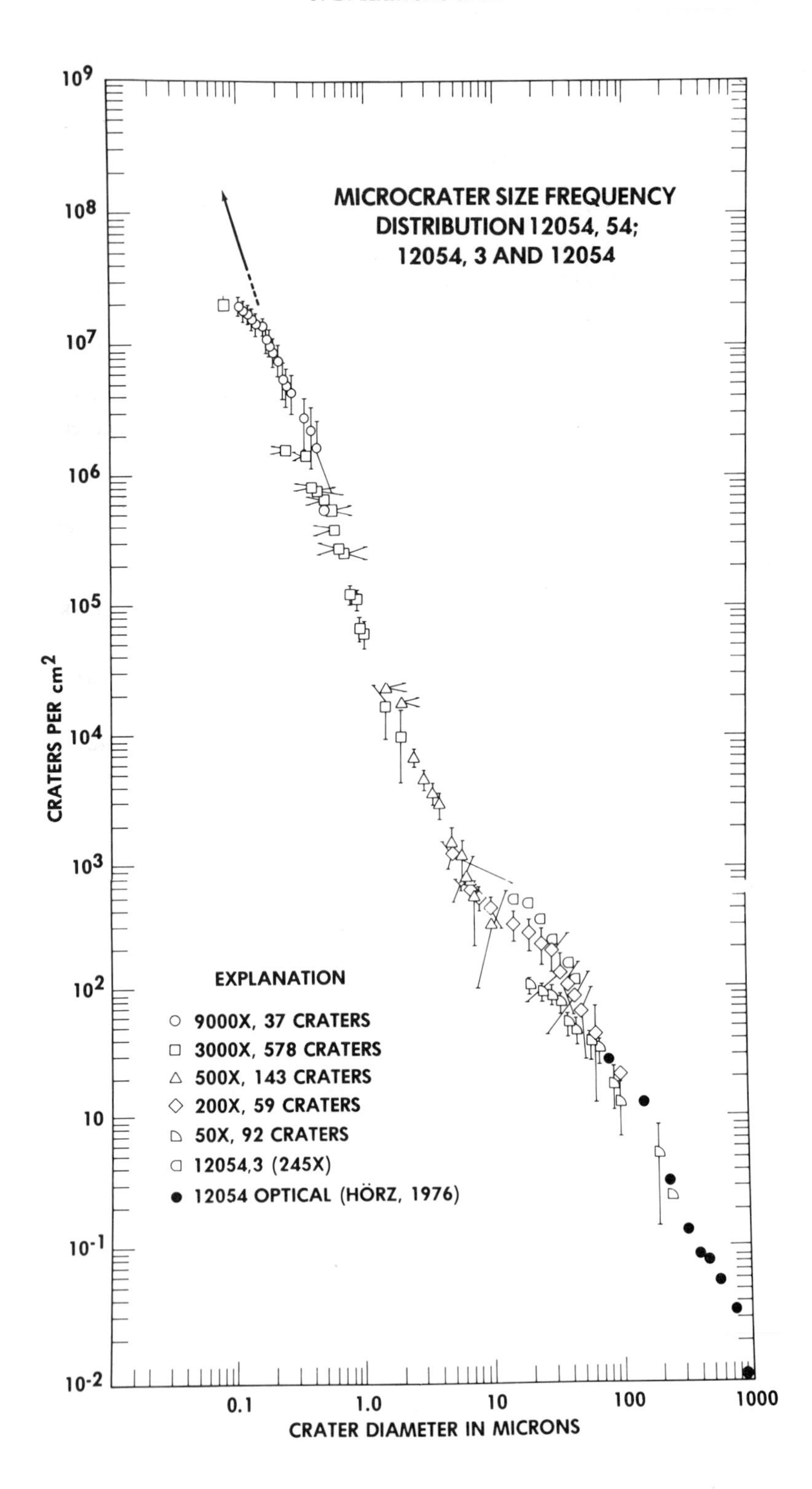
MICROCRATER SIZE FREQUENCY
DISTRIBUTION 12054, 54;
12054, 3 AND 12054
EXPLANATION
9000X, 37 CRATERS
3000X, 578 CRATERS
500X, 143 CRATERS
200X, 59 CRATERS
50X, 92 CRATERS
12054,3 (245X)
12054 OPTICAL (HÖRZ, 1976)
CRATERS PER cm2
CRATER DIAMETER IN MICRONS
10-2
10-1
10
10²
10³
10⁴
10⁵
10⁶
10⁷
10⁸
10⁹
0.1
1.0
10
100
1000

μm) and glassy accreta. The glass surface of 12054,59 formed a bowl-shaped depression which apparently collected dust and shielded the surface. Cumulative size distribution data for samples 12054,54 and 12054,3 together with optical microscope data for the north face of 12054 are presented in Fig. 13. These data have not been normalized with respect to solid angle of exposure.

There is excellent agreement between data using various samples for larger craters, but divergence at diameters less than a few μm. Rocks 12054,54 and 12054,55, for example, show identical distributions for craters greater than one μm diameter but 12054,55 shows a flatter slope in the sub-μm diameter range. Rock 12054,58 (Mandeville, 1978) begins to diverge, showing fewer craters than 12054,54 at a diameter of about 5 μm. Although on different faces, the exposure age and solid angles of exposure for these samples are the same, which is consistent with the same density of large craters on both surfaces. Which of these curves is correct for μm and sub-μm diameters?

The divergence at smaller crater diameters can be attributed most plausibly to variations in processes which renew or shield the host surface, such as accumulation of accretionary particles, or accreta, and dust particles. Variations in local microtopography may also shield surfaces or cause differences in the effectiveness of these processes. Support for this hypothesis comes from several sources. Zook (1978b) has documented an area on 12054,54 which is totally lacking in craters and accretionary particles, and therefore must have been shielded from the impact process by either nearby topographic features or loosely adhering lunar dust. Rock 12054,59 has a population of large craters but a complete absence of μm-sized craters and glassy accreta. This surface must also have been shielded. These observations show that suppression of smaller-sized craters does occur. Consequently, the distribution with the steepest slope and showing the highest density in the smallest diameter ranges (12054,54) is most likely to yield the most accurate record of the impacting particle flux (Morrison and Zinner, 1975).

The surfaces of almost all of the samples examined by SEM techniques are overprinted by accretionary particles, principally various forms of glass and angular fragments (Morrison *et al.*, 1973; McDonnell, 1977, 1978). Complete shielding by loosely adhering dust may cover existing pits or prevent the formation of pits on the host surface. These effects are particularly important for craters of sub-μm dimensions. For example, the microcrater size frequency distribution on 12054,54 (Fig. 13) bends over at a diameter of around 0.2 μm, whereas observations of much cleaner surfaces (Brownlee *et al.*, 1975; Blanford *et al.*, 1975; Morrison and Zinner, 1976, 1977) show a constant steep slope extending to craters of <0.1 μm diameter. The solid line in Fig. 13, accordingly, extrapolates the observed sub-μm crater densities to a 0.1 μm crater diameter to

Fig. 13. Standard cumulative size distribution of impact pit diameters. Measurements were made on the north face of 12054 and are not normalized with respect to solid angle of exposure. Areas measured were selected to give the highest areal densities of microcraters in a given size range.

account for the depletion of craters in this size range caused by covering or shielding processes.

Comparison of 12054,54 with samples which show no evidence of shielding and which have no appreciable accreta population, such as 76215 (Morrison and Zinner, 1976, 1977) and 60095 (Brownlee *et al.*, 1975), shows agreement in slope in the sub-μm range. The ratio of sub-μm crater densities at a given diameter, 0.3 μm, for example, to densities of larger craters, 100 or 500 μm in diameter, is the same for 12054,54 and 76215,77 (Morrison and Zinner, 1977). These samples differ in crater density by an order of magnitude and 76215,77, the less cratered sample, is essentially free of accreta (about 1% coverage). This agreement corroborates the hypothesis put forward earlier, and justifies the extrapolation to smaller crater sizes shown in Fig. 13.

We conclude that the distribution extending over four orders of magnitude of crater diameter shown in Fig. 13, including the extrapolation, is the best available estimate of the lunar impact crater size frequency distribution.

All of the distributions measured show complexities at a crater diameter around 10 μm, where a relative deficiency of craters is apparent. We emphasize that these distributions are not normalized from one magnification to another. Further, at a single magnification of 500× shown in Fig. 13, there is a relative deficiency of craters about 7 μm diameter but an increase in smaller craters. Because the latter were recognized and counted, and verified by observation at 3000×, the deficiency cannot be an observational artifact. Although the inflection in the size distribution curve in this size range appears to be real, its magnitude seems to be different for different rocks and even for different areas on the same rock. It has been suggested that this bimodal character of the crater size distribution is a reflection of the bimodal character of the meteoroid complex. According to this suggestion, most larger craters are formed by larger particles on elliptical orbits and spiralling into the sun due to the Poynting-Robertson effect. The smallest craters are thought to be formed by very small, fast-moving grains leaving the solar system due to acceleration by radiation pressure. Our contribution to this discussion is to show that perhaps an order of magnitude more sub-μm-sized particles exist than assumed in the model calculations of Dohnanyi (1976), which were based on the size distribution of Fechtig *et al.* (1974). We also recognize the possibility that inflections in the measured microcrater size frequency distributions may be due only to shielding by lunar dust, which may accumulate to preferred thicknesses on different samples.

A detailed study of microcraters with pit diameters between about 20 and 200 μm on sample 12054,54 (Hauser, 1978) has led to the characterization of three types of microcraters: (1) those with spallation zones displaying lobate concoidal fracture surfaces; (2) those with spallation zones dominated by radial and concentric fracture patterns ending with a scarp at the radial limit of the spallation; and (3) those without a spallation zone. These classes are correlated with microcrater pit size. The larger pits tend to be of the first type, although some craters with pits as small as 20 μm in diameter fall into this class. The second type of crater has pits no larger than about 50 μm in diameter. Only pits smaller than about 5 μm in diameter have no surrounding spall zones (Hartung *et al.*, 1972a). In the size range where the first two types of pits both occur, between about 20 and 50 μm pit diameter, the first type of crater has a significantly larger spall-to-pit-diameter ratio, tends to have pits with wavy interiors, and probably is formed by the impact of relatively larger, slow-moving meteoroids. In this range the second type of crater has a smaller spall-to-pit-diameter ratio, smooth pit bottoms, and probably was formed by the impact of tiny, high-speed, meteoroids. Craters in this limited size range are formed by impacts having similar total energies, and apparently the character of the spallation is an indicator of the relative specific energies or velocities of the impacts.

Densities of μm-sized impact pits and accretionary disks were measured on the surfaces of these larger microcrater pits to find their relative exposure ages. The distribution of relative exposure ages of individual pits on 12054,54 had an exponential decay character similar to that found previously for other microcrater populations (Hartung *et al.*, 1975; Hartung and Comstock, 1978). This type of age distribution apparently arises because the larger craters are randomly removed from the aging process not by superposition of later craters, but by the shielding of pit surfaces by the accumulation of loose dust inside the pits.

Solar flare tracks

Solar flare track data were difficult to obtain for 12054 because of the relatively high level of shock experienced by the entire sample. Track density versus depth profiles for 76215 and 12054 are shown in Fig. 14. The data for 76215, which was exposed a much shorter time than 12054, are shown for comparative purposes. Track densities at depths less than about 70 μm in 12054,3 could not be measured because of the glass coating. The track density versus depth profile for 12054 shows a steep slope, characteristic of solar flare produced tracks, from about 70 to 300 μm. At greater depths the background track density produced by galactic cosmic rays is present. The solid lines for both samples correct for the galactic track background. With this correction the slopes of the solar flare track profiles are -1.86 and -1.97 for 12054 and 76215, respectively. No track gradient exists at the bottom of the sample, thus indicating an absence of solar flare tracks and no exposure to space of the bottom of the rock.

Accreta

Using an uncleaned, south-facing sample, 12054,58, McDonnell (1977) studied the component of the accreta showing no evidence of flow or fracturing upon contact with the host surface. This important component includes accreta which were judged to have been sputter-degraded. The size distribution of this material was measured on the host surface, within impact crater pits, on impact crater spallation zones and within voids exposed only to the post lunar retrieval environment. Levels of contamination during transport and sectioning were

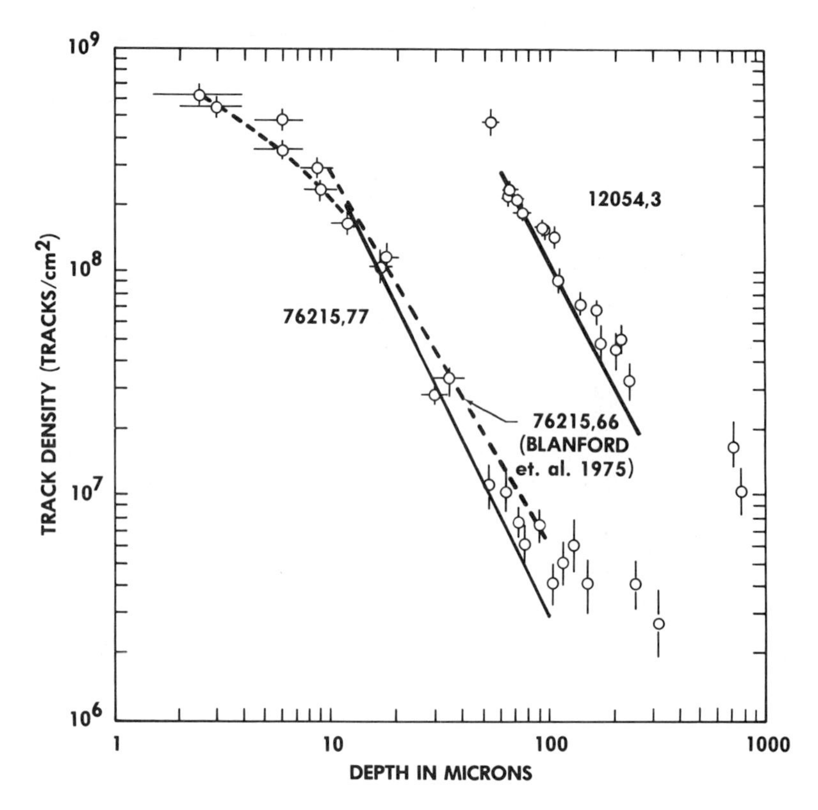

Fig. 14. Solar flare track density profiles for 12054,3 and 76215,66. Galactic cosmic ray track densities are subtracted to yield the straight lines which correspond to the actual solar flare track profiles. A solar flare track exposure age of $(1.75 \pm 0.5) \times 10^5$ yr is derived from these data.

deduced from the amount of accreta-like material present on the cleanest of the microcrater pit surfaces and on fracture surfaces exposed during sample preparation. An accreta production distribution was obtained based on the upper envelope of accreta size distributions measured on a large number of different crater surfaces. This approach assumes at least a few surfaces retain accreta efficiently. This accreta distribution is shown in Fig. 15 and forms an estimate for the average flux of low-velocity lunar microparticles. These data have been used in numerical calculations describing the relative importance of different surface processes by Ashworth *et al.* (1978) and Ashworth (unpublished data, 1978). The accreta production curve derived from 12054,58 has been used in a Monte Carlo simulation of accreta deposition by Carey and McDonnell (1978), where areal coverage, crater obscuration, and mass balances can be studied both with and without solar wind sputter. Different sides of a single, large impact pit were studied and accreta populations evaluated by McDonnell (1978). Populations showed differences attributed to the effects of solar wind sputter but were not

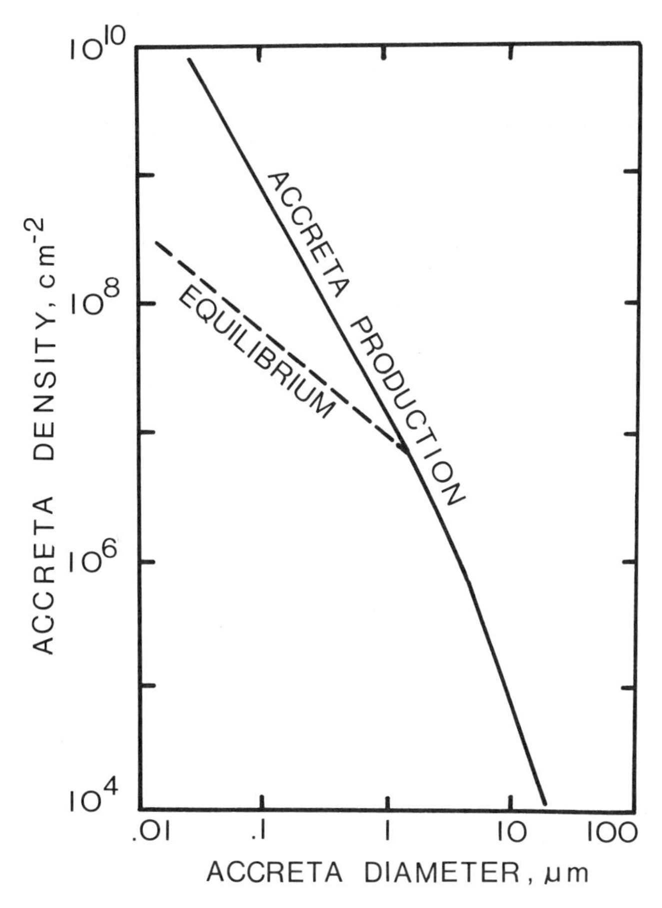

Fig. 15. Production accreta cumulative size distribution derived from measurements on 12054,58. The equilibrium curve reflects the loss of small grains by sputter erosion.

strongly anisotropic, a factor of 2 or 3 being typical. An exception is the south, upward-facing side of the pit, which showed a clear excess of accreta particles larger than about 0.1 μm in diameter, which may have been caused by particles rolling off vertical faces and collecting on more horizontal surfaces.

Solar wind sputter

Since solar wind sputter has not been measured as a real-time *in-situ* process, this is an effect which must be inferred rather than directly evaluated. Solar wind sputter was studied using sample 12054,58 by McDonnell (1977), who observed sputter effects on accreta populations. Using these measurements, Carey and McDonnell (1978) have extended the study by Monte Carlo simulations of accreta deposition under solar wind sputter and also by the computer simulation and laboratory measurement of the sputter process itself. The sputter process acting on the lunar surface should permit a flat surface to maintain its smoothness, not due to the intrinsic effect of the sputter process, since this leads to a roughening and needle formation aligned towards the incident ion direction, but instead due to the rotation of the surface during the lunar day. Topographic features, such as accreta and microcraters, will be rounded and smoothed due again to rotation of the surfaces. Fillets will be formed by sputter redeposition in protected areas. Impact crater pits will also retain preferentially such redeposited atoms. Characteristics of accreta on 12054,58 offer evidence of the occurrence of these processes. The shape of accreta grains varies from sharp, angular grains to mounds with a somewhat Gaussian profile. This variation can be interpreted in terms of solar wind sputter. More convincing than any such individual features is the effect of sputtering on the accreta size distribution. Smaller accreta grains are relatively depleted on "older" surfaces, and their areal densities may reach equilibrium levels, as is shown in Fig. 15.

Vapor deposition

Vapor deposition has been suggested as a process operating on the lunar surface. At present, strong, directly observable evidence showing visible effects of vapor deposition is lacking. Its effects, if any are found to exist, might be similar to those of sputter erosion and redeposition, such as rounding and smoothing and, in general, feature removal. In the absence of a strong case for observable vapor deposition, we attribute the removal of sub-μm-sized features, accreta and microcraters, entirely to solar wind sputtering.

Solar wind noble gases

Areal concentrations and isotopic compositions of solar wind He, Ne, and Ar were measured at 75-μm-diameter sites on exposed and shaded surfaces of sample 12054,31 using laser probe mass spectrometry techniques (Hartung *et al.*, 1977). Gas concentrations were roughly correlated with exposure geometry in the dominantly shaded region. A striking difference between $^{40}Ar/^{36}Ar$ ratios on nominally shaded ($^{40}Ar/^{36}Ar > 2$) and sunlit ($^{40}Ar/^{36}Ar < 0.5$) surfaces was observed. This difference may be explained by a low retention probability for ^{40}Ar on sunlit surfaces, due to its lower implantation energy and more probable loss by sputtering, and eventual collection of ^{40}Ar on less exposed surfaces.

The range in the amount of gas released at each analysis site was about a factor of five for Ne and Ar and a factor of 10 for He. No differences in the sites were apparent from optical microscope examinations. The sample has not been cleaned in a fluid, and the variations are well outside of any experimental uncertainty. The precise reason for these wide variations is not known, but the effect may be related in some way to the dynamic character of processes acting on scales of much less than .1 mm.

Gas concentrations were measured for microcrater spall zone surfaces which were formed at different times. Amounts of gas were similar in quantity and range to those obtained for the host surface, thus indicating some equilibrium concentration of gas was present on both spall zone and host surfaces. Because equilibrium levels of solar wind ^{36}Ar, the most retentive gas analyzed, are reached after times which are short compared to the total exposure lifetime of 12054, the accumulation of ^{36}Ar cannot be used as an exposure time clock for such long times. Areal concentrations of ^{36}Ar measured were about a factor of ten less than the ^{36}Ar incident upon the sample, based on the results of the Apollo solar wind composition experiments (J. Geiss, pers. comm.). The loss mechanism for ^{36}Ar may be sputtering. If the sputter erosion rate of .031 Å/yr (McDonnell, 1977) is accepted and all other processes are ignored, then sputter erosion would start to erode ^{36}Ar atoms implanted at a depth of 300 Å after 10^4 years. However, the effects of sputter redeposition and accreta accumulation cannot be ignored and these may be important factors relative to ^{36}Ar accumulation and retention.

Discussion

Relative importance of surface processes

Our primary objective was to determine the relative importance of as many surface processes as possible. Before the 12054 Consortium effort it was understood that microcratering was the dominant process occurring in the size range capable of being studied using an optical microscope ~10 μm to several cm or larger. Erosion of lunar rocks, mixing of lunar soil, and formation of melt droplets are caused mainly by meteoroid impacts that produce craters in the mm-size range or larger. Careful SEM studies showed that deposition of accretionary material is a much more important process than microcratering at the μm scale. These accretionary materials, or accreta, consist of rounded, angular and disk-shaped objects and are welded or otherwise firmly attached to essentially all lunar surfaces exposed directly to space. Bonded accreta are apparently produced by nearby mm-sized impact events because the phenomenon is observed on all exposed surfaces and the μm-sized melt droplets could not go far without solidifying. At the sub-μm scale, sputter erosion and deposition appear to compete with accreta accumulation as a dominant surface modifying process. Only now, with increasing use of high resolution electron microscope techniques, are the sub-μm-scale processes being studied carefully. Our appreciation and understanding of processes and phenomena on this and smaller scales, such as solar wind implantation, vapor deposition, and radiation damage are clearly incomplete.

The effects of the processes described so far are directly observable; material is removed or deposited. During the course of the study of 12054 we have come to appreciate the importance of loose dust or unwelded accreta. The accumulation of loose dust on a surface serves to inhibit the effects of other processes which are more readily observable. Processes occurring at all size ranges may be affected, but the effect of loose dust on the μm scale is particularly important because this material may act as a kind of filter, permitting 10 μm-scale processes to be recorded continuously while inhibiting sub-μm scale processes.

That fine-grained lunar dust would have extreme mobility on the lunar surface and be supported by electrostatic forces was suggested first by Gold (1955). Singer and Walker (1962) concluded that the initial elevation of material may be the result of meteoroid impact. Observations of "horizon glow" (Rennilson and Criswell, 1974) and the large number of charged particle events detected by the lunar ejecta and meteorite experiments only at the times of terminator passage (Berg *et al.*, 1976) have confirmed the mobility of μm-sized grains above the lunar surface.

Although the occurrence of loose dust on a variety of lunar surfaces is not firmly established, its presence may explain a variety of observations. In studies by McDonnell (1977) of the accreta age distribution of crater pits, the distribution was found to be uniform for crater diameters between 5 μm and 50 μm, indicating a uniform impact crater production rate and efficient loose accreta retention. Larger craters, of diameter >50 μm, showed a strong bias towards "young" ages, similar to that found for the 60015 craters studied by Hartung and Comstock (1978). This effect may be the result of accreta shielding of these larger craters. Studies on the opposite face of the rock using sample 12054,54 by Zook (1978b) have distinguished between the populations of rounded, angular and disk accreta. On the same areas, microcrater populations were studied. Results show a clear correlation between disk accreta and microcrater densities, less of a correlation between rounded accreta and microcrater densities, and no correlation between angular accreta and microcrater densities. Apparently, microcraters, disk accreta, and rounded accreta accumulate more or less together until total shielding by the loose accreta population is established.

The concept of a transitory population of dust, which was invoked by Hartung and Comstock (1978), may refer only to a population which does not survive sample cleaning procedures in the laboratory. That such a population may well remain intact on the lunar surface is shown by the much higher accreta densities observed by McDonnell (1977) on sample 12054,58, which was not cleaned. Total populations of firmly bonded material on photographs by Morrison and Zinner (1977) on 12054,54 show 20% of the area is covered; on 12054,58 accreta saturation is observed where particle retention is good. These results may reflect sample cleaning differences.

The presence of loose accreta in microcrater pits offers another explanation of solar flare track age measurements in individual pits which showed an apparent excess of "young" craters, which has been interpreted in terms of either a higher meteoroid flux or lower solar flare particle flux in recent times (Hartung and Storzer, 1974; Zook *et al.*, 1977). And finally, different thicknesses of dust on surfaces might explain the relatively wide variety of shapes of lunar microcrater size frequency distributions measured in the past, especially the relative depletion of microcraters with pits about 10 μm in diameter (Fig. 13).

Accreta studies on 12054 are incomplete, but have highlighted the fact that one of the most significant processes affecting the top few μm of the lunar surface is loose accreta deposition. The mass of accreta deposited on an area is probably quite substantially less than that removed by meteoroid impacts, causing craters of diameter 20 μm and over, but in the periods between the arrival of such particles full areal coverage by accreta is entirely possible, and the effects of such coverage must be considered.

Exposure times and process rates

Another objective of the 12054 Consortium effort was to determine the rates of surface processes, both relative to one another and absolutely. This task has been more difficult than expected. To determine absolute rates of surface processes it is essential that the exposure time for the surface be known. In principle, any one of the processes studied, accumulation of microcraters, solar flare tracks, solar-flare-produced radioactive ^{26}Al, and solar wind noble gases, and sputter erosion, can be used as an exposure time "clock" against which the other processes can be "timed." We have shown that solar wind ^{36}Ar accumulation is not an acceptable "clock" because it begins to be lost from surfaces after relatively short times, possibly through the action of sputter erosion of surface material.

Morrison and Zinner (1977) have applied the solar flare track "clock," which uses a track production rate of Blanford *et al.* (1975), based ultimately on concordant galactic track and Kr-Kr exposure age measurements. The exposure time obtained for 12054 using this method is $(1.75 \pm 0.5) \times 10^5$ yr.

The solar wind sputter "clock" relies on the fact that, as time goes by, populations of increasingly larger accreta grains are brought into equilibrium with respect to loss by sputtering (McDonnell, 1977). The characteristic grain size increases linearly with time and is identified by the point of transition between the accreta production size distribution at large accreta sizes and the equilibrium size distribution at smaller accreta sizes. The diameter at the transition is approximately RT, where R is the sputter rate and T is the exposure time. For a measured transition diameter of 0.6 μm and an exposure age of 1.75×10^5 yr (Morrison and Zinner, 1977) and after correcting for geometry and redeposition effects, the lunar sputter rate is deduced to be .031 A yr^{-1} by McDonnell (1977). Laboratory measurements support this value and thereby also imply that the solar flare track age is not in serious error. However, widespread agreement on the lunar sputter erosion rate does not exist. See also Zinner *et al.* (1977) for a discussion of this area of investigation.

The production rate of microcraters formed on recently exposed spacecraft parts has been determined by Zook (1978a). Applying this production rate of craters over 1.75×10^5 yr results in more craters than are observed on 12054 by about a factor of six. One way these data can be reconciled is for either microcrater or solar flare track production to have varied significantly with time. The argument has been made by Zook *et al.* (1977) that the sun may have been more active in the past. In contrast, Morrison and Zinner (1977) point out that the meteoroid detection experiment aboard the HEOS II spacecraft (Hoffman *et al.*, 1975) shows a near-earth enhancement of meteoroid flux which, if accounted for, would reconcile the apparent discrepancy without the need to appeal to time-varying fluxes. They also find a constant ratio of track densities at a depth of 100 μm to 0.1-μm-diameter impact pit densities exists for surfaces with greatly different exposure ages. This shows that either track and pit production vary with time together, which is not likely, or they are both constant. In view of our discussion of the importance of loose accreta, it should be pointed out that the surfaces used for these measurements were selected to yield the highest densities of pits, which corresponds to the least amount of undetected shielding by accreta.

Lunar surface "clocks" may be susceptible to interference from loose accreta accumulation. Assume for the moment a 30-μm-thick loose accreta layer on a surface. A "clock" based on the accumulation of tracks at a depth of 100 μm would continue to operate, albeit more slowly, than if no accreta were present, but a 20-μm-diameter impact pit "clock" would stop altogether. In this situation a discrepancy would arise in the track and crater "ages" of the surface similar to that described above.

A Monte Carlo study by Comstock (1978) was designed to model solar flare track accumulation in lunar rocks which are subjected to impact erosion and loose dust accumulation. He finds observed track gradients and densities are more easily obtained if dust is allowed to accumulate on surfaces at a uniform rate until removed by an impact event.

Finally, it is known that corresponding to each type of exposure age measurement there is, on the average, a characteristic depth which is correlated with the magnitude of the exposure age measured. Galactic cosmic rays penetrate tens of cm and yield cosmogenic noble gas exposure ages of the order $\sim 10^8$ yr; galactic tracks register only in the upper $\sim$10 cm and yield exposure ages of the order 10^7 yr; and so on. Exposure ages obtained using these methods are dependent upon the exact depth assumed for the sample during its residence in the regolith. We must now face the fact that because of the prospect of unknown amounts of loose accreta on sample surfaces, solar flare track and microcrater exposure ages may suffer the same type of uncertainty as do the other methods mentioned. We cannot necessarily expect solar flare track, microcrater, and solar wind accumulation exposure ages to be concordant, even if the same grain or surface is analyzed for the different effects. Concordant ages may be expected only for those surfaces where accreta build-up, amount of erosion, and mixing depths are small compared to the characteristic depths of the measurement method. According to this line of reasoning, the following sequence of processes, listed in descending order of average ages and corresponding characteristic depths, should prevail:

1. Cosmogenic noble gases.
2. Galactic cosmic ray tracks.

3. sized microcraters.
4. Solar flare tracks and solar cosmic ray effects.
5. μm-sized microcraters.
6. Solar wind sputtering of accreta.
7. Accumulation of solar wind atoms.

Concordant ages may be the exception—brought about by very careful sample selection—and not the rule.

Acknowledgments—We gratefully acknowledge the cooperation of the lunar sample curatorial staff, and especially Claire, who was responsible for the sampling and documentation activities, the efforts of G. Heiken, who arranged for excellent close-up color photography of 12054 during the Apollo 12 preliminary examination period, and the manuscript preparation assistance of M. O'Dowd.

Active participants in 12054 activities whose work on 12054 is already published elsewhere, is still in progress, was not directly related to surface processes, or was not done for scientific reasons are: J. Adams, J. Arnold, D. Blanchard, G. Comstock, A. Dollfus, J. Hartung, E. Hauser, F. Hörz, C. Hohenberg, R. Housley, C. Kohl, D. Lal, J. McDonnell, J.-C. Mandeville, D. Morrison, M. Rhodes, R. Schaal, O. Schaeffer, E. Schonfeld, R. Walker, E. Zinner and H. Zook.

The work of JBH was supported by NASA grant NSG 9013.

References

Ashworth D. G., McDonnell J. A. M. and Carey W. C. (1978) The role of accretionary particles in the approach to lunar equilibrium topology. *COSPAR: Space Research XVIII*. Pergamon, N.Y. In press.

Bailey N. G. and Ulrich G. E. (1975) Apollo 12 voice transcript pertaining to the geology of the landing site. U.S. Geol. Survey, Branch of Astrogeology, Flagstaff, Arizona, 1975. PB 241811, p. 12. Nat'l Tech. Info. Serv.

Berg O. E., Wolf H. and Rhee J. (1976) Lunar soil movement registered by the Apollo 17 cosmic dust experiment. In *Interplanetary Dust and Zodiacal Light* (H. Elsässer and H. Fechtig, eds.), p. 233–237. Springer-Verlag, N.Y.

Blanford G. E., Fruland R. M., McKay D. S. and Morrison D. A. (1974) Lunar surface phenomena: Solar flare track gradients, microcraters, and accretionary particles. *Proc. Lunar Sci. Conf. 5th*, p. 2501–2526.

Blanford G. E., Fruland R. M. and Morrison D. A. (1975) Long term differential energy spectrum for solar flare iron-group particles. *Proc. Lunar Sci. Conf. 6th*, p. 3557–3576.

Bogard D. D., Funkhouser J. G., Schaeffer O. A. and Zähringer J. (1971) Noble gas abundances in lunar material—cosmic ray spallation products and radiation ages from the Sea of Tranquility and Ocean of Storms. *J. Geophys. Res.* **76**, 2757–2779.

Brownlee D. E., Hörz F., Hartung J. B. and Gault D. E. (1975) Density, chemistry and size distribution of interplanetary dust. *Proc. Lunar Sci. Conf. 6th*, p. 3409–3416.

Carey W. C. and McDonnell J. A. M. (1978) Monte Carlo sputter simulations and laboratory ion sputter measurements on lunar surfaces. *Proc. Lunar Planet. Sci. Conf. 9th*. This volume.

Chao E. C. T. (1976) Mineral produced high pressure striae and clay polish: Key evidence for nonballistic transport of ejecta from the Ries Crater, South Germany. *Science* **194**, 615–618.

Comstock G. M. (1978) How dusty are lunar rocks? (abstract). In *Lunar and Planetary Science IX*, p. 189–191. Lunar and Planetary Institute, Houston.

Dohnanyi J. (1976) Flux of hyperbolic meteoroids. In *Interplanetary Dust and Zodiacal Light* (H. Elsässer and H. Fechtig, eds.), p. 170–180. Springer-Verlag, N.Y.

Dungan M. A. and Brown R. W. (1977) The petrology of the Apollo 12 ilmenite basalt suite. *Proc. Lunar Sci. Conf. 8th*, p. 1339–1381.

Fechtig H., Hartung J. B., Nagel K. and Neukum G. (1974) Lunar microcrater studies, derived meteoroid fluxes, and comparison with satellite-borne experiments. *Proc. Lunar Sci. Conf. 5th*, p. 2463–2474.

Frondel C., Klein C. and Ito J. (1971) Mineralogical and chemical data on Apollo 12 lunar fines. *Proc. Lunar Sci. Conf. 2nd*, p. 719–726.

Gold T. (1955) The lunar surface. *Monthly Notices Roy. Astron. Soc.* **115**, 585.

Hartung J. B. and Comstock G. M. (1978) New lunar microcrater evidence against a time varying meteoroid flux. *COSPAR: Space Research XVIII*. Pergamon, N.Y. In press.

Hartung J. B., Hodges F., Hörz F. and Storzer D. (1975) Microcrater investigations on lunar rock 12002. *Proc. Lunar Sci. Conf. 6th*, p. 3351–3371.

Hartung J. B., Hörz F. and Gault D. E. (1972) Lunar rocks as meteoroid detectors. In *Evolutionary and Physical Properties of Meteoroids* (C. L. Hemenway, P. M. Millman and A. F. Cook, eds.), p. 227–239. NASA SP-319.

Hartung J. B., Hörz F., McKay D. S. and Baiamonte F. L. (1972a) Surface features on glass spherules from the Luna 16 sample. *The Moon* **5**, 436–446.

Hartung J. B., Plieninger T., Müller H. W. and Schaeffer O. A. (1977) Helium, neon, and argon on sunlit and shaded surfaces of lunar rock 12054. *Proc. Lunar Sci. Conf. 8th*, p. 865–881.

Hartung J. B. and Storzer D. (1974) Lunar microcraters and their solar flare track record. *Proc. Lunar Sci. Conf. 5th*, p. 2527–2541.

Hauser E. E. (1978) Microcraters on lunar sample 12054,54. Masters thesis, State Univ. of New York at Stony Brook.

Hoffman H. S., Fechtig H., Grün E. and Kissel J. (1975) First results of the micrometeoroid experiment S-215 on the HEOS 2 satellite. *Planet. Space Sci.* **23**, 215–224.

Hörz F. and Hartung J. B. (1971) The lunar surface orientation of some Apollo 12 rocks. *Proc. Lunar Sci. Conf. 2nd*, p. 2629–2638.

Mandeville J.-C. (1977) Impact microcraters on 12054 rock. *Proc. Lunar Sci. Conf. 8th*, p. 883–888.

Mandeville J.-C. (1978) 12054 Consortium: More on microcraters (abstract). In *Lunar and Planetary Science IX*, p. 690–692. Lunar and Planetary Institute, Houston.

McDonnell J. A. M. (1977) Accretionary particle studies on Apollo 12054,58: In-situ lunar surface microparticle flux rate and solar wind sputter rate defined. *Proc. Lunar Sci. Conf. 8th*, p. 3835–3857.

McDonnell J. A. M. (1978) Whence low velocity lunar accreta? Directional anisotropic measurements from ancient and modern lunar surface sensors (abstract). In *Lunar and Planetary Science IX*, p. 717–719. Lunar and Planetary Institute, Houston.

Morrison D. A., McKay D. S., Fruland R. M. and Moore H. J. (1973) Microcraters on Apollo 15 and 16 rocks. *Proc. Lunar Sci. Conf. 4th*, p. 3235–3253.

Morrison D. A., McKay D. S., Heiken G. H. and Moore H. J. (1972) Microcraters on lunar rocks. *Proc. Lunar Sci. Conf. 3rd*, p. 2767–2791.

Morrison D. A. and Zinner E. (1975) Studies of solar flares and impact craters in partially protected crystals. *Proc. Lunar Sci. Conf. 6th*, p. 3373–3390.

Morrison D. A. and Zinner E. (1976) The size frequency distribution and rate of production of microcraters. In *Interplanetary Dust and Zodiacal Light* (H. Elsässer and H. Fechtig, eds.), p. 227–231. Springer-Verlag, N.Y.

Morrison D. A. and Zinner E. (1977) 12054 and 76215: New measurements of interplanetary dust and solar flare fluxes. *Proc. Lunar Sci. Conf. 8th*, p. 841–863.

Papike J. J., Hodges F. N., Bence A. E., Cameron M. and Rhodes J. M. (1976) Mare basalts: Crystal chemistry, mineralogy, petrology. *Rev. Geophys. Space Phys.* **14**, 475–540.

Rennilson J. J. and Criswell D. R. (1974) Surveyor observations of lunar horizon-glow. *The Moon* **10**, 121–142.

Rhodes J. M., Blanchard D. P., Dungan M. A., Brannon J. C. and Rodgers K. V. (1977) Chemistry of Apollo 12 mare basalts: Magma types and fractionation processes. *Proc. Lunar Sci. Conf. 8th*, p. 1305–1338.

Schaal R. B. and Hörz F. (1977) Shock metamorphism of lunar and terrestrial basalts. *Proc. Lunar Sci. Conf. 8th*, p. 1697–1729.

Shoemaker E. M., Batson R. M., Bean A. L., Conrad C. Jr., Dahlem D. H., Goddard E. N., Hait M. H., Larson K. B., Schaber G. G., Schleicher D. L., Sutton R. L., Swann G. A. and Waters A. C. (1970) Preliminary geologic investigation of the Apollo 12 landing site, part A. In *Apollo 12 Prelim. Sci. Rep.* NASA SP-235, p. 113–156.

Singer S. F. and Walker E. H. (1962) Electrostatic dust transport on the lunar surface. *Icarus* **1**, 112–120.

Stephenson W. K. (1970) Mission description. In *Apollo 12 Prelim. Sci. Rep.* NASA SP-235, p. xi–xii.

Wade L. C. (1970) Photographic summary of the Apollo 12 mission. In *Apollo 12 Prelim. Sci. Rep.* NASA SP-235, p. 7–27.

Zinner E., Walker R. M., Chaumont J. and Dran J. C. (1977) Ion probe surface concentration measurements of Mg and Fe and microcraters in crystals from lunar rock and soil samples. *Proc. Lunar Sci. Conf. 8th*, p. 3859–3883.

Zook H. A. (1978a) Temporal and spatial variations of the interplanetary dust flux. In *COSPAR: Space Research XVIII*, p. 411–422. Pergamon, N.Y.

Zook H. A. (1978b) Dust, impact pits, and accreta on lunar rock 12054. *Proc. Lunar Planet. Sci. Conf. 9th*. This volume.

Zook H. A., Hartung J. B. and Storzer D. (1977) Solar flare activity: Evidence for large-scale changes in the past. *Icarus* **32**, 106–126.

Proc. Lunar Planet. Sci. Conf. 9th (1978), p. 2539–2556.
Printed in the United States of America

Primary, secondary and tertiary microcrater populations on lunar rocks: Effects of hypervelocity impact microejecta on primary populations

R. P. FLAVILL, R. J. ALLISON and J. A. M. MCDONNELL

Space Sciences Laboratory, University of Kent, Canterbury, Kent, England.

Abstract—Computer models have been developed to simulate recapture of microscale hypervelocity ejecta from micron sized meteoroid impacts on the lunar regolith. The use of an experimentally measured secondary crater distribution from hypervelocity ejecta generated from impacts of 10^{-11}g iron microspheres at 4–6 km s^{-1} impact velocity on lunar sample 62235 has allowed partial decoding of the probable true primary components of measured microcrater populations on lunar samples 12054 and 60015. This study of line-of-sight ejecta recapture indicates the importance of secondary and, in some cases, even tertiary ejecta microcrater contributions for certain lunar surface exposure geometries and size ranges. The presence of directional anisotropies in the micrometeorite flux therefore requires that care be taken in the interpretation of observed microcrater distributions due to the highly variable contribution of ejecta components, and geometrical dilution of the incident primary micrometeorite flux.

1. INTRODUCTION

Telescopic lunar observations have provided evidence in the form of ray structures and some apparently linked crater chains, which has suggested a secondary impact contribution to lunar crater populations.

Since the Apollo age, which has provided lunar regolith samples for terrestrial laboratory examination, extension of populations measured for craters of undoubted hypervelocity impact origin has been achieved down to submicron sizes. Notable examples of this microcrater distribution have been published by Fechtig *et al.* (1974) and more recently by Morrison and Zinner (1977). We now ask how valid and accurate such model microcrater distributions are as indicators of the interplanetary micrometeoroid influx, particularly in relation to superimposed variable secondary microcrater distributions on these sample surfaces.

Attempts have been made to estimate the proportion of lunar impact microcraters, which are of secondary origin. Previously these have been based on measurements of secondary ejecta from relatively large scale laboratory hypervelocity impacts. A recent example is the measurement of micron-sized secondary microcraters due to ejecta from the centimetre scale primary impact crater caused by impact of a light gas gun accelerated 1.58 mm steel sphere at 4.1 km.s^{-1} on Duran glass (Schneider 1975). Downward extrapolation from these types of experiment to represent secondary contributions to micron and submicron observed lunar crater distribution (e.g., Stöffler *et al.*, 1975) have indicated these to be negligible.

We now present new measurements and calculations from laboratory *micron* scale primary impacts on lunar crystalline rock which show that, for favourable exposure geometries, secondary and even tertiary impact craters may be comparable to the primary micron and submicron lunar microcrater population over restricted size ranges.

A requirement for the production of secondary impacts considered here is that of an impact velocity in excess of some critical lower limit. This threshold velocity is of the order of kilometres per second for micron scale impacts on silicates. *Submicron* impact craters observed on lunar glasses, however, invariably lack spallation zones, removing that ready indicator of impact velocity. Observations within a 240 μm glass impact pit on 60015 (McDonnell *et al.*, 1975) have shown submicron impact features ranging from shallow depressions clearly of low velocity glass splash impact origin to well developed hemispherical impact pits. The presence of a range of intermediate features firmly demonstrates that on this surface at least some well formed impact pits *were* of secondary origin.

'Line of sight' trajectories prior to impact, imply that lunar secondary microcrater populations are expected to be highly dependent on the exposure configuration—in general, therefore, highly variable. Optimum exposure geometry for intercepting interplanetary particles corresponds to a flat horizontal surface and secondary crater populations should be zero; by contrast for a boulder side, the primary influx is reduced by a factor of two, and the recapture of secondary ejecta from line-of-sight trajectories is far more efficient.

A major difficulty affecting the resolution of the primary and secondary collection efficiencies is the generally scant knowledge of *in situ* lunar sample orientation. This difficulty is magnified further by the strongly non-isotropic distribution of the micrometeoroid flux as measured by recent space probes (e.g., Fechtig, 1976).

We therefore investigate primarily the exposure of the well documented consortium lunar sample 12054; we use as a basis the microcrater population measured on this rock by Morrison and Zinner (1977). Comparisons are also made with the rather different microcrater distribution characterised on Apollo 60015 by Fechtig *et al.* (1974) and for submicron craters by McDonnell *et al.* (1975). Microcrater distributions on both these samples are particularly well documented, and important differences between them are studied with regard to the possibility that:

(i) A similar primary particle flux incident on both surfaces resulted in different crater distributions due to (a) appreciable numbers of non-primary impacts and (b) differences in geometrical collection probability—both for primary and non-primary impact events—for these two samples.

(ii) Each surface has been an indicator of a different primary crater size distribution because of differences in time integration of surface geometry limited portions of a spatially anisotropic micrometeorite flux.

(iii) A third possibility—that observational biasses or errors are responsible—must also be considered. To a large extent this factor is not easily quantified, but it may well be that restricted areas, particularly for counting large microcraters on 60015, have caused an artificial steepening of the distribution of craters in the hundreds of micron size range. This selection effect is described in detail by Hartung *et al.* (1972).

2. Experimental Measurement of Secondary Microcraters

A series of microscale hypervelocity impact experiments was performed using the 2MV van de Graaff dust accelerator at the Max-Planck Institute, Heidelberg, Federal Republic of Germany. Approximately 31,000 iron particles of mass 10^{-11}g accelerated to 4–6 km.s^{-1} were allowed to impact an unpolished fracture surface of lunar crystalline rock 62235,28. An experimental arrangement as shown in Fig. 1a was used (Flavill and McDonnell, 1977) and primary microcraters were generated on 62235 similar to those shown in Fig. 1b.

The escaping fraction of ejecta from impacts in the unpolished lunar sample surface was collected on a quartz disc diameter approximately 0.8 cm which had been partly vacuum coated with aluminium.

Of the order of tens of secondary ejecta submicron craters were observed (Fig. 2) on aluminium areas together with much low velocity granular debris over most of the collector.

Although there must be *some* doubt about the velocity of the ejecta particles which caused these, the secondary microcraters typically have well developed lip and pit form. The causative particles have similar density to aluminium, which suggests that their velocity may be high enough for hypervelocity cratering to occur, particularly on a submicron size scale, on lunar rocks and glasses.

An evenly spread areal distribution of secondary microcraters was observed throughout the aluminium collecting area. Very much smaller numbers of microcraters were observed on quartz areas. A ratio of 3.5:1 between pit diameters for similar velocity and mass micron sized craters on aluminium and quartz as measured by McDonnell *et al.* (1972) suggests that the majority of quartz submicron secondary microcraters in this experiment are approximately 0.1 μm, which is approximately the limits of detection using our current SEM. Intensive study of limited areas of collector regions revealed very small numbers of microcraters on quartz collector areas. These numbers, while although not in themselves being very significant statistically, do improve our confidence in this explanation; although due to the small size of these craters, spallation does not, in this case, provide a ready test of impact velocity.

Details of the measurement of numbers of primary impacting particles in this type of experiment have been described by McDonnell *et al.* (1976) in connection with the measurement of the comminution distribution of low velocity cratering ejecta from micron scale lunar impacts. Figure 3 shows the measured size distributions of low velocity spall ejecta (McDonnell *et al.*, 1976) and secondary microcraters (Flavill and McDonnell, 1977).

These measurements relate to a narrow velocity and mass range of causative particles and therefore have been normalised for one primary microcrater.

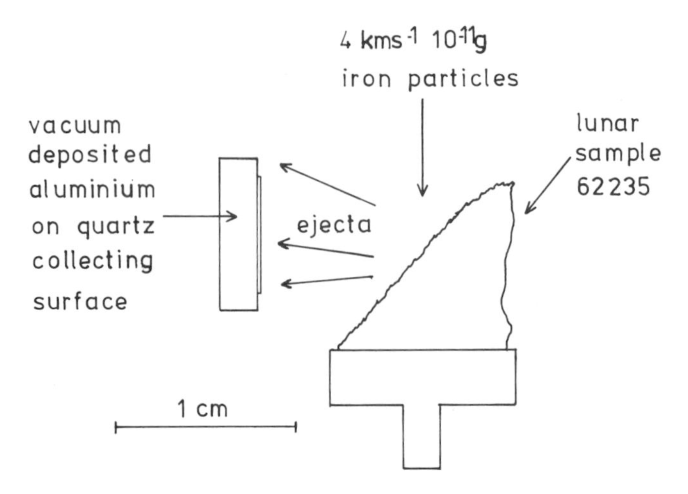

Fig. 1a. Experimental erosion of an unpolished fracture surface of 62235,28 by hypervelocity impact of iron dust particles. The escaping fraction of ejecta from impact sites is sampled by a test surface of aluminium deposited on quartz.

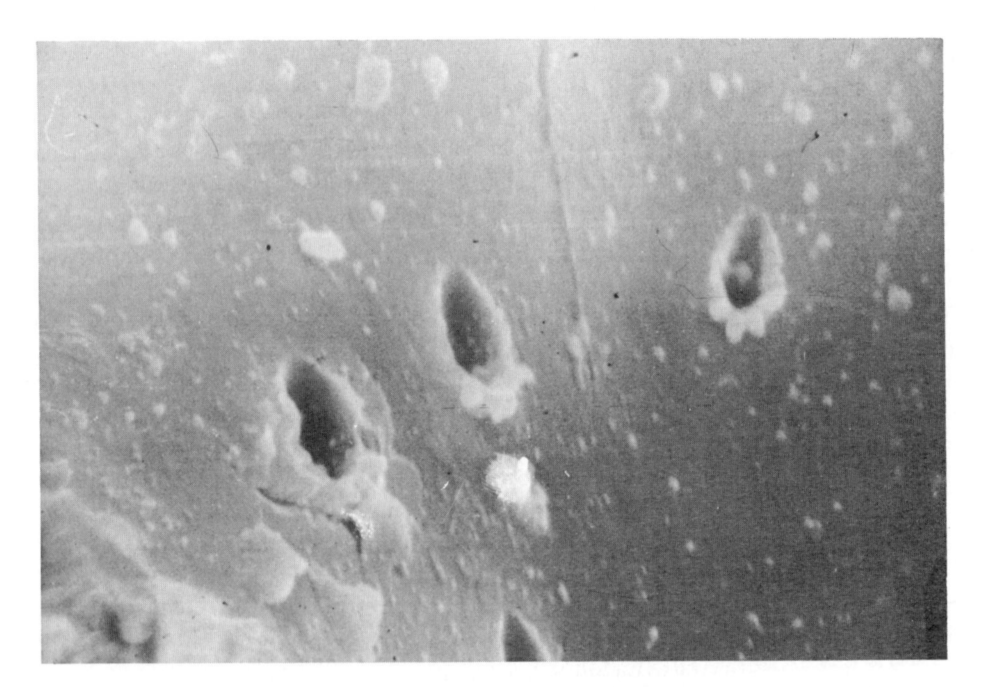

Fig. 1b. Typical microcraters formed on 62235 by iron dust particles. Microcrater lips show features representing almost-detached glass droplets.

Fig. 2. A submicron hypervelocity impact crater typical of those observed on the aluminium secondary collection target.

Even distribution of secondary microcraters throughout the collecting surfaces results very probably from the randomly oriented rough nature of the face of sample 62235,28 rather than representing the typical angular distribution from any particular impact situation. We observe in this experiment the spatial distribution of ejecta which is *not being recaptured locally within the target surface roughness*, and therefore escapes the immediate sample area. That this *measured* ejecta is the fraction escaping from the immediate impact area underlines an important difference between micron scale hypervelocity impact simulations and much larger (e.g., centimetre) scale impact ejecta measurements. In the latter case, very little of the ejecta is recaptured locally—because the target surface roughness size scale is small compared to the impact crater size. Thus, these new measurements from *micron* sized laboratory generated hypervelocity impact craters are particularly relevant to the *in-situ* microscale lunar conditions, in addition to better dimensional comparability.

For impact crater size regimes comparable with surface roughness sizes, local recapture is expected to account for sizeable fractions of ejecta. Comparison with an extrapolation from measured spatial distributions of hypervelocity impact ejecta (e.g., Gault and Heitowit, 1963; McDonnell *et al.*, 1976), and computer modelling by Carey and McDonnell (1976) of the geometrically similar processes associated with recapture of solar wind sputtered atoms from an atomic scale surface, leads us to suggest that to a first order approximation, 50% of hypervelocity ejecta from micrometeorite impact cratering will be locally recap-

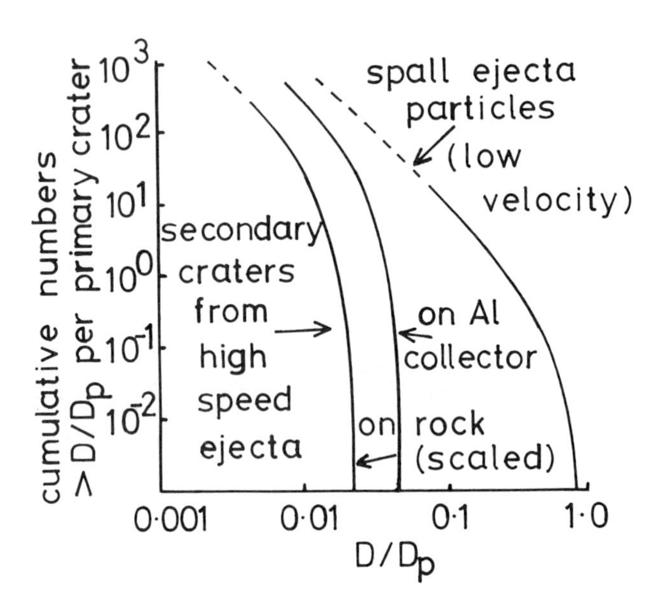

Fig. 3. Comminution distributions for: (a) low velocity spall ejecta particles, and (b) hypervelocity secondary microcraters measured from micron scale impacts on lunar rock. These distributions are normalised to one primary impact event pit diameter D.

tured within distances of the order of microns on a rough surface typical of many returned lunar sample areas. This scaling is important for both the laboratory experiments described and also in application of these measurements to an actual microscale lunar surface impact scenario.

3. SECONDARY AND TERTIARY MICROCRATER POPULATIONS RELATED TO PRIMARY MICROCRATER DISTRIBUTIONS

With knowledge of a purely primary microcrater population, and of geometry for emission and recapture of ejecta as a function of size, calculation of a total microcrater population on any sample surface of well defined exposure could be simply achieved. The basic method would be of convolution of the primary distribution with the measured distribution of secondary microcraters formed by each primary crater, with suitable weighting for acceptance angles.

We also include an upper size limit for the size of the generating crater. This describes the situation within a surface with partially enclosed geometry. For example, the surface of a hemispherical pit on a horizontal rock surface only recaptures line-of-sight ejecta from impacts within itself. The restricted area × time product for such an environment implies that corresponding hypervelocity ejecta generation can only originate from microcraters up to and including the largest meteoroid impact observed within the host pit.

The exposure of a lunar boulder side to hypervelocity ejecta is, however, quite different. This is exposed to line-of-sight trajectories for ejecta generated from meteoroid impacts from nearby sites to the lunar horizon—i.e., a very large area × time product. For this secondary microcrater calculation, the upper limit for primary craters, therefore, will be very large for typical sample exposure ages.

The geometry of exposure however does not enhance tertiary recapture on the boulder face. This is determined by boulder surface roughness.

Figures 4a and 4b respectively show calculated convolutions assuming an entirely primary origin for the measured model microcrater distributions of Apollo 60015 and 12054. Convolutions of each of these distributions with the secondary microcrater production profile measured by Flavill and McDonnell (1977) are shown which yield secondary lunar microcrater distributions of 100% recapture environments: a second convolution yields the tertiary distribution.

Figure 5 is the result of the convolution of the 12054 model distribution with the production distribution for the low velocity spall ejecta as depicted in Fig. 3. This corresponds to the distribution of freshly generated lunar soil and granular lunar sample accreta. For comparison, we also show this same convolution calculated for the 60015 model microcrater distribution as first presented by McDonnell *et al.* (1976). The size distribution measured by Heywood (1971) for the more mature lunar soil of 12057,72 is also shown for an arbitrary total number of particles together with accreta populations measured on 12054,58 by McDonnell (1977).

Heywood notes an increasing trend towards a majority of granular fragmented particles below 100 μm, supporting the idea of spallation as a major source: McDonnell observed much angular accreta between 100 μm and 10 μm, but below this size rounded spheroids became predominant.

Calculations of ejecta numbers presented in this publication are proposed as being valid for rock and glass surfaces.

On lunar terrain many impacts will occur on loosely bonded or noncohesive dust particle assemblies. For grain to meteoroid diameter ratios small enough to preclude semi-infinite target impact conditions, catastrophic rupture can result from meteoroid impact.

The critical destruction limit for grain rupture is a function of the size scale. At cm scale impacts destruction occurs for grain diameters less than ten times the primary crater diameter (Hörz *et al.*, 1975b): for micron scale impacts a maximum limit of four times the primary crater diameter is inferred by McDonnell (1976) from impacts on a raised lunar sample chip.

McDonnell (1976) considers impacts on individual lunar soil grains and near free edges, particularly on target material previously weakened by shock damage. For these circumstances, he calculates that ejecta mass may be enhanced by up to an order of magnitude compared to impacts on solid rock. This again is a function of size scale, which is not well understood.

Figures 6a and 6b show the calculated secondary and tertiary microcrater distributions of Figs. 4a and 4b normalised to the respective 'primary' distributions.

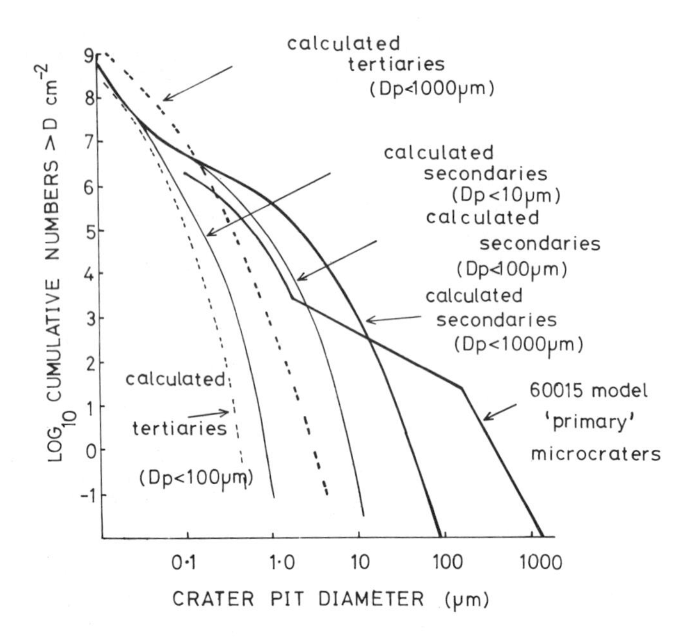

Fig. 4a. The microcrater distribution observed on 60015 (Fechtig *et al.*, 1974) together with secondary and tertiary microcrater distributions calculated for various upper size limits (Dp) of primary microcrater. For these calculations 100% recapture of secondary and tertiary ejecta has been assumed. The calculations show the integration of the effects of ejecta from primary craters of several chosen limits, of diameter Dp.

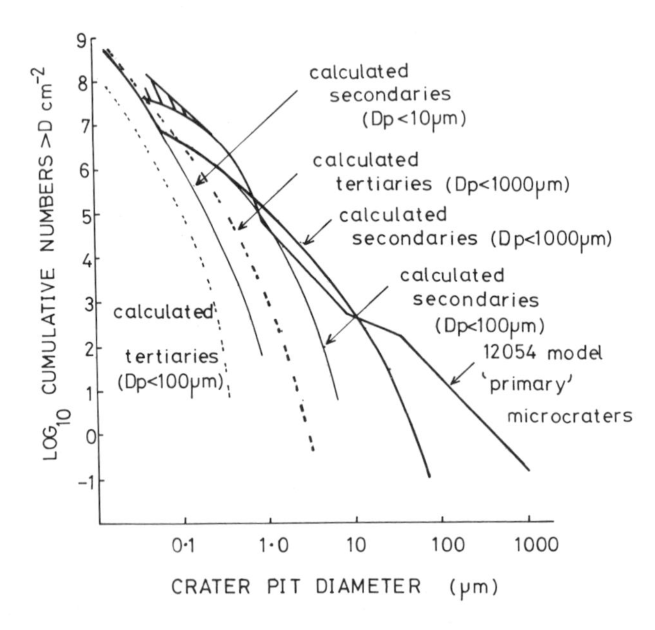

Fig. 4b. As Fig. 4a, using observations on lunar sample 12054 (Morrison and Zinner, 1977).

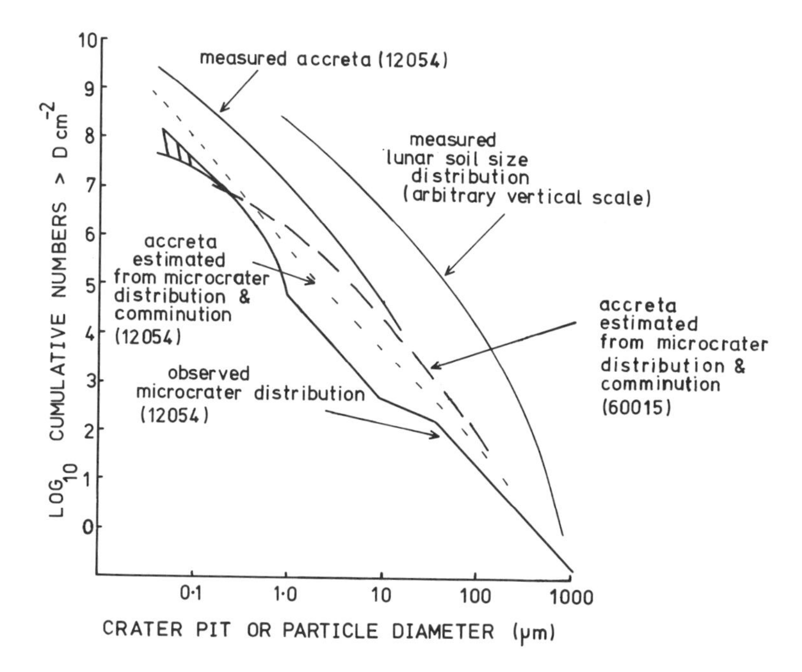

Fig. 5. 12054 microcrater distribution (Morrison & Zinner, 1977) together with the distribution of spall generated accreta which was calculated from this distribution and the microscale hypervelocity impact comminution distribution measured by McDonnell *et al.* (1976). The accreta population measured on 12054 by McDonnell (1977) and the lunar soil size distribution measured for 12057 by Heywood (1971) are shown for comparison. Freshly generated impact ejecta is insufficient to account for observed accreta populations.

Surface roughness on an initially smooth glass lining of a micrometeorite impact pit will develop due to collection of accreta from other meteorite impact events, and perhaps the accumulation of grains transported by electrostatic forces. This internal particle population may become bonded to the host surface with varying efficiencies due to partial impact melting and subsequent resolidification, or more gradually by solar wind sputter welding (Flavill *et al.*, 1977).

Enhanced *local* recapture of ejecta from hypervelocity impacts within the host pit occurs due to these developing microfeatures. For a well exposed hemispherical pit in lunar equatorial regions (e.g., on a boulder top face), the internal microcrater population caused by the micrometeorite flux is reduced by a factor of approximately two compared with a corresponding flat surface. This reduction is due to the restricted solid angle exposed. Of the secondary hypervelocity particles produced from each micrometeorite impact we estimate approximately 50% will be locally recaptured. Of the 50% escaping the immediate impact surroundings, on average only 50% is expected to leave the host hemispherical pit. Thus, a total of 75% of all high velocity ejecta produced from microimpacts within a host microcrater pit is calculated to form secondary microcraters within

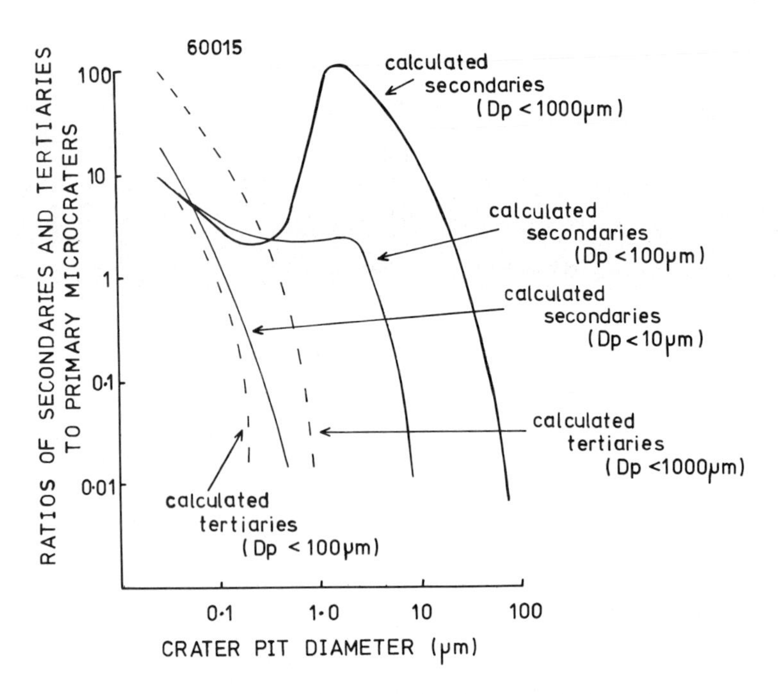

Fig. 6a. Secondary and tertiary microcrater distributions calculated for 100% recapture environments and normalised to an assumed 'primary' observed 60015 model microcrater population. Calculations are shown for several upper size limits (Dp) of the input primary microcrater distribution.

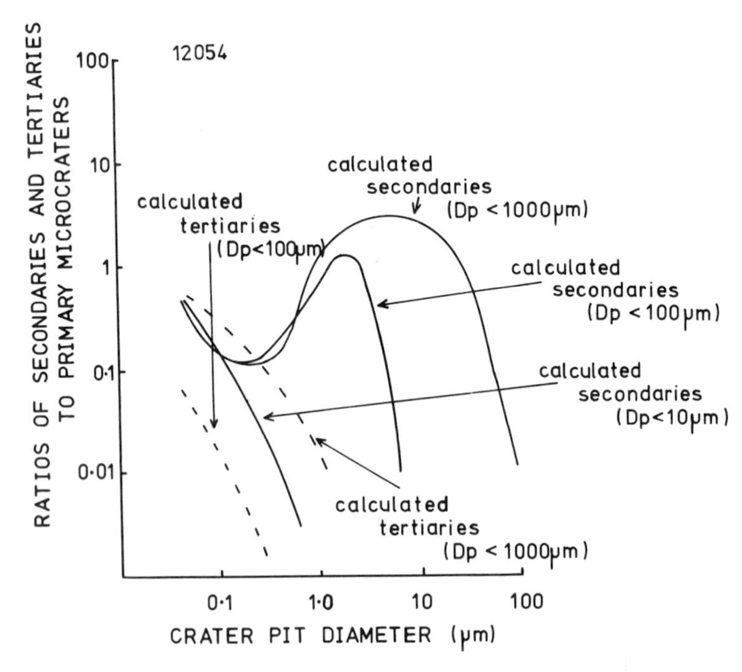

Fig. 6b. As Fig. 6a, using observations on lunar sample 12054.

that host pit. Therefore, the numbers of secondary microcraters calculated in Fig. 4a and 4b corresponding to 100% recapture should be reduced by a factor of ¾ to describe observed microcrater distributions within a host pit as described above relative to the primary populations within the same pits. Similarly, a reduction of the calculated tertiary populations by a factor of 9/16 is required.

For submicron sized impacts on silicates evaluation of the particle velocity is often very difficult. Observation of a range of submicron features intermediate between obviously low velocity glass splash impacts and well developed pits on 60015 (McDonnell *et al.*, 1975) suggests that some inclusion of secondary impacts in crater counts on this size scale has occurred.

Model calculations based on the 12054 crater distributions show possible importance of secondaries in the size range 1–10 μm. This is more difficult to evaluate because of the uncertainty in velocities of the secondary particles which formed our laboratory microcraters, mostly measured on aluminium. The development of clearly defined central pit and spallation zones for hypervelocity impact microcraters of this size on silicates would provide clear definition of a lower particle velocity limit. The application of our results to the 12054 distribution must therefore be regarded as tentative awaiting measurement of larger numbers of secondary microcraters on a test surface capable of giving impact velocity definition.

Results of these model simulatons suggest that secondary microcrater populations can become comparable to primary microcrater populations for sizes less than 0.1 μm pit diameter using the 60015 crater distribution. Surprisingly, calculated *tertiary* populations can exceed secondary ejecta populations on a submicron size scale for both input models. It is, therefore, suggested that both these observed microcrater distributions could contain significant non-primary components. The morphology of the hypervelocity secondary and tertiary craters here discussed would not differ in any yet known way from primary impacts.

Unfortunately, this expected similarity means that we cannot as yet see a method of separating primary and secondary microcrater components using observations of *in situ* (i.e., lunar surface) bombarded lunar samples.

3.1 Effect of exposure geometry on the ratio of primary to non-primary microcrater populations

The relative contributions to any observed lunar sample microcrater population from primary, secondary and tertiary impact sources depend on the magnitude of the solid angle subtended by the lunar surface. Orientation of the axis of this subtended solid angle relative to the local surface horizontal and to the lunar equator is also important. Selection of time and angle integrated portions of the probably spatially anisotropic incoming lunar micrometeoroid flux and also line-of-sight trajectory secondary ejecta from lunar regolith impact events are both functions of these exposure parameters.

Spatial anisotropies in the interplanetary micrometeorite flux have been measured by recent interplanetary probes such as Pioneers 8 and 9 (McDonnell and Berg, 1975) and HEOS II (Hoffman *et al.*, 1975), and such results invoked to reconcile lunar microcrater measurements with current micrometeoroid fluxes (e.g., Fechtig, 1976). HEOS II data is interpreted by Fechtig (1976) to represent two main spatial concentrations of micrometeorites. These comprise submicron very high (100 km.s^{-1}) velocity micrometeoroids ejected radially from near the Sun, and micron sized particles impacting the moon as 'swept up' by the heliocentric orbital motion of the earth-moon system. Directional anisotropies in the time integrated flux of micrometeoroids impacting lunar samples also are implied by comparison of lunar microcrater measurements made by Morrison and Zinner (1977) and Mandeville (1978). These measurements are made of populations on the NW and S faces respectively of 12054. Both these sample areas are able to yield good observational conditions due to the existence of relatively large numbers of microcraters on a glass covered surface. An excess of craters smaller than approximately 5 μm pit diameter was measured on the NW face compared to the S face. Zodiacal light observations (Weinberg, 1978) also show a concentration of micrometeoroids centred on the plane of the ecliptic. Particle sizes are those inferred from models of optical scattering mechanisms and are generally accepted to be in the tens of micron size range. To date no space probes have performed *in situ* micrometeoroid detection at orbital inclinations of more than ±15° from the ecliptic plane. Direct spacecraft measurement of these particles therefore awaits NASA's out of ecliptic mission.

The existence of meteoroid flux anisotropies with maxima aligned differently for different particle sizes creates variations in the incident microparticle flux size composition for partly enclosed sample surfaces which depend on surface inclination to the plane of the ecliptic: lunar orbital motion causes integration of the incident flux over 360° referred to the heliocentric radial direction. Sides of lunar boulders may be exposed substantially higher than the local surface and, therefore, efficient collectors of line-of-sight secondary ejecta from nearby to the lunar horizon. These surfaces are expected to possess singularly variable proportions of primary, secondary and tertiary microcrater populations. Consider, for example, a lunar north (or south) facing vertical equatorial boulder face. Both N and S faces are particularly inefficient collectors of meteorite flux components integrated over the lunar equatorial plane.

Assuming the micrometeorite flux anisotropies described previously, only 'background', i.e., out of ecliptic meteorite flux components impact these faces, diluted by a factor of two due to the restricted exposure solid angle. Secondary ejecta from cratering integrated from sites nearby up to the lunar horizon can be produced from surfaces very efficiently exposed to the primary flux. Ratios of secondary and tertiary to primary microparticle populations could be enhanced relative to those shown in Fig. 6a and 6b by the ratio of maxima to minima influx anisotropies. This is particularly relevant to lunar submicron craters where (i) the absence of spallation as a ready indicator of impact velocity creates difficulties in differentiating between primary and secondary origins and (ii) an important primary incident micrometeorite flux anisotropy maximum is expected.

3.2 Derivation of the primary component of observed 12054 and 60015 microcrater populations

Our main effort has been directed towards delineating the origins of components of the 12054 model microcrater distribution. Work by Morrison and Zinner (1977), Hörz *et al.* (1975a) and Mandeville (1977) has provided a cumulative crater size frequency distribution over approximately four orders of magnitude in size, and 12054 is the first lunar rock measured over such a wide size range for which adequate statistics exist for all sizes.

Whilst 60015 and associated rock surface studies (Fechtig *et al.*, 1974) represent an extensive and carefully executed observational achievement, it may be that restricted available areas, giving small numbers counted for larger microcrater sizes, produced an artificial steepening of the distribution of craters in the hundreds of micron size range due to a selection effect as defined by Hartung *et al.* (1972).

For the 60015 microcrater distribution smaller than 10 μm pit diameter, we have used measurements by McDonnell *et al.* (1975). These were made within a single 240 μm glass lined impact pit on 60015,6 which was dated relative to the surrounding area by relative numbers of intermediate sized craters distributed over the spall zone. Measurements extend to a lower limit of 0.2 μm.

3.2(a) The 12054 primary microcrater distribution Decoding the primary component of the 12054 microcrater population has been achieved by applying negative feedback concepts to the previously described computer convolution. By considering the observed distribution as the sum of a primary distribution and a secondary distribution due to it, a set of simultaneous linear equations was obtained relating values of these quantities over the range considered. Computer solution of this system yielded the appropriate primary distribution. The method was extended to consider exposure geometry and to investigate tertiary effects.

Secondary microcraters of any particular size are generated from a range of primary crater sizes. Relative numbers contributed from each primary size are a function of the assumed primary distribution and also the form of secondary crater production curve. Dimensional scaling of cratering parameters is not expected to be very important over the size range considered in this calculation because all ejecta considered is generated from craters having a central glass pit and surrounding spall zone, although it is an effect which will have to be studied in further detail. Table 1 shows these relative contributions calculated for the 12054 model distribution limited to pit diameters smaller than 1000 μm.

Figure 7 shows the primary microcrater distribution derived from the observed population on 12054.

The observed microcrater distribution and the calculated secondary-plus-primary microcrater distributions are also shown for comparison. For the *in situ* exposure environment of 12054 using geometry as defined by Apollo photography, secondary microcraters (mostly formed from primary impact pits of some hundreds of micron diameter) could be important in the size range 1–10 μm.

Table 1.

Primary Crater Median Pit Diameter in μm.	Secondary Crater Median Pit Diameter in μm							
	31.62	10.00	3.16	1.00	0.32	0.10	0.03	0.01
1,000.00	—	—	—	—	—	—	—	—
316.23	—	1.00	0.53	0.26	0.04	0.02	—	—
100.00	—	—	0.47	0.39	0.36	0.06	0.03	—
31.62	—	—	—	0.35	0.54	0.58	0.09	0.01
10.00	—	—	—	—	0.06	0.11	0.11	0.00
3.16	—	—	—	—	—	0.23	0.39	0.05
1.00	—	—	—	—	—	—	0.38	0.08
0.32	—	—	—	—	—	—	—	0.86
0.10	—	—	—	—	—	—	—	—

Relative numbers of secondary craters of various pit diameters due to primary craters of pit diameters less than 1000 μm, based on 12054 data.

Contributions from these would then result in smoothing of the changes in slope of the underlying primary microcrater distribution within this size range.

3.2(b) The 60015 primary microcrater distribution Obvious and important differences in slope are apparent between the 12054 and 60015 model microcrater populations. Calculation of the secondary ejecta microcrater component in the submicron and micron range is particularly sensitive to the observed crater distribution slope for microcraters larger than some tens of microns. Because of the better defined geometrical parameters for the 12054 exposure, this model was chosen as a standard from which derivation of the 60015 distribution was to be attempted. We have investigated the idea that purely geometrical exposure factors might be responsible for essentially the same incoming micrometeorite flux causing different observed microcrater populations on each sample surface.

The 12054 primary microcrater model was used then as an initial assumed primary microcrater distribution and convolved with the measured secondary ejecta microcrater comminution distribution measured by Flavill and McDonnell (1977). The calculated secondary microcrater distribution was linearly scaled to account for exposure acceptance angle and summed with the assumed primary model. This total was used in comparison with the observed 60015 model.

Figure 8 shows calculated convolutions of the previously derived 12054 primary microcrater distribution with the secondary microcrater production distribution for a range of acceptance angles referred to the local horizontal, and the observed 60015 microcrater model distribution for comparison. These calculations show that purely geometrical factors are not responsible for differences between the two models, if we assume an isotropic micrometeorite flux. It is unlikely that purely local anomalies could be responsible for these differences. The slope measured for the 60015 distribution for large sizes is suspect due to probable influence of a selection effect, but we suggest that disparities over

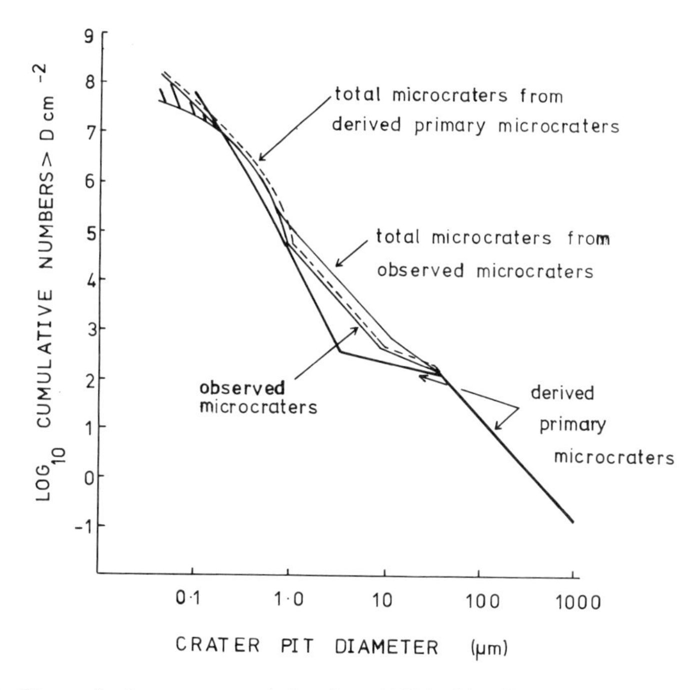

Fig. 7. Observed microcrater population from 12054 with primary microcrater population derived from it. Also shown are total (primary + secondary + tertiary) microcrater populations obtained by taking each of these as the actual primary microcrater population. Exposure geometry of the sample has been taken into account.

smaller size ranges could result from different averaged incident micrometeorite flux size distribution impacting each surface due to flux spatial anisotropies.

The importance of a secondary component in composition of the 60015 microcrater distribution for small sizes is not doubted, but lack of geometrical exposure data severely limits interpretation of this.

Conclusion

(1) Practical lunar regolith exposure geometries are considered for recapture of the hypervelocity fraction of micrometeorite impact ejecta (a) within hemispherical pits, and (b) on the sides of raised boulders.

For surfaces with partially enclosed solid geometry, secondary and even tertiary microcrater populations may become comparable with primary micrometeorite impact crater distributions for sizes less than 10 μm. For the 60015 distribution, secondary *and* tertiary populations may exceed the primary populations at $<.1$ μm crater diameter: for the 12054 distribution, secondary populations only could exceed the primary distribution over the range 1–10 μm.

(2) Scaling due to the solid angle acceptance for the numbers of impacts of

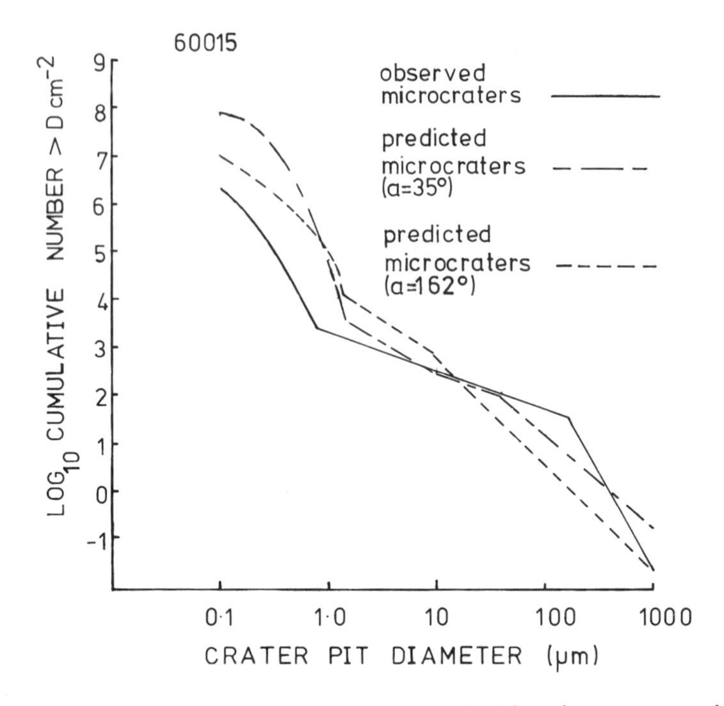

Fig. 8. Observed microcrater population on 60015. Also shown are populations predicted by the primary microcrater population derived from 12054, assuming two arbitrarily chosen different exposure geometries—angles of elevation of 35° and 162°. Neither agrees with the observations, and it appears unlikely that intermediate angles would offer better agreement.

(a) primary micrometeorites, and (b) secondary and tertiary hypervelocity ejecta, respectively, on a lunar boulder face may provide wide variations in the relative contributions of these types of event to observed microcrater populations.

Optimum exposure to the micrometeorite flux corresponds to a horizontal flat, unshielded equatorial lunar rock. This provides very poor capture efficiency for hypervelocity secondary ejecta from meteorite cratering events, except within the pits of craters already formed. These areas have often been the areas selected for study, particularly on crystalline rocks.

Conversely, a rock overhang surface has poor exposure to the micrometeorite flux, but the solid angle included at the surface by relatively efficiently exposed regolith areas can lead to good recapture geometry for secondaries from line of sight trajectories.

(3) Directional anisotropies in the incoming lunar surface micrometeorite flux have been proposed, e.g., (i) radial heliocentric submicron sized 100 km s^{-1} particles, (ii) orbitally swept micron sized 10 km s^{-1} particles.

These mechanisms imply variations in the *time-integrated* lunar surface micrometeorite flux as a function of inclination of a sample surface to the plane of the ecliptic, i.e., the lunar equator.

Assuming such anisotropies exist, then the importance of secondary microcraters due to ejecta from line-of-sight trajectories also varies with inclination of boulder side to the lunar equator.

(4) Secondary impacts are suggested to contribute to both the Apollo 12054 and 60015 observed microcrater populations. Calculated numbers of secondary and tertiary microcraters are *not* sufficient to explain both observed distributions in terms of the same micrometeorite flux with differences purely due to different acceptance angles both for primary and secondary ejecta fluxes. It may well be that 60015 was in fact exposed to a different micrometeorite flux than 12054. If 60015 pointed out of the ecliptic, a different form of flux may well apply. Lack of geometrical information about 60015 is a severe restriction in analysis of its cratering record, but it seems likely that differences in slopes of these two models for craters in the hundreds of microns size range are due to selection effects.

Acknowledgments—We acknowledge support offered by the Science Research Council (U.K.) and thank Professor R. C. Jennison for the use of laboratory facilities. We thank Mrs. J. Elmes and Miss D. Paine for the preparation of this manuscript.

REFERENCES

Carey W. C. and McDonnell J. A. M. (1976) Lunar surface sputter erosion: A Monte Carlo approach to microcrater erosion and sputter redeposition. *Proc. Lunar Sci. Conf. 7th*, p. 913–926.

Fechtig H., Gentner W., Hartung J. B., Nagel K., Neukum G., Schneider E. and Storzer D. (1974) Microcraters on lunar samples. In *Proc. Soviet-American Conference on Cosmochemistry of the Moon and Planets*, p. 585–603. NASA SP-370, Washington, D.C.

Fechtig H., (1976) *In situ* records of interplanetary dust particles: Methods and results. In *Interplanetary Dust and Zodiacal Light* (H. Elsässer and H. Fechtig, eds.), p. 143–158. Springer-Verlag, N.Y.

Flavill R. P., Carey W. C. and McDonnell J. A. M. (1977) Lunar surface microscale transportation phenomena: II *Space Research XVII*, p. 617–622.

Flavill R. P. and McDonnell J. A. M. (1977) Laboratory simulations of secondary lunar microcraters from micron scale hypervelocity impacts on lunar rock. *Meteoritics* **12**, 220–225.

Gault D. E. and Heitowit E. D. (1963) The partition of energy for hypervelocity impact craters formed in rocks. *Proc. 2nd Hypervelocity Impact Symposium*, p. 419. Clearning house for Federal Scientific and Technical Information, U.S.A.

Heywood H. (1971) Particle size and shape distribution for lunar fines sample 12057,72. *Proc. Lunar Sci. Conf. 2nd*, p. 1989–2001.

Hoffmann H. J., Fechtig H., Grün E. and Kissel J. (1975) First results of the micrometeoroid experiment S-215 on the HEOS 2 satellite. *Planet. Space Sci.* **23**, 215–224.

Hörz F., Brownlee D. E., Fechtig H., Hartung J. B., Morrison D. A., Neukum G., Schneider E., Vedder J. F. and Gault D. E. (1975a) Lunar microcraters: Implications for the micrometeoroid complex. *Planet. Space Sci.* **23**, 151–172.

Hörz F., Schneider E., Gault D. E., Hartung J. B. and Brownlee D. E. (1975b) Catastrophic rupture of lunar rocks: A Monte Carlo simulation. *The Moon* **13**, 235–258.
Mandeville J.-C. (1977) Microcraters of 12054 rock (abstract). In *Lunar Sci. VIII*, p. 613–615. The Lunar Science Institute, Houston.
Mandeville J.-C. (1978) 12054 Consortium: More on microcraters (abstract). In Lunar and *Planetary Science IX*, p. 690–692. Lunar and Planetary Institute, Houston.
McDonnell J. A. M. (1976) Finite target hypervelocity impact measurements at microscale dimensions—implications for regolith impacts (abstract). In *Symposium on Planetery Cratering Mechanics*, p. 73–75. The Lunar Science Institute, Houston.
McDonnell J. A. M. (1977) Accretionary particle studies on Apollo 12054,58: *In situ* lunar surface microparticle flux rate and solar wind sputter rate defined. *Proc. Lunar Sci. Conf. 8th*, p. 3835-3857.
McDonnell J. A. M., Ashworth D. G. and Flavill R. P. (1975) Lunar crater distributions under solar wind erosion. *Space Research XVI*, p. 951–958. Akademie-Verlag, Berlin.
McDonnell J. A. M., Ashworth D. G., Flavill R. P. and Jennison R. C. (1972) *Proc. Lunar Sci. Conf. 3rd*, p. 2755–2765.
McDonnell J. A. M. and Berg O. E. (1974) Bounds for the interstellar to solar system microparticle flux ratio deduced from 5 years of Pioneers 8 and 9 data over the mass range 10^{-11} to 10^{-13}g. *Space Research XV*, p. 553–563. Akademie-Verlag, Berlin.
McDonnell J. A. M., Flavill R. P. and Carey W. C. (1976) The micrometeoroid impact crater comminution distribution and accretionary populations on lunar rocks: Experimental measurements. *Proc. Lunar Sci. Conf. 7th*, p. 1055–1072.
Morrison D. A., McKay D. S., Fruland R. M. and Moore H. J. (1973) Microcraters on Apollo 15 and 16 rocks. *Proc. Lunar Sci. Conf. 4th*, p. 3235–3253.
Morrison D. A. and Zinner E. (1977) 12054 and 76215: New measurements of interplanetary dust and solar flare fluxes. *Proc. Lunar Sci. Conf. 8th*, p. 841–863.
Stöffler D., Gault D. E., Wedekind J. and Palkowski G. (1975) Experimental hypervelocity impact into quartz sand: Distribution and shock metamorphism of ejecta. *J. Geophys. Res.* **80**, 4062–4077.
Weinberg J. L. and Sparrow J. G. (1978) Zodiacal light as an indicator of interplanetary dust. In *Cosmic Dust* (J. A. M. McDonnell, ed.), p. 75–122. Wiley, N.Y.

Proc. Lunar Planet. Sci. Conf. 9th (1978), p. 2557–2577.
Printed in the United States of America

Miniregoliths I: Dusty lunar rocks and lunar soil layers

G. M. COMSTOCK

CosmoScience Associates
21 Erland Road, Stony Brook, New York 11790

Abstract—Using a detailed Monte-Carlo model for rock surface evolution we have verified that erosion processes alone cannot account for the shapes of the solar flare particle track profiles generally observed at depths of about 100 μm and less in rocks. We demonstrate that the observed profiles are easily explained by a steady accumulation of fine dust at a rate of 0.3 to 3 mm/10^6 years, depending on the micrometeoroid impact rate which controls the dust cover and results in maximum dust thicknesses on the order of 100 μm to 1 mm.

We have derived the commonly used lunar soil track parameters (ρ_{min}, ρ_q, ρ_{med}, N_H/N) in terms of parameters characterizing the exposure of soil grains in the few-millimeter-thick surface mixing and maturation zone which is one form of miniregolith. We present ρ_q vs. ρ_{min} and N_H/N vs. ρ_q plots which allow us to determine the degree of mixing in soil samples and the amount of processing (maturation) in surface miniregoliths. The ratio ρ_q/ρ_{min} is particularly sensitive to the mixing of soils of different maturities and we use it to show that the sampling process often artificially mixes together finer distinct layers, and that ancient miniregolith layers on the order of a millimeter thick are probably common in the lunar soil.

1. INTRODUCTION

The history of exposure of material on any space-exposed surface determines the extent of and correlation among many maturation processes, such as microcrater formation, accreta collection, solar wind implantation and chemical alterations, solar flare particle track accumulation, and glassy agglutinate formation in soil. The details on this exposure history are governed primarily by impact events that affect material within the first few millimeters of the surface (Gault *et al.*, 1974 for soil; Hörz *et al.*, 1974 for rock) and possibly by other surface transport processes such as electrostatic effects (Criswell, 1972; Pelizzari and Criswell, 1978.) We may define, therefore, a zone of a few millimeters thick at space-exposed surfaces in which maturation actively takes place. Except for rock surfaces constantly kept clean, this maturing zone will involve a layer of loose, impacted material which we call a miniregolith. As part of a general study of such miniregoliths we have been investigating the two types of space-exposed lunar surfaces, rocks and soil, using Monte-Carlo calculations together with primary particle track data.

In the case of rock surfaces we seek in this paper to establish the general existence of thin, variable dust coatings. The possibility of occasional loose dust on rocks has usually been accepted, but not studied systematically. The term dust refers to rather fine-grained material that has been kicked, blown, levitated, or otherwise transported from one surface to another and settles loosely or sticks to the surface.

The generally pitted and rounded aspect of rock surfaces exposed on the order of a million years or more indicates that over a long time mass wastage by primary impacts dominates the rock surface morphology on a scale of greater than about a millimeter. Comparison of the density of etchable tracks left by galactic cosmic ray iron nuclei with the concentration of cosmogenic nuclides produced by cosmic ray protons yields an upper limit of 0.5–2 mm/10^6 years for the mass wastage rate (Crozaz *et al.*, 1974.)

Surface exposure history on a scale less than about 1 mm can be studied using the depth profiles of solar flare particle track density. Some examples of depth profiles in rocks and a review of the mechanisms involved can be found in Fleischer *et al.* (1975). These profiles generally show an inverse power law exponent of about 1 to within 10 μm of the present surface, and have been explained as representing an equilibrium between rock erosion and track production (e.g., Crozaz *et al.*, 1971.) However, such shallow equilibrium profiles can be maintained down to 10 μm or less only if the erosion mechanism removes chips on a very fine size scale of less than about 10 μm. Moreover, the track density attained by the equilibrium profile depends on the rate of this erosion and measured profiles imply about 0.2–0.8 mm/10^6 years (see review by Fleischer *et al.*, 1975.)

A difficulty arises when we look for an erosion mechanism acting on a scale of 10 μm or less with this high rate. We will show in the next section that erosion by micrometeoroid impacts alone cannot account for the observed track profiles at depths $\lesssim$100 μm; it generally removes chips that are too large. Atomic sputtering (e.g., McDonnell *et al.*, 1972; McDonnell and Carey, 1975), thermal flaking (Seitz and Wittels, 1971), and other fine-scale mechanisms are at least two orders of magnitude too weak. Moreover, we find no direct evidence in the form of gradual degradation of microcraters for any fine-scale surface erosion of the required rate. These considerations force us to investigate other mechanisms which affect the surface exposure history.

Fine-scale erosion is not the only mechanism which will produce the observed solar flare track profiles. Any process which leads to a distribution of exposure depths roughly uniform on a scale of about 10 μm increments will have the same effect. The most likely such process would be a continual accumulation of small dust grains (note that a single "slab" layer will not work). We know that dust generated by relatively distant events is continually falling on the soil surrounding the rock. It is reasonable to assume that some of it will stick to the rock, however loosely, especially the smaller size fraction. This accumulation of dust would be modified and occasionally knocked off by micrometeoroid impacts and possibly by elecrostatic effects. A fraction of this accreta sticks well enough to be observed after sampling and cleaning (McDonnell, 1977; Zook *et al.*, 1978.)

We have modeled these processes with a Monte-Carlo computer program called MESS (Model for the Evolution of Space-exposed Surfaces) and report the first results in this paper. We shall show that for reasonable values of dust accumulation rates and micrometeoroid fluxes we can reproduce the observed solar flare particle track profiles and densities and account for a wide variation from rock to rock while maintaining intermittent, variable dust coatings of ~100 μm to ~1 mm thickness.

In the case of lunar soil, previous model calculations (e.g., Gault *et al.*, 1974; Duraud *et al.*, 1975; Comstock, 1977a) have included a surface mixing zone, or lunar skin, but this has not been related in a generally useful way to the various statistical track parameters (e.g., minimum, quartile, and median track densities) often used to characterize the measurements in individual soil layers (see, for example, Crozaz *et al.*, 1970; Arrhenius *et al.*, 1971; Bhandari *et al.*, 1973; Fleischer *et al.*, 1974; Crozaz and Dust, 1977; and others.) Price *et al.* (1975) and Goswami *et al.* (1976a) have used track parameters to identify microstratigraphy in impregnated core tubes, which could represent ancient surface maturation zones or miniregoliths. In the third part of this paper we derive typical track parameters in terms of model parameters governing surface exposure. The results allow us to establish quantitative criteria for determining the degree of mixing in individual soil samples and the amount of processing through surface miniregoliths. The evidence suggests that ancient miniregolith layers are common in the lunar soil.

2. How Dusty are Lunar Rocks? Evidence from Track Profiles

Rock model

The rock surface calculations in this paper were made using a new Monte-Carlo computer model which we have developed to serve as a general vehicle for the simulation of surface-correlated processes in any size range. The model uses a bootstrap technique which quickly generates an impacted surface of any desired scale and resolution while faithfully simulating topography and the distribution over the surface of exposure ages and erosion rates, explicitly giving the size-scale dependence of each. This model was first reported by Comstock (1977b).

Our method is to divide the crater population into consecutive size regimes; in the present case we use a factor of 10 in pit diameters: 0.4–4 μm, 4–40 μm, 40–400 μm, and so on. Starting with the smallest physically significant regime, the model is run for each regime long enough to establish the effects of discrete events in that size regime, that is until an equilibrium surface is well established. Each such run tells us how the distribution of surface exposure times, surface erosion, number of surviving craters, and so on develop with time for that regime. At each step the erosion information can be used as input for the next larger regime in order to establish cumulative effects as follows. There is some variable interval of time between simulated events for a given size regime. During that time interval the cumulative effects of smaller-scale regimes are included by assigning to each cell a depth change, or no change, dependent on that time interval and chosen from the distribution of erosion depths found by earlier runs for smaller-scale regimes.

Clearly this procedure can be extended to any size scale without either consuming computer time, needlessly processing the much more numerous smaller events, or ignoring them altogether. The time thus saved can be used to compute other information, such as track accumulation. The limited size range of each regime allows us to reduce the number of cells and to set a uniformly high resolution; in each regime the smallest event considered can cover many cells; 91 in the present model. The total simulated surface area thus scales with the size regime. This bootstrap procedure has an advantage over previous rock surface calculations (Hartung *et al.*, 1973; Hörz *et al.*, 1974) which require the processing of a great many small events before the effects of large events become statistically meaningful. Time limitations then prevent larger events from being considered and computer size limitations prevent much small-scale resolution. Hörz *et al.* (1974) use a cell size of 400 μm square; our smallest cell size used is 0.2 μm diameter and smaller size ranges could easily be considered.

We employ the nominal crater shape shown in Fig. 1 which has a central pit (stippled) and spall zone (shaded). For a pit diameter D_p, the pit depth $H_p = 2D_p/3$, the spall zone diameter $D_s = 4D_p$ at the original surface sloping down to $3D_p$ at the bottom, which has a depth $H_s = 0.4D_p$. The pit is formed by a spherical bottom tangent to a cone with opening angle $\theta_O = 40°$. This shape is based on observed microcraters (Hörz *et al.*, 1974, 1975; Brownlee *et al.*, 1975). Although the central pit generally represents a small fraction of the microcrater volume, we define it carefully because the model records which cells represent pit glass. In this way surviving microcraters may be identified by the same criterion as is used experimentally, that is preservation of the central pit (Hartung *et al.*, 1973; Hartung *et al.*, 1975). This idealized crater shape is translated to the quantized surface grid in such a way as to preserve volume, as shown in Fig. 2 for the smallest event (n = 1), the largest event (n = 11), and an intermediate size (n = 6).

Other innovations are included in the model which we believe enhance its faithfulness is simulating rock surface processes. We use a hexagonal close-packed grid of cells, not the usual rectangular grid. A hexagonal grid possesses

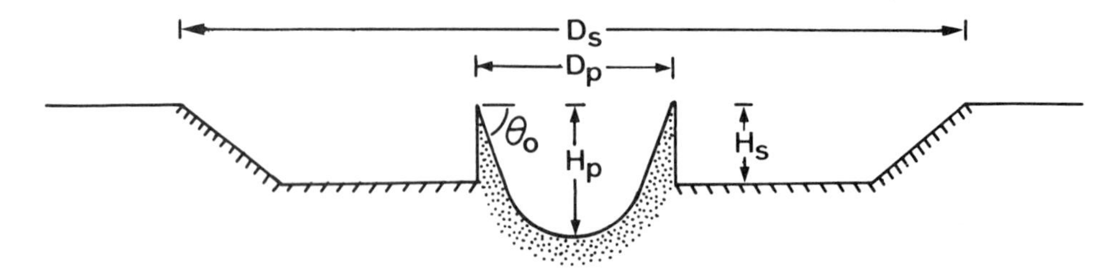

Fig. 1. Idealized cross-sectional shape of microcraters used in rock surface model. Stippled area is glassy pit, shaded area is spall zone which tends to be absent for D_p less than ~10 μm.

the unique property that it has nearly circular symmetry, at least locally, about any cell in the grid (and the noncircularity is easily corrected). This is very efficient for reproducing the circular symmetries inherent in impact events. It also conveniently enables us to divide the spall zone of each event into six independent angular sectors each of which we allow to conform to the average radial slope of the pre-spall surface in that sector while removing local high spots. The result is more convincing topography and reliable results. Studies of microcraters have shown (Hörz *et al.*, 1975) that pits of diameter on the order of 10 μm or less tend to lack spall zones, which decreases their erosion efficiency. In the model all pits of less than 4 μm diameter have no spall zones, all pits of greater than 40 μm have normal spall zones, and between 4 μm and 40 μm pits have a probability ranging from 0 to 1, respectively, of having a spall zone in a given angular sector.

Some impact pit production rates used with the model are shown in Fig. 3. The distribution marked *lunar rock data* is essentially the same curve preferred by Hörz *et al.* (1974), extended to smaller pit diameters based on the measurements of Morrison and Zinner (1977); we use an analytic form given by: $N(>D) = 8.3 \times 10^{-7}D^{-3} + 8.3 \times 10^{-4}(D + 0.005)^{-3}$ where D is the pit diameter in cm and N is the cumulative crater production rate per $cm^2 - 10^6$ years. Some satellite measurements are indicated by solid points for comparison; see Hörz *et al.* (1975) for a complete discussion of satellite measurements and a compilation of references.

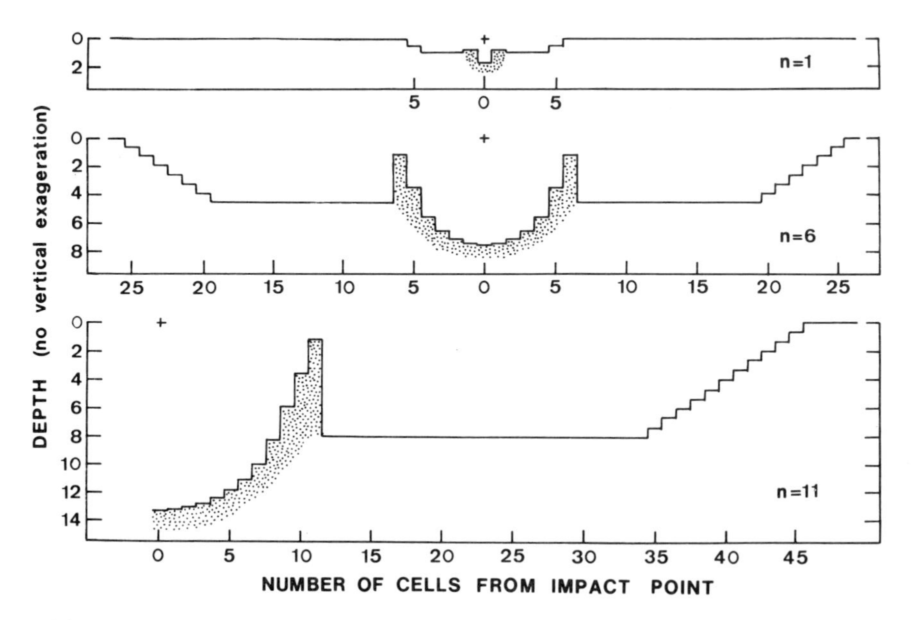

Fig. 2. Examples of microcrater shapes on actual rock model grid. Within a given size-scale regime, n = 1 is the smallest crater considered and n = 11 is the largest, with 11 size steps altogether. The n = 1 crater has an area of 91 cells on the hexagonal grid used.

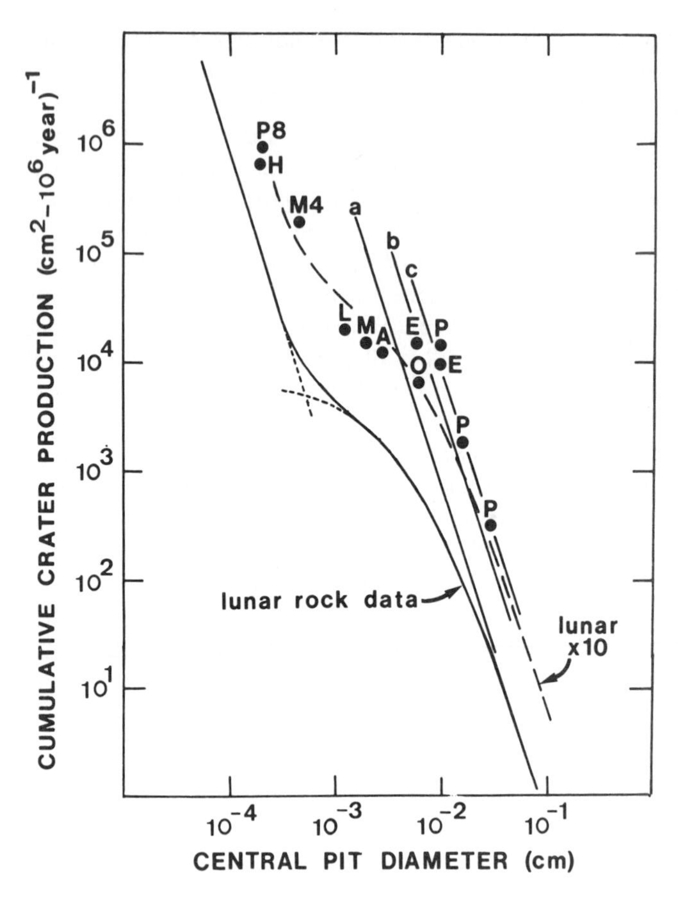

Fig. 3. Cumulative microcrater production curves used in rock surface model. Curve marked *lunar rock data* is from Hörz *et al.* (1974) and Morrison and Zinner (1977). Solid points are satellite measurements: A—Apollo windows; E—Explorer XVI and XXIII; H—HEOS 2; L—Lunar Explorer 35; M—Mariner II; M4—Mariner IV; O—Lunar Orbiter 1–5; P—Pegasus I, II, III; P8—Pioneer 8 (see Hörz *et al.*, 1975).

The model accumulates track densities at sample points assigned on a logarithmic depth scale; if these sample points are eroded away during the course of a simulation run, the computer generates new points by interpolation to maintain the logarithmic grid. In this way the track density profile is accumulated efficiently over a wide depth range. Some track production profiles that have been suggested are shown in Fig. 4 for a semi-infinite medium. Curves *a*, *b*, and *d* are for solar flare particles, curve *c* is due to galactic cosmic rays.

Track profile data

A large number of particle track density vs. depth profiles have been measured in lunar rocks by several groups since the first samples were returned. In order to

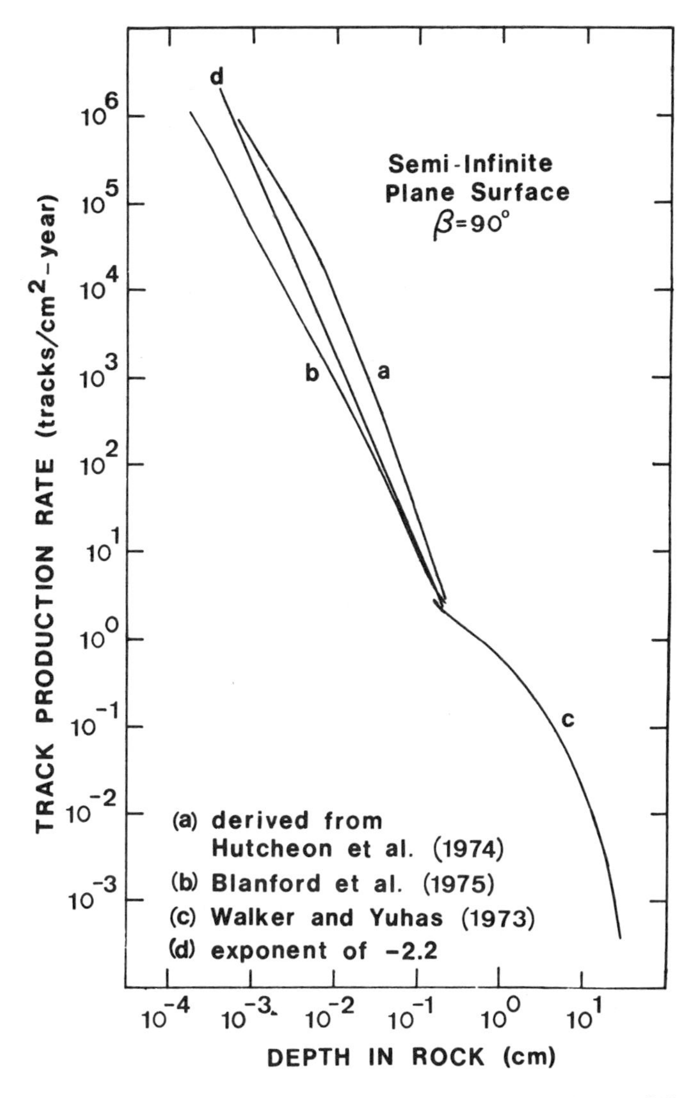

Fig. 4. Iron-group track production rates suggested by lunar samples carefully chosen to reduce the complicating effects of erosion and dust. All of these rates were tried in the rock surface model.

present clearly the relevant features of these data we have plotted in Fig. 5 the track density at a depth of 100 μm vs. the negative exponent (log-log slope) of the depth profile at 100 μm for 31 rocks reported in 21 papers (references are given in Table 1). We have attempted to include in Fig. 5 most of the published data measured in rock crystals; no measurements in glass are included. All data points shown are based on our own fits to published data before authors' corrections for any presumed recent chipping or dust shielding. Errors in track density are generally about 10% and errors in exponent are 10%–20% depending on the smoothness of the profile. Open circles refer to special samples originally chosen to yield the production profile, with exponents of 1.8–2.5. Most rocks show much shallower profiles, varying over a wide range within the parallelogram marked in Fig. 5.

Model results: Clean rocks

Some results of model calculations for impact-saturated rock surfaces kept continuously clean are shown in Fig. 6 by circular points. The most important

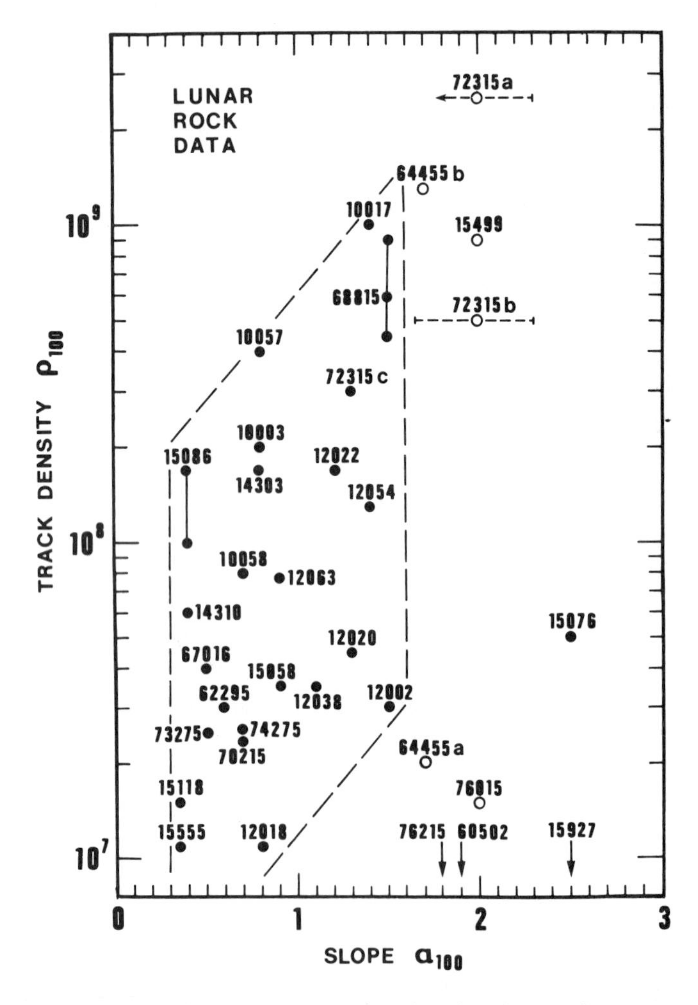

Fig. 5. Measured track density at 100 μm plotted against the negative exponent of the depth profile at 100 μm for 31 rocks reported in 21 papers. References are given in Table 1. Open circles refer to special samples chosen to yield the track production profile.

model parameters are α, the negative exponent of the solar flare track production profile, and ϕ, a meteoroid flux factor equal to the number of impacts per cm^2-10^6 years with central pit diameter greater than 500 μm. The parameter η is a dust accumulation factor which is 0 for clean rocks. The rate $\phi = 5$ refers to the crater production rate in Fig. 3 marked *lunar rock data* and $\phi = 50$ refers to *lunar* × *10*, which fits the satellite data better. The label Model I refers to the shape of these two crater production curves. Exponent $\alpha = 1.8$ represents the shallowest track production profile suggested by lunar data (curve *b* in Fig. 4, from Blanford *et al.*, 1975) while $\alpha = 2.2$ is a more representative model. The dashed parallelogram outlines the normal rock data in Fig. 5. The "error bars" show the total spread obtained in several trial computer runs for a given set of parameter values and represents the variation expected for clean rocks. The dotted arrows connect models of increasing meteoroid flux.

Figure 6 shows that only by taking the shallowest possible track production profile, $\alpha = 1.8$, and a high impact rate can we obtain results falling just within the range of normal rock data, and even this case does not explain most of the impact-saturated rock surfaces measured. Models with $\alpha \geq 2$ lie outside the data range. All clean rock models are found to have the following properties.

Most of the erosion by microcratering is due to the formation of pits larger than about 100 μm diameter. As a result, microcratering can produce enough mass wastage to explain the track profiles at depths $\gtrsim 1$ mm, but does not yield appreciable erosion on a finer size scale. We find that this effect is caused by two independent conditions, either one of which is sufficient to reduce the effect of erosion on the track profile shape at depths $\lesssim 100$ μm. These are: 1) the bend-over in crater production below about 100 μm pit diameter (Fig. 3), and 2) the tendency of pits smaller than ~10 μm diameter to lack full spall zones. The occasional removal of large chips of $\gtrsim 100$ μm size temporarily flattens that track profile, but in the absence of fine-scale erosion the profile relatively quickly regains nearly the production slope at 100 μm, with a smooth transition to an erosion-shaped profile at depths greater than 100 μm. The occurrence of a recent, large chip is too rare an event to explain the abundance of shallow track profiles observed (Fig. 5), especially in view of the fact that track workers generally attempt to measure the steepest track profile on each sample. During most of a computer simulation run the calculated track profile has an exponent at 100 μm depth, α_{100}, of 0–0.5 unit less than that of the track production profile, instead of the 1 unit less predicted for fine-scale erosion equilibrium. Increasing the cratering rate causes a decrease in track density as expected and also a slight decrease in exponent because the profile has less time to recover between chipping events.

Model results: Dusty rocks

In order to produce the kind of universal degradation shown in Fig. 5 we need a mechanism which continually changes the exposure depth in increments of much less than 100 μm. Impacts apparently do not do this efficiently, but fine

Table 1. Track profile measurements in lunar rock crystals.

Rock	Reference	Remarks
10003	Price and O'Sullivan (1970)	
10017	Fleischer *et al.* (1970)	
10057	Crozaz *et al.* (1971)	Average of different regions
10058	Crozaz *et al.* (1970)	Extrapolated from 30 μm to 100 μm
12002	Bhandari *et al.* (1971)	
12018	Bhandari *et al.* (1971)	
12020	Bhandari *et al.* (1971)	
12022	Barber *et al.* (1971)	
12038	Bhandari *et al.* (1971)	
12054	Morrison and Zinner (1977)	Feldspar crystal in glass coating
12063	Crozaz *et al.* (1971)	
14303	Bhandari *et al.* (1972)	
14310	Crozaz *et al.* (1972)	
15058	Bhandari *et al.* (1973)	
15076	Schneider *et al.* (1972)	
15086	Goswami *et al.* (1976b)	
15118	Bhandari *et al.* (1973)	
15499	Hutcheon *et al.* (1972)	Bottom of a surface vug
15555	Bhandari *et al.* (1973)	
15927	Schneider *et al.* (1972)	
60502	Storzer *et al.* (1973)	
62295	Bhandari *et al.* (1973)	
64455(a)	Blanford *et al.* (1975)	Quench crystal, raw data
64455(b)	Blanford *et al.* (1975)	Quench crystal, normalized
67016	Bhandari *et al.* (1973)	
68815	Dust and Crozaz (1977)	Three locations measured
70215	Goswami and Lal (1974)	
72315(a)	Hutcheon *et al.* (1974)	Inside crevice, uncorrected
72315(b)	Hutcheon *et al.* (1974)	Fresh crystal inside crevice
72315(c)	Hutcheon *et al.* (1974)	Outside of boulder
73275	Goswami and Lal (1974)	
74275	Goswami and Lal (1974)	
76015	Crozaz *et al.* (1974)	Inside narrow crevice
76215	Morrison and Zinner (1977)	

dust accumulation can. Lunar rocks reside in a dusty environment which generally experiences a fine-scale accumulation rate of about 4 mm/10^6 years (e.g., Comstock, 1977a). Assuming that only some fine fraction of this will stick, however loosely, to rock surfaces, we may expect an accumulation on rocks of 0.3 to 3 mm/10^6 years. This accumulation would be modulated and occasionally cleaned off by impacts. To evaluate this effect we included in the model a steady accumulation of fine dust described by an efficiency parameter η, defined to be

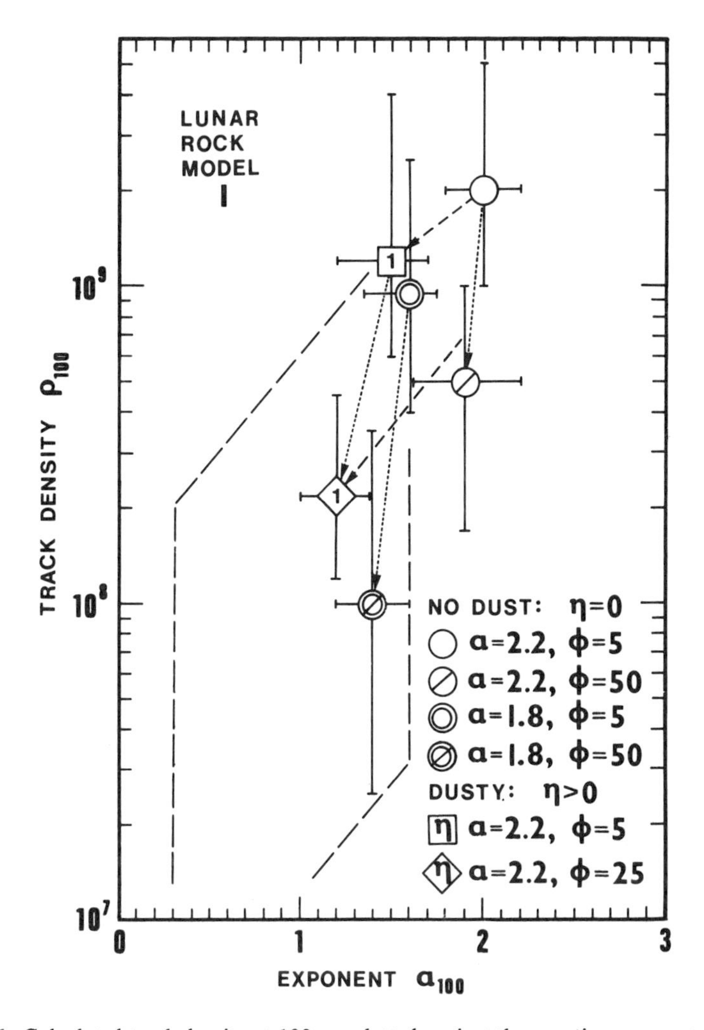

Fig. 6. Calculated track density at 100 μm plotted against the negative exponent of the depth profile at 100 μm for models of impact-saturated rock surfaces which are assumed to be continuously clean (no dust), along with two dusty rock models. η is a dust accumulation rate parameter defined in the text. α is the track production profile exponent. ϕ is the number of impact pits per cm^2-10^6 years with pit diameter >500 μm. The parallelogram encloses the normal rock data in Fig. 5. Dashed arrows show the effect of adding dust accumulation and dotted arrows show the effect of increasing the meteoroid flux level.

the rate of dust accumulation divided by the rate of removal (erosion) due to impacts with central pits of less than 4 mm diameter.

For $\eta = 1$ we find that the dust is kept well in check by impact cleaning, yielding intermittent, variable coverings less than 100 μm thick on a flat surface. Two typical examples of this case are shown in Fig. 6, by the square for a dust accumulation rate of 0.16 mm/10^6 years, and by the diamond for a 10-

times-higher impact rate and a dust accumulation rate of 1.6 mm/10^6 years. The dashed arrows indicate the effect of adding even a relatively small amount of dust with other parameters constant.

If the rock surfaces are indeed usually covered with a dust layer, then this layer will also partially shield the surface from micrometeoroids resulting in a reduction in the apparent rate of smaller impacts. This reduction, or shielding

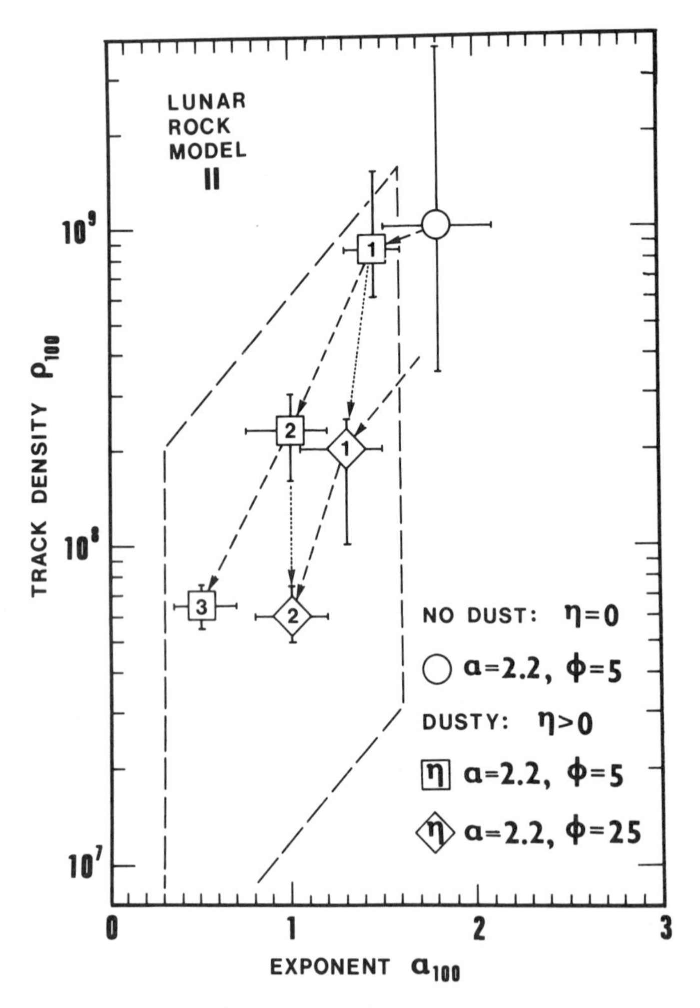

Fig. 7. Calculated track density at 100 μm plotted against the negative exponent of the depth profile at 100 μm for self-consistent models of rock surfaces subjected to a steady accumulation of dust. η is a dust accumulation rate parameter defined in the text. The parallelogram encloses the normal rock data in Fig. 5. Dashed arrows show the effect of increasing the dust accumulation efficiency with fixed impact rate and dotted arrows show the effect of increasing impact and dust accumulation rates proportionally.

factor, should be dependent on crater size for ~10 μm to ~1 mm diameter pits, and dependent on the actual dust-free exposure time for micron and submicron pits. Tracks profiles alone do not help much in determining this factor; a more detailed analysis of the dust layer, the cratering process in a thin dust layer, and the effect of surface condition is needed. A self-consistent model will therefore involve some higher micrometeoroid flux. This could eliminate any need to invoke a time-dependent meteoroid flux as has been suggested by satellite measurements (Fig. 3); but the reader is reminded that there are serious difficulties in interpreting the satellite data (Hörz *et al.*, 1975), and it is not clear how much disagreement there actually is between the lunar and satellite micrometeoroid flux determinations. Finally, since dust accumulation is in competition with removal by impacts, the strongest test case for the dust hypothesis will assume the steepest plausible crater production size distribution.

In order to investigate these effects we have also tried in the model the steeper crater production curves in Fig. 3 marked *a*, *b*, and *c*, which we refer to as Model II. Curve *a* has the form $N(>D) = 8.3 \times 10^{-4} D^{-3}$, where D is the pit diameter in cm; curve *b* is 5 times higher; and curve *c* is 10 times higher. Considering the satellite data in Fig. 3 it is likely that the true crater production curve does have an inflection at 10–100 μm pit diameter, not primarily due to shielding. In fact the model results show that we may tailor curves *a*, *b*, and *c* in Fig. 3 to bend over below 50–100 μm, in order to follow the satellite data, without strongly altering calculated track profiles at 100 μm depth, with or without dust. For example curves *b* and *c* give results similar to the *lunar ×10* production curve.

Figure 7 shows some results for models assuming crater production curves *a* ($\phi = 5$) or *b* ($\phi = 25$), with or without a bend-over below 50–100 μm. These results may be more appropriate for dusty rocks than those shown in Fig. 6. To emphasize the effect of dust we have included in Fig. 7 a computer run with $\eta = 0$ (clean rock) shown by the circle; this point also shows that the bend-over in crater production is not the only factor preventing clean rock models from matching observed track profiles at 100 μm depth. The dust accumulation rates, before modulation by impacts, for the models shown in Fig. 7 are as follows. With $\phi = 5$: 0.3, 0.6, and 0.9 mm/10^6 years for $\eta = 1$, 2, and 3, respectively; with $\phi = 25$: 1.5 and 3.0 mm/10^6 years for $\eta = 1$ and 2, respectively.

As before, $\eta = 1$ yields dust layers only up to ~100 μm thick. For higher η the dust begins to get the upper hand; for $\eta = 3$ only the larger impacts are efficient in cleaning dust off, and the covering can build to 1 mm on a flat surface. The efficiency with which dust will stick to a rock surface should depend on the orientation, roughness, and other surface conditions, so the rate of dust accumulation will vary over a rock and from rock to rock, thus producing a wide variation in track profiles as is observed. Rocks in the lower part of the parallelogram have clearly had a history of partial burial.

We have included in the present dust model only the effects of impacts on removing accumulated dust. Electrostatic effects also may play an important role in both the deposition and the removal of dust from rocks (see Pelizzari and Criswell, 1978, and references therein).

3. Model Derivation of Track Parameters in Lunar Soil Layers

Track parameters

In order to usefully characterize the great amount of particle track density data obtained from many soil samples we need to employ statistical track parameters which are both convenient to measure and readily interpretable. The most widely used parameters are ρ_{min}, the minimum track density; ρ_q, quartile density; ρ_{med}, median density; N_H/N, the fraction of grains with either $\rho > 10^8$ cm^{-2} or a track density gradient; and the fraction with $\rho > 10^8$ cm^{-2} which we denote N_8/N. ρ_{min} clearly can be used to estimate the maximum exposure time for units deposited as a single layer and buried all at once (Crozaz *et al.*, 1970). On the other hand, much of the soil appears to have been deposited and buried in finer increments of less than about 1 mm, so a single irradiation depth cannot be identified and a fine-scale burial rate together with surface mixing have more significance than a single exposure time of a static layer (Comstock, 1977a). The possibility of multiple episodes near the surface (pre-irradiation) also complicates the interpretation of track parameters (see Gault *et al.*, 1974 for further discussion).

In order to interpret the track parameters more quantitatively in such cases we have derived them with a Monte-Carlo soil model developed by Comstock (1976, 1977a) and used there to interpret track density gradient parameters. The model characterizes soil layers in terms of a fine-scale surface burial rate L, generally about 4 mm/10^6 years, and the number of surface exposure episodes, N_{SEE}, during which the soil has been exposed to solar flare particles within 1 mm of the surface while subjected to micrometeoroid gardening. $N_{SEE} = 0$ refers to soil deposited in a layer much thicker than 1 mm and never exposed to solar flare particles; for these cases the track parameters depend simply on layer thickness. The severe depth dependence of impact gardening can result in the upper part of a soil unit, the miniregolith, having $N_{SEE} > 1$ while the lower part has $N_{SEE} = 0$. Hence pre-irradiated soil samples and disturbed units may be complex mixtures of these "pure" models, no longer characterized by a single value of N_{SEE} or L.

Soil model

Our model treats individually all events with a depth greater than 100 μm whether layering or excavation, which roughly balance. The net accumulation of material on a scale smaller than 100 μm is treated as a continuous process which will depend on local topography and hence may vary from sample to sample. Hence a range of values is tried for the fine-scale burial rate L from 1 to 10 mm/10^6 years. A detailed description of this model is given by Comstock (1977a). The cratering event rate distribution used in the model is given by Comstock (1977a, Fig. 6) and is based primarily on the meteoroid flux used by Gault *et al.* (1974). The track production rate used is given by Comstock (1977a, Fig. 8) and is based on the revised energy spectra given by Walker and Yuhas (1973) and by Hutcheon *et al.* (1974). No time variations in either micrometeoroid flux or charged particle flux are assumed. The present model also includes the contribution of galactic cosmic rays as a layer is continuously buried through several centimeters depth. This contribution will vary if a layer is catastrophically buried or re-exposed by relatively larger events; in any case, it is generally much less than the solar flare particle track accumulation, except for soil never exposed within 1 mm of the surface.

The history of emplacement of a particular sample near the surface is governed by one (for fresh soil) or a few relatively rare, large events, hence we expect any distribution and mixture of N_{SEE} values from 0 for immature samples to <50 for mature samples since a purely statistical model can place no strong constraints on what sequence of characteristics we might expect to find in a particular series of

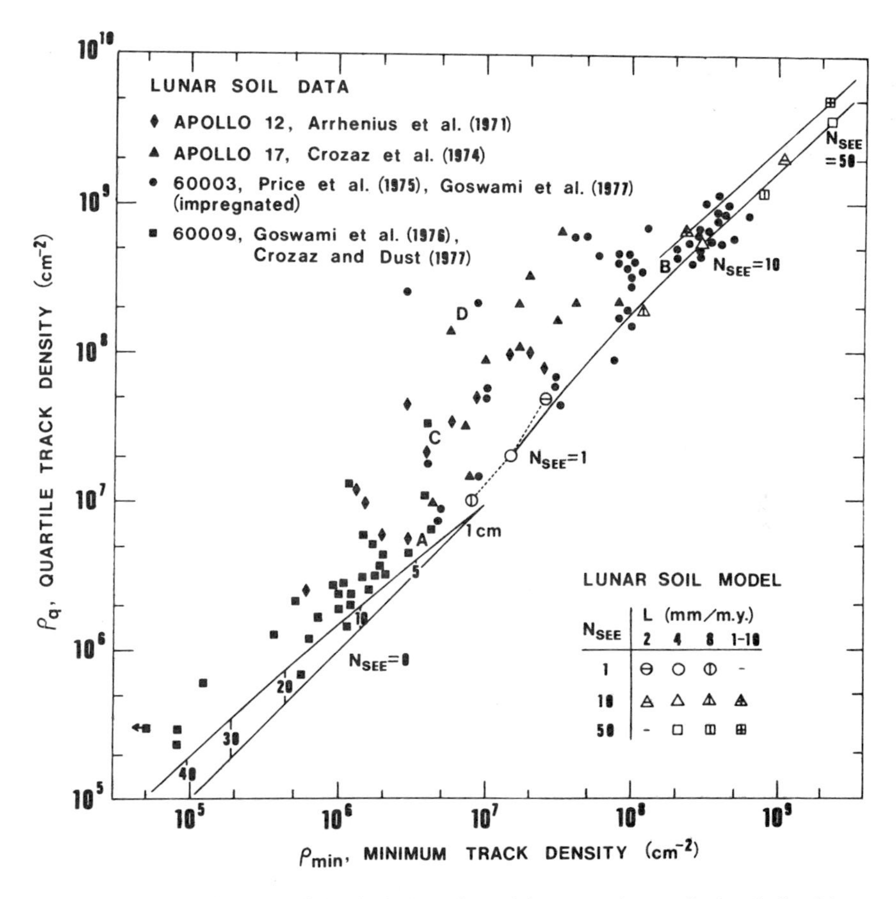

Fig. 8. Results of soil model calculations for minimum and quartile track densities plotted together with selected soil data. Pure models described by single values of surface burial rate L and number of surface exposure episodes N_{SEE} lie near the solid lines, mixtures of these lie off the line. For example, component soils at A and B will have mixtures along line A-C-D. N_{SEE} = O means no solar flare particle exposure, for units of the thickness shown, in cm.

soil samples. Our approach therefore is not to generate possible sequences of simulated soil layers, but to use N_{SEE} as a meaningful track maturation parameter derived with the help of the model from measured track parameters such as ρ_{min}, ρ_q, ρ_{med}, and N_H/N.

Results

The results of these calculations are most usefully presented as correlation plots between measured parameters. Figure 8 shows a plot of ρ_q vs. ρ_{min} calculated by the soil model for the various parameter values indicated, along with some soil data selected to show the wide variation observed among real soil layers. The model points are based on 100 simulated grains of 100 μm diameter; data points

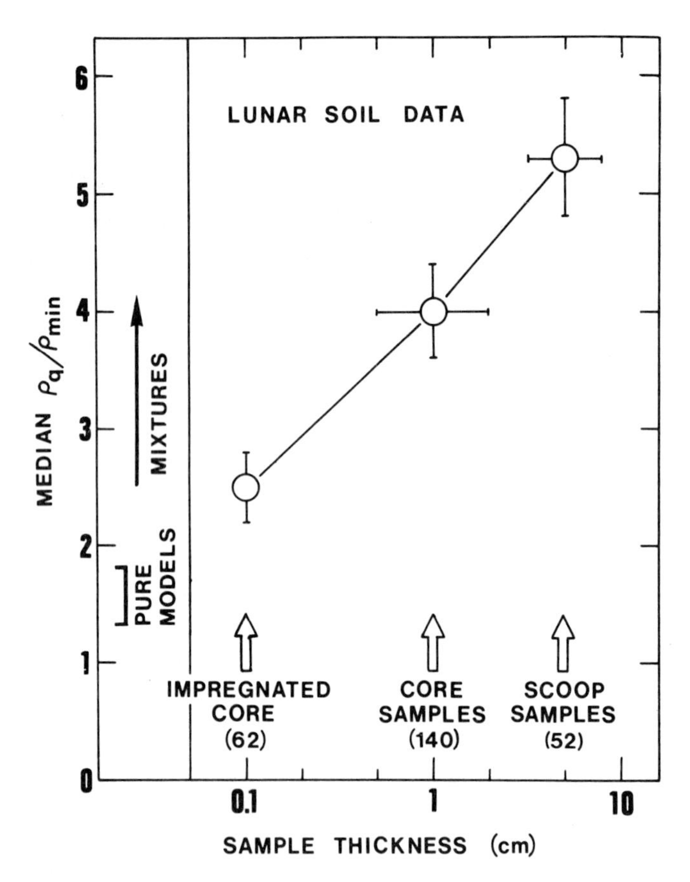

Fig. 9. A plot of the median values of the measured ratios ρ_q/ρ_{min} for most published soil data collected according to sample type, corresponding roughly to thickness sampled. The number of samples in each category is given in parentheses. The range of ρ_q/ρ_{min} values calculated for pure models and for mixtures are indicated for comparison.

are generally based on about 25 measured grains of similar size. Pure models, those represented by single values of L and N_{SEE}, all lie near the solid lines in Fig. 8. The two models marked L = 1–10 mm/10^6 years incorporated a different, random L value for each surface exposure episode simulated. The dart marked N_{SEE} = 0 refers to units of the thickness indicated, in cm, and is calculated for a representative burial rate of L = 4 mm/10^6 years; lower or higher values of L will slide the dart proportionally up or down the diagonal, respectively. The width of the dart represents the difference between sampling the whole thick unit or only the bottom part of it.

In general a soil sample will be some mixture of components with different N_{SEE} values, or maturity, and will lie on a mixing line with ρ_{min} similar to that of the least-irradiated component measured and ρ_q less than that of the most-irradiated component. For example, when a sample with no solar flare exposure from a unit a few cm thick, indicated by the letter A in Fig. 8, is mixed with a highly irradiated sample at point B the mixture will lie somewhere along the line A-C-D and have a much broader distribution of track densities. The data shown

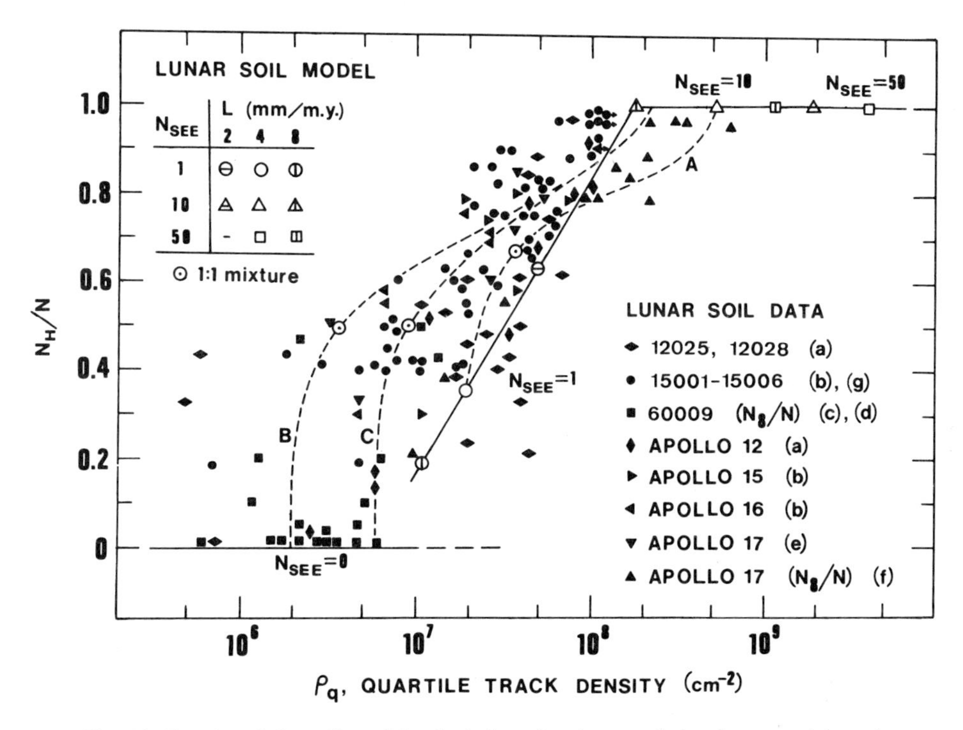

Fig. 10. Results of the soil model calculations for the correlation between the track parameters N_H/N and ρ_q plotted together with soil data. Pure models lie along the solid lines; three examples of mixing lines between two pure model components are shown by dashed lines. (a) Arrhenius *et al.* (1971), (b) Bhandari *et al.* (1973), (c) Goswami *et al.* (1976a), (d) Crozaz and Dust (1977), (e) Goswami and Lal (1974), (f) Crozaz *et al.* (1974), (g) Goswami and Lal (1977).

in Fig. 8 imply that material from thick units are less involved in mixtures.

Individual soil samples are well characterized by their position on a ρ_q vs. ρ_{min} plot, and it is clear which samples are more homogeneous and which are mixtures, and approximately what the components are. Only a fraction of the available data is shown in Fig. 8; we find that each sample site tends to have a particular signature on a ρ_q vs. ρ_{min} plot, so we cannot define an average soil. Plotting ρ_q vs. ρ_{min} emphasizes the lightly-irradiated component of a mixture; we could also plot ρ_{med} vs. ρ_{min} which brings out the more irradiated components.

One interesting result complicates the interpretation of soil samples. When many different samples are plotted on a ρ_q vs. ρ_{min} diagram we find that scoop samples tend to lie further from the pure model line and millimeter-by-millimeter samples from impregnated cores tend to lie the closest to this line. This effect is illustrated best in Fig. 9, where we have separated all available soil data into three categories according to the type of sample, corresponding roughly to thickness sampled: impregnated cores (1 mm intervals), regular core samples (0.5–2 cm intervals), and scoop samples (~5 cm thick) and determined for each category the median value of the ratios ρ_q/ρ_{min} obtained from individual samples. The ratios ρ_q/ρ_{min} calculated for pure models and for mixtures are also indicated for comparison. The range of ρ_q/ρ_{min} for pure models is very small, $1.2 \lesssim \rho_q/\rho_{min} \lesssim 2.0$, because ρ_q/ρ_{min} is nearly independent of L and N_{SEE}. It is also independent of time variations in the track production rate and nearly independent of time variations in the impact rate. ρ_q/ρ_{min} is strongly increased only by mixing. The trend evident in Fig. 9 suggests that the sampling procedure itself has artificially mixed thin layers with different surface exposure histories, implying the general existence of microstratigraphy on a size scale similar to that of the surface maturation zone or miniregolith and with track properties similar to those derived for miniregoliths.

Goswami and Lal (1974) introduced the N_H/N vs. ρ_q diagram in an attempt to isolate evidence for a time variation in the micrometeoroid flux. The result of our calculations deriving N_H/N vs. ρ_q are shown in Fig. 10. Pure models are connected by the solid line and three examples of mixing lines between pure models are shown. This type of plot is somewhat less useful than ρ_q vs. ρ_{min} for two reasons. First, the definition of N_H/N forces the data and models to follow a general trend similar to that shown because the region with both $\rho_q > 10^8$ cm^{-2} and $N_H/N < 0.75$ is forbidden. Secondly, the mixing lines connecting pure models tend to lie close to the pure model line so it is difficult to distinguish mixtures unless one component has had no solar flare irradiation, in which case N_H/N is a good indicator of a small admixture of highly irradiated soil. The spread of data in Fig. 10 is consistent with calculated models and mixing lines, without the need to invoke time variations. Time variations are not ruled out. However, their record is obscured by the complicating effects of mixing.

4. CONCLUSIONS

Our analysis of solar flare particle track profiles in lunar rocks indicates that erosion mechanisms alone are insufficient to account for the observed profiles at depths less than a few hundred microns. A Monte-Carlo model which includes the steady accumulation of dust at rates of 0.3 to 3 mm/10^6 years, depending on sticking efficiency, readily explains the magnitude, slope, and wide variation of observed track profiles. Variable dust layers with thicknesses from 100 μm to 1 mm are expected, depending on sticking efficiency and impact rates. The details of this dust layer and its effects on rock surface exposure need to be studied further; the dust cover is capable of drastically reducing the surface exposure time.

Using a Monte-Carlo soil model we have derived common statistical particle track parameters such as ρ_{min}, ρ_q, and N_H/N in terms of physical parameters characterizing the exposure of soil grains in surface mixing and maturation zones, a form of miniregolith, a few millimeters thick. A plot of ρ_q vs. ρ_{min} is very useful for determining the degree of mixing and surface exposure for individual soil samples; the ratio ρ_q/ρ_{min} is a good indicator of mixing. It is concluded that remnants of ancient surface miniregoliths are common in lunar soil, and often have been artificially mixed together upon sampling.

These results are important for the general study of miniregoliths on space-exposed surfaces, including the correlations among the various maturation indices and the evolution of asteroidal and meteorite parent body surfaces.

Acknowledgments—This work was primarily supported by NASA grant NSG-9013 under the P.I.-ship of Dr. J. B. Hartung, and is being continued under NASA contract NAS9-15582.

References

Arrhenius G., Liang S., Macdougall D., Wilkening L., Bhandari N., Bhat S., Lal D., Rajagopalan G., Tamhane A. S. and Venkatavaradan V. S. (1971) The exposure history of the Apollo 12 regolith. *Proc. Lunar Sci. Conf. 2nd*, p. 2583–2598.

Barber D. J., Cowsik R., Hutcheon I. D., Price P. B. and Rajan R. S. (1971) Solar flares, the lunar surface, and gas-rich meteorites. *Proc. Lunar Sci. Conf. 2nd*, p. 2705–2714.

Bhandari N., Bhat S., Lal D., Rajagopalan G., Tamhane A. S. and Venkatavaradan V. S. (1971) High resolution time averaged (millions of years) energy spectrum and chemical composition of iron-group cosmic ray nuclei at 1 A.U. based on fossil tracks in Apollo samples. *Proc. Lunar Sci. Conf. 2nd*, p. 2611–2619.

Bhandari N., Goswami J. N., Gupta S. K., Lal D., Tamhane A. S. and Venkatavaradan V. S. (1972) Collision controlled radiation history of the lunar regolith. *Proc. Lunar Sci. Conf. 3rd*, p. 2811–2829.

Bhandari N., Goswami J. N. and Lal D. (1973) Surface irradiation and evolution of the lunar regolith. *Proc. Lunar Sci. Conf. 4th*, p. 2275–2290.

Blanford G. E., Fruland R. M. and Morrison D. A. (1975) Long-term differential energy spectrum for solar-flare iron-group particles. *Proc. Lunar Sci. Conf. 6th*, p. 3557–3576.

Brownlee D. E., Hörz F., Hartung J. B. and Gault D. E. (1975) Density, chemistry, and size distribution of interplanetary dust. *Proc. Lunar Sci. Conf. 6th*, p. 3409–3416.

Comstock G. M. (1976) Particle tracks and lunar soil exposure history (abstract). In *Lunar Science VII*, p. 169–171. The Lunar Science Institute, Houston.

Comstock G. M. (1977a) On deciphering the particle track record of lunar regolith history. *J. Geophys. Res.* **82**, 357–367.

Comstock G. M. (1977b) New techniques for simulation of lunar surface processes (abstract). In *Lunar Science VIII*, p. 202–204. The Lunar Science Institute, Houston.
Criswell D. R. (1972) Lunar dust motion. *Proc. Lunar Sci. Conf. 3rd*, p. 2671–2680.
Crozaz G., Drozd R., Hohenberg C. M., Hoyt H. P., Jr., Ragan D., Walker R. M. and Yuhas D. (1972) Solar flare and galactic cosmic ray studies of Apollo 14 and 15 samples. *Proc. Lunar Sci. Conf. 3rd*, p. 2917–2931.
Crozaz G., Drozd R., Hohenberg C., Morgan C., Ralston C., Walker R. and Yuhas D. (1974) Lunar surface dynamics: Some general conclusions and new results from Apollo 16 and 17. *Proc. Lunar Sci. Conf. 5th*, p. 2475–2499.
Crozaz G. and Dust S. (1977) Irradiation history of lunar cores and the development of the regolith. *Proc. Lunar Sci. Conf. 8th*, p. 3001–3016.
Crozaz G., Haack U., Hair M., Maurette M., Walker R. M. and Woolum D. (1970) Nuclear track studies of ancient solar radiations and dynamic lunar surface processes. *Proc. Apollo 11 Lunar Sci. Conf.*, p. 2051–2080.
Crozaz G., Walker R. and Woolum D. (1971) Nuclear track studies of dynamic surface processes on the moon and the constancy of solar activity. *Proc. Lunar Sci. Conf. 2nd*, p. 2543–2558.
Duraud J. P., Langevin Y., Maurette M., Comstock G. M. and Burlingame A. L. (1975) The simulated depth history of dust grains in the lunar regolith. *Proc. Lunar Sci. Conf. 6th*, p. 2397–2415.
Dust S. and Crozaz G. (1977) 68815 revisited. *Proc. Lunar Sci. Conf. 8th*, p. 2315–2319.
Fleischer R. L., Haines E. L., Hart H. R., Jr., Woods R. T. and Comstock G. M. (1970) The particle track record of the Sea of Tranquility. *Proc. Apollo 11 Lunar Sci. Conf.*, p. 2103–2120.
Fleischer R. L., Hart H. R., Jr. and Giard W. R. (1974) Surface history of lunar soil and soil columns. *Geochim. Cosmochim. Acta* **38**, 365–380.
Fleischer R. L., Price P. B. and Walker R. M. (1975) Ancient energetic particles in space. In *Nuclear Tracks in Solids*, p. 307–431. University of California Press, Berkeley, Calif.
Gault D. E., Hörz F., Brownlee D. E. and Hartung J. B. (1974) Mixing of the lunar regolith. *Proc. Lunar Sci. Conf. 5th*, p. 2365–2386.
Goswami J. N., Borg J., Langevin Y., Maurette M. and Price P. B. (1977) Microstratification in Apollo 15 and 16 core tubes: Implications to regolith dynamics (abstract). In *Lunar Science VIII*, p. 365–367. The Lunar Science Institute, Houston.
Goswami J. N., Braddy D. and Price P. B. (1976a) Microstratigraphy of the lunar regolith and compaction ages of lunar breccias. *Proc. Lunar Sci. Conf. 7th*, p. 55–74.
Goswami J. N., Hutcheon I. D. and Macdougall J. D. (1976b) Particle track and microcrater records in lunar samples and meteorites. *Proc. Lunar Sci. Conf. 7th*, p. 543–562.
Goswami J. N. and Lal D. (1974) Cosmic ray irradiation pattern at the Apollo 17 site: Implications to lunar regolith dynamics. *Proc. Lunar Sci. Conf. 5th*, p. 2643–2662.
Goswami J. N. and Lal D. (1977) Particle track correlation studies in lunar soils: Possible long-term periodic fluctuations in ancient meteoritic flux at 1 A.U. *Proc. Lunar Sci. Conf. 8th*, p. 813–824.
Hartung J. B., Hodges F. and Hörz F., Storzer D. (1975) Microcrater investigations on lunar rock 12002. *Proc. Lunar Sci. Conf. 6th*, p. 3351–3371.
Hartung J. B., Hörz F., Aitken F. K., Gault D. E. and Brownlee D. E. (1973) The development of microcrater populations on lunar rocks. *Proc. Lunar Sci. Conf. 4th*, p. 3213–3234.
Hörz F., Brownlee D. E., Fechtig H., Hartung J. B., Morrison D. A., Neukum G., Schneider E., Vedder J. F. and Gault D. E. (1975) Lunar microcraters: Implications for the micrometeoroid complex. *Planet. Space Sci.* **23**, 151–172.
Hörz F., Schneider E. and Hill R. E. (1974) Micrometeoroid abrasion of lunar rocks: A Monte-Carlo simulation. *Proc. Lunar Sci. Conf. 5th*, p. 2397-2412.
Hutcheon I. D., Braddy D., Phakey P. P. and Price P. B. (1972) Study of solar flares, cosmic dust and lunar erosion with vesicular basalts (abstract). In *The Apollo 15 Lunar Samples* (J. W. Chamberlain and C. Watkins, eds.), p. 412–414. The Lunar Science Institute, Houston.
Hutcheon I. D., Macdougall J. D. and Price P. B. (1974) Improved determination of the long-term average Fe spectrum from ~1 to ~460 MeV/amu. *Proc. Lunar Sci. Conf. 5th*, p. 2561–2576.

McDonnell J. A. M. (1977) Accretionary particles: Production and equilibrium of 12054. *Proc. Lunar Sci. Conf. 8th*, p. 3835–3857.
McDonnell J. A. M., Ashworth D. G., Flavill R. P. and Jennison R. C. (1972) Simulated microscale erosion on the lunar surface by hypervelocity impact, solar wind sputtering, and thermal cycling. *Proc. Lunar Sci. Conf. 3rd*, p. 2755–2765.
McDonnell J. A. M. and Carey W. C. (1975) Solar-wind sputter erosion of microcrater populations on the lunar surface. *Proc. Lunar Sci. Conf. 6th*, p. 3391–3402.
Morrison D. and Zinner E. (1977) Microcraters and solar cosmic ray tracks (abstract). In *Lunar Science VIII*, p. 691–693. The Lunar Science Institute, Houston.
Pelizzari M. A. and Criswell D. R. (1978) Lunar dust transport by photoelectric charging (abstract). In *Lunar and Planetary Science IX*, p. 876–878. Lunar and Planetary Institute, Houston.
Price P. B., Hutcheon I. D., Braddy D. and Macdougall D. (1975) Track studies bearing on solar-system regoliths. *Proc. Lunar Sci. Conf. 6th*, p. 3449–3469.
Price P. B. and O'Sullivan D. (1970) Lunar erosion rate and solar flare paleontology. *Proc. Apollo 11 Lunar Sci. Conf.*, p. 2351–2359.
Schneider E., Storzer D. and Fechtig H. (1972) Exposure ages of Apollo 15 samples by means of microcrater statistics and solar flare particle tracks (abstract). In *The Apollo 15 Lunar Samples* (J. W. Chamberlain and C. Watkins, eds.), p. 415–419. The Lunar Science Institute, Houston.
Seitz M. G. and Wittels M. C. (1971) A limit on the radiation erosion in lunar surface material. *Earth Planet. Sci. Lett.* **10**, 268–270.
Storzer D., Poupeau G. and Krätschmer W. (1973) Track-exposure and formation ages of some lunar samples. *Proc. Lunar Sci. Conf. 4th*, p. 2363–2377.
Walker R. and Yuhas D. (1973) Cosmic ray track production rates in lunar materials. *Proc. Lunar Sci. Conf. 4th*, p. 2379–2389.
Zook H., Hartung J. B. and Hauser E. (1978) Loosely bound dust, impact pits, and accreta on lunar rock 12054,54 (abstract). In *Lunar and Planetary Science IX*, p. 1300–1302. Lunar and Planetary Institute, Houston.

Proc. Lunar Planet. Sci. Conf. 9th (1978), p. 2579–2608.
Printed in the United States of America

The melt rocks at Brent Crater, Ontario, Canada*

RICHARD A. F. GRIEVE

Earth Physics Branch, Department of Energy, Mines and Resources
Ottawa, Ontario, Canada K1A 0Y3

Abstract—The impact melt at Brent, 46°05'N; 78°29'W, a simple crater with an original diameter of 3.8 km, occurs as altered glass clasts in allochthonous mixed breccias and as a 34 m thick melt-zone at depths of 823–857 m at the base of the breccia lens. The melt-bearing mixed breccias are concentrated in the upper 160 m of the breccia lens and are similar to the suevitic breccias in the Nördlingen drill hole at the Ries. They, and underlying slightly mixed clastic breccias, are interpreted as material slumped into the cavity following excavation. The melt-zone is inclusion-rich at the top and bottom and the matrix has a modal composition: 70% feldspar, 13% mesostasis sheet silicate, 7% pyroxene, and 3% amphibole, opaques and quartz. Feldspars generally have cores of altered and intergrown albite (>Ab_{92}) and K-feldspar (>Or_{80}), and rims of sanidine (Or_{45-65}). In the coarser grained melt rocks, the cores decrease in size relative to the rims and also contain unaltered potassic-oligoclase (Or_{10} Ab_{65}) to anorthoclase. These complex feldspars are interpreted as shocked xenocrystal country rock feldspars (the albite-K-feldspar cores) that have been partly digested and mantled by melt feldspars (potassic-oligoclase and sanidine).

The melt rocks contain up to 575 ppm Ni and 120 ppm Cr. These elements vary sympathetically and reflect meteoritic contamination. The average Ni/Cr ratio of 2.76 and other siderophile data suggest the bolide was an L-chondrite. Mixing models, with the country rocks and an L-chondrite as components, demonstrate that the melt composition is best matched by a mix of approximately 98.5% mesoperthite gneiss and 1.5% L-chondrite. This agrees with the structural and petrographic data which indicate that the crater was excavated principally in a mesoperthite gneiss unit of the Grenville granodioritic gneiss complex of the area.

High potassium contents relative to the country rocks are associated with samples of gneiss shocked to >20 GPa that occur in the autochthonous basement and as clasts in the breccias and melt-zone. Although these gneisses are spatially associated with melt-bearing lithologies, the melt matrix itself is not enriched in potassium. Potassium enrichment is interpreted as the result of alkali exchange between structurally damaged, shocked feldspars and aqueous solutions during post-impact cooling. The solutions were probably saline, as the Brent event occurred on the edge of a transgressive Ordovician sea.

The principal difference between the melt at Brent and at complex structures is that only a small portion, 1–2%, of the melt occurs in a coherent basal zone at Brent. This has implications for variations in the relative timing of melt motion and cavity collapse with crater size. Other differences, such as alteration, potassium enrichment and siderophile content, are specific to the local conditions of the Brent impact event.

INTRODUCTION

The Brent crater is visible as a circular depression, approximately 3 km in diameter and 60 m deep (Fig. 1a), located at 46°05'N, 78°29'W near the

*Contribution from the Earth Physics Branch No. 726.

northern boundary of Algonquin Provincial Park in south-eastern Ontario. It is considered to have had a pre-erosional rim diameter of the order of 3.8 km and is generally regarded as a type example of a simple hypervelocity impact crater formed in a crystalline target (Dence, 1972). The Brent crater was excavated in an igneous-metamorphic basement complex of mesoperthite and microcline-bearing gneisses of granodioritic composition and minor amphibolites of the Grenville structural province of the Canadian Shield (Fig. 1b). The regional setting of the crater is described in detail in Currie (1971) and Dence and Guy-Bray (1972).

The sub-surface cross-section of the crater has been determined through drilling as roughly parabolic (Dence, 1968, 1973). The drilling was undertaken

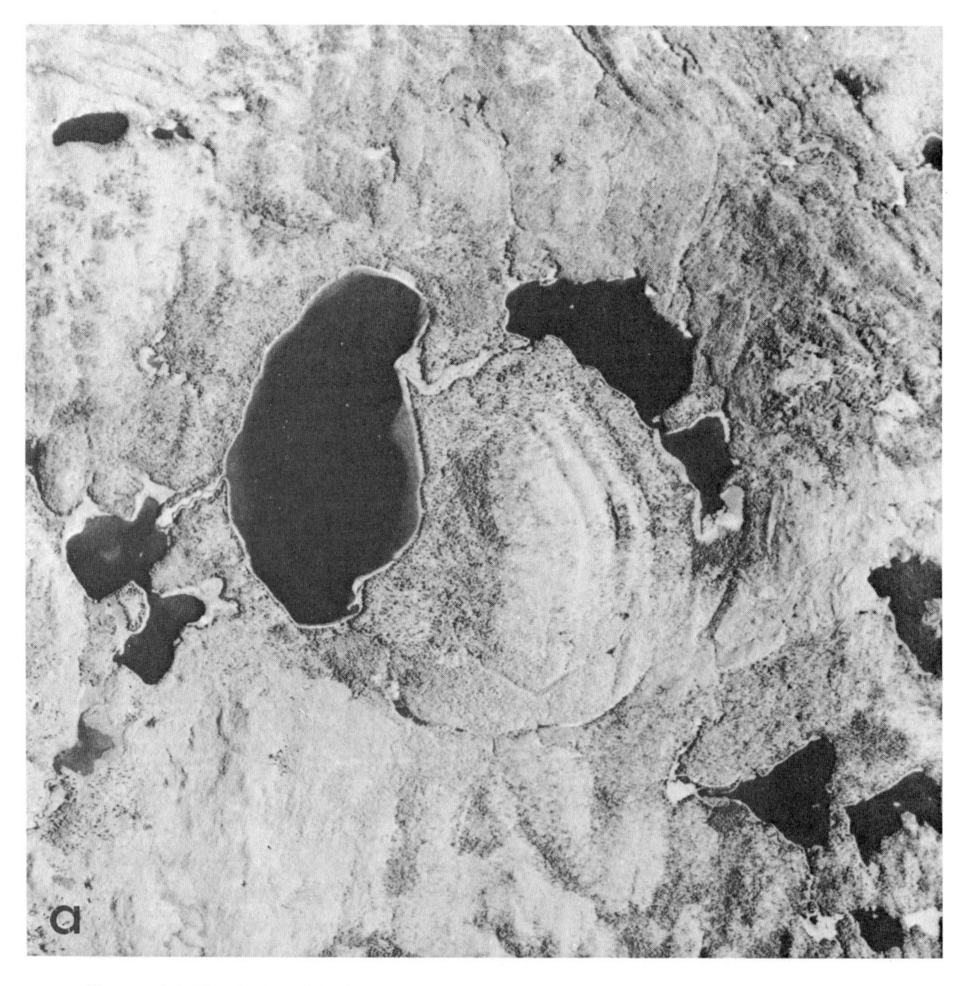

Fig. 1. (a) Vertical aerial photograph of Brent crater. The crater depression contains the arcuate lakes, Gilmour Lake on the west and Tecumseh Lake on the east, and

by the Dominion Observatory (now the Earth Physics Branch) and a total of 5035 m of core was recovered from 12 holes drilled between 1955 and 1967. The central holes, B1-59 and B1-67, penetrated superficial Pleistocene drift, a sequence up to 263 m thick of undeformed Middle Ordovician post-crater sediments (Lozej and Beales, 1975) and a lens of allochthonous breccias with a maximum thickness of 630 m before passing into fractured and shocked autochthonous basement (Dence, 1968). A 34 m section of inclusion-bearing, igneous-textured rock is present at the base of the breccia lens in hole B1-59. This section is referred to as the 'melt-zone' (Dence, 1968, 1971). Concentrations of glassy melt rocks are also encountered in altered, mixed allochthonous breccias, which have similarities to the suevitic breccias of the Ries (Hörz, 1965). Conventional K-Ar analyses of the melt-zone indicate a minimum age for Brent of 414 ± 20 m.y. (Hartung *et al.*, 1971), while an Ar^{40}-Ar^{39} age of 450 ± 9 m.y. is quoted by Lozej and Beales (1975).

Previous studies at Brent have included general topographical, geological and geophysical surveys (Millman *et al.*, 1960) and petrographic analysis of the breccias and underlying basement (Dence, 1964, 1968). Detailed investigations have also been carried out on the quartz and feldspar petrofabrics of the country rocks (Gold, 1968; Aitken and Gold, 1968), the correlation of thermal resistivity

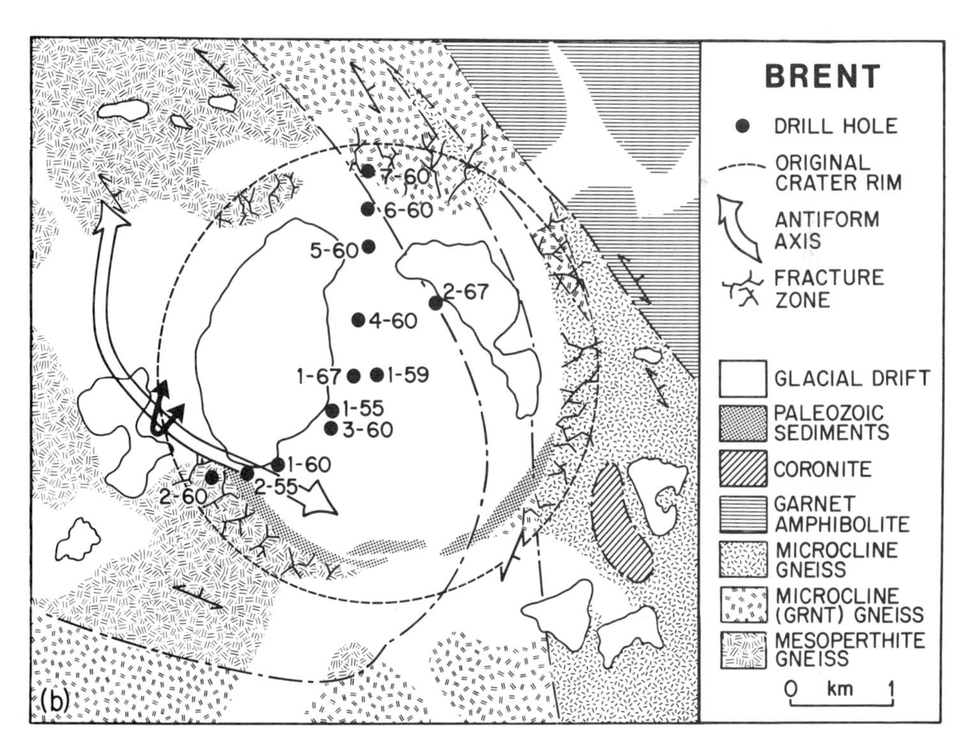

transects the trend of the basement gneisses.(b) Simplified geologic map of the area of Brent crater, after Dence and Guy-Bray (1972). Locations of drill holes are indicated.

with degree of shock deformation (Beck *et al.*, 1976), the shock metamorphism of potassium feldspars (Robertson, 1973) and the effect of shock grade on apparent K-Ar age (Hartung *et al.*, 1971). Data from the Brent crater have been used to define the radial attenuation rate of shock deformation at simple craters (Robertson and Grieve, 1977) and have been applied to the development of general energy-scaling relationships and cratering models of hypervelocity impact events (Dence *et al.*, 1977). Geochemical studies have been undertaken by Currie and Shafiqullah (1967) and Currie (1971). However, unlike the workers quoted above, they interpret the Brent crater as the product of explosive alkaline igneous activity related to the Cambrian igneous center at Callander Bay 80 km to the west.

The melt rocks at Brent are interpreted here as the products of impact melting. As such, they represent a sample of that part of the target shocked to pressures of the order of 100 GPa and greater (Schaal and Hörz, 1977; Hörz, pers. comm.). In recent years, attention has focussed on impact melt rocks because of their importance as a surface lunar lithology and the constraints they provide for the energetic and physical aspects of crater formation (Dence, 1971; Grieve *et al.*, 1974; O'Keefe and Ahrens, 1975; Phinney and Simonds, 1977; and others). Although vestiges of impact melted material are present at virtually all hypervelocity impact structures formed in crystalline targets (Simonds *et al.*, 1977), recent investigations have concentrated on the relatively well-exposed coherent melt sheets found at the larger central-uplift and ring structures (Floran *et al.*, 1978; Grieve, 1975; Marchand and Crocket, 1977; Masaitis *et al.*, 1975).

Impact events on the scale of Brent, with an impact energy of approximately 10^{18} joules (Dence *et al.*, 1977), are at least an order of magnitude more common than those which produce large complex craters. Consequently, although from a relatively small structure, the characteristics of the Brent impactites are important in establishing a comparative base for the interpretation of samples from the cratered terrestrial planets and developing models of the surface evolution of these planets. The melt rocks at Brent constitute one of the best preserved examples of impact melting at a simple crater. However, although the Brent crater is much referenced in the literature, the melt rocks at Brent have not been the specific subject of study. The purpose of this work is to detail the characteristics of the melt rocks, particularly in the melt-zone.

OCCURRENCE

The major sample of the melt rocks available for study is from the core of hole B1-59, which intersects both the melt-bearing mixed breccias of the upper portion of the breccia lens and the basal melt-zone (Fig. 2). In the core from B1-59, approximately 50% by volume of the sampled melt rocks are encountered in the melt-zone. However, the absence of a basal melt-zone in the other holes and the considerably larger radial extent of the mixed breccias at higher levels in the breccia lens (Fig. 2) results in an estimate for the crater as a whole of only

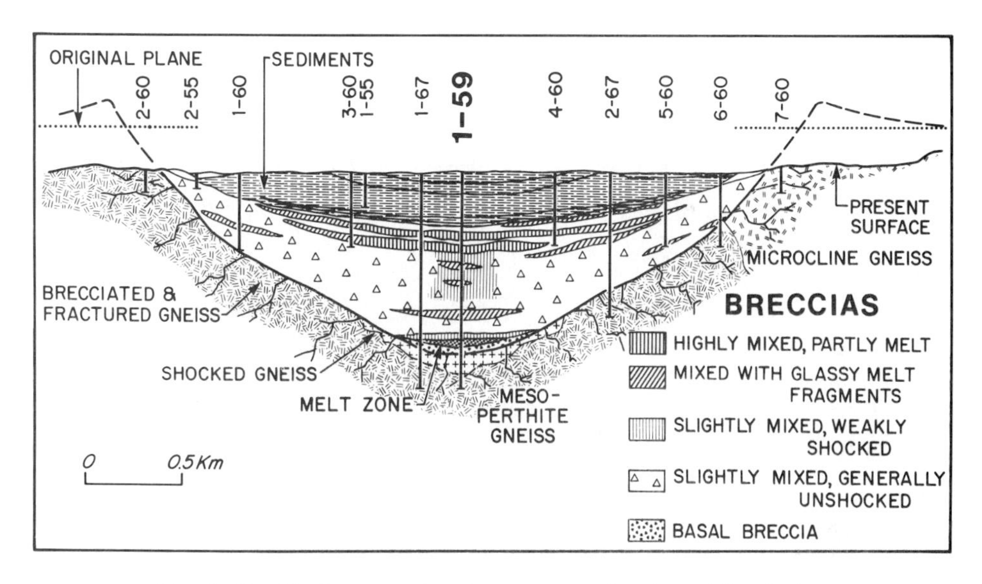

Fig. 2. Simplified cross-section of Brent crater, based on drill hole data.

2% of the melt being present in the discrete basal melt-zone. This figure is comparable with an earlier estimate of 1% by Dence (1971).

The melt rocks in the mixed breccias are generally in the form of inclusion-rich, altered glass clasts (Fig. 3a), ranging in size up to 10 cm with the majority <5 cm. The glass clasts often have contorted shapes and occasionally form a chilled rind to clasts of basement gneiss. These relationships indicate that the clasts were in at least a plastic state when incorporated into the mixed breccias. However, the glass clasts lack distinctive aerodynamic 'fladen' shapes such as found in melt clasts from suevite ejecta at the Ries and Popigai structures (Hörz, 1965; Masaitis *et al.*, 1975). It is believed that the bulk of the melt clasts in the Brent mixed breccias never achieved free-flight (Grieve *et al.*, 1977).

The percentage of glass clasts in the mixed breccias is variable, but is generally higher towards the top of the breccia lens (Fig. 4). In the core from B1-59, the average glass clast content in the mixed breccias is approximately 13% at depths of 264–427 m, decreasing to an average of 7% at 427–606 m*. Associated with this decrease in glass content of the mixed breccias is a decrease in the thickness of individual mixed breccia layers and an increase in the amount of intercalated slightly mixed allochthonous clastic breccias (Fig. 4), which contain only crystalline basement rocks as clasts (Fig. 3b). A similar relationship, of decreasing absolute and relative glass content with depth, has been noted in the 'suevitic' breccias in the core from the Nördlingen 1973 hole drilled within the trace of the transient cavity at the Ries structure (Stöffler *et al.*, 1977). These breccias at the

*A detailed log of hole B1-59 is available from the author on request.

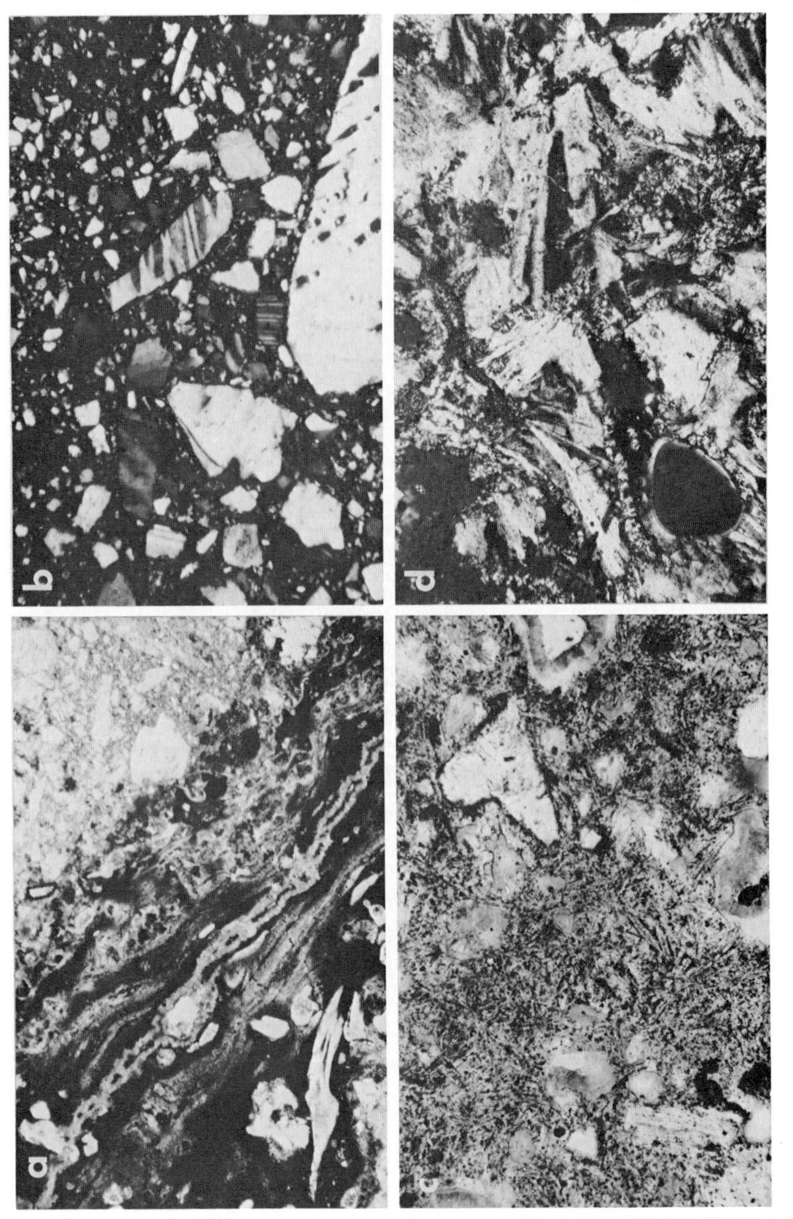

Fig. 3 (a) Photomicrograph of altered glass clast in mixed breccia, 371.6 m Bl-59. Plane light. Width of field of view 4.58 mm. (b) Photomicrograph of slightly mixed clastic breccia, 277.7 m Bl-59. Crossed nichols. Width of field of view 1.33 mm. (c) Photomicrograph of vesiculated melt-rich zone in mixed breccia, 282.5 m Bl-59. Plane light. Width of field of view 1.33 mm. (d) Photomicrograph of vesiculated, highly shocked gneiss clast in mixed breccia, 378.6 m Bl-59. Crossed nicols. Width of field of view 1.33 mm. Clasts, such as illustrated, contain only a single alkali feldspar and show potassium enrichment, see text for details.

Ries and Brent craters have similar glass contents, glass-shapes and structural positions. At the Ries, the lower melt-poor mixed breccias are considered to represent material slumped off the transient cavity wall and the upper melt-rich mixed breccias are fall back material which had insufficient velocity for ejection beyond the rim of the transient cavity (Stöffler, 1977). At Brent less emphasis is placed on the distinction between the melt-poor and melt-rich mixed breccias and the bulk of the mixed breccias is considered to represent slumped material (Grieve *et al.*, 1977). This interpretation is favoured by the observations that the uppermost melt-rich mixed breccia layer is not continuous throughout the crater and the melt-rich mixed breccias are thickest in the center of the breccia lens (Fig. 2). These observations are considered to be consistent with slumping and not with the interpretation that these breccias were simply fall back material.

Locally the mixed breccias have a vesiculated matrix. Six such vesiculated zones, ranging up to 5 m in thickness, occur in the melt-rich mixed breccias in the upper 160 m of the breccia lens in the core of B1-59 (Fig. 4). The central portions of these zones consist of <25% recrystallized feldspar inclusions in a fine-grained matrix of felted feldspar laths up to 0.45 mm in length (Fig. 3c). The vesicles are up to 1.25 cm in diameter and are locally filled with zeolites or less commonly barite.

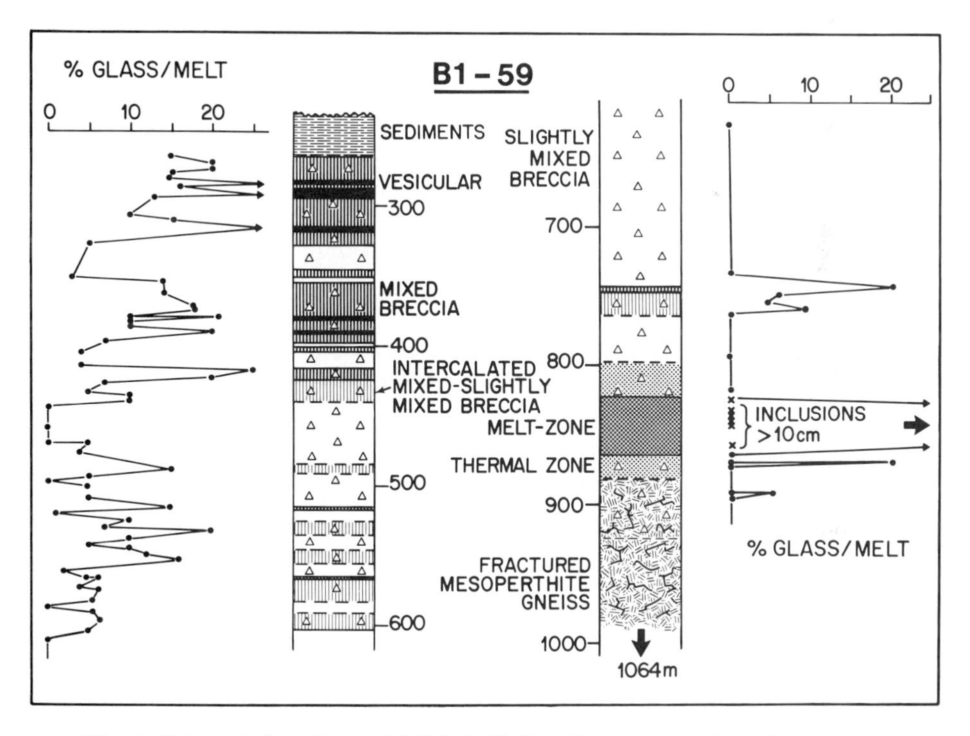

Fig. 4. Schematic log of central drill hole Bl-59 at Brent crater. The variation in melt content in the various lithologies, as determined by modal analysis of thin sections and estimation of percentage glass in the core, is indicated.

Clasts of vesiculated and partially melted gneiss (Fig. 3d) also occur in the upper 160 m and are commonly, but not exclusively, associated with areas of vesiculated matrix. These vesiculated gneiss clasts are heavily altered and recrystallized and are readily distinguishable from the vesiculated matrix by their brick-red color and the absence of inclusions. Textural equivalents to these vesiculated gneisses have been observed at other hypervelocity impact structures (Chao, 1967), nuclear explosion craters (Borg, 1972) and have been produced experimentally at shock pressures of between 4.5 GPa and 6.0 GPa (Stöffler, 1971; Stöffler and Hornemann, 1972).

The melt-zone is intersected at 823–857 m in B1-59 and is separated from the overlying melt-bearing mixed breccias by 213 m of allochthonous clastic gneiss breccia (Fig. 4). The exact radial extent of the melt-zone is not known. It does not occur in the core from B1-67, drilled 200 m west of B1-59 (Fig. 2), its place having been taken by a hornfels zone at depths of 824–840 m. Assuming a maximum radius of 200 m and a parabolic cross-section, the volume of melt in the melt-zone, exclusive of lithic inclusions, is estimated at 1.1×10^{-3} km^3.

PETROGRAPHIC CHARACTERISTICS OF THE MELT-ZONE

Lithic inclusion content of the melt rock in the melt-zone averages approximately 45%. At the top of the melt-zone microscopic lithic inclusions are >50% but by 5 m into the melt-zone have dropped to <18%. Within the bulk of the melt-zone microscopic lithic inclusion content ranges from 16 to <1%. The basal 8 m of melt-zone commonly contains more inclusions than melt, with some sections having as high as 95% inclusions. In addition to microscopic lithic inclusions, a number of megascopic recrystallized xenoliths are included in the melt-zone. They range up to 1 m in length and are generally altered gneisses. However, mafic inclusions of recrystallized amphibolite also occur in the lower 6 m of the inclusion-rich base of the melt-zone.

The melt matrix is microcrystalline near the upper and lower contacts of the melt-zone. Grain size increases rapidly away from the contacts and the basic fabric of the melt is established by tabular feldspars >0.35 mm within 4 m of the upper contact. The average modal composition of the melt matrix is 70% feldspar, 7% pyroxene, 3% amphibole, 3% quartz, 3% opaques, 13% interstitial mesostasis sheet silicate and accessory apatite.

The melt matrix feldspars are complex and have well developed cores and rims. Above 843 m and below 850 m in Bl-59, the feldspar cores consist of a patchy intergrowth of K-feldspar, > Or_{80}, and Na-feldspar, > Ab_{92} (Figs. 5a and 6a). Both feldspars are heavily altered to clay minerals, the albite occasionally has fine polysynthetic twinning, and the entire grain may be a simple twin. Patches of brown mesostasis sheet silicate occur in these cores, particularly in the coarser grained portions of the melt. Averaging individual microprobe analyses of the feldspar phases in the core yields a compositional range for the bulk cores of $Or_{40\text{-}60}Ab_{38\text{-}57}An_{1\text{-}3}$ (Fig. 6a).

At depths between 843 and 850 m the feldspar cores contain, in addition to

K- and Na-feldspar, areas of clear feldspar which range in composition from potassic-oligoclase, approximately $Or_{10}Ab_{65}An_{25}$, to anorthoclase (Figs. 5b and 6b). In the coarsest grained sample, at 844.6 m, the cores are virtually all potassic-oligoclase with only rare core analyses yielding alkali feldspar compositions. Some of the plagioclase cores have polysynthetic twinning, and although relatively unaltered, generally contain patches of brown mesostasis. In some cases the plagioclase cores have a poorly developed 'hopper' structure and may also be outlined by mesostasis.

Throughout the melt zone the rims on the feldspar grains are of sanidine, $Or_{45-65}Ab_{35-55}An_{1-2}$ (Figs. 6a and 6b), with film perthite developed on the outer margins. The rims are free from alteration, except in the areas of film perthite, and contain numerous needles of apatite. The proportion of core to rim is variable, generally decreasing with increasing grain size. At 843 m cores average 75% of the area of individual grains. However, this decreases rapidly as the core develops a potassic-oligoclase phase and by 844.6 m, where the cores contain little or no alkali feldspar, only 36% of the area of individual feldspar laths is core. Below 844.6 m, the percentage of core increases again and by 850 m an intergrown K- and Na-feldspar core with no potassic-oligoclase forms 90% of individual feldspars.

The following hypothesis is presented as a possible explanation for the petrogenesis of these complex melt feldspars. The origin of the altered and intergrown K- and Na-feldspar cores is related to the incorporation of xenocrystal feldspar from the country rock gneisses. Analyses of feldspars in the mesoperthite gneisses below the melt-zone (Fig. 2) indicate that as the melt-zone is approached the composition of both phases in the mesoperthite grains and the plagioclase grains changes towards values of $Or_{40-60}Ab_{40-60}An_{1-2}$ (Robertson, 1973). In the completely recrystallized gneiss at 866.3 m, directly below the melt-zone, the feldspars are microscopically homogeneous and have a composition $Or_{44}Ab_{55}An_{1}$ (Robertson, 1973). This is within the range of the average composition of the two-phase feldspar cores in the melt.

In addition to the compositional correspondence, the variation in the abundance and form of the two-phase cores with increasing melt grain size is considered to support the xenocrystal hypothesis. As melt matrix grain size increases, the two-phase cores become less abundant relative to the feldspar rims and contain more melt mesostasis and more potassic-oligoclase. This is taken as indicating increasing digestion and assimilation, with the mesostasis and potassic-oligoclase cores being melt products. These feldspars are therefore the textural equivalent of the sieved or checkerboard plagioclase xenocrysts partially filled with melt products observed at other impact structures (Carstens, 1975; Grieve, 1975).

In what crystallizes as the coarsest grained melt, virtually all xenocrystal feldspar is digested and the cores are formed by potassic-oligoclase. Growth of this melt phase was rapid; hopper-type crystals developed and some interstitial liquid, now observed as mesostasis sheet silicate, was trapped. These plagioclase cores were mantled by sanidine and the plagioclase was no longer in equilibrium

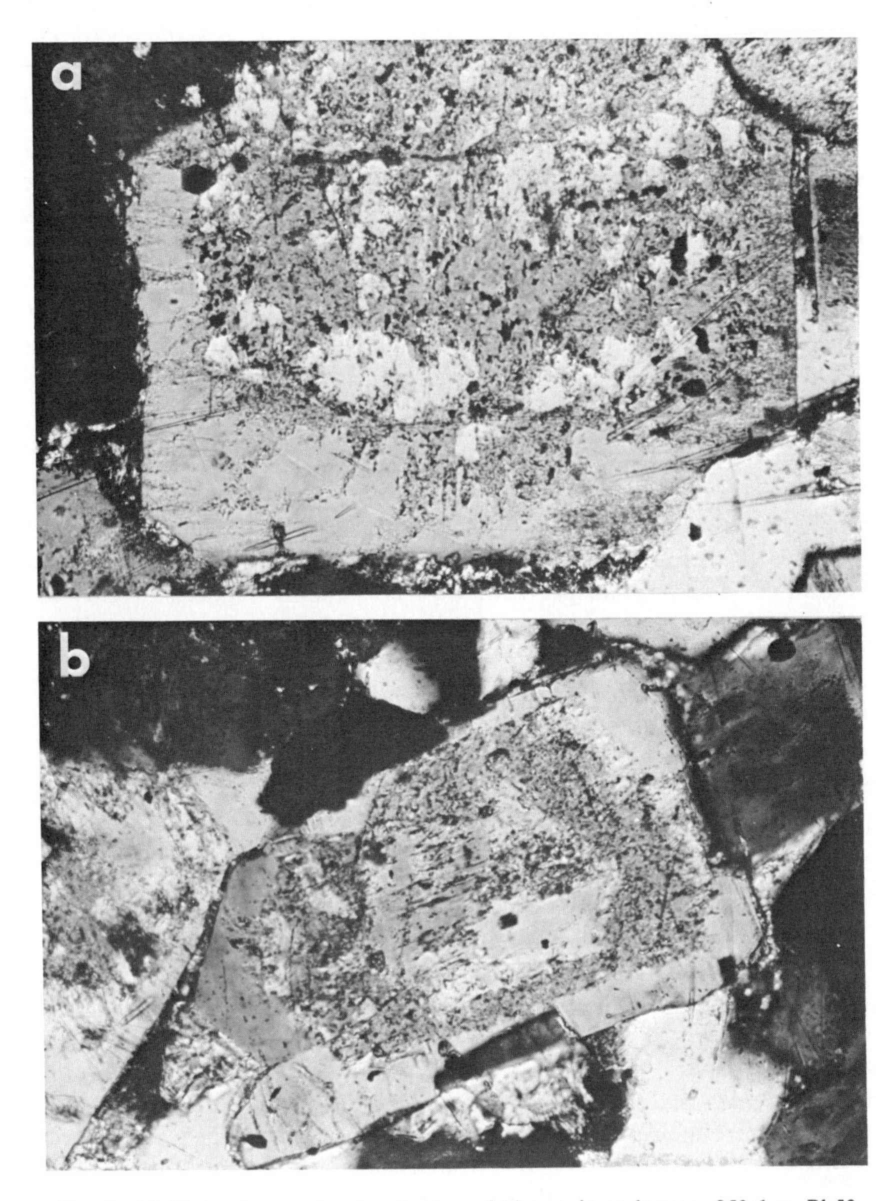

Fig. 5. (a) Photomicrograph of melt phase feldspars in melt-zone, 852.6 m Bl-59. Crossed nicols. Width of field of view 0.53 mm. Feldspars have a dusty, altered core of intergrown albite and K-feldspar and a rim of clear, unaltered sanidine. (b) Photomicrograph of melt phase feldspars in coarser grained portion of melt-zone, 848.5 m in Bl-59. Crossed nicols. Width of field of view 0.53 mm. Feldspars have a core of clear, unaltered potassic-oligoclase to anorthoclase and remanents of the dusty, altered albite and K-feldspar core. Cores are rimmed by sanidine, which may develop marginal film perthite. In coarsest grained melt, 844.6 m in Bl-59, only rare patches of albite and K-feldspar core remain, and core is essentially potassic-oligoclase.

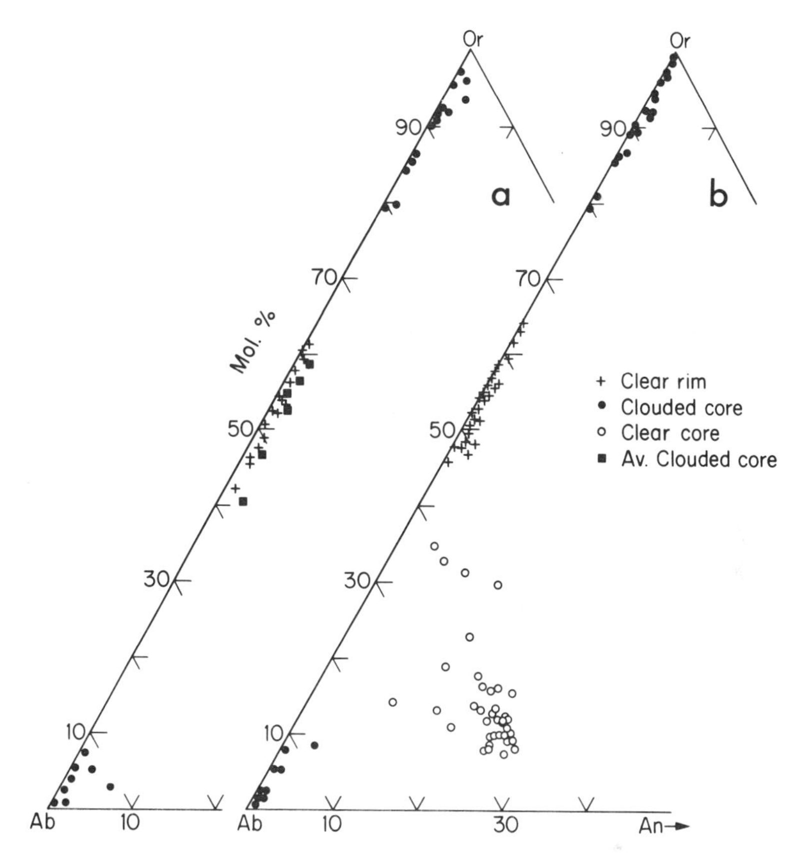

Fig. 6. (a) Composition of intergrown albite and K-feldspar core and sanidine rim in melt feldspars from 841.2 and 842.9 m in melt-zone. (b) Composition of potassic-oligoclase to anorthoclase core, with remanents of intergrown albite and K-feldspar core, and sanidine rims from 844.6 m, 847.6 m and 849.6 m in melt-zone. Albite and K-feldspar are very rare (only 5 analyses) in 844.6 m. See text for details.

with the melt as alkali feldspar continued to crystallize. This situation is similar to that observed in some acid volcanics (Rahman and MacKenzie, 1969). An equivalent mantling by sanidine took place throughout the melt-zone, with sanidine also overgrowing the altered xenocrystal two-feldspar cores.

The origin of the patchy intergrowth of K- and Na-feldspar in the two-phase cores is a further problem. If it is accepted that these cores have a xenocrystal origin, then the original mesoperthite texture has been destroyed. This may have been accomplished by shock before incorporation in the melt, similar to that observed in anti-perthites from shocked anorthosites with incipient maskelynite development at Mistastin and Manicouagan (Grieve, 1975 and unpublished data), or by heat during residence in the melt or by a combination of the two. On cooling of the melt rocks, these homogenized xenocrysts reordered. However, they did not revert to mesoperthite. They probably still contained sub-microscopic areas or domains of original K- or Na-feldspar, as suggested by X-ray studies of experimentally shocked feldspars which indicate internal fragmentation but retention of some short range order to shock pressures of 30 GPa (Hörz and Quaide, 1973). These domains acted as nucleation sites for the reordering of the K- and Na-feldspar phases in the homogenized xenocryst and thus a patchy intergrowth of alkali feldspars developed. Extensive reordering did not take place in the sanidine rims because of the absence of suitable nucleation sites and only marginal film perthite developed.

Pyroxene is generally interstitial to the melt feldspar and is often associated with and altered to mesostasis sheet silicate. The dominant pyroxene is a ferro-augite, $En_{19-27}Fs_{29-37}Wo_{41-44}$ [Table 1 (i)]. A second pyroxene with bright green pleochroism was detected in one sample at 849.6 m. It is a minor constituent, 1.3% of the mode, and is sodic, containing approximately 12.5% of the acmite molecule [Table 1 (ii)].

Table 1. Microprobe analyses of mafic minerals in the melt-zone.

	(i)	(ii)	(iii)	(iv)	(v)
SiO_2	49.36	47.79	47.24	47.05	44.54
TiO_2	0.80	0.41	2.19	1.18	0.02
Al_2O_3	1.84	0.52	0.65	0.65	4.80
Cr_2O_3	0.30	0.12	0.16	0.06	0.04
FeO	17.40	28.55	34.04	38.67	35.93
MnO	0.61	0.76	0.74	0.55	0.55
MgO	8.85	1.93	1.94	0.00	1.12
CaO	20.36	16.16	3.64	0.22	10.54
Na_2O	0.00	2.64	6.67	8.69	0.33
K_2O	0.00	0.00	1.46	0.85	0.76
	99.52	98.86	98.63	97.92	98.63
		Cation proportions			
O=	24	24	23	23	23
Si	7.71	7.94	7.64	7.78	7.22
Al(iv)	0.29	0.06	0.13	0.13	0.78
Al(vi)	0.05	0.04	0.00	0.00	0.13
Ti	0.09	0.05	0.27	0.15	0.00
Cr	0.04	0.02	0.01	0.01	0.00
Fe	2.27	3.97	4.60	5.35	4.87
Mn	0.08	0.11	0.10	0.08	0.07
Mg	2.06	0.48	0.47	0.00	0.27
Ca	3.41	2.88	0.63	0.04	1.83
Na	0.00	0.85	2.09	2.79	0.10
K	0.00	0.00	0.30	0.18	0.16

(i) Pyroxene (En_{26}, Fs_{30}, Wo_{44}), 849.6 m Bl–59; (ii) Sodic pyroxene, 849.6 m Bl–59; (iii) Arfvedsonitic amphibole, 844.6 m Bl–59; (iv) Riebeckitic amphibole rimming arfvedsonite, 844.6 m Bl–59; (v) Fibrous aluminous ferro-actinolite, 849.6 m Bl–59.

The melt rocks also contain two sodic amphiboles and a calcic amphibole. The amphiboles occur in association with each other and areas of mesostasis. Individual amphibole grains have a core of arfvedsonite, pleochroic from violet to yellowish green, and are overgrown by a dark green-blue riebeckite. The calcic amphibole is pale green and fibrous and rims the riebeckite or occurs separately in areas of mesostasis. It is generally less than 1% of the mode and is an aluminous ferro-actinolite. Analyses of the amphiboles are listed in Table 1 (iii)-(v). There are, however, uncertainties in the stoichiometry due to the lack of discrimination between Fe^{+2} and Fe^{+3} in the microprobe technique.

Quartz is anhedral and interstitial. Where in contact with the mesostasis it has a reaction rim and occasionally occurs in a graphic intergrowth with sanidine. The mesostasis is an olive-brown microcrystalline mass of sheet silicate of unknown compositon. The mesostasis sheet silicate may represent altered interstitial glass.

Chemical Characteristics

A large body of published and unpublished analytical data is available from the Brent structure (Currie, 1971; Dence, 1971; Hartung *et al.*, 1971; Hamza, 1973 and others). From these data it is apparent that a number of geochemical anomalies are associated with Brent. Among the problems which are addressed here are: the behaviour of potassium, the anomalous values of Ni and Cr and their variation in the melt-zone and the overall compositional relationship between the melt phase and the country rocks.

(i) *Potassium behaviour*

In drill core from Bl-59, high potassium values are encountered in the upper melt-bearing mixed breccias and in the area of the melt-zone (Fig. 7). Fewer analyses of the mixed breccias are available from the core of Bl-67; however, the same general relationship as in Bl-59 is apparent. In addition, a few high potassium values are encountered above and below the hornfels zone in Bl-67 (Fig. 3, Beck *et al.*, 1976).

On close inspection, it is apparent that the highest potassium values are not associated with the mixed breccias with the highest melt content, that is, the mixed breccias with vesiculated matrices (Fig. 7). Furthermore, high potassium values do not necessarily correspond to sections with relatively high melt content, as measured by the percentage of altered glass in the mixed breccias. The highest potassium values are related to individual samples of highly shocked gneiss fragments in the mixed breccias [Table 2 (i)]. These fragments are generally highly oxidised, with an overall pink to brick-red colour, and are at least partially recrystallized. The potassium content of samples of melt (glass)-bearing mixed breccia may be low [Table 2(ii)] or relatively high. The relatively high values result from the presence of fragments of shocked gneiss in the analysed samples

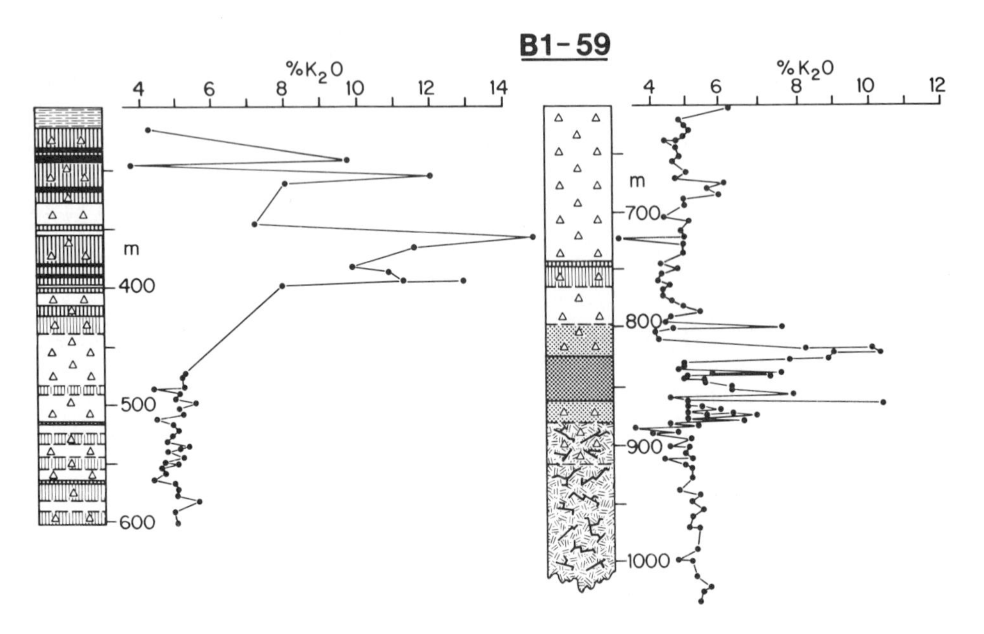

Fig. 7. Variation in potassium content with lithology in Bl-59. Symbols in schematic log are as in Fig. 4. High potassium values are generally associated with highly-shocked, melt-bearing lithologies but do not necessarily correspond to samples with high melt contents. See text for details.

of melt-bearing mixed breccia and from microscopic inclusions in the altered glass or melt phase itself.

Figure 7 indicates that the allochthonous clastic gneiss breccias, which lie between the melt-bearing mixed breccias and the melt-zone (Fig. 4), do not show anomalous potassium contents [Table 2(iii)]. These breccias are only weakly to moderately shocked (Dence, 1968; Hartung *et al.*, 1971) and have experienced pressures of less than 16 GPa, as estimated from the development of planar features in quartz (Robertson and Grieve, 1977). The relationship between potassium content of the gneisses as fragments in the breccias and from the autochthonous basement beneath the melt zone, and shock pressure is shown in Fig. 8. Gneiss samples shocked to less than 20 GPa have approximately 5% K_2O. However, above 20 GPa potassium content rises dramatically and in general increases with increasing recorded shock pressure (Fig. 8).

As stated previously, the general area of the melt-zone is also high in potassium, although the range of potassium values is less than in the breccias (Fig. 7). Petrographic examination of those samples with high potassium indicates that the highest potassium contents are correlated with the presence of crystalline gneiss fragments [Table 2(iv)], which occur as megascopic xenoliths or microscopic inclusions. The variation of potassium in the melt-zone is shown in Fig. 8, where the samples from the melt-zone have been subdivided on the basis

Table 2. Major element composition of specific impact-derived lithologies.

	(i)	(ii)	(iii)	(iv)	(v)	(vi)
SiO_2	64.3	65.0	64.0	64.5	61.6	25.0
TiO_2	0.49	0.83	0.45	0.42	0.73	4.0
Al_2O_3	13.9	14.7	15.1	15.1	15.0	6.9
Cr_2O_3	0.00	0.00	0.00	0.00	0.01	0.07
Fe_2O_3	1.6	–	–	1.9	2.2	3.1
FeO	4.1	7.02*	6.30*	2.4	6.2	9.3
MnO	0.06	0.04	0.09	0.06	0.25	0.50
MgO	2.55	2.95	1.3	1.2	1.0	8.0
CaO	0.20	0.70	2.0	0.9	2.8	20.7
Na_2O	0.9	3.21	3.20	3.0	4.3	0.74
K_2O	9.87	4.2	4.64	8.7	4.8	1.55
H_2O	1.5	n.d.	n.d.	1.5	1.5	n.d.
	99.49	98.65	97.08	99.68	100.39	79.86

*Total iron as FeO.

(i) Shocked and vesiculated gneiss fragment in mixed breccia, 381.6 m Bl–59; (ii) Glass-rich breccia, 264.0 m Bl–59; (iii) Weakly shocked clastic gneiss breccia, 640.0 m Bl–59; (iv) Pink xenolith in melt, 857.1 m Bl–59; (v) Med. grnd. melt with approximately 2% microscopic inclusions, 844.6 m Bl-59; (vi) Alnoite clast in breccia, 458.9 m Bl–59. Data sources: (i), (iv), (v) Earth Physics Branch unpublished data; (ii), (iii), (vi) Currie, pers. comm., 1972; individual analyses used in averages given in Currie and Shafiqullah (1967).

of modal analyses into melt with <10% inclusions, melt with >10% inclusions, and inclusions. These subdivisions have considerable overlap in terms of potassium; however, they do form a general series of increasing potassium content with increasing inclusion content.

All the melt-zone samples have been arbitrarily assigned a shock pressure of 70 GPa in Fig. 8. This pressure value is slightly higher than the minimum required for the onset of melting by shock (Stöffler, 1971) and does not necessarily reflect the shock pressures experienced by the inclusions. The level of original shock damage in the inclusions is unknown due to recrystallization by the thermal effects of the melt. The potassium content of the inclusions is variable and covers the range of values for gneiss fragments in the breccias shocked to pressures between approximately 25 GPa and 45 GPa (Fig. 8). It is a matter of speculation to what extent these pressures mirror the actual pressures to which the inclusions were shocked.

Details of the behaviour of potassium in the melt-zone are given in Fig. 9, which illustrates the general principle that potassium varies sympathetically with percentage of gneiss inclusions. Figure 9 does not indicate a regular numerical relationship. This can be ascribed to differences in the potassium content of individual gneiss inclusions, as indicated in Fig. 9.

It should be noted that there is no extensive concentration of high potassium

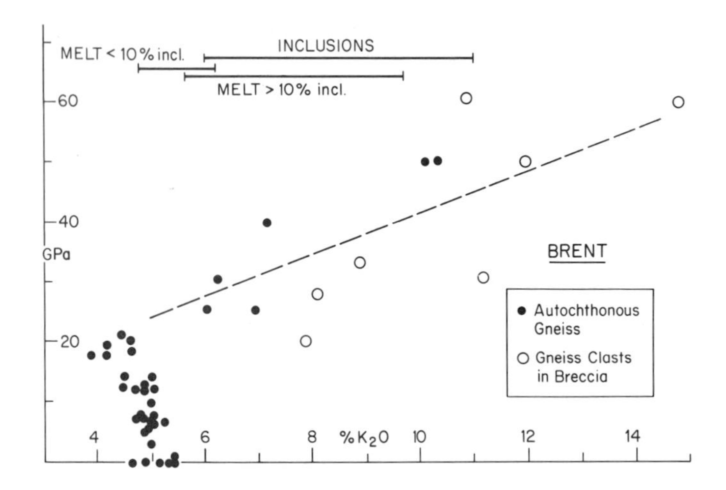

Fig. 8. Variation in potassium content of gneiss clasts and autochthonous basement gneisses with recorded shock pressure, as estimated from the extent of shock deformation (Hartung *et al.*, 1971; Robertson and Grieve, 1977). Also shown is range in potassium content of melt rock and inclusions from melt-zone.

values at Brent. High potassium is in detail a local phenomenon. There is no obvious "igneous" source for the fenitizing fluids required in Currie's (1971) endogenic model for the origin of Brent crater. The melt, a melted potash metasomatite according to Currie (1971), cannot be the source, because where it is relatively free from lithic inclusions, it does not have abnormally high potassium concentrations (Table 2(v); Fig. 9). In addition, the potassium distribution shows no regular gradients in alkali concentration (Fig. 7), such as is observed in areas of established fenitization (Currie and Ferguson, 1971; McKie, 1966).

The question remains as to the source and mechanism of the potassium enrichment. Dence (1971), who did not chemically differentiate between the melt phase and its inclusions, suggested the potassic character of the melt rocks resulted from the action of solutions heated by the residual thermal effects of the impact. This hypothesis has been criticized by Currie (1971), who compared the nature of the mineralogical and chemical changes at Brent with the effects of circulating groundwater in thermal springs and contended that the results were not equivalent. However, the thermal spring analogy is not exact and it is believed that the differences cited by Currie (1971) do not necessitate the rejection of chemical changes through interaction with aqueous solutions.

The general high level of alteration and extensive development of sheet silicates in the impact-produced lithologies at Brent indicate that post-impact cooling took place in an aqueous environment. From a detailed examination of the petrologic character of the sedimentary fill, Lozej and Beales (1975) concluded that the Brent impact event occurred on the edge of a transgressive

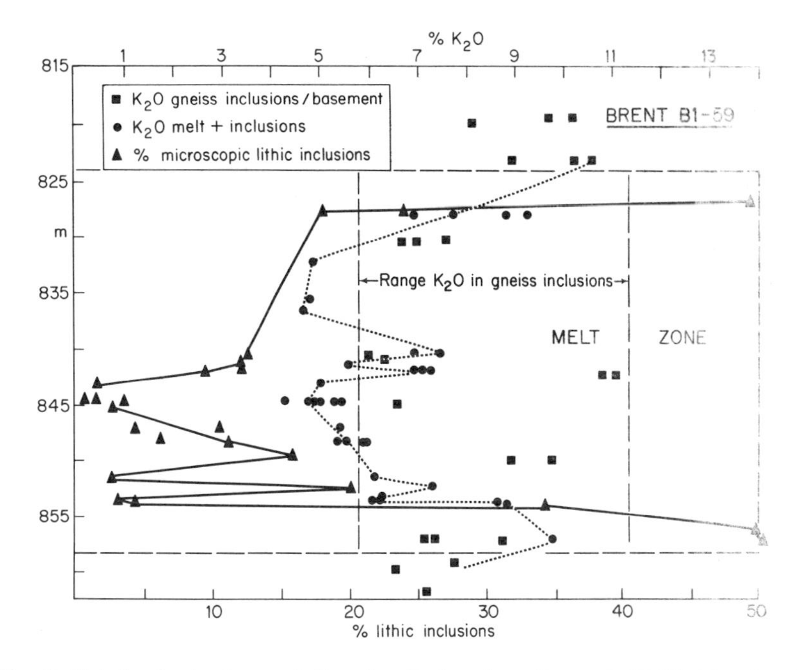

Fig. 9. Details of potassium variation of melt phase in melt-zone in Bl-59. Also plotted are potassium content of megascopic inclusions and percentage microscopic lithic inclusions in melt-zone. Note general sympathetic variation in potassium and percentage inclusions.

tropical sea and the crater was inundated almost immediately. Thus, it is likely that fluids present during post-impact cooling were not simply vadose waters but were at least in part saline. These waters would contain alkali chlorides and thus provide a source of potassium.

The enrichment of oceanic basalts in potassium upon hydration by sea water is a well documented phenomenon (Hart, 1969, 1970). A similar process through the development of clay minerals may account for some of the potassium enrichment at Brent (Hart, 1969; Dasch *et al.*, 1973). However, unlike the case of oceanic basalts there is an increase in the K/Rb ratio at Brent with increasing potassium content (630–1059 in the altered gneisses compared to 260–450 in the relatively unaltered autochthonous basement). It is also obvious from the petrographic data that simple alteration is insufficient to account for the extreme levels of potassium enrichment.

Thermodynamic and experimental data indicate that potassium will replace sodium in feldspar in the presence of alkali-bearing solutions at metamorphic temperatures (Orville, 1963; Helgeson, 1969; Lagache and Weisbord, 1977) and it is suggested here that the principal mechanism of potassium metasomatism involves alkali ion exchange between the feldspar phases and aqueous solutions. Such a process with potassium replacing sodium in plagioclase, resulting in a complementary enrichment in bulk rock potassium and reduction in sodium and calcium, is indicated by microprobe analyses of the feldspars from the autochthonous gneiss below the melt zone. There is a progressive change, as the melt zone is approached, in the composition of the albite and K-feldspar phases in the mesoperthite feldspar to compositions in the range Or_{40-60} and equivalent changes in the composition of discrete plagioclase grains, originally with compositions $Or_{1-2}Ab_{93-98}An_{2-8}$(Robertson, 1973). For example, in the gneiss at 880.3 m the discrete plagioclase grains have been replaced by a feldspar with Or_{51} and the two phases in the mesoperthite have compositions of Or_{59} and Or_{54}, respectively (Robertson, 1973). The gneiss itself contains 6.9% K_2O. The original modal percentage of plagioclase in the mesoperthite gneisses is 15–20% (Currie, 1971) and a crude calculation indicates that an increase of approximately 7.8% K_2O (equivalent to Or_{51}) in what were discrete plagioclase grains is equivalent to an increase of 1.5% K_2O in bulk rock potassium content. This is comparable to the potassium enrichment of this gneiss sample over the average autochthonous basement. There are no data on the bulk composition of the mesoperthites and it is not known what changes in bulk composition, if any, took place in the alkali feldspars.

If potassium enrichment is due to plagioclase replacement then the bulk composition of the gneiss fragments with extremely high potassium contents in the mixed breccias requires that replacement in these fragments is virtually complete. As an example, the high-potassium, vesiculated gneiss fragment listed in Table 2(i) contains 60.5% normative K-feldspar and only 8.9% albite plus anorthite. This is in keeping with petrographic and preliminary x-ray data, which indicate that these high potassium gneisses contain only a single alkali feldspar and no plagioclase.

Unlike the experimental case (Orville, 1963), the degree of replacement by potassium at Brent was not simply a function of decreasing tempertaure. The initial post-shock temperature of the vesiculated gneiss fragments in the mixed breccias was as high as 1000°C, whereas the autochthonous gneisses below the melt had post-shock temperatures of less than 250°C (Stöffler, 1971). As the gneisses cooled, their feldspars reacted with aqueous solutions. However, although all the gneisses cooled to approximately the same ambient temperature, the end products were not the same. The highly shocked gneisses underwent more complete replacement suggesting that reaction continued to lower temperatures (Orville, 1963) and/or the replacement reaction was not equivalent for all levels of shock damage.

It is not clear whether the threshold of potassium enrichment is controlled by shock pressures in excess of 20 GPa or post-shock temperatures above approximately 200°C (Fig. 8). However, the inclusions in the melt-zone all had the same thermal history after initial rapid thermal equilibration with the melt phase (Simonds *et al.*, 1978a), but have an extensive range of potassium values. Some of this variation may be due to differences in mineralogy with some inclusions containing less feldspar than others. However, it is a well documented observation that impact melts contain lithic inclusions with a wide range of shock damage (Dence, 1971; Grieve, 1975; Floran *et al.*, 1978). Therefore, it is suggested that this range in potassium values is a reflection of the original variation in the degree of shock damage.

In summary, it is believed that potassium enrichment arose from alkali exchange between feldspars and saline aqueous solutions during post-impact cooling. The onset and degree of potassium replacement is a function of the degree of shock damage (Fig. 8), with the lower limit of detectable enrichment corresponding to pressures of approximately 20 GPa. Presumably, as structural damage increased with increasing shock pressure, the degree of replacement increased, with the reaction rate being the most rapid in those gneisses where the feldspar was converted to glass (Orville, 1972). Potassium enrichment occurs in the inclusions in the melt-zone but not in the melt matrix itself. The feldspars which crystallized from the melt had no structural damage and were therefore relatively unreactive, as were the feldspars in the unshocked or weakly shocked autochthonous gneisses and clastic breccias.

(ii) *Ni and Cr anomalies*

Samples from the melt-bearing mixed breccias and the melt-zone have up to 575 ppm Ni and 120 ppm Cr. These values are well in excess of those encountered in endogenic igneous rocks with equivalent major element compositions (Krauskopf, 1967). These melt-bearing samples are also enriched in Ni and Cr compared to the allochthonous clastic gneiss breccias and the autochthonous gneiss basement (Fig. 10). Amongst the country rocks at Brent, only the relatively rare alnoite dikes (Currie, 1971) have comparable Ni and Cr values. The intersection in the drill core of a fragment of alnoite is responsible for the peak in Ni (327–360 ppm) and Cr (350–500 ppm) values at 459 m in Bl-59 (Fig. 10). However, the alnoite is chemically distinct from the high Ni and Cr melt-bearing lithologies in both its major [Table 2 (vi)] and other trace element contents. For example, the alnoite contains 330 ppm V as opposed to an average content of 18 ppm V in the melt-zone.

Details of the Ni-Cr distribution in the melt-zone are given in Fig. 11, where it is apparent that there is a considerable range in values and that Ni and Cr vary sympathetically. Although less dramatically, total iron as FeO also tends to mirror the distribution of Ni and Cr. Neutron activation analyses of two country rock samples and three samples from the melt-zone indicate that the melt is also enriched in other siderophile trace elements, in particular those which are depleted in terrestrial rocks but enriched in most meteorites (Palme *et al.*, 1978). The source of the enrichment in Ni, the other siderophiles and Cr is therefore considered to be meteoritic in origin (Grieve and Dence, 1978). The concentration of these elements in the melt-bearing lithologies at Brent is consistent with general models of crater and melt formation in a hypervelocity impact event, in which the melt represents a preserved portion of that part of the target which was closest to the impacting body (Grieve *et al.*, 1977). That is, the melt rocks are the most likely candidates to have an appreciable amount of meteoritic contamina-

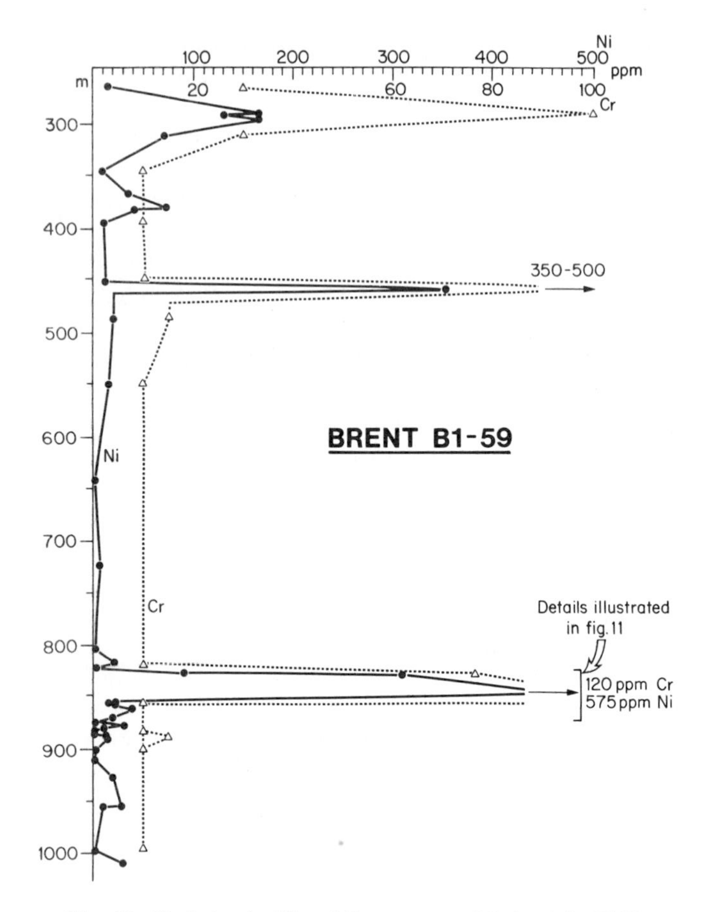

Fig. 10. Variation in Ni and Cr contents with depth in Bl-59.

tion. This is in keeping with data on siderophile enrichment in the impactites at craters where the character of the meteorite is known from recovered fragments (Morgan *et al.*, 1975).

Some of the variation in Ni and Cr in the melt-zone can be ascribed to the gneiss inclusions. The autochthonous gneisses average 10 ± 8 (1 s.d.) ppm Ni, whereas the gneiss inclusions in the melt-zone have slightly higher values, averaging 35 ± 14 ppm. This may reflect a small percentage of the melt phase in the analyzed samples of gneiss inclusions or represent the actual Ni content of the shocked inclusions, with the higher than average Ni due to direct contamination by meteoritic "vapor" condensed on the relatively cool inclusions or due to introduction into the gneiss inclusions of Ni from the melt. Assuming that the gneiss inclusions originally contained 10 ppm Ni, the Ni content of the melt rocks exclusive of inclusions can be adjusted for the effect of megascopic and microscopic inclusions. This helps to reduce but does not remove the Ni variation

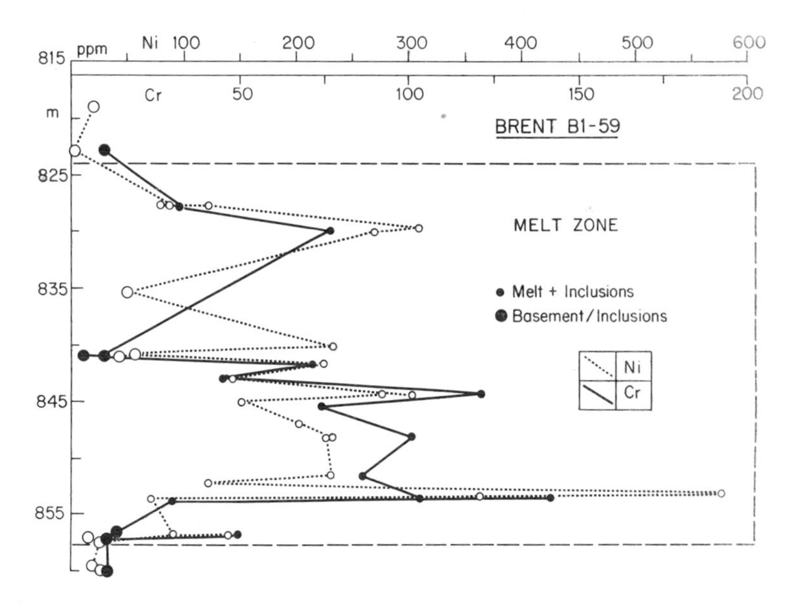

Fig. 11. Details of Ni and Cr variation in melt-zone of Bl-59. Note association of low Ni and Cr with gneiss inclusions and mutual variation of Ni and Cr in melt-phase.

within the melt. The "adjusted" average Ni content of the melt matrix is 253 ± 133 ppm, with the bulk of the melt rocks having values in the range 150–300 ppm. Towards the base of the melt-zone, Ni values increase to a maximum of 592 ppm. However, local variations of 200 ppm, occurring within <25 cm, remain (Fig. 11).

The "adjusted" Ni/Cr ratio of the melt matrix is 2.76 ± 0.36, within the range of values given for chondritic meteorites (Mason, 1971). If it is assumed that the Ni/Cr ratio of the melt rocks is dominated by a meteoritic component then this value suggests that the impacting body was an L-chondrite. This is consistent with the siderophile trace element data, which indicate a relative siderophile abundance pattern for the melt rocks equivalent to that of L-chondrites (Palme *et al.*, 1978).

(iii) *Compositional relationship with country rocks*

In general, the composition of impact melts is similar to the average composition of the country or target rocks in the area of the impact structure (Dence, 1971; Masaitis *et al.*, 1975). In some cases it has been shown that the melt rocks correspond, within the compositional uncertainty of both their major and trace elements, to mixtures of specific country rock units (Grieve *et al.*, 1977 and references listed therein). Accordingly, the compositional correspondence between the melt and country rocks at Brent has been tested. The procedure followed utilizes least-squares mixing models and has been described in detail in Grieve and Floran (1978) and Grieve (1978).

Of primary concern in the mixing models is the choice of geologically and chemically representative rock components. In the case of the country rocks at Brent this is a relatively straightforward procedure. Unlike structures such as Manicouagan, Quebec (Floran *et al.*, 1978; Grieve and Floran, 1978) and Mistastin, Labrador (Grieve, 1975; Marchand and Crocket, 1977) extensive drilling has provided petrographic data on the third dimension of the Brent structure. From these data it is apparent that the bulk of the crater was excavated in the mesoperthite gneiss unit (Figs. 1b and 2). This is confirmed by the overwhelming dominance of this lithology as lithic clasts in the allochthonous breccias. According to the structural data the mesoperthite gneiss was in the center of the target area at the time of impact and therefore should dominate the composition of the melt (Dence, 1971; Grieve *et al.*, 1977).

There is some interlayering of the various gneiss lithologies (Currie, 1971), so for completeness the other gneiss units were also considered as potential melt components. These units, following the petrographic subdivisions given by Dence and Guy-Bray (1972), are microcline-biotite ± garnet gneiss and garnet-amphibolite gneiss. The microcline-bearing gneiss occurs on the east side of the crater (Figs. 1b and 2) and contains a few lenses of garnet-amphibolite (Dence and Guy-Bray, 1972). Representatives of the alnoite dikes of Cambrian age (Hartung *et al.*, 1971) occur as rare fragments in the breccias and are also a possible component of the melt. The final component considered is the meteorite. From the Ni/Cr data presented previously and the siderophile data of Palme *et al.* (1978) an L-chondrite was chosen as best representing the composition of the impacting body.

A number of analyses of the country rocks are given in Currie (1971). However, the petrographic subdivisions are not equivalent to those in Dence and Guy-Bray (1972) and the exact locations of the samples are not known. In an effort to resolve these problems, Currie's (1971) data were subjected to cluster analysis (Davis, 1973) in the same manner as previously employed for the country rocks at the Manicouagan structure (Grieve and Floran, 1978). It was found that in the case of the quartz-feldspar gneisses the chemical groupings did not correspond exactly to the petrographic subdivisions of Currie (1971) or the subdivisions of Dence and Guy-Bray (1972), when the best estimate of the locations of Currie's samples was plotted on the Dence and Guy-Bray geologic map. Therefore, the analytical data used to define the country rock compositions were restricted to samples where, if possible, exact locations were known and petrographic thin sections were available. This essentially limited the data base to drill hole samples. Care was taken to ensure that the gneiss samples selected did not exhibit shock deformation equivalent to recorded shock pressures in excess of 20 GPa; that is, recorded shock pressures were below those apparently required for the onset of potassium enrichment (Fig. 8).

Only in the cases of the mafic gneiss and alnoite units, which were chemically distinct in the cluster analysis, was it necessary to supplement the data with analyses of samples which did not have corresponding petrography. The average compositions of the country rocks, grouped according to their petrography into Dence and Guy-Bray's (1972) units, are given in Table 3 along with an average composition for an L-chondrite, as calculated from the data in Mason (1971). An average composition for the melt matrix is also listed in Table 3. In order to minimize the effects of alteration and abnormal potassium values, the melt analyses were restricted to samples with <10% recognizable lithic inclusions from the coarsest-grained section of the melt-zone in B1-59.

The results of the mixing model calculations, for both the major and trace elements and the trace elements only, are given in Table 4. The closest compositional correspondence between the melt matrix and potential components at Brent is achieved in both cases by a mix of mesoperthite gneiss and a small amount of L-chondrite (Table 4). Other mixes involving other lithologies were a poorer match and all had higher reduced chi-squares than those listed in Table 4. The mixing models are consistent with the composition of the impacting body, as suggested from the siderophile data of Palme *et al.* (1978) and the structural and petrographic data on the type of country rock in the center of the target area at the time of impact (Figs. 1b and 2).

The reduced chi-square for the mixing model involving both major and trace elements is higher than those calculated for the model impact melts at the Lac à l'Eau Claire and Manicouagan structures (Grieve, 1978; Grieve and Floran, 1978). As can be seen from the trace element-only model (Table 4), which requires the same components in essentially the same proportions, the imbalance is principally in the major oxides. This may be due to the natural variability of the mesoperthite gneiss unit, which is weakly layered with variations in mafic content, particularly garnet, on the scale of metres to tens of metres. The differences in the major oxides of the model and observed melt matrix are, with the exception of CaO, within one standard deviation of the average composition of the mesoperthite gneiss used in the mixing calculation (Table 3). That is, they can be accommodated within the natural variability of the mesoperthite gneiss unit. The sense of the differences is consistent with the addition of garnet, and it may be that the portion of the target melted on impact was slighty more garnet-rich than the mesoperthite gneiss composition used in the model. It should be noted that there are no major differences in the alkalis between the model and observed melt compositions (Table 4). This is inconsistent with Currie's (1971) hypothesis involving fenitization. An alternative or additional reason for the slight differences in the major oxides may be the alteration of the melt rocks, particularly the extensive development of mesostasis sheet silicate, which may have resulted from the alteration of an original interstitial glass phase.

Summary and Concluding Remarks

The impact melt rocks at the Brent simple crater have the same general compositional relationships to their target lithologies as the extensive melt sheets at recently studied large complex structures. They differ principally in their

Table 3. Composition of lithologies used in Brent mixing models.

	Mesoperthite gneiss		Microcline gneiss		Mafic gneiss	
	$\bar{X}$	s.d.	$\bar{X}$	s.d.	$\bar{X}$	s.d.
SiO_2	63.30	4.43	69.83	1.46	42.20	2.52
TiO_2	0.91	0.81	0.44	0.12	1.84	1.60
Al_2O_3	15.01	1.44	15.03	1.90	15.83	4.28
FeO*	6.52	2.60	3.80	0.69	14.79	4.67
MnO	0.20	0.15	0.06	0.03	0.27	0.20
MgO	1.18	0.88	0.80	0.50	8.32	3.73
CaO	1.78	0.81	1.15	0.22	6.31	1.90
Na_2O	4.61	1.36	3.94	1.24	2.27	0.81
K_2O	4.92	0.20	4.20	1.09	2.20	1.57
	98.43		99.25		94.03	
			Trace elements, ppm			
Cr	12	5	10	0	62	30
Ni	10	9	7	4	17	6
Zr	900	660	570	234	271	214
Sr	136	59	90	—	297	265
Rb	136	48	122	—	47	30
Zn	120	38	67	31	140	28
Ba	938	280	1133	573	460	412
V	20	18	13	6	290	195
Cu	16	14	9	6	7	7
Co	8	3	7	3	50	17

*Total iron as FeO.

Mesoperthite gneiss from ≥911 m in Bl–59, 592 m and 599 m in Bl–67 and surface samples DB–78A–61 and DB–186–66; n = 10 for major oxides, except K_2O where n = 45; n ranges from 7 to 20 for trace elements.

Microcline gneiss from 754 m and 902 m in Bl–59 and DB–5B–61 on surface.

Mafic gneiss from 1013 m in Bl–67 and 1297 and 1293 on surface (Currie, 1971, Table 1).

higher degree of alteration. Enrichment in potassium over normal target rock compositions is present at Brent, but it is concluded here that this enrichment is not directly related to the occurrence of the melt or fenitization due to alkaline igneous activity. Potassium enrichment is confined to gneiss clasts and autochthonous basement, which have been shocked to pressures in excess of 20 GPa, and is the result of post-impact alkali-exchange between feldspars with a high degree of structural damage and aqueous solutions. These solutions are the source of potassium and are responsible for the general high degree of alteration of the Brent impacities. The dramatic enrichment in potassium and alteration at Brent relative to other Canadian impact structures is the result of the physical environment at the time of impact, with the Brent structure being formed on the edge of a transgressive Ordovician sea. The significance of an apparent threshold shock pressure of 20 GPa for potassium enrichment is not known and requires

Table 3. *(cont'd.)*

	Alnoite		Impact melt		L-chondrite
	$\overline{X}$	s.d.	$\overline{X}$	s.d.	
SiO_2	37.80	0.14	60.15	1.90	38.06
TiO_2	3.37	0.74	0.65	0.11	0.11
Al_2O_3	11.25	0.35	15.65	0.87	1.91
FeO*	10.70	0.67	7.97	0.46	28.06
MnO	0.24	0.02	0.31	0.21	0.32
MgO	10.50	2.82	0.88	0.26	25.18
CaO	11.35	3.04	2.80	0.30	1.79
Na_2O	1.80	0.03	4.58	0.66	0.87
K_2O	4.02	0.98	5.07	0.22	0.10
	91.03		98.06		96.40
		Trace elements, ppm			
Cr	424	107	79	38	3800
Ni	308	81	219	85	11767
Zr	300	—	825	275	13
Sr	664	223	161	55	10
Rb	40	15	161	47	3
Zn	1200	—	375	361	58
Ba	500	—	854	386	4
V	300	—	18	7	65
Cu	70	42	13	7	90
Co	62	22	14	4	550

Alnoite from 485.9 m in Bl-59, 316.7 m in B7-60 and 1519 and 1520 on surface (Currie, 1971, Table 9).

Melt from 843–845.3 m in Bl-59, n = 4 for major oxides, except K_2O n = 7 and Na_2O n = 5. Values for Ni, Cr, Co from 843–845.3 m with n = 4, all other trace elements are averages of all melt analyses in melt-zone with n ranging from 4 to 11.

Individual analyses and data sources are available on request from author.

further investigation. However, it may be related to the onset of the breakdown of short-range order and the conversion of feldspar to diapletic glass at pressures in excess of 20 GPa (Hörz and Quaide, 1973; Stöffler and Hornemann, 1972). Data from other impact structures with alkali enrichment or structures believed to have been formed in a "wet" environment should be examined in order to test the validity and general application of the conclusions presented here for the Brent structure.

If the physical characteristics of the impactites at Brent are typical of a simple crater, then major differences exist between the occurrence of melt rocks within simple and complex craters formed in a crystalline target. At Brent the melt rocks are not concentrated in a coherent basal sheet, but occur principally as glass clasts in the breccias. Only 1–2% of the melt rocks within the cavity remain as a coherent basal body. A possible explanation for this difference in the degree

Table 4. Mixing model calculations.

	Observed melt	Model 1	Model 2
SiO_2 (%)	60.15	62.97	–
TiO_2	0.65	0.90	–
Al_2O_3	15.65	14.84	–
FeO*	7.97	6.80	–
MnO	0.31	0.20	–
MgO	0.88	1.49	–
CaO	2.80	1.78	–
Na_2O	4.58	4.56	–
K_2O	5.07	4.86	–
Cr (ppm)	79	61	71
Ni	219	165	194
Zr	825	888	886
Sr	161	134	134
Rb	161	134	134
Zn	375	119	119
Ba	854	925	923
V	18	20	21
Cu	13	17	17
Co	14	15	16
	Reduced χ^2	2.07	0.27
	Components, %		
Mesoperthite gneiss		98.7	98.4
L-chondrite		1.3	1.6

*All iron as FeO.
Model 1—major and trace elements.
Model 2—trace elements only.

of melt concentration has been outlined in Grieve *et al.* (1977) and involves variations in the type and timing of the collapse of the excavated cavity between simple and complex craters. It is believed that a simple crater is a relatively more stable structure and that cavity collapse through wall slumping does not occur until the bulk of the melt is in motion up and outward along the cavity wall. In a complex structure, collapse is by base failure and occurs earlier in the cratering sequence with a larger portion of the melt remaining within the cavity and spreading out as a basal sheet along the already collapsing and flattened cavity floor. Further development and testing of this hypothesis is anticipated through theoretical modelling and on-going geological and geophysical studies of the relationships between shocked autochthonous basement, breccias, and melt rocks at complex structures in crystalline targets such as Lac à l'Eau Claire (Clearwater Lake) and Lake St. Martin (Simonds *et al.*, 1978b).

The melt-bearing mixed breccias at Brent are concentrated towards the top of the breccia lens and are essentially identical in character to the suevitic breccias penetrated by the Nördlingen 1973 drill hole within the inner crystalline ring at the Ries structure. At the Reis the upper melt-rich suevitic breccias are interpreted as fall back material. At Brent the equivalent lithologies are considered to be material slumped off the cavity wall during collapse. This conclusion is reached in part through the observations that the melt-bearing mixed breccias at Brent are not laterally continuous across the crater and that they are thickest in the center of the breccia lens and virtually absent close to the apparent crater rim. Although the differences in the interpretation of the physical conditions of formation of these breccias at the Ries and Brent structures may be partially semantic, the conclusions from Brent, if correct, relegate to a minor role, material falling back into the excavated cavity from a central ejecta cloud as a mechanism for providing crater-fill products at simple craters.

Although not a direct result of this study, the identification of the impacting body as an L-chondrite (Palme *et al.*, 1978) is significant. It is supported by the data presented here and is further evidence for Brent as an impact structure and for the general relationship between hypervelocity meteoritic impact and the production of shock metamorphic effects. The variability in the degree of meteoritic contamination within the melt-zone, even when allowance is made for dilution by target rock xenoliths, is appreciable and contamination generally increases towards the base. This observation has implications for models of the relative motions and mixing of the meteorite and target materials during a cratering event on the scale of Brent. Whether it also applies in events of a larger scale, which produce complex craters, is an open question. However, the variability in the degree of meteoritic contamination does emphasize the need for well-planned and adequate sampling in studies specificially undertaken to identify the nature of the bolide at terrestrial impact structures.

Acknowledgments—I would like to thank M. R. Dence and P. B. Robertson for useful discussions and the assistance of their experience on the Brent crater. Much of the collation of the basic chemical data was performed by P. B. Robertson. The paper has profited from reviews by F. Hörz, O. B. James and C. H. Simonds.

REFERENCES

Aitken F. K. and Gold D. P. (1968) The structural state of potash feldspar: A possible criterion for meteorite impact? In *Shock Metamorphism of Natural Materials* (B. M. French and N. M. Short, eds.), p. 519–530. Mono Book Corp., Baltimore.

Beck A. E., Hamza V. M. and Chang C. C. (1976) Analysis of heat flow data - correlation of thermal resistivity and shock metamorphic grade and its use as evidence for an impact origin of the Brent Crater. *Can. J. Earth Sci.* **13**, 929–936.

Borg I. Y. (1972) Some shock effects in granodiorite to 270 kilobars at the Piledriver site. Amer. Geophys. Union Monograph No. 16, p. 293–312.

Carstens H. (1975) Thermal history of impact melt rocks in the Fennoscandian Shield. *Contrib. Mineral. Petrol.* **50**, 145–155.

Chao E. C. T. (1967) Impact metamorphism. In *Researches in Geochemistry* (P. H. Abelson, ed.), p. 204–233. Wiley, N. Y.

Currie K. L. (1971) A study of potash fenitization around the Brent Crater, Ontario, a Paleozoic alkaline complex. *Can. J. Earth Sci.* **8**, 481–497.

Currie K. L. and Ferguson J. (1971) A study of fenitization around the alkaline carbonatite complex at Callander Bay, Ontario, Canada. *Can. J. Earth Sci.* **8**, 498–517.

Currie K. L. and Shafiqullah M. (1967) Carbonatite and alkaline igneous rocks in the Brent Crater, Ontario. *Nature* **215**, 725–726.

Dasch E. J., Hedge C. E. and Dymond J. (1973) Effect of sea water interaction on strontium isotope composition of deep sea basalts. *Earth Planet. Sci. Lett.* **19**, 177–183.

Davis J. C. (1973) *Statistics and Data Analysis in Geology*. Wiley, N.Y. 550 pp.

Dence M. R. (1964) A comparative structural and petrographic study of probable Canadian meteorite craters. *Meteoritics* **2**, 249–270.

Dence M. R. (1968) Shock zoning at Canadian craters: Petrography and structural implications. In *Shock Metamorphism of Natural Materials* (B. M. French and N. M. Short, eds.), p. 339–362. Mono Book Corp., Baltimore.

Dence M. R. (1971) Impact melts. *J. Geophys. Res.* **76**, 5552–5565.

Dence M. R. (1972) The nature and significance of terrestrial impact structures. *Proc. 24th Int. Geol. Congr.*, Section 15, p. 77–89.

Dence M. R. (1973) Dimensional analysis of impact structures (abstract). *Meteoritics* **8**, 343–344.

Dence M. R., Grieve R. A. F. and Robertson P. B. (1977) Terrestrial impact structures: Principal characteristics and energy considerations. In *Impact and Explosion Cratering* (D. J. Roddy, R. O. Pepin and R. B. Merrill, eds.), p. 247–276. Pergamon, N.Y.

Dence M. R. and Guy-Bray J. V. (1972) Some astroblemes, craters and cryptovolcanic structures in Ontario and Quebec. *Proc. 24th Int. Geol. Congr.*, Excursion A65 Guidebook.

Floran R. J., Grieve R. A. F., Phinney W. C., Warner J. L., Simonds C. H., Blanchard D. P. and Dence M. R. (1978) Manicouagan impact melt, Quebec. Part I: Stratigraphy, petrology and chemistry. *J. Geophys. Res.* **83**, 2737–2759.

Gold D. P. (1968) A study of quartz subfabrics from the Brent Crater, Ontario. In *Shock Metamorphism of Natural Materials* (B. M. French and N. M. Short, eds.), p. 495–508. Mono Book Corp., Baltimore.

Grieve R. A. F. (1975) Petrology of the impact melt at Mistastin Lake crater, Labrador. *Bull. Geol. Soc. Amer.* **86**, 1617–1629.

Grieve R. A. F. (1978) Meteoritic component and impact melt composition at the Lac à l'Eau Claire (Clearwater) impact structures, Quebec. *Geochim. Cosmochim. Acta* **42**, 429–431.

Grieve R. A. F. and Dence M. R. (1978) Principal characteristics of the impactites at Brent Crater, Ontario, Canada (abstract). In *Lunar and Planetary Science IX*, p. 416–418. Lunar and Planetary Institute, Houston.

Grieve R. A. F., Dence M. R. and Robertson P. B. (1977) Cratering processes: As interpreted from the occurrence of impact melts. In *Impact and Explosion Cratering* (D. J. Roddy, R. O. Pepin and R. B. Merrill, eds.), p. 791–814. Pergamon, N.Y.

Grieve R. A. F. and Floran R. J. (1978) Manicouagan impact melt, Quebec. Part II: Chemical inter-relations with basement and formational processes. *J. Geophys. Res.* **83**, 2761–2771.

Grieve R. A. F., Plant A. J. and Dence M. R. (1974) Lunar impact melts and terrestrial analogs: Their characteristics, formation and implications for lunar crustal evolution. *Proc. Lunar Sci. Conf. 5th*, p. 261–273.

Hamza V. M. (1973) Vertical distribution of radioactive heat production in the Grenville Province and the sedimentary sections overlying it. Ph.D. Thesis, University of Western Ontario, London, Ontario, Canada.

Hart R. (1970) Chemical exchange between sea water and deep ocean basalts. *Earth Planet. Sci. Lett.* **9**, 269–279.

Hart S. R. (1969) K, Rb, Cs contents and K/Rb, K/Cs ratios of fresh and altered submarine basalts. *Earth Planet. Sci. Lett.* **6**, 295–303.

Hartung J. B., Dence M. R. and Adams J. A. (1971) Potassium-argon dating of shock metamorphosed rocks from the Brent impact crater, Ontario, Canada. *J. Geophys. Res.* **76**, 5437–5448.

Helgeson H. C. (1969) Thermodynamics of hydrothermal systems at elevated temperature and pressure. *Amer. J. Sci.* **267**, 729–804.

Hörz F. (1965) Untersuchungen an Riesglasern. *Bietr. Mineral. Petrogr.* **11**, 621–661.

Hörz F. and Quaide W. L. (1973) Debye-Scherrer investigations of experimentally shocked silicates. *The Moon* **6**, 45–82.

Krauskopf K. (1967) *Introduction to Geochemistry*. McGraw-Hill, N.Y. 721 pp.

Lagache M. and Weisbord A. (1977) The system: Two alkali felspars-KCl-NaCl-H_2O at moderate to high temperatures and low pressures. *Contrib. Mineral. Petrol.* **62**, 77–101.

Lozej G. P. and Beales F. W. (1975) The unmetamorphosed sedimentary fill of the Brent meteorite crater, south-eastern Ontario. *Can. J. Earth Sci.* **12**, 606–628.

Marchand M. and Crocket J. H. (1977) Sr isotopes and trace element geochemistry of the impact melt and target rocks at Mistastin Lake crater, Labrador. *Geochim. Cosmochim. Acta* **41**, 1487–1496.

Masaitis V. L., Mikhaylov M. V. and Selivanouskaya T. V. (1975) *Popigayskiy Meteoritnyy Krater*. Nauk Press, Moscow. 124 pp.

Mason B. (1971) *Handbook of Elemental Abundances in Meteorites*. Gordon and Breach, N.Y. 555 pp.

McKie D. (1966) Fenitization. In *Carbonatites* (O. F. Tuttle and J. Gittins, eds.), p. 261–294. Wiley, N.Y.

Millman P. M., Liberty B. A., Clark J. F., Willmore P. L. and Innes M. J. S. (1960) The Brent Crater. *Publ. Dom. Obs.*, XXIV, no. 1, Ottawa. 43 pp.

Morgan J. W., Higuchi H., Ganapathy R. and Anders E. (1975) Meteoritic material in four terrestrial meteorite craters. *Proc. Lunar Sci. Conf. 6th*, p. 1609–1623.

O'Keefe J. D. and Ahrens T. J. (1975) Shock effects from a large impact on the Moon. *Proc. Lunar Sci. Conf. 6th*, p. 2831–2844.

Orville P. M. (1963) Alkali ion exchange between vapor and feldspar phases. *Amer. J. Sci.* **261**, 201–237.

Orville P. M. (1972) Plagioclase cation exchange equilibria with aqueous chloride solutions: Results at 700°C and 2000 bars in the presence of quartz. *Amer. J. Sci.* **272**, 234–272.

Palme H., Rainer W. and Grieve R. A. F. (1978) New data on meteoritic material at terrestrial impact craters (abstract). In *Lunar and Planetary Science IX*, p. 856–858. Lunar and Planetary Institute, Houston.

Phinney W. C. and Simonds C. H. (1977) Dynamical implications of the petrology and distribution of impact melt rocks. In *Impact and Explosion Cratering* (D. J. Roddy, R. O. Pepin and R. B. Merrill, eds.). p. 771–790. Pergamon, N.Y.

Rahman S. and MacKenzie W. S. (1969) The crystallization of ternary feldspars: A study from natural rocks. *Amer. J. Sci.* **267A**, 391–406.

Robertson P. B. (1973) Shock metamorphism of potassic feldspars. Ph.D. Thesis, Durham University, England.

Robertson P. B. and Grieve R. A. F. (1977) Shock attenuation at terrestrial impact structures. In *Impact and Explosion Cratering* (D. J. Roddy, R. O. Pepin and R. B. Merrill, eds.). p. 687–706. Pergamon, N.Y.

Schaal R. B. and Hörz F. (1977) Shock metamorphism of lunar and terrestrial basalts. *Proc. Lunar Sci. Conf. 8th*, p. 1677–1730.

Simonds C. H., Floran R. J., McGee P. M., Phinney W. C. and Warner J. L. (1978a) Petrogenesis of melt rocks, Manicouagan impact structure, Quebec. *J. Geophys. Res.* **83**, 2773–2788.

Simonds C. H., Phinney W. C., McGee P. E. and Cochran A. (1978b) Geology of West Clearwater, Quebec, Impact structure, Part I: Structure and field geology (abstract). In *Lunar and Planetary Science IX*, p. 1059–1061. Lunar and Planetary Institute, Houston.

Simonds C. H., Warner J. L. and Phinney W. C. (1977) Effect of water on cratering (abstract). In *EOS* (*Trans. Amer. Geophys. Union*) **58**, 425.

Stöffler D. (1971) Progressive metamorphism and classification of shocked and brecciated crystalline rocks at impact craters. *J. Geophys. Res.* **76**, 5541–5551.

Stöffler D. (1977) Research drilling Nördlingen 1973: Polymict breccias, crater basement, and cratering model of the Ries impact structure. *Geol. Bavarica* **75**, 443–458.
Stöffler D., Ewald V., Ostertag R. and Reimold W.-U. (1977) Ries deep drilling: I. Composition and texture of polymict impact breccias. *Geol. Bavarica* **75**, 163–189.
Stöffler D. and Hornemann U. (1972) Quartz and feldspar glasses produced by natural and experimental shock. *Meteoritics* **7**, 371–394.

Proc. Lunar Planet. Sci. Conf. 9th (1978), p. 2609–2632.
Printed in the United States of America

A structural study of the Kentland, Indiana impact site

R. T. LANEY* and W. R. VAN SCHMUS

Department of Geology, University of Kansas, Lawrence, Kansas 60045

Abstract—The structurally disturbed area near Kentland, Indiana is a typical example of a so-called 'cryptoexplosion' structure which we attribute to be the result of a large natural impact event. The structure includes a central uplifted area approximately 4 km in diameter with the most intense deformation confined to a core slightly greater than 1 km in diameter. Stratigraphic uplift in the core region is probably greater than 600 m. The central uplifted area is surrounded by a ring depression 1.5 to 2.0 km wide, situated 3.2 km from the center of the uplifted area and within which are preserved soft Pennsylvanian coals. At a radius of approximately 6.2 km from the center of the site, a subdued structural high is present and is believed to be the limit of the disturbance.

Although most of the site is covered with a thin veneer of glacial till, a large quarry (0.4 km^2 in area, 70–80 m deep) located on the northern flank of the central uplift exposes in detail some of the complexities which occur as a result of structural modifications of an original transient cavity. Quarry exposures are dominated by a large NNW plunging synclinal fold and numerous high angle normal and reverse faults. Megabreccia zones containing blocks 100 m or more in length bound this synclinal fold on three sides. These zones are structurally very complex and characteristically seem to involve the more mobile sandstone and shale formations along with the competent carbonates. Recent drilling indicates that this pattern of large carbonate blocks and megabreccia zones also characterizes the structural pattern of the remaining covered portions surrounding the core of the central uplift.

Shock deformational features produced by the impact event, including monomict and polymict breccias and shatter cones, are exposed in the quarry. Petrographic studies of quartz grains from the breccias and other sandy units show that brittle failure was common with irregular fractures, microfaulting, and cleavage being the dominant features in almost all grains. Less than one percent of the grains studied contained basal deformation lamellae; multiple sets of other planar features in quartz grains have not been observed. These results suggest shock pressures in the range of 50 kb for the material in the quarry.

Information gathered in this study indicates that the probable mechanism producing central uplift formation in impact craters in sedimentary rocks is some type of rebound of the impact-compressed material. Although most terrestrial impact structures in sedimentary rock are in various states of degradation and information generally is lacking in regard to the structural details at deeper levels in these impact sites, comparative studies indicate several variables may be important during a cratering event. Comparison of similar-size impact sites indicate that various factors, including target characteristics, projectile characteristics, and basement depth, could play important roles in determining final crater morphology.

INTRODUCTION

The area of disturbed rock known as the Kentland, Indiana structure is the fourth largest known impact site in the United States. Quarry operations over the

*Present Address: Department of Earth and Space Sciences, State University of New York, Stony Brook, New York 11794

last 70 years have exposed a complex assemblage of Ordovician and Silurian rocks near the center of the disturbance. Though quarry operations have accelerated recently, only about 0.3 percent of the entire structure has been exposed. For almost 100 years geologists who have visited Kentland have speculated as to the origin of this disturbance, which is in an area that is otherwise regionally tectonically stable. Only within the last 15 years has more intensive investigation brought to light the extent and nature of the structure (e.g., Gutschick, 1976; Tudor, 1971). Although various endogenetic origins have been proposed to account for the disturbance (Shrock, 1937; Gutschick, 1976; Tudor, 1971) these hypotheses still encounter the same objections as when they were applied to explain other similar impact sites (e.g., Roddy, 1968; Wilshire *et al.*, 1972; French, 1968). On the other hand, interpretation and integration of available geophysical data, along with information gathered from quarry exposures and core drilling, allow for a consistent picture to be drawn of the Kentland site that compares favorably with other well-studied impact sites, such as Flynn Creek (Roddy, 1968, 1977b), Sierra Madera (Wilshire *et al.*, 1972), Wells Creek (Wilson and Stearns, 1968), and Decaturville (Offield and Pohn, 1977). Because of the excellent, although limited, exposures of a portion of the Kentland central uplift, important information regarding late-stage crater modification in well-stratified rocks has been obtained from a study of this structure.

GEOLOGIC SETTING

The Kentland structure lies approximately 4 km east of the town of Kentland, Newton County, Indiana (Fig. 1). More than 700 m of Lower Ordovician through Pennsylvanian strata are known to be involved in the 12.5 km diameter disturbance. Almost all of the bedrock in this region is veneered with Pleistocene glacial till, reaching a thickness of over 40 m in the ring depression. Several minor structural sags occur in the Kentland region (Tudor, 1971), but for the most part, the regional bedrock is essentially flat-lying and consists of Lower Mississippian and Upper Devonian strata which dip gently to the southwest toward the Illinois Basin. The Kentland area lies approximately 30 km SW of the crest of the Wisconsin (Kankakee) Arch, and the nearest known basement fault occurs approximately 80 km to the east.

STRATIGRAPHY

In the vicinity of the Kentland structure, the Precambrian basement occurs at a depth of approximagely 1680 m, 1460 m below sea level (Bond *et al.*, 1971). Unconformably overlying this terrain of granite and rhyolite is a 1645 m thick succession of Paleozoic marine sandstone, carbonate, and shale formations which make up the local stratigraphic sequence. The majority of the Cambrian sediments are represented by the 1015 m thick Potsdam Sandstone Megagroup. The Cambrian-Ordovician Knox Dolomite Megagroup is composed of 266 m of dolomite and sandstone which conformably overlie the Potsdam Megagroup. The

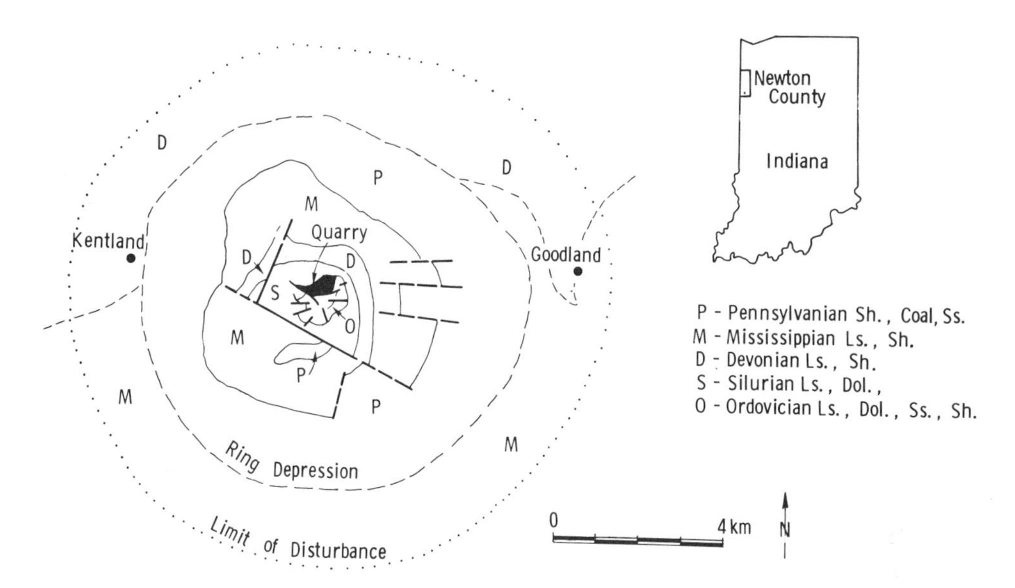

Fig. 1. Bedrock geology map of the Kentland, Indiana impact site. Inset indicates location of the disturbance. Quarry site is indicated by blackened section near center of the structure. Diagram is based on the work of Tudor (1971), Gutschick (1976), and this study.

upper member of the Knox, the Shakopee Dolomite, is the oldest unit exposed in the Kentland quarry. Unconformably overlying the Knox are 123 m of the Ordovician Ottawa Limestone Megagroup. The relatively pure Platteville and Galena carbonates that comprise this Megagroup are the rocks presently being quarried for crushed stone purposes. 80 m of Upper Ordovician Maquoketa shale disconformably overlie the Middle-Ordovician carbonates and conformably underlie 146 m of Silurian dolomite. Approximately 50–60 m of limestone and shale of Devonian and Mississippian age comprise the bedrock in the Kentland area. Disturbed sandstone, shale, and coal presumably of Early Pennsylvanian age are preserved in the ring syncline of the Kentland disturbance. Unconformable and undisturbed Pennsylvanian sediments occur in structural lows elsewhere in the surrounding region (Gutschick, 1976).

The age of the Kentland disturbance has been estimated as between post-Early-Pennsylvanian and Pre-Pleistocene. Several periods of erosion and deposition of sediments have occurred in the mid-continent region during late Paleozoic, Mesozoic, and Cenozoic time (Willman *et al.*, 1975) making it difficult to estimate the thickness of sediments at Kentland at the time of impact. For instance, the St. Peter sandstone outside the area of disturbance has a present regional depth of 450 m. Assuming that the maximum amount of known post-Mississippian sediments found in the Midwest was present at the time of impact, the St. Peter sandstone would have been at a maximum depth of 2400 m below the surface (Willman *et al.*, 1975). This additional 1950 m of overlying sediments at the time of impact should be considered an upper limit only. It appears probable that most of these overlying sediments were already eroded away at the time of impact and that post-event erosion has been on the order of 300 m or less. However, this point is still a matter of contention and more field data will be needed in order to resolve this question.

Geophysical Studies

Several geophysical studies of the Kentland disturbance have been conducted, and these have shed considerable light on the form and areal extent of the structure. An aeromagnetic survey flown in 1948 by the U.S. Geological Survey in the vicinity of the Kentland site showed that no unusual susceptibility contrast is associated with the structure (Joesting and Henderson, 1948), and this finding was confirmed by a later localized ground magnetic survey (Lucas, unpublished material, 1952). However, the 1952 survey indicated that a positive gravity anomaly did coincide with the disturbance, and this finding was the basis for the more thorough and detailed geophysical study undertaken by Tudor (1971).

This latter geophysical survey consisted of shallow seismic reflection and refraction studies, a shallow bedrock drilling program in conjunction with the seismic survey and a detailed gravity study involving 2600 metered stations covering an area of about 184 km^2 (an area approximately 13.5 km on a side). The results of the refraction survey and drilling program provided information on the bedrock topography and bedrock geology from which the map in Fig. 1 was drawn. Although this map is based on limited information, it is to represent the most general type of structure present. The best geologic control exists for the Silurian and Ordovician strata on the flanks of the central uplift due to quarry exposures and more drilling information.

The gravity survey and subsequent modelling done by Tudor (1971) provides information on the extent and form of the total disturbed area. A 3.5 to 4.0 milligal positive gravity anomaly is centered near the middle of the disturbance. An encircling 1.0 milligal negative anomaly, at a radius of 3.2 km from the center of the structure, is believed to represent a ring depression or moat fault system within which Pennsylvania strata are preserved. Two dimensional gravity modelling of this feature suggests that, where the anomaly is a maximum, up to 90 m of Pennsylvanian rock could be present (Tudor, 1971). An outer encircling positive gravity anomaly of 0.5 milligals occurs at a radius of 6.2 km and is thought to be a low relief, structurally high area that is a ring anticline and which represents the outer limit of the disturbance.

From this study, Tudor (1971) was not convinced of an exogenetic origin for the Kentland structure. He was concerned with the problem of explaining the origin of the 3.5 to 4.0 milligal anomaly coincident with the central uplift. He considered that the structure did not conform to the geophysical model of simple meteorite impact craters, such as the Canadian structures described by Innes (1961). Instead he invoked a ". . . subterranean source for the deformation . . ."

involving the uplift of 600 to 900 m of the basement rock (Tudor, 1971). Presumably this elevator-type uplift would have affected the overlying 1680 m of Paleozoic cover and produced the 550 to 600 m uplift that is present in the center of the Kentland structure. The mechanism to produce such an uplift was not stated, but instead the reader was referred to Jamieson's (1963) work on natural analogues that could produce local pressures of tens of kilobars.

Apparently Tudor was not aware that small positive gravity anomalies exist at other suspected impact sites in sedimentary rock (e.g., Wells Creek; Wilson and Stearns, 1968), and basement uplift is not indicated at this structure either. The positive gravity feature described by Tudor can, we believe, be more readily explained by the uplift of the more dense Upper Cambrian to Middle Ordovician carbonates into the core of the structure. Assuming realistic density contrasts, the 3.5 to 4.0 milligal positive anomaly can be accounted for in the sedimentary section alone without postulating basement involvement (Laney, 1978). In fact, little, if any, basement involvement should be present if the results from other impact structures can be applied to Kentland. At the 6.0 km diameter Decaturville, Missouri structure, Offield and Pohn (1977) have postulated that the transient cavity penetrated approximately 550 m of Paleozoic cover to briefly expose the basement surface. They note that isolated Precambrian ejecta blocks lying on the present ground surface, and other structural evidence supports this hypothesis. However, deep drilling near the center of the structure indicated little (50 m or less) general uplift of the basement surface. At Kentland, where the Paleozoic cover is some three times thicker, it seems improbable that the transient cavity would have excavated down to the level of the Precambrian basement. This is consistent with our contention that the positive gravity anomaly at Kentland did not originate due to 600–900 m of basement uplift as proposed by Tudor (1971).

Structure

A) Ring anticline

Very limited information exists for this part of the Kentland structure. Two water wells (total depth, 72 and 328 m) near the townsite of Goodland, Indiana indicate a structurally high area (approximately 15 m). This area is coincident with the small positive (0.5) milligal anomaly determined from Tudor's (1971) gravity survey. Presumably a structural high of this type surrounds the Kentland disturbance at a radius of 6.2 km from the center of the structure and separates the disturbed area from the surrounding flat-lying regional bedrock.

B) Ring depression

Most of the information derived from geologic studies about the overall configuration of this element of the Kentland structure has been presented above. This information was derived from several deep holes (total depth, 73, 93, and

158 m) drilled by the Indiana Geological Survey in the area surrounding the central uplift (Gutschick, 1976) and the shallow (generally 10–25 m) bedrock drilling and gravity survey of Tudor (1971). Compared to better exposed impact sites (e.g., Sierra Madera and Wells Creek), the bedrock geology map presented in Fig. 1 is lacking in detail, particularly for the area surrounding the Ordovician core. A hint of the possible true complexity of the area comes from the negative results of Tudor's seismic reflection study. Continuous reflection profiles were attempted along four lines at distances up to 3 km from the center of the structure. No usable seismic information was recorded, indicating that faulting and other subsurface complexities are more abundant than indicated on the bedrock map (Tudor, 1971). A more comprehensive drilling program would bring to light additional information about the area surrounding the central uplift.

The origin of the structural depression is believed to be the result of inward displacement of materials from deeper levels to support central uplift formation. Structural depressions surrounding central uplifts have been reported at many suspected impact sites in the United States (Wilshire and Howard, 1972; Wilson and Stearns, 1968; Offield and Pohn, 1977; Hendricks, 1965; Black, 1964). Such depressions also occur in large-scale, high-explosive cratering experiments (Roddy, 1976). Although no calculation has been made for Kentland, at other sites it has been shown that the amount of material brought into the center of the structure is approximately accounted for by the amount of material downfaulted into the ring depression (e.g., at Decaturville: Offield and Pohn, 1977). Roddy (1976, 1977a,b) argues that this inward flow beneath crater floors is common at both flat-floored impact (e.g., Flynn Creek) and flat-floored experimental explosion craters with central uplifts.

C) *Central uplift*

A somewhat simplified bedrock geology map and cross section of the Kentland central uplift are presented in Fig. 2. A more detailed geologic map and cross sections of the quarry are presented in Fig. 3. These are based on information from quarry exposures and core drilling by the Newton County Stone Company (pers. comm.), the Indiana Geological Survey, and work by Tudor (1971) and Gutschick (1976). Although field data are limited for the southern half of the uplift, where most of the geology is indicated by dashed lines, our interpretation presented in Fig. 2 is consistent with the presently available geologic information. We believe this interpretation represents at least the gross structure of the central uplift. More drilling information will greatly improve our knowledge about the structural details of the uplift.

Proceeding from the outer Silurian strata inward, successively older formations are exposed as the dips of the beds and structural complexity increase toward the center. Silurian dolomites have been uplifted approximately 275 m above their regional stratigraphic level, whereas the Ordovician Shakopee Dolomite (the oldest unit exposed in the quarry) is some 550 m above its normal

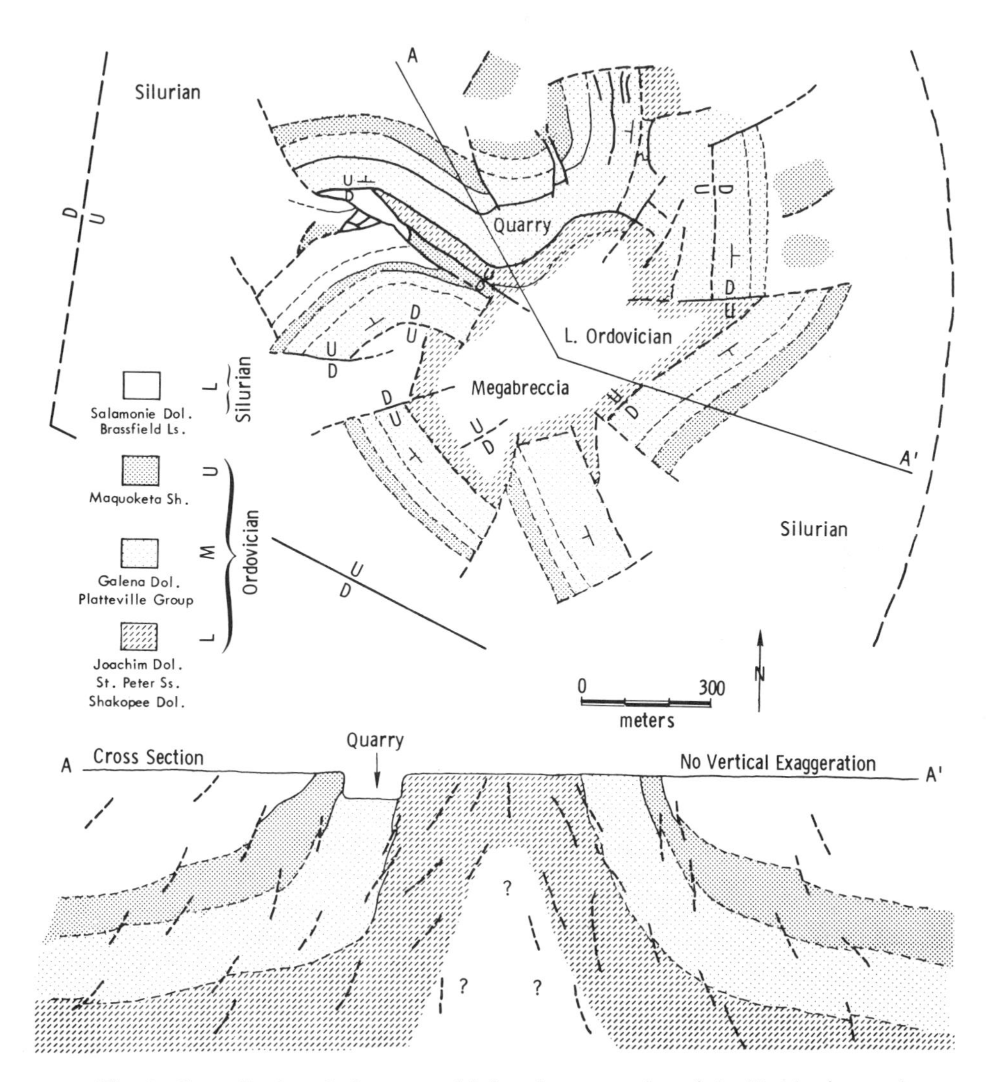

Fig. 2. Generalized geologic map and inferred cross section of the Kentland central uplift. Note that at the present erosional level most of the intense structural dislocations are confined to a relatively small area (1.0–1.2 km^2) compared to the area occupied by the total disturbance (123 km^2). Information for these diagrams comes from quarry exposures and drilling by Tudor (1971) and the Newton County Stone Company (pers. comm.).

position. Dips on the Silurian strata are about 20–30° and increase abruptly to 75–90° for the Middle-Ordovician carbonates. Intensity of faulting and folding also increase from the flanks of the central uplift towards the center, where large-scale structure is still recognizable, inward to an area of about 0.25 km^2 which is thought to be composed of blocks of lower Ordovician units in a complex structural arrangement or megabreccia.

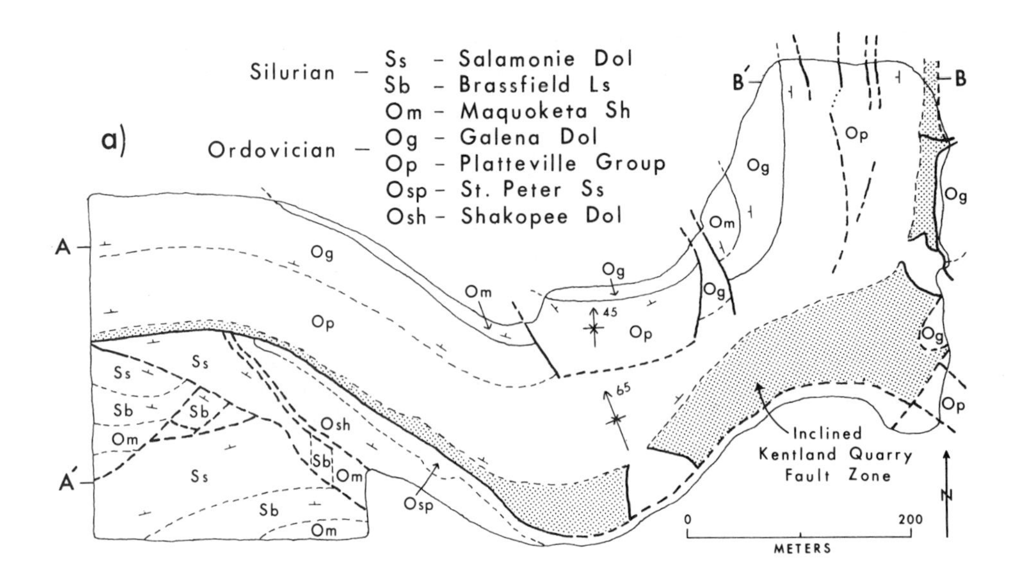
a)
Silurian — Ss – Salamonie Dol
Sb – Brassfield Ls
Om – Maquoketa Sh
Ordovician — Og – Galena Dol
Op – Platteville Group
Osp – St. Peter Ss
Osh – Shakopee Dol
A
A′
B′
B
Og
Op
Om
Ss
Sb
Osh
Osp
45
65
Inclined
Kentland Quarry
Fault Zone
0
200
METERS
N

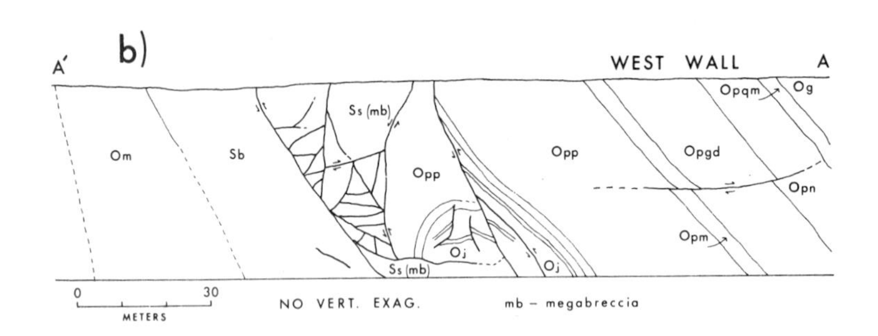
b)
A′
WEST WALL
A
Om
Sb
Ss (mb)
Opp
Opgd
Opqm
Og
Opn
Opm
Oj
0
30
METERS
NO VERT. EXAG.
mb – megabreccia

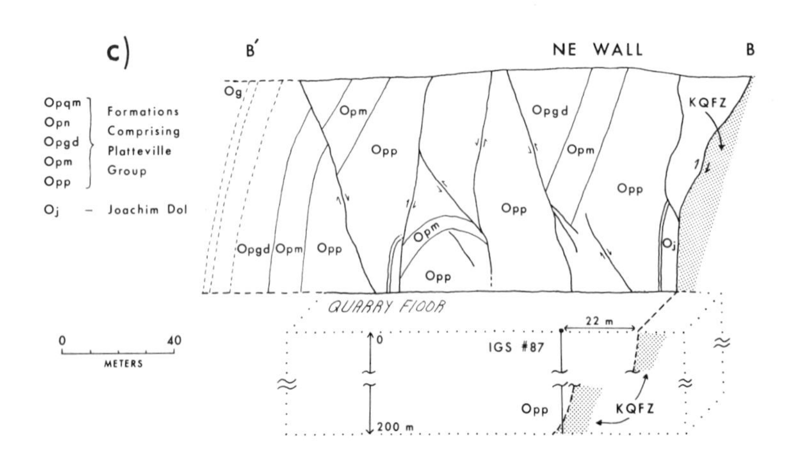
c)
B′
NE WALL
B
Opqm
Opn
Opgd
Opm
Opp
Formations Comprising Platteville Group
Oj – Joachim Dol
Og
Opm
Opp
Opgd
KQFZ
Oj
QUARRY FLOOR
22 m
0
IGS #87
200 m
Opp
KQFZ
0
40
METERS

For the area outside of the quarry, core drilling is the main source of information. It appears from the presently available data that relationships similar to those in the quarry characterize the remaining portions of the flanks of the uplift. There also appears to be correlation between lithology and deformational style, with the Middle-Ordovician carbonates deforming as competent blocks 400–800 m in length; whereas, the more mobile sandstone and shale units are involved in complex faulting between these blocks. Circumferential faulting, presumably dip-slip or closely subparallel to bedding, separates these blocks from the megabreccia core and radial faults cut these blocks perpendicular to their strike,

A cross section for the Kentland central uplift is given in the lower half of Fig. 2. The steeply dipping attitude of the strata adjacent to the inferred megabreccia core is indicated from drilling in the quarry by the Indiana Survey. A drill hole placed near the Kentland Quarry fault (Fig. 3c) has shown that the steeply dipping fault surface extends at least 280 m below the present ground surface. It is not certain what strata are present deeper in the megabreccia core, but Cambrian carbonates could be at a relatively shallow depth below the ground surface. Drilling in the core will be needed to clarify this.

In the sequence of rocks at the Kentland site, several good marker beds are present, and these allow elucidation of detailed structural relationships. This has been done at the quarry site located on the northern flank of the central uplift (Fig. 3). The first comprehensive studies of the structures in the quarry were done by Shrock and Malott (1933) and Shrock (1937). More recent investigations by Gutschick (1961, 1971, 1972, and 1976) laid the groundwork for a large part of this study.

The Newton County Stone Company has been quarrying continuously at this site since 1906 (Gutschick, 1976), and its recently increased activity has exposed much new information concerning structural relationships. The main quarry is about 70 m deep and quarry extraction has followed the Platteville and Galena carbonates westward. This has dramatically exposed the main structure of the quarry, in which the Middle-Ordovician carbonates are synclinally folded with the axis of the fold plunging about 65° NNW and oriented radially to the center of the structure (Fig. 3a). The westward limb of this synclinal fold appears to

---

Fig. 3. Plan map and geologic profiles of the West and NE walls of the Kentland quarry. a) Major structure present in quarry is a NNW plunging synclinal fold. Stippled pattern indicates where quarry extraction has excavated down to the surface of the Kentland Quarry fault zone. Map modified after Gutschick (1976). b) Geologic profile of West wall of quarry. Center portion indicates steep reverse fault that has placed the Ordovician Platteville Group (Op) over younger Silurian strata (Ss) which have been jumbled into a megabreccia. c) Geologic profile of NE wall of quarry. IGS #87 denotes Indiana Geological Survey drill #87. Drilling started on the level of the quarry floor (approximately 68 m below the present ground surface) and continued for a total depth of 198 m. At approximately 190 m in the drilling, the Kentland Quarry fault zone (KQFZ) was encountered indicating the steeply dipping nature of the fault plane.

begin to take on an anticlinal aspect but is broken by a fault, as is the east limb. A major fault, the Kentland Quarry fault (KQF) dips steeply northward and it bounds this synclinal fold throughout the quarry on the south. This fault involves the friable St. Peter sandstone and the massive Joachim dolomite. As pointed out by Gutschick (1976), this fault is actually a wide zone of deformation (i.e., a fault zone). Strata on the hanging wall are severely distorted and broken up by many smaller faults up to 15 m or more from the main fault surface. Movement along the main fault has been essentially dip-slip near the axis, whereas hanging wall material farther out on the limbs of the fold has moved obliquely toward the axis. Mullion, mylonite, and slickensides are well exposed where quarrying has exposed the main fault surface. Intimately associated with this fault zone are monomict and polymict breccias.

The west wall of the quarry (Fig. 3b) displays a steep reverse fault that has placed Middle-Ordovician carbonates over younger Silurian dolomites, with a zone approximately 30–40 m wide of Silurian strata that has been jumbled into a megabreccia. Other structures present in the quarry include a small nested syncline involving Middle-Ordovician carbonates near the axis of the main synclinal fold, and several high angle faults have duplicated sections of the Platteville group in the NE section of the quarry (Fig. 3c). These faults, along with several others present in the quarry, appear to be clean separations with no breccia, while others (e.g., the Kentland Quarry fault) are associated with polymict injection breccias or intensely shattered rock near the fault surface.

SHOCK DEFORMATION AT KENTLAND

Perhaps the most persuasive arguments supporting an impact origin for structures such as Kentland are recorded in the rocks themselves. Features such as shock deformed mineral grains, deformational breccias, and shatter cones have been noted to occur at almost all suspected impact sites and an extensive literature exists on these subjects (e.g., French and Short, 1968; Roddy *et al.*, 1977). Similar features occur at Kentland and these will be discussed below.

A) Shocked quartz

A petrographic study of quartz grains from the Kentland quarry was undertaken in order to determine the approximate peak shock pressures experienced by the grains. Similar studies of quartz grains from other impact sites have shown that the grains exhibit features that are rare or less well-developed in quartz from normal tectonic environments (Bunch, 1968, Carter, 1968). When this information is compared to data on experimentally shocked quartz, approximate ranges of shock pressures at various levels in the impact structure can be determined (Dence, 1968; Stoffler, 1966; Engelhardt and Bertsch, 1969).

Our data were obtained from universal stage measurements on 21 thin sections of rock from mixed breccias and from sandstone and sandy dolomite units, namely the basal Platteville, St. Peter, and Joachim units, which are situated approximately 50 m apart stratigraphically. Cleavage and microfaults are the most common shock features in these grains. Cleavage as used in this study conforms to the definition of Carter (1968) as being fairly wide (2 to 10 μ) planar elements that form as the result of brittle fracturing along crystallographic planes of low indices. A typical grain displaying rhombohedral cleavage is shown in Fig. 4b. Microfaulting may be related to crystallographically controlled cleavage, but offsets along the plane of discontinuity are visible.

The only other shock feature noted in these quartz grains was deformation lamellae in the basal (0001) orientation (Fig. 4c). These lamellae are similar to those described by Carter (1968) as being sharp, narrow (1 to 2 μ wide) planar features that occur in sets of 5 or more individual lamellae. These lamellae generally do not extend across an entire grain in the Kentland material, but occur either in one section of the grain or are distributed in sets across the grain. Multiple sets of planar features in other crystallographic orientations as described by Carter (1968) have not been observed. However, our sampling is limited to the quarry area, which occupies only a small part of the total central uplift.

The results of measurements of the orientations of 2421 planar fractures or cleavages in 875 grains from 21 thin sections are presented in Fig. 4e–f. From the histograms, it is apparent that crystallographic control of some fracturing has occurred with the most frequent orientations involving the positive and negative rhombohedra (r,z), the unit prism (m,a), and the basal pinacoid (c). Similar results have been reported from a study of experimentally shock-loaded quartz by Hörz (1968) and of quartz from other impact sites (Robertson *et al.*, 1968, Robertson, 1975; Bunch, 1968). Hörz (1968) has shown that cleavage of this type usually begins to develop at about 50 kb and that microfaults begin to develop as low as 40 kb. Basal deformation lamellae occur very infrequently in quartz grains from the Kentland rocks. They occur in slightly greater numbers in the mixed breccia samples but are still present in less than one percent of all the grains surveyed. Carter (1968) places a shock range of from 23–76 kb for the formation of these lamellae. We conclude from this study that the shock pressures experienced by these quartz grains are probably in the range of 30–60 kb and did not exceed 80 kb.

An unusual feature noted in some of the quartz grains are fractures that appear to be radiating from the contact surfaces with other grains. A typical example is shown in Fig. 4d. These features appear to be almost exact analogues of the concussion fractures described by Kieffer (1971) which occur in some quartz grains from the Coconino Sandstone at Meteor Crater, Arizona. These features are believed to be the result of tensile fracturing as neighboring grains collide during the passage of the initial compressional shock wave through the rock.

B) Breccias

Two types of breccias occur at the Kentland quarry: monomict breccias and

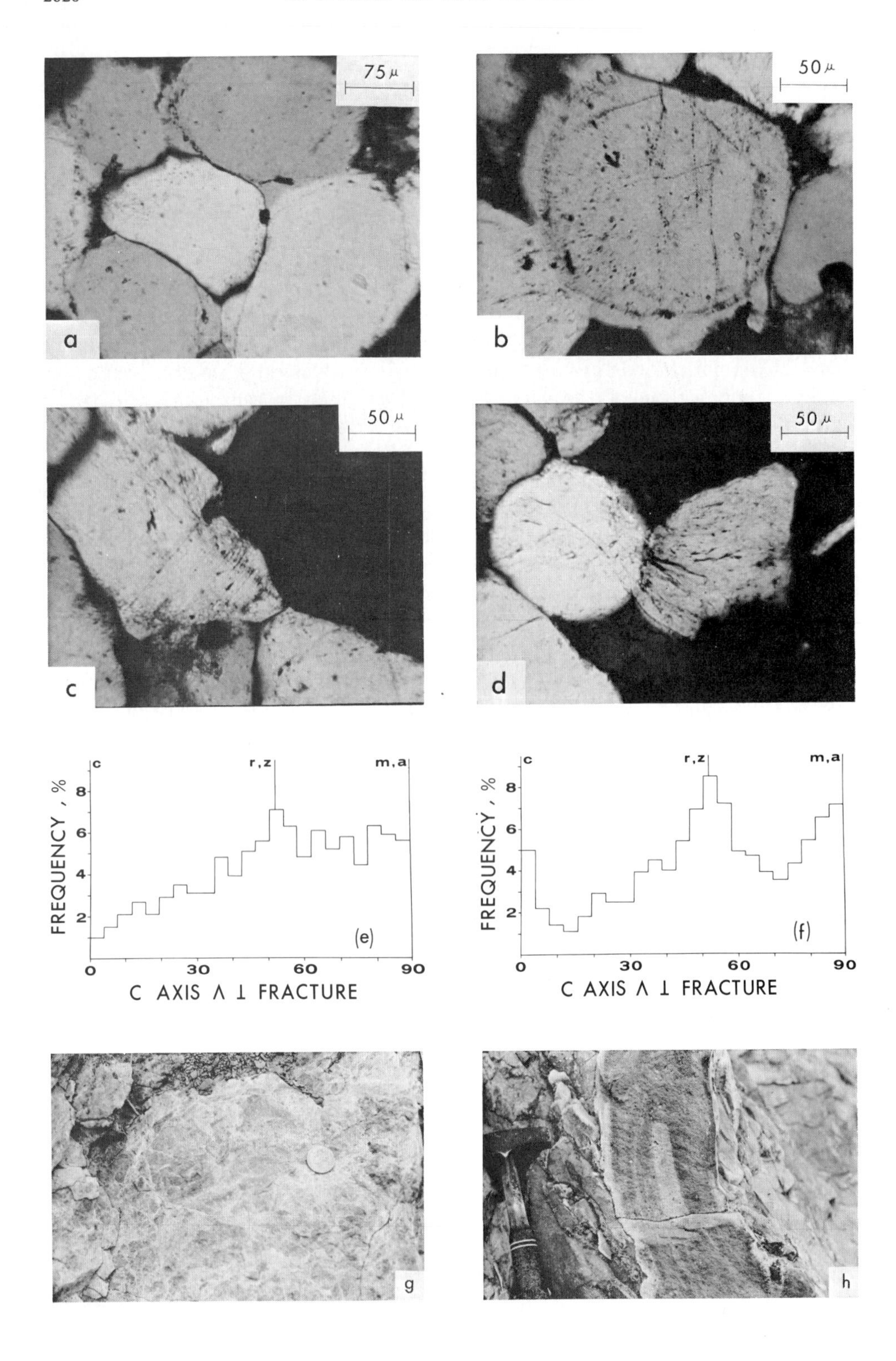
75 μ
a
50 μ
b
50 μ
c
50 μ
d
FREQUENCY, %
c
r, z
m, a
(e)
C AXIS ∧ ⊥ FRACTURE
0
30
60
90
2
4
6
8
(f)
g
h

the more commonly occurring polymict, or multiple lithology, breccias. The monomict breccias are generally recognizable in the carbonate units but may also be developed in some of the sandy units. In the carbonate rocks this type of breccia consists of angular to subangular clasts in an extremely fine-grained or mylonitic matrix. The clasts are variable in size (0.5–5 cm), and the larger ones are usually shattered. The matrix occurs as irregular veinlets or stringers variously oriented around the clasts; the clasts generally appear to have undergone little or no rotation. The lighter colored, mylonitic material contrasts with the darker clasts and these breccias can easily be recognized in the field (Fig. 4g). Presumably analogous brecciation has occurred in some of the sandy units, particularly in some outcrops of the St. Peter Sandstone. Granulation and pulverization has been intense enough to turn the well-rounded quartz grains into a fine grained rock powder which is only weakly cemented by silica cement.

Monomict breccias in carbonate rocks described from other impact sites (Roddy, 1968; Wilson and Stearns, 1968; Wilshire *et al.*, 1971; Offield and Pohn, 1977) are believed to form as the result of shattering, dilation, and subsequent recompaction of individual units with little or no mixing between units. A similar set of events is believed to have operated to produce the monomict breccias at the Kentland site.

Polymict breccias are abundantly exposed in the quarry. They usually occur as irregular to somewhat tabular bodies emplaced along fault surfaces and bedding planes, grouted into any available opening, or even injected as cross-cutting dikes into some sandstone beds (Fig. 4h). Clasts for these breccias are apparently derived from all lithologies present in the quarry and consist of quartz grains, stringers of quartz grains, chert, dolomite, calcite, shale, monomict breccia fragments, and occasional shatter cone segments. Clast sizes typically range more or less continuously from a few millimeters up to blocks 0.5 m across; blocks tens of meters or more across have been observed in some of the megabreccia zones. The matrix of these breccias is usually light gray, fine grained carbonate material that is quite hard and dense. This material could have originally been finely comminuted carbonate rock powder that has been recrystallized to form the

---

Fig. 4. Examples of unshocked and shock-deformed quartz. a) Unshocked St. Peter quartz sandstone from the La Salle Anticline, Ill. b) Shocked quartz displaying rhombohedral cleavage. c) Deformation lamellae in the basal (0001) orientation. d) Concussion fractures developed as the result of neighboring grains colliding during shock compression. e) Frequency histogram of the angles between the quartz optic axes and poles to cleavage fractures. Grains measured are from *in situ* sandstone formations (principally the St. Peter Sandstone). f) Frequency histogram for the cleavage and fracture orientations developed in quartz grains from polymict breccia occurrences collected from quarry exposures. Figs. b-f suggest shock pressures in the range 30–60 kb. g) Example of a field occurrence of monomict brecciation from quarry. White stringers are of the same composition as the darker clasts, but have a mylonitic appearance due to intense granulation. h) Example of a polymict breccia showing flow foliation texture. Sample is from an 'injection dike' intruded into a St. Peter Sandstone bed.

matrix. In some occurrences of this type breccia, material that was originally separated 250 m stratigraphically can now be found as clasts occurring side-by-side. This indicates extensive mixing has occurred over relatively large distances during uplift formation. Some breccias are also flow-foliated, displaying preferred orientation of elongate clasts and distinct color banding of matrix material. This type of feature has been reported at other sites (e.g., Sierra Madera), and Wilshire *et al.* (1971) postulate that such breccias were emplaced ". . . as dense suspensions of clasts . . . probably in CO_2-rich water or water vapor." This also implies that during rotational uplift of the strata there was transient dilation, presumably along bedding surfaces and faults. Roddy (1968) describes large sill-like injections of finely brecciated rock along the flanks of the central uplift at Flynn Creek and indicates that injection normally occurs during the rarefaction stage of the cratering event. During this stage much of the sub-crater floor and walls are in a tensional condition (Roddy, 1976). During this dilation phase, mixing of various lithologies took place and subsequent settling and compaction of the strata forcibly injected the breccia material into available openings and spaces. Such a sequence of events would account for many of the features of the breccias described above.

C) Shatter cones

Shatter cones are an unusual type of mesoscopic shock deformation feature present at a large number of impact sites (Dietz, 1947, 1963, 1968, 1972). Shatter cones usually display a conical fracture surface with striae, grooves, and small parasitic half-cones radiating away from the master cone apex. Shatter cones have now been reported to occur at over 37 confirmed or suspected impact sites including several which are associated with meteoritic material (Dietz, 1963, 1968; Milton, 1977). Recently, Roddy and Davis (1969, 1977) reported on a detailed study of shatter cones formed by several large-scale TNT explosion experiments that firmly establish that shatter cones can be formed by shock wave processes. Formational pressures suggested by Roddy and Davis (1977) and from other studies (Dence *et al.*, 1977; Hörz, 1971; Robertson, 1975) are in the range of 20–80 kb.

At Kentland, shatter cones apparently have formed in all lithologies; however, they are most abundant in the denser carbonate rocks, which is typical at other impact sites. Especially good development has been noted in Silurian dolomite in the south-west section of the quarry (Fig. 5a). Kentland shatter cones generally range in size from 0.01 to 0.2 m, but Dietz (1963) has reported one gigantic cone surface that reached a length of 15 m. Apical angles for Kentland shatter cones generally range from 83° to 105°.

As noted by several other workers (Milton, 1977; Dietz, 1963) shatter cones are often developed in association with mesoscopic inhomogeneities in the rock (small clastic grains, voids, bedding planes). At Kentland several excellent examples have been noted of shatter cones developing in association with fossil fragments. Figure 5b shows a Silurian cephalopod with multiple shatter cone surfaces radiating away from the shell.

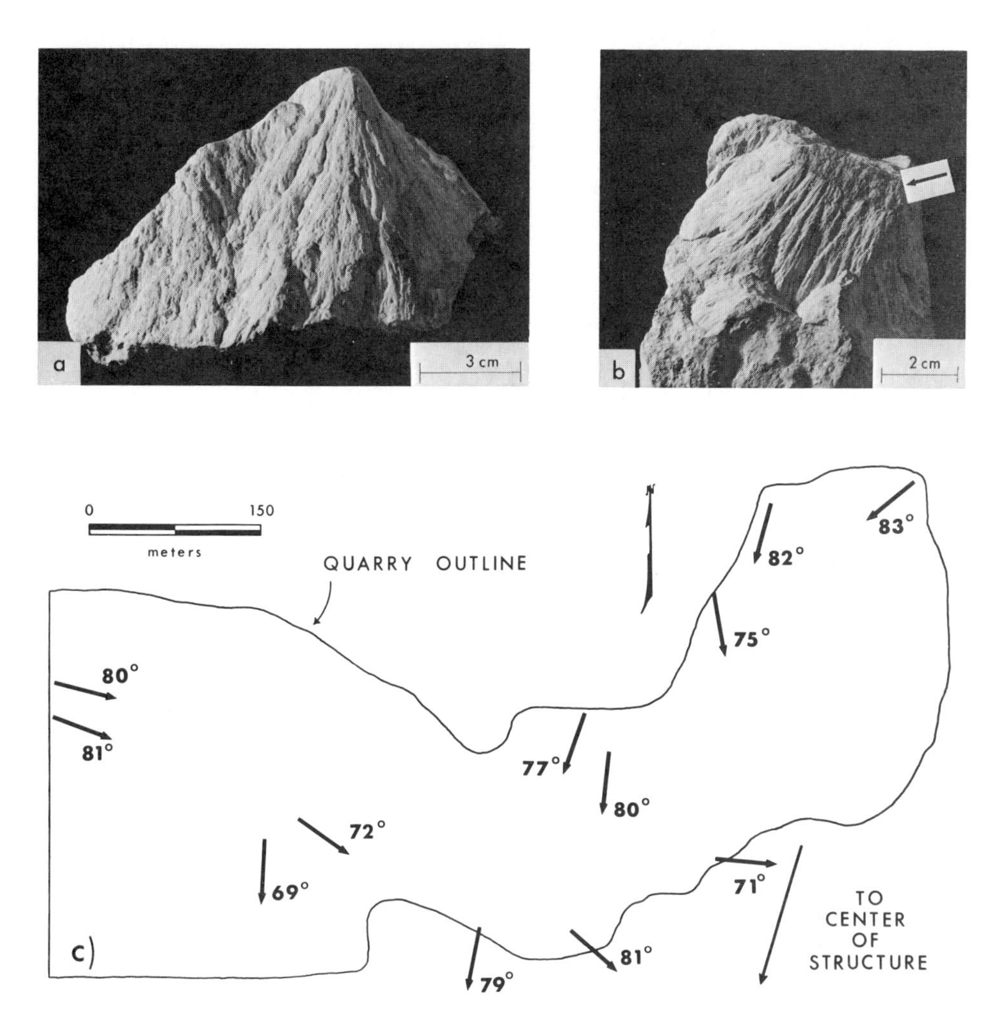

Fig. 5. a) Example of a complete shatter cone recovered from the SW quarry section. Shatter cone developed in the Silurian Salamonie dolomite unit. b) Multiple shatter cone surfaces radiating away from a Silurian cephalopod shell (arrow), recovered from the SW quarry section. c) Sketch of quarry limits and orientation of shatter cone axes restored to their position before major structural displacements have taken place. Note that most axes are oriented at high angles to the horizontal, in accord with a surface origin for formational shock wave.

Orientations of shatter cone striations have been measured from quarry exposures and have been plotted stereographically in order to determine the orientation of the full cone axes. These plots were then rotated to bring the host strata back to a horizontal attitude in order to determine the orientation of the cone axes before structural displacements had taken place (Manton, 1965). As can be seen (Fig. 5c), the orientations of the restored cone axes are generally at high angles to the horizontal. This orientation is in accord with a surface origin for the shock wave. Since only a small portion of the central uplift has shatter cones accessible for measurement, analysis similar to those performed by Manton (1965) and Milton *et al.* (1972), which have shown a general convergence of all restored cone axes toward the center of the structure, was not possible. As a larger portion of the flanks of the central uplift are exposed by quarry operations, an analysis of this type may be feasible.

DISCUSSION

Much effort has been expended in recent years in an attempt to understand and characterize the mechanical factors which influcence the formation of various structural features in impact craters (Dence, 1968; Dence *et al.*, 1977; Milton and Roddy, 1972; Roddy, 1968, 1976, 1977 a,b; Roddy *et al.*, 1977; Ullrich *et al.*, 1977). Central uplifts are an example of one such structural feature and have been shown to occur in certain size ranges of craters on the terrestrial planets and their satellites (Pike, 1971; Cintala *et al.*, 1976; Wood, 1973; Cintala *et al.*, 1977). On Earth, central uplifts commonly occur in impact structures with diameters ranging from 3.5 to 30 km (Dence *et al.*, 1977). The apparent morphological evolution, with increasing crater diameter from simple bowl-shape craters to flat floor-central peak craters to multi-ring basins, implies that central uplift formation is an important phase in the repositioning of material during a medium-size impact cratering event. The detailed sequence of events and driving forces leading to central uplift formation in medium-sized craters, however, remains unclear (Roddy, 1976; Dence *et al.*, 1977; Ullrich *et al.*, 1977).

Central uplift formation resulting from some type of rebound mechanism was first proposed by Boon and Albritton (1937) and has been expounded on and modified by other workers (Short, 1965; Baldwin, 1963; Ullrich *et al.*, 1977). Recently, numerical simulations of high-explosive cratering experiments (Ullrich, 1976; Ullrich *et al.*, 1977) have indicated that central uplift formation *might* be influenced by either a rebound mechanism following compressional shock or late-stage gravitational collapse of the crater walls or a combination of the two. An alternative mechanism involving gravitational deep sliding has also been proposed by Dence (1968), Gault *et al.* (1968), and Shoemaker (1963). Extensive photomorphometric studies by Cintala *et al.* (1977) and Wood (1973) suggest that a gravitational slumping mechanism for central uplift formation may play only a minor role at best.

Field studies of terrestrial impact structures with central uplifts (e.g., Wilshire and Howard, 1968; Wilson and Stearns, 1968; Milton *et al.*, 1972; Offield and Pohn, 1977) have provided basic data on some of the structural relationships observed in the uplifted material at each of these sites. A common problem encountered in the field study of these impact structures, including the Kentland structure, is the lack of detailed surface exposures, a serious lack of adequate deep drilling information in the central uplifted region, and the different erosional level present at each site. However, there are several impact structures with central uplifts which exhibit very little erosion (e.g., Steiheim Basin, West Germany and Flynn Creek crater, Tennessee,) and from these, important information regarding the form and extent of the central uplifted region has been obtained. Roddy's study of the Flynn Creek crater (1968, 1976, 1977 a,b,c) has shed considerable light on the processes and sequence of events leading to the formation of the central uplift at this structure. Roddy (1977b) argues that the Flynn Creek central uplift was formed as a late-stage material response to the initial compression and rarefaction phases of the impact event. The inferred configuration of the central uplift (a steeply conical, disrupted zone) resulted from a volumetric rebound phenomenon and inward displacement of sub-crater floor material. The displacements of greatest extent were presumably confined to a relatively limited volume centered near the crater axis. Roddy states that due to the steeply conical subsurface shape of the central uplift, a large bowl-shaped transient cavity was probably not formed.

Our study of the Kentland impact site also supports a model of abrupt inward and upward displacement of material to form the central uplift. However, it must be stated that exposures and drilling information are limited, much more so than, for example, at Flynn Creek. At the present erosional level of the Kentland site, the most intense structural dislocations are confined to an area approximately 1.0 km in diameter at the center of the uplift. In this region stratigraphic uplift of 600 m or more has occurred, with most of the fault planes dipping steeply and interpreted to extend to relatively deep levels (e.g., the Kentland quarry fault). From these relationships, observable only near the center of the structure, deep gravitational collapse and sliding of the crater walls seems to be an insufficient mechanism to produce the Kentland uplift.

During the early phases of uplift at Kentland it appears that, initially, steep, normal faulting occurred and was confined to a relatively limited area centered under the original cavity. Upward displacements continued, with inward displacements at a deep level occurring to provide support for the uplifted material. Finally, late stage settling of this uplifted material occurred, accounting for the reverse fault relationships seen on the flanks of the uplift. Another possiblity is that early inward movement dominated, producing thrust-reverse type faulting with lesser initial dips which were steepened during upward displacement. Later normal faults were produced as the rapidly rising core separated from the flanks of the uplift. Similar inward and upward displacements of both types have been reported from other impact sites and from craters produced by high-explosive devices (Wilshire *et al.*, 1972; Wilson and Stearns, 1968; Milton *et al.*, 1972; Roddy, 1968, 1976).

Evidence for inward displacements at Kentland comes from the fact that the length of the segments at the stratigraphic level of the base of the Platteville Group is approximately 10 percent greater than the length of the perimeter on which they now lie. This crowding or shortening of perimeter is an indication that inward movement took place as well as upward displacement. Similar horizontal movement has been reported from five other impact sites (Wilshire and Howard, 1968; Offield and Pohn, 1977). Inward crowding of material also is indicated by repeated faulting that duplicates portions of the Middle Ordovician carbonates in the quarry (for example, Fig. 3c).

The central uplift of the Kentland impact site displays most of the features that have been reported from uplifts of other impact structures occurring in well-stratified sedimentary rock sequences. The value of a study of the exposures at Kentland is high because an ever increasing portion of the central uplift is being systematically exposed by quarry operations. Similar detailed three-dimensional views of other central uplifts are obtainable only through an extensive drilling program. Continued quarry expansion and subsequent study of the exposures will do much to improve our understanding of processes operating during central uplift formation.

COMPARISON WITH OTHER IMPACT STRUCTURES

In recent years investigators have identified various factors which could influence impact crater morphology. These factors have included: a) properties of the projectile, such as velocity and strength; b) energy coupling to the target material; c) properties of the target material, such as strength and layering; and d) gravity effects (Baldwin, 1963; Oberbeck and Quaide, 1967; Gault *et al.*, 1968; Cintala *et al.*, 1977; Head, 1976; Roddy, 1968, 1977a,b). Since the Earth's surface only preserves the vestiges of an impact event, information concerning projectile type, mass, and velocity, and the distribution of energy into the target rock, is not usually available for study. However, if erosion has not been too severe, information regarding the nature of the target material is available and this allows for comparisons to be made of the possible variations in impact structures due to differences in the target rock, if such differences do indeed exist.

Gault *et al.* (1968) have shown that variations in target rock layering have influenced final crater shape during laboratory experimental impact studies.

Fig. 6. Cross sections for three impact structures of approximately the same size developed in sedimentary rock. a) Kentland, Indiana: P-Pennsylvanian; DM-Devonian and Mississippian; S-Silurian; Opm-Ordovician Platteville Group to Maquoketa Sh.; Oshj-Ord. Shakopee Dolomite to Joachim Dol.; O€po-Ord and Cambrian Potosi Dol. to Oneota Dol.; €ecf-Cambrian Eau Claire to Franconia Ss.; €ms-Cambrian Mt. Simon Ss. b) Wells Creek: DM-Devonian, Mississippian, S-Silurian, Ofh-Ord. Fernvale Ls. and Hermitage Formation, Osr-Ord. Stones River Group, O€k-Ord. Knox Dol. Diagram modified after Wilson and Stearns, 1968. c) Sierra Madera: K-Lower Cretaceous, Pg-Permian Gilliam Ls., Pw+Pwv-Permain Word Formation, Ph-Permian Hess Fm., Pwc-Permian Wolfcamp Fm. Diagram modified after Wilshire *et al.*, 1972. Erosional estimates are uncertain but probably do not exceed the values listed. Details are not evident on these cross sections, but at their present erosional levels, structural differences do occur in the central uplifts of each of these impact sites. Although erosional level could account for some of these differences, it appears that other impact parameters, such as target material properties, could play a role in determining impact crater structure.

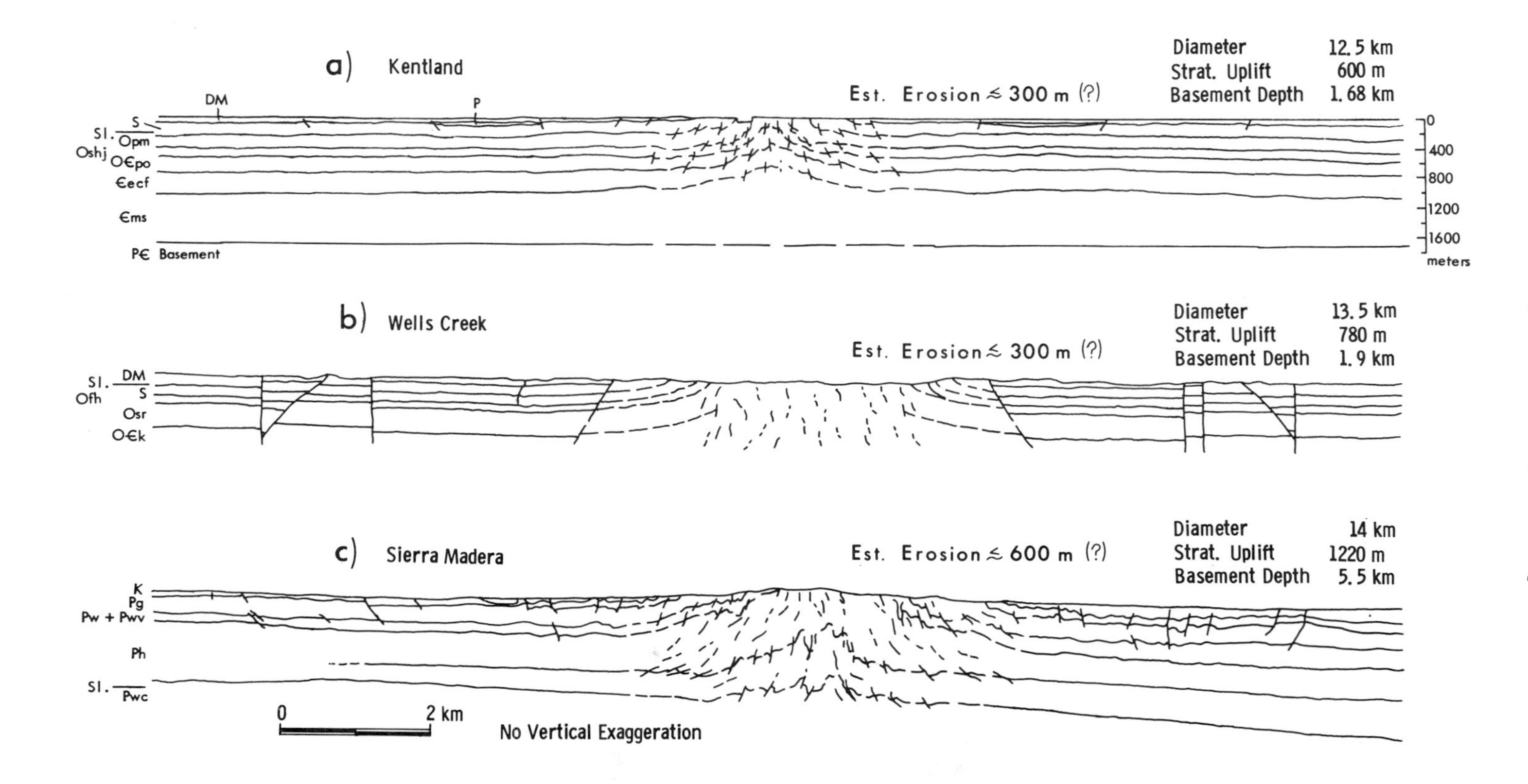
a) Kentland
Est. Erosion ≲ 300 m (?)
Diameter 12.5 km
Strat. Uplift 600 m
Basement Depth 1.68 km
DM
P
S
Sl.
Opm
Oshj
O€po
€ecf
€ms
P€ Basement
0
400
800
1200
1600
meters
b) Wells Creek
Est. Erosion ≲ 300 m (?)
Diameter 13.5 km
Strat. Uplift 780 m
Basement Depth 1.9 km
Sl.
DM
S
Ofh
Osr
O€k
c) Sierra Madera
Est. Erosion ≲ 600 m (?)
Diameter 14 km
Strat. Uplift 1220 m
Basement Depth 5.5 km
K
Pg
Pw + Pwv
Ph
Sl.
Pwc
0
2 km
No Vertical Exaggeration

Interplanetary comparisons of fresh crater morphology by Cintala *et al.* (1977), Head (1976), and others have indicated that substrate characteristics may play a role in determining initial diameters for various crater features, thus indicating that target properties may be a significant determining factor of final crater morphology. From studies of high-explosive produced craters (Roddy, 1976) and the numerical simulation of these events (Ullrich *et al.*, 1977), it has been pointed out that: a) compactibility of target materials, b) strength of target materials, c) layering of the target, and d) the presence of a low strength or a fluid layer underlying a more competent, dry layer may significantly influence or enhance upward motions in the sub-crater floor region. These motions ultimately combined to form the central uplifts of the experimental and numerically simulated craters.

It has been pointed out by Roddy (1977c), and others, that most terrestrial impact structures are in various stages of degradation, and if structural comparisons are to be meaningful the amount of erosion that has taken place at each impact site must be known. Cross sections are presented in Fig. 6 for the Kentland, Wells Creek, and Sierra Madera structures, which are all of comparable diameter. For each structure an estimated amount of erosional leveling has been given; these are based on estimates given by the various authors. However, recently Offield and Pohn (1977) have questioned the accuracy of the erosional estimate of 600 m given for the Sierra Madera structure. Based on their finding that relatively little erosion has occurred at the Decaturville structure, they argue that erosion at Sierra Madera may be much less than previously estimated (Wilshire *et al.*, 1972). In fact, until more field evidence is available, the amount of erosion at Kentland is still a matter of conjecture (the same probably can be said for Wells Creek, also). However, based on scaling from structures such as Flynn Creek (Roddy, 1977c) it appears that the *relative* amount of erosional leveling has been greatest at the Kentland structure [i.e., the Kentland site possibly displays material from a deeper level in the impact structure than either Wells Creek or Sierra Madera (Laney, 1968)]. Referring again to Fig. 6, it is to be noted that although these structures involve similar type sequences of carbonate, sandstone, and shaley units and have approximately the same overall diameter of disturbance, some structural differences do occur in the central uplifted regions of each. Very similar target rocks occur at the Kentland and Wells Creek sites, except for the Upper Cambrian sequence (at Kentland it consists of 1015 m of sandstone; whereas, at Wells Creek the Upper Cambrian is composed of dolomite). At their present erosional levels, stratigraphic uplifts at both sites are of the same order of magnitude, but the central uplift at Wells Creek occupies an area more than nine times larger than the equivalent one at Kentland. The area occupied by the uplifts at Wells Creek and Sierra Madera are similar, but the amount of uplift at Sierra Madera is significantly larger. The criteria used for determining the diameters of equivalent uplifted areas were points where the styles and intensity of deformation were similar in each structure.

The cause of these differences could be the result of differences in the properties of the layered substrate at each impact site. As the initial cavity expanded after impact, it encountered the differing sedimentary rock types (e.g., more mobile sand or shale units as compared to the more competent carbonates) at different levels in each structure. Alternatively, projectile characteristics, such as strength and impact velocity and the resulting changes in the energy imparted to the rock, could have been different at each site; however, this does not strictly conform to the ideas presented by Baldwin (1963) and others relating crater diameter to the energy of impact. Another factor which could have been an influence is the depth at which the shock wave encountered a significant physical (seismic) discontinuity (e.g., basement surface). This surface occurs at different levels for each of these structures. Various implications of physical discontinuities, large and small, have been discussed by Short (1965), Dence *et al.* (1965) and Ullrich (1975). However, due to the complex nature of the impact process, all three of the above features could play a role in determining differences in the uplift structure. More work will be needed on the mechanics of central uplift formation before specific differences in projectile and target media properties can be related to differences in impact crater structures.

Acknowledgments—We would like to express our sincere thanks to the management and personnel of the Newton County Stone Company and especially to M. Scott of the Bloomington Crushed Stone Company, Bloomington, Indiana. We also wish to thank Dr. R. C. Gutschick and acknowledge his important contributions concerning quarry structure and stratigraphy. We also appreciate the help of the Indiana Geological Survey and the Indiana Highway Commission. Special thanks go to Dr. D. J. Roddy who first suggested this study to us and to G. E. King and K. E. Kerdolff who assisted in the data preparation. Comments and suggestions by Drs. Roddy, T. W. Offield, F. Hörz, A. J. Rowell, and J. A. Peoples materially improved this paper. We gratefully acknowledge the National Aeronautics and Space Administration who funded this study through grant NSG-7351.

References

Baldwin R. B. (1963) *The Measure of the Moon.* Univ. Chicago Press, Chicago. 488 pp.

Black D. F. (1964) Cryptoexplosive structure near Versailles, Kentucky. *U. S. Geol. Survey Prof. Paper 501-B*, p. B9–B12.

Bond D. C., Atherton E., Bristol H. M., Buschbach T. C., Stevenson D. L., Becker L. E., Dawson T. A., Fernalld E. C., Schwalb H., Wilson E. N., Statler A. T., Stearns R. G. and Buehner J. H. (1971) Possible future petroleum potential of Region 9—Illinois Basin, Cincinnati Arch, and Northern Mississippi Embayment. In *Future Petroleum Provinces of the United States—Their Geology and Potential, 1971*, Vol. 2 (I. H. Cram, ed.), p. 1165–1218. Mem., Amer. Assoc. Pet. Geol. 15.

Boon J. D. and Albritton C. C., Jr. (1937) Meteorite scars in ancient rocks. *Field and Lab* **5**, 53–64.

Bunch T. E. (1968) Some characteristics of selected minerals from craters. In *Shock Metamorphism of Natural Materials* (B. M. French and N. M. Short, eds.), p. 413–432. Mono, Baltimore.

Carter N. L. (1968) Dynamic deformation of quartz. In *Shock Metamorphism of Natural Materials* (B. M. French and N. M. Short, eds.), p. 453–474. Mono, Baltimore.

Cintala M. J., Head J. W. and Mutch T. A. (1976) Characteristics of fresh martian craters as a function of diameter: Comparison with the Moon and Mercury. *Geophys. Res. Lett.* **3**, 117–120.

Cintala M. J., Wood C. A. and Head J. W. (1977) The effects of target characteristics on fresh crater morphology: Preliminary results for the Moon and Mercury. *Proc. Lunar Sci. Conf. 8th*, p. 3409–3425.

Dence M. R. (1968) Shock zoning at Canadian craters: Petrography and structural implications. In *Shock Metamorphism of Natural Materials* (B. M. French and N. M. Short, eds.), p. 169–184. Mono, Baltimore.

Dence M. R., Grieve R. A. F. and Robertson P. B. (1977) Terrestrial impact structures: Principal characteristics and energy considerations. In *Impact and Explosion Cratering* (D. J. Roddy, R. O. Pepin and R. B. Merrill, eds.), p. 247–275. Pergamon, N.Y.

Dence M. R., Innes M. J. S. and Beals C. S. (1965) On the probable meteorite origin of the Clearwater Lakes, Quebec. *Roy. Astron. Soc. Canada Jour.* **59**, p. 13–22.

Dietz R. S. (1947) Meteorite impact suggested by the orientation of shatter-cones at the Kentland, Indiana disturbance. *Science* **105**, 42.

Dietz R. S. (1963) Astroblemes: Ancient meteorite-impact structures on the earth. In *The Moon, Meteorites and Comets* (B. M. Middlehurst and G. P. Kuiper, eds.) p. 285–300. Univ. Chicago Press, Chicago.

Dietz R. S. (1968) Shatter cones in cryptoexplosion structures. In *Shock Metamorphism of Natural Materials* (B. M. French and N. M. Short, eds.), p. 267–285. Mono, Baltimore.

Dietz R. S. (1972) Shatter cones (shock fractures) in astroblemes. *Proc. 24th Int. Geol. Congr.*, Section 15, p. 112–118.

Engelhardt W. V. and Bertsch W. (1969) Shock induced planar deformation structures in quartz from the Ries crater, Germany. *Contrib. Mineral. Petrol.* **20**, 203–234.

French B. M. (1968) Shock metamorphism as a geological process. In *Shock Metamorphism of Natural Materials* (B. M. French and N. M. Short, eds.), p. 1–17. Mono, Baltimore.

French B. M. and Short N. M. (eds.) (1968) *Shock Metamorphism of Natural Materials*, Mono, Baltimore. 644 pp.

Gault D. E., Quaide W. L. and Oberbeck V. R. (1968) Impact cratering mechanics and structures. In *Shock Metamorphism of Natural Materials* (B. M. French and N. M. Short, eds.), p. 87–99. Mono, Baltimore.

Gutschick R. C. (1961) The Kentland structural anomaly. In *Guidebook for Field Trips, Cincinnati Meeting,* Geol. Soc. Amer., 12–17.

Gutschick R. C. (1971) Geology of the Kentland impact structural anomaly, north-western Indiana. *Field Guide for Nat. Assoc. Geology Teachers East-Central Section,* Apr. 17, 1971, 20 pp.

Gutschick R. C. (1972) Geology of the Kentland impact structural anomaly. *Guideboook for 35th Ann. Mtg. Meteor. Soc. Field Trip*, Nov. 15, 1972.

Gutschick R. C. (1976) Geology of the Kentland structural anomaly, northwestern Indiana. *Guidebook for the 10th Ann. Mtg., North Central Section,* Geol. Soc. Amer., April 28, 1976.

Head J. W. (1976) The significance of substrate characteristics in determining morphology and morphometry of lunar craters. *Proc. Lunar Sci. Conf. 7th*, p. 2913–2929.

Hendricks H. E. (1965) The Crooked Creek structure. In *Cryptoexplosive Structures in Missouri, Geol. Soc. Amer. Field Trip Guidebook, Report of Investigation* No. 30.

Hörz F. (1968) Statistical measurements of deformation structures and refractive indices in experimentally shock loaded quartz. In *Shock Metamorphism of Natural Materials.* (B. M. French and N. M. Short, eds.), p. 243–253. Mono, Baltimore.

Hörz F. (ed.) (1971) Meteorite impact and volcanism. *J. Geophys. Res.* **76**, 5381–5798.

Innes M. J. S. (1961) The use of gravity methods to study the underground structure and impact energy of meteorite craters. *J. Geophys. Res.* **66**, 2225–2239.

Jamieson J. C. (1963) Possible occurrence of exceedingly high pressure in geological processes. *Bull. Geol. Soc. Amer.* **74**, 1067–1070.

Joesting H. R. and Henderson J. R. (1948) Preliminary report on an experimental aeromagnetic survey in Northwestern Indiana. *U. S. Geol. Survey.*

Kieffer S. W. (1971) Shock metamorphism of the Coconino sandstone at Meteor Crater, Arizona. *J. Geophys. Res.* **76**, 5449–5473.

Laney R. T. (1978) A structural and petrographic study of the Kentland, Indiana impact site. M. S. thesis, Univ. of Kansas.

Manton W. I. (1965) The orientation and origin of shatter cones in the Vredefort Ring. *Ann. N. Y. Acad. Sci.* **123**, 1017–1049.

Milton D. J. (1977) Shatter cones—An outstanding problem in shock mechanics. In *Impact and Explosion Cratering* (D. J. Roddy, R. O. Pepin and R. B. Merrill, eds.), p. 703–714. Pergamon, N.Y.

Milton D. J., Barlow B. C., Brett R., Brown A. R., Glikson A. Y., Manwaring E. A., Moss F. J., Sedmik E. C. E., Van Son J., Young G. A. (1972) "Gosses Bluff impact structure, Australia", *Science* **175**, 1199–1207.

Milton D. J. and Roddy D. J. (1972) Diplacement within impact craters. *Proc. 24th Int. Geol. Congr.*, Section 15, p. 119–124.

Oberbeck V. and Quaide W. L. (1967) Estimated thickness of a fragmental surface layer of Oceanus Procellarum. *J. Geophys. Res.* **73**, 4697-4704.

Offield T. W. and Pohn H. A. (1977) Deformation at the Decaturville impact structure, Missouri. In *Impact and Explosion Cratering* (D. J. Roddy, R. O. Pepin and R. B. Merrill, eds.), p. 321–341. Pergamon, N.Y.

Pike R. J. (1971) Genetic implications of the shapes of martian and lunar craters. *Icarus* **15**, 384–395.

Robertson P. B. (1975) Zones of shock metamorphism at the Chalevoix impact structure, Quebec. *Bull. Geol. Soc. Amer.* **86**, 1630–1638.

Robertson P. B., Dence M. R. and Vos M. A. (1968) Deformation in rock-forming minerals from Canadian craters. In *Shock Metamorphism of Natural Materials* (B. M. French and N. M. Short, eds.), p. 433–452. Mono, Baltimore.

Roddy D. J. (1968) The Flynn Creek Crater, Tennessee. In *Shock Metamorphism of Natural Materials* (B. M. French and N. M. Short, eds.), p. 291–332. Mono, Baltimore.

Roddy D. J. (1976) High-explosive cratering analogs for bowl-shaped, central uplift, and multi-ring impact craters. *Proc. Lunar Sci. Conf. 7th*, p. 3027–3056.

Roddy D. J. (1977a) Large-scale impact and explosion craters: Comparisons of morphological and structural analogs. In *Impact and Explosion Cratering* (D. J. Roddy, R. O. Pepin and R. B. Merrill, eds.), p. 185–246. Pergamon, N.Y.

Roddy D. J. (1977b) Initial pre-impact conditions and cratering at the Flynn Creek Crater, Tennessee. In *Impact and Explosion Cratering* (D. J. Roddy, R. O. Pepin and R. B. Merrill, eds.), p. 277–309. Pergamon, N.Y.

Roddy D. J. (1977c) Tabular comparisons of the Flynn Creek impact crater, United States, Steinheim impact crater, Germany, and Snowball explosion crater, Canada. In *Impact and Explosion Cratering* (D. J. Roddy, R. O. Pepin and R. B. Merrill, eds.), p. 125–161. Pergamon, N.Y.

Roddy D. J. and Davis L. K. (1969) Shatter cones at TNT explosion craters (abstract). In *EOS (Trans. Amer. Geophys. Union)* **50**, 220.

Roddy D. J. and Davis L. K. (1977) Shatter cones formed in large-scale experimental explosion craters. In *Impact and Explosion Craters* (D. J. Roddy, R. O. Pepin and R. B. Merrill, eds.), p. 715–750. Pergamon, N.Y.

Roddy D. J., Pepin R. O. and Merrill R. B. (eds.) (1977) *Impact and Explosion Cratering*, Pergamon, N.Y. 1301 pp.

Shoemaker E. M. (1963) Impact mechanics at Meteor Crater, Arizona. *The Moon, Meteorites, and Comets* (B. M. Middlehurst and G. P. Kuiper, eds.), p. 301–336. Univ. Chicago Press, Chicago.

Short N. M. (1965) "A comparison of features characteristic of nuclear explosion craters and astroblemes. *Ann. N.Y. Acad. Sci.* **132**, p. 573–616.

Shrock R. R. (1937) Stratigraphy and structure of the area of disturbed Ordovician rocks near Kentland, Indiana. *Amer. Midland Naturalist* **18**, 471–531.

Shrock R. R. and Malott C. A. (1933) The Kentland area of disturbed Ordovician rocks in northwestern Indiana. *J. Geol.* **41**, 337–370.

Stöffler D. (1966) Zones of impact metamorphism in the crystalline rocks of the Nordlinger Ries crater. *Contrib. Mineral. Petrol.* **12**, 15–24.

Tudor D. S. (1971) A geophysical study of the Kentland disturbed area. Ph.D. thesis, Indiana University, Bloomington. 111 pp.

Ullrich G. W. (1976) *The Mechanics of Central Peak Formation in Shock Wave Cratering Events.* AFWL-TR-75-88, Air Force Weapons Laboratory, Kirtland AFB, New Mexico 87117.

Ullrich G. W., Roddy D. J. and Simmons G. (1977) Numerical simulation of a 20-ton TNT detonation on the earth's surface and implications concerning the mechanics of central uplift formation. In *Impact and Explosion Cratering* (D. J. Roddy, R. O. Pepin and R. B. Merrill, eds.), p. 959–982. Pergamon, N.Y.

Willman H. B., Atherton E., Bushbach T. C., Collson C., Frye J. C., Hopkins M. E., Lineback J. A. and Simon J. A. (1975) Handbook of Illinois Stratigraphy. *Illinois State Geol. Survey Bull. 95*, 261 pp.

Wilshire H. G. and Howard K. A. (1968) Structural pattern in central uplifts of cryptoexplosion structures as typified by Sierra Madera. *Science* **162**, 258–261.

Wilshire H. G., Howard K. A. and Offield T. W. (1971) Impact breccias in carbonate rocks, Sierra Madera, Texas, *Bull. Geol. Soc. Amer.* **82**, 1009–1018.

Wilshire H. G., Offield T. W., Howard K. A. and Cummings D. (1972) Geology of the Sierra Madera cryptoexplosion structure, Pecos County, Texas. *U. S. Geol. Survey Prof. Paper 599-H*, p. H1–42.

Wilson C. W. Jr. and Stearns R. G. (1968) Geology of the Wells Creek structure, Tennessee. *Tenn. Div. Geol. Bull. 68*, 236 pp.

Wood C. A. (1973) Moon: Central peak heights and crater origins. *Icarus* **20**, 503–506.

Proc. Lunar Planet. Sci. Conf. 9th (1978), p. 2633–2658.
Printed in the United States of America

West Clearwater, Quebec impact structure, Part I: Field geology, structure and bulk chemistry

CHARLES H. SIMONDS

Lunar Curatorial Laboratory, Northrop Services, Inc.
Box 34416, Houston, Texas 77034

WILLIAM C. PHINNEY

Geology Branch, NASA Johnson Space Center, Houston, Texas 77058

PATRICIA E. MCGEE and ANN COCHRAN

Lockheed Electronics Company, Inc.
1800 Nasa Road 1, Houston, Texas 77058

Abstract—The 30 km diameter West Clearwater structure was formed in relatively anhydrous, upper amphibolite to granulitic facies, Archean, metamorphic rocks of tonalite composition, covered by a thin layer of Middle to Upper Ordovician limestone. A well defined stratigraphy of impact-melt-bearing units is preserved on islands located 6–10 km from the center of the structure. The total thickness of the melt-bearing units is about 100 m and they dip at 0.5° toward the center of the structure. The exposures on the islands are believed to be part of the bottom of the crater cavity subsequent to modification, however, unambiguous definition of the transient cavity is virtually impossible. Extending at least 300 m below the melt-bearing units is Archean basement fractured on a submillimeter scale. Only rare fractures spaced several meters apart show signs of movement across them; many of those fractures are filled with holocrystalline pseudotachylite. The basement is virtually free of maskelynite, planar features or other evidence of shock pressures over 100 kb. Too little is known about the structure of the Archean rocks to determine whether the fractured basement is actually a mega-breccia or simply cut by numerous faults. Overlying the fragmented basement is a red, friable, impact-melt-bearing fragmental breccia 0–20 m thick, with over 50% fragments larger than 1 mm. Planar features and cryptocrystalline feldspar are abundant in the clasts suggesting that a fraction of the clastic debris was shocked to pressures well over 100 kb. In the breccia, melt rock forms rinds around clasts of basement rock and the ratio of melt to fragmental debris varies laterally and vertically on scales ranging from mm to 10 m. The melt rinds and small scale heterogeneities in clast-melt mixtures resemble features of lunar impactites. The clast-rich melt rock with 25–60% clastic debris is separated from the underlying fragmental breccia by a chilled contact. The red, well jointed, massive melt rock is uniformly about 15 m thick. Dikes of glass intrude the fragmental breccia from the clast-rich melt rock, a geometry reminiscent of the Apollo 17 Station 7 boulder. Clast-poor melt rock with less than 25% (generally <15%) clasts is separated from the underlying clast-rich melt rock by a sharp, locally knife edge, contact. The contact resembles one in the Apollo 17 Station 6 boulder. The massive, poorly jointed rock is red at the base and tan or blue-grey at the top of the 85 m of preserved section. Grain size increases to the top of the clast-poor melt and by analogy with other terrestrial impact melt sheets; the upper portion of the melt sheet and overlying fragmental debris has been removed by erosion. The center of the structure has maskelynite-bearing crystalline rocks, considerably less fractured than the basement underlying the melt rocks. Poorly formed shatter cones are observed in the central islands and in the rocks underlying the melt on the islands 6–10 km from the center. The sequence of units 6–10 km from the center of West Clearwater is similar to that preserved at 65 km diameter Manicouagan and 28 km diameter Mistastin. The contrast in maskelynite formation and extent of fracturing of the basement between the center and the region further out under the melt rocks is similar to that observed at Lake St. Martin and, to a lesser degree, Manicouagan.

The impact produced over 24 km^3 of impact melt which incorporated the solid debris as it moved outward from the center of the excavation. The flow of melt was turbulent as indicated by the chemical homogeneity of the melt, the intimate mixture of debris subjected to a wide range of peak shock pressures, and the high Reynolds number expected for impact induced flows. Debris was initially incorporated into the melt as large blocks of finely shattered basement. Further flow disaggregated these blocks and mixed the finer fragments with the melt.

INTRODUCTION

A field study at the 30 km diameter West Clearwater impact crater (Fig. 1) was carried out during the summer of 1977 to identify the nature of the flow of impact melt and less shocked debris during the excavation stage of the cratering process. This paper and the companion petrographic and mineral chemistry study (Phinney *et al.*, 1978) address:

1) Comparisons with other impact craters and lunar samples.
2) Generation and emplacement mechanisms of impact melts and associated clastic debris.
3) The geometry of the transient cavity and the pristine morphology of the crater.

Structural features discussed in this paper are on a scale ranging from that of a LANDSAT image down to individual hand specimens. Those in the paper by Phinney *et al.* (1978) range in scale down to observations made by scanning electron microscope.

Previous studies

Geologic mapping, major element analyses of basement and melt rocks, and petrologic observations at West Clearwater Lake were reported by Kranck and Sinclair (1963) and Bostock (1969). Both papers conclude that the structure is of volcanic origin. However, maskelynite, shock-vitrified feldspar formed only at peak pressures in excess of 250 kb, occurs in the center of the structure (Dence, 1964) and numerous shatter cones are observed both in the central islands and the ring of islands 6–10 km from the center. Such shock features are not associated with any known type of volcanic activity.

Five holes were drilled for the Dominion Observatory (Fig. 2), now the Earth Physics Branch, Department of Energy, Mines and Resources of the Canadian government. Reports of that drilling and a gravity study are given in Dence (1964) and Dence *et al.* (1965) and are interpreted in favor of the structure as a crater of meteorite impact origin.

Grieve (1978) demonstrated that a plausible mixture of basement rocks (Bostock's 1969 analyses) yielded the average composition of the melt rocks, which form the bulk of the impact-generated units within the structure. No volcanic, sedimentary, or meteoritic additions were necessary in the mix. Palme *et al.* (1978) reported siderophile trace element analyses for two melt rocks, and found no indications of the projectile, although both Grieve (1978) and Palme *et al.* (1978) inferred about 6% carbonaceous chondrite contamination in the companion 22 km eastern structure (Fig. 1). Considerable nickel enrichment in the melt rocks of the East Clearwater structure was also reported by Currie (1971).

The melt rocks yielded a K-Ar whole rock age of 300 ± 24 m.y. and 285 ± 23 m.y. (work of R. K. Wanless, J. A. Loudon and R. D. Stevens reported in Bostock, 1969). A fission track age of 34 m.y. was determined by Fleischer and Price (1963) on melt glass. The young age probably results from annealing of the glass. The only stratigraphic control on the age is offered by the occurrence of fossiliferous middle-upper Ordovician limestone as inclusions in the melt. Paleomagnetic and Rb-Sr isotopic studies are currently underway by R. Coles (Earth Physics Branch, Ottawa) and J. Wooden (Lockheed Electronics Co., Inc., Houston) respectively, which may help define more precisely the age of the structures.

Location—access

The West Clearwater Lake structure centered at 56°14′ N 74°30′ W, is about 125 km east of Hudson Bay. The structure forms the western portion of a twin circular lake known collectively as Lac à l'Eau Claire. The structure lies south of the perma-frost line and has trees (<5 m high) only in protected depressions. The prominent E-W topographic lineations in Figs. 1 and 2 reflect the westward flow of the Laurentide ice sheet. Outcrops typically occur on N, E and S facing slopes. The gradual incline of west facing slopes marks the regions covered with till. Access for the 1977 field season was by chartered flying boat from Great Whale River, Quebec, the nearest settlement with scheduled air service and that was frequently interrupted by fog. A 22-foot long canvas and wood freighter canoe was left on Drillers Island during the winter of 1964 by Dence and associates and remains serviceable. Future users are urged to bring paint and caulking to reduce leakage, and a 5–10 HP outboard motor with a long shaft to reduce water inflow over the stern.

Our field activity during 1977 was confined to the islands at the center and 6–10 km from the center (Figs. 1 and 2), and focused on studies of the melt rocks, fragmented breccia, and their contacts with the basement. The field party consisted of Phinney and Simonds as well as M. R. Dence, R. Coles and R. Wirthlin of the Earth Physics Branch.

Country rock geology

Country rocks consist of massive to slightly foliated granodiorite, quartz monzonite, minor granite and amphibolite compositions with a variable propor-

Fig. 1. Landsat Band 5 image of the twin (Lac a' l' Eau Claire) impact structures. The field of view is 185 km, West Clearwater is 30 km in diameter. Hudson Bay and Richmond Gulf are on the west side of the image. The lighter colored land areas are those free of trees, (possibly burn scars), and the darker are tree covered. Dence *et al.* (1977) ascribe structural significance to the unforested ring extending about 15 km around the Clearwater Lake. The drainage pattern in that area is more closely spaced suggesting greater fracturing.

tion of more mafic inclusions. Mineral assemblages indicate up to granulite facies metamorphism for at least some of the country rocks. The average water content is low (0.76 ± .27%), as expected for such high grade rocks. Both metamorphosed and unmetamorphosed diabase dikes cut the Archean basement. The regional geology has been mapped only in reconnaissance mode. The basement rock samples discussed by Bostock (1969) come from 16 helicopter landing sites evenly spaced around the periphery of the western portion of the lake.

Middle-Upper Ordovician limestone blocks occur in the melt. None is observed

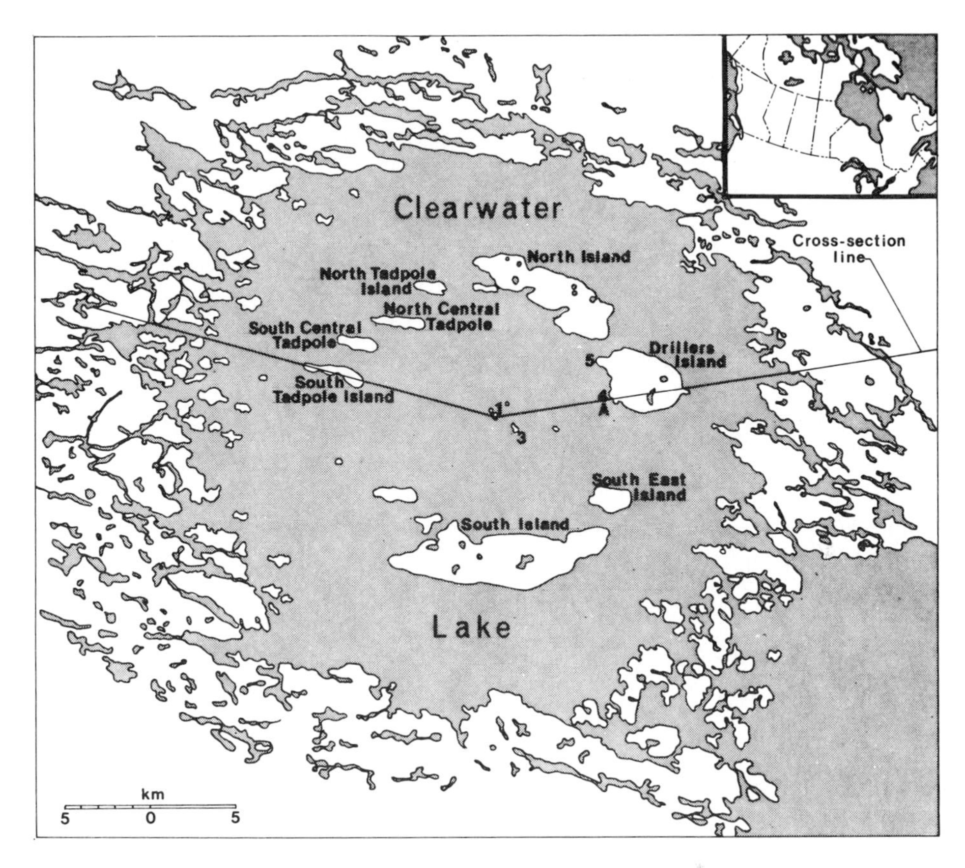

Fig. 2. Location map for islands and drill holes referred to in the text. "1" marks location of drill hole 1-63, "3" marks 3-63, "4" marks 4-63 and "A" marks 4a-63. Hole 2-63 was drilled in the East Structure.

in depositional contact with the Archean rocks. The largest limestone block, on the east end of Drillers Island, forms a 10 × 100 m outcrop. The thickness of the sedimentary cover at the time of impact is unknown, but probably was only a small fraction of the depth penetrated during crater excavation.

General observations

Figure 3 presents a summary cross-section of the geologic relations. It is essentially identical to that prepared by Dence (1964) and Dence *et al.* (1965). It differs from that of Bostock (1969) in that there is no indication of the feeder dikes for the melt sheet. Small dikes of melt do cut the basement. However, such dikes are either pseudotachylite formed *in situ* in the basement or are injected impact melt. There is no evidence for the genetic connection between large diabase dikes cutting basement and the melt sheet, particularly considering the

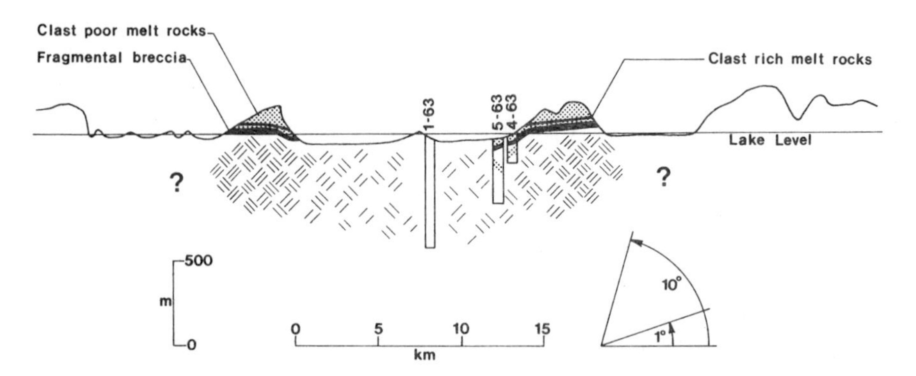

Fig. 3. Generalized and vertically exaggerated cross section of West Clearwater structure. The density of the hatches intended to schematically illustrate the density of fractures in the basement.

gabbroic composition and mode of those dikes and the feldspathic composition and mode of the melt.

Although the thickness of the three major units; the friable breccia, the clast-rich melt rocks and the clast-poor melt rocks, varies, the sequence of units is everywhere the same on the island ring 6–10 km from the center. Each of these three units is a mixture of two components: superheated impact melt from near the point of impact and less shocked debris from nearer the final edge of the excavation. This concept was put forward by Simonds (1975), Simonds *et al.*, (1978) and the papers referenced therein. Petrographically the prime difference between units is in the ratio of the two components, the fragmental breccia containing the largest proportion of the cold, less shocked debris, and the clast-poor melt rocks containing the least debris.

The three main contacts: fractured basement against fragmental breccia, fragmental breccia against clast-rich melt and clast-rich melt against clast-poor melt all dip consistently toward the center of the structure at about 0.5°. This general pattern is readily apparent on each of the islands; the melt occurs at least 50 m above the lake level on the outer edge of each island and at or below the water level on the inside. The melt also occurs below water level in drill holes 4-63, 4A-63 and 5-63 inside the main island ring (Fig. 2). On this general pattern of inward dipping units is imposed a smaller scale relief on the contacts and variations in thickness of each of the units.

There are trends of decreasing clast content and increasing grain size upward in the clast-poor and clast-rich impact melt rock units. However, no such consistent trend is noted within the fragmental breccia.

The preserved volume of the impact melt rocks is estimated by Bostock (1969) as only about 2 km^3 and he estimates that most of the melt must be preserved since the melt rocks are clearly less easily eroded than the basement rocks. He sets 8 km^3 as an upper limit in the volume. Grieve *et al.* (1977) estimate 24 km^3 assuming that the melt sheet was a continuous annular sheet or washer-shaped mass. Our estimate for a melt sheet extending for about 6 to 10 km from the center with a thickness not significantly greater than the maximum present thickness of about 100 m is the same as Grieve *et al.*'s. The thickness could be considerably greater since grain size in the sheet increases upward to the highest outcrops.

Studies at four large impact structures less eroded than West Clearwater including East Clearwater (Dence, 1964 and Dence *et al.*, 1965), Boltysh (Yurk, 1975), Popagai (Masaitis *et al.*, 1975) and Lake St. Martin (McCabe and Bannatyne, 1970) suggest that impact melt sheets are coarsest grained and least clast-rich in their interior, with decreased grain size of the melt and increased clast content at both the upper and lower margins of the sheets. A variety of melt-bearing breccias are preserved above the melt sheets in those structures. By inference, similar melt-bearing fragmental breccias must have overlain the West Clearwater melt sheet and the sheet may have been thicker than the preserved 100 m.

The energy required to melt 24 km^3 of rock is about 1.3×10^{27} ergs and the total energy of the impact may have been on the order of at least 10^{28} ergs (Dence *et al.*, 1977; O'Keefe and Ahrens, 1975). For an average terrestrial impact of 17 km/sec (Zook, 1975) and a carbonaceous chondrite composition and density as inferred from the meteorite contamination at East Clearwater, the bolide is estimated to be 1.9 km in diameter.

Fractured basement

A typical occurrence of the melt-bearing rocks is shown in Fig. 4. The fractured basement does not crop out except where exposed by wave and ice action along the shore of the lake. In most locations it is a red, oxidized mass of obviously sheared rock. The most resistant basement outcrops are the diabase dikes. The contact between the fragmental breccia and the basement was not observed either in outcrop or core. In the field the bottom of the breccia is typically covered by talus and core recovery of these friable rocks is poor.

The basement underlying the islands 6–10 km from the center, and penetrated by drill holes 4-63, 9A-63, and 5-63 (Fig. 2), is fractured on a submillimeter scale. However, mafic dikes and pegmatite pods extend continuously for several meters in each core, suggesting that the basement has not been mixed on a meter scale. Numerous dikes of pseudotachylite or clast-rich melt rock up to a few tens of centimeters thick occur in the cores. These dikes are holocrystalline, unlike the glass dikes originating in the melt sheet and penetrating the underlying fragmental breccia. Many of these holocrystalline dikes have subhedral to euhedral feldspar up to 0.1 mm long, unlike the feldspars in clast-rich melt rocks within the melt sheet. The preferred interpretation of the dikes is that they are pseudotachylite resulting from frictional heating. The gneiss on either side of a dike typically differs in color, mode, degree of foliation or grain size suggesting that some displacement has taken place across these dikes. On a thin section scale, the basement consists in large part of subrounded to angular fragments, 0.1–1 mm across, which lack shock features. The fractures define anastomosing

Fig. 4. An exposure of the melt bearing units about 30 m high on the south side of South Tadpole Island (Fig. 2). The top unit forming the rounded slopes above the cliff is the clast-poor impact melt unit. The cliff-forming unit is the clast-rich impact melt rock. The fragmental breccia included the white highly fractured gneiss block seen on the left side of the picture. The fragmental breccia extends from below the water line up to the base of the cliff.

surfaces which cut through all thin sections of basement from the cores. There are few zones even a cm across that are free of such fractures. Fracturing is sufficiently wide-spread that it has apparently lowered the density of the rocks in the crater producing the 10 mgal residual gravity anomaly over the structure (Dence *et al.*, 1965). The gravity data suggest that the fracturing extends out beyond the island ring to the margin of the lake. Since the topographic highs are the melt rocks, glacial excavation, apparently, was easier in these fractured rocks than in the impact melt rocks. Fracturing in the cores 4-63, 4A-63 and 5-63 correlates with mineralogy. Pegmatites and mica-bearing gneisses are finely comminuted; the mica-free gneisses and mafic dike rocks are less fractured with millimeter-wide spacings between fractures.

The degree of visible fracturing in outcrops on the central islands and in drill holes 1-63 and 3-63 is less than that on the island ring or in drill holes 4-63, 4A-63, or 5-63.

There are some shatter cones preserved in the fractured basement on the island ring 6–10 km from the center. Poorly formed shatter cones also occur on the central islands with axes that appear to plunge at more than 45°. However, no detailed measurements were made in the field.

The samples from the central islands display partial to complete transformation of feldspar to maskelynite. Planar features abound in the quartz. Pyroxenes are twinned and biotite has kinkbands. Peak shock pressures appear to be 200–250 kb (Stöffler, 1972). Thin sections from core 1-63, 36 m below lake level, show plagioclase with no reduction in birefringence, suggesting rapid attenuation of shock pressure with depth, provided the original geometrical relationships between rocks on the islands and the cores have not been disturbed during crater modification.

Fragmental breccia

The fragmental breccia is a red friable rock with over 50% fragments coarser than 1 mm. As discussed by Phinney *et al.* (1978), the matrix may have contained glass, in which case the breccia is not simply crushed basement. This unit corresponds to Bostock's (1969) "friable volcanic breccia", and is readily differentiated in the field from the underlying fractured basement by the latter's lack of a fine matrix and general preservation of mafic dikes, pegmatite pods and compositional banding. The fragmental breccia can be differentiated from the overlying clast-laden melt rock by the latter's tough, coherent structure, abundance of columnar cooling joints and content of less than 20% clasts over 1 mm.

The thickness of the unit varies from only a few meters to a maximum of over 20 m (Fig. 5). The unit is typically less than 5 m thick. As stated previously, the lower contact of the breccia with the underlying fractured basement was not observed either in the field or drill core.

On the outcrops on the south side of Drillers Island and east sides of North and South East Islands (Fig. 2), there is at least 15 m of local relief on the contact between breccia of basement within horizontal distances of 100–200 m. The relief is greater than the thickness of the fragmental breccia and the entire stratigraphy (fragmental breccia, clast-rich and clast-poor melt rock) swings up over protrusions in the basement. There is some suggestion that there is more relief on exposures (cliffs or steep slopes) which are radial to the structure than those which are concentric.

Blocks up to 10 m are recognizable as clasts in the fragmental breccia, but blocks >1 m across make up only about 1% of the total deposit. Blocks in the centimeter-meter range make up 20–40%. Detailed size distributions for material less than 1 cm are given by Phinney *et al.* (1978). It should be emphasized that the latter measurements are made on samples collected between clasts larger than 10 cm. The larger blocks are angular to subangular. Comminution of these blocks seems to occur by breaking along fractures rather than by rounding-off of corners.

The breccia on the whole is massive with flow foliations developed only around some of the larger clasts. The distribution of clasts and melt-rich matrix is horizontally and vertically irregular (Fig. 5). In Fig. 5, the melt-rich regions can be seen protruding from the cliff face because they are more coherent than the clast-rich zones surrounding them.

Fig. 5. Fragmental breccia exposed on cliff on south side of North Central Tadpole Island (Fig. 2). The field of view is about 15 m high. The massive crystalline looking zone in the right half of the picture has a greater amount of melt than the more friable rocks on the left half. Irregular distribution of melt seems to be characteristic of the fragmental breccia.

Smaller scale melt-rich zones occur as: 1) glass dikes, 2) crystallized rinds around individual fragments, and 3) millimeter to centimeter thick zones of crystallized melt with irregular morphology. A photo of well preserved examples of glass dikes is shown in Fig. 6. At least one of the dikes can be traced into the overlying melt (Fig. 6), indicating that the dike was intruded from above. The intrusion of the dikes into the underlying unit requires that the fragmental breccia existed as a unit prior to emplacement of the dikes.

The geometry of these dikes is reminiscent of that observed in the Apollo 17 Station 7 boulder [Apollo Field Geology Investigation Team (AFGIT), 1973] (Fig. 7). The rinds of melt (Fig. 8) surrounding clasts are also reminiscent of features observed on Apollo 17, in this case sample 72275 (Ryder *et al.*, 1975) (Fig. 9). The irregular distribution of clasts and melt shown in Fig. 10 also has lunar analogs, particularly in an Apollo 14 breccia, as discussed by Stöffler *et al.* (1978). The general lack of clast digestion in the fragmental breccia (Phinney *et al.*, 1978) is similar to that observed in Apollo 14 by Simonds *et al.* (1977).

Clast-rich melt

The clast-rich rock is red to purple coherent rock with up to 25% fragments coarser than 1 mm. The subhedral morphology of feldspar in the rock indicates

Fig. 6. Exposure of clast-rich melt rock overlying fragmental breccia on north east end of North Island (Fig. 2). The glass dike appears to extend nearly continuously down from the clast-rich impact melt into the underlying breccia. Other dikes of glass can not be directly connected with the overlying melt. The glass shows strong foliation parallel to the long dimension of the dikes. The foliation is defined by both lines of clasts and schlieren in the glass.

Fig. 7. Lunar equivalent of outcrop shown in Fig. 6. The glass dikelet sampled as 77075-77 intruded from clast-laden melt into a larger clast in a fragmental breccia. Exposure is the Apollo 17 Station 7 Boulder, photo number AS 17-146-22305. Diagram taken from Apollo Field Geology Investigation Team (1973).

that the matrix is crystallized from a melt (Phinney *et al.*, 1978; Bostock, 1969). The unit was referred to by Bostock (1969) as the "coherent volcanic breccia". It has much greater coherence and fewer clasts than the underlying fragmental breccia, and is finer grained and has more clasts than the overlying clast-poor melt rocks. Jointing normal to the basal contact is well developed; however, the jointing dies out near the top of the unit resulting in better exposures of the clast-rich unit compared to the clast-poor unit.

The thickness of the clast-rich melt zone is relatively constant at about 15 m,

Fig. 8. Sample of melt rind surrounding granite inclusion in sample 77-149 from the outcrop illustrated in Fig. 5. Cube is 1 cm on a side. The inner dark part of the rind is glass and glass altered to clay minerals. The outer lighter part of the rind is finely crystalline with a grain size of about 1 um.

and persists over the local relief on the basal contact. The lower contact of the unit with the underlying fragmental breccia is sharp, and is a chilled margin.

Typically sub-angular blocks over 1 m across are concentrated at the base of the melt sheet, where they make up a total of over 1%. The only vesicular zones in the clast-rich or clast-poor melt rocks are adjacent to large blocks of gneiss. Debris in the 1 cm–1 m range makes up about 10% in the lower few meters falling off to about 1% 15 m above the base. As discussed by Phinney *et al.* (1978), this trend for decreasing clast content upward in the sheet is observed in each traverse as a whole, although variations in the pattern are observed locally.

The clast-rich melt rock is massive. The only foliations recognizable in the field or hand specimen are around the larger clasts.

The extreme oxidation of the clast-rich melt has resulted in virtually total removal of mafic silicates (augite and low calcium pyroxene) from the matrix. There is no obvious tendency for the degree of oxidation to decrease upwards in the unit.

Fig. 9. Finely crystalline melt rind surrounding clast in lunar sample 72275 from the Apollo 17 Station 2 Boulder. The sample as a whole is a fragmental breccia with small amounts of melt binding the matrix together.

Clast-poor melt rocks

The clast-poor melt rock is a purple, tan or grey coherent unit with less than 15% fragments coarser than 1 mm. The unit corresponds to Bostock's (1969) "massive quartz latite". In the field the coarser grain size, lower clast content and generally poorer jointing of clast-poor melt all allow unambiguous identification of this unit.

The unit occurs on the highest points on all of the islands 6–10 km from the center. The clast content decreases and grain size continuously increases upwards to the top of the unit. As indicated previously, some has been removed by erosion. The preserved portion of the unit is 85 m thick and probably made up most of the volume of the impact melt sheet prior to erosion. The contact between clast-rich and clast-poor melt, at least insofar as it is reflected by the change in jointing

Fig. 10. Fragmental breccia sample 77-141 from outcrop in Fig. 5 illustrating the variable distribution of clasts.

pattern, mimics the relief on the base of the fragmental breccia, and the clast-poor, clast-rich contact displays the 0.5° dip to the center of the structure shown by the lower contacts.

The base of the clast-poor melt locally displays a knife edge contact with the underlying clast-rich melt (Fig. 11). Throughout the sheet the transition always takes place over a vertical distance of less than 3 m (Phinney *et al.*, 1978). Similar sharp contacts were noted in the Apollo 17 Station 6 boulder, sample 76215 (Fig. 12).

Blocks over 1 m across occur in the clast-poor melt but they are rare. Blocks 1 cm–1 m make up about 1% of the volume. The blocks are made of gneiss as well as the Ordovician limestone mentioned earlier. Blocks of such large size should settle with velocities of up to cm/sec through the melt if it were totally liquid. The blocks have subangular to subrounded outlines and in general, seem more rounded than in the lower units.

Only the lower part of the clast-poor unit is oxidized to give the red, or purple color to the rocks. The least weathered upper rocks are blue-grey in color. The

Fig. 11. Sample 77-84D from the center of South Island (Fig. 2) illustrating the nature of the contact between very fine-grained, dark colored clast-rich melt, and the coarser-grained, lighter colored clast-poor melt. Top of the sample is located on the left. The coarser clast-poor melt has a spherulitic texture. The cube is 1 cm on a side.

height of the highest red or purple rocks varies from 15–20 m to nearly 40 m above the clast-poor clast-rich contact. Petrographically some oxidation is visible in even the freshest and highest samples, but not enough to strongly color the rocks.

Bulk chemistry

The bulk chemical analyses of melt rocks and basement (Bostock, 1969) have been supplemented by microprobe analyses of glass from dikes in the fragmental breccia, and pulverized and fused chips of glassy and crystalline melt rock prepared according to the procedures outlined by Brown (1977). Results are given in Fig. 13 and 14. The plots reveal that the glass and crystalline melt rocks have a more restricted range of composition than the target rocks, a fact

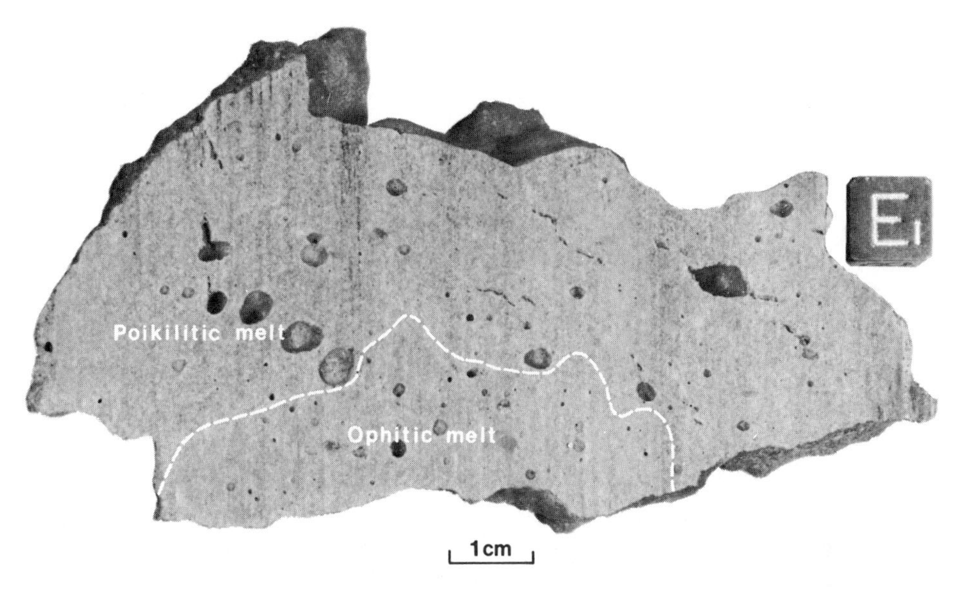

Fig. 12. Lunar equivalent to material illustrated in Fig. 11. The slab is of lunar sample 76215 from the Apollo 17 station 6 boulder. Petrographic studies of thin sections allow textural classification of the two textures of the melt rock. The poikilitic rock has a clast content of about 9%, the ophitic part about 2%.

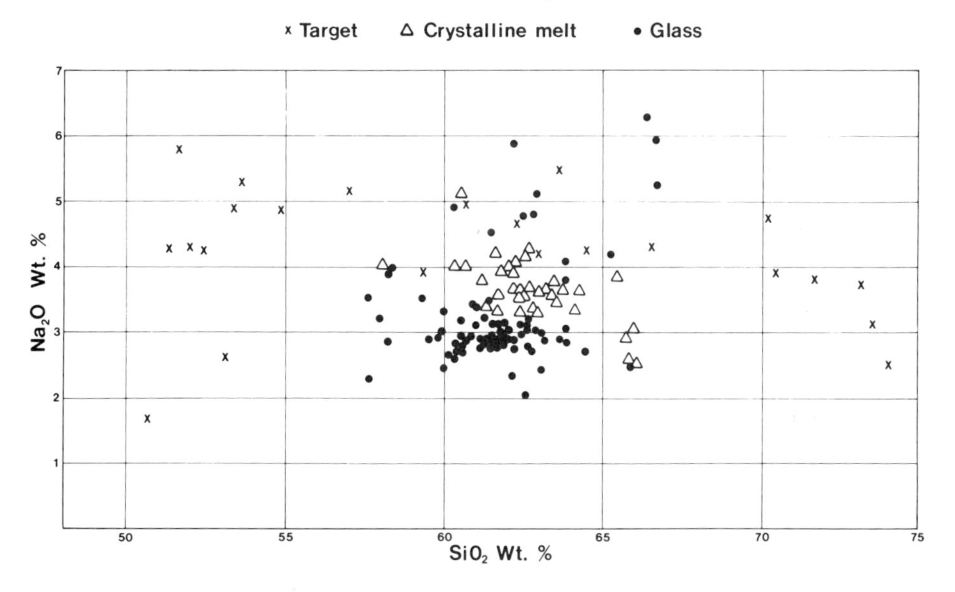

Fig. 13. Water-free silica and soda contents with major element analyses normalized to 100% (from both Bostock 1969 and new analyses). The silica and soda-rich glass analyses are by microprobe of light colored schlieren in rocks from glass dikes such as shown in Fig. 6.

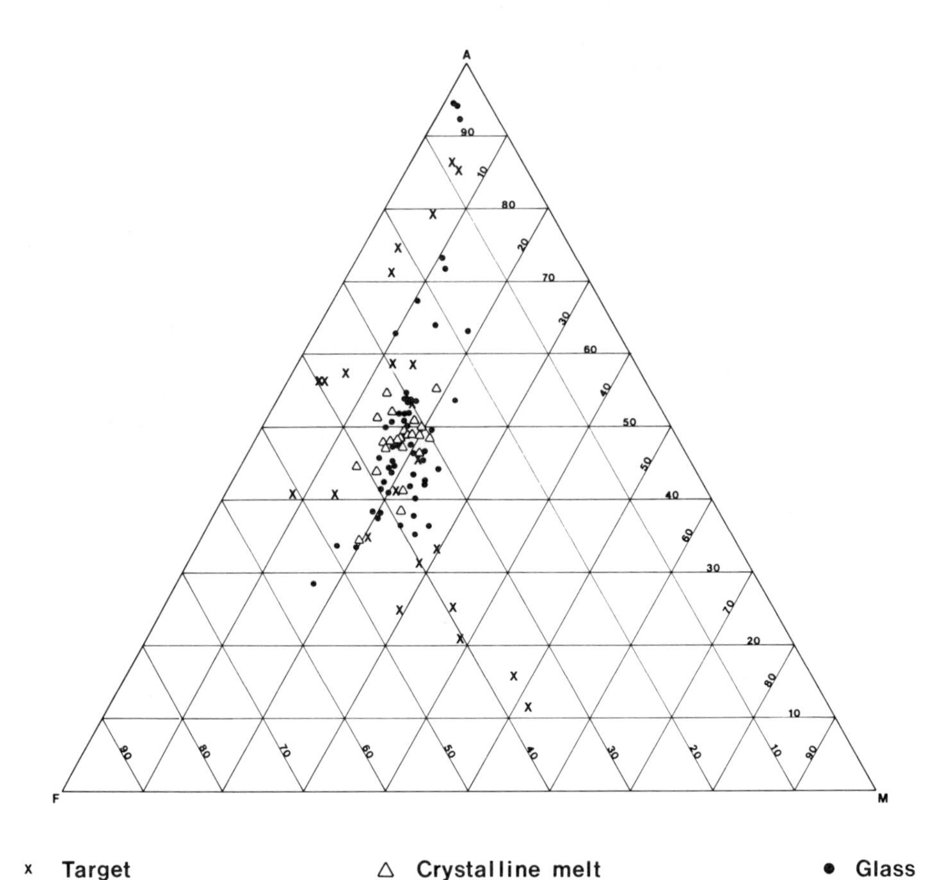

Fig. 14. Al-Fe-Mg plot (weight%) for West Clearwater constitutents with the Fe component including both FeO and Fe_2O_3. Note that most of the melt rock analyses form a tight cluster in the center of a much broader distribution of the country rocks. The few Al rich glasses are microprobe analyses of light colored schlieren.

previously discussed by Grieve *et al.* (1977). The few extremely aluminous, alkali-rich glasses tend toward feldspar compositions and are microprobe analyses of light colored schlieren in the glasses. The AFM plot (Fig. 14) does not show any obvious fractionation trend in the melt.

The only systematic compositional differences between melt and target rocks are in Fe_2O_3/FeO and H_2O, both of which are higher in the melt rocks. Fe_2O_3/FeO is 0.61 for target rocks and 3.5 for melt. H_2O is 0.76 ± .27% for target and 1.33 ± .29% melt rocks. Two lines of evidence suggest that oxidation took place after crystallization. First, primary titano-magnetite is replaced by ilmenite-hematite. Second, the average pyroxene has molar MgO/MgO + FeO = .72 ± .05; such pyroxenes could crystallize from a melt with MgO/MgO + FeO of 0.57 characteristic of the country rock. However, the melt rock analyses have a MgO/MgO + FeO of 0.80. Since a melt must crystallize a pyroxene more magnesian than itself, the observed oxidation state of the bulk melt analyses must have been attained after pyroxene crystallization.

The SiO_2-Na_2O (Fig. 13) plot suggests that the glass compositions are depleted in Na_2O. The one glass chip sample reported by Bostock (1969) had 3.1% Na_2O in contrast to all other melt rocks which ranged from 3.6–4.9% (average 3.8%). Precautions were taken to minimize loss due to electron beam effects during the analyses and included large spot size, short counting time, and analyzing Na_2O in the first triad of elements. If Na_2O depletion is real, the Na_2O is believed to have been leached during the same events which oxidized and hydrated the melt rocks. Phinney *et al.* (1978) point out that the material which, in hand specimen and thin section, looks like glass is, in fact, largely montmorillonite. Montmorillonite readily exchanges Na.

Discussion

Comparisons with other terrestrial craters

Stratigraphy and structure: The succession of units recognized on the island 6–10 km from the center of the structure, is identical to that observed 10–23 km from the center of the 65 km diameter Manicouagan impact structure (Floran *et al.*, 1978), and to outcrops in a similar structural position at the 28 km Mistastin structure (Grieve 1975). The upward progression from fractured basement (mega-breccia?) through variable amounts of melt-bearing, polymict, friable, fragmental breccia into a sequence of clast-rich and then clast-poor impact melt rocks appears to be generally characteristic of large terrestrial craters formed in crystalline rocks.

Rocks under the Ries crater and near the present topographic rim show similar fracturing to that observed under the islands 6–10 km from the center at Clearwater (Pohl *et al.*, 1977). Because of the well established stratigraphy at the Ries Crater, it is possible to demonstrate that the fractured rocks are a mega-breccia with blocks tens of meters across. The only evidence for displacements of large blocks at Clearwater are the mismatches of lithology across the pseudotachylite dikes.

It is a characteristic of larger craters that the level of shock effects observed in crystalline clasts of the fragmental breccia [abundant cryptocrystalline feldspar (crystallized maskelynite?), planar features, undulose extinction] is greater than in the fractured basement underlying the melt and breccia away from the center of the structure (Phinney *et al.*, 1978). The shock effects preserved in the breccia are greater than in the melt rocks because clasts with high levels of shock are relatively hot and the samples have considerable stored strain energy and, therefore, are more completely digested by the melt (Grieve, 1975; Simonds, 1975; Phinney *et al.*, 1978).

West Clearwater is like other large craters in crystalline targets in having maskelynite-bearing rocks in the central uplift, and having central uplift rocks showing less fracturing and granulation than the basement rocks further from the center of the structure. Maskelynite, or at least feldspar with reduced birefringence, has been observed in central uplift rocks from Manicouagan (Bunch *et al.*, 1967) and Lake St. Martin (McCabe and Bannatyne, 1970). At Manicouagan a central uplift of shocked garnet anorthosite forms two prominent mountains that stand 300 m above the surrounding impact melt sheet (Murtaugh, 1975; Currie, 1972). At Manicouagan, Paleozoic limestone occurs locally at the base of the 200 m thick melt sheet (possibly marking the former ground surface) suggesting that the central mountains are uplifted at least 500 m relative to the base of the periphery of the sheet where the limestone occurs. Sweeney (1978) argues that the center has been uplifted over 5 km based on interpretation of gravity data. The central peaks are north of the center of the structure and may be an eccentric central peak or part of a central peak ring. Traverses up to 2 km long across the central peaks indicate that the rocks are much less deformed than several uplifts of sedimentary target craters such as Sierra Madera (Wilshire *et al.*, 1972) or Gosses Bluff (Milton *et al.*, 1972). Brick-red dikes (pseudotachylite?), up to 10 cm thick, cut the anorthosite on a spacing of about 100 m. Dikes vary greatly in thickness and strike along their length. In most cases metamorphic features are not offset by the dikes. However, a small percentage of the dikes are straighter, have well-developed shear zones along their contacts and flow banding along their margins, and offset features in the anorthosite. Although there has been differential movement across these latter dikes, foliation within the anorthosite in the entire central uplift is remarkably constant, ranging over less than 60° of strike and conforming to lithologic contacts mapped within the anorthosite. The amount of rotation thus seems to be less than that observed in craters with targets of sedimentary rock where 90° rotations are common (Wilshire *et al.*, 1972; Milton *et al.*, 1972).

The samples from the central uplift at Manicouagan are garnet anorthosites with variable amounts of biotite, scapolite, and hornblende. The plagioclase in all samples has reduced birefringence, particularly at garnet-plagioclase grain boundaries, suggesting partial conversion to maskelynite. Samples with complete transformation of plagioclase to maskelynite are rare, suggesting that the peak shock pressure in the anorthosite was only 250 kb (Gibbons and Ahrens, 1977). Hand specimens and thin sections of the basement under the melt sheet more than 15 km from the center are free of the fractures which are abundant under the melt sheet at West Clearwater. In general, the samples are also free of planar elements, maskelynite, or reduced birefringence which could be due to shock deformation.

Studies of the Lake St. Martin structure have been limited to examination of drill cores obtained by the Manitoba Dept. of Mines, Resources and Environmental Management (McCabe and Bannatyne, 1970). The quartz-biotite gneisses of the deepest core from the center of the structure seem more disrupted than the outcrops and cores from the center of West Clearwater. The Lake St. Martin core is cut by several red dikes with sheared boundaries, some of which separate lithologies. In spite of the evidence for at least localized deformation, the dip of the metamorphic foliation is always about 45°. As at Manicouagan, there has been minimal rotation of even the blocks separated by the pseudotachylite dikes. Net upward movement of the central gneiss by at least 215 m is required to lift the crystalline Precambrian basement above the level inferred from drill holes outside the structure.

Only about 1 m of core of Precambrian basement from the region flanking the central uplift is available from Lake St. Martin, and that basement could be simply a large block in a breccia. The granite in that core is fractured to about the same degree as the gneisses in the cores 6 km from the center of West Clearwater.

We have only briefly examined thin sections from Lake St. Martin, however, McCabe and Bannatyne (1970) report for the central hole extensive formation of maskelynite, some of which is recrystallized to a mosaic texture. Peak shock pressures are, therefore, over 250 kb (Gibbons and Ahrens, 1977).

Composition and mixing of melt: The most comprehensive suite of analyses of impact melt rocks and target rocks comparable to the set available for West Clearwater come from Manicouagan (Floran *et al.*, 1978 and Grieve and Floran, 1978). Extensive data sets are also available from Popagai (Masaitis *et al.*, 1975) and the Ries (Graup, 1975; von Engelhardt, 1967; and Stähle, 1972); however, those two craters had thick sedimentary covers at the time of impact. Table 1 shows that the range in composition of chips of melt rocks from West Clearwater is greater than for chips from the melt rocks at Manicouagan. However, the Manicouagan target rocks have a more diverse compositional range (SiO_2 40–74%) than those at West Clearwater (SiO_2 47–73%). The 1 σ variations of the published analyses of melt rocks and target rocks for the two structures show the same trend (Table 1).

The greater compositional range of West Clearwater samples may be related to the higher clast content in equivalent lithologies than was the case at Manicouagan. However, the reason for the greater clast content at West Clearwater is not understood. It may be related to differences in the melt's liquidus temperatures. However, precise estimates of even the relative temperature of the liquidi and solidi cannot be made. West Clearwater has a more silicic melt but the target water content is lower than at Manicouagan and the two features should produce off-setting effects on the melting temperature. The combined compositional effects give West Clearwater a significantly more viscous melt, 28,000 poise at 1100°C, compared to the Manicouagan melt's viscosity of 3500 poise at the same temperature. Possibly the greater melt viscosity in some way enhances clast incorporation at West Clearwater.

Limits on transient cavity size and pre-erosion crater shape

Trying to remove the effects of 250–300 million years of terrestrial erosion to reconstruct the mechanics of crater deformation is full of ambiguities, particularly for craters in crystalline rocks where the target lacks obvious stratigraphic markers. However, some attempt is necessary to facilitate the use of terrestrial

Table 1.

	Manicouagan			West Clearwater		
	Average Melt[1]	Melt 1σ	Country Rock 1σ[2]	Average Melt[3]	Melt 1σ	Country Rock 1σ[4]
SiO_2	57.75	1.21	9.18	59.09	2.64	6.95
TiO_2	.77	.06	.67	.97	.60	.41
Al_2O_3	16.57	.64	5.67	16.15	1.56	1.19
Cr_2O_3	?			.15	.08	
Total Fe as FeO	5.87	.26	4.43	5.22	1.49	2.28
MnO	.11	.02	.09	.20	.095	.05
MgO	3.50	.54	2.86	2.77	.78	1.85
CaO	5.92	.97	3.66	3.70	1.07	1.69
Na_2O	3.82	.29	1.22	3.90	.87	.53
K_2O	3.03	.38	1.77	4.08	.79	.82
P_2O_5	.22	.03	.39	.29	.08	.18
H_2O	1.73	.37	1.12	1.58	.68	.27

(1) Average and standard deviation of all analyses from Floran *et al.* 1978.
(2) Unweighted average of all country rock analyses, from Currie (1972) and Grieve and Floran (1978).
(3) Unweighted and standard deviation average of 84 analyses for all elements except H_2O (10 analyses) and Cr_2O_3 (67 analyses) for melt rocks from Bostock (1969) and new analyses.
(4) Unweighted average of all analyses of Bostock (1969).

crater studies in unraveling the geology of lunar and martian craters where surficial morphology comprises much of the data base. Figure 15 shows schematically some of the limits on the geometry of the West Clearwater event. The peak shock pressures were inferred assuming that the meteorite plus 24 km^3 of melt (a minimum estimate) represented the total volume inside the 600 kb peak shock pressure isobar, and that peak shock pressure fell off as $R^{-2.5}$ (Dence *et al.*, 1977). Recent studies give rates from R^{-2} (Ahrens and O'Keefe, 1977) to R^{-5} and higher (Robertson and Grieve, 1977), and it is not obvious what is the cause of the discrepancy. The 600 kb peak pressure is probably a good estimate for the beginning of melting but is an underestimate of the pressure required to produce a total melt. The work of Schaal and Hörz (1977) suggests that the range in peak shock pressure over which melting occurs is over 400 kb for basalt with total melting at well over 1000 kb. The value for granitic rocks is probably somewhat less, 900–1000 kb, but relevant experimental data is not available. The maskelynite-bearing rocks located at the center limit the depth of excavation to less than the depth of the 250 kb isobar. The outer margin of the excavated cavity is not well documented by field observations; the one shown in Fig. 15 is along the lines favored by Dence *et al.* (1977). They propose that the initial excavation extended out no further than the island ring and that the 30 km diameter structure forming the circular lake is the result of slumping of transient cavity

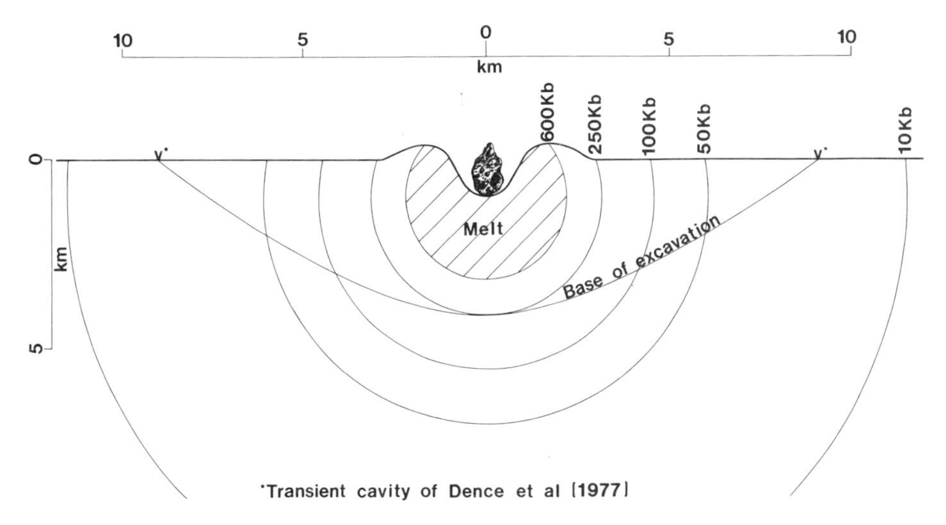

Fig. 15. Relationships between excavated cavity and shock contours for Clearwater crater at peak shock pressure.

walls extending from 9 to 15 km from the center. Critical to their arguments are the limestone outcrops which they interpret as part of the slumped walls but which also may be interpreted as clasts turbulently incorporated into the melt and carried downward by the impact-induced flow. The evidence is inconclusive that the limestone outcrops moved down during the modification stage rather than being driven down by the excavation because the limestone is not in sedimentary depositional contact with the Archean rocks. The great degree of fracturing under the region from 6–15 km from the center is implied by the 10 mgal negative gravity anomaly which extends beyond the 10 km radius of the island ring (Dence *et al.*, 1965), and the fact that glacial erosion has been able to remove the basement rocks inside the 15 km radius more easily than outside that circle. Dence *et al.* (1977) argue that these fractured rocks are slumped in a manner similar to the "mega-block zone" of the Ries (Pohl *et al.*, 1977). Studies of explosion craters such as the review by Cooper (1977) or the calculations of Orphal (1977) point out that nuclear explosion craters excavate rocks near the surface where peak shock pressures range from 1–2 kb for shallow explosions to only a few hundred bars for more deeply buried explosives. Using the 2.5 attenuation law, the 15 km radius surface corresponds to a peak pressure of about 5 kb. Therefore, if the flow during the excavation stage for a large impact is as effective as for moderately buried small nuclear events, then rocks far outside the 15 km radius may be excavated, or they may lie outside the transient cavity.

Nature of impact induced flow

Turbulence: The chemical homogenization of the melt rocks implies that individual analyzed aliquots of melt rock contain material from a range of target lithologies. Such intense mixing implies very turbulent flow. This turbulent flow is also implied by the very high initial velocities to which the melt must be accelerated (Cooper, 1977; O'Keefe and Ahrens, 1975). Characteristic Reynolds numbers for impact induced flows should be 10^6 for 1 km/sec velocities across distances of kilometers. Such high Reynolds numbers are characteristic of turbulent flow.

Boundaries between flow units: The sequence of units (fragmented basement, fragmental breccia, clast-poor melt rocks and clast-rich melt rocks) observed on the islands 6–10 km from the center of the structure represent at least three types of flow units.

As indicated above, the fragmented basement may be a breccia with blocks several meters across. But the amount and nature of flow of those rocks is not known. Dence *et al.* (1977) argue that the rocks are part of the mass which moved downward and inward during the modification stage of the event.

Some flow must have taken place in the fragmental breccia because it contains both melted material from inside the 1000 kb peak shock pressure isobar and less shocked debris from outside this isobar. The melt which is observed is texturally different and finer-grained than that in either the pseudotachylite dikes in the basement or that in the overlying melt sheet. If the flow had been simply laminar flow along the floor of the cavity margin it should have developed a better foliation than we observed. The greater shock effects in the fragmental breccia indicate that it came from nearer the center of the crater than the underlying fractured basement.

The clast-rich melt rock appears to have a chilled margin against the underlying fragmental breccia, implying that the breccia was emplaced prior to the clast-rich melt. Furthermore, dikes of the melt were injected into the fragmental breccia implying that the melt was still fluid after the breccia was in place.

The irregular detailed distribution of clasts in the the overall pattern of an upward decreasing clast content suggests that the clast incorporation process is one of first combining larger masses of melt and fragmental debris and then having the two become progressively better mixed. It is interesting to note that the most evenly distributed clast populations are in the uppermost rocks with the fewest clasts. The preservation of zones with the spherulitic texture discussed by Phinney *et al.* (1978) suggests that some virtually clast-free melt from inside the 600 kb isobar must remain intact throughout the cratering process.

Acknowledgments—The manuscript was improved greatly by reviews of Fred Hörz, Richard A. F. Grieve and Thomas Offield.

Portions of this work performed by C. H. Simonds, Principal Investigator, were supported by the Lunar Support Research and Technology program through the Universities Space Research Association, which is under contract No. NSR-09-051-001 with the National Aeronautics and Space Administration; the Lunar Curatorial Laboratory support task order contract NAS 9-15425 between the Johnson Space Center and Northrop Services, Inc.; and contract NAS 9-15200 between the Johnson Space Center and Lockheed Electronics Co., Inc. This paper constitutes the Lunar and Planetary Institute Contribution No. 339.

Field equipment was supplied by the Canadian government, and the cooperation of M. R. Dence and R. Coles of the Earth Physics Branch is greatly acknowledged.

References

AFGIT (Apollo Field Geology Investigation Team) (1975) Documentation and environments of Apollo 17 samples. In *Astrogeology #73, Apollo 17 Landing Site Geology*, Part I, p. 1–316. NASA-JSC, Houston.

Ahrens T. J. and O'Keefe J. D. (1977) Equations of state and impact-induced shock wave attenuation on the moon. In *Impact and Explosion Cratering* (D. J. Roddy, R. O. Pepin and R. B. Merrill, eds.), p. 639–656. Pergamon, N.Y.

Bostock H. H. (1969) The Clearwater Complex, New Quebec. *Geol. Survey Can. Bull.* **178**.

Brown R. W. (1977) A sample fusion technique for whole rock analysis with the electron microprobe. *Geochim. Cosmochim. Acta* **41**, 435–438.

Bunch T. E., Cohen A. J. and Dence M. R. (1967) Natural terrestrial maskelynite. *Amer. Mineral.* **52**, 244–253.

Cooper H. F. Jr. (1977) A summary of explosion cratering phenomena relevant to meteor impact events. In *Impact and Explosion Cratering* (D. J. Roddy, R. O. Pepin and R. B. Merrill, eds.). p. 11–44. Pergamon, N.Y.

Currie K. L. (1971) Origin of igneous rocks associated with shock metamorphism as suggested by geochemical investigations of Canadian craters. *J. Geophys. Res.* **76**, 5575–5585.

Currie K. L. (1972) Geology and petrology of Manicouagan resurgent caldera, Quebec. *Geol. Survey Can. Bull.* **198**, 153 pp.

Dence M. R. (1964) A comparative structural and petrographic study of probable Canadian meteorite craters. *Meteoritics* **2**, 249–269.

Dence M. R., Grieve R. A. F. and Robertson P. B. (1977) Terrestrial impact structures: Principal characteristics and energy considerations. In *Impact and Explosion Cratering* (D. J. Roddy, R. O. Pepin and R. B. Merrill, eds.), p. 247–276. Pergamon, N.Y.

Dence M. R., Innes M. J. S. and Beals C. S. (1965) On the probable meteorite origin of the Clearwater Lakes, Quebec. *J. Roy. Astron. Soc. Can.* **59**, 13–22.

Engelhardt W. v. (1967) Chemical composition of Ries glass bombs. *Geochim. Cosmochim. Acta* **31**, 1667–1689.

Fleisher R. L. and Price P. B. (1963) Chunks of the moon. *Time M.* (Can. Ed.) **8**, #22, 44.

Floran R. J., Grieve R. A. F., Phinney W. C., Warner J. L., Simonds C. H., Blanchard D. P. and Dence M. R. (1978) Manicouagan Impact Melt, Quebec, Part I: Stratigraphy, Petrology and Chemistry. *J. Geophys. Res.* In press.

Gibbons R. V. and Ahrens T. J. (1977) Effects of shock pressures on calcic plagioclase. *Phys. Chem. Min.* **1**, 95–107.

Graup G. (1975) Das Kristallin in Nordlinger Ries. Diss. Univ. Tubingen, Germany.

Grieve R. A. F. (1975) Petrology of the impact melt at Mistastin Lake Crater, Labrador. *Bull. Geol. Soc. Amer.* **86**, 1617–1629.

Grieve R. A. F. (1978) Meteoritic component and impact melt composition at the Lac a' l' Eau Claire (Clearwater) impact structures, Quebec. *Geochim. Cosmochim. Acta* **42**, 429–431.

Grieve R. A. F., Dence M. R. and Robertson P. B. (1977) Cratering process: As interpreted from the occurrence of impact melts. In *Impact and Explosion Cratering* (D. J. Roddy, R. O. Pepin, and R. B. Merrill, eds.), p. 791–814. Pergamon, N.Y.

Grieve R. A. F. and Floran R. J. (1978) Manicouagan impact melt Quebec. Part II: Chemical inter-relations with basement and formational processes. *J. Geophys. Res.* **83**. In press.

Kranck S. H. and Sinclair G. W. (1963) Clearwater Lake, New Quebec. *Geol. Survey Can. Bull.* **100**.

McCabe H. R. and Bannatyne B. B. (1970) Lake St. Martin crypto-explosion crater and geology of surrounding area. *Geol. Survey Manitoba, Geol. Paper 3/70*. 79 pp.

Masaitis V. L., Mikhaylov M. V. and Selivanouskaya T. V. (1975) Popagai Meteorite Crater. *Publ. of All Union Sci. Res. Geol. Inst. Nauk, Moscow*. 124 pp. In Russian.

Milton O. J., Barlow B. C., Brett R., Brown A. R., Glikson A. Y., Manwaring E. A., Moss F. J., Sedmik E. C. E., VanSoon J. and Young G. A. (1972) Gosses Bluff Impact Structure, Australia. *Science* **175**, 1199–1207.

Murtaugh J. G. (1975) Geology of the Manicouagan crypto-explosion structure. Unpublished Ph.D. thesis, Ohio State University, Colombus.
O'Keefe J. D. and Ahrens T. J. (1975) Shock effects from a loose impact on the moon. *Proc. Lunar Sci. Conf. 6th*, p. 2831–2844.
Orphal D. L. (1977) Calculations of Explosion Cratering—I. The shallow-buried nuclear detonation JOHNIE BOY. In *Impact and Explosion Cratering* (D. J. Roddy, R. O. Pepin, and R. B. Merrill, eds.), p. 897–906. Pergamon, N.Y.
Palme H., Janssen M. J., Takahashi H., Anders E. and Hertogen J. (1978) Meteoritic material at five large impact craters. *Geochim. Cosmochim. Acta* **42**, 313–323.
Pohl J., Stöffler D., Gall H. and Ernstson K. (1977) The Ries impact crater. In *Impact and Explosion Cratering* (D. J. Roddy, R. O. Pepin and R. B. Merrill, eds.), p. 343–404. Pergamon, N.Y.
Phinney W. C., Simonds C. H., Cochran A. and McGee P. E. (1978) West Clearwater, Quebec impact structure, Part II: Petrology. *Proc. Lunar Planet. Sci. Conf. 9th*. This volume.
Robertson P. B. and Grieve R. A. F. (1977) Shock attenuation at terrestrial impact structures. In *Impact and Explosion Cratering* (D. J. Roddy, R. O. Pepin and R. B. Merrill, eds.), p. 687–702. Pergamon, N.Y.
Ryder G., Stoeser O. B., Marvin U. B., Bower J. F. and Wood J. A. (1975) Boulder 1, Station 2, Apollo 17: Petrology and petrogensis. *The Moon* **14**, 327–357.
Schaal R. B. and Hörz F. (1977) Shock metamorphism of lunar and terrestrial basalts. *Proc. Lunar Sci. Conf. 8th*, p. 1697–1729.
Simonds C. H. (1975) Thermal regimes in impact melts and the petrology of the Apollo 17 Station 6 boulder. *Proc. Lunar Sci. Conf. 6th*, p. 641–672.
Simonds C. H., Phinney W. C., Warner J. L., McGee P. E., Geeslin J., Brown R. W. and Rhodes J. M. (1977) Apollo 14 revisited, or breccias aren't so bad after all. *Proc. Lunar Sci. Conf. 8th*, p. 1869–1893.
Simonds C. H., Floran R. J., McGee P. E., Phinney W. C. and Warner J. L. (1978) Petrogenesis of melt rocks, Manicouagan impact structure, Quebec. *J. Geophys. Res.* **83**. In press.
Stähle V. (1972) Impact glasses from suevite of the Nördlinger Ries. *Earth Planet. Sci. Lett.* **17**, 275–293.
Stöffler D. (1972) Deformation and transformation of rock forming minerals by natural and experimental shock processes I. Behavior of minerals under shock compression. *Fortschr. Mineral.* **49**, 50–113.
Stöffler D., Knoll H. D., Stähle V. and Ottemann J. (1978) Textural variations of crystalline matrix of Fra Mauro breccias and a model of breccia formation (abstract). In *Lunar and Planetary Science IX*, p. 1116–1118. Lunar and Planetary Institute, Houston.
Sweeney J. F. (1978) Gravity study of a great impact. *J. Geophys. Res.* **83**. In press.
Wilshire H. G., Offield T. W., Howard K. A. and Cummings O. (1972) Geology of the Sierra Madera crypto-explosion structure, Pecos County, Texas. *U.S. Geol. Survey Prof. Paper 599-H*, 42 pp.
Yurk Yu. Yu., Yeremenko G. K. and Polkanov Yu. A. (1975) The Boltysh depression: A fossil meteorite crater. *Internat. Geol. Rev.* **18**, 196–202.
Zook H. A. (1975) The state of meteoritic material on the moon. *Proc. Lunar Sci. Conf. 6th*, p. 1653–1672.

Proc. Lunar Planet. Sci. Conf. 9th (1978), p. 2659–2693.
Printed in the United States of America

West Clearwater, Quebec impact structure, Part II: Petrology

WILLIAM C. PHINNEY
NASA Johnson Space Center, Houston, Texas

CHARLES H. SIMONDS
Northrop Services, Inc., Houston, Texas

ANN COCHRAN and PATRICIA E. MCGEE
Lockheed Electronics Company, Inc., Houston, Texas

Abstract—Impactites lining the base of the crater at West Clearwater Lake consist of a 0 to 20 m thickness of fragmental breccia overlain by about 18 m of clast-rich melt rock which in turn is overlain by at least 85 m of clast-poor melt rock. Grain size of matrix material increases upward throughout the section by over two orders of magnitude. Total clast contents decrease upward by factors of 3 to 5. Clasts display features indicating shock pressures from <50 kb to >800 kb. Irregularities in grain size and clast content occur within the clast-rich rocks but within a given vertical traverse these irregularities stabilize over a few meters within the clast-poor rocks. The matrix grain-size is clearly related to clast-content: the larger the clast-content, the smaller the grain-size. Distribution of clast-contents suggests that two types of material flowed outward from the center in the lower part of the crater: a basal fragmental material with up to 20% melt and an overriding melted material with up to 25% included fragments. Turbulence throughout the entire mass produced a complex mixing of melt and fragments throughout a 15 to 20 m zone between the two types of material. A pervasive red to purple alteration occurs throughout the breccia and the clast-rich melt rocks as is observed at many other terrestrial craters. The alteration results from hydrothermal formation of montmorillonitic clays (Mg-rich saponite at West Clearwater) together with hematite or hydrous ferric oxides from glass and mafic minerals. These reactions take place shortly after impact by the influx of groundwater while the rocks are still relatively warm but below solidus temperatures.

INTRODUCTION

Purpose of study

Clearwater Lake, Quebec (officially Lac à l'Eau Claire) consists of two circular bodies of water commonly referred to as West Clearwater Lake and East Clearwater Lake. The West Clearwater Lake structure forms a circle whose diameter is approximately 35 km. In an accompanying paper, Simonds *et al.* (1978b) describe the field and structural relations on scales ranging from the structure as a whole, down to hand specimens. Although the circular nature of the lake and an inner ring of islands led to the inclusion of the structure in a list of possible meteorite impact craters (Beals *et al.*, 1960), the first published geologic and petrologic description of the unusual breccias and igneous rocks of the inner islands and central islands was by Kranck and Sinclair (1963) who interpreted the structure as a volcanic complex. Following a core drilling program at both East and West Clearwater Lakes, Dence (1964) and Dence *et al.* (1965), on the basis of structure and lithology, ascribed the pair of circular features to a meteorite impact origin. Dence suggested three broad divisions of the impact-produced allochthonous blanket: a basal breccia, a fragment-rich unit with a crystalline matrix, and a crystalline unit with few fragments and a coarser-grained texture resembling that of a diabase sill. In support of an impact origin McIntyre (1968), Bunch (1968), and Robertson *et al.* (1968) provided petrographic descriptions of shock metamorphism in fragments from the breccias. The most detailed prior study of West Clearwater Lake is that of Bostock (1969) who favored a volcanic origin for the structure. Bostock was the first to provide a systematic, detailed description of the unusual rocks which he divided into three units from bottom to top: friable volcanic breccia, coherent volcanic breccia, and massive quartz latite. The correlative terms in the present paper are fragmental breccia, clast-rich melt rock, and clast-poor melt rock.

Previous studies of impact melt sheets at Lake St. Martin (McCabe and Bannatyne, 1970), Popigay (Masaitis *et al.*, 1975), Mistastin (Grieve, 1975) and Manicouagan (Floran *et al.*, 1978; Simonds *et al.*, 1978) have been limited by scarcity of outcrops, extensive cover of vegetation, or incomplete sections through the strata. Clearwater Lake is north of the tree line, displays numerous outcrops on cliffs, terraces and knobs, and provides several sections of the three major units on the ring of islands. In addition, there is a greater abundance of clasts in the units at West Clearwater thus allowing more detailed determinations of the distribution and shock levels of clasts. Observation of fragmental breccia is limited to three small outcrops below the melt sheet at Manicouagan in sharp contrast to about a dozen, well-displayed, thick exposures of the equivalent unit at West Clearwater. Therefore, detailed petrologic variations at West Clearwater can be interpreted within a detailed structural and stratigraphic context. The first purpose of this study is to provide detailed petrologic data for the three major stratigraphic units.

Alteration of the rocks at West Clearwater was a major concern in the work of Bostock (1969). The development of clay in the breccias, glass, and pyroxenes is widespread. Oxidation is apparent from both the abundance of hematite and the high Fe_2O_3/FeO ratio. The reddish color of the fine-grained rocks and the alternation of reddish purple and gray zones even in the coarser-grained rocks were noted for all vertical sections studied. Determining the origin and time of formation of such extensive alteration is a second purpose of this study.

The overall similarity between the observations at Manicouagan and West Clearwater are striking. Common to both structures are fragmental breccia at the base, a melt sheet grading from fine-grained and clast-rich near the base to coarser-grained and clast-poor higher in the section, extensive alteration involving oxidation and clays especially in the lower parts of the section, and shock features in clasts and the central uplift. Further comparison of these features and implications for the dynamic behavior of melt and clasts during cratering as derived from detailed petrologic observations are a further purpose of this study.

ANALYTICAL METHODS

Sampling was conducted on all of the four largest islands of the island ring, on two of the Tads, from three of the low-lying central islands (Fig. 1) and from the five drill cores previously drilled at West Clearwater Lake (Dence, 1964). Detailed petrologic observations were conducted on samples collected along six vertical traverses: two each on Driller's Island, Southeast Island, and South Island (Fig. 1). Samples were collected at vertical intervals typically between ½ meter and 10 meters depending on the scale of the variations (see Fig. 2). In most cases segments of a vertical traverse had to be laterally offset by hundreds of meters to obtain a complete vertical section.

Clast counting was performed in different ways for the different size ranges: binocular microscope counts of slabbed, wet, hand specimens for the 1 mm–3 cm range; petrographic microscope counts on polished thin sections for the 40 μm–1 mm and 10 μm–40 μm range; and SEM photo counts of etched thin sections for some 10 μm–40 μm and all <10 μm ranges. Care was taken to insure that the areas counted were representative. A line intercept technique was used in all clast counting. Clast

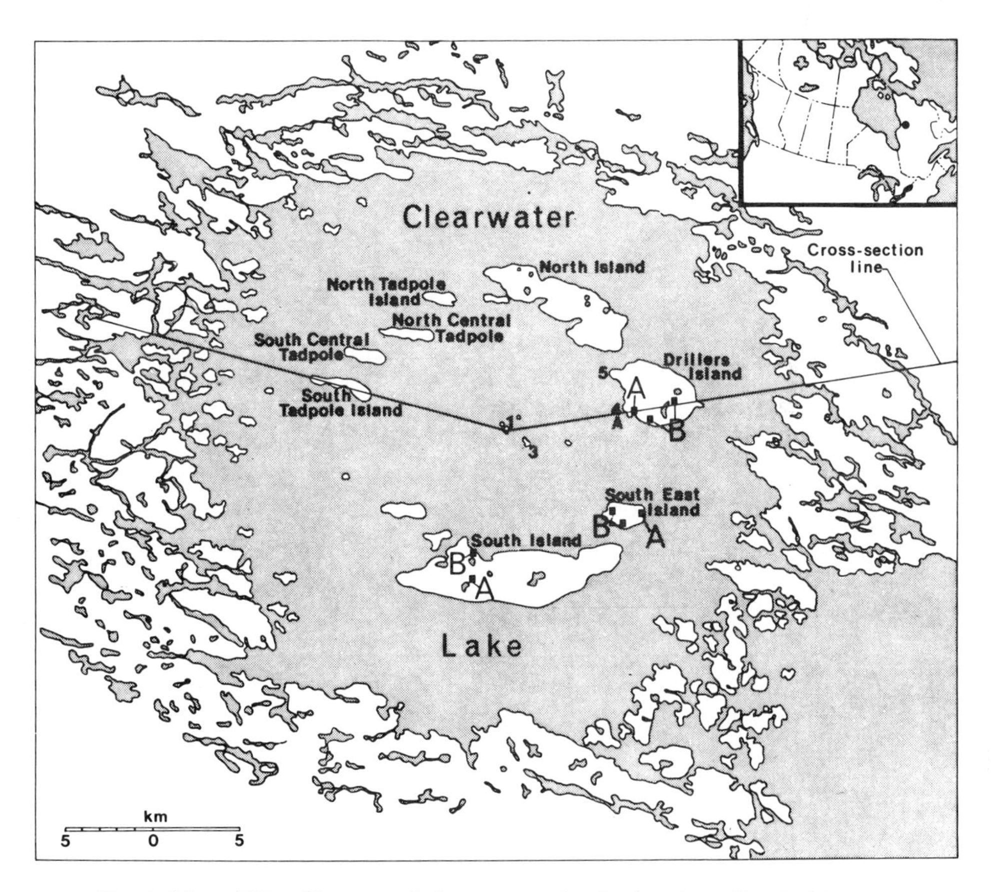

Fig. 1. Map of West Clearwater Lake structure showing locations of vertical traverses (A and B).

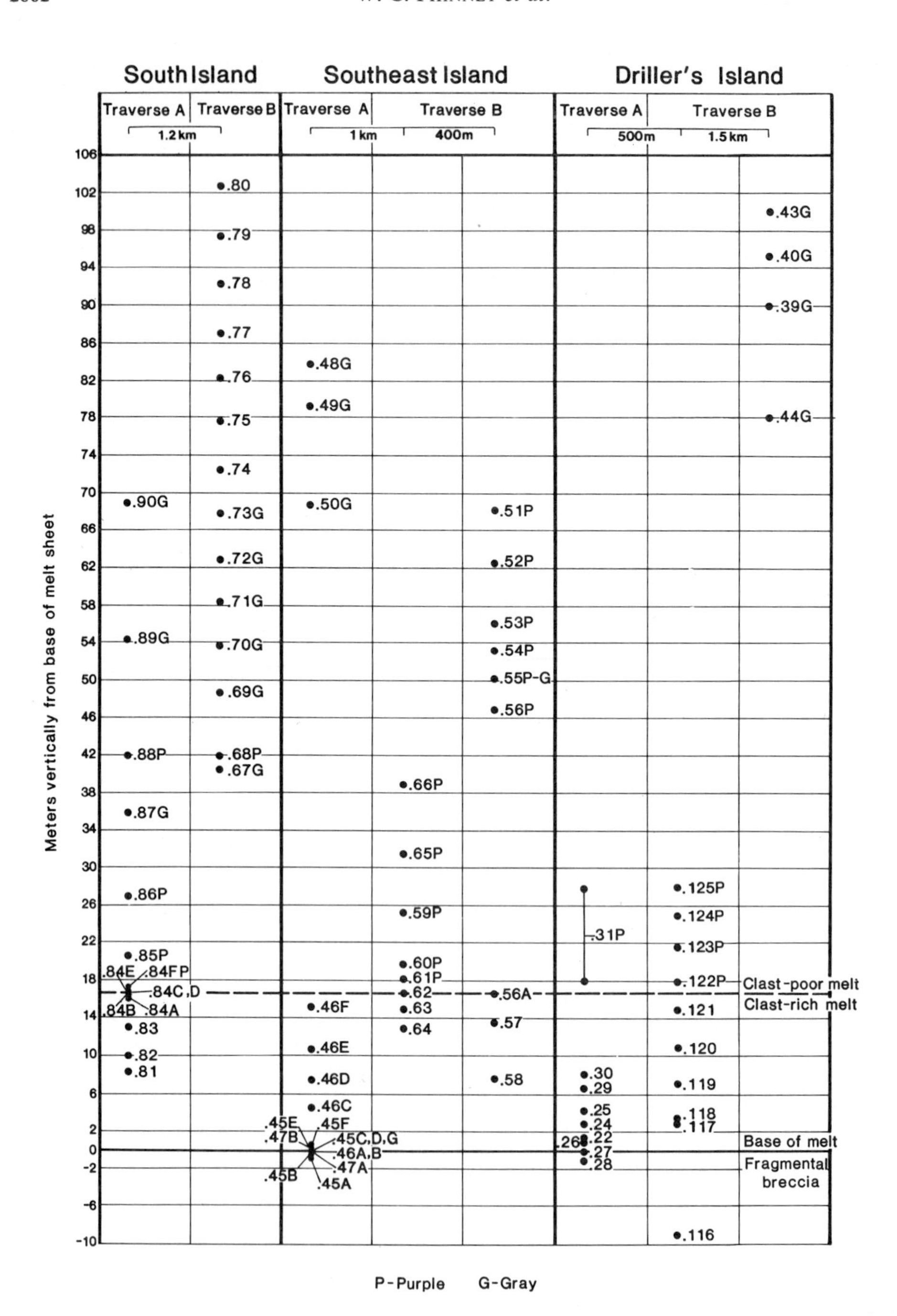

Fig. 2. Locations of samples in the vertical traverses selected for study.

percentages were calculated by integrating the lengths of clast intercepts over each of a series of parallel, linear traverses. The total distance of clasts traversed divided by the total distance traversed equals the average clast percentage for that sample. The clast percentages were also totalled at the end of each traverse, to test the homogeneity of the clast distributions. Because the clasts are randomly oriented, no bias should be introduced from preferred orientations. All clast counts for a given size range were performed by one person to insure consistency in clast identification within that size range.

Matrix grain size determinations were made for feldspars, pyroxenes, and opaques on polished thin sections. Dimensions of 15 to 20 grains of each mineral were measured in each section, excluding the long dimension of laths.

The petrographic descriptions were generally based on a transmitted and reflected light study of etched or unetched, polished thin sections. Larger-scale features, such as interfingering contacts, were studied from slabbed hand specimens under the binocular microscope and the fine-grained matrices and alteration products were studied with the SEM.

Percentages of clasts displaying shock features were determined by identifying clasts with characteristic shock textures (Stöffler, 1971) and calculating the percentage of shocked clasts in the total clast population by the intercept technique described previously.

The ARL and MAC microprobes were used to determine chemical compositions for feldspars, pyroxenes, and opaques in the matrices of various samples. Because of the often fine-grained nature of the matrix, a small electron beam ($<5\ \mu$) was used. Mineral standards were used on both the MAC and the ARL. The opaque minerals were analyzed for nine elements on the MAC using matrix and Bence-Albee corrections. Most feldspar and pyroxene analyses were done for three elements on both the MAC and the ARL. Three-element analyses (K, Ca, and Na for feldspar and Ca, Fe, and Mg for pyroxene), were corrected for beam drift, background, and dead time.

Series of SEM, 100-second, polaroid photos (from 100× to 15,000×) were taken at 25 KV of ion-etched, gold-palladium coated thin sections of fine-grained samples from several traverses. The ion-etching technique, which enhances mineralogical differences and textural relationships is discussed in Phinney *et al.* (1976). The gold-palladium coating technique was acquired from U. S. Clanton (pers. comm., 1977). On the photographed areas energy dispersive (EDAX) spectra were run at 15 KV to permit identification of phases.

X-ray powder diffraction studies were run on the matrix clays of two fragmental breccias. Pieces of the samples were hand-ground then stirred into deionized water, and decanted twice: once after 15 minutes and again after two hours to separate out the coarser clasts. The decanted liquid containing the suspended clay-sized fraction was settled overnight. A slurry from the settled clay was placed on a glass slide, allowed to dry, and run on the X-ray powder diffractometer from 4° to $40°2\theta$. The same sections were than glycolated in two ways which produced the same expansion of the clays: a watch glass of ethylene glycol was placed on a hot plate under a jar with the sample and evaporated at a low temperature for several hours; a drop of ethylene glycol was placed directly on the sample and allowed to dry. The glycolated samples were again run from 4° to $40°2\theta$. Distinct feldspar peaks were used to correlate the diffractometer scans.

Samples of the separated clay-sized fraction were dispersed onto a nucleapore filter by vacuuming a mixture of a small amount of the sample, dispersing agent, and water, down onto the filter. The filter was then cemented at the corners onto a carbon planchette and gold-paladium coated for the SEM. In this way the EDAX could be used on individual clay grains, reducing feldspar and opaque contamination of the spectra.

Description of Units

Fragmental breccia

The nature of fragmental breccia occurrences is highly varied. On North Island there are blocks 3 to 10 m across of disrupted basement with infillings of

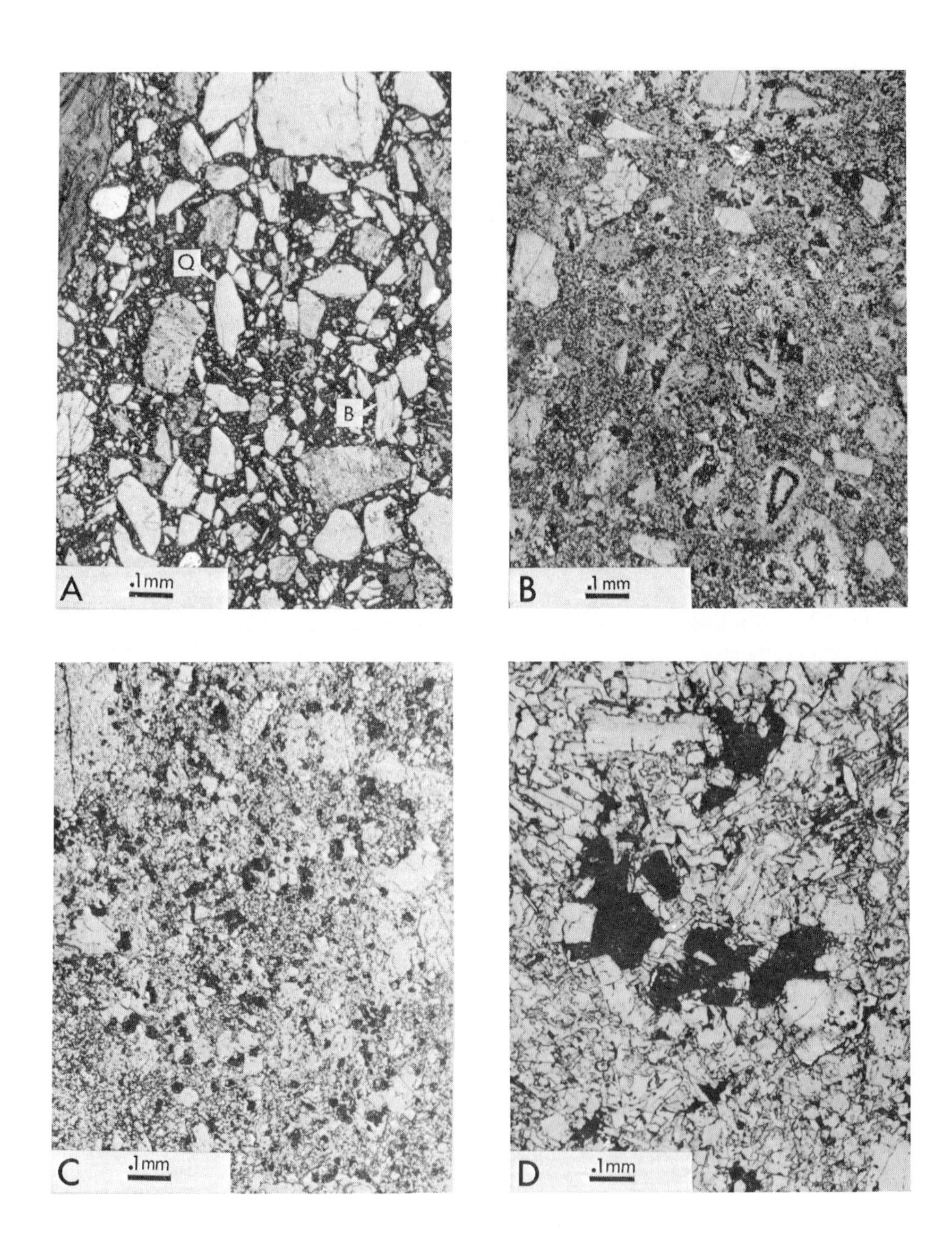
Q
B
A
.1mm
B
.1mm
C
.1mm
D
.1mm

finer-grained fragmental breccia. On one of the Tads a 20 meter cliff is composed entirely of fragmental breccia containing meter-sized lenses of clast-laden melt rocks. On the east shore of Southeast Island the melt sheet seems to rest directly on basement rocks although the basement rocks may be large disrupted blocks as discussed by Simonds et al. (1978b). On Driller's Island there is a 15 to 20 meter thickness of fragmental breccia below a reasonably good section of melt rocks. The samples of fragmental breccia chosen for detailed petrologic studies are from Driller's Island, although additional samples from the Tads were also studied by SEM for comparison.

Fragmental breccias display overall a clast-matrix texture (Fig. 3a) in thin section. The breccia as a whole contains many large lithic clasts: at least one percent are larger than one meter across and between 20 and 40% are between one centimenter and one meter across. Samples selected for thin sections and studies of clast distribution (Table 2 and Fig. 12) represent the finer-grained portions of the breccia as it occurs between the cobbles and boulders. Therefore, the clast proportions given in Table 2 and Fig. 12 do not represent the breccia as a whole but rather the 50 to 80% of the breccia which exists between the large clasts.

Clasts in the thin sections include various mineral fragments from the

---

Fig. 3. (A) Reflected light photo of a clast-laden fragmental breccia (WCW-77-116) located 9 meters below the base of the melt sheet on Driller's Island. Most of the clasts are unfractured, twinned, angular feldspars with undulatory extinction. Some of the feldspars are altered and have a grey mottled appearance in the picture. Other clasts include undulatory quartz (Q), opaques (rounded, white) and deformed biotites (B). The clasts are set in a fine-grained clay matrix (see Fig. 6). The breccia also contains a high percentage of amber-colored glass lenses (upper left corner). Note the flow lines in the glass and the way it has molded around an opaque clast, indicating that the glass was soft when the breccia formed. (B) Reflected light photo of a clast-rich melt rock (WCW-77-23) located 1.5 meters above the base of the melt sheet on Driller's Island. The abundant mono- and polycrystalline undulatory quartz clasts are rimmed by fine-grained pyroxenes which are, in turn, enclosed in a medium-grained interlocking quartz and feldspar rim. The twinned undulatory feldspar (light grey), altered feldspar (mottled), opaque (white) and pyroxene clasts do not have these necklaces. The fine-grained matrix contains interlocking plagioclase and alkali feldspar laths and blocks, quartz grains, interspersed opaques and minor interstitial clays. (C) Reflected light photo of a red, clast-poor melt (WCW-77-124), collected 8.5 meters above the clast-rich, clast-poor boundary on Driller's Island. The clasts are indulatory twinned feldspars, mottled feldspars, undulatory quartz, opaques (white, top center) and pyroxenes. The ragged edges of some clasts (mottled feldspar in upper left corner) indicate that they are partially digested. The matrix is composed of medium-grained granular feldspars, opaques, pyroxenes (grey-black), and quartz. (D) Reflected light photo of very large-grained, subophitic, clast-poor melt (WCW-77-41) which was collected 78.5 meters above the clast-rich, clast-poor contact on Driller's Island. The large, dark grey, altered Ca-rich pyroxene oikocryst sits in a matrix of twinned, zoned, seriate feldspar blocks and laths, interstitial quartz fingers, and white opaques. The feldspar clast in the lower right corner has a remnant core which is visible under transmitted polarized light.

basement rocks: rounded to angular fragments of polymict breccias with frothy to glassy matrices; angular to rounded fragments of glass, some of which contain flow structure; lithic fragments of basement rocks; and a few angular fragments of breccias with very fine-grained crystalline matrices. The quartz and feldspar fragments exhibit a range of shock features (see Table 1) suggesting that most of the debris was exposed to shock pressures of greater than 50 kb (Stöffler, 1971) and that some was exposed to pressures greater than 800 kb (Schaal and Hörz, 1977). The lithic clasts display undulatory extinction in their feldspars and some of the mafic mineral clasts display fracturing and chemical alteration to clay and hydrous Fe-oxides. The clasts of breccias with frothy or glassy matrices (Fig. 4a,b and c) have less distinct boundaries between clast and matrix than do the typical fragmental breccias (Fig. 6). This suggests that reaction between clasts and matrix was more extensive in the frothy breccias because of higher temperatures or greater porportions of melt within the forthy breccias. This interpretation would also account for the lack of micrometer-sized clasts which appear to have been digested in the matrix of the frothy breccias. The fragments of glass with flow structure may be either angular, indicating that they were brittle and were fractured before incorporation into the breccia, or they may be rounded or elongate stringers which are molded around other clasts (Figs. 3a and 5a) indicating that they were hot and soft when incorporated into the breccias. The occurrence of clasts containing glass and crystalline matrices further verifies the extreme range of shock pressures and temperatures undergone by the debris

Table 1. Distribution of Shock Features in Quartz and Feldspar Clasts.

		Fragmental Breccias			Clast-Rich	Melt Rock
		145	142	141	23	46c
FELDSPAR	Devitrified	42%	35%	35%	30%	26%
	Undulatory extinction	24	33	27	33	33
	Mosaic texture	20	9	24	0	0
	Maskelynite	0	0	0	0	4
	No shock features	14	22	14	37	37
QUARTZ	Planar features	64	9	87	17	22
	Perlitic cracks	21	91	13	69	37
	No shock features	15	0	0	14	41

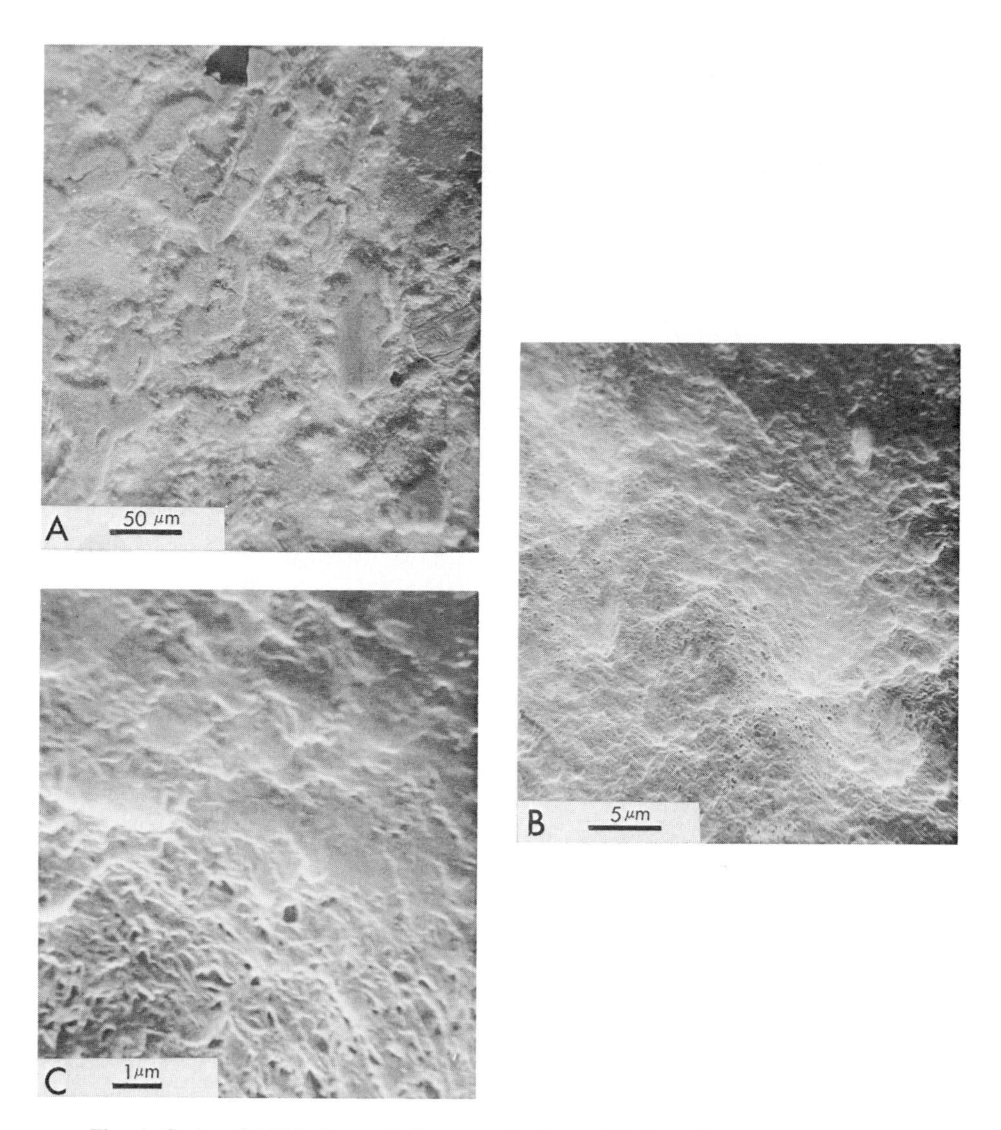

Fig. 4. Series of SEM photos of glassy matrix clasts. (A) Overall texture showing the partly digested rounded clasts present in the frothy matrix. Note the light-grey plagioclase lath at the top center, the striated, digested pyroxene at the center right-hand edge, and the medium-grey quartz clast just to the left of the pyroxene. (B) Contact of a digested clast in the frothy glassy matrix. Compare the lumpy surface of this clast (center right) with the smoother more distinct clasts in the fragmental breccias of Fig. 6. (C) Clay-rich matrix at high magnification. It is similar to the fragmental matrix (Fig. 6) in composition and texture except for a somewhat tighter network of clays.

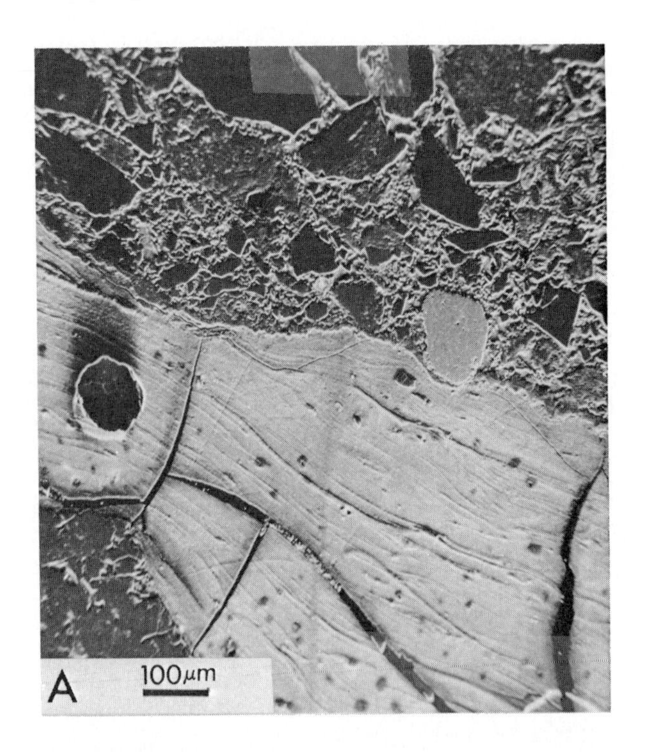
A
100μm

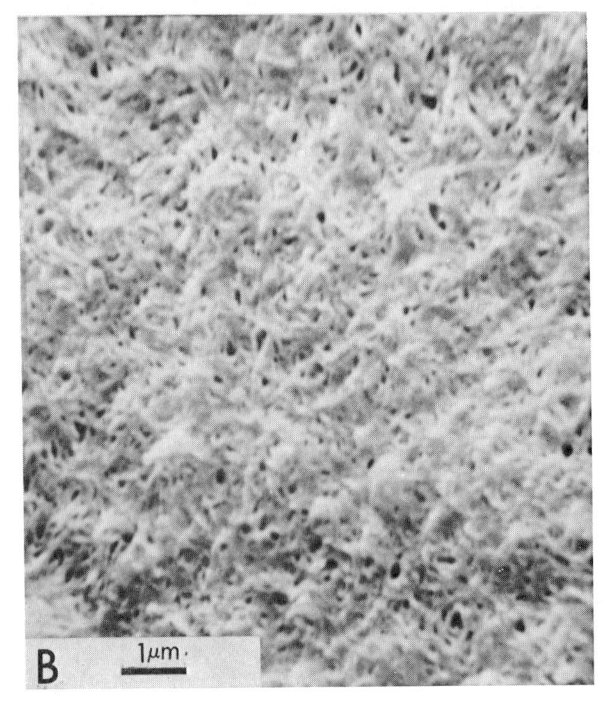
B
1μm

in the fragmental breccias. Clearly this debris must represent a mixture of materials whose sources represent a wide range of distances from the point of impact.

The matrix of the fragmental breccias is defined as any material that is not clearly clastic debris. Figure 6 displays two sequences of fragmental breccias. Although the clastic debris is similar in both of them the texture of the matrix differs considerably. In sample 28B the matrix displays a honeycomb or open boxwork pattern (Fig. 6a,b and c) whereas sample 116 displays a more closely woven wormy intergrowth (Fig. 6d,e, and f). Photomicrographs of the matrices of frothy breccia clasts (Fig. 4c) and glass fragments (Fig. 5b) at similar high magnification show textures that resemble the intergrowths of 116 but are even more densely interwoven. On the assumption that these textures were of clay-minerals, an unsuccessful search of the clay literature was made for SEM photomicrographs showing similar morphologies.

X-ray diffraction studies of concentrates from the matrices of 28B and 116 indicated expandable lattice clays. Sample 28B expanded from 13.96 Å to 16.67 Å after glycolation and 116 expanded from 14.03 Å to 16.52 Å. Energy dispersive X-ray analyses of these clays while in the SEM indicate a very Mg-rich composition (Fig. 7) with variable amounts of Al and Fe and small amounts of Ca and K. These data suggest clay minerals close to the Mg end-member, stevensite, of the saponitic clay series (Carroll, 1970; Millot, 1970). The nearly pure Mg product is probably the result of nearly complete oxidation of iron in the original matrix component to Fe^{+++} oxides which seems confirmed by the presence of hematite in the fragmental breccias (see section on alteration).

The initial material of the matrix can only be inferred. Inasmuch as the alteration of once glassy clasts with flow structure (Fig. 5) and glassy breccia clasts (Fig. 4) has resulted in a similar composition and a texture very similar to but more dense than the matrix of 116, it is reasonable to infer an initially glassy matrix for 116. The fact that matrices of both 116 and 28B contain a similar clay mineral with a similar range of compositions suggests that 28B and 116 contained similar initial matrices. Perhaps the differences in morphology of the clay are related to initial differences in porosity. Alternatively the initial matrices may have been very fine-grained, sub-micron fragments of glass and minerals that were small enough to react with circulating fluids to form clays. Although

---

Fig. 5. (A) The amber-colored, glassy clast shows flow lines and the formerly plastic nature of the glass, as it has conformed in shape to an opaque clast. The glass also is fractured, a result of its rapid cooling history. Most of the rounded, small, dark spots in the glass are vesicles; there are some small, angular feldspar clasts present. The large dark clast in the lower left is a feldspar which has been engulfed by the glass. (B) Close-up of the glass, which appears to be altered in the same manner as the glassy matrix in the fragmental breccias (Fig. 6). It has lower Mg, higher K, and higher Fe than the glassy matrix, and it contains some Ti which was not noted in the fragmental and glassy matrices.

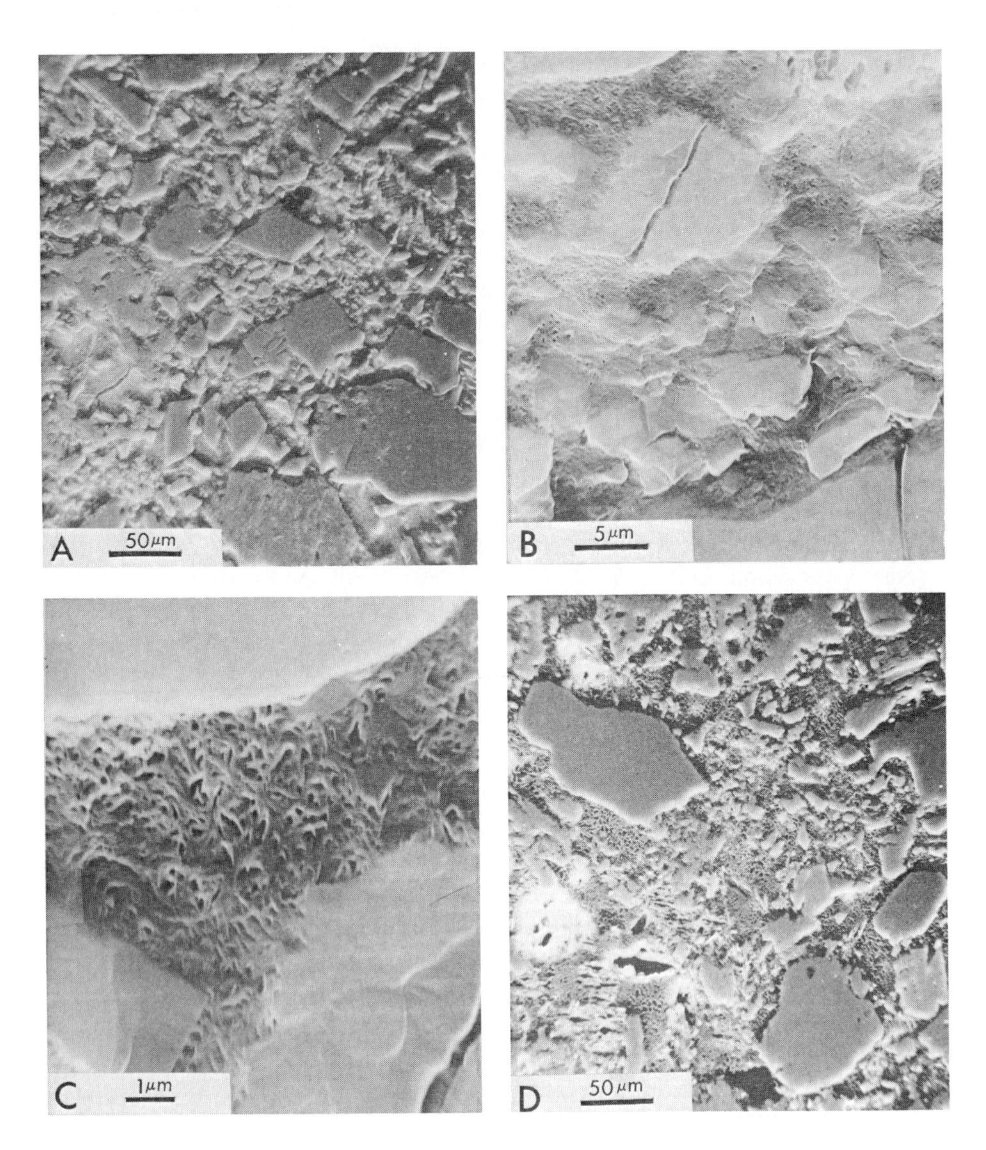

the nature of the initial material remains uncertain, the nature of the alteration is similar to that of hydrothermal alteration (Millot, 1970, p. 44; Faust *et al.*, 1959; Deer *et al.*, 1962; Eberl *et al.*, 1978) associated with volcanism and formation of ores. This suggests that the alteration occurred while the rocks were still relatively warm.

Clast-rich melt rock

Occurrences of clast-rich melt rock were sampled on all of the major islands and two of the Tads. The base of the melt sheet rests with a sharp contact on the

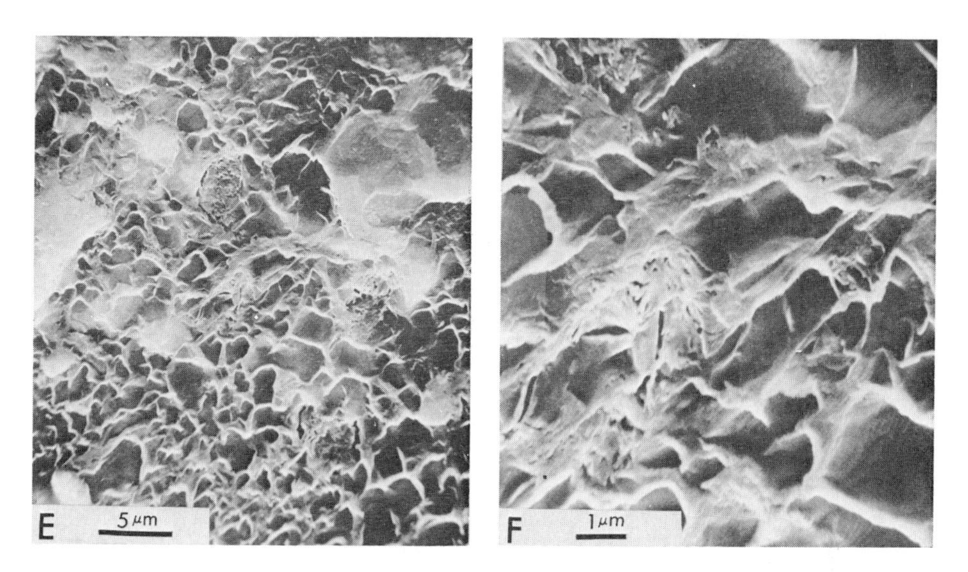

Fig. 6. Series of SEM photos of the fragmental breccias. (A) WCW-77-116 from Driller's Island. Note the wide range of sizes of the clasts and their angularity. A large, rounded, ragged, light-grey, Ca-rich pyroxene clast, located at the center of the left edge, appears to have been partially melted, probably by shock. The mottled, medium-grey clast in the center of the bottom edge is an altered feldspar. Just above and to the right of the mottled clast is a large, fresh, medium-grey plagioclase clast. In the center of the photo is a larger, medium-grey, fractured quartz clast. Along the upper part of the right edge is a totally altered oblong-shaped clast with a light-grey, opaque reaction rim. (B) Higher magnification view of center of A. The quartz clast of A appears along the bottom right edge. The cracked, circular, light-grey clast in the upper center of the photo is K-feldspar. Contrast this with the lathy shape of the plagioclase located directly above the quartz clast. Part of a ragged, light-grey, probably melted Ca-rich pyroxene, is visible in the upper right corner. (C) High magnification of the fragmental breccia matrix of 116 (upper left in B) showing the interwoven wormy texture of the matrix. It is a low Al, Mg-rich montmorillonite, probably an alteration of an originally glassy matrix. (D) Fragmental breccia WCW-77-28B. Note the wide range of sizes and shapes of the clasts. Some of the clasts are as angular and distinct (large, smooth, medium-grey quartz in upper left) as those in WCW-77-116. Other clasts are altered, such as the light-colored, perforated and ragged, potassium feldspar located along the top edge, and the striated pyroxene located just below the upper right-hand corner. (E) Higher magnification view of center of D. There are fewer small clasts compared to WCW-77-116. A more open boxwork-patterned matrix of clay occurs in contrast to the light, wormy-textured matrix of 116. (F) High magnification of the matrix shows wide range of sizes and shapes in the boxwork pattern plus a few patches of the more interwoven network seen in 116. This matrix contains less Al and more Mg than the breccia matrix of 116.

fragmental breccia at all of the several locations where it was observed. On Southeast Island and Driller's Island there were continuous exposures of the clast-rich melt rock upward from exposed contacts with the fragmental breccia. On these islands as well as on South Island there were also nearly continuous exposures across the upper contact into the clast-poor melt rock. Therefore, the

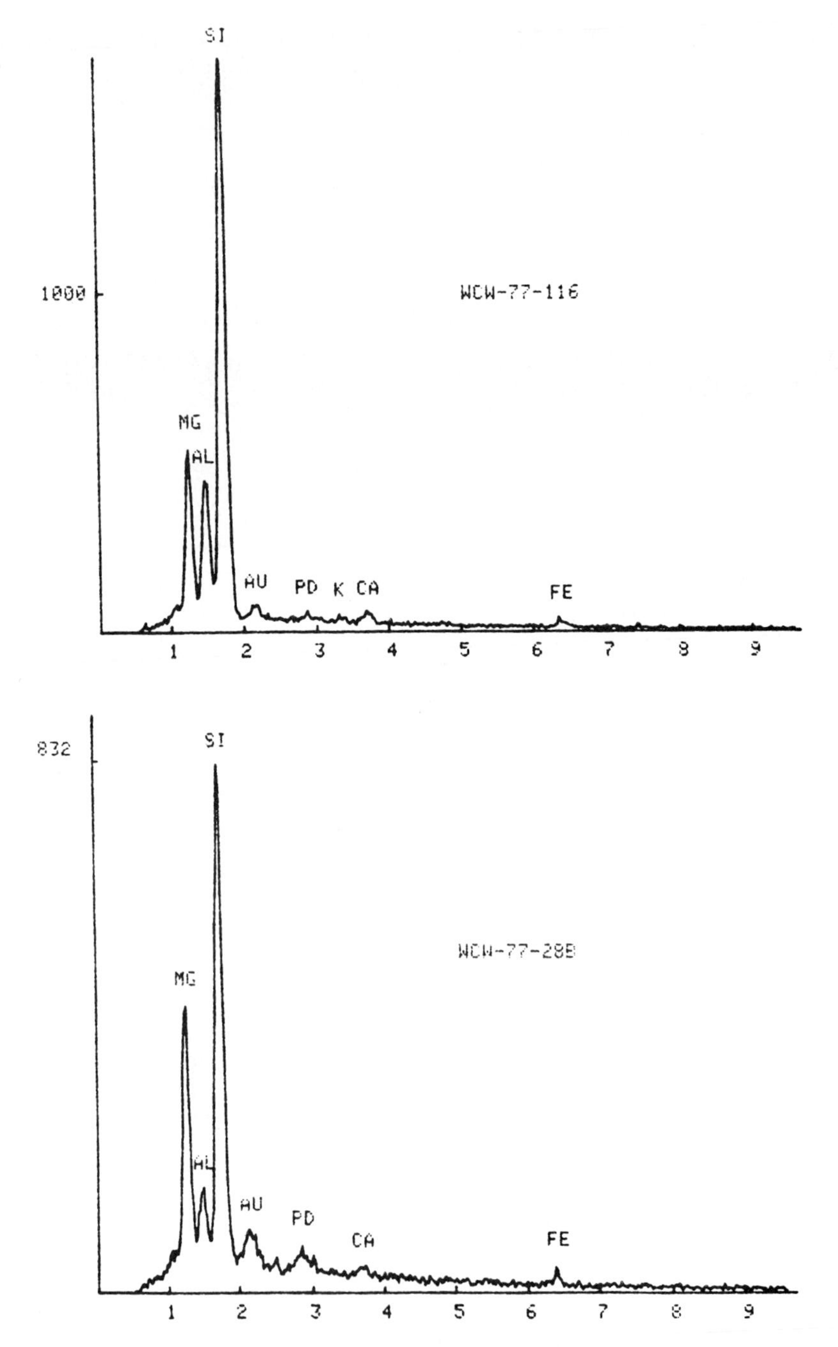

Fig. 7. Energy dispersive X-ray spectra of clay alteration in matrix of fragmental breccias.

samples chosen for detailed petrologic studies are from these three islands (Fig. 2). The upper contact of the clast-rich unit is defined as the location of a sharp decrease in clast content that persists into the upper unit. There may be local sharp decreases in clast content within the clast-rich unit but the clast content will increase again within a meter above the decrease in such cases.

Most clast-rich melt rocks display, overall, a clast-matrix texture in thin section (Fig. 3b), but a few have relatively coarse-grained spherulitic to intersertal textures (Fig. 8f) which will be discussed later. Within a few centimeters above the contact on Southeast Island the melt-rocks contained about 3% by volume of amygdules about one millimeter in diameter. The clast-rich unit contains few large clasts. Less than one percent by volume are larger than one meter across. About 10% are between one centimeter and one meter across near the base of the melt and this decreases upwards to about one percent at the contact with clast-poor melt rocks. Samples selected from this unit for thin sections and clast distributions (Table 2 and Fig. 12) do not include the large clasts and, therefore, are representative of 90 to 99% of the volume of the unit.

Clasts in thin sections of the clast-rich melt rocks consist largely of mineral fragments from the basement rocks. No glassy clasts occur. Many of the mineral clasts show indications of reaction with the melt. Clasts of biotite and pyroxene occur rarely and only near the base of this unit. They are not found higher then a few meters above the base and apparently are easily digested. Quartz fragments near the base of the clast-rich unit develop a rim of augite and above a few meters from the base of the unit they develop concentric rims of augite and alkali feldspar-quartz (Fig. 8a). Feldspar clasts develop zoning or patchy intergrowths of alkali feldspar and plagioclase at the rims. Similar patterns of clast reactions have been reported from the impact melt rocks at Mistastin (Grieve, 1975), Lake St. Martin (McCabe and Bannatyne, 1970), Manicouagan (Simonds *et al.*, 1978 and Floran *et al.*, 1978) and at several Scandinavian craters (Carstens, 1975). Another common feature in the clasts of this unit is a perlitic fracturing in polycrystalline quartz grains (Fig. 8b). This feature was encountered by Carstens (1975) who suggested that these were originally quartz clasts that were shocked and heated sufficiently to form cristobalite or lechatelierite and then, on cooling, transformed rapidly back to quartz resulting in enough strain from shrinkage to cause the perlitic fracturing. We have seen such perlitic quartz clasts in our thin sections from Lake St. Martin. It is not clear why this feature was not noted at Manicouagan. In the upper half of the clast-rich melt rocks there are numerous small (100's of μm) patches of material that are slightly finer or coarser grained than the matrix. These patches may also have slightly different colorations or mineral proportions than the matrix and are presumed to be patches representing partly digested clasts. Several samples from the clast-rich rocks contain irregular veins and patches usually a few millimeters across of relatively coarse, crushed, clastic material. These veins and patches form an anastomosing network similar to the coarse tan material in the lunar melt rocks 76275 and 76295 from the Apollo 17 Station 6 boulder (Simonds, 1975). Such networks apparently result from incomplete mixing between crushed clasts and melt.

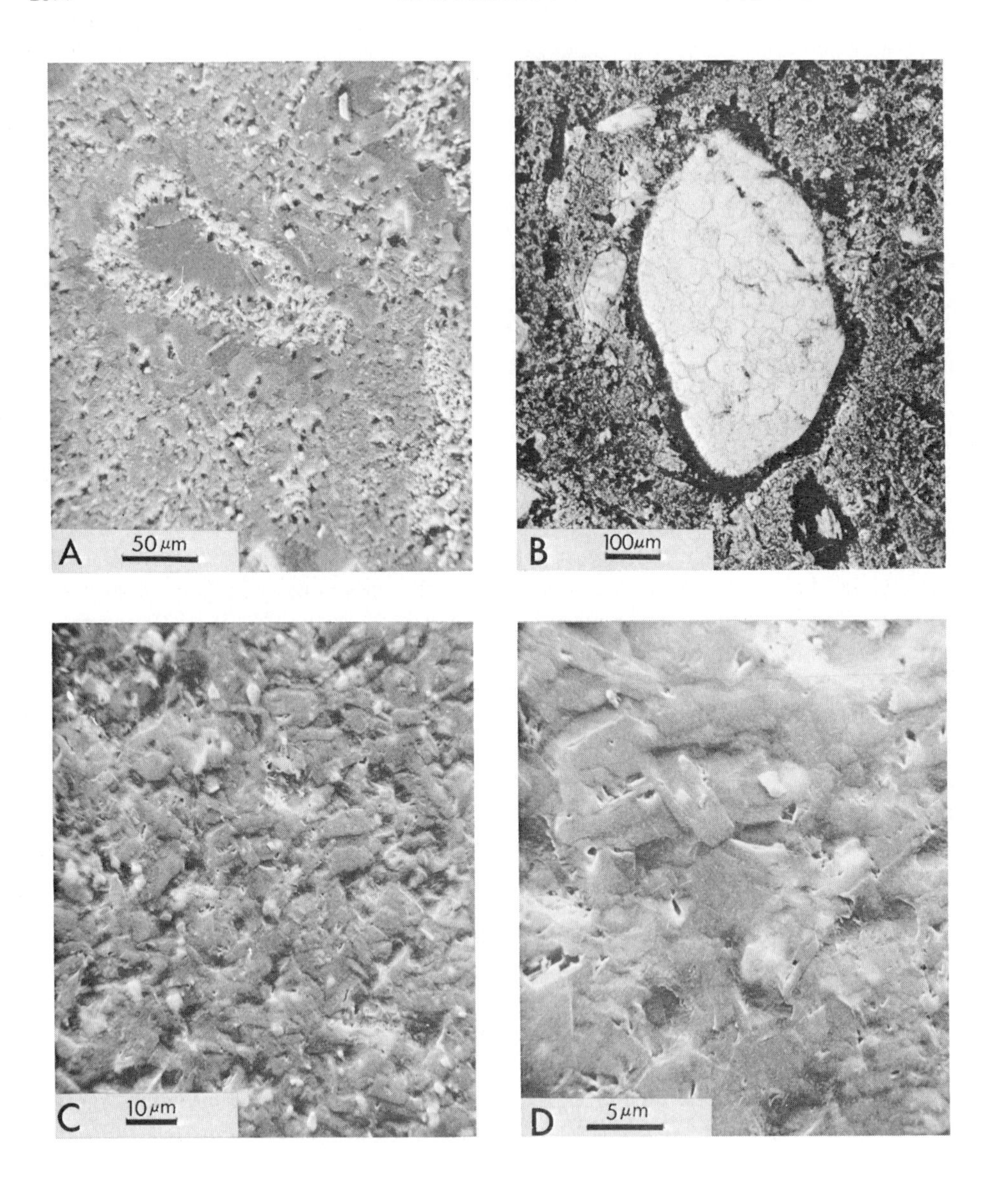

A smaller proportion of quartz and feldspar fragments in the clast-rich melt unit exhibit shock features compared with those in the underlying fragmental breccia (see Table 1). This suggests that the more highly shocked debris was preferentially assimilated in the melt unit because it attained a higher temperature during the initial shock event and, therefore, was assimilated more easily than the cooler, less shocked debris. Most lithic clasts contain feldspars with undulatory extinction or the plumose, feathery texture presumed to form by devitrification of maskelynite. Mafic mineral clasts all show some stage of digestion or reaction with the melt.

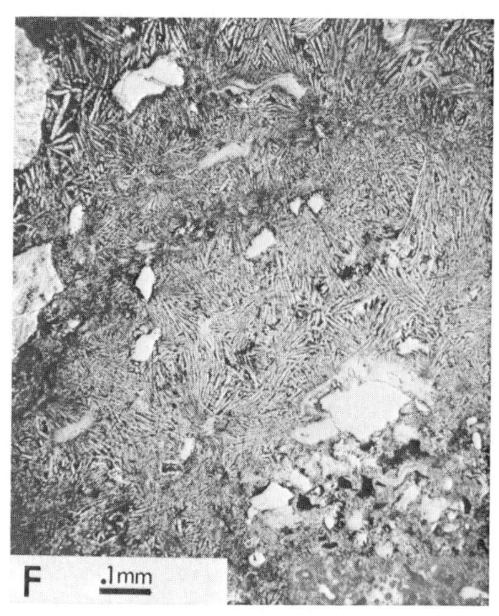

Fig. 8. (A) SEM photo of a quartz clast (center left) in a clast-rich melt rock (WCW-77-23). The quartz clast is surrounded by a fine-grained, Ca-rich pyroxene necklace (white), which, in turn, is enclosed in a rim of medium-grained, interlocking quartz (medium-grey) and feldspar (light-grey). Part of another similar clast is visible at the upper right. Note the sharp break in grain size between the medium-grained outer rim of the clast and the fine-grained matrix surrounding it. (B) Transmitted light photo of a pyroxene-rimmed quartz clast with perlitic fracturing located in a clast-rich melt rock (WCW-77-16A). It is probable that this clast was converted to high temperature cristobalite or lechatelierite by the heat of the melt, and then transformed back to quartz upon cooling. The shrinkage caused perlitic fracturing. (C) SEM view of matrix in a clast-rich melt rock (WCW-77-26B) shows the even, fine-grained interlocking texture. Laths and blocky to rounded grains of feldspar occur as mounds while interstitial quartz is lower in relief and darker in shade. Most of the small white bumps are ilmenite, some are Fe oxides. (D) Higher magnification SEM photo of center of C showing matrix of clast-rich melt rock (WCW-77-26B). In the center of the photo is a medium-relief potassium feldspar grain containing numerous white needles. Directly above the feldspar is some interstitial quartz recognized by its lower relief and darker shade. To the left of the quartz is a plagioclase lath which is higher in relief than both the quartz and potassium-feldspar. In the middle of the plagioclase lath is a small, square, white, opaque. (E) SEM photo of altered Ca-poor pyroxene blade in the matrix of the clast-rich melt rock WCW-77-83. The interwoven wormy alteration is comparable to the clay texture of the clay matrix in fragmental breccia 116 (Fig. 6). (F) Transmitted light view of a partly spherulitic clast-rich melt (WCW-77-46E) from 10.5 meters above the base of the melt sheet on Southeast Island. The spherulites are composed of radiating feldspar needles which often have a remnant clast core. The spherulitic patch grades from coarse to fine grained, from top to bottom. Where the clast population is greatest (bottom of photo), the spherulitic texture grades to a typical fine-grained, reddish, clast-rich matrix, composed of feldspar, quartz, Fe-oxides, and minor clays. The clasts include feldspars, quartz, opaques and pyroxenes.

The matrix of the clast-rich melt rocks is quite fine-grained, generally in the 3 to 10 μm range of grain sizes, and is best studied by the SEM petrographic techniques discussed by Phinney *et al.* (1976). The texture is diabasic to subophitic (Fig. 8c and d) and includes blocky to lathy plagioclase, blocky alkali feldspar, altered blade-shaped pyroxene probably of the Ca-poor variety (Fig. 8e), elongate ilmenite, platy iron oxides and interstitial quartz. In addition, the matrix contains numerous interstitial, 5 to 10 μm patches of the same wormy intergrowths of Mg-rich clay that formed most of the matrix of the fragmental breccias. The paragenetic sequence is not clear, but the feldspars, pyroxene, and opaques clearly precede the quartz and altered interstitial material. The pyroxene is elongate and of larger grain size than the other phases presenting a somewhat porphyritic appearance suggesting that it may have preceded the feldspars.

Within the clast-rich unit of each vertical traverse there occur zones of spherulitic to intersertal textures (Fig. 8f) typically as an interfingering set of thin (up to a few centimeters) layers or as an irregular anastomosing network. This texture is associated with abrupt decreases in clast contents to 20% or less by volume. Where this texture occurs in layers a few centimeters thick, the grain size increases by nearly an order of magnitude to include large needles of plagioclase, interstitial Ca-rich pyroxene, and some interstitial brown glass. In the more irregular networks (Fig. 8f), the grain size is clearly related inversely to the abundance of clasts which in some instances form nuclei for radiating plagioclase needles. Similar spherulitic textures and relations with clast proportions were observed at Manicouagan (Floran *et al.*, 1978). As the clast population increases the spherulitic texture becomes finer grained until it merges with the more typical matrix grain size and texture. Attempts to duplicate experimentally the textures of impactites (Lofgren and Smith, 1978) suggest that such spherulitic to intersertal textures result from rapid cooling of melts with very little included debris. In such cases nearly all residual submicroscopic nuclei which promote crystallization were lacking. Alternatively, the lack of clasts might cause a slightly slower cooling rate and fewer nucleation centers. Compositions of the feldspar and pyroxene in the spherulitic texture are shown in Fig. 9. Because of their extremely small grain size (4 to 8 μm) the feldspar and pyroxene compositions could not be determined in the typical clast-rich melt matrix. Notice the much broader compositional range of these rapidly cooled phases compared with their counterparts in the more slowly cooled clast-poor melt rocks. The Fe and Mg contents of the feldspars show similar relations. The total Fe and Mg is less than 1% (usually less than 0.5%) in the clast-poor rocks but ranges from 1 to 3% in the clast-rich unit. Although compositions of the feldspars in the non-spherulitic clast-rich matrices were not measured quantitatively, the qualitative EDAX spectra taken during SEM studies show a similar wide range of Ca, Na, and K contents. This lack of included debris within a rock unit that generally displays thoroughly mixed melt and debris is thought to represent a rare, small zone of melt that somehow managed to survive the mixing process without significant addition of clastic fragments.

Clast-poor melt rock

Occurrences of clast-poor melt rock were sampled on all of the major islands and the southernmost Tad. The lower contact is marked by a sharp increase in

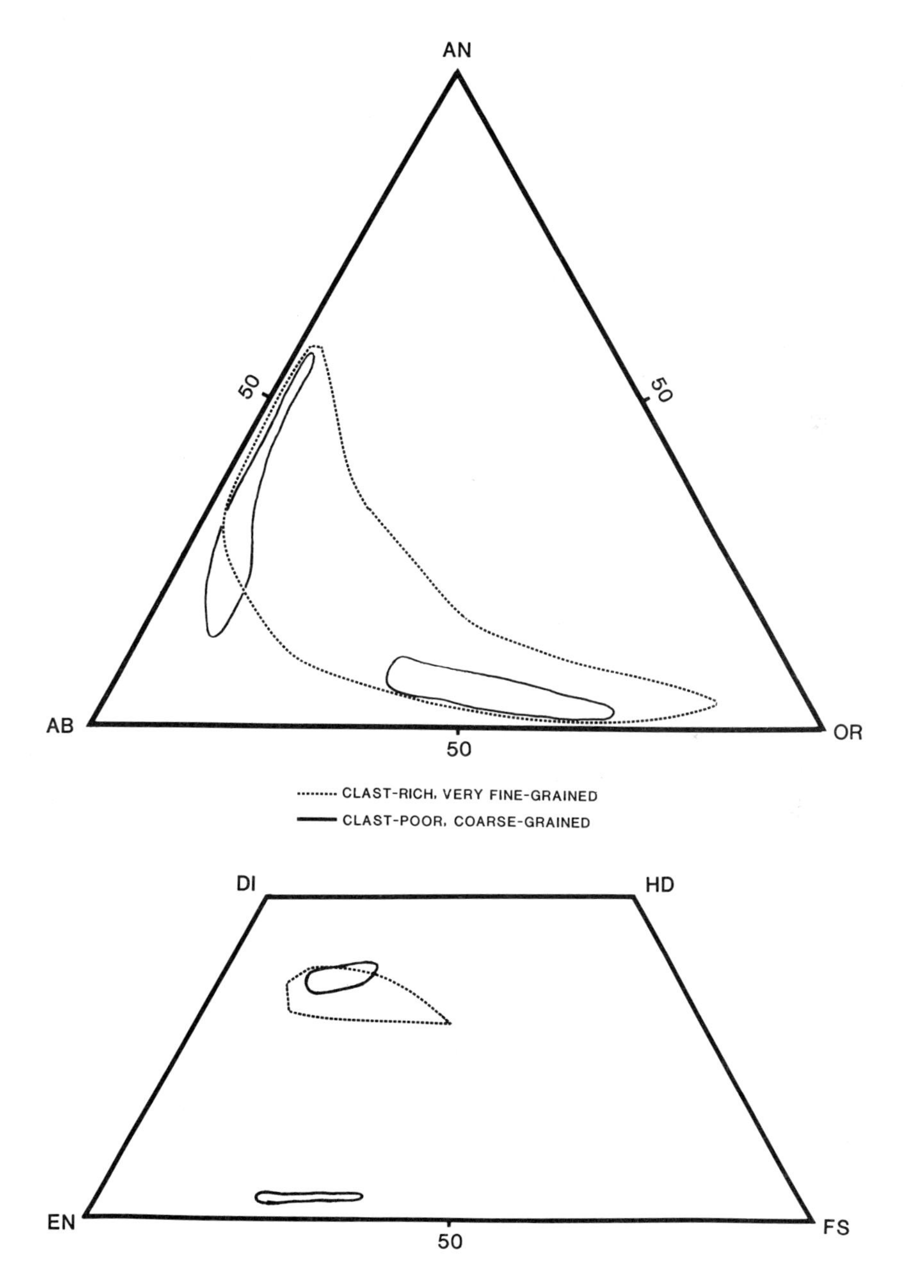

Fig. 9. Compositions of feldspars and pyroxenes in clast-rich and clast-poor melt rocks.

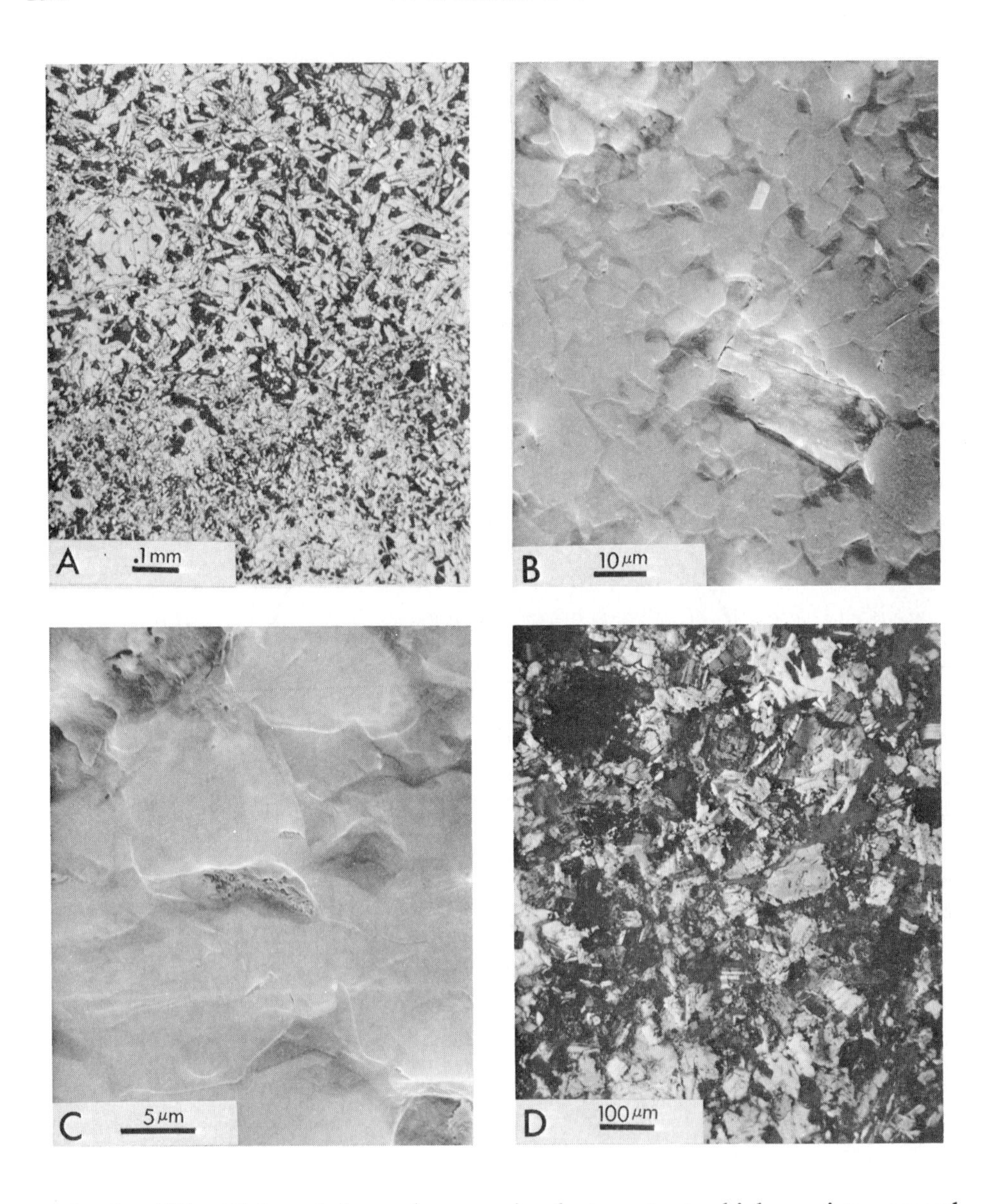

grain size (Fig. 10a) and sharp decrease in clast content which persists upward into the clast-poor unit where the clast content continuously decreases (see Fig. 12) and the grain size increases (see Fig. 11). The upper contact is above the highest elevations of any of the islands and has not been observed. The grain size is coarsest at the highest exposed levels suggesting that the finer-grained upper zone typical of igneous units and present in the drill core at East Clearwater Lake (Dence *et al.*, 1965) was well above the present level of the hilltops. On Driller's Island, Southeast Island and South Island there are good exposures of the contact between the clast-rich and clast-poor rocks, and the latter have exposures over 85 m thick. The two samples chosen for detailed studies (WCW-77-90 and 44) are from these islands (Fig. 2).

The clast-poor melt rocks range from an overall fine-grained, clast-matrix texture (Fig. 3c) at the base of the unit to a relatively coarse-grained, diabasic to subophitic texture (Fig. 3d) about 15 to 20 meters into the unit. The unit contains very few large clasts: less than one percent by volume are larger than one meter and about one percent are between one centimeter and one meter. Samples selected from this unit for thin sections and clast distributions (Table 2 and Fig. 12), therefore, are representative of about 99% of the volume of the unit.

Clasts in thin section always show increasing degrees of recrystallization or reaction with the matrix upward into the unit. Quartz fragments form cores with extensive concentric rings of augite inside the alkali feldspar-quartz outside. Feldspar fragments show extensive zoning and patchy checkerboard intergrowths. Some large polycrystalline feldspars occur. There occur numerous submillimeter to millimeter-size patches of material that is finer-grained or coarser-grained than the typical matrix and commonly of different mineral proportions as was seen in the clast-rich melt rocks. Again these are interpreted as partly-digested clasts. Recognizable shock features are not present throughout

---

Fig. 10. (A) Reflected light photo of the sharp clast-rich, clast-poor contact (WCW-77-84C) on South Island. The contact separates the clast-poor, subophitic, large-grained matrix (top half of photo) containing lathy feldspars and altered pyroxene oikocrysts (dark grey; top of picture) from the clast-rich, opaque-rich, reddish, finer-grained, granular matrix. Note the kidney-shaped, partly digested, pyroxene clast with a dark grey rim located in the center of the photo along the contact. The large, rounded, light-colored patch in the upper left center of the photo is a polycrystalline quartz clast. (B) SEM photo of a clast-poor melt (WCW-77-122) located 1.5 meters above the clast-rich, clast-poor boundary. It has a granular texture consisting of intergrown, rounded to blocky feldspar (light grey), interstitial quartz (medium grey), and some opaques (white) and interstitial clays. The rectangular, light grey area near the center is an altered low-Ca pyroxene. This and the following photo can be compared with photos of the clast-rich matrix (Figs. 8C and D) for texture and grain size. (C) Close-up of the inter-grain boundaries in the clast-poor matrix. In the middle is a lens-shaped patch of interstitial clay. The light grey, blocky to irregular grains are feldspars of various compositions. Generally, the more elevated, lighter grey parts of the feldspars are more Ca-rich. Between the feldspars are darker interstitial quartz patches. In the upper left and bottom right corner are parts of altered clasts. Note that white needles are not present in these feldspars as is the case in altered feldspars lower in the section (Fig. 13A). (D) Polarized transmitted light view of a typical, large-grained, clast-poor melt rock (WCW-77-89) located 39.5 meters above the clast-rich clast-poor contact. The twinned and zoned feldspars are lathy or blocky and display straight extinction. Compare the fresh medium-grey pyroxene wedge (left of and above center) with the altered, dark grey, granular pyroxene patches (lower left of photo). The opaques appear as black blocks, such as the large rounded opaque in the upper left. There is a large, bright-white, interstitial quartz patch enclosing medium-grey feldspar blocks in the upper right. A small patch of bright-white, interstitial apatite needles within medium-grey quartz is located at the bottom left. The large light-grey, blocky feldspar, just to the right of center is a partly digested clast with a remnant core.

most of this unit although traces of such features occur in the lowest few meters of the unit where clasts of feldspar rarely display undulatory extinction and, even more rarely, devitrified textures. Quartz is much less abundant in these lower few meters than in the underlying units and very rarely displays planar features or perlitic fractures.

The matrix of the clast-poor melt rocks contains the same mineral assemblage as the matrix of the clast-rich melt rocks although there is a change in texture from fine-grained and blocky at the base of the unit (Fig. 10b and c) to coarser-grained and subophitic higher in the unit (Fig. 10d). In the coarser rocks the Ca-rich pyroxenes enclose a few plagioclase grains as the texture becomes somewhat ophitic. Also there are intergrowths of Ca-rich and Ca-poor pyroxenes, some of which may be exsolution lamellae. However, the extensive alteration of Ca-poor pyroxene (Fig. 13) makes interpretation difficult. Compositions of the feldspars and pyroxenes in the clast-poor unit are shown in Fig. 9. The range of compositions shown for the minerals are similar in each of two analyzed samples (44 and 90) from widely separated localities. Crystallization commenced with plagioclase. Small plagioclase laths are enclosed poikilitically in Ca-rich pyroxenes indicating that the plagioclase was soon joined by pyroxene. Alkali feldspar then crystallized with pyroxene and plagioclase. Bacause of extensive alteration in the pyroxenes, it is not possible to determine the paragenetic relations between Ca-poor and Ca-rich pyroxenes. Large grains of ilmenite and titanomagnetite formed after the pyroxenes, but it is not clear how they relate temporally to the alkali feldspar. Large oikocrysts of interstitial quartz form the major late-stage mineral. Intergrown with the quartz are long needles of apatite, small laths of plagioclase, small blocky grains of alkali feldspar, and very rarely, biotite.

Grain sizes

Grain-size variations within the melt sheet are relatively constant throughout the clast-rich unit but the rate of increase varies from traverse to traverse in the clast-poor unit (Fig. 11). Within the clast-rich melt rocks the feldspars, pyroxenes, and opaque minerals are generally less than 10 μm across: the average feldspar grains range down to 3 μm and the average opaque minerals down to 1 μm. Of the six traverses measured only traverse A on Driller's Island departs from this pattern. The consistently larger grains in this traverse appear to be correlated with a significantly lower clast content in this traverse (Fig. 12a). Total clast contents in the unit are generally between 40 and 80% by volume, whereas in this traverse the volume of clasts is between 20 and 30%. In other cases where the grain sizes are anomalously large within the clast-rich rocks there is a spherulitic texture and very few clasts. Such is the case for the large spikes shown near the base in traverse A on Driller's Island and at the upper contact in Traverse A on South Island. Coarse spherulitic textures also occur (not plotted) about halfway up the unit on Southeast Island and near the top of the unit on North Island. Similar coarse spherulitic textures were associated with volumes of low clast content in the clast-rich unit near the base of the

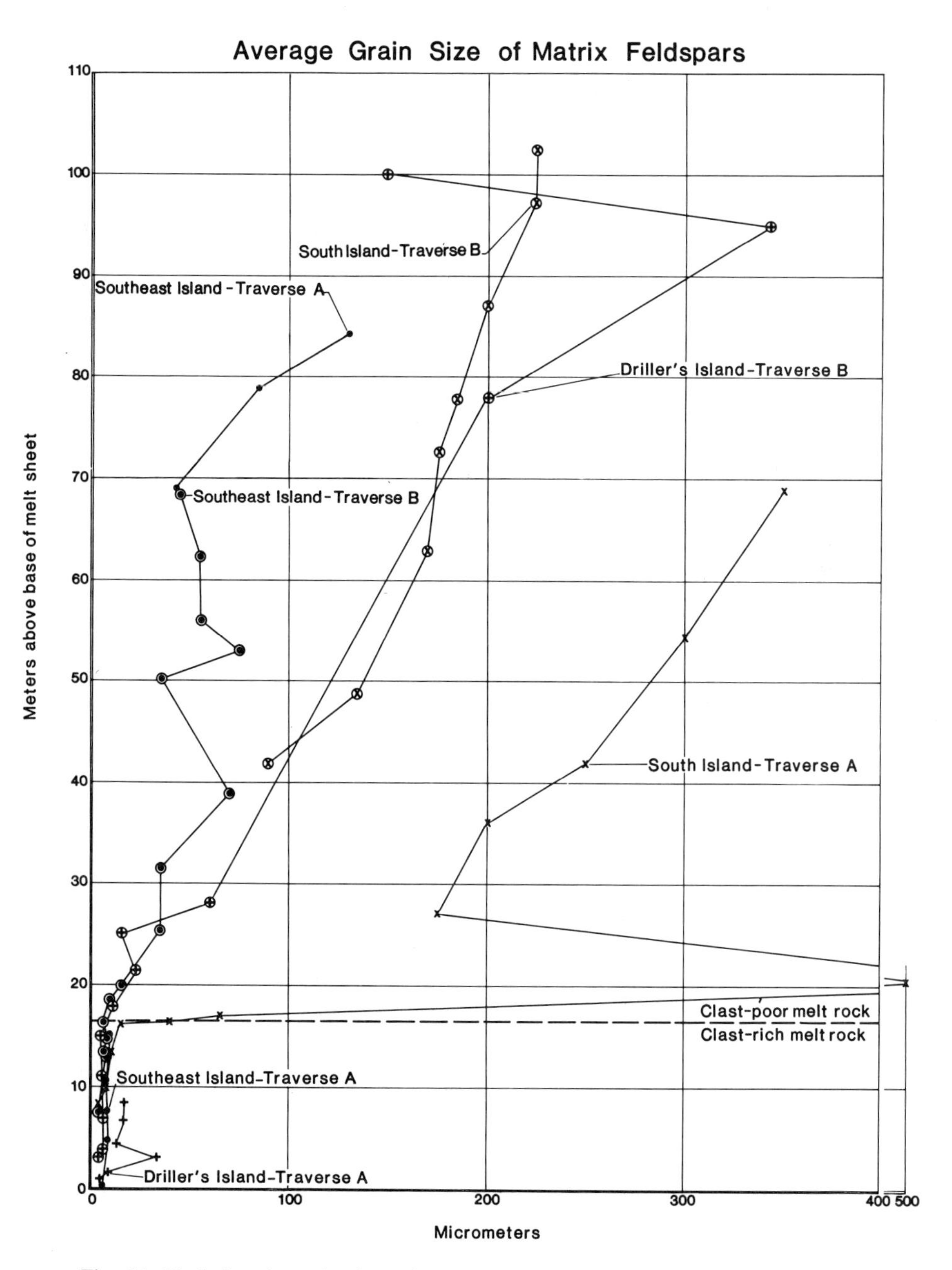

Fig. 11. Variations in grain sizes of matrix in the vertical traverses shown in Figs. 1 and 2.

Table 2. Clast Populations, Driller's Island.

	Matrix	<10 μm	10 μm–40 μm	40 μm–1 mm	1 mm–3 cm
Traverse A					
31	74.5%	0%	2.9%	7.3%	15.3%
30	65.4	0	1.7	22.7	11.9
29	66.8	0	1.3	24.6	8.6
25	70.3	0.2	2.4	18.8	8.2
24	75.9	0	0.5	12.3	11.3
23	52.2	0.2	4.8	32.3	10.4
26B	45.5	0.2	25.5	20.2	8.6
28B	28.2	5.0	10.6	19.6	36.6
Traverse B					
43	86.3	0	0	7.9	5.8
40	83.5	0	0	14.0	2.5
44	92.3	0	0	6.2	1.5
125	58.5	0	0	33.3	8.2
124	53.6	0	2.0	29.5	14.9
123	70.1	0	2.1	18.8	8.9
122	49.4	0.1	2.5	31.7	16.4
121	32.2	0.1	3.2	39.0	25.5
120	29.9	0.2	4.4	46.2	19.4
119	28.3	0.3	3.5	47.5	20.5
118	30.0	0.2	3.7	44.2	21.9
117	24.0	0.4	6.6	40.5	28.5
116	17.9	14.0	14.0	19.5	34.6

Manicouagan melt sheet (Floran *et al.*, 1978).

In the clast-poor unit the grain sizes display significantly different trends, although the grain sizes generally increase upward. At a height of 70 meters into the melt sheet, however, the grain sizes vary from traverse to traverse by over an order of magnitude. Again these variations may be associated with total clast contents which would have controlled the rate of heat loss during initial stages of cooling (Onorato *et al.*, 1978). The smallest grain-sizes are in traverse B on Southeast Island where the total clast populations are about 25%, whereas the largest grain-sizes are in traverse A on South Island where the total clast population is about 10%. Intermediate grain-sizes occur in traverse B on South Island where the total clast population is about 15 to 20% (Table 2). The unusually sharp decrease in grain size displayed by the uppermost sample in traverse B on Driller's Island is almost certainly a result of its proximity to a huge (10's of meters) inclusion of limestone which must have been a significant heat sink during cooling in this zone of melt.

Clast contents

The distribution of clast sizes and total clast populations varies significantly from traverse to traverse although there is a general decrease of clasts upward in

Table 2. *(cont'd.)* Clast Populations, SE Island.

	Matrix	<10 μm	10 μm–40 μm	40 μm–1 mm	1 mm–3 cm
Traverse A					
48	82.4	0	0	15.1	2.5
49	81.2	0	0	12.9	5.9
50	81.4	0	0	13.8	4.8
46F	65.5	0.3	3.1	22.9	8.3
46E	59.7	0.1	2.6	29.5	8.1
46D	74.2	0.2	2.7	13.4	9.5
46C	50.8	0.8	5.3	29.7	13.4
45F	72.4	0.2	2.0	13.2	9.3
Traverse B					
51	74.3	0	0	21.4	4.3
52	75.0	0	0	19.7	5.3
53	76.1	0	0	21.8	2.1
54	74.0	0	0	22.3	3.4
55	78.3	0	0	16.6	5.1
66	73.3	0	0	22.2	4.5
65	67.1	0	0	27.0	5.9
59	74.2	0	0	20.3	5.5
60	66.7	0	1.26	24.5	7.5
61	58.8	0.1	4.5	22.6	14.0
62	48.5	0.4	4.7	24.4	22.0
63	40.4	0.2	3.9	32.4	23.1
64	42.8	0.2	3.9	32.6	20.5
56A	49.4	0.2	2.7	23.4	24.4
57	37.8	0.5	5.9	24.6	31.2
58	63.6	0.3	4.5	13.5	18.1

Table 2. *(cont'd.)* Clast Populations, South Island.

	Matrix	<10 μm	10 μm–40 μm	40 μm–1 mm	1 mm–3 cm
Traverse A					
90	88.9	0	0	6.5	4.6
89	90.4	0	0	4.1	5.5
88	89.4	0	0	6.9	3.7
87	89.9	0	0	6.9	3.2
86	84.0	0	0	9.2	6.7
85	82.9	0	0	12.3	4.8
84E	76.2	0	0	16.2	7.7
84C	60.9	0	0	25.4	13.7
84B	56.4	0.1	2.4	26.7	14.5
83	52.8	0.1	2.9	27.5	16.7
82B	48.9	0.1	2.0	27.9	21.1
81	28.7	0.2	5.0	33.3	32.7
Traverse B					
80	86.2	0	0	10.4	3.4
77	84.0	0	0	13.5	2.5
69	80.5	0	0	13.3	6.2

each section (Table 2 and Fig. 12). The fragmental breccias contain 70 to 80% by volume of clastic debris ranging down to one micrometer and less in size. Although the two samples whose clast populations were studied in detail are from Driller's Island, the samples from other islands appear very similar.

As the matrix grain size increases, the small clasts are no longer observed. Apparently they become assimilated into the melt. Although clasts between one and ten micrometers are present in quantities of 5 to 10% in the fragmental breccia they are present in quantities of only a few tenths of a percent in the clast-rich melt rock, and are not recognized at all above the contact of the clast-rich and clast-poor melt rocks. Similarly the 10 to 40 μm clasts are not recognized above 8 meters into the clast-poor melt rocks.

At the contact of fragmental breccia and clast-rich melt rock there is a sharp discontinuity in clast populations (Fig. 12a and b). A similar discontinuity occurs at the contact between clast-rich and clast-poor melt rocks on each of the three islands where continuous sampling was possible across this contact (Fig. 12b,d and e). Within the clast-rich rocks sharp changes in clast distribution occur without any regular predictable pattern. The sharp decrease at 3 meters in traverse A on Driller's Island occurs in a spherulitic unit. Also, the overall total clast contents within the clast-rich unit may vary from about 70% in traverse B on Driller's Island (Fig. 12b) to about 35 to 40% 400 meters away in traverse A on Driller's Island (Fig. 12a). The clast populations in the clast-poor melt rocks, although a bit irregular in the lowest 5 to 10 meters, seem to stabilize at rather constant values throughout most of the unit. The highest content of clasts in the clast-rich rocks occurs in traverse B on Driller's Island (Fig. 12b) yet the clast content of the upper part of the clast-poor rocks in this traverse is relatively low. The highest content of clasts in the clast-poor rocks occurs in traverse B on Southeast Island (Fig. 12d), yet the clast content of the clast-rich rocks in this traverse is not drastically different from that of traverse A on South Island which has a low clast content in the upper unit (Fig. 12e). Thus there does not seem to be any relationship betwen the clast content in the clast-poor rocks and the clast contents of the underlying clast-rich rocks.

Discussion

Alteration

The most obvious evidence of alteration in the sequence of units at West Clearwater Lake is the red color of the fragmental breccia, the clast-rich melt rocks, and the lower part of the clast-poor melt rocks. The red coloration is pervasive and apparently independent of weathering. In the upper parts of the clast-poor unit the rocks are normally gray and have passed through an alternating sequence of purple and gray coloration lower in the melt sheet (Fig. 2). The gray clast-poor rocks when weathered display a typical yellow-brown color rather than the red to purplish red color common to the breccias and clast-rich melt rocks. The red color results from the formation of various ferric oxides which formed with the magnesium-rich clay minerals during alteration of mafic minerals and glass. Micrometer-sized hematite and ilmenite grains are abundant throughout the matrices. The alteration of feldspars commonly includes the formation of micrometer- to submicrometer-sized needles of an opaque mineral, which is too fine-grained to be identified with certainty (Fig. 13a), but which contains iron as one of its chief constituents. The Ca-poor pyroxenes are extremely susceptible to alteration during which they develop both the Mg-rich saponitic clays (Fig. 8e) that are found in the fragmental breccias and frothy patches of iron oxides (Fig. 13b, c, d). Similar frothy patches rich in iron occur throughout the matrices. As noted previously, such clays are normally associated with late-stage deuteric or hydrothermal alteration. Experimental studies suggest that these clays would begin to form mixed layers with talc above about 500°C (Eberl *et al.*, 1978) thus providing an approximate upper limit on their temperature of formation.

A second set of iron oxides in the rocks from West Clearwater indicates that the oxide assemblages result from two separate origins. Within the gray, coarser-grained rocks the major oxide assemblage is a three-phase mixture consisting of intergrowths of hematite and ilmenite within partial pseudomorphs of primary titanomagnetite. Within the red, finer-grained rocks the major oxide is very fine-grained ilmenite, hematite and, perhaps, hydrous ferric oxide. In samples 44 and 90 from the gray, coarse-grained rocks from the upper parts of traverses on Driller's Island and South Island there are blocky oxide grains up to 0.5 mm across. The major phase in these grains is a titaniferous hematite with about 10% TiO_2. The second most abundant phase is nearly stoichiometric ilmenite. Only small amounts of the initial titanomagnetite are preserved, invariably near the centers of the grains. The titanomagnetite seems clearly to be part of the initial igneous assemblage. Such replacement of magnetite by more oxidized subolidus assemblages is characteristic during cooling of basalts (Haggerty, 1976). Also within these rocks are sparse, wispy hematite fillings of small fractures up to one micrometer wide. Qualitative analysis of this hematite suggests that it is very low in TiO_2. This difference in composition from the other hematite and its occurrence in fractures suggests that this hematite is of a different origin and a later generation than the assemblages in the blocky grains. In samples from the red fragmental breccia, the few hematite grains large enough for even qualitative analysis were essentially TiO_2-free, suggesting an origin more closely associated with the late stage formation of the Ti-poor fracture fillings of the gray rocks. Thus the pervasive red color is associated with the formation of fine-grained hematite (or hydrous ferric oxides) during a late-stage alteration, almost certainly the same hydrothermal alteration that produced the Mg-rich saponitic clays.

In a companion paper (Simonds *et al.*, 1978b) the lower content of H_2O in the target rocks compared with the impact-generated rocks (especially the permeable fragmental breccias) is used as an argument for an influx of groundwater into the impactites shortly after their formation while they were still relatively warm. The relatively permeable fragmental breccias and fractured basement would provide

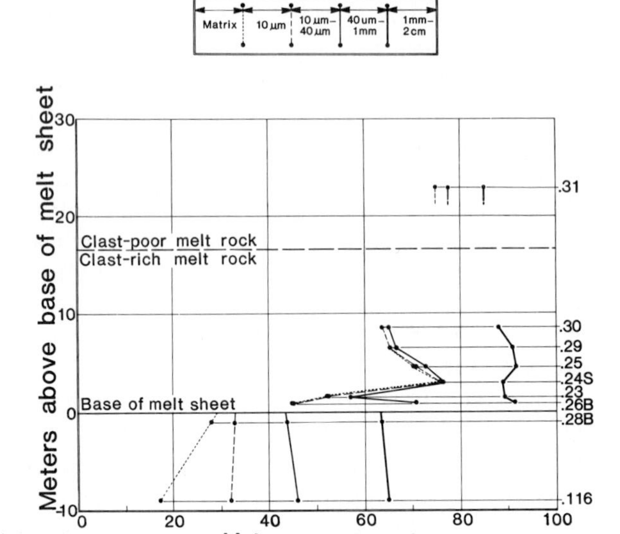
Clast-size Distribution, Driller's Island, Traverse A
Matrix
10 μm
10 μm–40 μm
40 μm–1 mm
1 mm–2 cm
Meters above base of melt sheet
Clast-poor melt rock
Clast-rich melt rock
Base of melt sheet
.31
.30
.29
.25
.24S
.23
.26B
.28B
.116
Volume percent

(A)

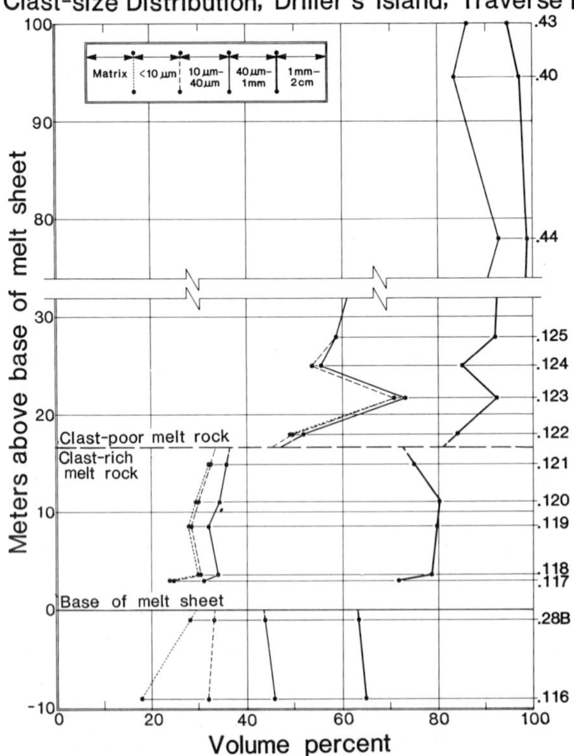
Clast-size Distribution, Driller's Island, Traverse B
Matrix
<10 μm
10 μm–40 μm
40 μm–1 mm
1 mm–2 cm
Meters above base of melt sheet
Clast-poor melt rock
Clast-rich melt rock
Base of melt sheet
.43
.40
.44
.125
.124
.123
.122
.121
.120
.119
.118
.117
.28B
.116
Volume percent

(B)

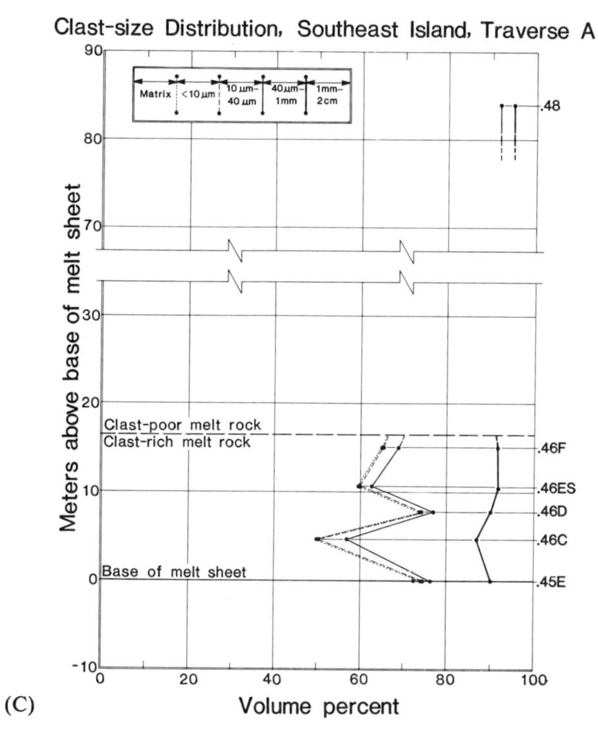
Clast-size Distribution, Southeast Island, Traverse A
Matrix
<10 µm
10 µm–40 µm
40 µm–1 mm
1 mm–2 cm
Meters above base of melt sheet
Volume percent
Clast-poor melt rock
Clast-rich melt rock
Base of melt sheet
.48
.46F
.46ES
.46D
.46C
.45E

(C)

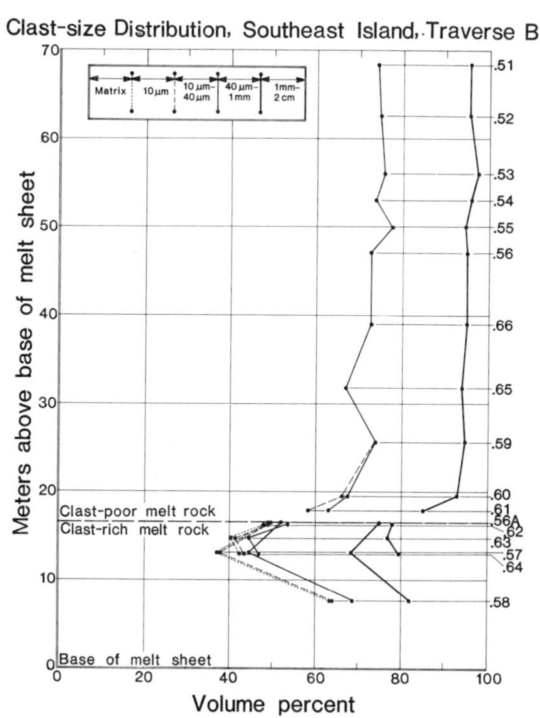
Clast-size Distribution, Southeast Island, Traverse B
Matrix
10 µm
10 µm–40 µm
40 µm–1 mm
1 mm–2 cm
Meters above base of melt sheet
Volume percent
Clast-poor melt rock
Clast-rich melt rock
Base of melt sheet
.51
.52
.53
.54
.55
.56
.66
.65
.59
.60
.61
.56A
.62
.63
.57
.64
.58

(D)

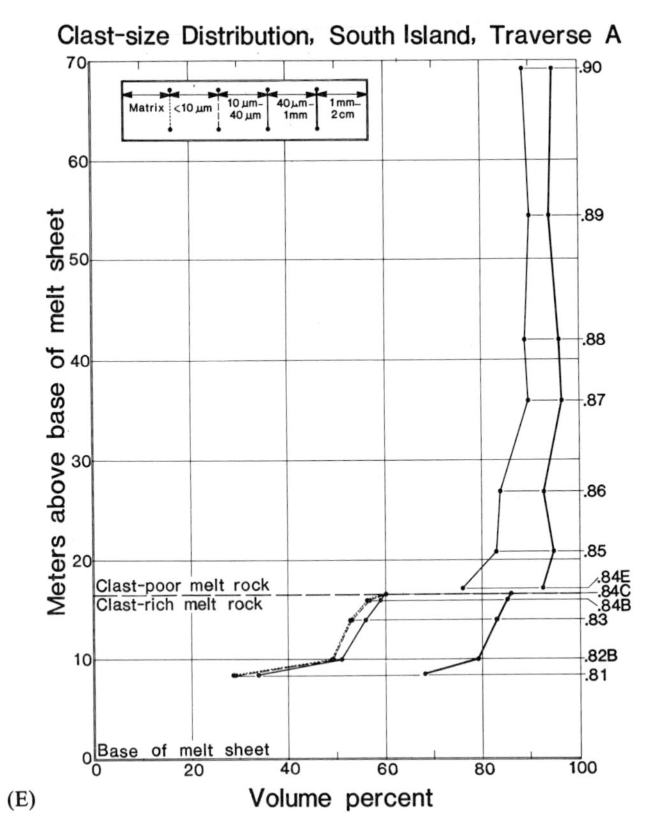

Fig. 12. Distributions of clasts and matrix in rocks selected from five of the vertical traverses shown in Figs. 1 and 2.

a conduit for the expected influx from the hydrostatic head developed between the bottom of the excavation and the groundwater level in the surrounding rocks. This process is postulated as the cause of the hydrothermal alteration reactions producing the red color in the permeable lower parts of the crater fill.

Comparison with Manicouagan

In addition to the overall structural and chemical similarities between West Clearwater Lake and Manicouagan, there are many similarities between the observed petrologic and stratigraphic features (Floran *et al.*, 1978 and Simonds *et al.*, 1978b). Both show decreasing clast populations and increasing grain sizes upward through the melt rocks although irregularities may occur in any vertical traverse. Both show sharp contacts at the base of the melt sheet with either fragmental breccias or basement rocks. Within the melt sheet both contain a sharply gradational contact between clast-rich and clast-poor melt rocks,

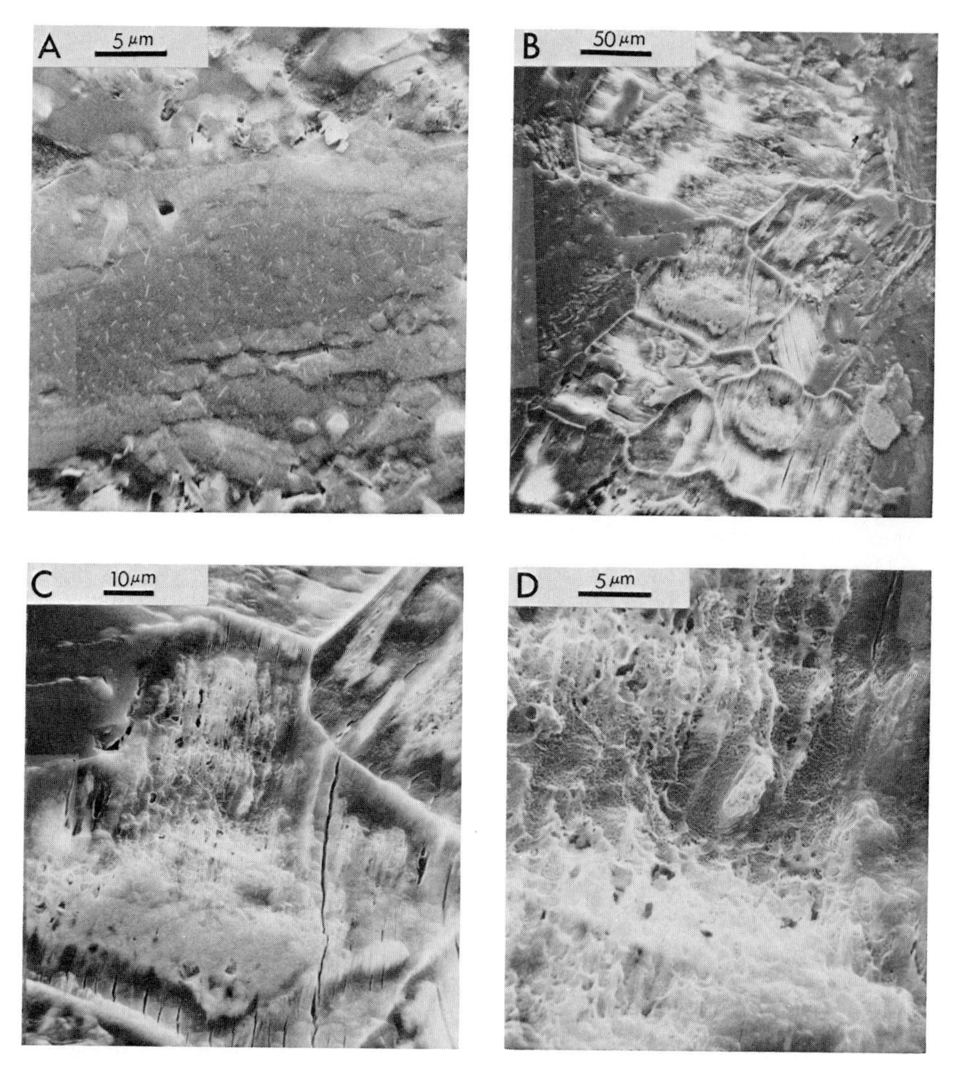

Fig. 13. (A) White needles of undetermined compositon in a plagioclase clast of clast-rich melt rock WCW-77-26B. These needles are abundant in feldspar clasts and matrix grains of the clast-rich melt. They are of varying lengths and have a random orientation. These white needles probably contain iron and represent an alteration of the feldspars. (B) SEM photo showing a large, altered, low-Ca pyroxene, located 4 meters above the clast-rich clast-poor boundary (WCW-77-85). Along the left side is plagioclase and on the right is Ca-rich pyroxene. Note the polygonal divisioning and the striated pattern caused by alteration. (C) Higher magnification of center of B. The vertically cracked strip along the bottom has a composition similar to low-Ca pyroxene. Directly above this is a lighter-colored plateau containing mostly iron plus minor silica. (D) High magnification of center of C showing the frothy nature of the Fe-rich alteration. This is probably a hydrous form of iron oxide.

although this contact is significantly higher in the melt sheet at Manicouagan. Perhaps this is due to Manicouagan's much larger size which results in a greater scale of mixing. Both structures have irregular development of spherulitic to intersertal textures in the clast-rich unit. Both show the same alteration to bright red colors in the breccias and clast-rich melt rocks and to saponitic clays in the mafic minerals, especially Ca-poor pyroxenes. Both display a similar widespread range of shock features in clasts. Both show the same types of reactions between clasts and matrices. Both have the same sequence of types of igneous textures in the matrices of melt rocks.

There have been several recent papers on the origin of and processes associated with the Manicouagan structure (Floran *et al.*, 1976; Floran *et al.*, 1978; Grieve and Floran, 1978; Simonds *et al.*, 1978a; Onorato *et al.*, 1978). All of these papers provide arguments for an origin by meteorite impact. Essentially all of the arguments can be applied to the observations at West Clearwater Lake. In addition, recent analyses of trace siderophile elements from the melt rocks at the companion East Clearwater Lake crater are two to three orders of magnitude higher than in the target rocks (Palme *et al.*, 1978). Such values indicate significant contamination by a stony meteorite. Similar studies have been used successfully as indicators of meteoritic material in lunar samples and other terrestrial impact craters (Morgan *et al.*, 1974; Janssens *et al.*, 1977). Thus, there is no question remaining about the origin of the Clearwater and Manicouagan structures. The problem is now one of applying the available data from these structures to a more detailed understanding of the dynamics which produced the observed stratigraphy, structure, and petrography.

Dynamics of mixing

The clast distributions suggest several important interpretations about the nature of mixing in the debris formed during impact events. Within the rapidly outward-moving debris in the lower part of the crater there were two different types of material. Along the base was a mass of fragmental debris that included clasts displaying no shock features to low, moderate and high shock features and ranging in size from less than 1 μm to tens of meters. Overriding the fragmental debris was a mass of melt-rich material containing some clastic fragments whose proportions varied as indicated by differing values in various traverses. It is suggested that the clast-rich melt rock with its extremely irregular clast distributions represents a zone of turbulent mixing of the underlying fragmental debris and the overriding melt-rich material. The occurrence of very clast-poor spherulitic layers and veins irregularly interfingered with very clast-rich zones seems to require such mixing. Sporadic zones showing foliation of elongate clasts further amplify the occurrence of such a flow. Although the upper parts of the clast-poor melt rocks have a relatively constant clast distribution within each traverse, a thin zone above the contact with clast-rich melt rocks displays a somewhat erratic clast distribution. Within this narrow zone the clast content generally decreases and stabilizes to the nearly constant value of the overlying clast-poor part of the section. The typical 15 to 18 m distance from the contact between the clast-rich and clast-poor melt rocks to the base of the melt sheet may be an indication of the approximate maximum size of turbulent eddies in the melt sheet. Such eddies would provide the mechanism for erratically stirring the debris from the underlying fragmental debris into the lower part of the melt sheet. There was probably turbulent mixing throughout the entire melt sheet but the initial melt probably contained a relatively low but variable content of clasts which were well mixed with the melt as presently preserved in the upper part of the clast-poor rocks. The thin zone above the contact of clast-rich and clast-poor melt, where the clast-content is somewhat irregular and the clast content decreases, would represent the transition where a few eddies overlapped between the clast-rich and clast-poor zones of melt. The presence of a sharp contact at the base of the clast-rich melt rocks and dikes of melt into the fragmental breccias suggest that the final solidification of the melt rocks did not occur until after the flowage and mixing had ceased and the structure of the units had stabilized.

The model of Grieve *et al.* (1977) for movement of melt in a complex crater has a clast-poor melt from the center of the cavity below the impact, overriding a clast-rich melt from the outer margin of the melt zone as it moves outward along the fragmental debris at the base of the excavation. In the present interpretation, as the melted mass, which may have an increase in clasts near its margin, moves outward overriding the fragmental debris, the lower part of the melt will incorporate debris to produce cooling and greater viscosity. The upper clast-poor part of the melt will maintain a higher temperature and lower viscosity allowing it to override the lower clast-rich melt as in Greive's model. There appears to be reasonable agreement, therefore, between this part of Grieve's model and the present interpretation. Enough observational data on radial variations in thickness and clast contents of units at Clearwater Lake do not exist for a more detailed comparison with the additional lithologic variations predicted by the model of Grieve *et al.* (1977).

Acknowledgments—One of the authors (C. S.), performed a portion of this work while at the Lunar and Planetary Institute, which is operated by the Universities Space Research Association under contract No. NSR-09-051-001 with the National Aeronautics and Space Administration. This paper constitutes the Lunar and Planetary Institute Contribution No. 340.

References

Beals C. S., Innes M. J. S. and Rottenberg J. A. (1960) The search for fossil meteorite craters. *Contrib. Dominion Observatory* **4**, no. 4, 1–3.

Bostock H. H. (1969) The Clearwater Complex, New Quebec. *Geol. Survey Can., Bull.* **178**, 63 pp.

Bunch T. E. (1968) Some characteristics of selected minerals from craters. In *Shock Metamorphism of Natural Materials* (B. M. French and N. M. Short, eds.), p. 413–432. Mono, Baltimore.

Carroll D. (1970) Clay minerals: A guide to their x-ray identification. *Geol. Soc. Amer. Spec. Paper* **126**, 80 pp.

Carstens H. (1975) Thermal history of impact melt rocks in the Fennoscandian Shield. *Contrib. Mineral. Petrol.* **50**, 145–155.

Deer W. A., Howie R. A. and Zussman J. (1962) *Rock-forming Minerals*, vol. 3, p. 241. Longmans, Green and Co. Ltd., London

Dence M. R. (1964) A comparative structural and petrographic study of probable Canadian meteorite craters. *Meteoritics* **2**, 249–270.

Dence M. R., Innes M. J. S. and Beals C. A. (1965) On the probable meteorite origin of the Clearwater Lakes, Quebec. *Roy. Astron. Soc. Can. J.* **59**, 13–22.

Eberl D., Whitney G. and Khoury H. (1978) Hydrothermal reactivity of smectite. *Amer. Mineral.* **63**, 401–409.

Faust G. T., Hathaway J. C. and Millot G. (1959) A restudy of stevensite and allied minerals. *Amer. Mineral.* **44**, 342–370.

Floran R. J., Grieve R. A. F., Phinney W. C., Warner J. L., Simonds C. H., Blanchard D. P. and Dence M. R. (1978) Manicouagan impact melt, Quebec, 1, stratigraphy, petrology, and chemistry. *J. Geophys. Res.* **83**, 2737–2759.

Floran R. J., Simonds C. H., Grieve R. A. F., Phinney W. C., Warner J. L., Rhodes M. J. and Dence M. R. (1976) Petrology, structure and origin of the Manicouagan melt sheet, Quebec, Canada: A preliminary report. *Geophys. Res. Lett.* **3**, 49–52.

Grieve R. A. F. (1975) Petrology and chemistry of the impact melt at Mistastin Lake Crater, Labrador. *Bull. Geol. Soc. Amer.* **86**, 1617–1629.

Grieve R. A. F., Dence M. R. and Robertson P. B. (1977) Cratering processes: As interpreted from the occurrence of impact melts. In *Impact and explosion cratering* (D. J. Roddy, R. O. Pepin, and R. B. Merrill, eds.), p. 791–814. Pergamon, N.Y.

Grieve R. A. F. and Floran R. J. (1978) Manicouagan impact melt, 2, chemical interrelations with basement and formational processes. *J. Geophys. Res.* **83**, 2761–2771.

Haggerty S. E. (1976) Oxidation of opaque mineral oxides in basalts. In *Mineral. Soc. Amer. Short Course Notes* **3**, Hg 1–Hg 100.

Janssens M-J., Hertogen J., Takahashi H., Anders E. and Lambert P. (1977) Rochechouart meteorite crater: Identification of projectile *J. Geophys. Res.* **82**, 750–758.

Kranck S. H. and Sinclair G. W. (1963) Clearwater Lake, New Quebec. *Geol. Survey Can., Bull.* **100**, 25 pp.

Lofgren G. and Smith D. (1978) Dynamic melting and crystallization studies on a lunar soil (abstract). *In Lunar and Planetary Science IX*, p. 654–656. Lunar and Planetary Institute, Houston.

Masaitis V. L., Mikhaylov M. V. and Selivanovskaya T. V. (1975) *The Popigay meteorite crater.* Nauka Press, Moscow. 124 pp.

McCabe H. R. and Bannatyne B. B. (1970) Lake St. Martin crypto-explosion crater and geology of the surrounding area. Manit. Dept. Min. Nat. Resour., Mines Br., Geol. Pap. 3/70, 79 pp.

McIntyre D. B. (1968) Impact metamorphism at Clearwater Lake, Quebec. In *Shock Metamorphism of Natural Materials* (B. M. French and N. M. Short, eds.), p. 363–366. Mono, Baltimore.

Millot G. (1970) *Geology of clays: Weathering, sedimentology, geochemistry.* Springer-Verlag, N. Y. 425 pp.

Morgan J. W., Ganapathy R., Higuchi H., Krähenbühl R. and Anders E. (1974) Lunar basins: Tentative characterization of projectiles from meteoritic elements in Apollo 17 boulders. *Proc. Lunar Sci. Conf. 5th*, p. 1703–1736.

Onorato P. I. K., Uhlmann D. R. and Simonds C. H. (1978) Thermal history of the Manicouagan impact melt sheet, Quebec. *J. Geophys. Res.* **83**, 2789–2798.

Palme H., Rainer W. and Grieve R. A. F. (1978) New data on meteoritic material at terrestrial impact craters (abstract). In *Lunar and Planetary Science IX*, p. 856–858. Lunar and Planetary Institute, Houston.

Phinney W. C., McKay D. S., Simonds C. H. and Warner J. L. (1976) Lithification of vitric- and clastic-matrix breccias: SEM petrography. *Proc. Lunar Sci. Conf. 7th*, p. 2469–2492.

Robertson P. B., Dence M. R. and Vos M. A. (1968) Deformation in rock-forming minerals from Canadian craters. In *Shock Metamorphism of Natural Materials* (B. M. French and N. M. Short, eds.), p. 433–452. Mono, Baltimore.

Schaal R. B. and Hörz F. (1977) Shock metamorphism of lunar and terrestrial basalts. *Proc. Lunar Sci. Conf. 8th*, p. 1697–1729.

Simonds C. H. (1975) Thermal regimes in impact melts and the petrology of the Apollo 17 Station 6 boulder. *Proc. Lunar Sci. Conf. 6th*, p. 641–672.

Simonds C. H., Floran R. J., McGee P. E., Phinney W. C. and Warner J. L. (1978a) Petrogenesis of melt rocks, Manicougan impact structure. Quebec. *J. Geophys. Res.*, **83**, 2773–2788.

Simonds C. H., Phinney W. C., McGee P. E. and Cochran A. (1978b) West Clearwater, Quebec impact structure, Part I: Field geology, structure and bulk chemistry. *Proc. Lunar Planet. Sci. Conf. 9th*. This volume.

Stöffler D. (1971) Progressive metamorphism and classification of shocked and brecciated crystalline rocks at impact craters. *J. Geophys. Res.* **76**, 5541–5551.

Proc. Lunar Planet. Sci. Conf. 9th (1978), p. 2695–2712.
Printed in the United States of America

An alternative model for the Manicouagan impact structure

DENNIS L. ORPHAL

California Research and Technology, 4049 First Street, Livermore, California 94550

PETER H. SCHULTZ

Lunar and Planetary Institute, 3303 NASA Road One, Houston, Texas 77058

Abstract—The 210 m.y. old Manicouagan impact structure of eastern Quebec has been previously interpreted as an analog of large (400 km diameter) multi-ringed basins on the Moon. However, existing geologic evidence also supports an alternative proposal that Manicouagan is an endogenically modified crater analogous to many lunar floor-fractured craters.

In the proposed model of endogenic modification, the impact event at Manicouagan initially formed a transient cavity 80 km in diameter generating about 270–455 km^3 of melted target material, forming a pool about 65 km in diameter at the bottom of the crater. Shortly after formation, collapse of the transient crater rim increased the crater diameter to about 100 km, corresponding to an existing change in joint pattern and drainage direction. Highly fractured limestone blocks currently found in contact with the outer portion of the melt sheet are interpreted as debris slumped to the floor during this adjustment. Inward motion of mass associated with the failure of the transient cavity resulted in the formation of a central peak-ring complex about 15 km in diameter, corresponding closely to the present diameter of the crescent-shaped elevated region in the central portion of the structure.

At some time, probably before the most recent periods of glaciation, the highly fractured and brecciated rock below the crater was intruded by a tabular magmatic body. This intrusion gradually uplifted the crater floor forming the existing annular moat surrounding a 55 km diameter floor plate. This adjustment of the crater floor resulted in downdropping part of the central peak-ring along steeply dipping normal faults. Dike intrusion along these faults could account for the high intensity, ring-like magnetic anomaly now associated with the central portion of the crater. The intrusion also produced an outer ring fracture with a diameter about 1.5 times the crater diameter in a manner analogous to those surrounding several lunar floor-fractured craters such as those found in Mare Smythii.

INTRODUCTION

The Manicouagan impact structure is located within the Canadian Shield of central Quebec (51° 23′N, 68° 42′W). The generally circular geometry of the structure is made prominent by a 65 km diameter annular trough or moat. Portions of this moat were originally occupied by the arcuate Manicouagan and Mouchalagane Lakes, but construction of the Daniel Johnson Dam has connected the water-filled moat by formation of the Manicouagan reservoir (Fig. 1).

Manicouagan is one of the largest and better preserved terrestrial impact structures. Because of its size, state of preservation and relatively well known geological setting, Manicouagan represents an excellent opportunity to test and improve models of impact crater formation and, thus, interpretations of the

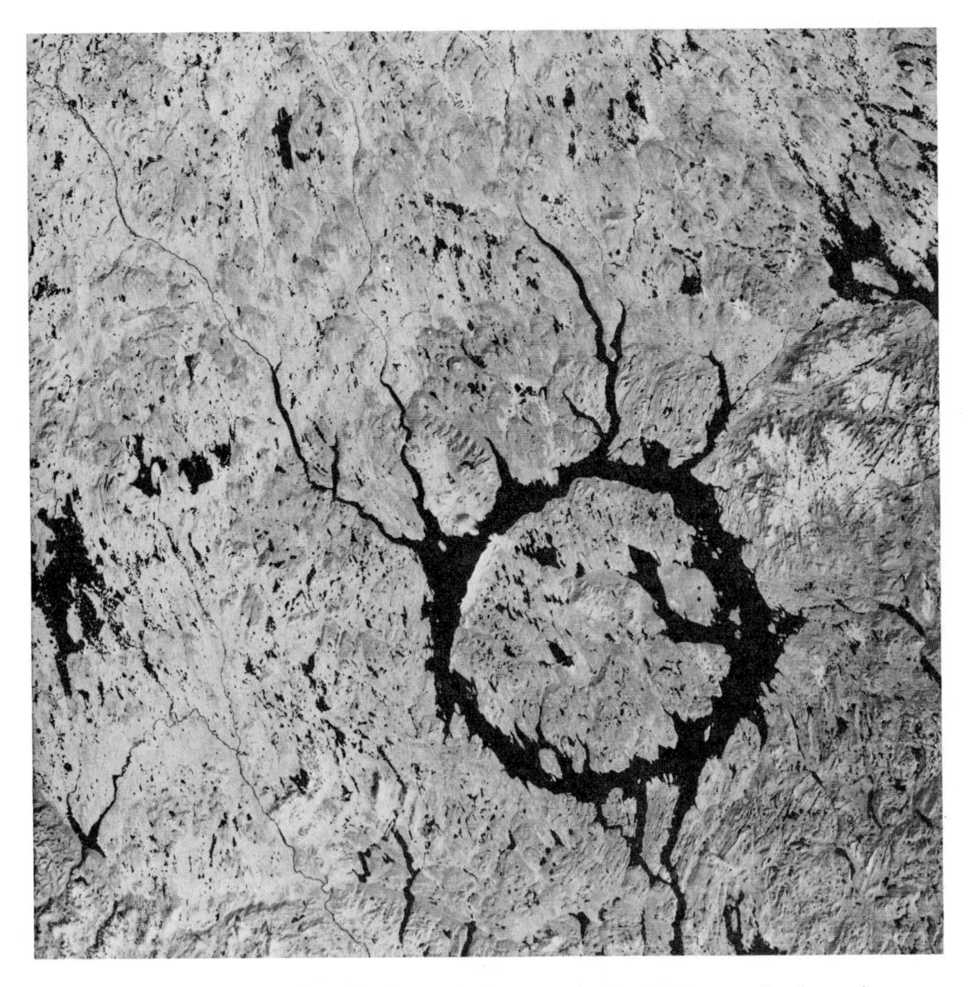

Fig. 1. ERTS image of the Manicouagan structure clearly showing annular depression (ERTS-81438150245N000, Band 6).

surface features and geological evolution of the Moon, Mars and Mercury. Consequently, Manicouagan has been extensively studied (e.g., Floran and Dence, 1976; Currie, 1972; Murtaugh, 1972 and 1975; Wolfe, 1972; Floran *et al.*, 1976; Coles and Clark, 1978; Phinney *et al.*, 1978; and Sweeney, 1978).

An important extrapolation from studies of Manicouagan to date is the interpretation of the structure as a multi-ringed basin analogous to 400 km-size lunar basins such as Moscoviense (Floran and Dence, 1976) and 200 km ring structures on Mars and Mercury (Dence, 1977). In this paper we propose an alternate interpretation wherein the Manicouagan impact structure represents an endogenically modified crater analogous to many floor-fractured craters found on the Moon (Schultz 1972, 1976a, 1976b).

GEOLOGIC SETTING

There have been a number of field investigations of the geology of the Manicouagan structure (Berard, 1961; Murtaugh and Currie, 1969; Currie, 1972; Murtaugh 1972 and 1975; and Wolfe, 1972). Figure 2 is a simplified geologic map of the inner portion of the structure.

The Manicouagan structure occurs in the northeastern part of the Grenville province of the Canadian Shield. This province is characterized by high grade metamorphic rocks with K-Ar ages of about 1 b.y. In the area of Manicouagan,

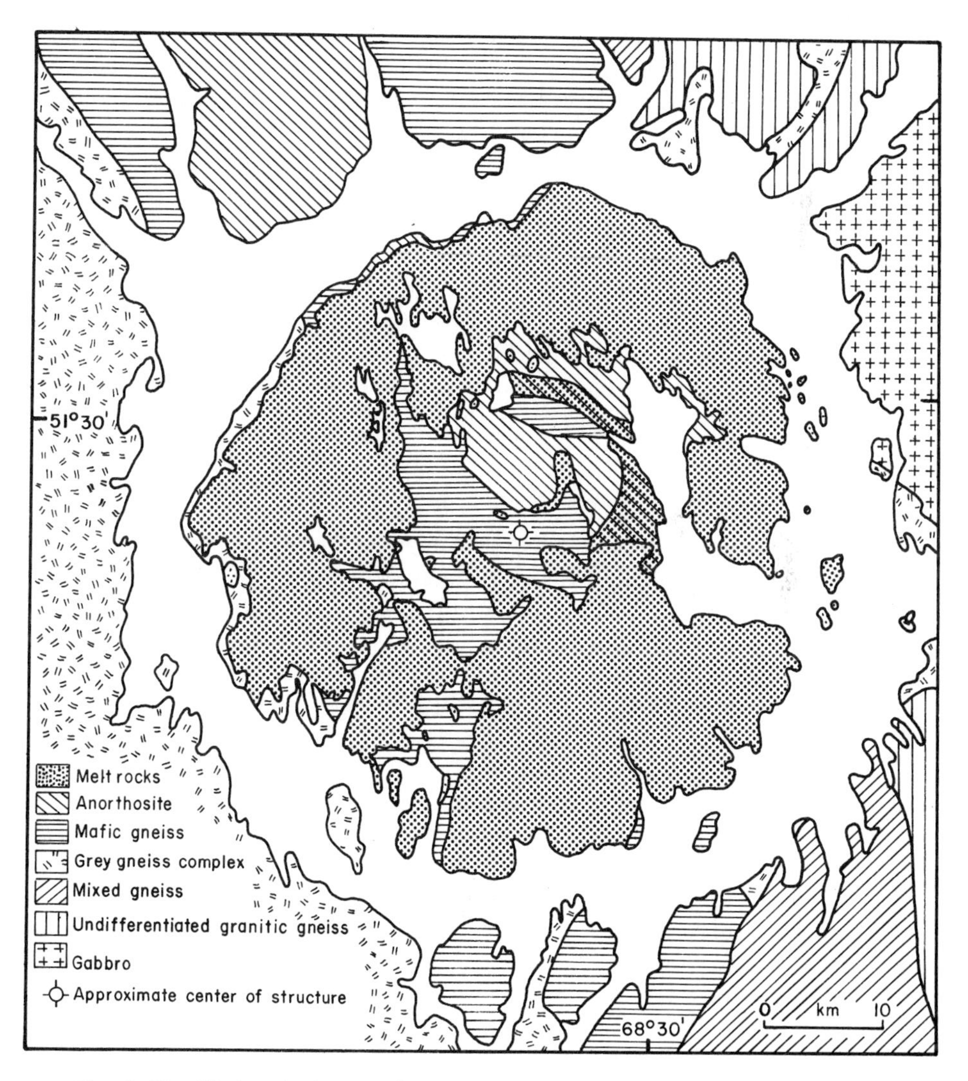

Fig. 2. Simplified geologic map of the Manicouagan structure. Modified from Grieve and Floran (1978).

Murtaugh (1975) recognized seven major Grenville units; Currie (1972) recognized eight. These units include meta-gabbros, meta-anorthosites and granitic gneisses. Mafic and grey gneisses are the primary country rock. In the vicinity of Manicouagan, this primarily granitic terrain was intruded by a mafic to anorthosite complex.

At the time of impact, about 210 m.y. ago (Jahn *et al.*, 1978), the Manicouagan area was overlain by a thin veneer of Ordovician limestone. Blocks of this limestone, up to 1 km in size, are preserved within the present crater at a few locations along the inner shore of the annular moat.

An extensive melt sheet is partly preserved inside the moat perimeter. The melt sheet is essentially continuous and forms an annulus: the outer boundary defined by the moat and an inner boundary, approximately 25 km in diameter, defined by a central topographic high. The melt sheet ranges up to 200 m thick in the east central portion of the structure, where the melt extends from the current water level in the moat to the tops of nearby hills. The preserved melt volume is about 150 km^3 (Dence, 1977). Simonds *et al.* (1978) estimate that the pre-erosional thickness of the melt sheet was 130 to 556 m with a volume of 270–475 km^3. The chemical composition of the melt is consistent with derivation from a mixture of basement rocks near the impact point (Grieve and Floran, 1978; Floran *et al.*, 1978). The melt rocks have been dated by a variety of radiometric techniques, all resulting in an estimated age for the melt rocks of about 210 m.y. (Wolfe, 1972; Jahn *et al.*, 1978).

MORPHOLOGY OF THE MANICOUAGAN STRUCTURE

The Manicouagan structure may be subdivided into six (6) morphological elements based on topography and structural style. These elements, their corresponding approximate outer diameters (in parentheses) and their characteristics are listed below and illustrated in Fig. 3:

1. Central region (25 km); a structurally complex zone of uplifted, shocked and metamorphosed basement rocks with small tabular bodies of impact melt and pseudotachylite veins.
2. Inner plateau (55 km); bounded by the annular moat, overlain by melt sheet, underlain by shocked basement rock.
3. Annular moat (65 km); ring graben; highly deformed limestone blocks near inner boundary.
4. Inner fractured zone (~100 km); defined by distinct change in joint pattern and spacing and drainage divide; within zone, joints increase in number and variety; drainage is towards (former) Lake Mouchalagane and (former) Lake Manicouagan.
5. Outer fractured zone (~150 km); fractured basement; drainage away from Manicouagan.
6. Outer circumferential depression (~150 km); circular depression.

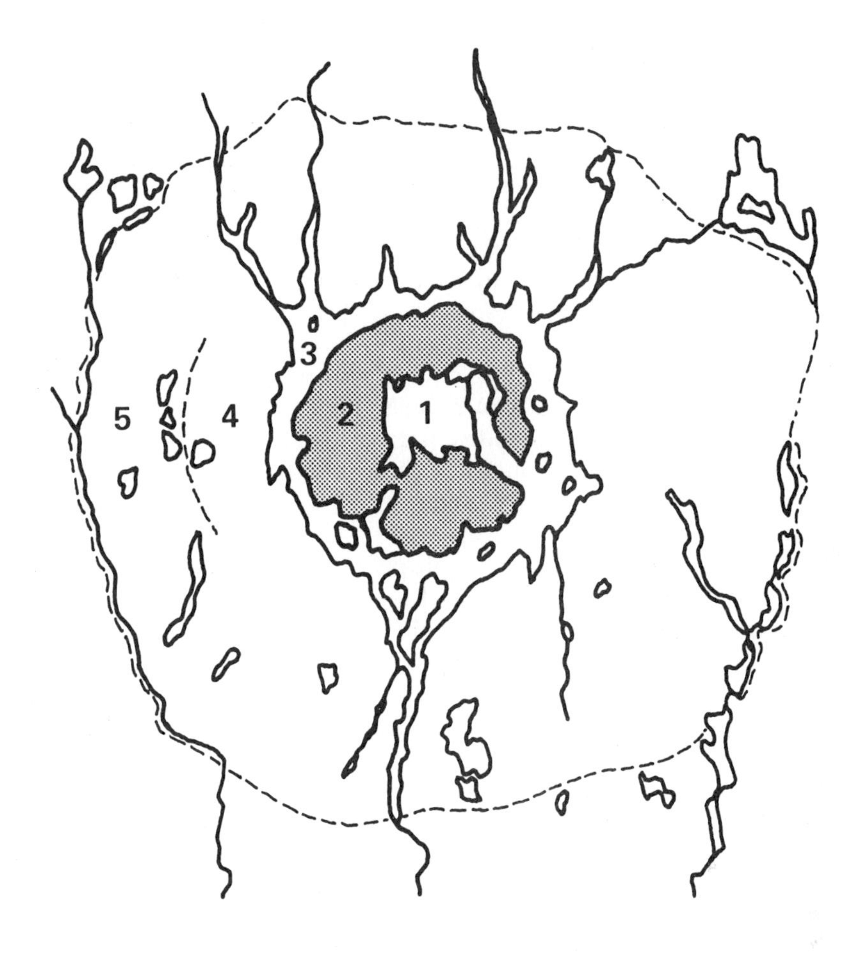

(ADAPTED FROM FLORAN AND DENCE, 1976)

Fig. 3. Major morphological elements of the Manicouagan structure as described by Floran and Dence (1976): central region (1); inner plateau (2); annular moat (3); inner fractured zone (4); outer fractured zone (5); and outer circumferential depression indicated by broken line.

Dence (1977) and Floran and Dence (1976) recognize five of these structural elements. They combine elements 4 and 5 above into a single element denoted as the "outer disturbed zone." The importance of the distinction between zones 4 and 5 will be discussed in some detail below.

The *central region* of the Manicouagan structure varies topographically but is dominated by two broad upland areas generally north and south of the geometric center of the structure. The northern upland area includes Mont de Babel, which exhibits the highest relief associated with the inner portion of the Manicouagan structure (~300 m). Mont de Babel is an eroded rectangular horst composed of shocked and fractured anorthosite basement rocks. In addition to Mont de Babel, several smaller isolated peaks of anorthosite occur in the area. The existence of maskelynite on several peaks of Mont de Babel (Floran and Dence, 1976) suggests that the higher elevations experienced shock pressures of ~200–400 kb (Stöffler, 1971) and were never covered by melt. Patches of melt are present, however, at the base of Mont de Babel (Floran and Dence, 1976). Mont de Babel and the associated smaller peaks are often referred to as the central peak(s) of the structure, although they are displaced 5–7 km from the geometric center of the structure.

Based on existing regional gravity data, Sweeney (1978) has constructed the residual Bouguer anomaly associated with the Manicouagan structure. The residual gravity field exhibits a relative gravity high of 0 mGal over the exposed central gneisses near the geometric center of the structure. The uplifted anorthositic rocks of Mont de Babel and nearby peaks are associated with a −10 mGal low.

Aeromagnetic surveys by the Geological Survey of Canada and the Quebec Department of Natural Resources reveal a prominent 2000 γ anomaly over the geometric center of the structure (Geological Survey of Canada, 1969). Mont de Babel exhibits a low magnetic field. The magnetic field near the center of the structure displays very high horizontal gradients, thereby implying a shallow source. The depth of the causative body is not well defined, primarily because of uncertainty in the assignment of a magnetization intensity for the body. However, the observed intense peaks and horizontal gradients require shallow, highly magnetic sections. Modeling suggests that the causative body is almost at the geometric center of the structure, extends over an area of about 12 km × 8 km and has several peaks or extensions that form a crude ring (~6–10 km diameter) around the center of the structure (Coles and Clark, 1978).

The transition from the central region to the *inner plateau* is marked by a distinct rise in elevation beginning generally at a distance of about 10–15 km from the center of the structure (Floran and Dence, 1976). The inner plateau is overlain by a nearly continuous sheet of impact melt. The outer boundary of the inner plateau is marked by the inner boundary of the annular moat.

The *annular moat* is the most prominent feature of the present Manicouagan structure. The 5–10 km-wide moat is formed by a 65 km ring graben bounded on the outside by steeply dipping normal faults (Floran and Dence, 1976). Before flooding of the reservoir isolated outcrops of tilted and deformed limestone, siltstone and shale were found on the inner edges of the moat (Murtaugh, 1975). The residual Bouguer gravity field constructed by Sweeney (1978) reveals a −4 to −10 mGal low associated with the moat that grades gently upward to the 0 mGal high over the geometric center of the impact structure. Figure 4 shows a generalized cross section of the Manicouagan structure within the circular moat.

Beyond the outer shore of the circular moat and extending to a diameter of about 100 km is a region we have called the *inner fractured zone*. The

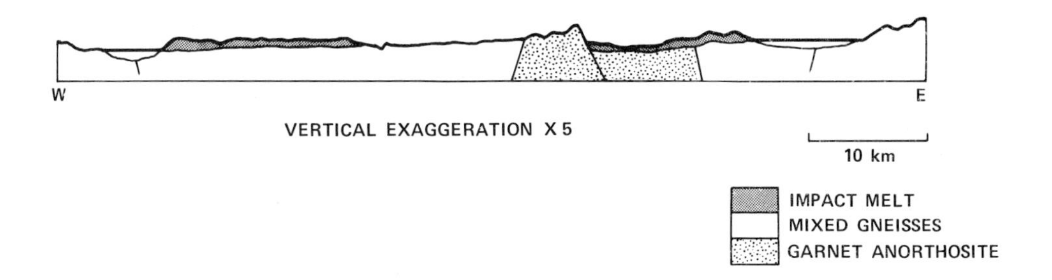

Fig. 4. Generalized cross section of the inner portion of the Manicouagan structure showing depressed annular moat, cap of impact melt, and garnet anorthosite central relief. Adapted from Floran and Dence (1976).

characteristic features of this zone are much more subtle than those for the zones previously described. The inner fractured zone is distinguished principally on the basis of the number of variety of joints and the drainage direction. Within the inner fractured zone, drainage is towards the Manicouagan structure. Murtaugh (1975) reports that the drainage divide is most easily observed in the low hills to the west of the Manicouagan structure. The eastern side exhibits a similar divide, but the transition is more difficult to define. To the west the drainage divide is about 15 km beyond the outer shore of former Lake Mouchalagane and is coincident with a distinct change in joint pattern. Joints within the inner fractured zone are much more numerous and complex than beyond the region (Murtaugh, 1975).

The *outer fractured zone* is the region between the inner fractured zone and the outer circumferential depression. The inner and outer diameters of the outer zone are about 100 and 150 km, respectively. Drainage within this zone is away from the Manicouagan structure. Basement rocks in this zone are fractured and jointed but not so extensively jointed as the basement rocks of the inner fractured zone (Murtaugh, 1975; Floran and Dence, 1976).

The presence of a subdued *outer circumferential depression* of about 150 km diameter associated with the Manicouagan structure has only recently been revealed (Floran and Dence, 1976; Dence, 1977) as a result of satellite photography (Fig. 5).

Current Models of the Manicouagan Impact Structure

Dence (1977) and Floran and Dence (1976) have proposed a model for the Manicouagan impact structure. Briefly, Floran and Dence propose that the impact formed a transient cavity of radius 19 ± 4 km. Floran and Dence fix the outer limit of 23 km for the transient cavity radius on the basis of the limestone

Fig. 5. Outer circumferential depression (arrows) controlling stream flow at several locations and corresponding to a slight change in albedo. Skylab Photo No. SL3-122-2628.

exposures occurring at that distance from the center of the existing structure. Such a small radius requires that considerable volumes of impact melt overrode the rim of the transient cavity (Floran and Dence, 1976).

During the last stages of crater formation, the transient cavity collapsed. Floran and Dence propose that the rim diameter of the collapsed transient cavity was 35–40 km, corresponding to a subdued topographic ring in the existing structure. In the Floran and Dence model, the downfaulted circular moat (called the peripheral trough by Floran and Dence) represents the collapsed outer margin of the transient cavity. They estimate the pre-erosional crater diameter to be 75 km. The collapse of the transient cavity results in an uplift of material from below the crater to form discontinuous, asymmetrically located central peaks or a peak-ring structure.

The depth of the transient cavity is uncertain. If the transient cavity had a parabolic shape (Dence, 1973; Orphal, 1977; and Dence *et al.*, 1977), a transient crater radius of ~20 km would correspond to a depth of ~14 km. Based on studies of the gravity data, however, Sweeney (1978) suggests a depth of $\lesssim$8 km. Phinney *et al.* (1978) suggest a depth of 6–9 km for the transient cavity.

Floran and Dence (1976) and Dence (1977) propose that Manicouagan is a multi-ring structure with ring diameters of 30–45 km, 75 km and 145–160 km and a ring spacing ratio of ~2. They further propose that Manicouagan is a structural analog of 400–500 km basins on the Moon and 200 km-diameter ringed basins on Mars and Mercury and may, in fact, be transitional to the larger lunar multi-ringed basins such as Orientale.

There are several difficulties with the model proposed by Floran and Dence (1976) and Dence (1977). First, unusual cratering phenomena must be imposed in order to reconcile the transient cavity diameter and the extent of impact melt. Specifically, large volumes of impact melt must override the rim. Lunar analogs of such phenomena occur only where the crater rim has intersected an adjacent topographic low (e.g., another crater), and the resulting deposition of melt is concentrated in this low-lying region (Schultz, 1972, 1976a; Hawke and Head, 1977). In these cases, the distribution of melt deposits on the crater rim are highly asymmetric, in contrast to the relatively symmetric melt distribution observed around Manicouagan. Moreover, the total volume of impact melt estimated for Manicouagan corresponds to 3–17 percent of the transient crater volume. This value for the ratio of the volume of impact melt to the transient crater volume is considerably greater than is normally estimated for impact craters (Gault and Heitowit, 1963; Dence, 1965; O'Keefe and Ahrens, 1975). While it is true that there is a great deal of uncertainty concerning the details of energy partitioning in large impact events, the simplest explanation would seem to be that the transient cavity proposed by Floran and Dence is too small.

Finally, the multi-ring model seems to offer no obvious explanation for the well-defined and distinctive ring-like magnetic anomaly associated with the central portion of the structure. We feel an acceptable model for the formation of the Manicouagan crater must offer a reasonable explanation for such a distinctive geophysical feature.

Floran and Dence (1976) suggest that Manicouagan is a terrestrial analog of the lunar multi-ring basins. However, multi-ring plans also occur in certain modified lunar craters that approach the same dimensions of well-preserved terrestrial impact craters. These modified lunar craters provide an alternative framework for the interpretation of the Manicouagan structure.

ENDOGENICALLY MODIFIED LUNAR IMPACT CRATERS AS AN ANALOGY TO THE MANICOUAGAN IMPACT STRUCTURE

Endogenically modified lunar craters

Over 200 craters on the Moon exhibit tectonic and volcanic modifications believed to be related to the epochs of mare basalt flooding (Schultz, 1976b). Similar endogenically modified craters exist on Mars (Schultz, 1976c, 1978) and Mercury (Schultz, 1977). There are six different types of lunar floor-fractured craters. These different types represent combinations of 1) degrees of modification, 2) initial appearance (time interval between crater formation and modification), 3) size of initial impact structure, and 4) possibly different crater origins. Craters that have been extensively modified develop a multi-ring plan (Class III in Schultz, 1976b) that exhibits the same structural elements described for Manicouagan.

Figure 6 illustrates the crater Haldane in Mare Smythii. The accompanying sketch identifies features characteristic of extensively modified lunar craters. The crater floor is shallow and approaches the same elevation as the exterior terrain. Separating the crater rim from the elevated crater floor is an annular depression (moat). In this example, the moat is filled with mare basalts believed to have been extruded from vents along a ring fracture beneath the moat.

The outer margin of the floor commonly exhibits remnants of rim material slumped during crater formation and uplifted with the floor during modification. The central peak complex also is commonly uplifted such that the summit approaches the elevation of the rim. Central peaks of fresh lunar impact craters approaching 100 km in diameter typically extend 1.5 km to 2.5 km above the crater floor, yet remain 1.5 to 2.5 km below the rim crest. Central peaks of extensively endogenically modified craters in the same size range extend from 0.5 km to 1.5 km above the floor but as close as 0.2 km to 1.0 km to the rim. In several examples, the central peak complex defines a faulted block, which is separate from the rest of the crater floor and has undergone a combination of uplift and subsidence. The old crater rim also commonly exhibits additional faulting. A feature common to several extensively modified craters is a ring fracture approximately 1.5 crater radii from the crater rim. Limited pyroclastic eruptions have occurred along this fracture around several craters, particularly in Mare Smythii (Schultz, 1976b; Wolfe and El Baz, 1976).

Comparisons of craters with different degrees of modification have led to an interpreted sequence of development where a tabular intrusion accumulates beneath the crater and uplifts the floor largely as a block (Schultz, 1976b). A depth of intrusion is believed to be approximately one-third of the transient

Fig. 6. (a) The 40 km-diameter lunar floor-fractured crater Haldane on the margin of Mare Smythii. Multi-ring appearance results from annular moat between old crater floor and wall as well as additional scarp to the east and concentric fractures (AS-17-M-2641). (b) Photogeologic sketch map modified from Wolfe and El Baz (1976) showing major geologic units and structural elements.

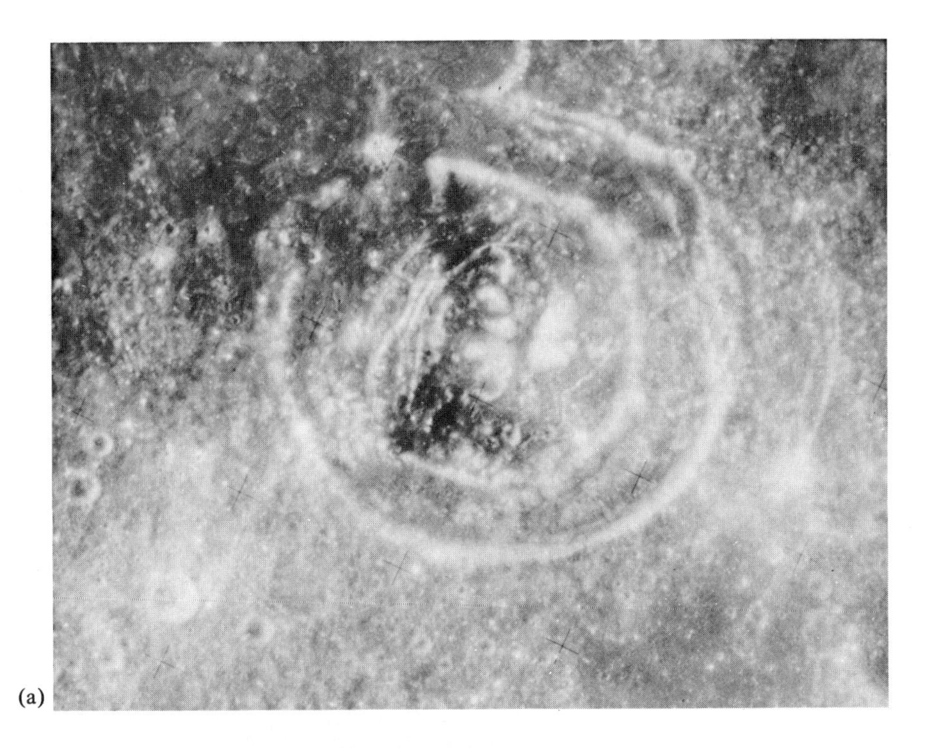
(a)

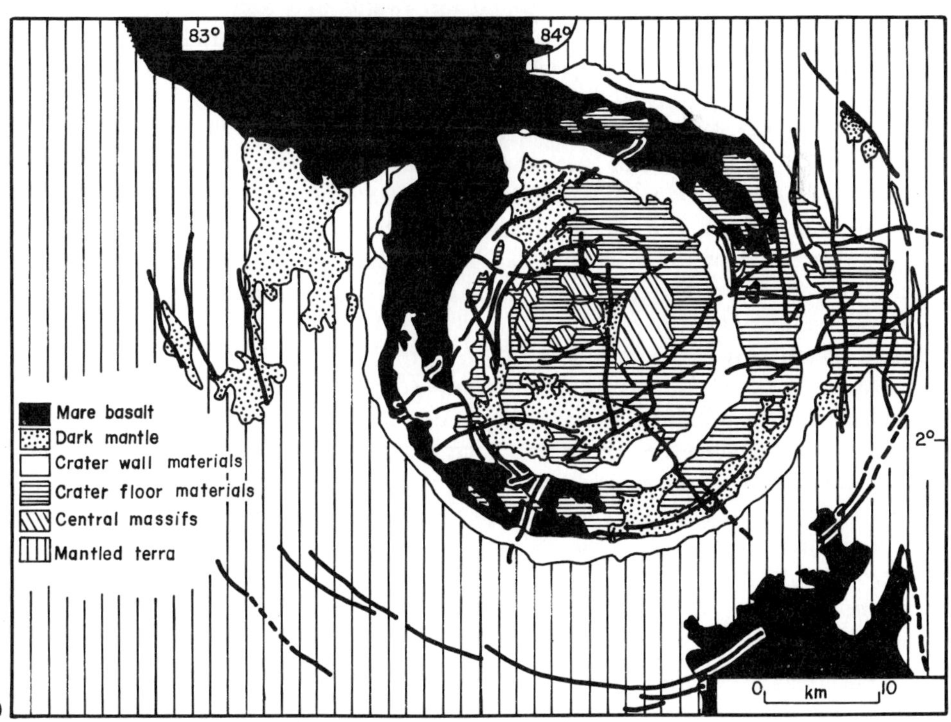
83°
84°
2°
Mare basalt
Dark mantle
Crater wall materials
Crater floor materials
Central massifs
Mantled terra
0 km 10
(b)

crater diameter near the inferred transition from the shock-induced breccia zone to the fractured country rock. At such depths, it can be shown that the expected gravity anomaly associated with an intrusion with low density contrast is effectively masked. Most floor-fractured lunar craters exhibit weak gravity lows.

Application to Manicouagan

The preceding general descriptions have many similarities with the description of Manicouagan crater. Subsequent discussion compares such similarities and offers an alternative interpretation for the development of terrestrial multi-ring craters.

The sequence of events corresponding to the proposed endogenically modified crater model for Manicouagan is illustrated in Fig. 7.

We suggest that a meteor impact about 210 m.y. ago formed a transient cavity about 80 km in diamter. The depth of the transient crater is not certain but was probably not less than 15 km and not more than about 20 km. The diameter-to-depth ratio for the transient cavity was thus about 4–5. The impact produced a volume of melted target rock greater than 270–475 km^3. Some melted material was ejected from the transient crater during formation. The 270–475 km^3 of impact melt estimated to remain in the crater corresponds to 0.5–1 percent of the transient cavity volume, assuming a depth of 15 km and an elliptical geometry. In contrast, a transient cavity of about 20 km radius and about 9 km depth, as proposed by Floran and Dence (1976), implies an estimated melt volume *remaining* in the crater corresponding to 4–6 percent of the transient cavity volume. In further contrast to the model by Floran and Dence (1976), large volumes of melt rock do not overflow the transient cavity rim. Melted target rock that was not ejected formed a pool about 65 km in diameter at the bottom of the crater. If one assumes an ellipsoidal geometry for the transient crater, the melt pool would have been 120–215 m deep at the center.

Very shortly after crater formation, the transient cavity collapsed owing to deep-seated lateral redistribution of mass. In addition, slumping and slope failure occurred along the crater wall. The failure of the transient cavity and crater walls resulted in an increase in rim to rim diameter of at least 20 percent (Quaide *et al.*, 1965). The post failure crater diameter is taken to be about 100 km. This position for the post-collapse, pre-erosional crater rim corresponds approximately to the division between the "inner fractured zone" and the "outer fractured zone" of the Manicouagan structure. The massive slope failure and final position of the crater rim could account for the corresponding change in drainage direction and abrupt change in jointing style. Slumping of the transient cavity walls also resulted in transport of blocks of tilted and folded limestone from near the transient crater rim onto the crater floor and, in cases, resulted in contact with the solidifying melt pool.

Along with failure of the transient cavity wall, material from beneath the crater is uplifted to form a central peak complex. For a crater this size, the

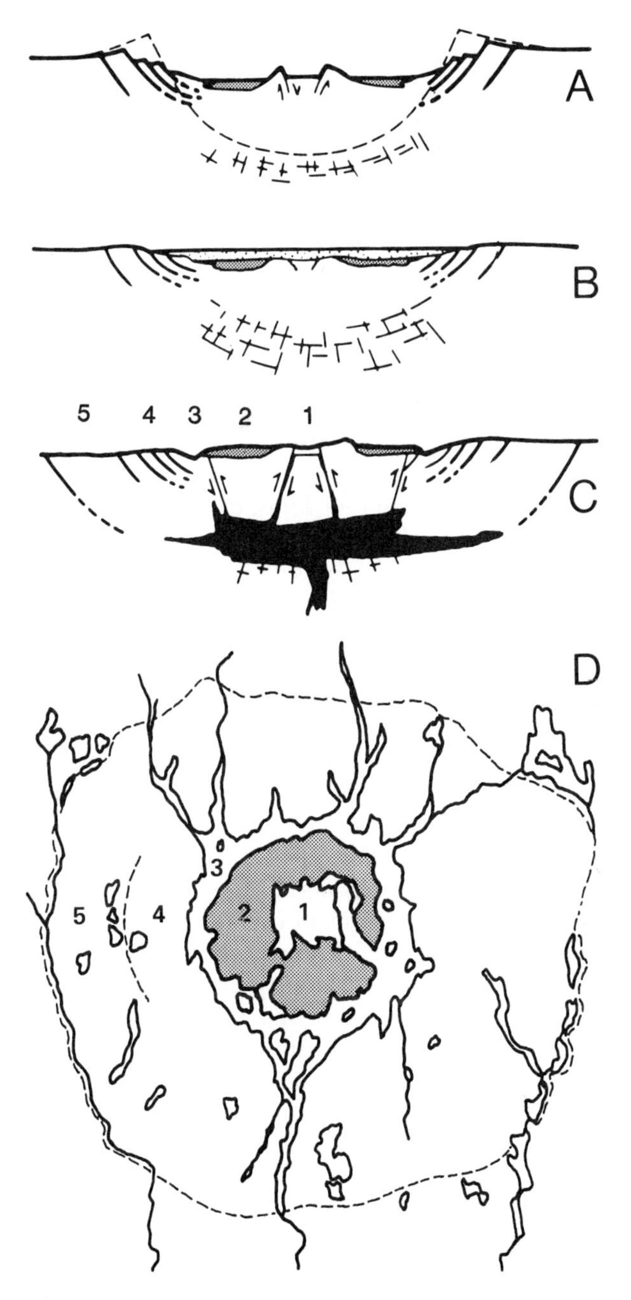

Fig. 7. Proposed geologic history of the Manicouagan structure. Initial impact produced an 80 km transient crater (broken line in A) that subsequently collapsed to produce a 100 km rim-rim diameter cavity. Sediments rapidly filled the crater (B). Subsequent tabular intrusion uplifted the old crater floor resulting in stripping away of overlying sediments and portions of the impact melt (C). Uplift also produced annular moat and intrusion beneath the central peak, a sequence of events common to floor-fractured craters on the Moon.

central portion of the uplifted peak often collapses forming a central peak-ring (Baldwin, 1963; Schultz, 1972). The diameter of central peaks with central depressions and peak-rings within craters smaller than 200 km increases linearly with crater diameter (Schultz, 1976b). For a 100 km crater, the diameter of such a central peak-ring is expected to be about 15 km. This corresponds very closely with the diameter of the crescent-shaped elevated region in the central portion of the existing Manicouagan structure.

As in numerous lunar craters, the central peak-ring may not have been continuous since the southern uplands of the existing Manicouagan structure is covered with the melt sheet. Portions of the central uplift ring complex originally may have been draped and interfingered with a thin layer of melt and breccia. The present maximum relief of the central peak complex is approximately 0.3 km; whereas, central peaks of lunar craters of comparable size typically approach 2 km. Drawing on such well-preserved lunar craters, and noting the inferred depths and shock history exhibited by Mont de Babel, we suggest that Mont de Babel represents only the core of one component of the central peak-ring complex.

Over time, as much as 1.5 km of the central peak-ring and most of the crater rim were eroded, thereby depositing on the crater floor a thick protective deposit of sediments.

At some time, probably before the most recent periods of glaciation, the highly fractured and brecciated rock beneath the crater was intruded by a magmatic body forming a sill-like or tabular body. This intrusion resulted in uplift of the crater floor forming a circular moat surrounding a floor "plate". It is envisioned that this massive readjustment of the crater floor was accompanied by downdropping of most of the central portion of the ring peak structure along steeply dipping normal faults, thereby leaving Mont de Babel and the immediately surrounding peaks as the only prominent surface remnant of the original central peak complex. Dike intrusion along these steeply dipping faults could account for the high intensity, ring-like magnetic anomaly associated with the central portion of the Manicouagan structure (Coles and Clark, 1978). The presence of an intrusive body is not inconsistent with the small relative gravity high associated with the center of the Manicouagan structure (Sweeney, 1978).

During the stages of extensive crater adjustment, the 150 km-diameter circumferential depression developed. The diameter of this depression may have been controlled by the shock-weakened state of the bedrock surrounding the impact crater and was enhanced by stresses resulting from the intruding mass.

DISCUSSION

Interpretation of Manicouagan as an endogenically modified crater potentially explains a number of important features of the impact structure. Perhaps most important, the distinctive ring-like magnetic anomaly associated with the center of the structure is interpreted as the result of dike intrusion along fault structures associated with the central ring-peak complex.

A number of endogenically modified craters on the moon display multi-ring plans remarkably similar to Manicouagan. In particular, many modified lunar craters display a circular moat and central floor plate structure very similar to that found at Manicouagan. Moreover, several such craters exhibit a circumferential fracture surrounding the crater that is similar in structure to the outer circumferential depression found at Manicouagan. For modified lunar craters, the diameter of this outer circumferential fracture is about 1.5 times the crater diameter. If the diameter of the Manicouagan crater after failure of the transient cavity walls is taken to be about 100 km as proposed here, the outer circumferential depression at Manicouagan also has a diameter of about 1.5 times the crater diameter. A crater diameter of 100 km seems consistent with the change in joint pattern and drainage direction found between the "inner" and "outer fractured zones" at Manicouagan.

The endogenically modified crater model proposed for Manicouagan has additional appealing features. In particular, it is not required that significant volumes of impact melt override the transient crater rim, a phenomenon not observed in any experiments or calculations performed to date. Also, the ratio of the volume of impact melt to the transient crater volume for the modified crater model is about 0.5 to 1 percent as compared to the 4–6 percent or more for previously proposed models. Although it must be admitted that the details of energy partitioning in impacts the size of that at Manicouagan are not well understood at present, the data and evidence that is available suggests that the volume of impact melt is about 1 percent of the transient crater volume.

Although there is no evidence of volcanic activity for the past 500 million years or so in the portion of the Canadian Shield containing the Manicouagan structure, such a nonvolcanic regional history does not necessarily weaken the model proposed here. It must be recalled that the depth of the Manicouagan transient cavity may well have been 15–20 km, a depth representing a significant fraction of the thickness of the crust of this region. Such a deep but localized disturbance and weakening of the crust could have provided a ready path for the local intrusion of magma, whether immediately after or, more probably, long after the impact event. The intrusion was highly localized due to the finite extent of the disruption of the crustal material by the impact, just as the modification process of lunar basalt-filled and floor-fractured craters is highly localized to the crater interior. The intrusion did not reach the surface; rather, it simply filled the highly fractured and brecciated region immediately beneath the crater. In these respects, the volcano/tectonic modification of the Manicouagan crater may be analogous to the isolated floor-fractured craters found in the lunar highlands. For example, Fig. 8 shows the lunar crater Kostinskij in the eastern farside hemisphere far removed from a major mare-filled region. The structural similarity between this lunar crater and Manicouagan is obvious. Moreover, the lunar crater Tsiolkovskij occurs deep in the farside cratered highlands nearly 1000 km from a major mare inundated basin, yet its floor is clearly filled with mare

Fig. 8. The 75 km-diameter crater Kostinskij in the eastern farside lunar highlands (+ 14°, 118°). This floor-fractured crater exhibits floor modification but is spatially far removed from a major mare-filled basin (AS-16-M-3000).

basalts. Thus, the isolation of Manicouagan from a major volcanic region is no more mysterious than several isolated mare-filled and floor-fractured craters on the Moon.

The evolution of surface features on the Moon and Mars is the result of a complex interaction of impact and volcanic processes. If, as proposed here, the Manicouagan impact structure is a terrestrial example of a volcanically modified crater, it offers a valuable opportunity to better understand the response of a cratered surface to volcanic processes. Also raised is the possibility that a number of additional terrestrial impact structures may be endogenically modified. Only one other terrestrial impact crater is commonly accepted as volcanically modified, the Sudbury structure in Ontario, Canada (French, 1970). The Sudbury structure, however, is much more poorly preserved than Manicouagan.

Although we believe the existing data for Manicouagan support the endogenically modified crater model proposed here, we also freely admit that the data are not sufficiently definitive to select this model to the exclusion of all others, in particular the multi-ring-structure model offered by Floran and Dence (1976). Since Manicouagan represents a potentially excellent opportunity to improve our understanding of the mechanics of formation of complex craters on the Moon andMars (and possibly Mercury), it is important to determine the detailed structure of Manicouagan more completely. A detailed seismic survey and deep drilling should provide the data necessary to determine whether an intrusive mass is present at depth at Manicouagan. Such data would also provide additional details of the subsurface structure of the crater essential for formulating a mechanically consistent model of the formation and perhaps modification of the crater.

Acknowledgments—The majority of this work was performed in September, 1977 during a Visiting Scientist appointment to Dennis L. Orphal at the Lunar and Planetary Institute. The receipt of this appointment and the assistance of the Institute staff is greatfully acknowledged. This work was originally stimulated by discussions with Chuck Simonds; his help and patience in answering many of our questions are appreciated. We also wish to thank Mike Dence and Jeffrey Warner for their very constructive reviews of the draft manuscript.

The Lunar and Planetary Institute is operated by the Universities Space Research Association under Contract No. NSR 09-051-001 with the National Aeronautics and Space Administration. This paper constitutes the Lunar and Planetary Institute Contribution No. 341.

REFERENCES

Baldwin R. B. (1963) *The Measure of the Moon.* Univ. Chicago Press. 488 pp.

Berard J. L. (1961) Geology of the Manicouagan-Mouchalagane Lakes Area: Quebec. Quebec Dept., Nat. Res. Prelim. Rept. 849.

Coles R. L. and Clark J. F. (1978) The central magnetic anomaly, Manicouagan structure, Quebec. *J. Geophys. Res.* **83**, 2805–2808.

Currie K. L. (1972) Geology and petrology of the Manicouagan resurgent caldera, Quebec. *Geol. Survey Can., Bull.* **198**, 153 pp.

Dence M. R. (1965) The extraterrestrial origin of Canadian craters. *Ann. N.Y. Acad. Sci.* **123**, 941–969.

Dence M. R. (1973) Dimensional analysis of impact structures (abstract). *Meteoritics* **8**, 343–344.

Dence M. R. (1977) The Manicouagan impact structure, Quebec, from Skylab. In *Skylab Explores the Earth*, p. 175–190. NASA SP-380. Washington, D.C.

Dence M. R., Grieve R. A. F. and Robertson P. B. (1977) Terrestrial impact structures: Principal characteristics and energy considerations. In *Impact and Explosion Cratering* (D. J. Roddy, R. O. Pepin and R. B. Merrill, eds.), p. 247-275. Pergamon, N.Y.

Floran R. J. and Dence M. R. (1976) Morphology of the Manicouagan ring-structure, Quebec, and some comparisons with lunar basins and craters. *Proc. Lunar Sci. Conf. 7th*, p. 2845–2865.

Floran R. J., Grieve R. A. F., Phinney W. C., Warner J. L., Simonds C. H., Blanchard D. P. and Dence M. R. (1978) Manicouagan impact melt, Quebec, Part I: Stratigraphy, petrology and chemistry. *J. Geophys. Res.* **83**, 2737–2759.

Floran R. J., Simonds C. H., Grieve R. A. F., Phinney W. C., Warner J. L., Rhodes M. J. and Dence M. R. (1976) Petrology, structure and origin of the Manicouagan melt sheet, Quebec, Canada: A preliminary report. *Geophys. Res. Lett.* **3**, 49–52.

French B. M. (1970) Possible relations between meteorite impact and igneous petrogenesis, as indicated by the Sudbury Structure, Ontario, Canada. *Bull. Volc.* **34**, 466–517.

Gault D. E. and Heitowit E. D. (1963) The partition of energy for hypervelocity impact craters formed in rock. *Proc. 6th Hypervelocity Impact Symposium*, p. 419–456. Clearinghouse for Federal Scientific and Technical Information.

Greive R. A. F. and Floran R. J. (1978) Manicouagan impact melt, Quebec, Part II: Chemical inter-relations with basement and formational processes. *J. Geophys. Res.* **83**, 2761–2771.

Hawke B. R. and Head J. W. (1977) Impact melt on lunar crater rims. In *Impact and Explosion Cratering* (D. J. Roddy, R. O. Pepin, and R. B. Merrill, eds.), p. 815–841. Pergamon, N.Y.

Jahn B. M., Floran R. J. and Simonds C. H. (1978) Rb-S_r isochron age of the Manicouagan melt sheet, Quebec, Canada. *J. Geophys. Res.* **83**, 2799–2803.

Murtaugh J. G. (1972) Shock metamorphism in the Manicouagan cryptoexplosion structure, Quebec. *Proc. 24th Int. Geol. Congr.*, Montreal.

Murtaugh J. G. (1975) Geology of the Manicouagan cryptoexplosion structure. Ph.D. thesis, Ohio State University.

Murtaugh J. G. and Currie K. L. (1969) Preliminary Study of the Manicouagan Structure. Quebec Dept. Nat. Res. Prelim. Rep. 536.

O'Keefe J. D. and Ahrens T. J. (1975) Shock effects from a large impact on the moon. *Proc. Lunar Sci. Conf. 6th*, p. 2831–2844.

Orphal D. L. (1977) Calculations of explosion cratering, Part II: Cratering mechanics and phenomenology. In *Impact and Explosion Cratering* (D. J. Roddy, R. O. Pepin and R. B. Merrill, eds.), p. 907–917. Pergamon Press, N.Y.

Phinney W. C., Dence M. R. and Grieve R. A. F. (1978) Investigation of the Manicouagan impact crater, Quebec: An introduction. *J. Geophys. Res.* **83**, 2729–2735.

Quaide W. L., Gault D. E. and Schmidt R. A. (1965) Gravitative effects on lunar impact structures. *Ann. N.Y. Acad. Sci.* **123**, 563–572.

Schultz P. H. (1972) A preliminary morphologic study of lunar surface features. Ph.D. thesis, University of Texas, Austin, Texas.

Schultz P. H. (1976a) *Moon Morphology*. University of Texas Press, Austin. 626 pp.

Schultz P. H. (1976b) Floor-fractured lunar craters. *The Moon* **15**, 241–273.

Schultz P. H. (1976c) Floor fracture craters on the Moon, Mars and Mercury. NASA TNX-3364, p. 159–160.

Schultz P. H. (1977) Endogenic modification of impact craters on Mercury. *Phys. Earth Planet. Inter.* **15**, 202–219.

Schultz P. H. (1978) Martian intrusions: Possible sites and implications. *Geophys. Res. Lett.* In press.

Simonds C. H., Floran R. J., McGee P. E., Phinney W. C. and Warner J. L. (1978) Petrogenesis of melt rocks, Manicouagan impact structure, Quebec. *J. Geophys. Res.* **83**, 2773–2788.

Stöffler D. (1971) Progressive metamorphism and classification of shocked and brecciated crystalline rocks at impact craters. *J. Geophys. Res.* **76**, 5541–5551.

Sweeney J. F. (1978) Gravity study of great impact. *J. Geophys. Res.* **83**, 2809–2815.

Wolfe R. W. and El Baz F. (1976) Photogeology of the multi-ringed crater Haldane in Mare Smythii. *Proc. Lunar Sci. Conf. 7th*, p. 2903–2912.

Wolfe S. H. (1972) Part I: Geology of the Manicouagan-Mushalagan Lakes Structure and Part II: Geochronology of the Manicouagan Mushalagan-Lakes Structure. Ph.D. thesis, California Institute of Technology, Pasadena.

Proc. Lunar Planet. Sci. Conf. 9th (1978), p. 2713–2730.
Printed in the United States of America

Lonar crater glasses and high-magnesium australites: Trace element volatilization and meteoritic contamination

JOHN W. MORGAN

U. S. Geological Survey, Mail Stop 923, National Center, Reston, Virginia 22092

Abstract—A suite of 6 siderophile elements (Ni, Pd, Re, Os, Ir, Au) and three volatile elements (Zn, Se, Cd) was measured by radiochemical neutron activation analysis in two basalts, a "pumice" and three impact glasses from Lonar Crater, India, and in six high-magnesium (H-Mg) australites from South Australia.

Lonar glasses are significantly depleted relative to parent basalts in Re (by a factor of 7) and Se (by a factor of 2.5); depletion and degree of shock appear to be correlated. The loss seems to be due to volatilization under oxidizing conditions and is probably peculiar to terrestrial impacts. Slight depletions of Au (27%) and Zn (14%) may merely reflect variations in target composition. Slight meteoritic enrichment of Ir (4×10^{-5} Cl-chondrite equivalent) is present only in the most heavily shocked glass and is too low to characterize.

Australite #46 is substantially and uniformly enriched in siderophile elements ($\sim 4 \times 10^{-3}$ Cl-chondrite equivalent) relative to the quite constant base levels in the other five tektites; Se enrichment is very low ($\sim 4 \times 10^{-5}$ Cl-chondrite equivalent), however. Of the known meteorites, only some IA irons resemble this pattern; silicate spherules from deep-Pacific sediments ("Brownlee particles") are also a reasonable match. Enrichments of Zn and Cd, if related to the projectile, spoil the similarity. The pattern for all nine elements qualitatively resembles the "carbonaceous" or "cometary" component in lunar soil 61220 and slightly favors an origin for australites by cometary impact.

INTRODUCTION

Terrestrial craters allow the study of the intermediate-sized objects that have crossed the earth-moon orbit in the last 2 b.y. The trace-element signature of these bodies is obscured on the moon by the ancient meteoritic components found in breccias older than 3.8 b.y. and by micrometeorites in mature soils. On the tectonically active earth, however, each new impact is imprinted on a *tabula rasa*.

The Lonar crater, Maharashtra, India, is a 1.8 km diameter, almost perfectly circular feature in the Deccan Traps basalt flows; it is uniquely interesting as an analogue of lunar craters because it seems to be the only terrestrial crater excavated in basalt (Fredriksson *et al.*, 1973; Kieffer *et al.*, 1976). For trace-element studies, this crater has the added advantage that it is very young, perhaps as little as 5×10^4 years (Fredriksson *et al.*, 1973).

Australites are clearly impact glasses, although no parent crater has been unequivocally identified. Attempts to detect excess siderophile elements of probable meteoritic origin in tektites have not been very rewarding. Rare metallic spherules in Southeast Asian tektites are undoubtedly Ni-rich (Chao *et al.*, 1964) and contain ppm amounts of Au and Ir (Ehmann and Baedecker, 1968).

Measurements of "whole-rock" tektites, however, generally show little or no enrichment of siderophile trace elements (Ehmann, 1963; Morgan *et al.*, 1975; Palme *et al.*, 1978a). Lovering and Morgan (1964) found one tektite (a philippinite) out of ten analyzed that contained $\sim 2 \times 10^{-3}$ Cl-chondrite abundance of Re and Os in approximately cosmic proportions, but the measurements were relatively insensitive. Palme *et al.* (1978a,b) made measurements of greater sensitivity on two Ivory Coast tektites, which contained 2×10^{-4} of Cl-chondrite equivalent of siderophiles. Gold abundances show large variations (Ehmann and Showalter, 1971, and pers. comm., 1970), but these are apparently mainly indigenous. Clearly, the search for meteoritic material in tektites has not been particularly rewarding, and, in the present work, I attempted to load the odds in favor of success by selecting the high-magnesium (H-Mg) australites (Chapman and Scheiber, 1969). The reasons for this were twofold. First, the H-Mg australasian tektites are high in Ni [91–390 ppm (Chapman and Scheiber, 1969)] and *could* contain up to several percent Cl-chondrite equivalent of meteoritic material, whereas the low Ni tektite groups (for example, high-Ca tektites; 14–42 ppm) demonstrably contain very little. Second, the Mg enrichment itself might conceivably be a result of meteoritic contamination (Taylor, 1962, 1964).

A suite of six siderophile elements (Ni, Pd, Re, Os, Ir, Au) was determined by radiochemical neutron activation in six H-Mg australites (2.4 to 3.5% MgO), two basalts, three vesicular glasses and a very vesicular "pumice" from the Lonar crater. In addition, three volatile elements (Zn, Se, Cd) were determined to detect any effects of volatile transfer.

METHODS AND MATERIALS

Samples and preparation

In order to minimize contamination during preparation, samples were not finely ground and sieved but merely were coarsely crushed until the fragments were less than 2 mm in the largest dimension.

Lonar Crater

The six Lonar crater samples were generously provided by Kurt Fredriksson of the Smithsonian Institution and have been described by Fredriksson *et al.* (1978).

Exterior and sawn surfaces were avoided as far as possible. The basalt pieces were shattered in a steel percussion mortar; clean fragments were selected and coarsely crushed in a Coors* alundum mortar. Basalt #16 was coherent and very fresh, #5 was slightly rotten and more friable. The glasses were broken up directly in the alundum mortar, and fragments having shiny conchoidal surfaces were chosen. The pumice was very friable and an interior sample was broken out using gloved fingers.

---

*Any trade names in this publication are used for descriptive purposes and do not constitute endorsement by the U.S. Geological Survey.

H-Mg australites

The six H-Mg australite samples were kindly donated by Brian Mason of the Smithsonian Institution; descriptions and analyses were given by Mason (1978). Samples #46 and #48 were received as small clean chips, and the whole material was coarsely crushed in an alundum mortar. The remaining samples were provided as prepared powders and were used without further treatment.

PROCEDURE

Irradiations

Samples were sealed in 3 mm internal diameter "Suprasil" vitreous silica vials and, together with appropriate duplicate standards, were irradiated by a thermal neutron flux of $1{\cdot}1 \times 10^{14}$ neutrons cm^{-2} sec^{-1} for five days in the G2 position of the National Bureau of Standards reactor, Gaithersburg, Maryland.

Radiochemical yields of Ir were determined by re-irradiation for five minutes in the RT3 position of the same reactor in a thermal flux of 5×10^{13} neutrons cm^{-2} sec^{-1}.

Separations

Three days after the end of irradiation the samples were returned to the U.S. Geological Survey, Reston, Virginia, for analysis. Individual elements were separated broadly following the procedure of Keays *et al.* (1974) as supplemented by Gros *et al.* (1976) for the addition of Os and Re. Because only nine elements were analyzed, several steps were omitted; in the initial ion exchange separation, the Bi and Tl elutions and the subsequent *aqua regia* treatment were not required; in the treatment of supernate from the sulfide-hydroxide precipitation, after removal of excess sulfide with CdS it was possible to acidify directly for precipitation of the Se-Re fraction. Radiochemical recoveries tended to be somewhat better than for the full procedure except for some Au fractions.

Counting

All Cd^{115}-In^{115m}, Au^{198} sources and samples containing high level of Se^{75} were measured by NaI (Tl) γ-ray spectrometry; although this technique has a high intrinsic photopeak efficiency, its sensitivity for low activities is limited by poor resolution and structured background.

Low level Se^{75} and all Zn^{65} and Ir^{192} sources were radio-assayed by use of a 13% efficiency Ge(Li) detector; Ni^{59}, Pd^{103}, and Os^{191} were counted by using a 13 mm thick intrinsic Ge detector. For the tektites, it soon became apparent (to anticipate the "results" section) that only one australite contained significant siderophiles and it became important to have a reliable estimate of the average Pd and Os contents in the "uncontaminated" australites For Os, low level beta counting enabled much lower limits to be determined, as discussed below. To determine the mean Pd base level the precipitates of all five meteorite-free tektites were combined to improve the signal-to-noise ratio, and counted for several days.

A Beckman WIDEBETA automatic beta counter was used for Re^{186} and Os^{191} assay. The background of two counts per minute was two to three times higher than optimum. This was partly due to the high ambient background in the counting room in which the equipment was temporarily located; and partly because Al planchets and fiber glass filters were used instead of stainless steel discs and paper filters, which were not immediately available.

RESULTS

The analytical results are given in Table 1. The Lonar crater glasses are arranged in approximate decreasing order of shock metamorphism, beginning

Table 1. Siderophile and volatile elements in Lonar crater rocks and H-Mg australites (Abundances in ppb, except Ni and Zn in ppm and MgO in percent).

	Ir	Re	Os	Pd	Au	Ni	Se	Zn	Cd	MgO*	Mass, mg
Lonar crater											
Vesicular glass LN10	0.026b	0.070a	≤0.008	11.5a	3.4a	63b	17a	120a	165a		115
" " LN1	0.004c	0.080a	≤0.009	9.7b	3.3a	60b	18a	113a	69a		119
" " LN2	0.012b	0.112a	≤0.013	11.7a	2.6a	63b	22a	121a	38a		112
"Pumice" LN2	0.005c	0.49 a	≤0.008	9.5b	3.4a	86b	38a	119a	96a		107
Basalt #5 (altered)	0.004c	0.61 a	0.013d	13.3a	4.4a	66b	161a	138a	361a		132
" #16 (fresh)	0.006c	0.61 a	≤0.006	6.9a	4.3a	69b	48a	137a	92a		142
H-Mg Australites											
Lake Wilson, #46	1.95 a	0.21 a	2.3 a	1.33c	1.84a	260b	1.36b	17a	90a	3.53	58
" " #47	0.022b	0.005a	0.020d	≤0.60	2.6 a	156b	0.64b	11a	19b	3.33	102
Mount Davies, #53	0.018d	0.003c	≤0.034	≤0.84	0.56a	200b	0.92c	11a	11c	3.13	43
Lake Wilson, #48	0.022c	0.004b	0.059d	≤1.02	0.37a	220b	0.33d	15a	6d	2.92	45
" " #49	0.019b	0.007b	≤0.009	≤0.35	2.6 a	176a	0.76b	12a	8d	2.70	130
" " #51	0.024b	0.004b	0.017d	≤0.37	0.55a	116b	0.64b	14a	11b	2.41	104
Composite of #47 thru #53				≤0.08							
Mean Cl	514	35.2	480	490	152	10300	19500	303	639		

*Mason (1978).

Estimated errors: a, <5%; b, 5%–10%; c, 10%–25%; d, 25%–50%.

with the high-temperature dense glass LN10 through the most vesicular glass LN2 to the frothy "pumice" from LN2 which is essentially sintered basalt and is of relatively low temperature origin. The australites are in decreasing order of MgO content.

Lonar Crater

Meteoritic component

Iridium is probably the most sensitive indicator element for many types of Fe-rich meteoritic material, because it is rather easily determined down to 10^{-12} g/g, and indigenous Ir abundances in most crustal rocks are in the 10^{-11} to 10^{-12} g/g range. As shown in Fig. 1, glasses LN10 and perhaps LN2 have marginal Ir enrichments, equivalent to 4×10^{-5} and 1×10^{-5} Cl chondrite,

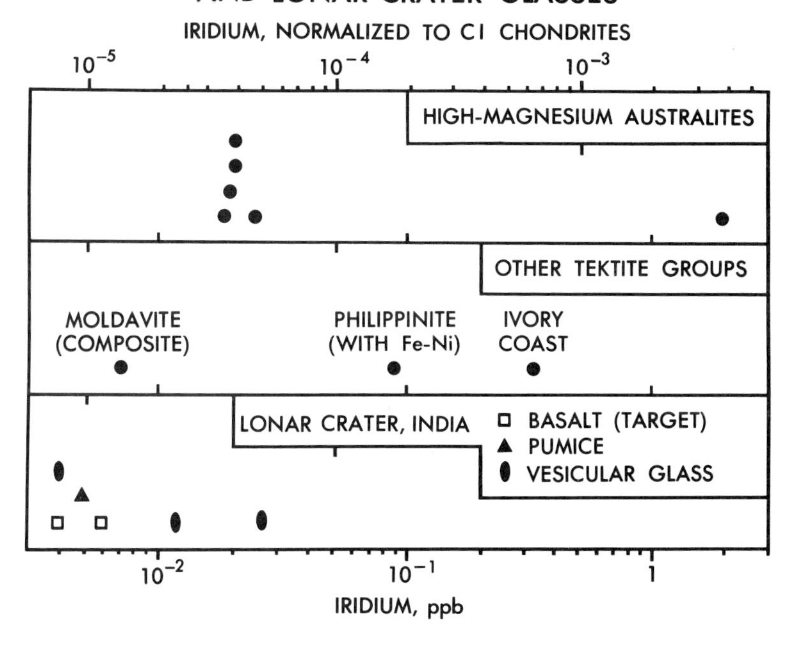

Fig. 1. Iridium is a good indicator of meteoritic contamination. Lonar crater glasses show little or no meteoritic enrichment. The highest Ir enhancement ($\sim 4 \times 10^{-5}$ Cl) is found in the most heavily shocked glass LN10. H-Mg australite #46 is enriched in Ir by almost two orders of magnitude (4×10^{-3} Cl) over the almost constant levels in the other H-Mg australites. Even the unenriched australites have surprisingly high Ir for silicic igneous rocks or sediments. Some other tektite groups are shown for comparison [moldavite composite (Palme *et al.*, 1978a); philippinite (Morgan *et al.*, 1975); Ivory Coast (Palme *et al.*, 1978b)].

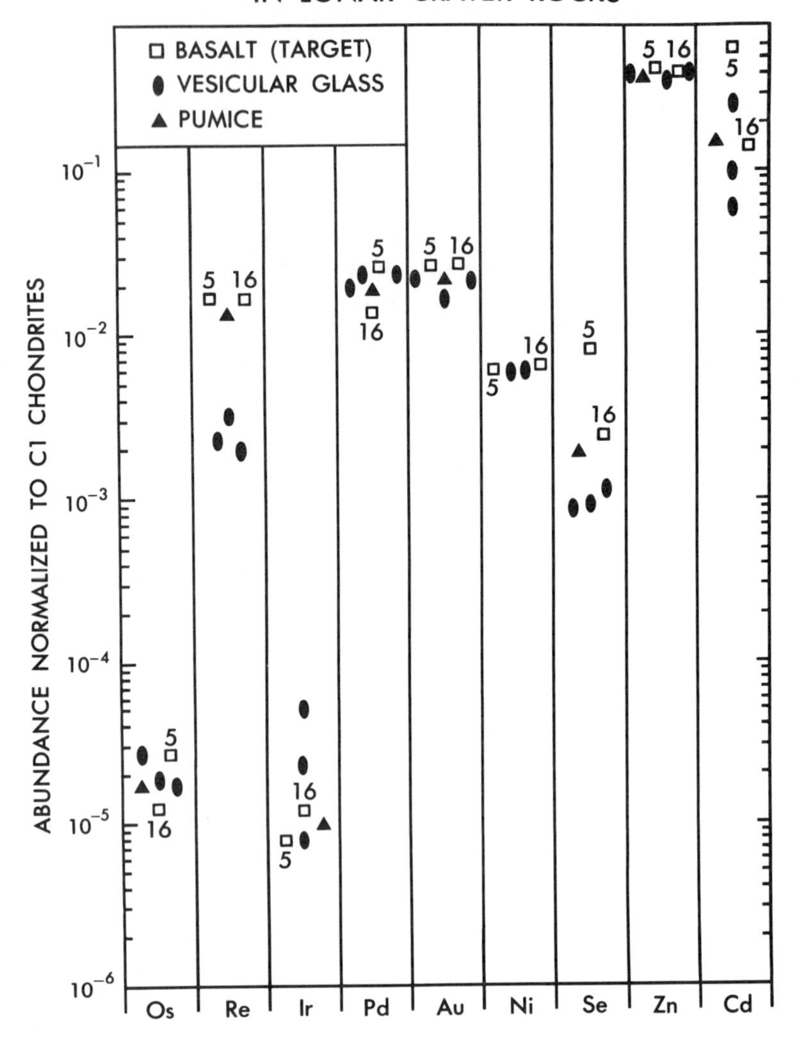

Fig. 2. In Lonar glasses, only Ir shows any hint of meteoritic enrichment; Os is too poorly determined; indigenous abundances of Re, Pd, Au and Ni are too high. Glasses are depleted in Re and Se, and to a lesser extent in Au and Zn. Altered, vesicular basalt #5 is enriched in Cd and Se over fresh compact basalt #16.

respectively. Although these are very parsimonious excesses, the Ir content of LN10 seems to be significantly above the rather constant indigenous levels of 0.005 ± 0.001 ppb Ir. It is not surprising that LN10 has the highest meteoritic contamination, because it is the densest, least vesicular of the glasses analysed, and is probably representative of the most heavily shocked "fladle" type glasses found at Lonar crater (Fredriksson *et al.*, 1973; Fredriksson, pers. comm., 1978). The small (≤3 mm diameter) glass spherules currently being studied may be more promising material, if degree of shock is a major criterion; these have apparently been shocked up to 1 Mbar (Kieffer *et al.*, 1976).

None of the other elements determined are sufficiently sensitive indicators to confirm these very small enrichments (Fig. 2). The contribution of several siderophiles anticipated from the suspected meteoritic component is greatly exceeded by the indigenous levels in, for example, the fresh nonvesicular Lonar basalt #16: Ni = 0.7×10^{-2}, Pd = 1.4×10^{-2}, Re = 1.7×10^{-2}, Au = 2.8×10^{-2}, all relative to Cl chondrites. (Incidentally, it is remarkable that these elements are present in proportions so close to cosmic.) The indigenous levels of Os and Ir are probably very similar to each other, but Os is not well enough determined to detect very small enrichments. In addition, Os may just possibly have been lost by volatilization of OsO_4; evidence discussed in a later section suggests that the conditions may have been mildly oxidizing during formation of the glasses.

Volatile transfer

Although the Lonar crater glasses have proven to be unpromising material for the characterization of a meteoritic component, they show abundance changes that appear to be related to the impact process (Fig. 2). In Fig. 3, the Lonar crater glasses are arranged in order of increasing shock metamorphism and elements are normalized to mean basalt abundances (except Se and Cd which are normalized only to compact basalt #16; the vesicular slightly weathered basalt #5 is three to four times higher in these elements). The elements divide into four groups: Ni and Pd show essentially no variation; Zn and Au are systematically lower in pumice and glasses than in basalts, but no effects of the degree of shock are obvious within the group of glasses; Re and Se show large depletions in glasses and vary with degree of shock; Cd is odd.

Nickel and palladium

Relative to the basalts, Ni and Pd are not significantly depleted in pumice and glasses. Even if the pumice is excluded, the glasses show only a small apparent depletion of about 8%, whereas the counting error for each Ni determination is between 6% and 10%. The basalt values for Pd differ by almost a factor of two; the proximity to two immediately raised the spectre of experimental error, but careful checking of the raw data appears to have lain it again to rest. The Pd contents of the pumice and glasses are remarkably constant, and all are close to the mean Pd content of the basalts.

Zinc and gold

The pumice and glasses are lower than the basalts in both Zn and Au by rather constant factors; 14 ± 3% and 27 ± 9% respectively. Within the glasses them-

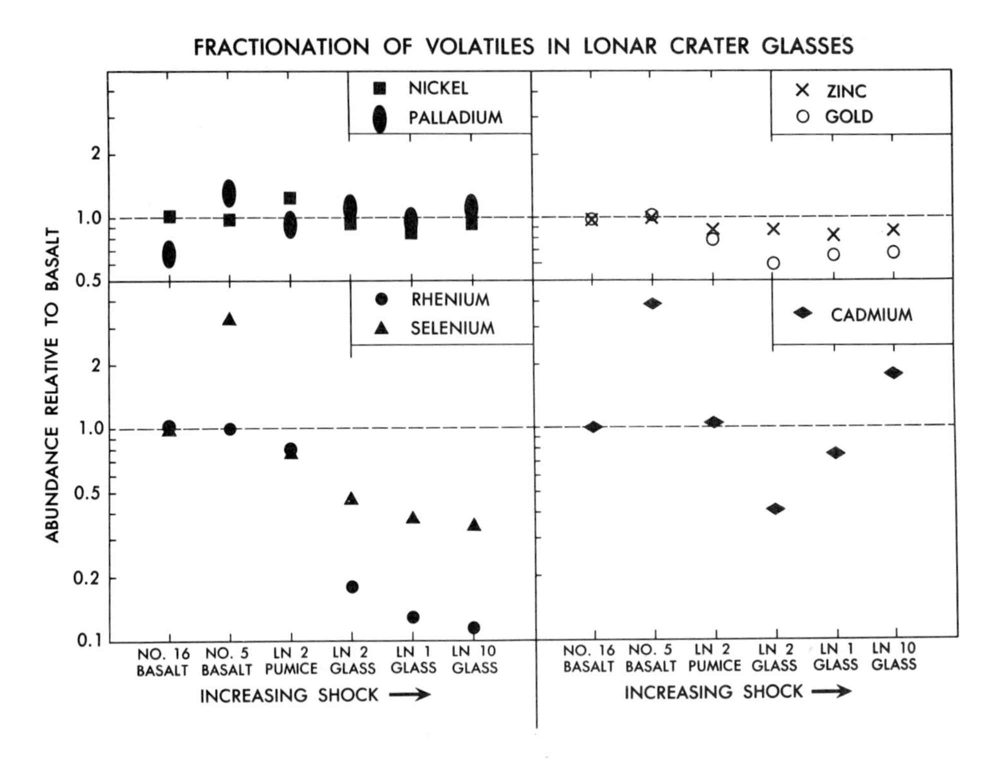

Fig. 3. Depletion of Re and Se in Lonar glasses is apparently correlated with degree of shock and probably results from oxidizing conditions during impact. Zn and Au are slightly lower in glasses than basalts, but Ni, Pd, and Cd show no systematic variation.

selves, Zn and Au do not decrease systematically with shock metamorphism; in fact, for Au there is a hint of an apparent increase in the more heavily shocked glasses (the same trend, but much more marked, is seen for Cd). There are two likely explanations for the Zn and Au depletion. First, the two basalt samples analyzed may not be entirely representative of the composition of the glasses. The flows sampled by the impact may have included basalts different from the samples analyzed. Although the original flows were relatively uniform in composition, the observed differences in degree of weathering may be responsible for the slightly depleted Zn and Au in the glasses. On the other hand, the compact fresh basalt #16 and vesicular, somewhat altered basalt #5 have virtually identical Zn and Au contents suggesting that mobilization by weathering may not have been very significant, at least for these elements. [Fractionation of plagioclase during the shock process is known to have affected the composition of some Lonar glasses (see Fig. 9 of Kieffer *et al.*, 1976), and in principle dilution by plagioclase could account for the Zn and Au depletion in the glasses; one would then expect Ni and Au to be depleted by similar amounts, since plagioclase is not particularly hospitable to either, judging from data for anorthosites (Morgan *et al.*, 1976). Nickel and Au are demonstrably not as depleted as Zn and Au.]

The second, and by default, more probable explanation is that Zn and Au were depleted by volatile loss, perhaps enhanced by the presence of a large, water-rich volatile phase. If the loss of Zn and Au is a shock-related effect, intuitively a trend with degree of shock in the glasses might be expected, but is not observed. The reason may be simple; to produce the basaltic glass required 800 kb to 1 Mb shock pressure, whereas the variations within the glasses may represent perhaps only 100–200 kb.

Rhenium and selenium

The effect of volatile loss is pronounced in Re; the Re contents of the glasses as a group (excluding the pumice) are lower than those of the basalts by a factor of 7. As Re is generally regarded as a refractory element under reducing conditions, the loss of Re is probably a consequence of heating in an oxidizing environment. A corresponding though lesser depletion in Se exists; the Se contents of the glasses are 2.5 times lower than that of basalt #16 and are about 8 times lower than that of vesicular basalt #5. The inference is that Re and Se were present in a sulfide phase and were lost by conversion to their volatile oxides; Re is commonly found to be highly enriched in the flue dusts resulting from the roasting in air of sulfide ores, particularly molybdenite. The observed depletions are in qualitative agreement with the volatility of the stable oxides; Re_2O_7 sublimes at 250°C, and SeO_2 at 340°–350°.

The erratic distribution of Re of largely meteoritic origin in weathered impact melts and breccias from Rochechouart (170–200 m.y. old) and Brent (450 m.y. old) craters has been attributed to the solubility of ReO_4 in aqueous solutions (Janssens *et al.*, 1977; Palme *et al.*, 1978a,b). [Impact melts from Clearwater East crater (285 m.y.), however, show a very well defined chondritic pattern that has been altered only slightly if at all (Palme *et al.*, 1978a,b).] The Lonar crater is much younger (~50,000 y. old) than Rochechouart and Clearwater East, and Lonar glasses are so fresh that it is very unlikely that weathering has been a major factor in the depletion of predominantly indigenous Re.

Cadmium

Cosmochemically, Cd is generally regarded as a rather volatile element, but pyrolysis curves indicate that it is considerably less volatile under oxidizing conditions (Duval, 1953; 1963). The sulfide is relatively volatile, but depending on the degree of oxidation, can convert to $CdSO_4$ or CdO, both of which are relatively stable to temperatures ≥900°C. In the basalts, Cd abundances differ by a factor of 4; a similar variation in Se content suggests that Cd is enriched in the vesicular basalt #5 as the sulfide either as an alteration product or with the vesiculating phase. Because of the variability in the target rocks, the apparent increase of Cd with shock in the glasses may be no more than a reflection of target material sampling. Nevertheless, a similar but more subdued variation in Au would tend to suggest that it is a real effect. Possibly the vesiculation in glasses LN1 and LN2 is a manifestation of the evolution and escape of a Cd-bearing vapor phase.

HIGH-MAGNESIUM AUSTRALITES

Meteoritic component

Iridium enrichment One of the six H-Mg australites (Lake Wilson #46) contains 2 ppb Ir and is enriched about 2 orders of magnitude over the rather constant level of 0.021 ± 0.002 ppb Ir found in the other five; as shown in Fig. 1, this appears to be the highest value found in a true tektite (Baedecker and Ehmann, 1965; Morgan *et al.*, 1975; Palme *et al.*, 1978a,b). [Ehmann and Baedecker (1968) reported 2 ppm Ir in a metallic spherule from an unspecified tektite, however; and Palme *et al.* (1978b) find 2.4 ppb in irghizites from the Zhamanshin structure.] Although no sample of country rock has been analyzed to establish an indigenous Ir level in the australite target rocks, the Ir levels in the five uncontaminated H-Mg australites suggest a rather low and constant upper limit. Making the appropriate correction, the Ir content of #46 corresponds to a meteoritic component equivalent to $\sim 4 \times 10^{-2}$ Cl chondrite. The Ir content of the "unenriched" H-Mg australites is suprisingly high for acidic rocks and, unlike Ni (see below), does not correlate with MgO; in other words, Ir has not been added along with varying amounts of a Mg, Ni-rich component.

Enrichment of other siderophile elements The probable meteoritic origin of the Ir enrichment in #46 is clearly confirmed by significant enrichments of Os, Re and Pd greatly in excess of the mean abundances in the unenriched australites (Fig. 4). The indigenous Os is less well determined than Ir but appears to be relatively constant, and the indigenous Ir/Os ratio seems quite close to cosmic. The very low Re (0.005 ± 0.002 ppb) in the "unenriched" H-Mg australites is an unexpected bonus; terrestrial acidic igneous rocks have substantial Re (0.2 to 1.1 ppb) and even sandstones (0.03–0.05 ppb) and shale (0.5 ppb) have Re contents an order of magnitude higher than those of the H-Mg australites (Morris and Fifield, 1961; Lovering and Morgan, 1964; Morgan *et al.*, 1975). Volatilization may not be an acceptable explanation for the low abundance of Re in australites, because the extremely low Fe^{+3}/Fe^{+2} ratio in these glasses indicates reducing rather than oxidizing conditions. Additionally, Ir/Re in the enriched australite is so close to the cosmic ratio that Re loss must have been very small. The individual indigenous Pd values were too low to be determined, but the composite average is probably a reasonable estimate in view of the uniform abundances of other platinum metals in the unenriched australites.

The meteoritic enrichment of Ni is considerably less clear cut. All the H-Mg australites have significant Ni, and abundances in the five unenriched australites

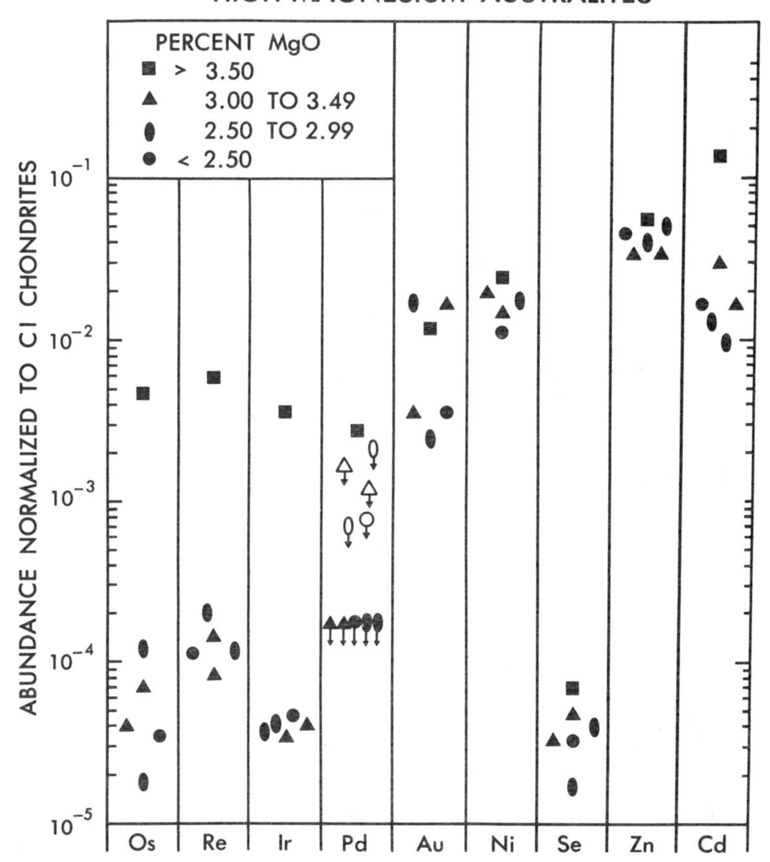

Fig. 4. H-Mg australite #46 is higher than the other five australites in all elements determined except Au (where contamination may be a problem). Base levels of most elements are rather constant, but Ni shows a variation of a factor of ~2 which is weakly correlated with MgO content. Downward arrows represent upper limits; open symbols for Pd are values derived from counting individual sources, filled symbols for Pd indicate the upper limit obtained from pooled sources.

are feebly correlated with MgO content. Although the correlation coefficient ($r = 0.47$) is not significantly different from zero, the value of 207 ppm Ni derived from the MgO-Ni regression is nevertheless the best estimate of the indigenous Ni in H-Mg australite #46; accordingly, the meteoritic Ni is 53 ± 20 ppm. Indigenous Au in H-Mg australites is very variable, and in fact the "unenriched" australites #47 and #49 have higher Au than #46 [Au values reported by Ehmann and Baedecker (1968) suggest that the variability of Au in australites as a whole may be even greater than that seen here in H-Mg group].

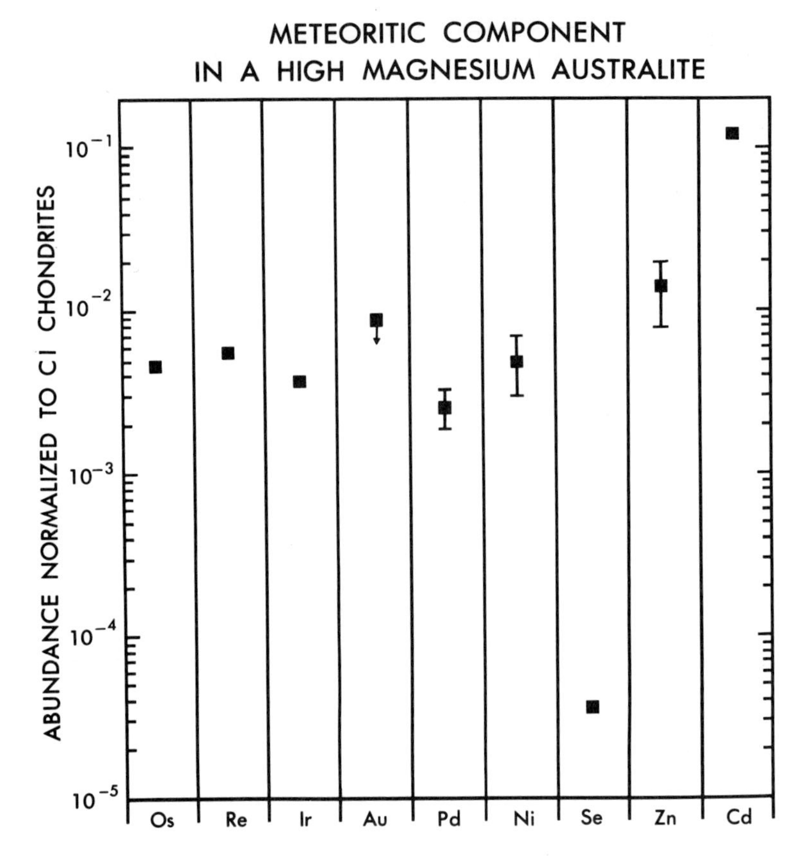

Fig. 5. The meteoritic component in H-Mg australite #46 shows a flat siderophile pattern with very depleted Se. Some IA iron meteorites could produce this pattern; the parent material of the "Brownlee particles" (Ganapathy *et al.*, 1978) is also a possible source. Neither can account for the large apparent enrichment of Zn and Cd.

The Au seems to be bimodally distributed, a "base" level is at 0.5 ppb, and a higher level is at 2.6 ppb. There is no way to be sure that the high values are not the result of contamination; because of the prevalence of Au jewelry, the "Midas touch" is no myth to the trace element analyst. In any event, an upper limit in Au enrichment can be estimated by using the "base" level values. The net siderophile meteoritic component of australite #46 is shown in Fig. 5.

Meteoritic enrichment of volatiles The enriched australite #46 shows enhanced abundances of the three volatile elements Se, Zn and Cd. Abundances in the five unenriched australites are low and quite uniform; Se, 0.7 ± 0.2 ppb; Zn, 13 ± 2 ppm; Cd, 11 ± 5 ppb. These values for Zn agree well with previous spectrophotometric results of 10 ± 5 ppm in 21 australites reported by Greenland and

Lovering (1963). The apparent enrichments of volatiles in #46 are also shown in Fig. 5.

Characterization of meteoritic component

Siderophile pattern The refractory siderophile enrichments are well defined and show essentially cosmic or chondritic ratios. The Os/Re weight ratio of ~11 is about 10% higher than that reported earlier (Morgan 1978); the change results from an improved determination of Os using both low-energy photon counting and beta counting. Because Os and Re are both very refractory in the solar nebula, they mutually fractionated very little and the Os/Re ratio is not very definitive. Nevertheless, this ratio has been determined in a large number of chondrites (Morgan and Lovering, 1967) and the ratio of 11 observed in the present work is marginally more compatible with enstatite or ordinary chondrites than with carbonaceous chondrites. Members of several iron meteorite groups also have appropriate Os/Re ratios (Herr *et al.*, 1961; Herpers *et al.*, 1969).

More important in terms of characterization is the ratio of the refractory siderophiles Os, Re, Ir as a group to the more volatile siderophiles Ni, Au, Pd. The latter group is not well determined, but the similarity of Pd and Ni enrichments (when normalized to Cl chondrites) to each other and to those of the refractory siderophiles suggests that the meteoritic component in the australite was relatively unfractionated material. Enstatite chondrites can therefore be excluded as possible contaminants, as can most iron meteorite groups except IA [and most members of this group have too low Ir/Ni and Ir/Au (Scott, 1972)].

Volatile elements The volatile elements do not present such a tidy picture as the siderophiles. The very marked depletion of Se cannot be dismissed lightly; even if no indigenous correction were made in #46, the Cl-normalized abundance of Se would still be 10^{-4}. The low Se content in the meteoritic component cannot be attributed to volatile loss under oxidizing conditions (as observed in the Lonar glasses) because a corresponding deficiency of Re would exist and there is no evidence whatever of this; in addition the tektites appear to have been formed under reducing conditions. On its own, the low Se would tend to eliminate a chondritic object, and would seem to identify the projectile as a IA iron meteorite. Such a pattern need not necessarily be carried by metallic material, however. In a remarkable piece of analytical work, Ganapathy *et al.* (1978) have shown that a 78 μg silicate spherule of extraterrestrial origin recovered from deep-Pacific sediments contains Os, Ir, Ni and a number of non-volatile and moderately volatile elements in approximate Cl absolute abundances, but Se is depleted by a factor of 500. Given the limited number of elements for comparison, the match with the H-Mg australite component is very close (except for Zn).

The situation is complicated by the apparent enrichment of Zn and Cd in H-Mg australite #46; the net Zn enhancement is in approximately Cl chondrite proportions to the siderophile elements (although the error limits are rather large), but Cd is higher by a factor of 20. In lunar soils, Zn and Cd are often overabundant relative to siderophile and volatile elements of micrometeorite origin; the overabundance has been variously ascribed to a greater probability of contamination, occasionally high indigenous contribution and higher mobility in the soil (Morgan *et al.*, 1977). In view of the rather constant abundances in the H-Mg australites [and indeed australites in general for Zn at least (Greenland and Lovering, 1963)], these reasons for Zn-Cd variability may be "inoperative" for tektites, and the enrichment of these elements in H-Mg australite #46 might be considered as part of the meteoritic component. As a corollary, if one accepts that Cd and/or Zn may be overabundant in the australite projectile, some of the apparent overenrichments in lunar soil may reflect cosmic rather than lunar effects. In particular, the enigmatic lunar soil 61221 contains a component which bears a general resemblance to that in #46.

Meteoritic components in 61220 The meteoritic contributions to most Apollo 16 soils are dominated by the 1H ancient meteorite component in breccia. This can be seen very clearly in 61220 particularly since dark crystalline separates from 61222, la contain a rather pure example of the 1H component corresponding to ~9% Cl (based on Au)(Krähenbühl *et al.*, 1973). The micrometeorite component in most Apollo 16 soils, as indicated particularly by Bi, Se and Te, correlates with cosmogenic Ne^{21} exposure ages (Walton *et al.*, 1973; Ganapathy *et al.*, 1974). Soil 61221 apparently falls on the Bi-Ne_c^{21} correlation line, but Se and Te are markedly depleted relative to other Apollo 16 soils. The depletion was initially interpreted as loss of Se and Te as hydride, since it has been suggested that 61221 has been involved in a cometary impact (Gibson and Moore, 1973). It seems now, however, that 61221 is a very young soil in terms of indicators of surface exposure; N content, I_s/FeO, and agglutinate content (Kerridge *et al.*, 1977; Morris, 1977; Heiken *et al.*, 1973), and Bi could not have been introduced in substantial amounts by micrometeorite impact at the lunar surface [although there does seem to have been a small admixture of surface-exposed material (Morris, 1977; Wszolek *et al.*, 1973)]. Instead, Bi and other volatiles appear to have been introduced by a carbonaceous material judging from the unusual light-element isotopic compositions (Epstein and Taylor, 1973; Becker and Clayton, 1977); the projectile may have been cometary because such volatile compounds as HCN have been reported in 61221 (Gibson and Moore, 1973, Wszolek *et al.*, 1973; for an opposing view see Epstein and Taylor, 1973; Kerridge *et al.*, 1978).

Comparison of components in 61221 and #46 australite In Fig. 6, the enrichment pattern in H-Mg australite #46 is compared with the gross meteorite contribution in 61221; also with the net enrichment after subtraction of the 1H ancient meteorite component and a small contribution from surface-exposed material (assumed to be a mature soil resembling 64801). The corrections are substantial and introduce considerable uncertainty into the residual values, nevertheless the H-Mg australite and the net "carbonaceous" component are qualitatively similar; siderophiles exhibit an essentially flat distribution, Se is underabundant, and Cd is greatly enriched. The disagreement in Zn may be due to the large error in the australite component. The greatest discrepancy, however, is in Se, which is far more depleted in the #46 australite. Part of the problem lies in the large uncertainty in the appropriate corrections made to derive the net "carbonaceous" component in 61221. Nevertheless, it is clear that the 61221 and australite components differ significantly from any carbonaceous chondrite studied in a terrestrial laboratory. If these components reflect projectile composition, then known types of carbonaceous chondrites can be excluded, particularly because of the Se depletion.

In view of the suggestions that cometary impact may have been involved in the

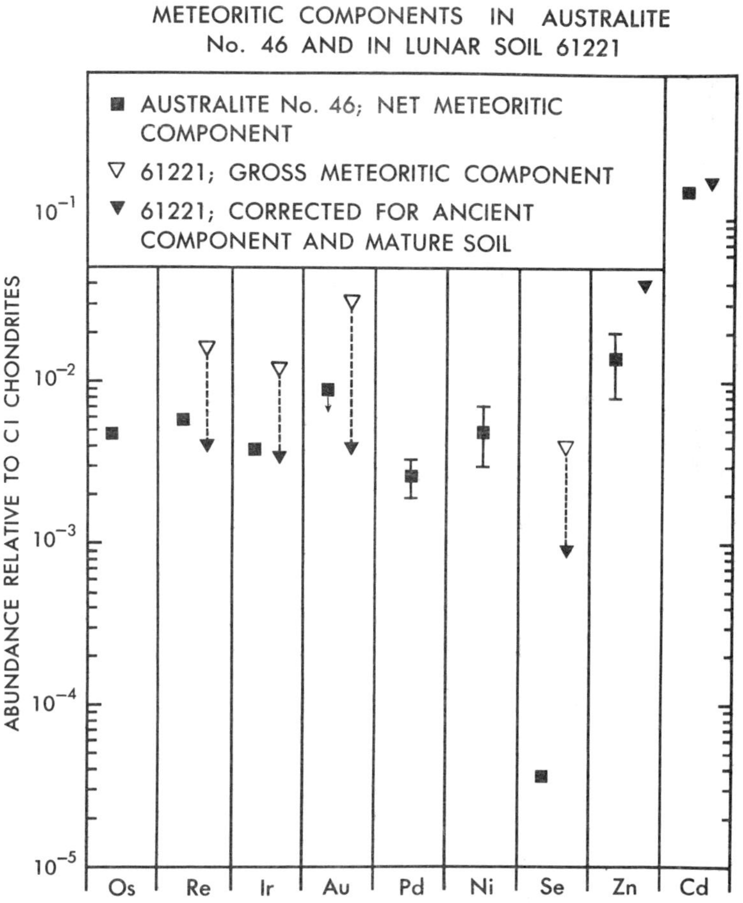

Fig. 6. If the Zn and Cd enrichments in H-Mg australites reflect the projectile composition, the best match may be the "carbonaceous" or "cometary" component in lunar soil 61221. Error bars represent one standard deviation.

genesis of both lunar soil 61221 (Gibson and Moore, 1973) and the australites (Urey, 1963; Taylor, 1968), it is very tempting to consider the meteoritic pattern in H-Mg australite #46 in terms of this hypothesis. Current opinion holds that cometary material resembles carbonaceous chondrites, and is supported by the presence of a volatile-rich micrometeorite component in lunar soils (Morgan *et al.*, 1977 and references summarized therein) and by meteor spectra (Millman, 1972). The flat siderophile pattern of the H-Mg australite projectile is clearly compatible with a carbonaceous chondrite type of material. The enrichment of Zn and particularly Cd relative to siderophiles is apparently characteristic of a late condensate which collected volatiles remaining after the formation of previous generations of dust (Higuchi *et al.*, 1977). The fractionation of Se from other volatiles does occur in carbonaceous interplanetary debris; Kapoeta (dark) contains a component essentially complementary to the australite projectile in that it resembles C2 chondrites, but for Se which is approximately 3 times higher then other volatiles (Anders, 1978). An australite origin by cometary impact may discomfort William of Ockham, however, because moldavites appear to have been ejected from the Ries crater by an aubritic object (Morgan *et al.*, unpublished material) and Ivory Coast tektites bear traces of an iron meteorite (Palme *et al.*, 1978a,b).

Acknowledgments—I thank Greg Wandless for assistance with data reduction. This work was carried out under NASA contract T4089F.

REFERENCES

Anders E. (1978) Most stony meteorites come from the asteroid belt. In *Asteroids: An Exploration Assessment*. NASA Conference Publication 2053, p. 57–75.

Baedecker P. A. and Ehmann W. D. (1965) The distribution of some noble metals in meteorites and natural materials. *Geochim. Cosmochim. Acta* **29**, 329–342.

Becker R. H. and Clayton R. N. (1977) Nitrogen isotopes in lunar soils as a measure of cosmic-ray exposure and regolith history. *Proc. Lunar Sci. Conf. 8th*, p. 3685–3704.

Chao E. C. T., Dwornik E. J. and Littler J. (1964) New data on the nickel-iron spherules from Southeast Asian tektites and their implications. *Geochim. Cosmochim. Acta* **28**, 971–980.

Chapman D. R. and Scheiber L. C. (1969) Chemical investigation of Australasian tektites. *J. Geophys Res.* **74**, 6737–6776.

Duval C. (1953) *Inorganic Thermogravimetric Analysis*, p. 302–303. Elsevier, Amsterdam.

Duval C. (1963) *Inorganic Thermogravimetric Analysis*, p. 430–431. Elsevier, Amsterdam.

Ehmann W. D. (1963) New determinations of iridium and tantalum in meteoritic materials (an interim report). *Meteoritics* **2**, 30–35.

Ehmann W. D. and Baedecker P. A. (1968) The distribution of gold and iridium in meteoritic and terrestrial materials. In *Origin and Distribution of the Elements* (L. H. Ahrens, ed.), p. 301–311. Pergamon, Oxford.

Ehmann W. D. and Showalter D. L. (1971) Elemental abundance trends in the Australite strewn field by non-destructive neutron activation. In *Activation Analysis in Geochemistry and Cosmochemistry* (A. O. Brunfelt and E. Steinnes, eds.), p. 253–260. Universitetsforlaget, Oslo.

Epstein S. and Taylor H. P. (1973) The isotopic composition and concentration of water, hydrogen and carbon in some Apollo 15 and 16 soils and in the Apollo 17 orange soil: *Proc. Lunar Sci. Conf. 4th*, p. 1559–1575.

Fredriksson D., Brenner P., Dube A., Milton D., Mooring O. and Nelson J. (1978) Petrology, mineralogy and distribution of Lonar (India) and lunar impact breccias and glasses. *Smithsonian Contrib. Earth Sci.* Smithsonian Press, Washington, D.C. In press.

Fredriksson D., Dube A., Milton D. J. and Balasundaram M. S. (1973) Lonar Lake, India; an impact crater in basalt. *Science* **180**, 862–864.

Ganapathy R., Brownlee D. E. and Hodge P. W. (1978) Silicate spherules from deep sea sediments; confirmation of extraterrestrial origin. *Science* **201**, 1119–1121.

Ganapathy R., Morgan J. W., Higuchi H., Anders E. and Anderson A. T. (1974) Meteoritic and volatile elements in Apollo 16 rocks and in separated phases from 14306. *Proc. Lunar Sci. Conf. 5th*, p 1659–1683.

Gibson E. K. and Moore G. W. (1973) Volatile-rich lunar soil; evidence of possible cometary impact. *Science* **179**, 69–71.

Greenland L. P. and Lovering J. F. (1963) The evolution of tektites: Elemental volatilization in tektites. *Geochim. Cosmochim. Acta* **27**, 249–259.

Gros J., Takahashi H., Hertogen J., Morgan J. W. and Anders E. (1976) Composition of the projectiles that bombarded the lunar highlands. *Proc. Lunar Sci. Conf. 7th*, p. 2403–2425.

Heiken G. H., McKay D. S. and Fruland R. M. (1973) Apollo 16 soils: Grain size analyses and petrography. *Proc. Lunar Sci. Conf. 4th*, p. 251–265.

Herpers U., Herr W. and Wölfle R. (1969) Evaluation of ^{53}Mn by (n,γ) activation, ^{26}Al and special trace elements in meteorites by γ-coincidence techniques. In *Meteorite Research* (P. M. Millman, ed.), p. 387–396. Springer, N.Y.

Herr W., Hoffmeister W., Hirt B., Geiss J. and Houtermans F. G. (1961) Versuch zur Datierung von Eisenmeteoriten nach der Rhenium-Osmium-Methode. *Z. Naturforsch* **16a**, 1053–1058.

Higuchi H., Ganapathy R., Morgan J. W. and Anders E. (1977) "Mysterite": A late condensate from the solar nebula. *Geochim. Cosmochim. Acta* **41**, 843–852.

Janssens M.-J., Hertogen J., Takahashi H., Anders E. and Lambert P. (1977) Rochechouart meteorite crater: Identification of projectile. *J. Geophys. Res.* **82**, 750–758.

Keays R. R., Ganapathy R., Laul J. C., Krahenbuhl U. and Morgan J. W. (1974) The simultaneous determination of 20 trace elements in terrestrial, lunar and meteoritic material by radiochemical neutron activation analysis. *Anal. Chim. Acta* **72**, 1–29.

Kerridge J. F., Kaplan I. R., Lingenfelter R. E. and Boynton W. V. (1977) Solar wind nitrogen: Mechanisms for isotopic evolution. *Proc. Lunar Sci. Conf. 8th*, p. 3773–3789.

Kerridge J. F., Kaplan I. R. and Petrowski C. (1978) Carbon isotope systematics in the Apollo 16 regolith. *Lunar and Planetary Science IX*, p. 618–620. Lunar and Planetary Institute, Houston.

Keiffer S. W., Schaal R. B., Gibbons R., Hörz F., Milton D. J. and Dube A. (1976) Shocked basalt from Lonar Crater, India, and experimental analogues. *Proc. Lunar Sci. Conf. 7th*, p. 1391–1412.

Krähenbühl U., Ganapathy R., Morgan J. W. and Anders E. (1973) Volatile elements in Apollo 16 samples: Implications for highland volcanism accretion history of the moon. *Proc. Lunar Sci. Conf. 4th*, p. 1325–1348.

Lovering J. F. and Morgan J. W. (1964) Rhenium and osmium abundances in tektites. *Geochim. Cosmochim. Acta* **28**, 761–768.

Mason B. (1978) Chemical variations among australasian tektites. *Smithsonian Contrib. Earth Sci.* Smithsonian Press, Washington, D. C. In press.

Millman P. M. (1972) Giacobinid meteor spectra. *J. Roy. Astron. Soc. Can.* **66**, 209–211.

Morgan J. W. (1978) Siderophile and volatile trace elements in high-magnesium australites and in glasses from Lonar Crater, India (abstract). In *Lunar and Planetary Science IX*, p. 754–756. Lunar and Planetary Institute, Houston.

Morgan J. W., Ganapathy R., Higuchi H. and Anders E. (1977) Meteoritic material on the Moon. In *The Soviet-American Conference on Cosmochemistry of the Moon and Planets* (J. H. Pomeroy and N. J. Hubbard, eds.), p. 659–689. NASA SP-370. Washington, D. C.

Morgan J. W., Ganapathy R., Higuchi H. and Krahenbuhl U. (1976) Volatile and siderophile trace elements in anorthositic rocks from Fiskenaesset, West Greenland: Comparison with lunar and meteoritic analogues. *Geochim. Cosmochim. Acta* **40**, 861–887.

Morgan J. W., Higuchi H., Ganapathy R. and Anders E. (1975) Meteoritic material in four terrestrial meteorite craters. *Proc. Lunar Sci. Conf. 6th*, p. 1609–1623.

Morgan J. W. and Lovering J. F. (1967) Rhenium and osmium abundances in chondritic meteorites. *Geochim. Cosmochim. Acta* **31**, 1893–1909.

Morris D. F. C. and Fifield F. W. (1961) Rhenium content of rocks. *Geochim. Cosmochim. Acta* **25**, 232–233.

Morris R. V. (1977) Origin and evolution of the grain-size dependence of the concentration of fine-grained metal in lunar soils: The maturation of lunar soils to a steady-state stage. *Proc. Lunar Sci. Conf. 8th*, p. 3719–3747.

Palme H., Janssens M.-J., Takahashi H., Anders E. and Hertogen J. (1978a) Meteoritic material of five large impact craters. *Geochim. Cosmochim. Acta* **42**, 313–323.

Palme H., Wolf R. and Grieve R. A. F. (1978b) New data on meteoritic material at terrestrial impact craters (abstract). In *Lunar and Planetary Science IX*, p. 856–858. Lunar and Planetary Institute, Houston.

Scott E. R. D. (1972) Chemical fractionation in iron meteorites and its intepretation. *Geochim. Cosmochim. Acta* **36**, 1205–1236.

Taylor S. R. (1962) Consequences for tektite composition of an origin by meteoritic splash. *Geochim. Cosmochim. Acta* **26**, 915–920.

Taylor S. R. (1964) Nickel-rich tektites from Australia. *Nature* **201**, 281–282.

Taylor S. R. (1968) Criteria for the source of australites. *Chem. Geol.* **4**, 451–459.

Urey H. C. (1963) Cometary collisions and tektites. *Nature* **179**, 228–239.

Walton J. R., Lakatos S. and Heymann D. (1973) Distribution of inert gases in fines from the Cayley-Descartes region. *Proc. Lunar Sci. Conf. 4th*, p. 2079–2095.

Wszolek P. C., Simoneit B. R. and Burlinghame A. L. (1973) Studies of magnetic fines and volatile-rich soils: Possible meteoritic and volcanic contributions to lunar carbon and light element chemistry. *Proc. Lunar Sci. Conf. 4th*, p. 1693–1706.

Proc. Lunar Planet. Sci. Conf. 9th (1978), p. 2731–2748.
Printed in the United States of America

Characteristics of microcracks in samples from the drill hole Nördlingen 1973 in the Ries crater, Germany

ELAINE R. PADOVANI, MICHAEL L. BATZLE and GENE SIMMONS

Department of Earth and Planetary Sciences, Massachusetts Institute of Technology
Cambridge, Massachusetts 02139

Abstract—Samples from the Nördlingen 1973 drill core contain abundant shock-induced microfractures which exhibit varying amounts of healing and sealing. Many of the microcracks resemble morphologically the open microcracks present in returned lunar samples. Data derived from petrography, scanning electron microscopy, and differential strain analysis indicate that fewer microcracks were formed at greater depths and crack sealing processes are more effective for cracks associated with planar elements. Although the microcracks in the Ries core are now sealed, they are valid analogues of the open shock-induced cracks of lunar rocks and demonstrate that open cracks do form in rocks at depth during a naturally-occurring shock event. The healing of the shock-induced cracks in the Ries core also precludes their use for laboratory measurements of physical properties intended to be used as analogue measurements of lunar samples *in situ*.

INTRODUCTION

An important assumption that is necessary for characterizing the physical properties of rocks of the lunar crust on the basis of data obtained on returned lunar samples is that the microcracks present in the lunar samples loose on the surface are analogous to those cracks present in rocks *in situ* in bedrock. Simmons *et al.* (1975) have discussed the validity of this assumption and the importance of differentiating between the types of microcracks produced by various processes active on the lunar surface (thermal cycling, changes in pressure and/or temperature, phase changes, or shock). Subsequent studies (Richter *et al.*, 1976; Siegfried *et al.*, 1977) have shown that microcracks in lunar rocks differ considerably from those observed in experimentally and naturally shocked terrestrial analogues. Lunar rocks exhibit a broader crack spectrum. Simmons *et al.* (1975) and Feves *et al.* (1977) attributed these differences to either multiple shock effects in the same sample or to differences in shock effects on confined versus unconfined samples. Siegfried *et al.* (1977) examined the crack spectra of shock-induced cracks. They found this distribution to be a function of both the duration of the shock pulse and of the initial crack distribution of the rock.

The 1973 drill hole at Nördlingen in the Ries Crater, Germany (described by Gudden, 1974; Bauberger *et al.*, 1974; Engelhardt and Graup, 1977; Graup, 1977; Stöffler *et al.*, 1977; and Engelhardt, 1974, 1975, for example) has provided the opportunity to study the effects of a naturally-occurring shock event on the distribution of microcracks as a function of depth in the crater material.

Five samples of drill core were provided for our study by the Bavarian Geological Survey. Sample depths are shown in Fig. 1, which includes a partial stratigraphic column. The samples are representative of the crystalline rocks present in the lower 600 meters of the hole. The samples consist of a variety of rock types, all of which exhibit microscopic indicators of weak shock, stages 0–1 of Stöffler (1972), although cross-cutting breccia dikelets contain shock features indicative of stages 2 and 3. By examining the samples with a variety of

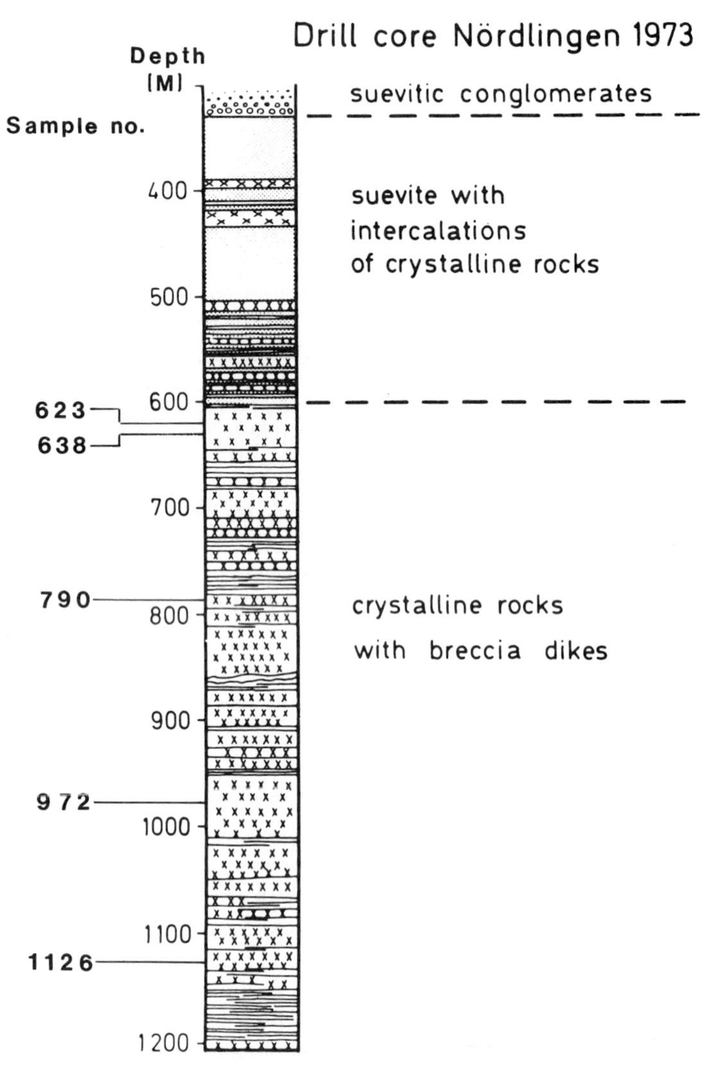

Fig. 1. Diagrammatic profile of the core from the 1973 drill hole at Nördlingen, Ries, Germany (Stöffler *et al.*, 1977). The locations of our samples are indicated by lines.

techniques, we can obtain data with which to test the assumption of whether or not the microcracks produced *in situ* during a shock event on earth are analogous to microcracks in the returned lunar samples.

The drill hole penetrated 1206 meters of Tertiary lake sediments, suevite, and crystalline basement. Several authors (for example, Stettner, 1974; Engelhardt and Graup, 1977; Stöffler, 1977; Stöffler *et al.*, 1977) believe that only the rock present in the bottom 20 meters of core is autochthonous. They suggest that the crystalline rocks below 670 meters are disconnected slices which originated in the crater rim and were emplaced beneath the more intensely shocked material in the center of the crater. However, Hüttner (1977) and Chao (1977) disagree with this view because (1) shock effects are gradually attenuated with depth, (2) shatter cone fractures diminish in number with depth, (3) only small movements are indicated on fracture planes, and (4) injection of breccia dikes and offsets in the basement associated with the impact event occurred along pre-existing fractures, joints, and crushed bands in the older, tectonically deformed basement. A petrographic description of the core from 623 to 1206 meters can be found in Bauberger *et al.* (1974), Engelhardt and Graup (1977) and Graup (1977).

Techniques

The sample preparation and the techniques used to study microcracks in the Ries samples are similar to those described by Simmons *et al.* (1975). In brief, the techniques include the use of petrographic microscopy (PM) and scanning electron microscopy (SEM) for examination of crack morphology and mineralogy, energy dispersive X-ray analyses (EDX) for semiquantitative chemical analysis, and differential strain analysis (DSA) for measuring the characteristics of open microcracks. Polished thin sections (PTS), crack sections (CS), and 2.5 × 2.5 × 5 cm DSA blocks have been made of each sample studied and used in the analyses mentioned above. Strain gages were applied in such a way as to avoid any observed breccia dikes in the DSA blocks.

Observations and Discussion

Petrography

The petrographic description of Bauberger *et al.* (1974) is adequate for most features in our samples that can be seen with the petrographic microscope. Therefore, we shall restrict our petrographic discussion to those features in our samples that differ significantly from Bauberger *et al.* (1974) or were not discussed by them. A summary of the features present in our samples is given in Table 1.

Alteration and oxidation is common and, in some samples, extensive and pervasive. Amphibole alters to chlorite plus magnetite. Chlorite, in turn, is oxidized to magnetite and/or hematite along grain boundaries and cleavage traces. An opaque mineral (probably magnetite) has also been oxidized to hematite along grain boundaries. Such oxidation effects in both naturally and artificially occurring shocked mineral assemblages have been reported previously (e.g., Chao, 1968 and El Goresy, 1968). Stöffler (1972) has reported similar

Table 1. Summary of sample descriptions and textural observations.

Sample Depth (Meters)	Rock Types	Mineralogy	Alteration	Shock Features	Breccia Dikes	SEM Observations
623	Amphibolite with some interbedding of hornblende gneiss and no aploid dikes.	Quartz Plagioclase Amphibole Chlorite Oxides	Amphibole to chlorite plus oxides. Plagioclase sausseritized. Minor oxidation of chlorite.	Planar elements in: Quartz (decorated) Plagioclase (nondecorated) Amphibole (decorated)	No.	Microcracks (planar elements) in amphibole partly sealed with chlorite and oxides. Some short, wide intragranular cracks in plagioclase sealed with chlorite. Grain boundary cracks (amphibole and chlorite) partially sealed with calcite, chlorite or sulfide phase.
638	Predominantly amphibolite with about 30 percent of cross-cutting hornblende gneiss.	Quartz Plagioclase Amphibole Chlorite Oxides	Amphibole to chlorite plus oxides. Plagioclase sausseritized.	Planar elements in: Quartz (decorated) Plagioclase (nondecorated) Amphibole (decorated)	No.	Microcracks (planar elements) in amphibole partly sealed with chlorite and oxides. Quartz has bubble planes (planar elements) and partially sealed, bridged cracks. Grain boundary cracks associated with amphibole and chlorite.

790	Chlorite gneiss with elongate porphyroblasts of quartz and feldspar.	Quartz Plagioclase Chlorite Oxides Sulfides	Extensive oxidation of chlorite and magnetite along cracks and grain boundaries.	Planar elements in: Quartz (decorated) Plagioclase (nondecorated) Glass in microbreccia dike	Yes: one microbreccia dike (1–3 mm wide) cuts through PTS.	Some microcracks (planar elements) in plagioclase coincide with planes of fluid inclusions. Other microcracks in plagioclase are sealed with either microbreccia (locally derived mineral debris) or calcite. Quartz has bubble planes (planar elements) and partially sealed bridged cracks.
972	Medium-grained gneiss with aploid schlieren of plagioclase and chlorite. Sample is cut by breccia-filled dikelets and calcite-filled cracks.	Quartz Plagioclase Chlorite Oxides	Oxidation of chlorite. Plagioclase sausseritized.	Planar elements in: Quartz (bubble planes often parallel to transgranular cracks) Plagioclase (nondecorated) Glass in microbreccia dike	Yes: one branching microbreccia dike (2 mm wide) cuts PTS.	Partially sealed microcracks in quartz are either bubble planes (planar elements) or bridged cracks. Microcracks in plagioclase are partially sealed with bridges of plagioclase. Grain boundary cracks (chlorite).
1126	Aplite, characterized by intense fracturing and injection of microbreccia along fractures.	Quartz Plagioclase K-feldspar Chlorite	Oxidation of chlorite. Feldspars sausseritized.	Planar elements in: Quartz (decorated) Plagioclase (nondecorated)	Yes: several reddish breccia dikes (2–4 mm wide) cut pre-existing partially sealed fractures.	Veins of quartz and breccia-sealed fractures are cross-cut and offset by reddish breccia dikelets. Quartz and plagioclase in host and microbreccia dikelet contain partially sealed, bridged cracks. Quartz contains bubble planes (planar elements) as well.

oxidation of amphibole from amphibolite included in the Otting suevite from Ries, Germany.

There seems to be little difference in the intensity of shock effects with depth in our samples from the Ries drill core from a petrographic viewpoint. Although the relative amount of fracturing and injection of breccia dikes increases with depth from petrographic and SEM observations, microscopic indicators of shock are comparable in all samples. In addition to the healed and sealed cracks listed in Table 1, a distinct set of open microcracks with abundant straight segments exists and cuts shock-induced features in all samples. We infer that these cracks are due to stress release during drilling and recovery of the samples.

SEM observations

The scanning electron microscope (SEM), because of its extreme magnification (≤100,000×), is a useful adjunct to the PM. With the SEM, we can study the extent of healing and sealing of microcracks, the nature of mechanical twin lamellae, and the daughter crystals in fluid inclusions. With an energy dispersive X-ray system combined with the SEM, we can estimate elemental compositions of phases. The SEM/EDX has much greater spatial resolution ($\simeq 0.01\ \mu$) than the electron microprobe. In the following sections we shall describe SEM/EDX observations of planar shock features, microcracks, and the sealing of microcracks.

Planar shock features

In our set of Ries core samples, planar shock features occur in quartz, plagioclase, and amphibole. A typical view with the petrographic microscope of shock-induced planar features in quartz is shown in Fig. 2. At the magnifications available with the PM we cannot determine whether the decorations are fluid or solid inclusions. However, the SEM observations reveal that the planar features in quartz are now 'planes' (or smoothly varying 'surfaces') of fluid inclusions approximately ½ to 1 micron in diameter (Fig. 3). At one time, the surfaces of fluid inclusions were probably microcracks with widths approximately the diameter of the fluid inclusions; they have become almost completely sealed.

The planar features in plagioclase, shown in Figs. 4 and 5, are microcracks that coincide with 'surfaces' of irregularly shaped fluid inclusions. No solid phase is observed in these microcracks. Thus, the decorations of shock-induced planar features in plagioclase appear to be only fluid inclusions. Most of the microcracks (of the planar features) in Fig. 5 are also spanned by crystal regrowth, evidence that they are partially sealed and unequivocal evidence that they were open when the rock was *in situ*.

In Figs. 6 and 7, we show typical planar features in amphibole. They consist of partially sealed microcracks that are a few microns wide. In every case they were observed to terminate on other microcracks or grain boundaries. The planes of the cracks coincide with twin planes and are normal to the c-axis. We have

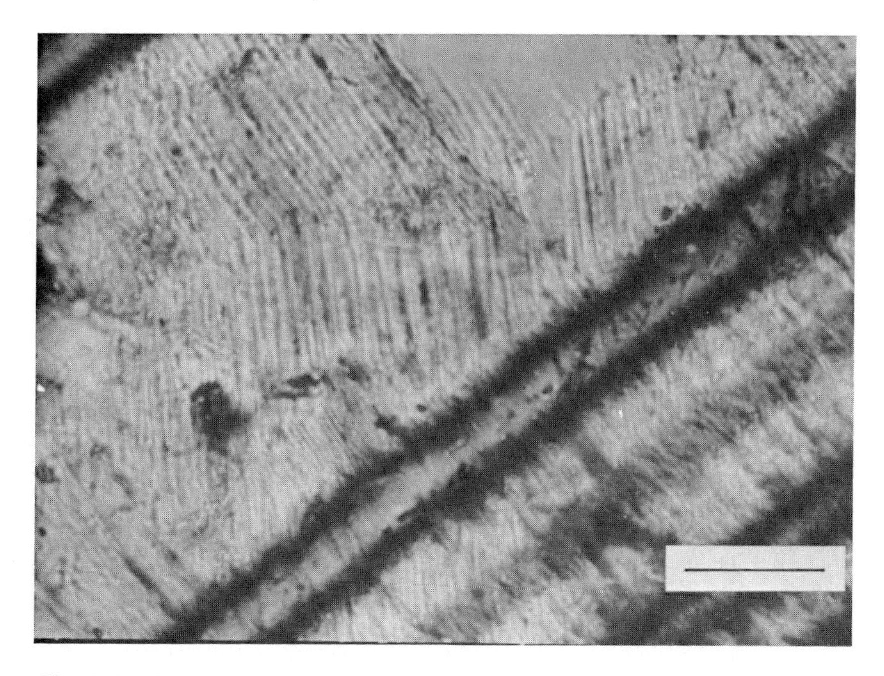

Fig. 2. Mechanical twin lamellae in a plagioclase from sample 638. Scale bar is 30 microns. Crossed polars.

identified the sealing minerals as chlorite and magnetite on the basis of EDX spectra and crystal morphology and habit. Figure 8 shows a portion of a chlorite sealed microcrack, which is identical to a shock-induced planar feature, in amphibole. The fibrous chlorite forms a network that extends between each crack wall.

Sealed microcracks

Abundant sealed microcracks (other than those microcracks identical to or associated with the planar features) are present in each sample. For these cracks, the sealing materials are chlorite, quartz, amphibole, plagioclase, oxide phases, calcite, and microbreccia. With the exception of the breccia sealed cracks, only one or two minerals seal a given crack. Chlorite sealed cracks occur in amphibole and plagioclase (see Fig. 9) but not quartz; at high magnification, they appear identical to the chlorite sealed microcracks of shock-induced planar features as shown in Fig. 8. A calcite sealed crack is shown in Fig. 5.

Several microbreccia sealed cracks observed in sample 790 appear to differ significantly from the breccia dikelets described by Bauberger *et al.* (1974) in origin as well as size. The material sealing the microcracks is derived from the host rock whereas the material of the breccia dikelets is not. See Fig. 10. In sample 790, we observe abundant fragmentation and displacement of alloch-

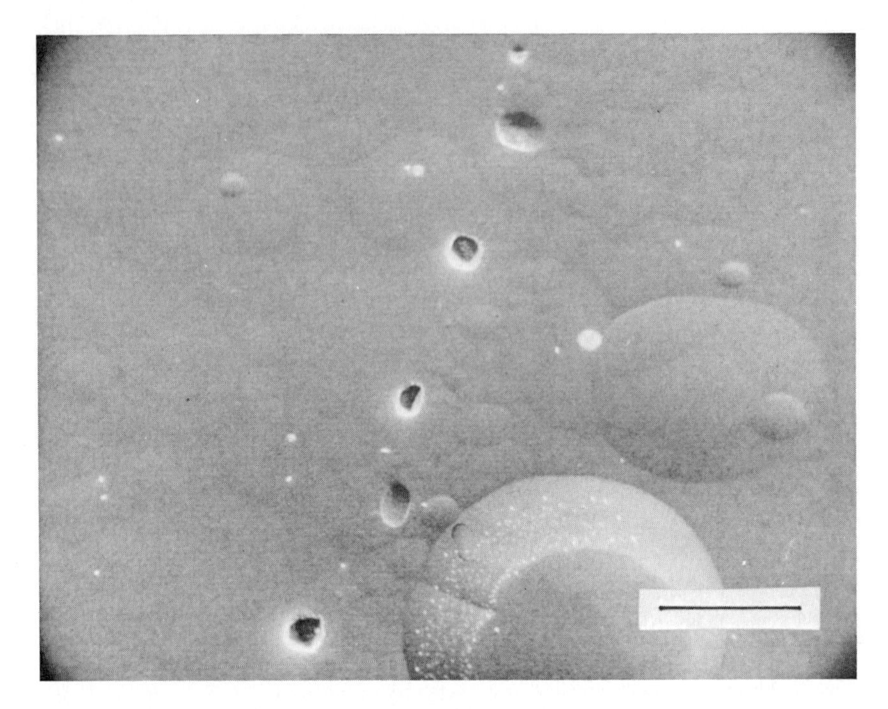

Fig. 3. SEM photomicrograph of the surface expression of a bubble plane in quartz, sample 638. The pores can be rounded like the ones illustrated, or irregular in shape. These features are ubiquitous in quartz from all our samples. Scale bar is 4 microns.

thonous grains adjacent to the breccia dikes and breccia sealed microcracks. A noteworthy characteristic of the microbreccia filled cracks is the abundance of hematite, which reaches as much as 50% in some regions. Other mineral debris in such cracks include quartz and plagioclase. The breccia filled cracks cut and displace some of the cracks sealed with calcite.

Short, bluntly terminated cracks are abundant in quartz and feldspar in our deepest sample (1126 meters). Most of these cracks are partially sealed. Less frequently observed are short, wide, partially-healed cracks with bridging structures such as those seen in Fig. 11 and 12. These cracks are similar morphologically to some cracks seen in returned lunar samples by Richter *et al.* (1976); however, we suggest that the mechanisms of sealing are different. For the Ries samples, the sealing material was probably transported in water. For the lunar samples, the cracks closed by thermal annealing.

Variation of sealed crack density with depth

Even though the number of samples examined by us with the SEM is small, we can see the trend of crack density with depth. The number of sealed microcracks associated with planar shock features per unit volume appears to decrease with

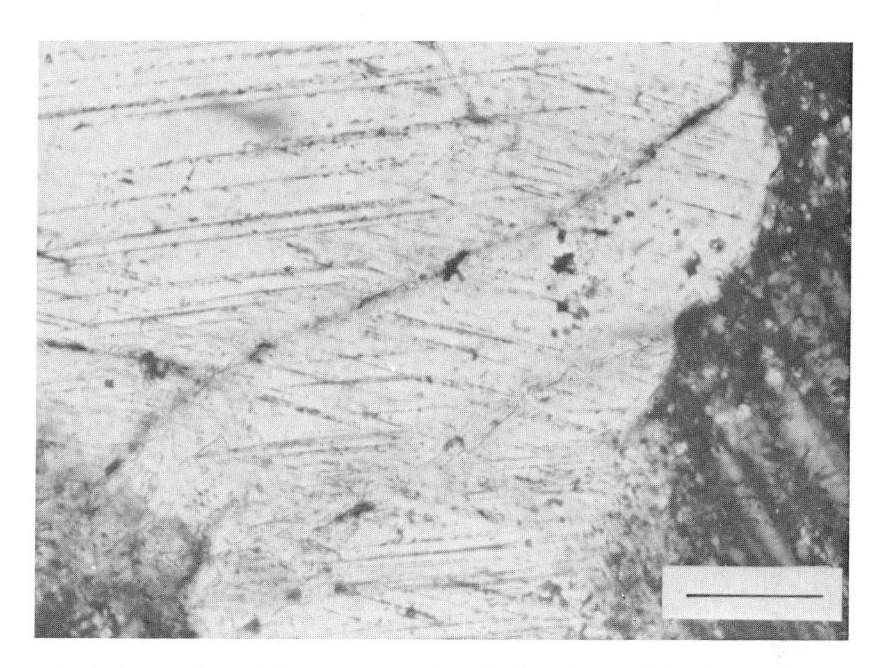

Fig. 4. Planar elements in quartz, sample 638. Scale bar is 150 microns. Crossed polars.

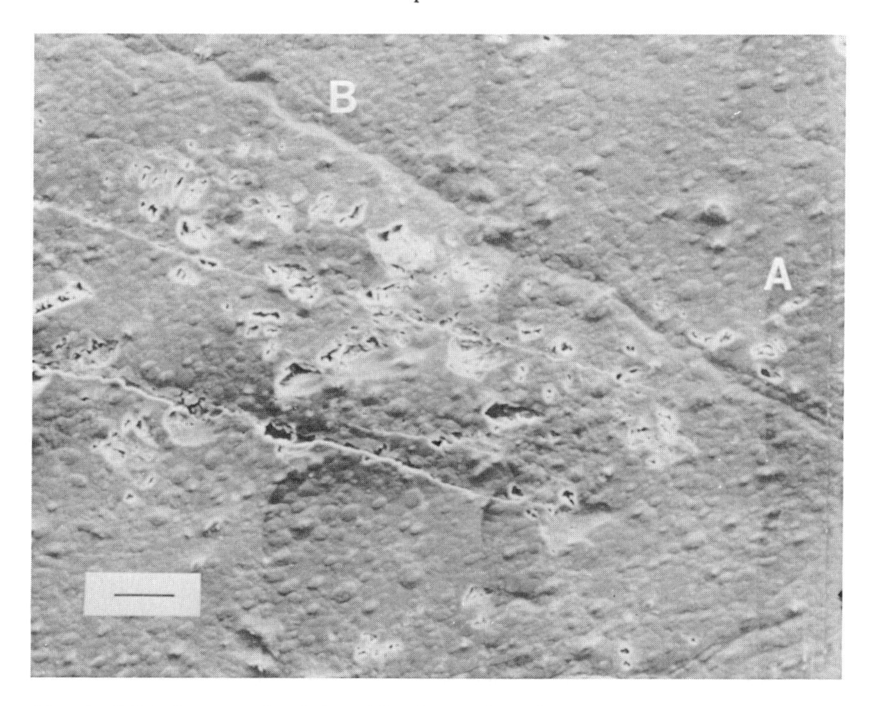

Fig. 5. SEM photomosaic of short, bluntly-terminated microcracks trending from NW to SE and associated with fluid inclusions. Note the trail of fluid inclusions below A trending NE to SW and indicating a partially-sealed crack. The broad crack at B above the cluster of fluid inclusions is a transgranular crack filled with calcite. Scale bar is 40 microns. Sample 790.

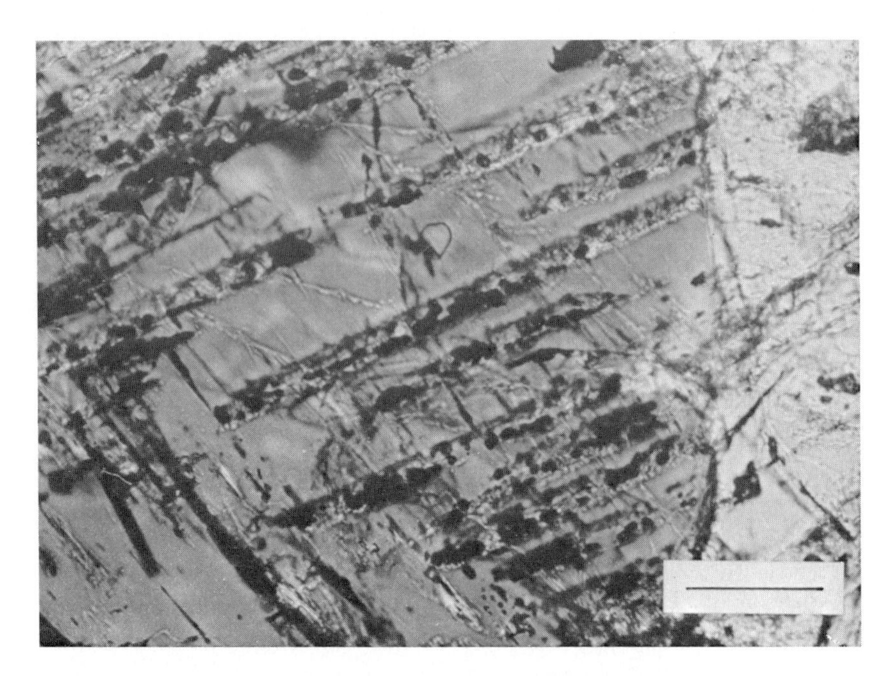

Fig. 6. Planar features in amphibole from sample 623. Scale bar is 60 microns. Plane light. Note that there are two phases associated with these planar features, an opaque (iron oxide) and a non-opaque (chlorite).

Fig. 7. SEM mosaic of sample 623. The field of view is dominated by a single amphibole grain, which exhibits well-developed planar features. Extensive chlorite alteration is visible at A. A plagioclase grain occurs at B. Scale bar is 20 microns.

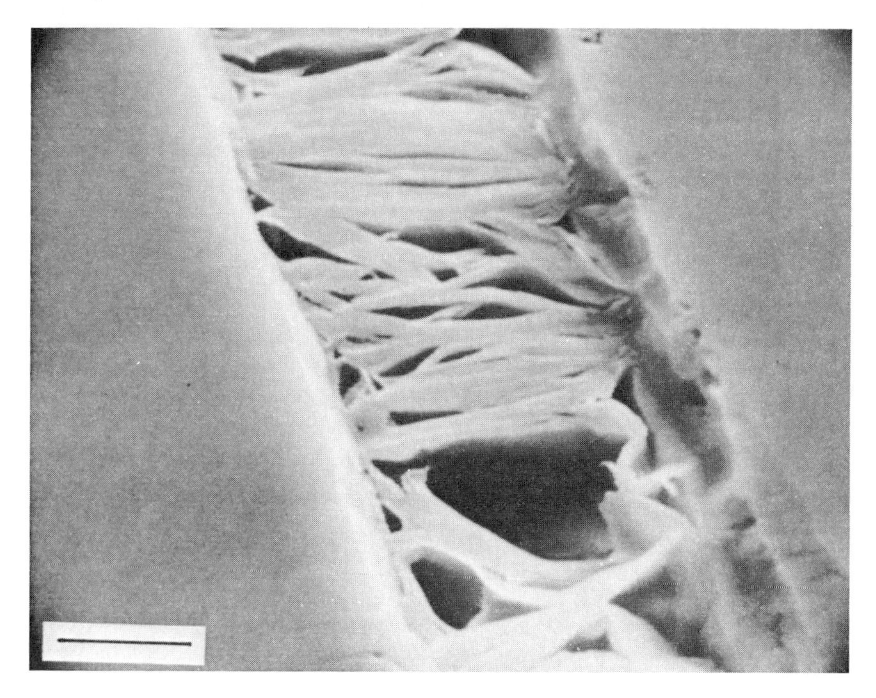

Fig. 8. SEM micrograph of a microcrack adjacent to shock lamellae in amphibole. The microcrack is the planar feature seen with an optical microscope. The host is amphibole and chlorite is the sealing material. No iron oxide phase is present in this field of view, although it has been observed elsewhere in similar cracks with the PM as small (1 micron) grains. Scale bar is 4 microns. Sample 623.

depth. The density of sealed microcracks not associated with planar features, approximately 30 to 40 cracks per millimeter, is roughly constant over the depth interval 790–1126 meters.

DSA Results

To characterize the open microcracks in these rocks we use a high precision strain technique termed differential strain analysis (DSA) which has been described in detail by Simmons *et al.* (1975), Siegfried and Simmons (1978), and Feves *et al.* (1977). With the DSA technique we measure the linear strain in various directions on a sample as a function of increasing hydrostatic confining pressure to obtain the crack porosity. This measured open crack porosity can be compared with the observed apparent volume of healed and sealed fractures in order to examine the effectiveness of crack sealing with depth. Figures 13 and 14 show the DSA data for samples 623 and 1126, respectively.

The amount of strain due to crack closure is the zero pressure intercept of the tangent to the strain curve at high pressure. This zero intercept is graphed as a

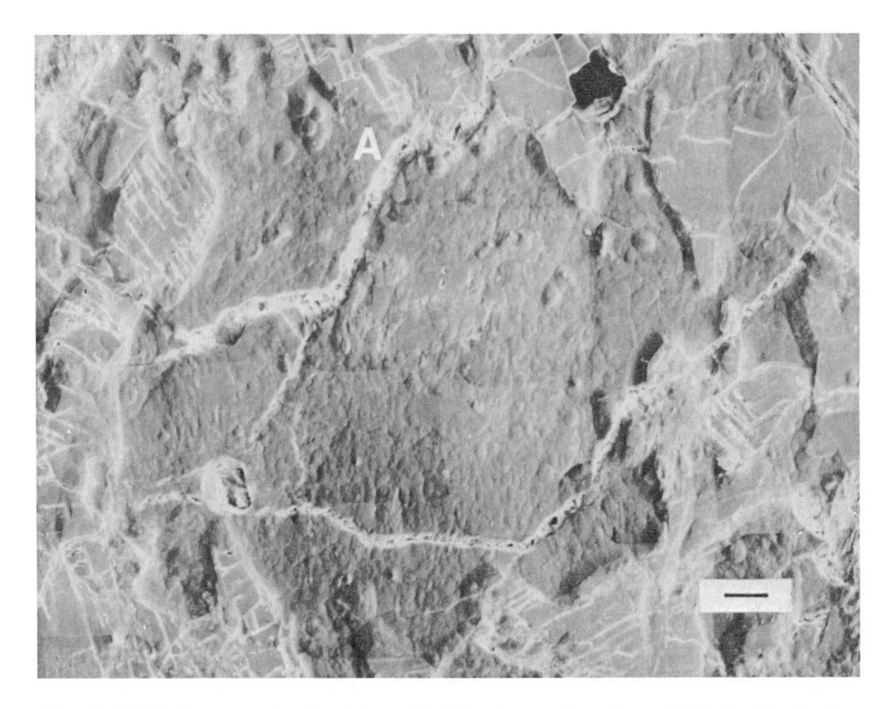

Fig. 9. SEM photomosaic of a short, wide, bluntly-terminated crack filled with chlorite and cutting a plagioclase grain in the center of the field of view. Note that the upper crack terminates at an amphibole grain boundary (A). Scale bar is 40 microns. Sample 623.

function of pressure and indicates the strain due to cracks that have been closed by some specific pressure. The total crack porosity is the sum of the values of this linear crack strain in any three orthogonal directions. As indicated in Fig. 14, multiple gages were often mounted in one direction to check reproducibility and sample inhomogeneity.

The total crack porosity indicated in the curves of Figs. 13 and 14 includes contributions from cracks that were produced during drilling and sample handling as well as from cracks that were open *in situ*. The contribution from drilling cracks can be recognized by their having closure pressures less than the pressure *in situ* for any given sample. We estimate the *in situ* pressure from ρgh, where ρ is an assumed dry rock density, g is gravitational acceleration, and h is thickness of overburden. In the results from our shallowest sample, shown in Fig. 13, the measured crack porosity continues to increase with increasing pressure. Fractures are closed over a wide range of pressures and most are undoubtedly open *in situ*. However, in our deepest sample, shown in Fig. 14, there is an abrupt decrease in the rate of crack closure at the approximate *in situ* pressure. Most of the total crack porosity in this sample is therefore due to cracks that would be closed *in situ*. The measured crack porosity, estimated *in situ* pressure, crack

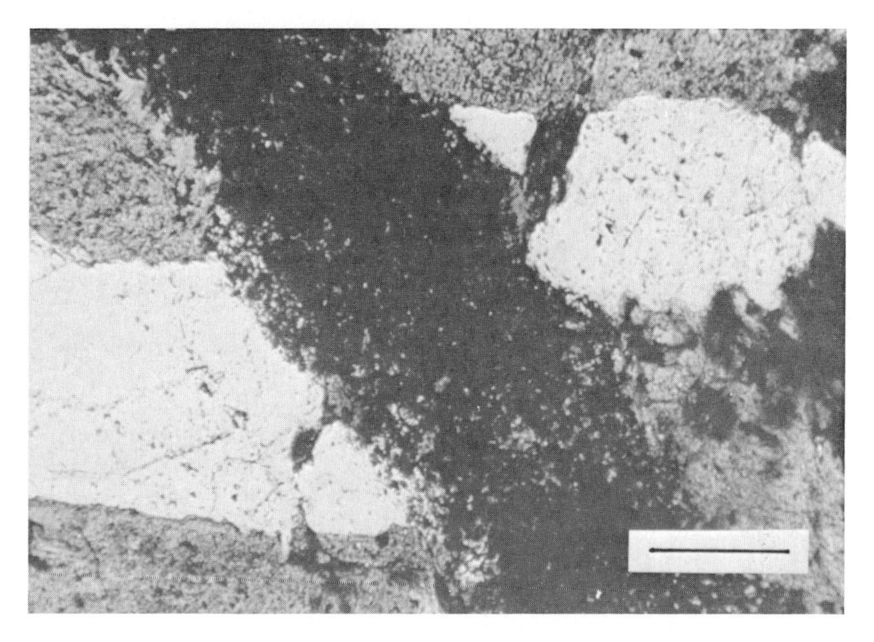

Fig. 10. Microcrack sealed with a breccia composed predominantly of fine-grained iron oxide phases and chlorite. Plane light. The scale bar is 60 microns. Note the offset in the quartz grain in the center of the photomicrograph. Sample 1126.

porosity of the cracks that must have been produced in drilling or sample handling, and the crack porosity of those cracks that may have been open *in situ* are given in Table 2. Figure 15 shows the distribution of crack porosity with respect to the depth for each sample.

The data for *in situ* cracks are not the values of crack porosity present in the sample *in situ*, rather they are the values of crack porosity at atmospheric pressure due to cracks that were present and open in the sample *in situ*, an important distinction. The density of open microcracks in our samples appears to be roughly constant with sample depth. Of course, the volume of open microcracks present in the rocks *in situ* decreases with depth, an inference drawn from each DSA curve.

Conclusions

(1) The microcracks in the Ries samples are morphologically similar to the cracks seen in returned lunar samples. The cracks in the Ries crater persist to depth. We conclude that the microcracks in the Ries samples are valid analogues of the cracks produced in lunar samples as a result of meteorite impact and that impact produces cracks in rock *in situ* and to at least 1.2 km depth.

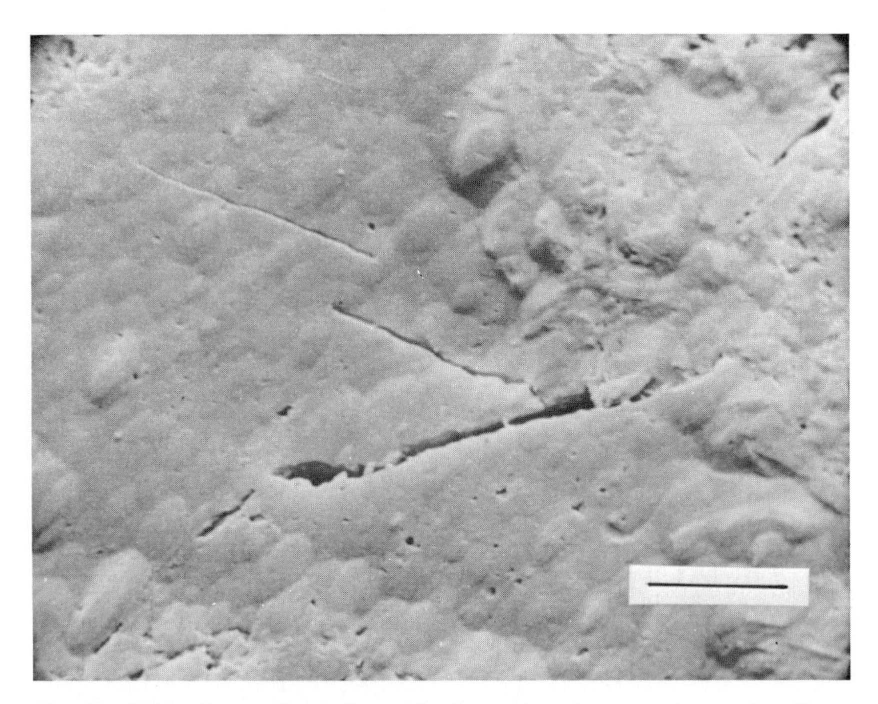

Fig. 11. SEM micrograph of short, bluntly-terminated, open microcracks. These cracks are partially healed, have bridging structures, and are in a potassium feldspar. Scale bar is 10 microns. Sample 1126.

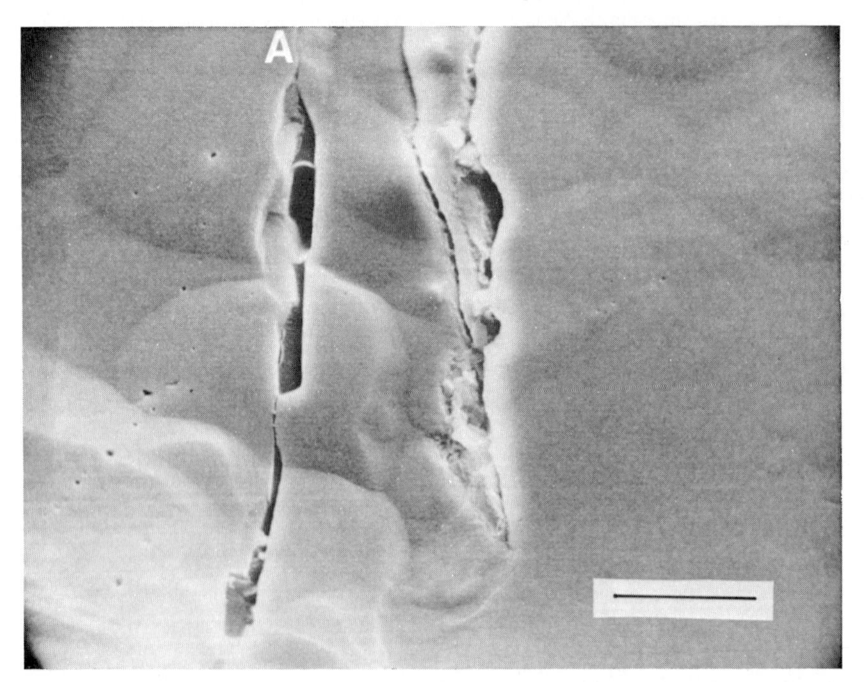

Fig. 12. Partially-healed microcracks and bubble planes in quartz. Sample 1126. At A, the width of the original crack has been preserved by quartz filling that has a texture that differs from the texture of the host quartz grain. The scale bar is 20 microns.

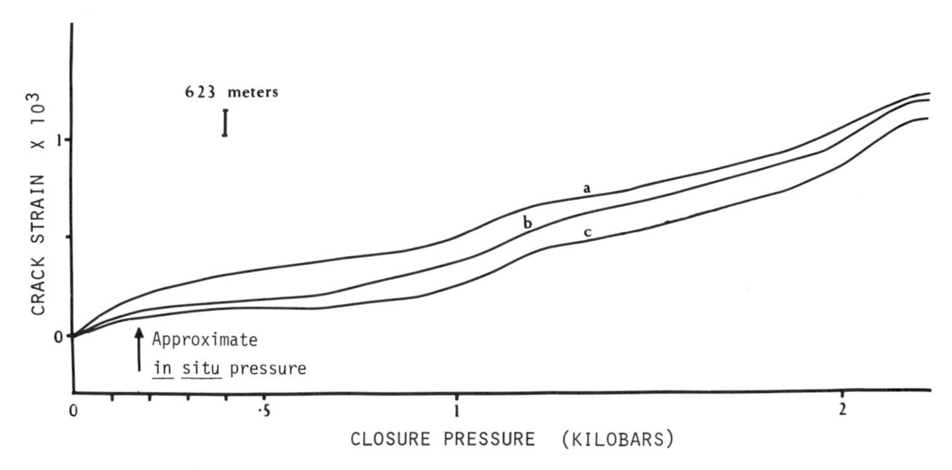

Fig. 13. Plot of the linear strain due to crack closure versus closure pressure for sample 623. Scale bar indicates the average error over the pressure range. The gages are in mutually orthogonal directions (a,b,c) with the b direction parallel to the core axis.

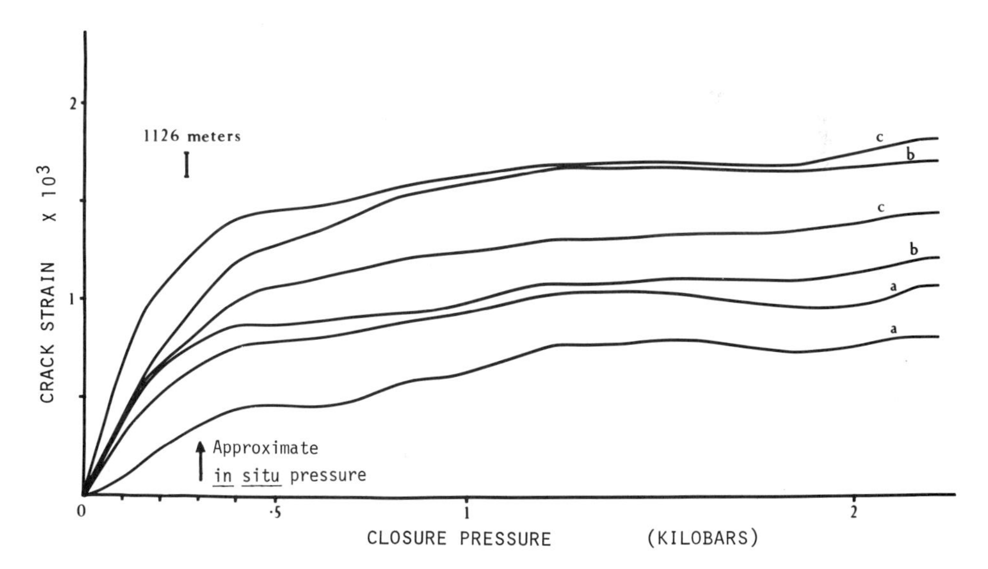

Fig. 14. Plot of the linear strain due to crack closure versus closure pressure for sample 1126. Note that all three directions have redundant gages. The spread of the data suggests considerable inhomogeneity in the sample. Note also that the major contribution to the total crack porosity comes from fractures that would be closed by the *in situ* pressure.

Table 2. Zero pressure crack porosities determined with differential strain analysis (DSA) and the estimated *in situ* pressures for each sample.

Sample (depth, m)	Total Crack Porosity to 2.5 kbar ($\times 10^6$)	Approximate *in situ* Pressure (in bars)	Crack Porosity Due to Drilling ($\times 10^6$)	Crack Porosity for Cracks Open in situ ($\times 10^6$)
623	3440	171	380	3060
638	1260	175	480	780
790	4420	217	1780	2640
972	2650	267	1250	1400
1126	4200	309	2460	1740

(2) From SEM and PM examination, the total fracture content (open, healed, and sealed cracks) decreases with depth. The content of fractures *not* along planar features, however, remains roughly constant (i.e., the fractures along planar features are more abundant in our shallower samples). The DSA data indicated that the porosity (at P = 0) of fractures which were open *in situ* remains approximately constant with

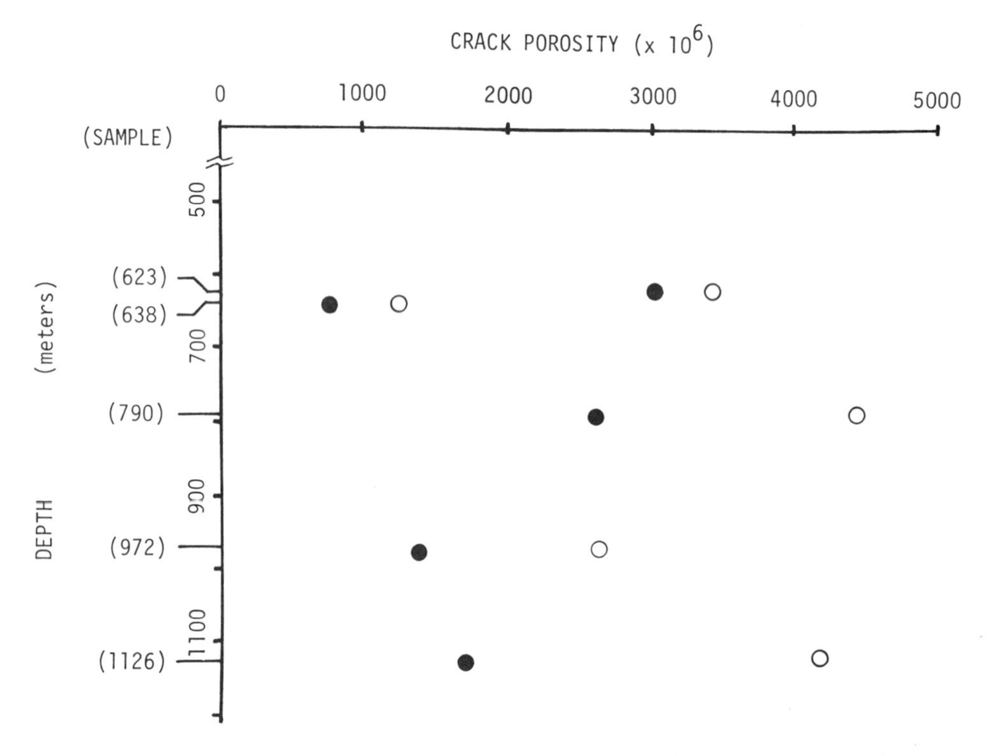

Fig. 15. Crack porosity versus depth as measured to 2.5 kbar closure pressure by differential strain analysis (DSA). The open circles represent the total measured open crack porosity at atmospheric pressure. The solid circles represent the crack porosity with the porosity due to cracks closed by the calculated *in situ* pressure removed.

depth. We conclude that the sealing processes are more effective in the fractures associated with planar features. These fractures are more completely sealed.

(3) The shallower samples (623, 638) show a continuous increase in fracture porosity with closure pressure. The deeper samples show a rapid increase at low closure pressure and very little, if any, increase at higher closure pressure. We conclude that the shallow samples contain broad spectra of open fracture shapes. This mixutre of both high and low closure pressure cracks (large and small aspect ratio cracks) could be due to either the original impact event or to subsequent modification of the open cracks.

(4) The sealing mechanisms are predominantly oxidation and hydration and these processes have sealed most of the fractures produced in impacted terrestrial rocks. They are unimportant for lunar samples. We conclude that such sealing mechanisms are uniquely terrestrial and preclude direct comparisons between physical properties of lunar rocks and those measured on the naturally impacted terrestrial rocks from the Ries crater.

(5) The shock-induced planar lamellae in amphibole are surrounded on all sides by either microcracks or grain boundaries. We conclude that the 'twinning planes' in amphiboles are actually fractures (or are coincident with fractures), the 'twin lamellae' form only with bounding fractures, or that possibly the 'twin lamellae' are not twins but are rotated portions of grains bounded by open or partially healed microcracks.

(6) Most of the planar features in amphibole and plagioclase consist of microcracks. These microcracks are usually open or only partially sealed with other minerals. We conclude that the distinction of decorated from undecorated planar features is a matter of scale and a complete gradation exists from macroscopic to submicroscopic decoration, a suggestion of Engelhardt and Bertsch (1969) on the basis of optical microscopy.

(7) The microcracks that are shock-induced planar features in quartz are almost completely healed. In amphibole, they are still almost completely open. In plagioclase, the amount of filling is intermediate between that of quartz and that of amphibole. We conclude that the rate of healing varies with the mineralogy of the host grain and is lowest, intermediate, and highest in amphibole, plagioclase, and quartz, respectively.

Acknowledgments—This work was supported by National Aeronautics and Space Administration contract NGR 22-009-540. Samples were provided by the Bavarian Geological Survey.

References

Bauberger W. v., Mielke H., Schmeer D. and Stettner G. (1974) Petrographische Profildarstellung der Forschungsbohrung Nördlingen 1973. *Geol. Bavarica* **72**, 33–34.

Chao E. C. T. (1968) Pressure and temperature histories of impact metamorphosed rocks—based on petrographic observations. In *Shock Metamorphism of Natural Materials* (B. M. French and N. M. Short, eds.), p. 135–158. Mono, Baltimore.

Chao E. C. T. (1977) Preliminary interpretation of the 1973 Ries research deep drill core and a new Ries cratering model. *Geol. Bavarica* **75**, 421–441.

El Goresy A. (1968) The opaque minerals in impactite glasses. In *Shock Metamorphism of Natural Materials*. (B. M. French and N. M. Short, eds.), p. 531–553. Mono, Baltimore.

Engelhardt W. v. (1974) Ries meteorite crater, Germany. *Fortschr. Mineral.* **52**, 103–122.

Engelhardt W. v. (1975) Some new results and suggestions on the origin of the Ries basin. *Fortschr. Mineral.* **52**, 375–384.

Engelhardt W. v. and Bertsch W. (1969) Shock induced planar deformation structures in quartz from the Ries Crater, Germany. *Contrib. Mineral. Petrol.* **20**, 203–234.

Engelhardt W. v. and Graup G. (1977) Stosswellenmetamorphose im Kristallin der Forschungsbohrung Nördlingen 1973. *Geol. Bavarica* **75**, 255–271.

Feves M., Simmons G. and Siegfried R. W. (1977) Microcracks in crustal igneous rocks: Physical properties. In *The Earth's Crust: Its Nature and Physical Properties*, Geophys. Monogr. Ser., **20**, (J. G. Heacock, ed.), p. 95–117. AGU, Washington, DC.

Graup G. v. (1977) Die Petrographie der kristallinen Gesteine der Forschungsbohrung Nördlingen 1973. *Geol. Bavarica* **75**, 219–229.

Gudden H. v. (1974) Die Forschungsbohrung Nördlingen 1973 Durchführung und erste Befunde. *Geol. Bavarica* **72**, 11–31.

Hüttner R. v. (1977) Makroskopische Beobachtungen zur Deformation des Kristallins in der Forschungsbohrung Nördlingen 1973. *Geol. Bavarica* **75**, 273–283.

Richter D., Simmons G. and Siegfried R. (1976) Microcracks, micropores, and their petrologic interpretation for 72415 and 15418. *Proc. Lunar Sci. Conf. 7th*, p. 1901–1923.

Siegfried R. and Simmons G. (1978) Characterization of oriented cracks with differential strain analysis. *J. Geophys. Res.* **83**, 1269–1278.

Siegfried R., Simmons G., Richter D. and Hörz F. (1977) Microfractures produced by a laboratory scale hypervelocity impact into granite. *Proc. Lunar Sci. Conf. 8th*, p. 1249–1270.

Simmons G., Siegfried R. and Richter D. (1975) Characteristics of microcracks in lunar samples. *Proc. Lunar Sci. Conf. 6th*, p. 3227–3254.

Stettner G. v. (1974) Das Grundgebirge in der Forschungsbohrung Nördlingen 1973 im regionalen Rahmen und seine Veränderungen durch den Impakt. *Geol. Bavarica* **72**, 35–51.

Stöffler D. (1972) Deformation and transformation of rock-forming minerals by natural and experimental shock processes. *Fortschr. Mineral.* **49**, 50–113.

Stöffler D. (1977) Research drilling Nördlingen 1973: Polymict breccias, crater basement, and cratering model of the Ries impact structure. *Geol. Bavarica* **75**, 443–458.

Stöffler D., Ewald U., Ostertag R. and Reimold W.-U. (1977) Research drilling Nördlingen 1973 (Ries): Composition and texture of polymict impact breccias. *Geol. Bavarica* **75**, 163–189.

Proc. Lunar Planet. Sci. Conf. 9th (1978), p. 2749–2771.
Printed in the United States of America

Is there any record of meteorite impact in the Archean rocks of North America?

PAUL W. WEIBLEN

Department of Geology and Geophysics, University of Minnesota, Minneapolis, Minnesota 55455

KLAUS J. SCHULZ

Johnson Space Center, Houston, Texas 77058

Abstract—Current speculations on the possibilities for recognizing a record of the early impact history of the earth between 4.5 and 3.0 b.y. ago range from simple analogies between mare basalts and Archean terrestrial greenstones to causal dismissal. The latter is based on the view that the high rate of convection in the Archean would have obliterated the effects of impact via plate tectonic processes. We suggest that the impact process may have been the prime control on *localizing* the convection process. The Superior Province and basin-in-basin structure within it may be a record of such localized convection.

The higher heat budget, difference in composition, and greater mobility of the earth's crust and mantle compared to the moon would result in development of a more complex terrestrial crater filling than on the moon. In the terrestrial case a mafic through felsic suite of igneous rocks is metamorphosed at depth and laterally to intermediate composition gneiss. The end product is a topographically high, low density, isoclinally folded crater-filling compared to the topographically low, high density, horizontally stratified lunar maria.

The stratigraphic succession in the postulated terrestrial crater-filling is more complex than the moon but the present data suggest that a major terrestrial subdivision into old gneiss terranes at >3.0 b.y. and greenstone-granite terranes at 2.5–2.7 b.y. may be equivalent to the lunar stratigraphic subdivision of KREEP basalts and mare basalts. We discuss the possible role of impact in terrestrial crustal evolution which is consistent with these interpretations of current Archean age and paleomagnetic data, structure, magmatic processes, and stratigraphy of the North American craton.

INTRODUCTION

The question of whether any record of an early meteoritic impact history (4.5–3.0 b.y.) on earth may have been preserved and is presently recognizable has not been seriously evaluated (Windley, 1977, p. 58). Considerations of the high heat budget, the relatively high intensity of convection processes required to dissipate the heat, and the resulting deformation of the thin lithosphere of early earth models have led to the view that all evidence of early impact would be obliterated.

Despite these considerations or perhaps by ignoring them a number of workers have suggested analogies between certain aspects of lunar and Archean (4.0–2.5 b.y.) terrestrial geology. Green (1972) and Glickson (1976) suggested that Archean greenstone belts may include equivalents of lunar maria. Wilson and Morrice (1977) have proposed that entire shield areas consisting of greenstone-granite terranes could be the result of single asteroidal impacts. Goodwin (1976) has suggested that the pre-Pangean distribution of continents may record the terrestrial equivalents of the large impact structures on the moon. The latter two suggestions are based solely on a similarity between the overall size and configuration of the lunar maria and the terrestrial continents in the Pangean reconstruction. Frey (1977) has used a scaling model to relate the observed lunar impact history to the earth. His assumptions and method lead him to the suggestion that the terrestrial dichotomy of continents and oceans may have been established by early terrestrial impact processes, and he has proposed that the Archean equivalents of present day ocean basins were the sites of large impact events. Thus Frey's analogy that the lunar highlands and maria are equivalent to the terrestrial continents and oceans respectively is just the reverse of Goodwin's.

Does the present divergence of opinion concerning the early impact history of the earth reflect the futility of the search for an early terrestrial impact record? Is it simply the result of the fact that to date the evidence has been sought in simple analogies with the moon, rather than in detailed reevaluation of Archean terrestrial geology in light of lunar analogies? We believe that the latter is the case and suggest that the time for such a reevaluation is particularly opportune in view of the scope and quality of the relevant data base. This includes: 1) A large body of experimental and interpretative data on impact processes (Roddy *et al.*, 1977); 2) high quality lunar mapping and interpretation (Head, 1977); 3) sophisticated models of the early thermal history of planets (Solomon and Chaiken, 1976; Toksöz and Johnston, 1977); 4) comprehensive summaries of Archean geology (Windley, 1976, 1977); and 5) the data generated by a serious debate in the current literature on the extent to which plate tectonic concepts can explain Archean geology (McElhinny, 1977; Kröner, 1977; Kerr, 1978). The data on impact phenomena and the mapping suggest the nature of impact related effects that should be looked for in the Archean record. The thermal models place constraints on the convection and its change with time in the Archean, and the plate tectonic debate provides a guide to applications of the concept of uniformitarianism to studies of the Archean.

It is interesting to note that the oldest impact structure recognized on the basis of shatter cones is the Proterozoic Sudbury structure (French, 1970). The direct evidence of shatter cones, ejecta blankets, melt sheets and actual craters will all have been obliterated or modified beyond easy recognition in the Archean record. It is necessary, therefore, to consider the indirect effects of impact processes such as the temporal and spatial distribution of igneous rocks and crustal deformation associated with the response of the Archean crust to impact events. A great deal has been learned about this on the moon and herein lies the need for attempting to draw valid analogies between lunar and Archean terrestrial geology. It should be pointed out that drawing such analogies does not require de-emphasizing the fundamental differences between the moon and earth. The analogies should be viewed as an attempt to establish criteria for recognizing the effects of a common process (impact) on two distinctly different planetary bodies (Warner and

Morrison, 1978; Morrison and Warner, 1978; Kaula, 1975). The differences in size, composition, and thermal history are reflected in the basic contrasts between the two crusts. Simply stated, lunar crustal evolution occurred in a "mechanically rigid and chemically dry regime" (Wood, 1977, p. 44–45), and in a thermal regime which permitted early formation of a thick crust (Solomon and Chaiken, 1976; Kaula, 1975). This has left a record of both the early cooling and impact history of the moon in the form of remnants of a primordial feldspathic crust modified by impact and overlain by ejecta and, largely on the front side, by a mafic volcanic succession. The emplacement and surface distribution of the lavas were controlled at least in part by the impact process (Head, 1975).

In contrast, terrestrial Archean crustal evolution occurred in a more mobile and chemically wet regime, and a thermal regime which has maintained mantle convection to the present (McKenzie and Weiss, 1975; Kaula, 1975). There is no recognized vestige of a primordial terrestrial crust and the exposed Archean crust consists essentially of two distinct geologic terranes: High-grade gneiss terranes and low grade greenstone-granite terranes (Windley, 1977), both of which have experienced extensive reworking. Critical attempts to apply plate tectonic concepts to the origin of the Archean crust (Burke and Dewey, 1973) have recognized the shorter temporal and smaller spatial scale of crustal processes in the Archean compared to the present day. The term "permobile regime" was used by Burke and Dewey (1973) perhaps to indicate that a uniformitarian application of plate tectonics to Archean geology is inappropriate. In a similar vein it should be emphasized that the attempt to recognize the effects of terrestrial impact processes in a "permobile regime" cannot be based on simple direct analogies with the "rigid" lunar record.

In this paper we shall examine analogies between certain aspects of the following facets of the Archean and lunar crusts: geochronology, paleomagnetism, structure, magmatic processes, and stratigraphy. Comparison of pertinent terrestrial data from the North American craton with relevant lunar data will be made in support of a serious search for a record of an early terrestrial impact history. Although the Proterozoic (2.5–1.5 b.y.) crustal record can be more readily interpreted in terms of plate tectonic processes (Burke and Dewey, 1973), we will discuss the possibility that an imprint of impact processes may be recognizable in these Precambrian rocks as well as in the Archean.

Geochronology

Mare volcanism may be temporarily decoupled by as much as 0.5 billion years from the early impact history of the moon (Tatsumoto *et al.*, 1977, p. 519). Hence, arguments that Archean greenstone volcanism is too young to be in any way related to early impact related processes are not convincing. Furthermore the number of reasonably well-constrained ages on terrestrial rocks which formed or were modified during the period of probably significant impact (4.5–3.0 b.y.) (Wetherill, 1977, p. 554; Neukum *et al.*, 1975) continues to grow (Table 1).

The earliest date of publication of a pre-3 b.y. terrestrial age is from the

Transvaal Craton in South Africa (Allsopp, 1961) and to date there are only a total of ~12 localities (Table 1) where rocks have been determined to be older than the ubiquitous 2.7 b.y., which was the oldest reported age prior to 1961. This explains why the prejudice persists that Archean rocks are too young for serious analogies to be drawn with lunar rocks.

It should be noted that in the old Archean gneiss terrane in Minnesota a much more complete Precambrian record of thermal and tectonic events is recorded than in the younger greenstone-granite terrane (Morey *et al.*, 1978, manuscript in preparation). This means that the potential for obtaining reset ages is greatest in the old gneiss terrane. It should be emphasized, therefore, that the present radiometric age data from gneiss terranes are biased toward minimum ages. In view of this, it is a reasonable expectation that as more data become available, more rather than fewer Archean rocks will be considered to have originated prior to 3 b.y. during the period of significant terrestrial impact. However, the first comprehensive application of the Sm-Nd method to the geochronology of the Archean (McCulloch and Wasserburg, 1978) has not turned up any new ages older than 3.5 b.y. on the North American craton. The Sm-Nd data give a 3.5 b.y. age for certain samples of Minneosta River Valley gneisses. The data also support the conclusion that the Superior, Slave, and Churchill provinces formed within the period 2.5–2.7 b.y. These new data suggest that the Minnesota River Valley and West Greenland gneisses (Table 1) are isolated, small fragments of early crust included in the Churchill and Central Provinces of North America (Fig. 1). However, the sampling is not adequate to rule out the possibility that these provinces may consist dominantly of gneisses older than 3 b.y. and that the 2.5–2.7 b.y. ages from the Churchill Province (McCulloch and Wasserburg, 1978) and the Proterozoic ages from the Central province (Van Schmus, 1976) are from younger supracrustal or intrusive rocks. Thus the present age data do not preclude the idea that a significant part of the North American Continent may have existed during the period of early impact.

PALEOMAGNETISM

The search for a record of impact in the Archean raises questions concerning the relative mobility of the Archean crust, the extent of Archean cratons, and whether the geologic provinces within Precambrian cratons were assembled by plate motion (Sutton, 1977), or possibly developed in contiguous crust by impact processes. Paleomagnetic data could potentially place strong constraints on answers to these questions.

With regard to the relative mobility of the Archean crust, Irving and Naldrett (1977) have recently suggested that paleomagnetic data do not indicate any greater horizontal mobility of the crust in the late Archean than in the Proterozoic.

A major problem in the Precambrian geology of North America is to determine the size of continuous Archean crust. Did it include all the Precambrian provinces as suggested above in the interpretation of the age data, or were all the provinces isolated cratons (Fig. 1)? Piper (1976) has postulated the existence of supercontinents in Early Proterozoic times. McElhinny and McWilliams (1977), Embelton (1978), and Nairn and Ressetar (1978) all conclude that the Proterozoic paleomagnetic data do not suggest differential movements of various tectonic provinces within Precambrian cratons. This would suggest that the North American craton within the confines of the Paleozoic troughs (Fig. 1) may have existed as a single crustal block during Late Archean times. The paleomagnetic data, however, do not preclude assembly of such a craton at that time (McElhinny and McWilliams, 1977; Cavanaugh and Seyfert, 1977). In effect, the present data elude definitive interpretation (Nairn and Ressetar, 1978).

Structure

Current general models for crustal evolution during the Archean make extensive use of analogies with the present-day earth. Because, at present, new additions to the crust and deformation of the crust are fundamentally controlled by lateral plate motion (Watson, 1976), the role of vertical tectonics in crustal processes is minimized or relegated to second order effects. Although, according to Bridgewater *et al.* (1974), there is evidence of lateral tectonics in the 3.7 b.y. gneiss terrane in West Greenland, we see no *a priori* reason for biasing the interpretations of the Archean structural record, commonly thought to be characterized by vertical tectonics (Barragar and McGlynn, 1978, p. 13), in favor of a lateral tectonic regime (Burke *et al.*, 1976). As movement on nearly vertical fractures would be characteristic of an impact dominated tectonic regime (Solomon and Head, 1978), minimization of vertical movements is tantamount to ignoring the potential structural evidence for impact.

The boundary between high-grade gneiss and low-grade greenstone-granite terranes is the most significant structural feature of the Archean crust. In North America the most striking example of such a boundary (Morey and Sims, 1976; Sims, 1976) is the locus of the Proterozoic section which forms a discontinuous annulus around the Superior Province (Fig. 1). The boundary has been interpreted by Burke and Dewey (1973) to represent a Proterozoic example of a plate collision regime. Recent gravity data from traverses across the Labrador Trough (Keary and Halliday, 1976) and the Huronian section in Michigan (Klasner and Bomke, 1977) indicate the presence of a positive gravity anomaly within the Proterozoic troughs. This feature is interpreted by Keary (1976) to represent a foundered section of greenstone analogous to foundered oceanic lithosphere in a modern subduction zone. This interpretation lends additional support to Burke and Dewey's (1973) plate tectonic model. Furthermore, the similarities between the structural style of the Proterozoic in North America and the classic Himalayan continent-continent plate collision boundary appear to be strengthened (Burke, pers. comm.) as new compilations of the structural deformation of the Proterozoic rocks become available (Dimroth, 1972; Gibb, 1975).

The plate tectonic interpretation of North American Proterozoic geology does

Table 1. Archean ages older than three billion years.

Location	Rock Types	Age[1] b.y.	Initial Sr^{87}/Sr^{86}	Reference	Date	Comments
		±2	±			
AUSTRALIA						
Pilbara Block	Duffer Formation	3.45 02	(U/Pb)	Pidgeon	1978	Zircons from dacite.
INDIA						
Dharwar Craton	gneiss and granite	3.25 15	0.7020 30	Venkatasubramanian	1974	Cobbles in conglomerate.
NORTH AMERICA						
Central Province	Minnesota River Valley gneisses	3.30	(U/Pb)	Catanzaro	1963	Continuous diffusion model.
	" "	3.55	"	"		Episodic Pb loss model.
	" "	3.80	0.700	Goldich and Hedge	1974	Visual fit to six whole rock data points.
Churchill Province	Uviak gneiss	3.62 07	0.7014 08	Hurst *et al.*	1975	Whole rock isochron
Superior Province	Lac Seul gneiss	3.04	(U/Pb)	Krogh *et al.*	1976	
Wyoming Province	Beartooth Mt. gneiss	3.20	"	Catanzaro and Kulp	1964	
SCANDINAVIA						
Norway	Vesteralen gneiss	3.46 07	0.7020 30	Taylor	1975	Whole rock isochron

SOUTH AFRICA						
Rhodesian Craton						
Mushandike	granite	3.52 26		Hickman	1974	Whole rock isochron
Limpopo	gneisses	3.86 12	0.7012 02	Barton *et al.*	1977	" " "
Transvaal Craton						
Barberton Area	Komatiite	3.50 20	0.7005 1	Jahn and Shih	1974	Mineral isochron
	Middle Marker Unit[3]	3.36 07	0.7015 18	Hurley *et al.*	1972	Whole rock isochron
Forbes Reef	granite	3.10 08		Allsopp *et al.*	1962	" " "
Milba	"	3.58 51		" " "		" " "
South Transvaal[4]	"	3.20 07		Allsopp	1961	" " "
WEST GREENLAND						
Isua Area	Iron Formation	3.76 70	(Pb/Pb)	Moorbath *et al.*	1973	
	Amitsoq Gneiss	3.70 14	0.7011 20	Moorbath *et al.*	1972	" " "
Narssaq Area	" "	3.75 09	0.7015 08	" " "	"	" " "
Praestefjord	" "	3.69 23	0.7001 17	" " "	"	" " "
Qilangarssuit	" "	3.74 10	0.7009 11	" " "	"	" " "

Notes

[1]See Jahn and Nyquist (1976) for a tabulation of younger Archean ages. Only first reference to 3.0 b.y. ages are given. Ages taken from original reference and not corrected for new decay constants.

[2]Error given in references for last two digits.

[3]Onverwacht System.

[4]Five samples of "Old Granite" between Johannesburg and Pretoria.

not address, however, the question of the temporal and spatial origin of the unusual annular configuration of the boundary (Fig. 1) between the Superior Province and the surrounding gneiss terrane. This is somewhat of an embarrassment to the plate tectonic interpretation because it requires the unusual circumstance of the near-simultaneous collision of continental blocks with the Superior Province from at least three directions (Fig. 1) (Keary and Halliday, 1976). Furthermore, evidence for deep troughs roughly perpendicular to the inferred collision boundary which would be analogous to the Bengal Fan of India, are not known in the North American crust surrounding the Superior Province (Condie, 1976a). Rather than developing *ad hoc* explanations for these problems which arise from the application of only plate tectonic models to the interpreta-

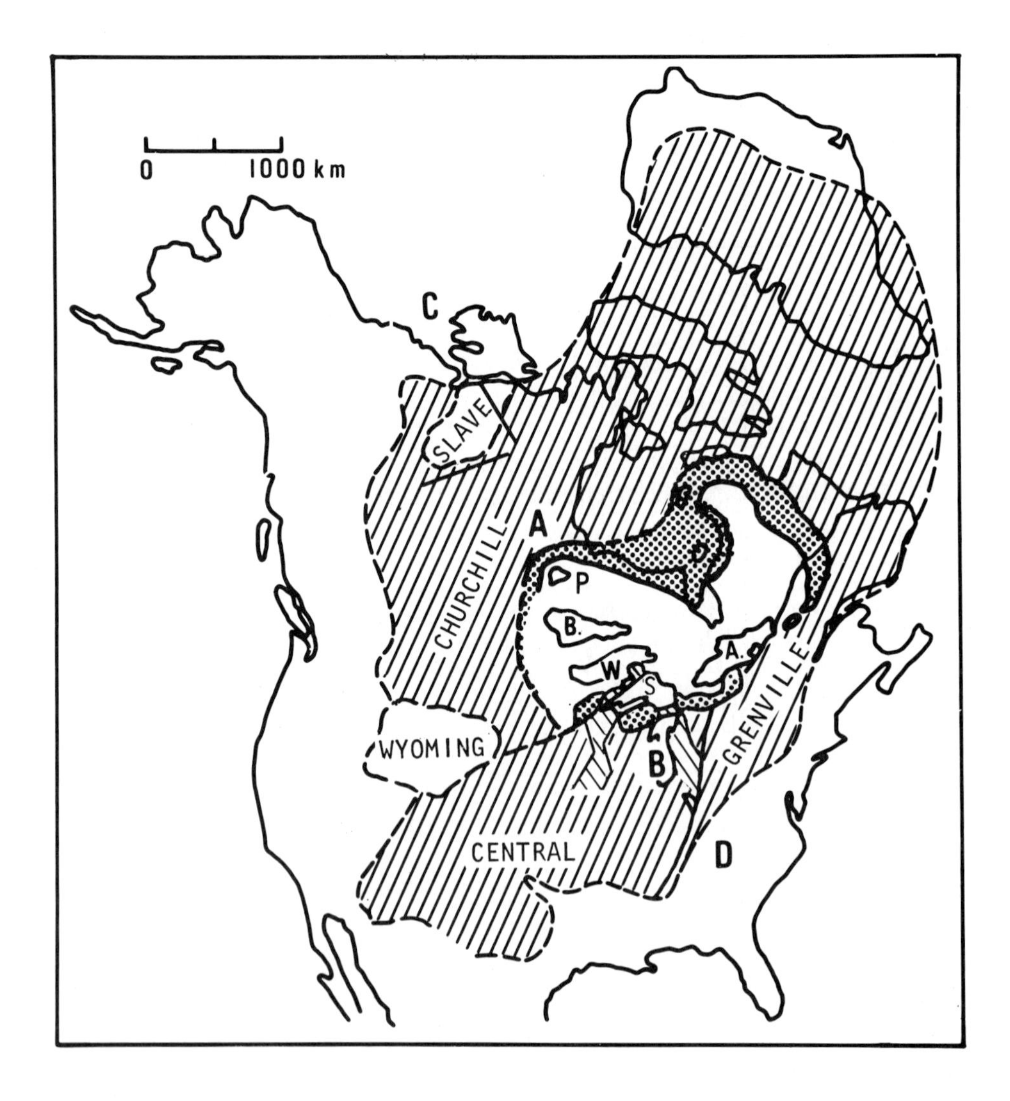

tion of this old crustal boundary, we suggest that it might be profitable to entertain the idea that meteorite impact may have established the boundary very early in the earth's history and that it was subsequently modified by plate tectonic processes.

It is an accepted dictum that pre-existing crustal structures affect tectonic processes (e.g., Baars, 1966; Nur, 1978; Zamarayev and Ruzhich, 1978; Lindholm, 1978). Thus it is as probable that impact fractures influenced Archean magmatic and tectonic processes as that the impact fractures may have been obliterated by subsequent igneous and/or tectonic events. In view of this it seems a reasonable assumption that the fracture systems and multiple basin geometry of the early impact history of the earth may have provided a configurational control on Proterozoic as well as Archean crustal evolution. (This idea has been used by Saul (1978) to support the view that upward propagation of fractures from old impact structures may explain certain circular topographic features in Arizona.)

A recent compilation of Superior Province geology by Goodwin (1977) suggests that in addition to the general delineation of the province into

---

Fig. 1. Generalized geologic map of the North American craton. Outer dashed line—Approximate locus of Paleozoic troughs (Adapted from Stewart, 1976). Right diagonal pattern—Speculative extent of contiguous Archean crust in North America. Only three of the many post-Archean geologic features that are known in this terrane are indicated: The Proterozoic fault-bounded aulacogens (thick straight lines) which separate the Slave and Churchill Provinces (Windley, 1977); the Late Precambrian Midcontinent Rift (left diagonal pattern) (Morey *et al.*, 1978) and the continental plate collision boundary (thick line) which separates the Grenville Province from the Central, Superior and Churchill Provinces. S—Lake Superior. Dotted pattern—Proterozoic troughs. These troughs define a discontinuous annulus around the Superior Province: on the north by the Labrador Trough (Dimroth, 1972), on the southeast by extension of the Labrador Trough in the Grenville Province (Windley, 1977) and the Huronian Supergroup (Windley, 1977), on the south by the Animikie Supergroup in Wisconsin and Minnesota (Sims, 1976), and on the west by geophysical anomalies (Lididak, 1971 and Douglas, 1973). The Wyoming, Slave, and Superior Provinces are Archean (>2.5 b.y.) greenstone-granite terranes. The Central and Churchill Provinces together and the Grenville Province have been thought to represent Proterozoic (<2.5 b.y.) and Late-Precambrian (<1.5 b.y.) additions to the North American craton respectively (Hurley *et al.*, 1962 and Condie, 1976a,b). The northeast boundary of the Churchill Province is not shown.

Goodwin (1977) has defined a number of elliptical-shaped basins within the Superior Province: A—Abitibi Basin, W—Wabigoon Basin B—Berens Craton, and P—Pikwitonei region. These basins are developed in gneissic terrane within the Superior Province, and the basins consist of increasing amounts of gneiss of increasing metamorphic grade in the order listed above. Goodwin (1977) suggests that this reflects deeper intersections of the basins from southeast to northeast across the Superior Province. We suggest that the disposition of these basins within the Superior Province as well as the Superior, Slave, and Wyoming Provinces within the confines of the Paleozoic troughs may be analogous to the basin-in-basin geometry of lunar geology. Hypothetical cross sections which illustrate this interpretation of the North American craton are depicted in Figs. 2 and 5.

alternating linear belts of greenstone-granite and gneiss, the province may contain a series of elliptical-shaped basins (Fig. 1). Goodwin (1977) suggests that from the southeast to the northwest across the Province successively deeper levels of these basins are exposed (Fig. 2). The Abitibi basin represents the shallowest level of exposures and provides one of the best studied and most complete examples of a greenstone-granite terrane. Goodwin implies that this basin was developed on an older gneiss terrane which he refers to as infrastructure (Goodwin, 1977, p. 2744). Within the basin, however, Goodwin points out that the granitic rocks become more gneissic and the amount of gneiss of intermediate composition increases toward the edges of the basin. Similarly the deeper intersections of basins become progressively more gneissic and of higher metamorphic grade across the Province (Fig. 2).

Although the possible occurrence of elliptical basins within the Superior Province has important implications with respect to an impact record, there are a number of problems associated with the delineation of such basins within the Superior province: 1) The delineation is based on interpretation; 2) The relationships of the basin structures to the previously mapped linear belts of gneiss and greenstone-granite are not clear; 3) The nature of the tectonic regime responsible

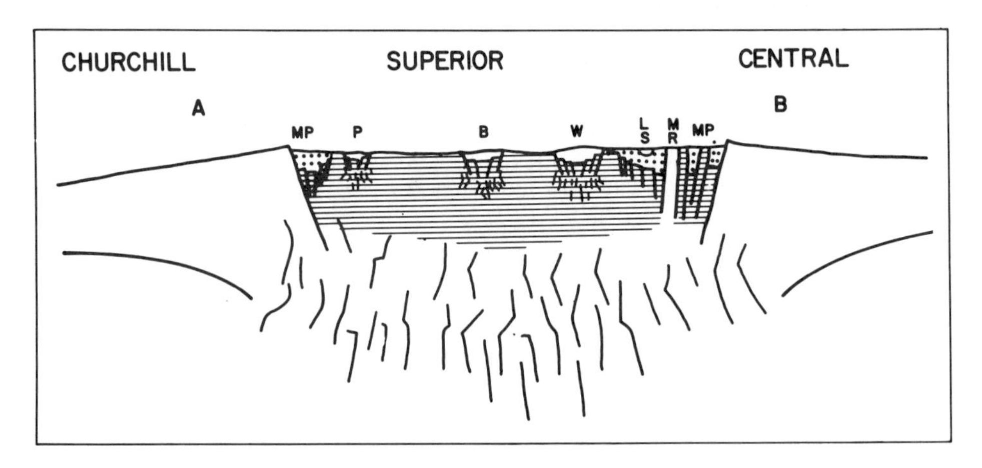

Fig. 2. Hypothetical N.W.—S.E. cross section across the Superior Province. See Fig. 1 for location and explanation of letters. This cross section illustrates relationships between postulated impact structures and the geology of the Superior Province and the surrounding Proterozoic troughs (MP, dotted pattern). A gneissic terrane, Goodwin's (1977) infrastructure, (horizontal lines) would have developed in a crater or multi-crater basin now outlined by the Proterozoic troughs. The process is discussed in the text and illustrated in Fig. 4. Smaller and later craters would have localized the basins within the Superior Province. Still later crustal adjustments along the boundary of the Superior Province would have produced the Proterozoic troughs. Such graben-like features have been recognized on the moon (Lucchitta and Watkins, 1978). The location of the Midcontinent Rift (MR) may have been in part controlled by the impact fracture system. This diagramatic fracture system (thin lines) is patterned after the much smaller Ries crater (Pohl *et al.*, 1977). LS—Lake Superior.

for development of the basin structures has not been established; 4) The processes which produced the progressive change from greenstone-granite lithologies to intermediate composition gneisses laterally and with depth in the basins have not been satisfactorily explained. The results of the ongoing studies of the Superior Province must be continually applied, of course, to items 1 and 2 rather than accepting Goodwin's model of basin structures at face value. We present in the following paragraph a brief outline of the relevance of the impact process to item 3, and the discussion on Archean magmatic processes below provides an indication of the present level of understanding of item 4.

The basin-in-basin structural relations suggested by Goodwin's model of the Superior Province (Fig. 1) resembles the basin geometry of the moon (Fig. 3) more closely than modern-day plate tectonic environments. We will pursue this analogy further after discussing Archean magmatic processes and stratigraphy.

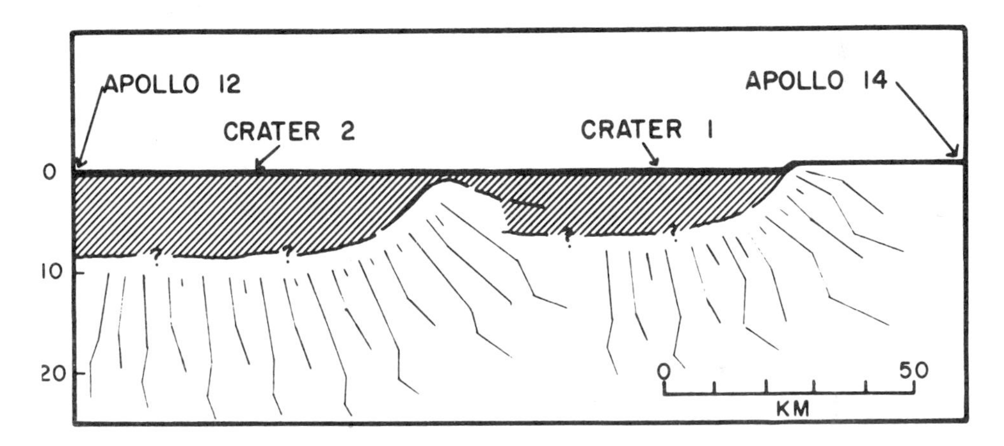

Fig. 3. Schematic cross section of the upper 25 km of the lunar crust in the Apollo 12–14 region. (From Head, 1977). Thin, radiating lines define gradational zone of shock effects adjacent to craters. At the Apollo 12 site the vertical stratigraphy downward would be: local regolith, Low-Ti basalt, Fra Mauro Formation (all indicated by the thick horizontal line), filling of the older, larger crater 2 and probably including equivalents of KREEP basalt (slanted lines), and finally repetitions of the above unit or primordial crust. Crater 2 is younger and superimposed on crater 1. This basin-in-basin geometry is characteristic of the lunar crust.

Magmatic Processes

As outlined in the introduction, comparisons of Archean greenstones and mare basalts have been central to all analogies made between early lunar and terrestrial history. Most authors (Green, 1972; Glickson, 1976; Goodwin, 1976) have referred to a fundamental difference between the products of lunar and terrestrial volcanism, i.e., that the lunar record consists of flat lying, relatively undifferentiated lavas predominantly distributed on the front side of the moon, whereas the terrestrial greenstones are part of complex isoclinally folded sequences of ultramafic through mafic and mafic through felsic volcanic rocks.

The chemical complexity of the early terrestrial crust can be taken as the expected record of intense convection (Burke and Kidd, 1978). However, this complexity by itself does not invalidate analogies between lunar and terrestrial magmatism. The REE constrained model of Arth and Hanson (1975) for the petrogenesis of the igneous rocks in greenstone-granite terranes provides a framework for contrasting possible modes of formation of lunar and terrestrial igneous rocks. Arth and Hanson's model consists of a three stage fractionation process involving ~20% partial melting of mantle peridotite to produce tholeiitic magma; ~25% partial melting of "basalt" to produce dacitic magma and ~20% to 50% partial melting of "graywacke" to produce rhyolitic magma. This scheme can be accomplished by extensive recycling of the products of the first two magmas as outlined in Table 2 and illustrated in Fig. 4. Recycling must take place within a short time span (~40 m.y.) to meet the age constraints on the rocks in typical greenstone sequences (Schulz, 1977). The recycling is consistent with the high convection rates postulated for the Archean.

Cycle	Rock Type	Magma Sources
I	Basalt	Mantle Periodotite
II	Dacite Na-Granite	Recycled "basalt"*
III	Rhyolite K-granite	Recycled "graywacke"*

*The exact metamorphic equivalent of these rocks is not specified in the model (Arth and Hanson, 1975).

Arth and Hanson's model also provides a mechanism for developing the increase in intermediate composition rocks laterally and with depth in Goodwin's basins. The plutonic products of the dacitic and rhyolitic magmas would occur as synkinematic, tonalitic and postkinematic, granitic intrusions as envisioned by Archibald *et al.*, (1978) (Fig. 4). The greenstones would essentially be remnants which escaped the recycling process (Fig. 4).

In the plate tectonic interpretation of Archean greenstone-granite terranes, the recycling implied by Arth and Hanson's model would be accomplished in an island arc-like environment and the Superior Province would represent a fortuitous product of a lateral assembling of such environments during a permobile regime of earth history (Langford and Morin, 1976).

We suggest that if the basins in the Superior Province are possible relicts of impact structures in a gneiss terrane, the fracture systems associated with such

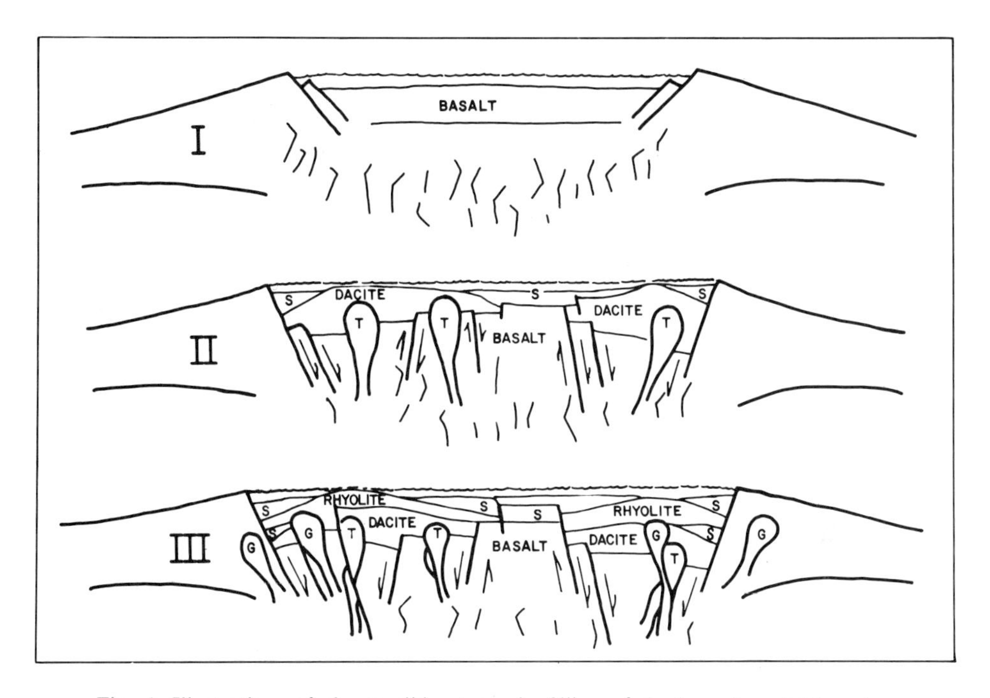

Fig. 4. Illustrations of the possible stages in filling of Archean terrestrial craters. Initiation of the Arth and Hanson igneous cycle outlined in Table 2 would produce a high density, basaltic crater filling (I). Its emplacement would be localized by the crater fracture system and its surface distribution by the crater walls. Isostatic adjustments within the crater would result in foundering of the basalt and triggering of dacitic volcanism and tonalitic (T) plutonism (II). The foundering would be most intense at the edges of the crater due to the local tensional stresses that would develop there (Solomon and Head, 1978). This would tend to preserve basalt in the center of the basin and develop annular, half-graben sedimentary troughs at the edge of the basin (S). Continued adjustments would lead to generation of late granitic and rhyolitic-volcanoclastic magmatism (III). The deformation and metamorphism associated with the intense faulting and plutonism postulated here are not illustrated in the cross sections. Isoclinal folding of the volcanic-sedimentary succession would begin in stage II. With the concentration of magmatism and deformation around the edges of the crater it can be seen that metamorphism would be most intense here. The end result of stage III would be a crater filling as described in the section on stratigraphy. Note the substantial thickening of the crust from stage I through III. Because the new section of crust within the basin is of overall intermediate composition, a gravity low and topographic high will be associated with the crater in contrast to the relations found in lunar craters (Fig. 3).

structures would have localized emplacement of tholeiitic magma analogous to mare basalt volcanism on the moon. The process of recycling of the products of this magmatism could have been confined to the impact structure giving rise to a feature like the Abitibi basin (Fig. 1). Foundering rather than subduction of the greenstones would perhaps be a more apt description of this recycling (Fig. 4).

Current studies of greenstone terranes are just beginning to shed light on possible detailed models for greenstone magmatism and structure (Archibald *et al.*, 1978; Groves *et al.*, 1978; Condie and Hunter, 1976). Regardless of the exact mechanisms, magmatic activity and deformation no doubt would be more intense in the vicinity of crater rims (Fig. 4) in response to crater modification processes that occur here after impact (Lucchitta and Watkins, 1978). The net result of crater (or multi-crater) confined Archean igneous activity in view of Arth and Hanson's model would be to produce crater-filled basins with rocks of predominantly intermediate composition. Metamorphism up to intermediate and granulite grade would be associated with plutonism which would be best developed in the deeper parts and at the edges of basins. The impact craters would thus be eventually converted to topographic highs and gravity lows (Fig. 4) in contrast to the opposite profiles of lunar craters (Fig. 3).

The simple filling of single craters would be only one manifestation of the process outlined above. A wide variety of multi-crater filling would no doubt occur as recognized in the lunar crust by Head (1975). This raises important questions of the timing, scale, and extent of Archean igneous activity that may have been influenced by the early impact history of the earth. Valid answers to these questions must await the constraints of new thermal and structural models of the Archean crust which take into account the impact process (Solomon and Chaiken, 1976; Solomon and Head, 1978).

STRATIGRAPHY

The true nature of the general stratigraphic succession in Archean terranes has always been highly speculative, but recent syntheses of the data base provide a new basis for comparison with lunar stratigraphy (Bridgewater *et al.*, 1974; Goodwin, 1977; Morey *et al.*, 1978).

A fundamental difference between the lunar and the Archean terrestrial crust is the fact that an unequivocal record of a primordial crust is preserved in the feldspathic rocks of the lunar highlands whereas the nature of a terrestrial equivalent is a matter of conjecture (Hargraves, 1976; Glickson, 1978; Barragar and McGlynn, 1978; Warner and Morrison, 1978). In fact the nature of the precursors of the oldest gneisses has not been established. One example of the oldest rocks in West Greenland, the Isua Iron Formation (Allaart, 1976), is part of a supracrustal metasedimentary sequence developed on a stratigraphically older gneiss terrane with the same radiometric age. The oldest gneisses in West Greenland (Amitsoq gneiss) and the Minnesota River Valley (Montevidio and Morton gneiss) (Table 1) are characterized by complex structural and stratigraphic relationships of amphibolitic and Na- and K-rich quartzofeldspathic rocks. It has not been possible to unequivocally deduce the stratigraphy of these contrasting lithologies. However, in both West Greenland and Minnesota the Na-rich phase of quartzofeldspathic gneiss appears, in general, to predate the K-rich phase (Bridgewater *et al.*, 1974; Goldich and Hedge, 1974).

The Archean record is complicated further by the fact that greenstone-granite terranes contain extensive metasedimentary sequences and the combined igneous-metasedimentary succession gives way laterally and with depth to intermediate to high grade metamorphic rocks. This succession is the natural consequence of the Arth and Hanson igneous cycle described above (Fig. 4). The stratigraphy of an Archean greenstone-granite terrane in this interpretation may be viewed in a number of ways: (From top to bottom or late to early.)

1) The near-surface vertical succession is volcanoclastic rocks, graywackes, dacites, and tholeiites.
2) The late to early lateral variation from the edges to the centers of basins is equivalent to the vertical succession described above. Interbedded oxide facies iron formation appears to be concentrated in the sedimentary sections at the edges of basins (Goodwin, 1977, p. 2743). This suggests a similarity in locus of occurrence to the Proterozoic iron formation around the perimeter of the Superior Province (Fig. 1).
3) The intrusive sequence late to early is granite, tonalite, and mafic dikes and sills.
4) A vertical and lateral stratigraphy may also be defined on the basis of metamorphic grade with late, high-grade rocks at the edges and deeper parts of basins, and early, low-grade rocks in the centers and upper levels of basins.
5) Overall bulk composition of the metamorphic rocks also defines a vertical and lateral compositional stratigraphy of intermediate to mafic compositions—late to early.

A significant large scale aspect of the Archean stratigraphy of North America is the fact that the oldest terranes (3 b.y.) consist of high grade gneisses of intermediate composition (Table 1). These rocks are the last to form in the stratigraphy outlined above. On the other hand the low-grade greenstones which constitute the oldest rocks in the greenstone-granite stratigraphy are all younger (2.5–2.7 b.y.) (McCulloch and Wasserburg, 1978). This emphasizes that the stratigraphic succession represents the products of a cyclic but directional process. The process is fundamentally a mechanism of energy dissipation which is still operating today and producing similar stratigraphic successions in plate collision environments. However, the magmatic and structural response to impact in the Archean crust outlined above and summarized in Fig. 4 also could have produced the same stratigraphic succession. Is it possible that in the Archean, the loci and mode of operation of this process was controlled by a deep crustal fracture system and a surface topography produced by meteorite impact? We examine below some ramifications of a positive answer to this question.

The Possible Role of Impact in Archean Crustal Evolution

If impact played a significant role in localizing the igneous and metamorphic cycles outlined above, one might expect to find a variety of styles of "crater filling" as recognized by Head (1975) on the moon. These would range from single, nearly circular crater fillings to irregular-shaped multicrater fillings. In contrast to the relatively simple general vertical stratigraphy in the lunar maria of regolith, mare basalt, KREEP basalt, ejecta, and primordial crust (Fig. 3), the impact basins on earth might be expected to have developed a series of superimposed igneous and metamorphic successions whose internal stratigraphy would have been basically controlled by the Arth and Hanson igneous cycle (Table 2 and Fig. 4).

The nature of the primodial crust is crucial to models of the role of impact in crustal evolution because the density contrast between the primordial crust and crater filling would have had a strong effect on the tectonics of the early crust. If the primordial crust is assumed to have been a global feature of mafic composition (Glickson, 1978), the effects of impacts would have produced a thickened, intermediate composition-mafic crust (Fig. 4). We suggest that the asymmetric distribution of Archean continental crust (Goodwin, 1976) could be a result of asymmetric filling of impact structures (Evensen, 1977), which originally may have been more or less uniformly distributed as on the moon (Florensky *et al.*, 1977, p. 572). Any unfilled part of the terrestrial primordial mafic crust even though highly cratered could have been subducted in post-Archean plate tectonic processes.

If the primordial crust were of intermediate or felsic composition (Hargraves, 1976; Barragar and McGlynn, 1978), impact may again have localized the igneous and metamorphic processes envisioned by Barragar and McGlynn (1978) to have produced the greenstone-granite and associated gneiss terranes. The cratered but unfilled part of a thin intermediate to felsic composition primordial crust would not have been subducted but could have been completely reworked and welded to the more stable greenstone-granite terranes in subsequent plate tectonic collision environments. Ocean basins floored by basalt would have first formed by rifting during this time. In this case one might envision preservation of evidence of the early impact processes in greenstone-granite terranes and early plate tectonic processes in the intervening gneiss terranes. However, because the evolution within greenstone-granite terranes leads to development of high grade gneisses with overall intermediate composition as outlined in the section on stratigraphy, it would be difficult to unravel such a record.

We have not considered the behavior of an early anorthositic crust (Warner and Morrison, 1978; Morrison and Warner, 1978). At this point it can only be emphasized that the ultimate composition and distribution of Archean continental crust would not only be a function of the composition of the primordial crust and the early thermal history of the earth but also of the details of the impact process and the nature of the impact crater filling. All of these are presently poorly defined (Cloud, 1977).

We have suggested above that the record of crater filling may be the igneous products of the Arth and Hanson igneous cycle and their metamorphic equivalents. If our premise is accepted that early impact fracture systems were not obliterated by convection, then the evidence for impact in the Archean may be found in large scale structural features of the continental crust. These would include features such as the Proterozoic and Paleozoic troughs (Fig. 1), which we suggest are prime candidates for the fracture systems related to old multi-crater boundaries. This conjecture carries with it the assumption that the deep crust between these troughs is basically >3.0 b.y. gneiss rather than solely the product of post-Archean continental accretion (Hurley *et al.*, 1962). In this view of North American geology, the gneisses of the Minnesota River Valley and West Greenland are considered to represent isolated exposures of more extensive Archean continental crust.

The Archean crust is clearly not now continuous within the confines of the Paleozoic troughs. The structures between the Slave and Churchill Provinces (Hoffman, 1973), the Midcontinent Rift System (Morey *et al.*, 1978), the Grenville Front (Windley, 1977), the Colorado Lineament (Warner, 1978), and the geology of the Belt Supergroup (Stewart, 1976) are some of the examples of a variety of post-Archean crustal additions and modifications found in the North American craton. All can be explained in terms of plate tectonic processes which involve, at one stage or another, fragmentation of preexisting crust. However, none require, nor provide evidence of, large-scale differential movements of plates (e.g., McWilliams and Dunlop, 1978 and references therein). We conclude, therefore, that there is no definitive evidence for lateral accretion of the North American craton between the Proterozoic and Paleozoic troughs (Fig. 1). We suggest, rather, that serious consideration should be given to the idea that this part of the North American craton may have formed in an early (>3.5 b.y.) response to cratering of this part of the primordial crust (Fig. 5). Such an episode

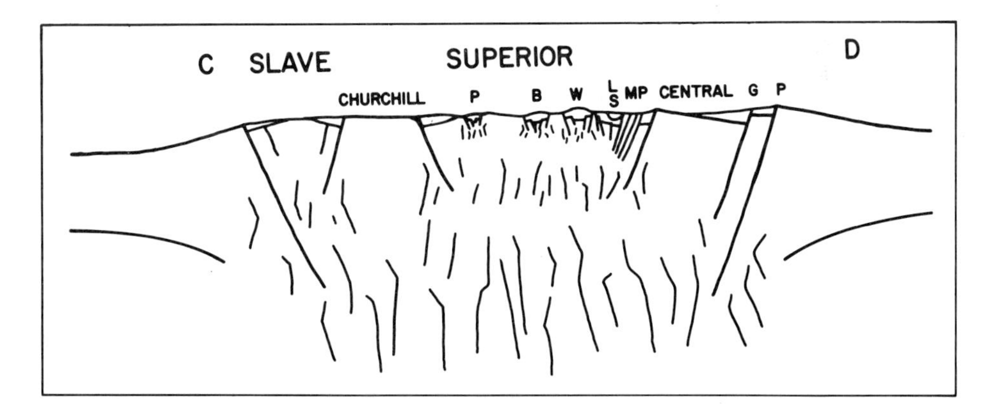

Fig. 5. N.W.—S.E. cross section across North America illustrating the geometrical relationships of a hypothetical impact related crustal evolution. See Fig. 1 for location and explanation of some letters. MP—Proterozoic trough in Wisconsin. G—Boundary between the Central and Grenville Provinces. P—Paleozoic trough. LS—Lake Superior. This cross section implies that the Churchill and Central Provinces are inherently older than the Superior Province, and that old crater fracture systems may have localized the plate boundaries developed around the North American craton during the Paleozoic (Thomas, 1978).

of crustal formation would be analogous to KREEP basalt volcanism on the moon. The formation of new crust within the Slave, Wyoming, and Superior Provinces and the basins within the latter (Fig. 1) would be equivalent to the lunar Low- and High-Ti mare basalt volcanism (Table 3). These analogies are highly speculative, but they present an intuitively satisfying picture of progressively smaller crater filling with time in the Archean as found on the moon.

We hypothesize that the crater or multicrater boundaries illustrated in Figs. 2 and 5 may have subsequently become plate boundaries. The ultimate value of this view and the conjectures made above may lie in calling attention to a possible explanation for the location and configuration of certain plate boundaries which are currently implied to be a more or less fortuitous aspect of mantle convection (Windley, 1977; Burke *et al.*, 1977; McKenzie and Weiss, 1975).

We are unable to suggest any unequivocal terrestrial equivalents for lunar highlands rocks (Table 3). It remains to be seen if these will be found in the old gneiss terranes. The search for such remnants of terrestrial primordial crust will require a program of persistent high-quality analysis and continual review of working hypotheses to document real differences between the contrasting lithologies in the old gneiss terranes. This is the approach initiated over 40 years ago

Table 3. Speculative comparison of the early crustal evolution of the moon and the North American continent.

		MOON[1]			
AGE	4.5	4.0		3.5	3.0
EVENTS	Accretion	Early, Final Bombardment		Sporadic Events	
MARIA ROCKS	Magma Ocean	KREEP Basalts	High-Ti Basalts	Low-Ti Basalts	
HIGHLANDS ROCKS		Crystalline Matrix Breccias		Vitric Matrix Breccias	
		EARTH			
AGE	4.5	4.0	3.5	3.0	
EVENTS[2]	North American Continent[3]		Superior Province	Abitibi Basin	
ROCKS			3.7 b.y. Gneisses	Greenstone-Granite Terranes	

Notes

[1]Summary of lunar crustal evolution from Warner *et al.* (1977).

[2]Speculative time of formation of the impact fracture systems which may have controlled the subsequent evolution of the crustal features indicated.

[3]The area within the confines of the Paleozoic Troughs shown in Fig. 1.

by S. S. Goldich (1938) in his studies of the gneiss terrane in the Minnesota River Valley and it is essentially the approach which led Wood *et al.* (1970) to the recognition of highlands material in the lunar regolith.

Summary

The cursory review of Archean geology outlined here has been presented with a view to encouraging a more serious search for a record of the early impact history of the earth. We tentatively conclude that although the age and paleomagnetic data are equivocal they may be construed to suggest that crustal blocks such as the North American craton were in existence during the period of early impact. The principal clue to recognition of an impact history in this craton may lie in a consideration of the role of impact in *localizing* Archean structural and magmatic processes. This contrasts with plate tectonic interpretations which do not elucidate this aspect of the crustal record.

Understanding the distinctive igneous cycle of interrelated tholeiitic through felsic magmatism found in greenstone-granite terranes and the associated deformation and metamorphism are essential to comparative studies or model analogies of lunar maria and possible terrestrial equivalents.

Despite the fundamental differences between the moon and the earth, a common impact history might be expected to develop parallel though contrasting stratigraphic successions. Current syntheses of the lunar and terrestrial Archean data base may be interpreted to show a parallel between the two periods of lunar KREEP and Low- and High-Ti volcanism and two episodes of terrestrial magmatism which led to formation of an old gneiss terrane and younger greenstone-granite terranes in North America.

Acknowledgments—This paper benefitted from reviews by R. A. F. Grieve, W. C. Phinney and M. Settle. However, the views expressed and conjectures made remain the responsibility of the authors.

References

Allsopp H. L. (1961) Rb-Sr age measurements on total rock and separated mineral fractions from the old granite of the Central Transvaal. *J. Geophys. Res.* **66**, 1499–1961.

Allsopp H. L., Roberts H. R., Schreiner G. D. L. and Hunter D. R. (1962) Rb-Sr age measurements on various Swaziland granites. *J. Geophys. Res.* **67**, 5307–5313.

Allsopp H. L., Viljoen M. J. and Viljoen R. P. (1973) Strontium isotopic studies of the mafic and felsic rocks of the Onverwacht Group of the Swaziland Sequence. *Geologisch Rundschau* **62**, 902–917.

Allaart J. H. (1976) The pre-3760 old supracrustal rocks of the Isua area, central west Greenland, and the associated occurrence of quartz-banded ironstone. In *The Early History of the Earth.* (B. F. Windley, ed.), p. 177–189, Wiley, N.Y.

Archibald N. J., Bettenay L. F., Binns R. A., Groves D. I. and Gunthorpe R. J. (1978) The evolution of Archean greenstone terrains, Eastern Goldfields Province, Western Australia. *Precambrian Research* **6**, 103–132.

Arth J. G. and Hanson G. N. (1975) Geochemistry and origin of the early Precambrian crust of north-eastern Minnesota. *Geochim. Cosmochim. Acta* **39**, 325–362.

Baars D. L. (1966) Pre-Pennsylvanian paleotectonics—Key to basin evolution and petroleum occurrences in Paradox Basin, Utah and Colorado. *Amer. Assoc. Pet. Geol. Bull.* **50**, 2082–2111.

Barragar W. R. A. and McGlynn J. C. (1978) On the basement of Canadian greenstone belts: Discussion. *Geoscience Canada* **5**, 13–15.

Barton J. H., Fripp R. E. P. and Ryan B. (1977) Rb-Sr ages aid geological setting of ancient dikes in the Sand River area Limpopo mobile belt S. Africa. *Nature* **267**, 487–490.

Bridgewater D., McGregor V. R. and Myers J. S. (1974) A horizontal tectonic regime in the Archaean of Greenland and its implications for early crustal thickening. *Precambrian Research* **1**, 179–197.

Burke K. and Dewey J. F. (1973) An outline of Precambrian Plate Development. In *Implications of Continental Drift to the Earth Sciences* **2**, (D. H. Tarling and S. K. Runcorn, eds.), p. 1035–1045, Academic Press, N.Y.

Burke K., Dewey J. F. and Kidd W. S. F. (1976) Dominance of horizontal movements, arc and microcontinental collisions during the later permobile regime. In *The Early History of the Earth*, (B. F. Windley, ed.), p. 113–129, John Wiley & Sons, N.Y.

Burke K., Dewey J. F. and Kidd W. S. F. (1977) World distribution of sutures—the sites of former oceans. *Tectonophys.* **40**, 69–99.

Burke K. and Kidd W. S. F. (1978) Were Archean Continental geothermal gradients much steeper than those of today? *Nature* **272**, 240–241.

Catanzaro E. J. (1963) Zircon ages in southwestern Minnesota. *J. Geophys. Res.* **68**, 2045–2048.

Catanzaro E. J. and Kulp J. L. (1964) Discordant zircons from the Little Belt (Montana), Beartooth (Montana) and Santa Catalina (Arizona) Mountains. *Geochim. Cosmochim. Acta* **28**, 87–124.

Cloud P. (1977) The veils of Gaia. In *The Encyclopedia of Ignorance* (R. Duncan and M. Weston-Smith, eds.), p. 387–390. Pergamon, N.Y.

Condie K. C. (1976a) *Plate Tectonics and Crustal Evolution.* Pergamon, N.Y. 288 pp.

Condie K. C. (1976b) The Wyoming Archean Province in the western United States. In *The Early History of the Earth* (B. F. Windley, ed.), p. 499–510. Wiley, N.Y.

Condie K. C. and Hunter D. R. (1976) Trace element geochemistry of Archean granite rocks from the Barberton region South Africa, *Earth Planet. Sci. Lett.* **29**, 389–400.

Cavanaugh M. D. and Seyfert C. K. (1977) Apparent polar wander paths and the joining of the Superior and Slave provinces during early Proterozoic time. *Geology* **5**, 207–211.

Dimroth E. (1972) The Labrador geosyncline revisited. *Amer. J. Sci.* **272**, 487–506.

Douglas R. J. W. (1973) Geologic provinces (map). In *The National Atlas of Canada*, p. 27–28, Dept. of Energy, Mines and Resources, Ottawa, Canada.

Embleton B. J. J. (1978) The paleomagnetism of 2400 M. Y. old rocks from the Australian Pilbara craton and its relation to Archean-Proterozoic tectonics. *Precambrian Research* **6**, 275–291.

Evensen N. (1977) The nature of the exotic component in Lunar soils and its implication for Lunar differentiation history. Phd. thesis, Univ. Minnesota, Minneapolis. 222 pp.

Florensky K. P., Bazilevsky A. T. and Ivanov A. V. (1977) The role of exogenic factors in the formation of the lunar surface. In *The Soviet-American Conference on Cosmochemistry of the Moon and Planets.* p. 571–584, NASA SP-370, Washington, D. C.

French B. M. (1970) Possible relations between meteorite impact and igneous petrogenesis, as indicated by the Sudbury structure, Ontario. *Bull. Volcanol.* **34**, 455–517.

Frey H. (1977) Origin of the Earth's Ocean Basins. *Icarus* **32**, 235–270.

Gibb R. A. (1975) Collision tectonics in the Canadian Shield. *Earth Planet. Sci. Lett.* **27**, 378–382.

Glickson A. Y. (1976) Earliest Precambrian ultramafic-mafic volcanic rocks; ancient oceanic crust or relic terrestrial maria? *Geology* **4**, 201–205.

Glickson A. Y. (1978) On the basement of Canadian greenstone belts. *Geoscience Canada* **5**, 3–12.

Goldich S. S. (1938) A study in rock weathering. *J. Geol.* **46**, 17–58.

Goldich S. S. and Hedge C. E. (1974) 3800 myr. gneiss in southwestern Minnesota. *Nature* **252**, 467–468.

Goodwin A. M. (1976) Giant impacting and the development of continental crust. In *The Early History of the Earth*, (B. F. Windley, ed.), p. 77–95, John Wiley & Sons, N.Y.

Goodwin A. M. (1977) Archean basin-craton complexes and the growth of Precambrian shields. *Can. J. Earth Sci.* **14**, 2737–2759.

Green D. H. (1972) Archaen greenstone belts may include terrestrial equivalents of lunar maria? *Earth Planet. Sci. Letter* **15**, 263–270.

Groves D. I., Archibald N. J., Bettenay L. F. and Binns R. A. (1978) Greenstone belts as ancient marginal basins or ensialic rift zones. *Nature* **273**, 460–461.

Hargraves R. B. (1976) Precambrian geologic history. *Science* **193**, 363–371.

Head J. W. (1975) Mode of occurrence and style of emplacement of lunar mare deposits. In *Conf. on Origins of Mare Basalts and Their Implication for Lunar Evolution*, p. 61–65. The Lunar Science Institute, Houston.

Head J. W. (1977) Some geological observations concerning lunar geophysical models. In *The Soviet-American Conference on Cosmochemistry of the Moon and Planets.* p. 407–416. NASA SP-370, Washington, D.C.

Hickman M. H. (1974) 3500-Myr.-old granite in Southern Africa. *Nature* **251**, 295–296.

Hoffman P. (1973) Evolution of an early proterozoic continental margin: the Coronation Geosyncline and associated anlasogens of the northwestern Canadian Shield. *Phil. Trans. Roy. Soc. Lond.* **A273**, 547–581.

Hurley P. M., Hughes H., Faure G., Fairbairn H. W. and Pinson W. H. (1962) Radiogenic Strontium-87 Model of Continent Formation. *J. Geophys. Res.* **67**, 5315–5334.

Hurley P. M., Pinson W. H., Jr., Nagy B., Teska T. M. (1972) Ancient age of the Middle Marker Horizon: Onverwacht Group, Swaziland Sequence, South Africa. *Earth Planet. Sci. Lett.* **14**, 360–366.

Hurst R. W., Bridgewater D., Collerson K. D. and Wetherwill G. W. (1975) 3600 M.Y. Rb-Sr ages from very early Archean gneisses from Saglek Bay, Labrador. *Earth Planet. Sci. Lett.* **27**, 393-403.

Irving E. and Naldrett A. J. (1977) Paleomagnetism in Abitibi greenstone belt, and Abitibi and Matachewan diabase dikes: Evidence of the Archean geomagnetic field. *J. Geol.* **85**, 157–176.

Jahn B. M. and Nyquist L. E. (1976) Crustal evolution in the early earth-moon system: Constraints from Rb-Sr studies. In *The Early History of the Earth*, (B. F. Windley, ed.), p. 55–76. Wiley, N.Y.

Jahn B. and Shih C. (1974) On the age of the Onverwacht Groupt, Swaziland Sequence, South Africa. *Geochim. Cosmochim. Acta* **38**, 873–885.

Kaula W. M. (1975) The seven ages of a planet. *Icarus* **26**, 1–15.

Keary P. (1976) A regional structural model of the Labrador trough, northern Quebec from gravity studies, and its relevance to continental collision in The Precambrian. *Earth Planet. Sci. Lett.* **28**, 371–378.

Keary P. and Halliday D. W. (1976) The gravity field of the central Labrador trough, Northern Quebec. Gravity Map Series 162, Gravity Service of Canada, Ottawa, Ontario.

Kerr R. A. (1978) Precambrian tectonics: Is the present the key to the past? *Science* **199**, 282–285, and 330.

Klasner J. S. and Bomke D. (1977) Crustal model studies of a regional gravity anomaly in northern Michigan and Wisconsin, extent of anomaly and its relationship to near surface geology (abstract). In *Inst. on Lake Superior Geology, 23rd, Abs. and Proc.* p. 21. Lakehead University, Thunder Bay, Ontario.

Krogh T. E., Harris N. B. W. and Davis G. L. (1976) Archean rocks from the Eastern Lac Seul region of the English River gneiss belt, northwestern Ontario, part 2. Geochronology. *Can. J. Earth Sci.* **13**, 1212–1215.

Kröner A. (1977) Precambrian mobile belts of southern and eastern Africa—ancient sutures or sites of ensialic mobility? A case for crustal evolution towards plate tectonics. *Tectonophys.* **40**, 101-135.

Langford F. F. and Morin J. A. (1976) The development of the Superior Province of Northwestern Ontario by merging island arcs. *Amer. J. Sci.* **276**, 1023–1034.

Lindholm R. C. (1978) Triassic-Jurassic faulting in eastern North America—a model based on pre-Triassic structures. *Geology* **6**, 365–368.

Lidiak E. G. (1971) Buried Precambrian rocks of South Dakota. *Bull. Geol. Soc. Amer.* **82**, 1411–1420.

Lucchitta B. K. and Watkins J. A. (1978) Large grabens and lunar tectonism (abstract). In *Lunar and Planet. Science IX*, p. 666–668. Lunar and Planetary Institute, Houston.

McCulloch M. T. and Wasserburg G. J. (1978) Sm-Nd and Rb-Sr chronology of continental crust formation. *Science* **200**, 1003-1011.

McElhinny M. W. (ed.) (1977) Past distribution of continents. *Tectonophys.* **40**, 181 pp.

McElhinny M. W. and McWilliams M. O. (1977) Precambrian geodynamics—a paleomagnetic view, *Tectonophys.* **40**, 137–160.

McKenzie D. P. and Weiss N. (1975) Speculations on the thermal and tectonic history of the earth. *Geophys. J. R. Astron. Soc.* **42**, 131–174.

McWilliams M. O. and Dunlop D. J. (1978) Grenville paleomagnetism and tectonics. *Can. J. Earth Sci.* **15**, 687–695.

Moorbath S., O'Nions R. K. and Pankhurst R. J. (1973) Early Archaean age for the Isua iron formation, West Greenland. *Nature* **245**, 138–139.

Moorbath S., O'Nions R. K., Parkhurst R. J., Gale N. H. and McGregor V. R. (1972) Further Rb-Sr age determinations on the very early Precambrian rocks of the Godthaab District, West Greenland. *Nature* **240**, 78–82.

Morey G. B. and Sims P. K. (1976) Boundary between two Precambrian W terranes in Minnesota and its geological significance. *Bull. Geol. Soc. Amer.* **87**, 141–152.

Morrison D. A. and Warner J. L. (1978) Planetary tectonics II: A gravitational effect (abstract). In *Lunar and Planetary Science IX*, p. 769–771. Lunar and Planetary Institute, Houston.

Nairn A. E. M. and Ressetar R. (1978) Paleomagnetism of the peri-Atlantic Precambrian. *Ann. Rev. Earth Planet. Sci.* **6**, 75–91.

Neukum G., König B. and Fechtig H. (1975) Cratering in the earth-moon system: Consequences for age determination by crater counting. *Proc. Lunar Sci. Conf. 6th*, p. 2597–2620.

Nur A. (1978) The origin of lineaments and their depth, spacing and directions (abstract). In *3rd Internat. Conf. on Basement Tectonics*, p. 35. Fort Lewis College, Durango, Colorado.

Pidgeon R. T. (1978) 3450-m.y.-old volcanics in the Archaean layered greenstone succession of the Pilbara Block, Western Australia. *Earth Planet. Sci. Lett.* **37**, 421–428.

Piper J. D. A. (1976) Paleomagnetic evidence for a Proterozoic super-continent. *Philos. Trans. Roy. Soc. London* **A280**, 469–490.

Pohl J., Stöffler D., Gall H. and Ernstson K. (1977) The Ries impact crater. In *Impact and Explosion Cratering.* (D. J. Roddy, R. O. Pepin and R. B. Merrill, eds.) p. 343–404. Pergamon, N.Y.

Roddy D. J., Pepin R. O. and Merrill R. B. (eds.) (1977) *Impact and Explosion Cratering.* Pergamon, N.Y. 1301 pp.

Saul J. M. (1978) Circular structures of large scale and great age on the Earth's surface. *Nature* **271**, 345–349.

Sims P. K. (1976) Precambrian tectonics and mineral deposits, Lake Superior Region. *Econ. Geol.* **71**, 1092-1127.

Schulz K. (1977) The petrology and geochemistry of archean volcanics, western Vermilion District, northeastern Minnesota. Ph.D. thesis, Univ. Minnesota. 349 pp.

Solomon S. C. and Chaiken J. (1976) Thermal expansion and thermal stress in the moon and terrestrial planets: Clues to early thermal history. *Proc. Lunar Sci. Conf. 7th*, p. 3229–3243.

Solomon S. C. and Head J. W. (1978) Vertical movement in mare basins: Relations to mare emplacement, basic tectonics, and lunar thermal history (abstract). In *Lunar and Planetary Science IX*, p. 1083–1085. Lunar and Planetary Institute, Houston.

Stewart J. H. (1976) Late Precambrian evolution of North America: Plate tectonics implication. *Geology* **4**, 11–15.

Sutton J. (1977) Some consequences of horizontal displacements in the Precambrian. *Tectonophys.* **40**, 161–181.

Tatsumoto M., Nunes P. D. and Unruh D. M. (1977) Early history of the Moon: Implications from U-Th-Pb and Rb-Sr systematics. In *The Soviet-American Conference on Cosmochemistry of the Moon and Planets*. p. 507–524. NASA SP-370, Washington, D. C.

Taylor P. N. (1975) An early Precambrian age for migmatitic gneisses Vikan I Bø Vesteralen, North Norway. *Earth Planet Sci. Lett.* **27**, 35–42.

Thomas G. E. (1978) Basement structures along the Appalachian-Ouachita continental margin (abstract) *3rd. Internat. Conf. on Basement Tectonics*, p. 49. Fort Lewis College, Durango, Colorado.

Toksöz M. N. and Johnston D. H. (1977) The evolution of the Moon and the terrestrial planets. In *The Soviet-American Conference on Cosmochemistry of the Moon and Planets*. p. 295–327. NASA SP-370, Washington, D.C.

Van Schmus W. R. (1976) Early and middle proterozoic history of the Great Lakes area, North America. *Phil. Trans. Roy. Soc. Lond.* **A280**, 605–628.

Venkatasubramanian V. S. (1974) Geochronology of the Dharwar Craton, a review. *J. Geol. India* **15**, 463–468.

Warner J. L. and Morrison D. A. (1978) Planetary tectonics I: the role of water (abstract). In *Lunar and Planetary Science IX*, p. 1217–1219. Lunar and Planetary Institute, Houston.

Warner J. L., Phinny W. C., Bickel C. E. and Simonds C. H. (1977) Feldspathic granulitic impactites and prefinal bombardment lunar evolution. *Proc. Lunar Sci. Conf. 8th*, p. 2051–2066.

Warner L. A. (1978) The Colorado lineament: a Middle Precambrian wrench fault system. *Bull. Geol. Soc. Amer.* **89**, 161–171.

Watson J. (1976) Vertical movement in Proterozoic structural provinces, *Phil. Trans. Roy. Soc. Lond.* **A280**, 629–640.

Wetherill G. W. (1977) Pre-mare cratering and early solar system history. In *The Soviet American Conf. on Cosmochemistry of Moon and Planets*. p. 553–567, NASA SP-370, Washington, D.C.

Wilson H. D. B. and Morrice M. G. (1977) The volcanic sequence in Archean Shields. In *Volcanic Regimes in Canada* (W. R. A. Baragar, L. C. Coleman and J. M. Hall, eds.), p. 355–374, *Geol. Assoc. Canada*, Univ. of Waterloo, Waterloo, Ontario.

Windley B. F. (1976) *The Early History of the Earth*. Wiley, N.Y., 619 pp.

Windley B. F. (1977) *The Evolving Continents*. Wiley, N.Y., 385 pp.

Wood J. A. (1977) A survey of lunar rock types and comparison of the crusts of earth and moon. In *The Soviet-American Conference on Cosmochemistry of the Moon and Planets*. p. 35–54. NASA SP-370, Washington, D.C.

Wood J. A., Dickey J. S., Marvin U. B. and Powell B. N. (1970) Lunar anorthosites and a geophysical model of the moon. *Proc. Apollo 11 Lunar Sci. Conf.*, p. 965–988.

Zamarayev S. M. and Ruzhich V. V. (1978) On relationships between the Baikal rift and ancient structures. *Tectonophys.* **45**, 41–48.

Proc. Lunar Planet. Sci. Conf. 9th (1978), p. 2773–2787.
Printed in the United States of America

X-ray diffractometer studies of shocked materials

R. E. HANSS, B. R. MONTAGUE, M. K. DAVIS, C. GALINDO
St. Mary's University, San Antonio, Texas 78284

FRIEDRICH HÖRZ
NASA Johnson Space Center, Houston, Texas 77058

Abstract—This study explores the utility of X-ray diffractometer scans for determination of shock pressure histories of geological materials from meteorite impact sites. The technique is based on quantification of increased crystal lattice disorder with increasing shock pressure as expressed by decreasing diffraction peak amplitude and pronounced line-broadening caused by decreasing mosaic domain size and increasing strain. The ratio of peak height (PH) to half width (HW) decreases systematically with increasing shock pressure. Data are given for experimentally shocked quartz, feldspars, pyroxene and olivine along with data for granitic materials from the Piledriver nuclear event and from the Ries Crater, Germany. Although the technique in principle may be capable of yielding relatively accurate pressure determinations, its application to naturally shocked materials may be severely limited.

INTRODUCTION

A variety of techniques exist to determine the solid state shock effects in geologic materials associated with natural impact events. Most commonly petrographic criteria such as specific deformation features and/or phase transitions are employed (e.g.; Stöffler, 1972; Robertson and Grieve, 1977; Schaal and Hörz, 1977). The loss of noble gases and structural-metallurgical features form the basis for inferred shock histories of meteorites (Heymann, 1967; Taylor and Heymann, 1969). Lipschutz (1968) suggested the usefulness of X-ray diffraction techniques. Jeanloz (1977) demonstrated that the degree of electron damage in shocked plagioclases depends on peak shock pressure. The application of SEM and TEM techniques to shocked materials is becoming increasingly more useful (e.g., Nord *et al.*, 1975; Kieffer *et al.*, 1976). Accordingly, a variety of techniques exist to address various aspects of a rock's shock history, especially if supported by controlled shock experiments for accurate peak pressure calibrations.

However, no single technique exists which accurately determines the total amount of shock energy deposited into a given rock. Even a cursory survey under the optical microscope indicates that solid state shock effects are distributed heterogeneously throughout a thin section, e.g., diaplectic feldspar glasses (maskelynite) may coexist with crystalline feldspar, or a quartz grain with numerous sets of planar deformation features may coexist with undeformed quartz grains. Such heterogeneous distribution of shock effects is caused by rarefaction and reflection of the shock wave at grain boundaries of variable acoustical impedance (e.g., Kieffer, 1971).

Because the above techniques generally rely on the characterization of a selected number of component grains and/or even one specific mineral species only, they do not quantitatively address the shock pressure history of the entire phase assemblage; shock pressures are largely inferred for the entire specimen based on the selected number of grains studied. While this is permissible and extremely useful to establish relative shock histories, a statistically valid approach is necessary for more quantitative information. Such an approach must characterize a large number of component grains. While this is possible in principle for most of the above techniques, it is in reality so time consuming that it becomes impractical.

The necessity for a technique which is statistically more valid is illustrated in Fig. 1. This graph presents the results of Debye-Scherrer studies of a total of 205 individual quartz and feldspar grains dislodged from ≈ 1 cm^3 volume of a naturally shocked granite from the Ries Crater, Germany (kindly provided by D. Stöffler). Using the Debye-Scherrer criteria developed by Hörz and Quaide (1972), each grain was assigned a specific peak pressure; these pressures are plotted in cumulative fashion (Fig. 1). It can readily be seen that a rather wide distribution of shock damage is contained even in a volume as small as 1 cm^3 from the 26 km diameter Ries Crater. This attests to extremely localized stress environments on the scales of individual grains and grain boundaries.

The basic criteria used in X-ray techniques for assignment of peak pressures rest on the gradual breakdown of the crystal lattice into smaller and smaller mosaic domains; such breakdown may proceed to domain-sizes beyond the limits of coherent X-ray diffraction as detailed by Hörz and Quaide (1972). The major effects controlling "single grain" X-ray patterns (e.g., Debye-Scherrer) will, therefore, be internal strain and intergranular domain size distribution resulting in line broadening, streaking arcs, etc.

In this paper we investigate the shock induced lattice breakdown via X-ray diffractometer techniques. The major objective is to evaluate the usefulness of the diffractometer method as a rapid technique for characterizing the shock histories of a statistically significant number of component grains. We use naturally and experimentally shocked single crystal and polycrystalline materials.

Experimental Methods

1. *Shocked materials*

Single crystal quartz (Arkansas), oligoclase (Muskwa Lake, Canada), enstatite (Bamle, Norway) and orthoclase (Itirongai, Madagascar) along with granodiorite (Climax Stock, Nevada) and dunite (Twin Sisters, Washington) were cored and sliced into discs 6 mm across and 0.5 mm thick. These were shock loaded in the JSC 20 mm Flat Plate Accelerator according to the procedures described by Gibbons *et al.* (1975). The metal capsules that contained the cylindrical target discs were carefully machined open upon recovery and the shocked silicates were removed. The shocked samples range

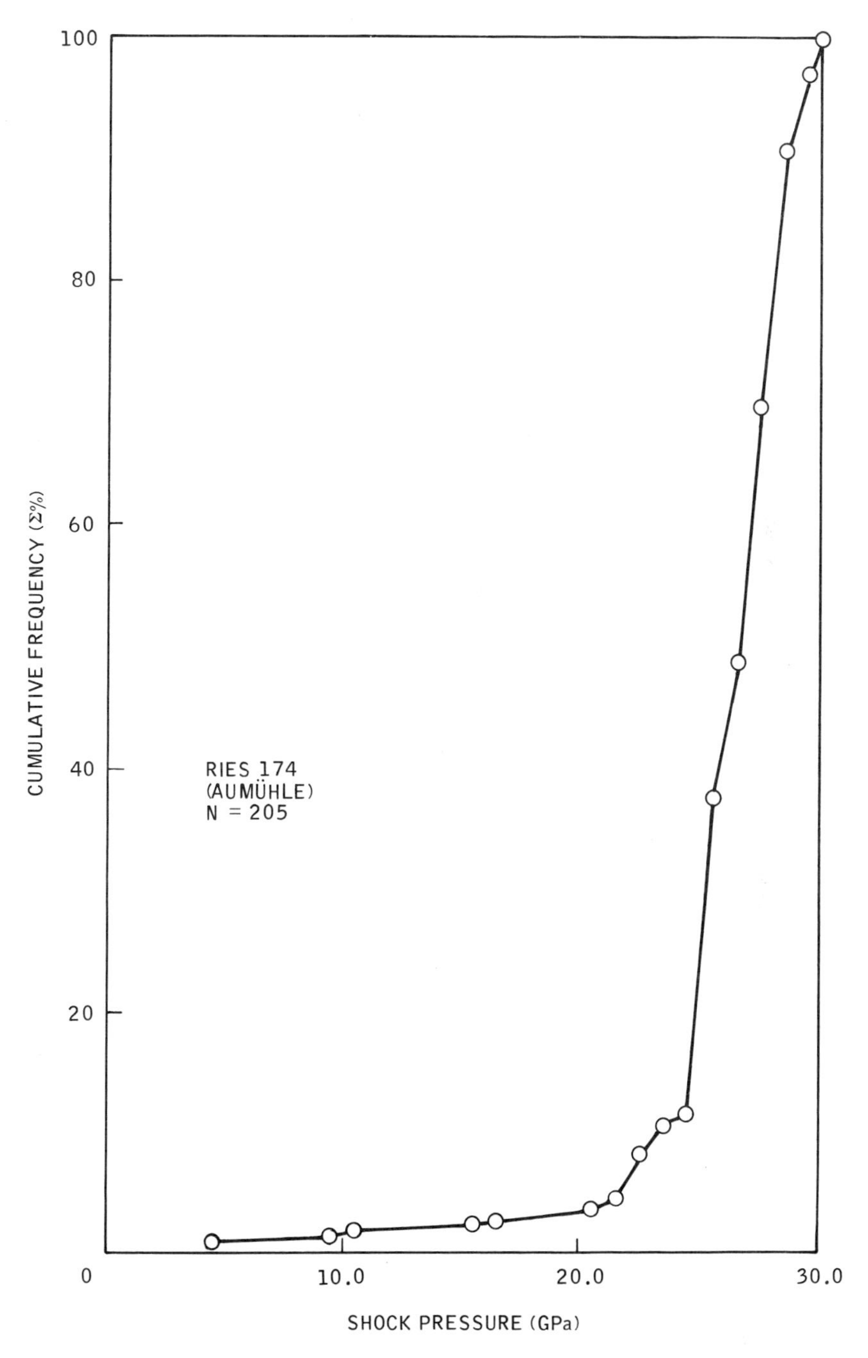

Fig. 1. Distribution of peak shock pressures of individual quartz and feldspar grains dislodged from ≃1 cm^3 volume of a naturally shocked Ries granite (summary of 205 individual Debye-Scherrer strips).

from fractured, coherent discs to loose friable masses. Furthermore, a series of naturally shocked granites from the Ries Crater, Germany was also studied.

The above suite of materials, therefore, allows us to proceed from controlled laboratory shock experiments on single crystals and polycrystalline materials to rocks shocked in large scale natural cratering events.

2. *X-ray diffractometry*

Because absolute grain size is a crucial factor that affects reproducibility of diffraction intensity measurements, care was taken to produce powders of consistent grain size for all samples. Klug and Alexander (1954) note that the mean deviation of integrated intensity for ten samples of a 15 to 50 μm fraction of quartz is approximately 18% under their sample preparation conditions. Our samples were ground to $\leq$44 μm; deviations from the mean peak intensity were $\approx$10–15%. A comparison of three size fractions 37 μm, 37 to 44 μm, 44 to $\sim$100 μm) for unshocked quartz and for oligoclase yielded absolute peak intensity differences of no more than 10%. In order to obviate extreme fineness ($<$1 μm) that might contribute to line broadening, samples were not ground longer than necessary to pass through the 44 μm screen. A fixed amount of each sample (usually 5 mg) was mixed with 5 μm Al_2O_3 or CaF_2 in a definite mass ratio selected for optimum peak height of the internal standard. This sample mixture was then placed inside a 10 mm (ID) washer, set on a suitable substrate (glass, aluminum or lead). The powder was then stirred with a few drops of ethanol and left to dry before the washer was removed. This leaves a circular sample area of constant surface dimension and uniform sample thickness, ensuring identical sample geometries.

The actual X-ray scans were performed under variable conditions (voltage, amperage, count rates) selected individually for each suite of materials (see below); however, conditions were kept identical during analysis of a specific suite. Slow scan rates of 0.2 to 0.5°/min were employed for high spectral resolution.

3. *Data reduction*

The diffractometer scans were analyzed in the following way: The peak height (PH) of the major diffractometer peaks was measured along with the width of the peak at ½ PH, i.e., "half width" (HW). Both amplitude (PH) and half width (HW) contain information related to total diffracting volume and grain size distribution. Therefore, the ratio of PH/HW may be taken as a measure of shock-induced lattice breakdown, provided all other parameters are kept constant. The use of an internal standard, such as Al_2O_3 or CaF_2, aims to ascertain the constancy of sample volume, grain orientation and operating conditions. PH/HW ratios were normalized to a constant internal standard intensity value. Most PH/HW ratios reported here represent a mean of three sample preparations. In some cases enough sample material was recovered to prepare only one slide; this sample was re-dispersed on a new substrate and a second diffraction scan was made. A comparison of different samples and different X-ray operating conditions for quartz are presented by Hanss *et al.* (1977a).

RESULTS

The basic diffractometer data are illustrated in Figs. 2 through 8. It can be seen that the diffractometer peaks decrease in amplitude and increase in linewidth with increasing peak shock pressure. However, normalization of PH/HW ratios according to internal standard intensities is necessary to visualize this decrease in some sample data (e.g., oligoclase). Furthermore, in agreement with Debye-Scherrer studies, the higher angle reflections tend to fade out at relatively modest pressure (see Fig. 2). Following the interpretations of Hörz and Quaide (1972) and Hanss *et al.* (1977a), we conclude that there is a continuous

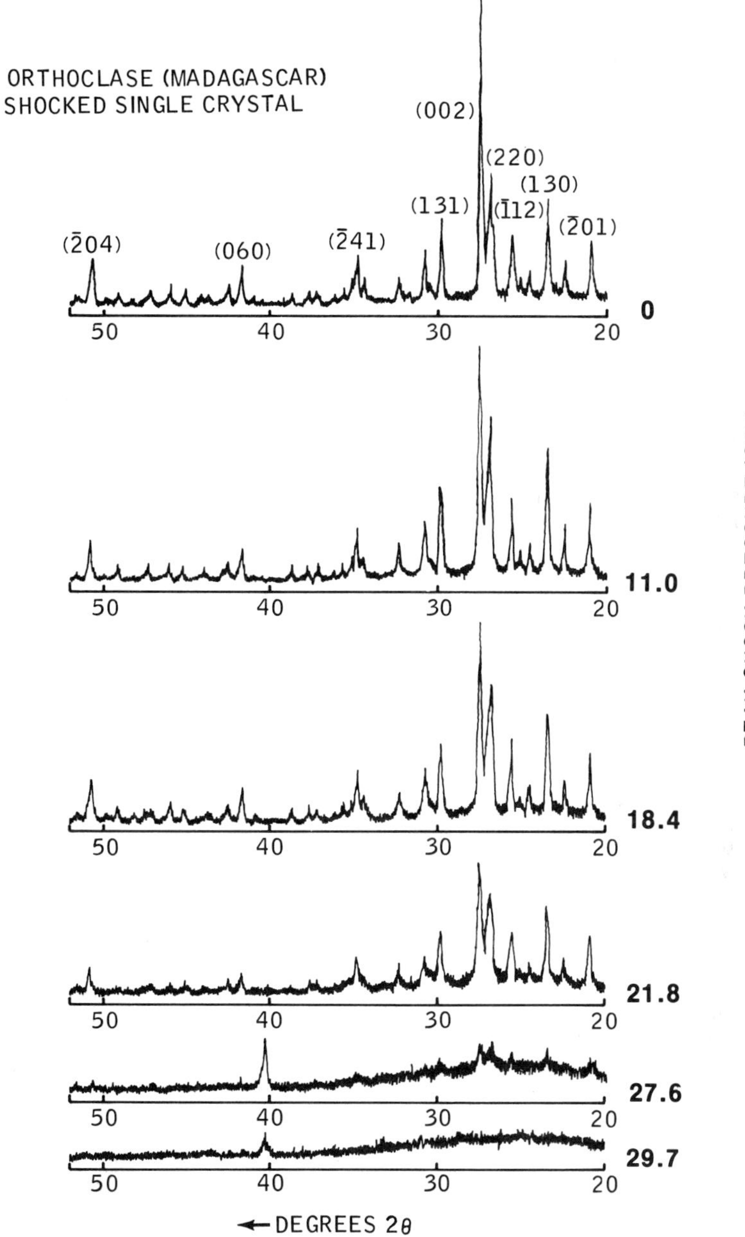

Fig. 2. Rapid diffractometer scans ($1°2\theta$/min) of experimentally shocked orthoclase from Itirongai, Madagascar. Scans were made over a large range in 2θ to illustrate general behavior of X-ray patterns. Anomalous 40° peak is from tungsten projectile (CuKα, monochromatized, 40 Kvp, 20 mA).

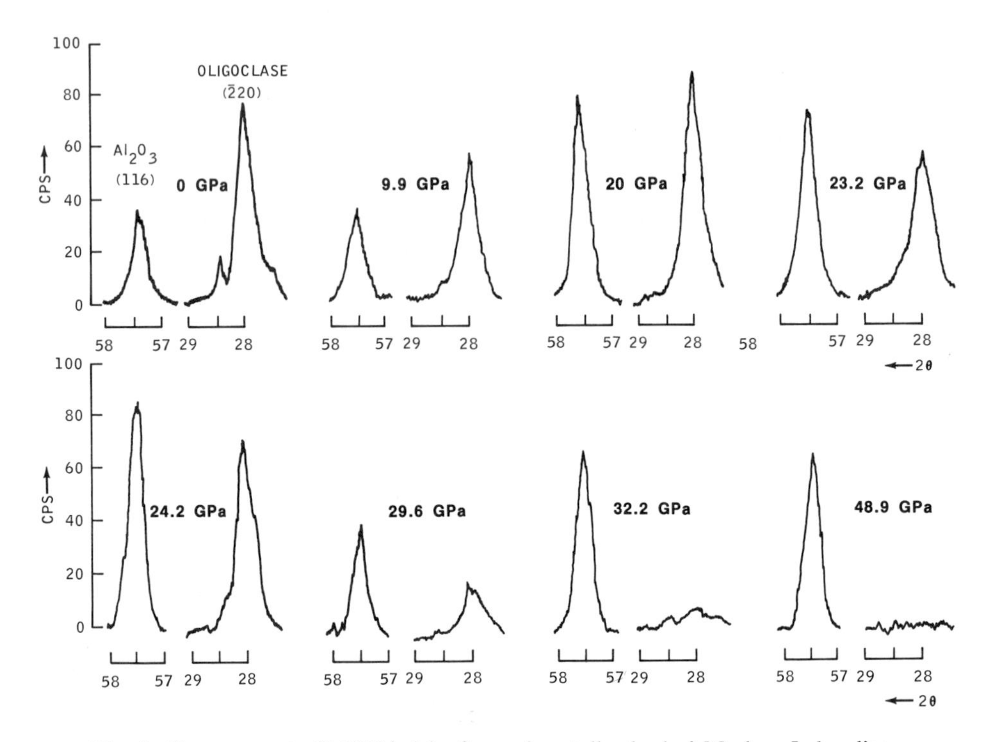

Fig. 3. Slow scan rate (0.2°2θ/min) of experimentally shocked Muskwa Lake oligoclase and Al_2O_3 internal standard (CuKα, monochromatized, 50 Kvp, 11 mA).

decrease in median domain size and possibly an increase in internal strain. The progressively decreasing domain size causes the coherently diffracting volume to decrease until a certain threshold pressure is reached beyond which coherent X-ray diffraction is precluded, i.e., where diaplectic phases (e.g., in quartz and feldspar) have formed. The observed line broadening may be due both to decreasing domain size, leading to increase in X-ray scatter, and to increasing internal strain, leading to genuine, small variations in the d-spacing which diffract about the mean value. Such lattice changes in shocked quartz are described by Schneider and Hornemann (1976).

As described above, we may take the PH/HW ratio of specific diffractometer peaks as a measure for lattice breakdown. If plotted against peak pressure for the experimentally shocked materials, "shock disorder" curves as illustrated in Fig. 9 may be obtained. Actual data points and corresponding least square fit lines are illustrated. Figure 9d shows data for quartz from experimentally shocked Climax Stock granodiorite: the quartz data from this polycrystalline target are in good agreement with the single crystal quartz experiments (Figs. 4; 9c). The agreement between the Climax Stock feldspars (Borg, 1972) and Muskwa Lake oligoclase is fair (not illustrated), attesting to their different resistance to shock

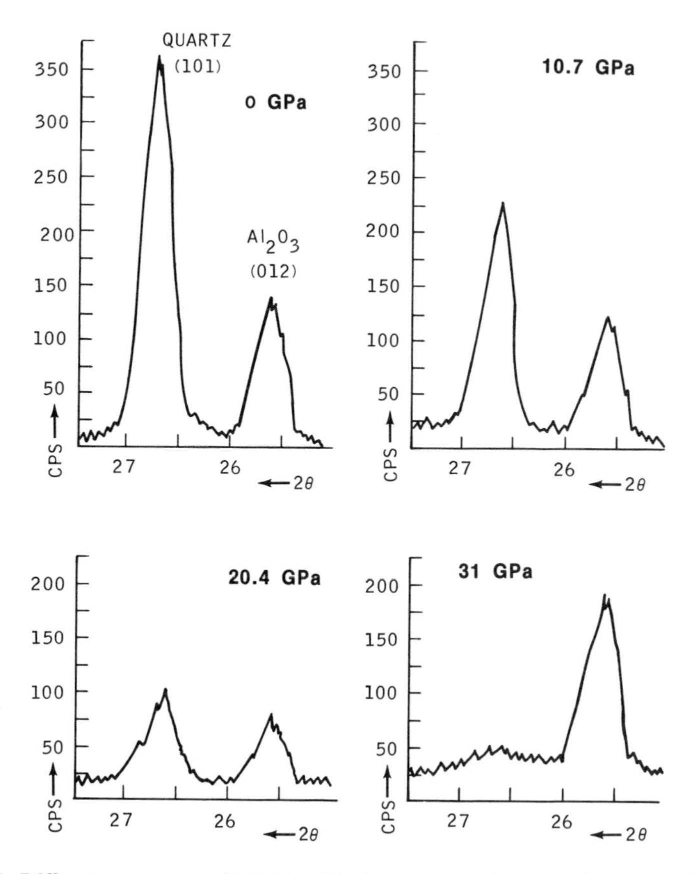

Fig. 4. Diffractometer scans (0.2°2θ/min) of experimentally shocked quartz and Al_2O_3 internal standard (CuKα, monochromatized, 50 Kvp, 5 mA).

disordering. However, the important finding is that component grains in polycrystalline targets appear to suffer the same amount of shock damage as their single crystal equivalents. The two data sets are internally self consistent which implies that our shock disorder curves (Fig. 9) may, in principle, be applicable in evaluating the shock history of polycrystalline rocks. Thus, if experimentally shocked materials or component phases are available as "standards", a relatively accurate (±10%) pressure calibration for the entire rock appears possible.

As illustrated in Fig. 8, the naturally shocked Ries granites display identical diffraction characteristics as the experimentally shock loaded targets. It is relatively easy to order the "unknown" pressure histories in a *relative* sequence. However, since we did not have an unshocked standard and because the mode of these granitic rocks varies slightly from specimen to specimen, precise absolute peak pressures cannot be estimated with great confidence. However, all specimens had grain sizes >0.5 mm, and thus, the lattice breakdown observed in

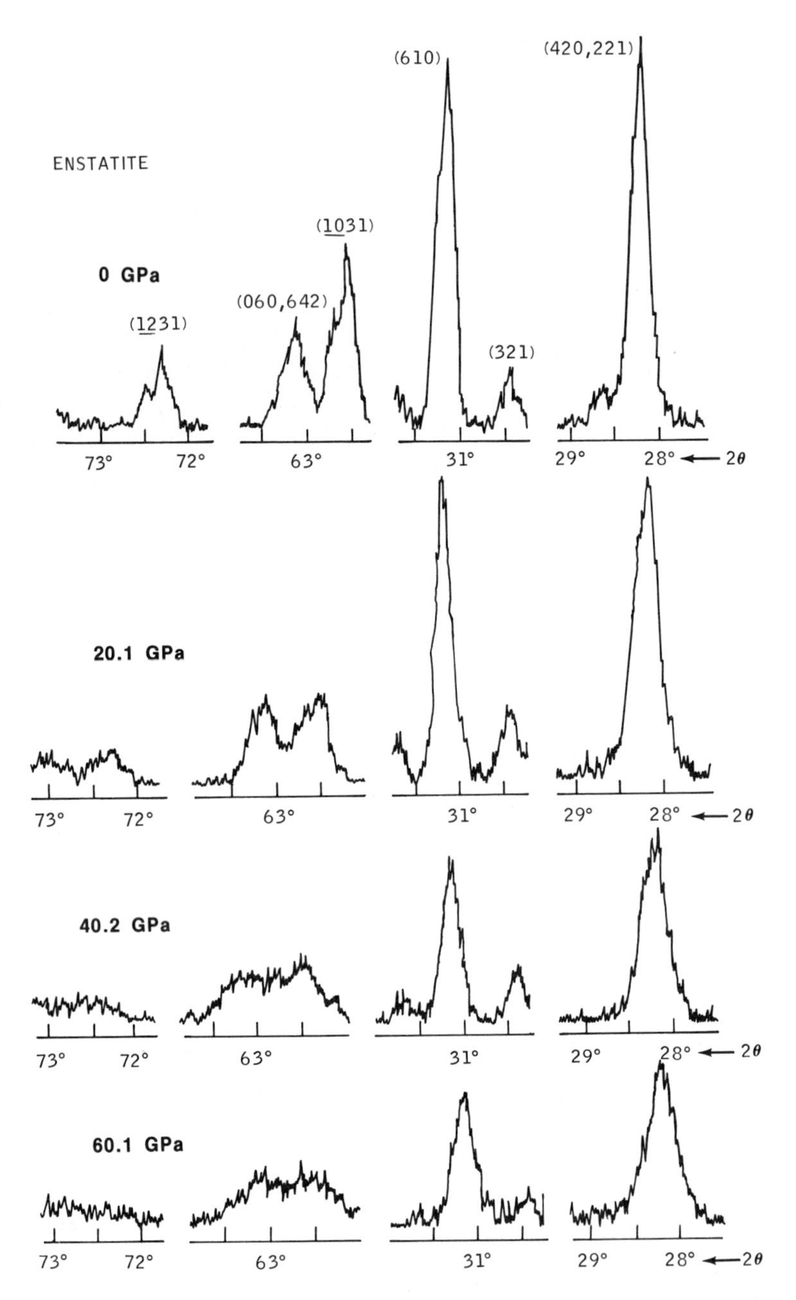

Fig. 5. Diffractometer scans (0.2°2θ/min) of experimentally shocked enstatite from Bamble, Norway. Note rapid decline of higher-angle peaks (CuKα, monochromatized, 50 Kvp, 5 mA).

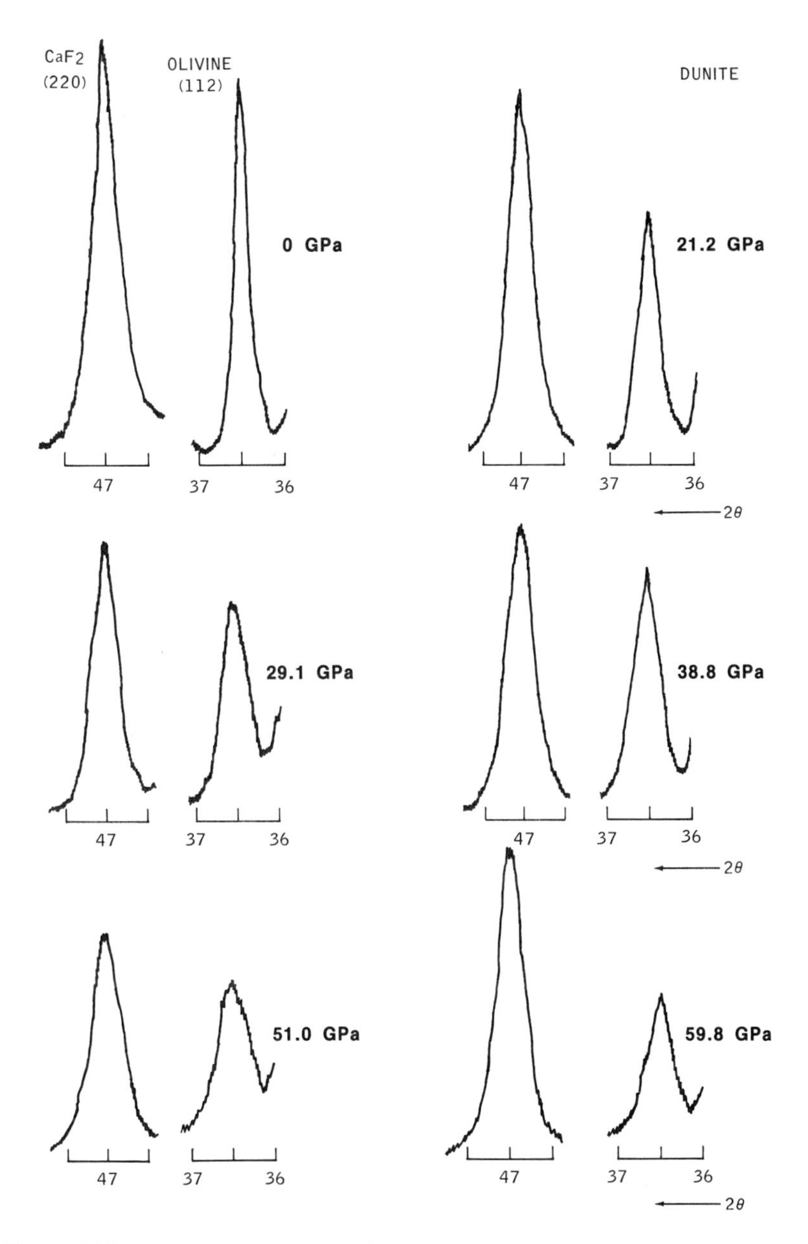

Fig. 6. Diffractometer scans (0.2°2θ/min) of experimentally shocked dunite from Twin Sisters, Washington, and CaF_2 internal standard (CuKα, monochromatized, 50 Kvp, 5 mA).

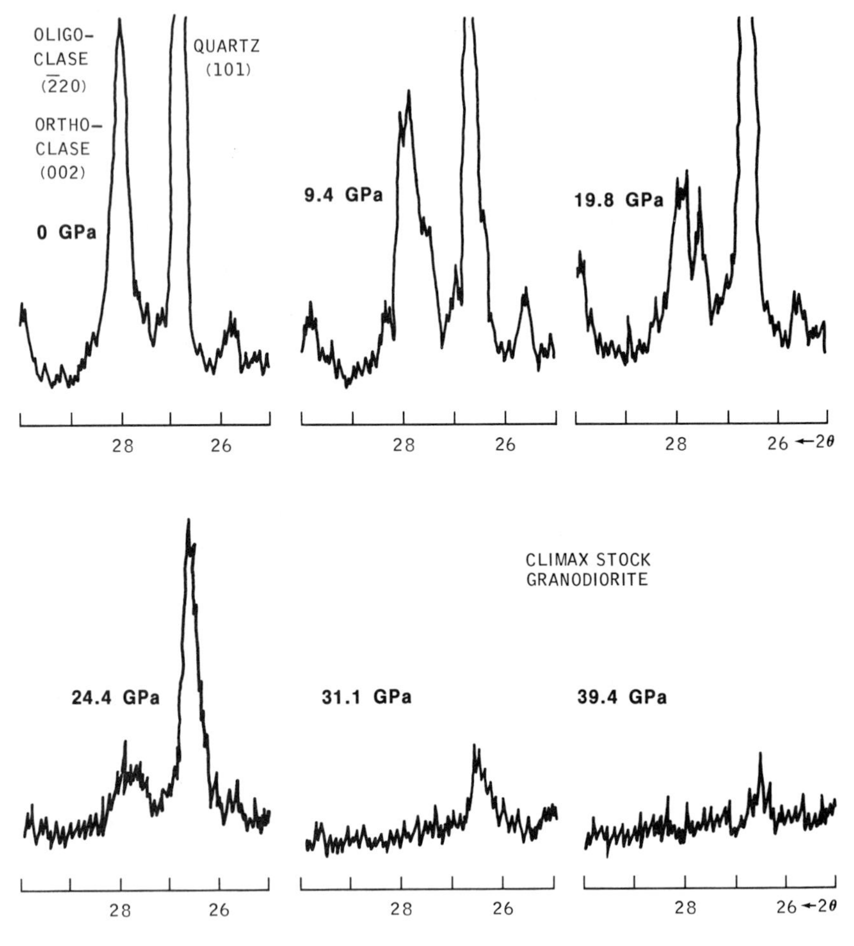

Fig. 7. Diffractometer scans (0.4°2θ/min) of experimentally shocked granodiorite from the Climax Stock, Nevada (CuKα, monochromatized, 40 Kvp, 20 mA).

Fig. 8 is unquestionably caused by shock. We estimate the peak pressures experienced by these samples by directly applying the single crystal quartz data of Figs. 4 and 9c, respectively. The shock pressures obtained are as follows: R 604; 9.4 GPa; R 352; 22.0 GPa; R 174; 28.3 GPa. The utility of the feldspar diffractions is extremely limited both because the feldspar mode and its composition were variable for the Ries sample. Therefore, the most reliable pressure estimates are based on the behavior of quartz only.

The three Ries granites were also used to test the sensitivity of the diffractometer technique with regard to variable X-ray excitation voltage. Three runs at 40, 45 and 50 Kvp were performed for each sample as well as an artificial plagioclase/quartz mixture (66%/33%). The results are summarized in Table 1.

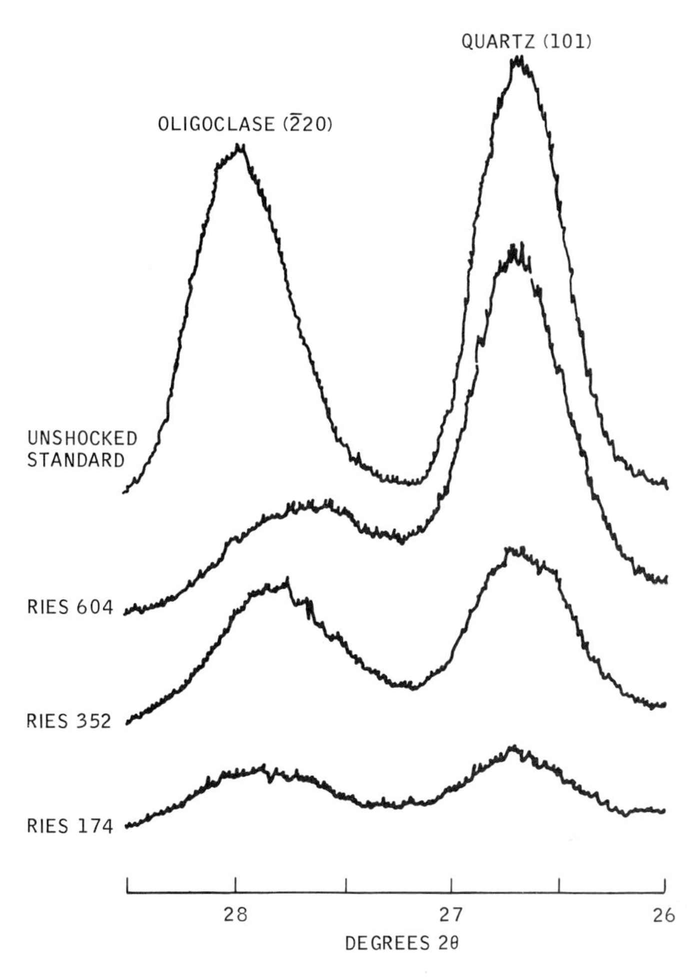

Fig. 8. Diffractometer scans (0.2°2θ/min) of three naturally shocked Ries granites and unshocked standard (CuKα, monochromatized, 50 Kvp, 5 mA).

The resulting pressure difference may be as much as 2 GPa illustrating that a standard set of conditions must be applied rigorously to any given suite of materials and associated internal standards.

Comparison Between Optical, Debye-Scherrer and Diffractometer Pressure Calibrations

We have reported previously on data presented in Fig. 1 obtained via Debye-Scherrer rotating grain techniques (Hanss *et al.*, 1977b; Hanss *et al.*, 1978). The data indicate that the distribution of peak pressures experienced by individual grains is highly heterogeneous. If plotted in a cumulative fashion (see

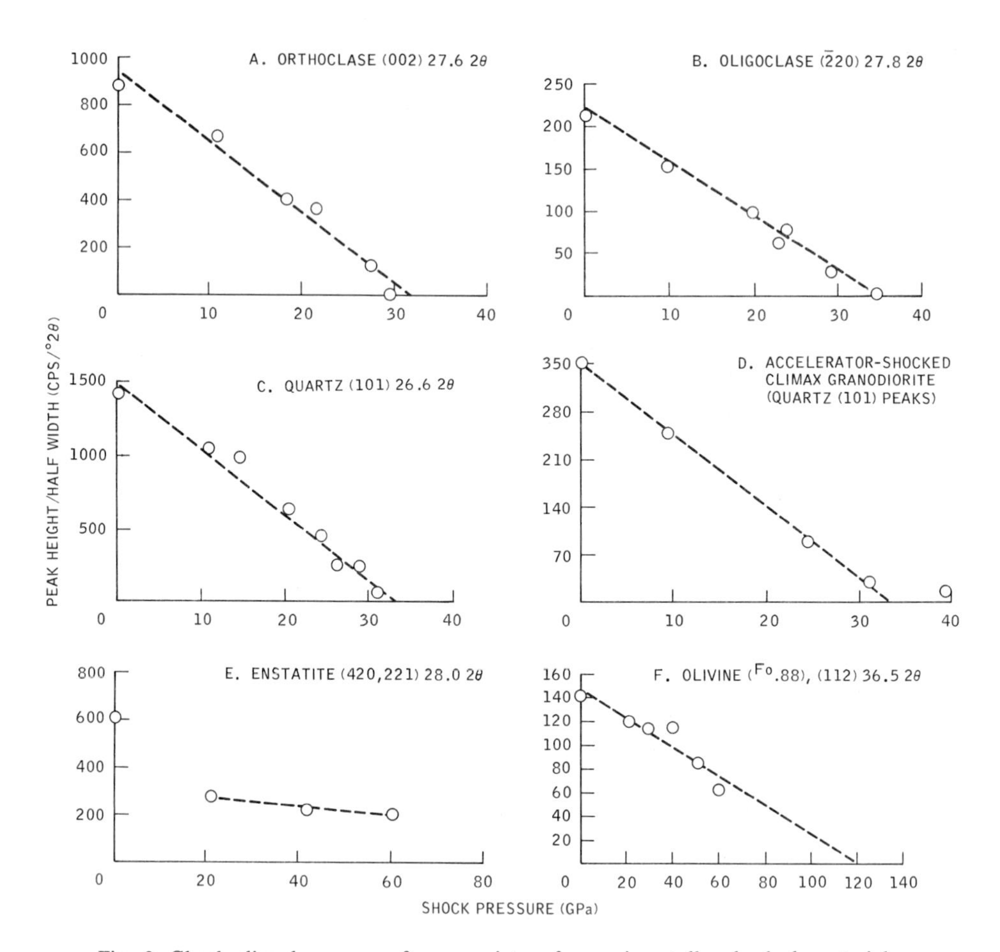

Fig. 9. Shock disorder curves for a variety of experimentally shocked materials. (A) Single crystal orthoclase. (B) Single crystal oligoclase. (C) Single crystal quartz. (D) Quartz in Climax Stock granodiorite. (E) Single crystal enstatite. (F) Olivine in Twin Sisters dunite.

Fig. 1) a median pressure for the entire rock may be obtained. These median pressures (DS) will be compared to optically obtained pressures (Stöffler, 1972). The Ries samples investigated were characterized by D. Stöffler and represented types I and II respectively on his scale of shock metamorphic classes. These pressure determinations along with the newly obtained diffractometer data are summarized in Table 2.

In the case of the three Ries samples the diffractometer data appear to compare favorably with the pressure ranges obtained by optical studies, although the new diffractometer pressures hopefully provide a more refined estimate. The discrepancy between the median DS pressure and the diffractometer value for the Ries 352 sample, however, is alarmingly large and not well understood at present. The most plausible explanation is that we failed to obtain suitable "0

Table 1. Effects of various X-ray conditions on PH/HW ratios of naturally shocked Ries granites (the unshocked "standard" is a mixture of 2 parts oligoclase and 1 part quartz). X-ray conditions: (1) 40KV/5mA (2) 45KV/5mA (3) 50KV/5mA.

QUARTZ Sample	PH/HW			Percentage			Estimated GPa			Mean GPa
	(1)	(2)	(3)	(1)	(2)	(3)	(1)	(2)	(3)	
Unshocked standard	821	974	1181	100	100	100	0	0	0	0
Ries 604	585	670	856	71	69	72	9.3	9.9	9.0	9.4
Ries 352	238	347	404	29	36	34	23.3	21.0	21.6	22.0
Ries 174	111	124	174	14	13	13	27.8	28.6	28.6	28.3

Table 2. Comparison of various pressure calibration techniques for three Ries samples.

Sample	Shock Pressure (GPa)				
	A	B	R_B	C	R_C
Ries 604	9.4	9	5–20	I	10–30
Ries 352	22.0	13	5–25	II	25–55
Ries 174	28.3	26	18–27	II	25–55

A Diffractometer technique.
B Debye-Scherrer technique (R_B = range observed).
C Optical classification (Stöffler, 1971) and associated range of pressures.

pressure" calibration for the Ries quartz, i.e., constant grain size distribution and mode of quartz in the unshocked parent material. The modal compositions of the Ries granites are variable, and optical investigation indicated quartz contents between 30–50%. At present we tend to rely more on the DS pressures, because they are independent of mode, albeit strongly dependent on original, possibly submicroscopic, "grain size." Thus, extreme care is required if absolute pressure data are to be obtained from natural materials without having unshocked parent materials for comparison.

However, in the case of Climax Stock granodiorite, both X-ray techniques yield internally consistent data for experimentally shocked samples (see Table 3). As reported previously, there is a systematic difference between *in situ* pressure determinations and our X-ray calibrations for this material shocked in the Piledriver nuclear event (Hanss *et al.*, 1977b). However, the internal consistency of the X-ray techniques is taken as evidence that the diffractometer technique alone can in principle provide a rapid means of estimating shock pressure histories if unshocked materials are available for "0 pressure" calibration.

Table 3. Comparison of various pressure calibration techniques for six experimentally shocked Climax Stock granodiorite samples.

Sample	Shock Pressure (GPa)			
	A	B	R_B	C
CS-1	0	0	0	0
CS-2	9.5	10.0	9.0–11.0	9.4
CS-3	7.0	20.0	14.0–25.0	19.8
CS-4	25.0	25.0	23.0–32.0	24.4
CS-5	30.5	30.5	29.0–32.0	31.1
CS-6	>32.0	>32.0	>32.0–?	39.4

A Diffractometer technique.
B Debye-Scherrer technique (R_B = range observed).
C Experimental pressure (based on projectile velocity and equation of state).

Since both X-ray techniques, in a strict sense, yield information only on domain size distribution and internal strain and the diffractometer technique is, furthermore, extremely sensitive to total coherently diffracting volume; both techniques can potentially yield reliable results only if unshocked parent material is available for comparison. Without such a comparison, the utility of diffractometer technique appears limited and requires further work. For the same reasons, and as already pointed out by Hörz and Quaide (1972), the X-ray techniques also will fail in naturally shocked samples which were annealed or recrystallized subsequent to the impact event, i.e., where original shock induced domain sizes and internal strains have been modified.

Acknowledgments—T. D. Thompson and R. B. Schaal assisted in preparing the shocked samples at the Johnson Space Center flat-plate accelerator; J. E. Wainwright offered many valuable discussions of X-ray theory and technique. This work was performed under NASA Grant NSG 9039.

REFERENCES

Borg I. Y. (1972) Some shock effects in granodiorite to 270 kb at the Piledriver site. *Amer. Geophys. Union Monograph 16*, p. 293–211.

Gibbons R. V., Morris R. V., Hörz F., and Thompson T. D. (1975) Petrographic and ferromagnetic resonance studies of experimentally shocked regolith analogs. *Proc. Lunar Sci. Conf. 6th*, p. 3143–3171.

Hanss R. E., Montague B. R., and Galindo C. (1977a) X-ray diffraction studies of shocked lunar analogs. *Advances in X-Ray Analysis* **20**, 337–344.

Hanss R. E., Montague B. R., Galindo C. and Hörz F. (1977b) X-ray diffraction studies of shocked materials (abstract). In *Lunar Science VIII*, p. 395–397. The Lunar Science Institute, Houston.

Hanss R. E., Montague B. R., Davis M. K., Galindo C., and Hörz F. (1978) Pressure distribution in naturally and experimentally shocked granodiorites (abstract). In *Lunar and Planetary Science IX*, p. 451–453. Lunar and Planetary Institute, Houston.

Heymann D. (1967) On the origin of hypersthene chondrites: Ages and shock effects of black chondrites. *Icarus* **6**, 189.

Hörz F. and Quaide W. L. (1972) Debye-Scherrer investigations of experimentally shocked silicates. *The Moon* **6**, 45–82.

Jeanloz R. (1977) Electron damage: A new analytical technique applied to plagioclase in shocked chondrites and basalt (abstract). In *Lunar Science VIII*, p. 508–510. The Lunar Science Institute, Houston.

Kieffer S. W. (1971) Shock metamorphism of Coconino sandstone at Meteor Crater, Arizona. *J. Geophys. Res.* **76**, 5449-5473.

Kieffer S. W., Phakey P. P. and Christie J. M. (1976) Shock processes in porous quartzite: Transmission electron microscope observations and theory. *Contrib. Mineral. Petrol.* **59**, p. 41–53.

Klug H. P. and Alexander L. E. (1954) *X-ray Diffraction Procedures*. Wiley, N.Y. 716pp.

Lipschutz M. E. (1968) Shock effects in iron meteorites: A review. In *Shock Metamorphism of Natural Materials* (B. M. French and N. M. Short, eds.), p. 571–583. Mono, Baltimore.

Nord G. L., Christie J. M., Heuer A. H. and Lang J. S. (1975) North Ray Crater breccias: An electron petrographic study. *Proc. Lunar Sci. Conf. 6th*, p. 779–797.

Robertson P. B. and Grieve R. A. F. (1977) Shock attenuation at terrestrial impact structures. In *Impact and Explosion Cratering* (D. J. Roddy, R. O. Pepin, and R. B. Merrill, eds.), p. 687–702. Pergamon, N.Y.

Schaal R. B. and Hörz F. (1977) Shock metamorphism of lunar and terrestrial basalts. *Proc. Lunar Sci. Conf. 8th*, p. 1697–1729.

Schneider H. and Hornemann U. (1976) X-ray investigations on the deformation of experimentally shock-loaded quartz. *Contrib. Mineral. Petrol.* **55**, 205–215.

Stöffler D. (1972) Deformation and transformation of rock forming minerals by natural and experimental shock processes I: Behavior of minerals under shock compression. *Fortschr. Mineral.* **49**, 50–113.

Taylor G. Y. and Heymann D. (1969) Shock, reheating and gas retention ages of chondrites. *Earth Planet. Sci. Lett.* **7**, 151.

Proc. Lunar Planet. Sci. Conf. 9th (1978), p. 2789–2803.
Printed in the United States of America

The equation of state of a lunar anorthosite: 60025

RAYMOND JEANLOZ and THOMAS J. AHRENS

Seismological Laboratory (252-21), Division of Geological and Planetary Sciences
California Institute of Technology, Pasadena, California 91125

Abstract—New, shock-wave, equation-of-state measurements of lunar anorthosite 60025 (~18% initial porosity) and single crystal anorthite in the 40 to 120 GPa (0.4–1.2 Mbar) pressure range are presented and compared, along with previous results on nonporous anorthosite and lunar samples. The porous lunar anorthosite exhibits a lower shock impedance than nonporous anorthosite which, in turn, has a lower impedance than either nonporous gabbroic anorthosite (15418) or high-titanium, mare basalt (70215). This suggests that crater statistics (and, hence, apparent cratering ages) for different lunar terranes are biased by the properties of the different target rocks: for a given set of impacts, systematically smaller craters will tend to be formed in (nonporous) high-Ti mare basalt, gabbroic anorthosite and anorthosite, respectively. The effect of initial porosity in anorthosite 60025 is significant: for a given impact, peak stresses are distinctly lower in the porous anorthosite (typically by about 20% in the 40–100 GPa range) than in the nonporous equivalent, whereas both shock and post-shock temperatures are considerably higher. For example, shock temperatures (at a given pressure) differ by about 10^3 to 4×10^3 K over the range 40–100 GPa. Thus, maturing of a planetary surface by repeated impact (resulting in even mild brecciation and, hence, porosity) strongly enhances the thermal energy partitioning into the planet during meteoritic bombardment. The coupling of kinetic energy on impact is, however, decreased leading to a decrease in cratering efficiency due to increased porosity. These results imply that, as a result of successive impact events, the dynamic properties of a given rock unit as well as the entire planetary surface will evolve such that the efficiency in the trapping of thermal energy associated with impact will tend to increase with time.

INTRODUCTION

We present the first high-pressure equation-of-state data for calcic anorthosite, an important lunar rock-type. Our new data help to illustrate the range of dynamic (shock-wave) properties among the major lithologies found on the moon and, in particular, the profound effects of mechanical properties (e.g., porosity) on the response of target rocks to impact.

An anorthositic component is thought to make up some two-thirds of the lunar highland crust, as the result of early lunar differentiation (e.g., Adler *et al.*, 1973; Walker *et al.*, 1973a; Taylor and Jakeš, 1974, 1977; Taylor and Bence, 1975, 1978; *Bielefeld et al.*, 1977; Schonfeld, 1977). Indeed, anorthosite 60025, the subject of the present study, is a likely representative of the primitive lunar crust (as suggested, for example, by its low $^{87}Sr/^{86}Sr$ values: Papanastassiou and Wasserburg, 1972; see also, Taylor and Bence, 1975). It is, therefore, a particularly interesting rock to subject to shock-wave studies, in view of the current interest in modelling (and understanding) the early evolution of the moon and the response of its surface to meteoritic bombardment associated with the late stages of accretion. In fact, sample 60025 itself has apparently been affected by a rather ancient (basin-forming?) impact (about 4.2 AE ago, according to $^{40}Ar/^{39}Ar$ data: Schaefer and Husain, 1974), which has left a textural imprint of fracturing and mild brecciation.

Finally, we are interested in acquiring Hugoniot (shock-wave) equation-of-state data on Ca- and Al-bearing oxides and silicates (in the present case, anorthite): i.e., phases of early, high-temperature condensates from the primitive solar nebula (Grossman, 1972). Such data, in conjunction with component mixing models (e.g., Birch, 1952; McQueen, 1968; McQueen *et al.*, 1970; Al'tshuler and Sharipdzhanov, 1971), help to constrain physical models of the growth and evolution of planetary bodies, and may be of direct interest in the study of the deep interiors of terrestrial planets (e.g., Anderson, 1972, 1973; Taylor and Bence, 1975; Ruff and Anderson, 1977).

EXPERIMENTAL

Although most of the experimental procedures used in this study have been previously described (Jeanloz and Ahrens, 1977; Ahrens *et al.*, 1977; Jackson and Ahrens, unpublished material), details specific to this study, as well as changes and improvements in techniques, are discussed in this section.

Samples approximately 15 mm × 10 mm × 5 mm (0.9–1.2 gm) were cut from a block of specimen 60025.36, and ground and lapped into the shape of a rectangular solid to within dimensional tolerances of 0.1–0.3% (thickness was controlled to 0.04–0.17%, i.e., ±1–6 μm). After thorough

Table 1. Shock-wave, equation-of-state data for Lunar Anorthosite, 60025.

Shot Number	026	029	036	042
Initial Density-Bulk (Mg/m³)	2.234(±0.001)	2.244(±0.002)	2.199(±0.001)	2.239(±0.009)
Initial Porosity	18%	18%	20%	18½%
Impact Velocity (km/s)	5.762(±0.012)	4.467(±0.003)	5.668(±0.003)	4.971(±0.002)
Flyer/Driver Material	Ta	Ta	Ta	A1-2024
Hugoniot State				
Shock Wave Velocity (km/s)	9.17(±0.13)	7.73(±0.05)	9.15(±0.07)	6.75(±0.04)
Particle Velocity (km/s)	4.591(±0.015)	3.617(±0.004)	4.525(±0.006)	2.964(±0.009)
Pressure (GPa)	94.0(±1.1)	62.7(±0.3)	91.1(±0.6)	44.8(±0.2)
Density (Mg/m³)	4.473(±0.075)	4.216(±0.027)	4.351(±0.039)	3.990(±0.025)
Release State				
Particle Velocity (km/s)	3.682(±0.164)	3.355(±0.069)	4.417(±0.054)	2.783(±0.023)
Pressure (GPa)	56.4(±4.6)	47.5(±1.8)	79.0(±1.8)	33.9(±0.5)
Density (Mg/m³)	4.074(±0.096)	4.137(±0.039)	4.333(±0.039)	3.943(±0.025)

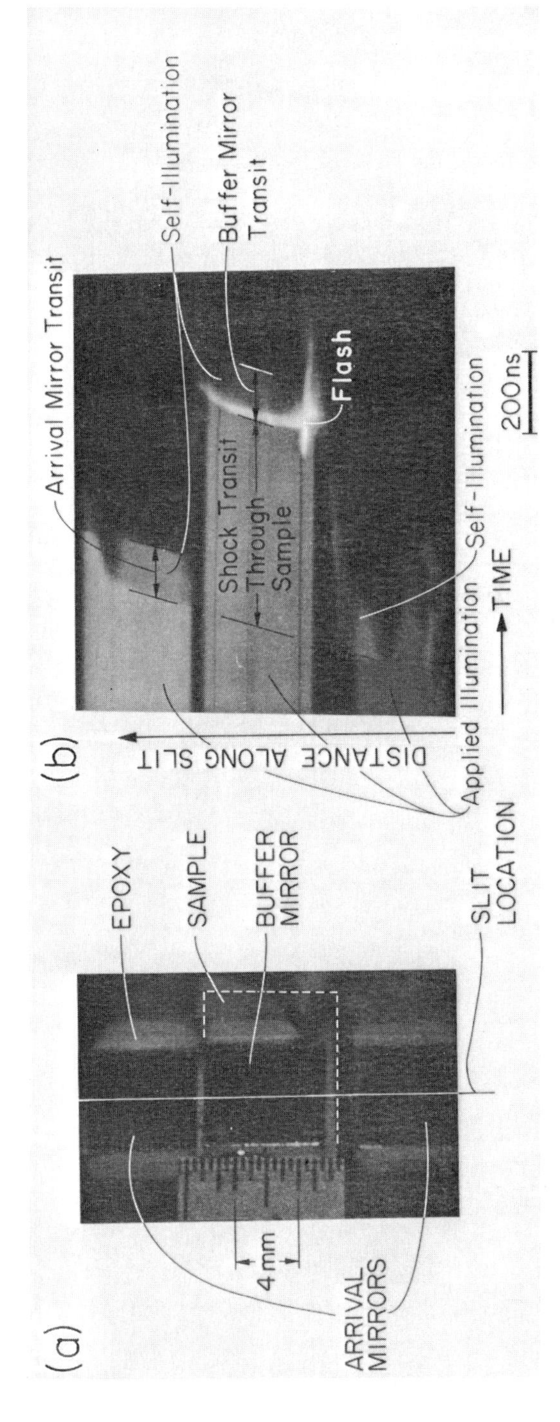

Fig. 1. (a) Still photograph taken through image converter camera, viewing the back-side of the target (#029): the sample and mirrors can be seen, as well as the location of the slit. (b) Streak photograph for shot #029 taken by sweeping the image of the slit towards the right (at nearly constant velocity) as the projectile impacted the target. Note the intense flash of light as the shock-wave crosses the sample/buffer mirror interface.

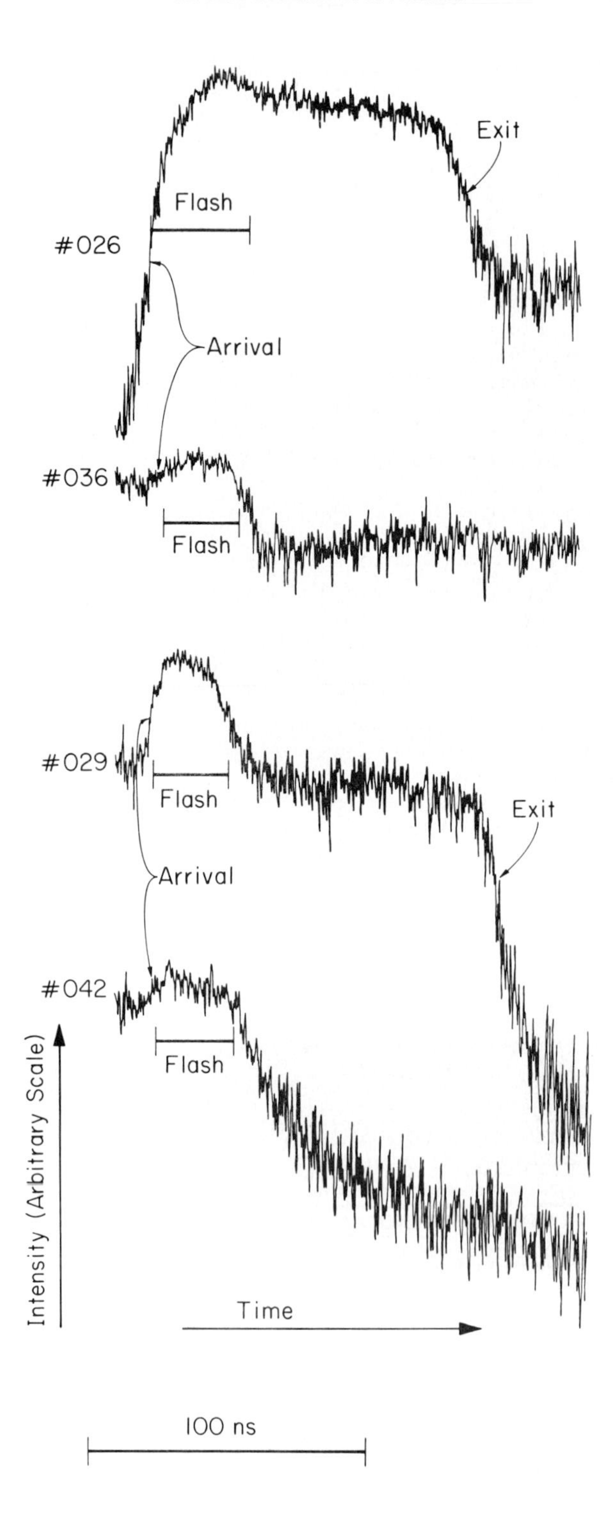
Exit
Flash
#026
Arrival
#036
Flash
#029
Flash
Exit
Arrival
#042
Flash
Intensity (Arbitrary Scale)
Time
100 ns

cleaning and washing in acetone, each sample was heated to 105°C in a partial vacuum ($<10^{-1}$ torr) for 12 hours or more, and then weighed. Values of the bulk density were calculated from the sample dimensions and mass (see Table 1); these are accurate to better than 0.5% (based on repeated measurements).

Archimedean ("intrinsic" or crystal) density was measured using toluene (cf., Ahrens *et al.*, 1977). Rather than placing the sample in a toluene bath as we have previously done, the sample was evacuated (to $<10^{-1}$ torr, at 20°C) inside a flask and then the toluene let in. In this way, vaporized toluene soaks into the sample very effectively, yielding highly reproducible density determinations. We tried to exclude pyroxene aggregates from our samples; however, despite these precautions, our values of crystal density may be low by as much as 1–2% (indicating less than complete saturation by the toluene) for a pyroxene content of up to 2–3% in our samples (cf., Dixon and Papike, 1975). For the samples listed in Table 1, crystal densities range between 2.709 and 2.751 (±0.003) Mg/m^3, using the correction factors of Berman (1939). These values are very consistent with the composition of 60025 (>98% plagioclase of composition An_{96} and density 2.75 Mg/m^3; Walker *et al.*, 1973b; Dixon and Papike, 1975; Hodges and Kushiro, 1973). Finally, the crystal and bulk densities of each sample were used to calculate the porosities given in Table 1 (listed to the nearest 1/2%, their probable accuracy).

The samples were mounted along with fused quartz arrival and buffer mirrors on pure tantalum or 2024 aluminum driver plates, and impacted by projectiles launched from a two-stage light-gas gun at velocities ranging between 4.5 and 5.8 km/s (see Jeanloz and Ahrens, 1977, and Ahrens *et al.*, 1977). The impacts were produced in a chamber evacuated to about 2×10^{-2} torr. Data, in the form of a streak record from an image-converter streak camera (Fig. 1), were read both visually and with a scanning microdensitometer (see Fig. 2). Boundaries were picked at 1/2 amplitude of the maximum gradient in light intensity, and the sharpness of the boundary was taken from the slope of the gradient in light intensity. The boundaries were fit by straight lines (all forced to be parallel for a given record) using a least-squares criterion weighted by the sharpness of each boundary crossing. These fits were then used in a statistical reduction of the shock and partial release data similar to that discussed by Jackson and Ahrens (unpublished material), using the Hugoniot equations-of-state of McQueen *et al.* (1970) for Ta and 2024 Al, and of Wackerle (1962) for fused SiO_2.

We have confirmed the reliability of using shock-induced self-illumination to measure shock-wave travel times through the mirrors on the target (cf., Jeanloz and Ahrens, 1977) by mounting small, supplementary arrival mirrors on the main mirrors (as described above) in several experiments (six to date) and comparing the extinction of applied illumination to the onset in self-illumination.

Results

The Hugoniot and release data for anorthosite 60025 are shown in three alternative representations in Figs. 3, 4, and 5, and are listed in Table 1. For comparison, data on nonporous plagioclase are also shown in these figures. These latter data represent a baseline against which the effects of porosity on the high-pressure equation-of-state of anorthite are dramatically evident.

Two datum points, the early results of a separate study which is underway on

Fig. 2. Microdensitometer scans taken from the streak records of experiments on 60025, showing the region of the flash (sample/buffer mirror interface) only. Intensity of light is shown as a function of time (10 μm spatial resolution on the film), at constant position along the slit (approximately at the center of the sample). Note the nearly constant flash duration. Arrival of the shock-wave into the buffer mirror and (in two cases) exit of the shock-wave from the back surface of the buffer mirror is also indicated. The noise in these traces is due wholly to the grain of the film.

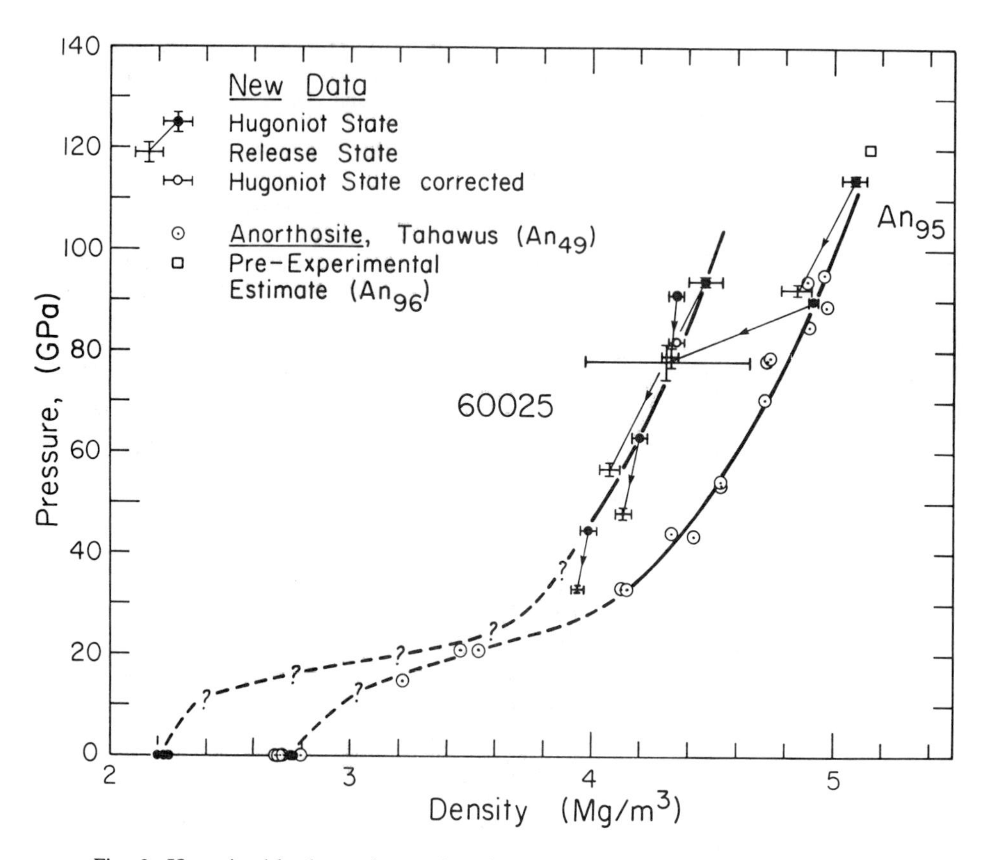

Fig. 3. Hugoniot (shock-wave) equation-of-state data for sample 60025 (including the datum from shot #036 at 91 GPa, corrected for extra porosity: see text) compared with single crystal anorthite (labelled An_{95}; source: Miyake-zima, Izu Islands, JAPAN), a terrestrial (intermediate plagioclase) anorthosite (from Tahawus, NY: McQueen *et al.*, 1967), and a theoretical estimate at 120 GPa (cf., Ahrens *et al.*, 1977), in the pressure-density plane. Note the steep release adiabat slopes for the three, low-pressure experiments on 60025.

nonporous anorthite (composition of An_{95}; results to be reported elsewhere), are shown in these figures. These data are consistent with previous results for Tahawus Anorthosite (McQueen *et al.*, 1967). Although the Tahawus Anorthosite consists mainly of intermediate plagioclase (An_{49}), it also contains about 10% augite (density ~3.4 Mg/m^3) such that its initial density and, hence, its impedance are fortuitously very close to that of single-crystal anorthite (see Fig. 3; McQueen *et al.*, 1967).

The Tahawus Anorthosite data above 30 GPa have been interpreted by McQueen *et al.* (1967) as representing a high-density polymorph (or assemblage) formed at high pressure; this is suggested by the anomalous compressibility below 30 GPa (Fig. 3) and the linear trend of the shock-wave velocity (U_S) *versus* particle velocity (u_p) data with an intercept below the bulk sound speed of

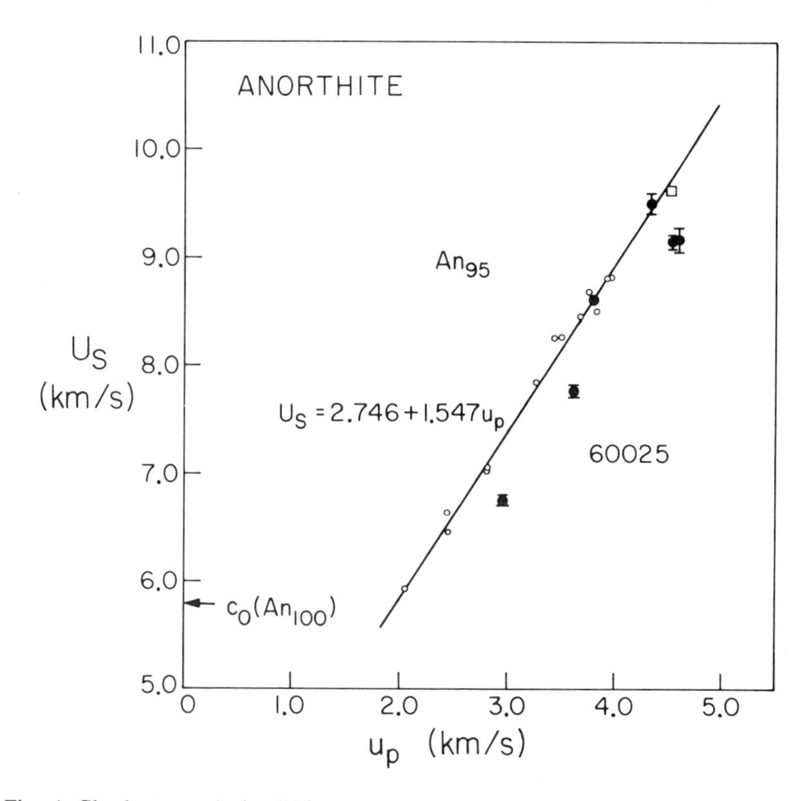

Fig. 4. Shock-wave velocity (U_S) *versus* particle velocity (u_p) data pertaining to calcic plagioclase (symbols are defined in Fig. 3). Only data above 30 GPa are shown. The bulk sound speed (c_0) from Liebermann and Ringwood (1976) is indicated for comparison. The porous (60025) and non-porous data define two, linear and essentially parallel trends, and the consistency of the combined results on non-porous samples is evident (despite significant variations in chemical composition: see text).

anorthite (Fig. 4, McQueen *et al.*, 1967; Liebermann and Ringwood, 1976; cf., also Ahrens *et al.*, 1969a and b). McQueen *et al.* (1967) inferred a zero-pressure density of about 3.5 Mg/m^3 corresponding to these high-pressure data; this is also consistent with one release adiabat measurement of Ahrens and Rosenberg (1968) but see Ahrens *et al.* (1969a and b) as well.

Also shown in Figs. 3 and 4 is a pre-experimental estimate of the Hugoniot state of anorthite (An_{96}) at 120 GPa (1.2 Mbar) based on the inversion of Hugoniot data for various rocks, using an additive (noninteracting continua) mixture theory (Ahrens *et al.*, 1973; 1977). This theoretical estimate is consistent with the Hugoniot data shown in these figures, thus lending further confidence to the validity and usefulness of the mixture model in extracting the properties of components or in predicting the behavior of aggregates which have not been (and perhaps cannot be) studied at high pressures.

The equation-of-state of nonporous anorthite, then, is well described by (i) the new single-crystal data, (ii) existing data on anorthosite (fortuitously), and (iii) a theoretical estimate.

It is clear from the figures, however, that the data for porous anorthite (anorthosite 60025) are dramatically offset from the nonporous Hugoniot. Presumably, this is due to the large amount of thermal energy imparted to the porous samples in the form of irreversible work done in compacting void space (see, e.g., Ze'ldovich and Raizer, 1967; McQueen *et al.*, 1970). It is worth noting that sample #036 (91 GPa shock state) with an initial porosity which was approximately 2% higher than that of the other samples (about 18 ± ½%; Table 1), has a slightly larger off-set from the nonporous Hugoniot than do the other porous samples, thus supporting the present interpretation. Correcting for the extra 2% in initial porosity (based on the effect of 18% porosity on the other three samples, when compared with the nonporous data) brings the Hugoniot state of sample #036 down in pressure (to about 83 GPa; see Fig. 3) and well in line with the three other Hugoniot points for 60025.

The streak record shown in Fig. 1b demonstrates an unusual feature. There is a bright flash of light at the back surface of the sample as the shock-wave emerges. We have seen this feature in all four experiments on 60025 (Fig. 2), but it has not been seen in experiments either on single-crystal anorthite or other nonporous samples (despite the high pressures to which these have been shocked). This seems to indicate that a considerable quantity of thermal energy resides in the porous samples (compared to the nonporous samples) at high pressures. We do not believe that this flash is associated with a second wave because its width does not correlate with shock pressure (see Fig. 2), and because preliminary results of an experiment at about 25 GPa with an inclined mirror assembly (cf., Ahrens *et al.*, 1969b) do not show a multiple-wave structure.

In the shock-wave velocity *versus* particle velocity plane (Fig. 4), the lunar anorthosite data lie on a straight line which is remarkably parallel to the line defined by the high-pressure Hugoniot points for nonporous anorthite, but offset towards lower U_S. Using the conventional, hydrodynamic interpretation of the linear U_S-u_p equation-of-state (e.g., Ruoff, 1967; McQueen *et al.*, 1967), this observation would imply a slightly lower initial bulk sound speed (but similar pressure derivative of the bulk modulus) for the porous anorthite when compared to the nonporous anorthite. Assuming no further complications due to phase transformations (e.g., kinetic effects: Jeanloz and Ahrens, 1977; see also Davies, 1972; and McQueen *et al.*, 1967), this is exactly what might be expected as a thermal effect: a lower, "effective" initial bulk modulus for the porous (much hotter) anorthite. The lower initial density tends to counteract the thermal effect, raising the initial sound speed somewhat.

Release adiabat data are also shown in Fig. 3, but their interpretation is problematical. The data are reduced from buffer-mirror, travel times (see Jeanloz and Ahrens, 1977, and Ahrens *et al.*, 1977), and exhibit the steep initial release slopes that have been documented for several silicates and rocks (Ahrens and Rosenberg, 1968; Ahrens *et al.*, 1969b; Grady *et al.*, 1974; Grady and Murri, 1976; Grady, 1977; Jeanloz and Ahrens, 1977; Ahrens *et al.*, 1977). Unfortunately, there are several possible explanations for such steep slopes.

In the simplest (nontransforming, hydrodynamic) case, the release path is expected to be isentropic (Courant and Friedrichs, 1948; Ze'ldovich and Raizer, 1967). This "normal" release path, then, is nearly parallel to but somewhat less steep than the Hugoniot. However, deviations from these conditions can lead to steeper release adiabats. Grady (1977), for example, believes that material strength (nonhydrodynamic behavior) can be regained by a sample *while* it is at peak pressure, based on a specific model in which yielding and transformation under shock are associated with hot shear zones. Material strength will, in general, result in a steep, initial release path (corresponding to elastic unloading: e.g., Courant and Friedrichs, 1948; Fowles, 1961; Al'tshuler, 1965; Murri *et al.*, 1974). On the other hand, one could argue that a material which has lost virtually all of its strength during shock compression due to its dynamic strength being significantly surpassed (see Rice *et al.*, 1958; Ahrens and Duvall, 1966; van Thiel and Kusubov, 1968; Fowles and Williams, 1970; however, cf., also Ahrens *et al.*, 1968, and Gust and Royce, 1971), could retain enough strength to affect the release slope (Ahrens and Rosenberg, 1968, and Ahrens *et al.*, 1969b, document a Hugoniot elastic limit of about 5 GPa for plagioclases). Although such an interpretation seems consistent with the fact that the highest pressure experiment on 60025 (#026: 94 GPa shock state) resulted in an apparently "normal" release path, the observed, steep release adiabats for the other experiments carried out on porous anorthite (for which the shock temperatures were presumably very high) make it appear quite unlikely that the steep release paths are due to strength effects. In particular, the steep slope for #036 (91 GPa shock pressure) compared with that for the single-crystal anorthite (at slightly lower pressure and a temperature which is less by about 4×10^3 K, according to preliminary analyses), make it virtually inconceivable that material strength is significantly affecting the release paths.

A more likely class of effects are those involving transformations (reactions, phase transformations, changes in material properties) or, in general, entropy production during release. A simple case is that in which partial phase transformation has occurred on shock-loading so that the release path is a better representation of a true compression curve than the Hugoniot itself (Ahrens and Rosenberg, 1968; Grady *et al.*, 1974; Grady and Murri, 1976; Jeanloz and Ahrens, 1977). Since the present data do not fall in what would normally be called a "mixed-phase region" (McQueen *et al.*, 1967; Ahrens *et al.*, 1969a), such phase transformation effects may not seem likely. However, it is possible that: i) phase transformation is still not complete in states beyond the "mixed-phase region" (Jeanloz and Ahrens, 1977; Jeanloz, 1977; Jackson and Ahrens, unpublished material); or ii) other kinetic effects are involved. For example, further phase transformation may still be underway even as the sample is beginning to unload, so that density could even increase (momentarily) as unloading begins. (Given the large amount of "overdriving" that typically occurs under shock conditions, such continued transformation is quite conceivable (e.g., McQueen *et al.*, 1967).

Also, we note that an additional factor affecting the slope of the release adiabats is the use of a finite difference in evaluating the Riemann integral (by necessity, since only one release state is measured in each experiment). As shown by Lyzenga and Ahrens (1978), this yields an upper bound to the actual density on release. Thus, a number of factors may be invoked to explain the steep release adiabats for 60025. Unfortunately, it is not clear what role each of these factors plays (quantitatively) in the apparent behavior observed on release.

DISCUSSION

The present results, when combined with previous data on the shock properties of lunar rocks, provide several interesting implications for the dynamics of impact phenomena on the lunar surface.

Our work to date suggests that essentially all of the major lunar rock types will have Hugoniots (see Fig. 5) bracketed by the Hugoniots of 70215 (a high-titanium, mare basalt; Ahrens *et al.*, 1977) and of nonporous anorthite (labelled An_{95}) for *solid* (nonporous) target materials. For example, a point for gabbroic anorthosite (15418), calculated via the additive mixing model (see above) and shown in Fig. 5, has an impedance ($\rho_0 U_S$ = slope in this Fig.), intermediate between that of the basalt and anorthite. Highland rocks appear, then, to have somewhat lower impedances than mare rocks implying the generation of higher pressures in the latter for a given impact velocity. As an example, a curve indicating the peak pressure generated by an iron projectile impacting at 6.0 km/s is shown in Fig. 5, showing that a given phase (a plagioclase, for example) might experience peak stresses different by about 20% for the same impact velocity, depending on whether it was located in a mare or in a highland terrane. These differences will increase with increasing pressure (or impact velocity). However, most solid lunar rock types will probably fall between the bounds set by the anorthite and the mare basalt in Fig. 5.

More dramatic, though, is the substantially lower impedance exhibited by 60025, compared to the solid rocks (Fig. 5). For a given impact event, this porous rock clearly attains much lower pressures than the solid rocks, but much higher shock and post-shock temperatures than the nonporous samples. For the 6.0 km/s iron impact, although peak pressures in the porous anorthosite are about 20% lower than for the nonporous equivalent, a preliminary analysis indicates peak temperatures on the order of 4,000 K *higher* than in the single-crystal anorthite. This same analysis yields a difference of about $1\text{–}4 \times 10^3$ K in shock temperature (at a given pressure) between 60025 and nonporous anorthite over the pressure range 40–100 GPa. Thus, porosity can have a very substantial effect on the thermal energy partitioning and the shock pressures reached in an impact process, ultimately affecting the nature and degree of shock-metamorphism as well as the residual ("post-shock") heating.

The porosity in 60025 is due to mild brecciation and fracturing (Walker *et al.*, 1973a,b; Hodges and Kushiro, 1973; Dixon and Papike, 1975) resulting from light shocking after crystallization. In general, a planetary surface is expected to

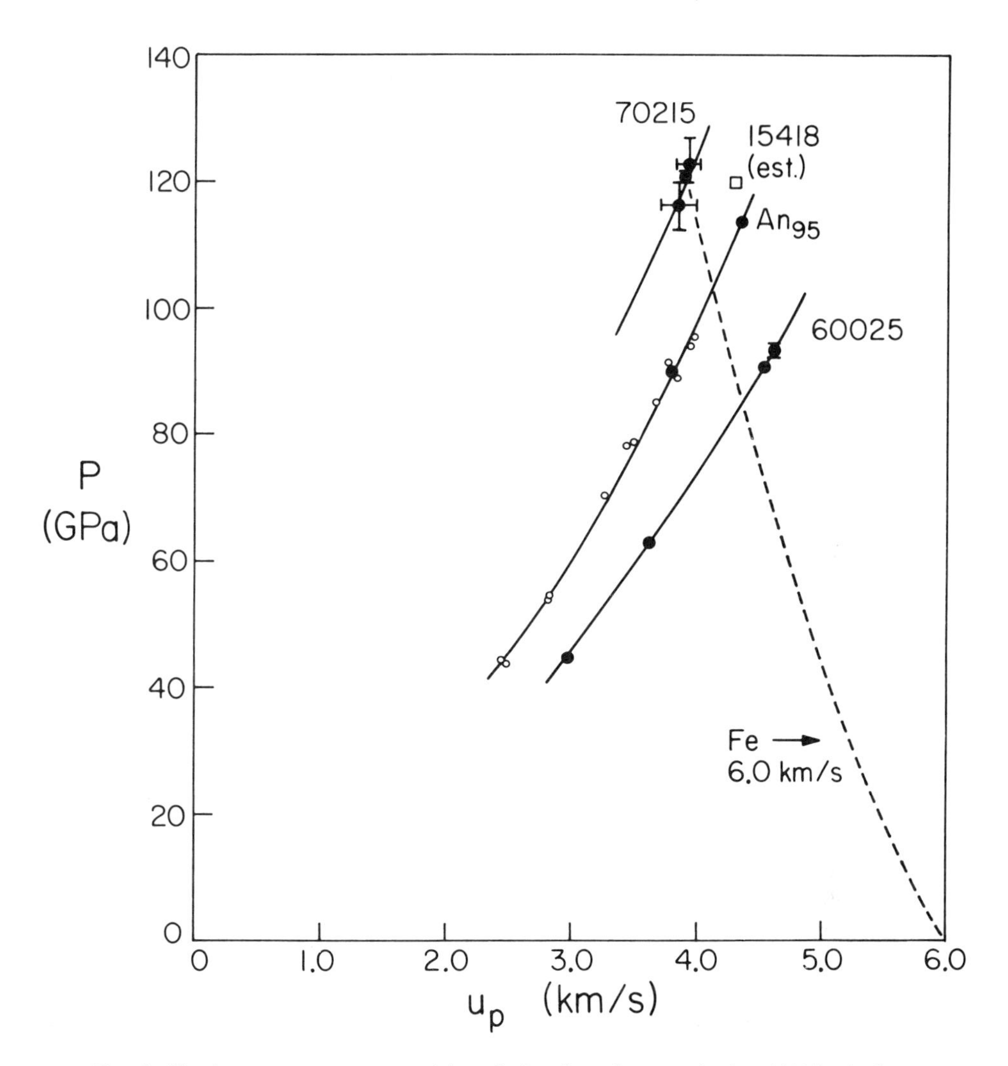

Fig. 5. Shock pressure *versus* particle velocity data for anorthosite 60025, single-crystal anorthite, terrestrial anorthosite (Tahawus, NY) and mare basalt 70215, and a theoretical estimate for gabbroic anorthosite 15418 (*Ahrens et al.*, 1977). Only data above 30 GPa are shown. The Hugoniot of iron is shown to illustrate the impedance-matching conditions for the 6.0 km/s impact of an iron projectile onto a lunar target.

"mature" with time in that continued bombardment tends to fracture and brecciate the surface, forming a regolith-like layer. Safronov (1972, 1978) and Kaula (1978) have discussed the importance of such a regolith in the thermal evolution of young planetary bodies in that it acts both as an insulating layer which traps impact-produced thermal energy, and also controls the loss of this energy by stirring due to subsequent overturn. Our data emphasize a separate (perhaps equally important) effect of such a layer indicating that, with its higher initial porosity (lower impedance), it could allow substantially increased partitioning of thermal energy into a planet during accretion or bombardment of its surface.

Therefore, with time, the dynamic properties of the pristine planetary surface will tend to mature (with the appropriate Hugoniots sweeping downward, to the right in Fig. 5), resulting in a more substantial shock-heating associated with any given impact. The porosity needed for such dramatic changes can result from very mild shocking (as in the case of 60025). Of course, this effect can be counterbalanced by several processes: for example, the eruption of fresh volcanic flows onto a planetary surface.

CONCLUSION

We have presented the first high-pressure, equation-of-state data for a calcic, aluminous silicate: anorthite. Our shock-wave results on lunar anorthosite 60025, which has an average initial porosity of about 18% contrast sharply with the equation-of-state of nonporous anorthite due to the large amount of thermal energy associated with pore closure under shock-loading. This thermal energy results in a substantial offset (e.g., as a significant thermal pressure) of the high-pressure states achieved by the porous anorthosite (compared to the nonporous anorthite), illustrating the important role of porosity in determining the response of a target material. For a given impact, although peak pressures are less in a porous than a nonporous target, both shock and post-shock temperatures can be considerably larger, as in the case of 60025. Cratering efficiency (per mass of target) is therefore significantly reduced for the porous case as a given projectile will impart a higher proportion of its kinetic energy to a smaller mass of target.

The present data provide a shock-wave equation-of-state for an important lunar rock-type and a prime candidate for material representative of the primordial lunar crust. These data can be used to describe the dynamic response of much of the lunar surface to meteoritic impact (e.g., during the late stages of accretion). The combined results on lunar samples to date indicate significant variations in the dynamic properties among the major lunar lithologies. These imply that for a given distribution of high impedance meteorites, the resulting crater sizes will tend to be systematically larger for lunar anorthosite, gabbroic anorthosite, and high-Ti mare basalt respectively. Hence, such lateral variations in crater-forming processes will be reflected as a systematic bias in apparent cratering ages (e.g., in maria vs. terrae). More significantly, however, the present study of a porous highland rock illustrates quite clearly the time-dependence of the dynamic properties of a petrologically well-defined rock unit or planetary surface due to its continual "maturing" under meteoritic bombardment. This results in a general (and potentially very substantial) enhancement, with time, in the production of thermal energy due to the impact of this planetary surface.

Acknowledgments—We thank E. Gelle and R. Smith for their dedicated assistance with the experiments, and Dr. Y. Syono for providing samples of anorthite. We also appreciate the helpful comments of Malcolm Somerville and James Long. The critical reviews of Rex V. Gibbons and Jon F. Bauer also improved the paper. This work was supported by NASA Grant NSG 9019. Contribution No. 3065, Division of Geological and Planetary Sciences, California Institute of Technology, Pasadena, California 91125.

REFERENCES

Adler I., Trombka J. I., Schmadeback R., Lowman P., Blodget H., Yin L., Eller E., Podwysocki M., Weidner J. R., Bickel A. L., Lum R. K. L., Gerard J., Gorenstein P., Rjorkholm P. and Harris B. (1973) Results of the Apollo 15 and 16 X-ray experiment. *Proc. Lunar Sci. Conf. 4th*, p. 2783–2791.

Ahrens T. J., Anderson D. L. and Ringwood A. E. (1969a) The equation of state and crystal structures of high-pressure phases of silicates and oxides. *Rev. Geophys. Space Phys.* **7**, 667–707.

Ahrens T. J. and Duvall G. E. (1966) Stress relaxation behind elastic shock waves in rock. *J. Geophys. Res.* **71**, 4349–4360.

Ahrens T. J., Gust W. H. and Royce E. B. (1968) Material strength effect in the shock compression of alumina. *J. Appl. Phys.* **39**, 4610–4616.

Ahrens T. J., Jackson I. and Jeanloz R. (1977) Shock compression and adiabatic release of a titaniferous mare basalt. *Proc. Lunar Sci. Conf. 8th*, p. 3437–3455.

Ahrens T. J., O'Keefe J. D. and Gibbons R. V. (1973) Shock compression of a recrystallized anorthositic rock from Apollo 15. *Proc. Lunar Sci. Conf. 4th*, p. 2575–2590.

Ahrens T. J., Petersen C. F. and Rosenberg J. T. (1969b) Shock compression of feldspars. *J. Geophys. Res.* **74**, 2727–2746.

Ahrens T. J. and Rosenberg J. T. (1968) Shock metamorphism: Experiments on quartz and plagioclase. In *Shock Metamorphism of Natural Materials* (B. M. French and N. M. Short, eds.), p. 59–81. Mono, Baltimore.

Al'tshuler L. V. (1965) Use of shock waves in high-pressure physics. *Sov. Phys. Uspekhi* **8**, 52–91.

Al'tshuler L. V. and Sharipdzhanov I. I. (1971) Additive equations of state of silicates at high pressures. *Izv. Earth Phys.* **3**, 11–28.

Anderson D. L. (1972) Implications of the inhomogeneous planetary accretion hypothesis, Comments on Earth sciences. *Geophysics* **2**, 93–98.

Anderson D. L. (1973) The formation of the moon (abstract). In *Lunar Science IV*, p. 40–42. The Lunar Science Institute, Houston.

Berman H. (1939) A torsional microbalance for the determination of specific gravities of minerals. *Amer. Mineral.* **24**, 434–440.

Bielefeld M. J., Andre C. G., Eliason E. M., Clark P. E., Adler I. and Trombka J. I. (1977) Imaging of lunar surface chemistry from orbital X-ray data. *Proc. Lunar Sci. Conf. 8th*, p. 901–908.

Birch F. (1952) Elasticity and constitution of the Earth's interior. *J. Geophys. Res.* **57**, 227–286.

Courant R. and Friedrichs K. (1948) *Supersonic Flow and Shock Waves.* Interscience, N.Y. 464 pp.

Davies G. F. (1972) Equations of state and phase equilibria of stishovite and a coesitelike phase from shock-wave and other data. *J. Geophys. Res.* **77**, 4920–4933.

Dixon J. R. and Papike J. J. (1975) Petrology of anorthosites from the Descartes region of the moon: Apollo 16. *Proc. Lunar Sci. Conf. 6th*, p. 263–291.

Fowles G. R. (1961) Shock wave compression of hardened and annealed 2024 aluminum. *J. Appl. Phys.* **32**, 1475–1487.

Fowles G. R. and Williams R. F. (1970) Plane stress wave propagation in solids. *J. Appl. Phys.* **41**, 360–363.

Grady D. E. (1977) Processes occurring in shock wave compression of rocks and minerals. In *High-Pressure Research: Applications to Geophysics* (M. H. Manghnani and S. Akimoto, eds.), p. 389–438. Academic Press, N.Y.

Grady D. E. and Murri W. J. (1976) Dynamic unloading in shock compressed feldspar. *Geophys. Res. Lett.* **3**, 472–474.

Grady D. E., Murri W. J. and Fowles G. R. (1974) Quartz to stishovite: Wave propagation in the mixed phase region. *J. Geophys. Res.* **79**, 332–338.

Grossman L. (1972) Condensation in the primitive solar nebula. *Geochim. Cosmochim. Acta* **36**, 597–619.

Gust W. H. and Royce E. B. (1971) Dynamic yield strengths of B_4C, BeO, and Al_2O_3 ceramics. *J. Appl. Phys.* **42**, 276–295.

Hodges F. N. and Kushiro I. (1973) Petrology of Apollo 16 lunar highland rocks. *Proc. Lunar Sci. Conf. 4th*, p. 1033–1048.

Jeanloz R. (1977) Shocked olivines and the spinel phase: IR spectra, observations and implications (abstract). *Geol. Soc. Amer. Abstracts with Programs* **9**, 1036.

Jeanloz R. and Ahrens T. J. (1977) Pyroxenes and olivines: Structural implications of shock-wave data for high-pressure phases. In *High-Pressure Research: Applications to Geophysics* (M. H. Manghnani and S. Akimoto, eds.), p. 439–461. Academic Press, N.Y.

Kaula W. M. (1978) Planetary thermal evolution during accretion (abstract). In *Lunar and Planetary Science IX*, p. 615–617. Lunar and Planetary Institute, Houston.

Liebermann R. C. and Ringwood A. E. (1976) Elastic properties of anorthite and the nature of the lunar crust. *Earth Planet. Sci. Lett.* **31**, 69–74.

Lyzenga G. A. and Ahrens T. J. (1978) The relation between the shock-induced free-surface velocity and the post-shock specific volume of solids. *J. Appl. Phys.* **49**, 201–204.

McQueen R. G. (1968) Shock-wave data and equations of state. In *Seismic Coupling*, p. 53–106. National Technical Information Service No. AD 848 608. Springfield, Virginia 22161.

McQueen R. G., Marsh S. P. and Fritz J. N. (1967) Hugoniot equation of state of twelve rocks. *J. Geophys. Res.* **72**, 4999–5036.

McQueen R. G., Marsh S. P., Taylor J. W., Fritz J. N. and Carter W. J. (1970) The equation of state of solids from shock wave studies. In *High Velocity Impact Phenomena* (R. Kinslow, ed.), p. 294–419. Academic Press, N.Y.

Murri W. J., Curran D. R., Petersen C. F. and Crewdson R. C. (1974) Response of solids to shock waves. In *Advances in High-Pressure Research* (R. H. Wentorf, Jr., ed.), V. 4, p. 1–163. Academic Press, N. Y.

Papanastassiou D. A. and Wasserburg G. J. (1972) Rb-Sr systematics of Luna 20 and Apollo 16 samples. *Earth Planet. Sci. Lett.* **17**, 52–63.

Rice M. H., McQueen R. G. and Walsh J. M. (1958) Compression of solids by strong shock waves. *Solid State Phys.* **6**, 1–63.

Ruoff A. L. (1967) Linear shock-velocity—particle-velocity relationship. *J. Appl. Phys.* **38**, 4976–4980.

Ruff L. and Anderson D. L. (1977) A geophysical model for core convection due to lateral heat sources (abstract). *EOS (Trans. Amer. Geophys. Union)* **58**, 1129.

Safronov V. S. (1972) Evolution of the Protoplanetary Cloud and Formation of the Earth and the Planets. NASA TT F-677 (trans. from Nauka, Moscow, 1969).

Safronov V. S. (1978) The heating of the Earth during its formation. *Icarus* **33**, 3–12.

Schaeffer O. A. and Husain L. (1974) Chronology of lunar basin formation. *Proc. Lunar Sci. Conf. 5th*, p. 1541–1555.

Schonfeld E. (1977) Comparison of orbital chemistry with crustal thickness and lunar sample chemistry. *Proc. Lunar Sci. Conf. 8th*, p. 1149–1162.

Taylor S. R. and Bence A. E. (1975) Evolution of the lunar highland crust. *Proc. Lunar Sci. Conf. 6th*, p. 1121–1141.

Taylor S. R. and Bence A. E. (1978) Chemical constraints on lunar highland petrogenesis (abstract). In *Lunar and Planetary Science IX*, p. 1155–1157. Lunar and Planetary Institute, Houston.

Taylor S. R. and Jakeš P. (1974) The geochemical evolution of the moon. *Proc. Lunar Sci. Conf. 5th*, p. 1287–1305.

Taylor S. R. and Jakeš P. (1977) Geochemical evolution of the moon revisited. *Proc. Lunar Sci. Conf. 8th*, p. 433–446.

van Thiel M. and Kusubov A. (1968) Effect of 2024 aluminum alloy strength on high-pressure shock measurements. In *Accurate Characterization of the High-Pressure Environment* (E. C. Lloyd, ed.), p. 125–130. National Bureau of Standards Special Publication 326.

Wackerle J. (1962) Shock-wave compression of quartz. *J. Appl. Phys.* **33**, 922–930.

Walker D., Grove T. L., Longhi J., Stolper E. M. and Hays J. F. (1973a) Origin of lunar feldspathic rocks. *Earth Planet. Sci. Lett.* **20**, 325–336.

Walker D., Longhi J., Grove T. L., Stolper E. and Hays J. F. (1973b) Experimental petrology and origin of rocks from the Descartes Highlands. *Proc. Lunar Sci. Conf. 4th*, p. 1013–1032.

Ze'ldovich Y. B. and Raizer Y. P. (1966) *Physics of Shock Waves and High-Temperature Hydrodynamic Phenomena.* Academic Press, N.Y. 916 pp.

Proc. Lunar Planet. Sci. Conf. 9th (1978), p. 2805–2824.
Printed in the United States of America

Experimental shock metamorphism of dunite

W. U. REIMOLD and D. STÖFFLER

Institute of Mineralogy, University of Münster
D-44 Münster, Germany

Abstract—Samples of dunite from Åheim were shock loaded to pressures of 5, 15.5, 20.0, 29.3, 38.5, 45 and 59 GPa using compressed air gun and high explosive accelerators. Additional shots at 29.3 GPa were performed with dunites of variable composition and texture (Twin Sisters, Balsam Gap, Corundum Hill).

The physical state of the olivine and the texture of the dunite change progressively with increasing peak pressure due to mechanical deformations and phase transformations including incipient melting at the highest pressure.

By means of quantitative textural analyses, measurement of the degree of X-ray asterism, and of the fracture density the following progressive stages of shock metamorphism can be established for the Åheim dunite: Irregular and planar fracturing at 5 GPa; mosaicism and planar fracturing at 15.5, 20 and 29.3 GPa; planar fracturing, mosaicism and intergranular brecciation at 38.5 and 45 GPa; planar fracturing, intragranular solid state recrystallization, and intergranular melting with subsequent crystallization at 59 GPa. To a certain degree the intensity of those shock effects which could be measured quantitatively was found to be dependent on the textural and compositional properties of the starting material. The frequency distribution of the crystallographic orientation of planar fractures is largely independent of peak pressure in the 15–45 GPa pressure range. It seems to be influenced not only by textural parameters of the starting material but also by the direction of shock wave propagation. The intergranular brecciation textures of dunites shocked in the 29–45 GPa pressure regime closely resemble the texture of the lunar dunite 72415.

1. INTRODUCTION

Shock-induced deformation and transformation features are widespread in olivine of basaltic and dunitic rocks from lunar and terrestrial impact formations and in olivine of meteorites as well (e.g., Carter *et al.*, 1968; Müller and Hornemann, 1969; Engelhardt *et al.*, 1970; Sclar and Morzenti, 1971; Levi-Donati, 1971; Stöffler, 1972; Snee and Ahrens, 1975; Ashworth and Barber, 1975; Lally *et al.*, 1976). The knowledge of the pressure and temperature conditions of shock effects in olivine-bearing lunar and meteoritic rocks, which usually have been displaced from the primary geological and physicochemical system in which they crystallized, may help to unravel their complex history involving impact metamorphism, ballistic transportation and relocation into breccia or impact melt formations. Shock recovery experiments on olivine and olivine-bearing rocks have been performed by several laboratories in order to deduce a pressure calibration for a number of characteristic shock metamorphic effects (Carter *et al.*, 1968; Müller and Hornemann, 1969; Sclar, 1969; Short, 1969; James, 1969; Hörz and Quaide, 1972; Snee and Ahrens, 1975; Ahrens *et al.*, 1976; Schneider and Hornemann, 1977; Jeanloz *et al.*, 1977). In many of the experiments single crystal olivine was used. Since the early shock experiments with Twin Sisters Peaks dunite (Carter *et al.*, 1968) little work has been devoted to polycrystalline dunitic rocks until recently (Bauer, 1978; Reimold and Stöffler, 1978). In particular, the influence of textural parameters (grain size, texture, porosity) on the intra- and intergranular shock deformation of olivine has not been considered in sufficient detail. There also remain some unexplained discrepancies between the results of various laboratories with respect to the occurrence and type of phase transitions (solid state transformations or melting with subsequent recrystallization) of olivine shocked to pressures in excess of about 50 GPa.

The present paper attempts to achieve a better and more quantitative understanding of shock metamorphism of polycrystalline olivine-rocks as a function of the textural parameters of the starting material and of the experimental peak shock pressure.

2. SAMPLE MATERIAL

Dunites from 4 different localities were selected for shock recovery experiments. Their composi-

Table 1. Origin, modal analysis, physical and textural characteristics of the shocked dunites; grain size statistics: GM = graphic mean in phi units, IGSD = inclusive graphic standard deviation in phi units, IGS = inclusive graphic skewness, GK = graphic kurtosis (parameters after Folk and Ward, 1957); mol% Fo determined by x-ray powder techniques, porosity by Archimedes principle using CCl_4 as liquid.

Sample No.	1	2	3	4	5	6	9	12	17
Locality*			Corundun Hill				Balsam Gap	Twin Sisters	Åheim
Olivine	90.80	86.20	90.50	74.40	91.90	87.00	94.00	94.25	83.90
Ortho-pyroxene	0.30	0.20	0.10	0.10	0.20	—	0.10	2.30	11.40
Spinel	1.30	1.20	1.10	0.90	1.30	3.20	1.00	1.05	0.50
Serpentine	5.30	8.00	6.60	12.40	4.70	7.50	4.70	1.60	1.50
Talc	1.90	3.20	1.30	10.50	1.30	1.20	—	0.35	0.40
Chlorite	0.40	1.20	0.40	1.90	0.60	1.10	0.20	0.15	2.20
Iddingsite	—	—	—	—	—	—	—	0.30	—
mol% Fo in olivine	94.50	94.50	94.75	91.00	92.50	93.50	95.50	94.50	94.75
ρ [g/cm³]	3.221	3.214	3.249	3.017	3.212	3.253	3.289	3.323	3.243
Porosity [Vol %]	0.85	1.94	1.18	0.99	1.50	0.88	0.94	0.81	1.44
GM	2.454	2.351	1.759	2.691	2.862	2.592	1.443	0.287	1.435
IGSD	0.243	0.391	0.367	0.273	0.292	0.322	0.588	1.190	0.711
IGS	−0.316	−0.465	−0.301	−0.411	−0.211	−0.322	−0.331	−0.157	−0.403
GK	1.015	1.280	0.937	1.145	1.010	1.045	0.950	1.013	0.906

*Corundum Hill and Balsam Gap: North Carolina, USA; Twin Sisters Peaks: Washington, USA; Åheim: Norway.

tional, textural, and physical properties are summarized in Table 1. The composition of olivine is rather constant within 4.5% in all samples. However, the modal abundance of olivine and sheet silicates and the grain size vary considerably (Reimold, 1977).

3. Experimental Techniques

Two different experimental set-ups were used for shock loading cylindrical specimens of 25 mm diameter and 0.5–2.5 mm thickness: 1) a compressed air gun accelerator for the pressure range below 10 GPa (Fig. 1) and 2) a high explosive device using TNT explosives (composition B) for pressures in the 10–59 GPa range (for details see Müller and Hornemann, 1969). A flyer plate of stainless steel of variable thickness was impacted in both cases into a steel sample holder which contained the specimen. The sample holder of the high explosive device was surrounded by loose steel plates (momentum trap technique).

In the high explosive device, pressures of 15.5 and 20 GPa were achieved by the impedance matching method using dunitic impedance cylinders (Fig. 1). Peak pressures >29.3 GPa were produced by multiple shock reverberations using steel impedance cylinders (Fig. 1, Table 2). Details of the experimental devices are given by Reimold (1977), Hornemann *et al.* (1975), and Müller and Hornemann (1969). The peak pressure achieved in the specimen was calculated on the basis of the Hugoniot data for steel (McQueen *et al.*, 1970) and Twin Sisters dunite (McQueen *et al.*, 1967). For the high explosive devices, separate calibration runs using the electrical pin method were made in order to determine the peak shock pressure as described by Müller and Hornemann (1969). The error in the calculated peak pressure is estimated to be on the order of ±1% for the air gun and ±3–4% for the high explosive device.

Two series of recovery shock experiments were performed:

1) Shock loading of the Åheim dunite at 5, 15.5, 20, 29.3, 38.5, 45 and 59 GPa, and
2) Shock loading of 5 different samples of the Corundum Hill dunite, Balsam Gap dunite, and Twin Sisters dunite at 29.3 GPa.

All samples were recovered as complete, solid disks except for some Corundum Hill runs and the 59 GPa shots, which were largely pulverized. The recovered specimen disks were split into two parts. One half was impregnated with epoxy, and polished thin sections were made of that part of the disks which was in direct contact with the flyer plate. The remaining specimen was divided into three sieve fractions (>0.15, 0.15-0.075, and <0.075 mm) from which olivine grains and olivine fractions were

Table 2. Experimental conditions for the shock recovery runs on dunites.

Accelerator	Shock pressure achieved by	Thickness of steel target D [mm]	Thickness of flyer plate d [mm]	Pressure [GPa]
Compressed air gun	shock wave reverberations	2	10	5.0 ± 0.05
High explosive	impedance matching	17.5	4	15.5 ± 0.6
		14.5	4	20.0 ± 0.8
High explosive	shock wave reverberations	10.8	4	29.3 ± 1.2
		5	3.45	38.5 ± 1.5
		5	2.44	45.0 ± 1.8
		5	2	59.0 ± 2.4

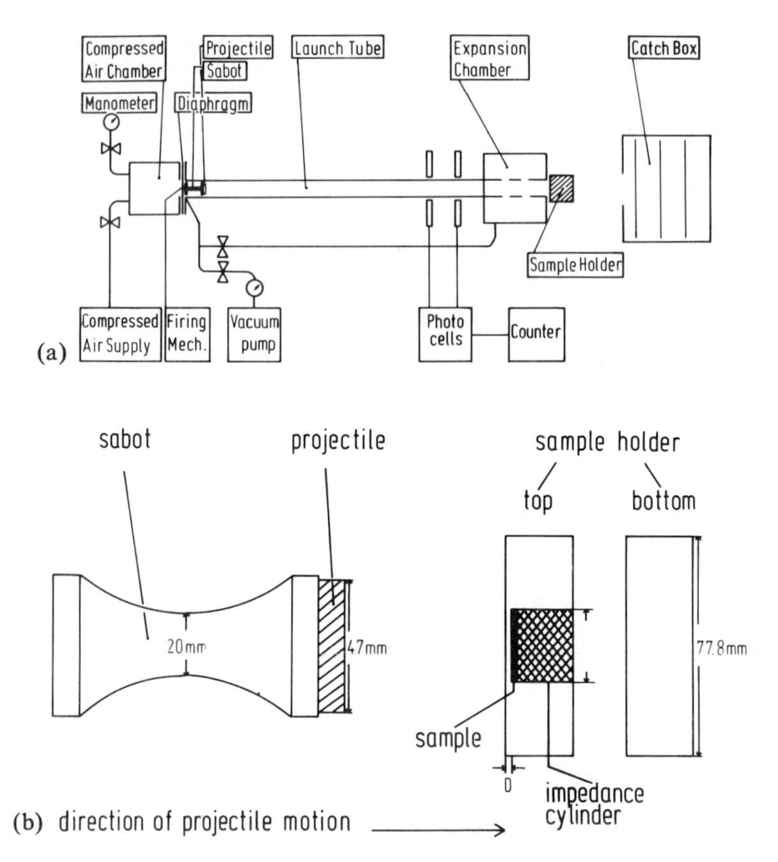

Fig. 1. (a) Compressed air gun accelerator; (b) Enlarged portion of (a) with projectile and sample holder; D = see Table 2.

obtained by hand-picking under the stereo microscope or by magnetic separation. Subsequent investigations included optical microscopy and universal stage methods, scanning electron microscopy, x-ray diffraction, microprobe analysis, infrared and electron spin resonance spectroscopy.

4. RESULTS

4.1 *Inter- and intragranular deformation*

The mechanical shock effects observed in the recovered specimens may be classified into two categories:

1) Intragranular deformation which includes irregular and planar fractures and mosaicism of single crystals,
2) Intergranular deformation which leads to fractures across grain boundaries and to brecciation and displacement of mineral fragments along grain boundaries.

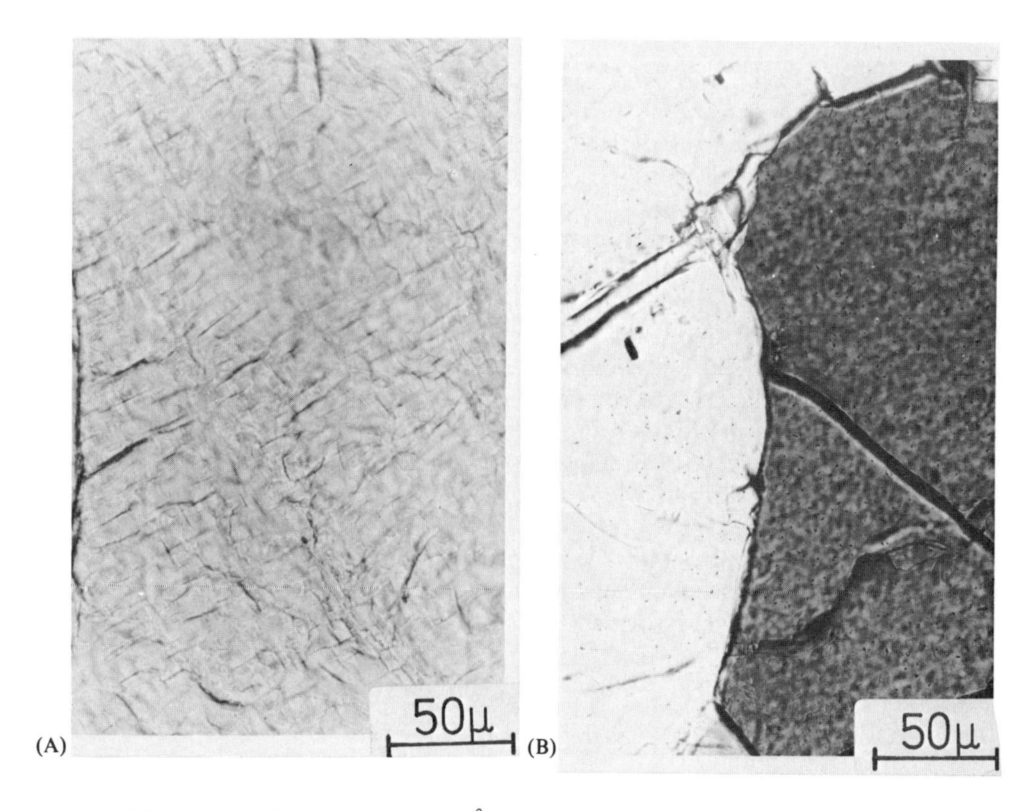

Fig. 2. Typical fracture pattern of Åheim dunite shocked at 38.5 (a), and olivine in unshocked Åheim dunite (b). (a) crossed polarizers; (b) plane polarized light.

In the first category only *fractures* were observed microscopically. Planar elements as defined by Stöffler (1972) which are most common in tectosilicates are not detectable. In the scanning electron microscope all observed structures appear as open fractures.

4.1.1 Intragranular fracturing All dunite samples used in the shock experiments display a certain degree of pre-shock fracturing. These fractures are predominantly irregular fractures which in many cases cut through grain boundaries. Their density was recorded in thin sections of unshocked samples (see below). Shocked samples show additional fractures which typically form a pattern of subparallel fractures in sets of variable orientation. These features are in most cases much shorter than the grain diameter (Fig. 2) and their most common separation is less than 10 μm. In some preferred crystallographic orientations parallel sets of more perfectly planar fractures with a spacing of about 10 to 50 μm occur (compare Fig. 6 of Müller and Hornemann, 1969). In most cases they are restricted to particular regions of a single crystal. The type and occurrence of the various kinds of fractures is in good agreement with previous observations made by several authors (Müller and Hornemann, 1969; Sclar, 1969; Snee and Ahrens, 1975).

The crystallographic orientation of planar fractures as measured with a universal stage is plotted in Fig. 3 as a function of peak shock pressure. These plots contain a large number of fracture sets more or less subparallel to each other which were measured separately in various regions of a single grain. All fracture sets were measured in eight to ten grains of each sample (Fig. 3). Therefore the crystallographically equivalent forms are not equally represented in the stereoplots. The frequency distribution of pinacoids, prisms and bipyramids as given in Table 3 and Fig. 4 does not show a systematic change with increasing peak shock pressure as described in single crystal shock experiments by Snee and Ahrens (1975). The frequency distribution of the orientation of planar structures observed in different dunites at a constant pressure of 29.3 GPa (Table 3) shows a variation which is larger than that of the Åheim dunite shocked at variable peak pressures from 0.5–45 GPa. This demonstrates clearly that the frequency statistics of planar deformation structures cannot be used for pressure calibration of shocked olivine-bearing rocks.

The orientation of planar fracture sets with respect to the direction of shock

Table 3. Frequency statistics of the crystallographic orientation of planar fractures in shocked olivine from various dunites.

Sample	Pressure [GPa]	Number of grains	Percentage of pinacoids	Percentage of prisms	Percentage of bipyramids
Åheim	5	8	37.7	36.0	26.3
	15.5	10	41.0	38.0	21.0
	20.0	9	48.0	30.5	21.5
	29.3	9	34.0	29.7	36.3
	38.5	9	34.8	25.8	39.4
	45.0	10	43.7	26.0	30.3
Corundum Hill					
1	29.3	7	20.7	48.3	31.0
2	29.3	9	27.7	52.3	20.0
4	29.3	8	28.3	40.3	21.3
6	29.3	7	24.3	43.3	41.3
Balsam Gap					
9a	29.3	3	26.0	51.0	23.0
9b	29.3	5	24.0	33.0	43.0
Twin Sisters					
12a	29.3	5	21.6	26.9	51.6
12b	29.3	4	22.1	29.9	48.0

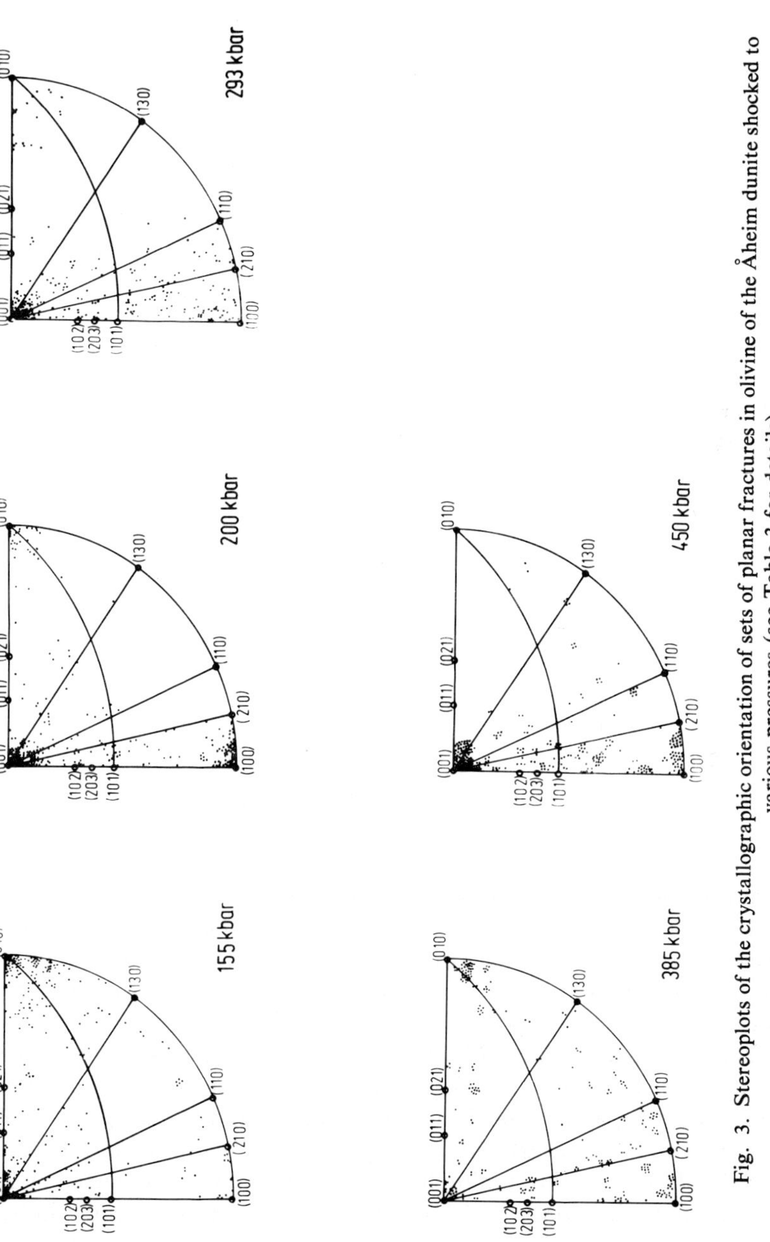

Fig. 3. Stereoplots of the crystallographic orientation of sets of planar fractures in olivine of the Åheim dunite shocked to various pressures (see Table 3 for details).

wave propagation was recorded in thin sections which were cut perpendicular to this direction. Examples for the Åheim dunite are given in Fig. 5. The fractures present in the unshocked dunite are almost randomly orientated. At 20 GPa two preferred orientations appear which form an angle of about 90°. At 45 GPa only one broad maximum of the frequency distribution of fracture orientations remains. This non-random distribution of shock induced fractures is also observed in all other dunite samples shocked at 29.3 GPa. The weak anisotropy of the orientation of the olivine crystals (as measured by the orientation of their optical indicatrix) observed in the unshocked dunite samples is not sufficient to explain this effect. Consequently the observed frequency distribution of planar fractures within the olivine crystals (Table 3) appears to be not only controlled by the crystallographic properties of olivine but also by the direction of the shock wave propagation (see also Müller and Hornemann, 1969).

An increasing intensity of shock induced fracturing of olivine was found with increasing shock pressure up to 45 GPa (Fig. 6). As a measure of this progressive deformation a "fracture index" was determined in each shocked sample by counting the fracture density microscopically in thin sections, according to the method of Short (1966). The index is defined as the number of fractures per unit length of a scanning traverse. Results are plotted in Fig. 7. For a given rock the fracture index increases linearly with increasing shock pressure. For different dunites the fracture index varies by a factor of almost two at constant shock pressure, most probably due to differences in texture, grain size, and modal composition. In comparing the degree of shock of such rocks the fracture index appears to be a rather useless indicator of the shock pressure.

4.1.2 Intergranular brecciation The primary crystallization texture of the dunite samples which is unaffected by shock pressure up to 20 GPa, is changed by intergranular brecciation upon shock compression above about 29 GPa. The large primary olivine crystals are brecciated along grain boundaries by a comminution process which produces small olivine fragments. The fragments which cover a broad grain size range down to the submicron region were displaced and rotated from the primary orientation of their parent crystals within narrow and irregularly shaped intergranular zones. Based on the grain size distribution these zones appear to be a fine grained matrix in which somewhat larger fragments are floating (Fig. 6d). Brecciation textures of this type are very weakly developed in the 29.3 GPa shots of the Corundum Hill dunite (only sample 2, Table 3), and of the Twin Sisters Peaks dunite (sample 12b, Table 3). They appear much more strongly developed in the Balsam Gap samples shocked to 29.3 GPa. In the Åheim dunite, brecciation was not yet produced at 29.3 GPa but very strongly developed at 38.5 GPa. There is no microscopic indication of any solid state recrystallization within the breccia zones. In all dunite samples shocked to 45 GPa, intergranular brecciation is the most conspicuous textural feature. These samples closely resemble the texture of the shocked lunar dunite 72415 (see Dymek *et al.*, 1975; Snee and Ahrens, 1975). In agreement with the

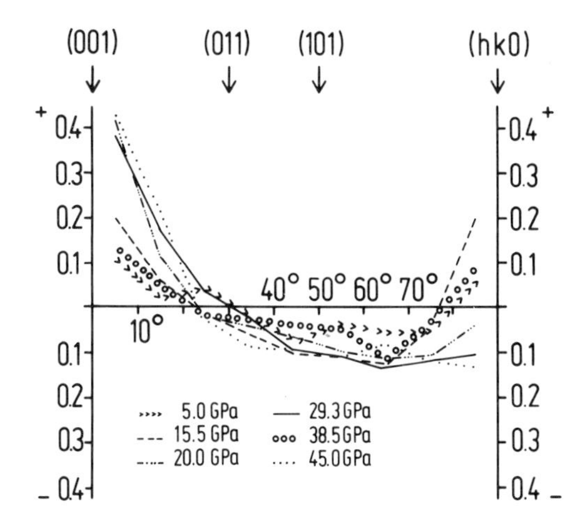

Fig. 4. Deviation of the frequency distribution of planar fractures from the isotropic distribution of fractures in olivine (Åheim dunite) shocked to various peak pressures. Isotropy of fracturing is defined by the zero line as calculated by Engelhardt (1944). The ordinate gives the number of measured fracture sets out of 100 isotropically distributed fractures which are in excess or deficient at a particular angle to the c-axis of olivine.

estimates of these authors we conclude that 72415 was shocked to pressures of at least 30 GPa, most probably to 40-45 GPa.

The increasing intensity of brecciation with increasing shock pressure was measured quantitatively with an electronic computerized scanning microscope. The size of olivine grains and of their fragments was recorded along a scanning traverse on thin sections of the Åheim dunite. Results are given in Fig. 8. The median grain size, the skewness, and the sorting change drastically from the unshocked sample to the sample shocked at 38.5 GPa. Consequently, the specific surface area of all grains of a given unit volume increases. These effects are even more pronounced in the 45 GPa sample.

4.1.3 Plastic deformation Plastic deformation of the crystal lattice as revealed by strong mosaicism is a most characteristic shock effect in silicate minerals (see review in Stöffler, 1972; Hörz and Quaide, 1972). As seen from the optical extinction pattern of single crystals, olivine in all of the shocked dunites shows an increasing degree of mosaicism with increasing shock pressure above peak pressures between 5 and 15 GPa (Fig. 9). The dunite shocked at 5 GPa is not affected by mosaicism and is very similar to the unshocked counterpart. Hörz and Quaide (1972) measured the degree of mosaicism quantitatively by Debye-Scherrer x-ray diffraction of single crystal grains. Using the same technique, four selected olivine grains of each shot were measured and averaged. Results are plotted in Fig. 10. The intensity of x-ray asterism increases with pressure but a

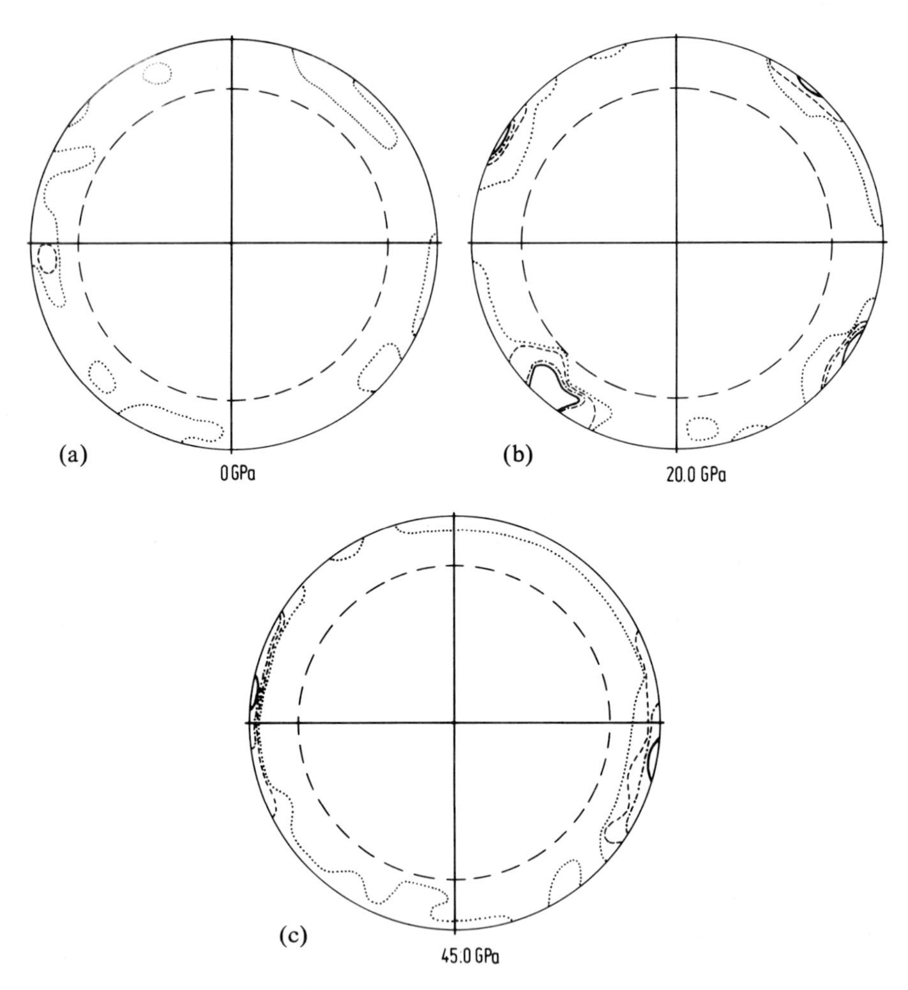

Fig. 5. Stereoplots of the orientation of 155 fracture sets of shocked and unshocked Åheim dunite. The direction of the shock wave propagation is perpendicular to the plane of the drawing (see text).

complete misorientation of the mosaic blocks resulting in complete diffraction rings is not achieved. This observation differs from the results of Hörz and Quaide (1972) who found complete diffraction rings of the front reflections of shocked olivine single crystals at pressures above 35 GPa. Evidently, the degree of shock-induced plastic deformation is less severe in polycrystalline material.

Fig. 6. Photomicrographs of progressively shock metamorphosed Åheim dunite (a) unshocked; (b) 5 GPa, irregular fractures; (c) 15.5 GPa, intragranular sets of planar fractures; (d) 38.5 GPa, intergranular brecciation. (a)- (c) plane polarized light; (d) crossed polarizers.

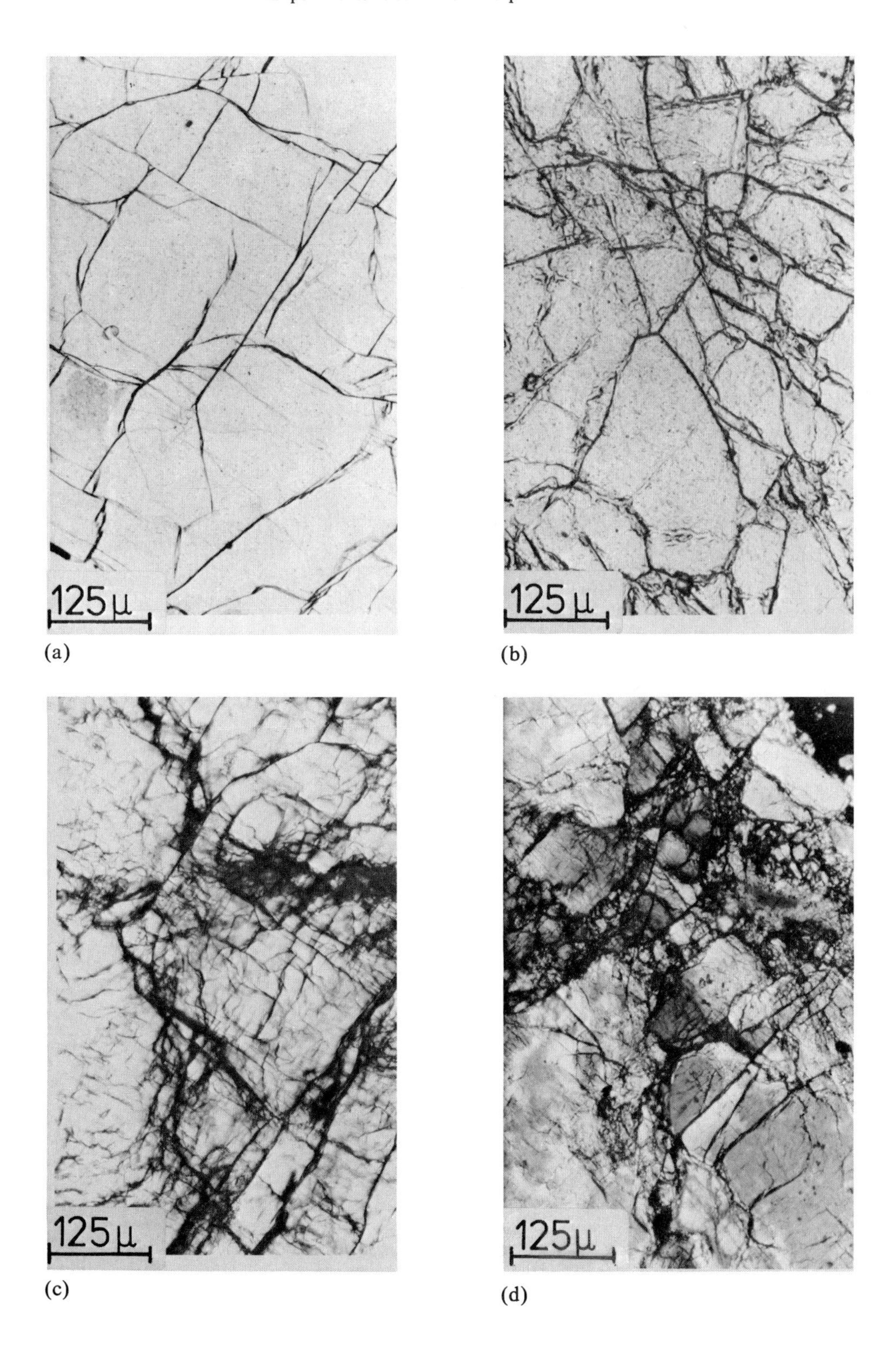

(a)
(b)
(c)
(d)

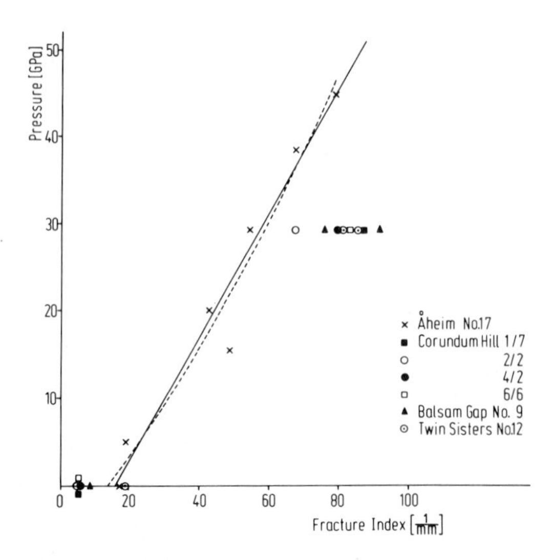

Fig. 7. Fracture index of shocked dunite as a function of peak shock pressure (see text); solid line = linear least-squares regression curve, broken line = polynomial least-squares regression curve.

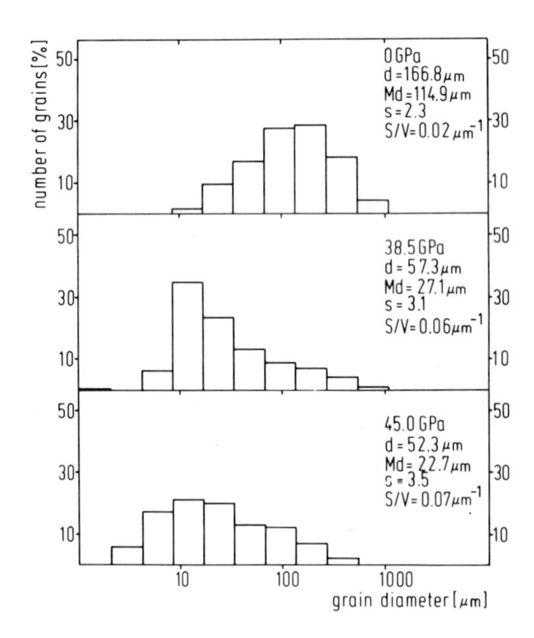

Fig. 8. Grain size statistics of unshocked and shock brecciated Åheim dunite measured by the Zeiss electronic scanning microscope; number of measured grain traverses per sample: 500–1000; d = Heyn diameter (average length of chord of all scanned grains); Md = median diameter; S = standard deviation of the grain size distribution; S/V = specific surface of all measured grains per unit volume.

Fig. 9. Mosaicism of olivine in Åheim dunite shocked at 29.3 GPa (a), and unshocked sample (b); crossed polarizers.

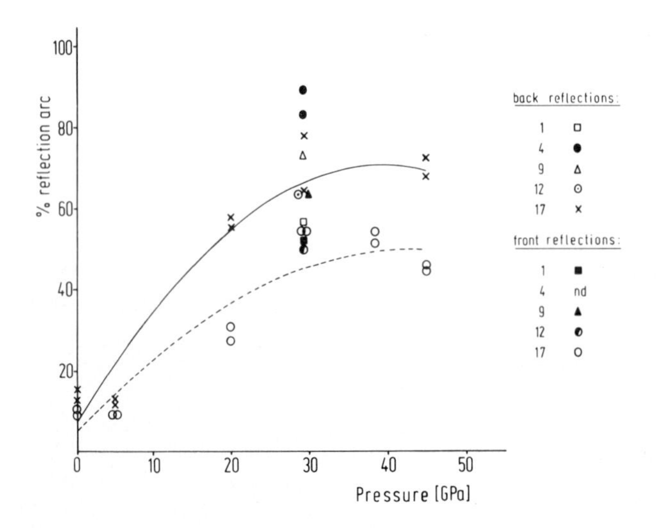

Fig. 10. Intensity of x-ray asterism measured according to the method of Hörz and Quaide (1972) for olivine (average of 4 grains) of various dunites as a function of peak shock pressure; 1,4 = Corundum Hill dunite, 9 = Balsam Gap dunite, 12 = Twin Sisters Peaks dunite, 17 = Åheim dunite.

Moreover, it seems to reach a maximum at pressures of about 30–40 GPa and to decrease at higher pressures, probably due to thermal recovery processes (Lally *et al.*, 1976).

4.1.4 Phase transitions Jeanloz *et al.* (1977) and Bauer (1978) found indications of high pressure solid state transformations in olivine single crystals recovered from shock experiments at pressures between 54 and 57.5 GPa, respectively. These authors reported traces of diaplectic olivine glass whereas Schneider and Hornemann (1977) observed complete reaction of olivine to new oxide phases as indicated by infrared absorption spectra. None of our samples which were analyzed by scanning electron microscopy, infrared absorption spectroscopy and x-ray diffraction gives any evidence for high pressure phase transitions in the solid state, although it cannot be excluded that small amounts of isotropic olivine might be found by transmission electron microscopy. However, evidence of melting and subsequent recrystallization was found along grain boundaries in two experiments at 59 GPa. Both samples were severely pulverized and most of the material was lost during recovery operations. In thin sections made from some of the remaining grain aggregates the primary dunite texture was still discernible. Finer grained prismatic olivine with completely random orientation was formed in relatively broad zones of several tens of μm along former grain boundaries. This olivine lacks mosaicism and fracturing. The occurrence of vesicles and small spherules of iron, which are derived from the adjacent steel sample holder and are incorporated in the newly formed olivine, indicate that these zones of polycrystalline olivine result from crystallized melt (Fig. 11). The "melt zones" are brownish in color and may be similar to the "brown fracture-free regions" of Bauer (1978) but no Fe^{3+} was found by electron spin resonance spectroscopy. In the regions between melt zones the primary large olivine grains display strong fracturing and, in some cases, are partially recrystallized, but the type of mosaicism typical of lower shock pressure (see Fig. 9) is generally lacking. In the recrystallized parts inside the primary olivine crystals a fine-grained texture is developed in which the newly formed equant olivine crystals have a totally random orientation (Fig. 12). This texture is typically different from the low pressure mosaicism described in section 4.1.3 which results from plastic deformation.

4.2 *Progressive Shock Metamorphism*

From the results of shock recovery experiments on dunites over a pressure range of about 60 GPa a system of progressive shock metamorphism can be deduced. It may be based on a number of characteristic deformational and thermal effects which are successively produced by the increasing peak shock pressure. We propose to classify this metamorphic sequence into four main stages (Table 4). Each stage may be typified by one main type of shock effect which is accompanied by other shock metamorphic features less characteristic of the shock stage in question. Irregular *fracturing* is observed in Stage 1 which covers

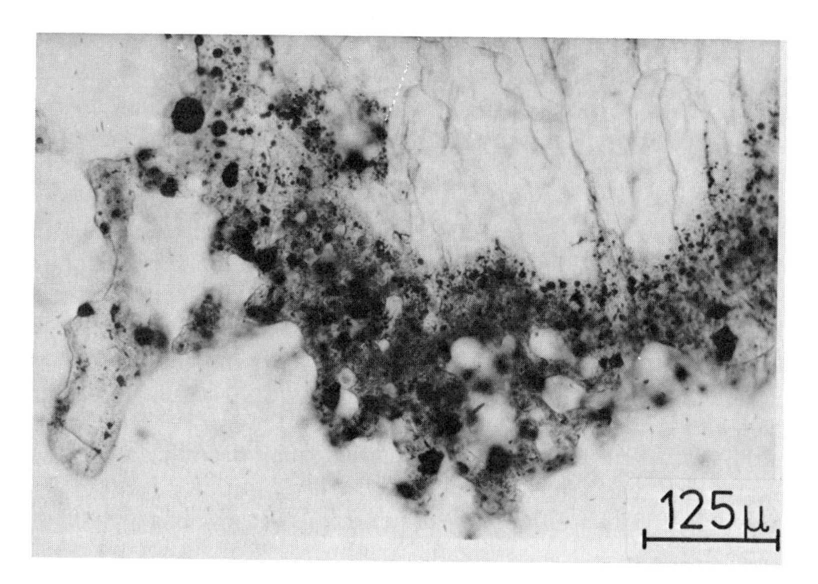

Fig. 11. Intergranular zone of brownish olivine aggregates crystallized from melt surrounding a primary olivine grain with sets of subparallel fractures; Åheim dunite, 59 GPa shock pressure; note vesicles and inclusions of black spherules of iron; plane polarized light.

Fig. 12. Region of solid state recrystallization in the core of olivine which has been marginally shock melted (see Fig. 11); Åheim dunite, 59 GPa; crossed polarizers.

the pressure range from 0 to 5 GPa (lower limit). At pressures between 5 and 15.5 GPa *mosaicism* of olivine develops as the typical shock effect of Stage 2 accompanied by planar fractures. Shock pressures in excess of about 29 GPa produce intergranular *brecciation* of olivine characteristic of Stage 3. In Stage 4, probably above some 55 GPa intergranular *melting* and *recrystallization* occurs. The proposed pressure limits between the four stages of metamorphism are not yet very accurate. Most probably, they will vary if texture and composition of the dunite is changed.

The proposed classification of progressive shock metamorphism of dunite may be compared with other experimentally deduced classifications by Carter *et al.* (1968) and by Bauer (1978). These studies were made on samples of the Twin Sisters Peaks dunite which is the coarsest grained dunite among the samples studied in this work. As seen in Table 4 the sequence and types of shock effects observed in all three sets of experiments are in reasonable agreement in the pressure range below about 55 GPa. The main differences in the results of our experiments and those of Carter *et al.* (1968) and Bauer (1978) in this pressure regime is the fact that intergranular brecciation is not described or considered as a major and typical mechanical shock effect by those authors. It is not clear from either paper whether this difference results from different interpretations of the textures or from different experimental methods and different ways of sample recovery from the sample holder. In our experiments the samples are larger than in those of the other authors. Also the grain size of the Åheim dunite is much smaller than that of the Twin Sisters Peaks dunite. Intergranular brecciation can only be observed if large sample disks are recovered intact for the preparation of

Table 4. Comparison of various classifications of progressive shock metamorphism of dunite (Bauer, 1978; Carter *et al.,* 1968; this work); *boxes* indicate experimentally observed pressure ranges for a given shock effect; *arrows in the boxes:* direction of increase or decrease of a particular shock effect; *other arrows* indicate the most probable or expected pressure range of a given shock effect.

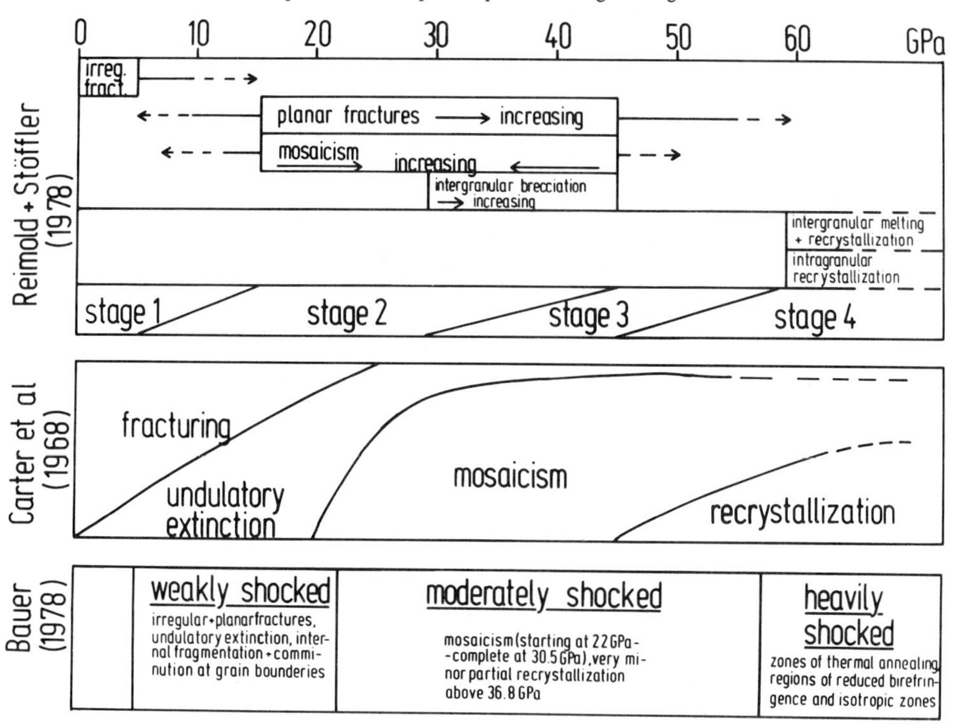

thin sections. Shock effects observed in samples from the shock pressure regime above about 55 GPa by the three working groups (Table 4) are in less agreement. In part, this may be due to different methods of experimentation: powder gun (Bauer, 1978); high explosive in contact with the sample holder (Carter *et al.*, 1968); and explosively accelerated flyer plate (this paper) and different methods of sample investigation. Thermochemical energy generated in high explosive experiments may create higher post shock sample temperatures than those in flyer plate ballistic impact experiments (Bauer, 1978). Quench times are correspondingly longer in explosion experiments, thus, explaining why recrystallization is more pronounced in high explosive experiments.

Notwithstanding the differences in experimentation, a major problem in understanding the thermally induced effects at high shock pressures arises from the uncertainties in determining the magnitude of the post-shock temperature achieved in the shock experiments. Shock temperatures in olivine have been calculated from Hugoniot data (McQueen *et al.*, 1967; Ahrens *et al.*, 1976). Raikes and Ahrens (1978) published post-shock temperatures of forsterite measured directly during shock compression up to a pressure of 30 GPa. Shock pressures above 59 GPa are accompanied by shock temperatures in excess of some 890°C (Ahrens *et al.*, 1976) which is far below the melting point of Mg-rich olivine. Ahrens *et al.* (1976) indicate that higher temperatures may be achieved when the reversion of a high pressure phase to olivine is taken into account upon shock unloading. For 44 GPa they estimate an additional temperature gain of bout 700° which raises the post-shock temperature to about 1140°C, still too low for melting of olivine.

These data and the rather low residual temperatures of 150°C at 30 GPa reported from direct temperature measurements (Raikes and Ahrens, 1978) readily explain the observed lack of any thermal effects in shocked dunite up to 45 GPa. They do not explain the pronounced thermal effects of our 59 GPa shots. We believe that the observed melting must be due to local stress concentrations caused by shock wave reflections at the interfaces to minerals of different shock impedance or at small air spaces between the interface of the sample holder and the sample disk which may lead to very localized heat concentrations. There might be other explanations, e.g., local heating due to "heterogeneous yielding" of the olivine crystals (Grady, 1977).

The solid state recrystallization in the cores of large olivine grains which remained unaffected by melting, most probably results from the recrystallization of a disordered phase which was produced upon pressure release from a high pressure polymorph of olivine as proposed by Ahrens *et al.* (1976). At 59 GPa the calculated mass fraction of the high pressure phase would be at about 55% (Ahrens *et al.*, 1976) which is in reasonable agreement with the petrographic observation concerning the abundance of solid state recrystallization. At 45 GPa, where the mass fraction of the high pressure phase is still some 25%, the post-shock temperature seems to be so low that recrystallization of the type observed at 59 GPa (Fig. 12) does not occur. It seems highly improbable that mosaicism may be the result of the reversion of a high pressure phase as discussed by Carter *et al.* (1968), because it forms at pressures well below the onset of the mixed phase regime of Mg-rich olivine which, according to Ahrens *et al.* (1976), extends from 26 to 75 GPa. Instead, mosaicism seems to be the characteristic shock effect resulting from shock states which lie on the plastic Hugoniot curve and probably on the lower part of the Hugoniot of the mixed phase regime where an appreciable amount of low-pressure-phase is still present (compare Table 4).

In conclusion, dunite of Stage 1 results from shock states on the elastic Hugoniot, Stage 2 and 3 are roughly related to the plastic Hugoniot and the mixed phase regime of the Hugoniot, and Stage 4 can be correlated presumably with the upper region of the mixed phase regime and with the high pressure phase Hugoniot as defined by Ahrens *et al.* (1976).

Acknowledgments—We are indebted to Dr. U. Hornemann, H. Hornauer and D. Franz (Ernst-Mach-Institut, Arbeitsgruppe für Ballistische Forschung, Weil a. Rh.) for their excellent assistance with the experimental techniques. Thanks are also due to all colleagues of the Institute of Mineralogy, Münster, especially to Mrs. G. Grant, Miss F. Möllers, H. Kreikenbaum and W. Ortjohann for their technical help. Prof. R. Blaschke, Münster, and Dr. Jakubith provided analyses with the scanning electron microscope and the electron spin resonance spectroscope, respectively. We thank Drs. J. F. Bauer, Houston, and R. Jeanloz, Pasadena, for their careful review of the manuscript.

REFERENCES

Ahrens T. J., Tsay F.-D. and Live D. H. (1976) Shock-induced fine-grained recrystallization of olivine: Evidence against subsolidus reduction of Fe^{2+}. *Proc. Lunar Sci. Conf. 7th*, p. 1143–1156.

Ashworth J. R. and Barber D. J. (1975) Electron petrography of shock-deformed olivine in stony meteorites. *Earth Planet. Sci. Lett.* **27**, p. 43–50.

Bauer J. F. (1978) Shock-induced deformation in olivine from polycrystalline dunites and particulate samples (abstract). In *Lunar and Planetary Science IX*, p. 55–57. Lunar and Planetary Institute, Houston.

Carter N. L., Raleigh C. B. and DeCarli P. S. (1968) Deformation of olivine in stony meteorites. *J. Geophys. Res.* **73**, 5439–5461.

Dymek R. F., Albee A. L. and Chodos A. A. (1975) Comparative petrology of lunar cumulate rocks of possible primary origin: Dunite 72415, troctolite 76535, norite 78235, and anorthosite 62237. *Proc. Lunar Sci. Conf. 6th*, p. 301–341.

Engelhardt W. v. (1944) Die Anisotropie der Teilbarkeit des Quarzes. *Nachr. d. Akad.d.Wiss. Göttingen* **3**, 43–56.

Engelhardt W. v., Arndt J., Müller W. F. and Stöffler D. (1970) Shock metamorphism of lunar rocks and origin of the regolith at the Apollo 11 landing site. *Proc. Apollo 11 Lunar Sci. Conf.* p. 363–384.

Folk R. L. and Ward W. C. (1957) Brazos river bar: A study of the significance of grain size parameters. *J. Sediment. Petrol.* **24**, 3–26.

Grady D. E. (1977) Processes occurring in shock wave compression of rocks and minerals. In *High-Pressure Research—Applications in Geophysics* (H. H. Manghnani and S.-I. Akimoto, eds.), p. 389–438. Acad. Press, N. Y.

Hornemann U., Pohl J. and Bleil U. (1975) A compressed air gun accelerator for shock magnetization and demagnetization—Experiments up to 20 Kbar. *J. Geophys.* **41**, 13–22.

Hörz F. and Quaide W. L. (1972) Debye-Scherrer investigations of experimentally shocked silicates. *The Moon* **6**, 45–82.

James O. B. (1969) Shock and thermal metamorphism of basalt by nuclear explosion, Nevada Test Site. *Science* **166**, 1615–1620.

Jeanloz R., Ahrens T. J., Lally J. S., Nord G. L. Jr., Christie J. M. and Heuer A. H. (1977) Shock-produced olivine glass: First observation. *Science* **197**, 457–459

Lally J. S., Christie J. M., Nord G. L. Jr. and Heuer A. H. (1976) Deformation, recovery and recrystallization of lunar dunite 72417. *Proc. Lunar Sci. Conf. 7th*, p. 1845–1863.

Levi-Donati G. R. (1971) Petrological features of shock metamorphism in chondrites: Alfionello. *Meteoritics* **6**, 225–235.

McQueen R. G., Marsh S. P. and Fritz J. N. (1967) Hugoniot equation of state of twelve rocks. *J. Geophys. Res.* **72**, 4999–5036.

McQueen R. G., Marsh S. P., Taylor J. W., Fritz J. N. and Carter W. J. (1970) Equation of state of solids from shock wave studies. In *High Velocity Impact Phenomena* (R. Kinslow, ed.), p. 293–417. Acad. Press, N. Y.

Müller W. F. and Hornemann U. (1969) Shock-induced planar deformation structures in experimentally shock-loaded olivines and in olivines from chondritic meteorites. *Earth Planet. Sci Lett.* **7** p. 251–264.

Raikes S. A. and Ahrens T. J. (1978) Measurement of post-shock temperatures in silicates (abstract). In *Lunar and Planetary Science IX*, p. 922–924. Lunar and Planetary Institute, Houston.

Reimold W. U. (1977) Stoßwellendeformation von dunitischen Gesteinen. Diplomarbeit, Münster. Diploma thesis.

Reimold W. U. and Stöffler D. (1978) Experimental shock metamorphism of dunitic rocks (abstract). In *Lunar and Planetary Science IX*, p. 955–957. Lunar and Planetary Institute, Houston.

Schneider H. and Hornemann U. (1977) Preliminary data on the shock-induced high pressure transformation of olivine. *Earth Planet. Sci Lett.* **36**, 322–324.

Sclar C. B. (1969) Shock wave damage in olivine. *Trans. Amer. Geophys. Union* **50**, 219.

Sclar C. B. and Morzenti S. P. (1971) Shock-induced planar deformation structures in olivine from the Chassigny meteorite. *Meteoritics* **6**, 310–311.

Short N. M. (1966) Effects of shock pressures from a nuclear explosion on mechanical and optical properties of granodiorite. *J. Geophys. Res.* **71**, 1195–1215.

Short N. M. (1969) Shock metamorphism of basalt. *Mod. Geol.* **1**, 81–95.

Snee L. W. and Ahrens T. J. (1975) Shock-induced deformation features in terrestrial peridot and lunar dunite. *Proc. Lunar Sci. Conf. 6th*, p. 833–842.

Stöffler D. (1972) Deformation and transformation of rock-forming minerals by natural and experimental processes. I. Behavior of minerals under shock compression. *Fortschr. Mineral.* **49**, 50–113.

Lunar Sample Index

Pages 1–1447: Volume 1, Petrogenetic Studies: The Moon and Meteorites
Pages 1449–2824: Volume 2, Lunar and Planetary Surfaces
Pages 2825–3973: Volume 3, The Moon and the Inner Solar System

(Index entries were compiled from information supplied by the authors, and refer only to opening pages of articles.)

Heavenly Bodies Index

Pages 1-1447: Volume 1, Petrogenetic Studies: The Moon and Meteorites
Pages 1449-2824: Volume 2, Lunar and Planetary Surfaces
Pages 2825-3973: Volume 3, The Moon and the Inner Solar System

(Index entries were compiled from lists of meteorites and localities supplied by the authors, and refer only to opening pages of articles.)

Subject Index

Pages 1-1477: Volume 1, Petrogenetic Studies: The Moon and Meteorites
Pages 1449-2824: Volume 2, Lunar and Planetary Surfaces
Pages 2825-3973: Volume 3, The Moon and the Inner Solar System

(Index entries were compiled from key words supplied by the authors, and refer only to opening pages of articles.)

Author Index

Pages 1-1477: Volume 1, Petrogenetic Studies: The Moon and Meteorites
Pages 1449-2824: Volume 2, Lunar and Planetary Surfaces
Pages 2825-3973: Volume 3, The Moon and the Inner Solar System

COMETS

Compiled by Ray Newburn.

Selected Short-period Comets

Comet**	Period# (yrs)	Perihelion Dist.# (AU)	Inclination# (deg.)	Absolute Magnitude* H_{10}	Remarks
Encke (1974 V) (1977 —)	3.30	0.34	12.0	9.7 (Pre-perihelion)	Shortest period comet known. Smallest perihelion distance of any comet with P < 100 years. Fan shaped coma. Sometimes shows weak ion tail. No continuum in spectra. Often mentioned as space probe target.
Tempel 2 (1977d)	5.26	1.36	12.5	8.4	Third shortest period. Fan shaped coma. No tails. Strong continuum in spectra. Possible space probe target. Brightest about three weeks after perihelion.
Tuttle-Giacobini-Kresak (1973 VI)	5.56	1.15	13.6	11.5	A few days before perihelion and 35-8 days after perihelion in 1973 this comet flared up in brightness by nine magnitudes (a factor of 4000).
d'Arrest (1976 IX)	6.23	1.17	16.7	9.5	Visible to naked eye briefly in 1976, close approach to Jupiter in 1979 will increase perihelion to 1.29 AU. Large, fan-shaped coma. Brightness usually increases 3 magnitudes *after* perihelion.
Giacobini-Zinner (1978h)	6.52	0.99	31.7	10.0	Striking tail for so faint a comet. Meteoroids apparently very low density, "fluffy" objects.
Schwassmann-Wachmann 1 (1974 II)	15.03	5.45	9.7	~7½	Smallest eccentricity (.105) and largest perihelion of any short period comet. Shows periodic flares in brightness of ~5 magnitude.
Halley (1910 II)	76.09	0.59	162.2	5.0	Brightest comet with P < 100 years. Well developed ion and dust tails. Strong continuum in spectra. Poor apparition as seen from Earth in 1986. Possible target for space probe flyby.